The Elements

Element	Symbol	Atomic Number	Atomic Mass*	Element	Symbol	Atomic Number	Atomic Mass*
Actinium	Ac	89	(227)	Mercury	Hg	80	200.6
Aluminum	Al	13	26.98	Molybdenum	Mo	42	95.94
Americium	Am	95	(243)	Neodymium	Nd	60	144.2
Antimony	Sb	51	121.8	Neon	Ne	10	20.18
Argon	Ar	18	39.95	Neptunium	Np	93	(244)
Arsenic	As	33	74.92	Nickel	Ni	28	58.70
Astatine	At	85	(210)	Niobium	Nb	41	92.91
Barium	Ba	56	137.3	Nitrogen	N	7	14.01
Berkelium	Bk	97	(247)	Nobelium	No	102	(253)
Beryllium	Be	4	9.012	Osmium	Os	76	190.2
Bismuth	Bi	83	209.0	Oxygen	O	8	16.00
Bohrium	Bh	107	(267)	Palladium	Pd	46	106.4
Boron	B	5	10.81	Phosphorus	P	15	30.97
Bromine	Br	35	79.90	Platinum	Pt	78	195.1
Cadmium	Cd	48	112.4	Plutonium	Pu	94	(242)
Calcium	Ca	20	40.08	Polonium	Po	84	(209)
Californium	Cf	98	(249)	Potassium	K	19	39.10
Carbon	C	6	12.01	Praseodymium	Pr	59	140.9
Cerium	Ce	58	140.1	Promethium	Pm	61	(145)
Cesium	Cs	55	132.9	Protactinium	Pa	91	(231)
Chlorine	Cl	17	35.45	Radium	Ra	88	(226)
Chromium	Cr	24	52.00	Radon	Rn	86	(222)
Cobalt	Co	27	58.93	Rhenium	Re	75	186.2
Copper	Cu	29	63.55	Rhodium	Rh	45	102.9
Curium	Cm	96	(247)	Roentgenium	Rg	111	(272)
Darmstadtium	Ds	110	(281)	Rubidium	Rb	37	85.47
Dubnium	Db	105	(262)	Ruthenium	Ru	44	101.1
Dysprosium	Dy	66	162.5	Rutherfordium	Rf	104	(263)
Einsteinium	Es	99	(254)	Samarium	Sm	62	150.4
Erbium	Er	68	167.3	Scandium	Sc	21	44.96
Europium	Eu	63	152.0	Seaborgium	Sg	106	(266)
Fermium	Fm	100	(253)	Selenium	Se	34	78.96
Fluorine	F	9	19.00	Silicon	Si	14	28.09
Francium	Fr	87	(223)	Silver	Ag	47	107.9
Gadolinium	Gd	64	157.3	Sodium	Na	11	22.99
Gallium	Ga	31	69.72	Strontium	Sr	38	87.62
Germanium	Ge	32	72.61	Sulfur	S	16	32.07
Gold	Au	79	197.0	Tantalum	Ta	73	180.9
Hafnium	Hf	72	178.5	Technetium	Tc	43	(98)
Hassium	Hs	108	(277)	Tellurium	Te	52	127.6
Helium	He	2	4.003	Terbium	Tb	65	158.9
Holmium	Ho	67	164.9	Thallium	Tl	81	204.4
Hydrogen	H	1	1.008	Thorium	Th	90	232.0
Indium	In	49	114.8	Thulium	Tm	69	168.9
Iodine	I	53	126.9	Tin	Sn	50	118.7
Iridium	Ir	77	192.2	Titanium	Ti	22	47.88
Iron	Fe	26	55.85	Tungsten	W	74	183.9
Krypton	Kr	36	83.80	Uranium	U	92	238.0
Lanthanum	La	57	138.9	Vanadium	V	23	50.94
Lawrencium	Lr	103	(257)	Xenon	Xe	54	131.3
Lead	Pb	82	207.2	Ytterbium	Yb	70	173.0
Lithium	Li	3	6.941	Yttrium	Y	39	88.91
Lutetium	Lu	71	175.0	Zinc	Zn	30	65.41
Magnesium	Mg	12	24.31	Zirconium	Zr	40	91.22
Manganese	Mn	25	54.94			112**	(285)
Meitnerium	Mt	109	(268)			114	(289)
Mendelevium	Md	101	(256)			116	(292)

*All atomic masses are given to four significant figures. Values in parentheses represent the mass number of the most stable isotope.

**The names and symbols for elements 112, 114, and 116 have not been chosen.

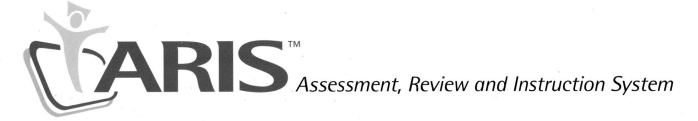

ONLINE STUDY & HOMEWORK MATERIALS

I M P O R T A N T : Following are instructions to access online resources to support your McGraw-Hill textbook

The URL associated with your text is:
http://www.mhhe.com/silberberg

Option 1: ARIS LOGIN. Your instructor may use **ARIS** as a homework and assessment tool. If so, you must register in order to ensure your assignments are recorded into your instructor's gradebook.

Option 2: If your instructor is **NOT** using **ARIS** as a homework and assessment tool, you are welcome to access the material on the site without registering. Simply go to the URL listed above. You are free to access these materials for your own self-study.

ARIS LOGIN.   *To Register you need:*

1. Section Code: Provided by your instructor.

2. Registration Code: Provided in the gray scratch-off area below.

3. URL: Go to the URL listed at the top of this card and follow the directions for creating an ARIS account.

VM4W-WUWM-QMYP-KUU4-NAUC

Scratch off for registration code
This registration code can be used by one individual and is not transferable.

I M P O R T A N T : The registration code printed above can only be used once to create a unique student account. Students do not need a registration code to access the content of the site. Students choosing "ARIS Login" must login each time they visit the site in order for their grades to be saved to their instructor's gradebook.

Silberberg: Principles of General Chemistry
ISBN-13: 978-0-07-326291-8
ISBN-10: 0-07-326291-9

Principles of
GENERAL CHEMISTRY

Martin S. Silberberg

Principles of
GENERAL CHEMISTRY

 **Higher Education**

Boston   Burr Ridge, IL   Dubuque, IA   Madison, WI   New York   San Francisco   St. Louis
Bangkok   Bogotá   Caracas   Kuala Lumpur   Lisbon   London   Madrid   Mexico City
Milan   Montreal   New Delhi   Santiago   Seoul   Singapore   Sydney   Taipei   Toronto

Higher Education

PRINCIPLES OF GENERAL CHEMISTRY

Published by McGraw-Hill, a business unit of The McGraw-Hill Companies, Inc., 1221 Avenue of the Americas, New York, NY 10020. Copyright © 2007 by The McGraw-Hill Companies, Inc. All rights reserved. No part of this publication may be reproduced or distributed in any form or by any means, or stored in a database or retrieval system, without the prior written consent of The McGraw-Hill Companies, Inc., including, but not limited to, in any network or other electronic storage or transmission, or broadcast for distance learning.

Some ancillaries, including electronic and print components, may not be available to customers outside the United States.

This book is printed on acid-free paper.

1 2 3 4 5 6 7 8 9 0 DOW/DOW 0 9 8 7 6

ISBN-13   978-0-07-310720-2
ISBN-10   0-07-310720-4

Publisher: *Thomas D. Timp*
Managing Developmental Editor: *Shirley R. Oberbroeckling*
Senior Developmental Editor: *Donna Nemmers*
Outside Developmental Services: *Karen Pluemer*
Senior Marketing Manager: *Tamara L. Good-Hodge*
Lead Project Manager: *Peggy J. Selle*
Lead Media Project Manager: *Judi David*
Lead Media Producer: *John J. Theobald*
Senior Designer: *David W. Hash*
Interior Designer: *Jamie O'Neal*
Cover Illustration: *Michael Goodman*
Cover Photo: *©Richard Megna/Fundamental Photographs*
Senior Photo Research Coordinator: *Lori Hancock*
Photo Research: *Chris Hammond/PhotoFind, LLC*
Supplement Producer: *Tracy L. Konrardy*
Compositor: *The GTS Companies/Los Angeles, CA Campus*
Typeface: *10.5/12 Times*
Printer: *R. R. Donnelley Willard, OH*

The credits section for this book begins on page C-1 and is considered an extension of the copyright page.

Library of Congress Cataloging-in-Publication Data

Silberberg, Martin S. (Martin Stuart), 1945–
    Principles of general chemistry / Martin S. Silberberg. — 1st ed.
      p.   cm.
    Includes index.
    ISBN   978-0-07-310720-2 — 0-07-310720-4 (acid-free paper)
      1. Chemistry—Textbooks. I. Title.

QD31.3.S55   2007
540—dc22   2005054377
                                                  CIP

To Ruth and Daniel, with all my love.
I can't even imagine doing this without
the two of you behind me.

Brief Contents

Contents

4 CHAPTER

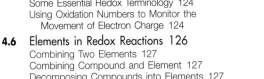

The Major Classes of Chemical Reactions 108

5 CHAPTER

Gases and the Kinetic-Molecular Theory 138

6 CHAPTER

Thermochemistry: Energy Flow and Chemical Change 177

7 CHAPTER

Quantum Theory and Atomic Structure 205

8 CHAPTER

Electron Configuration and Chemical Periodicity 235

9 CHAPTER

Models of Chemical Bonding 268

10 CHAPTER

The Shapes of Molecules 296

11 CHAPTER

Theories of Covalent Bonding 323

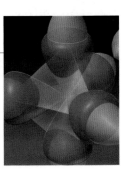

12 CHAPTER

Intermolecular Forces: Liquids, Solids, and Phase Changes 347

13 CHAPTER

The Properties of Solutions 389

14 CHAPTER

The Main-Group Elements: Applying Principles of Bonding and Structure 423

15 CHAPTER

Organic Compounds and the Atomic Properties of Carbon 457

16 CHAPTER

Kinetics: Rates and Mechanisms of Chemical Reactions 498

17 CHAPTER

Equilibrium: The Extent of Chemical Reactions 540

18 CHAPTER

Acid-Base Equilibria 577

19 CHAPTER

Ionic Equilibria in Aqueous Systems 615

20 CHAPTER

Thermodynamics: Entropy, Free Energy, and the Direction of Chemical Reactions 650

21 CHAPTER

Electrochemistry: Chemical Change and Electrical Work 681

22 CHAPTER

The Transition Elements and Their Coordination Compounds 734

23 CHAPTER

Nuclear Reactions and Their Applications 762

About the Author

Martin S. Silberberg received a B.S. in Chemistry from the City University of New York and a Ph.D. in Chemistry from the University of Oklahoma. He then accepted a research position in analytical biochemistry at the Albert Einstein College of Medicine in New York City, where he developed advanced methods to study fundamental brain mechanisms as well as neurotransmitter metabolism in Parkinson's disease. Following his years in research, Dr. Silberberg joined the faculty of Simon's Rock College of Bard, a liberal arts college known for excellence in teaching small classes of highly motivated students. As Head of the Natural Sciences Major and Director of Premedical Studies, he has taught courses in general chemistry, organic chemistry, biochemistry, and liberal arts chemistry. The close student contact has afforded him insights into how students learn chemistry, where they have difficulties, and what strategies can help them succeed. Dr. Silberberg has applied these insights in a broader context by establishing a text writing, editing, and consulting company. Before writing his own text, he worked as a consulting and developmental editor on chemistry, biochemistry, and physics texts for several major college publishers. He resides with his wife and son in the Pioneer Valley near Amherst, Massachusetts, where he enjoys the rich cultural and academic life of the area and relaxes by cooking, gardening, and hiking.

CREATING A NEW TEXT

Like the science of chemistry itself, the teaching of chemistry is evolving, and the course texts that professors and students rely on must do so as well. The large, thousand-page or more books that most courses use provide a complete survey of the field, with a richness of relevance and content. *Chemistry: The Molecular Nature of Matter and Change,* the parent of this new text, stands at the forefront of dynamic, modern texts. Yet, extensive market research demonstrates that some professors prefer a less expansive treatment, with coverage confined to the core principles and skills. Such a text would allow professors to enrich their course with topics relevant to their own students. And, most importantly, it would allow the entire book to be more easily covered in one year—all the essential material a science major needs to go on to other courses in chemistry and related disciplines.

Sensing the need for a more succinct text, we created *Principles of General Chemistry.* This new text retains the molecular artwork, problem-solving approach, and student-friendly pedagogy so admired in its parent, *Chemistry: The Molecular Nature of Matter and Change.* This new text is leaner and more concise, targeting only the topics that a general chemistry course at this level should include and that instructors expect to see.

Crafting the content of the new text involved assessing which topics constituted the core of a general chemistry course and distilling them from the parent text. To confirm my assessment, we invited three professors to serve as content editors and review my suggested changes. Using their experience and my detailed outline, the content editors pruned the parent text to generate a rough draft, which I then reworked into the final manuscript. It was very gratifying, even remarkable, to find that the four of us defined the essential content of the modern general chemistry course in virtually identical terms.

HOW *CHEMISTRY* AND THE NEW *PRINCIPLES OF GENERAL CHEMISTRY* ARE ALIKE

Both *Chemistry: The Molecular Nature of Matter and Change* and *Principles of General Chemistry* maintain the same high standards of accuracy, depth, clarity, and rigor and have the same three distinguishing hallmarks:

1. *Visualizing chemical models.* In many discussions, concepts are explained first at the macroscopic level and then from a molecular point of view. Placed near the related discussion, the text's celebrated graphics bring the point home for today's visually oriented students—depicting the change at the observable level in the lab, at the molecular level, and, when appropriate, at the symbolic level with the balanced equation.

2. *Thinking logically to solve problems.* The problem-solving approach, based on a four-step method widely approved by chemical educators, is introduced in Chapter 1 and employed consistently throughout the text. It encourages students to *first* plan a logical approach and *then* proceed to the arithmetic solution. A check step, universally recommended by instructors, fosters the habit of considering the reasonableness and magnitude of the answer. For practice and reinforcement, each worked problem has a matched follow-up problem, for which an abbreviated, multi-step solution—not just a brief answer—appears at the end of the chapter.

3. *Applying ideas to the real world.* For today's students, who may enter one of numerous chemistry-related fields, real-world applications are woven into the worked in-text sample problems and the chapter problem sets.

HOW *CHEMISTRY* AND *PRINCIPLES OF GENERAL CHEMISTRY* ARE DIFFERENT

Principles of General Chemistry achieves authoritative topic coverage in 300 fewer pages than its parent text, thereby appealing to today's efficiency-minded instructors and value-conscious students. To accomplish this shortening, most of the material in the boxed applications essays and margin notes was removed, thereby allowing instructors to include their own favorite examples.

The content editors and I also felt that several other topics, while constituting important fields of modern research, were not central to the core subject matter of general chemistry; these include colloids, green chemistry, and much of advanced materials. The chapters on descriptive chemistry, organic chemistry, and transition elements were tightened extensively, and the chapter on the industrial isolation of the elements was removed (except for a few topics that were blended into the chapter on electrochemistry).

The new text includes all the worked sample problems of the parent text but has about one-third fewer end-of-chapter problems. Nevertheless, there are more than enough representative problems for every topic, and they are packed with relevance and real-world applications.

Principles of General Chemistry is a powerhouse of pedagogy. All the learning aids that students find so useful in the parent text have been retained—Concepts and Skills to Review, Section Summaries, Key Terms, Key Equations, and Brief Solutions to Follow-up Problems. In addition, two new aids help students further focus their efforts:

1. *Key Principles.* At the beginning of each chapter, short paragraphs state the main concepts concisely, using many of the same phrases and terms that will appear in the pages that follow. A student can preview these principles before reading the chapter and then review them afterward.

2. *Problem-Based Learning Objectives.* At the end of each chapter, the list of learning objectives now includes the numbers of homework problems that relate to each objective. Thus, a student, or an instructor, can select problems that apply specifically to a given topic.

The new text is a lean and direct introduction to chemistry for science majors. Unlike its parent, which offers almost any topic that *any* instructor could want, *Principles of General Chemistry* offers every topic that *every* instructor would need.

Acknowledgments

Principles of General Chemistry and its author are fortunate to have supplement authors so committed to accuracy and clarity for student and instructor. Patricia Amateis of Virginia Tech diligently prepared the *Instructors' Solutions Manual* and *Student Solutions Manual*. Libby Weberg has prepared the *Student Study Guide*. S. Walter Orchard of Tacoma Community College updated the *Test Bank*. Christina Bailey of California Polytechnic University provided the excellent *PowerPoint Lecture Outlines* that appear on the Digital Content Manager CD.

It was a great pleasure to work closely with the three content editors, Patricia Amateis of Virginia Tech, Ramesh Arasasingham of the University of California–Irvine, and Steven Keller of the University of Missouri–Columbia. All three are superb professors dedicated to making general chemistry an enriching experience for their students. Their help and insight has ensured that this first edition contains all the essential principles necessary for the science major, two-semester, general chemistry course.

Special thanks go to Professor Dorothy B. Kurland for her exceptionally thorough accuracy check of the entire text. And I extend my gratitude to all the other professors who reviewed portions of this first edition or participated in our developmental survey process to assess the content needs for the text:

Edwin H. Abbott, *Montana State University*
James P. Birk, *Arizona State University*
Bob Blake, *Texas Tech University*
Jeffrey O. Boles, *Tennessee Tech University*
Wayne B. Bosma, *Bradley University*
Brian Buffin, *Western Michigan University*
Paul Chirik, *Cornell University–Ithaca*
Ramon Lopez De La Vega, *Florida International University*
Milagros Delgado, *Florida International University*
Stephen C. Foster, *Mississippi State University*
Phil Franklin, *Johnson County Community College*
Nancy Gardner, *California State University–Long Beach*
Graeme C. Gerrans, *University of Virginia*

Thomas Greenbowe, *Iowa State University*
Greg Hale, *University of Texas at Arlington*
C. Alton Hassell, *Baylor University*
Narayan S. Hosmane, *Northern Illinois University*
Andy Jorgenson, *University of Toledo*
Philip C. Keller, *University of Arizona*
Laurence Lavelle, *UCLA*
Michael M. Lerner, *Oregon State University*
Rudy L. Luck, *Michigan Technological University*
Pamela Marks, *Arizona State University*
Scott H. Northrup, *Tennessee Tech University*
Cortlandt Pierpont, *University of Colorado–Boulder*
Helen Place, *Washington State University*
John R. Pollard, *University of Arizona*

Daniel Rabinovich, *The University of North Carolina at Charlotte*
Cathrine E. Reck, *Indiana University*
Barbara Reisner, *James Madison University*
Suzanne F. Rottman, *University of Maryland–Baltimore County*
Jadwiga Sipowska, *University of Michigan–Ann Arbor*
Lothar Stahl, *University of North Dakota*
Alan M. Stolzenberg, *West Virginia University*
Thomas D. Tullius, *Boston University*
Thomas Webb, *Auburn University*
Steven M. Wietstock, *University of Notre Dame–Notre Dame*
Charles A. Wilkie, *Marquette University*
Frank Woodruff, *University of Southern Mississippi*

The superb publishing team at McGraw-Hill Higher Education has done a terrific job once again in the development and production of this new text, and they have my deepest appreciation. Heading the team with guidance, friendship, and support were Michael Lange, Vice President—New Product Launches, Director of Marketing Kent Peterson, and Publisher Thomas Timp. Senior Developmental Editor Donna Nemmers was in charge throughout the project overseeing innumerable text and supplement details; Lead Project Manager Peggy Selle handled the complex production expertly; Senior Designer David Hash supervised the modern interior design by freelancer Jamie O'Neal; and Marketing Manager Tami Hodge applied her enthusiasm and skill to presenting this new book and its supplements to the academic community.

A wonderful group of expert freelancers made indispensable contributions as well. I never could have finished this project on time without the hard work and remarkable organizational and personal skills of Freelance Developmental Editor Karen Pluemer. Jane Hoover performed a masterful copyediting job once again, and Katie Aiken and Janelle Pregler followed with superb proofreading. Chris Hammond of Photofind found some striking new photos. And my friend Michael Goodman created the exciting new cover.

As always, my wife Ruth was there every step of the way, from helping to set up the project to checking and correcting manuscript and proofs. I rely daily on her devoted support. And my son Daniel not only contributed his artistic skill in helping to design artwork, but, as a recent chemistry student, he also provided valuable input on the clarity of explanations.

A Guide to Student Success

This guided tour of **Principles of General Chemistry** will show you how the special features of this text can help you be successful in this course.

Chapter Opener

The opener provides a thought-provoking figure and legend that relate to a main topic of the chapter.

Key Principles

The main principles from the chapter are presented in a few sentences so that you can keep them in mind as you study. You can also use this list for review when you finish the chapter.

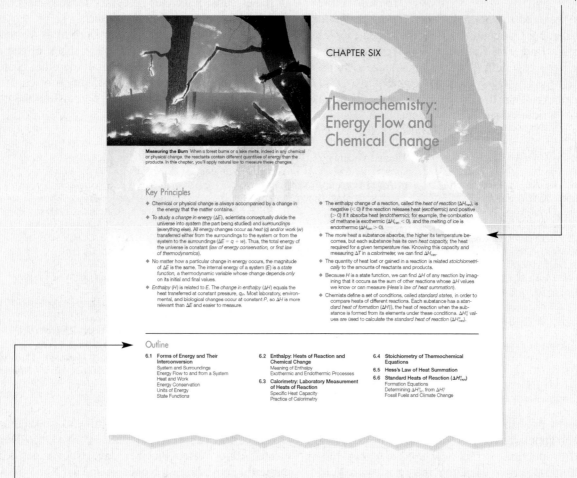

CHAPTER SIX

Thermochemistry: Energy Flow and Chemical Change

Measuring the Burn When a forest burns or a lake melts, indeed in any chemical or physical change, the reactants contain different quantities of energy than the products. In this chapter, you'll apply natural law to measure these changes.

Key Principles

◆ Chemical or physical change is *always* accompanied by a change in the energy that the matter contains.

◆ To study a *change in energy* (ΔE), scientists conceptually divide the universe into *system* (the part being studied) and *surroundings* (everything else). All energy changes occur as *heat* (q) and/or *work* (w) transferred either from the surroundings to the system or from the system to the surroundings ($\Delta E = q + w$). Thus, the total energy of the universe is constant (*law of energy conservation*, or *first law of thermodynamics*).

◆ No matter how a particular change in energy occurs, the magnitude of ΔE is the same. The internal energy of a system (E) is a *state function*, a thermodynamic variable whose change depends *only* on its initial and final values.

◆ *Enthalpy* (H) is related to E. The *change in enthalpy* (ΔH) equals the heat transferred at constant pressure, q_p. Most laboratory, environmental, and biological changes occur at constant P, so ΔH is more relevant than ΔE and easier to measure.

◆ The enthalpy change of a reaction, called the *heat of reaction* (ΔH_{rxn}), is negative (< 0) if the reaction releases heat (*exothermic*) and positive (> 0) if it absorbs heat (*endothermic*); for example, the combustion of methane is exothermic ($\Delta H_{rxn} < 0$), and the melting of ice is endothermic ($\Delta H_{fus} > 0$).

◆ The more heat a substance absorbs, the higher its temperature becomes, but each substance has its own *heat capacity*, the heat required for a given temperature rise. Knowing this capacity and measuring ΔT in a *calorimeter*, we can find ΔH_{rxn}.

◆ The quantity of heat lost or gained in a reaction is related *stoichiometrically* to the amounts of reactants and products.

◆ Because H is a state function, we can find ΔH of any reaction by imagining that it occurs as the sum of other reactions whose ΔH values we know or can measure (*Hess's law of heat summation*).

◆ Chemists define a set of conditions, called *standard states*, in order to compare heats of different reactions. Each substance has a *standard heat of formation* (ΔH_f°), the heat of reaction when the substance is formed from its elements under these conditions. ΔH_f° values are used to calculate the *standard heat of reaction* (ΔH_{rxn}°).

Outline

6.1 Forms of Energy and Their Interconversion
System and Surroundings
Energy Flow to and from a System
Heat and Work
Energy Conservation
Units of Energy
State Functions

6.2 Enthalpy: Heats of Reaction and Chemical Change
Meaning of Enthalpy
Exothermic and Endothermic Processes

6.3 Calorimetry: Laboratory Measurement of Heats of Reaction
Specific Heat Capacity
Practice of Calorimetry

6.4 Stoichiometry of Thermochemical Equations

6.5 Hess's Law of Heat Summation

6.6 Standard Heats of Reaction (ΔH_{rxn}°)
Formation Equations
Determining ΔH_{rxn}° from ΔH_f°
Fossil Fuels and Climate Change

Chapter Outline

The outline shows the sequence of topics and subtopics.

Concepts and Skills to Review

This unique feature helps you prepare for the upcoming chapter by referring to key material from earlier chapters that you should understand *before* you start reading this one.

Concepts & Skills to Review Before You Study This Chapter

- classification of mixtures (Section 2.9)
- calculations involving mass percent (Section 3.1) and molarity (Section 3.5)
- electrolytes; water as a solvent (Sections 4.1 and 12.5)
- mole fraction and Dalton's law of partial pressures (Section 5.4)
- types of intermolecular forces and the concept of polarizability (Section 12.3)
- vapor pressure of liquids (Section 12.2)

Nearly all the gases, liquids, and solids that make up our world are *mixtures*—two or more substances physically mixed together but not chemically combined. Synthetic mixtures, such as glass and soap, usually contain relatively few components, whereas natural mixtures, such as seawater and soil, are more complex, often containing more than 50 different substances. Living mixtures, such as trees and students, are the most complex—even a simple bacterial cell contains well over 5000 different compounds (Table 13.1).

Table 13.1 **Approximate Composition of a Bacterium**

Substance	Mass % of Cell	Number of Types	Number of Molecules

SOLVING PROBLEMS STEP-BY-STEP

Sample Problems

A worked-out problem appears whenever an important new concept or skill is introduced. The step-by-step approach is shown consistently for every sample problem in the text.

- **Plan** analyzes the problem, showing how you can use what is known to find what is unknown. This approach develops the habit of thinking through the solution *before* performing calculations.
- **Problem-solving roadmaps** specific to the problem lead you visually through the calculation steps.
- **Solution** shows the calculation steps *in the same order* as they are discussed in the plan and shown in the roadmap.
- **Check** fosters the habit of going over your work quickly to make sure that the answer is reasonable, both chemically and mathematically—a great way to avoid careless errors.
- **Comment** provides an additional insight or an alternative approach or notes a common mistake to avoid.
- **Follow-up Problem** gives you immediate practice by presenting a similar problem.

Molecular-view Sample Problems unique to Silberberg texts, conceptual (picture) problems apply this stepwise strategy to help you interpret molecular scenes and solve problems based on them.

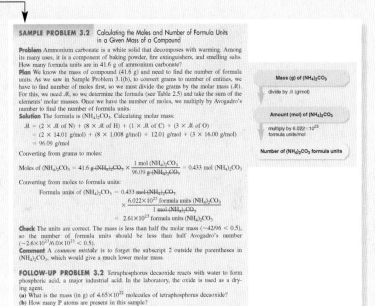

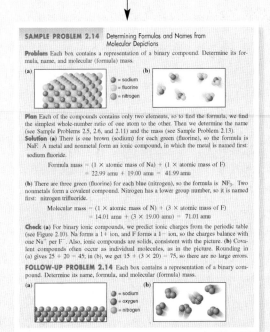

Brief Solutions to Follow-up Problems

These provide multistep solutions at the end of each chapter, not just an answer at the back of the book. This fuller treatment is an excellent way for you to reinforce your problem-solving skills.

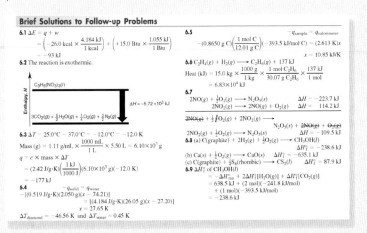

Three-Level Illustrations

A Silberberg hallmark, these illustrations provide macroscopic and molecular views of a process that help you connect these two levels of reality with each other and with the chemical equation that describes the process in symbols.

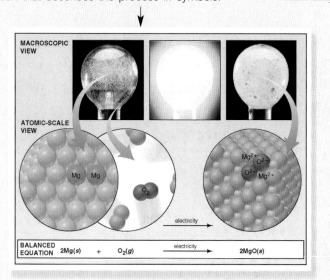

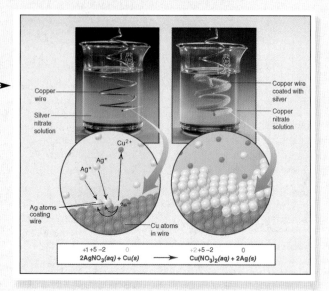

Cutting-Edge Molecular Models

Author and artist worked side by side and employed the most advanced computer-graphic software to provide accurate molecular-scale models and vivid scenes.

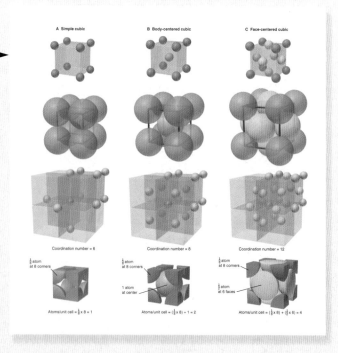

For Review and Reference

A rich catalog of study aids ends each chapter to help you review its content:

- **Learning Objectives** are listed, with section, sample problem, and end-of-chapter problem numbers, to focus you on key concepts and skills.
- **Key Terms** are boldfaced within the chapter and listed here by section (with page numbers); they are defined again in the Glossary.
- **Key Equations and Relationships** are screened and numbered within the chapter and listed here with page numbers.

[Reproduced page 100 excerpt:]

100   CHAPTER 3 Stoichiometry of Formulas and Equations

For Review and Reference [Numbers in parentheses refer to pages, unless noted otherwise.]

Learning Objectives

To help you review these learning objectives, the numbers of related sections (§), sample problems (SP), and upcoming end-of-chapter problems (EP) are listed in parentheses.

1. Realize the usefulness of the mole concept, and use the relation between molecular (or formula) mass and molar mass to calculate the molar mass of any substance (§ 3.1) (EPs 3.1–3.5, 3.7–3.10)
2. Understand the relationships among amount of substance (in moles), mass (in grams), and number of chemical entities and convert from one to any other (§ 3.1) (SPs 3.1, 3.2) (EPs 3.6, 3.11–3.16, 3.19)
3. Use mass percent to find the mass of element in a given mass of compound (§ 3.3) (SP 3.3) (EPs 3.17, 3.18, 3.20–3.23)
4. Determine the empirical and molecular formulas of a compound from mass analysis of its elements (§ 3.2) (SPs 3.4–3.6) (EPs 3.24–3.34)

5. Balance an equation given formulas or names, and use molar ratios to calculate amounts of reactants and products for reactions of pure or dissolved substances (§ 3.3 and 3.5) (SPs 3.7, 3.8, 3.15) (EPs 3.35–3.46, 3.62, 3.71, 3.72)
6. Understand why one reactant limits the yield of product, and solve limiting-reactant problems for reactions of pure or dissolved substances (§ 3.4, 3.5) (SPs 3.9, 3.10, 3.16) (EPs 3.47–3.54, 3.61, 3.73, 3.74)
7. Explain the reasons for lower-than-expected yields and the distinction between theoretical and actual yields, and calculate percent yield (§ 3.4) (SP 3.11) (EPs 3.55–3.60, 3.63)
8. Understand the meaning of concentration and the effect of dilution, and calculate molarity or mass of dissolved solute (§ 3.5) (SPs 3.12–3.14) (EPs 3.64–3.70, 3.75)

Key Terms

stoichiometry (70)	**Section 3.3**	**Section 3.4**	**Section 3.5**
Section 3.1	chemical equation (83)	limiting reactant (90)	solute (95)
mole (mol) (70)	reactant (83)	theoretical yield (93)	solvent (95)
Avogadro's number (70)	product (83)	side reaction (93)	concentration (95)
molar mass (M) (72)	balancing (stoichiometric)	actual yield (93)	molarity (M) (95)
Section 3.2	coefficient (83)	percent yield (% yield) (94)	
combustion analysis (80)			
isomer (81)			

Key Equations and Relationships

3.1 Number of entities in one mole (70):
1 mole contains 6.022×10^{23} entities (to 4 sf)

3.2 Converting amount (mol) to mass using M (73):
$$\text{Mass (g)} = \text{no. of moles} \times \frac{\text{no. of grams}}{1 \text{ mol}}$$

3.3 Converting mass to amount (mol) using 1/M (73):
$$\text{No. of moles} = \text{mass (g)} \times \frac{1 \text{ mol}}{\text{no. of grams}}$$

3.4 Converting amount (mol) to number of entities (73):
$$\text{No. of entities} = \text{no. of moles} \times \frac{6.022 \times 10^{23} \text{ entities}}{1 \text{ mol}}$$

3.5 Converting number of entities to amount (mol) (73):
$$\text{No. of moles} = \text{no. of entities} \times \frac{1 \text{ mol}}{6.022 \times 10^{23} \text{ entities}}$$

3.6 Calculating mass % (75):
Mass % of element X
$$= \frac{\text{moles of X in formula} \times \text{molar mass of X (g/mol)}}{\text{mass (g) of 1 mol of compound}} \times 100$$

3.7 Calculating percent yield (94):
$$\% \text{ yield} = \frac{\text{actual yield}}{\text{theoretical yield}} \times 100$$

3.8 Defining molarity (95):
$$\text{Molarity} = \frac{\text{moles of solute}}{\text{liters of solution}} \quad \text{or} \quad M = \frac{\text{mol solute}}{\text{L soln}}$$

3.9 Diluting a concentrated solution (97):
$$M_{dil} \times V_{dil} = \text{number of moles} = M_{conc} \times V_{conc}$$

Brief Solutions to Follow-up Problems

3.1 (a) Moles of C = 315 mg C $\times \frac{1 \text{ g}}{10^3 \text{ mg}} \times \frac{1 \text{ mol C}}{12.01 \text{ g C}}$
= 2.62×10^{-2} mol C
(b) Mass (g) of Mn = 3.22×10^{20} Mn atoms
$\times \frac{1 \text{ mol Mn}}{6.022 \times 10^{23} \text{ Mn atoms}} \times \frac{54.94 \text{ g Mn}}{1 \text{ mol Mn}}$
= 2.94×10^{-2} g Mn

3.2 (a) Mass (g) of P_4O_{10}
= 4.65×10^{22} molecules P_4O_{10}
$\times \frac{1 \text{ mol } P_4O_{10}}{6.022 \times 10^{23} \text{ molecules } P_4O_{10}} \times \frac{283.88 \text{ g } P_4O_{10}}{1 \text{ mol } P_4O_{10}}$
= 21.9 g P_4O_{10}

[Reproduced page 387 excerpt:]

Problems   387

12.66 Of the five major types of crystalline solid, which does each of the following form: (a) Sn; (b) Si; (c) Xe?
12.67 Of the five major types of crystalline solid, which does each of the following form: (a) cholesterol ($C_{27}H_{45}OH$); (b) KCl; (c) BN?
12.68 Zinc oxide adopts the zinc blende crystal structure (Figure P12.68). How many Zn^{2+} ions are in the ZnO unit cell?

O^{2-}
Zn^{2+}

12.69 Calcium sulfide adopts the sodium chloride crystal structure (Figure P12.69). How many S^{2-} ions are in the CaS unit cell?

S^{2-}
Ca^{2+}

12.70 Zinc selenide (ZnSe) crystallizes in the zinc blende structure and has a density of 5.42 g/cm³ (see Figure P12.68).
(a) How many Zn and Se ions are in each unit cell?
(b) What is the mass of a unit cell?
(c) What is the volume of a unit cell?
(d) What is the edge length of a unit cell?
12.71 An element crystallizes in a face-centered cubic lattice and has a density of 1.45 g/cm³. The edge of its unit cell is 4.52×10^{-8} cm.
(a) How many atoms are in each unit cell?
(b) What is the volume of a unit cell?
(c) What is the mass of a unit cell?
(d) Calculate an approximate atomic mass for the element.
12.72 Classify each of the following as a conductor, insulator, or semiconductor: (a) phosphorus; (b) mercury; (c) germanium.
12.73 Predict the effect (if any) of an increase in temperature on the electrical conductivity of (a) antimony, Sb; (b) tellurium, Te; (c) bismuth, Bi.
12.74 Use condensed electron configurations to predict the relative hardnesses and melting points of rubidium (Z = 37), vanadium (Z = 23), and cadmium (Z = 48).
12.75 One of the most important enzymes in the world—nitrogenase, the plant protein that catalyzes nitrogen fixation—contains active clusters of iron, sulfur, and molybdenum atoms. Crystalline molybdenum (Mo) has a body-centered cubic unit cell (d of Mo = 10.28 g/cm³). (a) Determine the edge length of the unit cell. (b) Calculate the atomic radius of Mo.

Comprehensive Problems
Problems with an asterisk (*) are more challenging.

12.76 Because bismuth has several well-characterized solid, crystalline phases, it is used to calibrate instruments employed in high-pressure studies. The following phase diagram for bismuth shows the liquid phase and five different solid phases stable

above 1 katm (1000 atm) and up to 300°C. (a) Which solid phases are stable at 25°C? (b) Which phase is stable at 50 katm and 175°C? (c) Identify the phase transitions that bismuth undergoes at 200°C as the pressure is reduced from 100 to 1 katm. (d) What phases are present at each of the triple points?

12.77 Mercury (Hg) vapor is toxic and readily absorbed from the lungs. At 20.°C, mercury ($\Delta H_{vap} = 59.1$ kJ/mol) has a vapor pressure of 1.20×10^{-3} torr, which is high enough to be hazardous. To reduce the danger to workers in processing plants, Hg is cooled to lower its vapor pressure. At what temperature would the vapor pressure of Hg be at the safer level of 5.0×10^{-5} torr?
12.78 Consider the phase diagram shown for substance X. (a) What phase(s) is (are) present at point A? E? F? H? B? C? (b) Which point corresponds to the critical point? Which point corresponds to the triple point? (c) What curve corresponds to conditions at which the solid and gas are in equilibrium? (d) Describe what happens when you start at point A and increase the temperature at constant pressure. (e) Describe what happens when you start at point H and decrease the pressure at constant temperature. (f) Is liquid X more or less dense than solid X?

12.79 Some high-temperature superconductors adopt a crystal structure similar to that of *perovskite* (CaTiO₃). The unit cell is cubic with a Ti^{4+} ion in each corner, a Ca^{2+} ion in the body center, and O^{2-} ions at the midpoint of each edge. (a) Is this unit cell simple, body-centered, or face-centered? (b) If the unit cell edge length is 3.84 Å, what is the density of perovskite (in g/cm³)?
12.80 The only alkali metal halides that do not adopt the NaCl structure are CsCl, CsBr, and CsI, formed from the largest alkali metal cation and the three largest halide ions. These crystallize in the *cesium chloride structure* (shown here for CsCl). This structure has been used as an example of how dispersion forces can dominate in the presence of ionic forces. Use the ideas of coordination number and polarizability to explain why the CsCl structure exists.

Cl^-
Cs^+

End-of-Chapter Problems

The numerous problems that end each chapter are sorted by section. Many are grouped in similar pairs, and the answer to one of each pair appears in Appendix E. Following these section-based problems is a large group of comprehensive problems, which are based on concepts and skills from any section and/or earlier chapter and are filled with applications from related sciences. Especially challenging problems are indicated with an asterisk.

Section Summaries

Concise summary paragraphs conclude each section, immediately restating the major ideas just covered.

SECTION SUMMARY
Surface tension is a measure of the energy required to increase a liquid's surface area. Greater intermolecular forces within a liquid create higher surface tension. Capillary action, the rising of a liquid through a narrow space, occurs when the forces between a liquid and a solid surface (adhesive) are greater than those within the liquid itself (cohesive). Viscosity, the resistance to flow, depends on molecular shape and decreases with temperature. Stronger intermolecular forces create higher viscosity.

Supplements for the Instructor

MULTIMEDIA SUPPLEMENTS

Digital Content Manager

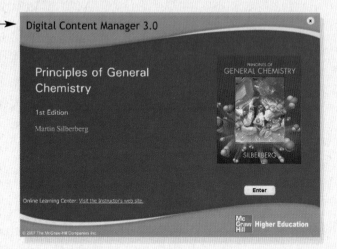

Electronic art at your fingertips! This cross-platform product provides you with artwork from the text in multiple formats. You can easily create customized classroom presentations, visually based tests and quizzes, dynamic content for a course website, or attractive printed support materials. Available on CD-ROM or DVD are the following resources:

- **Active Art Library** These key art pieces—formatted as PowerPoint slides—illustrate difficult concepts in a step-by-step manner. The artwork is broken into small, incremental frames, allowing you to incorporate the pieces into your lecture in whatever sequence or format you desire.
- **PowerPoint Lecture Outlines** Ready-made presentations—combining art and lecture notes—cover all of the chapters in the text. These lectures can be used as is or customized by you to meet your specific needs.
- **Art and Photo Library** Full-color digital files of all of the illustrations and many of the photos in the text can be readily incorporated into lecture presentations, exams, or custom-made classroom materials.
- **Worked Example Library** and **Table Library** Access the worked examples and visual tables from the text in electronic format for inclusion in your classroom presentations or materials.

Chemistry Animations DVD

This DVD contains more than 300 animations, several authored by Martin Silberberg. This easy-to-use DVD allows you to view the animations quickly and import them into PowerPoint to create multimedia presentations.

ARIS

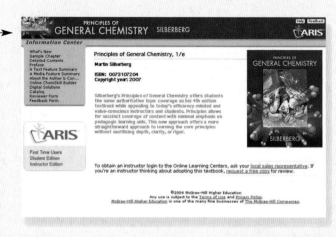

McGraw-Hill's Assessment, Review, and Instruction System for *Principles of General Chemistry* is a complete electronic homework and course management system, designed for greater ease of use than any other system available. Instructors can create and share course materials and assignments with colleagues with a few clicks of the mouse. Instructors can edit questions, import their own content, and create announcements and due dates for assignments. ARIS has automatic grading and reporting of easy-to-assign homework, quizzing, and testing. Once a student is registered in the course, all student activity within ARIS is automatically recorded and available to the instructor through a fully integrated grade book that can be downloaded to Excel. This book-specific website is found at www.mhhe.com/silberberg, and contains many useful tools for empowering both students and instructors.

- Most assignments and questions are directly tied to text-specific materials in *Principles of General Chemistry,* but you can edit questions and algorithms, import your own content, and create announcements and due dates for assignments.

- A secured Instructor Center stores your course materials, saving you preparation time.
- ARIS provides you with access to these essential instructor resources for each chapter: PowerPoint lecture outlines, *Instructor's Manual,* and animations.
- *ChemSkill Builder* McGraw-Hill's powerful electronic homework system, gives students the tutorial practice they need to master concepts covered in your general chemistry course. *ChemSkill Builder* contains more than 1500 algorithmically generated questions as well as interactive exercises, quizzes, animations, and study tools that correlate directly with each chapter of the text. A record of student work is maintained in an online gradebook so that homework can be easily assigned and factored into the course syllabus.

Course Management Software

With help from **Blackboard** or **WebAssign,** you can take complete control over your course content. These course cartridges also feature online testing and powerful student tracking. The *Principles of General Chemistry* Online Learning Center is available within either of these platforms. Contact your McGraw-Hill sales representative for more details.

Instructor's Testing and Resource CD-ROM

This cross-platform CD-ROM includes the *Instructor's Solutions Manual,* which provides all answers for the textbook's end-of-chapter problems, and the Test Bank, which offers additional questions that can be used for homework assignments and/or exams; both are available in Word and PDF formats. The computerized Test Bank utilizes testing software to allow you to quickly create customized exams by sorting questions by format, editing existing questions, adding new ones, and scrambling questions for multiple versions of the same test.

Instructor's Solutions Manual

By Patricia Amateis of Virginia Tech
This supplement contains complete, worked-out solutions for all the end-of-chapter problems in the text. It can be found within the secure Instructor's Center, within the Online Learning Center.

PRINTED SUPPLEMENTS

Transparencies

This boxed set of 300 full-color transparency acetates features images from the text that are modified to ensure maximum readability in both small and large classroom settings.

Primis LabBase

By Joseph Lagowski of University of Texas at Austin
More than 40 general chemistry lab experiments are available in this database collection, some from the *Journal of Chemical Education* and others provided by Professor Lagowski, enabling you to create your own custom laboratory manual.

General Chemistry Laboratory Manual

By Petra A. M. van Koppen of University of California, Santa Barbara
This definitive lab manual for the two-semester general chemistry course contains 21 experiments that cover the most commonly assigned experiments for the introductory level.

Cooperative Chemistry Laboratory Manual

By Melanie Cooper of Clemson University
This innovative guide features open-ended problems designed to simulate experience in a research lab. Working in groups, students investigate one problem over a period of several weeks, thus completing three or four projects during the semester, rather than one preprogrammed experiment per class. The emphasis here is on experimental design, analysis, problem solving, and communication.

Learning Aids for Students

MULTIMEDIA SUPPLEMENTS

ARIS for *Principles of General Chemistry* is your online source page for help. Text-specific features to complement and solidify lecture concepts include:

- Online homework and quizzes (which are automatically graded and recorded for your instructor)
- Study tools that relate directly to each chapter of the text

This book-specific website is found at www.mhhe.com/silberberg.

ChemSkill Builder, McGraw-Hill's powerful electronic homework system, gives you the tutorial practice you need to master concepts covered in your general chemistry course. *ChemSkill Builder* contains more than 1500 algorithmically generated questions as well as interactive exercises, quizzes, animations, and study tools matched to each chapter of the text. A record of your work is maintained in an online gradebook so that your homework scores can be easily viewed.

PRINTED SUPPLEMENTS

Student Solutions Manual

By Patricia Amateis of Virginia Tech
This supplement contains detailed solutions and explanations for all even-numbered problems in the main text.

Chemistry Resource Card

The resource card is a quick and easy source of information on general chemistry. Without having to consult the text, you have right at hand the periodic table and list of elements, tables for conversion factors, equilibrium and thermodynamic data, nomenclature, and key equations.

CHAPTER ONE

Keys to the Study of Chemistry

Unlocking the Door Learning the principles of chemistry opens your mind to an amazing world a billion times smaller than the one you see every day. This chapter introduces some key ideas and skills that prepare you to enter this new level of reality.

Key Principles

◆ *Matter* can undergo two kinds of change: a *physical change* involves a change in *state*—gas, liquid, or solid—but not in ultimate makeup (*composition*); a *chemical change* (*reaction*) is more fundamental because it does involve a change in composition. The changes we observe result from changes too small to observe.

◆ *Energy* occurs in two forms that are interconvertible. When oppositely charged particles are pulled apart, their *potential energy* increases; when they are released, that increase in potential energy is converted to *kinetic energy* as the particles move together again. Matter consists of charged particles, so changes in energy accompany changes in matter.

◆ *Scientific thinking* involves making *observations* and gathering *data* to develop *hypotheses*. *Controlled experiments* then test hypotheses until enough results are obtained to create a *model* (*theory*) that explains the observed phenomena. A sound theory can predict events but must be changed if new experimental results conflict with it.

◆ *Conversion factors* are ratios of equivalent measured quantities expressed in different units; they are used in calculations to change the units of a quantity.

◆ *Measured quantities,* such as mass, volume, and temperature, are expressed in numbers and *units*. *Decimal prefixes* and *exponential notation* are used to handle very large or very small quantities. Every measurement has some uncertainty, which is indicated by the number of *significant figures*. We *round* the final answer of a calculation to the same number of digits as the least certain measurement.

◆ *Extensive properties,* such as mass, depend on sample size; *intensive properties,* such as temperature, do not.

◆ *Accuracy* refers to how close a measurement is to the true value; *precision* refers to how close one measurement is to another.

Outline

Concepts & Skills to Review
Before You Study This Chapter

- exponential (scientific) notation
  (Appendix A)

The science of chemistry stands at the forefront of change in the 21st century, creating "greener" energy sources to power society *and* sustain the environment, using breakthrough knowledge of the human genetic makeup to understand disease and design medicines, and even researching the origin of life as we investigate our and nearby solar systems. Addressing these and countless other challenges and opportunities in biology, engineering, and environmental science depends on an understanding of the concepts you will learn in this course.

The impact of chemistry on your daily life is mind-boggling. Consider the beginning of a typical day—perhaps this one—from a chemical point of view. Molecules align in the liquid crystal display of your clock, electrons flow through its circuitry to create a noise, and you throw off a thermal insulator of manufactured polymer. You jump in the shower to emulsify fatty substances on your skin and hair with treated water and formulated detergents. You adorn yourself in an array of processed chemicals—pleasant-smelling pigmented materials suspended in cosmetic gels, dyed polymeric fibers, synthetic footwear, and metal-alloyed jewelry. Breakfast is a bowl of nutrient-enriched, spoilage-retarded cereal and milk, a piece of fertilizer-grown, pesticide-treated fruit, and a cup of a hot aqueous solution of neurally stimulating alkaloid. After brushing your teeth with artificially flavored, dental-hardening agents dispersed in a colloidal abrasive, you're ready to leave. You grab your laptop, an electronic device based on ultrathin, microetched semiconductor layers powered by a series of voltaic cells, collect some books—processed cellulose and plastic, electronically printed with light- and oxygen-resistant inks—hop in your hydrocarbon-fueled, metal-vinyl-ceramic vehicle, electrically ignite a synchronized series of controlled gaseous explosions, and you're off to class!

This course comes with a bonus—the development of two mental skills you can apply to any science-related field. The first is the ability to solve quantitative problems systematically. The second is specific to chemistry: as you comprehend its ideas, your mind's eye will learn to see a hidden level of the universe, one filled with incredibly minute particles hurtling at fantastic speeds, colliding billions of times a second, and interacting in ways that determine everything inside and outside of you. The first chapter holds the keys to help you enter this new world.

1.1   SOME FUNDAMENTAL DEFINITIONS

The science of chemistry deals with the makeup of the entire physical universe. A good place to begin our discussion is with the definition of a few central ideas, some of which may already be familiar to you. **Chemistry** is *the study of matter and its properties, the changes that matter undergoes, and the energy associated with those changes.*

The Properties of Matter

Matter is the "stuff" of the universe: air, glass, planets, students—*anything that has mass and volume.* (In Section 1.4, we discuss the meanings of mass and volume in terms of how they are measured.) Chemists are particularly interested in the **composition** of matter, *the types and amounts of simpler substances that make it up.* A *substance* is a type of matter that has a defined, fixed composition.

We learn about matter by observing its **properties,** *the characteristics that give each substance its unique identity.* To identify a person, we observe such properties as height, weight, eye color, race, fingerprints, and, now, even a DNA fingerprint, until we arrive at a unique identification. To identify a substance, chemists observe two types of properties, physical and chemical, which are closely related to two types of change that matter undergoes. **Physical properties**

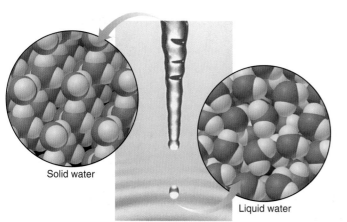

A Physical change:
Solid form of water becomes liquid form; composition does not change because particles are the same.

B Chemical change:
Electric current decomposes water into different substances (hydrogen and oxygen); composition does change because particles are different.

Figure 1.1 The distinction between physical and chemical change.

Animation: The Three States of Matter
Online Learning Center

are those that a substance shows *by itself, without changing into or interacting with another substance.* Some physical properties are color, melting point, electrical conductivity, and density.

A **physical change** occurs when a substance *alters its physical form, **not** its composition.* Thus, a physical change results in different physical properties. For example, when ice melts, several physical properties have changed, such as hardness, density, and ability to flow. But the sample has *not* changed its composition: it is still water. The photo in Figure 1.1A shows this change the way you would see it in everyday life. In your imagination, try to see the magnified view that appears in the "blow-up" circles. Here we see the particles that make up the sample; note that the same particles appear in solid and liquid water.
Physical change (same substance before and after):

$$\text{Water (solid form)} \longrightarrow \text{water (liquid form)}$$

On the other hand, **chemical properties** are those that a substance shows *as it changes into or interacts with another substance (or substances).* Some examples of chemical properties are flammability, corrosiveness, and reactivity with acids. A **chemical change,** also called a **chemical reaction,** occurs when *a substance (or substances) is converted into a different substance (or substances).*

Figure 1.1B shows the chemical change (reaction) that occurs when you pass an electric current through water: the water decomposes (breaks down) into two other substances, hydrogen and oxygen, each with physical and chemical properties different from each other *and* from water. The sample *has* changed its composition: it is no longer water, as you can see from the different particles in the magnified view.
Chemical change (different substances before and after):

$$\text{Water} \xrightarrow{\text{electric current}} \text{hydrogen gas} + \text{oxygen gas}$$

The Three States of Matter

Matter occurs commonly in three physical forms called **states:** solid, liquid, and gas. As shown in Figure 1.2 (on the next page) for a general substance, each state is defined by the way it fills a container. A **solid** *has a fixed shape that does not conform to the container shape.* A **liquid** *conforms to the container shape but fills the container only to the extent of the liquid's volume;* thus, a liquid forms a

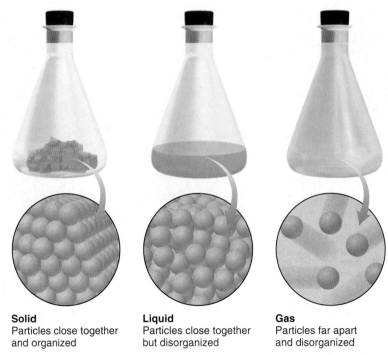

Solid
Particles close together
and organized

Liquid
Particles close together
but disorganized

Gas
Particles far apart
and disorganized

Figure 1.2 The physical states of matter. The magnified (blow-up) views show the atomic-scale arrangement of the particles in the three states of matter.

surface. A **gas** *conforms to the container shape also, but it fills the entire container,* and thus, does *not* form a surface. Now, look at the views within the blow-up circles of the figure. The particles in the solid lie next to each other in a regular, three-dimensional array with a definite pattern. Particles in the liquid also lie together but are jumbled and move randomly around one another. Particles in the gas usually have great distances between them, as they move randomly throughout the entire container.

Depending on the temperature and pressure of the surroundings, many substances can exist in each of the three physical states and undergo changes in state as well. For example, as the temperature increases, solid water melts to liquid water, which boils to gaseous water (also called *water vapor*). Similarly, with decreasing temperature, water vapor condenses to liquid water, and with further cooling, the liquid freezes to ice. Benzene, iron, nitrogen, and many other substances behave similarly.

Thus, a physical change caused by heating can generally be reversed by cooling, and vice versa. This is *not* generally true for a chemical change. For example, heating iron in moist air causes a chemical reaction that yields the brown, crumbly substance known as rust. Cooling does not reverse this change; rather, another chemical change (or series of them) is required.

To summarize the key distinctions:

- A physical change leads to a different form of the same substance (same composition), whereas a chemical change leads to a different substance (different composition).
- A physical change caused by a temperature change can generally be reversed by the opposite temperature change, but this is not generally true of a chemical change.

The following sample problem provides practice identifying these types of changes.

SAMPLE PROBLEM 1.1 Distinguishing Between Physical and Chemical Change

Problem Decide whether each of the following processes is primarily a physical or a chemical change, and explain briefly:
(a) Frost forms as the temperature drops on a humid winter night.
(b) A cornstalk grows from a seed that is watered and fertilized.
(c) Dynamite explodes to form a mixture of gases.
(d) Perspiration evaporates when you relax after jogging.
(e) A silver fork tarnishes slowly in air.
Plan The basic question we ask to decide whether a change is chemical or physical is, "Does the substance change composition or just change form?"
Solution (a) Frost forming is a physical change: the drop in temperature changes water vapor (gaseous water) in humid air to ice crystals (solid water).
(b) A seed growing involves chemical change: the seed uses substances from air, fertilizer, soil, and water, and energy from sunlight to make complex changes in composition.
(c) Dynamite exploding is a chemical change: the dynamite is converted into other substances.
(d) Perspiration evaporating is a physical change: the water in sweat changes its form, from liquid to gas, but not its composition.
(e) Tarnishing is a chemical change: silver changes to silver sulfide by reacting with sulfur-containing substances in the air.

FOLLOW-UP PROBLEM 1.1 Decide whether each of the following processes is primarily a physical or a chemical change, and explain briefly:
(a) Purple iodine vapor appears when solid iodine is warmed.
(b) Gasoline fumes are ignited by a spark in an automobile engine cylinder.
(c) A scab forms over an open cut.

The Central Theme in Chemistry

Understanding the properties of a substance and the changes it undergoes leads to the central theme in chemistry: *macroscopic* properties and behavior, those we can see, are the results of *submicroscopic* properties and behavior that we cannot see. The distinction between chemical and physical change is defined by composition, which we study macroscopically. But it ultimately depends on the makeup of substances at the atomic scale, as the magnified views of Figure 1.1 show. Similarly, the defining properties of the three states of matter are macroscopic, but they arise from the submicroscopic behavior shown in the magnified views of Figure 1.2. Picturing a chemical scene on the molecular scale clarifies what is taking place. What is happening when water boils or copper melts? What events occur in the invisible world of minute particles that cause a seed to grow, a neon light to glow, or a nail to rust? Throughout the text, we return to this central idea: we study *observable* changes in matter to understand their *unobservable* causes.

The Importance of Energy in the Study of Matter

In general, physical and chemical changes are accompanied by energy changes. **Energy** is often defined as *the ability to do work.* Essentially, all work involves moving something. Work is done when your arm lifts a book, when an engine moves a car's wheels, or when a falling rock moves the ground as it lands. The object doing the work (arm, engine, rock) transfers some of the energy it possesses to the object on which the work is done (book, wheels, ground).

The total energy an object possesses is the sum of its potential energy and its kinetic energy. **Potential energy** is the *energy due to the **position** of the object.* **Kinetic energy** is the *energy due to the **motion** of the object.* Let's examine four systems that illustrate the relationship between these two forms of energy: (1) a weight raised above the ground, (2) two balls attached by a spring, (3) two electrically charged particles, and (4) a fuel and its waste products. A key concept

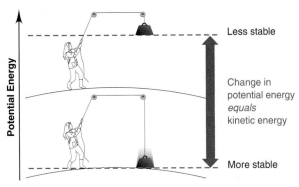

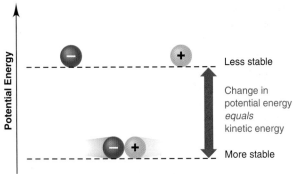

A **A gravitational system.** The potential energy gained when a weight is lifted is converted to kinetic energy as the weight falls.

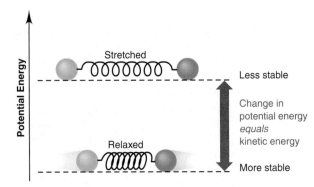

B **A system of two balls attached by a spring.** The potential energy gained when the spring is stretched is converted to the kinetic energy of the moving balls when it is released.

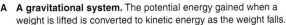

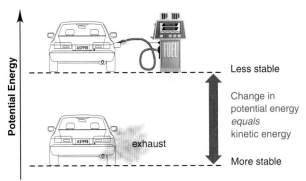

C **A system of oppositely charged particles.** The potential energy gained when the charges are separated is converted to kinetic energy as the attraction pulls them together.

D **A system of fuel and exhaust.** A fuel is higher in chemical potential energy than the exhaust. As the fuel burns, some of its potential energy is converted to the kinetic energy of the moving car.

Figure 1.3 Potential energy is converted to kinetic energy. In all four parts of the figure, the dashed horizontal lines indicate the potential energy of the system in each situation.

illustrated by all four cases is that *energy is conserved: it may be converted from one form to the other, but it is not destroyed.*

Suppose you lift a weight off the ground, as in Figure 1.3A. The energy you use to move the weight against the gravitational attraction of Earth increases the weight's potential energy (energy due to its position). When the weight is dropped, this additional potential energy is converted to kinetic energy (energy due to motion). Some of this kinetic energy is transferred to the ground as the weight does work, such as driving a stake or simply moving dirt and pebbles. As you can see, the added potential energy does not disappear: it is converted to kinetic energy.

In nature, situations of lower energy are typically favored over those of higher energy: because the weight has less potential energy (and thus less total energy) at rest on the ground than held in the air, it will fall when released. Therefore, the situation with the weight elevated and higher in potential energy is *less stable,* and the situation after the weight has fallen and is lower in potential energy is *more stable.*

Next, consider the two balls attached by a relaxed spring in Figure 1.3B. When you pull the balls apart, the energy you exert to stretch the spring increases its potential energy. This change in potential energy is converted to kinetic energy when you release the balls and they move closer together. The system of balls and spring is less stable (has more potential energy) when the spring is stretched than when it is relaxed.

There are no springs in a chemical substance, of course, but the following situation is similar in terms of energy. Much of the matter in the universe is composed of positively and negatively charged particles. A well-known behavior of charged particles (similar to the behavior of the poles of magnets) results from interactions known as *electrostatic forces: opposite charges attract each other, and like charges repel each other.* When work is done to separate a positive particle from a negative one, the potential energy of the particles increases. As Figure 1.3C shows, that increase in potential energy is converted to kinetic energy when the particles move together again. Also, when two positive (or two negative) particles are pushed toward each other, their potential energy increases, and when they are allowed to move apart, that increase in potential energy is changed into kinetic energy. Like the weight above the ground and the balls connected by a spring, charged particles move naturally toward a position of lower energy, which is more stable.

The chemical potential energy of a substance results from the relative positions and the attractions and repulsions among all its particles. Some substances are richer in this chemical potential energy than others. Fuels and foods, for example, contain more potential energy than the waste products they form. Figure 1.3D shows that when gasoline burns in a car engine, substances with higher chemical potential energy (gasoline and air) form substances with lower potential energy (exhaust gases). This difference in potential energy is converted into the kinetic energy that moves the car, heats the passenger compartment, makes the lights shine, and so forth. Similarly, the difference in potential energy between the food and air we take in and the waste products we excrete is used to move, grow, keep warm, study chemistry, and so on. Note again the essential point: *energy is neither created nor destroyed—it is always conserved as it is converted from one form to the other.*

SECTION SUMMARY
Chemists study the composition and properties of matter and how they change. Each substance has a unique set of physical properties (attributes of the substance itself) and chemical properties (attributes of the substance as it interacts with or changes to other substances). Changes in matter can be physical (different form of the same substance) or chemical (different substance). Matter exists in three physical states—solid, liquid, and gas. The observable features that distinguish these states reflect the arrangement of their particles. A change in physical state brought about by heating may be reversed by cooling. A chemical change can be reversed only by other chemical changes. Macroscopic changes result from submicroscopic changes.

Changes in matter are accompanied by changes in energy. An object's potential energy is due to its position; an object's kinetic energy is due to its motion. Chemical potential energy arises from the positions and interactions of the particles in a substance. Higher energy substances are less stable than lower energy substances. When a less stable substance is converted into a more stable substance, some potential energy is converted into kinetic energy, which can do work.

1.2 THE SCIENTIFIC APPROACH: DEVELOPING A MODEL

The principles of chemistry have been modified over time and are *still* evolving. At the dawn of human experience, our ancestors survived through knowledge acquired by *trial and error:* which types of stone were hard enough to shape others, which plants were edible, and so forth. Today, the science of chemistry, with its powerful *quantitative theories,* helps us understand the essential nature of materials to make better use of them and create new ones: specialized drugs, advanced composites, synthetic polymers, and countless other new materials (Figure 1.4, on the next page).

Is there something special about the way scientists think? If we could break down a "typical" modern scientist's thought processes, we could organize them into an approach called the **scientific method.** This approach is not a stepwise

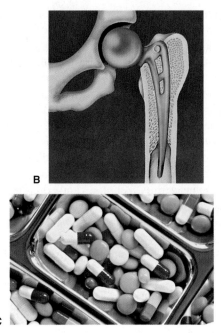

Figure 1.4 Modern materials in a variety of applications. A, Specialized steels in bicycles; synthetic polymers in clothing and helmets. **B,** High-tension polymers in synthetic hip joints. **C,** Medicinal agents in pills. **D,** Liquid crystal displays in electronic devices.

checklist, but rather a flexible process of creative thinking and testing aimed at objective, verifiable discoveries about how nature works. It is very important to realize that there is no typical scientist and no single method, and that luck can and often has played a key role in scientific discovery. In general terms, the scientific approach includes the following parts (Figure 1.5):

1. *Observations.* These are the facts that our ideas must explain. Observation is basic to scientific thinking. The most useful observations are quantitative because they can be compared and allow trends to be seen. Pieces of quantitative information are **data.** When the same observation is made by many investigators in many situations with no clear exceptions, it is summarized, often in mathematical terms, and called a **natural law.**

2. *Hypothesis.* Whether derived from actual observation or from a "spark of intuition," a hypothesis is a proposal made to explain an observation. A valid hypothesis need not be correct, but it must be *testable.* Thus, a hypothesis is often the reason for performing an experiment. If the hypothesis is inconsistent with the experimental results, it must be revised or discarded.

3. *Experiment.* An experiment is a clear set of procedural steps that tests a hypothesis. Often, hypothesis leads to experiment, which leads to revised

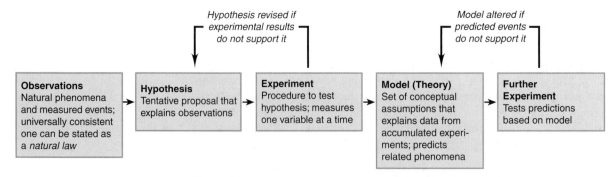

Figure 1.5 The scientific approach to understanding nature. Note that hypotheses and models are mental pictures that are changed to match observations and experimental results, *not* the other way around.

hypothesis, and so forth. Hypotheses can be altered, but the results of an experiment cannot.

An experiment typically contains at least two **variables,** quantities that can have more than a single value. A well-designed experiment is **controlled** in that it measures the effect of one variable on another while keeping all others constant. For experimental results to be accepted, they must be *reproducible,* not only by the person who designed the experiment, but also by others. Both skill and creativity play a part in experimental design.

4. *Model.* Formulating conceptual models, or **theories,** *based on experiments* is what distinguishes scientific thinking from speculation. As hypotheses are revised according to experimental results, a model gradually emerges that describes how the observed phenomenon occurs. A model is not an exact representation of nature, but rather a simplified version of it that can be used to make *predictions* about related phenomena. Further investigation refines a model by testing its predictions and altering it to account for new facts.

In an informal way, we often use a scientific approach in daily life. Consider this familiar scenario. While listening to an FM broadcast on your stereo system, you notice the sound is garbled (observation) and assume it is caused by poor reception (hypothesis). To isolate this variable, you play a CD (experiment): the sound is still garbled. If the problem is not poor reception, perhaps the speakers are at fault (new hypothesis). To isolate this variable, you play the CD and listen with headphones (experiment): the sound is clear. You conclude that the speakers need to be repaired (model). The repair shop says the speakers check out fine (new observation), but the power amplifier may be at fault (new hypothesis). Replacing a transistor in the amplifier corrects the garbled sound (new experiment), so the power amplifier was the problem (revised model). Approaching a problem scientifically is a common practice, even if you're not aware of it.

SECTION SUMMARY
The scientific method is not a rigid sequence of steps, but rather a dynamic process designed to explain and predict real phenomena. Observations (sometimes expressed as natural laws) lead to hypotheses about how or why something occurs. Hypotheses are tested in controlled experiments and adjusted if necessary. If all the data collected support a hypothesis, a model (theory) can be developed to explain the observations. A good model is useful in predicting related phenomena but must be refined if conflicting data appear.

1.3 CHEMICAL PROBLEM SOLVING

In many ways, learning chemistry is learning how to solve chemistry problems, not only those in exams or homework, but also more complex ones in professional life and society. In this section, we discuss the problem-solving approach. Most problems include calculations, so let's first go over some important ideas about measured quantities.

Units and Conversion Factors in Calculations

All measured quantities consist of a number *and* a unit; a person's height is "6 feet," not "6." Ratios of quantities have ratios of units, such as miles/hour. (We discuss the most important units in chemistry in the next section.) To minimize errors, try to make a habit of including units in all calculations.

The arithmetic operations used with measured quantities are the same as those used with pure numbers; in other words, units can be multiplied, divided, and canceled:

• A carpet measuring 3 feet (ft) by 4 ft has an area of

$$\text{Area} = 3 \text{ ft} \times 4 \text{ ft} = (3 \times 4)(\text{ft} \times \text{ft}) = 12 \text{ ft}^2$$

- A car traveling 350 miles (mi) in 7 hours (h) has a speed of

$$\text{Speed} = \frac{350 \text{ mi}}{7 \text{ h}} = \frac{50 \text{ mi}}{1 \text{ h}} \text{ (often written 50 mi·h}^{-1})$$

- In 3 hours, the car travels a distance of

$$\text{Distance} = 3 \text{ h} \times \frac{50 \text{ mi}}{1 \text{ h}} = 150 \text{ mi}$$

Conversion factors are ratios used to express a measured quantity in different units. Suppose we want to know the distance of that 150-mile car trip in feet. To convert the distance between miles and feet, we use equivalent quantities to construct the desired conversion factor. The equivalent quantities in this case are 1 mile and the number of feet in 1 mile:

$$1 \text{ mi} = 5280 \text{ ft}$$

We can construct two conversion factors from this equivalency. Dividing both sides by 5280 ft gives one conversion factor (shown in blue):

$$\frac{1 \text{ mi}}{5280 \text{ ft}} = \frac{5280 \text{ ft}}{5280 \text{ ft}} = 1$$

And, dividing both sides by 1 mi gives the other conversion factor (the inverse):

$$\frac{1 \text{ mi}}{1 \text{ mi}} = \frac{5280 \text{ ft}}{1 \text{ mi}} = 1$$

It's very important to see that, since the numerator and denominator of a conversion factor are equal, multiplying by a conversion factor is the same as multiplying by 1. Therefore, *even though the number and unit of the quantity change, the size of the quantity remains the same.*

In our example, we want to convert the distance in miles to the equivalent distance in feet. Therefore, we choose the conversion factor with units of feet in the numerator, because it cancels units of miles and gives units of feet:

$$\text{Distance (ft)} = 150 \text{ mi} \times \frac{5280 \text{ ft}}{1 \text{ mi}} = 792,000 \text{ ft}$$

$$\text{mi} \implies \text{ft}$$

Choosing the correct conversion factor is made much easier if you think through the calculation to decide whether the answer expressed in the new units should have a larger or smaller number. In the previous case, we know that a foot is *smaller* than a mile, so the distance in feet should have a *larger* number (792,000) than the distance in miles (150). The conversion factor has the larger number (5280) in the numerator, so it gave a larger number in the answer. The main goal is that *the chosen conversion factor cancels all units except those required for the answer.* Set up the calculation so that the unit you are converting *from* (beginning unit) is in the *opposite position in the conversion factor* (numerator or denominator). It will then cancel and leave the unit you are converting *to* (final unit):

$$\text{beginning unit} \times \frac{\text{final unit}}{\text{beginning unit}} = \text{final unit} \qquad \text{as in} \qquad \text{mi} \times \frac{\text{ft}}{\text{mi}} = \text{ft}$$

Or, in cases that involve units raised to a power,

$$(\text{beginning unit} \times \text{beginning unit}) \times \frac{\text{final unit}^2}{\text{beginning unit}^2} = \text{final unit}^2$$

$$\text{as in} \qquad (\text{ft} \times \text{ft}) \times \frac{\text{mi}^2}{\text{ft}^2} = \text{mi}^2$$

Or, in cases that involve a ratio of units,

$$\frac{\text{beginning unit}}{\text{final unit}_1} \times \frac{\text{final unit}_2}{\text{beginning unit}} = \frac{\text{final unit}_2}{\text{final unit}_1} \qquad \text{as in} \qquad \frac{\text{mi}}{\text{h}} \times \frac{\text{ft}}{\text{mi}} = \frac{\text{ft}}{\text{h}}$$

We use the same procedure to convert between systems of units, for example, between the English (or American) unit system and the International System (a revised metric system discussed fully in the next section). Suppose we know the height of Angel Falls in Venezuela to be 3212 ft, and we find its height in miles as

$$\text{Height (mi)} = 3212 \; \cancel{\text{ft}} \times \frac{1 \; \text{mi}}{5280 \; \cancel{\text{ft}}} = 0.6083 \; \text{mi}$$

$$\text{ft} \implies \text{mi}$$

Now, we want its height in kilometers (km). The equivalent quantities are

$$1.609 \; \text{km} = 1 \; \text{mi}$$

Because we are converting from miles to kilometers, we use the conversion factor with kilometers in the numerator in order to cancel miles:

$$\text{Height (km)} = 0.6083 \; \cancel{\text{mi}} \times \frac{1.609 \; \text{km}}{1 \; \cancel{\text{mi}}} = 0.9788 \; \text{km}$$

$$\text{mi} \implies \text{km}$$

Notice that, because kilometers are *smaller* than miles, this conversion factor gave us a *larger* number (0.9788 is larger than 0.6083).

If we want the height of Angel Falls in meters (m), we use the equivalent quantities 1 km = 1000 m to construct the conversion factor:

$$\text{Height (m)} = 0.9788 \; \cancel{\text{km}} \times \frac{1000 \; \text{m}}{1 \; \cancel{\text{km}}} = 978.8 \; \text{m}$$

$$\text{km} \implies \text{m}$$

In longer calculations, we often string together several conversion steps:

$$\text{Height (m)} = 3212 \; \cancel{\text{ft}} \times \frac{1 \; \cancel{\text{mi}}}{5280 \; \cancel{\text{ft}}} \times \frac{1.609 \; \cancel{\text{km}}}{1 \; \cancel{\text{mi}}} \times \frac{1000 \; \text{m}}{1 \; \cancel{\text{km}}} = 978.8 \; \text{m}$$

$$\text{ft} \implies \text{mi} \implies \text{km} \implies \text{m}$$

The use of conversion factors in calculations is known by various names, such as the factor-label method or **dimensional analysis** (because units represent physical dimensions). We use this method in quantitative problems throughout the text.

A Systematic Approach to Solving Chemistry Problems

The approach we use in this text provides a systematic way to work through a problem. It emphasizes reasoning, not memorizing, and is based on a very simple idea: plan how to solve the problem *before* you go on to solve it, and then check your answer. Try to develop a similar approach on homework and exams. In general, the sample problems consist of several parts:

1. **Problem.** This part states all the information you need to solve the problem (usually framed in some interesting context).
2. **Plan.** The overall solution is broken up into two parts, *plan* and *solution,* to make a point: *think* about how to solve the problem *before* juggling numbers. There is often more than one way to solve a problem, and the plan shown in a given problem is just one possibility; develop a plan that suits you best. The plan will
   - Clarify the known and unknown. (What information do you have, and what are you trying to find?)
   - Suggest the steps from known to unknown. (What ideas, conversions, or equations are needed to solve the problem?)
   - Present a "roadmap" of the solution for many problems in early chapters (and in some later ones). The roadmap is a visual summary of the planned steps. Each step is shown by an arrow labeled with information about the conversion factor or operation needed.

3. **Solution.** In this part, the steps appear in the same order as in the plan.

4. **Check.** In most cases, a quick check is provided to see if the results make sense: Are the units correct? Does the answer seem to be the right size? Did the change occur in the expected direction? Is it reasonable chemically? We often do a rough calculation to see if the answer is "in the same ballpark" as the calculated result, just to make sure we didn't make a large error. Here's a typical "ballpark" calculation. You are at the music store and buy three CDs at $14.97 each. With a 5% sales tax, the bill comes to $47.16. In your mind, you quickly check that 3 times approximately $15 is $45, and, given the sales tax, the cost should be a bit more. So, the amount of the bill *is* in the right ballpark. *Always* check your answers, especially in a multipart problem, where an error in an early step can affect all later steps.

5. **Comment.** This part is included occasionally to provide additional information, such as an application, an alternative approach, a common mistake to avoid, or an overview.

6. **Follow-up Problem.** This part consists of a problem statement only and provides practice by applying the same ideas as the sample problem. Try to solve it *before* you look at the brief worked-out solution at the end of the chapter.

Of course, you can't learn to solve chemistry problems, any more than you can learn to swim, by reading about doing it. Practice is the key. Try to:

- Follow along in the sample problem with pencil, paper, and calculator.
- Do the follow-up problem as soon as you finish studying the sample problem. Check your answer against the solution at the end of the chapter.
- Read the sample problem and text explanations again if you have trouble.
- Work on as many of the problems at the end of the chapter as you can. They review and extend the concepts and skills in the text. Answers are given in the back of the book for problems with a colored number, but try to solve them yourself first. Let's apply this approach in a unit-conversion problem.

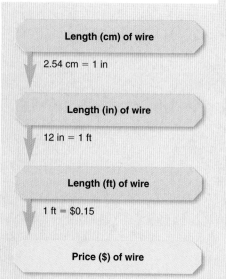

SAMPLE PROBLEM 1.2   Converting Units of Length

Problem To wire your stereo equipment, you need 325 centimeters (cm) of speaker wire that sells for $0.15/ft. What is the price of the wire?

Plan We know the length of wire in centimeters and the cost in dollars per foot ($/ft). We can find the unknown price of the wire by converting the length from centimeters to inches (in) and from inches to feet. Then the cost (1 ft = $0.15) gives us the equivalent quantities to construct the factor that converts feet of wire to price in dollars. The roadmap starts with the known and moves through the calculation steps to the unknown.

Solution Converting the known length from centimeters to inches: The equivalent quantities alongside the roadmap arrow are the ones needed to construct the conversion factor. We choose 1 in/2.54 cm, rather than the inverse, because it gives an answer in inches:

$$\text{Length (in)} = \text{length (cm)} \times \text{conversion factor} = 325 \, \text{cm} \times \frac{1 \, \text{in}}{2.54 \, \text{cm}} = 128 \, \text{in}$$

Converting the length from inches to feet:

$$\text{Length (ft)} = \text{length (in)} \times \text{conversion factor} = 128 \, \text{in} \times \frac{1 \, \text{ft}}{12 \, \text{in}} = 10.7 \, \text{ft}$$

Converting the length in feet to price in dollars:

$$\text{Price (\$)} = \text{length (ft)} \times \text{conversion factor} = 10.7 \, \text{ft} \times \frac{\$0.15}{1 \, \text{ft}} = \$1.60$$

Check The units are correct for each step. The conversion factors make sense in terms of the relative unit sizes: the number of inches is *smaller* than the number of centimeters (an inch is *larger* than a centimeter), and the number of feet is *smaller* than the number of inches. The total price seems reasonable: a little more than 10 ft of wire at $0.15/ft should cost a little more than $1.50.

Comment 1. We could also have strung the three steps together:

$$\text{Price ($)} = 325 \text{ cm} \times \frac{1 \text{ in}}{2.54 \text{ cm}} \times \frac{1 \text{ ft}}{12 \text{ in}} \times \frac{\$0.15}{1 \text{ ft}} = \$1.60$$

2. There are usually alternative sequences in unit-conversion problems. Here, for example, we would get the same answer if we first converted the cost of wire from $/ft to $/cm and kept the wire length in cm. Try it yourself.

FOLLOW-UP PROBLEM 1.2 A furniture factory needs 31.5 ft^2 of fabric to upholster one chair. Its Dutch supplier sends the fabric in bolts of exactly 200 m^2. What is the maximum number of chairs that can be upholstered by 3 bolts of fabric (1 m = 3.281 ft)?

SECTION SUMMARY
A measured quantity consists of a number and a unit. Conversion factors are used to express a quantity in different units and are constructed as a ratio of equivalent quantities. The problem-solving approach used in this text usually has four parts: (1) devise a plan for the solution, (2) put the plan into effect in the calculations, (3) check to see if the answer makes sense, and (4) practice with similar problems.

1.4 MEASUREMENT IN SCIENTIFIC STUDY

Almost everything we own—clothes, house, food, vehicle—is manufactured with measured parts, sold in measured amounts, and paid for with measured currency. Measurement has a history characterized by the search for *exact, invariable standards.* Our current system of measurement began in 1790, when the newly formed National Assembly of France set up a committee to establish consistent unit standards. This effort led to the development of the *metric system.* In 1960, another international committee met in France to establish the International System of Units, a revised metric system now accepted by scientists throughout the world. The units of this system are called **SI units,** from the French **S**ystème **I**nternational d'Unités.

General Features of SI Units

As Table 1.1 shows, the SI system is based on a set of seven **fundamental units,** or **base units,** each of which is identified with a physical quantity. All other units, called **derived units,** are combinations of these seven base units. For example, the derived unit for speed, meters per second (m/s), is the base unit for length (m) divided by the base unit for time (s). (Derived units that occur as a ratio of

Table 1.1 SI Base Units		
Physical Quantity (Dimension)	**Unit Name**	**Unit Abbreviation**
Mass	kilogram	kg
Length	meter	m
Time	second	s
Temperature	kelvin	K
Electric current	ampere	A
Amount of substance	mole	mol
Luminous intensity	candela	cd

Table 1.2 Common Decimal Prefixes Used with SI Units

Prefix*	Prefix Symbol	Word	Conventional Notation	Exponential Notation
tera	T	trillion	1,000,000,000,000	1×10^{12}
giga	G	billion	1,000,000,000	1×10^{9}
mega	M	million	1,000,000	1×10^{6}
kilo	k	thousand	1,000	1×10^{3}
hecto	h	hundred	100	1×10^{2}
deka	da	ten	10	1×10^{1}
—	—	one	1	1×10^{0}
deci	d	tenth	0.1	1×10^{-1}
centi	c	hundredth	0.01	1×10^{-2}
milli	m	thousandth	0.001	1×10^{-3}
micro	μ	millionth	0.000001	1×10^{-6}
nano	n	billionth	0.000000001	1×10^{-9}
pico	p	trillionth	0.000000000001	1×10^{-12}
femto	f	quadrillionth	0.000000000000001	1×10^{-15}

*The prefixes most frequently used by chemists appear in bold type.

two or more base units can be used as conversion factors.) For quantities that are much smaller or much larger than the base unit, we use decimal prefixes and exponential (scientific) notation. Table 1.2 shows the most important prefixes. (If you need a review of exponential notation, read Appendix A.) Because these prefixes are based on powers of 10, SI units are easier to use in calculations than are English units such as pounds and inches.

Some Important SI Units in Chemistry

Let's discuss some of the SI units for quantities that we use early in the text: length, volume, mass, density, temperature, and time. (Units for other quantities are presented in later chapters, as they are used.) Table 1.3 shows some useful SI quantities for length, volume, and mass, along with their equivalents in the English system.

Table 1.3 Common SI-English Equivalent Quantities

Quantity	SI	SI Equivalents	English Equivalents	English to SI Equivalent
Length	1 kilometer (km)	1000 (10^3) meters	0.6214 mile (mi)	1 mile = 1.609 km
	1 meter (m)	100 (10^2) centimeters	1.094 yards (yd)	1 yard = 0.9144 m
		1000 millimeters (mm)	39.37 inches (in)	1 foot (ft) = 0.3048 m
	1 centimeter (cm)	0.01 (10^{-2}) meter	0.3937 inch	1 inch = 2.54 cm (exactly)
Volume	1 cubic meter (m^3)	1,000,000 (10^6) cubic centimeters	35.31 cubic feet (ft^3)	1 cubic foot = 0.02832 m^3
	1 cubic decimeter (dm^3)	1000 cubic centimeters	0.2642 gallon (gal)	1 gallon = 3.785 dm^3
			1.057 quarts (qt)	1 quart = 0.9464 dm^3
				1 quart = 946.4 cm^3
	1 cubic centimeter (cm^3)	0.001 dm^3	0.03381 fluid ounce	1 fluid ounce = 29.57 cm^3
Mass	1 kilogram (kg)	1000 grams	2.205 pounds (lb)	1 pound = 0.4536 kg
	1 gram (g)	1000 milligrams (mg)	0.03527 ounce (oz)	1 ounce = 28.35 g

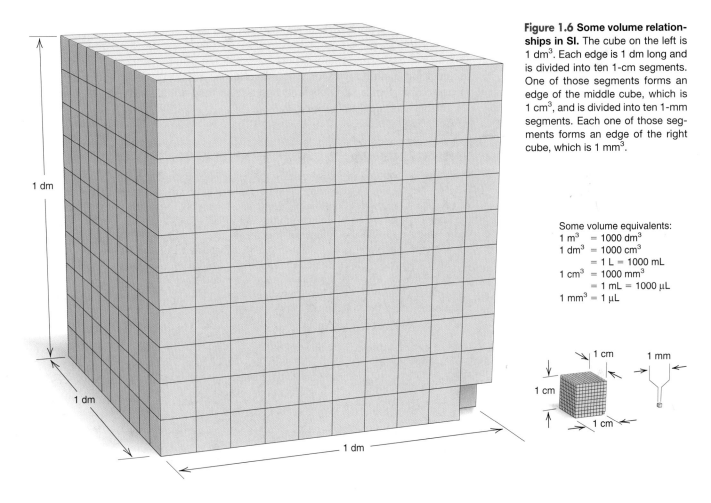

Figure 1.6 Some volume relation-ships in SI. The cube on the left is 1 dm³. Each edge is 1 dm long and is divided into ten 1-cm segments. One of those segments forms an edge of the middle cube, which is 1 cm³, and is divided into ten 1-mm segments. Each one of those segments forms an edge of the right cube, which is 1 mm³.

Some volume equivalents:
$1\ m^3 = 1000\ dm^3$
$1\ dm^3 = 1000\ cm^3$
$= 1\ L = 1000\ mL$
$1\ cm^3 = 1000\ mm^3$
$= 1\ mL = 1000\ \mu L$
$1\ mm^3 = 1\ \mu L$

Length The SI base unit of length is the **meter (m).** The standard meter is defined as the distance light travels in a vacuum in 1/299,792,458 second. Biological cells are often measured in micrometers (1 μm $= 10^{-6}$ m). On the atomic-size scale, nanometers and picometers are used (1 nm $= 10^{-9}$ m; 1 pm $= 10^{-12}$ m). Many proteins have diameters of around 2 nm; atomic diameters are around 200 pm (0.2 nm). An older unit still in use is the angstrom (1 Å $= 10^{-10}$ m $= 0.1$ nm $= 100$ pm).

Volume Any sample of matter has a certain **volume (V),** the amount of space that the sample occupies. The SI unit of volume is the **cubic meter (m³).** In chemistry, the most important volume units are non-SI units, the **liter (L)** and the **milliliter (mL)** (note the uppercase L). A liter is slightly larger than a quart (qt) (1 L = 1.057 qt; 1 qt = 946.4 mL). Physicians and other medical practitioners measure body fluids in cubic decimeters (dm³), which is equivalent to liters:

$$1\ L = 1\ dm^3 = 10^{-3}\ m^3$$

As the prefix *milli-* indicates, 1 mL is $\frac{1}{1000}$ of a liter, and it is equal to exactly 1 cubic centimeter (cm³):

$$1\ mL = 1\ cm^3 = 10^{-3}\ dm^3 = 10^{-3}\ L = 10^{-6}\ m^3$$

Figure 1.6 is a life-size depiction of the two 1000-fold decreases in volume from the cubic decimeter to the cubic millimeter. The edge of a cubic meter would be about 2.5 times the width of this textbook when open.

Figure 1.7 shows some of the types of laboratory glassware designed to contain liquids or measure their volumes. Many come in sizes from a few milliliters

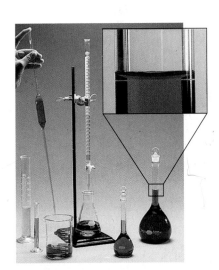

Figure 1.7 Common laboratory volumetric glassware. From left to right are two graduated cylinders, a pipet being emptied into a beaker, a buret delivering liquid to an Erlenmeyer flask, and two volumetric flasks. **Inset,** In contact with glass, this liquid forms a concave meniscus (curved surface).

to a few liters. Erlenmeyer flasks and beakers are used to contain liquids. Graduated cylinders, pipets, and burets are used to measure and transfer liquids. Volumetric flasks and many pipets have a fixed volume indicated by a mark on the neck. In quantitative work, liquid solutions are prepared in volumetric flasks, measured in cylinders, pipets, and burets, and then transferred to beakers or flasks for further chemical operations.

SAMPLE PROBLEM 1.3   Converting Units of Volume

Problem The volume of an irregularly shaped solid can be determined from the volume of water it displaces. A graduated cylinder contains 19.9 mL of water. When a small piece of galena, an ore of lead, is added, it sinks and the volume increases to 24.5 mL. What is the volume of the piece of galena in cm^3 and in L?

Plan We have to find the volume of the galena from the change in volume of the cylinder contents. The volume of galena in mL is the difference in the known volumes before and after adding it. The units mL and cm^3 represent identical volumes, so the volume of the galena in mL equals the volume in cm^3. We construct a conversion factor to convert the volume from mL to L. The calculation steps are shown in the roadmap.

Solution Finding the volume of galena:

$$\text{Volume (mL)} = \text{volume after} - \text{volume before} = 24.5 \text{ mL} - 19.9 \text{ mL} = 4.6 \text{ mL}$$

Converting the volume from mL to cm^3:

$$\text{Volume (cm}^3) = 4.6 \text{ mL} \times \frac{1 \text{ cm}^3}{1 \text{ mL}} = 4.6 \text{ cm}^3$$

Converting the volume from mL to L:

$$\text{Volume (L)} = 4.6 \text{ mL} \times \frac{10^{-3} \text{ L}}{1 \text{ mL}} = 4.6 \times 10^{-3} \text{ L}$$

Check The units and magnitudes of the answers seem correct. It makes sense that the volume expressed in mL would have a number 1000 times larger than the volume expressed in L, because a milliliter is $\frac{1}{1000}$ of a liter.

FOLLOW-UP PROBLEM 1.3 Within a cell, proteins are synthesized on particles called ribosomes. Assuming ribosomes are generally spherical, what is the volume (in dm^3 and μL) of a ribosome whose average diameter is 21.4 nm (V of a sphere $= \frac{4}{3}\pi r^3$)?

Roadmap (margin)

Volume (mL) before and after addition

↓ subtract

Volume (mL) of galena

1 mL = 1 cm³   1 mL = 10⁻³ L

Volume (cm³) of galena   Volume (L) of galena

Mass The **mass** of an object refers to the quantity of matter it contains. The SI unit of mass is the **kilogram (kg),** the only base unit whose standard is a physical object—a platinum-iridium cylinder kept in France. It is also the only base unit whose name has a prefix. (In contrast to the practice with other base units, however, we attach prefixes to the word "gram," as in "microgram," rather than to the word "kilogram"; thus, we never say "microkilogram.")

The terms *mass* and *weight* have distinct meanings. Because a given object's quantity of matter cannot change, its *mass is constant.* Its **weight,** on the other hand, depends on its mass *and* the strength of the local gravitational field pulling on it. Because the strength of this field varies with height above the Earth's surface, the object's weight also varies. For instance, you actually weigh slightly less on a high mountaintop than at sea level.

SAMPLE PROBLEM 1.4   Converting Units of Mass

Problem International computer communications are often carried by optical fibers in cables laid along the ocean floor. If one strand of optical fiber weighs 1.19×10^{-3} lb/m, what is the mass (in kg) of a cable made of six strands of optical fiber, each long enough to link New York and Paris (8.84×10^3 km)?

Plan We have to find the mass of cable (in kg) from the given mass/length of fiber, number of fibers/cable, and the length (distance from New York to Paris). One way to do this (as shown in the roadmap) is to first find the mass of one fiber and then find the mass of cable. We convert the length of one fiber from km to m and then find its mass (in lb) by using the lb/m factor. The cable mass is six times the fiber mass, and finally we convert lb to kg.

Solution Converting the fiber length from km to m:

$$\text{Length (m) of fiber} = 8.84 \times 10^3 \text{ km} \times \frac{10^3 \text{ m}}{1 \text{ km}} = 8.84 \times 10^6 \text{ m}$$

Converting the length of one fiber to mass (lb):

$$\text{Mass (lb) of fiber} = 8.84 \times 10^6 \text{ m} \times \frac{1.19 \times 10^{-3} \text{ lb}}{1 \text{ m}} = 1.05 \times 10^4 \text{ lb}$$

Finding the mass of the cable (lb):

$$\text{Mass (lb) of cable} = \frac{1.05 \times 10^4 \text{ lb}}{1 \text{ fiber}} \times \frac{6 \text{ fibers}}{1 \text{ cable}} = 6.30 \times 10^4 \text{ lb/cable}$$

Converting the mass of cable from lb to kg:

$$\text{Mass (kg) of cable} = \frac{6.30 \times 10^4 \text{ lb}}{1 \text{ cable}} \times \frac{1 \text{ kg}}{2.205 \text{ lb}} = 2.86 \times 10^4 \text{ kg/cable}$$

Check The units are correct. Let's think through the relative sizes of the answers to see if they make sense: The number of m should be 10^3 larger than the number of km. If 1 m of fiber weighs about 10^{-3} lb, about 10^7 m should weigh about 10^4 lb. The cable mass should be six times as much, or about 6×10^4 lb. Since 1 lb is about $\frac{1}{2}$ kg, the number of kg should be about half the number of lb.

FOLLOW-UP PROBLEM 1.4 An intravenous bag delivers a nutrient solution to a hospital patient at a rate of 1.5 drops per second. If a drop weighs 65 mg on average, how many kilograms of solution are delivered in 8.0 h?

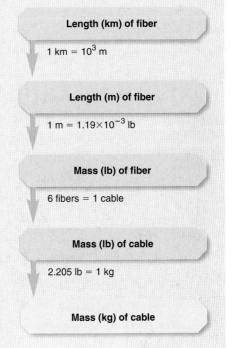

Length (km) of fiber

$1 \text{ km} = 10^3 \text{ m}$

Length (m) of fiber

$1 \text{ m} = 1.19 \times 10^{-3} \text{ lb}$

Mass (lb) of fiber

6 fibers = 1 cable

Mass (lb) of cable

2.205 lb = 1 kg

Mass (kg) of cable

Density The **density** (*d*) of an object is its mass divided by its volume:

$$\text{Density} = \frac{\text{mass}}{\text{volume}} \qquad (1.1)$$

Whenever needed, you can isolate mathematically each of the component variables by treating density as a conversion factor:

$$\text{Mass} = \text{volume} \times \text{density} = \text{volume} \times \frac{\text{mass}}{\text{volume}}$$

Or,

$$\text{Volume} = \text{mass} \times \frac{1}{\text{density}} = \text{mass} \times \frac{\text{volume}}{\text{mass}}$$

Because volume may change with temperature, density may change also. But, under given conditions of temperature and pressure, *density is a characteristic physical property of a substance* and has a specific value. Mass and volume are examples of **extensive properties,** those dependent on the amount of substance present. Density, on the other hand, is an **intensive property,** one that is independent of the amount of substance. For example, the mass of a gallon of water is four times the mass of a quart of water, but its volume is also four times greater; therefore, the density of the water, the *ratio* of its mass to its volume, is constant at a particular temperature and pressure, regardless of the sample size.

The SI unit of density is the kilogram per cubic meter (kg/m^3), but in chemistry, density is typically given in units of g/L (g/dm^3) or g/mL (g/cm^3). As you might expect from the magnified views in Figure 1.2, at ordinary pressure and 20°C, the densities of gases (for example, 0.0000899 g/cm^3 for hydrogen) are much lower than the densities of either liquids (1.00 g/cm^3 for water) or solids (2.17 g/cm^3 for aluminum).

SAMPLE PROBLEM 1.5   Calculating Density from Mass and Length

Problem Lithium is a soft, gray solid that has the lowest density of any metal. It is an essential component of some advanced batteries, such as the one in your laptop. If a small rectangular slab of lithium weighs 1.49×10^3 mg and has sides that measure 20.9 mm by 11.1 mm by 11.9 mm, what is the density of lithium in g/cm^3?

Plan To find the density in g/cm^3, we need the mass of lithium in g and the volume in cm^3. The mass is given in mg, so we convert mg to g. Volume data are not given, but we can convert the given side lengths from mm to cm, and then multiply them to find the volume in cm^3. Finally, we divide mass by volume to get density. The steps are shown in the roadmap.

Solution Converting the mass from mg to g:

$$\text{Mass (g) of lithium} = 1.49 \times 10^3 \text{ mg} \left(\frac{10^{-3} \text{ g}}{1 \text{ mg}} \right) = 1.49 \text{ g}$$

Converting side lengths from mm to cm:

$$\text{Length (cm) of one side} = 20.9 \text{ mm} \times \frac{1 \text{ cm}}{10 \text{ mm}} = 2.09 \text{ cm}$$

Similarly, the other side lengths are 1.11 cm and 1.19 cm.
Finding the volume:

$$\text{Volume (cm}^3\text{)} = 2.09 \text{ cm} \times 1.11 \text{ cm} \times 1.19 \text{ cm} = 2.76 \text{ cm}^3$$

Calculating the density:

$$\text{Density of lithium} = \frac{\text{mass}}{\text{volume}} = \frac{1.49 \text{ g}}{2.76 \text{ cm}^3} = 0.540 \text{ g/cm}^3$$

Check Since 1 cm = 10 mm, the number of cm in each length should be $\frac{1}{10}$ the number of mm. The units for density are correct, and the size of the answer (~ 0.5 g/cm^3) seems correct since the number of g (1.49) is about half the number of cm^3 (2.76). Since the problem states that lithium has a very low density, this answer makes sense.

FOLLOW-UP PROBLEM 1.5 The piece of galena in Sample Problem 1.3 has a volume of 4.6 cm^3. If the density of galena is 7.5 g/cm^3, what is the mass (in kg) of that piece of galena?

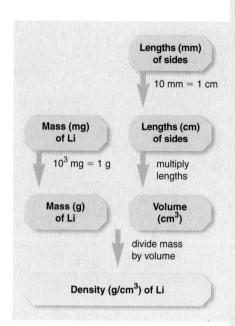

Temperature There is a common misunderstanding about heat and temperature. **Temperature (T)** is a measure of how hot or cold a substance is *relative to another substance.* **Heat** is the energy that flows between objects that are at different temperatures. Temperature is related to the *direction* of that energy flow: when two objects at different temperatures touch, energy flows *from* the one with the higher temperature *to* the one with the lower temperature until their temperatures are equal. When you hold an ice cube, its "cold" seems to flow *into* your hand; actually, heat flows *from* your hand into the ice. (In Chapter 6, we will see how heat is measured and how it is related to chemical and physical change.) Energy is an *extensive* property (as is volume), but temperature is an *intensive* property (as is density): a vat of boiling water has more energy than a cup of boiling water, but the temperatures of the two water samples are the same.

In the laboratory, the most common means for measuring temperature is the **thermometer,** a device that contains a fluid that expands when it is heated. When the thermometer's fluid-filled bulb is immersed in a substance hotter than itself, heat flows from the substance through the glass and into the fluid, which expands and rises in the thermometer tube. If a substance is colder than the thermometer, heat flows outward from the fluid, which contracts and falls within the tube.

The three temperature scales most important for us to consider are the Celsius (°C, formerly called centigrade), the Kelvin (K), and the Fahrenheit (°F) scales. The SI base unit of temperature is the **kelvin (K);** note that the kelvin has

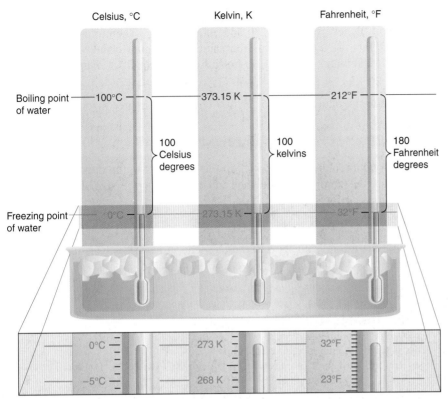

Figure 1.8 The freezing point and the boiling point of water in the Celsius, Kelvin (absolute), and Fahrenheit temperature scales. As you can see, this range consists of 100 degrees on the Celsius and Kelvin scales, but 180 degrees on the Fahrenheit scale. At the bottom of the figure, a portion of each of the three thermometer scales is expanded to show the sizes of the units. A Celsius degree (°C; *left*) and a kelvin (K; *center*) are the same size, and each is $\frac{9}{5}$ the size of a Fahrenheit degree (°F; *right*).

no degree sign (°). The Kelvin scale, also known as the *absolute scale,* is preferred in all scientific work, although the Celsius scale is used frequently. In the United States, the Fahrenheit scale is still used for weather reporting, body temperature, and other everyday purposes. *The three scales differ in the size of the unit and/or the temperature of the zero point.* Figure 1.8 shows the freezing and boiling points of water in the three scales.

The **Celsius scale,** devised in the 18th century by the Swedish astronomer Anders Celsius, is based on changes in the physical state of water: 0°C is set at water's freezing point, and 100°C is set at its boiling point (at normal atmospheric pressure). The **Kelvin (absolute) scale** was devised by the English physicist William Thomson, known as Lord Kelvin, in 1854 during his experiments on the expansion and contraction of gases. *The Kelvin scale uses the same size degree unit as the Celsius scale—$\frac{1}{100}$ of the difference between the freezing and boiling points of water—but it differs in zero point.* The zero point in the Kelvin scale, 0 K, is called *absolute zero* and equals -273.15°C. In the Kelvin scale, *all temperatures have positive values.* Water freezes at $+273.15$ K (0°C) and boils at $+373.15$ K (100°C).

We can convert between the Celsius and Kelvin scales by remembering the difference in zero points: since 0°C = 273.15 K,

$$T \text{ (in K)} = T \text{ (in °C)} + 273.15 \tag{1.2}$$

Solving Equation 1.2 for T (in °C) gives

$$T \text{ (in °C)} = T \text{ (in K)} - 273.15 \tag{1.3}$$

The Fahrenheit scale differs from the other scales in its zero point *and* in the size of its unit. Water freezes at 32°F and boils at 212°F. Therefore, 180 Fahrenheit degrees (212°F − 32°F) represents the same temperature change as 100 Celsius degrees (or 100 kelvins). Because 100 Celsius degrees equal 180 Fahrenheit degrees,

$$1 \text{ Celsius degree} = \tfrac{180}{100} \text{ Fahrenheit degrees} = \tfrac{9}{5} \text{ Fahrenheit degrees}$$

To convert a temperature in °C to °F, first change the degree size and then adjust the zero point:

$$T \text{ (in °F)} = \tfrac{9}{5}T \text{ (in °C)} + 32 \qquad (1.4)$$

To convert a temperature in °F to °C, do the two steps in the opposite order; that is, first adjust the zero point and then change the degree size. In other words, solve Equation 1.4 for T (in °C):

$$T \text{ (in °C)} = [T \text{ (in °F)} - 32]\tfrac{5}{9} \qquad (1.5)$$

(The only temperature with the same numerical value in the Celsius and Fahrenheit scales is −40°; that is, −40°F = −40°C.)

SAMPLE PROBLEM 1.6   Converting Units of Temperature

Problem A child has a body temperature of 38.7°C.
(a) If normal body temperature is 98.6°F, does the child have a fever?
(b) What is the child's temperature in kelvins?
Plan (a) To find out if the child has a fever, we convert from °C to °F (Equation 1.4) and see whether 38.7°C is higher than 98.6°F.
(b) We use Equation 1.2 to convert the temperature in °C to K.
Solution (a) Converting the temperature from °C to °F:

$$T \text{ (in °F)} = \tfrac{9}{5}T \text{ (in °C)} + 32 = \tfrac{9}{5}(38.7°C) + 32 = \boxed{101.7°F}; \text{ yes, the child has a fever.}$$

(b) Converting the temperature from °C to K:

$$T \text{ (in K)} = T \text{ (in °C)} + 273.15 = 38.7°C + 273.15 = \boxed{311.8 \text{ K}}$$

Check (a) From everyday experience, you know that 101.7°F is a reasonable temperature for someone with a fever.
(b) We know that a Celsius degree and a kelvin are the same size. Therefore, we can check the math by approximating the Celsius value as 40°C and adding 273: 40 + 273 = 313, which is close to our calculation, so there is no large error.

FOLLOW-UP PROBLEM 1.6 Mercury melts at 234 K, lower than any other pure metal. What is its melting point in °C and °F?

Time The SI base unit of time is the **second (s).** The standard second is defined by the number of oscillations of microwave radiation absorbed by cooled gaseous cesium atoms in an atomic clock; precisely 9,192,631,770 of these oscillations are absorbed in 1 second.

In the laboratory, we study the speed (or *rate*) of a reaction by measuring the time it takes a fixed amount of substance to undergo a chemical change. The range of reaction rates is enormous: a fast reaction may be over in less than a nanosecond (10^{-9} s), whereas slow ones, such as rusting or aging, take years. Chemists now use lasers to study changes that occur in a few picoseconds (10^{-12} s) or femtoseconds (10^{-15} s).

SECTION SUMMARY

SI units consist of seven base units and numerous derived units. Exponential notation and prefixes based on powers of 10 are used to express very small and very large numbers. The SI base unit of length is the meter (m). Length units on the atomic scale are the nanometer (nm) and picometer (pm). Volume units are derived from length units; the most important volume units are the cubic meter (m^3) and the liter

(L). The mass of an object, a measure of the quantity of matter present in it, is constant. The SI unit of mass is the kilogram (kg). The weight of an object varies with the gravitational field influencing it. Density (d) is the ratio of mass to volume of a substance and is one of its characteristic physical properties. Temperature (T) is a measure of the relative hotness of an object. Heat is energy that flows from an object at higher temperature to one at lower temperature. Temperature scales differ in the size of the degree unit and/or the zero point. In chemistry, temperature is measured in kelvins (K) or degrees Celsius (°C). Extensive properties, such as mass, volume, and energy, depend on the amount of a substance. Intensive properties, such as density and temperature, are independent of amount.

1.5 UNCERTAINTY IN MEASUREMENT: SIGNIFICANT FIGURES

We can never measure a quantity exactly, because measuring devices are made to limited specifications and we use our imperfect senses and skills to read them. Therefore, every measurement includes some **uncertainty.**

The measuring device we choose in a given situation depends on how much uncertainty we are willing to accept. When you buy potatoes, a supermarket scale that measures in 0.1-kg increments is perfectly acceptable; it tells you that the mass is, for example, 2.0 ± 0.1 kg. The term "± 0.1 kg" expresses the uncertainty in the measurement: the potatoes weigh between 1.9 and 2.1 kg. For a large-scale reaction, a chemist uses a lab balance that measures in 0.001-kg increments in order to obtain 2.036 ± 0.001 kg of a chemical, that is, between 2.035 and 2.037 kg. The greater number of digits in the mass of the chemical indicates that we know its mass with *more certainty* than we know the mass of the potatoes. The uncertainty of a measured quantity can be expressed with the ± sign, but generally we drop the sign and *assume an uncertainty of one unit in the rightmost digit.*

The digits we record in a measurement, both the certain and the uncertain ones, are called **significant figures.** There are four significant figures in 2.036 kg and two in 2.0 kg. *The greater the number of significant figures in a measurement, the greater is the certainty.* Figure 1.9 shows this point for two thermometers.

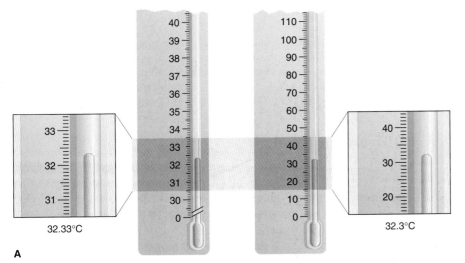

32.33°C

A

32.3°C

B

Figure 1.9 The number of significant figures in a measurement depends on the measuring device. A, Two thermometers measuring the same temperature are shown with expanded views. The thermometer on the left is graduated in 0.1°C and reads 32.33°C; the one on the right is graduated in 1°C and reads 32.3°C. Therefore, a reading with more significant figures (more certainty) can be made with the thermometer on the left. **B,** This modern electronic thermometer measures the resistance through a fine platinum wire in the probe to determine temperatures to the nearest microkelvin (10^{-6} K).

Determining Which Digits Are Significant

When you take measurements or use them in calculations, you must know the number of digits that are significant. In general, *all digits are significant, except zeros that are not measured but are used only to position the decimal point.* Here is a simple procedure that applies this general point:

1. Make sure that the measured quantity has a decimal point.
2. Start at the left of the number and move right to the first nonzero digit.
3. Count that digit and every digit to its right as significant.

A complication may arise with zeros that *end* a number. Zeros that end a number and lie either after or before the decimal point *are* significant; thus, 1.030 mL has four significant figures, and 5300. L has four significant figures also. If there is no decimal point, as in 5300 L, we assume that the zeros are *not* significant; exponential notation is needed to show which of the zeros, if any, were measured and therefore are significant. Thus, 5.300×10^3 L has four significant figures, 5.30×10^3 L has three, and 5.3×10^3 L has only two. A terminal decimal point is used to clarify the number of significant figures; thus, 500 mL has one significant figure, but 5.00×10^2 mL, 500. mL, and 0.500 L have three.

SAMPLE PROBLEM 1.7 Determining the Number of Significant Figures

Problem For each of the following quantities, underline the zeros that are significant figures (sf), and determine the number of significant figures in each quantity. For (d) to (f), express each in exponential notation first.
(a) 0.0030 L   **(b)** 0.1044 g   **(c)** 53,069 mL
(d) 0.00004715 m   **(e)** 57,600. s   **(f)** 0.0000007160 cm^3
Plan We determine the number of significant figures by counting digits, as just presented, paying particular attention to the position of zeros in relation to the decimal point.
Solution (a) 0.0030 L has 2 sf
(b) 0.1044 g has 4 sf
(c) 53,069 mL has 5 sf
(d) 0.00004715 m, or 4.715×10^{-5} m, has 4 sf
(e) 57,600. s, or 5.7600×10^4 s, has 5 sf
(f) 0.0000007160 cm^3, or 7.160×10^{-7} cm^3, has 4 sf
Check Be sure that every zero counted as significant comes after nonzero digit(s) in the number.

FOLLOW-UP PROBLEM 1.7 For each of the following quantities, underline the zeros that are significant figures and determine the number of significant figures (sf) in each quantity. For (d) to (f), express each in exponential notation first.
(a) 31.070 mg   **(b)** 0.06060 g   **(c)** 850.°C
(d) 200.0 mL   **(e)** 0.0000039 m   **(f)** 0.000401 L

Significant Figures in Calculations

Measurements often contain differing numbers of significant figures. In a calculation, we keep track of the number of significant figures in each quantity so that we don't claim more significant figures (more certainty) in the answer than in the original data. If we have too many significant figures, we **round off** the answer to obtain the proper number of them.

The general rule for rounding is that *the least certain measurement sets the limit on certainty for the entire calculation and determines the number of significant figures in the final answer.* Suppose you want to find the density of a new ceramic. You measure the mass of a piece on a precise laboratory balance and obtain 3.8056 g; you measure its volume as 2.5 mL by displacement of water in a graduated cylinder. The mass has five significant figures, but the volume has only two. Should you report the density as 3.8056 g/2.5 mL = 1.5222 g/mL or as 1.5 g/mL? The answer with five significant figures implies more certainty than the answer with two. But you didn't measure the volume to five significant figures, so you can't possibly know the density with that much certainty. Therefore, you report the answer as 1.5 g/mL.

Significant Figures and Arithmetic Operations The following two rules tell how many significant figures to show based on the arithmetic operation:

    1. *For multiplication and division.* The answer contains the same number of *significant figures* as in the measurement with the fewest significant figures. Suppose you want to find the volume of a sheet of a new graphite composite. The length (9.2 cm) and width (6.8 cm) are obtained with a meterstick and the thickness (0.3744 cm) with a set of fine calipers. The volume calculation is

$$\text{Volume (cm}^3) = 9.2 \text{ cm} \times 6.8 \text{ cm} \times 0.3744 \text{ cm} = 23 \text{ cm}^3$$

The calculator shows 23.4225 cm^3, but you should report the answer as 23 cm^3, with two significant figures, because the length and width measurements determine the overall certainty, and they contain only two significant figures.

    2. *For addition and subtraction.* The answer has the same number of *decimal places* as there are in the measurement with the fewest decimal places. Suppose you measure 83.5 mL of water in a graduated cylinder and add 23.28 mL of protein solution from a buret. The total volume is

$$\text{Volume (mL)} = 83.5 \text{ mL} + 23.28 \text{ mL} = 106.8 \text{ mL}$$

Here the calculator shows 106.78 mL, but you report the volume as 106.8 mL, with one decimal place, because the measurement with fewer decimal places (83.5 mL) has one decimal place.

Rules for Rounding Off In most calculations, you need to round off the answer to obtain the proper number of significant figures or decimal places. Notice that in calculating the volume of the graphite composite above, we removed the extra digits, but in calculating the total protein solution volume, we removed the extra digit and increased the last digit by one. Here are rules for rounding off:

    1. If the digit removed is *more than 5,* the preceding number is increased by 1: 5.379 rounds to 5.38 if three significant figures are retained and to 5.4 if two significant figures are retained.

    2. If the digit removed is *less than 5,* the preceding number is unchanged: 0.2413 rounds to 0.241 if three significant figures are retained and to 0.24 if two significant figures are retained.

    3. If the digit removed *is 5,* the preceding number is increased by 1 if it is odd and remains unchanged if it is even: 17.75 rounds to 17.8, but 17.65 rounds to 17.6. If the 5 is followed only by zeros, rule 3 is followed; if the 5 is followed by nonzeros, rule 1 is followed: 17.6500 rounds to 17.6, but 17.6513 rounds to 17.7.

    4. *Always carry one or two additional significant figures through a multistep calculation and round off the final answer **only**.* Don't be concerned if you string together a calculation to check a sample or follow-up problem and find that your answer differs in the last decimal place from the one in the book. To show you the correct number of significant figures in text calculations, *we round off intermediate steps,* and this process may sometimes change the last digit.

    A calculator usually gives answers with too many significant figures. For example, if your calculator displays ten digits and you divide 15.6 by 9.1, it will show 1.714285714. Obviously, most of these digits are not significant; the answer should be rounded off to 1.7 so that it has two significant figures, the same as in 9.1.

Exact Numbers Some numbers are called **exact numbers** because they have no uncertainty associated with them. Some exact numbers are part of a unit definition: there are 60 minutes in 1 hour, 1000 micrograms in 1 milligram, and

2.54 centimeters in 1 inch. Other exact numbers result from actually counting individual items: there are exactly 3 quarters in my hand, 26 letters in the English alphabet, and so forth. Because they have no uncertainty, *exact numbers do not limit the number of significant figures in the answer.* Put another way, exact numbers have as many significant figures as a calculation requires.

SAMPLE PROBLEM 1.8 Significant Figures and Rounding

Problem Perform the following calculations and round the answer to the correct number of significant figures:

(a) $\dfrac{16.3521 \text{ cm}^2 - 1.448 \text{ cm}^2}{7.085 \text{ cm}}$

(b) $\dfrac{(4.80 \times 10^4 \text{ mg}) \left(\dfrac{1 \text{ g}}{1000 \text{ mg}} \right)}{11.55 \text{ cm}^3}$

Plan We use the rules just presented in the text. In (a), we subtract before we divide. In (b), we note that the unit conversion involves an exact number.

Solution (a) $\dfrac{16.3521 \text{ cm}^2 - 1.448 \text{ cm}^2}{7.085 \text{ cm}} = \dfrac{14.904 \text{ cm}^2}{7.085 \text{ cm}} = 2.104 \text{ cm}$

(b) $\dfrac{(4.80 \times 10^4 \text{ mg}) \left(\dfrac{1 \text{ g}}{1000 \text{ mg}} \right)}{11.55 \text{ cm}^3} = \dfrac{48.0 \text{ g}}{11.55 \text{ cm}^3} = 4.16 \text{ g/cm}^3$

Check Note that in (a) we lose a decimal place in the numerator, and in (b) we retain 3 sf in the answer because there are 3 sf in 4.80. Rounding to the nearest whole number is always a good way to check: **(a)** $(16 - 1)/7 \approx 2$; **(b)** $(5 \times 10^4 / 1 \times 10^3)/12 \approx 4$.

FOLLOW-UP PROBLEM 1.8 Perform the following calculation and round the answer to the correct number of significant figures: $\dfrac{25.65 \text{ mL} + 37.4 \text{ mL}}{73.55 \text{ s} \left(\dfrac{1 \text{ min}}{60 \text{ s}} \right)}$

Precision, Accuracy, and Instrument Calibration

Precision and accuracy are two aspects of certainty. We often use these terms interchangeably in everyday speech, but in scientific measurements they have distinct meanings. **Precision,** or *reproducibility,* refers to how close the measurements in a series are to each other. **Accuracy** refers to how close a measurement is to the actual value.

Precision and accuracy are linked with two common types of error:

1. **Systematic error** produces values that are *either* all higher or all lower than the actual value. Such error is part of the experimental system, often caused by a faulty measuring device or by a consistent mistake in taking a reading.
2. **Random error,** in the absence of systematic error, produces values that are higher *and* lower than the actual value. Random error *always* occurs, but its size depends on the measurer's skill and the instrument's precision.

Precise measurements have low random error, that is, small deviations from the average. *Accurate measurements have low systematic error and, generally, low random error as well.* In some cases, when many measurements are taken that have a high random error, the *average* may still be accurate.

Suppose each of four students measures 25.0 mL of water in a pre-weighed graduated cylinder and then weighs the water *plus* cylinder on a balance. If the density of water is 1.00 g/mL at the temperature of the experiment, the actual mass of 25.0 mL of water is 25.0 g. Each student performs the operation four times, subtracts the mass of the empty cylinder, and obtains one of the four graphs shown in Figure 1.10. In graphs A and B, the random error is small; that is, the precision is high (the weighings are reproducible). In A, however, the accuracy is high as well (all the values are close to 25.0 g), whereas in B

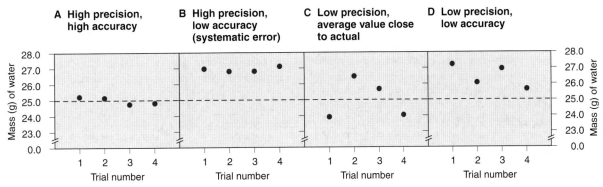

Figure 1.10 Precision and accuracy in a laboratory calibration. Each graph represents four measurements made with a graduated cylinder that is being calibrated (see text for details).

the accuracy is low (there is a systematic error). In graphs C and D, there is a large random error; that is, the precision is low. Large random error is often called large *scatter*. Note, however, that in D there is also a systematic error (all the values are high), whereas in C the average of the values is close to the actual value.

Systematic error can be avoided, or at least taken into account, through **calibration** of the measuring device, that is, by comparing it with a known standard. The systematic error in graph B, for example, might be caused by a poorly manufactured cylinder that reads "25.0" when it actually contains about 27 mL. If you detect such an error by means of a calibration procedure, you could adjust all volumes measured with that cylinder. Instrument calibration is an essential part of careful measurement.

SECTION SUMMARY

Because the final digit of a measurement is estimated, all measurements have a limit to their certainty, which is expressed by the number of significant figures. The certainty of a calculated result depends on the certainty of the data, so the answer has as many significant figures as in the least certain measurement. Excess digits are rounded off in the final answer. Exact numbers have as many significant figures as the calculation requires.

Precision (how close values are to each other) and accuracy (how close values are to the actual value) are two aspects of certainty. Systematic errors result in values that are either all higher or all lower than the actual value. Random errors result in some values that are higher and some values that are lower than the actual value. Precise measurements have low random error; accurate measurements have low systematic error and often low random error. The size of random errors depends on the skill of the measurer and the precision of the instrument. A systematic error, however, is often caused by faulty equipment and can be compensated for by calibration.

For Review and Reference (Numbers in parentheses refer to pages, unless noted otherwise.)

Learning Objectives

To help you review these learning objectives, the numbers of related sections (§), sample problems (SP), and upcoming end-of-chapter problems (EP) are listed in parentheses.

1. Distinguish between physical and chemical properties and changes (§ 1.1) (SP 1.1) (EPs 1.1, 1.3–1.6)
2. Define the features of the states of matter (§ 1.1) (EP 1.2)
3. Understand the nature of potential and kinetic energy and their interconversion (§ 1.1) (EPs 1.7–1.8)

4. Understand the scientific approach to studying phenomena and distinguish between observation, hypothesis, experiment, and model (§ 1.2) (EPs 1.9–1.12)
5. Use conversion factors in calculations (§ 1.3) (SP 1.2) (EPs 1.13–1.15)
6. Distinguish between mass and weight, heat and temperature, and intensive and extensive properties (§ 1.4) (EPs 1.16, 1.17, 1.19)

Learning Objectives *(continued)*

7. Use numerical prefixes and common units of length, mass, volume, and temperature in unit-conversion calculations (§ 1.4) (SPs 1.3–1.6) (EPs 1.21–1.39)

8. Understand scientific notation and the meaning of uncertainty; determine the number of significant figures and the number of digits after rounding (§ 1.5) (SP 1.7, 1.8) (EPs 1.40–1.54)

9. Distinguish between accuracy and precision and between systematic and random error (§ 1.5) (EPs 1.55–1.57)

Key Terms

Section 1.1
chemistry (2)
matter (2)
composition (2)
property (2)
physical property (2)
physical change (3)
chemical property (3)
chemical change (chemical reaction) (3)
state of matter (3)
solid (3)
liquid (3)
gas (3)
energy (5)
potential energy (5)
kinetic energy (5)

Section 1.2
scientific method (7)
observation (8)
data (8)
natural law (8)
hypothesis (8)
experiment (8)
variable (9)
controlled experiment (9)
model (theory) (9)

Section 1.3
conversion factor (10)
dimensional analysis (11)

Section 1.4
SI units (13)
base (fundamental) unit (13)

derived unit (13)
meter (m) (15)
volume (V) (15)
cubic meter (m^3) (15)
liter (L) (15)
milliliter (mL) (15)
mass (16)
kilogram (kg) (16)
weight (16)
density (d) (17)
extensive property (17)
intensive property (17)
temperature (T) (18)
heat (18)
thermometer (18)

kelvin (K) (18)
Celsius scale (19)
Kelvin (absolute) scale (19)
second (s) (20)

Section 1.5
uncertainty (21)
significant figures (21)
round off (22)
exact number (23)
precision (24)
accuracy (24)
systematic error (24)
random error (24)
calibration (25)

Key Equations and Relationships

1.1 Calculating density from mass and volume (17):

$$\text{Density} = \frac{\text{mass}}{\text{volume}}$$

1.2 Converting temperature from °C to K (19):

$$T \text{ (in K)} = T \text{ (in °C)} + 273.15$$

1.3 Converting temperature from K to °C (19):

$$T \text{ (in °C)} = T \text{ (in K)} - 273.15$$

1.4 Converting temperature from °C to °F (20):

$$T \text{ (in °F)} = \tfrac{9}{5}T \text{ (in °C)} + 32$$

1.5 Converting temperature from °F to °C (20):

$$T \text{ (in °C)} = [T \text{ (in °F)} - 32]\tfrac{5}{9}$$

Brief Solutions to Follow-up Problems

1.1 (a) Physical. Solid iodine changes to gaseous iodine.
(b) Chemical. Gasoline burns in air to form different substances.
(c) Chemical. In contact with air, torn skin and blood react to form different substances.

1.2 No. of chairs

$$= 3 \text{ bolts} \times \frac{200 \text{ m}^2}{1 \text{ bolt}} \times \frac{3.281 \text{ ft}}{1 \text{ m}} \times \frac{3.281 \text{ ft}}{1 \text{ m}} \times \frac{1 \text{ chair}}{31.5 \text{ ft}^2}$$

$$= 205 \text{ chairs}$$

1.3 Radius of ribosome (dm) $= \dfrac{21.4 \text{ nm}}{2} \times \dfrac{1 \text{ dm}}{10^8 \text{ nm}}$

$$= 1.07 \times 10^{-7} \text{ dm}$$

Volume of ribosome (dm³) $= \tfrac{4}{3}\pi r^3 = \tfrac{4}{3}(3.14)(1.07 \times 10^{-7} \text{ dm})^3$

$$= 5.13 \times 10^{-21} \text{ dm}^3$$

Volume of ribosome (μL) $= (5.13 \times 10^{-21} \text{ dm}^3)\left(\dfrac{1 \text{ L}}{1 \text{ dm}^3}\right)\left(\dfrac{10^6 \text{ μL}}{1 \text{ L}}\right)$

$$= 5.13 \times 10^{-15} \text{ μL}$$

1.4 Mass (kg) of solution $= 8.0 \text{ h} \times \dfrac{60 \text{ min}}{1 \text{ h}} \times \dfrac{60 \text{ s}}{1 \text{ min}} \times \dfrac{1.5 \text{ drops}}{1 \text{ s}}$

$$\times \dfrac{65 \text{ mg}}{1 \text{ drop}} \times \dfrac{1 \text{ g}}{10^3 \text{ mg}} \times \dfrac{1 \text{ kg}}{10^3 \text{ g}}$$

$$= 2.8 \text{ kg}$$

1.5 Mass (kg) of sample $= 4.6 \text{ cm}^3 \times \dfrac{7.5 \text{ g}}{1 \text{ cm}^3} \times \dfrac{1 \text{ kg}}{10^3 \text{ g}}$

$$= 0.034 \text{ kg}$$

1.6 T (in °C) $= 234 \text{ K} - 273.15 = -39°\text{C}$
T (in °F) $= \tfrac{9}{5}(-39°\text{C}) + 32 = -38°\text{F}$
Answer contains two significant figures (see Section 1.5).

1.7 (a) 31.0̲70 mg, 5 sf   (b) 0.060̲60 g, 4 sf
(c) 850̲.°C, 3 sf   (d) $2.00̲0 \times 10^2$ mL, 4 sf
(e) 3.9×10^{-6} m, 2 sf   (f) $4.0̲1 \times 10^{-4}$ L, 3 sf

1.8 $\dfrac{25.65 \text{ mL} + 37.4 \text{ mL}}{73.55 \text{ s} \left(\dfrac{1 \text{ min}}{60 \text{ s}}\right)} = 51.4 \text{ mL/min}$

Problems

Problems with **colored** numbers are answered in Appendix E. Sections match the text and provide the numbers of relevant sample problems. Bracketed problems are grouped in pairs (indicated by a short rule) that cover the same concept. Comprehensive Problems are based on material from any section.

Some Fundamental Definitions
(Sample Problem 1.1)

1.1 Scenes A and B depict changes in matter at the atomic scale:

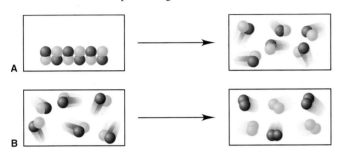

(a) Which show(s) a physical change?
(b) Which show(s) a chemical change?
(c) Which result(s) in different physical properties?
(d) Which result(s) in different chemical properties?
(e) Which result(s) in a change in state?

1.2 Describe solids, liquids, and gases in terms of how they fill a container. Use your descriptions to identify the physical state (at room temperature) of the following: (a) helium in a toy balloon; (b) mercury in a thermometer; (c) soup in a bowl.

1.3 Define *physical property* and *chemical property*. Identify each type of property in the following statements:
(a) Yellow-green chlorine gas attacks silvery sodium metal to form white crystals of sodium chloride (table salt).
(b) A magnet separates a mixture of black iron shavings and white sand.

1.4 Define *physical change* and *chemical change*. State which type of change occurs in each of the following statements:
(a) Passing an electric current through molten magnesium chloride yields molten magnesium and gaseous chlorine.
(b) The iron in discarded automobiles slowly forms reddish brown, crumbly rust.

1.5 Which of the following is a chemical change? Explain your reasoning: (a) boiling canned soup; (b) toasting a slice of bread; (c) chopping a log; (d) burning a log.

1.6 Which of the following changes can be reversed by changing the temperature (that is, which are physical changes): (a) dew condensing on a leaf; (b) an egg turning hard when it is boiled; (c) ice cream melting; (d) a spoonful of batter cooking on a hot griddle?

1.7 For each pair, which has higher potential energy?
(a) The fuel in your car or the products in its exhaust
(b) Wood in a fireplace or the ashes in the fireplace after the wood burns

1.8 For each pair, which has higher kinetic energy?
(a) A sled resting at the top of a hill or a sled sliding down the hill
(b) Water above a dam or water falling over the dam

The Scientific Approach: Developing a Model

1.9 How are the key elements of scientific thinking used in the following scenario? While making your breakfast toast, you notice it fails to pop out of the toaster. Thinking the spring mechanism is stuck, you notice that the bread is unchanged. Assuming you forgot to plug in the toaster, you check and find it is plugged in. When you take the toaster into the dining room and plug it into a different outlet, you find the toaster works. Returning to the kitchen, you turn on the switch for the overhead light and nothing happens.

1.10 Why is a quantitative observation more useful than a non-quantitative one? Which of the following are quantitative?
(a) The Sun rises in the east.
(b) An astronaut weighs one-sixth as much on the Moon as on Earth.
(c) Ice floats on water.
(d) An old-fashioned hand pump cannot draw water from a well more than 34 ft deep.

1.11 Describe the essential features of a well-designed experiment.

1.12 Describe the essential features of a scientific model.

Chemical Problem Solving
(Sample Problem 1.2)

1.13 When you convert feet to inches, how do you decide which portion of the conversion factor should be in the numerator and which in the denominator?

1.14 Write the conversion factor(s) for (a) in^2 to cm^2; (b) km^2 to m^2; (c) mi/h to m/s; (d) lb/ft^3 to g/cm^3.

1.15 Write the conversion factor(s) for (a) cm/min to in/min; (b) m^3 to in^3; (c) m/s^2 to km/h^2; (d) gallons/h to L/s.

Measurement in Scientific Study
(Sample Problems 1.3 to 1.6)

1.16 Describe the difference between intensive and extensive properties. Which of the following properties are intensive: (a) mass; (b) density; (c) volume; (d) melting point?

1.17 Explain the difference between mass and weight. Why is your weight on the Moon one-sixth that on Earth?

1.18 For each of the following cases, state whether the density of the object increases, decreases, or remains the same:
(a) A sample of chlorine gas is compressed.
(b) A lead weight is carried from sea level to the top of a high mountain.
(c) A sample of water is frozen.
(d) An iron bar is cooled.
(e) A diamond is submerged in water.

1.19 Explain the difference between heat and temperature. Does 1 L of water at 65°F have more, less, or the same quantity of energy as 1 L of water at 65°C?

1.20 A one-step conversion is sufficient to convert a temperature in the Celsius scale into the Kelvin scale, but not into the Fahrenheit scale. Explain.

1.21 The average radius of a molecule of lysozyme, an enzyme in tears, is 1430 pm. What is its radius in nanometers (nm)?

1.22 The radius of a barium atom is 2.22×10^{-10} m. What is its radius in angstroms (Å)?

1.23 A small hole in the wing of a space shuttle requires a 17.7-cm^2 patch. (a) What is the patch's area in square kilometers (km^2)? (b) If the patching material costs NASA \$3.25/in^2, what is the cost of the patch?

1.24 The area of a telescope lens is 6322 mm^2. (a) What is the area of the lens in square feet (ft^2)? (b) If it takes a technician 45 s to polish 135 mm^2, how long does it take her to polish the entire lens?

1.25 The average density of Earth is 5.52 g/cm^3. What is its density in (a) kg/m^3; (b) lb/ft^3?

1.26 The speed of light in a vacuum is 2.998×10^8 m/s. What is its speed in (a) km/h; (b) mi/min?

1.27 The volume of a certain bacterial cell is 1.72 μm^3. (a) What is its volume in cubic millimeters (mm^3)? (b) What is the volume of 10^5 cells in liters (L)?

1.28 (a) How many cubic meters of milk are in 1 qt (946.4 mL)? (b) How many liters of milk are in 835 gallons (1 gal = 4 qt)?

1.29 An empty vial weighs 55.32 g. (a) If the vial weighs 185.56 g when filled with liquid mercury ($d = 13.53$ g/cm^3), what is its volume? (b) How much would the vial weigh if it were filled with water ($d = 0.997$ g/cm^3 at 25°C)?

1.30 An empty Erlenmeyer flask weighs 241.3 g. When filled with water ($d = 1.00$ g/cm^3), the flask and its contents weigh 489.1 g. (a) What is the flask's volume? (b) How much does the flask weigh when filled with chloroform ($d = 1.48$ g/cm^3)?

1.31 A small cube of aluminum measures 15.6 mm on a side and weighs 10.25 g. What is the density of aluminum in g/cm^3?

1.32 A steel ball-bearing with a circumference of 32.5 mm weighs 4.20 g. What is the density of the steel in g/cm^3 (V of a sphere = $\frac{4}{3}\pi r^3$; circumference of a circle = $2\pi r$)?

1.33 Perform the following conversions:
(a) 72°F (a pleasant spring day) to °C and K
(b) −164°C (the boiling point of methane, the main component of natural gas) to K and °F
(c) 0 K (absolute zero, theoretically the coldest possible temperature) to °C and °F

1.34 Perform the following conversions:
(a) 106°F (the body temperature of many birds) to K and °C
(b) 3410°C (the melting point of tungsten, the highest for any element) to K and °F
(c) 6.1×10^3 K (the surface temperature of the Sun) to °F and °C

1.35 Anton van Leeuwenhoek, a 17th-century pioneer in the use of the microscope, described the microorganisms he saw as "animalcules" whose length was "25 thousandths of an inch." How long were the animalcules in meters?

1.36 The distance between two adjacent peaks on a wave is called the *wavelength*.
(a) The wavelength of a beam of ultraviolet light is 255 nanometers (nm). What is its wavelength in meters?
(b) The wavelength of a beam of red light is 683 nm. What is its wavelength in angstroms (Å)?

1.37 In the early 20th century, thin metal foils were used to study atomic structure. (a) How many in^2 of gold foil with a thickness of 1.6×10^{-5} in could have been made from 2.0 troy oz? (b) If gold cost \$20.00/troy oz at that time, how many cm^2 of gold foil could have been made from \$75.00 worth of gold (1 troy oz = 31.1 g; d of gold = 19.3 g/cm^3)?

1.38 A cylindrical tube 7.8 cm high and 0.85 cm in diameter is used to collect blood samples. How many cubic decimeters (dm^3) of blood can it hold (V of a cylinder = $\pi r^2 h$)?

1.39 Copper can be drawn into thin wires. How many meters of 34-gauge wire (diameter = 6.304×10^{-3} in) can be produced from the copper in 5.01 lb of covellite, an ore of copper that is 66% copper by mass? (*Hint:* Treat the wire as a cylinder: V of cylinder = $\pi r^2 h$; d of copper = 8.95 g/cm^3.)

Uncertainty in Measurement: Significant Figures
(Sample Problems 1.7 and 1.8)

1.40 What is an exact number? How are exact numbers treated differently from other numbers in a calculation?

1.41 All nonzero digits are significant. State a rule that tells which zeros are significant.

1.42 Underline the significant zeros in the following numbers:
(a) 0.39   (b) 0.039   (c) 0.0390   (d) 3.0900×10^4

1.43 Underline the significant zeros in the following numbers:
(a) 5.08   (b) 508   (c) 5.080×10^3 (d) 0.05080

1.44 Carry out the following calculations, making sure that your answer has the correct number of significant figures:
(a) $\dfrac{2.795 \text{ m} \times 3.10 \text{ m}}{6.48 \text{ m}}$
(b) $V = \frac{4}{3}\pi r^3$, where $r = 9.282$ cm
(c) 1.110 cm + 17.3 cm + 108.2 cm + 316 cm

1.45 Carry out the following calculations, making sure that your answer has the correct number of significant figures:
(a) $\dfrac{2.420 \text{ g} + 15.6 \text{ g}}{4.8 \text{ g}}$ (b) $\dfrac{7.87 \text{ mL}}{16.1 \text{ mL} - 8.44 \text{ mL}}$
(c) $V = \pi r^2 h$, where $r = 6.23$ cm and $h = 4.630$ cm

1.46 Write the following numbers in scientific notation:
(a) 131,000.0   (b) 0.00047   (c) 210,006   (d) 2160.5

1.47 Write the following numbers in scientific notation:
(a) 281.0   (b) 0.00380   (c) 4270.8   (d) 58,200.9

1.48 Write the following numbers in standard notation. Use a terminal decimal point when needed:
(a) 5.55×10^3 (b) 1.0070×10^4
(c) 8.85×10^{-7} (d) 3.004×10^{-3}

1.49 Write the following numbers in standard notation. Use a terminal decimal point when needed:
(a) 6.500×10^3 (b) 3.46×10^{-5} (c) 7.5×10^2 (d) 1.8856×10^2

1.50 Carry out each of the following calculations, paying special attention to significant figures, rounding, and units (J = joule, the SI unit of energy; mol = mole, the SI unit for amount of substance):
(a) $\dfrac{(6.626 \times 10^{-34} \text{ J·s})(2.9979 \times 10^8 \text{ m/s})}{489 \times 10^{-9} \text{ m}}$
(b) $\dfrac{(6.022 \times 10^{23} \text{ molecules/mol})(1.19 \times 10^2 \text{ g})}{46.07 \text{ g/mol}}$
(c) $(6.022 \times 10^{23} \text{ atoms/mol})(2.18 \times 10^{-18} \text{ J/atom})\left(\dfrac{1}{2^2} - \dfrac{1}{3^2}\right)$,

where the numbers 2 and 3 in the last term are exact.

1.51 Carry out each of the following calculations, paying special attention to significant figures, rounding, and units:
(a) $\dfrac{8.32 \times 10^7 \text{ g}}{\frac{4}{3}(3.1416)(1.95 \times 10^2 \text{ cm})^3}$ (The term $\frac{4}{3}$ is exact.)

(b) $\dfrac{(1.84\times10^2 \text{ g})(44.7 \text{ m/s})^2}{2}$ (The term 2 is exact.)

(c) $\dfrac{(1.07\times10^{-4} \text{ mol/L})^2(2.6\times10^{-3} \text{ mol/L})}{(8.35\times10^{-5} \text{ mol/L})(1.48\times10^{-2} \text{ mol/L})^3}$

1.52 Which statements include exact numbers?
(a) Angel Falls in Venezuela is 3212 ft high.
(b) There are nine known planets in the Solar System.
(c) There are 453.59 g in 1 lb.
(d) There are 1000 mm in 1 m.

1.53 Which of the following include exact numbers?
(a) The speed of light in a vacuum is a physical constant; to six significant figures, it is 2.99792×10^8 m/s.
(b) The density of mercury at 25°C is 13.53 g/mL.
(c) There are 3600 s in 1 h.
(d) In 2003, the United States had 50 states.

1.54 How long is the metal strip shown below? Be sure to answer with the correct number of significant figures.

1.55 These organic solvents are used to clean compact discs:

Solvent	Density (g/mL) at 20°C
Chloroform	1.492
Diethyl ether	0.714
Ethanol	0.789
Isopropanol	0.785
Toluene	0.867

(a) If a 15.00-mL sample of CD cleaner weighs 11.775 g at 20°C, which solvent is most likely to be present?
(b) The chemist analyzing the cleaner calibrates her equipment and finds that the pipet is accurate to ±0.02 mL, and the balance is accurate to ±0.003 g. Is this equipment precise enough to distinguish between ethanol and isopropanol?

1.56 A laboratory instructor gives a sample of amino-acid powder to each of four students, I, II, III, and IV, and they weigh the samples. The true value is 8.72 g. Their results for three trials are
I: 8.72 g, 8.74 g, 8.70 g   II: 8.56 g, 8.77 g, 8.83 g
III: 8.50 g, 8.48 g, 8.51 g   IV: 8.41 g, 8.72 g, 8.55 g
(a) Calculate the average mass from each set of data, and tell which set is the most accurate.
(b) Precision is a measure of the average of the deviations of each piece of data from the average value. Which set of data is the most precise? Is this set also the most accurate?
(c) Which set of data is both the most accurate and most precise?
(d) Which set of data is both the least accurate and least precise?

1.57 The following dartboards illustrate the types of errors often seen in measurements. The bull's-eye represents the actual value, and the darts represent the data.

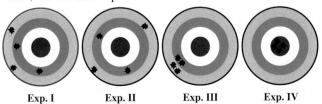

| Exp. I | Exp. II | Exp. III | Exp. IV |

(a) Which experiments yield the same average result?
(b) Which experiment(s) display(s) high precision?
(c) Which experiment(s) display(s) high accuracy?
(d) Which experiment(s) show(s) a systematic error?

Comprehensive Problems

Problems with an asterisk (*) are more challenging.

1.58 To make 2.000 gal of a powdered sports drink, a group of students measure out 2.000 gal of water with 500.-mL, 50.-mL, and 5-mL graduated cylinders. Show how they could get closest to 2.000 gal of water, using these cylinders the fewest times.

1.59 Two blank potential energy diagrams (see Figure 1.3) appear below. Beneath each diagram are objects to place in the diagram. Draw the objects on the dashed lines to indicate higher or lower potential energy and label each case as more or less stable:

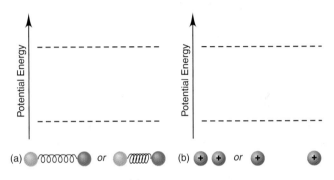

(a) Two balls attached to a relaxed *or* a compressed spring
(b) Two positive charges near *or* apart from each other

1.60 Soft drinks are about as dense as water (1.0 g/cm^3); many common metals, including iron, copper, and silver, have densities around 9.5 g/cm^3. (a) What is the mass of the liquid in a standard 12-oz bottle of diet cola? (b) What is the mass of a dime? (*Hint:* A stack of five dimes has a volume of about 1 cm^3.)

1.61 Suppose your dorm room is 11 ft wide by 12 ft long by 8.5 ft high and has an air conditioner that exchanges air at a rate of 1200 L/min. How long would it take the air conditioner to exchange the air in your room once?

1.62 In 1933, the United States went off the international gold standard, and the price of gold increased from $20.00 to $35.00/troy oz. The twenty-dollar gold piece, known as the double eagle, weighed 33.436 g and was 90.0% gold by mass. (a) What was the value of the gold in the double eagle before and after the price change? (b) How many coins could be made from 50.0 troy oz of gold? (c) How many coins could be made from 2.00 in^3 of gold (1 troy oz = 31.1 g; d of gold = 19.3 g/cm^3)?

1.63 An Olympic-size pool is 50.0 m long and 25.0 m wide. (a) How many gallons of water (d = 1.0 g/mL) are needed to fill the pool to an average depth of 4.8 ft? (b) What is the mass (in kg) of water in the pool?

* **1.64** At room temperature (20°C) and pressure, the density of air is 1.189 g/L. An object will float in air if its density is less than that of air. In a buoyancy experiment with a new plastic, a chemist creates a rigid, thin-walled ball that weighs 0.12 g and has a volume of 560 cm^3.
(a) Will the ball float if it is evacuated?
(b) Will it float if filled with carbon dioxide (d = 1.830 g/L)?

(c) Will it float if filled with hydrogen ($d = 0.0899$ g/L)?

(d) Will it float if filled with oxygen ($d = 1.330$ g/L)?

(e) Will it float if filled with nitrogen ($d = 1.165$ g/L)?

(f) For any case that will float, how much weight must be added to make the ball sink?

1.65 Asbestos is a fibrous silicate mineral with remarkably high tensile strength. But it is no longer used because airborne asbestos particles can cause lung cancer. Grunerite, a type of asbestos, has a *tensile strength* of 3.5×10^4 kg/cm^2 (thus, a bar of grunerite with a 1-cm^2 cross-sectional area can hold up to 3.5×10^4 kg). The tensile strengths of aluminum and Steel No. 5137 are 2.5×10^4 lb/in^2 and 5.0×10^4 lb/in^2, respectively. Calculate the cross-sectional area (in cm^2) of bars of aluminum and Steel No. 5137 that have the same tensile strength as a bar of grunerite with a cross-sectional area of 25 mm^2.

1.66 According to the lore of ancient Greece, Archimedes discovered the displacement method of density determination while bathing and used it to find the composition of the king's crown. If a crown weighing 4 lb 13 oz displaces 186 mL of water, is the crown made of pure gold ($d = 19.3$ g/cm^3)?

1.67 Earth's oceans have an average depth of 3800 m, a total area of 3.63×10^8 km^2, and an average concentration of dissolved gold of 5.8×10^{-9} g/L. (a) How many grams of gold are in the oceans? (b) How many m^3 of gold are in the oceans? (c) If a recent price of gold was \$370.00/troy oz, what is the value of gold in the oceans (1 troy oz = 31.1 g; d of gold = 19.3 g/cm^3)?

1.68 For the year 2002, worldwide production of aluminum was 24.4 million metric tons (t). (a) How many pounds of aluminum were produced? (b) What was its volume in cubic feet (1 t = 1000 kg; d of aluminum = 2.70 g/cm^3)?

1.69 Liquid nitrogen is obtained from liquefied air and is used industrially to prepare frozen foods. It boils at 77.36 K. (a) What is this temperature in °C? (b) What is this temperature in °F? (c) At the boiling point, the density of the liquid is 809 g/L and that of the gas is 4.566 g/L. How many liters of liquid nitrogen are produced when 895.0 L of nitrogen gas is liquefied at 77.36 K?

1.70 The speed of sound varies according to the material through which it travels. Sound travels at 5.4×10^3 cm/s through rubber and at 1.97×10^4 ft/s through granite. Calculate each of these speeds in m/s.

1.71 If a raindrop weighs 65 mg on average and 5.1×10^5 raindrops fall on a lawn every minute, what mass (in kg) of rain falls on the lawn in 1.5 h?

* 1.72 The Environmental Protection Agency (EPA) has proposed a new safety standard for microparticulates in air: for particles up to 2.5 μm in diameter, the maximum allowable amount is 50. μg/m^3. If your 10.0 ft × 8.25 ft × 12.5 ft dorm room just meets the new EPA standard, how many of these particles are in your room? How many are in each 0.500-L breath you take? (Assume the particles are spheres of diameter 2.5 μm and made primarily of soot, a form of carbon with a density of 2.5 g/cm^3.)

* 1.73 Earth's surface area is 5.10×10^8 km^2, and its crust has a mean thickness of 35 km and mean density of 2.8 g/cm^3. The two most abundant elements in the crust are oxygen (4.55×10^5 g/metric ton, t) and silicon (2.72×10^5 g/t), and the two rarest non-radioactive elements are ruthenium and rhodium, each with an abundance of 1×10^{-4} g/t. What is the total mass of each of these elements in Earth's crust (1 t = 1000 kg)?

* 1.74 The three states of matter differ greatly in their viscosity, a measure of their resistance to flow. Rank the three states from highest to lowest viscosity. Explain in submicroscopic terms.

* 1.75 If a temperature scale were based on the freezing point (5.5°C) and boiling point (80.1°C) of benzene and the temperature difference between these points was divided into 50 units (called °X), what would be the freezing and boiling points of water in °X? (See Figure 1.8.)

CHAPTER TWO

The Components of Matter

Taking It Apart Like this machine transmission, every day matter consists of simpler components that are themselves made of even simpler parts. In this chapter, you'll learn their properties and discover how chemists identify the components to see how they combine.

Key Principles

◆ A *substance* is matter with a *fixed composition:* an *element* consists of a single type of *atom;* a *compound* consists of *molecules* (or formula units) made up of two or more atoms combined in a specific ratio. A *mixture* consists of two or more substances intermingled physically and, thus, has a *variable composition.*

◆ According to *Dalton's atomic theory,* atoms of a given element have a unique mass and other properties. Mass is *conserved* during a chemical reaction because all the atoms of the reacting substances are just rearranged into different substances.

◆ Atoms have a structure made of three types of *subatomic particles:* positively charged *protons* and uncharged *neutrons* make up the *nucleus,* which contains nearly all the mass of the atom; negatively charged *electrons* move continuously around the nucleus. All the atoms of a given element have the same number of protons (*atomic number, Z*). Atoms are *neutral* because the number of protons equals the number of electrons.

◆ *Isotopes* of an element are atoms of different masses because they have different numbers of neutrons, but they behave the same way chemically. The atomic *mass* of an element is the weighted *average* of the masses of its naturally occurring isotopes.

◆ In the *periodic table,* the elements are arranged by increasing atomic number into a grid of horizontal rows (*periods*) and vertical columns (*groups*). *Metals* occupy most of the lower left portion, and *non-metals* are found in the upper right corner, with *metalloids* in between. Elements in a group have similar properties.

◆ The electrons of atoms are involved in forming compounds. In *ionic bonding,* metal atoms *transfer* electrons to nonmetal atoms, and the resulting charged particles (*ions*) attract each other into solid arrays. In *covalent bonding,* nonmetal atoms *share* electrons and usually form individual molecules. Each compound has a unique name, formula, and mass based on its component elements.

◆ Unlike compounds, mixtures can be *separated by physical means* into their components. A *heterogeneous mixture* has a non-uniform composition with visible boundaries between the components. A *homogeneous mixture* (*solution*) has a uniform composition because the components (elements and/or compounds) are mixed as individual atoms, ions, or molecules.

Outline

Questioning what something is made of was a common practice even among the philosophers of ancient Greece. They believed that everything was made of one or, at most, a few elemental substances (elements). Some believed this substance to be water, others thought it was air, and still others believed there were four elements—fire, air, water, and earth.

Democritus (c. 460–370 BC), the father of atomism, focused on the ultimate components of *all* substances, and his reasoning went something like this: If you cut a piece of, say, copper smaller and smaller, you must eventually reach a particle of copper so small that it can no longer be cut. Therefore, matter is ultimately composed of indivisible particles, with nothing between them but empty space. He called the particles *atoms* (Greek *atomos,* "uncuttable"). However, Aristotle (384–322 BC) held that it was impossible for "nothing" to exist, and his influence suppressed the concept of atoms for 2000 years.

Finally, in the 17th century, the great English scientist Robert Boyle argued that an element is composed of "simple Bodies, . . . of which all mixed Bodies are compounded." Boyle's hypothesis is remarkably similar to today's idea of an element, in which the "simple Bodies" are atoms. Further studies in the 18th century gave rise to laws concerning the relative masses of substances that react with each other. Then, at the beginning of the 19th century, John Dalton proposed an atomic model that explained these mass laws. By that century's close, however, further observation exposed the need to revise Dalton's model. A burst of creativity in the early 20th century gave rise to a picture of the atom with a complex internal structure, which led to our current model.

2.1 ELEMENTS, COMPOUNDS, AND MIXTURES: AN ATOMIC OVERVIEW

Matter can be broadly classified into three types—elements, compounds, and mixtures. An **element** is the simplest type of matter with unique physical and chemical properties. *An element consists of only one kind of atom.* Therefore, it cannot be broken down into a simpler type of matter by any physical or chemical methods. An element is one kind of **pure substance** (or just **substance**), matter whose composition is fixed. Each element has a name, such as silicon, oxygen, or copper. A sample of silicon contains only silicon atoms. A key point to remember is that the *macroscopic* properties of a piece of silicon, such as color, density, and combustibility, are different from those of a piece of copper because silicon atoms are different from copper atoms; in other words, *each element is unique because the properties of its atoms are unique.*

Most elements exist in nature as populations of atoms. Figure 2.1A shows atoms of a gaseous element such as neon. Several elements occur naturally as molecules: a **molecule** is an independent structural unit consisting of two or more

Figure 2.1 Elements, compounds, and mixtures on the atomic scale. A, Most elements consist of a large collection of identical atoms. **B,** Some elements occur as molecules. **C,** A molecule of a compound consists of characteristic numbers of atoms of two or more elements chemically bound together. **D,** A mixture contains the individual units of two or more elements and/or compounds that are physically intermingled. The samples shown here are gases, but elements, compounds, and mixtures occur as liquids and solids also.

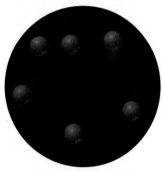

A Atoms of an element

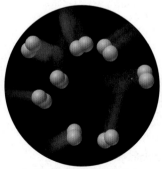

B Molecules of an element

C Molecules of a compound

D Mixture of two elements and a compound

Table 2.1 Some Properties of Sodium, Chlorine, and Sodium Chloride							
Property	Sodium	+		Chlorine	$\longrightarrow$		Sodium Chloride
Melting point	97.8°C			−101°C			801°C
Boiling point	881.4°C			−34°C			1413°C
Color	Silvery			Yellow-green			Colorless (white)
Density	0.97 g/cm^3			0.0032 g/cm^3			2.16 g/cm^3
Behavior in water	Reacts			Dissolves slightly			Dissolves freely

atoms chemically bound together (Figure 2.1B). Elemental oxygen, for example, occurs in air as *diatomic* (two-atom) molecules.

A **compound** is a type of matter composed of *two or more different elements that are chemically bound together* (Figure 2.1C). Ammonia, water, and carbon dioxide are some common compounds. One defining feature of a compound is that *the elements are present in fixed parts by mass* (fixed mass ratio). Because of this fixed composition, *a compound is also considered a substance.* Any molecule of the compound has the same fixed parts by mass because it consists of *fixed numbers* of atoms of the component elements. For example, any sample of ammonia is 14 parts nitrogen by mass plus 3 parts hydrogen by mass. Since 1 nitrogen atom has 14 times the mass of 1 hydrogen atom, ammonia must consist of 1 nitrogen atom for every 3 hydrogen atoms:

Ammonia is 14 parts N and 3 parts H by mass

1 N atom has 14 times the mass of 1 H atom

Therefore, ammonia has 1 N atom for every 3 H atoms.

Another defining feature of a compound is that *its properties are different from those of its component elements.* Table 2.1 shows a striking example. Soft, silvery sodium metal and yellow-green, poisonous chlorine gas have very different properties from the compound they form—white, crystalline sodium chloride, or common table salt! Unlike an element, a compound *can* be broken down into simpler substances—its component elements. For example, an electric current breaks down molten sodium chloride into metallic sodium and chlorine gas. Note that this breakdown is a *chemical change,* not a physical one.

Figure 2.1D depicts a **mixture,** a group of two or more substances (elements and/or compounds) that are physically intermingled. In contrast to a compound, *the components of a mixture **can** vary in their parts by mass.* Because its composition is not fixed, a mixture is *not* a substance. A mixture of the two compounds sodium chloride and water, for example, can have many different parts by mass of salt to water. At the atomic scale, a mixture is merely a group of the individual units that make up its component elements and/or compounds. Therefore, *a mixture retains many of the properties of its components.* Saltwater, for instance, is colorless like water and tastes salty like sodium chloride. Unlike compounds, mixtures can be separated into their components by *physical changes;* chemical changes are not needed. For example, the water in saltwater can be boiled off, a physical process that leaves behind the sodium chloride.

SECTION SUMMARY

All matter exists as either elements, compounds, or mixtures. An element consists of only one type of atom. A compound contains two or more elements in chemical combination; it exhibits different properties from its component elements. The elements of a compound occur in fixed parts by mass because each unit of the compound has fixed numbers of each type of atom. Elements and compounds are referred to as substances because their compositions are fixed. A mixture consists of two or more substances mixed together, not chemically combined. The components retain their individual properties and can be present in any proportion.

2.2   THE OBSERVATIONS THAT LED TO AN ATOMIC VIEW OF MATTER

Any model of the composition of matter had to explain two extremely important chemical observations that were well established by the end of the 18th century: the *law of mass conservation* and the *law of definite (or constant) composition*. As you'll see, John Dalton's atomic theory explained these laws and another observation now known as the *law of multiple proportions*.

Mass Conservation

The most fundamental chemical observation of the 18th century was the **law of mass conservation:** *the total mass of substances does not change during a chemical reaction.* The *number* of substances may change and, by definition, their properties must, but the *total amount* of matter remains constant. Antoine Lavoisier (1743–1794), the great French chemist and statesman, had first stated this law on the basis of experiments in which he reacted mercury with oxygen. He found the mass of oxygen plus the mass of mercury always equaled the mass of mercuric oxide that formed.

Even in a complex biochemical change within an organism, such as the metabolism of the sugar glucose, which involves many reactions, mass is conserved:

$$180 \text{ g glucose} + 192 \text{ g oxygen gas} \longrightarrow 264 \text{ g carbon dioxide} + 108 \text{ g water}$$
$$372 \text{ g material before change} \longrightarrow 372 \text{ g material after change}$$

Mass conservation means that, based on all chemical experience, *matter cannot be created or destroyed.* (As you'll see later, however, mass *does* change in nuclear reactions, although *not* in chemical reactions.)

Definite Composition

Another fundamental chemical observation is summarized as the **law of definite (or constant) composition:** *no matter what its source, a particular compound is composed of the same elements in the same parts (fractions) by mass.* The **fraction by mass (mass fraction)** is that part of the compound's mass contributed by the element. It is obtained by dividing the mass of each element by the total mass of compound. The **percent by mass (mass percent, mass %)** is the fraction by mass expressed as a percentage.

Consider calcium carbonate, the major compound in marble. It is composed of three elements—calcium, carbon, and oxygen—and each is present in a fixed fraction (or percent) by mass. The following results are obtained for the elemental mass composition of 20.0 g of calcium carbonate (for example, 8.0 g of calcium/20.0 g = 0.40 parts of calcium):

Analysis by Mass (grams/20.0 g)	Mass Fraction (parts/1.00 part)	Percent by Mass (parts/100 parts)
8.0 g calcium	0.40 calcium	40% calcium
2.4 g carbon	0.12 carbon	12% carbon
9.6 g oxygen	0.48 oxygen	48% oxygen
20.0 g	1.00 part by mass	100% by mass

As you can see, the sum of the mass fractions (or mass percents) equals 1.00 part (or 100%) by mass. The law of definite composition tells us that pure samples of calcium carbonate, no matter where they come from, always contain these elements in the same percents by mass (Figure 2.2).

Because a given element always constitutes the same mass fraction of a given compound, we can use that mass fraction to find the actual mass of the element in any sample of the compound:

$$\text{Mass of element} = \text{mass of compound} \times \frac{\text{part by mass of element}}{\text{one part by mass of compound}}$$

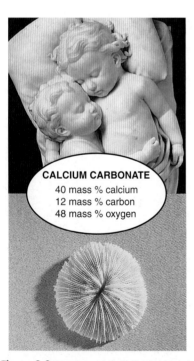

CALCIUM CARBONATE
40 mass % calcium
12 mass % carbon
48 mass % oxygen

Figure 2.2 The law of definite composition. Calcium carbonate is found naturally in many forms, including marble (*top*), coral (*bottom*), chalk, and seashells. The mass percents of its component elements do not change regardless of the compound's source.

Or, more simply, mass analysis tells us the parts by mass, so we can use that directly with *any* mass unit and skip the need to find the mass fraction first:

Mass of element in sample

$$= \text{mass of compound in sample} \times \frac{\text{mass of element in compound}}{\text{mass of compound}} \quad \textbf{(2.1)}$$

SAMPLE PROBLEM 2.1 Calculating the Mass of an Element in a Compound

Problem Pitchblende is the most commercially important compound of uranium. Analysis shows that 84.2 g of pitchblende contains 71.4 g of uranium, with oxygen as the only other element. How many grams of uranium can be obtained from 102 kg of pitchblende?

Plan We have to find the mass of uranium in a known mass of pitchblende, given the mass of uranium in a different mass of pitchblende. The mass ratio of uranium/pitchblende is the same for any sample of pitchblende. Therefore, as shown by Equation 2.1, we multiply the mass (in kg) of pitchblende by the ratio of uranium to pitchblende that we construct from the mass analysis. This gives the mass (in kg) of uranium, and we just convert kilograms to grams.

Solution Finding the mass (kg) of uranium in 102 kg of pitchblende:

$$\text{Mass (kg) of uranium} = \text{mass (kg) of pitchblende} \times \frac{\text{mass (kg) of uranium in pitchblende}}{\text{mass (kg) of pitchblende}}$$

$$\text{Mass (kg) of uranium} = 102 \text{ kg pitchblende} \times \frac{71.4 \text{ kg uranium}}{84.2 \text{ kg pitchblende}} = 86.5 \text{ kg uranium}$$

Converting the mass of uranium from kg to g:

$$\text{Mass (g) of uranium} = 86.5 \text{ kg uranium} \times \frac{1000 \text{ g}}{1 \text{ kg}}$$

$$= 8.65 \times 10^4 \text{ g uranium}$$

Check The analysis showed that most of the mass of pitchblende is due to uranium, so the large mass of uranium makes sense. Rounding off to check the math gives:

$$\sim 100 \text{ kg pitchblende} \times \frac{70}{85} = 82 \text{ kg uranium}$$

FOLLOW-UP PROBLEM 2.1 How many metric tons (t) of oxygen are combined in a sample of pitchblende that contains 2.3 t of uranium? (*Hint:* Remember that oxygen is the only other element present.)

Mass (kg) of pitchblende

multiply by mass ratio of uranium to pitchblende from analysis

Mass (kg) of uranium

1 kg = 1000 g

Mass (g) of uranium

Multiple Proportions

Dalton described a phenomenon that occurs when two elements form more than one compound. His observation is now called the **law of multiple proportions:** *if elements A and B react to form two compounds, the different masses of B that combine with a fixed mass of A can be expressed as a ratio of small whole numbers.* Consider two compounds that form from carbon and oxygen; for now, let's call them carbon oxides I and II. They have very different properties. For example, measured at the same temperature and pressure, the density of carbon oxide I is 1.25 g/L, whereas that of II is 1.98 g/L. Moreover, I is poisonous and flammable, but II is not. Analysis shows that their compositions by mass are

        Carbon oxide I: 57.1 mass % oxygen and 42.9 mass % carbon
        Carbon oxide II: 72.7 mass % oxygen and 27.3 mass % carbon

To see the phenomenon of multiple proportions, we use the mass percents of oxygen and of carbon in each compound to find the masses of these elements in a given mass, for example, 100 g, of each compound. Then we divide the mass of oxygen by the mass of carbon in each compound to obtain the mass of oxygen that combines with a fixed mass of carbon:

	Carbon Oxide I	Carbon Oxide II
g oxygen/100 g compound	57.1	72.7
g carbon/100 g compound	42.9	27.3
g oxygen/g carbon	$\dfrac{57.1}{42.9} = 1.33$	$\dfrac{72.7}{27.3} = 2.66$

If we then divide the grams of oxygen per gram of carbon in II by that in I, we obtain a ratio of small whole numbers:

$$\frac{2.66 \text{ g oxygen/g carbon in II}}{1.33 \text{ g oxygen/g carbon in I}} = \frac{2}{1}$$

The law of multiple proportions tells us that in two compounds of the same elements, the mass fraction of one element relative to the other element changes in *increments based on ratios of small whole numbers.* In this case, the ratio is 2:1 for a given mass of carbon, II contains *2 times* as much oxygen as I, not 1.583 times, 1.716 times, or any other intermediate amount. As you'll see next, Dalton's theory allows us to explain the composition of carbon oxides I and II on the atomic scale.

SECTION SUMMARY
Three fundamental observations known as the mass laws state that (1) the total mass remains constant during a chemical reaction; (2) any sample of a given compound has the same elements present in the same parts by mass; and (3) in different compounds of the same elements, the masses of one element that combine with a fixed mass of the other can be expressed as a ratio of small whole numbers.

2.3   DALTON'S ATOMIC THEORY

With almost 200 years of hindsight, it may be easy to see how the mass laws could be explained by an atomic model—matter existing in indestructible units, each with a particular mass—but it was a major breakthrough in 1808 when John Dalton (1766–1844) presented his atomic theory of matter in *A New System of Chemical Philosophy.*

Postulates of the Atomic Theory

Dalton expressed his theory in a series of postulates. Like most great thinkers, Dalton incorporated the ideas of others into his own to create the new theory. As we go through the postulates, which are presented here in modern terms, let's see which were original and which came from others.

1. All matter consists of **atoms,** tiny indivisible particles of an element that cannot be created or destroyed. (Derives from the "eternal, indestructible atoms" proposed by Democritus more than 2000 years earlier and conforms to mass conservation as stated by Lavoisier.)
2. Atoms of one element *cannot* be converted into atoms of another element. In chemical reactions, the atoms of the original substances recombine to form different substances. (Rejects the earlier belief by alchemists that one element could be magically transformed into another, such as lead into gold.)
3. Atoms of an element are identical in mass and other properties and are different from atoms of any other element. (Contains Dalton's major new ideas: unique mass and properties for all the atoms of a given element.)
4. Compounds result from the chemical combination of a specific ratio of atoms of different elements. (Follows directly from the fact of definite composition.)

How the Theory Explains the Mass Laws

Let's see how Dalton's postulates explain the mass laws:

- *Mass conservation.* Atoms cannot be created or destroyed (postulate 1) or converted into other types of atoms (postulate 2). Since each type of atom has a fixed mass (postulate 3), a chemical reaction, in which atoms are just combined differently with each other, cannot possibly result in a mass change.
- *Definite composition.* A compound is a combination of a *specific* ratio of different atoms (postulate 4), each of which has a particular mass (postulate 3). Thus, each element in a compound constitutes a fixed fraction of the total mass.
- *Multiple proportions.* Atoms of an element have the same mass (postulate 3) and are indivisible (postulate 1). The masses of element B that combine with a fixed mass of element A give a small, whole-number ratio because different numbers of B atoms combine with each A atom in different compounds.

The *simplest* arrangement consistent with the mass data for carbon oxides I and II in our earlier example is that one atom of oxygen combines with one atom of carbon in compound I (carbon monoxide) and that two atoms of oxygen combine with one atom of carbon in compound II (carbon dioxide) (Figure 2.3).

Dalton's atomic model was crucial to the idea that masses of reacting elements could be explained in terms of atoms, and it led to experiments to learn the relative masses of atoms in compounds. However, the model did not explain why atoms bond as they do: for example, why do two, and not three, hydrogen atoms bond with one oxygen atom in water? Also, Dalton's model did not account for the charged particles that were being observed in experiments. Clearly, a more complex atomic model was needed.

Carbon oxide I (carbon monoxide)   Carbon oxide II (carbon dioxide)

Figure 2.3 The atomic basis of the law of multiple proportions. Carbon and oxygen combine to form carbon oxide I (carbon monoxide) and carbon oxide II (carbon dioxide). The masses of oxygen in the two compounds relative to a fixed mass of carbon are in a ratio of small whole numbers.

SECTION SUMMARY
Dalton's atomic theory explained the mass laws by proposing that all matter consists of indivisible, unchangeable atoms of fixed, unique mass. Mass is constant during a reaction because atoms form new combinations; each compound has a fixed mass fraction of each of its elements because it is composed of a fixed number of each type of atom; and different compounds of the same elements exhibit multiple proportions because they each consist of whole atoms.

2.4 THE OBSERVATIONS THAT LED TO THE NUCLEAR ATOM MODEL

The path of discovery is often winding and unpredictable. Basic research into the nature of electricity eventually led to the discovery of *electrons,* negatively charged particles that are part of all atoms. Soon thereafter, other experiments revealed that the atom has a *nucleus*—a tiny, central core of mass and positive charge. In this section, we examine some key experiments that led to our current model of the atom.

Discovery of the Electron and Its Properties

Nineteenth-century investigators of electricity knew that matter and electric charge were somehow related. What they did not know, however, was what an electric current itself might consist of. Some investigators tried passing current from a high-voltage source through nearly evacuated glass tubes fitted with metal electrodes that were sealed in place and connected to an external source of electricity. When the power was turned on, a "ray" could be seen striking the phosphor-coated end of the tube and emitting a glowing spot of light. The rays were called **cathode rays** because they originated at the negative electrode (cathode) and moved

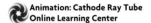
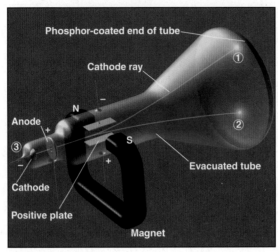

OBSERVATION	CONCLUSION
1. Ray bends in magnetic field	Consists of charged particles
2. Ray bends toward positive plate in electric field	Consists of negative particles
3. Ray is identical for any cathode	Particles found in all matter

Figure 2.4 Experiments to determine the properties of cathode rays. A cathode ray forms when high voltage is applied to a partially evacuated tube. The ray passes through a hole in the anode and hits the coated end of the tube to produce a glow.

to the positive electrode (anode). Cathode rays typically travel in a straight line, but in a magnetic field the path is bent, indicating that the particles are charged, and in an electric field the path bends toward the positive plate. The ray is identical no matter what metal is used as the cathode (Figure 2.4). It was concluded that cathode rays consist of negatively charged particles found in all matter. The rays appear when these particles collide with the few remaining gas molecules in the evacuated tube. Cathode ray particles were later named *electrons.*

In 1897, the British physicist J. J. Thomson (1856–1940) used magnetic and electric fields to measure the ratio of the cathode ray particle's mass to its charge. By comparing this value with the mass/charge ratio for the lightest charged particle in solution, Thomson estimated that the cathode ray particle weighed less than $\frac{1}{1000}$ as much as hydrogen, the lightest atom! He was shocked because this implied that, contrary to Dalton's atomic theory, *atoms are divisible into even smaller particles.* Fellow scientists reacted at first with disbelief, and some even thought Thomson was joking.

In 1909, the American physicist Robert Millikan (1868–1953) measured the *charge* of the electron. He did so by observing the movement of tiny droplets of the "highest grade clock oil" in an apparatus that contained electrically charged plates and an x-ray source (Figure 2.5). X-rays knocked electrons from gas molecules in the air, and as an oil droplet fell through a hole in the positive (upper) plate, the electrons stuck to the drop, giving it a negative charge. With the electric field off, Millikan measured the mass of the droplet from its rate of fall. By turning on the field and varying its strength, he could make the drop fall more slowly, rise, or pause suspended. From these data, Millikan calculated the total charge of the droplet.

After studying many droplets, Millikan calculated that the various charges of the droplets were always some *whole-number multiple of a minimum charge.* He reasoned that different oil droplets picked up different numbers of electrons, so this minimum charge must be that of the electron itself. The value, which he calculated almost 100 years ago, is within 1% of the modern value of the electron's charge, -1.602×10^{-19} C (C stands for *coulomb,* the SI unit of charge). Using the electron's

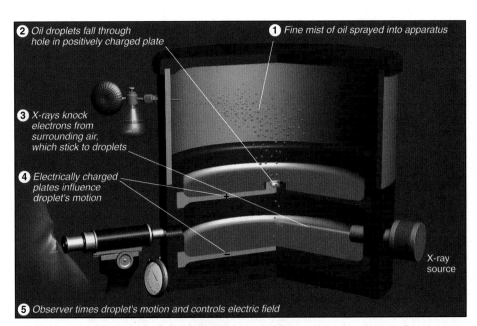

② Oil droplets fall through hole in positively charged plate

① Fine mist of oil sprayed into apparatus

③ X-rays knock electrons from surrounding air, which stick to droplets

④ Electrically charged plates influence droplet's motion

X-ray source

⑤ Observer times droplet's motion and controls electric field

Figure 2.5 Millikan's oil-drop experiment for measuring an electron's charge. The motion of a given oil droplet depends on the variation in electric field and the total charge on the droplet, which depends in turn on the number of attached electrons. Millikan reasoned that the total charge must be some whole-number multiple of the charge of the electron.

mass/charge ratio from work by Thomson and others and this value for the electron's charge, let's calculate the electron's *extremely* small mass the way Millikan did:

$$\text{Mass of electron} = \frac{\text{mass}}{\text{charge}} \times \text{charge} = \left(-5.686 \times 10^{-12} \frac{\text{kg}}{\text{C}}\right)(-1.602 \times 10^{-19} \text{ C})$$

$$= 9.109 \times 10^{-31} \text{ kg} = 9.109 \times 10^{-28} \text{ g}$$

Discovery of the Atomic Nucleus

Clearly, the properties of the electron posed problems about the inner structure of atoms. If everyday matter is electrically neutral, the atoms that make it up must be neutral also. But if atoms contain negatively charged electrons, what positive charges balance them? And if an electron has such an incredibly tiny mass, what accounts for an atom's much larger mass? To address these issues, Thomson proposed a model of a spherical atom composed of diffuse, positively charged matter, in which electrons were embedded like "raisins in a plum pudding."

Near the turn of the 20th century, French scientists discovered radioactivity, the emission of particles and/or radiation from atoms of certain elements. Just a few years later, in 1910, the New Zealand–born physicist Ernest Rutherford (1871–1937) used one type of radioactive particle in a series of experiments that solved this dilemma of atomic structure.

Figure 2.6 (on the next page) is a three-part representation of Rutherford's experiment. Tiny, dense, positively charged alpha (α) particles emitted from radium were aimed, like minute projectiles, at thin gold foil. The figure illustrates (A) the "plum pudding" hypothesis, (B) the apparatus used to measure the deflection (scattering) of the α particles from the light flashes created when the particles struck a circular, coated screen, and (C) the actual result.

A Hypothesis: Expected result based on "plum pudding" model

B Experiment

C Actual Result

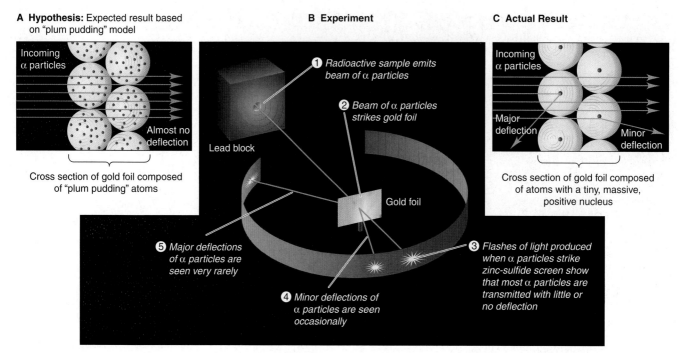

Figure 2.6 Rutherford's α-scattering experiment and discovery of the atomic nucleus. A, HYPOTHESIS: Atoms consist of electrons embedded in diffuse, positively charged matter, so the speeding α particles should pass through the gold foil with, at most, minor deflections. **B,** EXPERIMENT: α particles emit a flash of light when they pass through the gold atoms and hit a phosphor-coated screen. **C,** RESULTS: Occasional minor deflections and very infrequent major deflections are seen. This means very high mass and positive charge are concentrated in a small region within the atom, the nucleus.

 **Animation: Rutherford's Experiment**
Online Learning Center

With Thomson's model in mind, Rutherford expected only minor, if any, deflections of the α particles because they should act as tiny, dense, positively charged "bullets" and go right through the gold atoms. According to the model, the embedded electrons could not deflect the α particles any more than a Ping-Pong ball could deflect a speeding baseball. Initial results confirmed this, but soon the unexpected happened. As Rutherford recalled: "Then I remember two or three days later Geiger [one of his coworkers] coming to me in great excitement and saying, 'We have been able to get some of the α particles coming backwards . . .' It was quite the most incredible event that has ever happened to me in my life. It was almost as incredible as if you fired a 15-inch shell at a piece of tissue paper and it came back and hit you."

The data showed that very few α particles were deflected at all, and that only 1 in 20,000 was deflected by more than 90° ("coming backwards"). It seemed that these few α particles were being repelled by something small, dense, and positive within the gold atoms. From the data, Rutherford calculated that *an atom is mostly space occupied by electrons,* but in the center of that space is a tiny region, which he called the **nucleus,** that contains *all the positive charge and essentially all the mass of the atom.* He proposed that positive particles lay within the nucleus and called them *protons.* Rutherford's model explained the charged nature of matter, but it could not account for all the atom's mass. After more than 20 years, this issue was resolved when, in 1932, James Chadwick discovered the *neutron,* an uncharged dense particle that also resides in the nucleus.

SECTION SUMMARY

Several major discoveries at the turn of the 20th century led to our current model of atomic structure. Cathode rays were shown to consist of negative particles (electrons) that exist in all matter. J. J. Thomson measured their mass/charge ratio and concluded that they are much smaller and lighter than atoms. Robert Millikan determined the charge of the electron, which he combined with other data to calculate its mass. Ernest Rutherford proposed that atoms consist of a tiny, massive, positive nucleus surrounded by electrons.

2.5 THE ATOMIC THEORY TODAY

For over 200 years, scientists have known that all matter consists of atoms, and they have learned astonishing things about them. Dalton's tiny indivisible particles have given way to atoms with "fuzzy," indistinct boundaries and an elaborate internal architecture of subatomic particles. In this section, we examine our current model and begin to see how the properties of subatomic particles affect the properties of atoms.

Structure of the Atom

An *atom* is an electrically neutral, spherical entity composed of a positively charged central nucleus surrounded by one or more negatively charged electrons (Figure 2.7). The electrons move rapidly within the available atomic volume, held there by the attraction of the nucleus. The nucleus is incredibly dense: it contributes 99.97% of the atom's mass but occupies only about 1 ten-trillionth of its volume. (A nucleus the size of a period on this page would weigh about 100 tons, as much as 50 cars!) An atom's diameter ($\sim 10^{-10}$ m) is about 10,000 times the diameter of its nucleus ($\sim 10^{-14}$ m).

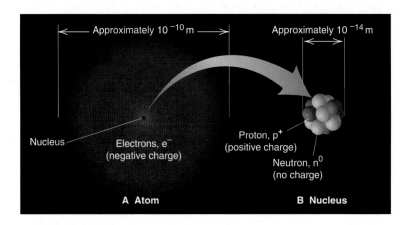

A Atom   **B Nucleus**

Figure 2.7 General features of the atom. A, A "cloud" of rapidly moving, negatively charged electrons occupies virtually all the atomic volume and surrounds the tiny, central nucleus. **B,** The nucleus contains virtually all the mass of the atom and consists of positively charged protons and uncharged neutrons. If the nucleus were actually the size in the figure (~ 1 cm across), the atom would be about 100 m across—slightly more than the length of a football field!

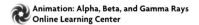
Animation: Alpha, Beta, and Gamma Rays
Online Learning Center

   An atomic nucleus consists of protons and neutrons (the only exception is the simplest hydrogen nucleus, which is a single proton). The **proton (p^+)** has a positive charge, and the **neutron (n^0)** has no charge; thus, the positive charge of the nucleus results from its protons. The *magnitude* of charge possessed by a proton is equal to that of an **electron (e^-),** but the *signs* of the charges are opposite. *An atom is neutral because the number of protons in the nucleus equals the number of electrons surrounding the nucleus.* Some properties of these three subatomic particles are listed in Table 2.2.

Table 2.2 Properties of the Three Key Subatomic Particles					
	Charge		Mass		
Name (Symbol)	Relative	Absolute (C)*	Relative (amu)[†]	Absolute (g)	Location in Atom
Proton (p^+)	1+	$+1.60218 \times 10^{-19}$	1.00727	1.67262×10^{-24}	Nucleus
Neutron (n^0)	0	0	1.00866	1.67493×10^{-24}	Nucleus
Electron (e^-)	1−	-1.60218×10^{-19}	0.00054858	9.10939×10^{-28}	Outside nucleus

*The coulomb (C) is the SI unit of charge.
[†]The atomic mass unit (amu) equals 1.66054×10^{-24} g; discussed later in this section.

Atomic Number, Mass Number, and Atomic Symbol

The **atomic number (Z)** of an element equals the number of protons in the nucleus of each of its atoms. *All atoms of a particular element have the same atomic number, and each element has a different atomic number from that of any other element.* All carbon atoms ($Z = 6$) have 6 protons, all oxygen atoms ($Z = 8$) have 8 protons, and all uranium atoms ($Z = 92$) have 92 protons. There are currently 114 known elements, of which 90 occur in nature; the remaining 24 have been synthesized by nuclear scientists.

The total number of protons and neutrons in the nucleus of an atom is its **mass number (A).** Each proton and each neutron contributes one unit to the mass number. Thus, a carbon atom with 6 protons and 6 neutrons in its nucleus has a mass number of 12, and a uranium atom with 92 protons and 146 neutrons in its nucleus has a mass number of 238.

The nuclear mass number and charge are often written with the **atomic symbol** (or *element symbol*). Every element has a symbol based on its English, Latin, or Greek name, such as C for carbon, O for oxygen, S for sulfur, and Na for sodium (Latin *natrium*). The atomic number (Z) is written as a left *sub*script and the mass number (A) as a left *super*script to the symbol, so element X would be $^{A}_{Z}X$. The mass number is the sum of protons and neutrons, so the number of neutrons (N) equals the mass number minus the atomic number:

$$\text{Number of neutrons} = \text{mass number} - \text{atomic number}, \quad \text{or} \quad N = A - Z \quad \textbf{(2.2)}$$

Thus, a chlorine atom, which is symbolized as $^{35}_{17}Cl$, has $A = 35$, $Z = 17$, and $N = 35 - 17 = 18$. Each element has its own atomic number, so we know the atomic number from the symbol. For example, every carbon atom has 6 protons. Therefore, instead of writing $^{12}_{6}C$ for carbon with mass number 12, we can write ^{12}C (spoken "carbon twelve"), with $Z = 6$ understood. Another way to write this atom is carbon-12.

Isotopes and Atomic Masses of the Elements

All atoms of an element are identical in atomic number but not in mass number. **Isotopes** of an element are atoms that have *different numbers of neutrons* and therefore different mass numbers. For example, all carbon atoms ($Z = 6$) have 6 protons and 6 electrons, but only 98.89% of naturally occurring carbon atoms have 6 neutrons in the nucleus ($A = 12$). A small percentage (1.11%) have 7 neutrons in the nucleus ($A = 13$), and even fewer (less than 0.01%) have 8 ($A = 14$). These are carbon's three naturally occurring isotopes—^{12}C, ^{13}C, and ^{14}C. Five other carbon isotopes—^{9}C, ^{10}C, ^{11}C, ^{15}C, and ^{16}C—have been created in the laboratory. Figure 2.8 depicts the atomic number, mass number, and symbol for four atoms, two of which are isotopes of the same element.

A key point is that the chemical properties of an element are primarily determined by the number of electrons, so *all isotopes of an element have nearly identical chemical behavior,* even though they have different masses.

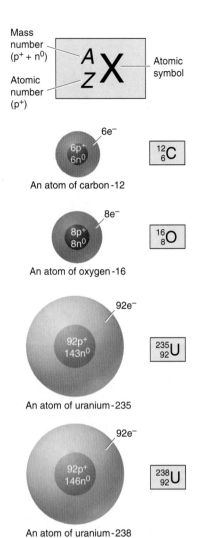

Figure 2.8 Depicting the atom. Atoms of carbon-12, oxygen-16, uranium-235, and uranium-238 are shown (nuclei not drawn to scale) with their symbolic representations. The sum of the number of protons (Z) and the number of neutrons (N) equals the mass number (A). An atom is neutral, so the number of protons in the nucleus equals the number of electrons around the nucleus. The two uranium atoms are isotopes of the element.

SAMPLE PROBLEM 2.2 Determining the Number of Subatomic Particles in the Isotopes of an Element

Problem Silicon (Si) is essential to the computer industry as a major component of semiconductor chips. It has three naturally occurring isotopes: ^{28}Si, ^{29}Si, and ^{30}Si. Determine the numbers of protons, neutrons, and electrons in each silicon isotope.

Plan The mass number (A) of each of the three isotopes is given, so we know the sum of protons and neutrons. From the elements list on the text's inside front cover, we find the atomic number (Z, number of protons), which equals the number of electrons. We obtain the number of neutrons from Equation 2.2.

Solution From the elements list, the atomic number of silicon is 14. Therefore,

$$^{28}\text{Si has } 14\text{p}^+, 14\text{e}^-, \text{ and } 14\text{n}^0 \ (28 - 14)$$
$$^{29}\text{Si has } 14\text{p}^+, 14\text{e}^-, \text{ and } 15\text{n}^0 \ (29 - 14)$$
$$^{30}\text{Si has } 14\text{p}^+, 14\text{e}^-, \text{ and } 16\text{n}^0 \ (30 - 14)$$

FOLLOW-UP PROBLEM 2.2 How many protons, neutrons, and electrons are in **(a)** $^{11}_{5}\text{Q}$? **(b)** $^{41}_{20}\text{X}$? **(c)** $^{131}_{53}\text{Y}$? What element symbols do Q, X, and Y represent?

The mass of an atom is measured *relative* to the mass of an atomic standard. The modern atomic mass standard is the carbon-12 atom. Its mass is defined as *exactly* 12 atomic mass units. Thus, the **atomic mass unit (amu)** is $\frac{1}{12}$ the mass of a carbon-12 atom. Based on this standard, the ^{1}H atom has a mass of 1.008 amu; in other words, a ^{12}C atom has almost 12 times the mass of an ^{1}H atom. We will continue to use the term *atomic mass unit* in the text, although the name of the unit has been changed to the **dalton (Da);** thus, one ^{12}C atom has a mass of 12 daltons (12 Da, or 12 amu). The atomic mass unit, which is a unit of relative mass, has an absolute mass of 1.66054×10^{-24} g.

The isotopic makeup of an element is determined by **mass spectrometry,** a method for measuring the relative masses and abundances of atomic-scale particles very precisely. In one type of mass spectrometer, atoms of a sample of, say, elemental neon are bombarded by a high-energy electron beam (Figure 2.9A). As a result, one electron is knocked off each Ne atom, and each resulting particle has one positive charge. Thus, its mass/charge ratio (*m/e*) equals the mass of an Ne atom divided by $1+$. The *m/e* values can be measured in the mass spectrometer to identify the masses of different isotopes of the element. The positively charged Ne particles are attracted toward a series of negatively charged plates with slits in them, and some of the particles pass through into an evacuated tube exposed to a magnetic field. As the particles zoom through this region, they are deflected (their paths are bent) according to their *m/e* values: the lightest particles are deflected most and the heaviest particles least. At the end of the magnetic region, the particles strike a detector, which records their relative positions and abundances (Figure 2.9B).

Figure 2.9 The mass spectrometer and its data. A, Charged particles are separated on the basis of their *m/e* values. Ne is the sample here. **B,** Data show the percent abundance of each Ne isotope.

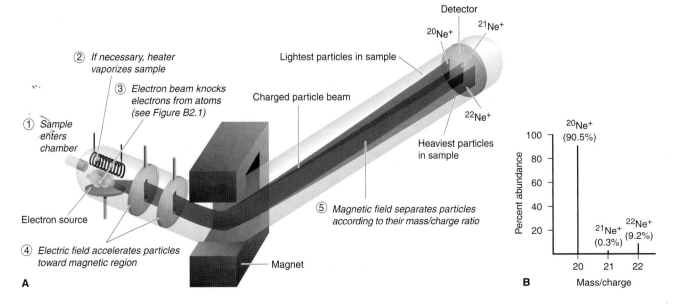

Mass spectrometry is now used to measure the mass of virtually any atom or molecule. In 2002, the Nobel Prize in chemistry was awarded for the study of proteins by mass spectrometry.

Let's see how data obtained with this instrument give us key information. Using a mass spectrometer, we measure the mass ratio of, say, ^{28}Si to ^{12}C as

$$\frac{\text{Mass of }^{28}\text{Si atom}}{\text{Mass of }^{12}\text{C standard}} = 2.331411$$

From this mass ratio, we find the **isotopic mass** of the ^{28}Si atom, the mass of the isotope relative to the mass of the standard carbon-12 isotope:

$$\text{Isotopic mass of }^{28}\text{Si} = \text{measured mass ratio} \times \text{mass of }^{12}\text{C}$$
$$= 2.331411 \times 12 \text{ amu} = 27.97693 \text{ amu}$$

Along with the isotopic mass, the mass spectrometer gives the relative abundance (fraction) of each isotope in a sample of the element. For example, the percent abundance of ^{28}Si is 92.23%. Such data allow us to calculate the **atomic mass** (also called *atomic weight*) of an element, the *average* of the masses of its naturally occurring isotopes weighted according to their abundances.

Each naturally occurring isotope of an element contributes a certain portion to the atomic mass. For instance, as just noted, 92.23% of Si atoms are ^{28}Si. Using this percent abundance as a fraction and multiplying by the isotopic mass of ^{28}Si gives the portion of the atomic mass of Si contributed by ^{28}Si:

$$\text{Portion of Si atomic mass from }^{28}\text{Si} = 27.97693 \text{ amu} \times 0.9223 = 25.8031 \text{ amu}$$
(retaining two additional significant figures)

Similar calculations give the portions contributed by ^{29}Si (28.976495 amu × 0.0467 = 1.3532 amu) and by ^{30}Si (29.973770 amu × 0.0310 = 0.9292 amu), and adding the three portions together (rounding to two decimal places at the end) gives the atomic mass of silicon:

$$\text{Atomic mass of Si} = 25.8031 \text{ amu} + 1.3532 \text{ amu} + 0.9292 \text{ amu}$$
$$= 28.0855 \text{ amu} = 28.09 \text{ amu}$$

Note that this atomic mass is an average value, and averages must be interpreted carefully. Although the average number of children in an American family in 1985 was 2.4, no family actually had 2.4 children; similarly, no individual silicon atom has a mass of 28.09 amu. But for most laboratory purposes, we consider a sample of silicon to consist of atoms with this average mass.

SAMPLE PROBLEM 2.3   Calculating the Atomic Mass of an Element

Problem Silver (Ag; Z = 47) has 46 known isotopes, but only two occur naturally, ^{107}Ag and ^{109}Ag. Given the following mass spectrometric data, calculate the atomic mass of Ag:

Isotope	Mass (amu)	Abundance (%)
^{107}Ag	106.90509	51.84
^{109}Ag	108.90476	48.16

Plan From the mass and abundance of the two Ag isotopes, we have to find the atomic mass of Ag (weighted average of the isotopic masses). We multiply each isotopic mass by its fractional abundance to find the portion of the atomic mass contributed by each isotope. The sum of the isotopic portions is the atomic mass.

Solution Finding the portion of the atomic mass from each isotope:

$$\text{Portion of atomic mass from }^{107}\text{Ag:} = \text{isotopic mass} \times \text{fractional abundance}$$
$$= 106.90509 \text{ amu} \times 0.5184 = 55.42 \text{ amu}$$

$$\text{Portion of atomic mass from }^{109}\text{Ag:} = 108.90476 \text{ amu} \times 0.4816 = 52.45 \text{ amu}$$

Finding the atomic mass of silver:

$$\text{Atomic mass of Ag} = 55.42 \text{ amu} + 52.45 \text{ amu} = 107.87 \text{ amu}$$

Mass (g) of each isotope

multiply by fractional abundance of each isotope

Portion of atomic mass from each isotope

add isotopic portions

Atomic mass

Check The individual portions seem right: ~100 amu $\times$ 0.50 = 50 amu. The portions should be almost the same because the two isotopic abundances are almost the same. We rounded each portion to four significant figures because that is the number of significant figures in the abundance values. This is the correct atomic mass (to two decimal places), as shown in the list of elements (*inside front cover*).

FOLLOW-UP PROBLEM 2.3 Boron (B; $Z = 5$) has two naturally occurring isotopes. Find the percent abundances of ^{10}B and ^{11}B given the atomic mass of B = 10.81 amu, the isotopic mass of ^{10}B = 10.0129 amu, and the isotopic mass of ^{11}B = 11.0093 amu. (*Hint:* The sum of the fractional abundances is 1. If x = abundance of ^{10}B, then $1 - x$ = abundance of ^{11}B.)

SECTION SUMMARY
An atom has a central nucleus, which contains positively charged protons and uncharged neutrons and is surrounded by negatively charged electrons. An atom is neutral because the number of electrons equals the number of protons. An atom is represented by the notation $^{A}_{Z}X$, in which Z is the atomic number (number of protons), A the mass number (sum of protons and neutrons), and X the atomic symbol. An element occurs naturally as a mixture of isotopes, atoms with the same number of protons but different numbers of neutrons. Each isotope has a mass relative to the ^{12}C mass standard. The atomic mass of an element is the average of its isotopic masses weighted according to their natural abundances and is determined by mass spectrometry.

2.6 ELEMENTS: A FIRST LOOK AT THE PERIODIC TABLE

At the end of the 18th century, Lavoisier compiled a list of the 23 elements known at that time; by 1870, 65 were known; by 1925, 88; today, there are 114 and still counting! These elements combine to form millions of compounds, so we clearly need some way to organize what we know about their behavior. By the mid-19th century, enormous amounts of information concerning reactions, properties, and atomic masses of the elements had been accumulated. Several researchers noted recurring, or *periodic,* patterns of behavior and proposed schemes to organize the elements according to some fundamental property.

In 1871, the Russian chemist Dmitri Mendeleev published the most successful of these organizing schemes in the form of a table that listed the elements by increasing atomic mass, arranged so that elements with similar chemical properties fell in the same column. The modern **periodic table of the elements,** based on Mendeleev's earlier version (but arranged by atomic number, not mass), is one of the great classifying schemes in science and has become an indispensable tool to chemists. Throughout your study of chemistry, the periodic table will guide you through an otherwise dizzying amount of chemical and physical behavior.

Organization of the Periodic Table A modern version of the periodic table appears in Figure 2.10 (on the next page) and inside the front cover. It is formatted as follows:

1. Each element has a box that contains its atomic number, atomic symbol, and atomic mass. The boxes lie in order of *increasing atomic number* (number of protons) as you move from left to right.
2. The boxes are arranged into a grid of **periods** (horizontal rows) and **groups** (vertical columns). Each period has a number from 1 to 7. Each group has a number from 1 to 8 *and* either the letter A or B. A new system, with group numbers from 1 to 18 but no letters, appears in parentheses under the number-letter designations. (Most chemists still use the number-letter system, so the text retains it, but shows the new numbering system in parentheses.)

3. The eight A groups (two on the left and six on the right) contain the *main-group,* or *representative, elements.* The ten B groups, located between Groups 2A(2) and 3A(13), contain the *transition elements.* Two horizontal series of *inner transition elements,* the lanthanides and the actinides, fit *between* the elements in Group 3B(3) and Group 4B(4) and are usually placed below the main body of the table.

At this point in the text, the clearest distinction among the elements is their classification as metals, nonmetals, or metalloids. The "staircase" line that runs from the top of Group 3A(13) to the bottom of Group 6A(16) in Period 6 is a dividing line for this classification. The **metals** (three shades of blue) appear in the large lower-left portion of the table. About three-quarters of the elements are metals, including many main-group elements and all the transition and inner

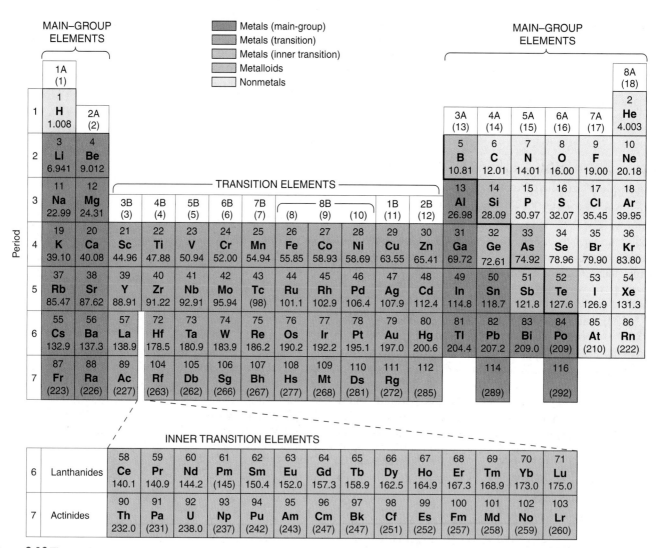

Figure 2.10 The modern periodic table. The table consists of element boxes arranged by *increasing* atomic number into groups (vertical columns) and periods (horizontal rows). Each box contains the atomic number, atomic symbol, and atomic mass. (A mass in parentheses is the mass number of the most stable isotope of that element.) The periods are numbered 1 to 7. The groups (sometimes called *families*) have a number-letter designation and a new group number in parentheses. The A groups are the main-group elements; the B groups are the transition elements. Two series of inner transition elements are placed below the main body of the table but actually fit between the elements indicated. Metals lie below and to the left of the thick "staircase" line [top of 3A(13) to bottom of 6A(16) in Period 6] and include main-group metals (*purple-blue*), transition elements (*blue*), and inner transition elements (*gray-blue*). Nonmetals (*yellow*) lie to the right of the line. Metalloids (*green*) lie along the line. We discuss the placement of hydrogen in Chapter 14. As of late 2005, elements 112, 114, and 116 had not yet been named.

transition elements. They are generally shiny solids at room temperature (mercury is the only liquid) that conduct heat and electricity well and can be tooled into sheets (malleable) and wires (ductile). The **nonmetals** (yellow) appear in the small upper-right portion of the table. They are generally gases or dull, brittle solids at room temperature (bromine is the only liquid) and conduct heat and electricity poorly. Along the staircase line lie the **metalloids** (green; also called **semi-metals**), elements that have properties between those of metals and nonmetals. Several metalloids, such as silicon (Si) and germanium (Ge), play major roles in modern electronics.

Two of the major branches of chemistry have traditionally been defined by the elements that each studies. *Organic chemistry* studies the compounds of carbon, specifically those that contain hydrogen and often oxygen, nitrogen, and a few other elements. This branch is concerned with fuels, drugs, dyes, polymers, and the like. *Inorganic chemistry,* on the other hand, focuses mainly on the compounds of all the other elements. It is concerned with catalysts, electronic materials, metal alloys, mineral salts, and the like. With the explosive growth in biomedical and materials research, the line between these branches has all but disappeared.

It is important to learn some of the group (family) names. Group 1A(1), except for hydrogen, consists of the *alkali metals,* and Group 2A(2) consists of the *alkaline earth metals.* Both groups consist of highly reactive elements. The *halogens,* Group 7A(17), are highly reactive nonmetals, whereas the *noble gases,* Group 8A(18), are relatively unreactive nonmetals. Other main groups [3A(13) to 6A(16)] are often named for the first element in the group; for example, Group 6A is the *oxygen family.*

A key point that we return to many times is that, in general, *elements in a group have* **similar** *chemical properties and elements in a period have* **different** *chemical properties.* We begin applying the organizing power of the periodic table in the next section, where we discuss how elements combine to form compounds.

SECTION SUMMARY

In the periodic table, the elements are arranged by atomic number into horizontal periods and vertical groups. Because of the periodic recurrence of certain key properties, elements within a group have similar behavior, whereas elements in a period have dissimilar behavior. Nonmetals appear in the upper-right portion of the table, metalloids lie along a staircase line, and metals fill the rest of the table.

2.7 COMPOUNDS: INTRODUCTION TO BONDING

The overwhelming majority of elements occur in chemical combination with other elements. In fact, only a few elements occur free in nature. The noble gases—helium (He), neon (Ne), argon (Ar), krypton (Kr), xenon (Xe), and radon (Rn)—occur in air as separate atoms. In addition to occurring in compounds, oxygen (O), nitrogen (N), and sulfur (S) occur in the most common elemental form as the molecules O_2, N_2, and S_8, and carbon (C) occurs in vast, nearly pure deposits of coal. Some of the metals, such as copper (Cu), silver (Ag), gold (Au), and platinum (Pt), may also occur uncombined with other elements. But these few exceptions reinforce the general rule that elements occur combined in compounds.

It is the electrons of the atoms of interacting elements that are involved in compound formation. Elements combine in two general ways:

1. *Transferring electrons* from the atoms of one element to those of another to form **ionic compounds**
2. *Sharing electrons* between atoms of different elements to form **covalent compounds**

These processes generate **chemical bonds,** the forces that hold the atoms of elements together in a compound. We'll introduce compound formation next and have much more to say about it in later chapters.

The Formation of Ionic Compounds

Ionic compounds are composed of **ions,** charged particles that form when an atom (or small group of atoms) gains or loses one or more electrons. The simplest type of ionic compound is a **binary ionic compound,** one composed of just two elements. It typically forms *when a metal reacts with a nonmetal.* Each metal atom loses a certain number of its electrons and becomes a **cation,** a positively charged ion. The nonmetal atoms gain the electrons lost by the metal atoms and become **anions,** negatively charged ions. In effect, the metal atoms *transfer electrons* to the nonmetal atoms. The resulting cations and anions attract each other through electrostatic forces and form the ionic compound. *All binary ionic compounds are solids.* A cation or anion derived from a single atom is called a **monatomic ion;** we'll discuss polyatomic ions, those derived from a small group of atoms, later.

The formation of the binary ionic compound sodium chloride, common table salt, is depicted in Figure 2.11, from the elements through the atomic-scale electron transfer to the compound. In the electron transfer, a sodium atom, which is neutral because it has the same number of protons as electrons, *loses* 1 electron and forms a sodium cation, Na^+. (The charge on the ion is written as a *right superscript.*) A chlorine atom *gains* the electron and becomes a chloride anion, Cl^-. (The name change from the nonmetal atom to the ion is discussed in the next section.) Even the tiniest visible grain of table salt contains an enormous number of sodium and chloride ions. The oppositely charged ions (Na^+ and Cl^-) attract each other, and the similarly charged ions (Na^+ and Na^+, or Cl^- and Cl^-) repel each other. The resulting solid aggregation is a regular array of alternating Na^+ and Cl^- ions that extends in all three dimensions.

Animation: Formation of an Ionic Compound
Online Learning Center

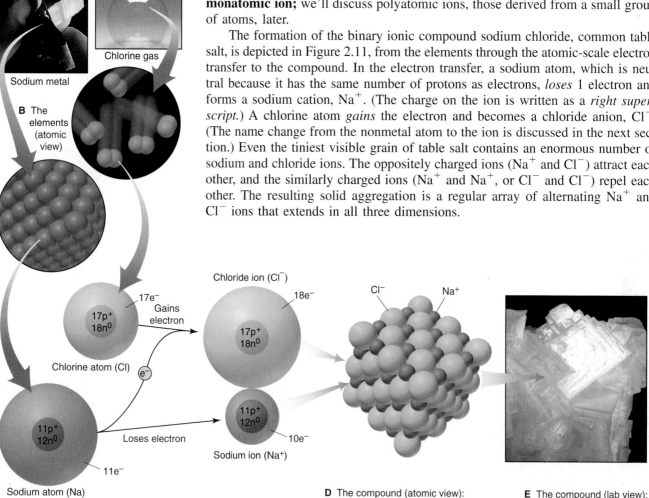

Figure 2.11 The formation of an ionic compound. A, The two elements as seen in the laboratory. **B,** The elements as they might appear on the atomic scale. **C,** The neutral sodium atom loses one electron to become a sodium cation (Na^+), and the chlorine atom gains one electron to become a chloride anion (Cl^-). (Note that when atoms lose electrons, they become ions that are smaller, and when they gain electrons, they become ions that are larger.) **D,** Na^+ and Cl^- ions attract each other and form a regular three-dimensional array. **E,** This arrangement of the ions is reflected in the structure of crystalline NaCl, which occurs naturally as the mineral halite, hence the name *halogens* for the Group 7A(17) elements.

The strength of the ionic bonding depends to a great extent on the net strength of these attractions and repulsions and is described by *Coulomb's law,* which can be expressed as follows: *the energy of attraction (or repulsion) between two particles is directly proportional to the product of the charges and inversely proportional to the distance between them.*

$$\text{Energy} \propto \frac{\text{charge 1} \times \text{charge 2}}{\text{distance}}$$

In other words, ions with higher charges attract (or repel) each other more strongly than ions with lower charges. Likewise, smaller ions attract (or repel) each other more strongly than larger ions, because their charges are closer together. These effects are summarized in Figure 2.12.

Ionic compounds are neutral; that is, they possess no net charge. For this to occur, they must contain equal numbers of positive and negative *charges*—not necessarily equal numbers of positive and negative *ions.* Because Na^+ and Cl^- each bear a unit charge ($1+$ or $1-$), equal numbers of these ions are present in sodium chloride; but in sodium oxide, for example, there are two Na^+ ions needed to balance the $2-$ charge of each oxide ion, O^{2-}.

Can we predict the number of electrons a given atom will lose or gain when it forms an ion? In the formation of sodium chloride, for example, why does each sodium atom give up only 1 of its 11 electrons? Why doesn't each chlorine atom gain two electrons, instead of just one? For A-group elements, the periodic table provides an answer. We generally find that metals lose electrons and nonmetals gain electrons to *form ions with the same number of electrons as in an atom of the nearest noble gas* [Group 8A(18)]. Noble gases have a stability (low reactivity) that is related to their number (and arrangement) of electrons. A sodium atom ($11e^-$) can attain the stability of neon ($10e^-$), the nearest noble gas, by losing one electron. Similarly, by gaining one electron, a chlorine atom ($17e^-$) attains the stability of argon ($18e^-$), its nearest noble gas. Thus, when an element located near a noble gas forms a monatomic ion, *it gains or loses enough electrons to attain the same number as that noble gas.* Specifically, the elements in Group 1A(1) lose one electron, those in Group 2A(2) lose two, and aluminum in Group 3A(13) loses three; the elements in Group 7A(17) gain one electron, oxygen and sulfur in Group 6A(16) gain two, and nitrogen in Group 5A(15) gains three.

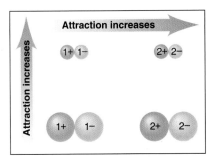

Figure 2.12 Factors that influence the strength of ionic bonding. For ions of a given size, strength of attraction (*arrows*) increases with higher ionic charge (*left to right*). For ions of a given charge, strength of attraction increases with smaller ionic size (*bottom to top*).

SAMPLE PROBLEM 2.4   Predicting the Ion an Element Forms

Problem What monatomic ions do the following elements form?
(a) Iodine ($Z = 53$)   **(b)** Calcium ($Z = 20$)   **(c)** Aluminum ($Z = 13$)
Plan We use the given Z value to find the element in the periodic table and see where its group lies relative to the noble gases. Elements in Groups 1A, 2A, and 3A *lose* electrons to attain the same number as the nearest noble gas and become positive ions; those in Groups 5A, 6A, and 7A *gain* electrons and become negative ions.
Solution (a) I^- Iodine ($_{53}I$) is a nonmetal in Group 7A(17), one of the halogens. Like any member of this group, it gains 1 electron to have the same number as the nearest Group 8A(18) member, in this case $_{54}Xe$.
(b) Ca^{2+} Calcium ($_{20}Ca$) is a member of Group 2A(2), the alkaline earth metals. Like any Group 2A member, it loses 2 electrons to attain the same number as the nearest noble gas, in this case, $_{18}Ar$.
(c) Al^{3+} Aluminum ($_{13}Al$) is a metal in the boron family [Group 3A(13)] and thus loses 3 electrons to attain the same number as its nearest noble gas, $_{10}Ne$.

FOLLOW-UP PROBLEM 2.4 What monatomic ion does each of the following elements form: **(a)** $_{16}S$; **(b)** $_{37}Rb$; **(c)** $_{56}Ba$?

A No interaction

B Attraction begins

C Covalent bond

D Combination of forces

Figure 2.13 Formation of a covalent bond between two H atoms. A, The distance is too great for the atoms to affect each other. **B,** As the distance decreases, the nucleus of each atom begins to attract the electron of the other. **C,** The covalent bond forms when the two nuclei mutually attract the pair of electrons at some optimum distance. **D,** The H_2 molecule is more stable than the separate atoms because the attractive forces (*black arrows*) between each nucleus and the two electrons are greater than the repulsive forces (*red arrows*) between the electrons and between the nuclei.

The Formation of Covalent Compounds

Covalent compounds form when elements share electrons, which usually occurs between nonmetals. Even though relatively few nonmetals exist, they interact in many combinations to form a very large number of covalent compounds.

The simplest case of electron sharing occurs not in a compound but between two hydrogen atoms (H; $Z = 1$). Imagine two separated H atoms approaching each other, as in Figure 2.13. As they get closer, the nucleus of each atom attracts the electron of the other atom more and more strongly, and the separated atoms begin to interpenetrate each other. At some optimum distance between the nuclei, the two atoms form a **covalent bond,** a pair of electrons mutually attracted by the two nuclei. The result is a hydrogen molecule, in which each electron no longer "belongs" to a particular H atom: the two electrons are *shared* by the two nuclei. Repulsions between the nuclei and between the electrons also occur, but the net attraction is greater than the net repulsion. (We discuss the properties of covalent bonds in great detail in Chapter 9.)

A sample of hydrogen gas consists of these diatomic molecules (H_2)—pairs of atoms that are chemically bound and behave as an independent unit—*not* separate H atoms. Other nonmetals that exist as diatomic molecules at room temperature are nitrogen (N_2), oxygen (O_2), and the halogens [fluorine (F_2), chlorine (Cl_2), bromine (Br_2), and iodine (I_2)]. Phosphorus exists as tetratomic molecules (P_4), and sulfur and selenium as octatomic molecules (S_8 and Se_8) (Figure 2.14). At room temperature, covalent substances may be gases, liquids, or solids.

Atoms of different elements share electrons to form the molecules of a covalent compound. A sample of hydrogen fluoride, for example, consists of molecules in which one H atom forms a covalent bond with one F atom; water consists of molecules in which one O atom forms covalent bonds with two H atoms:

Hydrogen fluoride, HF   Water, H_2O

(As you'll see in Chapter 9, covalent bonding provides another way for atoms to attain the same number of electrons as the nearest noble gas.)

Distinguishing the Entities in Covalent and Ionic Substances There is a key distinction between the chemical entities in covalent and ionic substances. *Most covalent substances consist of molecules.* A cup of water, for example, consists of individual water molecules lying near each other. In contrast, under ordinary

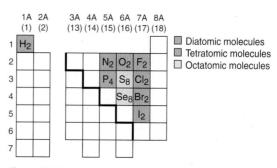

Figure 2.14 Elements that occur as molecules.

conditions, *no molecules exist in a sample of an ionic compound.* A piece of sodium chloride, for example, is a continuous array of oppositely charged sodium and chloride ions, *not* a collection of individual "sodium chloride molecules."

Another key distinction exists between the particles attracting each other. Covalent bonding involves the mutual attraction between two (positively charged) nuclei and the two (negatively charged) electrons that reside between them. Ionic bonding involves the mutual attraction between positive and negative ions.

Polyatomic Ions: Covalent Bonds Within Ions Many ionic compounds contain **polyatomic ions,** which consist of two or more atoms bonded *covalently* and have a net positive or negative charge. For example, the ionic compound calcium carbonate is an array of polyatomic carbonate anions and monatomic calcium cations attracted to each other. The carbonate ion consists of a carbon atom covalently bonded to three oxygen atoms, and two additional electrons give the ion its 2− charge (Figure 2.15). In many reactions, a polyatomic ion stays together as a unit.

SECTION SUMMARY

Although a few elements occur uncombined in nature, the great majority exist in compounds. Ionic compounds form when a metal *transfers electrons* to a nonmetal, and the resulting positive and negative ions attract each other to form a three-dimensional array. In many cases, metal atoms lose and nonmetal atoms gain enough electrons to attain the same number of electrons as in atoms of the nearest noble gas. Covalent compounds form when elements, usually nonmetals, *share electrons.* Each covalent bond is an electron pair mutually attracted by two atomic nuclei. Monatomic ions are derived from single atoms. Polyatomic ions consist of two or more covalently bonded atoms that have a net positive or negative charge due to a deficit or excess of electrons.

2.8 COMPOUNDS: FORMULAS, NAMES, AND MASSES

Names and formulas of compounds form the vocabulary of the chemical language. In this discussion, you'll learn the names and formulas of ionic and simple covalent compounds and how to calculate the mass of a unit of a compound from its formula.

Types of Chemical Formulas

In a **chemical formula,** element symbols and numerical subscripts show the type and number of each atom present in the smallest unit of the substance. There are several types of chemical formulas for a compound:

1. The **empirical formula** shows the *relative* number of atoms of each element in the compound. It is the simplest type of formula and is derived from the masses of the component elements. For example, in hydrogen peroxide, there is 1 part by mass of hydrogen for every 16 parts by mass of oxygen. Therefore, the empirical formula of hydrogen peroxide is HO: one H atom for every O atom.
2. The **molecular formula** shows the *actual* number of atoms of each element in a molecule of the compound. The molecular formula of hydrogen peroxide is H_2O_2; there are two H atoms and two O atoms in each molecule.
3. A **structural formula** shows the number of atoms and *the bonds between them;* that is, the relative placement and connections of atoms in the molecule. The structural formula of hydrogen peroxide is H—O—O—H; each H is bonded to an O, and the O's are bonded to each other.

Carbonate ion
CO_3^{2-}

Figure 2.15 A polyatomic ion. Calcium carbonate is a three-dimensional array of monatomic calcium cations (*purple spheres*) and polyatomic carbonate anions. As the bottom structure shows, each carbonate ion consists of four covalently bonded atoms.

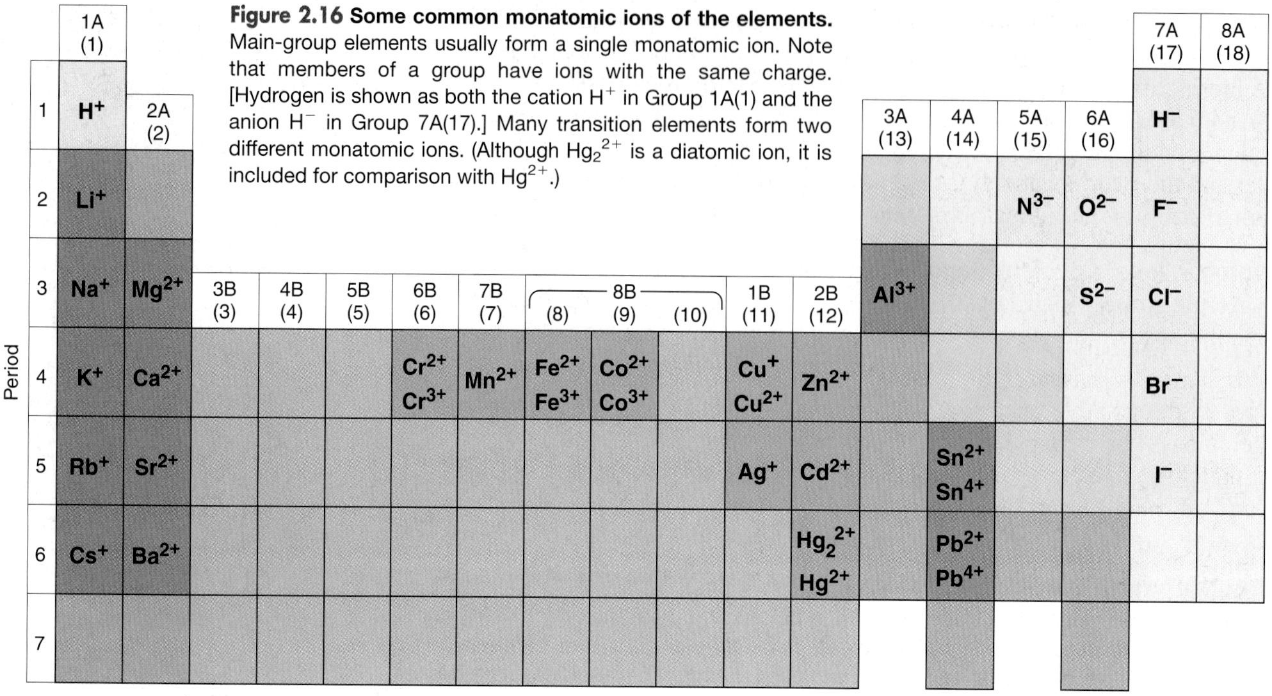

Figure 2.16 Some common monatomic ions of the elements. Main-group elements usually form a single monatomic ion. Note that members of a group have ions with the same charge. [Hydrogen is shown as both the cation H^+ in Group 1A(1) and the anion H^- in Group 7A(17).] Many transition elements form two different monatomic ions. (Although Hg_2^{2+} is a diatomic ion, it is included for comparison with Hg^{2+}.)

Names and Formulas of Ionic Compounds

All ionic compound names give the positive ion (cation) first and the negative ion (anion) second. Here are some points to note about ion charges:

- Members of a periodic table group have the same ionic charge; for example, Li, Na, and K are all in Group 1A and all have a 1+ charge.
- For A-group cations, ion charge = group number: for example, Na^+ is in Group 1A, Ba^{2+} in Group 2A. (Exceptions in Figure 2.16 are Sn^{2+} and Pb^{2+}.)
- For anions, ion charge = group number minus 8: for example, S is in Group 6A $(6 - 8 = -2)$, so the ion is S^{2-}. Table 2.3 lists some of the more common monatomic ions.

Compounds Formed from Monatomic Ions Let's first consider how to name binary ionic compounds, those composed of ions of two elements.

- *The name of the cation is the same as the name of the metal.* Many metal names end in *-ium.*
- *The name of the anion takes the root of the nonmetal name and adds the suffix -ide.*

For example, the anion formed from brom*ine* is named brom*ide* (brom+ide). Therefore, the compound formed from the metal calcium and the nonmetal bromine is named *calcium bromide.*

SAMPLE PROBLEM 2.5   Naming Binary Ionic Compounds

Problem Name the ionic compound formed from the following pairs of elements:
(a) Magnesium and nitrogen   (b) Iodine and cadmium
(c) Strontium and fluorine   (d) Sulfur and cesium

Plan The key to naming a binary ionic compound is to recognize which element is the metal and which is the nonmetal. When in doubt, check the periodic table. We place the cation name first, add the suffix *-ide* to the nonmetal root, and place the anion name last.

Solution (a) Magnesium is the metal; *nitr-* is the nonmetal root: magnesium nitride

(b) Cadmium is the metal; *iod-* is the nonmetal root: cadmium iodide

---

Table 2.3 Common Monatomic Ions*

Charge	Formula	Name
Cations		
1+	H^+	hydrogen
	Li^+	**lithium**
	Na^+	**sodium**
	K^+	**potassium**
	Cs^+	cesium
	Ag^+	**silver**
2+	Mg^{2+}	**magnesium**
	Ca^{2+}	**calcium**
	Sr^{2+}	strontium
	Ba^{2+}	**barium**
	Zn^{2+}	**zinc**
	Cd^{2+}	cadmium
3+	Al^{3+}	**aluminum**
Anions		
1−	H^-	hydride
	F^-	**fluoride**
	Cl^-	**chloride**
	Br^-	**bromide**
	I^-	**iodide**
2−	O^{2-}	**oxide**
	S^{2-}	**sulfide**
3−	N^{3-}	nitride

*Listed by charge; those in **boldface** are most common.

(c) Strontium is the metal; *fluor-* is the nonmetal root: strontium fluoride (Note the spelling is fl*u*oride, not fl*o*uride.)
(d) Cesium is the metal; *sulf-* is the nonmetal root: cesium sulfide

FOLLOW-UP PROBLEM 2.5 For the following ionic compounds, give the name and periodic table group number of each of the elements present: **(a)** zinc oxide; **(b)** silver bromide; **(c)** lithium chloride; **(d)** aluminum sulfide.

Ionic compounds are arrays of oppositely charged ions rather than separate molecular units. Therefore, we write a formula for the **formula unit,** which gives the *relative* numbers of cations and anions in the compound. Thus, ionic compounds generally have only empirical formulas.* The compound has zero net charge, so the positive charges of the cations must balance the negative charges of the anions. For example, calcium bromide is composed of Ca^{2+} ions and Br^- ions; therefore, two Br^- balance each Ca^{2+}. The formula is $CaBr_2$, not Ca_2Br. In this and all other formulas,

- The subscript refers to the element *preceding* it.
- The *subscript 1 is understood* from the presence of the element symbol alone (that is, we do not write Ca_1Br_2).
- The charge (without the sign) of one ion becomes the subscript of the other:

$$Ca^{2+} \quad Br^{1-} \qquad \text{gives} \qquad Ca_1Br_2 \quad \text{or} \quad CaBr_2$$

Reduce the subscripts to the smallest whole numbers that retain the ratio of ions. Thus, for example, from the ions Ca^{2+} and O^{2-} we have Ca_2O_2, which we reduce to the formula CaO (but see the footnote).

SAMPLE PROBLEM 2.6 Determining Formulas of Binary Ionic Compounds

Problem Write empirical formulas for the compounds named in Sample Problem 2.5.
Plan We write the empirical formula by finding the smallest number of each ion that gives the neutral compound. These numbers appear as *right subscripts* to the element symbol.
Solution
(a) Mg^{2+} and N^{3-}; three Mg^{2+} ions (6+) balance two N^{3-} ions (6−): Mg_3N_2
(b) Cd^{2+} and I^-; one Cd^{2+} ion (2+) balances two I^- ions (2−): CdI_2
(c) Sr^{2+} and F^-; one Sr^{2+} ion (2+) balances two F^- ions (2−): SrF_2
(d) Cs^+ and S^{2-}; two Cs^+ ions (2+) balance one S^{2-} ion (2−): Cs_2S
Comment Note that ion charges do *not* appear in the compound formula. That is, for cadmium iodide, we do *not* write $Cd^{2+}I_2^-$.

FOLLOW-UP PROBLEM 2.6 Write the formulas of the compounds named in Follow-up Problem 2.5.

Compounds with Metals That Can Form More Than One Ion

Many metals, particularly the transition elements (B groups), can form more than one ion, each with its own particular charge. Table 2.4 (on the next page) lists some examples, and Figure 2.16 shows their placement in the periodic table. Names of compounds containing these elements include a *Roman numeral within parentheses* immediately after the metal ion's name to indicate its ionic charge. For example, iron can form Fe^{2+} and Fe^{3+} ions. The two compounds that iron forms with

---

*Compounds of the mercury(I) ion, such as Hg_2Cl_2, and peroxides of the alkali metals, such as Na_2O_2, are the only two common exceptions. Their empirical formulas are HgCl and NaO, respectively.

Table 2.4   Some Metals That Form More Than One Monatomic Ion*

Element	Ion Formula	Systematic Name	Common (Trivial) Name
Chromium	Cr^{2+}	chromium(II)	chromous
	Cr^{3+}	**chromium(III)**	chromic
Cobalt	Co^{2+}	cobalt(II)	
	Co^{3+}	cobalt(III)	
Copper	**Cu^+**	**copper(I)**	cuprous
	Cu^{2+}	**copper(II)**	cupric
Iron	**Fe^{2+}**	**iron(II)**	ferrous
	Fe^{3+}	**iron(III)**	ferric
Lead	**Pb^{2+}**	**lead(II)**	
	Pb^{4+}	lead(IV)	
Mercury	Hg_2^{2+}	mercury(I)	mercurous
	Hg^{2+}	**mercury(II)**	mercuric
Tin	**Sn^{2+}**	**tin(II)**	stannous
	Sn^{4+}	tin(IV)	stannic

*Listed alphabetically by metal name; those in **boldface** are most common.

Table 2.5   Common Polyatomic Ions*

Formula	Name
Cations	
NH_4^+	**ammonium**
H_3O^+	**hydronium**
Anions	
CH_3COO^- (or **$C_2H_3O_2^-$**)	**acetate**
CN^-	cyanide
OH^-	**hydroxide**
ClO^-	hypochlorite
ClO_2^-	chlorite
ClO_3^-	**chlorate**
ClO_4^-	**perchlorate**
NO_2^-	nitrite
NO_3^-	**nitrate**
MnO_4^-	**permanganate**
CO_3^{2-}	**carbonate**
HCO_3^-	**hydrogen carbonate** (or **bicarbonate**)
CrO_4^{2-}	chromate
$Cr_2O_7^{2-}$	**dichromate**
O_2^{2-}	peroxide
PO_4^{3-}	**phosphate**
HPO_4^{2-}	hydrogen phosphate
$H_2PO_4^-$	dihydrogen phosphate
SO_3^{2-}	sulfite
SO_4^{2-}	**sulfate**
HSO_4^-	hydrogen sulfate (or bisulfate)

*Boldface ions are most common.

chlorine are $FeCl_2$, named iron(II) chloride (spoken "iron two chloride"), and $FeCl_3$, named iron(III) chloride.

In common names, the Latin root of the metal is followed by either of two suffixes:

- The suffix *-ous* for the ion with the lower charge
- The suffix *-ic* for the ion with the higher charge

Thus, iron(II) chloride is also called ferr*ous* chloride and iron(III) chloride is ferr*ic* chloride. (You can easily remember this naming relationship because there is an *o* in *-ous* and *lower,* and an *i* in *-ic* and *higher.*)

SAMPLE PROBLEM 2.7   Determining Names and Formulas of Ionic Compounds of Elements That Form More Than One Ion

Problem Give the systematic names for the formulas or the formulas for the names of the following compounds: (**a**) tin(II) fluoride; (**b**) CrI_3; (**c**) ferric oxide; (**d**) CoS.

Solution (**a**) Tin(II) is Sn^{2+}; fluoride is F^-. Two F^- ions balance one Sn^{2+} ion: tin(II) fluoride is SnF_2. (The common name is stannous fluoride.)

(**b**) The anion is I^-, iodide, and the formula shows three I^-. Therefore, the cation must be Cr^{3+}, chromium(III): CrI_3 is chromium(III) iodide. (The common name is chromic iodide.)

(**c**) Ferric is the common name for iron(III), Fe^{3+}; oxide ion is O^{2-}. To balance the ionic charges, the formula of ferric oxide is Fe_2O_3. [The systematic name is iron(III) oxide.]

(**d**) The anion is sulfide, S^{2-}, which requires that the cation be Co^{2+}. The name is cobalt(II) sulfide.

FOLLOW-UP PROBLEM 2.7 Give the systematic names for the formulas or the formulas for the names of the following compounds: (**a**) lead(IV) oxide; (**b**) Cu_2S; (**c**) $FeBr_2$; (**d**) mercuric chloride.

Compounds Formed from Polyatomic Ions Ionic compounds in which one or both of the ions are polyatomic are very common. Table 2.5 gives the formulas and the names of some common polyatomic ions. Remember that *the polyatomic ion stays together as a charged unit.* The formula for potassium nitrate is KNO_3: each K^+ balances one NO_3^-. The formula for sodium carbonate is Na_2CO_3: two Na^+ balance one CO_3^{2-}. *When two or more of the same polyatomic ion are* *present in the formula unit, that ion appears in parentheses with the subscript written outside.* For example, calcium nitrate, which contains one Ca^{2+} and two NO_3^- ions, has the formula $Ca(NO_3)_2$. Parentheses and a subscript are *not* used unless *more than one* of the polyatomic ions is present; thus, sodium nitrate is $NaNO_3$, *not* $Na(NO_3)$.

Families of Oxoanions As Table 2.5 shows, most polyatomic ions are **oxoanions,** those in which an element, usually a nonmetal, is bonded to one or more oxygen atoms. There are several families of two or four oxoanions that differ only in the number of oxygen atoms. A simple naming convention is used with these ions.

With two oxoanions in the family:

- The ion with *more* O atoms takes the nonmetal root and the suffix *-ate.*
- The ion with *fewer* O atoms takes the nonmetal root and the suffix *-ite.*

For example, SO_4^{2-} is the sulf*ate* ion; SO_3^{2-} is the sulf*ite* ion; similarly, NO_3^- is nitr*ate*, and NO_2^- is nitr*ite*.

With four oxoanions in the family (usually a halogen bonded to O), as Figure 2.17 shows:

- The ion with *most* O atoms has the prefix *per-*, the nonmetal root, and the suffix *-ate.*
- The ion with *one fewer* O atom has just the root and the suffix *-ate.*
- The ion with *two fewer* O atoms has just the root and the suffix *-ite.*
- The ion with *least (three fewer)* O atoms has the prefix *hypo-*, the root, and the suffix *-ite.*

For example, for the four chlorine oxoanions,

ClO_4^- is *per*chlor*ate*, ClO_3^- is chlor*ate*, ClO_2^- is chlor*ite*, ClO^- is *hypo*chlor*ite*

Hydrated Ionic Compounds Ionic compounds called **hydrates** have a specific number of water molecules associated with each formula unit. In their formulas, this number is shown after a centered dot. It is indicated in the systematic name by a Greek numerical prefix before the word *hydrate*. Table 2.6 shows these prefixes. For example, Epsom salt has the formula $MgSO_4 \cdot 7H_2O$ and the name magnesium sulfate *hepta*hydrate. Similarly, the mineral gypsum has the formula $CaSO_4 \cdot 2H_2O$ and the name calcium sulfate *di*hydrate. The water molecules, referred to as "waters of hydration," are part of the hydrate's structure. Heating can remove some or all of them, leading to a different substance. For example, when heated strongly, blue copper(II) sulfate pentahydrate ($CuSO_4 \cdot 5H_2O$) is converted to white copper(II) sulfate ($CuSO_4$).

	Prefix	Root	Suffix
	per	*root*	ate
		root	ate
		root	ite
	hypo	*root*	ite

No. of O atoms ↑

Figure 2.17 Naming oxoanions. Prefixes and suffixes indicate the number of O atoms in the anion.

Table 2.6 Numerical Prefixes for Hydrates and Binary Covalent Compounds

Number	Prefix
1	mono-
2	di-
3	tri-
4	tetra-
5	penta-
6	hexa-
7	hepta-
8	octa-
9	nona-
10	deca-

SAMPLE PROBLEM 2.8   Determining Names and Formulas of Ionic Compounds Containing Polyatomic Ions

Problem Give the systematic names for the formulas or the formulas for the names of the following compounds:
(a) $Fe(ClO_4)_2$ **(b)** Sodium sulfite   **(c)** $Ba(OH)_2 \cdot 8H_2O$
Solution (a) ClO_4^- is perchlorate, which has a 1− charge, so the cation must be Fe^{2+}. The name is iron(II) perchlorate. (The common name is ferrous perchlorate.)
(b) Sodium is Na^+; sulfite is SO_3^{2-}. Therefore, two Na^+ ions balance one SO_3^{2-} ion. The formula is Na_2SO_3. **(c)** Ba^{2+} is barium; OH^- is hydroxide. There are eight (*octa-*) water molecules in each formula unit. The name is barium hydroxide octahydrate.

FOLLOW-UP PROBLEM 2.8 Give the systematic names for the formulas or the formulas for the names of the following compounds:
(a) Cupric nitrate trihydrate   **(b)** Zinc hydroxide   **(c)** LiCN

SAMPLE PROBLEM 2.9 Recognizing Incorrect Names and Formulas of Ionic Compounds

Problem Something is wrong with the second part of each statement. Provide the correct name or formula.
(a) $Ba(C_2H_3O_2)_2$ is called barium diacetate.
(b) Sodium sulfide has the formula $(Na)_2SO_3$.
(c) Iron(II) sulfate has the formula $Fe_2(SO_4)_3$.
(d) Cesium carbonate has the formula $Cs_2(CO_3)$.

Solution (a) The charge of the Ba^{2+} ion *must* be balanced by *two* $C_2H_3O_2^-$ ions, so the prefix *di-* is unnecessary. For ionic compounds, we do not indicate the number of ions with numerical prefixes. The correct name is barium acetate.
(b) Two mistakes occur here. The sodium ion is monatomic, so it does *not* require parentheses. The sulfide ion is S^{2-}, *not* SO_3^{2-} (called "sulfite"). The correct formula is Na_2S.
(c) The Roman numeral refers to the charge of the ion, *not* the number of ions in the formula. Fe^{2+} is the cation, so it requires one SO_4^{2-} to balance its charge. The correct formula is $FeSO_4$.
(d) Parentheses are *not* required when only one polyatomic ion of a kind is present. The correct formula is Cs_2CO_3.

FOLLOW-UP PROBLEM 2.9 State why the second part of each statement is incorrect, and correct it:
(a) Ammonium phosphate is $(NH_3)_4PO_4$.
(b) Aluminum hydroxide is $AlOH_3$.
(c) $Mg(HCO_3)_2$ is manganese(II) carbonate.
(d) $Cr(NO_3)_3$ is chromic(III) nitride.
(e) $Ca(NO_2)_2$ is cadmium nitrate.

Acid Names from Anion Names Acids are an important group of hydrogen-containing compounds that have been used in chemical reactions for centuries. In the laboratory, acids are typically used in water solution. When naming them and writing their formulas, we consider them as anions connected to the number of hydrogen ions (H^+) needed for charge neutrality. The two common types of acids are binary acids and oxoacids:

1. *Binary acid* solutions form when certain gaseous compounds dissolve in water. For example, when gaseous hydrogen chloride (HCl) dissolves in water, it forms a solution whose name consists of the following parts:

    Prefix *hydro-* + nonmetal *root* + suffix *-ic* + separate word *acid*

    hydro   +   chlor   +   ic   +   acid

    or *hydrochloric acid.* This naming pattern holds for many compounds in which hydrogen combines with an anion that has an *-ide* suffix.

2. *Oxoacid* names are similar to those of the oxoanions, except for two suffix changes:
    • *-ate* in the anion becomes *-ic* in the acid
    • *-ite* in the anion becomes *-ous* in the acid
    The oxoanion prefixes *hypo-* and *per-* are kept. Thus,

    BrO_4^- is *per*brom*ate*, and $HBrO_4$ is *per*brom*ic* acid

    IO_2^- is iod*ite*, and HIO_2 is iod*ous* acid

SAMPLE PROBLEM 2.10 Determining Names and Formulas of Anions and Acids

Problem Name the following anions and give the names and formulas of the acids derived from them: (a) Br^-; (b) IO_3^-; (c) CN^-; (d) SO_4^{2-}; (e) NO_2^-.
Solution (a) The anion is bromide; the acid is hydrobromic acid, HBr.
(b) The anion is iodate; the acid is iodic acid, HIO_3.
(c) The anion is cyanide; the acid is hydrocyanic acid, HCN.

(d) The anion is sulfate; the acid is sulfuric acid, H_2SO_4. (In this case, the suffix is added to the element name *sulfur*, not to the root, *sulf-*.)

(e) The anion is nitrite; the acid is nitrous acid, HNO_2.

Comment We added *two* H^+ ions to the sulfate ion to obtain sulfuric acid because it has a 2− charge.

FOLLOW-UP PROBLEM 2.10 Write the formula for the name or name for the formula of each acid: **(a)** chloric acid; **(b)** HF; **(c)** acetic acid; **(d)** sulfurous acid; **(e)** HBrO.

Names and Formulas of Binary Covalent Compounds

Binary covalent compounds are formed by the combination of two elements, usually nonmetals. Several are so familiar, such as ammonia (NH_3), methane (CH_4), and water (H_2O), that we use their common names, but most are named in a systematic way:

1. The element with the lower group number in the periodic table is the first word in the name; the element with the higher group number is the second word. (*Exception:* When the compound contains oxygen and any of the halogens chlorine, bromine, and iodine, the halogen is named first.)
2. If both elements are in the same group, the one with the higher period number is named first.
3. The second element is named with its root and the suffix *-ide*.
4. Covalent compounds have Greek numerical prefixes (see Table 2.6) to indicate the number of atoms of each element in the compound. The first word has a prefix *only* when more than one atom of the element is present; the second word *usually* has a numerical prefix.

SAMPLE PROBLEM 2.11 Determining Names and Formulas of Binary Covalent Compounds

Problem (a) What is the formula of carbon disulfide? **(b)** What is the name of PCl_5? **(c)** Give the name and formula of the compound whose molecules each consist of two N atoms and four O atoms.
Solution (a) The prefix *di-* means "two." The formula is CS_2.
(b) P is the symbol for phosphorus; there are five chlorine atoms, which is indicated by the prefix *penta-*. The name is phosphorus pentachloride.
(c) Nitrogen (N) comes first in the name (lower group number). The compound is dinitrogen tetraoxide, N_2O_4.

FOLLOW-UP PROBLEM 2.11 Give the name or formula for **(a)** SO_3; **(b)** SiO_2; **(c)** dinitrogen monoxide; **(d)** selenium hexafluoride.

SAMPLE PROBLEM 2.12 Recognizing Incorrect Names and Formulas of Binary Covalent Compounds

Problem Explain what is wrong with the name or formula in the second part of each statement and correct it: **(a)** SF_4 is monosulfur pentafluoride. **(b)** Dichlorine heptaoxide is Cl_2O_6. **(c)** N_2O_3 is dinitrotrioxide.
Solution (a) There are two mistakes. *Mono-* is not needed if there is only one atom of the first element, and the prefix for four is *tetra-*, not *penta-*. The correct name is sulfur tetrafluoride.
(b) The prefix *hepta-* indicates seven, not six. The correct formula is Cl_2O_7.
(c) The full name of the first element is needed, and a space separates the two element names. The correct name is dinitrogen trioxide.

Naming Alkanes

Organic compounds typically have complex structural formulas that consist of chains, branches, and/or rings of carbon atoms bonded to hydrogen atoms and, often, to atoms of oxygen, nitrogen, and a few other elements. At this point, we'll see how the simplest organic compounds are named. Much more on the rules of organic nomenclature appears in Chapter 15.

Hydrocarbons, the simplest type of organic compound, contain *only* carbon and hydrogen. *Alkanes* are the simplest type of hydrocarbon; many function as important fuels, such as methane, propane, butane, and the mixture of alkanes in gasoline. The simplest alkanes to name are the *straight-chain alkanes* because the carbon chains have no branches. Alkanes are named with a *root,* based on the number of C atoms in the chain, followed by the suffix *-ane.* Table 2.7 gives the names, molecular formulas, and space-filling models (discussed shortly) of the first 10 straight-chain alkanes. Note that the roots of the four smallest ones are new, but those for the larger ones are the same as the Greek prefixes in Table 2.6.

Molecular Masses from Chemical Formulas

In Section 2.5, we calculated the atomic mass of an element. Using the periodic table and the formula of a compound to see the number of atoms of each element present, we calculate the **molecular mass** (also called *molecular weight*) of a formula unit of the compound as the sum of the atomic masses:

$$\text{Molecular mass} = \text{sum of atomic masses} \qquad (2.3)$$

The molecular mass of a water molecule (using atomic masses to four significant figures from the periodic table) is

$$\text{Molecular mass of } H_2O = (2 \times \text{atomic mass of H}) + (1 \times \text{atomic mass of O})$$
$$= (2 \times 1.008 \text{ amu}) + 16.00 \text{ amu} = 18.02 \text{ amu}$$

Ionic compounds are treated the same, but because they do not consist of molecules, we use the term **formula mass** for an ionic compound. To calculate its formula mass, *the number of atoms of each element inside the parentheses is multiplied by the subscript outside the parentheses.* For barium nitrate, $Ba(NO_3)_2$,

Formula mass of $Ba(NO_3)_2$

$$= (1 \times \text{atomic mass of Ba}) + (2 \times \text{atomic mass of N}) + (6 \times \text{atomic mass of O})$$
$$= 137.3 \text{ amu} + (2 \times 14.01 \text{ amu}) + (6 \times 16.00 \text{ amu}) = 261.3 \text{ amu}$$

Note that atomic, not ionic, masses are used. Although masses of ions differ from those of their atoms by the masses of the electrons, electron loss equals electron gain in the compound, so electron mass is balanced.

Table 2.7 The First 10 Straight-Chain Alkanes

Name (Formula)	Model
Methane (CH_4)	
Ethane (C_2H_6)	
Propane (C_3H_8)	
Butane (C_4H_{10})	
Pentane (C_5H_{12})	
Hexane (C_6H_{14})	
Heptane (C_7H_{16})	
Octane (C_8H_{18})	
Nonane (C_9H_{20})	
Decane ($C_{10}H_{22}$)	

SAMPLE PROBLEM 2.13 Calculating the Molecular Mass of a Compound

Problem Using data in the periodic table, calculate the molecular (or formula) mass of:
(a) Tetraphosphorus trisulfide   **(b)** Ammonium nitrate
Plan We first write the formula, then multiply the number of atoms (or ions) of each element by its atomic mass, and find the sum.
Solution (a) The formula is P_4S_3.

$$\text{Molecular mass} = (4 \times \text{atomic mass of P}) + (3 \times \text{atomic mass of S})$$
$$= (4 \times 30.97 \text{ amu}) + (3 \times 32.07 \text{ amu}) = \boxed{220.09 \text{ amu}}$$

(b) The formula is NH_4NO_3. We count the total number of N atoms even though they belong to different ions:

Formula mass

$$= (2 \times \text{atomic mass of N}) + (4 \times \text{atomic mass of H}) + (3 \times \text{atomic mass of O})$$
$$= (2 \times 14.01 \text{ amu}) + (4 \times 1.008 \text{ amu}) + (3 \times 16.00 \text{ amu}) = \boxed{80.05 \text{ amu}}$$

Check You can often find large errors by rounding atomic masses to the nearest 5 and adding: **(a)** $(4 \times 30) + (3 \times 30) = 210 \approx 220.09$. The sum has two decimal places because the atomic masses have two. **(b)** $(2 \times 15) + 4 + (3 \times 15) = 79 \approx 80.05$.

FOLLOW-UP PROBLEM 2.13 What is the formula and molecular (or formula) mass of each of the following compounds: **(a)** hydrogen peroxide; **(b)** cesium chloride; **(c)** sulfuric acid; **(d)** potassium sulfate?

In the next sample problem, we use molecular depictions to find the formula, name, and mass.

SAMPLE PROBLEM 2.14   Determining Formulas and Names from Molecular Depictions

Problem Each box contains a representation of a binary compound. Determine its formula, name, and molecular (formula) mass.

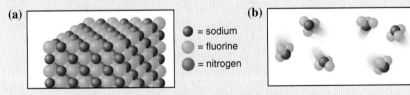

(a)   **(b)**

= sodium
= fluorine
= nitrogen

Plan Each of the compounds contains only two elements, so to find the formula, we find the simplest whole-number ratio of one atom to the other. Then we determine the name (see Sample Problems 2.5, 2.6, and 2.11) and the mass (see Sample Problem 2.13).

Solution (a) There is one brown (sodium) for each green (fluorine), so the formula is NaF. A metal and nonmetal form an ionic compound, in which the metal is named first: sodium fluoride.

$$\text{Formula mass} = (1 \times \text{atomic mass of Na}) + (1 \times \text{atomic mass of F})$$
$$= 22.99 \text{ amu} + 19.00 \text{ amu} = \boxed{41.99 \text{ amu}}$$

(b) There are three green (fluorine) for each blue (nitrogen), so the formula is $\boxed{NF_3}$. Two nonmetals form a covalent compound. Nitrogen has a lower group number, so it is named first: nitrogen trifluoride.

$$\text{Molecular mass} = (1 \times \text{atomic mass of N}) + (3 \times \text{atomic mass of F})$$
$$= 14.01 \text{ amu} + (3 \times 19.00 \text{ amu}) = \boxed{71.01 \text{ amu}}$$

Check (a) For binary ionic compounds, we predict ionic charges from the periodic table (see Figure 2.10). Na forms a $1+$ ion, and F forms a $1-$ ion, so the charges balance with one Na^+ per F^-. Also, ionic compounds are solids, consistent with the picture. **(b)** Covalent compounds often occur as individual molecules, as in the picture. Rounding in (a) gives $25 + 20 = 45$; in (b), we get $15 + (3 \times 20) = 75$, so there are no large errors.

FOLLOW-UP PROBLEM 2.14 Each box contains a representation of a binary compound. Determine its name, formula, and molecular (formula) mass.

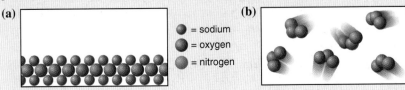

(a)   **(b)**

= sodium
= oxygen
= nitrogen

Figure 2.18 Representations of a water molecule.

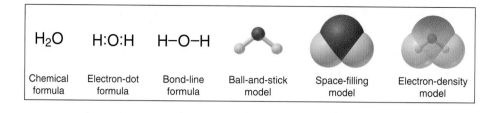

H_2O	H:O:H	H–O–H			
Chemical formula	Electron-dot formula	Bond-line formula	Ball-and-stick model	Space-filling model	Electron-density model

Picturing Molecules

Molecules are depicted in a variety of useful ways, as shown in Figure 2.18 for the water molecule. A *chemical formula* shows only the relative number of atoms. *Electron-dot* and *bond-line formulas* show a bond between atoms as either a pair of dots or a line. A *ball-and-stick model* shows atoms as spheres and bonds as sticks, with accurate angles and relative sizes, but distances are exaggerated. A *space-filling model* is an accurately scaled-up version of a molecule, but it does not show bonds. An *electron-density model* shows the ball-and-stick model within the space-filling shape and colors the regions of high (*red*) and low (*blue*) electron charge.

SECTION SUMMARY
Chemical formulas describe the simplest atom ratio (empirical formula), actual atom number (molecular formula), and atom arrangement (structural formula) of one unit of a compound. An ionic compound is named with cation first and anion second. For metals that can form more than one ion, the charge is shown with a Roman numeral. Oxoanions have suffixes, and sometimes prefixes, attached to the element root name to indicate the number of oxygen atoms. Names of hydrates give the number of associated water molecules with a numerical prefix. Acid names are based on anion names. Covalent compounds have the first word of the name for the element that is leftmost or lower down in the periodic table, and prefixes show the number of each atom. The molecular (or formula) mass of a compound is the sum of the atomic masses in the formula. Molecules are depicted by various types of formulas and models.

2.9   CLASSIFICATION OF MIXTURES

Although chemists pay a great deal of attention to pure substances, this form of matter almost never occurs around us. In the natural world, *matter usually occurs as mixtures*, such as air, seawater, soil, and organisms.

There are two broad classes of mixtures. A **heterogeneous mixture** has one or more visible boundaries between the components. Thus, its composition is *not* uniform; it varies from one region to another. Many rocks are heterogeneous, showing individual grains and flecks of different minerals. In some cases, as in milk and blood, the boundaries can be seen only with a microscope. A **homogeneous mixture** has no visible boundaries because the components are mixed as individual atoms, ions, and molecules. Thus, its composition *is* uniform. A mixture of sugar dissolved in water is homogeneous, for example, because the sugar molecules and water molecules are uniformly intermingled on the molecular level. We have no way to tell visually whether an object is a substance (element or compound) or a homogeneous mixture.

A homogeneous mixture is also called a **solution.** Although we usually think of solutions as liquid, they can exist in all three physical states. For example, air is a gaseous solution of mostly oxygen and nitrogen molecules, and wax is a solid solution of several fatty substances. Solutions in water, called **aqueous solutions,** are especially important in chemistry and comprise a major portion of the environment and of all organisms.

Recall that mixtures differ fundamentally from compounds in three ways: (1) the proportions of the components can vary; (2) the individual properties of the components are observable; and (3) the components can be separated by physical means. In some cases, as in a mixture of iron and sulfur (Figure 2.19),

A

S_8

Fe

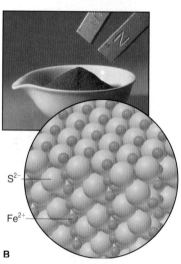

B

S^{2-}

Fe^{2+}

Figure 2.19 The distinction between mixtures and compounds. A, A *mixture* of iron and sulfur can be separated with a magnet because only the iron is magnetic. The blow-up shows separate regions of the two elements. **B,** After strong heating, the *compound* iron(II) sulfide forms, which is no longer magnetic. The blow-up shows the structure of the compound.

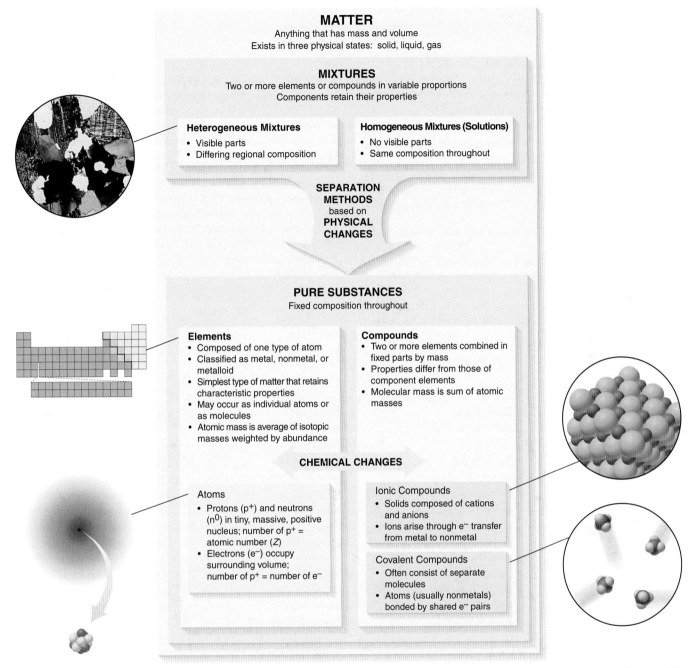

Figure 2.20 The classification of matter from a chemical point of view. Mixtures are separated by physical changes into elements and compounds. Chemical changes are required to convert elements into compounds, and vice versa.

if we apply enough energy to the components, they react with each other and form a compound, after which their individual properties are no longer observable. The characteristics of mixtures and pure substances that we covered in this chapter are summarized in Figure 2.20.

SECTION SUMMARY

Heterogeneous mixtures have visible boundaries between the components. Homogeneous mixtures have no visible boundaries because mixing occurs at the molecular level. A solution is a homogeneous mixture and can occur in any physical state. Mixtures (not compounds) can have variable proportions, can be separated physically, and retain their components' properties.

For Review and Reference (Numbers in parentheses refer to pages, unless noted otherwise.)

Learning Objectives

To help you review these learning objectives, the numbers of related sections (§), sample problems (SP), and upcoming end-of-chapter problems (EP) are listed in parentheses.

1. Define the characteristics of the three types of matter—element, compound, and mixture—on the macroscopic and atomic levels (§ 2.1) (EPs 2.1–2.5)
2. Understand the laws of mass conservation, definite composition, and multiple proportions; use the mass ratio of element-to-compound to find the mass of an element in a compound (§ 2.1) (SP 2.1) (EPs 2.6–2.18)
3. Understand Dalton's atomic theory and how it explains the mass laws (§ 2.3) (EP 2.19)
4. Describe the results of the key experiments by Thomson, Millikan, and Rutherford concerning atomic structure (§ 2.4) (EPs 2.20, 2.21)
5. Explain the structure of the atom, the main features of the subatomic particles, and the significance of isotopes; use atomic

notation to express the subatomic makeup of an isotope; calculate the atomic mass of an element from its isotopic composition (§ 2.5) (SP 2.2, 2.3) (EPs 2.22–2.34)
6. Describe the format of the periodic table and the general location and characteristics of metals, metalloids, and nonmetals (§ 2.6) (EPs 2.35–2.41)
7. Explain the essential features of ionic and covalent bonding and distinguish between them; predict the monatomic ion formed from a main-group element (§ 2.7) (SP 2.4) (EPs 2.42–2.54)
8. Name, write the formula, and calculate the molecule (or formula) mass of ionic and binary covalent compounds (§ 2.8) (SP 2.5–2.14) (EPs 2.55–2.78)
9. Describe the types of mixtures and their properties (§ 2.9) (EPs 2.79–2.83)

Key Terms

Section 2.1
element (32)
pure substance (32)
molecule (32)
compound (33)
mixture (33)

Section 2.2
law of mass
    conservation (34)
law of definite (or constant)
    composition (34)
fraction by mass (mass
    fraction) (34)
percent by mass (mass
    percent, mass %) (34)
law of multiple
    proportions (35)

Section 2.3
atom (36)

Section 2.4
cathode ray (37)
nucleus (40)

Section 2.5
proton (p^+) (41)
neutron (n^0) (41)
electron (e^-) (41)
atomic number (Z) (42)
mass number (A) (42)
atomic symbol (42)
isotope (42)
atomic mass unit (amu) (43)
dalton (Da) (43)
mass spectrometry (43)
isotopic mass (44)
atomic mass (44)

Section 2.6
periodic table of the
    elements (45)

period (45)
group (45)
metal (46)
nonmetal (47)
metalloid (semimetal) (47)

Section 2.7
ionic compound (47)
covalent compound (47)
chemical bond (48)
ion (48)
binary ionic compound (48)
cation (48)
anion (48)
monatomic ion (48)
covalent bond (50)
polyatomic ion (51)

Section 2.8
chemical formula (51)
empirical formula (51)

molecular formula (51)
structural formula (51)
formula unit (53)
oxoanion (55)
hydrate (55)
binary covalent compound (57)
molecular mass (58)
formula mass (58)

Section 2.9
heterogeneous mixture (60)
homogeneous mixture (60)
solution (60)
aqueous solution (60)

Key Equations and Relationships

2.1 Finding the mass of an element in a given mass of compound (35):

Mass of element in sample

$$= \text{mass of compound in sample} \times \frac{\text{mass of element}}{\text{mass of compound}}$$

2.2 Calculating the number of neutrons in an atom (42):

Number of neutrons = mass number − atomic number

or $N = A - Z$

2.3 Determining the molecular mass of a formula unit of a compound (58):

Molecular mass = sum of atomic masses

Brief Solutions to Follow-up Problems

2.1 Mass (t) of pitchblende

$$= 2.3 \text{ t uranium} \times \frac{84.2 \text{ t pitchblende}}{71.4 \text{ t uranium}} = 2.7 \text{ t pitchblende}$$

Mass (t) of oxygen

$$= 2.7 \text{ t pitchblende} \times \frac{(84.2 - 71.4 \text{ t oxygen})}{84.2 \text{ t pitchblende}} = 0.41 \text{ t oxygen}$$

2.2 (a) Q = B; $5p^+$, $6n^0$, $5e^-$
(b) X = Ca; $20p^+$, $21n^0$, $20e^-$
(c) Y = I; $53p^+$, $78n^0$, $53e^-$
2.3 $10.0129x + [11.0093(1 - x)] = 10.81$; $0.9964x = 0.1993$;
$x = 0.2000$ and $1 - x = 0.8000$; % abundance of $^{10}B = 20.00\%$;
% abundance of $^{11}B = 80.00\%$
2.4 (a) S^{2-}; (b) Rb^+; (c) Ba^{2+}
2.5 (a) Zinc [Group 2B(12)] and oxygen [Group 6A(16)]
(b) Silver [Group 1B(11)] and bromine [Group 7A(17)]
(c) Lithium [Group 1A(1)] and chlorine [Group 7A(17)]
(d) Aluminum [Group 3A(13)] and sulfur [Group 6A(16)]
2.6 (a) ZnO; (b) AgBr; (c) LiCl; (d) Al_2S_3
2.7 (a) PbO_2; (b) copper(I) sulfide (cuprous sulfide); (c) iron(II) bromide (ferrous bromide); (d) $HgCl_2$
2.8 (a) $Cu(NO_3)_2 \cdot 3H_2O$; (b) $Zn(OH)_2$; (c) lithium cyanide
2.9 (a) $(NH_4)_3PO_4$; ammonium is NH_4^+ and phosphate is PO_4^{3-}.
(b) $Al(OH)_3$; parentheses are needed around the polyatomic ion OH^-.

(c) Magnesium hydrogen carbonate; Mg^{2+} is magnesium and can have only a 2+ charge, so it does not need (II); HCO_3^- is hydrogen carbonate (or bicarbonate).
(d) Chromium(III) nitrate; the *-ic* ending is not used with Roman numerals; NO_3^- is nit*rate*.
(e) Calcium nitrite; Ca^{2+} is calcium and NO_2^- is nit*rite*.
2.10 (a) $HClO_3$; (b) hydrofluoric acid; (c) CH_3COOH (or $HC_2H_3O_2$); (d) H_2SO_3; (e) hypobromous acid
2.11 (a) Sulfur trioxide; (b) silicon dioxide; (c) N_2O; (d) SeF_6
2.12 (a) Disulfur dichloride; the *-ous* suffix is not used.
(b) NO; the name indicates one nitrogen.
(c) Bromine trichloride; Br is in a higher period in Group 7A(17), so it is named first.
2.13 (a) H_2O_2, 34.02 amu; (b) CsCl, 168.4 amu; (c) H_2SO_4, 98.09 amu; (d) K_2SO_4, 174.27 amu
2.14 (a) Na_2O. This is an ionic compound, so the name is sodium oxide.
Formula mass
$$= (2 \times \text{atomic mass of Na}) + (1 \times \text{atomic mass of O})$$
$$= (2 \times 22.99 \text{ amu}) + 16.00 \text{ amu} = 61.98 \text{ amu}$$
(b) NO_2. This is a covalent compound, and N has the lower group number, so the name is nitrogen dioxide.
Molecular mass
$$= (1 \times \text{atomic mass of N}) + (2 \times \text{atomic mass of O})$$
$$= 14.01 \text{ amu} + (2 \times 16.00 \text{ amu}) = 46.01 \text{ amu}$$

Problems

Problems with **colored** numbers are answered in Appendix E. Sections match the text and provide the numbers of relevant sample problems. Bracketed problems are grouped in pairs (indicated by a short rule) that cover the same concept. Comprehensive Problems are based on material from any section or previous chapter.

Elements, Compounds, and Mixtures: An Atomic Overview

2.1 What is the key difference between an element and a compound?
2.2 List two differences between a compound and a mixture.
2.3 Which of the following are pure substances? Explain.
(a) Calcium chloride, used to melt ice on roads, consists of two elements, calcium and chlorine, in a fixed mass ratio.
(b) Sulfur consists of sulfur atoms combined into octatomic molecules.
(c) Baking powder, a leavening agent, contains 26% to 30% sodium hydrogen carbonate and 30% to 35% calcium dihydrogen phosphate by mass.
(d) Cytosine, a component of DNA, consists of H, C, N, and O atoms bonded in a specific arrangement.
2.4 Classify each substance in Problem 2.3 as an element, compound, or mixture, and explain your answers.
2.5 Samples of illicit "street" drugs often contain an inactive component, such as ascorbic acid (vitamin C). After obtaining a sample of cocaine, government chemists calculate the mass of vitamin C per gram of drug sample, and use it to track the drug's distribution. For example, if different samples of cocaine obtained on the streets of New York, Los Angeles, and Paris all contain 0.6384 g of vitamin C per gram of sample, they very likely come from a common source. Is this street sample a compound, element, or mixture? Explain.

The Observations That Led to an Atomic View of Matter
(Sample Problem 2.1)

2.6 To which classes of matter—element, compound, and/or mixture—do the following apply: (a) law of mass conservation; (b) law of definite composition; (c) law of multiple proportions?
2.7 Identify the mass law that each of the following observations demonstrates, and explain your reasoning:
(a) A sample of potassium chloride from Chile contains the same percent by mass of potassium as one from Poland.
(b) A flashbulb contains magnesium and oxygen before use and magnesium oxide afterward, but its mass does not change.
(c) Arsenic and oxygen form one compound that is 65.2 mass % arsenic and another that is 75.8 mass % arsenic.
2.8 (a) Does the percent by mass of each element in a compound depend on the amount of compound? Explain. (b) Does the mass of each element in a compound depend on the amount of compound? Explain. (c) Does the percent by mass of each element in a compound depend on the amount of that element used to make the compound? Explain.
2.9 State the mass law(s) demonstrated by the following experimental results, and explain your reasoning:
Experiment 1: A student heats 1.00 g of a blue compound and obtains 0.64 g of a white compound and 0.36 g of a colorless gas.

Experiment 2: A second student heats 3.25 g of the same blue compound and obtains 2.08 g of a white compound and 1.17 g of a colorless gas.

2.10 State the mass law(s) demonstrated by the following experimental results, and explain your reasoning:
Experiment 1: A student heats 1.27 g of copper and 3.50 g of iodine to produce 3.81 g of a white compound, and 0.96 g of iodine remains.
Experiment 2: A second student heats 2.55 g of copper and 3.50 g of iodine to form 5.25 g of a white compound, and 0.80 g of copper remains.

2.11 Fluorite, a mineral of calcium, is a compound of the metal with fluorine. Analysis shows that a 2.76-g sample of fluorite contains 1.42 g of calcium. Calculate the (a) mass of fluorine in the sample; (b) mass fractions of calcium and fluorine in fluorite; (c) mass percents of calcium and fluorine in fluorite.

2.12 Galena, a mineral of lead, is a compound of the metal with sulfur. Analysis shows that a 2.34-g sample of galena contains 2.03 g of lead. Calculate the (a) mass of sulfur in the sample; (b) mass fractions of lead and sulfur in galena; (c) mass percents of lead and sulfur in galena.

2.13 A compound of copper and sulfur contains 88.39 g of metal and 44.61 g of nonmetal. How many grams of copper are in 5264 kg of compound? How many grams of sulfur?

2.14 A compound of iodine and cesium contains 63.94 g of metal and 61.06 g of nonmetal. How many grams of cesium are in 38.77 g of compound? How many grams of iodine?

2.15 Show, with calculations, how the following data illustrate the law of multiple proportions:
Compound 1: 47.5 mass % sulfur and 52.5 mass % chlorine
Compound 2: 31.1 mass % sulfur and 68.9 mass % chlorine

2.16 Show, with calculations, how the following data illustrate the law of multiple proportions:
Compound 1: 77.6 mass % xenon and 22.4 mass % fluorine
Compound 2: 63.3 mass % xenon and 36.7 mass % fluorine

2.17 Dolomite is a carbonate of magnesium and calcium. Analysis shows that 7.81 g of dolomite contains 1.70 g of Ca. Calculate the mass percent of Ca in dolomite. On the basis of the mass percent of Ca, and neglecting all other factors, which is the richer source of Ca, dolomite or fluorite (see Problem 2.11)?

2.18 The mass percent of sulfur in a sample of coal is a key factor in the environmental impact of the coal because the sulfur combines with oxygen when the coal is burned and the oxide can then be incorporated into acid rain. Which of the following coals would have the smallest environmental impact?

	Mass (g) of Sample	Mass (g) of Sulfur in Sample
Coal A	378	11.3
Coal B	495	19.0
Coal C	675	20.6

Dalton's Atomic Theory

2.19 Use Dalton's theory to explain why potassium nitrate from India or Italy has the same mass percents of K, N, and O.

The Observations That Led to the Nuclear Atom Model

2.20 Thomson was able to determine the mass/charge ratio of the electron but not its mass. (a) How did Millikan's experiment allow determination of the electron's mass? (b) The following charges on individual oil droplets were obtained during an experiment similar to Millikan's. Determine a charge for the electron (in C, coulombs), and explain your answer: -3.204×10^{-19} C; -4.806×10^{-19} C; -8.010×10^{-19} C; -1.442×10^{-18} C.

2.21 When Rutherford's coworkers bombarded gold foil with α particles, they obtained results that overturned the existing (Thomson) model of the atom. Explain.

The Atomic Theory Today
(Sample Problems 2.2 and 2.3)

2.22 Choose the correct answer. The difference between the mass number of an isotope and its atomic number is (a) directly related to the identity of the element; (b) the number of electrons; (c) the number of neutrons; (d) the number of isotopes.

2.23 Argon has three naturally occurring isotopes, ^{36}Ar, ^{38}Ar, and ^{40}Ar. What is the mass number of each? How many protons, neutrons, and electrons are present in each?

2.24 Chlorine has two naturally occurring isotopes, ^{35}Cl and ^{37}Cl. What is the mass number of each isotope? How many protons, neutrons, and electrons are present in each?

2.25 Do both members of the following pairs have the same number of protons? Neutrons? Electrons?
(a) $^{16}_{8}$O and $^{17}_{8}$O   (b) $^{40}_{18}$Ar and $^{41}_{19}$K   (c) $^{60}_{27}$Co and $^{60}_{28}$Ni
Which pair(s) consist(s) of atoms with the same Z value? N value? A value?

2.26 Do both members of the following pairs have the same number of protons? Neutrons? Electrons?
(a) $^{3}_{1}$H and $^{3}_{2}$He   (b) $^{14}_{6}$C and $^{15}_{7}$N   (c) $^{19}_{9}$F and $^{18}_{9}$F
Which pair(s) consist(s) of atoms with the same Z value? N value? A value?

2.27 Write the $^{A}_{Z}$X notation for each atomic depiction:

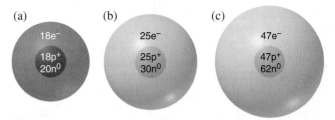

2.28 Write the $^{A}_{Z}$X notation for each atomic depiction:

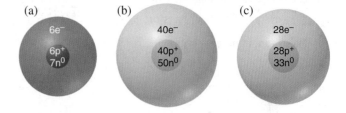

2.29 Draw atomic depictions similar to those in Problem 2.27 for (a) $^{48}_{22}$Ti; (b) $^{79}_{34}$Se; (c) $^{11}_{5}$B.

2.30 Draw atomic depictions similar to those in Problem 2.27 for (a) $^{207}_{82}$Pb; (b) $^{9}_{4}$Be; (c) $^{75}_{33}$As.

2.31 Gallium has two naturally occurring isotopes, ^{69}Ga (isotopic mass 68.9256 amu, abundance 60.11%) and ^{71}Ga (isotopic mass 70.9247 amu, abundance 39.89%). Calculate the atomic mass of gallium.

2.32 Magnesium has three naturally occurring isotopes, ^{24}Mg (isotopic mass 23.9850 amu, abundance 78.99%), ^{25}Mg (isotopic mass 24.9858 amu, abundance 10.00%), and ^{26}Mg (isotopic mass 25.9826 amu, abundance 11.01%). Calculate the atomic mass of magnesium.

2.33 Chlorine has two naturally occurring isotopes, ^{35}Cl (isotopic mass 34.9689 amu) and ^{37}Cl (isotopic mass 36.9659 amu). If chlorine has an atomic mass of 35.4527 amu, what is the percent abundance of each isotope?

2.34 Copper has two naturally occurring isotopes, ^{63}Cu (isotopic mass 62.9396 amu) and ^{65}Cu (isotopic mass 64.9278 amu). If copper has an atomic mass of 63.546 amu, what is the percent abundance of each isotope?

Elements: A First Look at the Periodic Table

2.35 Correct each of the following statements:
(a) In the modern periodic table, the elements are arranged in order of increasing atomic mass.
(b) Elements in a period have similar chemical properties.
(c) Elements can be classified as either metalloids or nonmetals.

2.36 What class of elements lies along the "staircase" line in the periodic table? How do their properties compare with those of metals and nonmetals?

2.37 What are some characteristic properties of elements to the left of the elements along the "staircase"? To the right?

2.38 Give the name, atomic symbol, and group number of the element with the following Z value, and classify it as a metal, metalloid, or nonmetal:
(a) $Z = 32$ (b) $Z = 16$ (c) $Z = 2$ (d) $Z = 3$ (e) $Z = 42$

2.39 Give the name, atomic symbol, and group number of the element with the following Z value, and classify it as a metal, metalloid, or nonmetal:
(a) $Z = 33$ (b) $Z = 20$ (c) $Z = 35$ (d) $Z = 19$ (e) $Z = 12$

2.40 Fill in the blanks:
(a) The symbol and atomic number of the heaviest alkaline earth metal are _____ and _____.
(b) The symbol and atomic number of the lightest metalloid in Group 5A(15) are _____ and _____.
(c) Group 1B(11) consists of the *coinage metals*. The symbol and atomic mass of the coinage metal whose atoms have the fewest electrons are _____ and _____.
(d) The symbol and atomic mass of the halogen in Period 4 are _____ and _____.

2.41 Fill in the blanks:
(a) The symbol and atomic number of the heaviest noble gas are _____ and _____.
(b) The symbol and group number of the Period 5 transition element whose atoms have the fewest protons are _____ and _____.
(c) The elements in Group 6A(16) are sometimes called the *chalcogens*. The symbol and atomic number of the only metallic chalcogen are _____ and _____.
(d) The symbol and number of protons of the Period 4 alkali metal atom are _____ and _____.

Compounds: Introduction to Bonding
(Sample Problem 2.4)

2.42 Describe the type and nature of the bonding that occurs between reactive metals and nonmetals.

2.43 Describe the type and nature of the bonding that often occurs between two nonmetals.

2.44 Given that the ions in LiF and in MgO are of similar size, which compound has stronger ionic bonding? Use Coulomb's law in your explanation.

2.45 Describe the formation of solid magnesium chloride ($MgCl_2$) from large numbers of magnesium and chlorine atoms.

2.46 Does potassium nitrate (KNO_3) incorporate ionic bonding, covalent bonding, or both? Explain.

2.47 What monatomic ions do potassium ($Z = 19$) and iodine ($Z = 53$) form?

2.48 What monatomic ions do barium ($Z = 56$) and selenium ($Z = 34$) form?

2.49 For each ionic depiction, give the name of the parent atom, its mass number, and its group and period numbers:

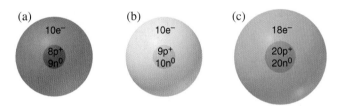

2.50 For each ionic depiction, give the name of the parent atom, its mass number, and its group and period numbers:

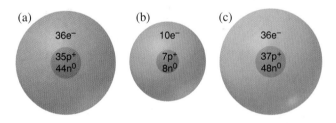

2.51 An ionic compound forms when lithium ($Z = 3$) reacts with oxygen ($Z = 8$). If a sample of the compound contains 5.3×10^{20} lithium ions, how many oxide ions does it contain?

2.52 An ionic compound forms when calcium ($Z = 20$) reacts with iodine ($Z = 53$). If a sample of the compound contains 7.4×10^{21} calcium ions, how many iodide ions does it contain?

2.53 The radii of the sodium and potassium ions are 102 pm and 138 pm, respectively. Which compound has stronger ionic attractions, sodium chloride or potassium chloride?

2.54 The radii of the lithium and magnesium ions are 76 pm and 72 pm, respectively. Which compound has stronger ionic attractions, lithium oxide or magnesium oxide?

Compounds: Formulas, Names, and Masses
(Sample Problems 2.5 to 2.14)

2.55 What is the difference between an empirical formula and a molecular formula? Can they ever be the same?

2.56 Consider a mixture of 10 billion O_2 molecules and 10 billion H_2 molecules. In what way is this mixture similar to a sample containing 10 billion hydrogen peroxide (H_2O_2) molecules? In what way is it different?

2.57 Write an empirical formula for each of the following:
(a) Hydrazine, a rocket fuel, molecular formula N_2H_4
(b) Glucose, a sugar, molecular formula $C_6H_{12}O_6$
2.58 Write an empirical formula for each of the following:
(a) Ethylene glycol, car antifreeze, molecular formula $C_2H_6O_2$
(b) Peroxodisulfuric acid, a compound used to make bleaching agents, molecular formula $H_2S_2O_8$

2.59 Give the name and formula of the compound formed from the following elements: (a) lithium and nitrogen; (b) oxygen and strontium; (c) aluminum and chlorine.
2.60 Give the name and formula of the compound formed from the following elements: (a) rubidium and bromine; (b) sulfur and barium; (c) calcium and fluorine.

2.61 Give the name and formula of the compound formed from the following elements:
(a) $_{12}$L and $_9$M   (b) $_{11}$L and $_{16}$M   (c) $_{17}$L and $_{38}$M
2.62 Give the name and formula of the compound formed from the following elements:
(a) $_3$Q and $_{35}$R   (b) $_8$Q and $_{13}$R   (c) $_{19}$Q and $_{53}$R

2.63 Give the systematic names for the formulas or the formulas for the names: (a) tin(IV) chloride; (b) $FeBr_3$; (c) cuprous bromide; (d) Mn_2O_3.
2.64 Give the systematic names for the formulas or the formulas for the names: (a) Na_2HPO_4; (b) potassium carbonate dihydrate; (c) $NaNO_2$; (d) ammonium perchlorate.

2.65 Correct each of the following formulas:
(a) Barium oxide is BaO_2.
(b) Iron(II) nitrate is $Fe(NO_3)_3$.
(c) Magnesium sulfide is $MnSO_3$.
2.66 Correct each of the following names:
(a) CuI is cobalt(II) iodide.
(b) $Fe(HSO_4)_3$ is iron(II) sulfate.
(c) $MgCr_2O_7$ is magnesium dichromium heptaoxide.

2.67 Give the name and formula for the acid derived from each of the following anions:
(a) hydrogen sulfate   (b) IO_3^- (c) cyanide   (d) HS^-
2.68 Give the name and formula for the acid derived from each of the following anions:
(a) perchlorate   (b) NO_3^- (c) bromite   (d) F^-

2.69 Give the name and formula of the compound whose molecules consist of two sulfur atoms and four fluorine atoms.
2.70 Give the name and formula of the compound whose molecules consist of two iodine atoms and seven oxygen atoms.

2.71 Give the number of atoms of the specified element in a formula unit of each of the following compounds, and calculate the molecular (formula) mass:
(a) Oxygen in aluminum sulfate, $Al_2(SO_4)_3$
(b) Hydrogen in ammonium hydrogen phosphate, $(NH_4)_2HPO_4$
(c) Oxygen in the mineral azurite, $Cu_3(OH)_2(CO_3)_2$
2.72 Give the number of atoms of the specified element in a formula unit of each of the following compounds, and calculate the molecular (formula) mass:
(a) Hydrogen in ammonium benzoate, $C_6H_5COONH_4$
(b) Nitrogen in hydrazinium sulfate, $N_2H_6SO_4$
(c) Oxygen in the mineral leadhillite, $Pb_4SO_4(CO_3)_2(OH)_2$

2.73 Write the formula of each compound, and determine its molecular (formula) mass: (a) ammonium sulfate; (b) sodium dihydrogen phosphate; (c) potassium bicarbonate.
2.74 Write the formula of each compound, and determine its molecular (formula) mass: (a) sodium dichromate; (b) ammonium perchlorate; (c) magnesium nitrite trihydrate.

2.75 Give the name, empirical formula, and molecular mass of the molecule depicted in Figure P2.75.
2.76 Give the name, empirical formula, and molecular mass of the molecule depicted in Figure P2.76.

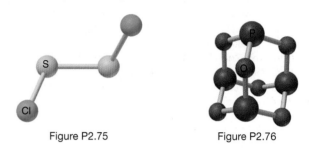

Figure P2.75   Figure P2.76

2.77 Give the formula, name, and molecular mass of the following molecules:

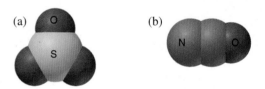

2.78 Before the use of systematic names, many compounds had common names. Give the systematic name for each of the following: (a) blue vitriol, $CuSO_4 \cdot 5H_2O$; (b) slaked lime, $Ca(OH)_2$; (c) oil of vitriol, H_2SO_4; (d) washing soda, Na_2CO_3; (e) muriatic acid, HCl; (f) Epsom salts, $MgSO_4 \cdot 7H_2O$; (g) chalk, $CaCO_3$; (h) dry ice, CO_2; (i) baking soda, $NaHCO_3$; (j) lye, NaOH.

Mixtures: Classification and Separation

2.79 In what main way is separating the components of a mixture different from separating the components of a compound?
2.80 What is the difference between a homogeneous and a heterogeneous mixture?
2.81 Is a solution a homogeneous or a heterogeneous mixture? Give an example of an aqueous solution.

2.82 Classify each of the following as a compound, a homogeneous mixture, or a heterogeneous mixture: (a) distilled water; (b) gasoline; (c) beach sand; (d) wine; (e) air.
2.83 Classify each of the following as a compound, a homogeneous mixture, or a heterogeneous mixture: (a) orange juice; (b) vegetable soup; (c) cement; (d) calcium sulfate; (e) tea.

Comprehensive Problems

Problems with an asterisk (*) are more challenging.

2.84 Helium is the lightest noble gas and the second most abundant element (after hydrogen) in the universe.
(a) The radius of a helium atom is 3.1×10^{-11} m; the radius of its nucleus is 2.5×10^{-15} m. What fraction of the spherical

atomic volume is occupied by the nucleus (V of a sphere = $\frac{4}{3}\pi r^3$)?

(b) The mass of a helium-4 atom is 6.64648×10^{-24} g, and each of its two electrons has a mass of 9.10939×10^{-28} g. What fraction of this atom's mass is contributed by its nucleus?

2.85 Scenes A–I depict various types of matter on the atomic scale. Choose the correct scene(s) for each of the following:

(a) A mixture that fills its container

(b) A substance that cannot be broken down into simpler ones

(c) An element with a very high resistance to flow

(d) A homogeneous mixture

(e) An element that conforms to the walls of its container and displays a surface

(f) A gas consisting of diatomic particles

(g) A gas that can be broken down into simpler substances

(h) A substance with a 2:1 number ratio of its component atoms

(i) Matter that can be separated into its component substances by physical means

(j) A heterogeneous mixture

(k) Matter that obeys the law of definite composition

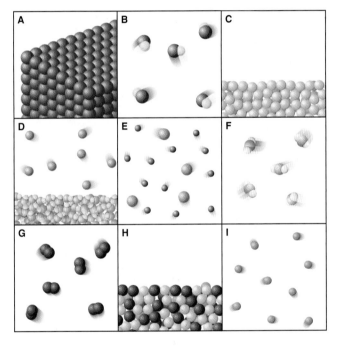

2.86 Nitrogen forms more oxides than any other element. The percents by mass of N in three different nitrogen oxides are (I) 46.69%; (II) 36.85%; (III) 25.94%. (a) Determine the empirical formula of each compound. (b) How many grams of oxygen per 1.00 g of nitrogen are in each compound?

2.87 Which of the following pictures is (are) consistent with the fact that compounds of nitrogen *(blue)* and oxygen *(red)* exhibit the law of multiple proportions? Explain.

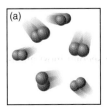
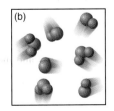
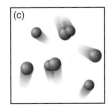

2.88 The seven most abundant ions in seawater make up more than 99% by mass of the dissolved compounds. They are listed in units of mg ion/kg seawater: chloride 18,980; sodium 10,560; sulfate 2650; magnesium 1270; calcium 400; potassium 380; hydrogen carbonate 140.

(a) What is the mass % of each ion in seawater?

(b) What percent of the total mass of ions is sodium ion?

(c) How does the total mass % of alkaline earth metal ions compare with the total mass % of alkali metal ions?

(d) Which makes up the larger mass fraction of dissolved components, anions or cations?

* **2.89** When barium (Ba) reacts with sulfur (S) to form barium sulfide (BaS), each Ba atom reacts with an S atom. If 2.00 cm^3 of Ba reacts with 1.50 cm^3 of S, are there enough Ba atoms to react with the S atoms (d of Ba = 3.51 g/cm^3; d of S = 2.07 g/cm^3)?

2.90 Succinic acid *(below)* is an important metabolite in biological energy production. Give the molecular formula, empirical formula, and molecular mass of succinic acid, and calculate the mass percent of each element.

2.91 Antimony has many uses—for example, in semiconductor infrared devices and as part of an alloy in lead storage batteries. The element has two naturally occurring isotopes, one with mass 120.904 amu, the other with mass 122.904 amu. (a) Write the $_Z^A$X notation for each isotope. (b) Use the atomic mass of antimony from the periodic table to calculate the natural abundance of each isotope.

2.92 The two isotopes of potassium with significant abundance in nature are ^{39}K (isotopic mass 38.9637 amu, 93.258%) and ^{41}K (isotopic mass 40.9618 amu, 6.730%). Fluorine has only one naturally occurring isotope, ^{19}F (isotopic mass 18.9984 amu). Calculate the formula mass of potassium fluoride.

2.93 Nuclei differ in their stability, and some are so unstable that they undergo radioactive decay. The ratio of the number of neutrons to number of protons (N/Z) in a nucleus correlates with its stability. Calculate the N/Z ratio for (a) ^{144}Sm; (b) ^{56}Fe; (c) ^{20}Ne; (d) ^{107}Ag. (e) The radioactive isotope ^{238}U decays in a series of nuclear reactions that includes another uranium isotope, ^{234}U, and three lead isotopes, ^{214}Pb, ^{210}Pb, and ^{206}Pb. How many neutrons, protons, and electrons are in each of these five isotopes?

* **2.94** TNT (trinitrotoluene; *below*) is used as an explosive in construction. Calculate the mass of each element in 1.00 lb of TNT.

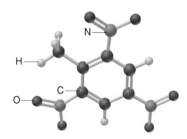

2.95 The anticancer drug Platinol (Cisplatin), $Pt(NH_3)_2Cl_2$, reacts with the cancer cell's DNA and interferes with its growth. (a) What is the mass % of platinum (Pt) in Platinol? (b) If Pt costs $19/g, how many grams of Platinol can be made for $1.00 million (assume that the cost of Pt determines the cost of the drug)?

2.96 Dimercaprol ($HSCH_2CHSHCH_2OH$) is a complexing agent developed during World War I as an antidote to arsenic-based poison gas and used today to treat heavy-metal poisoning. Such an agent binds and removes the toxic element from the body.
(a) If each molecule binds one arsenic (As) atom, how many atoms of As could be removed by 250. mg of dimercaprol?
(b) If one molecule binds one metal atom, calculate the mass % of each of the following metals in a metal-dimercaprol complex: mercury, thallium, chromium.

2.97 Choose the box color(s) in the periodic table below that match(es) the following:

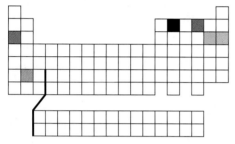

(a) Four elements that are nonmetals
(b) Two elements that are metals
(c) Three elements that are gases at room temperature
(d) Three elements that are solid at room temperature
(e) One pair of elements likely to form a covalent compound
(f) Another pair of elements likely to form a covalent compound
(g) One pair of elements likely to form an ionic compound with formula MX
(h) Another pair of elements likely to form an ionic compound with formula MX
(i) Two elements likely to form an ionic compound with formula M_2X
(j) Two elements likely to form an ionic compound with formula MX_2
(k) An element that forms no compounds
(l) A pair of elements whose compounds exhibit the law of multiple proportions

2.98 From the following ions and their radii (in pm), choose a pair that gives the strongest ionic bonding and a pair that gives the weakest: Mg^{2+} 72; K^+ 138; Rb^+ 152; Ba^{2+} 135; Cl^- 181; O^{2-} 140; I^- 220.

2.99 Transition metals, located in the center of the periodic table, have many essential uses as elements and form many important compounds as well. Calculate the molecular mass of the following transition metal compounds:
(a) $[Co(NH_3)_6]Cl_3$ (b) $[Pt(NH_3)_4BrCl]Cl_2$
(b) $K_4[V(CN)_6]$ (d) $[Ce(NH_3)_6][FeCl_4]_3$

2.100 A rock is 5.0% by mass fayalite (Fe_2SiO_4), 7.0% by mass forsterite (Mg_2SiO_4), and the remainder silicon dioxide. What is the mass percent of each element in the rock?

2.101 Fluoride ion is poisonous in relatively low amounts 0.2 g of F^- per 70 kg of body weight can cause death. Nevertheless, in order to prevent tooth decay, F^- ions are added to drinking water at a concentration of 1 mg of F^- ion per L of water. How many liters of fluoridated drinking water would a 70-kg person have to consume in one day to reach this toxic level? How many kilograms of sodium fluoride would be needed to treat a 7.00×10^7-gal reservoir?

2.102 The scene below represents a chemical reaction in a container. Which of the mass laws—mass conservation, definite composition, or multiple proportions—is (are) illustrated?

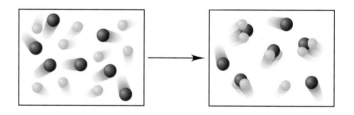

2.103 Nitrogen monoxide (NO) is a bioactive molecule in blood. Low NO concentrations cause respiratory distress and the formation of blood clots. Doctors prescribe nitroglycerin, $C_3H_5N_3O_9$, and isoamyl nitrate, $(CH_3)_2CHCH_2CH_2ONO_2$, to increase NO. If each compound releases one molecule of NO per atom of N, calculate the mass percent of NO in each medicine.

2.104 Which of the following steps in an overall process involve(s) a physical change and which involve(s) a chemical change?

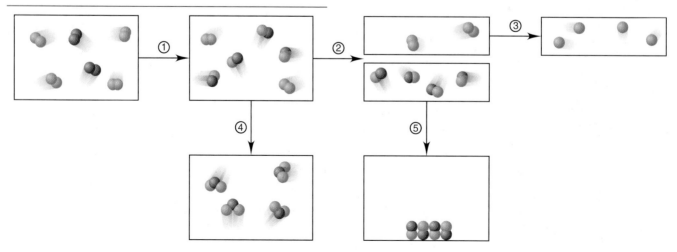

CHAPTER THREE

Stoichiometry of Formulas and Equations

Weighing the Matter By knowing the masses of aluminum and iron(III) oxide reacting in this spectacular chemical change, you also know the number of atoms of each and can predict the masses of aluminum oxide and iron metal that form. In this chapter, you'll learn the arithmetic of chemical formulas and reactions.

Key Principles

◆ The *mole* (*mol*) is the standard unit for *amount of substance* and consists of *Avogadro's number* (6.022×10^{23}) of atoms, molecules, or ions. It has the same numerical value in grams as a single entity of the substance has in atomic mass units; for example, 1 molecule of H_2O weighs 18.02 amu and 1 mol of H_2O weighs 18.02 g. Therefore, if the amount of a substance is expressed in moles, we know the number of entities in a given mass of it.

◆ The subscripts in a *chemical formula* provide quantitative information about the amounts of each element in a mole of compound. In an *empirical formula,* the subscripts show the *relative* number of moles of each element; in a *molecular formula,* they show the *actual* number. Isomers are different compounds with the same molecular formula.

◆ In a *balanced equation,* formulas preceded by integer *balancing coefficients* are used to show the same numbers of each kind of atom on the left (*reactants*) as on the right (*products*) but in different combinations; we can therefore use the amount of one substance to calculate the amount of any other.

◆ During a typical reaction, one substance (the *limiting reactant*) is used up, so it limits the amount of product that can form; the other reactant(s) are in excess. The *theoretical yield,* the amount based on the balanced equation, is never obtained in the lab because of competing *side reactions* and physical losses.

◆ For reactions in solution, we determine amounts of substances from their *concentration* (*molarity*) and volume. To dilute a solution, we add *solvent,* which lowers the amount of *solute* dissolved in each unit volume.

Outline

Concepts & Skills to Review Before You Study This Chapter

- isotopes and atomic mass (Section 2.5)
- names and formulas of compounds (Section 2.8)
- molecular mass of a compound (Section 2.8)
- empirical and molecular formulas (Section 2.8)
- mass laws in chemical reactions (Section 2.2)

Chemistry is a practical science. Just imagine how useful it could be to determine the formula of a compound from the masses of its elements or to predict the amounts of substances consumed and produced in a reaction. Suppose you are a polymer chemist preparing a new plastic: how much of this new material will a given polymerization reaction yield? Or suppose you're a chemical engineer studying rocket engine thrust: what amount of exhaust gases will a test of this fuel mixture produce? Perhaps you are on a team of environmental chemists examining coal samples: what quantity of air pollutants will this sample release when burned? Or, maybe you're a biomedical researcher who has extracted a new cancer-preventing substance from a tropical plant: what is its formula, and what quantity of metabolic products will establish a safe dosage level? You can answer such questions and countless others like them with a knowledge of **stoichiometry** (pronounced "stoy-key-AHM-uh-tree"; from the Greek *stoicheion,* "element or part," and *metron,* "measure"), the study of the quantitative aspects of chemical formulas and reactions.

3.1 THE MOLE

All the ideas and skills discussed in this chapter depend on an understanding of the *mole* concept, so let's begin there. In daily life, we typically measure things out by counting or by weighing, with the choice based on convenience. It is more convenient to weigh beans or rice than to count individual pieces, and it is more convenient to count eggs or pencils than to weigh them. To measure such things, we use mass units (a kilogram of rice) or counting units (a dozen pencils). Similarly, daily life in the laboratory involves measuring substances. However, an obvious problem arises when we try to do this. The atoms, ions, molecules, or formula units are the entities that react with one another, so we would like to know the numbers of them that we mix together. But, how can we possibly count entities that are so small? To do this, chemists have devised a unit called the mole to *count chemical entities by weighing them.*

Defining the Mole

The **mole** (abbreviated **mol**) is the SI unit for amount of substance. It is defined as *the amount of a substance that contains the same number of entities as there are atoms in exactly 12 g of carbon-12.* This number is called **Avogadro's number,** in honor of the 19th-century Italian physicist Amedeo Avogadro, and as you can tell from the definition, it is enormous:*

One mole (1 mol) contains 6.022×10^{23} entities (to four significant figures) **(3.1)**

Thus,

1 mol of carbon-12	contains	6.022×10^{23} carbon-12 atoms
1 mol of H_2O	contains	6.022×10^{23} H_2O molecules
1 mol of NaCl	contains	6.022×10^{23} NaCl formula units

However, the mole is not just a counting unit like the dozen, which specifies only the *number* of objects. The definition of the mole specifies the *number* of objects in a fixed *mass* of substance. Therefore, *1 mole of a substance represents a fixed number of chemical entities* ***and*** *has a fixed mass.* To see why this is important, consider the marbles in Figure 3.1A, which we'll use as

*A mole of any ordinary object is a staggering amount: a mole of periods (.) lined up side by side would equal the radius of our galaxy; a mole of marbles stacked tightly together would cover the United States 70 miles deep. However, atoms and molecules are not ordinary objects: a mole of water molecules (about 18 mL) can be swallowed in one gulp!

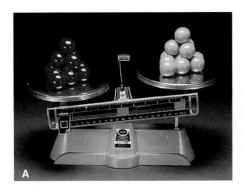

Figure 3.1 Counting objects of fixed relative mass. A, If marbles had a fixed mass, we could count them by weighing them. Each red marble weighs 7 g, and each yellow marble weighs 4 g, so 84 g of red marbles and 48 g of yellow marbles each contains 12 marbles. Equal numbers of the two types of marbles always have a 7:4 mass ratio of red:yellow marbles. **B,** Because atoms of a substance have a fixed mass, we can weigh the substance to count the atoms; 55.85 g of Fe (left pan) and 32.07 g of S (right pan) each contains 6.022×10^{23} atoms (1 mol of atoms). Any two samples of Fe and S that contain equal numbers of atoms have a 55.85:32.07 mass ratio of Fe:S.

an analogy for atoms. Suppose you have large groups of red marbles and yellow marbles; each red marble weighs 7 g and each yellow marble weighs 4 g. Right away you know that there are 12 marbles in 84 g of red marbles or in 48 g of yellow marbles. Moreover, because one red marble weighs $\frac{7}{4}$ as much as one yellow marble, any given *number* of red and of yellow marbles always has this 7:4 *mass* ratio. By the same token, any given *mass* of red and of yellow marbles always has a 4:7 *number* ratio. For example, 280 g of red marbles contains 40 marbles, and 280 g of yellow marbles contains 70 marbles. As you can see, the fixed masses of the marbles allow you to count marbles by weighing them.

Atoms have fixed masses, too. Let's recall a key point from Chapter 2: the atomic mass of an element (which appears on the periodic table) is the weighted average of the masses of its naturally occurring isotopes. That is, all iron (Fe) atoms have an atomic mass of 55.85 amu, all sulfur (S) atoms have an atomic mass of 32.07 amu, and so forth.

The central relationship between the mass of one atom and the mass of 1 mole of those atoms is that *the atomic mass of an element expressed in* **amu** *is numerically the same as the mass of 1 mole of atoms of the element expressed in* **grams**. You can see this from the definition of the mole, which referred to the number of atoms in "**12** g of carbon-**12**." Thus,

1 Fe atom	has a mass of	55.85 amu	and	1 mol of Fe atoms	has a mass of	55.85 g
1 S atom	has a mass of	32.07 amu	and	1 mol of S atoms	has a mass of	32.07 g
1 O atom	has a mass of	16.00 amu	and	1 mol of O atoms	has a mass of	16.00 g
1 O_2 molecule	has a mass of	32.00 amu	and	1 mol of O_2 molecules	has a mass of	32.00 g

Moreover, because of their fixed atomic masses, we know that 55.85 g of Fe atoms and 32.07 g of S atoms each contains 6.022×10^{23} atoms. As with marbles of fixed mass, one Fe atom weighs $\frac{55.85}{32.07}$ as much as one S atom, and 1 mol of Fe atoms weighs $\frac{55.85}{32.07}$ as much as 1 mol of S atoms (Figure 3.1B).

A similar relationship holds for compounds: *the molecular mass (or formula mass) of a compound expressed in* **amu** *is numerically the same as the mass of 1 mole of the compound expressed in* **grams**. Thus, for example,

1 molecule of H_2O	has a mass of	18.02 amu	and	1 mol of H_2O (6.022×10^{23} molecules)	has a mass of	18.02 g
1 formula unit of NaCl	has a mass of	58.44 amu	and	1 mol of NaCl (6.022×10^{23} formula units)	has a mass of	58.44 g

To summarize the two key points about the usefulness of the mole concept:

- The mole maintains the *same mass relationship* between macroscopic samples as exists between individual chemical entities.
- The mole relates the *number* of chemical entities to the *mass* of a sample of those entities.

A grocer cannot obtain 1 dozen eggs by weighing them because eggs vary in mass. But a chemist *can* obtain 1 mol of copper atoms (6.022×10^{23} atoms) simply

Figure 3.2 One mole of some familiar substances. One mole of a substance is the amount that contains 6.022×10^{23} atoms, molecules, or formula units. From left to right: 1 mol (172.19 g) of writing chalk (calcium sulfate dihydrate), 1 mol (32.00 g) of gaseous O_2, 1 mol (63.55 g) of copper, and 1 mol (18.02 g) of liquid H_2O.

by weighing 63.55 g of copper. Figure 3.2 shows 1 mol of some familiar elements and compounds.

Molar Mass

The **molar mass** ($\mathcal{M}$) of a substance is the mass per mole of its entities (atoms, molecules, or formula units). Thus, molar mass has units of grams per mole (g/mol). The periodic table is indispensable for calculating the molar mass of a substance. Here's how the calculations are done:

1. **Elements.** You find the molar mass of an element simply by looking up its atomic mass in the periodic table and then noting whether the element occurs naturally as individual atoms or as molecules.

- *Monatomic elements.* For elements that occur as individual atoms, the molar mass is the numerical value from the periodic table expressed in units of grams per mole.* Thus, the molar mass of neon is 20.18 g/mol, the molar mass of iron is 55.85 g/mol, and the molar mass of gold is 197.0 g/mol.

- *Molecular elements.* For elements that occur as molecules, you must know the molecular formula to determine the molar mass. For example, oxygen exists normally in air as diatomic molecules, so the molar mass of O_2 molecules is twice that of O atoms:

$$\text{Molar mass } (\mathcal{M}) \text{ of } O_2 = 2 \times \mathcal{M} \text{ of O} = 2 \times 16.00 \text{ g/mol} = 32.00 \text{ g/mol}$$

The most common form of sulfur exists as octatomic molecules, S_8:

$$\mathcal{M} \text{ of } S_8 = 8 \times \mathcal{M} \text{ of S} = 8 \times 32.07 \text{ g/mol} = 256.6 \text{ g/mol}$$

2. **Compounds.** *The molar mass of a compound is the sum of the molar masses of the atoms of the elements in the formula.* For example, the formula of sulfur dioxide (SO_2) tells us that 1 mol of SO_2 molecules contains 1 mol of S atoms and 2 mol of O atoms:

$$\mathcal{M} \text{ of } SO_2 = \mathcal{M} \text{ of S} + (2 \times \mathcal{M} \text{ of O}) = 32.07 \text{ g/mol} + (2 \times 16.00 \text{ g/mol})$$
$$= 64.07 \text{ g/mol}$$

Similarly, for ionic compounds, such as potassium sulfide (K_2S), we have

$$\mathcal{M} \text{ of } K_2S = (2 \times \mathcal{M} \text{ of K}) + \mathcal{M} \text{ of S} = (2 \times 39.10 \text{ g/mol}) + 32.07 \text{ g/mol}$$
$$= 110.27 \text{ g/mol}$$

A key point to note is that *the subscripts in a formula refer to individual atoms (or ions), as well as to moles of atoms (or ions).* Table 3.1 presents this idea for glucose ($C_6H_{12}O_6$), the essential sugar in energy metabolism.

---

*The mass value in the periodic table has no units because it is a *relative* atomic mass, given by the atomic mass (in amu) divided by 1 amu ($\frac{1}{12}$ mass of one ^{12}C atom in amu):

$$\text{Relative atomic mass} = \frac{\text{atomic mass (amu)}}{\frac{1}{12} \text{ mass of } ^{12}C \text{ (amu)}}$$

Therefore, you use the same number for the atomic mass (weighted average mass of one atom in amu) and the molar mass (mass of 1 mole of atoms in grams).

Table 3.1	Information Contained in the Chemical Formula of Glucose, $C_6H_{12}O_6$ ($\mathcal{M}$ = 180.16 g/mol)		
	Carbon (C)	**Hydrogen (H)**	**Oxygen (O)**
Atoms/molecule of compound	6 atoms	12 atoms	6 atoms
Moles of atoms/mole of compound	6 mol of atoms	12 mol of atoms	6 mol of atoms
Atoms/mole of compound	$6(6.022 \times 10^{23})$ atoms	$12(6.022 \times 10^{23})$ atoms	$6(6.022 \times 10^{23})$ atoms
Mass/molecule of compound	6(12.01 amu) = 72.06 amu	12(1.008 amu) = 12.10 amu	6(16.00 amu) = 96.00 amu
Mass/mole of compound	72.06 g	12.10 g	96.00 g

Interconverting Moles, Mass, and Number of Chemical Entities

One of the reasons the mole is such a convenient unit for laboratory work is that it allows you to calculate the mass or number of entities of a substance in a sample if you know the amount (number of moles) of the substance. Conversely, if you know the mass or number of entities of a substance, you can calculate the number of moles.

The molar mass, which expresses the equivalent relationship between 1 mole of a substance and its mass in grams, can be used as a conversion factor. We multiply by the molar mass of an element or compound ($\mathcal{M}$, in g/mol) to convert a given amount (in moles) to mass (in grams):

$$\text{Mass (g)} = \text{no. of moles} \times \frac{\text{no. of grams}}{1 \text{ mol}} \qquad (3.2)$$

Or, we divide by the molar mass (multiply by $1/\mathcal{M}$) to convert a given mass (in grams) to amount (in moles):

$$\text{No. of moles} = \text{mass (g)} \times \frac{1 \text{ mol}}{\text{no. of grams}} \qquad (3.3)$$

In a similar way, we use Avogadro's number, which expresses the equivalent relationship between 1 mole of a substance and the number of entities it contains, as a conversion factor. We multiply by Avogadro's number to convert amount of substance (in moles) to the number of entities (atoms, molecules, or formula units):

$$\text{No. of entities} = \text{no. of moles} \times \frac{6.022 \times 10^{23} \text{ entities}}{1 \text{ mol}} \qquad (3.4)$$

Or, we divide by Avogadro's number to do the reverse:

$$\text{No. of moles} = \text{no. of entities} \times \frac{1 \text{ mol}}{6.022 \times 10^{23} \text{ entities}} \qquad (3.5)$$

Converting Moles of Elements For problems involving mass-mole-number relationships of elements, keep these points in mind:

- To convert between amount (mol) and mass (g), use the molar mass ($\mathcal{M}$ in g/mol).
- To convert between amount (mol) and number of entities, use Avogadro's number (6.022×10^{23} entities/mol). For elements that occur as molecules, use the molecular formula to find atoms/mol.
- Mass and number of entities relate directly to number of moles, *not* to each other. Therefore, to convert between number of entities and mass, *first convert to number of moles*. For example, to find the number of atoms in a given mass,

$$\text{No. of atoms} = \text{mass (g)} \times \frac{1 \text{ mol}}{\text{no. of grams}} \times \frac{6.022 \times 10^{23} \text{ atoms}}{1 \text{ mol}}$$

These relationships are summarized in Figure 3.3 and demonstrated in Sample Problem 3.1.

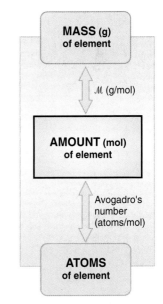

Figure 3.3 Summary of the mass-mole-number relationships for elements. The amount (mol) of an element is related to its mass (g) through the molar mass ($\mathcal{M}$ in g/mol) and to its number of atoms through Avogadro's number (6.022×10^{23} atoms/mol). For elements that occur as molecules, Avogadro's number gives *molecules* per mole.

SAMPLE PROBLEM 3.1   Calculating the Mass and Number of Atoms in a Given Number of Moles of an Element

Problem (a) Silver (Ag) is used in jewelry and tableware but no longer in U.S. coins. How many grams of Ag are in 0.0342 mol of Ag?
(b) Iron (Fe), the main component of steel, is the most important metal in industrial society. How many Fe atoms are in 95.8 g of Fe?

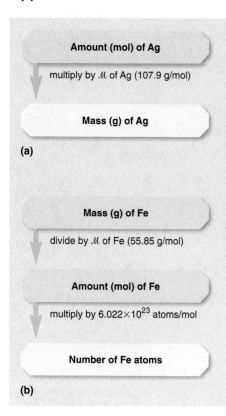

(a)

(b)

(a) Determining the mass (g) of Ag

Plan We know the number of moles of Ag (0.0342 mol) and have to find the mass (in g). To convert *moles* of Ag to *grams* of Ag, we multiply by the *molar mass* of Ag, which we find in the periodic table (see roadmap a).

Solution Converting from moles of Ag to grams:

$$\text{Mass (g) of Ag} = 0.0342 \, \cancel{\text{mol Ag}} \times \frac{107.9 \text{ g Ag}}{1 \, \cancel{\text{mol Ag}}} = 3.69 \text{ g Ag}$$

Check We rounded the mass to three significant figures because the number of moles has three. The units are correct. About 0.03 mol × 100 g/mol gives 3 g; the small mass makes sense because 0.0342 is a small fraction of a mole.

(b) Determining the number of Fe atoms

Plan We know the grams of Fe (95.8 g) and need the number of Fe atoms. We cannot convert directly from grams to atoms, so we first convert to moles by dividing grams of Fe by its molar mass. [This is the reverse of the step in part (a).] Then, we multiply number of moles by Avogadro's number to find number of atoms (see roadmap b).

Solution Converting from grams of Fe to moles:

$$\text{Moles of Fe} = 95.8 \, \cancel{\text{g Fe}} \times \frac{1 \text{ mol Fe}}{55.85 \, \cancel{\text{g Fe}}} = 1.72 \text{ mol Fe}$$

Converting from moles of Fe to number of atoms:

$$\text{No. of Fe atoms} = 1.72 \, \cancel{\text{mol Fe}} \times \frac{6.022 \times 10^{23} \text{ atoms Fe}}{1 \, \cancel{\text{mol Fe}}}$$
$$= 10.4 \times 10^{23} \text{ atoms Fe} = 1.04 \times 10^{24} \text{ atoms Fe}$$

Check When we approximate the mass of Fe and the molar mass of Fe, we have ~100 g/(~50 g/mol) = 2 mol. Therefore, the number of atoms should be about twice Avogadro's number: $2(6 \times 10^{23}) = 1.2 \times 10^{24}$.

FOLLOW-UP PROBLEM 3.1 (a) Graphite is the crystalline form of carbon used in "lead" pencils. How many moles of carbon are in 315 mg of graphite?
(b) Manganese (Mn) is a transition element essential for the growth of bones. What is the mass in grams of 3.22×10^{20} Mn atoms, the number found in 1 kg of bone?

Converting Moles of Compounds Solving mass-mole-number problems involving compounds requires a very similar approach to the one for elements. We need the chemical formula to find the molar mass and to determine the moles of a given element in the compound. These relationships are shown in Figure 3.4, and an example is worked through in Sample Problem 3.2.

Figure 3.4 Summary of the mass-mole-number relationships for compounds. Moles of a compound are related to grams of the compound through the molar mass ($\mathcal{M}$ in g/mol) and to the number of molecules (or formula units) through Avogadro's number (6.022×10^{23} molecules/mol). To find the number of molecules (or formula units) in a given mass, or vice versa, convert the information to moles first. With the chemical formula, you can calculate mass-mole-number information about each component element.

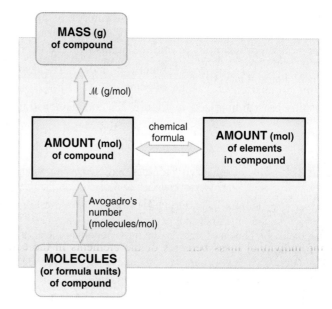

SAMPLE PROBLEM 3.2 Calculating the Moles and Number of Formula Units in a Given Mass of a Compound

Problem Ammonium carbonate is a white solid that decomposes with warming. Among its many uses, it is a component of baking powder, fire extinguishers, and smelling salts. How many formula units are in 41.6 g of ammonium carbonate?

Plan We know the mass of compound (41.6 g) and need to find the number of formula units. As we saw in Sample Problem 3.1(b), to convert grams to number of entities, we have to find number of moles first, so we must divide the grams by the molar mass ($\mathcal{M}$). For this, we need $\mathcal{M}$, so we determine the formula (see Table 2.5) and take the sum of the elements' molar masses. Once we have the number of moles, we multiply by Avogadro's number to find the number of formula units.

Solution The formula is $(NH_4)_2CO_3$. Calculating molar mass:

$$\mathcal{M} = (2 \times \mathcal{M} \text{ of N}) + (8 \times \mathcal{M} \text{ of H}) + (1 \times \mathcal{M} \text{ of C}) + (3 \times \mathcal{M} \text{ of O})$$
$$= (2 \times 14.01 \text{ g/mol}) + (8 \times 1.008 \text{ g/mol}) + 12.01 \text{ g/mol} + (3 \times 16.00 \text{ g/mol})$$
$$= 96.09 \text{ g/mol}$$

Converting from grams to moles:

$$\text{Moles of } (NH_4)_2CO_3 = 41.6 \text{ g } (NH_4)_2CO_3 \times \frac{1 \text{ mol } (NH_4)_2CO_3}{96.09 \text{ g } (NH_4)_2CO_3} = 0.433 \text{ mol } (NH_4)_2CO_3$$

Converting from moles to formula units:

$$\text{Formula units of } (NH_4)_2CO_3 = 0.433 \text{ mol } (NH_4)_2CO_3$$
$$\times \frac{6.022 \times 10^{23} \text{ formula units } (NH_4)_2CO_3}{1 \text{ mol } (NH_4)_2CO_3}$$
$$= 2.61 \times 10^{23} \text{ formula units } (NH_4)_2CO_3$$

Check The units are correct. The mass is less than half the molar mass ($\sim 42/96 < 0.5$), so the number of formula units should be less than half Avogadro's number ($\sim 2.6 \times 10^{23}/6.0 \times 10^{23} < 0.5$).

Comment A *common mistake* is to forget the subscript 2 outside the parentheses in $(NH_4)_2CO_3$, which would give a much lower molar mass.

FOLLOW-UP PROBLEM 3.2 Tetraphosphorus decaoxide reacts with water to form phosphoric acid, a major industrial acid. In the laboratory, the oxide is used as a drying agent.
(a) What is the mass (in g) of 4.65×10^{22} molecules of tetraphosphorus decaoxide?
(b) How many P atoms are present in this sample?

Mass (g) of $(NH_4)_2CO_3$

divide by $\mathcal{M}$ (g/mol)

Amount (mol) of $(NH_4)_2CO_3$

multiply by 6.022×10^{23} formula units/mol

Number of $(NH_4)_2CO_3$ formula units

Mass Percent from the Chemical Formula

Each element in a compound constitutes its own particular portion of the compound's mass. For an individual molecule (or formula unit), we use the molecular (or formula) mass and chemical formula to find the mass percent of any element X in the compound:

$$\text{Mass \% of element X} = \frac{\text{atoms of X in formula} \times \text{atomic mass of X (amu)}}{\text{molecular (or formula) mass of compound (amu)}} \times 100$$

The formula also tells the number of *moles* of each element in the compound, so we can use the molar mass to find the mass percent of each element on a mole basis:

$$\text{Mass \% of element X} = \frac{\text{moles of X in formula} \times \text{molar mass of X (g/mol)}}{\text{mass (g) of 1 mol of compound}} \times 100 \quad \textbf{(3.6)}$$

As always, the individual mass percents of the elements in the compound must add up to 100% (within rounding). As Sample Problem 3.3 demonstrates, an important practical use of mass percent is to determine the amount of an element in any size sample of a compound.

SAMPLE PROBLEM 3.3   Calculating Mass Percents and Masses of Elements in a Sample of a Compound

Problem In mammals, lactose (milk sugar) is metabolized to glucose ($C_6H_{12}O_6$), the key nutrient for generating chemical potential energy.
(a) What is the mass percent of each element in glucose?
(b) How many grams of carbon are in 16.55 g of glucose?

(a) Determining the mass percent of each element

Plan We know the relative numbers of moles of the elements in glucose from the formula (6 C, 12 H, 6 O). We multiply the number of moles of each element by its molar mass to find grams. Dividing each element's mass by the mass of 1 mol of glucose gives the mass fraction of each element, and multiplying each fraction by 100 gives the mass percent. The calculation steps for any element X are shown in the roadmap.

Solution Calculating the mass of 1 mol of $C_6H_{12}O_6$:

$$\mathcal{M} = (6 \times \mathcal{M} \text{ of C}) + (12 \times \mathcal{M} \text{ of H}) + (6 \times \mathcal{M} \text{ of O})$$
$$= (6 \times 12.01 \text{ g/mol}) + (12 \times 1.008 \text{ g/mol}) + (6 \times 16.00 \text{ g/mol})$$
$$= 180.16 \text{ g/mol}$$

Converting moles of C to grams: There are 6 mol of C in 1 mol of glucose, so

$$\text{Mass (g) of C} = 6 \text{ mol C} \times \frac{12.01 \text{ g C}}{1 \text{ mol C}} = 72.06 \text{ g C}$$

Finding the mass fraction of C in glucose:

$$\text{Mass fraction of C} = \frac{\text{total mass C}}{\text{mass of 1 mol glucose}} = \frac{72.06 \text{ g}}{180.16 \text{ g}} = 0.4000$$

Finding the mass percent of C:

$$\text{Mass \% of C} = \text{mass fraction of C} \times 100 = 0.4000 \times 100 = 40.00 \text{ mass \% C}$$

Combining the steps for each of the other two elements in glucose:

$$\text{Mass \% of H} = \frac{\text{mol H} \times \mathcal{M} \text{ of H}}{\text{mass of 1 mol glucose}} \times 100 = \frac{12 \text{ mol H} \times \frac{1.008 \text{ g H}}{1 \text{ mol H}}}{180.16 \text{ g}} \times 100$$
$$= 6.714 \text{ mass \% H}$$

$$\text{Mass \% of O} = \frac{\text{mol O} \times \mathcal{M} \text{ of O}}{\text{mass of 1 mol glucose}} \times 100 = \frac{6 \text{ mol O} \times \frac{16.00 \text{ g O}}{1 \text{ mol O}}}{180.16 \text{ g}} \times 100$$
$$= 53.29 \text{ mass \% O}$$

Check The answers make sense: even though there are equal numbers of moles of O and C in the compound, the mass % of O is greater than the mass % of C because the molar mass of O is greater than the molar mass of C. The mass % of H is small because the molar mass of H is small. The total of the mass percents is 100.00%.

(b) Determining the mass (g) of carbon

Plan To find the mass of C in the glucose sample, we multiply the mass of the sample by the mass fraction of C from part (a).

Solution Finding the mass of C in a given mass of glucose (with units for mass fraction):

$$\text{Mass (g) of C} = \text{mass of glucose} \times \text{mass fraction of C} = 16.55 \text{ g glucose} \times \frac{0.4000 \text{ g C}}{1 \text{ g glucose}}$$
$$= 6.620 \text{ g C}$$

Check Rounding shows that the answer is "in the right ballpark": 16 g times less than 0.5 parts by mass should be less than 8 g.

Comment 1. A *more direct approach* to finding the mass of element in any mass of compound is similar to the approach we used in Sample Problem 2.1 and eliminates the need

---

Roadmap (side panel):

Amount (mol) of element X in 1 mol of compound

↓ multiply by $\mathcal{M}$ (g/mol) of X

Mass (g) of X in 1 mol of compound

↓ divide by mass (g) of 1 mol of compound

Mass fraction of X

↓ multiply by 100

Mass % of X

to calculate the mass fraction. Just multiply the given mass of compound by the ratio of the total mass of element to the mass of 1 mol of compound:

$$\text{Mass (g) of C} = 16.55 \ \text{g glucose} \times \frac{72.06 \ \text{g C}}{180.16 \ \text{g glucose}} = 6.620 \ \text{g C}$$

2. From here on, you should be able to determine the molar mass of a compound, so that calculation will no longer be shown.

FOLLOW-UP PROBLEM 3.3 Ammonium nitrate is a common fertilizer. Agronomists base the effectiveness of fertilizers on their nitrogen content.
(a) Calculate the mass percent of N in ammonium nitrate.
(b) How many grams of N are in 35.8 kg of ammonium nitrate?

SECTION SUMMARY

A mole of substance is the amount that contains Avogadro's number (6.022×10^{23}) of chemical entities (atoms, molecules, or formula units). The mass (in grams) of a mole has the same numerical value as the mass (in amu) of the entity. Thus, the mole allows us to count entities by weighing them. Using the molar mass ($\mathcal{M}$, g/mol) of an element (or compound) and Avogadro's number as conversion factors, we can convert among amount (mol), mass (g), and number of entities. The mass fraction of element X in a compound is used to find the mass of X in any amount of the compound.

3.2 DETERMINING THE FORMULA OF AN UNKNOWN COMPOUND

In Sample Problem 3.3, we knew the formula and used it to find the mass percent (or mass fraction) of an element in a compound *and* the mass of the element in a given mass of the compound. In this section, we do the reverse: use the masses of elements in a compound to find its formula. We'll present the mass data in several ways and then look briefly at molecular structures.

Empirical Formulas

An analytical chemist investigating a compound decomposes it into simpler substances, finds the mass of each component element, converts these masses to numbers of moles, and then arithmetically converts the moles to whole-number (integer) subscripts. This procedure yields the empirical formula, the *simplest whole-number ratio* of moles of each element in the compound (see Section 2.8). Let's see how to obtain the subscripts from the moles of each element.

Analysis of an unknown compound shows that the sample contains 0.21 mol of zinc, 0.14 mol of phosphorus, and 0.56 mol of oxygen. Because the subscripts in a formula represent individual atoms or moles of atoms, we write a preliminary formula that contains fractional subscripts: $Zn_{0.21}P_{0.14}O_{0.56}$. Next, we convert these fractional subscripts to whole numbers using one or two simple arithmetic steps (rounding when needed):

1. Divide each subscript by the smallest subscript:

$$Zn_{\frac{0.21}{0.14}}P_{\frac{0.14}{0.14}}O_{\frac{0.56}{0.14}} \longrightarrow Zn_{1.5}P_{1.0}O_{4.0}$$

This step alone often gives integer subscripts.

2. If any of the subscripts is still not an integer, multiply through by the *smallest integer* that will turn all subscripts into integers. Here, we multiply by 2, the smallest integer that will make 1.5 (the subscript for Zn) into an integer:

$$Zn_{(1.5 \times 2)}P_{(1.0 \times 2)}O_{(4.0 \times 2)} \longrightarrow Zn_{3.0}P_{2.0}O_{8.0}, \ \text{or} \ Zn_3P_2O_8$$

Notice that the *relative* number of moles has not changed because we multiplied *all* the subscripts by 2.

Always check that the subscripts are the smallest set of integers with the same ratio as the original numbers of moles; that is, 3:2:8 is *in the same ratio* as 0.21:0.14:0.56. A more conventional way to write this formula is $Zn_3(PO_4)_2$; the compound is zinc phosphate, a dental cement.

The following three sample problems (3.4, 3.5, and 3.6) demonstrate how other types of compositional data are used to determine chemical formulas. In the first problem, the empirical formula is found from data given as grams of each element rather than as moles.

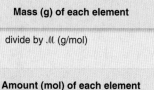

Mass (g) of each element

divide by $\mathcal{M}$ (g/mol)

Amount (mol) of each element

use nos. of moles as subscripts

Preliminary formula

change to integer subscripts

Empirical formula

SAMPLE PROBLEM 3.4   Determining an Empirical Formula from Masses of Elements

Problem Elemental analysis of a sample of an ionic compound showed 2.82 g of Na, 4.35 g of Cl, and 7.83 g of O. What is the empirical formula and name of the compound?
Plan This problem is similar to the one we just discussed, except that we are given element *masses*, so we must convert the masses into integer subscripts. We first divide each mass by the element's molar mass to find *number of moles*. Then we construct a preliminary formula and convert the numbers of moles to integers.
Solution Finding moles of elements:

$$\text{Moles of Na} = 2.82 \text{ g Na} \times \frac{1 \text{ mol Na}}{22.99 \text{ g Na}} = 0.123 \text{ mol Na}$$

$$\text{Moles of Cl} = 4.35 \text{ g Cl} \times \frac{1 \text{ mol Cl}}{35.45 \text{ g Cl}} = 0.123 \text{ mol Cl}$$

$$\text{Moles of O} = 7.83 \text{ g O} \times \frac{1 \text{ mol O}}{16.00 \text{ g O}} = 0.489 \text{ mol O}$$

Constructing a preliminary formula: $Na_{0.123}Cl_{0.123}O_{0.489}$
Converting to integer subscripts (dividing all by the smallest subscript):

$$Na_{\frac{0.123}{0.123}}Cl_{\frac{0.123}{0.123}}O_{\frac{0.489}{0.123}} \longrightarrow Na_{1.00}Cl_{1.00}O_{3.98} \approx Na_1Cl_1O_4, \text{ or } NaClO_4$$

We rounded the subscript of O from 3.98 to 4. The empirical formula is $NaClO_4$; the name is sodium perchlorate.
Check The moles seem correct because the masses of Na and Cl are slightly more than 0.1 of their molar masses. The mass of O is greatest and its molar mass is smallest, so it should have the greatest number of moles. The ratio of subscripts, 1:1:4, is the same as the ratio of moles, 0.123:0.123:0.489 (within rounding).

FOLLOW-UP PROBLEM 3.4 An unknown metal M reacts with sulfur to form a compound with the formula M_2S_3. If 3.12 g of M reacts with 2.88 g of S, what are the names of M and M_2S_3? (*Hint:* Determine the number of moles of S and use the formula to find the number of moles of M.)

Molecular Formulas

If we know the molar mass of a compound, we can use the empirical formula to obtain the molecular formula, the *actual* number of moles of each element in 1 mol of compound. In some cases, such as water (H_2O), ammonia (NH_3), and methane (CH_4), the empirical and molecular formulas are identical, but in many others the molecular formula is a *whole-number multiple* of the empirical formula. Hydrogen peroxide, for example, has the empirical formula HO and the molecular formula H_2O_2. Dividing the molar mass of H_2O_2 (34.02 g/mol) by the empirical formula mass (17.01 g/mol) gives the whole-number multiple:

$$\text{Whole-number multiple} = \frac{\text{molar mass (g/mol)}}{\text{empirical formula mass (g/mol)}}$$

$$= \frac{34.02 \text{ g/mol}}{17.01 \text{ g/mol}} = 2.000 = 2$$

Instead of giving compositional data in terms of masses of each element, analytical laboratories provide it as mass percents. From this, we determine the empirical formula by (1) assuming 100.0 g of compound, which allows us to express mass percent directly as mass, (2) converting the mass to number of moles, and (3) constructing the empirical formula. With the molar mass, we can also find the whole-number multiple and then the molecular formula.

SAMPLE PROBLEM 3.5 Determining a Molecular Formula from Elemental Analysis and Molar Mass

Problem During excessive physical activity, lactic acid ($\mathcal{M}$ = 90.08 g/mol) forms in muscle tissue and is responsible for muscle soreness. Elemental analysis shows that this compound contains 40.0 mass % C, 6.71 mass % H, and 53.3 mass % O.
(a) Determine the empirical formula of lactic acid.
(b) Determine the molecular formula.

(a) Determining the empirical formula
Plan We know the mass % of each element and must convert each to an integer subscript. Although the mass of lactic acid is not given, mass % is the same for any mass of compound, so we can assume 100.0 g of lactic acid and express each mass % directly as grams. Then, we convert grams to moles and construct the empirical formula as we did in Sample Problem 3.4.
Solution Expressing mass % as grams, assuming 100.0 g of lactic acid:

$$\text{Mass (g) of C} = \frac{40.0 \text{ parts C by mass}}{100 \text{ parts by mass}} \times 100.0 \text{ g} = 40.0 \text{ g C}$$

Similarly, we have 6.71 g of H and 53.3 g of O.
Converting from grams of each element to moles:

$$\text{Moles of C} = \text{mass of C} \times \frac{1}{\mathcal{M} \text{ of C}} = 40.0 \text{ g C} \times \frac{1 \text{ mol C}}{12.01 \text{ g C}} = 3.33 \text{ mol C}$$

Similarly, we have 6.66 mol of H and 3.33 mol of O.
Constructing the preliminary formula: $C_{3.33}H_{6.66}O_{3.33}$
Converting to integer subscripts:

$$C_{\frac{3.33}{3.33}}H_{\frac{6.66}{3.33}}O_{\frac{3.33}{3.33}} \longrightarrow C_{1.00}H_{2.00}O_{1.00} \approx C_1H_2O_1; \text{ the empirical formula is } \boxed{CH_2O}$$

Check The numbers of moles seem correct: the masses of C and O are each slightly more than 3 times their molar masses (e.g., for C, 40 g/(12 g/mol) > 3 mol), and the mass of H is over 6 times its molar mass.

(b) Determining the molecular formula
Plan The molecular formula subscripts are whole-number multiples of the empirical formula subscripts. To find this whole number, we divide the given molar mass (90.08 g/mol) by the empirical formula mass, which we find from the sum of the elements' molar masses. Then we multiply the whole number by each subscript in the empirical formula.
Solution The empirical-formula molar mass is 30.03 g/mol. Finding the whole-number multiple:

$$\text{Whole-number multiple} = \frac{\mathcal{M} \text{ of lactic acid}}{\mathcal{M} \text{ of empirical formula}} = \frac{90.08 \text{ g/mol}}{30.03 \text{ g/mol}}$$
$$= 3.000 = 3$$

Determining the molecular formula:

$$C_{(1\times3)}H_{(2\times3)}O_{(1\times3)} = \boxed{C_3H_6O_3}$$

Check The calculated molecular formula has the same ratio of moles of elements (3:6:3) as the empirical formula (1:2:1) and corresponds to the given molar mass:

$$\mathcal{M} \text{ of lactic acid} = (3 \times \mathcal{M} \text{ of C}) + (6 \times \mathcal{M} \text{ of H}) + (3 \times \mathcal{M} \text{ of O})$$
$$= (3 \times 12.01) + (6 \times 1.008) + (3 \times 16.00) = 90.08 \text{ g/mol}$$

FOLLOW-UP PROBLEM 3.5 One of the most widespread environmental carcinogens (cancer-causing agents) is benzo[a]pyrene ($\mathcal{M}$ = 252.30 g/mol). It is found in coal dust, in cigarette smoke, and even in charcoal-grilled meat. Analysis of this hydrocarbon shows 95.21 mass % C and 4.79 mass % H. What is the molecular formula of benzo[a]pyrene?

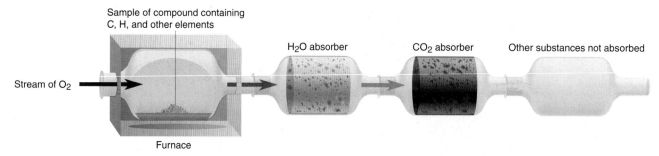

Sample of compound containing
C, H, and other elements

H₂O absorber   CO₂ absorber   Other substances not absorbed

Stream of O₂

Furnace

Figure 3.5 Combustion apparatus for determining formulas of organic compounds. A sample of compound that contains C and H (and perhaps other elements) is burned in a stream of O_2 gas. The CO_2 and H_2O formed are absorbed separately, while any other element oxides are carried through by the O_2 gas stream. H_2O is absorbed by $Mg(ClO_4)_2$; CO_2 is absorbed by NaOH. The increases in mass of the absorbers are used to calculate the amounts (mol) of C and H in the sample.

Combustion Analysis of Organic Compounds Still another type of compositional data is obtained through **combustion analysis,** a method used to measure the amounts of carbon and hydrogen in a combustible organic compound. The unknown compound is burned in pure O_2 in an apparatus that consists of a combustion chamber and chambers containing compounds that absorb either H_2O or CO_2 (Figure 3.5). All the H in the unknown is converted to H_2O, which is absorbed in the first chamber, and all the C is converted to CO_2, which is absorbed in the second. By weighing the contents of the chambers before and after combustion, we find the masses of CO_2 and H_2O and use them to calculate the masses of C and H in the compound, from which we find the empirical formula.

Many organic compounds also contain at least one other element, such as oxygen, nitrogen, or a halogen. As long as a third element doesn't interfere with the absorption of CO_2 and H_2O, we calculate its mass by subtracting the masses of C and H from the original mass of the compound.

SAMPLE PROBLEM 3.6   Determining a Molecular Formula from Combustion Analysis

Problem Vitamin C ($\mathcal{M} = 176.12$ g/mol) is a compound of C, H, and O found in many natural sources, especially citrus fruits. When a 1.000-g sample of vitamin C is placed in a combustion chamber and burned, the following data are obtained:

$$\text{Mass of } CO_2 \text{ absorber after combustion} = 85.35 \text{ g}$$
$$\text{Mass of } CO_2 \text{ absorber before combustion} = 83.85 \text{ g}$$
$$\text{Mass of } H_2O \text{ absorber after combustion} = 37.96 \text{ g}$$
$$\text{Mass of } H_2O \text{ absorber before combustion} = 37.55 \text{ g}$$

What is the molecular formula of vitamin C?

Plan We find the masses of CO_2 and H_2O by subtracting the masses of the absorbers before the reaction from the masses after. From the mass of CO_2, we use the mass fraction of C in CO_2 to find the mass of C (see Comment in Sample Problem 3.3). Similarly, we find the mass of H from the mass of H_2O. The mass of vitamin C (1.000 g) minus the sum of the C and H masses gives the mass of O, the third element present. Then, we proceed as in Sample Problem 3.5: calculate numbers of moles using the elements' molar masses, construct the empirical formula, determine the whole-number multiple from the given molar mass, and construct the molecular formula.

Solution Finding the masses of combustion products:

$$\text{Mass (g) of } CO_2 = \text{mass of } CO_2 \text{ absorber after} - \text{mass before}$$
$$= 85.35 \text{ g} - 83.85 \text{ g} = 1.50 \text{ g } CO_2$$
$$\text{Mass (g) of } H_2O = \text{mass of } H_2O \text{ absorber after} - \text{mass before}$$
$$= 37.96 \text{ g} - 37.55 \text{ g} = 0.41 \text{ g } H_2O$$

Calculating masses of C and H using their mass fractions:

$$\text{Mass of element} = \text{mass of compound} \times \frac{\text{mass of element in compound}}{\text{mass of 1 mol of compound}}$$

$$\text{Mass (g) of C} = \text{mass of CO}_2 \times \frac{1 \text{ mol C} \times \mathcal{M} \text{ of C}}{\text{mass of 1 mol CO}_2} = 1.50 \text{ g CO}_2 \times \frac{12.01 \text{ g C}}{44.01 \text{ g CO}_2}$$

$$= 0.409 \text{ g C}$$

$$\text{Mass (g) of H} = \text{mass of H}_2\text{O} \times \frac{2 \text{ mol H} \times \mathcal{M} \text{ of H}}{\text{mass of 1 mol H}_2\text{O}} = 0.41 \text{ g H}_2\text{O} \times \frac{2.016 \text{ g H}}{18.02 \text{ g H}_2\text{O}}$$

$$= 0.046 \text{ g H}$$

Calculating the mass of O:

$$\text{Mass (g) of O} = \text{mass of vitamin C sample} - (\text{mass of C} + \text{mass of H})$$
$$= 1.000 \text{ g} - (0.409 \text{ g} + 0.046 \text{ g}) = 0.545 \text{ g O}$$

Finding the amounts (mol) of elements: Dividing the mass in grams of each element by its molar mass gives 0.0341 mol of C, 0.046 mol of H, and 0.0341 mol of O.
Constructing the preliminary formula: $C_{0.0341}H_{0.046}O_{0.0341}$
Determining the empirical formula: Dividing through by the smallest subscript gives

$$C_{\frac{0.0341}{0.0341}}H_{\frac{0.046}{0.0341}}O_{\frac{0.0341}{0.0341}} = C_{1.00}H_{1.3}O_{1.00}$$

By trial and error, we find that 3 is the smallest integer that will make all subscripts approximately into integers:

$$C_{(1.00\times3)}H_{(1.3\times3)}O_{(1.00\times3)} = C_{3.00}H_{3.9}O_{3.00} \approx C_3H_4O_3$$

Determining the molecular formula:

$$\text{Whole-number multiple} = \frac{\mathcal{M} \text{ of vitamin C}}{\mathcal{M} \text{ of empirical formula}} = \frac{176.12 \text{ g/mol}}{88.06 \text{ g/mol}} = 2.000 = 2$$

$$C_{(3\times2)}H_{(4\times2)}O_{(3\times2)} = C_6H_8O_6$$

Check The element masses seem correct: carbon makes up slightly more than 0.25 of the mass of CO_2 (12 g/44 g > 0.25), as do the masses in the problem (0.409 g/1.50 g > 0.25). Hydrogen makes up slightly more than 0.10 of the mass of H_2O (2 g/18 g > 0.10), as do the masses in the problem (0.046 g/0.41 g > 0.10). The molecular formula has the same ratio of subscripts (6:8:6) as the empirical formula (3:4:3) and adds up to the given molar mass:

$$(6 \times \mathcal{M} \text{ of C}) + (8 \times \mathcal{M} \text{ of H}) + (6 \times \mathcal{M} \text{ of O}) = \mathcal{M} \text{ of vitamin C}$$
$$(6 \times 12.01) + (8 \times 1.008) + (6 \times 16.00) = 176.12 \text{ g/mol}$$

Comment In determining the subscript for H, if we string the calculation steps together, we obtain the subscript 4.0, rather than 3.9, and don't need to round:

$$\text{Subscript of H} = 0.41 \text{ g H}_2\text{O} \times \frac{2.016 \text{ g H}}{18.02 \text{ g H}_2\text{O}} \times \frac{1 \text{ mol H}}{1.008 \text{ g H}} \times \frac{1}{0.0341 \text{ mol}} \times 3 = 4.0$$

FOLLOW-UP PROBLEM 3.6 A dry-cleaning solvent ($\mathcal{M} = 146.99$ g/mol) that contains C, H, and Cl is suspected to be a cancer-causing agent. When a 0.250-g sample was studied by combustion analysis, 0.451 g of CO_2 and 0.0617 g of H_2O formed. Find the molecular formula.

Isomers A molecular formula tells the *actual* number of each type of atom, providing as much information as possible from mass analysis. Yet *different compounds can have the same molecular formula* because the atoms can bond to each other in different arrangements to give more than one *structural formula*. **Isomers** are compounds with the same molecular formula but different properties. The simplest type of isomerism, called *constitutional*, or *structural*, *isomerism*, occurs when the atoms link together in different arrangements. The pair of constitutional isomers shown in Table 3.2 (on the next page) share the molecular formula C_2H_6O

Table 3.2	**Constitutional Isomers of C_2H_6O**	
Property	**Ethanol**	**Dimethyl Ether**
$\mathcal{M}$ (g/mol)	46.07	46.07
Boiling point	78.5°C	−25°C
Density (at 20°C)	0.789 g/mL (liquid)	0.00195 g/mL (gas)
Structural formula		
Space-filling model		

but have very different properties because they are different compounds. In this case, they are even different *types* of compounds—one is an alcohol, and the other an ether.

As the number and kinds of atoms increase, the number of constitutional isomers—that is, the number of structural formulas that can be written for a given molecular formula—also increases: C_2H_6O has the two that you've seen, C_3H_8O has three, and $C_4H_{10}O$, seven. (We'll discuss this and other types of isomerism fully later in the text.)

SECTION SUMMARY
From the masses of elements in an unknown compound, the relative amounts (in moles) can be found and the empirical formula determined. If the molar mass is known, the molecular formula can also be determined. Methods such as combustion analysis provide data on the masses of elements in a compound, which can be used to obtain the formula. Because atoms can bond in different arrangements, more than one compound may have the same molecular formula (constitutional isomers).

3.3 WRITING AND BALANCING CHEMICAL EQUATIONS

Perhaps the most important reason for thinking in terms of moles is because it greatly clarifies the amounts of substances taking part in a reaction. Comparing masses doesn't tell the ratio of substances reacting but comparing numbers of moles does. It allows us to view substances as large populations of interacting particles rather than as grams of material. To clarify this idea, consider the formation of hydrogen fluoride gas from H_2 and F_2, a reaction that occurs explosively at room temperature. If we weigh the gases, we find that

2.016 g of H_2 and 38.00 g of F_2 react to form 40.02 g of HF

This information tells us little except that mass is conserved. However, if we convert these masses (in grams) to amounts (in moles), we find that

1 mol of H_2 and 1 mol of F_2 react to form 2 mol of HF

This information reveals that equal-size populations of H_2 and F_2 molecules combine to form twice as large a population of HF molecules. Dividing through by Avogadro's number shows us the chemical event that occurs between individual molecules:

1 H_2 molecule and 1 F_2 molecule react to form 2 HF molecules

Figure 3.6 shows that when we express the reaction in terms of moles, *the macroscopic (molar) change corresponds to the submicroscopic (molecular) change.* As you'll see, a balanced chemical equation shows both changes.

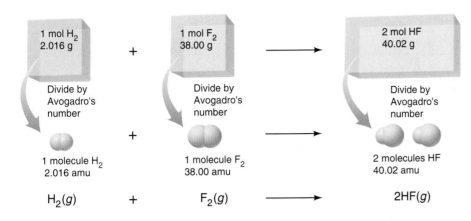

Figure 3.6 The formation of HF gas on the macroscopic and molecular levels. When 1 mol of H_2 (2.016 g) and 1 mol of F_2 (38.00 g) react, 2 mol of HF (40.02 g) forms. Dividing by Avogadro's number shows the change at the molecular level.

A **chemical equation** is a statement in formulas that expresses the identities and quantities of the substances involved in a chemical or physical change. Equations are the "sentences" of chemistry, just as chemical formulas are the "words" and atomic symbols the "letters." The left side of an equation shows the amount of each substance present before the change, and the right side shows the amounts present afterward. *For an equation to depict these amounts accurately, it must be balanced; that is, the same number of each type of atom must appear on both sides of the equation.* This requirement follows directly from the mass laws and the atomic theory:

- In a chemical process, atoms cannot be created, destroyed, or changed, only rearranged into different combinations.
- A formula represents a fixed ratio of the elements in a compound, so a different ratio represents a different compound.

Consider the chemical change that occurs in an old-fashioned photographic flashbulb: magnesium wire and oxygen gas yield powdery magnesium oxide. (Light and heat are produced as well, but here we're concerned only with the substances involved.) Let's convert this chemical statement into a balanced equation through the following steps:

1. *Translating the statement.* We first translate the chemical statement into a "skeleton" equation: chemical formulas arranged in an equation format. All the substances that react during the change, called **reactants,** are placed to the left of a "yield" arrow, which points to all the substances produced, called **products:**

$$\underbrace{\_\_Mg \quad + \quad \_\_O_2}_{\text{reactants}} \quad \overset{\text{yield}}{\longrightarrow} \quad \overset{\text{product}}{\_\_MgO}$$

$$\underbrace{\text{magnesium and oxygen}} \quad \text{yield} \quad \text{magnesium oxide}$$

At the beginning of the balancing process, we put a blank in front of each substance to remind us that we have to account for its atoms.

2. *Balancing the atoms.* The next step involves shifting our attention back and forth from right to left in order to *match the number of each type of atom on each side.* At the end of this step, each blank will contain a **balancing (stoichiometric) coefficient,** a numerical multiplier of *all the atoms* in the formula that follows it. In general, balancing is easiest when we

- Start with the most complex substance, the one with the largest number of atoms or different types of atoms.
- End with the least complex substance, such as an element by itself.

In this case, MgO is the most complex, so we place a coefficient 1 *in front of* the compound:

$$\_\_Mg + \_\_O_2 \longrightarrow \underline{1}\,MgO$$

To balance the Mg in MgO on the right, we place a 1 in front of Mg on the left:

$$\underline{1}\,Mg + \underline{}\,O_2 \longrightarrow \underline{1}\,MgO$$

The O atom on the right must be balanced by one O atom on the left. One-half an O_2 molecule provides one O atom:

$$\underline{1}\,Mg + \underline{\tfrac{1}{2}}\,O_2 \longrightarrow \underline{1}\,MgO$$

In terms of number and type of atom, the equation is balanced.

3. *Adjusting the coefficients.* There are several conventions about the final form of the coefficients:

- In most cases, *the smallest whole-number coefficients are preferred.* Whole numbers allow entities such as O_2 molecules to be treated as intact particles. One-half of an O_2 molecule cannot exist, so we multiply the equation by 2:

$$2Mg + 1O_2 \longrightarrow 2MgO$$

- We used the coefficient 1 to remind us to balance each substance. In the final form, a coefficient of 1 is implied just by the presence of the formula of the substance, so we don't need to write it:

$$2Mg + O_2 \longrightarrow 2MgO$$

(This convention is similar to not writing a subscript 1 in a formula.)

4. *Checking.* After balancing and adjusting the coefficients, always check that the equation is balanced:

$$\text{Reactants (2 Mg, 2 O)} \longrightarrow \text{products (2 Mg, 2 O)}$$

5. *Specifying the states of matter.* The final equation also indicates the physical state of each substance or whether it is dissolved in water. The abbreviations that are used for these states are solid (*s*), liquid (*l*), gas (*g*), and aqueous solution (*aq*). From the original statement, we know that the Mg "wire" is solid, the O_2 is a gas, and the "powdery" MgO is also solid. The balanced equation, therefore, is

$$2Mg(s) + O_2(g) \longrightarrow 2MgO(s)$$

Of course, the key point to realize is, as was pointed out in Figure 3.6, *the balancing coefficients refer to both individual chemical entities and moles of chemical entities.* Thus, 2 mol of Mg and 1 mol of O_2 yield 2 mol of MgO. Figure 3.7 shows this reaction from three points of view—as you see it on the macroscopic level, as chemists (and you!) can imagine it on the atomic level (darker colored atoms represent the stoichiometry), and on the symbolic level of the chemical equation.

Keep in mind these other key points about the balancing process:

- A coefficient operates on *all the atoms in the formula* that follows it: 2MgO means 2 × (MgO), or 2 Mg atoms and 2 O atoms; $2Ca(NO_3)_2$ means 2 × $[Ca(NO_3)_2]$, or 2 Ca atoms, 4 N atoms, and 12 O atoms.
- In balancing an equation, *chemical formulas cannot be altered.* In step 2 of the example, we *cannot* balance the O atoms by changing MgO to MgO_2 because MgO_2 has a different elemental composition and thus is a different compound.
- We *cannot add other reactants or products* to balance the equation because this would represent a different reaction. For example, we *cannot* balance the O atoms by changing O_2 to O or by adding one O atom to the products, because the chemical statement does not say that the reaction involves O atoms.

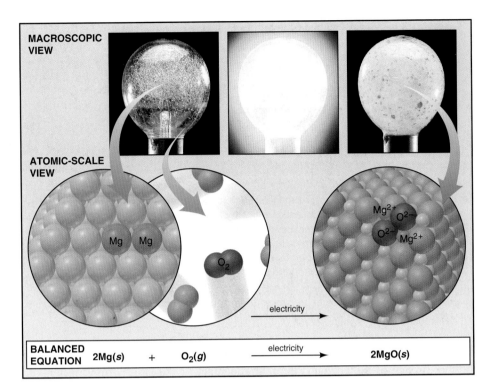

MACROSCOPIC VIEW

ATOMIC-SCALE VIEW

Mg   Mg

O_2

Mg^{2+} O^{2-}
O^{2-} Mg^{2+}

electricity

BALANCED EQUATION $2Mg(s)$ + $O_2(g)$ $\xrightarrow{\text{electricity}}$ $2MgO(s)$

Figure 3.7 A three-level view of the chemical reaction in a flashbulb. The photos present the macroscopic view that you see. Before the reaction occurs, a fine magnesium filament is surrounded by oxygen (*left*). After the reaction, white, powdery magnesium oxide coats the bulb's inner surface (*right*). The blow-up arrows lead to an atomic-scale view, a representation of the chemist's mental picture of the reaction. The darker colored spheres show the stoichiometry. By knowing the substances before and after a reaction, we can write a balanced equation (*bottom*), the chemist's symbolic shorthand for the change.

- A balanced equation remains balanced even if you multiply all the coefficients by the same number. For example,

$$4Mg(s) + 2O_2(g) \longrightarrow 4MgO(s)$$

is also balanced: it is just the same balanced equation multiplied by 2. However, we balance an equation with the *smallest* whole-number coefficients.

SAMPLE PROBLEM 3.7   Balancing Chemical Equations

Problem Within the cylinders of a car's engine, the hydrocarbon octane (C_8H_{18}), one of many components of gasoline, mixes with oxygen from the air and burns to form carbon dioxide and water vapor. Write a balanced equation for this reaction.

Solution

1. *Translate* the statement into a skeleton equation (with coefficient blanks). Octane and oxygen are reactants; "oxygen from the air" implies molecular oxygen, O_2. Carbon dioxide and water vapor are products:

$$\_C_8H_{18} + \_O_2 \longrightarrow \_CO_2 + \_H_2O$$

2. *Balance the atoms.* We start with the most complex substance, C_8H_{18}, and balance O_2 last:

$$\underline{1}\,C_8H_{18} + \_O_2 \longrightarrow \_CO_2 + \_H_2O$$

The C atoms in C_8H_{18} end up in CO_2. Each CO_2 contains one C atom, so 8 molecules of CO_2 are needed to balance the 8 C atoms in each C_8H_{18}:

$$\underline{1}\,C_8H_{18} + \_O_2 \longrightarrow \underline{8}\,CO_2 + \_H_2O$$

The H atoms in C_8H_{18} end up in H_2O. The 18 H atoms in C_8H_{18} require a coefficient 9 in front of H_2O:

$$\underline{1}\,C_8H_{18} + \_O_2 \longrightarrow \underline{8}\,CO_2 + \underline{9}\,H_2O$$

There are 25 atoms of O on the right (16 in $8CO_2$ plus 9 in $9H_2O$), so we place the coefficient $\frac{25}{2}$ in front of O_2:

$$\underline{1}\,C_8H_{18} + \underline{\tfrac{25}{2}}\,O_2 \longrightarrow \underline{8}\,CO_2 + \underline{9}\,H_2O$$

3. *Adjust the coefficients.* Multiply through by 2 to obtain whole numbers:

$$2C_8H_{18} + 25O_2 \longrightarrow 16CO_2 + 18H_2O$$

4. *Check* that the equation is balanced:

Reactants (16 C, 36 H, 50 O) $\longrightarrow$ products (16 C, 36 H, 50 O)

5. *Specify* states of matter. C_8H_{18} is liquid; O_2, CO_2, and H_2O vapor are gases:

$$2C_8H_{18}(l) + 25O_2(g) \longrightarrow 16CO_2(g) + 18H_2O(g)$$

Comment This is an example of a combustion reaction. *Any* compound containing C and H that burns in an excess of air produces CO_2 and H_2O.

FOLLOW-UP PROBLEM 3.7 Write a balanced equation for each chemical statement:
(a) A characteristic reaction of Group 1A(1) elements: chunks of sodium react violently with water to form hydrogen gas and sodium hydroxide solution.
(b) The destruction of marble statuary by acid rain: aqueous nitric acid reacts with calcium carbonate to form carbon dioxide, water, and aqueous calcium nitrate.
(c) Halogen compounds exchanging bonding partners: phosphorus trifluoride is prepared by the reaction of phosphorus trichloride and hydrogen fluoride; hydrogen chloride is the other product. The reaction involves gases only.
(d) Explosive decomposition of dynamite: liquid nitroglycerine ($C_3H_5N_3O_9$) explodes to produce a mixture of gases—carbon dioxide, water vapor, nitrogen, and oxygen.

Viewing an equation in a schematic molecular scene is a great way to focus on the essence of the change—the rearrangement of the atoms from reactants to products. Here's a simple schematic for the combustion of octane:

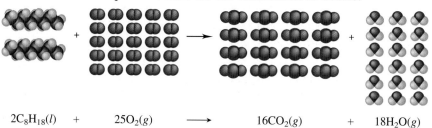

$$2C_8H_{18}(l) \quad + \quad 25O_2(g) \quad \longrightarrow \quad 16CO_2(g) \quad + \quad 18H_2O(g)$$

We can also derive a balanced equation from a molecular scene. Let's do this for the following change (black = carbon, red = oxygen):

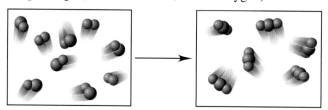

We can see from the colors that the reactant box includes carbon monoxide (CO) and oxygen (O_2) molecules, and the product box contains carbon dioxide (CO_2). Once we identify the substances with formulas, we count the number of each kind of molecule and put the numbers and formulas in equation format. There are six CO, three O_2, and six CO_2, that is, twice as many CO as O_2 and the same number as CO_2. So, given that they are all gases, we adjust the coefficients and have

$$2CO(g) + O_2(g) \longrightarrow 2CO_2(g)$$

SECTION SUMMARY

To conserve mass and maintain the fixed composition of compounds, a chemical equation must be balanced in terms of number and type of each atom. A balanced equation has reactant formulas on the left of a yield arrow and product formulas on the right. Balancing coefficients are integer multipliers for *all* the atoms in a formula and apply to the individual entity or to moles of entities.

3.4 CALCULATING AMOUNTS OF REACTANT AND PRODUCT

A balanced equation contains a wealth of quantitative information relating individual chemical entities, amounts of chemical entities, and masses of substances. It is essential for all calculations involving amounts of reactants and products: *if you know the number of moles of one substance, the balanced equation tells you the number of moles of all the others in the reaction.*

Stoichiometrically Equivalent Molar Ratios from the Balanced Equation

In a balanced equation, *the number of moles of one substance is stoichiometrically equivalent to the number of moles of any other substance.* The term *stoichiometrically equivalent* means that a definite amount of one substance is formed from, produces, or reacts with a definite amount of the other. These quantitative relationships are expressed as *stoichiometrically equivalent molar ratios* that we use as conversion factors to calculate these amounts. Table 3.3 presents the quantitative information contained in the equation for the combustion of propane, a hydrocarbon fuel used in cooking and water heating:

$$C_3H_8(g) + 5O_2(g) \longrightarrow 3CO_2(g) + 4H_2O(g)$$

If we view the reaction quantitatively in terms of C_3H_8, we see that

1 mol of C_3H_8 reacts with 5 mol of O_2
1 mol of C_3H_8 produces   3 mol of CO_2
1 mol of C_3H_8 produces   4 mol of H_2O

Therefore, in this reaction,

1 mol of C_3H_8 is stoichiometrically equivalent to 5 mol of O_2
1 mol of C_3H_8 is stoichiometrically equivalent to 3 mol of CO_2
1 mol of C_3H_8 is stoichiometrically equivalent to 4 mol of H_2O

Table 3.3   Information Contained in a Balanced Equation

Viewed in Terms of	Reactants $C_3H_8(g) + 5O_2(g)$	$\longrightarrow$	Products $3CO_2(g) + 4H_2O(g)$
Molecules	1 molecule C_3H_8 + 5 molecules O_2	$\longrightarrow$	3 molecules CO_2 + 4 molecules H_2O
Amount (mol)	1 mol C_3H_8 + 5 mol O_2	$\longrightarrow$	3 mol CO_2 + 4 mol H_2O
Mass (amu)	44.09 amu C_3H_8 + 160.00 amu O_2	$\longrightarrow$	132.03 amu CO_2 + 72.06 amu H_2O
Mass (g)	44.09 g C_3H_8 + 160.00 g O_2	$\longrightarrow$	132.03 g CO_2 + 72.06 g H_2O
Total mass (g)	204.09 g	$\longrightarrow$	204.09 g

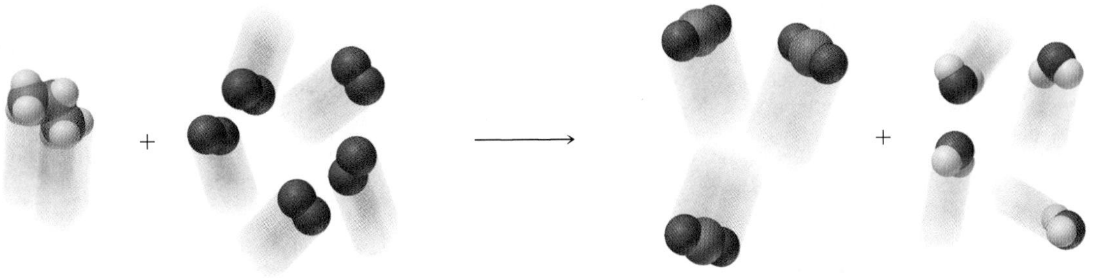

We chose to look at C_3H_8, but any two of the substances are stoichiometrically equivalent to each other. Thus,

3 mol of CO_2 is stoichiometrically equivalent to 4 mol of H_2O

5 mol of O_2 is stoichiometrically equivalent to 3 mol of CO_2

and so on.

Here's a typical problem that shows how stoichiometric equivalence is used to create conversion factors: in the combustion of propane, how many moles of O_2 are consumed when 10.0 mol of H_2O are produced? To solve this problem, we have to find the molar ratio between O_2 and H_2O. From the balanced equation, we see that for every 5 mol of O_2 consumed, 4 mol of H_2O is formed:

5 mol of O_2 is stoichiometrically equivalent to 4 mol of H_2O

We can construct two conversion factors from this equivalence, depending on the quantity we want to find:

$$\frac{5 \text{ mol } O_2}{4 \text{ mol } H_2O} \quad \text{or} \quad \frac{4 \text{ mol } H_2O}{5 \text{ mol } O_2}$$

Because we want to find moles of O_2 and we know moles of H_2O, we choose "5 mol O_2/4 mol H_2O" to cancel "mol H_2O":

$$\text{Moles of } O_2 \text{ consumed} = 10.0 \, \cancel{\text{mol } H_2O} \times \frac{5 \text{ mol } O_2}{4 \, \cancel{\text{mol } H_2O}} = 12.5 \text{ mol } O_2$$

$$\text{mol } H_2O \xrightarrow[\substack{\text{molar ratio as} \\ \text{conversion factor}}]{} \text{mol } O_2$$

Obviously, we could not have solved this problem without the balanced equation. Here is a general approach for solving *any* stoichiometry problem that involves a reaction:

1. Write a balanced equation for the reaction.
2. Convert the given mass (or number of entities) of the first substance to amount (mol).
3. Use the appropriate molar ratio from the balanced equation to calculate the amount (mol) of the second substance.
4. Convert the amount of the second substance to the desired mass (or number of entities).

This approach is shown in Figure 3.8 and demonstrated in the following sample problem.

Figure 3.8 Summary of the mass-mole-number relationships in a chemical reaction. The amount of one substance in a reaction is related to that of any other. Quantities are expressed in terms of grams, moles, or number of entities (atoms, molecules, or formula units). Start at any box in the diagram (known) and move to any other box (unknown) by using the information on the arrows as conversion factors. As an example, if you know the mass (in g) of A and want to know the number of molecules of B, the path involves three calculation steps:
1. Grams of A to moles of A, using the molar mass ($\mathcal{M}$) of A
2. Moles of A to moles of B, using the molar ratio from the balanced equation
3. Moles of B to molecules of B, using Avogadro's number
Steps 1 and 3 refer to calculations discussed in Section 3.1 (see Figure 3.4).

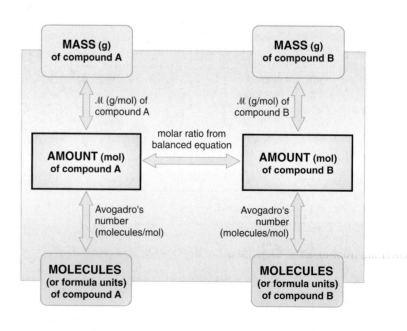

SAMPLE PROBLEM 3.8 Calculating Amounts of Reactants and Products

Problem In a lifetime, the average American uses 1750 lb (794 kg) of copper in coins, plumbing, and wiring. Copper is obtained from sulfide ores, such as chalcocite, or copper(I) sulfide, by a multistep process. After an initial grinding, the first step is to "roast" the ore (heat it strongly with oxygen gas) to form powdered copper(I) oxide and gaseous sulfur dioxide.

(a) How many moles of oxygen are required to roast 10.0 mol of copper(I) sulfide?
(b) How many grams of sulfur dioxide are formed when 10.0 mol of copper(I) sulfide is roasted?
(c) How many kilograms of oxygen are required to form 2.86 kg of copper(I) oxide?

(a) Determining the moles of O_2 needed to roast 10.0 mol of Cu_2S
Plan We *always* write the balanced equation first. The formulas of the reactants are Cu_2S and O_2, and the formulas of the products are Cu_2O and SO_2, so we have

$$2Cu_2S(s) + 3O_2(g) \longrightarrow 2Cu_2O(s) + 2SO_2(g)$$

We are given the *moles* of Cu_2S and need to find the *moles* of O_2. The balanced equation shows that 3 mol of O_2 is needed for every 2 mol of Cu_2S consumed, so the conversion factor is "3 mol O_2/2 mol Cu_2S" (see roadmap a).
Solution Calculating number of moles of O_2:

$$\text{Moles of } O_2 = 10.0 \text{ mol } Cu_2S \times \frac{3 \text{ mol } O_2}{2 \text{ mol } Cu_2S} = 15.0 \text{ mol } O_2$$

Check The units are correct, and the answer is reasonable because this O_2/Cu_2S molar ratio (15:10) is equivalent to the ratio in the balanced equation (3:2).
Comment A *common mistake* is to use the incorrect conversion factor; the calculation would then be

$$\text{Moles of } O_2 = 10.0 \text{ mol } Cu_2S \times \frac{2 \text{ mol } Cu_2S}{3 \text{ mol } O_2} = \frac{6.67 \text{ mol}^2 Cu_2S}{1 \text{ mol } O_2}$$

Such strange units should signal that you made an error in setting up the conversion factor. In addition, the answer, 6.67, is *less* than 10.0, whereas the balanced equation shows that *more* moles of O_2 than of Cu_2S are needed. Be sure to think through the calculation when setting up the conversion factor and canceling units.

(b) Determining the mass (g) of SO_2 formed from 10.0 mol of Cu_2S
Plan Here we need the *grams* of product (SO_2) that form from the given *moles* of reactant (Cu_2S). We first find the moles of SO_2 using the molar ratio from the balanced equation (2 mol SO_2/2 mol Cu_2S) and then multiply by its molar mass (64.07 g/mol) to find grams of SO_2. The steps appear in roadmap b.
Solution Combining the two conversion steps into one calculation, we have

$$\text{Mass (g) of } SO_2 = 10.0 \text{ mol } Cu_2S \times \frac{2 \text{ mol } SO_2}{2 \text{ mol } Cu_2S} \times \frac{64.07 \text{ g } SO_2}{1 \text{ mol } SO_2} = 641 \text{ g } SO_2$$

Check The answer makes sense, since the molar ratio shows that 10.0 mol of SO_2 are formed and each mole weighs about 64 g. We rounded to three significant figures.

(c) Determining the mass (kg) of O_2 that yields 2.86 kg of Cu_2O
Plan Here the mass of product (Cu_2O) is known, and we need the mass of reactant (O_2) that reacts to form it. We first convert the quantity of Cu_2O from *kilograms* to *moles* (in two steps, as shown in roadmap c). Then, we use the molar ratio (3 mol O_2/2 mol Cu_2O) to find the *moles* of O_2 required. Finally, we convert *moles* of O_2 to *kilograms* (in two steps).
Solution Converting from kilograms of Cu_2O to moles of Cu_2O: Combining the mass unit conversion with the mass-to-mole conversion gives

$$\text{Moles of } Cu_2O = 2.86 \text{ kg } Cu_2O \times \frac{10^3 \text{ g}}{1 \text{ kg}} \times \frac{1 \text{ mol } Cu_2O}{143.10 \text{ g } Cu_2O} = 20.0 \text{ mol } Cu_2O$$

Converting from moles of Cu_2O to moles of O_2:

$$\text{Moles of } O_2 = 20.0 \text{ mol } Cu_2O \times \frac{3 \text{ mol } O_2}{2 \text{ mol } Cu_2O} = 30.0 \text{ mol } O_2$$

(a)

Amount (mol) of Cu_2S
↓ molar ratio
Amount (mol) of O_2

(b)

Amount (mol) of Cu_2S
↓ molar ratio
Amount (mol) of SO_2
↓ multiply by $\mathcal{M}$ (g/mol)
Mass (g) of SO_2

(c)

Mass (kg) of Cu_2O
↓ 1 kg = 10^3 g
Mass (g) of Cu_2O
↓ divide by $\mathcal{M}$ (g/mol)
Amount (mol) of Cu_2O
↓ molar ratio
Amount (mol) of O_2
↓ multiply by $\mathcal{M}$ (g/mol)
Mass (g) of O_2
↓ 10^3 g = 1 kg
Mass (kg) of O_2

Converting from moles of O_2 to kilograms of O_2: Combining the mole-to-mass conversion with the mass unit conversion gives

$$\text{Mass (kg) of } O_2 = 30.0 \; \cancel{\text{mol } O_2} \times \frac{32.00 \text{ g } O_2}{1 \; \cancel{\text{mol } O_2}} \times \frac{1 \text{ kg}}{10^3 \text{ g}} = 0.960 \text{ kg } O_2$$

Check The units are correct. Round off to check the math: for example, in the final step, $\sim$30 mol $\times$ 30 g/mol $\times$ 1 kg/10^3 g = 0.90 kg. The answer seems reasonable: even though the amount (mol) of O_2 is greater than the amount (mol) of Cu_2O, the mass of O_2 is less than the mass of Cu_2O because $\mathcal{M}$ of O_2 is less than $\mathcal{M}$ of Cu_2O.

Comment This problem highlights a key point for solving stoichiometry problems: *convert the information given into moles.* Then, use the appropriate molar ratio and any other conversion factors to complete the solution.

FOLLOW-UP PROBLEM 3.8 Thermite is a mixture of iron(III) oxide and aluminum powders that was once used to weld railroad tracks. It undergoes a spectacular reaction to yield solid aluminum oxide and molten iron.
(a) How many grams of iron form when 135 g of aluminum reacts?
(b) How many atoms of aluminum react for every 1.00 g of aluminum oxide formed?

Chemical Reactions That Involve a Limiting Reactant

In the problems we've considered up to now, the amount of *one* reactant was given, and we assumed there was enough of any other reactant for the first reactant to be completely used up. For example, to find the amount of SO_2 that forms when 100 g of Cu_2S reacts, we convert the grams of Cu_2S to moles and assume that the Cu_2S reacts with as much O_2 as needed. Because all the Cu_2S is used up, its initial amount determines, or limits, how much SO_2 can form. We call Cu_2S the **limiting reactant** (or *limiting reagent*) because the product stops forming once the Cu_2S is gone, no matter how much O_2 is present.

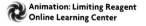
Animation: Limiting Reagent
Online Learning Center

Suppose, however, that the amounts of both Cu_2S *and* O_2 are given in the problem, and we need to find out how much SO_2 forms. We first have to determine whether Cu_2S *or* O_2 is the limiting reactant (that is, which one is completely used up) because the amount of that reactant limits how much SO_2 can form. The other reactant is *in excess,* and whatever amount of it is not used is left over.

To clarify the idea of a limiting reactant, let's consider a situation from real life. A car assembly plant has 1500 car bodies and 4000 tires. How many cars can be made with the supplies on hand? Does the plant manager need to order more car bodies or more tires? Obviously, 4 tires are required for each car body, so the "balanced equation" is

$$1 \text{ car body} + 4 \text{ tires} \longrightarrow 1 \text{ car}$$

How much "product" (cars) can we make from the amount of each "reactant"?

$$1500 \; \cancel{\text{car bodies}} \times \frac{1 \text{ car}}{1 \; \cancel{\text{car body}}} = 1500 \text{ cars}$$

$$4000 \; \cancel{\text{tires}} \times \frac{1 \text{ car}}{4 \; \cancel{\text{tires}}} = 1000 \text{ cars}$$

The number of tires limits the number of cars because less "product" (fewer cars) can be produced from the available tires. There will be 1500 − 1000 = 500 car bodies in excess, and they cannot be turned into cars until more tires are delivered.

Now let's apply these ideas to solving chemical problems. In limiting-reactant problems, the amounts of two (or more) reactants are given, and we must first determine which is limiting. To do this, just as we did with the cars, we first note how much of each reactant *should* be present to completely use up the other, and then we compare it with the amount that is *actually* present. Simply put, the limiting reactant is the one there is not enough of; that is, *it is the reactant that limits the amount of the other reactant that can react, and thus the amount of product that can form.* In mathematical terms, *the limiting reactant is the one that yields the **lower** amount of product.*

We'll examine limiting reactants in the following two sample problems. Sample Problem 3.9 has two parts, and in both we have to identify the limiting reactant. In the first part, we look at a simple molecular view of a reaction and compare the number of molecules to find the limiting reactant; in the second part, we start with the amounts (mol) of two reactants and perform two calculations, each of which assumes an excess of one of the reactants, to see which reactant forms less product. Then, in Sample Problem 3.10, we go through a similar process but start with the masses of the two reactants.

SAMPLE PROBLEM 3.9   Using Molecular Depictions to Solve a Limiting-Reactant Problem

Problem Nuclear engineers use chlorine trifluoride in the processing of uranium fuel for power plants. This extremely reactive substance is formed as a gas in special metal containers by the reaction of elemental chlorine and fluorine.
(a) Suppose the box shown at right represents a container of the reactant mixture before the reaction occurs (with chlorine colored green). Name the limiting reactant and draw the container contents after the reaction is complete.
(b) When the reaction is run again with 0.750 mol of Cl_2 and 3.00 mol of F_2, what mass of chlorine trifluoride will be prepared?

(a) Determining the limiting reactant and drawing the container contents
Plan We first write the balanced equation. From its name, we know that chlorine trifluoride consists of one Cl atom bonded to three F atoms, ClF_3. Elemental chlorine and fluorine refer to the diatomic molecules Cl_2 and F_2. All the substances are gases. To find the limiting reactant, we compare the number of molecules we have of each reactant, with the number we need for the other to react completely. The limiting reactant limits the amount of the other reactant that can react and the amount of product that will form.
Solution The balanced equation is

$$Cl_2(g) + 3F_2(g) \longrightarrow 2ClF_3(g)$$

The equation shows that two ClF_3 molecules are formed for every one Cl_2 molecule and three F_2 molecules that react. Before the reaction, there are three Cl_2 molecules (six Cl atoms). For all the Cl_2 to react, we need three times three, or nine, F_2 molecules (18 F atoms). But there are only six F_2 molecules (12 F atoms). Therefore, F_2 is the limiting reactant because it limits the amount of Cl_2 that can react, and thus the amount of ClF_3 that can form. After the reaction, as the box at right depicts, all 12 F atoms and four of the six Cl atoms make four ClF_3 molecules, and one Cl_2 molecule remains in excess.
Check The equation is balanced: reactants (2 Cl, 6 F) $\longrightarrow$ products (2 Cl, 6 F), and, in the boxes, the number of each type of atom before the reaction equals the number after the reaction. You can check the choice of limiting reactant by examining the reaction from the perspective of Cl_2: Two Cl_2 molecules are enough to react with the six F_2 molecules in the container. But there are three Cl_2 molecules, so there is not enough F_2.

(b) Calculating the mass of ClF_3 formed
Plan We first determine the limiting reactant by using the molar ratios from the balanced equation to convert the moles of each reactant to moles of ClF_3 formed, assuming an excess of the other reactant. Whichever reactant forms fewer moles of ClF_3 is limiting. Then we use the molar mass of ClF_3 to convert this lower number of moles to grams.
Solution Determining the limiting reactant:
Finding moles of ClF_3 from moles of Cl_2 (assuming F_2 is in excess):

$$\text{Moles of } ClF_3 = 0.750 \text{ mol } Cl_2 \times \frac{2 \text{ mol } ClF_3}{1 \text{ mol } Cl_2} = 1.50 \text{ mol } ClF_3$$

Finding moles of ClF_3 from moles of F_2 (assuming Cl_2 is in excess):

$$\text{Moles of } ClF_3 = 3.00 \text{ mol } F_2 \times \frac{2 \text{ mol } ClF_3}{3 \text{ mol } F_2} = 2.00 \text{ mol } ClF_3$$

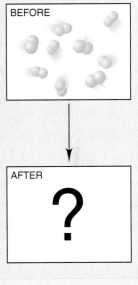

BEFORE

AFTER

?

In this experiment, Cl_2 is limiting because it forms fewer moles of ClF_3. Calculating grams of ClF_3 formed:

$$\text{Mass (g) of ClF}_3 = 1.50 \cancel{\text{ mol ClF}_3} \times \frac{92.45 \text{ g ClF}_3}{1 \cancel{\text{ mol ClF}_3}} = 139 \text{ g ClF}_3$$

Check Let's check our reasoning that Cl_2 is the limiting reactant by assuming, for the moment, that F_2 is limiting. In that case, all 3.00 mol of F_2 would react to form 2.00 mol of ClF_3. Based on the balanced equation, however, that amount of product would require that 1.00 mol of Cl_2 reacted. But that is impossible because only 0.750 mol of Cl_2 is present.
Comment Note that a reactant can be limiting even though it is present in the greater amount. It is the *reactant molar ratio in the balanced equation* that is the determining factor. In both parts (a) and (b), F_2 is present in greater amount than Cl_2. However, in (a), the F_2/Cl_2 ratio is 6/3, or 2/1, which is less than the required molar ratio of 3/1, so F_2 is limiting; in (b), the F_2/Cl_2 ratio is 3.00/0.750, greater than the required 3/1, so F_2 is in excess.

FOLLOW-UP PROBLEM 3.9 B_2 (red spheres) reacts with AB as shown below:

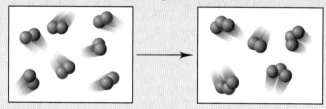

(a) Write a balanced equation for the reaction, and determine the limiting reactant.
(b) How many moles of product can form from the reaction of 1.5 mol of each reactant?

SAMPLE PROBLEM 3.10   Calculating Amounts of Reactant and Product in a Limiting-Reactant Problem

Problem A fuel mixture used in the early days of rocketry is composed of two liquids, hydrazine (N_2H_4) and dinitrogen tetraoxide (N_2O_4), which ignite on contact to form nitrogen gas and water vapor. How many grams of nitrogen gas form when 1.00×10^2 g of N_2H_4 and 2.00×10^2 g of N_2O_4 are mixed?
Plan We first write the balanced equation. *Because the amounts of two reactants are given, we know this is a limiting-reactant problem.* To determine which reactant is limiting, we calculate the mass of N_2 formed from each reactant *assuming an excess of the other*. We convert the grams of each reactant to moles and use the appropriate molar ratio to find the moles of N_2 each forms. Whichever yields *less* N_2 is the limiting reactant. Then, we convert this lower number of moles of N_2 to mass. The roadmap shows the steps.
Solution Writing the balanced equation:

$$2N_2H_4(l) + N_2O_4(l) \longrightarrow 3N_2(g) + 4H_2O(g)$$

Finding the moles of N_2 from the moles of N_2H_4 (if N_2H_4 is limiting):

$$\text{Moles of N}_2\text{H}_4 = 1.00\times10^2 \cancel{\text{ g N}_2\text{H}_4} \times \frac{1 \text{ mol N}_2\text{H}_4}{32.05 \cancel{\text{ g N}_2\text{H}_4}} = 3.12 \text{ mol N}_2\text{H}_4$$

$$\text{Moles of N}_2 = 3.12 \cancel{\text{ mol N}_2\text{H}_4} \times \frac{3 \text{ mol N}_2}{2 \cancel{\text{ mol N}_2\text{H}_4}} = \textbf{4.68 mol N}_2$$

Finding the moles of N_2 from the moles of N_2O_4 (if N_2O_4 is limiting):

$$\text{Moles of N}_2\text{O}_4 = 2.00\times10^2 \cancel{\text{ g N}_2\text{O}_4} \times \frac{1 \text{ mol N}_2\text{O}_4}{92.02 \cancel{\text{ g N}_2\text{O}_4}} = 2.17 \text{ mol N}_2\text{O}_4$$

$$\text{Moles of N}_2 = 2.17 \cancel{\text{ mol N}_2\text{O}_4} \times \frac{3 \text{ mol N}_2}{1 \cancel{\text{ mol N}_2\text{O}_4}} = \textbf{6.51 mol N}_2$$

Thus, N_2H_4 is the limiting reactant because it yields fewer moles of N_2.
Converting from moles of N_2 to grams:

$$\text{Mass (g) of N}_2 = 4.68 \cancel{\text{ mol N}_2} \times \frac{28.02 \text{ g N}_2}{1 \cancel{\text{ mol N}_2}} = 131 \text{ g N}_2$$

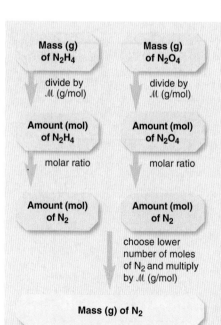

Check The mass of N_2O_4 is greater than that of N_2H_4, but there are fewer moles of N_2O_4 because its $\mathcal{M}$ is much higher. Round off to check the math: for N_2H_4, 100 g N_2H_4 × 1 mol/32 g ≈ 3 mol; ~3 mol × $\frac{3}{2}$ ≈ 4.5 mol N_2; ~4.5 mol × 30 g/mol ≈ 135 g N_2.
Comment 1. Here are two *common mistakes* in solving limiting-reactant problems:
- The limiting reactant is not the *reactant* present in fewer moles (2.17 mol of N_2O_4 vs. 3.12 mol of N_2H_4). Rather, it is the reactant that forms fewer moles of *product*.
- Similarly, the limiting reactant is not the *reactant* present in lower mass. Rather, it is the reactant that forms the lower mass of *product*.

2. Here is an *alternative approach* to finding the limiting reactant. Find the moles of each reactant that would be needed to react with the other reactant. Then see which amount actually given in the problem is sufficient. That substance is in excess, and the other substance is limiting. For example, the balanced equation shows that 2 mol of N_2H_4 reacts with 1 mol of N_2O_4. The moles of N_2O_4 needed to react with the given moles of N_2H_4 are

$$\text{Moles of } N_2O_4 \text{ needed} = 3.12 \text{ mol } N_2H_4 \times \frac{1 \text{ mol } N_2O_4}{2 \text{ mol } N_2H_4} = 1.56 \text{ mol } N_2O_4$$

The moles of N_2H_4 needed to react with the given moles of N_2O_4 are

$$\text{Moles of } N_2H_4 \text{ needed} = 2.17 \text{ mol } N_2O_4 \times \frac{2 \text{ mol } N_2H_4}{1 \text{ mol } N_2O_4} = 4.34 \text{ mol } N_2H_4$$

We are given 2.17 mol of N_2O_4, which is *more* than the amount of N_2O_4 needed (1.56 mol) to react with the given amount of N_2H_4, and we are given 3.12 mol of N_2H_4, which is *less* than the amount of N_2H_4 needed (4.34 mol) to react with the given amount of N_2O_4. Therefore, N_2H_4 is limiting, and N_2O_4 is in excess. Once we determine this, we continue with the final calculation to find the amount of N_2.

FOLLOW-UP PROBLEM 3.10 How many grams of solid aluminum sulfide can be prepared by the reaction of 10.0 g of aluminum and 15.0 g of sulfur? How much of the nonlimiting reactant is in excess?

Chemical Reactions in Practice: Theoretical, Actual, and Percent Yields

Up until now, we've been optimistic about the amount of product obtained from a reaction. We have assumed that 100% of the limiting reactant becomes product, that ideal separation and purification methods exist for isolating the product, and that we use perfect lab technique to collect all the product formed. In other words, we have assumed that we obtain the **theoretical yield,** the amount indicated by the stoichiometrically equivalent molar ratio in the balanced equation.

It's time to face reality. The theoretical yield is *never* obtained, for reasons that are largely uncontrollable. For one thing, although the major reaction predominates, many reactant mixtures also proceed through one or more **side reactions** that form smaller amounts of different products (Figure 3.9). In the rocket fuel reaction in Sample Problem 3.10, for example, the reactants might form some NO in the following side reaction:

$$2N_2O_4(l) + N_2H_4(l) \longrightarrow 6NO(g) + 2H_2O(g)$$

This reaction decreases the amounts of reactants available for N_2 production (see Problem 3.81 at the end of the chapter). Even more important, as we'll discuss in later chapters, many reactions seem to stop before they are complete, which leaves some limiting reactant unused. But, even when a reaction does go completely to product, losses occur in virtually every step of the separation procedure used to isolate the product from the reaction mixture. With careful technique, you can minimize these losses but never eliminate them.

The amount of product that you actually obtain is the **actual yield.** Theoretical and actual yields are expressed in units of amount (moles) or mass (grams).

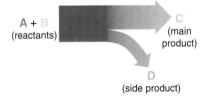

Figure 3.9 The effect of side reactions on yield. One reason the theoretical yield is never obtained is that other reactions lead some of the reactants along side paths to form undesired products.

The **percent yield (% yield)** is the actual yield expressed as a percentage of the theoretical yield:

$$\% \text{ yield} = \frac{\text{actual yield}}{\text{theoretical yield}} \times 100 \qquad (3.7)$$

Because the actual yield *must* be less than the theoretical yield, the percent yield is *always* less than 100%. In multistep reaction sequences, the percent yield of each step is expressed as a fraction and multiplied by the others to find the overall yield. The result may sometimes be surprising. For example, suppose a six-step reaction sequence has a 90.0% yield for each step, which is quite high. Even so, the overall percent yield would be

Overall % yield = $(0.900 \times 0.900 \times 0.900 \times 0.900 \times 0.900 \times 0.900) \times 100 = 53.1\%$

SAMPLE PROBLEM 3.11   Calculating Percent Yield

Problem Silicon carbide (SiC) is an important ceramic material that is made by allowing sand (silicon dioxide, SiO_2) to react with powdered carbon at high temperature. Carbon monoxide is also formed. When 100.0 kg of sand is processed, 51.4 kg of SiC is recovered. What is the percent yield of SiC from this process?

Plan We are given the actual yield of SiC (51.4 kg), so we need the theoretical yield to calculate the percent yield. After writing the balanced equation, we convert the given mass of SiO_2 (100.0 kg) to amount (mol). We use the molar ratio to find the amount of SiC formed and convert that amount to mass (kg) to obtain the theoretical yield [see Sample Problem 3.8(c)]. Then, we use Equation 3.7 to find the percent yield (see the roadmap).

Solution Writing the balanced equation:

$$SiO_2(s) + 3C(s) \longrightarrow SiC(s) + 2CO(g)$$

Converting from kilograms of SiO_2 to moles:

$$\text{Moles of } SiO_2 = 100.0 \text{ kg } SiO_2 \times \frac{1000 \text{ g}}{1 \text{ kg}} \times \frac{1 \text{ mol } SiO_2}{60.09 \text{ g } SiO_2} = 1664 \text{ mol } SiO_2$$

Converting from moles of SiO_2 to moles of SiC: The molar ratio is 1 mol SiC/1 mol SiO_2, so

$$\text{Moles of } SiO_2 = \text{moles of SiC} = 1664 \text{ mol SiC}$$

Converting from moles of SiC to kilograms:

$$\text{Mass (kg) of SiC} = 1664 \text{ mol SiC} \times \frac{40.10 \text{ g SiC}}{1 \text{ mol SiC}} \times \frac{1 \text{ kg}}{1000 \text{ g}} = 66.73 \text{ kg SiC}$$

Calculating the percent yield:

$$\% \text{ yield of SiC} = \frac{\text{actual yield}}{\text{theoretical yield}} \times 100 = \frac{51.4 \text{ kg}}{66.73 \text{ kg}} \times 100 = \boxed{77.0\%}$$

Check Rounding shows that the mass of SiC seems correct: ~1500 mol × 40 g/mol × 1 kg/1000 g = 60 kg. The molar ratio of SiC:SiO_2 is 1:1, and the M of SiC is about two-thirds ($\sim\frac{40}{60}$) the M of SiO_2, so 100 kg of SiO_2 should form about 66 kg of SiC.

FOLLOW-UP PROBLEM 3.11 Marble (calcium carbonate) reacts with hydrochloric acid solution to form calcium chloride solution, water, and carbon dioxide. What is the percent yield of carbon dioxide if 3.65 g of the gas is collected when 10.0 g of marble reacts?

Roadmap (left margin):

Mass (kg) of SiO_2

1. multiply by 10^3
2. divide by M (g/mol)

Amount (mol) of SiO_2

molar ratio

Amount (mol) of SiC

1. multiply by M (g/mol)
2. divide by 10^3

Mass (kg) of SiC

Eq. 3.7

% Yield of SiC

SECTION SUMMARY

The substances in a balanced equation are related to each other by stoichiometrically equivalent molar ratios, which can be used as conversion factors to find the moles of one substance given the moles of another. In limiting-reactant problems, the amounts of two (or more) reactants are given, and one of them limits the amount of product that forms. The limiting reactant is the one that forms the lower amount of product. In practice, side reactions, incomplete reactions, and physical losses result in an actual yield of product that is less than the theoretical yield, the amount based solely on the molar ratio. The percent yield is the actual yield expressed as a percentage of the theoretical yield. In multistep reaction sequences, the overall yield is found by multiplying the percent yields for each step.

3.5 FUNDAMENTALS OF SOLUTION STOICHIOMETRY

Many environmental reactions and almost all biochemical reactions occur in solution, so an understanding of reactions in solution is extremely important in chemistry and related sciences. We'll discuss solution chemistry at many places in the text, but here we focus on solution stoichiometry. Only one aspect of the stoichiometry of dissolved substances is different from what we've seen so far. We know the amounts of pure substances by converting their masses directly into moles. For dissolved substances, we must know the *concentration*—the number of moles present in a certain volume of solution—to find the volume that contains a given number of moles. Of the various ways to express concentration, the most important is *molarity,* so we discuss it here (and wait until Chapter 13 to discuss the other ways). Then, we see how to prepare a solution of a specific molarity and how to use solutions in stoichiometric calculations.

Expressing Concentration in Terms of Molarity

A typical solution consists of a smaller amount of one substance, the **solute,** dissolved in a larger amount of another substance, the **solvent.** When a solution forms, the solute's individual chemical entities become evenly dispersed throughout the available volume and surrounded by solvent molecules. The **concentration** of a solution is usually expressed as *the amount of solute dissolved in a given amount of solution.* Concentration is an *intensive* quantity (like density or temperature) and thus independent of the volume of solution: a 50-L tank of a given solution has the *same concentration* (solute amount/solution amount) as a 50-mL beaker of the solution. **Molarity *(M)*** expresses the concentration in units of *moles of solute per liter of solution:*

$$\text{Molarity} = \frac{\text{moles of solute}}{\text{liters of solution}} \quad \text{or} \quad M = \frac{\text{mol solute}}{\text{L soln}} \qquad (3.8)$$

SAMPLE PROBLEM 3.12   Calculating the Molarity of a Solution

Problem Glycine (H_2NCH_2COOH) is the simplest amino acid. What is the molarity of an aqueous solution that contains 0.715 mol of glycine in 495 mL?
Plan The molarity is the number of moles of solute in each liter of solution. We are given the number of moles (0.715 mol) and the volume (495 mL), so we divide moles by volume and convert the volume to liters to find the molarity (see the roadmap).
Solution

$$\text{Molarity} = \frac{0.715 \text{ mol glycine}}{495 \text{ mL soln}} \times \frac{1000 \text{ mL}}{1 \text{ L}} = 1.44 \text{ } M \text{ glycine}$$

Check A quick look at the math shows about 0.7 mol of glycine in about 0.5 L of solution, so the concentration should be about 1.4 mol/L, or 1.4 *M.*

FOLLOW-UP PROBLEM 3.12 How many moles of KI are in 84 mL of 0.50 *M* KI?

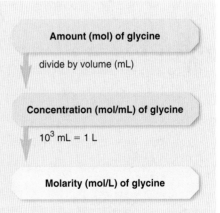

Amount (mol) of glycine

divide by volume (mL)

Concentration (mol/mL) of glycine

10^3 mL = 1 L

Molarity (mol/L) of glycine

Mole-Mass-Number Conversions Involving Solutions

Molarity can be thought of as a conversion factor used to convert between volume of solution and amount (mol) of solute, from which we then find the mass or the number of entities of solute. Figure 3.10 (on the next page) shows this new stoichiometric relationship, and Sample Problem 3.13 applies it.

Figure 3.10 Summary of mass-mole-number-volume relationships in solution. The amount (in moles) of a compound in solution is related to the volume of solution in liters through the molarity (*M*) in moles per liter. The other relationships shown are identical to those in Figure 3.4, except that here they refer to the quantities *in solution*. As in previous cases, to find the quantity of substance expressed in one form or another, convert the given information to moles first.

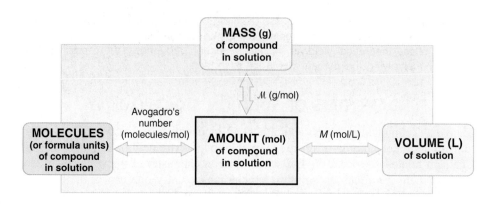

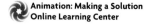

Volume (L) of solution

↓ multiply by *M* (mol/L)

Amount (mol) of solute

↓ multiply by $\mathcal{M}$ (g/mol)

Mass (g) of solute

SAMPLE PROBLEM 3.13 Calculating Mass of Solute in a Given Volume of Solution

Problem A *buffered* solution maintains acidity as a reaction occurs. In living cells, phosphate ions play a key buffering role, so biochemists often study reactions in such solutions. How many grams of solute are in 1.75 L of 0.460 *M* sodium monohydrogen phosphate?
Plan We know the solution volume (1.75 L) and molarity (0.460 *M*), and we need the mass of solute. We use the known quantities to find the amount (mol) of solute and then convert moles to grams with the solute molar mass, as shown in the roadmap.
Solution Calculating moles of solute in solution:

$$\text{Moles of Na}_2\text{HPO}_4 = 1.75 \ \cancel{\text{L soln}} \times \frac{0.460 \ \text{mol Na}_2\text{HPO}_4}{1 \ \cancel{\text{L soln}}} = 0.805 \ \text{mol Na}_2\text{HPO}_4$$

Converting from moles of solute to grams:

$$\text{Mass (g) Na}_2\text{HPO}_4 = 0.805 \ \cancel{\text{mol Na}_2\text{HPO}_4} \times \frac{141.96 \ \text{g Na}_2\text{HPO}_4}{1 \ \cancel{\text{mol Na}_2\text{HPO}_4}} = 114 \ \text{g Na}_2\text{HPO}_4$$

Check The answer seems to be correct: ~1.8 L of 0.5 mol/L contains 0.9 mol, and 150 g/mol × 0.9 mol = 135 g, which is close to 114 g of solute.

FOLLOW-UP PROBLEM 3.13 In biochemistry laboratories, solutions of sucrose (table sugar, $C_{12}H_{22}O_{11}$) are used in high-speed centrifuges to separate the parts of a biological cell. How many liters of 3.30 *M* sucrose contain 135 g of solute?

Dilution of Molar Solutions

A concentrated solution (higher molarity) can be converted to a dilute solution (lower molarity) by *adding solvent* to it. The solution volume increases while the number of moles of solute remains the same. Thus, a given volume of the final (dilute) solution contains fewer solute particles and has a lower concentration than the original (concentrated) solution did (Figure 3.11). If various low concentrations of a solution are used frequently, it is common practice to prepare a more concentrated solution (called a *stock solution*), which is stored and diluted as needed.

SAMPLE PROBLEM 3.14 Preparing a Dilute Solution from a Concentrated Solution

Problem Isotonic saline is a 0.15 *M* aqueous solution of NaCl that simulates the total concentration of ions found in many cellular fluids. Its uses range from a cleansing rinse for contact lenses to a washing medium for red blood cells. How would you prepare 0.80 L of isotonic saline from a 6.0 *M* stock solution?

Plan To dilute a concentrated solution, we add only solvent, so the *moles of solute are the same in both solutions*. We know the volume (0.80 L) and molarity (0.15 M) of the dilute (dil) NaCl solution we need, so we find the moles of NaCl it contains and then find the volume of concentrated (conc; 6.0 M) NaCl solution that contains the same number of moles. Then, we dilute this volume with solvent *up to* the final volume (see roadmap).

Solution Finding moles of solute in dilute solution:

$$\text{Moles of NaCl in dil soln} = 0.80 \text{ L soln} \times \frac{0.15 \text{ mol NaCl}}{1 \text{ L soln}} = 0.12 \text{ mol NaCl}$$

Finding moles of solute in concentrated solution: Because we add only solvent to dilute the solution,

$$\text{Moles of NaCl in dil soln} = \text{moles of NaCl in conc soln} = 0.12 \text{ mol NaCl}$$

Finding the volume of concentrated solution that contains 0.12 mol of NaCl:

$$\text{Volume (L) of conc NaCl soln} = 0.12 \text{ mol NaCl} \times \frac{1 \text{ L soln}}{6.0 \text{ mol NaCl}} = 0.020 \text{ L soln}$$

To prepare 0.80 L of dilute solution, place 0.020 L of 6.0 M NaCl in a 1.0-L cylinder, add distilled water (~780 mL) to the 0.80-L mark, and stir thoroughly.

Check The answer seems reasonable because a small volume of concentrated solution is used to prepare a large volume of dilute solution. Also, the ratio of volumes (0.020 L:0.80 L) is the same as the ratio of concentrations (0.15 M:6.0 M).

Comment An *alternative approach* to solving dilution problems uses the formula

$$M_{\text{dil}} \times V_{\text{dil}} = \text{number of moles} = M_{\text{conc}} \times V_{\text{conc}} \qquad (3.9)$$

where the M and V terms are the molarity and volume of the *dil*ute and *conc*entrated solutions. In this problem, we need the volume of concentrated solution, V_{conc}:

$$V_{\text{conc}} = \frac{M_{\text{dil}} \times V_{\text{dil}}}{M_{\text{conc}}} = \frac{0.15 \, M \times 0.80 \text{ L}}{6.0 \, M} = 0.020 \text{ L}$$

The method worked out in the Solution (above) is actually the same calculation broken into two parts to emphasize the thinking process:

$$V_{\text{conc}} = 0.80 \text{ L} \times \frac{0.15 \text{ mol NaCl}}{1 \text{ L}} \times \frac{1 \text{ L}}{6.0 \text{ mol NaCl}} = 0.020 \text{ L}$$

FOLLOW-UP PROBLEM 3.14 To prepare a fertilizer, an engineer dilutes a stock solution of sulfuric acid by adding 25.0 m³ of 7.50 M acid to enough water to make 500. m³. What is the mass (in g) of sulfuric acid per milliliter of the diluted solution?

Volume (L) of dilute solution

multiply by M (mol/L) of dilute solution

Amount (mol) of NaCl in dilute solution = Amount (mol) of NaCl in concentrated solution

divide by M (mol/L) of concentrated solution

Volume (L) of concentrated solution

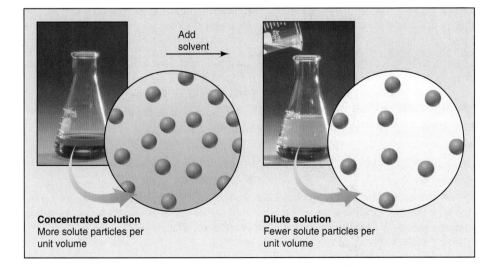

Concentrated solution
More solute particles per unit volume

Add solvent

Dilute solution
Fewer solute particles per unit volume

Figure 3.11 Converting a concentrated solution to a dilute solution. When a solution is diluted, only solvent is added. The solution volume increases while the total number of moles of solute remains the same. Therefore, as shown in the blow-up views, a unit volume of concentrated solution contains more solute particles than the same unit volume of dilute solution.

Stoichiometry of Chemical Reactions in Solution

Solving stoichiometry problems for reactions in solution requires the same approach as before, with the additional step of converting the volume of reactant or product to moles: (1) balance the equation, (2) find the number of moles of one substance, (3) relate it to the stoichiometrically equivalent number of moles of another substance, and (4) convert to the desired units.

SAMPLE PROBLEM 3.15 Calculating Amounts of Reactants and Products for a Reaction in Solution

Problem Specialized cells in the stomach release HCl to aid digestion. If they release too much, the excess can be neutralized with an antacid to avoid discomfort. A common antacid contains magnesium hydroxide, $Mg(OH)_2$, which reacts with the acid to form water and magnesium chloride solution. As a government chemist testing commercial antacids, you use 0.10 M HCl to simulate the acid concentration in the stomach. How many liters of "stomach acid" react with a tablet containing 0.10 g of $Mg(OH)_2$?

Plan We know the mass of $Mg(OH)_2$ (0.10 g) that reacts and the acid concentration (0.10 M), and we must find the acid volume. After writing the balanced equation, we convert the grams of $Mg(OH)_2$ to moles, use the molar ratio to find the moles of HCl that react with these moles of $Mg(OH)_2$, and then use the molarity of HCl to find the volume that contains this number of moles. The steps appear in the roadmap.

Solution Writing the balanced equation:

$$Mg(OH)_2(s) + 2HCl(aq) \longrightarrow MgCl_2(aq) + 2H_2O(l)$$

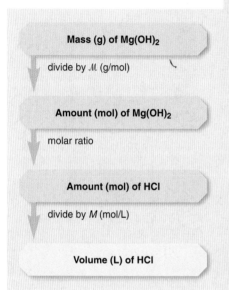

Converting from grams of $Mg(OH)_2$ to moles:

$$\text{Moles of } Mg(OH)_2 = 0.10 \text{ g } Mg(OH)_2 \times \frac{1 \text{ mol } Mg(OH)_2}{58.33 \text{ g } Mg(OH)_2} = 1.7 \times 10^{-3} \text{ mol } Mg(OH)_2$$

Converting from moles of $Mg(OH)_2$ to moles of HCl:

$$\text{Moles of HCl} = 1.7 \times 10^{-3} \text{ mol } Mg(OH)_2 \times \frac{2 \text{ mol HCl}}{1 \text{ mol } Mg(OH)_2} = 3.4 \times 10^{-3} \text{ mol HCl}$$

Converting from moles of HCl to liters:

$$\text{Volume (L) of HCl} = 3.4 \times 10^{-3} \text{ mol HCl} \times \frac{1 \text{ L}}{0.10 \text{ mol HCl}}$$
$$= 3.4 \times 10^{-2} \text{ L}$$

Check The size of the answer seems reasonable: a small volume of dilute acid (0.034 L of 0.10 M) reacts with a small amount of antacid (0.0017 mol).

Comment The reaction as written is an oversimplification; in reality, HCl and $MgCl_2$ exist as separated ions in solution. This point will be covered in great detail in Chapters 4 and 18.

FOLLOW-UP PROBLEM 3.15 Another active ingredient found in some antacids is aluminum hydroxide. Which is more effective at neutralizing stomach acid, magnesium hydroxide or aluminum hydroxide? [*Hint:* Effectiveness refers to the amount of acid that reacts with a given mass of antacid. You already know the effectiveness of 0.10 g of $Mg(OH)_2$.]

In limiting-reactant problems for reactions in solution, we first determine which reactant is limiting and then determine the yield, as demonstrated in the next sample problem.

SAMPLE PROBLEM 3.16 Solving Limiting-Reactant Problems for Reactions in Solution

Problem Mercury and its compounds have many uses, from fillings for teeth (as a mixture with silver, copper, and tin) to the industrial production of chlorine. Because of their toxicity, however, soluble mercury compounds, such as mercury(II) nitrate, must be removed from industrial wastewater. One removal method reacts the wastewater with sodium sulfide solution to produce solid mercury(II) sulfide and sodium nitrate solution. In a laboratory simulation, 0.050 L of 0.010 M mercury(II) nitrate reacts with 0.020 L of 0.10 M sodium sulfide. How many grams of mercury(II) sulfide form?

Plan This is a limiting-reactant problem because *the amounts of two reactants are given*. After balancing the equation, we must determine the limiting reactant. The molarity (0.010 M) and volume (0.050 L) of the mercury(II) nitrate solution tell us the moles of one reactant, and the molarity (0.10 M) and volume (0.020 L) of the sodium sulfide solution tell us the moles of the other. Then, as in Sample Problem 3.10, we use the molar ratio to find the moles of HgS that form from each reactant, *assuming the other reactant is present in excess*. The limiting reactant is the one that forms fewer moles of HgS, which we convert to mass using the HgS molar mass. The roadmap shows the process.

Solution Writing the balanced equation:

$$Hg(NO_3)_2(aq) + Na_2S(aq) \longrightarrow HgS(s) + 2NaNO_3(aq)$$

Finding moles of HgS assuming $Hg(NO_3)_2$ is limiting: Combining the steps gives

$$\text{Moles of HgS} = 0.050 \text{ L soln} \times \frac{0.010 \text{ mol Hg(NO}_3)_2}{1 \text{ L soln}} \times \frac{1 \text{ mol HgS}}{1 \text{ mol Hg(NO}_3)_2}$$

$$= 5.0 \times 10^{-4} \text{ mol HgS}$$

Finding moles of HgS assuming Na_2S is limiting: Combining the steps gives

$$\text{Moles of HgS} = 0.020 \text{ L soln} \times \frac{0.10 \text{ mol Na}_2S}{1 \text{ L soln}} \times \frac{1 \text{ mol HgS}}{1 \text{ mol Na}_2S}$$

$$= 2.0 \times 10^{-3} \text{ mol HgS}$$

$Hg(NO_3)_2$ is the limiting reactant because it forms fewer moles of HgS. Converting the moles of HgS formed from $Hg(NO_3)_2$ to grams:

$$\text{Mass (g) of HgS} = 5.0 \times 10^{-4} \text{ mol HgS} \times \frac{232.7 \text{ g HgS}}{1 \text{ mol HgS}}$$

$$= 0.12 \text{ g HgS}$$

Check As a check, let's use the alternative method for finding the limiting reactant (see Comment in Sample Problem 3.10). Finding moles of reactants available:

$$\text{Moles of Hg(NO}_3)_2 = 0.050 \text{ L soln} \times \frac{0.010 \text{ mol Hg(NO}_3)_2}{1 \text{ L soln}} = 5.0 \times 10^{-4} \text{ mol Hg(NO}_3)_2$$

$$\text{Moles of Na}_2S = 0.020 \text{ L soln} \times \frac{0.10 \text{ mol Na}_2S}{1 \text{ L soln}} = 2.0 \times 10^{-3} \text{ mol Na}_2S$$

The molar ratio of the reactants is 1 $Hg(NO_3)_2$/1 Na_2S. Therefore, $Hg(NO_3)_2$ is limiting because there are fewer moles of it than are needed to react with the moles of Na_2S. Finding grams of product from moles of limiting reactant and the molar ratio:

$$\text{Mass (g) of HgS} = 5.0 \times 10^{-4} \text{ mol Hg(NO}_3)_2 \times \frac{1 \text{ mol HgS}}{1 \text{ mol Hg(NO}_3)_2} \times \frac{232.7 \text{ g HgS}}{1 \text{ mol HgS}}$$

$$= 0.12 \text{ g HgS}$$

FOLLOW-UP PROBLEM 3.16 Even though gasoline sold in the United States no longer contains lead, this metal persists in the environment as a poison. Despite their toxicity, many compounds of lead are still used to make pigments.
(a) What volume of 1.50 M lead(II) acetate contains 0.400 mol of Pb^{2+} ions?
(b) When this volume reacts with 125 mL of 3.40 M sodium chloride, how many grams of solid lead(II) chloride can form? (Sodium acetate solution also forms.)

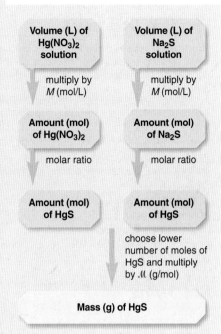

SECTION SUMMARY
When reactions occur in solution, reactant and product amounts are given in terms of concentration and volume. Molarity is the number of moles of solute dissolved in one liter of solution. Using molarity as a conversion factor, we apply the principles of stoichiometry to all aspects of reactions in solution.

For Review and Reference (Numbers in parentheses refer to pages, unless noted otherwise.)

Learning Objectives

To help you review these learning objectives, the numbers of related sections (§), sample problems (SP), and upcoming end-of-chapter problems (EP) are listed in parentheses.

1. Realize the usefulness of the mole concept, and use the relation between molecular (or formula) mass and molar mass to calculate the molar mass of any substance (§ 3.1) (EPs 3.1–3.5, 3.7–3.10)
2. Understand the relationships among amount of substance (in moles), mass (in grams), and number of chemical entities and convert from one to any other (§ 3.1) (SPs 3.1, 3.2) (EPs 3.6, 3.11–3.16, 3.19)
3. Use mass percent to find the mass of element in a given mass of compound (§ 3.1) (SP 3.3) (EPs 3.17, 3.18, 3.20–3.23)
4. Determine the empirical and molecular formulas of a compound from mass analysis of its elements (§ 3.2) (SPs 3.4–3.6) (EPs 3.24–3.34)

5. Balance an equation given formulas or names, and use molar ratios to calculate amounts of reactants and products for reactions of pure or dissolved substances (§ 3.3 and 3.5) (SPs 3.7, 3.8, 3.15) (EPs 3.35–3.46, 3.62, 3.71, 3.72)
6. Understand why one reactant limits the yield of product, and solve limiting-reactant problems for reactions of pure or dissolved substances (§ 3.4, 3.5) (SPs 3.9, 3.10, 3.16) (EPs 3.47–3.54, 3.61, 3.73, 3.74)
7. Explain the reasons for lower-than-expected yields and the distinction between theoretical and actual yields, and calculate percent yield (§ 3.4) (SP 3.11) (EPs 3.55–3.60, 3.63)
8. Understand the meaning of concentration and the effect of dilution, and calculate molarity or mass of dissolved solute (§ 3.5) (SPs 3.12–3.14) (EPs 3.64–3.70, 3.75)

Key Terms

stoichiometry (70)

Section 3.1
mole (mol) (70)
Avogadro's number (70)
molar mass ($\mathcal{M}$) (72)

Section 3.2
combustion analysis (80)
isomer (81)

Section 3.3
chemical equation (83)
reactant (83)
product (83)
balancing (stoichiometric) coefficient (83)

Section 3.4
limiting reactant (90)
theoretical yield (93)
side reaction (93)
actual yield (93)
percent yield (% yield) (94)

Section 3.5
solute (95)
solvent (95)
concentration (95)
molarity (M) (95)

Key Equations and Relationships

3.1 Number of entities in one mole (70):
$$\text{1 mole contains } 6.022\times10^{23} \text{ entities (to 4 sf)}$$

3.2 Converting amount (mol) to mass using $\mathcal{M}$ (73):
$$\text{Mass (g)} = \text{no. of moles} \times \frac{\text{no. of grams}}{1 \text{ mol}}$$

3.3 Converting mass to amount (mol) using $1/\mathcal{M}$ (73):
$$\text{No. of moles} = \text{mass (g)} \times \frac{1 \text{ mol}}{\text{no. of grams}}$$

3.4 Converting amount (mol) to number of entities (73):
$$\text{No. of entities} = \text{no. of moles} \times \frac{6.022\times10^{23} \text{ entities}}{1 \text{ mol}}$$

3.5 Converting number of entities to amount (mol) (73):
$$\text{No. of moles} = \text{no. of entities} \times \frac{1 \text{ mol}}{6.022\times10^{23} \text{ entities}}$$

3.6 Calculating mass % (75):
Mass % of element X
$$= \frac{\text{moles of X in formula} \times \text{molar mass of X (g/mol)}}{\text{mass (g) of 1 mol of compound}} \times 100$$

3.7 Calculating percent yield (94):
$$\text{\% yield} = \frac{\text{actual yield}}{\text{theoretical yield}} \times 100$$

3.8 Defining molarity (95):
$$\text{Molarity} = \frac{\text{moles of solute}}{\text{liters of solution}} \quad \text{or} \quad M = \frac{\text{mol solute}}{\text{L soln}}$$

3.9 Diluting a concentrated solution (97):
$$M_{\text{dil}} \times V_{\text{dil}} = \text{number of moles} = M_{\text{conc}} \times V_{\text{conc}}$$

Brief Solutions to Follow-up Problems

3.1 (a) Moles of C = $315 \text{ mg C} \times \dfrac{1 \text{ g}}{10^3 \text{ mg}} \times \dfrac{1 \text{ mol C}}{12.01 \text{ g C}}$
$\qquad = 2.62\times10^{-2} \text{ mol C}$
(b) Mass (g) of Mn = $3.22\times10^{20} \text{ Mn atoms}$
$\qquad \times \dfrac{1 \text{ mol Mn}}{6.022\times10^{23} \text{ Mn atoms}} \times \dfrac{54.94 \text{ g Mn}}{1 \text{ mol Mn}}$
$\qquad = 2.94\times10^{-2} \text{ g Mn}$

3.2 (a) Mass (g) of P_4O_{10}
$\qquad = 4.65\times10^{22} \text{ molecules } P_4O_{10}$
$\qquad \times \dfrac{1 \text{ mol } P_4O_{10}}{6.022\times10^{23} \text{ molecules } P_4O_{10}} \times \dfrac{283.88 \text{ g } P_4O_{10}}{1 \text{ mol } P_4O_{10}}$
$\qquad = 21.9 \text{ g } P_4O_{10}$

(b) No. of P atoms = 4.65×10^{22} ~~molecules P₄O₁₀~~

$$\times \frac{4 \text{ atoms P}}{1 \text{ molecule P}_4\text{O}_{10}}$$

$$= 1.86 \times 10^{23} \text{ P atoms}$$

3.3 (a) Mass % of N = $\dfrac{2 \text{ mol N} \times \dfrac{14.01 \text{ g N}}{1 \text{ mol N}}}{80.05 \text{ g NH}_4\text{NO}_3} \times 100$

$$= 35.00 \text{ mass \% N}$$

(b) Mass (g) of N = $35.8 \text{ kg NH}_4\text{NO}_3 \times \dfrac{10^3 \text{ g}}{1 \text{ kg}} \times \dfrac{0.3500 \text{ g N}}{1 \text{ g NH}_4\text{NO}_3}$

$$= 1.25 \times 10^4 \text{ g N}$$

3.4 Moles of S = $2.88 \text{ g S} \times \dfrac{1 \text{ mol S}}{32.07 \text{ g S}} = 0.0898 \text{ mol S}$

Moles of M = $0.0898 \text{ mol S} \times \dfrac{2 \text{ mol M}}{3 \text{ mol S}} = 0.0599 \text{ mol M}$

Molar mass of M = $\dfrac{3.12 \text{ g M}}{0.0599 \text{ mol M}} = 52.1 \text{ g/mol}$

M is chromium, and M_2S_3 is chromium(III) sulfide.

3.5 Assuming 100.00 g of compound, we have 95.21 g of C and 4.79 g of H:

$$\text{Moles of C} = 95.21 \text{ g C} \times \frac{1 \text{ mol C}}{12.01 \text{ g C}}$$

$$= 7.928 \text{ mol C}$$

Also, 4.75 mol H

Preliminary formula = $C_{7.928}H_{4.75} \approx C_{1.67}H_{1.00}$

Empirical formula = C_5H_3

Whole-number multiple = $\dfrac{252.30 \text{ g/mol}}{63.07 \text{ g/mol}} = 4$

Molecular formula = $C_{20}H_{12}$

3.6 Mass (g) of C = $0.451 \text{ g CO}_2 \times \dfrac{12.01 \text{ g C}}{44.01 \text{ g CO}_2}$

$$= 0.123 \text{ g C}$$

Also, 0.00690 g H

Mass (g) of Cl = $0.250 \text{ g} - (0.123 \text{ g} + 0.00690 \text{ g}) = 0.120 \text{ g Cl}$

Moles of elements

$$= 0.0102 \text{ mol C}; \ 0.00685 \text{ mol H}; \ 0.00339 \text{ mol Cl}$$

Empirical formula = C_3H_2Cl; multiple = 2

Molecular formula = $C_6H_4Cl_2$

3.7 (a) $2Na(s) + 2H_2O(l) \longrightarrow H_2(g) + 2NaOH(aq)$

(b) $2HNO_3(aq) + CaCO_3(s) \longrightarrow$

$$H_2O(l) + CO_2(g) + Ca(NO_3)_2(aq)$$

(c) $PCl_3(g) + 3HF(g) \longrightarrow PF_3(g) + 3HCl(g)$

(d) $4C_3H_5N_3O_9(l) \longrightarrow$

$$12CO_2(g) + 10H_2O(g) + 6N_2(g) + O_2(g)$$

3.8 $Fe_2O_3(s) + 2Al(s) \longrightarrow Al_2O_3(s) + 2Fe(l)$

(a) Mass (g) of Fe

$$= 135 \text{ g Al} \times \frac{1 \text{ mol Al}}{26.98 \text{ g Al}} \times \frac{2 \text{ mol Fe}}{2 \text{ mol Al}} \times \frac{55.85 \text{ g Fe}}{1 \text{ mol Fe}}$$

$$= 279 \text{ g Fe}$$

(b) No. of Al atoms = $1.00 \text{ g Al}_2\text{O}_3 \times \dfrac{1 \text{ mol Al}_2\text{O}_3}{101.96 \text{ g Al}_2\text{O}_3}$

$$\times \frac{2 \text{ mol Al}}{1 \text{ mol Al}_2\text{O}_3} \times \frac{6.022 \times 10^{23} \text{ Al atoms}}{1 \text{ mol Al}}$$

$$= 1.18 \times 10^{22} \text{ Al atoms}$$

3.9 (a) $2AB + B_2 \longrightarrow 2AB_2$

In the boxes, the AB/B_2 ratio is 4/3, which is less than the 2/1 ratio in the equation. Thus, there is not enough AB, so it is the limiting reactant; note that one B_2 is in excess.

(b) Moles of AB_2 = $1.5 \text{ mol AB} \times \dfrac{2 \text{ mol AB}_2}{2 \text{ mol AB}} = 1.5 \text{ mol AB}_2$

Moles of AB_2 = $1.5 \text{ mol B}_2 \times \dfrac{2 \text{ mol AB}_2}{1 \text{ mol B}_2} = 3.0 \text{ mol AB}_2$

3.10 $2Al(s) + 3S(s) \longrightarrow Al_2S_3(s)$

Mass (g) of Al_2S_3 formed from Al

$$= 10.0 \text{ g Al} \times \frac{1 \text{ mol Al}}{26.98 \text{ g Al}} \times \frac{1 \text{ mol Al}_2\text{S}_3}{2 \text{ mol Al}} \times \frac{150.17 \text{ g Al}_2\text{S}_3}{1 \text{ mol Al}_2\text{S}_3}$$

$$= 27.8 \text{ g Al}_2\text{S}_3$$

Similarly, mass (g) of Al_2S_3 formed from S = 23.4 g Al_2S_3. Therefore, S is limiting reactant, and 23.4 g of Al_2S_3 can form.

Mass (g) of Al in excess

$$= \text{total mass of Al} - \text{mass of Al used}$$

$$= 10.0 \text{ g Al}$$

$$- \left(15.0 \text{ g S} \times \frac{1 \text{ mol S}}{32.07 \text{ g S}} \times \frac{2 \text{ mol Al}}{3 \text{ mol S}} \times \frac{26.98 \text{ g Al}}{1 \text{ mol Al}}\right)$$

$$= 1.6 \text{ g Al}$$

(We would obtain the same answer if sulfur were shown more correctly as S_8.)

3.11 $CaCO_3(s) + 2HCl(aq) \longrightarrow CaCl_2(aq) + H_2O(l) + CO_2(g)$

Theoretical yield (g) of CO_2

$$= 10.0 \text{ g CaCO}_3 \times \frac{1 \text{ mol CaCO}_3}{100.09 \text{ g CaCO}_3} \times \frac{1 \text{ mol CO}_2}{1 \text{ mol CaCO}_3}$$

$$\times \frac{44.01 \text{ g CO}_2}{1 \text{ mol CO}_2} = 4.40 \text{ g CO}_2$$

% yield = $\dfrac{3.65 \text{ g CO}_2}{4.40 \text{ g CO}_2} \times 100 = 83.0\%$

3.12 Moles of KI = $84 \text{ mL soln} \times \dfrac{1 \text{ L}}{10^3 \text{ mL}} \times \dfrac{0.50 \text{ mol KI}}{1 \text{ L soln}}$

$$= 0.042 \text{ mol KI}$$

3.13 Volume (L) of soln

$$= 135 \text{ g sucrose} \times \frac{1 \text{ mol sucrose}}{342.30 \text{ g sucrose}} \times \frac{1 \text{ L soln}}{3.30 \text{ mol sucrose}}$$

$$= 0.120 \text{ L soln}$$

3.14 M_{dil} of H_2SO_4 = $\dfrac{7.50 \, M \times 25.0 \text{ m}^3}{500. \text{ m}^3} = 0.375 \, M \text{ H}_2\text{SO}_4$

Mass (g) of H_2SO_4/mL soln

$$= \frac{0.375 \text{ mol H}_2\text{SO}_4}{1 \text{ L soln}} \times \frac{1 \text{ L}}{10^3 \text{ mL}} \times \frac{98.09 \text{ g H}_2\text{SO}_4}{1 \text{ mol H}_2\text{SO}_4}$$

$$= 3.68 \times 10^{-2} \text{ g/mL soln}$$

3.15 $Al(OH)_3(s) + 3HCl(aq) \longrightarrow AlCl_3(aq) + 3H_2O(l)$

Volume (L) of HCl consumed

$$= 0.10 \text{ g Al(OH)}_3 \times \frac{1 \text{ mol Al(OH)}_3}{78.00 \text{ g Al(OH)}_3}$$

$$\times \frac{3 \text{ mol HCl}}{1 \text{ mol Al(OH)}_3} \times \frac{1 \text{ L soln}}{0.10 \text{ mol HCl}}$$

$$= 3.8 \times 10^{-2} \text{ L soln}$$

Therefore, $Al(OH)_3$ is more effective than $Mg(OH)_2$.

3.16 (a) Volume (L) of soln

$$= 0.400 \text{ mol Pb}^{2+}$$

$$\times \frac{1 \text{ mol Pb(C}_2\text{H}_3\text{O}_2)_2}{1 \text{ mol Pb}^{2+}} \times \frac{1 \text{ L soln}}{1.50 \text{ mol Pb(C}_2\text{H}_3\text{O}_2)_2}$$

$$= 0.267 \text{ L soln}$$

(b) $Pb(C_2H_3O_2)_2(aq) + 2NaCl(aq) \longrightarrow$

$$PbCl_2(s) + 2NaC_2H_3O_2(aq)$$

Mass (g) of $PbCl_2$ from $Pb(C_2H_3O_2)_2$ soln = 111 g $PbCl_2$

Mass (g) of $PbCl_2$ from NaCl soln = 59.1 g $PbCl_2$

Thus, NaCl is the limiting reactant, and 59.1 g of $PbCl_2$ can form.

Problems

Problems with **colored** numbers are answered in Appendix E. Sections match the text and provide the numbers of relevant sample problems. Bracketed problems are grouped in pairs (indicated by a short rule) that cover the same concept. Comprehensive Problems are based on material from any section or previous chapter.

The Mole

(Sample Problems 3.1 to 3.3)

3.1 The atomic mass of Cl is 35.45 amu, and the atomic mass of Al is 26.98 amu. What are the masses in grams of 2 mol of Al atoms and of 3 mol of Cl atoms?

3.2 (a) How many moles of C atoms are in 1 mol of sucrose $(C_{12}H_{22}O_{11})$?
(b) How many C atoms are in 1 mol of sucrose?

3.3 Why might the expression "1 mol of nitrogen" be confusing? What change would remove any uncertainty? For what other elements might a similar confusion exist? Why?

3.4 How is the molecular mass of a compound the same as the molar mass, and how is it different?

3.5 What advantage is there to using a counting unit (the mole) in chemistry rather than a mass unit?

3.6 Each of the following balances weighs the indicated numbers of atoms of two elements:

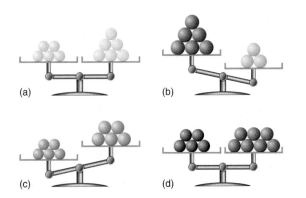

(a)   (b)

(c)   (d)

Which element—left, right, or neither,
(a) Has the higher molar mass?
(b) Has more atoms per gram?
(c) Has fewer atoms per gram?
(d) Has more atoms per mole?

3.7 Calculate the molar mass of each of the following:
(a) $Sr(OH)_2$ (b) N_2O (c) $NaClO_3$ (d) Cr_2O_3

3.8 Calculate the molar mass of each of the following:
(a) $(NH_4)_3PO_4$ (b) CH_2Cl_2 (c) $CuSO_4 \cdot 5H_2O$ (d) BrF_5

3.9 Calculate the molar mass of each of the following:
(a) SnO_2 (b) BaF_2 (c) $Al_2(SO_4)_3$ (d) $MnCl_2$

3.10 Calculate the molar mass of each of the following:
(a) N_2O_4 (b) C_8H_{10} (c) $MgSO_4 \cdot 7H_2O$ (d) $Ca(C_2H_3O_2)_2$

3.11 Calculate each of the following quantities:
(a) Mass in grams of 0.57 mol of $KMnO_4$
(b) Moles of O atoms in 8.18 g of $Mg(NO_3)_2$
(c) Number of O atoms in 8.1×10^{-3} g of $CuSO_4 \cdot 5H_2O$

3.12 Calculate each of the following quantities:
(a) Mass in kilograms of 3.8×10^{20} molecules of NO_2
(b) Moles of Cl atoms in 0.0425 g of $C_2H_4Cl_2$
(c) Number of H^- ions in 4.92 g of SrH_2

3.13 Calculate each of the following quantities:
(a) Mass in grams of 0.64 mol of $MnSO_4$
(b) Moles of compound in 15.8 g of $Fe(ClO_4)_3$
(c) Number of N atoms in 92.6 g of NH_4NO_2

3.14 Calculate each of the following quantities:
(a) Total number of ions in 38.1 g of CaF_2
(b) Mass in milligrams of 3.58 mol of $CuCl_2 \cdot 2H_2O$
(c) Mass in kilograms of 2.88×10^{22} formula units of $Bi(NO_3)_3 \cdot 5H_2O$

3.15 Calculate each of the following quantities:
(a) Mass in grams of 8.41 mol of copper(I) carbonate
(b) Mass in grams of 2.04×10^{21} molecules of dinitrogen pentaoxide
(c) Number of moles and formula units in 57.9 g of sodium perchlorate
(d) Number of sodium ions, perchlorate ions, Cl atoms, and O atoms in the mass of compound in part (c)

3.16 Calculate each of the following quantities:
(a) Mass in grams of 3.52 mol of chromium(III) sulfate decahydrate
(b) Mass in grams of 9.64×10^{24} molecules of dichlorine heptaoxide
(c) Number of moles and formula units in 56.2 g of lithium sulfate
(d) Number of lithium ions, sulfate ions, S atoms, and O atoms in the mass of compound in part (c)

3.17 Calculate each of the following:
(a) Mass % of H in ammonium bicarbonate
(b) Mass % of O in sodium dihydrogen phosphate heptahydrate

3.18 Calculate each of the following:
(a) Mass % of I in strontium periodate
(b) Mass % of Mn in potassium permanganate

3.19 Cisplatin (*right*), or Platinol, is a powerful drug used in the treatment of certain cancers. Calculate (a) the moles of compound in 285.3 g of cisplatin; (b) the number of hydrogen atoms in 0.98 mol of cisplatin.

3.20 Propane is widely used in liquid form as a fuel for barbecue grills and camp stoves. For 75.3 g of propane, calculate (a) the moles of compound; (b) the grams of carbon.

3.21 The effectiveness of a nitrogen fertilizer is determined mainly by its mass % N. Rank the following fertilizers in terms of their effectiveness: potassium nitrate; ammonium nitrate; ammonium sulfate; urea, $CO(NH_2)_2$.

3.22 The mineral galena is composed of lead(II) sulfide and has an average density of 7.46 g/cm^3. (a) How many moles of lead(II) sulfide are in 1.00 ft^3 of galena? (b) How many lead atoms are in 1.00 dm^3 of galena?

3.23 Hemoglobin, a protein in red blood cells, carries O_2 from the lungs to the body's cells. Iron (as ferrous ion, Fe^{2+}) makes up

0.33 mass % of hemoglobin. If the molar mass of hemoglobin is 6.8×10^4 g/mol, how many Fe^{2+} ions are in one molecule?

Determining the Formula of an Unknown Compound
(Sample Problems 3.4 to 3.6)

3.24 Which of the following sets of information allows you to obtain the molecular formula of a covalent compound? In each case that allows it, explain how you would proceed (write a solution "Plan").
(a) Number of moles of each type of atom in a given sample of the compound
(b) Mass % of each element and the total number of atoms in a molecule of the compound
(c) Mass % of each element and the number of atoms of one element in a molecule of the compound
(d) Empirical formula and the mass % of each element in the compound
(e) Structural formula of the compound

3.25 What is the empirical formula and empirical formula mass for each of the following compounds?
(a) C_2H_4 (b) $C_2H_6O_2$ (c) N_2O_5 (d) $Ba_3(PO_4)_2$ (e) Te_4I_{16}
3.26 What is the empirical formula and empirical formula mass for each of the following compounds?
(a) C_4H_8 (b) $C_3H_6O_3$ (c) P_4O_{10} (d) $Ga_2(SO_4)_3$ (e) Al_2Br_6

3.27 What is the molecular formula of each compound?
(a) Empirical formula CH_2 ($\mathcal{M}$ = 42.08 g/mol)
(b) Empirical formula NH_2 ($\mathcal{M}$ = 32.05 g/mol)
(c) Empirical formula NO_2 ($\mathcal{M}$ = 92.02 g/mol)
(d) Empirical formula CHN ($\mathcal{M}$ = 135.14 g/mol)
3.28 What is the molecular formula of each compound?
(a) Empirical formula CH ($\mathcal{M}$ = 78.11 g/mol)
(b) Empirical formula $C_3H_6O_2$ ($\mathcal{M}$ = 74.08 g/mol)
(c) Empirical formula $HgCl$ ($\mathcal{M}$ = 472.1 g/mol)
(d) Empirical formula $C_7H_4O_2$ ($\mathcal{M}$ = 240.20 g/mol)

3.29 Determine the empirical formula of each of the following compounds:
(a) 0.063 mol of chlorine atoms combined with 0.22 mol of oxygen atoms
(b) 2.45 g of silicon combined with 12.4 g of chlorine
(c) 27.3 mass % carbon and 72.7 mass % oxygen
3.30 Determine the empirical formula of each of the following compounds:
(a) 0.039 mol of iron atoms combined with 0.052 mol of oxygen atoms
(b) 0.903 g of phosphorus combined with 6.99 g of bromine
(c) A hydrocarbon with 79.9 mass % carbon

3.31 A sample of 0.600 mol of a metal M reacts completely with excess fluorine to form 46.8 g of MF_2.
(a) How many moles of F are in the sample of MF_2 that forms?
(b) How many grams of M are in this sample of MF_2?
(c) What element is represented by the symbol M?
3.32 A sample of 0.370 mol of a metal oxide (M_2O_3) weighs 55.4 g.
(a) How many moles of O are in the sample?
(b) How many grams of M are in the sample?
(c) What element is represented by the symbol M?

3.33 Cortisol ($\mathcal{M}$ = 362.47 g/mol), one of the major steroid hormones, is a key factor in the synthesis of protein. Its profound effect on the reduction of inflammation explains its use in the treatment of rheumatoid arthritis. Cortisol is 69.6% C, 8.34% H, and 22.1% O by mass. What is its molecular formula?
3.34 Menthol ($\mathcal{M}$ = 156.3 g/mol), a strong-smelling substance used in cough drops, is a compound of carbon, hydrogen, and oxygen. When 0.1595 g of menthol was subjected to combustion analysis, it produced 0.449 g of CO_2 and 0.184 g of H_2O. What is menthol's molecular formula?

Writing and Balancing Chemical Equations
(Sample Problem 3.7)

3.35 In the process of balancing the equation
$$Al + Cl_2 \longrightarrow AlCl_3$$
Student I writes: $Al + Cl_2 \longrightarrow AlCl_2$
Student II writes: $Al + Cl_2 + Cl \longrightarrow AlCl_3$
Student III writes: $2Al + 3Cl_2 \longrightarrow 2AlCl_3$
Is the approach of Student I valid? Student II? Student III? Explain.
3.36 The boxes below represent a chemical reaction between elements A (red) and B (green):

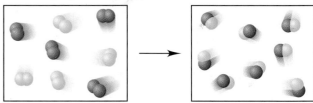

Which of the following best represents the balanced equation for the reaction?
(a) $2A + 2B \longrightarrow A_2 + B_2$ (b) $A_2 + B_2 \longrightarrow 2AB$
(c) $B_2 + 2AB \longrightarrow 2B_2 + A_2$ (d) $4A_2 + 4B_2 \longrightarrow 8AB$

3.37 Write balanced equations for each of the following by inserting the correct coefficients in the blanks:
(a) __$Cu(s)$ + __$S_8(s) \longrightarrow$ __$Cu_2S(s)$
(b) __$P_4O_{10}(s)$ + __$H_2O(l) \longrightarrow$ __$H_3PO_4(l)$
(c) __$B_2O_3(s)$ + __$NaOH(aq) \longrightarrow$
$\qquad$ __$Na_3BO_3(aq)$ + __$H_2O(l)$
(d) __$CH_3NH_2(g)$ + __$O_2(g) \longrightarrow$
$\qquad$ __$CO_2(g)$ + __$H_2O(g)$ + __$N_2(g)$
3.38 Write balanced equations for each of the following by inserting the correct coefficients in the blanks:
(a) __$Cu(NO_3)_2(aq)$ + __$KOH(aq) \longrightarrow$
$\qquad$ __$Cu(OH)_2(s)$ + __$KNO_3(aq)$
(b) __$BCl_3(g)$ + __$H_2O(l) \longrightarrow$ __$H_3BO_3(s)$ + __$HCl(g)$
(c) __$CaSiO_3(s)$ + __$HF(g) \longrightarrow$
$\qquad$ __$SiF_4(g)$ + __$CaF_2(s)$ + __$H_2O(l)$
(d) __$(CN)_2(g)$ + __$H_2O(l) \longrightarrow$ __$H_2C_2O_4(aq)$ + __$NH_3(g)$

3.39 Convert the following into balanced equations:
(a) When gallium metal is heated in oxygen gas, it melts and forms solid gallium(III) oxide.
(b) Liquid hexane burns in oxygen gas to form carbon dioxide gas and water vapor.
(c) When solutions of calcium chloride and sodium phosphate are mixed, solid calcium phosphate forms and sodium chloride remains in solution.
3.40 Convert the following into balanced equations:
(a) When lead(II) nitrate solution is added to potassium iodide solution, solid lead(II) iodide forms and potassium nitrate solution remains.

(b) Liquid disilicon hexachloride reacts with water to form solid silicon dioxide, hydrogen chloride gas, and hydrogen gas.
(c) When nitrogen dioxide is bubbled into water, a solution of nitric acid forms and gaseous nitrogen monoxide is released.

Calculating Amounts of Reactant and Product
(Sample Problems 3.8 to 3.11)

3.41 Potassium nitrate decomposes on heating, producing potassium oxide and gaseous nitrogen and oxygen:

$$4KNO_3(s) \longrightarrow 2K_2O(s) + 2N_2(g) + 5O_2(g)$$

To produce 88.6 kg of oxygen, how many (a) moles of KNO_3 must be heated? (b) Grams of KNO_3 must be heated?

3.42 Chromium(III) oxide reacts with hydrogen sulfide (H_2S) gas to form chromium(III) sulfide and water:

$$Cr_2O_3(s) + 3H_2S(g) \longrightarrow Cr_2S_3(s) + 3H_2O(l)$$

To produce 421 g of Cr_2S_3, (a) how many moles of Cr_2O_3 are required? (b) How many grams of Cr_2O_3 are required?

3.43 Calculate the mass of each product formed when 33.61 g of diborane (B_2H_6) reacts with excess water:

$$B_2H_6(g) + H_2O(l) \longrightarrow H_3BO_3(s) + H_2(g) \text{ [unbalanced]}$$

3.44 Calculate the mass of each product formed when 174 g of silver sulfide reacts with excess hydrochloric acid:

$$Ag_2S(s) + HCl(aq) \longrightarrow AgCl(s) + H_2S(g) \text{ [unbalanced]}$$

3.45 Elemental phosphorus occurs as tetratomic molecules, P_4. What mass of chlorine gas is needed to react completely with 355 g of phosphorus to form phosphorus pentachloride?

3.46 Elemental sulfur occurs as octatomic molecules, S_8. What mass of fluorine gas is needed to react completely with 17.8 g of sulfur to form sulfur hexafluoride?

3.47 Many metals react with oxygen gas to form the metal oxide. For example, calcium reacts as follows:

$$2Ca(s) + O_2(g) \longrightarrow 2CaO(s)$$

You wish to calculate the mass of calcium oxide that can be prepared from 4.20 g of Ca and 2.80 g of O_2.
(a) How many moles of CaO can form from the given mass of Ca?
(b) How many moles of CaO can form from the given mass of O_2?
(c) Which is the limiting reactant?
(d) How many grams of CaO can form?

3.48 Metal hydrides react with water to form hydrogen gas and the metal hydroxide. For example,

$$SrH_2(s) + 2H_2O(l) \longrightarrow Sr(OH)_2(s) + 2H_2(g)$$

You wish to calculate the mass of hydrogen gas that can be prepared from 5.63 g of SrH_2 and 4.80 g of H_2O.
(a) How many moles of H_2 can form from the given mass of SrH_2?
(b) How many moles of H_2 can form from the given mass of H_2O?
(c) Which is the limiting reactant?
(d) How many grams of H_2 can form?

3.49 Calculate the maximum numbers of moles and grams of iodic acid (HIO_3) that can form when 685 g of iodine trichloride reacts with 117.4 g of water:

$$ICl_3 + H_2O \longrightarrow ICl + HIO_3 + HCl \text{ [unbalanced]}$$

What mass of the excess reactant remains?

3.50 Calculate the maximum numbers of moles and grams of H_2S that can form when 158 g of aluminum sulfide reacts with 131 g of water:

$$Al_2S_3 + H_2O \longrightarrow Al(OH)_3 + H_2S \text{ [unbalanced]}$$

What mass of the excess reactant remains?

3.51 When 0.100 mol of carbon is burned in a closed vessel with 8.00 g of oxygen, how many grams of carbon dioxide can form? Which reactant is in excess, and how many grams of it remain after the reaction?

3.52 A mixture of 0.0359 g of hydrogen and 0.0175 mol of oxygen in a closed container is sparked to initiate a reaction. How many grams of water can form? Which reactant is in excess, and how many grams of it remain after the reaction?

3.53 Aluminum nitrite and ammonium chloride react to form aluminum chloride, nitrogen, and water. What mass of each substance is present after 62.5 g of aluminum nitrite and 54.6 g of ammonium chloride react completely?

3.54 Calcium nitrate and ammonium fluoride react to form calcium fluoride, dinitrogen monoxide, and water vapor. What mass of each substance is present after 16.8 g of calcium nitrate and 17.50 g of ammonium fluoride react completely?

3.55 Two successive reactions, A $\longrightarrow$ B and B $\longrightarrow$ C, have yields of 82% and 65%, respectively. What is the overall percent yield for conversion of A to C?

3.56 Two successive reactions, D $\longrightarrow$ E and E $\longrightarrow$ F, have yields of 48% and 73%, respectively. What is the overall percent yield for conversion of D to F?

3.57 What is the percent yield of a reaction in which 41.5 g of tungsten(VI) oxide (WO_3) reacts with excess hydrogen gas to produce metallic tungsten and 9.50 mL of water ($d = 1.00$ g/mL)?

3.58 What is the percent yield of a reaction in which 200. g of phosphorus trichloride reacts with excess water to form 128 g of HCl and aqueous phosphorous acid (H_3PO_3)?

3.59 When 18.5 g of methane and 43.0 g of chlorine gas undergo a reaction that has an 80.0% yield, what mass of chloromethane (CH_3Cl) forms? Hydrogen chloride also forms.

3.60 When 56.6 g of calcium and 30.5 g of nitrogen gas undergo a reaction that has a 93.0% yield, what mass of calcium nitride forms?

3.61 Cyanogen, $(CN)_2$, has been observed in the atmosphere of Titan, Saturn's largest moon, and in the gases of interstellar nebulas. On Earth, it is used as a welding gas and a fumigant. In its reaction with fluorine gas, carbon tetrafluoride and nitrogen trifluoride gases are produced. What mass of carbon tetrafluoride forms when 80.0 g of each reactant is used?

3.62 Gaseous butane is compressed and used as a liquid fuel in disposable cigarette lighters and lightweight camping stoves. Suppose a lighter contains 6.50 mL of butane ($d = 0.579$ g/mL).
(a) How many grams of oxygen are needed to burn the butane completely?
(b) How many moles of CO_2 form when all the butane burns?
(c) How many total molecules of gas form when the butane burns completely?

3.63 Sodium borohydride ($NaBH_4$) is used industrially in many organic syntheses. One way to prepare it is by reacting sodium hydride with gaseous diborane (B_2H_6). Assuming a 95.5% yield, how many grams of $NaBH_4$ can be prepared by reacting 7.88 g of sodium hydride and 8.12 g of diborane?

Fundamentals of Solution Stoichiometry
(Sample Problems 3.12 to 3.16)

3.64 Box A represents a unit volume of a solution. Choose from boxes B and C the one representing the same unit volume of solution that has (a) more solute added; (b) more solvent added; (c) higher molarity; (d) lower concentration.

A   **B**   **C**

3.65 Calculate each of the following quantities:
(a) Grams of solute in 175.8 mL of 0.207 M calcium acetate
(b) Molarity of 500. mL of solution containing 21.1 g of potassium iodide
(c) Moles of solute in 145.6 L of 0.850 M sodium cyanide

3.66 Calculate each of the following quantities:
(a) Volume in liters of 2.26 M potassium hydroxide that contains 8.42 g of solute
(b) Number of Cu^{2+} ions in 52 L of 2.3 M copper(II) chloride
(c) Molarity of 275 mL of solution containing 135 mmol of glucose

3.67 Calculate each of the following quantities:
(a) Molarity of a solution prepared by diluting 37.00 mL of 0.250 M potassium chloride to 150.00 mL
(b) Molarity of a solution prepared by diluting 25.71 mL of 0.0706 M ammonium sulfate to 500.00 mL
(c) Molarity of sodium ion in a solution made by mixing 3.58 mL of 0.288 M sodium chloride with 500. mL of 6.51×10^{-3} M sodium sulfate (assume volumes are additive)

3.68 Calculate each of the following quantities:
(a) Volume of 2.050 M copper(II) nitrate that must be diluted with water to prepare 750.0 mL of a 0.8543 M solution
(b) Volume of 1.03 M calcium chloride that must be diluted with water to prepare 350. mL of a 2.66×10^{-2} M chloride ion solution
(c) Final volume of a 0.0700 M solution prepared by diluting 18.0 mL of 0.155 M lithium carbonate with water

3.69 A sample of concentrated nitric acid has a density of 1.41 g/mL and contains 70.0% HNO_3 by mass.
(a) What mass of HNO_3 is present per liter of solution?
(b) What is the molarity of the solution?

3.70 Concentrated sulfuric acid (18.3 M) has a density of 1.84 g/mL.
(a) How many moles of sulfuric acid are present per milliliter of solution?
(b) What is the mass % of H_2SO_4 in the solution?

3.71 How many milliliters of 0.383 M HCl are needed to react with 16.2 g of $CaCO_3$?

$$2HCl(aq) + CaCO_3(s) \longrightarrow CaCl_2(aq) + CO_2(g) + H_2O(l)$$

3.72 How many grams of NaH_2PO_4 are needed to react with 38.74 mL of 0.275 M NaOH?

$$NaH_2PO_4(s) + 2NaOH(aq) \longrightarrow Na_3PO_4(aq) + 2H_2O(l)$$

3.73 How many grams of solid barium sulfate form when 25.0 mL of 0.160 M barium chloride reacts with 68.0 mL of 0.055 M sodium sulfate? Aqueous sodium chloride is the other product.

3.74 Which reactant is in excess and by how many moles when 350.0 mL of 0.210 M sulfuric acid reacts with 0.500 L of 0.196 M sodium hydroxide to form water and aqueous sodium sulfate?

3.75 Muriatic acid, an industrial grade of concentrated HCl, is used to clean masonry and etch cement for painting. Its concentration is 11.7 M.
(a) Write instructions for diluting the concentrated acid to make 5.0 gallons of 3.5 M acid for routine use (1 gal = 4 qt; 1 qt = 0.946 L).
(b) How many milliliters of the muriatic acid solution contain 9.55 g of HCl?

Comprehensive Problems
Problems with an asterisk (*) are more challenging.

3.76 Narceine is a narcotic in opium. It crystallizes from water solution as a hydrate that contains 10.8 mass % water. If the molar mass of narceine hydrate is 499.52 g/mol, determine x in narceine·xH_2O.

3.77 Hydrogen-containing fuels have a "fuel value" based on their mass % H. Rank the following compounds from highest mass % H to lowest: ethane, propane, benzene, ethanol, cetyl palmitate (whale oil, $C_{32}H_{64}O_2$).

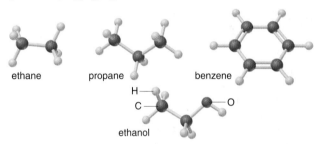

ethane   propane   benzene

ethanol

3.78 Convert the following descriptions of reactions into balanced equations:
(a) In a gaseous reaction, hydrogen sulfide burns in oxygen to form sulfur dioxide and water vapor.
(b) When crystalline potassium chlorate is heated to just above its melting point, it reacts to form two different crystalline compounds, potassium chloride and potassium perchlorate.
(c) When hydrogen gas is passed over powdered iron(III) oxide, iron metal and water vapor form.
(d) The combustion of gaseous ethane in air forms carbon dioxide and water vapor.
(e) Iron(II) chloride is converted to iron(III) fluoride by treatment with chlorine trifluoride gas. Chlorine gas is also formed.

3.79 Isobutylene is a hydrocarbon used in the manufacture of synthetic rubber. When 0.847 g of isobutylene was analyzed by combustion (using an apparatus similar to that in Figure 3.5), the gain in mass of the CO_2 absorber was 2.657 g and that of the H_2O absorber was 1.089 g. What is the empirical formula of isobutylene?

3.80 One of the compounds used to increase the octane rating of gasoline is toluene (*right*). Suppose 15.0 mL of toluene ($d = 0.867$ g/mL) is consumed when a sample of gasoline burns in air.
(a) How many grams of oxygen are needed for complete combustion of the toluene?

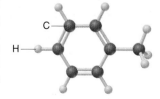

(b) How many total moles of gaseous products form?

(c) How many molecules of water vapor form?

3.81 During studies of the reaction in Sample Problem 3.10,

$$N_2O_4(l) + 2N_2H_4(l) \longrightarrow 3N_2(g) + 4H_2O(g)$$

a chemical engineer measured a less-than-expected yield of N_2 and discovered that the following side reaction occurs:

$$2N_2O_4(l) + N_2H_4(l) \longrightarrow 6NO(g) + 2H_2O(g)$$

In one experiment, 10.0 g of NO formed when 100.0 g of each reactant was used. What is the highest percent yield of N_2 that can be expected?

3.82 The following boxes represent a chemical reaction between AB_2 and B_2:

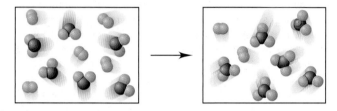

(a) Write a balanced equation for the reaction.

(b) What is the limiting reactant in this reaction?

(c) How many moles of product can be made from 3.0 mol of B_2 and 5.0 mol of AB_2?

(d) How many moles of excess reactant remain after the reaction in part (c)?

3.83 Seawater is approximately 4.0% by mass dissolved ions. About 85% of the mass of the dissolved ions is from NaCl.

(a) Calculate the mass percent of NaCl in seawater.

(b) Calculate the mass percent of Na^+ ions and of Cl^- ions in seawater.

(c) Calculate the molarity of NaCl in seawater at 15°C (*d* of seawater at 15°C = 1.025 g/mL).

3.84 Is each of the following statements true or false? Correct any that are false:

(a) A mole of one substance has the same number of atoms as a mole of any other substance.

(b) The theoretical yield for a reaction is based on the balanced chemical equation.

(c) A limiting-reactant problem is presented when the quantity of available material is given in moles for one of the reactants.

(d) The concentration of a solution is an intensive property, but the amount of solute in a solution is an extensive property.

3.85 Box A represents one unit volume of solution A. Which box—B, C, or D—represents one unit volume after adding enough solvent to solution A to (a) triple its volume; (b) double its volume; (c) quadruple its volume?

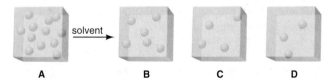

3.86 In each pair, choose the larger of the indicated quantities or state that the samples are equal:

(a) Entities: 0.4 mol of O_3 molecules or 0.4 mol of O atoms

(b) Grams: 0.4 mol of O_3 molecules or 0.4 mol of O atoms

(c) Moles: 4.0 g of N_2O_4 or 3.3 g of SO_2

(d) Grams: 0.6 mol of C_2H_4 or 0.6 mol of F_2

(e) Total ions: 2.3 mol of sodium chlorate or 2.2 mol of magnesium chloride

(f) Molecules: 1.0 g of H_2O or 1.0 g of H_2O_2

(g) Na^+ ions: 0.500 L of 0.500 *M* NaBr or 0.0146 kg of NaCl

(h) Mass: 6.02×10^{23} atoms of ^{235}U or 6.02×10^{23} atoms of ^{238}U

3.87 Balance the equation for the reaction between solid tetraphosphorus trisulfide and oxygen gas to form solid tetraphosphorus decaoxide and gaseous sulfur dioxide. Tabulate the equation (see Table 3.2) in terms of (a) molecules, (b) moles, and (c) grams.

3.88 Hydrogen gas has been suggested as a clean fuel because it produces only water vapor when it burns. If the reaction has a 98.8% yield, what mass of hydrogen forms 85.0 kg of water?

3.89 Assuming that the volumes are additive, what is the concentration of KBr in a solution prepared by mixing 0.200 L of 0.053 *M* KBr with 0.550 L of 0.078 *M* KBr?

3.90 Calculate each of the following quantities:

(a) Moles of compound in 0.588 g of ammonium bromide

(b) Number of potassium ions in 68.5 g of potassium nitrate

(c) Mass in grams of 5.85 mol of glycerol ($C_3H_8O_3$)

(d) Volume of 2.55 mol of chloroform ($CHCl_3$; *d* = 1.48 g/mL)

(e) Number of sodium ions in 2.11 mol of sodium carbonate

(f) Number of atoms in 10.0 µg of cadmium

(g) Number of atoms in 0.0015 mol of fluorine gas

3.91 Elements X (green) and Y (purple) react according to the following equation: $X_2 + 3Y_2 \longrightarrow 2XY_3$. Which molecular scene represents the product of the reaction?

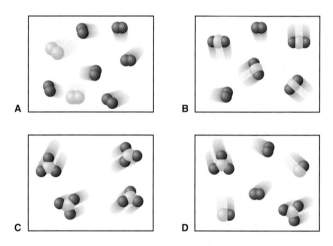

3.92 Hydrocarbon mixtures are used as fuels. How many grams of $CO_2(g)$ are produced by the combustion of 200. g of a mixture that is 25.0% CH_4 and 75.0% C_3H_8 by mass?

* **3.93** Nitrogen (N), phosphorus (P), and potassium (K) are the main nutrients in plant fertilizers. According to an industry convention, the numbers on the label refer to the mass percents of N, P_2O_5, and K_2O, in that order. Calculate the N:P:K ratio of a 30:10:10 fertilizer in terms of moles of each element, and express it as *x*:*y*:1.0.

3.94 A 0.652-g sample of a pure strontium halide reacts with excess sulfuric acid, and the solid strontium sulfate formed is separated, dried, and found to weigh 0.755 g. What is the formula of the original halide?

* **3.95** When carbon-containing compounds are burned in a limited amount of air, some $CO(g)$ as well as $CO_2(g)$ is produced. A gaseous product mixture is 35.0 mass % CO and 65.0 mass % CO_2. What is the mass % C in the mixture?

3.96 Ferrocene, first synthesized in 1951, was the first organic iron compound with Fe—C bonds. An understanding of the structure of ferrocene gave rise to new ideas about chemical bonding and led to the preparation of many useful compounds. In the combustion analysis of ferrocene, which contains only Fe, C, and H, a 0.9437-g sample produced 2.233 g of CO_2 and 0.457 g of H_2O. What is the empirical formula of ferrocene?

* 3.97 Citric acid *(right)* is concentrated in citrus fruits and plays a central metabolic role in nearly every animal and plant cell.
(a) What are the molar mass and formula of citric acid?
(b) How many moles of citric acid are in 1.50 qt of lemon juice (d = 1.09 g/mL) that is 6.82% citric acid by mass?

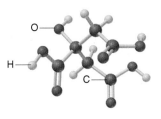

* 3.98 Fluorine is so reactive that it forms compounds with materials inert to other treatments.
(a) When 0.327 g of platinum is heated in fluorine, 0.519 g of a dark red, volatile solid forms. What is its empirical formula?
(b) When 0.265 g of this red solid reacts with excess xenon gas, 0.378 g of an orange-yellow solid forms. What is the empirical formula of this compound, the first noble gas compound formed?
(c) Fluorides of xenon can be formed by direct reaction of the elements at high pressure and temperature. Depending on conditions, the product mixture may include the difluoride, the tetrafluoride, and the hexafluoride. Under conditions that produce only the tetra- and hexafluorides, 1.85×10^{-4} mol of xenon reacted with 5.00×10^{-4} mol of fluorine, and 9.00×10^{-6} mol of xenon was found in excess. What are the mass percents of each xenon fluoride in the product mixture?

3.99 Hemoglobin is 6.0% heme ($C_{34}H_{32}FeN_4O_4$) by mass. To remove the heme, hemoglobin is treated with acetic acid and NaCl to form hemin ($C_{34}H_{32}N_4O_4FeCl$). At a crime scene, a blood sample contains 0.45 g of hemoglobin.
(a) How many grams of heme are in the sample?
(b) How many moles of heme?
(c) How many grams of Fe?
(d) How many grams of hemin could be formed for a forensic chemist to measure?

* 3.100 Manganese is a key component of extremely hard steel. The element occurs naturally in many oxides. A 542.3-g sample of a manganese oxide has an Mn:O ratio of 1.00:1.42 and consists of braunite (Mn_2O_3) and manganosite (MnO).
(a) What masses of braunite and manganosite are in the ore?
(b) What is the ratio $Mn^{3+}:Mn^{2+}$ in the ore?

3.101 Sulfur dioxide is a major industrial gas used primarily for the production of sulfuric acid, but also as a bleach and food preservative. One way to produce it is by roasting iron pyrite (iron disulfide, FeS_2) in oxygen, which yields the gas and solid iron(III) oxide. What mass of each of the other three substances are involved in producing 1.00 kg of sulfur dioxide?

3.102 The human body excretes nitrogen in the form of urea, NH_2CONH_2. The key biochemical step in urea formation is the reaction of water with arginine to produce urea and ornithine:

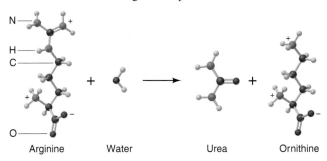

| Arginine | Water | Urea | Ornithine |

(a) What is the mass percent of nitrogen in urea, arginine, and ornithine? (b) How many grams of nitrogen can be excreted as urea when 143.2 g of ornithine is produced?

3.103 Aspirin (acetylsalicylic acid, $C_9H_8O_4$) can be made by reacting salicylic acid ($C_7H_6O_3$) with acetic anhydride [$(CH_3CO)_2O$]:
$$C_7H_6O_3(s) + (CH_3CO)_2O(l) \longrightarrow C_9H_8O_4(s) + CH_3COOH(l)$$
In one reaction, 3.027 g of salicylic acid and 6.00 mL of acetic anhydride react to form 3.261 g of aspirin.
(a) Which is the limiting reactant (d of acetic anhydride = 1.080 g/mL)?
(b) What is the percent yield of this reaction?

* 3.104 When powdered zinc is heated with sulfur, a violent reaction occurs, and zinc sulfide forms:
$$Zn(s) + S_8(s) \longrightarrow ZnS(s) \text{ [unbalanced]}$$
Some of the reactants also combine with oxygen in air to form zinc oxide and sulfur dioxide. When 85.2 g of Zn reacts with 52.4 g of S_8, 105.4 g of ZnS forms. What is the percent yield of ZnS? (b) If all the remaining reactants combine with oxygen, how many grams of each of the two oxides form?

* 3.105 High-temperature superconducting oxides hold great promise in the utility, transportation, and computer industries. (a) One superconductor is $La_{2-x}Sr_xCuO_4$. Calculate the molar mass of this oxide when $x = 0$, $x = 1$, and $x = 0.163$ (the last characterizes the compound with optimum superconducting properties).
(b) Another common superconducting oxide is made by heating a mixture of barium carbonate, copper(II) oxide, and yttrium(III) oxide, followed by further heating in O_2:
$$4BaCO_3(s) + 6CuO(s) + Y_2O_3(s) \longrightarrow$$
$$2YBa_2Cu_3O_{6.5}(s) + 4CO_2(g)$$
$$2YBa_2Cu_3O_{6.5}(s) + \tfrac{1}{2}O_2(g) \longrightarrow 2YBa_2Cu_3O_7(s)$$
When equal masses of the three reactants are heated, which reactant is limiting?
(c) After the product in part (b) is removed, what is the mass percent of each reactant in the solid mixture remaining?

* 3.106 The zirconium oxalate $K_2Zr(C_2O_4)_3(H_2C_2O_4) \cdot H_2O$ was synthesized by mixing 1.60 g of $ZrOCl_2 \cdot 8H_2O$ with 5.20 g of $H_2C_2O_4 \cdot 2H_2O$ and an excess of aqueous KOH. After 2 months, 1.20 g of crystalline product was obtained, as well as aqueous KCl and water. Calculate the percent yield.

CHAPTER FOUR

The Major Classes of Chemical Reactions

Classifying the Countless There are countless chemical reactions but only a few reaction classes. These swirling particles of silver chromate form through one of the three major classes discussed in this chapter.

Key Principles

◆ Many chemical reactions occur in water. Because of the *shape* of its molecules and their *distribution of electrons,* water is a *polar* molecule. In water, many ionic compounds and a few simple covalent compounds, such as HCl, *dissociate* into ions.

◆ To describe an aqueous ionic reaction, chemists use a *net ionic equation* because it eliminates *spectator ions* (those not involved in the reaction) and shows the *actual chemical change* taking place.

◆ *Precipitation reactions* occur when soluble ionic compounds exchange ions (*metathesis*) and form an insoluble product (*precipitate*) in which the ions attract each other so strongly that water cannot pull them apart.

◆ In solution, an *acid* produces H^+ ions, and a *base* produces OH^- ions. In an *acid-base (neutralization) reaction,* the H^+ and the OH^- ions form water. Another way to view such a reaction is to see an acid as a substance that *transfers a proton* (H^+) to a base.

◆ *Oxidation* is defined as *electron loss,* and *reduction* as *electron gain*. In an *oxidation-reduction (redox) reaction,* electrons move from one reactant to the other: the *reducing agent* is oxidized (loses electrons), and the *oxidizing agent* is reduced (gains electrons). Chemists use *oxidation numbers,* the numbers of electrons "owned" by the atoms in reactants and products, to follow these changes.

◆ Many common redox reactions involve *elements* as reactants or products. In an *activity series,* metals are ranked according to their ability to *displace* H_2 (reduce H^+) from water or acid or reduce the cation of another metal from solution.

Outline

Rapid chemical changes occur among gas molecules as sunlight bathes the atmosphere or lightning rips through a stormy sky. Aqueous reactions go on unceasingly in the gigantic containers we know as oceans. And, in every cell of your body, thousands of reactions taking place right now allow you to function. Indeed, the amazing variety in nature is largely a consequence of the amazing variety of chemical reactions.

Of the millions of chemical reactions occurring in and around you, we have examined only a tiny fraction so far, and it would be impossible to examine them all. Fortunately, it isn't necessary to catalog every reaction, because when we survey even a small percentage of reactions, a few major patterns emerge. In this chapter, we examine the underlying nature of the three most common reaction processes. Many reactions occur in aqueous solution, so first we'll highlight the importance of water.

Concepts & Skills to Review
Before You Study This Chapter

- names and formulas of compounds (Section 2.8)
- nature of ionic and covalent bonding (Section 2.7)
- mole-mass-number conversions (Section 3.1)
- molarity and mole-volume conversions (Section 3.5)
- balancing chemical equations (Section 3.3)
- calculating amounts of reactants and products (Section 3.4)

4.1 THE ROLE OF WATER AS A SOLVENT

Our first step toward comprehending aqueous reactions is to understand how water acts as a solvent. The role a solvent plays in a reaction depends on its chemical nature. Some solvents play a passive role, dispersing the dissolved substances into individual molecules but doing nothing further. Water plays a much more active role, interacting strongly with the substances and, in some cases, even reacting with them. To understand this active role, we'll examine the structure of water and how it interacts with ionic and covalent solutes.

The Polar Nature of Water

Of the many thousands of reactions that occur in the environment and in organisms, nearly all take place in water. Water's remarkable power as a solvent results from two features of its molecules: *the distribution of the bonding electrons* and *the overall shape.*

Recall from Section 2.7 that the electrons in a covalent bond are shared between the bonded atoms. In a covalent bond between identical atoms (as in H_2, Cl_2, O_2, etc.), the sharing is equal, so no imbalance of charge appears (Figure 4.1A). On the other hand, in covalent bonds between nonidentical atoms, the sharing is unequal: one atom attracts the electron pair more strongly than the other. For reasons discussed in Chapter 9, an O atom attracts electrons more strongly than an H atom. Therefore, in each O—H bond in water, the shared electrons spend more time closer to the O atom.

This unequal distribution of negative charge creates partially charged "poles" at the ends of each O—H bond (Figure 4.1B). The O end acts as a slightly negative pole (represented by the red shading and the $\delta-$), and the H end acts as a slightly positive pole (represented by the blue shading and the $\delta+$). Figure 4.1C indicates the bond's polarity with a *polar arrow* (the arrowhead points to the negative pole and the tail is crossed to make a "plus").

The H—O—H arrangement forms an angle, so the water molecule is bent. The combined effects of its bent shape and its polar bonds make water a **polar molecule:** the O portion of the molecule is the partially negative pole, and the region midway between the H atoms is the partially positive pole (Figure 4.1D).

Ionic Compounds in Water

In an ionic solid, the oppositely charged ions are held next to each other by electrostatic attraction (see Figure 1.3C). *Water separates the ions by replacing that attraction with one between the water molecules and the ions.* Imagine a granule

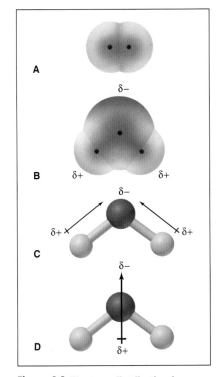

Figure 4.1 Electron distribution in molecules of H_2 and H_2O. A, In H_2, the identical nuclei attract the electrons equally. The central region of higher electron density (red) is balanced by the two outer regions of lower electron density (blue). **B,** In H_2O, the O nucleus attracts the shared electrons more strongly than the H nucleus. **C,** In this ball-and-stick model, a polar arrow points to the negative end of each O—H bond. **D,** The two polar O—H bonds and the bent shape give rise to the polar H_2O molecule.

Figure 4.2 The dissolution of an ionic compound. When an ionic compound dissolves in water, H_2O molecules separate, surround, and disperse the ions into the liquid. The negative ends of the H_2O molecules face the positive ions and the positive ends face the negative ions.

Animation: Dissolution of an Ionic and a Covalent Compound
Online Learning Center

of an ionic compound surrounded by bent, polar water molecules. The negative ends of some water molecules are attracted to the cations, and the positive ends of others are attracted to the anions (Figure 4.2). Gradually, the attraction between each ion and the nearby water molecules outweighs the attraction of the ions for each other. By this process, the ions separate (dissociate) and become **solvated,** surrounded tightly by solvent molecules, as they move randomly in the solution. A similar scene occurs whenever an ionic compound dissolves in water.

Although many ionic compounds dissolve in water, many others do not. In the latter cases, the electrostatic attraction among ions in the compound remains greater than the attraction between ions and water molecules, so the solid stays largely intact. Actually, these so-called insoluble substances *do* dissolve to a very small extent, usually several orders of magnitude less than so-called soluble substances. Compare, for example, the solubilities of NaCl (a "soluble" compound) and AgCl (an "insoluble" compound):

Solubility of NaCl in H_2O at 20°C = 365 g/L

Solubility of AgCl in H_2O at 20°C = 0.009 g/L

When an ionic compound dissolves, an important change occurs in the solution. Figure 4.3 shows this change with a simple apparatus that demonstrates *electrical conductivity,* the flow of electric current. When the electrodes are immersed in pure water or pushed into an ionic solid, such as potassium bromide (KBr), no current flows. In an aqueous KBr solution, however, a significant current flows, as shown by the brightly lit bulb. This current flow implies the *movement of charged particles:* when KBr dissolves in water, the K^+ and Br^- ions dissociate, become solvated, and move toward the electrode of opposite charge. A substance that conducts a current when dissolved in water is an **electrolyte.** Soluble ionic compounds are called *strong* electrolytes because they dissociate completely into ions and create a large current. We express the dissociation of KBr into solvated ions in water as follows:

$$KBr(s) \xrightarrow{\text{H}_2\text{O}} K^+(aq) + Br^-(aq)$$

The "H_2O" above the arrow indicates that water is required as the solvent but is not a reactant in the usual sense.

The formula of the compound tells the number of moles of different ions that result when the compound dissolves. Thus, 1 mol of KBr dissociates into 2 mol of ions—1 mol of K^+ and 1 mol of Br^-. The upcoming sample problem goes over this idea.

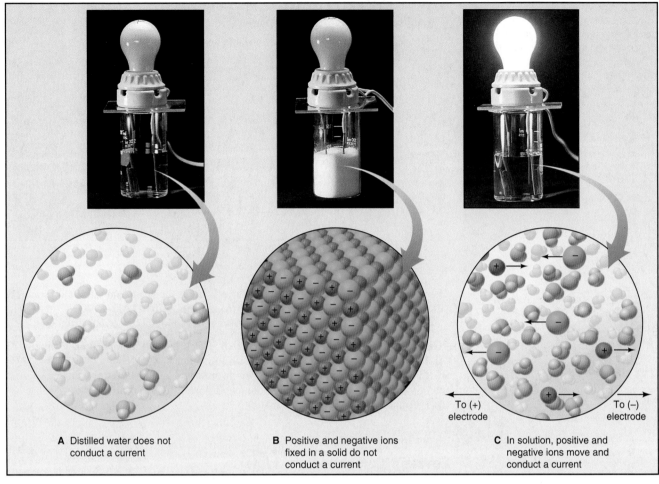

A Distilled water does not conduct a current

B Positive and negative ions fixed in a solid do not conduct a current

C In solution, positive and negative ions move and conduct a current

Figure 4.3 The electrical conductivity of ionic solutions. A, When electrodes connected to a power source are placed in distilled water, no current flows and the bulb is unlit. **B,** A solid ionic compound, such as KBr, conducts no current because the ions are bound tightly together. **C,** When KBr dissolves in H_2O, the ions separate and move through the solution toward the oppositely charged electrodes, thereby conducting a current.

Animation: Strong Electrolytes, Weak Electrolytes, and Nonelectrolytes Online Learning Center

SAMPLE PROBLEM 4.1 Determining Moles of Ions in Aqueous Ionic Solutions

Problem How many moles of each ion are in each solution?
(a) 5.0 mol of ammonium sulfate dissolved in water
(b) 78.5 g of cesium bromide dissolved in water
(c) 7.42×10^{22} formula units of copper(II) nitrate dissolved in water
(d) 35 mL of 0.84 M zinc chloride

Plan We write an equation that shows 1 mol of compound dissociating into ions. In (a), we multiply the moles of ions by 5.0. In (b), we first convert grams to moles. In (c), we first convert formula units to moles. In (d), we first convert molarity and volume to moles.

Solution (a) $(NH_4)_2SO_4(s) \xrightarrow{H_2O} 2NH_4^+(aq) + SO_4^{2-}(aq)$

Remember that, in general, *polyatomic ions remain as intact units in solution.*
Calculating moles of NH_4^+ ions:

$$\text{Moles of } NH_4^+ = 5.0 \text{ mol } (NH_4)_2SO_4 \times \frac{2 \text{ mol } NH_4^+}{1 \text{ mol } (NH_4)_2SO_4} = \boxed{10. \text{ mol } NH_4^+}$$

$\boxed{5.0 \text{ mol of } SO_4^{2-}}$ is also present.

(b) $CsBr(s) \xrightarrow{H_2O} Cs^+(aq) + Br^-(aq)$
Converting from grams to moles:

$$\text{Moles of } CsBr = 78.5 \text{ g } CsBr \times \frac{1 \text{ mol } CsBr}{212.8 \text{ g } CsBr} = 0.369 \text{ mol } CsBr$$

Thus, $\boxed{0.369 \text{ mol of } Cs^+ \text{ and } 0.369 \text{ mol of } Br^-}$ are present.

(c) $Cu(NO_3)_2(s) \xrightarrow{H_2O} Cu^{2+}(aq) + 2NO_3^-(aq)$
Converting from formula units to moles:

$$\text{Moles of } Cu(NO_3)_2 = 7.42 \times 10^{22} \text{ formula units } Cu(NO_3)_2$$

$$\times \frac{1 \text{ mol } Cu(NO_3)_2}{6.022 \times 10^{23} \text{ formula units } Cu(NO_3)_2}$$

$$= 0.123 \text{ mol } Cu(NO_3)_2$$

$$\text{Moles of } NO_3^- = 0.123 \text{ mol } Cu(NO_3)_2 \times \frac{2 \text{ mol } NO_3^-}{1 \text{ mol } Cu(NO_3)_2} = 0.246 \text{ mol } NO_3^-$$

0.123 mol of Cu^{2+} is also present.

(d) $ZnCl_2(aq) \longrightarrow Zn^{2+}(aq) + 2Cl^-(aq)$
Converting from liters to moles:

$$\text{Moles of } ZnCl_2 = 35 \text{ mL} \times \frac{1 \text{ L}}{10^3 \text{ mL}} \times \frac{0.84 \text{ mol } ZnCl_2}{1 \text{ L}} = 2.9 \times 10^{-2} \text{ mol } ZnCl_2$$

$$\text{Moles of } Cl^- = 2.9 \times 10^{-2} \text{ mol } ZnCl_2 \times \frac{2 \text{ mol } Cl^-}{1 \text{ mol } ZnCl_2} = 5.8 \times 10^{-2} \text{ mol } Cl^-$$

2.9×10^{-2} mol of Zn^{2+} is also present.

Check After you round off to check the math, see if the relative moles of ions are consistent with the formula. For instance, in **(a)**, 10 mol NH_4^+/5.0 mol SO_4^{2-} = 2 NH_4^+/1 SO_4^{2-}, or $(NH_4)_2SO_4$. In **(d)**, 0.029 mol Zn^{2+}/0.058 mol Cl^- = 1 Zn^{2+}/2 Cl^-, or $ZnCl_2$.

FOLLOW-UP PROBLEM 4.1 How many moles of each ion are in each solution?
(a) 2 mol of potassium perchlorate dissolved in water
(b) 354 g of magnesium acetate dissolved in water
(c) 1.88×10^{24} formula units of ammonium chromate dissolved in water
(d) 1.32 L of 0.55 M sodium bisulfate

Covalent Compounds in Water

Water dissolves many covalent compounds also. Table sugar (sucrose, $C_{12}H_{22}O_{11}$), beverage (grain) alcohol (ethanol, CH_3CH_2OH), and automobile antifreeze (ethylene glycol, $HOCH_2CH_2OH$) are some familiar examples. All contain their own polar bonds, which interact with those of water. However, even though these substances dissolve, they do not dissociate into ions but remain as intact molecules. As a result, their aqueous solutions do not conduct an electric current, and these substances are called **nonelectrolytes.** Many other covalent substances, such as benzene (C_6H_6) and octane (C_8H_{18}), do not contain polar bonds, and these substances do not dissolve appreciably in water.

Acids and the Solvated Proton A small, but extremely important, group of H-containing covalent compounds interacts so strongly with water that their molecules *do* dissociate into ions. In aqueous solution, these substances are *acids,* as you'll see shortly. The molecules contain polar bonds to hydrogen, in which the atom bonded to H pulls more strongly on the shared electron pair. A good example is hydrogen chloride gas. The Cl end of the HCl molecule is partially negative, and the H end is partially positive. When HCl dissolves in water, the partially charged poles of H_2O molecules are attracted to the oppositely charged poles of HCl. The H—Cl bond breaks, with the H becoming the solvated cation $H^+(aq)$ (but see the discussion following the next sample problem) and the Cl becoming the solvated anion $Cl^-(aq)$. Hydrogen bromide behaves similarly when it dissolves in water:

$$HBr(g) \xrightarrow{H_2O} H^+(aq) + Br^-(aq)$$

SAMPLE PROBLEM 4.2   Determining the Molarity of H$^+$ Ions in an Aqueous Solution of an Acid

Problem Nitric acid is a major chemical in the fertilizer and explosives industries. In aqueous solution, each molecule dissociates and the H becomes a solvated H$^+$ ion. What is the molarity of H$^+$(aq) in 1.4 M nitric acid?

Plan We know the molarity of acid (1.4 M), so we just need the formula to find the number of moles of H$^+$(aq) present in 1 L of solution.

Solution Nitrate ion is NO$_3$$^-$, so nitric acid is HNO$_3$. Thus, 1 mol of H$^+$(aq) is released per mole of acid:

$$HNO_3(l) \xrightarrow{H_2O} H^+(aq) + NO_3^-(aq)$$

Therefore, 1.4 M HNO$_3$ contains 1.4 mol of H$^+$(aq) per liter and is 1.4 M H$^+$(aq).

FOLLOW-UP PROBLEM 4.2 How many moles of H$^+$(aq) are present in 451 mL of 3.20 M hydrobromic acid?

Water interacts strongly with many ions, but most strongly with the hydrogen cation, H$^+$, a very unusual species. The H atom is a proton surrounded by an electron, so the H$^+$ ion is just a proton. Because its full positive charge is concentrated in such a tiny volume, H$^+$ attracts the negative pole of surrounding water molecules so strongly that it actually forms a covalent bond to one of them. We usually show this interaction by writing the aqueous H$^+$ ion as H$_3$O$^+$ (hydronium ion). Thus, to show more accurately what takes place when HBr(g) dissolves, we should write

$$HBr(g) + H_2O(l) \longrightarrow H_3O^+(aq) + Br^-(aq)$$

SECTION SUMMARY
Water plays an active role in dissolving ionic compounds because it consists of polar molecules that are attracted to the ions. When an ionic compound dissolves in water, the ions dissociate from each other and become solvated by water molecules. Because the ions are free to move, their solutions conduct electricity. Water also dissolves many covalent substances with polar bonds. It interacts with some H-containing molecules so strongly it breaks their bonds and dissociates them into H$^+$(aq) ions and anions. In water, the H$^+$ ion is bonded to an H$_2$O, forming H$_3$O$^+$.

4.2   WRITING EQUATIONS FOR AQUEOUS IONIC REACTIONS

Chemists use three types of equations to represent aqueous ionic reactions: molecular, total ionic, and net ionic equations. As you'll see in the two types of ionic equations, by balancing the atoms, we also balance the charges.

Let's examine a reaction to see what each of these equations shows. When solutions of silver nitrate and sodium chromate are mixed, the brick-red solid silver chromate (Ag$_2$CrO$_4$) forms. Figure 4.4 (on the next page) depicts three views of this reaction: the change you would see if you mixed these solutions in the lab, how you might imagine the change at the atomic level among the ions, and how you can symbolize the change with the three types of equations. (The ions that are reacting are shown in red type.)

The **molecular equation** (*top*) reveals the least about the species in solution and is actually somewhat misleading because *it shows all the reactants and products as if they were intact, undissociated compounds*:

$$2AgNO_3(aq) + Na_2CrO_4(aq) \longrightarrow Ag_2CrO_4(s) + 2NaNO_3(aq)$$

Only by examining the state-of-matter designations (*s*) and (*aq*), can you tell the change that has occurred.

Figure 4.4 An aqueous ionic reaction and its equations. When silver nitrate and sodium chromate solutions are mixed, a reaction occurs that forms solid silver chromate and a solution of sodium nitrate. The photos present the macroscopic view of the reaction, the view the chemist sees in the lab. The blow-up arrows lead to an atomic-scale view, a representation of the chemist's mental picture of the reactants and products. (The pale ions are spectator ions, present for electrical neutrality, but not involved in the reaction.) Three equations represent the reaction in symbols. (The ions that are reacting are shown in red type.) The *molecular equation* shows all substances intact. The *total ionic equation* shows all *soluble* substances as separate, solvated ions. The *net ionic equation* eliminates the spectator ions to show only the reacting species.

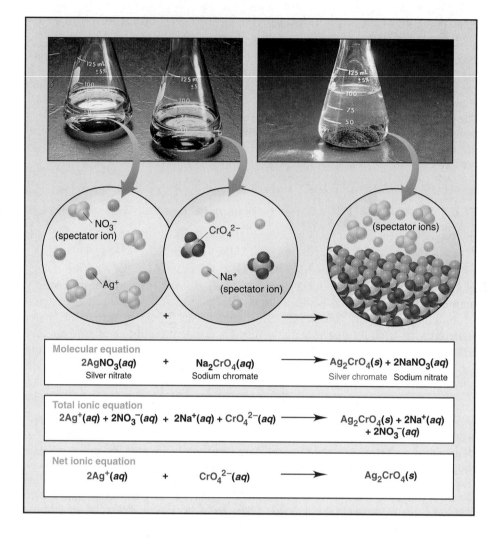

The **total ionic equation** (*middle*) is a much more accurate representation of the reaction because *it shows all the soluble ionic substances dissociated into ions.* Now the $Ag_2CrO_4(s)$ stands out as the only undissociated substance:

$$2Ag^+(aq) + 2NO_3^-(aq) + 2Na^+(aq) + CrO_4^{2-}(aq) \longrightarrow$$
$$Ag_2CrO_4(s) + 2Na^+(aq) + 2NO_3^-(aq)$$

Notice that charges balance: there are four positive and four negative charges on the left for a net zero charge, and there are two positive and two negative charges on the right for a net zero charge.

Note that $Na^+(aq)$ and $NO_3^-(aq)$ appear in the same form on both sides of the equation. They are called **spectator ions** because they are not involved in the actual chemical change. These ions are present as part of the reactants to balance the charge. That is, we can't add an Ag^+ ion without also adding an anion, in this case, NO_3^- ion.

The **net ionic equation** (*bottom*) is the most useful because *it eliminates the spectator ions and shows the actual chemical change taking place:*

$$2Ag^+(aq) + CrO_4^{2-}(aq) \longrightarrow Ag_2CrO_4(s)$$

The formation of solid silver chromate from silver ions and chromate ions *is* the only change. In fact, if we had originally mixed solutions of potassium chromate, $K_2CrO_4(aq)$, and silver acetate, $AgC_2H_3O_2(aq)$, instead of sodium chromate and silver nitrate, the same change would have occurred. Only the spectator ions would differ—$K^+(aq)$ and $C_2H_3O_2^-(aq)$ instead of $Na^+(aq)$ and $NO_3^-(aq)$.

Now, let's apply these types of equations to the three most important types of chemical reactions—precipitation, acid-base, and oxidation-reduction.

SECTION SUMMARY

A molecular equation for an aqueous ionic reaction shows undissociated substances. A total ionic equation shows all soluble ionic compounds as separate, solvated ions. Spectator ions appear unchanged on both sides of the equation. By eliminating them, you see the actual chemical change in a net ionic equation.

4.3 PRECIPITATION REACTIONS

Precipitation reactions are common in both nature and commerce. Many geological formations, including coral reefs, some gems and minerals, and deep-sea structures form, in part, through this type of chemical process. And the chemical industry employs precipitation methods to produce several important inorganic compounds.

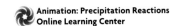
Animation: Precipitation Reactions
Online Learning Center

The Key Event: Formation of a Solid from Dissolved Ions

In **precipitation reactions,** two soluble ionic compounds react to form an insoluble product, a **precipitate.** The reaction you just saw between silver nitrate and sodium chromate is an example. Precipitates form for the same reason that some ionic compounds do not dissolve: the electrostatic attraction between the ions outweighs the tendency of the ions to remain solvated and move randomly throughout the solution. When solutions of such ions are mixed, the ions collide and stay together, and the resulting substance "comes out of solution" as a solid, as shown in Figure 4.5 for calcium fluoride. Thus, the key event in a precipitation reaction is *the formation of an insoluble product through the net removal of solvated ions from solution.*

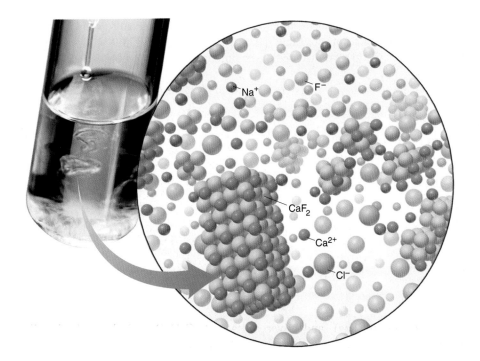

Figure 4.5 The precipitation of calcium fluoride. When an aqueous solution of NaF is added to a solution of $CaCl_2$, Ca^{2+} and F^- ions form particles of solid CaF_2.

Predicting Whether a Precipitate Will Form

If you mix aqueous solutions of two ionic compounds, can you predict if a precipitate will form? Consider this example. When solid sodium iodide and potassium nitrate are each dissolved in water, each solution consists of separated ions dispersed throughout the solution:

$$NaI(s) \xrightarrow{H_2O} Na^+(aq) + I^-(aq)$$
$$KNO_3(s) \xrightarrow{H_2O} K^+(aq) + NO_3^-(aq)$$

Let's follow three steps to predict whether a precipitate will form:

1. *Note the ions present in the reactants.* The reactant ions are

$$Na^+(aq) + I^-(aq) + K^+(aq) + NO_3^-(aq) \longrightarrow ?$$

2. *Consider the possible cation-anion combinations.* In addition to the two original ones, NaI and KNO_3, which you know are soluble, the other possible cation-anion combinations are $NaNO_3$ and KI.

3. *Decide whether any of the combinations is insoluble.* A reaction does *not* occur when you mix these starting solutions because all the combinations—NaI, KNO_3, $NaNO_3$, and KI—are soluble. All the ions remain in solution. (You'll see shortly a set of rules for deciding if a product is soluble or not.)

Now, what happens if you substitute a solution of lead(II) nitrate, $Pb(NO_3)_2$, for the KNO_3? The reactant ions are Na^+, I^-, Pb^{2+}, and NO_3^-. In addition to the two soluble reactants, NaI and $Pb(NO_3)_2$, the other two possible cation-anion combinations are $NaNO_3$ and PbI_2. Lead(II) iodide is insoluble, so a reaction *does* occur because ions are removed from solution (Figure 4.6):

$$2Na^+(aq) + 2I^-(aq) + Pb^{2+}(aq) + 2NO_3^-(aq) \longrightarrow 2Na^+(aq) + 2NO_3^-(aq) + PbI_2(s)$$

A close look (with color) at the molecular equation shows that *the ions are exchanging partners*:

$$2NaI(aq) + Pb(NO_3)_2(aq) \longrightarrow PbI_2(s) + 2NaNO_3(aq)$$

Such reactions are called double-displacement, or **metathesis** (pronounced *meh-TA-thuh-sis*) **reactions.** Several are important in industry, such as the preparation of silver bromide for the manufacture of black-and-white film:

$$AgNO_3(aq) + KBr(aq) \longrightarrow AgBr(s) + KNO_3(aq)$$

As we said, there is no simple way to decide whether any given ion combination is soluble or not, so Table 4.1 provides a short list of solubility rules to memorize. They allow you to predict the outcome of many precipitation reactions.

Figure 4.6 The reaction of $Pb(NO_3)_2$ and NaI. When aqueous solutions of these ionic compounds are mixed, the yellow solid PbI_2 forms.

Table 4.1	Solubility Rules for Ionic Compounds in Water
Soluble Ionic Compounds	**Insoluble Ionic Compounds**
1. All common compounds of Group 1A(1) ions (Li$^+$, Na$^+$, K$^+$, etc.) and ammonium ion (NH$_4^+$) are soluble.	1. All common metal hydroxides are insoluble, *except* those of Group 1A(1) and the larger members of Group 2A(2) (beginning with Ca^{2+}).
2. All common nitrates (NO$_3^-$), acetates (CH$_3$COO$^-$ or C$_2$H$_3$O$_2^-$), and most perchlorates (ClO$_4^-$) are soluble.	2. All common carbonates (CO$_3^{2-}$) and phosphates (PO$_4^{3-}$) are insoluble, *except* those of Group 1A(1) and NH$_4^+$.
3. All common chlorides (Cl$^-$), bromides (Br$^-$), and iodides (I$^-$) are soluble, *except* those of Ag$^+$, Pb^{2+}, Cu$^+$, and Hg$_2^{2+}$. All common fluorides (F$^-$) are soluble, *except* those of Pb^{2+} and Group 2A(2).	3. All common sulfides are insoluble *except* those of Group 1A(1), Group 2A(2), and NH$_4^+$.
4. All common sulfates (SO$_4^{2-}$) are soluble, *except* those of Ca^{2+}, Sr^{2+}, Ba^{2+}, Ag$^+$, and Pb^{2+}.	

SAMPLE PROBLEM 4.3 Predicting Whether a Precipitation Reaction Occurs; Writing Ionic Equations

Problem Predict whether a reaction occurs when each of the following pairs of solutions are mixed. If a reaction does occur, write balanced molecular, total ionic, and net ionic equations, and identify the spectator ions.

(a) Sodium sulfate(*aq*) + strontium nitrate(*aq*) $\longrightarrow$

(b) Ammonium perchlorate(*aq*) + sodium bromide(*aq*) $\longrightarrow$

Plan For each pair of solutions, we note the ions present in the reactants, write the cation-anion combinations, and refer to Table 4.1 to see if any are insoluble. For the molecular equation, we predict the products. For the total ionic equation, we write the soluble compounds as separate ions. For the net ionic equation, we eliminate the spectator ions.

Solution (a) In addition to the reactants, the two other ion combinations are strontium sulfate and sodium nitrate. Table 4.1 shows that strontium sulfate is insoluble, so a reaction *does* occur. Writing the molecular equation:

$$Na_2SO_4(aq) + Sr(NO_3)_2(aq) \longrightarrow SrSO_4(s) + 2NaNO_3(aq)$$

Writing the total ionic equation:

$$2Na^+(aq) + SO_4^{2-}(aq) + Sr^{2+}(aq) + 2NO_3^-(aq) \longrightarrow$$
$$SrSO_4(s) + 2Na^+(aq) + 2NO_3^-(aq)$$

Writing the net ionic equation:

$$Sr^{2+}(aq) + SO_4^{2-}(aq) \longrightarrow SrSO_4(s)$$

The spectator ions are Na^+ and NO_3^-.

(b) The other ion combinations are ammonium bromide and sodium perchlorate. Table 4.1 shows that all ammonium, sodium, and most perchlorate compounds are soluble, and all bromides are soluble except those of Ag^+, Pb^{2+}, Cu^+, and Hg_2^{2+}. Therefore, *no* reaction occurs. The compounds remain dissociated in solution as solvated ions.

FOLLOW-UP PROBLEM 4.3 Predict whether a reaction occurs, and write balanced total and net ionic equations:

(a) Iron(III) chloride(*aq*) + cesium phosphate(*aq*) $\longrightarrow$

(b) Sodium hydroxide(*aq*) + cadmium nitrate(*aq*) $\longrightarrow$

(c) Magnesium bromide(*aq*) + potassium acetate(*aq*) $\longrightarrow$

(d) Silver sulfate(*aq*) + barium chloride(*aq*) $\longrightarrow$

SECTION SUMMARY

Precipitation reactions involve the formation of an insoluble ionic compound from two soluble ones. They occur because electrostatic attractions among certain pairs of solvated ions are strong enough to cause their removal from solution. Such reactions can be predicted by noting whether any possible ion combinations are insoluble, based on a set of solubility rules.

4.4 ACID-BASE REACTIONS

Aqueous acid-base reactions involve water not only as solvent but also in the more active roles of reactant and product. These reactions occur in processes as diverse as the biochemical synthesis of proteins, the industrial production of fertilizer, and some of the methods for revitalizing lakes damaged by acid rain.

Obviously, an **acid-base reaction** (also called a **neutralization reaction**) occurs when an acid reacts with a base, but the definitions of these terms and the scope of this reaction class have changed considerably over the years. For our purposes at this point, we'll use definitions that apply to chemicals you commonly encounter in the lab:

- *An **acid** is a substance that produces H^+ ions when dissolved in water.*

$$HX \xrightarrow{H_2O} H^+(aq) + X^-(aq)$$

Table 4.2	Strong and Weak Acids and Bases

Acids

Strong

Hydrochloric acid, HCl
Hydrobromic acid, HBr
Hydriodic acid, HI
Nitric acid, HNO_3
Sulfuric acid, H_2SO_4
Perchloric acid, $HClO_4$

Weak

Hydrofluoric acid, HF
Phosphoric acid, H_3PO_4
Acetic acid, CH_3COOH
(or $HC_2H_3O_2$)

Bases

Strong

Sodium hydroxide, NaOH
Potassium hydroxide, KOH
Calcium hydroxide, $Ca(OH)_2$
Strontium hydroxide, $Sr(OH)_2$
Barium hydroxide, $Ba(OH)_2$

Weak

Ammonia, NH_3

Strong acids and bases are strong electrolytes.

Weak acids and bases are weak electrolytes.

• A **base** is a substance that produces OH^- ions when dissolved in water.

$$MOH \xrightarrow{H_2O} M^+(aq) + OH^-(aq)$$

(Other definitions of *acid* and *base* are presented later in this section and again in Chapter 18, along with a fuller meaning of *neutralization*.)

Acids and bases are electrolytes. Table 4.2 lists some acids and bases categorized in terms of their "strength"—the degree to which they dissociate into ions in aqueous solution. In water, *strong acids and strong bases dissociate completely* into ions. Therefore, like soluble ionic compounds, they are *strong* electrolytes and conduct a current well (see left photo in margin). In contrast, *weak acids and weak bases dissociate into ions very little, and most of their molecules remain intact.* As a result, they conduct only a small current and are *weak* electrolytes (see right photo).

Strong and weak acids have one or more H atoms as part of their structure. Strong bases have either the OH^- or the O^{2-} ion as part of their structure. Soluble ionic oxides, such as K_2O, are strong bases because the oxide ion is not stable in water and reacts immediately to form hydroxide ion:

$$K_2O(s) + H_2O(l) \longrightarrow 2K^+(aq) + 2OH^-(aq)$$

Weak bases, such as ammonia, do not contain OH^- ions, but they produce them in a reaction with water that occurs to a small extent:

$$NH_3(g) + H_2O(l) \rightleftharpoons NH_4^+(aq) + OH^-(aq)$$

(Note the reaction arrow in the preceding equation. This type of arrow indicates that the reaction proceeds in both directions; we'll discuss this important idea further in Chapter 17.)

The Key Event: Formation of H_2O from H^+ and OH^-

Let's use ionic equations to see what occurs in acid-base reactions. We begin with the molecular equation for the reaction between the strong acid HCl and the strong base $Ba(OH)_2$:

$$2HCl(aq) + Ba(OH)_2(aq) \longrightarrow BaCl_2(aq) + 2H_2O(l)$$

Because HCl and $Ba(OH)_2$ dissociate completely and H_2O remains undissociated, the total ionic equation is

$$2H^+(aq) + 2Cl^-(aq) + Ba^{2+}(aq) + 2OH^-(aq) \longrightarrow Ba^{2+}(aq) + 2Cl^-(aq) + 2H_2O(l)$$

In the net ionic equation, we eliminate the spectator ions $Ba^{2+}(aq)$ and $Cl^-(aq)$ and see the actual reaction:

$$2H^+(aq) + 2OH^-(aq) \longrightarrow 2H_2O(l)$$

Or

$$H^+(aq) + OH^-(aq) \longrightarrow H_2O(l)$$

Thus, *the essential change in all aqueous reactions between a strong acid and a strong base is that an H^+ ion from the acid and an OH^- ion from the base form a water molecule. In fact,* only the spectator ions differ from one strong acid–strong base reaction to another.

Now it's easy to understand how these reactions take place: like precipitation reactions, acid-base reactions occur through *the electrostatic attraction of ions and their removal from solution* in the formation of the product. In this case, the ions are H^+ and OH^- and the product is H_2O, which consists almost entirely of undissociated molecules. (Actually, water molecules *do* dissociate, but *very* slightly. As you'll see in Chapter 18, this slight dissociation is very important, but the formation of water in a neutralization reaction nevertheless represents an enormous net removal of H^+ and OH^- ions.)

Evaporate the water from the above reaction mixture, and the ionic solid barium chloride remains. An ionic compound that results from the reaction of an acid and a base is called a **salt.** Thus, in a typical aqueous neutralization reaction, *the reactants are an acid and a base, and the products are a salt solution and water:*

$$HX(aq) + MOH(aq) \longrightarrow MX(aq) + H_2O(l)$$

$$\quad\text{acid}\qquad\quad\text{base}\qquad\qquad\text{salt}\qquad\text{water}$$

The color shows that *the cation of the salt comes from the base and the anion comes from the acid.*

As you can see, acid-base reactions, like precipitation reactions, are metathesis (double-displacement) reactions. The molecular equation for the reaction of aluminum hydroxide, the active ingredient in some antacid tablets, with HCl, the major component of stomach acid, shows this clearly:

$$3HCl(aq) + Al(OH)_3(s) \longrightarrow AlCl_3(aq) + 3H_2O(l)$$

Acid-base reactions occur frequently in the synthesis and breakdown of large biological molecules.

SAMPLE PROBLEM 4.4   Writing Ionic Equations for Acid-Base Reactions

Problem Write balanced molecular, total ionic, and net ionic equations for each of the following acid-base reactions and identify the spectator ions:
(a) Strontium hydroxide(*aq*) + perchloric acid(*aq*) $\longrightarrow$
(b) Barium hydroxide(*aq*) + sulfuric acid(*aq*) $\longrightarrow$
Plan All are strong acids and bases (see Table 4.2), so the essential reaction is between H^+ and OH^-. The products are H_2O and a salt solution consisting of the spectator ions. Note that in (b), the salt ($BaSO_4$) is insoluble (see Table 4.1), so virtually all ions are removed from solution.
Solution (a) Writing the molecular equation:

$$Sr(OH)_2(aq) + 2HClO_4(aq) \longrightarrow Sr(ClO_4)_2(aq) + 2H_2O(l)$$

Writing the total ionic equation:

$$Sr^{2+}(aq) + 2OH^-(aq) + 2H^+(aq) + 2ClO_4^-(aq) \longrightarrow$$
$$Sr^{2+}(aq) + 2ClO_4^-(aq) + 2H_2O(l)$$

Writing the net ionic equation:

$$2OH^-(aq) + 2H^+(aq) \longrightarrow 2H_2O(l) \quad \text{or} \quad OH^-(aq) + H^+(aq) \longrightarrow H_2O(l)$$

$Sr^{2+}(aq)$ and $ClO_4^-(aq)$ are the spectator ions.
(b) Writing the molecular equation:

$$Ba(OH)_2(aq) + H_2SO_4(aq) \longrightarrow BaSO_4(s) + 2H_2O(l)$$

Writing the total ionic equation:

$$Ba^{2+}(aq) + 2OH^-(aq) + 2H^+(aq) + SO_4^{2-}(aq) \longrightarrow BaSO_4(s) + 2H_2O(l)$$

The net ionic equation is the same as the total ionic equation. This is a precipitation *and* a neutralization reaction. There are no spectator ions because all the ions are used to form the two products.

FOLLOW-UP PROBLEM 4.4 Write balanced molecular, total ionic, and net ionic equations for the reaction between aqueous solutions of calcium hydroxide and nitric acid.

Acid-Base Titrations

Chemists study acid-base reactions quantitatively through titrations. In any **titration,** *one solution of known concentration is used to determine the concentration of another solution through a monitored reaction.*

Figure 4.7 An acid-base titration.
A, In this procedure, a measured volume of the unknown acid solution is placed in a flask beneath a buret containing the known (standardized) base solution. A few drops of indicator are added to the flask; the indicator used here is phenolphthalein, which is colorless in acid and pink in base. After an initial buret reading, base (OH$^-$ ions) is added slowly to the acid (H$^+$ ions). **B,** Near the end of the titration, the indicator momentarily changes to its base color but reverts to its acid color with swirling. **C,** When the end point is reached, a tiny excess of OH$^-$ is present, shown by the permanent change in color of the indicator. The difference between the final buret reading and the initial buret reading gives the volume of base used.

$$H^+(aq) + X^-(aq) + M^+(aq) + OH^-(aq) \longrightarrow H_2O(l) + M^+(aq) + X^-(aq)$$

In a typical acid-base titration, a *standardized* solution of base, one whose concentration is *known,* is added slowly to an acid solution of *unknown* concentration (Figure 4.7). A known volume of the acid solution is placed in a flask, and a few drops of indicator solution are added. An *acid-base indicator* is a substance whose color is different in acid than in base. (We examine indicators in Chapters 18 and 19.) The standardized solution of base is added slowly to the flask from a buret. As the titration is close to its end, indicator molecules near a drop of added base change color due to the temporary excess of OH$^-$ ions there. As soon as the solution is swirled, however, the indicator's acidic color returns. The **equivalence point** in the titration occurs when *all the moles of H$^+$ ions present in the original volume of acid solution have reacted with an equivalent number of moles of OH$^-$ ions added from the buret:*

Moles of H$^+$ (originally in flask) = moles of OH$^-$ (added from buret)

The **end point** of the titration occurs when a tiny excess of OH$^-$ ions changes the indicator permanently to its color in base. In calculations, we assume this tiny excess is insignificant, and therefore *the amount of base needed to reach the end point is the same as the amount needed to reach the equivalence point.*

SAMPLE PROBLEM 4.5   Finding the Concentration of Acid
from an Acid-Base Titration

Problem You perform an acid-base titration to standardize an HCl solution by placing 50.00 mL of HCl in a flask with a few drops of indicator solution. You put 0.1524 *M* NaOH into the buret, and the initial reading is 0.55 mL. At the end point, the buret reading is 33.87 mL. What is the concentration of the HCl solution?
Plan We must find the molarity of acid from the volume of acid (50.00 mL), the initial (0.55 mL) and final (33.87 mL) volumes of base, and the molarity of base (0.1524 *M*). First, we balance the equation. We find the volume of base added from the difference in buret readings and use the base's molarity to calculate the amount (mol) of base added. Then, we use the molar ratio from the balanced equation to find the amount (mol) of acid originally present and divide by the acid's original volume to find the molarity.
Solution Writing the balanced equation:

$$NaOH(aq) + HCl(aq) \longrightarrow NaCl(aq) + H_2O(l)$$

Finding volume (L) of NaOH solution added:

$$\text{Volume (L) of solution} = (33.87 \text{ mL soln} - 0.55 \text{ mL soln}) \times \frac{1 \text{ L}}{1000 \text{ mL}}$$

$$= 0.03332 \text{ L soln}$$

Finding amount (mol) of NaOH added:

$$\text{Moles of NaOH} = 0.03332 \text{ L soln} \times \frac{0.1524 \text{ mol NaOH}}{1 \text{ L soln}}$$

$$= 5.078 \times 10^{-3} \text{ mol NaOH}$$

Finding amount (mol) of HCl originally present: Because the molar ratio is 1:1,

$$\text{Moles of HCl} = 5.078 \times 10^{-3} \text{ mol NaOH} \times \frac{1 \text{ mol HCl}}{1 \text{ mol NaOH}} = 5.078 \times 10^{-3} \text{ mol HCl}$$

Calculating molarity of HCl:

$$\text{Molarity of HCl} = \frac{5.078 \times 10^{-3} \text{ mol HCl}}{50.00 \text{ mL}} \times \frac{1000 \text{ mL}}{1 \text{ L}}$$

$$= 0.1016 \text{ } M \text{ HCl}$$

Check The answer makes sense: a larger volume of less concentrated acid neutralized a smaller volume of more concentrated base. Rounding shows that the moles of H^+ and OH^- are about equal: 50 mL $\times$ 0.1 M H^+ = 0.005 mol = 33 mL $\times$ 0.15 M OH^-.

FOLLOW-UP PROBLEM 4.5 What volume of 0.1292 M $Ba(OH)_2$ would neutralize 50.00 mL of the HCl solution standardized in the preceding sample problem?

> **Volume (L) of base**
> (difference in buret readings)
>
> multiply by M (mol/L) of base
>
> **Amount (mol) of base**
>
> molar ratio
>
> **Amount (mol) of acid**
>
> divide by volume (L) of acid
>
> **M (mol/L) of acid**

Proton Transfer: A Closer Look at Acid-Base Reactions

We gain deeper insight into acid-base reactions if we look closely at the species in solution. Let's see what takes place when HCl gas dissolves in water. Polar water molecules pull apart each HCl molecule, and the H^+ ion ends up bonded to a water molecule. In essence, HCl *transfers its proton* to H_2O:

$$\overset{\frown}{\text{H}^+ \text{ transfer}}$$
$$\text{HCl}(g) + \text{H}_2\text{O}(l) \longrightarrow \text{H}_3\text{O}^+(aq) + \text{Cl}^-(aq)$$

Thus, hydrochloric acid (an aqueous solution of HCl gas) actually consists of solvated H_3O^+ and Cl^- ions.

When sodium hydroxide solution is added, the H_3O^+ ion transfers a proton to the OH^- ion of the base (with the product water shown here as HOH):

$$\overset{\frown}{\text{H}^+ \text{ transfer}}$$
$$[\text{H}_3\text{O}^+(aq) + \text{Cl}^-(aq)] + [\text{Na}^+(aq) + \text{OH}^-(aq)] \longrightarrow$$
$$\text{H}_2\text{O}(l) + \text{Cl}^-(aq) + \text{Na}^+(aq) + \text{HOH}(l)$$

Without the spectator ions, the transfer of a proton from H_3O^+ to OH^- is obvious:

$$\overset{\frown}{\text{H}^+ \text{ transfer}}$$
$$\text{H}_3\text{O}^+(aq) + \text{OH}^-(aq) \longrightarrow \text{H}_2\text{O}(l) + \text{HOH}(l) \quad [\text{or } 2\text{H}_2\text{O}(l)]$$

This net ionic equation is identical with the one we saw earlier (see p. 118),

$$\text{H}^+(aq) + \text{OH}^-(aq) \longrightarrow \text{H}_2\text{O}(l)$$

with the additional H_2O molecule coming from the H_3O^+. Thus, *an acid-base reaction is a proton-transfer process.* In this case, the Cl^- and Na^+ ions remain in solution, and if the water is evaporated, they crystallize as the salt NaCl. Figure 4.8 (on the next page) shows this process on the atomic level.

In the early 20th century, the chemists Johannes Brønsted and Thomas Lowry realized the proton-transfer nature of acid-base reactions. They defined *an acid as a molecule (or ion) that donates a proton, and a base as a molecule (or ion) that accepts a proton.* Therefore, in the aqueous reaction between strong acid and

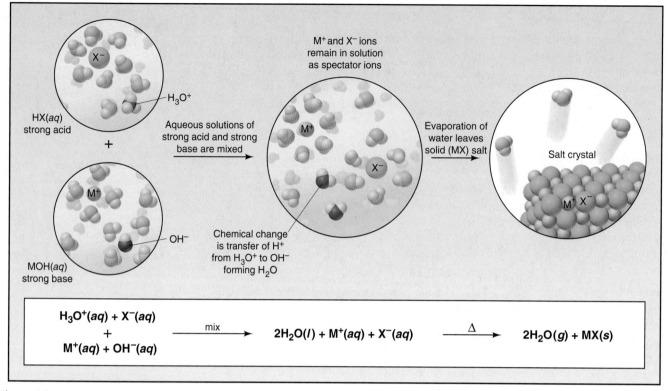

Figure 4.8 An aqueous strong acid–strong base reaction on the atomic scale. When solutions of a strong acid (HX) and a strong base (MOH) are mixed, the H_3O^+ from the acid transfers a proton to the OH^- from the base to form an H_2O molecule. Evaporation of the water leaves the spectator ions, X^- and M^+, as a solid ionic compound called a *salt*.

strong base, H_3O^+ *ion acts as the acid and* OH^- *ion acts as the base.* Because it ionizes completely, a given amount of strong acid (or strong base) creates an equivalent amount of H_3O^+ (or OH^-) when it dissolves in water. (We discuss the Brønsted-Lowry concept thoroughly in Chapter 18.)

Reactions of Weak Acids Ionic equations are written differently for the reactions of weak acids. When solutions of sodium hydroxide and acetic acid (CH_3COOH) are mixed, the molecular, total ionic, and net ionic equations are

Molecular equation:
$$CH_3COOH(aq) + NaOH(aq) \longrightarrow CH_3COONa(aq) + H_2O(l)$$

Total ionic equation:
$$CH_3COOH(aq) + Na^+(aq) + OH^-(aq) \longrightarrow CH_3COO^-(aq) + Na^+(aq) + H_2O(l)$$

Net ionic equation:
$$\overset{\frown\ \text{H}^+ \text{ transfer}\ \searrow}{CH_3COOH(aq)\ +\ OH^-(aq)} \longrightarrow CH_3COO^-(aq) + H_2O(l)$$

Acetic acid dissociates very little because it is a weak acid (see Table 4.2). To show this, *it appears undissociated in both ionic equations.* Note that H_3O^+ does not appear; rather, the proton is transferred from CH_3COOH. Therefore, only $Na^+(aq)$ is a spectator ion; $CH_3COO^-(aq)$ is not.

SECTION SUMMARY

Acid-base (neutralization) reactions occur when an acid (an H^+-yielding substance) and a base (an OH^--yielding substance) react and the H^+ and OH^- ions form a water molecule. Strong acids and bases dissociate completely in water; weak acids and bases dissociate slightly. In a titration, a known concentration of one reactant is used to determine the concentration of the other. An acid-base reaction can also be viewed as the transfer of a proton from an acid to a base. Because weak acids dissociate very little, equations involving them show the acid as an intact molecule.

4.5 OXIDATION-REDUCTION (REDOX) REACTIONS

Redox reactions are the third and, perhaps, most important type of chemical process. They include the formation of a compound from its elements (and vice versa), all combustion reactions, the reactions that generate electricity in batteries, the reactions that produce cellular energy, and many others. In this section, we examine the process and introduce some essential terminology.

The Key Event: Movement of Electrons Between Reactants

In **oxidation-reduction** (or **redox**) **reactions,** the key chemical event is the *net movement of electrons from one reactant to the other.* This movement of electrons occurs *from the reactant (or atom in the reactant) with less attraction for electrons to the reactant (or atom) with more attraction for electrons.*

Such movement of electron charge occurs in the formation of both ionic and covalent compounds. As an example, let's reconsider the flashbulb reaction (see Figure 3.7), in which an ionic compound, MgO, forms from its elements:

$$2Mg(s) + O_2(g) \longrightarrow 2MgO(s)$$

Figure 4.9A shows that during the reaction, each Mg atom loses two electrons and each O atom gains them; that is, two electrons move from each Mg atom to each O atom. This change represents a **transfer** of electron charge away from each Mg atom toward each O atom, resulting in the formation of Mg^{2+} and O^{2-} ions. The ions aggregate and form an ionic solid.

During the formation of a covalent compound from its elements, there is again a net movement of electrons, but it is more of a *shift* in electron charge than a full transfer. Thus, *ions do not form.* Consider the formation of HCl gas:

$$H_2(g) + Cl_2(g) \longrightarrow 2HCl(g)$$

To see the electron movement here, compare the electron charge distributions in the reactant bonds and in the product bonds. As Figure 4.9B shows, H_2 and Cl_2 molecules are each held together by covalent bonds in which the electrons are shared equally between the atoms (the tan shading is symmetrical). In the HCl molecule, the electrons are shared unequally because the Cl atom attracts them more strongly than the H atom does. Thus, in HCl, the H has less electron charge (*blue shading*) than it had in H_2, and the Cl has more charge (*red shading*) than it had in Cl_2. In other words, in the formation of HCl, there has been a relative **shift** of electron charge away from the H atom toward the Cl atom. This electron

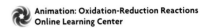
**Animation: Oxidation-Reduction Reactions
Online Learning Center**

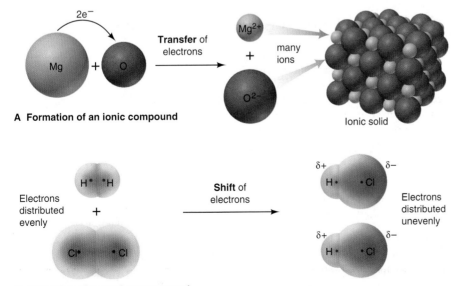

A Formation of an ionic compound

B Formation of a covalent compound

Figure 4.9 The redox process in compound formation. A, In forming the ionic compound MgO, each Mg atom transfers two electrons to each O atom. (Note that atoms become smaller when they lose electrons and larger when they gain electrons.) The resulting Mg^{2+} and O^{2-} ions aggregate with many others to form an ionic solid. **B,** In the reactants H_2 and Cl_2, the electron pairs are shared equally (indicated by even electron density shading). In the covalent product HCl, Cl attracts the shared electrons more strongly than H does. In effect, the H electron shifts toward Cl, as shown by higher electron density (*red*) near the Cl end of the molecule and lower electron density (*blue*) near the H end.

shift is not nearly as extreme as the electron *transfer* during MgO formation. In fact, in some reactions, the net movement of electrons may be very slight, but the reaction is still a redox process.

Some Essential Redox Terminology

Chemists use some important terminology to describe the movement of electrons in oxidation-reduction reactions. **Oxidation** is the *loss* of electrons, and **reduction** is the *gain* of electrons. (The original meaning of *reduction* comes from the process of reducing large amounts of metal ore to smaller amounts of metal, but you'll see shortly why we use the term "reduction" for the act of gaining.)

For example, during the formation of magnesium oxide, Mg undergoes oxidation (electron loss) and O_2 undergoes reduction (electron gain). The loss and gain are simultaneous, but we can imagine them occurring in separate steps:

$$\text{Oxidation (electron loss by Mg):} \qquad Mg \longrightarrow Mg^{2+} + 2e^-$$
$$\text{Reduction (electron gain by } O_2\text{):} \quad \tfrac{1}{2}O_2 + 2e^- \longrightarrow O^{2-}$$

One reactant acts on the other. Thus, we say that *O_2 oxidizes Mg,* and that O_2 is the **oxidizing agent,** the species doing the oxidizing. Similarly, *Mg reduces O_2,* so Mg is the **reducing agent,** the species doing the reducing.

Note especially that O_2 takes the electrons that Mg loses or, put the other way around, Mg gives up the electrons that O_2 gains. This give-and-take of electrons means that *the oxidizing agent is reduced* because it takes the electrons (and thus gains them), and *the reducing agent is oxidized* because it gives up the electrons (and thus loses them). In the formation of HCl, Cl_2 oxidizes H_2 (H loses some electron charge and Cl gains it), which is the same as saying that H_2 reduces Cl_2. The reducing agent, H_2, is oxidized and the oxidizing agent, Cl_2, is reduced.

Using Oxidation Numbers to Monitor the Movement of Electron Charge

Chemists have devised a useful "bookkeeping" system to monitor which atom loses electron charge and which atom gains it. Each atom in a molecule (or ionic compound) is assigned an **oxidation number (O.N.),** or *oxidation state,* the charge the atom would have *if* electrons were not shared but were transferred completely. Oxidation numbers are determined by the set of rules in Table 4.3. [Note that an oxidation number has the sign *before* the number (+2), whereas an ionic charge has the sign *after* the number (2+).]

Table 4.3   **Rules for Assigning an Oxidation Number (O.N.)**

General Rules

1. For an atom in its elemental form (Na, O_2, Cl_2, etc.):   O.N. = 0
2. For a monatomic ion: O.N. = ion charge
3. The sum of O.N. values for the atoms in a molecule or formula unit of a compound equals zero. The sum of O.N. values for the atoms in a polyatomic ion equals the ion's charge.

Rules for Specific Atoms or Periodic Table Groups

1. For Group 1A(1):	O.N. = +1 in all compounds
2. For Group 2A(2):	O.N. = +2 in all compounds
3. For hydrogen:	O.N. = +1 in combination with nonmetals
	O.N. = −1 in combination with metals and boron
4. For fluorine:	O.N. = −1 in all compounds
5. For oxygen:	O.N. = −1 in peroxides
	O.N. = −2 in all other compounds (except with F)
6. For Group 7A(17):	O.N. = −1 in combination with metals, nonmetals (except O), and other halogens lower in the group

SAMPLE PROBLEM 4.6 Determining the Oxidation Number of an Element

Problem Determine the oxidation number (O.N.) of each element in these compounds:
(a) Zinc chloride   **(b)** Sulfur trioxide   **(c)** Nitric acid
Plan We apply Table 4.3, noting the general rules that the O.N. values in a compound add up to zero, and the O.N. values in a polyatomic ion add up to the ion's charge.
Solution (a) $ZnCl_2$. The sum of O.N.s for the monatomic ions in the compound must equal zero. The O.N. of the Zn^{2+} ion is $+2$. The O.N. of each Cl^- ion is -1, for a total of -2. The sum of O.N.s is $+2 + (-2)$, or 0.
(b) SO_3. The O.N. of each oxygen is -2, for a total of -6. The O.N.s must add up to zero, so the O.N. of S is $+6$.
(c) HNO_3. The O.N. of H is $+1$, so the O.N.s of the NO_3 group must add up to -1 to give zero for the compound. The O.N. of each O is -2 for a total of -6. Therefore, the O.N. of N is $+5$.

FOLLOW-UP PROBLEM 4.6 Determine the O.N. of each element in the following:
(a) Scandium oxide (Sc_2O_3)   **(b)** Gallium chloride ($GaCl_3$)
(c) Hydrogen phosphate ion   **(d)** Iodine trifluoride

The periodic table is a great help in learning the highest and lowest oxidation numbers of most main-group elements, as Figure 4.10 shows:

• For most main-group elements, the A-group number (1A, 2A, and so on) is the *highest* oxidation number (always positive) of any element in the group. The exceptions are O and F (see Table 4.3).
• For main-group nonmetals and some metalloids, the A-group number minus 8 is the *lowest* oxidation number (always negative) of any element in the group.

For example, the highest oxidation number of S (Group 6A) is $+6$, as in SF_6, and the lowest is $(6 - 8)$, or -2, as in FeS and other metal sulfides.

Thus, another way to define a redox reaction is one in which *the oxidation numbers of the species change,* and the most important use of oxidation numbers is to monitor these changes:

• If a given atom has a higher (more positive or less negative) oxidation number in the product than it had in the reactant, the reactant species that contains the atom was oxidized (lost electrons). Thus, *oxidation is represented by an increase in oxidation number.*
• If an atom has a lower (more negative or less positive) oxidation number in the product than it had in the reactant, the reactant species that contains the atom was reduced (gained electrons). Thus, *the gain of electrons is represented by a decrease (a "reduction") in oxidation number.*

Figure 4.11 summarizes redox terminology. Oxidation numbers are assigned according to the relative attraction of an atom for electrons, so they are ultimately based on atomic properties, as you'll see in Chapters 8 and 9. (For the remainder of this section, blue oxidation numbers represent oxidation, and red oxidation numbers indicate reduction.)

SAMPLE PROBLEM 4.7 Recognizing Oxidizing and Reducing Agents

Problem Identify the oxidizing agent and reducing agent in each of the following:
(a) $2Al(s) + 3H_2SO_4(aq) \longrightarrow Al_2(SO_4)_3(aq) + 3H_2(g)$
(b) $PbO(s) + CO(g) \longrightarrow Pb(s) + CO_2(g)$
(c) $2H_2(g) + O_2(g) \longrightarrow 2H_2O(g)$
Plan We first assign an oxidation number (O.N.) to each atom (or ion) based on the rules in Table 4.3. The reactant is the reducing agent if it contains an atom that is oxidized (O.N. increased from left side to right side of the equation). The reactant is the oxidizing agent if it contains an atom that is reduced (O.N. decreased).

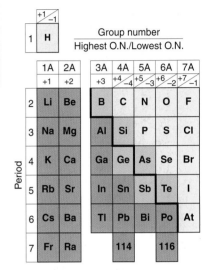

Figure 4.10 Highest and lowest oxidation numbers of reactive main-group elements. The A-group number shows the highest possible oxidation number (O.N.) for a main-group element. (Two important exceptions are O, which never has an O.N. of $+6$, and F, which never has an O.N. of $+7$.) For nonmetals (*yellow*) and metalloids (*green*), the A-group number minus 8 gives the lowest possible oxidation number.

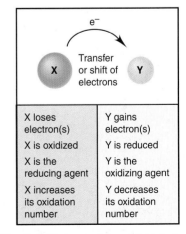

Figure 4.11 A summary of terminology for oxidation-reduction (redox) reactions.

Solution (a) Assigning oxidation numbers:

$$\overset{0}{2Al}(s) + 3\overset{+1}{\underset{+6}{\overset{-2}{H_2SO_4}}}(aq) \longrightarrow \overset{+3}{\underset{+6}{\overset{-2}{Al_2(SO_4)_3}}}(aq) + 3\overset{0}{H_2}(g)$$

The O.N. of Al increased from 0 to +3 (Al lost electrons), so Al was oxidized; Al is the reducing agent.

The O.N. of H decreased from +1 to 0 (H gained electrons), so H^+ was reduced; H_2SO_4 is the oxidizing agent.

(b) Assigning oxidation numbers:

$$\overset{+2}{\underset{}{\overset{-2}{PbO}}}(s) + \overset{+2}{\underset{}{\overset{-2}{CO}}}(g) \longrightarrow \overset{0}{Pb}(s) + \overset{+4}{\underset{}{\overset{-2}{CO_2}}}(g)$$

Pb decreased its O.N. from +2 to 0, so PbO was reduced; PbO is the oxidizing agent.

C increased its O.N. from +2 to +4, so CO was oxidized; CO is the reducing agent.

In general, when a substance (such as CO) becomes one with more O atoms (as in CO_2), it is oxidized; and when a substance (such as PbO) becomes one with fewer O atoms (as in Pb), it is reduced.

(c) Assigning oxidation numbers:

$$\overset{0}{2H_2}(g) + \overset{0}{O_2}(g) \longrightarrow \overset{+1}{\underset{}{\overset{-2}{2H_2O}}}(g)$$

O_2 was reduced (O.N. of O decreased from 0 to −2); O_2 is the oxidizing agent.

H_2 was oxidized (O.N. of H increased from 0 to +1); H_2 is the reducing agent. Oxygen is always the oxidizing agent in a combustion reaction.

Comment 1. Compare the O.N. values in (c) with those in another common reaction that forms water—the net ionic equation for an acid-base reaction:

$$\overset{+1}{H^+}(aq) + \overset{-2}{\underset{}{\overset{+1}{OH^-}}}(aq) \longrightarrow \overset{+1}{\underset{}{\overset{-2}{H_2O}}}(l)$$

Note that the O.N. values remain the same on both sides of the acid-base equation. Therefore, *an acid-base reaction is **not** a redox reaction.*

2. If a substance occurs in its elemental form on one side of an equation, it can't possibly be in its elemental form on the other side, so the reaction must be a redox process. Notice that elements appear in all three parts of this problem.

FOLLOW-UP PROBLEM 4.7 Identify each oxidizing agent and each reducing agent:
(a) $2Fe(s) + 3Cl_2(g) \longrightarrow 2FeCl_3(s)$
(b) $2C_2H_6(g) + 7O_2(g) \longrightarrow 4CO_2(g) + 6H_2O(g)$
(c) $5CO(g) + I_2O_5(s) \longrightarrow I_2(s) + 5CO_2(g)$

SECTION SUMMARY

When one reactant has a greater attraction for electrons than another, there is a net movement of electron charge, and a redox reaction takes place. Electron gain (reduction) and electron loss (oxidation) occur simultaneously. The redox process is tracked by assigning oxidation numbers to each atom in a reaction. The species that is oxidized (contains an atom that increases in oxidation number) is the reducing agent; the species that is reduced (contains an atom that decreases in oxidation number) is the oxidizing agent.

4.6 ELEMENTS IN REDOX REACTIONS

As we saw in Sample Problem 4.7, *whenever atoms appear in the form of a free element on one side of an equation and as part of a compound on the other, there must have been a change in oxidation state, and the reaction is a redox process.*

And, while there are many redox reactions that do *not* involve free elements, we'll focus here on the many others that do. One way to classify these is by comparing the numbers of reactants and products. By that approach, we have three types:

- *Combination reactions:* two or more reactants form one product:

$$X + Y \longrightarrow Z$$

- *Decomposition reactions:* one reactant forms two or more products:

$$Z \longrightarrow X + Y$$

- *Displacement reactions:* the number of substances is the same but atoms (or ions) exchange places:

$$X + YZ \longrightarrow XZ + Y$$

Combining Two Elements Two elements may react to form binary ionic or covalent compounds. Here are some important examples:

1. *Metal and nonmetal form an ionic compound.* A metal, such as aluminum, reacts with a nonmetal, such as oxygen. The change in O.N.'s shows that the metal is oxidized, so it is the reducing agent; the nonmetal is reduced, so it is the oxidizing agent.

$$\overset{0}{4Al(s)} + \overset{0}{3O_2(g)} \longrightarrow \overset{+3\ -2}{2Al_2O_3(s)}$$

2. *Two nonmetals form a covalent compound.* In one of thousands of examples, ammonia forms from nitrogen and hydrogen in a reaction that occurs in industry on an enormous scale:

$$\overset{0}{N_2(g)} + \overset{0}{3H_2(g)} \longrightarrow \overset{-3\ +1}{2NH_3(g)}$$

Combining Compound and Element Many binary covalent compounds react with nonmetals to form larger compounds. Many nonmetal oxides react with additional O_2 to form "higher" oxides (those with more O atoms in each molecule). For example,

$$\overset{+2\ -2}{2NO(g)} + \overset{0}{O_2(g)} \longrightarrow \overset{+4\ -2}{2NO_2(g)}$$

Similarly, many nonmetal halides combine with additional halogen:

$$\overset{+3\ -1}{PCl_3(l)} + \overset{0}{Cl_2(g)} \longrightarrow \overset{+5\ -1}{PCl_5(s)}$$

Decomposing Compounds into Elements A decomposition reaction occurs when a reactant absorbs enough energy for one or more of its bonds to break. The energy can take many forms; we'll focus in this discussion on heat and electricity. The products are either elements or elements and smaller compounds. Following are several common examples:

1. *Thermal decomposition.* When the energy absorbed is heat, the reaction is a thermal decomposition. (A Greek *delta,* Δ, above a reaction arrow indicates that heat is required for the reaction.) Many metal oxides, chlorates, and perchlorates release oxygen when strongly heated. Heating potassium chlorate is a method for forming small amounts of oxygen in the laboratory; the same reaction occurs in some explosives and fireworks:

$$\overset{+1\ +5\ -2}{2KClO_3(s)} \overset{\Delta}{\longrightarrow} \overset{+1\ -1}{2KCl(s)} + \overset{0}{3O_2(g)}$$

Notice that the lone reactant is the oxidizing *and* the reducing agent.

2. *Electrolytic decomposition.* In the process of *electrolysis,* a compound absorbs electrical energy and decomposes into its elements. Observing the electrolysis of water was crucial in the establishment of atomic masses:

$$\overset{+1\,-2}{2H_2O(l)} \xrightarrow{\text{electricity}} \overset{0}{2H_2(g)} + \overset{0}{O_2(g)}$$

Many active metals, such as sodium, magnesium, and calcium, are produced industrially by electrolysis of their molten halides:

$$\overset{+2\,-1}{MgCl_2(l)} \xrightarrow{\text{electricity}} \overset{0}{Mg(l)} + \overset{0}{Cl_2(g)}$$

(We'll examine the details of electrolysis in Chapter 21.)

Displacing One Element by Another; Activity Series As we said, displacement reactions have the same number of reactants as products. We mentioned double-displacement (metathesis) reactions in discussing precipitation and acid-base reactions. The other type, *single-displacement* reactions, are all oxidation-reduction processes. They occur when one atom displaces the ion of a different atom from solution. When the reaction involves metals, the atom reduces the ion; when it involves nonmetals (specifically halogens), the atom oxidizes the ion. Chemists rank various elements into activity series—one for metals and one for halogens—in order of their ability to displace one another.

1. *The activity series of the metals.* Metals can be ranked by their ability to displace H_2 (actually reduce H^+) from various sources or by their ability to displace one another from solution.

- *A metal displaces H_2 from water or acid.* The most reactive metals, such as those from Group 1A(1) and Ca, Sr, and Ba from Group 2A(2), displace H_2 from water, and they do so vigorously. Figure 4.12 shows this reaction for

Figure 4.12 An active metal displacing hydrogen from water. Lithium displaces hydrogen from water in a vigorous reaction that yields an aqueous solution of lithium hydroxide and hydrogen gas, as shown on the macroscopic scale (*top*), at the atomic scale (*middle*), and as a balanced equation (*bottom*). (For clarity, the atomic-scale view of water has been greatly simplified, and only water molecules involved in the reaction are colored red and blue.)

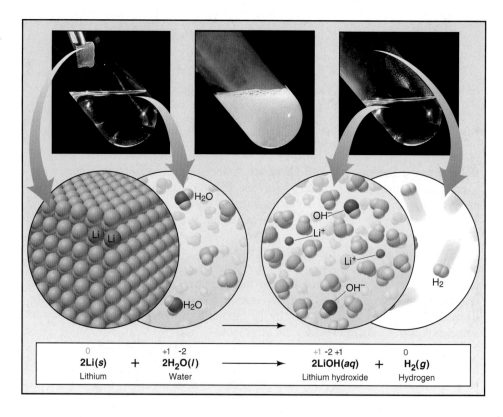

lithium. Heat is needed to speed the reaction of slightly less reactive metals, such as Al and Zn, so these metals displace H_2 from steam:

$$\overset{0}{2Al(s)} + \overset{+1-2}{6H_2O(g)} \longrightarrow \overset{+3-2\overset{+1}{|}}{2Al(OH)_3(s)} + \overset{0}{3H_2(g)}$$

Still less reactive metals, such as nickel and tin, do not react with water but *do* react with acids. Because the concentration of H^+ is higher in acid solutions than in water, H_2 is displaced more easily. Here is the net ionic equation:

$$\overset{0}{Ni(s)} + \overset{+1}{2H^+(aq)} \longrightarrow \overset{+2}{Ni^{2+}(aq)} + \overset{0}{H_2(g)}$$

Notice that in all such reactions, the metal is the reducing agent (O.N. of metal increases), and water or acid is the oxidizing agent (O.N. of H decreases). The least reactive metals, such as silver and gold, cannot displace H_2 from any source.

- *A metal displaces another metal ion from solution.* Direct comparisons of metal reactivity are clearest in these reactions. For example, zinc metal displaces copper(II) ion from (actually reduces Cu^{2+} in) copper(II) sulfate solution, as the total ionic equation shows:

$$\overset{+2}{Cu^{2+}(aq)} + \overset{+6}{\underset{-2}{SO_4^{2-}}(aq)} + \overset{0}{Zn(s)} \longrightarrow \overset{0}{Cu(s)} + \overset{+2}{Zn^{2+}(aq)} + \overset{+6}{\underset{-2}{SO_4^{2-}}(aq)}$$

Figure 4.13 demonstrates in atomic detail that copper metal can displace silver ion from solution. Thus, zinc is more reactive than copper, which is more reactive than silver.

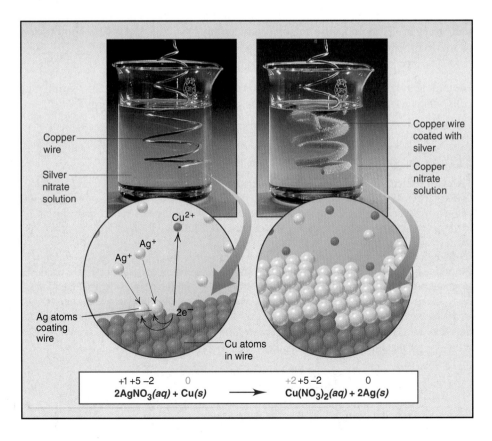

Copper wire

Silver nitrate solution

Copper wire coated with silver

Copper nitrate solution

Cu^{2+}

Ag^+

Ag^+

Ag atoms coating wire

$2e^-$

Cu atoms in wire

$$\overset{+1+5-2}{2AgNO_3(aq)} + \overset{0}{Cu(s)} \longrightarrow \overset{+2+5-2}{Cu(NO_3)_2(aq)} + \overset{0}{2Ag(s)}$$

Figure 4.13 Displacing one metal by another. More reactive metals displace less reactive metals from solution. In this reaction, Cu atoms each give up two electrons as they become Cu^{2+} ions and leave the wire. The electrons are transferred to two Ag^+ ions that become Ag atoms and deposit on the wire. With time, a coating of crystalline silver coats the wire. Thus, copper has displaced silver (reduced silver ion) from solution. The reaction is depicted as the laboratory view (*top*), the atomic-scale view (*middle*), and the balanced redox equation (*bottom*).

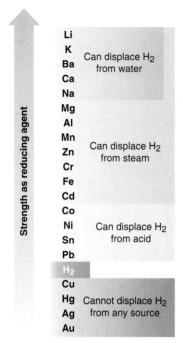

Figure 4.14 The activity series of the metals. This list of metals (and H_2) is arranged with the most active metal (strongest reducing agent) at the top and the least active metal (weakest reducing agent) at the bottom. The four metals below H_2 cannot displace it from any source.

The results of many such reactions between metals and water, aqueous acids, and metal-ion solutions form the basis of the **activity series of the metals.** In Figure 4.14 elements higher on the list are stronger reducing agents than elements lower down; that is, for those that are stable in water, elements higher on the list can reduce aqueous ions of elements lower down. The list also shows whether the metal can displace H_2 (reduce H^+) and, if so, from which source. Look at the metals in the equations we've just discussed. Note that Li, Al, and Ni lie above H_2, while Ag lies below it; also, Zn lies above Cu, which lies above Ag. The most reactive metals on the list are in Groups 1A(1) and 2A(2) of the periodic table, and the least reactive lie at the right of the transition elements in Groups 1B(11) and 2B(12).

2. *The activity series of the halogens.* Reactivity decreases down Group 7A(17), so we can arrange the halogens into their own activity series:

$$F_2 > Cl_2 > Br_2 > I_2$$

A halogen higher in the periodic table is a stronger oxidizing agent than one lower down. Thus, chlorine can oxidize bromide ions or iodide ions from solution, and bromine can oxidize iodide ions. Here, chlorine displaces bromine:

$$\overset{-1}{2Br^-}(aq) + \overset{0}{Cl_2}(aq) \longrightarrow \overset{0}{Br_2}(aq) + \overset{-1}{2Cl^-}(aq)$$

Combustion Reactions *Combustion* is the process of combining with oxygen, often with the release of heat and light, as in a flame. Combustion reactions do not fall neatly into classes based on the number of reactants and products, but *all are redox processes* because elemental oxygen is a reactant:

$$2CO(g) + O_2(g) \longrightarrow 2CO_2(g)$$

The combustion reactions that we commonly use to produce energy involve organic mixtures such as coal, gasoline, and natural gas as reactants. These mixtures consist of substances with many carbon-carbon and carbon-hydrogen bonds. During the reaction, these bonds break, and each C and H atom combines with oxygen. Therefore, the major products are CO_2 and H_2O. The combustion of the hydrocarbon butane, which is used in camp stoves, is typical:

$$2C_4H_{10}(g) + 13O_2(g) \longrightarrow 8CO_2(g) + 10H_2O(g)$$

Biological *respiration* is a multistep combustion process that occurs within our cells when we "burn" organic foodstuffs, such as glucose, for energy:

$$C_6H_{12}O_6(s) + 6O_2(g) \longrightarrow 6CO_2(g) + 6H_2O(g)$$

SAMPLE PROBLEM 4.8   Identifying the Type of Redox Reaction

Problem Classify each of the following redox reactions as a combination, decomposition, or displacement reaction, write a balanced molecular equation for each, as well as total and net ionic equations for part (c), and identify the oxidizing and reducing agents:
(a) Magnesium(*s*) + nitrogen(*g*) $\longrightarrow$ magnesium nitride(*s*)
(b) Hydrogen peroxide(*l*) $\longrightarrow$ water + oxygen gas
(c) Aluminum(*s*) + lead(II) nitrate(*aq*) $\longrightarrow$ aluminum nitrate(*aq*) + lead(*s*)
Plan To decide on reaction type, recall that combination reactions produce fewer products than reactants, decomposition reactions produce more products, and displacement reactions have the same number of reactants and products. The oxidation number (O.N.) becomes more positive for the reducing agent and less positive for the oxidizing agent.
Solution (a) Combination: two substances form one. This reaction occurs, along with formation of magnesium oxide, when magnesium burns in air:

$$\overset{0}{3Mg}(s) + \overset{0}{N_2}(g) \longrightarrow \overset{+2\ -3}{Mg_3N_2}(s)$$

Mg is the reducing agent; N_2 is the oxidizing agent.

(b) Decomposition: one substance forms two. This reaction occurs within every bottle of this common household antiseptic. Hydrogen peroxide is very unstable and breaks down from heat, light, or just shaking:

$$\overset{\substack{-1\\ +1\ -1}}{2H_2O_2(l)} \longrightarrow \overset{+1\ -2}{2H_2O(l)} + \overset{0}{O_2(g)}$$

H_2O_2 is both the oxidizing *and* the reducing agent. The O.N. of O in peroxides is -1.
It increases to 0 in O_2 and decreases to -2 in H_2O.

(c) Displacement: two substances form two others. As Figure 4.14 shows, Al is more active than Pb and, thus, displaces it from aqueous solution:

$$\overset{0}{2Al(s)} + \overset{\substack{-2\\ +2\ +5}}{3Pb(NO_3)_2(aq)} \longrightarrow \overset{\substack{-2\\ +3\ +5}}{2Al(NO_3)_3(aq)} + \overset{0}{3Pb(s)}$$

Al is the reducing agent; $Pb(NO_3)_2$ is the oxidizing agent.
The total ionic equation is

$$2Al(s) + 3Pb^{2+}(aq) + 6NO_3^-(aq) \longrightarrow 2Al^{3+}(aq) + 6NO_3^-(aq) + 3Pb(s)$$

The net ionic equation is

$$2Al(s) + 3Pb^{2+}(aq) \longrightarrow 2Al^{3+}(aq) + 3Pb(s)$$

FOLLOW-UP PROBLEM 4.8 Classify each of the following redox reactions as a combination, decomposition, or displacement reaction, write a balanced molecular equation for each, as well as total and net ionic equations for parts (b) and (c), and identify the oxidizing and reducing agents:
(a) $S_8(s) + F_2(g) \longrightarrow SF_4(g)$ **(b)** $CsI(aq) + Cl_2(aq) \longrightarrow CsCl(aq) + I_2(aq)$
(c) $Ni(NO_3)_2(aq) + Cr(s) \longrightarrow Ni(s) + Cr(NO_3)_3(aq)$

SECTION SUMMARY

Any reaction that includes a free element as reactant or product is a redox reaction. In combination reactions, elements combine to form a compound, or a compound and an element combine. Decomposition of compounds by absorption of heat or electricity can form elements or a compound and an element. In displacement reactions, one element displaces another from solution. Activity series rank elements in order of reactivity. The activity series of the metals ranks metals by their ability to displace H_2 from water, steam, or acid, or to displace one another from solution. Combustion typically releases heat and light energy through reaction of a substance with O_2.

For Review and Reference (Numbers in parentheses refer to pages, unless noted otherwise.)

Learning Objectives

To help you review these learning objectives, the numbers of related sections (§), sample problems (SP), and upcoming end-of-chapter problems (EP) are listed in parentheses.

1. Understand how water dissolves an ionic compound compared to a covalent compound and which solution contains an electrolyte; use a compound's formula to find moles of ions (including H^+) in solution (§ 4.1) (SPs 4.1, 4.2) (EPs 4.1–4.19)
2. Understand the key events in precipitation and acid-base reactions and use ionic equations to describe them; distinguish between strong and weak acids and bases and calculate an unknown concentration from a titration (§ 4.3, 4.4) (SPs 4.3–4.5) (EPs 4.20–4.43)
3. Understand the key event in the redox process; determine the oxidation number of any element in a compound; identify the oxidizing and reducing agents in a reaction (§ 4.5) (SPs 4.6, 4.7) (EPs 4.44–4.58)
4. Identify three important types of redox reactions that include elements: combination, decomposition, displacement (§ 4.6) (SP 4.8) (EPs 4.59–4.73)

Key Terms

Section 4.1
polar molecule (109)
solvated (110)
electrolyte (110)
nonelectrolyte (112)

Section 4.2
molecular equation (113)
total ionic equation (114)
spectator ion (114)

net ionic equation (114)
Section 4.3
precipitation reaction (115)
precipitate (115)
metathesis reaction (116)

Section 4.4
acid-base reaction (117)
neutralization reaction (117)
acid (117)

base (118)
salt (119)
titration (119)
equivalence point (120)
end point (120)

Section 4.5
oxidation-reduction (redox)
  reaction (123)
oxidation (124)

reduction (124)
oxidizing agent (124)
reducing agent (124)
oxidation number (O.N.)
  (or oxidation state) (124)

Section 4.6
activity series of the
  metals (130)

Brief Solutions to Follow-up Problems

4.1 (a) $KClO_4(s) \xrightarrow{H_2O} K^+(aq) + ClO_4^-(aq)$;
2 mol of K^+ and 2 mol of ClO_4^-

(b) $Mg(C_2H_3O_2)_2(s) \xrightarrow{H_2O} Mg^{2+}(aq) + 2C_2H_3O_2^-(aq)$;
2.49 mol of Mg^{2+} and 4.97 mol of $C_2H_3O_2^-$

(c) $(NH_4)_2CrO_4(s) \xrightarrow{H_2O} 2NH_4^+(aq) + CrO_4^{2-}(aq)$;
6.24 mol of NH_4^+ and 3.12 mol of CrO_4^{2-}

(d) $NaHSO_4(s) \xrightarrow{H_2O} Na^+(aq) + HSO_4^-(aq)$;
0.73 mol of Na^+ and 0.73 mol of HSO_4^-

4.2 Moles of $H^+ = 451 \text{ mL} \times \dfrac{1 \text{ L}}{10^3 \text{ mL}}$

$\times \dfrac{3.20 \text{ mol HBr}}{1 \text{ L soln}} \times \dfrac{1 \text{ mol } H^+}{1 \text{ mol HBr}}$

$= 1.44 \text{ mol } H^+$

4.3 (a) $Fe^{3+}(aq) + 3Cl^-(aq) + 3Cs^+(aq) + PO_4^{3-}(aq) \longrightarrow$
$ FePO_4(s) + 3Cl^-(aq) + 3Cs^+(aq)$
$Fe^{3+}(aq) + PO_4^{3-}(aq) \longrightarrow FePO_4(s)$
(b) $2Na^+(aq) + 2OH^-(aq) + Cd^{2+}(aq) + 2NO_3^-(aq) \longrightarrow$
$ 2Na^+(aq) + 2NO_3^-(aq) + Cd(OH)_2(s)$
$2OH^-(aq) + Cd^{2+}(aq) \longrightarrow Cd(OH)_2(s)$
(c) No reaction occurs
(d) $2Ag^+(aq) + SO_4^{2-}(aq) + Ba^{2+}(aq) + 2Cl^-(aq) \longrightarrow$
$ 2AgCl(s) + BaSO_4(s)$
Total and net ionic equations are identical.
4.4 $Ca(OH)_2(aq) + 2HNO_3(aq) \longrightarrow Ca(NO_3)_2(aq) + 2H_2O(l)$
$Ca^{2+}(aq) + 2OH^-(aq) + 2H^+(aq) + 2NO_3^-(aq) \longrightarrow$
$ Ca^{2+}(aq) + 2NO_3^-(aq) + 2H_2O(l)$
$H^+(aq) + OH^-(aq) \longrightarrow H_2O(l)$

4.5 $Ba(OH)_2(aq) + 2HCl(aq) \longrightarrow BaCl_2(aq) + 2H_2O(l)$
Volume (L) of soln

$= 50.00 \text{ mL HCl soln} \times \dfrac{1 \text{ L}}{10^3 \text{ mL}} \times \dfrac{0.1016 \text{ mol HCl}}{1 \text{ L soln}}$

$\times \dfrac{1 \text{ mol Ba(OH)}_2}{2 \text{ mol HCl}} \times \dfrac{1 \text{ L soln}}{0.1292 \text{ mol Ba(OH)}_2}$

$= 0.01966 \text{ L}$

4.6 (a) O.N. of Sc $= +3$; O.N. of O $= -2$
(b) O.N. of Ga $= +3$; O.N. of Cl $= -1$
(c) O.N. of H $= +1$; O.N. of P $= +5$; O.N. of O $= -2$
(d) O.N. of I $= +3$; O.N. of F $= -1$
4.7 (a) Fe is the reducing agent; Cl_2 is the oxidizing agent.
(b) C_2H_6 is the reducing agent; O_2 is the oxidizing agent.
(c) CO is the reducing agent; I_2O_5 is the oxidizing agent.
4.8 (a) Combination:
$S_8(s) + 16F_2(g) \longrightarrow 8SF_4(g)$
S_8 is the reducing agent; F_2 is the oxidizing agent.
(b) Displacement:
$2CsI(aq) + Cl_2(aq) \longrightarrow 2CsCl(aq) + I_2(aq)$
Cl_2 is the oxidizing agent; CsI is the reducing agent.
$2Cs^+(aq) + 2I^-(aq) + Cl_2(aq) \longrightarrow$
$ 2Cs^+(aq) + 2Cl^-(aq) + I_2(aq)$
$2I^-(aq) + Cl_2(aq) \longrightarrow 2Cl^-(aq) + I_2(aq)$
(c) Displacement:
$3Ni(NO_3)_2(aq) + 2Cr(s) \longrightarrow 3Ni(s) + 2Cr(NO_3)_3(aq)$
$3Ni^{2+}(aq) + 6NO_3^-(aq) + 2Cr(s) \longrightarrow$
$ 3Ni(s) + 2Cr^{3+}(aq) + 6NO_3^-(aq)$
$3Ni^{2+}(aq) + 2Cr(s) \longrightarrow 3Ni(s) + 2Cr^{3+}(aq)$
Cr is the reducing agent; $Ni(NO_3)_2$ is the oxidizing agent.

Problems

Problems with **colored** numbers are answered in Appendix E. Sections match the text and provide the numbers of relevant sample problems. Bracketed problems are grouped in pairs (indicated by a short rule) that cover the same concept. Comprehensive Problems are based on material from any section or previous chapter.

The Role of Water as a Solvent
(Sample Problems 4.1 and 4.2)

4.1 What two factors cause water to be polar?
4.2 What must be present in an aqueous solution for it to conduct an electric current? What general classes of compounds form solutions that conduct?

4.3 What occurs on the molecular level when an ionic compound dissolves in water?

4.4 Which of the following best represents how the ions occur in aqueous solutions of (a) $CaCl_2$, (b) Li_2SO_4, and (c) NH_4Br?

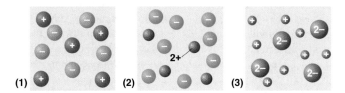

(1)   **(2)**   **(3)**

4.5 Which of the following best represents a volume from a solution of magnesium nitrate?

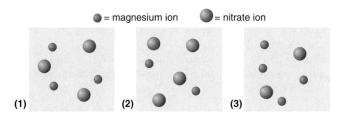

● = magnesium ion   ● = nitrate ion

(1)   **(2)**   **(3)**

4.6 Why are some ionic compounds soluble in water and others are not?

4.7 Some covalent compounds dissociate into ions when they dissolve in water. What atom do these compounds have in their structures? What type of aqueous solution do they form? Name three examples of such an aqueous solution.

4.8 State whether each of the following substances is likely to be very soluble in water. Explain.
(a) Benzene, C_6H_6 (b) Sodium hydroxide
(c) Ethanol, CH_3CH_2OH (d) Potassium acetate

4.9 State whether each of the following substances is likely to be very soluble in water. Explain.
(a) Lithium nitrate   (b) Glycine, H_2NCH_2COOH
(c) Pentane   (d) Ethylene glycol, $HOCH_2CH_2OH$

4.10 State whether an aqueous solution of each of the following substances conducts an electric current. Explain your reasoning.
(a) Cesium iodide   (b) Hydrogen bromide

4.11 State whether an aqueous solution of each of the following substances conducts an electric current. Explain your reasoning.
(a) Potassium hydroxide   (b) Glucose, $C_6H_{12}O_6$

4.12 How many total moles of ions are released when each of the following samples dissolves completely in water?
(a) 0.83 mol of K_3PO_4 (b) 8.11×10^{-3} g of $NiBr_2 \cdot 3H_2O$
(c) 1.23×10^{21} formula units of $FeCl_3$

4.13 How many total moles of ions are released when each of the following samples dissolves completely in water?
(a) 0.734 mol of Na_2HPO_4 (b) 3.86 g of $CuSO_4 \cdot 5H_2O$
(c) 8.66×10^{20} formula units of $NiCl_2$

4.14 How many moles and numbers of ions of each type are present in the following aqueous solutions?
(a) 100 mL of 2.45 M aluminum chloride
(b) 1.80 L of a solution containing 2.59 g lithium sulfate/L
(c) 225 mL of a solution containing 1.68×10^{22} formula units of potassium bromide per liter

4.15 How many moles and numbers of ions of each type are present in the following aqueous solutions?
(a) 88 mL of 1.75 M magnesium chloride
(b) 321 mL of a solution containing 0.22 g aluminum sulfate/L
(c) 1.65 L of a solution containing 8.83×10^{21} formula units of cesium nitrate per liter

4.16 How many moles of H^+ ions are present in the following aqueous solutions?
(a) 1.40 L of 0.25 M perchloric acid
(b) 1.8 mL of 0.72 M nitric acid
(c) 7.6 L of 0.056 M hydrochloric acid

4.17 How many moles of H^+ ions are present in the following aqueous solutions?
(a) 1.4 mL of 0.75 M hydrobromic acid
(b) 2.47 mL of 1.98 M hydriodic acid
(c) 395 mL of 0.270 M nitric acid

4.18 To study a marine organism, a biologist prepares a 1.00-kg sample to simulate the ion concentrations in seawater. She mixes 26.5 g of NaCl, 2.40 g of $MgCl_2$, 3.35 g of $MgSO_4$, 1.20 g of $CaCl_2$, 1.05 g of KCl, 0.315 g of $NaHCO_3$, and 0.098 g of NaBr in distilled water. (a) If the density of this solution is 1.04 g/cm^3, what is the molarity of each ion? (b) What is the total molarity of alkali metal ions? (c) What is the total molarity of alkaline earth metal ions? (d) What is the total molarity of anions?

4.19 Water "softeners" remove metal ions such as Ca^{2+} and Fe^{3+} by replacing them with enough Na^+ ions to maintain the same number of positive charges in the solution. If 1.0×10^3 L of "hard" water is 0.015 M Ca^{2+} and 0.0010 M Fe^{3+}, how many moles of Na^+ are needed to replace these ions?

Writing Equations for Aqueous Ionic Reactions

4.20 Write two sets of equations (both molecular and total ionic) with different reactants that have the same net ionic equation as the following equation:
$$Ba(NO_3)_2(aq) + Na_2CO_3(aq) \longrightarrow BaCO_3(s) + 2NaNO_3(aq)$$

Precipitation Reactions
(Sample Problem 4.3)

4.21 Why do some pairs of ions precipitate and others do not?

4.22 Use Table 4.1 to determine which of the following combinations leads to a reaction. How can you identify the spectator ions in the reaction?
(a) Calcium nitrate(aq) + sodium chloride(aq) $\longrightarrow$
(b) Potassium chloride(aq) + lead(II) nitrate(aq) $\longrightarrow$

4.23 The beakers represent the aqueous reaction of $AgNO_3$ and NaCl. Silver ions are gray. What colors are used to represent NO_3^-, Na^+, and Cl^-? Write molecular, total ionic, and net ionic equations for the reaction.

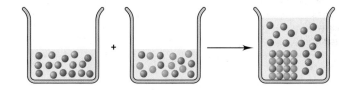

4.24 Complete the following precipitation reactions with balanced molecular, total ionic, and net ionic equations:
(a) $Hg_2(NO_3)_2(aq) + KI(aq) \longrightarrow$
(b) $FeSO_4(aq) + Ba(OH)_2(aq) \longrightarrow$

4.25 Complete the following precipitation reactions with balanced molecular, total ionic, and net ionic equations:
(a) $CaCl_2(aq) + Cs_3PO_4(aq) \longrightarrow$
(b) $Na_2S(aq) + ZnSO_4(aq) \longrightarrow$

4.26 When each of the following pairs of aqueous solutions is mixed, does a precipitation reaction occur? If so, write balanced molecular, total ionic, and net ionic equations:
(a) Sodium nitrate + copper(II) sulfate
(b) Ammonium iodide + silver nitrate

4.27 When each of the following pairs of aqueous solutions is mixed, does a precipitation reaction occur? If so, write balanced molecular, total ionic, and net ionic equations:
(a) Potassium carbonate + barium hydroxide
(b) Aluminum nitrate + sodium phosphate

4.28 If 35.0 mL of lead(II) nitrate solution reacts completely with excess sodium iodide solution to yield 0.628 g of precipitate, what is the molarity of lead(II) ion in the original solution?

4.29 If 25.0 mL of silver nitrate solution reacts with excess potassium chloride solution to yield 0.842 g of precipitate, what is the molarity of silver ion in the original solution?

4.30 The mass percent of Cl^- in a seawater sample is determined by titrating 25.00 mL of seawater with $AgNO_3$ solution, causing a precipitation reaction. An indicator is used to detect the end point, which occurs when free Ag^+ ion is present in solution after all the Cl^- has reacted. If 43.63 mL of 0.3020 M $AgNO_3$ is required to reach the end point, what is the mass percent of Cl^- in the seawater (d of seawater = 1.04 g/mL)?

4.31 Aluminum sulfate, known as *cake alum,* has a wide range of uses, from dyeing leather and cloth to purifying sewage. In aqueous solution, it reacts with base to form a white precipitate. (a) Write balanced total and net ionic equations for its reaction with aqueous NaOH. (b) What mass of precipitate forms when 185.5 mL of 0.533 M NaOH is added to 627 mL of a solution that contains 15.8 g of aluminum sulfate per liter?

Acid-Base Reactions
(Sample Problems 4.4 and 4.5)

4.32 Is the total ionic equation the same as the net ionic equation when $Sr(OH)_2(aq)$ and $H_2SO_4(aq)$ react? Explain.

4.33 (a) Name three common strong acids. (b) Name three common strong bases. (c) What is a characteristic behavior of a strong acid or a strong base?

4.34 (a) Name three common weak acids. (b) Name one common weak base. (c) What is the major difference between a weak acid and a strong acid or between a weak base and a strong base, and what experiment would you perform to observe it?

4.35 The net ionic equation for the aqueous neutralization reaction between acetic acid and sodium hydroxide is different from that for the reaction between hydrochloric acid and sodium hydroxide. Explain by writing balanced net ionic equations.

4.36 Complete the following acid-base reactions with balanced molecular, total ionic, and net ionic equations:
(a) Potassium hydroxide(*aq*) + hydriodic acid(*aq*) $\longrightarrow$
(b) Ammonia(*aq*) + hydrochloric acid(*aq*) $\longrightarrow$

4.37 Complete the following acid-base reactions with balanced molecular, total ionic, and net ionic equations:
(a) Cesium hydroxide(*aq*) + nitric acid(*aq*) $\longrightarrow$
(b) Calcium hydroxide(*aq*) + acetic acid(*aq*) $\longrightarrow$

4.38 Limestone (calcium carbonate) is insoluble in water but dissolves when a hydrochloric acid solution is added. Write balanced total ionic and net ionic equations, showing hydrochloric acid as it actually exists in water and the reaction as a proton-transfer process. [*Hint:* The H_2CO_3 that forms decomposes to H_2O and gaseous CO_2.]

4.39 Zinc hydroxide is insoluble in water but dissolves when a nitric acid solution is added. Write balanced total ionic and net ionic equations, showing nitric acid as it actually exists in water and the reaction as a proton-transfer process.

4.40 If 25.98 mL of a standard 0.1180 M KOH solution reacts with 52.50 mL of CH_3COOH solution, what is the molarity of the acid solution?

4.41 If 36.25 mL of a standard 0.1750 M NaOH solution is required to neutralize 25.00 mL of H_2SO_4, what is the molarity of the acid solution?

4.42 An auto mechanic spills 78 mL of 2.6 M H_2SO_4 solution from a rebuilt auto battery. How many milliliters of 1.5 M $NaHCO_3$ must be poured on the spill to react completely with the sulfuric acid? [*Hint:* H_2O and CO_2 are among the products.]

4.43 One of the first steps in the enrichment of uranium for use in nuclear power plants involves a displacement reaction between UO_2 and aqueous HF:
$$UO_2(s) + HF(aq) \longrightarrow UF_4(s) + 2H_2O(l) \text{ [unbalanced]}$$
How many liters of 2.50 M HF will react with 2.25 kg of UO_2?

Oxidation-Reduction (Redox) Reactions
(Sample Problems 4.6 and 4.7)

4.44 Why must every redox reaction involve an oxidizing agent and a reducing agent?

4.45 In which of the following equations does sulfuric acid act as an oxidizing agent? In which does it act as an acid? Explain.
(a) $4H^+(aq) + SO_4^{2-}(aq) + 2NaI(s) \longrightarrow$
$$2Na^+(aq) + I_2(s) + SO_2(g) + 2H_2O(l)$$
(b) $BaF_2(s) + 2H^+(aq) + SO_4^{2-}(aq) \longrightarrow$
$$2HF(aq) + BaSO_4(s)$$

4.46 Give the oxidation number of nitrogen in the following:
(a) NH_2OH (b) N_2H_4 (c) NH_4^+ (d) HNO_2

4.47 Give the oxidation number of sulfur in the following:
(a) $SOCl_2$ (b) H_2S_2 (c) H_2SO_3 (d) Na_2S

4.48 Give the oxidation number of arsenic in the following:
(a) AsH_3 (b) H_3AsO_4 (c) $AsCl_3$

4.49 Give the oxidation number of phosphorus in the following:
(a) $H_2P_2O_7^{2-}$ (b) PH_4^+ (c) PCl_5

4.50 Give the oxidation number of manganese in the following:
(a) MnO_4^{2-} (b) Mn_2O_3 (c) $KMnO_4$

4.51 Give the oxidation number of chromium in the following:
(a) CrO_3 (b) $Cr_2O_7^{2-}$ (c) $Cr_2(SO_4)_3$

4.52 Identify the oxidizing and reducing agents in the following:
(a) $5H_2C_2O_4(aq) + 2MnO_4^-(aq) + 6H^+(aq) \longrightarrow$
$$2Mn^{2+}(aq) + 10CO_2(g) + 8H_2O(l)$$
(b) $3Cu(s) + 8H^+(aq) + 2NO_3^-(aq) \longrightarrow$
$$3Cu^{2+}(aq) + 2NO(g) + 4H_2O(l)$$

4.53 Identify the oxidizing and reducing agents in the following:
(a) $Sn(s) + 2H^+(aq) \longrightarrow Sn^{2+}(aq) + H_2(g)$
(b) $2H^+(aq) + H_2O_2(aq) + 2Fe^{2+}(aq) \longrightarrow$
$$2Fe^{3+}(aq) + 2H_2O(l)$$

4.54 Identify the oxidizing and reducing agents in the following:
(a) $8H^+(aq) + 6Cl^-(aq) + Sn(s) + 4NO_3^-(aq) \longrightarrow$
$$SnCl_6^{2-}(aq) + 4NO_2(g) + 4H_2O(l)$$
(b) $2MnO_4^-(aq) + 10Cl^-(aq) + 16H^+(aq) \longrightarrow$
$$5Cl_2(g) + 2Mn^{2+}(aq) + 8H_2O(l)$$

4.55 Identify the oxidizing and reducing agents in the following:
(a) $8H^+(aq) + Cr_2O_7^{2-}(aq) + 3SO_3^{2-}(aq) \longrightarrow$
$$2Cr^{3+}(aq) + 3SO_4^{2-}(aq) + 4H_2O(l)$$
(b) $NO_3^-(aq) + 4Zn(s) + 7OH^-(aq) + 6H_2O(l) \longrightarrow$
$$4Zn(OH)_4^{2-}(aq) + NH_3(aq)$$

4.56 Discuss each conclusion from a study of redox reactions:
(a) The sulfide ion functions only as a reducing agent.
(b) The sulfate ion functions only as an oxidizing agent.
(c) Sulfur dioxide functions as an oxidizing or a reducing agent.

4.57 Discuss each conclusion from a study of redox reactions:
(a) The nitride ion functions only as a reducing agent.
(b) The nitrate ion functions only as an oxidizing agent.
(c) The nitrite ion functions as an oxidizing or a reducing agent.

4.58 A person's blood alcohol (C_2H_5OH) level can be determined by titrating a sample of blood plasma with a potassium dichromate solution. The balanced equation is
$$16H^+(aq) + 2Cr_2O_7^{2-}(aq) + C_2H_5OH(aq) \longrightarrow$$
$$4Cr^{3+}(aq) + 2CO_2(g) + 11H_2O(l)$$
If 35.46 mL of 0.05961 M $Cr_2O_7^{2-}$ is required to titrate 28.00 g of plasma, what is the mass percent of alcohol in the blood?

Elements in Redox Reactions
(Sample Problem 4.8)

4.59 Which type of redox reaction leads to the following?
(a) An increase in the number of substances
(b) A decrease in the number of substances
(c) No change in the number of substances

4.60 Why do decomposition reactions typically have compounds as reactants, whereas combination and displacement reactions have one or more elements?

4.61 Which of the three types of reactions discussed in this section commonly produce one or more compounds?

4.62 Balance each of the following redox reactions and classify it as a combination, decomposition, or displacement reaction:
(a) $Sb(s) + Cl_2(g) \longrightarrow SbCl_3(s)$
(b) $AsH_3(g) \longrightarrow As(s) + H_2(g)$
(c) $Mn(s) + Fe(NO_3)_3(aq) \longrightarrow Mn(NO_3)_2(aq) + Fe(s)$

4.63 Balance each of the following redox reactions and classify it as a combination, decomposition, or displacement reaction:
(a) $Mg(s) + H_2O(g) \longrightarrow Mg(OH)_2(s) + H_2(g)$
(b) $Cr(NO_3)_3(aq) + Al(s) \longrightarrow Al(NO_3)_3(aq) + Cr(s)$
(c) $PF_3(g) + F_2(g) \longrightarrow PF_5(g)$

4.64 Predict the product(s) and write a balanced equation for each of the following redox reactions:
(a) $N_2(g) + H_2(g) \longrightarrow$
(b) $NaClO_3(s) \xrightarrow{\Delta}$
(c) $Ba(s) + H_2O(l) \longrightarrow$

4.65 Predict the product(s) and write a balanced equation for each of the following redox reactions:
(a) $Fe(s) + HClO_4(aq) \longrightarrow$
(b) $S_8(s) + O_2(g) \longrightarrow$
(c) $BaCl_2(l) \xrightarrow{electricity}$

4.66 Predict the product(s) and write a balanced equation for each of the following redox reactions:
(a) Cesium + iodine $\longrightarrow$
(b) Aluminum + aqueous manganese(II) sulfate $\longrightarrow$
(c) Sulfur dioxide + oxygen $\longrightarrow$
(d) Propane and oxygen $\longrightarrow$
(e) Write a balanced net ionic equation for (b).

4.67 Predict the product(s) and write a balanced equation for each of the following redox reactions:
(a) Pentane (C_5H_{12}) + oxygen $\longrightarrow$
(b) Phosphorus trichloride + chlorine $\longrightarrow$
(c) Zinc + hydrobromic acid $\longrightarrow$
(d) Aqueous potassium iodide + bromine $\longrightarrow$
(e) Write a balanced net ionic equation for (d).

4.68 How many grams of O_2 can be prepared from the thermal decomposition of 4.27 kg of HgO? Name and calculate the mass (in kg) of the other product.

4.69 How many grams of chlorine gas can be produced from the electrolytic decomposition of 874 g of calcium chloride? Name and calculate the mass (in g) of the other product.

4.70 In a combination reaction, 1.62 g of lithium is mixed with 6.00 g of oxygen. (a) Which reactant is present in excess? (b) How many moles of product are formed? (c) After reaction, how many grams of each reactant and product are present?

4.71 In a combination reaction, 2.22 g of magnesium is heated with 3.75 g of nitrogen. (a) Which reactant is present in excess? (b) How many moles of product are formed? (c) After reaction, how many grams of each reactant and product are present?

4.72 A mixture of $CaCO_3$ and CaO weighing 0.693 g was heated to produce gaseous CO_2. After heating, the remaining solid weighed 0.508 g. Assuming all the $CaCO_3$ broke down to CaO and CO_2, calculate the mass percent of $CaCO_3$ in the original mixture.

4.73 Before arc welding was developed, a displacement reaction involving aluminum and iron(III) oxide was commonly used to produce molten iron (the thermite process). This reaction was used, for example, to connect sections of iron railroad track. Calculate the mass of molten iron produced when 1.00 kg of aluminum reacts with 2.00 mol of iron(III) oxide.

Comprehensive Problems
Problems with an asterisk (*) are more challenging.

4.74 Nutritional biochemists have known for decades that acidic foods cooked in cast-iron cookware can supply significant amounts of dietary iron (ferrous ion).
(a) Write a balanced net ionic equation, with oxidation numbers, that supports this fact.
(b) Measurements show an increase from 3.3 mg of iron to 49 mg of iron per $\frac{1}{2}$-cup (125-g) serving during the slow preparation of tomato sauce in a cast-iron pot. How many ferrous ions are present in a 26-oz (737-g) jar of the tomato sauce?

4.75 The brewing industry uses yeast microorganisms to convert glucose to ethanol for wine and beer. The baking industry uses the carbon dioxide produced to make bread rise:

$$C_6H_{12}O_6(s) \xrightarrow{\text{yeast}} 2C_2H_5OH(l) + 2CO_2(g)$$

How many grams of ethanol can be produced from 10.0 g of glucose? What volume of CO_2 is produced? (Assume 1 mol of gas occupies 22.4 L at the conditions used.)

* 4.76 A chemical engineer determines the mass percent of iron in an ore sample by converting the Fe to Fe^{2+} in acid and then titrating the Fe^{2+} with MnO_4^-. A 1.1081-g sample was dissolved in acid and then titrated with 39.32 mL of 0.03190 M KMnO₄. The balanced equation is

$$8H^+(aq) + 5Fe^{2+}(aq) + MnO_4^-(aq) \longrightarrow$$
$$5Fe^{3+}(aq) + Mn^{2+}(aq) + 4H_2O(l)$$

Calculate the mass percent of iron in the ore.

4.77 You are given solutions of HCl and NaOH and must determine their concentrations. You use 27.5 mL of NaOH to titrate 100. mL of HCl and 18.4 mL of NaOH to titrate 50.0 mL of 0.0782 M H_2SO_4. Find the unknown concentrations.

4.78 The flask (*right*) represents the products of the titration of 25 mL of sulfuric acid with 25 mL of sodium hydroxide.
(a) Write balanced molecular, total ionic, and net ionic equations for the reaction.
(b) If each orange sphere represents 0.010 mol of sulfate ion, how many moles of acid and of base reacted?
(c) What are the molarities of the acid and the base?

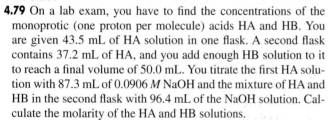

4.79 On a lab exam, you have to find the concentrations of the monoprotic (one proton per molecule) acids HA and HB. You are given 43.5 mL of HA solution in one flask. A second flask contains 37.2 mL of HA, and you add enough HB solution to it to reach a final volume of 50.0 mL. You titrate the first HA solution with 87.3 mL of 0.0906 M NaOH and the mixture of HA and HB in the second flask with 96.4 mL of the NaOH solution. Calculate the molarity of the HA and HB solutions.

4.80 Nitric acid, a major industrial and laboratory acid, is produced commercially by the multistep Ostwald process, which begins with the oxidation of ammonia:

Step 1. $4NH_3(g) + 5O_2(g) \longrightarrow 4NO(g) + 6H_2O(l)$
Step 2. $2NO(g) + O_2(g) \longrightarrow 2NO_2(g)$
Step 3. $3NO_2(g) + H_2O(l) \longrightarrow 2HNO_3(l) + NO(g)$

(a) What are the oxidizing and reducing agents in each step?
(b) Assuming 100% yield in each step, what mass (in kg) of ammonia must be used to produce 3.0×10^4 kg of HNO₃?

4.81 For the following aqueous reactions, complete and balance the molecular equation and write a net ionic equation:
(a) Manganese(II) sulfide + hydrobromic acid
(b) Potassium carbonate + strontium nitrate
(c) Potassium nitrite + hydrochloric acid
(d) Calcium hydroxide + nitric acid
(e) Barium acetate + iron(II) sulfate
(f) Barium hydroxide + hydrocyanic acid
(g) Copper(II) nitrate + hydrosulfuric acid
(h) Magnesium hydroxide + chloric acid
(i) Potassium chloride + ammonium phosphate

4.82 Sodium peroxide (Na_2O_2) is often used in self-contained breathing devices, such as those used in fire emergencies, because it reacts with exhaled CO_2 to form Na_2CO_3 and O_2. How many liters of respired air can react with 80.0 g of Na_2O_2 if each liter of respired air contains 0.0720 g of CO_2?

4.83 Magnesium is used in airplane bodies and other lightweight alloys. The metal is obtained from seawater in a process that includes precipitation, neutralization, evaporation, and electrolysis. How many kilograms of magnesium can be obtained from 1.00 km³ of seawater if the initial Mg^{2+} concentration is 0.13% by mass (d of seawater = 1.04 g/mL)?

4.84 Physicians who specialize in sports medicine routinely treat athletes and dancers. Ethyl chloride, a local anesthetic commonly used for simple injuries, is the product of the combination of ethylene with hydrogen chloride:

$$C_2H_4(g) + HCl(g) \longrightarrow C_2H_5Cl(g)$$

If 0.100 kg of C_2H_4 and 0.100 kg of HCl react:
(a) How many molecules of gas (reactants plus products) are present when the reaction is complete?
(b) How many moles of gas are present when half the product forms?

4.85 Carbon dioxide is removed from the atmosphere of space capsules by reaction with a solid metal hydroxide. The products are water and the metal carbonate.
(a) Calculate the mass of CO_2 that can be removed by reaction with 3.50 kg of lithium hydroxide.
(b) How many grams of CO_2 can be removed by 1.00 g of each of the following: lithium hydroxide, magnesium hydroxide, and aluminum hydroxide?

* 4.86 Calcium dihydrogen phosphate, $Ca(H_2PO_4)_2$, and sodium hydrogen carbonate, $NaHCO_3$, are ingredients of baking powder that react with each other to produce CO_2, which causes dough or batter to rise:

$$Ca(H_2PO_4)_2(s) + NaHCO_3(s) \longrightarrow$$
$$CO_2(g) + H_2O(g) + CaHPO_4(s) + Na_2HPO_4(s) \text{ [unbalanced]}$$

If the baking powder contains 31% $NaHCO_3$ and 35% $Ca(H_2PO_4)_2$ by mass:
(a) How many moles of CO_2 are produced from 1.00 g of baking powder?
(b) If 1 mol of CO_2 occupies 37.0 L at 350°F (a typical baking temperature), what volume of CO_2 is produced from 1.00 g of baking powder?

4.87 In a titration of HNO₃, you add a few drops of phenolphthalein indicator to 50.00 mL of acid in a flask. You quickly add 20.00 mL of 0.0502 M NaOH but overshoot the end point, and the solution turns deep pink. Instead of starting over, you add 30.00 mL of the acid, and the solution turns colorless. Then, it takes 3.22 mL of the NaOH to reach the end point. (a) What is the concentration of the HNO₃ solution? (b) How many moles of NaOH were in excess after the first addition?

* 4.88 The active compound in Pepto-Bismol contains C, H, O, and Bi.
(a) When 0.22105 g of it was burned in excess O_2, 0.1422 g of bismuth(III) oxide, 0.1880 g of carbon dioxide, and 0.02750 g of water were formed. What is the empirical formula of this compound?
(b) Given a molar mass of 1086 g/mol, determine the molecular formula.

(c) Complete and balance the acid-base reaction between bismuth(III) hydroxide and salicylic acid ($HC_7H_5O_3$), which is used to form this compound.

(d) A dose of Pepto-Bismol contains 0.600 mg of the active ingredient. If the yield of the reaction in part (c) is 88.0%, what mass (in mg) of bismuth(III) hydroxide is required to prepare one dose?

4.89 Two aqueous solutions contain the ions indicated below.

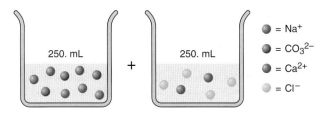

(a) Write balanced molecular, total ionic, and net ionic equations for the reaction that occurs when the solutions are mixed. (b) If each sphere represents 0.050 mol of ion, what mass (in g) of precipitate forms, assuming 100% reaction. (c) What is the concentration of each ion in solution after reaction?

* 4.90 In 1997, at the United Nations Conference on Climate Change, the major industrial nations agreed to expand their research efforts to develop renewable sources of carbon-based fuels. For more than a decade, Brazil has been engaged in a program to replace gasoline with ethanol derived from the root crop manioc (cassava). (a) Write separate balanced equations for the complete combustion of ethanol (C_2H_5OH) and of gasoline (represented by the formula C_8H_{18}).

(b) What mass of oxygen is required to burn completely 1.00 L of a mixture that is 90.0% gasoline ($d = 0.742$ g/mL) and 10.0% ethanol ($d = 0.789$ g/mL) by volume?

(c) If 1.00 mol of O_2 occupies 22.4 L, what volume of O_2 is needed to burn 1.00 L of the mixture?

(d) Air is 20.9% O_2 by volume. What volume of air is needed to burn 1.00 L of the mixture?

* 4.91 In a car engine, gasoline (represented by C_8H_{18}) does not burn completely, and some CO, a toxic pollutant, forms along with CO_2 and H_2O. If 5.0% of the gasoline forms CO:

(a) What is the ratio of CO_2 to CO molecules in the exhaust?

(b) What is the mass ratio of CO_2 to CO?

(c) What percentage of the gasoline must form CO for the mass ratio of CO_2 to CO to be exactly 1:1?

* 4.92 The amount of ascorbic acid (vitamin C; $C_6H_8O_6$) in tablets is determined by reaction with bromine and then titration of the hydrobromic acid with standard base:

$$C_6H_8O_6 + Br_2 \longrightarrow C_6H_6O_6 + 2HBr$$
$$HBr + NaOH \longrightarrow NaBr + H_2O$$

A certain tablet is advertised as containing 500 mg of vitamin C. One tablet was dissolved in water and reacted with Br_2. The solution was then titrated with 43.20 mL of 0.1350 M NaOH. Did the tablet contain the advertised quantity of vitamin C?

4.93 In the process of *pickling,* rust is removed from newly produced steel by washing the steel in hydrochloric acid:

(1) $6HCl(aq) + Fe_2O_3(s) \longrightarrow 2FeCl_3(aq) + 3H_2O(l)$

During the process, some iron is lost as well:

(2) $2HCl(aq) + Fe(s) \longrightarrow FeCl_2(aq) + H_2(g)$

(a) Which reaction, if either, is a redox process? (b) If reaction 2 did not occur and all the HCl were used, how many grams of Fe_2O_3 could be removed and $FeCl_3$ produced in a 2.50×10^3-L bath of 3.00 M HCl? (c) If reaction 1 did not occur and all the HCl were used, how many grams of Fe could be lost and $FeCl_2$ produced in a 2.50×10^3-L bath of 3.00 M HCl? (d) If 0.280 g of Fe is lost per gram of Fe_2O_3 removed, what is the mass ratio of $FeCl_2$ to $FeCl_3$?

4.94 At liftoff, the space shuttle uses a solid mixture of ammonium perchlorate and aluminum powder to obtain great thrust from the volume change of solid to gas. In the presence of a catalyst, the mixture forms solid aluminum oxide and aluminum trichloride and gaseous water and nitrogen monoxide. (a) Write a balanced equation for the reaction, and identify the reducing and oxidizing agents. (b) How many total moles of gas (water vapor and nitrogen monoxide) are produced when 50.0 kg of ammonium perchlorate reacts with a stoichiometric amount of Al? (c) What is the volume change from this reaction? (d of $NH_4ClO_4 = 1.95$ g/cc, Al $= 2.70$ g/cc, $Al_2O_3 = 3.97$ g/cc, and $AlCl_3 = 2.44$ g/cc; assume 1 mol of gas occupies 22.4 L.)

CHAPTER FIVE

Gases and the Kinetic-Molecular Theory

Floating on a Gas Law Ballooning, bread-making, breathing, and many other familiar phenomena operate through a few simple behaviors of gases that you'll learn in this chapter.

Key Principles

◆ Properties of gases differ significantly from those of liquids and solids because gas particles are very far apart.

◆ *Pressure* is a force acting on an area; atmospheric gases exert a pressure on Earth's surface that is measured with a *barometer*.

◆ Measurable gas variables—*volume (V), pressure (P), temperature (T),* and *amount (n)*—are *interdependent*. For a hypothetical *ideal gas,* volume changes *linearly* with a change in any one of the other variables, as long as the remaining two are held constant. These behaviors are described by gas laws (*Boyle's, Charles's,* and *Avogadro's*), which are combined into the *ideal gas law (PV = nRT)*. Most simple gases behave ideally at ordinary conditions (low pressure and high temperature).

◆ Rearrangements of the ideal gas law are used to calculate the *molar mass* of a gas, the *density* of a gas, the *partial pressure* of each gas in a mixture of gases, and the amounts of gaseous reactants or products in a reaction.

◆ To explain the gas laws, the *kinetic-molecular theory* postulates that an ideal gas consists of points of mass *moving randomly in straight lines* between *elastic collisions* (no loss of energy). A key result is that all gases have the same average kinetic energy at a given temperature. Thus, temperature is a measure of energy of molecular motion.

◆ Heavier gas molecules move more slowly than lighter ones, which helps explain why gases *effuse* (move out of a container through a tiny hole into a vacuum) or *diffuse* (move through one another) at rates inversely proportional to the square root of their molar mass (*Graham's law*).

◆ At extreme conditions (low temperature and high pressure), *real gas behavior deviates* from ideal behavior because the volume of the gas molecules and the attractions (and repulsions) they experience during collisions become important factors. The *van der Waals equation,* an adjusted version of the ideal gas law, accounts for these effects.

Outline

Gases are everywhere. People have been observing their behavior, and that of matter in other states, throughout history—three of the four "elements" of the ancients were air (gas), water (liquid), and earth (solid). However, many questions remain. In this chapter and its companion, Chapter 12, we examine these states and their interrelations. Here, we highlight the gaseous state, the one we understand best.

Although the *chemical* behavior of a gas depends on its composition, all gases have remarkably similar *physical* behavior, which is the focus of this chapter. For instance, although the particular gases differ, the same physical behaviors are at work in the operation of a car and in the baking of bread, in the thrust of a rocket engine and in the explosion of a kernel of popcorn, in the process of breathing and in the creation of thunder.

Concepts & Skills to Review
Before You Study This Chapter

- physical states of matter (Section 1.1)
- SI unit conversions (Section 1.4)
- mole-mass-number conversions (Section 3.1)

5.1 AN OVERVIEW OF THE PHYSICAL STATES OF MATTER

Under appropriate conditions of pressure and temperature, most substances can exist as a solid, a liquid, or a gas. In Chapter 1 we described these physical states in terms of how each fills a container and began to develop a molecular view that explains this macroscopic behavior: a solid has a fixed shape regardless of the container shape because its particles are held rigidly in place; a liquid conforms to the container shape but has a definite volume and a surface because its particles are close together but free to move around each other; and a gas fills the container because its particles are far apart and moving randomly. Several other aspects of their behavior distinguish gases from liquids and solids:

1. *Gas volume changes greatly with pressure.* When a sample of gas is confined to a container of variable volume, such as the piston-cylinder assembly of a car engine, an external force can compress the gas. Removing the external force allows the gas volume to increase again. In contrast, a liquid or solid resists significant changes in volume.
2. *Gas volume changes greatly with temperature.* When a gas sample at constant pressure is heated, its volume increases; when it is cooled, its volume decreases. This volume change is 50 to 100 times greater for gases than for liquids or solids.
3. *Gases have relatively low viscosity.* Gases flow much more freely than liquids and solids. Low viscosity allows gases to be transported through pipes over long distances but also to leak rapidly out of small holes.
4. *Most gases have relatively low densities* under normal conditions. Gas density is usually tabulated in units of grams per *liter*, whereas liquid and solid densities are in grams per *milliliter*, about 1000 times as dense. For example, at 20°C and normal atmospheric pressure, the density of $O_2(g)$ is 1.3 g/**L**, whereas the density of $H_2O(l)$ is 1.0 g/**mL** and that of NaCl(s) is 2.2 g/**mL.** When a gas is cooled, its density increases because its volume decreases: at 0°C, the density of $O_2(g)$ increases to 1.4 g/L.
5. *Gases are miscible. Miscible* substances mix with one another in any proportion to form a solution. Dry air, for example, is a solution of about 18 gases. Two liquids, however, may or may not be miscible: water and ethanol are, but water and gasoline are not. Two solids generally do not form a solution unless they are mixed as molten liquids and then allowed to solidify.

Each of these observable properties offers a clue to the molecular properties of gases. For example, consider these density data. At 20°C and normal atmospheric pressure, gaseous N_2 has a density of 1.25 g/L. If cooled below −196°C, it condenses to liquid N_2 and its density becomes 0.808 g/mL. (Note the change

Figure 5.1 The three states of matter. Many pure substances, such as bromine (Br_2), can exist under appropriate conditions of pressure and temperature as **A,** a gas; **B,** a liquid; or **C,** a solid. The atomic-scale views show that molecules are much farther apart in a gas than in a liquid or solid.

A Gas: Molecules are far apart, move freely, and fill the available space

B Liquid: Molecules are close together but move around one another

C Solid: Molecules are close together in a regular array and do not move around one another

in units.) The same amount of nitrogen occupies less than $\frac{1}{600}$ as much space! Further cooling to below $-210°C$ yields solid N_2 ($d = 1.03$ g/mL), which is only slightly more dense than the liquid. These values show again that *the molecules are much farther apart in the gas than in either the liquid or the solid.* Moreover, a large amount of space between molecules is consistent with gases' miscibility, low viscosity, and compressibility. Figure 5.1 compares macroscopic and atomic-scale views of the physical states of a real substance.

SECTION SUMMARY
The volume of a gas can be altered significantly by changing the applied external force or the temperature. The corresponding changes for liquids and solids are much smaller. Gases flow more freely and have lower densities than liquids and solids, and they mix in any proportion to form solutions. The reason for these differences is the greater distance between particles in a gas than in a liquid or a solid.

5.2   GAS PRESSURE AND ITS MEASUREMENT

Blowing up a balloon provides clear evidence that a gas exerts pressure on the walls of its container. **Pressure (P)** is defined as the force exerted per unit of surface area:

$$\text{Pressure} = \frac{\text{force}}{\text{area}}$$

Earth's gravitational attraction pulls the atmospheric gases toward its surface, where they exert a force on all objects. The force, or weight, of these gases creates a pressure of about 14.7 pounds per square inch (lb/in^2; psi) of surface.

The molecules in a gas are moving in every direction, so the pressure of the atmosphere is exerted uniformly on the floor, walls, ceiling, and every object in a room. The pressure on the outside of your body is equalized by the pressure on the inside, so there is no net pressure on your body's outer surface. What would happen if this were not the case? As an analogy, consider the empty metal can attached to a vacuum pump in Figure 5.2. With the pump off, the can maintains its shape because the pressure on the outside is equal to the pressure on the inside. With the pump on, the internal pressure decreases greatly, and the ever-present external pressure easily crushes the can. A vacuum-filtration flask (and tubing), which you may have used in the lab, has thick walls that can withstand the external pressure when the flask is evacuated.

Measuring Atmospheric Pressure

The **barometer** is a common device used to measure atmospheric pressure. Invented in 1643 by Evangelista Torricelli, the barometer is still basically just a tube about 1 m long, closed at one end, filled with mercury, and inverted into a dish containing more mercury. When the tube is inverted, some of the mercury flows out into the dish, and a vacuum forms above the mercury remaining in the tube, as shown in Figure 5.3. At sea level under ordinary atmospheric conditions, the outward flow of mercury stops when the surface of the mercury in the tube is about 760 mm above the surface of the mercury in the dish. It stops at 760 mm because at that point the column of mercury in the tube exerts the same pressure (weight/area) on the mercury surface in the dish as does the column of air that extends from the dish to the outer reaches of the atmosphere. All that air pushing down keeps any more of the mercury in the tube from flowing out. Likewise, if you place an evacuated tube into a dish filled with mercury, the mercury rises about 760 mm into the tube because the atmosphere pushes the mercury up to that height.

Notice that we did not specify the diameter of the barometer tube. If the mercury in a 1-cm diameter tube rises to a height of 760 mm, the mercury in a 2-cm diameter tube will rise to that height also. The *weight* of mercury is greater in the wider tube, but the area is larger also; thus the *pressure,* the *ratio* of weight to area, is the same.

Since the pressure of the mercury column is directly proportional to its height, a unit commonly used for pressure is mmHg, the height of the column in millimeters (mm). At sea level and 0°C, normal atmospheric pressure is 760 mmHg; at the top of Mt. Everest (29,028 ft, or 8848 m), the atmospheric pressure is only about 270 mmHg. Thus, *pressure decreases with altitude:* the column of air above the sea is taller and weighs more than the column of air above Mt. Everest.

Laboratory barometers contain mercury because its high density allows the barometer to be a convenient size. For example, the pressure of the atmosphere would equal the pressure of a column of water about 10,300 mm, almost 34 ft, high. Note that, for a given pressure, the ratio of heights (h) of the liquid columns is inversely related to the ratio of the densities (d) of the liquids:

$$\frac{h_{H_2O}}{h_{Hg}} = \frac{d_{Hg}}{d_{H_2O}}$$

Units of Pressure

Pressure results from a force exerted on an area. The SI unit of force is the newton (N): $1\ N = 1\ kg \cdot m/s^2$. The SI unit of pressure is the **pascal (Pa),** which equals a force of one newton exerted on an area of one square meter:

$$1\ Pa = 1\ N/m^2$$

A much larger unit is the **standard atmosphere (atm),** the average atmospheric pressure measured at sea level and 0°C. It is defined in terms of the pascal:

$$1\ atm = 101.325\ kilopascals\ (kPa) = 1.01325 \times 10^5\ Pa$$

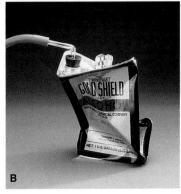

Figure 5.2 Effect of atmospheric pressure on objects at Earth's surface. A, A metal can filled with air has equal pressure on the inside and outside. **B,** When the air inside the can is removed, the atmospheric pressure crushes the can.

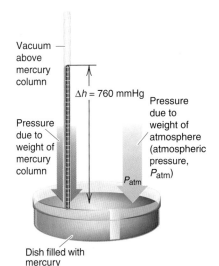

Figure 5.3 A mercury barometer.

Table 5.1	**Common Units of Pressure**	
Unit	Atmospheric Pressure	Scientific Field
pascal (Pa); kilopascal (kPa)	1.01325×10^5 Pa; 101.325 kPa	SI unit; physics, chemistry
atmosphere (atm)	1 atm*	Chemistry
millimeters of mercury (mmHg)	760 mmHg*	Chemistry, medicine, biology
torr	760 torr*	Chemistry
pounds per square inch (lb/in² or psi)	14.7 lb/in²	Engineering
bar	1.01325 bar	Meteorology, chemistry, physics

*This is an exact quantity; in calculations, we use as many significant figures as necessary.

Another common unit is the **millimeter of mercury (mmHg),** mentioned above, which is based on measurement with a barometer. In honor of Torricelli, this unit has been named the **torr:**

$$1 \text{ torr} = 1 \text{ mmHg} = \frac{1}{760} \text{ atm} = \frac{101.325}{760} \text{ kPa} = 133.322 \text{ Pa}$$

The *bar* is coming into more common use in chemistry:

$$1 \text{ bar} = 1 \times 10^2 \text{ kPa} = 1 \times 10^5 \text{ Pa}$$

Despite a gradual change to SI units, many chemists still express pressure in torrs and atmospheres. Table 5.1 lists some important pressure units used in various scientific fields.

SAMPLE PROBLEM 5.1   Converting Units of Pressure

Problem A geochemist heats a limestone ($CaCO_3$) sample and collects the CO_2 released in an evacuated flask. The CO_2 pressure is 291.4 mmHg. Calculate the CO_2 pressure in torrs, atmospheres, and kilopascals.

Plan The CO_2 pressure is given in units of mmHg, so we construct conversion factors from Table 5.1 to find the pressure in the other units.

Solution Converting from mmHg to torr:

$$P_{CO_2}(\text{torr}) = 291.4 \text{ mmHg} \times \frac{1 \text{ torr}}{1 \text{ mmHg}} = 291.4 \text{ torr}$$

Converting from torr to atm:

$$P_{CO_2}(\text{atm}) = 291.4 \text{ torr} \times \frac{1 \text{ atm}}{760 \text{ torr}} = 0.3834 \text{ atm}$$

Converting from atm to kPa:

$$P_{CO_2}(\text{kPa}) = 0.3834 \text{ atm} \times \frac{101.325 \text{ kPa}}{1 \text{ atm}} = 38.85 \text{ kPa}$$

Check There are 760 torr in 1 atm, so ~300 torr should be <0.5 atm. There are ~100 kPa in 1 atm, so <0.5 atm should be <50 kPa.

Comment 1. In the conversion from torr to atm, we retained four significant figures because this unit conversion factor involves *exact* numbers; that is, 760 torr has as many significant figures as the calculation requires.

2. From here on, except in particularly complex situations, *the canceling of units in calculations is no longer shown.*

FOLLOW-UP PROBLEM 5.1 The CO_2 released from another mineral sample was collected in an evacuated flask. If the pressure of the CO_2 is 579.6 torr, calculate P_{CO_2} in mmHg, pascals, and lb/in².

SECTION SUMMARY
Gases exert pressure (force/area) on all surfaces with which they make contact. A barometer measures atmospheric pressure in terms of the height of the mercury column that the atmosphere can support (760 mmHg at sea level and 0°C). Chemists measure pressure in units of atmospheres (atm), torrs (equivalent to mmHg), and pascals (Pa, the SI unit).

5.3 THE GAS LAWS AND THEIR EXPERIMENTAL FOUNDATIONS

The physical behavior of a sample of gas can be described completely by four variables: pressure (P), volume (V), temperature (T), and amount (number of moles, n). The variables are interdependent: *any one of them can be determined by measuring the other three.* We know now that this quantitatively predictable behavior is a direct outcome of the structure of gases on the molecular level. Yet, it was discovered, for the most part, before Dalton's atomic theory was published!

Three key relationships exist among the four gas variables—Boyle's, Charles's, and Avogadro's laws. Each of these gas laws expresses *the effect of one variable on another, with the remaining two variables held constant.* Because gas volume is so easy to measure, the laws are expressed as the effect on gas volume of a change in pressure, temperature, or amount of gas.

These three laws are special cases of an all-encompassing relationship among gas variables called the *ideal gas law.* This unifying observation quantitatively describes the state of a so-called **ideal gas,** one that exhibits simple linear relationships among volume, pressure, temperature, and amount. Although no ideal gas actually exists, most simple gases, such as N_2, O_2, H_2, and the noble gases, show nearly ideal behavior at ordinary temperatures and pressures. We discuss the ideal gas law after the three special cases.

The Relationship Between Volume and Pressure: Boyle's Law

The great 17th-century English chemist Robert Boyle performed a series of experiments that led him to conclude that at a given temperature, *the volume occupied by a gas is inversely related to its pressure.* Figure 5.4 shows Boyle's experiment

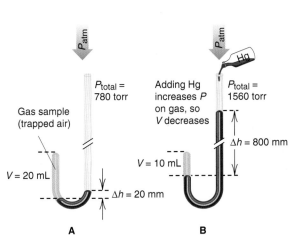

V (mL)	P (torr)			$\dfrac{1}{P_{total}}$	PV (torr•mL)
	Δh	+ P_{atm}	= P_{total}		
20.0	20.0	760	780	0.00128	1.56×10^4
15.0	278	760	1038	0.000963	1.56×10^4
10.0	800	760	1560	0.000641	1.56×10^4
5.0	2352	760	3112	0.000321	1.56×10^4

Figure 5.4 The relationship between the volume and pressure of a gas. A, A small amount of air (the gas) is trapped in the short arm of a J tube; n and T are fixed. The total pressure on the gas (P_{total}) is the sum of the pressure due to the difference in heights of the mercury columns (Δh) plus the pressure of the atmosphere (P_{atm}). If P_{atm} = 760 torr, P_{total} = 780 torr. **B,** As mercury is added, the total pressure on the gas increases and its volume (V) decreases. Note that if P_{total} is doubled (to 1560 torr), V is halved (not drawn to scale). **C,** Some typical pressure-volume data from the experiment. **D,** A plot of V vs. P_{total} shows that V is inversely proportional to P. **E,** A plot of V vs. $1/P_{total}$ is a straight line whose slope is a constant characteristic of *any* gas that behaves ideally.

and some typical data he might have collected. Boyle fashioned a J-shaped glass tube, sealed the shorter end, and poured mercury into the longer end, thereby trapping some air, the gas in the experiment. From the height of the trapped air column and the diameter of the tube, he calculated the air volume. The total pressure applied to the trapped air was the pressure of the atmosphere (measured with a barometer) plus that of the mercury column (part A). By adding mercury, Boyle increased the total pressure exerted on the air, and the air volume decreased (part B). With the temperature and amount of air constant, Boyle could directly measure the effect of the applied pressure on the volume of air.

Note the following results, shown in Figure 5.4:

- The product of corresponding P and V values is a constant (part C, rightmost column).
- V is *inversely* proportional to P (part D).
- V is *directly* proportional to $1/P$ (part E) and generates a linear plot of V against $1/P$. This *linear relationship between two gas variables* is a hallmark of ideal gas behavior.

The generalization of Boyle's observations is known as **Boyle's law:** *at constant temperature, the volume occupied by a fixed amount of gas is inversely proportional to the applied (external) pressure,* or

$$V \propto \frac{1}{P} \qquad [T \text{ and } n \text{ fixed}] \qquad (5.1)$$

This relationship can also be expressed as

$$PV = \text{constant} \qquad \text{or} \qquad V = \frac{\text{constant}}{P} \qquad [T \text{ and } n \text{ fixed}]$$

The constant is the same for the great majority of gases. Thus, tripling the external pressure reduces the volume to one-third its initial value; halving the external pressure doubles the volume; and so forth.

The wording of Boyle's law focuses on *external* pressure. In his experiment, however, adding more mercury caused the mercury level to rise until the pressure of the trapped air stopped the rise at some new level. At that point, the pressure exerted *on* the gas equaled the pressure exerted *by* the gas. In other words, by measuring the applied pressure, Boyle was also measuring the gas pressure. Thus, when gas volume doubles, gas pressure is halved. In general, if V_{gas} increases, P_{gas} decreases, and vice versa.

The Relationship Between Volume and Temperature: Charles's Law

One question raised by Boyle's work was why the pressure-volume relationship holds only at constant temperature. It was not until the early 19th century, through the separate work of the French scientists J. A. C. Charles and J. L. Gay-Lussac, that the relationship between gas volume and temperature was clearly understood.

Let's examine this relationship by measuring the volume of a fixed amount of a gas under constant pressure but at different temperatures. A straight tube, closed at one end, traps a fixed amount of air under a small mercury plug. The tube is immersed in a water bath that can be warmed with a heater or cooled with ice. After each change of water temperature, we measure the length of the air column, which is proportional to its volume. The pressure exerted on the gas is constant because the mercury plug and the atmospheric pressure do not change (Figure 5.5A and B).

Some typical data are shown for different amounts and pressures of gas in Figure 5.5C. Again, note the linear relationships, but this time the variables are

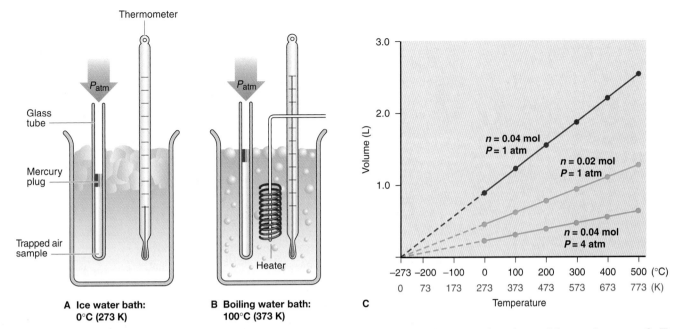

Figure 5.5 The relationship between the volume and temperature of a gas. At constant P, the volume of a given amount of gas is *directly* proportional to the absolute temperature. A fixed amount of gas (air) is trapped under a small plug of mercury at a fixed pressure. **A,** The sample is in an ice water bath. **B,** The sample is in a boiling water bath. As the temperature increases, the volume of the gas increases. **C,** The three lines show the effect of amount (n) of gas (compare red and green) and pressure (P) of gas (compare red and blue). The dashed lines extrapolate the data to lower temperatures. For any amount of an ideal gas at any pressure, the volume is theoretically zero at −273.15°C (0 K).

directly proportional: for a given amount of gas at a given pressure, *volume increases as temperature increases.* For example, the red line shows how the volume of 0.04 mol of gas at 1 atm pressure changes as the temperature changes. Extending (*extrapolating*) the line to lower temperatures (dashed portion) shows that the volume shrinks until the gas occupies a theoretical zero volume at −273.15°C (the intercept on the temperature axis). Similar plots for a different amount of gas (green) and a different gas pressure (blue) show lines with different slopes, but they all converge at this temperature, called *absolute zero* (0 K or −273.15°C). Of course, no sample of matter can have zero volume, and every real gas condenses to a liquid at some temperature higher than 0 K. Nevertheless, *this linear relationship between volume and absolute temperature holds for most common gases over a wide temperature range.*

The modern statement of the volume-temperature relationship is known as **Charles's law:** *at constant pressure, the volume occupied by a fixed amount of gas is directly proportional to its absolute (Kelvin) temperature,* or

$$V \propto T \qquad [P \text{ and } n \text{ fixed}] \qquad (5.2)$$

This relationship can also be expressed as

$$\frac{V}{T} = \text{constant} \qquad \text{or} \qquad V = \text{constant} \times T \qquad [P \text{ and } n \text{ fixed}]$$

If T increases, V increases, and vice versa. Once again, for any given P and n, the constant is the same for the great majority of gases.

The dependence of gas volume on *absolute* temperature means that you must *use the Kelvin scale in gas law calculations.* For instance, if the temperature changes from 200 K to 400 K, the volume of 1 mol of gas doubles. But, if the temperature changes from 200°C to 400°C, the volume increases by a factor of 1.42; that is, $\left(\dfrac{400°C + 273.15}{200°C + 273.15}\right) = \dfrac{673}{473} = 1.42$.

Other Relationships Based on Boyle's and Charles's Laws Two other important relationships in gas behavior emerge from an understanding of Boyle's and Charles's laws:

1. *The pressure-temperature relationship.* Charles's law is expressed as the effect of a temperature change on gas *volume.* However, volume and pressure are interdependent, so the effect of temperature on volume is closely related to its effect on pressure (sometimes referred to as *Amontons's law*). Measure the pressure in your car's tires before and after a long drive, and you will find that it has increased. Frictional heating between the tire and the road increases the air temperature inside the tire, but since the tire volume doesn't change appreciably, the air exerts more pressure. Thus, *at constant volume, the pressure exerted by a fixed amount of gas is directly proportional to the absolute temperature:*

$$P \propto T \quad [V \text{ and } n \text{ fixed}] \tag{5.3}$$

or

$$\frac{P}{T} = \text{constant} \qquad \text{or} \qquad P = \text{constant} \times T$$

2. *The combined gas law.* A simple combination of Boyle's and Charles's laws gives the *combined gas law,* which applies to situations when two of the three variables (*V, P, T*) change and you must find the effect on the third:

$$V \propto \frac{T}{P} \qquad \text{or} \qquad V = \text{constant} \times \frac{T}{P} \qquad \text{or} \qquad \frac{PV}{T} = \text{constant}$$

The Relationship Between Volume and Amount: Avogadro's Law

Boyle's and Charles's laws both specify a fixed amount of gas. Let's see why. Figure 5.6 shows an experiment that involves two small test tubes, each fitted with a piston-cylinder assembly. We add 0.10 mol (4.4 g) of dry ice (frozen CO_2) to the first (tube A) and 0.20 mol (8.8 g) to the second (tube B). As the solid warms, it changes directly to gaseous CO_2, which expands into the cylinder and pushes up the piston. When all the solid has changed to gas and the temperature is constant, we find that cylinder A has half the volume of cylinder B. (We can neglect the volume of the tube because it is so much smaller than the volume of the cylinder.)

This experimental result shows that twice the amount (mol) of gas occupies twice the volume. Notice that, for both cylinders, the *T* of the gas equals room temperature and the *P* of the gas equals atmospheric pressure. Thus, *at fixed*

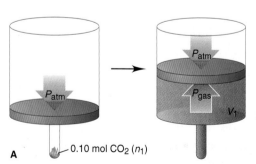

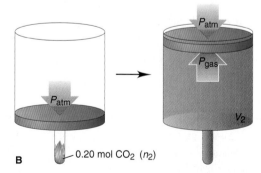

Figure 5.6 An experiment to study the relationship between the volume and amount of a gas. A, At a given external *P* and *T*, a given amount (n_1) of $CO_2(s)$ is put into the tube. When the CO_2 changes from solid to gas, it pushes up the piston until $P_{gas} = P_{atm}$, at which point it occupies a given volume of the cylinder. **B,** When twice the amount (n_2) of $CO_2(s)$ is used, twice the volume of the cylinder becomes occupied. Thus, at fixed *P* and *T*, the volume (*V*) of a gas is directly proportional to the amount of gas (*n*).

temperature and pressure, the volume occupied by a gas is directly proportional to the amount (mol) of gas:

$$V \propto n \qquad [P \text{ and } T \text{ fixed}] \tag{5.4}$$

As n increases, V increases, and vice versa. This relationship is also expressed as

$$\frac{V}{n} = \text{constant} \qquad \text{or} \qquad V = \text{constant} \times n$$

The constant is the same for all gases at a given temperature and pressure. This relationship is another way of expressing **Avogadro's law,** which states that *at fixed temperature and pressure, equal volumes of **any** ideal gas contain equal numbers of particles (or moles).*

The gas laws apply to many familiar phenomena: gasoline burning in a car engine, dough rising—even the act of breathing. When you inhale, the downward movement of your diaphragm and the expansion of your rib cage increase your lung volume, which decreases the air pressure inside, so air rushes in (Boyle's law). The greater amount of air stretches the lung tissue, which expands the volume further (Avogadro's law), and the air expands slightly as it warms to body temperature (Charles's law). When you exhale, these steps occur in reverse.

Gas Behavior at Standard Conditions

To better understand the factors that influence gas behavior, chemists use a set of *standard conditions* called **standard temperature and pressure (STP):**

$$\text{STP:} \qquad 0°\text{C (273.15 K) and 1 atm (760 torr)} \tag{5.5}$$

Under these conditions, the volume of 1 mol of an ideal gas is called the **standard molar volume:**

$$\text{Standard molar volume} = 22.4141 \text{ L} \quad \text{or} \quad 22.4 \text{ L} \quad [\text{to 3 sf}] \tag{5.6}$$

Figure 5.7 compares the properties of three simple gases at STP.

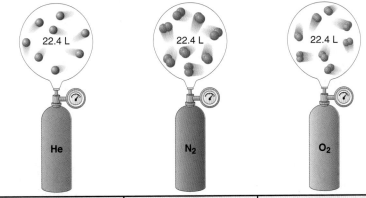

$n = 1$ mol	$n = 1$ mol	$n = 1$ mol
$P = 1$ atm (760 torr)	$P = 1$ atm (760 torr)	$P = 1$ atm (760 torr)
$T = 0°$C (273 K)	$T = 0°$C (273 K)	$T = 0°$C (273 K)
$V = 22.4$ L	$V = 22.4$ L	$V = 22.4$ L
Number of gas particles $= 6.022 \times 10^{23}$	Number of gas particles $= 6.022 \times 10^{23}$	Number of gas particles $= 6.022 \times 10^{23}$
Mass = 4.003 g	Mass = 28.02 g	Mass = 32.00 g
$d = 0.179$ g/L	$d = 1.25$ g/L	$d = 1.43$ g/L

Figure 5.7 Standard molar volume. One mole of an ideal gas occupies 22.4 L at STP (0°C and 1 atm). At STP, helium, nitrogen, oxygen, and most other simple gases behave ideally. Note that the *mass* of a gas, and thus its density (*d*), depends on its molar mass.

Figure 5.8 The volumes of 1 mol of an ideal gas and some familiar objects. A basketball (7.5 L), 5-gal fish tank (18.9 L), 13-in television (21.6 L), and 22.4 L of He gas in a balloon.

Figure 5.8 compares the volumes of some familiar objects with the standard molar volume of an ideal gas.

The Ideal Gas Law

Each of the gas laws focuses on the effect that changes in one variable have on gas volume:

- Boyle's law focuses on pressure ($V \propto 1/P$).
- Charles's law focuses on temperature ($V \propto T$).
- Avogadro's law focuses on amount (mol) of gas ($V \propto n$).

We can combine these individual effects into one relationship, called the **ideal gas law** (or *ideal gas equation*):

$$V \propto \frac{nT}{P} \quad \text{or} \quad PV \propto nT \quad \text{or} \quad \frac{PV}{nT} = R$$

where R is a proportionality constant known as the **universal gas constant.** Rearranging gives the most common form of the ideal gas law:

$$PV = nRT \tag{5.7}$$

We can obtain a value of R by measuring the volume, temperature, and pressure of a given amount of gas and substituting the values into the ideal gas law. For example, using standard conditions for the gas variables, we have

$$R = \frac{PV}{nT} = \frac{1 \text{ atm} \times 22.4141 \text{ L}}{1 \text{ mol} \times 273.15 \text{ K}} = 0.082058 \, \frac{\text{atm·L}}{\text{mol·K}} = 0.0821 \, \frac{\text{atm·L}}{\text{mol·K}} \quad \text{[3 sf]} \tag{5.8}$$

This numerical value of R corresponds to the gas variables P, V, and T expressed *in these units*. R has a different numerical value when different units are used. For example, later in this chapter, R has the value 8.314 J/mol·K (J stands for joule, the SI unit of energy).

Figure 5.9 makes a central point: the ideal gas law *becomes* one of the individual gas laws when two of the four variables are kept constant. When initial conditions (subscript $_1$) change to final conditions (subscript $_2$), we have

$$P_1V_1 = n_1RT_1 \quad \text{and} \quad P_2V_2 = n_2RT_2$$

Thus,

$$\frac{P_1V_1}{n_1T_1} = R \quad \text{and} \quad \frac{P_2V_2}{n_2T_2} = R, \quad \text{so} \quad \frac{P_1V_1}{n_1T_1} = \frac{P_2V_2}{n_2T_2}$$

Notice that if two of the variables remain constant, say P and T, then $P_1 = P_2$ and $T_1 = T_2$, and we obtain an expression for Avogadro's law:

$$\frac{P_{\cancel{1}}V_1}{n_1T_{\cancel{1}}} = \frac{P_{\cancel{2}}V_2}{n_2T_{\cancel{2}}} \quad \text{or} \quad \frac{V_1}{n_1} = \frac{V_2}{n_2}$$

We use rearrangements of the ideal gas law such as this one to solve gas law problems, as you'll see next. The point to remember is that there is no need to memorize the individual gas laws.

Figure 5.9 Relationship between the ideal gas law and the individual gas laws. Boyle's, Charles's, and Avogadro's laws are contained within the ideal gas law.

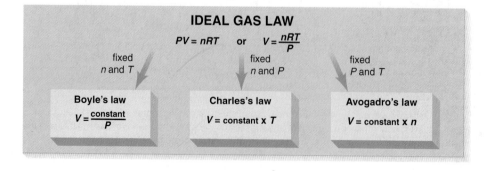

Solving Gas Law Problems

Gas law problems are phrased in many ways, but they can usually be grouped into two main types:

1. *A change in one of the four variables causes a change in another, while the two remaining variables remain constant.* In this type, the ideal gas law reduces to one of the individual gas laws, and you solve for the new value of the variable. Units must be consistent, T must always be in kelvins, but R is not involved. Sample Problems 5.2 to 5.4 and 5.6 are of this type. (A variation on this type involves the combined gas law for simultaneous changes in two of the variables that cause a change in a third.)

2. *One variable is unknown, but the other three are known and no change occurs.* In this type, exemplified by Sample Problem 5.5, the ideal gas law is applied directly to find the unknown, and the units must conform to those in R.

These problems are far easier to solve if you follow a systematic approach:

- Summarize the information: identify the changing gas variables—knowns and unknown—and those held constant.
- Predict the direction of the change, and later check your answer against the prediction.
- Perform any necessary unit conversions.
- Rearrange the ideal gas law to obtain the appropriate relationship of gas variables, and solve for the unknown variable.

The following series of sample problems applies the various gas behaviors.

SAMPLE PROBLEM 5.2 Applying the Volume-Pressure Relationship

Problem Boyle's apprentice finds that the air trapped in a J tube occupies 24.8 cm³ at 1.12 atm. By adding mercury to the tube, he increases the pressure on the trapped air to 2.64 atm. Assuming constant temperature, what is the new volume of air (in L)?

Plan We must find the final volume (V_2) in liters, given the initial volume (V_1), initial pressure (P_1), and final pressure (P_2). The temperature and amount of gas are fixed. We convert the units of V_1 from cm³ to mL and then to L, rearrange the ideal gas law to the appropriate form, and solve for V_2. We can predict the direction of the change: we know that P increases, so V will decrease; thus, $V_2 < V_1$. (Note that the roadmap has two parts.)

Solution Summarizing the gas variables:

$$P_1 = 1.12 \text{ atm} \qquad P_2 = 2.64 \text{ atm}$$
$$V_1 = 24.8 \text{ cm}^3 \text{ (convert to L)} \qquad V_2 = \text{unknown} \qquad T \text{ and } n \text{ remain constant}$$

Converting V_1 from cm³ to L:

$$V_1 = 24.8 \text{ cm}^3 \times \frac{1 \text{ mL}}{1 \text{ cm}^3} \times \frac{1 \text{ L}}{1000 \text{ mL}} = 0.0248 \text{ L}$$

Arranging the ideal gas law and solving for V_2: At fixed n and T, we have

$$\frac{P_1 V_1}{n_1 T_1} = \frac{P_2 V_2}{n_2 T_2} \qquad \text{or} \qquad P_1 V_1 = P_2 V_2$$

$$V_2 = V_1 \times \frac{P_1}{P_2} = 0.0248 \text{ L} \times \frac{1.12 \text{ atm}}{2.64 \text{ atm}} = 0.0105 \text{ L}$$

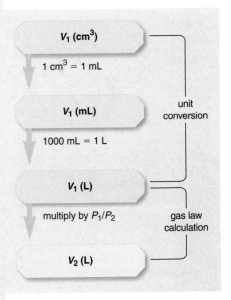

V_1 (cm³)

1 cm³ = 1 mL

V_1 (mL)

1000 mL = 1 L

V_1 (L)

multiply by P_1/P_2

V_2 (L)

unit conversion

gas law calculation

Check As we predicted, $V_2 < V_1$. Let's think about the relative values of P and V as we check the math. P more than doubled, so V_2 should be less than $\frac{1}{2}V_1$ (0.0105/0.0248 < $\frac{1}{2}$).
Comment Predicting the direction of the change provides another check on the problem setup: To make $V_2 < V_1$, we must multiply V_1 by a number *less than* 1. This means the ratio of pressures must be *less than* 1, so the larger pressure (P_2) must be in the denominator, P_1/P_2.

FOLLOW-UP PROBLEM 5.2 A sample of argon gas occupies 105 mL at 0.871 atm. If the temperature remains constant, what is the volume (in L) at 26.3 kPa?

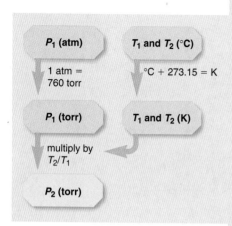

SAMPLE PROBLEM 5.3 Applying the Pressure-Temperature Relationship

Problem A steel tank used for fuel delivery is fitted with a safety valve that opens if the internal pressure exceeds 1.00×10^3 torr. It is filled with methane at 23°C and 0.991 atm and placed in boiling water at exactly 100°C. Will the safety valve open?

Plan The question "Will the safety valve open?" translates into "Is P_2 greater than 1.00×10^3 torr at T_2?" Thus, P_2 is the unknown, and T_1, T_2, and P_1 are given, with V (steel tank) and n fixed. We convert both T values to kelvins and P_1 to torrs in order to compare P_2 with the safety-limit pressure. We rearrange the ideal gas law to the appropriate form and solve for P_2. We predict that $P_2 > P_1$ because $T_2 > T_1$.

Solution Summary of gas variables:

$$P_1 = 0.991 \text{ atm (convert to torr)} \qquad P_2 = \text{unknown}$$
$$T_1 = 23°C \text{ (convert to K)} \qquad T_2 = 100°C \text{ (convert to K)}$$
$$V \text{ and } n \text{ remain constant}$$

Converting T from °C to K:

$$T_1 \text{ (K)} = 23°C + 273.15 = 296 \text{ K} \qquad T_2 \text{ (K)} = 100°C + 273.15 = 373 \text{ K}$$

Converting P from atm to torr:

$$P_1 \text{ (torr)} = 0.991 \text{ atm} \times \frac{760 \text{ torr}}{1 \text{ atm}} = 753 \text{ torr}$$

Arranging the ideal gas law and solving for P_2: At fixed n and V, we have

$$\frac{P_1 V_1}{n_1 T_1} = \frac{P_2 V_2}{n_2 T_2} \qquad \text{or} \qquad \frac{P_1}{T_1} = \frac{P_2}{T_2}$$

$$P_2 = P_1 \times \frac{T_2}{T_1} = 753 \text{ torr} \times \frac{373 \text{ K}}{296 \text{ K}} = 949 \text{ torr}$$

P_2 is less than 1.00×10^3 torr, so the valve will *not* open.

Check Our prediction is correct: because $T_2 > T_1$, we have $P_2 > P_1$. Thus, the temperature ratio should be >1 (T_2 in the numerator). The T ratio is about 1.25 (373/296), so the P ratio should also be about 1.25 (950/750 ≈ 1.25).

FOLLOW-UP PROBLEM 5.3 An engineer pumps air at 0°C into a newly designed piston-cylinder assembly. The volume measures 6.83 cm³. At what temperature (in K) will the volume be 9.75 cm³?

SAMPLE PROBLEM 5.4 Applying the Volume-Amount Relationship

Problem A scale model of a blimp rises when it is filled with helium to a volume of 55.0 dm³. When 1.10 mol of He is added to the blimp, the volume is 26.2 dm³. How many more grams of He must be added to make it rise? Assume constant T and P.

Plan We are given the initial amount of helium (n_1), the initial volume of the blimp (V_1), and the volume needed for it to rise (V_2), and we need the additional mass of helium to make it rise. So we first need to find n_2. We rearrange the ideal gas law to the appropriate form, solve for n_2, subtract n_1 to find the additional amount ($n_{add'l}$), and then convert moles to grams. We predict that $n_2 > n_1$ because $V_2 > V_1$.

Solution Summary of gas variables:

$$n_1 = 1.10 \text{ mol} \qquad n_2 = \text{unknown (find, and then subtract } n_1)$$
$$V_1 = 26.2 \text{ dm}^3 \qquad V_2 = 55.0 \text{ dm}^3$$
$$P \text{ and } T \text{ remain constant}$$

Arranging the ideal gas law and solving for n_2: At fixed P and T, we have

$$\frac{P_1 V_1}{n_1 T_1} = \frac{P_2 V_2}{n_2 T_2} \qquad \text{or} \qquad \frac{V_1}{n_1} = \frac{V_2}{n_2}$$

$$n_2 = n_1 \times \frac{V_2}{V_1} = 1.10 \text{ mol He} \times \frac{55.0 \text{ dm}^3}{26.2 \text{ dm}^3} = 2.31 \text{ mol He}$$

Finding the additional amount of He:

$$n_{add'l} = n_2 - n_1 = 2.31 \text{ mol He} - 1.10 \text{ mol He} = 1.21 \text{ mol He}$$

Converting moles of He to grams:

$$\text{Mass (g) of He} = 1.21 \text{ mol He} \times \frac{4.003 \text{ g He}}{1 \text{ mol He}} = \boxed{4.84 \text{ g He}}$$

Check V_2 is about twice V_1 ($55/26 \approx 2$), so n_2 should be about twice n_1 ($2.3/1.1 \approx 2$). Because $n_2 > n_1$, we were right to multiply n_1 by a number >1 (that is, V_2/V_1). About $1.2 \text{ mol} \times 4 \text{ g/mol} \approx 4.8 \text{ g}$.

Comment **1.** A different sequence of steps will give you the same answer: first find the additional volume ($V_{add'l} = V_2 - V_1$), and then solve directly for $n_{add'l}$. Try it for yourself. **2.** You saw that Charles's law ($V \propto T$ at fixed P and n) translates into a similar relationship between P and T at fixed V and n. The follow-up problem demonstrates that Avogadro's law ($V \propto n$ at fixed P and T) translates into an analogous relationship at fixed V and T.

FOLLOW-UP PROBLEM 5.4 A rigid plastic container holds 35.0 g of ethylene gas (C_2H_4) at a pressure of 793 torr. What is the pressure if 5.0 g of ethylene is removed at constant temperature?

SAMPLE PROBLEM 5.5   Solving for an Unknown Gas Variable at Fixed Conditions

Problem A steel tank has a volume of 438 L and is filled with 0.885 kg of O_2. Calculate the pressure of O_2 at 21°C.

Plan We are given V, T, and the mass of O_2, and we must find P. Conditions are not changing, so we apply the ideal gas law without rearranging it. We use the given V in liters, convert T to kelvins and mass of O_2 to moles, and solve for P.

Solution Summary of gas variables:

$$V = 438 \text{ L} \qquad\qquad\qquad T = 21°C \text{ (convert to K)}$$
$$n = 0.885 \text{ kg } O_2 \text{ (convert to mol)} \qquad P = \text{unknown}$$

Converting T from °C to K:

$$T \text{ (K)} = 21°C + 273.15 = 294 \text{ K}$$

Converting from mass of O_2 to moles:

$$n = \text{mol of } O_2 = 0.885 \text{ kg } O_2 \times \frac{1000 \text{ g}}{1 \text{ kg}} \times \frac{1 \text{ mol } O_2}{32.00 \text{ g } O_2} = 27.7 \text{ mol } O_2$$

Solving for P (note the unit canceling here):

$$P = \frac{nRT}{V} = \frac{27.7 \text{ mol} \times 0.0821 \frac{\text{atm·L}}{\text{mol·K}} \times 294 \text{ K}}{438 \text{ L}} = \boxed{1.53 \text{ atm}}$$

Check The amount of O_2 seems correct: $\sim 900 \text{ g}/(30 \text{ g/mol}) = 30 \text{ mol}$. To check the approximate size of the final calculation, round off the values, including that for R:

$$P = \frac{30 \text{ mol } O_2 \times 0.1 \frac{\text{atm·L}}{\text{mol·K}} \times 300 \text{ K}}{450 \text{ L}} = 2 \text{ atm}$$

which is reasonably close to 1.53 atm.

FOLLOW-UP PROBLEM 5.5 The tank in the sample problem develops a slow leak that is discovered and sealed. The new measured pressure is 1.37 atm. How many grams of O_2 remain?

Finally, in a slightly different type of problem that depicts a simple laboratory scene, we apply the gas laws to determine the correct balanced equation for a process.

SAMPLE PROBLEM 5.6   Using Gas Laws to Determine a Balanced Equation

Problem The piston-cylinders below depict a gaseous reaction carried out at constant pressure. Before the reaction, the temperature is 150 K; when it is complete, the temperature is 300 K.

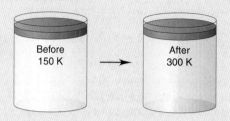

Which of the following balanced equations describes the reaction?
(1) $A_2(g) + B_2(g) \longrightarrow 2AB(g)$ (2) $2AB(g) + B_2(g) \longrightarrow 2AB_2(g)$
(3) $A(g) + B_2(g) \longrightarrow AB_2(g)$ (4) $2AB_2(g) \longrightarrow A_2(g) + 2B_2(g)$

Plan We are shown a depiction of a gaseous reaction and must choose the balanced equation. The problem says that P is constant, and the pictures show that T doubles and V stays the same. If n were also constant, the gas laws tell us that V should double when T doubles. Therefore, n cannot be constant, and the only way to maintain V with P constant and T doubling is for n to be halved. So we examine the four balanced equations and count the number of moles on each side to see in which equation n is halved.

Solution In equation (1), n does not change, so doubling T would double V.

In equation (2), n decreases from 3 mol to 2 mol, so doubling T would increase V by one-third.

In equation (3), n decreases from 2 mol to 1 mol. Doubling T would exactly balance the decrease from halving n, so V would stay the same.

In equation (4), n increases, so doubling T would more than double V.

Equation (3) is correct:

$$A(g) + B_2(g) \longrightarrow AB_2(g)$$

FOLLOW-UP PROBLEM 5.6 The gaseous reaction depicted below is carried out at constant pressure and an initial temperature of $-73°C$:

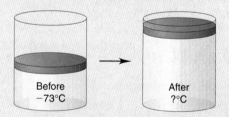

The *unbalanced* equation is $CD \longrightarrow C_2 + D_2$. What is the final temperature (in °C)?

SECTION SUMMARY

Four variables define the physical behavior of an ideal gas: volume (V), pressure (P), temperature (T), and amount (number of moles, n). Most simple gases display nearly ideal behavior at ordinary temperatures and pressures. Boyle's, Charles's, and Avogadro's laws relate volume to pressure, to temperature, and to amount of gas, respectively. At STP (0°C and 1 atm), 1 mol of an ideal gas occupies 22.4 L. The ideal gas law incorporates the individual gas laws into one equation: $PV = nRT$, where R is the universal gas constant.

5.4   FURTHER APPLICATIONS OF THE IDEAL GAS LAW

The ideal gas law can be recast in additional ways to determine other properties of gases. In this section, we use it to find gas density, molar mass, and the partial pressure of each gas in a mixture.

The Density of a Gas

One mole of any gas occupies nearly the same volume at a given temperature and pressure, so differences in gas density ($d = m/V$) depend on differences in molar mass (see Figure 5.7). For example, at STP, 1 mol of O_2 occupies the same volume as 1 mol of N_2, but each O_2 molecule has a greater mass than each N_2 molecule, so O_2 is denser.

All gases are miscible, but if two or more gases are not thoroughly mixed, a more dense gas will sink below the less dense one(s). A common example of this behavior occurs with carbon dioxide fire extinguishers. Carbon dioxide is denser than air, so it sinks onto the fire and prevents air (containing oxygen) from feeding the fire. By the same principle, dense gases in smog blanket urban centers, contributing to respiratory illnesses.

We can rearrange the ideal gas law to calculate the density of a gas from its molar mass. Recall that the number of moles (n) is the mass (m) divided by the molar mass ($\mathcal{M}$), $n = m/\mathcal{M}$. Substituting for n in the ideal gas law gives

$$PV = \frac{m}{\mathcal{M}}RT$$

Rearranging to isolate m/V gives

$$\frac{m}{V} = d = \frac{\mathcal{M} \times P}{RT} \tag{5.9}$$

Two important ideas are expressed by Equation 5.9:

- *The density of a gas is directly proportional to its molar mass* because a given amount of a heavier gas occupies the same volume as that amount of a lighter gas (Avogadro's law).
- *The density of a gas is inversely proportional to the temperature.* As the volume of a gas increases with temperature (Charles's law), the same mass occupies more space; thus, the density is lower.

The second of these relationships explains why, for example, safety experts recommend staying near the floor when escaping from a fire to avoid the hot, and therefore less dense, noxious gases. We use Equation 5.9 to find the density of a gas at any temperature and pressure near standard conditions.

SAMPLE PROBLEM 5.7   Calculating Gas Density

Problem A chemical engineer uses waste CO_2 from a manufacturing process, instead of chlorofluorocarbons, as a "blowing agent" in the production of polystyrene containers. Find the density (in g/L) of CO_2 and the number of molecules per liter **(a)** at STP (0°C and 1 atm) and **(b)** at room conditions (20.°C and 1.00 atm).

Plan We must find the density (d) and number of molecules of CO_2, given the two sets of P and T data. We find $\mathcal{M}$, convert T to kelvins, and calculate d with Equation 5.9. Then we convert the mass per liter to molecules per liter with Avogadro's number.

Solution (a) Density and molecules per liter of CO_2 at STP. Summary of gas properties:

$$T = 0°C + 273.15 = 273 \text{ K} \qquad P = 1 \text{ atm} \qquad \mathcal{M} \text{ of } CO_2 = 44.01 \text{ g/mol}$$

Calculating density (note the unit canceling here):

$$d = \frac{\mathcal{M} \times P}{RT} = \frac{44.01 \text{ g/mol} \times 1.00 \text{ atm}}{0.0821 \dfrac{\text{atm·L}}{\text{mol·K}} \times 273 \text{ K}} = 1.96 \text{ g/L}$$

Converting from mass/L to molecules/L:

$$\text{Molecules } CO_2/L = \frac{1.96 \text{ g } CO_2}{1 \text{ L}} \times \frac{1 \text{ mol } CO_2}{44.01 \text{ g } CO_2} \times \frac{6.022 \times 10^{23} \text{ molecules } CO_2}{1 \text{ mol } CO_2}$$

$$= 2.68 \times 10^{22} \text{ molecules } CO_2/L$$

(b) Density and molecules of CO_2 per liter at room conditions. Summary of gas properties:

$$T = 20.°C + 273.15 = 293 \text{ K} \qquad P = 1.00 \text{ atm} \qquad \mathcal{M} \text{ of } CO_2 = 44.01 \text{ g/mol}$$

Calculating density:

$$d = \frac{\mathcal{M} \times P}{RT} = \frac{44.01 \text{ g/mol} \times 1.00 \text{ atm}}{0.0821 \dfrac{\text{atm·L}}{\text{mol·K}} \times 293 \text{ K}} = \boxed{1.83 \text{ g/L}}$$

Converting from mass/L to molecules/L:

$$\text{Molecules } CO_2/L = \frac{1.83 \text{ g } CO_2}{1 \text{ L}} \times \frac{1 \text{ mol } CO_2}{44.01 \text{ g } CO_2} \times \frac{6.022 \times 10^{23} \text{ molecules } CO_2}{1 \text{ mol } CO_2}$$

$$= 2.50 \times 10^{22} \text{ molecules } CO_2/L$$

Check Round off to check the density values; for example, in (a), at STP:

$$\frac{50 \text{ g/mol} \times 1 \text{ atm}}{0.1 \dfrac{\text{atm·L}}{\text{mol·K}} \times 250 \text{ K}} = 2 \text{ g/L} \approx 1.96 \text{ g/L}$$

At the higher temperature in (b), the density should decrease, which can happen only if there are fewer molecules per liter, so the answer is reasonable.

Comment 1. An *alternative approach* for finding the density of most simple gases, but *at STP only*, is to divide the molar mass by the standard molar volume, 22.4 L:

$$d = \frac{\mathcal{M}}{V} = \frac{44.01 \text{ g/mol}}{22.4 \text{ L/mol}} = 1.96 \text{ g/L}$$

Once you know the density at one temperature (0°C), you can find it at any other temperature with the following relationship: $d_1/d_2 = T_2/T_1$.

2. Note that we have different numbers of significant figures for the pressure values. In (a), "1 atm" is part of the definition of STP, so it is an exact number. In (b), we specified "1.00 atm" to allow three significant figures in the answer.

FOLLOW-UP PROBLEM 5.7 Compare the density of CO_2 at 0°C and 380 torr with its density at STP.

In Sample Problem 5.7, we knew the identity of the gas. Finding the density of an unknown gas is a straightforward experimental procedure. You evacuate a flask of known volume, weigh the empty flask, then fill it with the gas at known temperature and pressure, and weigh it again. The difference in weights is the mass of the gas, and dividing this mass by the flask's volume gives the density.

The Molar Mass of a Gas

Through another simple rearrangement of the ideal gas law, we can determine the molar mass of an unknown gas or volatile liquid (one that is easily vaporized):

$$n = \frac{m}{\mathcal{M}} = \frac{PV}{RT} \qquad \text{so} \qquad \mathcal{M} = \frac{mRT}{PV} \qquad \text{or} \qquad \mathcal{M} = \frac{dRT}{P} \tag{5.10}$$

Notice that this equation is just a rearrangement of Equation 5.9.

SAMPLE PROBLEM 5.8   Finding the Molar Mass of a Volatile Liquid

Problem An organic chemist isolates a colorless liquid from a petroleum sample. She places the liquid in a flask and puts the flask in a boiling water bath, which vaporizes the liquid and fills the flask with gas. She closes the flask, reweighs it, and obtains the following data:

Volume (V) of flask = 213 mL $T = 100.0°C$ $P = 754$ torr

Mass of flask + gas = 78.416 g   Mass of empty flask = 77.834 g

Calculate the molar mass of the liquid.

Plan We are given V, T, P, and mass data and must find the molar mass ($\mathcal{M}$) of the liquid, which is the same as $\mathcal{M}$ of the gas. The flask was closed when the gas had reached the surrounding temperature and pressure, so we convert V to liters, T to kelvins, and P to atmospheres, find the mass of gas by subtracting the mass of the empty flask, and use Equation 5.10 to solve for $\mathcal{M}$.

Solution Summary of gas variables:

$$m = 78.416 \text{ g} - 77.834 \text{ g} = 0.582 \text{ g} \qquad P \text{ (atm)} = 754 \text{ torr} \times \frac{1 \text{ atm}}{760 \text{ torr}} = 0.992 \text{ atm}$$

$$V \text{ (L)} = 213 \text{ mL} \times \frac{1 \text{ L}}{1000 \text{ mL}} = 0.213 \text{ L} \qquad T \text{ (K)} = 100.0°\text{C} + 273.15 = 373.2 \text{ K}$$

Solving for $\mathcal{M}$:

$$\mathcal{M} = \frac{mRT}{PV} = \frac{0.582 \text{ g} \times 0.0821 \dfrac{\text{atm·L}}{\text{mol·K}} \times 373.2 \text{ K}}{0.992 \text{ atm} \times 0.213 \text{ L}} = 84.4 \text{ g/mol}$$

Check Rounding to check the arithmetic, we have

$$\frac{0.6 \text{ g} \times 0.08 \dfrac{\text{atm·L}}{\text{mol·K}} \times 375 \text{ K}}{1 \text{ atm} \times 0.2 \text{ L}} = 90 \text{ g/mol} \qquad \text{which is close to 84.4 g/mol}$$

FOLLOW-UP PROBLEM 5.8 At 10.0°C and 102.5 kPa, the density of dry air is 1.26 g/L. What is the average "molar mass" of dry air at these conditions?

The Partial Pressure of a Gas in a Mixture of Gases

All of the behaviors we've discussed so far were observed from experiments with air, which is a complex mixture of gases. The ideal gas law holds for virtually any gas, whether pure or a mixture, at ordinary conditions for two reasons:

- Gases mix homogeneously (form a solution) in any proportions.
- Each gas in a mixture behaves as if it were the only gas present (assuming no chemical interactions).

Dalton's Law of Partial Pressures The second point above was discovered by John Dalton in his life-long study of humidity. He observed that when water vapor is added to dry air, the total air pressure increases by an increment equal to the pressure of the water vapor:

$$P_{\text{humid air}} = P_{\text{dry air}} + P_{\text{added water vapor}}$$

In other words, each gas in the mixture exerts a **partial pressure,** a portion of the total pressure of the mixture, that is the same as the pressure it would exert *by itself.* This observation is formulated as **Dalton's law of partial pressures:** *in a mixture of unreacting gases, the total pressure is the sum of the partial pressures of the individual gases:*

$$P_{\text{total}} = P_1 + P_2 + P_3 + \cdots \tag{5.11}$$

As an example, suppose you have a tank of fixed volume that contains nitrogen gas at a certain pressure, and you introduce a sample of hydrogen gas into the tank. Each gas behaves independently, so we can write an ideal gas law expression for each:

$$P_{N_2} = \frac{n_{N_2}RT}{V} \qquad \text{and} \qquad P_{H_2} = \frac{n_{H_2}RT}{V}$$

Because each gas occupies the same total volume and is at the same temperature, the pressure of each gas depends only on its amount, n. Thus, the total pressure is

$$P_{\text{total}} = P_{N_2} + P_{H_2} = \frac{n_{N_2}RT}{V} + \frac{n_{H_2}RT}{V} = \frac{(n_{N_2} + n_{H_2})RT}{V} = \frac{n_{\text{total}}RT}{V}$$

where $n_{\text{total}} = n_{N_2} + n_{H_2}$.

Each component in a mixture contributes a fraction of the total number of moles in the mixture, which is the **mole fraction (X)** of that component. Multiplying X by 100 gives the mole percent. Keep in mind that the sum of the mole fractions of all components in any mixture must be 1, and the sum of the mole percents must be 100%. For N_2, the mole fraction is

$$X_{N_2} = \frac{n_{N_2}}{n_{total}} = \frac{n_{N_2}}{n_{N_2} + n_{H_2}}$$

The total pressure is due to the total number of moles, so the partial pressure of gas A is the total pressure multiplied by the mole fraction of A, X_A:

$$P_A = X_A \times P_{total} \tag{5.12}$$

Equation 5.12 is a very important result. To see that it is valid for the mixture of N_2 and H_2, we recall that $X_{N_2} + X_{H_2} = 1$ and obtain

$$P_{total} = P_{N_2} + P_{H_2} = (X_{N_2} \times P_{total}) + (X_{H_2} \times P_{total}) = (X_{N_2} + X_{H_2})P_{total} = 1 \times P_{total}$$

SAMPLE PROBLEM 5.9   Applying Dalton's Law of Partial Pressures

Problem In a study of O_2 uptake by muscle at high altitude, a physiologist prepares an atmosphere consisting of 79 mole % N_2, 17 mole % $^{16}O_2$, and 4.0 mole % $^{18}O_2$. (The isotope ^{18}O will be measured to determine O_2 uptake.) The total pressure is 0.75 atm to simulate high altitude. Calculate the mole fraction and partial pressure of $^{18}O_2$ in the mixture.

Plan We must find $X_{^{18}O_2}$ and $P_{^{18}O_2}$ from P_{total} (0.75 atm) and the mole % of $^{18}O_2$ (4.0). Dividing the mole % by 100 gives the mole fraction, $X_{^{18}O_2}$. Then, using Equation 5.12, we multiply $X_{^{18}O_2}$ by P_{total} to find $P_{^{18}O_2}$.

Solution Calculating the mole fraction of $^{18}O_2$:

$$X_{^{18}O_2} = \frac{4.0 \text{ mol } \% \ ^{18}O_2}{100} = 0.040$$

Solving for the partial pressure of $^{18}O_2$:

$$P_{^{18}O_2} = X_{^{18}O_2} \times P_{total} = 0.040 \times 0.75 \text{ atm} = 0.030 \text{ atm}$$

Check $X_{^{18}O_2}$ is small because the mole % is small, so $P_{^{18}O_2}$ should be small also.

Comment At high altitudes, specialized brain cells that are sensitive to O_2 and CO_2 levels in the blood trigger an increase in rate and depth of breathing for several days, until a person becomes acclimated.

FOLLOW-UP PROBLEM 5.9 To prevent the presence of air, noble gases are placed over highly reactive chemicals to act as inert "blanketing" gases. A chemical engineer places a mixture of noble gases consisting of 5.50 g of He, 15.0 g of Ne, and 35.0 g of Kr in a piston-cylinder assembly at STP. Calculate the partial pressure of each gas.

Mole % of $^{18}O_2$

↓ divide by 100

Mole fraction, $X^{18}O_2$

↓ multiply by P_{total}

Partial pressure, $P^{18}O_2$

Collecting a Gas over Water The law of partial pressures is frequently used to determine the yield of a water-insoluble gas formed in a reaction. The gaseous product bubbles through water and is collected into an inverted container, as shown in Figure 5.10. The water vapor that mixes with the gas contributes a portion of the total pressure, called the *vapor pressure,* which depends only on the water temperature.

In order to determine the yield of gaseous product, we find the appropriate vapor pressure value from a list, such as the one in Table 5.2, and subtract it from the total gas pressure (corrected to barometric pressure) to get the partial pressure of the gaseous product. With V and T known, we can calculate the amount of product.

Animation: Collecting a Gas over Water
Online Learning Center

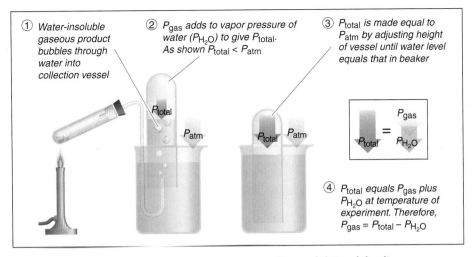

Figure 5.10 Collecting a water-insoluble gaseous product and determining its pressure.

Table 5.2 Vapor Pressure of Water (P_{H_2O}) at Different T	
T (°C)	P (torr)
0	4.6
5	6.5
10	9.2
12	10.5
14	12.0
16	13.6
18	15.5
20	17.5
22	19.8
24	22.4
26	25.2
28	28.3
30	31.8
35	42.2
40	55.3
45	71.9
50	92.5
55	118.0
60	149.4
65	187.5
70	233.7
75	289.1
80	355.1
85	433.6
90	525.8
95	633.9
100	760.0

SAMPLE PROBLEM 5.10   Calculating the Amount of Gas Collected over Water

Problem Acetylene (C_2H_2), an important fuel in welding, is produced in the laboratory when calcium carbide (CaC_2) reacts with water:

$$CaC_2(s) + 2H_2O(l) \longrightarrow C_2H_2(g) + Ca(OH)_2(aq)$$

For a sample of acetylene collected over water, total gas pressure (adjusted to barometric pressure) is 738 torr and the volume is 523 mL. At the temperature of the gas (23°C), the vapor pressure of water is 21 torr. How many grams of acetylene are collected?

Plan In order to find the mass of C_2H_2, we first need to find the number of moles of C_2H_2, $n_{C_2H_2}$, which we can obtain from the ideal gas law by calculating $P_{C_2H_2}$. The barometer reading gives us P_{total}, which is the sum of $P_{C_2H_2}$ and P_{H_2O}, and we are given P_{H_2O}, so we subtract to find $P_{C_2H_2}$. We are also given V and T, so we convert to consistent units, and find $n_{C_2H_2}$ from the ideal gas law. Then we convert moles to grams using the molar mass from the formula.

Solution Summary of gas variables:

$$P_{C_2H_2} \text{ (torr)} = P_{total} - P_{H_2O} = 738 \text{ torr} - 21 \text{ torr} = 717 \text{ torr}$$

$$P_{C_2H_2} \text{ (atm)} = 717 \text{ torr} \times \frac{1 \text{ atm}}{760 \text{ torr}} = 0.943 \text{ atm}$$

$$V \text{ (L)} = 523 \text{ mL} \times \frac{1 \text{ L}}{1000 \text{ mL}} = 0.523 \text{ L}$$

$$T \text{ (K)} = 23°C + 273.15 = 296 \text{ K}$$

$$n_{C_2H_2} = \text{unknown}$$

Solving for $n_{C_2H_2}$:

$$n_{C_2H_2} = \frac{PV}{RT} = \frac{0.943 \text{ atm} \times 0.523 \text{ L}}{0.0821 \frac{\text{atm·L}}{\text{mol·K}} \times 296 \text{ K}} = 0.0203 \text{ mol}$$

Converting $n_{C_2H_2}$ to mass:

$$\text{Mass (g) of } C_2H_2 = 0.0203 \text{ mol } C_2H_2 \times \frac{26.04 \text{ g } C_2H_2}{1 \text{ mol } C_2H_2} = 0.529 \text{ g } C_2H_2$$

Check Rounding to one significant figure, a quick arithmetic check for n gives

$$n \approx \frac{1 \text{ atm} \times 0.5 \text{ L}}{0.08 \frac{\text{atm·L}}{\text{mol·K}} \times 300 \text{ K}} = 0.02 \text{ mol} \approx 0.0203 \text{ mol}$$

P_{total}

subtract P_{H_2O}

$P_{C_2H_2}$

$n = \frac{PV}{RT}$

$n_{C_2H_2}$

multiply by $\mathcal{M}$ (g/mol)

Mass (g) of C_2H_2

FOLLOW-UP PROBLEM 5.10 A small piece of zinc reacts with dilute HCl to form H_2, which is collected over water at 16°C into a large flask. The total pressure is adjusted to barometric pressure (752 torr), and the volume is 1495 mL. Use Table 5.2 to help calculate the partial pressure and mass of H_2.

SECTION SUMMARY

The ideal gas law can be rearranged to calculate the density and molar mass of a gas. In a mixture of gases, each component contributes its own partial pressure to the total pressure (Dalton's law of partial pressures). The mole fraction of each component is the ratio of its partial pressure to the total pressure. When a gas is in contact with water, the total pressure is the sum of the gas pressure and the vapor pressure of water at the given temperature.

5.5 THE IDEAL GAS LAW AND REACTION STOICHIOMETRY

In Chapters 3 and 4, we encountered many reactions that involved gases as reactants (e.g., combustion with O_2) or as products (e.g., a metal displacing H_2 from acid). From the balanced equation, we used stoichiometrically equivalent molar ratios to calculate the amounts (moles) of reactants and products and converted these quantities into masses, numbers of molecules, or solution volumes (see Figure 3.10). Figure 5.11 shows how you can expand your problem-solving repertoire by using the ideal gas law to convert between gas variables (P, T, and V) and amounts (moles) of gaseous reactants and products. In effect, you combine a gas law problem with a stoichiometry problem; it is more realistic to measure the volume, pressure, and temperature of a gas than its mass.

Figure 5.11 Summary of the stoichiometric relationships among the amount (mol, n) of gaseous reactant or product and the gas variables pressure (P), volume (V), and temperature (T).

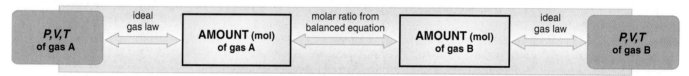

SAMPLE PROBLEM 5.11   Using Gas Variables to Find Amounts of Reactants or Products

Problem Hot H_2 can reduce copper(II) oxide, forming the pure metal and H_2O. What volume of H_2 at 765 torr and 225°C is needed to reduce 35.5 g of copper(II) oxide?
Plan This is a stoichiometry *and* gas law problem. To find V_{H_2}, we first need n_{H_2}. We write and balance the equation. Next, we convert the given mass of CuO (35.5 g) to amount (mol) and use the molar ratio to find moles of H_2 needed (stoichiometry portion). Then, we use the ideal gas law to convert moles of H_2 to liters (gas law portion). A roadmap is shown, but you are familiar with all the steps.
Solution Writing the balanced equation:

$$CuO(s) + H_2(g) \longrightarrow Cu(s) + H_2O(g)$$

Calculating n_{H_2}:

$$n_{H_2} = 35.5 \text{ g CuO} \times \frac{1 \text{ mol CuO}}{79.55 \text{ g CuO}} \times \frac{1 \text{ mol } H_2}{1 \text{ mol CuO}} = 0.446 \text{ mol } H_2$$

Summary of other gas variables:

$$V = \text{unknown} \qquad P \text{ (atm)} = 765 \text{ torr} \times \frac{1 \text{ atm}}{760 \text{ torr}} = 1.01 \text{ atm}$$

$$T \text{ (K)} = 225°C + 273.15 = 498 \text{ K}$$

Solving for V_{H_2}:

$$V = \frac{nRT}{P} = \frac{0.446 \text{ mol} \times 0.0821 \dfrac{\text{atm·L}}{\text{mol·K}} \times 498 \text{ K}}{1.01 \text{ atm}} = 18.1 \text{ L}$$

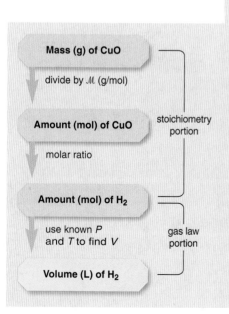

Check One way to check the answer is to compare it with the molar volume of an ideal gas at STP (22.4 L at 273.15 K and 1 atm). One mole of H_2 at STP occupies about 22 L, so less than 0.5 mol occupies less than 11 L. T is less than twice 273 K, so V should be less than twice 11 L.

Comment The main point here is that the stoichiometry provides one gas variable (n), two more are given, and the ideal gas law is used to find the fourth.

FOLLOW-UP PROBLEM 5.11 Sulfuric acid reacts with sodium chloride to form aqueous sodium sulfate and hydrogen chloride gas. How many milliliters of gas form at STP when 0.117 kg of sodium chloride reacts with excess sulfuric acid?

SAMPLE PROBLEM 5.12   Using the Ideal Gas Law in a Limiting-Reactant Problem

Problem What mass of potassium chloride forms when 5.25 L of chlorine gas at 0.950 atm and 293 K reacts with 17.0 g of potassium?

Plan The only difference between this and previous limiting-reactant problems (see Sample Problem 3.10) is that here we use the ideal gas law to find the amount (n) of gaseous reactant from the known V, P, and T. We first write the balanced equation and then use it to find the limiting reactant and the amount of product.

Solution Writing the balanced equation:

$$2K(s) + Cl_2(g) \longrightarrow 2KCl(s)$$

Summary of gas variables:

$$P = 0.950 \text{ atm} \qquad V = 5.25 \text{ L}$$
$$T = 293 \text{ K} \qquad n = \text{unknown}$$

Solving for n_{Cl_2}:

$$n_{Cl_2} = \frac{PV}{RT} = \frac{0.950 \text{ atm} \times 5.25 \text{ L}}{0.0821 \dfrac{\text{atm·L}}{\text{mol·K}} \times 293 \text{ K}} = 0.207 \text{ mol}$$

Converting from grams of potassium (K) to moles:

$$\text{Moles of K} = 17.0 \text{ g K} \times \frac{1 \text{ mol K}}{39.10 \text{ g K}} = 0.435 \text{ mol K}$$

Determining the limiting reactant: If Cl_2 is limiting,

$$\text{Moles of KCl} = 0.207 \text{ mol Cl}_2 \times \frac{2 \text{ mol KCl}}{1 \text{ mol Cl}_2} = 0.414 \text{ mol KCl}$$

If K is limiting,

$$\text{Moles of KCl} = 0.435 \text{ mol K} \times \frac{2 \text{ mol KCl}}{2 \text{ mol K}} = 0.435 \text{ mol KCl}$$

Cl_2 is the limiting reactant because it forms less KCl.

Converting from moles of KCl to grams:

$$\text{Mass (g) of KCl} = 0.414 \text{ mol KCl} \times \frac{74.55 \text{ g KCl}}{1 \text{ mol KCl}} = 30.9 \text{ g KCl}$$

Check The gas law calculation seems correct. At STP, 22 L of Cl_2 gas contains about 1 mol, so a 5-L volume would contain a bit less than 0.25 mol of Cl_2. Moreover, the fact that P (in numerator) is slightly lower than STP and T (in denominator) is slightly higher than STP should lower the calculated n further below the ideal value. The mass of KCl seems correct: less than 0.5 mol of KCl gives $<0.5 \times \mathcal{M}$ (30.9 g $< 0.5 \times 75$ g).

FOLLOW-UP PROBLEM 5.12 Ammonia and hydrogen chloride gases react to form solid ammonium chloride. A 10.0-L reaction flask contains ammonia at 0.452 atm and 22°C, and 155 mL of hydrogen chloride gas at 7.50 atm and 271 K is introduced. After the reaction occurs and the temperature returns to 22°C, what is the pressure inside the flask? (Neglect the volume of the solid product.)

By converting the variables P, V, and T of gaseous reactants (or products) to amount (n, mol), we can solve stoichiometry problems for gaseous reactions.

5.6 THE KINETIC-MOLECULAR THEORY: A MODEL FOR GAS BEHAVIOR

So far we have discussed observations of macroscopic samples of gas: decreasing cylinder volume, increasing tank pressure, and so forth. This section presents the central model that explains macroscopic gas behavior at the level of individual particles: the **kinetic-molecular theory.** The theory draws conclusions through mathematical derivations, but here our discussion will be largely qualitative.

How the Kinetic-Molecular Theory Explains the Gas Laws

Developed by some of the great scientists of the 19th century, most notably James Clerk Maxwell and Ludwig Boltzmann, the kinetic-molecular theory was able to explain the gas laws that scientists of the century before had arrived at empirically.

Questions Concerning Gas Behavior To model gas behavior, we must rationalize certain questions at the molecular level:

1. *Origin of pressure.* Pressure is a measure of the force a gas exerts on a surface. How do individual gas particles create this force?
2. *Boyle's law ($V \propto 1/P$).* A change in gas pressure in one direction causes a change in gas volume in the other. What happens to the particles when external pressure compresses the gas volume? And why aren't liquids and solids compressible?
3. *Dalton's law ($P_{total} = P_1 + P_2 + P_3$).* The pressure of a gas mixture is the sum of the pressures of the individual gases. Why does each gas contribute to the total pressure in proportion to its mole fraction?
4. *Charles's law ($V \propto T$).* A change in temperature is accompanied by a corresponding change in volume. What effect does higher temperature have on gas particles that increases the volume—or increases the pressure if volume is fixed? This question raises a more fundamental one: what does temperature measure on the molecular scale?
5. *Avogadro's law ($V \propto n$).* Gas volume (or pressure) depends on the number of moles present, not on the nature of the particular gas. But shouldn't 1 mol of larger molecules occupy more space than 1 mol of smaller molecules? And why doesn't 1 mol of heavier molecules exert more pressure than 1 mol of lighter molecules?

Postulates of the Kinetic-Molecular Theory The theory is based on three postulates (assumptions):

Postulate 1. *Particle volume.* A gas consists of a large collection of individual particles. The volume of an individual particle is *extremely* small compared with the volume of the container. In essence, the model pictures gas particles as points of mass with empty space between them.

Postulate 2. *Particle motion.* Gas particles are in constant, random, straight-line motion, except when they collide with the container walls or with each other.

Postulate 3. *Particle collisions.* Collisions are *elastic,* which means that the colliding molecules exchange energy but they do not lose any energy through friction. Thus, their total kinetic energy (E_k) is constant. Between collisions, the molecules do not influence each other by attractive or repulsive forces.

Gas behavior that conforms to these postulates is called *ideal*. As you'll see in the next section, most real gases behave almost ideally at ordinary temperatures and pressures.

Picture the scene envisioned by the postulates: Countless particles moving in every direction, smashing into the container walls and one another. Any given particle changes its speed with each collision, perhaps one instant standing nearly still from a head-on crash and the next instant zooming away from a smash on the side. Thus, *the particles have an average speed,* with most moving near the average speed, some moving faster, and some slower.

Figure 5.12 depicts the distribution of molecular speeds (u) for N_2 gas at three temperatures. The curves flatten and spread at higher temperatures. Note especially that the *most probable speed (the peak of each curve) increases as the temperature increases.* This increase occurs because the average kinetic energy of the molecules ($\overline{E_k}$: the overbar indicates the average value of a quantity), which incorporates the most probable speed, is proportional to the absolute temperature: $\overline{E_k} \propto T$, or $\overline{E_k} = c \times T$, where c is a constant that is the same for any gas. (We'll return to this equation shortly.) Thus, a major conclusion based on the distribution of speeds, which arises directly from postulate 3, is that *at a given temperature, all gases have the same average kinetic energy.*

A Molecular View of the Gas Laws Let's continue visualizing the particles to see how the theory explains the macroscopic behavior of gases and answers the questions posed above:

1. *Origin of pressure.* When a moving object collides with a surface, it exerts a force. We conclude from postulate 2, which describes particle motion, that when a particle collides with the container wall, it too exerts a force. Many such collisions result in the observed pressure. The greater the number of molecules in a given container, the more frequently they collide with the walls, and the greater the pressure is.

2. *Boyle's law* ($V \propto 1/P$). Gas molecules are points of mass with empty space between them (postulate 1), so as the pressure exerted *on* the sample increases at constant temperature, the distance between molecules decreases, and the sample volume decreases. The pressure exerted *by* the gas increases simultaneously because in a smaller volume of gas, there are shorter distances between gas molecules and the walls and between the walls themselves; thus, collisions are more frequent (Figure 5.13). The fact that liquids and solids cannot be compressed means there is little, if any, free space between the molecules.

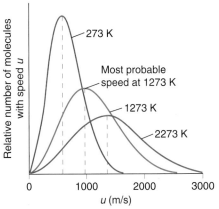

Figure 5.12 Distribution of molecular speeds at three temperatures. At a given temperature, a plot of the relative number of N_2 molecules vs. molecular speed (u) results in a skewed bell-shaped curve, with the most probable speed at the peak. Note that the curves spread at higher temperatures and the most probable speed is directly proportional to the temperature.

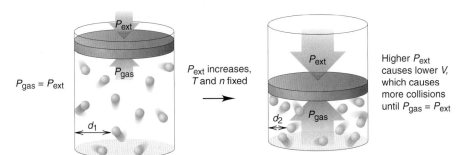

Figure 5.13 A molecular description of Boyle's law. At a given T, gas molecules collide with the walls across an average distance (d_1) and give rise to a pressure (P_{gas}) that equals the external pressure (P_{ext}). If P_{ext} increases, V decreases, and so the average distance between a molecule and the walls is shorter ($d_2 < d_1$). Molecules strike the walls more often, and P_{gas} increases until it again equals P_{ext}. Thus, V decreases when P increases.

3. *Dalton's law of partial pressures* ($P_{total} = P_A + P_B$). Adding a given amount of gas A to a given amount of gas B causes an increase in the total number of molecules in proportion to the amount of A that is added. This increase causes a corresponding increase in the number of collisions per second with the walls (postulate 2), which causes a corresponding increase in the pressure (Figure 5.14, next page). Thus, each gas exerts a fraction of the total pressure based on the fraction of molecules (or fraction of moles; that is, the mole fraction) of that gas in the mixture.

Figure 5.14 A molecular description of Dalton's law of partial pressures. A piston-cylinder assembly containing 0.30 mol of gas A at 0.50 atm is connected to a tank of fixed volume containing 0.60 mol of gas B at 1.0 atm. When the piston is depressed at fixed temperature, gas A is forced into the tank of gas B and the gases mix. The new total pressure, 1.5 atm, equals the sum of the partial pressures, which is related to the new total amount of gas, 0.90 mol. Thus, each gas undergoes a fraction of the total collisions related to its fraction of the total number of molecules (moles), which is equal to its mole fraction.

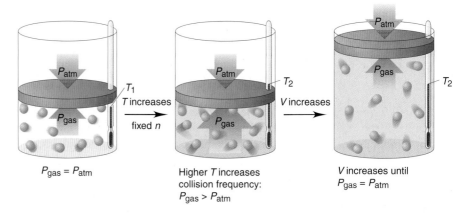

$P_A = P_{total}$
$= 0.50$ atm
$n_A = 0.30$ mol

$P_B = P_{total}$
$= 1.0$ atm
$n_B = 0.60$ mol

$P_{total} = P_A + P_B = 1.5$ atm
$n_{total} = 0.90$ mol
$X_A = 0.33$ mol $X_B = 0.67$ mol

4. *Charles's law* ($V \propto T$). As the temperature increases, the most probable molecular speed and the average kinetic energy increase (postulate 3). Thus, the molecules hit the walls more frequently *and* more energetically. A higher frequency of collisions causes higher internal pressure. As a result, the walls move outward, which increases the volume and restores the starting pressure (Figure 5.15).

Figure 5.15 A molecular description of Charles's law. At a given temperature (T_1), $P_{gas} = P_{atm}$. When the gas is heated to T_2, the molecules move faster and collide with the walls more often, which increases P_{gas}. This increases V, and so the molecules collide less often until P_{gas} again equals P_{atm}. Thus, V increases when T increases.

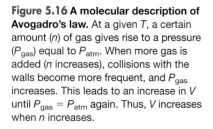

$P_{gas} = P_{atm}$

Higher T increases collision frequency:
$P_{gas} > P_{atm}$

V increases until
$P_{gas} = P_{atm}$

5. *Avogadro's law* ($V \propto n$). Adding more molecules to a container increases the total number of collisions with the walls and, therefore, the internal pressure. As a result, the volume expands until the number of collisions per unit of wall area is the same as it was before the addition (Figure 5.16).

Figure 5.16 A molecular description of Avogadro's law. At a given T, a certain amount (n) of gas gives rise to a pressure (P_{gas}) equal to P_{atm}. When more gas is added (n increases), collisions with the walls become more frequent, and P_{gas} increases. This leads to an increase in V until $P_{gas} = P_{atm}$ again. Thus, V increases when n increases.

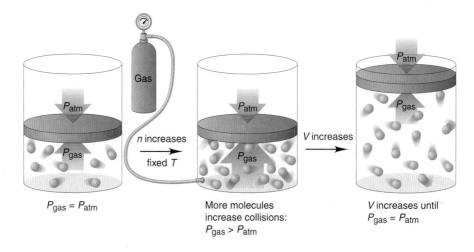

$P_{gas} = P_{atm}$

More molecules increase collisions:
$P_{gas} > P_{atm}$

V increases until
$P_{gas} = P_{atm}$

Relation Between Molecular Speed and Mass We still need to explain why equal numbers of molecules of two different gases, such as O_2 and H_2, occupy the same volume. Let's first see why heavier O_2 particles *do not* hit the container walls with more energy than lighter H_2 particles. Recall that the kinetic energy of an object is the energy associated with its motion (Chapter 1). This energy is related to the mass *and* the speed of the object:

$$E_k = \tfrac{1}{2}\,\text{mass} \times \text{speed}^2$$

This equation shows that if a heavy object and a light object have the same kinetic energy, *the heavy object must be moving more slowly*. For a large population of molecules, the average kinetic energy is

$$\overline{E_k} = \tfrac{1}{2}m\overline{u^2}$$

where m is the molecular mass and $\overline{u^2}$ is the average of the squares of the molecular speeds. The square root of $\overline{u^2}$ is called the root-mean-square speed, or **rms speed (u_{rms})**. *A molecule moving at this speed has the average kinetic energy.* The rms speed is somewhat higher than the most probable speed, but the speeds are proportional to each other and we will use them interchangeably. The rms speed is related to the temperature and the molar mass as follows:

$$u_{rms} = \sqrt{\frac{3RT}{\mathcal{M}}} \qquad (5.13)$$

where R is the gas constant, T is the absolute temperature, and $\mathcal{M}$ is the molar mass. (Because we want u in m/s, and R includes the joule, which has units of $kg \cdot m^2/s^2$, we use the value 8.314 J/mol·K for R and express $\mathcal{M}$ in kg/mol.)

Postulate 3 leads to the conclusion that different gases at the same temperature have the same average kinetic energy. Therefore, Avogadro's law requires that, on average, *molecules with a higher mass have a lower speed*. In other words, at the same temperature, O_2 molecules move more slowly, on average, than H_2 molecules. Figure 5.17 shows that, in general, at the same temperature, lighter gases have higher speeds. This means that H_2 molecules collide with the walls more often than do O_2 molecules, but each collision has less force. Because, at the same T, lighter and heavier molecules hit the walls with the same average kinetic energy, lighter and heavier gases have the same pressure and, thus, the same volume.

The Meaning of Temperature Earlier we said that the average kinetic energy of a particle was equal to the absolute temperature times a constant, that is, $\overline{E_k} = \text{constant} \times T$. A derivation of the full relationship gives the following equation:

$$\overline{E_k} = \tfrac{3}{2}\left(\frac{R}{N_A}\right)T$$

where R is the gas constant and N_A is Avogadro's number. This equation expresses the important point that *temperature is related to the average energy of molecular motion*. Note that it is not related to the *total* energy, which depends on the size of the sample, but to the *average* energy: as T increases, $\overline{E_k}$ increases.

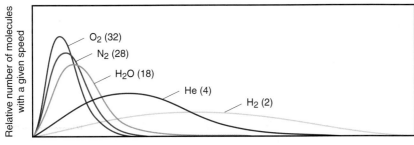

Relative number of molecules with a given speed

O_2 (32)
N_2 (28)
H_2O (18)
He (4)
H_2 (2)

Molecular speed at a given T

Figure 5.17 Relationship between molar mass and molecular speed. At a given temperature, gases with lower molar masses (numbers in parentheses) have higher most probable speeds (peak of each curve).

Effusion and Diffusion

The movement of gases, either through one another or into regions of very low pressure, has many important applications.

The Process of Effusion One of the early triumphs of the kinetic-molecular theory was an explanation of **effusion,** the process by which a gas escapes from its container through a tiny hole into an evacuated space. In 1846, Thomas Graham studied this process and concluded that the effusion rate was inversely proportional to the square root of the gas density. The effusion rate is the number of moles (or molecules) of gas effusing per unit time. Because density is directly proportional to molar mass, we can state **Graham's law of effusion** as follows: *the rate of effusion of a gas is inversely proportional to the square root of its molar mass,*

$$\text{Rate of effusion} \propto \frac{1}{\sqrt{\mathcal{M}}}$$

Argon (Ar) is lighter than krypton (Kr), so it effuses faster, assuming equal pressures of the two gases. Thus, the ratio of the rates is

$$\frac{\text{Rate}_{Ar}}{\text{Rate}_{Kr}} = \frac{\sqrt{\mathcal{M}_{Kr}}}{\sqrt{\mathcal{M}_{Ar}}} \quad \text{or, in general,} \quad \frac{\text{rate}_{A}}{\text{rate}_{B}} = \frac{\sqrt{\mathcal{M}_{B}}}{\sqrt{\mathcal{M}_{A}}} = \sqrt{\frac{\mathcal{M}_{B}}{\mathcal{M}_{A}}} \quad \text{(5.14)}$$

The kinetic-molecular theory explains that, at a given temperature and pressure, *the gas with the lower molar mass effuses faster because the most probable speed of its molecules is higher; therefore, more molecules escape per unit time.*

One of the most important applications of Graham's law is the *enrichment* of nuclear reactor fuel: separating nonfissionable, more abundant ^{238}U from fissionable ^{235}U to increase the proportion of ^{235}U in the mixture. The two isotopes have identical chemical properties, so they are separated by differences in physical properties—the effusion rates of their gaseous compounds. Uranium ore is converted to gaseous UF_6 (a mixture of $^{238}UF_6$ and $^{235}UF_6$), which is pumped through a series of chambers with porous barriers. Because they are a bit lighter and thus move very slightly faster, molecules of $^{235}UF_6$ ($\mathcal{M} = 349.03$) effuse through each barrier 1.0043 times faster than do molecules of $^{238}UF_6$ ($\mathcal{M} = 352.04$):

$$\frac{\text{Rate}_{^{235}UF_6}}{\text{Rate}_{^{238}UF_6}} = \sqrt{\frac{\mathcal{M}_{^{238}UF_6}}{\mathcal{M}_{^{235}UF_6}}} = \sqrt{\frac{352.04 \text{ g/mol}}{349.03 \text{ g/mol}}} = 1.0043$$

Many passes are made, each increasing the fraction of $^{235}UF_6$ until a mixture is obtained that contains enough $^{235}UF_6$. This isotope-enrichment process was developed during the latter years of World War II and produced enough ^{235}U for two of the world's first three atomic bombs. The principle is still used to prepare nuclear fuel for power plants.

SAMPLE PROBLEM 5.13 Applying Graham's Law of Effusion

Problem Calculate the ratio of the effusion rates of helium and methane (CH_4).
Plan The effusion rate is inversely proportional to $\sqrt{\mathcal{M}}$, so we find the molar mass of each substance from the formula and take its square root. The inverse of the ratio of the square roots is the ratio of the effusion rates.
Solution

$$\mathcal{M} \text{ of } CH_4 = 16.04 \text{ g/mol} \qquad \mathcal{M} \text{ of He} = 4.003 \text{ g/mol}$$

Calculating the ratio of the effusion rates:

$$\frac{\text{Rate}_{He}}{\text{Rate}_{CH_4}} = \sqrt{\frac{\mathcal{M}_{CH_4}}{\mathcal{M}_{He}}} = \sqrt{\frac{16.04 \text{ g/mol}}{4.003 \text{ g/mol}}} = \sqrt{4.007} = 2.002$$

Check A ratio >1 makes sense because the lighter He should effuse faster than the heavier CH_4. Because the molar mass of He is about one-fourth that of CH_4, He should effuse about twice as fast (the inverse of $\sqrt{\frac{1}{4}}$).

FOLLOW-UP PROBLEM 5.13 If it takes 1.25 min for 0.010 mol of He to effuse, how long will it take for the same amount of ethane (C_2H_6) to effuse?

Graham's law is also used to determine the molar mass of an unknown gas, X, by comparing its effusion rate with that of a known gas, such as He:

$$\frac{\text{Rate}_X}{\text{Rate}_{He}} = \sqrt{\frac{\mathcal{M}_{He}}{\mathcal{M}_X}}$$

Squaring both sides and solving for the molar mass of X gives

$$\mathcal{M}_X = \mathcal{M}_{He} \times \left(\frac{\text{rate}_{He}}{\text{rate}_X}\right)^2$$

The Process of Diffusion Closely related to effusion is the process of gaseous **diffusion,** the movement of one gas through another. Diffusion rates are also described generally by Graham's law:

$$\text{Rate of diffusion} \propto \frac{1}{\sqrt{\mathcal{M}}}$$

For two gases at equal pressures, such as NH_3 and HCl, moving through another gas or a mixture of gases, such as air, we find

$$\frac{\text{Rate}_{NH_3}}{\text{Rate}_{HCl}} = \sqrt{\frac{\mathcal{M}_{HCl}}{\mathcal{M}_{NH_3}}}$$

The reason for this dependence on molar mass is the same as for effusion rates: lighter molecules have higher molecular speeds than heavier molecules, so they move farther in a given amount of time.

If gas molecules move at hundreds of meters per second at ordinary temperatures (see Figure 5.12), why does it take a second or two after you open a bottle of perfume to smell the fragrance? Although convection plays an important role, a molecule moving by diffusion does not travel very far before it collides with a molecule in the air. As you can see from Figure 5.18, the path of each molecule is tortuous. Imagine your walking speed through an empty room and then through a room crowded with other moving people.

Diffusion also occurs in liquids (and even to a small extent in solids). However, because the distances between molecules are much shorter in a liquid than in a gas, collisions are much more frequent; thus, diffusion is *much* slower.

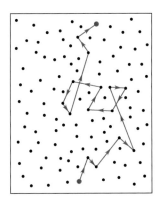

Figure 5.18 Diffusion of a gas particle through a space filled with other particles. In traversing a space, a gas molecule collides with many other molecules, which gives it a tortuous path. For clarity, the path of only one particle (*red dot*) is shown (*red lines*).

SECTION SUMMARY
The kinetic-molecular theory postulates that gas molecules take up a negligible portion of the gas volume, move in straight-line paths between elastic collisions, and have average kinetic energies proportional to the absolute temperature of the gas. This theory explains the gas laws in terms of changes in distances between molecules and the container walls and changes in molecular speed. Temperature is a measure of the average kinetic energy of molecules. Effusion and diffusion rates are inversely proportional to the square root of the molar mass (Graham's law) because they are directly proportional to molecular speed.

5.7 REAL GASES: DEVIATIONS FROM IDEAL BEHAVIOR

A fundamental principle of science is that simpler models are more useful than complex ones—as long as they explain the data. You can certainly appreciate the usefulness of the kinetic-molecular theory. With simple postulates, it explains the behavior of the ideal gases in terms of particles acting like infinitesimal billiard balls, moving at speeds governed by the absolute temperature, and experiencing only perfectly elastic collisions.

In reality, however, you know that *molecules are not points of mass.* They have volumes determined by the sizes of their atoms and the lengths and directions of their bonds. You also know that atoms contain charged particles and many bonds are polar, giving rise to *attractive and repulsive forces among molecules.* Therefore, we expect these properties of real gases to cause deviations from ideal behavior under some conditions, and this is indeed the case. We must alter the simple model and the ideal gas law to predict gas behavior at low temperatures and very high pressures.

Effects of Extreme Conditions on Gas Behavior

At ordinary conditions—relatively high temperatures and low pressures—most real gases exhibit nearly ideal behavior. Even at STP (0°C and 1 atm), however, real gases deviate *slightly* from ideal behavior. Table 5.3 shows the standard molar volumes of several gases to five significant figures. Note that they do not quite equal the ideal value. The phenomena that cause these slight deviations under standard conditions exert more influence as the temperature decreases toward the condensation point of the gas, the temperature at which it liquefies. As you can see, the largest deviations from ideal behavior in Table 5.3 are for Cl_2 and NH_3, because, at the standard temperature of 0°C, they are already close to their condensation points.

At pressures greater than 10 atm, we begin to see significant deviations from ideal behavior in many gases. Figure 5.19 shows a plot of *PV/RT* versus P_{ext} for *1 mol* of several real gases and an ideal gas. For 1 mol of an *ideal* gas, the ratio *PV/RT* is equal to 1 at any pressure. The values on the horizontal axis are the external pressures at which the *PV/RT* ratios are calculated. The pressures range from normal (at 1 atm, *PV/RT* = 1) to very high (at ~1000 atm, *PV/RT* ≈ 1.6 to 2.3).

The *PV/RT* curve shown in Figure 5.19 for 1 mol of methane (CH_4) is typical of that for most real gases: it decreases below the ideal value at moderately high pressures and then rises above it as pressure increases further. This shape arises from two overlapping effects of the two characteristics of real molecules just mentioned:

1. At moderately high pressure, values of *PV/RT* lower than ideal (less than 1) are due predominantly to *intermolecular attractions.*
2. At very high pressure, values of *PV/RT* greater than ideal (more than 1) are due predominantly to *molecular volume.*

Table 5.3	Molar Volume of Some Common Gases at STP (0°C and 1 atm)	
Gas	**Molar Volume (L/mol)**	**Condensation Point (°C)**
He	22.435	−268.9
H₂	22.432	−252.8
Ne	22.422	−246.1
Ideal gas	**22.414**	—
Ar	22.397	−185.9
N₂	22.396	−195.8
O₂	22.390	−183.0
CO	22.388	−191.5
Cl₂	22.184	−34.0
NH₃	22.079	−33.4

Figure 5.19 The behavior of several real gases with increasing external pressure. The horizontal line shows the behavior of 1 mol of ideal gas: *PV/RT* = 1 at all P_{ext}. At very high pressures, all real gases deviate significantly from such ideal behavior. Even at ordinary pressures, these deviations begin to appear (expanded portion).

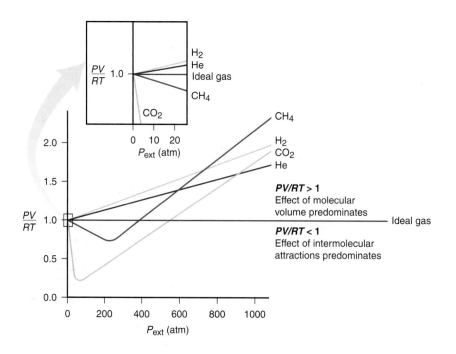

Let's examine these effects on the molecular level:

1. *Intermolecular attractions.* Attractive forces between molecules are *much* weaker than the covalent bonding forces that hold a molecule together. Most intermolecular attractions are caused by slight imbalances in electron distributions and are important only over relatively short distances. At normal pressures, the spaces between the molecules of any real gas are so large that attractions are negligible and the gas behaves nearly ideally. As the pressure rises and the volume of the sample decreases, however, the average intermolecular distance becomes smaller and attractions have a greater effect.

Picture a molecule at these higher pressures (Figure 5.20). As it approaches the container wall, nearby molecules attract it, which lessens the force of its impact. *Repeated throughout the sample, this effect results in decreased gas pressure and, thus, a smaller numerator in the PV/RT ratio.* Lowering the temperature has the same effect because it slows the molecules, so attractive forces exert an influence for a longer time. At a low enough temperature, the attractions among molecules become overwhelming, and the gas condenses to a liquid.

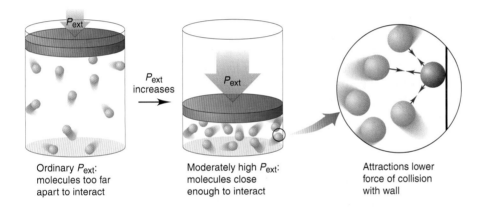

Ordinary P_{ext}: molecules too far apart to interact

Moderately high P_{ext}: molecules close enough to interact

Attractions lower force of collision with wall

Figure 5.20 The effect of intermolecular attractions on measured gas pressure. At ordinary pressures, the volume is large and gas molecules are too far apart to experience significant attractions. At moderately high external pressures, the volume decreases enough for the molecules to influence each other. As the close-up shows, a gas molecule approaching the container wall experiences intermolecular attractions from neighboring molecules that reduce the force of its impact. As a result, real gases exert *less* pressure than the ideal gas law predicts.

2. *Molecular volume.* At normal pressures, the space between molecules of a real gas (free volume) is enormous compared with the volume of the molecules *themselves* (molecular volume), so the free volume is essentially equal to the container volume. As the applied pressure increases, however, and the free volume decreases, the molecular volume makes up a greater proportion of the container volume, which you can see in Figure 5.21. Thus, at very high pressures, the free volume becomes significantly *less* than the container volume. However, we continue to use the container volume as the *V* in the *PV/RT* ratio, so the ratio is artificially high. This makes the numerator artificially high. The molecular volume effect becomes more important as the pressure increases, eventually outweighing the effect of the intermolecular attractions and causing *PV/RT* to rise above the ideal value.

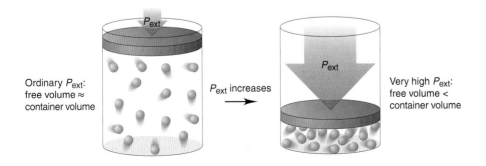

Ordinary P_{ext}: free volume ≈ container volume

Very high P_{ext}: free volume < container volume

Figure 5.21 The effect of molecular volume on measured gas volume. At ordinary pressures, the volume *between* molecules of a real gas (free volume) is essentially equal to the container volume because the molecules occupy only a tiny fraction of the available space. At very high external pressures, however, the free volume is significantly *less* than the container volume because of the volume of the molecules themselves.

In Figure 5.19, the H_2 and He curves do not show the typical dip at moderate pressures. These gases consist of particles with such weak intermolecular attractions that the molecular volume effect predominates at all pressures.

The van der Waals Equation: The Ideal Gas Law Redesigned

To describe real gas behavior more accurately, we need to "redesign" the ideal gas equation to do two things:

1. Adjust the measured pressure *up* by adding a factor that accounts for intermolecular attractions, and
2. Adjust the measured volume *down* by subtracting a factor from the entire container volume that accounts for the molecular volume.

In 1873, Johannes van der Waals realized the limitations of the ideal gas law and proposed an equation that accounts for the behavior of real gases. The **van der Waals equation** for *n* moles of a real gas is

$$\left(P + \frac{n^2a}{V^2}\right)(V - nb) = nRT$$

adjusts *P* up   adjusts *V* down

where *P* is the measured pressure, *V* is the container volume, *n* and *T* have their usual meanings, and *a* and *b* are *van der Waals constants,* experimentally determined positive numbers specific for a given gas. Values of these constants for several gases are given in Table 5.4. The constant *a* relates to the number of electrons, which in turn relates to the complexity of a molecule and the strength of its intermolecular attractions. The constant *b* relates to molecular volume.

Consider this typical application of the van der Waals equation to calculate a gas variable. A 1.98-L vessel contains 215 g (4.89 mol) of dry ice. After standing at 26°C (299 K), the $CO_2(s)$ changes to $CO_2(g)$. The pressure is measured (P_{real}) and calculated by the ideal gas law (P_{IGL}) and, using the appropriate values of *a* and *b,* by the van der Waals equation (P_{VDW}). The results are revealing:

$$P_{real} = 44.8 \text{ atm} \qquad P_{IGL} = 60.6 \text{ atm} \qquad P_{VDW} = 45.9 \text{ atm}$$

Comparing the real with each calculated value shows that P_{IGL} is 35.3% greater than P_{real}, but P_{VDW} is only 2.5% greater than P_{real}. At these conditions, CO_2 deviates so much from ideal behavior that the ideal gas law is not very useful.

Table 5.4	Van der Waals Constants for Some Common Gases	
Gas	$a \left(\dfrac{atm \cdot L^2}{mol^2}\right)$	$b \left(\dfrac{L}{mol}\right)$
He	0.034	0.0237
Ne	0.211	0.0171
Ar	1.35	0.0322
Kr	2.32	0.0398
Xe	4.19	0.0511
H_2	0.244	0.0266
N_2	1.39	0.0391
O_2	1.36	0.0318
Cl_2	6.49	0.0562
CH_4	2.25	0.0428
CO	1.45	0.0395
CO_2	3.59	0.0427
NH_3	4.17	0.0371
H_2O	5.46	0.0305

SECTION SUMMARY

At very high pressures or low temperatures, all gases deviate greatly from ideal behavior. As pressure increases, most real gases exhibit first a lower and then a higher *PV/RT* ratio than the value for the same amount (1 mol) of an ideal gas. These deviations are due to attractions between molecules, which lower the pressure (and the ratio), and to the larger fraction of the container volume occupied by the molecules, which increases the ratio. By including parameters characteristic of each gas, the van der Waals equation corrects for these deviations.

For Review and Reference (Numbers in parentheses refer to pages, unless noted otherwise.)

Learning Objectives

To help you review these learning objectives, the numbers of related sections (§), sample problems (SP), and end-of-chapter problems (EP) are listed in parentheses.

1. Explain how gases differ from liquids and solids (§ 5.1) (EPs 5.1, 5.2)
2. Understand how a barometer works and interconvert units of pressure (§ 5.2) (SP 5.1) (EPs 5.3–5.11)

3. Describe Boyle's, Charles's, and Avogadro's laws, understand how they relate to the ideal gas law, and apply them in calculations (§ 5.3) (SPs 5.2–5.6) (EPs 5.12–5.26)
4. Apply the ideal gas law to determine the molar mass of a gas, the density of a gas at different temperatures, and the partial pressure (or mole fraction) of each gas in a mixture (Dalton's law) (§ 5.4) (SPs 5.7–5.10) (EPs 5.27–5.43)

5. Use stoichiometry and the gas laws to calculate amounts of reactants and products (§ 5.5) (SPs 5.11, 5.12) (EPs 5.44–5.52)

6. Understand the kinetic-molecular theory and how it explains the gas laws, average molecular speed and kinetic energy, and the processes of effusion and diffusion (§ 5.6) (SP 5.13) (EPs 5.53–5.65)

7. Explain why intermolecular attractions and molecular volume cause real gases to deviate from ideal behavior and how the van der Waals equation corrects for the deviations (§ 5.7) (EPs 5.66–5.69)

Key Terms

Section 5.2
pressure (P) (140)
barometer (141)
pascal (Pa) (141)
standard atmosphere
  (atm) (141)
millimeter of mercury
  (mmHg) (142)
torr (142)

Section 5.3
ideal gas (143)
Boyle's law (144)
Charles's law (145)
Avogadro's law (147)
standard temperature and
  pressure (STP) (147)
standard molar
  volume (147)

ideal gas law (148)
universal gas constant
  (R) (148)

Section 5.4
partial pressure (155)
Dalton's law of partial
  pressures (155)
mole fraction (X) (156)

Section 5.6
kinetic-molecular theory (160)
rms speed (u_{rms}) (163)
effusion (164)
Graham's law of effusion (164)
diffusion (165)

Section 5.7
van der Waals equation (168)

Key Equations and Relationships

5.1 Expressing the volume-pressure relationship (Boyle's law) (144):
$$V \propto \frac{1}{P} \quad \text{or} \quad PV = \text{constant} \quad [T \text{ and } n \text{ fixed}]$$

5.2 Expressing the volume-temperature relationship (Charles's law) (145):
$$V \propto T \quad \text{or} \quad \frac{V}{T} = \text{constant} \quad [P \text{ and } n \text{ fixed}]$$

5.3 Expressing the pressure-temperature relationship (Amontons's law) (146):
$$P \propto T \quad \text{or} \quad \frac{P}{T} = \text{constant} \quad [V \text{ and } n \text{ fixed}]$$

5.4 Expressing the volume-amount relationship (Avogadro's law) (147):
$$V \propto n \quad \text{or} \quad \frac{V}{n} = \text{constant} \quad [P \text{ and } T \text{ fixed}]$$

5.5 Defining standard temperature and pressure (147):
STP: 0°C (273.15 K) and 1 atm (760 torr)

5.6 Defining the volume of 1 mol of an ideal gas at STP (147):
Standard molar volume = 22.4141 L = 22.4 L   [3 sf]

5.7 Relating volume to pressure, temperature, and amount (ideal gas law) (148):
$$PV = nRT \quad \text{and} \quad \frac{P_1 V_1}{n_1 T_1} = \frac{P_2 V_2}{n_2 T_2}$$

5.8 Calculating the value of R (148):
$$R = \frac{PV}{nT} = \frac{1 \text{ atm} \times 22.4141 \text{ L}}{1 \text{ mol} \times 273.15 \text{ K}}$$
$$= 0.082058 \frac{\text{atm·L}}{\text{mol·K}} = 0.0821 \frac{\text{atm·L}}{\text{mol·K}} \quad [3 \text{ sf}]$$

5.9 Rearranging the ideal gas law to find gas density (153):
$$PV = \frac{m}{\mathcal{M}} RT \quad \text{so} \quad \frac{m}{V} = d = \frac{\mathcal{M} \times P}{RT}$$

5.10 Rearranging the ideal gas law to find molar mass (154):
$$n = \frac{m}{\mathcal{M}} = \frac{PV}{RT} \quad \text{so} \quad \mathcal{M} = \frac{mRT}{PV} \quad \text{or} \quad \mathcal{M} = \frac{dRT}{P}$$

5.11 Relating the total pressure of a gas mixture to the partial pressures of the components (Dalton's law of partial pressures) (155):
$$P_{total} = P_1 + P_2 + P_3 + \cdots$$

5.12 Relating partial pressure to mole fraction (156):
$$P_A = X_A \times P_{total}$$

5.13 Defining rms speed as a function of molar mass and temperature (163):
$$u_{rms} = \sqrt{\frac{3RT}{\mathcal{M}}}$$

5.14 Applying Graham's law of effusion (164):
$$\frac{\text{Rate}_A}{\text{Rate}_B} = \frac{\sqrt{\mathcal{M}_B}}{\sqrt{\mathcal{M}_A}} = \sqrt{\frac{\mathcal{M}_B}{\mathcal{M}_A}}$$

Brief Solutions to Follow-up Problems

5.1 P_{CO_2} (mmHg) = 579.6 torr $\times \dfrac{1 \text{ mmHg}}{1 \text{ torr}}$
  = 579.6 mmHg

P_{CO_2} (Pa) = 579.6 torr $\times \dfrac{1 \text{ atm}}{760 \text{ torr}} \times \dfrac{1.01325 \times 10^5 \text{ Pa}}{1 \text{ atm}}$
  = 7.727×10^4 Pa

P_{CO_2} (lb/in^2) = 579.6 torr $\times \dfrac{1 \text{ atm}}{760 \text{ torr}} \times \dfrac{14.7 \text{ lb/in}^2}{1 \text{ atm}}$
  = 11.2 lb/in^2

5.2 P_2 (atm) = 26.3 kPa $\times \dfrac{1 \text{ atm}}{101.325 \text{ kPa}}$ = 0.260 atm

V_2 (L) = 105 mL $\times \dfrac{1 \text{ L}}{1000 \text{ mL}} \times \dfrac{0.871 \text{ atm}}{0.260 \text{ atm}}$ = 0.352 L

5.3 T_2 (K) = 273 K $\times \dfrac{9.75 \text{ cm}^3}{6.83 \text{ cm}^3}$ = 390. K

5.4 P_2 (torr) = 793 torr $\times \dfrac{35.0 \text{ g} - 5.0 \text{ g}}{35.0 \text{ g}}$ = 680. torr

(There is no need to convert mass to moles because the ratio of masses equals the ratio of moles.)

5.5 $n = \dfrac{PV}{RT} = \dfrac{1.37 \text{ atm} \times 438 \text{ L}}{0.0821 \dfrac{\text{atm·L}}{\text{mol·K}} \times 294 \text{ K}} = 24.9 \text{ mol O}_2$

Mass (g) of $O_2 = 24.9 \text{ mol O}_2 \times \dfrac{32.00 \text{ g O}_2}{1 \text{ mol O}_2} = 7.97 \times 10^2 \text{ g O}_2$

5.6 The balanced equation is $2CD(g) \longrightarrow C_2(g) + D_2(g)$, so n does not change. Therefore, given constant P, the absolute temperature T must double: $T_1 = -73°C + 273.15 = 200 \text{ K}$; so $T_2 = 400 \text{ K}$, or $400 \text{ K} - 273.15 = 127°C$.

5.7 d (at 0°C and 380 torr) $= \dfrac{44.01 \text{ g/mol} \times \dfrac{380 \text{ torr}}{760 \text{ torr/atm}}}{0.0821 \dfrac{\text{atm·L}}{\text{mol·K}} \times 273 \text{ K}}$

$= 0.982 \text{ g/L}$

The density is lower at the smaller P because V is larger. In this case, d is lowered by one-half because P is one-half as much.

5.8 $\mathcal{M} = \dfrac{1.26 \text{ g} \times 0.0821 \dfrac{\text{atm·L}}{\text{mol·K}} \times 283.2 \text{ K}}{\dfrac{102.5 \text{ kPa}}{101.325 \text{ kPa/1 atm}} \times 1.00 \text{ L}} = 29.0 \text{ g/mol}$

5.9 $n_{\text{total}} = \left(5.50 \text{ g He} \times \dfrac{1 \text{ mol He}}{4.003 \text{ g He}}\right)$

$+ \left(15.0 \text{ g Ne} \times \dfrac{1 \text{ mol Ne}}{20.18 \text{ g Ne}}\right)$

$+ \left(35.0 \text{ g Kr} \times \dfrac{1 \text{ mol Kr}}{83.80 \text{ g Kr}}\right)$

$= 2.53 \text{ mol}$

$P_{\text{He}} = \left(\dfrac{5.50 \text{ g He} \times \dfrac{1 \text{ mol He}}{4.003 \text{ g He}}}{2.53 \text{ mol}}\right) \times 1 \text{ atm} = 0.543 \text{ atm}$

$P_{\text{Ne}} = 0.294 \text{ atm} \qquad P_{\text{Kr}} = 0.165 \text{ atm}$

5.10 $P_{\text{H}_2} = 752 \text{ torr} - 13.6 \text{ torr} = 738 \text{ torr}$

Mass (g) of $H_2 = \left(\dfrac{\dfrac{738 \text{ torr}}{760 \text{ torr/atm}} \times 1.495 \text{ L}}{0.0821 \dfrac{\text{atm·L}}{\text{mol·K}} \times 289 \text{ K}}\right) \times \dfrac{2.016 \text{ g H}_2}{1 \text{ mol H}_2}$

$= 0.123 \text{ g H}_2$

5.11 $H_2SO_4(aq) + 2NaCl(s) \longrightarrow Na_2SO_4(aq) + 2HCl(g)$

$n_{\text{HCl}} = 0.117 \text{ kg NaCl} \times \dfrac{10^3 \text{ g}}{1 \text{ kg}} \times \dfrac{1 \text{ mol NaCl}}{58.44 \text{ g NaCl}} \times \dfrac{2 \text{ mol HCl}}{2 \text{ mol NaCl}}$

$= 2.00 \text{ mol HCl}$

At STP, V (mL) $= 2.00 \text{ mol} \times \dfrac{22.4 \text{ L}}{1 \text{ mol}} \times \dfrac{10^3 \text{ mL}}{1 \text{ L}}$

$= 4.48 \times 10^4 \text{ mL}$

5.12 $NH_3(g) + HCl(g) \longrightarrow NH_4Cl(s)$

$n_{\text{NH}_3} = 0.187 \text{ mol} \qquad n_{\text{HCl}} = 0.0522 \text{ mol}$

n_{NH_3} after reaction

$= 0.187 \text{ mol NH}_3 - \left(0.0522 \text{ mol HCl} \times \dfrac{1 \text{ mol NH}_3}{1 \text{ mol HCl}}\right)$

$= 0.135 \text{ mol NH}_3$

$P = \dfrac{0.135 \text{ mol} \times 0.0821 \dfrac{\text{atm·L}}{\text{mol·K}} \times 295 \text{ K}}{10.0 \text{ L}} = 0.327 \text{ atm}$

5.13 $\dfrac{\text{Rate of He}}{\text{Rate of C}_2\text{H}_6} = \sqrt{\dfrac{30.07 \text{ g/mol}}{4.003 \text{ g/mol}}} = 2.741$

Time for C_2H_6 to effuse $= 1.25 \text{ min} \times 2.741 = 3.43 \text{ min}$

Problems

Problems with **colored** numbers are answered in Appendix E. Sections match the text and provide the numbers of relevant sample problems. Bracketed problems are grouped in pairs (indicated by a short rule) that cover the same concept. Comprehensive Problems are based on material from any section or previous chapter.

An Overview of the Physical States of Matter

5.1 How does a sample of gas differ in its behavior from a sample of liquid in each of the following situations?
(a) The sample is transferred from one container to a larger one.
(b) The sample is heated in an expandable container, but no change of state occurs.
(c) The sample is placed in a cylinder with a piston, and an external force is applied.

5.2 Are the particles in a gas farther apart or closer together than the particles in a liquid? Use your answer to this question in order to explain each of the following general observations:
(a) Gases are more compressible than liquids.
(b) Gases have lower viscosities than liquids.

(c) After thorough stirring, all gas mixtures are solutions.
(d) The density of a substance in the gas state is lower than in the liquid state.

Gas Pressure and Its Measurement

(Sample Problem 5.1)

5.3 How does a barometer work?

5.4 How can a unit of length such as millimeter of mercury (mmHg) be used as a unit of pressure, which has the dimensions of force per unit area?

5.5 Is the column of mercury in a barometer shorter when it is on a mountaintop or at sea level? Explain.

5.6 On a cool, rainy day, the barometric pressure is 725 mmHg. Calculate the barometric pressure in centimeters of water (cmH$_2$O) (d of Hg = 13.5 g/mL; d of H$_2$O = 1.00 g/mL).

5.7 A long glass tube, sealed at one end, has an inner diameter of 10.0 mm. The tube is filled with water and inverted into a pail of water. If the atmospheric pressure is 755 mmHg, how high (in mmH$_2$O) is the column of water in the tube (d of Hg = 13.5 g/mL; d of H$_2$O = 1.00 g/mL)?

5.8 Convert the following:
(a) 0.745 atm to mmHg
(b) 992 torr to bar
(c) 365 kPa to atm
(d) 804 mmHg to kPa

5.9 Convert the following:
(a) 74.8 cmHg to atm
(b) 27.0 atm to kPa
(c) 8.50 atm to bar
(d) 0.907 kPa to torr

5.10 Convert each of the pressures described below into atmospheres:
(a) At the peak of Mt. Everest, atmospheric pressure is only 2.75×10^2 mmHg.
(b) A cyclist fills her bike tires to 91 psi.
(c) The surface of Venus has an atmospheric pressure of 9.15×10^6 Pa.
(d) At 100 ft below sea level, a scuba diver experiences a pressure of 2.44×10^4 torr.

5.11 The gravitational force exerted by an object is given by $F = mg$, where F is the force in newtons, m is the mass in kilograms, and g is the acceleration due to gravity (9.81 m/s^2).
(a) Use the definition of the pascal to calculate the mass (in kg) of the atmosphere on 1 m^2 of ocean.
(b) Osmium ($Z = 76$) has the highest density of any element (22.6 g/mL). If an osmium column is 1 m^2 in area, how high must it be for its pressure to equal atmospheric pressure? [Use the answer from part (a) in your calculation.]

The Gas Laws and Their Experimental Foundations
(Sample Problems 5.2 to 5.6)

5.12 When asked to state Boyle's law, a student replies, "The volume of a gas is inversely proportional to its pressure." How is this statement incomplete? Give a correct statement of Boyle's law.

5.13 Which quantities are variables and which are fixed in (a) Charles's law; (b) Avogadro's law?

5.14 Boyle's law relates gas volume to pressure, and Avogadro's law relates gas volume to number of moles. State a relationship between gas pressure and number of moles.

5.15 Each of the following processes caused the gas volume to double, as shown. For each process, state how the remaining gas variable changed or that it remained fixed:
(a) T doubles at fixed P.
(b) T and n are fixed.
(c) At fixed T, the reaction is $CD_2(g) \longrightarrow C(g) + D_2(g)$.
(d) At fixed P, the reaction is $A_2(g) + B_2(g) \longrightarrow 2AB(g)$.

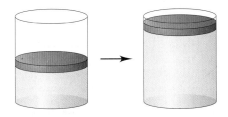

5.16 What is the effect of the following on the volume of 1 mol of an ideal gas?
(a) The pressure is tripled (at constant T).
(b) The absolute temperature is increased by a factor of 2.5 (at constant P).
(c) Two more moles of the gas are added (at constant P and T).

5.17 What is the effect of the following on the volume of 1 mol of an ideal gas?
(a) The pressure is reduced by a factor of 4 (at constant T).
(b) The pressure changes from 760 torr to 202 kPa, and the temperature changes from 37°C to 155 K.
(c) The temperature changes from 305 K to 32°C, and the pressure changes from 2 atm to 101 kPa.

5.18 A sample of sulfur hexafluoride gas occupies a volume of 5.10 L at 198°C. Assuming that the pressure remains constant, what temperature (in °C) is needed to reduce the volume to 2.50 L?

5.19 A 93-L sample of dry air is cooled from 145°C to −22°C while the pressure is maintained at 2.85 atm. What is the final volume?

5.20 A sample of Freon-12 (CF_2Cl_2) occupies 25.5 L at 298 K and 153.3 kPa. Find its volume at STP.

5.21 Calculate the volume of a sample of carbon monoxide that is at −14°C and 367 torr if it occupies 3.65 L at 298 K and 745 torr.

5.22 A sample of chlorine gas is confined in a 5.0-L container at 228 torr and 27°C. How many moles of gas are in the sample?

5.23 If 1.47×10^{-3} mol of argon occupies a 75.0-mL container at 26°C, what is the pressure (in torr)?

5.24 You have 207 mL of chlorine trifluoride gas at 699 mmHg and 45°C. What is the mass (in g) of the sample?

5.25 A 75.0-g sample of dinitrogen monoxide is confined in a 3.1-L vessel. What is the pressure (in atm) at 115°C?

5.26 In preparation for a demonstration, your professor brings a 1.5-L bottle of sulfur dioxide into the lecture hall before class to allow the gas to reach room temperature. If the pressure gauge reads 85 psi and the lecture hall is 23°C, how many moles of sulfur dioxide are in the bottle? (*Hint:* The gauge reads zero when 14.7 psi of gas remains.)

Further Applications of the Ideal Gas Law
(Sample Problems 5.7 to 5.10)

5.27 Why is moist air less dense than dry air?

5.28 To collect a beaker of H$_2$ gas by displacing the air already in the beaker, would you hold the beaker upright or inverted? Why? How would you hold the beaker to collect CO$_2$?

5.29 Why can we use a gas mixture, such as air, to study the general behavior of an ideal gas under ordinary conditions?

5.30 How does the partial pressure of gas A in a mixture compare to its mole fraction in the mixture? Explain.

5.31 What is the density of Xe gas at STP?

5.32 What is the density of Freon-11 ($CFCl_3$) at 120°C and 1.5 atm?

5.33 How many moles of gaseous arsine (AsH_3) will occupy 0.0400 L at STP? What is the density of gaseous arsine?

5.34 The density of a noble gas is 2.71 g/L at 3.00 atm and 0°C. Identify the gas.

5.35 Calculate the molar mass of a gas at 388 torr and 45°C if 206 ng occupies 0.206 μL.

5.36 When an evacuated 63.8-mL glass bulb is filled with a gas at 22°C and 747 mmHg, the bulb gains 0.103 g in mass. Is the gas N$_2$, Ne, or Ar?

5.37 When 0.600 L of Ar at 1.20 atm and 227°C is mixed with 0.200 L of O_2 at 501 torr and 127°C in a 400-mL flask at 27°C, what is the pressure in the flask?

5.38 A 355-mL container holds 0.146 g of Ne and an unknown amount of Ar at 35°C and a total pressure of 626 mmHg. Calculate the moles of Ar present.

5.39 The air in a hot-air balloon at 744 torr is heated from 17°C to 60.0°C. Assuming that the moles of air and the pressure remain constant, what is the density of the air at each temperature? (The average molar mass of air is 28.8 g/mol.)

5.40 On a certain winter day in Utah, the average atmospheric pressure is 650. torr. What is the molar density (in mol/L) of air if the temperature is −25°C?

5.41 A sample of a liquid hydrocarbon known to consist of molecules with five carbon atoms is vaporized in a 0.204-L flask by immersion in a water bath at 101°C. The barometric pressure is 767 torr, and the remaining gas weighs 0.482 g. What is the molecular formula of the hydrocarbon?

5.42 A sample of air contains 78.08% nitrogen, 20.94% oxygen, 0.05% carbon dioxide, and 0.93% argon, by volume. How many molecules of each gas are present in 1.00 L of the sample at 25°C and 1.00 atm?

5.43 An environmental chemist sampling industrial exhaust gases from a coal-burning plant collects a CO_2-SO_2-H_2O mixture in a 21-L steel tank until the pressure reaches 850. torr at 45°C.
(a) How many moles of gas are collected?
(b) If the SO_2 concentration in the mixture is 7.95×10^3 parts per million by volume (ppmv), what is its partial pressure? [*Hint:* ppmv = (volume of component/volume of mixture) $\times 10^6$.]

The Ideal Gas Law and Reaction Stoichiometry
(Sample Problems 5.11 and 5.12)

5.44 How many grams of phosphorus react with 35.5 L of O_2 at STP to form tetraphosphorus decaoxide?
$$P_4(s) + 5O_2(g) \longrightarrow P_4O_{10}(s)$$

5.45 How many grams of potassium chlorate decompose to potassium chloride and 638 mL of O_2 at 128°C and 752 torr?
$$2KClO_3(s) \longrightarrow 2KCl(s) + 3O_2(g)$$

5.46 How many grams of phosphine (PH_3) can form when 37.5 g of phosphorus and 83.0 L of hydrogen gas react at STP?
$$P_4(s) + H_2(g) \longrightarrow PH_3(g) \quad \text{[unbalanced]}$$

5.47 When 35.6 L of ammonia and 40.5 L of oxygen gas at STP burn, nitrogen monoxide and water are produced. After the products return to STP, how many grams of nitrogen monoxide are present?
$$NH_3(g) + O_2(g) \longrightarrow NO(g) + H_2O(l) \quad \text{[unbalanced]}$$

5.48 Aluminum reacts with excess hydrochloric acid to form aqueous aluminum chloride and 35.8 mL of hydrogen gas over water at 27°C and 751 mmHg. How many grams of aluminum reacted?

5.49 How many liters of hydrogen gas are collected over water at 18°C and 725 mmHg when 0.84 g of lithium reacts with water? Aqueous lithium hydroxide also forms.

5.50 "Strike anywhere" matches contain the compound tetraphosphorus trisulfide, which burns to form tetraphosphorus decaoxide

and sulfur dioxide gas. How many milliliters of sulfur dioxide, measured at 725 torr and 32°C, can be produced from burning 0.800 g of tetraphosphorus trisulfide?

5.51 Xenon hexafluoride was one of the first noble gas compounds synthesized. The solid reacts rapidly with the silicon dioxide in glass or quartz containers to form liquid $XeOF_4$ and gaseous silicon tetrafluoride. What is the pressure in a 1.00-L container at 25°C after 2.00 g of xenon hexafluoride reacts? (Assume that silicon tetrafluoride is the only gas present and that it occupies the entire volume.)

5.52 Roasting galena [lead(II) sulfide] is an early step in the industrial isolation of lead. How many liters of sulfur dioxide, measured at STP, are produced by the reaction of 3.75 kg of galena with 228 L of oxygen gas at 220°C and 2.0 atm? Lead(II) oxide also forms.

The Kinetic-Molecular Theory: A Model for Gas Behavior
(Sample Problem 5.13)

5.53 Use the kinetic-molecular theory to explain the change in gas pressure that results from warming a sample of gas.

5.54 How does the kinetic-molecular theory explain why 1 mol of krypton and 1 mol of helium have the same volume at STP?

5.55 Is the rate of effusion of a gas higher than, lower than, or equal to its rate of diffusion? Explain. For two gases with molecules of approximately the same size, is the ratio of their effusion rates higher than, lower than, or equal to the ratio of their diffusion rates? Explain.

5.56 Consider two 1-L samples of gas: one is H_2 and the other is O_2. Both are at 1 atm and 25°C. How do the samples compare in terms of (a) mass, (b) density, (c) average molecular kinetic energy, (d) average molecular speed, and (e) time for a given fraction of molecules to effuse?

5.57 Three 5-L flasks, fixed with pressure gauges and small valves, each contain 4 g of gas at 273 K. Flask A contains H_2, flask B contains He, and flask C contains CH_4. Rank the flask contents in terms of (a) pressure, (b) average molecular kinetic energy, (c) diffusion rate after the valve is opened, (d) total kinetic energy of the molecules, and (e) density.

5.58 What is the ratio of effusion rates for the lightest gas, H_2, and the heaviest known gas, UF_6?

5.59 What is the ratio of effusion rates for O_2 and Kr?

5.60 The graph below shows the distribution of molecular speeds for argon and helium at the same temperature.

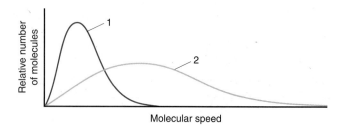

(a) Does curve 1 or 2 better represent the behavior of argon?
(b) Which curve represents the gas that effuses more slowly?
(c) Which curve more closely represents the behavior of fluorine gas? Explain.

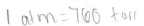

5.61 The graph below shows the distribution of molecular speeds for a gas at two different temperatures.

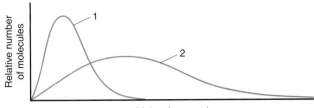

(a) Does curve 1 or 2 better represent the behavior of the gas at the lower temperature?
(b) Which curve represents the sample with the higher $\overline{E_k}$?
(c) Which curve represents the sample that diffuses more quickly?

5.62 At a given pressure and temperature, it takes 4.55 min for a 1.5-L sample of He to effuse through a membrane. How long does it take for 1.5 L of F_2 to effuse under the same conditions?

5.63 A sample of an unknown gas effuses in 11.1 min. An equal volume of H_2 in the same apparatus at the same temperature and pressure effuses in 2.42 min. What is the molar mass of the unknown gas?

5.64 Solid white phosphorus melts and then vaporizes at high temperature. Gaseous white phosphorus effuses at a rate that is 0.404 times that of neon in the same apparatus under the same conditions. How many atoms are in a molecule of gaseous white phosphorus?

5.65 Helium is the lightest noble gas component of air, and xenon is the heaviest. [For this problem, use $R = 8.314$ J/(mol·K) and M in kg/mol.]
(a) Calculate the rms speed of helium in winter (0.°C) and in summer (30.°C).
(b) Compare the rms speed of helium with that of xenon at 30.°C.
(c) Calculate the average kinetic energy per mole of helium and of xenon at 30.°C.
(d) Calculate the average kinetic energy per molecule of helium at 30.°C.

Real Gases: Deviations from Ideal Behavior

5.66 Do intermolecular attractions cause negative or positive deviations from the PV/RT ratio of an ideal gas? Use data from Table 5.4 to rank Kr, CO_2, and N_2 in order of increasing magnitude of these deviations.

5.67 Does molecular size cause negative or positive deviations from the PV/RT ratio of an ideal gas? Use data from Table 5.4 to rank Cl_2, H_2, and O_2 in order of increasing magnitude of these deviations.

5.68 Does N_2 behave more ideally at 1 atm or at 500 atm? Explain.
5.69 Does SF_6 (boiling point = 16°C at 1 atm) behave more ideally at 150°C or at 20°C? Explain.

Comprehensive Problems

Problems with an asterisk (*) are more challenging.

5.70 An "empty" gasoline can with dimensions 15.0 cm by 40.0 cm by 12.5 cm is attached to a vacuum pump and evacuated. If the atmospheric pressure is 14.7 lb/in², what is the total force (in pounds) on the outside of the can?

5.71 Hemoglobin is the protein that transports O_2 through the blood from the lungs to the rest of the body. In doing so, each molecule of hemoglobin combines with four molecules of O_2. If 1.00 g of hemoglobin combines with 1.53 mL of O_2 at 37°C and 743 torr, what is the molar mass of hemoglobin?

5.72 A baker uses sodium hydrogen carbonate (baking soda) as the leavening agent in a banana-nut quickbread. The baking soda decomposes according to two possible reactions:
Reaction 1. $2NaHCO_3(s) \longrightarrow Na_2CO_3(s) + H_2O(l) + CO_2(g)$
Reaction 2. $NaHCO_3(s) + H^+(aq) \longrightarrow$
$$H_2O(l) + CO_2(g) + Na^+(aq)$$
Calculate the volume (in mL) of CO_2 that forms at 200.°C and 0.975 atm per gram of $NaHCO_3$ by each of the reaction processes.

* **5.73** Chlorine is produced from concentrated seawater by the electrochemical chlor-alkali process. During the process, the chlorine is collected in a container that is isolated from the other products to prevent unwanted (and explosive) reactions. If a 15.00-L container holds 0.5850 kg of Cl_2 gas at 225°C, calculate
(a) P_{IGL} (b) P_{VDW} $\left(\text{use } R = 0.08206 \dfrac{\text{atm·L}}{\text{mol·K}}\right)$

* **5.74** In a certain experiment, magnesium boride (Mg_3B_2) reacted with acid to form a mixture of four boron hydrides (B_xH_y), three as liquids (labeled I, II, and III) and one as a gas (IV).
(a) When a 0.1000-g sample of each liquid was transferred to an evacuated 750.0-mL container and volatilized at 70.00°C, sample I had a pressure of 0.05951 atm, sample II 0.07045 atm, and sample III 0.05767 atm. What is the molar mass of each liquid?
(b) The mass of boron was found to be 85.63% in sample I, 81.10% in II, and 82.98% in III. What is the molecular formula of each sample?
(c) Sample IV was found to be 78.14% boron. Its rate of effusion was compared to that of sulfur dioxide and under identical conditions, 350.0 mL of sample IV effused in 12.00 min and 250.0 mL of sulfur dioxide effused in 13.04 min. What is the molecular formula of sample IV?

5.75 When air is inhaled, it enters the alveoli of the lungs, and varying amounts of the component gases exchange with dissolved gases in the blood. The resulting alveolar gas mixture is quite different from the atmospheric mixture. The following table presents selected data on the composition and partial pressure of four gases in the atmosphere and in the alveoli:

	Atmosphere (sea level)		Alveoli	
Gas	Mole %	Partial Pressure (torr)	Mole %	Partial Pressure (torr)
N_2	78.6	—	—	569
O_2	20.9	—	—	104
CO_2	0.04	—	—	40
H_2O	0.46	—	—	47

If the total pressure of each gas mixture is 1.00 atm, calculate the following:
(a) The partial pressure (in torr) of each gas in the atmosphere
(b) The mole % of each gas in the alveoli
(c) The number of O_2 molecules in 0.50 L of alveolar air (volume of an average breath at rest) at 37°C

5.76 Radon (Rn) is the heaviest, and only radioactive, member of Group 8A(18) (noble gases). It is a product of the disintegration of heavier radioactive nuclei found in minute concentrations in many common rocks used for building and construction. In recent years, health concerns about the cancers caused from inhaled residential radon have grown. If 1.0×10^{15} atoms of radium (Ra) produce an average of 1.373×10^4 atoms of Rn per second, how many liters of Rn, measured at STP, are produced per day by 1.0 g of Ra?

5.77 At 1400. mmHg and 286 K, a skin diver exhales a 208-mL bubble of air that is 77% N_2, 17% O_2, and 6.0% CO_2 by volume. (a) How many milliliters would the volume of the bubble be if it were exhaled at the surface at 1 atm and 298 K? (b) How many moles of N_2 are in the bubble?

5.78 Nitrogen dioxide is used industrially to produce nitric acid, but it contributes to acid rain and photochemical smog. What volume of nitrogen dioxide is formed at 735 torr and 28.2°C by reacting 4.95 cm³ of copper ($d = 8.95$ g/cm³) with 230.0 mL of nitric acid ($d = 1.42$ g/cm³, 68.0% HNO_3 by mass):

$$Cu(s) + 4HNO_3(aq) \longrightarrow Cu(NO_3)_2(aq) + 2NO_2(g) + 2H_2O(l)$$

5.79 In the average adult male, the residual volume (RV) of the lungs, the volume of air remaining after a forced exhalation, is 1200 mL. (a) How many moles of air are present in the RV at 1.0 atm and 37°C? (b) How many molecules of gas are present under these conditions?

5.80 In a collision of sufficient force, automobile air bags respond by electrically triggering the explosive decomposition of sodium azide (NaN_3) to its elements. A 50.0-g sample of sodium azide was decomposed, and the nitrogen gas generated was collected over water at 26°C. The total pressure was 745.5 mmHg. How many liters of dry N_2 were generated?

5.81 An anesthetic gas contains 64.81% carbon, 13.60% hydrogen, and 21.59% oxygen, by mass. If 2.00 L of the gas at 25°C and 0.420 atm weighs 2.57 g, what is the molecular formula of the anesthetic?

5.82 Aluminum chloride is easily vaporized at temperatures above 180°C. The gas escapes through a pinhole 0.122 times as fast as helium at the same conditions of temperature and pressure in the same apparatus. What is the molecular formula of gaseous aluminum chloride?

* **5.83** (a) What is the total volume of gaseous *products,* measured at 350°C and 735 torr, when an automobile engine burns 100. g of C_8H_{18} (a typical component of gasoline)?
(b) For part (a), the source of O_2 is air, which is about 78% N_2, 21% O_2, and 1.0% Ar by volume. Assuming all the O_2 reacts, but none of the N_2 or Ar does, what is the total volume of gaseous *exhaust?*

* **5.84** An atmospheric chemist studying the reactions of the pollutant SO_2 places a mixture of SO_2 and O_2 in a 2.00-L container at 900. K and an initial pressure of 1.95 atm. When the reaction occurs, gaseous SO_3 forms, and the pressure eventually falls to 1.65 atm. How many moles of SO_3 form?

5.85 Liquid nitrogen trichloride is heated in a 2.50-L closed reaction vessel until it decomposes completely to gaseous elements. The resulting mixture exerts a pressure of 754 mmHg at 95°C. (a) What is the partial pressure of each gas in the container? (b) What is the mass of the original sample?

5.86 Ammonium nitrate, a common fertilizer, is used as an explosive in fireworks and by terrorists. It was the material used in the devastating and tragic explosion of the Oklahoma City federal building in 1995. How many liters of gas at 307°C and 1.00 atm are formed by the explosive decomposition of 15.0 kg of ammonium nitrate to nitrogen, oxygen, and water vapor?

5.87 Analysis of a newly discovered gaseous silicon-fluorine compound shows that it contains 33.01 mass % silicon. At 27°C, 2.60 g of the compound exerts a pressure of 1.50 atm in a 0.250-L vessel. What is the molecular formula of the compound?

5.88 A gaseous organic compound containing only carbon, hydrogen, and nitrogen is burned in oxygen gas, and the individual volume of each reactant and product is measured under the same conditions of temperature and pressure. Reaction of four volumes of the compound produces four volumes of CO_2, two volumes of N_2, and ten volumes of water vapor. (a) What volume of oxygen gas was required? (b) What is the empirical formula of the compound?

5.89 Containers A, B, and C are attached by closed stopcocks of negligible volume.

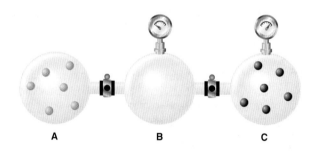

If each particle shown in the picture represents 10^6 particles, (a) How many blue particles and black particles are in B after the stopcocks are opened and the pressure stops changing?
(b) How many blue particles and black particles are in A after the stopcocks are opened and the pressure stops changing?
(c) If the pressure in C, P_C, is 750 torr before the stopcocks are opened, what is P_C afterward? (d) What is P_B afterward?

* **5.90** By what factor would a scuba diver's lungs expand if she ascended rapidly to the surface from a depth of 125 ft without inhaling or exhaling? If an expansion factor greater than 1.5 causes lung rupture, how far could she safely ascend from 125 ft without breathing? Assume constant temperature (d of seawater = 1.04 g/mL; d of Hg = 13.5 g/mL).

5.91 When 15.0 g of fluorite (CaF_2) reacts with excess sulfuric acid, hydrogen fluoride gas is collected at 744 torr and 25.5°C. Solid calcium sulfate is the other product. What gas temperature is required to store the gas in an 8.63-L container at 875 torr?

5.92 At a height of 300 km above Earth's surface, an astronaut finds that the atmospheric pressure is about 10^{-8} mmHg and the temperature is 500 K. How many molecules of gas are there per milliliter at this altitude?

5.93 What is the rms speed of O_2 molecules at STP? [Use $R = 8.314$ J/(mol·K) and $\mathcal{M}$ in kg/mol.]

5.94 Standard conditions are based on relevant environmental conditions. If normal average surface temperature and pressure on Venus are 730. K and 90 atm, respectively, what is the standard molar volume of an ideal gas on Venus?

5.95 The Hawaiian volcano Kilauea emits an average of 1.5×10^3 m^3 of gas each day, when corrected to 298 K and 1.00 atm. The mixture contains gases that contribute to global warming and acid rain, and some are toxic. An atmospheric chemist analyzes a sample and finds the following mole fractions: 0.4896 CO_2, 0.0146 CO, 0.3710 H_2O, 0.1185 SO_2, 0.0003 S_2, 0.0047 H_2, 0.0008 HCl, and 0.0003 H_2S. How many metric tons (t) of each gas is emitted per year (1 t = 1000 kg)?

5.96 To study a key fuel-cell reaction, a chemical engineer has 20.0-L tanks of H_2 and of O_2 and wants to use up both tanks to form 28.0 mol of water at 23.8°C. (a) Use the ideal gas law to find the pressure needed in each tank. (b) Use the van der Waals equation to find the pressure needed in each tank. (c) Compare the results from the two equations.

5.97 For each of the following, which shows the greater deviation from ideal behavior at the same set of conditions? Explain your choice.
(a) Argon or xenon   (b) Water vapor or neon
(c) Mercury vapor or radon   (d) Water vapor or methane

5.98 How many liters of gaseous hydrogen bromide at 27°C and 0.975 atm will a chemist need if she wishes to prepare 3.50 L of 1.20 M hydrobromic acid?

5.99 Sulfur dioxide is used primarily to make sulfuric acid. One method of producing it is by roasting mineral sulfides, for example,

$$FeS_2(s) + O_2(g) \longrightarrow SO_2(g) + Fe_2O_3(s) \quad \text{[unbalanced]}$$

A production error leads to the sulfide being placed in a 950-L vessel with insufficient oxygen. The partial pressure of O_2 is 0.64 atm and the total pressure is initially 1.05 atm, with the balance N_2. The reaction is run until 85% of the O_2 is consumed, and the vessel is then cooled to its initial temperature. What is the total pressure and partial pressure of each gas in the vessel?

* 5.100 A mixture of CO_2 and Kr weighs 35.0 g and exerts a pressure of 0.708 atm in its container. Since Kr is expensive, you wish to recover it from the mixture. After the CO_2 is completely removed by absorption with NaOH(s), the pressure in the container is 0.250 atm. How many grams of CO_2 were originally present? How many grams of Kr can you recover?

* 5.101 Aqueous sulfurous acid (H_2SO_3) was made by dissolving 0.200 L of sulfur dioxide gas at 20.°C and 740. mmHg in water to yield 500.0 mL of solution. The acid solution required 10.0 mL of sodium hydroxide solution to reach the titration end point. What was the molarity of the sodium hydroxide solution?

* 5.102 A person inhales air richer in O_2 and exhales air richer in CO_2 and water vapor. During each hour of sleep, a person exhales a total of about 300 L of this CO_2-enriched and H_2O-enriched air. (a) If the partial pressures of CO_2 and H_2O in exhaled air are each 30.0 torr at 37.0°C, calculate the masses of

CO_2 and of H_2O exhaled in 1 h of sleep. (b) How many grams of body mass does the person lose in an 8-h sleep if all the CO_2 and H_2O exhaled come from the metabolism of glucose?

$$C_6H_{12}O_6(s) + 6O_2(g) \longrightarrow 6CO_2(g) + 6H_2O(g)$$

5.103 Given these relationships for average kinetic energy,

$$\overline{E_k} = \tfrac{1}{2}m\overline{u^2} \qquad \text{and} \qquad \overline{E_k} = \tfrac{3}{2}\left(\frac{R}{N_A}\right)T$$

where m is molecular mass, u is rms speed, R is the gas constant [in J/(mol·K)], N_A is Avogadro's number, and T is absolute temperature: (a) derive Equation 5.13; (b) derive Equation 5.14.

5.104 Cylinder A in the picture below contains 0.1 mol of a gas that behaves ideally. Choose the cylinder (B, C, or D) that correctly represents the volume of the gas after each of the following changes. If none of the cylinders is correct, specify "none":
(a) P is doubled at fixed n and T.
(b) T is reduced from 400 K to 200 K at fixed n and P.
(c) T is increased from 100°C to 200°C at fixed n and P.
(d) 0.1 mol of gas is added at fixed P and T.
(e) 0.1 mol of gas is added and P is doubled at fixed T.

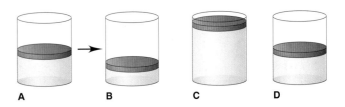

A   B   C   D

5.105 Ammonia is essential to so many industries that, on a molar basis, it is the most heavily produced substance in the world. Calculate P_{IGL} and P_{VDW} (in atm) of 51.1 g of ammonia in a 3.000-L container at 0°C and 400.°C, the industrial temperature. (From Table 5.4, for NH_3, $a = 4.17$ atm·L^2/mol^2 and $b = 0.0371$ L/mol.)

5.106 A 6.0-L flask contains a mixture of methane (CH_4), argon, and helium at 45°C and 1.75 atm. If the mole fractions of helium and argon are 0.25 and 0.35, respectively, how many molecules of methane are present?

5.107 A large portion of metabolic energy arises from the biological combustion of glucose:

$$C_6H_{12}O_6(s) + 6O_2(g) \longrightarrow 6CO_2(g) + 6H_2O(g)$$

(a) If this reaction is carried out in an expandable container at 35°C and 780. torr, what volume of CO_2 is produced from 18.0 g of glucose and excess O_2? (b) If the reaction is carried out at the same conditions with the stoichiometric amount of O_2, what is the partial pressure of each gas when the reaction is 50% complete (9.0 g of glucose remains)?

5.108 According to the American Conference of Governmental Industrial Hygienists, the *8-h threshold limit value* is 5000 ppmv for CO_2 and 0.1 ppmv for Br_2 (1 ppmv is 1 part by volume in 10^6 parts by volume). Exposure to either gas for 8 h above these limits is unsafe. At STP, which of the following would be unsafe for 8 h of exposure?
(a) Air with a partial pressure of 0.2 torr of Br_2
(b) Air with a partial pressure of 0.2 torr of CO_2
(c) 1000 L of air containing 0.0004 g of Br_2 gas
(d) 1000 L of air containing 2.8×10^{22} molecules of CO_2

5.109 One way to prevent emission of the pollutant NO from industrial plants is to react it with NH_3:

$$4NH_3(g) + 4NO(g) + O_2(g) \longrightarrow 4N_2(g) + 6H_2O(g)$$

(a) If the NO has a partial pressure of 4.5×10^{-5} atm in the flue gas, how many liters of NH_3 are needed per liter of flue gas at 1.00 atm? (b) If the reaction takes place at 1.00 atm and 150°C, how many grams of NH_3 are needed per kL of flue gas?

* 5.110 An equimolar mixture of Ne and Xe is accidentally placed in a container that has a tiny leak. After a short while, a very small proportion of the mixture has escaped. What is the mole fraction of Ne in the effusing gas?

* 5.111 One way to utilize naturally occurring uranium (0.72% ^{235}U and 99.27% ^{238}U) as a nuclear fuel is to enrich it (increase its ^{235}U content) by allowing gaseous UF_6 to effuse through a porous membrane (see p. 164). From the relative rates of effusion of $^{235}UF_6$ and $^{238}UF_6$, find the number of steps needed to produce uranium that is 3.0 mole % ^{235}U, the enriched fuel used in many nuclear reactors.

* 5.112 In preparation for a combustion demonstration, a professor's assistant fills a balloon with equal molar amounts of H_2 and O_2, but the demonstration has to be postponed until the next day. During the night, both gases leak through pores in the balloon. If 45% of the H_2 leaks, what is the O_2/H_2 ratio in the balloon the next day?

* 5.113 Phosphorus trichloride is important in the manufacture of insecticides, fuel additives, and flame retardants. Phosphorus has only one naturally occurring isotope, ^{31}P, whereas chlorine has two, 75% ^{35}Cl and 25% ^{37}Cl. (a) What different molecular masses (amu) can be found for PCl_3? (b) What is the ratio of the effusion rates for the heaviest and the lightest PCl_3 molecules?

5.114 A truck tire has a volume of 208 L and is filled with air to 35.0 psi at 295 K. After a drive, the air heats up to 319 K. (a) If the tire volume is constant, what is the pressure? (b) If the tire volume increases 2.0%, what is the pressure? (c) If the tire leaks 1.5 g of air per minute and the temperature is constant, how many minutes will it take for the tire to reach the original pressure of 35.0 psi ($\mathcal{M}$ of air = 28.8 g/mol)?

* 5.115 Many water treatment plants use chlorine gas to kill microorganisms before the water is released for residential use. A plant engineer has to maintain the chlorine pressure in a tank below the 85.0-atm rating and, to be safe, decides to fill the tank to 80.0% of this maximum pressure. (a) How many moles of Cl_2 gas can be kept in the 850.-L tank at 298 K if she uses the ideal gas law in the calculation? (b) What is the tank pressure if she uses the van der Waals equation for this amount of gas? (c) Did the engineer fill the tank to the desired pressure?

CHAPTER SIX

Thermochemistry: Energy Flow and Chemical Change

Measuring the Burn When a forest burns or a lake melts, indeed in any chemical or physical change, the reactants contain different quantities of energy than the products. In this chapter, you'll apply natural law to measure these changes.

Key Principles

♦ Chemical or physical change is *always* accompanied by a change in the energy that the matter contains.

♦ To study a *change in energy* (ΔE), scientists conceptually divide the universe into *system* (the part being studied) and *surroundings* (everything else). All energy changes occur as *heat* (q) and/or *work* (w) transferred either from the surroundings to the system or from the system to the surroundings ($\Delta E = q + w$). Thus, the total energy of the universe is constant (*law of energy conservation,* or *first law of thermodynamics*).

♦ No matter how a particular change in energy occurs, the magnitude of ΔE is the same. The internal energy of a system (E) is a *state function,* a thermodynamic variable whose change depends *only* on its initial and final values.

♦ *Enthalpy* (H) is related to E. The *change in enthalpy* (ΔH) equals the heat transferred at constant pressure, q_P. Most laboratory, environmental, and biological changes occur at constant P, so ΔH is more relevant than ΔE and easier to measure.

♦ The enthalpy change of a reaction, called the *heat of reaction* (ΔH_{rxn}), is negative (< 0) if the reaction releases heat (*exothermic*) and positive (> 0) if it absorbs heat (*endothermic*); for example, the combustion of methane is exothermic ($\Delta H_{rxn} < 0$), and the melting of ice is endothermic ($\Delta H_{rxn} > 0$).

♦ The more heat a substance absorbs, the higher its temperature becomes, but each substance has its own *heat capacity,* the heat required for a given temperature rise. Knowing this capacity and measuring ΔT in a *calorimeter,* we can find ΔH_{rxn}.

♦ The quantity of heat lost or gained in a reaction is related *stoichiometrically* to the amounts of reactants and products.

♦ Because H is a state function, we can find ΔH of any reaction by imagining that it occurs as the sum of other reactions whose ΔH values we know or can measure (*Hess's law of heat summation*).

♦ Chemists define a set of conditions, called *standard states,* in order to compare heats of different reactions. Each substance has a *standard heat of formation* (ΔH_f°), the heat of reaction when the substance is formed from its elements under these conditions. ΔH_f° values are used to calculate the *standard heat of reaction* (ΔH_{rxn}°).

Outline

Concepts & Skills to Review
Before You Study This Chapter

- energy and its interconversion (Section 1.1)
- distinction between heat and temperature (Section 1.4)
- nature of chemical bonding (Section 2.7)
- calculations of reaction stoichiometry (Section 3.4)
- properties of the gaseous state (Section 5.1)
- relation between kinetic energy and temperature (Section 5.6)

Whenever matter changes, whether chemically or physically, the energy content of the matter changes also. In the inferno of a forest fire, the wood and oxygen reactants contain more energy than the ash and gas products, and this difference in energy is *released* as heat and light. In contrast, some of the energy in a flash of lightning is *absorbed* when lower energy N_2 and O_2 in the air react to form higher energy NO. Energy is *absorbed* when snow melts and is *released* when water vapor condenses.

The production and utilization of energy in its many forms have an enormous impact on society. Some of the largest industries manufacture products that release, absorb, or limit the flow of energy. Common fuels—oil, wood, coal, and natural gas—release energy for heating and for powering combustion engines and steam turbines. Fertilizers help crops absorb solar energy and convert it to the chemical energy of food, which our bodies convert into other forms. Numerous plastic, fiberglass, and ceramic materials serve as insulators that limit the flow of energy.

Thermodynamics is the branch of physical science concerned with heat and its transformations to and from other forms of energy. It will be our focus here and again in Chapter 20. In this chapter, we highlight **thermochemistry,** which deals with the heat involved in chemical and physical changes.

6.1 FORMS OF ENERGY AND THEIR INTERCONVERSION

As we discussed in Chapter 1, all energy is either potential or kinetic, and these forms are convertible from one to the other. An object has potential energy by virtue of its position and kinetic energy by virtue of its motion. The potential energy of a weight raised above the ground is converted to kinetic energy as it falls (see Figure 1.3). When the weight hits the ground, it transfers some of that kinetic energy to the soil and pebbles, causing them to move, and thereby doing *work.* In addition, some of the transferred kinetic energy appears as *heat,* as it slightly warms the soil and pebbles. Thus, the potential energy of the weight is converted to kinetic energy, which is transferred to the ground as work and as heat.

Modern atomic theory allows us to consider other forms of energy—solar, electrical, nuclear, and chemical—as examples of potential and kinetic energy on the atomic and molecular scales. No matter what the details of the situation, *when energy is transferred from one object to another, it appears as work and/or as heat.* In this section, we examine this idea in terms of the loss or gain of energy that takes place during a chemical or physical change.

The System and Its Surroundings

In order to observe and measure a change in energy, we must first define the **system,** that part of the universe that we are going to focus on. The moment we define the system, everything else relevant to the change is defined as the **surroundings.** For example, in a flask containing a solution, the system is the contents of the flask; the flask itself, the other equipment, and perhaps the rest of the laboratory are the surroundings. In principle, the rest of the universe is the surroundings, but in practice, we need to consider only the portions of the universe relevant to the system. That is, it's not likely that a thunderstorm in central Asia or a methane blizzard on Neptune will affect the contents of the flask, but the temperature, pressure, and humidity of the lab might.

Energy Flow to and from a System

Each particle in a system has potential and kinetic energy, and the sum of these energies for all the particles is the **internal energy, E,** of the system (some texts use the symbol U). When a chemical system changes from reactants to products

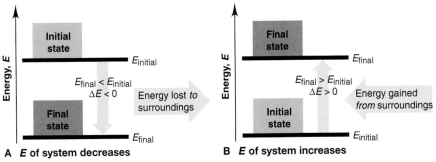

Figure 6.1 Energy diagrams for the transfer of internal energy (E) between a system and its surroundings. A, When the internal energy of a system *decreases,* the change in energy (ΔE) is lost *to* the surroundings; therefore, ΔE of the system ($E_{final} - E_{initial}$) is negative. **B,** When the system's internal energy *increases,* ΔE is gained *from* the surroundings and is positive. Note that the vertical yellow arrow, which signifies the direction of the change in energy, *always* has its tail at the initial state and its head at the final state.

and the products return to the starting temperature, the internal energy has changed. To determine this change, ΔE, we measure the difference between the system's internal energy *after* the change (E_{final}) and *before* the change ($E_{initial}$):

$$\Delta E = E_{final} - E_{initial} = E_{products} - E_{reactants} \qquad (6.1)$$

where Δ (Greek *delta*) means "change (or difference) in." Note especially that Δ refers to the *final state of the system **minus** the initial state.*

Because the total energy must be conserved, *a change in the energy of the system is always accompanied by an **opposite** change in the energy of the surroundings.* We often represent this change with an *energy diagram* in which the final and initial states are horizontal lines on a vertical energy axis. The change in internal energy, ΔE, is the difference between the heights of the two lines. A system can change its internal energy in one of two ways:

1. By losing some energy *to* the surroundings, as shown in Figure 6.1A:

$$E_{final} < E_{initial} \qquad \Delta E < 0 \qquad (\Delta E \text{ is negative})$$

2. By gaining some energy *from* the surroundings, as shown in Figure 6.1B:

$$E_{final} > E_{initial} \qquad \Delta E > 0 \qquad (\Delta E \text{ is positive})$$

Note that the change in energy is always an energy *transfer* from system to surroundings, or vice versa.

Heat and Work: Two Forms of Energy Transfer

Just as we saw when a weight hits the ground, energy transfer outward from the system or inward from the surroundings can appear in two forms, heat and work. **Heat** (or *thermal energy,* symbol *q*) is the energy transferred between a system and its surroundings as a result of a difference in their temperatures only. Energy in the form of heat is transferred from hot soup (system) to the bowl, air, and table (surroundings) because the surroundings have a lower temperature. All other forms of energy transfer (mechanical, electrical, and so on) involve some type of **work (*w*),** the energy transferred when an object is moved by a force. When you (system) kick a football (surroundings), energy is transferred as work to move the ball. When you inflate the ball, the inside air (system) exerts a force on the inner wall of the ball and nearby air (surroundings) and does work to move them outward.

The total change in a system's internal energy is the sum of the energy transferred as heat and/or work:

$$\Delta E = q + w \qquad (6.2)$$

The numerical values of q and w (and thus that of ΔE) can be either positive or negative, depending on the change the *system* undergoes. In other words, *we define the sign of the energy transfer from the system's perspective.* Energy coming *into* the system is *positive.* Energy going *out from* the system is *negative.* Of the innumerable changes possible in the system's internal energy, we'll examine the four simplest—two that involve only heat and two that involve only work.

Energy Transfer as Heat Only For a system that does no work but transfers energy only as heat (q), we know that $w = 0$. Therefore, from Equation 6.2, we have $\Delta E = q + 0 = q$. The two possibilities are:

1. *Heat flowing **out from** a system.* Suppose a sample of hot water is the system; then, the beaker containing it and the rest of the lab are the surroundings. The water transfers energy as heat to the surroundings until the temperature of the water equals that of the surroundings. The system's energy decreases as heat flows *out from* the system, so the final energy of the system is less than its initial energy. Heat was lost by the system, so q *is negative,* and therefore ΔE *is negative* (Figure 6.2A). This situation occurs in a refrigerator. The air (surroundings) has a lower temperature than a newly added piece of food (system), so the food loses energy as heat to the refrigerator air, $q < 0$.

2. *Heat flowing **into** a system.* If the system consists of ice water, it gains energy as heat from the surroundings until the temperature of the water equals that of the surroundings. In this case, energy is transferred *into* the system, so the final energy of the system is higher than its initial energy. Heat was gained by the system, so q *is positive,* and therefore ΔE *is positive* (Figure 6.2B). A common example of this situation occurs in a hot oven. In this case, the air (surroundings) has a higher temperature than a newly added piece of food (system), so the food gains energy as heat from the oven air, $q > 0$.

Animation: Energy Flow
Online Learning Center

Energy Transfer as Work Only For a system that transfers energy only as work (w), $q = 0$; therefore, $\Delta E = 0 + w = w$. The possibilities are

1. *Work done **by** a system.* Consider the reaction between zinc and hydrochloric acid as it takes place in an insulated container attached to a piston-cylinder assembly. We define the system as the atoms that make up the substances. In the initial state, the system's internal energy is that of the atoms in the form of the reactants, metallic Zn and aqueous H^+ and Cl^- ions. In the final state, the system's

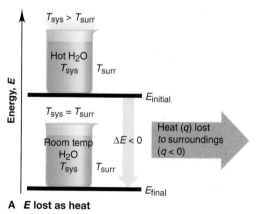

A *E* lost as heat

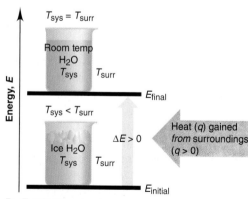

B *E* gained as heat

Figure 6.2 A system transferring energy as heat only. A, Hot water (the system, sys) transfers energy as heat (q) *to* the surroundings (surr) until $T_{sys} = T_{surr}$. Here $E_{initial} > E_{final}$ and $w = 0$, so $\Delta E < 0$ and the sign of q is negative. **B,** Ice water gains energy as heat (q) *from* the surroundings until $T_{sys} = T_{surr}$. Here $E_{initial} < E_{final}$ and $w = 0$, so $\Delta E > 0$ and the sign of q is positive.

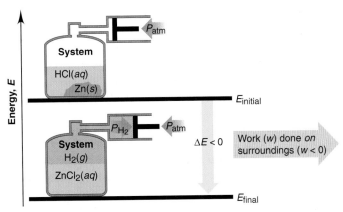

Figure 6.3 A system losing energy as work only. The internal energy of the system decreases as the reactants form products because the $H_2(g)$ does work (w) *on* the surroundings by pushing back the piston. The reaction vessel is insulated, so $q = 0$. Here $E_{initial} > E_{final}$, so $\Delta E < 0$ and the sign of w is negative.

internal energy is that of the same atoms in the form of the products, H_2 gas and aqueous Zn^{2+} and Cl^- ions:

$$Zn(s) + 2H^+(aq) + 2Cl^-(aq) \longrightarrow H_2(g) + Zn^{2+}(aq) + 2Cl^-(aq)$$

As the H_2 gas forms, some of the internal energy is used *by* the system to do work *on* the surroundings and push the piston outward. Energy is lost by the system as work, so *w is negative* and *ΔE is negative,* as you see in Figure 6.3. The H_2 gas is doing **pressure-volume work (PV work),** the type of work in which a volume changes against an external pressure. The work done here is not very useful because it simply pushes back the piston and outside air. But, if the system is a ton of burning coal and O_2, and the surroundings are a locomotive engine, much of the internal energy lost from the system does the work of moving a train.

2. *Work done **on** a system.* If we increase the external pressure on the piston in Figure 6.3, the system gains energy because work is done *on* the system *by* the surroundings: *w is positive, so ΔE is positive.*

Table 6.1 summarizes the sign conventions for q and w and their effect on the sign of ΔE.

The Law of Energy Conservation

As you've seen, when a system gains energy, the surroundings lose it, and when a system loses energy, the surroundings gain it. Energy can be converted from one form to another as these transfers take place, but it cannot simply appear or disappear—it cannot be created or destroyed. The **law of conservation of energy** restates this basic observation as follows: *the total energy of the universe is constant.* This law is also known as the **first law of thermodynamics.**

Table 6.1	**The Sign Conventions* for q, w, and ΔE**		
q	**+** **w**	**=**	**ΔE**
+	+		+
+	−		Depends on *sizes* of q and w
−	+		Depends on *sizes* of q and w
−	−		−

*For *q:* + means system *gains* heat; − means system *loses* heat.
  For *w:* + means work done *on* system; − means work done *by* system.

Conservation of energy applies everywhere. As gasoline burns in a car engine, the released energy appears as an equivalent amount of heat and work. The heat warms the car parts, passenger compartment, and surrounding air. The work appears as mechanical energy to turn the car's wheels and belts. That energy is converted further into the electrical energy of the clock and radio, the radiant energy of the headlights, the chemical energy of the battery, the heat due to friction, and so forth. If you took the sum of all these energy forms, you would find that it equals the change in energy between the reactants and products as the gasoline is burned. Complex biological processes also obey energy conservation. Through photosynthesis, green plants convert radiant energy from the Sun into chemical energy, as low-energy CO_2 and H_2O are used to make high-energy carbohydrates (such as wood) and O_2. When the wood is burned in air, those low-energy compounds form again, and the energy difference is released to the surroundings.

Thus, energy transfers between system and surroundings can be in the forms of heat and/or various types of work—mechanical, electrical, radiant, chemical—but *the energy of the system plus the energy of the surroundings remains constant: energy is conserved.* A mathematical expression of the law of conservation of energy (first law of thermodynamics) is

$$\Delta E_{universe} = \Delta E_{system} + \Delta E_{surroundings} = 0 \qquad (6.3)$$

This profound idea pertains to all systems, from a burning match to the movement of continents, from the inner workings of your heart to the formation of the Solar System.

Units of Energy

The SI unit of energy is the **joule (J),** a derived unit composed of three base units:

$$1 \text{ J} = 1 \text{ kg} \cdot \text{m}^2/\text{s}^2$$

Both heat and work are expressed in joules. Let's see how these units arise in the case of work. The work (w) done on a mass is the force (F) times the distance (d) that the mass moves: $w = F \times d$. A *force* changes the velocity of (accelerates) a mass. Velocity has units of meters per second (m/s), so acceleration (a) has units of m/s^2. Force, therefore, has units of mass (m, in kilograms) times acceleration:

$$F = m \times a \quad \text{in units of} \quad \text{kg} \cdot \text{m/s}^2$$

Therefore, $\quad w = F \times d \quad$ has units of $\quad (\text{kg} \cdot \text{m/s}^2) \times \text{m} = \text{kg} \cdot \text{m}^2/\text{s}^2 = \text{J}$

Potential energy, kinetic energy, and PV work are combinations of the same physical quantities and are also expressed in joules.

The **calorie (cal)** is an older unit that was defined originally as the quantity of energy needed to raise the temperature of 1 g of water by 1°C (from 14.5°C to 15.5°C). The calorie is now defined in terms of the joule:

$$1 \text{ cal} \equiv 4.184 \text{ J} \quad \text{or} \quad 1 \text{ J} = \frac{1}{4.184} \text{ cal} = 0.2390 \text{ cal}$$

Because the quantities of energy involved in chemical reactions are usually quite large, chemists use the kilojoule (kJ), or sometimes the kilocalorie (kcal):

$$1 \text{ kJ} = 1000 \text{ J} = 0.2390 \text{ kcal} = 239.0 \text{ cal}$$

The nutritional Calorie (note the capital C), the unit that diet tables use to show the energy available from food, is actually a kilocalorie. The *British thermal unit (Btu),* a unit in engineering that you may have seen used to indicate energy output of appliances, is the quantity of energy required to raise the temperature of 1 lb of water by 1°F and is equivalent to 1055 J. In general, the SI unit (J or kJ) is used throughout this text.

SAMPLE PROBLEM 6.1 Determining the Change in Internal Energy of a System

Problem When gasoline burns in a car engine, the heat released causes the products CO_2 and H_2O to expand, which pushes the pistons outward. Excess heat is removed by the car's cooling system. If the expanding gases do 451 J of work on the pistons and the system loses 325 J to the surroundings as heat, calculate the change in energy (ΔE) in J, kJ, and kcal.

Plan We must define system and surroundings, assign signs to q and w, and then calculate ΔE with Equation 6.2. The system is the reactants and products, and the surroundings are the pistons, the cooling system, and the rest of the car. Heat is released by the system, so q is negative. Work is done by the system to push the pistons outward, so w is also negative. We obtain the answer in J and then convert it to kJ and kcal.

Solution Calculating ΔE (from Equation 6.2) in J:

$$q = -325 \text{ J}$$
$$w = -451 \text{ J}$$
$$\Delta E = q + w = -325 \text{ J} + (-451 \text{ J})$$
$$= -776 \text{ J}$$

Converting from J to kJ:

$$\Delta E = -776 \text{ J} \times \frac{1 \text{ kJ}}{1000 \text{ J}}$$
$$= \boxed{-0.776 \text{ kJ}}$$

Converting from kJ to kcal:

$$\Delta E = -0.776 \text{ kJ} \times \frac{1 \text{ kcal}}{4.184 \text{ kJ}}$$
$$= \boxed{-0.185 \text{ kcal}}$$

Check The answer is reasonable: combustion releases energy from the system, so $E_{\text{final}} < E_{\text{initial}}$ and ΔE should be negative. Given that 4 kJ ≈ 1 kcal with rounding, nearly 0.8 kJ should be nearly 0.2 kcal.

FOLLOW-UP PROBLEM 6.1 In a reaction, gaseous reactants form a liquid product. The heat absorbed by the surroundings is 26.0 kcal, and the work done on the system is 15.0 Btu. Calculate ΔE (in kJ).

State Functions and the Path Independence of the Energy Change

An important point to understand is that there is no particular sequence by which the internal energy (E) of a system must change. This is because E is a **state function,** a property dependent only on the *current* state of the system (its composition, volume, pressure, and temperature), *not* on the path the system took to reach that state; the current state depends only on the *difference* between the final and initial states. As an analogy, the *balance* in your checkbook is a state function of your personal financial system. You can open a new account with a birthday gift of $50, or you can open a new account with a deposit of a $100 paycheck and then write two $25 checks. The two paths to the balance are different, but the balance (current state) is the same. And you can imagine countless other paths to the same balance.

In the same sense, the energy change of a system can occur by countless combinations of heat (q) and work (w). No matter what the combination, however, the same overall energy change occurs, because ΔE does **not** *depend on how the change takes place.* As an example, let's define a system in its initial state as 1 mol of octane (a component of gasoline) together with enough O_2 to burn it. In its final state, the system is the CO_2 and H_2O that form (a fractional coefficient is needed for O_2 because we specified 1 mol of octane):

$$C_8H_{18}(l) + \tfrac{25}{2} O_2(g) \longrightarrow 8CO_2(g) + 9H_2O(g)$$

initial state (E_{initial})   final state (E_{final})

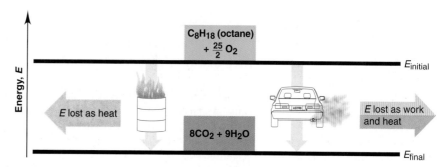

Figure 6.4 Two different paths for the energy change of a system. The change in internal energy when a given amount of octane burns in air is the same no matter how the energy is transferred. On the left, the fuel is burned in an open can, and the energy is lost almost entirely as heat. On the right, it is burned in a car engine; thus, a portion of the energy is lost as work to move the car, and less is lost as heat.

Energy is released to warm the surroundings and/or do work on them, so ΔE is negative. Two of the ways the change can occur are shown in Figure 6.4. If we burn the octane in an open container, ΔE appears almost completely as heat (with a small amount of work done to push back the atmosphere). If we burn it in a car engine, a much larger portion ($\sim 30\%$) of ΔE appears as work that moves the car, with the rest used to heat the car, exhaust gases, and surrounding air. If we burn the octane in a lawn mower or a plane, ΔE appears as other combinations of work and heat.

Thus, even though the separate quantities of work and heat available from the change *do* depend on how the change occurs, the change in internal energy (the *sum* of the heat and work) *does not*. In other words, for a given change, ΔE *(sum of q and w) is constant, even though q and w can vary.* Thus, heat and work are *not* state functions because their values *do* depend on the path the system takes in undergoing the energy change.

The pressure (P) of an ideal gas or the volume (V) of water in a beaker are other examples of state functions. This path independence means that *changes in state functions—ΔE, ΔP, and ΔV—depend only on their initial and final states.* (Note that symbols for state functions, such as E, P, and V, are capitalized.)

SECTION SUMMARY
Energy is transferred as heat (q) when the system and surroundings are at different temperatures; energy is transferred as work (w) when an object is moved by a force. Heat or work gained by a system ($q > 0$; $w > 0$) increases its internal energy (E); heat or work lost by the system ($q < 0$; $w < 0$) decreases E. The total change in the system's internal energy is the sum of the heat and work: $\Delta E = q + w$. Heat and work are measured in joules (J). Energy is always conserved: it changes from one form into another, moving into or out of the system, but the total quantity of energy in the universe (system *plus* surroundings) is constant. Energy is a state function; therefore, the same ΔE can occur through any combination of q and w.

6.2 ENTHALPY: HEATS OF REACTION AND CHEMICAL CHANGE

Most physical and chemical changes occur at virtually constant atmospheric pressure—a reaction in an open flask, the freezing of a lake, a drug response in an organism. In this section, we define a thermodynamic variable that makes it much easier to measure energy changes at constant pressure.

The Meaning of Enthalpy

To determine ΔE, we must measure both heat and work. The two most important types of chemical work are electrical work, the work done by moving charged particles (Chapter 21), and PV work, the work done by an expanding gas. We find the quantity of PV work done by multiplying the external pressure (P) by the change in volume of the gas (ΔV, or $V_{final} - V_{initial}$). In an open flask (or a cylinder with a weightless, frictionless piston), a gas does work by pushing back the atmosphere (Figure 6.5). Work done *on the surroundings*, is a negative quantity; work done *on the system* is a positive quantity:

$$w = -P\Delta V \qquad (6.4)$$

For reactions at constant pressure, a thermodynamic variable called **enthalpy (*H*)** eliminates the need to consider PV work separately. The enthalpy of a system is defined as the internal energy *plus* the product of the pressure and volume:

$$H = E + PV$$

The **change in enthalpy (*ΔH*)** is the change in internal energy *plus* the product of the constant pressure and the change in volume:

$$\Delta H = \Delta E + P\Delta V \qquad (6.5)$$

Combining Equations 6.2 ($\Delta E = q + w$) and 6.4 leads to a key point about ΔH:

$$\Delta E = q + w = q + (-P\Delta V) = q - P\Delta V$$

At constant pressure, we denote q as q_P and solve for it:

$$q_P = \Delta E + P\Delta V$$

Notice the right side of this equation is identical to the right side of Equation 6.5:

$$q_P = \Delta E + P\Delta V = \Delta H \qquad (6.6)$$

Thus, *the change in enthalpy equals the heat gained or lost at constant pressure.* With most changes occurring at constant pressure, ΔH is more relevant than ΔE *and* easier to find: *to find ΔH, measure q_P.*

We discuss the laboratory method for measuring this heat in Section 6.3.

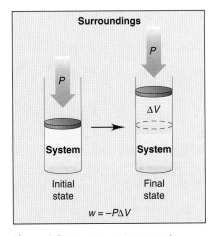

Figure 6.5 Pressure-volume work. When the volume (V) of a system increases by an amount ΔV against an external pressure (P), the system pushes back, and thus does PV work *on the* surroundings ($w = -P\Delta V$).

Exothermic and Endothermic Processes

Because E, P, and V are state functions, H is also a state function, which means that ΔH depends only on the *difference* between H_{final} and $H_{initial}$. The enthalpy change of a reaction, also called the **heat of reaction, ΔH_{rxn}**, *always refers to* H_{final} *minus* $H_{initial}$:

$$\Delta H = H_{final} - H_{initial} = H_{products} - H_{reactants}$$

Therefore, because $H_{products}$ can be either more or less than $H_{reactants}$, the sign of ΔH indicates whether heat is absorbed or released in the change. We determine the sign of ΔH by *imagining the heat as a "reactant" or "product."* When methane burns in air, for example, we know that heat is produced, so we show it as a product (on the right):

$$CH_4(g) + 2O_2(g) \longrightarrow CO_2(g) + 2H_2O(g) + heat$$

Because heat is released to the surroundings, the products (1 mol of CO_2 and 2 mol of H_2O) must have less enthalpy than the reactants (1 mol of CH_4 and 2 mol of O_2). Therefore, ΔH ($H_{final} - H_{initial}$) is negative, as the **enthalpy diagram** in Figure 6.6A (on the next page) shows. An **exothermic** ("heat out") **process** *releases* heat and results in a *decrease* in the enthalpy of the system:

Exothermic: $\quad H_{final} < H_{initial} \qquad \Delta H < 0 \qquad$ (ΔH is negative)

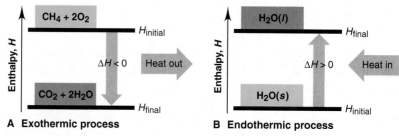

Figure 6.6 Enthalpy diagrams for exothermic and endothermic processes. A, Methane burns with a decrease in enthalpy because heat *leaves* the system. Therefore, $H_{final} < H_{initial}$, and the process is exothermic: $\Delta H < 0$. **B,** Ice melts with an increase in enthalpy because heat *enters* the system. Therefore, $H_{final} > H_{initial}$, and the process is endothermic: $\Delta H > 0$.

An **endothermic** ("heat in") **process** *absorbs* heat and results in an *increase* in the enthalpy of the system. When ice melts, for instance, heat flows *into* the ice from the surroundings, so we show the heat as a reactant (on the left):

$$heat + H_2O(s) \longrightarrow H_2O(l)$$

Because heat is absorbed, the enthalpy of the liquid water is higher than that of the solid water, as Figure 6.6B shows. Therefore, ΔH ($H_{water} - H_{ice}$) is positive:

$$\text{Endothermic:} \quad H_{final} > H_{initial} \quad \Delta H > 0 \quad (\Delta H \text{ is positive})$$

In general, the value of an enthalpy change refers to reactants and products at the same temperature.

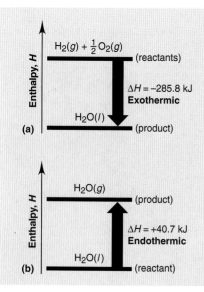

SAMPLE PROBLEM 6.2 Drawing Enthalpy Diagrams and Determining the Sign of ΔH

Problem In each of the following cases, determine the sign of ΔH, state whether the reaction is exothermic or endothermic, and draw an enthalpy diagram:
(a) $H_2(g) + \frac{1}{2}O_2(g) \longrightarrow H_2O(l) + 285.8 \text{ kJ}$
(b) $40.7 \text{ kJ} + H_2O(l) \longrightarrow H_2O(g)$
Plan From each equation, we see whether heat is a "product" (exothermic; $\Delta H < 0$) or a "reactant" (endothermic; $\Delta H > 0$). For exothermic reactions, reactants are above products on the enthalpy diagram; for endothermic reactions, reactants are below products. The ΔH arrow *always* points from reactants to products.
Solution (a) Heat is a product (on the right), so $\Delta H < 0$ and the reaction is exothermic. The enthalpy diagram appears in the margin (*top*).

(b) Heat is a reactant (on the left), so $\Delta H > 0$ and the reaction is endothermic. The enthalpy diagram appears in the margin (*bottom*).
Check Substances that are on the same side of the equation as the heat have less enthalpy than substances on the other side, so make sure they are placed on the lower line of the diagram.
Comment ΔH values depend on conditions. In (b), for instance, $\Delta H = 40.7 \text{ kJ}$ at 1 atm and 100°C; at 1 atm and 25°C, $\Delta H = 44.0 \text{ kJ}$.

FOLLOW-UP PROBLEM 6.2 When 1 mol of nitroglycerine decomposes, it causes a violent explosion and releases 5.72×10^3 kJ of heat:

$$C_3H_5(NO_3)_3(l) \longrightarrow 3CO_2(g) + \frac{5}{2}H_2O(g) + \frac{1}{4}O_2(g) + \frac{3}{2}N_2(g)$$

Is the reaction exothermic or endothermic? Draw an enthalpy diagram for it.

SECTION SUMMARY

The change in enthalpy, ΔH, is equal to the heat lost or gained during a chemical or physical change that occurs at constant pressure, q_P. A change that releases heat is exothermic ($\Delta H < 0$); a change that absorbs heat is endothermic ($\Delta H > 0$).

6.3 CALORIMETRY: LABORATORY MEASUREMENT OF HEATS OF REACTION

Data about energy content and use is everywhere—the calories per serving of a slice of bread, the energy efficiency rating of a washing machine, or the city/highway mileage of a new car. How do we measure the heat released (or absorbed) by a change? To determine the energy content of a teaspoon of sugar, for example, you might think we can simply measure the enthalpies of the reactants (sucrose and O_2) and subtract them from the enthalpies of the products (CO_2 and H_2O). The problem is that the enthalpy (H) of a system in a given state cannot be measured because we have no starting point with which to compare it, no zero enthalpy. However, we *can* measure the *change* in enthalpy (ΔH) of a system. In this section, we'll see how ΔH values are determined.

To measure q_P, which is equal to ΔH, we construct "surroundings" that retain the heat, and we observe the temperature change. Then, we relate the quantity of heat released (or absorbed) to that temperature change through a physical property called the *specific heat capacity*.

Specific Heat Capacity

You know from everyday experience that the more you heat an object, the higher its temperature; that is, the quantity of heat (q) absorbed by an object is proportional to its temperature change:

$$q \propto \Delta T \quad \text{or} \quad q = \text{constant} \times \Delta T \quad \text{or} \quad \frac{q}{\Delta T} = \text{constant}$$

Every object has its own **heat capacity,** the quantity of heat required to change its temperature by 1 K. Heat capacity is the proportionality constant in the preceding equation:

$$\text{Heat capacity} = \frac{q}{\Delta T} \quad \text{[in units of J/K]}$$

A related property is **specific heat capacity (c),** the quantity of heat required to change the temperature of 1 *gram* of a substance by 1 K:*

$$\text{Specific heat capacity (c)} = \frac{q}{\text{mass} \times \Delta T} \quad \text{[in units of J/g·K]}$$

If we know c of the substance being heated (or cooled), we can measure its mass and temperature change and calculate the heat absorbed or released:

$$q = c \times \text{mass} \times \Delta T \tag{6.7}$$

Notice that when an object gets hotter, ΔT (that is, $T_{\text{final}} - T_{\text{initial}}$) is positive. The object gains heat, so $q > 0$, as we expect. Similarly, when an object gets cooler, ΔT is negative; so $q < 0$ because heat is lost. Table 6.2 lists the specific heat capacities of some representative substances and materials.

Closely related to the specific heat capacity is the **molar heat capacity (C;** note capital letter), the quantity of heat required to change the temperature of 1 *mole* of a substance by 1 K:

$$\text{Molar heat capacity (C)} = \frac{q}{\text{moles} \times \Delta T} \quad \text{[in units of J/mol·K]}$$

Table 6.2 **Specific Heat Capacities of Some Elements, Compounds, and Materials**	
Substance	**Specific Heat Capacity (J/g·K)***
Elements	
Aluminum, Al	0.900
Graphite, C	0.711
Iron, Fe	0.450
Copper, Cu	0.387
Gold, Au	0.129
Compounds	
Water, $H_2O(l)$	4.184
Ethyl alcohol, $C_2H_5OH(l)$	2.46
Ethylene glycol, $(CH_2OH)_2(l)$	2.42
Carbon tetrachloride, $CCl_4(l)$	0.862
Solid materials	
Wood	1.76
Cement	0.88
Glass	0.84
Granite	0.79
Steel	0.45
*At 298 K (25°C).	

*Some texts use the term *specific heat* in place of *specific heat capacity.* This usage is very common but somewhat incorrect. *Specific heat* is the ratio of the heat capacity of 1 g of a substance to the heat capacity of 1 g of H_2O and therefore has no units.

The specific heat capacity of liquid water is 4.184 J/g·K, so

$$C \text{ of } H_2O(l) = 4.184 \frac{J}{g \cdot K} \times \frac{18.02 \text{ g}}{1 \text{ mol}} = 75.40 \frac{J}{mol \cdot K}$$

SAMPLE PROBLEM 6.3   Finding Quantity of Heat from Specific Heat Capacity

Problem A layer of copper welded to the bottom of a skillet weighs 125 g. How much heat is needed to raise the temperature of the copper layer from 25°C to 300.°C? The specific heat capacity (c) of Cu is 0.387 J/g·K.

Plan We know the mass and c of Cu and can find ΔT in °C, which equals ΔT in K. We use this ΔT and Equation 6.7 to solve for the heat.

Solution Calculating ΔT and q:

$$\Delta T = T_{final} - T_{initial} = 300.°C - 25°C = 275°C = 275 \text{ K}$$
$$q = c \times \text{mass (g)} \times \Delta T = 0.387 \text{ J/g·K} \times 125 \text{ g} \times 275 \text{ K} = 1.33 \times 10^4 \text{ J}$$

Check Heat is absorbed by the copper bottom (system), so q is positive. Rounding shows that the arithmetic seems reasonable: $q \approx 0.4$ J/g·K $\times$ 100 g $\times$ 300 K = 1.2×10^4 J.

FOLLOW-UP PROBLEM 6.3 Find the heat transferred (in kJ) when 5.50 L of ethylene glycol ($d = 1.11$ g/mL; see Table 6.2 for c) in a car radiator cools from 37.0°C to 25.0°C.

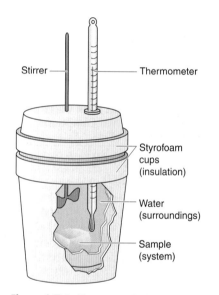

Stirrer ——
Thermometer ——
Styrofoam cups (insulation) ——
Water (surroundings) ——
Sample (system) ——

Figure 6.7 Coffee-cup calorimeter. This apparatus is used to measure the heat at constant pressure (q_P).

The Practice of Calorimetry

The **calorimeter** is used to measure the heat released (or absorbed) by a physical or chemical process. This apparatus is the "surroundings" that change temperature when heat is transferred to or from the system. Two common types are the constant-pressure and constant-volume calorimeters.

Constant-Pressure Calorimetry A "coffee-cup" calorimeter (Figure 6.7) is often used to measure the heat transferred (q_P) in processes open to the atmosphere. One common use is to find the specific heat capacity of a solid that does not react with or dissolve in water. The solid (system) is weighed, heated to some known temperature, and added to a sample of water (surroundings) of known temperature and mass in the calorimeter. With stirring, the final water temperature, which is also the final temperature of the solid, is measured.

The heat lost by the system ($-q_{sys}$, or $-q_{solid}$) is equal in magnitude but opposite in sign to the heat gained by the surroundings ($+q_{surr}$, or $+q_{H_2O}$):

$$-q_{solid} = q_{H_2O}$$

Substituting Equation 6.7 for each side of this equality gives

$$-(c_{solid} \times \text{mass}_{solid} \times \Delta T_{solid}) = c_{H_2O} \times \text{mass}_{H_2O} \times \Delta T_{H_2O}$$

All the quantities are known or measured except c_{solid}:

$$c_{solid} = -\frac{c_{H_2O} \times \text{mass}_{H_2O} \times \Delta T_{H_2O}}{\text{mass}_{solid} \times \Delta T_{solid}}$$

For example, suppose you heat a 25.64-g solid in a test tube to 100.00°C and carefully add it to 50.00 g of water in a coffee-cup calorimeter. The water temperature changes from 25.10°C to 28.49°C, and you want to find the specific heat capacity of the solid. Converting ΔT directly from °C to K, we know $\Delta T_{H_2O} = 3.39$ K (28.49°C − 25.10°C) and $\Delta T_{solid} = -71.51$ K (28.49°C − 100.00°C). Then, assuming all the heat lost by the solid is gained by the water, we have

$$c_{solid} = -\frac{c_{H_2O} \times \text{mass}_{H_2O} \times \Delta T_{H_2O}}{\text{mass}_{solid} \times \Delta T_{solid}} = -\frac{4.184 \text{ J/g·K} \times 50.00 \text{ g} \times 3.39 \text{ K}}{25.64 \text{ g} \times (-71.51 \text{ K})} = 0.387 \text{ J/g·K}$$

The next follow-up problem applies this calculation, but the sample problem first shows how to find the heat of a reaction that takes place in the calorimeter.

SAMPLE PROBLEM 6.4 Determining the Heat of a Reaction

Problem You place 50.0 mL of 0.500 M NaOH in a coffee-cup calorimeter at 25.00°C and carefully add 25.0 mL of 0.500 M HCl, also at 25.00°C. After stirring, the final temperature is 27.21°C. Calculate q_{soln} (in J) and ΔH_{rxn} (in kJ/mol). (Assume the total volume is the sum of the individual volumes and that the final solution has the same density and specific heat capacity as water: $d = 1.00$ g/mL and $c = 4.184$ J/g·K.)

Plan We first find the heat given off to the solution (q_{soln}) for the amounts given and then use the equation to find the heat per mole of reaction. We know the solution volumes (25.0 mL and 50.0 mL), so we can find their masses from the given density (1.00 g/mL). Multiplying their total mass by the change in T and the given c, we can find q_{soln}. Then, writing the balanced net ionic equation for the acid-base reaction, we use the volumes and the concentrations (0.500 M) to find moles of reactants (H^+ and OH^-) and, thus, product (H_2O). Dividing q_{soln} by the moles of water formed gives ΔH_{rxn} per mole.

Solution Finding mass$_{soln}$ and ΔT_{soln}:

$$\text{Total mass (g) of solution} = (25.0 \text{ mL} + 50.0 \text{ mL}) \times 1.00 \text{ g/mL} = 75.0 \text{ g}$$
$$\Delta T = 27.21°C - 25.00°C = 2.21°C = 2.21 \text{ K}$$

Finding q_{soln}:

$$q_{soln} = c_{soln} \times \text{mass}_{soln} \times \Delta T_{soln} = (4.184 \text{ J/g·K})(75.0 \text{ g})(2.21 \text{ K}) = 693 \text{ J}$$

Writing the net ionic equation: $HCl(aq) + NaOH(aq) \longrightarrow H_2O(l) + NaCl(aq)$
$$H^+(aq) + OH^-(aq) \longrightarrow H_2O(l)$$

Finding moles of reactants and products:

$$\text{Moles of } H^+ = 0.500 \text{ mol/L} \times 0.0250 \text{ L} = 0.0125 \text{ mol } H^+$$
$$\text{Moles of } OH^- = 0.500 \text{ mol/L} \times 0.0500 \text{ L} = 0.0250 \text{ mol } OH^-$$

Therefore, H^+ is limiting, so 0.0125 mol of H_2O is formed.

Finding ΔH_{rxn}: Heat gained by the water was lost by the reaction; that is,

$$q_{soln} = -q_{rxn} = 693 \text{ J} \qquad \text{so} \qquad q_{rxn} = -693 \text{ J}$$
$$\Delta H_{rxn} \text{ (kJ/mol)} = \frac{q_{rxn}}{\text{mol } H_2O} \times \frac{1 \text{ kJ}}{1000 \text{ J}} = \frac{-693 \text{ J}}{0.0125 \text{ mol}} \times \frac{1 \text{ kJ}}{1000 \text{ J}} = -55.4 \text{ kJ/mol}$$

Check Rounding to check q_{soln} gives 4 J/g·K × 75 g × 2 K = 600 J. The volume of H^+ is half the volume of OH^-, so moles of H^+ determines moles of product. Taking the negative of q_{soln} to find ΔH_{rxn} gives −600 J/0.012 mol = -5×10^4 J/mol, or −50 kJ/mol.

FOLLOW-UP PROBLEM 6.4 In a purity check for industrial diamonds, a 10.25-carat (1 carat = 0.2000 g) diamond is heated to 74.21°C and immersed in 26.05 g of water in a constant-pressure calorimeter. The initial temperature of the water is 27.20°C. Calculate ΔT of the water and of the diamond ($c_{diamond} = 0.519$ J/g·K).

Constant-Volume Calorimetry In the coffee-cup calorimeter, we assume all the heat is gained by the water, but some must be gained by the stirrer, thermometer, and so forth. For more precise work, as in constant-volume calorimetry, the *heat capacity of the entire calorimeter* must be known. One type of constant-volume apparatus is the *bomb calorimeter*, designed to measure very precisely the heat released in a combustion reaction. As Sample Problem 6.5 will show, this need for greater precision requires that we know (or determine) the heat capacity of the calorimeter.

Figure 6.8 (on the next page) depicts the preweighed combustible sample in a metal-walled chamber (the bomb), which is filled with oxygen gas and immersed in an insulated water bath fitted with motorized stirrer and thermometer. A heating coil connected to an electrical source ignites the sample, and the heat evolved raises the temperature of the bomb, water, and other calorimeter parts. Because we know the mass of the sample and the heat capacity of the entire calorimeter, we can use the measured ΔT to calculate the heat released.

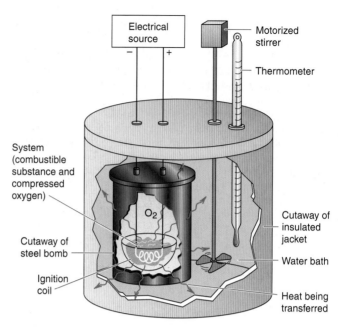

Figure 6.8 A bomb calorimeter. This device (not drawn to scale) is used to measure heat of combustion at constant volume (q_V).

SAMPLE PROBLEM 6.5   Calculating the Heat of Combustion

Problem A manufacturer claims that its new dietetic dessert has "fewer than 10 Calories per serving." To test the claim, a chemist at the Department of Consumer Affairs places one serving in a bomb calorimeter and burns it in O_2 (heat capacity of the calorimeter = 8.151 kJ/K). The temperature increases 4.937°C. Is the manufacturer's claim correct?

Plan When the dessert burns, the heat released is gained by the calorimeter:

$$-q_{sample} = q_{calorimeter}$$

To find the heat, we multiply the given heat capacity of the calorimeter (8.151 kJ/K) by ΔT (4.937°C).

Solution Calculating the heat gained by the calorimeter:

$$q_{calorimeter} = \text{heat capacity} \times \Delta T = 8.151 \text{ kJ/K} \times 4.937 \text{ K} = 40.24 \text{ kJ}$$

Recall that 1 Calorie = 1 kcal = 4.184 kJ. Therefore, 10 Calories = 41.84 kJ, so the claim is correct.

Check A quick math check shows that the answer is reasonable: 8 kJ/K × 5 K = 40 kJ.

Comment With the volume of the steel bomb fixed, $\Delta V = 0$, and thus $P\Delta V = 0$. Thus, the energy change measured is the *heat at constant volume* (q_V), which equals ΔE, not ΔH:

$$\Delta E = q + w = q_V + 0 = q_V$$

However, in most cases, ΔH is usually very close to ΔE. For example, ΔH is only 0.5% larger than ΔE for the combustion of H_2 and only 0.2% smaller for the combustion of octane.

FOLLOW-UP PROBLEM 6.5 A chemist burns 0.8650 g of graphite (a form of carbon) in a new bomb calorimeter, and CO_2 forms. If 393.5 kJ of heat is released per mole of graphite and T increases 2.613 K, what is the heat capacity of the bomb calorimeter?

SECTION SUMMARY

We calculate ΔH of a process by measuring the heat at constant pressure (q_P). To do this, we determine ΔT of the surroundings and relate it to q_P through the mass of the substance and its specific heat capacity (c), the quantity of energy needed to raise the temperature of 1 g of the substance by 1 K. Calorimeters measure the heat released from a system either at constant pressure ($q_P = \Delta H$) or at constant volume ($q_V = \Delta E$).

6.4 STOICHIOMETRY OF THERMOCHEMICAL EQUATIONS

A **thermochemical equation** is a balanced equation that includes the heat of reaction (ΔH_{rxn}). Keep in mind that the ΔH_{rxn} value shown refers to the *amounts (moles) of substances **and** their states of matter in that specific equation.* The enthalpy change of any process has two aspects:

1. *Sign.* The sign of ΔH depends on whether the reaction is exothermic ($-$) or endothermic ($+$). A forward reaction has the *opposite* sign of the reverse reaction.
   Decomposition of 2 mol of water to its elements (endothermic):

   $$2H_2O(l) \longrightarrow 2H_2(g) + O_2(g) \quad \Delta H_{rxn} = 572 \text{ kJ}$$

   Formation of 2 mol of water from its elements (exothermic):

   $$2H_2(g) + O_2(g) \longrightarrow 2H_2O(l) \quad \Delta H_{rxn} = -572 \text{ kJ}$$

2. *Magnitude.* The magnitude of ΔH is *proportional to the amount of substance* reacting.
   Formation of 1 mol of water from its elements (half the amount in the preceding equation):

   $$H_2(g) + \tfrac{1}{2}O_2(g) \longrightarrow H_2O(l) \quad \Delta H_{rxn} = -286 \text{ kJ}$$

Note that, in thermochemical equations, we often use fractional coefficients to specify the magnitude of ΔH_{rxn} for a *particular amount of substance.* Moreover, *in a particular reaction,* a certain amount of substance is thermochemically equivalent to a certain quantity of energy. In the reaction just shown,

> 286 kJ is thermochemically equivalent to 1 mol of $H_2(g)$
> 286 kJ is thermochemically equivalent to $\tfrac{1}{2}$ mol of $O_2(g)$
> 286 kJ is thermochemically equivalent to 1 mol of $H_2O(l)$

Just as we use stoichiometrically equivalent molar ratios to find amounts of substances, we use thermochemically equivalent quantities to find the heat of reaction for a given amount of substance. Also, just as we use molar mass (in g/mol of substance) to convert moles of a substance to grams, we use the heat of reaction (in kJ/mol of substance) to convert moles of a substance to an equivalent quantity of heat (in kJ). Figure 6.9 shows this new relationship, and the next sample problem applies it.

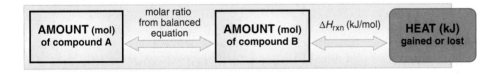

Figure 6.9 Summary of the relationship between amount (mol) of substance and the heat (kJ) transferred during a reaction.

SAMPLE PROBLEM 6.6   Using the Heat of Reaction (ΔH_{rxn}) to Find Amounts

Problem The major source of aluminum in the world is bauxite (mostly aluminum oxide). Its thermal decomposition can be represented by

$$Al_2O_3(s) \xrightarrow{\Delta} 2Al(s) + \tfrac{3}{2}O_2(g) \quad \Delta H_{rxn} = 1676 \text{ kJ}$$

If aluminum is produced this way (see Comment), how many grams of aluminum can form when 1.000×10^3 kJ of heat is transferred?

Plan From the balanced equation and the enthalpy change, we see that 2 mol of Al forms when 1676 kJ of heat is absorbed. With this equivalent quantity, we convert the given kJ transferred to moles formed and then convert moles to grams.

Solution Combining steps to convert from heat transferred to mass of Al:

$$\text{Mass (g) of Al} = (1.000\times10^3 \text{ kJ}) \times \frac{2 \text{ mol Al}}{1676 \text{ kJ}} \times \frac{26.98 \text{ g Al}}{1 \text{ mol Al}} = 32.20 \text{ g Al}$$

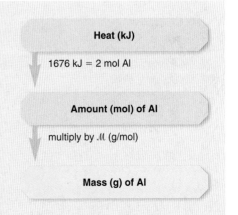

Check The mass of aluminum seems correct: ~1700 kJ forms about 2 mol of Al (54 g), so 1000 kJ should form a bit more than half that amount (27 g).

Comment In practice, aluminum is not obtained by heating but by supplying electrical energy (Chapter 21). Because ΔH is a state function, however, the total energy required for this chemical change is the same no matter how it occurs.

FOLLOW-UP PROBLEM 6.6 Organic hydrogenation reactions, in which H_2 and an "unsaturated" organic compound combine, are used in the food, fuel, and polymer industries. In the simplest case, ethene (C_2H_4) and H_2 form ethane (C_2H_6). If 137 kJ is given off per mole of C_2H_4 reacting, how much heat is released when 15.0 kg of C_2H_6 forms?

SECTION SUMMARY

A thermochemical equation shows the balanced equation *and* its ΔH_{rxn}. The sign of ΔH for a forward reaction is opposite that for the reverse reaction. The magnitude of ΔH depends on the amount and physical state of the substance reacting and the ΔH per mole of substance. We use the thermochemically equivalent amounts of substance and heat from the balanced equation as conversion factors to find the quantity of heat when a given amount of substance reacts.

6.5 HESS'S LAW OF HEAT SUMMATION

Many reactions are difficult, even impossible, to carry out separately. A reaction may be part of a complex biochemical process; or it may take place only under extreme environmental conditions; or it may require a change in conditions while it is occurring. Even if we can't run a reaction in the lab, it is still possible to find its enthalpy change. One of the most powerful applications of the state-function property of enthalpy (H) allows us to find the ΔH of *any* reaction for which we can write an equation.

This application is based on **Hess's law of heat summation:** *the enthalpy change of an overall process is the sum of the enthalpy changes of its individual steps.* To use Hess's law, we imagine an overall reaction as the sum of a series of reaction steps, whether or not it really occurs that way. Each step is chosen because its ΔH is known. Because the overall ΔH depends only on the initial and final states, Hess's law says that we add together the known ΔH values for the steps to get the unknown ΔH of the overall reaction. Similarly, if we know the ΔH values for the overall reaction and all but one of the steps, we can find the unknown ΔH of that step.

Let's see how we apply Hess's law in the case of the oxidation of sulfur to sulfur trioxide, the central process in the industrial production of sulfuric acid and in the formation of acid rain. (To introduce the approach, we'll simplify the equations by using S as the formula for sulfur, rather than the more correct S_8.) When we burn S in an excess of O_2, sulfur dioxide (SO_2) forms, *not* sulfur trioxide (SO_3). Equation 1 shows this step and its ΔH. If we change conditions and then add more O_2, we can oxidize SO_2 to SO_3 (Equation 2). In other words, we cannot put S and O_2 in a calorimeter and find ΔH for the overall reaction of S to SO_3 (Equation 3). But, we *can* find it with Hess's law. The three equations are

$$\text{Equation 1:} \qquad S(s) + O_2(g) \longrightarrow SO_2(g) \qquad \Delta H_1 = -296.8 \text{ kJ}$$

$$\text{Equation 2:} \quad 2SO_2(g) + O_2(g) \longrightarrow 2SO_3(g) \qquad \Delta H_2 = -198.4 \text{ kJ}$$

$$\text{Equation 3:} \qquad S(s) + \tfrac{3}{2}O_2(g) \longrightarrow SO_3(g) \qquad \Delta H_3 = ?$$

Hess's law tells us that if we manipulate Equations 1 and/or 2 so that they add up to Equation 3, then ΔH_3 is the sum of the manipulated ΔH values of Equations 1 and 2.

First, we identify Equation 3 as our "target" equation, the one whose ΔH we want to find, and we carefully note the number of moles of each reactant and product in it. We also note that ΔH_1 and ΔH_2 are the values for Equations 1 and 2 *as written*. Now we manipulate Equations 1 and/or 2 as follows to make them add up to Equation 3:

- Equations 1 and 3 contain the same amount of S, so we leave Equation 1 unchanged.
- Equation 2 has twice as much SO_3 as Equation 3, so we multiply it by $\frac{1}{2}$, being sure to halve ΔH_2 as well.
- With the targeted amounts of reactants and products now present, we add Equation 1 to the halved Equation 2 and cancel terms that appear on both sides:

Equation 1: $\qquad\qquad\qquad$ $S(s) + O_2(g) \longrightarrow SO_2(g)$ $\qquad$ $\Delta H_1 = -296.8$ kJ

$\frac{1}{2}$(Equation 2): $\qquad\qquad$ $SO_2(g) + \frac{1}{2}O_2(g) \longrightarrow SO_3(g)$ $\qquad$ $\frac{1}{2}(\Delta H_2) = -99.2$ kJ

Equation 3: $S(s) + O_2(g) + \cancel{SO_2(g)} + \frac{1}{2}O_2(g) \longrightarrow \cancel{SO_2(g)} + SO_3(g)$ $\quad$ $\Delta H_3 = ?$

or, $\qquad\qquad\qquad\qquad$ $S(s) + \frac{3}{2}O_2(g) \longrightarrow SO_3(g)$

Adding the ΔH values gives

$$\Delta H_3 = \Delta H_1 + \tfrac{1}{2}(\Delta H_2) = -296.8 \text{ kJ} + (-99.2 \text{ kJ}) = -396.0 \text{ kJ}$$

Once again, the key point is that H is a state function, so the overall ΔH depends on the difference between the initial and final enthalpies only. Hess's law tells us that the difference between the enthalpies of the reactants (1 mol of S and $\frac{3}{2}$ mol of O_2) and that of the product (1 mol of SO_3) is the same, whether S is oxidized directly to SO_3 (impossible) or through the formation of SO_2 (actual).

To summarize, calculating an unknown ΔH involves three steps:

1. Identify the target equation, the step whose ΔH is unknown, and note the number of moles of each reactant and product.
2. Manipulate the equations with known ΔH values so that the target numbers of moles of reactants and products are on the correct sides. Remember to:
   - Change the sign of ΔH when you reverse an equation.
   - Multiply numbers of moles and ΔH by the same factor.
3. Add the manipulated equations to obtain the target equation. All substances except those in the target equation must cancel. Add their ΔH values to obtain the unknown ΔH.

SAMPLE PROBLEM 6.7 Using Hess's Law to Calculate an Unknown ΔH

Problem Two gaseous pollutants that form in auto exhaust are CO and NO. An environmental chemist is studying ways to convert them to less harmful gases through the following equation:

$$CO(g) + NO(g) \longrightarrow CO_2(g) + \tfrac{1}{2}N_2(g) \quad \Delta H = ?$$

Given the following information, calculate the unknown ΔH:

Equation A: $\quad$ $CO(g) + \frac{1}{2}O_2(g) \longrightarrow CO_2(g)$ $\quad$ $\Delta H_A = -283.0$ kJ

Equation B: $\quad$ $N_2(g) + O_2(g) \longrightarrow 2NO(g)$ $\quad$ $\Delta H_B = 180.6$ kJ

Plan We note the numbers of moles of each substance in the target equation, manipulate Equations A and/or B *and* their ΔH values, and then add them together to obtain the target equation and the unknown ΔH.

Solution Noting moles of substances in the target equation: There are 1 mol each of reactants CO and NO, 1 mol of product CO_2, and $\frac{1}{2}$ mol of product N_2.

Manipulating the given equations: Equation A has the same number of moles of CO and CO_2 as the target, so we leave it as written. Equation B has twice the needed amounts of N_2 and NO, and they are on the opposite sides from the target; therefore, we reverse Equation B, change the sign of ΔH_B, and multiply both by $\frac{1}{2}$:

$$\tfrac{1}{2}[2NO(g) \longrightarrow N_2(g) + O_2(g)] \quad \Delta H = -\tfrac{1}{2}(\Delta H_B) = -\tfrac{1}{2}(180.6 \text{ kJ})$$

Or, $\qquad\qquad$ $NO(g) \longrightarrow \frac{1}{2}N_2(g) + \frac{1}{2}O_2(g)$ $\quad$ $\Delta H = -90.3$ kJ

Adding the manipulated equations to obtain the target equation:

Equation A:	$CO(g) + \frac{1}{2}O_2(g) \longrightarrow CO_2(g)$	$\Delta H = -283.0$ kJ
$\frac{1}{2}$(Equation B reversed):	$NO(g) \longrightarrow \frac{1}{2}N_2(g) + \frac{1}{2}O_2(g)$	$\Delta H = -90.3$ kJ

Target:	$CO(g) + NO(g) \longrightarrow CO_2(g) + \frac{1}{2}N_2(g)$	$\Delta H = -373.3$ kJ

Check Obtaining the desired target equation is its own check. Be sure to remember to change the *sign* of ΔH for any equation you reverse.

FOLLOW-UP PROBLEM 6.7 Nitrogen oxides undergo many interesting reactions in the environment and in industry. Given the following information, calculate ΔH for the overall equation $2NO_2(g) + \frac{1}{2}O_2(g) \longrightarrow N_2O_5(s)$:

$$N_2O_5(s) \longrightarrow 2NO(g) + \frac{3}{2}O_2(g) \quad \Delta H = 223.7 \text{ kJ}$$
$$NO(g) + \frac{1}{2}O_2(g) \longrightarrow NO_2(g) \quad \Delta H = -57.1 \text{ kJ}$$

SECTION SUMMARY

Because H is a state function, $\Delta H = H_{final} - H_{initial}$ and does not depend on how the reaction takes place. Using Hess's law ($\Delta H_{total} = \Delta H_1 + \Delta H_2 + \cdots + \Delta H_n$), we can determine ΔH of any equation by manipulating the coefficients of other appropriate equations and their known ΔH values.

6.6 STANDARD HEATS OF REACTION (ΔH°_{rxn})

In this section, we see how Hess's law is used to determine the ΔH values of an enormous number of reactions. To begin we must take into account that thermodynamic variables, such as ΔH, vary somewhat with conditions. Therefore, to use heats of reaction, as well as other thermodynamic data that we will encounter in later chapters, chemists have established **standard states,** a set of specified conditions and concentrations:

- For a *gas,* the standard state is 1 atm* with the gas behaving ideally.
- For a substance in *aqueous solution,* the standard state is 1 M concentration.
- For a *pure substance* (element or compound), the standard state is usually the most stable form of the substance at 1 atm and the temperature of interest. In this text, that temperature is usually 25°C (298 K).[†]

We use a degree sign to indicate these standard states. In other words, when the heat of reaction, ΔH_{rxn}, has been measured with all the reactants and products in their standard states, it is referred to as the **standard heat of reaction, ΔH°_{rxn}.**

Formation Equations and Their Standard Enthalpy Changes

In a **formation equation,** 1 mole of a compound forms from its elements. The **standard heat of formation (ΔH°_f)** is the enthalpy change for the formation equation when all the substances are in their standard states. For instance, the formation equation for methane (CH_4) is

$$C(graphite) + 2H_2(g) \longrightarrow CH_4(g) \quad \Delta H^\circ_f = -74.9 \text{ kJ}$$

---

*The definition of the standard state for gases has been changed to 1 bar, a slightly lower pressure than the 1 atm standard on which the data in this book are based (1 atm = 101.3 kPa = 1.013 bar). For most purposes, this makes *very* little difference in the standard enthalpy values.

[†]In the case of phosphorus, the most *common* form, white phosphorus (P_4), is chosen as the standard state, even though red phosphorus is more stable at 1 atm and 298 K.

Thus, the standard heat of formation of methane is -74.9 kJ/mol. Some other examples are

$$Na(s) + \tfrac{1}{2}Cl_2(g) \longrightarrow NaCl(s) \qquad \Delta H^{\circ}_f = -411.1 \text{ kJ}$$

$$2C(\text{graphite}) + 3H_2(g) + \tfrac{1}{2}O_2(g) \longrightarrow C_2H_5OH(l) \qquad \Delta H^{\circ}_f = -277.6 \text{ kJ}$$

Standard heats of formation have been tabulated for many substances. Table 6.3 shows ΔH°_f values for several, and a much more extensive table appears in Appendix B.

The values in Table 6.3 were selected to make two points:

1. *An element in its standard state is assigned a ΔH°_f of zero.* For example, note that $\Delta H^{\circ}_f = 0$ for Na(s), but $\Delta H^{\circ}_f = 107.8$ kJ/mol for Na(g). These values mean that the gaseous state is *not* the most stable state of sodium at 1 atm and 298.15 K, and that heat is required to form Na(g). Note also that the standard state of chlorine is Cl_2 molecules, not Cl atoms. Several elements exist in different forms, only one of which is the standard state. Thus, the standard state of carbon is graphite, not diamond, so ΔH°_f of C(graphite) $= 0$. Similarly, the standard state of oxygen is dioxygen (O_2), not ozone (O_3), and the standard state of sulfur is S_8 in its rhombic crystal form, rather than its monoclinic form.
2. *Most compounds have a negative ΔH°_f.* That is, most compounds have exothermic formation reactions under standard conditions: *heat is given off when the compound forms.*

Table 6.3 Selected Standard Heats of Formation at 25°C (298 K)

Formula	ΔH°_f (kJ/mol)
Calcium	
Ca(s)	0
CaO(s)	-635.1
$CaCO_3(s)$	-1206.9
Carbon	
C(graphite)	0
C(diamond)	1.9
CO(g)	-110.5
$CO_2(g)$	-393.5
$CH_4(g)$	-74.9
$CH_3OH(l)$	-238.6
HCN(g)	135
$CS_2(l)$	87.9
Chlorine	
Cl(g)	121.0
$Cl_2(g)$	0
HCl(g)	-92.3
Hydrogen	
H(g)	218.0
$H_2(g)$	0
Nitrogen	
$N_2(g)$	0
$NH_3(g)$	-45.9
NO(g)	90.3
Oxygen	
$O_2(g)$	0
$O_3(g)$	143
$H_2O(g)$	-241.8
$H_2O(l)$	-285.8
Silver	
Ag(s)	0
AgCl(s)	-127.0
Sodium	
Na(s)	0
Na(g)	107.8
NaCl(s)	-411.1
Sulfur	
S_8(rhombic)	0
S_8(monoclinic)	0.3
$SO_2(g)$	-296.8
$SO_3(g)$	-396.0

SAMPLE PROBLEM 6.8   Writing Formation Equations

Problem Write balanced equations for the formation of 1 mole of each of the following compounds from their elements in their standard states, and include ΔH°_f.
(a) Silver chloride, AgCl, a solid at standard conditions
(b) Calcium carbonate, $CaCO_3$, a solid at standard conditions
(c) Hydrogen cyanide, HCN, a gas at standard conditions
Plan We write the elements as the reactants and 1 mol of the compound as the product, being sure all substances are in their standard states. Then, we balance the atoms and obtain the ΔH°_f values from Table 6.3 or Appendix B.
Solution (a) $\quad Ag(s) + \tfrac{1}{2}Cl_2(g) \longrightarrow AgCl(s) \qquad \Delta H^{\circ}_f = -127.0 \text{ kJ}$

(b) $\quad Ca(s) + C(\text{graphite}) + \tfrac{3}{2}O_2(g) \longrightarrow CaCO_3(s) \qquad \Delta H^{\circ}_f = -1206.9 \text{ kJ}$

(c) $\quad \tfrac{1}{2}H_2(g) + C(\text{graphite}) + \tfrac{1}{2}N_2(g) \longrightarrow HCN(g) \qquad \Delta H^{\circ}_f = 135 \text{ kJ}$

FOLLOW-UP PROBLEM 6.8 Write balanced equations for the formation of 1 mol of **(a)** $CH_3OH(l)$, **(b)** CaO(s), and **(c)** $CS_2(l)$ from their elements in their standard states. Include ΔH°_f for each reaction.

Determining ΔH°_{rxn} from ΔH°_f Values of Reactants and Products

By applying Hess's law, we can use ΔH°_f values to determine ΔH°_{rxn} for any reaction. All we have to do is view the reaction as an imaginary two-step process.

Step 1. Each reactant decomposes to its elements. This is the *reverse* of the formation reaction for each *reactant,* so each standard enthalpy change is $-\Delta H^{\circ}_f$.
Step 2. Each product forms from its elements. This step *is* the formation reaction for each *product,* so each standard enthalpy change is ΔH°_f.

According to Hess's law, we add the enthalpy changes for these steps to obtain the overall enthalpy change for the reaction (ΔH°_{rxn}). Figure 6.10 (on the next page) depicts the conceptual process. Suppose we want ΔH°_{rxn} for

$$TiCl_4(l) + 2H_2O(g) \longrightarrow TiO_2(s) + 4HCl(g)$$

We write this equation as though it were the sum of four individual equations, one for each compound. The first two of these equations show the decomposition

Figure 6.10 The general process for determining ΔH°_{rxn} from ΔH°_{f} values. For any reaction, ΔH°_{rxn} can be considered as the sum of the enthalpy changes for the decomposition of reactants to their elements $[-\Sigma n\Delta H^{\circ}_{f(reactants)}]$ and the formation of products from their elements $[+\Sigma m\Delta H^{\circ}_{f(products)}]$. [The factors m and n are the amounts (mol) of the products and reactants and equal the coefficients in the balanced equation, and Σ is the symbol for "sum of."]

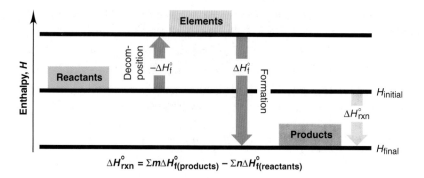

$$\Delta H^{\circ}_{rxn} = \Sigma m\Delta H^{\circ}_{f(products)} - \Sigma n\Delta H^{\circ}_{f(reactants)}$$

of the reactants to their elements (*reverse* of their formation), and the second two show the formation of the products from their elements:

$$TiCl_4(l) \longrightarrow Ti(s) + 2Cl_2(g) \qquad -\Delta H^{\circ}_{f}[TiCl_4(l)]$$
$$2H_2O(g) \longrightarrow 2H_2(g) + O_2(g) \qquad -2\Delta H^{\circ}_{f}[H_2O(g)]$$
$$Ti(s) + O_2(g) \longrightarrow TiO_2(s) \qquad \Delta H^{\circ}_{f}[TiO_2(s)]$$
$$2H_2(g) + 2Cl_2(g) \longrightarrow 4HCl(g) \qquad 4\Delta H^{\circ}_{f}[HCl(g)]$$

$$TiCl_4(l) + 2H_2O(g) + \cancel{Ti(s)} + O_2(g) + 2H_2(g) + \cancel{2Cl_2(g)} \longrightarrow$$
$$\cancel{Ti(s)} + \cancel{2Cl_2(g)} + 2H_2(g) + O_2(g) + TiO_2(s) + 4HCl(g)$$

Or, $\qquad TiCl_4(l) + 2H_2O(g) \longrightarrow TiO_2(s) + 4HCl(g)$

It's important to realize that when titanium(IV) chloride and water react, the reactants don't *actually* decompose to their elements, which then recombine to form the products. But that is the great usefulness of Hess's law and the state-function concept. Because ΔH°_{rxn} is the difference between two state functions, $\Delta H^{\circ}_{products}$ minus $\Delta H^{\circ}_{reactants}$, it doesn't matter *how* the change actually occurs. We simply add the individual enthalpy changes to find ΔH°_{rxn}:

$$\Delta H^{\circ}_{rxn} = \Delta H^{\circ}_{f}[TiO_2(s)] + 4\Delta H^{\circ}_{f}[HCl(g)] + \{-\Delta H^{\circ}_{f}[TiCl_4(l)]\} + \{-2\Delta H^{\circ}_{f}[H_2O(g)]\}$$
$$= \underbrace{\{\Delta H^{\circ}_{f}[TiO_2(s)] + 4\Delta H^{\circ}_{f}[HCl(g)]\}}_{\text{Products}} - \underbrace{\{\Delta H^{\circ}_{f}[TiCl_4(l)] + 2\Delta H^{\circ}_{f}[H_2O(g)]\}}_{\text{Reactants}}$$

Notice the pattern here. By generalizing it, we see that *the standard heat of reaction is the sum of the standard heats of formation of the **products** minus the sum of the standard heats of formation of the **reactants*** (see Figure 6.10):

$$\Delta H^{\circ}_{rxn} = \Sigma m\Delta H^{\circ}_{f(products)} - \Sigma n\Delta H^{\circ}_{f(reactants)} \qquad \textbf{(6.8)}$$

where the symbol Σ means "sum of," and m and n are the amounts (mol) of the products and reactants indicated by the coefficients from the balanced equation.

SAMPLE PROBLEM 6.9 Calculating the Heat of Reaction from Heats of Formation

Problem Nitric acid, whose worldwide annual production is nearly 10 billion kilograms, is used to make many products, including fertilizers, dyes, and explosives. The first step in the production process is the oxidation of ammonia:

$$4NH_3(g) + 5O_2(g) \longrightarrow 4NO(g) + 6H_2O(g)$$

Calculate ΔH°_{rxn} from ΔH°_{f} values.
Plan We use values from Table 6.3 (or Appendix B) and apply Equation 6.8 to find ΔH°_{rxn}.
Solution Calculating ΔH°_{rxn}:

$$\Delta H^{\circ}_{rxn} = \Sigma m\Delta H^{\circ}_{f(products)} - \Sigma n\Delta H^{\circ}_{f(reactants)}$$
$$= \{4\Delta H^{\circ}_{f}[NO(g)] + 6\Delta H^{\circ}_{f}[H_2O(g)]\} - \{4\Delta H^{\circ}_{f}[NH_3(g)] + 5\Delta H^{\circ}_{f}[O_2(g)]\}$$
$$= (4 \text{ mol})(90.3 \text{ kJ/mol}) + (6 \text{ mol})(-241.8 \text{ kJ/mol})$$
$$\qquad -[(4 \text{ mol})(-45.9 \text{ kJ/mol}) + (5 \text{ mol})(0 \text{ kJ/mol})]$$
$$= 361 \text{ kJ} - 1451 \text{ kJ} + 184 \text{ kJ} - 0 \text{ kJ} = -906 \text{ kJ}$$

Check One way to check is to write formation equations for the amounts of individual compounds in the correct direction and take their sum:

$$4NH_3(g) \longrightarrow 2N_2(g) + 6H_2(g) \quad -4(-45.9\ kJ) = \quad 184\ kJ$$
$$2N_2(g) + 2O_2(g) \longrightarrow 4NO(g) \qquad\qquad 4(90.3\ kJ) = \quad 361\ kJ$$
$$6H_2(g) + 3O_2(g) \longrightarrow 6H_2O(g) \qquad 6(-241.8\ kJ) = -1451\ kJ$$
$$\overline{4NH_3(g) + 5O_2(g) \longrightarrow 4NO(g) + 6H_2O(g) \qquad\qquad\quad -906\ kJ}$$

Comment In this problem, we know the individual ΔH_f° values and find the sum, ΔH_{rxn}°. In the follow-up problem, we know the sum and want to find an individual value.

FOLLOW-UP PROBLEM 6.9 Use the following information to find ΔH_f° of methanol [CH$_3$OH(l)]:

$$CH_3OH(l) + \tfrac{3}{2}O_2(g) \longrightarrow CO_2(g) + 2H_2O(g) \quad \Delta H_{rxn}^\circ = -638.5\ kJ$$
$$\Delta H_f^\circ\ of\ CO_2(g) = -393.5\ kJ/mol \quad \Delta H_f^\circ\ of\ H_2O(g) = -241.8\ kJ/mol$$

Fossil Fuels and Climate Change

Out of necessity, modern societies are finally beginning to rethink the issue of energy use. No scientific challenge today is greater than reversing the climatic effects of our increasing dependence on the combustion of **fossil fuels**—coal, petroleum, and natural gas. Because these fuels form so much more slowly than we consume them, they are *nonrenewable*. In contrast, wood and other fuels derived from plant and animal matter are *renewable*.

The combustion of all carbon-based fuels releases carbon dioxide, which plays a key temperature-regulating role in the atmosphere. Much of the sunlight that shines on Earth is absorbed by the surface and converted to heat (Figure 6.11). Because atmospheric CO_2 absorbs heat, some of the heat reflected back from Earth's surface is trapped by this CO_2.

The large amount of CO_2 present in Earth's early atmosphere decreased to a relatively constant 0.028% by volume as a result of the spread of plants, which use CO_2 in photosynthesis. However, for the past 150 years, this amount has been increasing as we burn fossil fuels, and today it exceeds 0.036%. (Trees and other renewable fuels also release CO_2 when they burn, but new trees take it up as they grow.) Thus, although the same amount of solar energy passes into the atmosphere, more is being trapped as heat in a process called the *greenhouse effect*, which is changing Earth's climate through *global warming*. Based on current fossil fuel use, CO_2 concentrations are predicted to reach 0.070% to 0.075% by 2100.

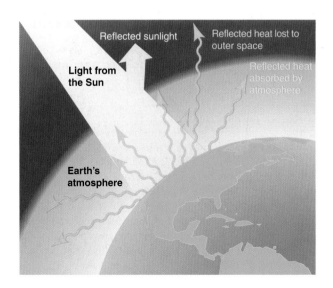

Figure 6.11 The trapping of heat by the atmosphere. About 25% of sunlight is reflected by the atmosphere. Much of the remaining sunlight is converted into heat by the air and Earth's surface. Some of the heat emanating from the surface is trapped by atmospheric gases, mostly CO_2, and some is lost to outer space.

Many groups of scientists have attempted to model the effects of CO_2 and other *greenhouse gases* (most importantly, CH_4) on the temperature of the atmosphere, but the system is extremely complex. There are natural fluctuations in temperature, as well as cyclic changes in solar activity that may provide additional warming through 2010 or so. Moreover, as fuels are burned and CO_2 concentration increases, so does the amount of soot, which may block sunlight and have a cooling effect. As temperatures rise, more water is vaporized, which may thicken the cloud cover and also cause cooling. But recent data show that less volcanic activity coupled with power plant improvements may reduce the particulates that now cool the atmosphere.

Despite these complexities, most models predict a net warming. Indeed, scientists are already observing it and documenting the predicted climate disruptions. The average temperature has increased by 0.6°C over the past 50 years. When extra heat is trapped, 20% of it warms the air and 80% evaporates water. Globally, the 10 warmest years on record have occurred since 1980, with 1997–1999 among the hottest. The decade from 1987 to 1997 included 10 times as many catastrophic floods from storms as did the previous decade, a trend that continued in 1998 and 1999. Very recent evidence shows a 10% decrease in global snow cover, a 40% thinning of Arctic Sea ice, a 20% decrease in glacier extent, and nearly a 10% increase in rainfall in the northern hemisphere.

Today, the most accepted models predict a temperature rise of 1.4–5.8°C (2.5–10.4°F). An average increase of 5°C would disrupt rainfall patterns and crop yields and melt large parts of the polar ice caps, flooding low-lying regions such as the Netherlands, half of Florida, and much of India, and inundating some Pacific island nations. To make matters worse, as we burn fuels that *release* CO_2, we cut down the forests that naturally *absorb* CO_2. As a solution, some have proposed large-scale tree planting. Others suggest liquefying CO_2 and burying it underground or injecting it into the deep oceans.

International conferences, such as the 1997 Conference on Climate Change in Kyoto, Japan, provided a forum for politicians, business interests, and scientists to seek ways to cut greenhouse emissions. But the largest producer of CO_2, the United States, refused to ratify the Kyoto Protocol and, as a result, many expect global warming and its associated effects on Earth's climate to increase.

SECTION SUMMARY

Standard states are chosen conditions for substances. When 1 mol of a compound forms from its elements with all substances in their standard states, the enthalpy change is ΔH_f°. Hess's law allows us to picture a reaction as the decomposition of reactants to their elements, followed by the formation of products from their elements. We use tabulated ΔH_f° values to find ΔH_{rxn}° or use known ΔH_{rxn}° and ΔH_f° values to find an unknown ΔH_f°. As a result of increased fossil-fuel combustion, the amount of atmospheric CO_2 is climbing, which is seriously affecting Earth's climate.

For Review and Reference (Numbers in parentheses refer to pages, unless noted otherwise.)

Learning Objectives

To help you review these learning objectives, the numbers of related sections (§), sample problems (SP), and end-of-chapter problems (EP) are listed in parentheses.

1. Interconvert energy units; understand that ΔE of a system appears as the total heat and/or work transferred to or from its surroundings; understand the meaning of a state function (§ 6.1) (SP 6.1) (EPs 6.1–6.9)

2. Understand the meaning of H, why we measure ΔH, and the distinction between exothermic and endothermic reactions; draw enthalpy diagrams for chemical and physical changes (§ 6.2) (SP 6.2) (EPs 6.10–6.16)

3. Understand the relation between specific heat capacity and heat transferred in both constant-pressure (coffee-cup) and constant-volume (bomb) calorimeters (§ 6.3) (SPs 6.3–6.5) (EPs 6.17–6.30)

4. Understand the relation between heat of reaction and amount of substance (§ 6.4) (SP 6.6) (EPs 6.31–6.40)

5. Explain the importance of Hess's law and use it to find an unknown ΔH (§ 6.5) (SP 6.7) (EPs 6.41–6.46)

6. View a reaction as the decomposition of reactants followed by the formation of products; understand formation equations and how to use ΔH_f° values to find ΔH_{rxn}° (§ 6.6) (SPs 6.8, 6.9) (EPs 6.47–6.56)

Key Terms

thermodynamics (178)
thermochemistry (178)

Section 6.1
system (178)
surroundings (178)
internal energy (E) (178)
heat (q) (179)
work (w) (179)
pressure-volume work
   (PV work) (181)

law of conservation of energy
  (first law of
   thermodynamics) (181)
joule (J) (182)
calorie (cal) (182)
state function (183)

Section 6.2
enthalpy (H) (185)
change in enthalpy (ΔH) (185)
heat of reaction (ΔH_{rxn}) (185)

enthalpy diagram (185)
exothermic process (185)
endothermic process (186)

Section 6.3
heat capacity (187)
specific heat capacity (c) (187)
molar heat capacity (C) (187)
calorimeter (188)

Section 6.4
thermochemical equation (191)

Section 6.5
Hess's law of heat
  summation (192)

Section 6.6
standard states (194)
standard heat of reaction
  (ΔH°_{rxn}) (194)
formation equation (194)
standard heat of formation
  (ΔH°_{f}) (194)
fossil fuel (197)

Key Equations and Relationships

6.1 Defining the change in internal energy (179):
$$\Delta E = E_{final} - E_{initial} = E_{products} - E_{reactants}$$
6.2 Expressing the change in internal energy in terms of heat and work (179):
$$\Delta E = q + w$$
6.3 Stating the first law of thermodynamics (law of conservation of energy) (182):
$$\Delta E_{universe} = \Delta E_{system} + \Delta E_{surroundings} = 0$$
6.4 Determining the work due to a change in volume at constant pressure (PV work) (185):
$$w = -P\Delta V$$

6.5 Relating the enthalpy change to the internal energy change at constant pressure (185):
$$\Delta H = \Delta E + P\Delta V$$
6.6 Identifying the enthalpy change with the heat gained or lost at constant pressure (185):
$$q_P = \Delta E + P\Delta V = \Delta H$$
6.7 Calculating the heat absorbed or released when a substance undergoes a temperature change (187):
$$q = c \times \text{mass} \times \Delta T$$
6.8 Calculating the standard heat of reaction (196):
$$\Delta H^{\circ}_{rxn} = \Sigma m\Delta H^{\circ}_{f(products)} - \Sigma n\Delta H^{\circ}_{f(reactants)}$$

Brief Solutions to Follow-up Problems

6.1 $\Delta E = q + w$
$$= \left(-26.0 \text{ kcal} \times \frac{4.184 \text{ kJ}}{1 \text{ kcal}}\right) + \left(+15.0 \text{ Btu} \times \frac{1.055 \text{ kJ}}{1 \text{ Btu}}\right)$$
$$= -93 \text{ kJ}$$

6.2 The reaction is exothermic.

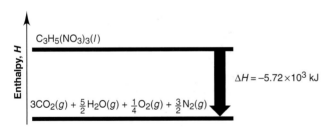

6.3 $\Delta T = 25.0°C - 37.0°C = -12.0°C = -12.0 \text{ K}$
$$\text{Mass (g)} = 1.11 \text{ g/mL} \times \frac{1000 \text{ mL}}{1 \text{ L}} \times 5.50 \text{ L} = 6.10 \times 10^3 \text{ g}$$
$q = c \times \text{mass} \times \Delta T$
$$= (2.42 \text{ J/g·K})\left(\frac{1 \text{ kJ}}{1000 \text{ J}}\right)(6.10 \times 10^3 \text{ g})(-12.0 \text{ K})$$
$$= -177 \text{ kJ}$$

6.4 $\qquad -q_{solid} = q_{water}$
$-[(0.519 \text{ J/g·K})(2.050 \text{ g})(x - 74.21)]$
$$= [(4.184 \text{ J/g·K})(26.05 \text{ g})(x - 27.20)]$$
$$x = 27.65 \text{ K}$$
$\Delta T_{diamond} = -46.56 \text{ K}$ and $\Delta T_{water} = 0.45 \text{ K}$

6.5 $\qquad\qquad\qquad -q_{sample} = q_{calorimeter}$
$$-(0.8650 \text{ g C})\left(\frac{1 \text{ mol C}}{12.01 \text{ g C}}\right)(-393.5 \text{ kJ/mol C}) = (2.613 \text{ K})x$$
$$x = 10.85 \text{ kJ/K}$$

6.6 $C_2H_4(g) + H_2(g) \longrightarrow C_2H_6(g) + 137 \text{ kJ}$
$$\text{Heat (kJ)} = 15.0 \text{ kg} \times \frac{1000 \text{ g}}{1 \text{ kg}} \times \frac{1 \text{ mol } C_2H_6}{30.07 \text{ g } C_2H_6} \times \frac{137 \text{ kJ}}{1 \text{ mol}}$$
$$= 6.83 \times 10^4 \text{ kJ}$$

6.7
$$2NO(g) + \tfrac{3}{2}O_2(g) \longrightarrow N_2O_5(s) \qquad \Delta H = -223.7 \text{ kJ}$$
$$2NO_2(g) \longrightarrow 2NO(g) + O_2(g) \qquad \Delta H = 114.2 \text{ kJ}$$

$\overline{2\text{NO}(g) + \tfrac{1}{2}\tfrac{3}{2}O_2(g) + 2NO_2(g) \longrightarrow}$
$$\qquad\qquad\qquad N_2O_5(s) + 2\text{NO}(g) + \text{O}_2(g)$$
$$2NO_2(g) + \tfrac{1}{2}O_2(g) \longrightarrow N_2O_5(s) \qquad \Delta H = -109.5 \text{ kJ}$$

6.8 (a) $C(\text{graphite}) + 2H_2(g) + \tfrac{1}{2}O_2(g) \longrightarrow CH_3OH(l)$
$$\Delta H^{\circ}_{f} = -238.6 \text{ kJ}$$
(b) $Ca(s) + \tfrac{1}{2}O_2(g) \longrightarrow CaO(s) \quad \Delta H^{\circ}_{f} = -635.1 \text{ kJ}$
(c) $C(\text{graphite}) + \tfrac{1}{4}S_8(\text{rhombic}) \longrightarrow CS_2(l) \quad \Delta H^{\circ}_{f} = 87.9 \text{ kJ}$

6.9 ΔH°_{f} of $CH_3OH(l)$
$$= -\Delta H^{\circ}_{rxn} + 2\Delta H^{\circ}_{f}[H_2O(g)] + \Delta H^{\circ}_{f}[CO_2(g)]$$
$$= 638.5 \text{ kJ} + (2 \text{ mol})(-241.8 \text{ kJ/mol})$$
$$+ (1 \text{ mol})(-393.5 \text{ kJ/mol})$$
$$= -238.6 \text{ kJ}$$

Problems

Problems with **colored** numbers are answered in Appendix E. Sections match the text and provide the numbers of relevant sample problems. Bracketed problems are grouped in pairs (indicated by a short rule) that cover the same concept. Comprehensive Problems are based on material from any section or previous chapter.

Forms of Energy and Their Interconversion
(Sample Problem 6.1)

6.1 If you feel warm after exercising, have you increased the internal energy of your body? Explain.

6.2 An *adiabatic* process is one that involves no heat transfer. What is the relationship between work and the change in internal energy in an adiabatic process?

6.3 Name a common device that is used to accomplish each energy change:
(a) Electrical energy to thermal energy
(b) Electrical energy to sound energy
(c) Electrical energy to light energy
(d) Mechanical energy to electrical energy
(e) Chemical energy to electrical energy

6.4 Imagine lifting your textbook into the air and dropping it onto a desktop. Describe all the energy transformations (from one form to another) that occur, moving backward in time from a moment after impact.

6.5 A system receives 425 J of heat and delivers 425 J of work to its surroundings. What is the change in internal energy of the system (in J)?

6.6 A system conducts 255 cal of heat to the surroundings while delivering 428 cal of work. What is the change in internal energy of the system (in cal)?

6.7 Complete combustion of 1.0 metric ton of coal (assuming pure carbon) to gaseous carbon dioxide releases 3.3×10^{10} J of heat. Convert this energy to (a) kilojoules; (b) kilocalories; (c) British thermal units.

6.8 Thermal decomposition of 5.0 metric tons of limestone to lime and carbon dioxide requires 9.0×10^6 kJ of heat. Convert this energy to (a) joules; (b) calories; (c) British thermal units.

6.9 The nutritional calorie (Calorie) is equivalent to 1 kcal. One pound of body fat is equivalent to about 4.1×10^3 Calories. Express this quantity of energy in joules and kilojoules.

Enthalpy: Heats of Reaction and Chemical Change
(Sample Problem 6.2)

6.10 Classify the following processes as exothermic or endothermic: (a) freezing of water; (b) boiling of water; (c) digestion of food; (d) a person running; (e) a person growing; (f) wood being chopped; (g) heating with a furnace.

6.11 Draw an enthalpy diagram for a general exothermic reaction; label axis, reactants, products, and ΔH with its sign.

6.12 Draw an enthalpy diagram for a general endothermic reaction; label axis, reactants, products, and ΔH with its sign.

6.13 Write a balanced equation and draw an approximate enthalpy diagram for each of the following: (a) the combustion of 1 mol of methane in oxygen; (b) the freezing of liquid water.

6.14 Write a balanced equation and draw an approximate enthalpy diagram for each of the following: (a) the formation of 1 mol of sodium chloride from its elements (heat is released); (b) the conversion of liquid benzene to gaseous benzene.

6.15 Write a balanced equation and draw an approximate enthalpy diagram for each of the following changes: (a) the combustion of 1 mol of liquid ethanol (C_2H_5OH); (b) the formation of 1 mol of nitrogen dioxide from its elements (heat is absorbed).

6.16 Write a balanced equation and draw an approximate enthalpy diagram for each of the following changes: (a) the sublimation of dry ice [conversion of $CO_2(s)$ directly to $CO_2(g)$]; (b) the reaction of 1 mol of sulfur dioxide with oxygen.

Calorimetry: Laboratory Measurement of Heats of Reaction
(Sample Problems 6.3 to 6.5)

6.17 Why can we measure only *changes* in enthalpy, not absolute enthalpy values?

6.18 What data do you need to determine the specific heat capacity of a substance?

6.19 Is the specific heat capacity of a substance an intensive or extensive property? Explain.

6.20 Calculate q when 12.0 g of water is heated from 20.°C to 100.°C.

6.21 Calculate q when 0.10 g of ice is cooled from 10.°C to -75°C ($c_{ice} = 2.087$ J/g·K).

6.22 A 295-g aluminum engine part at an initial temperature of 3.00°C absorbs 85.0 kJ of heat. What is the final temperature of the part (c of Al = 0.900 J/g·K)?

6.23 A 27.7-g sample of ethylene glycol, a car radiator coolant, loses 688 J of heat. What was the initial temperature of the ethylene glycol if the final temperature is 32.5°C (c of ethylene glycol = 2.42 J/g·K)?

6.24 Two iron bolts of equal mass—one at 100.°C, the other at 55°C—are placed in an insulated container. Assuming the heat capacity of the container is negligible, what is the final temperature inside the container (c of iron = 0.450 J/g·K)?

6.25 One piece of copper jewelry at 105°C has exactly twice the mass of another piece, which is at 45°C. Both pieces are placed inside a calorimeter whose heat capacity is negligible. What is the final temperature inside the calorimeter (c of copper = 0.387 J/g·K)?

6.26 When 165 mL of water at 22°C is mixed with 85 mL of water at 82°C, what is the final temperature? (Assume that no heat is lost to the surroundings; d of water is 1.00 g/mL.)

6.27 An unknown volume of water at 18.2°C is added to 24.4 mL of water at 35.0°C. If the final temperature is 23.5°C, what was the unknown volume? (Assume that no heat is lost to the surroundings; d of water is 1.00 g/mL.)

6.28 High-purity benzoic acid (C_6H_5COOH; ΔH_{rxn} for combustion = -3227 kJ/mol) is used as a standard for calibrating bomb calorimeters. A 1.221-g sample burns in a calorimeter (heat capacity = 1365 J/°C) that contains exactly 1.200 kg of water. What temperature change is observed?

6.29 Two aircraft rivets, one of iron and the other of copper, are placed in a calorimeter that has an initial temperature of 20.°C. The data for the metals are as follows:

	Iron	Copper
Mass (g)	30.0	20.0
Initial T (°C)	0.0	100.0
c (J/g·K)	0.450	0.387

(a) Will heat flow from Fe to Cu or from Cu to Fe?
(b) What other information is needed to correct any measurements that would be made in an actual experiment?
(c) What is the maximum final temperature of the system (assuming the heat capacity of the calorimeter is negligible)?

6.30 When 25.0 mL of 0.500 M H_2SO_4 is added to 25.0 mL of 1.00 M KOH in a coffee-cup calorimeter at 23.50°C, the temperature rises to 30.17°C. Calculate ΔH of this reaction. (Assume that the total volume is the sum of the individual volumes and that the density and specific heat capacity of the solution are the same as for pure water.)

Stoichiometry of Thermochemical Equations
(Sample Problem 6.6)

6.31 Would you expect $O_2(g) \longrightarrow 2O(g)$ to have a positive or a negative ΔH_{rxn}? Explain.

6.32 Is ΔH positive or negative when 1 mol of water vapor condenses to liquid water? Why? How does this value compare with that for the conversion of 2 mol of liquid water to water vapor?

6.33 Consider the following balanced thermochemical equation for a reaction sometimes used for H_2S production:
$$\tfrac{1}{8}S_8(s) + H_2(g) \longrightarrow H_2S(g) \quad \Delta H_{rxn} = -20.2 \text{ kJ}$$
(a) Is this an exothermic or endothermic reaction?
(b) What is ΔH_{rxn} for the reverse reaction?
(c) What is ΔH when 3.2 mol of S_8 reacts?
(d) What is ΔH when 20.0 g of S_8 reacts?

6.34 Consider the following balanced thermochemical equation for the decomposition of the mineral magnesite:
$$MgCO_3(s) \longrightarrow MgO(s) + CO_2(g) \quad \Delta H_{rxn} = 117.3 \text{ kJ}$$
(a) Is heat absorbed or released in the reaction?
(b) What is ΔH_{rxn} for the reverse reaction?
(c) What is ΔH when 5.35 mol of CO_2 reacts with excess MgO?
(d) What is ΔH when 35.5 g of CO_2 reacts with excess MgO?

6.35 When 1 mol of NO(g) forms from its elements, 90.29 kJ of heat is absorbed. (a) Write a balanced thermochemical equation for this reaction. (b) How much heat is involved when 1.50 g of NO decomposes to its elements?

6.36 When 1 mol of KBr(s) decomposes to its elements, 394 kJ of heat is absorbed. (a) Write a balanced thermochemical equation for this reaction. (b) How much heat is released when 10.0 kg of KBr forms from its elements?

6.37 Liquid hydrogen peroxide, an oxidizing agent in many rocket fuel mixtures, releases oxygen gas on decomposition:
$$2H_2O_2(l) \longrightarrow 2H_2O(l) + O_2(g) \quad \Delta H_{rxn} = -196.1 \text{ kJ}$$
How much heat is released when 732 kg of H_2O_2 decomposes?

6.38 Compounds of boron and hydrogen are remarkable for their unusual bonding (described in Section 14.5) and also for their reactivity. With the more reactive halogens, for example, diborane (B_2H_6) forms trihalides even at low temperatures:
$$B_2H_6(g) + 6Cl_2(g) \longrightarrow 2BCl_3(g) + 6HCl(g)$$
$$\Delta H_{rxn} = -755.4 \text{ kJ}$$
How much heat is released per kilogram of diborane that reacts?

6.39 Ethylene (C_2H_4) is the starting material for the preparation of polyethylene. Although typically made during the processing of petroleum, ethylene occurs naturally as a fruit-ripening hormone and as a component of natural gas.
(a) The heat of reaction for the combustion of C_2H_4 is -1411 kJ/mol. Write a balanced thermochemical equation for the combustion of C_2H_4.
(b) How many grams of C_2H_4 must burn to give 70.0 kJ of heat?

6.40 Sucrose ($C_{12}H_{22}O_{11}$, table sugar) is oxidized in the body by O_2 via a complex set of reactions that ultimately produces $CO_2(g)$ and $H_2O(g)$ and releases 5.64×10^3 kJ/mol sucrose.
(a) Write a balanced thermochemical equation for this reaction.
(b) How much heat is released per gram of sucrose oxidized?

Hess's Law of Heat Summation
(Sample Problem 6.7)

6.41 Express Hess's law in your own words.

6.42 Calculate ΔH_{rxn} for
$$Ca(s) + \tfrac{1}{2}O_2(g) + CO_2(g) \longrightarrow CaCO_3(s)$$
given the following set of reactions:
$$Ca(s) + \tfrac{1}{2}O_2(g) \longrightarrow CaO(s) \qquad \Delta H = -635.1 \text{ kJ}$$
$$CaCO_3(s) \longrightarrow CaO(s) + CO_2(g) \quad \Delta H = 178.3 \text{ kJ}$$

6.43 Calculate ΔH_{rxn} for
$$2NOCl(g) \longrightarrow N_2(g) + O_2(g) + Cl_2(g)$$
given the following set of reactions:
$$\tfrac{1}{2}N_2(g) + \tfrac{1}{2}O_2(g) \longrightarrow NO(g) \qquad \Delta H = 90.3 \text{ kJ}$$
$$NO(g) + \tfrac{1}{2}Cl_2(g) \longrightarrow NOCl(g) \quad \Delta H = -38.6 \text{ kJ}$$

6.44 Write the balanced overall equation for the following process (equation 3), calculate $\Delta H_{overall}$, and match the number of each equation with the letter of the appropriate arrow in Figure P6.44:

(1) $N_2(g) + O_2(g) \longrightarrow 2NO(g)$	$\Delta H = 180.6 \text{ kJ}$
(2) $2NO(g) + O_2(g) \longrightarrow 2NO_2(g)$	$\Delta H = -114.2 \text{ kJ}$
(3)	$\Delta H_{overall} = ?$

6.45 Write the balanced overall equation for the following process (equation 3), calculate $\Delta H_{overall}$, and match the number of each equation with the letter of the appropriate arrow in Figure P6.45:

(1) $P_4(s) + 6Cl_2(g) \longrightarrow 4PCl_3(g)$	$\Delta H = -1148 \text{ kJ}$
(2) $4PCl_3(g) + 4Cl_2(g) \longrightarrow 4PCl_5(g)$	$\Delta H = -460 \text{ kJ}$
(3)	$\Delta H_{overall} = ?$

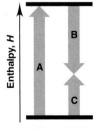

Figure P6.44

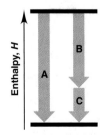

Figure P6.45

6.46 Diamond and graphite are two crystalline forms of carbon. At 1 atm and 25°C, diamond changes to graphite so slowly that the enthalpy change of the process must be obtained indirectly. Determine ΔH_{rxn} for

$$C(diamond) \longrightarrow C(graphite)$$

with equations from the following list:

(1) $C(diamond) + O_2(g) \longrightarrow CO_2(g)$	$\Delta H =$	-395.4 kJ
(2) $2CO_2(g) \longrightarrow 2CO(g) + O_2(g)$	$\Delta H =$	566.0 kJ
(3) $C(graphite) + O_2(g) \longrightarrow CO_2(g)$	$\Delta H =$	-393.5 kJ
(4) $2CO(g) \longrightarrow C(graphite) + CO_2(g)$	$\Delta H =$	-172.5 kJ

Standard Heats of Reaction (ΔH°_{rxn})
(Sample Problems 6.8 and 6.9)

6.47 What is the difference between the standard heat of formation and the standard heat of reaction?

6.48 Make any changes needed in each of the following equations to make ΔH°_{rxn} equal to ΔH°_f for the compound present:
(a) $Cl(g) + Na(s) \longrightarrow NaCl(s)$
(b) $H_2O(g) \longrightarrow 2H(g) + \frac{1}{2}O_2(g)$
(c) $\frac{1}{2}N_2(g) + \frac{3}{2}H_2(g) \longrightarrow NH_3(g)$

6.49 Write balanced formation equations at standard conditions for each of the following compounds: (a) $CaCl_2$; (b) $NaHCO_3$; (c) CCl_4; (d) HNO_3.

6.50 Write balanced formation equations at standard conditions for each of the following compounds: (a) HI; (b) SiF_4; (c) O_3; (d) $Ca_3(PO_4)_2$.

6.51 Calculate ΔH°_{rxn} for each of the following:
(a) $2H_2S(g) + 3O_2(g) \longrightarrow 2SO_2(g) + 2H_2O(g)$
(b) $CH_4(g) + Cl_2(g) \longrightarrow CCl_4(l) + HCl(g)$ [unbalanced]

6.52 Calculate ΔH°_{rxn} for each of the following:
(a) $SiO_2(s) + 4HF(g) \longrightarrow SiF_4(g) + 2H_2O(l)$
(b) $C_2H_6(g) + O_2(g) \longrightarrow CO_2(g) + H_2O(g)$ [unbalanced]

6.53 Copper(I) oxide can be oxidized to copper(II) oxide:
$$Cu_2O(s) + \tfrac{1}{2}O_2(g) \longrightarrow 2CuO(s) \quad \Delta H^\circ_{rxn} = -146.0 \text{ kJ}$$
Given that ΔH°_f of $Cu_2O(s) = -168.6$ kJ/mol, what is ΔH°_f of $CuO(s)$?

6.54 Acetylene burns in air according to the following equation:
$$C_2H_2(g) + \tfrac{5}{2}O_2(g) \longrightarrow 2CO_2(g) + H_2O(g)$$
$$\Delta H^\circ_{rxn} = -1255.8 \text{ kJ}$$
Given that ΔH°_f of $CO_2(g) = -393.5$ kJ/mol and that ΔH°_f of $H_2O(g) = -241.8$ kJ/mol, what is ΔH°_f of $C_2H_2(g)$?

6.55 Nitroglycerine, $C_3H_5(NO_3)_3(l)$, a powerful explosive used in mining, detonates to produce a hot gaseous mixture of nitrogen, water, carbon dioxide, and oxygen.
(a) Write a balanced equation for this reaction using the smallest whole-number coefficients.
(b) If $\Delta H^\circ_{rxn} = -2.29\times10^4$ kJ for the equation as written in part (a), calculate ΔH°_f of nitroglycerine.

6.56 The common lead-acid car battery produces a large burst of current, even at low temperatures, and is rechargeable. The reaction that occurs while recharging a "dead" battery is
$$2PbSO_4(s) + 2H_2O(l) \longrightarrow Pb(s) + PbO_2(s) + 2H_2SO_4(l)$$
(a) Use ΔH°_f values from Appendix B to calculate ΔH°_{rxn}.
(b) Use the following equations to check your answer in part (a):
(1) $Pb(s) + PbO_2(s) + 2SO_3(g) \longrightarrow 2PbSO_4(s)$
$$\Delta H^\circ = -768 \text{ kJ}$$
(2) $SO_3(g) + H_2O(l) \longrightarrow H_2SO_4(l)$ $\Delta H^\circ = -132$ kJ

Comprehensive Problems
Problems with an asterisk (*) are more challenging.

6.57 Stearic acid ($C_{18}H_{36}O_2$) is a typical fatty acid, a molecule with a long hydrocarbon chain and an organic acid group (COOH) at the end. It is used to make cosmetics, ointments, soaps, and candles and is found in animal tissue as part of many saturated fats. In fact, when you eat meat, chances are that you are ingesting some fats that contain stearic acid.
(a) Write a balanced equation for the complete combustion of stearic acid to gaseous products.
(b) Calculate ΔH°_{rxn} for this combustion ($\Delta H^\circ_f = -948$ kJ/mol).
(c) Calculate the heat (q) in kJ and kcal when 1.00 g of stearic acid is burned completely.
(d) The nutritional information for a candy bar states that one serving contains 11.0 g of fat and 100. Cal from fat (1 Cal = 1 kcal). Is this information consistent with your answer for part (c)?

6.58 A balloonist is preparing to make a trip in a helium-filled balloon. The trip begins in early morning at a temperature of 15°C. By midafternoon, the temperature has increased to 30.°C. Assuming the pressure remains constant at 1.00 atm, for each mole of helium, calculate:
(a) The initial and final volumes
(b) The change in internal energy, ΔE [*Hint:* Helium behaves like an ideal gas, so $E = \frac{3}{2}nRT$. Be sure the units of R are consistent with those of E.]
(c) The work (w) done by the helium (in J)
(d) The heat (q) transferred (in J)
(e) ΔH for the process (in J)
(f) Explain the relationship between the answers to (d) and (e).

6.59 In winemaking, the sugars in grapes undergo *fermentation* by yeast to yield CH_3CH_2OH and CO_2. During cellular *respiration,* sugar and ethanol are "burned" to water vapor and CO_2.
(a) Using $C_6H_{12}O_6$ for sugar, calculate ΔH°_{rxn} of fermentation and of respiration (combustion).
(b) Write a combustion reaction for ethanol. Which has a higher ΔH°_{rxn} for the combustion per mol of C, sugar or ethanol?

6.60 Iron metal is produced in a blast furnace through a complex series of reactions that involve reduction of iron(III) oxide with carbon monoxide.
(a) Write a balanced overall equation for the process, including the other product.
(b) Use the equations below to calculate ΔH°_{rxn} for the overall equation:
(1) $3Fe_2O_3(s) + CO(g) \longrightarrow 2Fe_3O_4(s) + CO_2(g)$
$$\Delta H^\circ = -48.5 \text{ kJ}$$
(2) $Fe(s) + CO_2(g) \longrightarrow FeO(s) + CO(g)$ $\Delta H^\circ = -11.0$ kJ
(3) $Fe_3O_4(s) + CO(g) \longrightarrow 3FeO(s) + CO_2(g)$ $\Delta H^\circ = 22$ kJ

6.61 Pure liquid octane (C_8H_{18}; $d = 0.702$ g/mL) is used as the fuel in a test of a new automobile drive train.
(a) How much energy (in kJ) is released by complete combustion of the octane in a 20.4-gal fuel tank to gases ($\Delta H^\circ_{rxn} = -5.45\times10^3$ kJ/mol)?
(b) The energy delivered to the wheels at 65 mph is 5.5×10^4 kJ/h. Assuming all the energy is transferred to the wheels, what is the cruising range (in km) of the car on a full tank?
(c) If the actual cruising range is 455 miles, explain your answer to part (b).

6.62 When simple sugars, called *monosaccharides,* link together, they form a variety of complex sugars and, ultimately, *polysaccharides,* such as starch, glycogen, and cellulose. Glucose and fructose have the same formula, $C_6H_{12}O_6$, but different arrangements of atoms. They link together to form a molecule of sucrose (table sugar) and a molecule of liquid water. The ΔH_f° values of glucose, fructose, and sucrose are -1273 kJ/mol, -1266 kJ/mol, and -2226 kJ/mol, respectively. Write a balanced equation for this reaction and calculate ΔH_{rxn}°.

6.63 Oxidation of gaseous ClF by F_2 yields liquid ClF_3, an important fluorinating agent. Use the following thermochemical equations to calculate ΔH_{rxn}° for the production of ClF_3:

(1) $2ClF(g) + O_2(g) \longrightarrow Cl_2O(g) + OF_2(g)$ $\Delta H^\circ = 167.5$ kJ
(2) $2F_2(g) + O_2(g) \longrightarrow 2OF_2(g)$ $\Delta H^\circ = -43.5$ kJ
(3) $2ClF_3(l) + 2O_2(g) \longrightarrow Cl_2O(g) + 3OF_2(g)$
$\Delta H^\circ = 394.1$ kJ

6.64 Silver bromide is used to coat ordinary black-and-white photographic film, while high-speed film uses silver iodide.
(a) When 50.0 mL of 5.0 g/L $AgNO_3$ is added to a coffee-cup calorimeter containing 50.0 mL of 5.0 g/L NaI, with both solutions at 25°C, what mass of AgI forms?
(b) Use Appendix B to find ΔH_{rxn}°.
(c) What is ΔT_{soln} (assume the volumes are additive and the solution has the density and specific heat capacity of water)?

* 6.65 Whenever organic matter is decomposed under oxygen-free (anaerobic) conditions, methane is one of the products. Thus, enormous deposits of natural gas, which is almost entirely methane, exist as a major source of fuel for home and industry.
(a) It is estimated that known sources of natural gas can produce 5600 EJ of energy (1 EJ = 10^{18} J). Current total global energy usage is 4.0×10^2 EJ per year. Find the mass (in kg) of known sources of natural gas (ΔH_{rxn}° for combustion of $CH_4 = -802$ kJ/mol).
(b) For how many years could these sources supply the world's total energy needs?
(c) What volume (in ft^3) of natural gas is required to heat 1.00 qt of water from 20.0°C to 100.0°C (d of $H_2O = 1.00$ g/mL; d of CH_4 at STP = 0.72 g/L)?
(d) The fission of 1 mol of uranium (about 4×10^{-4} ft^3) in a nuclear reactor produces 2×10^{13} J. What volume (in ft^3) of natural gas would produce the same amount of energy?

6.66 The heat of atomization (ΔH_{atom}°) is the heat needed to form separated gaseous atoms from a substance in its standard state. The equation for the atomization of graphite is

$$C(graphite) \longrightarrow C(g)$$

Use Hess's law to calculate ΔH_{atom}° of graphite from these data:
(1) ΔH_f° of $CH_4 = -74.9$ kJ/mol
(2) ΔH_{atom}° of $CH_4 = 1660$ kJ/mol
(3) ΔH_{atom}° of $H_2 = 432$ kJ/mol

6.67 A reaction is carried out in a steel vessel within a chamber filled with argon gas. Below are molecular views of the argon adjacent to the surface of the reaction vessel before and after the reaction. Was the reaction exothermic or endothermic? Explain.

6.68 Benzene (C_6H_6) and acetylene (C_2H_2) have the same empirical formula, CH. Which releases more energy per mole of CH (ΔH_f° of gaseous $C_6H_6 = 82.9$ kJ/mol)?

6.69 Kerosene, a common space-heater fuel, is a mixture of hydrocarbons whose "average" formula is $C_{12}H_{26}$.
(a) Write a balanced equation, using the simplest whole-number coefficients, for the complete combustion of kerosene to gases.
(b) If $\Delta H_{rxn}^\circ = -1.50 \times 10^4$ kJ for the combustion equation as written in part (a), determine ΔH_f° of kerosene.
(c) Calculate the heat produced by combustion of 0.50 gal of kerosene (d of kerosene = 0.749 g/mL).
(d) How many gallons of kerosene must be burned for a kerosene furnace to produce 1250. Btu (1 Btu = 1.055 kJ)?

* 6.70 Coal gasification is a multistep process to convert coal into cleaner-burning gaseous fuels. In one step, a certain coal sample reacts with superheated steam:
$C(coal) + H_2O(g) \longrightarrow CO(g) + H_2(g)$ $\Delta H_{rxn}^\circ = 129.7$ kJ
(a) Combine this reaction with the following two to write an overall reaction for the production of methane:
$CO(g) + H_2O(g) \longrightarrow CO_2(g) + H_2(g)$ $\Delta H_{rxn}^\circ = -41$ kJ
$CO(g) + 3H_2(g) \longrightarrow CH_4(g) + H_2O(g)$ $\Delta H_{rxn}^\circ = -206$ kJ
(b) Calculate ΔH_{rxn}° for this overall change.
(c) Using the value in (b) and calculating the ΔH_{rxn}° for combustion of methane, find the total heat for gasifying 1.00 kg of coal and burning the methane formed (assume water forms as a gas and $\mathcal{M}$ of coal = 12.00 g/mol).

6.71 Phosphorus pentachloride is used in the industrial preparation of organic phosphorus compounds. Equation 1 shows its preparation from PCl_3 and Cl_2:
(1) $PCl_3(l) + Cl_2(g) \longrightarrow PCl_5(s)$
Use equations 2 and 3 to calculate ΔH_{rxn} of equation 1:
(2) $P_4(s) + 6Cl_2(g) \longrightarrow 4PCl_3(l)$ $\Delta H = -1280$ kJ
(3) $P_4(s) + 10Cl_2(g) \longrightarrow 4PCl_5(s)$ $\Delta H = -1774$ kJ

6.72 A typical candy bar weighs about 2 oz (1.00 oz = 28.4 g).
(a) Assuming that a candy bar is 100% sugar and that 1.0 g of sugar is equivalent to about 4.0 Calories of energy, calculate the energy (in kJ) contained in a typical candy bar.
(b) Assuming that your mass is 58 kg and you convert chemical potential energy to work with 100% efficiency, how high would you have to climb to work off the energy in a candy bar? (Potential energy = mass $\times g \times$ height, where $g = 9.8$ m/s^2.)
(c) Why is your actual conversion of potential energy to work less than 100% efficient?

6.73 Silicon tetrachloride is produced annually on the multikiloton scale for making transistor-grade silicon. It can be made directly from the elements (reaction 1) or, more cheaply, by heating sand and graphite with chlorine gas (reaction 2). If water is present in reaction 2, some tetrachloride may be lost in an unwanted side reaction (reaction 3):
(1) $Si(s) + 2Cl_2(g) \longrightarrow SiCl_4(g)$
(2) $SiO_2(s) + 2C(graphite) + 2Cl_2(g) \longrightarrow SiCl_4(g) + 2CO(g)$
(3) $SiCl_4(g) + 2H_2O(g) \longrightarrow SiO_2(s) + 4HCl(g)$
$\Delta H_{rxn}^\circ = -139.5$ kJ
(a) Use reaction 3 to calculate the heats of reaction of reactions 1 and 2. (b) What is the heat of reaction for the new reaction that is the sum of reactions 2 and 3?

6.74 Use the following information to find ΔH_f° of gaseous HCl:
$N_2(g) + 3H_2(g) \longrightarrow 2NH_3(g)$ $\Delta H_{rxn}^\circ = -91.8$ kJ
$N_2(g) + 4H_2(g) + Cl_2(g) \longrightarrow 2NH_4Cl(s)$ $\Delta H_{rxn}^\circ = -628.8$ kJ
$NH_3(g) + HCl(g) \longrightarrow NH_4Cl(s)$ $\Delta H_{rxn}^\circ = -176.2$ kJ

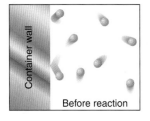

Before reaction

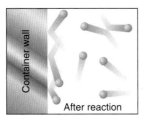

After reaction

* 6.75 You want to determine $\Delta H°$ for the reaction

$$Zn(s) + 2HCl(aq) \longrightarrow ZnCl_2(aq) + H_2(g)$$

(a) To do so, you first determine the heat capacity of a calorimeter using the following reaction, whose ΔH is known:

$$NaOH(aq) + HCl(aq) \longrightarrow NaCl(aq) + H_2O(l)$$
$$\Delta H° = -57.32 \text{ kJ}$$

Calculate the heat capacity of the calorimeter from these data:

Amounts used: 50.0 mL of 2.00 M HCl and 50.0 mL of 2.00 M NaOH

Initial T of both solutions: 16.9°C

Maximum T recorded during reaction: 30.4°C

Density of resulting NaCl solution: 1.04 g/mL

c of 1.00 M NaCl(aq) = 3.93 J/g·K

(b) Use the result from part (a) and the following data to determine $\Delta H°_{rxn}$ for the reaction between zinc and HCl(aq):

Amounts used: 100.0 mL of 1.00 M HCl and 1.3078 g of Zn

Initial T of HCl solution and Zn: 16.8°C

Maximum T recorded during reaction: 24.1°C

Density of 1.0 M HCl solution = 1.015 g/mL

c of resulting ZnCl$_2(aq)$ = 3.95 J/g·K

(c) Given the values below, what is the error in your experiment?

$$\Delta H°_f \text{ of HCl}(aq) = -1.652 \times 10^2 \text{ kJ/mol}$$

$$\Delta H°_f \text{ of ZnCl}_2(aq) = -4.822 \times 10^2 \text{ kJ/mol}$$

* 6.76 One mole of nitrogen gas confined within a cylinder by a piston is heated from 0°C to 819°C at 1.00 atm.

(a) Calculate the work of expansion of the gas in joules (1 J = 9.87×10^{-3} atm·L). Assume all the energy is used to do work.

(b) What would be the temperature change if the gas were heated with the same amount of energy in a container of fixed volume? (Assume the specific heat capacity of N$_2$ is 1.00 J/g·K.)

6.77 The chemistry of nitrogen oxides is very versatile. Given the following reactions and their standard enthalpy changes,

(1) $NO(g) + NO_2(g) \longrightarrow N_2O_3(g)$ $\quad \Delta H°_{rxn} = -39.8 \text{ kJ}$

(2) $NO(g) + NO_2(g) + O_2(g) \longrightarrow N_2O_5(g)$ $\Delta H°_{rxn} = -112.5 \text{ kJ}$

(3) $2NO_2(g) \longrightarrow N_2O_4(g)$ $\quad \Delta H°_{rxn} = -57.2 \text{ kJ}$

(4) $2NO(g) + O_2(g) \longrightarrow 2NO_2(g)$ $\quad \Delta H°_{rxn} = -114.2 \text{ kJ}$

(5) $N_2O_5(g) \longrightarrow N_2O_5(s)$ $\quad \Delta H°_{rxn} = -54.1 \text{ kJ}$

calculate the heat of reaction for

$$N_2O_3(g) + N_2O_5(s) \longrightarrow 2N_2O_4(g)$$

* 6.78 Liquid methanol (CH$_3$OH) is used as an alternative fuel in truck engines. An industrial method for preparing it uses the catalytic hydrogenation of carbon monoxide:

$$CO(g) + 2H_2(g) \xrightarrow{\text{catalyst}} CH_3OH(l)$$

How much heat (in kJ) is released when 15.0 L of CO at 85°C and 112 kPa reacts with 18.5 L of H$_2$ at 75°C and 744 torr?

* 6.79 (a) How much heat is released when 25.0 g of methane burns in excess O$_2$ to form gaseous products?

(b) Calculate the temperature of the product mixture if the methane and air are both at an initial temperature of 0.0°C. Assume a stoichiometric ratio of methane to oxygen from the air, with air being 21% O$_2$ by volume (c of CO$_2$ = 57.2 J/mol·K; c of H$_2$O(g) = 36.0 J/mol·K; c of N$_2$ = 30.5 J/mol·K).

Explaining the Spectacular The light from breathtaking firework and aurora displays, as well as from the more common neon signs and TV screens, occur through changes in atomic energy levels, which you'll examine in this chapter.

CHAPTER SEVEN

Quantum Theory and Atomic Structure

Key Principles

◆ All forms of *electromagnetic radiation* can be characterized as waves traveling at the *speed of light (c)*. The properties of a wave are its *wavelength* (λ, distance between corresponding points on adjacent waves), *frequency* (ν, number of cycles the wave undergoes per second), and *amplitude* (height of the crest or trough), which is related to the *intensity* (brightness) of the light. A region of the *electromagnetic spectrum* includes a range of wavelengths.

◆ Although light seems to have a wavelike nature, certain phenomena— *blackbody radiation* (light emitted by very hot objects), the *photoelectric effect* (the flow of current when light strikes a metal surface), and *atomic spectra* (separate lines of different colors observed when a substance is excited)—can only be explained if light energy consists of "packets" (*quanta*) of energy. The energy of a quantum is related to the light's frequency.

◆ According to the *Bohr model,* an atomic spectrum consists of separate lines because an atom has only certain, allowable energy levels (*states*). The energy of the atom changes when an electron moves from one *orbit* to another as the atom absorbs (or emits) light of a specific frequency.

◆ *Wave-particle duality* means that matter has wavelike properties and energy has particle-like properties. These properties become observable only at the atomic scale. Because of wave-particle duality, we can never know the exact position *and* momentum of an electron simultaneously (*uncertainty principle*).

◆ According to the *quantum-mechanical model,* each energy state of an atom is associated with an *atomic orbital,* a mathematical function describing an electron's motion in three dimensions. We can know the *probability* that the electron is within a particular tiny volume of space, but *not* its exact location. This probability decreases quickly with distance from the nucleus.

◆ *Quantum numbers* denote the size (n), shape (l), and spatial orientation (m_l) of each atomic orbital. An *energy level* consists of *sublevels,* which consist of *orbitals*. There is a *hierarchy* among the quantum numbers: n determines the values of l, which determines the values of m_l.

◆ In the H atom, there is only one electron. For this simplest case, the energy state of the H atom depends *only* on the principal quantum number (n).

Outline

Concepts & Skills to Review
Before You Study This Chapter

- discovery of the electron and atomic nucleus (Section 2.4)
- major features of atomic structure (Section 2.5)
- changes in energy state of a system (Section 6.1)

Over a few remarkable decades—from around 1890 to 1930—a revolution took place in how we view the makeup of the universe. But revolutions in science are not the violent upheavals of political overthrow. Rather, flaws appear in an established model as conflicting evidence mounts, a startling discovery or two widens the flaws into cracks, and the conceptual structure crumbles gradually from its inconsistencies. New insight, verified by experiment, then guides the building of a model more consistent with reality. So it was when Dalton's atomic theory established the idea of individual units of matter, and when Rutherford's nuclear model substituted atoms with rich internal structure for "plum puddings." In this chapter, you will see this process unfold again with the development of modern atomic theory.

Almost as soon as Rutherford proposed his nuclear model, a major problem arose. A nucleus and an electron attract each other, so if they are to remain apart, the energy of the electron's motion (kinetic energy) must balance the energy of attraction (potential energy). However, the laws of classical physics had established that a negative particle moving in a curved path around a positive one *must* emit radiation and thus lose energy. If this requirement applied to atoms, why didn't the orbiting electron lose energy continuously and spiral into the nucleus? Clearly, if electrons behaved the way classical physics predicted, all atoms would have collapsed eons ago! The behavior of subatomic matter seemed to violate real-world experience and accepted principles.

The breakthroughs that soon followed Rutherford's model forced a complete rethinking of the classical picture of matter and energy. In the macroscopic world, the two are distinct. Matter occurs in chunks you can hold and weigh, and you can change the amount of matter in a sample piece by piece. In contrast, energy is "massless," and its quantity changes in a continuous manner. Matter moves in specific paths, whereas light and other types of energy travel in diffuse waves. As you'll see in this chapter, however, as soon as 20th-century scientists probed the subatomic world, these clear distinctions between particulate matter and wavelike energy began to fade, revealing a much more amazing reality.

7.1 THE NATURE OF LIGHT

Visible light is one type of **electromagnetic radiation** (also called *electromagnetic energy* or *radiant energy*). Other familiar types include x-rays, microwaves, and radio waves. All electromagnetic radiation consists of energy propagated by means of electric and magnetic fields that alternately increase and decrease in intensity as they move through space. This classical wave model distinguishes clearly between waves and particles; it is essential for understanding why rainbows form, how magnifying glasses work, why objects look distorted under water, and many other everyday observations. But, it cannot explain observations on the atomic scale because, in that unfamiliar realm, energy behaves as though it consists of particles!

The Wave Nature of Light

The wave properties of electromagnetic radiation are described by two interdependent variables, as Figure 7.1 shows:

- **Frequency** (ν, Greek *nu*) is the number of cycles the wave undergoes per second and is expressed in units of 1/second [s^{-1}; also called *hertz* (Hz)].
- **Wavelength** (λ, Greek *lambda*) is the distance between any point on a wave and the corresponding point on the next crest (or trough) of the wave, that is, the distance the wave travels during one cycle. Wavelength is expressed in meters and often, for very short wavelengths, in nanometers (nm, 10^{-9} m), picometers (pm, 10^{-12} m), or the non-SI unit angstroms (Å, 10^{-10} m).

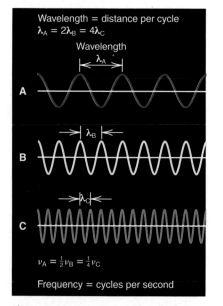

Wavelength = distance per cycle
$\lambda_A = 2\lambda_B = 4\lambda_C$

Wavelength

λ_A

A

λ_B

B

λ_C

C

$\nu_A = \frac{1}{2}\nu_B = \frac{1}{4}\nu_C$

Frequency = cycles per second

Figure 7.1 Frequency and wavelength. Three waves with different wavelengths (λ) and thus different frequencies (ν) are shown. Note that *as the wavelength decreases, the frequency increases, and vice versa.*

The speed of the wave, the distance traveled per unit time (in units of meters per second), is the product of its frequency (cycles per second) and its wavelength (meters per cycle):

$$\text{Units for speed of wave:} \quad \frac{\cancel{\text{cycles}}}{\text{s}} \times \frac{\text{m}}{\cancel{\text{cycle}}} = \frac{\text{m}}{\text{s}}$$

In a vacuum, all types of electromagnetic radiation travel at 2.99792458×10^8 m/s (3.00×10^8 m/s to three significant figures), which is a physical constant called the **speed of light (c):**

$$c = \nu \times \lambda \tag{7.1}$$

As Equation 7.1 shows, the product of ν and λ is a constant. Thus, the individual terms have a reciprocal relationship to each other: *radiation with a high frequency has a short wavelength, and vice versa.*

Another characteristic of a wave is its **amplitude,** the height of the crest (or depth of the trough) of each wave (Figure 7.2). The amplitude of an electromagnetic wave is a measure of the strength of its electric and magnetic fields. Thus, amplitude is related to the *intensity* of the radiation, which we perceive as brightness in the case of visible light. Light of a particular color—fire-engine red, for instance—has a specific frequency and wavelength, but it can be dimmer (lower amplitude) or brighter (higher amplitude).

The Electromagnetic Spectrum Visible light represents a small portion of the continuum of radiant energy known as the **electromagnetic spectrum** (Figure 7.3). *All the waves in the spectrum travel at the same speed through a vacuum but differ in frequency and, therefore, wavelength.* Some regions of the spectrum are utilized by particular devices; for example, the long-wavelength, low-frequency radiation is used by microwave ovens and radios. Note that each region meets the next. For instance, the **infrared (IR)** region meets the microwave region on one end and the visible region on the other.

We perceive different wavelengths (or frequencies) of *visible* light as different colors, from red ($\lambda \approx 750$ nm) to violet ($\lambda \approx 400$ nm). Light of a single wavelength is called *monochromatic* (Greek, "one color"), whereas light of many wavelengths is *polychromatic*. White light is polychromatic. The region adjacent to visible light on the short-wavelength end consists of **ultraviolet (UV)** radiation (also called *ultraviolet light*). Still shorter wavelengths (higher frequencies) make up the x-ray and gamma (γ) ray regions. Thus, a TV signal, the green light from a traffic signal, and a gamma ray emitted by a radioactive element all travel at the same speed but differ in their frequency (and wavelength).

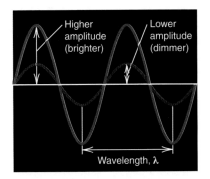

Figure 7.2 Amplitude (intensity) of a wave. Amplitude is represented by the height of the crest (or depth of the trough) of the wave. The two waves shown have the same wavelength (color) but different amplitudes and, therefore, different brightnesses (intensities).

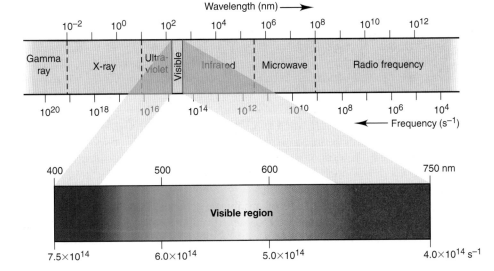

Figure 7.3 Regions of the electromagnetic spectrum. The electromagnetic spectrum extends from the very short wavelengths (very high frequencies) of gamma rays through the very long wavelengths (very low frequencies) of radio waves. The relatively narrow visible region is expanded (and the scale made linear) to show the component colors.

SAMPLE PROBLEM 7.1 Interconverting Wavelength and Frequency

Problem A dental hygienist uses x-rays ($\lambda = 1.00$ Å) to take a series of dental radiographs while the patient listens to a radio station ($\lambda = 325$ cm) and looks out the window at the blue sky ($\lambda = 473$ nm). What is the frequency (in s^{-1}) of the electromagnetic radiation from each source? (Assume that the radiation travels at the speed of light, 3.00×10^8 m/s.)

Plan We are given the wavelengths, so we use Equation 7.1 to find the frequencies. However, we must first convert the wavelengths to meters because c has units of m/s.

Solution For the x-rays: Converting from angstroms to meters,

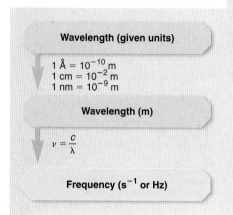

Wavelength (given units)

1 Å $= 10^{-10}$ m
1 cm $= 10^{-2}$ m
1 nm $= 10^{-9}$ m

Wavelength (m)

$\nu = \dfrac{c}{\lambda}$

Frequency (s^{-1} or Hz)

$$\lambda = 1.00 \text{ Å} \times \frac{10^{-10} \text{ m}}{1 \text{ Å}} = 1.00 \times 10^{-10} \text{ m}$$

Calculating the frequency:

$$\nu = \frac{c}{\lambda} = \frac{3.00 \times 10^8 \text{ m/s}}{1.00 \times 10^{-10} \text{ m}} = 3.00 \times 10^{18} \text{ s}^{-1}$$

For the radio signal: Combining steps to calculate the frequency,

$$\nu = \frac{c}{\lambda} = \frac{3.00 \times 10^8 \text{ m/s}}{325 \text{ cm} \times \dfrac{10^{-2} \text{ m}}{1 \text{ cm}}} = 9.23 \times 10^7 \text{ s}^{-1}$$

For the blue sky: Combining steps to calculate the frequency,

$$\nu = \frac{c}{\lambda} = \frac{3.00 \times 10^8 \text{ m/s}}{473 \text{ nm} \times \dfrac{10^{-9} \text{ m}}{1 \text{ nm}}} = 6.34 \times 10^{14} \text{ s}^{-1}$$

Check The orders of magnitude are correct for the regions of the electromagnetic spectrum (see Figure 7.3): x-rays (10^{19} to 10^{16} s^{-1}), radio waves (10^9 to 10^4 s^{-1}), and visible light (7.5×10^{14} to 4.0×10^{14} s^{-1}).

Comment The radio station here is broadcasting at 92.3×10^6 s^{-1}, or 92.3 million Hz (92.3 MHz), about midway in the FM range.

FOLLOW-UP PROBLEM 7.1 Some diamonds appear yellow because they contain nitrogen compounds that absorb purple light of frequency 7.23×10^{14} Hz. Calculate the wavelength (in nm and Å) of the absorbed light.

The Distinction Between Energy and Matter In the everyday world around us, energy and matter behave very differently. Let's examine some important differences. Light of a given wavelength travels at different speeds through different transparent media—vacuum, air, water, quartz, and so forth. Therefore, when a light wave passes from one medium into another, say, from air to water, the speed of the wave changes. Figure 7.4A shows the phenomenon known as **refraction**. If the wave strikes the boundary between air and water, at an angle other than 90°, the change in speed causes a change in direction, and the wave continues at a different angle. The new angle (angle of refraction) depends on the materials on either side of the boundary and the wavelength of the light. In the process of *dispersion,* white light separates (disperses) into its component colors, as when it passes through a prism, because each incoming wave is refracted at a slightly different angle. Rainbows result when sunlight is dispersed through water droplets.

In contrast, a particle, like a pebble, does not undergo refraction when passing from one medium to another. Figure 7.4B shows that if you throw a pebble through the air into a pond, its speed changes abruptly and then it continues to slow down gradually in a curved path.

When a wave strikes the edge of an object, it bends around it in a phenomenon called **diffraction**. If the wave passes through a slit about as wide as its wavelength, it bends around both edges of the slit and forms a semicircular wave on the other side of the opening, as shown in Figure 7.4C.

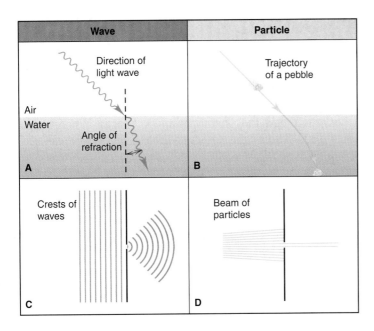

Figure 7.4 Different behaviors of waves and particles. A, A wave passing from air into water is *refracted* (bent at an angle). **B,** In contrast, a particle of matter (such as a pebble) entering a pond moves in a curved path, because gravity and the greater resistance (drag) of the water slow it down gradually. **C,** A wave is *diffracted* through a small opening, which gives rise to a circular wave on the other side. (The lines represent the crests of water waves as seen from above.) **D,** In contrast, when a collection of moving particles encounters a small opening, as when a handful of sand is thrown at a hole in a fence, some particles move through the opening and continue along their individual paths.

Once again, particles act very differently. Figure 7.4D shows that if you throw a collection of particles, like a handful of sand, at a small opening, some particles hit the edge, while others go through the opening and continue linearly in a narrower group.

If waves of light pass through two adjacent slits, the emerging circular waves interact with each other through the process of *interference.* If the crests of the waves coincide (*in phase*), they interfere *constructively* and the amplitudes add together. If the crests coincide with troughs (*out of phase*), they interfere *destructively* and the amplitudes cancel. The result is a diffraction pattern of brighter and darker regions (Figure 7.5). In contrast, particles passing through adjacent openings continue in straight paths, some colliding with each other and moving at different angles.

At the end of the 19th century, all everyday and laboratory experience seemed to confirm these classical distinctions between the wave nature of energy and the particle nature of matter.

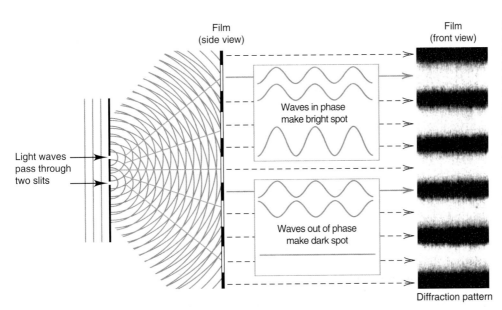

Diffraction pattern

Figure 7.5 The diffraction pattern caused by light passing through two adjacent slits. As light waves pass through two closely spaced slits, they emerge as circular waves and interfere with each other. They create a diffraction (interference) pattern of bright and dark regions on a sheet of film. Bright regions appear where crests coincide and the amplitudes combine with each other (in phase); dark regions appear where crests meet troughs and the amplitudes cancel each other (out of phase).

The Particle Nature of Light

Three phenomena involving matter and light confounded physicists at the turn of the 20th century: (1) blackbody radiation, (2) the photoelectric effect, and (3) atomic spectra. Explaining these phenomena required a radically new picture of energy. We discuss the first two here and the third in the next section.

Blackbody Radiation and the Quantization of Energy When a solid object is heated to about 1000 K, it begins to emit visible light, as you can see in the soft red glow of smoldering coal. At about 1500 K, the light is brighter and more orange, like that from an electric heating coil. At temperatures greater than 2000 K, the light is still brighter and whiter, as from the filament of a lightbulb. These changes in intensity and wavelength of emitted light as an object is heated are characteristic of *blackbody radiation,* light given off by a hot *blackbody.** All attempts to account for these observed changes by applying classical electromagnetic theory failed. Then, in 1900, the German physicist Max Planck (1858–1947) made a radical assumption that eventually led to an entirely new view of energy. He proposed that the hot, glowing object could emit (or absorb) only *certain quantities* of energy:

$$E = nh\nu$$

where E is the energy of the radiation, ν is its frequency, n is a positive integer (1, 2, 3, and so on) called a **quantum number,** and h is a proportionality constant now known very precisely and called **Planck's constant.** With energy in joules (J) and frequency in s^{-1}, h has units of J·s:

$$h = 6.62606876{\times}10^{-34}\ \text{J·s} = 6.626{\times}10^{-34}\ \text{J·s} \qquad \text{(4 sf)}$$

Later interpretations of Planck's proposal stated that the hot object's radiation is emitted by the atoms contained within it. If an atom can *emit* only certain quantities of energy, it follows that *the atom itself can have only certain quantities of energy.* Thus, the energy of an atom is *quantized:* it exists only in certain fixed quantities, rather than being continuous. Each change in the atom's energy results from the gain or loss of one or more "packets," definite amounts, of energy. Each energy packet is called a **quantum** ("fixed quantity"; plural, *quanta*), and its energy is equal to $h\nu$. Thus, *an atom changes its energy state by emitting (or absorbing) one or more quanta,* and the energy of the emitted (or absorbed) radiation is equal to the *difference in the atom's energy states:*

$$\Delta E_{\text{atom}} = E_{\text{emitted (or absorbed) radiation}} = \Delta nh\nu$$

Because the atom can change its energy only by integer multiples of $h\nu$, the smallest change occurs when an atom in a given energy state changes to an adjacent state, that is, when $\Delta n = 1$:

$$\Delta E = h\nu \qquad\qquad (7.2)$$

The Photoelectric Effect and the Photon Theory of Light Despite the idea that energy is quantized, Planck and other physicists continued to picture the emitted energy as traveling in waves. However, the wave model could not explain the **photoelectric effect,** the flow of current when monochromatic light of sufficient frequency shines on a metal plate (Figure 7.6). The existence of the current was not puzzling: it could be understood as arising when the light transfers energy to the electrons at the metal surface, which break free and are collected by the positive electrode. However, the photoelectric effect had certain confusing features, in particular, the *presence of a threshold frequency* and the *absence of a time lag:*

1. *Presence of a threshold frequency.* Light shining on the metal must have a minimum *frequency,* or no current flows. (Different metals have different minimum

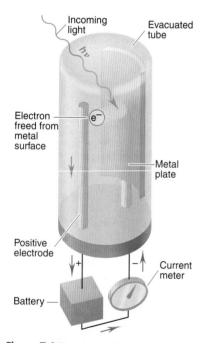

Figure 7.6 Demonstration of the photoelectric effect. When monochromatic light of high enough frequency strikes the metal plate, electrons are freed from the plate and travel to the positive electrode, creating a current.

---

*A blackbody is an idealized object that absorbs all the radiation incident on it. A hollow cube with a small hole in one wall approximates a blackbody.

frequencies.) The wave theory, however, associates the energy of the light with the *amplitude* (intensity) of the wave, not with its frequency (color). Thus, the wave theory predicts that an electron would break free when it absorbed enough energy from light of *any* color.

2. *Absence of a time lag.* Current flows the moment light of this minimum frequency shines on the metal, regardless of the light's intensity. The wave theory, however, predicts that in dim light there would be a time lag before the current flowed, because the electrons had to absorb enough energy to break free.

Carrying Planck's idea of quantized energy further, the great physicist Albert Einstein proposed that light itself is particulate, that is, quantized into small "bundles" of electromagnetic energy, which were later called **photons.** In terms of Planck's work, we can say that each atom changes its energy whenever it absorbs or emits one photon, one "particle" of light, whose energy is fixed by its *frequency:*

$$E_{\text{photon}} = h\nu = \Delta E_{\text{atom}}$$

Let's see how Einstein's photon theory explains the photoelectric effect:

1. *Explanation of the threshold frequency.* According to the photon theory, a beam of light consists of an enormous number of photons. Light intensity (brightness) is related to the number of photons striking the surface per unit time, but *not* to their energy. Therefore, a photon of a certain minimum *energy* must be absorbed for an electron to be freed. Because energy depends on frequency ($h\nu$), the theory predicts a threshold frequency.

2. *Explanation of the time lag.* An electron cannot "save up" energy from several photons below the minimum energy until it has enough to break free. Rather, one electron breaks free the moment it absorbs one photon of *enough* energy. The current is weaker in dim light than in bright light because fewer photons of enough energy are present, so fewer electrons break free per unit time. But *some* current flows the moment photons reach the metal plate.

SAMPLE PROBLEM 7.2 Calculating the Energy of Radiation from Its Wavelength

Problem A cook uses a microwave oven to heat a meal. The wavelength of the radiation is 1.20 cm. What is the energy of one photon of this microwave radiation?

Plan We know λ in centimeters (1.20 cm) so we convert to meters, find the frequency with Equation 7.1, and then find the energy of one photon with Equation 7.2.

Solution Combining steps to find the energy:

$$E = h\nu = \frac{hc}{\lambda} = \frac{(6.626\times10^{-34}\text{ J·s})(3.00\times10^{8}\text{ m/s})}{(1.20\text{ cm})\left(\dfrac{10^{-2}\text{ m}}{1\text{ cm}}\right)} = 1.66\times10^{-23}\text{ J}$$

Check Checking the order of magnitude gives $\dfrac{10^{-33}\text{ J·s}\times10^{8}\text{ m/s}}{10^{-2}\text{ m}} = 10^{-23}\text{ J}$.

FOLLOW-UP PROBLEM 7.2 Calculate the energies of one photon of ultraviolet ($\lambda = 1\times10^{-8}$ m), visible ($\lambda = 5\times10^{-7}$ m), and infrared ($\lambda = 1\times10^{-4}$ m) light. What do the answers indicate about the relationship between the wavelength and energy of light?

Planck's quantum theory and Einstein's photon theory assigned properties to energy that, until then, had always been reserved for matter: fixed quantity and discrete particles. These properties have since proved essential to explaining the interactions of matter and energy at the atomic level. But how can a particulate model of energy be made to fit the facts of diffraction and refraction, phenomena explained only in terms of waves? As you'll see shortly, the photon model does not *replace* the wave model. Rather, we have to accept *both* to understand reality. Before we discuss this astonishing notion, however, let's see how the new idea of quantized energy led to a key understanding about atomic behavior.

SECTION SUMMARY

Electromagnetic radiation travels in waves of specific wavelength (λ) and frequency (ν). All electromagnetic waves travel through a vacuum at the speed of light, c (3.00×10^8 m/s), which is equal to $\nu \times \lambda$. The intensity (brightness) of a light wave is related to its amplitude. The electromagnetic spectrum ranges from very long radio waves to very short gamma rays and includes the visible region [750 nm (red) to 400 nm (violet)]. Refraction and diffraction indicate that electromagnetic radiation is wavelike, but blackbody radiation and the photoelectric effect indicate that it is particle-like. Light exists as photons (quanta) that have an energy proportional to the frequency. According to quantum theory, an atom has only certain quantities of energy ($E = nh\nu$), which it can change only by absorbing or emitting a photon.

7.2 ATOMIC SPECTRA

The third key observation about matter and energy that late 19th-century physicists could not explain involved the light emitted when an element is vaporized and then thermally or electrically excited, as you see in a neon sign. Figure 7.7A shows the result when light from excited hydrogen atoms passes through a narrow slit and is then refracted by a prism. Note that this light does not create a *continuous spectrum,* or rainbow, as sunlight does. Instead, it creates a **line spectrum,** a series of fine lines of individual colors separated by colorless (black) spaces.* The wavelengths of these spectral lines are characteristic of the element producing them (Figure 7.7B).

---

*The appearance of the spectrum as a series of lines results from the construction of the apparatus. If the light passed through a small hole, rather than a narrow slit, the spectrum would appear as a circular field of dots rather than a horizontal series of lines. The key point is that *the spectrum is discrete, rather than continuous.*

Figure 7.7 The line spectra of several elements. A, A sample of gaseous H_2 is dissociated into atoms and excited by an electric discharge. The emitted light passes through a slit and a prism, which disperses the light into individual wavelengths. The line spectrum of atomic H is shown (*top*). **B,** The continuous spectrum of white light is compared with the line spectra of mercury and strontium. Note that each line spectrum is different from the others.

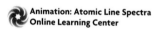
Animation: Atomic Line Spectra
Online Learning Center

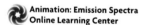
Animation: Emission Spectra
Online Learning Center

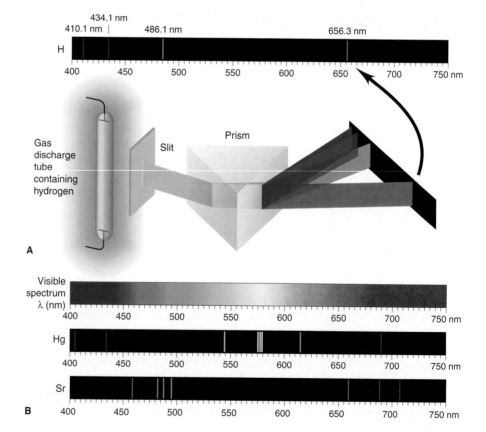

Spectroscopists studying the spectrum of atomic hydrogen had identified several series of such lines in different regions of the electromagnetic spectrum. Figure 7.8 shows three of these series of lines. Equations of the following general form, called the *Rydberg equation,* were found to predict the position and wavelength of any line in a given series:

$$\frac{1}{\lambda} = R\left(\frac{1}{n_1^2} - \frac{1}{n_2^2}\right) \tag{7.3}$$

where λ is the wavelength of a spectral line, n_1 and n_2 are positive integers with $n_2 > n_1$, and R is the Rydberg constant (1.096776×10^7 m^{-1}). For the visible series of lines, $n_1 = 2$:

$$\frac{1}{\lambda} = R\left(\frac{1}{2^2} - \frac{1}{n_2^2}\right), \quad \text{with } n_2 = 3, 4, 5, \ldots$$

The Rydberg equation and the value of the constant are based on data rather than theory. No one knew *why* the spectral lines of hydrogen appear in this pattern. (Problems 7.18 and 7.19 are two of several that apply the Rydberg equation.)

The observation of line spectra did not correlate with classical theory for one major reason. As was mentioned in the chapter introduction, if an electron spiraled closer to the nucleus, it should emit radiation. Moreover, the frequency of the radiation should be related to the time of revolution. On the spiral path inward, that time should decrease smoothly, so the frequency of the radiation should change smoothly and create a continuous spectrum. Rutherford's nuclear model seemed totally at odds with atomic line spectra.

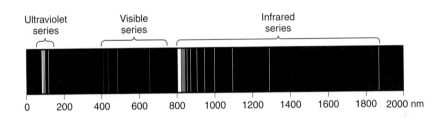

Figure 7.8 Three series of spectral lines of atomic hydrogen. These series appear in different regions of the electromagnetic spectrum. The hydrogen spectrum shown in Figure 7.7A is the visible series.

The Bohr Model of the Hydrogen Atom

Soon after the nuclear model was proposed, Niels Bohr (1885–1962), a young Danish physicist working in Rutherford's laboratory, suggested a model for the H atom that predicted the existence of line spectra. In his model, Bohr used Planck's and Einstein's ideas about quantized energy and proposed three postulates:

1. *The H atom has only certain allowable energy levels,* which Bohr called **stationary states.** Each of these states is associated with a fixed circular orbit of the electron around the nucleus.

2. *The atom does **not** radiate energy while in one of its stationary states.* That is, even though it violates the ideas of classical physics, the atom does not change energy while the electron moves *within* an orbit.

3. *The atom changes to another stationary state* (the electron moves to another orbit) *only by absorbing or emitting a photon whose energy equals the difference in energy between the two states:*

$$E_{\text{photon}} = E_{\text{state A}} - E_{\text{state B}} = h\nu$$

where the energy of state A is higher than that of state B. A spectral line results when a photon of specific energy (and thus specific frequency) is *emitted* as the electron moves from a higher energy state to a lower one. Therefore, Bohr's model explains that an atomic spectrum is not continuous because *the atom's energy has only certain discrete levels, or states.*

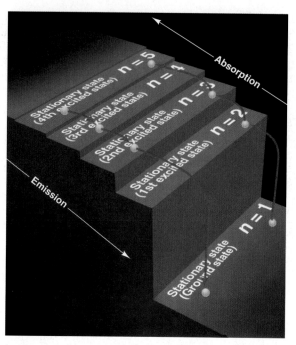

Figure 7.9 Quantum staircase. In this analogy for the energy levels of the hydrogen atom, an electron can absorb a photon and jump up to a higher "step" (stationary state) or emit a photon and jump down to a lower one. But the electron cannot lie between two steps.

In Bohr's model, the quantum number n (1, 2, 3, . . .) is associated with the radius of an electron orbit, which is directly related to the electron's energy: *the lower the n value, the smaller the radius of the orbit, and the lower the energy level.* When the electron is in the first orbit ($n = 1$), the orbit closest to the nucleus, the H atom is in its lowest (first) energy level, called the **ground state.** If the H atom absorbs a photon whose energy equals the *difference* between the first and second energy levels, the electron moves to the second orbit ($n = 2$), the next orbit out from the nucleus. When the electron is in the second or any higher orbit, the atom is said to be in an **excited state.** If the H atom in the first excited state (the electron in the second orbit) emits a photon of that same energy, it returns to the ground state. Figure 7.9 shows a staircase analogy for this behavior.

Figure 7.10A shows how Bohr's model accounts for the three line spectra of hydrogen. When a sample of gaseous H atoms is excited, different atoms absorb different quantities of energy. Each atom has one electron, but so many atoms are present that all the energy levels (orbits) are populated by electrons. When the electrons drop from outer orbits to the $n = 3$ orbit (second excited state), the emitted photons create the infrared series of lines. The visible series arises when electrons drop to the $n = 2$ orbit (first excited state). Figure 7.10B shows that the ultraviolet series arises when electrons drop to the $n = 1$ orbit (ground state).

Despite its great success in accounting for the spectral lines of the H atom, the Bohr model failed to predict the spectrum of any other atom, even that of helium, the next simplest element. In essence, the Bohr model predicts spectral lines for the H atom and other one-electron species, such as He^+ ($Z = 2$), Li^{2+} ($Z = 3$), and Be^{3+} ($Z = 4$). But, it fails for atoms with more than one electron because in these systems, electron-electron repulsions and additional nucleus-electron attractions are present as well. Nevertheless, we still use the terms "ground state" and "excited state" and retain one of Bohr's central ideas in our current model: *the energy of an atom occurs in discrete levels.*

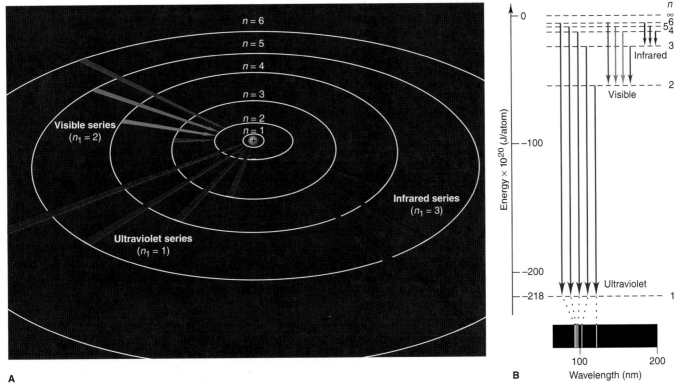

A

B

Figure 7.10 The Bohr explanation of three series of spectral lines. A, According to the Bohr model, when an electron drops from an outer orbit to an inner one, the atom emits a photon of specific energy that gives rise to a spectral line. In a given spectral series, each electron drop has the same inner orbit, that is, the same value of n_1 in the Rydberg equation (see Equation 7.3). (The orbit radius is proportional to n^2. Only the first six orbits are shown.) **B,** An energy diagram shows how the ultraviolet series arises. Within each series, the greater the *difference* in orbit radii, the greater the difference in energy levels (depicted as a downward arrow), and the higher the energy of the photon emitted. For example, in the ultraviolet series, in which $n_1 = 1$, a drop from $n = 5$ to $n = 1$ emits a photon with more energy (shorter λ, higher ν) than a drop from $n = 2$ to $n = 1$. [The axis shows negative values because $n = \infty$ (the electron completely separated from the nucleus) is *defined* as the atom with zero energy.]

The Energy States of the Hydrogen Atom

A very useful result from Bohr's work is an equation for calculating the energy levels of an atom, which he derived from the classical principles of electrostatic attraction and circular motion:

$$E = -2.18 \times 10^{-18} \text{ J}\left(\frac{Z^2}{n^2}\right)$$

where Z is the charge of the nucleus. For the H atom, $Z = 1$, so we have

$$E = -2.18 \times 10^{-18} \text{ J}\left(\frac{1^2}{n^2}\right) = -2.18 \times 10^{-18} \text{ J}\left(\frac{1}{n^2}\right)$$

Therefore, the energy of the ground state ($n = 1$) is

$$E = -2.18 \times 10^{-18} \text{ J}\left(\frac{1}{1^2}\right) = -2.18 \times 10^{-18} \text{ J}$$

Don't be confused by the negative sign for the energy values (see the axis in Figure 7.10B). It appears because we *define* the zero point of the atom's energy when *the electron is completely removed from the nucleus*. Thus, $E = 0$ when $n = \infty$, so $E < 0$ for any smaller n. As an analogy, consider a book resting on the floor. You can define the zero point of the book's potential energy in many ways. If you define zero when the book is on the floor, the energy is positive when the book

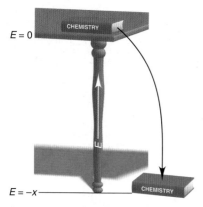

$E = 0$

CHEMISTRY

$E = -x$

CHEMISTRY

Figure 7.11 A tabletop analogy for the H atom's energy.

is on a tabletop. But, if you define zero when the book is on a tabletop, the energy is negative when the book lies on the floor; the latter case is analogous to the energy of the H atom (Figure 7.11).

With n in the denominator of the energy equation, as the electron moves closer to the nucleus (n decreases), the atom becomes more stable (less energetic) and its energy becomes a *larger negative number*. As the electron moves away from the nucleus (n increases), the atom's energy increases (becomes a smaller negative number).

This equation is easily adapted to find the energy difference between any two levels:

$$\Delta E = E_{\text{final}} - E_{\text{initial}} = -2.18 \times 10^{-18} \text{ J} \left(\frac{1}{n_{\text{final}}^2} - \frac{1}{n_{\text{initial}}^2} \right) \quad (7.4)$$

Using Equation 7.4, we can predict the wavelengths of the spectral lines of the H atom. Note that if we combine Equation 7.4 with Planck's expression for the change in an atom's energy (Equation 7.2), we obtain the Rydberg equation (Equation 7.3):

$$\Delta E = h\nu = \frac{hc}{\lambda} = -2.18 \times 10^{-18} \text{ J} \left(\frac{1}{n_{\text{final}}^2} - \frac{1}{n_{\text{initial}}^2} \right)$$

Therefore,
$$\frac{1}{\lambda} = -\frac{2.18 \times 10^{-18} \text{ J}}{hc} \left(\frac{1}{n_{\text{final}}^2} - \frac{1}{n_{\text{initial}}^2} \right)$$

$$= -\frac{2.18 \times 10^{-18} \text{ J}}{(6.626 \times 10^{-34} \text{ J·s})(3.00 \times 10^8 \text{ m/s})} \left(\frac{1}{n_{\text{final}}^2} - \frac{1}{n_{\text{initial}}^2} \right)$$

$$= -1.10 \times 10^7 \text{ m}^{-1} \left(\frac{1}{n_{\text{final}}^2} - \frac{1}{n_{\text{initial}}^2} \right)$$

where $n_{\text{final}} = n_2$, $n_{\text{initial}} = n_1$, and $1.10 \times 10^7 \text{ m}^{-1}$ is the Rydberg constant ($1.096776 \times 10^7 \text{ m}^{-1}$) to three significant figures. Thus, from classical relationships of charge and of motion combined with the idea that the H atom can have only certain values of energy, we obtain an equation from theory that leads directly to the empirical one! (In fact, Bohr obtained a value for the Rydberg constant that differed from the spectroscopists' value by only 0.05%!) We can use Equation 7.4 to find the quantity of energy needed to completely remove the electron from an H atom. In other words, what is ΔE for the following change?

$$\text{H}(g) \longrightarrow \text{H}^+(g) + \text{e}^-$$

We substitute $n_{\text{final}} = \infty$ and $n_{\text{initial}} = 1$ and obtain

$$\Delta E = E_{\text{final}} - E_{\text{initial}} = -2.18 \times 10^{-18} \text{ J} \left(\frac{1}{\infty^2} - \frac{1}{1^2} \right)$$

$$= -2.18 \times 10^{-18} \text{ J}(0 - 1) = 2.18 \times 10^{-18} \text{ J}$$

ΔE is positive because energy is *absorbed* to remove the electron from the vicinity of the nucleus. For 1 mol of H atoms,

$$\Delta E = \left(2.18 \times 10^{-18} \frac{\text{J}}{\text{atom}} \right) \left(6.022 \times 10^{23} \frac{\text{atoms}}{\text{mol}} \right) \left(\frac{1 \text{ kJ}}{10^3 \text{ J}} \right) = 1.31 \times 10^3 \text{ kJ/mol}$$

This is the *ionization energy* of the H atom, the quantity of energy required to form 1 mol of gaseous H$^+$ ions from 1 mol of gaseous H atoms. We return to this idea in Chapter 8.

Spectral Analysis in the Laboratory

Analysis of the spectrum of the H atom led to the Bohr model, the first step toward our current model of the atom. From its use by 19th-century chemists as a means of identifying elements and compounds, spectrometry has developed into a major

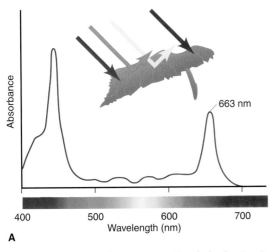

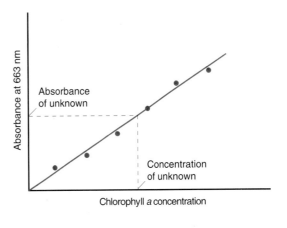

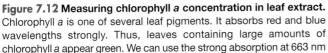

A

B

Figure 7.12 Measuring chlorophyll *a* concentration in leaf extract. Chlorophyll *a* is one of several leaf pigments. It absorbs red and blue wavelengths strongly. Thus, leaves containing large amounts of chlorophyll *a* appear green. We can use the strong absorption at 663 nm in the spectrum **(A)** to quantify the amount of chlorophyll *a* present in a plant extract by comparing that absorbance to a series of known standards **(B)**.

tool of modern chemistry. The terms *spectroscopy, spectrometry,* and **spectrophotometry** refer to a large group of instrumental techniques that obtain spectra corresponding to a substance's atomic or molecular energy levels. The two types of spectra most often obtained are emission and absorption spectra. An **emission spectrum** is produced when atoms in an excited state emit photons characteristic of the element as they return to lower energy states. The characteristic colors of fireworks and sodium-vapor streetlights are due to one or a few prominent lines in the emission spectra of the atoms present.

An **absorption spectrum** is produced when atoms absorb photons of certain wavelengths and become excited from lower to higher energy states. Therefore, the absorption spectrum of an element appears as dark lines against a bright background. Both the emission and absorption spectra of a substance are characteristic of that substance and used to identify it. A spectrometer can also be used to measure the concentration of a substance in a solution because *the absorbance, the amount of light of a given wavelength absorbed by a substance, is proportional to the number of molecules.* Suppose, for example, you want to determine the concentration of chlorophyll in an ether solution of leaf extract. You select a strongly absorbed wavelength from the chlorophyll spectrum (such as 663 nm in Figure 7.12A), measure the absorbance of the leaf-extract solution, and compare it with the absorbances of a series of ether solutions with known chlorophyll concentrations (Figure 7.12B).

SECTION SUMMARY

To explain the line spectrum of atomic hydrogen, Bohr proposed that the atom's energy is quantized because the electron's motion is restricted to fixed orbits. The electron can move from one orbit to another only if the atom absorbs or emits a photon whose energy equals the difference in energy levels (orbits). Line spectra are produced because these energy changes correspond to photons of specific wavelength. Bohr's model predicted the hydrogen atomic spectrum but cannot predict that of any species with more than one electron. Despite this, Bohr's idea that atoms have quantized energy levels is a cornerstone of our current atomic model. Spectrophotometry is an instrumental technique in which emission and absorption spectra are used to identify substances and/or measure their concentrations.

7.3 THE WAVE-PARTICLE DUALITY OF MATTER AND ENERGY

The early proponents of quantum theory demonstrated that *energy is particle-like*. Physicists who developed the theory turned this proposition upside down and showed that *matter is wavelike*. The sharp divisions we perceive between matter (chunky and massive) and energy (diffuse and massless) have been completely blurred. Strange as this idea may seem, it is the key to our modern atomic model.

The Wave Nature of Electrons and the Particle Nature of Photons

Bohr's efforts were a perfect case of fitting theory to data: he *assumed* that an atom has only certain allowable energy levels in order to *explain* the observed line spectrum. However, his assumption had no basis in physical theory. Then, in the early 1920s, a young French physics student named Louis de Broglie proposed a startling reason for fixed energy levels: *if energy is particle-like, perhaps matter is wavelike*. De Broglie had been thinking of other systems that display only certain allowed motions, such as the wave of a plucked guitar string. Figure 7.13 shows that, because the ends of the string are fixed, only certain vibrational frequencies (and wavelengths) are possible. De Broglie reasoned that *if electrons have wavelike motion* and are restricted to orbits of fixed radii, that would explain why they have only certain possible frequencies and energies.

Combining Einstein's famous equation for the quantity of energy equivalent to a given amount of mass ($E = mc^2$) with the equation for the energy of a photon ($E = h\nu = hc/\lambda$), de Broglie derived an equation for the wavelength of any particle of mass m—whether planet, baseball, or electron—moving at speed u:

$$\lambda = \frac{h}{mu} \tag{7.5}$$

Figure 7.13 Wave motion in restricted systems. A, In a musical analogy to electron waves, one half-wavelength ($\lambda/2$) is the "quantum" of the guitar string's vibration. The string length L is fixed, so the only allowed vibrations occur when L is a whole-number multiple (n) of $\lambda/2$. **B,** If an electron occupies a circular orbit, only whole numbers of wavelengths are allowed ($n = 3$ and $n = 5$ are shown). A wave with a fractional number of wavelengths (such as $n = 3\frac{1}{3}$) is "forbidden" because it rapidly dies out through overlap of crests and troughs.

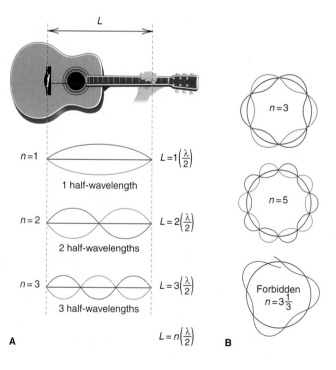

Table 7.1 **The de Broglie Wavelengths of Several Objects**			
Substance	Mass (g)	Speed (m/s)	λ (m)
Slow electron	9×10^{-28}	1.0	7×10^{-4}
Fast electron	9×10^{-28}	5.9×10^{6}	1×10^{-10}
Alpha particle	6.6×10^{-24}	1.5×10^{7}	7×10^{-15}
One-gram mass	1.0	0.01	7×10^{-29}
Baseball	142	25.0	2×10^{-34}
Earth	6.0×10^{27}	3.0×10^{4}	4×10^{-63}

According to this equation for the **de Broglie wavelength,** *matter behaves as though it moves in a wave.* Note also that an object's wavelength is *inversely* proportional to its mass, so heavy objects such as planets and baseballs have wavelengths that are *many* orders of magnitude smaller than the object itself, as you can see in Table 7.1.

SAMPLE PROBLEM 7.3 Calculating the de Broglie Wavelength of an Electron

Problem Find the de Broglie wavelength of an electron with a speed of 1.00×10^{6} m/s (electron mass = 9.11×10^{-31} kg; $h = 6.626\times10^{-34}$ kg·m^2/s).
Plan We know the speed (1.00×10^{6} m/s) and mass (9.11×10^{-31} kg) of the electron, so we substitute these into Equation 7.5 to find λ.
Solution

$$\lambda = \frac{h}{mu} = \frac{6.626\times10^{-34}\ \text{kg·m}^2\text{/s}}{(9.11\times10^{-31}\ \text{kg})(1.00\times10^{6}\ \text{m/s})} = 7.27\times10^{-10}\ \text{m}$$

Check The order of magnitude and units seem correct:

$$\lambda \approx \frac{10^{-33}\ \text{kg·m}^2\text{/s}}{(10^{-30}\ \text{kg})(10^{6}\ \text{m/s})} = 10^{-9}\ \text{m}$$

FOLLOW-UP PROBLEM 7.3 What is the speed of an electron that has a de Broglie wavelength of 100. nm?

If particles travel in waves, electrons should exhibit diffraction and interference (see Section 7.1). A fast-moving electron has a wavelength of about 10^{-10} m, so perhaps a beam of electrons would be diffracted by the spaces about this size between atoms in a crystal. Indeed, in 1927, C. Davisson and L. Germer guided a beam of electrons at a nickel crystal and obtained a diffraction pattern. Figure 7.14 shows the diffraction patterns obtained when either x-rays or electrons impinge on aluminum foil. Apparently, electrons—particles with mass and charge—create diffraction patterns, just as electromagnetic waves do! (Indeed, the electron microscope—and its revolutionary impact on modern biology—depends on the wavelike behavior of electrons.) Even though electrons do not have orbits of fixed radius, as de Broglie thought, the energy levels of the atom *are* related to the wave nature of the electron.

If electrons have properties of energy, do photons have properties of matter? The de Broglie equation suggests that we can calculate the momentum (p), the product of mass and speed, for a photon of a given wavelength. Substituting the speed of light (c) for speed u in Equation 7.5 and solving for p gives

$$\lambda = \frac{h}{mc} = \frac{h}{p} \quad \text{and} \quad p = \frac{h}{\lambda}$$

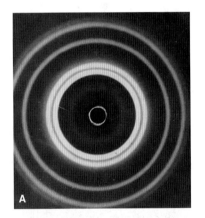

A

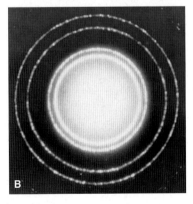

B

Figure 7.14 Comparing diffraction patterns of x-rays and electrons. A, X-ray diffraction pattern of aluminum. **B,** Electron diffraction pattern of aluminum. This behavior implies that both x-rays, which are electromagnetic radiation, and electrons, which are particles, travel in waves.

Notice the inverse relationship between p and λ. This means that shorter wavelength (higher energy) photons have greater momentum. Thus, a decrease in a photon's momentum should appear as an increase in its wavelength. In 1923, Arthur Compton directed a beam of x-ray photons at a sample of graphite and observed that the wavelength of the reflected photons increased. This result means that the photons transferred some of their momentum to the electrons in the carbon atoms of the graphite, just as colliding billiard balls transfer momentum to one another. In this experiment, photons behave as particles with momentum!

To scientists of the time, these results were very unsettling. Classical experiments had shown matter to be particle-like and energy to be wavelike, but these new studies showed that, on the atomic scale, every characteristic trait used to define the one now also defined the other. Figure 7.15 summarizes the conceptual and experimental breakthroughs that led to this juncture.

The truth is that *both* matter and energy show *both* behaviors: each possesses both "faces." In some experiments, we observe one face; in other experiments, we observe the other face. The distinction between a particle and a wave is meaningful only in the macroscopic world, *not* in the atomic world. The distinction between matter and energy is in our minds and our limiting definitions, not inherent in nature. This dual character of matter and energy is known as the **wave-particle duality.**

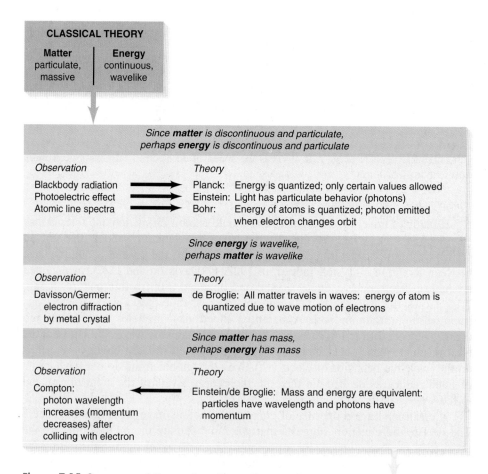

Figure 7.15 Summary of the major observations and theories leading from classical theory to quantum theory. As often happens in science, an observation (experiment) stimulates the need for an explanation (theory), and/or a theoretical insight provides the impetus for an experimental test.

The Heisenberg Uncertainty Principle

In the classical view of the world, a moving particle has a definite location at any instant, whereas a wave is spread out in space. If an electron has the properties of *both* a particle and a wave, what can we determine about its position in the atom? In 1927, the German physicist Werner Heisenberg postulated the **uncertainty principle,** which states that it is impossible to know the exact position *and* momentum (mass times speed) of a particle simultaneously. For a particle with constant mass m, the principle is expressed mathematically as

$$\Delta x \cdot m\Delta u \geq \frac{h}{4\pi} \tag{7.6}$$

where Δx is the uncertainty in position and Δu is the uncertainty in speed. The more accurately we know the position of the particle (smaller Δx), the less accurately we know its speed (larger Δu), and vice versa.

By knowing the position and speed of a pitched baseball and using the classical laws of motion, we can predict its trajectory and whether it will be a strike or a ball. For a baseball, Δx and Δu are insignificant because its mass is enormous compared with $h/4\pi$. Knowing the position and speed of an electron, and from them its trajectory, is another situation entirely. For example, if we take an electron's speed as 6×10^6 m/s $\pm$ 1%, then Δu in Equation 7.6 is 6×10^4 m/s, and the uncertainty in the electron's position (Δx) is 10^{-9} m, which is about 10 times greater than the diameter of the entire atom (10^{-10} m)! Therefore, we have no precise idea where in the atom the electron is located.

The uncertainty principle has profound implications for an atomic model. It means that *we cannot assign fixed paths for electrons,* such as the circular orbits of Bohr's model. As you'll see next, the most we can ever hope to know is the *probability*—the odds—of finding an electron in a given region of space; but we are not *sure* it is there any more than a gambler is sure of the next roll of the dice.

SECTION SUMMARY
As a result of Planck's quantum theory and Einstein's relativity theory, we do not view matter and energy as distinct entities. The de Broglie wavelength expresses the idea that electrons (and all matter) have wavelike motion. Allowed atomic energy levels are related to allowed wavelengths of the electron's motion. Electrons exhibit diffraction patterns, as do waves of energy, and photons exhibit transfer of momentum, as do particles of mass. The wave-particle duality of matter and energy is observable only on the atomic scale. According to the uncertainty principle, we cannot know simultaneously the exact position and speed of an electron.

7.4   THE QUANTUM-MECHANICAL MODEL OF THE ATOM

Acceptance of the dual nature of matter and energy and of the uncertainty principle culminated in the field of **quantum mechanics,** which examines the wave nature of objects on the atomic scale. In 1926, Erwin Schrödinger derived an equation that is the basis for the quantum-mechanical model of the hydrogen atom. The model describes an atom that has certain allowed quantities of energy due to the allowed frequencies of an electron whose behavior is wavelike and whose exact location is impossible to know.

The Atomic Orbital and the Probable Location of the Electron

The electron's matter-wave occupies the three-dimensional space near the nucleus and experiences a continuous, but varying, influence from the nuclear charge. The **Schrödinger equation** is quite complex but is represented as

$$\mathcal{H}\psi = E\psi$$

where E is the energy of the atom. The symbol ψ (Greek *psi*, pronounced "sigh") is called a **wave function,** a mathematical description of the electron's matter-wave in terms of position in three dimensions. The symbol $\mathscr{H}$, called the Hamiltonian operator, represents a set of mathematical operations that, when carried out on a particular ψ, yields an allowed energy value.*

Each solution to the equation (that is, each energy state of the atom) is associated with a given wave function, also called an **atomic orbital.** It's important to keep in mind that an "orbital" in the quantum-mechanical model *bears no resemblance* to an "orbit" in the Bohr model: an *orbit* was, supposedly, an electron's path around the nucleus, whereas an *orbital* is a mathematical function with no direct physical meaning.

We cannot know precisely where the electron is at any moment, but we can describe where it *probably* is, that is, where it is most likely to be found, or where it spends most of its time. Although the wave function (atomic orbital) has no direct physical meaning, the square of the wave function, ψ^2, is the *probability density,* a measure of the probability that the electron can be found within a particular tiny volume of the atom. (Whereas ψ can have positive or negative values, ψ^2 is always positive, which makes sense for a value that expresses a probability.) For a given energy level, we can depict this probability with an *electron probability density diagram,* or more simply, an **electron density diagram.** In Figure 7.16A, the value of ψ^2 for a given volume is represented pictorially by a certain density of dots: the greater the density of dots, the higher the probability of finding the electron within that volume.

Electron density diagrams are sometimes called **electron cloud** representations. If we *could* take a time-exposure photograph of the electron in wavelike motion around the nucleus, it would appear as a "cloud" of electron positions. The electron cloud is an *imaginary* picture of the electron changing its position rapidly over time; it does *not* mean that an electron is a diffuse cloud of charge. Note that *the electron probability density decreases with distance from the nucleus* along a line, r. The same concept is shown graphically in the plot of ψ^2 vs. r in Figure 7.16B. Note that due to the thickness of the printed line, the curve touches the axis; nevertheless, *the probability of the electron being far from the nucleus is very small, but not zero.*

The *total* probability of finding the electron at any distance r from the nucleus is also important. To find this, we mentally divide the volume around the nucleus into thin, concentric, spherical layers, like the layers of an onion (shown in cross section in Figure 7.16C), and ask in which *spherical layer* we are most likely to find the electron. This is the same as asking for the *sum of ψ^2 values* within each spherical layer. The steep falloff in probability density with distance (see Figure 7.16B) has an important effect. Near the nucleus, the volume of each layer increases faster than its probability density decreases. As a result, the *total* probability of finding the electron in the second layer is higher than in the first. Electron density drops off so quickly, however, that this effect soon diminishes with greater distance. Thus, even though the volume of each layer continues to increase, the total probability for a given layer gradually decreases. Because of these opposing effects of decreasing probability density and increasing layer volume, the total probability peaks in a layer some distance from the nucleus. Figure 7.16D shows this as a **radial probability distribution plot.**

---

*The complete form of the Schrödinger equation in terms of the three linear axes is

$$\left[-\frac{h^2}{8\pi^2 m_e}\left(\frac{d^2}{dx^2} + \frac{d^2}{dy^2} + \frac{d^2}{dz^2}\right) + V(x,y,z)\right]\psi(x,y,z) = E\psi(x,y,z)$$

where ψ is the wave function; m_e is the electron's mass; E is the total quantized energy of the atomic system; and V is the potential energy at point (x,y,z).

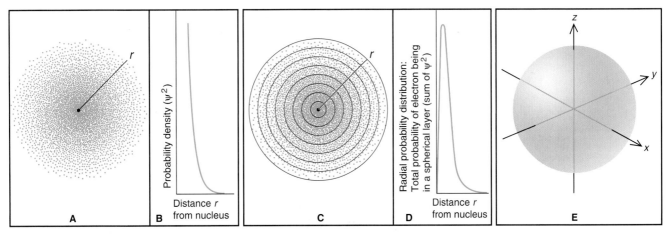

Figure 7.16 Electron probability density in the ground-state H atom. A, An electron density diagram shows a cross section of the H atom. The dots, each representing the probability of the electron being within a tiny volume, decrease along a line outward from the nucleus. **B,** A plot of the data in **A** shows that the probability density (ψ^2) decreases with distance from the nucleus but does not reach zero (the thickness of the line makes it appear to do so). **C,** Dividing the atom's volume into thin, concentric, spherical layers (shown in cross section) and counting the dots within each layer gives the total probability of finding the electron within that layer. **D,** A radial probability distribution plot shows total electron density in each spherical layer vs. r. Because electron density decreases more slowly than the volume of each concentric layer increases, the plot shows a peak. **E,** A 90% probability contour shows the ground state of the H atom (orbital of lowest energy) and represents the volume in which the electron spends 90% of its time.

As an analogy for the behavior shown in Figure 7.16D, picture fallen apples around the base of an apple tree: the density of apples is greatest near the trunk and decreases with distance (Figure 7.17A). Divide the ground under the tree into foot-wide concentric rings and collect the apples within each ring. Apple density is greatest in the first ring, but the area of the second ring is larger, and so it contains a greater *total* number of apples. Farther out near the edge of the tree, rings have more area but lower apple "density," so the total number of apples decreases. In Figure 7.17B, a plot of "number of apples within a ring" vs. "distance of ring from trunk" shows a peak at some distance from the trunk, as in Figure 7.16D.

The peak of the radial probability distribution for the ground-state H atom appears at the same distance from the nucleus (0.529Å, or 5.29×10^{-11} m) as Bohr postulated for the closest orbit. Thus, at least for the ground state, the Schrödinger model predicts that the electron spends *most* of its time at the same distance that the Bohr model predicted it spent *all* of its time. The difference between "most" and "all" reflects the uncertainty of the electron's location in the Schrödinger model.

How far away from the nucleus can we find the electron? This is the same as asking "How large is the atom?" Recall from Figure 7.16B that the probability of finding the electron far from the nucleus is not zero. Therefore, we *cannot* assign a definite volume to an atom. However, we often visualize atoms with a 90% **probability contour,** such as in Figure 7.16E, which shows the volume within which the electron of the hydrogen atom spends 90% of its time.

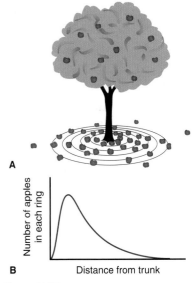

Figure 7.17 A radial probability distribution of apples.

Quantum Numbers of an Atomic Orbital

So far we have discussed the electron density for the *ground* state of the H atom. When the atom absorbs energy, it exists in an *excited* state and the region of space occupied by the electron is described by a different atomic orbital (wave function). As you'll see, each atomic orbital has a distinctive radial probability distribution and 90% probability contour.

An atomic orbital is specified by three quantum numbers. One is related to the orbital's size, another to its shape, and the third to its orientation in space.* The quantum numbers have a hierarchical relationship: the size-related number limits the shape-related number, which limits the orientation-related number. Let's examine this hierarchy and then look at the shapes and orientations.

1. The **principal quantum number (n)** *is a positive integer* (1, 2, 3, and so forth). It indicates the relative *size* of the orbital and therefore the relative *distance from the nucleus* of the peak in the radial probability distribution plot. The principal quantum number specifies the *energy level* of the H atom: *the higher the n value, the higher the energy level.* When the electron occupies an orbital with $n = 1$, the H atom is in its ground state and has lower energy than when the electron occupies the $n = 2$ orbital (first excited state).

2. The **angular momentum quantum number (l)** *is an integer from 0 to $n - 1$.* It is related to the *shape* of the orbital and is sometimes called the *orbital-shape (or azimuthal) quantum number.* Note that the principal quantum number sets a limit on the values for the angular momentum quantum number; that is, n limits l. For an orbital with $n = 1$, l can have a value of only 0. For orbitals with $n = 2$, l can have a value of 0 or 1; for those with $n = 3$, l can be 0, 1, or 2; and so forth. Note that the number of possible l values equals the value of n.

3. The **magnetic quantum number (m_l)** *is an integer from $-l$ through 0 to $+l$.* It prescribes the *orientation* of the orbital in the space around the nucleus and is sometimes called the *orbital-orientation quantum number.* The possible values of an orbital's magnetic quantum number are set by its angular momentum quantum number; that is, l sets the possible values of m_l. An orbital with $l = 0$ can have only $m_l = 0$. However, an orbital with $l = 1$ can have any one of three m_l values, -1, 0, or $+1$; thus, there are three possible orbitals with $l = 1$, each with its own orientation. Note that the number of possible m_l values *equals* the number of orbitals, which is $2l + 1$ for a given l value.

Table 7.2 summarizes the hierarchy among the three quantum numbers. (In Chapter 8, we'll discuss a fourth quantum number that relates to a property of the electron itself.) The total number of orbitals for a given n value is n^2.

---

*For ease in discussion, we refer to the size, shape, and orientation of an "atomic orbital," although we really mean the size, shape, and orientation of an "atomic orbital's radial probability distribution." This usage is common in both introductory and advanced texts.

Table 7.2 **The Hierarchy of Quantum Numbers for Atomic Orbitals**

Name, Symbol (Property)	Allowed Values	Quantum Numbers									
Principal, n (size, energy)	Positive integer (1, 2, 3, . . .)	1	2			3					
Angular momentum, l (shape)	0 to $n - 1$	0	0	1		0	1		2		
Magnetic, m_l (orientation)	$-l, \ldots, 0, \ldots, +l$	0	0	$-1\ 0\ +1$		0	$-1\ 0\ +1$		$-2\ -1\ 0\ +1\ +2$		

SAMPLE PROBLEM 7.4 Determining Quantum Numbers for an Energy Level

Problem What values of the angular momentum (l) and magnetic (m_l) quantum numbers are allowed for a principal quantum number (n) of 3? How many orbitals exist for $n = 3$?

Plan We determine allowable quantum numbers with the rules from the text: l values are integers from 0 to $n - 1$, and m_l values are integers from $-l$ to 0 to $+l$. One m_l value is assigned to each orbital, so the number of m_l values gives the number of orbitals.

Solution Determining l values: for $n = 3$, $l = 0, 1, 2$

Determining m_l for each l value:

$$\text{For } l = 0, \quad m_l = 0$$
$$\text{For } l = 1, \quad m_l = -1, 0, +1$$
$$\text{For } l = 2, \quad m_l = -2, -1, 0, +1, +2$$

There are nine m_l values, so there are nine orbitals with $n = 3$.

Check Table 7.2 shows that we are correct. The total number of orbitals for a given n value is n^2, and for $n = 3$, $n^2 = 9$.

FOLLOW-UP PROBLEM 7.4 Specify the l and m_l values for $n = 4$.

The energy states and orbitals of the atom are described with specific terms and associated with one or more quantum numbers:

1. *Level.* The atom's energy **levels,** or *shells,* are given by the n value: the smaller the n value, the lower the energy level and the greater the probability of the electron being closer to the nucleus.

2. *Sublevel.* The atom's levels contain **sublevels,** or *subshells,* which designate the orbital shape. Each sublevel has a letter designation:

$l = 0$ is an *s* sublevel.

$l = 1$ is a *p* sublevel.

$l = 2$ is a *d* sublevel.

$l = 3$ is an *f* sublevel.

(The letters derive from the names of spectroscopic lines: *s*harp, *p*rincipal, *dif*fuse, and *f*undamental. Sublevels with l values greater than 3 are designated alphabetically: *g* sublevel, *h* sublevel, etc.) Sublevels are named by joining the n value and the letter designation. For example, the sublevel (subshell) with $n = 2$ and $l = 0$ is called the 2*s* sublevel.

3. *Orbital.* Each allowed combination of n, l, and m_l values specifies one of the atom's *orbitals.* Thus, the three quantum numbers that describe an orbital express its size (energy), shape, and spatial orientation. You can easily give the quantum numbers of the orbitals in any sublevel if you know the sublevel letter designation and the quantum number hierarchy. For example, the 2*s* sublevel has only one orbital, and its quantum numbers are $n = 2$, $l = 0$, and $m_l = 0$. The 3*p* sublevel has three orbitals: one with $n = 3$, $l = 1$, and $m_l = -1$; another with $n = 3$, $l = 1$, and $m_l = 0$; and a third with $n = 3$, $l = 1$, and $m_l = +1$.

SAMPLE PROBLEM 7.5 Determining Sublevel Names and Orbital Quantum Numbers

Problem Give the name, magnetic quantum numbers, and number of orbitals for each sublevel with the given quantum numbers:

(a) $n = 3, l = 2$ **(b)** $n = 2, l = 0$ **(c)** $n = 5, l = 1$ **(d)** $n = 4, l = 3$

Plan To name the sublevel (subshell), we combine the n value and l letter designation. We know l, so we can find the possible m_l values, whose total number equals the number of orbitals.

Solution

	n	l	Sublevel Name	Possible m_l Values	No. of Orbitals
(a)	3	2	3d	$-2, -1, 0, +1, +2$	5
(b)	2	0	2s	0	1
(c)	5	1	5p	$-1, 0, +1$	3
(d)	4	3	4f	$-3, -2, -1, 0, +1, +2, +3$	7

Check Check the number of orbitals in each sublevel using

$$\text{No. of orbitals} = \text{no. of } m_l \text{ values} = 2l + 1$$

FOLLOW-UP PROBLEM 7.5 What are the n, l, and possible m_l values for the 2p and 5f sublevels?

SAMPLE PROBLEM 7.6   Identifying Incorrect Quantum Numbers

Problem What is wrong with each of the following quantum number designations and/or sublevel names?

	n	l	m_l	Name
(a)	1	1	0	1p
(b)	4	3	+1	4d
(c)	3	1	-2	3p

Solution **(a)** A sublevel with $n = 1$ can have only $l = 0$, not $l = 1$. The only possible sublevel name is 1s.
(b) A sublevel with $l = 3$ is an f sublevel, not a d sublevel. The name should be 4f.
(c) A sublevel with $l = 1$ can have only m_l of $-1, 0, +1$, not -2.
Check Check that l is always less than n, and m_l is always $\geq -l$ and $\leq +l$.

FOLLOW-UP PROBLEM 7.6 Supply the missing quantum numbers and sublevel names.

	n	l	m_l	Name
(a)	?	?	0	4p
(b)	2	1	0	?
(c)	3	2	-2	?
(d)	?	?	?	2s

Shapes of Atomic Orbitals

Each sublevel of the H atom consists of a set of orbitals with characteristic shapes. As you'll see in Chapter 8, orbitals for the other atoms have similar shapes.

The s Orbital An orbital with $l = 0$ has a *spherical* shape with the nucleus at its center and is called an *s* **orbital.** The H atom's ground state, for example, has the electron in the 1s orbital, and *the electron probability density is highest at the nucleus.* Figure 7.18A shows this fact graphically *(top)*, and an electron density *relief map (inset)* depicts this curve in three dimensions. The quarter-section of an electron cloud representation *(middle)* has the darkest shading at the nucleus. On the other hand, the radial probability distribution plot *(bottom)*, which represents the probability of finding the electron (that is, locates where the electron spends most of its time), is highest slightly out from the nucleus. Both plots fall off smoothly with distance.

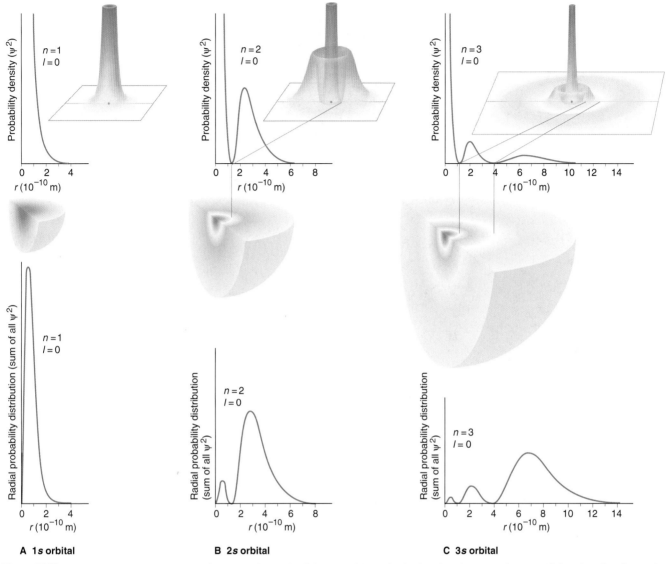

A 1s orbital

B 2s orbital

C 3s orbital

Figure 7.18 The 1s, 2s, and 3s orbitals. Information for each of the s orbitals is shown as a plot of probability density vs. distance (*top,* with the relief map (*inset*) showing the plot in three dimensions); as an electron cloud representation (*middle*), in which shading coincides with peaks in the plot above; and as a radial probability distribution (*bottom*) that shows where the electron spends its time. **A,** The 1s orbital. **B,** The 2s orbital. **C,** The 3s orbital. Nodes (regions of zero probability) appear in the 2s and 3s orbitals.

The 2s orbital (Figure 7.18B) has two regions of higher electron density. The radial probability distribution (Figure 7.18B, *bottom*) of the more distant region is *higher* than that of the closer one because the sum of its ψ^2 is taken over a much larger volume. Between the two regions is a spherical **node,** a shell-like region where the probability drops to zero ($\psi^2 = 0$ at the node, analogous to zero amplitude of a wave). Because the 2s orbital is larger than the 1s, an electron in the 2s spends more time *farther* from the nucleus than when it occupies the 1s.

The 3s orbital, shown in Figure 7.18C, has three regions of high electron density and two nodes. Here again, the highest radial probability is at the greatest distance from the nucleus because the sum of all ψ^2 is taken over a larger volume. This pattern of more nodes and higher probability with distance continues

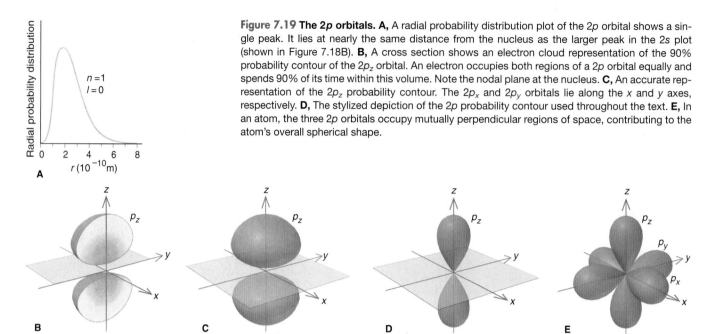

Figure 7.19 The 2p orbitals. A, A radial probability distribution plot of the 2p orbital shows a single peak. It lies at nearly the same distance from the nucleus as the larger peak in the 2s plot (shown in Figure 7.18B). **B,** A cross section shows an electron cloud representation of the 90% probability contour of the $2p_z$ orbital. An electron occupies both regions of a 2p orbital equally and spends 90% of its time within this volume. Note the nodal plane at the nucleus. **C,** An accurate representation of the $2p_z$ probability contour. The $2p_x$ and $2p_y$ orbitals lie along the x and y axes, respectively. **D,** The stylized depiction of the 2p probability contour used throughout the text. **E,** In an atom, the three 2p orbitals occupy mutually perpendicular regions of space, contributing to the atom's overall spherical shape.

for s orbitals of higher n value. An s orbital has a spherical shape, so it can have only one orientation and, thus, only one value for the magnetic quantum number: for any s orbital, $m_l = 0$.

The p Orbital An orbital with $l = 1$, called a **p orbital,** has two regions (lobes) of high probability, one on *either side* of the nucleus. Thus, as you can see in Figure 7.19, the *nucleus lies at the nodal plane* of this dumbbell-shaped orbital. The maximum value of l is $n - 1$, so only levels with $n = 2$ or higher can have a p orbital. Therefore, the lowest energy p orbital (the one closest to the nucleus) is the 2p. Keep in mind that *one p orbital consists of both lobes* and that the electron spends *equal* time in both. Similar to the pattern for s orbitals, a 3p orbital is larger than a 2p orbital, a 4p orbital is larger than a 3p orbital, and so forth.

Unlike an s orbital, each p orbital *does* have a specific orientation in space. The $l = 1$ value has three possible m_l values: -1, 0, and $+1$, which refer to three *mutually perpendicular* p orbitals. They are identical in size, shape, and energy, differing only in orientation. For convenience, we associate p orbitals with the x, y, and z axes (but there is no necessary relation between a spatial axis and a given m_l value): the p_x orbital lies along the x axis, the p_y along the y axis, and the p_z along the z axis.

The d Orbital An orbital with $l = 2$ is called a **d orbital.** There are five possible m_l values for the $l = 2$ value: -2, -1, 0, $+1$, and $+2$. Thus, a d orbital can have any one of five different orientations, as shown in Figure 7.20. Four of the five d orbitals have four lobes (a cloverleaf shape) prescribed by two mutually perpendicular nodal planes, with the nucleus lying at the junction of the lobes. Three of these orbitals lie in the mutually perpendicular xy, xz, and yz planes, with their lobes *between* the axes, and are called the d_{xy}, d_{xz}, and d_{yz} orbitals. A fourth, the $d_{x^2-y^2}$ orbital, also lies in the xy plane, but its lobes are directed *along* the axes. The fifth d orbital, the d_{z^2}, has a different shape: two major lobes lie along the z axis, and a donut-shaped region girdles the center. An electron associated with a given d orbital has equal probability of being in any of the orbital's lobes.

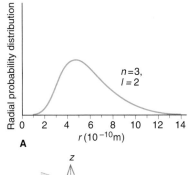

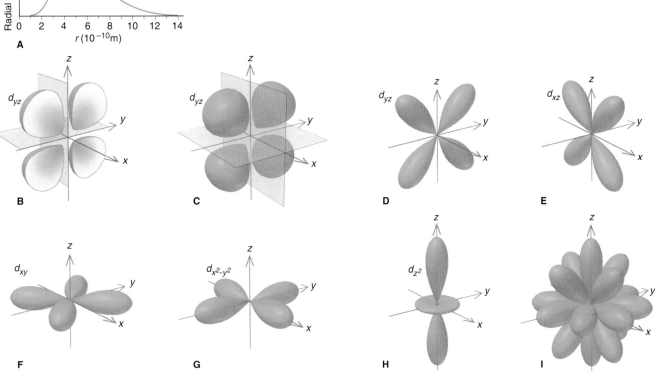

Figure 7.20 The 3*d* orbitals. A, A radial probability distribution plot. **B,** An electron cloud representation of the $3d_{yz}$ orbital in cross section. Note the mutually perpendicular nodal planes and the lobes lying *between* the axes. **C,** An accurate representation of the $3d_{yz}$ orbital probability contour. **D,** The stylized depiction of the $3d_{yz}$ orbital used throughout the text. **E,** The $3d_{xz}$ orbital. **F,** The $3d_{xy}$ orbital. **G,** The lobes of the $3d_{x^2-y^2}$ orbital lie *on* the x and y axes. **H,** The $3d_{z^2}$ orbital has two lobes and a central, donut-shaped region. **I,** A composite of the five 3*d* orbitals, which again contributes to an atom's overall spherical shape.

As we said for the *p* orbitals, an axis designation for a *d* orbital is *not* associated with a given m_l value. In keeping with the quantum number rules, a *d* orbital ($l = 2$) must have a principal quantum number of $n = 3$ or greater. The 4*d* orbitals extend farther from the nucleus than the 3*d* orbitals, and the 5*d* orbitals extend still farther.

Orbitals with Higher *l* Values Orbitals with $l = 3$ are *f* orbitals and must have a principal quantum number of at least $n = 4$. There are seven *f* orbitals ($2l + 1 = 7$), each with a complex, multilobed shape; Figure 7.21 shows one of them. Orbitals with $l = 4$ are *g* orbitals, but we will not discuss them further because they play no known role in chemical bonding.

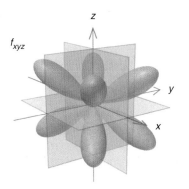

Figure 7.21 One of the seven possible 4*f* orbitals. The $4f_{xyz}$ orbital has eight lobes and three nodal planes. The other six 4*f* orbitals also have multilobed contours.

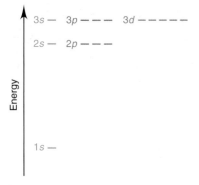

Figure 7.22 The energy levels in the H atom. The H atom is the only atom in which the energy level depends *only* on the *n* value of the sublevels. For example, the 2*s* and the three 2*p* sublevels (shown as short lines) all have the same energy.

The Special Case of the Hydrogen Atom

The energy state of the H atom depends only on the principal quantum number n because there is only one electron. If the electron occupies an orbital with a higher *n* value, it occurs (on average) farther from the nucleus, so it is higher in energy. In other words, *in the special case of the H atom,* all four *n* = 2 orbitals (one 2*s* and three 2*p*) have the same energy, and all nine *n* = 3 orbitals (one 3*s*, three 3*p*, and five 3*d*) have the same energy (Figure 7.22). Of course, atoms of all other elements have more than one electron; as you'll see in Chapter 8, the additional attractions and repulsions that result make their energy states depend on *both* the *n* and *l* values of the occupied orbitals.

SECTION SUMMARY

The electron's wave function (ψ, atomic orbital) is a mathematical description of the electron's wavelike behavior in an atom. Each wave function is associated with one of the atom's allowed energy states. The probability density of finding the electron at a particular location is represented by ψ^2. An electron density diagram and a radial probability distribution plot show how the electron occupies the space near the nucleus for a particular energy level. Three features of an atomic orbital are described by quantum numbers: size (*n*), shape (*l*), and orientation (m_l). Orbitals with the same *n* and *l* values constitute a sublevel; sublevels with the same *n* value constitute an energy level. A sublevel with *l* = 0 has a spherical (*s*) orbital; a sublevel with *l* = 1 has three, two-lobed (*p*) orbitals; and a sublevel with *l* = 2 has five, multilobed (*d*) orbitals. In the special case of the H atom, the energy levels depend on the *n* value only.

For Review and Reference (Numbers in parentheses refer to pages, unless noted otherwise.)

Learning Objectives

To help you review these learning objectives, the numbers of related sections (§), sample problems (SP), and end-of-chapter problems (EP) are listed in parentheses.

1. Describe the relationships among frequency, wavelength, and energy of light, and know the meaning of amplitude; have a general understanding of the electromagnetic spectrum (§ 7.1) (SPs 7.1, 7.2) (EPs 7.1, 7.2, 7.6–7.13)
2. Understand how particles and waves differ and how the work of Planck (quantization of energy) and Einstein (photon theory) changed thinking about it (§ 7.1) (EPs 7.3–7.5)
3. Explain the Bohr theory and the importance of discrete atomic energy levels (§ 7.2) (EPs 7.14–7.26)

4. Describe the wave-particle duality of matter and energy and the theories and experiments that led to it (particle wavelength, electron diffraction, photon momentum, uncertainty principle) (§ 7.3) (SP 7.3) (EPs 7.27–7.34)
5. Distinguish between ψ (wave function) and ψ^2 (probability density); understand the meaning of electron density diagrams and radial probability distribution plots; describe the hierarchy of quantum numbers, the hierarchy of levels, sublevels, and orbitals, and the shapes and nodes of *s*, *p*, and *d* orbitals; and determine quantum numbers and sublevel designations (§ 7.4) (SPs 7.4–7.6) (EPs 7.35–7.47)

Key Terms

Section 7.1
electromagnetic radiation (206)
frequency (ν) (206)
wavelength (λ) (206)
speed of light (*c*) (207)
amplitude (207)
electromagnetic spectrum (207)
infrared (IR) (207)
ultraviolet (UV) (207)
refraction (208)
diffraction (208)
quantum number (210)
Planck's constant (*h*) (210)

quantum (210)
photoelectric effect (210)
photon (211)

Section 7.2
line spectrum (212)
stationary state (213)
ground state (214)
excited state (214)

spectrophotometry (217)
emission spectrum (217)
absorption spectrum (217)

Section 7.3
de Broglie wavelength (219)
wave-particle duality (220)
uncertainty principle (221)

Section 7.4

quantum mechanics (221)
Schrödinger equation (221)
wave function (atomic orbital) (222)

electron density diagram (222)
electron cloud (222)
radial probability distribution plot (222)
probability contour (223)

principal quantum number (n) (224)
angular momentum quantum number (l) (224)
magnetic quantum number (m_l) (224)

level (shell) (225)
sublevel (subshell) (225)
s orbital (226)
node (227)
p orbital (228)
d orbital (228)

Key Equations and Relationships

7.1 Relating the speed of light to its frequency and wavelength (207):
$$c = \nu \times \lambda$$
7.2 Determining the smallest change in an atom's energy (210):
$$\Delta E = h\nu$$
7.3 Calculating the wavelength of any line in the H atom spectrum (Rydberg equation) (213):
$$\frac{1}{\lambda} = R\left(\frac{1}{n_1^2} - \frac{1}{n_2^2}\right)$$
where n is a positive integer and $n_2 > n_1$

7.4 Finding the difference between two energy levels in the H atom (216):
$$\Delta E = E_{\text{final}} - E_{\text{initial}} = -2.18\times10^{-18} \text{ J}\left(\frac{1}{n_{\text{final}}^2} - \frac{1}{n_{\text{initial}}^2}\right)$$
7.5 Calculating the wavelength of any moving particle (de Broglie wavelength) (218):
$$\lambda = \frac{h}{mu}$$
7.6 Finding the uncertainty in position or speed of a particle (Heisenberg uncertainty principle) (221):
$$\Delta x \cdot m\Delta u \geq \frac{h}{4\pi}$$

Brief Solutions to Follow-up Problems

7.1 λ (nm) $= \dfrac{3.00\times10^8 \text{ m/s}}{7.23\times10^{14} \text{ s}^{-1}} \times \dfrac{10^9 \text{ nm}}{1 \text{ m}} = 415$ nm

λ (Å) $= 415$ nm $\times \dfrac{10 \text{ Å}}{1 \text{ nm}} = 4150$ Å

7.2 UV: $E = hc/\lambda$
$= \dfrac{(6.626\times10^{-34} \text{ J·s})(3.00\times10^8 \text{ m/s})}{1\times10^{-8} \text{ m}} = 2\times10^{-17}$ J

Visible: $E = 4\times10^{-19}$ J; IR: $E = 2\times10^{-21}$ J
As λ increases, E decreases.

7.3 $u = \dfrac{h}{m\lambda} = \dfrac{6.626\times10^{-34} \text{ kg·m}^2/\text{s}}{(9.11\times10^{-31} \text{ kg})\left(100 \text{ nm} \times \dfrac{1 \text{ m}}{10^9 \text{ nm}}\right)}$
$= 7.27\times10^3$ m/s

7.4 $n = 4$, so $l = 0, 1, 2, 3$. In addition to the m_l values in Sample Problem 7.4, we have those for $l = 3$:
$$m_l = -3, -2, -1, 0, +1, +2, +3$$
7.5 For $2p$: $n = 2, l = 1, m_l = -1, 0, +1$
For $5f$: $n = 5, l = 3, m_l = -3, -2, -1, 0, +1, +2, +3$
7.6 (a) $n = 4, l = 1$; (b) name is $2p$; (c) name is $3d$;
(d) $n = 2, l = 0, m_l = 0$

Problems

Problems with **colored** numbers are answered in Appendix E. Sections match the text and provide the numbers of relevant sample problems. Bracketed problems are grouped in pairs (indicated by a short rule) that cover the same concept. Comprehensive Problems are based on material from any section or previous chapter.

The Nature of Light
(Sample Problems 7.1 and 7.2)

7.1 In what ways are microwave and ultraviolet radiation the same? In what ways are they different?

7.2 Consider the following types of electromagnetic radiation:
(1) microwave   (2) ultraviolet   (3) radio waves
(4) infrared   (5) x-ray   (6) visible
(a) Arrange them in order of increasing wavelength.
(b) Arrange them in order of increasing frequency.
(c) Arrange them in order of increasing energy.

7.3 In the mid-17th century, Isaac Newton proposed that light existed as a stream of particles, and the wave-particle debate continued for over 250 years until Planck and Einstein presented their revolutionary ideas. Give two pieces of evidence for the wave model and two for the particle model.

7.4 What new idea about energy did Planck use to explain blackbody radiation?

7.5 What new idea about light did Einstein use to explain the photoelectric effect? Why does the photoelectric effect exhibit a threshold frequency? Why does it *not* exhibit a time lag?

7.6 An AM station broadcasts rock music at "960 on your radio dial." Units for AM frequencies are given in kilohertz (kHz). Find the wavelength of the station's radio waves in meters (m), nanometers (nm), and angstroms (Å).

7.7 An FM station broadcasts classical music at 93.5 MHz (megahertz, or 10^6 Hz). Find the wavelength (in m, nm, and Å) of these radio waves.

7.8 A radio wave has a frequency of 3.6×10^{10} Hz. What is the energy (in J) of one photon of this radiation?

7.9 An x-ray has a wavelength of 1.3 Å. Calculate the energy (in J) of one photon of this radiation.

7.10 Rank the following photons in terms of increasing energy: (a) blue ($\lambda = 453$ nm); (b) red ($\lambda = 660$ nm); (c) yellow ($\lambda = 595$ nm).

7.11 Rank the following photons in terms of decreasing energy: (a) IR ($\nu = 6.5 \times 10^{13}$ s^{-1}); (b) microwave ($\nu = 9.8 \times 10^{11}$ s^{-1}); (c) UV ($\nu = 8.0 \times 10^{15}$ s^{-1}).

7.12 Cobalt-60 is a radioactive isotope used to treat cancers of the brain and other tissues. A gamma ray emitted by an atom of this isotope has an energy of 1.33 MeV (million electron volts; 1 eV = 1.602×10^{-19} J). What is the frequency (in Hz) and the wavelength (in m) of this gamma ray?

7.13 (a) The first step in the formation of ozone in the upper atmosphere occurs when oxygen molecules absorb UV radiation of wavelengths ≤ 242 nm. Calculate the frequency and energy of the least energetic of these photons.
(b) Ozone absorbs light having wavelengths of 2200 to 2900 Å, thus protecting organisms on Earth's surface from this high-energy UV radiation. What are the frequency and energy of the most energetic of these photons?

Atomic Spectra

7.14 How is n_1 in the Rydberg equation (Equation 7.3) related to the quantum number n in the Bohr model?

7.15 Distinguish between an absorption spectrum and an emission spectrum. With which did Bohr work?

7.16 Which of these electron transitions correspond to absorption of energy and which to emission?
(a) $n = 2$ to $n = 4$ 　　　　(b) $n = 3$ to $n = 1$
(c) $n = 5$ to $n = 2$ 　　　　(d) $n = 3$ to $n = 4$

7.17 Why could the Bohr model not predict line spectra for atoms other than hydrogen?

7.18 Use the Rydberg equation (Equation 7.3) to calculate the wavelength (in nm) of the photon emitted when a hydrogen atom undergoes a transition from $n = 5$ to $n = 2$.

7.19 Use the Rydberg equation to calculate the wavelength (in Å) of the photon absorbed when a hydrogen atom undergoes a transition from $n = 1$ to $n = 3$.

7.20 Calculate the energy difference (ΔE) for the transition in Problem 7.18 for 1 mol of H atoms.

7.21 Calculate the energy difference (ΔE) for the transition in Problem 7.19 for 1 mol of H atoms.

7.22 Arrange the following H atom electron transitions in order of *increasing* frequency of the photon absorbed or emitted:
(a) $n = 2$ to $n = 4$ 　　　　(b) $n = 2$ to $n = 1$
(c) $n = 2$ to $n = 5$ 　　　　(d) $n = 4$ to $n = 3$

7.23 Arrange the following H atom electron transitions in order of *decreasing* wavelength of the photon absorbed or emitted:
(a) $n = 2$ to $n = \infty$ 　　　(b) $n = 4$ to $n = 20$
(c) $n = 3$ to $n = 10$ 　　　(d) $n = 2$ to $n = 1$

7.24 The electron in a ground-state H atom absorbs a photon of wavelength 97.20 nm. To what energy level does the electron move?

7.25 An electron in the $n = 5$ level of an H atom emits a photon of wavelength 1281 nm. To what energy level does the electron move?

7.26 In addition to continuous radiation, fluorescent lamps emit sharp lines in the visible region from a mercury discharge within the tube. Much of this light has a wavelength of 436 nm. What is the energy (in J) of one photon of this light?

The Wave-Particle Duality of Matter and Energy
(Sample Problem 7.3)

7.27 If particles have wavelike motion, why don't we observe that motion in the macroscopic world?

7.28 Why can't we overcome the uncertainty predicted by Heisenberg's principle by building more precise devices to reduce the error in measurements below the $h/4\pi$ limit?

7.29 A 220-lb fullback runs the 40-yd dash at a speed of 19.6 ± 0.1 mi/h. What is his de Broglie wavelength (in meters)?

7.30 An alpha particle (mass = 6.6×10^{-24} g) emitted by radium travels at $3.4 \times 10^7 \pm 0.1 \times 10^7$ mi/h. What is its de Broglie wavelength (in meters)?

7.31 How fast must a 56.5-g tennis ball travel in order to have a de Broglie wavelength that is equal to that of a photon of green light (5400 Å)?

7.32 How fast must a 142-g baseball travel in order to have a de Broglie wavelength that is equal to that of an x-ray photon with $\lambda = 100.$ pm?

7.33 A sodium flame has a characteristic yellow color due to emissions of wavelength 589 nm. What is the mass equivalence of one photon of this wavelength (1 J = 1 kg·m^2/s^2)?

7.34 A lithium flame has a characteristic red color due to emissions of wavelength 671 nm. What is the mass equivalence of 1 mol of photons of this wavelength (1 J = 1 kg·m^2/s^2)?

The Quantum-Mechanical Model of the Atom
(Sample Problems 7.4 to 7.6)

7.35 What physical meaning is attributed to the square of the wave function, ψ^2?

7.36 Explain in your own words what the "electron density" in a particular tiny volume of space means.

7.37 What feature of an orbital is related to each of the following quantum numbers?
(a) Principal quantum number (n)
(b) Angular momentum quantum number (l)
(c) Magnetic quantum number (m_l)

7.38 How many orbitals in an atom can have each of the following designations: (a) 1s; (b) 4d; (c) 3p; (d) $n = 3$?

7.39 How many orbitals in an atom can have each of the following designations: (a) 5f; (b) 4p; (c) 5d; (d) $n = 2$?

7.40 Give all possible m_l values for orbitals that have each of the following: (a) $l = 2$; (b) $n = 1$; (c) $n = 4$, $l = 3$.

7.41 Give all possible m_l values for orbitals that have each of the following: (a) $l = 3$; (b) $n = 2$; (c) $n = 6$, $l = 1$.

7.42 For each of the following, give the sublevel designation, the allowable m_l values, and the number of orbitals:
(a) $n = 4$, $l = 2$ 　　(b) $n = 5$, $l = 1$ 　　(c) $n = 6$, $l = 3$

7.43 For each of the following, give the sublevel designation, the allowable m_l values, and the number of orbitals:
(a) $n = 2, l = 0$ (b) $n = 3, l = 2$ (c) $n = 5, l = 1$

7.44 For each of the following sublevels, give the n and l values and the number of orbitals: (a) $5s$; (b) $3p$; (c) $4f$.

7.45 For each of the following sublevels, give the n and l values and the number of orbitals: (a) $6g$; (b) $4s$; (c) $3d$.

7.46 Are the following quantum number combinations allowed? If not, show two ways to correct them:
(a) $n = 2; l = 0; m_l = -1$ (b) $n = 4; l = 3; m_l = -1$
(c) $n = 3; l = 1; m_l = 0$ (d) $n = 5; l = 2; m_l = +3$

7.47 Are the following quantum number combinations allowed? If not, show two ways to correct them:
(a) $n = 1; l = 0; m_l = 0$ (b) $n = 2; l = 2; m_l = +1$
(c) $n = 7; l = 1; m_l = +2$ (d) $n = 3; l = 1; m_l = -2$

Comprehensive Problems

Problems with an asterisk (*) are more challenging.

7.48 The quantum-mechanical treatment of the hydrogen atom gives the energy, E, of the electron as a function of the principal quantum number, n:

$$E = -\frac{h^2}{8\pi^2 m_e a_0^2 n^2} \qquad (n = 1, 2, 3, \ldots)$$

where h is Planck's constant, m_e is the electron mass, and a_0 is 52.92×10^{-12} m.

(a) Write the expression in the form $E = -(\text{constant})\dfrac{1}{n^2}$, evaluate the constant (in J), and compare it with the corresponding expression from Bohr's theory.
(b) Use the expression to find ΔE between $n = 2$ and $n = 3$.
(c) Calculate the wavelength of the photon that corresponds to this energy change. Is this photon seen in the hydrogen spectrum obtained from experiment (see Figure 7.7)?

7.49 The photoelectric effect is illustrated in a plot of the kinetic energies of electrons ejected from the surface of potassium metal or silver metal at different frequencies of incident light.

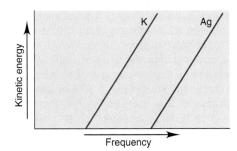

(a) Why don't the lines begin at the origin?
(b) Why don't the lines begin at the same point?
(c) From which metal will light of a shorter wavelength eject an electron?
(d) Why are the slopes of the lines equal?

7.50 The human eye is a complex sensing device for visible light. The optic nerve needs a minimum of 2.0×10^{-17} J of energy to trigger a series of impulses that eventually reach the brain.
(a) How many photons of red light (700. nm) are needed?
(b) How many photons of blue light (475 nm)?

7.51 One reason carbon monoxide (CO) is toxic is that it binds to the blood protein hemoglobin more strongly than oxygen does. The bond between hemoglobin and CO absorbs radiation of 1953 cm^{-1}. (The units are the reciprocal of the wavelength in centimeters.) Calculate the wavelength (in nm and Å) and the frequency (in Hz) of the absorbed radiation.

7.52 A metal ion M^{n+} has a single electron. The highest energy line in its emission spectrum occurs at a frequency of 2.961×10^{16} Hz. Identify the ion.

7.53 TV and radio stations transmit in specific frequency bands of the radio region of the electromagnetic spectrum.
(a) TV channels 2 to 13 (VHF) broadcast signals between the frequencies of 59.5 and 215.8 MHz, whereas FM radio stations broadcast signals with wavelengths between 2.78 and 3.41 m. Do these bands of signals overlap?
(b) AM radio signals have frequencies between 550 and 1600 kHz. Which has a broader transmission band, AM or FM?

7.54 In his explanation of the threshold frequency in the photoelectric effect, Einstein reasoned that the absorbed photon must have the minimum energy required to dislodge an electron from the metal surface. This energy is called the *work function* (ϕ) of that metal. What is the longest wavelength of radiation (in nm) that could cause the photoelectric effect in each of these metals?
(a) Calcium, $\phi = 4.60 \times 10^{-19}$ J
(b) Titanium, $\phi = 6.94 \times 10^{-19}$ J
(c) Sodium, $\phi = 4.41 \times 10^{-19}$ J

7.55 You have three metal samples—A, B, and C—that are tantalum (Ta), barium (Ba), and tungsten (W), but you don't know which is which. Metal A emits electrons in response to visible light; metals B and C require UV light. (a) Identify metal A, and find the longest wavelength that removes an electron. (b) What range of wavelengths would distinguish B and C? [The work functions are Ta (6.81×10^{-19} J), Ba (4.30×10^{-19} J), and W (7.16×10^{-19} J); work function is explained in Problem 7.54.]

7.56 A laser (*light amplification by stimulated emission of radiation*) provides a coherent (in-phase) nearly monochromatic source of high-intensity light. Lasers are used in eye surgery, CD/DVD players, basic research, etc. Some modern dye lasers can be "tuned" to emit a desired wavelength. Fill in the blanks in the following table of the properties of some common lasers:

Type	λ (nm)	ν (s^{-1})	E (J)	Color
He-Ne	632.8	?	?	?
Ar	?	6.148×10^{14}	?	?
Ar-Kr	?	?	3.499×10^{-19}	?
Dye	663.7	?	?	?

7.57 As space exploration increases, means of communication with humans and probes on other planets are being developed.
(a) How much time (in s) does it take for a radio wave of frequency 8.93×10^7 s^{-1} to reach Mars, which is 8.1×10^7 km from Earth? (b) If it takes this radiation 1.2 s to reach the Moon, how far (in m) is the Moon from Earth?

* 7.58 A ground-state H atom absorbs a photon of wavelength 94.91 nm, and its electron attains a higher energy level. The atom then emits two photons: one of wavelength 1281 nm to reach an intermediate level, and a second to reach the ground state.
(a) What higher level did the electron reach?

(b) What intermediate level did the electron reach?

(c) What was the wavelength of the second photon emitted?

* **7.59** Why do the spaces between spectral lines within a series decrease as the wavelength becomes shorter?

7.60 Enormous numbers of microwave photons are needed to warm macroscopic samples of matter. A portion of soup containing 252 g of water is heated in a microwave oven from 20.°C to 98°C, with radiation of wavelength 1.55×10^{-2} m. How many photons are absorbed by the water in the soup?

* **7.61** The quantum-mechanical treatment of the hydrogen atom gives an expression for the wave function, ψ, of the 1s orbital:

$$\psi = \frac{1}{\sqrt{\pi}}\left(\frac{1}{a_0}\right)^{3/2} e^{-r/a_0}$$

where r is the distance from the nucleus and a_0 is 52.92 pm. The electron probability density is the probability of finding the electron in a tiny volume at distance r from the nucleus and is proportional to ψ^2. The radial probability distribution is the total probability of finding the electron at all points at distance r from the nucleus and is proportional to $4\pi r^2\psi^2$. Calculate the values (to three significant figures) of ψ, ψ^2, and $4\pi r^2\psi^2$ to fill in the following table, and sketch plots of these quantities versus r. Compare the latter two plots with those in Figure 7.18A:

r (pm)	ψ (pm$^{-3/2}$)	ψ^2 (pm^{-3})	$4\pi^2\psi^2$ (pm^{-1})
0			
50			
100			
200			

7.62 In fireworks displays, light of a given wavelength indicates the presence of a particular element. What are the frequency and color of the light associated with each of the following?

(a) Li^+, $\lambda = 671$ nm   (b) Cs^+, $\lambda = 456$ nm

(c) Ca^{2+}, $\lambda = 649$ nm   (d) Na^+, $\lambda = 589$ nm

7.63 Photoelectron spectroscopy applies the principle of the photoelectric effect to study orbital energies of atoms and molecules. High-energy radiation (usually UV or x-ray) is absorbed by a sample and an electron is ejected. By knowing the energy of the radiation and measuring the energy of the electron lost, the orbital energy can be calculated. The following energy differences were determined for several electron transitions:

$$\Delta E_{2 \longrightarrow 1} = 4.088 \times 10^{-17} \text{ J}$$
$$\Delta E_{3 \longrightarrow 1} = 4.844 \times 10^{-17} \text{ J}$$
$$\Delta E_{5 \longrightarrow 1} = 5.232 \times 10^{-17} \text{ J}$$
$$\Delta E_{4 \longrightarrow 2} = 1.022 \times 10^{-17} \text{ J}$$

Calculate the energy change and the wavelength of a photon emitted in the following transitions.

(a) Level 3 $\longrightarrow$ 2   (b) Level 4 $\longrightarrow$ 1   (c) Level 5 $\longrightarrow$ 4

7.64 An electron microscope focuses electrons through magnetic lenses to observe objects at higher magnification than is possible with a light microscope. For any microscope, the smallest object that can be observed is one-half the wavelength of the radiation used. Thus, for example, the smallest object that can be observed with light of 400 nm is 2×10^{-7} m. (a) What is the smallest object observable with an electron microscope using electrons moving at 5.5×10^4 m/s? (b) At 3.0×10^7 m/s?

7.65 In a typical fireworks device, the heat of the reaction between a strong oxidizing agent, such as $KClO_4$, and an organic compound excites certain salts, which emit specific colors. Strontium salts have an intense emission at 641 nm, and barium salts have one at 493 nm. (a) What colors do these emissions produce? (b) What is the energy (in kJ) of these emissions for 1.00 g each of the chloride salts of Sr and Ba? (Assume that all the heat released is converted to light emitted.)

* **7.66** Atomic hydrogen produces well-known series of spectral lines in several regions of the electromagnetic spectrum. Each series fits the Rydberg equation with its own particular n_1 value. Calculate the value of n_1 (by trial and error if necessary) that would produce a series of lines in which:

(a) The *highest* energy line has a wavelength of 3282 nm.

(b) The *lowest* energy line has a wavelength of 7460 nm.

7.67 Fish-liver oil is a good source of vitamin A, which is measured spectrophotometrically at a wavelength of 329 nm.

(a) Suggest a reason for using this wavelength.

(b) In what region of the spectrum does this wavelength lie?

(c) When 0.1232 g of fish-liver oil is dissolved in 500. mL of solvent, the absorbance is 0.724 units. When 1.67×10^{-3} g of vitamin A is dissolved in 250. mL of solvent, the absorbance is 1.018 units. Calculate the vitamin A concentration in the fish-liver oil.

7.68 Many calculators use photocells to provide their energy. Find the maximum wavelength needed to remove an electron from silver ($\phi = 7.59 \times 10^{-19}$ J). Is silver a good choice for a photocell that uses visible light?

7.69 In a game of "Clue," Ms. White is killed in the conservatory. You have a device in each room to help you find the murderer—a spectrometer that emits the entire visible spectrum to indicate who is in that room. For example, if someone wearing yellow is in a room, light at 580 nm is reflected. The suspects are Col. Mustard, Prof. Plum, Mr. Green, Ms. Peacock (blue), and Ms. Scarlet. At the time of the murder, the spectrometer in the dining room recorded a reflection at 520 nm, those in the lounge and study recorded reflections of lower frequencies, and the one in the library recorded a reflection of the shortest possible wavelength. Who killed Ms. White? Explain.

7.70 Technetium (Tc; $Z = 43$) is a synthetic element used as a radioactive tracer in medical studies. A Tc atom emits a beta particle (electron) with a kinetic energy (E_k) of 4.71×10^{-15} J. What is the de Broglie wavelength of this electron ($E_k = \frac{1}{2}mv^2$)?

7.71 Electric power is typically given in units of watts (1 W = 1 J/s). About 95% of the power output of an incandescent bulb is converted to heat and 5% to light. If 10% of that light shines on your chemistry text, how many photons per second shine on the book from a 75-W bulb? (Assume the photons have a wavelength of 550 nm.)

* **7.72** The net change in the multistep biochemical process of photosynthesis is that CO_2 and H_2O form glucose ($C_6H_{12}O_6$) and O_2. Chlorophyll absorbs light in the 600 to 700 nm region. (a) Write a balanced thermochemical equation for formation of 1.00 mol of glucose. (b) What is the minimum number of photons with $\lambda = 680$. nm needed to prepare 1.00 mol of glucose?

7.73 In contrast to the situation for an electron, calculate the uncertainty in the position of a 142-g baseball, pitched during the 2004 Red Sox–Cardinals World Series, that moved at 100.0 mi/h $\pm$ 1.00%.

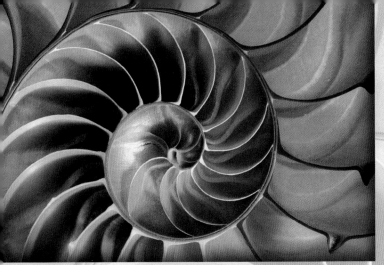

CHAPTER EIGHT

Electron Configuration and Chemical Periodicity

Observing the Trend Many natural phenomena, such as the increasing spiral of a *Nautilus* shell and, as you'll see in this chapter, the arrangement of electrons in atoms, recur with such periodic regularity that they allow us to predict properties and behavior.

Key Principles

◆ The *electron configuration* of an atom, the distribution of electrons into its energy levels and sublevels, ultimately determines the behavior of the element.

◆ Three new features are important for atoms with more than one electron (all elements except hydrogen): a fourth quantum number (m_s) specifies *electron spin;* an orbital holds no more than two electrons (*exclusion principle*); and interactions between electrons and between nucleus and electrons (*shielding* and *penetration*) cause energy levels to split into sublevels of different energy.

◆ The periodic table is "*built up*" by adding one electron (and one proton and one or more neutrons) to each preceding atom. This method

results in vertical groups of elements having *identical outer electron configurations* and, thus, similar behavior.

◆ Three atomic properties—*atomic size, ionization energy* (energy involved in removing an electron from an atom), and *electron affinity* (energy involved in adding an electron to an atom)—exhibit recurring trends throughout the periodic table.

◆ These atomic properties have a profound effect on many macroscopic properties, including *metallic behavior, acid-base behavior of oxides, ionic behavior,* and *magnetic behavior* of the elements and their compounds.

Outline

In Chapter 7, you saw how an outpouring of scientific creativity by early 20th-century physicists led to a new understanding of matter and energy, which in turn led to the quantum-mechanical model of the atom. But you can be sure that late 19th-century chemists were not sitting idly by, waiting for their colleagues in physics to develop that model. They were exploring the nature of electrolytes, establishing the kinetic-molecular theory, and developing chemical thermodynamics. The fields of organic chemistry and biochemistry were born, as were the fertilizer, explosives, glassmaking, soapmaking, bleaching, and dyestuff industries. And, for the first time, chemistry became a university subject in Europe and America. Superimposed on this activity was the accumulation of an enormous body of facts about the elements, which became organized into the periodic table.

The goal of this chapter is to show how the organization of the table, condensed from countless hours of laboratory work, was explained perfectly by the new quantum-mechanical atomic model. This model answers one of the central questions in chemistry: why do the elements behave as they do? Or, rephrasing the question to fit the main topic of this chapter: how does the **electron configuration** of an element—*the distribution of electrons within the orbitals of its atoms*—relate to its chemical and physical properties?

8.1 DEVELOPMENT OF THE PERIODIC TABLE

An essential requirement for the amazing growth in theoretical and practical chemistry in the second half of the 19th century was the ability to organize the facts known about element behavior. In Chapter 2, you saw the organizing scheme created by the Russian chemist Dmitri Mendeleev. In 1870, he arranged the 65 elements then known into a *periodic table* and summarized their behavior in the **periodic law:** when arranged by atomic mass, the elements exhibit a periodic recurrence of similar properties. It is a curious quirk of history that Mendeleev and the German chemist Julius Lothar Meyer arrived at virtually the same organization simultaneously, yet independently. Mendeleev focused on chemical properties and Meyer on physical properties. The greater credit has gone to Mendeleev because he *predicted* the properties of several as-yet-undiscovered elements, for which he had left blank spaces in his table.

Today's periodic table, which appears on the inside front cover of the text, resembles Mendeleev's in most details, although it includes 49 (and still counting) elements that were unknown in 1870. The only substantive change is that the elements are now arranged in order of *atomic number* (number of protons) rather than atomic mass.

8.2 CHARACTERISTICS OF MANY-ELECTRON ATOMS

Like the Bohr model, the Schrödinger equation does not give *exact* solutions for many-electron atoms. However, unlike the Bohr model, the Schrödinger equation gives very good *approximate* solutions. These solutions show that the atomic orbitals of many-electron atoms resemble those of the H atom, which means we can use the same quantum numbers that we used for the H atom to describe the orbitals of other atoms.

Nevertheless, the existence of more than one electron in an atom requires us to consider three features that were not relevant in the case of hydrogen: (1) the need for a fourth quantum number, (2) a limit on the number of electrons allowed in a given orbital, and (3) a more complex set of orbital energy levels. Let's examine these new features and then go on to determine the electron configuration for each element.

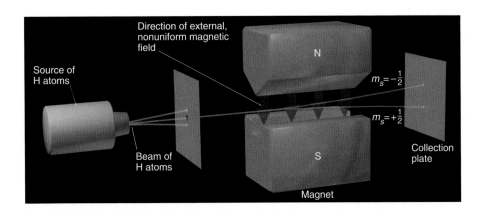

Source of H atoms

Direction of external, nonuniform magnetic field

N

$m_s = -\frac{1}{2}$

$m_s = +\frac{1}{2}$

Beam of H atoms

S

Collection plate

Magnet

Figure 8.1 Observing the effect of electron spin. A nonuniform magnetic field, created by magnet faces with different shapes, splits a beam of hydrogen atoms in two. The split beam results from the two possible values of electron spin within each atom.

The Electron-Spin Quantum Number

Recall from Chapter 7 that the three quantum numbers n, l, and m_l describe the size (energy), shape, and orientation, respectively, of an atomic orbital. However, an additional quantum number is needed to describe a property of the electron itself, called *spin,* which is not a property of the orbital. Electron spin becomes important when more than one electron is present.

When a beam of H atoms passes through a nonuniform magnetic field, as shown in Figure 8.1, it splits into two beams that bend away from each other. The explanation of the split beam is that the electron generates a tiny magnetic field, as though it were a spinning charge. The single electron in each H atom can have one of two possible values of *spin,* each of which generates a tiny magnetic field. These two fields have opposing directions, so half of the electrons are attracted into the large external magnetic field and the other half are repelled by it. As a result, the beam of H atoms splits.

Like its charge, spin is an intrinsic property of the electron, and the **spin quantum number (m_s)** has values of either $+\frac{1}{2}$ or $-\frac{1}{2}$. Thus, *each electron in an atom is described completely by a set of **four** quantum numbers: the first three describe its orbital, and the fourth describes its spin.* The quantum numbers are summarized in Table 8.1.

Table 8.1	Summary of Quantum Numbers of Electrons in Atoms		
Name	Symbol	Permitted Values	Property
Principal	n	Positive integers (1, 2, 3, ...)	Orbital energy (size)
Angular momentum	l	Integers from 0 to $n-1$	Orbital shape (The l values 0, 1, 2, and 3 correspond to s, p, d, and f orbitals, respectively.)
Magnetic	m_l	Integers from $-l$ to 0 to $+l$	Orbital orientation
Spin	m_s	$+\frac{1}{2}$ or $-\frac{1}{2}$	Direction of e^- spin

Now we can write a set of four quantum numbers for any electron in the ground state of any atom. For example, the set of quantum numbers for the lone electron in hydrogen (H; $Z = 1$) is $n = 1$, $l = 0$, $m_l = 0$, and $m_s = +\frac{1}{2}$. (The spin quantum number for this electron could just as well have been $-\frac{1}{2}$, but by convention, we assign $+\frac{1}{2}$ for the first electron in an orbital.)

The Exclusion Principle

The element after hydrogen is helium (He; $Z = 2$), the first with atoms having more than one electron. The first electron in the He ground state has the same set of quantum numbers as the electron in the H atom, but the second He electron does not. Based on observations of the excited states of atoms, the Austrian physicist Wolfgang Pauli formulated the **exclusion principle:** *no two electrons in the same atom can have the same four quantum numbers.* That is, each electron must have a unique "identity" as expressed by its set of quantum numbers. Therefore, the second He electron occupies the same orbital as the first but has an opposite spin: $n = 1$, $l = 0$, $m_l = 0$, and $m_s = -\frac{1}{2}$. You can think of the unique set of quantum numbers for an electron as analogous to the unique location of a box seat at a baseball game. The stadium (atom) is divided into section (n, level), box (l, sublevel), row (m_l, orbital), and seat (m_s, spin). Only one person (electron) can have this particular set of stadium "quantum numbers."

Because the spin quantum number (m_s) can have only two values, the major consequence of the exclusion principle is that *an atomic orbital can hold a maximum of two electrons and they must have opposing spins.* We say that the 1s orbital in He is *filled* and that the electrons have *paired spins.* Thus, a beam of He atoms is not split in an experiment like that in Figure 8.1.

Electrostatic Effects and Energy-Level Splitting

Electrostatic effects play a major role in determining the energy states of many-electron atoms. Recall that the energy state of the H atom is determined *only* by the n value of the occupied orbital. In other words, in the H atom, all sublevels of a given level, such as the 2s and 2p, have the same energy. The reason is that the only electrostatic interaction is the attraction between nucleus and electron. On the other hand, the energy states of many-electron atoms arise not only from nucleus-electron attractions, but also electron-electron repulsions. One major consequence of these additional interactions is *the splitting of energy levels into sublevels of differing energies: the energy of an orbital in a many-electron atom depends mostly on its n value (size) and to a lesser extent on its l value (shape).* Thus, a more complex set of energy states exists for a many-electron atom than we saw in the H atom.

For example, with lithium (Li; $Z = 3$) in its ground state, the first two electrons fill the 1s orbital because the electrons of an atom in its ground state occupy the orbitals of lowest energy. Then the third Li electron must go into the $n = 2$ level, which has 2s and 2p sublevels. As you'll see, the 2s is lower in energy than the 2p. The reasons for this energy difference are based on three factors—*nuclear charge, electron repulsions,* and *orbital shape* (more specifically, radial probability distribution). Their interplay leads to the phenomena of *shielding* and *penetration,* which occur in all the atoms in the periodic table—except hydrogen.

The Effect of Nuclear Charge (Z) on Orbital Energy
Nuclear protons create an ever-present pull on the electrons. You know that higher charges attract each other more strongly than lower charges (Coulomb's law, Section 2.7). Therefore, *higher nuclear charge lowers orbital energy (stabilizes the system) by increasing nucleus-electron attractions.* We can see this effect clearly by comparing the 1s orbital energies of three species with one electron—the H atom ($Z = 1$), He^+ ion ($Z = 2$), and Li^{2+} ion ($Z = 3$). The H 1s orbital is the least stable (highest energy), and the Li^{2+} 1s orbital is the most stable.

Shielding: The Effect of Electron Repulsions on Orbital Energy
In many-electron atoms, each electron "feels" not only the attraction to the nucleus but also the repulsion from other electrons. This repulsion counteracts the nuclear attraction somewhat, making each electron easier to remove by, in effect, helping to push it away. We speak of each electron "shielding" the other electrons somewhat from the nucleus. **Shielding** (also called *screening*) reduces the full nuclear charge to

an **effective nuclear charge (Z_{eff})**, the nuclear charge an electron *actually experiences. This lower nuclear charge makes the electron easier to remove.* We see the effect of shielding by electrons in the *same* orbital when we compare the He atom and He^+ ion: both have a 2+ nuclear charge, but He has two electrons in the 1s orbital and He^+ has only one. It takes less than half as much energy to remove an electron from He (2372 kJ/mol) than from He^+ (5250 kJ/mol) because the second electron in He repels the first, in effect shielding the first electron from the full nuclear charge (lowering Z_{eff}).

Much greater shielding is provided by inner electrons. Because they spend nearly all their time *between* the outer electrons and the nucleus, inner electrons shield outer electrons very effectively, in fact, *much more effectively* than do electrons in the same sublevel. *Shielding by inner electrons greatly lowers the Z_{eff} felt by outer electrons.*

Penetration: The Effect of Orbital Shape on Orbital Energy

Why does the third electron occupy the 2s orbital in the Li ground state, rather than the 2p? To answer this, we have to consider orbital shapes, that is, radial probability distributions (Figure 8.2). At first, we might expect that the electron would enter the 2p orbital (*orange curve*) because it is slightly closer to the nucleus than the major portion of the 2s orbital (*blue curve*). But note that a minor portion of the 2s radial probability distribution appears within the 1s region. As a result, an electron in the 2s orbital spends part of its time "penetrating" very close to the nucleus. Charges attract more strongly if they are near each other than far apart (Coulomb's law, Section 2.7). Therefore, **penetration** by the 2s electron increases its overall attraction to the nucleus relative to that for a 2p electron. At the same time, penetration into the 1s region decreases the shielding of the 2s electron by the 1s electrons. Evidence shows that, indeed, the 2s orbital of Li *is* lower in energy than the 2p orbital, because it takes more energy to remove a 2s electron (520 kJ/mol) than a 2p (341 kJ/mol).

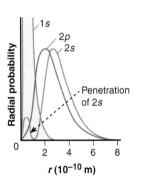

Figure 8.2 Penetration and orbital energy. Radial probability distributions show that a 2s electron spends most of its time slightly farther from the nucleus than does a 2p electron but penetrates near the nucleus for a small part of the time. Penetration by the 2s electron increases its overall attraction to the nucleus; thus, the 2s orbital is more stable (lower in energy) than the 2p.

In general, *penetration and the resulting effects on shielding cause an energy level to split into sublevels of differing energy.* The lower the *l* value of an orbital, the more its electrons penetrate, and so the greater their attraction to the nucleus. Therefore, *for a given n value, the lower the l value, the lower the sublevel energy:*

$$\text{Order of sublevel energies: } s < p < d < f \qquad (8.1)$$

Thus, the 2s ($l = 0$) is lower in energy than the 2p ($l = 1$), the 3p ($l = 1$) is lower than the 3d ($l = 2$), and so forth.

Figure 8.3 shows the general energy order of levels (*n* value) and how they are split into sublevels (*l* values) of differing energies. (Compare this with the H atom energy levels in Figure 7.22.) Next, we use this energy order to construct a periodic table of ground-state atoms.

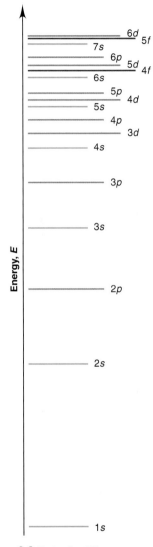

Figure 8.3 Order for filling energy sublevels with electrons. In many-electron atoms, energy levels split into sublevels. The relative energies of sublevels increase with principal quantum number *n* (1 < 2 < 3, etc.) and angular momentum quantum number *l* (s < p < d < f). As *n* increases, the energies become closer together. The penetration effect, together with this narrowing of energy differences, results in the overlap of some sublevels; for example, the 4s sublevel is slightly lower in energy than the 3d, so it is filled first. (Line color is by sublevel type; line lengths differ for ease in labeling.)

SECTION SUMMARY
Identifying electrons in many-electron atoms requires four quantum numbers: three (n, l, m_l) describe the orbital, and a fourth (m_s) describes electron spin. The exclusion principle requires each electron to have a unique set of four quantum numbers; therefore, an orbital can hold no more than two electrons, and their spins must be paired (opposite). Electrostatic interactions determine orbital energies as follows:
1. Greater nuclear charge lowers orbital energy and makes electrons harder to remove.
2. Electron-electron repulsions raise orbital energy and make electrons easier to remove. Repulsions have the effect of *shielding* electrons from the full nuclear charge, reducing it to an effective nuclear charge, Z_{eff}. Inner electrons shield outer electrons most effectively.
3. Greater radial probability distribution near the nucleus (greater *penetration*) makes an electron harder to remove because it is attracted more strongly and shielded less effectively. As a result, an energy level (shell) is split into sublevels (subshells) with the energy order $s < p < d < f$.

8.3 THE QUANTUM-MECHANICAL MODEL AND THE PERIODIC TABLE

Quantum mechanics provides the theoretical foundation for the experimentally based periodic table. In this section, we fill the table with elements and determine their electron configurations—the distributions of electrons within their atoms' orbitals. Note especially the *recurring pattern in electron configurations, which is the basis for the recurring pattern in chemical behavior.*

Building Up Periods 1 and 2

A useful way to determine the electron configurations of the elements is to start at the beginning of the periodic table and add one electron per element to the *lowest energy orbital available.* (Of course, one proton and one or more neutrons are also added to the nucleus.) This approach is based on the **aufbau principle** (German *aufbauen,* "to build up"), and it results in *ground-state* electron configurations. Let's assign sets of quantum numbers to the electrons in the ground state of the first 10 elements, those in the first two periods (horizontal rows).

For the electron in H, as you've seen, the set of quantum numbers is

$$\text{H } (Z = 1): \ n = 1, l = 0, m_l = 0, m_s = +\tfrac{1}{2}$$

You also saw that the first electron in He has the same set as the electron in H, but the second He electron has opposing spin (exclusion principle):

$$\text{He } (Z = 2): \ n = 1, l = 0, m_l = 0, m_s = -\tfrac{1}{2}$$

(As we go through each element in this discussion, the quantum numbers that follow refer to the element's *last added* electron.)

Here are two common ways to designate the orbital and its electrons:

1. *The electron configuration.* This shorthand notation consists of the principal energy level (n value), the letter designation of the sublevel (l value), and the number of electrons (#) in the sublevel, written as a superscript: $nl^{\#}$. The electron configuration of H is $1s^1$ (spoken "one-ess-one"); that of He is $1s^2$ (spoken "one-ess-two," *not* "one-ess-squared"). This notation does *not* indicate electron spin but assumes you know that the two $1s$ electrons have paired (opposite) spins.

2. *The orbital diagram.* An **orbital diagram** consists of a box (or circle, or just a line) for each orbital in a given energy level, grouped by sublevel, with an arrow indicating an electron *and* its spin. (Traditionally, ↑ is $+\tfrac{1}{2}$ and ↓ is $-\tfrac{1}{2}$, but these are arbitrary; it is necessary only to be consistent. Throughout the text, orbital occupancy is also indicated by color intensity: an orbital with no color is empty, pale color means half-filled, and full color means filled.) The electron

configurations and orbital diagrams for the first two elements are

H ($Z = 1$) $1s^1$ [↑]   He ($Z = 2$) $1s^2$ [↑↓]
               $1s$ $1s$

The exclusion principle tells us that an orbital can hold only two electrons, so the $1s$ orbital in He is filled, and the $n = 1$ level is also filled. The $n = 2$ level is filled next, beginning with the $2s$ orbital, the next lowest in energy. As we said earlier, the first two electrons in Li fill the $1s$ orbital, and the last added Li electron has quantum numbers $n = 2$, $l = 0$, $m_l = 0$, $m_s = +\frac{1}{2}$. The electron configuration for Li is $1s^2 2s^1$. Note that the orbital diagram shows all the orbitals for $n = 2$, whether or not they are occupied:

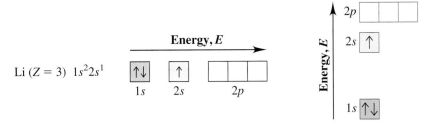

To save space on a page, orbital diagrams are often written horizontally, as shown at left. Note that *the energy of the sublevels increases from left to right.* The orbital diagram at right shows the sublevels vertically.

With the $2s$ orbital only half-filled in Li, the fourth electron of beryllium fills it with the electron's spin paired: $n = 2$, $l = 0$, $m_l = 0$, $m_s = -\frac{1}{2}$.

Be ($Z = 4$) $1s^2 2s^2$ [↑↓] [↑↓] [| |]
                             $1s$ $2s$ $2p$

The next lowest energy sublevel is the $2p$. A p sublevel has $l = 1$, so the m_l (orientation) values can be -1, 0, or $+1$. The three orbitals in the $2p$ sublevel have *equal energy* (same n and l values), which means that the fifth electron of boron can go into *any one of the 2p orbitals*. For convenience, let's label the boxes from left to right, -1, 0, $+1$. By convention, we place the electron in the $m_l = -1$ orbital: $n = 2$, $l = 1$, $m_l = -1$, $m_s = +\frac{1}{2}$.

                                                 -1 0 $+1$

B ($Z = 5$) $1s^2 2s^2 2p^1$ [↑↓] [↑↓] [↑ | |]
                                 $1s$ $2s$ $2p$

To minimize electron-electron repulsions, the last added (sixth) electron of carbon enters one of the *unoccupied 2p orbitals*; by convention, we place it in the $m_l = 0$ orbital. Experiment shows that the spin of this electron is *parallel* to (the same as) the spin of the other $2p$ electron: $n = 2$, $l = 1$, $m_l = 0$, $m_s = +\frac{1}{2}$.

C ($Z = 6$) $1s^2 2s^2 2p^2$ [↑↓] [↑↓] [↑ | ↑ |]
                                 $1s$ $2s$ $2p$

This placement of electrons for carbon exemplifies **Hund's rule:** *when orbitals of equal energy are available, the electron configuration of lowest energy has the maximum number of unpaired electrons with parallel spins.* Based on Hund's rule, nitrogen's seventh electron enters the last empty $2p$ orbital, with its spin parallel to the two other $2p$ electrons: $n = 2$, $l = 1$, $m_l = +1$, $m_s = +\frac{1}{2}$.

N ($Z = 7$) $1s^2 2s^2 2p^3$ [↑↓] [↑↓] [↑ | ↑ | ↑]
                                 $1s$ $2s$ $2p$

The eighth electron in oxygen must enter one of these three half-filled $2p$ orbitals and "pair up" with (have opposing spin to) the electron already present. With the $2p$ orbitals all having the same energy, we place the electron in the orbital previously designated $m_l = -1$. The quantum numbers are $n = 2$, $l = 1$, $m_l = -1$, $m_s = -\frac{1}{2}$.

O $(Z = 8)$ $1s^2 2s^2 2p^4$ ⬆⬇ ⬆⬇ ⬆⬇ ⬆ ⬆
                                    $1s$ $2s$ $2p$

Fluorine's ninth electron enters either of the two remaining half-filled $2p$ orbitals: $n = 2$, $l = 1$, $m_l = 0$, $m_s = -\frac{1}{2}$.

F $(Z = 9)$ $1s^2 2s^2 2p^5$ ⬆⬇ ⬆⬇ ⬆⬇ ⬆⬇ ⬆
                                    $1s$ $2s$ $2p$

Only one unfilled orbital remains in the $2p$ sublevel, so the tenth electron of neon occupies it: $n = 2$, $l = 1$, $m_l = +1$, $m_s = -\frac{1}{2}$. With neon, the $n = 2$ level is filled.

Ne $(Z = 10)$ $1s^2 2s^2 2p^6$ ⬆⬇ ⬆⬇ ⬆⬇ ⬆⬇ ⬆⬇
                                      $1s$ $2s$ $2p$

SAMPLE PROBLEM 8.1   Determining Quantum Numbers from Orbital Diagrams

Problem Write a set of quantum numbers for the third electron and a set for the eighth electron of the F atom.

Plan Referring to the orbital diagram, we count to the electron of interest and note its level (n), sublevel (l), orbital (m_l), and spin (m_s).

Solution The third electron is in the $2s$ orbital. The upward arrow indicates a spin of $+\frac{1}{2}$:

$$n = 2, \; l = 0, \; m_l = 0, \; m_s = +\frac{1}{2}$$

The eighth electron is in the first $2p$ orbital, which is designated $m_l = -1$, and has a downward arrow:

$$n = 2, \; l = 1, \; m_l = -1, \; m_s = -\frac{1}{2}$$

FOLLOW-UP PROBLEM 8.1 Use the periodic table to identify the element with the electron configuration $1s^2 2s^2 2p^4$. Write its orbital diagram, and give the quantum numbers of its sixth electron.

Building Up Period 3

The Period 3 elements, sodium through argon, lie directly under the Period 2 elements, lithium through neon. The sublevels of the $n = 3$ level are filled in the order $3s$, $3p$, $3d$. Table 8.2 presents *partial* orbital diagrams ($3s$ and $3p$ sublevels only) and electron configurations for the eight elements in Period 3 (with *filled inner levels* in brackets and the sublevel to which the last electron is added in colored type). Note the *group similarities in outer electron configuration* with the elements in Period 2 (refer to Figure 8.4).

In sodium (the second alkali metal) and magnesium (the second alkaline earth metal), electrons are added to the $3s$ sublevel, which contains the $3s$ orbital only, just as they filled the $2s$ sublevel in lithium and beryllium in Period 2. Then, just as for boron, carbon, and nitrogen in Period 2, the last electrons added to aluminum, silicon, and phosphorus in Period 3 half-fill the three $3p$ orbitals with spins parallel (Hund's rule). The last electrons added to sulfur, chlorine, and argon then successively enter the three half-filled $3p$ orbitals, thereby filling the $3p$ sublevel. With argon, the next noble gas after helium and neon, we arrive at the end of Period 3. (As you'll see shortly, the $3d$ orbitals are filled in Period 4.)

Table 8.2 | **Partial Orbital Diagrams and Electron Configurations* for the Elements in Period 3**

Atomic Number	Element	Partial Orbital Diagram (3s and 3p Sublevels Only)		Full Electron Configuration	Condensed Electron Configuration
		3s	**3p**		
11	Na	↑	▯▯▯	$[1s^2 2s^2 2p^6]\, 3s^1$	$[Ne]\, 3s^1$
12	Mg	↑↓	▯▯▯	$[1s^2 2s^2 2p^6]\, 3s^2$	$[Ne]\, 3s^2$
13	Al	↑↓	↑ ▯▯	$[1s^2 2s^2 2p^6]\, 3s^2 3p^1$	$[Ne]\, 3s^2 3p^1$
14	Si	↑↓	↑ ↑ ▯	$[1s^2 2s^2 2p^6]\, 3s^2 3p^2$	$[Ne]\, 3s^2 3p^2$
15	P	↑↓	↑ ↑ ↑	$[1s^2 2s^2 2p^6]\, 3s^2 3p^3$	$[Ne]\, 3s^2 3p^3$
16	S	↑↓	↑↓ ↑ ↑	$[1s^2 2s^2 2p^6]\, 3s^2 3p^4$	$[Ne]\, 3s^2 3p^4$
17	Cl	↑↓	↑↓ ↑↓ ↑	$[1s^2 2s^2 2p^6]\, 3s^2 3p^5$	$[Ne]\, 3s^2 3p^5$
18	Ar	↑↓	↑↓ ↑↓ ↑↓	$[1s^2 2s^2 2p^6]\, 3s^2 3p^6$	$[Ne]\, 3s^2 3p^6$

*Colored type indicates the sublevel to which the last electron is added.

The rightmost column of Table 8.2 shows the *condensed electron configuration*. In this simplified notation, the electron configuration of the previous noble gas is shown by its element symbol in brackets, and it is followed by the electron configuration of the energy level being filled. The condensed electron configuration of sulfur, for example, is $[Ne]\, 3s^2 3p^4$, where $[Ne]$ stands for $1s^2 2s^2 2p^6$.

Electron Configurations Within Groups

One of the central points in all chemistry is that *similar outer electron configurations correlate with similar chemical behavior.* Figure 8.4 shows the condensed electron configurations of the first 18 elements. Note the similarities within each group. Here are some examples from just three groups:

- In Group 1A(1), lithium and sodium have the condensed electron configuration [noble gas] ns^1 (where n is the quantum number of the outermost energy level), as do all the other alkali metals (K, Rb, Cs, Fr). All are highly reactive metals that form ionic compounds with nonmetals with formulas such as MCl, M_2O, and M_2S (where M represents the alkali metal), and all react vigorously with water to displace H_2.

	1A (1)							8A (18)
	1 **H** $1s^1$	2A (2)	3A (13)	4A (14)	5A (15)	6A (16)	7A (17)	2 **He** $1s^2$
Period 2	3 **Li** $[He]\, 2s^1$	4 **Be** $[He]\, 2s^2$	5 **B** $[He]\, 2s^2 2p^1$	6 **C** $[He]\, 2s^2 2p^2$	7 **N** $[He]\, 2s^2 2p^3$	8 **O** $[He]\, 2s^2 2p^4$	9 **F** $[He]\, 2s^2 2p^5$	10 **Ne** $[He]\, 2s^2 2p^6$
Period 3	11 **Na** $[Ne]\, 3s^1$	12 **Mg** $[Ne]\, 3s^2$	13 **Al** $[Ne]\, 3s^2 3p^1$	14 **Si** $[Ne]\, 3s^2 3p^2$	15 **P** $[Ne]\, 3s^2 3p^3$	16 **S** $[Ne]\, 3s^2 3p^4$	17 **Cl** $[Ne]\, 3s^2 3p^5$	18 **Ar** $[Ne]\, 3s^2 3p^6$

Figure 8.4 Condensed ground-state electron configurations in the first three periods. The first 18 elements, H through Ar, are arranged in three periods containing two, eight, and eight elements. Each box shows the atomic number, atomic symbol, and condensed ground-state electron configuration. Note that elements in a group have similar outer electron configurations (*color*).

- In Group 7A(17), fluorine and chlorine have the condensed electron configuration [noble gas] ns^2np^5, as do the other halogens (Br, I, At). Little is known about rare, radioactive astatine (At), but all the others are reactive nonmetals that occur as diatomic molecules, X_2 (where X represents the halogen). All form ionic compounds with metals (KX, MgX_2), covalent compounds with hydrogen (HX) that yield acidic solutions in water, and covalent compounds with carbon (CX_4).
- In Group 8A(18), all of the elements have the condensed configuration [noble gas] ns^2np^6. Consistent with their *filled* energy levels, all of these elements are monatomic gases that are extremely unreactive.

To summarize the major connection between quantum mechanics and chemical periodicity: *orbitals are filled in order of increasing energy, which leads to outer electron configurations that recur periodically, which leads to chemical properties that recur periodically.*

The First *d*-Orbital Transition Series: Building Up Period 4

The 3*d* orbitals are filled in Period 4. Note, however, that *the 4s orbital is filled before the 3d.* This switch in filling order is due to the shielding and penetration effects that we discussed in Section 8.2. The radial probability distribution of the 3*d* orbital is greater outside the filled, inner $n = 1$ and $n = 2$ levels, so a 3*d* electron is shielded very effectively from the nuclear charge. In contrast, penetration by the 4*s* electron means that it spends a significant part of its time near the nucleus and feels a greater nuclear attraction. Thus, the 4*s* orbital is slightly *lower* in energy than the 3*d*, and so fills first. Similarly, the 5*s* orbital fills before the 4*d*, and the 6*s* fills before the 5*d*. In general, *the ns sublevel fills before the (n − 1)d sublevel.* As we proceed through the transition series, however, you'll see several exceptions to this pattern because the energies of the *ns* and $(n − 1)d$ sublevels become extremely close at higher values of *n*.

Table 8.3 shows the partial orbital diagrams and ground-state electron configurations for the 18 elements in Period 4 (again with filled inner levels in brackets and the sublevel to which the last electron has been added in colored type). The first two elements of the period, potassium and calcium, are the next alkali and alkaline earth metals, respectively, and their electrons fill the 4*s* sublevel. The third element, scandium ($Z = 21$), is the first of the **transition elements,** those in which *d* orbitals are being filled. The last electron in scandium occupies any one of the five 3*d* orbitals because they are equal in energy. Scandium has the electron configuration [Ar] $4s^23d^1$.

The filling of 3*d* orbitals proceeds one at a time, as with *p* orbitals, except in two cases: chromium ($Z = 24$) and copper ($Z = 29$). Vanadium ($Z = 23$), the element before chromium, has three half-filled *d* orbitals ([Ar] $4s^23d^3$). Rather than having its last electron enter a fourth empty *d* orbital to give [Ar] $4s^23d^4$, chromium has one electron in the 4*s* sublevel and five in the 3*d* sublevel. Thus, both the 4*s* and the 3*d* sublevels are *half-filled* (see margin). In manganese ($Z = 25$), the 4*s* sublevel is filled again ([Ar] $4s^23d^5$).

The other anomalous filling pattern occurs with copper. Following nickel ([Ar] $4s^23d^8$), copper would be expected to have the [Ar] $4s^23d^9$ configuration. Instead, the 4*s* orbital of copper is half-filled (1 electron), and the 3*d* orbitals are *filled* with 10 electrons (see margin). The anomalous filling patterns in Cr and Cu lead us to conclude that *half-filled and filled sublevels are unexpectedly stable.* These are the first two cases of a pattern seen with many other elements.

Cr ($Z = 24$) [Ar] $4s^13d^5$

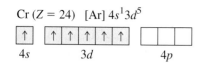

Cu ($Z = 29$) [Ar] $4s^13d^{10}$

Table 8.3 | **Partial Orbital Diagrams and Electron Configurations* for the Elements in Period 4**

Atomic Number	Element	Partial Orbital Diagram (4s, 3d, and 4p Sublevels Only)			Full Electron Configuration	Condensed Electron Configuration
		4s	3d	4p		
19	K	↑			$[1s^22s^22p^63s^23p^6]\,4s^1$	$[Ar]\,4s^1$
20	Ca	↑↓			$[1s^22s^22p^63s^23p^6]\,4s^2$	$[Ar]\,4s^2$
21	Sc	↑↓	↑		$[1s^22s^22p^63s^23p^6]\,4s^23d^1$	$[Ar]\,4s^23d^1$
22	Ti	↑↓	↑ ↑		$[1s^22s^22p^63s^23p^6]\,4s^23d^2$	$[Ar]\,4s^23d^2$
23	V	↑↓	↑ ↑ ↑		$[1s^22s^22p^63s^23p^6]\,4s^23d^3$	$[Ar]\,4s^23d^3$
24	Cr	↑	↑ ↑ ↑ ↑ ↑		$[1s^22s^22p^63s^23p^6]\,4s^13d^5$	$[Ar]\,4s^13d^5$
25	Mn	↑↓	↑ ↑ ↑ ↑ ↑		$[1s^22s^22p^63s^23p^6]\,4s^23d^5$	$[Ar]\,4s^23d^5$
26	Fe	↑↓	↑↓ ↑ ↑ ↑ ↑		$[1s^22s^22p^63s^23p^6]\,4s^23d^6$	$[Ar]\,4s^23d^6$
27	Co	↑↓	↑↓ ↑↓ ↑ ↑ ↑		$[1s^22s^22p^63s^23p^6]\,4s^23d^7$	$[Ar]\,4s^23d^7$
28	Ni	↑↓	↑↓ ↑↓ ↑↓ ↑ ↑		$[1s^22s^22p^63s^23p^6]\,4s^23d^8$	$[Ar]\,4s^23d^8$
29	Cu	↑	↑↓ ↑↓ ↑↓ ↑↓ ↑↓		$[1s^22s^22p^63s^23p^6]\,4s^13d^{10}$	$[Ar]\,4s^13d^{10}$
30	Zn	↑↓	↑↓ ↑↓ ↑↓ ↑↓ ↑↓		$[1s^22s^22p^63s^23p^6]\,4s^23d^{10}$	$[Ar]\,4s^23d^{10}$
31	Ga	↑↓	↑↓ ↑↓ ↑↓ ↑↓ ↑↓	↑	$[1s^22s^22p^63s^23p^6]\,4s^23d^{10}4p^1$	$[Ar]\,4s^23d^{10}4p^1$
32	Ge	↑↓	↑↓ ↑↓ ↑↓ ↑↓ ↑↓	↑ ↑	$[1s^22s^22p^63s^23p^6]\,4s^23d^{10}4p^2$	$[Ar]\,4s^23d^{10}4p^2$
33	As	↑↓	↑↓ ↑↓ ↑↓ ↑↓ ↑↓	↑ ↑ ↑	$[1s^22s^22p^63s^23p^6]\,4s^23d^{10}4p^3$	$[Ar]\,4s^23d^{10}4p^3$
34	Se	↑↓	↑↓ ↑↓ ↑↓ ↑↓ ↑↓	↑↓ ↑ ↑	$[1s^22s^22p^63s^23p^6]\,4s^23d^{10}4p^4$	$[Ar]\,4s^23d^{10}4p^4$
35	Br	↑↓	↑↓ ↑↓ ↑↓ ↑↓ ↑↓	↑↓ ↑↓ ↑	$[1s^22s^22p^63s^23p^6]\,4s^23d^{10}4p^5$	$[Ar]\,4s^23d^{10}4p^5$
36	Kr	↑↓	↑↓ ↑↓ ↑↓ ↑↓ ↑↓	↑↓ ↑↓ ↑↓	$[1s^22s^22p^63s^23p^6]\,4s^23d^{10}4p^6$	$[Ar]\,4s^23d^{10}4p^6$

*Colored type indicates sublevel(s) whose occupancy changes when the last electron is added.

In zinc, both the $4s$ and $3d$ sublevels are completely filled, and the first transition series ends. As Table 8.3 shows, the $4p$ sublevel is then filled by the next six elements. Period 4 ends with krypton, the next noble gas.

General Principles of Electron Configurations

There are 78 known elements beyond the 36 we have considered. Let's survey the ground-state electron configurations to highlight some key ideas.

Similar Outer Electron Configurations Within a Group To repeat one of chemistry's central themes and the key to the usefulness of the periodic table, *elements in a group have similar chemical properties because they have similar outer electron configurations* (Figure 8.5 on the next page). Among the main-group elements (A groups)—the *s*-block and *p*-block elements—outer electron configurations within a group are essentially identical, as shown by the group headings in Figure 8.5. Some variations in the transition elements (B groups, *d* block) and inner transition elements (*f* block) occur, as you'll see.

Figure 8.5 A periodic table of partial ground-state electron configurations. These ground-state electron configurations show the electrons beyond the previous noble gas in the sublevel block being filled (excluding filled inner sublevels). For main-group elements, the group heading identifies the general outer configuration. Anomalous electron configurations occur often among the d-block and f-block elements, with the first two appearing for Cr ($Z = 24$) and Cu ($Z = 29$). Helium is colored as an s-block element but placed with the other members of Group 8A(18). Configurations for elements 110 to 112, 114, and 116 have not yet been confirmed.

Main-Group Elements (s block)

Main-Group Elements (p block)

Transition Elements (d block)

Period number: highest occupied energy level

1A (1) ns^1	2A (2) ns^2	3B (3)	4B (4)	5B (5)	6B (6)	7B (7)	8B (8)	8B (9)	8B (10)	1B (11)	2B (12)	3A (13) ns^2np^1	4A (14) ns^2np^2	5A (15) ns^2np^3	6A (16) ns^2np^4	7A (17) ns^2np^5	8A (18) ns^2np^6
1 **H** $1s^1$																	2 **He** $1s^2$
3 **Li** $2s^1$	4 **Be** $2s^2$											5 **B** $2s^22p^1$	6 **C** $2s^22p^2$	7 **N** $2s^22p^3$	8 **O** $2s^22p^4$	9 **F** $2s^22p^5$	10 **Ne** $2s^22p^6$
11 **Na** $3s^1$	12 **Mg** $3s^2$											13 **Al** $3s^23p^1$	14 **Si** $3s^23p^2$	15 **P** $3s^23p^3$	16 **S** $3s^23p^4$	17 **Cl** $3s^23p^5$	18 **Ar** $3s^23p^6$
19 **K** $4s^1$	20 **Ca** $4s^2$	21 **Sc** $4s^23d^1$	22 **Ti** $4s^23d^2$	23 **V** $4s^23d^3$	24 **Cr** $4s^13d^5$	25 **Mn** $4s^23d^5$	26 **Fe** $4s^23d^6$	27 **Co** $4s^23d^7$	28 **Ni** $4s^23d^8$	29 **Cu** $4s^13d^{10}$	30 **Zn** $4s^23d^{10}$	31 **Ga** $4s^24p^1$	32 **Ge** $4s^24p^2$	33 **As** $4s^24p^3$	34 **Se** $4s^24p^4$	35 **Br** $4s^24p^5$	36 **Kr** $4s^24p^6$
37 **Rb** $5s^1$	38 **Sr** $5s^2$	39 **Y** $5s^24d^1$	40 **Zr** $5s^24d^2$	41 **Nb** $5s^14d^4$	42 **Mo** $5s^14d^5$	43 **Tc** $5s^24d^5$	44 **Ru** $5s^14d^7$	45 **Rh** $5s^14d^8$	46 **Pd** $4d^{10}$	47 **Ag** $5s^14d^{10}$	48 **Cd** $5s^24d^{10}$	49 **In** $5s^25p^1$	50 **Sn** $5s^25p^2$	51 **Sb** $5s^25p^3$	52 **Te** $5s^25p^4$	53 **I** $5s^25p^5$	54 **Xe** $5s^25p^6$
55 **Cs** $6s^1$	56 **Ba** $6s^2$	57 **La*** $6s^25d^1$	72 **Hf** $6s^25d^2$	73 **Ta** $6s^25d^3$	74 **W** $6s^25d^4$	75 **Re** $6s^25d^5$	76 **Os** $6s^25d^6$	77 **Ir** $6s^25d^7$	78 **Pt** $6s^15d^9$	79 **Au** $6s^15d^{10}$	80 **Hg** $6s^25d^{10}$	81 **Tl** $6s^26p^1$	82 **Pb** $6s^26p^2$	83 **Bi** $6s^26p^3$	84 **Po** $6s^26p^4$	85 **At** $6s^26p^5$	86 **Rn** $6s^26p^6$
87 **Fr** $7s^1$	88 **Ra** $7s^2$	89 **Ac*** $7s^26d^1$	104 **Rf** $7s^26d^2$	105 **Db** $7s^26d^3$	106 **Sg** $7s^26d^4$	107 **Bh** $7s^26d^5$	108 **Hs** $7s^26d^6$	109 **Mt** $7s^26d^7$	110 **Ds** $7s^26d^8$	111 **Rg** $7s^26d^9$	112 $7s^26d^{10}$		114 $7s^27p^2$		116 $7s^27p^4$		

Inner Transition Elements (f block)

6 *Lanthanides	58 **Ce** $6s^24f^15d^1$	59 **Pr** $6s^24f^3$	60 **Nd** $6s^24f^4$	61 **Pm** $6s^24f^5$	62 **Sm** $6s^24f^6$	63 **Eu** $6s^24f^7$	64 **Gd** $6s^24f^75d^1$	65 **Tb** $6s^24f^9$	66 **Dy** $6s^24f^{10}$	67 **Ho** $6s^24f^{11}$	68 **Er** $6s^24f^{12}$	69 **Tm** $6s^24f^{13}$	70 **Yb** $6s^24f^{14}$	71 **Lu** $6s^24f^{14}5d^1$
7 **Actinides	90 **Th** $7s^26d^2$	91 **Pa** $7s^25f^26d^1$	92 **U** $7s^25f^36d^1$	93 **Np** $7s^25f^46d^1$	94 **Pu** $7s^25f^6$	95 **Am** $7s^25f^7$	96 **Cm** $7s^25f^76d^1$	97 **Bk** $7s^25f^9$	98 **Cf** $7s^25f^{10}$	99 **Es** $7s^25f^{11}$	100 **Fm** $7s^25f^{12}$	101 **Md** $7s^25f^{13}$	102 **No** $7s^25f^{14}$	103 **Lr** $7s^25f^{14}6d^1$

Orbital Filling Order When the elements are "built up" by filling their levels and sublevels in order of increasing energy, we obtain the actual sequence of elements in the periodic table. Thus, reading the table from left to right, as you read words on a page, gives the energy order of levels and sublevels, shown in Figure 8.6. The arrangement of the periodic table is the best way to learn the orbital filling order of the elements, but a useful memory aid is shown in Figure 8.7.

Categories of Electrons The elements have three categories of electrons:

1. **Inner (core) electrons** are those seen in the previous noble gas and any completed transition series. They fill all the *lower energy levels* of an atom.
2. **Outer electrons** are those in the *highest energy level* (highest *n* value). They spend most of their time farthest from the nucleus.
3. **Valence electrons** are those involved in forming compounds. *Among the main-group elements, the valence electrons* ***are*** *the outer electrons.* Among the transition elements, the $(n - 1)d$ electrons are counted among the valence electrons because some or all of them are often involved in bonding.

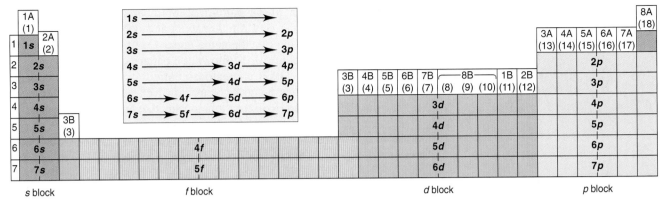

Figure 8.6 The relation between orbital filling and the periodic table. If we "read" the periods like the words on a page, the elements are arranged into sublevel blocks that occur in the order of increasing energy. This form of the periodic table shows the sublevel blocks. (The *f* blocks fit between the first and second elements of the *d* blocks in Periods 6 and 7.) *Inset:* A simple version of sublevel order.

Group and Period Numbers Key information is embedded in the periodic table:

1. Among the main-group elements (A groups), *the group number equals the number of outer electrons* (those with the highest *n*): chlorine (Cl; Group **7**A) has 7 outer electrons, tellurium (Te; Group **6**A) has 6, and so forth.
2. *The period number is the n value of the highest energy level.* Thus, in Period 2, the $n = 2$ level has the highest energy; in Period 5, it is the $n = 5$ level.
3. The *n* value squared (n^2) gives the total number of *orbitals* in that energy level. Because an orbital can hold no more than two electrons (exclusion principle), $2n^2$ gives the maximum number of *electrons* (or elements) in the energy level. For example, for the $n = 3$ level, the number of orbitals is $n^2 = 9$: one 3*s*, three 3*p*, and five 3*d*. The number of electrons is $2n^2$, or 18: two 3*s* and six 3*p* electrons occur in the eight elements of Period 3, and ten 3*d* electrons are added in the ten transition elements of Period 4.

Unusual Configurations: Transition and Inner Transition Elements

Periods 4, 5, 6, and 7 incorporate the *d*-block transition elements. The general pattern, as you've seen, is that the $(n - 1)d$ orbitals are filled between the *ns* and *np* orbitals. Thus, Period 5 follows the same general pattern as Period 4. In Period 6, the 6*s* sublevel is filled in cesium (Cs) and barium (Ba), and then lanthanum (La; $Z = 57$), the first member of the 5*d* transition series, occurs. At this point, the first series of **inner transition elements,** those in which *f* orbitals are being filled, intervenes (Figure 8.6). The *f* orbitals have $l = 3$, so the possible m_l values are $-3, -2, -1, 0, +1, +2,$ and $+3$; that is, there are seven *f* orbitals, for a total of 14 elements in *each* of the two inner transition series.

The Period 6 inner transition series fills the 4*f* orbitals and consists of the **lanthanides** (or *rare earths*), so called because they occur after and are similar to lanthanum. The other inner transition series holds the **actinides,** which fill the 5*f* orbitals that appear in Period 7 after actinium (Ac; $Z = 89$). In both series, the $(n - 2)f$ orbitals are filled, after which filling of the $(n - 1)d$ orbitals proceeds. Period 6 ends with the filling of the 6*p* orbitals as in other *p*-block elements. Period 7 is incomplete because only two elements with 7*p* electrons have been confirmed at this time.

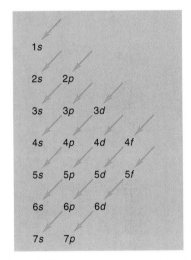

Figure 8.7 Aid to memorizing sublevel filling order. List the sublevels as shown, and read from 1*s*, following the direction of the arrows. Note that the
• *n* value is constant horizontally
• *l* value is constant vertically
• $n + l$ sum is constant diagonally.

Several irregularities in filling pattern occur in both the d and f blocks. Two already mentioned occur in chromium (Cr) and copper (Cu) in Period 4. Silver (Ag) and gold (Au), the two elements under Cu in Group 1B(11), follow copper's pattern. Molybdenum (Mo) follows the pattern of Cr in Group 6B(6), but tungsten (W) does not. Other anomalous configurations appear among the transition elements in Periods 5 and 6. Note, however, that even though minor variations from the expected configurations occur, the sum of ns electrons and $(n - 1)d$ electrons always equals the *new* group number. For instance, despite variations in the electron configurations in Group 6B(**6**)—Cr, Mo, W, and Sg— the sum of ns and $(n - 1)d$ electrons is 6; in Group 8B(**10**)—Ni, Pd, and Pt— the sum is 10.

Whenever our observations differ from our expectations, remember that the fact always takes precedence over the model; in other words, the electrons don't "care" what orbitals *we* think they should occupy. As the atomic orbitals in larger atoms fill with electrons, sublevel energies differ very little, which results in these variations from the expected pattern.

Animation: Electron Configurations/Orbital Diagrams
Online Learning Center

SAMPLE PROBLEM 8.2 Determining Electron Configurations

Problem Using the periodic table on the inside front cover of the text (not Figure 8.5 or Table 8.3), give the full and condensed electron configurations, partial orbital diagrams showing valence electrons, and number of inner electrons for the following elements:

(a) Potassium (K; $Z = 19$)   **(b)** Molybdenum (Mo; $Z = 42$)   **(c)** Lead (Pb; $Z = 82$)

Plan The atomic number tells us the number of electrons, and the periodic table shows the order for filling sublevels. In the partial orbital diagrams, we include all electrons after those of the previous noble gas *except* those in *filled* inner sublevels. The number of inner electrons is the sum of those in the previous noble gas and in filled d and f sublevels.

Solution (a) For K ($Z = 19$), the full electron configuration is $1s^2 2s^2 2p^6 3s^2 3p^6 4s^1$. The condensed configuration is $[Ar] 4s^1$.

The partial orbital diagram for valence electrons is

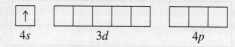

K is a main-group element in Group 1A(1) of Period 4, so there are 18 inner electrons.

(b) For Mo ($Z = 42$), we would expect the full electron configuration to be $1s^2 2s^2 2p^6 3s^2 3p^6 4s^2 3d^{10} 4p^6 5s^2 4d^4$. However, Mo lies under Cr in Group 6B(6) and exhibits the same variation in filling pattern in the ns and $(n - 1)d$ sublevels:
$1s^2 2s^2 2p^6 3s^2 3p^6 4s^2 3d^{10} 4p^6 5s^1 4d^5$.

The condensed electron configuration is $[Kr] 5s^1 4d^5$.

The partial orbital diagram for valence electrons is

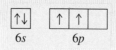

Mo is a transition element in Group 6B(6) of Period 5, so there are 36 inner electrons.

(c) For Pb ($Z = 82$), the full electron configuration is
$1s^2 2s^2 2p^6 3s^2 3p^6 4s^2 3d^{10} 4p^6 5s^2 4d^{10} 5p^6 6s^2 4f^{14} 5d^{10} 6p^2$.

The condensed electron configuration is $[Xe] 6s^2 4f^{14} 5d^{10} 6p^2$.

The partial orbital diagram for valence electrons (no filled inner sublevels) is

Pb is a main-group element in Group 4A(14) of Period 6, so there are 54 (in Xe) + 14 (in 4f series) + 10 (in 5d series) = 78 inner electrons.

Check Be sure the sum of the superscripts (electrons) in the full electron configuration equals the atomic number, and that the number of *valence* electrons in the condensed configuration equals the number of electrons in the partial orbital diagram.

FOLLOW-UP PROBLEM 8.2 Without referring to Table 8.3 or Figure 8.5, give full and condensed electron configurations, partial orbital diagrams showing valence electrons, and the number of inner electrons for the following elements:
(a) Ni (Z = 28)   **(b)** Sr (Z = 38)   **(c)** Po (Z = 84)

SECTION SUMMARY

In the aufbau method, one electron is added to an atom of each successive element in accord with Pauli's exclusion principle (no two electrons can have the same set of quantum numbers) and Hund's rule (orbitals of equal energy become half-filled, with electron spins parallel, before any pairing occurs). The elements of a group have similar outer electron configurations and similar chemical behavior. For the main-group elements, valence electrons (those involved in reactions) are in the outer (highest energy) level only. For transition elements, (n − 1)d electrons are also involved in reactions. In general, (n − 1)d orbitals fill after ns and before np orbitals. In Periods 6 and 7, (n − 2)f orbitals fill between the first and second (n − 1)d orbitals.

8.4 TRENDS IN THREE KEY ATOMIC PROPERTIES

All physical and chemical behavior of the elements is based ultimately on the electron configurations of their atoms. In this section, we focus on three properties of atoms that are directly influenced by electron configuration and, thus, effective nuclear charge: atomic size, ionization energy (the energy required to remove an electron from a gaseous atom), and electron affinity (the energy change involved in adding an electron to a gaseous atom). These properties are *periodic:* they generally increase and decrease in a recurring manner throughout the periodic table. As a result, their relative magnitudes can often be predicted, and they often exhibit consistent changes, or *trends,* within a group or period that correlate with element behavior.

Trends in Atomic Size

In Chapter 7, we noted that an electron in an atom can lie relatively far from the nucleus, so we commonly represent atoms as spheres in which the electrons spend 90% of their time. However, we often *define* **atomic size** in terms of how closely one atom lies next to another. In practice, we measure the distance between identical, adjacent atomic nuclei in a sample of an element and divide that distance in half. (The technique is discussed in Chapter 12.) Because atoms do not have hard surfaces, the size of an atom in a compound depends somewhat on the atoms near it. In other words, *atomic size varies slightly from substance to substance.*

Figure 8.8 shows two common definitions of atomic size. The **metallic radius** is one-half the distance between nuclei of adjacent atoms in a crystal of the element; we typically use this definition for metals. For elements commonly occurring as molecules, mostly nonmetals, we define atomic size by the **covalent radius,** one-half the distance between nuclei of identical covalently bonded atoms.

Trends Among the Main-Group Elements
Atomic size greatly influences other atomic properties and is critical to understanding element behavior. Figure 8.9 (on the next page) shows the atomic radii of the main-group elements and most of the transition elements. Among the main-group elements, note that atomic size

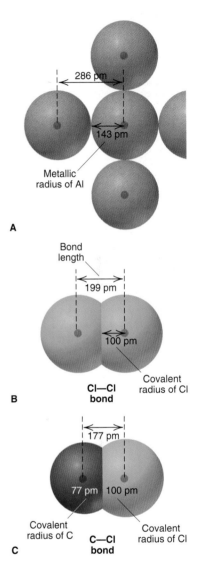

Figure 8.8 Defining metallic and covalent radii. A, The metallic radius is one-half the distance between nuclei of adjacent atoms in a crystal of the element, as shown here for aluminum. **B,** The covalent radius is one-half the distance between bonded nuclei in a molecule of the element, as shown here for chlorine. In effect, it is one-half the bond length. **C,** In a covalent compound, the bond length and known covalent radii are used to determine other radii. Here the C—Cl bond length (177 pm) and the covalent radius of Cl (100 pm) are used to find a value for the covalent radius of C (177 pm − 100 pm = 77 pm).

Figure 8.9 Atomic radii of the main-group and transition elements. Atomic radii (in picometers) are shown as half-spheres of proportional size for the main-group elements (*tan*) and the transition elements (*blue*). Among the main-group elements, atomic radius generally *increases* from top to bottom and *decreases* from left to right. The transition elements do not exhibit these trends as consistently. (Values in parentheses have only two significant figures; values for the noble gases are based on quantum-mechanical calculations.)

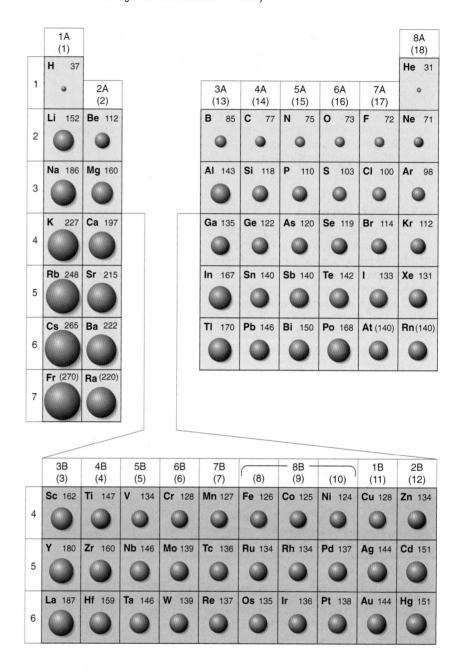

varies within both a group and a period. These variations in atomic size are the result of two opposing influences:

1. *Changes in n.* As the principal quantum number (n) increases, the probability that the outer electrons will spend more time farther from the nucleus increases as well; thus, the atoms are larger.
2. *Changes in Z_{eff}.* As the effective nuclear charge (Z_{eff})—the positive charge "felt" by an electron—increases, outer electrons are pulled closer to the nucleus; thus, the atoms are smaller.

The net effect of these influences depends on shielding of the increasing nuclear charge by inner electrons:

1. *Down a group, n dominates.* As we move down a main group, each member has one more level of *inner* electrons that shield the *outer* electrons very effectively. Even though calculations show Z_{eff} on the outer electrons rising moderately for

each element in the group, the atoms get larger as a result of the increasing n value. *Atomic radius generally **increases** in a group from top to bottom.*

2. *Across a period, Z_{eff} dominates.* As we move across a period of main-group elements, electrons are added to the same outer level, so the shielding by inner electrons does not change. Because outer electrons shield each other poorly, Z_{eff} on the outer electrons rises significantly, and so they are pulled closer to the nucleus. *Atomic radius generally **decreases** in a period from left to right.*

Trends Among the Transition Elements As Figure 8.9 shows, these trends hold well for the main-group elements but *not* as consistently for the transition elements. As we move from left to right, size shrinks through the first two or three transition elements because of the increasing nuclear charge. But, from then on, *the size remains relatively constant* because shielding by the inner d electrons counteracts the usual increase in Z_{eff}. For instance, vanadium (V; $Z = 23$), the third Period 4 transition metal, has the same atomic radius as zinc (Zn; $Z = 30$), the last Period 4 transition metal. This pattern of atomic size shrinking also appears in Periods 5 and 6 in the d-block transition series and in both series of inner transition elements. The *lack* of a vertical size increase from the Period 5 to 6 transition metal is especially obvious.

This shielding by d electrons causes a major size decrease from Group 2A(2) to Group 3A(13), the two main groups that flank the transition series. The size decrease in Periods 4, 5, and 6 (*with* a transition series) is much greater than in Period 3 (*without* a transition series). Because electrons in the np orbitals penetrate more than those in the $(n - 1)d$ orbitals, the first np electron [Group 3A(13)] "feels" a Z_{eff} that has been increased by the protons added to all the intervening transition elements. The greatest change in size occurs in Period 4, in which calcium (Ca; $Z = 20$) is nearly 50% larger than gallium (Ga; $Z = 31$). In fact, shielding by the d orbitals in the transition series causes such a major size contraction that gallium is slightly *smaller* than aluminum (Al; $Z = 13$), even though Ga is below Al in the same group!

SAMPLE PROBLEM 8.3 Ranking Elements by Atomic Size

Problem Using only the periodic table (not Figure 8.9), rank each set of main-group elements in order of *decreasing* atomic size:
(a) Ca, Mg, Sr   (b) K, Ga, Ca
(c) Br, Rb, Kr   (d) Sr, Ca, Rb
Plan To rank the elements by atomic size, we find them in the periodic table. They are main-group elements, so size increases down a group and decreases across a period.
Solution (a) Sr > Ca > Mg. These three elements are in Group 2A(2), and size decreases up the group.
(b) K > Ca > Ga. These three elements are in Period 4, and size decreases across a period.
(c) Rb > Br > Kr. Rb is largest because it has one more energy level and is farthest to the left. Kr is smaller than Br because Kr is farther to the right in Period 4.
(d) Rb > Sr > Ca. Ca is smallest because it has one fewer energy level. Sr is smaller than Rb because it is farther to the right.
Check From Figure 8.9, we see that the rankings are correct.

FOLLOW-UP PROBLEM 8.3 Using only the periodic table, rank the elements in each set in order of *increasing* size: (a) Se, Br, Cl; (b) I, Xe, Ba.

Figure 8.10 (on the next page) shows the overall variation in atomic size with increasing atomic number. Note the recurring up-and-down pattern as size drops across a period to the noble gas and then leaps up to the alkali metal that begins the next period. Also note how each transition series, beginning with that in Period 4 (K to Kr), throws off the smooth size decrease.

Figure 8.10 Periodicity of atomic radius. A plot of atomic radius vs. atomic number for the elements in Periods 1 through 6 shows a periodic change: the radius generally decreases through a period to the noble gas [Group 8A(18); *purple*] and then increases suddenly to the next alkali metal [Group 1A(1); *brown*]. Deviation from the general decrease occurs among the transition elements.

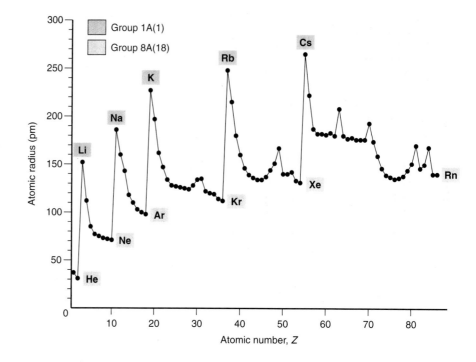

Trends in Ionization Energy

The **ionization energy (IE)** is the energy (in kJ) required for the *complete removal* of 1 mol of electrons from 1 mol of gaseous atoms or ions. Pulling an electron away from a nucleus *requires* energy to overcome the attraction. Because energy flows *into* the system, the ionization energy is always positive (like ΔH of an endothermic reaction).

In Chapter 7, you saw that the ionization energy of the H atom is the energy difference between $n = 1$ and $n = \infty$, the point at which the electron is completely removed. Many-electron atoms can lose more than one electron. The first ionization energy (IE_1) removes an outermost electron (highest energy sublevel) from the gaseous atom:

$$\text{Atom}(g) \longrightarrow \text{ion}^+(g) + e^- \qquad \Delta E = IE_1 > 0 \qquad (8.2)$$

The second ionization energy (IE_2) removes a second electron. This electron is pulled away from a positively charged ion, so IE_2 is always larger than IE_1:

$$\text{Ion}^+(g) \longrightarrow \text{ion}^{2+}(g) + e^- \qquad \Delta E = IE_2 \text{ (always} > IE_1)$$

The first ionization energy is a key factor in an element's chemical reactivity because, as you'll see, *atoms with a low IE_1 tend to form cations during reactions, whereas those with a high IE_1 (except the noble gases) often form anions.*

Variations in First Ionization Energy The elements exhibit a periodic pattern in first ionization energy, as shown in Figure 8.11. By comparing this figure with Figure 8.10, you can see a roughly *inverse* relationship between IE_1 and atomic size: *as size decreases, it takes more energy to remove an electron.* This inverse relationship appears throughout the groups and periods of the table.

Let's examine the two trends and their exceptions.

1. *Down a group.* As we move *down* a main group, the orbital's n value increases and so does atomic size. As the distance from nucleus to outermost electron increases, the attraction between them lessens, which makes the electron easier to remove. Figure 8.12 shows that *ionization energy generally decreases down a group:* it is easier to remove an outer electron from an element in Period 6 than Period 2.

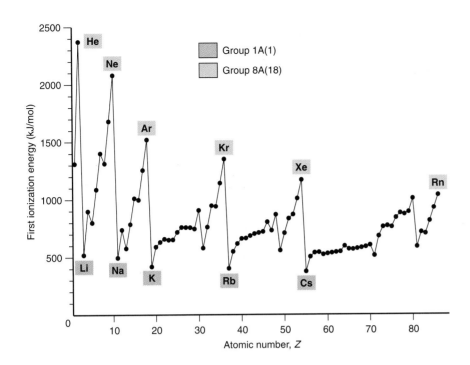

Figure 8.11 Periodicity of first ionization energy (IE₁). A plot of IE_1 vs. atomic number for the elements in Periods 1 through 6 shows a periodic pattern: the lowest values occur for the alkali metals (*brown*) and the highest for the noble gases (*purple*). This is the *inverse* of the trend in atomic size (see Figure 8.10).

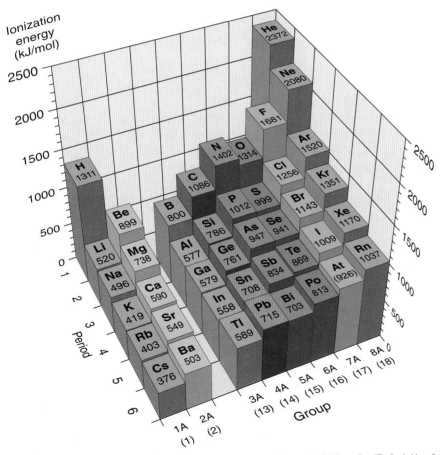

Figure 8.12 First ionization energies of the main-group elements. Values for IE_1 (in kJ/mol) of the main-group elements are shown as posts of varying height. Note the general increase within a period and decrease within a group. Thus, the lowest value is at the bottom of Group 1A(1), and the highest is at the top of Group 8A(18).

The only significant exception to this pattern occurs in Group 3A(13), right after the transition series, and is due to the effect of the series on atomic size: IE_1 decreases from boron (B) to aluminum (Al), but not for the rest of the group. Filling the d sublevels in Periods 4, 5, and 6 causes a greater-than-expected Z_{eff}, which holds the outer electrons more tightly in the larger Group 3A members.

2. *Across a period.* As we move left to right across a period, the orbital's n value stays the same, so Z_{eff} increases and atomic size decreases. As a result, the attraction between nucleus and outer electrons increases, which makes an electron harder to remove. *Ionization energy generally **increases across** a period:* it is easier to remove an outer electron from an alkali metal than from a noble gas.

There are several small "dips" in the otherwise smooth increase in ionization energy. These occur in Group 3A(13) for B and Al and in Group 6A(16) for O and S. The dips in Group 3A occur because these electrons are the first in the np sublevel. This sublevel is higher in energy than the ns, so the electron in it is pulled off more easily, leaving a stable, filled ns sublevel. The dips in Group 6A occur because the np^4 electron is the first to pair up with another np electron, and electron-electron repulsions raise the orbital energy. Removing the np^4 electron relieves the repulsions and leaves a stable, half-filled np sublevel; so the fourth p electron comes off more easily than the third one does.

SAMPLE PROBLEM 8.4   Ranking Elements by First Ionization Energy

Problem Using the periodic table only, rank the elements in each of the following sets in order of *decreasing* IE_1:
(a) Kr, He, Ar   **(b)** Sb, Te, Sn   **(c)** K, Ca, Rb   **(d)** I, Xe, Cs
Plan As in Sample Problem 8.3, we first find the elements in the periodic table and then apply the general trends of decreasing IE_1 down a group and increasing IE_1 across a period.
Solution (a) He > Ar > Kr. These three are all in Group 8A(18), and IE_1 decreases down a group.
(b) Te > Sb > Sn. These three are all in Period 5, and IE_1 increases across a period.
(c) Ca > K > Rb. IE_1 of K is larger than IE_1 of Rb because K is higher in Group 1A(1). IE_1 of Ca is larger than IE_1 of K because Ca is farther to the right in Period 4.
(d) Xe > I > Cs. IE_1 of I is smaller than IE_1 of Xe because I is farther to the left. IE_1 of I is larger than IE_1 of Cs because I is farther to the right and in the previous period.
Check Because trends in IE_1 are generally the opposite of the trends in size, you can rank the elements by size and check that you obtain the reverse order.

FOLLOW-UP PROBLEM 8.4 Rank the elements in each of the following sets in order of increasing IE_1:
(a) Sb, Sn, I   **(b)** Sr, Ca, Ba

Variations in Successive Ionization Energies Successive ionization energies (IE_1, IE_2, and so on) of a given element increase because each electron is pulled away from an ion with a progressively higher positive charge. Note from Figure 8.13, however, that this increase is not smooth, but includes an enormous jump.

A more complete picture is presented in Table 8.4, which shows successive ionization energies for the elements in Period 2 and the first element in Period 3. Move horizontally through the values for a given element, and you reach a point that separates relatively low from relatively high IE values (shaded area to right of line). This jump appears *after* the outer (valence) electrons have been removed and, thus, reflects the much greater energy needed to remove an inner (core) electron. For example, follow the values for boron (B): IE_1 is lower than IE_2, which is lower than IE_3, which is *much* lower than IE_4. Thus, boron has three electrons in the highest energy level ($1s^2 2s^2 2p^1$). Because of the significantly greater energy needed to remove core electrons, they are *not* involved in chemical reactions.

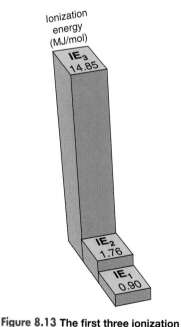

Ionization energy (MJ/mol)

IE_3 14.85

IE_2 1.76

IE_1 0.90

Figure 8.13 The first three ionization energies of beryllium (in MJ/mol). Successive ionization energies always increase, but an exceptionally large increase occurs when the first core electron is removed. For Be, this occurs with the third electron (IE_3). (Also see Table 8.4.)

Table 8.4	Successive Ionization Energies of the Elements Lithium Through Sodium											
		Number of Valence	Ionization Energy (MJ/mol)*									
Z	Element	Electrons	IE$_1$	IE$_2$	IE$_3$	IE$_4$	IE$_5$	IE$_6$	IE$_7$	IE$_8$	IE$_9$	IE$_{10}$
3	Li	1	0.52	7.30	11.81							
4	Be	2	0.90	1.76	14.85	21.01		CORE ELECTRONS				
5	B	3	0.80	2.43	3.66	25.02	32.82					
6	C	4	1.09	2.35	4.62	6.22	37.83	47.28				
7	N	5	1.40	2.86	4.58	7.48	9.44	53.27	64.36			
8	O	6	1.31	3.39	5.30	7.47	10.98	13.33	71.33	84.08		
9	F	7	1.68	3.37	6.05	8.41	11.02	15.16	17.87	92.04	106.43	
10	Ne	8	2.08	3.95	6.12	9.37	12.18	15.24	20.00	23.07	115.38	131.43
11	Na	1	0.50	4.56	6.91	9.54	13.35	16.61	20.11	25.49	28.93	141.37

*MJ/mol, or megajoules per mole = 10^3 kJ/mol.

SAMPLE PROBLEM 8.5   Identifying an Element from Successive Ionization Energies

Problem Name the Period 3 element with the following ionization energies (in kJ/mol), and write its electron configuration:

IE$_1$	IE$_2$	IE$_3$	IE$_4$	IE$_5$	IE$_6$
1012	1903	2910	4956	6278	22,230

Plan We look for a large jump in the IE values, which occurs after all valence electrons have been removed. Then we refer to the periodic table to find the Period 3 element with this number of valence electrons and write its electron configuration.

Solution The exceptionally large jump occurs after IE$_5$, indicating that the element has five valence electrons and, thus, is in Group 5A(15). This Period 3 element is phosphorus (P; $Z = 15$). Its electron configuration is $1s^2 2s^2 2p^6 3s^2 3p^3$.

FOLLOW-UP PROBLEM 8.5 Element Q is in Period 3 and has the following ionization energies (in kJ/mol):

IE$_1$	IE$_2$	IE$_3$	IE$_4$	IE$_5$	IE$_6$
577	1816	2744	11,576	14,829	18,375

Name element Q and write its electron configuration.

Trends in Electron Affinity

The **electron affinity (EA)** is the energy change (in kJ) accompanying the *addition* of 1 mol of electrons to 1 mol of gaseous atoms or ions. As with ionization energy, there is a first electron affinity, a second, and so forth. The *first electron affinity* (EA$_1$) refers to the formation of 1 mol of monovalent (1−) gaseous anions:

$$\text{Atom}(g) + \text{e}^- \longrightarrow \text{ion}^-(g) \quad \Delta E = \text{EA}_1$$

In most cases, *energy is released when the first electron is added* because it is attracted to the atom's nuclear charge. Thus, EA$_1$ is usually negative (just as ΔH for an exothermic reaction is negative).* The second electron affinity (EA$_2$), on the other hand, is always positive because energy must be *absorbed*

---

*Tables of first electron affinity often list them as positive if energy is *absorbed* to remove an electron from the anion. Keep this convention in mind when researching these values in reference texts. Electron affinities are difficult to measure, so values are frequently updated with more accurate data. Values for Group 2A(2) reflect recent changes.

Figure 8.14 Electron affinities of the main-group elements. The electron affinities (in kJ/mol) of the main-group elements are shown. Negative values indicate that energy is released when the anion forms. Positive values, which occur in Group 8A(18), indicate that energy is absorbed to form the anion; in fact, these anions are unstable and the values are estimated.

1A (1)			3A (13)	4A (14)	5A (15)	6A (16)	7A (17)	8A (18)
H −72.8	**2A** (2)							**He** (0.0)
Li −59.6	**Be** ≤0		**B** −26.7	**C** −122	**N** +7	**O** −141	**F** −328	**Ne** (+29)
Na −52.9	**Mg** ≤0		**Al** −42.5	**Si** −134	**P** −72.0	**S** −200	**Cl** −349	**Ar** (+35)
K −48.4	**Ca** −2.37		**Ga** −28.9	**Ge** −119	**As** −78.2	**Se** −195	**Br** −325	**Kr** (+39)
Rb −46.9	**Sr** −5.03		**In** −28.9	**Sn** −107	**Sb** −103	**Te** −190	**I** −295	**Xe** (+41)
Cs −45.5	**Ba** −13.95		**Tl** −19.3	**Pb** −35.1	**Bi** −91.3	**Po** −183	**At** −270	**Rn** (+41)

in order to overcome electrostatic repulsions and add another electron to a negative ion.

Factors other than Z_{eff} and atomic size affect electron affinities, so trends are not as regular as those for the previous two properties. For instance, we might expect electron affinities to decrease smoothly down a group (smaller negative number) because the nucleus is farther away from an electron being added. But, as Figure 8.14 shows, only Group 1A(1) exhibits this behavior. We might also expect a regular increase in electron affinities across a period (larger negative number) because size decreases and the increasing Z_{eff} should attract the electron being added more strongly. An overall left-to-right increase in magnitude is there, but we certainly cannot say that it is a regular increase. These exceptions arise from changes in sublevel energy and in electron-electron repulsion.

Despite the irregularities, three key points emerge when we examine the relative values of ionization energy and electron affinity:

1. *Reactive nonmetals.* The elements in Groups 6A(16) and especially those in Group 7A(17) (halogens) have high ionization energies and highly negative (exothermic) electron affinities. These elements lose electrons with difficulty but attract them strongly. Therefore, *in their ionic compounds, they form negative ions.*

2. *Reactive metals.* The elements in Groups 1A(1) and 2A(2) have low ionization energies and slightly negative (exothermic) electron affinities. Both groups lose electrons readily but attract them only weakly, if at all. Therefore, *in their ionic compounds, they form positive ions.*

3. *Noble gases.* The elements in Group 8A(18) have very high ionization energies and slightly positive (endothermic) electron affinities. Therefore, *these elements tend **not** to lose or gain electrons.* In fact, only the larger members of the group (Kr, Xe, Rn) form any compounds at all.

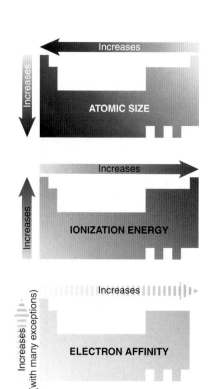

Figure 8.15 Trends in three atomic properties. Periodic trends are depicted as gradations in shading on miniature periodic tables, with arrows indicating the direction of general *increase* in a group or period. For electron affinity, Group 8A(18) is not shown, and the dashed arrows indicate the numerous exceptions to expected trends.

SECTION SUMMARY

Trends in three atomic properties are summarized in Figure 8.15. Atomic size increases down a main group and decreases across a period. Across a transition series, size remains relatively constant. First ionization energy (the energy required to remove the outermost electron from a mole of gaseous atoms) is inversely related to atomic size: IE_1 decreases down a main group and increases across a period. An element's successive ionization energies show a very large increase when the first inner (core) electron is removed. Electron affinity (the energy involved in adding an electron to a mole of gaseous atoms) shows many variations from expected trends. Based on the relative sizes of IEs and EAs, in their ionic compounds, the Group 1A(1) and 2A(2) elements tend to form cations, and the Group 6A(16) and 7A(17) elements tend to form anions.

8.5 ATOMIC STRUCTURE AND CHEMICAL REACTIVITY

Our main purpose for discussing atomic properties is, of course, to see how they affect element behavior. In this section, you'll see how the properties we just examined influence metallic behavior and determine the type of ion an element can form, as well as how electron configuration relates to magnetic properties.

Trends in Metallic Behavior

Metals are located in the left and lower three-quarters of the periodic table. They are typically shiny solids with moderate to high melting points, are good thermal and electrical conductors, can be drawn into wires and rolled into sheets, and tend to lose electrons to nonmetals. *Nonmetals* are located in the upper right quarter of the table. They are typically not shiny, have relatively low melting points, are poor thermal and electrical conductors, are mostly crumbly solids or gases, and tend to gain electrons from metals. *Metalloids* are located in the region between the other two classes and have properties between them as well. Thus, *metallic behavior decreases left to right and increases top to bottom* in the periodic table (Figure 8.16).

It's important to realize, however, that an element's properties may not fall neatly into our categories. For instance, the nonmetal carbon in the form of graphite is a good electrical conductor. Iodine, another nonmetal, is a shiny solid. Gallium and cesium are metals that melt at temperatures below body temperature, and mercury is a liquid at room temperature. And iron is quite brittle. Despite such exceptions, we can make several generalizations about metallic behavior.

Relative Tendency to Lose Electrons Metals tend to lose electrons during chemical reactions because they have low ionization energies compared to nonmetals. The increase in metallic behavior down a group is most obvious in the physical and chemical behavior of the elements in Groups 3A(13) through 6A(16), which contain more than one class of element. For example, consider the elements in Group 5A(15). Here, the change is so great that, with regard to monatomic ions, *elements at the top tend to form anions and those at the bottom tend to form cations.* Nitrogen (N) is a gaseous nonmetal, and phosphorus (P) is a solid nonmetal. Both occur occasionally as 3− anions in their compounds. Arsenic (As) and antimony (Sb) are metalloids, with Sb the more metallic of the two; neither forms ions readily. Bismuth (Bi), the largest member, is a typical metal, forming mostly ionic compounds in which it appears as a 3+ cation. Even in Group 2A(2), which consists entirely of metals, the tendency to form cations increases down the group. Thus, beryllium (Be) forms covalent compounds with nonmetals, whereas the compounds of barium (Ba) are ionic.

As we move across a period, it becomes more difficult to lose an electron (IE increases) and easier to gain one (EA becomes more negative). Therefore, with regard to monatomic ions, *elements at the left tend to form cations and those at the right tend to form anions.* The typical decrease in metallic behavior across a period is clear among the elements in Period 3. Sodium and magnesium are metals. Sodium is shiny when freshly cut under mineral oil, but it loses an electron so readily to O_2 that, if cut in air, its surface is coated immediately with a dull oxide. These metals exist naturally as Na^+ and Mg^{2+} ions in oceans, minerals, and

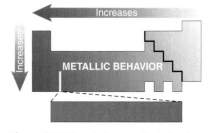

Figure 8.16 Trends in metallic behavior. The gradation in metallic behavior among the elements is depicted as a gradation in shading from bottom left to top right, with arrows showing the direction of increase. (Hydrogen appears next to helium in this periodic table.)

organisms. Aluminum is metallic in its physical properties and forms the Al^{3+} ion in some compounds, but it bonds covalently in most others. Silicon (Si) is a shiny metalloid that does not occur as a monatomic ion. The most common form of phosphorus is a white, waxy nonmetal that, as noted above, forms the P^{3-} ion in a few compounds. Sulfur is a crumbly yellow nonmetal that forms the sulfide ion (S^{2-}) in many compounds. Diatomic chlorine (Cl_2) is a yellow-green, gaseous nonmetal that attracts electrons avidly and exists in nature as the Cl^- ion.

Acid-Base Behavior of the Element Oxides Metals are also distinguished from nonmetals by the acid-base behavior of their oxides in water:

- Most main-group metals *transfer* electrons to oxygen, so their *oxides are ionic. In water, the oxides act as bases,* producing OH^- ions from O^{2-} and reacting with acids.
- Nonmetals *share* electrons with oxygen, so *nonmetal oxides are covalent. In water, they act as acids,* producing H^+ ions and reacting with bases.

Some metals and many metalloids form oxides that are **amphoteric:** they can act as acids *or* as bases in water.

Figure 8.17 classifies the acid-base behavior of some common oxides, focusing on the elements in Group 5A(15) and Period 3. Note that *as the elements become more metallic down a group, their oxides become more basic.* Among oxides of Group 5A(15) elements, dinitrogen pentaoxide (N_2O_5) forms nitric acid, HNO_3, a strong acid, while tetraphosphorus decaoxide (P_4O_{10}) forms a weaker acid, H_3PO_4. The oxide of the metalloid arsenic (As_2O_5) is weakly acidic, whereas that of the metalloid antimony (Sb_2O_5) is weakly basic. Bismuth, the most metallic element of the group, forms a basic oxide (Bi_2O_3).

Figure 8.17 The trend in acid-base behavior of element oxides. The trend in acid-base behavior for some common oxides of Group 5A(15) and Period 3 elements is shown as a gradation in color (*red* = acidic; *blue* = basic). Note that the metals form basic oxides and the nonmetals form acidic oxides. Aluminum forms an oxide (*purple*) that can act as an acid or as a base. Thus, as atomic size increases, ionization energy decreases, and oxide basicity increases.

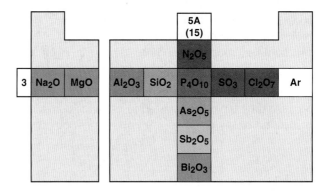

Note that *as the elements become less metallic across a period, their oxides become more acidic.* In Period 3, sodium and magnesium form the strongly basic oxides Na_2O and MgO. Metallic aluminum forms amphoteric aluminum oxide (Al_2O_3), which reacts with acid or with base, whereas silicon dioxide is weakly acidic. The common oxides of phosphorus, sulfur, and chlorine form acids of increasing strength: H_3PO_4, H_2SO_4, and $HClO_4$.

Properties of Monatomic Ions

So far we have focused on the reactants—the atoms—in the process of electron loss and gain. Now we focus on the products—the ions. We examine electron configurations, magnetic properties, and ionic radius relative to atomic radius.

Electron Configurations of Main-Group Ions In Chapter 2, you learned the symbols and charges of many monatomic ions. But *why* does an ion have a particular charge in its compounds? Why is a sodium ion Na^+ and not Na^{2+}, and why is a fluoride ion F^- and not F^{2-}? For elements at the left and right ends of the periodic table, the explanation concerns the very low reactivity of the noble gases. As we said earlier, because they have high IEs and positive (endothermic) EAs, the noble gases typically do not form ions but remain chemically stable with a *filled* outer energy level (ns^2np^6). *Elements in Groups 1A(1), 2A(2), 6A(16), and 7A(17) that readily form ions either lose or gain electrons to attain a filled outer level and thus a noble gas configuration.* Their ions are said to be **isoelectronic** (Greek *iso,* "same") with the nearest noble gas. Figure 8.18 shows this relationship.

When an alkali metal atom [Group 1A(1)] loses its single valence electron, it becomes isoelectronic with the *previous* noble gas. The Na^+ ion, for example, is isoelectronic with neon (Ne):

$$Na\ (1s^22s^22p^63s^1) \longrightarrow Na^+\ (1s^22s^22p^6)\ \text{[isoelectronic with Ne } (1s^22s^22p^6)] + e^-$$

When a halogen atom [Group 7A(17)] adds a single electron to the five in its *np* sublevel, it becomes isoelectronic with the *next* noble gas. Bromide ion, for example, is isoelectronic with krypton (Kr):

$$Br\ ([Ar]\ 4s^23d^{10}4p^5) + e^- \longrightarrow Br^-\ ([Ar]\ 4s^23d^{10}4p^6)\ \text{[isoelectronic with Kr } ([Ar]\ 4s^23d^{10}4p^6)]$$

The energy needed to remove the electrons from metals to attain the previous noble gas configuration is supplied during their exothermic reactions with nonmetals. Removing more than one electron from Na to form Na^{2+} or more than two from Mg to form Mg^{3+} means removing core electrons, which requires more energy than is available in a reaction. This is the reason that $NaCl_2$ and MgF_3 do *not* exist. Similarly, adding two electrons to F to form F^{2-} or three to O to form O^{3-} means placing the extra electron into the next energy level. With 18 electrons acting as inner electrons and shielding the nuclear charge very effectively, adding an electron to the negative ion, F^- or O^{2-}, requires too much energy. Thus, we never see Na_2F or Mg_3O_2.

The larger metals of Groups 3A(13), 4A(14), and 5A(15) form cations through a different process, because it would be energetically impossible for them to lose enough electrons to attain a noble gas configuration. For example, tin (Sn; $Z = 50$) would have to lose 14 electrons—two 5*p*, ten 4*d*, and two 5*s*—to be isoelectronic with krypton (Kr; $Z = 36$), the previous noble gas. Instead, tin loses far fewer electrons and attains two different stable configurations. In the tin(IV) ion (Sn^{4+}), the metal atom empties its outer energy level and attains the stability of empty 5*s* and 5*p* sublevels and a filled inner 4*d* sublevel. This $(n - 1)d^{10}$ configuration is called a **pseudo–noble gas configuration:**

$$Sn\ ([Kr]\ 5s^24d^{10}5p^2) \longrightarrow Sn^{4+}\ ([Kr]\ 4d^{10}) + 4e^-$$

Alternatively, in the more common tin(II) ion (Sn^{2+}), the atom loses the two 5*p* electrons only and attains the stability of filled 5*s* and 4*d* sublevels:

$$Sn\ ([Kr]\ 5s^24d^{10}5p^2) \longrightarrow Sn^{2+}\ ([Kr]\ 5s^24d^{10}) + 2e^-$$

The retained ns^2 electrons are sometimes called an *inert pair* because they seem difficult to remove. Thallium, lead, and bismuth, the largest and most metallic members of Groups 3A(13) to 5A(15), commonly form ions that retain the ns^2 pair of electrons: Tl^+, Pb^{2+}, and Bi^{3+}.

Excessively high energy cost is also the reason that some elements do not form monatomic ions in any of their reactions. For instance, carbon would have to lose four electrons to form C^{4+} and attain the He configuration, or gain four to form C^{4-} and attain the Ne configuration, but neither ion forms. (Such multivalent ions *are* observed in the spectra of stars, however, where temperatures exceed 10^6 K.) As you'll see in Chapter 9, carbon and other atoms that do not form ions attain a filled shell by *sharing* electrons through covalent bonding.

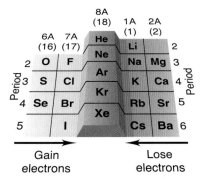

Figure 8.18 Main-group ions and the noble gas electron configurations. Most of the elements that form monatomic ions that are isoelectronic with a noble gas lie in the four groups that flank Group 8A(18), two on either side.

SAMPLE PROBLEM 8.6 Writing Electron Configurations of Main-Group Ions

Problem Using condensed electron configurations, write reactions for the formation of the common ions of the following elements:
(a) Iodine ($Z = 53$)   **(b)** Potassium ($Z = 19$)   **(c)** Indium ($Z = 49$)

Plan We identify the element's position in the periodic table and recall two general points:
• Ions of elements in Groups 1A(1), 2A(2), 6A(16), and 7A(17) are isoelectronic with the nearest noble gas.
• Metals in Groups 3A(13) to 5A(15) lose the ns and np electrons or just the np electrons.

Solution (a) Iodine is in Group 7A(17), so it gains one electron and is isoelectronic with xenon:

$$I\ ([Kr]\ 5s^2 4d^{10} 5p^5) + e^- \longrightarrow I^-\ ([Kr]\ 5s^2 4d^{10} 5p^6)\quad \text{(same as Xe)}$$

(b) Potassium is in Group 1A(1), so it loses one electron and is isoelectronic with argon:

$$K\ ([Ar]\ 4s^1) \longrightarrow K^+\ ([Ar]) + e^-$$

(c) Indium is in Group 3A(13), so it loses either three electrons to form In^{3+} (pseudo–noble gas configuration) or one to form In^+ (inert pair):

$$In\ ([Kr]\ 5s^2 4d^{10} 5p^1) \longrightarrow In^{3+}\ ([Kr]\ 4d^{10}) + 3e^-$$

$$In\ ([Kr]\ 5s^2 4d^{10} 5p^1) \longrightarrow In^+\ ([Kr]\ 5s^2 4d^{10}) + e^-$$

Check Be sure that the number of electrons in the ion's electron configuration, plus those gained or lost to form the ion, equals Z.

FOLLOW-UP PROBLEM 8.6 Using condensed electron configurations, write reactions showing the formation of the common ions of the following elements:
(a) Ba ($Z = 56$)   **(b)** O ($Z = 8$)   **(c)** Pb ($Z = 82$)

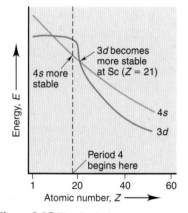

Figure 8.19 The Period 4 crossover in sublevel energies. The 3*d* orbitals are empty in elements at the beginning of Period 4. Because the 4*s* electron penetrates closer to the nucleus, the energy of the 4*s* orbital is lower in K and Ca; thus, the 4*s* fills before the 3*d*. But as the 3*d* orbitals fill, beginning with $Z = 21$, these inner electrons are attracted by the increasing nuclear charge, and they also shield the 4*s* electrons. As a result, there is an energy crossover, with the 3*d* sublevel becoming lower in energy than the 4*s*. For this reason, the 4*s* electrons are removed first when the transition metal ion forms. In other words,
• For a main-group metal ion, the highest *n* level of electrons is "last-in, first-out."
• For a transition metal ion, the highest *n* level of electrons is "first-in, first out."

Electron Configurations of Transition Metal Ions In contrast to most main-group ions, *transition metal ions rarely attain a noble gas configuration,* and the reason, once again, is that energy costs are too high. The exceptions in Period 4 are scandium, which forms Sc^{3+}, and titanium, which occasionally forms Ti^{4+} in some compounds. The typical behavior of a transition element is to *form more than one cation by losing all of its ns and some of its (n − 1)d electrons.* (We focus here on the Period 4 series, but these points hold for Periods 5 and 6 also.)

In the aufbau process of building up the ground-state atoms, Period 3 ends with the noble gas argon. At the beginning of Period 4, the radial probability distribution of the 4*s* orbital near the nucleus makes it more stable than the empty 3*d*. Therefore, the first and second electrons added in the period enter the 4*s* in K and Ca. But, as soon as we reach the transition elements and the 3*d* orbitals begin to fill, the increasing nuclear charge attracts their electrons more and more strongly. Moreover, the added 3*d* electrons fill inner orbitals, so they are not very well shielded from the increasing nuclear charge by the 4*s* electrons. As a result, the 3*d* orbital becomes *more stable* than the 4*s*. In effect, a *crossover in orbital energy* occurs as we enter the transition series (Figure 8.19). The effect on ion formation is critical: because the 3*d* orbitals are more stable, *the 4s electrons are lost before the 3d electrons to form the Period 4 transition metal ions.* Thus, the 4*s* electrons are added before the 3*d* to form the *atom* but lost before the 3*d* to form the *ion:* "first-in, first-out."

To summarize, *electrons with the highest n value are removed first.* Here are a few simple rules for forming the ion of any main-group or transition element:

• For main-group, *s*-block metals, remove all electrons with the highest *n* value.
• For main-group, *p*-block metals, remove *np* electrons before *ns* electrons.
• For transition (*d*-block) metals, remove *ns* electrons before $(n - 1)d$ electrons.
• For nonmetals, add electrons to the *p* orbitals of highest *n* value.

Magnetic Properties of Transition Metal Ions If we can't see electrons in orbitals, how do we know that a particular electron configuration is correct? Although analysis of atomic spectra is the most important method for determining configuration, the magnetic properties of an element and its compounds can support or refute conclusions from spectra. Recall that electron spin generates a tiny magnetic field, which causes a beam of H atoms to split in an external magnetic field (see Figure 8.1). Only chemical species (atoms, ions, or molecules) with one or more *unpaired* electrons are affected by the external field. The species used in the original 1921 split-beam experiment was the silver atom:

Ag ($Z = 47$)   [Kr] $5s^1 4d^{10}$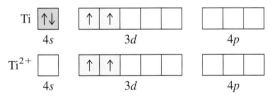

Note the unpaired $5s$ electron. A beam of cadmium atoms, the element after silver, is not split because their $5s$ electrons are *paired* (Cd: [Kr] $5s^2 4d^{10}$).

A species with unpaired electrons exhibits **paramagnetism:** it is attracted by an external magnetic field. A species with all electrons paired exhibits **diamagnetism:** it is not attracted (and, in fact, is slightly repelled) by a magnetic field. Figure 8.20 shows how this magnetic behavior is studied. Many transition metals and their compounds are paramagnetic because their atoms and ions have unpaired electrons.

Studies of paramagnetism can be used to provide additional evidence for a proposed electron configuration. Spectral analysis of the titanium atom yields the configuration [Ar] $4s^2 3d^2$. Experiment shows that Ti metal is paramagnetic, which is consistent with the presence of unpaired electrons in its atoms. Spectral analysis of the Ti^{2+} ion yields the configuration [Ar] $3d^2$, indicating loss of the two $4s$ electrons. Once again, experiment supports these findings by showing that Ti^{2+} compounds are paramagnetic. If Ti had lost its two $3d$ electrons during ion formation, its compounds would be diamagnetic because the $4s$ electrons are paired. Thus, the [Ar] $3d^2$ configuration supports the conclusion that electrons of highest n value are lost first:

$$Ti\ ([Ar]\ 4s^2 3d^2) \longrightarrow Ti^{2+}\ ([Ar]\ 3d^2) + 2e^-$$

The partial orbital diagrams are

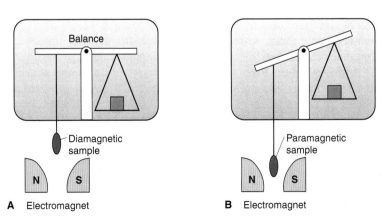

Figure 8.20 Apparatus for measuring the magnetic behavior of a sample. The substance is weighed on a very sensitive balance in the absence of an external magnetic field. **A,** If the substance is diamagnetic (has all *paired* electrons), its apparent mass is unaffected (or slightly reduced) when the magnetic field is "on." **B,** If the substance is paramagnetic (has *unpaired* electrons), its apparent mass increases when the field is "on" because the balance arm feels an additional force. This method is used to estimate the number of unpaired electrons in transition metal compounds.

SAMPLE PROBLEM 8.7 Writing Electron Configurations and Predicting Magnetic Behavior of Transition Metal Ions

Problem Use condensed electron configurations to write the reaction for the formation of each transition metal ion, and predict whether the ion is paramagnetic:
(a) Mn^{2+} $(Z = 25)$ **(b)** Cr^{3+} $(Z = 24)$ **(c)** Hg^{2+} $(Z = 80)$

Plan We first write the condensed electron configuration of the atom, noting the irregularity for Cr in (b). Then we remove electrons, beginning with ns electrons, to attain the ion charge. If unpaired electrons are present, the ion is paramagnetic.

Solution (a) Mn ([Ar] $4s^23d^5$) $\longrightarrow$ Mn^{2+} ([Ar] $3d^5$) $+ 2e^-$

There are five unpaired e^-, so Mn^{2+} is paramagnetic.

(b) Cr ([Ar] $4s^13d^5$) $\longrightarrow$ Cr^{3+} ([Ar] $3d^3$) $+ 3e^-$

There are three unpaired e^-, so Cr^{3+} is paramagnetic.

(c) Hg ([Xe] $6s^24f^{14}5d^{10}$) $\longrightarrow$ Hg^{2+} ([Xe] $4f^{14}5d^{10}$) $+ 2e^-$

There are no unpaired e^-, so Hg^{2+} is *not* paramagnetic (is diamagnetic).

Check We removed the ns electrons first, and the sum of the lost electrons and those in the electron configuration of the ion equals Z.

FOLLOW-UP PROBLEM 8.7 Write the condensed electron configuration of each transition metal ion, and predict whether it is paramagnetic:
(a) V^{3+} $(Z = 23)$ **(b)** Ni^{2+} $(Z = 28)$ **(c)** La^{3+} $(Z = 57)$

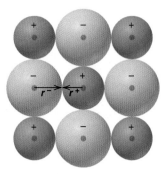

Figure 8.21 Depicting ionic radius. The cation radius (r^+) and the anion radius (r^-) each make up a portion of the total distance between the nuclei of adjacent ions in a crystalline ionic compound.

Ionic Size vs. Atomic Size The **ionic radius** is an estimate of the size of an ion in a crystalline ionic compound. You can picture it as one ion's portion of the distance between the nuclei of neighboring ions in the solid (Figure 8.21). From the relation between effective nuclear charge and atomic size, we can predict the size of an ion relative to its parent atom:

• *Cations are smaller than their parent atoms.* When a cation forms, electrons are *removed from* the outer level. The resulting decrease in electron repulsions allows the nuclear charge to pull the remaining electrons closer.

• *Anions are larger than their parent atoms.* When an anion forms, electrons are *added to* the outer level. The increase in repulsions causes the electrons to occupy more space.

Figure 8.22 shows the radii of some common main-group monatomic ions relative to their parent atoms. As you can see, *ionic size increases down a group* because the number of energy levels increases. Across a period, however, the pattern is more complex. Size decreases among the cations, then increases tremendously with the first of the anions, and finally decreases again among the anions.

This pattern results from changes in effective nuclear charge and electron-electron repulsions. In Period 3 (Na through Cl), for example, increasing Z_{eff} from left to right makes Na^+ larger than Mg^{2+}, which in turn is larger than Al^{3+}. The great jump in size from cations to anions occurs because we are *adding* electrons rather than removing them, so repulsions increase sharply. For instance, P^{3-} has eight more electrons than Al^{3+}. Then, the ongoing rise in Z_{eff} makes P^{3-} larger than S^{2-}, which is larger than Cl^-. These factors lead to some striking effects even among ions with the same number of electrons. Look at the ions within the dashed outline in Figure 8.22, which are all isoelectronic with neon. Even though the cations form from elements in the next period, the anions are still much larger. The pattern is

$$3- > 2- > 1- > 1+ > 2+ > 3+$$

When an element forms more than one cation, *the greater the ionic charge, the smaller the ionic radius.* Consider Fe^{2+} and Fe^{3+}. The number of protons is the same, but Fe^{3+} has one fewer electron, so electron repulsions are reduced somewhat. As a result, Z_{eff} increases, which pulls all the electrons closer, so Fe^{3+} is smaller than Fe^{2+}.

GROUP

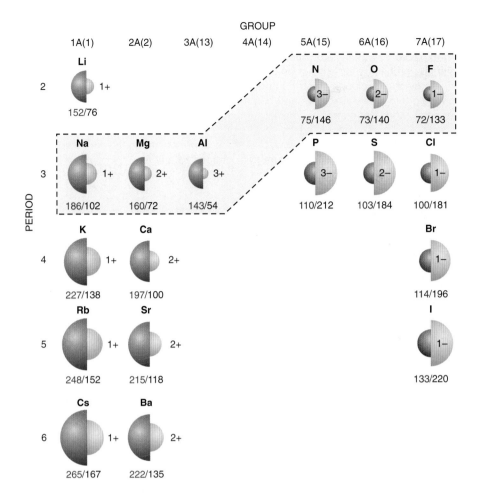

Figure 8.22 Ionic vs. atomic radii. The atomic radii (*colored half-spheres*) and ionic radii (*gray half-spheres*) of some main-group elements are arranged in periodic table format (with all radii values in picometers). Note that metal atoms (*blue*) form *smaller* positive ions, whereas nonmetal atoms (*red*) form *larger* negative ions. The dashed outline sets off ions of Period 2 nonmetals and Period 3 metals that are *isoelectronic* with neon. Note the size decrease from anions to cations.

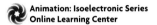
**Animation: Isoelectronic Series
Online Learning Center**

To summarize the main points,

- Ionic size increases down a group.
- Ionic size decreases across a period but increases from cations to anions.
- Ionic size decreases with increasing positive (or decreasing negative) charge in an isoelectronic series.
- Ionic size decreases as charge increases for different cations of a given element.

SAMPLE PROBLEM 8.8 Ranking Ions by Size

Problem Rank each set of ions in order of *decreasing* size, and explain your ranking:
(a) Ca^{2+}, Sr^{2+}, Mg^{2+} **(b)** K^+, S^{2-}, Cl^- **(c)** Au^+, Au^{3+}
Plan We find the position of each element in the periodic table and apply the ideas presented in the text.
Solution (a) Because Mg^{2+}, Ca^{2+}, and Sr^{2+} are all from Group 2A(2), they decrease in size up the group: $Sr^{2+} > Ca^{2+} > Mg^{2+}$.
(b) The ions K^+, S^{2-}, and Cl^- are isoelectronic. S^{2-} has a lower Z_{eff} than Cl^-, so it is larger. K^+ is a cation, and has the highest Z_{eff}, so it is smallest: $S^{2-} > Cl^- > K^+$.
(c) Au^+ has a lower charge than Au^{3+}, so it is larger: $Au^+ > Au^{3+}$.

FOLLOW-UP PROBLEM 8.8 Rank the ions in each set in order of *increasing* size:
(a) Cl^-, Br^-, F^- **(b)** Na^+, Mg^{2+}, F^- **(c)** Cr^{2+}, Cr^{3+}

SECTION SUMMARY
Metallic behavior correlates with large atomic size and low ionization energy. Thus, metallic behavior increases down a group and decreases across a period. Within the main groups, metal oxides are basic and nonmetal oxides acidic. Thus, oxides become

more acidic across a period and more basic down a group. Many main-group elements form ions that are isoelectronic with the nearest noble gas. Removing (or adding) more electrons than needed to attain the previous noble gas configuration requires a prohibitive amount of energy. Metals in Groups 3A(13) to 5A(15) lose either their *np* electrons or both their *ns* and *np* electrons. Transition metals lose *ns* electrons before $(n - 1)d$ electrons and commonly form more than one ion. Many transition metals and their compounds are paramagnetic because their atoms (or ions) have unpaired electrons. Cations are smaller and anions larger than their parent atoms. Ionic radius increases down a group. Across a period, cationic and anionic radii decrease, but a large increase occurs from cations to anions.

For Review and Reference (Numbers in parentheses refer to pages, unless noted otherwise.)

Learning Objectives

To help you review these learning objectives, the numbers of related sections (§), sample problems (SP), and end-of-chapter problems (EP) are listed in parentheses.

1. Understand the periodic law and the arrangement of elements by atomic number (§ 8.1) (EPs 8.1–8.3)
2. Describe the importance of the spin quantum number (m_s) and the exclusion principle for populating an orbital; understand how shielding and penetration lead to the splitting of energy levels into sublevels (§ 8.2) (EPs 8.4–8.13)
3. Understand orbital filling order, how outer configuration correlates with chemical behavior, and the distinction among inner, outer, and valence electrons; write the set of quantum numbers for any electron in an atom as well as full and condensed electron configurations and orbital diagrams for the atoms of any element (§ 8.3) (SPs 8.1, 8.2) (EPs 8.14–8.32)
4. Describe atomic radius, ionization energy, and electron affinity and their periodic trends; explain patterns in successive ionization energies and identify which electrons are involved in ion formation (to yield a noble gas or pseudo–noble gas electron configuration) (§ 8.4) (SPs 8.3–8.5) (EPs 8.33–8.46)
5. Describe the general properties of metals and nonmetals and understand how trends in metallic behavior relate to ion formation, oxide acidity, and magnetic behavior; understand the relation between atomic and ionic size and write ion electron configurations (§ 8.5) (SPs 8.6–8.8) (EPs 8.47–8.65)

Key Terms

electron configuration (236)

Section 8.1
periodic law (236)

Section 8.2
spin quantum number
  (m_s) (237)
exclusion principle (238)
shielding (238)

effective nuclear charge
  (Z_{eff}) (239)
penetration (239)

Section 8.3
aufbau principle (240)
orbital diagram (240)
Hund's rule (241)
transition elements (244)
inner (core) electrons (246)
outer electrons (246)

valence electrons (246)
inner transition elements (247)
lanthanides (247)
actinides (247)

Section 8.4
atomic size (249)
metallic radius (249)
covalent radius (249)
ionization energy (IE) (252)
electron affinity (EA) (255)

Section 8.5
amphoteric (258)
isoelectronic (259)
pseudo–noble gas
  configuration (259)
paramagnetism (261)
diamagnetism (261)
ionic radius (262)

Key Equations and Relationships

8.1 Defining the energy order of sublevels in terms of the angular momentum quantum number (*l* value) (239):

  Order of sublevel energies: $s < p < d < f$

8.2 Meaning of the first ionization energy (252):

$$\text{Atom}(g) \longrightarrow \text{ion}^+(g) + e^- \qquad \Delta E = \text{IE}_1 > 0$$

Brief Solutions to Follow-up Problems

8.1 The element has eight electrons, so $Z = 8$: oxygen.
Sixth electron: $n = 2$, $l = 1$, $m_l = 0$, $m_s = +\frac{1}{2}$

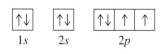

$\quad$ 1*s* $\qquad$ 2*s* $\qquad$ 2*p*

8.2 (a) For Ni, $1s^2 2s^2 2p^6 3s^2 3p^6 4s^2 3d^8$; [Ar] $4s^2 3d^8$

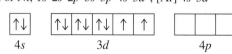

$\quad$ 4*s* $\qquad$ 3*d* $\qquad\qquad$ 4*p*

Ni has 18 inner electrons.

(b) For Sr, $1s^22s^22p^63s^23p^64s^23d^{10}4p^65s^2$; [Kr] $5s^2$

↑↓									

5s   4d   5p

Sr has 36 inner electrons.

(c) For Po, $1s^22s^22p^63s^23p^64s^23d^{10}4p^65s^24d^{10}5p^66s^24f^{14}5d^{10}6p^4$; [Xe] $6s^24f^{14}5d^{10}6p^4$

↑↓		↑↓	↑	↑

6s   6p

Po has 78 inner electrons.

8.3 (a) Cl < Br < Se; (b) Xe < I < Ba
8.4 (a) Sn < Sb < I; (b) Ba < Sr < Ca
8.5 Q is aluminum: $1s^22s^22p^63s^23p^1$
8.6 (a) Ba ([Xe] $6s^2$) $\longrightarrow$ Ba^{2+} ([Xe]) $+$ 2e$^-$
(b) O ([He] $2s^22p^4$) $+$ 2e$^-$ $\longrightarrow$ O^{2-} ([He] $2s^22p^6$) (same as Ne)
(c) Pb ([Xe] $6s^24f^{14}5d^{10}6p^2$) $\longrightarrow$ Pb^{2+} ([Xe] $6s^24f^{14}5d^{10}$) $+$ 2e$^-$
Pb ([Xe] $6s^24f^{14}5d^{10}6p^2$) $\longrightarrow$ Pb^{4+} ([Xe] $4f^{14}5d^{10}$) $+$ 4e$^-$
8.7 (a) V^{3+}: [Ar] $3d^2$; paramagnetic
(b) Ni^{2+}: [Ar] $3d^8$; paramagnetic
(c) La^{3+}: [Xe]; not paramagnetic (diamagnetic)
8.8 (a) F$^-$ < Cl$^-$ < Br$^-$; (b) Mg^{2+} < Na$^+$ < F$^-$;
(c) Cr^{3+} < Cr^{2+}

Problems

Problems with **colored** numbers are answered in Appendix E. Sections match the text and provide the numbers of relevant sample problems. Bracketed problems are grouped in pairs (indicated by a short rule) that cover the same concept. Comprehensive Problems are based on material from any section or previous chapter.

Development of the Periodic Table

8.1 What would be your reaction to a claim that a new element had been discovered, and it fit between tin (Sn) and antimony (Sb) in the periodic table?

8.2 Mendeleev arranged the elements in his periodic table by atomic mass. By what property are the elements now ordered in the periodic table? Give an example of a sequence of element order that would change if mass were still used.

8.3 Before Mendeleev published his periodic table, Johann Döbereiner grouped elements with similar properties into "triads," in which the unknown properties of one member could be predicted by averaging known values of the properties of the others. Predict the values of the following quantities:
(a) The atomic mass of K from the atomic masses of Na and Rb
(b) The melting point of Br$_2$ from the melting points of Cl$_2$ ($-101.0°C$) and I$_2$ ($113.6°C$) (actual value $= -7.2°C$)
(c) The boiling point of HBr from the boiling points of HCl ($-84.9°C$) and HI ($-35.4°C$) (actual value $= -67.0°C$)

Characteristics of Many-Electron Atoms

8.4 Summarize the rules for the allowable values of the four quantum numbers of an electron in an atom.

8.5 Which of the quantum numbers relate(s) to the electron only? Which relate(s) to the orbital?

8.6 State the exclusion principle. What does it imply about the number and spin of electrons in an atomic orbital?

8.7 What is the key distinction between sublevel energies in one-electron species, such as the H atom, and those in many-electron species, such as the C atom? What factors lead to this distinction? Would you expect the pattern of sublevel energies in Be^{3+} to be more like that in H or that in C? Explain.

8.8 Define *shielding* and *effective nuclear charge*. What is the connection between the two?

8.9 What is penetration? How is it related to shielding? Use the penetration effect to explain the difference in relative orbital energies of a 3p and a 3d electron in the same atom.

8.10 How many electrons in an atom can have each of the following quantum number or sublevel designations?
(a) $n = 2, l = 1$ (b) $3d$ (c) $4s$

8.11 How many electrons in an atom can have each of the following quantum number or sublevel designations?
(a) $n = 2, l = 1, m_l = 0$ (b) $5p$ (c) $n = 4, l = 3$

8.12 How many electrons in an atom can have each of the following quantum number or sublevel designations?
(a) $4p$ (b) $n = 3, l = 1, m_l = +1$ (c) $n = 5, l = 3$

8.13 How many electrons in an atom can have each of the following quantum number or sublevel designations?
(a) $2s$ (b) $n = 3, l = 2$ (c) $6d$

The Quantum-Mechanical Model and the Periodic Table
(Sample Problems 8.1 and 8.2)

8.14 State the periodic law, and explain its relation to electron configuration. (Use Na and K in your explanation.)

8.15 State Hund's rule in your own words, and show its application in the orbital diagram of the nitrogen atom.

8.16 How does the aufbau principle, in connection with the periodic law, lead to the format of the periodic table?

8.17 For main-group elements, are outer electron configurations similar or different within a group? Within a period? Explain.

8.18 Write a full set of quantum numbers for the following:
(a) The outermost electron in an Rb atom
(b) The electron gained when an S$^-$ ion becomes an S^{2-} ion
(c) The electron lost when an Ag atom ionizes
(d) The electron gained when an F$^-$ ion forms from an F atom

8.19 Write a full set of quantum numbers for the following:
(a) The outermost electron in an Li atom
(b) The electron gained when a Br atom becomes a Br$^-$ ion
(c) The electron lost when a Cs atom ionizes
(d) The highest energy electron in the ground-state B atom

8.20 Write the full ground-state electron configuration for each:
(a) Rb   (b) Ge   (c) Ar

8.21 Write the full ground-state electron configuration for each:
(a) Br   (b) Mg   (c) Se

8.22 Draw an orbital diagram showing valence electrons, and write the condensed ground-state electron configuration for each:
(a) Ti   (b) Cl   (c) V

8.23 Draw an orbital diagram showing valence electrons, and write the condensed ground-state electron configuration for each:
(a) Ba   (b) Co   (c) Ag

8.24 Draw the partial (valence-level) orbital diagram, and write the symbol, group number, and period number of the element:
(a) [He] $2s^2 2p^4$ (b) [Ne] $3s^2 3p^3$

8.25 Draw the partial (valence-level) orbital diagram, and write the symbol, group number, and period number of the element:
(a) [Kr] $5s^2 4d^{10}$ (b) [Ar] $4s^2 3d^8$

8.26 From each partial (valence-level) orbital diagram, write the ground-state electron configuration and group number:

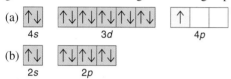

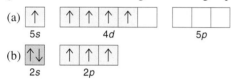

8.27 From each partial (valence-level) orbital diagram, write the ground-state electron configuration and group number:

8.28 How many inner, outer, and valence electrons are present in an atom of each of the following elements?
(a) O   (b) Sn   (c) Ca   (d) Fe   (e) Se

8.29 How many inner, outer, and valence electrons are present in an atom of each of the following elements?
(a) Br   (b) Cs   (c) Cr   (d) Sr   (e) F

8.30 Identify each element below, and give the symbols of the other elements in its group:
(a) [He] $2s^2 2p^1$ (b) [Ne] $3s^2 3p^4$ (c) [Xe] $6s^2 5d^1$

8.31 Identify each element below, and give the symbols of the other elements in its group:
(a) [Ar] $4s^2 3d^{10} 4p^4$ (b) [Xe] $6s^2 4f^{14} 5d^2$ (c) [Ar] $4s^2 3d^5$

8.32 One reason spectroscopists study excited states is to gain information about the energies of orbitals that are unoccupied in an atom's ground state. Each of the following electron configurations represents an atom in an excited state. Identify the element, and write its condensed ground-state configuration:
(a) $1s^2 2s^2 2p^6 3s^1 3p^1$ (b) $1s^2 2s^2 2p^6 3s^2 3p^3 4s^1$
(c) $1s^2 2s^2 2p^6 3s^2 3p^6 4s^2 3d^4 4p^1$ (d) $1s^2 2s^2 2p^5 3s^1$

Trends in Three Key Atomic Properties
(Sample Problems 8.3 to 8.5)

8.33 Explain the relationship between the trends in atomic size and in ionization energy within the main groups.

8.34 In what region of the periodic table will you find elements with relatively high IEs? With relatively low IEs?

8.35 Why do successive IEs of a given element always increase? When the difference between successive IEs of a given element is exceptionally large (for example, between IE_1 and IE_2 of K), what do we learn about its electron configuration?

8.36 In a plot of IE_1 for the Period 3 elements (see Figure 8.11), why do the values for elements in Groups 3A(13) and 6A(16) drop slightly below the generally increasing trend?

8.37 Which group in the periodic table has elements with high (endothermic) IE_1 and very negative (exothermic) first electron affinities (EA_1)? Give the charge on the ions these atoms form.

8.38 How does d-electron shielding influence atomic size among the Period 4 transition elements?

8.39 Arrange each set in order of *increasing* atomic size:
(a) Rb, K, Cs   (b) C, O, Be   (c) Cl, K, S   (d) Mg, K, Ca

8.40 Arrange each set in order of *decreasing* atomic size:
(a) Ge, Pb, Sn   (b) Sn, Te, Sr   (c) F, Ne, Na   (d) Be, Mg, Na

8.41 Arrange each set of atoms in order of *increasing* IE_1:
(a) Sr, Ca, Ba   (b) N, B, Ne   (c) Br, Rb, Se   (d) As, Sb, Sn

8.42 Arrange each set of atoms in order of *decreasing* IE_1:
(a) Na, Li, K   (b) Be, F, C   (c) Cl, Ar, Na   (d) Cl, Br, Se

8.43 Write the full electron configuration of the Period 2 element with the following successive IEs (in kJ/mol):
$IE_1 = 801$ $IE_2 = 2427$ $IE_3 = 3659$
$IE_4 = 25{,}022$ $IE_5 = 32{,}822$

8.44 Write the full electron configuration of the Period 3 element with the following successive IEs (in kJ/mol):
$IE_1 = 738$ $IE_2 = 1450$ $IE_3 = 7732$
$IE_4 = 10{,}539$ $IE_5 = 13{,}628$

8.45 Which element in each of the following sets would you expect to have the *highest* IE_2?
(a) Na, Mg, Al   (b) Na, K, Fe   (c) Sc, Be, Mg

8.46 Which element in each of the following sets would you expect to have the *lowest* IE_3?
(a) Na, Mg, Al   (b) K, Ca, Sc   (c) Li, Al, B

Atomic Structure and Chemical Reactivity
(Sample Problems 8.6 to 8.8)

8.47 List three ways in which metals and nonmetals differ.

8.48 Summarize the trend in metallic character as a function of position in the periodic table. Is it the same as the trend in atomic size? Ionization energy?

8.49 Summarize the acid-base behavior of the main-group metal and nonmetal oxides in water. How does oxide acidity in water change down a group and across a period?

8.50 What is a pseudo–noble gas configuration? Give an example of one ion from Group 3A(13) that has it.

8.51 The charges of a set of isoelectronic ions vary from 3+ to 3−. Place the ions in order of increasing size.

8.52 Which element would you expect to be *more* metallic?
(a) Ca or Rb   (b) Mg or Ra   (c) Br or I

8.53 Which element would you expect to be *less* metallic?
(a) S or Cl   (b) In or Al   (c) As or Br

8.54 Does the reaction of a main-group nonmetal oxide in water produce an acidic or a basic solution? Write a balanced equation for the reaction of a Group 6A(16) nonmetal oxide with water.

8.55 Does the reaction of a main-group metal oxide in water produce an acidic solution or a basic solution? Write a balanced equation for the reaction of a Group 2A(2) oxide with water.

8.56 Write the charge and full ground-state electron configuration of the monatomic ion most likely to be formed by each:
(a) Cl   (b) Na   (c) Ca

8.57 Write the charge and full ground-state electron configuration of the monatomic ion most likely to be formed by each:
(a) Rb   (b) N   (c) Br

8.58 How many unpaired electrons are present in the ground state of an atom from each of the following groups?
(a) 2A(2)   (b) 5A(15)   (c) 8A(18)   (d) 3A(13)

8.59 How many unpaired electrons are present in the ground state of an atom from each of the following groups?
(a) 4A(14)   (b) 7A(17)   (c) 1A(1)   (d) 6A(16)

8.60 Write the condensed ground-state electron configurations of these transition metal ions, and state which are paramagnetic:
(a) V^{3+} (b) Cd^{2+} (c) Co^{3+} (d) Ag^+

8.61 Write the condensed ground-state electron configurations of these transition metal ions, and state which are paramagnetic:
(a) Mo^{3+} (b) Au^+ (c) Mn^{2+} (d) Hf^{2+}

8.62 Palladium (Pd; $Z = 46$) is diamagnetic. Draw partial orbital diagrams to show which of the following electron configurations is consistent with this fact:
(a) $[Kr]\,5s^2 4d^8$ (b) $[Kr]\,4d^{10}$ (c) $[Kr]\,5s^1 4d^9$

8.63 Niobium (Nb; $Z = 41$) has an anomalous ground-state electron configuration for a Group 5B(5) element: $[Kr]\,5s^1 4d^4$. What is the expected electron configuration for elements in this group? Draw partial orbital diagrams to show how paramagnetic measurements could support niobium's actual configuration.

8.64 Rank the ions in each set in order of *increasing* size, and explain your ranking:
(a) Li^+, K^+, Na^+ (b) Se^{2-}, Rb^+, Br^- (c) O^{2-}, F^-, N^{3-}

8.65 Rank the ions in each set in order of *decreasing* size, and explain your ranking:
(a) Se^{2-}, S^{2-}, O^{2-} (b) Te^{2-}, Cs^+, I^- (c) Sr^{2+}, Ba^{2+}, Cs^+

Comprehensive Problems

Problems with an asterisk (*) are more challenging.

8.66 Name the element described in each of the following:
(a) Smallest atomic radius in Group 6A
(b) Largest atomic radius in Period 6
(c) Smallest metal in Period 3
(d) Highest IE_1 in Group 14
(e) Lowest IE_1 in Period 5
(f) Most metallic in Group 15
(g) Group 3A element that forms the most basic oxide
(h) Period 4 element with filled outer level
(i) Condensed ground-state electron configuration is $[Ne]\,3s^2 3p^2$
(j) Condensed ground-state electron configuration is $[Kr]\,5s^2 4d^6$
(k) Forms 2+ ion with electron configuration $[Ar]\,3d^3$
(l) Period 5 element that forms 3+ ion with pseudo–noble gas configuration
(m) Period 4 transition element that forms 3+ diamagnetic ion
(n) Period 4 transition element that forms 2+ ion with a half-filled d sublevel
(o) Heaviest lanthanide
(p) Period 3 element whose 2− ion is isoelectronic with Ar
(q) Alkaline earth metal whose cation is isoelectronic with Kr
(r) Group 5A(15) metalloid with the most acidic oxide

* 8.67 When a nonmetal oxide reacts with water, it forms an oxoacid with the same nonmetal oxidation state. Give the name and formula of the oxide used to prepare each of these oxoacids: (a) hypochlorous acid; (b) chlorous acid; (c) chloric acid; (d) perchloric acid; (e) sulfuric acid; (f) sulfurous acid; (g) nitric acid; (h) nitrous acid; (i) carbonic acid; (j) phosphoric acid.

* 8.68 The energy difference between the $5d$ and $6s$ sublevels in gold accounts for its color. Assuming this energy difference is about 2.7 eV [1 electron volt (eV) $= 1.602 \times 10^{-19}$ J], explain why gold has a warm yellow color.

8.69 Write the formula and name of the compound formed from the following ionic interactions: (a) The 2+ ion and the 1− ion are both isoelectronic with the atoms of a chemically unreactive Period 4 element. (b) The 2+ ion and the 2− ion are both isoelectronic with the Period 3 noble gas. (c) The 2+ ion is the smallest with a filled d subshell; the anion forms from the smallest halogen. (d) The ions form from the largest and smallest ionizable atoms in Period 2.

8.70 The hot glowing gases around the Sun, the *corona,* can reach millions of degrees Celsius, high enough to remove many electrons from gaseous atoms. Iron ions with charges as high as 14+ have been observed in the corona. Which ions from Fe^+ to Fe^{14+} are paramagnetic? Which would be most attracted to a magnetic field?

* 8.71 There are some exceptions to the trends of first and successive ionization energies. For each of the following pairs, explain which ionization energy would be higher:
(a) IE_1 of Ga or IE_1 of Ge   (b) IE_2 of Ga or IE_2 of Ge
(c) IE_3 of Ga or IE_3 of Ge   (d) IE_4 of Ga or IE_4 of Ge

8.72 Half of the first 18 elements have an odd number of electrons, and half have an even number. Show why these elements aren't half paramagnetic and half diamagnetic.

8.73 Draw the partial (valence-level) orbital diagram and write the electron configuration of the atom and monatomic ion of the element with the following ionization energies (in kJ/mol):

IE_1	IE_2	IE_3	IE_4	IE_5	IE_6	IE_7	IE_8
999	2251	3361	4564	7013	8495	27,106	31,669

* 8.74 On the planet Zog in the Andromeda galaxy, all of the stable elements have been studied. Data for some main-group elements are shown below (Zoggian units are unknown on Earth and, therefore, not shown). Limited communications with the Zoggians have indicated that balloonium is a monatomic gas with two positive charges in its nucleus. Use the data to deduce the names that Earthlings give to these elements:

Name	Atomic Radius	IE_1	EA_1
Balloonium	10	339	0
Inertium	24	297	+4.1
Allotropium	34	143	−28.6
Brinium	63	70.9	−7.6
Canium	47	101	−15.3
Fertilium	25	200	0
Liquidium	38	163	−46.4
Utilium	48	82.4	−6.1
Crimsonium	72	78.4	−2.9

CHAPTER NINE

Models of Chemical Bonding

Binding Atoms Together The properties of substances, such as the low conductivity and high melting point of sodium chloride crystals, depend on the properties of their atoms and how they bind together, as you'll see in this chapter.

Key Principles

◆ Two classes of elements, metals and nonmetals, combine through three types of bonding: metal and nonmetal through *ionic bonding*, nonmetal and nonmetal through *covalent bonding,* and metal and metal through *metallic bonding*.

◆ Ionic bonding is the attraction among the ions that are created when metal atoms *transfer* electrons to nonmetal atoms. Even though energy is required to form the ions, more energy is released when the ions attract each other and form a solid.

◆ The strong attractions among their ions make ionic compounds hard, high-melting solids that conduct a current only when melted or dissolved.

◆ A covalent bond is the attraction between the nuclei of two nonmetal atoms and the electron pair they share. Each covalent bond has specific *energy* and *length* that depend on the bonded atoms and an *order* that depends on the number of electron pairs shared.

◆ Most covalent compounds are molecular substances with low melting and boiling points, because these physical changes disrupt the weak attractions *between* the molecules while leaving the strong covalent bonds *within* the molecules intact.

◆ During a reaction, energy is absorbed to break certain bonds in the reactant molecules and is released to form other bonds that create the product molecules; the heat of reaction is the *difference* between the energy absorbed and the energy released.

◆ Each atom in a covalent bond attracts the shared electron pair according to its *electronegativity (EN)*. A covalent bond is *polar* if the two atoms have different EN values. The *ionic character* of a bond—from highly ionic to nonpolar covalent—varies with the difference in EN values of the atoms.

Outline

W hy is table salt (or any other ionic substance) a hard, brittle, high-melting solid that conducts a current only when molten or dissolved in water? Why is candle wax (along with most covalent substances) low melting, soft, and non-conducting, while diamond (and some other exceptions) is high melting and extremely hard? And why is copper (and most other metallic substances) shiny, malleable, and able to conduct a current whether molten or solid? The answers lie in the *type of bonding within the substance*. In Chapter 8, we examined the properties of individual atoms and ions. Yet, in virtually all the substances in and around you, these particles are bonded to one another. As you'll see in this chapter, deeper insight comes as we discover how the properties of atoms influence the types of chemical bonds they form, because these are ultimately responsible for the behavior of substances.

*Concepts & Skills to Review
Before You Study This Chapter*

- characteristics of ionic and covalent bonding (Section 2.7)
- polar covalent bonds and the polarity of water (Section 4.1)
- Hess's law, ΔH°_{rxn}, and ΔH°_f (Sections 6.5 and 6.6)
- atomic and ionic electron configurations (Sections 8.3 and 8.5)
- trends in atomic properties and metallic behavior (Sections 8.4 and 8.5)

9.1 ATOMIC PROPERTIES AND CHEMICAL BONDS

Before we examine the types of chemical bonding, we should ask why atoms bond at all. In general terms, they do so for one overriding reason: *bonding lowers the potential energy between positive and negative particles*, whether those particles are oppositely charged ions or atomic nuclei and the electrons between them. Just as the electron configuration and the strength of the nucleus-electron attraction determine the properties of an atom, the type and strength of chemical bonds determine the properties of a substance.

The Three Types of Chemical Bonding

On the atomic level, we distinguish a metal from a nonmetal on the basis of several properties that correlate with position in the periodic table (Figure 9.1 and inside the front cover). Recall from Chapter 8 that, in general, there is a gradation from more metal-like to more nonmetal-like behavior from left to right across a period and from bottom to top within most groups. Three types of bonding result from the three ways these two types of atoms can combine—metal with nonmetal, nonmetal with nonmetal, and metal with metal:

1. *Metal with nonmetal: electron transfer and ionic bonding* (Figure 9.2A, on the next page). We typically observe **ionic bonding** between atoms with large differences in their tendencies to lose or gain electrons. Such differences occur between reactive metals [Groups 1A(1) and 2A(2)] and nonmetals [Group 7A(17) and the top of Group 6A(16)]. The metal atom (low IE) loses its one or two

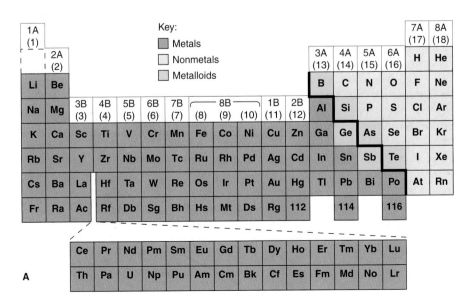

A

Figure 9.1 A general comparison of metals and nonmetals. A, The positions of metals, nonmetals, and metalloids within the periodic table. **B,** The relative magnitudes of some key atomic properties vary from left to right within a period and correlate with whether an element is metallic or nonmetallic.

PROPERTY	METAL ATOM	NONMETAL ATOM
Atomic size	Larger	Smaller
Z_{eff}	Lower	Higher
IE	Lower	Higher
EA	Less negative	More negative

B Relative magnitudes of atomic properties within a period

Figure 9.2 The three models of chemical bonding. A, In ionic bonding, metal atoms transfer electron(s) to nonmetal atoms, forming oppositely charged ions that attract each other to form a solid. **B,** In covalent bonding, two atoms share an electron pair localized between their nuclei (shown here as a bond line). Most covalent substances consist of individual molecules, each made from two or more atoms. **C,** In metallic bonding, many metal atoms pool their valence electrons to form a delocalized electron "sea" that holds the metal-ion cores together.

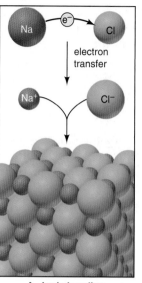

A   Ionic bonding

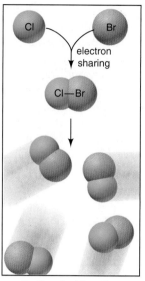

B   Covalent bonding

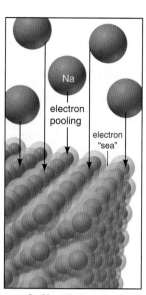

C   Metallic bonding

valence electrons, whereas the nonmetal atom (highly negative EA) gains the electron(s). *Electron transfer* from metal to nonmetal occurs, and each atom forms an ion with a noble gas electron configuration. The electrostatic attraction between these positive and negative ions draws them into the three-dimensional array of an ionic solid, whose chemical formula represents the cation-to-anion ratio (empirical formula).

2. *Nonmetal with nonmetal: electron sharing and covalent bonding* (Figure 9.2B). When two atoms have a small difference in their tendencies to lose or gain electrons, we observe *electron sharing* and **covalent bonding.** This type of bonding most commonly occurs between nonmetal atoms (although a pair of metal atoms can sometimes form a covalent bond). Each nonmetal atom holds onto its own electrons tightly (high IE) and tends to attract other electrons as well (highly negative EA). The attraction of each nucleus for the valence electrons of the other draws the atoms together. A shared electron pair is considered to be *localized* between the two atoms because it spends most of its time there, linking them in a covalent bond of a particular length and strength. In most cases, separate molecules form when covalent bonding occurs, and the chemical formula reflects the actual numbers of atoms in the molecule (molecular formula).

3. *Metal with metal: electron pooling and metallic bonding* (Figure 9.2C). In general, metal atoms are relatively large, and their few outer electrons are well shielded by filled inner levels. Thus, they lose outer electrons comparatively easily (low IE) but do not gain them very readily (slightly negative or positive EA). These properties lead large numbers of metal atoms to share their valence electrons, but in a way that differs from covalent bonding. In the simplest model of **metallic bonding,** all the metal atoms in a sample *pool* their valence electrons into an evenly distributed "sea" of electrons that "flows" between and around the metal-ion cores (nucleus plus inner electrons), attracting them and holding them together. Unlike the localized electrons in covalent bonding, electrons in metallic bonding are *delocalized,* moving freely throughout the piece of metal. (For the remainder of this chapter, we'll focus on ionic and covalent bonding. We discuss electron delocalization in Chapter 11 and the structures of solids, including metallic solids, in Chapter 12. So we'll postpone the coverage of metallic bonding until then.)

It's important to remember that there are exceptions to these idealized bonding models in the world of real substances. For instance, all binary ionic compounds contain a metal and a nonmetal, but all metals do not form binary ionic

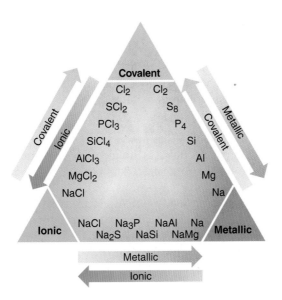

Figure 9.3 Gradation in bond type among the Period 3 elements. Along *the left side* of the triangle, compounds of each element with chlorine display a gradual change from ionic to covalent bonding. Along *the right side,* the elements themselves display a gradual change from covalent to metallic bonding. Along *the base,* compounds of each element with sodium display a gradual change from ionic to metallic bonding.

compounds with all nonmetals. As just one example, when the metal beryllium [Group 2A(2)] combines with the nonmetal chlorine [Group 7A(17)], the bonding fits the covalent model better than the ionic model. In other words, just as we see a gradation in metallic behavior within groups and periods, we also see a gradation in bonding from one type to another (Figure 9.3).

Lewis Electron-Dot Symbols: Depicting Atoms in Chemical Bonding

Before turning to the two bonding models, let's discuss a method for depicting the valence electrons of interacting atoms. In the **Lewis electron-dot symbol** (named for the American chemist G. N. Lewis), the element symbol represents the nucleus *and* inner electrons, and the surrounding dots represent the valence electrons (Figure 9.4). Note that the pattern of dots is the same for elements within a group.

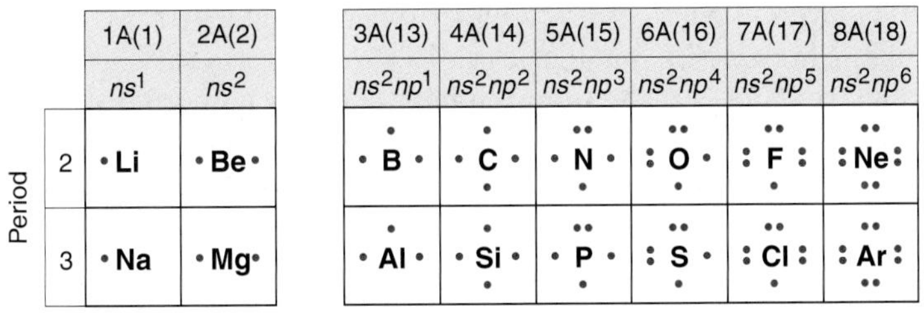

	1A(1)	2A(2)		3A(13)	4A(14)	5A(15)	6A(16)	7A(17)	8A(18)
	ns^1	ns^2		ns^2np^1	ns^2np^2	ns^2np^3	ns^2np^4	ns^2np^5	ns^2np^6
2	·Li	·Be·		·B·	·C·	·N·	:O·	:F:	:Ne:
3	·Na	·Mg·		·Al·	·Si·	·P·	:S·	:Cl:	:Ar:

(Period)

Figure 9.4 Lewis electron-dot symbols for elements in Periods 2 and 3. The element symbol represents the nucleus and inner electrons, and the dots around it represent valence electrons, either paired or unpaired. The number of unpaired dots indicates the number of electrons a metal atom loses, or the number a nonmetal atom gains, or the number of covalent bonds a nonmetal atom usually forms.

It's easy to write the Lewis symbol for any main-group element:

1. Note its A-group number (1A to 8A), which gives the number of valence electrons.

2. Place one dot at a time on the four sides (top, right, bottom, left) of the element symbol.

3. Keep adding dots, pairing the dots until all are used up.

The specific placement of dots is not important; that is, in addition to the one shown in Figure 9.4, the Lewis symbol for nitrogen can *also* be written as

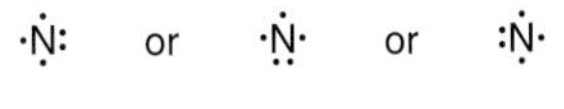

$$\cdot\ddot{N}: \quad \text{or} \quad \cdot\dot{\ddot{N}}\cdot \quad \text{or} \quad :\ddot{N}\cdot$$

The Lewis symbol provides information about an element's bonding behavior:

- For a metal, the *total* number of dots is the maximum number of electrons an atom loses to form a cation.
- For a nonmetal, the number of *unpaired* dots equals either the number of electrons an atom gains in becoming an anion or the number it shares in forming covalent bonds.

To illustrate the last point, look at the Lewis symbol for carbon. Rather than one pair of dots and two unpaired dots, as its electron configuration ([He] $2s^2 2p^2$) would indicate, carbon has four unpaired dots because it forms four bonds. That is, in its compounds, carbon's four electrons are paired with four more electrons from its bonding partners for a total of eight electrons around carbon. (In Chapter 10, we'll see that larger nonmetals can form as many bonds as they have dots in the Lewis symbol.)

In his studies of bonding, Lewis generalized much of bonding behavior into the **octet rule:** *when atoms bond, they lose, gain, or share electrons to attain a filled outer level of eight (or two) electrons.* The octet rule holds for nearly all of the compounds of Period 2 elements and a large number of others as well.

SECTION SUMMARY
Nearly all naturally occurring substances consist of atoms or ions bonded to others. Chemical bonding allows atoms to lower their energy. Ionic bonding occurs when metal atoms transfer electrons to nonmetal atoms, and the resulting ions attract each other and form an ionic solid. Covalent bonding most commonly occurs between nonmetal atoms and usually results in molecules. The bonded atoms share a pair of electrons, which remain localized between them. Metallic bonding occurs when many metal atoms pool their valence electrons in a delocalized electron "sea" that holds all the atoms together. The Lewis electron-dot symbol of an atom depicts the number of valence electrons for a main-group element. In bonding, many atoms lose, gain, or share electrons to attain a filled outer level of eight (or two).

9.2 THE IONIC BONDING MODEL

The central idea of the ionic bonding model is the *transfer of electrons from metal atoms to nonmetal atoms to form ions that come together in a solid ionic compound.* For nearly every monatomic ion of a main-group element, the electron configuration has a filled outer level: either two or eight electrons, the same number as in the nearest noble gas (octet rule).

The transfer of an electron from a lithium atom to a fluorine atom is depicted in three ways in Figure 9.5. In each, Li loses its single outer electron and is left with a filled $n = 1$ level, while F gains a single electron to fill its $n = 2$ level. In

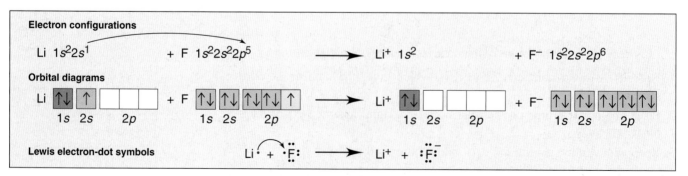

Figure 9.5 Three ways to represent the formation of Li$^+$ and F$^-$ through electron transfer. The electron being transferred is indicated in red.

this case, each atom is one electron away from its nearest noble gas—He for Li and Ne for F—so the number of electrons lost by each Li equals the number gained by each F. Therefore, equal numbers of Li^+ and F^- ions form, as the formula LiF indicates. That is, in ionic bonding, *the total number of electrons lost by the metal atoms equals the total number of electrons gained by the nonmetal atoms.*

SAMPLE PROBLEM 9.1   Depicting Ion Formation

Problem Use partial orbital diagrams and Lewis symbols to depict the formation of Na^+ and O^{2-} ions from the atoms, and determine the formula of the compound the ions form.
Plan First we draw the orbital diagrams and Lewis symbols for the Na and O atoms. To attain filled outer levels, Na loses one electron and O gains two. Thus, to make the number of electrons lost equal the number gained, two Na atoms are needed for each O atom.
Solution

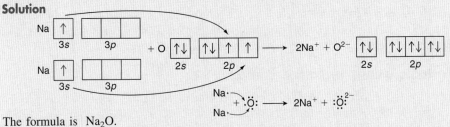

The formula is Na_2O.

FOLLOW-UP PROBLEM 9.1 Use condensed electron configurations and Lewis symbols to depict the formation of Mg^{2+} and Cl^- ions from the atoms, and write the formula of the ionic compound.

Energy Considerations in Ionic Bonding: The Importance of Lattice Energy

You may be surprised to learn that the electron-transfer process by itself actually *absorbs* energy! Consider just the electron-transfer process for the formation of lithium fluoride, which involves two steps—a gaseous Li atom loses an electron, and a gaseous F atom gains it:

- The first ionization energy (IE_1) of Li is the energy change that occurs when 1 mol of gaseous Li atoms loses 1 mol of outer electrons:

$$Li(g) \longrightarrow Li^+(g) + e^- \qquad IE_1 = 520 \text{ kJ}$$

- The electron affinity (EA) of F is the energy change that occurs when 1 mol of gaseous F atoms gains 1 mol of electrons:

$$F(g) + e^- \longrightarrow F^-(g) \qquad EA = -328 \text{ kJ}$$

Note that the two-step electron-transfer process *by itself* requires energy:

$$Li(g) + F(g) \longrightarrow Li^+(g) + F^-(g) \qquad IE_1 + EA = 192 \text{ kJ}$$

The total energy needed for ion formation is even greater than this because metallic lithium and diatomic fluorine must first be converted to separate gaseous atoms, which also requires energy. Despite this, the standard heat of formation (ΔH_f°) of solid LiF is -617 kJ/mol; that is, 617 kJ is *released* when 1 mol of LiF(s) forms from its elements. The case of LiF is typical of many reactions between active metals and nonmetals: despite the endothermic electron transfer, ionic solids form readily, often vigorously. Figure 9.6 shows another example, the formation of NaBr.

Clearly, if the overall reaction of Li(s) and $F_2(g)$ to form LiF(s) releases energy, there must be some exothermic energy component large enough to overcome the endothermic steps. This component arises from the strong *attraction*

A

B

Figure 9.6 The reaction between sodium and bromine. A, Despite the endothermic electron-transfer process, all the Group 1A(1) metals react exothermically with any of the Group 7A(17) nonmetals to form solid alkali-metal halides. The reactants in the example shown are sodium (in beaker under mineral oil) and bromine. **B,** The reaction is usually rapid and vigorous.

between oppositely charged ions. When 1 mol of $Li^+(g)$ and 1 mol of $F^-(g)$ form 1 mol of gaseous LiF molecules, a large quantity of heat is released:

$$Li^+(g) + F^-(g) \longrightarrow LiF(g) \qquad \Delta H^\circ = -755 \text{ kJ}$$

Of course, under ordinary conditions, LiF does not consist of gaseous molecules because *much more energy is released when the gaseous ions coalesce into a crystalline solid.* That occurs because each ion attracts others of opposite charge:

$$Li^+(g) + F^-(g) \longrightarrow LiF(s) \qquad \Delta H^\circ = -1050 \text{ kJ}$$

The negative of this value, 1050 kJ, is the lattice energy of LiF. The **lattice energy** ($\Delta H^\circ_{\text{lattice}}$) is the enthalpy change that occurs when 1 mol of ionic solid separates into gaseous ions. It indicates the strength of ionic interactions and influences melting point, hardness, solubility, and other properties.

A key point to keep in mind is that *ionic solids exist only because the lattice energy exceeds the energy required for the electron transfer.* In other words, the energy *required* for elements to lose or gain electrons is *supplied* by the attraction between the ions they form: energy is expended to form the ions, but it is more than regained when they attract each other and form a solid.

Periodic Trends in Lattice Energy

Because the lattice energy is the result of electrostatic interactions among ions, we expect its magnitude to depend on several factors, including ionic size, ionic charge, and ionic arrangement in the solid. In Chapter 2, you were introduced to **Coulomb's law,** which states that the electrostatic energy between two charges is directly proportional to the product of their magnitudes and inversely proportional to the distance between them:

$$\text{Electrostatic energy} \propto \frac{\text{charge A} \times \text{charge B}}{\text{distance}}$$

The lattice energy ($\Delta H^\circ_{\text{lattice}}$) is directly proportional to the electrostatic energy. In an ionic solid, cations and anions lie as close to each other as possible; the distance between them is the distance between their centers, which equals the sum of their radii (see Figure 8.21):

$$\text{Electrostatic energy} \propto \frac{\text{cation charge} \times \text{anion charge}}{\text{cation radius} + \text{anion radius}} \propto \Delta H^\circ_{\text{lattice}} \qquad \textbf{(9.1)}$$

This relationship helps us predict trends in lattice energy and explain the effects of ionic size and charge:

1. *Effect of ionic size.* As we move down a group in the periodic table, the ionic radius increases. Therefore, the electrostatic energy between cations and anions should decrease because the interionic distance is greater; thus, the lattice energies of their compounds should decrease as well. This prediction is borne out by the lattice energies of the alkali-metal halides shown in Figure 9.7. A smooth decrease in lattice energy occurs down a group whether we hold the cation constant (LiF to LiI) or the anion constant (LiF to RbF).

2. *Effect of ionic charge.* When we compare lithium fluoride with magnesium oxide, we find cations of about equal radii ($Li^+ = 76$ pm and $Mg^{2+} = 72$ pm) and anions of about equal radii ($F^- = 133$ pm and $O^{2-} = 140$ pm). Thus, the only significant difference is the ionic charge: LiF contains the singly charged Li^+ and F^- ions, whereas MgO contains the doubly charged Mg^{2+} and O^{2-} ions. The difference in their lattice energies is striking:

$$\Delta H^\circ_{\text{lattice}} \text{ of LiF} = 1050 \text{ kJ/mol} \qquad \text{and} \qquad \Delta H^\circ_{\text{lattice}} \text{ of MgO} = 3923 \text{ kJ/mol}$$

This nearly fourfold increase in $\Delta H^\circ_{\text{lattice}}$ reflects the fourfold increase in the product of the charges (1×1 vs. 2×2) in the numerator of Equation 9.1. The very large lattice energy of MgO more than compensates for the energy required to form the Mg^{2+} and O^{2-} ions. In fact, the lattice energy is the reason that compounds with 2+ cations and 2− anions even exist.

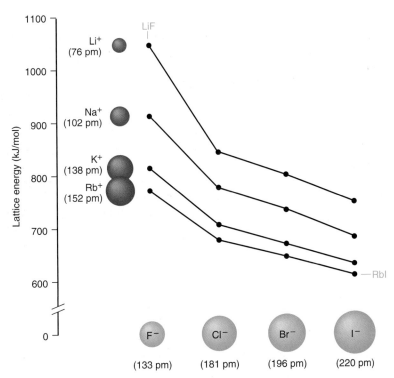

Figure 9.7 Trends in lattice energy. The lattice energies for many of the alkali-metal halides are shown. Each series of four points represents a given Group 1A(1) cation (*left side*) combining with each of the Group 7A(17) anions (*bottom*). As ionic radii increase, the electrostatic attractions decrease, so the lattice energies of the compounds decrease as well. Thus, LiF (smallest ions shown) has the largest lattice energy, and RbI (largest ions) has the smallest.

How the Model Explains the Properties of Ionic Compounds

The first and most important job of any model is to explain the facts. By magnifying our view, we can see how the ionic bonding model accounts for the properties of ionic solids. You may have seen a piece of rock salt (NaCl). It is *hard* (does not dent), *rigid* (does not bend), and *brittle* (cracks without deforming). These properties are due to the powerful attractive forces that hold the ions *in specific positions* throughout the crystal. Moving the ions out of position requires overcoming these forces, so the sample resists denting and bending. If *enough* pressure is applied, ions of like charge are brought next to each other, and repulsive forces crack the sample suddenly (Figure 9.8).

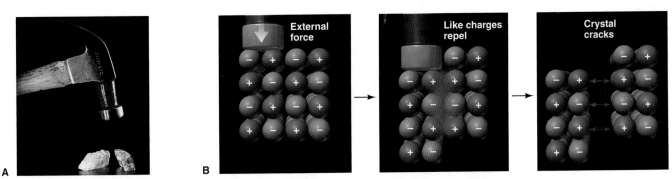

Figure 9.8 Electrostatic forces and the reason ionic compounds crack. A, Ionic compounds are hard and will crack, rather than bend, when struck with enough force. **B,** The positive and negative ions in the crystal are arranged to maximize their attractions. When an external force is applied, like charges move near each other, and the repulsions crack the piece apart.

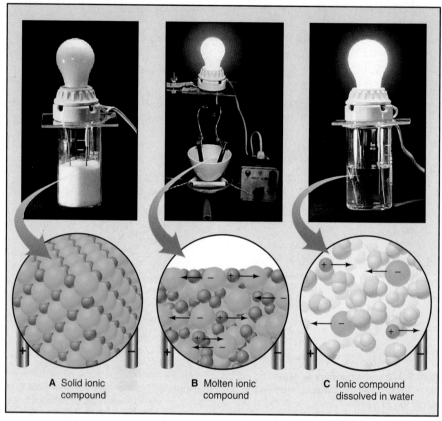

Figure 9.9 Electrical conductance and ion mobility. A, No current flows in the ionic solid because ions are immobile. **B,** In the molten compound, mobile ions flow toward the oppositely charged electrodes and carry a current. **C,** In an aqueous solution of the compound, mobile solvated ions carry a current.

Table 9.1	Melting and Boiling Points of Some Ionic Compounds	
Compound	mp (°C)	bp (°C)
CsBr	636	1300
NaI	661	1304
KBr	734	1435
NaCl	801	1413
LiF	845	1676
KF	858	1505
MgO	2852	3600

Most ionic compounds *do not* conduct electricity in the solid state but *do* conduct it when melted or when dissolved in water. According to the ionic bonding model, the solid consists of immobilized ions. When it melts or dissolves, however, the ions are free to move and carry an electric current, as shown in Figure 9.9.

The model also explains that high temperatures are needed to melt and boil an ionic compound (Table 9.1) because freeing the ions from their positions (melting) requires large amounts of energy, and vaporizing them requires even more. In fact, the interionic attraction is so strong that the vapor consists of **ion pairs,** gaseous ionic molecules rather than individual ions. But keep in mind that in their ordinary (solid) state, ionic compounds consist of arrays of alternating ions that extend in all directions, and *no separate molecules exist.*

SECTION SUMMARY

In ionic bonding, a metal transfers electrons to a nonmetal, and the resulting ions attract each other strongly to form a solid. Main-group elements often attain a filled outer level of electrons (either eight or two) by forming ions with the electron configuration of the nearest noble gas. Ion formation by itself *requires* energy. However, the lattice energy, the energy *absorbed* when the solid separates into gaseous ions, is large and is the major reason ionic solids exist. The lattice energy depends on ionic size and charge. The ionic bonding model pictures oppositely charged ions held rigidly in position by strong electrostatic attractions and explains why ionic solids crack rather than bend and why they conduct electric current only when melted or dissolved. Gaseous ion pairs form when an ionic compound vaporizes, which requires very high temperatures.

9.3 THE COVALENT BONDING MODEL

Look through any large reference source of chemical compounds, such as the *Handbook of Chemistry and Physics,* and you'll find that the number of known covalent compounds dwarfs the number of known ionic compounds. Molecules held together by covalent bonds range from diatomic hydrogen to biological and synthetic macromolecules consisting of many hundreds or even thousands of atoms. We also find covalent bonds in many polyatomic ions. Without doubt, *sharing electrons is the principal way that atoms interact chemically.*

The Formation of a Covalent Bond

A sample of hydrogen gas consists of H_2 molecules. But why *do* the atoms bond to one another in pairs? Look at Figure 9.10 and imagine what happens to two isolated H atoms that approach each other from a distance (move *right* to *left* on the graph). When the atoms are far apart, each behaves as though the other were not present (point 1). As the distance between the nuclei decreases, each nucleus starts to attract the other atom's electron, which lowers the potential energy of the system. Attractions continue to draw the atoms closer, and the system becomes progressively lower in energy (point 2). As attractions increase, however, so do repulsions between the nuclei and between the electrons. At some internuclear distance, maximum attraction is achieved in the face of the increasing repulsion, and the system has its minimum energy (point 3, at the bottom of the energy "well"). Any shorter distance would increase repulsions and cause a rise in potential energy (point 4). Thus a **covalent bond,** such as the one that holds the atoms together in the H_2 molecule, arises from the balance between nucleus-electron attractions and electron-electron and nucleus-nucleus repulsions. Formation of a bond always results in *greater electron density between the nuclei.*

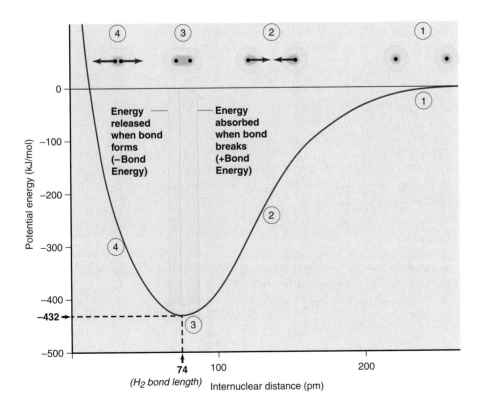

Figure 9.10 Covalent bond formation in H_2. The potential energy of a system of two H atoms is plotted against the distance between the nuclei, with a depiction of the atomic systems above. At point 1, the atoms are too far apart to attract each other. At 2, each nucleus attracts the other atom's electron. At 3, the combination of nucleus-electron attractions and electron-electron and nucleus-nucleus repulsions gives the minimum energy of the system. The energy difference between points 1 and 3 is the H_2 bond energy (432 kJ/mol). It is released when the bond forms and must be absorbed to break the bond. The internuclear distance at point 3 is the H_2 bond length (74 pm). If the atoms move closer, as at point 4, repulsions increase the system's energy and force the atoms apart to point 3 again.

Figure 9.11 Distribution of electron density in H_2. A, At some optimum distance (bond length), attractions balance repulsions. Electron density (*blue shading*) is highest around and between the nuclei. **B,** This *contour map* shows a doubling of electron densities with each contour line; the dots represent the nuclei. **C,** This *relief map* depicts the varying electron densities of the contour map as peaks. The densest regions, by far, are around the nuclei (black dots on the "floor"), but the region between the nuclei—the bonding region—also has higher electron density.

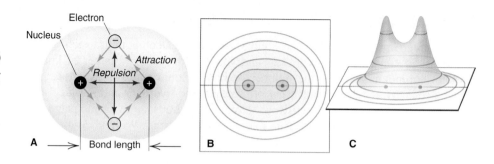

Animation: Formation of a Covalent Bond Online Learning Center

Figure 9.11 depicts this fact in three ways: a cross-section of a space-filling model; an *electron density contour map,* with lines representing regular increments in electron density; and an *electron density relief map,* which portrays the contour map three-dimensionally as peaks of electron density.

Bonding Pairs and Lone Pairs In covalent bonding, as in ionic bonding, each atom achieves a full outer (valence) level of electrons, but this is accomplished by different means. *Each atom in a covalent bond "counts" the shared electrons as belonging entirely to itself.* Thus, the two electrons in the shared electron pair of H_2 simultaneously fill the outer level of *both* H atoms. The **shared pair,** or **bonding pair,** is represented by either a pair of dots or a line, H:H or H—H.

An outer-level electron pair that is *not* involved in bonding is called a **lone pair,** or **unshared pair.** The bonding pair in HF fills the outer level of the H atom *and,* together with three lone pairs, fills the outer level of the F atom as well:

$$\text{bonding pair} \quad H \!:\! \ddot{F} \!: \quad \text{lone pairs} \qquad \text{or} \qquad H\!-\!\ddot{F}\!:$$

In F_2 the bonding pair and three lone pairs fill the outer level of *each* F atom:

$$:\!\ddot{F}\!:\!\ddot{F}\!: \qquad \text{or} \qquad :\!\ddot{F}\!-\!\ddot{F}\!:$$

(This text generally shows bonding pairs as lines and lone pairs as dots.)

Types of Bonds and Bond Order The **bond order** is the number of electron pairs being shared by any pair of bonded atoms. The covalent bond in H_2, HF, or F_2 is a **single bond,** one that consists of a single bonding pair of electrons. A *single bond has a bond order of 1.*

Single bonds are the most common type of bond, but many molecules (and ions) contain multiple bonds. Multiple bonds most frequently involve C, O, N, and/or S atoms. A **double bond** consists of two bonding electron pairs, four electrons shared between two atoms, so *the bond order is 2.* Ethylene (C_2H_4) is a simple hydrocarbon that contains a carbon-carbon double bond and four carbon-hydrogen single bonds:

$$\begin{array}{cc} H & H \\ \diagdown C :: C \diagup \\ H \diagup \diagdown H \end{array} \quad \text{or} \quad \begin{array}{cc} H H \\ \diagdown C = C \diagup \\ H \diagup \diagdown H \end{array}$$

Each carbon "counts" the four electrons in the double bond and the four in its two single bonds to hydrogens to attain an octet.

A **triple bond** consists of three bonding pairs; two atoms share six electrons, so *the bond order is 3.* In the N_2 molecule, the atoms are held together by a triple bond, and each N atom also has a lone pair:

$$:\!N\!:::\!N\!: \qquad \text{or} \qquad :\!N\!\equiv\!N\!:$$

Six shared and two unshared electrons give *each* N atom an octet.

Properties of a Covalent Bond: Bond Energy and Bond Length

The strength of a covalent bond depends on the magnitude of the mutual attraction between bonded nuclei and shared electrons. The **bond energy (BE)** (also called *bond enthalpy* or *bond strength*) is the energy required to overcome this attraction and is defined as the standard enthalpy change for breaking the bond in 1 mol of gaseous molecules. *Bond breakage is an endothermic process, so the bond energy is always positive:*

$$A\text{—}B(g) \longrightarrow A(g) + B(g) \qquad \Delta H^{\circ}_{\text{bond breaking}} = BE_{A\text{—}B} \text{ (always > 0)}$$

Stated in another way, the bond energy is the difference in energy between the separated atoms and the bonded atoms (the potential energy difference between points 1 and 3 in Figure 9.10; the depth of the energy well). The same amount of energy that is absorbed to break the bond is released when it forms. *Bond formation is an exothermic process, so the sign of the enthalpy change is negative:*

$$A(g) + B(g) \longrightarrow A\text{—}B(g) \qquad \Delta H^{\circ}_{\text{bond forming}} = -BE_{A\text{—}B} \text{ (always < 0)}$$

Because bond energies depend on characteristics of the bonded atoms—their electron configurations, nuclear charges, and atomic radii—each type of bond has its own bond energy. Energies for some common bonds are listed in Table 9.2, along with bond lengths, which we discuss below. *Stronger bonds are lower in energy (have a deeper energy well); weaker bonds are higher in energy (have a shallower energy well).* The energy of a given type of bond varies slightly from molecule to molecule, and even within the same molecule, so each tabulated value is an *average* bond energy.

A covalent bond has a **bond length,** the distance between the nuclei of two bonded atoms. In Figure 9.10, bond length is shown as the distance between the

Table 9.2 Average Bond Energies (kJ/mol) and Bond Lengths (pm)

Bond	Energy	Length	Bond	Energy	Length	Bond	Energy	Length	Bond	Energy	Length
Single Bonds											
H—H	432	74	N—H	391	101	Si—H	323	148	S—H	347	134
H—F	565	92	N—N	160	146	Si—Si	226	234	S—S	266	204
H—Cl	427	127	N—P	209	177	Si—O	368	161	S—F	327	158
H—Br	363	141	N—O	201	144	Si—S	226	210	S—Cl	271	201
H—I	295	161	N—F	272	139	Si—F	565	156	S—Br	218	225
			N—Cl	200	191	Si—Cl	381	204	S—I	~170	234
C—H	413	109	N—Br	243	214	Si—Br	310	216			
C—C	347	154	N—I	159	222	Si—I	234	240	F—F	159	143
C—Si	301	186							F—Cl	193	166
C—N	305	147	O—H	467	96	P—H	320	142	F—Br	212	178
C—O	358	143	O—P	351	160	P—Si	213	227	F—I	263	187
C—P	264	187	O—O	204	148	P—P	200	221	Cl—Cl	243	199
C—S	259	181	O—S	265	151	P—F	490	156	Cl—Br	215	214
C—F	453	133	O—F	190	142	P—Cl	331	204	Cl—I	208	243
C—Cl	339	177	O—Cl	203	164	P—Br	272	222	Br—Br	193	228
C—Br	276	194	O—Br	234	172	P—I	184	243	Br—I	175	248
C—I	216	213	O—I	234	194				I—I	151	266
Multiple Bonds											
C=C	614	134	N=N	418	122	C≡C	839	121	N≡N	945	110
C=N	615	127	N=O	607	120	C≡N	891	115	N≡O	631	106
C=O	745	123	O_2	498	121	C≡O	1070	113			
	(799 in CO_2)										

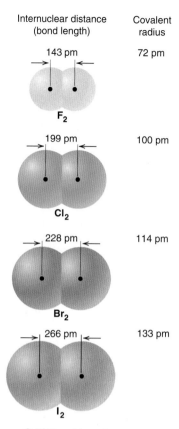

Internuclear distance (bond length)   Covalent radius

143 pm   72 pm

F₂

199 pm   100 pm

Cl₂

228 pm   114 pm

Br₂

266 pm   133 pm

I₂

Figure 9.12 Bond length and covalent radius. Within a series of similar molecules, such as the diatomic halogen molecules, bond length increases as covalent radius increases.

nuclei at the point of minimum energy. The values given in Table 9.2 represent *average* bond lengths for the given bond in different substances. Bond length is related to the sum of the radii of the bonded atoms. In fact, most atomic radii are calculated from measured bond lengths (see Figure 8.8C). Bond lengths for a series of similar bonds increase with atomic size, as shown in Figure 9.12 for the halogens.

A close relationship exists among bond order, bond length, and bond energy. Two nuclei are more strongly attracted to two shared electron pairs than to one: the atoms are drawn closer together *and* are more difficult to pull apart. Therefore, *for a given pair of atoms, a higher bond order results in a shorter bond length and a higher bond energy.* So, as Table 9.3 shows, for a given pair of atoms, *a shorter bond is a stronger bond.*

Table 9.3 The Relation of Bond Order, Bond Length, and Bond Energy

Bond	Bond Order	Average Bond Length (pm)	Average Bond Energy (kJ/mol)
C—O	1	143	358
C=O	2	123	745
C≡O	3	113	1070
C—C	1	154	347
C=C	2	134	614
C≡C	3	121	839
N—N	1	146	160
N=N	2	122	418
N≡N	3	110	945

In some cases, we can extend this relationship among atomic size, bond length, and bond strength by holding one atom in the bond constant and varying the other atom within a group or period. For example, the trend in carbon-halogen single bond lengths, C—I > C—Br > C—Cl, parallels the trend in atomic size, I > Br > Cl, and is opposite to the trend in bond energy, C—Cl > C—Br > C—I. Thus, for *single bonds,* longer bonds are usually weaker, and you can see many other examples of this relationship in Table 9.2.

SAMPLE PROBLEM 9.2   *Comparing Bond Length and Bond Strength*

Problem Without referring to Tables 9.2 and 9.3, rank the bonds in each set in order of *decreasing* bond length and bond strength:
(a) S—F, S—Br, S—Cl   **(b)** C=O, C—O, C≡O
Plan In part (a), S is singly bonded to three different halogen atoms, so all members of the set have a bond order of 1. Bond length increases and bond strength decreases as the halogen's atomic radius increases, and that size trend is clear from the periodic table. In all the bonds in part (b), the same two atoms are involved, but the bond orders differ. In this case, bond strength increases and bond length decreases as bond order increases.
Solution (a) Atomic size increases down a group, so F < Cl < Br.

Bond length: S—Br > S—Cl > S—F

Bond strength: S—F > S—Cl > S—Br

(b) By ranking the bond orders, C≡O > C=O > C—O, we obtain

Bond length: C—O > C=O > C≡O

Bond strength: C≡O > C=O > C—O

Check From Tables 9.2 and 9.3, we see that the rankings are correct.
Comment Remember that for bonds involving pairs of different atoms, as in part (a), *the relationship between length and strength holds* only *for single bonds* and not in every case, so apply it carefully.

FOLLOW-UP PROBLEM 9.2 Rank the bonds in each set in order of *increasing* bond length and bond strength: **(a)** Si—F, Si—C, Si—O; **(b)** N=N, N—N, N≡N.

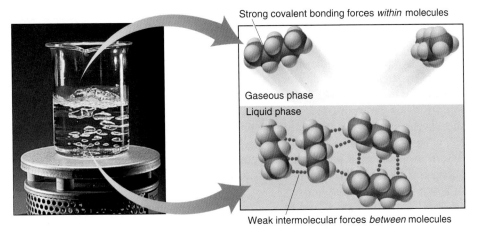

Strong covalent bonding forces *within* molecules

Gaseous phase

Liquid phase

Weak intermolecular forces *between* molecules

Figure 9.13 Strong forces within molecules and weak forces between them. When pentane boils, weak forces *between* molecules (intermolecular forces) are overcome, but the strong covalent bonds holding the atoms together within each molecule remain unaffected. Thus, the pentane molecules leave the liquid phase as intact units.

How the Model Explains the Properties of Covalent Substances

The covalent bonding model proposes that electron sharing between pairs of atoms leads to strong, localized bonds, usually within individual molecules. At first glance, however, it seems that the model is inconsistent with some of the familiar physical properties of covalent substances. After all, most are gases (such as methane and ammonia), liquids (such as benzene and water), or low-melting solids (such as sulfur and paraffin wax). Covalent bonds are strong (~200 to 500 kJ/mol), so why do covalent substances melt and boil at such low temperatures?

To answer this question, we must distinguish between two different sets of forces: (1) the *strong covalent bonding forces* holding the atoms together within the molecule (those we have been discussing), and (2) the *weak intermolecular forces* holding the molecules near each other in the macroscopic sample. It is these weak forces *between* the molecules, not the strong covalent bonds *within* each molecule, that are responsible for the physical properties of covalent substances. Consider, for example, what happens when pentane (C_5H_{12}) boils. As Figure 9.13 shows, the weak interactions *between* the pentane molecules are affected, not the strong C—C and C—H covalent bonds *within* each molecule.

Some covalent substances, called *network covalent solids,* do not consist of separate molecules. Rather, they are held together by covalent bonds that extend in three dimensions *throughout* the sample. The properties of these substances *do* reflect the strength of their covalent bonds. Two examples, quartz and diamond, are shown in Figure 9.14. Quartz (SiO_2) has silicon-oxygen covalent bonds that extend throughout the sample; no separate SiO_2 molecules exist. Quartz is very hard and melts at 1550°C. Diamond has covalent bonds connecting each of its carbon atoms to four others throughout the sample. It is the hardest substance known and melts at around 3550°C. Clearly, covalent bonds *are* strong, but because most covalent substances consist of separate molecules with weak forces between them, their physical properties do not reflect this bond strength. (We discuss intermolecular forces in detail in Chapter 12.)

Unlike ionic compounds, most covalent substances are poor electrical conductors, even when melted or when dissolved in water. An electric current is carried by either mobile electrons or mobile ions. In covalent substances the electrons are localized as either shared or unshared pairs, so they are not free to move, and no ions are present.

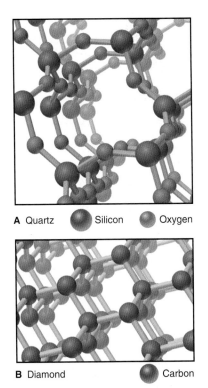

A Quartz ● Silicon ● Oxygen

B Diamond ● Carbon

Figure 9.14 Covalent bonds of network covalent solids. A, In quartz (SiO_2), each Si atom is bonded covalently to four O atoms and each O atom is bonded to two Si atoms in a pattern that extends throughout the sample. Because no separate SiO_2 molecules are present, the melting point of quartz is very high, and it is very hard. **B,** In diamond, each C atom is covalently bonded to four other C atoms throughout the crystal. Diamond is the hardest natural substance known and has an extremely high melting point.

Chemists often study the types of covalent bonds in a molecule using a technique called **infrared (IR) spectroscopy.** All molecules, whether occurring as a gas, a liquid, or a solid, undergo continual vibrations. We can think of any covalent bond between two atoms, say, the C—C bond in ethane (H_3C—CH_3), as a spring that is continually stretching, twisting, and bending. Each motion occurs at a frequency that depends on the "stiffness" of the spring (the bond energy), the type of motion, and the masses of the atoms. The frequencies of these vibrational motions correspond to the wavelengths of photons that lie within the IR region of the electromagnetic spectrum. Thus, the energies of these motions are quantized. And, just as an atom can absorb a photon of a particular energy and attain a different electron energy level (Chapter 7), a molecule can absorb an IR photon of a particular energy and attain a different vibrational energy level.

Each kind of bond (C—C, C=C, C—O, etc.) absorbs a characteristic range of IR wavelengths and quantity of radiation, which depends on the molecule's overall structure. The absorptions by all the bonds in a given molecule create a unique pattern of downward pointing peaks of varying depth and sharpness. Thus, *each compound has a characteristic IR spectrum* that can be used to identify it, much like a fingerprint is used to identify a person. As an example, consider the compounds 2-butanol and diethyl ether. These compounds have the same molecular formula ($C_4H_{10}O$) but different structural formulas and, therefore, are constitutional (structural) isomers. Figure 9.15 shows that they have very different IR spectra.

Figure 9.15 The infrared spectra of 2-butanol (*green*) and diethyl ether (*red*).

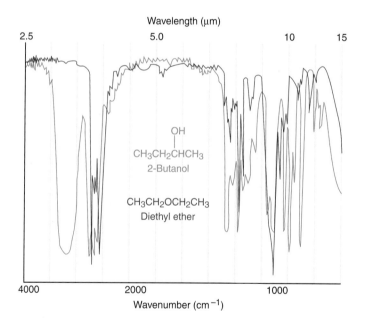

SECTION SUMMARY

A shared pair of valence electrons attracts the nuclei of two atoms and holds them together in a covalent bond while filling each atom's outer level. The number of shared pairs between the two atoms is the bond order. For a given type of bond, the bond energy is the average energy required to completely separate the bonded atoms; the bond length is the average distance between their nuclei. For a given pair of bonded atoms, bond order is directly related to bond energy and inversely related to bond length. Substances that consist of separate molecules are generally soft and low melting because of the weak forces between molecules. Solids held together by covalent bonds extending in three dimensions throughout the sample are extremely hard and high melting. Most covalent substances have low electrical conductivity because electrons are localized and ions are absent. The atoms in a covalent bond vibrate, and the energies of these vibrations can be studied with IR spectroscopy.

9.4   BOND ENERGY AND CHEMICAL CHANGE

The relative strengths of the bonds in reactants and products of a chemical change determine whether heat is released or absorbed. In fact, as you'll see in Chapter 20, bond strength is one of two essential factors determining whether the change occurs at all. In this section, we'll discuss the importance of bond energy in chemical change, especially in the combustion of fuels and foods.

Changes in Bond Strengths: Where Does ΔH°_{rxn} Come From?

In Chapter 6, we discussed the heat involved in a chemical change (ΔH°_{rxn}), but we never stopped to ask a central question. When, for example, 1 mol of H_2 and 1 mol of F_2 react at 298 K, 2 mol of HF forms and 546 kJ of heat is released:

$$H_2(g) + F_2(g) \longrightarrow 2HF(g) + 546 \text{ kJ}$$

Where does this heat come from? We find the answer through a very close-up view of the molecules and their energy components.

A system's internal energy has kinetic energy (E_k) and potential energy (E_p) components. Let's examine the contributions to these components to see which one changes during the reaction of H_2 and F_2 to form HF.

Of the various contributions to the kinetic energy, the most important come from the molecules moving through space, rotating, and vibrating and, of course, from the electrons moving within the atoms. Of the various contributions to the potential energy, the most important are electrostatic forces between the vibrating atoms, between nucleus and electrons (and between electrons) in each atom, between protons and neutrons in each nucleus, and, of course, between nuclei and shared electron pair in each bond.

The kinetic energy doesn't change during the reaction because the molecules' motions in space, rotations, and vibrations are proportional to the temperature, which is constant at 298 K; and electron motion is not affected by a reaction. Of the potential energy contributions, those within the atoms and nuclei don't change, and vibrational forces vary only slightly as the bonded atoms change. The only significant change in potential energy is in the strength of attraction of the nuclei for the shared electron pair, that is, in the bond energy.

In other words, the answer to "Where does the heat come from?" is that it doesn't really "come from" anywhere: *the energy released or absorbed during a chemical change is due to differences between the reactant bond energies and the product bond energies.*

Using Bond Energies to Calculate ΔH°_{rxn}

We can think of a reaction as a two-step process in which *heat is absorbed (ΔH° is positive) to break reactant bonds and form separate atoms and then is released (ΔH° is negative) when the atoms rearrange to form product bonds.* The sum (symbolized by Σ) of these enthalpy changes is the heat of reaction, ΔH°_{rxn}:

$$\Delta H^{\circ}_{rxn} = \Sigma \Delta H^{\circ}_{\text{reactant bonds broken}} + \Sigma \Delta H^{\circ}_{\text{product bonds formed}} \qquad (9.2)$$

- In an exothermic reaction, the total ΔH° for product bonds formed is *greater* than that for reactant bonds broken, so the sum, ΔH°_{rxn}, is *negative*.
- In an endothermic reaction, the total ΔH° for product bonds formed is *smaller* than that for reactant bonds broken, so the sum, ΔH°_{rxn}, is *positive*.

An equivalent form of Equation 9.2 uses bond energies:

$$\Delta H^{\circ}_{rxn} = \Sigma BE_{\text{reactant bonds broken}} - \Sigma BE_{\text{product bonds formed}}$$

The minus sign is needed because all bond energies are positive values (see Table 9.2).

Figure 9.16 Using bond energies to calculate ΔH°_{rxn}. Any chemical reaction can be divided conceptually into two hypothetical steps: (1) reactant bonds break to yield separate atoms in a step that absorbs heat (+ sum of BE), and (2) the atoms combine to form product bonds in a step that releases heat (− sum of BE). When the total bond energy of the products is greater than that of the reactants, more energy is released than is absorbed, and the reaction is exothermic (as shown); ΔH°_{rxn} is negative. When the total bond energy of the products is less than that of the reactants, the reaction is endothermic; ΔH°_{rxn} is positive.

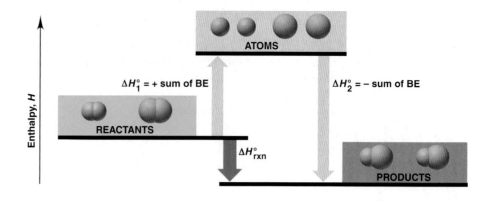

When 1 mol of H—H bonds and 1 mol of F—F bonds absorb energy and break, the 2 mol each of H and F atoms form 2 mol of H—F bonds, which releases energy (Figure 9.16). Recall that *weaker bonds (less stable, more reactive) are easier to break than stronger bonds (more stable, less reactive) because they are higher in energy.* Heat is released when HF forms because the bonds in H_2 and F_2 are weaker (less stable) than the bonds in HF (more stable). Put another way, the sum of the bond energies in 1 mol of H_2 and 1 mol of F_2 is *smaller* than the sum of the bond energies in 2 mol of HF.

We use bond energies to calculate ΔH°_{rxn} by assuming that **all** *the reactant bonds break to give individual atoms, from which* **all** *the product bonds form.* Even though the actual reaction may not occur this way—typically, only certain bonds break and form—Hess's law (see Section 6.5) allows us to sum the bond energies (with their appropriate signs) to arrive at the overall heat of reaction. (This method assumes that all reactants and products are in the same physical state. When phase changes occur, additional heat must be taken into account. We address this topic in Chapter 12.)

Let's use bond energies to calculate ΔH°_{rxn} for the combustion of methane. Figure 9.17 shows that all the bonds in CH_4 and O_2 break, and the atoms form the bonds in CO_2 and H_2O. We find the bond energy values in Table 9.2, and use a positive sign for bonds broken and a negative sign for bonds formed:

Bonds broken
$$4 \times C—H = (4 \text{ mol})(413 \text{ kJ/mol}) = 1652 \text{ kJ}$$
$$\underline{2 \times O_2 = (2 \text{ mol})(498 \text{ kJ/mol}) = 996 \text{ kJ}}$$
$$\Sigma\Delta H^\circ_{\text{reactant bonds broken}} = 2648 \text{ kJ}$$

Bonds formed
$$2 \times C{=}O = (2 \text{ mol})(-799 \text{ kJ/mol}) = -1598 \text{ kJ}$$
$$\underline{4 \times O—H = (4 \text{ mol})(-467 \text{ kJ/mol}) = -1868 \text{ kJ}}$$
$$\Sigma\Delta H^\circ_{\text{product bonds formed}} = -3466 \text{ kJ}$$

Applying Equation 9.2 gives

$$\Delta H^\circ_{rxn} = \Sigma\Delta H^\circ_{\text{reactant bonds broken}} + \Sigma\Delta H^\circ_{\text{product bonds formed}}$$
$$= 2648 \text{ kJ} + (-3466 \text{ kJ}) = -818 \text{ kJ}$$

It is interesting to compare this value with the value obtained by calorimetry (Section 6.3), which is

$$CH_4(g) + 2O_2(g) \longrightarrow CO_2(g) + 2H_2O(g) \qquad \Delta H^\circ_{rxn} = -802 \text{ kJ}$$

Why is there a discrepancy between the bond energy value (−818 kJ) and the calorimetric value (−802 kJ)? Variations in experimental method always introduce small discrepancies, but there is a more basic reason in this case. Because bond energies are *average* values obtained from many different compounds, the energy of the bond *in a particular substance* is usually close, but not equal, to this average. For example, the tabulated C—H bond energy of 413 kJ/mol is the

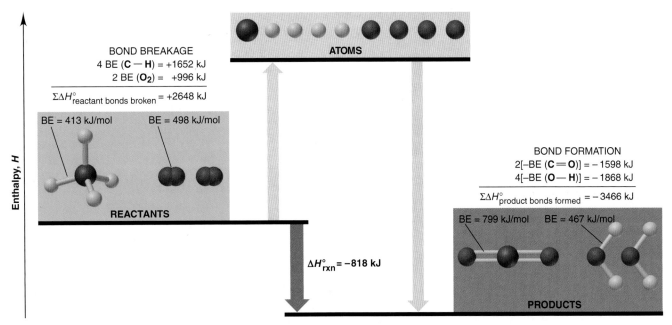

Figure 9.17 Using bond energies to calculate ΔH°_{rxn} of methane. Treating the combustion of methane as a hypothetical two-step process (see Figure 9.16) means breaking all the bonds in the reactants and forming all the bonds in the products.

average value of C—H bonds in many different molecules. In fact, 415 kJ is actually required to break 1 mol of C—H bonds in methane, or 1660 kJ for 4 mol of these bonds, which gives a ΔH°_{rxn} even closer to the calorimetric value. Thus, it isn't surprising to find a discrepancy between the two ΔH°_{rxn} values. What is surprising—and satisfying in its confirmation of bond theory—is that the values are so close.

SAMPLE PROBLEM 9.3 Using Bond Energies to Calculate ΔH°_{rxn}

Problem Calculate ΔH°_{rxn} for the chlorination of methane to form chloroform:

$$\underset{\underset{H}{|}}{\overset{\overset{H}{|}}{H-C-H}} + 3\,Cl-Cl \longrightarrow \underset{\underset{Cl}{|}}{\overset{\overset{H}{|}}{Cl-C-Cl}} + 3\,H-Cl$$

Plan We assume that, in the reaction, *all* the reactant bonds break and *all* the product bonds form. We find the bond energies in Table 9.2 and substitute the two sums, with correct signs, into Equation 9.2.

Solution Finding the standard enthalpy changes for bonds broken and for bonds formed: For bonds broken, the bond energy values are

$$4 \times C-H = (4\ \text{mol})(413\ \text{kJ/mol}) = 1652\ \text{kJ}$$
$$\underline{3 \times Cl-Cl = (3\ \text{mol})(243\ \text{kJ/mol}) = \ \ 729\ \text{kJ}}$$
$$\Sigma\Delta H^{\circ}_{\text{bonds broken}} = 2381\ \text{kJ}$$

For bonds formed, the values are

$$3 \times C-Cl = (3\ \text{mol})(-339\ \text{kJ/mol}) = -1017\ \text{kJ}$$
$$1 \times C-H = (1\ \text{mol})(-413\ \text{kJ/mol}) = \ \ -413\ \text{kJ}$$
$$\underline{3 \times H-Cl = (3\ \text{mol})(-427\ \text{kJ/mol}) = -1281\ \text{kJ}}$$
$$\Sigma\Delta H^{\circ}_{\text{bonds formed}} = -2711\ \text{kJ}$$

Calculating ΔH°_{rxn}:

$$\Delta H^{\circ}_{rxn} = \Sigma\Delta H^{\circ}_{\text{bonds broken}} + \Sigma\Delta H^{\circ}_{\text{bonds formed}} = 2381\ \text{kJ} + (-2711\ \text{kJ}) = -330\ \text{kJ}$$

Check The signs of the enthalpy changes are correct: $\Sigma\Delta H^\circ_{\text{bonds broken}}$ should be >0, and $\Sigma\Delta H^\circ_{\text{bonds formed}} <0$. More energy is released than absorbed, so $\Delta H^\circ_{\text{rxn}}$ is negative:

$$\sim2400 \text{ kJ} + [\sim(-2700 \text{ kJ})] = -300 \text{ kJ}$$

FOLLOW-UP PROBLEM 9.3 One of the most important industrial reactions is the formation of ammonia from its elements:

$$\text{N}\equiv\text{N} + 3\,\text{H}—\text{H} \longrightarrow 2\,\text{H}—\overset{\displaystyle |}{\underset{\displaystyle \underset{\textstyle \text{H}}{|}}{\text{N}}}—\text{H}$$

Use bond energies to calculate $\Delta H^\circ_{\text{rxn}}$.

Relative Bond Strengths in Fuels and Foods

A *fuel* is a material that reacts with atmospheric oxygen to release energy *and* is available at a reasonable cost. The most common fuels for machines are hydrocarbons and coal, and the most common ones for organisms are the fats and carbohydrates in foods. Both types of fuels are composed of large organic molecules with mostly C—C and C—H bonds; the foods also contain some C—O and O—H bonds. As with any reaction, the energy released from the combustion of a fuel arises from differences in bond energies between the reactants (fuel plus O_2) and the products (CO_2 and H_2O). When the fuel reacts with O_2, the bonds break, and the C, H, and O atoms form C=O and O—H bonds in the products. Because the reaction is exothermic, we know that the total strength of the bonds in the products is greater than that of the bonds in the reactants. In other words, the bonds in CO_2 and H_2O are stronger (lower in energy, more stable) than those in gasoline (or salad oil) and O_2 (weaker, higher in energy, less stable).

Fuels with more weak bonds yield more energy than fuels with fewer weak bonds. When an alkane burns, C—C and C—H bonds break; when an alcohol burns, C—O and O—H bonds break as well. A look at the bond energy values in Table 9.2 shows that the sum for C—C and C—H bonds (760 kJ/mol) is less than the sum for C—O and O—H bonds (825 kJ/mol). Fuels with relatively more of the weaker bonds will release more energy because it takes less energy to break them apart before the atoms rearrange to form the products (Figure 9.18). Thus, we would expect that as the number of C—C and C—H bonds decreases and/or the number of C—O and O—H bonds increases, less energy is released from combustion; that is, ΔH_{rxn} for the combustion reaction is less negative. As a generalization, we can say that *the fewer bonds to O in a fuel, the more energy it releases when burned.*

Both fats and carbohydrates serve as high-energy foods—carbohydrates for shorter term release and fats for longer term storage. Fats consist of chains of carbon atoms (C—C bonds) attached to hydrogen atoms (C—H bonds), with a few C—O and O—H bonds (shown in red in the structures). In contrast, carbohydrates have far fewer C—C and C—H bonds and many more C—O and O—H bonds:

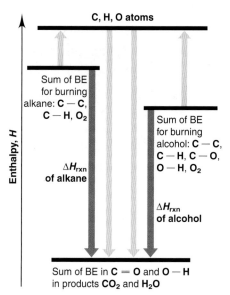

Figure 9.18 Bond strength and the energy from fuels. Fuels with weaker bonds release more energy during combustion than do fuels with stronger bonds.

Sucrose, a carbohydrate

Triolein, a fat

Both kinds of foods are metabolized in the body to CO_2 and H_2O. Fats "contain more Calories" per gram than carbohydrates because fats have more weaker bonds and fewer stronger bonds—that is, fewer bonds to O. Thus, fats should release more energy than carbohydrates, which Table 9.4 confirms.

SECTION SUMMARY
The only component of internal energy that changes significantly during a reaction is the energy of the bonds in reactants and products; this change in bond energy appears as the heat of reaction, ΔH°_{rxn}. A reaction involves breaking reactant bonds and forming product bonds. Applying Hess's law, we use tabulated bond energies to calculate ΔH°_{rxn}. Bonds in fuels and foods are weaker (less stable, higher energy) than bonds in waste products. Fuels with more weak bonds release more energy than fuels with fewer.

9.5 BETWEEN THE EXTREMES: ELECTRONEGATIVITY AND BOND POLARITY

Scientific models are idealized descriptions of reality. As we've discussed them so far, the ionic and covalent bonding models portray compounds as being formed by *either* complete electron transfer *or* complete electron sharing. In most real substances, however, the type of bonding lies somewhere between these extremes. Thus, the great majority of compounds have bonds that are more accurately thought of as "polar covalent," that is, partially ionic and partially covalent (Figure 9.19). Let's consider this important feature of covalent bonding.

Electronegativity

One of the most important concepts in chemical bonding is **electronegativity (EN),** the relative ability of a bonded atom to attract the shared electrons.* More than 50 years ago, the American chemist Linus Pauling developed the most common scale of relative EN values for the elements. Here is an example to show the basis of Pauling's approach. We might expect the bond energy of the HF bond to be the average of the energies of an H—H bond (432 kJ/mol) and an F—F bond (159 kJ/mol), or 296 kJ/mol. However, the actual bond energy of H—F is 565 kJ/mol, or 269 kJ/mol *higher* than the average. Pauling reasoned that this difference is due to an *electrostatic (charge) contribution* to the H—F bond energy. If F attracts the shared electron pair more strongly than H, that is, if F is more electronegative than H, the electrons will spend more time closer to F. This unequal sharing of electrons makes the F end of the bond partially negative and the H end partially positive, and the attraction between these partial charges *increases* the energy required to break the bond.

From similar studies with the remaining hydrogen halides and many other compounds, Pauling arrived at the scale of *relative EN values* shown in Figure 9.20 (on the next page). These values are not measured quantities but are based on Pauling's assignment of the highest EN value, 4.0, to fluorine.

Trends in Electronegativity Because the nucleus of a smaller atom is closer to the shared pair than that of a larger atom, it attracts the bonding electrons more strongly. So, in general, electronegativity is inversely related to atomic size. Thus, for the main-group elements, *electronegativity generally increases up a group and across a period.*

Table 9.4 Heats of Reaction for the Combustion of Some Foods	
Substance	ΔH_{rxn} **(kJ/g)**
Fats	
Vegetable oil	−37.0
Margarine	−30.1
Butter	−30.0
Carbohydrates	
Table sugar (sucrose)	−16.2
Brown rice	−14.9
Maple syrup	−10.4

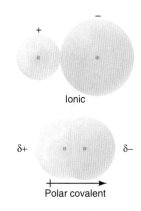

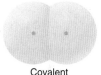

Figure 9.19 The prevalence of electron sharing. The extremes of pure ionic (*top*) and pure covalent (*bottom*) bonding occur rarely, if ever. Much more common is the unequal electron sharing seen in polar covalent bonding.

---

*Electronegativity is not the same as electron affinity (EA), although many elements with a high EN also have a highly negative EA. Electronegativity refers to a bonded atom attracting the shared electron pair; electron affinity refers to a separate atom in the gas phase gaining an electron to form a gaseous anion.

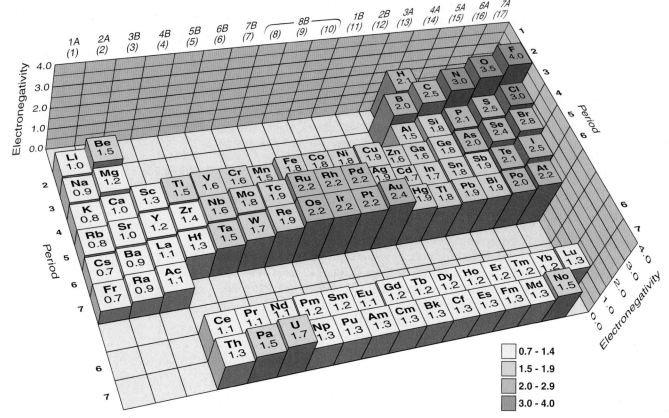

Figure 9.20 The Pauling electronegativity (EN) scale. The EN is shown by the height of the post with the value on top. The key indicates arbitrary EN cutoffs. In the main groups, EN generally *increases* from left to right and *decreases* from top to bottom. The noble gases are not shown. The transition and inner transition elements show relatively little change in EN. Hydrogen is shown near elements of similar EN.

Electronegativity and Oxidation Number One important use of electronegativity is in determining an atom's oxidation number (O.N.; see Section 4.5):

1. The more electronegative atom in a bond is assigned *all* the *shared* electrons; the less electronegative atom is assigned *none*.
2. Each atom in a bond is assigned *all* of its *unshared* electrons.
3. The oxidation number is given by

$$\text{O.N.} = \text{no. of valence } e^- - (\text{no. of shared } e^- + \text{no. of unshared } e^-)$$

In HCl, for example, Cl is more electronegative than H. It has 7 valence electrons but is assigned 8 (2 shared + 6 unshared), so its oxidation number is $7 - 8 = -1$. The H atom has 1 valence electron and is assigned none, so its oxidation number is $1 - 0 = +1$.

Polar Covalent Bonds and Bond Polarity

Whenever atoms of different electronegativities form a bond, as in HF, the bonding pair is shared *un*equally. This unequal distribution of electron density gives the bond partially negative and positive poles. Such a **polar covalent bond** is depicted by a polar arrow ($\longmapsto$) pointing toward the negative pole or by $\delta+$ and $\delta-$ symbols, where the lowercase Greek letter delta (δ) represents a partial charge (see also the discussion of O—H bonds in water, Section 4.1):

$$\overset{\longmapsto}{\text{H}-\ddot{\text{F}}:} \quad \text{or} \quad \overset{\delta+ \quad \delta-}{\text{H}-\ddot{\text{F}}:}$$

In the H—H and F—F bonds, the atoms are identical, so the bonding pair is shared equally, and a **nonpolar covalent bond** results. By knowing the EN values of the atoms in a bond, we can find the direction of the bond polarity. Figure 9.21 compares the distribution of electron density in H_2, F_2, and HF.

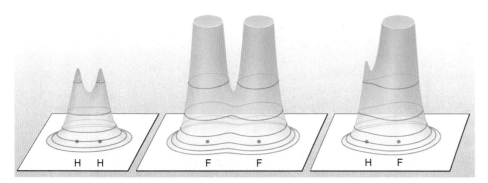

Figure 9.21 Electron density distributions in H_2, F_2, and HF. As these relief maps show, electron density is distributed equally around the two nuclei in the nonpolar covalent molecules H_2 and F_2. (The electron density around the F nuclei is so great that the peaks must be "cut off" to fit within the figure.) But, in polar covalent HF, the electron density is shifted away from H and toward F.

SAMPLE PROBLEM 9.4 Determining Bond Polarity from EN Values

Problem (a) Use a polar arrow to indicate the polarity of each bond: N—H, F—N, I—Cl.
(b) Rank the following bonds in order of increasing polarity: H—N, H—O, H—C.
Plan (a) We use Figure 9.20 to find the EN values of the bonded atoms and point the polar arrow toward the negative end. (b) Each choice has H bonded to an atom from Period 2. EN increases across a period, so the polarity is greatest for the bond whose Period 2 atom is farthest to the right.
Solution (a) The EN of N = 3.0 and the EN of H = 2.1, so N is more electronegative than H: $\overset{\longleftarrow+}{\text{N—H}}$
The EN of F = 4.0 and the EN of N = 3.0, so F is more electronegative: $\overset{\longleftarrow+}{\text{F—N}}$

The EN of I = 2.5 and the EN of Cl = 3.0, so I is less electronegative: $\overset{+\longrightarrow}{\text{I—Cl}}$
(b) The order of increasing EN is C < N < O, and each has a higher EN than H does. Therefore, O pulls most on the electron pair shared with H, and C pulls least; so the order of bond polarity is H—C < H—N < H—O.
Comment In Chapter 10, you'll see that the polarity of the bonds in a molecule contributes to the overall polarity of the molecule, which is a major factor determining the magnitudes of several physical properties.

FOLLOW-UP PROBLEM 9.4 Arrange each set of bonds in order of increasing polarity, and indicate bond polarity with $\delta+$ and $\delta-$ symbols:
(a) Cl—F, Br—Cl, Cl—Cl   (b) Si—Cl, P—Cl, S—Cl, Si—Si

The Partial Ionic Character of Polar Covalent Bonds

As we've just seen, if you ask "Is an X—Y bond ionic or covalent?" the answer in almost every case is "Both, partially!" A better question is *"To what extent* is the bond ionic or covalent?" The **partial ionic character** of a bond is related directly to the **electronegativity difference (ΔEN),** the difference between the EN values of the bonded atoms: *a greater ΔEN results in larger partial charges and a higher partial ionic character.* Consider these three chlorine-containing molecules: ΔEN for LiCl(g) is 3.0 − 1.0 = 2.0; for HCl(g), it is 3.0 − 2.1 = 0.9; and for $Cl_2(g)$, it is 3.0 − 3.0 = 0. Thus, the bond in LiCl has more ionic character than the H—Cl bond, which has more than the Cl—Cl bond.

Various attempts have been made to classify the ionic character of bonds, but they all use arbitrary cutoff values, which is inconsistent with the gradation of ionic character observed experimentally. One approach uses ΔEN values to divide bonds into ionic, polar covalent, and nonpolar covalent. Based on a range of ΔEN values from 0 (completely nonpolar) to 3.3 (highly ionic), some *approximate* guidelines are given in Figure 9.22A and appear within a gradient in Figure 9.22B (on the next page). Keep in mind that these ΔEN cutoff values are only useful for gaining a *general* idea of a compound's ionic character.

ΔEN	IONIC CHARACTER
>1.7	Mostly ionic
0.4–1.7	Polar covalent
<0.4	Mostly covalent
0	Nonpolar covalent

A

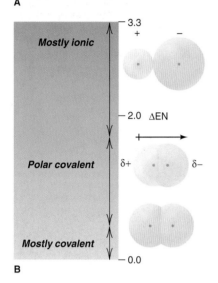

B

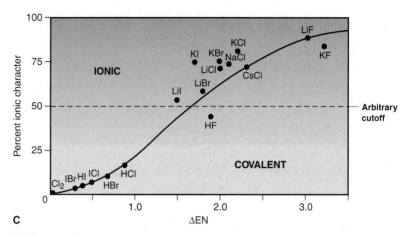

C

Figure 9.22 The ionic character of chemical bonds. A, The electronegativity difference (ΔEN) between bonded atoms shows cutoff values that act as a very general guide to a bond's ionic character. **B,** The gradation in ionic character across the entire bonding range is shown as shading from ionic (*green*) to covalent (*yellow*). **C,** The percent ionic character is plotted against ΔEN for some gaseous diatomic molecules. Note that, in general, ΔEN correlates with ionic character. (The arbitrary cutoff for an ionic compound is >50% ionic character.)

Another approach calculates the *percent ionic character* of a bond by observing the actual behavior of a polar molecule in an electric field. A value of 50% ionic character divides substances we call "ionic" from those we call "covalent." Such methods show 43% ionic character for the H—F bond and expected decreases for the other hydrogen halides: H—Cl is 19% ionic, H—Br 11%, and H—I 4%. A plot of percent ionic character vs. ΔEN for a variety of gaseous diatomic molecules is shown in Figure 9.22C. The specific values are not important, but note that *percent ionic character generally increases with ΔEN.* Another point to note is that whereas some molecules, such as $Cl_2(g)$, have 0% ionic character, none has 100% ionic character. Thus, *electron sharing occurs to some extent in every bond,* even one between an alkali metal and a halogen.

The Continuum of Bonding Across a Period

A metal and a nonmetal—elements from opposite sides of the periodic table—have a relatively large ΔEN and typically interact by electron transfer to form an ionic compound. Two nonmetals—elements from the same side of the table—have a small ΔEN and interact by electron sharing to form a covalent compound. When we combine the nonmetal chlorine with each of the other elements in Period 3, starting with sodium, we should observe a steady decrease in ΔEN and a gradation in bond type from ionic through polar covalent to nonpolar covalent.

Figure 9.23 shows samples of the common Period 3 chlorides—NaCl, $MgCl_2$, $AlCl_3$, $SiCl_4$, PCl_3, and SCl_2, as well as Cl_2—with graphs showing their ΔEN values and two key macroscopic properties. With a ΔEN of 2.1, sodium chloride is the most ionic of this group, displaying the physical properties characteristic of an ionic solid: high melting point and high electrical conductivity when molten. $MgCl_2$, also a white crystalline solid, is also ionic, with ΔEN of 1.8, but it has a lower melting point and a slightly lower conductivity in the molten state.

The structure of $AlCl_3$ reflects its smaller ΔEN value of 1.5, which lies near the upper cutoff for polar covalent bonding. Instead of a three-dimensional lattice of Al^{3+} and Cl^- ions, $AlCl_3$ consists of extended layers of Al and Cl atoms connected through highly polar covalent bonds. Much weaker forces between the layers result in a significantly lower melting point for $AlCl_3$, and its very low conductivity when molten is consistent with a scarcity of ions.

The tendency toward less ionic and more covalent character in the bonds continues through $SiCl_4$, PCl_3, and SCl_2. Each of these compounds occurs as separate

Figure 9.23 Properties of the Period 3 chlorides. Samples of the compounds formed from each of the Period 3 elements with chlorine are shown in periodic table sequence in the photo. Note the trend in properties displayed in the bar graphs: as ΔEN decreases, both melting point and electrical conductivity (at the melting point) decrease. These trends are consistent with a change in bond type from ionic through polar covalent to nonpolar covalent.

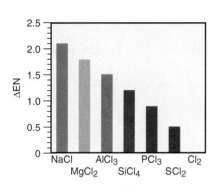
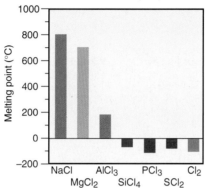
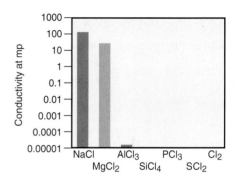

molecules, as indicated by the absence of any measurable conductivity. Moreover, the forces *between* the molecules are weak enough to lower the melting points for these compounds below room temperature. In Cl_2, the bond is nonpolar, and this is the only substance in this group that is a gas at room temperature.

Thus, *as ΔEN becomes smaller, the bond becomes more covalent,* and the macroscopic properties of the Period 3 chlorides and Cl_2 change from those of a solid consisting of ions to those of a gas consisting of individual molecules.

SECTION SUMMARY
An atom's electronegativity refers to its ability to pull bonded electrons toward it, which generates partial charges at the ends of a bond. Electronegativity increases across a period and decreases down a group, the reverse of the trends in atomic size. The greater the ΔEN for the two atoms in a bond, the more polar the bond is and the greater its ionic character. For chlorides of Period 3 elements, there is a gradation of bond type from ionic to polar covalent to nonpolar covalent.

For Review and Reference

Learning Objectives

To help you review these learning objectives, the numbers of related sections (§), sample problems (SP), and end-of-chapter problems (EP) are listed in parentheses.

1. Explain how differences in atomic properties lead to the three types of chemical bonding (§ 9.1) (EPs 9.1–9.7)
2. Depict main-group atoms with Lewis electron-dot symbols (§ 9.1) (EPs 9.8–9.11)
3. Understand the key features of ionic bonding, the significance of the lattice energy, and how the model explains the properties of ionic compounds (§ 9.2) (EPs 9.12–9.15, 9.20–9.22)
4. Depict the formation of binary ionic compounds with electron configurations, partial orbital diagrams, and Lewis electron-dot symbols (§ 9.2) (SP 9.1) (EPs 9.16–9.19)

5. Describe the formation of a covalent bond, the interrelationship among bond length, strength, and order, and how the model explains the properties of covalent compounds (§ 9.3) (SP 9.2) (EPs 9.23–9.31)
6. Understand how changes in bond energy account for ΔH°_{rxn} and be able to divide a reaction into bond-breaking and bond-forming steps (§ 9.4) (SP 9.3) (EPs 9.32–9.40)
7. Describe the trends in electronegativity, and understand how the polarity of a bond and the partial ionic character of a compound relate to ΔEN of the bonded atoms (§ 9.5) (SP 9.4) (EPs 9.41–9.56)

Key Terms

Section 9.1
ionic bonding (269)
covalent bonding (270)
metallic bonding (270)
Lewis electron-dot
  symbol (271)
octet rule (272)

Section 9.2
lattice energy (274)
Coulomb's law (274)
ion pair (276)

Section 9.3
covalent bond (277)
bonding (shared) pair (278)
lone (unshared) pair (278)

bond order (278)
single bond (278)
double bond (278)
triple bond (278)
bond energy (BE) (279)
bond length (279)
infrared (IR) spectroscopy (282)

Section 9.5
electronegativity (EN) (287)
polar covalent bond (288)
nonpolar covalent bond (288)
partial ionic character (289)
electronegativity difference
  (ΔEN) (289)

Key Equations and Relationships

9.1 Relating the energy of attraction to the lattice energy (274):

$$\text{Electrostatic energy} \propto \frac{\text{cation charge} \times \text{anion charge}}{\text{cation radius} + \text{anion radius}} \propto \Delta H^\circ_{\text{lattice}}$$

9.2 Calculating heat of reaction from enthalpy changes or bond energies (283):

$$\Delta H^\circ_{\text{rxn}} = \Sigma \Delta H^\circ_{\text{reactant bonds broken}} + \Sigma \Delta H^\circ_{\text{product bonds formed}}$$

or $\quad \Delta H^\circ_{\text{rxn}} = \Sigma BE_{\text{reactant bonds broken}} - \Sigma BE_{\text{product bonds formed}}$

Brief Solutions to Follow-up Problems

9.1 Mg ([Ne] $3s^2$) + 2Cl ([Ne] $3s^23p^5$) $\longrightarrow$
$\quad\quad\quad\quad\quad\quad\quad\quad$ Mg^{2+} ([Ne]) + 2Cl$^-$ ([Ne] $3s^23p^6$)

$\quad\quad\quad\quad\quad\quad\quad\quad\quad\quad\quad\quad$ Formula: MgCl$_2$

9.2 (a) Bond length: Si—F < Si—O < Si—C
Bond strength: Si—C < Si—O < Si—F
(b) Bond length: N≡N < N=N < N—N
Bond strength: N—N < N=N < N≡N

9.3 N≡N + 3 H—H $\longrightarrow$ 2 H—N—H
$\quad\quad\quad\quad\quad\quad\quad\quad\quad\quad\quad\quad\quad\quad$ |
$\quad\quad\quad\quad\quad\quad\quad\quad\quad\quad\quad\quad\quad\quad$ H

$\Sigma \Delta H^\circ_{\text{bonds broken}}$ = 1 N≡N + 3 H—H
$\quad\quad\quad\quad\quad\quad$ = 945 kJ + 1296 kJ = 2241 kJ
$\Sigma \Delta H^\circ_{\text{bonds formed}}$ = 6 N—H = −2346 kJ
$\Delta H^\circ_{\text{rxn}}$ = −105 kJ

9.4 (a) Cl—Cl < $\overset{\delta+}{\text{Br}}$—$\overset{\delta-}{\text{Cl}}$ < $\overset{\delta+}{\text{Cl}}$—$\overset{\delta-}{\text{F}}$
(b) Si—Si < $\overset{\delta+}{\text{S}}$—$\overset{\delta-}{\text{Cl}}$ < $\overset{\delta+}{\text{P}}$—$\overset{\delta-}{\text{Cl}}$ < $\overset{\delta+}{\text{Si}}$—$\overset{\delta-}{\text{Cl}}$

Problems

Problems with **colored** numbers are answered in Appendix E. Sections match the text and provide the numbers of relevant sample problems. Bracketed problems are grouped in pairs (indicated by a short rule) that cover the same concept. Comprehensive Problems are based on material from any section or previous chapter.

Atomic Properties and Chemical Bonds

9.1 In general terms, how does each of the following atomic properties influence the metallic character of the main-group elements in a period?
(a) Ionization energy
(b) Atomic radius
(c) Number of outer electrons
(d) Effective nuclear charge
9.2 Both nitrogen and bismuth are members of Group 5A(15). Which is more metallic? Explain your answer in terms of atomic properties.
9.3 What is the relationship between the tendency of a main-group element to form a monatomic ion and its position in the periodic

table? In what part of the table are the main-group elements that typically form cations? Anions?

9.4 Which member of each pair is *more* metallic?
(a) Na or Cs   (b) Mg or Rb   (c) As or N
9.5 Which member of each pair is *less* metallic?
(a) I or O   (b) Be or Ba   (c) Se or Ge

9.6 State whether the type of bonding in the following compounds is best described as ionic or covalent: (a) CsF(s); (b) N$_2$(g); (c) H$_2$S(g).
9.7 State whether the type of bonding in the following compounds is best described as ionic or covalent: (a) ICl$_3$(g); (b) N$_2$O(g); (c) LiCl(s).

9.8 Draw a Lewis electron-dot symbol for each atom:
(a) Rb   (b) Si   (c) I
9.9 Draw a Lewis electron-dot symbol for each atom:
(a) Ba   (b) Kr   (c) Br

9.10 Give the group number and general electron configuration of an element with each electron-dot symbol:
(a) ·Ẍ: (b) Ẍ·

9.11 Give the group number and general electron configuration of an element with each electron-dot symbol:
(a) ·Ẍ: (b) ·Ẍ·

The Ionic Bonding Model
(Sample Problem 9.1)

9.12 If energy is required to form monatomic ions from metals and nonmetals, why do ionic compounds exist?

9.13 In general, how does the lattice energy of an ionic compound depend on the charges and sizes of the ions?

9.14 When gaseous Na^+ and Cl^- ions form gaseous NaCl ion pairs, 548 kJ/mol of energy is released. Why, then, does NaCl occur as a solid under ordinary conditions?

9.15 To form S^{2-} ions from gaseous sulfur atoms requires 214 kJ/mol, but these ions exist in solids such as K_2S. Explain.

9.16 Use condensed electron configurations and Lewis electron-dot symbols to depict the monatomic ions formed from each of the following atoms, and predict the formula of the compound the ions produce:
(a) Ba and Cl   (b) Sr and O   (c) Al and F   (d) Rb and O

9.17 Use condensed electron configurations and Lewis electron-dot symbols to depict the monatomic ions formed from each of the following atoms, and predict the formula of the compound the ions produce:
(a) Cs and S   (b) O and Ga   (c) N and Mg   (d) Br and Li

9.18 Identify the main group to which X belongs in each ionic compound formula: (a) X_2O_3; (b) XCO_3; (c) Na_2X.

9.19 Identify the main group to which X belongs in each ionic compound formula: (a) CaX_2; (b) Al_2X_3; (c) XPO_4.

9.20 For each pair, choose the compound with the lower lattice energy, and explain your choice:
(a) CaS or BaS   (b) NaF or MgO

9.21 For each pair, choose the compound with the lower lattice energy, and explain your choice:
(a) NaF or NaCl   (b) K_2O or K_2S

9.22 Aluminum oxide (Al_2O_3) is a widely used industrial abrasive (emery, corundum), for which the specific application depends on the hardness of the crystal. What does this hardness imply about the magnitude of the lattice energy? Would you have predicted from the chemical formula that Al_2O_3 is hard? Explain.

The Covalent Bonding Model
(Sample Problem 9.2)

9.23 Describe the interactions that occur between individual chlorine atoms as they approach each other and form Cl_2. What combination of forces gives rise to the energy holding the atoms together and to the final internuclear distance?

9.24 Define bond energy using the H—Cl bond as an example. When this bond breaks, is energy absorbed or released? Is the accompanying ΔH value positive or negative? How do the magnitude and sign of this ΔH value relate to the value that accompanies H—Cl bond formation?

9.25 For single bonds between similar types of atoms, how does the strength of the bond relate to the sizes of the atoms? Explain.

9.26 How does the energy of the bond between a given pair of atoms relate to the bond order? Why?

9.27 When liquid benzene (C_6H_6) boils, does the gas consist of molecules, ions, or separate atoms? Explain.

9.28 Using the periodic table only, arrange the members of each of the following sets in order of increasing bond *strength:*
(a) Br—Br, Cl—Cl, I—I   (b) S—H, S—Br, S—Cl
(c) C=N, C—N, C≡N

9.29 Using the periodic table only, arrange the members of each of the following sets in order of increasing bond *length:*
(a) H—F, H—I, H—Cl   (b) C—S, C=O, C—O
(c) N—H, N—S, N—O

9.30 Formic acid (HCOOH) has the structural formula shown at right. It is secreted by certain species of ants when they bite. Rank the relative strengths of (a) the C—O and C=O bonds, and (b) the H—C and H—O bonds. Explain these rankings.

$$\overset{\displaystyle O}{\underset{}{\overset{\|}{H—C—O—H}}}$$

9.31 In IR spectra, the stretching of a C=C bond appears at a shorter wavelength than that of a C—C bond. Would you expect the wavelength for the stretching of a C≡C bond to be shorter or longer than that for a C=C bond? Explain.

Bond Energy and Chemical Change
(Sample Problem 9.3)

9.32 Write a solution plan (without actual numbers, but including the bond energies you would use and how you would combine them algebraically) for calculating the total enthalpy change of the following reaction:
$$H_2(g) + O_2(g) \longrightarrow H_2O_2(g) \text{ (H—O—O—H)}$$

9.33 The text points out that, for similar types of substances, one with weaker bonds is usually more reactive than one with stronger bonds. Why is this generally true?

9.34 Why is there often a discrepancy between a heat of reaction obtained from calorimetry and one obtained from bond energies?

9.35 Which gas has the greater heat of reaction per mole from combustion? Why?

methane   or   formaldehyde

9.36 Which liquid fuel has the greater heat of reaction per mole from combustion? Why?

ethanol   or   methanol

9.37 Use bond energies to calculate the heat of reaction:

9.38 Use bond energies to calculate the heat of reaction:

$$O=C=O \ + \ 2\underset{\underset{H}{|}}{\overset{\overset{H}{|}}{N}}-H \ \longrightarrow \ H-\underset{\underset{H}{|}}{\overset{\overset{H}{|}}{N}}-\overset{\overset{O}{||}}{C}-\underset{\underset{H}{|}}{\overset{\overset{H}{|}}{N}}-H \ + \ H-O-H$$

9.39 An important industrial route to extremely pure acetic acid is the reaction of methanol with carbon monoxide:

$$H-\underset{\underset{H}{|}}{\overset{\overset{H}{|}}{C}}-O-H \ + \ C\equiv O \ \longrightarrow \ H-\underset{\underset{H}{|}}{\overset{\overset{H}{|}}{C}}-\overset{\overset{O}{||}}{C}-O-H$$

Use bond energies to calculate the heat of reaction.

9.40 Sports trainers treat sprains and soreness with ethyl bromide. It is manufactured by reacting ethylene with hydrogen bromide:

$$\underset{\underset{H}{}}{\overset{\overset{H}{}}{}}C=C\underset{\underset{H}{}}{\overset{\overset{H}{}}{}} \ + \ H-Br \ \longrightarrow \ H-\underset{\underset{H}{|}}{\overset{\overset{H}{|}}{C}}-\underset{\underset{H}{|}}{\overset{\overset{H}{|}}{C}}-Br$$

Use bond energies to find the enthalpy change for this reaction.

Between the Extremes: Electronegativity and Bond Polarity
(Sample Problem 9.4)

9.41 Describe the vertical and horizontal trends in electronegativity (EN) among the main-group elements. According to Pauling's scale, what are the two most electronegative elements? The two least electronegative elements?

9.42 What is the general relationship between IE_1 and EN for the elements? Why?

9.43 Is the H—O bond in water nonpolar covalent, polar covalent, or ionic? Define each term, and explain your choice.

9.44 How does electronegativity differ from electron affinity?

9.45 How is the partial ionic character of a bond in a diatomic molecule related to ΔEN for the bonded atoms? Why?

9.46 Using the periodic table only, arrange the elements in each set in order of *increasing* EN: (a) S, O, Si; (b) Mg, P, As.

9.47 Using the periodic table only, arrange the elements in each set in order of *decreasing* EN: (a) I, Br, N; (b) Ca, H, F.

9.48 Use Figure 9.20 to indicate the polarity of each bond with a *polar arrow:* (a) N—B; (b) N—O; (c) C—S; (d) S—O; (e) N—H; (f) Cl—O.

9.49 Use Figure 9.20 to indicate the polarity of each bond with *partial charges:* (a) Br—Cl; (b) F—Cl; (c) H—O; (d) Se—H; (e) As—H; (f) S—N.

9.50 Which is the more polar bond in each of the following pairs from Problem 9.48: (a) or (b); (c) or (d); (e) or (f)?

9.51 Which is the more polar bond in each of the following pairs from Problem 9.49: (a) or (b); (c) or (d); (e) or (f)?

9.52 Are the bonds in each of the following substances ionic, non-polar covalent, or polar covalent? Arrange the substances with polar covalent bonds in order of increasing bond polarity:
(a) S_8 (b) RbCl   (c) PF_3 (d) SCl_2 (e) F_2 (f) SF_2

9.53 Are the bonds in each of the following substances ionic, non-polar covalent, or polar covalent? Arrange the substances with polar covalent bonds in order of increasing bond polarity:
(a) KCl   (b) P_4 (c) BF_3 (d) SO_2 (e) Br_2 (f) NO_2

9.54 Rank the members of each set of compounds in order of *increasing* ionic character of their bonds. Use a *polar arrow* to indicate the bond polarity of each:
(a) HBr, HCl, HI   (b) H_2O, CH_4, HF   (c) SCl_2, PCl_3, $SiCl_4$

9.55 Rank the members of each set of compounds in order of *decreasing* ionic character of their bonds. Use *partial charges* to indicate the bond polarity of each:
(a) PCl_3, PBr_3, PF_3 (b) BF_3, NF_3, CF_4 (c) SeF_4, TeF_4, BrF_3

9.56 The energy of the C—C bond is 347 kJ/mol, and that of the Cl—Cl bond is 243 kJ/mol. Which of the following values might you expect for the C—Cl bond energy? Explain.
(a) 590 kJ/mol (sum of the values given)
(b) 104 kJ/mol (difference of the values given)
(c) 295 kJ/mol (average of the values given)
(d) 339 kJ/mol (greater than the average of the values given)

Comprehensive Problems
Problems with an asterisk (*) are more challenging.

9.57 Geologists have a rule of thumb: when molten rock cools and solidifies, crystals of compounds with the smallest lattice energies appear at the bottom of the mass. Suggest a reason for this.

9.58 Acetylene gas (ethyne; $HC\equiv CH$) burns with oxygen in an oxyacetylene torch to produce carbon dioxide, water vapor, and the heat needed to weld metals. The heat of reaction for the combustion of acetylene is 1259 kJ/mol. (a) Calculate the $C\equiv C$ bond energy, and compare your value with that in Table 9.2. (b) When 500.0 g of acetylene burns, how many kilojoules of heat are given off? (c) How many grams of CO_2 are produced? (d) How many liters of O_2 at 298 K and 18.0 atm are consumed?

* 9.59 Even though so much energy is required to form a metal cation with a 2+ charge, the alkaline earth metals form halides with general formula MX_2, rather than MX.
(a) Use the following data to calculate the ΔH_f° of MgCl:

$Mg(s) \longrightarrow Mg(g)$	$\Delta H^\circ =$	148 kJ
$Cl_2(g) \longrightarrow 2Cl(g)$	$\Delta H^\circ =$	243 kJ
$Mg(g) \longrightarrow Mg^+(g) + e^-$	$\Delta H^\circ =$	738 kJ
$Cl(g) + e^- \longrightarrow Cl^-(g)$	$\Delta H^\circ =$	-349 kJ

$$\Delta H_{lattice}^\circ \text{ of MgCl} = 783.5 \text{ kJ/mol}$$

(b) Is MgCl favored energetically relative to its elements? Explain.
(c) Use Hess's law to calculate ΔH° for the conversion of MgCl to $MgCl_2$ and Mg (ΔH_f° of $MgCl_2 = -641.6$ kJ/mol).
(d) Is MgCl favored energetically relative to $MgCl_2$? Explain.

* 9.60 By using photons of specific wavelengths, chemists can dissociate gaseous HI to produce H atoms with certain speeds. When HI dissociates, the H atoms move away rapidly, whereas the relatively heavy I atoms move more slowly.
(a) What is the longest wavelength (in nm) that can dissociate a molecule of HI?
(b) If a photon of 254 nm is used, what is the excess energy (in J) over that needed for the dissociation?
(c) If all this excess energy is carried away by the H atom as kinetic energy, what is its speed (in m/s)?

9.61 Carbon dioxide is a linear molecule. Its vibrational motions include symmetrical stretching, bending, and asymmetrical stretching, and their frequencies are 4.02×10^{13} s^{-1}, 2.00×10^{13} s^{-1}, and

7.05×10^{13} s^{-1}, respectively. (a) In what region of the electromagnetic spectrum are these frequencies? (b) Calculate the energy (in J) of each vibration. Which occurs most readily (takes the least energy)?

9.62 In developing the concept of electronegativity, Pauling used the term *excess bond energy* for the difference between the actual bond energy of X—Y and the average bond energies of X—X and Y—Y (see text discussion for the case of HF). Based on the values in Figure 9.20, which of the following substances contains bonds with *no* excess bond energy?

(a) PH_3 (b) CS_2 (c) BrCl   (d) BH_3 (e) Se_8

9.63 Without stratospheric ozone (O_3), harmful solar radiation would cause gene alterations. Ozone forms when O_2 breaks and each O atom reacts with another O_2 molecule. It is destroyed by reaction with Cl atoms that are formed when the C—Cl bond in synthetic chemicals breaks. Find the wavelengths of light that can break the C—Cl bond and the bond in O_2.

* 9.64 "Inert" xenon actually forms many compounds, especially with highly electronegative fluorine. The ΔH_f° values for xenon difluoride, tetrafluoride, and hexafluoride are -105, -284, and -402 kJ/mol, respectively. Find the average bond energy of the Xe—F bonds in each fluoride.

* 9.65 The HF bond length is 92 pm, 16% shorter than the sum of the covalent radii of H (37 pm) and F (72 pm). Suggest a reason for this difference. Similar calculations show that the difference becomes smaller down the group from HF to HI. Explain.

* 9.66 There are two main types of covalent bond breakage. In homolytic breakage (as in Table 9.2), each atom in the bond gets one of the shared electrons. In some cases, the electronegativity of adjacent atoms affects the bond energy. In heterolytic breakage, one atom gets both electrons and the other gets none; thus, a cation and an anion form.

(a) Why is the C—C bond in H_3C—CF_3 (423 kJ/mol) stronger than the bond in H_3C—CH_3 (376 kJ/mol)?

(b) Use bond energy and any other data to calculate the heat of reaction for the heterolytic cleavage of O_2.

9.67 Find the longest wavelengths of light that can cleave the bonds in elemental nitrogen, oxygen, and fluorine.

9.68 We can write equations for the formation of methane from ethane (C_2H_6) with its C—C bond, from ethene (C_2H_4) with its C=C bond, and from ethyne (C_2H_2) with its C≡C bond:

$C_2H_6(g) + H_2(g) \longrightarrow 2CH_4(g)$ $\Delta H_{rxn}^\circ = -65.07$ kJ/mol

$C_2H_4(g) + 2H_2(g) \longrightarrow 2CH_4(g)$ $\Delta H_{rxn}^\circ = -202.21$ kJ/mol

$C_2H_2(g) + 3H_2(g) \longrightarrow 2CH_4(g)$ $\Delta H_{rxn}^\circ = -376.74$ kJ/mol

Given that the average C—H bond energy in CH_4 is 415 kJ/mol, use Table 9.2 values to calculate the average C—H bond energy in ethane, in ethene, and in ethyne.

9.69 Carbon-carbon bonds form the "backbone" of nearly every organic and biological molecule. The average bond energy of the C—C bond is 347 kJ/mol. Calculate the frequency and wavelength of the least energetic photon that can break an average C—C bond. In what region of the electromagnetic spectrum is this radiation?

9.70 In a future hydrogen-fuel economy, the cheapest source of H_2 will certainly be water. It takes 467 kJ to produce 1 mol of H atoms from water. What is the frequency, wavelength, and minimum energy of a photon that can free an H atom from water?

9.71 Dimethyl ether (CH_3OCH_3) and ethanol (CH_3CH_2OH) are constitutional isomers. (a) Use Table 9.2 to calculate ΔH_{rxn}° for the formation of each compound as a gas from methane and oxygen; water vapor also forms. (b) State which reaction is more exothermic. (c) Calculate ΔH_{rxn}° for the conversion of ethanol to dimethyl ether.

CHAPTER TEN

The Shapes of Molecules

Fitting Together The tiles in this wall detail from the Alhambra in Granada, Spain interconnect to form a whole. Similarly, biomolecules fit together to trigger reactions within organisms. In this chapter, you'll view a molecule as an object with a specific shape and learn the importance of shape in reactivity.

Key Principles

◆ A *Lewis structure* shows the relative positions of the atoms in the molecule or a polyatomic ion, as well as the placement of all the shared and unshared electron pairs.

◆ In many molecules or ions one electron pair in a double bond spreads over an adjacent single bond, thereby *delocalizing* its charge, which stabilizes the system. In such cases, more than one Lewis structure, each called a *resonance form,* can be drawn, and the species actually exists as a *resonance hybrid,* a mixture of the resonance forms.

◆ By assigning to each atom a *formal charge* based on the electrons belonging to the atom and shared by it, we can select the most important of the various resonance forms.

◆ According to VSEPR theory, each group of valence electrons, whether a bonding pair or a lone pair, around a central atom repels the others. These repulsions give rise to five geometric arrangements—linear, trigonal planar, tetrahedral, trigonal bipyramidal, and octahedral. Various molecular shapes, with distinctive bond angles, arise from these arrangements.

◆ A whole molecule may be polar, depending on its shape and the polarities of its bonds.

Outline

The printed page, covered with atomic symbols, lines, and pairs of dots, makes it easy to forget the amazing, three-dimensional reality of molecular shape. In any molecule, each atom, bonding pair, and lone pair has its own position in space relative to the others, determined by the attractive and repulsive forces that govern all matter. With definite angles and distances between the nuclei, a molecule has a characteristic minute architecture, extending throughout its tiny volume of space. Whether we consider the details of simple reactions, the properties of synthetic materials, or the intricate life-sustaining processes of living cells, molecular shape is a crucial factor. In this chapter, we see how to depict molecules, first as two-dimensional drawings and then as three-dimensional objects.

10.1 DEPICTING MOLECULES AND IONS WITH LEWIS STRUCTURES

The first step toward visualizing what a molecule looks like is to convert its molecular formula to its **Lewis structure** (or **Lewis formula**).* This two-dimensional structural formula consists of electron-dot symbols that depict each atom and its neighbors, the bonding pairs that hold them together, and the lone pairs that fill each atom's outer level (valence shell). In many cases, the octet rule (Section 9.1) guides us in allotting electrons to the atoms in a Lewis structure; in many other cases, however, we set the rule aside.

Using the Octet Rule to Write Lewis Structures

To write a Lewis structure, we decide on the relative placement of the atoms in the molecule (or polyatomic ion)—that is, which atoms are adjacent and become bonded to each other—and distribute the total number of valence electrons as bonding and lone pairs. Let's begin by examining Lewis structures for species that "obey" the octet rule—those in which each atom fills its outer level with eight electrons (or two for hydrogen).

Lewis Structures for Molecules with Single Bonds First, we discuss the steps for writing Lewis structures for molecules that have only single bonds, using nitrogen trifluoride, NF_3, as an example. Figure 10.1 lays out the steps.

Step 1. Place the atoms relative to each other. For compounds of molecular formula AB_n, place the atom with *lower group number* in the center because it needs more electrons to attain an octet; usually, this is also the atom with the *lower electronegativity*. In NF_3, the N (Group 5A; EN = 3.0) has five electrons and so needs three, whereas each F (Group 7A; EN = 4.0) has seven and so needs only one; thus, N goes in the center with the three F atoms around it:

$$\begin{array}{c} F \\ N \\ F \quad F \end{array}$$

- electron configurations of main-group elements (Section 8.3)
- electron-dot symbols (Section 9.1)
- the octet rule (Section 9.1)
- bond order, bond length, and bond energy (Sections 9.3 and 9.4)
- polar covalent bonds and bond polarity (Section 9.5)

---

*A Lewis *structure* may be more correctly called a Lewis *formula* because it provides information about the relative placement of atoms in a molecule or ion and shows which atoms are bonded to each other, but it does **not** indicate the three-dimensional shape. Nevertheless, use of the term Lewis *structure* is a convention that we follow.

Figure 10.1 The steps in converting a molecular formula into a Lewis structure.

If the atoms have the same group number, as in SO_3 or ClF_3, place the atom with the *higher period number* in the center. H can form only one bond, so it is *never* a central atom.

Step 2. Determine the total number of valence electrons available. For molecules, add up the valence electrons of all the atoms. (Recall that the number of valence electrons equals the A-group number.) In NF_3, N has five valence electrons, and each F has seven:

$$[1 \times N(5e^-)] + [3 \times F(7e^-)] = 5e^- + 21e^- = 26 \text{ valence } e^-$$

For polyatomic ions, *add* one e^- for each negative charge of the ion, or *subtract* one e^- for each positive charge.

Step 3. Draw a single bond from each surrounding atom to the central atom, and subtract two valence electrons for each bond. There must be at least a single bond between bonded atoms:

$$
\begin{array}{c}
\text{F} \\
| \\
\text{F}\diagdown\text{N}\diagup\text{F}
\end{array}
$$

Subtract $2e^-$ for each single bond from the total number of valence electrons available (from step 2) to find the number remaining:

$$3 \text{ N—F bonds} \times 2e^- = 6e^- \quad \text{so} \quad 26e^- - 6e^- = 20e^- \text{ remaining}$$

Step 4. Distribute the remaining electrons in pairs so that each atom ends up with eight electrons (or two for H). First, place lone pairs on the *surrounding (more electronegative) atoms* to give each an octet. If any electrons remain, place them around the central atom. Then check that each atom has $8e^-$:

$$
\begin{array}{c}
\text{:}\ddot{\text{F}}\text{:} \\
| \\
\text{:}\ddot{\text{F}}\diagdown\overset{..}{\text{N}}\diagup\ddot{\text{F}}\text{:}
\end{array}
$$

This is the Lewis structure for NF_3. Always check that the total number of electrons (bonds plus lone pairs) equals the sum of the valence electrons: $6e^-$ in three bonds plus $20e^-$ in ten lone pairs equals 26 valence electrons.

This particular arrangement of F atoms around an N atom resembles the molecular shape of NF_3, as you'll see when we discuss shapes in Section 10.2. But, Lewis structures do not indicate shape, so an equally correct depiction of NF_3 is

$$
\begin{array}{c}
\text{:}\ddot{\text{F}}\text{:} \\
| \\
\text{:}\ddot{\text{F}}\text{—N—}\ddot{\text{F}}\text{:}
\end{array}
$$

or any other that retains the *same connections among the atoms*—a central N atom connected by single bonds to three surrounding F atoms.

Using these four steps, you can write a Lewis structure for any singly bonded molecule whose central atom is C, N, or O, as well as for some molecules with central atoms from higher periods. Remember that, in nearly all their compounds,

- Hydrogen atoms form one bond.
- Carbon atoms form four bonds.
- Nitrogen atoms form three bonds.
- Oxygen atoms form two bonds.
- Halogens form one bond when they are surrounding atoms; fluorine is always a surrounding atom.

SAMPLE PROBLEM 10.1   Writing Lewis Structures for Molecules with One Central Atom

Problem Write a Lewis structure for CCl_2F_2, one of the compounds responsible for the depletion of stratospheric ozone.

Solution *Step 1.* Place the atoms relative to each other. In CCl_2F_2, carbon has the lowest group number and EN, so it is the central atom. The halogen atoms surround it, but their specific positions are not important (see margin).

Step 2. Determine the total number of valence electrons (from A-group numbers): C is in Group 4A, F is in Group 7A, and Cl is in Group 7A, too. Therefore, we have

$$[1 \times C(4e^-)] + [2 \times F(7e^-)] + [2 \times Cl(7e^-)] = 32 \text{ valence } e^-$$

Step 3. Draw single bonds to the central atom and subtract $2e^-$ for each bond (see margin). Four single bonds use $8e^-$, so $32e^- - 8e^-$ leaves $24e^-$ remaining.

Step 4. Distribute the remaining electrons in pairs, beginning with the surrounding atoms, so that each atom has an octet (see margin).

Check Counting the electrons shows that each atom has an octet. Remember that bonding electrons are counted as belonging to each atom in the bond. The total number of electrons in bonds (8) and lone pairs (24) equals 32 valence electrons. Note that, as expected, C has four bonds and the surrounding halogens have one each.

FOLLOW-UP PROBLEM 10.1 Write a Lewis structure for each of the following:
(a) H_2S **(b)** OF_2 **(c)** $SOCl_2$

A slightly more complex situation occurs when molecules have two or more central atoms bonded to each other, with the other atoms around them.

SAMPLE PROBLEM 10.2 Writing Lewis Structures for Molecules with More Than One Central Atom

Problem Write the Lewis structure for methanol, (molecular formula CH_4O), an industrial alcohol that is also being considered as a gasoline alternative.

Solution *Step 1.* Place the atoms relative to each other. The H atoms can have only one bond, so C and O must be adjacent to each other. Recall that C has four bonds and O has two, so we arrange the H atoms to show this (see margin).

Step 2. Find the sum of valence electrons:

$$[1 \times C(4e^-)] + [1 \times O(6e^-)] + [4 \times H(1e^-)] = 14e^-$$

Step 3. Add single bonds and subtract $2e^-$ for each bond (see margin). Five bonds use $10e^-$, so $14e^- - 10e^-$ leaves $4e^-$ remaining.

Step 4. Add the remaining electrons in pairs. Carbon already has an octet, and each H shares two electrons with the C; so the four remaining valence electrons form two lone pairs on O. We now have the Lewis structure for methanol (see margin).

Check Each H atom has $2e^-$, and the C and O each have $8e^-$. The total number of valence electrons is $14e^-$, which equals $10e^-$ in bonds plus $4e^-$ in lone pairs. Also note that each H has one bond, C has four, and O has two.

FOLLOW-UP PROBLEM 10.2 Write a Lewis structure for each of the following:
(a) hydroxylamine (NH_3O) **(b)** dimethyl ether (C_2H_6O; no O—H bonds)

Lewis Structures for Molecules with Multiple Bonds Sometimes, you'll find that, after steps 1 to 4, there are not enough electrons for the central atom (or one of the central atoms) to attain an octet. This usually means that a multiple bond is present, and the following additional step is needed:

Step 5. Cases involving multiple bonds. If, after step 4, a central atom still does not have an octet, make a multiple bond by changing a lone pair from one of the surrounding atoms into a bonding pair to the central atom.

SAMPLE PROBLEM 10.3 Writing Lewis Structures for Molecules with Multiple Bonds

Problem Write Lewis structures for the following:
(a) Ethylene (C_2H_4), the most important reactant in the manufacture of polymers
(b) Nitrogen (N_2), the most abundant atmospheric gas

Plan We show the solution following steps 1 to 4: placing the atoms, counting the total valence electrons, making single bonds, and distributing the remaining valence electrons in pairs to attain octets. Then we continue with step 5, if needed.

Solution (a) For C_2H_4. After steps 1 to 4, we have

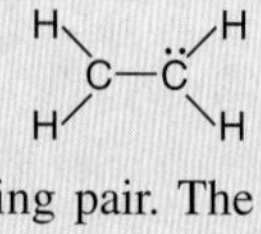

Step 5. Change a lone pair to a bonding pair. The C on the right has an octet, but the C on the left has only 6e⁻, so we convert the lone pair to another bonding pair between the two C atoms:

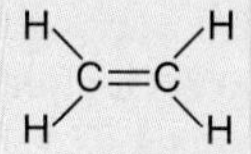

(b) For N_2. After steps 1 to 4, we have :N̈—N̈:

Step 5. Neither N has an octet, so we change a lone pair to a bonding pair: :N̈=N:
In this case, moving one lone pair to make a double bond still does not give the N on the right an octet, so we move a lone pair from the left N to make a triple bond: :N≡N:

Check In part (a), each C has four bonds and counts the 4e⁻ in the double bond as part of its own octet. The valence electron total is 12e⁻, all in six bonds. In part (b), each N has three bonds and counts the 6e⁻ in the triple bond as part of its own octet. The valence electron total is 10e⁻, which equals the electrons in three bonds and two lone pairs.

FOLLOW-UP PROBLEM 10.3 Write Lewis structures for each of the following:
(a) CO (the only common molecule in which C has only three bonds) **(b)** HCN **(c)** CO_2

Resonance: Delocalized Electron-Pair Bonding

We can often write more than one Lewis structure, each with the same relative placement of atoms, for a molecule or ion with *double bonds next to single bonds.* Consider ozone (O_3), a serious air pollutant at ground level but a life-sustaining absorber of harmful ultraviolet (UV) radiation in the stratosphere. Two valid Lewis structures (with lettered O atoms for clarity) are

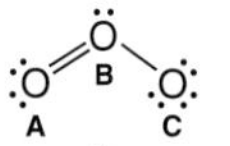

In structure I, oxygen B has a double bond to oxygen A and a single bond to oxygen C. In structure II, the single and double bonds are reversed. These are *not* two different O_3 molecules, just different Lewis structures for the same molecule.

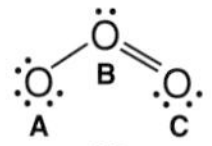

In fact, *neither* Lewis structure depicts O_3 accurately. Bond length and bond energy measurements indicate that the two oxygen-oxygen bonds in O_3 are identical, with properties that lie between those of an O—O bond and an O=O bond, something like a "one-and-a-half" bond. The molecule is shown more correctly with two Lewis structures, called **resonance structures** (or **resonance forms**), and a two-headed resonance arrow (⟷) between them. Resonance structures *have the same relative placement of atoms but different locations of bonding and lone electron pairs.* You can convert one resonance form to another by moving lone pairs to bonding positions, and vice versa:

Resonance structures are not real bonding depictions: O_3 does *not* change back and forth from structure I at one instant to structure II the next. The actual molecule is a **resonance hybrid,** an average of the resonance forms.

Consider these analogies. A mule is a genetic mix, a hybrid, of a horse and a donkey; it is not a horse one instant and a donkey the next. Similarly, the color purple is a mix of two other colors, red and blue, not red one instant and blue the next (Figure 10.2). In the same sense, a resonance hybrid is one molecular species, not one resonance form this instant and another resonance form the next. The problem is that we cannot depict the hybrid accurately with a single Lewis structure.

Our need for more than one Lewis structure to depict the ozone molecule is the result of **electron-pair delocalization.** In a single, double, or triple bond, each electron pair is attracted by the nuclei of the two bonded atoms, and the electron density is greatest in the region between the nuclei: each electron pair is *localized.* In the resonance hybrid for O_3, however, two of the electron pairs (one bonding and one lone pair) are *delocalized:* their density is "spread" over the entire molecule. In O_3, this results in two identical bonds, each consisting of a single bond (the localized electron pair) and a *partial bond* (the contribution from one of the delocalized electron pairs). We draw the resonance hybrid with a curved dashed line to show the delocalized pairs:

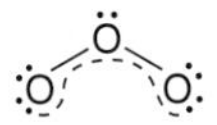

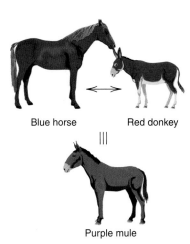

Electron delocalization diffuses electron density over a greater volume, which reduces electron-electron repulsions and thus stabilizes the molecule. Resonance is very common, and many molecules (and ions) are best depicted as resonance hybrids. Benzene (C_6H_6), for example, has two important resonance forms in which alternating single and double bonds have different positions. The actual molecule has six identical carbon-carbon bonds because there are six C—C bonds and three electron pairs delocalized over all six C atoms, often shown as a dashed circle (or simply a circle):

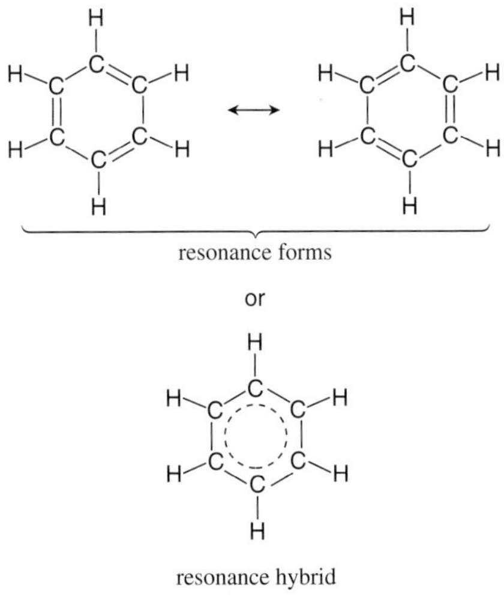

resonance forms

or

resonance hybrid

Partial bonding, such as that occurring in resonance hybrids, often leads to fractional bond orders. For O_3, we have

$$\text{Bond order} = \frac{3 \text{ electron pairs}}{2 \text{ bonded-atom pairs}} = 1\frac{1}{2}$$

The carbon-to-carbon bond order in benzene is 9 electron pairs/6 bonded-atom pairs, or $1\frac{1}{2}$ also. For the carbonate ion, CO_3^{2-}, three resonance structures can be drawn. Each has 4 electron pairs shared among 3 bonded-atom pairs, so the bond order is 4/3, or $1\frac{1}{3}$. One of the three resonance structures for CO_3^{2-} is

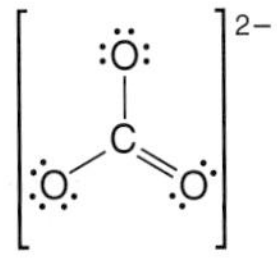

Note that *the Lewis structure of a polyatomic ion is shown in square brackets, with its charge as a right superscript outside the brackets.*

SAMPLE PROBLEM 10.4   Writing Resonance Structures

Problem Write resonance structures for the nitrate ion, NO_3^-.

Plan We write a Lewis structure, remembering to add $1e^-$ to the total number of valence electrons because of the $1-$ ionic charge. Then we move lone and bonding pairs to write other resonance forms and connect them with the resonance arrow.

Solution After steps 1 to 4, we have

$$\left[\ \ddot{\overset{\displaystyle :\ddot{O}:}{\underset{:\ddot{O}.\quad.\ddot{O}:}{N}}}\ \right]^-$$

Step 5. Because N has only $6e^-$, we change one lone pair on an O atom to a bonding pair and form a double bond, which gives each atom an octet. All the O atoms are equivalent, however, so we can move a lone pair from any of the three O atoms and obtain three resonance structures:

$$\left[\ \ddot{N}\ \right]^- \longleftrightarrow \left[\ \ddot{N}\ \right]^- \longleftrightarrow \left[\ \ddot{N}\ \right]^-$$

Check Each structure has the same relative placement of atoms, an octet around each atom, and $24e^-$ (the sum of the valence electron total and $1e^-$ from the ionic charge distributed in four bonds and eight lone pairs).

Comment Remember that no double bond actually exists in the NO_3^- ion. The ion is a resonance hybrid of these three structures with a bond order of $1\frac{1}{3}$. (You'll see in the upcoming discussion why N can have four bonds here.)

FOLLOW-UP PROBLEM 10.4 One of the three resonance structures for CO_3^{2-} was shown just before Sample Problem 10.4. Draw the other two.

Formal Charge: Selecting the More Important Resonance Structure

In the previous examples, the resonance forms were mixed equally to form the resonance hybrid because the central atoms had surrounding atoms that were all the same. Often this is not the case, and one resonance form may look more like the hybrid than the others. In other words, because the resonance hybrid is an average of the resonance forms, one form may contribute more and "weight" the average in its favor. One way to select the more important resonance form is to determine each atom's **formal charge,** the charge it would have *if the bonding electrons were shared equally.*

An atom's formal charge is its total number of valence electrons *minus* the number of valence electrons it "owns" in the molecule: it owns *all* of its unshared valence electrons and *half* of its shared valence electrons. Thus,

Formal charge of atom $=$
no. of valence e^- $-$ (no. of unshared valence e^- $+ \frac{1}{2}$ no. of shared valence e^-)   **(10.1)**

For example, in O_3, the formal charge of oxygen A in resonance form I is

6 valence e^- $-$ (4 unshared e^- $+ \frac{1}{2}$ of 4 shared e^-) $= 6 - 4 - 2 = 0$

The formal charges of all the atoms in the two O_3 resonance forms are

$O_A[6 - 4 - \frac{1}{2}(4)] = 0$
$O_B[6 - 2 - \frac{1}{2}(6)] = +1$
$O_C[6 - 6 - \frac{1}{2}(2)] = -1$

I

$\longleftrightarrow$

II

$O_A[6 - 6 - \frac{1}{2}(2)] = -1$
$O_B[6 - 2 - \frac{1}{2}(6)] = +1$
$O_C[6 - 4 - \frac{1}{2}(4)] = 0$

Forms I and II have the same formal charges but on different O atoms, so they contribute equally to the resonance hybrid. *Formal charges must sum to the actual charge on the species:* zero for a molecule and the ionic charge for an ion.

Note that, in form I, instead of the usual two bonds for oxygen, O_B has three bonds and O_C has one. Only when an atom has a zero formal charge does it have its usual number of bonds; the same holds for C in CO, N in NO_3^-, and so forth.

Three criteria help us choose the more important resonance structures:

- Smaller formal charges (positive or negative) are preferable to larger ones.
- Like formal charges on adjacent atoms are not desirable.
- A more negative formal charge should reside on a more electronegative atom.

Let's apply these criteria next to the cyanate ion, NCO^-, which has two different atoms around the central one. Three resonance forms with formal charges are

Formal charges: (−2) (0) (+1)   (−1) (0) (0)   (0) (0) (−1)

Resonance forms: $[:\ddot{N}-C\equiv O:]^- \longleftrightarrow [\ddot{N}=C=\ddot{O}]^- \longleftrightarrow [:N\equiv C-\ddot{O}:]^-$
                        I   **II**   **III**

We eliminate form I because it has larger formal charges than the others and a positive formal charge on O, which is more electronegative than N. Forms II and III have the same magnitude of formal charges, but form III has a −1 charge on the more electronegative atom, O. Therefore, II and III are significant contributors to the resonance hybrid of the cyanate ion, but III is the more important.

Formal charge (used to examine resonance structures) is *not* the same as oxidation number (used to monitor redox reactions):

- For a formal charge, bonding electrons are assigned *equally* to the atoms (as if the bonding were *nonpolar covalent*), so each atom has half of them:

$$\text{Formal charge} = \text{valence } e^- - (\text{lone pair } e^- + \tfrac{1}{2} \text{ bonding } e^-)$$

- For an oxidation number, bonding electrons are assigned *completely* to the more electronegative atom (as if the bonding were *ionic*):

$$\text{Oxidation number} = \text{valence } e^- - (\text{lone pair } e^- + \text{ bonding } e^-)$$

Here are the formal charges and oxidation numbers for the three cyanate ion resonance structures:

Formal charges: (−2) (0) (+1)   (−1) (0) (0)   (0) (0) (−1)

$[:\ddot{N}-C\equiv O:]^- \longleftrightarrow [\ddot{N}=C=\ddot{O}]^- \longleftrightarrow [:N\equiv C-\ddot{O}:]^-$

Oxidation numbers: −3 +4 −2   −3 +4 −2   −3 +4 −2

Note that the oxidation numbers *do not* change from one resonance form to another (because the electronegativities *do not* change), but the formal charges *do* change (because the numbers of bonding and lone pairs *do* change).

Lewis Structures for Exceptions to the Octet Rule

The octet rule is a useful guide for most molecules with Period 2 central atoms, but not for every one. Also, many molecules have central atoms from higher periods. As you'll see, some central atoms have fewer than eight electrons around them, and others have more. The most significant octet rule exceptions are for molecules containing electron-deficient atoms, odd-electron atoms, and especially atoms with expanded valence shells.

Electron-Deficient Molecules Gaseous molecules containing either beryllium or boron as the central atom are often **electron deficient;** that is, they have *fewer*

Animation: Formal Charge Calculations
Online Learning Center

than eight electrons around the Be or B atom. The Lewis structures of gaseous beryllium chloride* and boron trifluoride are

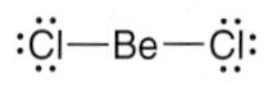

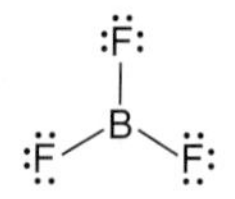

$$:\ddot{C}l\!-\!Be\!-\!\ddot{C}l: \qquad\qquad \ddot{F}\!-\!B\!\!\begin{smallmatrix}\ddot{F}:\\\\ \ddot{F}:\end{smallmatrix}$$

There are only four electrons around beryllium and six around boron. Why don't lone pairs from the surrounding halogen atoms form multiple bonds to the central atoms, thereby satisfying the octet rule? Halogens are much more electronegative than beryllium or boron, and formal charges show that the following are unlikely structures:

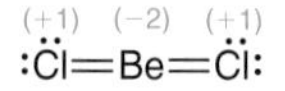

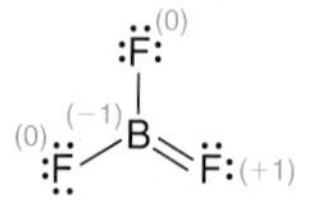

$$\overset{(+1)\;\;(-2)\;\;(+1)}{:\ddot{C}l\!=\!Be\!=\!\ddot{C}l:} \qquad\qquad \text{(structure with B=F bond)}$$

(Some data for BF_3 show a shorter than expected B—F bond. Shorter bonds indicate double-bond character, so the structure with the B=F bond may be a minor contributor to a resonance hybrid.) The main way electron-deficient atoms attain an octet is by forming additional bonds in reactions. When BF_3 reacts with ammonia, for instance, a compound forms in which boron attains its octet:

$$\text{(reaction of } BF_3 \text{ with } NH_3 \text{)}$$

Odd-Electron Molecules A few molecules contain a central atom with an odd number of valence electrons, so they cannot possibly have all their electrons in pairs. Such species, called **free radicals,** contain a lone (unpaired) electron, which makes them paramagnetic (Section 8.5) and extremely reactive. Let's use formal charges to decide where the lone electron resides. Most odd-electron molecules have a central atom from an odd-numbered group, such as N [Group 5A(15)] or Cl [Group 7A(17)].

Consider nitrogen dioxide (NO_2) as an example. A major contributor to urban smog is formed when the NO in auto exhaust is oxidized. NO_2 has several resonance forms. Two involve the O atom that is doubly bonded, as in the case of ozone. Several others involve the location of the lone electron. Two of these resonance forms are shown below. The form with the lone electron on the singly bonded O has zero formal charges (right):

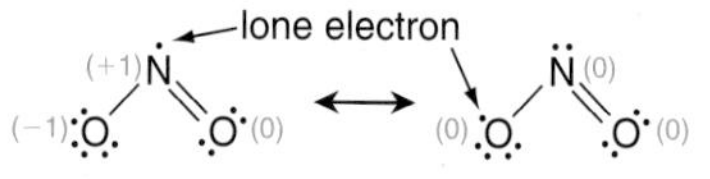

$$\text{(resonance forms of } NO_2 \text{ showing lone electron)}$$

But the form with the lone electron on N (left) must be important also because of the way NO_2 reacts. Free radicals react with each other to pair up their lone electrons. When two NO_2 molecules collide, the lone electrons pair up to form the N—N bond in dinitrogen tetraoxide (N_2O_4) and each N attains an octet:

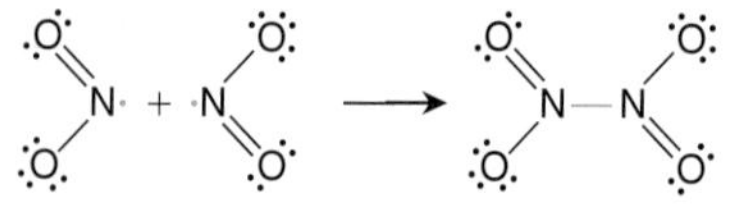

$$\text{(reaction of two } NO_2 \text{ to form } N_2O_4 \text{)}$$

Expanded Valence Shells Many molecules and ions have more than eight valence electrons around the central atom. *An atom expands its valence shell to form more bonds,* a process that releases energy. A central atom can accommodate additional pairs by using empty *outer d* orbitals in addition to occupied *s* and *p* orbitals.

---

*Even though beryllium is an alkaline earth metal [Group 2A(2)], most of its compounds have properties consistent with covalent, rather than ionic, bonding (Chapter 14). For example, molten $BeCl_2$ does not conduct electricity, indicating the absence of ions.

Therefore, **expanded valence shells** occur only with a *central nonmetal atom in which d orbitals are available, that is, one from Period 3 or higher.*

One example is sulfur hexafluoride, SF_6, a remarkably dense and inert gas used as an insulator in electrical equipment. The central sulfur is surrounded by six single bonds, one to each fluorine, for a total of 12 electrons:

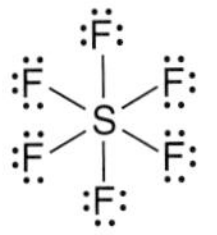

Another example is phosphorus pentachloride, PCl_5, a fuming yellow-white solid used in the manufacture of lacquers and films. PCl_5 is formed when phosphorus trichloride, PCl_3, reacts with chlorine gas. The P in PCl_3 has an octet, but it uses the lone pair to form two more bonds to chlorine and expands its valence shell in PCl_5 to a total of 10 electrons. Note that when PCl_5 forms, *one* Cl—Cl bond breaks (left side of the equation), and *two* P—Cl bonds form (right side), for a net increase of one bond:

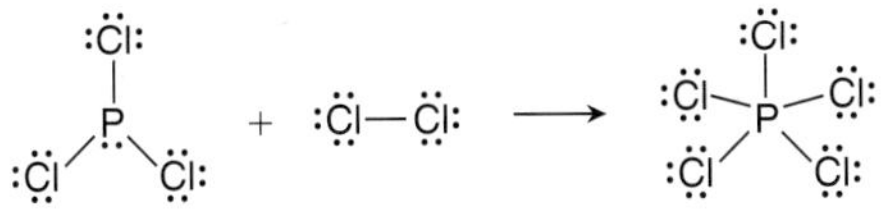

In SF_6 and PCl_5, the central atom forms bonds to *more than four* atoms. But there are many cases of expanded valence shells in which the central atom bonds to *four or even fewer* atoms. Consider sulfuric acid, the industrial chemical produced in the greatest quantity. Two of the resonance forms for H_2SO_4, with formal charges, are

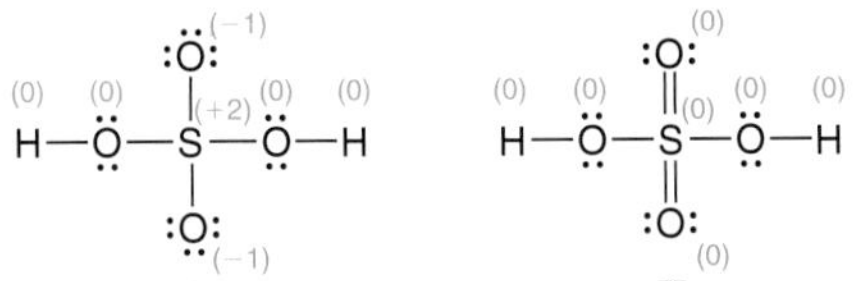

In form II, sulfur has an expanded valence shell of 12 electrons. Based on the formal charge rules, II contributes more than I to the resonance hybrid. More importantly, form II is consistent with observed bond lengths. In gaseous H_2SO_4, the two sulfur-oxygen bonds *with* H atoms attached to O are 157 pm long, whereas the two sulfur-oxygen bonds *without* H atoms attached to O are 142 pm long. This shorter bond length indicates double-bond character, which is shown in form II.

It's important to realize that formal charge is a useful, but not perfect, tool for assessing the importance of contributions to a resonance hybrid. You've already seen that it does not predict an important resonance form of NO_2. In fact, recent theoretical calculations indicate that, for many species with central atoms from Period 3 or higher, forms with expanded valence shells and zero formal charges may be less important than forms with higher formal charges. But we will continue to apply the formal charge rules because it is usually the simplest approach consistent with experimental data.

SAMPLE PROBLEM 10.5 Writing Lewis Structures for Octet-Rule Exceptions

Problem Write Lewis structures for (a) H_3PO_4 (pick the most likely structure); (b) $BFCl_2$.
Plan We write each Lewis structure and examine it for exceptions to the octet rule. In (a), the central atom is P, which is in Period 3, so it can use *d* orbitals to have more than an octet. Therefore, we can write more than one Lewis structure. We use formal charges to decide if one resonance form is more important. In (b), the central atom is B, which can have fewer than an octet of electrons.

Solution (a) For H_3PO_4, two possible Lewis structures, with formal charges, are

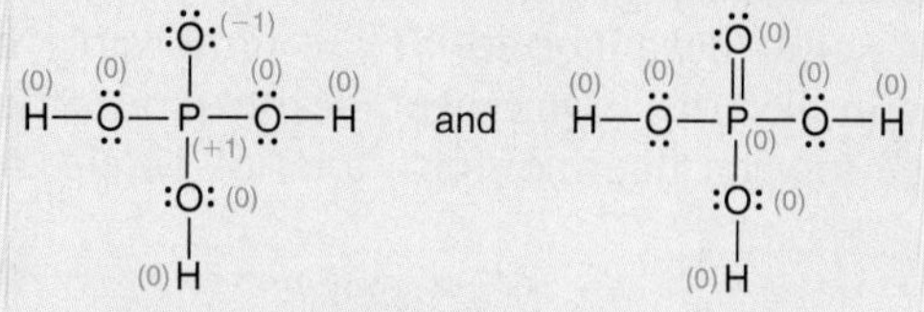

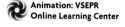

Structure **II** has lower formal charges, so it is the more important resonance form.
(b) For $BFCl_2$, the Lewis structure leaves B with only six electrons surrounding it:

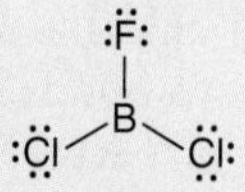

Comment In (a), structure **II** is also consistent with bond length measurements, which show one shorter (152 pm) phosphorus-oxygen bond and three longer (157 pm) ones.

FOLLOW-UP PROBLEM 10.5 Write the most likely Lewis structure for (a) $POCl_3$; (b) ClO_2; (c) XeF_4.

SECTION SUMMARY

A stepwise process is used to convert a molecular formula into a Lewis structure, a *two-dimensional* representation of a molecule (or ion) that shows the relative placement of atoms and distribution of valence electrons among bonding and lone pairs. When two or more Lewis structures can be drawn for the same relative placement of atoms, the actual structure is a hybrid of those resonance forms. Formal charges are often useful for determining the most important contributor to the hybrid. Electron-deficient molecules (central Be or B) and odd-electron species (free radicals) have less than an octet around the central atom but often attain an octet in reactions. In a molecule (or ion) with a central atom from Period 3 or higher, the atom can hold more than eight electrons by using *d* orbitals to expand its valence shell.

10.2 VALENCE-SHELL ELECTRON-PAIR REPULSION (VSEPR) THEORY AND MOLECULAR SHAPE

Virtually every biochemical process hinges to a great extent on the shapes of interacting molecules. Every medicine you take, odor you smell, or flavor you taste depends on part or all of one molecule fitting physically together with another. This universal importance of molecular shape in the functioning of each organism carries over to the ecosystem. Biologists have learned of complex interactions regulating behaviors (such as mating, defense, navigation, and feeding) that depend on one molecule's shape matching up with that of another. In this section, we discuss a model for understanding and predicting molecular shape.

The Lewis structure of a molecule is something like the blueprint of a building: a flat drawing showing the relative placement of parts (atom cores), structural connections (groups of bonding valence electrons), and various attachments (pairs of nonbonding valence electrons). To construct the molecular shape from the Lewis structure, chemists employ **valence-shell electron-pair repulsion (VSEPR) theory.** Its basic principle is that *each group of valence electrons around a central atom is located as far away as possible from the others in order to minimize repulsions.* We define a "group" of electrons as any number of electrons that occupy a localized region around an atom. Thus, an electron group may consist of a single bond, a double bond, a triple bond, a lone pair, or even a lone electron. [The two electron pairs in a double bond (or the three pairs in a triple bond) occupy separate orbitals, so they remain near each other and act as one electron group, as you'll see in Chapter 11.] Each group of valence electrons around an atom repels the other groups to maximize the angles between them. It is the three-dimensional arrangement of nuclei joined by these groups that gives rise to the molecular shape.

Electron-Group Arrangements and Molecular Shapes

When two, three, four, five, or six objects attached to a central point maximize the space that each can occupy around that point, five geometric patterns result. Figure 10.3A depicts these patterns with balloons. If the objects are the valence-electron groups of a central atom, their repulsions maximize the space each occupies and give rise to the five *electron-group arrangements* of minimum energy seen in the great majority of molecules and polyatomic ions.

The electron-group arrangement is defined by the valence-electron groups, both bonding and nonbonding, around the central atom. On the other hand, the **molecular shape** is defined by the relative positions of the atomic nuclei. Figure 10.3B shows the molecular shapes that occur when *all* the surrounding electron groups are *bonding* groups. When some are *nonbonding* groups, different molecular shapes occur. Thus, *the same electron-group arrangement can give rise to different molecular shapes:* some with all bonding groups (as in Figure 10.3B) and others with bonding and nonbonding groups. To classify molecular shapes, we assign each a specific AX_mE_n designation, where m and n are integers, A is the central atom, X is a surrounding atom, and E is a nonbonding valence-electron group (usually a lone pair).

The **bond angle** is the angle formed by the nuclei of two surrounding atoms with the nucleus of the central atom at the vertex. The angles shown for the shapes in Figure 10.3B are *ideal* bond angles, those predicted by simple geometry alone. These are observed when all the bonding electron groups around a central atom are identical and are connected to atoms of the same element. When this is not the case, the bond angles deviate from the ideal angles, as you'll see shortly.

It's important to realize that we use the VSEPR model to account for the molecular shapes observed by means of various laboratory instruments. In almost

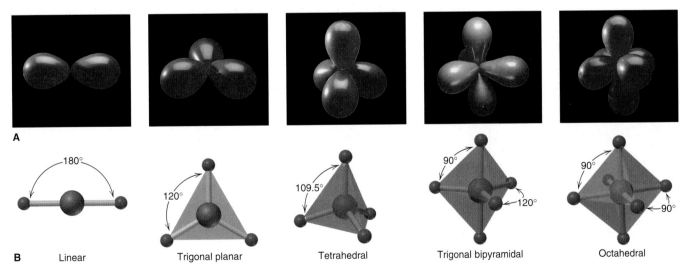

Figure 10.3 Electron-group repulsions and the five basic molecular shapes. A, As an analogy for electron-group arrangements, two to six attached balloons form five geometric orientations such that each balloon occupies as much space as possible. **B,** Mutually repelling electron groups attached to a central atom (*red*) occupy as much space as possible. If each is a bonding group to a surrounding atom (*dark gray*), these molecular shapes and bond angles are observed. The shape has the same name as the electron-group arrangement.

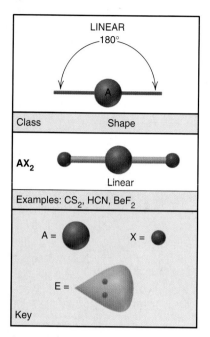

Figure 10.4 The single molecular shape of the linear electron-group arrangement. The key (*bottom*) for A, X, and E also refers to Figures 10.5, 10.6, 10.8, and 10.9.

Animation: **VSEPR Theory and the Shape of Molecules**
Online Learning Center

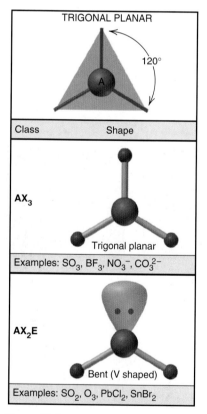

Figure 10.5 The two molecular shapes of the trigonal planar electron-group arrangement.

every case, VSEPR predictions are in accord with actual observations. (We discuss some of these observational methods in Chapter 12.)

The Molecular Shape with Two Electron Groups (Linear Arrangement)

When two electron groups attached to a central atom are oriented as far apart as possible, they point in opposite directions. The **linear arrangement** of electron groups results in a molecule with a **linear shape** and a bond angle of 180°. Figure 10.4 shows the general form (top) and shape (middle) with VSEPR shape class (AX_2), and the formulas of some linear molecules.

Gaseous beryllium chloride ($BeCl_2$) is a linear molecule (AX_2). Gaseous beryllium compounds are electron deficient, with only two electron pairs around the central Be atom:

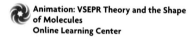

In carbon dioxide, the central C atom forms two double bonds with the O atoms:

Each double bond acts as one electron group and is oriented 180° away from the other, so CO_2 is linear. Notice that the lone pairs on the O atoms of CO_2 or on the Cl atoms of $BeCl_2$ are not involved in the molecular shape: only electron groups around the *central* atom influence shape.

Molecular Shapes with Three Electron Groups (Trigonal Planar Arrangement)

Three electron groups around the central atom repel each other to the corners of an equilateral triangle, which gives the **trigonal planar arrangement,** shown in Figure 10.5, and an ideal bond angle of 120°. This arrangement has two possible molecular shapes, one with three surrounding atoms and the other with two atoms and one lone pair. It provides our first opportunity to see the effects of double bonds and lone pairs on bond angles.

When the three electron groups are bonding groups, the molecular shape is *trigonal planar* (AX_3). Boron trifluoride (BF_3), another electron-deficient molecule, is an example. It has six electrons around the central B atom in three single bonds to F atoms. The nuclei lie in a plane, and each F—B—F angle is 120°:

The nitrate ion (NO_3^-) is one of several polyatomic ions with the trigonal planar shape. One of three resonance forms of the nitrate ion (Sample Problem 10.4) is

The resonance hybrid has three identical bonds of bond order $1\frac{1}{3}$, so the ideal bond angle is observed.

Effect of Double Bonds How do bond angles deviate from the ideal angles when the surrounding atoms and electron groups are not identical? Consider formaldehyde (CH_2O), a substance with many uses, including the manufacture of Formica countertops, the production of methanol, and the preservation of cadavers. Its trigonal planar shape is due to two types of surrounding atoms (O and H) and two types of electron groups (single and double bonds):

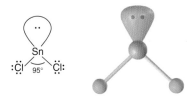

The actual bond angles deviate from the ideal because *the double bond, with its greater electron density, repels the two single bonds more strongly than they repel each other.* Note that the H—C—H bond angle is less than 120°.

Effect of Lone Pairs The molecular shape is defined *only* by the positions of the nuclei, so when one of the three electron groups is a lone pair (AX_2E), the shape is **bent**, or **V shaped**, not trigonal planar. Gaseous tin(II) chloride is an example, with the three electron groups in a trigonal plane and the lone pair at one of the triangle's corners. A lone pair can have a major effect on bond angle. Because a lone pair is held by only one nucleus, it is less confined and exerts stronger repulsions than a bonding pair. Thus, *a lone pair repels bonding pairs more strongly than bonding pairs repel each other.* This stronger repulsion *decreases* the angle between bonding pairs. Note the decrease from the ideal 120° angle in $SnCl_2$:

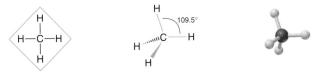

Molecular Shapes with Four Electron Groups (Tetrahedral Arrangement)

The shapes described so far have all been easy to depict in two dimensions, but four electron groups must use three dimensions to achieve maximal separation. Recall that *Lewis structures do **not** depict shape.* Consider methane. The Lewis structure shown below (left) indicates four bonds pointing to the corners of a square, which suggests a 90° bond angle. However, in three dimensions, the four electron groups move farther apart and point to the vertices of a tetrahedron, a polyhedron with four faces made of identical equilateral triangles. Methane has a bond angle of 109.5°. Perspective drawings, such as the one shown below (middle) for methane, indicate depth by using solid and dashed wedges for some of the bonds:

In a *perspective drawing* (middle), the normal bond lines (blue) represent bonds in the plane of the page; the solid wedge (green) is the bond between the atom in the plane of the page and a group lying toward you above the page; and the dashed wedge (red) is the bond to a group lying away from you below the page. The ball-and-stick model (right) shows the tetrahedral shape clearly.

*All molecules or ions with four electron groups around a central atom adopt the **tetrahedral arrangement*** (Figure 10.6). When all four electron groups are bonding groups, as in the case of methane, the molecular shape is also *tetrahedral* (AX_4), a very common geometry in organic molecules. In Sample Problem 10.1, we drew the Lewis structure for the tetrahedral molecule dichlorodifluoromethane

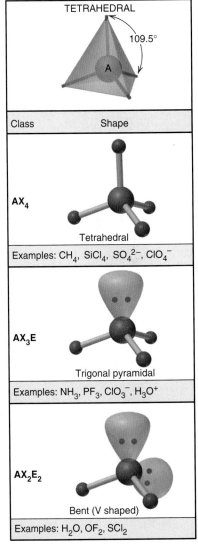

Figure 10.6 The three molecular shapes of the tetrahedral electron-group arrangement.

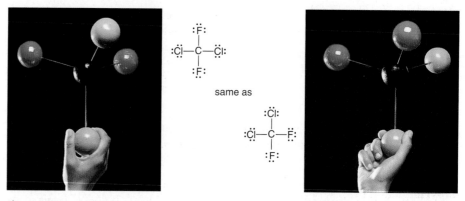

Figure 10.7 Lewis structures and molecular shapes. Lewis structures do not indicate geometry. For example, it may seem as if two different Lewis structures can be written for CCl_2F_2, but a twist of the model (Cl, *green;* F, *yellow*) shows that they represent the same molecule.

(CCl_2F_2), without regard to how the halogen atoms surround the carbon atom. Because Lewis structures are flat, it may seem as if we can write two different structures for CCl_2F_2, but these actually represent the same molecule, as Figure 10.7 makes clear.

When one of the four electron groups in the tetrahedral arrangement is a lone pair, the molecular shape is that of a **trigonal pyramid** (AX_3E), a tetrahedron with one vertex "missing." As we would expect from the stronger repulsions due to the lone pair, the measured bond angle is slightly less than the ideal 109.5°. In ammonia (NH_3), for example, the lone pair forces the N—H bonding pairs closer, and the H—N—H bond angle is 107.3°.

Picturing molecular shapes is a great way to visualize what happens during a reaction. For instance, when ammonia accepts the proton from an acid, the lone pair on the N atom of trigonal pyramidal NH_3 forms a covalent bond to the H^+ and yields the ammonium ion (NH_4^+), one of many tetrahedral polyatomic ions. Note how the H—N—H bond angle expands from 107.3° in NH_3 to 109.5° in NH_4^+, as the lone pair becomes another bonding pair:

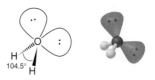

When the four electron groups around the central atom include two bonding and two nonbonding groups, the molecular shape is *bent, or V shaped* (AX_2E_2). [In the trigonal planar arrangement, the shape with two bonding groups and one lone pair (AX_2E) is also called bent, but its ideal bond angle is 120°, not 109.5°.] Water is the most important V-shaped molecule with the tetrahedral arrangement. We might expect the repulsions from its two lone pairs to have a *greater* effect on the bond angle than the repulsions from the single lone pair in NH_3. Indeed, the H—O—H bond angle is 104.5°, even less than the H—N—H angle in NH_3:

Thus, for similar molecules within a given electron-group arrangement, electron-pair repulsions cause deviations from ideal bond angles in the following order:

$$\text{Lone pair–lone pair} > \text{lone pair–bonding pair} > \text{bonding pair–bonding pair} \quad \text{(10.2)}$$

Molecular Shapes with Five Electron Groups (Trigonal Bipyramidal Arrangement)

All molecules with five or six electron groups have a central atom from Period 3 or higher because only these atoms have the *d* orbitals available to expand the valence shell beyond eight electrons.

When five electron groups maximize their separation, they form the **trigonal bipyramidal arrangement.** In a trigonal bipyramid, two trigonal pyramids share a common base, as shown in Figure 10.8. Note that, in a molecule with this arrangement, *there are two types of positions for surrounding electron groups and two ideal bond angles.* Three **equatorial groups** lie in a trigonal plane that includes the central atom, and two **axial groups** lie above and below this plane. Therefore, a 120° bond angle separates equatorial groups, and a 90° angle separates axial from equatorial groups. In general, the greater the bond angle, the weaker the repulsions, so *equatorial-equatorial (120°) repulsions are weaker than axial-equatorial (90°) repulsions.* The tendency of the electron groups to occupy *equatorial* positions, and thus minimize the stronger axial-equatorial repulsions, governs the four shapes of the trigonal bipyramidal arrangement.

With all five positions occupied by bonded atoms, the molecule has the *trigonal bipyramidal* shape (AX_5), as in phosphorus pentachloride (PCl_5):

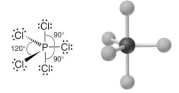

Three other shapes arise for molecules with lone pairs. Lone pairs exert stronger repulsions than bonding pairs, so *lone pairs occupy equatorial positions.* With one lone pair present at an equatorial position, the molecule has a **seesaw shape** (AX_4E). Sulfur tetrafluoride (SF_4), a powerful fluorinating agent, has this shape, shown here and in Figure 10.8 with the "seesaw" tipped up on an end. Note how the equatorial lone pair repels all four bonding pairs to reduce the bond angles:

The tendency of lone pairs to occupy equatorial positions causes molecules with three bonding groups and two lone pairs to have a **T shape** (AX_3E_2). Bromine trifluoride (BrF_3), one of many compounds with fluorine bonded to a larger halogen, has this shape. Note the predicted decrease from the ideal 90° F—Br—F bond angle:

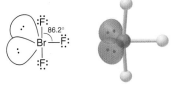

Molecules with three lone pairs in equatorial positions must have the two bonding groups in axial positions, which gives the molecule a *linear* shape (AX_2E_3) and

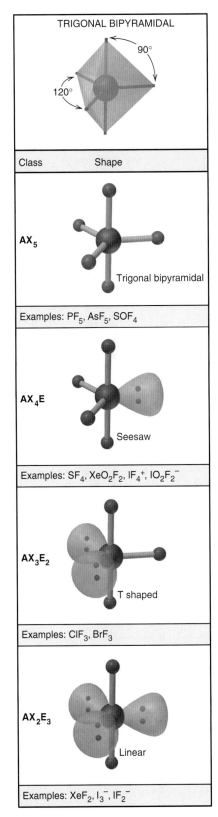

Figure 10.8 The four molecular shapes of the trigonal bipyramidal electron-group arrangement.

a 180° axial-to-central-to-axial (X—A—X) bond angle. For example, the triiodide ion (I_3^-), which forms when I_2 dissolves in aqueous I^- solution, is linear:

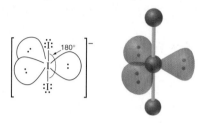

Molecular Shapes with Six Electron Groups (Octahedral Arrangement)

The last of the five major electron-group arrangements is the **octahedral arrangement.** An octahedron is a polyhedron with eight faces made of identical equilateral triangles and six identical vertices, as shown in Figure 10.9. In a molecule (or ion) with this arrangement, six electron groups surround the central atom and each points to one of the six vertices, which gives all the groups a 90° ideal bond angle. Three important molecular shapes occur with this arrangement.

With six bonding groups, the molecular shape is *octahedral* (AX_6), as in sulfur hexafluoride (SF_6):

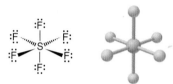

Because all six electron groups have the same ideal bond angle, it makes no difference which position one lone pair occupies. Five bonded atoms and one lone pair define the **square pyramidal shape** (AX_5E), as in iodine pentafluoride (IF_5):

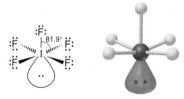

When a molecule has four bonded atoms and two lone pairs, however, the lone pairs always lie at *opposite vertices* to avoid the stronger 90° lone pair–lone pair repulsions. This positioning gives the **square planar shape** (AX_4E_2), as in xenon tetrafluoride (XeF_4):

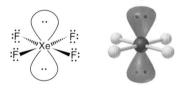

Using VSEPR Theory to Determine Molecular Shape

Let's apply a stepwise method for using the VSEPR theory to determine a molecular shape from a molecular formula:

Step 1. Write the Lewis structure from the molecular formula (Figure 10.1) to see the relative placement of atoms and the number of electron groups.

Step 2. Assign an electron-group arrangement by counting *all* electron groups around the central atom, bonding *plus* nonbonding.

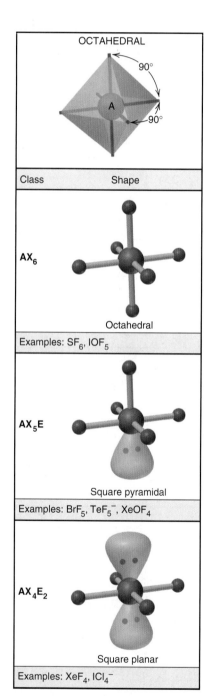

Figure 10.9 The three molecular shapes of the octahedral electron-group arrangement.

Step 3. Predict the ideal bond angle from the electron-group arrangement and *the direction of any deviation* caused by lone pairs or double bonds.

Step 4. Draw and name the molecular shape by counting bonding groups and nonbonding groups separately.

Figure 10.10 summarizes these steps, and the next two sample problems apply them.

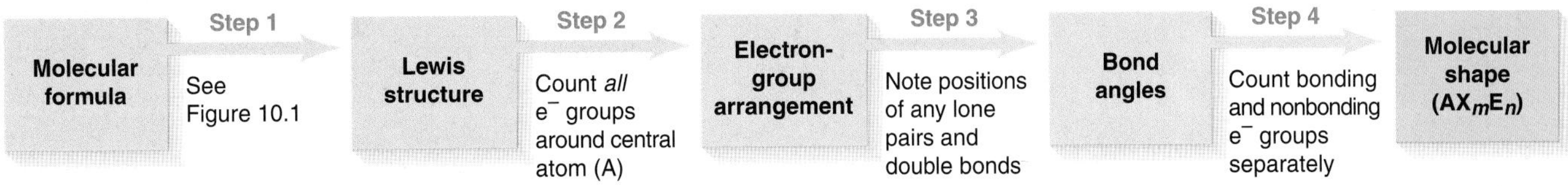

Figure 10.10 The steps in determining a molecular shape. Four steps are needed to convert a molecular formula to a molecular shape.

SAMPLE PROBLEM 10.6   Predicting Molecular Shapes with Two, Three, or Four Electron Groups

Problem Draw the molecular shapes and predict the bond angles (relative to the ideal angles) of **(a)** PF_3 and **(b)** $COCl_2$.

Solution (a) For PF_3.

Step 1. Write the Lewis structure from the formula (see below left).

Step 2. Assign the electron-group arrangement: Three bonding groups plus one lone pair give four electron groups around P and the *tetrahedral arrangement*.

Step 3. Predict the bond angle: For the tetrahedral electron-group arrangement, the ideal bond angle is $109.5°$. There is one lone pair, so the actual bond angle should be less than $109.5°$.

Step 4. Draw and name the molecular shape: With four electron groups, one of them a lone pair, PF_3 has a trigonal pyramidal shape (AX_3E):

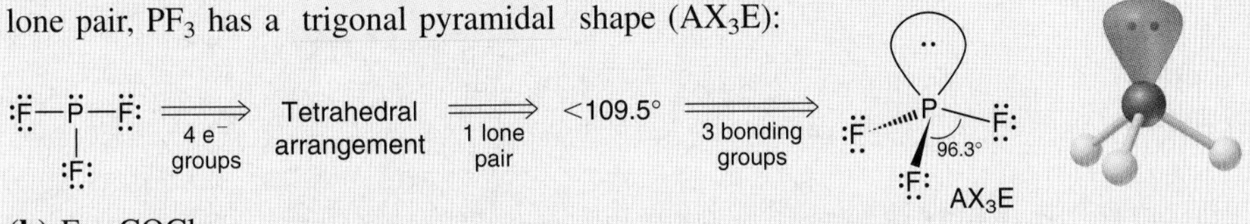

(b) For $COCl_2$.

Step 1. Write the Lewis structure from the formula (see below left).

Step 2. Assign the electron-group arrangement: Two single bonds plus one double bond give three electron groups around C and the *trigonal planar arrangement*.

Step 3. Predict the bond angles: The ideal bond angle is $120°$, but the double bond between C and O should compress the Cl—C—Cl angle to less than $120°$.

Step 4. Draw and name the molecular shape: With three electron groups and no lone pairs, $COCl_2$ has a trigonal planar shape (AX_3):

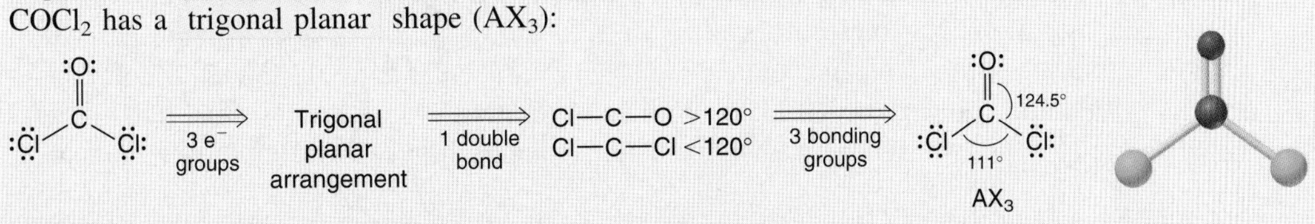

Check We compare the answers with the information in Figures 10.5 and 10.6.

Comment Be sure the Lewis structure is correct because it determines the other steps.

FOLLOW-UP PROBLEM 10.6 Draw the molecular shapes and predict the bond angles (relative to the ideal angles) of **(a)** CS_2; **(b)** $PbCl_2$; **(c)** CBr_4; **(d)** SF_2.

SAMPLE PROBLEM 10.7   Predicting Molecular Shapes with Five or Six Electron Groups

Problem Determine the molecular shapes and predict the bond angles (relative to the ideal angles) of **(a)** SbF_5 and **(b)** BrF_5.

Plan We proceed as in Sample Problem 10.6, keeping in mind the need to minimize the number of $90°$ repulsions.

Solution (a) For SbF$_5$.

Step 1. Lewis structure (see below left).

Step 2. Electron-group arrangement: With five electron groups, this is the *trigonal bipyramidal* arrangement.

Step 3. Bond angles: All the groups and surrounding atoms are identical, so the bond angles are ideal: 120° between equatorial groups and 90° between axial and equatorial groups.

Step 4. Molecular shape: Five electron groups and no lone pairs give the trigonal bipyramidal shape (AX$_5$):

(b) For BrF$_5$.

Step 1. Lewis structure (see below left).

Step 2. Electron-group arrangement: Six electron groups give the *octahedral* arrangement.

Step 3. Bond angles: The lone pair should make all bond angles less than the ideal 90°.

Step 4. Molecular shape: With six electron groups and one of them a lone pair, BrF$_5$ has the square pyramidal shape (AX$_5$E):

Check We compare our answers with Figures 10.8 and 10.9.

Comment We will encounter the linear, tetrahedral, square planar, and octahedral shapes in *coordination compounds,* which we discuss in Chapter 22.

FOLLOW-UP PROBLEM 10.7 Draw the molecular shapes and predict the bond angles (relative to the ideal angles) of **(a)** ICl$_2^-$; **(b)** ClF$_3$; **(c)** SOF$_4$.

Molecular Shapes with More Than One Central Atom

Many molecules, especially those in living systems, have more than one central atom. The shapes of these molecules are combinations of the molecular shapes for each central atom. For these molecules, we find the molecular shape around one central atom at a time. Consider ethane (CH$_3$CH$_3$; molecular formula C$_2$H$_6$), a component of natural gas (Figure 10.11A). With four bonding groups and no lone pairs around each of the two central carbons, ethane is shaped like two overlapping tetrahedra.

Figure 10.11 The tetrahedral centers of ethane and of ethanol. When a molecule has more than one central atom, the overall shape is a composite of the shape around each center. **A,** Ethane's shape can be viewed as two overlapping tetrahedra. **B,** Ethanol's shape can be viewed as three overlapping tetrahedral arrangements, with the shape around the O atom bent (V shaped) because of its two lone pairs.

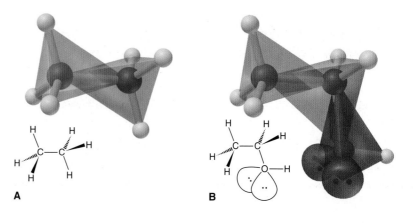

A   **B**

Ethanol (CH_3CH_2OH; molecular formula C_2H_6O), the intoxicating substance in beer and wine, has three central atoms (Figure 10.11B). The CH_3— group is tetrahedrally shaped, and the —CH_2— group has four bonding groups around its central C atom, so it is tetrahedrally shaped also. The O atom has four electron groups and two lone pairs around it, which gives the V shape (AX_2E_2).

SAMPLE PROBLEM 10.8 Predicting Molecular Shapes with More Than One Central Atom

Problem Determine the shape around each of the central atoms in acetone, $(CH_3)_2C{=}O$.
Plan There are three central atoms, all C, two of which are in CH_3— groups. We determine the shape around one central atom at a time.
Solution
Step 1. Lewis structure (see below left).
Step 2. Electron-group arrangement: Each CH_3— group has four electron groups around its central C, so its electron-group arrangement is *tetrahedral*. The third C atom has three electron groups around it, so it has the *trigonal planar* arrangement.
Step 3. Bond angles: The H—C—H angle in the CH_3— groups should be near the ideal tetrahedral angle of 109.5°. The C=O double bond should compress the C—C—C angle to less than the ideal 120°.
Step 4. Shapes around central atoms: With four electron groups and no lone pairs, the shapes around the two C atoms in the CH_3— groups are tetrahedral (AX_4). With three electron groups and no lone pairs, the shape around the middle C atom is trigonal planar (AX_3):

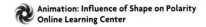

FOLLOW-UP PROBLEM 10.8 Determine the shape around each central atom and predict any deviations from ideal bond angles in the following: **(a)** H_2SO_4; **(b)** propyne (C_3H_4; there is one C≡C bond); **(c)** S_2F_2.

SECTION SUMMARY

The VSEPR theory proposes that each group of electrons (single bond, multiple bond, lone pair, or lone electron) around a central atom remains as far away from the others as possible. One of five electron-group arrangements results when two, three, four, five, or six electron groups surround a central atom. Each arrangement is associated with one or more molecular shapes. Ideal bond angles are prescribed by the regular geometric shapes; deviations from these angles occur when the surrounding atoms or electron groups are not identical. Lone pairs and double bonds exert greater repulsions than single bonds. Larger molecules have shapes that are composites of the shapes around each central atom.

10.3 MOLECULAR SHAPE AND MOLECULAR POLARITY

Knowing the shape of a substance's molecules is a key to understanding its physical and chemical behavior. One of the most important and far-reaching effects of molecular shape is molecular polarity, which can influence melting and boiling points, solubility, chemical reactivity, and even biological function.

In Chapter 9, you learned that a covalent bond is *polar* when it joins atoms of different electronegativities because the atoms share the electrons unequally. In diatomic molecules, such as HF, where there is only one bond, the bond polarity causes the molecule itself to be polar. Molecules with a net imbalance of charge have a **molecular polarity.** In molecules with more than two atoms, *both shape and bond polarity determine molecular polarity.* In an electric field, polar molecules become

Animation: Influence of Shape on Polarity
Online Learning Center

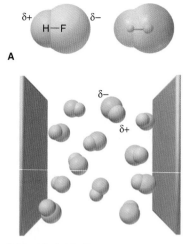

A

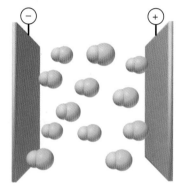

B Electric field off

C Electric field on

Figure 10.12 The orientation of polar molecules in an electric field. A, A space-filling model of HF (*left*) shows the partial charges of this polar molecule. The electron density model (*right*) shows high electron density (*red*) associated with the F end and low electron density (*blue*) with the H end. **B,** In the absence of an external electric field, HF molecules are oriented randomly. **C,** In the presence of an electric field, the molecules, on average, become oriented with their partial charges pointing toward the oppositely charged plates.

oriented, on average, with their partial charges pointing toward the oppositely charged electric plates, as shown for HF in Figure 10.12. The **dipole moment (μ)** is the product of these partial charges and the distance between them. It is typically measured in *debye* (D) units; using the SI units of charge (coulomb, C) and length (meter, m), $1 \text{ D} = 3.34 \times 10^{-30}$ C·m. [The unit is named for Peter Debye (1884–1966), the Dutch American chemist and physicist who won a Nobel Prize in 1936 for his major contributions to our understanding of molecular structure and solution behavior.]

To determine molecular polarity, we also consider shape because the presence of polar bonds does not *always* lead to a polar molecule. In carbon dioxide, for example, the large electronegativity difference between C (EN = 2.5) and O (EN = 3.5) makes each C=O bond quite polar. However, CO_2 is linear, so its bonds point 180° from each other. As a result, the two identical bond polarities are counterbalanced and give the molecule *no net dipole moment* (μ = 0 D). The electron density model shows regions of high negative charge (*red*) distributed equally on either side of the central region of high positive charge (*blue*):

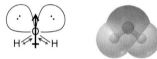

Water also has identical atoms bonded to the central atom, but it *does* have a significant dipole moment (μ = 1.85 D). In each O—H bond, electron density is pulled toward the more electronegative O atom. Here, the bond polarities are *not* counterbalanced, because the water molecule is V shaped (also see Figure 4.1). Instead, the bond polarities are partially reinforced, and the O end of the molecule is more negative than the other end (the region between the H atoms), which the electron density model shows clearly:

(The molecular polarity of water has some amazing effects, from determining the composition of the oceans to supporting life itself, as you'll see in Chapter 12.)

In the two previous examples, molecular shape influences polarity. When different molecules have the same shape, the nature of the atoms surrounding the central atom can have a major effect on polarity. Consider carbon tetrachloride (CCl_4) and chloroform ($CHCl_3$), two tetrahedral molecules with very different polarities. In CCl_4, the surrounding atoms are all Cl atoms. Although each C—Cl bond is polar (ΔEN = 0.5), the molecule is nonpolar (μ = 0 D) because the individual bond polarities counterbalance each other. In $CHCl_3$, H substitutes for one Cl atom, disrupting the balance and giving chloroform a significant dipole moment (μ = 1.01 D):

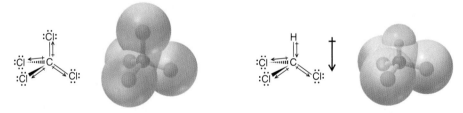

SAMPLE PROBLEM 10.9 Predicting the Polarity of Molecules

Problem From electronegativity (EN) values and their periodic trends (see Figure 9.20), predict whether each of the following molecules is polar and show the direction of bond dipoles and the overall molecular dipole when applicable:
(a) Ammonia, NH_3 **(b)** Boron trifluoride, BF_3
(c) Carbonyl sulfide, COS (atom sequence SCO)

Plan First, we draw and name the molecular shape. Then, using relative EN values, we decide on the direction of each bond dipole. Finally, we see if the bond dipoles balance or reinforce each other in the molecule as a whole.

Solution (a) For NH_3. The molecular shape is trigonal pyramidal. From Figure 9.20, we see that N (EN = 3.0) is more electronegative than H (EN = 2.1), so the bond dipoles point toward N. The bond dipoles partially reinforce each other, and thus the molecular dipole points toward N:

molecular shape

bond dipoles

molecular dipole

Therefore, ammonia is polar.

(b) For BF_3. The molecular shape is trigonal planar. Because F (EN = 4.0) is farther to the right in Period 2 than B (EN = 2.0), it is more electronegative; thus, each bond dipole points toward F. However, the bond angle is 120°, so the three bond dipoles counterbalance each other, and BF_3 has no molecular dipole:

molecular shape

bond dipoles

no molecular dipole

Therefore, boron trifluoride is nonpolar.

(c) For COS. The molecular shape is linear. With C and S having the same EN, the C=S bond is nonpolar, but the C=O bond is quite polar (ΔEN = 1.0), so there is a net molecular dipole toward the O:

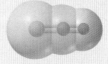

molecular shape   bond dipole   molecular dipole

Therefore, carbonyl sulfide is polar.

FOLLOW-UP PROBLEM 10.9 Show the bond dipoles and molecular dipole, if any, for **(a)** dichloromethane (CH_2Cl_2); **(b)** iodine oxide pentafluoride (IOF_5); **(c)** nitrogen tribromide (NBr_3).

SECTION SUMMARY
Bond polarity and molecular shape determine molecular polarity, which is measured as a dipole moment. When bond polarities counterbalance each other, the molecule is nonpolar; when they reinforce each other, even partially, the molecule is polar.

For Review and Reference (Numbers in parentheses refer to pages, unless noted otherwise.)

Learning Objectives

To help you review these learning objectives, the numbers of related sections (§), sample problems (SP), and end-of-chapter problems (EP) are listed in parentheses.

1. Use the octet rule to draw a Lewis structure from a molecular formula (§ 10.1) (SPs 10.1–10.3) (EPs 10.1, 10.5–10.8)
2. Understand how electron delocalization explains bond properties, and draw resonance structures (§ 10.1) (SP 10.4) (EPs 10.2, 10.9–10.12)
3. Describe the three exceptions to the octet rule, draw Lewis structures for them, and use formal charges to select the most

important resonance structure (§ 10.1) (SP 10.5) (EPs 10.3, 10.4, 10.13–10.24)
4. Describe the five electron-group arrangements and associated molecular shapes, predict molecular shapes from Lewis structures, and explain deviations from ideal bond angles (§ 10.2) (SPs 10.6–10.8) (EPs 10.25–10.49)
5. Understand how a molecule's polarity arises, and use molecular shape and EN values to predict the direction of a dipole (§ 10.3) (SP 10.9) (EPs 10.50–10.55)

Key Terms

Section 10.1

Lewis structure (Lewis formula) (297)

resonance structure (resonance form) (300)

resonance hybrid (300)

electron-pair delocalization (301)

formal charge (302)

electron deficient (303)

free radical (304)

expanded valence shell (305)

Section 10.2

valence-shell electron-pair repulsion (VSEPR) theory (306)

molecular shape (307)

bond angle (307)

linear arrangement (308)

linear shape (308)

trigonal planar arrangement (308)

bent shape (V shape) (309)

tetrahedral arrangement (309)

trigonal pyramidal shape (310)

trigonal bipyramidal arrangement (311)

equatorial group (311)

axial group (311)

seesaw shape (311)

T shape (311)

octahedral arrangement (312)

square pyramidal shape (312)

square planar shape (312)

Section 10.3

molecular polarity (315)

dipole moment (μ) (316)

Key Equations and Relationships

10.1 Calculating the formal charge on an atom (302):

Formal charge of atom

$$= \text{no. of valence e}^- - (\text{no. of unshared valence e}^- + \tfrac{1}{2}\text{ no. of shared valence e}^-)$$

10.2 Ranking the effect of electron-pair repulsions on bond angle (311):

Lone pair–lone pair > lone pair–bonding pair

> bonding pair–bonding pair

Brief Solutions to Follow-up Problems

10.1 (a) 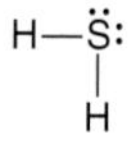 H—S̈—H (b) 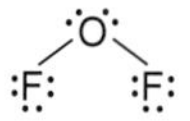 :Ö: with :F̈: :F̈: (c) 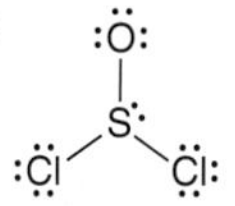 :Ö: S with :C̈l: :C̈l:

10.2 (a) 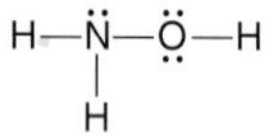 H—N̈—Ö—H with H (b) 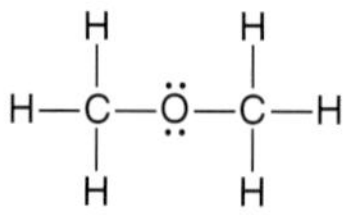 H—C—Ö—C—H with H H / H H

10.3 (a) :C≡O: (b) H—C≡N: (c) :Ö=C=Ö:

10.4 $\left[\begin{array}{c}:Ö: \\ C \\ :Ö: \ :Ö:\end{array}\right]^{2-} \longleftrightarrow \left[\begin{array}{c}:Ö: \\ C \\ :Ö: \ :Ö:\end{array}\right]^{2-}$

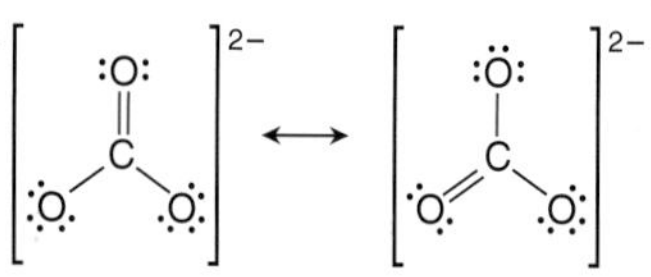

10.5 (a) 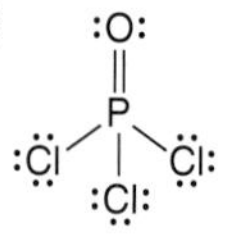 :O: P with :C̈l: :C̈l: :C̈l: (b) :Ö:C̈l:Ö: (c) :F̈: Xe :F̈: / :F̈: :F̈:

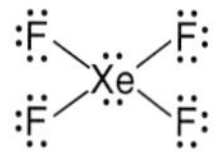

10.6 (a) 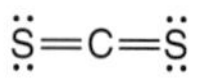 S̈=C=S̈   Linear, 180°

(b) 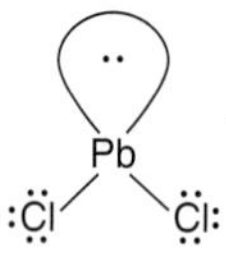 Pb with :C̈l: :C̈l: V shaped, <120°

(c) 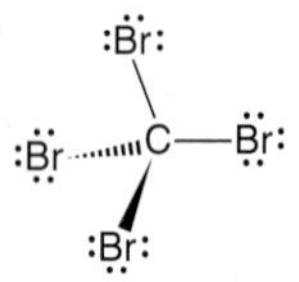 :B̈r: :B̈r:—C—B̈r: :B̈r: Tetrahedral, 109.5°

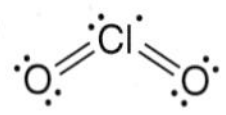

(d) 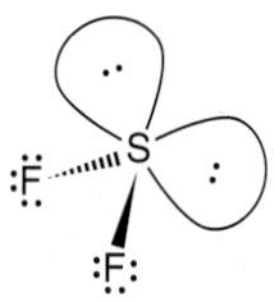 :F̈: S :F̈:   V shaped, <109.5°

10.7

(a) 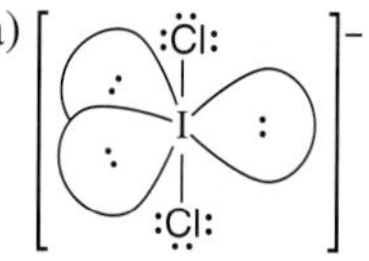 $\left[\ :Cl: \ I \ :Cl:\ \right]^-$ Linear, 180°

(b) 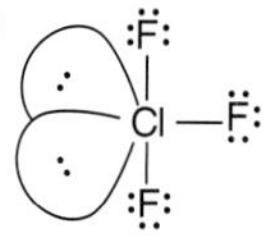 :F̈: Cl—F̈: :F̈:   T shaped, <90°

(c) 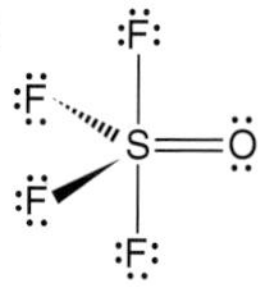 :F̈: F—S=Ö :F̈: :F̈:   Trigonal bipyramidal
F_{eq}—S—F_{eq} angle <120°
F_{ax}—S—F_{eq} angle <90°

10.8 (a) 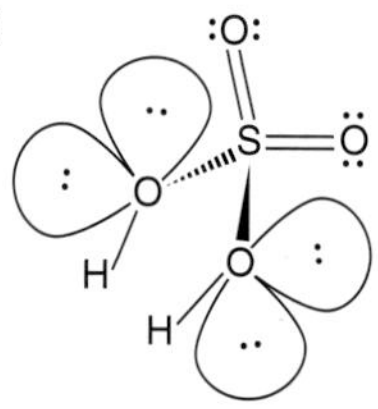 :O: S=Ö O with H / H

S is tetrahedral; double bonds compress O—S—O angle to <109.5°. Each central O is V shaped; lone pairs compress H—O—S angle to <109.5°.

(b) 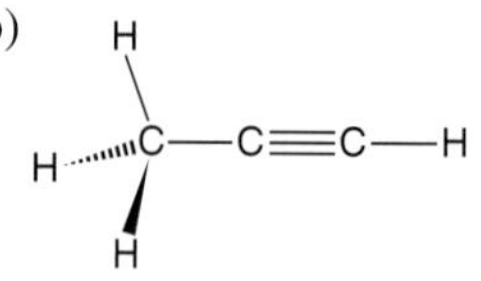 H H—C—C≡C—H H

C in CH₃— is tetrahedral ~109.5°; other C atoms are linear, 180°.

(c) 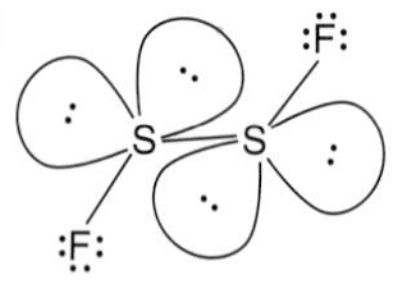 :F̈: S—S :F̈:

Each S is V shaped; F—S—S angle <109.5°.

10.9

(a) 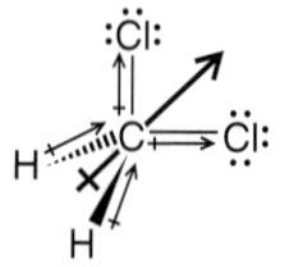 :C̈l: C—C̈l: H H

(b) :O: F—I—F :F̈: :F̈:

(c) N with :B̈r: :B̈r: :B̈r:

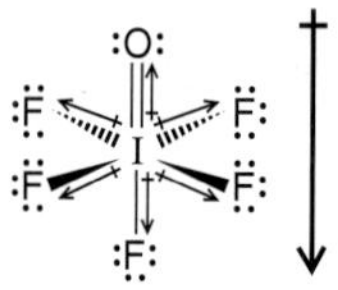

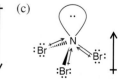

Problems

Problems with **colored** numbers are answered in Appendix E. Sections match the text and provide the numbers of relevant sample problems. Bracketed problems are grouped in pairs (indicated by a short rule) that cover the same concept. Comprehensive Problems are based on material from any section or previous chapter.

Depicting Molecules and Ions with Lewis Structures
(Sample Problems 10.1 to 10.5)

10.1 Which of these atoms *cannot* serve as a central atom in a Lewis structure: (a) O; (b) He; (c) F; (d) H; (e) P? Explain.

10.2 When is a resonance hybrid needed to adequately depict the bonding in a molecule? Using NO_2 as an example, explain how a resonance hybrid is consistent with the actual bond length, bond strength, and bond order.

10.3 In which of these bonding patterns does X obey the octet rule?

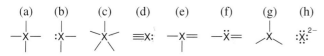

10.4 What is required for an atom to expand its valence shell? Which of the following atoms can expand its valence shell: F, S, H, Al, Se, Cl?

10.5 Draw a Lewis structure for (a) SiF_4; (b) $SeCl_2$; (c) COF_2 (C is central).

10.6 Draw a Lewis structure for (a) PH_4^+; (b) C_2F_4; (c) SbH_3.

10.7 Draw a Lewis structure for (a) PF_3; (b) H_2CO_3 (both H atoms are attached to O atoms); (c) CS_2.

10.8 Draw a Lewis structure for (a) CH_4S; (b) S_2Cl_2; (c) $CHCl_3$.

10.9 Draw Lewis structures of all the important resonance forms of (a) NO_2^+; (b) NO_2F (N is central).

10.10 Draw Lewis structures of all the important resonance forms of (a) HNO_3 ($HONO_2$); (b) $HAsO_4^{2-}$ ($HOAsO_3^{2-}$).

10.11 Draw Lewis structures of all the important resonance forms of (a) N_3^-; (b) NO_2^-.

10.12 Draw Lewis structures of all the important resonance forms of (a) HCO_2^- (H is attached to C); (b) $HBrO_4$ ($HOBrO_3$).

10.13 Draw a Lewis structure and calculate the formal charge of each atom in (a) IF_5; (b) AlH_4^-.

10.14 Draw a Lewis structure and calculate the formal charge of each atom in (a) COS (C is central); (b) NO.

10.15 Draw a Lewis structure for the most important resonance form of each ion, showing formal charges and oxidation numbers of the atoms: (a) BrO_3^-; (b) SO_3^{2-}.

10.16 Draw a Lewis structure for the most important resonance form of each ion, showing formal charges and oxidation numbers of the atoms: (a) AsO_4^{3-}; (b) ClO_2^-.

10.17 These species do not obey the octet rule. Draw a Lewis structure for each, and state the type of octet-rule exception:
(a) BH_3 (b) AsF_4^- (c) $SeCl_4$

10.18 These species do not obey the octet rule. Draw a Lewis structure for each, and state the type of octet-rule exception:
(a) PF_6^- (b) ClO_3 (c) H_3PO_3

10.19 These species do not obey the octet rule. Draw a Lewis structure for each, and state the type of octet-rule exception:
(a) BrF_3 (b) ICl_2^- (c) BeF_2

10.20 These species do not obey the octet rule. Draw a Lewis structure for each, and state the type of octet-rule exception:
(a) O_3^- (b) XeF_2 (c) SbF_4^-

10.21 Molten beryllium chloride reacts with chloride ion from molten NaCl to form the $BeCl_4^{2-}$ ion, in which the Be atom attains an octet. Show the net ionic reaction with Lewis structures.

10.22 Despite many attempts, the perbromate ion (BrO_4^-) was not prepared in the laboratory until about 1970. (Indeed, articles were published explaining theoretically why it could never be prepared!) Draw a Lewis structure for BrO_4^- in which all atoms have the lowest formal charges.

10.23 Cryolite (Na_3AlF_6) is an indispensable component in the electrochemical manufacture of aluminum. Draw a Lewis structure for the AlF_6^{3-} ion.

10.24 Phosgene is a colorless, highly toxic gas employed against troops in World War I and used today as a key reactant in organic syntheses. Use formal charges to select the most important of the following resonance structures:

$$\begin{array}{ccc} :\!\overset{\displaystyle :O:}{\underset{\displaystyle :\ddot{C}l:}{C}}\!\!\ddot{C}l: & :\!\overset{\displaystyle :\ddot{O}:}{\underset{\displaystyle :\ddot{C}l:}{C}}\!\!\ddot{C}l: & :\!\overset{\displaystyle :\ddot{O}:}{\underset{\displaystyle :\ddot{C}l}{C}}\!\!=\!\!\ddot{C}l: \\ \mathbf{A} & \mathbf{B} & \mathbf{C} \end{array}$$

Valence-Shell Electron-Pair Repulsion (VSEPR) Theory and Molecular Shape
(Sample Problems 10.6 to 10.8)

10.25 If you know the formula of a molecule or ion, what is the first step in predicting its shape?

10.26 In what situation is the name of the molecular shape the same as the name of the electron-group arrangement?

10.27 Which of the following numbers of electron groups can give rise to a bent (V-shaped) molecule: two, three, four, five, six? Draw an example for each case, showing the shape classification (AX_mE_n) and the ideal bond angle.

10.28 Name all the molecular shapes that have a tetrahedral electron-group arrangement.

10.29 Why aren't lone pairs considered along with surrounding bonding groups when describing the molecular shape?

10.30 Use wedge-bond perspective drawings (if necessary) to sketch the atom positions in a general molecule of formula (not shape class) AX_n that has each of the following shapes:
(a) V shaped   (b) trigonal planar   (c) trigonal bipyramidal
(d) T shaped   (e) trigonal pyramidal   (f) square pyramidal

10.31 What would you expect to be the electron-group arrangement around atom A in each of the following cases? For each arrangement, give the ideal bond angle and the direction of any expected deviation:

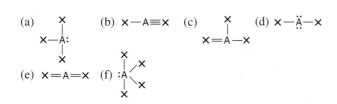

10.32 Determine the electron-group arrangement, molecular shape, and ideal bond angle(s) for each of the following:
(a) O_3 (b) H_3O^+ (c) NF_3

10.33 Determine the electron-group arrangement, molecular shape, and ideal bond angle(s) for each of the following:
(a) SO_4^{2-} (b) NO_2^- (c) PH_3

10.34 Determine the electron-group arrangement, molecular shape, and ideal bond angle(s) for each of the following:
(a) CO_3^{2-} (b) SO_2 (c) CF_4

10.35 Determine the electron-group arrangement, molecular shape, and ideal bond angle(s) for each of the following:
(a) SO_3 (b) N_2O (N is central)   (c) CH_2Cl_2

10.36 Name the shape and give the AX_mE_n classification and ideal bond angle(s) for each of the following general molecules:

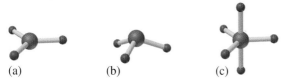

10.37 Name the shape and give the AX_mE_n classification and ideal bond angle(s) for each of the following general molecules:

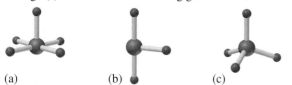

10.38 Determine the shape, ideal bond angle(s), and the direction of any deviation from these angles for each of the following:
(a) ClO_2^- (b) PF_5 (c) SeF_4 (d) KrF_2

10.39 Determine the shape, ideal bond angle(s), and the direction of any deviation from these angles for each of the following:
(a) ClO_3^- (b) IF_4^- (c) $SeOF_2$ (d) TeF_5^-

10.40 Determine the shape around each central atom in each molecule, and explain any deviation from ideal bond angles:
(a) CH_3OH (b) N_2O_4 (O_2NNO_2)

10.41 Determine the shape around each central atom in each molecule, and explain any deviation from ideal bond angles:
(a) H_3PO_4 (no H—P bond)   (b) $CH_3—O—CH_2CH_3$

10.42 Determine the shape around each central atom in each molecule, and explain any deviation from ideal bond angles:
(a) CH_3COOH (b) H_2O_2

10.43 Determine the shape around each central atom in each molecule, and explain any deviation from ideal bond angles:
(a) H_2SO_3 (no H—S bond)   (b) N_2O_3 ($ONNO_2$)

10.44 Arrange the following AF_n species in order of *increasing* F—A—F bond angles: BF_3, BeF_2, CF_4, NF_3, OF_2.

10.45 Arrange the following ACl_n species in order of *decreasing* Cl—A—Cl bond angles: SCl_2, OCl_2, PCl_3, $SiCl_4$, $SiCl_6^{2-}$.

10.46 State an ideal value for each of the bond angles in each molecule, and note where you expect deviations:

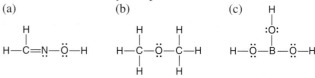

10.47 State an ideal value for each of the bond angles in each molecule, and note where you expect deviations:
(a)

$$\ddot{O}=N—\ddot{O}—H$$
$$\quad\quad\quad :\ddot{O}:$$

(b)
H
|
H—C—C=Ö
|
H

(c)
:O:
‖
H—C—Ö—H

10.48 Because both tin and carbon are members of Group 4A(14), they form structurally similar compounds. However, tin exhibits a greater variety of structures because it forms several ionic species. Predict the shapes and ideal bond angles, including any deviations, for the following:
(a) $Sn(CH_3)_2$ (b) $SnCl_3^-$ (c) $Sn(CH_3)_4$
(d) SnF_5^- (e) SnF_6^{2-}

10.49 In the gas phase, phosphorus pentachloride exists as separate molecules. In the solid phase, however, the compound is composed of alternating PCl_4^+ and PCl_6^- ions. What change(s) in molecular shape occur(s) as PCl_5 solidifies? How does the Cl—P—Cl angle change?

Molecular Shape and Molecular Polarity
(Sample Problem 10.9)

10.50 How can a molecule with polar covalent bonds not be polar? Give an example.

10.51 Consider the molecules SCl_2, F_2, CS_2, CF_4, and BrCl.
(a) Which has bonds that are the most polar?
(b) Which have a molecular dipole moment?

10.52 Consider the molecules BF_3, PF_3, BrF_3, SF_4, and SF_6.
(a) Which has bonds that are the most polar?
(b) Which have a molecular dipole moment?

10.53 Which molecule in each pair has the greater dipole moment? Give the reason for your choice.
(a) SO_2 or SO_3 (b) ICl or IF
(c) SiF_4 or SF_4 (d) H_2O or H_2S

10.54 Which molecule in each pair has the greater dipole moment? Give the reason for your choice.
(a) ClO_2 or SO_2 (b) HBr or HCl
(c) $BeCl_2$ or SCl_2 (d) AsF_3 or AsF_5

Comprehensive Problems
Problems with an asterisk (*) are more challenging.

* 10.55 There are three different dichloroethylenes (molecular formula $C_2H_2Cl_2$), which we can designate X, Y, and Z. Compound X has no dipole moment, but compound Z does. Compounds X and Z each combine with hydrogen to give the same product:

$$C_2H_2Cl_2 \text{ (X or Z)} + H_2 \longrightarrow ClCH_2—CH_2Cl$$

What are the structures of X, Y, and Z? Would you expect compound Y to have a dipole moment?

10.56 In addition to ammonia, nitrogen forms three other hydrides: hydrazine (N_2H_4), diazene (N_2H_2), and tetrazene (N_4H_4).

(a) Use Lewis structures to compare the strength, length, and order of nitrogen-nitrogen bonds in hydrazine, diazene, and N_2.

(b) Tetrazene (atom sequence H_2NNNNH_2) decomposes above 0°C to hydrazine and nitrogen gas. Draw a Lewis structure for tetrazene, and calculate ΔH°_{rxn} for this decomposition.

10.57 Both aluminum and iodine form chlorides, Al_2Cl_6 and I_2Cl_6, with "bridging" Cl atoms. The Lewis structures are

(a) What is the formal charge on each atom? (b) Which of these molecules has a planar shape? Explain.

10.58 Nitrosyl fluoride (NOF) has an atom sequence in which all atoms have formal charges of zero. Write the Lewis structure consistent with this fact.

10.59 The VSEPR model was developed in the 1950s, before any xenon compounds had been prepared. Thus, in the early 1960s, these compounds provided an excellent test of the model's predictive power. What would you have predicted for the shapes of XeF_2, XeF_4, and XeF_6?

10.60 The actual bond angle in NO_2 is 134.3°, and in NO_2^- it is 115.4°, although the ideal bond angle is 120° in both. Explain.

10.61 "Inert" xenon actually forms several compounds, especially with the highly electronegative elements oxygen and fluorine. The simple fluorides XeF_2, XeF_4, and XeF_6 are all formed by direct reaction of the elements. As you might expect from the size of the xenon atom, the Xe—F bond is not a strong one. Calculate the Xe—F bond energy in XeF_6, given that the heat of formation is -402 kJ/mol.

10.62 Chloral, Cl_3C—CH=O, reacts with water to form the sedative and hypnotic agent chloral hydrate, Cl_3C—$CH(OH)_2$. Draw Lewis structures for these substances, and describe the change in molecular shape, if any, that occurs around each of the carbon atoms during the reaction.

* **10.63** Like several other bonds, carbon-oxygen bonds have lengths and strengths that depend on the bond order. Draw Lewis structures for the following species, and arrange them in order of increasing carbon-oxygen bond length and then by increasing carbon-oxygen bond strength: (a) CO; (b) CO_3^{2-}; (c) H_2CO; (d) CH_4O; (e) HCO_3^- (H attached to O).

10.64 The four bonds of carbon tetrachloride (CCl_4) are polar, but the molecule is nonpolar because the bond polarity is canceled by the symmetric tetrahedral shape. When other atoms substitute for some of the Cl atoms, the symmetry is broken and the molecule becomes polar. Use Figure 9.20 to rank the following molecules from the least polar to the most polar: CH_2Br_2, CF_2Cl_2, CH_2F_2, CH_2Cl_2, CBr_4, CF_2Br_2.

* **10.65** Ethanol (CH_3CH_2OH) is being used as a gasoline additive or alternative in many parts of the world.

(a) Use bond energies to find the ΔH°_{rxn} for the combustion of gaseous ethanol. (Assume H_2O forms as a gas.)

(b) In its standard state at 25°C, ethanol is a liquid. Its vaporization requires 40.5 kJ/mol. Correct the value from part (a) to find the heat of reaction for the combustion of liquid ethanol.

(c) How does the value from part (b) compare with the value you calculate from standard heats of formation (Appendix B)?

(d) Methods for sustainable energy produce ethanol from corn and other plant material, but the main industrial method still involves hydrating ethylene from petroleum. Use Lewis structures

and bond energies to calculate ΔH°_{rxn} for the formation of gaseous ethanol from ethylene gas with water vapor.

10.66 An oxide of nitrogen is 25.9% N by mass, has a molar mass of 108 g/mol, and contains no nitrogen-nitrogen or oxygen-oxygen bonds. Draw its Lewis structure, and name it.

* **10.67** An experiment requires 50.0 mL of 0.040 M NaOH for the titration of 1.00 mmol of acid. Mass analysis of the acid shows 2.24% hydrogen, 26.7% carbon, and 71.1% oxygen. Draw the Lewis structure of the acid.

* **10.68** A major short-lived, neutral species in flames is OH.

(a) What is unusual about the electronic structure of OH?

(b) Use the standard heat of formation of OH(g) and bond energies to calculate the O—H bond energy in OH(g) [ΔH°_f of OH(g) = 39.0 kJ/mol].

(c) From the average value for the O—H bond energy in Table 9.2 and your value for the O—H bond energy in OH(g), find the energy needed to break the first O—H bond in water.

10.69 Pure HN_3 (atom sequence HNNN) is a very explosive compound. In aqueous solution, it is a weak acid (comparable to acetic acid) that yields the azide ion, N_3^-. Draw resonance structures to explain why the nitrogen-nitrogen bond lengths are equal in N_3^- but unequal in HN_3.

10.70 Except for nitrogen, the elements of Group 5A(15) all form pentafluorides, and most form pentachlorides. The chlorine atoms of PCl_5 can be replaced with fluorine atoms one at a time to give, successively, PCl_4F, PCl_3F_2, ..., PF_5.

(a) Given the sizes of F and Cl, would you expect the first two F substitutions to be at axial or equatorial positions? Explain.

(b) Which of the five fluorine-containing molecules have no dipole moment?

10.71 Dinitrogen monoxide (N_2O), used as the anesthetic "laughing gas" in dental surgery, supports combustion in a manner similar to oxygen, with the nitrogen atoms forming N_2. Draw three resonance structures for N_2O (one N is central), and use formal charges to decide the relative importance of each. What correlation can you suggest between the most important structure and the observation that N_2O supports combustion?

10.72 Some scientists speculate that many organic molecules required for life on the young Earth arrived on meteorites. The Murchison meteorite that landed in Australia in 1969 contained 92 different amino acids, including 21 found in Earth organisms. A skeleton structure (single bonds only) of one of these extraterrestrial amino acids is

Draw a Lewis structure, and identify any atoms with a nonzero formal charge.

10.73 When gaseous sulfur trioxide is dissolved in concentrated sulfuric acid, disulfuric acid forms:

$$SO_3(g) + H_2SO_4(l) \longrightarrow H_2S_2O_7(l)$$

Use bond energies (Table 9.2) to determine ΔH°_{rxn}. (The S atoms in $H_2S_2O_7$ are bonded through an O atom. Assume Lewis structures with zero formal charges; BE of S=O is 552 kJ/mol.)

10.74 In addition to propyne (see Follow-up Problem 10.8), there are two other constitutional isomers of formula C_3H_4. Draw a

Lewis structure for each, determine the shape around each carbon, and predict any deviations from ideal bond angles.

10.75 A molecule of formula AY_3 is found experimentally to be polar. Which molecular shapes are possible and which impossible for AY_3?

* 10.76 In contrast to the cyanate ion (NCO^-), which is stable and found in many compounds, the fulminate ion (CNO^-), with its different atom sequence, is unstable and forms compounds with heavy metal ions, such as Ag^+ and Hg^{2+}, that are explosive. Like the cyanate ion, the fulminate ion has three resonance structures. Which is the most important contributor to the resonance hybrid? Suggest a reason for the instability of fulminate.

10.77 Hydrogen cyanide (and organic nitriles, which contain the cyano group) can be catalytically reduced with hydrogen to form amines. Use Lewis structures and bond energies to determine ΔH°_{rxn} for

$$HCN(g) + 2H_2(g) \longrightarrow CH_3NH_2(g)$$

10.78 Ethylene, C_2H_4, and tetrafluoroethylene, C_2F_4, are used to make the polymers polyethylene and polytetrafluoroethylene (Teflon), respectively.
(a) Draw the Lewis structures for C_2H_4 and C_2F_4, and give the ideal H—C—H and F—C—F bond angles.

(b) The actual H—C—H and F—C—F bond angles are $117.4°$ and $112.4°$, respectively. Explain these deviations.

10.79 Lewis structures of mescaline, a hallucinogenic compound in peyote cactus, and dopamine, a neurotransmitter in the mammalian brain, appear below. Suggest a reason for mescaline's ability to disrupt nerve impulses.

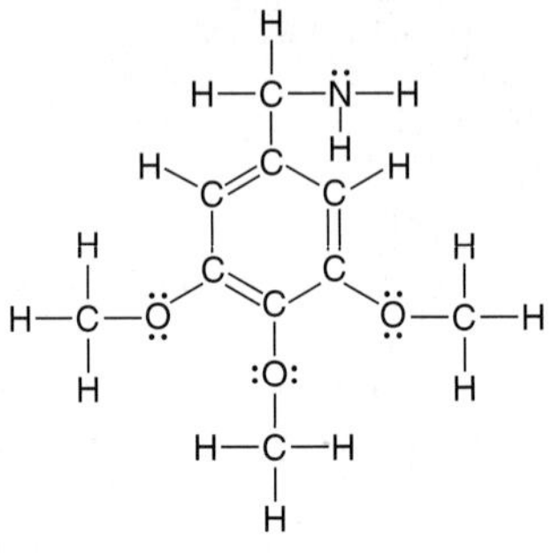

mescaline 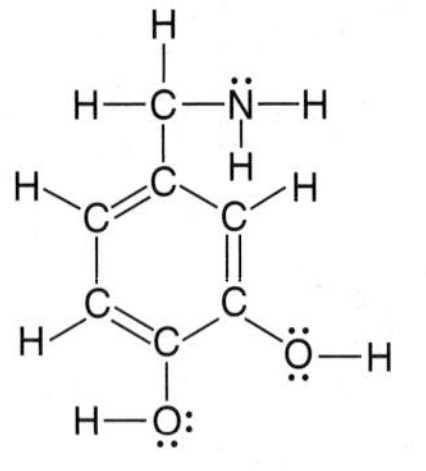   dopamine

10.80 Phosphorus pentachloride, a key industrial compound with annual world production of about 2×10^7 kg, is used to make other compounds. It reacts with sulfur dioxide to produce phosphorus oxychloride ($POCl_3$) and thionyl chloride ($SOCl_2$). Draw a Lewis structure and name the molecular shape of each product.

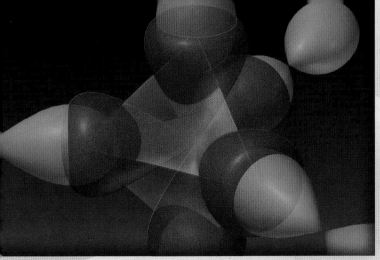

CHAPTER ELEVEN

Theories of Covalent Bonding

Rationalizing the Bond What is a chemical bond, and how do molecular shapes, like this depiction of phosphorus pentachloride, emerge from interacting atoms? In this chapter, you'll explore two models that rationalize the rich properties of the covalent bond.

Key Principles

◆ According to *valence bond (VB) theory,* a covalent bond forms when an electron pair is localized in the region where orbitals of the bonding atoms *overlap.*

◆ To account for observed molecular shapes, VB theory proposes that the orbitals of an isolated atom mix together and become *hybrid orbitals* that have the same orientation in space as the electron-group arrangements of VSEPR theory.

◆ The *mode of orbital overlap* determines the type of covalent bond: a *sigma* (σ) bond results from *end-to-end* overlap, and a *pi* (π) bond results from *side-to-side* overlap. A single bond is a σ bond, a double bond consists of a σ bond and a π bond, and a triple bond consists of a σ bond and two π bonds.

◆ According to *molecular orbital (MO) theory,* atomic orbitals (AOs) combine mathematically to form *molecular orbitals (MOs),* which are spread over the whole molecule. Adding AOs together gives a *bonding MO;* subtracting them gives an *antibonding MO.* Each type of MO has its own shape and energy and can hold a maximum of two electrons. The molecule is stabilized when electrons occupy a bonding MO.

◆ Like AOs, MOs become occupied one electron at a time and have specific energy levels. MO theory proposes that the magnetic properties of molecules depend on the number of unpaired electrons, and the spectral properties arise from electrons moving to different energy levels as a molecule absorbs or emits energy.

Outline

All scientific models have limitations because they are simplifications of reality. The VSEPR model we discussed in Chapter 10 accounts for molecular shapes by assuming that electron groups minimize their repulsions, and thus occupy as much space as possible around a central atom. But it does *not* explain how the shapes arise from interactions of atomic orbitals. After all, the orbitals we examined in Chapter 7 aren't oriented toward the corners of a tetrahedron or a trigonal bipyramid, to mention just two of the common molecular shapes. Moreover, knowing the shape doesn't help us explain the magnetic and spectral properties of molecules; only an understanding of their orbitals and energy levels can do that.

In this chapter, we discuss two theories of bonding in molecules, both of which are based on quantum mechanics. Valence bond (VB) theory rationalizes observed molecular shapes through interactions of atomic orbitals; molecular orbital (MO) theory explains molecular energy levels and properties.

11.1 VALENCE BOND (VB) THEORY AND ORBITAL HYBRIDIZATION

What *is* a covalent bond, and what characteristic gives it strength? And how can we explain *molecular* shapes based on the interactions of *atomic* orbitals? The most useful approach for answering these questions is **valence bond (VB) theory.**

The Central Themes of VB Theory

The basic principle of VB theory is that *a covalent bond forms when orbitals of two atoms overlap and the overlap region, which is between the nuclei, is occupied by a pair of electrons.* ("Orbital overlap" is another way of saying that the two wave functions are *in phase,* so the amplitude increases between the nuclei.) The central themes of VB theory derive from this principle:

1. *Opposing spins of the electron pair.* As the exclusion principle (Section 8.2) prescribes, the space formed by the overlapping orbitals *has a maximum capacity of two electrons that must have opposite spins.* When a molecule of H_2 forms, for instance, the two 1s electrons of two H atoms occupy the overlapping 1s orbitals and have opposite spins (Figure 11.1A).

2. *Maximum overlap of bonding orbitals.* The bond strength depends on the attraction of the nuclei for the shared electrons, so *the greater the orbital overlap, the stronger (more stable) the bond.* The extent of overlap depends on the shapes and directions of the orbitals. An *s* orbital is spherical, but *p* and *d* orbitals have more electron density in one direction than in another. Thus, whenever possible, a bond involving *p* or *d* orbitals will be oriented in the direction that maximizes overlap. In the HF bond, for example, the 1s orbital of H overlaps the half-filled 2p orbital of F *along the long axis* of that orbital (Figure 11.1B). Any

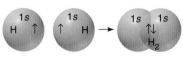

A Hydrogen, H_2

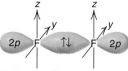

B Hydrogen fluoride, HF   **C** Fluorine, F_2

Figure 11.1 Orbital overlap and spin pairing in three diatomic molecules. A, In the H_2 molecule, the two overlapping 1s orbitals are occupied by the two 1s electrons with opposite spins. (The electrons, shown as arrows, spend the most time between the nuclei but move throughout the overlapping orbitals.) **B,** To maximize overlap in HF, half-filled H 1s and F 2p orbitals overlap along the long axis of the 2p orbital involved in the bonding. (The $2p_x$ orbital is shown bonding; the other two 2p orbitals of F are not shown.) **C,** In F_2, the half-filled $2p_x$ orbital on one F points end to end toward the similar orbital on the other F to maximize overlap.

other direction would result in less overlap and, thus, a weaker bond. Similarly, in the F—F bond of F_2, the two half-filled $2p$ orbitals interact end to end, that is, *along the long axes* of the orbitals, to maximize overlap (Figure 11.1C).

3. *Hybridization of atomic orbitals.* To account for the bonding in simple diatomic molecules like HF, we picture the direct overlap of s and p orbitals of isolated atoms. But how can we account for the shapes of so many molecules and polyatomic ions through the overlap of spherical s orbitals, dumbbell-shaped p orbitals, and cloverleaf-shaped d orbitals?

Consider a methane molecule, CH_4. It has four H atoms bonded to a central C atom. An isolated ground-state C atom ([He] $2s^2 2p^2$) has four valence electrons: two in the $2s$ orbital and one each in two of the three $2p$ orbitals. We might easily see how the two half-filled p orbitals of C could overlap with the $1s$ orbitals of two H atoms to form *two* C—H bonds with a 90° H—C—H bond angle. But methane is not CH_2 and doesn't have a bond angle of 90°. It's not as easy to see how the orbitals overlap to form the *four* C—H bonds with the 109.5° bond angle that occurs in methane.

To explain such facts, Linus Pauling proposed that *the valence atomic orbitals in the molecule are **different** from those in the isolated atoms.* Indeed, quantum-mechanical calculations show that if we "mix" specific combinations of orbitals mathematically, we obtain *new atomic orbitals.* The spatial orientations of these new orbitals lead to more stable bonds and are consistent with observed molecular shapes. The process of orbital mixing is called **hybridization,** and the new atomic orbitals are called **hybrid orbitals.** Two key points about the number and type of hybrid orbitals are that

- The *number* of hybrid orbitals obtained *equals* the number of atomic orbitals mixed.
- The *type* of hybrid orbitals obtained *varies* with the types of atomic orbitals mixed.

You can imagine hybridization as a process in which atomic orbitals mix, hybrid orbitals form, and electrons enter them with spins parallel (Hund's rule) to create stable bonds. In truth, though, hybridization is a mathematically derived result from quantum mechanics that accounts for the molecular shapes we observe.

Types of Hybrid Orbitals

We postulate the presence of a certain type of hybrid orbital *after* we observe the molecular shape. As we discuss the five common types of hybridization, notice that the spatial orientation of each type of hybrid orbital corresponds with one of the five common electron-group arrangements predicted by VSEPR theory.

***sp* Hybridization** When two electron groups surround the central atom, we observe a linear shape, which means that the bonding orbitals must have a linear orientation. VB theory explains this by proposing that mixing two *nonequivalent* orbitals of a central atom, one s and one p, gives rise to two *equivalent sp* **hybrid orbitals** that lie 180° apart (Figure 11.2A on the next page). Note the shape of the hybrid orbital: with one large and one small lobe, it differs markedly from the shapes of the atomic orbitals that were mixed. The orientations of hybrid orbitals extend electron density in the bonding direction and minimize repulsions between the electrons that occupy them. Thus, *both shape and orientation maximize overlap with the orbital of the other atom in the bond.*

In gaseous $BeCl_2$, for example, the paired $2s$ electrons in the isolated Be atom are distributed into two sp hybrid orbitals, which form two Be—Cl bonds by overlapping with the $3p$ orbitals of two Cl atoms (Figure 11.2B–D).

Animation: Molecular Shapes and Orbital Hybridization
Online Learning Center

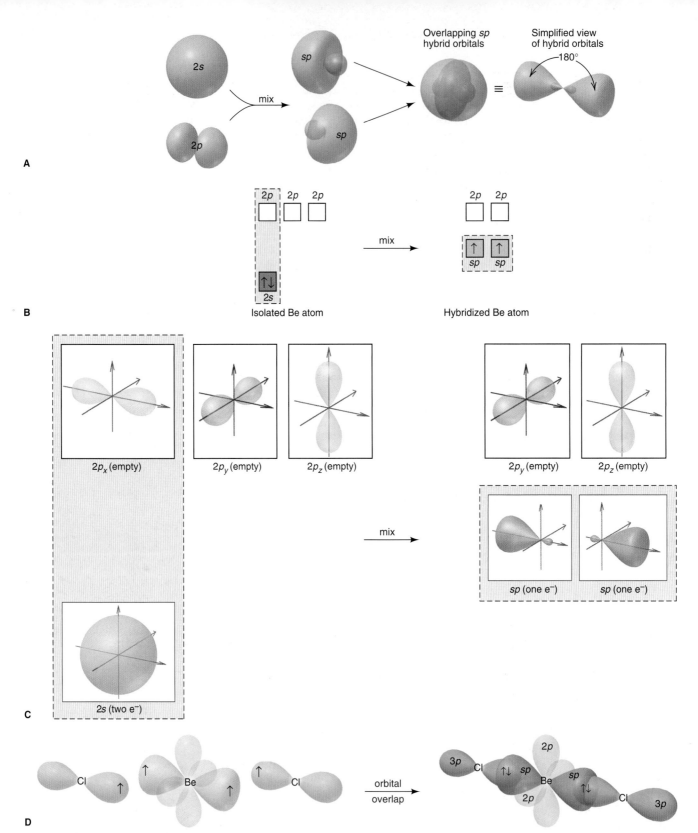

Figure 11.2 The *sp* hybrid orbitals in gaseous BeCl₂. A, One 2*s* and one 2*p* atomic orbital mix to form two *sp* hybrid orbitals (shown above and below one another and slightly to the side for ease of viewing). Note the large and small lobes of the hybrid orbitals. In the molecule, the two *sp* orbitals of Be are oriented in opposite directions. For clarity, the simplified hybrid orbitals will be used throughout the text, usually without the small lobe. **B,** The orbital diagram for hybridization in Be is drawn vertically and shows that the 2*s* and one of the three 2*p* orbitals form two *sp* hybrid orbitals, and the two other 2*p* orbitals remain unhybridized. Electrons half-fill the *sp* hybrid orbitals. During bonding, each *sp* orbital fills by sharing an electron from Cl (not shown). **C,** The orbital diagram is shown with orbital contours instead of electron arrows. **D,** BeCl₂ forms by overlap of the two *sp* hybrids with the 3*p* orbitals of two Cl atoms; the two unhybridized Be 2*p* orbitals lie perpendicular to the *sp* hybrids. (For clarity, only the 3*p* orbital involved in bonding is shown for each Cl.)

sp^2 Hybridization To rationalize the trigonal planar electron-group arrangement and the shapes of molecules based on it, we introduce the mixing of one s and two p orbitals of the central atom to give three hybrid orbitals that point toward the vertices of an equilateral triangle, their axes 120° apart. These are called **sp^2 hybrid orbitals.** (Note that, unlike electron configuration notation, hybrid orbital notation uses superscripts for the *number* of atomic orbitals of a given type that are mixed, *not* for the number of electrons in the orbital: here, one s and two p orbitals were mixed, so we have s^1p^2, or sp^2.)

For example, VB theory proposes that the central B atom in the BF_3 molecule is sp^2 hybridized. Figure 11.3 shows the three sp^2 orbitals in the trigonal plane, with the third $2p$ orbital unhybridized and perpendicular to this plane. Each sp^2 orbital overlaps the $2p$ orbital of an F atom, and the six valence electrons—three from B and one from each of the three F atoms—form three bonding pairs.

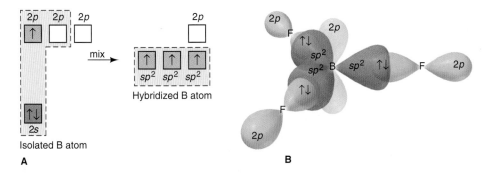

Figure 11.3 The sp^2 hybrid orbitals in BF_3. A, The orbital diagram shows that the $2s$ and two of the three $2p$ orbitals of the B atom mix to make three sp^2 hybrid orbitals. Three electrons (*up arrows*) half-fill the sp^2 hybrids. The third $2p$ orbital remains empty and unhybridized. **B,** BF_3 forms through overlap of $2p$ orbitals on three F atoms with the sp^2 hybrids. During bonding, each sp^2 orbital fills with an electron from one F (*down arrow*). The three sp^2 hybrids of B lie 120° apart, and the unhybridized $2p$ orbital is perpendicular to the trigonal bonding plane.

To account for other molecular shapes within a given electron-group arrangement, we postulate that one or more of the hybrid orbitals contains lone pairs. In ozone (O_3), for example, the central O is sp^2 hybridized and a lone pair fills one of its three sp^2 orbitals, so ozone has a bent molecular shape.

sp^3 Hybridization Now let's return to the question posed earlier about the orbitals in methane, the same question that arises for any species with a tetrahedral electron-group arrangement. VB theory proposes that the one s and all three p orbitals of the central atom mix and form four **sp^3 hybrid orbitals,** which point toward the vertices of a tetrahedron. As shown in Figure 11.4, the C atom in methane is sp^3 hybridized. Its four valence electrons half-fill the four sp^3 hybrids, which overlap the half-filled $1s$ orbitals of four H atoms and form four C—H bonds.

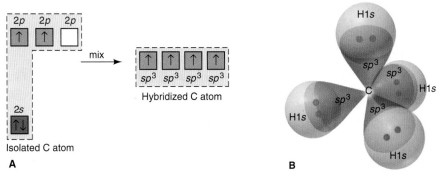

Figure 11.4 The sp^3 hybrid orbitals in CH_4. A, The $2s$ and all three $2p$ orbitals of C are mixed to form four sp^3 hybrids. Carbon's four valence electrons half-fill the sp^3 hybrids. **B,** In methane, the four sp^3 orbitals of C point toward the corners of a tetrahedron and overlap the $1s$ orbitals of four H atoms. Each sp^3 orbital fills by addition of an electron from one H (electrons are shown as dots).

Figure 11.5 The sp^3 hybrid orbitals in NH$_3$ and H$_2$O. A, The orbital diagrams show sp^3 hybridization, as in CH$_4$. In NH$_3$ (*top*), one sp^3 orbital is filled with a lone pair. In H$_2$O (*bottom*), two sp^3 orbitals are filled with lone pairs. **B,** Contour diagrams show the tetrahedral orientation of the sp^3 orbitals and the overlap of the bonded H atoms. Each half-filled sp^3 orbital fills by addition of an electron from one H. (Shared pairs and lone pairs are shown as dots.)

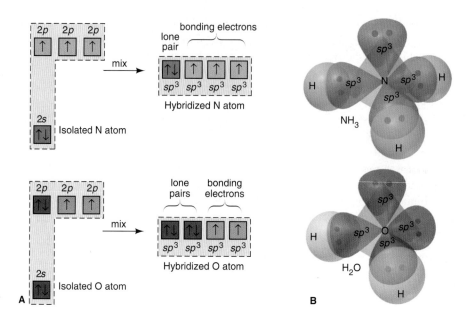

Figure 11.5 shows the bonding in other molecular shapes with the tetrahedral electron-group arrangement. The trigonal pyramidal shape of NH$_3$ arises when a lone pair fills one of the four sp^3 orbitals of N, and the bent shape of H$_2$O arises when lone pairs fill two of the sp^3 orbitals of O.

sp^3d Hybridization The shapes of molecules with trigonal bipyramidal or octahedral electron-group arrangements are rationalized with VB theory through similar arguments. The only new point is that such molecules have central atoms from Period 3 or higher, so atomic d orbitals, as well as s and p orbitals, are mixed to form the hybrid orbitals.

For the trigonal bipyramidal shape of the PCl$_5$ molecule, for example, the VB model proposes that the one $3s$, the three $3p$, and one of the five $3d$ orbitals of the central P atom mix and form five **sp^3d hybrid orbitals,** which point to the vertices of a trigonal bipyramid (Figure 11.6). Seesaw, T-shaped, and linear molecules have this electron-group arrangement with lone pairs in one, two, or three of the central atom's sp^3d orbitals, respectively.

Figure 11.6 The sp^3d hybrid orbitals in PCl$_5$. A, The orbital diagram shows that one $3s$, three $3p$, and one of the five $3d$ orbitals of P mix to form five sp^3d orbitals that are half-filled. Four $3d$ orbitals are unhybridized and empty. **B,** The trigonal bipyramidal PCl$_5$ molecule forms by the overlap of a $3p$ orbital from each of the five Cl atoms with the sp^3d hybrid orbitals of P (unhybridized, empty $3d$ orbitals not shown). Each sp^3d orbital fills by addition of an electron from one Cl. (The five bonding pairs are not shown.)

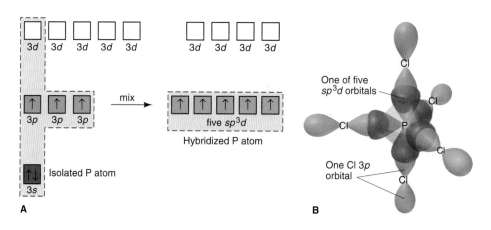

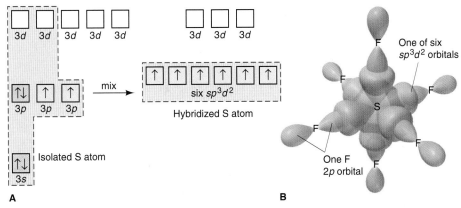

A   **B**

Figure 11.7 The sp^3d^2 hybrid orbitals in SF$_6$. A, The orbital diagram shows that one 3s, three 3p, and two 3d orbitals of S mix to form six sp^3d^2 orbitals that are half-filled. Three 3d orbitals are unhybridized and empty. **B,** The octahedral SF$_6$ molecule forms from overlap of a 2p orbital from each of six F atoms with the sp^3d^2 orbitals of S (unhybridized, empty 3d orbitals not shown). Each sp^3d^2 orbital fills by addition of an electron from one F. (The six bonding pairs are not shown.)

sp³d²* Hybridization** To rationalize the shape of SF$_6$, the VB model proposes that the one 3s, the three 3p, and two of the five 3d orbitals of the central S atom mix and form six ***sp³d² hybrid orbitals, which point to the vertices of an octahedron (Figure 11.7). Square pyramidal and square planar molecules have lone pairs in one and two of the central atom's sp^3d^2 orbitals, respectively.

Table 11.1 summarizes the numbers and types of atomic orbitals that are mixed to obtain the five types of hybrid orbitals. Once again, note the similarities between the orientations of the hybrid orbitals proposed by VB theory and the shapes predicted by VSEPR theory (see Figure 10.3). Figure 11.8 (on the next page) shows the three conceptual steps from molecular formula to postulating the hybrid orbitals in the molecule, and Sample Problem 11.1 details the end of that process.

Table 11.1   **Composition and Orientation of Hybrid Orbitals**

	Linear	Trigonal Planar	Tetrahedral	Trigonal Bipyramidal	Octahedral
Atomic orbitals mixed	one s one p	one s two p	one s three p	one s three p one d	one s three p two d
Hybrid orbitals formed	two sp	three sp^2	four sp^3	five sp^3d	six sp^3d^2
Unhybridized orbitals remaining	two p	one p	none	four d	three d
Orientation					

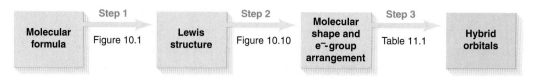

Figure 11.8 The conceptual steps from molecular formula to the hybrid orbitals used in bonding. (See Figures 10.1 and 10.10.)

SAMPLE PROBLEM 11.1 Postulating Hybrid Orbitals in a Molecule

Problem Use partial orbital diagrams to describe how mixing of the atomic orbitals of the central atom(s) leads to the hybrid orbitals in each of the following:
(a) Methanol, CH_3OH **(b)** Sulfur tetrafluoride, SF_4
Plan From the Lewis structure, we determine the number and arrangement of electron groups around each central atom, along with the molecular shape. From that, we postulate the type of hybrid orbitals involved. Then, we write the partial orbital diagram for each central atom before and after the orbitals are hybridized.
Solution (a) For CH_3OH. The electron-group arrangement is tetrahedral around both C and O atoms. Therefore, each central atom is sp^3 hybridized. The C atom has four half-filled sp^3 orbitals:

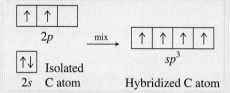

The O atom has two half-filled sp^3 orbitals and two filled with lone pairs:

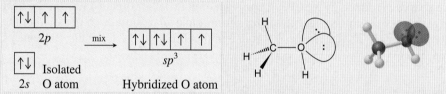

During bonding, each half-filled C or O orbital becomes filled. The second electron comes from an H atom, or in the C—O bond, from the overlapping C and O orbitals.
(b) For SF_4. The molecular shape is seesaw, which is based on the trigonal bipyramidal electron-group arrangement. Thus, the central S atom is surrounded by five electron groups, which implies sp^3d hybridization. One 3s orbital, three 3p orbitals, and one 3d orbital are mixed. One hybrid orbital is filled with a lone pair, and four are half-filled. Four unhybridized 3d orbitals remain empty:

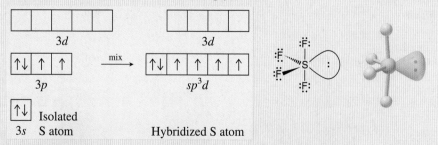

During bonding, each half-filled S orbital becomes filled, with the second electrons coming from F atoms.

FOLLOW-UP PROBLEM 11.1 Use partial orbital diagrams to show how the atomic orbitals of the central atoms mix to form hybrid orbitals in **(a)** beryllium fluoride, BeF_2; **(b)** silicon tetrachloride, $SiCl_4$; **(c)** xenon tetrafluoride, XeF_4.

SECTION SUMMARY
VB theory explains that a covalent bond forms when two atomic orbitals overlap and two electrons with paired (opposite) spins occupy the overlapped region. Orbital hybridization allows us to explain how atomic orbitals mix and change their characteristics during bonding. Based on the observed molecular shape (and the related electron-group arrangement), we postulate the type of hybrid orbital needed.

11.2 THE MODE OF ORBITAL OVERLAP AND THE TYPES OF COVALENT BONDS

In this section, we focus on the *mode* by which orbitals overlap—end to end or side to side—to see the detailed makeup of covalent bonds. These two modes give rise to the two types of covalent bonds—*sigma* bonds and *pi* bonds. We'll use valence bond theory to describe the two types here, but they are essential features of molecular orbital theory as well.

Orbital Overlap in Single and Multiple Bonds

The VSEPR model predicts, and measurements verify, different shapes for ethane (C_2H_6), ethylene (C_2H_4), and acetylene (C_2H_2). Ethane is tetrahedrally shaped at both carbons, with bond angles of about 109.5°. Ethylene is trigonal planar at both carbons, with the double bond acting as one electron group and bond angles near the ideal 120°. Acetylene has a linear shape, with the triple bond acting as one electron group and bond angles of 180°:

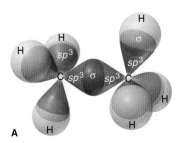

A close look at the bonds shows two modes of orbital overlap, which correspond to the two types of covalent bonds:

1. *End-to-end overlap and sigma (σ) bonding.* Both C atoms of ethane are sp^3 hybridized (Figure 11.9). The C—C bond involves overlap of one sp^3 orbital from each C, and each of the six C—H bonds involves overlap of a C sp^3 orbital with an H $1s$ orbital. The bonds in ethane are like all the others described so far in this chapter. Look closely at the C—C bond, for example. It involves the overlap of the end of one orbital with the end of the other. The bond resulting from such *end-to-end* overlap is called a **sigma (σ) bond.** It has its *highest electron density along the bond axis* (an imaginary line joining the nuclei) and is shaped like an ellipse rotated about its long axis (the shape resembles a football). All

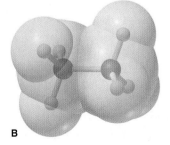
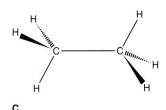

Figure 11.9 The σ bonds in ethane (C_2H_6). A, Two sp^3 hybridized C atoms and six H atoms in ethane form one C—C σ bond and six C—H σ bonds. **B,** Electron density is distributed relatively evenly among the σ bonds. **C,** Wedge-bond perspective drawing.

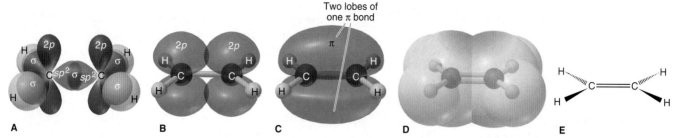

Figure 11.10 **The σ and π bonds in ethylene (C₂H₄). A,** Two *sp²* hybridized C atoms form one C—C σ bond and four C—H σ bonds. Half-filled unhybridized *2p* orbitals (shown in stylized form) lie perpendicular to the σ-bond axis. **B,** The *2p* orbitals (shown in accurate form) actually overlap side to side; the σ bonds are shown as ball-and-stick models. **C,** The two overlapping regions comprise *one* π bond, which is occupied by two electrons. **D,** With four electrons (one σ bond and one π bond) between the C atoms, electron density (*red shading*) is high. **E,** Wedge-bond perspective drawing.

single bonds, formed by any combination of overlapping hybrid, *s*, or *p* orbitals, have their electron density concentrated along the bond axis, and thus are σ bonds.

2. *Side-to-side overlap and pi (π) bonding.* A close look at Figure 11.10 reveals the double nature of the carbon-carbon bond in ethylene. Here, each C atom is *sp²* hybridized. Each C atom's four valence electrons half-fill its three *sp²* orbitals and its unhybridized *2p* orbital, which lies perpendicular to the *sp²* plane. Two *sp²* orbitals of each C form C—H σ bonds by overlapping the *1s* orbitals of two H atoms. The third *sp²* orbital forms a C—C σ bond with an *sp²* orbital of the other C because their orientation allows end-to-end overlap. With the σ-bonded C atoms near each other, their half-filled unhybridized *2p* orbitals are close enough to overlap *side to side.* Such overlap forms another type of covalent bond called a **pi (π) bond.** It has *two regions of electron density,* one above and one below the σ-bond axis. *One π bond holds two electrons that occupy both regions of the bond. A double bond always consists of one σ bond and one π bond.* As shown in Figure 11.10D, the double bond increases electron density between the C atoms.

Now we can answer a question brought up in our discussion of VSEPR theory in Chapter 10: why do the two electron pairs in a double bond act as one electron group; that is, why don't they push each other apart? The answer is that each electron pair occupies a distinct orbital, a specific region of electron density, so repulsions are reduced.

A triple bond, such as the C≡C bond in acetylene, *consists of one σ and two π bonds* (Figure 11.11). To maximize overlap in a linear shape, one *s* and one *p* orbital in *each* C atom form two *sp* hybrids, and two *2p* orbitals remain unhybridized. The four valence electrons half-fill all four orbitals. Each C uses one of its *sp* orbitals to form a σ bond with an H atom and uses the other to form the C—C σ bond. Side-to-side overlap of one pair of *2p* orbitals gives one π bond, with electron density above and below the σ bond. Side-to-side overlap of the other pair of *2p* orbitals gives the other π bond, 90° away from the first, with electron density in front and back of the σ bond. The result is a *cylindrically symmetrical* H—C≡C—H molecule. Note the greater electron density between the C atoms created by the six bonding electrons. Figure 11.12 shows electron density relief maps of the carbon-carbon bonds in these compounds; note the increasing electron density between the nuclei from single to double to triple bond.

Mode of Overlap and Molecular Properties

The way in which orbitals overlap affects several properties of molecules, such as bond strength and molecular rotation. The mode of overlap influences bond strength because side-to-side overlap is not as extensive as end-to-end overlap.

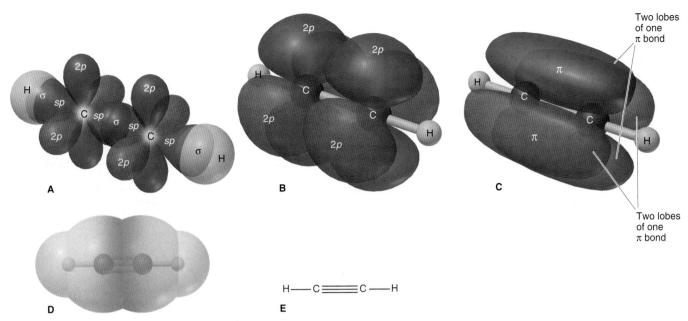

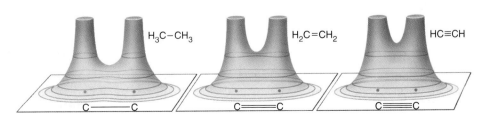

Figure 11.11 The σ bonds and π bonds in acetylene (C$_2$H$_2$). A, The C—C σ bond in acetylene forms when an *sp* orbital of each C overlaps. Two C—H σ bonds form from overlap of the other *sp* orbital of each C and an *s* orbital of an H. Two unhybridized *2p* orbitals on each C are shown in stylized form (one pair is in red for clarity). **B,** The *2p* orbitals (shown in accurate form) actually overlap side to side; the σ bonds are

shown as ball-and-stick models. **C,** Overlap of pairs of *2p* orbitals results in two π bonds, one (*black*) with its lobes above and below the C—C σ bond, and the other (*red*) with its lobes in front of and behind the σ bond. **D,** The two π bonds give the molecule cylindrical symmetry. Six electrons—one σ bond and two π bonds—create greater electron density between the C atoms than in ethylene. **E,** Bond-line drawing.

Therefore, we would expect a π bond to be weaker than a σ bond, and so a double bond should be less than twice as strong as a single bond. This expectation *is* borne out for carbon-carbon bonds. However, many factors, such as lone-pair repulsions, bond polarities, and other electrostatic contributions, affect overlap and the relative strength of σ and π bonds between other pairs of atoms. Thus, as a rough approximation, a double bond is about twice as strong as a single bond, and a triple bond is about three times as strong.

The mode of overlap also influences molecular rotation, the ability of one part of a molecule to rotate relative to another part. A *σ bond allows free rotation* of the parts of the molecule with respect to each other because the extent of overlap is not affected. If you could hold one CH$_3$ group of the ethane molecule, the other CH$_3$ group could spin like a pinwheel without affecting the C—C σ-bond overlap (see Figure 11.9).

However, *p* orbitals must be parallel to engage in side-to-side overlap, so *a π bond restricts rotation* around it. Rotating one CH$_2$ group in ethylene with respect to the other must decrease the side-to-side overlap and break the π bond (see Figure 11.10). (In Chapter 15, you'll see that restricted rotation leads to another type of isomerism.)

Figure 11.12 Electron density and bond order. Relief maps of the carbon-carbon bonding region in ethane, ethylene, and acetylene show a large increase in electron density as the bond order increases.

SAMPLE PROBLEM 11.2 Describing the Types of Bonds in Molecules

Problem Describe the types of bonds and orbitals in acetone, $(CH_3)_2CO$.

Plan We note, as in Sample Problem 11.1, the shape around each central atom to postulate the type of the hybrid orbitals, paying attention to the multiple bonding of the C and O bond.

Solution In Sample Problem 10.8, we determined the shapes around the three central atoms of acetone: tetrahedral around each C of the two CH_3 (methyl) groups and trigonal planar around the middle C atom. Thus, the middle C has three sp^2 orbitals and one unhybridized p orbital. Each of the two methyl C atoms has four sp^3 orbitals. Three of these sp^3 orbitals overlap the $1s$ orbitals of the H atoms to form σ bonds; the fourth overlaps an sp^2 orbital of the middle C atom. Thus, two of the three sp^2 orbitals of the middle C form σ bonds to the other two C atoms.

The O atom is also sp^2 hybridized and has an unhybridized p orbital that can form a π bond. Two of the O atom's sp^2 orbitals hold lone pairs, and the third forms a σ bond with the third sp^2 orbital of the middle C atom. The unhybridized, half-filled p orbitals of C and O form a π bond. The σ and π bonds constitute the C=O bond:

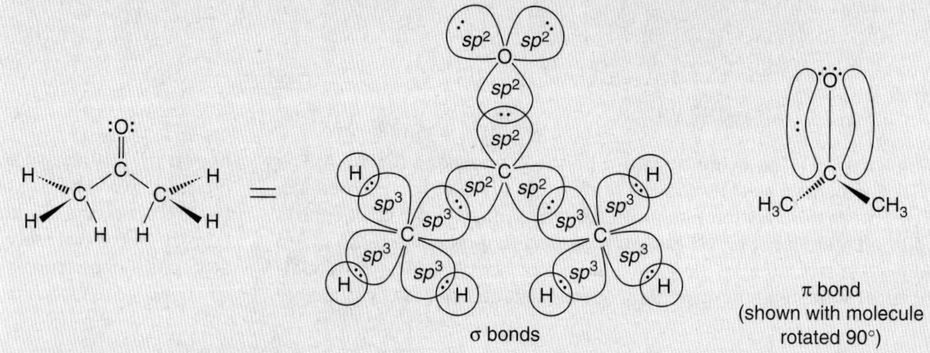

Comment How can we tell if a terminal atom, such as the O atom in acetone, is hybridized? After all, it could use two perpendicular p orbitals for the σ and π bonds with C and leave the other p and the s orbital to hold the two lone pairs. But, having each lone pair in an sp^2 orbital oriented away from the C=O bond, rather than in an s orbital, lowers repulsions.

FOLLOW-UP PROBLEM 11.2 Describe the types of bonds and orbitals in (a) hydrogen cyanide, HCN, and (b) carbon dioxide, CO_2.

SECTION SUMMARY

End-to-end overlap of atomic orbitals forms a σ bond and allows free rotation of the parts of the molecule. Side-to-side overlap forms a π bond, which restricts rotation. A multiple bond consists of a σ bond and either one π bond (double bond) or two π bonds (triple bond). Multiple bonds have greater electron density than single bonds.

11.3 MOLECULAR ORBITAL (MO) THEORY AND ELECTRON DELOCALIZATION

Scientists choose the model that best helps them answer a particular question. If the question concerns molecular shape, chemists choose the VSEPR model, followed by hybrid-orbital analysis with VB theory. But VB theory does not adequately explain magnetic and spectral properties, and it understates the importance of electron delocalization. In order to deal with these phenomena, which involve molecular energy levels, chemists choose **molecular orbital (MO) theory.**

In VB theory, a molecule is pictured as a group of atoms bound together through *localized* overlap of valence-shell atomic orbitals. In MO theory, a molecule is pictured as a collection of nuclei with the electron orbitals *delocalized* over the entire molecule. The MO model is a quantum-mechanical treatment for molecules similar to the one for atoms in Chapter 8. Just as an atom has atomic orbitals (AOs) with a given energy and shape that are occupied by the atom's electrons, a molecule has **molecular orbitals (MOs)** with a given energy and

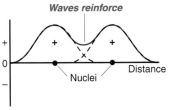 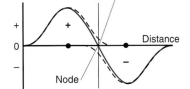

A Amplitudes of wave functions added

B Amplitudes of wave functions subtracted

Figure 11.13 An analogy between light waves and atomic wave functions. When light waves undergo interference, their amplitudes either add together or subtract. **A,** When the amplitudes of atomic wave functions (*dashed lines*) are added, a bonding molecular orbital (MO) results, and electron density (*red line*) increases between the nuclei. **B,** Conversely, when the amplitudes of the wave functions are subtracted, an antibonding MO results, which has a node (region of zero electron density) between the nuclei.

shape that are occupied by the molecule's electrons. Despite the great usefulness of MO theory, it too has a drawback: MOs are more difficult to visualize than the easily depicted shapes of VSEPR theory or the hybrid orbitals of VB theory.

The Central Themes of MO Theory

Several key ideas of MO theory appear in its description of the hydrogen molecule and other simple species. These ideas include the formation of MOs, their energy and shape, and how they fill with electrons.

Formation of Molecular Orbitals Because electron motion is so complex, we use approximations to solve the Schrödinger equation (see Section 7.4) for an atom with more than one electron. Similar complications arise even with H_2, the simplest molecule, so we use approximations to solve for the properties of MOs. The most common approximation mathematically *combines* (adds or subtracts) the atomic orbitals (atomic wave functions) of nearby atoms to form MOs (molecular wave functions).

When two H nuclei lie near each other, as in H_2, their AOs overlap. The two ways of combining the AOs are as follows:

- *Adding the wave functions together.* This combination forms a **bonding MO,** which has *a region of high electron density between the nuclei.* Additive overlap is analogous to light waves reinforcing each other, making the resulting amplitude higher and the light brighter. For electron waves, the overlap *increases* the probability that the electrons are between the nuclei (Figure 11.13A).
- *Subtracting the wave functions from each other.* This combination forms an **antibonding MO,** which has *a region of zero electron density (a node) between the nuclei* (Figure 11.13B). Subtractive overlap is analogous to light waves canceling each other, so that the light disappears. With electron waves, the probability that the electrons lie between the nuclei *decreases* to zero.

The two possible combinations for hydrogen atoms H_A and H_B are

AO of H_A + AO of H_B = bonding MO of H_2 (more e^- density between nuclei)
AO of H_A − AO of H_B = antibonding MO of H_2 (less e^- density between nuclei)

Notice that *the number of AOs combined always equals the number of MOs formed:* two H atomic orbitals combine to form two H_2 molecular orbitals.

Energy and Shape of H_2 Molecular Orbitals *The bonding MO is lower in energy and the antibonding MO higher in energy than the AOs that combined to form them.* Let's examine Figure 11.14 to see why this is so.

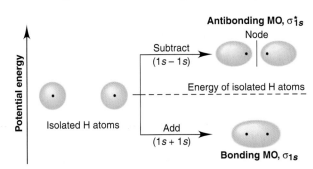

Figure 11.14 Contours and energies of the bonding and antibonding molecular orbitals (MOs) in H_2. When two H 1*s* atomic orbitals (AOs) combine, they form two H_2 MOs. The bonding MO (σ_{1s}) forms from addition of the AOs and is lower in energy than those AOs because most of its electron density lies *between* the nuclei (shown as dots). The antibonding MO (σ_{1s}^*) forms from subtraction of the AOs and is higher in energy because there is a *node* between the nuclei and most of the electron density lies outside the internuclear region.

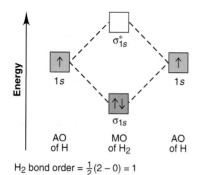

H_2 bond order $= \frac{1}{2}(2-0) = 1$

Figure 11.15 The MO diagram for H_2. The positions of the boxes indicate the relative energies and the arrows show the electron occupancy of the MOs and the AOs from which they formed. Two electrons, one from each H atom, fill the lower energy σ_{1s} MO, while the higher energy σ_{1s}^* MO remains empty. Orbital occupancy is also shown by color (darker = full, paler = half-filled, no color = empty).

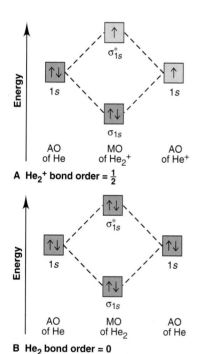

A He_2^+ bond order $= \frac{1}{2}$

B He_2 bond order $= 0$

Figure 11.16 MO diagrams for He_2^+ and He_2. A, In He_2^+, three electrons enter MOs in order of increasing energy to give a filled σ_{1s} MO and a half-filled σ_{1s}^* MO. The bond order of $\frac{1}{2}$ implies that He_2^+ exists. **B,** In He_2, the four electrons fill both the σ_{1s} and the σ_{1s}^* MOs, so there is no net stabilization (bond order = 0).

The bonding MO in H_2 is spread mostly *between* the nuclei, with the nuclei attracted to the intervening electrons. An electron in this MO can delocalize its charge over a much larger volume than is possible in an individual AO in either H. Because the electron-electron repulsions are reduced, the bonding MO is *lower* in energy than the isolated AOs. Therefore, when electrons occupy this orbital, the molecule is *more stable* than the separate atoms. In contrast, the antibonding MO has a node between the nuclei and most of its electron density *outside* the internuclear region. The electrons do not shield one nucleus from the other, which increases the nucleus-nucleus repulsion and makes the antibonding MO *higher* in energy than the isolated AOs. Therefore, when the antibonding orbital is occupied, the molecule is *less stable* than when this orbital is empty.

Both the bonding and antibonding MOs of H_2 are **sigma (σ) MOs** because they are cylindrically symmetrical about an imaginary line that runs through the two nuclei. The bonding MO is denoted by σ_{1s}, that is, a σ MO formed by combination of $1s$ AOs. Antibonding orbitals are denoted with a superscript star, so the orbital derived from the $1s$ AOs is σ_{1s}^* (spoken "sigma, one ess, star").

To interact effectively and form MOs, *atomic orbitals must have similar energy and orientation.* The $1s$ orbitals on two H atoms have identical energy and orientation, so they interact strongly.

Filling Molecular Orbitals with Electrons Electrons fill MOs just as they fill AOs:

- MOs are filled in order of increasing energy (aufbau principle).
- An MO has a maximum capacity of two electrons with opposite spins (exclusion principle).
- MOs of equal energy are half-filled, with spins parallel, before any of them is completely filled (Hund's rule).

A **molecular orbital (MO) diagram** shows the relative energy and number of electrons in each MO, as well as the AOs from which they formed. Figure 11.15 is the MO diagram for H_2.

MO theory redefines bond order. In a Lewis structure, bond order is the number of electron pairs per linkage. The **MO bond order** is the number of electrons in bonding MOs minus the number in antibonding MOs, divided by two:

Bond order $= \frac{1}{2}[$(no. of e^- in bonding MO) $-$ (no. of e^- in antibonding MO)$]$ (11.1)

Thus, for H_2, the bond order is $\frac{1}{2}(2-0) = 1$. A bond order greater than zero indicates that the molecular species is stable relative to the separate atoms, whereas a bond order of zero implies no net stability and, thus, no likelihood that the species will form. In general, *the higher the bond order, the stronger the bond.*

Another similarity of MO theory to the quantum-mechanical model for atoms is that we can write electron configurations for a molecule. The symbol of each occupied MO is shown in parentheses, and the number of electrons in it is written outside as a superscript. Thus, the electron configuration of H_2 is $(\sigma_{1s})^2$.

One of the early triumphs of MO theory was its ability to *predict* the existence of He_2^+, the dihelium molecule-ion, which is composed of two He nuclei and three electrons. Let's use MO theory to see why He_2^+ exists and, at the same time, why He_2 does not. In He_2^+, the $1s$ atomic orbitals form the molecular orbitals, so the MO diagram, shown in Figure 11.16A, is similar to that for H_2. The three electrons enter the MOs to give a pair in the σ_{1s} MO and a lone electron in the σ_{1s}^* MO. The bond order is $\frac{1}{2}(2-1) = \frac{1}{2}$. Thus, He_2^+ has a relatively weak bond, but it should exist. Indeed, this molecular ionic species has been observed frequently when He atoms collide with He^+ ions. Its electron configuration is $(\sigma_{1s})^2(\sigma_{1s}^*)^1$.

On the other hand, He_2 has four electrons to place in its σ_{1s} and σ_{1s}^* MOs. As Figure 11.16B shows, both the bonding and antibonding orbitals are filled. The stabilization arising from the electron pair in the bonding MO is canceled by the destabilization due to the electron pair in the antibonding MO. From its zero bond order $[\frac{1}{2}(2 - 2) = 0]$, we predict, and experiment has so far confirmed, that a covalent He_2 molecule does not exist.

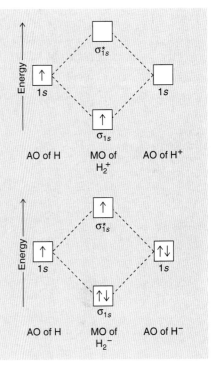

SAMPLE PROBLEM 11.3 Predicting Stability of Species Using MO Diagrams

Problem Use MO diagrams to predict whether H_2^+ and H_2^- exist. Determine their bond orders and electron configurations.

Plan In these species, the $1s$ orbitals form MOs, so the MO diagrams are similar to that for H_2. We determine the number of electrons in each species and distribute the electrons in pairs to the bonding and antibonding MOs in order of increasing energy. We obtain the bond order with Equation 11.1 and write the electron configuration as described in the text.

Solution For H_2^+. H_2 has two e^-, so H_2^+ has only one, as shown in the margin (top diagram). The bond order is $\frac{1}{2}(1 - 0) = \frac{1}{2}$, so we predict that H_2^+ does exist. The electron configuration is $(\sigma_{1s})^1$.

For H_2^-. H_2 has two e^-, so H_2^- has three, as shown in the margin (bottom diagram). The bond order is $\frac{1}{2}(2 - 1) = \frac{1}{2}$, so we predict that H_2^- does exist. The electron configuration is $(\sigma_{1s})^2(\sigma_{1s}^*)^1$.

Check The number of electrons in the MOs equals the number of electrons in the AOs, as it should.

Comment Both these species have been detected spectroscopically: H_2^+ occurs in the hydrogen-containing material around stars; H_2^- has been formed in the laboratory.

FOLLOW-UP PROBLEM 11.3 Use an MO diagram to predict whether two hydride ions (H^-) will form H_2^{2-}. Calculate the bond order of H_2^{2-}, and write its electron configuration.

Homonuclear Diatomic Molecules of the Period 2 Elements

Homonuclear diatomic molecules are those composed of two identical atoms. In addition to H_2 from Period 1, you're also familiar with several from Period 2—N_2, O_2, and F_2—as the elemental forms under standard conditions. Others in Period 2—Li_2, Be_2, B_2, C_2, and Ne_2—are observed, if at all, only in high-temperature gas-phase experiments. Molecular orbital descriptions of these species provide some interesting tests of the model. Let's look first at the molecules from the *s* block, Groups 1A(1) and 2A(2), and then at those from the *p* block, Groups 3A(13) through 8A(18).

Bonding in the *s*-Block Homonuclear Diatomic Molecules

Both Li and Be occur as metals under normal conditions, but let's see what MO theory predicts for their stability as the diatomic gases dilithium (Li_2) and diberyllium (Be_2).

These atoms have both inner ($1s$) and outer ($2s$) electrons, but the $1s$ orbitals interact negligibly. As we do when writing Lewis structures, we ignore the inner electrons here because, in general, *only outer (valence) orbitals interact enough to form molecular orbitals*. Like the MOs formed from $1s$ AOs, those formed from $2s$ AOs are σ orbitals, cylindrically symmetrical around the internuclear axis. Bonding (σ_{2s}) and antibonding (σ_{2s}^*) MOs form, and the two valence electrons fill the bonding MO, with opposing spins (Figure 11.17A, next page). Dilithium has two electrons in bonding MOs and none in antibonding MOs; therefore, its bond order is $\frac{1}{2}(2 - 0) = 1$. In fact, Li_2 *has* been observed, and the MO electron configuration is $(\sigma_{2s})^2$.

With its two additional electrons, the MO diagram for Be_2 has filled σ_{2s} and σ_{2s}^* MOs (Figure 11.17B, next page). This is similar to the case of He_2. The bond order is $\frac{1}{2}(2 - 2) = 0$. In keeping with a zero bond order, the ground state of Be_2 has never been observed.

Figure 11.17 Bonding in *s*-block homonuclear diatomic molecules. Only outer (valence) AOs interact enough to form MOs. **A,** Li$_2$. The two valence electrons from two Li atoms fill the bonding (σ_{2s}) MO, and the antibonding (σ_{2s}^*) remains empty. With a bond order of 1, Li$_2$ does form. **B,** Be$_2$. The four valence electrons from two Be atoms fill both MOs to give no net stabilization. Ground-state Be$_2$ has a zero bond order and has never been observed.

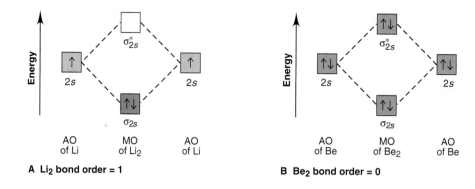

A Li$_2$ bond order = 1   **B Be$_2$ bond order = 0**

Molecular Orbitals from Atomic *p*-Orbital Combinations As we move to boron in the *p* block, atomic 2*p* orbitals become involved, so let's first consider the shapes and energies of the MOs that result from their combinations. Recall that *p* orbitals can overlap with each other in two different modes, as shown in Figure 11.18. End-to-end combination gives a pair of σ MOs, the σ_{2p} and σ_{2p}^*. Side-to-side combination gives a pair of **pi (π) MOs,** π_{2p} and π_{2p}^*. Similar to MOs formed from *s* orbitals, bonding MOs from *p*-orbital combinations have their greatest electron density *between* the nuclei, whereas antibonding MOs from *p*-orbital combinations have a node between the nuclei and most of their electron density *outside* the internuclear region.

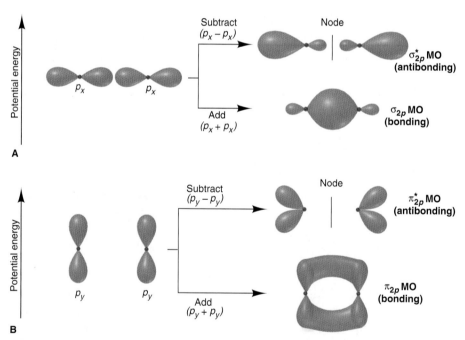

Figure 11.18 Contours and energies of σ and π MOs from combinations of 2*p* atomic orbitals. **A,** The *p* orbitals lying along the line between the atoms (usually designated *p$_x$*) undergo end-to-end overlap and form σ_{2p} and σ_{2p}^* MOs. Note the greater electron density between the nuclei for the bonding MO and the node between the nuclei for the antibonding MO. **B,** The *p* orbitals that lie perpendicular to the internuclear axis (*p$_y$* and *p$_z$*) undergo side-to-side overlap to form two π MOs. The *p$_z$* interactions are the same as those shown here for the *p$_y$* orbitals, so a total of four π MOs form. A π_{2p} is a bonding MO with its greatest density above and below the internuclear axis; a π_{2p}^* is an antibonding MO with a node between the nuclei and its electron density outside the internuclear region.

The order of MO energy levels, whether bonding or antibonding, is based on the order of AO energy levels *and* on the mode of the *p*-orbital combination:

- MOs formed from $2s$ orbitals are lower in energy than MOs formed from $2p$ orbitals because $2s$ AOs are lower in energy than $2p$ AOs.
- Bonding MOs are lower in energy than antibonding MOs, so σ_{2p} is lower in energy than σ_{2p}^* and π_{2p} is lower than π_{2p}^*.
- Atomic *p* orbitals can interact more extensively end to end than they can side to side. Thus, the σ_{2p} MO is lower in energy than the π_{2p} MO. Similarly, the destabilizing effect of the σ_{2p}^* MO is greater than that of the π_{2p}^* MO.

Thus, the energy order for MOs derived from $2p$ orbitals is

$$\sigma_{2p} < \pi_{2p} < \pi_{2p}^* < \sigma_{2p}^*$$

There are three mutually perpendicular $2p$ orbitals in each atom. When the six *p* orbitals in two atoms combine, the two orbitals that interact end to end form a σ and a σ^* MO, and the two pairs of orbitals that interact side to side form two π MOs of the same energy and two π^* MOs of the same energy. Combining these orientations with the energy order gives the *expected* MO diagram for the *p*-block Period 2 homonuclear diatomic molecules (Figure 11.19A).

One other factor influences the MO energy order. Recall that only AOs of similar energy interact to form MOs. The order in Figure 11.19A assumes that the *s* and *p* AOs are so different in energy that they do not interact with each other: the orbitals do not *mix*. This is true for O, F, and Ne. These atoms are small, so relatively strong repulsions occur as the $2p$ electrons pair up; these repulsions raise the energy of the $2p$ orbitals high enough above the energy of the $2s$ orbitals to minimize orbital mixing. In contrast, B, C, and N atoms are larger, and when the $2p$ AOs are half-filled, repulsions are relatively small; so the $2p$ energies are much closer to the $2s$ energy. As a result, some mixing occurs between the $2s$ orbital of one atom and the end-on $2p$ orbital of the other. This orbital mixing *lowers* the energy of the σ_{2s} and σ_{2s}^* MOs and raises

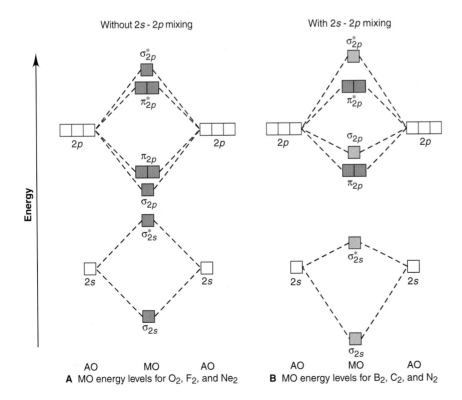

A MO energy levels for O_2, F_2, and Ne_2

B MO energy levels for B_2, C_2, and N_2

Figure 11.19 Relative MO energy levels for Period 2 homonuclear diatomic molecules. A, MO energy levels for O_2, F_2, and Ne_2. The six $2p$ orbitals of the two atoms form six MOs that are higher in energy than the two MOs formed from the two $2s$ orbitals. The AOs forming π orbitals give rise to two bonding MOs (π_{2p}) of equal energy and two antibonding MOs (π_{2p}^*) of equal energy. This sequence of energy levels arises from minimal $2s$-$2p$ orbital mixing. **B,** MO energy levels for B_2, C_2, and N_2. Because of significant $2s$-$2p$ orbital mixing, the energies of σ MOs formed from $2p$ orbitals increase and those formed from $2s$ orbitals decrease. The major effect of this orbital mixing on the MO sequence is that the σ_{2p} is higher in energy than the π_{2p}. (For clarity, those MOs affected by mixing are shown in purple.)

the energy of the σ_{2p} and σ_{2p}^* MOs; the π MOs are not affected. The MO diagram for B_2 through N_2 (Figure 11.19B) reflects this AO mixing. The only change that affects this discussion is the *reverse in energy order* of the σ_{2p} and π_{2p} MOs.

Bonding in the *p*-Block Homonuclear Diatomic Molecules Figure 11.20 shows the MOs, their electron occupancy, and some properties of B_2 through Ne_2. Note how *a higher bond order correlates with a greater bond energy and shorter bond length.* Also note how orbital occupancy correlates with magnetic properties. Recall from Chapter 8 that the spins of unpaired electrons in an atom (or ion) cause the substance to be *paramagnetic*, attracted to an external magnetic field. If all the electron spins are paired, the substance is *diamagnetic,* unaffected (or weakly repelled) by the magnetic field. The same observations apply to molecules. These properties are not addressed directly in VSEPR or VB theory.

The B_2 molecule has six outer electrons to place in its MOs. Four of these fill the σ_{2s} and σ_{2s}^* MOs. The remaining two electrons occupy the two π_{2p} MOs, one in each orbital, in keeping with Hund's rule. With four electrons in bonding MOs and two electrons in antibonding MOs, the bond order of B_2 is $\frac{1}{2}(4 - 2) = 1$. As expected from its MO diagram, B_2 is paramagnetic.

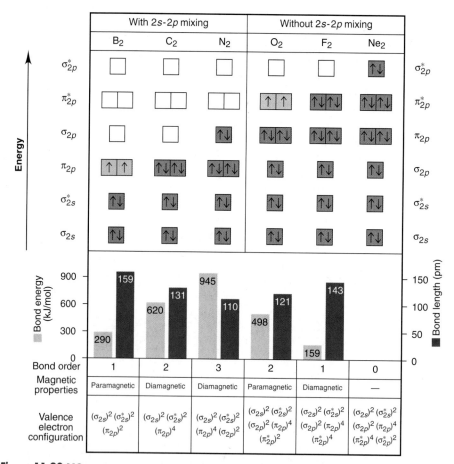

Figure 11.20 MO occupancy and molecular properties for B_2 through Ne_2. The sequence of MOs and their electron populations are shown for the homonuclear diatomic molecules in the *p* block of Period 2 [Groups 3A(13) to 8A(18)]. The bond energy, bond length, bond order, magnetic properties, and outer (valence) electron configuration appear below the orbital diagrams. Note the correlation between bond order and bond energy, both of which are inversely related to bond length.

The two additional electrons present in C_2 fill the two π_{2p} MOs. Because C_2 has two more bonding electrons than B_2, it has a bond order of 2 and the expected stronger, shorter bond. All the electrons are paired, and, as the model predicts, C_2 is diamagnetic.

In N_2, the two additional electrons fill the σ_{2p} MO. The resulting bond order is 3, which is consistent with the triple bond in the Lewis structure. As the model predicts, the bond energy is higher and the bond length shorter than for C_2, and N_2 is diamagnetic.

With O_2, we really see the power of MO theory compared to theories based on localized orbitals. For years, it seemed impossible to reconcile bonding theories with the bond strength and magnetic behavior of O_2. On the one hand, the data show a double-bonded molecule that is paramagnetic. On the other hand, we can write two possible Lewis structures for O_2, but neither gives such a molecule. One has a double bond and all electrons paired, the other a single bond and two electrons unpaired:

$$\ddot{\text{O}}=\ddot{\text{O}} \quad \text{or} \quad :\dot{\text{O}}-\dot{\text{O}}:$$

MO theory resolves this paradox beautifully. As Figure 11.20 shows, the bond order of O_2 is 2: eight electrons occupy bonding MOs and four occupy antibonding MOs $[\frac{1}{2}(8-4) = 2]$. Note the lower bond energy and greater bond length relative to N_2. The *two* electrons with highest energy occupy the *two* π^*_{2p} MOs with unpaired (parallel) spins, making the molecule paramagnetic. Figure 11.21 shows liquid O_2 suspended between the poles of a powerful magnet.

The two additional electrons in F_2 fill the π^*_{2p} orbitals, which decreases the bond order to 1, and the absence of unpaired electrons makes F_2 diamagnetic. As expected, the bond energy is lower and the bond distance longer than in O_2. Note that the bond energy for F_2 is only about half that for B_2, even though they have the same bond order. F is smaller than B, so we might expect a stronger bond. But the 18 electrons in the smaller volume of F_2 compared with the 10 electrons in B_2 cause greater repulsions and make the F_2 single bond easier to break.

The final member of the series, Ne_2, does not exist for the same reason that He_2 does not: all the MOs are filled, so the stabilization from bonding electrons cancels the destabilization from antibonding electrons, and the bond order is zero.

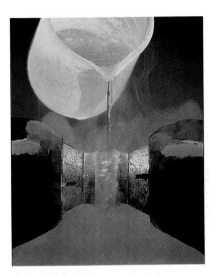

Figure 11.21 The paramagnetic properties of O_2. Liquid O_2 is attracted to the poles of a magnet because it is paramagnetic, as MO theory predicts. A diamagnetic substance would fall between the poles.

SAMPLE PROBLEM 11.4   Using MO Theory to Explain Bond Properties

Problem As the following data show, removing an electron from N_2 forms an ion with a weaker, longer bond than in the parent molecule, whereas the ion formed from O_2 has a stronger, shorter bond:

	N_2	N_2^+	O_2	O_2^+
Bond energy (kJ/mol)	945	841	498	623
Bond length (pm)	110	112	121	112

Explain these facts with diagrams that show the sequence and occupancy of MOs.
Plan We first determine the number of valence electrons in each species. Then, we draw the sequence of MO energy levels for the four species, recalling that they differ for N_2 and O_2 (see Figures 11.19 and 11.20), and fill them with electrons. Finally, we calculate bond orders and compare them with the data. Recall that bond order is related directly to bond energy and inversely to bond length.
Solution Determining the valence electrons:

N has 5 valence e^-, so N_2 has 10 and N_2^+ has 9
O has 6 valence e^-, so O_2 has 12 and O_2^+ has 11

Drawing and filling the MO diagrams:

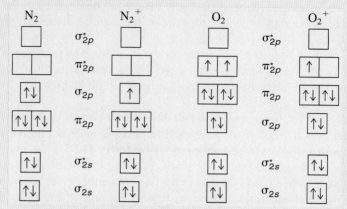

Calculating bond orders:

$$\tfrac{1}{2}(8-2)=3 \qquad \tfrac{1}{2}(7-2)=2.5 \qquad \tfrac{1}{2}(8-4)=2 \qquad \tfrac{1}{2}(8-3)=2.5$$

When N_2 forms N_2^+, a *bonding* electron is removed, so the bond order decreases. Thus, N_2^+ has a weaker, longer bond than N_2. When O_2 forms O_2^+, an *antibonding* electron is removed, so the bond order increases. Thus, O_2^+ has a stronger, shorter bond than O_2.
Check The answer makes sense in terms of the relationships among bond order, bond energy, and bond length. Check that the total number of bonding and antibonding electrons equals the number of valence electrons calculated.

FOLLOW-UP PROBLEM 11.4 Determine the bond orders for the following species: F_2^{2-}, F_2^-, F_2, F_2^+, F_2^{2+}. List the species in order of increasing bond energy and in order of increasing bond length.

SECTION SUMMARY

MO theory treats a molecule as a collection of nuclei with molecular orbitals delocalized over the entire structure. Atomic orbitals of comparable energy can be added and subtracted to obtain bonding and antibonding MOs, respectively. Bonding MOs have most of their electron density between the nuclei and are lower in energy than the atomic orbitals; most of the electron density of antibonding MOs does not lie between the nuclei, so these MOs are higher in energy. MOs are filled in order of their energy with paired electrons having opposing spins. MO diagrams show energy levels and orbital occupancy. The diagrams for homonuclear diatomic molecules of Period 2 explain observed bond energy, bond length, and magnetic behavior.

For Review and Reference (Numbers in parentheses refer to pages, unless noted otherwise.)

Learning Objectives

To help you review these learning objectives, the numbers of related sections (§), sample problems (SP), and upcoming end-of-chapter problems (EP) are listed in parentheses.

1. Describe how orbitals mix to form hybrid orbitals, and use the molecular shape from VSEPR theory to postulate the hybrid orbitals of a central atom (§ 11.1) (SP 11.1) (EPs 11.1–11.19)
2. Describe the modes of overlap that give sigma (σ) or pi (π) bonds, and explain the makeup of single and multiple bonds (§ 11.2) (SP 11.2) (EPs 11.20–11.24)

3. Understand how MOs arise from AOs; describe the shapes of MOs, and draw MO diagrams, with electron configurations and bond orders; and explain properties of homonuclear diatomic species from Periods 1 and 2 (§ 11.3) (SPs 11.3, 11.4) (EPs 11.25–11.36)

Key Terms

Section 11.1
valence bond (VB)
   theory (324)
hybridization (325)
hybrid orbital (325)
sp hybrid orbital (325)
sp^2 hybrid orbital (327)

sp^3 hybrid orbital (327)
sp^3d hybrid orbital (328)
sp^3d^2 hybrid orbital (329)
Section 11.2
sigma (σ) bond (331)
pi (π) bond (332)

Section 11.3
molecular orbital (MO)
   theory (334)
molecular orbital (MO) (334)
bonding MO (335)
antibonding MO (335)
sigma (σ) MO (336)

molecular orbital (MO)
   diagram (336)
MO bond order (336)
homonuclear diatomic
   molecule (337)
pi (π) MO (338)

Key Equations and Relationships

11.1 Calculating the MO bond order (336):   Bond order = $\frac{1}{2}$[(no. of e^- in bonding MO) − (no. of e^- in antibonding MO)]

Brief Solutions to Follow-up Problems

11.1 (a) Shape is linear, so Be is sp hybridized.

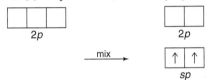

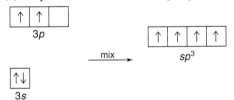

Isolated Be atom   Hybridized Be atom

(b) Shape is tetrahedral, so Si is sp^3 hybridized.

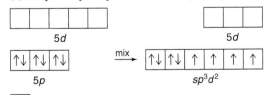

Isolated Si atom   Hybridized Si atom

(c) Shape is square planar, so Xe is sp^3d^2 hybridized.

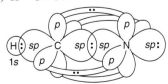

Isolated Xe atom   Hybridized Xe atom

11.2 (a) H—C≡N:

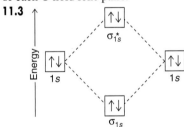

HCN is linear, so C is sp hybridized. N is also sp hybridized. One sp of C overlaps the $1s$ of H to form a σ bond. The other sp of C overlaps one sp of N to form a σ bond. The other sp of N holds a lone pair. Two unhybridized p orbitals of N and two of C overlap to form two π bonds.

(b) Ö=C=Ö

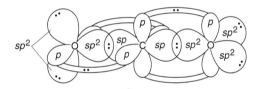

CO_2 is linear, so C is sp hybridized. Both O atoms are sp^2 hybridized. Each sp of C overlaps one sp^2 of an O to form two σ bonds. Each of the two unhybridized p orbitals of C forms a π bond with the unhybridized p of one of the two O atoms. Two sp^2 of each O hold lone pairs.

11.3

AO of H⁻   MO of $H_2{}^{2-}$ AO of H⁻

Does not exist: bond order = $\frac{1}{2}$(2 − 2) = 0; $(\sigma_{1s})^2(\sigma^*_{1s})^2$
11.4 Bond orders: $F_2{}^{2-}$ = 0; $F_2{}^-$ = $\frac{1}{2}$; F_2 = 1; $F_2{}^+$ = $1\frac{1}{2}$
            $F_2{}^{2+}$ = 2
Bond energy: $F_2{}^{2-}$ < $F_2{}^-$ < F_2 < $F_2{}^+$ < $F_2{}^{2+}$
Bond length: $F_2{}^{2+}$ < $F_2{}^+$ < F_2 < $F_2{}^-$; $F_2{}^{2-}$ does not exist

Problems

Problems with **colored** numbers are answered in Appendix E. Sections match the text and provide the numbers of relevant sample problems. Bracketed problems are grouped in pairs (indicated by a short rule) that cover the same concept. Comprehensive Problems are based on material from any section or previous chapter.

Valence Bond (VB) Theory and Orbital Hybridization
(Sample Problem 11.1)

11.1 What type of central-atom orbital hybridization corresponds to each electron-group arrangement: (a) trigonal planar; (b) octahedral; (c) linear; (d) tetrahedral; (e) trigonal bipyramidal?

11.2 What is the orbital hybridization of a central atom that has one lone pair and bonds to: (a) two other atoms; (b) three other atoms; (c) four other atoms; (d) five other atoms?

11.3 How do carbon and silicon differ with regard to the *types* of orbitals available for hybridization? Explain.

11.4 How many hybrid orbitals form when four atomic orbitals of a central atom mix? Explain.

11.5 Give the number and type of hybrid orbital that forms when each set of atomic orbitals mixes:
(a) two d, one s, and three p (b) three p and one s

11.6 Give the number and type of hybrid orbital that forms when each set of atomic orbitals mixes:
(a) one p and one s (b) three p, one d, and one s

11.7 What is the hybridization of nitrogen in each of the following: (a) NO; (b) NO_2; (c) NO_2^-?

11.8 What is the hybridization of carbon in each of the following: (a) CO_3^{2-}; (b) $C_2O_4^{2-}$; (c) NCO^-?

11.9 What is the hybridization of chlorine in each of the following: (a) ClO_2; (b) ClO_3^-; (c) ClO_4^-?

11.10 What is the hybridization of bromine in each of the following: (a) BrF_3; (b) BrO_2^-; (c) BrF_5?

11.11 Which types of atomic orbitals of the central atom mix to form hybrid orbitals in (a) $SiClH_3$; (b) CS_2?

11.12 Which types of atomic orbitals of the central atom mix to form hybrid orbitals in (a) Cl_2O; (b) $BrCl_3$?

11.13 Which types of atomic orbitals of the central atom mix to form hybrid orbitals in (a) SCl_3F; (b) NF_3?

11.14 Which types of atomic orbitals of the central atom mix to form hybrid orbitals in (a) PF_5; (b) SO_3^{2-}?

11.15 Use partial orbital diagrams to show how the atomic orbitals of the central atom lead to hybrid orbitals in (a) $GeCl_4$; (b) BCl_3; (c) CH_3^+.

11.16 Use partial orbital diagrams to show how the atomic orbitals of the central atom lead to hybrid orbitals in (a) BF_4^-; (b) PO_4^{3-}; (c) SO_3.

11.17 Use partial orbital diagrams to show how the atomic orbitals of the central atom lead to hybrid orbitals in (a) $SeCl_2$; (b) H_3O^+; (c) IF_4^-.

11.18 Use partial orbital diagrams to show how the atomic orbitals of the central atom lead to hybrid orbitals in (a) $AsCl_3$; (b) $SnCl_2$; (c) PF_6^-.

11.19 Methyl isocyanate, $CH_3\!-\!\ddot{N}\!=\!C\!=\!\ddot{O}:$, is an intermediate in the manufacture of many pesticides. It received notoriety in 1984 when a leak from a manufacturing plant resulted in the death of more than 2000 people in Bhopal, India. What are the hybridizations of the N atom and the two C atoms in methyl isocyanate? Sketch the molecular shape.

The Mode of Orbital Overlap and the Types of Covalent Bonds
(Sample Problem 11.2)

11.20 Are these statements true or false? Correct any that are false.
(a) Two σ bonds comprise a double bond.
(b) A triple bond consists of one π bond and two σ bonds.
(c) Bonds formed from atomic s orbitals are always σ bonds.
(d) A π bond restricts rotation about the σ-bond axis.
(e) A π bond consists of two pairs of electrons.
(f) End-to-end overlap results in a bond with electron density above and below the bond axis.

11.21 Describe the hybrid orbitals used by the central atom and the type(s) of bonds formed in (a) NO_3^-; (b) CS_2; (c) CH_2O.

11.22 Describe the hybrid orbitals used by the central atom and the type(s) of bonds formed in (a) O_3; (b) I_3^-; (c) $COCl_2$ (C is central).

11.23 Describe the hybrid orbitals used by the central atom(s) and the type(s) of bonds formed in (a) FNO; (b) C_2F_4; (c) $(CN)_2$.

11.24 Describe the hybrid orbitals used by the central atom(s) and the type(s) of bonds formed in (a) BrF_3; (b) $CH_3C\!\equiv\!CH$; (c) SO_2.

Molecular Orbital (MO) Theory and Electron Delocalization
(Sample Problems 11.3 and 11.4)

11.25 Two p orbitals from one atom and two p orbitals from another atom are combined to form molecular orbitals for the joined atoms. How many MOs will result from this combination? Explain.

11.26 Certain atomic orbitals on two atoms were combined to form the following MOs. Name the atomic orbitals used and the MOs formed, and explain which MO has higher energy:

11.27 How do the bonding and antibonding MOs formed from a given pair of AOs compare to each other with respect to (a) energy; (b) presence of nodes; (c) internuclear electron density?

11.28 Antibonding MOs always have at least one node. Can a bonding MO have a node? If so, draw an example.

11.29 How many electrons does it take to fill (a) a σ bonding MO; (b) a π antibonding MO; (c) the MOs formed from combination of the 1s orbitals of two atoms?

11.30 How many electrons does it take to fill (a) the MOs formed from combination of the 2p orbitals of two atoms; (b) a σ_{2p}^* MO; (c) the MOs formed from combination of the 2s orbitals of two atoms?

11.31 Show the shapes of bonding and antibonding MOs formed by combination of (a) an *s* orbital and a *p* orbital; (b) two *p* orbitals (end to end).

11.32 Show the shapes of bonding and antibonding MOs formed by combination of (a) two *s* orbitals; (b) two *p* orbitals (side to side).

11.33 Use MO diagrams and the bond orders you obtain from them to answer: (a) Is Be_2^+ stable? (b) Is Be_2^+ diamagnetic? (c) What is the outer (valence) electron configuration of Be_2^+?

11.34 Use MO diagrams and the bond orders you obtain from them to answer: (a) Is O_2^- stable? (b) Is O_2^- paramagnetic? (c) What is the outer (valence) electron configuration of O_2^-?

11.35 Use MO diagrams to place C_2^-, C_2, and C_2^+ in order of (a) increasing bond energy; (b) increasing bond length.

11.36 Use MO diagrams to place B_2^+, B_2, and B_2^- in order of (a) decreasing bond energy; (b) decreasing bond length.

Comprehensive Problems

Problems with an asterisk (*) are more challenging.

11.37 Predict the shape, state the hybridization of the central atom, and give the ideal bond angle(s) and any expected deviations for
(a) BrO_3^- (b) $AsCl_4^-$ (c) SeO_4^{2-} (d) BiF_5^{2-}
(e) SbF_4^+ (f) AlF_6^{3-} (g) IF_4^+

11.38 Butadiene (shown below) is a colorless gas used to make synthetic rubber and many other compounds:

How many σ bonds and π bonds does the molecule have?

11.39 Epinephrine (or adrenaline) is a naturally occurring hormone that is also manufactured commercially for use as a heart stimulant, a nasal decongestant, and to treat glaucoma. A valid Lewis structure is

(a) What is the hybridization of each C, O, and N atom?
(b) How many σ bonds does the molecule have?
(c) How many π electrons are delocalized in the ring?

11.40 Use partial orbital diagrams to show how the atomic orbitals of the central atom lead to the hybrid orbitals in
(a) IF_2^- (b) ICl_3 (c) $XeOF_4$ (d) BHF_2

11.41 Isoniazid is an antibacterial agent that is very useful against many common strains of tuberculosis. A valid Lewis structure is

(a) How many σ bonds are in the molecule?
(b) What is the hybridization of each C and N atom?

11.42 Hydrazine, N_2H_4, and carbon disulfide, CS_2, form a cyclic molecule with the following Lewis structure:

(a) Draw Lewis structures for N_2H_4 and CS_2.
(b) How do electron-group arrangement, molecular shape, and hybridization of N change when N_2H_4 reacts to form the product?
(c) How do electron-group arrangement, molecular shape, and hybridization of C change when CS_2 reacts to form the product?

11.43 In each of the following equations, what hybridization change, if any, occurs for the underlined atom?
(a) $\underline{B}F_3 + NaF \longrightarrow Na^+BF_4^-$
(b) $\underline{P}Cl_3 + Cl_2 \longrightarrow PCl_5$
(c) $H\underline{C}\equiv CH + H_2 \longrightarrow H_2C=CH_2$
(d) $\underline{Si}F_4 + 2F^- \longrightarrow SiF_6^{2-}$
(e) $\underline{S}O_2 + \frac{1}{2}O_2 \longrightarrow SO_3$

11.44 Glyphosate (*below*) is a common herbicide that is relatively harmless to animals but deadly to most plants. Describe the shape around and the hybridization of the P, N, and three numbered C atoms.

* **11.45** The sulfate ion can be represented with four S—O bonds or with two S—O and two S=O bonds. (a) Which representation is better from the standpoint of formal charges? (b) What is the shape of the sulfate ion, and what hybrid orbitals of S are postulated for the σ bonding? (c) In view of the answer to part (b), what orbitals of S must be used for the π bonds? What orbitals of O? (d) Draw a diagram to show how one atomic orbital from S and one from O overlap to form a π bond.

11.46 Tryptophan is one of the amino acids found in proteins:

(a) What is the hybridization of each of the numbered C, N, and O atoms?
(b) How many σ bonds are present in tryptophan?
(c) Predict the bond angles at points a, b, and c.

11.47 Sulfur forms oxides, oxoanions, and halides. What is the hybridization of the central S in SO_2, SO_3, SO_3^{2-}, SCl_4, SCl_6, and S_2Cl_2 (atom sequence Cl—S—S—Cl)?

* **11.48** The hydrocarbon allene, $H_2C=C=CH_2$, is obtained indirectly from petroleum and used as a precursor for several types of plastics. What is the hybridization of each C atom in allene? Draw a bonding picture for allene with lines for σ bonds, and show the arrangement of the π bonds. Be sure to represent the geometry of the molecule in three dimensions.

11.49 Some species with two oxygen atoms only are the oxygen molecule, O_2, the peroxide ion, O_2^{2-}, the superoxide ion, O_2^-, and the dioxygenyl ion, O_2^+. Draw an MO diagram for each, rank them in order of increasing bond length, and find the number of unpaired electrons in each.

11.50 There is concern in health-related government agencies that the American diet contains too much meat, and numerous recommendations have been made urging people to consume more fruit and vegetables. One of the richest sources of vegetable protein is soy, available in many forms. Among these is soybean curd, or tofu, which is a staple of many Asian diets. Chemists have isolated an anticancer agent called *genistein* from tofu, which may explain the much lower incidence of cancer among people in the Far East. A valid Lewis structure for genistein is

(a) Is the hybridization of each C in the right-hand ring the same? Explain.
(b) Is the hybridization of the O atom in the center ring the same as that of the O atoms in OH groups? Explain.
(c) How many carbon-oxygen σ bonds are there? How many carbon-oxygen π bonds?
(d) Do all the lone pairs on oxygens occupy the same type of hybrid orbital? Explain.

11.51 Silicon tetrafluoride reacts with F⁻ to produce the hexafluorosilicate ion, SiF_6^{2-}; GeF_4 behaves similarly, but CF_4 does

not. (a) Draw Lewis structures for SiF_4, GeF_6^{2-}, and CF_4.
(b) What is the hybridization of the central atom in each species?
(c) Why doesn't CF_4 react with F⁻ to form CF_6^{2-}?

11.52 The compound 2,6-dimethylpyrazine gives chocolate its odor and is used in flavorings. A valid Lewis structure is

(a) Which atomic orbitals mix to form the hybrid orbitals of N?
(b) In what type of hybrid orbital do the lone pairs of N reside?
(c) Is the hybridization of each C in a CH_3 group the same as that of each C in the ring? Explain.

11.53 Acetylsalicylic acid (aspirin), the most widely used medicine in the world, has the following Lewis structure:

(a) What is the hybridization of each C and each O atom?
(b) How many localized π bonds are present?
(c) How many C atoms have a trigonal planar shape around them? A tetrahedral shape?

Intermolecular Forces: Liquids, Solids, and Phase Changes

Holding Molecules Together Various combinations of attractive and repulsive forces among molecules create these spherical droplets and attach them to a stem. As you'll discover in this chapter, they also determine the properties of the three states of matter and their changes from one to another.

Key Principles

◆ The relative magnitudes of the energy of motion *(kinetic)* and the energy of attraction *(potential)* determine whether a substance exists as a gas, a liquid, or a solid under a given set of conditions: in a gas, kinetic energy is much greater than potential energy, whereas in a solid, the relative magnitudes are reversed.

◆ A *heating-cooling curve* shows a substance gaining or losing heat in a series of stages. Within a given state *(phase)*, adding or subtracting heat results in a change in temperature, but a change in state *(phase change)* occurs at constant temperature.

◆ Each type of phase change is *reversible* and has an enthalpy change that is positive in one direction and negative in the other. It takes more energy to convert a liquid to a gas than to convert a solid to a liquid, because the molecules are being more completely separated in the conversion to a gas.

◆ When a liquid vaporizes in a closed container at a given temperature, it reaches a state of *equilibrium* as the rates of *vaporization* and *condensation* become equal. At that point, the pressure of the gas *(vapor pressure)* becomes constant.

◆ A *phase diagram* for a substance shows the conditions of pressure and temperature at which each phase is stable and at which phase changes occur.

◆ Charged regions of one molecule can attract oppositely charged regions of another, and the strength of these nonbonding forces— *intermolecular forces*—determines many physical properties. *Hydrogen bonding* requires a specific arrangement of atoms; *dispersion forces* exist between all atoms, ions, and molecules.

◆ A combination of intermolecular forces determines the properties of liquids—*surface tension, capillarity,* and *viscosity.* The unique properties of water allow it to play vital roles in biology and the environment.

◆ Crystalline solids consist of particles tightly packed into a regular array called a *crystal lattice.* The *unit cell* is the simplest portion of the crystal that, when repeated, gives the crystal. Many substances crystallize in one of three types of cubic unit cells, which differ in the arrangement of the particles and, therefore, in the number of particles per unit cell and how efficiently they are packed.

◆ Solids have properties based on the type(s) of particles present— atoms, molecules, or ions—and the bonding or nonbonding forces among them. Properties of metals also depend on the number and *delocalization* of valence electrons.

Outline

Concepts & Skills to Review Before You Study This Chapter

- properties of gases, liquids, and solids (Section 5.1)
- kinetic-molecular theory of gases (Section 5.6)
- kinetic and potential energy (Section 6.1)
- enthalpy change, heat capacity, and Hess's law (Sections 6.2, 6.3, and 6.5)
- diffraction of light (Section 7.1)
- Coulomb's law (Section 9.2)
- chemical bonding models (Chapter 9)
- molecular polarity (Section 10.3)
- molecular orbital treatment of diatomic molecules (Section 11.3)

All the matter around you occurs in one of three physical states—gas, liquid, or solid. Under specific conditions, many pure substances can exist in any of the states. The three states were introduced in Chapter 1 and their properties compared when we examined gases in Chapter 5; now we turn our attention to liquids and solids. A physical state is one type of **phase,** any physically distinct, homogeneous part of a system. The water in a glass constitutes a single phase. Add some ice and you have two phases; or, if there are bubbles in the ice, you have three.

Liquids and solids are called *condensed* phases (or condensed states) because their particles are extremely close together. Electrostatic forces among the particles, called *interparticle forces* or, more commonly, **intermolecular forces,** combine with the particles' kinetic energy to create the properties of each phase as well as **phase changes,** the changes from one phase to another.

12.1 AN OVERVIEW OF PHYSICAL STATES AND PHASE CHANGES

Imagine yourself among the particles of a molecular substance, HF, and you'll discover two types of electrostatic forces at work:

- *Intra*molecular forces (bonding forces) exist *within* each molecule and influence the *chemical* properties of the substance.
- *Inter*molecular forces (nonbonding forces) exist *between* the molecules and influence the *physical* properties of the substance.

Now imagine a molecular view of the three states of water, as an example, and focus on just one molecule from each state. They look identical—bent, polar H—O—H molecules. In fact, the *chemical* behavior of the three states *is* identical because their molecules are held together by the same *intra*molecular bonding forces. However, the *physical* behavior of the states differs greatly because the strengths of the *inter*molecular nonbonding forces differ greatly.

A Kinetic-Molecular View of the Three States Whether a substance is a gas, liquid, or solid depends on the interplay of the potential energy of the intermolecular attractions, which tends to draw the molecules together, and the kinetic energy of the molecules, which tends to disperse them. According to Coulomb's law, the potential energy depends on the charges of the particles and the distances between them (see Section 9.2). The average kinetic energy, which is related to the particles' average speed, is proportional to the absolute temperature.

In Table 12.1, we distinguish among the three states by focusing on three properties—shape, compressibility, and ability to flow. The following kinetic-molecular view of the three states is an extension of the model we used to understand gases (see Section 5.6):

Table 12.1 A Macroscopic Comparison of Gases, Liquids, and Solids

State	Shape and Volume	Compressibility	Ability to Flow
Gas	Conforms to shape and volume of container	High	High
Liquid	Conforms to shape of container; volume limited by surface	Very low	Moderate
Solid	Maintains its own shape and volume	Almost none	Almost none

- *In a gas,* the energy of attraction is small relative to the energy of motion; so, on average, the particles are far apart. This large interparticle distance has several macroscopic consequences. A gas moves randomly throughout its container and fills it. Gases are highly compressible, and they flow and diffuse easily through one another.
- *In a liquid,* the attractions are stronger because the particles are in contact. But their kinetic energy still allows them to tumble randomly over and around each other. Therefore, a liquid conforms to the shape of its container but has a surface. With very little free space between the particles, liquids resist an applied external force and thus compress only very slightly. They flow and diffuse but *much* more slowly than gases.
- *In a solid,* the attractions dominate the motion so much that the particles remain in position relative to one another, jiggling in place. With the positions of the particles fixed, a solid has a specific shape and does not flow significantly. The particles are usually slightly closer together than in a liquid, so solids compress even less than liquids.

Types of Phase Changes Phase changes are also determined by the interplay between kinetic energy and intermolecular forces. As the temperature increases, the average kinetic energy increases as well, so the faster moving particles can overcome attractions more easily; conversely, lower temperatures allow the forces to draw the slower moving particles together.

What happens when gaseous water is cooled? A mist appears as the particles form tiny microdroplets that then collect into a bulk sample of liquid with a single surface. The process by which a gas changes into a liquid is called **condensation;** the opposite process, changing from a liquid into a gas, is called **vaporization.** With further cooling, the particles move even more slowly and become fixed in position as the liquid solidifies in the process of **freezing;** the opposite change is called **melting,** or **fusion.** In common speech, the term *freezing* implies low temperature because we typically think of water. But, most metals, for example, freeze (solidify) at much higher temperatures.

Enthalpy changes accompany phase changes. As the molecules of a gas attract each other and come closer together in the liquid, and then become fixed in the solid, the system of particles loses energy, which is released as heat. Thus, *condensing and freezing are exothermic changes.* On the other hand, energy must be absorbed by the system to overcome the attractive forces that keep the particles in a liquid together and those that keep them fixed in place in a solid. Thus, *melting and vaporizing are endothermic changes.* As a familiar example, the evaporation of sweat cools our bodies.

For a pure substance, each phase change has a specific enthalpy change *per mole* (measured at 1 atm and the temperature of the change). For vaporization, it is called the **heat of vaporization ($\Delta H^\circ_{\text{vap}}$),** and for fusion, it is the **heat of fusion ($\Delta H^\circ_{\text{fus}}$).** In the case of water, we have

$$H_2O(l) \longrightarrow H_2O(g) \qquad \Delta H = \Delta H^\circ_{\text{vap}} = 40.7 \text{ kJ/mol (at 100°C)}$$
$$H_2O(s) \longrightarrow H_2O(l) \qquad \Delta H = \Delta H^\circ_{\text{fus}} = 6.02 \text{ kJ/mol (at 0°C)}$$

The reverse processes, condensing and freezing, have enthalpy changes of the *same magnitude but opposite sign:*

$$H_2O(g) \longrightarrow H_2O(l) \qquad \Delta H = -\Delta H^\circ_{\text{vap}} = -40.7 \text{ kJ/mol}$$
$$H_2O(l) \longrightarrow H_2O(s) \qquad \Delta H = -\Delta H^\circ_{\text{fus}} = -6.02 \text{ kJ/mol}$$

Water is typical of most pure substances in that it takes less energy to melt 1 mol of solid water than to vaporize 1 mol of liquid water: $\Delta H^\circ_{\text{fus}} < \Delta H^\circ_{\text{vap}}$.

Figure 12.1 Heats of vaporization and fusion for several common substances. ΔH°_{vap} is always larger than ΔH°_{fus} because it takes more energy to separate particles completely than just to free them from their fixed positions in the solid.

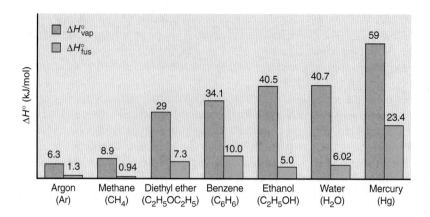

Figure 12.1 shows several examples. The reason is that a phase change is essentially a change in intermolecular distance and freedom of motion. Less energy is needed to overcome the forces holding the molecules in fixed positions (melt a solid) than to separate them completely from each other (vaporize a liquid).

The three states of water are so common because they all are stable under ordinary conditions. Carbon dioxide, on the other hand, is familiar as a gas and a solid (dry ice), but liquid CO_2 occurs only at external pressures greater than 5 atm. At ordinary conditions, solid CO_2 becomes a gas without first becoming a liquid. This process is called **sublimation.** Freeze-dried foods are prepared by sublimation. The opposite process, changing from a gas directly into a solid, is called **deposition**—you may have seen ice crystals form on a cold window from the deposition of water vapor. The **heat of sublimation (ΔH°_{subl})** is the enthalpy change when 1 mol of a substance sublimes. From Hess's law (Section 6.5), it equals the sum of the heats of fusion and vaporization:

$$
\begin{array}{ll}
\text{Solid} \longrightarrow \text{liquid} & \Delta H^{\circ}_{fus} \\
\underline{\text{Liquid} \longrightarrow \text{gas}} & \underline{\Delta H^{\circ}_{vap}} \\
\text{Solid} \longrightarrow \text{gas} & \Delta H^{\circ}_{subl}
\end{array}
$$

Figure 12.2 summarizes the terminology of the various phase changes and shows the enthalpy changes associated with them.

Figure 12.2 Phase changes and their enthalpy changes. Each type of phase change is shown with its associated enthalpy change. Fusion (or melting), vaporization, and sublimation are endothermic changes (positive ΔH°), whereas freezing, condensation, and deposition are exothermic changes (negative ΔH°).

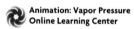
Animation: Vapor Pressure
Online Learning Center

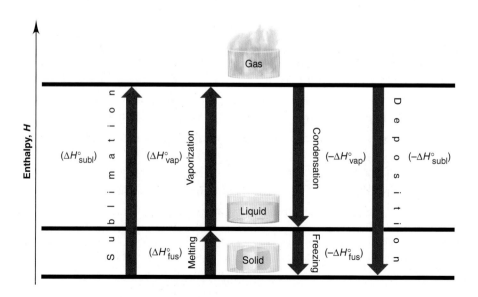

Because of the relative magnitudes of intermolecular forces and kinetic energy, the particles in a gas are far apart and moving randomly, those in a liquid are in contact but still moving relative to each other, and those in a solid are in contact and fixed relative to one another in a rigid structure. These molecular-level differences in the states of matter account for macroscopic differences in shape, compressibility, and ability to flow. When a solid becomes a liquid (melting, or fusion) or a liquid becomes a gas (vaporization), energy is absorbed to overcome intermolecular forces and increase the average distance between particles. When particles come closer together in the reverse changes (freezing and condensation), energy is released. Sublimation is the changing of a solid directly into a gas. Each phase change is associated with a given enthalpy change under specified conditions.

12.2 QUANTITATIVE ASPECTS OF PHASE CHANGES

In this section, we examine the heat absorbed or released in a phase change and the equilibrium nature of the process.

Heat Involved in Phase Changes: A Kinetic-Molecular Approach

We can apply the kinetic-molecular theory quantitatively to phase changes by means of a **heating-cooling curve,** which shows the changes that occur when heat is added to or removed from a particular sample of matter at a constant rate. As an example, the cooling process is depicted in Figure 12.3 for a 2.50-mol sample of gaseous water in a closed container, with the pressure kept at 1 atm and

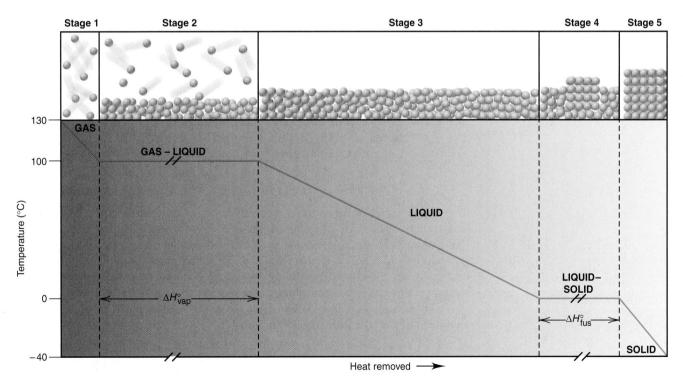

Figure 12.3 A cooling curve for the conversion of gaseous water to ice. A plot is shown of temperature vs. heat removed as gaseous water at 130°C changes to ice at −40°C. This process occurs in five stages, with a molecular-level depiction shown for each stage. Stage 1: Gaseous water cools. Stage 2: Gaseous water condenses. Stage 3: Liquid water cools. Stage 4: Liquid water freezes. Stage 5: Solid water cools. The slopes of the lines in stages 1, 3, and 5 reflect the magnitudes of the molar heat capacities of the phases. Although not drawn to scale, the line in stage 2 is longer than the line in stage 4 because ΔH°_{vap} of water is greater than ΔH°_{fus}. A plot of temperature vs. heat added would have the same steps but in reverse order.

the temperature changing from 130°C to −40°C. We can divide the process into the five heat-releasing (exothermic) stages of the curve:

Stage 1. Gaseous water cools. Picture a collection of water molecules behaving like a typical gas. At a high enough temperature, the most probable speed, and thus the average kinetic energy (E_k), of the molecules is high enough to overcome the potential energy (E_p) of attractions among them. As the temperature falls, the average E_k decreases, so the attractions become increasingly important.

The change is $H_2O(g)$ [130°C] ⟶ $H_2O(g)$ [100°C]. The heat (q) is the product of the amount (number of moles, n) of water, the molar heat capacity of *gaseous water,* $C_{water(g)}$, and the temperature change, ΔT ($T_{final} - T_{initial}$):

$$q = n \times C_{water(g)} \times \Delta T = (2.50 \text{ mol}) (33.1 \text{ J/mol·°C}) (100°C - 130°C)$$
$$= -2482 \text{ J} = -2.48 \text{ kJ}$$

The minus sign indicates that heat is released. (For purposes of canceling, the units for C include °C, rather than K, which represents the same temperature increment.)

Stage 2. Gaseous water condenses. At the condensation point, the slowest of the molecules are near each other long enough for intermolecular attractions to form groups of molecules, which aggregate into microdroplets and then a bulk liquid. Note that while the state is changing from gas to liquid, *the temperature remains constant,* so the average E_k is constant. However, removing heat from the system involves a decrease in the average E_p, as the molecules approach and attract each other more strongly. In other words, at 100°C, gaseous water and liquid water have the same average E_k, but the liquid has lower E_p.

The change is $H_2O(g)$ [100°C] ⟶ $H_2O(l)$ [100°C]. The heat released is the amount (n) times the negative of the heat of vaporization ($-\Delta H^\circ_{vap}$):

$$q = n(-\Delta H^\circ_{vap}) = (2.50 \text{ mol}) (-40.7 \text{ kJ/mol}) = -102 \text{ kJ}$$

This step contributes the *greatest portion of the total heat released* because of the decrease in potential energy that occurs with the enormous decrease in distance between molecules in a gas and those in a liquid.

Stage 3. Liquid water cools. The molecules are now in the liquid state. The continued loss of heat appears as a decrease in temperature, that is, as a decrease in the most probable molecular speed and, thus, the average E_k. The temperature decreases as long as the sample remains liquid.

The change is $H_2O(l)$ [100°C] ⟶ $H_2O(l)$ [0°C]. The heat depends on amount (n), the molar heat capacity of *liquid water,* and ΔT:

$$q = n \times C_{water(l)} \times \Delta T = (2.50 \text{ mol}) (75.4 \text{ J/mol·°C}) (0°C - 100°C)$$
$$= -18,850 \text{ J} = -18.8 \text{ kJ}$$

Stage 4. Liquid water freezes. At the freezing temperature of water, 0°C, intermolecular attractions overcome the motion of the molecules around one another. Beginning with the slowest, the molecules lose E_p and align themselves into the crystalline structure of ice. Molecular motion continues, but only as vibration of atoms about their fixed positions. As during condensation, the temperature and average E_k remain constant during freezing.

The change is $H_2O(l)$ [0°C] ⟶ $H_2O(s)$ [0°C]. The heat released is n times the negative of the heat of fusion ($-\Delta H^\circ_{fus}$):

$$q = n(-\Delta H^\circ_{fus}) = (2.50 \text{ mol}) (-6.02 \text{ kJ/mol}) = -15.0 \text{ kJ}$$

Stage 5. Solid water cools. With motion restricted to jiggling in place, further cooling merely reduces the average speed of this jiggling.

The change is $H_2O(s)$ [0°C] ⟶ $H_2O(s)$ [−40°C]. The heat released depends on n, the molar heat capacity of *solid water,* and ΔT:

$$q = n \times C_{water(s)} \times \Delta T = (2.50 \text{ mol}) (37.6 \text{ J/mol·°C}) (-40°C - 0°C)$$
$$= -3760 \text{ J} = -3.76 \text{ kJ}$$

According to Hess's law, the total heat released is the sum of the heats released for the individual stages. The sum of q for stages 1 to 5 is -142 kJ.

The Equilibrium Nature of Phase Changes

In everyday experience, phase changes take place in open containers—the out-doors, a pot on a stove, the freezer compartment of a refrigerator—so such a change is not reversible. In a closed container under controlled conditions, how-ever, *phase changes of many substances are reversible.*

Liquid-Gas Equilibria Picture an *open* flask containing a pure liquid at constant temperature and focus on the molecules at the surface. Within their range of molecular speeds, some are moving fast enough and in the right direction to over-come attractions, so they vaporize. Nearby molecules immediately fill the gap, and with energy supplied by the constant-temperature surroundings, the process continues until the entire liquid phase is gone.

Now picture starting with a *closed* flask at constant temperature, as in Figure 12.4A, and assume that a vacuum exists above the liquid. As before, some of the molecules at the surface have a high enough E_k to vaporize. As the number of molecules in the vapor phase increases, the pressure of the vapor increases. At the same time, some of the molecules in the vapor that collide with the surface have a low enough E_k to become attracted too strongly to leave the liquid and they condense. For a given surface area, the number of molecules that make up the surface is constant; therefore, the rate of vaporization—the number of mol-ecules leaving the surface per unit time—is also constant. On the other hand, as the vapor becomes more populated, molecules collide with the surface more often, so the rate of condensation slowly increases. As condensation continues to offset vaporization, the increase in the pressure of the vapor slows. Eventually, the rate of condensation equals the rate of vaporization, as depicted in Figure 12.4B. From this time onward, *the pressure of the vapor is constant at that temperature.* Macro-scopically, the situation seems static, but at the molecular level, molecules are entering and leaving the liquid surface at equal rates. The system has reached a state of **dynamic equilibrium:**

$$\text{Liquid} \rightleftharpoons \text{gas}$$

Figure 12.4C depicts the entire process graphically. The pressure exerted by the vapor at equilibrium is called the *equilibrium vapor pressure,* or just the **vapor pressure,** of the liquid at that temperature. (In later chapters, beginning with Chapter 17, you'll see that reactions also reach a state of equilibrium, in which reactants are changing into products and products into reactants at the same rate. Thus, the yield of product becomes constant, and it seems as if no changes are occurring.)

**Figure 12.4 Liquid-gas equilibrium.
A,** In a closed flask at constant tempera-ture with the air removed, the initial pressure is zero. As molecules leave the surface and enter the space above the liquid, the pressure of the vapor rises. **B,** At equilibrium, the same number of molecules leave as enter the liquid within a given time, so the pressure of the vapor reaches a constant value. **C,** A plot of pressure vs. time shows that the pressure of the vapor increases as long as the rate of vaporization is greater than the rate of condensation. At equilibrium, the rates are equal, so the pressure is constant. The pressure at this point is the *vapor pressure* of the liquid at that temperature.

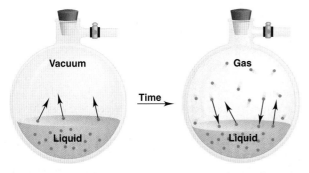

A Molecules in liquid vaporize.

B Molecules enter and leave liquid at same rate.

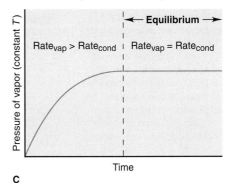

C

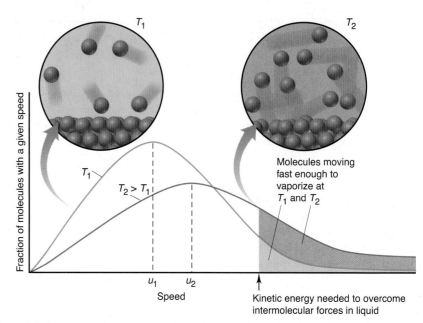

Figure 12.5 The effect of temperature on the distribution of molecular speeds in a liquid. With T_1 lower than T_2, the most probable molecular speed u_1 is less than u_2. (Note the similarity to Figure 5.12) The fraction of molecules with enough energy to escape the liquid *(shaded area)* is *greater at the higher temperature.* The molecular views show that at the higher T, equilibrium is reached with more gas molecules in the same volume and thus at a higher vapor pressure.

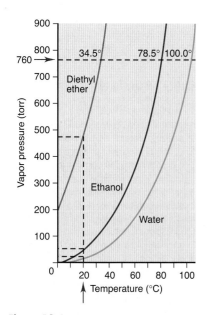

Figure 12.6 Vapor pressure as a function of temperature and intermolecular forces. The vapor pressures of three liquids are plotted against temperature. At any given temperature (see the vertical dashed line at 20°C), diethyl ether has the highest vapor pressure and water the lowest, because diethyl ether has the weakest intermolecular forces and water the strongest. The horizontal dashed line at 760 torr shows the normal boiling point of each liquid, the temperature at which the vapor pressure equals atmospheric pressure at sea level.

The Effects of Temperature and Intermolecular Forces on Vapor Pressure The vapor pressure of a substance depends on the temperature. Raising the temperature of a liquid *increases* the fraction of molecules moving fast enough to escape the liquid and *decreases* the fraction moving slowly enough to be recaptured. This important idea is shown in Figure 12.5. In general, *the higher the temperature is, the higher the vapor pressure.*

The vapor pressure also depends on the intermolecular forces present. The average E_k is the same for different substances at a given temperature. Therefore, molecules with weaker intermolecular forces vaporize more easily. In general, *the weaker the intermolecular forces are, the higher the vapor pressure.*

Figure 12.6 shows the vapor pressure of three liquids as a function of temperature. Notice that each curve rises more steeply as the temperature increases. Note also, at a given temperature, the substance with the weakest intermolecular forces has the highest vapor pressure: the intermolecular forces in diethyl ether are weaker than those in ethanol, which are weaker than those in water.

The nonlinear relationship between vapor pressure and temperature shown in Figure 12.6 can be expressed as a linear relationship between ln P and $1/T$:

$$\ln P = \frac{-\Delta H_{\text{vap}}}{R}\left(\frac{1}{T}\right) + C$$
$$y \quad = \quad m \quad x \quad + \quad b$$

where ln P is the natural logarithm of the vapor pressure, ΔH_{vap} is the heat of vaporization, R is the universal gas constant (8.31 J/mol·K), T is the absolute temperature, and C is a constant (not related to heat capacity). This is the **Clausius-Clapeyron equation,** which gives us a way of finding the heat of vaporization, the energy needed to vaporize 1 mol of molecules in the liquid state. The blue equation beneath the Clausius-Clapeyron equation is the equation for a straight line, where $y = \ln P$, $x = 1/T$, m (the slope) $= -\Delta H_{\text{vap}}/R$, and b (the y-axis intercept) $= C$. Figure 12.7 shows plots for diethyl ether and water. A two-point version of the equation allows a nongraphical determination of ΔH_{vap}:

$$\ln \frac{P_2}{P_1} = \frac{-\Delta H_{vap}}{R}\left(\frac{1}{T_2} - \frac{1}{T_1}\right) \tag{12.1}$$

If ΔH_{vap} and P_1 at T_1 are known, we can calculate the vapor pressure (P_2) at any other temperature (T_2) or the temperature at any other pressure.

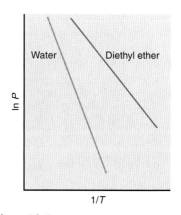

Figure 12.7 A linear plot of the relationship between vapor pressure and temperature. The Clausius-Clapeyron equation gives a straight line when the natural logarithm of the vapor pressure (ln P) is plotted against the inverse of the absolute temperature ($1/T$). The slopes ($-\Delta H_{vap}/R$) allow determination of the heats of vaporization of the two liquids. Note that the slope is steeper for water because its ΔH_{vap} is greater.

SAMPLE PROBLEM 12.1   Using the Clausius-Clapeyron Equation

Problem The vapor pressure of ethanol is 115 torr at 34.9°C. If ΔH_{vap} of ethanol is 40.5 kJ/mol, calculate the temperature (in °C) when the vapor pressure is 760 torr.

Plan We are given ΔH_{vap}, P_1, P_2, and T_1 and substitute them into Equation 12.1 to solve for T_2. The value of R here is 8.31 J/mol·K, so we must convert T_1 to K to obtain T_2, and then convert T_2 to °C.

Solution Substituting the values into Equation 12.1 and solving for T_2:

$$\ln \frac{P_2}{P_1} = \frac{-\Delta H_{vap}}{R}\left(\frac{1}{T_2} - \frac{1}{T_1}\right)$$

$$T_1 = 34.9°C + 273.15 = 308.0 \text{ K}$$

$$\ln \frac{760 \text{ torr}}{115 \text{ torr}} = \left(-\frac{40.5\times10^3 \text{ J/mol}}{8.314 \text{ J/mol·K}}\right)\left(\frac{1}{T_2} - \frac{1}{308.0 \text{ K}}\right)$$

$$1.888 = (-4.87\times10^3)\left[\frac{1}{T_2} - (3.247\times10^{-3})\right]$$

$$T_2 = 350. \text{ K}$$

Converting T_2 from K to °C:

$$T_2 = 350. \text{ K} - 273.15 = \boxed{77°C}$$

Check Round off to check the math. The change is in the right direction: higher P should occur at higher T. As we discuss next, a substance has a vapor pressure of 760 torr at its normal boiling point. Checking the *CRC Handbook of Chemistry and Physics* shows that the boiling point of ethanol is 78.5°C, very close to our answer.

FOLLOW-UP PROBLEM 12.1 At 34.1°C, the vapor pressure of water is 40.1 torr. What is the vapor pressure at 85.5°C? The ΔH_{vap} of water is 40.7 kJ/mol.

Vapor Pressure and Boiling Point In an *open* container, the atmosphere bears down on the liquid surface. As the temperature rises, molecules leave the surface more often, and they also move more quickly throughout the liquid. At some temperature, the average E_k of the molecules in the liquid is great enough for bubbles of vapor to form *in the interior,* and the liquid boils. At any lower temperature, the bubbles collapse as soon as they start to form because the external pressure is greater than the vapor pressure inside the bubbles. Thus, the **boiling point** is *the temperature at which the vapor pressure equals the external pressure,* usually that of the atmosphere. Once boiling begins, the temperature of the liquid remains constant until the liquid is gone, because the applied heat is used by the molecules to overcome attractions and enter the gas phase.

The boiling point varies with elevation because the atmospheric pressure does. At higher elevations, a lower pressure is exerted on the liquid surface, so molecules in the interior need less kinetic energy to form bubbles. At lower elevations, the opposite is true. Thus, *the boiling point depends on the applied pressure.* The *normal boiling point* is observed at standard atmospheric pressure (760 torr, or 101.3 kPa; see the horizontal dashed line in Figure 12.6).

The boiling point is the temperature at which the vapor pressure equals the external pressure, so we can also interpret the curves in Figure 12.6 as a plot of external pressure vs. boiling point. For instance, the H_2O curve shows that water boils at 100°C at 760 torr (sea level), at 94°C at 610 torr (Boulder, Colorado), and at about 72°C at 270 torr (top of Mt. Everest). People who live or hike in mountainous regions cook their meals under lower atmospheric pressure and the resulting lower boiling point of the liquid means that the food takes *more* time to cook.

Solid-Liquid Equilibria At the molecular level, the particles in a crystal are continually vibrating about their fixed positions. As the temperature rises, the particles vibrate more violently, until some have enough kinetic energy to break free of their positions, and melting begins. As more molecules enter the liquid (molten) phase, some collide with the solid and become fixed again. Because the phases remain in contact, a dynamic equilibrium is established when the melting rate equals the freezing rate. The temperature at which this occurs is the **melting point;** it is the same temperature as the freezing point, differing only in the direction of the energy flow. As with the boiling point, the temperature remains at the melting point as long as both phases are present.

Because liquids and solids are nearly incompressible, a change in pressure has little effect on the rate of movement to or from the solid. Therefore, in contrast to the boiling point, the melting point is affected by pressure only very slightly, and a plot of pressure (*y* axis) vs. temperature (*x* axis) for a solid-liquid phase change is typically a straight, *nearly* vertical line.

Solid-Gas Equilibria Solids have much lower vapor pressures than liquids. Sublimation, the process of a solid changing directly into a gas, is much less familiar than vaporization because the necessary conditions of pressure and temperature are uncommon for most substances. Some solids *do* have high enough vapor pressures to sublime at ordinary conditions, including dry ice (carbon dioxide), iodine, and solid room deodorizers. A substance sublimes rather than melts because the combination of intermolecular attractions and atmospheric pressure is not great enough to keep the particles near one another when they leave the solid state. The pressure vs. temperature plot for the solid-gas transition shows a large effect of temperature on the pressure of the vapor; thus, this curve resembles the liquid-gas line in curving upward at higher temperatures.

Phase Diagrams: Effect of Pressure and Temperature on Physical State

To describe the phase changes of a substance at various conditions of temperature and pressure, we construct a **phase diagram,** which combines the liquid-gas, solid-liquid, and solid-gas curves. The shape of the phase diagram for CO_2 is typical for most substances (Figure 12.8A). A phase diagram has these four features:

1. *Regions of the diagram.* Each region corresponds to one phase of the substance. A particular phase is stable for any combination of pressure and temperature within its region. If any of the other phases is placed under those conditions, it will change to the stable phase. In general, the solid is stable at low temperature and high pressure, the gas at high temperature and low pressure, and the liquid at intermediate conditions.

2. *Lines between regions.* The lines separating the regions represent the phase-transition curves discussed earlier. Any point along a line shows the pressure and temperature at which the two phases exist in equilibrium. Note that the solid-liquid line has a *positive* slope (slants to the *right* with increasing pressure) because, for most substances, the solid is more dense than the liquid. Because the liquid occupies slightly more space than the solid, an increase in pressure favors the solid phase. (Water is the major exception, as you'll soon see.)

3. *The critical point.* The liquid-gas line ends at the **critical point.** Picture a liquid in a closed container. As it is heated, it expands, so its density decreases.

Animation: Phase Diagrams of the States of Matter
Online Learning Center

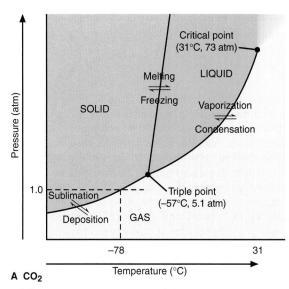

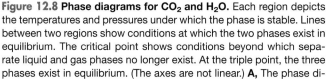

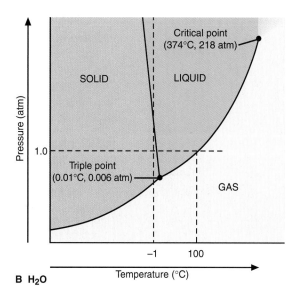

Figure 12.8 Phase diagrams for CO₂ and H₂O. Each region depicts the temperatures and pressures under which the phase is stable. Lines between two regions show conditions at which the two phases exist in equilibrium. The critical point shows conditions beyond which separate liquid and gas phases no longer exist. At the triple point, the three phases exist in equilibrium. (The axes are not linear.) **A,** The phase diagram for CO₂ is typical of most substances in that the solid-liquid line slopes to the right with increasing pressure: the solid is *more* dense than the liquid. **B,** Water is one of the few substances whose solid-liquid line slopes to the left with increasing pressure: the solid is *less* dense than the liquid. (The slopes of the solid-liquid lines in both diagrams are exaggerated.)

At the same time, more liquid vaporizes, so the density of the vapor increases. The liquid and vapor densities become closer and closer to each other until, at the *critical temperature* (T_c), the two densities are equal and the phase boundary disappears. The pressure at this temperature is the *critical pressure* (P_c). At this point, the average E_k of the molecules is so high that the vapor cannot be condensed no matter what pressure is applied.

4. *The triple point.* The three phase-transition curves meet at the **triple point:** the pressure and temperature at which three phases are in equilibrium. As strange as it sounds, at the triple point in Figure 12.8A, CO₂ is subliming and depositing, melting and freezing, and vaporizing and condensing simultaneously! Phase diagrams for substances with several solid forms, such as sulfur, have more than one triple point.

The CO₂ phase diagram explains why dry ice (solid CO₂) doesn't melt under ordinary conditions. The triple-point pressure for CO₂ is 5.1 atm; therefore, at around 1 atm, liquid CO₂ does not occur. By following the horizontal dashed line in Figure 12.8A, you can see that when solid CO₂ is heated at 1.0 atm, it sublimes at −78°C to gaseous CO₂ rather than melting. If normal atmospheric pressure were 5.2 atm, liquid CO₂ would be common.

The phase diagram for water differs in one key respect from the general case and reveals an extremely important property (Figure 12.8B). Unlike almost every other substance, solid water is *less dense* than liquid water. Because the solid occupies more space than the liquid, *water expands on freezing.* This behavior results from the unique open crystal structure of ice, which we discuss in a later section. As always, an increase in pressure favors the phase that occupies less space, but in the case of water, this is the *liquid* phase. Therefore, the solid-liquid line for water has a *negative* slope (slants to the *left* with increasing pressure): the higher the pressure, the lower the temperature at which water freezes. In Figure 12.8B, the vertical dashed line at −1°C crosses the solid-liquid line, which means that ice melts at that temperature with only an increase in pressure.

The triple point of water occurs at low pressure (0.006 atm). Therefore, when solid water is heated at 1.0 atm, the horizontal dashed line crosses the solid-liquid line (at 0°C, the normal melting point) and enters the liquid region. Thus, at ordinary pressures, ice melts rather than sublimes. As the temperature rises, the horizontal line crosses the liquid-gas curve (at 100°C, the normal boiling point) and enters the gas region.

SECTION SUMMARY

A heating-cooling curve depicts the change in temperature with heat gain or loss. Within a phase, temperature (and average E_k) changes as heat is added or removed. During a phase change, temperature (and average E_k) is constant, but E_p changes. The total heat change for the curve is calculated using Hess's law. In a closed container, equilibrium is established between the liquid and gas phases. Vapor pressure, the pressure of the gas at equilibrium, is related directly to temperature and inversely to the strength of the intermolecular forces. The Clausius-Clapeyron equation uses ΔH_{vap} to relate the vapor pressure to the temperature. A liquid in an open container boils when its vapor pressure equals the external pressure. Solid-liquid equilibrium occurs at the melting point. Many solids sublime at low pressures and high temperatures. A phase diagram shows the phase that exists at a given pressure and temperature and the conditions at the critical point and the triple point of a substance. Water differs from most substances in that its solid phase is less dense than its liquid phase, so its solid-liquid line has a negative slope.

12.3 TYPES OF INTERMOLECULAR FORCES

As we've seen, the nature of the phases and their changes are due primarily to forces among the molecules. Both bonding (intramolecular) forces and intermolecular forces arise from electrostatic attractions between opposite charges. Bonding forces are due to the attraction between cations and anions (ionic bonding), nuclei and electron pairs (covalent bonding), or metal cations and delocalized valence electrons (metallic bonding). Intermolecular forces, on the other hand, are due to the attraction between molecules as a result of partial charges, or the attraction between ions and molecules. The two types of forces differ in magnitude, and Coulomb's law explains why:

- *Bonding forces are relatively strong* because they involve larger charges that are closer together.
- *Intermolecular forces are relatively weak* because they typically involve smaller charges that are farther apart.

How far apart are the charges between molecules that give rise to intermolecular forces? Consider Cl_2 as an example. When we measure the distances between two Cl nuclei in a sample of solid Cl_2, we obtain two different values, as shown in Figure 12.9. The shorter distance is between *two **bonded** Cl atoms in the same*

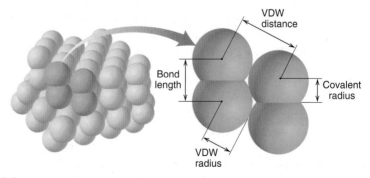

Figure 12.9 Covalent and van der Waals radii. As shown here for solid chlorine, the van der Waals (VDW) radius is one-half the distance between adjacent *nonbonded* atoms ($\frac{1}{2}$ × VDW distance), and the covalent radius is one-half the distance between *bonded* atoms ($\frac{1}{2}$ × bond length).

molecule. It is, as you know, called the *bond length,* and one-half this distance is the *covalent radius.* The longer distance is between *two **nonbonded** Cl atoms in adjacent molecules.* It is called the *van der Waals distance* (named after the Dutch physicist Johannes van der Waals, who studied the effects of intermolecular forces on the behavior of real gases). This distance is the closest one Cl_2 molecule can approach another, the point at which intermolecular attractions balance electron-cloud repulsions. One-half this distance is the **van der Waals radius,** one-half the closest distance between the nuclei of identical *nonbonded* Cl atoms. *The van der Waals radius of an atom is always larger than its covalent radius,* but van der Waals radii decrease across a period and increase down a group, just as covalent radii do. Figure 12.10 shows these relationships for many of the nonmetals.

There are several types of intermolecular forces: ion-dipole, dipole-dipole, hydrogen bonding, dipole–induced dipole, and dispersion forces. As we discuss these *intermolecular* forces (also called *van der Waals forces*), look at Table 12.2, which compares them with the stronger *intramolecular* (bonding) forces.

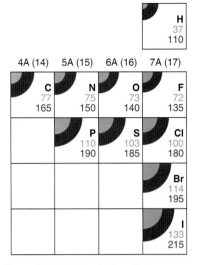

Figure 12.10 Periodic trends in covalent and van der Waals radii (in pm). Like covalent radii (*blue quarter-circles and top numbers*), van der Waals radii (*red quarter-circles and bottom numbers*) increase down a group and decrease across a period. The covalent radius of an element is *always* less than its van der Waals radius.

Table 12.2 **Comparison of Bonding and Nonbonding (Intermolecular) Forces**

Force	Model	Basis of Attraction	Energy (kJ/mol)	Example
Bonding				
Ionic		Cation–anion	400–4000	NaCl
Covalent		Nuclei–shared e^- pair	150–1100	H—H
Metallic		Cations–delocalized electrons	75–1000	Fe
Nonbonding (Intermolecular)				
Ion-dipole		Ion charge–dipole charge	40–600	$Na^+ \cdots O\begin{smallmatrix}H\\H\end{smallmatrix}$
H bond	$\delta-$ $\delta+$ $\delta-$ —A—H$\cdots$:B—	Polar bond to H–dipole charge (high EN of N, O, F)	10–40	:Ö—H$\cdots$:Ö—H
Dipole-dipole		Dipole charges	5–25	I—Cl$\cdots$I—Cl
Ion–induced dipole		Ion charge–polarizable e^- cloud	3–15	$Fe^{2+}\cdots O_2$
Dipole–induced dipole		Dipole charge–polarizable e^- cloud	2–10	H—Cl$\cdots$Cl—Cl
Dispersion (London)		Polarizable e^- clouds	0.05–40	F—F$\cdots$F—F

Ion-Dipole Forces

When an ion and a nearby polar molecule (dipole) attract each other, an **ion-dipole force** results. The most important example takes place when an ionic compound dissolves in water. The ions become separated because the attractions between the ions and the oppositely charged poles of the H_2O molecules overcome the attractions between the ions themselves. Ion-dipole forces in solutions and their associated energy are discussed fully in Chapter 13.

Dipole-Dipole Forces

In Figure 10.12, you saw how an external electric field can orient gaseous polar molecules. When polar molecules lie near one another, as in liquids and solids, their partial charges act as tiny electric fields that orient them and give rise to **dipole-dipole forces:** the positive pole of one molecule attracts the negative pole of another (Figure 12.11).

Figure 12.11 Polar molecules and dipole-dipole forces. In a solid or a liquid, the polar molecules are close enough for the partially positive pole of one molecule to attract the partially negative pole of a nearby molecule. The orientation is more orderly in the solid (*left*) than in the liquid (*right*) because, at the lower temperatures required for freezing, the average kinetic energy of the particles is lower. (Interparticle spaces are increased for clarity.)

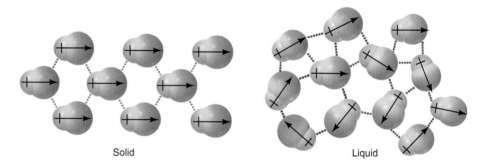

Solid   Liquid

These forces give polar molecules a higher boiling point than nonpolar molecules of similar molar mass. In fact, the greater the dipole moment, the greater the dipole-dipole forces between the molecules are, and so the more energy it takes to separate them. Consider the boiling points of the compounds in Figure 12.12, which have similar molar masses. Methyl chloride, for instance, has a smaller dipole moment than acetaldehyde, so less energy is needed to overcome the dipole-dipole forces between its molecules and it boils at a lower temperature.

Figure 12.12 Dipole moment and boiling point. For compounds of similar molar mass, the boiling point increases with increasing dipole moment. (Note also the increasing color intensities in the electron-density models.) The greater dipole moment creates stronger dipole-dipole forces, which require higher temperatures to overcome.

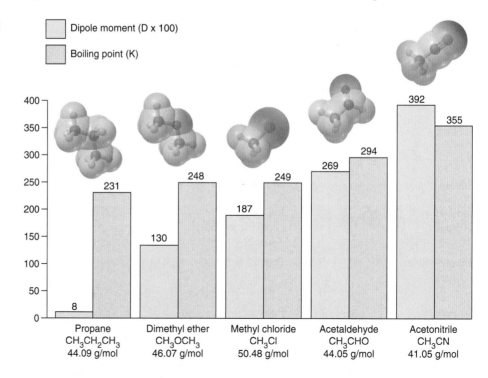

The Hydrogen Bond

A special type of dipole-dipole force arises between molecules that have *an H atom bonded to a small, highly electronegative atom with lone electron pairs.* The most important atoms that fit this description are N, O, and F. The H—N, H—O, and H—F bonds are very polar, so electron density is withdrawn from H. As a result, the partially positive H of one molecule is attracted to the partially negative lone pair on the N, O, or F of another molecule, and a **hydrogen bond (H bond)** forms. Thus, the atom sequence that allows an H bond (dotted line) to form is —B:····H—A—, where *both* A and B are N, O, or F. Three examples are

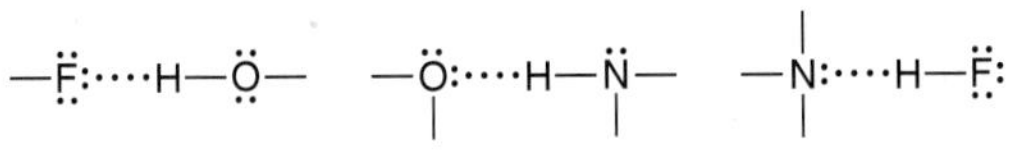

$$-\ddot{\text{F}}:\cdots\text{H}-\ddot{\text{O}}- \quad -\ddot{\text{O}}:\cdots\text{H}-\ddot{\text{N}}- \quad -\ddot{\text{N}}:\cdots\text{H}-\ddot{\text{F}}:$$

The small sizes of N, O, and F* are essential to H bonding for two reasons:

1. It makes these atoms so electronegative that their covalently bonded H is highly positive.
2. It allows the lone pair on the other N, O, or F to come close to the H.

SAMPLE PROBLEM 12.2   Drawing Hydrogen Bonds Between Molecules of a Substance

Problem Which of the following substances exhibits H bonding? For those that do, draw two molecules of the substance with the H bond(s) between them.

(a) C_2H_6 (b) CH_3OH (c) $CH_3\overset{\text{O}}{\overset{\|}{C}}-NH_2$

Plan We draw each structure to see if it contains N, O, or F covalently bonded to H. If it does, we draw two molecules of the substance in the —B:····H—A— pattern.

Solution (a) For C_2H_6. No H bonds are formed.

(b) For CH_3OH. The H covalently bonded to the O in one molecule forms an H bond to the lone pair on the O of an adjacent molecule:

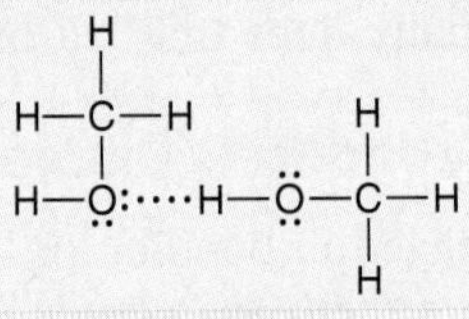

(c) For $CH_3\overset{\text{O}}{\overset{\|}{C}}-NH_2$. Two of these molecules can form one H bond between an H bonded to N and the O, or they can form two such H bonds:

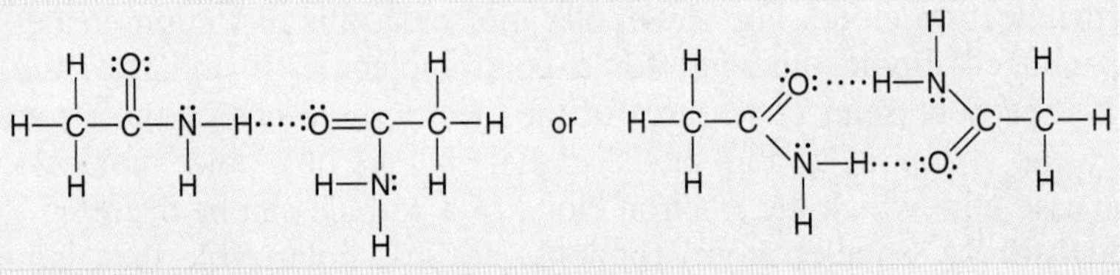

A third possibility (not shown) could be between an H attached to N in one molecule and the lone pair of N in another molecule.

Check The —B:····H—A— sequence (with A and B either N, O, or F) is present.

Comment Note that *H covalently bonded to C does not form H bonds* because carbon is not electronegative enough to make the C—H bond very polar.

FOLLOW-UP PROBLEM 12.2 Which of these substances exhibits H bonding? Draw the H bond(s) between two molecules of the substance where appropriate.

(a) $CH_3\overset{\text{O}}{\overset{\|}{C}}-OH$ (b) CH_3CH_2OH (c) $CH_3\overset{\text{O}}{\overset{\|}{C}}CH_3$

---

*H-bond–type interactions occur with the larger atoms P, S, and Cl, but those are so much weaker than the interactions with N, O, and F that we will not consider them.

Figure 12.13 Hydrogen bonding and boiling point. Boiling points of the binary hydrides of Groups 4A(14) to 7A(17) are plotted against period number. H bonds in NH_3, H_2O, and HF give them much higher boiling points than if the trend were based on molar mass, as it is for Group 4A. The red dashed line extrapolates to the boiling point H_2O would have if it formed no H bonds.

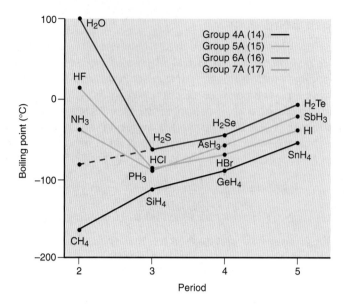

The Significance of Hydrogen Bonding

Hydrogen bonding has a profound impact in many systems. Here we'll examine one major effect it has on physical properties. Figure 12.13 shows the effect of H bonding on the boiling points of the binary hydrides of Groups 4A(14) through 7A(17). For reasons we'll discuss shortly, boiling points typically rise as molar mass increases, as you can see in the Group 4A hydrides, CH_4 through SnH_4. In the other groups, however, the first member in each series—NH_3, H_2O, and HF—deviates enormously from this expected increase. The H bonds between the molecules in these substances require additional energy to break before the molecules can separate and enter the gas phase. For example, on the basis of molar mass alone, we would expect water to boil about 200°C lower than it actually does (dashed line).

Polarizability and Charge-Induced Dipole Forces

Even though electrons are localized in bonding or lone pairs, they are in constant motion, and so we often picture them as "clouds" of negative charge. A nearby electric field can distort a cloud, pulling electron density toward a positive charge or pushing it away from a negative charge. In effect, the field *induces* a distortion in the electron cloud. For a nonpolar molecule, this distortion creates a temporary, induced dipole moment; for a polar molecule, it enhances the dipole moment already present. The source of the electric field can be the electrodes of a battery, the charge of an ion, or the partial charges of a polar molecule.

The ease with which the electron cloud of a particle can be distorted is called its **polarizability.** Smaller atoms (or ions) are less polarizable than larger ones because their electrons are closer to the nucleus and therefore are held more tightly. Thus, we observe several trends:

- *Polarizability increases down a group* because atomic size increases, so the larger electron clouds are farther from the nucleus and, thus, more easily distorted.
- *Polarizability decreases from left to right across a period* because the increasing Z_{eff} shrinks atomic size and holds the electrons more tightly.
- Cations are *less* polarizable than their parent atoms because they are smaller; anions are *more* polarizable because they are larger.

Ion–induced dipole and dipole–induced dipole forces are the two types of charge-induced dipole forces; they are most important in solution, so we'll focus on them in Chapter 13. Nevertheless, *polarizability affects all intermolecular forces.*

Dispersion (London) Forces

Polarizability plays the central role in the most universal intermolecular force. Up to this point, we've discussed forces that depend on an existing charge, of either an ion or a polar molecule. But what forces cause nonpolar substances like octane, chlorine, and argon to condense and solidify? Some force must be acting between the particles, or these substances would be gases under any conditions. *The intermolecular force primarily responsible for the condensed states of nonpolar substances is the* **dispersion force** (or **London force,** named for Fritz London, the physicist who explained the quantum-mechanical basis of the attraction).

Dispersion forces are caused by *momentary oscillations of electron charge in atoms and, therefore, are present between all particles (atoms, ions, and molecules).* Picture one atom in a sample of argon gas. Averaged over time, the 18 electrons are distributed uniformly around the nucleus, so the atom is nonpolar. But at any instant, there may be more electrons on one side of the nucleus than on the other, so the atom has an *instantaneous dipole.* When far apart, a pair of argon atoms do not influence each other. But when close together, *the instantaneous dipole in one atom induces a dipole in its neighbor.* The result is a synchronized motion of the electrons in the two atoms, which causes an attraction between them. This process occurs with other nearby atoms and, thus, throughout the sample. At low enough temperatures, the attractions among the dipoles keep all the atoms together. Thus, dispersion forces are *instantaneous dipole–induced dipole forces.* Figure 12.14 depicts the dispersion forces among nonpolar particles.

Figure 12.14 Dispersion forces among nonpolar particles. The dispersion force is responsible for the condensed states of noble gases and nonpolar molecules. **A,** Separated Ar atoms are nonpolar. **B,** An instantaneous dipole in one atom induces a dipole in its neighbor. These partial charges attract the atoms together. **C,** This process takes place among atoms throughout the sample.

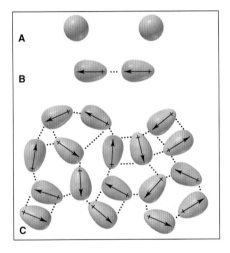

As we noted, the dispersion force is the *only* force existing between nonpolar particles. However, because they exist between *all* particles, *dispersion forces contribute to the overall energy of attraction of all substances.* In fact, except in cases involving small, polar molecules with large dipole moments or those forming strong H bonds, the *dispersion force is the dominant intermolecular force.* Calculations show, for example, that 85% of the total energy of attraction between HCl molecules is due to dispersion forces and only 15% to dipole-dipole forces. Even for water, estimates indicate that 75% of the total energy of attraction comes from the strong H bonds and nearly 25% from dispersion forces!

Dispersion forces are very weak for small particles, like H_2 and He, but much stronger for larger particles, like I_2 and Xe. The relative strength of the dispersion force depends on the polarizability of the particle. *Polarizability depends on the number of electrons, which correlates closely with molar mass* because heavier particles are either larger atoms or composed of more atoms and thus have more electrons. For example, molar mass increases down the halogens and the noble gases, so dispersion forces increase and so do boiling points (Figure 12.15).

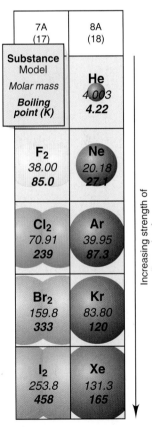

Figure 12.15 Molar mass and boiling point. The strength of dispersion forces increases with number of electrons, which usually correlates with molar mass. As a result, boiling points increase down the halogens and the noble gases.

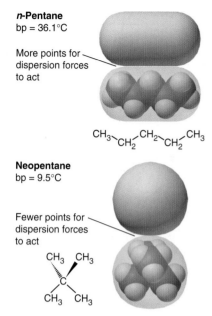

n-Pentane
bp = 36.1°C

More points for dispersion forces to act

CH₃—CH₂—CH₂—CH₂—CH₃

Neopentane
bp = 9.5°C

Fewer points for dispersion forces to act

Figure 12.16 Molecular shape and boiling point. Spherical neopentane molecules make less contact with each other than do cylindrical *n*-pentane molecules, so neopentane has a lower boiling point.

For nonpolar substances with the same molar mass, the strength of the dispersion forces is influenced by molecular shape. Shapes that allow more points of contact have more area over which electron clouds can be distorted, so stronger attractions result. Consider the two five-carbon alkanes, pentane (also called *n-pentane*) and 2,2-dimethylpropane (also called *neopentane*). These isomers have the same molecular formula (C_5H_{12}) but different shapes. *n*-Pentane is shaped like a cylinder, whereas neopentane has a more compact, spherical shape, as shown in Figure 12.16. Thus, two *n*-pentane molecules make more contact than do two neopentane molecules. Greater contact allows the dispersion forces to act at more points, so *n*-pentane has a higher boiling point. Figure 12.17 summarizes how to analyze the intermolecular forces in a sample.

SAMPLE PROBLEM 12.3 Predicting the Types of Intermolecular Force

Problem For each pair of substances, identify the key intermolecular force(s) in each substance, and select the substance with the higher boiling point:
(a) $MgCl_2$ or PCl_3
(b) CH_3NH_2 or CH_3F
(c) CH_3OH or CH_3CH_2OH
(d) Hexane ($CH_3CH_2CH_2CH_2CH_2CH_3$) or 2,2-dimethylbutane $\left(CH_3CCH_2CH_3 \text{ with } CH_3 \text{ groups} \right)$

Plan We examine the formulas and picture (or draw) the structures to identify key differences between members of each pair: Are ions present? Are molecules polar or nonpolar? Is N, O, or F bonded to H? Do the molecules have different masses or shapes? To rank the boiling points, we consult Table 12.2 and Figure 12.17 and remember that

- Bonding forces are stronger than intermolecular forces.
- Hydrogen bonding is a strong type of dipole-dipole force.
- Dispersion forces are always present, but they are decisive when the difference is molar mass or molecular shape.

Solution (a) $MgCl_2$ consists of Mg^{2+} and Cl^- ions held together by ionic bonding forces; PCl_3 consists of polar molecules, so intermolecular dipole-dipole forces are present. The forces in $MgCl_2$ are stronger, so it has a higher boiling point.
(b) CH_3NH_2 and CH_3F both consist of polar molecules of about the same molar mass. CH_3NH_2 has N—H bonds, so it can form H bonds (see margin). CH_3F contains a C—F bond but no H—F bond, so dipole-dipole forces occur but not H bonds. Therefore, CH_3NH_2 has the higher boiling point.
(c) CH_3OH and CH_3CH_2OH molecules both contain an O—H bond, so they can form H bonds (see margin). CH_3CH_2OH has an additional —CH_2— group and thus a larger molar mass, which correlates with stronger dispersion forces; therefore, it has a higher boiling point.
(d) Hexane and 2,2-dimethylbutane are nonpolar molecules of the same molar mass but different molecular shapes (see margin). Cylindrical hexane molecules can make more extensive intermolecular contact than the more spherical 2,2-dimethylbutane molecules can, so hexane should have greater dispersion forces and a higher boiling point.
Check The actual boiling points show that our predictions are correct:
(a) $MgCl_2$ (1412°C) and PCl_3 (76°C)
(b) CH_3NH_2 (−6.3°C) and CH_3F (−78.4°C)
(c) CH_3OH (64.7°C) and CH_3CH_2OH (78.5°C)
(d) Hexane (69°C) and 2,2-dimethylbutane (49.7°C)
Comment Dispersion forces are *always* present, but in parts (a) and (b), they are much less significant than the other forces involved.

FOLLOW-UP PROBLEM 12.3 In each pair, identify all the intermolecular forces present for each substance, and select the substance with the higher boiling point:
(a) CH_3Br or CH_3F (b) $CH_3CH_2CH_2OH$ or $CH_3CH_2OCH_3$ (c) C_2H_6 or C_3H_8

(b)

CH₃—N—H····N—CH₃ (with H atoms)

(c)

CH₃—O—H····O—CH₃ (with H atom)

CH₃CH₂—O—H····O—CH₂CH₃ (with H atom)

(d)

2,2-Dimethylbutane

Hexane

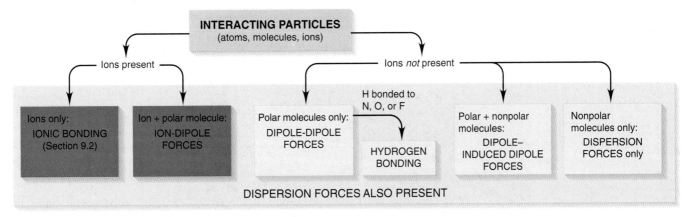

Figure 12.17 Summary diagram for analyzing the intermolecular forces in a sample.

SECTION SUMMARY

The van der Waals radius determines the shortest distance over which intermolecular forces operate; it is always larger than the covalent radius. Intermolecular forces are much weaker than bonding (intramolecular) forces. Ion-dipole forces occur between ions and polar molecules. Dipole-dipole forces occur between oppositely charged poles on polar molecules. Hydrogen bonding, a special type of dipole-dipole force, occurs when H bonded to N, O, or F is attracted to the lone pair of N, O, or F in another molecule. Electron clouds can be distorted (polarized) in an electric field. Dispersion (London) forces are instantaneous dipole–induced dipole forces that occur among all particles and increase with number of electrons (molar mass). Molecular shape determines the extent of contact between molecules and can be a factor in the strength of dispersion forces.

12.4 PROPERTIES OF THE LIQUID STATE

Of the three states of matter, the liquid is the least understood at the molecular level. Because of the *randomness* of the particles in a gas, any region of the sample is virtually identical to any other. As you'll see in Section 12.6, different regions of a crystalline solid are identical because of the *orderliness* of the particles. Liquids, however, have a combination of these attributes that changes continually: a region that is orderly one moment becomes random the next, and vice versa. Despite this complexity at the molecular level, the macroscopic properties of liquids are well understood. In this section, we discuss three liquid properties—surface tension, capillarity, and viscosity.

Surface Tension

In a liquid sample, intermolecular forces exert different effects on a molecule at the surface than on one in the interior (Figure 12.18). Interior molecules are attracted by others on all sides, whereas molecules at the surface have others only below and to the sides. As a result, molecules at the surface experience a *net attraction downward* and move toward the interior to increase attractions and become more stable. Therefore, *a liquid surface tends to have the smallest possible area,* that of a sphere, and behaves like a "taut skin" covering the interior.

To increase the surface area, molecules must move to the surface, thus breaking some attractions in the interior, which requires energy. The **surface tension** is the energy required to increase the surface area by a unit amount;

Figure 12.18 The molecular basis of surface tension. Molecules in the interior of a liquid experience intermolecular attractions in all directions. Molecules at the surface experience a net attraction downward (*red arrow*) and move toward the interior. Thus, a liquid tends to minimize the number of molecules at the surface, which results in surface tension.

Table 12.3	**Surface Tension and Forces Between Particles**		
Substance	Formula	Surface Tension (J/m^2) at 20°C	Major Force(s)
Diethyl ether	$CH_3CH_2OCH_2CH_3$	1.7×10^{-2}	Dipole-dipole; dispersion
Ethanol	CH_3CH_2OH	2.3×10^{-2}	H bonding
Butanol	$CH_3CH_2CH_2CH_2OH$	2.5×10^{-2}	H bonding; dispersion
Water	H_2O	7.3×10^{-2}	H bonding
Mercury	Hg	48×10^{-2}	Metallic bonding

units of J/m^2 are shown in Table 12.3. Comparing these values with those in Table 12.2 shows that, in general, *the stronger the forces are between the particles in a liquid, the greater the surface tension.* Water has a high surface tension because its molecules form multiple H bonds. *Surfactants (surface-active agents)*, such as soaps, petroleum recovery agents, and biological fat emulsifiers, decrease the surface tension of water by congregating at the surface and disrupting the H bonds.

Capillarity

The rising of a liquid through a narrow space against the pull of gravity is called *capillary action,* or **capillarity.** In blood-screening tests, a narrow *capillary tube* is held against the skin opening made by pricking a finger. Capillarity results from a competition between the intermolecular forces within the liquid (cohesive forces) and those between the liquid and the tube walls (adhesive forces).

Picture what occurs at the molecular level when you place a glass capillary tube in water. Glass is mostly silicon dioxide (SiO_2), so the water molecules form H bonds to the oxygen atoms of the tube's inner wall. Because the adhesive forces (H bonding) between the water and the wall are stronger than the cohesive forces (H bonding) within the water, a thin film of water creeps up the wall. At the same time, the cohesive forces that give rise to surface tension pull the liquid surface taut. These adhesive and cohesive forces combine to raise the water level and produce the familiar concave meniscus (Figure 12.19A). The liquid rises until gravity pulling down is balanced by the adhesive forces pulling up.

Figure 12.19 Shape of water or mercury meniscus in glass. A, Water displays a concave meniscus in a glass tube because the adhesive (H-bond) forces between the H_2O molecules and the O—Si—O groups of the glass are *stronger* than the cohesive (H-bond) forces within the water. **B,** Mercury displays a convex meniscus in a glass tube because the cohesive (metallic bonding) forces within the mercury are *stronger* than the adhesive (dispersion) forces between the mercury and the glass.

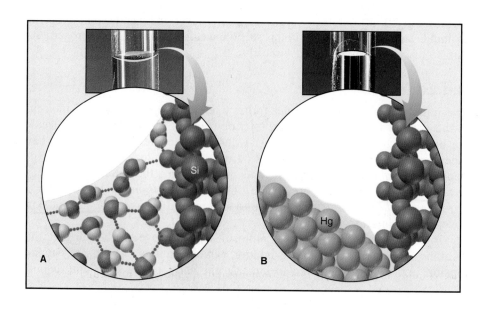

On the other hand, if you place a glass capillary tube in a dish of mercury, the mercury level in the tube drops below that in the dish. Mercury has a higher surface tension than water (see Table 12.3), which means it has stronger cohesive forces (metallic bonding). The cohesive forces among the mercury atoms are much stronger than the adhesive forces (mostly dispersion) between mercury and glass, so the liquid tends to pull away from the walls. At the same time, the surface atoms are being pulled toward the interior of the mercury by its high surface tension, so the level drops. These combined forces produce a convex meniscus (Figure 12.19B, seen in a laboratory barometer).

Viscosity

When a liquid flows, the molecules slide around and past each other. A liquid's **viscosity,** its resistance to flow, results from intermolecular attractions that impede this movement. Both gases and liquids flow, but liquid viscosities are *much* higher because the intermolecular forces operate over much shorter distances.

Viscosity decreases with heating, as Table 12.4 shows for water. When molecules move faster at higher temperatures, they can overcome intermolecular forces more easily, so the resistance to flow decreases. Next time you put cooking oil in a pan, watch the oil flow more easily and spread out in a thin layer as you heat it.

Molecular shape plays a key role in a liquid's viscosity. Small, spherical molecules make little contact and, like buckshot in a glass, pour easily. Long molecules make more contact and, like spaghetti in a glass, become entangled and pour slowly. Thus, given the same types of forces, liquids containing longer molecules have higher viscosities. A striking example of a change in viscosity occurs during the making of syrup. Even at room temperature, a concentrated aqueous sugar solution has a higher viscosity than water because of H bonding among the many hydroxyl (—OH) groups on the ring-shaped sugar molecules. When the solution is slowly heated to boiling, the sugar molecules react with each other and link covalently, gradually forming long chains. Hydrogen bonds and dispersion forces occur at many points along the chains, and the resulting syrup is a viscous liquid that pours slowly and clings to a spoon. When a viscous syrup is cooled, it may become stiff enough to be picked up and stretched—into taffy candy.

Table 12.4	Viscosity of Water at Several Temperatures
Temperature (°C)	**Viscosity $(N \cdot s/m^2)$***
20	1.00×10^{-3}
40	0.65×10^{-3}
60	0.47×10^{-3}
80	0.35×10^{-3}

*The units of viscosity are newton-seconds per square meter.

SECTION SUMMARY

Surface tension is a measure of the energy required to increase a liquid's surface area. Greater intermolecular forces within a liquid create higher surface tension. Capillary action, the rising of a liquid through a narrow space, occurs when the forces between a liquid and a solid surface (adhesive) are greater than those within the liquid itself (cohesive). Viscosity, the resistance to flow, depends on molecular shape and decreases with temperature. Stronger intermolecular forces create higher viscosity.

12.5 THE UNIQUENESS OF WATER

Water has some of the most unusual properties of any substance. These properties, which are vital to our very existence, arise inevitably from the nature of the H and O atoms that make up the molecule. With two bonding pairs and two lone pairs around the O atom and a large electronegativity difference in each O—H bond, the H_2O molecule is bent and highly polar. This arrangement is crucial because it allows each water molecule to engage in four H bonds with its neighbors (Figure 12.20). From these fundamental atomic and molecular properties emerges unique and remarkable macroscopic behavior.

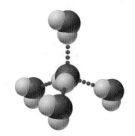

Figure 12.20 The H-bonding ability of the water molecule. Because it has two O—H bonds and two lone pairs, one H_2O molecule can engage in as many as four H bonds to surrounding H_2O molecules, which are arranged tetrahedrally.

Solvent Properties of Water

The *great solvent power* of water is the result of its polarity and exceptional H-bonding ability. It dissolves ionic compounds through ion-dipole forces that separate the ions from the solid and keep them in solution (see Figure 4.2). Water dissolves many polar nonionic substances, such as ethanol (CH_3CH_2OH) and glucose ($C_6H_{12}O_6$), by forming H bonds to them. It even dissolves nonpolar gases, such as those in the atmosphere, to a limited extent, through dipole–induced dipole and dispersion forces. Chapter 13 highlights these solvent properties.

Because it can dissolve so many substances, water is the environmental and biological solvent, forming the complex solutions we know as oceans, rivers, lakes, and cellular fluids. Indeed, from a chemical point of view, all organisms, from bacteria to humans, can be thought of as highly organized systems of membranes enclosing and compartmentalizing complex aqueous solutions.

Thermal Properties of Water

Water has an exceptionally *high specific heat capacity,* higher than almost any other liquid. Recall from Section 6.3 that heat capacity is a measure of the heat absorbed by a substance for a given temperature rise. When a substance is heated, some of the energy increases the average molecular speed, some increases molecular vibration and rotation, and some is used to overcome intermolecular forces. Because water has so many strong H bonds, it has a high specific heat capacity. With oceans and seas covering 70% of Earth's surface, daytime heat from the Sun causes relatively small changes in temperature, allowing our planet to support life.

Numerous strong H bonds also give water an exceptionally *high heat of vaporization.* A quick calculation shows how essential this property is to our existence. When 1000 g of water absorbs about 4 kJ of heat, its temperature rises 1°C. With a ΔH°_{vap} of 2.3 kJ/g, however, less than 2 g of water must evaporate to keep the temperature constant for the remaining 998 g. This high heat of vaporization results in a stable body temperature and minimal loss of body fluid despite the 10,000 kJ of heat that the average adult generates each day through metabolism. On a planetary scale, the Sun's energy supplies the heat of vaporization for ocean water. Water vapor, formed in warm latitudes, moves through the atmosphere, and its potential energy is released as heat to warm cooler regions when the vapor condenses to rain. The enormous energy involved in this cycling of water powers storms all over the planet.

Surface Properties of Water

The H bonds that give water its remarkable thermal properties are also responsible for its *high surface tension* and *high capillarity.* Except for some molten metals and salts, water has the highest surface tension of any liquid. This property is vital for surface aquatic life because it keeps plant debris resting on a pond surface, which provides shelter and nutrients for many fish, microorganisms, and insects. Water's high capillarity, a result of its high surface tension, is crucial to land plants. During dry periods, plant roots absorb deep groundwater, which rises by capillary action through the tiny spaces between soil particles.

The Density of Solid and Liquid Water

As you saw in Figure 12.20, through the H bonds of water, each O atom becomes connected to as many as four other O atoms via four H atoms. Continuing this tetrahedral pattern through many molecules in a fixed array leads to the hexagonal, *open structure* of ice shown in Figure 12.21A. The symmetrical beauty of snowflakes, as shown in Figure 12.21B, reflects this hexagonal organization.

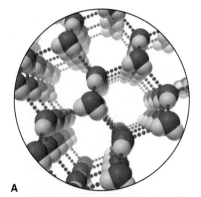

A

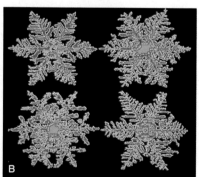

B

Figure 12.21 The hexagonal structure of ice. A, The geometric arrangement of the H bonds in H_2O leads to the open, hexagonally shaped crystal structure of ice. Thus, when liquid water freezes, the volume *increases.* **B,** The delicate six-pointed beauty of snowflakes reflects the hexagonal crystal structure of ice.

This organization explains the negative slope of the solid-liquid line in the phase diagram for water (see Figure 12.8B): As pressure is applied, some H bonds break; as a result, some water molecules enter the spaces. The crystal structure breaks down, and the sample liquefies.

The large spaces within ice give *the solid state a lower density than the liquid state.* When the surface of a lake freezes in winter, the ice floats on the liquid water below. If the solid were denser than the liquid, as is true for nearly every other substance, the surface of a lake would freeze and sink repeatedly until the entire lake was solid. As a result, aquatic plant and animal life would not survive from year to year.

The density of water changes in a complex way. When ice melts at 0°C, the tetrahedral arrangement around each O atom breaks down, and the loosened molecules pack much more closely, filling spaces in the collapsing solid structure. As a result, water is most dense (1.000 g/mL) at around 4°C (3.98°C). With more heating, the density decreases through normal thermal expansion.

This change in density is vital for freshwater life. As lake water becomes colder in the fall and early winter, it becomes more dense *before* it freezes. Similarly, in spring, less dense ice thaws to form more dense water *before* the water expands. During both of these seasonal density changes, the top layer of water reaches the high-density point first and sinks. The next layer of water rises because it is slightly less dense, reaches 4°C, and likewise sinks. This sinking and rising distribute nutrients and dissolved oxygen.

SECTION SUMMARY

The atomic properties of hydrogen and oxygen atoms result in the water molecule's bent shape, polarity, and H-bonding ability. These properties of water enable it to dissolve many ionic and polar compounds. Water's H bonding results in a high specific heat capacity and a high heat of vaporization, which combine to give Earth and its organisms a narrow temperature range. H bonds also confer high surface tension and capillarity, which are essential to plant and animal life. Water expands on freezing because of its H bonds, which lead to an open crystal structure in ice. Seasonal density changes foster nutrient mixing in lakes.

12.6 THE SOLID STATE: STRUCTURE, PROPERTIES, AND BONDING

Stroll through the mineral collection of any school or museum, and you'll be struck by the extraordinary variety and beauty of these solids. In this section, we first discuss the general structural features of crystalline solids and then examine a laboratory method for studying them. We survey the properties of the major types of solids and find the whole range of intermolecular forces at work. We then present a model for bonding in solids that explains many of their properties.

Structural Features of Solids

We can divide solids into two broad categories based on the orderliness of their shapes, which in turn is based on the orderliness of their particles. **Crystalline solids** generally have a well-defined shape, as shown in Figure 12.22, because their particles—atoms, molecules, or ions—occur in an orderly arrangement. **Amorphous solids** have poorly defined shapes because their particles lack long-range ordering throughout the sample. In this discussion, we focus, for the most part, on crystalline solids.

Figure 12.22 The striking beauty of crystalline solids. A, Pyrite. **B,** Beryl. **C,** Barite (*left*) on calcite (*right*).

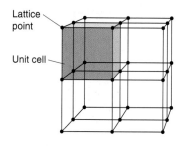

A Portion of a 3-D lattice

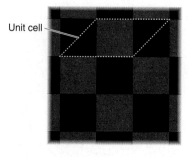

B 2-D analogy for unit cell and lattice

Figure 12.23 The crystal lattice and the unit cell. A, The lattice is an array of points that defines the positions of the particles in a crystal structure. It is shown here as points connected by lines. A unit cell *(colored)* is the simplest array of points that, when repeated in all directions, produces the lattice. A simple cubic unit cell, one of 14 types in nature, is shown. **B,** A checkerboard is a two-dimensional analogy for a lattice.

The Crystal Lattice and the Unit Cell If you could see the particles within a crystal, you would find them packed tightly together in an orderly, three-dimensional array. Consider the simplest case, in which all the particles are *identical* spheres, and imagine a point at the same location within each particle in this array, say, at the center. The points form a regular pattern throughout the crystal that is called the crystal **lattice.** Thus, *the lattice consists of all points with identical surroundings.* Put another way, if the rest of each particle were removed, leaving only the lattice point, and you were transported from one point to another, you would not be able to tell that you had moved. Keep in mind that there is no pre-existing array of lattice points; rather, *the arrangement of the points within the particles defines the lattice.*

Figure 12.23A shows a portion of a lattice and the **unit cell,** the *smallest* portion of the crystal that, if repeated in all three directions, gives the crystal. A two-dimensional analogy for a unit cell and a crystal lattice can be seen in a checkerboard (as shown in Figure 12.23B), a section of tiled floor, a strip of wallpaper, or any other pattern that is constructed from a repeating unit. The **coordination number** of a particle in a crystal is the number of nearest neighbors surrounding it.

There are 7 crystal systems and 14 types of unit cells that occur in nature, but we will be concerned primarily with the *cubic system,* which gives rise to the cubic lattice. The solid states of a majority of metallic elements, some covalent compounds, and many ionic compounds occur as cubic lattices. (We also describe the hexagonal unit cell a bit later.) There are three types of cubic unit cells within the cubic system:

1. In the **simple cubic unit cell,** shown in Figure 12.24A, the centers of eight identical particles define the corners of a cube. Attractions pull the particles together, so they touch along the cube's edges; but they do not touch diagonally along the cube's faces or through its center. The coordination number of each particle is 6: four in its own layer, one in the layer above, and one in the layer below.

2. In the **body-centered cubic unit cell,** shown in Figure 12.24B, identical particles lie at each corner *and* in the center of the cube. Those at the corners do not touch each other, but they all touch the one in the center. Each particle is surrounded by eight nearest neighbors, four above and four below, so the coordination number is 8.

3. In the **face-centered cubic unit cell,** shown in Figure 12.24C, identical particles lie at each corner *and* in the center of each face but not in the center of the cube. Those at the corners touch those in the faces but not each other. The coordination number is 12.

One unit cell lies adjacent to another throughout the crystal, with no gaps, so a particle at a corner or face is *shared* by adjacent unit cells. As you can see from Figure 12.24 (third row from the top), in the three cubic unit cells, the particle at each corner is part of eight adjacent cells, so one-eighth of each of these particles belongs to each unit cell (bottom row). There are eight corners in a cube, so each simple cubic unit cell contains $8 \times \frac{1}{8}$ particle = 1 particle. The body-centered cubic unit cell contains one particle from the eight corners and one in the center, for a total of two particles; and the face-centered cubic unit cell contains four particles, one from the eight corners and three from the half-particles in each of the six faces.

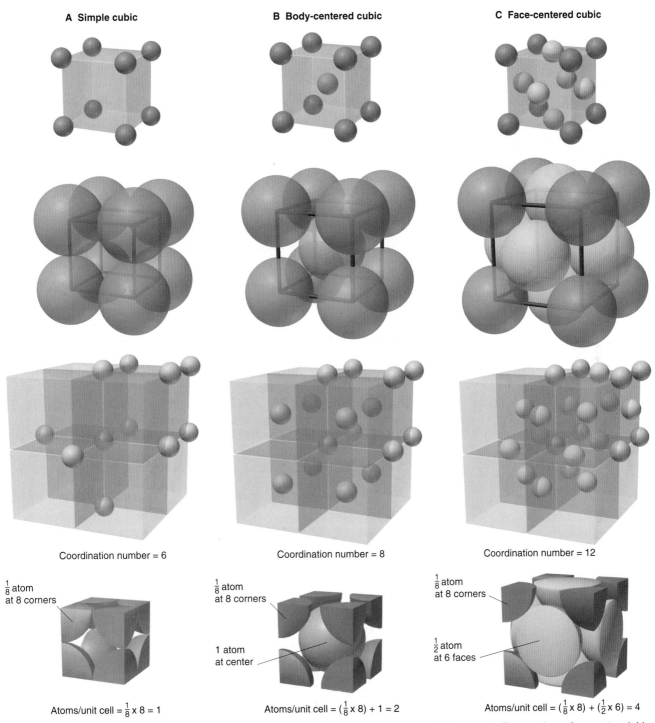

A Simple cubic

B Body-centered cubic

C Face-centered cubic

Coordination number = 6

Coordination number = 8

Coordination number = 12

$\frac{1}{8}$ atom at 8 corners

$\frac{1}{8}$ atom at 8 corners

1 atom at center

$\frac{1}{8}$ atom at 8 corners

$\frac{1}{2}$ atom at 6 faces

Atoms/unit cell = $\frac{1}{8} \times 8 = 1$

Atoms/unit cell = $(\frac{1}{8} \times 8) + 1 = 2$

Atoms/unit cell = $(\frac{1}{8} \times 8) + (\frac{1}{2} \times 6) = 4$

Figure 12.24 The three cubic unit cells. A, Simple cubic unit cell. **B,** Body-centered cubic unit cell. **C,** Face-centered cubic unit cell. *Top row:* Cubic arrangements of atoms in expanded view. *Second row:* Space-filling views of these cubic arrangements. All atoms are identical but, for clarity, corner atoms are blue, body-centered atoms pink, and face-centered atoms yellow. *Third row:* A unit cell (*shaded blue*) in a portion of the crystal. The number of nearest neighbors around one particle (*dark blue in center*) is the coordination number. *Bottom row:* The total numbers of atoms in the actual unit cell. The simple cubic has one atom; the body-centered has two; and the face-centered has four.

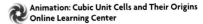 **Animation: Cubic Unit Cells and Their Origins Online Learning Center**

Figure 12.25 The efficient packing of fruit.

Packing Efficiency and the Creation of Unit Cells The unit cells found in nature result from the various ways atoms are packed together, which are similar to the ways macroscopic spheres—marbles, golf balls, fruit—are packed for shipping or display (Figure 12.25).

For particles of the same size, *the higher the coordination number of the crystal is, the greater the number of particles in a given volume.* Therefore, as the coordination numbers in Figure 12.24 indicate, a crystal structure based on the face-centered cubic unit cell has more particles packed into a given volume than one based on the body-centered cubic unit cell, which has more than one based on the simple cubic unit cell. Let's see how to pack *identical* spheres to create these unit cells and the hexagonal unit cell as well:

1. *The simple cubic unit cell.* Suppose we arrange the first layer of spheres as shown in Figure 12.26A. Note the large diamond-shaped spaces (cutaway portion). If we place the next layer of spheres *directly above* the first, as shown in Figure 12.26B, we obtain an arrangement based on the *simple* cubic unit cell. By calculating the **packing efficiency** of this arrangement—the percentage of the total volume occupied by the spheres themselves—we find that only 52% of the available unit-cell volume is occupied by spheres and 48% consists of the empty space between them. This is a very inefficient way to pack spheres, so that neither fruit nor atoms are usually packed this way.

2. *The body-centered cubic unit cell.* Rather than placing the spheres of the second layer directly above the first, we can use space more efficiently by placing the spheres (colored differently for clarity) on the diamond-shaped spaces in the first layer, as shown in Figure 12.26C. Then we pack the third layer onto the spaces in the second so that the first and third layers line up vertically. This arrangement is based on the *body-centered* cubic unit cell, and its packing efficiency is 68%—much higher than for the simple cubic unit cell. Several metallic elements, including chromium, iron, and all the Group 1A(1) elements, have a crystal structure based on the body-centered cubic unit cell.

3. *The hexagonal and face-centered cubic unit cells.* Spheres can be packed even more efficiently. First, we shift rows in the bottom layer so that the large diamond-shaped spaces become smaller triangular spaces. Then we place the second layer over these spaces. Figure 12.26D shows this arrangement, with the first layer labeled *a* (*orange*) and the second layer *b* (*green*).

We can place the third layer of spheres in two different ways, and how we do so gives rise to two different unit cells. If you look carefully at the spaces formed in layer *b* of Figure 12.26D, you'll see that some are orange because they lie above *spheres* in layer *a*, whereas others are white because they lie above *spaces* in layer *a*. If we place the third layer of spheres (*orange*) over the orange spaces (down and left to Figure 12.26E), they lie directly over spheres in layer *a*, and we obtain an *abab. . .* layering pattern because every other layer is placed identically. This gives **hexagonal closest packing,** which is based on the *hexagonal unit cell.*

On the other hand, if we place the third layer of spheres (*blue*) over the white spaces (down and right to Figure 12.26F), the spheres lie over spaces in layer *a*. This placement is different from both layers *a* and *b*, so we obtain an *abcabc. . .* pattern. This gives **cubic closest packing,** which is based on the *face-centered* cubic unit cell.

The packing efficiency of both hexagonal and cubic closest packing is 74%, and the coordination number of both is 12. There is no way to pack identical spheres more efficiently. Most metallic elements crystallize in either of these arrangements. Magnesium, titanium, and zinc are some elements that adopt the hexagonal structure; nickel, copper, and lead adopt the cubic structure, as do many

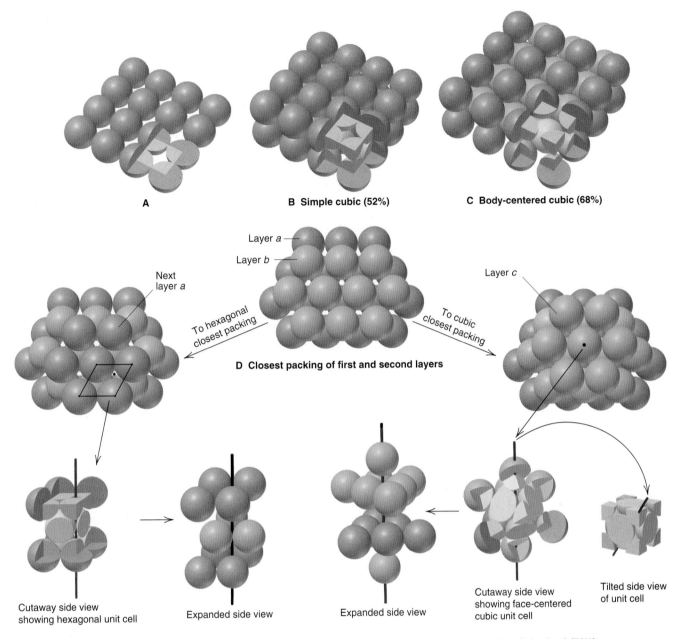

A, **B** Simple cubic (52%) **C** Body-centered cubic (68%)

Layer *a*
Layer *b*

Next layer *a*

To hexagonal closest packing

D Closest packing of first and second layers

To cubic closest packing

Layer *c*

Cutaway side view showing hexagonal unit cell

Expanded side view

Expanded side view

Cutaway side view showing face-centered cubic unit cell

Tilted side view of unit cell

E Hexagonal closest packing (*abab...*) (74%)

F Cubic closest packing (*abcabc...*) (74%)

Figure 12.26 Packing identical spheres. A, In the first layer, each sphere lies next to another horizontally and vertically; note the large diamond-shaped spaces (*see cutaway*). **B,** If the spheres in the next layer lie *directly* over those in the first, the packing is based on the *simple cubic* unit cell (*pale orange cube, lower right corner*). **C,** If the spheres in the next layer lie in the diamond-shaped spaces of the first layer, the packing is based on the *body-centered cubic* unit cell (*lower right corner*). **D,** The closest possible packing of the first layer (*layer a, orange*) is obtained by shifting every other row in part A, thus reducing the diamond-shaped spaces to smaller triangular spaces. The spheres of the second layer (*layer b, green*) are placed above these spaces; note the orange and white spaces that result. **E,** Follow the left arrow

from part D to obtain hexagonal closest packing. When the third layer (*next layer a, orange*) is placed directly over the first, that is, over the orange spaces, we obtain an *abab*. . . pattern. Rotating the layers 90° produces the side view, with the hexagonal unit cell shown as a cutaway segment, and the expanded side view. **F,** Follow the right arrow from part D to obtain cubic closest packing. When the third layer (*layer c, blue*) covers the white spaces, it lies in a different position from the first *and* second layers to give an *abcabc*. . . pattern. Rotating the layers 90° shows the side view, with the face-centered cubic unit cell as a cutaway, and a further tilt shows the unit cell clearly; finally, we see the expanded view. The packing efficiency for each type of unit cell is given in parentheses.

ionic compounds and other substances, such as frozen carbon dioxide, methane, and most noble gases.

In Sample Problem 12.4, we use the density of an element and the packing efficiency of its crystal structure to calculate its atomic radius. Variations of this approach are used to find the molar mass and as one of the ways to determine Avogadro's number.

SAMPLE PROBLEM 12.4 Determining Atomic Radius from Crystal Structure

Problem Barium is the largest nonradioactive alkaline earth metal. It has a body-centered cubic unit cell and a density of 3.62 g/cm³. What is the atomic radius of barium? (Volume of a sphere: $V = \frac{4}{3}\pi r^3$.)

Plan Because an atom is spherical, we can find its radius from its volume. If we multiply the reciprocal of density (volume/mass) by the molar mass (mass/mole), we find the volume of 1 mol of Ba metal. The metal crystallizes in the body-centered cubic structure, so 68% of this volume is occupied by 1 mol of the atoms themselves (see Figure 12.26C). Dividing by Avogadro's number gives the volume of one Ba atom, from which we find the radius.

Solution Combining steps to find the volume of 1 mol of Ba metal:

$$\text{Volume/mole of Ba metal} = \frac{1}{\text{density}} \times \mathcal{M}$$

$$= \frac{1 \text{ cm}^3}{3.62 \text{ g Ba}} \times \frac{137.3 \text{ g Ba}}{1 \text{ mol Ba}}$$

$$= 37.9 \text{ cm}^3/\text{mol Ba}$$

Finding the volume of 1 mol of Ba *atoms:*

$$\text{Volume/mole of Ba atoms} = \text{volume/mol Ba} \times \text{packing efficiency}$$

$$= 37.9 \text{ cm}^3/\text{mol Ba} \times 0.68 = 26 \text{ cm}^3/\text{mol Ba atoms}$$

Finding the volume of one Ba atom:

$$\text{Volume of Ba atom} = \frac{26 \text{ cm}^3}{1 \text{ mol Ba atoms}} \times \frac{1 \text{ mol Ba atoms}}{6.022 \times 10^{23} \text{ Ba atoms}}$$

$$= 4.3 \times 10^{-23} \text{ cm}^3/\text{Ba atom}$$

Finding the atomic radius of Ba from the volume of a sphere:

$$V \text{ of Ba atom} = \frac{4}{3}\pi r^3$$

So,

$$r^3 = \frac{3V}{4\pi}$$

Thus,

$$r = \sqrt[3]{\frac{3V}{4\pi}} = \sqrt[3]{\frac{3(4.3 \times 10^{-23} \text{ cm}^3)}{4 \times 3.14}}$$

$$= 2.2 \times 10^{-8} \text{ cm}$$

Check The order of magnitude is correct for an atom ($\sim 10^{-8}$ cm $\approx 10^{-10}$ m). The actual value for barium is, in fact, 2.22×10^{-8} cm (see Figure 8.9).

FOLLOW-UP PROBLEM 12.4 Iron crystallizes in a body-centered cubic structure. The volume of one Fe atom is 8.38×10^{-24} cm³, and the density of Fe is 7.874 g/cm³. Calculate an approximate value for Avogadro's number.

Flowchart (left margin):

Density (g/cm³) of Ba metal

↓ find reciprocal and multiply by $\mathcal{M}$ (g/mol)

Volume (cm³) per mole of Ba metal

↓ multiply by packing efficiency

Volume (cm³) per mole of Ba atoms

↓ divide by Avogadro's number

Volume (cm³) of Ba atom

↓ $V = \frac{4}{3}\pi r^3$

Radius (cm) of Ba atom

Observing Crystal Structures: X-Ray Diffraction Analysis Our understanding of solids is based on the ability to "see" their crystal structures. One of the most powerful tools for doing this is **x-ray diffraction analysis.** In Chapter 7, we discussed wave diffraction and saw how interference patterns of bright and dark regions appear when light passes through slits that are spaced at the distance of the light's wavelength (see Figure 7.5). Because x-ray wavelengths are about the same size as the spaces between layers of particles in many solids, the layers diffract x-rays.

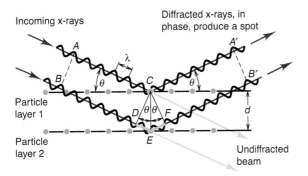

Figure 12.27 Diffraction of x-rays by crystal planes. As in-phase x-ray beams *A* and *B* pass into a crystal at angle θ, they are diffracted by interaction with the particles. Beam *B* travels the distance *DE* + *EF* farther than beam *A*. If this additional distance is equal to a whole number of wavelengths, the beams remain in phase and create a spot on a screen or photographic plate. From the pattern of spots and the Bragg equation, $n\lambda = 2d \sin \theta$, the distance *d* between layers of particles can be calculated.

Let's see how this technique is used to measure a key parameter in a crystal structure: the distance (*d*) between layers of atoms. Figure 12.27 depicts a side view of two layers in a simplified lattice. Two waves impinge on the crystal at an angle θ and are diffracted at the same angle by adjacent layers. When the first wave strikes the top layer and the second strikes the next layer, the waves are *in phase* (peaks aligned with peaks and troughs with troughs). If they are still in phase after being diffracted, a spot appears on a nearby photographic plate. Note that this will occur only if the additional distance traveled by the second wave (*DE* + *EF* in the figure) is a whole number of wavelengths, $n\lambda$, where *n* is an integer (1, 2, 3, and so on). From trigonometry, we find that

$$n\lambda = 2d \sin \theta$$

where θ is the known angle of incoming light, λ is its known wavelength, and *d* is the unknown distance between the layers in the crystal. This relationship is the *Bragg equation,* named for W. H. Bragg and his son W. L. Bragg, who shared the Nobel Prize in physics in 1915 for their work on crystal structure analysis.

Rotating the crystal changes the angle of incoming radiation and produces a different set of spots. Modern x-ray diffraction equipment automatically rotates the crystal and measures thousands of diffractions, and a computer calculates the distances and angles within the lattice (Figure 12.28).

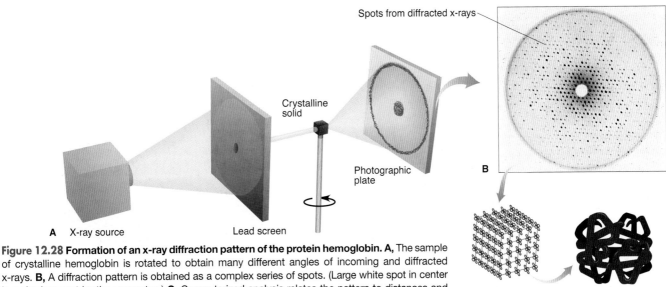

Figure 12.28 Formation of an x-ray diffraction pattern of the protein hemoglobin. A, The sample of crystalline hemoglobin is rotated to obtain many different angles of incoming and diffracted x-rays. **B,** A diffraction pattern is obtained as a complex series of spots. (Large white spot in center is a shadow cast by the apparatus.) **C,** Computerized analysis relates the pattern to distances and angles within the crystal. The data are used to generate a picture of the hemoglobin molecule.

Table 12.5	Characteristics of the Major Types of Crystalline Solids			
Type	**Particle(s)**	**Interparticle Forces**	**Physical Properties**	**Examples [mp, °C]**
Atomic	Atoms	Dispersion	Soft, very low mp, poor thermal and electrical conductors	Group 8A(18) [Ne (-249) to Rn (-71)]
Molecular	Molecules	Dispersion, dipole-dipole, H bonds	Fairly soft, low to moderate mp, poor thermal and electrical conductors	*Nonpolar** O_2 [-219], C_4H_{10} [-138] Cl_2 [-101], C_6H_{14} [-95], P_4 [44.1] *Polar* SO_2 [-73], $CHCl_3$ [-64], HNO_3 [-42], H_2O [0.0], CH_3COOH [17]
Ionic	Positive and negative ions	Ion-ion attraction	Hard and brittle, high mp, good thermal and electrical conductors when molten	NaCl [801] CaF_2 [1423] MgO [2852]
Metallic	Atoms	Metallic bond	Soft to hard, low to very high mp, excellent thermal and electrical conductors, malleable and ductile	Na [97.8] Zn [420] Fe [1535]
Network covalent	Atoms	Covalent bond	Very hard, very high mp, usually poor thermal and electrical conductors	SiO_2 (quartz) [1610] C (diamond) [~4000]

*Nonpolar molecular solids are arranged in order of increasing molar mass. Note the correlation with increasing melting point (mp).

Types and Properties of Crystalline Solids

Now we can turn to the five most important types of solids, which are summarized in Table 12.5. Each is defined by the type(s) of particle(s) in the crystal, which determines the interparticle forces. You may want to review the bonding models (Chapter 9) to clarify how they relate to the properties of different solids.

Atomic Solids Individual atoms held together by dispersion forces form an **atomic solid.** The noble gases [Group 8A(18)] are the only examples, and their physical properties reflect the very weak forces among the atoms. Melting and boiling points and heats of vaporization and fusion are all very low, rising smoothly with increasing molar mass. As shown in Figure 12.29, argon crystallizes in a cubic closest packed structure. The other atomic solids do so as well.

Figure 12.29 Cubic closest packing of frozen argon (face-centered cubic unit cell).

Molecular Solids In the many thousands of **molecular solids,** the lattice points are occupied by individual molecules. For example, methane crystallizes in a face-centered cubic structure, shown with the center of each carbon as the lattice point (Figure 12.30).

Various combinations of dipole-dipole, dispersion, and H-bonding forces are at work in molecular solids, which accounts for their wide range of physical properties. Dispersion forces are the principal force acting in nonpolar substances, so melting points generally increase with molar mass (Table 12.5). Among polar molecules, dipole-dipole forces and, where possible, H bonding dominate. Except for those substances consisting of the simplest molecules, molecular solids have higher melting points than the atomic solids (noble gases). Nevertheless, intermolecular forces are still relatively weak, so the melting points are much lower than those of ionic, metallic, and network covalent solids.

Ionic Solids In crystalline **ionic solids,** the unit cell contains particles with whole, rather than partial, charges. As a result, the interparticle forces (ionic bonds) are *much* stronger than the van der Waals forces in atomic or molecular solids. To maximize attractions, cations are surrounded by as many anions as possible, and vice versa, with *the smaller of the two ions lying in the spaces (holes) formed by the packing of the larger.* Because the unit cell is the smallest portion of the crystal that maintains the overall spatial arrangement, it is also the smallest portion that maintains the overall chemical composition. In other words, *the unit cell has the same cation:anion ratio as the empirical formula.*

Ionic compounds adopt several different crystal structures, but many use cubic closest packing. As an example, let's consider a structure that has a 1:1 ratio of ions. The *sodium chloride structure* is found in many compounds, including most of the alkali metal [Group 1A(1)] halides and hydrides, the alkaline earth metal [Group 2A(2)] oxides and sulfides, several transition-metal oxides and sulfides, and most of the silver halides. To visualize this structure, first imagine Cl^- anions and Na^+ cations organized separately in face-centered cubic (cubic closest packing) arrays. The crystal structure arises when these two arrays penetrate each other such that the smaller Na^+ ions end up in the holes between the larger Cl^- ions, as shown in Figure 12.31A. Thus, each Na^+ is surrounded by six Cl^-, and vice versa (coordination number = 6). Figure 12.31B is a space-filling depiction of the unit cell showing a face-centered cube of Cl^- ions with Na^+ ions between them. Note the four Cl^- $[(8 \times \frac{1}{8}) + (6 \times \frac{1}{2}) = 4\ Cl^-]$ and four Na^+ $[(12 \times \frac{1}{4}) + 1$ in the center $= 4\ Na^+]$, giving a 1:1 ion ratio.

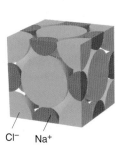

Figure 12.30 Cubic closest packing of frozen methane. Only one CH_4 molecule is shown.

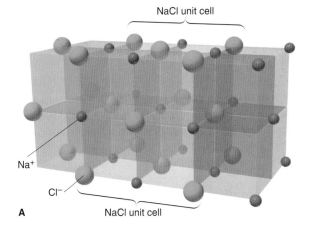

NaCl unit cell

Na$^+$

Cl$^-$

NaCl unit cell

A

Cl$^-$ Na$^+$

B

Figure 12.31 The sodium chloride structure. A, In an expanded view, the sodium chloride structure is pictured as resulting from the interpenetration of two face-centered cubic arrangements, one of Na$^+$ ions (*brown*) and the other of Cl$^-$ ions (*green*). **B,** A space-filling view of the NaCl unit cell (central overlapped portion in part A), which consists of four Cl$^-$ ions and four Na$^+$ ions.

The properties of ionic solids are a direct consequence of the *fixed ion positions* and *very strong interionic forces,* which create a high lattice energy. Thus, ionic solids typically have high melting points and low electrical conductivities. When a large amount of heat is supplied and the ions gain enough kinetic energy to break free of their positions, the solid melts and the mobile ions conduct a current. Ionic compounds are hard because only a strong external force can change the relative positions of many trillions of interacting ions. If enough force *is* applied to move them, ions of like charge are brought near each other, and their repulsions crack the crystal (see Figure 9.8).

Metallic Solids In contrast to the weak dispersion forces between the atoms in atomic solids, powerful metallic bonding forces hold individual atoms together in **metallic solids.** The properties of metals—high electrical and thermal conductivity, luster, and malleability—result from the presence of delocalized electrons, the essential feature of metallic bonding. Metals have a wide range of melting points and hardness, which are related to the packing efficiency of the crystal structure and the number of valence electrons available for bonding. Most metallic elements crystallize in one of the two closest packed structures (Figure 12.32). In an upcoming subsection in this chapter, we'll discuss two bonding models— one specific to metals and the other for metals and other solids.

Figure 12.32 Crystal structures of metals. Most metals crystallize in one of the two closest packed arrangements. **A,** Copper adopts cubic closest packing. **B,** Magnesium adopts hexagonal closest packing.

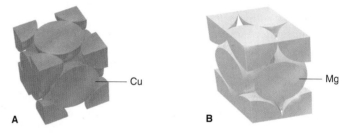

A — Cu

B — Mg

Network Covalent Solids In the final type of crystalline solid, separate particles are not present. Instead, strong covalent bonds link the atoms together throughout a **network covalent solid.** As a consequence of the strong bonding, all these substances have extremely high melting and boiling points, but their conductivity and hardness depend on the details of their bonding.

The two common crystalline forms of elemental carbon are examples of network covalent solids. Although graphite and diamond have the same composition, their properties are strikingly different, as Table 12.6 shows. Graphite occurs as stacked flat sheets of hexagonal carbon rings with a strong σ-bond framework and delocalized π bonds, reminiscent of benzene. The arrangement of hexagons looks like chicken wire or honeycomb. Whereas the π-bonding electrons of benzene are delocalized over one ring, those of graphite are delocalized over the entire sheet. These mobile electrons allow graphite to conduct electricity, but only in the plane of the sheets. Graphite is a common electrode material and was once used for lightbulb filaments. The sheets interact via dispersion forces. Common impurities, such as O_2, that lodge between the sheets allow them to slide past each other easily, which explains why graphite is so soft. Diamond crystallizes in a face-centered cubic unit cell, with each carbon atom tetrahedrally surrounded by four others in one virtually endless array. Strong, single bonds throughout the crystal make diamond the hardest substance known. Because of its

Table 12.6 **Comparison of the Properties of Diamond and Graphite**

Property	Graphite		Diamond	
Density (g/cm^3)	2.27		3.51	
Hardness	<1 (very soft)		10 (hardest)	
Melting point (K)	4100		4100	
Color	Shiny black		Colorless transparent	
Electrical conductivity	High (along sheet)		None	
ΔH°_{comb} (kJ/mol)	−393.5		−395.4	
ΔH°_f (kJ/mol)	0 (standard state)		1.90	

localized bonding electrons, diamond (like most network covalent solids) does not conduct electricity.

By far the most important network covalent solids are the *silicates.* They utilize a variety of bonding patterns, but nearly all consist of extended arrays of covalently bonded silicon and oxygen atoms. Quartz (SiO_2) is a common example. We'll discuss silicates, which form the structure of clays, rocks, and many minerals, when we consider the chemistry of silicon in Chapter 14.

Amorphous Solids

Amorphous solids are noncrystalline. Many have small, somewhat ordered regions connected by large disordered regions. Charcoal, rubber, and glass are some familiar examples of amorphous solids.

The process that forms quartz glass is typical of that for many amorphous solids. Crystalline quartz (SiO_2) has a cubic closest packed structure. The crystalline form is melted, and the viscous liquid is cooled rapidly to prevent it from recrystallizing. The chains of silicon and oxygen atoms cannot orient themselves quickly enough into an orderly structure, so they solidify in a distorted jumble containing many gaps and misaligned rows (Figure 12.33). The absence of regularity in the structure confers some properties of a liquid; in fact, glasses are sometimes referred to as *supercooled liquids.*

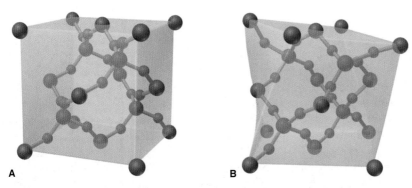

A   **B**

Figure 12.33 Crystalline and amorphous silicon dioxide. A, The atomic arrangement of cristobalite, one of the many crystalline forms of silica (SiO_2), shows the regularity of cubic closest packing. **B,** The atomic arrangement of a quartz glass is amorphous with a generally disordered structure.

Bonding in Solids

In this section, we discuss two models of bonding in solids. The first is a simple, qualitative model for metals; the second is more quantitative and therefore more useful. It explains not only the properties of metals but also differences in electrical conductivity of metals, metalloids, and nonmetals.

The Electron-Sea Model of Metallic Bonding The simplest model that accounts for the properties of metals is the **electron-sea model.** It proposes that all the metal atoms in a sample pool their valence electrons to form an electron "sea" that is delocalized throughout the piece. The metal ions (nuclei plus core electrons) are submerged within this electron sea in an orderly array (see Figure 9.2C). They are not held in place as rigidly as the ions in an ionic solid, and no two metal atoms are bonded through a localized pair of electrons as in a covalent bond. Rather, *the valence electrons are shared among all the atoms in the sample,* and the piece of metal is held together by the mutual attraction of the metal cations for the mobile, highly delocalized valence electrons.

The *regularity,* but not rigidity, of the metal-ion array and the *mobility* of the valence electrons account for the physical properties of metals. Metals have moderate to high melting points because the attractions between the cations and the delocalized electrons are not broken during melting, but boiling points are very high because each cation and its electron(s) must break away from the others. Gallium provides a striking example: it melts in your hand (mp 29.8°C) but doesn't boil until 2403°C. The alkaline earth metals [Group 2A(2)] have higher melting points than the alkali metals [Group 1A(1)] because of greater attraction between their 2+ cations and twice the number of valence electrons.

When struck by a hammer, metals usually bend or dent rather than crack or shatter. Instead of repelling each other, the metal cations slide past each other through the electron sea and end up in new positions (Figure 12.34). Compare this behavior with that of an ionic solid (see Figure 9.8). As a result, many metals can be flattened into sheets (malleable) and pulled into wires (ductile). Gold is in a class by itself: 1 g of gold (a cube 0.37 cm on a side) can be hammered into a 1.0-m^2 sheet that is only 230 atoms (50 nm) thick or drawn into a wire 165 m long and 20 μm thick!

Metals are good electrical conductors because the mobile electrons carry the current, and they conduct heat well because the mobile electrons disperse heat more quickly than do the localized electron pairs or fixed ions in other materials.

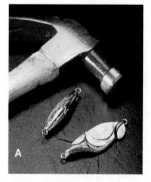
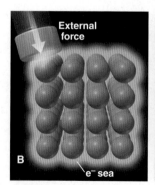
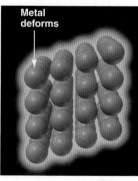

Figure 12.34 The reason metals deform. A, An external force applied to a piece of metal deforms but doesn't break it. **B,** The external force merely moves metal ions past each other through the electron sea.

Molecular Orbital Band Theory Quantum mechanics offers the second model of bonding in solids—an extension of molecular orbital (MO) theory called **band theory.** Recall that when two atoms form a diatomic molecule, their atomic orbitals (AOs) combine to form an equal number of molecular orbitals (MOs). Let's consider lithium as an example. Figure 12.35 shows the formation of MOs in lithium. In dilithium, Li_2, each atom has four valence orbitals (one 2s and three 2p). (Recall that in Section 11.3, we focused primarily on the 2s orbitals.) They combine to form eight MOs, four bonding and four antibonding, spread over both atoms. If two more Li atoms combine, they form Li_4, a slightly larger aggregate, with 16 delocalized MOs. As more Li atoms join the cluster, more MOs are created, their energy levels lying closer and closer together. Extending this process to a 7-g sample of lithium metal (the molar mass) results in 1 mol of Li atoms (Li_{N_A}, where N_A is Avogadro's number) combining to form an extremely large number (4 × Avogadro's number) of delocalized MOs, with *energies so closely spaced that they form a continuum, or band, of MOs.* It is almost as though the entire piece of metal were one enormous Li molecule.

The band model proposes that the lower energy MOs are occupied by the valence electrons and make up the **valence band.** The empty MOs that are higher in energy make up the **conduction band.** In Li metal, the valence band is derived from the 2s AOs, and the conduction band is derived mostly from an intermingling of the 2s and 2p AOs. In Li_2, two valence electrons fill the lowest energy MO and leave the antibonding MO empty. Similarly, in Li metal, 1 mol of valence electrons fills the valence band and leaves the conduction band empty.

The key to understanding metallic properties is that *in metals, the valence and conduction bands are contiguous,* which means that electrons can jump from the filled valence band to the unfilled conduction band if they receive even an infinitesimally small quantity of energy. In other words, the electrons are completely delocalized: *they are free to move throughout the piece of metal.* Thus, metals conduct electricity so well because an applied electric field easily excites the highest energy electrons into empty orbitals, and they move through the sample.

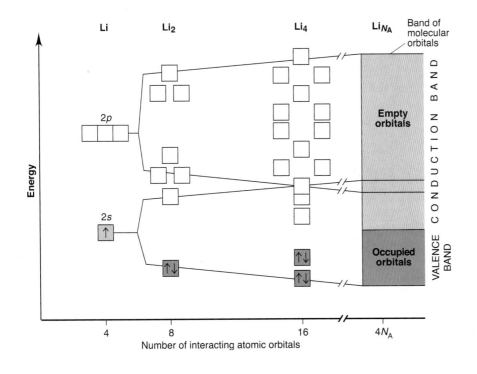

Figure 12.35 The band of molecular orbitals in lithium metal. Lithium atoms contain four valence orbitals, one 2s and three 2p *(left).* When two lithium atoms combine (Li_2), their AOs form eight MOs within a certain range of energy. Four Li atoms (Li_4) form 16 MOs. A mole of Li atoms forms $4N_A$ MOs (N_A = Avogadro's number). The orbital energies are so close together that they form a continuous band. The valence electrons enter the lower energy portion (valence band), while the higher energy portion (conduction band) remains empty. In lithium (and other metals), the valence and conduction bands have no gap between them.

Metallic luster (shininess) is another effect of the continuous band of MO energy levels. With so many closely spaced levels available, electrons can absorb and release photons of many frequencies as they move between the valence and conduction bands. Malleability and thermal conductivity also result from the completely delocalized electrons. Under an externally applied force, layers of positive metal ions simply move past each other, always protected from mutual repulsions by the presence of the delocalized electrons (see Figure 12.34B). When a metal wire is heated, the highest energy electrons are excited and their extra energy is transferred as kinetic energy along the wire's length.

Large numbers of nonmetal or metalloid atoms can also combine to form bands of MOs. Metals conduct a current well (conductors), whereas most nonmetals do not (insulators), and the conductivity of metalloids lies somewhere in between (semiconductors). Band theory explains these differences in terms of the size of the energy gaps between the valence and conduction bands, as shown in Figure 12.36:

1. *Conductors (metals).* The valence and conduction bands of a **conductor** have no gap between them, so electrons flow when even a tiny electrical potential difference is applied. When the temperature is raised, greater random motion of the atoms hinders electron movement, which *decreases* the conductivity of a metal.
2. *Semiconductors (metalloids).* In a **semiconductor,** a relatively small energy gap exists between the valence and conduction bands. Thermally excited electrons can cross the gap, allowing a small current to flow. Thus, in contrast to a conductor, the conductivity of a semiconductor *increases* when it is heated.
3. *Insulators (nonmetals).* In an **insulator,** the gap between the bands is too large for electrons to jump even when the substance is heated, so no current is observed.

Another type of electrical conductivity, called **superconductivity,** has been generating intense interest for more than two decades. When metals conduct at ordinary temperatures, electron flow is restricted by collisions with atoms vibrating in their lattice sites. Such restricted flow appears as resistive heating and represents a loss of energy. To conduct with no energy loss—to superconduct—requires extreme cooling to minimize atom movement. This remarkable phenomenon had been observed in metals only by cooling them to near absolute zero, which can be done only with liquid helium (bp = 4 K; price = $11/L).

In 1986, all this changed with the synthesis of certain ionic oxides that superconduct near the boiling point of liquid nitrogen (bp = 77 K; price = $0.25/L).

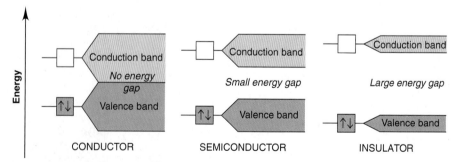

Figure 12.36 Electrical conductivity in a conductor, a semiconductor, and an insulator. Band theory explains differences in electrical conductivity in terms of the size of the energy gap between the material's valence and conduction bands. In conductors (metals), there is no gap. In semiconductors (many metalloids), electrons can jump the small gap if they are given energy, as when the sample is heated. In insulators (most nonmetals), electrons cannot jump the large energy gap.

Like metal conductors, oxide superconductors have no band gap. In the case of $YBa_2Cu_3O_7$, x-ray analysis shows that the Cu ions in the oxide lattice are aligned, which may be related to the superconducting property. In 1989, oxides with Bi and Tl instead of Y and Ba were synthesized and found to superconduct at 125 K; and in 1993, an oxide with Hg, Ba, and Ca, in addition to Cu and O, was shown to superconduct at 133 K. Engineering dreams for these materials include storage and transmission of electricity with no loss of energy (allowing power plants to be located far from cities), ultrasmall microchips for ultrafast computers, electromagnets to levitate superfast railway trains (Figure 12.37), and inexpensive medical diagnostic equipment with remarkable image clarity.

SECTION SUMMARY

The particles in crystalline solids lie at points that form a structure of repeating unit cells. The three types of unit cells in the cubic system are simple, body-centered, and face-centered. The most efficient packing arrangements are cubic closest packing and hexagonal closest packing. Bond angles and distances in a crystal structure can be determined with x-ray diffraction analysis. Atomic solids have a closest packed structure, with atoms held together by very weak dispersion forces. Molecular solids have molecules at the lattice points, often in a cubic closest packed structure. Their intermolecular forces (dispersion, dipole-dipole, H bonding) and resulting physical properties vary greatly. Ionic solids often crystallize with one type of ion filling holes in a cubic closest packed structure of the other. The high melting points, hardness, and low conductivity of these solids arise from strong ionic attractions. Most metals have a closest packed structure. The atoms of network covalent solids are covalently bonded throughout the sample. Amorphous solids have very little regularity among their particles. The electron-sea model proposes that metals consist of positive ions submerged in an electron "sea" of highly delocalized valence electrons. These electrons are shared by all the atoms in the sample. Band theory proposes that orbitals in the atoms of solids combine to form a continuum, or band, of molecular orbitals. Metals are electrical conductors because electrons move freely from the filled (valence band) to the empty (conduction band) portions of this energy continuum. Insulators have a large energy gap between the two portions; semiconductors have a small gap.

Figure 12.37 The levitating power of a superconducting oxide. A magnet is suspended above a cooled high-temperature superconductor. Someday, this phenomenon may be used to levitate trains above their tracks for quiet, fast travel.

For Review and Reference (Numbers in parentheses refer to pages, unless noted otherwise.)

Learning Objectives

To help you review these learning objectives, the numbers of related sections (§), sample problems (SP), and upcoming end-of-chapter problems (EP) are listed in parentheses.

1. Explain how kinetic energy and potential energy determine the properties of the three states and phase changes, what occurs when heat is added or removed from a substance, and how to calculate the enthalpy change (§ 12.1, 12.2) (EPs 12.1–12.9, 12.14, 12.15)
2. Understand that phase changes are equilibrium processes and how vapor pressure, temperature, and boiling point are related (§ 12.2) (SP 12.1) (EPs 12.10, 12.11, 12.16, 12.17, 12.20)
3. Use a phase diagram to show the phases and phase changes of a substance at different conditions of pressure and temperature (§ 12.2) (EPs 12.12, 12.13, 12.18, 12.19, 12.21)
4. Distinguish between bonding and molecular forces, predict the relative strengths of intermolecular forces acting in a substance,

and understand the impact of H bonding on physical properties (§ 12.3) (SPs 12.2, 12.3) (EPs 12.22–12.43)
5. Define surface tension, capillarity, and viscosity, and describe how intermolecular forces influence their magnitudes (§ 12.4) (EPs 12.44–12.50)
6. Understand how the macroscopic properties of water arise from its molecular properties (§ 12.5) (EPs 12.51–12.56)
7. Describe the three types of cubic unit cells and explain how to find the number of particles in each and how packing of spheres gives rise to each; calculate the atomic radius of an element from its density and crystal structure; distinguish the types of crystalline solids; explain how the electron-sea model and band theory account for the properties of metals and how the size of the energy gap explains the conductivity of substances (§ 12.6) (SP 12.4) (EPs 12.57–12.75)

Key Terms

phase (348)
intermolecular forces (348)
phase change (348)

Section 12.1
condensation (349)
vaporization (349)
freezing (349)
melting (fusion) (349)
heat of vaporization (ΔH°_{vap}) (349)
heat of fusion (ΔH°_{fus}) (349)
sublimation (350)
deposition (350)
heat of sublimation (ΔH°_{subl}) (350)

Section 12.2
heating-cooling curve (351)

dynamic equilibrium (353)
vapor pressure (353)
Clausius-Clapeyron equation (354)
boiling point (355)
melting point (356)
phase diagram (356)
critical point (356)
triple point (357)

Section 12.3
van der Waals radius (359)
ion-dipole force (360)
dipole-dipole force (360)
hydrogen bond (H bond) (361)
polarizability (362)
dispersion (London) force (363)

Section 12.4
surface tension (365)
capillarity (366)
viscosity (367)

Section 12.6
crystalline solid (369)
amorphous solid (369)
lattice (370)
unit cell (370)
coordination number (370)
simple cubic unit cell (370)
body-centered cubic unit cell (370)
face-centered cubic unit cell (370)
packing efficiency (372)

hexagonal closest packing (372)
cubic closest packing (372)
x-ray diffraction analysis (374)
atomic solid (376)
molecular solid (377)
ionic solid (377)
metallic solid (378)
network covalent solid (378)
electron-sea model (380)
band theory (381)
valence band (381)
conduction band (381)
conductor (382)
semiconductor (382)
insulator (382)
superconductivity (382)

Key Equations and Relationships

12.1 Using the vapor pressure at one temperature to find the vapor pressure at another temperature (two-point form of the Clausius-Clapeyron equation) (355):

$$\ln \frac{P_2}{P_1} = \frac{-\Delta H_{vap}}{R} \left(\frac{1}{T_2} - \frac{1}{T_1} \right)$$

Brief Solutions to Follow-up Problems

12.1 $\ln \dfrac{P_2}{P_1} = \left(\dfrac{-40.7 \times 10^3 \text{ J/mol}}{8.314 \text{ J/mol·K}} \right)$

$\times \left(\dfrac{1}{273.15 + 85.5 \text{ K}} - \dfrac{1}{273.15 + 34.1 \text{ K}} \right)$

$= (-4.90 \times 10^3 \text{ K})(-4.66 \times 10^{-4} \text{ K}^{-1}) = 2.28$

$\dfrac{P_2}{P_1} = 9.8$; thus, $P_2 = 40.1 \text{ torr} \times 9.8 = 3.9 \times 10^2 \text{ torr}$

12.2 (a)

(c) No H bonding

(b)

12.3 (a) Dipole-dipole, dispersion; CH_3Br
(b) H bonds, dipole-dipole, dispersion; $CH_3CH_2CH_2OH$
(c) Dispersion; C_3H_8

12.4 Avogadro's no. $= \dfrac{1 \text{ cm}^3}{7.874 \text{ g Fe}} \times \dfrac{55.85 \text{ g Fe}}{1 \text{ mol Fe}} \times 0.68$

$\times \dfrac{1 \text{ Fe atom}}{8.38 \times 10^{-24} \text{ cm}^3}$

$= 5.8 \times 10^{23} \text{ Fe atoms/mol Fe}$

Problems

Problems with **colored** numbers are answered in Appendix E. Sections match the text and provide the numbers of relevant sample problems. Bracketed problems are grouped in pairs (indicated by a short rule) that cover the same concept. Comprehensive Problems are based on material from any section or previous chapter.

An Overview of Physical States and Phase Changes

12.1 How does the energy of attraction between particles compare with their energy of motion in a gas and in a solid? As part of your answer, identify two macroscopic properties that differ between a gas and a solid.

12.2 What types of forces, intramolecular or intermolecular,
(a) prevent ice cubes from adopting the shape of their container?
(b) are overcome when ice melts?
(c) are overcome when liquid water is vaporized?
(d) are overcome when gaseous water is converted to hydrogen gas and oxygen gas?

12.3 (a) Why are gases more easily compressed than liquids?
(b) Why do liquids have a greater ability to flow than solids?

12.4 (a) Why is the heat of fusion (ΔH_{fus}) of a substance smaller than its heat of vaporization (ΔH_{vap})?
(b) Why is the heat of sublimation (ΔH_{subl}) of a substance greater than its ΔH_{vap}?
(c) At a given temperature and pressure, how does the magnitude of the heat of vaporization of a substance compare with that of its heat of condensation?

12.5 Name the phase change in each of these events:
(a) Dew appears on a lawn in the morning.
(b) Icicles change into liquid water.
(c) Wet clothes dry on a summer day.

12.6 Name the phase change in each of these events:
(a) A diamond film forms on a surface from gaseous carbon atoms in a vacuum.
(b) Mothballs in a bureau drawer disappear over time.
(c) Molten iron from a blast furnace is cast into ingots ("pigs").

12.7 Liquid propane, a widely used fuel, is produced by compressing gaseous propane at 20°C. During the process, approximately 15 kJ of energy is released for each mole of gas liquefied. Where does this energy come from?

12.8 Many heat-sensitive and oxygen-sensitive solids, such as camphor, are purified by warming under vacuum. The solid vaporizes directly, and the vapor crystallizes on a cool surface. What phase changes are involved in this method?

Quantitative Aspects of Phase Changes
(Sample Problem 12.1)

12.9 Describe the changes (if any) in potential energy and in kinetic energy among the molecules when gaseous PCl_3 condenses to a liquid at a fixed temperature.

12.10 When benzene is at its melting point, two processes occur simultaneously and balance each other. Describe these processes on the macroscopic and molecular levels.

12.11 Liquid hexane (bp = 69°C) is placed in a closed container at room temperature. At first, the pressure of the vapor phase increases, but after a short time, it stops changing. Why?

12.12 At 1.1 atm, will water boil at 100.°C? Explain.

12.13 The phase diagram for substance A has a solid-liquid line with a positive slope, and that for substance B has a solid-liquid line with a negative slope. What macroscopic property can distinguish A from B?

12.14 From the data below, calculate the total heat (in J) needed to convert 12.00 g of ice at −5.00°C to liquid water at 0.500°C:

mp at 1 atm:	0.0°C	ΔH_{fus}°: 6.02 kJ/mol
c_{liquid}:	4.21 J/g·°C	c_{solid}: 2.09 J/g·°C

12.15 From the data below, calculate the total heat (in J) needed to convert 0.333 mol of gaseous ethanol at 300°C and 1 atm to liquid ethanol at 25.0°C and 1 atm:

bp at 1 atm:	78.5°C	ΔH_{vap}°: 40.5 kJ/mol
c_{gas}:	1.43 J/g·°C	c_{liquid}: 2.45 J/g·°C

12.16 A liquid has a ΔH_{vap}° of 35.5 kJ/mol and a boiling point of 122°C at 1.00 atm. What is its vapor pressure at 109°C?

12.17 What is the ΔH_{vap}° of a liquid that has a vapor pressure of 641 torr at 85.2°C and a boiling point of 95.6°C at 1 atm?

12.18 Use these data to draw a qualitative phase diagram for ethylene (C_2H_4). Is $C_2H_4(s)$ more or less dense than $C_2H_4(l)$?

bp at 1 atm:	−103.7°C
mp at 1 atm:	−169.16°C
Critical point:	9.9°C and 50.5 atm
Triple point:	−169.17°C and 1.20×10^{-3} atm

12.19 Use these data to draw a qualitative phase diagram for H_2. Does H_2 sublime at 0.05 atm? Explain.

mp at 1 atm:	13.96 K
bp at 1 atm:	20.39 K
Triple point:	13.95 K and 0.07 atm
Critical point:	33.2 K and 13.0 atm
Vapor pressure of solid at 10 K:	0.001 atm

12.20 Butane is a common fuel used in cigarette lighters and camping stoves. Normally supplied in metal containers under pressure, the fuel exists as a mixture of liquid and gas, so high temperatures may cause the container to explode. At 25.0°C, the vapor pressure of butane is 2.3 atm. What is the pressure in the container at 150.°C (ΔH_{vap} = 24.3 kJ/mol)?

12.21 Use Figure 12.8A, to answer the following:
(a) Carbon dioxide is sold in steel cylinders under pressures of approximately 20 atm. Is there liquid CO_2 in the cylinder at room temperature (~20°C)? At 40°C? At −40°C? At −120°C?
(b) Carbon dioxide is also sold as solid chunks, called *dry ice*, in insulated containers. If the chunks are warmed by leaving them in an open container at room temperature, will they melt?
(c) If a container is nearly filled with dry ice and then sealed and warmed to room temperature, will the dry ice melt?
(d) If dry ice is compressed at a temperature below its triple point, will it melt?
(e) Will liquid CO_2 placed in a beaker at room temperature boil?

Types of Intermolecular Forces
(Sample Problems 12.2 and 12.3)

12.22 Why are covalent bonds typically much stronger than intermolecular forces?

12.23 Even though molecules are neutral, the dipole-dipole force is an important interparticle force that exists among them. Explain.

12.24 Oxygen and selenium are members of Group 6A(16). Water forms H bonds, but H_2Se does not. Explain.

12.25 Polar molecules exhibit dipole-dipole forces. Do they also exhibit dispersion forces? Explain.

12.26 Distinguish between *polarizability* and *polarity*. How does each influence intermolecular forces?

12.27 How can one nonpolar molecule induce a dipole in a nearby nonpolar molecule?

12.28 What is the strongest interparticle force in each substance?
(a) CH_3OH (b) CCl_4 (c) Cl_2

12.29 What is the strongest interparticle force in each substance?
(a) H_3PO_4 (b) SO_2 (c) $MgCl_2$

12.30 What is the strongest interparticle force in each substance?
(a) CH_3Cl (b) CH_3CH_3 (c) NH_3

12.31 What is the strongest interparticle force in each substance?
(a) Kr   (b) BrF   (c) H_2SO_4

12.32 Which member of each pair of compounds forms intermolecular H bonds? Draw the H-bonded structures in each case:
(a) CH_3CHCH_3 or CH_3SCH_3 (b) HF or HBr
       |
      OH

12.33 Which member of each pair of compounds forms intermolecular H bonds? Draw the H-bonded structures in each case:
(a) $(CH_3)_2NH$ or $(CH_3)_3N$ (b) $HOCH_2CH_2OH$ or FCH_2CH_2F

12.34 Which has the greater polarizability? Explain.
(a) Br^- or I^- (b) $CH_2=CH_2$ or CH_3-CH_3 (c) H_2O or H_2Se

12.35 Which has the greater polarizability? Explain.
(a) Ca^{2+} or Ca   (b) CH_3CH_3 or $CH_3CH_2CH_3$ (c) CCl_4 or CF_4

12.36 Which member in each pair of liquids has the *higher* vapor pressure at a given temperature? Explain.
(a) C_2H_6 or C_4H_{10} (b) CH_3CH_2OH or CH_3CH_2F
(c) NH_3 or PH_3

12.37 Which member in each pair of liquids has the *lower* vapor pressure at a given temperature? Explain.
(a) $HOCH_2CH_2OH$ or $CH_3CH_2CH_2OH$
(b) CH_3COOH or $(CH_3)_2C=O$ (c) HF or HCl

12.38 Which substance has the *higher* boiling point? Explain.
(a) LiCl or HCl   (b) NH_3 or PH_3 (c) Xe or I_2

12.39 Which substance has the *higher* boiling point? Explain.
(a) CH_3CH_2OH or $CH_3CH_2CH_3$ (b) NO or N_2 (c) H_2S or H_2Te

12.40 Which substance has the *lower* boiling point? Explain.
(a) $CH_3CH_2CH_2CH_3$ or CH_2-CH_2 (b) NaBr or PBr_3
                      | |
                     CH_2-CH_2
(c) H_2O or HBr

12.41 Which substance has the *lower* boiling point? Explain.
(a) CH_3OH or CH_3CH_3 (b) FNO or ClNO
(c)
$$\underset{H}{\overset{F}{}}C=C\underset{H}{\overset{F}{}} \quad or \quad \underset{F}{\overset{H}{}}C=C\underset{H}{\overset{F}{}}$$

12.42 Dispersion forces are the only intermolecular forces present in motor oil, yet it has a high boiling point. Explain.

12.43 Why does the antifreeze ingredient ethylene glycol ($HOCH_2CH_2OH$; $\mathcal{M}$ = 62.07 g/mol) have a boiling point of 197.6°C, whereas propanol ($CH_3CH_2CH_2OH$; $\mathcal{M}$ = 60.09 g/mol), a compound with a similar molar mass, has a boiling point of only 97.4°C?

Properties of the Liquid State

12.44 Before the phenomenon of surface tension was understood, physicists described the surface of water as being covered with a "skin." What causes this skinlike phenomenon?

12.45 Small, equal-sized drops of oil, water, and mercury lie on a waxed floor. How does each liquid behave? Explain.

12.46 Does the *strength* of the intermolecular forces in a liquid change as the liquid is heated? Explain. Why does liquid viscosity decrease with rising temperature?

12.47 Rank the following in order of *increasing* surface tension at a given temperature, and explain your ranking:
(a) $CH_3CH_2CH_2OH$ (b) $HOCH_2CH(OH)CH_2OH$
(c) $HOCH_2CH_2OH$

12.48 Rank the following in order of *decreasing* surface tension at a given temperature, and explain your ranking:
(a) CH_3OH (b) CH_3CH_3 (c) $H_2C=O$

12.49 Use Figure 12.1 to answer the following: (a) Does it take more heat to melt 12.0 g of CH_4 or 12.0 g of Hg? (b) Does it take more heat to vaporize 12.0 g of CH_4 or 12.0 g of Hg? (c) What is the principal intermolecular force in each sample?

12.50 Pentanol ($C_5H_{11}OH$; $\mathcal{M}$ = 88.15 g/mol) has nearly the same molar mass as hexane (C_6H_{14}; $\mathcal{M}$ = 86.17 g/mol) but is more than 12 times as viscous at 20°C. Explain.

The Uniqueness of Water

12.51 For what types of substances is water a good solvent? For what types is it a poor solvent? Explain.

12.52 A water molecule can engage in as many as four H bonds. Explain.

12.53 Warm-blooded animals have a narrow range of body temperature because their bodies have a high water content. Explain.

12.54 A drooping plant can be made upright by watering the ground around it. Explain.

12.55 Describe the molecular basis of the property of water responsible for the presence of ice on the surface of a frozen lake.

12.56 Describe in molecular terms what occurs when ice melts.

The Solid State: Structure, Properties, and Bonding
(Sample Problem 12.4)

12.57 What is the difference between an amorphous solid and a crystalline solid on the macroscopic and molecular levels? Give an example of each.

12.58 How are a solid's unit cell and crystal structure related?

12.59 For structures consisting of identical atoms, how many atoms are contained in the simple, body-centered, and face-centered cubic unit cells? Explain how you obtained the values.

12.60 List four physical characteristics of a solid metal.

12.61 Briefly account for the following relative values:
(a) The melting point of sodium is 89°C, whereas that of potassium is 63°C.
(b) The melting points of Li and Be are 180°C and 1287°C, respectively.
(c) Lithium boils more than 1100°C higher than it melts.

12.62 Magnesium metal is easily deformed by an applied force, whereas magnesium fluoride is shattered. Why do these two solids behave so differently?

12.63 What is the energy gap in band theory? Compare its size in superconductors, conductors, semiconductors, and insulators.

12.64 What type of crystal lattice does each metal form? (The number of atoms per unit cell is given in parentheses.)
(a) Ni (4)   (b) Cr (2)   (c) Ca (4)

12.65 What is the number of atoms per unit cell for each metal?
(a) Polonium, Po   (b) Iron, Fe   (c) Silver, Ag

12.66 Of the five major types of crystalline solid, which does each of the following form: (a) Sn; (b) Si; (c) Xe?

12.67 Of the five major types of crystalline solid, which does each of the following form: (a) cholesterol ($C_{27}H_{45}OH$); (b) KCl; (c) BN?

12.68 Zinc oxide adopts the zinc blende crystal structure (Figure P12.68). How many Zn^{2+} ions are in the ZnO unit cell?

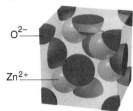

12.69 Calcium sulfide adopts the sodium chloride crystal structure (Figure P12.69). How many S^{2-} ions are in the CaS unit cell?

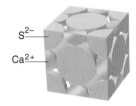

12.70 Zinc selenide (ZnSe) crystallizes in the zinc blende structure and has a density of 5.42 g/cm³ (see Figure P12.68).
(a) How many Zn and Se ions are in each unit cell?
(b) What is the mass of a unit cell?
(c) What is the volume of a unit cell?
(d) What is the edge length of a unit cell?

12.71 An element crystallizes in a face-centered cubic lattice and has a density of 1.45 g/cm³. The edge of its unit cell is 4.52×10^{-8} cm.
(a) How many atoms are in each unit cell?
(b) What is the volume of a unit cell?
(c) What is the mass of a unit cell?
(d) Calculate an approximate atomic mass for the element.

12.72 Classify each of the following as a conductor, insulator, or semiconductor: (a) phosphorus; (b) mercury; (c) germanium.

12.73 Predict the effect (if any) of an increase in temperature on the electrical conductivity of (a) antimony, Sb; (b) tellurium, Te; (c) bismuth, Bi.

12.74 Use condensed electron configurations to predict the relative hardnesses and melting points of rubidium ($Z = 37$), vanadium ($Z = 23$), and cadmium ($Z = 48$).

12.75 One of the most important enzymes in the world—nitrogenase, the plant protein that catalyzes nitrogen fixation—contains active clusters of iron, sulfur, and molybdenum atoms. Crystalline molybdenum (Mo) has a body-centered cubic unit cell (d of Mo = 10.28 g/cm³). (a) Determine the edge length of the unit cell. (b) Calculate the atomic radius of Mo.

Comprehensive Problems

Problems with an asterisk (*) are more challenging.

12.76 Because bismuth has several well-characterized solid, crystalline phases, it is used to calibrate instruments employed in high-pressure studies. The following phase diagram for bismuth shows the liquid phase and five different solid phases stable above 1 katm (1000 atm) and up to 300°C. (a) Which solid phases are stable at 25°C? (b) Which phase is stable at 50 katm and 175°C? (c) Identify the phase transitions that bismuth undergoes at 200°C as the pressure is reduced from 100 to 1 katm. (d) What phases are present at each of the triple points?

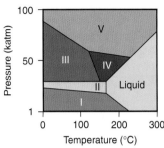

12.77 Mercury (Hg) vapor is toxic and readily absorbed from the lungs. At 20.°C, mercury ($\Delta H_{vap} = 59.1$ kJ/mol) has a vapor pressure of 1.20×10^{-3} torr, which is high enough to be hazardous. To reduce the danger to workers in processing plants, Hg is cooled to lower its vapor pressure. At what temperature would the vapor pressure of Hg be at the safer level of 5.0×10^{-5} torr?

12.78 Consider the phase diagram shown for substance X. (a) What phase(s) is (are) present at point A? E? F? H? B? C? (b) Which point corresponds to the critical point? Which point corresponds to the triple point? (c) What curve corresponds to conditions at which the solid and gas are in equilibrium? (d) Describe what happens when you start at point A and increase the temperature at constant pressure. (e) Describe what happens when you start at point H and decrease the pressure at constant temperature. (f) Is liquid X more or less dense than solid X?

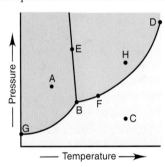

12.79 Some high-temperature superconductors adopt a crystal structure similar to that of *perovskite* ($CaTiO_3$). The unit cell is cubic with a Ti^{4+} ion in each corner, a Ca^{2+} ion in the body center, and O^{2-} ions at the midpoint of each edge. (a) Is this unit cell simple, body-centered, or face-centered? (b) If the unit cell edge length is 3.84 Å, what is the density of perovskite (in g/cm³)?

12.80 The only alkali metal halides that do not adopt the NaCl structure are CsCl, CsBr, and CsI, formed from the largest alkali metal cation and the three largest halide ions. These crystallize in the *cesium chloride structure* (shown here for CsCl). This structure has been used as an example of how dispersion forces can dominate in the presence of ionic forces. Use the ideas of coordination number and polarizability to explain why the CsCl structure exists.

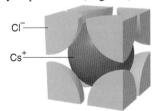

* **12.81** Corn is a valuable source of industrial chemicals. For example, furfural is prepared from corncobs. It is an important reactant in plastics manufacturing and a key solvent for the production of cellulose acetate, which is used to make everything from videotape to waterproof fabric. It can be reduced to furfuryl alcohol or oxidized to 2-furoic acid.

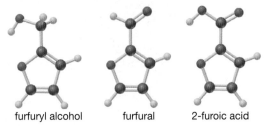

furfuryl alcohol   furfural   2-furoic acid

(a) Which of these compounds can form H bonds? Draw structures in each case.

(b) The molecules of some substances can form an "internal" H bond, that is, an H bond *within* a molecule. This takes the form of a polygon with atoms as corners and bonds as sides and an H bond as one of the sides. Which of these molecules is (are) likely to form a stable internal H bond? Draw the structure. (*Hint:* Structures with 5 or 6 atoms as corners are most stable.)

* **12.82** A 4.7-L sealed bottle containing 0.33 g of liquid ethanol, C_2H_6O, is placed in a refrigerator and reaches equilibrium with its vapor at $-11°C$. (a) What mass of ethanol is present in the vapor? (b) When the container is removed and warmed to room temperature, 20.°C, will all the ethanol vaporize? (c) How much liquid ethanol would be present at 0.0°C? The vapor pressure of ethanol is 10. torr at $-2.3°C$ and 40. torr at 19°C.

12.83 A cubic unit cell contains atoms of element A at each corner and atoms of element Z on each face. What is the empirical formula of the compound?

12.84 Is it possible for a salt of formula AB_3 to have a face-centered cubic unit cell of anions with cations in all the eight available holes? Explain.

* **12.85** In a body-centered cubic unit cell, the central atom lies on an internal diagonal of the cell and touches the corner atoms. (a) Find the length of the diagonal in terms of r, the atomic radius. (b) If the edge length of the cube is a, what is the length of a *face* diagonal? (c) Derive an expression for a in terms of r. (d) How many atoms are in this unit cell? (e) What fraction of the unit cell volume is filled with spheres?

* **12.86** KF has the same type of crystal structure as NaCl. The unit cell of KF has an edge length of 5.39 Å. Find the density of KF.

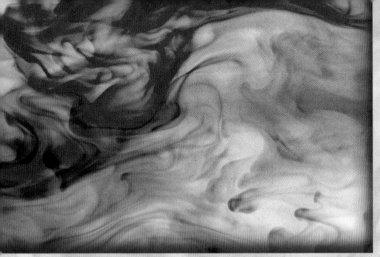

CHAPTER THIRTEEN

The Properties of Solutions

Mixing It Up Solutions are everywhere, both inside and outside us. Here, dyes swirl as they dissolve in an aqueous solvent. As you'll see in this chapter, calculating solution concentrations and predicting their properties are crucial skills in science and industry.

Key Principles

◆ *Solubility* refers to the amount of solute that can dissolve in a solvent at a given temperature. The *like-dissolves-like rule* summarizes the fact that solutions form when solute and solvent have similar types of *intermolecular forces.*

◆ Dissolving involves *enthalpy* changes: heat is absorbed to separate solute and solvent particles and is released when they mix. The relative magnitudes of these quantities of heat determine whether the overall *heat of solution* is positive or negative, that is, whether the process of dissolving is endothermic or exothermic. Dissolving also involves changes in particles' freedom of motion and thus in the distribution of their energy, which are related to changes in a quantity called *entropy.*

◆ A solution is *saturated* when the maximum amount of solute has dissolved at a given temperature; at that point, excess solute and dissolved solute are in *equilibrium.* Solubility is related to conditions:

most solids are more soluble in water at higher temperatures; all gases are less soluble in water at higher temperatures; and the solubility of a gas is directly proportional to its pressure *(Henry's law).*

◆ The *concentration* of a solution can be expressed in different terms *(molarity, molality, parts by mass, parts by volume,* and *mole fraction)*. Because a concentration is a ratio involving mass, volume, and/or amount (mol), the various terms are *interconvertible.*

◆ The physical properties of a solution differ from those of the solvent. These properties *(vapor-pressure lowering, boiling-point elevation, freezing-point depression,* and *osmotic pressure)* are called *colligative* because they depend on the number, not the chemical nature, of the dissolved particles. In salt solutions, interactions among ions cause deviations from expected properties.

Outline

Concepts & Skills to Review Before You Study This Chapter

- classification of mixtures (Section 2.9)
- calculations involving mass percent (Section 3.1) and molarity (Section 3.5)
- electrolytes; water as a solvent (Sections 4.1 and 12.5)
- mole fraction and Dalton's law of partial pressures (Section 5.4)
- types of intermolecular forces and the concept of polarizability (Section 12.3)
- vapor pressure of liquids (Section 12.2)

Nearly all the gases, liquids, and solids that make up our world are *mixtures*—two or more substances physically mixed together but not chemically combined. Synthetic mixtures, such as glass and soap, usually contain relatively few components, whereas natural mixtures, such as seawater and soil, are more complex, often containing more than 50 different substances. Living mixtures, such as trees and students, are the most complex—even a simple bacterial cell contains well over 5000 different compounds (Table 13.1).

Table 13.1 Approximate Composition of a Bacterium

Substance	Mass % of Cell	Number of Types	Number of Molecules
Water	~70	1	5×10^{10}
Ions	1	20	?
Sugars*	3	200	3×10^{8}
Amino acids*	0.4	100	5×10^{7}
Lipids*	2	50	3×10^{7}
Nucleotides*	0.4	200	1×10^{7}
Other small molecules	0.2	~200	?
Macromolecules (proteins, nucleic acids, polysaccharides)	23	~5000	6×10^{6}

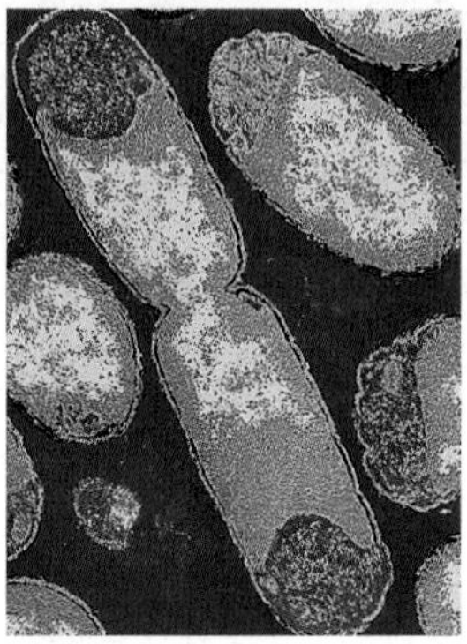

*Includes precursors and metabolites.

Recall from Chapter 2 that a mixture has two defining characteristics: *its composition is variable,* and *it retains some properties of its components.* In this chapter, we focus on solutions, the most common type of mixture. *A solution is a homogeneous mixture,* one with no boundaries separating its components; thus, a solution exists as one phase. A *heterogeneous mixture* has two or more phases. The pebbles in concrete or the bubbles in champagne are visible indications that these are heterogeneous mixtures. In some cases, the particles of one or more components may be very small, so distinct phases are not easy to see. Smoke and milk are heterogeneous mixtures with very small component particles and thus no visible distinct phases. The essential distinction is that in a solution all the particles are individual atoms, ions, or small molecules.

In this chapter, we discuss the types of solutions, why they form, the different concentration units that describe them, and how their properties differ from those of pure substances.

13.1 TYPES OF SOLUTIONS: INTERMOLECULAR FORCES AND SOLUBILITY

We often describe solutions in terms of one substance dissolving in another: the **solute** dissolves in the **solvent.** Usually, *the solvent is the most abundant component* of a given solution. In some cases, however, the substances are **miscible,** that is, soluble in each other in any proportion; in such cases, it may not be meaningful to call one the solute and the other the solvent.

The **solubility** (*S*) of a solute is the maximum amount that dissolves in a fixed quantity of a particular solvent at a specified temperature, given that excess solute is present. Different solutes have different solubilities. For example, for sodium chloride (NaCl), S = 39.12 g/100. mL water at 100.°C, whereas for silver chloride (AgCl), S = 0.0021 g/100. mL water at 100.°C. Obviously, NaCl is much more soluble in water than AgCl is. (Solubility is also expressed in other units, as you'll see later.) Although solubility has a quantitative meaning, *dilute* and *concentrated* are qualitative terms that refer to the *relative amounts of solute:* a dilute solution contains much less dissolved solute than a concentrated one.

From everyday experience, you know that some solvents can dissolve a given solute, whereas others can't. For example, butter doesn't dissolve in water, but it does in cooking oil. A major factor determining whether a solution forms is the relative strength of the intermolecular forces within and between solute and solvent. Experience shows that *substances with similar types of intermolecular forces dissolve in each other*. This fact is summarized in the rule-of-thumb **like dissolves like.** From a knowledge of these forces, we can often predict which solutes will dissolve in which solvents.

Intermolecular Forces in Solution

All the intermolecular forces we discussed in Section 12.3 for pure substances also occur between solute and solvent in solutions (Figure 13.1):

1. *Ion-dipole forces* are the principal force involved in the solubility of ionic compounds in water. When a salt dissolves, each ion on the crystal's surface attracts the oppositely charged end of the water dipole. These attractive forces overcome those between the ions and break down the crystal structure. As each ion becomes separated, more water molecules cluster around it in **hydration shells** (Figure 13.2). Recent evidence shows that water's normal H bonding is disrupted only among molecules in the closest hydration shell. These molecules are H bonded to others slightly farther away, which are in turn H bonded to other molecules in the bulk solvent. It's important to remember that this process does not lead to a jumble of ions and water molecules; rather, there is less freedom of motion for molecules in the closest hydration shell. For monatomic ions, the number of water molecules in the closest shell depends on the ion's size. Four water molecules can fit tetrahedrally around small ions, such as Li^+, while larger ions, such as Na^+ and F^-, have six water molecules surrounding them octahedrally.

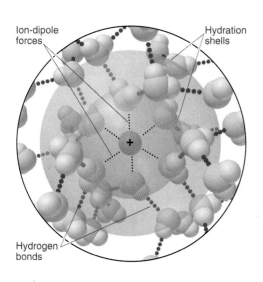

Ion-dipole forces

Hydration shells

Hydrogen bonds

Figure 13.2 Hydration shells around an aqueous ion. When an ionic compound dissolves in water, ion-dipole forces orient water molecules around the separated ions to form hydration shells. The cation shown here is octahedrally surrounded by six water molecules, which form H bonds with water molecules in the next hydration shell, and those form H bonds with others farther away.

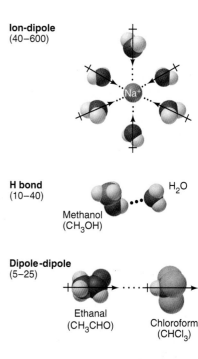

Ion-dipole
(40–600)

H bond
(10–40)

Methanol
(CH_3OH)

H_2O

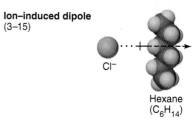

Dipole-dipole
(5–25)

Ethanal
(CH_3CHO)

Chloroform
($CHCl_3$)

Ion–induced dipole
(3–15)

Cl^-

Hexane
(C_6H_{14})

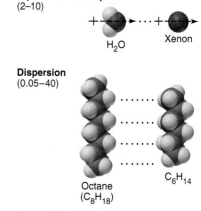

Dipole–induced dipole
(2–10)

H_2O

Xenon

Dispersion
(0.05–40)

Octane
(C_8H_{18})

C_6H_{14}

Figure 13.1 The major types of intermolecular forces in solutions. Forces are listed in decreasing order of strength (with values in kJ/mol), and an example of each is shown with space-filling models.

2. *Hydrogen bonding* is especially important in aqueous solution. In fact, it is a primary factor in the solubility in water—and, thus, cell fluid—of many oxygen- and nitrogen-containing organic and biological compounds, such as alcohols, sugars, amines, and amino acids. (Recall that O and N are small and electronegative, so their bound H atoms are partially positive and can get very close to the O of H_2O.)

3. *Dipole-dipole forces,* in the absence of H bonding, account for the solubility of polar organic molecules, such as ethanal (acetaldehyde, CH_3CHO), in polar, nonaqueous solvents like chloroform ($CHCl_3$).

4. **Ion–induced dipole forces** are one of two types of *charge-induced dipole forces*, which rely on the polarizability of the components. They result when an ion's charge distorts the electron cloud of a nearby nonpolar molecule. This type of force plays an essential biological role that initiates the binding of the Fe^{2+} ion in hemoglobin and an O_2 molecule in the bloodstream. Because an ion increases the magnitude of any nearby dipole, ion–induced dipole forces also contribute to the solubility of salts in less polar solvents, such as LiCl in ethanol.

5. **Dipole–induced dipole forces,** also based on polarizability, are weaker than the ion–induced dipole forces because the magnitude of the charge is smaller (Coulomb's law). They arise when a polar molecule distorts the electron cloud of a nearby nonpolar molecule. The solubility in water, limited though it is, of atmospheric O_2, N_2, and the noble gases is due in large part to these forces. Paint thinners and grease solvents also function through dipole–induced dipole forces.

6. *Dispersion forces* contribute to the solubility of all solutes in all solvents, but they are the *principal* type of intermolecular force in solutions of nonpolar substances, such as petroleum and gasoline.

As you'll see in Chapter 15, these intermolecular forces are also responsible for keeping cellular macromolecules in their biologically active shapes.

Liquid Solutions and the Role of Molecular Polarity

Solutions can be gaseous, liquid, or solid. In general, *the physical state of the solvent determines the physical state of the solution.* We focus on liquid solutions because they are by far the most common and important.

From cytoplasm to tree sap, gasoline to cleaning fluid, iced tea to urine, solutions in which the solvent is a liquid are familiar in everyday life. Water is the most prominent solvent because it is so common and dissolves so many ionic and polar substances. But there are many other liquid solvents, and their polarity ranges from very polar to nonpolar.

Liquid-Liquid and Solid-Liquid Solutions Many salts dissolve in water because the strong ion-dipole attractions that water molecules form with the ions are very similar to the strong attractions between the ions themselves and, therefore, can *substitute* for them. The same salts are insoluble in hexane (C_6H_{14}) because the weak ion–induced dipole forces their ions could form with the nonpolar molecules of this solvent *cannot* substitute for attractions between the ions. Similarly, oil does not dissolve in water because the weak dipole–induced dipole forces between oil and water molecules cannot substitute for the strong H bonds between water molecules. Oil *does* dissolve in hexane, however, because the dispersion forces in one substitute readily for the dispersion forces in the other. Thus, for a solution to form, "like dissolves like" means that the forces created between solute and solvent must be *comparable in strength* to the forces destroyed within both the solute and the solvent.

To examine this idea further, let's compare the solubilities of a series of alcohols in two solvents that act through very different intermolecular forces—water and hexane. Alcohols are organic molecules with a hydroxyl (—OH) group bound to a hydrocarbon group. The simplest type of alcohol has the general formula $CH_3(CH_2)_nOH$; we'll consider alcohols with $n = 0$ to 5. We can view an alcohol molecule as consisting of two portions: the polar —OH group and the nonpolar hydrocarbon chain. The —OH portion forms strong H bonds with water and weak dipole–induced dipole forces with hexane. The hydrocarbon portion interacts through dispersion forces with hexane and through dipole–induced dipole forces with water.

In Table 13.2, the models show the relative change in size of the polar and nonpolar portions of the alcohol molecules. In the smaller alcohols (one to three

Table 13.2   Solubility* of a Series of Alcohols in Water and Hexane

Alcohol	Model	Solubility in Water	Solubility in Hexane
CH_3OH (methanol)		∞	1.2
CH_3CH_2OH (ethanol)		∞	∞
$CH_3(CH_2)_2OH$ (1-propanol)		∞	∞
$CH_3(CH_2)_3OH$ (1-butanol)		1.1	∞
$CH_3(CH_2)_4OH$ (1-pentanol)		0.30	∞
$CH_3(CH_2)_5OH$ (1-hexanol)		0.058	∞

*Expressed in mol alcohol/1000 g solvent at 20°C.

carbons), the hydroxyl group is a relatively large portion, so the molecules interact with each other through H bonding, just as water molecules do. When the smaller alcohols mix with water, H bonding within solute and within solvent is replaced by H bonding *between* solute and solvent (Figure 13.3). As a result, these smaller alcohols are miscible with water.

Water solubility decreases dramatically for alcohols larger than three carbons, and those with chains longer than six carbons are insoluble in water. For these larger alcohols to dissolve, the nonpolar chains have to move between the water molecules, substituting their weak attractions with those water molecules for strong H bonds among the water molecules themselves. The —OH portion of the

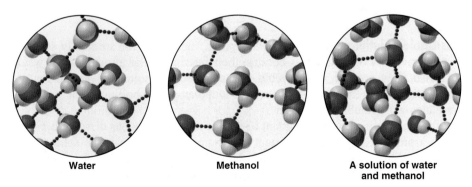

Water   **Methanol**   **A solution of water and methanol**

Figure 13.3 Like dissolves like: solubility of methanol in water. The H bonds in water and in methanol are similar in type and strength, so they can substitute for one another. Thus, methanol is soluble in water; in fact, the two substances are miscible.

alcohol does form H bonds to water, but these don't outweigh the H bonds among water molecules that break to make room for the hydrocarbon portion.

Table 13.2 shows that the opposite trend occurs with hexane. Now the major solute-solvent *and* solvent-solvent interactions are dispersion forces. The weak forces between the —OH group of methanol (CH_3OH) and hexane cannot substitute for the strong H bonding among CH_3OH molecules, so the solubility of methanol in hexane is relatively low. In any larger alcohol, however, dispersion forces become increasingly more important, and these *can* substitute for dispersion forces in pure hexane; thus, solubility increases. With no strong solvent-solvent forces to be replaced by weak solute-solvent forces, even the two-carbon chain of ethanol has strong enough solute-solvent attractions to be miscible in hexane.

Many other organic molecules have polar and nonpolar portions, and the predominance of one portion or the other determines their solubility in different solvents. For example, carboxylic acids and amines behave very much like alcohols. Thus, methanoic acid (HCOOH, formic acid) and methanamine (CH_3NH_2) are miscible with water and much less soluble in hexane, whereas hexanoic acid [$CH_3(CH_2)_4COOH$] and 1-hexanamine [$CH_3(CH_2)_5NH_2$] are slightly soluble in water and very soluble in hexane.

SAMPLE PROBLEM 13.1 Predicting Relative Solubilities of Substances

Problem Predict which solvent will dissolve more of the given solute:
(a) Sodium chloride in methanol (CH_3OH) or in 1-propanol ($CH_3CH_2CH_2OH$)
(b) Ethylene glycol ($HOCH_2CH_2OH$) in hexane ($CH_3CH_2CH_2CH_2CH_2CH_3$) or in water
(c) Diethyl ether ($CH_3CH_2OCH_2CH_3$) in water or in ethanol (CH_3CH_2OH)
Plan We examine the formulas of the solute and each solvent to determine the types of forces that could occur. A solute tends to be more soluble in a solvent whose intermolecular forces are similar to, and therefore can substitute for, its own.
Solution (a) Methanol. NaCl is an ionic solid that dissolves through ion-dipole forces. Both methanol and 1-propanol contain a polar —OH group, but 1-propanol's longer hydrocarbon chain can form only weak forces with the ions, so it is less effective at substituting for the ionic attractions in the solute.
(b) Water. Ethylene glycol molecules have two —OH groups, so the molecules interact with each other through H bonding. They are more soluble in H_2O, whose H bonds can substitute for their own H bonds better than the dispersion forces in hexane can.
(c) Ethanol. Diethyl ether molecules interact with each other through dipole-dipole and dispersion forces and can form H bonds to both H_2O and ethanol. The ether is more soluble in ethanol because that solvent can form H bonds *and* substitute for the ether's dispersion forces. Water, on the other hand, can form H bonds with the ether, but it lacks any hydrocarbon portion, so it forms much weaker dispersion forces with that solute.

FOLLOW-UP PROBLEM 13.1 Which solute is more soluble in the given solvent?
(a) 1-Butanol ($CH_3CH_2CH_2CH_2OH$) or 1,4-butanediol ($HOCH_2CH_2CH_2CH_2OH$) in water
(b) Chloroform ($CHCl_3$) or carbon tetrachloride (CCl_4) in water

---

Table 13.3	Correlation Between Boiling Point and Solubility in Water	
Gas	Solubility (M)*	bp (K)
He	4.2×10^{-4}	4.2
Ne	6.6×10^{-4}	27.1
N_2	10.4×10^{-4}	77.4
CO	15.6×10^{-4}	81.6
O_2	21.8×10^{-4}	90.2
NO	32.7×10^{-4}	121.4

*At 273 K and 1 atm.

Gas-Liquid Solutions Gases that are nonpolar, such as N_2, or are nearly so, such as NO, have low boiling points because their intermolecular attractions are weak. Likewise, they are not very soluble in water because solute-solvent forces are weak. In fact, as Table 13.3 shows, for nonpolar gases, boiling point generally correlates with solubility in water.

In some cases, the small amount of a nonpolar gas that does dissolve is essential to a process. The most important environmental example is the solubility of O_2 in water. At 25°C and 1 atm, the solubility of O_2 is only 3.2 mL/100. mL of water, but aquatic animal life would die without this small amount. In other cases, the solubility of a gas may *seem* high because the gas is not only dissolving but also reacting with the solvent or another component. Oxygen seems much more soluble in blood than in water because O_2 molecules are continually bonding with hemoglobin molecules in red blood cells. Similarly, carbon dioxide, which is essential for aquatic plants and coral-reef growth, seems very soluble in water (~81 mL of CO_2/100. mL of H_2O at 25°C and 1 atm) because it is reacting, in addition to simply dissolving:

$$CO_2(g) + H_2O(l) \longrightarrow H^+(aq) + HCO_3^-(aq)$$

Gas Solutions and Solid Solutions

Despite the central place of liquid solutions in chemistry, gaseous solutions and solid solutions also have vital importance and numerous applications.

Gas-Gas Solutions *All gases are infinitely soluble in one another.* Air is the classic example of a gaseous solution, consisting of about 18 gases in widely differing proportions. Anesthetic gas proportions are finely adjusted to the needs of the patient and the length of the surgical procedure.

Gas-Solid Solutions *When a gas dissolves in a solid, it occupies the spaces between the closely packed particles.* Hydrogen gas can be purified by passing an impure sample through a solid metal such as palladium. Only the H_2 molecules are small enough to enter the spaces between the Pd atoms, where they form Pd—H covalent bonds. Under high H_2 pressure, the H atoms are passed along the Pd crystal structure and emerge from the solid as H_2 molecules.

Solid-Solid Solutions Because solids diffuse so little, their mixtures are usually heterogeneous, as in gravel mixed with sand. Some solid-solid solutions can be formed by melting the solids and then mixing them and allowing them to freeze. Many **alloys,** mixtures of elements that have an overall metallic character, are examples of solid-solid solutions (although several common alloys have microscopic heterogeneous regions). Brass, a familiar example of an alloy, is a mixture of zinc and copper.

Waxes are another familiar type of solid-solid solution. Most waxes are amorphous solids that may contain small regions of crystalline regularity. A natural *wax* is a solid of biological origin that is insoluble in water but dissolves in nonpolar solvents.

SECTION SUMMARY
Solutions are homogeneous mixtures consisting of a solute dissolved in a solvent through the action of intermolecular forces. The solubility of a solute in a given amount of solvent is the maximum amount that can dissolve at a specified temperature. (For gaseous solutes, the pressure must be specified also.) In addition to the intermolecular forces that exist in pure substances, ion-dipole, ion–induced dipole, and dipole–induced dipole forces occur in solution. If similar intermolecular forces occur in solute and solvent, a solution will likely form ("like dissolves like"). When ionic compounds dissolve in water, the ions become surrounded by hydration shells of H-bonded water molecules. Solubility of organic molecules in various solvents depends on their polarity and the extent of their polar and nonpolar portions. The solubility of nonpolar gases in water is low because of weak intermolecular forces. Gases are miscible with one another and dissolve in solids by fitting into spaces in the crystal structure. Solid-solid solutions, such as alloys and waxes, form when the components are mixed while molten.

13.2 WHY SUBSTANCES DISSOLVE: UNDERSTANDING THE SOLUTION PROCESS

As a qualitative predictive tool, "like dissolves like" is helpful in many cases. As you might expect, this handy *macroscopic* rule is based on the *molecular* interactions that occur between solute and solvent particles. To see *why* like dissolves like, let's break down the solution process conceptually into steps and examine them in terms of changes in *enthalpy* and *entropy* of the system. We discussed enthalpy in Chapter 6 and focus on it first here. The concept of entropy is introduced at the end of this section and treated quantitatively in Chapter 20.

Heats of Solution and Solution Cycles

Picture a general solute and solvent about to form a solution. Both consist of particles attracting each other. For one substance to dissolve in another, three events must occur: (1) solute particles must separate from each other, (2) some solvent particles must separate to make room for the solute particles, and (3) solute and solvent particles must mix together. No matter what the nature of the attractions within the solute and within the solvent, some energy must be *absorbed* for particles to separate, and some energy is *released* when they mix and attract each other. As a result of these changes, the solution process is typically accompanied by a change in enthalpy. We can divide the process into these three steps, each with its own enthalpy term:

Step 1. Solute particles separate from each other. This step involves overcoming intermolecular attractions, so it is *endothermic:*

$$\text{Solute (aggregated)} + heat \longrightarrow \text{solute (separated)} \qquad \Delta H_{\text{solute}} > 0$$

Step 2. Solvent particles separate from each other. This step also involves overcoming attractions, so it is *endothermic,* too:

$$\text{Solvent (aggregated)} + heat \longrightarrow \text{solvent (separated)} \qquad \Delta H_{\text{solvent}} > 0$$

Step 3. Solute and solvent particles mix. The particles attract each other, so this step is *exothermic:*

$$\text{Solute (separated)} + \text{solvent (separated)} \longrightarrow \text{solution} + heat \qquad \Delta H_{\text{mix}} < 0$$

The total enthalpy change that occurs when a solution forms from solute and solvent is the **heat of solution (ΔH_{soln}),** and we combine the three individual enthalpy changes to find it. The overall process is called a *thermochemical solution cycle* and is yet another application of Hess's law:

$$\Delta H_{\text{soln}} = \Delta H_{\text{solute}} + \Delta H_{\text{solvent}} + \Delta H_{\text{mix}} \qquad (13.1)$$

If the sum of the endothermic terms ($\Delta H_{\text{solute}} + \Delta H_{\text{solvent}}$) is smaller than the exothermic term (ΔH_{mix}), ΔH_{soln} is negative; that is, the process is exothermic. Figure 13.4A is an enthalpy diagram for the formation of such a solution. If the sum of the endothermic terms is larger than the exothermic term, ΔH_{soln} is positive; that is, the process is endothermic (Figure 13.4B). However, if ΔH_{soln} is highly positive, the solute may not dissolve to any significant extent in that solvent.

Figure 13.4 Solution cycles and the enthalpy components of the heat of solution. ΔH_{soln} can be thought of as the sum of three enthalpy changes: $\Delta H_{\text{solvent}}$ (separating the solvent; always >0), ΔH_{solute} (separating the solute; always >0), and ΔH_{mix} (mixing solute and solvent; always <0). **A,** ΔH_{mix} is larger than the sum of ΔH_{solute} and $\Delta H_{\text{solvent}}$, so ΔH_{soln} is negative (exothermic process). **B,** ΔH_{mix} is smaller than the sum of the others, so ΔH_{soln} is positive (endothermic process).

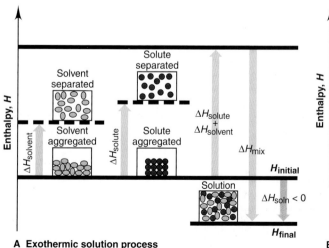

A Exothermic solution process

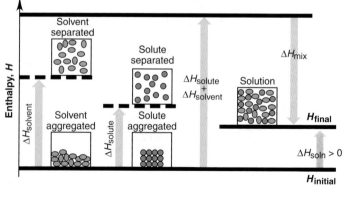

B Endothermic solution process

Heats of Hydration: Ionic Solids in Water

We can simplify the solution cycle for aqueous systems. The $\Delta H_{solvent}$ and ΔH_{mix} components of the heat of solution are difficult to measure individually. Combined, these terms represent the enthalpy change during **solvation,** the process of surrounding a solute particle with solvent particles. Solvation in water is called **hydration.** Thus, enthalpy changes for separating the water molecules ($\Delta H_{solvent}$) and mixing the solute with them (ΔH_{mix}) are combined into the **heat of hydration (ΔH_{hydr}).** In water, Equation 13.1 becomes

$$\Delta H_{soln} = \Delta H_{solute} + \Delta H_{hydr}$$

The heat of hydration is a key factor in dissolving an ionic solid. Breaking H bonds in water is more than compensated for by forming strong ion-dipole forces, so hydration of an ion is *always* exothermic. The ΔH_{hydr} of an ion is defined as the enthalpy change for the hydration of 1 mol of separated (gaseous) ions:

$$M^+(g) \text{ [or } X^-(g)] \xrightarrow{H_2O} M^+(aq) \text{ [or } X^-(aq)] \qquad \Delta H_{hydr \text{ of the ion}} \text{ (always } <0)$$

Heats of hydration exhibit trends based on the **charge density** of the ion, the ratio of the ion's charge to its volume. In general, the higher the charge density is, the more negative ΔH_{hydr} is. According to Coulomb's law, the greater the ion's charge is and the closer the ion can approach the oppositely charged end of the water molecule's dipole, the stronger the attraction. Therefore,

- A 2+ ion attracts H_2O molecules more strongly than a 1+ ion of similar size.
- A small 1+ ion attracts H_2O molecules more strongly than a large 1+ ion.

The energy required to separate an ionic solute (ΔH_{solute}) into gaseous ions is its lattice energy ($\Delta H_{lattice}$), and ΔH_{solute} is highly positive:

$$M^+X^-(s) \longrightarrow M^+(g) + X^-(g) \qquad \Delta H_{solute} \text{ (always } >0) = \Delta H_{lattice}$$

Thus, the heat of solution for ionic compounds in water combines the lattice energy (always positive) and the combined heats of hydration of cation and anion (always negative),

$$\Delta H_{soln} = \Delta H_{lattice} + \Delta H_{hydr \text{ of the ions}} \tag{13.2}$$

The sizes of the individual terms determine the sign of the heat of solution.

Figure 13.5 shows enthalpy diagrams for dissolving three ionic solutes in water. The first, NaCl, has a small positive heat of solution ($\Delta H_{soln} = 3.9$ kJ/mol).

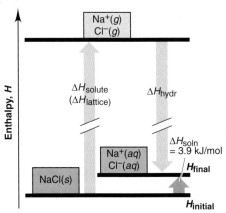

A NaCl. $\Delta H_{lattice}$ is slightly larger than ΔH_{hydr}: ΔH_{soln} is small and positive.

Figure 13.5 Dissolving ionic compounds in water. The enthalpy diagram for an ionic compound in water includes $\Delta H_{lattice}$ (ΔH_{solute}; always positive) and the combined ionic heats of hydration (ΔH_{hydr}; always negative).

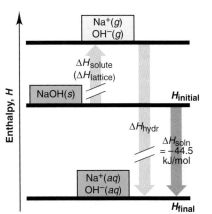

B NaOH. ΔH_{hydr} dominates: ΔH_{soln} is large and negative.

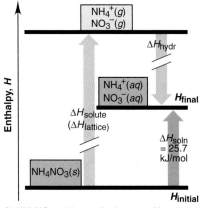

C NH4NO3. $\Delta H_{lattice}$ dominates: ΔH_{soln} is large and positive.

Its lattice energy is only slightly greater than the combined ionic heats of hydration, so if you dissolve NaCl in water in a flask, you do not notice any temperature change. However, if you dissolve NaOH in water, the flask feels hot. The lattice energy for NaOH is much smaller than the combined ionic heats of hydration, so dissolving NaOH is highly exothermic ($\Delta H_{soln} = -44.5$ kJ/mol). Finally, if you dissolve NH_4NO_3 in water, the flask feels cold. In this case, the lattice energy is much larger than the combined ionic heats of hydration, so the process is quite endothermic ($\Delta H_{soln} = 25.7$ kJ/mol).

The Solution Process and the Change in Entropy

If it takes energy ($\Delta H_{soln} > 0$) for NaCl and NH_4NO_3 to dissolve, why do they? It turns out that the heat of solution is only one of two factors that determine whether a solution forms. The other concerns the natural tendency of any system to distribute its energy in as many ways as possible. A thermodynamic variable called **entropy (S)** is directly related to the number of ways that a system can distribute its energy, which in turn is closely related to the freedom of motion of the particles and the number of ways they can be arranged.

Let's see what it means for a system to "distribute its energy" by comparing the three physical states first and then solute and solvent with solution. In a solid, the particles are relatively fixed in their positions, but in a liquid, they are free to move around each other. This greater freedom of motion means the particles can distribute their kinetic energy in more ways; thus, the liquid has higher entropy than the solid ($S_{liquid} > S_{solid}$). The gas, in turn, has higher entropy than the liquid because the particles have much more freedom of motion ($S_{gas} > S_{liquid}$). Another way to state this is that the change in entropy when the liquid vaporizes to a gas (ΔS_{vap}) is positive; that is, $\Delta S_{vap} > 0$.

Similarly, *a solution usually has higher entropy than the pure solute and pure solvent;* in other words, solutions form naturally but pure solute and pure solvent do not. The number of ways to distribute the energy and the freedom of motion of the particles are related to the number of different interactions between the particles, and there are far more interactions possible when solute and solvent are mixed than when they are pure; thus, $S_{soln} > (S_{solute} + S_{solvent})$, or $\Delta S_{soln} > 0$. The solution process involves the interplay of the change in enthalpy *and* the change in entropy: *systems tend toward a state of lower enthalpy and higher entropy,* and the relative magnitudes of ΔH_{soln} and ΔS_{soln} determine whether a solution forms.

To see this interplay of enthalpy and entropy, let's consider three solute-solvent pairs in which different influences dominate. In our first example, sodium chloride does not dissolve in hexane (C_6H_{14}), as you would predict from the dissimilar intermolecular forces (Figure 13.6A). Separating the solvent is easy because the dispersion forces are relatively weak, but separating the ionic solute requires $\Delta H_{lattice}$. Mixing releases very little heat because the ion–induced dipole attractions between Na^+ (or Cl^-) ions and hexane molecules are weak. Therefore, ΔH_{soln} is highly positive. A solution does not form because the entropy increase from mixing NaCl and hexane is much smaller than $\Delta H_{lattice}$.

The second example is octane (C_8H_{18}) dissolving in hexane. Both consist of nonpolar molecules held together by dispersion forces, so we predict that octane is soluble in hexane; in fact, they are infinitely soluble. But ΔH_{soln} is around zero (Figure 13.6B). With no release of heat driving the process, octane dissolves in hexane because the entropy increases greatly when the pure substances mix.

In the case of NaCl and especially NH_4NO_3 dissolving in water, $\Delta H_{soln} > 0$, but the increase in entropy that occurs when the crystal breaks down and the ions mix with water more than compensates for the increase in enthalpy. (The enthalpy/entropy interplay in physical and chemical systems is covered in depth in Chapter 20.)

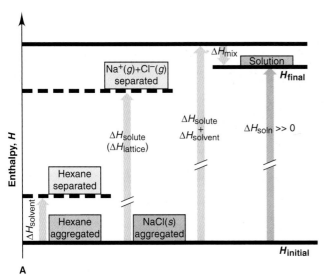

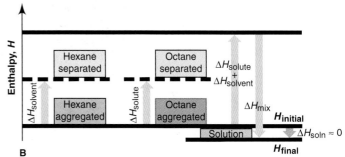

Figure 13.6 Enthalpy diagrams for dissolving NaCl and octane in hexane. A, Because attractions between Na$^+$ (or Cl$^-$) ions and hexane molecules are weak, ΔH_{mix} is *much* smaller than ΔH_{solute}. Thus, ΔH_{soln} is so positive that NaCl does *not* dissolve in hexane. **B,** Intermolecular forces in octane and in hexane are so similar that ΔH_{soln} is very small. Octane dissolves in hexane because the solution has greater entropy than the pure components.

SECTION SUMMARY

An overall heat of solution can be obtained from a thermochemical solution cycle as the sum of two endothermic steps (solute separation and solvent separation) and one exothermic step (solute-solvent mixing). In aqueous solutions, the combination of solvent separation and mixing is called *hydration.* Ionic heats of hydration are always negative because of strong ion-dipole forces. Most systems have a natural tendency to increase their entropy (distribute their energy in more ways), and a solution has greater entropy than the pure solute and solvent. The combination of enthalpy and entropy changes determines whether a solution forms. A substance with a positive ΔH_{soln} dissolves *only* if the entropy increase is large enough to outweigh it.

13.3 SOLUBILITY AS AN EQUILIBRIUM PROCESS

When an ionic solid dissolves, ions leave the solid and become dispersed in the solvent. Some dissolved ions collide occasionally with the undissolved solute and recrystallize. As long as the rate of dissolving is greater than the rate of recrystallizing, the concentration of ions rises. Eventually, given enough solid, ions are dissolving at the same rate as ions in the solution are recrystallizing (Figure 13.7).

At this point, even though dissolving and recrystallizing continue, there is no further change in the concentration with time. The system has reached equilibrium; that is, *excess undissolved solute is in equilibrium with the dissolved solute:*

$$\text{Solute (undissolved)} \rightleftharpoons \text{solute (dissolved)}$$

This solution is called **saturated:** it contains the maximum amount of dissolved solute at a given temperature in the presence of undissolved solute. Filter off the saturated solution and add more solute to it, and the added solute will not

Figure 13.7 Equilibrium in a saturated solution. In a saturated solution, equilibrium exists between excess solid solute and dissolved solute. At a particular temperature, the number of solute particles dissolving per unit time equals the number recrystallizing.

Figure 13.8 Sodium acetate crystallizing from a supersaturated solution. When a seed crystal of sodium acetate is added to a supersaturated solution of the compound (**A**), solute begins to crystallize out of solution (**B**) and continues to do so until the remaining solution is saturated (**C**).

dissolve. A solution that contains less than this concentration of dissolved solute is called **unsaturated:** add more solute, and more will dissolve until the solution becomes saturated.

In some cases, we can prepare a solution that contains more than the equilibrium concentration of dissolved solute. Such a solution is called **supersaturated.** It is unstable relative to the saturated solution: if you add a "seed" crystal of solute, or just tap the container, the excess solute crystallizes immediately, leaving a saturated solution (Figure 13.8).

Effect of Temperature on Solubility

Temperature affects the solubility of most substances. You may have noticed, for example, that not only does sugar dissolve more quickly in hot tea than in iced tea, but *more* sugar dissolves; in other words, the solubility of sugar in tea is greater at higher temperatures. Let's examine the effects of temperature on the solubility of solids and of gases.

Temperature and the Solubility of Solids Like sugar, *most solids are more soluble at higher temperatures.* Figure 13.9 shows the solubility of several ionic compounds in water as a function of temperature. Note that most of the graphed lines curve upward. Cerium sulfate is the only exception shown in the figure, but several other salts, mostly sulfates, behave similarly. Some salts exhibit

Figure 13.9 The relation between solubility and temperature for several ionic compounds. Most ionic compounds have higher solubilities at higher temperatures. Cerium sulfate is one of several exceptions.

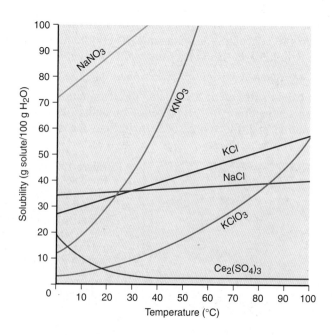

increasing solubility up to a certain temperature and then decreasing solubility at still higher temperatures. Unfortunately, the effect of temperature on the solubility of a solid solute is difficult to predict.

Temperature and Gas Solubility in Water The effect of temperature on gas solubility is much more predictable. When a solid dissolves in a liquid, the solute particles must separate, so energy must be added; thus, for a solid, $\Delta H_{solute} > 0$. In contrast, gas particles are already separated, so $\Delta H_{solute} \approx 0$. Because the hydration step is exothermic ($\Delta H_{hydr} < 0$), the sum of these two terms must be negative. Thus, for all gases in water, $\Delta H_{soln} < 0$:

$$\text{Solute}(g) + \text{water}(l) \rightleftharpoons \text{saturated solution}(aq) + heat$$

This equation means that *gas solubility in water **decreases** with rising temperature.* Gases have weak intermolecular forces, so there are relatively weak intermolecular forces between a gas and water. When the temperature rises, the average kinetic energy of the particles in solution increases, allowing the gas particles to easily overcome these weak forces and re-enter the gas phase.

This behavior can lead to an environmental problem known as *thermal pollution.* During many industrial processes, large amounts of water are taken from a nearby river or lake, pumped through the system to cool liquids, gases, and equipment, and then returned to the body of water at a higher temperature. The metabolic rates of fish and other aquatic animals increase in the warmer water released near the plant outlet; thus, their need for O_2 increases, but the concentration of dissolved O_2 is lower at the higher temperature. Farther from the plant, the water temperature returns to ambient levels and the O_2 solubility increases. One way to lessen the problem is with cooling towers, which lower the water's temperature before it exits the plant.

Effect of Pressure on Solubility

Because liquids and solids are almost incompressible, pressure has little effect on their solubility, but it has a major effect on gas solubility. Consider the piston-cylinder assembly in Figure 13.10, with a gas above a saturated aqueous solution of the gas. At a given pressure, the same number of gas molecules enter and leave the solution per unit time; that is, the system is at equilibrium:

$$\text{Gas} + \text{solvent} \rightleftharpoons \text{saturated solution}$$

If you push down on the piston, the gas volume decreases, its pressure increases, and gas particles collide with the liquid surface more often. Thus, more gas particles enter than leave the solution per unit time. In other words, higher gas pressure disturbs the balance at equilibrium. But, like any system at equilibrium,

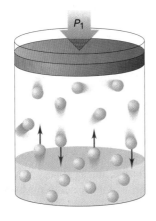
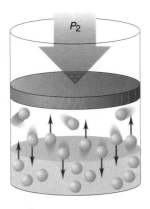

A **B**

Figure 13.10 The effect of pressure on gas solubility. A, A saturated solution of a gas is in equilibrium at pressure P_1. **B,** If the pressure is increased to P_2, the volume of the gas decreases. As a result, the frequency of collisions with the surface increases, and more gas is in solution when equilibrium is re-established.

the system adjusts to regain this balance: more gas dissolves (a shift to the right in the preceding equation) until the system re-establishes equilibrium.

   Henry's law expresses the quantitative relationship between gas pressure and solubility: *the solubility of a gas (S_{gas}) is directly proportional to the partial pressure of the gas (P_{gas}) above the solution:*

$$S_{gas} = k_H \times P_{gas} \qquad (13.3)$$

where k_H is the Henry's law constant and is specific for a given gas-solvent combination at a given temperature. With S_{gas} in mol/L and P_{gas} in atm, the units of k_H are mol/L·atm.

SAMPLE PROBLEM 13.2   Using Henry's Law to Calculate Gas Solubility

Problem The partial pressure of carbon dioxide gas inside a bottle of cola is 4 atm at 25°C. What is the solubility of CO_2? The Henry's law constant for CO_2 dissolved in water is 3.3×10^{-2} mol/L·atm at 25°C.

Plan We know P_{CO_2} (4 atm) and the value of k_H (3.3×10^{-2} mol/L·atm), so we substitute them into Equation 13.3 to find S_{CO_2}.

Solution $S_{CO_2} = k_H \times P_{CO_2} = (3.3 \times 10^{-2}$ mol/L·atm$)(4$ atm$) = 0.1$ mol/L

Check The units are correct for solubility. We rounded the answer to one significant figure because there is only one in the pressure value.

FOLLOW-UP PROBLEM 13.2 If air contains 78% N_2 by volume, what is the solubility of N_2 in water at 25°C and 1 atm (k_H for N_2 in H_2O at 25°C $= 7 \times 10^{-4}$ mol/L·atm)?

SECTION SUMMARY

A solution that contains the maximum amount of dissolved solute in the presence of excess undissolved solute is saturated. A state of equilibrium exists when a saturated solution is in contact with excess solute, because solute particles are entering and leaving the solution at the same rate. Most solids are more soluble at higher temperatures. All gases have a negative ΔH_{soln} in water, so heating lowers gas solubility in water. Henry's law says that the solubility of a gas is directly proportional to its partial pressure above the solution.

13.4 QUANTITATIVE WAYS OF EXPRESSING CONCENTRATION

Concentration is the *proportion* of a substance in a mixture, so it is an intensive property, one that does not depend on the quantity of mixture present: 1.0 L of 0.1 *M* NaCl has the same concentration as 1.0 mL of 0.1 *M* NaCl. Concentration is most often expressed as the ratio of the quantity of solute to the quantity of *solution,* but sometimes it is the ratio of solute to *solvent.* Because both parts of the ratio can be given in units of mass, volume, or amount (mol), chemists employ several concentration terms, including molarity, molality, and various expressions of "parts of solute per part of solution" (Table 13.4).

Molarity and Molality

Molarity (*M*) is the *number of moles of solute dissolved in 1 L of solution:*

$$\text{Molarity } (M) = \frac{\text{amount (mol) of solute}}{\text{volume (L) of solution}} \qquad (13.4)$$

In Chapter 3, you used molarity to convert liters of solution into moles of dissolved solute. Expressing concentration in terms of molarity may have drawbacks, however. Because volume is affected by temperature, so is molarity. A solution expands when heated, so a unit volume of hot solution contains slightly less solute than a unit volume of cold solution. This can be a source of error in very precise

Table 13.4	Concentration Definitions
Concentration Term	**Ratio**
Molarity (M)	$\dfrac{\text{amount (mol) of solute}}{\text{volume (L) of solution}}$
Molality (m)	$\dfrac{\text{amount (mol) of solute}}{\text{mass (kg) of solvent}}$
Parts by mass	$\dfrac{\text{mass of solute}}{\text{mass of solution}}$
Parts by volume	$\dfrac{\text{volume of solute}}{\text{volume of solution}}$
Mole fraction (X)	$\dfrac{\text{amount (mol) of solute}}{\text{amount (mol) of solute + amount (mol) of solvent}}$

work. More importantly, because of solute-solvent interactions that are difficult to predict, *solution volumes may not be additive;* that is, adding 500. mL of one solution to 500. mL of another may not give 1000. mL. Therefore, in precise work, a solution with a desired molarity may not be easy to prepare.

A concentration term that does not contain volume in its ratio is **molality (m),** *the number of moles of solute dissolved in 1000 g (1 kg) of solvent:*

$$\text{Molality } (m) = \frac{\text{amount (mol) of solute}}{\text{mass (kg) of solvent}} \qquad (13.5)$$

Note that molality includes the quantity of *solvent,* not solution. And, most important, molal solutions are prepared by measuring *masses* of solute and solvent, not solvent or solution *volume.* Mass does not change with temperature, so neither does molality. Moreover, unlike volumes, masses *are* additive: adding 500. g of one solution to 500. g of another *does* give 1000. g of final solution. For these reasons, molality is a preferred unit when temperature, and hence density, may change, as in the examination of solutions' physical properties. For the special case of water, 1 L has a mass of 1 kg, so *molality and molarity are nearly the same for dilute aqueous solutions.*

SAMPLE PROBLEM 13.3 Calculating Molality

Problem What is the molality of a solution prepared by dissolving 32.0 g of $CaCl_2$ in 271 g of water?
Plan To use Equation 13.5, we convert mass of $CaCl_2$ (32.0 g) to amount (mol) with the molar mass (g/mol) and then divide by the mass of water (271 g), being sure to convert from grams to kilograms.
Solution Converting from grams of solute to moles:

$$\text{Moles of } CaCl_2 = 32.0 \text{ g } CaCl_2 \times \frac{1 \text{ mol } CaCl_2}{110.98 \text{ g } CaCl_2} = 0.288 \text{ mol } CaCl_2$$

Finding molality:

$$\text{Molality} = \frac{\text{mol solute}}{\text{kg solvent}} = \frac{0.288 \text{ mol } CaCl_2}{271 \text{ g} \times \dfrac{1 \text{ kg}}{10^3 \text{ g}}} = 1.06 \text{ } m \text{ } CaCl_2$$

Mass (g) of $CaCl_2$

divide by $\mathcal{M}$ (g/mol)

Amount (mol) of $CaCl_2$

divide by kg of water

Molality (m) of $CaCl_2$ solution

Check The answer seems reasonable: the given numbers of moles of $CaCl_2$ and kilograms of H_2O are about the same, so their ratio is about 1.

FOLLOW-UP PROBLEM 13.3 How many grams of glucose ($C_6H_{12}O_6$) must be dissolved in 563 g of ethanol (C_2H_5OH) to prepare a 2.40×10^{-2} m solution?

Parts of Solute by Parts of Solution

Several concentration terms are based on the number of solute (or solvent) parts present in a specific number of *solution* parts. The solution parts can be expressed in terms of mass, volume, or amount (mol).

Parts by Mass   The most common of the parts-by-mass terms is **mass percent,** which you encountered in Chapter 3. The word *percent* means "per hundred," so mass percent of solute means the mass of solute dissolved in every 100. parts by mass of solution, or the mass fraction times 100:

$$\text{Mass percent} = \frac{\text{mass of solute}}{\text{mass of solute} + \text{mass of solvent}} \times 100$$

$$= \frac{\text{mass of solute}}{\text{mass of solution}} \times 100 \qquad (13.6)$$

Sometimes mass percent is symbolized as **% (w/w),** indicating that the percentage is a ratio of weights (more accurately, masses). You may have seen mass percent values on jars of solid chemicals to indicate the amounts of impurities present. Two very similar terms are parts per million (ppm) by mass and parts per billion (ppb) by mass: grams of solute per million or per billion grams of solution. For these quantities, in Equation 13.6 you multiply by 10^6 or by 10^9, respectively, instead of by 100.

Parts by Volume   The most common of the parts-by-volume terms is **volume percent,** the volume of solute in 100. volumes of solution:

$$\text{Volume percent} = \frac{\text{volume of solute}}{\text{volume of solution}} \times 100 \qquad (13.7)$$

A common symbol for volume percent is **% (v/v).** Commercial rubbing alcohol, for example, is an aqueous solution of isopropyl alcohol (a three-carbon alcohol) that contains 70 volumes of isopropyl alcohol per 100. volumes of solution, and the label indicates this as "70% (v/v)." Parts-by-volume concentrations are most often used for liquids and gases. Minor atmospheric components occur in parts per million by volume (ppmv). For example, there are about 0.05 ppmv of the toxic gas carbon monoxide (CO) in clean air, 1000 times as much (about 50 ppmv of CO) in air over urban traffic, and 10,000 times as much (about 500 ppmv of CO) in cigarette smoke.

A measure of concentration frequently used for aqueous solutions is % (w/v), a ratio of solute *weight* (actually mass) to solution *volume.* Thus, a 1.5% (w/v) NaCl solution contains 1.5 g of NaCl per 100. mL of *solution.* This way of expressing concentrations is particularly common in medical labs and other health-related facilities.

Mole Fraction   The **mole fraction** (X) of a solute is the ratio of number of solute moles to the total number of moles (solute plus solvent), that is, parts by mole. The *mole percent* is the mole fraction expressed as a percentage:

$$\text{Mole fraction} (X) = \frac{\text{amount (mol) of solute}}{\text{amount (mol) of solute} + \text{amount (mol) of solvent}} \qquad (13.8)$$

$$\text{Mole percent (mol \%)} = \text{mole fraction} \times 100$$

We discussed these terms in Chapter 5 in relation to Dalton's law of partial pressures for mixtures of gases, but they apply to liquids and solids as well. Concentrations given as mole fractions provide the clearest picture of the actual proportion of solute (or solvent) particles among all the particles in the solution.

SAMPLE PROBLEM 13.4 Expressing Concentrations in Parts by Mass, Parts by Volume, and Mole Fraction

Problem (a) Find the concentration of calcium (in ppm) in a 3.50-g pill that contains 40.5 mg of Ca.
(b) The label on a 0.750-L bottle of Italian chianti indicates "11.5% alcohol by volume." How many liters of alcohol does the wine contain?
(c) A sample of rubbing alcohol contains 142 g of isopropyl alcohol (C_3H_7OH) and 58.0 g of water. What are the mole fractions of alcohol and water?
Plan (a) We are given the masses of Ca (40.5 mg) and the pill (3.50 g). We convert the mass of Ca from mg to g, find the ratio of mass of Ca to mass of pill, and multiply by 10^6 to obtain ppm. **(b)** We know the volume % (11.5%, or 11.5 parts by volume of alcohol to 100 parts of chianti) and the total volume (0.750 mL), so we use Equation 13.7 to find the volume of alcohol. **(c)** We know the mass and formula of each component, so we convert masses to amounts (mol) and apply Equation 13.8 to find the mole fractions.
Solution (a) Finding parts per million by mass of Ca. Combining the steps, we have

$$\text{ppm Ca} = \frac{\text{mass of Ca}}{\text{mass of pill}} \times 10^6 = \frac{40.5 \text{ mg Ca} \times \dfrac{1 \text{ g}}{10^3 \text{ mg}}}{3.50 \text{ g}} \times 10^6 = 1.16 \times 10^4 \text{ ppm Ca}$$

(b) Finding volume (L) of alcohol:

$$\text{Volume (L) of alcohol} = 0.750 \text{ L chianti} \times \frac{11.5 \text{ L alcohol}}{100. \text{ L chianti}} = 0.0862 \text{ L}$$

(c) Finding mole fractions. Converting from grams to moles:

$$\text{Moles of } C_3H_7OH = 142 \text{ g } C_3H_7OH \times \frac{1 \text{ mol } C_3H_7OH}{60.09 \text{ g } C_3H_7OH} = 2.36 \text{ mol } C_3H_7OH$$

$$\text{Moles of } H_2O = 58.0 \text{ g } H_2O \times \frac{1 \text{ mol } H_2O}{18.02 \text{ g } H_2O} = 3.22 \text{ mol } H_2O$$

Calculating mole fractions:

$$X_{C_3H_7OH} = \frac{\text{moles of } C_3H_7OH}{\text{total moles}} = \frac{2.36 \text{ mol}}{2.36 \text{ mol} + 3.22 \text{ mol}} = 0.423$$

$$X_{H_2O} = \frac{\text{moles of } H_2O}{\text{total moles}} = \frac{3.22 \text{ mol}}{2.36 \text{ mol} + 3.22 \text{ mol}} = 0.577$$

Check (a) The mass ratio is about 0.04 g/4 g = 10^{-2}, and $10^{-2} \times 10^6 = 10^4$ ppm, so it seems correct. **(b)** The volume % is a bit more than 10%, so the volume of alcohol should be a bit more than 75 mL (0.075 L). **(c)** Always check that the *mole fractions add up to 1*; thus, in this case, 0.423 + 0.577 = 1.000.

FOLLOW-UP PROBLEM 13.4 An alcohol solution contains 35.0 g of 1-propanol (C_3H_7OH) and 150. g of ethanol (C_2H_5OH). Calculate the mass percent and the mole fraction of each alcohol.

Interconverting Concentration Terms

All the terms we just discussed represent different ways of expressing concentration, so they are interconvertible. Keep these points in mind:

- To convert a term based on amount (mol) to one based on mass, you need the molar mass. These conversions are similar to the mass-mole conversions you've done earlier.
- To convert a term based on mass to one based on volume, you need the solution *density*. Given the mass of a solution, the density (mass/volume) gives you the volume, or vice versa.
- Molality involves quantity of *solvent*, whereas the other concentration terms involve quantity of *solution*.

SAMPLE PROBLEM 13.5   Converting Concentration Terms

Problem Hydrogen peroxide is a powerful oxidizing agent used in concentrated solution in rocket fuels and in dilute solution as a hair bleach. An aqueous solution of H_2O_2 is 30.0% by mass and has a density of 1.11 g/mL. Calculate its
(a) Molality
(b) Mole fraction of H_2O_2
(c) Molarity
Plan We know the mass % (30.0) and the density (1.11 g/mL). **(a)** For molality, we need the amount (mol) of solute and the mass (kg) of *solvent*. Assuming 100.0 g of H_2O_2 solution allows us to express the mass % directly as grams of substance. We subtract the grams of H_2O_2 to obtain the grams of solvent. To find molality, we convert grams of H_2O_2 to moles and divide by mass of solvent (converting g to kg). **(b)** To find the mole fraction, we use the number of moles of H_2O_2 [from part (a)] and convert the grams of H_2O to moles. Then we divide the moles of H_2O_2 by the total moles. **(c)** To find molarity, we assume 100.0 g of solution and use the given solution density to find the volume. Then we divide the amount (mol) of H_2O_2 [from part (a)] by *solution volume* (in L).
Solution (a) From mass % to molality. Finding mass of solvent (assuming 100.0 g of solution):

$$\text{Mass (g) of } H_2O = 100.0 \text{ g solution} - 30.0 \text{ g } H_2O_2 = 70.0 \text{ g } H_2O$$

Converting from grams of H_2O_2 to moles:

$$\text{Moles of } H_2O_2 = 30.0 \text{ g } H_2O_2 \times \frac{1 \text{ mol } H_2O_2}{34.02 \text{ g } H_2O_2} = 0.882 \text{ mol } H_2O_2$$

Calculating molality:

$$\text{Molality of } H_2O_2 = \frac{0.882 \text{ mol } H_2O_2}{70.0 \text{ g} \times \dfrac{1 \text{ kg}}{10^3 \text{ g}}} = \boxed{12.6 \; m \; H_2O_2}$$

(b) From mass % to mole fraction:

$$\text{Moles of } H_2O_2 = 0.882 \text{ mol } H_2O_2 \text{ [from part (a)]}$$

$$\text{Moles of } H_2O = 70.0 \text{ g } H_2O \times \frac{1 \text{ mol } H_2O}{18.02 \text{ g } H_2O} = 3.88 \text{ mol } H_2O$$

$$X_{H_2O_2} = \frac{0.882 \text{ mol}}{0.882 \text{ mol} + 3.88 \text{ mol}} = \boxed{0.185}$$

(c) From mass % and density to molarity. Converting from solution mass to volume:

$$\text{Volume (mL) of solution} = 100.0 \text{ g} \times \frac{1 \text{ mL}}{1.11 \text{ g}} = 90.1 \text{ mL}$$

Calculating molarity:

$$\text{Molarity} = \frac{\text{mol } H_2O_2}{\text{L soln}} = \frac{0.882 \text{ mol } H_2O_2}{90.1 \text{ mL} \times \dfrac{1 \text{ L soln}}{10^3 \text{ mL}}} = \boxed{9.79 \; M \; H_2O_2}$$

Check Rounding shows the answers seem reasonable: **(a)** The ratio of ~0.9 mol/0.07 kg is greater than 10. **(b)** ~0.9 mol H_2O_2/(1 mol + 4 mol) ≈ 0.2. **(c)** The ratio of moles to liters (0.9/0.09) is around 10.

FOLLOW-UP PROBLEM 13.5 A sample of commercial concentrated hydrochloric acid is 11.8 *M* HCl and has a density of 1.190 g/mL. Calculate the mass %, molality, and mole fraction of HCl.

SECTION SUMMARY

The concentration of a solution is independent of the amount of solution and can be expressed as molarity (mol solute/L solution), molality (mol solute/kg solvent), parts by mass (mass solute/mass solution), parts by volume (volume solute/volume solution), or mole fraction [mol solute/(mol solute + mol solvent)]. The choice of units depends on convenience or the nature of the solution. If, in addition to the quantities of solute and solution, the solution density is also known, all ways of expressing concentration are interconvertible.

13.5 COLLIGATIVE PROPERTIES OF SOLUTIONS

We might expect the presence of solute particles to make the physical properties of a solution different from those of the pure solvent. However, what we might *not* expect is that, in the case of four important solution properties, the *number* of solute particles makes the difference, not their chemical identity. These properties, known as **colligative properties** (*colligative* means "collective"), are vapor pressure lowering, boiling point elevation, freezing point depression, and osmotic pressure. Even though most of these effects are small, they have many practical applications, including some that are vital to biological systems.

In Chapter 4, we classified solutes by their ability to conduct an electric current, which requires moving ions to be present. Recall that an **electrolyte** is a substance that dissociates into ions in aqueous solution: *strong electrolytes* dissociate completely, and *weak electrolytes* dissociate very little. **Nonelectrolytes** do not dissociate into ions at all. To predict the magnitude of a colligative property, we refer to the solute formula to find the number of particles in solution. Each mole of nonelectrolyte yields 1 mol of particles in the solution. For example, 0.35 M glucose contains 0.35 mol of solute particles per liter. In principle, each mole of strong electrolyte dissociates into the number of moles of ions in the formula unit: 0.4 M Na_2SO_4 contains 0.8 mol of Na^+ ions and 0.4 mol of SO_4^{2-} ions, or 1.2 mol of particles, per liter (see Sample Problem 4.1).

Colligative Properties of Nonvolatile Nonelectrolyte Solutions

In this section, we focus most of our attention on the simplest case, the colligative properties of solutes that do not dissociate into ions and have negligible vapor pressure even at the boiling point of the solvent. Such solutes are called *nonvolatile nonelectrolytes;* sucrose (table sugar) is an example. Later, we briefly explore the properties of volatile nonelectrolytes and of strong electrolytes.

Vapor Pressure Lowering The vapor pressure of a solution of a nonvolatile nonelectrolyte is always *lower* than the vapor pressure of the pure solvent. We can understand this **vapor pressure lowering (ΔP)** in terms of the opposing rates of vaporization (molecules leaving the liquid) and of condensation (molecules re-entering the liquid). At equilibrium, the two rates are equal. When we add some nonvolatile solute, the number of solvent molecules on the surface is lower, so fewer vaporize per unit time. To maintain equilibrium, fewer gas molecules can enter the liquid, which occurs only if the concentration of gas, that is, the vapor pressure, is lowered (Figure 13.11).

In quantitative terms, we find that the vapor pressure of solvent above the solution ($P_{solvent}$) equals the mole fraction of solvent in the solution ($X_{solvent}$) times the vapor pressure of the pure solvent ($P^\circ_{solvent}$). This relationship is expressed by **Raoult's law:**

$$P_{solvent} = X_{solvent} \times P^\circ_{solvent} \tag{13.9}$$

In a solution, $X_{solvent}$ is always less than 1, so $P_{solvent}$ is always less than $P^\circ_{solvent}$. An **ideal solution** is one that follows Raoult's law at any concentration. However, just as most gases deviate from ideality, so do most solutions. In practice, Raoult's law gives a good approximation of the behavior of *dilute* solutions only, and it becomes exact at infinite dilution.

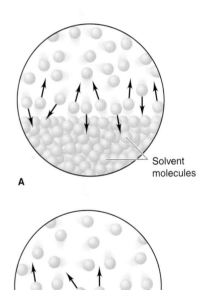

A

B

Solvent molecules

Nonvolatile solute molecules

Figure 13.11 The effect of the solute on the vapor pressure of a solution.
A, Equilibrium is established between a pure liquid and its vapor when the numbers of molecules vaporizing and condensing in a given time are equal. **B,** The presence of a dissolved solute decreases the number of solvent molecules at the surface so fewer solvent molecules vaporize in a given time. Therefore, fewer molecules need to condense to balance them, and equilibrium is established at a lower vapor pressure.

How does the *amount* of solute affect the *magnitude* of the vapor pressure lowering, ΔP? The solution consists of solvent and solute, so the sum of their mole fractions equals 1:

$$X_{solvent} + X_{solute} = 1; \quad \text{thus,} \quad X_{solvent} = 1 - X_{solute}$$

From Raoult's law, we have

$$P_{solvent} = X_{solvent} \times P^{\circ}_{solvent} = (1 - X_{solute}) \times P^{\circ}_{solvent}$$

Multiplying through on the right side gives

$$P_{solvent} = P^{\circ}_{solvent} - (X_{solute} \times P^{\circ}_{solvent})$$

Rearranging and introducing ΔP gives

$$P^{\circ}_{solvent} - P_{solvent} = \Delta P = X_{solute} \times P^{\circ}_{solvent} \tag{13.10}$$

Thus, the *magnitude* of ΔP equals the mole fraction of solute times the vapor pressure of the pure solvent—a relationship applied in the next sample problem.

SAMPLE PROBLEM 13.6   Using Raoult's Law to Find Vapor Pressure Lowering

Problem Calculate the vapor pressure lowering, ΔP, when 10.0 mL of glycerol ($C_3H_8O_3$) is added to 500. mL of water at 50.°C. At this temperature, the vapor pressure of pure water is 92.5 torr and its density is 0.988 g/mL. The density of glycerol is 1.26 g/mL.
Plan To calculate ΔP, we use Equation 13.10. We are given the vapor pressure of pure water ($P^{\circ}_{H_2O} = 92.5$ torr), so we just need the mole fraction of glycerol, $X_{glycerol}$. We convert the given volume of glycerol (10.0 mL) to mass using the given density (1.26 g/L), find the molar mass from the formula, and convert mass (g) to amount (mol). The same procedure gives amount of H_2O. From these, we find $X_{glycerol}$ and ΔP.
Solution Calculating the amount (mol) of glycerol and of water:

$$\text{Moles of glycerol} = 10.0 \text{ mL glycerol} \times \frac{1.26 \text{ g glycerol}}{1 \text{ mL glycerol}} \times \frac{1 \text{ mol glycerol}}{92.09 \text{ g glycerol}}$$

$$= 0.137 \text{ mol glycerol}$$

$$\text{Moles of } H_2O = 500. \text{ mL } H_2O \times \frac{0.988 \text{ g } H_2O}{1 \text{ mL } H_2O} \times \frac{1 \text{ mol } H_2O}{18.02 \text{ g } H_2O} = 27.4 \text{ mol } H_2O$$

Calculating the mole fraction of glycerol:

$$X_{glycerol} = \frac{0.137 \text{ mol}}{0.137 \text{ mol} + 27.4 \text{ mol}} = 0.00498$$

Finding the vapor pressure lowering:

$$\Delta P = X_{glycerol} \times P^{\circ}_{H_2O} = 0.00498 \times 92.5 \text{ torr} = 0.461 \text{ torr}$$

Check The amount of each component seems correct: for glycerol, ~10 mL × 1.25 g/mL ÷ 100 g/mol = 0.125 mol; for H_2O, ~500 mL × 1 g/mL ÷ 20 g/mol = 25 mol. The small ΔP is reasonable because the mole fraction of solute is small.
Comment The calculation assumes that glycerol is nonvolatile. At 1 atm, glycerol boils at 290.0°C, so the vapor pressure of glycerol at 50°C is so low it can be neglected.

FOLLOW-UP PROBLEM 13.6 Calculate the vapor pressure lowering of a solution of 2.00 g of aspirin ($\mathcal{M}$ = 180.15 g/mol) in 50.0 g of methanol (CH_3OH) at 21.2°C. Pure methanol has a vapor pressure of 101 torr at this temperature.

Sidebar flowchart:

Volume (mL) of glycerol (or H₂O)

multiply by density (g/mL)

Mass (g) of glycerol (or H₂O)

divide by $\mathcal{M}$ (g/mol)

Amount (mol) of glycerol (or H₂O)

divide by total number of moles

Mole fraction (X) of glycerol

multiply by $P^{\circ}_{H_2O}$

Vapor pressure lowering (ΔP)

Boiling Point Elevation *A solution boils at a higher temperature than the pure solvent.* Let's see why. The boiling point (boiling temperature, T_b) of a liquid is the temperature at which its vapor pressure equals the external pressure. The vapor pressure of a solution is lower than the external pressure at the solvent's boiling point because the vapor pressure of a solution is lower than that of the pure solvent at any temperature. Therefore, the solution does not yet boil. A higher temperature is needed to raise the solution's vapor pressure to equal the external pressure. We can see this **boiling point elevation (ΔT_b)** by superimposing a phase diagram for the solution on a phase diagram for the pure solvent, as shown in

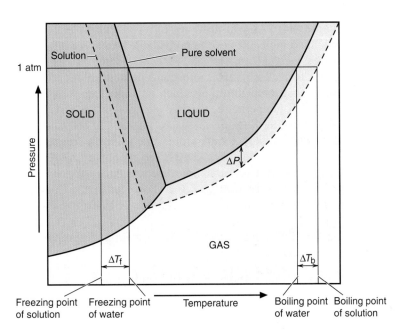

Figure 13.12 Phase diagrams of solvent and solution. Phase diagrams of an aqueous solution (*dashed lines*) and of pure water (*solid lines*) show that, by lowering the vapor pressure (ΔP), a dissolved solute elevates the boiling point (ΔT_b) and depresses the freezing point (ΔT_f).

Figure 13.12. Note that the gas-liquid line for the solution lies *below* that for the pure solvent at any temperature and to the right of it at any pressure.

Like the vapor pressure lowering, the magnitude of the boiling point elevation is proportional to the concentration of solute particles:

$$\Delta T_b \propto m \quad \text{or} \quad \Delta T_b = K_b m \qquad (13.11)$$

where m is the solution molality, K_b is the *molal boiling point elevation constant*, and ΔT_b is the boiling point elevation. We typically speak of ΔT_b as a positive value, so we subtract the lower temperature from the higher; that is, we subtract the solvent T_b from the solution T_b:

$$\Delta T_b = T_{b(\text{solution})} - T_{b(\text{solvent})}$$

Molality is the concentration term used because it is related to mole fraction and thus to particles of solute. It also involves mass rather than volume of solvent, so it is not affected by temperature changes. The constant K_b has units of degrees Celsius per molal unit (°C/m) and is specific for a given solvent (Table 13.5).

Notice that the K_b for water is only 0.512°C/m, so the changes in boiling point are quite small: if you dissolved 1.00 mol of glucose (180. g; 1.00 mol of particles) in 1.00 kg of water, or 0.500 mol of NaCl (29.2 g; a strong electrolyte, so also 1.00 mol of particles) in 1.00 kg of water, the boiling points of the resulting solutions at 1 atm would be only 100.512°C instead of 100.000°C.

Table 13.5	Molal Boiling Point Elevation and Freezing Point Depression Constants of Several Solvents			
Solvent	**Boiling Point (°C)***	**K_b (°C/m)**	**Melting Point (°C)**	**K_f (°C/m)**
Acetic acid	117.9	3.07	16.6	3.90
Benzene	80.1	2.53	5.5	4.90
Carbon disulfide	46.2	2.34	−111.5	3.83
Carbon tetrachloride	76.5	5.03	−23	30.
Chloroform	61.7	3.63	−63.5	4.70
Diethyl ether	34.5	2.02	−116.2	1.79
Ethanol	78.5	1.22	−117.3	1.99
Water	100.0	0.512	0.0	1.86

*At 1 atm.

Freezing Point Depression As you just saw, only solvent molecules can vaporize from the solution, so molecules of the nonvolatile solute are left behind. Similarly, in many cases, *only solvent molecules can solidify*, again leaving solute molecules behind to form a slightly more concentrated solution. The freezing point of a solution is that temperature at which its vapor pressure equals that of the pure solvent. At this temperature, the two phases—solid solvent and liquid solution—are in equilibrium. Because the vapor pressure of the solution is lower than that of the solvent at any temperature, the solution freezes at a lower temperature than the solvent. In other words, the numbers of solvent particles leaving and entering the solid per unit time become equal at a lower temperature. The **freezing point depression (ΔT_f)** is shown in Figure 13.12; note that the solid-liquid line for the solution lies to the left of that for the pure solvent at any pressure.

Like ΔT_b, the freezing point depression has a magnitude proportional to the molal concentration of solute:

$$\Delta T_f \propto m \quad \text{or} \quad \Delta T_f = K_f m \qquad (13.12)$$

where K_f is the *molal freezing point depression constant,* which also has units of °C/m (see Table 13.5). Also like ΔT_b, ΔT_f is considered a positive value, so we subtract the lower temperature from the higher; in this case, however, it is the solution T_f from the solvent T_f:

$$\Delta T_f = T_{f(\text{solvent})} - T_{f(\text{solution})}$$

Here, too, the overall effect in aqueous solution is quite small because the K_f value for water is small—only 1.86°C/m. Thus, 1 m glucose, 0.5 m NaCl, and 0.33 m K$_2$SO$_4$, all solutions with 1 mol of particles per kilogram of water, freeze at −1.86°C at 1 atm instead of at 0.00°C.

You have encountered practical applications of freezing point depression if you have added antifreeze—a solution of ethylene glycol in water—to your car's radiator or have seen airplane de-icers used before takeoff. Also, roads are "salted" with NaCl and CaCl$_2$ in winter to lower the freezing point of water, causing road ice to melt.

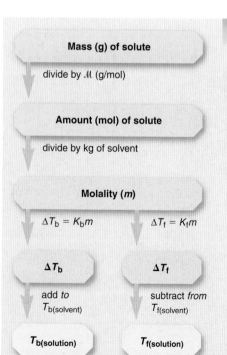

Mass (g) of solute

divide by $\mathcal{M}$ (g/mol)

Amount (mol) of solute

divide by kg of solvent

Molality (m)

$\Delta T_b = K_b m$ $\Delta T_f = K_f m$

ΔT_b ΔT_f

add *to*
$T_{b(\text{solvent})}$

subtract *from*
$T_{f(\text{solvent})}$

$T_{b(\text{solution})}$ $T_{f(\text{solution})}$

SAMPLE PROBLEM 13.7   Determining the Boiling Point Elevation and Freezing Point Depression of a Solution

Problem You add 1.00 kg of ethylene glycol (C$_2$H$_6$O$_2$) antifreeze to your car radiator, which contains 4450 g of water. What are the boiling and freezing points of the solution?

Plan To find the boiling and freezing points of the solution, we first find the molality by converting the given mass of solute (1.00 kg) to amount (mol) and dividing by mass of solvent (4450 g). Then we calculate ΔT_b and ΔT_f from Equations 13.11 and 13.12 (using constants from Table 13.5). We add ΔT_b to the solvent boiling point and subtract ΔT_f from the solvent freezing point. The roadmap shows the steps.

Solution Calculating the molality:

$$\text{Moles of C}_2\text{H}_6\text{O}_2 = 1.00 \text{ kg C}_2\text{H}_6\text{O}_2 \times \frac{10^3 \text{ g}}{1 \text{ kg}} \times \frac{1 \text{ mol C}_2\text{H}_6\text{O}_2}{62.07 \text{ g C}_2\text{H}_6\text{O}_2} = 16.1 \text{ mol C}_2\text{H}_6\text{O}_2$$

$$\text{Molality} = \frac{\text{mol solute}}{\text{kg solvent}} = \frac{16.1 \text{ mol C}_2\text{H}_6\text{O}_2}{4450 \text{ g H}_2\text{O} \times \frac{1 \text{ kg}}{10^3 \text{ g}}} = 3.62 \text{ } m \text{ C}_2\text{H}_6\text{O}_2$$

Finding the boiling point elevation and $T_{b(\text{solution})}$, with $K_b = 0.512$°C/m:

$$\Delta T_b = \frac{0.512\text{°C}}{m} \times 3.62 \text{ } m = 1.85\text{°C}$$

$$T_{b(\text{solution})} = T_{b(\text{solvent})} + \Delta T_b = 100.00\text{°C} + 1.85\text{°C} = 101.85\text{°C}$$

Finding the freezing point depression and $T_{f(\text{solution})}$, with $K_f = 1.86°C/m$:

$$\Delta T_f = \frac{1.86°C}{m} \times 3.62\ m = 6.73°C$$

$$T_{f(\text{solution})} = T_{f(\text{solvent})} - \Delta T_f = 0.00°C - 6.73°C = -6.73°C$$

Check The changes in boiling and freezing points should be in the same proportion as the constants used. That is, $\Delta T_b/\Delta T_f$ should equal K_b/K_f: $1.85/6.73 = 0.275 = 0.512/1.86$.
Comment These answers are only approximate because the concentration far exceeds that of a *dilute* solution, for which Raoult's law is most useful.

FOLLOW-UP PROBLEM 13.7 What is the minimum concentration (molality) of ethylene glycol solution that will protect the car's cooling system from freezing at $0.00°F$? (Assume the solution is ideal.)

Osmotic Pressure The fourth colligative property appears when two solutions of different concentrations are separated by a **semipermeable membrane,** one that allows solvent, but *not* solute, molecules to pass through. This process is called **osmosis.** Organisms have semipermeable membranes that regulate internal cellular concentrations by osmosis. You apply the principle of osmosis when you rinse your contact lenses, and your kidneys maintain fluid volume osmotically by controlling Na^+ concentration.

Consider a simple apparatus in which a semipermeable membrane lies at the curve of a U tube and separates an aqueous sugar solution from pure water. The membrane allows water molecules to pass in *either* direction, but not the larger sugar molecules. Because the solute molecules are present, fewer water molecules touch the membrane on the solution side, so fewer of them leave the solution in a given time than enter it (Figure 13.13A). This *net flow of water into the solution* increases the volume of the solution and thus decreases its concentration.

As the height of the solution rises and that of the solvent falls, the resulting pressure difference pushes some water molecules *from* the solution back through the membrane. At equilibrium, water is pushed out of the solution at the same rate it enters (Figure 13.13B). The pressure difference at equilibrium is the **osmotic pressure (Π),** which is defined as the applied pressure required to *prevent* the net movement of water from solvent to solution (Figure 13.13C).

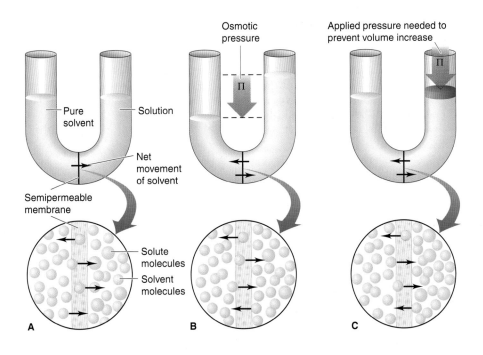

A Pure solvent / Solution / Semipermeable membrane / Net movement of solvent / Solute molecules / Solvent molecules

B Osmotic pressure / Π

C Applied pressure needed to prevent volume increase / Π

Figure 13.13 The development of osmotic pressure. A, In the process of osmosis, a solution and a solvent (or solutions of different concentrations) are separated by a semipermeable membrane, which allows only solvent molecules to pass through. The molecular-scale view (*below*) shows that more solvent molecules enter the solution than leave it in a given time. **B,** As a result, the solution volume increases, so its concentration decreases. At equilibrium, the difference in heights in the two compartments reflects the *osmotic pressure* (Π). The greater height in the solution compartment exerts a backward pressure that eventually equalizes the flow of solvent in both directions. **C,** Osmotic pressure is defined as the applied pressure required to *prevent* this volume change.

The osmotic pressure is proportional to the number of solute particles in a given *volume* of solution, that is, to the molarity (*M*):

$$\Pi \propto \frac{n_{solute}}{V_{soln}} \qquad \text{or} \qquad \Pi \propto M$$

The proportionality constant is *R* times the absolute temperature *T*. Thus,

$$\Pi = \frac{n_{solute}}{V_{soln}} RT = MRT \tag{13.13}$$

The similarity of Equation 13.13 to the ideal gas law ($P = nRT/V$) is not surprising, because both relate the pressure of a system to its concentration and temperature.

Using Colligative Properties to Find Solute Molar Mass

Each colligative property relates concentration to some measurable quantity—the number of degrees the freezing point is lowered, the magnitude of osmotic pressure created, and so forth. From these measurements, we can determine the amount (mol) of solute particles and, for a known mass of solute, the molar mass of the solute as well.

In principle, any of the colligative properties can be used to find the solute's molar mass, but in practice, some systems provide more precise data than others. For example, to determine the molar mass of an unknown solute by freezing point depression, you would select a solvent with as large a molal freezing point depression constant as possible (see Table 13.5). If the solute is soluble in acetic acid, for instance, a 1 *m* concentration of it depresses the freezing point of acetic acid by 3.90°C, more than twice the change in water (1.86°C).

Of the four colligative properties, osmotic pressure creates the largest changes and therefore the most precise measurements. Biological chemists estimate molar masses as great as 10^5 g/mol by measuring osmotic pressure. Because only a tiny fraction of a mole of a macromolecular solute dissolves, it would create too small a change in the other colligative properties.

SAMPLE PROBLEM 13.8   Determining Molar Mass from Osmotic Pressure

Problem Biochemists have discovered more than 400 mutant varieties of hemoglobin, the blood protein that carries oxygen throughout the body. A physician studying a variety associated with a fatal disease first finds its molar mass ($\mathcal{M}$). She dissolves 21.5 mg of the protein in water at 5.0°C to make 1.50 mL of solution and measures an osmotic pressure of 3.61 torr. What is the molar mass of this variety of hemoglobin?

Plan We know the osmotic pressure ($\Pi = 3.61$ torr), *R*, and *T* (5.0°C). We convert Π from torr to atm, and *T* from °C to K, and then use Equation 13.13 to solve for molarity (*M*). Then we calculate the amount (mol) of hemoglobin from the known volume (1.50 mL) and use the known mass (21.5 mg) to find $\mathcal{M}$.

Solution Combining unit conversion steps and solving for molarity from Equation 13.13:

$$M = \frac{\Pi}{RT} = \frac{\dfrac{3.61 \text{ torr}}{760 \text{ torr}/1 \text{ atm}}}{\left(0.0821 \dfrac{\text{atm·L}}{\text{mol·K}}\right)(273.15 \text{ K} + 5.0)} = 2.08 \times 10^{-4} M$$

Finding amount (mol) of solute (after changing mL to L):

$$\text{Moles of solute} = M \times V = \frac{2.08 \times 10^{-4} \text{ mol}}{1 \text{ L soln}} \times 0.00150 \text{ L soln} = 3.12 \times 10^{-7} \text{ mol}$$

Calculating molar mass of hemoglobin (after changing mg to g):

$$\mathcal{M} = \frac{0.0215 \text{ g}}{3.12 \times 10^{-7} \text{ mol}} = 6.89 \times 10^4 \text{ g/mol}$$

Π (atm)

$M = \Pi/RT$

M (mol/L)

multiply by volume (L) of solution

Amount (mol) of solute

divide *into* mass (g) of solute

$\mathcal{M}$ (g/mol)

Check The answers seem reasonable: The small osmotic pressure implies a very low molarity. Hemoglobin is a protein, a biological macromolecule, so we expect a small number of moles $[(\sim2\times10^{-4}\text{ mol/L})\,(1.5\times10^{-3}\text{ L}) = 3\times10^{-7}\text{ mol}]$ and a high molar mass $(\sim21\times10^{-3}\text{ g/}3\times10^{-7}\text{ mol} = 7\times10^{4}\text{ g/mol})$.

FOLLOW-UP PROBLEM 13.8 A 0.30 *M* solution of sucrose that is at 37°C has approximately the same osmotic pressure as blood does. What is the osmotic pressure of blood?

Colligative Properties of Volatile Nonelectrolyte Solutions

What is the effect on vapor pressure when the solute *is* volatile, that is, when the vapor consists of solute *and* solvent molecules? From Raoult's law (Equation 13.9), we know that

$$P_{\text{solvent}} = X_{\text{solvent}} \times P_{\text{solvent}}^{\circ} \quad\text{and}\quad P_{\text{solute}} = X_{\text{solute}} \times P_{\text{solute}}^{\circ}$$

where X_{solvent} and X_{solute} refer to the mole fractions in the liquid phase. According to Dalton's law of partial pressures, the total vapor pressure is the sum of the partial vapor pressures:

$$P_{\text{total}} = P_{\text{solvent}} + P_{\text{solute}} = (X_{\text{solvent}} \times P_{\text{solvent}}^{\circ}) + (X_{\text{solute}} \times P_{\text{solute}}^{\circ})$$

Just as a nonvolatile solute lowers the vapor pressure of the solvent by making the mole fraction of the solvent less than 1, *the presence of each volatile component lowers the vapor pressure of the other* by making each mole fraction less than 1.

Let's examine this effect in a solution that contains equal amounts (mol) of benzene (C_6H_6) and toluene (C_7H_8): $X_{\text{ben}} = X_{\text{tol}} = 0.500$. At 25°C, the vapor pressure of pure benzene (P_{ben}°) is 95.1 torr and that of pure toluene (P_{tol}°) is 28.4 torr; note that benzene is more volatile than toluene. We find the partial pressures from Raoult's law:

$$P_{\text{ben}} = X_{\text{ben}} \times P_{\text{ben}}^{\circ} = 0.500 \times 95.1\text{ torr} = 47.6\text{ torr}$$
$$P_{\text{tol}} = X_{\text{tol}} \times P_{\text{tol}}^{\circ} = 0.500 \times 28.4\text{ torr} = 14.2\text{ torr}$$

As you can see, the presence of benzene lowers the vapor pressure of toluene, and vice versa.

Does the composition of the vapor differ from that of the solution? To see, let's calculate the mole fraction of each substance *in the vapor* by applying Dalton's law. Recall from Section 5.4 that $X_A = P_A/P_{\text{total}}$. Therefore, for benzene and toluene in the vapor,

$$X_{\text{ben}} = \frac{P_{\text{ben}}}{P_{\text{total}}} = \frac{47.6\text{ torr}}{47.6\text{ torr} + 14.2\text{ torr}} = 0.770$$

$$X_{\text{tol}} = \frac{P_{\text{tol}}}{P_{\text{total}}} = \frac{14.2\text{ torr}}{47.6\text{ torr} + 14.2\text{ torr}} = 0.230$$

The vapor composition is very different from the solution composition. The essential point to notice is that *the vapor has a higher mole fraction of the **more** volatile solution component.* The 50:50 ratio of benzene:toluene in the liquid created a 77:23 ratio of benzene:toluene in the vapor. Condense this vapor into a separate container, and that new *solution* would have this 77:23 composition, and the new *vapor* above it would be enriched still further in benzene.

In the process of *fractional distillation*, this phenomenon is used to separate a mixture of volatile components. Numerous vaporization-condensation steps continually enrich the vapor, until the vapor consists solely of the most volatile component. Fractional distillation is used in the industrial process of petroleum refining to separate the hundreds of individual compounds in crude oil into a small number of "fractions" based on boiling point range.

Colligative Properties of Strong Electrolyte Solutions

When we consider colligative properties of strong electrolyte solutions, the solute formula tells us the number of particles. For instance, the boiling point elevation (ΔT_b) of 0.050 m NaCl should be twice that of 0.050 m glucose ($C_6H_{12}O_6$), because NaCl dissociates into two particles per formula unit. Thus, we include a multiplying factor in the equations for the colligative properties of electrolyte solutions. The *van't Hoff factor (i),* named after the Dutch chemist Jacobus van't Hoff (1852–1911), is the ratio of the *measured* value of the colligative property in the electrolyte solution to the *expected* value for a nonelectrolyte solution:

$$i = \frac{\text{measured value for electrolyte solution}}{\text{expected value for nonelectrolyte solution}}$$

To calculate the colligative properties of strong electrolyte solutions, we incorporate the van't Hoff factor into the equation:

For vapor pressure lowering: $\Delta P = i(X_{solute} \times P^\circ_{solvent})$
For boiling point elevation: $\Delta T_b = i(K_b m)$
For freezing point depression: $\Delta T_f = i(K_f m)$
For osmotic pressure: $\Pi = i(MRT)$

If strong electrolyte solutions behaved ideally, the factor i would be the amount (mol) of particles in solution divided by the amount (mol) of dissolved solute; that is, i would be 2 for NaCl, 3 for $Mg(NO_3)_2$, and so forth. Careful experiment shows, however, that *most strong electrolyte solutions are **not** ideal.* For example, comparing the boiling point elevation for 0.050 m NaCl solution with that for 0.050 m glucose solution gives a factor i of 1.9, not 2.0:

$$i = \frac{\Delta T_b \text{ of } 0.050\ m\ \text{NaCl}}{\Delta T_b \text{ of } 0.050\ m\ \text{glucose}} = \frac{0.049°C}{0.026°C} = 1.9$$

The measured value of the van't Hoff factor is typically *lower* than that expected from the formula. This deviation implies that the ions are not behaving as independent particles. However, we know from other evidence that soluble salts dissociate completely into ions. The fact that the deviation is greater with divalent and trivalent ions is a strong indication that the ionic charge is somehow involved (Figure 13.14).

To explain this nonideal behavior, we picture ions as separate but near each other. Clustered near a positive ion are, on average, more negative ions than positive ones, and vice versa. Figure 13.15 shows each ion surrounded by an **ionic atmosphere** of net opposite charge. Through these electrostatic associations, each type of ion behaves as if it were "tied up," so its concentration seems *lower* than it actually is. Thus, we often speak of an *effective* concentration, obtained by

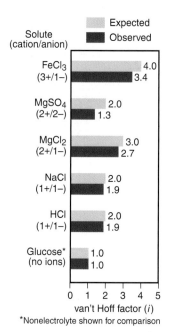

Figure 13.14 Nonideal behavior of strong electrolyte solutions. The van't Hoff factors *(i)* for various ionic solutes in dilute (0.05 m) aqueous solution show that the observed value *(dark blue)* is always *lower* than the expected value *(light blue)*. This deviation is due to ionic interactions that, in effect, reduce the number of free ions in solution. The deviation is greatest for multivalent ions.

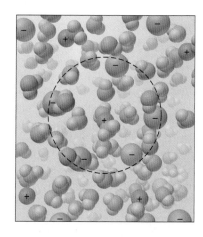

Figure 13.15 An ionic atmosphere model for nonideal behavior of electrolyte solutions. Hydrated anions cluster near cations, and vice versa, to form ionic atmospheres of net opposite charge. Because the ions do not act independently, their concentrations are effectively *less* than expected. Such interactions cause deviations from ideal behavior.

multiplying i by the *stoichiometric* concentration based on the formula. The greater the charge, the stronger the electrostatic associations, so the deviation from ideal behavior is greater for compounds that dissociate into multivalent ions.

SAMPLE PROBLEM 13.9 Depicting a Solution to Find Its Colligative Properties

Problem A 0.952-g sample of magnesium chloride is dissolved in 100. g of water.
(a) Which scene depicts the solution best?

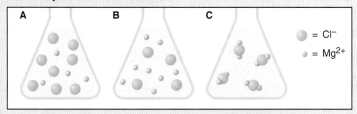

(b) What is the amount (mol) represented by each green sphere?
(c) Assuming the solution is ideal, what is its freezing point (at 1 atm)?
Plan (a) From the name, we recognize an ionic compound, so we determine the formula to find the numbers of cations and anions per formula unit and compare this result with the three scenes: there is 1 magnesium ion for every 2 chloride ions. **(b)** From the given mass of solute, we find the amount (mol); from part (a), there are twice as many moles of chloride ions (green spheres). Dividing by the total number of green spheres gives the moles/sphere. **(c)** From the moles of solute and the given mass (kg) of water, we find the molality (m). We use K_f for water from Table 13.5 and multiply by m to get ΔT_f, and then subtract that from 0.000°C to get the solution freezing point.
Solution (a) The formula is $MgCl_2$; only scene A has 1 Mg^{2+} for every 2 Cl^-.

(b)
$$\text{Moles of } MgCl_2 = \frac{0.952 \text{ g } MgCl_2}{95.21 \text{ g/mol } MgCl_2} = 0.0100 \text{ mol } MgCl_2$$

Therefore, $\quad \text{Moles of } Cl^- = 0.0100 \text{ mol } MgCl_2 \times \dfrac{2 \text{ Cl}^-}{1 \text{ MgCl}_2} = 0.0200 \text{ mol Cl}^-$

$$\text{Moles/sphere} = \frac{0.0200 \text{ mol } Cl^-}{8 \text{ spheres}} = 2.50 \times 10^{-3} \text{ mol/sphere}$$

(c) $\quad \text{Molality } (m) = \dfrac{\text{mol of solute}}{\text{kg of solvent}} = \dfrac{0.0100 \text{ mol } MgCl_2}{100. \text{ g} \times \dfrac{1 \text{ kg}}{1000 \text{ g}}} = 0.100 \; m \; MgCl_2$

Assuming an ideal solution, the van't Hoff factor, i, is 3 for $MgCl_2$ because there are 3 ions per formula unit, so we have

$$\Delta T_f = i(K_f m) = 3(1.86°C/m \times 0.100 \; m) = 0.558°C$$
And $\qquad\qquad T_f = 0.000°C - 0.558°C = -0.558°C$

Check Let's quickly check part (c): We have 0.01 mol dissolved in 0.1 kg, which gives 0.1 m. Then, rounding K_f, we have about $3(2°C/m \times 0.1 \; m) = 0.6°C$.

FOLLOW-UP PROBLEM 13.9 The $MgCl_2$ solution in the sample problem has a density of 1.006 g/mL at 20.0°C. **(a)** What is the osmotic pressure of the solution? **(b)** A U-tube fitted with a semipermeable membrane is filled with this $MgCl_2$ solution in the left arm and a glucose solution of equal concentration in the right arm. After time, which scene depicts the U-tube best?

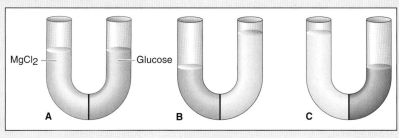

SECTION SUMMARY
Colligative properties are related to the number of dissolved solute particles, not their chemical nature. Compared with the pure solvent, a solution of a nonvolatile non-electrolyte has a lower vapor pressure (Raoult's law), an elevated boiling point, a depressed freezing point, and an osmotic pressure. Colligative properties can be used to determine the solute molar mass. When solute *and* solvent are volatile, the vapor pressure of each is lowered by the presence of the other. The vapor pressure of the more volatile component is always higher. Electrolyte solutions exhibit nonideal behavior because ionic interactions reduce the effective concentration of the ions.

For Review and Reference (Numbers in parentheses refer to pages, unless noted otherwise.)

Learning Objectives

To help you review these learning objectives, the numbers of related sections (§), sample problems (SP), and upcoming end-of-chapter problems (EP) are listed in parentheses.

1. Explain how solubility depends on the types of intermolecular forces ("like-dissolves-like" rule) and understand the characteristics of solutions consisting of gases, liquids, or solids (§ 13.1) (SP 13.1) (EPs 13.1–13.12)
2. Understand the enthalpy components of ΔH_{soln}, the dependence of ΔH_{hydr} on charge density, and why a solution process is exothermic or endothermic (§ 13.2) (EPs 13.13–13.15, 13.18–13.25, 13.28)
3. Comprehend the meaning of entropy and how the balance between ΔH and ΔS governs the solution process (§ 13.2) (EPs 13.16, 13.17, 13.26, 13.27)

4. Distinguish among saturated, unsaturated, and supersaturated solutions and explain the equilibrium nature of a saturated solution (§ 13.3) (EPs 13.29, 13.35)
5. Describe the effect of temperature on the solubility of solids and gases in water and the effect of pressure on gases (Henry's law) (§ 13.3) (SP 13.2) (EPs 13.30–13.34, 13.36)
6. Express concentration in terms of molarity, molality, mole fraction, and parts by mass or by volume and be able to interconvert these terms (§ 13.4) (SPs 13.3–13.5) (EPs 13.37–13.58)
7. Describe electrolyte behavior and the four colligative properties, explain the difference between phase diagrams for a solution and a pure solvent, explain vapor-pressure lowering for nonvolatile and volatile nonelectrolytes, and discuss the van't Hoff factor for colligative properties of electrolyte solutions (§ 13.5) (SPs 13.6–13.9) (EPs 13.59–13.83)

Key Terms

Section 13.1
solute (390)
solvent (390)
miscible (390)
solubility (S) (390)
like-dissolves-like rule (391)
hydration shell (391)
ion–induced dipole force (392)
dipole–induced dipole force (392)
alloy (395)

Section 13.2
heat of solution (ΔH_{soln}) (396)
solvation (397)
hydration (397)
heat of hydration (ΔH_{hydr}) (397)
charge density (397)
entropy (S) (398)

Section 13.3
saturated solution (399)
unsaturated solution (400)
supersaturated solution (400)
Henry's law (402)

Section 13.4
molality (m) (403)
mass percent [% (w/w)] (404)
volume percent [% (v/v)] (404)
mole fraction (X) (404)

Section 13.5
colligative property (407)
electrolyte (407)
nonelectrolyte (407)
vapor pressure lowering (ΔP) (407)
Raoult's law (407)

ideal solution (407)
boiling point elevation (ΔT_b) (408)
freezing point depression (ΔT_f) (410)
semipermeable membrane (411)
osmosis (411)
osmotic pressure (Π) (411)
ionic atmosphere (414)

Key Equations and Relationships

13.1 Dividing the general heat of solution into component enthalpies (396):

$$\Delta H_{soln} = \Delta H_{solute} + \Delta H_{solvent} + \Delta H_{mix}$$

13.2 Dividing the heat of solution of an ionic compound in water into component enthalpies (397):

$$\Delta H_{soln} = \Delta H_{lattice} + \Delta H_{hydr\ of\ the\ ions}$$

13.3 Relating gas solubility to its partial pressure (Henry's law) (402):

$$S_{gas} = k_H \times P_{gas}$$

13.4 Defining concentration in terms of molarity (402):

$$\text{Molarity } (M) = \frac{\text{amount (mol) of solute}}{\text{volume (L) of solution}}$$

13.5 Defining concentration in terms of molality (403):

$$\text{Molality } (m) = \frac{\text{amount (mol) of solute}}{\text{mass (kg) of solvent}}$$

13.6 Defining concentration in terms of mass percent (404):

$$\text{Mass percent [\% (w/w)]} = \frac{\text{mass of solute}}{\text{mass of solution}} \times 100$$

13.7 Defining concentration in terms of volume percent (404):

$$\text{Volume percent [\% (v/v)]} = \frac{\text{volume of solute}}{\text{volume of solution}} \times 100$$

13.8 Defining concentration in terms of mole fraction (404):

Mole fraction (X)

$$= \frac{\text{amount (mol) of solute}}{\text{amount (mol) of solute} + \text{amount (mol) of solvent}}$$

13.9 Expressing the relationship between the vapor pressure of solvent above a solution and its mole fraction in the solution (Raoult's law) (407):

$$P_{\text{solvent}} = X_{\text{solvent}} \times P^\circ_{\text{solvent}}$$

13.10 Calculating the vapor pressure lowering due to solute (408):

$$\Delta P = X_{\text{solute}} \times P^\circ_{\text{solvent}}$$

13.11 Calculating the boiling point elevation of a solution (409):

$$\Delta T_{\text{b}} = K_{\text{b}} m$$

13.12 Calculating the freezing point depression of a solution (410):

$$\Delta T_{\text{f}} = K_{\text{f}} m$$

13.13 Calculating the osmotic pressure of a solution (412):

$$\Pi = \frac{n_{\text{solute}}}{V_{\text{soln}}} RT = MRT$$

Brief Solutions to Follow-up Problems

13.1 (a) 1,4-Butanediol is more soluble in water because it can form more H bonds.
(b) Chloroform is more soluble in water because of dipole-dipole forces.

13.2 $S_{N_2} = (7 \times 10^{-4} \text{ mol/L·atm}) (0.78 \text{ atm})$
$= 5 \times 10^{-4} \text{ mol/L}$

13.3 Mass (g) of glucose $= 563 \text{ g ethanol}$
$$\times \frac{1 \text{ kg}}{10^3 \text{ g}} \times \frac{2.40 \times 10^{-2} \text{ mol glucose}}{1 \text{ kg ethanol}}$$
$$\times \frac{180.16 \text{ g glucose}}{1 \text{ mol glucose}}$$
$$= 2.43 \text{ g glucose}$$

13.4 Mass % $C_3H_7OH = \dfrac{35.0 \text{ g}}{35.0 \text{ g} + 150. \text{ g}} \times 100 = 18.9 \text{ mass \%}$

Mass % $C_2H_5OH = 100.0 - 18.9 = 81.1 \text{ mass \%}$

$$X_{C_3H_7OH} = \frac{35.0 \text{ g } C_3H_7OH \times \dfrac{1 \text{ mol } C_3H_7OH}{60.09 \text{ g } C_3H_7OH}}{\left(35.0 \text{ g } C_3H_7OH \times \dfrac{1 \text{ mol } C_3H_7OH}{60.09 \text{ g } C_3H_7OH}\right) + \left(150. \text{ g } C_2H_5OH \times \dfrac{1 \text{ mol } C_2H_5OH}{46.07 \text{ g } C_2H_5OH}\right)} = 0.152$$

$X_{C_2H_5OH} = 1.000 - 0.152 = 0.848$

13.5 Mass % HCl $= \dfrac{\text{mass of HCl}}{\text{mass of soln}} \times 100$

$$= \frac{\dfrac{11.8 \text{ mol HCl}}{1 \text{ L soln}} \times \dfrac{36.46 \text{ g HCl}}{1 \text{ mol HCl}}}{\dfrac{1.190 \text{ g}}{1 \text{ mL soln}} \times \dfrac{10^3 \text{ mL}}{1 \text{ L}}} \times 100$$

$$= 36.2 \text{ mass \%}$$

Mass (kg) of soln $= 1 \text{ L soln} \times \dfrac{1.190 \times 10^{-3} \text{ kg soln}}{1 \times 10^{-3} \text{ L soln}}$
$$= 1.190 \text{ kg soln}$$

Mass (kg) of HCl $= 11.8 \text{ mol HCl} \times \dfrac{36.46 \text{ g HCl}}{1 \text{ mol HCl}} \times \dfrac{1 \text{ kg}}{10^3 \text{ g}}$
$$= 0.430 \text{ kg HCl}$$

Molality of HCl $= \dfrac{\text{mol HCl}}{\text{kg water}} = \dfrac{\text{mol HCl}}{\text{kg soln} - \text{kg HCl}}$
$$= \dfrac{11.8 \text{ mol HCl}}{0.760 \text{ kg } H_2O} = 15.5 \; m \text{ HCl}$$

$$X_{\text{HCl}} = \frac{\text{mol HCl}}{\text{mol HCl} + \text{mol } H_2O}$$
$$= \frac{11.8 \text{ mol}}{11.8 \text{ mol} + \left(760 \text{ g } H_2O \times \dfrac{1 \text{ mol}}{18.02 \text{ g } H_2O}\right)} = 0.219$$

13.6 $\Delta P = X_{\text{aspirin}} \times P^\circ_{\text{methanol}}$

$$= \frac{\dfrac{2.00 \text{ g}}{180.15 \text{ g/mol}}}{\dfrac{2.00 \text{ g}}{180.15 \text{ g/mol}} + \dfrac{50.0 \text{ g}}{32.04 \text{ g/mol}}} \times 101 \text{ torr}$$

$$= 0.713 \text{ torr}$$

13.7 Molality of $C_2H_6O_2 = \dfrac{(0.00°F - 32°F)\left(\dfrac{5°C}{9°F}\right)}{1.86°C/m} = 9.56 \; m$

13.8 $\Pi = MRT = (0.30 \text{ mol/L}) \left(0.0821 \dfrac{\text{atm·L}}{\text{mol·K}}\right)(37°C + 273.15)$
$$= 7.6 \text{ atm}$$

13.9 (a) Mass of $0.100 \; m$ solution $= 1 \text{ kg water} + 0.100 \text{ mol MgCl}_2$
$$= 1000 \text{ g} + 9.52 \text{ g} = 1009.52 \text{ g}$$

Volume of solution $= 1009.52 \text{ g} \times \dfrac{1 \text{ mL}}{1.006 \text{ g}} = 1003 \text{ mL}$

Molarity $= \dfrac{9.52 \text{ g MgCl}_2}{1003 \text{ mL soln}} \times \dfrac{1 \text{ mol}}{9.52 \text{ g MgCl}_2} \times \dfrac{10^3 \text{ mL}}{1 \text{ L}}$
$$= 9.97 \times 10^{-2} \; M$$

Osmotic pressure (Π)
$$= i(MRT)$$
$$= 3(9.97 \times 10^{-2} \text{ mol/L})\left(0.0821 \dfrac{\text{atm·L}}{\text{mol·K}}\right)(293 \text{ K})$$
$$= 7.19 \text{ atm}$$

(b) Scene C

Problems

Problems with **colored** numbers are answered in Appendix E. Sections match the text and provide the numbers of relevant sample problems. Bracketed problems are grouped in pairs (indicated by a short rule) that cover the same concept. Comprehensive Problems are based on material from any section or previous chapter.

Types of Solutions: Intermolecular Forces and Solubility
(Sample Problem 13.1)

13.1 Describe how properties of seawater illustrate the two characteristics that define mixtures.

13.2 What types of intermolecular forces give rise to hydration shells in an aqueous solution of sodium chloride?

13.3 Acetic acid is miscible with water. Would you expect carboxylic acids, general formula $CH_3(CH_2)_nCOOH$, to become more or less water soluble as n increases? Explain.

13.4 Which gives the more concentrated solution, (a) KNO_3 in H_2O or (b) KNO_3 in carbon tetrachloride (CCl_4)? Explain.

13.5 Which gives the more concentrated solution, stearic acid [$CH_3(CH_2)_{16}COOH$] in (a) H_2O or (b) CCl_4? Explain.

13.6 What is the strongest type of intermolecular force between solute and solvent in each solution?

(a) $CsCl(s)$ in $H_2O(l)$ (b) $CH_3\overset{\overset{O}{\|}}{C}CH_3(l)$ in $H_2O(l)$
(c) $CH_3OH(l)$ in $CCl_4(l)$

13.7 What is the strongest type of intermolecular force between solute and solvent in each solution?
(a) $Cu(s)$ in $Ag(s)$ (b) $CH_3Cl(g)$ in $CH_3OCH_3(g)$
(c) $CH_3CH_3(g)$ in $CH_3CH_2CH_2NH_2(l)$

13.8 What is the strongest type of intermolecular force between solute and solvent in each solution?
(a) $CH_3OCH_3(g)$ in $H_2O(l)$ (b) $Ne(g)$ in $H_2O(l)$
(c) $N_2(g)$ in $C_4H_{10}(g)$

13.9 What is the strongest type of intermolecular force between solute and solvent in each solution?
(a) $C_6H_{14}(l)$ in $C_8H_{18}(l)$ (b) $H_2C{=}O(g)$ in $CH_3OH(l)$
(c) $Br_2(l)$ in $CCl_4(l)$

13.10 Which member of each pair is more soluble in diethyl ether, $CH_3CH_2{-}O{-}CH_2CH_3$? Why?
(a) $NaCl(s)$ or $HCl(g)$ (b) $H_2O(l)$ or $CH_3\overset{\overset{O}{\|}}{C}H(l)$
(c) $MgBr_2(s)$ or $CH_3CH_2MgBr(s)$

13.11 Which member of each pair is more soluble in water? Why?
(a) $CH_3CH_2OCH_2CH_3(l)$ or $CH_3CH_2OCH_3(g)$
(b) $CH_2Cl_2(l)$ or $CCl_4(l)$
(c)

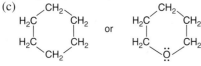

cyclohexane   tetrahydropyran

13.12 Gluconic acid is a derivative of glucose used in cleaners and in the dairy and brewing industries. Caproic acid is a carboxylic acid used in the flavoring industry. Although both are six-carbon

acids (see structures below), gluconic acid is soluble in water and nearly insoluble in hexane, whereas caproic acid has the opposite solubility behavior. Explain.

$$\underset{\text{gluconic acid}}{\overset{\displaystyle \overset{OH}{|}\;\overset{OH}{|}\;\overset{OH}{|}\;\overset{OH}{|}\;\overset{OH}{|}}{CH_2{-}CH{-}CH{-}CH{-}CH{-}COOH}}$$

$$\underset{\text{caproic acid}}{CH_3{-}CH_2{-}CH_2{-}CH_2{-}CH_2{-}COOH}$$

Why Substances Dissolve: Understanding the Solution Process

13.13 What is the relationship between solvation and hydration?

13.14 (a) What is the charge density of an ion, and what two properties of an ion affect it?
(b) Arrange the following in order of increasing charge density:

(c) How do the two properties in part (a) affect the ionic heat of hydration, ΔH_{hydr}?

13.15 For ΔH_{soln} to be very small, what quantities must be nearly equal in magnitude? Will their signs be the same or opposite?

13.16 Water is added to a flask containing solid NH_4Cl. As the salt dissolves, the solution becomes colder.
(a) Is the dissolving of NH_4Cl exothermic or endothermic?
(b) Is the magnitude of $\Delta H_{lattice}$ of NH_4Cl larger or smaller than the combined ΔH_{hydr} of the ions? Explain.
(c) Given the answer to (a), why does NH_4Cl dissolve in water?

13.17 An ionic compound has a highly negative ΔH_{soln} in water. Would you expect it to be very soluble or nearly insoluble in water? Explain in terms of enthalpy and entropy changes.

13.18 Sketch a qualitative enthalpy diagram for the process of dissolving $KCl(s)$ in H_2O (endothermic).

13.19 Sketch a qualitative enthalpy diagram for the process of dissolving $NaI(s)$ in H_2O (exothermic).

13.20 Which ion in each pair has greater charge density? Explain.
(a) Na^+ or Cs^+ (b) Sr^{2+} or Rb^+ (c) Na^+ or Cl^-
(d) O^{2-} or F^- (e) OH^- or SH^-

13.21 Which ion has the lower ratio of charge to volume? Explain.
(a) Br^- or I^- (b) Sc^{3+} or Ca^{2+} (c) Br^- or K^+
(d) S^{2-} or Cl^- (e) Sc^{3+} or Al^{3+}

13.22 Which has the *larger* ΔH_{hydr} in each pair of Problem 13.20?

13.23 Which has the *smaller* ΔH_{hydr} in each pair of Problem 13.21?

13.24 (a) Use the following data to calculate the combined heat of hydration for the ions in potassium bromate ($KBrO_3$):

$$\Delta H_{lattice} = 745 \text{ kJ/mol} \qquad \Delta H_{soln} = 41.1 \text{ kJ/mol}$$

(b) Which ion contributes more to the answer to part (a)? Why?

13.25 (a) Use the following data to calculate the combined heat of hydration for the ions in sodium acetate ($NaC_2H_3O_2$):

$$\Delta H_{lattice} = 763 \text{ kJ/mol} \qquad \Delta H_{soln} = 17.3 \text{ kJ/mol}$$

(b) Which ion contributes more to the answer to part (a)? Why?

13.26 State whether the entropy of the system increases or decreases in each of the following processes:
(a) Gasoline burns in a car engine.
(b) Gold is extracted and purified from its ore.
(c) Ethanol (CH_3CH_2OH) dissolves in 1-propanol ($CH_3CH_2CH_2OH$).

13.27 State whether the entropy of the system increases or decreases in each of the following processes:
(a) Pure gases are mixed to prepare an anesthetic.
(b) Electronic-grade silicon is prepared from sand.
(c) Dry ice (solid CO_2) sublimes.

13.28 Besides its use in making black-and-white film, silver nitrate ($AgNO_3$) is used similarly in forensic science. The NaCl left behind in the sweat of a fingerprint is treated with $AgNO_3$ solution to form AgCl. This precipitate is developed to show the black-and-white fingerprint pattern. Given $\Delta H_{lattice}$ of $AgNO_3$ = 822 kJ/mol and ΔH_{hydr} = -799 kJ/mol, calculate its ΔH_{soln}.

Solubility as an Equilibrium Process
(Sample Problem 13.2)

13.29 You are given a bottle of solid X and three aqueous solutions of X—one saturated, one unsaturated, and one supersaturated. How would you determine which solution is which?

13.30 Why does the solubility of any gas in water decrease with rising temperature?

13.31 For a saturated aqueous solution of each of the following at 20°C and 1 atm, will the solubility increase, decrease, or stay the same when the indicated change occurs?
(a) $O_2(g)$, increase P (b) $N_2(g)$, increase V

13.32 For a saturated aqueous solution of each of the following at 20°C and 1 atm, will the solubility increase, decrease, or stay the same when the indicated change occurs?
(a) $He(g)$, decrease T (b) $RbI(s)$, increase P

13.33 The Henry's law constant (k_H) for O_2 in water at 20°C is 1.28×10^{-3} mol/L·atm. (a) How many grams of O_2 will dissolve in 2.00 L of H_2O that is in contact with pure O_2 at 1.00 atm? (b) How many grams of O_2 will dissolve in 2.00 L of H_2O that is in contact with air, where the partial pressure of O_2 is 0.209 atm?

13.34 Argon makes up 0.93% by volume of air. Calculate its solubility (mol/L) in water at 20°C and 1.0 atm. The Henry's law constant for Ar under these conditions is 1.5×10^{-3} mol/L·atm.

13.35 Caffeine is about 10 times as soluble in hot water as in cold water. A chemist puts a hot-water extract of caffeine into an ice bath, and some caffeine crystallizes. Is the remaining solution saturated, unsaturated, or supersaturated?

13.36 The partial pressure of CO_2 gas above the liquid in a bottle of champagne at 20°C is 5.5 atm. What is the solubility of CO_2 in champagne? Assume Henry's law constant is the same for champagne as for water: at 20°C, $k_H = 3.7 \times 10^{-2}$ mol/L·atm.

Quantitative Ways of Expressing Concentration
(Sample Problems 13.3 to 13.5)

13.37 Explain the difference between molarity and molality. Under what circumstances would molality be a more accurate measure of the concentration of a prepared solution than molarity? Why?

13.38 A solute has a solubility in water of 21 g/kg solvent. Is this value the same as 21 g/kg solution? Explain.

13.39 You want to convert among molarity, molality, and mole fraction of a solution. You know the masses of solute and solvent and the volume of solution. Is this enough information to carry out all the conversions? Explain.

13.40 When a solution is heated, which ways of expressing concentration change in value? Which remain unchanged? Explain.

13.41 Calculate the molarity of each aqueous solution:
(a) 42.3 g of table sugar ($C_{12}H_{22}O_{11}$) in 100. mL of solution
(b) 5.50 g of $LiNO_3$ in 505 mL of solution

13.42 Calculate the molarity of each aqueous solution:
(a) 0.82 g of ethanol (C_2H_5OH) in 10.5 mL of solution
(b) 1.22 g of gaseous NH_3 in 33.5 mL of solution

13.43 Calculate the molarity of each aqueous solution:
(a) 75.0 mL of 0.250 M NaOH diluted to 0.250 L with water
(b) 35.5 mL of 1.3 M HNO_3 diluted to 0.150 L with water

13.44 Calculate the molarity of each aqueous solution:
(a) 25.0 mL of 6.15 M HCl diluted to 0.500 L with water
(b) 8.55 mL of 2.00×10^{-2} M KI diluted to 10.0 mL with water

13.45 How would you prepare the following aqueous solutions?
(a) 355 mL of 8.74×10^{-2} M KH_2PO_4 from solid KH_2PO_4
(b) 425 mL of 0.315 M NaOH from 1.25 M NaOH

13.46 How would you prepare the following aqueous solutions?
(a) 3.5 L of 0.55 M NaCl from solid NaCl
(b) 17.5 L of 0.3 M urea [($NH_2)_2C=O$] from 2.2 M urea

13.47 Calculate the molality of the following:
(a) A solution containing 88.4 g of glycine (NH_2CH_2COOH) dissolved in 1.250 kg of H_2O
(b) A solution containing 8.89 g of glycerol ($C_3H_8O_3$) in 75.0 g of ethanol (C_2H_5OH)

13.48 Calculate the molality of the following:
(a) A solution containing 164 g of HCl in 753 g of H_2O
(b) A solution containing 16.5 g of naphthalene ($C_{10}H_8$) in 53.3 g of benzene (C_6H_6)

13.49 What is the molality of a solution consisting of 34.0 mL of benzene (C_6H_6; d = 0.877 g/mL) in 187 mL of hexane (C_6H_{14}; d = 0.660 g/mL)?

13.50 What is the molality of a solution consisting of 2.77 mL of carbon tetrachloride (CCl_4; d = 1.59 g/mL) in 79.5 mL of methylene chloride (CH_2Cl_2; d = 1.33 g/mL)?

13.51 How would you prepare the following aqueous solutions?
(a) 3.00×10^2 g of 0.115 m ethylene glycol ($C_2H_6O_2$) from ethylene glycol and water
(b) 1.00 kg of 2.00 mass % HNO_3 from 62.0 mass % HNO_3

13.52 How would you prepare the following aqueous solutions?
(a) 1.00 kg of 0.0555 m ethanol (C_2H_5OH) from ethanol and water
(b) 475 g of 15.0 mass % HCl from 37.1 mass % HCl

13.53 A solution is made by dissolving 0.30 mol of isopropanol (C_3H_7OH) in 0.80 mol of water. (a) What is the mole fraction of isopropanol? (b) What is the mass percent of isopropanol? (c) What is the molality of isopropanol?

13.54 A solution is made by dissolving 0.100 mol of NaCl in 8.60 mol of water. (a) What is the mole fraction of NaCl? (b) What is the mass percent of NaCl? (c) What is the molality of NaCl?

13.55 Calculate the molality, molarity, and mole fraction of NH_3 in an 8.00 mass % aqueous solution (d = 0.9651 g/mL).

13.56 Calculate the molality, molarity, and mole fraction of $FeCl_3$ in a 28.8 mass % aqueous solution ($d = 1.280$ g/mL).

13.57 Wastewater from a cement factory contains 0.22 g of Ca^{2+} ion and 0.066 g of Mg^{2+} ion per 100.0 L of solution. The solution density is 1.001 g/mL. Calculate the Ca^{2+} and Mg^{2+} concentrations in ppm (by mass).

13.58 An automobile antifreeze mixture is made by mixing equal volumes of ethylene glycol ($d = 1.114$ g/mL; $\mathcal{M} = 62.07$ g/mol) and water ($d = 1.00$ g/mL) at 20°C. The density of the mixture is 1.070 g/mL. Express the concentration of ethylene glycol as
(a) volume percent   (b) mass percent   (c) molarity
(d) molality   (e) mole fraction

Colligative Properties of Solutions
(Sample Problems 13.6 to 13.9)

13.59 Express Raoult's law in words. Is Raoult's law valid for a solution of a volatile solute? Explain.

13.60 What are the most important differences between the phase diagram of a pure solvent and the phase diagram of a solution of that solvent?

13.61 Is the boiling point of 0.01 m KF(aq) higher or lower than that of 0.01 m glucose(aq)? Explain.

13.62 Which aqueous solution has a freezing point closer to its predicted value, 0.01 m NaBr or 0.01 m $MgCl_2$? Explain.

13.63 The freezing point depression constants of the solvents cyclohexane and naphthalene are 20.1°C/m and 6.94°C/m, respectively. Which solvent would give a more accurate result if you are using freezing point depression to determine the molar mass of a substance that is soluble in either one? Why?

13.64 Classify the following substances as strong electrolytes, weak electrolytes, or nonelectrolytes:
(a) hydrogen chloride (HCl)   (b) potassium nitrate (KNO_3)
(c) glucose ($C_6H_{12}O_6$)   (d) ammonia (NH_3)

13.65 Classify the following substances as strong electrolytes, weak electrolytes, or nonelectrolytes:
(a) sodium permanganate ($NaMnO_4$)
(b) acetic acid (CH_3COOH)
(c) methanol (CH_3OH)   (d) calcium acetate [$Ca(C_2H_3O_2)_2$]

13.66 How many moles of solute particles are present in 1 L of each of the following aqueous solutions?
(a) 0.2 M KI   (b) 0.070 M HNO_3
(c) 10^{-4} M K_2SO_4 (d) 0.07 M ethanol (C_2H_5OH)

13.67 How many moles of solute particles are present in 1 mL of each of the following aqueous solutions?
(a) 0.01 M $CuSO_4$ (b) 0.005 M $Ba(OH)_2$
(c) 0.06 M pyridine (C_5H_5N)   (d) 0.05 M $(NH_4)_2CO_3$

13.68 Which solution has the lower freezing point?
(a) 10.0 g of CH_3OH in 100. g of H_2O *or*
20.0 g of CH_3CH_2OH in 200. g of H_2O
(b) 10.0 g of H_2O in 1.00 kg of CH_3OH *or*
10.0 g of CH_3CH_2OH in 1.00 kg of CH_3OH

13.69 Which solution has the higher boiling point?
(a) 35.0 g of $C_3H_8O_3$ in 250. g of ethanol   *or*
35.0 g of $C_2H_6O_2$ in 250. g of ethanol

(b) 20. g of $C_2H_6O_2$ in 0.50 kg of H_2O *or*
20. g of NaCl in 0.50 kg of H_2O

13.70 Rank the following aqueous solutions in order of increasing (a) osmotic pressure; (b) boiling point; (c) freezing point; (d) vapor pressure at 50°C:
(I) 0.100 m $NaNO_3$ (II) 0.200 m glucose   (III) 0.100 m $CaCl_2$

13.71 Rank the following aqueous solutions in order of decreasing (a) osmotic pressure; (b) boiling point; (c) freezing point; (d) vapor pressure at 298 K:
(I) 0.04 m urea [$(NH_2)_2C{=}O$]
(II) 0.02 m $AgNO_3$
(III) 0.02 m $CuSO_4$

13.72 Calculate the vapor pressure of a solution of 44.0 g of glycerol ($C_3H_8O_3$) in 500.0 g of water at 25°C. The vapor pressure of water at 25°C is 23.76 torr. (Assume ideal behavior.)

13.73 Calculate the vapor pressure of a solution of 0.39 mol of cholesterol in 5.4 mol of toluene at 32°C. Pure toluene has a vapor pressure of 41 torr at 32°C. (Assume ideal behavior.)

13.74 What is the freezing point of 0.111 m urea in water?

13.75 What is the boiling point of 0.200 m lactose in water?

13.76 The boiling point of ethanol (C_2H_5OH) is 78.5°C. What is the boiling point of a solution of 3.4 g of vanillin ($\mathcal{M} = 152.14$ g/mol) in 50.0 g of ethanol (K_b of ethanol = 1.22°C/m)?

13.77 The freezing point of benzene is 5.5°C. What is the freezing point of a solution of 5.00 g of naphthalene ($C_{10}H_8$) in 444 g of benzene (K_f of benzene = 4.90°C/m)?

13.78 What is the minimum mass of ethylene glycol ($C_2H_6O_2$) that must be dissolved in 14.5 kg of water to prevent the solution from freezing at −10.0°F? (Assume ideal behavior.)

13.79 What is the minimum mass of glycerol ($C_3H_8O_3$) that must be dissolved in 11.0 mg of water to prevent the solution from freezing at −25°C? (Assume ideal behavior.)

13.80 Calculate the molality and van't Hoff factor (i) for the following aqueous solutions:
(a) 1.00 mass % NaCl, freezing point = −0.593°C
(b) 0.500 mass % CH_3COOH, freezing point = −0.159°C

13.81 Calculate the molality and van't Hoff factor (i) for the following aqueous solutions:
(a) 0.500 mass % KCl, freezing point = −0.234°C
(b) 1.00 mass % H_2SO_4, freezing point = −0.423°C

13.82 In a study designed to prepare new gasoline-resistant coatings, a polymer chemist dissolves 6.053 g of poly(vinyl alcohol) in enough water to make 100.0 mL of solution. At 25°C, the osmotic pressure of this solution is 0.272 atm. What is the molar mass of the polymer sample?

13.83 The U.S. Food and Drug Administration lists dichloromethane (CH_2Cl_2) and carbon tetrachloride (CCl_4) among the many chlorinated organic compounds that are carcinogenic (cancer-causing). What are the partial pressures of these substances in the vapor above a solution of 1.50 mol of CH_2Cl_2 and 1.00 mol of CCl_4 at 23.5°C? The vapor pressures of pure CH_2Cl_2 and CCl_4 at this temperature are 352 torr and 118 torr, respectively. (Assume ideal behavior.)

Comprehensive Problems

Problems with an asterisk (*) are more challenging.

13.84 Which of the following best represents a molecular-scale view of an ionic compound in aqueous solution? Explain.

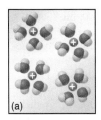  

(a) (b) (c)

13.85 Gold occurs in seawater at an average concentration of 1.1×10^{-2} ppb. How many liters of seawater must be processed to recover 1 troy ounce of gold, assuming 79.5% efficiency (*d* of seawater = 1.025 g/mL; 1 troy ounce = 31.1 g)?

13.86 Use atomic properties to explain why xenon is more than 25 times as soluble as helium in water at 0°C.

13.87 Thermal pollution from industrial wastewater causes the temperature of river or lake water to increase, which can affect fish survival as the concentration of dissolved O_2 decreases. Use the following data to find the molarity of O_2 at each temperature (assume the solution density is the same as water):

Temperature (°C)	Solubility of O_2 (mg/kg H_2O)	Density of H_2O (g/mL)
0.0	14.5	0.99987
20.0	9.07	0.99823
40.0	6.44	0.99224

13.88 "De-icing salt" is used to melt snow and ice on streets. The highway department of a small town is deciding whether to buy NaCl or $CaCl_2$ for the job. The town can obtain NaCl for $0.22/kg. What is the maximum the town should pay for $CaCl_2$ to be cost effective?

13.89 Is 50% by mass of methanol dissolved in ethanol different from 50% by mass of ethanol dissolved in methanol? Explain.

* **13.90** An industrial chemist is studying small organic compounds for their potential use as an automobile antifreeze. When 0.243 g of a compound is dissolved in 25.0 mL of water, the freezing point of the solution is −0.201°C.
(a) Calculate the molar mass of the compound (*d* of water = 1.00 g/mL at the temperature of the experiment).
(b) The compositional analysis of the compound shows that it is 53.31 mass % C and 11.18 mass % H, the remainder being O. Calculate the empirical and molecular formulas of the compound.
(c) Draw two possible Lewis structures for a compound with this formula, one that forms H bonds and one that does not.

13.91 β-Pinene ($C_{10}H_{16}$) and α-terpineol ($C_{10}H_{18}O$) are two of the many compounds used in perfumes and cosmetics to provide a "fresh pine" scent. At 367 K, the pure substances have vapor pressures of 100.3 torr and 9.8 torr, respectively. What is the composition of the vapor (in terms of mole fractions) above a solution containing equal masses of these compounds at 367 K? (Assume ideal behavior.)

* **13.92** A solution made by dissolving 1.50 g of solute in 25.0 mL of H_2O at 25°C has a boiling point of 100.45°C.
(a) What is the molar mass of the solute if it is a nonvolatile non-electrolyte and the solution behaves ideally (*d* of H_2O at 25°C = 0.997 g/mL)?
(b) Conductivity measurements indicate that the solute is actually ionic with general formula AB_2 or A_2B. What is the molar mass of the compound *if* the solution behaves ideally?
(c) Analysis indicates an empirical formula of CaN_2O_6. Explain the difference between the actual formula mass and that calculated from the boiling point elevation experiment.
(d) Calculate the van't Hoff factor (*i*) for this solution.

* **13.93** A pharmaceutical preparation made with ethanol (C_2H_5OH) is contaminated with methanol (CH_3OH). A sample of vapor above the liquid mixture contains a 97:1 mass ratio of C_2H_5OH:CH_3OH. What is the mass ratio of these alcohols in the liquid? At the temperature of the liquid, the vapor pressures of C_2H_5OH and CH_3OH are 60.5 torr and 126.0 torr, respectively.

13.94 Water-treatment plants commonly use chlorination to destroy bacteria. A by-product is chloroform ($CHCl_3$), a suspected carcinogen, produced when HOCl, formed by reaction of Cl_2 and water, reacts with dissolved organic matter. The United States, Canada, and the World Health Organization have set a limit of 100. ppb of $CHCl_3$ in drinking water. Convert this concentration into molarity, molality, mole fraction, and mass percent.

13.95 A biochemical engineer isolates a bacterial gene fragment and dissolves a 10.0-mg sample of the material in enough water to make 30.0 mL of solution. The osmotic pressure of the solution is 0.340 torr at 25°C.
(a) What is the molar mass of the gene fragment?
(b) If the solution density is 0.997 g/mL, how large is the freezing point depression for this solution (K_f of water = 1.86°C/*m*)?

* **13.96** Two beakers are placed in a closed container (*below, left*). One beaker contains water, the other a concentrated aqueous sugar solution. With time, the solution volume increases and the water volume decreases (*right*). Explain on the molecular level.

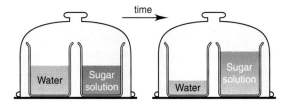

13.97 Glyphosate is the active ingredient in a common weed and grass killer. It is sold as an 18.0% by mass solution with a density of 8.94 lb/gal. (a) How many grams of Glyphosate are in a 16.0 fl oz container (1 gal = 128 fl oz)? (b) To treat a patio area of 300. ft^2, it is recommended that 3.00 fl oz be diluted with water to 1.00 gal. What is the mass percent of Glyphosate in the diluted solution (1 gal = 3.785 L)?

* **13.98** Although other solvents are available, dichloromethane (CH_2Cl_2) is still often used to "decaffeinate" foods because the solubility of caffeine in CH_2Cl_2 is 8.35 times that in water.

(a) A 100.0-mL sample of cola containing 10.0 mg of caffeine is extracted with 60.0 mL of CH_2Cl_2. What mass of caffeine remains in the aqueous phase?

(b) A second identical cola sample is extracted with two successive 30.0-mL portions of CH_2Cl_2. What mass of caffeine remains in the aqueous phase after each extraction?

(c) Which approach extracts more caffeine?

13.99 Tartaric acid can be produced from crystalline residues found in wine vats. It is used in baking powders and as an additive in foods. Analysis shows that it contains 32.3% by mass carbon and 3.97% by mass hydrogen; the balance is oxygen. When 0.981 g of tartaric acid is dissolved in 11.23 g of water, the solution freezes at $-1.26°C$. Use these data to find the empirical and molecular formulas of tartaric acid.

* **13.100** A florist prepares a solution of nitrogen-phosphorus fertilizer by dissolving 5.66 g of NH_4NO_3 and 4.42 g of $(NH_4)_3PO_4$ in enough water to make 20.0 L of solution. What are the molarities of NH_4^+ and of PO_4^{3-} in the solution?

13.101 Urea is a white crystalline solid used as a fertilizer, in the pharmaceutical industry, and in the manufacture of certain polymer resins. Analysis of urea reveals that, by mass, it is 20.1% carbon, 6.7% hydrogen, 46.5% nitrogen and the balance oxygen.

(a) Calculate the empirical formula of urea.

(b) A 5.0 g/L solution of urea in water has an osmotic pressure of 2.04 atm, measured at 25°C. What is the molar mass and molecular formula of urea?

13.102 The total concentration of dissolved particles in blood is 0.30 *M*. An intravenous (IV) solution must be *isotonic* with blood, which means it must have the same concentration.

(a) To relieve dehydration, a patient is given 100. mL/h of IV glucose ($C_6H_{12}O_6$) for 2.5 h. What mass (g) of glucose did she receive?

(b) If isotonic saline (NaCl) were used, what is the molarity of the solution?

(c) If the patient is given 150. mL/h of IV saline for 1.5 h, how many grams of NaCl did she receive?

13.103 In ice-cream making, the temperature of the ingredients is kept below 0.0°C in an ice-salt bath.

(a) Assuming that NaCl dissolves completely and forms an ideal solution, what mass of it is needed to lower the melting point of 5.5 kg of ice to $-5.0°C$?

(b) Given the same assumptions as in part (a), what mass of $CaCl_2$ is needed?

13.104 Carbonated soft drinks are canned under 4 atm of CO_2 and release much of it when opened. (a) How many moles of CO_2 are dissolved in a 355-mL can of soda before it is opened? (b) After it has gone flat? (c) What volume (in L) would the released CO_2 occupy at 1.00 atm and 25°C (k_H for CO_2 at 25°C is 3.3×10^{-2} mol/L·atm; P_{CO_2} in air is 3×10^{-4} atm)?

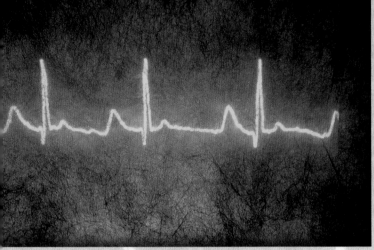

CHAPTER FOURTEEN

The Main-Group Elements: Applying Principles of Bonding and Structure

Recurring Patterns From the movement of the planets to the beat of a human heart, recurring patterns appear throughout nature. In this chapter, you'll discover the patterns of element behavior that arise from periodically recurring trends in atomic properties.

Key Principles

◆ *Hydrogen does not fit into any particular family (group)* because its tiny size and simple structure give it unique properties.

◆ Within a family of elements, *similar behavior* results from a *similar outer electron configuration.*

◆ Because the Period 2 elements have a small atomic size and only four outer-level orbitals, they exhibit some behavior that is *anomalous* within their groups.

◆ In Period 4 and higher, Group 3A and 4A elements deviate from expected trends because their nuclei attract outer *s* and *p* electrons very strongly due to poor shielding by their inner *d* and *f* electrons.

◆ Because atoms get larger down a group, *metallic behavior* (such as ability to form cations and basicity of oxides) increases, and this trend becomes especially apparent in Groups 3A to 6A.

◆ In Groups 3A to 6A, nearly every element exhibits *more than one oxidation state,* and the lower state becomes more common going down the group.

◆ Many elements occur in different forms (*allotropes*), each with its own properties.

◆ Group 1A and 7A elements are very reactive because each is one electron away from having a filled outer level; Group 8A elements have a filled outer level and thus are very unreactive.

Outline

In your study of chemistry so far, you've learned how to name compounds, balance equations, and calculate reaction yields. You've seen how heat is related to chemical and physical change, how electron configuration influences atomic properties, how elements bond to form compounds, and how the arrangement of bonding and lone pairs accounts for molecular shapes. You've learned modern theories of bonding and, most recently, seen how atomic and molecular properties give rise to the macroscopic properties of gases, liquids, solids, and solutions.

The purpose of this knowledge, of course, is to make sense of the magnificent diversity of chemical and physical behavior around you. The periodic table, which organizes much of this diversity, was derived from chemical facts observed in countless hours of 18th- and 19th-century research. One of the greatest achievements in science is 20th-century quantum theory, which provides a theoretical basis for the periodic table's arrangement. In this chapter, we apply general ideas of bonding and structure from earlier chapters to the main-group elements to see how their behavior correlates with their position in the periodic table.

14.1 HYDROGEN, THE SIMPLEST ATOM

A hydrogen (H) atom consists of a nucleus with a single proton, surrounded by a single electron. About 90% of all the atoms in the universe are H atoms, making it the most abundant element by far. On Earth, only tiny amounts of the free element (H_2) occur naturally because the molecules are so light that they escape Earth's gravity. However, hydrogen is abundant in combination with oxygen in water. Hydrogen's physical behavior results from its simple structure and low molar mass. Nonpolar gaseous H_2 is colorless and odorless, and its extremely weak dispersion forces result in very low melting ($-259°C$) and boiling points ($-253°C$).

Hydrogen's tiny size, low nuclear charge, and simple electron configuration make it difficult to place in the periodic table. We might think it belongs in Group 1A(1) because it has one valence electron. However, unlike the alkali metals, hydrogen *shares* its electron with nonmetals, and it has a much higher ionization energy and electronegativity than lithium, the highest of the alkali metals. Another possible placement might be with the halogens in Group 7A(17), because hydrogen occurs as a diatomic nonmetal that fills its outer shell either by sharing or by forming a monatomic anion (H^-). But hydrogen lacks the halogens' three valence electron pairs, and the H^- ion is rare and reactive, whereas halide ions are common and stable. Based on several atomic properties, a third possibility might be in Group 4A(14), because hydrogen has a half-filled valence level and ionization energy, electron affinity, electronegativity, and bond energy values close to those of Group 4A elements; but it shows little physical or chemical behavior that matches members of the carbon family. In this text, hydrogen will appear in either Group 1A(1) or 7A(17), depending on the property being considered.

Highlights of Hydrogen Chemistry

Beyond the enormous impact of hydrogen bonding on physical properties that we already discussed in Chapters 12 and 13, elemental hydrogen is very reactive, combining with nearly every other element to form ionic or covalent hydrides.

Ionic (Saltlike) Hydrides With very reactive metals, such as those in Group 1A(1) and the larger members of Group 2A(2) (Ca, Sr, and Ba), hydrogen forms *saltlike* *hydrides*—white, crystalline solids composed of the metal cation and the hydride ion:

$$2Li(s) + H_2(g) \longrightarrow 2LiH(s)$$
$$Ca(s) + H_2(g) \longrightarrow CaH_2(s)$$

In water, H^- reacts as a strong base to form H_2 and OH^-:

$$NaH(s) + H_2O(l) \longrightarrow Na^+(aq) + OH^-(aq) + H_2(g)$$

The H^- ion is also a strong reducing agent, as in this example:

$$TiCl_4(l) + 4LiH(s) \longrightarrow Ti(s) + 4LiCl(s) + 2H_2(g)$$

Covalent (Molecular) Hydrides Hydrogen reacts with nonmetals to form many *covalent hydrides*. In most of them, hydrogen has an oxidation number of $+1$ because the other nonmetal has a higher electronegativity.

Conditions for preparing covalent hydrides depend on the reactivity of the other nonmetal. For example, with stable, triple-bonded N_2, the reaction needs high temperatures ($\sim400°C$), high pressures (~250 atm), and a catalyst:

$$N_2(g) + 3H_2(g) \xrightarrow{\text{catalyst}} 2NH_3(g) \quad \Delta H°_{rxn} = -91.8 \text{ kJ}$$

Industrial facilities throughout the world use this reaction to produce millions of tons of ammonia each year for fertilizers, explosives, and synthetic fibers. On the other hand, hydrogen combines rapidly with reactive, single-bonded F_2, even at extremely low temperatures ($-196°C$):

$$F_2(g) + H_2(g) \longrightarrow 2HF(g) \quad \Delta H°_{rxn} = -546 \text{ kJ}$$

14.2 GROUP 1A(1): THE ALKALI METALS

The first group of elements is named for the alkaline (basic) nature of their oxides and for the basic solutions the elements form in water. Group 1A(1) provides the best example of regular trends with no significant exceptions. All the elements in the group—lithium (Li), sodium (Na), potassium (K), rubidium (Rb), cesium (Cs), and rare, radioactive francium (Fr)—are very reactive metals. The Family Portrait of Group 1A(1) on p. 426 is the first in a series that provides an overview of each of the main groups, summarizing key atomic, physical, and chemical properties.

The Unusual Physical Properties of the Alkali Metals

The alkali metals are softer and have lower melting and boiling points and lower densities than nearly any other metals. This unusual physical behavior can be traced to their atomic size, the largest in their respective periods, and to the ns^1 valence electron configuration. Because the single valence electron is relatively far from the nucleus, there is only weak metallic bonding, which results in a soft consistency (K can be squeezed like clay) and low melting point. And their low densities result from the lowest molar masses and largest atomic radii in their periods.

The High Reactivity of the Alkali Metals

The alkali metals are extremely reactive elements, acting as *powerful reducing agents*. Therefore, they always occur in nature as $1+$ cations rather than as free metals. (As we discuss in Section 21.7, highly endothermic reduction processes are needed to prepare the free metals industrially from their molten salts.)

The ns^1 configuration, which is the basis for their physical properties, is also the reason these metals form salts so readily. Their low ionization energies give rise to small cations, which allow them to lie close to anions, resulting in high

Family Portrait of Group 1A(1): The Alkali Metals

Key Atomic Properties, Physical Properties, and Reactions

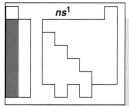

KEY
Atomic No.
Symbol
Atomic mass
Valence e⁻ configuration
Common oxidation states

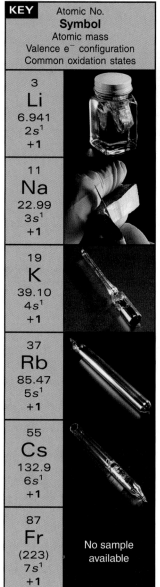

3 **Li** 6.941 $2s^1$ +1	
11 **Na** 22.99 $3s^1$ +1	
19 **K** 39.10 $4s^1$ +1	
37 **Rb** 85.47 $5s^1$ +1	
55 **Cs** 132.9 $6s^1$ +1	
87 **Fr** (223) $7s^1$ +1	No sample available

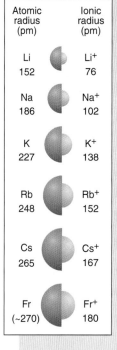

Atomic radius (pm)		Ionic radius (pm)
Li 152		Li⁺ 76
Na 186		Na⁺ 102
K 227		K⁺ 138
Rb 248		Rb⁺ 152
Cs 265		Cs⁺ 167
Fr (~270)		Fr⁺ 180

Atomic Properties

Group electron configuration is ns^1. All members have the +1 oxidation state and form an E^+ ion. Atoms have the largest size and lowest IE and EN in their periods. Down the group, atomic and ionic size increase, while IE and EN decrease.

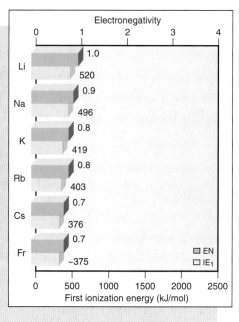

Physical Properties

Metallic bonding is relatively weak because there is only one valence electron. Therefore, these metals are soft with relatively low melting and boiling points. These values decrease down the group because larger atom cores attract delocalized electrons less strongly. Large atomic size and low atomic mass result in low density; thus density generally increases down the group because mass increases more than size.

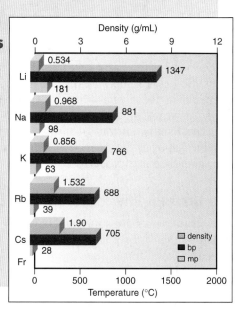

Reactions

1. The alkali metals reduce H in H_2O from the +1 to the 0 oxidation state:

$$2E(s) + 2H_2O(l) \longrightarrow 2E^+(aq) + 2OH^-(aq) + H_2(g)$$

The reaction becomes more vigorous down the group.

2. The alkali metals reduce oxygen, but the product depends on the metal. Li forms the oxide, Li_2O; Na forms the peroxide, Na_2O_2; K, Rb, and Cs form the superoxide, EO_2:

$$4Li(s) + O_2(g) \longrightarrow 2Li_2O(s)$$
$$K(s) + O_2(g) \longrightarrow KO_2(s)$$

In emergency breathing units, KO_2 reacts with H_2O and CO_2 in exhaled air to release O_2.

3. The alkali metals reduce hydrogen to form ionic (saltlike) hydrides:

$$2E(s) + H_2(g) \longrightarrow 2EH(s)$$

NaH is an industrial base and reducing agent that is used to prepare other reducing agents, such as $NaBH_4$.

4. The alkali metals reduce halogens to form ionic halides:

$$2E(s) + X_2 \longrightarrow 2EX(s) \qquad (X = F, Cl, Br, I)$$

lattice energies. Some examples of this reactivity occur with halogens, water, oxygen, and hydrogen. The alkali metals (E) reduce halogens to form ionic solids in highly exothermic reactions:

$$2E(s) + X_2 \longrightarrow 2EX(s) \quad (X = F, Cl, Br, I)$$

They reduce the hydrogen in water, reacting vigorously (Rb and Cs explosively) to form H_2 and a metal hydroxide solution:

$$2E(s) + 2H_2O(l) \longrightarrow 2E^+(aq) + 2OH^-(aq) + H_2(g)$$

They reduce O_2, but the product depends on the metal. Li forms the oxide, Li_2O; Na the peroxide (O.N. of O = -1), Na_2O_2; and K, Rb, and Cs the superoxide (O.N. of O = $-\frac{1}{2}$), EO_2:

$$4Li(s) + O_2(g) \longrightarrow 2Li_2O(s)$$
$$2Na(s) + O_2(g) \longrightarrow Na_2O_2(s)$$
$$K(s) + O_2(g) \longrightarrow KO_2(s)$$

Thus, in air, the metals tarnish rapidly, so Na and K are usually kept under mineral oil (an unreactive liquid) in the laboratory, and Rb and Cs are handled with gloves under an inert argon atmosphere. And, finally, the Group 1A(1) elements reduce molecular hydrogen to form ionic (saltlike) hydrides:

$$2E(s) + H_2(g) \longrightarrow 2EH(s)$$

For a given anion, the trend in lattice energy is the inverse of the trend in cation size: *as the cation becomes larger, the lattice energy becomes smaller.* Figure 14.1 shows this steady decrease in lattice energy within the Group 1A(1) and 2A(2) chlorides. Despite these strong ionic attractions in the solid, *nearly all Group 1A salts are water soluble* because the ions attract water molecules to create a highly exothermic heat of hydration (ΔH_{hydr}).

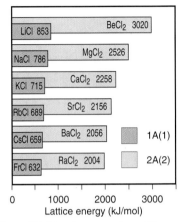

Figure 14.1 Lattice energies of the Group 1A(1) and 2A(2) chlorides. The lattice energy decreases regularly in both groups of metal chlorides as the cations become larger. Lattice energies for the 2A chlorides are greater because the 2A cations have higher charge and smaller size.

The Anomalous Behavior of Period 2 Members

A consistent feature within the main groups is that, as a result of their *small atomic size* and *small number of outer-level orbitals,* all the Period 2 members display some anomalous (unrepresentative) behavior within their groups.

In Group 1A(1), Li is the only member that forms a simple oxide and nitride, Li_2O and Li_3N, with O_2 and N_2 in air, and only Li forms molecular compounds with organic halides:

$$2Li(s) + CH_3CH_2Cl(g) \longrightarrow CH_3CH_2Li(s) + LiCl(s)$$

Because of the high charge density of Li^+, many lithium salts have significant covalent character. Thus, halides of Li are more soluble in polar organic solvents than the halides of Na and K.

In Group 2A(2), beryllium displays even more anomalous behavior than Li. Because of the extremely high charge density of Be^{2+}, the discrete ion does not exist, and all Be compounds exhibit covalent bonding. In Group 3A(13), boron is the only member to form a complex family of compounds with metals and covalent compounds with hydrogen (boranes). Carbon, in Group 4A(14), shows extremely unusual behavior: it bonds to itself (and a small number of other elements) so extensively and diversely that it gives rise to countless organic compounds. In Group 5A(15), triple-bonded, gaseous nitrogen is dramatically different from its reactive, solid family members. Oxygen, the only gas in Group 6A(16), is much more reactive than sulfur and the other members. In Group 7A(17), fluorine is so electronegative that it reacts violently with water, and it is the only member that forms a weak hydrohalic acid, HF. And, finally, helium, in Group 8A(18), has the lowest melting and boiling points and the smallest heats of phase change of any noble gas, indeed of any element.

14.3 GROUP 2A(2): THE ALKALINE EARTH METALS

The Group 2A(2) elements are called *alkaline earth metals* because their oxides give basic (alkaline) solutions and melt at such high temperatures that they remained as solids ("earths") in the alchemists' fires. The group includes rare beryllium (Be), common magnesium (Mg) and calcium (Ca), less familiar strontium (Sr) and barium (Ba), and radioactive radium (Ra). The Group 2A(2) Family Portrait presents an overview of these elements.

How Do the Physical Properties of the Alkaline Earth and Alkali Metals Compare?

In general, the elements in Groups 1A(1) and 2A(2) behave as close cousins. Whatever differences occur are due to an additional *s* electron: ns^2 vs. ns^1. Two valence electrons and a nucleus with one additional positive charge make for much stronger metallic bonding. Consequently, Group 2A melting and boiling points are much higher than those of the corresponding 1A metals. Compared with transition metals, such as iron and chromium, the alkaline earths are soft and lightweight, but they are much harder and more dense than the alkali metals. Magnesium is a particularly versatile member. Because it forms a tough oxide layer that prevents further reaction in air, it is alloyed with aluminum for camera bodies and luggage and with the lanthanides for auto engine blocks and missile parts.

How Do the Chemical Properties of the Alkaline Earth and Alkali Metals Compare?

The alkaline earth metals display a wider range of chemical behavior than the alkali metals, largely because of the unrepresentative covalent bonding of beryllium. The second valence electron lies in the same sublevel as the first, so it is poorly shielded and Z_{eff} is greater. Therefore, Group 2A(2) elements have smaller atomic radii and higher ionization energies than Group 1A(1) elements. Yet, despite the higher second IEs required to form the 2+ cations, *all the alkaline earths (except Be) form ionic compounds* because the resulting high lattice energies more than compensate for the large total IEs.

Like the alkali metals, the alkaline earth metals are *strong reducing agents*. They reduce O_2 in air. Except for Be and Mg, which form oxide coatings that adhere tightly to the sample's surface, they reduce H_2O at room temperature to form H_2, and, again except for Be, they reduce halogens, H_2, and N_2 to form the corresponding ionic halides, hydrides, and nitrides (see the Family Portrait).

The Group 2A oxides are very basic (except for amphoteric BeO) and react with acidic oxides to form salts, such as sulfites and carbonates; for example,

$$SrO(s) + CO_2(g) \longrightarrow SrCO_3(s)$$

The natural carbonates limestone and marble are major structural materials and the commercial sources for most 2A compounds.

One of the main differences between the two groups is the lower solubility of 2A salts. With such high lattice energies, most 2A fluorides, carbonates, phosphates, and sulfates are insoluble, unlike the corresponding 1A compounds.

Diagonal Relationships

Diagonal relationships are similarities between a Period 2 element and one diagonally down and to the right in Period 3. Three such relationships are of interest to us here (Figure 14.2). The first occurs between Li and Mg. Both form nitrides with N_2, hydroxides and carbonates that decompose easily with heat, organic compounds with a polar covalent metal-carbon bond, and salts with similar

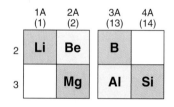

Figure 14.2 Three diagonal relationships in the periodic table. Certain Period 2 elements exhibit behaviors that are very similar to those of the Period 3 elements immediately below and to the right. Three such diagonal relationships exist: Li and Mg, Be and Al, and B and Si.

Key Atomic Properties, Physical Properties, and Reactions

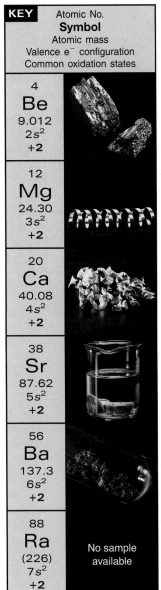

KEY
Atomic No.
Symbol
Atomic mass
Valence e⁻ configuration
Common oxidation states

| 4 |
| Be |
| 9.012 |
| $2s^2$ |
| +2 |

| 12 |
| Mg |
| 24.30 |
| $3s^2$ |
| +2 |

| 20 |
| Ca |
| 40.08 |
| $4s^2$ |
| +2 |

| 38 |
| Sr |
| 87.62 |
| $5s^2$ |
| +2 |

| 56 |
| Ba |
| 137.3 |
| $6s^2$ |
| +2 |

| 88 |
| Ra |
| (226) |
| $7s^2$ |
| +2 |

No sample available

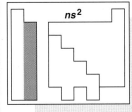

ns^2

	Atomic radius (pm)	Ionic radius (pm)
Be	112	
Mg	160	Mg^{2+} 72
Ca	197	Ca^{2+} 100
Sr	215	Sr^{2+} 118
Ba	222	Ba^{2+} 135
Ra	(~220)	Ra^{2+} 148

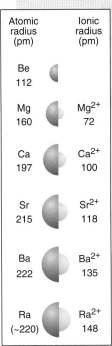

Atomic Properties

Group electron configuration is ns^2 (filled ns sublevel). All members have the +2 oxidation state and, except for Be, form compounds with an E^{2+} ion. Atomic and ionic sizes increase down the group but are smaller than for the corresponding 1A(1) elements. IE and EN decrease down the group but are higher than for the corresponding 1A(1) elements.

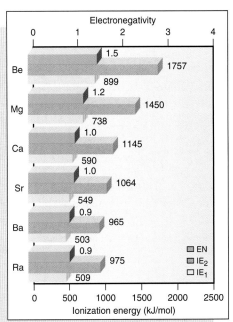

Electronegativity

	EN	IE_2	IE_1
Be	1.5	1757	899
Mg	1.2	1450	738
Ca	1.0	1145	590
Sr	1.0	1064	549
Ba	0.9	965	503
Ra	0.9	975	509

Ionization energy (kJ/mol)

Physical Properties

Metallic bonding involves two valence electrons. These metals are still relatively soft but are much harder than the 1A(1) metals. Melting and boiling points generally decrease, and densities generally increase down the group. These values are much higher than for 1A(1) elements, and the trend is not as regular.

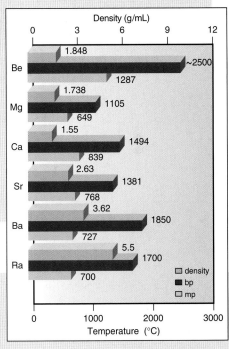

Density (g/mL)

	density	bp	mp
Be	1.848	~2500	1287
Mg	1.738	1105	649
Ca	1.55	1494	839
Sr	2.63	1381	768
Ba	3.62	1850	727
Ra	5.5	1700	700

Temperature (°C)

Reactions

1. The metals reduce O_2 to form the oxides:
$$2E(s) + O_2(g) \longrightarrow 2EO(s)$$
Ba also forms the peroxide, BaO_2.

2. The larger metals reduce water to form hydrogen gas:
$$E(s) + 2H_2O(l) \longrightarrow E^{2+}(aq) + 2OH^-(aq) + H_2(g)$$
$$(E = Ca, Sr, Ba)$$
Be and Mg form an oxide coating that allows only slight reaction.

3. The metals reduce halogens to form ionic halides:
$$E(s) + X_2 \longrightarrow EX_2(s) \qquad (X = F, Cl, Br, I)$$

4. Most of the elements reduce hydrogen to form ionic hydrides:
$$E(s) + H_2(g) \longrightarrow EH_2(s) \qquad (E = all \ except \ Be)$$

5. Most of the elements reduce nitrogen to form ionic nitrides:
$$3E(s) + N_2(g) \longrightarrow E_3N_2(s) \qquad (E = all \ except \ Be)$$

6. Except for amphoteric BeO, the element oxides are basic:
$$EO(s) + H_2O(l) \longrightarrow E^{2+}(aq) + 2OH^-(aq)$$
$Ca(OH)_2$ is a component of cement and mortar.

7. All carbonates undergo thermal decomposition to the oxide:
$$ECO_3(s) \xrightarrow{\Delta} EO(s) + CO_2(g)$$
This reaction is used to produce CaO (lime) in huge amounts from naturally occurring limestone.

solubilities. Beryllium in Group 2A(2) and aluminum in Group 3A(13) are another pair. Both metals form oxide coatings, so they don't react with water, and both form amphoteric, extremely hard, high-melting oxides. The third diagonal relationship occurs between the metalloids boron in Group 3A(13) and silicon in Group 4A(14). Both behave electrically as semiconductors and both form weakly acidic, solid oxoacids and flammable, low-melting, strongly reducing, covalent hydrides.

Looking Backward and Forward: Groups 1A(1), 2A(2), and 3A(13)

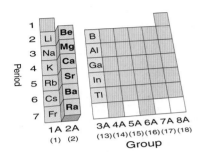

Figure 14.3 Standing in Group 2A(2), looking backward to 1A(1) and forward to 3A(13).

Throughout this chapter, comparing the previous, current, and upcoming groups (Figure 14.3) will help you keep horizontal trends in mind while examining vertical groups. Little changes from Group 1A to 2A, and all the elements behave as metals. With smaller atomic sizes and stronger metallic bonding, 2A elements are harder, higher melting, and denser than those in 1A. Nearly all 1A and most 2A compounds are ionic. The higher ionic charge in Group 2A (2+ vs. 1+) leads to higher lattice energies and less soluble salts. The range of behavior in 2A is wider than that in 1A because of Be, and the range widens much further in Group 3A, from metalloid boron to metallic thallium.

14.4 GROUP 3A(13): THE BORON FAMILY

Boron (B) heads the third family of main-group elements, but its properties are not representative, as the Group 3A(13) Family Portrait shows. Metallic aluminum (Al) is more typical of the group, but its great abundance and importance contrast with the rareness of gallium (Ga), indium (In), and thallium (Tl).

How Do Transition Elements Influence Group 3A(13) Properties?

Group 3A(13) is the first in the *p* block of the periodic table. In Periods 2 and 3, its members lie just one element away from those in Group 2A(2), but in Period 4 and higher, a large gap separates the two groups, with 10 transition elements (*d* block) each in Periods 4, 5, and 6 and an additional 14 inner transition elements (*f* block) in Period 6. Because *d* and *f* electrons spend very little time near the nucleus, they shield the outer (*s* and *p*) electrons in Ga, In, and Tl very little from the stronger nuclear attraction (greater Z_{eff}) (Sections 7.4 and 8.5). As a result, these elements deviate from the usual trends down a group, having smaller atomic radii and larger ionization energies and electronegativities than expected.

With respect to their physical properties, boron is a black, hard, very high-melting, network covalent metalloid, but the other 3A members are shiny, relatively soft, low-melting metals. Aluminum's low density and three valence electrons make it an exceptional conductor: for a given mass, it conducts a current twice as effectively as copper. Gallium has the largest liquid temperature range of any element: it melts in your hand but does not boil until 2403°C.

What New Features Appear in the Chemical Properties of Group 3A(13)?

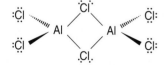

Figure 14.4 The dimeric structure of gaseous aluminum chloride. Despite its name, aluminum trichloride exists in the gas phase as the dimer, Al_2Cl_6.

Looking down Group 3A(13), we see a wide range of chemical behavior. Boron, the first metalloid we've encountered, is much less reactive at room temperature than the other members and forms covalent bonds exclusively. Although aluminum acts physically like a metal, its halides exist in the gas phase as covalent *dimers*—molecules formed by joining two identical smaller molecules (Figure 14.4)—and its oxide is amphoteric rather than basic. Most of the other 3A compounds are ionic, but with more covalent character than similar 2A compounds because the 3A cations can polarize nearby electron clouds more effectively.

Key Atomic Properties, Physical Properties, and Reactions

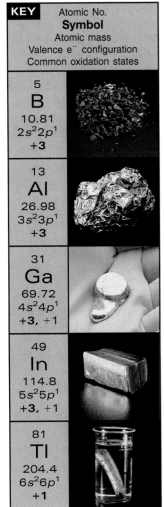

KEY
Atomic No.
Symbol
Atomic mass
Valence e^- configuration
Common oxidation states

5	
B	
10.81	
$2s^22p^1$	
+3	

13	
Al	
26.98	
$3s^23p^1$	
+3	

31	
Ga	
69.72	
$4s^24p^1$	
+3, +1	

49	
In	
114.8	
$5s^25p^1$	
+3, +1	

81	
Tl	
204.4	
$6s^26p^1$	
+1	

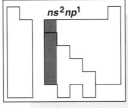

ns^2np^1

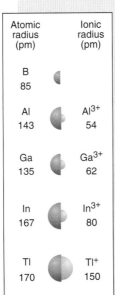

	Atomic radius (pm)	Ionic radius (pm)
B	85	
Al	143	Al^{3+} 54
Ga	135	Ga^{3+} 62
In	167	In^{3+} 80
Tl	170	Tl^+ 150

Atomic Properties

Group electron configuration is ns^2np^1. All except Tl commonly display the +3 oxidation state. The +1 state becomes more common down the group. Atomic size is smaller and EN is higher than for 2A(2) elements; IE is lower, however, because it is easier to remove an electron from the higher energy p sublevel. Atomic size, IE, and EN do not change as expected down the group because there are intervening transition and inner transition elements.

Physical Properties

Bonding changes from network covalent in B to metallic in the rest of the group. Thus, B has a much higher melting point than the others, but there is no overall trend. Boiling points decrease down the group. Densities increase down the group.

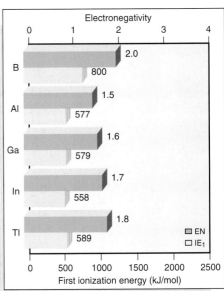

Electronegativity

	EN	IE₁
B	2.0	800
Al	1.5	577
Ga	1.6	579
In	1.7	558
Tl	1.8	589

First ionization energy (kJ/mol)

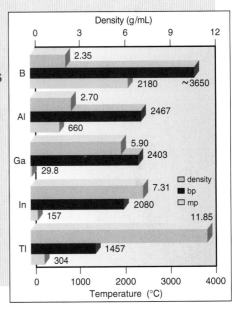

Density (g/mL)

	density	bp	mp
B	2.35	2180	~3650
Al	2.70	2467	660
Ga	5.90	2403	29.8
In	7.31	2080	157
Tl	11.85	1457	304

Temperature (°C)

Reactions

1. The elements react sluggishly, if at all, with water:

$$2Ga(s) + 6H_2O(hot) \longrightarrow 2Ga^{3+}(aq) + 6OH^-(aq) + 3H_2(g)$$
$$2Tl(s) + 2H_2O(steam) \longrightarrow 2Tl^+(aq) + 2OH^-(aq) + H_2(g)$$

Al becomes covered with a layer of Al_2O_3 that prevents further reaction.

2. When strongly heated in pure O_2, all members form oxides:

$$4E(s) + 3O_2(g) \xrightarrow{\Delta} 2E_2O_3(s) \qquad (E = B, Al, Ga, In)$$
$$4Tl(s) + O_2(g) \xrightarrow{\Delta} 2Tl_2O(s)$$

Oxide acidity decreases down the group: B_2O_3 (weakly acidic) > Al_2O_3 > Ga_2O_3 > In_2O_3 > Tl_2O (strongly basic), and the +1 oxide is more basic than the +3 oxide.

3. All members reduce halogens (X_2):

$$2E(s) + 3X_2 \longrightarrow 2EX_3 \qquad (E = B, Al, Ga, In)$$
$$2Tl(s) + X_2 \longrightarrow 2TlX(s)$$

The BX_3 compounds are volatile and covalent. Trihalides of Al, Ga, and In are (mostly) ionic solids.

Three features are common to the elements of Groups 3A(13) to 6A(16):

1. *Presence of multiple oxidation states.* Many of the larger elements in these groups also have an important oxidation state *two lower than the A-group number.* The lower state occurs when the atoms lose their *np* electrons but not their two *ns* electrons. This phenomenon is often called the *inert-pair effect* (Section 8.5).

2. *Increasing prominence of the lower oxidation state.* When a group exhibits more than one oxidation state, the lower state becomes more prominent going down the group. In Group 3A(13), for instance, all members exhibit the +3 state, but the +1 state first appears with some compounds of gallium and becomes the only important state of thallium.

3. *Relative basicity of oxides.* In general, oxides with the element in a lower oxidation state are more basic than oxides with the element in a higher oxidation state. For example, in Group 3A, In_2O is more basic than In_2O_3. The lower charge of In^+ does not polarize the O^{2-} ion as much as the higher charge of In^{3+} does, so the O^{2-} ion is more available to act as a base. In general, when an element has more than one oxidation state, *it acts more like a metal in its lower state.*

Highlights of Boron Chemistry

The chemical behavior of boron is strikingly different from that of the other Group 3A(13) members. Boron forms network covalent compounds or large molecules with metals, H, O, N, and C. Many boron compounds are *electron deficient,* but boron fills its outer level in two ways:

1. *Accepting a bonding pair from an electron-rich atom.* In gaseous boron trihalides, B has only six electrons around it (Section 10.1). To attain an octet, it accepts a lone pair from an electron-rich atom and forms a covalent bond:

$$BF_3(g) + :NH_3(g) \longrightarrow F_3B-NH_3(g)$$

Similarly, B has only six electrons in boric acid, $B(OH)_3$ (sometimes written as H_3BO_3). In water, the acid itself does not release a proton, but it bonds to the O of H_2O, which then releases an H^+ ion:

$$B(OH)_3(s) + H_2O(l) \rightleftharpoons B(OH)_4^-(aq) + H^+(aq)$$

2. *Forming bridge bonds with electron-poor atoms.* In elemental form and in the boron hydrides (boranes), boron attains an octet through an unusual type of bonding. In diborane (B_2H_6), for example, two types of B—H bonds exist. One is a normal electron-pair bond in which an sp^3 orbital of B overlaps the H 1*s* orbital in each of the four terminal B—H bonds (Figure 14.5). The other is a hydride *bridge bond* (or three-center, two-electron bond): each B—H—B grouping is held

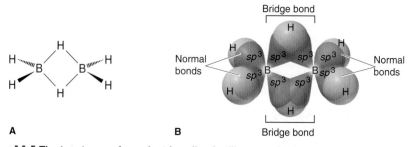

Figure 14.5 The two types of covalent bonding in diborane. A, A perspective diagram of B_2H_6 shows the unusual B—H—B bridge bond and the tetrahedral arrangement around each B atom. **B,** A valence bond depiction shows each sp^3-hybridized B forming normal covalent bonds with two hydrogens and two bridge bonds, in which two electrons bind three atoms, at the two central B—H—B groupings.

together by only two electrons, as two sp^3 orbitals, one from *each* B, overlap the H $1s$ orbital between them. Thus, each B atom is surrounded by four electrons from the two typical B—H bonds and four more from the two bridge bonds. In many other boranes and in elemental boron, we also see three-center, two-electron B—B—B bonds (Figure 14.6).

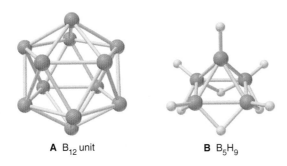

A B_{12} unit   **B** B_5H_9

Figure 14.6 The boron icosahedron and one of the boranes. A, The icosahedral structural unit of elemental boron. **B,** The structure of B_5H_9, one of many boranes.

14.5 GROUP 4A(14): THE CARBON FAMILY

All three categories of elements occur within Group 4A(14), from the nonmetal carbon (C) through the metalloids silicon (Si) and germanium (Ge) and down to the metals tin (Sn) and lead (Pb) [Group 4A(14) Family Portrait, p. 434].

How Does the Bonding in an Element Affect Physical Properties?

Trends among the elements of Group 4A(14) and their neighbors in Groups 3A(13) and 5A(15) illustrate how physical properties depend on the type of bonding in an element (Table 14.1). Within Group 4A, the large decrease in melting point between the network covalent solids C and Si is due to longer, weaker bonds in the Si structure; the large decrease between Ge and Sn is due to the change from network covalent to metallic bonding. Similarly, considering horizontal trends, the large increases in melting point and ΔH_{fus} across a period between Al and Si and between Ga and Ge reflect the change from metallic to network covalent bonding. Note the abrupt rises in these properties from metallic Al, Ga, and Sn to the network covalent metalloids Si, Ge, and Sb, and note the abrupt drops from the covalent networks of C and Si to the individual molecules of N and P.

Table 14.1 **Bond Type and the Melting Process in Groups 3A(13) to 5A(15)**

Period	Group 3A(13)				Group 4A(14)				Group 5A(15)			
	Element	Bond Type	Melting Point (°C)	ΔH_{fus} (kJ/mol)	Element	Bond Type	Melting Point (°C)	ΔH_{fus} (kJ/mol)	Element	Bond Type	Melting Point (°C)	ΔH_{fus} (kJ/mol)
2	B	covalent network	2180	23.6	C	covalent network	4100	Very high	N	covalent molecule	−210	0.7
3	Al	metallic	660	10.5	Si	covalent network	1420	50.6	P	covalent molecule	44.1	2.5
4	Ga	metallic	30	5.6	Ge	covalent network	945	36.8	As	covalent network	816	27.7
5	In	metallic	157	3.3	Sn	metallic	232	7.1	Sb	covalent network	631	20.0
6	Tl	metallic	304	4.3	Pb	metallic	327	4.8	Bi	metallic	271	10.5

Key:
☐ Metallic
⬡ Covalent network
∞ Covalent molecule
▢ Metal
▢ Metalloid
▢ Nonmetal

Family Portrait of Group 4A(14): The Carbon Family

Key Atomic Properties, Physical Properties, and Reactions

KEY

Atomic No.
Symbol
Atomic mass
Valence e⁻ configuration
Common oxidation states

6 **C** 12.01 $2s^2 2p^2$ (−4, **+4**, +2)	
14 **Si** 28.09 $3s^2 3p^2$ (−4, **+4**)	
32 **Ge** 72.61 $4s^2 4p^2$ (**+4**, +2)	
50 **Sn** 118.7 $5s^2 5p^2$ (**+4**, +2)	
82 **Pb** 207.2 $6s^2 6p^2$ (+4, **+2**)	
114 (285) $7s^2 7p^2$	Observed in experiments at Dubna, Russia, in 1998

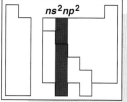

$ns^2 np^2$

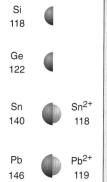

Atomic radius (pm)	Ionic radius (pm)
C 77	
Si 118	
Ge 122	
Sn 140	Sn²⁺ 118
Pb 146	Pb²⁺ 119

Atomic Properties

Group electron configuration is $ns^2 np^2$. Down the group, the number of oxidation states decreases, and the lower (+2) state becomes more common. Down the group, size increases. Because transition and inner transition elements intervene, IE and EN do not decrease smoothly.

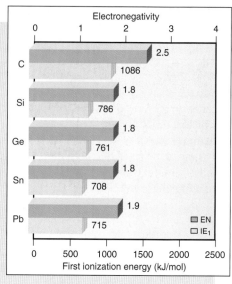

Electronegativity / First ionization energy (kJ/mol)

	EN	IE₁
C	2.5	1086
Si	1.8	786
Ge	1.8	761
Sn	1.8	708
Pb	1.9	715

Physical Properties

Trends in properties, such as decreasing hardness and melting point, are due to changes in types of bonding within the solid: covalent network in C, Si, and Ge; metallic in Sn and Pb (*see text*). Down the group, density increases because of several factors, including differences in crystal packing.

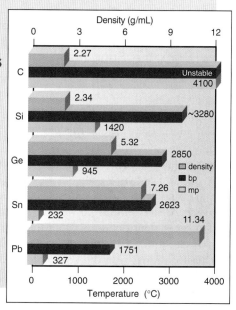

Density (g/mL) / Temperature (°C)

	density	bp	mp
C	2.27	Unstable	4100
Si	2.34	~3280	1420
Ge	5.32	2850	945
Sn	7.26	2623	232
Pb	11.34	1751	327

Reactions

1. The elements are oxidized by halogens:

$$E(s) + 2X_2 \longrightarrow EX_4 \quad (E = C, Si, Ge)$$

The +2 halides are more stable for tin and lead, SnX_2 and PbX_2.

2. The elements are oxidized by O_2:

$$E(s) + O_2(g) \longrightarrow EO_2 \quad (E = C, Si, Ge, Sn)$$

Pb forms the +2 oxide, PbO. Oxides become more basic down the group. The reaction of CO_2 and H_2O provides the weak acidity of natural unpolluted waters:

$$CO_2(g) + H_2O(l) \rightleftharpoons [H_2CO_3(aq)]$$
$$\rightleftharpoons H^+(aq) + HCO_3^-(aq)$$

3. Hydrocarbons react with O_2 to form CO_2 and H_2O. The reaction for methane is adapted to yield heat or electricity:

$$CH_4(g) + 2O_2(g) \longrightarrow CO_2(g) + 2H_2O(g)$$

4. Silica is reduced to form elemental silicon:

$$SiO_2(s) + 2C(s) \longrightarrow Si(s) + 2CO(g)$$

This crude silicon is made ultrapure through zone refining for use in the manufacture of computer chips.

Allotropism Striking variations in physical properties often appear among **allotropes,** different crystalline or molecular forms of a substance. One allotrope is usually more stable than another at a particular pressure and temperature. Group 4A(14) provides the first of many examples of allotropism. It is difficult to imagine two substances made entirely of the same atom that are more different than graphite and diamond. Graphite is a soft, black electrical conductor, whereas diamond is an extremely hard, colorless electrical insulator. Graphite is the more stable (standard state) form at ordinary temperatures and pressures (Figure 14.7). Fortunately for jewelry owners, diamond changes to graphite at a negligible rate under normal conditions.

In the mid-1980s, a newly discovered allotrope of carbon began generating great interest. Mass spectrometric analysis of soot showed evidence for a soccer ball–shaped molecule of formula C_{60}, dubbed *buckminsterfullerene* (informally called a "buckyball") (Figure 14.8A). Since 1990, when multigram quantities of C_{60} and related fullerenes were prepared, metal atoms have been incorporated into the structure and many different groups (fluorine, hydroxyl, sugars, etc.) have been attached. In 1991, extremely thin (~1 nm in diameter) graphite-like tubes with fullerene ends were prepared (Figure 14.8B). Along their length, such *nanotubes* are stronger than steel and conduct electricity. With potential applications in nanoscale electronics, catalysis, polymers, and medicine, fullerenes and nanotubes are receiving great attention from chemists and engineers.

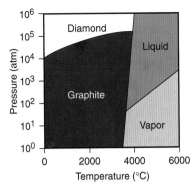

Figure 14.7 Phase diagram of carbon. Graphite is the more stable form of carbon at ordinary conditions (*small red circle at extreme lower left*). Diamond is more stable at very high pressure.

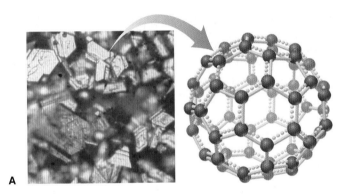

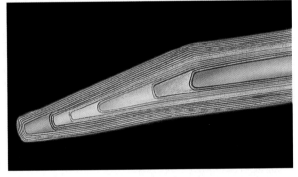

Figure 14.8 Buckyballs and nanotubes. A, Crystals of buckminsterfullerene (C_{60}) are shown leading to a ball-and-stick model. The parent of the fullerenes, the "buckyball," is a soccer ball–shaped molecule of 60 carbon atoms. **B,** Nanotubes are single or, as shown in this colorized transmission electron micrograph, concentric graphite-like tubes with fullerene ends.

How Does the Type of Bonding Change in Group 4A(14) Compounds?

The Group 4A(14) elements display a wide range of chemical behavior, from the covalent compounds of carbon to the ionic compounds of lead. Carbon's intermediate EN of 2.5 ensures that it virtually always forms covalent bonds, but the larger members of the group form bonds with increasing ionic character. With nonmetals, Si and Ge form strong polar covalent bonds. The most important is the Si—O bond, one of the strongest of any Period 3 element (BE = 368 kJ/mol), which is responsible for the physical and chemical stability of Earth's solid surface.

The pattern of elements having more than one oxidation state also appears here. Thus, compounds of Si(IV) are much more stable than those of Si(II), whereas compounds of Pb(II) are more stable than those of Pb(IV). The 4A elements also behave more like metals in the lower oxidation state. Thus, $SnCl_2$ and $PbCl_2$ are white, relatively high-melting, water-soluble crystals, typical properties of a salt, whereas $SnCl_4$ is a volatile, benzene-soluble liquid, and $PbCl_4$ is a thermally unstable oil. Similarly, SnO and PbO are more basic than SnO_2 and PbO_2.

Highlights of Carbon Chemistry

Carbon is not only an anomaly in its group, but its bonding ability makes it an anomaly throughout the periodic table. As a result of its small size and capacity for four bonds, carbon bonds to itself, a process known as *catenation,* to form chains, branches, and rings that lead to myriad structures. Add a lot of H, some O and N, a bit of S, P, halogens, and a few metals, and you have the whole organic world! Figure 14.9 shows three of the several million known organic compounds. Multiple bonds are common in these structures because the C—C bond is short enough for side-to-side overlap of two half-filled $2p$ orbitals to form π bonds. Because the other 4A members are larger, E—E bonds become longer and weaker down the group, and the presence of empty d orbitals of the larger atoms make their chains much more susceptible to chemical attack. Thus, none form molecules with stable chains.

In contrast to its organic compounds, carbon's inorganic compounds are simple. Metal carbonates in marble, limestone, chalk, and coral occur in enormous deposits throughout the world. Carbonates are used in some antacids because they react with the HCl in stomach acid:

$$CaCO_3(s) + 2HCl(aq) \longrightarrow CaCl_2(aq) + CO_2(g) + H_2O(l)$$

Identical net ionic reactions with sulfuric and nitric acids protect lakes bounded by limestone from the harmful effects of acid rain.

Carbon forms two common gaseous oxides. Carbon dioxide plays a vital role on Earth; through the process of photosynthesis, it is the primary source of carbon in plants, and thus animals as well. In solution, it is the cause of acidity in natural waters. However, its atmospheric buildup from deforestation and excessive use of fossil fuels is severely affecting the global climate. Carbon monoxide is a key component of fuel mixtures and is widely used in the production of methanol, formaldehyde, and other industrial compounds. Its toxicity arises from strong binding to the Fe(II) in hemoglobin, where it prevents the binding of O_2. The cyanide ion (CN^-), which is *isoelectronic* with CO,

$$[:C\equiv N:]^- \quad \text{same electronic structure as} \quad :C\equiv O:$$

is toxic because it binds to many other essential iron-containing proteins.

Monocarbon halides (or halomethanes) are tetrahedral molecules. The short, strong bonds in chlorofluorocarbons (CFCs, or Freons) make their long-term persistence in the upper atmosphere a major environmental problem (Chapter 16).

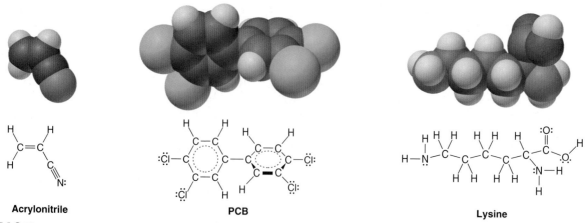

Acrylonitrile **PCB** **Lysine**

Figure 14.9 Three of the several million known organic compounds of carbon. Acrylonitrile is a precursor of acrylic fibers. PCB is an example of a polychlorinated biphenyl. Lysine is one of about 20 amino acids that occur in proteins.

Highlights of Silicon Chemistry

To a great extent, the chemistry of silicon is the chemistry of the *silicon-oxygen bond.* Just as carbon forms unending —C—C— chains, the —Si—O— grouping repeats itself endlessly in **silicates,** the most important minerals on the planet, and in **silicones,** extremely long synthetic molecules that have many applications:

1. *Silicate.* From common sand and clay to semiprecious amethyst and carnelian, silicate minerals dominate the nonliving world. In fact, oxygen and silicon account for four of every five atoms on Earth's surface! The silicate building unit is the *orthosilicate* grouping, —SiO$_4$—, a tetrahedral arrangement of four oxygens around a central silicon. Several well-known minerals, such as zircon and beryl, the natural source of beryllium, contain SiO$_4^{4-}$ ions or small groups of them linked together (Figure 14.10). In extended structures, one of the O atoms links the next Si—O group to form chains, a second one forms crosslinks to neighboring chains to form sheets, and the third forms more crosslinks to create three-dimensional frameworks. Chains of silicate groups compose the asbestos minerals, sheets give rise to talc and mica, and frameworks occur in quartz and feldspar.

2. *Silicone.* These compounds have two organic groups bonded to each Si atom in a very long Si—O chain, as in *poly(dimethyl siloxane):*

$$\cdots O-\underset{\underset{CH_3}{|}}{\overset{\overset{CH_3}{|}}{Si}}-O-\underset{\underset{CH_3}{|}}{\overset{\overset{CH_3}{|}}{Si}}-O-\underset{\underset{CH_3}{|}}{\overset{\overset{CH_3}{|}}{Si}}-O-\underset{\underset{CH_3}{|}}{\overset{\overset{CH_3}{|}}{Si}}-O-\underset{\underset{CH_3}{|}}{\overset{\overset{CH_3}{|}}{Si}}-O\cdots$$

The organic groups, with their weak intermolecular forces, give silicones flexibility, while the mineral-like —Si—O— backbone gives them thermal stability and inflammability.

Silicone chemists create structures similar to those of the silicates by adding various reactants to create silicone chains, sheets, and frameworks. Chains are oily liquids used as lubricants and as components of car polish and makeup. Sheets are components of gaskets, space suits, and contact lenses. Frameworks find uses as laminates on circuit boards, in nonstick cookware, and in artificial skin and bone.

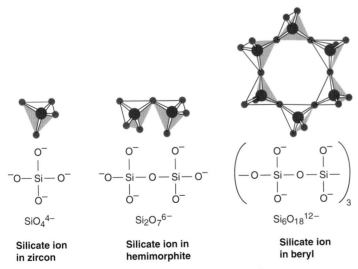

Figure 14.10 Structures of the silicate anions in some minerals.

438 CHAPTER 14 The Main-Group Elements

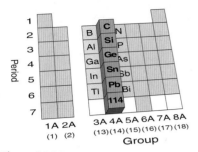

Figure 14.11 Standing in Group 4A(14), looking backward to 3A(13) and forward to 5A(15).

Looking Backward and Forward: Groups 3A(13), 4A(14), and 5A(15)

Standing in Group 4A(14), we see Group 3A(13) as the transition from the *s* block of metals to the *p* block of mostly metalloids and nonmetals (Figure 14.11). Changes occur in physical behavior, as we move from metals to covalent networks, and in chemical behavior, as cations give way to covalent tetrahedra. Looking ahead to Group 5A(15), we find many compounds with expanded valence shells and the first appearance of monatomic anions.

14.6 GROUP 5A(15): THE NITROGEN FAMILY

The first two elements of Group 5A(15), gaseous nonmetallic nitrogen (N) and solid nonmetallic phosphorus (P) have great industrial, environmental, and biological significance. Below these nonmetals are two metalloids, arsenic (As) and antimony (Sb), followed by the sole metal, bismuth (Bi), the last nonradioactive element in the periodic table [Group 5A(15) Family Portrait].

The Wide Range of Physical and Chemical Behavior in Group 5A(15)

Group 5A(15) displays the widest range of physical behavior we've seen so far. Nitrogen occurs as a gas consisting of N_2 molecules. Stronger dispersion forces due to heavier, more polarizable P atoms make phosphorus a solid. It has several allotropes. The white form consists of individual tetrahedral molecules (Figure 14.12A), making it low melting and soluble in nonpolar solvents; with a small 60° bond angle and, thus, weak P—P bonds, it is highly reactive (Figure 14.12B). In the red form, the P_4 units exist in chains, which make it much less reactive, high melting, and insoluble (Figure 14.12C). Arsenic consists of extended sheets, and a similar covalent network for Sb gives it a much higher melting point than metallic Bi.

Nearly all Group 5A(15) compounds have *covalent bonds*. A 5A element must *gain* three electrons to form an ion with a noble gas electron configuration. Enormous lattice energy results when 3− anions attract cations, but this occurs for N only with active metals, such as Li_3N and Mg_3N_2 (and perhaps with P in Na_3P).

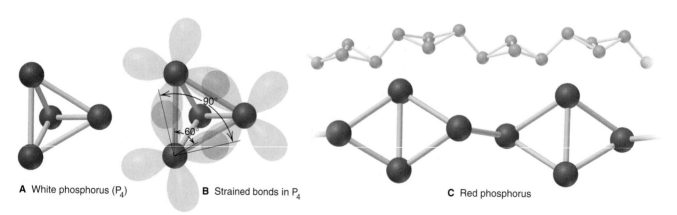

A White phosphorus (P_4)   **B** Strained bonds in P_4 **C** Red phosphorus

Figure 14.12 Two allotropes of phosphorus. A, White phosphorus exists as individual P_4 molecules, with the P—P bonds forming the edges of a tetrahedron. **B,** The reactivity of P_4 is due in part to the bond strain that arises from the 60° bond angle. Note how overlap of the 3*p* orbitals is decreased because they do not meet directly end to end (overlap is shown here for only three of the P—P bonds), which makes the bonds easier to break. **C,** In red phosphorus, one of the P—P bonds of the white form has broken and links the P_4 units together into long chains. Lone pairs (not shown) reside in *s* orbitals in both allotropes.

Key Atomic Properties, Physical Properties, and Reactions

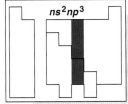

ns^2np^3

KEY
Atomic No.
Symbol
Atomic mass
Valence e⁻ configuration
Common oxidation states

7	**N** 14.01 $2s^2 2p^3$ (**−3, +5,** +4, +3, +2, +1)
15	**P** 30.97 $3s^2 3p^3$ (**−3, +5,** +3)
33	**As** 74.92 $4s^2 4p^3$ (**−3, +5,** +3)
51	**Sb** 121.8 $5s^2 5p^3$ (**−3, +5,** +3)
83	**Bi** 209.0 $6s^2 6p^3$ (**+3**)

Atomic radius (pm)		Ionic radius (pm)
N 75		N^{3-} 146
P 110		P^{3-} 212
As 120		
Sb 140		
Bi 150		Bi^{3+} 103

Atomic Properties

Group electron configuration is ns^2np^3. The np sublevel is half-filled, with each p orbital containing one electron. The number of oxidation states decreases down the group, and the lower (+3) state becomes more common. Atomic properties follow generally expected trends. The large (~50%) increase in size from N to P correlates with the much lower IE and EN of P.

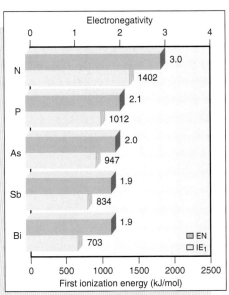

Physical Properties

Physical properties reflect the change from individual molecules (N, P) to network covalent solid (As, Sb) to metal (Bi). Thus, melting points increase and then decrease. Large atomic size and low atomic mass result in low density. Because mass increases more than size down the group, the density of the elements as solids increases. The dramatic increase from P to As is due to the intervening transition elements.

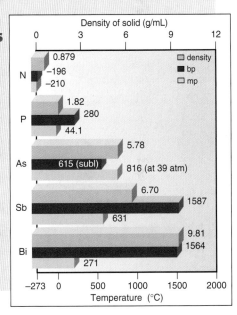

Reactions

1. Nitrogen is "fixed" industrially in the Haber process:

$$N_2(g) + 3H_2(g) \rightleftharpoons 2NH_3(g)$$

Further reactions convert NH_3 to NO, NO_2, and HNO_3 (*Highlights of Nitrogen Chemistry*). Hydrides of some other group members are formed from reaction in water (or with H_3O^+) of a metal phosphide, arsenide, and so forth:

$$Ca_3P_2(s) + 6H_2O(l) \longrightarrow 2PH_3(g) + 3Ca(OH)_2(aq)$$

2. Halides are formed by direct combination of the elements:

$$2E(s) + 3X_2 \longrightarrow 2EX_3 \quad \text{(E = all except N)}$$
$$EX_3 + X_2 \longrightarrow EX_5 \quad \text{(E = all except N and Bi with X = F and Cl, but no } BiCl_5; \text{ E = P for X = Br)}$$

3. Oxoacids are formed from the halides in a reaction with water that is common to many nonmetal halides:

$$EX_3 + 3H_2O(l) \longrightarrow H_3EO_3(aq) + 3HX(aq)$$
$$\text{(E = all except N)}$$
$$EX_5 + 4H_2O(l) \longrightarrow H_3EO_4(aq) + 5HX(aq)$$
$$\text{(E = all except N and Bi)}$$

Note that the oxidation number of E does *not* change.

As in Groups 3A and 4A, as we move down the group there are fewer oxidation states and the lower state becomes more prominent: N exhibits every state possible for a 5A element, from +5, as in HNO_3, to −3, as in NH_3; only the +5 and +3 states are common for P, As, and Sb; and +3 is the only common state of Bi. The oxides change from acidic to amphoteric to basic, and the lower oxide is more basic than the higher oxide because the lower oxide's E-to-O bond is more ionic.

Some characteristic reactions appear in the Family Portrait, but we point out some highlights here. All the elements form gaseous hydrides of formula EH_3. Ammonia is made industrially at high pressure and moderately high temperature:

$$N_2(g) + 3H_2(g) \rightleftharpoons 2NH_3(g)$$

The other hydrides are very poisonous and form by reaction in water of a metal phosphide, arsenide, and so forth, which acts as a strong base; for example,

$$Ca_3As_2(s) + 6H_2O(l) \longrightarrow 2AsH_3(g) + 3Ca(OH)_2(aq)$$

Molecular properties of the Group 5A(15) hydrides reveal some interesting patterns that appear again in Group 6A(16):

- Despite a much lower molar mass, NH_3 melts and boils at higher temperatures than the other 5A hydrides as a result of *H bonding*.
- Bond angles decrease from 107.3° for NH_3 to around 90° for the other hydrides, which suggests that the larger atoms use *un*hybridized *p* orbitals.
- E—H bond lengths increase down the group, so bond strength and thermal stability decrease: AsH_3 decomposes at 250°C, SbH_3 at 20°C, and BiH_3 at −45°C.

Through direct combination of the elements, the Group 5A(15) members form all possible trihalides (EX_3) and pentafluorides (EF_5), but few other pentahalides. As with the hydrides, stability of the halides decreases as the E—X bond becomes longer with larger halogens.

In an aqueous reaction pattern *typical of many nonmetal halides,* each 5A halide reacts with water to yield the hydrogen halide and the oxoacid, in which E has the *same* oxidation number as it had in the original halide. For example, PX_5 (O.N. of P = +5) produces HX and phosphoric acid (O.N. of P = +5):

$$PCl_5(s) + 4H_2O(l) \longrightarrow 5HCl(g) + H_3PO_4(l)$$

Highlights of Nitrogen Chemistry

The most striking highlight of nitrogen chemistry is the *inertness* of N_2. Even though the atmosphere consists of nearly four-fifths N_2 and one-fifth O_2, the searing temperature of a lightning bolt is needed to form significant amounts of nitrogen oxides. Indeed, N_2 reacts at high temperatures with H_2, Li, Group 2A(2) members, B, Al, C, Si, Ge, O_2, and many transition elements. Here we focus on the oxides and the oxoacids and their salts.

Nitrogen Oxides Nitrogen is remarkable for having six stable oxides, each with a *positive* heat of formation because of the great strength of the N≡N bond (Table 14.2). Unlike the hydrides and halides of nitrogen, the oxides are planar. Nitrogen displays all its positive oxidation states in these compounds, and in N_2O and N_2O_3, the two N atoms have different states. Of special interest are NO and NO_2.

Nitrogen monoxide (NO; also called nitric oxide) is an odd-electron molecule (see Section 10.1) with recently discovered biochemical functions ranging from neurotransmission to control of blood flow. Its commercial preparation occurs through the oxidation of ammonia during the production of nitric acid:

$$4NH_3(g) + 5O_2(g) \longrightarrow 4NO(g) + 6H_2O(g)$$

Table 14.2　Structures and Properties of the Nitrogen Oxides

Formula	Name	Space-filling Model	Lewis Structure	Oxidation State of N	ΔH_f° (kJ/mol) at 298 K	Comment
N_2O	Dinitrogen monoxide (dinitrogen oxide; nitrous oxide)		$:N\equiv N-\ddot{O}:$	+1 (0, +2)	82.0	Colorless gas; used as dental anesthetic ("laughing gas") and aerosol propellant
NO	Nitrogen monoxide (nitrogen oxide; nitric oxide)		$:\dot{N}=\ddot{O}:$	+2	90.3	Colorless, paramagnetic gas; biochemical messenger; air pollutant
N_2O_3	Dinitrogen trioxide			+3 (+2, +4)	83.7	Reddish brown gas (reversibly dissociates to NO and NO_2)
NO_2	Nitrogen dioxide			+4	33.2	Orange-brown, paramagnetic gas formed during HNO_3 manufacture; poisonous air pollutant
N_2O_4	Dinitrogen tetraoxide			+4	9.16	Colorless to yellow liquid (reversibly dissociates to NO_2)
N_2O_5	Dinitrogen pentaoxide			+5	11.3	Colorless, volatile solid consisting of NO_2^+ and NO_3^-; gas consists of N_2O_5 molecules

It also forms when air is heated to high temperatures in a car engine:

$$N_2(g) + O_2(g) \xrightarrow{\text{high } T} 2NO(g)$$

Heating converts NO to two other oxides:

$$3NO(g) \xrightarrow{\Delta} N_2O(g) + NO_2(g)$$

This redox reaction is called a **disproportionation,** one that involves a substance *acting as both an oxidizing and a reducing agent*. Thus, an atom in the reactant occurs in the products in both lower and higher states: the oxidation state of N in NO (+2) becomes +1 in N_2O and +4 in NO_2.

Nitrogen dioxide (NO_2), a brown poisonous gas, forms to a small extent when NO reacts with additional oxygen:

$$2NO(g) + O_2(g) \rightleftharpoons 2NO_2(g)$$

Like NO, NO_2 is an odd-electron molecule, but the unpaired electron is more localized on the N atom. Thus, NO_2 dimerizes reversibly to dinitrogen tetraoxide:

$$O_2N\cdot(g) + \cdot NO_2(g) \rightleftharpoons O_2N-NO_2(g) \quad (\text{or } N_2O_4)$$

In urban settings, a series of reactions involving sunlight, NO, NO_2, ozone (O_3), unburned gasoline, and various other species form photochemical smog.

Figure 14.13 The structures of nitric and nitrous acids and their oxoanions. A, Nitric acid loses a proton (H^+) to form the trigonal planar nitrate ion (one of three resonance forms is shown). **B,** Nitrous acid, a much weaker acid, forms the planar nitrite ion. Note the effect of nitrogen's lone pair in reducing the ideal 120° bond angle to 115° (one of two resonance forms is shown).

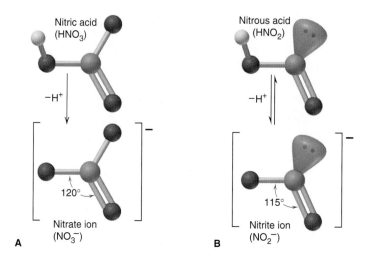

Nitrogen Oxoacids and Oxoanions The two common nitrogen oxoacids are nitric acid and nitrous acid. The first two steps in the *Ostwald process* for the production of nitric acid are the oxidations of NH_3 to NO and of NO to NO_2. The final step is a disproportionation, as the oxidation numbers show:

$$\overset{+4}{3NO_2}(g) + H_2O(l) \longrightarrow \overset{+5}{2HNO_3}(aq) + \overset{+2}{NO}(g)$$

The NO is recycled to make more NO_2.

In nitric acid, as in all oxoacids, *the acidic H is attached to one of the O atoms* (Figure 14.13A). In the laboratory, nitric acid is used as a strong oxidizing acid. The products of its reactions with metals vary with the metal's reactivity and the acid's concentration. In the following examples, notice from the net ionic equations that *the NO_3^- ion is the oxidizing agent.* Nitrate ion that is not reduced is a spectator ion and does not appear in the net ionic equations.

- With an active metal, such as Al, and dilute acid, N is reduced from the +5 state all the way to the −3 state in the ammonium ion, NH_4^+:

$$8Al(s) + 30HNO_3(aq; 1\ M) \longrightarrow 8Al(NO_3)_3(aq) + 3NH_4NO_3(aq) + 9H_2O(l)$$
$$8Al(s) + 30H^+(aq) + 3NO_3^-(aq) \longrightarrow 8Al^{3+}(aq) + 3NH_4^+(aq) + 9H_2O(l)$$

- With a less reactive metal, such as Cu, and more concentrated acid, N is reduced to the +2 state in NO:

$$3Cu(s) + 8HNO_3(aq; 3\ to\ 6\ M) \longrightarrow 3Cu(NO_3)_2(aq) + 4H_2O(l) + 2NO(g)$$
$$3Cu(s) + 8H^+(aq) + 2NO_3^-(aq) \longrightarrow 3Cu^{2+}(aq) + 4H_2O(l) + 2NO(g)$$

- With still more concentrated acid, N is reduced only to the +4 state in NO_2:

$$Cu(s) + 4HNO_3(aq; 12\ M) \longrightarrow Cu(NO_3)_2(aq) + 2H_2O(l) + 2NO_2(g)$$
$$Cu(s) + 4H^+(aq) + 2NO_3^-(aq) \longrightarrow Cu^{2+}(aq) + 2H_2O(l) + 2NO_2(g)$$

Nitrates form when HNO_3 reacts with metals or with their hydroxides, oxides, or carbonates. *All nitrates are soluble in water.*

Nitrous acid, HNO_2 (Figure 14.13B), a much weaker acid than HNO_3, forms when metal nitrites are treated with a strong acid:

$$NaNO_2(aq) + HCl(aq) \longrightarrow HNO_2(aq) + NaCl(aq)$$

These two acids reveal a *general pattern in relative acid strength among oxoacids:* the more O atoms bonded to the central nonmetal, the stronger the acid. We'll discuss the pattern quantitatively in Chapter 18.

Highlights of Phosphorus Chemistry: Oxides and Oxoacids

Phosphorus forms two important oxides, P_4O_6 and P_4O_{10}. Tetraphosphorus hexaoxide, P_4O_6, forms when white phosphorus, P_4, reacts with limited oxygen:

$$P_4(s) + 3O_2(g) \longrightarrow P_4O_6(s)$$

P_4O_6 has the tetrahedral orientation of the P atoms in P_4, with an O atom between each pair of P atoms (Figure 14.14A). It reacts with water to form phosphor*ous* acid (note the spelling):

$$P_4O_6(s) + 6H_2O(l) \longrightarrow 4H_3PO_3(l)$$

Despite the formula, H_3PO_3 has only two acidic H atoms; the third is bonded to the central P. It is a weak acid in water but reacts completely with strong base:

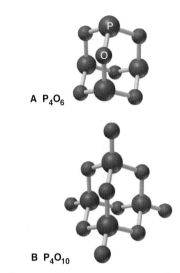

A P_4O_6

B P_4O_{10}

Figure 14.14 Important oxides of phosphorus. A, P_4O_6. **B,** P_4O_{10}.

Salts of phosphorous acid contain the phosphite ion, HPO_3^{2-}.

Commonly known as "phosphorus pentoxide" from the empirical formula (P_2O_5), tetraphosphorus decaoxide, P_4O_{10}, forms when P_4 burns in excess O_2:

$$P_4(s) + 5O_2(g) \longrightarrow P_4O_{10}(s)$$

It has the structure of P_4O_6 with another O atom bonded to each P atom (Figure 14.14B). P_4O_{10} is a powerful drying agent and, in a vigorous exothermic reaction with water, forms phosphoric acid (H_3PO_4), one of the "top-10" most important compounds in chemical manufacturing:

$$P_4O_{10}(s) + 6H_2O(l) \longrightarrow 4H_3PO_4(l)$$

The presence of many H bonds makes pure H_3PO_4 syrupy, more than 75 times as viscous as water. H_3PO_4 is a weak triprotic acid; in water, it loses one proton:

$$H_3PO_4(l) + H_2O(l) \rightleftharpoons H_2PO_4^-(aq) + H_3O^+(aq)$$

In strong base, however, it dissociates completely to give the three oxoanions:

dihydrogen phosphate ion   hydrogen phosphate ion   phosphate ion

Phosphoric acid has a central role in fertilizer production, and it is also used as a polishing agent for aluminum car trim and as an additive in soft drinks. The various phosphate salts have many essential applications, from paint stripper (Na_3PO_4) to rubber stabilizer (K_3PO_4) to fertilizer [$Ca(H_2PO_4)_2$ and $(NH_4)_2HPO_4$].

Polyphosphates are formed by heating hydrogen phosphates, which lose water as they form P—O—P linkages. This type of reaction, in which an H_2O molecule is lost for every pair of —OH groups that join, is called a **dehydration-condensation;** it occurs in the formation of polyoxoanion chains and other very large molecules, both synthetic and natural, made of repeating units.

14.7 GROUP 6A(16): THE OXYGEN FAMILY

Oxygen (O) and sulfur (S) are among the most important elements in industry, the environment, and organisms. Selenium (Se), tellurium (Te), radioactive polonium (Po), and newly synthesized element 116 lie beneath them in Group 6A(16) [Family Portrait, p. 444].

Family Portrait of Group 6A(16): The Oxygen Family

Key Atomic Properties, Physical Properties, and Reactions

KEY	Atomic No.
	Symbol
	Atomic mass
	Valence e⁻ configuration
	Common oxidation states

8
O
16.00
$2s^2 2p^4$
(−1, **−2**)

16
S
32.07
$3s^2 3p^4$
(**−2, +6,**
+4, +2)

34
Se
78.96
$4s^2 4p^4$
(**−2, +6,**
+4, +2)

52
Te
127.6
$5s^2 5p^4$
(**−2, +6,**
+4, +2)

84
Po
(209)
$6s^2 6p^4$
(**+4, +2**)

116
No sample
available
(292)
$7s^2 7p^4$

ns^2np^4

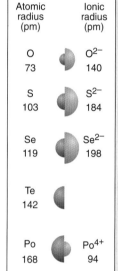

Atomic radius (pm)		Ionic radius (pm)
O 73		O²⁻ 140
S 103		S²⁻ 184
Se 119		Se²⁻ 198
Te 142		
Po 168		Po⁴⁺ 94

Atomic Properties

Group electron configuration is ns^2np^4. As in Groups 3A(13) and 5A(15), a lower (+4) oxidation state becomes more common down the group. Down the group, atomic and ionic size increase, and IE and EN decrease.

Physical Properties

Melting points increase through Te, which has covalent bonding, and then decrease for Po, which has metallic bonding. Densities of the elements as solids increase steadily.

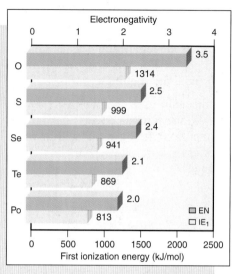

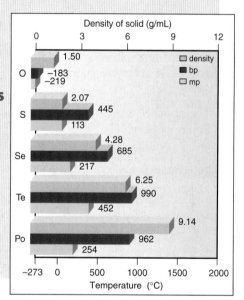

Reactions

1. Halides are formed by direct combination:

$$E(s) + X_2(g) \longrightarrow \text{various halides}$$
$$(E = S, Se, Te; X = F, Cl)$$

2. The other elements in the group are oxidized by O_2:

$$E(s) + O_2(g) \longrightarrow EO_2 \quad (E = S, Se, Te, Po)$$

SO_2 is oxidized further, and the product is used in the final step of H_2SO_4 manufacture (*Highlights of Sulfur Chemistry*):

$$2SO_2(g) + O_2(g) \longrightarrow 2SO_3(g)$$

How Do the Oxygen and Nitrogen Families Compare Physically?

Group 6A(16) resembles Group 5A(15) in many respects. Like nitrogen, oxygen occurs as a low-boiling diatomic gas. Like phosphorus, sulfur occurs as a polyatomic molecular solid. Like arsenic, selenium occurs as a gray metalloid. Like antimony, tellurium is slightly more metallic than the preceding member of its group but still displays network covalent bonding. Finally, like bismuth, polonium is a metal. Thus, as in Group 5A, electrical conductivity increases steadily as bonding changes from individual molecules (insulators) to metalloid networks (semiconductors) to a metallic solid (conductor).

Allotropism is common in Group 6A(16). Oxygen has two forms: life-giving dioxygen (O_2) and poisonous triatomic ozone (O_3). O_2 gas is colorless, odorless, paramagnetic, and thermally stable. In contrast, O_3 gas is bluish, has a pungent odor, is diamagnetic, and decomposes in heat and in ultraviolet (UV) light:

$$2O_3(g) \xrightarrow{\text{UV}} 3O_2(g)$$

This ability to absorb high-energy photons makes stratospheric ozone vital to life, as we'll discuss in Chapter 16.

The S atom's ability to bond to itself over a wide range of bond lengths and angles makes sulfur the allotrope "champion" of the periodic table, with more than 10 forms. The most stable is orthorhombic α-S_8, which consists of crown-shaped molecules of eight atoms, called *cyclo-S_8* (Figure 14.15); all other S allotropes eventually revert to this one. Some selenium allotropes consist of crown-shaped Se_8 molecules. Gray Se consists of layers of helical chains. Its ability to conduct a current when exposed to visible light gave birth to the photocopying industry.

How Do the Oxygen and Nitrogen Families Compare Chemically?

Groups 5A(15) and 6A(16) also have several chemical similarities. Like N and P, O and S bond covalently with almost every other nonmetal, even though they occur as anions much more often. Se and Te form mostly covalent compounds, as do As and Sb, and Po forms many saltlike compounds, as does Bi. In contrast to nitrogen, oxygen has few common oxidation states, but the earlier pattern returns with the other members: among the common positive (+6 and +4) states, the +4 state is seen more often in Te and Po.

Oxygen's high EN (3.5) and great oxidizing strength are second only to those of fluorine. But the other 6A members are much less electronegative, occur as anions much less often, and form hydrides that exhibit no H bonding.

All the elements except O form foul-smelling, poisonous, gaseous hydrides by treatment with acid of the metal sulfide, selenide, and so forth. For example,

$$FeSe(s) + 2HCl(aq) \longrightarrow H_2Se(g) + FeCl_2(aq)$$

In their bonding and stability, Group 6A hydrides are similar to 5A hydrides:

- Only water forms H bonds, so it melts and boils at much higher temperatures than the other hydrides (see Figure 12.13).
- Bond angles drop from 104.5° (nearly tetrahedral) in H_2O to around 90° in hydrides of larger group members, suggesting that the central atom uses *un*hybridized *p* orbitals.
- E—H bond length increases and bond energy decreases down the group. One result is that 6A hydrides are acids in water, as we discuss in Chapter 18.

Except for O, the Group 6A elements form a wide range of halides whose stability depends on *crowding between lone pairs and surrounding halogen (X) atoms.* Therefore, with increasing size of E and X, E—X bond length increases, electron repulsions between lone pairs and X atoms weaken,

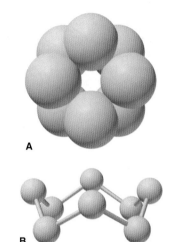

Figure 14.15 The cyclo-S_8 molecule. A, Top view of a space-filling model of the cyclo-S_8 molecule. **B,** Side view of a ball-and-stick model of the molecule; note the crownlike shape.

and a greater number of stable halides form. Thus, S, Se, and Te form hexa-fluorides; Se, Te, and Po form tetrachlorides and tetrabromides; and Te and Po form tetraiodides.

Highlights of Oxygen Chemistry

Oxygen is the most abundant element on Earth's surface, occurring in air as the free element, combined with hydrogen in water, and combined in innumerable oxides, silicates, carbonates, and phosphates. Virtually all O_2 has a biological origin, having been formed for billions of years by photosynthetic algae and multicellular plants in an overall equation that looks simple but involves many steps:

$$n\text{H}_2\text{O}(l) + n\text{CO}_2(g) \xrightarrow{\text{light}} n\text{O}_2(g) + (\text{CH}_2\text{O})_n \text{ (carbohydrates)}$$

The reverse process occurs during combustion and respiration.

Every element (except He, Ne, and Ar) forms at least one oxide, many by direct combination. For this reason, a useful way to classify elements is by the acid-base properties of their oxides. The oxides of Group 6A(16) exhibit expected trends down the group, with SO_3 the most acidic and PoO_2 the most basic.

Highlights of Sulfur Chemistry: Oxides and Oxoacids

Like phosphorus, sulfur forms two important oxides, sulfur dioxide (SO_2) and sulfur trioxide (SO_3). SO_2 is a colorless, choking gas that forms when S, H_2S, or a metal sulfide burns in air:

$$2\text{H}_2\text{S}(g) + 3\text{O}_2(g) \longrightarrow 2\text{H}_2\text{O}(g) + 2\text{SO}_2(g)$$
$$4\text{FeS}_2(s) + 11\text{O}_2(g) \longrightarrow 2\text{Fe}_2\text{O}_3(s) + 8\text{SO}_2(g)$$

In water, sulfur dioxide forms sulfurous acid, which exists in equilibrium with hydrated SO_2 rather than as independent H_2SO_3 molecules:

$$\text{SO}_2(aq) + \text{H}_2\text{O}(l) \rightleftharpoons [\text{H}_2\text{SO}_3(aq)] \rightleftharpoons \text{H}^+(aq) + \text{HSO}_3^-(aq)$$

(Similarly, carbonic acid occurs in equilibrium with hydrated CO_2 and cannot be isolated as H_2CO_3 molecules.) Sulfurous acid is weak and has two acidic protons, forming the hydrogen sulfite (bisulfite, HSO_3^-) and sulfite (SO_3^{2-}) ions with strong base. S is in the +4 state in SO_3^{2-}, so it can be oxidized easily to the +6 state; thus, sulfites are good reducing agents and are used to preserve foods and wine from air oxidation.

Most SO_2 produced industrially is used to make sulfuric acid. It is first oxidized to SO_3 by heating in O_2 over a catalyst:

$$\text{SO}_2(g) + \tfrac{1}{2}\text{O}_2(g) \xrightarrow{\text{V}_2\text{O}_5/\text{K}_2\text{O catalyst}} \text{SO}_3(g)$$

We discuss catalysts in Chapter 16 and H_2SO_4 production in Chapter 21. These two sulfur oxides also form when sulfur impurities in coal burn and then oxidize further. In contact with rain, they form H_2SO_3 and H_2SO_4 and contribute to a major pollution problem that we discuss in Chapter 19.

Sulfuric acid ranks first among all industrial chemicals in mass produced. The fertilizer, pigment, textile, and detergent industries are just a few that depend on it. The concentrated acid is a viscous, colorless liquid that is 98% H_2SO_4 by mass. It is a strong acid, but only the first proton dissociates completely. The hydrogen sulfate (or bisulfate) ion that results is a weak acid:

hydrogen sulfate ion   sulfate ion

Looking Backward and Forward: Groups 5A(15), 6A(16), and 7A(17)

Groups 5A(15) and 6A(16) are very similar in their physical and chemical trends (Figure 14.16). Their greatest difference is the sluggish behavior of N_2 compared with the striking reactivity of O_2. In both groups, metals appear only as the largest members. From here on, metals and metalloids are left behind: all Group 7A(17) elements are reactive nonmetals. Anion formation, which was rare in 5A and more common in 6A, is a dominant feature of 7A, as is the number of covalent compounds the elements form with oxygen *and* with each other.

14.8 GROUP 7A(17): THE HALOGENS

Our last chance to view very active elements occurs in Group 7A(17). The halogens begin with fluorine (F), the strongest electron "grabber" of all. Chlorine (Cl), bromine (Br), and iodine (I) also form compounds with most elements, and even rare astatine (At) is thought to be reactive [Group 7A(17) Family Portrait, p. 448].

What Accounts for the Regular Changes in the Halogens' Physical Properties?

Like the alkali metals, the halogens display regular trends in physical properties. But they display opposite trends because of differences in bonding. Alkali metal atoms are held together by metallic bonding, which *decreases* in strength down the group. The halogens, on the other hand, exist as diatomic molecules that interact through dispersion forces, which *increase* in strength as the atoms become larger. Thus, at room temperature, F_2 is a very pale yellow gas, Cl_2 a yellow-green gas, Br_2 a brown-orange liquid, and I_2 a purple-black solid.

Why Are the Halogens So Reactive?

The Group 7A(17) elements form many ionic and covalent compounds: metal and nonmetal halides, halogen oxides, and oxoacids. Like the alkali metals, the halogens have an electron configuration one electron away from that of a noble gas: whereas a 1A metal atom must *lose* one electron, a 7A nonmetal atom must *gain* one. It fills its outer level in either of two ways:

1. Gaining an electron from a metal atom, thus forming a negative ion as the metal forms a positive one
2. Sharing an electron pair with a nonmetal atom, thus forming a covalent bond

Down the group, reactivity reflects the decrease in electronegativity, but the exceptional reactivity of elemental F_2 is also related to the weakness of the F—F bond. The F—F bond is short, but F is so small that lone pairs on one atom repel those on the other, which weakens the bond (Figure 14.17). As a result, F_2 reacts with every element (except He, Ne, and Ar), in many cases, explosively.

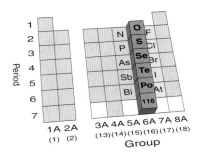

Figure 14.16 Standing in Group 6A(16), looking backward to Group 5A(15) and forward to Group 7A(17).

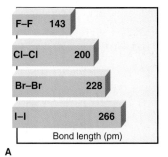

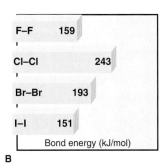

A   **B**

Figure 14.17 Bond energies and bond lengths of the halogens. A, In keeping with the increase in atomic size down the group, bond lengths increase steadily. **B,** The halogens show a general decrease in bond energy as bond length increases. However, F_2 deviates from this trend because its small, close, electron-rich atoms repel each other, thereby lowering its bond energy.

Family Portrait of Group 7A(17): The Halogens

Key Atomic Properties, Physical Properties, and Reactions

KEY	Atomic No.
	Symbol
	Atomic mass
	Valence e⁻ configuration
	Common oxidation states

9
F
19.00
$2s^2 2p^5$
(**−1**)

Photograph not available

17
Cl
35.45
$3s^2 3p^5$
(**−1**, +7, +5, +3, +1)

35
Br
79.90
$4s^2 4p^5$
(**−1**, +7, +5, +3, +1)

53
I
126.9
$5s^2 5p^5$
(**−1**, +7, +5, +3, +1)

85
At
(210)
$6s^2 6p^5$
(−1)

Extremely rare, no sample available

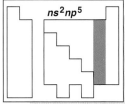

ns^2np^5

Atomic radius (pm)		Ionic radius (pm)
F 72		F⁻ 133
Cl 100		Cl⁻ 181
Br 114		Br⁻ 196
I 133		I⁻ 220
At (140)		no data

Atomic Properties

Group electron configuration is ns^2np^5; elements lack one electron to complete their outer level. The −1 oxidation state is the most common for all members. Except for F, the halogens exhibit all odd-numbered states (+7 through −1). Down the group, atomic and ionic size increase steadily, as IE and EN decrease.

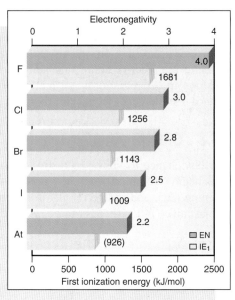

Physical Properties

Down the group, melting and boiling points increase smoothly as a result of stronger dispersion forces between larger molecules. The densities of the elements as liquids (at given *T*) increase steadily with molar mass.

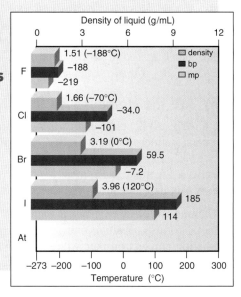

Reactions

1. The halogens (X_2) oxidize many metals and nonmetals. The reaction with hydrogen, although not used commercially for HX production (except for high-purity HCl), is characteristic of these strong oxidizing agents:

$$X_2 + H_2(g) \longrightarrow 2HX(g)$$

2. The halogens disproportionate in water:

$$X_2 + H_2O(l) \rightleftharpoons HX(aq) + HXO(aq)$$
$$(X = Cl, Br, I)$$

In aqueous base, the reaction goes to completion to form hypohalites (*see text*) and, at higher temperatures, halates; for example:

$$3Cl_2(g) + 6OH^-(aq) \xrightarrow{\Delta} ClO_3^-(aq) + 5Cl^-(aq) + 3H_2O(l)$$

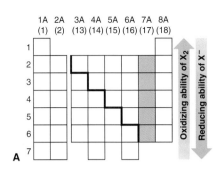

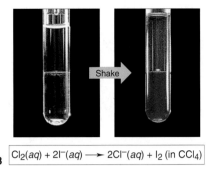

$$Cl_2(aq) + 2I^-(aq) \longrightarrow 2Cl^-(aq) + I_2 \text{ (in CCl}_4)$$

Figure 14.18 The relative oxidizing ability of the halogens. A, Halogen redox behavior is based on atomic properties such as electron affinity, ionic charge density, and electronegativity. A halogen (X_2) higher in the group can oxidize a halide ion (X^-) lower down. **B,** As an example, when aqueous Cl_2 is added to a solution of I^- (*top layer*), it oxidizes the I^- to I_2, which dissolves in the CCl_4 solvent (*bottom layer*) to give a purple solution.

The halogens act as *oxidizing agents* in the majority of their reactions, and halogens higher in the group can oxidize halide ions lower down:

$$F_2(g) + 2X^-(aq) \longrightarrow 2F^-(aq) + X_2(aq) \quad (X = Cl, Br, I)$$

Thus, the oxidizing ability of X_2 *decreases* down the group: the lower the EN, the less strongly each X atom pulls electrons. And the reducing ability of X^- *increases* down the group: the larger the ion, the more easily it gives up its electron (Figure 14.18).

Highlights of Halogen Chemistry

Let's examine the compounds the halogens form with hydrogen and with each other, as well as their oxides, oxoanions, and oxoacids.

The Hydrogen Halides The halogens form gaseous hydrogen halides (HX) through direct combination with H_2 or through the action of a concentrated acid on the metal halide. Commercially, HCl forms as a by-product during the chlorination of hydrocarbons to form useful materials, such as poly(vinyl chloride).

In water, gaseous HX molecules form *hydrohalic acids*. Only HF, with its relatively short, strong bond, forms a weak acid:

$$HF(g) + H_2O(l) \rightleftharpoons H_3O^+(aq) + F^-(aq)$$

The others dissociate completely to form the stoichiometric amount of H_3O^+ ions:

$$HBr(g) + H_2O(l) \longrightarrow H_3O^+(aq) + Br^-(aq)$$

(Recall from Chapter 4 that this type of transfer of a proton from acid to H_2O is a *Brønsted-Lowry acid-base reaction;* we discuss them in Chapter 18.)

Interhalogen Compounds: The "Halogen Halides" Halogens react exothermically with one another to form many *interhalogen compounds*. The simplest are diatomic molecules, such as ClF or BrCl. Every binary combination of the four common halogens is known. The more electronegative halogen is in the −1 oxidation state, and the less electronegative is in the +1 state; thus, for example, in BrCl, Br is +1 and Cl is −1. Interhalogens of general formula XY_n (n = 3, 5, 7) form when the larger members of the group (X) use *d* orbitals to expand their valence shells. In every case, the central atom has the *lower electronegativity* and a positive oxidation state; thus, for example, in BrF_3, Br is +3 and F is −1.

Halogen Oxides, Oxoacids, and Oxoanions The Group 7A(17) elements form many oxides that are *powerful oxidizing agents*. Dichlorine monoxide (Cl_2O), chlorine dioxide (ClO_2, with an unpaired electron and Cl in the unusual +4 oxidation state), and dichlorine heptaoxide (Cl_2O_7) are important examples.

The halogen oxoacids and oxoanions are produced by reaction of the halogens and their oxides with water. Most of the oxoacids are stable only in solution. Table 14.3 (next page) shows ball-and-stick models of the acids in which each atom has its lowest formal charge; note that H is bonded to O. (We'll discuss factors that determine the relative strengths of the halogen oxoacids in Chapter 18.)

Table 14.3 | **The Known Halogen Oxoacids***

Central Atom	Hypohalous Acid (HOX) 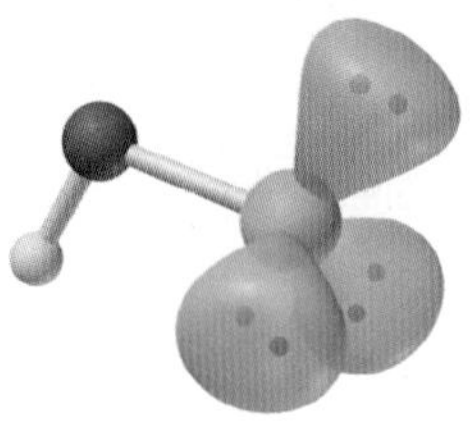	Halous Acid (HOXO) 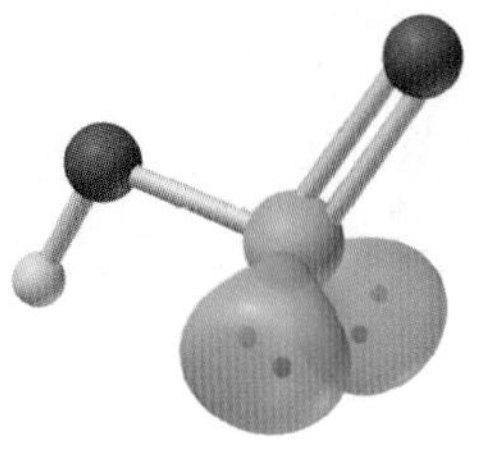	Halic Acid (HOXO$_2$) 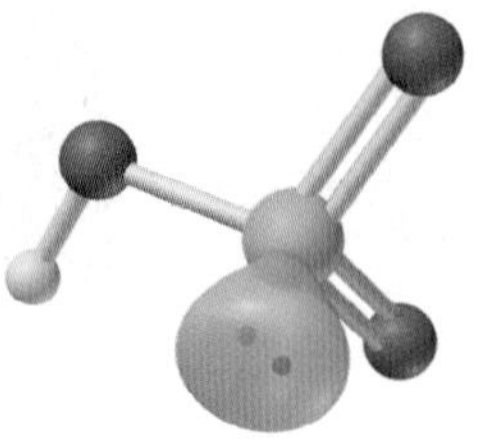	Perhalic Acid (HOXO$_3$) 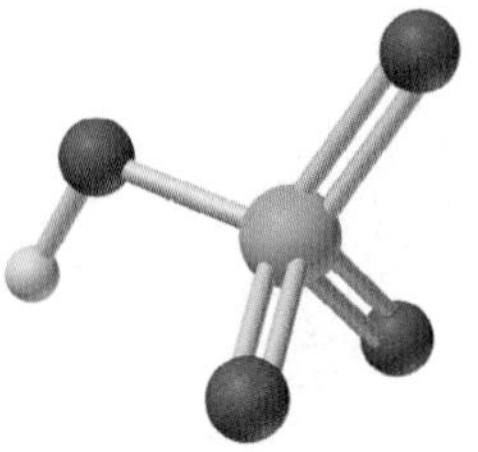
Fluorine	HOF	—	—	—
Chlorine	HOCl	HOClO	HOClO$_2$	HOClO$_3$
Bromine	HOBr	(HOBrO)?	HOBrO$_2$	HOBrO$_3$
Iodine	HOI	—	HOIO$_2$	HOIO$_3$, (HO)$_5$IO
Oxoanion	Hypohalite	Halite	Halate	Perhalate

*Lone pairs are shown only on the halogen atom.

The hypohalites (XO$^-$), halites (XO$_2^-$), and halates (XO$_3^-$) are oxidizing agents formed by aqueous disproportionation reactions [see Group 7A(17) Family Portrait]. Potassium chlorate is the oxidizer in "safety" matches. You may have heated it in the lab to form small amounts of O$_2$:

$$2KClO_3(s) \xrightarrow{\Delta} 2KCl(s) + 3O_2(g)$$

Some perhalates are especially strong oxidizing agents. Ammonium perchlorate, prepared from sodium perchlorate, is the oxidizing agent for the aluminum powder in the solid-fuel booster rocket of the space shuttle; each launch uses more than 700 tons of NH$_4$ClO$_4$:

$$10Al(s) + 6NH_4ClO_4(s) \longrightarrow 4Al_2O_3(s) + 12H_2O(g) + 3N_2(g) + 2AlCl_3(g)$$

14.9 GROUP 8A(18): THE NOBLE GASES

The Group 8A(18) elements are helium (He, the second most abundant element in the universe), neon (Ne), argon (Ar, which makes up about 0.93% of Earth's atmosphere), krypton (Kr), xenon (Xe), and radioactive radon (Rn). Only the last three form compounds [Group 8A(18) Family Portrait].

How Can Noble Gases Form Compounds?

Lying at the far right side of the periodic table, the Group 8A(18) elements consist of individual atoms with filled outer levels and the smallest radii in their periods: even Li, the smallest alkali metal (152 pm), is bigger than Rn, the largest noble gas (140 pm). These elements come as close to behaving as ideal gases as any substance. They condense and solidify at very low temperatures; in fact, He requires an increase in pressure to solidify, 25 atm at −272.2°C. With only dispersion forces at work, melting and boiling points increase with molar mass.

Ever since their discovery in the late 19th century, these elements were considered, and in fact, were generally referred to as, the "inert" gases. Atomic theory and, more important, all experiments had supported this idea. Then, in 1962, all this changed when the first noble gas compound was prepared.

The discovery of noble gas reactivity is a classic example of clear thinking in the face of an unexpected event. The young inorganic chemist Neil Bartlett was studying platinum fluorides. When he accidentally exposed PtF$_6$ to air, its deep-red color lightened slightly, and analysis showed that the PtF$_6$ had oxidized O$_2$ to form the ionic compound [O$_2$]$^+$[PtF$_6$]$^-$. Knowing that the IE$_1$ of O$_2$ to form

Family Portrait of Group 8A(18): The Noble Gases

Key Atomic Properties, Physical Properties, and Reactions

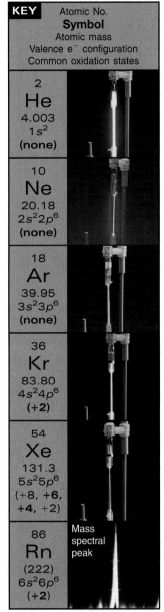

KEY
Atomic No.
Symbol
Atomic mass
Valence e⁻ configuration
Common oxidation states

| 2 |
| He |
| 4.003 |
| $1s^2$ |
| (none) |

| 10 |
| Ne |
| 20.18 |
| $2s^2 2p^6$ |
| (none) |

| 18 |
| Ar |
| 39.95 |
| $3s^2 3p^6$ |
| (none) |

| 36 |
| Kr |
| 83.80 |
| $4s^2 4p^6$ |
| (+2) |

| 54 |
| Xe |
| 131.3 |
| $5s^2 5p^6$ |
| (+8, +6, +4, +2) |

| 86 |
| Rn |
| (222) |
| $6s^2 6p^6$ |
| (+2) |

Mass spectral peak

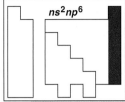

ns^2np^6

Atomic radius (pm)

He
31

Ne
71

Ar
98

Kr
112

Xe
131

Rn
(140)

Atomic Properties

Group electron configuration is $1s^2$ for He and ns^2np^6 for the others. The valence shell is filled. Only Kr and Xe (and perhaps Rn) are known to form compounds. The more reactive Xe exhibits all even oxidation states (+2 to +8). This group contains the smallest atoms with the highest IEs in their periods. Down the group, atomic size increases and IE decreases steadily. (EN values are given only for Kr and Xe.)

Physical Properties

Melting and boiling points of these gaseous elements are extremely low but increase down the group because of stronger dispersion forces. Note the extremely small liquid ranges. Densities (at STP) increase steadily, as expected.

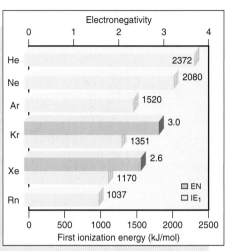

Electronegativity

He — 2372
Ne — 2080
Ar — 1520
Kr — 3.0, 1351
Xe — 2.6, 1170
Rn — 1037

☐ EN
☐ IE₁

First ionization energy (kJ/mol)

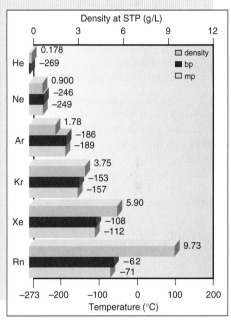

Density at STP (g/L)

☐ density
■ bp
☐ mp

He — 0.178, −269
Ne — 0.900, −246, −249
Ar — 1.78, −186, −189
Kr — 3.75, −153, −157
Xe — 5.90, −108, −112
Rn — 9.73, −62, −71

Temperature (°C)

O_2^+ (1175 kJ/mol) is very close to IE_1 of xenon (1170 kJ/mol), Bartlett reasoned that PtF_6 could oxidize xenon. He prepared $XePtF_6$, an orange-yellow solid, and within a few months, XeF_2 and XeF_4 were also prepared. In addition to its +2 and +4 oxidation states, Xe has the +6 state in several compounds, such as XeF_6 and XeO_3, and the +8 state in the unstable oxide, XeO_4. A few compounds of Kr and Rn have also been made.

Looking Backward and Forward: Groups 7A(17), 8A(18), and 1A(1)

The great reactivity of the Group 7A(17) elements, which form a host of anions, covalent oxides, and oxoanions, lies in stark contrast to the unreactivity of their 8A(18) neighbors. Filled outer levels render the noble gas atoms largely inert, with a limited ability to react directly only with highly electronegative fluorine. The least reactive family stands between the two most reactive: the halogens and the alkali metals (Figure 14.19), and atomic, physical, and chemical properties change dramatically from Group 8A(18) to Group 1A(1).

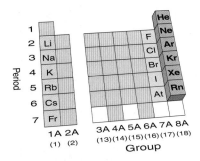

Figure 14.19 Standing in Group 8A(18), looking backward at the halogens, Group 7A(17), and ahead to the alkali metals, Group 1A(1).

For Review and Reference (Numbers in parentheses refer to pages, unless noted otherwise.)

Learning Objectives

To help you review these learning objectives, related sections (§) and upcoming end-of-chapter problems (EP) are listed in parentheses.

1. Compare hydrogen with alkali metals and halogens, and distinguish saltlike from covalent hydrides (§ 14.1) (EPs 14.1–14.5)
2. Discuss key features of Group 1A(1), and understand how the ns^1 configuration explains physical and chemical properties (§ 14.2) (EPs 14.7–14.13)
3. Understand the anomalous behaviors of the Period 2 elements (§ 14.2) (EPs 14.6, 14.14, 14.20–14.22, 14.26, 14.27)
4. Discuss key features of Group 2A(2), and understand how the ns^2 configuration explains differences between Groups 1A(1) and 2A(2) (§ 14.3) (EPs 14.15, 14.17–14.19)
5. Describe the three main diagonal relationships (§ 14.3) (EPs 14.16, 14.47)
6. Discuss key features of Group 3A(13), especially patterns in oxidation state and oxide acidity; understand how the presence of d and f electrons affects group properties; and describe major aspects of boron chemistry (§ 14.4) (EPs 14.25, 14.28–14.34)
7. Discuss key features of Group 4A(14), especially patterns in oxidation state and oxide acidity; describe how types of bonding affect physical behavior of Groups 4A to 6A; give examples of allotropism in these groups; and describe major aspects of carbon and silicon chemistry (§ 14.5) (EPs 14.35–14.47)
8. Discuss key features of Group 5A(15), especially patterns in oxidation state, oxide acidity, and hydride and halide structures; and describe the nitrogen and phosphorus oxides and oxoacids (§ 14.6) (EPs 14.48–14.58)
9. Discuss key features of Group 6A(16); compare the patterns in oxidation state, oxide acidity, and hydride and halide structures with those of Group 5A(15); and describe the sulfur oxides and oxoacids (§ 14.7) (EPs 14.59–14.67)
10. Discuss key features of Group 7A(17), understand how intermolecular forces and the ns^2np^5 configuration account for physical and chemical properties, and describe the halogen oxides and oxoacids (§ 14.8) (EPs 14.68–14.75)
11. Discuss key features of Group 8A(18), understand how intermolecular forces and the ns^2np^6 configuration account for physical and chemical properties, and explain how xenon can form compounds (§ 14.9) (EPs 14.76–14.78)

Key Terms

Section 14.3
diagonal relationship (428)

Section 14.5
allotrope (434)
silicate (437)
silicone (437)

Section 14.6
disproportionation reaction (441)

dehydration-condensation reaction (443)

Problems

Problems with **colored** numbers are answered in Appendix E. Sections match the text and provide the numbers of relevant sample problems. Bracketed problems are grouped in pairs (indicated by a short rule) that cover the same concept. Comprehensive Problems are based on material from any section or previous chapter.

Hydrogen, the Simplest Atom

14.1 Hydrogen has only one proton, but its IE_1 is much greater than that of lithium, which has three protons. Explain.

14.2 Complete and balance the following equations:
(a) An active metal reacting with acid,
$$Al(s) + HCl(aq) \longrightarrow$$
(b) A saltlike (alkali metal) hydride reacting with water,
$$LiH(s) + H_2O(l) \longrightarrow$$

14.3 Complete and balance the following equations:
(a) A saltlike (alkaline earth metal) hydride reacting with water,
$$CaH_2(s) + H_2O(l) \longrightarrow$$
(b) Reduction of a metal halide by hydrogen to form a metal,
$$PdCl_2(aq) + H_2(g) \longrightarrow$$

14.4 Compounds such as $NaBH_4$, $Al(BH_4)_3$, and $LiAlH_4$ are complex hydrides used as reducing agents in many syntheses.
(a) Give the oxidation state of each element in these compounds.
(b) Write a Lewis structure for the polyatomic anion in $NaBH_4$, and predict its shape.

14.5 Unlike the F^- ion, which has an ionic radius close to 133 pm in all alkali metal fluorides, the ionic radius of H^- varies from 137 pm in LiH to 152 pm in CsH. Suggest an explanation for the large variability in the size of H^- but not F^-.

Group 1A(1): The Alkali Metals

14.6 Lithium salts are often much less soluble in water than the corresponding salts of other alkali metals. For example, at 18°C, the concentration of a saturated LiF solution is 1.0×10^{-2} M, whereas that of a saturated KF solution is 1.6 M. How would you explain this behavior?

14.7 The alkali metals play virtually the same general chemical role in all their reactions.
(a) What is this role?
(b) How is it based on atomic properties?
(c) Using sodium, write two balanced equations that illustrate this role.

14.8 How do atomic properties account for the low densities of the Group 1A(1) elements?

14.9 Each of the following properties shows regular trends in Group 1A(1). Predict whether each increases or decreases *down* the group: (a) density; (b) ionic size; (c) E—E bond energy; (d) IE_1; (e) magnitude of ΔH_{hydr} of E^+ ion.

14.10 Each of the following properties shows regular trends in Group 1A(1). Predict whether each increases or decreases *up* the group: (a) melting point; (b) E—E bond length; (c) hardness; (d) molar volume; (e) lattice energy of EBr.

14.11 Write a balanced equation for the formation from its elements of sodium peroxide, an industrial bleach.

14.12 Write a balanced equation for the formation of rubidium bromide through a reaction of a strong acid and a strong base.

14.13 Although the alkali metal halides can be prepared directly from the elements, the far less expensive industrial route is treatment of the metal carbonate or hydroxide with aqueous hydrohalic acid (HX) followed by recrystallization. Balance the reaction between potassium carbonate and aqueous hydroiodic acid.

14.14 Lithium forms several useful organolithium compounds. Calculate the mass percent of Li in the following:
(a) Lithium stearate ($C_{17}H_{35}COOLi$), a water-resistant grease used in cars because it does not harden at cold temperatures
(b) Butyllithium (LiC_4H_9), a reagent in organic syntheses

Group 2A(2): The Alkaline Earth Metals

14.15 How do Groups 1A(1) and 2A(2) compare with respect to reaction of the metals with water?

14.16 Alkaline earth metals are involved in two key diagonal relationships in the periodic table. (a) Give the two pairs of elements in these diagonal relationships. (b) For each pair, cite two similarities that demonstrate the relationship. (c) Why are the members of each pair so similar in behavior?

14.17 The melting points of alkaline earth metals are many times higher than those of the alkali metals. Explain this difference on the basis of atomic properties. Name three other physical properties for which Group 2A(2) metals have higher values than the corresponding 1A(1) metals.

14.18 Write a balanced equation for each reaction:
(a) "Slaking" (treatment with water) of lime, CaO
(b) Combustion of calcium in air

14.19 Write a balanced equation for each reaction:
(a) Thermal decomposition of witherite (barium carbonate)
(b) Neutralization of stomach acid (HCl) by milk of magnesia (magnesium hydroxide)

14.20 In some reactions, Be behaves like a typical alkaline earth metal; in others, it does not. Complete and balance the following equations:
(a) $BeO(s) + H_2O(l) \longrightarrow$
(b) $BeCl_2(l) + Cl^-(l; from\ molten\ NaCl) \longrightarrow$
In which reaction does Be behave like the other Group 2A(2) members?

Group 3A(13): The Boron Family

14.21 How does the maximum oxidation number vary across a period in the main groups? Is the pattern in Period 2 different?

14.22 What correlation, if any, exists for the Period 2 elements between group number and the number of covalent bonds the element typically forms? How is the correlation different for elements in Periods 3 to 6?

14.23 How do the transition metals in Period 4 affect the pattern of ionization energies in Group 3A(13)?

14.24 How do the acidities of aqueous solutions of Tl_2O and Tl_2O_3 compare with each other? Explain.

14.25 Despite the expected decrease in atomic size, there is an unexpected drop in the first ionization energy between Groups

2A(2) and 3A(13) in Periods 2 through 4. Explain this pattern in terms of electron configurations and orbital energies.

14.26 Many compounds of Group 3A(13) elements have chemical behavior that reflects an electron deficiency.
(a) What is the meaning of *electron deficiency?*
(b) Give two reactions that illustrate this behavior.

14.27 Boron's chemistry is not typical of its group.
(a) Cite three ways in which boron and its compounds differ significantly from the other 3A(13) members and their compounds.
(b) What is the reason for these differences?

14.28 Rank the following oxides in order of increasing aqueous *acidity:* Ga_2O_3, Al_2O_3, In_2O_3.

14.29 Rank the following hydroxides in order of increasing aqueous *basicity:* $Al(OH)_3$, $B(OH)_3$, $In(OH)_3$.

14.30 Thallium forms the compound TlI_3. What is the apparent oxidation state of Tl in this compound? Given that the anion is I_3^-, what is the actual oxidation state of Tl? Draw the shape of the anion, giving its VSEPR class and bond angles. Propose a reason why the compound does not exist as $(Tl^{3+})(I^-)_3$.

14.31 Very stable dihalides of the Group 3A(13) metals are known. What is the apparent oxidation state of Ga in $GaCl_2$? Given that $GaCl_2$ consists of a Ga^+ cation and a $GaCl_4^-$ anion, what are the actual oxidation states of Ga? Draw the shape of the anion, giving its VSEPR class and bond angles.

14.32 Give the name and symbol or formula of a Group 3A(13) element or compound that fits each description or use:
(a) Largest temperature range for liquid state of an element
(b) Elementary substance with three-center, two-electron bonds
(c) Metal protected from oxidation by adherent oxide coat
(d) Toxic metal that lies between two other toxic metals

14.33 Indium (In) reacts with HCl to form a diamagnetic solid with the formula $InCl_2$.
(a) Write condensed electron configurations for In, In^+, In^{2+}, and In^{3+}.
(b) Which of these species is (are) diamagnetic and which paramagnetic?
(c) What is the apparent oxidation state of In in $InCl_2$?
(d) Given your answers to parts (b) and (c), explain how $InCl_2$ can be diamagnetic.

14.34 Use VSEPR theory to draw structures, with ideal bond angles, for boric acid and the anion it forms in reaction with water.

Group 4A(14): The Carbon Family

14.35 How do the physical properties of a network covalent solid and a molecular covalent solid differ? Why?

14.36 How does the basicity of SnO_2 in water compare with that of CO_2? Explain.

14.37 Nearly every compound of silicon has the element in the $+4$ oxidation state. In contrast, most compounds of lead have the element in the $+2$ state.
(a) What general observation do these facts illustrate?
(b) Explain in terms of atomic and molecular properties.
(c) Give an analogous example from Group 3A(13).

14.38 The sum of IE_1 through IE_4 for Group 4A(14) elements shows a decrease from C to Si, a slight increase from Si to Ge, a decrease from Ge to Sn, and an increase from Sn to Pb.
(a) What is the expected trend for ionization energy down a group?

(b) Suggest a reason for the deviations from the expected trend.
(c) Which group might show even greater deviations?

14.39 Give explanations for the large drops in melting point from C to Si and from Ge to Sn.

14.40 What is an allotrope? Name a Group 4A(14) element that exhibits allotropism, and name its three allotropes.

14.41 Even though EN values vary relatively little down Group 4A(14), the elements change from nonmetal to metal. Explain.

14.42 How do atomic properties account for the enormous number of carbon compounds? Why don't other Group 4A(14) elements behave similarly?

14.43 Draw a Lewis structure for
(a) The cyclic silicate ion $Si_4O_{12}^{8-}$
(b) A cyclic hydrocarbon with formula C_4H_8

14.44 Draw a Lewis structure for
(a) The cyclic silicate ion $Si_6O_{18}^{12-}$
(b) A cyclic hydrocarbon with formula C_6H_{12}

14.45 Zeolite A, $Na_{12}[(AlO_2)_{12}(SiO_2)_{12}]\cdot27H_2O$, is used to soften water by replacing Ca^{2+} and Mg^{2+} with Na^+. Hard water from a certain source is 4.5×10^{-3} M Ca^{2+} and 9.2×10^{-4} M Mg^{2+}, and a pipe delivers 25,000 L of this hard water per day. What mass (in kg) of zeolite A is needed to soften a week's supply of the water? (Assume zeolite A loses its capacity to exchange ions when 85 mol % of its Na^+ has been lost.)

14.46 Give the name and symbol or formula of a Group 4A(14) element or compound that fits each description or use:
(a) Hardest known substance
(b) Medicinal antacid
(c) Atmospheric gas implicated in the greenhouse effect
(d) Gas that binds to iron(II) in blood
(e) Toxic metal found in plumbing and paints

14.47 One similarity between B and Si is the explosive combustion of their hydrides in air. Write balanced equations for the combustion of B_2H_6 and of Si_4H_{10}.

Group 5A(15): The Nitrogen Family

14.48 The Group 5A(15) elements form all the trihalides but not pentahalides. Explain.

14.49 As you move down Group 5A(15), the melting points of the elements increase and then decrease. Explain.

14.50 (a) What is the range of oxidation states shown by the elements of Group 5A(15) as you move down the group?
(b) How does this range illustrate the general rule for the range of oxidation states in groups on the right side of the periodic table?

14.51 (a) How does the type of bonding in element oxides correlate with the electronegativity of the elements?
(b) How does the acid-base behavior of element oxides correlate with the electronegativity of the elements?

14.52 (a) How does the metallic character of an element correlate with the *acidity* of its oxide?
(b) What trends, if any, exist in oxide *basicity* across a period and down a group?

14.53 Rank the following oxides in order of increasing acidity in water: Sb_2O_3, Bi_2O_3, P_4O_{10}, Sb_2O_5.

14.54 Complete and balance the following:
(a) $As(s) + \text{excess } O_2(g) \longrightarrow$
(b) $Bi(s) + \text{excess } F_2(g) \longrightarrow$
(c) $Ca_3As_2(s) + H_2O(l) \longrightarrow$

14.55 Complete and balance the following:
(a) Excess $Sb(s)$ + $Br_2(l)$ $\longrightarrow$
(b) $HNO_3(aq)$ + $MgCO_3(s)$ $\longrightarrow$
(c) $PF_5(g)$ + $H_2O(l)$ $\longrightarrow$

14.56 The pentafluorides of the larger members of Group 5A(15) have been prepared, but N can have only eight electrons. A claim has been made that, at low temperatures, a compound with the empirical formula NF_5 forms. Draw a possible Lewis structure for this compound. (*Hint:* NF_5 is ionic.)

14.57 Give the name and symbol or formula of a Group 5A(15) element or compound that fits each description or use:
(a) Hydride produced at multimillion-ton level
(b) Element(s) essential in plant nutrition
(c) Oxide used as a laboratory drying agent
(d) Odd-electron molecule (two examples)
(e) Element that is an electrical conductor

14.58 Nitrous oxide (N_2O), the "laughing gas" used as an anesthetic by dentists, is made by thermal decomposition of solid NH_4NO_3. Write a balanced equation for this reaction. What are the oxidation states of N in NH_4NO_3 and in N_2O?

Group 6A(16): The Oxygen Family

14.59 Rank the following in order of increasing electrical conductivity, and explain your ranking: Po, S, Se.

14.60 The oxygen and nitrogen families have some obvious similarities and differences.
(a) State two general physical similarities between Group 5A(15) and 6A(16) elements.
(b) State two general chemical similarities between Group 5A(15) and 6A(16) elements.
(c) State two chemical similarities between P and S.
(d) State two physical similarities between N and O.
(e) State two chemical differences between N and O.

14.61 A molecular property of the Group 6A(16) hydrides changes abruptly down the group. This change has been explained in terms of a change in orbital hybridization.
(a) Between what periods does the change occur?
(b) What is the change in the molecular property?
(c) What is the change in hybridization?
(d) What other group displays a similar change?

14.62 Complete and balance the following:
(a) $NaHSO_4(aq)$ + $NaOH(aq)$ $\longrightarrow$
(b) $S_8(s)$ + excess $F_2(g)$ $\longrightarrow$
(c) $FeS(s)$ + $HCl(aq)$ $\longrightarrow$

14.63 Complete and balance the following:
(a) $H_2S(g)$ + $O_2(g)$ $\longrightarrow$
(b) $SO_3(g)$ + $H_2O(l)$ $\longrightarrow$
(c) $SF_4(g)$ + $H_2O(l)$ $\longrightarrow$

14.64 Is each oxide basic, acidic, or amphoteric in water:
(a) SeO_2; (b) N_2O_3; (c) K_2O; (d) BeO; (e) BaO?

14.65 Is each oxide basic, acidic, or amphoteric in water:
(a) MgO; (b) N_2O_5; (c) CaO; (d) CO_2; (e) TeO_2?

14.66 Give the name and symbol or formula of a Group 6A(16) element or compound that fits each description or use:
(a) Unstable allotrope of oxygen
(b) Oxide having sulfur in the same oxidation state as in sulfuric acid
(c) Air pollutant produced by burning sulfur-containing coal

14.67 Disulfur decafluoride is intermediate in reactivity between SF_4 and SF_6. It disproportionates at 150°C to these monosulfur fluorides. Write a balanced equation for this reaction, and give the oxidation state of S in each compound.

Group 7A(17): The Halogens

14.68 Iodine monochloride and elemental bromine have nearly the same molar mass and liquid density but very different boiling points.
(a) What molecular property is primarily responsible for this difference in boiling point? What atomic property gives rise to it? Explain.
(b) Which substance has a higher boiling point? Why?

14.69 Explain the change in physical state down Group 7A(17) in terms of molecular properties.

14.70 (a) What are the common oxidation states of the halogens?
(b) Give an explanation based on electron configuration for the range and values of the oxidation states of chlorine.
(c) Why is fluorine an exception to the pattern of oxidation states found for the other group members?

14.71 Select the stronger bond in each pair:
(a) Cl—Cl or Br—Br
(b) Br—Br or I—I
(c) F—F or Cl—Cl. Why doesn't the F—F bond strength follow the group trend?

14.72 A halogen (X_2) disproportionates in base in several steps to X^- and XO_3^-. Write the overall equation for the disproportionation of Br_2 to Br^- and BrO_3^-.

14.73 Complete and balance the following equations. If no reaction occurs, write NR:
(a) $I_2(s)$ + $H_2O(l)$ $\longrightarrow$
(b) $Br_2(l)$ + $I^-(aq)$ $\longrightarrow$
(c) $CaF_2(s)$ + $H_2SO_4(l)$ $\longrightarrow$

14.74 Complete and balance the following equations. If no reaction occurs, write NR:
(a) $Cl_2(g)$ + $I^-(aq)$ $\longrightarrow$
(b) $Br_2(l)$ + $Cl^-(aq)$ $\longrightarrow$
(c) $ClF(g)$ + $F_2(g)$ $\longrightarrow$

14.75 An industrial chemist treats solid NaCl with concentrated H_2SO_4 and obtains gaseous HCl and $NaHSO_4$. When she substitutes solid NaI for NaCl, gaseous H_2S, solid I_2, and S_8 are obtained but no HI.
(a) What type of reaction did the H_2SO_4 undergo with NaI?
(b) Why does NaI, but not NaCl, cause this type of reaction?
(c) To produce HI(g) by the reaction of NaI with an acid, how does the acid have to differ from sulfuric acid?

Group 8A(18): The Noble Gases

14.76 Which noble gas is the most abundant in the universe? In Earth's atmosphere?

14.77 Why do the noble gases have such low boiling points?

14.78 Explain why Xe and, to a limited extent, Kr form compounds, whereas He, Ne, and Ar do not.

Comprehensive Problems

Problems with an asterisk (*) are more challenging.

14.79 The main reason alkali metal dihalides (MX_2) do *not* form is the high IE_2 of the metal.
(a) Why is IE_2 so high for alkali metals?

(b) The IE_2 for Cs is 2255 kJ/mol, low enough for CsF_2 to form exothermically ($\Delta H_f^\circ = -125$ kJ/mol). This compound cannot be synthesized, however, because CsF forms with a much greater release of heat ($\Delta H_f^\circ = -530$ kJ/mol). Thus, the breakdown of CsF_2 to CsF happens readily. Write the equation for this breakdown, and calculate the heat of reaction per mole of CsF.

14.80 Semiconductors made from elements in Groups 3A(13) and 5A(15) are typically prepared by direct reaction of the elements at high temperature. An engineer treats 32.5 g of molten gallium with 20.4 L of white phosphorus vapor at 515 K and 195 kPa. If purification losses are 7.2% by mass, how many grams of gallium phosphide will be prepared?

* 14.81 Two substances with empirical formula HNO are hyponitrous acid ($\mathcal{M} = 62.04$ g/mol) and nitroxyl ($\mathcal{M} = 31.02$ g/mol).
(a) What is the molecular formula of each species?
(b) For each species, draw the Lewis structure having the lowest formal charges. (*Hint:* Hyponitrous acid has an N=N bond.)
(c) Predict the shape around the N atoms of each species.
(d) When hyponitrous acid loses two protons, it forms the hyponitrite ion. Because the double bond restricts rotation (Section 11.2), there are two possible structures for this ion; draw them.

14.82 For the species CO, CN^-, and C_2^{2-},
(a) Draw their Lewis structures.
(b) Draw their MO diagrams (assume $2s$-$2p$ mixing, as in N_2), and give the bond order and electron configuration for each.

* 14.83 The Ostwald process is a series of three reactions used for the industrial production of nitric acid from ammonia.
(a) Write a series of balanced equations for the Ostwald process.
(b) If NO is *not* recycled, how many moles of NH_3 are consumed per mole of HNO_3 produced?
(c) In a typical industrial unit, the process is very efficient, with a 96% yield for the first step. Assuming 100% yields for the subsequent steps, what volume of nitric acid (60.% by mass; $d = 1.37$ g/mL) can be prepared for each cubic meter of a gas mixture that is 90.% air and 10.% NH_3 by volume at the industrial conditions of 5.0 atm and 850.°C?

14.84 What is a disproportionation reaction, and which of the following fit the description?
(a) $I_2(s) + KI(aq) \longrightarrow KI_3(aq)$
(b) $2ClO_2(g) + H_2O(l) \longrightarrow HClO_3(aq) + HClO_2(aq)$
(c) $Cl_2(g) + 2NaOH(aq) \longrightarrow$
$$NaCl(aq) + NaClO(aq) + H_2O(l)$$
(d) $NH_4NO_2(s) \longrightarrow N_2(g) + 2H_2O(g)$
(e) $3MnO_4^{2-}(aq) + 2H_2O(l) \longrightarrow$
$$2MnO_4^-(aq) + MnO_2(s) + 4OH^-(aq)$$
(f) $3AuCl(s) \longrightarrow AuCl_3(s) + 2Au(s)$

14.85 Which group(s) of the periodic table is (are) described by each of the following general statements?

(a) The elements form neutral compounds of VSEPR class AX_3E.
(b) The free elements are strong oxidizing agents and form monatomic ions and oxoanions.
(c) The valence electron configuration allows the atoms to form compounds by combining with two atoms that donate one electron each.
(d) The free elements are strong reducing agents, show only one nonzero oxidation state, and form mainly ionic compounds.
(e) The elements can form stable compounds with only three bonds, but as a central atom, they can accept a pair of electrons from a fourth atom without expanding their valence shell.
(f) Only larger members of the group are chemically active.

14.86 Bromine monofluoride (BrF) disproportionates to bromine gas and bromine tri- and pentafluorides. Use the following to find ΔH_{rxn}° for the decomposition of BrF to its elements:
$$3BrF(g) \longrightarrow Br_2(g) + BrF_3(l) \qquad \Delta H_{rxn} = -125.3 \text{ kJ}$$
$$5BrF(g) \longrightarrow 2Br_2(g) + BrF_5(l) \qquad \Delta H_{rxn} = -166.1 \text{ kJ}$$
$$BrF_3(l) + F_2(g) \longrightarrow BrF_5(l) \qquad \Delta H_{rxn} = -158.0 \text{ kJ}$$

* 14.87 In addition to Al_2Cl_6, aluminum forms other species with bridging halide ions to two aluminum atoms. One such species is the ion $Al_2Cl_7^-$. The ion is symmetrical, with a 180° Al—Cl—Al bond angle.
(a) What orbitals does Al use to bond with the Cl atoms?
(b) What is the shape around each Al?
(c) What is the hybridization of the central Cl?
(d) What do the shape and hybridization suggest about the presence of lone pairs of electrons on the central Cl?

14.88 The bond angles in the nitrite ion, nitrogen dioxide, and the nitronium ion (NO_2^+) are 115°, 134°, and 180°, respectively. Explain these values using Lewis structures and VSEPR theory.

14.89 The triatomic molecular ion H_3^+ was first detected and characterized by J. J. Thomson using mass spectrometry. Use the bond energy of H_2 (432 kJ/mol) and the proton affinity of H_2 ($H_2 + H^+ \longrightarrow H_3^+$; $\Delta H = -337$ kJ/mol) to calculate the heat of reaction for
$$H + H + H^+ \longrightarrow H_3^+$$

* 14.90 Copper(II) hydrogen arsenite ($CuHAsO_3$) is a green pigment once used in wallpaper; in fact, forensic evidence suggests that Napoleon may have been poisoned by arsenic from his wallpaper. In damp conditions, mold metabolizes this compound to trimethylarsenic [$(CH_3)_3As$], a highly toxic gas.
(a) Calculate the mass percent of As in each compound.
(b) How much $CuHAsO_3$ must react to reach a toxic level in a room that measures 12.35 m × 7.52 m × 2.98 m (arsenic is toxic at 0.50 mg/m^3)?

CHAPTER FIFTEEN

Organic Compounds and the Atomic Properties of Carbon

Organic Beauty A polarized light micrograph of vitamin E crystals reveals a striking pattern. As you'll see in this chapter, molecular models of organic compounds are beautiful as well. These substances occur throughout every living thing and as countless industrial and medical products.

Key Principles

◆ Carbon's unusual ability to bond to other carbons and to many other nonmetals gives its compounds *structural complexity* and *chemical diversity*. The diversity arises from the presence of *functional groups,* specific combinations of bonded atoms that react in characteristic ways.

◆ *Hydrocarbons* (containing only C and H) are classified as *alkanes* (all single bonds), *alkenes* (at least one C=C bond), *alkynes* (at least one C≡C bond), and *aromatic* (at least one planar ring with delocalized π electrons). The C=C and C≡C bonds are functional groups.

◆ Two kinds of *isomers* are important in organic chemistry. *Constitutional (structural) isomers* have different arrangements of atoms. *Stereoisomers* have the same atom arrangement but different spatial orientations. There are two types: *optical isomers* are mirror images that cannot be superimposed, and *geometric (cis-trans) isomers* have different orientations of groups around a C=C bond.

◆ Three common types of organic reactions are *addition* (two atoms or groups are added and a C=C bond is converted to a C—C bond), *elimination* (two atoms or groups are removed and a C—C bond is converted to a C=C bond), and *substitution* (one atom or group replaces another).

◆ *Functional groups* undergo characteristic reactions: groups with only single bonds (*alcohol, amine,* and *alkyl halide*) undergo substitution or elimination; groups with double bonds (*alkene, aldehyde,* and *ketone*) and those with triple bonds (*alkyne* and *nitrile*) undergo addition; and groups with both single and double bonds (*acids, esters,* and *amides*) undergo substitution.

◆ *Polymers* are made by covalently linking many small repeat units (*monomers*). The monomer can be selected to give *synthetic polymers* desired properties. *Polysaccharides, proteins,* and *nucleic acids* are *natural polymers*. DNA occurs as a *double helix*, with bases in each strand H-bonded to specific bases in the other. The *base sequence* of an organism's DNA determines the *amino-acid sequences* of its proteins, which determine the protein's functions.

Outline

Is there any chemical system more remarkable than a living cell? Through delicately controlled mechanisms, it oxidizes food for energy, maintains the concentrations of thousands of aqueous components, interacts continuously with its environment, synthesizes both simple and complex molecules, and even reproduces itself! For all our technological prowess, no human-made system even approaches the cell in its complexity and sheer elegance of function.

This amazing chemical machine consumes, creates, and consists largely of *organic compounds*. Except for a few inorganic salts and ever-present water, everything you put into or on your body—food, medicine, cosmetics, and clothing—consists of organic compounds. Organic fuels warm our homes, cook our meals, and power our society. Major industries are devoted to producing organic compounds, such as polymers, pharmaceuticals, and insecticides.

What *is* an organic compound? Dictionaries define it as "a compound of carbon," but that definition is too general because it includes carbonates, cyanides, carbides, cyanates, and other carbon-containing ionic compounds that most chemists classify as inorganic. Here is a more specific definition: all **organic compounds** contain carbon, nearly always bonded to other carbons and hydrogen, and often to other elements.

In the early 19th century, organic compounds were thought to possess a spiritual "vital force" and to be impossible to synthesize. Today, we know that *the same chemical principles govern organic and inorganic systems* because the behavior of a compound arises from the properties of its elements, no matter how marvelous that behavior may be.

15.1 THE SPECIAL NATURE OF CARBON AND THE CHARACTERISTICS OF ORGANIC MOLECULES

Although there is nothing mystical about organic molecules, their indispensable role in biology and industry leads us to ask if carbon has some extraordinary attributes that give it a special chemical "personality." Of course, each element has its own specific properties, and carbon is no more unique than sodium, hafnium, or any other element. But the atomic properties of carbon do give it bonding capabilities beyond those of any other element, which in turn lead to the two obvious characteristics of organic molecules—structural complexity and chemical diversity.

The Structural Complexity of Organic Molecules

Most organic molecules have much more complex structures than most inorganic molecules, and a quick review of carbon's atomic properties and bonding behavior shows why:

1. *Electron configuration, electronegativity, and covalent bonding.* Carbon's ground-state electron configuration of [He] $2s^2 2p^2$—four electrons more than He and four fewer than Ne—means that the formation of carbon ions is energetically impossible under ordinary conditions. Carbon's position in the periodic table (Figure 15.1) and its electronegativity are midway between the most metallic and nonmetallic elements of Period 2: Li = 1.0, C = 2.5, F = 4.0. Therefore, *carbon shares electrons to attain a filled outer (valence) level*, bonding covalently in all its elemental forms and compounds.

2. *Bond properties, catenation, and molecular shape.* The *number* and *strength* of carbon's bonds lead to its outstanding ability to *catenate* (form chains of atoms), which allows it to form a multitude of chemically and thermally stable chain, ring, and branched compounds. Through the process of orbital hybridization (Section 11.1), *carbon forms four bonds in virtually all its compounds,* and they point in as many as four different directions. The small size of carbon allows close approach to another atom and thus greater orbital overlap, meaning that *carbon forms*

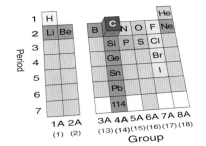

Figure 15.1 The position of carbon in the periodic table. Lying at the center of Period 2, carbon has an intermediate electronegativity (EN), and its position at the top of Group 4A(14) means it is relatively small. Other elements common in organic compounds are H, N, O, P, S, and the halogens.

relatively short, strong bonds. The C—C bond is short enough to allow side-to-side overlap of half-filled, unhybridized *p* orbitals and the formation of *multiple bonds,* which restrict rotation of attached groups (see Section 11.2). These features add more possibilities for the shapes of carbon compounds.

3. *Molecular stability.* Although silicon and several other elements also catenate, none can compete with carbon. Atomic and bonding properties confer three crucial differences between C and Si chains that explain why C chains are so stable and, therefore, so common:

- *Atomic size and bond strength.* As atomic size increases down Group 4A(14), bonds between identical atoms become longer and weaker. Thus, a C—C bond (347 kJ/mol) is much stronger than an Si—Si bond (226 kJ/mol).
- *Relative heats of reaction.* A C—C bond (347 kJ/mol) and a C—O bond (358 kJ/mol) have nearly the same energy, so relatively little heat is released when a C chain reacts and one bond replaces the other. In contrast, an Si—O bond (368 kJ/mol) is much stronger than an Si—Si bond (226 kJ/mol), so a large quantity of heat is released when an Si chain reacts.
- *Orbitals available for reaction.* Unlike C, Si has low-energy *d* orbitals that can be attacked (occupied) by the lone pairs of incoming reactants. Thus, for example, ethane (CH_3—CH_3) is stable in water and does not react in air unless sparked, whereas disilane (SiH_3—SiH_3) breaks down in water and ignites spontaneously in air.

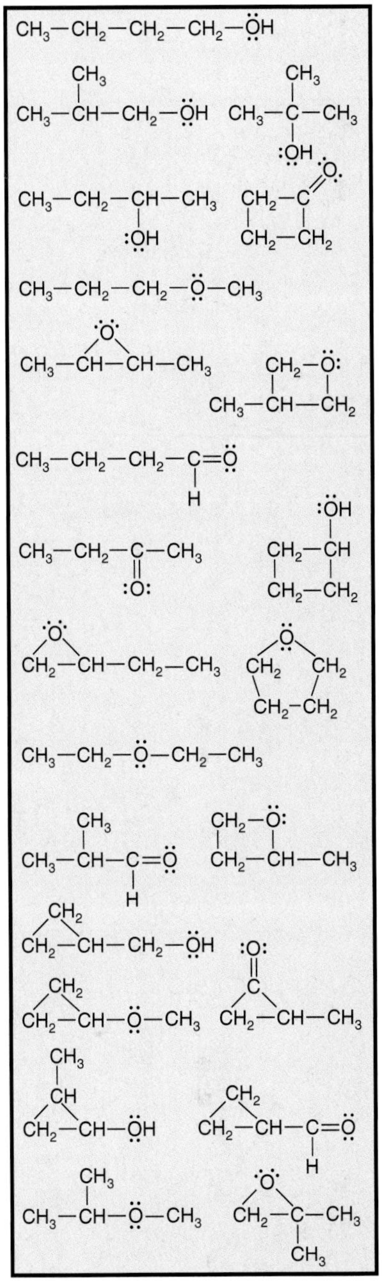

The Chemical Diversity of Organic Molecules

In addition to their elaborate geometries, organic compounds are noted for their sheer number and diverse chemical behavior. Several million of these compounds are known, and thousands more are discovered or synthesized each year. This incredible diversity is also founded on atomic and bonding behavior and is due to three interrelated factors:

1. *Bonding to heteroatoms.* Many organic compounds contain **heteroatoms,** atoms other than C or H. The most common heteroatoms are N and O, but S, P, and the halogens often occur as well. Figure 15.2 shows that 23 different molecular structures are possible from various arrangements of four C atoms singly bonded to each other, the necessary number of H atoms, and just one O atom (either singly or doubly bonded).

2. *Electron density and reactivity.* Most reactions start—that is, a new bond begins to form—*when a region of high electron density on one molecule meets a region of low electron density on another.* These regions may be due to the presence of a multiple bond or to the partial charges that occur in carbon-heteroatom bonds. For example, consider four bonds commonly found in organic molecules:

- *The C—C bond.* When C is singly bonded to another C, as occurs in portions of nearly every organic molecule, the EN values are equal and the bond is nonpolar. Therefore, in general, *C—C bonds are unreactive.*
- *The C—H bond.* This bond, which also occurs in nearly every organic molecule, is very nearly nonpolar because it is short (109 pm) and the EN values of H (2.1) and C (2.5) are close. Thus, *C—H bonds are largely unreactive* as well.
- *The C—O bond.* This bond, which occurs in many types of organic molecules, is highly polar ($\Delta EN = 1.0$), with the O end of the bond electron rich and the C end electron poor. As a result of this imbalance in electron density, *the C—O bond is reactive* and, given appropriate conditions, a reaction will occur there.
- *Bonds to other heteroatoms.* Even when a carbon-heteroatom bond has a small ΔEN, such as that for C—Br ($\Delta EN = 0.3$), or none at all, as for C—S ($\Delta EN = 0$), heteroatoms like these are large, and so their bonds to carbon are long, weak, and thus reactive.

Figure 15.2 The chemical diversity of organic compounds. Different arrangements of chains, branches, rings, and heteroatoms give rise to many structures. There are 23 different compounds possible from just four C atoms joined by single bonds, one O atom, and the necessary H atoms.

3. *Nature of functional groups.* One of the most important ideas in organic chemistry is that of the **functional group,** a specific combination of bonded atoms that reacts in a *characteristic* way, no matter what molecule it occurs in. In nearly every case, *the reaction of an organic compound takes place at the functional group.* Functional groups vary from carbon-carbon multiple bonds to several combinations of carbon-heteroatom bonds, and each has its own pattern of reactivity. A particular bond may be *part* of one or more functional groups. For example, the C—O bond occurs in four functional groups. We will discuss the reactivity of these three later in this chapter:

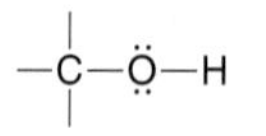

—C̈—Ö—H	—C̈—Ö—H	—C̈—Ö—C̈—
alcohol group	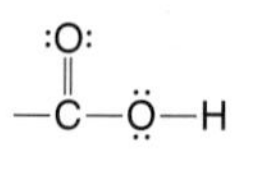 carboxylic acid group	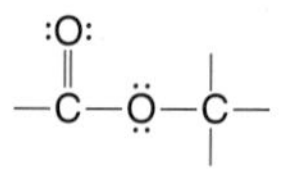 ester group

SECTION SUMMARY
The structural complexity of organic compounds arises from carbon's small size, intermediate EN, four valence electrons, ability to form multiple bonds, and absence of *d* orbitals in the valence level. These factors lead to chains, branches, and rings of C atoms joined by strong, chemically resistant bonds that point in as many as four directions from each C. The chemical diversity of organic compounds arises from carbon's ability to bond to many other elements, including O and N, which creates polar bonds and greater reactivity. These factors lead to compounds that contain functional groups, specific portions of molecules that react in characteristic ways.

15.2 THE STRUCTURES AND CLASSES OF HYDROCARBONS

A fanciful, anatomical analogy can be made between an organic molecule and an animal. The carbon-carbon bonds form the skeleton: the longest continual chain is the backbone, and any branches are the limbs. Covering the skeleton is a skin of hydrogen atoms, with functional groups protruding at specific locations, like chemical fingers ready to grab an incoming reactant.

In this section, we "dissect" one group of compounds down to their skeletons and see how to name and draw them. **Hydrocarbons,** the simplest type of organic compound, are a large group of substances containing only H and C atoms. Some common fuels, such as natural gas and gasoline, are hydrocarbon mixtures. Hydrocarbons are also important *feedstocks,* precursor reactants used to make other compounds. Ethylene, acetylene, and benzene, for example, are feedstocks for hundreds of other substances.

Carbon Skeletons and Hydrogen Skins

Let's begin by examining the possible bonding arrangements of C atoms only (we'll leave off the H atoms at first) in simple skeletons without multiple bonds or rings. To distinguish different skeletons, focus on the *arrangement* of C atoms (that is, the successive linkages of one to another) and keep in mind that *groups joined by single (sigma) bonds are relatively free to rotate* (Section 11.2).

Structures with one, two, or three carbons can be arranged in only one way. Whether you draw three C atoms in a line or with a bend, the arrangement is the same. Four C atoms, however, have two possible arrangements—a four-C chain or a three-C chain with a one-C branch at the central C:

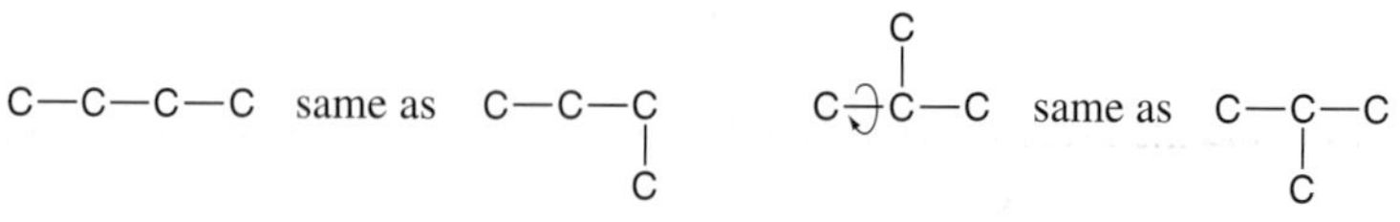

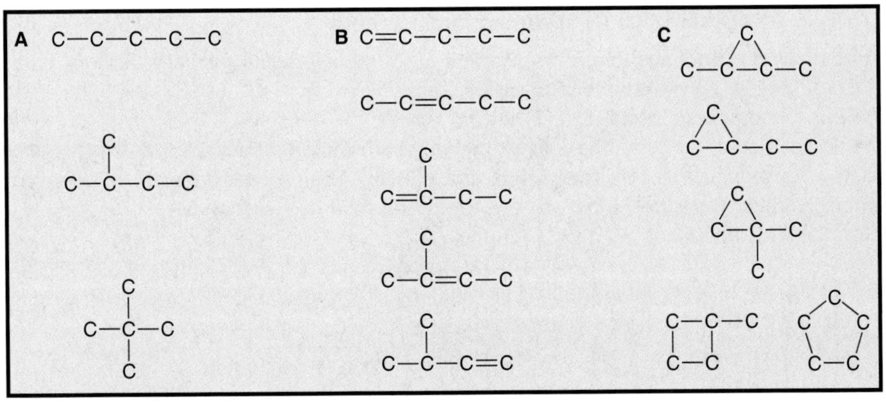

Figure 15.3 Some five-carbon skeletons. A, Three five-C skeletons are possible with only single bonds. **B,** Five more skeletons are possible with one C=C bond present. **C,** Five more skeletons are possible with one ring present. Even more would be possible with a ring *and* a double bond.

As the total number of C atoms increases, the number of different arrangements increases as well. Five C atoms have 3 possible arrangements; 6 C atoms can be arranged in 5 ways, 7 C atoms in 9 ways, 10 C atoms in 75 ways, and 20 C atoms in more than 300,000 ways! If we include multiple bonds and rings, the number of arrangements increases further. For example, including one C=C bond in the five-C skeletons creates 5 more arrangements, and including one ring creates 5 more (Figure 15.3).

When determining the number of different skeletons for a given number of C atoms, remember that

- Each C atom can form a *maximum* of four single bonds, or two single and one double bond, or one single and one triple bond.
- The *arrangement* of C atoms determines the skeleton, so a straight chain and a bent chain represent the same skeleton.
- Groups joined by single bonds can *rotate,* so a branch pointing down is the same as one pointing up. (Recall that a double bond restricts rotation.)

If we put a hydrogen "skin" on a carbon skeleton, we obtain a hydrocarbon. Figure 15.4 shows that the skeleton has the correct number of H atoms when each C has four bonds. Sample Problem 15.1 provides practice drawing hydrocarbons.

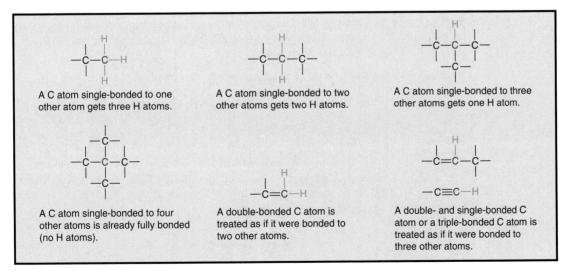

A C atom single-bonded to one other atom gets three H atoms.

A C atom single-bonded to two other atoms gets two H atoms.

A C atom single-bonded to three other atoms gets one H atom.

A C atom single-bonded to four other atoms is already fully bonded (no H atoms).

A double-bonded C atom is treated as if it were bonded to two other atoms.

A double- and single-bonded C atom or a triple-bonded C atom is treated as if it were bonded to three other atoms.

Figure 15.4 Adding the H-atom skin to the C-atom skeleton. In a hydrocarbon molecule, each carbon atom bonds to as many hydrogen atoms as needed to give the carbon a total of four bonds.

SAMPLE PROBLEM 15.1   Drawing Hydrocarbons

Problem Draw structures that have different atom arrangements for hydrocarbons with:
(a) Six C atoms, no multiple bonds, and no rings
(b) Four C atoms, one double bond, and no rings
Plan In each case, we draw the longest carbon chain first and then work down to smaller chains with branches at different points along them. The process typically involves trial and error. Then, we add H atoms to give each C a total of four bonds.
Solution (a) Compounds with six C atoms:

6-C chain:

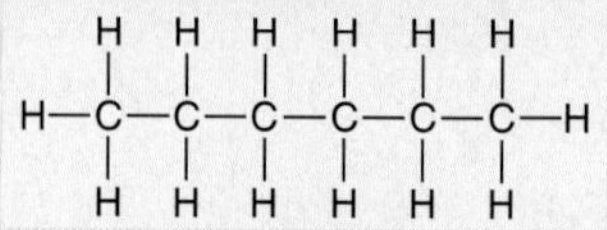

5-C chains:

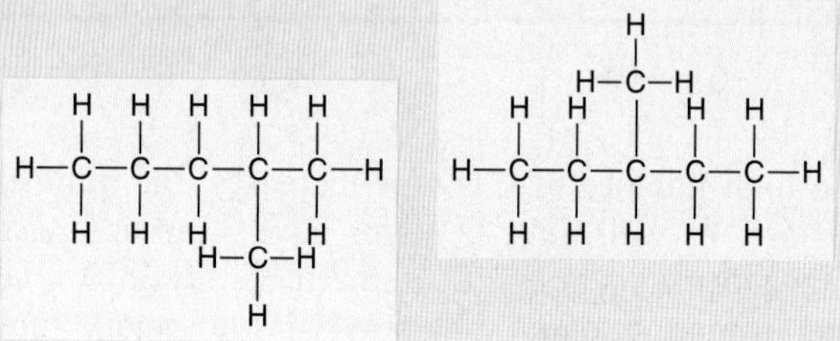

4-C chains:

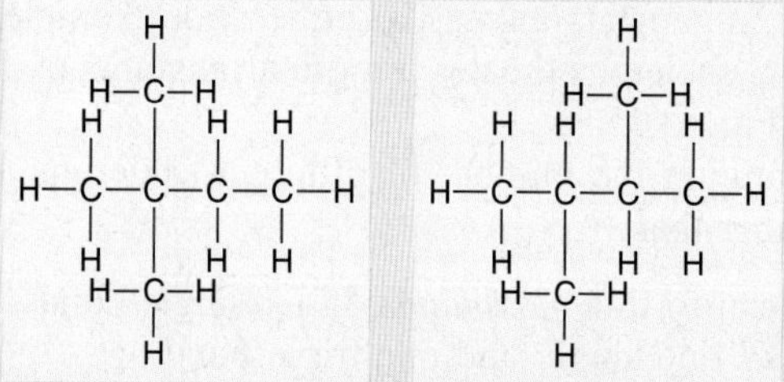

(b) Compounds with four C atoms and one double bond:

4-C chains:

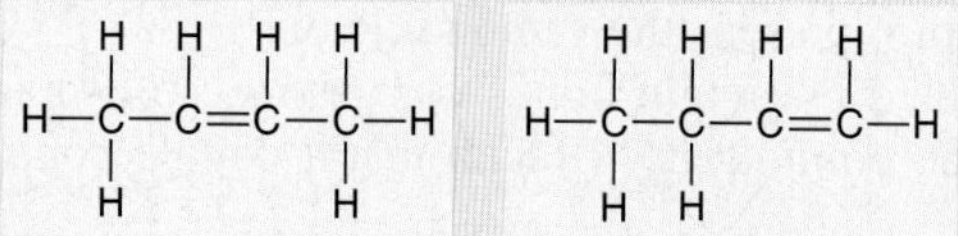

3-C chain:

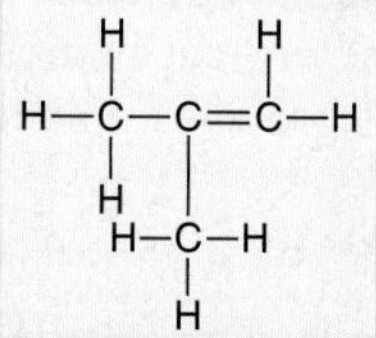

Check Be sure each skeleton has the correct number of C atoms and multiple bonds and no arrangements are repeated or omitted; remember a double bond counts as two bonds.
Comment Avoid some *common mistakes:*

In **(a)**: C—C—C—C—C is the same skeleton as C—C—C—C—C

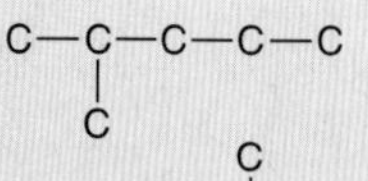

C—C—C—C is the same skeleton as C—C—C—C

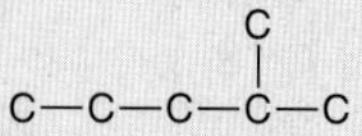

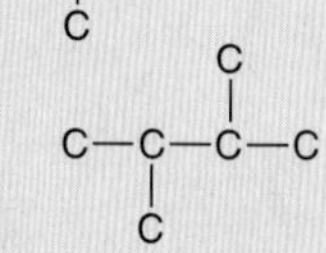

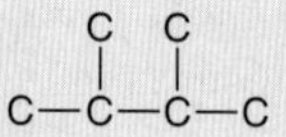

In **(b)**: C—C—C=C is the same skeleton as C=C—C—C
The double bond restricts rotation. Thus, in addition to the form shown in part (b), in which the H atoms are on the same side of the C=C bond, another possibility is the form in which the H atoms are on opposite sides:

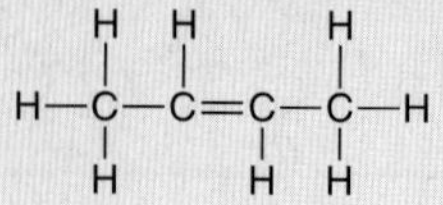

(We discuss these forms fully later in this section.)

Also, there are too many bonds to one C in

FOLLOW-UP PROBLEM 15.1 Draw all hydrocarbons that have five C atoms, one triple bond, and no rings (three arrangements).

Hydrocarbons can be classified into four main groups. In the remainder of this section, we examine the names and some structural features and physical properties of each group. Later, we discuss the chemical behavior of the hydrocarbons.

Alkanes: Hydrocarbons with Only Single Bonds

A hydrocarbon that contains only single bonds is an **alkane** (general formula C_nH_{2n+2}, where n is a positive integer). For example, if $n = 5$, the formula is $C_5H_{[(2\times5)+2]}$, or C_5H_{12}. The alkanes comprise a **homologous series,** one in which each member differs from the next by a $-CH_2-$ (methylene) group. In an alkane, each C is sp^3 hybridized. Because each C is bonded to the *maximum number of other atoms* (C or H), alkanes are referred to as **saturated hydrocarbons.**

Naming Alkanes You learned how to name simple alkanes in Section 2.8. Here we discuss general rules for naming any alkane and, by extension, other organic compounds as well. The key point is that *each chain, branch, or ring has a name based on the **number** of C atoms.* The name of a compound has three portions:

PREFIX + ROOT + SUFFIX

- *Root:* The root tells the number of C atoms in the longest *continuous* chain in the molecule. The roots for the ten smallest alkanes are shown in Table 15.1. Recall that there are special roots for compounds of one to four C atoms; roots of longer chains are based on Greek numbers.
- *Prefix:* Each prefix identifies a *group attached to the main chain* and the number of the carbon to which it is attached. Prefixes identifying hydrocarbon branches are the same as root names (Table 15.1) but have *-yl* as their ending. Each prefix is placed *before* the root.
- *Suffix:* The suffix tells the *type of organic compound* the molecule represents; that is, it identifies the key functional group the molecule possesses. The suffix is placed *after* the root.

For example, in the name 2-methylbutane, *2-methyl-* is the prefix (a one-carbon branch is attached to C-2 of the main chain), *-but-* is the root (the main chain has four C atoms), and *-ane* is the suffix (the compound is an alkane).

To obtain the systematic name of a compound,

1. Name the longest chain (root).
2. Add the compound type (suffix).
3. Name any branches (prefix).

Table 15.2 (next page) presents the rules for naming any organic compound and applies them to an alkane component of gasoline. Other organic compounds are named with a variety of other prefixes and suffixes (see Table 15.5). In addition to these *systematic* names, we'll also note important *common* names still in use.

Table 15.1 Numerical Roots for Carbon Chains and Branches

Roots	Number of C Atoms
meth-	1
eth-	2
prop-	3
but-	4
pent-	5
hex-	6
hept-	7
oct-	8
non-	9
dec-	10

Table 15.2 Rules for Naming an Organic Compound

1. Naming the longest chain (root)
   (a) Find the longest *continuous* chain of C atoms.
   (b) Select the root that corresponds to the number of C atoms in this chain.

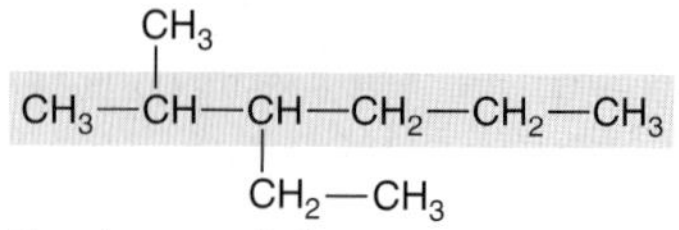

$$CH_3-CH-CH-CH_2-CH_2-CH_3$$
with CH_3 above the second carbon and CH_2-CH_3 below the third carbon

6 carbons $\Longrightarrow$ hex-

2. Naming the compound type (suffix)
   (a) For alkanes, add the suffix *-ane* to the chain root. (Other suffixes appear in Table 15.5 with their functional group and compound type.)
   (b) If the chain forms a ring, the name is preceded by *cyclo-*.

hex- + -ane $\Longrightarrow$ hexane

3. Naming the branches (prefixes) (If the compound has no branches, the name consists of the root and suffix.)
   (a) Each branch name consists of a subroot (number of C atoms) and the ending *-yl* to signify that it is not part of the main chain.
   (b) Branch names precede the chain name. When two or more branches are present, their names appear in *alphabetical* order.
   (c) To specify where the branch occurs along the chain, number the main-chain C atoms consecutively, starting at the end *closer* to a branch, to achieve the *lowest* numbers for the branches. Precede each branch name with the number of the main-chain C to which that branch is attached.

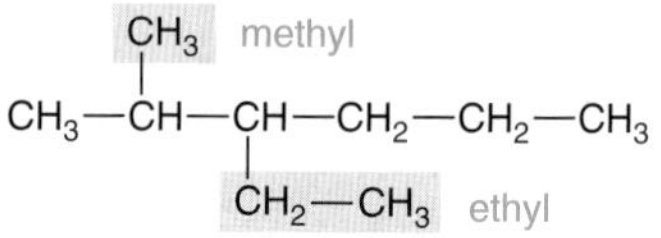

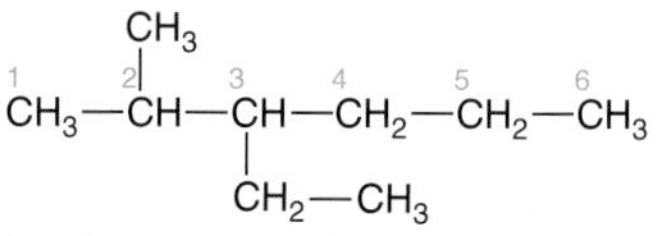

CH_3 methyl

$$CH_3-CH-CH-CH_2-CH_2-CH_3$$
with CH_2-CH_3 ethyl below

ethylmethylhexane

CH_3

$$\overset{1}{CH_3}-\overset{2}{CH}-\overset{3}{CH}-\overset{4}{CH_2}-\overset{5}{CH_2}-\overset{6}{CH_3}$$
with CH_2-CH_3 below

3-ethyl-2-methylhexane

Depicting Alkanes with Formulas and Models Chemists have several ways to depict organic compounds. Expanded, condensed, and carbon-skeleton formulas are easy to draw; ball-and-stick and space-filling models show the actual shapes.

The *expanded formula* shows each atom and bond. One type of *condensed formula* groups each C atom with its H atoms. *Carbon-skeleton formulas* show only carbon-carbon bonds and appear as zig-zag lines, often with branches. *Each end or bend of a zig-zag line or branch represents a C atom attached to the number of H atoms that gives it a total of four bonds:*

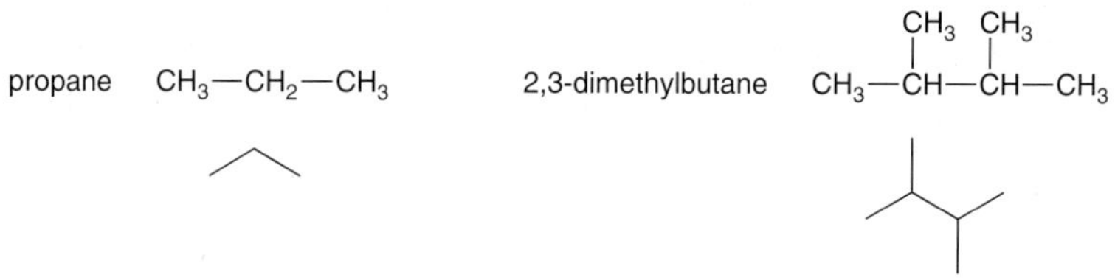

propane $CH_3-CH_2-CH_3$

2,3-dimethylbutane $CH_3-CH-CH-CH_3$ with CH_3 CH_3 above

Figure 15.5 Ways of depicting an alkane.

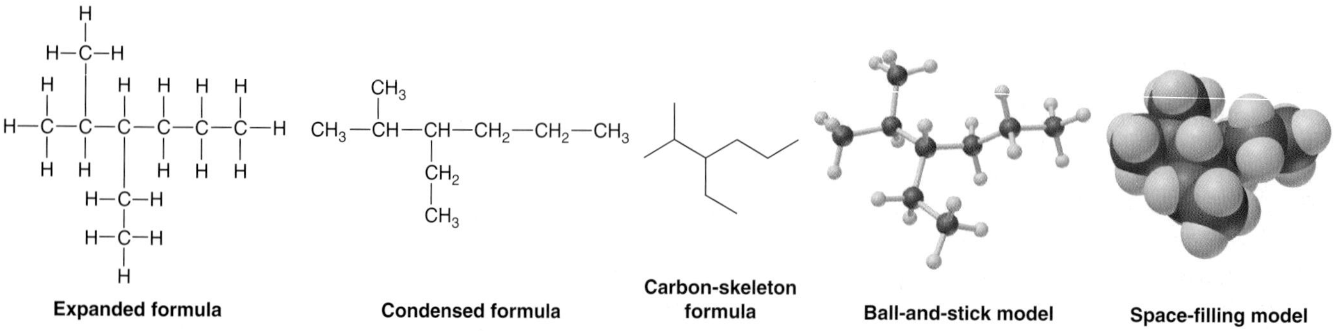

Figure 15.5 shows these types of formulas, together with ball-and-stick and space-filling models, of the compound named in Table 15.2.

Expanded formula

$CH_3-CH-CH_2-CH_2-CH_3$ with CH_3 above and CH_2, CH_3 below

Condensed formula

Carbon-skeleton formula

Ball-and-stick model

Space-filling model

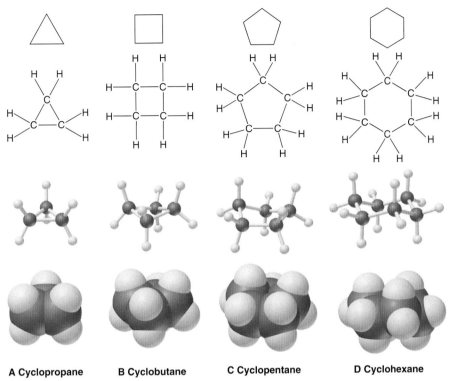

A Cyclopropane   B Cyclobutane   C Cyclopentane   D Cyclohexane

Figure 15.6 Depicting cycloalkanes. Cycloalkanes are usually drawn as regular polygons. Each side is a C—C bond, and each corner represents a C atom with its required number of H atoms. The expanded formulas show each bond in the molecule. The ball-and-stick and space-filling models show that, except for cyclopropane, the rings are not planar. These conformations minimize electron repulsions between adjacent H atoms. Cyclohexane **(D)** is shown in its more stable chair conformation.

Cyclic Hydrocarbons   A **cyclic hydrocarbon** contains one or more rings in its structure. When a straight-chain alkane (C_nH_{2n+2}) forms a ring, two H atoms are lost as the C—C bond forms to join the two ends of the chain. Thus, *cycloalkanes* have the general formula C_nH_{2n}. Cyclic hydrocarbons are often drawn with carbon-skeleton formulas, as shown at the top of Figure 15.6. Except for three-carbon rings, *cycloalkanes are nonplanar*. This structural feature arises from the tetrahedral shape around each C atom and the need to minimize electron repulsions between adjacent H atoms. As a result, orbital overlap of adjacent C atoms is maximized. The most stable form of cyclohexane, called the *chair conformation*, is shown in Figure 15.6D.

Constitutional Isomerism and the Physical Properties of Alkanes

Two or more compounds with the same molecular formula but different properties are called **isomers.** Isomers with *different arrangements of bonded atoms* are **constitutional** (or **structural**) **isomers;** alkanes with the same number of C atoms but different skeletons are examples. The smallest alkane to exhibit constitutional isomerism has four C atoms: two different compounds have the formula C_4H_{10}, as shown in Table 15.3 on the next page. The unbranched one is butane (common name, *n*-butane; *n*- stands for "normal," or having a straight chain), and the other is 2-methylpropane (common name, *iso*butane). Similarly, three compounds

Table 15.3 **The Constitutional Isomers of C_4H_{10} and C_5H_{12}**

Systematic Name (Common Name)	Expanded Formula	Condensed and Skeleton Formulas	Space-filling Model	Density (g/mL)	Boiling Point (°C)
Butane (*n*-butane)		CH_3—CH_2—CH_2—CH_3		0.579	−0.5
2-Methylpropane (isobutane)		CH_3—CH—CH_3 CH_3		0.549	−11.6
Pentane (*n*-pentane)		CH_3—CH_2—CH_2—CH_2—CH_3		0.626	36.1
2-Methylbutane (isopentane)		CH_3—CH—CH_2—CH_3 CH_3		0.620	27.8
2,2-Dimethylpropane (neopentane)		CH_3 CH_3—C—CH_3 CH_3		0.614	9.5

have the formula C_5H_{12} (shown in Table 15.3). The unbranched isomer is pentane (common name, *n*-pentane); the one with a methyl group at C-2 of a four-C chain is 2-methylbutane (common name, *iso*pentane). The third isomer has two methyl branches on C-2 of a three-C chain, so its name is 2,2-dimethylpropane (common name, *neo*pentane).

Because alkanes are nearly nonpolar, their physical properties are determined by dispersion forces. The four-C alkanes boil lower than the five-C compounds (Table 15.3). Moreover, within each group of isomers, the more spherical member (isobutane or neopentane) boils lower than the more elongated one (*n*-butane or *n*-pentane). As you saw in Chapter 12, this trend occurs because a spherical shape leads to less intermolecular contact, and thus lower total dispersion forces, than does an elongated shape.

A particularly clear example of the effect of dispersion forces on physical properties occurs among the unbranched alkanes (*n*-alkanes). Among these compounds, the longer the chain, the greater the intermolecular contact, the stronger the dispersion forces, and the higher the boiling point (Figure 15.7). The solubility of alkanes, and of all hydrocarbons, is easy to predict from the like-dissolves-like rule (Section 13.1). Alkanes are miscible in each other and in other nonpolar solvents, such as benzene, but are nearly insoluble in water. The solubility of pentane in water, for example, is only 0.36 g/L at room temperature.

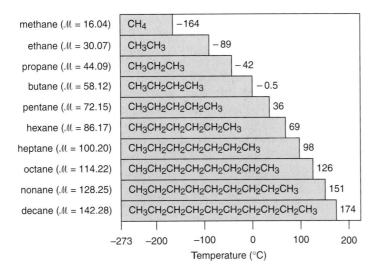

Figure 15.7 Boiling points of the first 10 unbranched alkanes. Boiling point increases smoothly with chain length because dispersion forces increase. Each entry includes the name, molar mass ($\mathcal{M}$, in g/mol), formula, and boiling point at 1 atm pressure.

Chiral Molecules and Optical Isomerism

Another type of isomerism exhibited by some alkanes and many other organic (as well as some inorganic) compounds is called *stereoisomerism.* **Stereoisomers** are molecules with the same arrangement of atoms *but different orientations of groups in space. Optical isomerism* is one type of stereoisomerism: *when two objects are mirror images of each other and cannot be superimposed, they are* **optical isomers,** also called *enantiomers.* To use a familiar example, your right hand is an optical isomer of your left. Look at your right hand in a mirror, and you will see that the *image* is identical to your left hand (Figure 15.8). No matter how you twist your arms around, however, your hands cannot lie on top of each other with all parts superimposed. They are not superimposable because each is *asymmetric:* there is no plane of symmetry that divides your hand into two identical parts.

An asymmetric molecule is called **chiral** (Greek *cheir,* "hand"). Typically, an organic molecule *is chiral if it contains a carbon atom that is bonded to four* **different** *groups.* This C atom is called a *chiral center* or an asymmetric carbon. In 3-methylhexane, for example, C-3 is a chiral center because it is bonded to four different groups: H—, CH_3—, CH_3—CH_2—, and CH_3—CH_2—CH_2— (Figure 15.9A). Like your two hands, the two forms are mirror images and cannot be superimposed on each other: when two of the groups are superimposed, the other two are opposite each other. Thus, the two forms are optical isomers. The central C atom in the amino acid alanine is also a chiral center (Figure 15.9B).

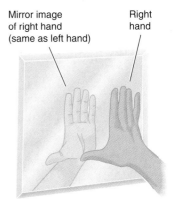

Figure 15.8 An analogy for optical isomers. The reflection of your right hand looks like your left hand. Each hand is asymmetric, so you cannot superimpose them with your palms facing in the same direction.

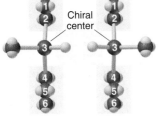 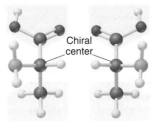

A Optical isomers of 3-methylhexane   **B Optical isomers of alanine**

Figure 15.9 Two chiral molecules. A, 3-Methylhexane is chiral because C-3 is bonded to four different groups. These two models are optical isomers (enantiomers). **B,** The central C in the amino acid alanine is also bonded to four different groups.

Unlike constitutional isomers, optical isomers are identical in all but two respects:

1. In their physical properties, *optical isomers differ only in the direction that each isomer rotates the plane of polarized light.* A *polarimeter* is used to measure the angle that the plane is rotated. A beam of light consists of waves that oscillate in all planes. A polarizing filter blocks all waves except those in one plane, so the light emerging through the filter is *plane-polarized.* An optical isomer is **optically active** because it rotates the plane of this polarized light, either in a clockwise direction (the *d* isomer) or in a counterclockwise direction (the *l* isomer). An equimolar mixture of the two isomers does not rotate the plane at all because the opposing rotations cancel.

2. In their chemical properties, *optical isomers differ only in a chiral (asymmetric) chemical environment,* one that distinguishes "right-handed" from "left-handed" molecules. As an analogy, your right hand fits well in your right glove but not in your left glove.

Optical isomerism plays a vital role in living cells. Nearly all carbohydrates and amino acids are optically active, but only one of the isomers is biologically usable. For example, *d*-glucose is metabolized for energy, but *l*-glucose is excreted unused. Similarly, *l*-alanine is incorporated naturally into proteins, but *d*-alanine is not. Many drugs are chiral molecules of which one optical isomer is biologically active and the other has either a different type of activity or none at all.

Alkenes: Hydrocarbons with Double Bonds

A hydrocarbon that contains at least one $C=C$ bond is called an **alkene** (general formula C_nH_{2n}). The double-bonded C atoms are sp^2 hybridized. Because their carbon atoms are bonded to fewer than the maximum of four atoms each, alkenes are considered **unsaturated hydrocarbons.**

Alkene names differ from those of alkanes in two respects:

1. The main chain (root) *must* contain both C atoms of the double bond, even if it is not the longest chain. The chain is numbered from the end *closer* to the $C=C$ bond, and the position of the bond is indicated by the number of the *first* C atom in it.
2. The suffix for alkenes is *-ene.*

For example, there are three four-C alkenes (C_4H_8), two unbranched and one branched (see Sample Problem 15.1b). The branched isomer is 2-methylpropene; the unbranched isomer with the $C=C$ bond between C-1 and C-2 is 1-butene; the unbranched isomer with the $C=C$ bond between C-2 and C-3 is 2-butene. As you'll see next, there are two isomers of 2-butene, but they are of a different sort.

The $C=C$ Bond and Geometric (*cis-trans*) Isomerism
There are two major structural differences between alkenes and alkanes. First, alkanes have a *tetrahedral* geometry (bond angles of ~109.5°) around each C atom, whereas the double-bonded C atoms in alkenes are *trigonal planar* (~120°). Second, the C—C bond *allows* rotation of bonded groups, so the atoms in an alkane continually change their relative positions. In contrast, the π bond of the $C=C$ bond *restricts* rotation, which fixes the relative positions of the atoms bonded to it.

This rotational restriction leads to another type of stereoisomerism. **Geometric isomers** (also called *cis-trans* isomers) have different orientations of groups around a double bond (or similar structural feature). Table 15.4 shows the two geometric isomers of 2-butene (also see Comment, Sample Problem

Table 15.4 **The Geometric Isomers of 2-Butene**

Systematic Name	Condensed and Skeleton Formulas	Space-filling Model	Density (g/mL)	Boiling Point (°C)
cis-2-Butene	(structure)	(model)	0.621	3.7
trans-2-Butene	(structure)	(model)	0.604	0.9

15.1), *cis*-2-butene and *trans*-2-butene. In general, the *cis* isomer has the larger portions of the main chain (in this case, two CH_3 groups) *on the same side* of the double bond, and the *trans* isomer has them on *opposite sides*. For a molecule to have geometric isomers, *each C atom in the C$=$C bond must be bonded to two different groups.* Like other isomers, geometric isomers have different properties. Note in Table 15.4 that the two 2-butenes differ in molecular shape *and* physical properties. The *cis* isomer has a bend in the chain that the *trans* isomer lacks.

Geometric Isomers and the Chemistry of Vision The first step in the remarkable sequence of events that allows us to see relies on the different shapes of a pair of geometric isomers. The molecule responsible for receiving the light energy is *retinal,* a 20-C compound with five C$=$C bonds in its structure. There are two biologically occurring isomers, which have very different shapes: the all-*trans* isomer is elongated, and the 11-*cis* isomer is sharply bent around the double bond between C-11 and C-12.

Certain cells of the retina are packed with rhodopsin, a large protein covalently bonded to 11-*cis*-retinal. The energies of photons of visible light have a range (165–293 kJ/mol) that includes the energy needed to break a C$=$C π bond (about 250 kJ/mol). Within a few millionths of a second after rhodopsin absorbs a photon, the 11-*cis* π bond breaks, the intact σ bond between C-11 and C-12 rotates, and the π bond re-forms to produce all-*trans* retinal.

This rapid and significant change in shape causes the attached protein to change its shape as well, which triggers a flow of ions into the retina's cells. This ion influx initiates electrical impulses to the optic nerve and brain. Because of the speed and efficiency with which light causes such a large structural change in retinal, natural selection has made it the photon absorber in organisms as different as purple bacteria, mollusks, insects, and vertebrates.

Alkynes: Hydrocarbons with Triple Bonds

Hydrocarbons that contain at least one C$\equiv$C bond are called **alkynes** (general formula C_nH_{2n-2}). Because a carbon in a C$\equiv$C bond can bond to only one other atom, the geometry around each C atom is linear (180°): each C is *sp* hybridized. Alkynes are named in the same way as alkenes, except that the suffix is *-yne.* Because of their localized π electrons, C$=$C and C$\equiv$C bonds are electron rich and act as functional groups. Thus, alkenes and alkynes are much more reactive than alkanes, as we'll discuss in Section 15.4.

SAMPLE PROBLEM 15.2 Naming Alkanes, Alkenes, and Alkynes

Problem Give the systematic name for each of the following, indicate the chiral center in part (d), and draw two geometric isomers for part (e):

(a)

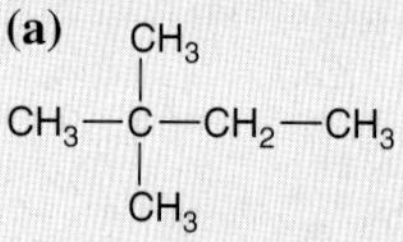

(b)

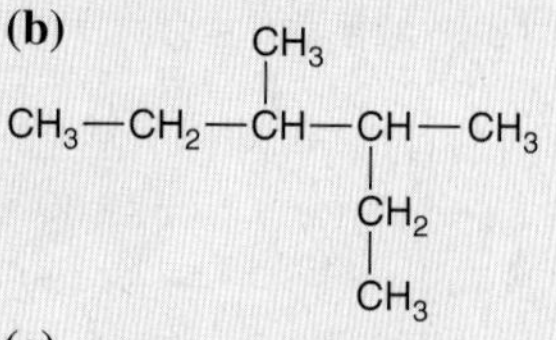

(c)

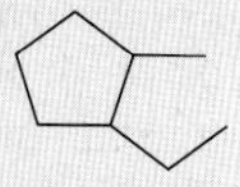

(d)

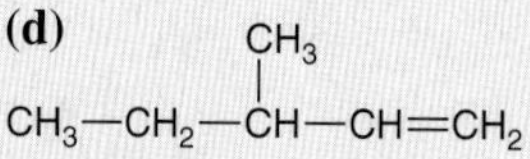

(e)

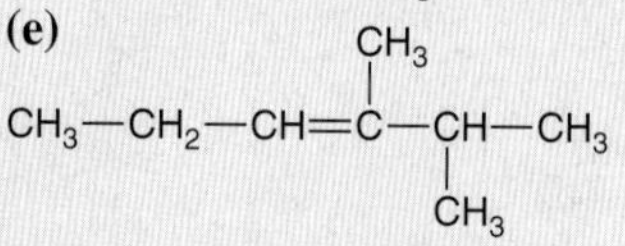

Plan For (a) to (c), we refer to Table 15.2. We first name the longest chain (*root-* + *-ane*). Then we find the *lowest* branch numbers by counting C atoms from the end *closer* to a branch. Finally, we name each branch (*root-* + *-yl*) and put the names alphabetically before the chain name. For (d) and (e), the longest chain that *includes* the multiple bond is numbered from the end closer to it. For (d), the chiral center is the C atom bonded to four different groups. In (e), the *cis* isomer has larger groups on the same side of the double bond, and the *trans* isomer has them on opposite sides.

Solution

(a)

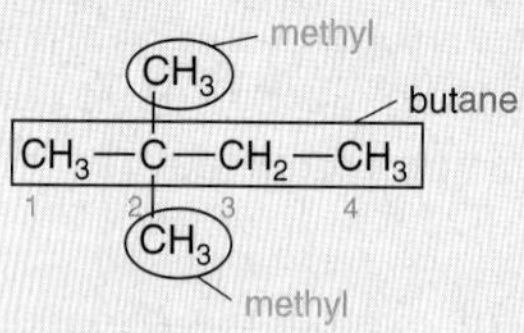

2,2-dimethylbutane

When a type of branch appears more than once, we group the chain numbers and indicate the number of branches with a numerical prefix, such as 2,2-*di*methyl.

(b)

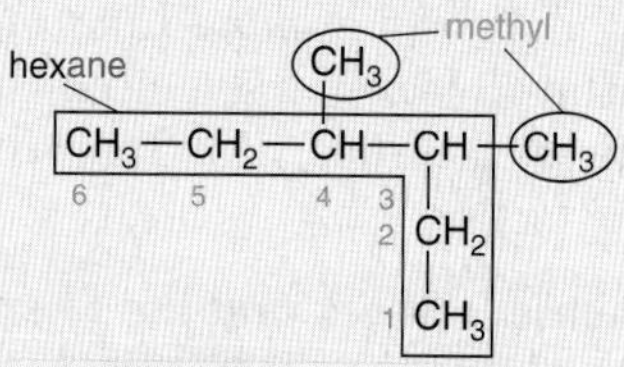

3,4-dimethylhexane

In this case, we can number the chain from either end because the branches are the same and are attached to the two central C atoms.

(c)

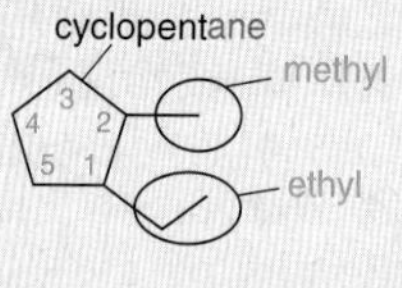

1-ethyl-2-methylcyclopentane

We number the ring C atoms so that a branch is attached to C-1.

(d)

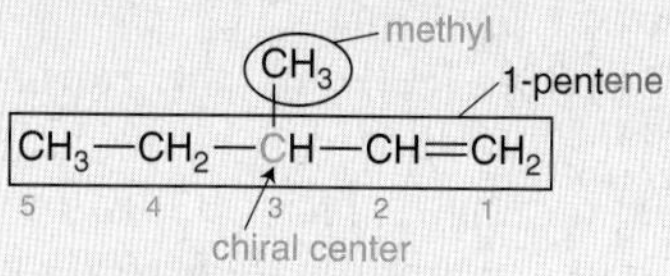

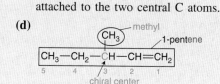

3-methyl-1-pentene

(e)

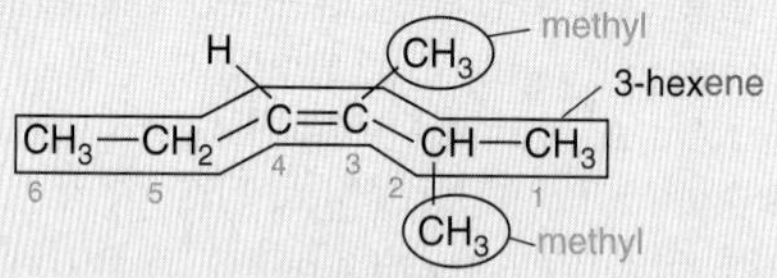

cis-2,3-dimethyl-3-hexene

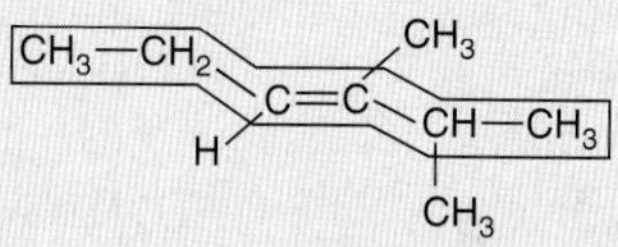

trans-2,3-dimethyl-3-hexene

Check A good check (and excellent practice) is to reverse the process by drawing structures for the names to see if you come up with the structures given in the problem.

Comment In part (b), C-3 and C-4 are chiral centers, as are C-1 and C-2 in part (c). However, in (b) the molecule is not chiral: it has a plane of symmetry between C-3 and C-4, so the molecule as a whole does not rotate the plane of polarized light. Avoid these common mistakes: In (b), 2-ethyl-3-methylpentane is wrong: the longest chain is *hexane.* In (c), 1-methyl-2-ethylcyclopentane is wrong: the branch names appear *alphabetically.*

FOLLOW-UP PROBLEM 15.2 Draw condensed formulas for the following compounds: **(a)** 3-ethyl-3-methyloctane; **(b)** 1-ethyl-3-propylcyclohexane (also draw a carbon-skeleton formula for this compound); **(c)** 3,3-diethyl-1-hexyne; **(d)** *trans*-3-methyl-3-heptene.

Aromatic Hydrocarbons: Cyclic Molecules with Delocalized π Electrons

Unlike the cycloalkanes, **aromatic hydrocarbons** are planar molecules, usually with one or more rings of six C atoms, and are often drawn with alternating single and double bonds. However, as you learned in Section 10.1, in benzene (C_6H_6), all the ring bonds are identical; their values of length and strength are *between* those of a C—C and a C=C bond. To indicate this, benzene is also shown as a resonance hybrid or with a circle (or dashed circle) representing the delocalized character of the π electrons:

The systematic naming of simple aromatic compounds is quite straightforward. Usually, benzene is the parent compound, and attached groups, or *substituents,* are named as prefixes. For example, benzene with one methyl group attached is systematically named *methylbenzene.* With only one substituent present, we do not number the ring C atoms; when two or more groups are attached, however, we number in such a way that one of the groups is attached to ring C-1. Thus, methylbenzene and the three structural isomers with two methyl groups attached are

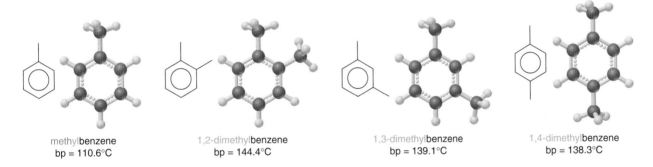

methylbenzene
bp = 110.6°C

1,2-dimethylbenzene
bp = 144.4°C

1,3-dimethylbenzene
bp = 139.1°C

1,4-dimethylbenzene
bp = 138.3°C

The dimethylbenzenes are important solvents and feedstocks for making polyester fibers and dyes. Benzene and many other aromatic hydrocarbons have carcinogenic (cancer-causing) properties.

SECTION SUMMARY

Hydrocarbons contain only C and H atoms, so their physical properties depend on the strength of their dispersion forces. The names of organic compounds have a root for the longest chain, a prefix for any attached group, and a suffix for the type of compound. Alkanes (C_nH_{2n+2}) have only single bonds. Cycloalkanes (C_nH_{2n}) have ring structures that are typically nonplanar. Alkenes (C_nH_{2n}) have at least one C=C bond. Alkynes (C_nH_{2n-2}) have at least one C≡C bond. Aromatic hydrocarbons have at least one planar ring with delocalized π electrons.

Isomers are compounds with the same molecular formula but different properties. Structural isomers have different atom arrangements. Stereoisomers (optical and geometric) have the same arrangement of atoms, but their atoms are oriented differently in space. Optical isomers cannot be superimposed on each other because they are asymmetric, with four different groups bonded to the C that is the chiral center. They have identical physical and chemical properties except in their rotation of plane-polarized light. Geometric (*cis-trans*) isomers have groups oriented differently around a C=C bond, which restricts rotation. Light converts a *cis* isomer of retinal to the all-*trans* form, which initiates the visual response.

15.3 SOME IMPORTANT CLASSES OF ORGANIC REACTIONS

Organic reactions are classified according to the chemical process involved. Three important classes are *addition, elimination,* and *substitution* reactions.

From here on, we use the notation of an uppercase R with a single bond, R—, to signify a general organic group attached to one of the atoms shown; you can usually picture R— as an **alkyl group,** a saturated hydrocarbon chain with one bond available to link to another atom. Thus, R—CH_2—Br has an alkyl group attached to a CH_2 group bearing a Br atom; R—CH=CH_2 is an alkene with an alkyl group attached to one of the carbons in the double bond; and so forth. (Often, when more than one R group is present, we write R, R′, R″, and so forth, to indicate that these groups may be different.)

The three classes of organic reactions we discuss here can be identified by comparing the *number of bonds to C* in reactants and products:

1. An **addition reaction** occurs when an unsaturated reactant becomes a saturated product:

$$R-CH=CH-R \ + \ X-Y \longrightarrow R-\underset{\underset{}{|}}{\overset{\overset{X}{|}}{C}}H-\underset{\underset{}{|}}{\overset{\overset{Y}{|}}{C}}H-R$$

Note the C atoms are bonded to *more* atoms in the product than in the reactant.

The C=C and C≡C bonds and the C=O bond commonly undergo addition reactions. In each case, the π bond breaks, leaving the σ bond intact. In the product, the two C atoms (or C and O) form two additional σ bonds. In the following addition reaction, H and Cl from HCl add to the double bond in ethylene:

$$CH_2=CH_2 \ + \ H-Cl \longrightarrow H-CH_2-CH_2-Cl$$

2. **Elimination reactions** are the opposite of addition reactions. They occur when a saturated reactant becomes an unsaturated product:

$$R-\underset{\underset{}{|}}{\overset{\overset{Y}{|}}{C}}H-\underset{\underset{}{|}}{\overset{\overset{X}{|}}{C}}H_2 \longrightarrow R-CH=CH_2 \ + \ X-Y$$

Note that the C atoms are bonded to *fewer* atoms in the product than in the reactant. A pair of halogen atoms, an H atom and a halogen atom, or an H atom and an —OH group are typically eliminated, but C atoms are not:

$$CH_3-\underset{\underset{}{|}}{\overset{\overset{OH}{|}}{C}}H-\underset{\underset{}{|}}{\overset{\overset{H}{|}}{C}}H_2 \xrightarrow{H_2SO_4} CH_3-CH=CH_2 \ + \ H-OH$$

3. A **substitution reaction** occurs when an atom (or group) from an added reagent substitutes for one in the organic reactant:

$$R-\underset{\underset{}{|}}{\overset{\overset{}{|}}{C}}-X \ + \ :Y \longrightarrow R-\underset{\underset{}{|}}{\overset{\overset{}{|}}{C}}-Y \ + \ :X$$

Note that the C atom is bonded to the *same number* of atoms in the product as in the reactant. The C atom may be saturated or unsaturated, and X and Y can be many different atoms, but generally *not* C. The main flavor ingredient in banana oil, for instance, forms through a substitution reaction; note that the O substitutes for the Cl:

$$CH_3-\overset{\overset{O}{||}}{C}-Cl \ + \ HO-CH_2-CH_2-\overset{\overset{CH_3}{|}}{C}H-CH_3 \longrightarrow$$

$$CH_3-\overset{\overset{O}{||}}{C}-O-CH_2-CH_2-\overset{\overset{CH_3}{|}}{C}H-CH_3 \ + \ H-Cl$$

SAMPLE PROBLEM 15.3 Recognizing the Type of Organic Reaction

Problem State whether each reaction is an addition, elimination, or substitution:

(a) $CH_3—CH_2—CH_2—Br \longrightarrow CH_3—CH=CH_2 + HBr$

(b) $+ H_2 \longrightarrow$

(c) $CH_3\overset{\overset{\displaystyle O}{\|}}{C}—Br + CH_3CH_2OH \longrightarrow CH_3\overset{\overset{\displaystyle O}{\|}}{C}—OCH_2CH_3 + HBr$

Plan We determine the type of reaction by looking for any change in the number of atoms bonded to C:
• More atoms bonded to C is an *addition*.
• Fewer atoms bonded to C is an *elimination*.
• Same number of atoms bonded to C is a *substitution*.

Solution
(a) Elimination: two bonds in the reactant, C—H and C—Br, are absent in the product, so fewer atoms are bonded to C.
(b) Addition: two more C—H bonds have formed in the product, so more atoms are bonded to C.
(c) Substitution: the reactant C—Br bond becomes a C—O bond in the product, so the same number of atoms are bonded to C.

FOLLOW-UP PROBLEM 15.3 Write a balanced equation for each of the following:
(a) An addition reaction between 2-butene and Cl_2
(b) A substitution reaction between $CH_3—CH_2—CH_2—Br$ and OH^-
(c) The elimination of H_2O from $(CH_3)_3C—OH$

SECTION SUMMARY

In an addition reaction, a π bond breaks and the two C atoms bond to more atoms. In an elimination reaction, a π bond forms and the two C atoms bond to fewer atoms. In a substitution reaction, one atom replaces another atom, but the total number of atoms bonded to C does not change.

15.4 PROPERTIES AND REACTIVITIES OF COMMON FUNCTIONAL GROUPS

The central organizing principle of organic reaction chemistry is the *functional group*. To predict how an organic compound might react, we narrow our focus to such groups because *the distribution of electron density in a functional group affects the reactivity.* The electron density can be high, as in the C=C and C≡C bonds, or it can be low at one end of a bond and high at the other, as in the C—Cl and C—O bonds. Such bond sites attract reactants that are charged or polar and begin a sequence of bond-forming and bond-breaking steps that lead to product(s). Table 15.5 on the next page lists some of the important functional groups in organic compounds.

When we classify functional groups by bond order (single, double, and so forth), they tend to follow certain patterns of reactivity:

• Functional groups with only single bonds undergo substitution or elimination.
• Functional groups with double or triple bonds undergo addition.
• Functional groups with both single and double bonds undergo substitution.

Table 15.5 Important Functional Groups in Organic Compounds

Functional Group	Compound Type	Prefix or *Suffix* of Name	Example		
			Lewis Structure	Ball-and-Stick Model	Systematic Name (Common Name)
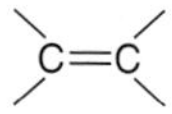 $\text{C}=\text{C}$	alkene	-ene	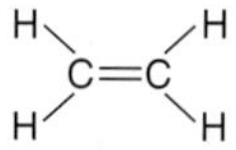	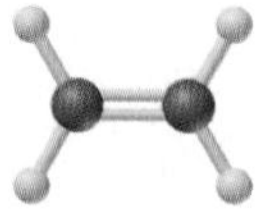	ethene (ethylene)
$-\text{C}\equiv\text{C}-$	alkyne	-yne	$\text{H}-\text{C}\equiv\text{C}-\text{H}$		ethyne (acetylene)
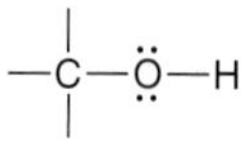 $-\overset{\vert}{\underset{\vert}{\text{C}}}-\ddot{\text{O}}-\text{H}$	alcohol	-ol	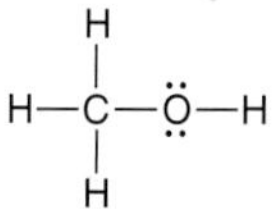	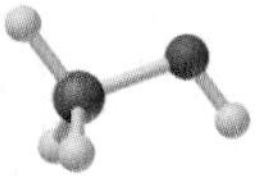	methanol (methyl alcohol)
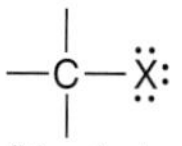 $-\overset{\vert}{\underset{\vert}{\text{C}}}-\ddot{\text{X}}:$ (X = halogen)	haloalkane	halo-	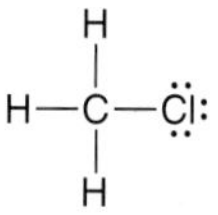	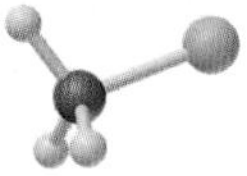	chloromethane (methyl chloride)
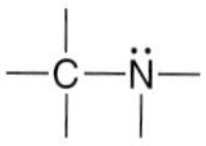 $-\overset{\vert}{\underset{\vert}{\text{C}}}-\overset{\vert}{\underset{\vert}{\text{N}}}-$	amine	-amine	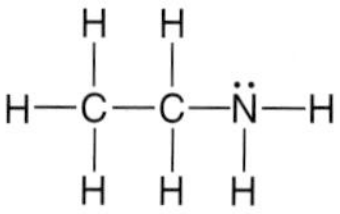	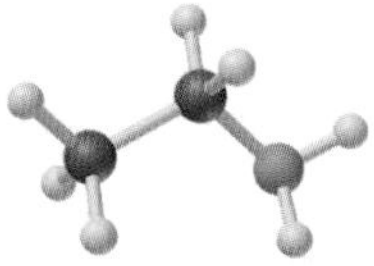	ethanamine (ethylamine)
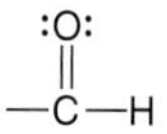 $:\text{O}:$ $-\text{C}-\text{H}$	aldehyde	-al	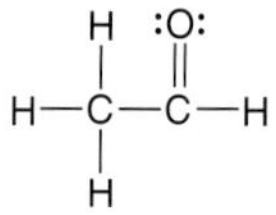	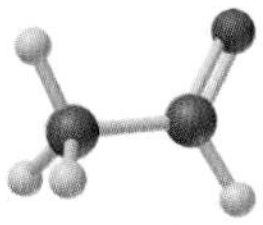	ethanal (acetaldehyde)
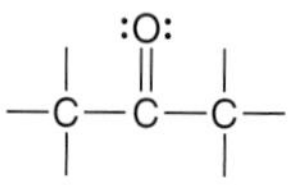 $:\text{O}:$ $-\text{C}-\text{C}-\text{C}-$	ketone	-one	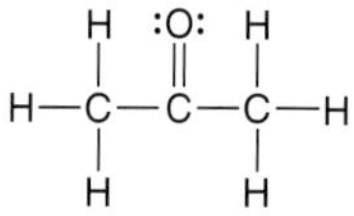	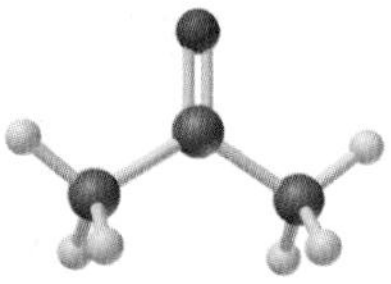	2-propanone (acetone)
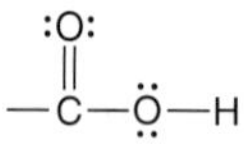 $:\text{O}:$ $-\text{C}-\ddot{\text{O}}-\text{H}$	carboxylic acid	-oic acid	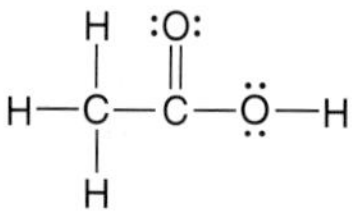	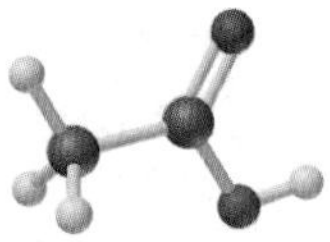	ethanoic acid (acetic acid)
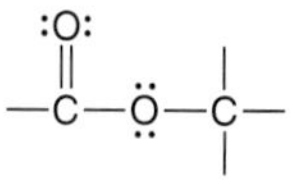 $:\text{O}:$ $-\text{C}-\ddot{\text{O}}-\text{C}-$	ester	-oate	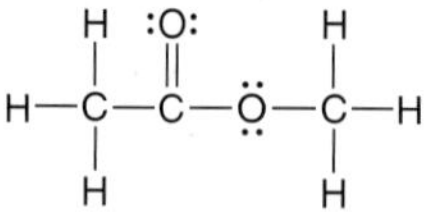	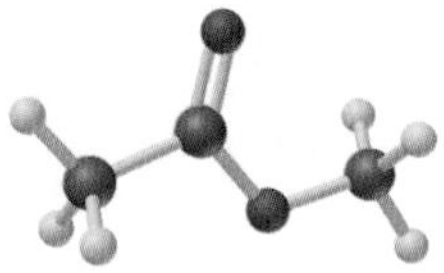	methyl ethanoate (methyl acetate)
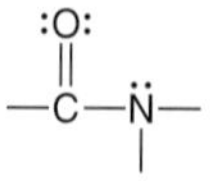 $:\text{O}:$ $-\text{C}-\overset{\vert}{\underset{\vert}{\text{N}}}-$	amide	-amide	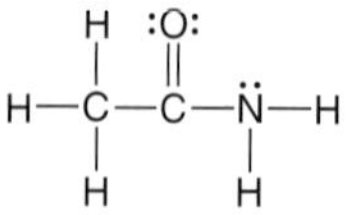	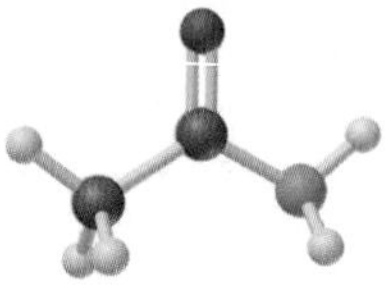	ethanamide (acetamide)
 $-\text{C}\equiv\text{N}:$	nitrile	-nitrile	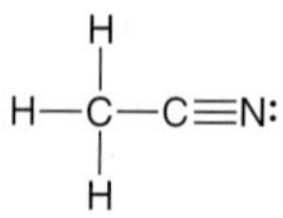	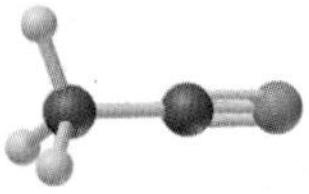	ethanenitrile (acetonitrile, methyl cyanide)

Functional Groups with Only Single Bonds

The most common functional groups with only single bonds are alcohols, haloalkanes, and amines.

Alcohols The **alcohol** functional group consists of carbon bonded to an —OH

group, $-\overset{|}{\underset{|}{C}}-\ddot{O}-H$, and the general formula of an alcohol is R—OH. Alcohols

are named by dropping the final -e from the parent hydrocarbon name and adding the suffix -ol. Thus, the two-carbon alcohol is ethanol (ethan- + -ol). The common name is the hydrocarbon root- + -yl, followed by "alcohol"; thus, the common name of ethanol is ethyl alcohol. (This substance has been consumed as an intoxicant since ancient times; today, it is recognized as the most abused drug in the world.) Alcohols are important laboratory reagents, and the functional group occurs in many biomolecules, including carbohydrates, sterols, and some amino acids. Figure 15.10 shows the names, structures, and uses of some important compounds that contain the alcohol group.

Alcohols undergo elimination and substitution reactions. Dehydration, the elimination of H and OH, requires acid and forms alkenes:

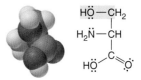

cyclohexanol   cyclohexene

Elimination of two H atoms requires inorganic oxidizing agents, such as $K_2Cr_2O_7$ in aqueous H_2SO_4. The reaction oxidizes the HC—OH group to the C=O group (shown with condensed and carbon-skeleton formulas):

$$CH_3-CH_2-\overset{OH}{\underset{|}{CH}}-CH_3 \xrightarrow[\text{H}_2\text{SO}_4]{\text{K}_2\text{Cr}_2\text{O}_7} CH_3-CH_2-\overset{O}{\overset{||}{C}}-CH_3$$

2-butanol   2-butanone

For alcohols with an OH group at the end of the chain (R—CH₂—OH), another oxidation occurs. Wine turns sour, for example, when the ethanol in contact with air is oxidized to acetic acid (vinegar):

$$CH_3-\overset{OH}{\underset{|}{CH_2}} \xrightarrow[-\text{H}_2\text{O}]{\text{½ O}_2} CH_3-\overset{O}{\overset{||}{CH}} \xrightarrow{\text{½ O}_2} CH_3-\overset{O}{\overset{||}{C}}-OH$$

Substitution yields products with other single-bonded functional groups. With hydrohalic acids, many alcohols give haloalkanes:

$$R_2CH-OH + HBr \longrightarrow R_2CH-Br + HOH$$

As you'll see below, *the C atom undergoing the change in a substitution is bonded to a more electronegative element,* which makes it partially positive and, thus, a target for a negatively charged or electron-rich group of an incoming reactant.

Haloalkanes A *halogen* atom (X) bonded to C gives the **haloalkane** functional

group, $-\overset{|}{\underset{|}{C}}-\ddot{\ddot{X}}:$, and compounds with the general formula R—X. Haloalkanes

(common name, **alkyl halides**) are named by adding the halogen as a prefix to the hydrocarbon name and numbering the C atom to which the halogen is attached, as in bromomethane, 2-chloropropane, or 1,3-diiodohexane.

Methanol (methyl alcohol)
By-product in coal gasification; de-icing agent; gasoline substitute; precursor of organic compounds

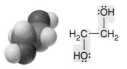

1,2-Ethanediol (ethylene glycol)
Main component of auto antifreeze

Serine
Amino acid found in most proteins

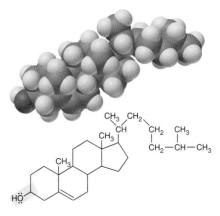

Cholesterol
Major sterol in animals; essential for cell membranes; precursor of steroid hormones

Figure 15.10 Some molecules with the alcohol functional group.

Just as many alcohols undergo substitution to alkyl halides when treated with halide ions in acid, many alkyl halides undergo substitution to alcohols in base. For example, OH^- attacks the positive C end of the C—X bond and displaces X^-:

$$CH_3—CH_2—CH_2—CH_2—Br \; + \; OH^- \; \longrightarrow \; CH_3—CH_2—CH_2—CH_2—OH \; + \; Br^-$$

1-bromobutane   1-butanol

Substitution by groups such as —CN, —SH, —OR, and —NH₂ allows chemists to convert alkyl halides to a host of other compounds.

Just as addition of HX *to* an alkene produces haloalkanes, elimination of HX *from* a haloalkane by reaction with a strong base, such as potassium ethoxide, produces an alkene:

CH₃

|

$$CH_3—C—CH_3 \; + \; CH_3—CH_2—OK \; \longrightarrow \; CH_3—C=CH_2 \; + \; KCl \; + \; CH_3—CH_2—OH$$

|

Cl

2-chloro-2-methylpropane   potassium ethoxide   2-methylpropene

Haloalkanes have many important uses, but many are carcinogenic in mammals, have severe neurological effects in humans, and, to make matters worse, are very stable and accumulate in the environment.

Amines The **amine** functional group is $—\overset{\displaystyle |}{\underset{\displaystyle |}{C}}—\overset{\displaystyle |}{N}{:}$. Chemists classify amines as derivatives of ammonia, with R groups in place of one or more H atoms. *Primary* (1°) amines are RNH₂, *secondary* (2°) amines are R₂NH, and *tertiary* (3°) amines are R₃N. Like ammonia, amines have trigonal pyramidal shapes and a lone pair of electrons on a partially negative N atom (Figure 15.11). Systematic names drop the final *-e* of the alkane and add the suffix *-amine,* as in ethanamine. However, there is still wide usage of common names, in which the suffix *-amine* follows the name of the alkyl group; thus, methylamine has one methyl group attached to N, diethylamine has two ethyl groups attached, and so forth. Figure 15.12 shows that the amine functional group occurs in many biomolecules.

Amines undergo substitution reactions in which the lone pair of N attacks the partially positive C in an alkyl halide to displace X^- and form a larger amine:

$$2CH_3—CH_2—\overset{..}{N}H_2 \; + \; CH_3—CH_2—Cl \; \longrightarrow \; CH_3—CH_2—\overset{..}{N}H \; + \; CH_3—CH_2—\overset{+}{N}H_3Cl^-$$

                                          |

                                    CH₃—CH₂

ethylamine   chloroethane   diethylamine   ethylammonium

                                                 chloride

(One molecule of ethylamine participates in the substitution, while the other binds the released H^+ and prevents it from remaining on the diethylamine product.)

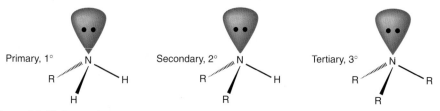

Figure 15.11 General structures of amines. Amines have a trigonal pyramidal shape and are classified by the number of R groups bonded to N. The lone pair of the nitrogen atom is the key to amine reactivity.

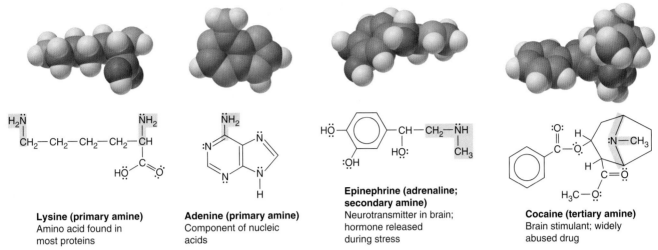

Figure 15.12 Some biomolecules with the amine functional group.

Lysine (primary amine)
Amino acid found in
most proteins

Adenine (primary amine)
Component of nucleic
acids

Epinephrine (adrenaline;
secondary amine)
Neurotransmitter in brain;
hormone released
during stress

Cocaine (tertiary amine)
Brain stimulant; widely
abused drug

SAMPLE PROBLEM 15.4 Predicting the Reactions of Alcohols, Alkyl Halides, and Amines

Problem Determine the reaction type and predict the product(s) for each of the following reactions:

(a) $CH_3-CH_2-CH_2-I + NaOH \longrightarrow$

(b) $CH_3-CH_2-Br + 2CH_3-CH_2-CH_2-NH_2 \longrightarrow$

(c) $CH_3-\overset{\displaystyle |}{\underset{\displaystyle OH}{CH}}-CH_3 \xrightarrow[H_2SO_4]{Cr_2O_7^{2-}}$

Plan We first determine the functional group(s) of the reactant(s) and then examine any inorganic reagent(s) to decide on the possible reaction type, keeping in mind that, in general, these functional groups undergo substitution or elimination. In **(a)**, the reactant is an alkyl halide, so the OH^- of the inorganic reagent substitutes for the $-I$. In **(b)**, the reactants are an amine and an alkyl halide, so the N: of the amine substitutes for the $-Br$. In **(c)**, the reactant is an alcohol, the inorganic reagents form a strong oxidizing agent, and the alcohol undergoes elimination to a carbonyl compound.

Solution (a) Substitution: The products are $CH_3-CH_2-CH_2-OH + NaI$

(b) Substitution: The products are $CH_3-CH_2-CH_2-NH + CH_3-CH_2-CH_2-\overset{+}{N}H_3Br^-$
$\qquad\qquad\qquad\qquad\qquad\qquad\quad |$
$\qquad\qquad\qquad\qquad\qquad\quad CH_2-CH_3$

(c) Elimination: (oxidation): The product is $CH_3-\overset{\displaystyle |\!|}{\underset{\displaystyle O}{C}}-CH_3$

Check The only changes should be at the functional group.

FOLLOW-UP PROBLEM 15.4 Fill in the blank in each reaction. (*Hint:* Examine any inorganic compounds and the organic product to determine the organic reactant.)

(a) _____ $+ CH_3-ONa \longrightarrow CH_3-CH=\overset{\displaystyle \overset{CH_3}{|}}{C}-CH_3 + NaCl + CH_3-OH$

(b) _____ $\xrightarrow[H_2SO_4]{Cr_2O_7^{2-}} CH_3-CH_2-\overset{\displaystyle \overset{O}{|\!|}}{C}-OH$

Functional Groups with Double Bonds

The most important functional groups with double bonds are the C=C of alkenes and the C=O of aldehydes and ketones. Both appear in many organic and biological molecules. *Their most common reaction type is addition.*

Alkenes Although the \C=C/ functional group in an alkene can be converted to the —C≡C— group of an alkyne, *alkenes typically undergo addition.* The electron-rich double bond is readily attracted to the partially positive H atoms of hydronium ions and hydrohalic acids, yielding alcohols and alkyl halides, respectively:

$$CH_3-\overset{\overset{\displaystyle CH_3}{|}}{C}=CH_2 \; + \; H_3O^+ \longrightarrow CH_3-\overset{\overset{\displaystyle CH_3}{|}}{\underset{\underset{\displaystyle OH}{|}}{C}}-CH_3 \; + \; H^+$$

2-methylpropene   2-methyl-2-propanol

$$CH_3-CH=CH_2 \; + \; HCl \longrightarrow CH_3-\overset{\overset{\displaystyle Cl}{|}}{C}H-CH_3$$

propene   2-chloropropane

Aldehydes and Ketones The C=O bond, or **carbonyl group,** is one of the most chemically versatile. In the **aldehyde** functional group, the carbonyl C is bonded to H (and often to another C), so it occurs *at the end of a chain,* $R-\overset{\overset{\displaystyle H}{|}}{C}=O$. Aldehyde names drop the final *-e* from the alkane name and add *-al.* For example, the three-C aldehyde is propanal. In the **ketone** functional group, the carbonyl C is bonded to two other C atoms, $-\overset{}{\underset{|}{C}}-\overset{\overset{\displaystyle :O:}{||}}{C}-\overset{}{\underset{|}{C}}-$, so it occurs *within the chain.*

Ketones, $R-\overset{\overset{\displaystyle O}{||}}{C}-R'$, are named by numbering the carbonyl C, dropping the final *-e* from the alkane name, and adding *-one.* For example, the unbranched, five-C ketone with the carbonyl C as C-2 in the chain is named 2-pentanone. Figure 15.13 shows some common carbonyl compounds.

Figure 15.13 Some common aldehydes and ketones.

Methanal (formaldehyde)
Used to make resins in plywood, dishware, countertops; biological preservative

Ethanal (acetaldehyde)
Narcotic product of ethanol metabolism; used to make perfumes, flavors, plastics, other chemicals

Benzaldehyde
Artificial almond flavoring

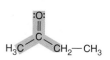

2-Propanone (acetone)
Solvent for fat, rubber, plastic, varnish, lacquer; chemical feedstock

2-Butanone (methyl ethyl ketone)
Important solvent

Like the C=C bond, the C=O bond is *electron rich;* unlike the C=C bond, it is *highly polar* (ΔEN = 1.0). Figure 15.14 emphasizes this polarity with an electron density model and a charged resonance form. Aldehydes and ketones are formed by the oxidation of alcohols:

CH₃—CH₂—OH $\xrightarrow{\text{oxidation}}$ CH₃—C(=O)—H
ethanol — ethanal (common name, acetaldehyde)

3-pentanol $\xrightarrow{\text{oxidation}}$ 3-pentanone (common name, diethylketone)

Conversely, as a result of their unsaturation, carbonyl compounds can undergo *addition* and be reduced to alcohols:

cyclobutanone $\xrightarrow{\text{reduction}}$ cyclobutanol

Functional Groups with Both Single and Double Bonds

A family of three functional groups contains a C double bonded to O (a carbonyl group) *and* single bonded to O or N. The parent of the family is the **carboxylic acid** group, —C(=O)—ÖH, also called the *carboxyl group* and written —COOH. *The most important reaction type of this family is substitution from one member to another.* Substitution for the —OH by the —OR of an alcohol gives the **ester** group, —C(=O)—Ö—R; substitution by the —N̈— of an amine gives the **amide** group, —C(=O)—N̈—.

Carboxylic Acids Carboxylic acids, R—C(=O)—OH, are named by dropping the *-e* from the alkane name and adding *-oic acid;* however, many common names are used. For example, the four-C acid is butanoic acid (the carboxyl C is counted when choosing the root); its common name is butyric acid. Figure 15.15 shows some important carboxylic acids. The carboxyl C already has three bonds, so it forms only one other. In formic acid (methanoic acid), the carboxyl C bonds to an H, but in all other carboxylic acids it bonds to a chain or ring.

Carboxylic acids are weak acids in water:

CH₃—C(=O)—OH(l) + H₂O(l) ⇌ CH₃—C(=O)—O⁻(aq) + H₃O⁺(aq)
ethanoic acid (acetic acid)

At equilibrium in acid solutions of typical concentration, more than 99% of the acid molecules are undissociated at any given moment. In strong base, however, they react completely to form a salt and water:

CH₃—C(=O)—OH(l) + NaOH(aq) ⟶ CH₃—C(=O)—O⁻(aq) + Na⁺(aq) + H₂O(l)

The anion is the *carboxylate ion,* named by dropping *-oic acid* and adding *-oate;* the sodium salt of butanoic acid, for instance, is sodium butanoate.

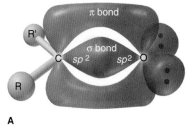

Figure 15.14 The carbonyl group. A, The σ and π bonds that make up the C=O bond of the carbonyl group. **B,** The charged resonance form shows that the C=O bond is polar (ΔEN = 1.0).

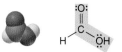

Methanoic acid (formic acid)
An irritating component of ant and bee stings

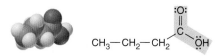

Butanoic acid (butyric acid)
Odor of rancid butter; suspected component of monkey sex attractant

Benzoic acid
Calorimetric standard; used in preserving food, dyeing fabric, curing tobacco

Octadecanoic acid (stearic acid)
Found in animal fats; used in making candles and soaps

Figure 15.15 Some molecules with the carboxylic acid functional group.

Carboxylic acids with long hydrocarbon chains are **fatty acids,** an essential group of compounds found in all cells. Animal fatty acids have saturated chains (see stearic acid, Figure 15.15, *bottom*), whereas many from vegetable sources are unsaturated. Fatty acid salts are soaps, with their cations usually from Group 1A(1) or 2A(2). When clothes with greasy spots are immersed in soapy water, the nonpolar "tails" of the soap molecules interact with the grease, while the ionic "heads" interact with the water. Agitation of the water rinses the grease away.

Substitution of carboxylic acids occurs through a two-step sequence: *addition plus elimination equals substitution.* Addition to the trigonal planar shape of the carbonyl group gives an unstable tetrahedral species, which immediately undergoes elimination to revert to a trigonal planar product:

Strong heating of carboxylic acids forms an **acid anhydride** through a type of substitution called a *dehydration-condensation reaction* (Section 14.6), in which two molecules condense into one with loss of water:

Esters *An alcohol and a carboxylic acid form an ester.* The first part of an ester name designates the alcohol portion and the second the acid portion (named in the same way as the carboxylate ion). For example, the ester formed between ethanol and ethanoic acid is ethyl ethanoate (common name, ethyl acetate), a solvent for nail polish and model glue.

The ester group occurs commonly in **lipids,** a large group of fatty biological substances. Most dietary fats are *triglycerides,* esters that are composed of three fatty acids linked to the alcohol 1,2,3-trihydroxypropane (common name, glycerol) and that function as energy stores. Some important lipids are shown in Figure 15.16; lecithin is one of several phospholipids that make up the lipid bilayer in all cell membranes.

Esters, like acid anhydrides, form through a dehydration-condensation reaction; in this case, it is called an *esterification:*

Note that the esterification reaction is reversible. The opposite of dehydration-condensation is called **hydrolysis,** in which the O atom of water is attracted to the

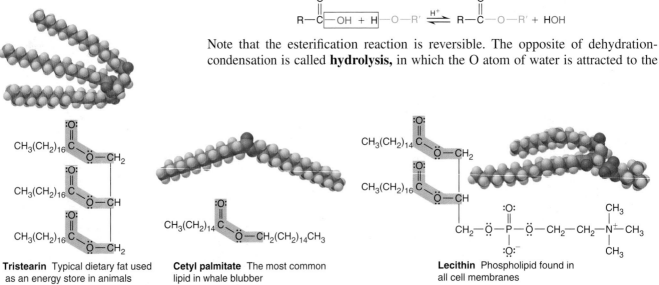

Tristearin Typical dietary fat used as an energy store in animals

Cetyl palmitate The most common lipid in whale blubber

Lecithin Phospholipid found in all cell membranes

Figure 15.16 Some lipid molecules with the ester functional group.

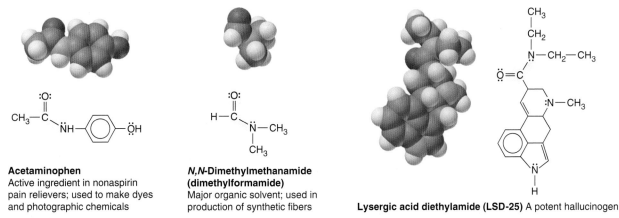

Acetaminophen
Active ingredient in nonaspirin
pain relievers; used to make dyes
and photographic chemicals

**N,N-Dimethylmethanamide
(dimethylformamide)**
Major organic solvent; used in
production of synthetic fibers

Lysergic acid diethylamide (LSD-25) A potent hallucinogen

Figure 15.17 Some molecules with the amide functional group.

partially positive C atom of the ester, cleaving (lysing) the molecule into two parts. One part receives water's —OH, and the other part receives water's other H.

Amides The product of a substitution between an amine (or NH_3) and an ester is an amide. The partially negative N of the amine is attracted to the partially positive C of the ester, an alcohol (ROH) is lost, and an amide forms:

methyl ethanoate   ethanamine   N-ethylethanamide   methanol
(methyl acetate)   (ethylamine)   (N-ethylacetamide)

Amides are named by denoting the amine portion with *N-* and replacing *-oic acid* from the parent carboxylic acid with *-amide*. In the amide from the previous reaction, the ethyl group comes from the amine, and the acid portion comes from ethanoic acid (acetic acid). Some amides are shown in Figure 15.17.

Amides are hydrolyzed in hot water (or base) to a carboxylic acid (or carboxylate ion) and an amine. Thus, even though amides are not normally formed in the following way, they can be viewed as the result of a reversible dehydration-condensation:

The most important example of the amide group is the *peptide bond* (discussed in Section 15.6), which links amino acids in a protein.

SAMPLE PROBLEM 15.5   Predicting Reactions of the Carboxylic Acid Family

Problem Predict the product(s) of the following reactions:

(a) CH_3—CH_2—CH_2—C(=O)—OH + CH_3—CH(OH)—CH_3 $\rightleftharpoons$ (H$^+$)

(b) CH_3—CH(CH_3)—CH_2—CH_2—C(=O)—NH—CH_2—CH_3 $\xrightarrow{\text{NaOH}}_{\text{H}_2\text{O}}$

Plan We discussed substitution reactions (including addition-elimination and dehydration-condensation) and hydrolysis. In **(a)**, a carboxylic acid and an alcohol react, so the reaction must be a substitution to form an ester and water. In **(b)**, an amide reacts with OH$^-$, so it is hydrolyzed to an amine and a sodium carboxylate.

Solution (a) Formation of an ester:

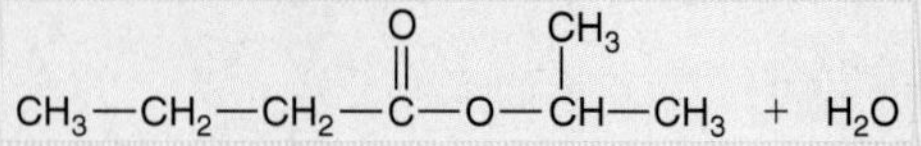

$$CH_3-CH_2-CH_2-\overset{\overset{\displaystyle O}{\|}}{C}-O-\overset{\overset{\displaystyle CH_3}{|}}{CH}-CH_3 \;+\; H_2O$$

(b) Basic hydrolysis of an amide:

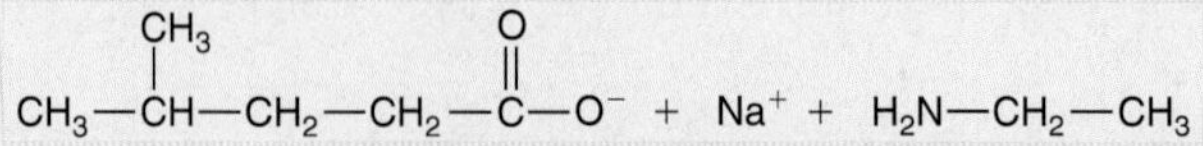

$$CH_3-\overset{\overset{\displaystyle CH_3}{|}}{CH}-CH_2-CH_2-\overset{\overset{\displaystyle O}{\|}}{C}-O^- \;+\; Na^+ \;+\; H_2N-CH_2-CH_3$$

Check Note that in part (b), the carboxylate ion forms, rather than the acid, because the aqueous NaOH that is present reacts with the carboxylic acid.

FOLLOW-UP PROBLEM 15.5 Fill in the blanks in the following reactions:

(a) 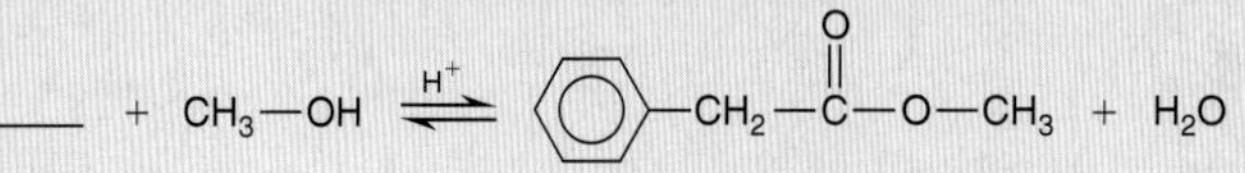 _____ + CH₃—OH $\overset{H^+}{\rightleftharpoons}$ ⬡—CH₂—$\overset{\overset{\displaystyle O}{\|}}{C}$—O—CH₃ + H₂O

(b) 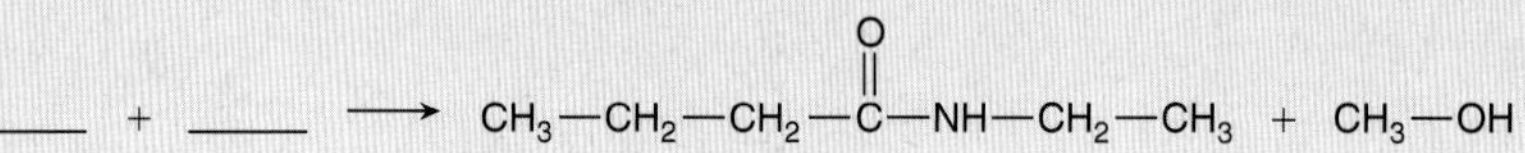 _____ + _____ ⟶ CH₃—CH₂—CH₂—$\overset{\overset{\displaystyle O}{\|}}{C}$—NH—CH₂—CH₃ + CH₃—OH

Functional Groups with Triple Bonds

There are only two important functional groups with triple bonds. *Alkynes,* with their electron-rich —C≡C— group, undergo addition (by H₂O, H₂, HX, X₂, and so forth) to form double-bonded or saturated compounds:

$$CH_3-C\equiv CH \xrightarrow{H_2} CH_3-CH=CH_2 \xrightarrow{H_2} CH_3-CH_2-CH_3$$
$$\text{propyne} \qquad\qquad \text{propene} \qquad\qquad \text{propane}$$

Nitriles (R—C≡N) contain the **nitrile** group (—C≡N:) and are made by substituting a CN⁻ (cyanide) ion for X⁻ in a reaction with an alkyl halide:

$$CH_3-CH_2-Cl \;+\; NaCN \longrightarrow CH_3-CH_2-C\equiv N \;+\; NaCl$$

This reaction is useful because it *increases the hydrocarbon chain by one C atom.* Nitriles are versatile because once they are formed, they can be reduced to amines or hydrolyzed to carboxylic acids:

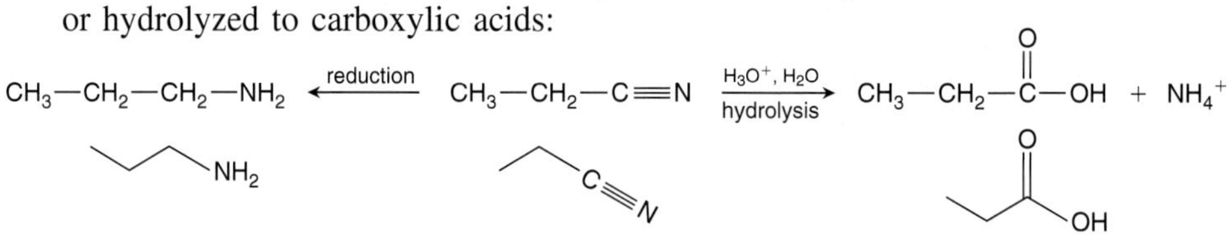

$$CH_3-CH_2-CH_2-NH_2 \xleftarrow{\text{reduction}} CH_3-CH_2-C\equiv N \xrightarrow[\text{hydrolysis}]{H_3O^+,\, H_2O} CH_3-CH_2-\overset{\overset{\displaystyle O}{\|}}{C}-OH \;+\; NH_4^+$$

SAMPLE PROBLEM 15.6 Recognizing Functional Groups

Problem Circle and name the functional groups in the following molecules:

(a)

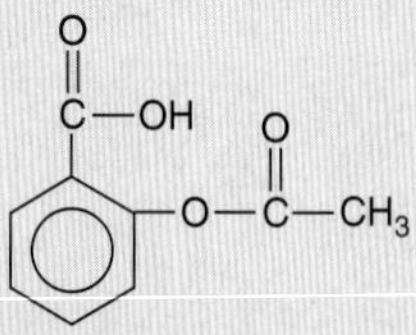

$$\overset{\overset{\displaystyle O}{\|}}{C}-OH \qquad \overset{\overset{\displaystyle O}{\|}}{O-C-CH_3}$$

(b)

$$\overset{\overset{\displaystyle OH}{|}}{CH}-CH_2-NH-CH_3$$

(c)

(ketone/haloalkane cyclohexenone structure with Cl)

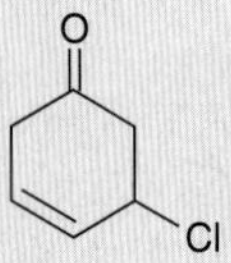

Plan We use Table 15.5 to identify the various functional groups.

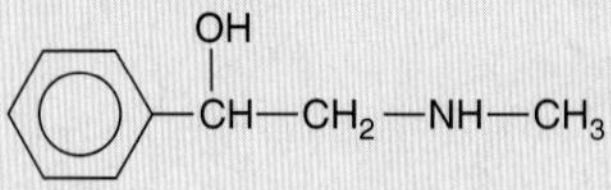

Solution

(a) 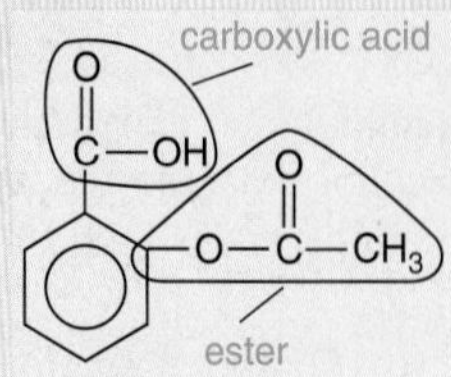 carboxylic acid, ester

(b) 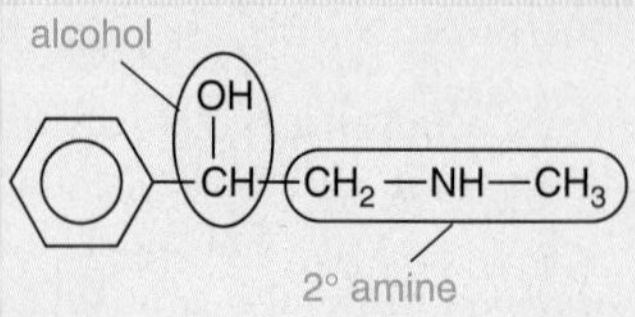 alcohol, 2° amine

(c) 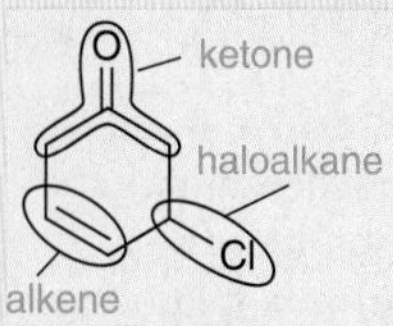 ketone, haloalkane, alkene

FOLLOW-UP PROBLEM 15.6 Circle and name the functional groups:

(a)

(b)
$$H_2N-\overset{\overset{\displaystyle O}{\|}}{C}-CH_2-\underset{\underset{\displaystyle Br}{|}}{CH}-CH_3$$

SECTION SUMMARY

Organic reactions are initiated when regions of high and low electron density of different reactant molecules attract each other. Groups containing only single bonds—alcohols, amines, and alkyl halides—take part in substitution and elimination reactions. Groups with double or triple bonds—alkenes, aldehydes, ketones, alkynes, and nitriles—generally take part in addition reactions. Groups with both double and single bonds—carboxylic acids, esters, and amides—generally take part in substitution reactions. Many reactions change one functional group to another, but some, such as those involving the cyanide ion, change the C skeleton.

15.5 THE MONOMER-POLYMER THEME I: SYNTHETIC MACROMOLECULES

In its simplest form, a **polymer** (Greek, "many parts") is an extremely large molecule, or **macromolecule**, consisting of a covalently linked chain of smaller molecules, called **monomers** (Greek, "one part"). The monomer is the *repeat unit* of the polymer, and a typical polymer may have from hundreds to hundreds of thousands of repeat units. There are many types of monomers, and their chemical structures allow for the complete repertoire of intermolecular forces. *Synthetic* polymers are created by chemical reactions in the laboratory; *natural* polymers (or *biopolymers*) are created by chemical reactions within organisms, and we'll discuss them in the next section.

Virtually every home, car, electronic device, and processed food contains synthetic polymers in its structure or packaging. You interact with dozens of these materials each day—from paints to floor coverings to clothing to the paper coating and adhesives in this textbook. Some of these materials, like those used in food containers, do not break down in the environment and have created a serious waste-disposal problem. Others are being actively recycled into numerous useful products, such as garbage bags, outdoor furniture, roofing tiles, and even marine pilings and roadside curbs. Still others, such as artificial skin, heart valve components, and hip joints, are designed to have as long a life as possible (see silicones in Section 14.5). In this section, we'll see how synthetic polymers are named and discuss the two types of reactions that link monomers covalently into a chain.

To name a polymer, just add the prefix *poly-* to the monomer name, as in *polyethylene* or *polystyrene*. When the monomer has a two-word name, parentheses are used, as in *poly(vinyl chloride)*.

The two major types of reaction processes that form synthetic polymers lend their names to the resulting polymer classes—addition and condensation.

Addition Polymers

Addition polymers form when monomers undergo an addition reaction with one another. These are also called *chain-reaction* (or *chain-growth*) *polymers* because as each monomer adds to the chain, it forms a new reactive site to continue the

process. The monomers of most addition polymers have the $\overset{\diagdown}{\underset{\diagup}{C}}=\overset{\diagup}{\underset{\diagdown}{C}}$ grouping.

Table 15.6   Some Major Addition Polymers

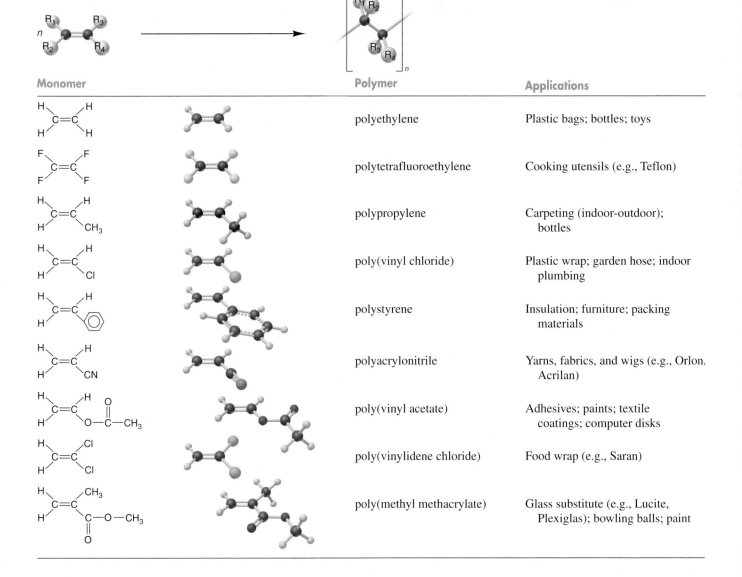

Monomer	Polymer	Applications
	polyethylene	Plastic bags; bottles; toys
	polytetrafluoroethylene	Cooking utensils (e.g., Teflon)
	polypropylene	Carpeting (indoor-outdoor); bottles
	poly(vinyl chloride)	Plastic wrap; garden hose; indoor plumbing
	polystyrene	Insulation; furniture; packing materials
	polyacrylonitrile	Yarns, fabrics, and wigs (e.g., Orlon. Acrilan)
	poly(vinyl acetate)	Adhesives; paints; textile coatings; computer disks
	poly(vinylidene chloride)	Food wrap (e.g., Saran)
	poly(methyl methacrylate)	Glass substitute (e.g., Lucite, Plexiglas); bowling balls; paint

As you can see from Table 15.6, the essential differences between an acrylic sweater, a plastic grocery bag, and a bowling ball are due to the different groups that are attached to the double-bonded C atoms of the monomer.

The *free-radical polymerization* of ethene (ethylene, $CH_2{=}CH_2$) to polyethylene is a simple example of the addition process. In Figure 15.18, the monomer reacts to form a *free radical,* a species with an unpaired electron, that seeks an electron from another monomer to form a covalent bond. The process begins when an *initiator,* usually a peroxide, generates a free radical that attacks the π bond of an ethylene unit, forming a σ bond with one of the electrons and leaving the other unpaired. This new free radical then attacks the π bond of another ethylene, joining it to the chain end, and the backbone of the polymer grows one unit longer. This process continues until two free radicals form a covalent bond or a very stable free radical is formed by addition of an *inhibitor* molecule.

The most important polymerization reactions are *stereoselective* to create polymers whose repeat units have groups spatially oriented in particular ways.

Through the use of these reactions, polyethylene chains with molar masses of 10^4 to 10^5 g/mol are made by varying conditions and reagents.

Similar methods are used to make polypropylenes, $\{\!-CH_2\!-\!CH\!-\!\}_n$, that have
$$\quad\quad\quad\quad\quad\quad\quad\quad\quad\quad\quad\quad\quad\quad\quad\quad\quad\quad\quad\quad\quad\quad\quad\quad\quad\quad\quad\quad CH_3$$
all the CH_3 groups of the repeat units oriented either on one side of the chain or on alternating sides. The different orientations lead to chains that pack differently, which leads to differences in such physical properties as density, rigidity, and elasticity.

Condensation Polymers

The monomers of **condensation polymers** must have *two functional groups;* we can designate such a monomer as A—**R**—B (where **R** is the rest of the molecule). Most commonly, the monomers link when an A group on one undergoes a *dehydration-condensation reaction* with a B group on another:

$$\tfrac{1}{2}n\text{H}\!-\!\text{A}\!-\!\text{R}\!-\!\text{B}\!-\!\text{OH} + \tfrac{1}{2}n\text{H}\!-\!\text{A}\!-\!\text{R}\!-\!\text{B}\!-\!\text{OH} \xrightarrow{-(n-1)\text{HOH}}$$
$$\text{H}\{\!-\!\text{A}\!-\!\text{R}\!-\!\text{B}\!-\!\}_n\,\text{OH}$$

Many condensation polymers are *copolymers,* those consisting of two or more different repeat units. For example, condensation of carboxylic acid and amine monomers forms *polyamides (nylons),* whereas carboxylic acid and alcohol monomers form *polyesters.*

One of the most common polyamides is *nylon-66,* manufactured by mixing equimolar amounts of a six-C diamine (1,6-diaminohexane) and a six-C diacid (1,6-hexanedioic acid). The basic amine reacts with the acid to form a "nylon salt." Heating drives off water and forms the amide bonds:

$$n\text{HO}\!-\!\overset{\displaystyle O}{\overset{\displaystyle \|}{\text{C}}}\!-\!(CH_2)_4\!-\!\overset{\displaystyle O}{\overset{\displaystyle \|}{\text{C}}}\!-\!\text{OH} + n\text{H}_2\text{N}\!-\!(CH_2)_6\!-\!\text{NH}_2 \xrightarrow[-(2n-1)\text{H}_2\text{O}]{\Delta}$$

$$\text{HO}\!\left[\!\overset{\displaystyle O}{\overset{\displaystyle \|}{\text{C}}}\!-\!(CH_2)_4\!-\!\overset{\displaystyle O}{\overset{\displaystyle \|}{\text{C}}}\!-\!\text{NH}\!-\!(CH_2)_6\!-\!\text{NH}\!\right]_{\!n}\!\text{H}$$

Covalent bonds within the chains give nylons great strength, and H bonds between chains give them great flexibility. About half of all nylons are made to reinforce automobile tires; the others are used for rugs, clothing, fishing line, and so forth.

Dacron, a popular polyester fiber, is woven from polymer strands formed when equimolar amounts of 1,4-benzenedicarboxylic acid and 1,2-ethanediol react. Blending these polyester fibers with various amounts of cotton gives fabrics that are durable, easily dyed, and crease resistant. Extremely thin Mylar films, used for recording tape and food packaging, are also made from this polymer.

SECTION SUMMARY
Polymers are extremely large molecules that are made of repeat units called monomers. Addition polymers are formed from unsaturated monomers that commonly link through free-radical reactions. Most condensation polymers are formed by linking two types of monomer through a dehydration-condensation reaction.

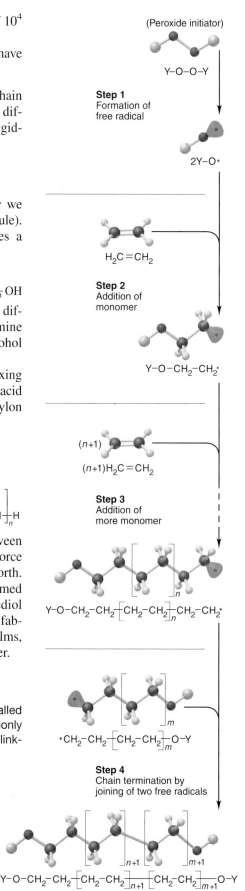

(Peroxide initiator)

Y—O—O—Y

Step 1
Formation of free radical

2Y—O•

$H_2C=CH_2$

Step 2
Addition of monomer

Y—O—CH$_2$—CH$_2$•

(n+1)
$(n+1)H_2C=CH_2$

Step 3
Addition of more monomer

Y—O—CH$_2$—CH$_2\!\{\!$CH$_2$—CH$_2\!\}_n\!$CH$_2$—CH$_2$•

•CH$_2$—CH$_2\!\{\!$CH$_2$—CH$_2\!\}_m\!$—O—Y

Step 4
Chain termination by joining of two free radicals

Y—O—CH$_2$—CH$_2\!\{\!$CH$_2$—CH$_2\!\}_{n+1}\!\{\!$CH$_2$—CH$_2\!\}_{m+1}\!$—O—Y

Figure 15.18 Steps in the free-radical polymerization of ethylene. In this polymerization method, free radicals initiate, propagate, and terminate the formation of an addition polymer. An initiator (Y—O—O—Y) is split to form two molecules of a free radical (Y—O•). The free radical attacks the π bond of a monomer and creates another free radical (Y—O—CH$_2$—CH$_2$•). The process continues, and the chain grows (propagates) until an inhibitor is added (not shown) or two free radicals combine.

15.6 THE MONOMER-POLYMER THEME II: BIOLOGICAL MACROMOLECULES

The monomer-polymer theme was being played out in nature eons before humans employed it to such great advantage. Biological macromolecules are nothing more than condensation polymers created by nature's reaction chemistry and improved through evolution. These remarkable molecules are the greatest proof of the versatility of carbon and its handful of atomic partners.

Natural polymers are the "stuff of life"—polysaccharides, proteins, and nucleic acids. Some have structures that make wood strong, hair curly, fingernails hard, and wool flexible. Others speed up the myriad reactions that occur in every cell or defend the body against infection. Still others possess the genetic information organisms need to forge other biomolecules. Remarkable as these giant molecules are, the functional groups of their monomers and the reactions that link them are identical to those for other, smaller organic molecules, and the same intermolecular forces that dissolve smaller molecules stabilize these giant molecules in the aqueous medium of the cell.

Sugars and Polysaccharides

In essence, the same chemical change occurs when you burn a piece of wood or eat a piece of bread. Wood and bread are mixtures of *carbohydrates,* substances that provide energy through oxidation.

Monomer Structure and Linkage Glucose and other simple sugars are called **monosaccharides** and consist of carbon chains with attached hydroxyl and carbonyl groups. In addition to their roles as individual molecules engaged in energy metabolism, they serve as the monomer units of **polysaccharides.** Most natural polysaccharides are formed from five- and six-C units. In aqueous solution, *an alcohol group and the aldehyde (or ketone) group of a given monosaccharide react with each other to form a cyclic molecule with either a five- or six-membered ring* (Figure 15.19A). When two monosaccharides undergo a dehydration-condensation reaction, a **disaccharide** forms. For example, sucrose (table sugar) is a disaccharide of glucose and fructose (Figure 15.19B).

A polysaccharide consists of *many* monosaccharide units linked covalently. The three major natural polysaccharides consist entirely of glucose units, but they differ in the details of how they are linked. *Cellulose* is the most abundant organic chemical on Earth. More than 50% of the carbon in plants occurs in the cellulose of stems and leaves; wood is largely cellulose, and cotton is more than 90% cellulose. It consists of long chains of glucose H-bonded to one another to form planes that H bond to planes above and below. Thus, the great strength of wood is due largely to H bonds. *Starch* serves as the energy storage molecule in plants. It occurs as a helical molecule of several thousand glucose units mixed with a highly

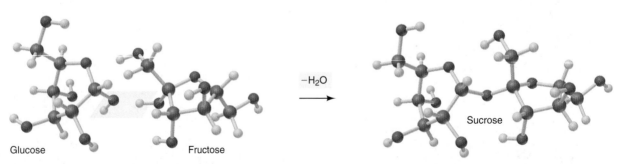

Figure 15.19 The structure of glucose in aqueous solution and the formation of a disaccharide. A, A molecule of glucose undergoes an internal addition reaction between the aldehyde group of C-1 and the alcohol group of C-5 to form a cyclic monosaccharide. **B,** In a dehydration-condensation reaction, the monosaccharides glucose and fructose form the disaccharide sucrose (table sugar) and a water molecule.

branched, bushlike molecule of up to a million glucose units. *Glycogen* functions as the energy store in animals. It occurs in liver and muscle cells as insoluble granules consisting of even more highly branched molecules made from 1000 to more than 500,000 glucose units. The bonds between glucose units in these polymers differ in their chirality. Humans lack the enzyme to break the particular link in cellulose, so we cannot digest it (unfortunately!), but we can break the links in starch and glycogen.

Amino Acids and Proteins

As you saw in Section 15.5, synthetic polyamides (such as nylon-66) are formed from two monomers, one with a carboxyl group at each end and the other with an amine group at each end. **Proteins,** the polyamides of nature, are unbranched polymers formed from monomers called **amino acids,** *each of which has a carboxyl group and an amine group.*

Monomer Structure and Linkage An amino acid has both its carboxyl and amine groups attached to the α-*carbon,* the second C atom in the chain. Proteins are made up of about 20 different types of amino acids, each with its own particular R group (Figure 15.20).

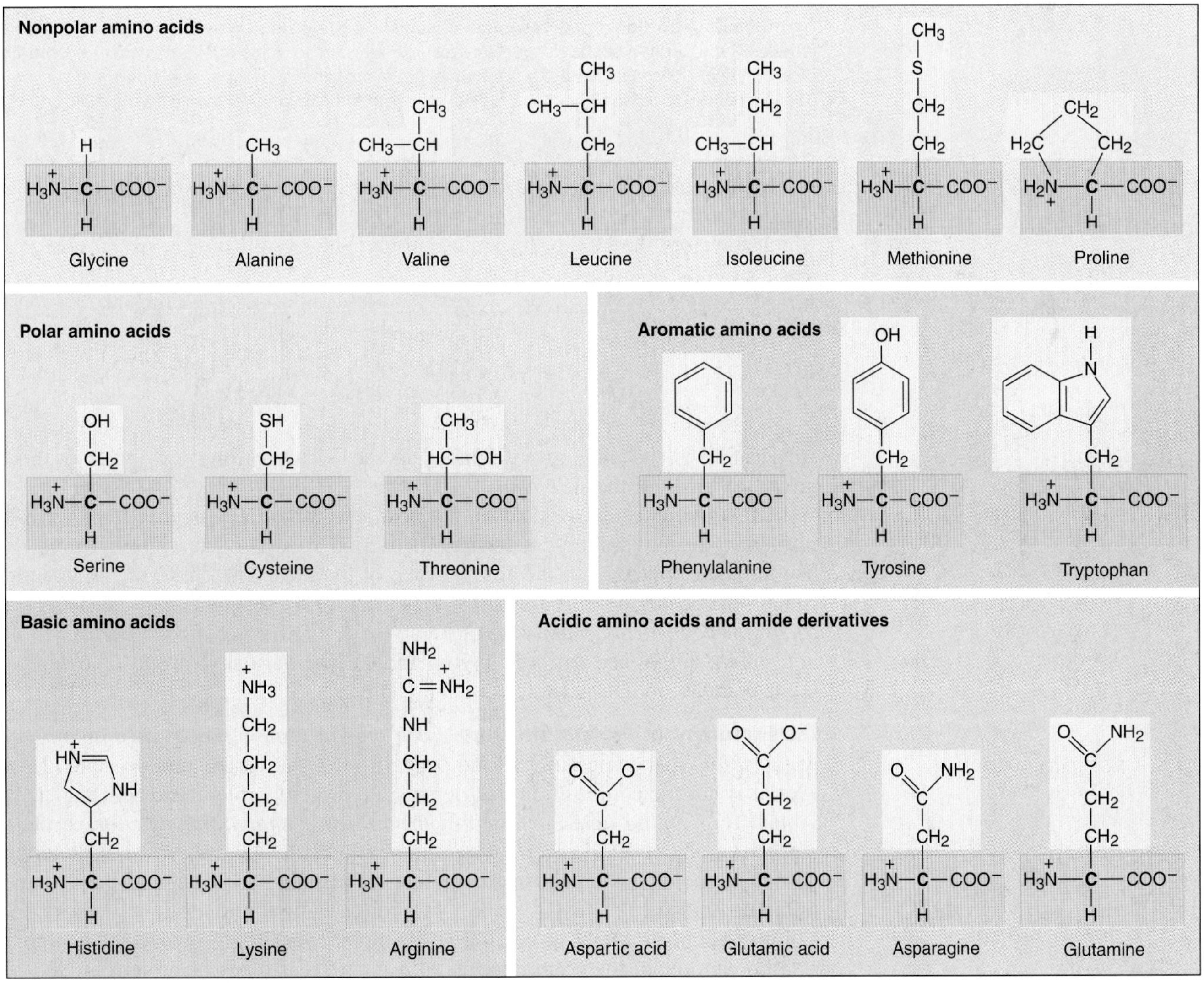

Figure 15.20 The common amino acids. About 20 different amino acids occur in proteins. The R groups are screened yellow, and the α-carbons (*boldface*), with carboxyl and amino groups, are screened gray. Here the amino acids are shown with the charges they have under physiological conditions. They are grouped by polarity, acid-base character, and presence of an aromatic ring. The R groups play a major role in the shape and function of the protein.

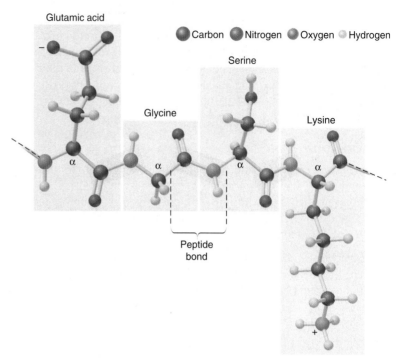

Figure 15.21 A portion of a polypeptide chain. The peptide bond holds monomers together in a protein. Three peptide bonds (*orange screens*) join four amino acids (*gray screens*) in this portion of a polypeptide chain. Note the repeating pattern of the chain: peptide bond—α-carbon—peptide bond—α-carbon—and so on. Also note that the side chains dangle off the main chain.

In the aqueous cell fluid, the NH_2 and COOH groups of amino acids are charged because the carboxyl group transfers an H^+ ion to H_2O to form H_3O^+, which transfers the H^+ to the amine group. The overall process is, in effect, an intramolecular acid-base reaction:

An H atom is the third group bonded to the α-carbon, and the fourth is the R group (also called the *side chain*).

Each amino acid is linked to the next one through a *peptide (amide) bond* formed by a dehydration-condensation reaction in which the carboxyl group of one monomer reacts with the amine group of the next. Therefore, the polypeptide chain—the backbone of the protein—has a repeating sequence that consists of an *α-carbon bonded to an amide group bonded to the next α-carbon bonded to the next amide group,* and so forth (Figure 15.21). The various R groups dangle from the α-carbons on alternate sides of the chain.

The Hierarchy of Protein Structure *Each type of protein has its own amino acid composition,* a specific number and proportion of the various amino acids. However, it is not the composition that defines the protein's role in the cell; rather, *the sequence of amino acids determines the protein's shape and function.* Proteins range from about 50 to several thousand amino acids, yet even a small protein of 100 amino acids has an enormous number of possible sequences of the 20 types of amino acids ($20^{100} \approx 10^{130}$). In fact, though, only a tiny fraction of these possibilities occur in actual proteins. For example, even in an organism as complex as a human being, there are only about 10^5 different types of protein.

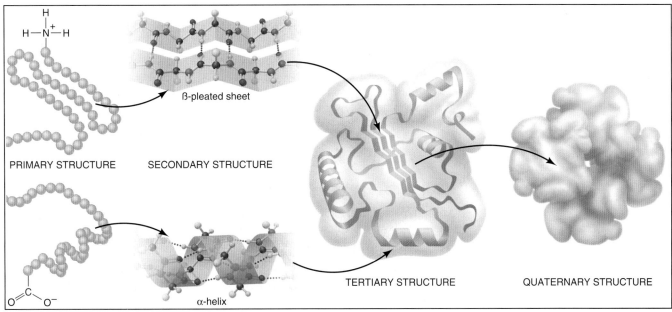

PRIMARY STRUCTURE

β-pleated sheet

SECONDARY STRUCTURE

α-helix

TERTIARY STRUCTURE

QUATERNARY STRUCTURE

Figure 15.22 The structural hierarchy of proteins. A typical protein's structure can be viewed at different levels. Primary structure (shown as a long string of balls leaving and returning to the picture frame) is the sequence of amino acids. Secondary structure consists of highly ordered regions that occur as an α-helix or a β-sheet. Tertiary structure combines these ordered regions with more random sections. In many proteins, several tertiary units interact to give the quaternary structure.

A protein folds into its *native shape* as it is being synthesized in the cell. Biochemists define a hierarchy for the overall structure of a protein (Figure 15.22):

1. *Primary (1°) structure,* the most basic level, refers to the sequence of covalently bonded amino acids in the polypeptide chain.
2. *Secondary (2°) structure* refers to sections of the chain that, as a result of H bonding between nearby peptide groupings, adopt shapes called α-helices and β-pleated sheets.
3. *Tertiary (3°) structure* refers to the three-dimensional folding of the whole polypeptide chain, which results from many forces. The —SH ends of two cysteine side chains form a covalent *disulfide* bridge (—S—S—) that brings together distant parts of the chain. Polar and ionic side chains interact with surrounding water through ion-dipole forces and H bonds. And nonpolar side chains interact through dispersion forces within the nonaqueous protein interior. Thus, *soluble proteins have polar-ionic exteriors and nonpolar interiors.*
4. *Quaternary (4°) structure,* the most complex level, occurs in proteins made up of several polypeptide chains (subunits) and refers to the way the chains assemble into the overall protein.

Note that *only the 1° structure involves covalent bonds; the 2°, 3°, and 4° structures rely primarily on intermolecular forces.*

The Relation Between Structure and Function Two broad classes of proteins differ in the complexity of their amino acid compositions and sequences and, therefore, in their structure and function:

1. *Fibrous proteins* are key components of materials that require strength and flexibility. They have simple amino acid compositions and repetitive structures. Consider collagen, the most common animal protein, which makes up as much as 40% of human body weight. More than 30% of its amino acids are glycine, and another 20% are proline. It exists as long, triple-helical cable in which the peptide C=O groups in one chain form H bonds to the peptide N—H groups in another. As the main component of tendons, skin, and blood vessels, collagen has a high tensile strength; in fact, a 1-mm thick strand can support a 10-kg weight!

2. *Globular proteins* have complex compositions, often containing all 20 common amino acids in varying proportions. They are typically compact, with a wide variety of shapes and functions—as antibodies, hormones, and enzymes, to name a few. The locations of particular amino-acid R groups are crucial to a globular protein's function. In enzymes, for example, these groups bring the reactants together through intermolecular forces and stretch their bonds to speed their reaction to products. Experiment shows that a slight change in a critical R group decreases function dramatically. This fact supports the essential idea that *the protein's amino acid sequence determines its structure, which in turn determines its function:*

$$\text{SEQUENCE} \Longrightarrow \text{STRUCTURE} \Longrightarrow \text{FUNCTION}$$

Nucleotides and Nucleic Acids

An organism's nucleic acids construct its proteins. And, given that the proteins determine how the organism looks and behaves, no job could be more essential.

Monomer Structure and Linkage **Nucleic acids** are *polynucleotides,* unbranched polymers that consist of **mononucleotides,** each of which consists of an N-containing base, a sugar, and a phosphate group. The two types of nucleic acid, *ribonucleic acid* (RNA) and *deoxyribonucleic acid* (DNA), differ in the sugar portions of their mononucleotides: RNA contains *ribose,* and DNA contains *deoxyribose,* in which —H substitutes for —OH on the second C of ribose.

The cellular precursors that form a nucleic acid are *nucleoside triphosphates* (Figure 15.23A). Dehydration-condensation reactions between them create a chain

Figure 15.23 Nucleic acid precursors and their linkage. A, In the cell, nucleic acids are constructed from nucleoside triphosphates, precursors of the mononucleotide units. Each one consists of an N-containing base (structure not shown), a sugar, and a triphosphate group. In RNA (*top*), the sugar is ribose; in DNA, it is 2′-deoxyribose (C atoms of the sugar are denoted by a number primed, e.g., 2′; note the absence of an —OH group on C-2 of the ring). **B,** A tiny segment of the polynucleotide chain of DNA shows the phosphodiester bonds that link the 5′-OH group of one sugar to the 3′-OH group of the next and are formed through dehydration-condensation reactions (which also release diphosphate ion). The bases dangle off the chain.

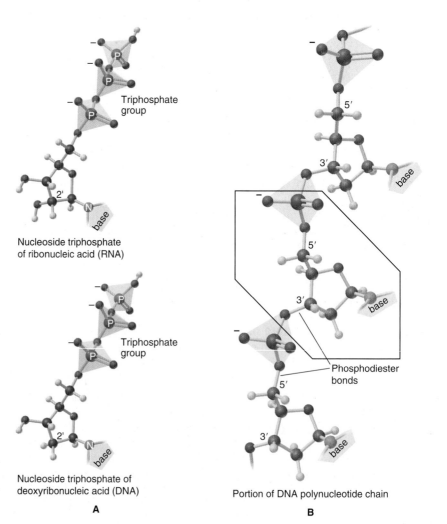

Triphosphate group

Nucleoside triphosphate of ribonucleic acid (RNA)

Triphosphate group

Nucleoside triphosphate of deoxyribonucleic acid (DNA)

Phosphodiester bonds

Portion of DNA polynucleotide chain

A

B

with the repeating pattern —*sugar—phosphate—sugar—phosphate—,* and so on (Figure 15.23B). Attached to each sugar is one of four N-containing bases—thymine (T), cytosine (C), guanine (G), and adenine (A). In RNA, uracil (U) substitutes for thymine. The bases dangle off the sugar-phosphate chain, much as R groups dangle off the polypeptide chain of a protein.

DNA Structure and Base Pairing In the cell nucleus, the many millions of nucleotides in DNA occur as two chains wrapped around each other in a **double helix** (Figure 15.24). Intermolecular forces play a central role in stabilizing this structure. On the exterior, negatively charged sugar-phosphate chains form ion-dipole and H bonds with the aqueous surroundings. In the interior, the flat, nitrogen-containing bases stack above each other, which allows extensive interaction through dispersion forces.

Most important, *each base in one chain "pairs" with a specific base in the other through H bonding.* The essential feature of these **base pairs,** which is crucial to the structure and function of DNA, is that *each type of base is always paired with the same partner:* A with T and G with C. Thus, *the base sequence of one chain is the complement of the sequence of the other.* For example, the sequence A—C—T on one chain is *always* paired with T—G—A on the other: A with T, C with G, and T with A.

Each DNA molecule is folded into a tangled mass that forms one of the cell's *chromosomes.* The DNA molecule is amazingly long and thin: if the largest human chromosome were stretched out, it would be 4 cm long; in the cell nucleus, however, it is wound into a rounded structure only 5 nm wide—8 million times shorter!

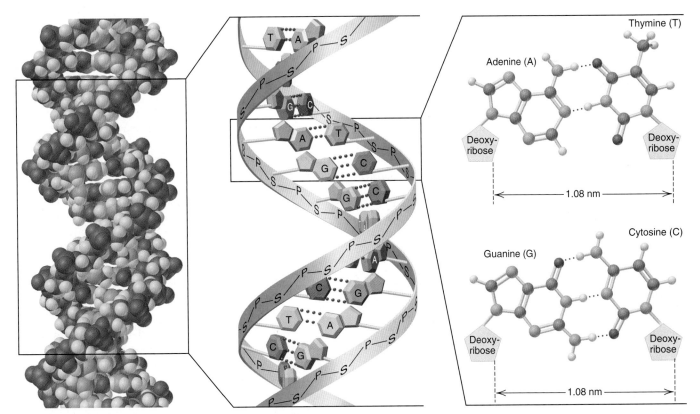

Figure 15.24 The double helix of DNA. A segment of DNA is shown as a space-filling model (*left*). The boxed area is expanded (*center*) to show how the polar sugar(S)-phosphate(P) backbone faces the watery outside, and the nonpolar bases form H bonds to each other in the DNA core. The boxed area is expanded (*right*) to show how a pyrimidine and a purine always form H-bonded base pairs to maintain the double helix width. The members of the pairs are always the same: A pairs with T, and G pairs with C.

From DNA to Protein Segments of the DNA chains are the *genes* that contain the chemical information for synthesizing the organism's proteins. In the **genetic code,** each base acts as a "letter," each three-base sequence as a "word," and *each word codes for a specific amino acid.* For example, the sequence C—A—C codes for the amino acid histidine, A—A—G codes for lysine, and so on. Through a complex process that occurs largely through *H bonding between base pairs,* the DNA message of three-base words is transcribed into an RNA message of three-base words, which is then translated into a sequence of amino acids that are linked to make a protein:

DNA BASE SEQUENCE $\Longrightarrow$ RNA BASE SEQUENCE $\Longrightarrow$ PROTEIN AMINO-ACID SEQUENCE

The biopolymers provide striking evidence that the same atomic properties that give rise to covalent bonds, molecular shape, and intermolecular forces provide the means for all life forms to flourish.

SECTION SUMMARY
Polysaccharides, proteins, and nucleic acids are formed by dehydration-condensation reactions. Polysaccharides are formed from cyclic monosaccharides, such as glucose. Cellulose, starch, and glycogen have structural or energy-storage roles. Proteins are polyamides formed from as many as 20 different types of amino acids. Fibrous proteins have extended shapes and play structural roles. Globular proteins have compact shapes and play metabolic, immunologic, and hormonal roles. The amino acid sequence of a protein determines its shape and function. Nucleic acids (DNA and RNA) are polynucleotides consisting of four different mononucleotides. Hydrogen bonding between specific base pairs is the key to DNA structure. The base sequence of DNA determines the amino-acid sequences of an organism's proteins.

For Review and Reference (Numbers in parentheses refer to pages, unless noted otherwise.)

Learning Objectives

To help you review these learning objectives, the numbers of related sections (§), sample problems (SP), and upcoming end-of-chapter problems (EP) are listed in parentheses.

1. Explain why carbon's atomic properties lead to formation of four strong bonds, multiple bonds, chains, and functional groups (§ 15.1) (EPs 15.1–15.4)
2. Name and draw alkanes, alkenes, and alkynes with expanded, condensed, and carbon-skeleton formulas (§ 15.2) (SPs 15.1, 15.2a–c) (EPs 15.5, 15.9–15.18, 15.29)
3. Distinguish among constitutional, optical, and geometric isomers (§ 15.2) (SP 15.2d, e) (EPs 15.6–15.8, 15.19–15.28, 15.30, 15.31)

4. Describe three types of organic reactions (addition, elimination, and substitution) and identify each type from reactants and products (§ 15.3) (SP 15.3) (EPs 15.32–15.36)
5. Understand the properties and reaction types of the various functional groups (§ 15.4) (SPs 15.4–15.6) (EPs 15.37–15.58)
6. Discuss the formation of addition and condensation polymers and draw abbreviated polymer structures (§ 15.5) (EPs 15.59–15.67)
7. Describe the three types of natural polymers, explain how amino-acid sequence determines protein shape, and thus function, draw small peptides, and use the sequence of one DNA strand to predict the sequence of the other (§ 15.6) (EPs 15.68–15.79)

Key Terms

organic compound (458)

Section 15.1
heteroatom (459)
functional group (460)

Section 15.2
hydrocarbon (460)
alkane (C_nH_{2n+2}) (463)
homologous series (463)
saturated hydrocarbon (463)

cyclic hydrocarbon (465)
isomers (465)
constitutional (structural) isomers (465)
stereoisomers (467)
optical isomers (467)
chiral molecule (467)
optically active (468)
alkene (C_nH_{2n}) (468)

unsaturated hydrocarbon (468)
geometric (*cis-trans*) isomers (468)
alkyne (C_nH_{2n-2}) (469)
aromatic hydrocarbon (471)

Section 15.3
alkyl group (472)
addition reaction (472)

elimination reaction (472)
substitution reaction (472)

Section 15.4
alcohol (475)
haloalkane (alkyl halide) (475)
amine (476)
carbonyl group (478)
aldehyde (478)
ketone (478)

Brief Solutions to Follow-up Problems

15.1 $H-C-C-C-C{\equiv}C-H$ (with H substituents); $H-C-C-C{\equiv}C-C-H$; and branched isomer with $H-C-H$ group: $H-C-C-C{\equiv}C-H$

15.2
(a) $CH_3-CH_2-CH_2-CH_2-CH_2-C(CH_2-CH_3)(CH_3)-CH_2-CH_3$

(b) cyclohexane with $CH_2-CH_2-CH_3$ and CH_2-CH_3 substituents, **and** cyclohexane with propyl and ethyl substituents

(c) $HC{\equiv}C-C(CH_2-CH_3)(CH_2-CH_3)-CH_2-CH_2-CH_3$

(d) CH_3-CH_2 and CH_3 on $C{=}C$ with H and $CH_2-CH_2-CH_3$

15.3
(a) $CH_3-CH{=}CH-CH_3 + Cl_2 \longrightarrow CH_3-CH(Cl)-CH(Cl)-CH_3$

(b) $CH_3-CH_2-CH_2-Br + OH^- \longrightarrow CH_3-CH_2-CH_2-OH + Br^-$

(c) $CH_3-C(CH_3)(OH)-CH_3 \xrightarrow{-H_2O} CH_3-C(CH_3){=}CH_2$

15.4
(a) $CH_3-CH(Cl)-CH(CH_3)-CH_3$ **or** $CH_3-CH_2-C(CH_3)(Cl)-CH_3$

(b) $CH_3-CH_2-CH_2-OH$

15.5
(a) phenyl$-CH_2-C({=}O)-OH$

(b) $CH_3-CH_2-CH_2-C({=}O)-O-CH_3 + CH_3-CH_2-NH_2$

15.6
(a) phenyl$-CH{=}CH-C({=}O)H$ — alkene, aldehyde

(b) $H_2N-C({=}O)-CH_2-CH(Br)-CH_3$ — amide, haloalkane

Problems

Problems with **colored** numbers are answered in Appendix E. Sections match the text and provide the numbers of relevant sample problems. Bracketed problems are grouped in pairs (indicated by a short rule) that cover the same concept. Comprehensive Problems are based on material from any section or previous chapter.

The Special Nature of Carbon and the Characteristics of Organic Molecules

15.1 Explain each of the following statements in terms of atomic properties:
(a) Carbon engages in covalent rather than ionic bonding.
(b) Carbon has four bonds in all its organic compounds.
(c) Carbon forms neither stable cations, like many metals, nor stable anions, like many nonmetals.
(d) Carbon bonds to itself more extensively than does any other element.
(e) Carbon forms stable multiple bonds.

15.2 Carbon bonds to many elements other than itself.
(a) Name six elements that commonly bond to carbon in organic compounds.
(b) Which of these elements are heteroatoms?
(c) Which of these elements are more electronegative than carbon? Less electronegative?

(d) How does bonding of carbon to heteroatoms increase the number of organic compounds?

15.3 Silicon lies just below carbon in Group 4A(14) and also forms four covalent bonds. Why aren't there as many silicon compounds as carbon compounds?

15.4 Which of these bonds to carbon would you expect to be relatively reactive: C—H, C—C, C—I, C=O, C—Li? Explain.

The Structures and Classes of Hydrocarbons
(Sample Problems 15.1 and 15.2)

15.5 (a) What structural feature is associated with each type of hydrocarbon: an alkane; a cycloalkane; an alkene; an alkyne?
(b) Give the general formula for each.
(c) Which hydrocarbons are considered saturated?

15.6 Define each type of isomer: (a) constitutional; (b) geometric; (c) optical. Which types of isomers are stereoisomers?

15.7 Among alkenes, alkynes, and aromatic hydrocarbons, only alkenes exhibit *cis-trans* isomerism. Why don't the others?

15.8 Which objects are asymmetric (have no plane of symmetry): (a) a circular clock face; (b) a football; (c) a dime; (d) a brick; (e) a hammer; (f) a spring?

15.9 Draw all possible skeletons for a 7-C compound with
(a) A 6-C chain and 1 double bond
(b) A 5-C chain and 1 double bond
(c) A 5-C ring and no double bonds

15.10 Draw all possible skeletons for a 6-C compound with
(a) A 5-C chain and 2 double bonds
(b) A 5-C chain and 1 triple bond
(c) A 4-C ring and no double bonds

15.11 Add the correct number of hydrogens to each of the skeletons in Problem 15.9.

15.12 Add the correct number of hydrogens to each of the skeletons in Problem 15.10.

15.13 Draw correct structures, by making a single change, for any that are incorrect:

(a) 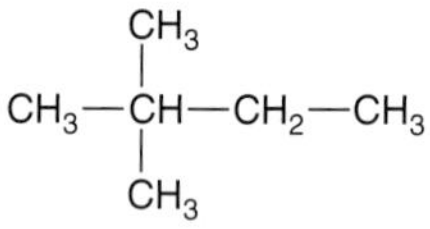 $CH_3-\overset{\overset{\displaystyle CH_3}{|}}{\underset{\underset{\displaystyle CH_3}{|}}{CH}}-CH_2-CH_3$ (b) $CH_3=CH-CH_2-CH_3$

(c) 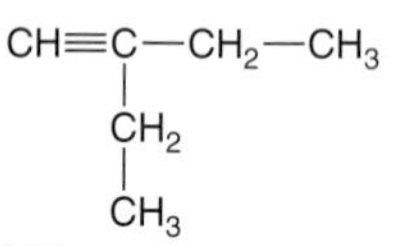 $CH\equiv C-\overset{\overset{\displaystyle |}{CH_2}}{\underset{\underset{\displaystyle CH_3}{|}}{\underset{\displaystyle |}{CH_2}}}-CH_3$ (d) 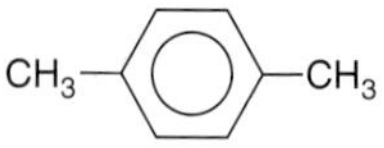 $CH_3-\langle\bigcirc\rangle-CH_3$

15.14 Draw correct structures, by making a single change, for any that are incorrect:

(a) $CH_3-CH=CH-CH_2-CH_3$ (b)

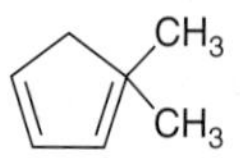

(c) $CH_3-C\equiv CH-CH_2-CH_3$ (d) 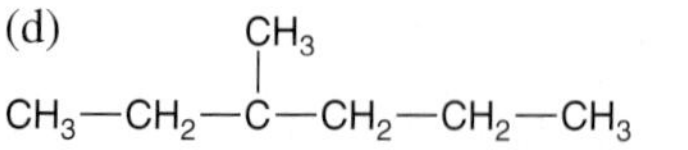
$CH_3-CH_2-\overset{\overset{\displaystyle CH_3}{|}}{\underset{\underset{\displaystyle}{}}{C}}-CH_2-CH_2-CH_3$

15.15 Draw the structure or give the name of each compound:
(a) 2,3-dimethyloctane
(b) 1-ethyl-3-methylcyclohexane

(c) 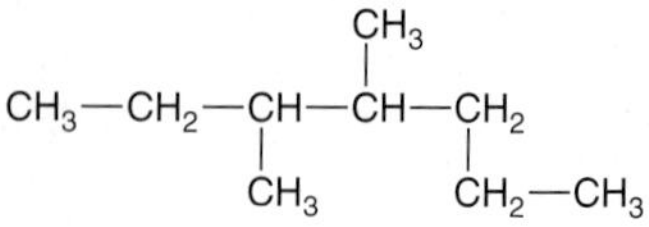 $CH_3-CH_2-\overset{}{CH}-\overset{\overset{\displaystyle CH_3}{|}}{CH}-\overset{}{CH_2}$ (d) 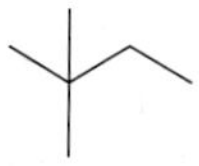
$\qquad\qquad\underset{\underset{\displaystyle CH_3}{|}}{}\quad\underset{\underset{\displaystyle CH_2-CH_3}{|}}{}$

15.16 Draw the structure or give the name of each compound:

(a) 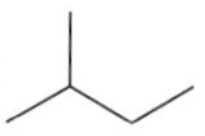   (b)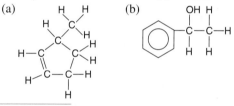

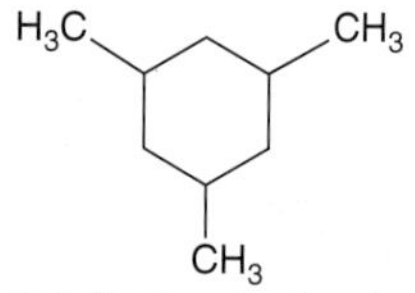

(c) 1,2-diethylcyclopentane   (d) 2,4,5-trimethylnonane

15.17 Each of the following names is wrong. Draw structures based on them, and correct the names:
(a) 4-methylhexane   (b) 2-ethylpentane
(c) 2-methylcyclohexane   (d) 3,3-methyl-4-ethyloctane

15.18 Each of the following names is wrong. Draw structures based on them, and correct the names:
(a) 3,3-dimethylbutane   (b) 1,1,1-trimethylheptane
(c) 1,4-diethylcyclopentane   (d) 1-propylcyclohexane

15.19 Each of the following compounds can exhibit optical activity. Circle the chiral center(s) in each:
(a) 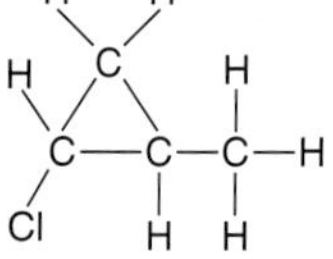   (b)

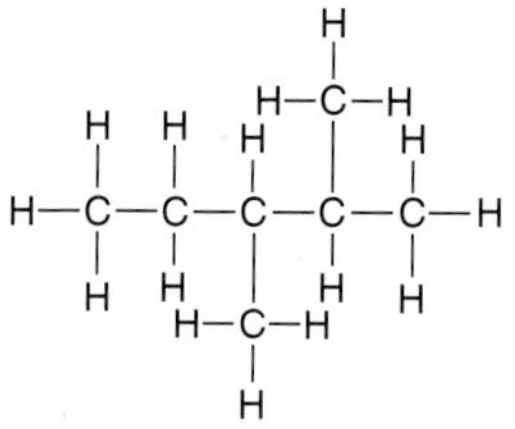

15.20 Each of the following compounds can exhibit optical activity. Circle the chiral center(s) in each:
(a)   (b)

15.21 Draw structures from the following names, and determine which compounds are optically active:
(a) 3-bromohexane
(b) 3-chloro-3-methylpentane
(c) 1,2-dibromo-2-methylbutane

15.22 Draw structures from the following names, and determine which compounds are optically active:
(a) 1,3-dichloropentane
(b) 3-chloro-2,2,5-trimethylhexane
(c) 1-bromo-1-chlorobutane

15.23 Which of the following structures exhibit geometric isomerism? Draw and name the two isomers in each case:
(a) $CH_3-CH_2-CH=CH-CH_3$ (b) 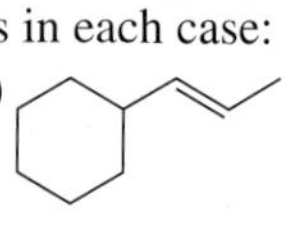
(c) 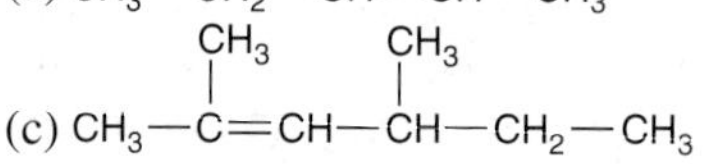 $CH_3-\overset{\overset{\displaystyle CH_3}{|}}{C}=CH-\overset{\overset{\displaystyle CH_3}{|}}{CH}-CH_2-CH_3$

15.24 Which of the following structures exhibit geometric isomerism? Draw and name the two isomers in each case:
(a) 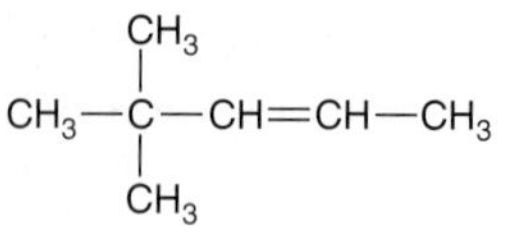 $CH_3-\overset{\overset{\displaystyle CH_3}{|}}{\underset{\underset{\displaystyle CH_3}{|}}{C}}-CH=CH-CH_3$ (b)

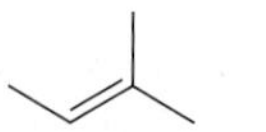

(c) 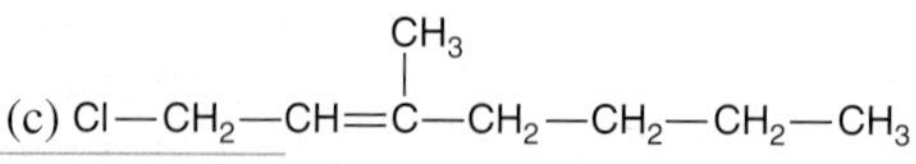 $Cl-CH_2-CH=\overset{\overset{\displaystyle CH_3}{|}}{C}-CH_2-CH_2-CH_2-CH_3$

15.25 Which compounds exhibit geometric isomerism? Draw and name the two isomers in each case:
(a) propene
(b) 3-hexene
(c) 1,1-dichloroethene
(d) 1,2-dichloroethene

15.26 Which compounds exhibit geometric isomerism? Draw and name the two isomers in each case:
(a) 1-pentene
(b) 2-pentene
(c) 1-chloropropene
(d) 2-chloropropene

15.27 Draw and name all the constitutional isomers of dichlorobenzene.

15.28 Draw and name all the constitutional isomers of trimethylbenzene.

15.29 Butylated hydroxytoluene (BHT) is a common preservative added to cereals and other dry foods. Its systematic name is 1-hydroxy-2,6-di-*tert*-butyl-4-methylbenzene (where "*tert*-butyl" is 1,1-dimethylethyl). Draw the structure of BHT.

15.30 There are two compounds with the name 2-methyl-3-hexene, but only one with the name 2-methyl-2-hexene. Explain with structures.

15.31 Any tetrahedral atom with four different groups attached can be a chiral center. Which of these compounds is optically active?
(a) CHClBrF   (b) $NBrCl_2H^+$ (c) $PFClBrI^+$ (d) SeFClBrH

Some Important Classes of Organic Reactions
(Sample Problem 15.3)

15.32 Determine the type of each of the following reactions:

(a) $CH_3-CH_2-\underset{\underset{Br}{|}}{CH}-CH_3 \xrightarrow[\Delta]{NaOH}$
$CH_3-CH=CH-CH_3 + NaBr + H_2O$

(b) $CH_3-CH=CH-CH_2-CH_3 + H_2 \xrightarrow{Pt}$
$CH_3-CH_2-CH_2-CH_2-CH_3$

15.33 Determine the type of each of the following reactions:

(a) $CH_3-\overset{\overset{O}{\|}}{CH} + HCN \longrightarrow CH_3-\underset{\underset{CN}{|}}{\overset{\overset{OH}{|}}{CH}}$

(b) $CH_3-\overset{\overset{O}{\|}}{C}-O-CH_3 + CH_3-NH_2 \xrightarrow{H^+}$
$CH_3-\overset{\overset{O}{\|}}{C}-NH-CH_3 + CH_3-OH$

15.34 Write equations for the following:
(a) An addition reaction between H_2O and 3-hexene (H^+ speeds the reaction but is not consumed)
(b) An elimination reaction between 2-bromopropane and hot potassium ethoxide, CH_3-CH_2-OK (KBr and ethanol are also products)
(c) A light-induced substitution reaction between Cl_2 and ethane to form 1,1-dichloroethane

15.35 Write equations for the following:
(a) A substitution reaction between 2-bromopropane and KI
(b) An addition reaction between cyclohexene and Cl_2
(c) An addition reaction between 2-propanone and H_2 (the reaction occurs on an Ni metal surface)

15.36 Phenylethylamine is a natural substance that is structurally similar to amphetamine. It is found in sources as diverse as almond oil and human urine, where it occurs at elevated concentrations as a result of stress and certain forms of schizophrenia. One method of synthesizing the compound for pharmacological and psychiatric studies involves two steps:

phenylethylamine

Classify each step as an addition, elimination, or substitution.

Properties and Reactivities of Common Functional Groups
(Sample Problems 15.4 to 15.6)

15.37 Compounds with nearly identical molar masses often have very different physical properties. Choose the compound with the higher value for each of the following properties, and explain your choice.
(a) Solubility in water: chloroethane or methylethylamine
(b) Melting point: diethyl ether or 1-butanol
(c) Boiling point: trimethylamine or propylamine

15.38 Fill in each blank with a general formula for the type of compound formed:

15.39 Why does the C=O group react differently from the C=C group? Show an example of the difference.

15.40 Many substitution reactions involve an initial electrostatic attraction between reactants. Show where this attraction arises in the formation of an amide from an amine and an ester.

15.41 What reaction type is common to the formation of esters and acid anhydrides? What is the other product?

15.42 Both alcohols and carboxylic acids undergo substitution, but the processes are very different. Explain.

15.43 Name the type of organic compound from the following description of its functional group:
(a) Polar group that has only single bonds and does not include O or N
(b) Group that is polar and has a triple bond
(c) Group that has single and double bonds and is acidic in water
(d) Group that has a double bond and must be at the end of a C chain

15.44 Name the type of organic compound from the following description of its functional group:
(a) N-containing group with single and double bonds
(b) Group that is not polar and has a double bond
(c) Polar group that has a double bond and cannot be at the end of a C chain
(d) Group that has only single bonds and is basic in water

15.45 Circle and name the functional group(s) in each compound:

(a) $CH_3-CH=CH-CH_2-OH$ (b)

(c) [structure: cyclohexene ring with C(=O)–N(H)–CH₃ amide group]

(d) N≡C—CH₂—C(=O)—CH₃

(e) [structure: cyclobutane ring with C(=O)—O—CH₂—CH₃ ester group]

15.46 Circle and name the functional group(s) in each compound:

(a) HO—[structure with two C=O]—H

(b) I—CH₂—CH₂—C≡CH

(c) CH₂=CH—CH₂—C(=O)—O—CH₃

(d) CH₃—NH—C(=O)—C(=O)—O—CH₃

(e) CH₃—CH(Br)—CH=CH—CH₂—NH—CH₃

15.47 Draw all possible alcohols with the formula $C_5H_{12}O$.
15.48 Draw all possible aldehydes and ketones with the formula $C_5H_{10}O$.

15.49 Draw all possible amines with the formula $C_4H_{11}N$.
15.50 Draw all possible carboxylic acids with the formula $C_5H_{10}O_2$.

15.51 Draw the organic product formed when the following compounds undergo a substitution reaction:
(a) Acetic acid and methylamine
(b) Butanoic acid and 2-propanol
(c) Formic acid and 2-methyl-1-propanol
15.52 Draw the organic product formed when the following compounds undergo a substitution reaction:
(a) Acetic acid and 1-hexanol
(b) Propanoic acid and dimethylamine
(c) Ethanoic acid and diethylamine

15.53 Draw condensed formulas for the carboxylic acid and alcohol portions of the following esters:

(a) [structure: pentanoic acid ethyl ester]

(b) [structure: benzoic acid C(=O)—O—CH₂—CH₂—CH₃ ester]

(c) CH₃—CH₂—O—C(=O)—CH₂—CH₂—[benzene ring]

15.54 Draw condensed formulas for the carboxylic acid and amine portions of the following amides:

(a) H₃C—[benzene ring]—CH₂—C(=O)—NH₂

(b) [structure: isopropyl C(=O)—N with ethyl groups]

(c) HC(=O)—NH—[benzene ring]

15.55 Fill in the expected organic substances:

(a) CH₃—CH₂—Br $\xrightarrow{OH^-}$ ____ $\xrightarrow[H^+]{CH_3-CH_2-C(=O)-OH}$ ____

(b) CH₃—CH₂—CH(Br)—CH₃ $\xrightarrow{CN^-}$ ____ $\xrightarrow{H_3O^+,\ H_2O}$ ____
15.56 Fill in the expected organic substances:

(a) CH₃—CH₂—CH=CH₂ $\xrightarrow{H^+,\ H_2O}$ ____ $\xrightarrow{Cr_2O_7^{2-},\ H^+}$ ____

(b) CH₃—CH₂—OH + ____ $\xrightarrow{H_3O^+,\ H_2O}$ CH₃—CH₂—O—C(=O)—CH₂—CH₃

15.57 (a) Draw the four isomers of $C_5H_{12}O$ that can be oxidized to an aldehyde. (b) Draw the three isomers of $C_5H_{12}O$ that can be oxidized to a ketone. (c) Draw the isomers of $C_5H_{12}O$ that cannot be easily oxidized to an aldehyde or ketone. (d) Name any isomer that is an alcohol.

15.58 Ethyl formate (HC(=O)—O—CH₂—CH₃) is added to foods to give them the flavor of rum. How would you synthesize ethyl formate from ethanol, methanol, and any inorganic reagents?

The Monomer-Polymer Theme I: Synthetic Macromolecules

15.59 Name the reaction processes that lead to the two types of synthetic polymers.
15.60 Which functional group is common to the monomers that make up addition polymers? What makes these polymers different from one another?
15.61 Which intermolecular force is primarily responsible for the different types of polyethylene? Explain.
15.62 Which of the two types of synthetic polymer is more similar chemically to biopolymers? Explain.
15.63 Which two functional groups react to form nylons? Polyesters?

15.64 Draw an abbreviated structure for the following polymers, with brackets around the repeat unit:

(a) Poly(vinyl chloride) (PVC) from [structure: H₂C=CHCl]

(b) Polypropylene from [structure: H₂C=CHCH₃]

15.65 Draw an abbreviated structure for the following polymers, with brackets around the repeat unit:
(a) Teflon from [structure: F₂C=CF₂]
(b) Polystyrene from [structure: H₂C=CH-phenyl]

15.66 Write a balanced equation for the reaction between 1,4-benzenedicarboxylic acid and 1,2-dihydroxyethane to form the polyester Dacron. Draw an abbreviated structure for the polymer, with brackets around the repeat unit.
15.67 Write a balanced equation for the reaction of the monomer dihydroxydimethylsilane (*below*) to form the condensation polymer known as Silly Putty.

[structure: HO—Si(CH₃)₂—OH]

The Monomer-Polymer Theme II: Biological Macromolecules

15.68 Which type of polymer is formed from each of the following monomers: (a) amino acids; (b) alkenes; (c) simple sugars; (d) mononucleotides?

15.69 What is the key structural difference between fibrous and globular proteins? How is it related, in general, to the proteins' amino acid composition?

15.70 Protein shape, function, and amino acid sequence are interrelated. Which determines which?

15.71 What is base pairing? How does it pertain to DNA structure?

15.72 Draw the structure of the R group of (a) alanine; (b) histidine; (c) methionine.

15.73 Draw the structure of the R group of (a) glycine; (b) isoleucine; (c) tyrosine.

15.74 Draw the structure of each of the following tripeptides:
(a) Aspartic acid-histidine-tryptophan
(b) Glycine-cysteine-tyrosine with the charges existing in cell fluid

15.75 Draw the structure of each of the following tripeptides:
(a) Lysine-phenylalanine-threonine
(b) Alanine-leucine-valine with the charges that exist in cell fluid

15.76 Write the sequence of the complementary DNA strand that pairs with each of the following DNA base sequences:
(a) TTAGCC   (b) AGACAT

15.77 Write the sequence of the complementary DNA strand that pairs with each of the following DNA base sequences:
(a) GGTTAC   (b) CCCGAA

15.78 Protein shapes are maintained by a variety of forces that arise from interactions between the amino-acid R groups. Name the amino acid that possesses each R group and the force that could arise in each of the following interactions:
(a) $-CH_2-SH$ with $HS-CH_2-$

(b) $-(CH_2)_4-NH_3^+$ with $^-O-\overset{\overset{\displaystyle O}{\|}}{C}-CH_2-$

(c) $-CH_2-\overset{\overset{\displaystyle O}{\|}}{C}-NH_2$ with $HO-CH_2-$

(d) $-\overset{\overset{\displaystyle CH_3}{|}}{CH}-CH_3$ with ⬡$-CH_2-$

15.79 Amino acids have an average molar mass of 100 g/mol. How many bases on a single strand of DNA are needed to code for a protein with a molar mass of 5×10^5 g/mol?

Comprehensive Problems

Problems with an asterisk (*) are more challenging.

15.80 A synthesis of 2-butanol was performed by treating 2-bromobutane with hot sodium hydroxide solution. The yield was 60%, indicating that a significant portion of the reactant was converted into a second product. Predict what this other product might be.

15.81 Pyrethrins, such as jasmolin II (*below*), are a group of natural compounds synthesized by flowers of the genus *Chrysanthemum* (known as pyrethrum flowers) to act as insecticides.
(a) Circle and name the functional groups in jasmolin II.
(b) What is the hybridization of the numbered carbons?
(c) Which, if any, of the numbered carbons are chiral centers?

* 15.82 Compound A is branched and optically active and contains C, H, and O. (a) A 0.500-g sample burns in excess O_2 to yield 1.25 g of CO_2 and 0.613 g of H_2O. Determine the empirical formula. (b) When 0.225 g of compound A vaporizes at 755 torr and 97°C, the vapor occupies 78.0 mL. Determine the molecular formula. (c) Careful oxidation of the compound yields a ketone. Name and draw compound A and circle the chiral center.

15.83 Vanillin (*right*) is a naturally occurring flavoring agent used in many food products. Name each functional group that contains oxygen. Which carbon-oxygen bond is shortest?

15.84 The genetic code consists of a series of three-base words that each code for a given amino acid.
(a) Using the selections from the genetic code shown below, determine the amino acid sequence coded by the following segment of RNA:

UCCACAGCCUAUAUGGCAAACUUGAAG

AUG = methionine	CCU = proline	CAU = histidine
UGG = tryptophan	AAG = lysine	UAU = tyrosine
GCC = alanine	UUG = leucine	CGG = arginine
UGU = cysteine	AAC = asparagine	ACA = threonine
UCC = serine	GCA = alanine	UCA = serine

(b) What is the complementary DNA sequence from which this RNA sequence was made?

* 15.85 Sodium propanoate ($CH_3-CH_2-\overset{\overset{\displaystyle O}{\|}}{C}-ONa$) is a common preservative found in breads, cheeses, and pies. How would you synthesize sodium propanoate from 1-propanol and any inorganic reagents?

15.86 Supply the missing organic and/or inorganic substances:

(a) $CH_3-\overset{\overset{\displaystyle Cl}{|}}{CH}-CH_3 \xrightarrow{?} CH_3-CH=CH_2 \xrightarrow{?} CH_3-\overset{\overset{\displaystyle Br}{|}}{CH}-\overset{\overset{\displaystyle Br}{|}}{CH_2}$

(b) $CH_3-CH_2-CH_2-OH \xrightarrow{?} CH_3-CH_2-\overset{\overset{\displaystyle O}{\|}}{C}-OH + \xrightarrow{?}\xrightarrow{?}$
$CH_3-CH_2-\overset{\overset{\displaystyle O}{\|}}{C}-O-CH_2-$⬡

CHAPTER SIXTEEN

Kinetics: Rates and Mechanisms of Chemical Reactions

Getting Things Moving The metabolic processes of cold-blooded animals like this baby Nile crocodile (*Crocodylus niloticus*) speed up as the temperatures rise toward midday. In this chapter, you'll see how temperature, as well as several other factors, influences the speed of a reaction.

Key Principles

◆ The *rate of reaction*—the change in concentration of reactant (or product) per unit time—varies as the reaction proceeds; it is fastest at the beginning of the reaction and slowest at the end.

◆ Rate depends on *concentration* and *temperature* because reactants must collide to react, and they must do so with enough kinetic energy.

◆ Kinetic studies often measure an *initial rate*, the rate at the instant the reactants are mixed. Because products are not yet present, only the forward reaction is taking place.

◆ Rate is expressed mathematically by a *rate law* (or *rate equation*), which consists of a temperature-dependent *rate constant* and concentration terms. Each concentration term has an exponent, called a *reaction order*, that defines how the concentration of that reactant affects the rate. The rate law must be determined through experiment, *not* from the balanced equation.

◆ An *integrated rate law* includes concentration *and* time as variables. In addition to reaction order, it allows determination of *half-life,* the time required for half of a reactant to be used up. The half-life of a first-order reaction does not depend on reactant concentration.

◆ *Collision theory* proposes that an *activation energy* (E_a) must be exceeded if a collision of reactants is to lead to products. Higher temperature increases the *fraction* of sufficiently energetic collisions. A collision must also be *effective*—that is, the colliding atoms must be *oriented* correctly to form a bond.

◆ *Transition state theory* explains that the activation energy is used to form a high-energy species with partially broken reactant bonds and partially formed product bonds exists momentarily. Every step in a reaction has such a *transition state (activated complex).*

◆ Chemists explain the rate law for an overall reaction by proposing a *reaction mechanism* that consists of several *elementary steps,* each with its own rate law. For a valid mechanism, the sum of these steps gives the balanced equation, and the rate law of the slowest step (the *rate-determining step*) corresponds to the overall rate law.

◆ A *catalyst* speeds a reaction but is not consumed. It functions by *lowering the E_a* of the rate-determining step of an alternative mechanism for the same overall reaction.

Outline

ntil now we've taken a rather simple approach to chemical change: reactants mix and products form. A balanced equation is an essential quantitative tool for calculating product yields from reactant amounts, but it tells us nothing about how fast the reaction proceeds. **Chemical kinetics** is the study of *reaction rates,* the changes in concentrations of reactants (or products) as a function of time (Figure 16.1).

Concepts & Skills to Review
Before You Study This Chapter

• influence of temperature on molecular speed (Section 5.6)

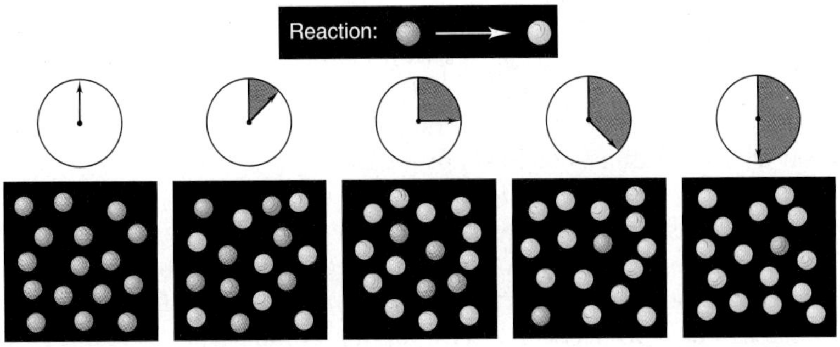

Figure 16.1 Reaction rate: the central focus of chemical kinetics. The rate at which reactant becomes product is the underlying theme of chemical kinetics. As time elapses, reactant *(purple)* decreases and product *(green)* increases.

Reactions occur at a wide range of rates. Some, like a neutralization, a precipitation, or an explosive redox process, seem to be over as soon as the reactants make contact—in a fraction of a second. Others, such as the reactions involved in cooking or rusting, take a moderate length of time, from minutes to months. Still others take much longer: the reactions that make up the human aging process continue for decades, and those involved in the formation of coal from dead plants take hundreds of millions of years.

Knowing how fast a chemical change occurs can be essential. How quickly a medicine acts or blood clots can make the difference between life and death. How long it takes for cement to harden or polyethylene to form can make the difference between profit and loss. In general, the rates of these diverse processes depend on the same variables, most of which chemists can manipulate to maximize yields within a given time or to slow down an unwanted reaction.

In this chapter, we first discuss reaction rate and then focus on the *reaction mechanism,* the steps a reaction goes through as reactant bonds are breaking and product bonds are forming.

16.1 FACTORS THAT INFLUENCE REACTION RATE

Let's begin our study of kinetics with a qualitative look at the key factors that affect how fast a reaction proceeds. Under any given set of conditions, *each reaction has its own characteristic rate,* which is determined by the chemical nature of the reactants. At room temperature, for example, hydrogen reacts explosively with fluorine but extremely slowly with nitrogen:

$$H_2(g) + F_2(g) \longrightarrow 2HF(g) \qquad \text{[very fast]}$$
$$3H_2(g) + N_2(g) \longrightarrow 2NH_3(g) \qquad \text{[very slow]}$$

We can control four factors that affect the rate of a given reaction: the concentrations of the reactants, the physical state of the reactants, the temperature at which the reaction occurs, and the use of a catalyst. We'll consider the first three factors here and discuss the fourth later in the chapter.

1. *Concentration: molecules must collide to react.* A major factor influencing the rate of a given reaction is reactant concentration. A reaction can occur only when the reactant molecules collide. The more molecules present in the

container, the more frequently they collide, and the more often a reaction between them occurs. Thus, *reaction rate is proportional to the concentration of reactants:*

Rate ∝ collision frequency ∝ concentration

2. *Physical state: molecules must mix to collide.* The frequency of collisions between molecules also depends on the physical states of the reactants. When the reactants are in the same phase, as in an aqueous solution, random thermal motion brings them into contact. When they are in different phases, contact occurs only at the interface, so vigorous stirring and grinding may be needed. In these cases, *the more finely divided a solid or liquid reactant, the greater its surface area per unit volume, the more contact it makes with the other reactant, and the faster the reaction occurs.* Thus, a steel nail heated in oxygen glows feebly, but the same mass of steel wool bursts into flame. For the same reason, you start a campfire with wood chips and thin branches, not logs.

3. *Temperature: molecules must collide with enough energy to react.* Temperature usually has a major effect on the speed of a reaction. Recall that molecules in a sample of gas have a range of speeds, with the most probable speed dependent on the temperature (see Figure 5.12). Thus, *at a higher temperature, more collisions occur in a given time.* Even more important, however, is the fact that temperature affects the kinetic energy of the molecules, and thus the *energy of the collisions.* Most collisions result in the molecules simply recoiling, somewhat like billiard balls, with no reaction taking place. However, some collisions occur with sufficient energy for the molecules to react. Figure 16.2 shows this occurring during a few collisions in the reaction between nitric oxide (NO) and ozone (O_3). At higher temperatures, more of these sufficiently energetic collisions occur. Thus, *raising the temperature increases the reaction rate by increasing the number and, especially, the energy of the collisions:*

Rate ∝ collision energy ∝ temperature

The *qualitative* idea that reaction rate is influenced by the frequency and energy of reactant collisions leads to several *quantitative* questions: How can we describe the dependence of rate on reactant concentration mathematically? Do all changes in concentration affect the rate to the same extent? Do all rates increase to the same extent with a given rise in temperature? How do reactant molecules use the collision energy to form product molecules, and is there a way to determine this energy? What do the reactants look like as they are turning into products? We address these questions in the following sections.

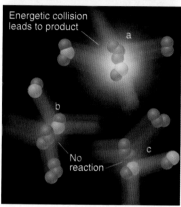

Figure 16.2 Collision energy and reaction rate. The reaction equation is shown in the panel. Although many collisions between NO and O_3 molecules occur, relatively few have enough energy to cause reaction. At this temperature, only collision *a* is energetic enough to lead to product; the reactant molecules in collisions *b* and *c* just bounce off each other.

SECTION SUMMARY
Chemical kinetics deals with reaction rates and the stepwise molecular events by which a reaction occurs. Under a given set of conditions, each reaction has its own rate. Concentration affects rate by influencing the frequency of collisions between reactant molecules. Physical state affects rate by determining the surface area per unit volume of reactant(s). Temperature affects rate by influencing the frequency and, even more importantly, the energy of the reactant collisions.

16.2 EXPRESSING THE REACTION RATE

A *rate* is a change in some variable per unit of time. The most common examples relate to the rate of motion (speed) of an object, which is the change in its position (that is, the distance it travels) divided by the change in time. Suppose, for instance, we measure a runner's starting position, x_1, at time t_1 and final position, x_2, at time t_2. The runner's average speed is

$$\text{Rate of motion} = \frac{\text{change in position}}{\text{change in time}} = \frac{x_2 - x_1}{t_2 - t_1} = \frac{\Delta x}{\Delta t}$$

In the case of a chemical change, we are concerned with the **reaction rate,** the changes in concentrations of reactants or products per unit time: *reactant concentrations decrease while product concentrations increase.* Consider a general reaction, A $\longrightarrow$ B. We quickly measure the starting reactant concentration (conc A$_1$) at t_1, allow the reaction to proceed, and then quickly measure the reactant concentration again (conc A$_2$) at t_2. The change in concentration divided by the change in time gives the *average* rate:

$$\text{Rate of reaction} = -\frac{\text{change in concentration of A}}{\text{change in time}} = -\frac{\text{conc A}_2 - \text{conc A}_1}{t_2 - t_1} = -\frac{\Delta(\text{conc A})}{\Delta t}$$

Note the minus sign. By convention, reaction rate is a *positive* number, but conc A$_2$ will always be *lower* than conc A$_1$, so the *change in (final − initial) concentration of reactant A is always negative.* We use the minus sign simply to convert the negative change in reactant concentration to a positive value for the rate. Suppose the concentration of A changes from 1.2 mol/L (conc A$_1$) to 0.75 mol/L (conc A$_2$) over a 125-s period. The average rate is

$$\text{Rate} = -\frac{0.75 \text{ mol/L} - 1.2 \text{ mol/L}}{125 \text{ s} - 0 \text{ s}} = 3.6 \times 10^{-3} \text{ mol/L·s}$$

We use *square brackets,* [ ], *to express concentration in moles per liter.* That is, [A] is the concentration of A in mol/L, so the rate expressed in terms of A is

$$\text{Rate} = -\frac{\Delta[A]}{\Delta t} \qquad (16.1)$$

The rate has units of moles per liter per second (mol L^{-1} s^{-1}, or mol/L·s), or any time unit convenient for the particular reaction (minutes, years, and so on).

If instead we measure the *product* to determine the reaction rate, we find its concentration *increasing* over time. That is, conc B$_2$ is always *higher* than conc B$_1$. Thus, the *change* in product concentration, $\Delta[B]$, is *positive,* and the reaction rate for A $\longrightarrow$ B expressed in terms of B is

$$\text{Rate} = \frac{\Delta[B]}{\Delta t}$$

Average, Instantaneous, and Initial Reaction Rates

Examining the rate of a real reaction reveals an important point: not only the concentration, but *the rate itself varies with time as the reaction proceeds.* Consider the reversible gas-phase reaction between ethylene and ozone, one of many reactions that can be involved in the formation of photochemical smog:

$$C_2H_4(g) + O_3(g) \rightleftharpoons C_2H_4O(g) + O_2(g)$$

For now, we consider only reactant concentrations. You can see from the equation coefficients that for every molecule of C_2H_4 that reacts, a molecule of O_3 reacts with it. In other words, the concentrations of both reactants decrease at the same rate in this particular reaction:

$$\text{Rate} = -\frac{\Delta[C_2H_4]}{\Delta t} = -\frac{\Delta[O_3]}{\Delta t}$$

By measuring the concentration of either reactant, we can follow the reaction rate.

Suppose we have a known concentration of O_3 in a closed reaction vessel kept at 30°C (303 K). Table 16.1 shows the concentration of O_3 at various times during the first minute after we introduce C_2H_4 gas. The rate over the entire 60.0 s is the total change in concentration divided by the change in time:

$$\text{Rate} = -\frac{\Delta[O_3]}{\Delta t} = -\frac{(1.10 \times 10^{-5} \text{ mol/L}) - (3.20 \times 10^{-5} \text{ mol/L})}{60.0 \text{ s} - 0.0 \text{ s}} = 3.50 \times 10^{-7} \text{ mol/L·s}$$

Table 16.1 Concentration of O_3 at Various Times in Its Reaction with C_2H_4 at 303 K

Time (s)	Concentration of O_3 (mol/L)
0.0	3.20×10^{-5}
10.0	2.42×10^{-5}
20.0	1.95×10^{-5}
30.0	1.63×10^{-5}
40.0	1.40×10^{-5}
50.0	1.23×10^{-5}
60.0	1.10×10^{-5}

This calculation gives us the **average rate** over that period; that is, during the first 60.0 s of the reaction, ozone concentration decreases an *average* of 3.50×10^{-7} mol/L each second. However, the average rate does not show that the rate is changing, and it tells us nothing about how fast the ozone concentration is decreasing *at any given instant.*

We can see the rate change during the reaction by calculating the average rate over two shorter periods—one earlier and one later. Between the starting time 0.0 s and 10.0 s, the average rate is

$$\text{Rate} = -\frac{\Delta[O_3]}{\Delta t} = -\frac{(2.42 \times 10^{-5}\ \text{mol/L}) - (3.20 \times 10^{-5}\ \text{mol/L})}{10.0\ \text{s} - 0.0\ \text{s}} = 7.80 \times 10^{-7}\ \text{mol/L·s}$$

During the last 10.0 s, between 50.0 s and 60.0 s, the average rate is

$$\text{Rate} = -\frac{\Delta[O_3]}{\Delta t} = -\frac{(1.10 \times 10^{-5}\ \text{mol/L}) - (1.23 \times 10^{-5}\ \text{mol/L})}{60.0\ \text{s} - 50.0\ \text{s}} = 1.30 \times 10^{-7}\ \text{mol/L·s}$$

The earlier rate is six times as fast as the later rate. Thus, *the rate decreases during the course of the reaction.* This makes perfect sense from a molecular point of view: as O_3 molecules are used up, fewer of them are present to collide with C_2H_4 molecules, so the rate decreases.

The change in rate can also be seen by plotting the concentrations vs. the times at which they were measured (Figure 16.3). A curve is obtained, which means that the rate changes. *The slope of the straight line ($\Delta y/\Delta x$, that is, $\Delta[O_3]/\Delta t$) joining any two points gives the average rate over that period.*

The shorter the time period we choose, the closer we come to the **instantaneous rate,** the rate at a particular instant during the reaction. *The slope of a line tangent to the curve at a particular point gives the instantaneous rate at that time.* For example, the rate of the reaction of C_2H_4 and O_3 at 35.0 s is 2.50×10^{-7} mol/L·s, the slope of the line drawn tangent to the curve through the point at which $t = 35.0$ s (line *d* in Figure 16.3). In general, we use the term *reaction rate* to mean the *instantaneous* reaction rate.

Figure 16.3 The concentration of O_3 vs. time during its reaction with C_2H_4. Plotting the data in Table 16.1 gives a curve because the rate changes during the reaction. The *average* rate over a given period is the slope of a line joining two points along the curve. The slope of line *b* is the average rate over the first 60.0 s of the reaction. The slopes of lines *c* and *e* give the average rate over the first and last 10.0-s intervals, respectively. Line *c* is steeper than line *e* because the average rate over the earlier period is higher. The *instantaneous* rate at 35.0 s is the slope of line *d*, the tangent to the curve at $t = 35.0$ s. The *initial* rate is the slope of line *a*, the tangent to the curve at $t = 0$ s.

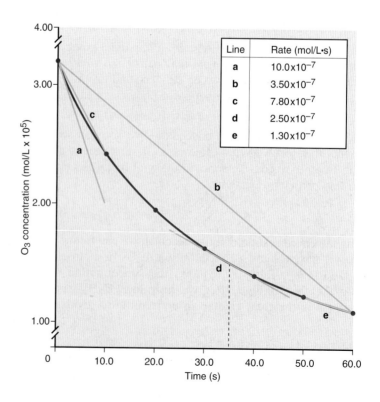

Line	Rate (mol/L·s)
a	10.0×10^{-7}
b	3.50×10^{-7}
c	7.80×10^{-7}
d	2.50×10^{-7}
e	1.30×10^{-7}

As a reaction continues, the product concentrations increase, and so the reverse reaction (reactants ⟵ products) proceeds more quickly. To find the overall (net) rate, we would have to take both forward and reverse reactions into account and calculate the difference between their rates. A common way to avoid this complication for many reactions is to measure the **initial rate,** the instantaneous rate at the moment the reactants are mixed. Under these conditions, the product concentrations are negligible, so the reverse rate is negligible. The initial rate is measured by determining the slope of the line tangent to the curve at $t = 0$ s. In Figure 16.3, the initial rate is 10.0×10^{-7} mol/L·s (line a). Unless stated otherwise, we will use initial rate data to determine other kinetic parameters.

Expressing Rate in Terms of Reactant and Product Concentrations

So far, in our discussion of the reaction of C_2H_4 and O_3, we've expressed the rate in terms of the decreasing concentration of O_3. The rate is the same in terms of C_2H_4, but it is exactly the opposite in terms of the products because their concentrations are *increasing*. From the balanced equation, we see that one molecule of C_2H_4O and one of O_2 appear for every molecule of C_2H_4 and of O_3 that disappear. We can express the rate in terms of any of the four substances involved:

$$\text{Rate} = -\frac{\Delta[C_2H_4]}{\Delta t} = -\frac{\Delta[O_3]}{\Delta t} = +\frac{\Delta[C_2H_4O]}{\Delta t} = +\frac{\Delta[O_2]}{\Delta t}$$

Again, note the negative values for the reactants and the positive values for the products (usually written without the plus sign). Figure 16.4 shows a plot of the simultaneous monitoring of one reactant and one product. Because, in this case, product concentration increases at the same rate that reactant concentration decreases, the curves have the same shapes but are inverted.

In the reaction between ethylene and ozone, the reactants disappear and the products appear at the same rate because all the coefficients in the balanced equation are equal. Consider next the reaction between hydrogen and iodine to form hydrogen iodide:

$$H_2(g) + I_2(g) \longrightarrow 2HI(g)$$

For every molecule of H_2 that disappears, one molecule of I_2 disappears and *two* molecules of HI appear. In other words, the rate of $[H_2]$ decrease is the same as the rate of $[I_2]$ decrease, but both are only half the rate of [HI] increase. By referring the change in $[I_2]$ and [HI] to the change in $[H_2]$, we have

$$\text{Rate} = -\frac{\Delta[H_2]}{\Delta t} = -\frac{\Delta[I_2]}{\Delta t} = \frac{1}{2}\frac{\Delta[HI]}{\Delta t}$$

If we refer the change in $[H_2]$ and $[I_2]$ to the change in [HI] instead, we obtain

$$\text{Rate} = \frac{\Delta[HI]}{\Delta t} = -2\frac{\Delta[H_2]}{\Delta t} = -2\frac{\Delta[I_2]}{\Delta t}$$

Notice that this expression is just a rearrangement of the previous one; also note that it gives a numerical value for the rate that is double the previous value. Thus, the mathematical expression for the rate of a particular reaction and *the numerical value of the rate depend on which substance serves as the reference.*

We can summarize these results for any reaction,

$$a\text{A} + b\text{B} \longrightarrow c\text{C} + d\text{D}$$

where a, b, c, and d are coefficients of the balanced equation. In general, the rate is related to reactant or product concentrations as follows:

$$\text{Rate} = -\frac{1}{a}\frac{\Delta[A]}{\Delta t} = -\frac{1}{b}\frac{\Delta[B]}{\Delta t} = \frac{1}{c}\frac{\Delta[C]}{\Delta t} = \frac{1}{d}\frac{\Delta[D]}{\Delta t} \qquad \textbf{(16.2)}$$

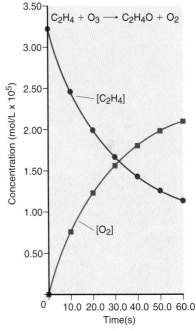

Figure 16.4 Plots of $[C_2H_4]$ and $[O_2]$ vs. time. Measuring reactant concentration, $[C_2H_4]$, and product concentration, $[O_2]$, gives curves of identical shapes but changing in opposite directions. The steep upward (positive) slope of $[O_2]$ early in the reaction mirrors the steep downward (negative) slope of $[C_2H_4]$ because the faster C_2H_4 is used up, the faster O_2 is formed. The curve shapes are identical in this case because the equation coefficients are identical.

SAMPLE PROBLEM 16.1 Expressing Rate in Terms of Changes in Concentration with Time

Problem Because it has a nonpolluting combustion product (water vapor), hydrogen gas is used for fuel aboard the space shuttle and in prototype cars with Earth-bound engines:

$$2H_2(g) + O_2(g) \longrightarrow 2H_2O(g)$$

(a) Express the rate in terms of changes in $[H_2]$, $[O_2]$, and $[H_2O]$ with time.
(b) When $[O_2]$ is decreasing at 0.23 mol/L·s, at what rate is $[H_2O]$ increasing?
Plan (a) Of the three substances in the equation, let's choose O_2 as the reference because its coefficient is 1. For every molecule of O_2 that disappears, two molecules of H_2 disappear, so the rate of $[O_2]$ decrease is one-half the rate of $[H_2]$ decrease. By similar reasoning, we see that the rate of $[O_2]$ decrease is one-half the rate of $[H_2O]$ increase. **(b)** Because $[O_2]$ is decreasing, the change in its concentration must be negative. We substitute the negative value into the expression and solve for $\Delta[H_2O]/\Delta t$.
Solution (a) Expressing the rate in terms of each component:

$$\text{Rate} = -\frac{1}{2}\frac{\Delta[H_2]}{\Delta t} = -\frac{\Delta[O_2]}{\Delta t} = \frac{1}{2}\frac{\Delta[H_2O]}{\Delta t}$$

(b) Calculating the rate of change of $[H_2O]$:

$$\frac{1}{2}\frac{\Delta[H_2O]}{\Delta t} = -\frac{\Delta[O_2]}{\Delta t} = -(-0.23 \text{ mol/L·s})$$

$$\frac{\Delta[H_2O]}{\Delta t} = 2(0.23 \text{ mol/L·s}) = 0.46 \text{ mol/L·s}$$

Check (a) A good check is to use the rate expression to obtain the balanced equation: $[H_2]$ changes twice as fast as $[O_2]$, so two H_2 molecules react for each O_2. $[H_2O]$ changes twice as fast as $[O_2]$, so two H_2O molecules form from each O_2. From this reasoning, we get $2H_2 + O_2 \longrightarrow 2H_2O$. The $[H_2]$ and $[O_2]$ decrease, so they take minus signs; $[H_2O]$ increases, so it takes a plus sign. Another check is to use Equation 16.2, with $A = H_2$, $a = 2$; $B = O_2$, $b = 1$; $C = H_2O$, $c = 2$. Thus,

$$\text{Rate} = -\frac{1}{a}\frac{\Delta[A]}{\Delta t} = -\frac{1}{b}\frac{\Delta[B]}{\Delta t} = \frac{1}{c}\frac{\Delta[C]}{\Delta t}$$

or

$$\text{Rate} = -\frac{1}{2}\frac{\Delta[H_2]}{\Delta t} = -\frac{\Delta[O_2]}{\Delta t} = \frac{1}{2}\frac{\Delta[H_2O]}{\Delta t}$$

(b) Given the rate expression, it makes sense that the numerical value of the rate of $[H_2O]$ increase is twice that of $[O_2]$ decrease.
Comment Thinking through this type of problem at the molecular level is the best approach, but use Equation 16.2 to confirm your answer.

FOLLOW-UP PROBLEM 16.1 (a) Balance the following equation and express the rate in terms of the change in concentration with time for each substance:

$$NO(g) + O_2(g) \longrightarrow N_2O_3(g)$$

(b) How fast is $[O_2]$ decreasing when $[NO]$ is decreasing at a rate of 1.60×10^{-4} mol/L·s?

SECTION SUMMARY

The average reaction rate is the change in reactant (or product) concentration over a change in time, Δt. The rate slows as reactants are used up. The instantaneous rate at time t is obtained from the slope of the tangent to a concentration vs. time curve at time t. The initial rate, the instantaneous rate at $t = 0$, occurs when reactants are just mixed and before any product accumulates. The expression for a reaction rate and its numerical value depend on which reaction component is being monitored.

16.3 THE RATE LAW AND ITS COMPONENTS

The centerpiece of any kinetic study is the **rate law** (or **rate equation**) for the reaction in question. The rate law expresses the rate as a function of reactant concentrations, product concentrations, and temperature. Any hypothesis we make about how the reaction occurs on the molecular level must conform to the rate law because it is based on experimental fact.

In this discussion, we generally consider reactions for which the products do not appear in the rate law. In these cases, *the reaction rate depends only on reactant concentrations and temperature.* First, we look at the effect of concentration on rate for reactions occurring at a fixed temperature. For a general reaction,

$$aA + bB + \cdots \longrightarrow cC + dD + \cdots$$

the rate law has the form

$$\text{Rate} = k[A]^m[B]^n \cdots \tag{16.3}$$

Aside from the concentration terms, [A] and [B], the other parameters in Equation 16.3 require some definition. The proportionality constant k, called the **rate constant,** is specific for a given reaction at a given temperature; it does *not* change as the reaction proceeds. (As you'll see in Section 16.5, k *does* change with temperature and therefore determines how temperature affects the rate.) The exponents m and n, called the **reaction orders,** define how the rate is affected by reactant concentration. Thus, if the rate doubles when [A] doubles, the rate depends on [A] raised to the first power, $[A]^1$, so $m = 1$. Similarly, if the rate quadruples when [B] doubles, the rate depends on [B] raised to the second power, $[B]^2$, so $n = 2$. In another reaction, the rate may not change at all when [A] doubles; in that case, the rate does *not* depend on [A], or, to put it another way, the rate depends on [A] raised to the zero power, $[A]^0$, so $m = 0$. Keep in mind that the coefficients a and b in the general balanced equation are *not* necessarily related in any way to these reaction orders m and n.

A key point to remember is that *the components of the rate law—rate, reaction orders, and rate constant—must be found by experiment;* they cannot be deduced from the reaction stoichiometry. Chemists take an experimental approach to finding these components by

1. Using concentration measurements to find the *initial rate*
2. Using initial rates from several experiments to find the *reaction orders*
3. Using these values to calculate the *rate constant*

Many experimental techniques have been developed to accomplish the first of these steps, the measurement of concentrations in order to find initial rates; here are three common approaches. For reactions that involve a colored substance, *spectroscopic methods* can be used. For example, in the oxidation of nitrogen monoxide, only the product, nitrogen dioxide, is colored:

$$2NO(g) + O_2(g) \longrightarrow 2NO_2(g;\ \text{brown})$$

As time proceeds, the brown color of the reaction mixture deepens.

For reactions that involve a change in number of moles of gas, the *change in pressure* can be monitored. Note that the above reaction could also be studied this way. Because 3 mol of gas becomes 2 mol of gas, the pressure in the reaction container decreases with time.

A third technique monitors a *change in conductivity*. In the reaction between an organic halide (2-bromo-2-methylpropane) and water,

$$(CH_3)_3C\text{—}Br(l) + H_2O(l) \longrightarrow (CH_3)_3C\text{—}OH(l) + H^+(aq) + Br^-(aq)$$

the HBr that forms is a strong acid and dissociates completely into ions; thus, the conductivity of the reaction mixture increases as time proceeds.

Once chemists have used initial rates to find reaction orders and have calculated the rate constant, they know the rate law and can then use it to predict the rate for any initial reactant concentrations. Let's proceed with finding the reaction orders and the rate constant.

Reaction Order Terminology

Before we see how reaction orders are determined from initial rate data, let's discuss the meaning of reaction order and some important terminology. We speak of a reaction as having an *individual* order "with respect to" or "in" each reactant as well as an *overall* order, which is simply the sum of the individual orders.

In the simplest case, a reaction with a single reactant A, the reaction is *first order* overall if the rate is directly proportional to [A]:

$$\text{Rate} = k[A]$$

It is *second order* overall if the rate is directly proportional to the square of [A]:

$$\text{Rate} = k[A]^2$$

And it is *zero order* overall if the rate is *not* dependent on [A] at all, a common situation in metal-catalyzed and biochemical processes, as you'll see later:

$$\text{Rate} = k[A]^0 = k(1) = k$$

Here are some real examples. For the reaction between nitrogen monoxide and ozone,

$$NO(g) + O_3(g) \longrightarrow NO_2(g) + O_2(g)$$

the rate law has been experimentally determined to be

$$\text{Rate} = k[NO][O_3]$$

This reaction is first order with respect to NO (or first order in NO), which means that the rate depends on NO concentration raised to the first power, that is, $[NO]^1$ (an exponent of 1 is generally omitted). It is also first order with respect to O_3, or $[O_3]^1$. This reaction is second order overall ($1 + 1 = 2$).

Now consider a different gas-phase reaction:

$$2NO(g) + 2H_2(g) \longrightarrow N_2(g) + 2H_2O(g)$$

The rate law for this reaction has been determined to be

$$\text{Rate} = k[NO]^2[H_2]$$

The reaction is second order in NO and first order in H_2, so it is third order overall.

Finally, for the reaction of 2-bromo-2-methylpropane and water that we considered earlier, the rate law has been found to be

$$\text{Rate} = k[(CH_3)_3CBr]$$

This reaction is first order in 2-bromo-2-methylpropane. Note that the concentration of H_2O does not even appear in the rate law. Thus, the reaction is zero order with respect to H_2O ($[H_2O]^0$). This means that the rate does not depend on the concentration of H_2O. We can also write the rate law as

$$\text{Rate} = k[(CH_3)_3CBr][H_2O]^0$$

Overall, this is a first-order reaction.

These examples demonstrate a major point: *reaction orders **cannot** be deduced from the balanced equation.* For the reaction between NO and H_2 and for the hydrolysis of 2-bromo-2-methylpropane, the reaction orders in the rate laws do *not* correspond to the coefficients of the balanced equations. Reaction orders *must* be determined from rate data.

Reaction orders are usually positive integers or zero, but they can also be fractional or negative. For the reaction

$$CHCl_3(g) + Cl_2(g) \longrightarrow CCl_4(g) + HCl(g)$$

a fractional order appears in the rate law:

$$Rate = k[CHCl_3][Cl_2]^{1/2}$$

This reaction order means that the rate depends on the square root of the Cl_2 concentration. For example, if the initial Cl_2 concentration is increased by a factor of 4, while the initial $CHCl_3$ concentration is kept the same, the rate increases by a factor of 2, the square root of the change in $[Cl_2]$. A negative exponent means that the rate *decreases* when the concentration of that component increases. Negative orders are often seen for reactions whose rate laws include products. For example, for the atmospheric reaction

$$2O_3(g) \rightleftharpoons 3O_2(g)$$

the rate law has been shown to be

$$Rate = k[O_3]^2[O_2]^{-1} = k\frac{[O_3]^2}{[O_2]}$$

If the O_2 concentration doubles, the reaction proceeds half as fast.

SAMPLE PROBLEM 16.2 Determining Reaction Order from Rate Laws

Problem For each of the following reactions, use the given rate law to determine the reaction order with respect to each reactant and the overall order:
(a) $2NO(g) + O_2(g) \longrightarrow 2NO_2(g)$; rate = $k[NO]^2[O_2]$
(b) $CH_3CHO(g) \longrightarrow CH_4(g) + CO(g)$; rate = $k[CH_3CHO]^{3/2}$
(c) $H_2O_2(aq) + 3I^-(aq) + 2H^+(aq) \longrightarrow I_3^-(aq) + 2H_2O(l)$; rate = $k[H_2O_2][I^-]$
Plan We inspect the exponents in the rate law, *not* the coefficients of the balanced equation, to find the individual orders, and then take their sum to find the overall reaction order.
Solution (a) The exponent of [NO] is 2, so the reaction is second order with respect to NO, first order with respect to O_2, and third order overall.

(b) The reaction is $\frac{3}{2}$ order in CH_3CHO and $\frac{3}{2}$ order overall.

(c) The reaction is first order in H_2O_2, first order in I^-, and second order overall.
The reactant H^+ does not appear in the rate law, so the reaction is zero order in H^+.
Check Be sure that each reactant has an order and that the sum of the individual orders gives the overall order.

FOLLOW-UP PROBLEM 16.2 Experiment shows that the reaction

$$5Br^-(aq) + BrO_3^-(aq) + 6H^+(aq) \longrightarrow 3Br_2(l) + 3H_2O(l)$$

obeys this rate law: rate = $k[Br^-][BrO_3^-][H^+]^2$. What are the reaction orders in each reactant and the overall reaction order?

Determining Reaction Orders Experimentally

Sample Problem 16.2 shows how to find the reaction orders from a known rate law. Now let's see how they are found from data *before* the rate law is known. Consider the reaction between oxygen and nitrogen monoxide, a key step in the formation of acid rain and in the industrial production of nitric acid:

$$O_2(g) + 2NO(g) \longrightarrow 2NO_2(g)$$

The rate law, expressed in general form, is

$$Rate = k[O_2]^m[NO]^n$$

To find the reaction orders, *we run a series of experiments, starting each one with a different set of reactant concentrations and obtaining an initial rate in each case.*

Table 16.2 **Initial Rates for a Series of Experiments with the Reaction Between O_2 and NO**

| Experiment | Initial Reactant Concentrations (mol/L) | | Initial Rate (mol/L·s) |
	O_2	NO	
1	1.10×10^{-2}	1.30×10^{-2}	3.21×10^{-3}
2	2.20×10^{-2}	1.30×10^{-2}	6.40×10^{-3}
3	1.10×10^{-2}	2.60×10^{-2}	12.8×10^{-3}
4	3.30×10^{-2}	1.30×10^{-2}	9.60×10^{-3}
5	1.10×10^{-2}	3.90×10^{-2}	28.8×10^{-3}

Table 16.2 shows experiments that change one reactant concentration while keeping the other constant. If we compare experiments 1 and 2, we see the effect of doubling $[O_2]$ on the rate. First, we take the ratio of their rate laws:

$$\frac{\text{Rate 2}}{\text{Rate 1}} = \frac{k[O_2]_2^m [NO]_2^n}{k[O_2]_1^m [NO]_1^n}$$

where $[O_2]_2$ is the O_2 concentration for experiment 2, $[NO]_1$ is the NO concentration for experiment 1, and so forth. Because k is a constant and $[NO]$ does not change between these two experiments, these quantities cancel:

$$\frac{\text{Rate 2}}{\text{Rate 1}} = \frac{[O_2]_2^m}{[O_2]_1^m} = \left(\frac{[O_2]_2}{[O_2]_1}\right)^m$$

Substituting the values from Table 16.2, we obtain

$$\frac{6.40 \times 10^{-3} \text{ mol/L·s}}{3.21 \times 10^{-3} \text{ mol/L·s}} = \left(\frac{2.20 \times 10^{-2} \text{ mol/L}}{1.10 \times 10^{-2} \text{ mol/L}}\right)^m$$

Dividing, we obtain

$$1.99 = (2.00)^m$$

Rounding to one significant figure gives

$$2 = 2^m; \quad \text{therefore,} \quad m = 1$$

The reaction is first order in O_2: when $[O_2]$ doubles, the rate doubles.

To find the order with respect to NO, we compare experiments 3 and 1, in which $[O_2]$ is held constant and $[NO]$ is doubled:

$$\frac{\text{Rate 3}}{\text{Rate 1}} = \frac{k[O_2]_3^m [NO]_3^n}{k[O_2]_1^m [NO]_1^n}$$

As before, k is constant, and in this pair of experiments $[O_2]$ does not change, so these quantities cancel:

$$\frac{\text{Rate 3}}{\text{Rate 1}} = \left(\frac{[NO]_3}{[NO]_1}\right)^n$$

The actual values give

$$\frac{12.8 \times 10^{-3} \text{ mol/L·s}}{3.21 \times 10^{-3} \text{ mol/L·s}} = \left(\frac{2.60 \times 10^{-2} \text{ mol/L}}{1.30 \times 10^{-2} \text{ mol/L}}\right)^n$$

Dividing, we obtain

$$3.99 = (2.00)^n$$

Rounding gives

$$4 = 2^n; \quad \text{therefore,} \quad n = 2$$

The reaction is second order in NO: when $[NO]$ doubles, the rate quadruples. Thus, the rate law is

$$\text{Rate} = k[O_2][NO]^2$$

You may want to use experiment 1 in combination with experiments 4 and 5 to check this result.

SAMPLE PROBLEM 16.3 Determining Reaction Orders from Initial Rate Data

Problem Many gaseous reactions occur in car engines and exhaust systems. One of these is

$$NO_2(g) + CO(g) \longrightarrow NO(g) + CO_2(g) \qquad \text{rate} = k[NO_2]^m[CO]^n$$

Use the following data to determine the individual and overall reaction orders:

Experiment	Initial Rate (mol/L·s)	Initial [NO₂] (mol/L)	Initial [CO] (mol/L)
1	0.0050	0.10	0.10
2	0.080	0.40	0.10
3	0.0050	0.10	0.20

Plan We need to solve the general rate law for the reaction orders m and n. To solve for each exponent, we proceed as in the text, taking the ratio of the rate laws for two experiments in which only the reactant in question changes.

Solution Calculating m in $[NO_2]^m$: We take the ratio of the rate laws for experiments 1 and 2, in which $[NO_2]$ varies but $[CO]$ is constant:

$$\frac{\text{Rate 2}}{\text{Rate 1}} = \frac{k[NO_2]_2^m\,[CO]_2^n}{k[NO_2]_1^m\,[CO]_1^n} = \left(\frac{[NO_2]_2}{[NO_2]_1}\right)^m \quad \text{or} \quad \frac{0.080 \text{ mol/L·s}}{0.0050 \text{ mol/L·s}} = \left(\frac{0.40 \text{ mol/L}}{0.10 \text{ mol/L}}\right)^m$$

This gives $16 = 4.0^m$, so $m = 2.0$. The reaction is second order in NO_2.

Calculating n in $[CO]^n$: We take the ratio of the rate laws for experiments 1 and 3, in which $[CO]$ varies but $[NO_2]$ is constant:

$$\frac{\text{Rate 3}}{\text{Rate 1}} = \frac{k[NO_2]_3^2[CO]_3^n}{k[NO_2]_1^2[CO]_1^n} = \left(\frac{[CO]_3}{[CO]_1}\right)^n \quad \text{or} \quad \frac{0.0050 \text{ mol/L·s}}{0.0050 \text{ mol/L·s}} = \left(\frac{0.20 \text{ mol/L}}{0.10 \text{ mol/L}}\right)^n$$

We have $1.0 = (2.0)^n$, so $n = 0$. The rate does not change when $[CO]$ varies, so the reaction is zero order in CO.

Therefore, the rate law is

$$\text{Rate} = k[NO_2]^2[CO]^0 = k[NO_2]^2(1) = k[NO_2]^2$$

The reaction is second order overall.

Check A good check is to reason through the orders. If $m = 1$, quadrupling $[NO_2]$ would quadruple the rate; but the rate *more* than quadruples, so $m > 1$. If $m = 2$, quadrupling $[NO_2]$ would increase the rate by a factor of 16 (4^2). The ratio of rates is $0.080/0.005 = 16$, so $m = 2$. In contrast, increasing $[CO]$ has no effect on the rate, which can happen only if $[CO]^n = 1$, so $n = 0$.

FOLLOW-UP PROBLEM 16.3 Find the rate law and the overall reaction order for the reaction $H_2 + I_2 \longrightarrow 2HI$ from the following data at 450°C:

Experiment	Initial Rate (mol/L·s)	Initial [H₂] (mol/L)	Initial [I₂] (mol/L)
1	1.9×10^{-23}	0.0113	0.0011
2	1.1×10^{-22}	0.0220	0.0033
3	9.3×10^{-23}	0.0550	0.0011
4	1.9×10^{-22}	0.0220	0.0056

Determining the Rate Constant

With the rate, reactant concentrations, and reaction orders known, the sole remaining unknown in the rate law is the rate constant, k. The rate constant is specific for a particular reaction *at a particular temperature*. The experiments with the reaction of O_2 and NO were run at the same temperature, so we can use data from any to solve for k. From experiment 1 in Table 16.2, for instance, we obtain

$$k = \frac{\text{rate 1}}{[O_2]_1[NO]_1^2} = \frac{3.21 \times 10^{-3} \text{ mol/L·s}}{(1.10 \times 10^{-2} \text{ mol/L})(1.30 \times 10^{-2} \text{ mol/L})^2}$$

$$= \frac{3.21 \times 10^{-3} \text{ mol/L·s}}{1.86 \times 10^{-6} \text{ mol}^3/\text{L}^3} = 1.73 \times 10^3 \text{ L}^2/\text{mol}^2\cdot\text{s}$$

Table 16.3 Units of the Rate Constant k for Several Overall Reaction Orders	
Overall Reaction Order	Units of k (t in seconds)
0	$mol/L \cdot s$ (or $mol \, L^{-1} \, s^{-1}$)
1	$1/s$ (or s^{-1})
2	$L/mol \cdot s$ (or $L \, mol^{-1} \, s^{-1}$)
3	$L^2/mol^2 \cdot s$ (or $L^2 \, mol^{-2} \, s^{-1}$)

General formula:

$$\text{Units of } k = \frac{\left(\dfrac{L}{mol}\right)^{order-1}}{\text{unit of } t}$$

Always check that the values of k for a series are constant within experimental error. To three significant figures, the average value of k for the five experiments in Table 16.2 is $1.72 \times 10^3 \, L^2/mol^2 \cdot s$.

Note the units for the rate constant. With concentrations in mol/L and the reaction rate in units of mol/L·time, the units for k depend on the order of the reaction and, of course, the time unit. The units for k in our example, $L^2/mol^2 \cdot s$, are required to give a rate with units of mol/L·s:

$$\frac{mol}{L \cdot s} = \frac{L^2}{mol^2 \cdot s} \times \frac{mol}{L} \times \left(\frac{mol}{L}\right)^2$$

The rate constant will *always* have these units for an overall third-order reaction with the time unit in seconds. Table 16.3 shows the units of k for some common overall reaction orders, but you can always determine the units mathematically.

SECTION SUMMARY

An experimentally determined rate law shows how the rate of a reaction depends on concentration. If we consider only initial rates, the rate law often takes this form: rate = $k[A]^m[B]^n \cdots$. With an accurate method for obtaining initial rates, reaction orders are determined experimentally by comparing rates for different initial concentrations, that is, by performing several experiments and varying the concentration of one reactant at a time to see its effect on the rate. With rate, concentrations, and reaction orders known, the rate constant is the only remaining unknown in the rate law and can be calculated.

16.4 INTEGRATED RATE LAWS: CONCENTRATION CHANGES OVER TIME

Notice that the rate laws we've developed so far do not include time as a variable. They tell us the rate or concentration at a given instant, allowing us to answer a critical question, "How fast is the reaction proceeding at the moment when y moles per liter of A are reacting with z moles per liter of B?" However, by employing different forms of the rate laws, called **integrated rate laws,** we can consider the time factor and answer other questions, such as "How long will it take for x moles per liter of A to be used up?" and "What is the concentration of A after y minutes of reaction?"

Integrated Rate Laws for First-Order, Second-Order, and Zero-Order Reactions

Consider a simple first-order reaction, A $\longrightarrow$ B. (Because first- and second-order reactions are more common, we'll discuss them before zero-order reactions.) As we discussed previously, the rate can be expressed as the change in the concentration of A divided by the change in time:

$$\text{Rate} = -\frac{\Delta[A]}{\Delta t}$$

It can also be expressed in terms of the rate law:

$$\text{Rate} = k[A]$$

Setting these different expressions equal to each other gives

$$-\frac{\Delta[A]}{\Delta t} = k[A]$$

Using calculus, this expression is integrated over time to obtain the integrated rate law for a first-order reaction:

$$\ln \frac{[A]_0}{[A]_t} = kt \quad \text{(first-order reaction; rate} = k[A]) \tag{16.4}$$

where ln is the natural logarithm, $[A]_0$ is the concentration of A at $t = 0$, and $[A]_t$ is the concentration of A at any time t during an experiment. In mathematical terms, $\ln \dfrac{a}{b} = \ln a - \ln b$, so we have

$$\ln [A]_0 - \ln [A]_t = kt$$

For a general second-order reaction, the expression including time is quite complex, so let's consider the case in which the rate law contains only one reactant. Setting the rate expressions equal to each other gives

$$\text{Rate} = -\frac{\Delta[A]}{\Delta t} = k[A]^2$$

Integrating over time gives the integrated rate law for a second-order reaction involving one reactant:

$$\frac{1}{[A]_t} - \frac{1}{[A]_0} = kt \quad \text{(second-order reaction; rate} = k[A]^2) \qquad \textbf{(16.5)}$$

For a zero-order reaction, we have

$$\text{Rate} = -\frac{\Delta[A]}{\Delta t} = k[A]^0$$

Integrating over time gives the integrated rate law for a zero-order reaction:

$$[A]_t - [A]_0 = -kt \quad \text{(zero-order reaction; rate} = k[A]^0 = k) \qquad \textbf{(16.6)}$$

Sample Problem 16.4 shows one way integrated rate laws are applied.

SAMPLE PROBLEM 16.4   Determining the Reactant Concentration at a Given Time

Problem At 1000°C, cyclobutane (C_4H_8) decomposes in a first-order reaction, with the very high rate constant of 87 s^{-1}, to two molecules of ethylene (C_2H_4).
(a) If the initial C_4H_8 concentration is 2.00 M, what is the concentration after 0.010 s?
(b) What fraction of C_4H_8 has decomposed in this time?
Plan (a) We must find the concentration of cyclobutane at time t, $[C_4H_8]_t$. The problem tells us this is a first-order reaction, so we use the integrated first-order rate law:

$$\ln \frac{[C_4H_8]_0}{[C_4H_8]_t} = kt$$

We know k (87 s^{-1}), t (0.010 s), and $[C_4H_8]_0$ (2.00 M), so we can solve for $[C_4H_8]_t$.
(b) The fraction decomposed is the concentration that has decomposed divided by the initial concentration:

$$\text{Fraction decomposed} = \frac{[C_4H_8]_0 - [C_4H_8]_t}{[C_4H_8]_0}$$

Solution (a) Substituting the data into the integrated rate law:

$$\ln \frac{2.00 \text{ mol/L}}{[C_4H_8]_t} = (87 \text{ s}^{-1})(0.010 \text{ s}) = 0.87$$

Taking the antilog of both sides:

$$\frac{2.00 \text{ mol/L}}{[C_4H_8]_t} = e^{0.87} = 2.4$$

Solving for $[C_4H_8]_t$:

$$[C_4H_8]_t = \frac{2.00 \text{ mol/L}}{2.4} = 0.83 \text{ mol/L}$$

(b) Finding the fraction that has decomposed after 0.010 s:

$$\frac{[C_4H_8]_0 - [C_4H_8]_t}{[C_4H_8]_0} = \frac{2.00 \text{ mol/L} - 0.83 \text{ mol/L}}{2.00 \text{ mol/L}} = 0.58$$

Check The concentration remaining after 0.010 s (0.83 mol/L) is less than the starting concentration (2.00 mol/L), which makes sense. Raising e to an exponent slightly less than 1 should give a number (2.4) slightly less than the value of e (2.718). Moreover, the final result makes sense: a high rate constant indicates a fast reaction, so it's not surprising that so much decomposes in such a short time.

Comment Integrated rate laws are also used to solve for the time it takes to reach a certain reactant concentration, as in the follow-up problem.

FOLLOW-UP PROBLEM 16.4 At 25°C, hydrogen iodide breaks down very slowly to hydrogen and iodine: rate $= k[\mathrm{HI}]^2$. The rate constant at 25°C is 2.4×10^{-21} L/mol·s. If 0.0100 mol of HI(g) is placed in a 1.0-L container, how long will it take for the concentration of HI to reach 0.00900 mol/L (10.0% reacted)?

Determining the Reaction Order from the Integrated Rate Law

Suppose you don't know the rate law for a reaction and don't have the initial rate data needed to determine the reaction orders (which we did have in Sample Problem 16.3). Another method for finding reaction orders is a graphical technique that uses concentration and time data directly.

An integrated rate law can be rearranged into the form of an equation for a straight line, $y = mx + b$, where m is the slope and b is the y-axis intercept. For a first-order reaction, we have

$$\ln [A]_0 - \ln [A]_t = kt$$

Rearranging and changing signs gives

$$\ln [A]_t = -kt + \ln [A]_0$$
$$y \quad = mx + \quad b$$

Therefore, a plot of $\ln [A]_t$ vs. time gives a straight line with slope $= -k$ and y intercept $= \ln [A]_0$ (Figure 16.5A).

For a simple second-order reaction, we have

$$\frac{1}{[A]_t} - \frac{1}{[A]_0} = kt$$

Rearranging gives

$$\frac{1}{[A]_t} = kt + \frac{1}{[A]_0}$$
$$y \quad = mx + \quad b$$

In this case, a plot of $1/[A]_t$ vs. time gives a straight line with slope $= k$ and y intercept $= 1/[A]_0$ (Figure 16.5B).

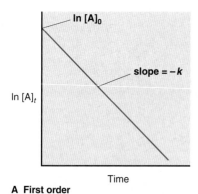

A First order

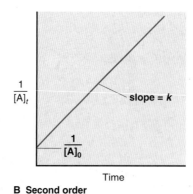

B Second order

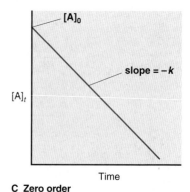

C Zero order

Figure 16.5 Integrated rate laws and reaction orders. A, Plot of $\ln [A]_t$ vs. time gives a straight line for a reaction that is first order in A. **B,** Plot of $1/[A]_t$ vs. time gives a straight line for a reaction that is second order in A. **C,** Plot of $[A]_t$ vs. time gives a straight line for a reaction that is zero order in A.

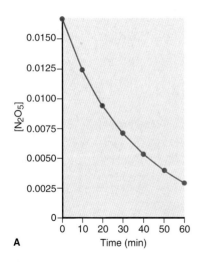

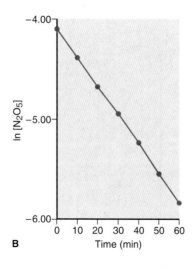

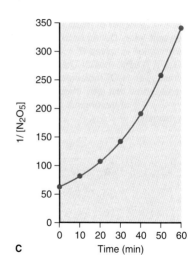

Time (min)	$[N_2O_5]$	$\ln [N_2O_5]$	$1/[N_2O_5]$
0	0.0165	−4.104	60.6
10	0.0124	−4.390	80.6
20	0.0093	−4.68	1.1×10^2
30	0.0071	−4.95	1.4×10^2
40	0.0053	−5.24	1.9×10^2
50	0.0039	−5.55	2.6×10^2
60	0.0029	−5.84	3.4×10^2

Figure 16.6 Graphical determination of the reaction order for the decomposition of N_2O_5. A table of time and concentration data for determining reaction order appears below the graphs. **A,** A plot of $[N_2O_5]$ vs. time is curved, indicating that the reaction is *not* zero order in N_2O_5. **B,** A plot of $\ln [N_2O_5]$ vs. time gives a straight line, indicating that the reaction *is* first order in N_2O_5. **C,** A plot of $1/[N_2O_5]$ vs. time is curved, indicating that the reaction is *not* second order in N_2O_5. Plots A and C support the conclusion from plot B.

For a zero-order reaction, we have

$$[A]_t - [A]_0 = -kt$$

Rearranging gives

$$[A]_t = -kt + [A]_0$$
$$y = mx + b$$

Thus, a plot of $[A]_t$ vs. time gives a straight line with slope $= -k$ and y intercept $= [A]_0$ (Figure 16.5C).

Therefore, some trial-and-error graphical plotting is required to find the reaction order from the concentration and time data:

- If you obtain a straight line when you plot $\ln$ [reactant] vs. time, the reaction is *first order* with respect to that reactant.
- If you obtain a straight line when you plot 1/[reactant] vs. time, the reaction is *second order* with respect to that reactant.
- If you obtain a straight line when you plot [reactant] vs. time, the reaction is *zero order* with respect to that reactant.

Figure 16.6 shows how this approach is used to determine the order for the decomposition of N_2O_5. Because the plot of $\ln [N_2O_5]$ *is* linear and the plot of $1/[N_2O_5]$ *is not,* the decomposition of N_2O_5 must be first order in N_2O_5.

Reaction Half-Life

The **half-life ($t_{1/2}$)** of a reaction is the time required for the reactant concentration to reach half its initial value. A half-life is expressed in time units appropriate for a given reaction and is characteristic of that reaction at a given temperature.

At fixed conditions, *the half-life of a first-order reaction is a constant, independent of reactant concentration.* For example, the half-life for the first-order decomposition of N_2O_5 at 45°C is 24.0 min. The meaning of this value is that if we start with, say, 0.0600 mol/L of N_2O_5 at 45°C, after 24 min (one half-life),

Figure 16.7 A plot of [N₂O₅] vs. time for three half-lives. During each half-life, the concentration is halved ($T = 45°C$ and $[N_2O_5]_0 = 0.0600$ mol/L). The blow-up volumes, with N_2O_5 molecules as colored spheres, show that after three half-lives, $\frac{1}{2} \times \frac{1}{2} \times \frac{1}{2} = \frac{1}{8}$ of the original concentration remains.

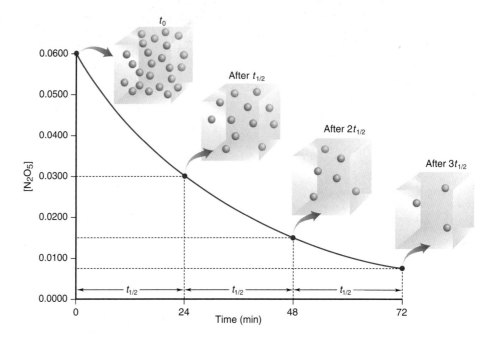

0.0300 mol/L has been consumed and 0.0300 mol/L remains; after 48 min (two half-lives), 0.0150 mol/L remains; after 72 min (three half-lives), 0.0075 mol/L remains, and so forth (Figure 16.7).

We can see from the integrated rate law why the half-life of a first-order reaction is independent of concentration:

$$\ln \frac{[A]_0}{[A]_t} = kt$$

After one half-life, $t = t_{1/2}$, and $[A]_t = \frac{1}{2}[A]_0$. Substituting, we obtain

$$\ln \frac{[A]_0}{\frac{1}{2}[A]_0} = kt_{1/2} \quad \text{or} \quad \ln 2 = kt_{1/2}$$

Then, solving for $t_{1/2}$, we have

$$t_{1/2} = \frac{\ln 2}{k} = \frac{0.693}{k} \quad \text{(first-order process; rate} = k[A]) \tag{16.7}$$

As you can see, *the time to reach one-half the starting concentration in a first-order reaction does not depend on what that starting concentration is.*

Radioactive decay of an unstable nucleus is another example of a first-order process. For example, the half-life for the decay of uranium-235 is 7.1×10^8 yr. After 710 million years, a 1-kg sample of uranium-235 will contain 0.5 kg of uranium-235, and a 1-mg sample of uranium-235 will contain 0.5 mg. (We discuss the kinetics of radioactive decay thoroughly in Chapter 23.) Whether we consider a molecule or a radioactive nucleus, the *decomposition of each particle in a first-order process is independent of the number of other particles present.*

SAMPLE PROBLEM 16.5 Determining the Half-Life of a First-Order Reaction

Problem Cyclopropane is the smallest cyclic hydrocarbon. Because its 60° bond angles reduce orbital overlap, its bonds are weak. As a result, it is thermally unstable and rearranges to propene at 1000°C via the following first-order reaction:

$$\underset{\displaystyle \overset{CH_2}{\diagup \ \diagdown}}{H_2C - CH_2}(g) \xrightarrow{\Delta} CH_3 - CH = CH_2(g)$$

The rate constant is 9.2 s^{-1}. (a) What is the half-life of the reaction? (b) How long does it take for the concentration of cyclopropane to reach one-quarter of the initial value?

Plan (a) The cyclopropane rearrangement is first order, so to find $t_{1/2}$ we use Equation 16.7 and substitute for k (9.2 s^{-1}). (b) Each half-life decreases the concentration to one-half of its initial value, so two half-lives decrease it to one-quarter.

Solution (a) Solving for $t_{1/2}$:

$$t_{1/2} = \frac{\ln 2}{k} = \frac{0.693}{9.2 \text{ s}^{-1}} = 0.075 \text{ s}$$

It takes 0.075 s for half the cyclopropane to form propene at this temperature.

(b) Finding the time to reach one-quarter of the initial concentration:

$$\text{Time} = 2(t_{1/2}) = 2(0.075 \text{ s}) = 0.15 \text{ s}$$

Check For (a), rounding gives 0.7/9 s^{-1} = 0.08 s, so the answer seems correct.

FOLLOW-UP PROBLEM 16.5 Iodine-123 is used to study thyroid gland function. This radioactive isotope breaks down in a first-order process with a half-life of 13.1 h. What is the rate constant for the process?

In contrast to the half-life of a first-order reaction, the half-life of a second-order reaction *does* depend on reactant concentration:

$$t_{1/2} = \frac{1}{k[A]_0} \quad \text{(second-order process; rate} = k[A]^2)$$

Note that here *the half-life is **inversely** proportional to the initial reactant concentration.* This relationship means that a second-order reaction with a high initial reactant concentration has a shorter half-life, and one with a low initial reactant concentration has a longer half-life. Therefore, *as a second-order reaction proceeds, the half-life increases.*

In contrast to the half-life of a second-order reaction, *the half-life of a zero-order reaction is **directly** proportional to the initial reactant concentration:*

$$t_{1/2} = \frac{[A]_0}{2k} \quad \text{(zero-order process; rate} = k)$$

Thus, if a zero-order reaction begins with a high reactant concentration, it has a longer half-life than if it begins with a low reactant concentration. Table 16.4 summarizes the essential features of zero-, first-, and second-order reactions.

Table 16.4 An Overview of Zero-Order, First-Order, and Simple Second-Order Reactions

	Zero Order	First Order	Second Order
Rate law	rate $= k$	rate $= k[A]$	rate $= k[A]^2$
Units for k	mol/L·s	1/s	L/mol·s
Integrated rate law in straight-line form	$[A]_t =$ $-kt + [A]_0$	$\ln [A]_t =$ $-kt + \ln [A]_0$	$1/[A]_t =$ $kt + 1/[A]_0$
Plot for straight line	$[A]_t$ vs. time	$\ln [A]_t$ vs. time	$1/[A]_t$ vs. time
Slope, y intercept	$-k$, $[A]_0$	$-k$, $\ln [A]_0$	k, $1/[A]_0$
Half-life	$[A]_0/2k$	$(\ln 2)/k$	$1/k[A]_0$

SECTION SUMMARY

Integrated rate laws are used to find either the time needed to reach a certain concentration of reactant or the concentration present after a given time. Rearrangements of the integrated rate laws allow us to determine reaction orders and rate constants graphically. The half-life is the time needed for the reaction to consume half the reactant; for first-order reactions, it is independent of concentration.

Figure 16.8 Dependence of the rate constant on temperature. A, In the hydrolysis of the ester ethyl acetate, $CH_3COOCH_2CH_3 + H_2O \rightleftharpoons CH_3COOH + CH_3CH_2OH$, when reactant concentrations are held constant and temperature increases, the rate and rate constant increase. Note the near doubling of k with each rise of 10 K (10°C). **B,** A plot of rate constant vs. temperature for this reaction shows an exponentially increasing curve.

Exp't	[Ester]	[H₂O]	T (K)	Rate (mol/L·s)	k (L/mol·s)
1	0.100	0.200	288	1.04×10^{-3}	0.0521
2	0.100	0.200	298	2.02×10^{-3}	0.101
3	0.100	0.200	308	3.68×10^{-3}	0.184
4	0.100	0.200	318	6.64×10^{-3}	0.332

A

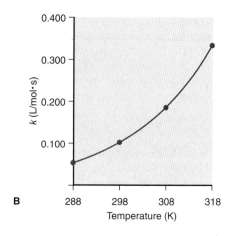

B

16.5 THE EFFECT OF TEMPERATURE ON REACTION RATE

Temperature often has a major effect on reaction rate. As Figure 16.8A shows for a common organic reaction—hydrolysis, or reaction with water, of an ester—when reactant concentrations are held constant, the rate nearly doubles with each rise in temperature of 10 K (or 10°C). In fact, for many reactions near room temperature, an increase of 10°C causes a doubling or tripling of the rate.

How does the rate law express this effect of temperature? If we collect concentration and time data for the same reaction run at *different* temperatures (T), and then solve each rate expression for k, we find that k increases as T increases. In other words, *temperature affects the rate by affecting the rate constant.* A plot of k vs. T gives a curve that increases exponentially (Figure 16.8B).

These results are consistent with studies made in 1889 by the Swedish chemist Svante Arrhenius, who discovered a key relationship between T and k. In its modern form, the **Arrhenius equation** is

$$k = Ae^{-E_a/RT} \tag{16.8}$$

where k is the rate constant, e is the base of natural logarithms, T is the absolute temperature, and R is the universal gas constant. We'll discuss the meaning of the constant A, which is related to the orientation of the colliding molecules, in the next section. The E_a term is the **activation energy** of the reaction, which Arrhenius considered the *minimum energy* the molecules must have to react; we'll explore its meaning in the next section as well. This negative exponential relationship between T and k means that *as T increases, the negative exponent becomes smaller, so the value of k becomes larger, which means that the rate increases:*

Higher $T \Longrightarrow$ larger $k \Longrightarrow$ increased rate

We can calculate E_a from the Arrhenius equation by taking the natural logarithm of both sides and recasting the equation into one for a straight line:

$$\ln k = \ln A - \frac{E_a}{R}\left(\frac{1}{T}\right)$$
$$y = b + mx$$

A plot of $\ln k$ vs. $1/T$ gives a straight line whose slope is $-E_a/R$ and whose y intercept is $\ln A$ (Figure 16.9). Therefore, with the constant R known, we can determine E_a graphically from a series of k values at different temperatures.

Because the relationship between $\ln k$ and $1/T$ is linear, we can use a simpler method to find E_a if we know the rate constants at two temperatures, T_2 and T_1:

$$\ln k_2 = \ln A - \frac{E_a}{R}\left(\frac{1}{T_2}\right) \qquad \ln k_1 = \ln A - \frac{E_a}{R}\left(\frac{1}{T_1}\right)$$

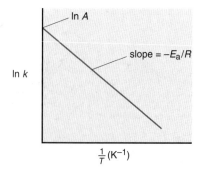

Figure 16.9 Graphical determination of the activation energy. A plot of $\ln k$ vs. $1/T$ gives a straight line with slope $= -E_a/R$.

When we subtract $\ln k_1$ from $\ln k_2$, the term $\ln A$ drops out and the other terms can be rearranged to give

$$\ln \frac{k_2}{k_1} = -\frac{E_a}{R}\left(\frac{1}{T_2} - \frac{1}{T_1}\right)$$ (16.9)

From this, we can solve for E_a.

SAMPLE PROBLEM 16.6   Determining the Energy of Activation

Problem The decomposition of hydrogen iodide,

$$2HI(g) \longrightarrow H_2(g) + I_2(g)$$

has rate constants of 9.51×10^{-9} L/mol·s at 500. K and 1.10×10^{-5} L/mol·s at 600. K. Find E_a.

Plan We are given the rate constants, k_1 and k_2, at two temperatures, T_1 and T_2, so we substitute into Equation 16.9 and solve for E_a.

Solution Rearranging Equation 16.9 to solve for E_a:

$$\ln \frac{k_2}{k_1} = -\frac{E_a}{R}\left(\frac{1}{T_2} - \frac{1}{T_1}\right)$$

$$E_a = -R\left(\ln \frac{k_2}{k_1}\right)\left(\frac{1}{T_2} - \frac{1}{T_1}\right)^{-1}$$

$$= -(8.314 \text{ J/mol·K})\left(\ln \frac{1.10 \times 10^{-5} \text{ L/mol·s}}{9.51 \times 10^{-9} \text{ L/mol·s}}\right)\left(\frac{1}{600. \text{ K}} - \frac{1}{500. \text{ K}}\right)^{-1}$$

$$= 1.76 \times 10^5 \text{ J/mol} = 1.76 \times 10^2 \text{ kJ/mol}$$

Comment Be sure to retain the same number of significant figures in $1/T$ as you have in T, or a significant error could be introduced. Round to the correct number of significant figures only at the final answer. On most pocket calculators, the expression $(1/T_2 - 1/T_1)$ is entered as follows: $(T_2)(1/x) - (T_1)(1/x) =$

FOLLOW-UP PROBLEM 16.6 The reaction $2NOCl(g) \longrightarrow 2NO(g) + Cl_2(g)$ has an E_a of 1.00×10^2 kJ/mol and a rate constant of 0.286 L/mol·s at 500. K. What is the rate constant at 490. K?

In this section and the previous two, we discussed a series of experimental and mathematical methods for the study of reaction kinetics. Figure 16.10 is a useful summary of this information. Note that the integrated rate law provides an alternative method for obtaining reaction orders and the rate constant.

SECTION SUMMARY

As the Arrhenius equation shows, rate increases with temperature because a temperature rise increases the rate constant. The activation energy, E_a, the minimum energy needed for a reaction to occur, can be determined graphically from k values at different T values.

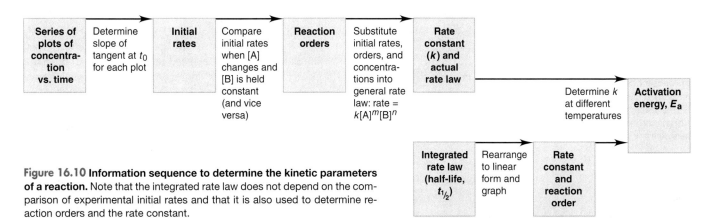

Figure 16.10 Information sequence to determine the kinetic parameters of a reaction. Note that the integrated rate law does not depend on the comparison of experimental initial rates and that it is also used to determine reaction orders and the rate constant.

16.6 EXPLAINING THE EFFECTS OF CONCENTRATION AND TEMPERATURE

The Arrhenius equation was developed empirically from the observations of many reactions. The two major models that explain the observed effects of concentration and temperature on reaction rate highlight different aspects of the reaction process but are completely compatible. *Collision theory* views the reaction rate as the result of particles colliding with a certain frequency and minimum energy. *Transition state theory* offers a close-up view of how the energy of a collision converts reactant to product.

Collision Theory: Basis of the Rate Law

The basic tenet of **collision theory** is that reactant particles—atoms, molecules, and ions—must collide with each other to react. Therefore, the number of collisions per unit time provides an upper limit on how fast a reaction can take place. The model restricts itself to simple one-step reactions in which two particles collide and form products: A + B $\longrightarrow$ products. With its emphasis on collisions between three-dimensional particles, this model explains why reactant concentrations are multiplied together in the rate law, how temperature affects the rate, and what influence molecular structure has on rate.

Why Concentrations Are Multiplied in the Rate Law If particles must collide to react, the laws of probability tell us why the rate depends on the *product* of the reactant concentrations, not their sum. Imagine that you have only two particles of A and two of B confined in a reaction vessel. Figure 16.11 shows that four A-B collisions are possible. If you add another particle of A, there can be six A-B collisions (3 × 2), not just five (3 + 2); add another particle of B, and there can be nine A-B collisions (3 × 3), not just six (3 + 3). Thus, collision theory is consistent with the observation that concentrations are *multiplied* in the rate law.

How Temperature Affects Rate: The Importance of Activation Energy Increasing the temperature of a reaction increases the average speed of particles and therefore their collision frequency. But collision frequency cannot be the only factor affecting rate. In fact, *in the vast majority of collisions, the molecules rebound without reacting.*

Arrhenius proposed that every reaction has an *energy threshold* that the colliding molecules must exceed in order to react. (An analogy might be an athlete who must exceed the height of the bar to accomplish a high jump.) This minimum collision energy is the *activation energy* (E_a), the energy required to activate the molecules into a state from which reactant bonds can change into product bonds. Recall that at any given temperature, molecules have a range of kinetic energies; thus, their collisions have a range of energies as well. According to collision theory, *only those collisions with enough energy to exceed E_a can lead to reaction.*

We noted earlier that many reactions near room temperature approximately double or triple their rates with a 10°C rise in temperature. Is the rate increase due to a higher number of collisions? Actually, this has only a minor effect. Calculations show that a 10°C rise increases the average molecular speed by only 2%. If an increase in speed is the only effect of temperature and if the speed of each colliding molecule increases by 2%, we should observe at most a 4% increase in rate. Far more important is that *the temperature rise enlarges the fraction of collisions with enough energy to exceed the activation energy.* This key point is shown in Figure 16.12 (next page).

At a given temperature, the fraction f of molecular collisions with energy greater than or equal to the activation energy E_a is given by

$$f = e^{-E_a/RT}$$

where e is the base of natural logarithms, T is the absolute temperature, and R is the universal gas constant. [Notice that the right side of this equation is the central

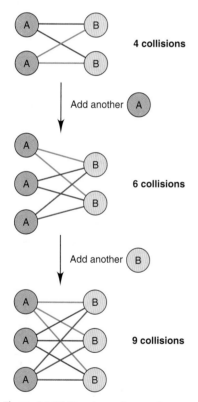

Figure 16.11 The dependence of number of possible collisions on the product of reactant concentrations. Concentrations are multiplied, not added, in the rate law because the number of possible collisions is the *product,* not the sum, of the numbers of particles present.

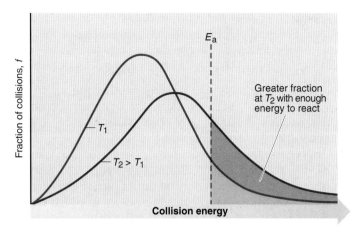

Figure 16.12 The effect of temperature on the distribution of collision energies. At the higher temperature, T_2, a larger fraction of collisions occur with enough energy to exceed E_a.

Table 16.5 The Effect of E_a and T on the Fraction (f) of Collisions with Sufficient Energy to Allow Reaction	
E_a (kJ/mol)	f (at T = 298 K)
50	1.70×10^{-9}
75	7.03×10^{-14}
100	2.90×10^{-18}
T	f (at E_a = 50 kJ/mol)
25°C (298 K)	1.70×10^{-9}
35°C (308 K)	3.29×10^{-9}
45°C (318 K)	6.12×10^{-9}

component in the Arrhenius equation (Equation 16.8).] *The magnitudes of both E_a and T affect the fraction of sufficiently energetic collisions.* In the top portion of Table 16.5, you can see the effect on this fraction of increasing E_a at a fixed temperature. Note how much the fraction shrinks with a 25-kJ/mol increase in activation energy. (As the height of the bar is raised, fewer athletes can accomplish the jump.) In the bottom portion of the table, you can see the effect on the fraction of increasing T at a fixed E_a of 50 kJ/mol, a typical value for many reactions. Note that the fraction nearly doubles for a 10°C increase. Doubling the fraction doubles the rate constant, which doubles the reaction rate.

A reversible reaction has two activation energies (Figure 16.13). The activation energy for the forward reaction, $E_{a(fwd)}$, is the energy difference between the activated state and the reactants; the activation energy for the reverse reaction, $E_{a(rev)}$, is the energy difference between the activated state and the products. The figure shows an energy-level diagram for an exothermic reaction, so the products are at a lower energy than the reactants, and $E_{a(fwd)}$ is less than $E_{a(rev)}$.

These observations are consistent with the Arrhenius equation; that is, *the smaller the E_a (or the higher the temperature), the larger the value of k, and the faster the reaction:*

$$\text{Smaller } E_a \text{ (or higher } T) \Longrightarrow \text{larger } k \Longrightarrow \text{increased rate}$$

Conversely, *the larger the E_a (or the lower the temperature), the smaller the value of k, and the slower the reaction:*

$$\text{Larger } E_a \text{ (or lower } T) \Longrightarrow \text{smaller } k \Longrightarrow \text{decreased rate}$$

How Molecular Structure Affects Rate

The enormous number of molecular collisions per second is greatly reduced when we count only those with enough energy to react. However, even this tiny fraction of the total collisions does not reveal the true number of **effective collisions,** those that actually lead to product. In addition to colliding with enough energy, *the molecules must collide so that the reacting atoms make contact.* In other words, to be effective, a collision must have enough energy *and* a particular *molecular orientation.*

In the Arrhenius equation, the effect of molecular orientation is contained in the term A:

$$k = Ae^{-E_a/RT}$$

This term is called the **frequency factor,** the product of the collision frequency Z and an *orientation probability factor, p,* which is specific for each reaction: $A = pZ$. The factor p is related to the structural complexity of the colliding

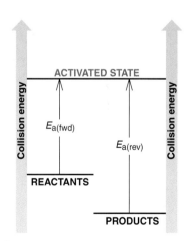

Figure 16.13 Energy-level diagram for a reaction. For molecules to react, they must collide with enough energy to reach an activated state. This minimum collision energy is the energy of activation, E_a. A reaction can occur in either direction, so the diagram shows two activation energies. Here, the forward reaction is exothermic and $E_{a(fwd)} < E_{a(rev)}$.

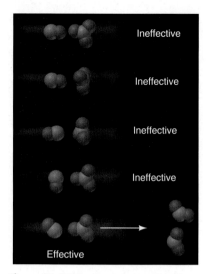

Figure 16.14 The importance of molecular orientation to an effective collision. Only one of the five orientations shown for the collision between NO and NO_3 has the correct orientation to lead to product. In the effective orientation, contact occurs between the atoms that will become bonded in the product.

particles. You can think of it as the ratio of effectively oriented collisions to all possible collisions. For example, Figure 16.14 shows a few of the possible collision orientations for the following simple gaseous reaction:

$$NO(g) + NO_3(g) \longrightarrow 2NO_2(g)$$

Of the five collisions shown, only one has an orientation in which the N of NO makes contact with an O of NO_3. Actually, the orientation probability factor (*p* value) for this reaction is 0.006: only 6 collisions in every 1000 (1 in 167) have an orientation that can lead to reaction.

Collisions between individual atoms have *p* values near 1: almost no matter how they hit, as long as the collision has enough energy, the particles react. In such cases, the rate constant depends only on the frequency and energy of the collisions. At the other extreme are biochemical reactions, in which the reactants are often two small molecules that can react only when they collide with a specific tiny region of a giant molecule—a protein or nucleic acid. The orientation probability factor for these reactions is often less than 10^{-6}: fewer than one in a million sufficiently energetic collisions leads to product. The fact that countless such biochemical reactions are occurring right now, as you read this sentence, helps make the point that the number of collisions per second is truly astounding.

Transition State Theory: Molecular Nature of the Activated Complex

Collision theory is a simple, easy-to-visualize model, but it provides no insight about why the activation energy is needed and how the activated molecules look. To understand these aspects of the process, we turn to **transition state theory,** which focuses on the high-energy species that forms through an effective collision.

Visualizing the Transition State Recall from our discussion of energy changes (Chapter 6) that the internal energy of a system is the sum of its kinetic and potential energies. When two molecules approach one another, some kinetic energy is converted to potential energy as the electron clouds repel each other. At the moment of a head-on collision, the molecules stop, and their kinetic energy is converted to the potential energy of the collision. *If this potential energy is less than the activation energy, the molecules recoil,* bouncing off each other like billiard balls; the molecules zoom apart without reacting.

The tiny fraction of molecules that are oriented effectively *and* moving at the highest speed behave differently. *Their kinetic energy pushes them together with enough force to overcome repulsions and react.* Nuclei in one atom attract electrons in another; atomic orbitals overlap and electron densities shift; some bonds lengthen and weaken while others start to form. At some point during this smooth transformation, what exists is *neither reactant nor product but a transitional species with partial bonds.* This extremely unstable species, which is called the **transition state,** or **activated complex,** exists only at the instant when the reacting system is highest in potential energy. Thus, *the activation energy is the quantity needed to stretch and deform bonds in order to reach the transition state.*

Consider the reaction between methyl bromide and hydroxide ion:

$$CH_3Br + OH^- \longrightarrow CH_3OH + Br^-$$

The electronegative bromine makes the carbon of methyl bromide partially positive. If the reactants are moving toward each other fast enough and are oriented effectively when they collide, the negatively charged oxygen in OH^- approaches the carbon with enough energy to begin forming a C—O bond, which causes the C—Br bond to weaken. In the transition state (Figure 16.15), C is surrounded by five atoms (trigonal bipyramidal), which never occurs in its stable compounds. This high-energy species has three normal C—H bonds and two partial bonds, one from C to O and the other from C to Br. Reaching this transition state is no guarantee that the reaction will proceed to products. A transition state can change

Figure 16.15 Nature of the transition state in the reaction between CH_3Br and OH^-. Note the partial (elongated) C—O and C—Br bonds and the trigonal bipyramidal shape of the transition state of this reaction.

in either direction: if the C—O bond continues to shorten and strengthen, products form; however, if the C—Br bond becomes shorter and stronger again, the transition state reverts to reactants.

Depicting the Change with Reaction Energy Diagrams A useful way to depict the events we just described is with a **reaction energy diagram,** which shows the potential energy of the system during the reaction as a smooth curve. Figure 16.16 shows the reaction energy diagram for the reaction of CH_3Br and OH^-, and also includes electron density relief maps, structural formulas, and molecular-scale views at various points during the change.

The horizontal axis, labeled "Reaction progress," means reactants change to products from left to right. This reaction is exothermic, so reactants are higher in energy than products. The diagram also shows activation energies for the forward and reverse reactions; in this case, $E_{a(fwd)}$ is less than $E_{a(rev)}$. This difference, which reflects the change in bond energies, equals the heat of reaction, ΔH_{rxn}:

$$\Delta H_{rxn} = E_{a(fwd)} - E_{a(rev)} \qquad (16.10)$$

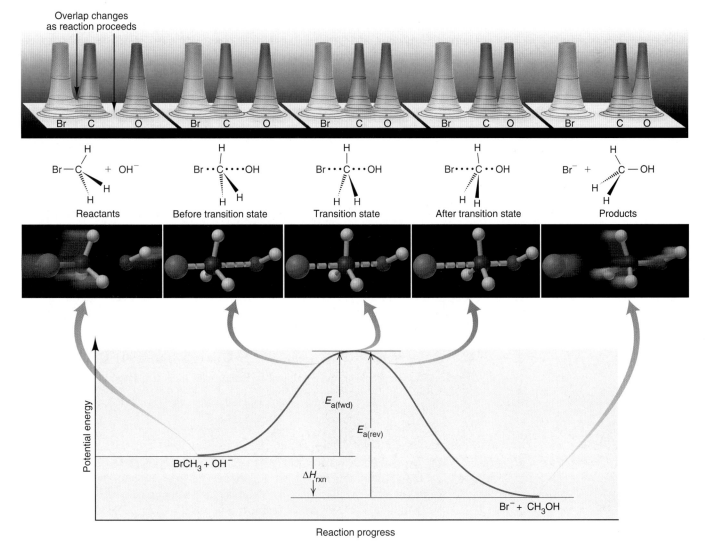

Figure 16.16 Reaction energy diagram for the reaction between CH_3Br and OH^-. A plot of potential energy vs. reaction progress shows the relative energy levels of reactants, products, and transition state joined by a curved line, as well as the activation energies of the forward and reverse steps and the heat of reaction. The electron density relief maps, structural formulas, and molecular-scale views depict the change at five points. Note the gradual bond forming and bond breaking as the system goes through the transition state.

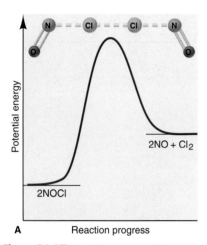

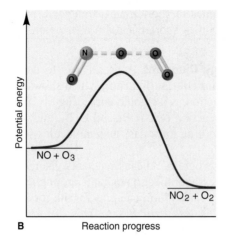

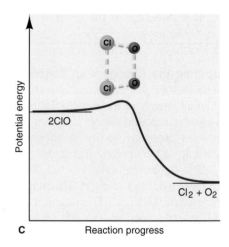

A   Reaction progress

B   Reaction progress

C   Reaction progress

Figure 16.17 Reaction energy diagrams and possible transition states for three reactions.
A, $2NOCl(g) \longrightarrow 2NO(g) + Cl_2(g)$ (Despite the formula NOCl, the atom sequence is ClNO.)
B, $NO(g) + O_3(g) \longrightarrow NO_2(g) + O_2(g)$
C, $2ClO(g) \longrightarrow Cl_2(g) + O_2(g)$
Note that reaction A is endothermic, B and C are exothermic, and C has a very small $E_{a(fwd)}$.

Transition state theory proposes that *every reaction (and every step in an overall reaction) goes through its own transition state,* from which it can continue in either direction. We imagine how a transition state might look by examining the reactant and product bonds that change. Figure 16.17 depicts reaction energy diagrams for three simple reactions. Note that the shape of the postulated transition state in each case is based on a specific collision orientation between the atoms that become bonded to form the product.

SAMPLE PROBLEM 16.7   Drawing Reaction Energy Diagrams and Transition States

Problem A key reaction in the upper atmosphere is

$$O_3(g) + O(g) \longrightarrow 2O_2(g)$$

The $E_{a(fwd)}$ is 19 kJ, and the ΔH_{rxn} for the reaction as written is -392 kJ. Draw a reaction energy diagram for this reaction, postulate a transition state, and calculate $E_{a(rev)}$.
Plan The reaction is highly exothermic ($\Delta H_{rxn} = -392$ kJ), so the products are much lower in energy than the reactants. The small $E_{a(fwd)}$ (19 kJ) means the energy of the reactants lies slightly below that of the transition state. We use Equation 16.10 to calculate $E_{a(rev)}$. To postulate the transition state, we sketch the species and note that one of the bonds in O_3 weakens, and this partially bonded O begins forming a bond to the separate O atom.
Solution Solving for $E_{a(rev)}$:

$$\Delta H_{rxn} = E_{a(fwd)} - E_{a(rev)}$$

So, $E_{a(rev)} = E_{a(fwd)} - \Delta H_{rxn} = 19 \text{ kJ} - (-392 \text{ kJ}) = 411 \text{ kJ}$

The reaction energy diagram (not drawn to scale), with transition state, is

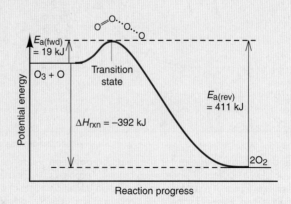

Check Rounding to find $E_{a(rev)}$ gives $\sim 20 + 390 = 410$.

FOLLOW-UP PROBLEM 16.7 The following reaction energy diagram depicts another key atmospheric reaction. Label the axes, identify $E_{a(fwd)}$, $E_{a(rev)}$, and ΔH_{rxn}, draw and label the transition state, and calculate $E_{a(rev)}$ for the reaction.

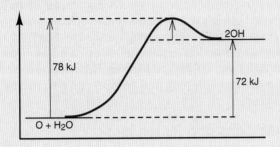

SECTION SUMMARY

According to collision theory, reactant particles must collide to react, and the number of collisions depends on the product of the reactant concentrations. At higher temperatures, more collisions have enough energy to exceed the activation energy (E_a). The relative E_a values for the forward and the reverse reactions depend on whether the overall reaction is exothermic or endothermic. Molecules must collide with an effective orientation for reaction to occur, so structural complexity decreases rate.

Transition state theory pictures the kinetic energy of the particles changing to potential energy during a collision. Given a sufficiently energetic collision and an effective molecular orientation, the reactant species become an unstable transition state, which either forms product(s) or reverts to reactant(s). Reaction energy diagrams depict the changing energy of the chemical system as it progresses from reactants through transition state(s) to products.

16.7 REACTION MECHANISMS: STEPS IN THE OVERALL REACTION

Imagine trying to figure out how a car works just by examining the body, wheels, and dashboard. It can't be done—you need to look under the hood and inside the engine to see how the parts fit together and function. Similarly, because our main purpose is to know how a reaction works *at the molecular level*, examining the overall balanced equation is not much help—we must "look under the yield arrow and inside the reaction" to see how reactants change into products.

When we do so, we find that most reactions occur through a **reaction mechanism,** a sequence of single reaction steps that sum to the overall reaction. For example, a possible mechanism for the overall reaction

$$2A + B \longrightarrow E + F$$

might involve these three simpler steps:

(1) $A + B \longrightarrow C$

(2) $C + A \longrightarrow D$

(3) $ D \longrightarrow E + F$

Adding them together and canceling common substances, we obtain the overall equation:

$$A + B + \cancel{C} + A + \cancel{D} \longrightarrow \cancel{C} + \cancel{D} + E + F \quad \text{or} \quad 2A + B \longrightarrow E + F$$

Note what happens to C and to D in this mechanism. C is a product in step 1 and a reactant in step 2, and D is a product in 2 and a reactant in 3. Each functions as a **reaction intermediate,** a substance that is formed and used up during the overall reaction. Reaction intermediates do not appear in the overall balanced equation but are absolutely essential for the reaction to occur. They are usually unstable relative to the reactants and products but are far more stable than transition states (activated complexes). Reaction intermediates are molecules with normal bonds and are sometimes stable enough to be isolated.

Chemists *propose* a reaction mechanism to explain how a particular reaction might occur, and then they *test* the mechanism. This section focuses on the nature of the individual steps and how they fit together to give a rate law consistent with experimental results.

Elementary Reactions and Molecularity

The individual steps, which together make up a proposed reaction mechanism, are called **elementary reactions** (or **elementary steps**). Each describes a *single molecular event,* such as one particle decomposing or two particles colliding and combining. *An elementary step is **not** made up of simpler steps.*

An elementary step is characterized by its **molecularity,** the number of *reactant* particles involved in the step. Consider the mechanism for the breakdown of ozone in the stratosphere. The overall reaction is

$$2O_3(g) \longrightarrow 3O_2(g)$$

A two-step mechanism has been proposed for this reaction. Notice that the two steps sum to the overall reaction. The first elementary step is a **unimolecular reaction,** one that involves the decomposition or rearrangement of a single particle:

(1) $O_3(g) \longrightarrow O_2(g) + O(g)$

The second step is a **bimolecular reaction,** one in which two particles react:

(2) $O_3(g) + O(g) \longrightarrow 2O_2(g)$

Some *termolecular* elementary steps occur, but they are extremely rare because the probability of three particles colliding simultaneously with enough energy and with an effective orientation is very small. Higher molecularities are not known. Unless evidence exists to the contrary, it makes good chemical sense to propose only unimolecular or bimolecular reactions as the elementary steps in a reaction mechanism.

The rate law for an elementary reaction, unlike that for an overall reaction, *can* be deduced from the reaction stoichiometry. An elementary reaction occurs in one step, so its rate must be proportional to the product of the reactant concentrations. Therefore, *we use the equation coefficients as the reaction orders in the rate law for an elementary step; that is, reaction order equals molecularity* (Table 16.6). Remember that this statement holds *only* when we know that the reaction is elementary; you've already seen that for an overall reaction, the reaction orders must be determined experimentally.

Table 16.6 **Rate Laws for General Elementary Steps**

Elementary Step	Molecularity	Rate Law
A $\longrightarrow$ product	Unimolecular	Rate = $k[A]$
2A $\longrightarrow$ product	Bimolecular	Rate = $k[A]^2$
A + B $\longrightarrow$ product	Bimolecular	Rate = $k[A][B]$
2A + B $\longrightarrow$ product	Termolecular	Rate = $k[A]^2[B]$

SAMPLE PROBLEM 16.8 Determining Molecularity and Rate Laws
for Elementary Steps

Problem The following two reactions are proposed as elementary steps in the mechanism
for an overall reaction:
(1) $\quad\quad\quad NO_2Cl(g) \longrightarrow NO_2(g) + Cl(g)$
(2) $NO_2Cl(g) + Cl(g) \longrightarrow NO_2(g) + Cl_2(g)$
(a) Write the overall balanced equation.
(b) Determine the molecularity of each step.
(c) Write the rate law for each step.
Plan We find the overall equation from the sum of the elementary steps. The molecularity of each step equals the total number of reactant particles. We write the rate law for
each step using the molecularities as reaction orders.
Solution (a) Writing the overall balanced equation:

$$NO_2Cl(g) \longrightarrow NO_2(g) + Cl(g)$$
$$NO_2Cl(g) + Cl(g) \longrightarrow NO_2(g) + Cl_2(g)$$
$$\overline{NO_2Cl(g) + NO_2Cl(g) + \cancel{Cl(g)} \longrightarrow NO_2(g) + \cancel{Cl(g)} + NO_2(g) + Cl_2(g)}$$
$$2NO_2Cl(g) \longrightarrow 2NO_2(g) + Cl_2(g)$$

(b) Determining the molecularity of each step: The first elementary step has only one reactant, NO_2Cl, so it is unimolecular. The second elementary step has two reactants, NO_2Cl
and Cl, so it is bimolecular.
(c) Writing rate laws for the elementary reactions:
(1) $Rate_1 = k_1[NO_2Cl]$
(2) $Rate_2 = k_2[NO_2Cl][Cl]$
Check In part (a), be sure the equation is balanced; in part (c), be sure the substances in
brackets are the reactants of each elementary step.

FOLLOW-UP PROBLEM 16.8 The following elementary steps constitute a proposed
mechanism for a reaction:
(1) $\quad\quad\quad\quad 2NO(g) \longrightarrow N_2O_2(g)$
(2) $\quad\quad\quad\quad 2H_2(g) \longrightarrow 4H(g)$
(3) $N_2O_2(g) + H(g) \longrightarrow N_2O(g) + HO(g)$
(4) $2HO(g) + 2H(g) \longrightarrow 2H_2O(g)$
(5) $\quad H(g) + N_2O(g) \longrightarrow HO(g) + N_2(g)$
(a) Write the balanced equation for the overall reaction.
(b) Determine the molecularity of each step.
(c) Write the rate law for each step.

The Rate-Determining Step of a Reaction Mechanism

All the elementary steps in a mechanism do not have the same rate. Usually, one
step is much slower than the others, so it limits how fast the overall reaction proceeds. This step is called the **rate-determining step** (or **rate-limiting step**).

*Because the rate-determining step limits the rate of the overall reaction, its
rate law represents the rate law for the overall reaction.* Consider the reaction
between nitrogen dioxide and carbon monoxide:

$$NO_2(g) + CO(g) \longrightarrow NO(g) + CO_2(g)$$

If the overall reaction were an elementary reaction—that is, if the mechanism
consisted of only one step—we could immediately write the overall rate law as

$$Rate = k[NO_2][CO]$$

However, as you saw in Sample Problem 16.3, experiment shows that the actual
rate law is

$$Rate = k[NO_2]^2$$

From this, we know immediately that the reaction shown cannot be elementary.

A proposed two-step mechanism is

(1) $NO_2(g) + NO_2(g) \longrightarrow NO_3(g) + NO(g)$ [slow; rate determining]
(2) $NO_3(g) + CO(g) \longrightarrow NO_2(g) + CO_2(g)$ [fast]

Note that NO_3 functions as a reaction intermediate in the mechanism. Rate laws for these elementary steps are

(1) $Rate_1 = k_1[NO_2][NO_2] = k_1[NO_2]^2$
(2) $Rate_2 = k_2[NO_3][CO]$

Note that if $k_1 = k$, *the rate law for the rate-determining step (step 1) is identical to the experimental rate law.* The first step is so slow compared with the second that the overall reaction takes essentially as long as the first step. Here you can see that one reason a reactant (in this case, CO) has a reaction order of zero is that it takes part in the reaction only *after* the rate-determining step.

Correlating the Mechanism with the Rate Law

Conjuring up a reasonable reaction mechanism can be a classic example of the use of the scientific method. We use observations and data from rate experiments to hypothesize what the individual steps might be and then test our hypothesis by gathering further evidence. If the evidence supports it, we continue to apply that mechanism; if not, we propose a new one. However, *we can never prove, just from data, that a particular mechanism represents the actual chemical change, only that it is consistent with it.*

Regardless of the elementary steps proposed for a mechanism, they must meet three criteria:

1. *The elementary steps must add up to the overall balanced equation.* We cannot wind up with more (or fewer) reactants or products than are present in the balanced equation.
2. *The elementary steps must be physically reasonable.* As we noted, most steps should involve one reactant particle (unimolecular) or two (bimolecular). Steps with three reactant particles (termolecular) are very unlikely.
3. *The mechanism must correlate with the rate law.* Most importantly, a mechanism must support the experimental facts shown by the rate law, not the other way around.

Let's see how the mechanisms of several reactions conform to these criteria and how the elementary steps fit together.

Mechanisms with a Slow Initial Step We've already seen one mechanism with a rate-determining first step—that for the reaction of NO_2 and CO. Another example is the reaction between nitrogen dioxide and fluorine gas:

$$2NO_2(g) + F_2(g) \longrightarrow 2NO_2F(g)$$

The experimental rate law is first order in NO_2 and in F_2:

$$Rate = k[NO_2][F_2]$$

The accepted mechanism for the reaction is

(1) $NO_2(g) + F_2(g) \longrightarrow NO_2F(g) + F(g)$ [slow; rate determining]
(2) $NO_2(g) + F(g) \longrightarrow NO_2F(g)$ [fast]

Molecules of reactant and product appear in both elementary steps. The free fluorine atom is a reaction intermediate.

Does this mechanism meet the three crucial criteria?

1. The elementary reactions sum to the balanced equation:

$$NO_2(g) + NO_2(g) + F_2(g) + \cancel{F(g)} \longrightarrow NO_2F(g) + NO_2F(g) + \cancel{F(g)}$$

or

$$2NO_2(g) + F_2(g) \longrightarrow 2NO_2F(g)$$

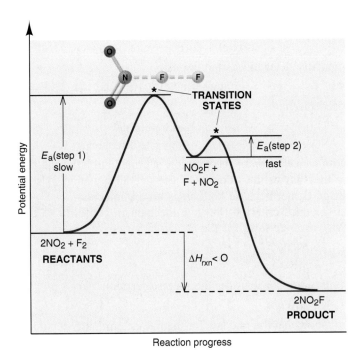

Figure 16.18 Reaction energy diagram for the two-step reaction of NO_2 and F_2. Each step in the mechanism has its own transition state. The proposed transition state is shown for step 1. Reactants for the second step are the F atom intermediate and the second molecule of NO_2. Note that the first step is slower (higher E_a). The overall reaction is exothermic ($\Delta H_{rxn} < 0$).

2. Both steps are bimolecular, so they are chemically reasonable.
3. The mechanism gives the rate law for the overall equation. To show this, we write the rate laws for the elementary steps:

  (1) $Rate_1 = k_1[NO_2][F_2]$
  (2) $Rate_2 = k_2[NO_2][F]$

Step 1 is the rate-determining step and therefore gives the overall rate law, with $k_1 = k$. Because the second molecule of NO_2 appears in the step that follows the rate-determining step, it does not appear in the overall rate law. Thus, we see that *the overall rate law includes only species active in the reaction up to and including those in the rate-determining step.* This point was also illustrated by the mechanism for NO_2 and CO shown earlier. Carbon monoxide was absent from the overall rate law because it appeared *after* the rate-determining step.

Figure 16.18 is a reaction energy diagram for the reaction of NO_2 and F_2. Note that

- *Each step in the mechanism has its own transition state.* (Note that only one molecule of NO_2 reacts in step 1, and only the first transition state is depicted.)
- The F atom intermediate is a reactive, unstable species (as you know from halogen chemistry), so it is higher in energy than the reactants or product.
- The first step is slower (rate limiting), so its activation energy is *larger* than that of the second step.
- The overall reaction is exothermic, so the product is lower in energy than the reactants.

Mechanisms with a Fast Initial Step If the rate-determining step in a mechanism is *not* the initial step, it acts as a bottleneck later in the reaction sequence. As a result, the product of a fast initial step builds up and starts reverting to reactant, while waiting for the slow step to remove it. With time, the product of the initial step is changing back to reactant as fast as it is forming. In other words, the *fast initial step reaches equilibrium.* As you'll see, this situation allows us to fit the mechanism to the overall rate law.

Consider once again the oxidation of nitrogen monoxide:

$$2NO(g) + O_2(g) \longrightarrow 2NO_2(g)$$

The experimentally determined rate law is

$$\text{Rate} = k[NO]^2[O_2]$$

and a proposed mechanism is

(1) $NO(g) + O_2(g) \rightleftharpoons NO_3(g)$ [fast, reversible]
(2) $NO_3(g) + NO(g) \longrightarrow 2NO_2(g)$ [slow; rate determining]

Note that, with cancellation of the reaction intermediate NO_3, the first criterion is met because the sum of the steps gives the overall equation. Also note that the second criterion is met because both steps are bimolecular.

To meet the third criterion (that the mechanism conforms to the overall rate law), we first write rate laws for the elementary steps:

(1) $\text{Rate}_{1(\text{fwd})} = k_1[NO][O_2]$
    $\text{Rate}_{1(\text{rev})} = k_{-1}[NO_3]$

where k_{-1} is the rate constant for the reverse reaction.

(2) $\text{Rate}_2 = k_2[NO_3][NO]$

Now we must show that the rate law for the rate-determining step (step 2) gives the overall rate law. As written, it does not, because it contains the intermediate NO_3, and *an overall rate law can include only reactants (and products)*. Therefore, we must eliminate $[NO_3]$ from the step 2 rate law. To do so, we express $[NO_3]$ in terms of reactants. Step 1 reaches equilibrium when the forward and reverse rates are equal:

$$\text{Rate}_{1(\text{fwd})} = \text{Rate}_{1(\text{rev})} \quad \text{or} \quad k_1[NO][O_2] = k_{-1}[NO_3]$$

To express $[NO_3]$ in terms of reactants, we isolate it algebraically:

$$[NO_3] = \frac{k_1}{k_{-1}}[NO][O_2]$$

Then, substituting for $[NO_3]$ in the rate law for step 2, we obtain

$$\text{Rate}_2 = k_2[NO_3][NO] = k_2\left(\frac{k_1}{k_{-1}}[NO][O_2]\right)[NO] = \frac{k_2 k_1}{k_{-1}}[NO]^2[O_2]$$

This rate law is identical to the overall rate law, with $k = \dfrac{k_2 k_1}{k_{-1}}$.

Thus, to test the validity of a mechanism with a fast initial, reversible step:

1. Write rate laws for both directions of the fast step and for the slow step.
2. Show the slow step's rate law is equivalent to the overall rate law, by *expressing [intermediate] in terms of [reactant]:* set the forward rate law of the fast, reversible step equal to the reverse rate law, and solve for [intermediate].
3. Substitute the expression for [intermediate] into the rate law for the slow step to obtain the overall rate law.

Several end-of-chapter problems, including 16.61 and 16.62, provide additional examples of this approach.

SECTION SUMMARY

The mechanisms of most common reactions consist of two or more elementary steps, reactions that occur in one step and depict a single chemical change. The molecularity of an elementary step equals the number of reactant particles and is the same as the reaction order of its rate law. Unimolecular and bimolecular steps are common. The rate-determining, or rate-limiting (slowest), step determines how fast the overall reaction occurs, and its rate law represents the overall rate law. Reaction intermediates are species that form in one step and react in a later one. The steps in a proposed mechanism must add up to the overall reaction, be physically reasonable, and conform to the overall rate law. If a fast step precedes a slow step, the fast step reaches equilibrium, and the concentrations of intermediates in the rate law of the slow step must be expressed in terms of reactants.

16.8 CATALYSIS: SPEEDING UP A CHEMICAL REACTION

There are many situations in which the rate of a reaction must be increased for it to be useful. In an industrial process, for example, a higher rate often determines whether a new product can be made economically. Sometimes, we can speed up a reaction sufficiently with a higher temperature, but energy is costly and many substances are heat sensitive and easily decomposed. Alternatively, we can often employ a **catalyst,** a substance that increases the rate *without* being consumed in the reaction. Because catalysts are not consumed, only very small, non-stoichiometric quantities are generally required. Nevertheless, these substances are employed in so many important processes that several million tons of industrial catalysts are produced annually in the United States alone! Nature is the master designer and user of catalysts. Even the simplest bacterium employs thousands of biological catalysts, known as *enzymes,* to speed up its cellular reactions. Every organism relies on enzymes to sustain life.

Each catalyst has its own specific way of functioning, but in general, *a catalyst causes a lower activation energy, which in turn makes the rate constant larger and the rate higher.* Two important points stand out in Figure 16.19:

- A catalyst speeds up the forward *and* reverse reactions. A reaction with a catalyst *does not yield more product* than one without a catalyst, but it yields the product *more quickly*.
- A catalyst causes a *lower activation energy* by providing a *different mechanism* for the reaction, and thus a new, lower energy pathway.

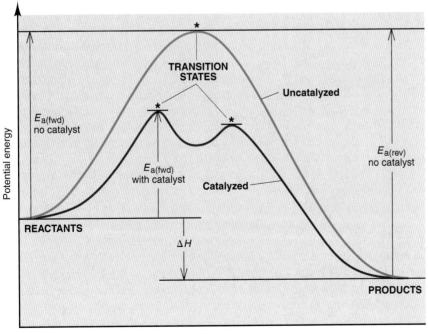

Figure 16.19 Reaction energy diagram for a catalyzed and an uncatalyzed process. A catalyst speeds a reaction by providing a new, lower energy pathway, in this case by replacing the one-step mechanism with a two-step mechanism. Both forward and reverse rates are increased to the same extent, so *a catalyst does not affect the overall reaction yield.* (The only activation energy shown for the catalyzed reaction is the larger one for the forward direction.)

Consider a general *uncatalyzed* reaction that proceeds by a one-step mechanism involving a bimolecular collision:

$$A + B \longrightarrow \text{product} \qquad \text{[slower]}$$

In the *catalyzed* reaction, a reactant molecule interacts with the catalyst, so the mechanism might involve a two-step pathway:

$$A + \text{catalyst} \longrightarrow C \qquad \text{[faster]}$$
$$C + B \longrightarrow \text{product} + \text{catalyst} \qquad \text{[faster]}$$

Note that *the catalyst is not consumed,* as its definition requires. Rather, it is used and then regenerated, and the activation energies of both steps are lower than the activation energy of the uncatalyzed pathway.

There are two general categories of catalyst—homogeneous catalysts and heterogeneous catalysts—based on whether the catalyst is in the same phase as the reactant and product.

Homogeneous Catalysis

A **homogeneous catalyst** exists in solution with the reaction mixture. All homogeneous catalysts are gases, liquids, or soluble solids.

A thoroughly studied example of homogeneous catalysis is the hydrolysis of an organic ester (RCOOR′), a reaction introduced in Section 15.4:

$$\underset{\text{O}}{R-\overset{\overset{\text{O}}{\|}}{C}-O-R'} + H_2O \rightleftharpoons R-\overset{\overset{\text{O}}{\|}}{C}-OH + R'-OH$$

Here R and R′ are hydrocarbon groups, $R-\overset{\overset{\text{O}}{\|}}{C}-OH$ is a carboxylic acid, and R′—OH is an alcohol. The reaction rate is low at room temperature but can be increased greatly by adding a small amount of strong acid, which provides H^+ ion, the catalyst in the reaction; strong bases, which supply OH^- ions, also speed ester hydrolysis, but by a slightly different mechanism.

In the first step of the acid-catalyzed reaction (Figure 16.20), the H^+ of a hydronium ion forms a bond to the double-bonded O atom. From the resonance forms, we see that the bonding of H^+ then makes the C atom more positive, which *increases its attraction* for the partially negative O atom of water. In effect, H^+ increases the likelihood that the bonding of water, which is the rate-determining step, will take place. Several steps later, a water molecule, acting as a base, removes the H^+ and returns it to solution. Thus, H^+ acts as a catalyst because it speeds up the reaction but is not itself consumed: it is used up in one step and re-formed in another.

Many digestive enzymes, which catalyze the hydrolysis of proteins, fats, and carbohydrates during the digestion of foods, employ very similar mechanisms. The difference is that the acids or bases that speed these reactions are not the strong inorganic reagents used in the lab, but rather specific amino-acid side chains of the enzymes that release or abstract H^+ ions.

Heterogeneous Catalysis

A **heterogeneous catalyst** speeds up a reaction that occurs in a separate phase. The catalyst is most often a solid interacting with gaseous or liquid reactants. Because reaction occurs on the solid's surface, heterogeneous catalysts usually have enormous surface areas for contact, between 1 and 500 m^2/g. Interestingly, many reactions that occur on a metal surface, such as the decomposition of HI on gold and the decomposition of N_2O on platinum, are zero order because the rate-determining step occurs on the surface itself. Thus, despite an enormous surface area, once the reactant gas covers the surface, increasing the reactant concentration cannot increase the rate.

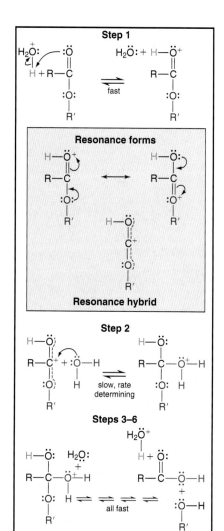

Figure 16.20 Mechanism for the catalyzed hydrolysis of an organic ester. In step 1, the catalytic H^+ ion binds to the electron-rich oxygen. The resonance hybrid of this product *(see gray panel)* shows the C atom is more positive than it would ordinarily be. The enhanced charge on C attracts the partially negative O of water more strongly, increasing the fraction of effective collisions and thus speeding up step 2, the rate-determining step. Loss of R′OH and removal of H^+ by water occur in a final series of fast steps.

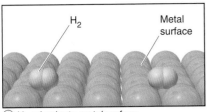

① H_2 adsorbs to metal surface

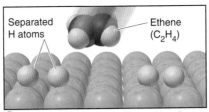

② Rate-limiting step is H—H bond breakage.

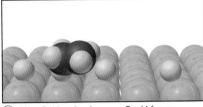

③ After C_2H_4 adsorbs, one C—H forms.

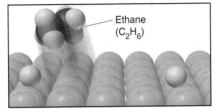

④ Another C—H bond forms; C_2H_6 leaves surface.

Figure 16.21 The metal-catalyzed hydrogenation of ethylene.

One of the most important examples of heterogeneous catalysis is the addition of H_2 to the C=C bonds of organic compounds to form C—C bonds. The petroleum, plastics, and food industries frequently use catalytic **hydrogenation.** The conversion of vegetable oil into margarine is one example.

The simplest hydrogenation converts ethylene to ethane:

$$H_2C=CH_2(g) + H_2(g) \longrightarrow H_3C-CH_3(g)$$

In the absence of a catalyst, the reaction occurs very slowly. At high H_2 pressure in the presence of finely divided Ni, Pd, or Pt, the reaction becomes rapid even at ordinary temperatures. These Group 8B(10) metals catalyze by *chemically adsorbing the reactants onto their surface* (Figure 16.21). The H_2 lands and splits into separate H atoms chemically bound to the solid catalyst's metal atoms (catM):

$$H-H(g) + 2catM(s) \longrightarrow 2catM-H \text{ (H atoms bound to metal surface)}$$

Then, C_2H_4 adsorbs and reacts with two H atoms, one at a time, to form C_2H_6. The H—H bond breakage is the rate-determining step in the overall process, and interaction with the catalyst's surface provides the low-E_a step as part of an alternative reaction mechanism.

Catalysis in Nature

Unlike the industrial examples we just discussed, catalytic processes occur in natural settings as well, and a brief description of two important systems follows. The first concerns the remarkable abilities of catalysts inside you, and the second focuses on how catalysis operates in the stratosphere.

Cellular Catalysis: The Function of Enzymes Within every living cell, thousands of individual reactions occur in dilute solution at ordinary temperatures and pressures. The rates of these reactions respond smoothly to various factors, including concentration changes, signals from other cells, and environmental stresses. Virtually every cell reaction is catalyzed by its own specific **enzyme,** a protein whose complex three-dimensional shape—and thus its function—has been perfected through natural selection (Section 15.6).

Every enzyme has an **active site,** a small region whose shape results from those of the side chains (R groups) of the amino acids that make it up. When reactant molecules, called the *substrates,* bind to an active site, usually through *intermolecular forces,* the chemical change begins. With molar masses ranging from 15,000 to 1,000,000 g/mol, most enzymes are enormous relative to their substrates, and they are often embedded within membranes. Thus, like a heterogeneous catalyst, an enzyme provides a surface on which a substrate is immobilized temporarily, waiting for another reactant to land nearby. Like a homogeneous catalyst, the active-site R groups interact directly with the substrates in multistep sequences.

Enzymes are incredibly *efficient* catalysts. Consider the hydrolysis of urea, a key component in amino-acid metabolism:

$$(NH_2)_2C{=}O(aq) + 2H_2O(l) + H^+(aq) \longrightarrow 2NH_4^+(aq) + HCO_3^-(aq)$$

In water at room temperature, the rate constant for the uncatalyzed reaction is $3{\times}10^{-10}$ s^{-1}. Under the same conditions in the presence of the enzyme urease (pronounced "*yur*-ee-ase"), the rate constant increases 10^{14}-fold, to $3{\times}10^4$ s^{-1}! Enzymes are also extremely *specific:* urease catalyzes *only* this hydrolysis reaction, and no other enzyme does so.

There are two main models of enzyme action. In the *lock-and-key model,* when the "key" (substrate) fits the "lock" (active site), the chemical change begins. However, experiments show that, in many cases, the enzyme changes shape when the substrate lands at its active site. Thus, rather than a rigidly shaped lock in which a particular key fits, the *induced-fit model* pictures a "hand" (substrate) entering a "glove" (active site), causing it to attain its functional shape.

Enzymes act through a variety of catalytic mechanisms. In some cases, the active-site R groups bring the reacting atoms of the substrates closer together. In other cases, the R groups stretch the substrate bond that is to be broken. Some R groups provide an H$^+$ ion that increases the speed of a rate-determining step; others remove an H$^+$ ion at a critical step. Regardless of their specific mode of action, *all enzymes function by binding to the reaction's transition state and thus stabilizing it.* In this way, the enzyme lowers the activation energy, which increases the reaction rate.

Atmospheric Catalysis: Depletion of the Ozone Layer Both homogeneous and heterogeneous catalysis play key roles in the depletion of ozone from the stratosphere. At Earth's surface, ozone is an air pollutant, contributing to smog and other problems. In the stratosphere, however, a natural layer of ozone absorbs UV radiation from the Sun. If this radiation reaches the surface, it can break bonds in DNA, promote skin cancer, and damage simple life forms at the base of the food chain.

Stratospheric ozone concentrations are maintained naturally by a simple sequence of reactions:

$$O_2 \xrightarrow{\text{UV}} 2O$$
$$O + O_2 \longrightarrow O_3 \qquad \text{[ozone formation]}$$
$$O + O_3 \longrightarrow 2O_2 \qquad \text{[ozone breakdown]}$$

In 1995, Paul J. Crutzen, Mario J. Molina, and F. Sherwood Rowland received the Nobel Prize in chemistry for showing that chlorofluorocarbons (CFCs), used as aerosol propellants and air-conditioning coolants, were disrupting this sequence by catalyzing the breakdown reaction. CFCs are unreactive in the lower atmosphere, but slowly rise to the stratosphere, where UV radiation cleaves them:

$$CF_2Cl_2 \xrightarrow{\text{UV}} CF_2Cl{\cdot} + Cl{\cdot}$$

(The dots are unpaired electrons resulting from bond cleavage.) Like many species with unpaired electrons (free radicals), atomic Cl is very reactive. It reacts with ozone to produce chlorine monoxide (ClO·), which then reacts to regenerate Cl atoms:

$$O_3 + Cl· \longrightarrow ClO· + O_2$$
$$ClO· + O \longrightarrow ·Cl + O_2$$

The sum of these steps is the ozone breakdown reaction:

$$O_3 + ·\cancel{Cl} + ·\cancel{ClO} + O \longrightarrow ·\cancel{ClO} + O_2 + ·\cancel{Cl} + O_2$$

or

$$O_3 + O \longrightarrow 2O_2$$

Thus, the Cl atom is a homogeneous catalyst: it exists in the same phase as the reactants, speeds the reaction by allowing a different mechanism, and is regenerated. During its stratospheric half-life of about 2 years, each Cl atom speeds the breakdown of about 100,000 ozone molecules.

High levels of chlorine monoxide over Antarctica have given rise to an *ozone hole,* an area of the stratosphere showing a severe reduction of ozone. The hole enlarges by heterogeneous catalysis, as stratospheric clouds and dust from volcanic activity provide a surface that speeds formation of Cl atoms by other mechanisms. Despite international agreements that are phasing out CFCs and similar compounds, full recovery of the ozone layer is likely to take the rest of this century! The good news is that halogen levels in the lower atmosphere have begun to fall.

SECTION SUMMARY

A catalyst is a substance that increases the rate of a reaction without being consumed. It accomplishes this by providing an alternative mechanism with a lower activation energy. Homogeneous catalysts function in the same phase as the reactants. Heterogeneous catalysts act in a different phase from the reactants. The hydrogenation of carbon-carbon double bonds takes place on a solid catalyst, which speeds the breakage of the H—H bond in H_2. Enzymes are biological catalysts with spectacular efficiency and specificity. Chlorine atoms derived from CFC molecules catalyze the breakdown of stratospheric ozone.

For Review and Reference (Numbers in parentheses refer to pages, unless noted otherwise.)

Learning Objectives

To help you review these learning objectives, the numbers of related sections (§), sample problems (SP), and upcoming end-of-chapter problems (EP) are listed in parentheses.

1. Explain why reaction rate depends on concentration, physical state, and temperature (§ 16.1) (EPs 16.1–16.6)
2. Understand how reaction rate is expressed in terms of changing reactant and product concentrations over time, and distinguish among average, instantaneous, and initial rates (§ 16.2) (SP 16.1) (EPs 16.7–16.19)
3. Describe the information needed to determine the rate law, and explain how to calculate reaction orders and rate constant (§ 16.3) (SPs 16.2, 16.3) (EPs 16.20–16.28)
4. Understand how to use integrated rate laws to find concentration at a given time (or vice versa) and reaction order, and explain the meaning of half-life (§ 16.4) (SPs 16.4, 16.5) (EPs 16.29–16.34)

5. Explain the importance of activation energy and the effect of temperature on the rate constant (Arrhenius equation) (§ 16.5) (SP 16.6) (EPs 16.35–16.40)
6. Understand collision theory (why concentrations are multiplied, how temperature affects the fraction of collisions exceeding E_a, and how rate depends on the number of effective collisions) and transition state theory (how E_a is used to form the transition state and how a reaction energy diagram depicts the progress of a reaction) (§ 16.6) (SP 16.7) (EPs 16.41–16.52)
7. Understand elementary steps and molecularity, and be able to construct a valid reaction mechanism with either a slow or a fast initial step (§ 16.7) (SP 16.8) (EPs 16.53–16.64)
8. Explain how a catalyst speeds a reaction by lowering E_a, and distinguish between homogeneous and heterogeneous catalysis (§ 16.8) (EPs 16.65, 16.66)

Key Terms

chemical kinetics (499)

Section 16.2
reaction rate (501)
average rate (502)
instantaneous rate (502)
initial rate (503)

Section 16.3
rate law (rate equation) (505)
rate constant (505)
reaction orders (505)

Section 16.4
integrated rate law (510)
half-life ($t_{1/2}$) (513)

Section 16.5
Arrhenius equation (516)
activation energy (E_a) (516)

Section 16.6
collision theory (518)
effective collision (519)
frequency factor (519)
transition state theory (520)

transition state (activated
  complex) (520)
reaction energy diagram (521)

Section 16.7
reaction mechanism (523)
reaction intermediate (524)
elementary reaction
  (elementary step) (524)
molecularity (524)
unimolecular reaction (524)
bimolecular reaction (524)

rate-determining (rate-
  limiting) step (525)

Section 16.8
catalyst (529)
homogeneous catalyst (530)
heterogeneous catalyst (530)
hydrogenation (531)
enzyme (531)
active site (531)

Key Equations and Relationships

16.1 Expressing reaction rate in terms of reactant A (501):

$$\text{Rate} = -\frac{\Delta[A]}{\Delta t}$$

16.2 Expressing the rate of a general reaction (503):

$$aA + bB \longrightarrow cC + dD$$

$$\text{Rate} = -\frac{1}{a}\frac{\Delta[A]}{\Delta t} = -\frac{1}{b}\frac{\Delta[B]}{\Delta t} = \frac{1}{c}\frac{\Delta[C]}{\Delta t} = \frac{1}{d}\frac{\Delta[D]}{\Delta t}$$

16.3 Writing a general rate law (for a case not involving products) (505):

$$\text{Rate} = k[A]^m[B]^n \cdots$$

16.4 Calculating the time to reach a given [A] in a first-order reaction (rate = k[A]) (510):

$$\ln\frac{[A]_0}{[A]_t} = kt$$

16.5 Calculating the time to reach a given [A] in a simple second-order reaction (rate = k[A]^2) (511):

$$\frac{1}{[A]_t} - \frac{1}{[A]_0} = kt$$

16.6 Calculating the time to reach a given [A] in a zero-order reaction (rate = k) (511):

$$[A]_t - [A]_0 = -kt$$

16.7 Finding the half-life of a first-order process (514):

$$t_{1/2} = \frac{\ln 2}{k} = \frac{0.693}{k}$$

16.8 Relating the rate constant to the temperature (Arrhenius equation) (516):

$$k = Ae^{-E_a/RT}$$

16.9 Calculating the activation energy (rearranged form of Arrhenius equation) (517):

$$\ln\frac{k_2}{k_1} = -\frac{E_a}{R}\left(\frac{1}{T_2} - \frac{1}{T_1}\right)$$

16.10 Relating the heat of reaction to the forward and reverse activation energies (521):

$$\Delta H_{rxn} = E_{a(fwd)} - E_{a(rev)}$$

Brief Solutions to Follow-up Problems

16.1 (a) $4NO(g) + O_2(g) \longrightarrow 2N_2O_3(g)$;

$$\text{rate} = -\frac{\Delta[O_2]}{\Delta t} = -\frac{1}{4}\frac{\Delta[NO]}{\Delta t} = \frac{1}{2}\frac{\Delta[N_2O_3]}{\Delta t}$$

(b) $-\dfrac{\Delta[O_2]}{\Delta t} = -\dfrac{1}{4}\dfrac{\Delta[NO]}{\Delta t} = -\dfrac{1}{4}(-1.60\times10^{-4}\ \text{mol/L·s})$

$= 4.00\times10^{-5}\ \text{mol/L·s}$

16.2 First order in Br^-, first order in BrO_3^-, second order in H^+, fourth order overall.

16.3 Rate = $k[H_2]^m[I_2]^n$. From experiments 1 and 3, $m = 1$. From experiments 2 and 4, $n = 1$.

Therefore, rate = $k[H_2][I_2]$; second order overall.

16.4 $1/[HI]_1 - 1/[HI]_0 = kt$;

$111\ \text{L/mol} - 100\ \text{L/mol} = (2.4\times10^{-21}\ \text{L/mol·s})(t)$

$t = 4.6\times10^{21}\ \text{s (or } 1.5\times10^{14}\ \text{yr})$

16.5 $t_{1/2} = (\ln 2)/k$; $k = 0.693/13.1\ \text{h} = 5.29\times10^{-2}\ \text{h}^{-1}$

16.6 $\ln\dfrac{0.286\ \text{L/mol·s}}{k_1} = -\dfrac{1.00\times10^5\ \text{J/mol}}{8.314\ \text{J/mol·K}}\times\left(\dfrac{1}{500.\ \text{K}} - \dfrac{1}{490.\ \text{K}}\right)$

$= 0.491$

$k_1 = 0.175\ \text{L/mol·s}$

16.7

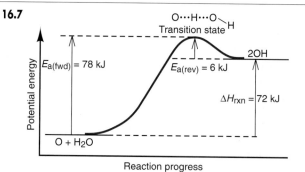

16.8 (a) Balanced equation:

$2NO(g) + 2H_2(g) \longrightarrow N_2(g) + 2H_2O(g)$

(b) Step 2 is unimolecular. All others are bimolecular.

(c) Rate$_1 = k_1[NO]^2$; rate$_2 = k_2[H_2]$; rate$_3 = k_3[N_2O_2][H]$; rate$_4 = k_4[HO][H]$; rate$_5 = k_5[H][N_2O]$.

Problems

Problems with **colored** numbers are answered in Appendix E. Sections match the text and provide the numbers of relevant sample problems. Bracketed problems are grouped in pairs (indicated by a short rule) that cover the same concept. Comprehensive Problems are based on material from any section or previous chapter.

Factors That Influence Reaction Rate

16.1 What variable of a chemical reaction is measured over time to obtain the reaction rate?

16.2 How does an increase in pressure affect the rate of a gas-phase reaction? Explain.

16.3 A reaction is carried out with water as the solvent. How does the addition of more water to the reaction vessel affect the rate of the reaction? Explain.

16.4 A gas reacts with a solid that is present in large chunks. Then the reaction is run again with the solid pulverized. How does the increase in the surface area of the solid affect the rate of its reaction with the gas? Explain.

16.5 How does an increase in temperature affect the rate of a reaction? Explain the two factors involved.

16.6 In a kinetics experiment, a chemist places crystals of iodine in a closed reaction vessel, introduces a given quantity of hydrogen gas, and obtains data to calculate the rate of hydrogen iodide formation. In a second experiment, she uses the same amounts of iodine and hydrogen, but she first warms the flask to 130°C, a temperature above the sublimation point of iodine. In which of these two experiments does the reaction proceed at a higher rate? Explain.

Expressing the Reaction Rate
(Sample Problem 16.1)

16.7 Define *reaction rate*. Assuming constant temperature and a closed reaction vessel, why does the rate change with time?

16.8 (a) What is the difference between an average rate and an instantaneous rate? (b) What is the difference between an initial rate and an instantaneous rate?

16.9 Give two reasons to measure initial rates in a kinetics study.

16.10 For the reaction $A(g) \longrightarrow B(g)$, sketch two curves on the same set of axes that show
(a) The formation of product as a function of time
(b) The consumption of reactant as a function of time

16.11 For the reaction $C(g) \longrightarrow D(g)$, [C] vs. time is plotted:

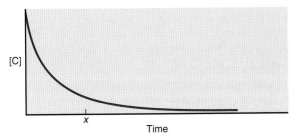

How do you determine each of the following?
(a) The average rate over the entire experiment
(b) The reaction rate at time x
(c) The initial reaction rate
(d) Would the values in parts (a), (b), and (c) be different if you plotted [D] vs. time? Explain.

16.12 The compound AX_2 decomposes according to the equation $2AX_2(g) \longrightarrow 2AX(g) + X_2(g)$. In one experiment, $[AX_2]$ was measured at various times and these data were obtained:

Time (s)	$[AX_2]$ (mol/L)
0.0	0.0500
2.0	0.0448
6.0	0.0300
8.0	0.0249
10.0	0.0209
20.0	0.0088

(a) Find the average rate over the entire experiment.
(b) Is the initial rate higher or lower than the rate in part (a)? Use graphical methods to estimate the initial rate.

16.13 (a) Use the data from Problem 16.12 to calculate the average rate from 8.0 to 20.0 s.
(b) Is the rate at exactly 5.0 s higher or lower than the rate in part (a)? Use graphical methods to estimate the rate at 5.0 s.

16.14 Express the rate of reaction in terms of the change in concentration of each of the reactants and products:
$$A(g) + 2B(g) \longrightarrow C(g)$$
When [B] is decreasing at 0.5 mol/L·s, how fast is [A] decreasing?

16.15 Express the rate of reaction in terms of the change in concentration of each of the reactants and products:
$$2D(g) + 3E(g) + F(g) \longrightarrow 2G(g) + H(g)$$
When [D] is decreasing at 0.1 mol/L·s, how fast is [H] increasing?

16.16 Reaction rate is expressed in terms of changes in concentration of reactants and products. Write a balanced equation for
$$\text{Rate} = -\frac{1}{2}\frac{\Delta[N_2O_5]}{\Delta t} = \frac{1}{4}\frac{\Delta[NO_2]}{\Delta t} = \frac{\Delta[O_2]}{\Delta t}$$

16.17 Reaction rate is expressed in terms of changes in concentration of reactants and products. Write a balanced equation for
$$\text{Rate} = -\frac{\Delta[CH_4]}{\Delta t} = -\frac{1}{2}\frac{\Delta[O_2]}{\Delta t} = \frac{1}{2}\frac{\Delta[H_2O]}{\Delta t} = \frac{\Delta[CO_2]}{\Delta t}$$

16.18 The decomposition of nitrosyl bromide is followed by measuring total pressure because the number of moles of gas changes; it cannot be followed colorimetrically because both NOBr and Br_2 are reddish brown:
$$2NOBr(g) \longrightarrow 2NO(g) + Br_2(g)$$
Use the data below to answer the following:
(a) Determine the average rate over the entire experiment.
(b) Determine the average rate between 2.00 and 4.00 s.
(c) Use graphical methods to estimate the initial reaction rate.
(d) Use graphical methods to estimate the rate at 7.00 s.
(e) At what time does the instantaneous rate equal the average rate over the entire experiment?

Time (s)	[NOBr] (mol/L)
0.00	0.0100
2.00	0.0071
4.00	0.0055
6.00	0.0045
8.00	0.0038
10.00	0.0033

16.19 Although the depletion of stratospheric ozone threatens life on Earth today, its accumulation was one of the crucial processes that allowed life to develop in prehistoric times:

$$3O_2(g) \longrightarrow 2O_3(g)$$

(a) Express the reaction rate in terms of $[O_2]$ and $[O_3]$.
(b) At a given instant, the reaction rate in terms of $[O_2]$ is 2.17×10^{-5} mol/L·s. What is it in terms of $[O_3]$?

The Rate Law and Its Components
(Sample Problems 16.2 and 16.3)

16.20 The rate law for the general reaction

$$aA + bB + \cdots \longrightarrow cC + dD + \cdots$$

is rate $= k[A]^m[B]^n \cdots$.
(a) Explain the meaning of k.
(b) Explain the meanings of m and n. Does $m = a$ and $n = b$? Explain.
(c) If the reaction is first order in A and second order in B, and time is measured in minutes (min), what are the units for k?

16.21 By what factor does the rate change in each of the following cases (assuming constant temperature)?
(a) A reaction is first order in reactant A, and [A] is doubled.
(b) A reaction is second order in reactant B, and [B] is halved.
(c) A reaction is second order in reactant C, and [C] is tripled.

16.22 Give the individual reaction orders for all substances and the overall reaction order from the following rate law:

$$\text{Rate} = k[BrO_3^-][Br^-][H^+]^2$$

16.23 Give the individual reaction orders for all substances and the overall reaction order from the following rate law:

$$\text{Rate} = k\frac{[O_3]^2}{[O_2]}$$

16.24 By what factor does the rate in Problem 16.22 change if each of the following changes occurs: (a) $[BrO_3^-]$ is doubled; (b) $[Br^-]$ is halved; (c) $[H^+]$ is quadrupled?

16.25 By what factor does the rate in Problem 16.23 change if each of the following changes occurs: (a) $[O_3]$ is doubled; (b) $[O_2]$ is doubled; (c) $[O_2]$ is halved?

16.26 For the reaction

$$4A(g) + 3B(g) \longrightarrow 2C(g)$$

the following data were obtained at constant temperature:

Experiment	Initial [A] (mol/L)	Initial [B] (mol/L)	Initial Rate (mol/L·min)
1	0.100	0.100	5.00
2	0.300	0.100	45.0
3	0.100	0.200	10.0
4	0.300	0.200	90.0

(a) What is the order with respect to each reactant? (b) Write the rate law. (c) Calculate k (using the data from experiment 1).

16.27 For the reaction

$$A(g) + B(g) + C(g) \longrightarrow D(g)$$

the following data were obtained at constant temperature:

Exp't	Initial [A] (mol/L)	Initial [B] (mol/L)	Initial [C] (mol/L)	Initial Rate (mol/L·s)
1	0.0500	0.0500	0.0100	6.25×10^{-3}
2	0.1000	0.0500	0.0100	1.25×10^{-2}
3	0.1000	0.1000	0.0100	5.00×10^{-2}
4	0.0500	0.0500	0.0200	6.25×10^{-3}

(a) What is the order with respect to each reactant? (b) Write the rate law. (c) Calculate k (using the data from experiment 1).

16.28 Phosgene is a toxic gas prepared by the reaction of carbon monoxide with chlorine:

$$CO(g) + Cl_2(g) \longrightarrow COCl_2(g)$$

These data were obtained in a kinetics study of its formation:

Experiment	Initial [CO] (mol/L)	Initial [Cl₂] (mol/L)	Initial Rate (mol/L·s)
1	1.00	0.100	1.29×10^{-29}
2	0.100	0.100	1.33×10^{-30}
3	0.100	1.00	1.30×10^{-29}
4	0.100	0.0100	1.32×10^{-31}

(a) Write the rate law for the formation of phosgene.
(b) Calculate the average value of the rate constant.

Integrated Rate Laws: Concentration Changes Over Time
(Sample Problems 16.4 and 16.5)

16.29 How are integrated rate laws used to determine reaction order? What is the order in reactant if a plot of
(a) The natural logarithm of [reactant] vs. time is linear?
(b) The inverse of [reactant] vs. time is linear?
(c) [Reactant] vs. time is linear?

16.30 Define the *half-life* of a reaction. Explain on the molecular level why the half-life of a first-order reaction is constant.

16.31 For the simple decomposition reaction

$$AB(g) \longrightarrow A(g) + B(g)$$

rate $= k[AB]^2$ and $k = 0.2$ L/mol·s. How long will it take for [AB] to reach $\frac{1}{3}$ of its initial concentration of 1.50 M?

16.32 For the reaction in Problem 16.31, what is [AB] after 10.0 s?

16.33 In a first-order decomposition reaction, 50.0% of a compound decomposes in 10.5 min. (a) What is the rate constant of the reaction? (b) How long does it take for 75.0% of the compound to decompose?

16.34 A decomposition reaction has a rate constant of 0.0012 yr^{-1}. (a) What is the half-life of the reaction? (b) How long does it take for [reactant] to reach 12.5% of its original value?

The Effect of Temperature on Reaction Rate
(Sample Problem 16.6)

16.35 Use the exponential term in the Arrhenius equation to explain how temperature affects reaction rate.

16.36 How is the activation energy determined from the Arrhenius equation?

16.37 (a) Graph the relationship between k (y axis) and T (x axis). (b) Graph the relationship between ln k (y axis) and $1/T$ (x axis). How is the activation energy determined from this graph?

16.38 The rate constant of a reaction is 4.7×10^{-3} s^{-1} at 25°C, and the activation energy is 33.6 kJ/mol. What is k at 75°C?

16.39 The rate constant of a reaction is 4.50×10^{-5} L/mol·s at 195°C and 3.20×10^{-3} L/mol·s at 258°C. What is the activation energy of the reaction?

16.40 Understanding the high-temperature formation and breakdown of the nitrogen oxides is essential for controlling the pollutants generated from power plants and cars. The first-order breakdown of dinitrogen monoxide to its elements has rate constants of 0.76/s at 727°C and 0.87/s at 757°C. What is the activation energy of this reaction?

Explaining the Effects of Concentration and Temperature
(Sample Problem 16.7)

16.41 What is the central idea of collision theory? How does this idea explain the effect of concentration on reaction rate?

16.42 Is collision frequency the only factor affecting rate? Explain.

16.43 Arrhenius proposed that each reaction has an energy threshold that must be reached for the particles to react. The kinetic theory of gases proposes that the average kinetic energy of the particles is proportional to the absolute temperature. How do these concepts relate to the effect of temperature on rate?

16.44 (a) For a reaction with a given E_a, how does an increase in T affect the rate? (b) For a reaction at a given T, how does a decrease in E_a affect the rate?

16.45 Assuming the activation energies are equal, which of the following reactions will occur at a higher rate at 50°C? Explain:

$$NH_3(g) + HCl(g) \longrightarrow NH_4Cl(s)$$
$$N(CH_3)_3(g) + HCl(g) \longrightarrow (CH_3)_3NHCl(s)$$

16.46 For the reaction $A(g) + B(g) \longrightarrow AB(g)$, how many unique collisions between A and B are possible if there are four particles of A and three particles of B present in the vessel?

16.47 For the reaction $A(g) + B(g) \longrightarrow AB(g)$, how many unique collisions between A and B are possible if 1.01 mol of $A(g)$ and 2.12 mol of $B(g)$ are present in the vessel?

16.48 At 25°C, what is the fraction of collisions with energy equal to or greater than an activation energy of 100. kJ/mol?

16.49 If the temperature in Problem 16.48 is increased to 50.°C, by what factor does the fraction of collisions with energy equal to or greater than the activation energy change?

16.50 For the reaction $ABC + D \rightleftharpoons AB + CD$, $\Delta H^{\circ}_{rxn} = -55$ kJ/mol and $E_{a(fwd)} = 215$ kJ/mol. Assuming a one-step reaction, (a) draw a reaction energy diagram; (b) calculate $E_{a(rev)}$; and (c) sketch a possible transition state if ABC is V-shaped.

16.51 For the reaction $A_2 + B_2 \longrightarrow 2AB$, $E_{a(fwd)} = 125$ kJ/mol and $E_{a(rev)} = 85$ kJ/mol. Assuming the reaction occurs in one step, (a) draw a reaction energy diagram; (b) calculate ΔH°_{rxn}; and (c) sketch a possible transition state.

16.52 Aqua regia, a mixture of HCl and HNO_3, has been used since alchemical times to dissolve many metals, including gold. Its orange color is due to the presence of nitrosyl chloride. Consider this one-step reaction for the formation of this compound:

$$NO(g) + Cl_2(g) \longrightarrow NOCl(g) + Cl(g) \qquad \Delta H^{\circ} = 83 \text{ kJ}$$

(a) Draw a reaction energy diagram, given $E_{a(fwd)}$ is 86 kJ/mol.
(b) Calculate $E_{a(rev)}$.
(c) Sketch a possible transition state for the reaction. (*Note:* The atom sequence of nitrosyl chloride is Cl—N—O.)

Reaction Mechanisms: Steps in the Overall Reaction
(Sample Problem 16.8)

16.53 Is the rate of an overall reaction lower, higher, or equal to the average rate of the individual steps? Explain.

16.54 Explain why the coefficients of an elementary step equal the reaction orders of its rate law but those of an overall reaction do not.

16.55 Is it possible for more than one mechanism to be consistent with the rate law of a given reaction? Explain.

16.56 What is the difference between a reaction intermediate and a transition state?

16.57 Why is a bimolecular step more reasonable physically than a termolecular step?

16.58 If a slow step precedes a fast step in a two-step mechanism, do the substances in the fast step appear in the rate law? Explain.

16.59 A proposed mechanism for the reaction of carbon dioxide with hydroxide ion in aqueous solution is
(1) $CO_2(aq) + OH^-(aq) \longrightarrow HCO_3^-(aq)$ [slow]
(2) $HCO_3^-(aq) + OH^-(aq) \longrightarrow CO_3^{2-}(aq) + H_2O(l)$ [fast]
(a) What is the overall reaction equation?
(b) Identify the intermediate(s), if any.
(c) What are the molecularity and the rate law for each step?
(d) Is the mechanism consistent with the actual rate law: rate $= k[CO_2][OH^-]$?

16.60 A proposed mechanism for the gas-phase reaction between chlorine and nitrogen dioxide is
(1) $Cl_2(g) + NO_2(g) \longrightarrow Cl(g) + NO_2Cl(g)$ [slow]
(1) $Cl(g) + NO_2(g) \longrightarrow NO_2Cl(g)$ [fast]
(a) What is the overall reaction equation?
(b) Identify the intermediate(s), if any.
(c) What are the molecularity and the rate law for each step?
(d) Is the mechanism consistent with the actual rate law: rate $= k[Cl_2][NO_2]$?

16.61 The proposed mechanism for a reaction is
(1) $A(g) + B(g) \rightleftharpoons X(g)$ [fast]
(2) $X(g) + C(g) \longrightarrow Y(g)$ [slow]
(3) $Y(g) \longrightarrow D(g)$ [fast]
(a) What is the overall equation?
(b) Identify the intermediate(s), if any.
(c) What are the molecularity and the rate law for each step?
(d) Is the mechanism consistent with the actual rate law: rate $= k[A][B][C]$?
(e) Is the following one-step mechanism equally valid: $A(g) + B(g) + C(g) \longrightarrow D(g)$?

16.62 Consider the following mechanism:
(1) $ClO^-(aq) + H_2O(l) \rightleftharpoons HClO(aq) + OH^-(aq)$ [fast]
(2) $I^-(aq) + HClO(aq) \longrightarrow HIO(aq) + Cl^-(aq)$ [slow]
(3) $OH^-(aq) + HIO(aq) \longrightarrow H_2O(l) + IO^-(aq)$ [fast]
(a) What is the overall equation?
(b) Identify the intermediate(s), if any.
(c) What are the molecularity and the rate law for each step?
(d) Is the mechanism consistent with the actual rate law: rate $= k[ClO^-][I^-]$?

16.63 In a study of nitrosyl halides, a chemist proposes the following mechanism for the synthesis of nitrosyl bromide:
$$NO(g) + Br_2(g) \rightleftharpoons NOBr_2(g) \qquad \text{[fast]}$$
$$NOBr_2(g) + NO(g) \longrightarrow 2NOBr(g) \qquad \text{[slow]}$$
If the rate law is rate $= k[NO]^2[Br_2]$, is the proposed mechanism valid? If so, show that it satisfies the three criteria for validity.

16.64 The rate law for $2NO(g) + O_2(g) \longrightarrow 2NO_2(g)$ is rate $= k[NO]^2[O_2]$. In addition to the mechanism in the text, the following ones have been proposed:
I $2NO(g) + O_2(g) \longrightarrow 2NO_2(g)$
II $2NO(g) \rightleftharpoons N_2O_2(g)$ [fast]
    $N_2O_2(g) + O_2(g) \longrightarrow 2NO_2(g)$ [slow]
III $2NO(g) \rightleftharpoons N_2(g) + O_2(g)$ [fast]
    $N_2(g) + 2O_2(g) \longrightarrow 2NO_2(g)$ [slow]
(a) Which of these mechanisms is consistent with the rate law?
(b) Which is most reasonable chemically? Why?

Catalysis: Speeding Up a Chemical Reaction

16.65 Consider the reaction $N_2O(g) \xrightarrow{Au} N_2(g) + \frac{1}{2}O_2(g)$.
(a) Is the gold a homogeneous or a heterogeneous catalyst?
(b) On the same set of axes, sketch the reaction energy diagrams for the catalyzed and the uncatalyzed reactions.

16.66 Does a catalyst increase reaction rate by the same means as a rise in temperature does? Explain.

Comprehensive Problems

Problems with an asterisk (*) are more challenging.

16.67 Consider the following reaction energy diagram:

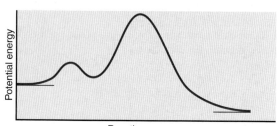

Reaction progress

(a) How many elementary steps are in the reaction mechanism?
(b) Which step is rate limiting?
(c) Is the overall reaction exothermic or endothermic?

16.68 The catalytic destruction of ozone occurs via a two-step mechanism, where X can be any of several species:
(1) $X + O_3 \longrightarrow XO + O_2$ [slow]
(2) $XO + O \longrightarrow X + O_2$ [fast]
(a) Write the overall reaction.
(b) Write the rate law for each step.
(c) X acts as _____, and XO acts as _____.
(d) High-flying aircraft release NO, which catalyzes this process, into the stratosphere. When O_3 and NO concentrations are 5×10^{12} molecule/cm^3 and 1.0×10^9 molecule/cm^3, respectively, what is the rate of O_3 depletion (k for the rate-determining step is 6×10^{-15} cm^3/molecule·s)?
(e) Is the O_3 concentration in part (d) reasonable for this reaction, given that stratospheric O_3 never exceeds 10 mg/L?

16.69 A slightly bruised apple will rot extensively in about 4 days at room temperature (20°C). If it is kept in the refrigerator at 0°C, the same extent of rotting takes about 16 days. What is the activation energy for the rotting reaction?

16.70 Benzoyl peroxide, the substance most widely used against acne, has a half-life of 9.8×10^3 days when refrigerated. How long will it take to lose 5% of its potency (95% remaining)?

16.71 The rate law for the reaction

$$NO_2(g) + CO(g) \longrightarrow NO(g) + CO_2(g)$$

is rate $= k[NO_2]^2$; one possible mechanism is shown on p. 526.
(a) Draw a reaction energy diagram for that mechanism, given that $\Delta H°_{\text{overall}} = -226$ kJ/mol.
(b) The following alternative mechanism has been proposed:
(1) $2NO_2(g) \longrightarrow N_2(g) + 2O_2(g)$ [slow]
(2) $2CO(g) + O_2(g) \longrightarrow 2CO_2(g)$ [fast]
(3) $N_2(g) + O_2(g) \longrightarrow 2NO(g)$ [fast]
Is the alternative mechanism consistent with the rate law? Is one mechanism more reasonable physically? Explain.

16.72 In acidic solution, the breakdown of sucrose into glucose and fructose has this rate law: rate $= k[H^+][\text{sucrose}]$. The initial rate of sucrose breakdown is measured in a solution that is 0.01 M

H^+, 1.0 M sucrose, 0.1 M fructose, and 0.1 M glucose. How does the rate change if
(a) [Sucrose] is changed to 2.5 M?
(b) [Sucrose], [fructose], and [glucose] are all changed to 0.5 M?
(c) [H^+] is changed to 0.0001 M?
(d) [Sucrose] and [H^+] are both changed to 0.1 M?

16.73 Biacetyl, the flavoring that makes margarine taste "just like butter," is extremely stable at room temperature, but at 200°C it undergoes a first-order breakdown with a half-life of 9.0 min. An industrial flavor-enhancing process requires that a biacetyl-flavored food be heated briefly at 200°C. How long can the food be heated and retain 85% of its buttery flavor?

16.74 At body temperature (37°C), k of an enzyme-catalyzed reaction is 2.3×10^{14} times greater than k of the uncatalyzed reaction. Assuming that the frequency factor A is the same for both reactions, by how much does the enzyme lower the E_a?

16.75 A biochemist studying breakdown of the insecticide DDT finds that it decomposes by a first-order reaction with a half-life of 12 yr. How long does it take DDT in a soil sample to decompose from 275 ppbm to 10. ppbm (parts per billion by mass)?

16.76 Proteins in the body undergo continual breakdown and synthesis. Insulin is a polypeptide hormone that stimulates fat and muscle to take up glucose. Once released from the pancreas, it has a first-order half-life in the blood of 8.0 min. To maintain an adequate blood concentration of insulin, it must be replenished in a time interval equal to $1/k$. How long is this interval?

16.77 The hydrolysis of sucrose occurs by this overall reaction:

$$\underset{\text{sucrose}}{C_{12}H_{22}O_{11}(s)} + H_2O(l) \longrightarrow \underset{\text{glucose}}{C_6H_{12}O_6(aq)} + \underset{\text{fructose}}{C_6H_{12}O_6(aq)}$$

A nutritional biochemist obtains the following kinetic data:

[Sucrose] (mol/L)	Time (h)
0.501	0.00
0.451	0.50
0.404	1.00
0.363	1.50
0.267	3.00

(a) Determine the rate constant and the half-life of the reaction.
(b) How long does it take to hydrolyze 75% of the sucrose?
(c) Other studies have shown that this reaction is actually second order overall but appears to follow first-order kinetics. (Such a reaction is called a *pseudo–first-order reaction*.) Suggest a reason for this apparent first-order behavior.

16.78 Is each of these statements true? If not, explain why.
(a) At a given T, all molecules have the same kinetic energy.
(b) Halving the P of a gaseous reaction doubles the rate.
(c) A higher activation energy gives a lower reaction rate.
(d) A temperature rise of 10°C doubles the rate of any reaction.
(e) If reactant molecules collide with greater energy than the activation energy, they change into product molecules.
(f) The activation energy of a reaction depends on temperature.
(g) The rate of a reaction increases as the reaction proceeds.
(h) Activation energy depends on collision frequency.
(i) A catalyst increases the rate by increasing collision frequency.
(j) Exothermic reactions are faster than endothermic reactions.
(k) Temperature has no effect on the frequency factor (A).
(l) The activation energy of a reaction is lowered by a catalyst.
(m) For most reactions, ΔH_{rxn} is lowered by a catalyst.

(n) The orientation probability factor (p) is near 1 for reactions between single atoms.

(o) The initial rate of a reaction is its maximum rate.

(p) A bimolecular reaction is generally twice as fast as a unimolecular reaction.

(q) The molecularity of an elementary reaction is proportional to the molecular complexity of the reactant(s).

16.79 Even when a mechanism is consistent with the rate law, later experimentation may show it to be incorrect or only one of several alternatives. As an example, the reaction between hydrogen and iodine has the following rate law: rate = $k[H_2][I_2]$. The long-accepted mechanism proposed a single bimolecular step; that is, the overall reaction was thought to be elementary:

$$H_2(g) + I_2(g) \longrightarrow 2HI(g)$$

In the 1960s, however, spectroscopic evidence showed the presence of free I atoms during the reaction. Kineticists have since proposed a three-step mechanism:

(1) $I_2(g) \rightleftharpoons 2I(g)$ [fast]
(2) $H_2(g) + I(g) \rightleftharpoons H_2I(g)$ [fast]
(3) $H_2I(g) + I(g) \longrightarrow 2HI(g)$ [slow]

Show that this mechanism is consistent with the rate law.

* 16.80 Many drugs decompose in blood by a first-order process. (a) Two tablets of aspirin supply 0.60 g of the active compound. After 30 min, this compound reaches a maximum concentration of 2 mg/100 mL of blood. If the half-life for its breakdown is 90 min, what is its concentration (in mg/100 mL) 2.5 h after it reaches its maximum concentration?

(b) In 8.0 h, secobarbital sodium, a common sedative, reaches a blood level that is 18% of its maximum. What is $t_{1/2}$ of the decomposition of secobarbital sodium in blood?

(c) The blood level of the sedative phenobarbital sodium drops to 59% of its maximum after 20. h. What is $t_{1/2}$ for its breakdown in blood?

(d) For the decomposition of an antibiotic in a person with a normal temperature (98.6°F), $k = 3.1\times10^{-5}$ s^{-1}; for a person with a fever at 101.9°F, $k = 3.9\times10^{-5}$ s^{-1}. If the sick person must take another pill when $\frac{2}{3}$ of the first pill has decomposed, how many hours should she wait to take a second pill? A third pill?

(e) Calculate E_a for decomposition of the antibiotic in part (d).

16.81 In Houston (near sea level), water boils at 100.0°C. In Cripple Creek, Colorado (near 9500 ft), it boils at 90.0°C. If it takes 4.8 min to cook an egg in Cripple Creek and 4.5 min in Houston, what is E_a for this process?

16.82 In the lower atmosphere, ozone is one of the components of photochemical smog. It is generated in air when nitrogen dioxide, formed by the oxidation of nitrogen monoxide from car exhaust, reacts by the following mechanism:

(1) $NO_2(g) \xrightarrow{k_1} NO(g) + O(g)$
(2) $O(g) + O_2(g) \xrightarrow{k_2} O_3(g)$

Assuming the rate of formation of atomic oxygen in step 1 equals the rate of its consumption in step 2, use the data below to calculate (a) the concentration of atomic oxygen [O]; (b) the rate of ozone formation.

$k_1 = 6.0\times10^{-3}$ s^{-1} $[NO_2] = 4.0\times10^{-9}$ M
$k_2 = 1.0\times10^6$ L/mol·s $[O_2] = 1.0\times10^{-2}$ M

16.83 Chlorine is commonly used to disinfect drinking water, and inactivation of pathogens by chlorine follows first-order kinetics. The following data show *E. coli* inactivation:

Contact time (min)	Percent (%) inactivation
0.00	0.0
0.50	68.3
1.00	90.0
1.50	96.8
2.00	99.0
2.50	99.7
3.00	99.9

(a) Determine the first-order inactivation constant, k. [*Hint:* % inactivation = $100 \times (1 - [A]_t/[A]_0)$.]

(b) How much contact time is required for 95% inactivation?

16.84 The reaction and rate law for the gas-phase decomposition of dinitrogen pentaoxide are

$$2N_2O_5(g) \longrightarrow 4NO_2(g) + O_2(g) \qquad \text{rate} = k[N_2O_5]$$

Which of the following can be considered valid mechanisms for the reaction?

I   One-step collision

II $2N_2O_5(g) \longrightarrow 2NO_3(g) + 2NO_2(g)$ [slow]
     $2NO_3(g) \longrightarrow 2NO_2(g) + 2O(g)$ [fast]
     $2O(g) \longrightarrow O_2(g)$ [fast]

III $N_2O_5(g) \rightleftharpoons NO_3(g) + NO_2(g)$ [fast]
     $NO_2(g) + N_2O_5(g) \longrightarrow 3NO_2(g) + O(g)$ [slow]
     $NO_3(g) + O(g) \longrightarrow NO_2(g) + O_2(g)$ [fast]

IV $2N_2O_5(g) \rightleftharpoons 2NO_2(g) + N_2O_3(g) + 3O(g)$ [fast]
     $N_2O_3(g) + O(g) \longrightarrow 2NO_2(g)$ [slow]
     $2O(g) \longrightarrow O_2(g)$ [fast]

V $2N_2O_5(g) \longrightarrow N_4O_{10}(g)$ [slow]
     $N_4O_{10}(g) \longrightarrow 4NO_2(g) + O_2(g)$ [fast]

* 16.85 Consider the following organic reaction, in which one halogen replaces another in an alkyl halide:

$$CH_3CH_2Br + KI \longrightarrow CH_3CH_2I + KBr$$

In acetone, this particular reaction goes to completion because KI is soluble in acetone but KBr is not. In the mechanism, I^- approaches the carbon *opposite* to the Br (see Figure 16.16, with I^- instead of OH^-). After Br^- has been replaced by I^- and precipitates as KBr, other I^- ions react with the ethyl iodide by the same mechanism.

(a) If we designate the carbon bonded to the halogen as C-1, what is the shape around C-1 and the hybridization of C-1 in ethyl iodide?

(b) In the transition state, one of the two lobes of the unhybridized $2p$ orbital of C-1 overlaps a p orbital of I, while the other lobe overlaps a p orbital of Br. What is the shape around C-1 and the hybridization of C-1 in the transition state?

(c) The deuterated reactant, CH_3CHDBr (where D is ^2H), has two optical isomers because C-1 is chiral. When the reaction is run with one of the isomers, the ethyl iodide is *not* optically active. Explain.

CHAPTER SEVENTEEN

Equilibrium: The Extent of Chemical Reactions

Balancing To and Fro As you'll learn in this chapter, the continual back and forth flow of students going to class mimics the forward and reverse steps of a chemical reaction in a state of dynamic equilibrium.

Key Principles

◆ The *extent (yield)* of a reaction is not necessarily related to its *rate*.

◆ When the rates of the forward and reverse steps of a reaction are *equal*, the system has reached *equilibrium*. After this point, there is no further observable change. The *ratio* of the rate constants equals the *equilibrium constant, K*. The size of *K* is directly related to the *extent* of the reaction at a given temperature.

◆ The *reaction quotient, Q*, is a particular ratio of product and reactant terms, written *directly* from the balanced equation. Its value *changes continually* until the system reaches equilibrium, where *Q = K*.

◆ The ideal gas law is used to relate an equilibrium constant based on concentrations, K_c, to one based on pressures, K_p.

◆ At any point, we can learn the direction of a reaction by comparing the values of *Q* and *K*.

◆ If the initial concentration of reactant is large relative to the change in concentration to reach equilibrium, we can make the *simplifying assumption* that the change can be neglected in performing calculations.

◆ If a system at equilibrium is *disturbed* by a change in conditions (concentration, pressure, or temperature), it will temporarily be *out of equilibrium*. However, it will then undergo a *net reaction* to reach equilibrium again (*Le Châtelier's principle*). A change in concentration or pressure does not affect *K*, but a change in temperature does.

Outline

Our study of kinetics in the last chapter addressed a different aspect of reaction chemistry than our upcoming study of equilibrium:

- Kinetics applies to the *speed* of a reaction, the concentration of product that appears (or of reactant that disappears) per unit time.
- Equilibrium applies to the *extent* of a reaction, the concentration of product that has appeared after an unlimited time, or once no further change occurs.

Just as reactions vary greatly in their speed, they also vary in their extent. A fast reaction may go almost completely or barely at all toward products. Consider the dissociation of an acid in water. In 1 M HCl, virtually all the hydrogen chloride molecules are dissociated into ions. In contrast, in 1 M CH_3COOH, fewer than 1% of the acetic acid molecules are dissociated at any given time. Yet both reactions take less than a second to reach completion. Similarly, some slow reactions eventually yield a large amount of product, whereas others yield very little. After a few years at ordinary temperatures, a steel water-storage tank will rust, and it will do so completely given enough time; but no matter how long you wait, the water inside will not decompose to hydrogen and oxygen.

Knowing the extent of a given reaction is crucial. How much product— medicine, polymer, or fuel—can you obtain from a particular reaction mixture? How can you adjust conditions to obtain more? If a reaction is slow but has a good yield, will a catalyst speed it up enough to make it useful?

In this chapter, we consider equilibrium principles in systems of gases and pure liquids and solids; we'll discuss various solution equilibria in the next two chapters.

17.1 THE EQUILIBRIUM STATE AND THE EQUILIBRIUM CONSTANT

Countless experiments with chemical systems have shown that, in a state of equilibrium, *the concentrations of reactants and products no longer change with time.* This apparent cessation of chemical activity occurs because *all reactions are reversible.* Let's examine a chemical system at the macroscopic and molecular levels to see how the equilibrium state arises. The system consists of two gases, colorless dinitrogen tetraoxide and brown nitrogen dioxide:

$$N_2O_4(g; \text{colorless}) \rightleftharpoons 2NO_2(g; \text{brown})$$

When we introduce some $N_2O_4(l)$ into a sealed flask kept at 200°C, a change occurs immediately. The liquid vaporizes (bp = 21°C) and the gas begins to turn pale brown. The color darkens, and after a few moments, the color stops changing (Figure 17.1 on the next page).

On the molecular level, a much more dynamic scene unfolds. The N_2O_4 molecules fly wildly throughout the flask, a few splitting into two NO_2 molecules. As time passes, more N_2O_4 molecules decompose and the concentration of NO_2 rises. As observers in the macroscopic world, we see the flask contents darken, because NO_2 is reddish brown. As the number of N_2O_4 molecules decreases, N_2O_4 decomposition slows. At the same time, increasing numbers of NO_2 molecules collide and combine, so re-formation of N_2O_4 speeds up. Eventually, N_2O_4 molecules decompose into NO_2 molecules as fast as the NO_2 molecules combine into N_2O_4. The system has reached equilibrium: *reactant and product concentrations stop changing because the forward and reverse rates have become equal:*

At equilibrium: $\text{rate}_{\text{fwd}} = \text{rate}_{\text{rev}}$ **(17.1)**

Concepts & Skills to Review
Before You Study This Chapter

- equilibrium vapor pressure (Section 12.2)
- equilibrium nature of a saturated solution (Section 13.3)
- dependence of rate on concentration (Sections 16.2 and 16.6)
- rate laws for elementary reactions (Section 16.7)
- function of a catalyst (Section 16.8)

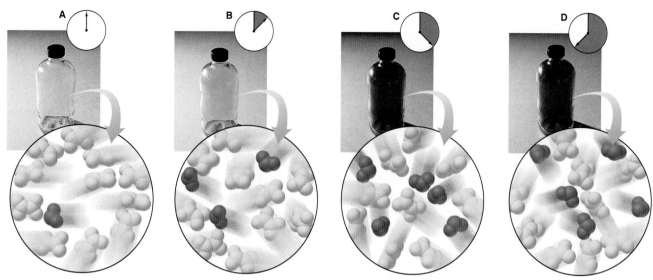

Figure 17.1 Reaching equilibrium on the macroscopic and molecular levels. A, When the experiment begins, the reaction mixture consists mostly of colorless N_2O_4. **B,** As N_2O_4 decomposes to reddish brown NO_2, the color of the mixture becomes pale brown. **C,** When equilibrium is reached, the concentrations of NO_2 and N_2O_4 are constant, and the color reaches its final intensity. **D,** Because the reaction continues in the forward and reverse directions at equal rates, the concentrations (and color) remain constant.

Thus, a system at equilibrium continues to be dynamic at the molecular level, but we observe *no further **net** change because changes in one direction are balanced by changes in the other.*

At a particular temperature, when the system reaches equilibrium, product and reactant concentrations are constant. Therefore, their ratio must be a constant. We'll use the N_2O_4-NO_2 system to derive this constant. At equilibrium, we have

$$\text{rate}_{\text{fwd}} = \text{rate}_{\text{rev}}$$

In this case, both forward and reverse reactions are elementary steps (Section 16.7), so we can write their rate laws directly from the balanced equation:

$$k_{\text{fwd}}[N_2O_4]_{\text{eq}} = k_{\text{rev}}[NO_2]^2_{\text{eq}}$$

where the subscript "eq" refers to concentrations at equilibrium. By rearranging, we set the ratio of the rate constants equal to the ratio of the concentration terms:

$$\frac{k_{\text{fwd}}}{k_{\text{rev}}} = \frac{[NO_2]^2_{\text{eq}}}{[N_2O_4]_{\text{eq}}}$$

The ratio of constants gives rise to a new overall constant called the **equilibrium constant (K):**

$$K = \frac{k_{\text{fwd}}}{k_{\text{rev}}} = \frac{[NO_2]^2_{\text{eq}}}{[N_2O_4]_{\text{eq}}} \tag{17.2}$$

The equilibrium constant K is a number equal to a particular ratio of equilibrium concentrations of product and reactant at a particular temperature. (We examine this idea closely in the next section.)

The magnitude of K is an indication of how far a reaction proceeds toward product at a given temperature. Remember, it is the opposing *rates* that are equal at equilibrium, not necessarily the concentrations. Indeed, different reactions, even at the same temperature, have a wide range of concentrations at equilibrium—from almost all reactant to almost all product—and, therefore, they have a wide

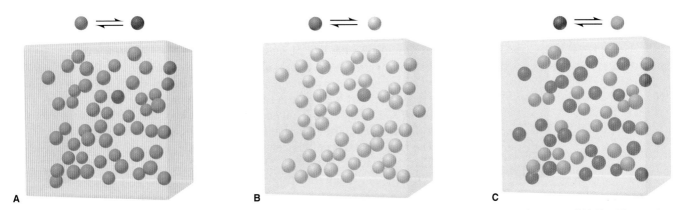

Figure 17.2 The range of equilibrium constants. A, A system that reaches equilibrium with very little product has a small K. For this reaction, $K = 1/49 = 0.020$. **B,** A system that reaches equilibrium with nearly all product has a large K. For this reaction, $K = 49/1 = 49$. **C,** A system that reaches equilibrium with significant concentrations of reactant and product has an intermediate K. For this reaction, $K = 25/25 = 1.0$.

range of equilibrium constants (Figure 17.2). Here are three examples of different magnitudes of K:

1. *Small K.* If a reaction yields very little product before reaching equilibrium, it has a small K, and we may even say there is "no reaction." For example, the oxidation of nitrogen barely proceeds at 1000 K:*

$$N_2(g) + O_2(g) \rightleftharpoons 2NO(g) \quad K = 1 \times 10^{-30}$$

2. *Large K.* Conversely, if a reaction reaches equilibrium with very little reactant remaining, it has a large K, and we say it "goes to completion." The oxidation of carbon monoxide goes to completion at 1000 K:

$$2CO(g) + O_2(g) \rightleftharpoons 2CO_2(g) \quad K = 2.2 \times 10^{22}$$

3. *Intermediate K.* When significant amounts of both reactant and product are present at equilibrium, K has an intermediate value, as when bromine monochloride breaks down to its elements at 1000 K:

$$2BrCl(g) \rightleftharpoons Br_2(g) + Cl_2(g) \quad K = 5$$

SECTION SUMMARY
The rate and extent of a reaction are not necessarily related. When the forward and reverse reactions occur at the same rate, the system has reached dynamic equilibrium and concentrations no longer change. The equilibrium constant (K) is a number based on a particular ratio of product and reactant concentrations. K is small for reactions that reach equilibrium with a high concentration of reactant(s) and large for reactions that reach equilibrium with a low concentration of reactant(s).

17.2 THE REACTION QUOTIENT AND THE EQUILIBRIUM CONSTANT

Our derivation of the equilibrium constant in Section 17.1 was based on kinetics. But the fundamental observation of equilibrium studies was stated many years before the principles of kinetics were developed. In 1864, two Norwegian chemists, Cato Guldberg and Peter Waage, observed that *at a given temperature, a chemical system reaches a state in which a particular ratio of reactant and*

---

*To distinguish the equilibrium constant from the Kelvin scale temperature unit, the equilibrium constant is an uppercase italic K, whereas the kelvin is an uppercase roman K. Also, because the kelvin is a unit, it always follows a number.

product concentrations has a constant value. This is one way of stating the **law of chemical equilibrium,** or the **law of mass action.** No mention of rates appears.

In formulating the law of mass action, Guldberg and Waage found that, *for a particular system and temperature, the same equilibrium state is attained regardless of how the reaction is run.* For example, in the N_2O_4-NO_2 system at 200°C, we can start with pure reactant (N_2O_4), pure product (NO_2), or any mixture of NO_2 and N_2O_4, and given enough time, the ratio of concentrations will attain the same equilibrium value (within experimental error).

The particular ratio of concentration terms that we write for a given reaction is called the **reaction quotient (Q).** For the reaction of N_2O_4 to form NO_2, the reaction quotient, based directly on the balanced equation as written, is

$$Q = \frac{[NO_2]^2}{[N_2O_4]}$$

As the reaction proceeds toward the equilibrium state, there is a continual, smooth change in the concentrations of reactants and products. Thus, the ratio of concentrations also changes: at the beginning of the reaction, the concentrations have initial values, and Q has an initial value; a moment later, after the reaction has proceeded a bit, the concentrations have slightly different values, and so does Q; another moment into the reaction, and there is more change in the concentrations and more change in Q; and on and on, *until the reacting system reaches equilibrium.* At that point, at a given temperature, the reactant and product concentrations have reached their equilibrium levels and no longer change. And the value of Q has reached its equilibrium value and no longer changes. It equals K at that temperature:

At equilibrium: $Q = K$ (17.3)

So, monitoring Q tells whether the system has reached equilibrium, how far away it is if it has not, and, as we discuss later, in which direction it is changing to reach equilibrium. Table 17.1 presents four experiments, each a different run of the N_2O_4-NO_2 reaction at 200°C. There are two essential points to note:

- The ratio of *initial* concentrations varies widely but always gives the same ratio of *equilibrium* concentrations.
- The *individual* equilibrium concentrations are different in each case, but the *ratio* of these equilibrium concentrations is constant.

The curves in Figure 17.3 show experiment 1 in Table 17.1. Note that $[N_2O_4]$ and $[NO_2]$ change smoothly during the course of the reaction and, thus, so does the value of Q. Once the system reaches equilibrium, as indicated by the constant brown color, the concentrations no longer change and Q equals K. In other words, for any given chemical system, K is a special value of Q that occurs when the reactant and product terms have their equilibrium values.

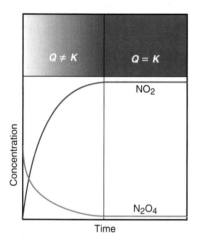

Figure 17.3 The change in Q during the N_2O_4-NO_2 reaction. The curved plots and the darkening brown screen above them show that $[N_2O_4]$ and $[NO_2]$, and therefore the value of Q, change with time. Before equilibrium is reached, the concentrations are changing continuously, so $Q \neq K$. Once equilibrium is reached (*vertical line*) and any time thereafter, $Q = K$.

Table 17.1	Initial and Equilibrium Concentration Ratios for the N_2O_4-NO_2 System at 200°C (473 K)					
	Initial		Ratio (Q)	Equilibrium		Ratio (K)
Exp't.	$[N_2O_4]$	$[NO_2]$	$[NO_2]^2/[N_2O_4]$	$[N_2O_4]_{eq}$	$[NO_2]_{eq}$	$[NO_2]^2_{eq}/[N_2O_4]_{eq}$
1	0.1000	0.0000	0.0000	0.00357	0.193	10.4
2	0.0000	0.1000	∞	9.24×10^{-4}	9.82×10^{-2}	10.4
3	0.0500	0.0500	0.0500	0.00204	0.146	10.4
4	0.0750	0.0250	0.00833	0.00275	0.170	10.5

Writing the Reaction Quotient

In Chapter 16, you saw that the rate law for an overall reaction cannot be written from the balanced equation, but must be determined from rate data. In contrast, the reaction quotient *can* be written directly from the balanced equation: *Q is a ratio made up of product concentration terms multiplied together and divided by reactant concentration terms multiplied together, with each term raised to the power of its stoichiometric coefficient in the balanced equation.*

The most common form of the reaction quotient shows reactant and product terms as molar concentrations, which are designated by square brackets, [ ]. In the cases you've seen so far, K is the *equilibrium constant based on concentrations,* designated from now on as K_c. Similarly, we designate the reaction quotient based on concentrations as Q_c. For the general balanced equation

$$aA + bB \rightleftharpoons cC + dD$$

where a, b, c, and d are the stoichiometric coefficients, the reaction quotient is

$$Q_c = \frac{[C]^c[D]^d}{[A]^a[B]^b} \tag{17.4}$$

(Another form of the reaction quotient that we discuss later shows gaseous reactant and product terms as pressures.)

To construct the reaction quotient for any reaction, write the balanced equation first. For the formation of ammonia from its elements, for example, the balanced equation (with colored coefficients for easy reference) is

$$N_2(g) + 3H_2(g) \rightleftharpoons 2NH_3(g)$$

To construct the reaction quotient, we place the product term in the numerator and the reactant terms in the denominator, multiplied by each other, and raise each term to the power of its balancing coefficient (colored as in the equation):

$$Q_c = \frac{[NH_3]^2}{[N_2][H_2]^3}$$

Let's practice this essential skill.

SAMPLE PROBLEM 17.1 Writing the Reaction Quotient from the Balanced Equation

Problem Write the reaction quotient, Q_c, for each of the following reactions:
(a) The decomposition of dinitrogen pentaoxide, $N_2O_5(g) \rightleftharpoons NO_2(g) + O_2(g)$
(b) The combustion of propane gas, $C_3H_8(g) + O_2(g) \rightleftharpoons CO_2(g) + H_2O(g)$
Plan We balance the equations and then construct the reaction quotient (Equation 17.4).

Solution (a) $2N_2O_5(g) \rightleftharpoons 4NO_2(g) + O_2(g)$ $Q_c = \dfrac{[NO_2]^4[O_2]}{[N_2O_5]^2}$

(b) $C_3H_8(g) + 5O_2(g) \rightleftharpoons 3CO_2(g) + 4H_2O(g)$ $Q_c = \dfrac{[CO_2]^3[H_2O]^4}{[C_3H_8][O_2]^5}$

Check Always be sure that the exponents in Q are the same as the balancing coefficients. A good check is to reverse the process: turn the numerator into products and the denominator into reactants, and change the exponents to coefficients.

FOLLOW-UP PROBLEM 17.1 Write a reaction quotient, Q_c, for each of the following reactions (unbalanced):
(a) The first step in nitric acid production, $NH_3(g) + O_2(g) \rightleftharpoons NO(g) + H_2O(g)$
(b) The disproportionation of nitric oxide, $NO(g) \rightleftharpoons N_2O(g) + NO_2(g)$

Variations in the Form of the Reaction Quotient

As you'll see in the upcoming discussion, the reaction quotient Q is a collection of terms based on the balanced equation *exactly as written* for a given reaction. Therefore, the value of Q, which varies during the reaction, and the value of K, which is the constant value that Q attains when the system has reached equilibrium, also depend on how the balanced equation is written.

A Word about Units for Q and K In this text (and most others), *the values of Q and K are shown as unitless numbers.* This is because each term in the reaction quotient represents the *ratio* of the measured quantity of the substance (molar concentration or pressure) to the thermodynamic standard-state quantity of the substance. Recall from Section 6.6 that these standard states are 1 M for a substance in solution, 1 atm for gases, and the pure substance for a liquid or solid. Thus, a concentration of 1.20 M becomes $\dfrac{1.20\ M}{1\ M} = 1.20$; similarly, a pressure of 0.53 atm becomes $\dfrac{0.53\ \text{atm}}{1\ \text{atm}} = 0.53$. With such quantity terms unitless, the ratio of terms we use to find the value of Q (or K) is also unitless.

Form of Q for an Overall Reaction Notice that we've been writing reaction quotients without knowing whether an equation represents an individual reaction step or an overall multistep reaction. We can do this because we obtain the same expression for the overall reaction as we do when we combine the expressions for the individual steps. That is, *if an overall reaction is the **sum** of two or more reactions, the overall reaction quotient (or equilibrium constant) is the **product** of the reaction quotients (or equilibrium constants) for the steps:*

$$Q_{\text{overall}} = Q_1 \times Q_2 \times Q_3 \times \cdots$$

and

$$K_{\text{overall}} = K_1 \times K_2 \times K_3 \times \cdots \tag{17.5}$$

For example, consider an overall equation for the formation of nitrogen dioxide, the toxic pollutant that contributes to photochemical smog and acid rain:

$$N_2(g) + 2O_2(g) \rightleftharpoons 2NO_2(g)$$

We can construct the reaction quotient directly:

$$Q_{c(\text{overall})} = \frac{[NO_2]^2}{[N_2][O_2]^2}$$

The overall reaction actually occurs in two steps with NO serving as the intermediate:

(1) $N_2(g) + O_2(g) \rightleftharpoons 2NO(g)$ $K_{c1} = 4.3 \times 10^{-25}$
(2) $O_2(g) + 2NO(g) \rightleftharpoons 2NO_2(g)$ $K_{c2} = 6.4 \times 10^{9}$

The reaction quotients for these steps are

$$Q_{c1} = \frac{[NO]^2}{[N_2][O_2]} \quad \text{and} \quad Q_{c2} = \frac{[NO_2]^2}{[O_2][NO]^2}$$

The overall reaction quotient is the product of Q_{c1} and Q_{c2}:

$$\frac{[\cancel{NO}]^2}{[N_2][O_2]} \times \frac{[NO_2]^2}{[O_2][\cancel{NO}]^2} = \frac{[NO_2]^2}{[N_2][O_2]^2} = Q_{c(\text{overall})}$$

Similarly, the equilibrium constant for the overall reaction is the product of the equilibrium constants for the elementary steps:

$$K_{c(\text{overall})} = K_{c1} \times K_{c2} = (4.3 \times 10^{-25})(6.4 \times 10^{9}) = 2.8 \times 10^{-15}$$

Form of Q for a Forward and Reverse Reaction The form of the reaction quotient depends on the *direction* in which the balanced equation is written. Consider, for example, the oxidation of sulfur dioxide to sulfur trioxide. This reaction is a key step in acid rain formation and sulfuric acid production. The balanced equation is

$$2SO_2(g) + O_2(g) \rightleftharpoons 2SO_3(g)$$

The reaction quotient for this equation *as written* is

$$Q_{c(fwd)} = \frac{[SO_3]^2}{[SO_2]^2[O_2]}$$

If we had written the reverse reaction, the decomposition of sulfur trioxide,

$$2SO_3(g) \rightleftharpoons 2SO_2(g) + O_2(g)$$

the reaction quotient would be the *reciprocal* of $Q_{c(fwd)}$:

$$Q_{c(rev)} = \frac{[SO_2]^2[O_2]}{[SO_3]^2} = \frac{1}{Q_{c(fwd)}}$$

Thus, *a reaction quotient (or equilibrium constant) for a forward reaction is the* **reciprocal** *of the reaction quotient (or equilibrium constant) for the reverse reaction:*

$$Q_{c(fwd)} = \frac{1}{Q_{c(rev)}} \quad \text{and} \quad K_{c(fwd)} = \frac{1}{K_{c(rev)}} \tag{17.6}$$

The K_c values for the forward and reverse reactions at 1000 K are

$$K_{c(fwd)} = 261 \quad \text{and} \quad K_{c(rev)} = \frac{1}{K_{c(fwd)}} = \frac{1}{261} = 3.83 \times 10^{-3}$$

These values make sense: if the forward reaction goes far to the right (high K_c), the reverse reaction does not (low K_c).

Form of Q for a Reaction with Coefficients Multiplied by a Common Factor Multiplying all the coefficients of the equation by some factor also changes the form of Q. For example, multiplying all the coefficients in the previous equation for the formation of SO_3 by $\frac{1}{2}$ gives

$$SO_2(g) + \tfrac{1}{2}O_2(g) \rightleftharpoons SO_3(g)$$

For this equation, the reaction quotient is

$$Q'_{c(fwd)} = \frac{[SO_3]}{[SO_2][O_2]^{1/2}}$$

Notice that Q_c for the halved equation equals Q_c for the original equation raised to the $\frac{1}{2}$ power:

$$Q'_{c(fwd)} = Q_{c(fwd)}^{1/2} = \left(\frac{[SO_3]^2}{[SO_2]^2[O_2]}\right)^{1/2} = \frac{[SO_3]}{[SO_2][O_2]^{1/2}}$$

Once again, the same property holds for the equilibrium constants. Relating the halved reaction to the original, we have

$$K'_{c(fwd)} = K_{c(fwd)}^{1/2} = (261)^{1/2} = 16.2$$

In general, *if all the coefficients of the balanced equation are multiplied by some factor, that factor becomes the exponent for relating the reaction quotients and the equilibrium constants.* For a multiplying factor n, which we can write as

$$n(aA + bB \rightleftharpoons cC + dD)$$

the reaction quotient and equilibrium constant are

$$Q' = Q^n = \left(\frac{[C]^c[D]^d}{[A]^a[B]^b}\right)^n \quad \text{and} \quad K' = K^n \tag{17.7}$$

Form of Q for a Reaction Involving Pure Liquids and Solids Until now, we've looked at *homogeneous* equilibria, systems in which all the components of the reaction are in the same phase, such as a system of reacting gases. When the components are in different phases, the system reaches *heterogeneous* equilibrium.

Consider the decomposition of limestone to lime and carbon dioxide, in which a gas and two solids make up the reaction components:

$$CaCO_3(s) \rightleftharpoons CaO(s) + CO_2(g)$$

Based on the rules for writing the reaction quotient, we have

$$Q_c = \frac{[CaO][CO_2]}{[CaCO_3]}$$

A pure solid, however, such as $CaCO_3$ or CaO, always has the same "concentration" at a given temperature, that is, the same number of moles per liter of its volume, just as it has the same density (g/cm^3) at a given temperature. Therefore, the concentration of a pure solid is constant, as is the concentration of a pure liquid.

Because we are concerned only with concentrations that *change* as they approach equilibrium, *we eliminate the terms for pure liquids and solids from the reaction quotient.* We do this by multiplying both sides of the above equation by $[CaCO_3]$ and dividing both sides by $[CaO]$ to get a new reaction quotient. Thus, the only substance whose concentration can change is the gas CO_2:

$$Q_c' = Q_c \frac{[CaCO_3]}{[CaO]} = [CO_2]$$

No matter how much CaO and $CaCO_3$ are in the reaction vessel, *as long as some of each is present,* the reaction quotient equals the CO_2 concentration.

Table 17.2 summarizes the ways of writing Q and calculating K.

Table 17.2 Ways of Expressing Q and Calculating K		
Form of Chemical Equation	**Form of Q**	**Value of K**
Reference reaction: A $\rightleftharpoons$ B	$Q_{(ref)} = \dfrac{[B]}{[A]}$	$K_{(ref)} = \dfrac{[B]_{eq}}{[A]_{eq}}$
Reverse reaction: B $\rightleftharpoons$ A	$Q = \dfrac{1}{Q_{(ref)}} = \dfrac{[A]}{[B]}$	$K = \dfrac{1}{K_{(ref)}}$
Reaction as sum of two steps: (1) A $\rightleftharpoons$ C (2) C $\rightleftharpoons$ B	$Q_1 = \dfrac{[C]}{[A]}; Q_2 = \dfrac{[B]}{[C]}$ $Q_{overall} = Q_1 \times Q_2 = Q_{(ref)}$ $= \dfrac{[C]}{[A]} \times \dfrac{[B]}{[C]} = \dfrac{[B]}{[A]}$	 $K_{overall} = K_1 \times K_2$ $= K_{(ref)}$
Coefficients multiplied by n	$Q = Q_{(ref)}^n$	$K = K_{(ref)}^n$
Reaction with pure solid or liquid component, such as A(s)	$Q = Q_{(ref)}[A] = [B]$	$K = K_{(ref)}[A] = [B]$

SECTION SUMMARY
The reaction quotient, Q, is a particular ratio of product to reactant terms. Substituting experimental values into this expression gives the value of Q, which changes as the reaction proceeds. When the system reaches equilibrium at a particular temperature, $Q = K$. If a reaction is the sum of two or more steps, the overall Q (or K) is the product of the individual Q's (or K's). The *form* of Q is based directly on the balanced equation for the reaction exactly as written, so it changes if the equation is reversed or multiplied by some factor, and K changes accordingly. Pure liquids or solids do not appear as terms in the expression for Q because their concentrations are constant.

17.3 EXPRESSING EQUILIBRIA WITH PRESSURE TERMS: RELATION BETWEEN K_c AND K_p

It is easier to measure the pressure of a gas than its concentration and, as long as the gas behaves nearly ideally under the conditions of the experiment, the ideal gas law (Section 5.3) allows us to relate these variables to each other:

$$PV = nRT, \quad \text{so} \quad P = \frac{n}{V}RT \quad \text{or} \quad \frac{P}{RT} = \frac{n}{V}$$

where P is the pressure of a gas and n/V is its molar concentration (M). Thus, at constant temperature, *pressure is directly proportional to molar concentration*. When the substances involved in the reaction are gases, we can express the reaction quotient and calculate its value in terms of partial pressures instead of concentrations. For example, in the reaction between gaseous NO and O_2,

$$2NO(g) + O_2(g) \rightleftharpoons 2NO_2(g)$$

the reaction quotient based on partial pressures, Q_p, is

$$Q_p = \frac{P_{NO_2}^2}{P_{NO}^2 \times P_{O_2}}$$

The equilibrium constant obtained when all components are present at their equilibrium partial pressures is designated K_p, the *equilibrium constant based on pressures*. In many cases, K_p has a value different from K_c, but if you know one, you can calculate the other by noting the *change in amount (mol) of gas*, Δn_{gas}, from the balanced equation. Let's see this relationship by converting the terms in Q_c for the reaction of NO and O_2 to those in Q_p:

$$2NO(g) + O_2(g) \rightleftharpoons 2NO_2(g)$$

As the balanced equation shows,

$$3 \text{ mol } (2 \text{ mol} + 1 \text{ mol}) \text{ gaseous reactants} \rightleftharpoons 2 \text{ mol gaseous products}$$

With Δ meaning final *minus* initial (products *minus* reactants), we have

$$\Delta n_{gas} = \text{moles of gaseous product} - \text{moles of gaseous reactant} = 2 - 3 = -1$$

Keep this value of Δn_{gas} in mind because it appears in the algebraic conversion that follows. The reaction quotient based on concentrations is

$$Q_c = \frac{[NO_2]^2}{[NO]^2[O_2]}$$

Rearranging the ideal gas law to $n/V = P/RT$, we express concentrations as n/V and convert them to partial pressures, P; then we collect the RT terms and cancel:

$$Q_c = \frac{\dfrac{n_{NO_2}^2}{V^2}}{\dfrac{n_{NO}^2}{V^2} \times \dfrac{n_{O_2}}{V}} = \frac{\dfrac{P_{NO_2}^2}{(RT)^2}}{\dfrac{P_{NO}^2}{(RT)^2} \times \dfrac{P_{O_2}}{RT}} = \frac{P_{NO_2}^2}{P_{NO}^2 \times P_{O_2}} \times \frac{\dfrac{1}{(RT)^2}}{\dfrac{1}{(RT)^2} \times \dfrac{1}{RT}} = \frac{P_{NO_2}^2}{P_{NO}^2 \times P_{O_2}} \times RT$$

The far right side of the previous expression is Q_p multiplied by RT: $Q_c = Q_p(RT)$. Also, at equilibrium, $K_c = K_p(RT)$; thus, $K_p = \dfrac{K_c}{RT}$, or $K_c(RT)^{-1}$.

Notice that *the exponent of the RT term equals the change in the amount (mol) of gas (Δn_{gas}) from the balanced equation*, -1. Thus, in general, we have

$$K_p = K_c(RT)^{\Delta n_{gas}} \tag{17.8}$$

The units for the partial pressure terms in K_p are generally atmospheres, pascals, or torr, raised to some power, and the units of R must be consistent with those units. As Equation 17.8 shows, for those reactions in which the amount (mol) of gas does not change, we have $\Delta n_{gas} = 0$, so the RT term drops out and $K_p = K_c$.

SAMPLE PROBLEM 17.2   Converting Between K_c and K_p

Problem A chemical engineer injects limestone ($CaCO_3$) into the hot flue gas of a coal-burning power plant to form lime (CaO), which scrubs SO_2 from the gas and forms gypsum ($CaSO_4 \cdot 2H_2O$). Find K_c for the following reaction, if CO_2 pressure is in atmospheres:

$$CaCO_3(s) \rightleftharpoons CaO(s) + CO_2(g) \qquad K_p = 2.1 \times 10^{-4} \text{ (at 1000. K)}$$

Plan We know K_p (2.1×10^{-4}), so to convert between K_p and K_c, we must first determine Δn_{gas} from the balanced equation. Then we rearrange Equation 17.8. With gas pressure in atmospheres, R is 0.0821 atm·L/mol·K.

Solution Determining Δn_{gas}: There is 1 mol of gaseous product and no gaseous reactant, so $\Delta n_{gas} = 1 - 0 = 1$.

Rearranging Equation 17.8 and calculating K_c:

$$K_p = K_c(RT)^1 \qquad \text{so} \qquad K_c = K_p(RT)^{-1}$$
$$K_c = (2.1 \times 10^{-4})(0.0821 \times 1000.)^{-1} = 2.6 \times 10^{-6}$$

Check Work backward to see whether you obtain the given K_p:

$$K_p = (2.6 \times 10^{-6})(0.0821 \times 1000.) = 2.1 \times 10^{-4}$$

FOLLOW-UP PROBLEM 17.2 Calculate K_p for the following reaction:

$$PCl_3(g) + Cl_2(g) \rightleftharpoons PCl_5(g) \qquad K_c = 1.67 \text{ (at 500. K)}$$

SECTION SUMMARY

The reaction quotient and the equilibrium constant can be expressed in terms of concentrations (Q_c and K_c); for gases, they are expressed in terms of partial pressures (Q_p and K_p). The values of K_p and K_c are related by using the ideal gas law: $K_p = K_c(RT)^{\Delta n_{gas}}$.

17.4   REACTION DIRECTION: COMPARING Q AND K

Suppose you start a reaction with a mixture of reactants and products and you know the equilibrium constant at the temperature of the reaction. Because the value of Q can change, depending on the initial concentrations, Q can be smaller than K, larger than K, or, when the system reaches equilibrium, equal to K. By comparing the value of Q at a particular time with the known K, you can tell whether the reaction has attained equilibrium or, if not, in which direction it is progressing. With product terms in the numerator of Q and reactant terms in the denominator, *more product makes Q larger, and more reactant makes Q smaller.*

The three possible relative sizes of Q and K are shown in Figure 17.4.

- $Q < K$. If the value of Q is smaller than K, the denominator (reactants) is large relative to the numerator (products). For Q to become equal to K, the reactants must decrease and the products increase. In other words, the reaction will progress to the right, toward products, until equilibrium is reached:

$$\text{If } Q < K, \text{ reactants} \longrightarrow \text{products}$$

- $Q > K$. If Q is larger than K, the numerator (products) will decrease and the denominator (reactants) increase until equilibrium is reached. Therefore, the reaction will progress to the left, toward reactants:

$$\text{If } Q > K, \text{ reactants} \longleftarrow \text{products}$$

- $Q = K$. This situation exists only when the reactant and product concentrations (or pressures) have attained their equilibrium values. Thus, no further net change occurs:

$$\text{If } Q = K, \text{ reactants} \rightleftharpoons \text{products}$$

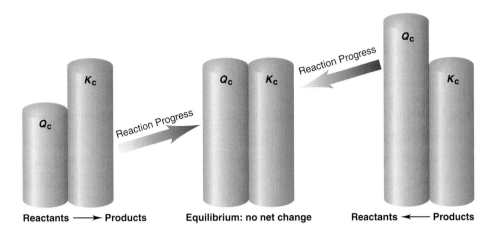

Figure 17.4 Reaction direction and the relative sizes of Q and K. When Q_c is smaller than K_c, the equilibrium of the reaction system shifts to the right, that is, toward products. When Q_c is larger than K_c, the equilibrium of the reaction system shifts to the left. Both shifts continue until $Q_c = K_c$. Note that the size of K_c remains the same throughout.

Reactants ⟶ Products   Equilibrium: no net change   Reactants ⟵ Products

---

SAMPLE PROBLEM 17.3   Comparing Q and K to Determine Reaction Direction

Problem For the reaction $N_2O_4(g) \rightleftharpoons 2NO_2(g)$, $K_c = 0.21$ at 100°C. At a point during the reaction, $[N_2O_4] = 0.12\ M$ and $[NO_2] = 0.55\ M$. Is the reaction at equilibrium? If not, in which direction is it progressing?

Plan We write the expression for Q_c, find its value by substituting the given concentrations, and then compare its value with the given K_c.

Solution Writing the reaction quotient and solving for Q_c:

$$Q_c = \frac{[NO_2]^2}{[N_2O_4]} = \frac{0.55^2}{0.12} = 2.5$$

With $Q_c > K_c$, the reaction is not at equilibrium and will proceed to the left until $Q_c = K_c$.

Check With $[NO_2] > [N_2O_4]$, we expect to obtain a value for Q_c that is greater than 0.21. If $Q_c > K_c$, the numerator will decrease and the denominator will increase until $Q_c = K_c$; that is, this reaction will proceed toward reactants.

FOLLOW-UP PROBLEM 17.3 Chloromethane forms by the reaction

$$CH_4(g) + Cl_2(g) \rightleftharpoons CH_3Cl(g) + HCl(g)$$

At 1500 K, $K_p = 1.6 \times 10^4$. In the reaction mixture, $P_{CH_4} = 0.13$ atm, $P_{Cl_2} = 0.035$ atm, $P_{CH_3Cl} = 0.24$ atm, and $P_{HCl} = 0.47$ atm. Is CH_3Cl or CH_4 forming?

SECTION SUMMARY

We compare the values of Q and K to determine the direction in which a reaction will proceed toward equilibrium.

- If $Q_c < K_c$, more product forms.
- If $Q_c > K_c$, more reactant forms.
- If $Q_c = K_c$, there is no net change.

17.5 HOW TO SOLVE EQUILIBRIUM PROBLEMS

As you've seen in the previous sections, three criteria define a system at equilibrium:

- Reactant and product concentrations are constant over time.
- The forward reaction rate equals the reverse reaction rate.
- The reaction quotient equals the equilibrium constant: $Q = K$.

Now, let's apply equilibrium principles to quantitative situations. Many kinds of equilibrium problems arise in the real world, as well as on chemistry exams, but we can group most of them into two types:

1. In one type, we are given equilibrium quantities (concentrations or partial pressures) and solve for K.
2. In the other type, we are given K and initial quantities and solve for the equilibrium quantities.

(From now on, the subscript "eq" is used only when it is not clear that a concentration is an equilibrium value.)

Using Quantities to Determine the Equilibrium Constant

There are two common variations on the type of equilibrium problem in which we solve for K: one involves a straightforward substitution of quantities, and the other requires first finding some of the quantities.

Substituting Given Equilibrium Quantities into Q to Find K In the straightforward case, we are given the equilibrium quantities and we must calculate K.

Suppose, for example, that equal amounts of gaseous hydrogen and iodine are injected into a 1.50-L reaction flask at a fixed temperature. In time, the following equilibrium is attained:

$$H_2(g) + I_2(g) \rightleftharpoons 2HI(g)$$

At equilibrium, analysis shows that the flask contains 1.80 mol of H_2, 1.80 mol of I_2, and 0.520 mol of HI. We calculate K_c by finding the concentrations and substituting them into the reaction quotient. Given the balanced equation, we then write the reaction quotient:

$$Q_c = \frac{[HI]^2}{[H_2][I_2]}$$

We first have to convert the amounts (mol) to concentrations (mol/L), using the flask volume of 1.50 L:

$$[H_2] = \frac{1.80 \text{ mol}}{1.50 \text{ L}} = 1.20 \ M$$

Similarly, $[I_2] = 1.20 \ M$, and $[HI] = 0.347 \ M$. Substituting these values into the expression for Q_c gives K_c:

$$K_c = \frac{(0.347)^2}{(1.20)(1.20)} = 8.36 \times 10^{-2}$$

Using a Reaction Table to Determine Equilibrium Quantities and Find K When some quantities are not given, we determine them first from the reaction stoichiometry and then find K. In the following example, pay close attention to a valuable tool being introduced: *the reaction table*.

In a study of carbon oxidation, an evacuated vessel containing a small amount of powdered graphite is heated to 1080 K, and then CO_2 is added to a pressure of 0.458 atm. Once the CO_2 is added, the system starts to produce CO. After equilibrium has been reached, the total pressure inside the vessel is 0.757 atm. Calculate K_p.

As always, we start to solve the problem by writing the balanced equation and the reaction quotient:

$$CO_2(g) + C(\text{graphite}) \rightleftharpoons 2CO(g)$$

The data are given in atmospheres and we must find K_p, so we write Q in terms of partial pressures (note the absence of a term for the solid, C):

$$Q_p = \frac{P_{CO}^2}{P_{CO_2}}$$

We are given the initial P_{CO_2} and P_{total} at equilibrium. To find K_p, we must find the equilibrium pressures of CO_2 and CO, which requires solving a stoichiometry problem, and then substitute them into the expression for Q_p.

Let's stop for a moment before we begin the calculations to think through what happened in the vessel. An unknown portion of the CO_2 reacted with graphite to form an unknown amount of CO. We already know the *relative* amounts of CO_2 and CO from the balanced equation: for each mole of CO_2 that reacts, 2 mol of CO forms, which means that when x atm of CO_2 reacts, $2x$ atm of CO forms:

$$x \text{ atm } CO_2 \longrightarrow 2x \text{ atm CO}$$

The pressure of CO_2 at equilibrium, P_{CO_2}, is the initial pressure, $P_{CO_2(init)}$, *minus* the change in CO_2 pressure, x, that is, the CO_2 that reacts:

$$P_{CO_2(init)} - x = P_{CO_2}$$

Similarly, the pressure of CO at equilibrium, P_{CO}, is the initial pressure, $P_{CO(init)}$, *plus* the change in CO pressure, which is the CO that forms, $2x$. Because $P_{CO(init)}$ is zero at the beginning of the reaction, we have

$$P_{CO(init)} + 2x = 0 + 2x = 2x = P_{CO}$$

A useful way to summarize this information is with a *reaction table* that shows the balanced equation and what we know about

- the *initial* quantities (concentrations or pressures) of reactants and products
- the *changes* in these quantities during the reaction
- the *equilibrium* quantities

Pressure (atm)	$CO_2(g)$	+	C(graphite)	⇌	$2CO(g)$
Initial	0.458		—		0
Change	$-x$		—		$+2x$
Equilibrium	$0.458 - x$		—		$2x$

Note that we *add the initial value to the change to obtain the equilibrium value* in each column. Note also that we include data *only for those substances whose concentrations change;* thus, in this case, the C(graphite) column is blank. We use reaction tables in many of the equilibrium problems in this and later chapters.

To solve for K_p, we substitute equilibrium values into the reaction quotient, so we have to find x. To do this, we use the other piece of data given, P_{total}. According to Dalton's law of partial pressures and using the quantities from the bottom (equilibrium) row of the reaction table,

$$P_{total} = 0.757 \text{ atm} = P_{CO_2} + P_{CO} = (0.458 \text{ atm} - x) + 2x$$

Thus,

$$0.757 \text{ atm} = 0.458 \text{ atm} + x \quad \text{and} \quad x = 0.299 \text{ atm}$$

With x known, we determine the equilibrium partial pressures:

$$P_{CO_2} = 0.458 \text{ atm} - x = 0.458 \text{ atm} - 0.299 \text{ atm} = 0.159 \text{ atm}$$
$$P_{CO} = 2x = 2(0.299 \text{ atm}) = 0.598 \text{ atm}$$

Now, we substitute these values into the expression for Q_p to find K_p:

$$Q_p = \frac{P_{CO}^2}{P_{CO_2}} = \frac{0.598^2}{0.159} = 2.25 = K_p$$

SAMPLE PROBLEM 17.4 Calculating K_c from Concentration Data

Problem In order to study hydrogen halide decomposition, a researcher fills an evacuated 2.00-L flask with 0.200 mol of HI gas and allows the reaction to proceed at 453°C:

$$2HI(g) \rightleftharpoons H_2(g) + I_2(g)$$

At equilibrium, [HI] = 0.078 M. Calculate K_c.

Plan To calculate K_c, we need the equilibrium concentrations. We can find the initial [HI] from the amount (0.200 mol) and the flask volume (2.00 L), and we are given [HI] at equilibrium (0.078 M). From the balanced equation, when $2x$ mol of HI reacts, x mol of H_2 and x mol of I_2 form. We set up a reaction table, use the known [HI] at equilibrium to solve for x (the change in [H_2] or [I_2]), and substitute the concentrations into Q_c.

Solution Calculating initial [HI]:

$$[HI] = \frac{0.200 \text{ mol}}{2.00 \text{ L}} = 0.100 \ M$$

Setting up the reaction table, with x = [H_2] or [I_2] that forms and $2x$ = [HI] that reacts:

Concentration (M)	2HI(g)	$\rightleftharpoons$	H₂(g)	+	I₂(g)
Initial	0.100		0		0
Change	$-2x$		$+x$		$+x$
Equilibrium	$0.100 - 2x$		x		x

Solving for x, using the known [HI] at equilibrium:

$$[HI] = 0.100 \ M - 2x = 0.078 \ M$$

$$x = 0.011 \ M$$

Therefore, the equilibrium concentrations are

$$[H_2] = [I_2] = 0.011 \ M \quad \text{and} \quad [HI] = 0.078 \ M$$

Substituting into the reaction quotient:

$$Q_c = \frac{[H_2][I_2]}{[HI]^2}$$

Thus,

$$K_c = \frac{(0.011)(0.011)}{0.078^2} = 0.020$$

Check Rounding gives $\sim 0.01^2/0.08^2 = 0.02$. Because the initial [HI] of 0.100 M fell slightly at equilibrium to 0.078 M, relatively little product formed; so we expect $K_c < 1$.

FOLLOW-UP PROBLEM 17.4 The atmospheric oxidation of nitrogen monoxide, $2NO(g) + O_2(g) \rightleftharpoons 2NO_2(g)$, was studied at 184°C with initial pressures of 1.000 atm of NO and 1.000 atm of O_2. At equilibrium, $P_{O_2} = 0.506$ atm. Calculate K_p.

Using the Equilibrium Constant to Determine Quantities

Like the type of problem that involves finding K, the type that involves finding equilibrium concentrations (or pressures) has several variations. Sample Problem 17.5 is one variation, in which we know K and some of the equilibrium concentrations and must find another equilibrium concentration.

SAMPLE PROBLEM 17.5 Determining Equilibrium Concentrations from K_c

Problem In a study of the conversion of methane to other fuels, a chemical engineer mixes gaseous CH_4 and H_2O in a 0.32-L flask at 1200 K. At equilibrium, the flask contains 0.26 mol of CO, 0.091 mol of H_2, and 0.041 mol of CH_4. What is [H_2O] at equilibrium? $K_c = 0.26$ for the equation

$$CH_4(g) + H_2O(g) \rightleftharpoons CO(g) + 3H_2(g)$$

Plan First, we use the balanced equation to write the reaction quotient. We can calculate the equilibrium concentrations from the given numbers of moles and the flask volume (0.32 L). Substituting these into Q_c and setting it equal to the given K_c (0.26), we solve for the unknown equilibrium concentration, [H₂O].

Solution Writing the reaction quotient:

$$CH_4(g) + H_2O(g) \rightleftharpoons CO(g) + 3H_2(g) \qquad Q_c = \frac{[CO][H_2]^3}{[CH_4][H_2O]}$$

Determining the equilibrium concentrations:

$$[CH_4] = \frac{0.041\ mol}{0.32\ L} = 0.13\ M$$

Similarly, $[CO] = 0.81\ M$ and $[H_2] = 0.28\ M$.

Calculating [H₂O] at equilibrium: Since $Q_c = K_c$, rearranging gives

$$[H_2O] = \frac{[CO][H_2]^3}{[CH_4]K_c} = \frac{(0.81)(0.28)^3}{(0.13)(0.26)} = 0.53\ M$$

Check Always check by substituting the concentrations into Q_c to confirm K_c:

$$Q_c = \frac{[CO][H_2]^3}{[CH_4][H_2O]} = \frac{(0.81)(0.28)^3}{(0.13)(0.53)} = 0.26 = K_c$$

FOLLOW-UP PROBLEM 17.5 Nitrogen monoxide, oxygen, and nitrogen react by the following equation: $2NO(g) \rightleftharpoons N_2(g) + O_2(g)$; $K_c = 2.3 \times 10^{30}$ at 298 K. In the atmosphere, $P_{O_2} = 0.209$ atm and $P_{N_2} = 0.781$ atm. What is the equilibrium partial pressure of NO in the air we breathe? [*Hint:* You need K_p to find the partial pressure.]

In a somewhat more involved variation, we know K and *initial* quantities and must find *equilibrium* quantities, for which we use a reaction table. In Sample Problem 17.6, the amounts were chosen to simplify the math, allowing us to focus more easily on the overall approach.

SAMPLE PROBLEM 17.6 Determining Equilibrium Concentrations from Initial Concentrations and K_c

Problem Fuel engineers use the extent of the change from CO and H₂O to CO₂ and H₂ to regulate the proportions of synthetic fuel mixtures. If 0.250 mol of CO and 0.250 mol of H₂O are placed in a 125-mL flask at 900 K, what is the composition of the equilibrium mixture? At this temperature, K_c is 1.56 for the equation

$$CO(g) + H_2O(g) \rightleftharpoons CO_2(g) + H_2(g)$$

Plan We have to find the "composition" of the equilibrium mixture, in other words, the equilibrium concentrations. As always, we use the balanced equation to write the reaction quotient. We find the initial [CO] and [H₂O] from the given amounts (0.250 mol of each) and volume (0.125 L), use the balanced equation to define x and set up a reaction table, substitute into Q_c, and solve for x, from which we calculate the concentrations.

Solution Writing the reaction quotient:

$$CO(g) + H_2O(g) \rightleftharpoons CO_2(g) + H_2(g) \qquad Q_c = \frac{[CO_2][H_2]}{[CO][H_2O]}$$

Calculating initial reactant concentrations:

$$[CO] = [H_2O] = \frac{0.250\ mol}{0.125\ L} = 2.00\ M$$

Setting up the reaction table, with $x = [CO]$ and [H₂O] that react:

Concentration (M)	CO(g)	+	H₂O(g)	⇌	CO₂(g)	+	H₂(g)
Initial	2.00		2.00		0		0
Change	$-x$		$-x$		$+x$		$+x$
Equilibrium	$2.00 - x$		$2.00 - x$		x		x

Substituting into the reaction quotient and solving for x:

$$Q_c = \frac{[CO_2][H_2]}{[CO][H_2O]} = \frac{(x)(x)}{(2.00-x)(2.00-x)} = \frac{x^2}{(2.00-x)^2}$$

At equilibrium, we have

$$Q_c = K_c = 1.56 = \frac{x^2}{(2.00-x)^2}$$

We can apply the following math shortcut in this case *but not in general:* Because the right side of the equation is a perfect square, we take the square root of both sides:

$$\sqrt{1.56} = \frac{x}{2.00-x} = \pm 1.25$$

A positive number (1.56) has a positive *and* a negative square root, but *only the positive root has any chemical meaning,* so we ignore the negative root:*

$$1.25 = \frac{x}{2.00-x} \quad \text{or} \quad 2.50 - 1.25x = x$$

So $2.50 = 2.25x;$ therefore, $x = 1.11\ M$

Calculating equilibrium concentrations:

$$[CO] = [H_2O] = 2.00\ M - x = 2.00\ M - 1.11\ M = 0.89\ M$$
$$[CO_2] = [H_2] = x = 1.11\ M$$

Check From the intermediate size of K_c, it makes sense that the changes in concentration are moderate. It's a good idea to check that the sign of x in the reaction table makes sense—only reactants were initially present, so the change had to proceed to the right: x is the change in concentration, so it has a negative sign for reactants and a positive sign for products. Also check that the equilibrium concentrations give the known K_c:
$\dfrac{(1.11)(1.11)}{(0.89)(0.89)} = 1.56$.

FOLLOW-UP PROBLEM 17.6 The decomposition of HI at low temperature was studied by injecting 2.50 mol of HI into a 10.32-L vessel at 25°C. What is $[H_2]$ at equilibrium for the reaction $2HI(g) \rightleftharpoons H_2(g) + I_2(g)$; $K_c = 1.26 \times 10^{-3}$?

---

*The negative root gives $-1.25 = \dfrac{x}{2.00-x}$, or $-2.50 + 1.25x = x$.

So $-2.50 = -0.25x$, and $x = 10.\ M$

This value has no chemical meaning because we started with 2.00 M of each reactant, so it is impossible for 10. M to react. Moreover, the square root of an equilibrium constant is another equilibrium constant, which cannot have a negative value.

Using the Quadratic Formula to Solve for the Unknown The shortcut that we used to simplify the math in Sample Problem 17.6 is a special case that occurs when both numerator and denominator of the reaction quotient are perfect squares. It worked out that way because we started with equal concentrations of the two reactants, but that is not ordinarily the case.

Suppose, for example, we instead start the reaction in the sample problem with 2.00 M CO and 1.00 M H$_2$O. The reaction table is

Concentration (M)	CO(g)	+	H₂O(g)	⇌	CO₂(g)	+	H₂(g)
Initial	2.00		1.00		0		0
Change	$-x$		$-x$		$+x$		$+x$
Equilibrium	$2.00 - x$		$1.00 - x$		x		x

Substituting these values into Q_c, we obtain

$$Q_c = \frac{[CO_2][H_2]}{[CO][H_2O]} = \frac{(x)(x)}{(2.00-x)(1.00-x)} = \frac{x^2}{x^2 - 3.00x + 2.00}$$

At equilibrium, we have

$$1.56 = \frac{x^2}{x^2 - 3.00x + 2.00}$$

To solve for x in this case, we rearrange the previous expression into the form of a *quadratic equation:*

$$a\ x^2 + \ b\ x + \ c \ = 0$$
$$0.56x^2 - 4.68x + 3.12 = 0$$

where $a = 0.56$, $b = -4.68$, and $c = 3.12$. Then we can find x with the quadratic formula (Appendix A):

$$x = \frac{-b \pm \sqrt{b^2 - 4ac}}{2a}$$

The $\pm$ sign means that we obtain two possible values for x:

$$x = \frac{4.68 \pm \sqrt{(-4.68)^2 - 4(0.56)(3.12)}}{2(0.56)}$$

$$x = 7.6\ M \quad \text{and} \quad x = 0.73\ M$$

Note that only one of the values for x makes sense chemically. The larger value gives negative concentrations at equilibrium (for example, $2.00\ M - 7.6\ M = -5.6\ M$), which have no meaning. Therefore, $x = 0.73\ M$, and we have

$$[CO] = 2.00\ M - x = 2.00\ M - 0.73\ M = 1.27\ M$$
$$[H_2O] = 1.00\ M - x = 0.27\ M$$
$$[CO_2] = [H_2] = x = 0.73\ M$$

Checking to see if these values give the known K_c, we have

$$K_c = \frac{(0.73)(0.73)}{(1.27)(0.27)} = 1.6 \ (\text{within rounding of } 1.56)$$

Simplifying Assumptions for Finding an Unknown Quantity In many cases, we can use chemical "common sense" to make an assumption that avoids the use of the quadratic formula to find x. In general, *if a reaction has a relatively small K and a relatively large initial reactant concentration, the concentration change (x) can often be neglected* without introducing significant error. This assumption does not mean that $x = 0$, because then there would be no reaction. It means that if a reaction proceeds very little (small K) and if there is a high initial reactant concentration, only a very small amount will be used up; therefore, at equilibrium, the reactant concentration will have hardly changed:

$$[\text{reactant}]_{init} - x = [\text{reactant}]_{eq} \approx [\text{reactant}]_{init}$$

You can imagine a similar situation in everyday life. On a bathroom scale, you weigh 158 lb. Take off your wristwatch, and you still weigh 158 lb. Within the precision of the measurement, the weight of the wristwatch is so small compared with your body weight that it can be neglected:

Initial body weight − weight of watch = final body weight ≈ initial body weight

Similarly, if the initial concentration of A is, for example, $0.500\ M$ and, because of a small K_c, the concentration of A that reacts is $0.002\ M$, we can assume that

$$0.500\ M - 0.002\ M = 0.498\ M \approx 0.500\ M$$

that is,
$$[A]_{init} - [A]_{reacting} = [A]_{eq} \approx [A]_{init} \qquad (17.9)$$

For the assumption that x is negligible to be justified, you must check that the error introduced is not significant. But how much is "significant"? One common criterion is the 5% rule: *if the assumption results in a change (error) in a concentration that is less than 5%, the error is not significant, and the assumption is justified.* Let's go through a sample problem and make this assumption to see how it simplifies the math, and then we'll see if the assumption is justified in the case of two different initial concentrations.

SAMPLE PROBLEM 17.7 Calculating Equilibrium Concentrations with Simplifying Assumptions

Problem Phosgene is a potent chemical warfare agent that is now outlawed by international agreement. It decomposes by the reaction

$$COCl_2(g) \rightleftharpoons CO(g) + Cl_2(g) \qquad K_c = 8.3 \times 10^{-4} \text{ (at } 360°C)$$

Calculate $[CO]$, $[Cl_2]$, and $[COCl_2]$ when each of the following amounts of phosgene decomposes and reaches equilibrium in a 10.0-L flask:
(a) 5.00 mol of $COCl_2$ **(b)** 0.100 mol of $COCl_2$

Plan We know from the balanced equation that when x mol of $COCl_2$ decomposes, x mol of CO and x mol of Cl_2 form. We convert amount (5.00 mol or 0.100 mol) to concentration, define x and set up the reaction table, and substitute the values into Q_c. Before using the quadratic formula, we simplify the calculation by assuming that x is negligibly small. After solving for x, we check the assumption and find the concentrations. If the assumption is not justified, we must use the quadratic formula to find x.

Solution (a) For 5.00 mol of $COCl_2$. Writing the reaction quotient:

$$Q_c = \frac{[CO][Cl_2]}{[COCl_2]}$$

Calculating initial $[COCl_2]$:

$$[COCl_2]_{init} = \frac{5.00 \text{ mol}}{10.0 \text{ L}} = 0.500 \, M$$

Setting up the reaction table, with $x = [COCl_2]_{reacting}$:

Concentration (M)	$COCl_2(g)$	$\rightleftharpoons$	$CO(g)$	+	$Cl_2(g)$
Initial	0.500		0		0
Change	$-x$		$+x$		$+x$
Equilibrium	$0.500 - x$		x		x

If we use the equilibrium values in Q_c, we obtain

$$Q_c = \frac{[CO][Cl_2]}{[COCl_2]} = \frac{x^2}{0.500 - x} = K_c = 8.3 \times 10^{-4}$$

Because K_c is small, the reaction does not proceed very far to the right, so let's assume that x (the $[COCl_2]$ that reacts) is so much smaller than the initial concentration, 0.500 M, that the equilibrium concentration is nearly the same. Therefore,

$$0.500 \, M - x \approx 0.500 \, M$$

Using this assumption, we substitute and solve for x:

$$K_c = 8.3 \times 10^{-4} \approx \frac{x^2}{0.500}$$

$$x^2 \approx (8.3 \times 10^{-4})(0.500) \qquad \text{so} \qquad x \approx 2.0 \times 10^{-2}$$

Checking the assumption by finding the percent error:

$$\frac{2.0 \times 10^{-2}}{0.500} \times 100 = 4\% \qquad \text{(less than 5\%, so the assumption is justified)}$$

Solving for the equilibrium concentrations:

$$[CO] = [Cl_2] = x = 2.0 \times 10^{-2} \, M$$
$$[COCl_2] = 0.500 \, M - x = 0.480 \, M$$

(b) For 0.100 mol of $COCl_2$. The calculation in this case is the same as the calculation in part (a), except that $[COCl_2]_{init} = 0.100 \text{ mol}/10.0 \text{ L} = 0.0100 \, M$. Thus, at equilibrium, we have

$$Q_c = \frac{[CO][Cl_2]}{[COCl_2]} = \frac{x^2}{0.0100 - x}$$

$$= K_c = 8.3 \times 10^{-4}$$

Making the assumption that $0.0100\ M - x \approx 0.0100\ M$ and solving for x:

$$K_c = 8.3 \times 10^{-4} \approx \frac{x^2}{0.0100}$$

$$x \approx 2.9 \times 10^{-3}$$

Checking the assumption:

$$\frac{2.9 \times 10^{-3}}{0.0100} \times 100 = 29\% \qquad \text{(more than 5\%, so the assumption is } not \text{ justified)}$$

We must solve the quadratic equation, $x^2 + (8.3 \times 10^{-4})x - (8.3 \times 10^{-6}) = 0$, for which the only meaningful value of x is 2.5×10^{-3} (see Appendix A).
Solving for the equilibrium concentrations:

$$[CO] = [Cl_2] = \boxed{2.5 \times 10^{-3}\ M}$$

$$[COCl_2] = 1.00 \times 10^{-2}\ M - x = \boxed{7.5 \times 10^{-3}\ M}$$

Check Once again, the best check is to use the calculated values to be sure you obtain the given K_c.
Comment Note that the assumption was justified at the high initial concentration, but *not* at the low initial concentration.

FOLLOW-UP PROBLEM 17.7 In a study of the effect of temperature on halogen decomposition, 0.50 mol of I_2 was heated in a 2.5-L vessel, and the following reaction occurred: $I_2(g) \rightleftharpoons 2I(g)$.
(a) Calculate $[I_2]$ and $[I]$ at equilibrium at 600 K; $K_c = 2.94 \times 10^{-10}$.
(b) Calculate $[I_2]$ and $[I]$ at equilibrium at 2000 K; $K_c = 0.209$.

Mixtures of Reactants and Products: Determining Reaction Direction

In the problems we've worked so far, the direction of the reaction was obvious: with only reactants present at the start, the reaction *had* to go toward products. Thus, in the reaction tables, we knew that the unknown change in reactant concentration had a negative sign $(-x)$ and the change in product concentration had a positive sign $(+x)$. Suppose, however, we start with a *mixture* of reactants and products. Whenever the reaction direction is not obvious, we first *compare the value of Q with K to find the direction* in which the reaction proceeds to reach equilibrium. This tells us the sign of x, the unknown change in concentration. (In order to focus on this idea, the next sample problem eliminates the need for the quadratic formula.)

SAMPLE PROBLEM 17.8   Predicting Reaction Direction and Calculating Equilibrium Concentrations

Problem The research and development unit of a chemical company is studying the reaction of CH_4 and H_2S, two components of natural gas:

$$CH_4(g) + 2H_2S(g) \rightleftharpoons CS_2(g) + 4H_2(g)$$

In one experiment, 1.00 mol of CH_4, 1.00 mol of CS_2, 2.00 mol of H_2S, and 2.00 mol of H_2 are mixed in a 250-mL vessel at 960°C. At this temperature, $K_c = 0.036$.
(a) In which direction will the reaction proceed to reach equilibrium?
(b) If $[CH_4] = 5.56\ M$ at equilibrium, what are the equilibrium concentrations of the other substances?
Plan **(a)** To find the direction, we convert the given initial amounts and volume (0.250 L) to concentrations, calculate Q_c, and compare it with K_c. **(b)** Based on the results from (a), we determine the sign of each concentration change for the reaction table and then use the known $[CH_4]$ at equilibrium (5.56 M) to determine x and the other equilibrium concentrations.

Solution (a) Calculating the initial concentrations:

$$[CH_4] = \frac{1.00 \text{ mol}}{0.250 \text{ L}} = 4.00 \ M$$

Similarly, $[H_2S] = 8.00 \ M$, $[CS_2] = 4.00 \ M$, and $[H_2] = 8.00 \ M$.
Calculating the value of Q_c:

$$Q_c = \frac{[CS_2][H_2]^4}{[CH_4][H_2S]^2} = \frac{(4.00)(8.00)^4}{(4.00)(8.00)^2} = 64.0$$

Comparing Q_c and K_c: $Q_c > K_c$ (64.0 > 0.036), so the reaction goes to the left. Therefore, concentrations of reactants increase and those of products decrease.
(b) Setting up a reaction table, with $x = [CS_2]$ that reacts, which equals $[CH_4]$ that forms:

Concentration (M)	$CH_4(g)$	+	$2H_2S(g)$	$\rightleftharpoons$	$CS_2(g)$	+	$4H_2(g)$
Initial	4.00		8.00		4.00		8.00
Change	$+x$		$+2x$		$-x$		$-4x$
Equilibrium	$4.00 + x$		$8.00 + 2x$		$4.00 - x$		$8.00 - 4x$

Solving for x: At equilibrium,

$$[CH_4] = 5.56 \ M = 4.00 \ M + x$$

So,

$$x = 1.56 \ M$$

Thus,

$$[H_2S] = 8.00 \ M + 2x = 8.00 \ M + 2(1.56 \ M) = 11.12 \ M$$
$$[CS_2] = 4.00 \ M - x = 2.44 \ M$$
$$[H_2] = 8.00 \ M - 4x = 1.76 \ M$$

Check The comparison of Q_c and K_c showed the reaction proceeding to the left. The given data from part (b) confirm this because $[CH_4]$ increases from 4.00 M to 5.56 M during the reaction. Check that the concentrations give the known K_c:

$$\frac{(2.44)(1.76)^4}{(5.56)(11.12)^2} = 0.0341, \text{ which is close to } 0.036$$

FOLLOW-UP PROBLEM 17.8 An inorganic chemist studying the reactions of phosphorus halides mixes 0.1050 mol of PCl_5 with 0.0450 mol of Cl_2 and 0.0450 mol of PCl_3 in a 0.5000-L flask at 250°C: $PCl_5(g) \rightleftharpoons PCl_3(g) + Cl_2(g)$; $K_c = 4.2 \times 10^{-2}$.
(a) In which direction will the reaction proceed?
(b) If $[PCl_5] = 0.2065 \ M$ at equilibrium, what are the equilibrium concentrations of the other components?

By this time, you've seen quite a few variations on the type of equilibrium problem in which you know K and some initial quantities and must find the equilibrium quantities. Figure 17.5 presents a useful summary of the steps involved in solving these types of equilibrium problems. A good way to organize the steps is to group them into three overall parts.

SECTION SUMMARY

In most equilibrium problems, we use quantities (concentrations or pressures) of reactants and products to find K, or we use K to find quantities. We use a reaction table to summarize the initial quantities, how they change, and the equilibrium quantities. When K is small and the initial quantity of reactant is large, we assume the unknown change in the quantity (x) is so much smaller than the initial quantity that it can be neglected. If this assumption is not justified (that is, if the error is greater than 5%), we use the quadratic formula to find x. To determine reaction direction, we compare the values of Q and K.

SOLVING EQUILIBRIUM PROBLEMS

PRELIMINARY SETTING UP

1. Write the balanced equation
2. Write the reaction quotient, Q
3. Convert all amounts into the correct units (M or atm)

WORKING ON THE REACTION TABLE

4. When reaction direction is not known, compare Q with K
5. Construct a reaction table

✓ Check the sign of x, the change in the quantity

SOLVING FOR x AND EQUILIBRIUM QUANTITIES

6. Substitute the quantities into Q
7. To simplify the math, assume that x is negligible
 ($[A]_{init} - x = [A]_{eq} \approx [A]_{init}$)
8. Solve for x

✓ Check that assumption is justified (< 5% error). If not, solve quadratic equation for x.

9. Find the equilibrium quantities

✓ Check to see that calculated values give the known K

Figure 17.5 Steps in solving equilibrium problems. These nine steps, grouped into three tasks, provide a useful approach to calculating equilibrium quantities, given initial quantities and K.

17.6 REACTION CONDITIONS AND THE EQUILIBRIUM STATE: LE CHÂTELIER'S PRINCIPLE

The most remarkable feature of a system at equilibrium is its ability to return to equilibrium after a change in conditions moves it away from that state. This drive to reattain equilibrium is stated in **Le Châtelier's principle:** when a chemical system at equilibrium is disturbed, it reattains equilibrium by undergoing a net reaction that reduces the effect of the disturbance.

Two phrases in this statement need further explanation. First, what does it mean to "disturb" a system? At equilibrium, Q equals K. When a change in conditions forces the system temporarily out of equilibrium ($Q \neq K$), we say that the system has been stressed, or disturbed. Three common disturbances are a change in concentration of a component (that appears in Q), a change in pressure (caused by a change in volume), or a change in temperature. We'll discuss each of these changes below.

The other phrase, "net reaction," is often referred to as a shift in the *equilibrium position* of the system to the right or left. The equilibrium position is just the specific equilibrium concentrations (or pressures). A shift in the equilibrium position to the right means that there is a net reaction to the right (reactant to product) until equilibrium is reattained; a shift to the left means that there is a net reaction to the left (product to reactant). Thus, when a disturbance occurs, we say that the equilibrium position shifts, which means that *concentrations (or pressures) change in a way that reduces the disturbance, and the system attains a new equilibrium position ($Q = K$ again).*

Le Châtelier's principle allows us to predict the direction of the shift in equilibrium position. Most importantly, it helps research and industrial chemists create conditions that maximize yields. For the remainder of this section, we examine each of the three kinds of disturbances—concentration, pressure (volume), and temperature—to see how a system at equilibrium responds; then, we'll note whether a catalyst has any effect.

In the following discussions, we focus on the reversible gaseous reaction between phosphorus trichloride and chlorine to produce phosphorus pentachloride:

$$PCl_3(g) + Cl_2(g) \rightleftharpoons PCl_5(g)$$

However, the basis of Le Châtelier's principle holds for any system at equilibrium.

The Effect of a Change in Concentration

When a system at equilibrium is disturbed by a change in concentration of one of the components, the system reacts in the direction that reduces the change:

- If the concentration increases, the system reacts to consume some of it.
- If the concentration decreases, the system reacts to produce some of it.

Of course, the component must be one that appears in Q; thus, pure liquids and solids, which do not appear in Q because their concentrations are constant, are not involved.

At 523 K, the PCl_3-Cl_2-PCl_5 system reaches equilibrium when

$$Q_c = \frac{[PCl_5]}{[PCl_3][Cl_2]} = 24.0 = K_c$$

What happens if we now inject some Cl_2 gas, one of the reactants? The system will always act to reduce the disturbance, so it will reduce the increase in reactant by proceeding toward the product side, thereby consuming some of the additional Cl_2. In terms of the reaction quotient, when we add Cl_2, the $[Cl_2]$ term increases, so the value of Q_c immediately falls as the denominator becomes larger; thus, the system is no longer at equilibrium. As some of the added Cl_2 reacts with some of the PCl_3 present and produces more PCl_5, the denominator becomes smaller once again and the numerator larger, until eventually Q_c again equals K_c. The concentrations of the components have changed, however: the concentrations of Cl_2 and PCl_5 are higher than in the original equilibrium position, and the concentration of PCl_3 is lower. Nevertheless, the ratio of values gives the same K_c. We describe this change by saying that *the equilibrium position shifts to the right when a component on the left is added:*

$$PCl_3 + Cl_2(\text{added}) \longrightarrow PCl_5$$

What happens if, instead of adding Cl_2, we remove some PCl_3, the other reactant? In this case, the system reduces the disturbance (the decrease in reactant), by proceeding toward the reactant side, thereby consuming some PCl_5. Once again, thinking in terms of Q_c, when we remove PCl_3, the $[PCl_3]$ term decreases, the denominator becomes smaller, and the value of Q_c rises above K_c. As some PCl_5 decomposes to PCl_3 and Cl_2, the numerator decreases and the denominator increases until Q_c equals K_c again. Here, too, the concentrations are different from those of the original equilibrium position, but K_c is not. We say that *the equilibrium position shifts to the left when a component on the left is removed:*

$$PCl_3(\text{removed}) + Cl_2 \longleftarrow PCl_5$$

The same points we just made for adding or removing a reactant also hold for adding or removing a product. If we add PCl_5, its concentration rises and the equilibrium position shifts to the left, just as it did when we removed some PCl_3; if we remove some PCl_5, the equilibrium position shifts to the right, just as it did when we added some Cl_2. In other words, no matter how the disturbance in concentration comes about, the system responds to make Q_c and K_c equal again. To summarize the effects of concentration changes (Figure 17.6):

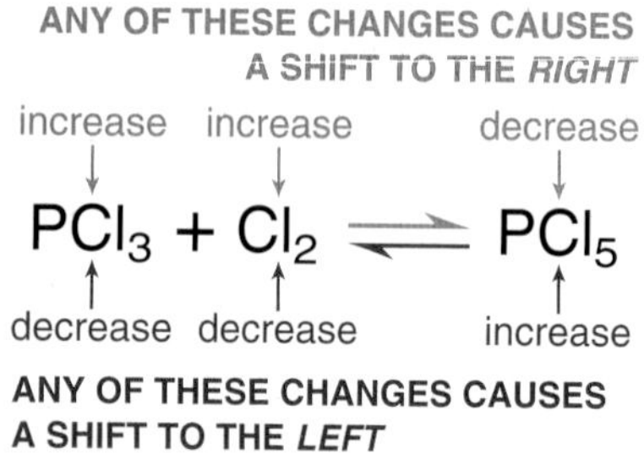

- The equilibrium position shifts to the *right* if a reactant is added or a product is removed: [reactant] increases or [product] decreases.
- The equilibrium position shifts to the *left* if a reactant is removed or a product is added: [reactant] decreases or [product] increases.

In general, whenever the concentration of a component changes, *the equilibrium system reacts to consume some of the added substance or produce some of the removed substance.* In this way, the system "reduces the effect of the disturbance." The effect is not completely eliminated, however, as we will see next from a quantitative comparison of original and new equilibrium positions.

Consider the case in which we added Cl_2 to the system at equilibrium. Suppose the original equilibrium position was established with the following concentrations: $[PCl_3] = 0.200\ M$, $[Cl_2] = 0.125\ M$, and $[PCl_5] = 0.600\ M$. Thus,

$$Q_c = \frac{[PCl_5]}{[PCl_3][Cl_2]} = \frac{0.600}{(0.200)(0.125)} = 24.0 = K_c$$

Now we add enough Cl_2 to increase its concentration by $0.075\ M$. Before any reaction occurs, this addition creates a new set of initial concentrations. Then the system reacts and comes to a new equilibrium position. From Le Châtelier's principle, we predict that adding more reactant will produce more product, that is, shift the equilibrium position to the right. Experiment shows that the new $[PCl_5]$ at equilibrium is $0.637\ M$.

Table 17.3 shows a reaction table of the entire process: the original equilibrium position, the disturbance, the (new) initial concentrations, the size and direction of the change needed to reattain equilibrium, and the new equilibrium position. Figure 17.7 depicts the process.

ANY OF THESE CHANGES CAUSES
A SHIFT TO THE *RIGHT*

increase   increase   decrease
$$PCl_3 + Cl_2 \rightleftharpoons PCl_5$$
decrease   decrease   increase

ANY OF THESE CHANGES CAUSES
A SHIFT TO THE *LEFT*

Figure 17.6 The effect of a change in concentration.

Table 17.3 **The Effect of Added Cl_2 on the PCl_3-Cl_2-PCl_5 System**				
Concentration (M)	$PCl_3(g)$	+	$Cl_2(g)$ $\rightleftharpoons$	$PCl_5(g)$
Original equilibrium	0.200		0.125	0.600
Disturbance			+0.075	
New initial	0.200		0.200	0.600
Change	$-x$		$-x$	$+x$
New equilibrium	$0.200 - x$		$0.200 - x$	$0.600 + x$ (0.637)*

*Experimentally determined value.

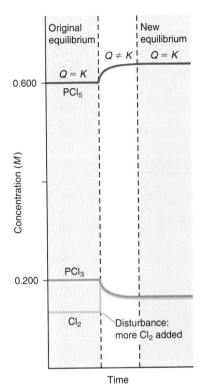

Figure 17.7 **The effect of added Cl_2 on the PCl_3-Cl_2-PCl_5 system.** In the original equilibrium (*gray region*), all concentrations are constant. When Cl_2 (*yellow curve*) is added, its concentration jumps and then starts to fall as Cl_2 reacts with some PCl_3 to form more PCl_5. After a period of time, equilibrium is re-established at new concentrations (*blue region*) but with the same K.

From Table 17.3,

$$[PCl_5] = 0.600 \ M + x = 0.637 \ M, \qquad \text{so } x = 0.037 \ M$$

Also, $\qquad [PCl_3] = [Cl_2] = 0.200 \ M - x = 0.163 \ M$

Therefore, at equilibrium,

$$K_{c(\text{original})} = \frac{0.600}{(0.200)(0.125)} = 24.0$$

$$K_{c(\text{new})} = \frac{0.637}{(0.163)(0.163)} = 24.0$$

There are several key points to notice about the new equilibrium concentrations that exist after Cl_2 is added:

- As we predicted, $[PCl_5]$ (0.637 M) is higher than its original concentration (0.600 M).
- $[Cl_2]$ (0.163 M) is higher than its original equilibrium concentration (0.125 M), but lower than its initial concentration just after the addition (0.200 M); thus, the disturbance (addition of Cl_2) is *reduced but not eliminated.*
- $[PCl_3]$ (0.163 M), the other left-side component, is lower than its original concentration (0.200 M) because some reacted with the added Cl_2.
- Most importantly, although the position of equilibrium shifted to the right, K_c *remains the same.*

Be sure to note that the system adjusts by changing concentrations, but the value of K_c at a given temperature does **not** change with a change in concentration.

SAMPLE PROBLEM 17.9 Predicting the Effect of a Change in Concentration on the Equilibrium Position

Problem To improve air quality and obtain a useful product, chemists often remove sulfur from coal and natural gas by treating the fuel contaminant hydrogen sulfide with O_2:

$$2H_2S(g) + O_2(g) \rightleftharpoons 2S(s) + 2H_2O(g)$$

What happens to
(a) $[H_2O]$ if O_2 is added?   (b) $[H_2S]$ if O_2 is added?
(c) $[O_2]$ if H_2S is removed?   (d) $[H_2S]$ if sulfur is added?

Plan We write the reaction quotient to see how Q_c is affected by each disturbance, relative to K_c. This effect tells us the direction in which the reaction proceeds for the system to reattain equilibrium and how each concentration changes.

Solution Writing the reaction quotient: $Q_c = \dfrac{[H_2O]^2}{[H_2S]^2[O_2]}$

(a) When O_2 is added, the denominator of Q_c increases, so $Q_c < K_c$. The reaction proceeds to the right until $Q_c = K_c$ again, so $[H_2O]$ increases.

(b) As in part (a), when O_2 is added, $Q_c < K_c$. Some H_2S reacts with the added O_2 as the reaction proceeds to the right, so $[H_2S]$ decreases.

(c) When H_2S is removed, the denominator of Q_c decreases, so $Q_c > K_c$. As the reaction proceeds to the left to re-form H_2S, more O_2 is produced as well, so $[O_2]$ increases.

(d) The concentration of solid S is unchanged as long as some is present, so it does not appear in the reaction quotient. Adding more S has no effect, so $[H_2S]$ is unchanged (but see Comment 2 below).

Check Apply Le Châtelier's principle to see that the reaction proceeds in the direction that lowers the increased concentration or raises the decreased concentration.

Comment 1. As you know, sulfur exists most commonly as S_8. How would this change in formula affect the answers? The balanced equation and Q_c would be

$$8H_2S(g) + 4O_2(g) \rightleftharpoons S_8(s) + 8H_2O(g) \qquad Q_c = \frac{[H_2O]^8}{[H_2S]^8[O_2]^4}$$

The value of K_c is different for this equation, but the changes described in the problem have the same effects. For example, in (a), if O_2 were added, the denominator of Q_c would increase, so $Q_c < K_c$. As above, the reaction would proceed to the right until $Q_c = K_c$ again. In other words, changes predicted by Le Châtelier's principle for a given reaction are not affected by a change in the balancing coefficients.

2. In (d), you saw that adding a solid has no effect on the concentrations of other components: because *the* **concentration** *of the solid cannot change*, it does not appear in Q. But *the* **amount** *of solid can change.* Adding H_2S shifts the reaction to the right, and more S forms.

FOLLOW-UP PROBLEM 17.9 In a study of the chemistry of glass etching, an inorganic chemist examines the reaction between sand (SiO_2) and hydrogen fluoride at a temperature above the boiling point of water:

$$SiO_2(s) + 4HF(g) \rightleftharpoons SiF_4(g) + 2H_2O(g)$$

Predict the effect on $[SiF_4]$ when (a) $H_2O(g)$ is removed; (b) some liquid water is added; (c) HF is removed; (d) some sand is removed.

The Effect of a Change in Pressure (Volume)

Changes in pressure have significant effects only on equilibrium systems with gaseous components. Aside from phase changes, a change in pressure has a negligible effect on liquids and solids because they are nearly incompressible. Pressure changes can occur in three ways:

- Changing the concentration of a gaseous component
- Adding an inert gas (one that does not take part in the reaction)
- Changing the volume of the reaction vessel

We just considered the effect of changing the concentration of a component, and that reasoning holds here. Next, let's see why *adding an inert gas has no effect on the equilibrium position.* Adding an inert gas does not change the volume, so all *reactant and product concentrations remain the same.* In other words, the volume and the number of moles of the reactant and product gases do not change, so their *partial pressures do not change.* Because we use these (unchanged) partial pressures in the reaction quotient, the equilibrium position cannot change. Moreover, the inert gas does not appear in Q, so it cannot have an effect.

On the other hand, changing the pressure by changing the volume often causes a large shift in the equilibrium position. Suppose we let the PCl_3-Cl_2-PCl_5 system come to equilibrium in a cylinder-piston assembly. Then, we press down on the piston to halve the volume: the gas pressure immediately doubles. To reduce this increase in gas pressure, the system responds by reducing the number of gas molecules. And it does so in the only possible way—by shifting the reaction toward the side with *fewer moles of gas,* in this case, toward the product side:

$$PCl_3(g) + Cl_2(g) \longrightarrow PCl_5(g)$$

2 mol gas $\longrightarrow$ 1 mol gas

Notice that *a change in volume results in a change in concentration:* a decrease in container volume raises the concentration, and an increase in volume lowers the concentration. Recall that $Q_c = \dfrac{[PCl_5]}{[PCl_3][Cl_2]}$. When the volume is halved, the concentrations double, but the denominator of Q_c is the product of two concentrations, so it quadruples while the numerator only doubles. Thus, Q_c becomes less than K_c. As a result, the system forms more PCl_5 and a new equilibrium position is reached. Because it is just another way to change the concentration, *a change in pressure due to a change in volume does **not** alter K_c.*

Thus, for a system that contains gases at equilibrium, in which the amount (mol) of gas, n_{gas}, changes during the reaction (Figure 17.8):

- If the volume becomes smaller (pressure is higher), the reaction shifts so that the total number of gas molecules decreases.
- If the volume becomes larger (pressure is lower), the reaction shifts so that the total number of gas molecules increases.

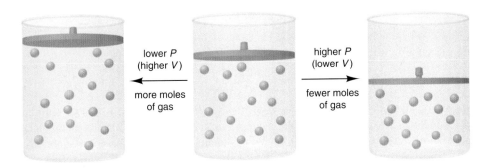

Figure 17.8 The effect of pressure (volume) on a system at equilibrium. The system of gases (*center*) is at equilibrium. For the reaction

⬤ + ○ ⇌ ⬤

an increase in pressure (*right*) decreases the volume, so the reaction shifts to the right to make fewer molecules. A decrease in pressure (*left*) increases the volume, so the reaction shifts to the left to make more molecules.

In many cases, however, n_{gas} does not change ($\Delta n_{gas} = 0$). For example,

$$H_2(g) + I_2(g) \rightleftharpoons 2HI(g)$$
$$\text{2 mol gas} \longrightarrow \text{2 mol gas}$$

Q_c has the same number of terms in the numerator and denominator:

$$Q_c = \frac{[HI]^2}{[H_2][I_2]} = \frac{[HI][HI]}{[H_2][I_2]}$$

Therefore, a change in volume has the same effect on the numerator and denominator. Thus, *if $\Delta n_{gas} = 0$, there is no effect on the equilibrium position.*

SAMPLE PROBLEM 17.10   Predicting the Effect of a Change in Volume (Pressure) on the Equilibrium Position

Problem How would you change the volume of each of the following reactions to *increase* the yield of the products?
(a) $CaCO_3(s) \rightleftharpoons CaO(s) + CO_2(g)$ **(b)** $S(s) + 3F_2(g) \rightleftharpoons SF_6(g)$
(c) $Cl_2(g) + I_2(g) \rightleftharpoons 2ICl(g)$
Plan Whenever gases are present, a change in volume causes a change in concentration. For reactions in which the number of moles of gas changes, if the volume decreases (pressure increases), the equilibrium position shifts to relieve the pressure by reducing the number of moles of gas. A volume increase (pressure decrease) has the opposite effect.
Solution (a) The only gas is the product CO_2. To make the system produce more CO_2, we increase the volume (decrease the pressure).
(b) With 3 mol of gas on the left and only 1 mol on the right, we decrease the volume (increase the pressure) to form more SF_6.
(c) The number of moles of gas is the same on both sides of the equation, so a change in volume (pressure) will have no effect on the yield of ICl.

Check Let's predict the relative values of Q_c and K_c. In (a), $Q_c = [CO_2]$, so increasing the volume will make $Q_c < K_c$, and the system will make more CO_2. In (b), $Q_c = [SF_6]/[F_2]^3$. Lowering the volume increases $[F_2]$ and $[SF_6]$ proportionately, but Q_c decreases because of the exponent 3 in the denominator. To make $Q_c = K_c$ again, $[SF_6]$ must increase. In (c), $Q_c = [ICl]^2/[Cl_2][I_2]$. A change in volume (pressure) affects the numerator (2 mol) and denominator (2 mol) equally, so it will have no effect.

FOLLOW-UP PROBLEM 17.10 Would you increase or decrease the pressure (via a volume change) of each of the following reaction mixtures to *decrease* the yield of products?
(a) $2SO_2(g) + O_2(g) \rightleftharpoons 2SO_3(g)$ **(b)** $4NH_3(g) + 5O_2(g) \rightleftharpoons 4NO(g) + 6H_2O(g)$
(c) $CaC_2O_4(s) \rightleftharpoons CaCO_3(s) + CO(g)$

The Effect of a Change in Temperature

Of the three types of disturbances—a change in concentration, in pressure, or in temperature—*only temperature changes alter K.* To see why, we must take the heat of reaction into account:

$$PCl_3(g) + Cl_2(g) \rightleftharpoons PCl_5(g) \quad \Delta H^\circ_{rxn} = -111 \text{ kJ}$$

The forward reaction is exothermic (releases heat; $\Delta H^\circ < 0$), so the reverse reaction is endothermic (absorbs heat; $\Delta H^\circ > 0$):

$$PCl_3(g) + Cl_2(g) \longrightarrow PCl_5(g) + \textbf{\textit{heat}} \text{ (exothermic)}$$
$$PCl_3(g) + Cl_2(g) \longleftarrow PCl_5(g) + \textbf{\textit{heat}} \text{ (endothermic)}$$

If we consider *heat as a component of the equilibrium system,* a rise in temperature means heat is "added" to the system and a drop in temperature means heat is "removed" from the system. As with a change in any other component, the system shifts to reduce the effect of the change. Therefore, *a temperature increase (adding heat) favors the endothermic (heat-absorbing) direction, and a temperature decrease (removing heat) favors the exothermic (heat-releasing) direction.*

If we start with the system at equilibrium, Q_c equals K_c. Increase the temperature, and the system responds by decomposing some PCl_5 to PCl_3 and Cl_2, which absorbs the added heat. The denominator of Q_c becomes larger and the numerator smaller, so the system reaches a new equilibrium position at a smaller ratio of concentration terms, that is, a lower K_c. Similarly, the system responds to a drop in temperature by forming more PCl_5 from some PCl_3 and Cl_2, which releases more heat. The numerator of Q_c becomes larger, the denominator smaller, and the new equilibrium position has a higher K_c. Thus,

- *A temperature rise will increase K_c for a system with a positive ΔH°_{rxn}.*
- *A temperature rise will decrease K_c for a system with a negative ΔH°_{rxn}.*

Let's review these ideas with a sample problem.

SAMPLE PROBLEM 17.11   Predicting the Effect of a Change in Temperature on the Equilibrium Position

Problem How does an *increase* in temperature affect the equilibrium concentration of the underlined substance and K_c for each of the following reactions?
(a) $CaO(s) + H_2O(l) \rightleftharpoons \underline{Ca(OH)_2}(aq)$ $\Delta H^\circ = -82 \text{ kJ}$
(b) $CaCO_3(s) \rightleftharpoons CaO(s) + \underline{CO_2}(g)$ $\Delta H^\circ = 178 \text{ kJ}$
(c) $\underline{SO_2}(g) \rightleftharpoons S(s) + O_2(g)$ $\Delta H^\circ = 297 \text{ kJ}$

Plan We write each equation to show heat as a reactant or product. Increasing the temperature adds heat, so the system shifts to absorb the heat; that is, the endothermic reaction occurs. K_c will increase if the forward reaction is endothermic and decrease if it is exothermic.
Solution (a) $CaO(s) + H_2O(l) \rightleftharpoons Ca(OH)_2(aq) + \textbf{\textit{heat}}$
Adding heat shifts the system to the left: $[Ca(OH)_2]$ and K_c will decrease.

(b) $CaCO_3(s) + \textbf{\textit{heat}} \rightleftharpoons CaO(s) + CO_2(g)$
Adding heat shifts the system to the right: $[CO_2]$ and K_c will increase.
(c) $SO_2(g) + \textbf{\textit{heat}} \rightleftharpoons S(s) + O_2(g)$
Adding heat shifts the system to the right: $[SO_2]$ will decrease and K_c will increase.
Check You can check your answers by going through the reasoning for a *decrease* in temperature: heat is removed and the exothermic direction is favored. All the answers should be opposite.

FOLLOW-UP PROBLEM 17.11 How does a *decrease* in temperature affect the partial pressure of the underlined substance and the value of K_p for each of the following reactions?
(a) $C(graphite) + 2H_2(g) \rightleftharpoons CH_4(g) \qquad \Delta H° = -75 \text{ kJ}$
(b) $\underline{N_2(g)} + O_2(g) \rightleftharpoons 2NO(g) \qquad \Delta H° = 181 \text{ kJ}$
(c) $\underline{P_4(s)} + 10Cl_2(g) \rightleftharpoons 4PCl_5(g) \qquad \Delta H° = -1528 \text{ kJ}$

The Lack of Effect of a Catalyst

Let's briefly consider a final external change to the reacting system: adding a catalyst. Recall from Chapter 16 that a catalyst speeds up a reaction by providing an alternative mechanism with a lower activation energy, thereby increasing the forward *and* reverse rates to the same extent. In other words, it shortens the time needed to attain the *final* concentrations. Thus, *a catalyst shortens the time it takes to reach equilibrium but has **no** effect on the equilibrium position.*

If, for instance, we add a catalyst to a mixture of PCl_3 and Cl_2 at 523 K, the system will attain the *same* equilibrium concentrations of PCl_3, Cl_2, and PCl_5 *more quickly* than it did without the catalyst. Nevertheless, a catalyst often plays a key role in optimizing the yield of a reaction system. The industrial production of ammonia, described in the following subsection, provides an example of a catalyzed improvement of yield.

Table 17.4 summarizes the effects of changing conditions on the position of equilibrium. Note that many changes alter the equilibrium *position,* but only temperature changes alter the value of the equilibrium *constant*. Sample Problem 17.12 shows how to visualize equilibrium at the molecular level.

Table 17.4 Effect of Various Disturbances on a System at Equilibrium

Disturbance	Net Direction of Reaction	Effect on Value of K
Concentration		
Increase [reactant]	Toward formation of product	None
Decrease [reactant]	Toward formation of reactant	None
Increase [product]	Toward formation of reactant	None
Decrease [product]	Toward formation of product	None
Pressure		
Increase P (decrease V)	Toward formation of fewer moles of gas	None
Decrease P (increase V)	Toward formation of more moles of gas	None
Increase P (add inert gas, no change in V)	None; concentrations unchanged	None
Temperature		
Increase T	Toward absorption of heat	Increases if $\Delta H°_{rxn} > 0$ Decreases if $\Delta H°_{rxn} < 0$
Decrease T	Toward release of heat	Increases if $\Delta H°_{rxn} < 0$ Decreases if $\Delta H°_{rxn} > 0$
Catalyst added	None; forward and reverse equilibrium attained sooner; rates increase equally	None

SAMPLE PROBLEM 17.12 Determining Equilibrium Parameters from Molecular Scenes

Problem For the reaction,

$$X(g) + Y_2(g) \rightleftharpoons XY(g) + Y(g) \qquad \Delta H > 0$$

the following molecular scenes depict different reaction mixtures (X = green, Y = purple):

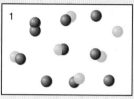
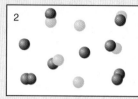

(a) If $K_c = 2$ at the temperature of the reaction, which scene represents the mixture at equilibrium?

(b) Will the reaction mixtures in the other two scenes proceed toward reactants or toward products to reach equilibrium?

(c) For the mixture at equilibrium, how will a rise in temperature affect $[Y_2]$?

Plan (a) We are given the balanced equation and the value of K_c and must choose the scene representing the mixture at equilibrium. We write the expression for Q_c, and for each scene, count particles and plug in the numbers to solve for the value of Q_c. Whichever scene gives a Q_c equal to K_c represents the mixture at equilibrium. **(b)** To determine the direction each reaction proceeds in the other two scenes, we compare the value of Q_c with the given K_c. If $Q_c > K_c$, the numerator (product side) is too high, so the reaction proceeds toward reactants; if $Q_c < K_c$, the reaction proceeds toward products. **(c)** We are given the sign of ΔH and must see whether a rise in T (corresponding to supplying heat) will increase or decrease the amount of the reactant Y_2. We treat heat as a reactant or product and see whether adding heat shifts the reaction right or left.

Solution (a) For the reaction, we have

$$Q_c = \frac{[XY][Y]}{[X][Y_2]}$$

scene 1: $Q_c = \dfrac{5 \times 3}{1 \times 1} = 15$ scene 2: $Q_c = \dfrac{4 \times 2}{2 \times 2} = 2$ scene 3: $Q_c = \dfrac{3 \times 1}{3 \times 3} = \dfrac{1}{3}$

For scene 2, $Q_c = K_c$, so it represents the mixture at equilibrium.

(b) For scene 1, Q_c (15) $> K_c$ (2), so the reaction proceeds toward reactants. For scene 3, Q_c ($\frac{1}{3}$) $< K_c$ (2), so the reaction proceeds toward products.

(c) The reaction is endothermic, so heat acts as a reactant:

$$X(g) + Y_2(g) + heat \rightleftharpoons XY(g) + Y(g)$$

Therefore, adding heat to the left shifts the reaction to the right, so $[Y_2]$ decreases.

Check (a) Remember that quantities in the numerator (or denominator) of Q_c are multiplied, not added. For example, the denominator for scene 1 is $1 \times 1 = 1$, not $1 + 1 = 2$.

(c) A good check is to imagine that $\Delta H < 0$ and see if you get the opposite result:

$$X(g) + Y_2(g) \rightleftharpoons XY(g) + Y(g) + heat$$

If $\Delta H < 0$, adding heat would shift the reaction to the left and increase $[Y_2]$.

FOLLOW-UP PROBLEM 17.12 For the reaction

$$C_2(g) + D_2(g) \rightleftharpoons 2CD(g) \qquad \Delta H < 0$$

these molecular scenes depict different reaction mixtures (C = red, D = blue):

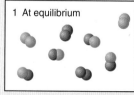
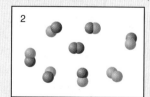
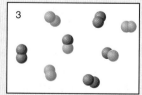

(a) Calculate the value of K_p. **(b)** In which direction will the reaction proceed for the mixtures *not* at equilibrium? **(c)** For the mixture at equilibrium, what effect will a rise in T have on the total moles of gas (increase, decrease, no effect)? Explain.

The Industrial Production of Ammonia

Nitrogen occurs in countless compounds, and four-fifths of the air we breathe is N_2. Yet, the use of N_2 in producing biological and industrial compounds is limited by its low chemical reactivity, the result of its strong triple bond. Thus, elemental nitrogen first must be *fixed*—combined with other elements in some usable form. The great majority of nitrogen fixation occurs naturally, either caused by lightning or carried out by enzymes found in certain bacteria on plant roots. However, nearly 13% of nitrogen fixation is accomplished industrially, through the **Haber process** for the synthesis of ammonia:

$$N_2(g) + 3H_2(g) \rightleftharpoons 2NH_3(g) \qquad \Delta H^\circ_{rxn} = -91.8 \text{ kJ}$$

The process was developed in 1913 by the German chemist Fritz Haber, and the first plant to use it produced 12,000 tons a year. Today, more than 110 million tons are produced each year—the highest production level of any compound (on a mole basis). Over 80% of this ammonia is used in fertilizers; other uses include the manufacture of explosives, nylons, and other polymers.

By inspecting the balanced equation and applying equilibrium principles, we can see three ways to maximize the yield of ammonia:

1. *Decrease concentration of ammonia.* NH_3 is the product, so removing it will shift the equilibrium position toward producing more.
2. *Decrease volume (increase pressure).* Because four moles of gas react to form two moles of gas, decreasing the volume will shift the equilibrium position toward fewer moles of gas, that is, toward forming more NH_3.
3. *Decrease temperature.* Because the formation of ammonia is exothermic, decreasing the temperature (removing heat) will shift the equilibrium position toward formation of product, thereby increasing K_c (Table 17.5).

Table 17.5	**Effect of Temperature on K_c for Ammonia Synthesis**
T (K)	**K_c**
200.	7.17×10^{15}
300.	2.69×10^{8}
400.	3.94×10^{4}
500.	1.72×10^{2}
600.	4.53×10^{0}
700.	2.96×10^{-1}
800.	3.96×10^{-2}

Therefore, the yield of NH_3 is maximized by removing it continuously as it forms while maintaining high pressure and low temperature. Figure 17.9 on the next page shows the very high yield (98.3%) attained at 1000 atm and 473 K (200.°C). Unfortunately, these conditions lead to a problem that highlights the distinction between equilibrium and kinetics. Although the *yield* is favored at this low temperature, the *rate* of formation is too slow to be economical. In practice, a compromise optimizes yield *and* rate. High pressure and continuous removal of product increase yield, and a somewhat higher temperature and a catalyst are used to increase rate. Achieving the same rate increase without a catalyst would require even higher temperatures, which would reduce the yield.

Modern ammonia plants operate at 200–300 atm and around 673 K (400.°C), and the catalyst consists of small iron crystals fused into a mixture of MgO, Al_2O_3, and SiO_2. The reactant gases in stoichiometric ratio (N_2:$H_2 = 1$:3) are injected into the reaction chamber and over the catalyst beds. The emerging equilibrium mixture, which contains about 35% NH_3 by volume, is cooled to condense and remove the NH_3; the remaining N_2 and H_2, which are still gaseous, are recycled into the reaction chamber.

Figure 17.9 Percent yield of ammonia vs. temperature (°C) at five different operating pressures. At very high pressure and low temperature (*top left*), the yield is high, but the rate of formation is low. Industrial conditions (*circle*) are between 200 and 300 atm at about 400°C.

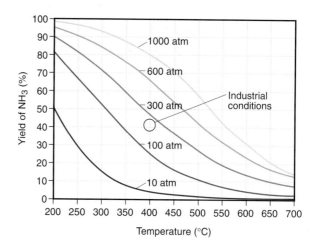

SECTION SUMMARY

Le Châtelier's principle states that if a system at equilibrium is disturbed, the system undergoes a net reaction that reduces the disturbance and allows equilibrium to be reattained. Changes in concentration cause a net reaction away from the added component or toward the removed component. For a reaction that involves a change in number of moles of gas, an increase in pressure (decrease in volume) causes a net reaction toward fewer moles of gas, and a decrease in pressure causes the opposite change. Although the equilibrium concentrations of components change as a result of concentration and volume changes, K does not change. A temperature change *does* change K: higher T increases K for an endothermic reaction (positive ΔH°_{rxn}) and decreases K for an exothermic reaction (negative ΔH°_{rxn}). A catalyst causes the system to reach equilibrium more quickly by speeding forward and reverse reactions equally, but it does not affect the equilibrium position. Ammonia production is favored by high pressure, low temperature, and continual removal of product. To make the process economical, an intermediate temperature and a catalyst are used.

For Review and Reference (Numbers in parentheses refer to pages, unless noted otherwise.)

Learning Objectives

To help you review these learning objectives, the numbers of related sections (§), sample problems (SP), and upcoming end-of-chapter problems (EP) are listed in parentheses.

1. Distinguish between the rate and the extent of a reaction; understand that the equilibrium constant (K) is a number whose magnitude is related to the extent of the reaction (§ 17.1) (EPs 17.1–17.4)

2. Understand that the reaction quotient (Q) changes until the system reaches equilibrium, when it equals K; write Q for any balanced equation, and calculate K given concentrations (§ 17.2) (SP 17.1) (EPs 17.5–17.18)

3. Use the ideal gas law and Δn_{gas} to convert between K_c and K_p (§ 17.3) (SP 17.2) (EPs 17.19–17.25)

4. Explain how the reaction direction depends on the relative values of Q and K (§ 17.4) (SP 17.3) (EPs 17.26–17.29)

5. Solve different types of equilibrium problems; calculate K given unknown quantities (concentrations or pressures), or unknown quantities given K, set up and use a reaction table, apply the quadratic equation, and make an assumption to simplify the calculations (§ 17.5) (SPs 17.4–17.8) (EPs 17.30–17.45)

6. Understand Le Châtelier's principle, and predict the effects of concentration, pressure (volume), temperature, and a catalyst on equilibrium position and on K (§ 17.6) (SPs 17.9–17.12) (EPs 17.46–17.60)

Key Terms

Section 17.1
equilibrium constant (K) (542)

Section 17.2
law of chemical equilibrium (law of mass action) (543)
reaction quotient (Q) (544)

Section 17.6
Le Châtelier's principle (561)
Haber process (569)

Key Equations and Relationships

17.1 Defining equilibrium in terms of reaction rates (541):

At equilibrium: $\text{rate}_{fwd} = \text{rate}_{rev}$

17.2 Defining the equilibrium constant for the reaction
A $\rightleftharpoons$ 2B (542):

$$K = \frac{k_{fwd}}{k_{rev}} = \frac{[B]_{eq}^2}{[A]_{eq}}$$

17.3 Defining the equilibrium constant in terms of the reaction quotient (544):

At equilibrium: $Q = K$

17.4 Expressing Q_c for the reaction aA + bB $\rightleftharpoons$ cC + dD (545):

$$Q_c = \frac{[C]^c[D]^d}{[A]^a[B]^b}$$

17.5 Finding the overall K for a reaction sequence (546):

$$K_{overall} = K_1 \times K_2 \times K_3 \times \cdots$$

17.6 Finding K of a reaction from K of the reverse reaction (547):

$$K_{fwd} = \frac{1}{K_{rev}}$$

17.7 Finding K of a reaction multiplied by a factor n (547):

$$K' = K^n$$

17.8 Relating K based on pressures to K based on concentrations (549):

$$K_p = K_c(RT)^{\Delta n_{gas}}$$

17.9 Assuming that ignoring the concentration that reacts introduces no significant error (557):

$$[A]_{init} - [A]_{reacting} = [A]_{eq} \approx [A]_{init}$$

Brief Solutions to Follow-up Problems

17.1 (a) $Q_c = \dfrac{[NO]^4[H_2O]^6}{[NH_3]^4[O_2]^5}$ (b) $Q_c = \dfrac{[N_2O][NO_2]}{[NO]^3}$

17.2 $K_p = K_c(RT)^{-1} = 1.67\left(0.0821\dfrac{atm\cdot L}{mol\cdot K} \times 500.\,K\right)^{-1}$

$= 4.07\times10^{-2}$

17.3 $Q_p = \dfrac{(P_{CH_3Cl})(P_{HCl})}{(P_{CH_4})(P_{Cl_2})} = \dfrac{(0.24)(0.47)}{(0.13)(0.035)} = 25;$

$Q_p < K_p$, so CH_3Cl is forming.

17.4 From the reaction table for $2NO + O_2 \rightleftharpoons 2NO_2$,

$P_{O_2} = 1.000\ atm - x = 0.506\ atm; x = 0.494\ atm$

Also, $P_{NO} = 0.012\ atm$ and $P_{NO_2} = 0.988\ atm$, so

$$K_p = \frac{0.988^2}{0.012^2(0.506)} = 1.3\times10^4$$

17.5 Since $\Delta n_{gas} = 0$, $K_p = K_c = 2.3\times10^{30} = \dfrac{(0.781)(0.209)}{P_{NO}^2}$

Thus, $P_{NO} = 2.7\times10^{-16}\ atm$

17.6 From the reaction table, $[H_2] = [I_2] = x$; $[HI] = 0.242 - 2x$.

Thus, $K_c = 1.26\times10^{-3} = \dfrac{x^2}{(0.242 - 2x)^2}$

Taking the square root of both sides, ignoring the negative root, and solving gives $x = [H_2] = 8.02\times10^{-3}\ M$.

17.7 (a) Based on the reaction table, and assuming that $0.20\ M - x \approx 0.20\ M$,

$$K_c = 2.94\times10^{-10} \approx \frac{4x^2}{0.20} \qquad x \approx 3.8\times10^{-6}$$

Error = $1.9\times10^{-3}\%$, so assumption is justified; therefore, at equilibrium, $[I_2] = 0.20\ M$ and $[I] = 7.6\times10^{-6}\ M$.

(b) Based on the same reaction table and assumption, $x \approx 0.10$; error is 50%, so assumption is *not* justified. Solve equation:

$$4x^2 + 0.209x - 0.042 = 0 \qquad x = 0.080\ M$$

Therefore, at equilibrium, $[I_2] = 0.12\ M$ and $[I] = 0.16\ M$.

17.8 (a) $Q_c = \dfrac{(0.0900)(0.0900)}{0.2100} = 3.86\times10^{-2}$

$Q_c < K_c$, so reaction proceeds to the right.

(b) From the reaction table,

$[PCl_5] = 0.2100\ M - x = 0.2065\ M \qquad x = 0.0035\ M$

So, $[Cl_2] = [PCl_3] = 0.0900\ M + x = 0.0935\ M$.

17.9 (a) $[SiF_4]$ increases; (b) decreases; (c) decreases; (d) no effect.

17.10 (a) Decrease P; (b) increase P; (c) increase P.

17.11 (a) P_{H_2} will decrease; K_p will increase; (b) P_{N_2} will increase; K_p will decrease; (c) P_{PCl_5} will increase; K_p will increase.

17.12 (a) Since $P = \dfrac{n}{V}RT$ and, in this case, V, R, and T cancel,

$$K_p = \frac{n_{CD}^2}{n_{C_2} \times n_{D_2}} = \frac{16}{(2)(2)} = 4$$

(b) Scene 2 to the left; scene 3 to the right. (c) There are 2 mol of gas on each side of the balanced equation, so there is no effect on total moles of gas.

Problems

Problems with **colored** numbers are answered in Appendix E. Sections match the text and provide the numbers of relevant sample problems. Bracketed problems are grouped in pairs (indicated by a short rule) that cover the same concept. Comprehensive Problems are based on material from any section or previous chapter.

The Equilibrium State and the Equilibrium Constant

17.1 A change in reaction conditions increases the rate of a certain forward reaction more than that of the reverse reaction. What is the effect on the equilibrium constant and the concentrations of reactants and products at equilibrium?

17.2 When a chemical company employs a new reaction to manufacture a product, the chemists consider its rate (kinetics) and yield (equilibrium). How do each of these affect the usefulness of a manufacturing process?

17.3 If there is no change in concentrations, why is the equilibrium state considered dynamic?

17.4 (a) Is K very large or very small for a reaction that goes essentially to completion? Explain. (b) White phosphorus, P_4, is produced by the reduction of phosphate rock, $Ca_3(PO_4)_2$. If exposed to oxygen, the waxy, white solid smokes, bursts into flames, and releases a large quantity of heat. Does the reaction

$$P_4(g) + 5O_2(g) \rightleftharpoons P_4O_{10}(s)$$

have a large or small equilibrium constant? Explain.

The Reaction Quotient and the Equilibrium Constant
(Sample Problem 17.1)

17.5 For a given reaction at a given temperature, the value of K is constant. Is the value of Q also constant? Explain.

17.6 In a series of experiments on the thermal decomposition of lithium peroxide,

$$2Li_2O_2(s) \rightleftharpoons 2Li_2O(s) + O_2(g)$$

a chemist finds that, as long as some Li_2O_2 is present at the end of the experiment, the amount of O_2 obtained in a given container at a given T is the same. Explain.

17.7 In a study of the formation of HI from its elements,

$$H_2(g) + I_2(g) \rightleftharpoons 2HI(g)$$

equal amounts of H_2 and I_2 were placed in a container, which was then sealed and heated.
(a) On one set of axes, sketch concentration vs. time curves for H_2 and HI, and explain how Q changes as a function of time.
(b) Is the value of Q different if $[I_2]$ is plotted instead of $[H_2]$?

17.8 Explain the difference between a heterogeneous and a homogeneous equilibrium. Give an example of each.

17.9 Does Q for the formation of 1 mol of NO from its elements differ from Q for the decomposition of 1 mol of NO to its elements? Explain and give the relationship between the two Q's.

17.10 Does Q for the formation of 1 mol of NH_3 from H_2 and N_2 differ from Q for the formation of NH_3 from H_2 and 1 mol of N_2? Explain and give the relationship between the two Q's.

17.11 Balance each reaction and write its reaction quotient, Q_c:
(a) $NO(g) + O_2(g) \rightleftharpoons N_2O_3(g)$
(b) $SF_6(g) + SO_3(g) \rightleftharpoons SO_2F_2(g)$
(c) $SClF_5(g) + H_2(g) \rightleftharpoons S_2F_{10}(g) + HCl(g)$

17.12 Balance each reaction and write its reaction quotient, Q_c:
(a) $C_2H_6(g) + O_2(g) \rightleftharpoons CO_2(g) + H_2O(g)$
(b) $CH_4(g) + F_2(g) \rightleftharpoons CF_4(g) + HF(g)$
(c) $SO_3(g) \rightleftharpoons SO_2(g) + O_2(g)$

17.13 At a particular temperature, $K_c = 1.6 \times 10^{-2}$ for

$$2H_2S(g) \rightleftharpoons 2H_2(g) + S_2(g)$$

Calculate K_c for each of the following reactions:
(a) $\frac{1}{2}S_2(g) + H_2(g) \rightleftharpoons H_2S(g)$
(b) $5H_2S(g) \rightleftharpoons 5H_2(g) + \frac{5}{2}S_2(g)$

17.14 At a particular temperature, $K_c = 6.5 \times 10^2$ for

$$2NO(g) + 2H_2(g) \rightleftharpoons N_2(g) + 2H_2O(g)$$

Calculate K_c for each of the following reactions:
(a) $NO(g) + H_2(g) \rightleftharpoons \frac{1}{2}N_2(g) + H_2O(g)$
(b) $2N_2(g) + 4H_2O(g) \rightleftharpoons 4NO(g) + 4H_2(g)$

17.15 Balance each of the following examples of heterogeneous equilibria and write its reaction quotient, Q_c:
(a) $Na_2O_2(s) + CO_2(g) \rightleftharpoons Na_2CO_3(s) + O_2(g)$
(b) $H_2O(l) \rightleftharpoons H_2O(g)$
(c) $NH_4Cl(s) \rightleftharpoons NH_3(g) + HCl(g)$

17.16 Balance each of the following examples of heterogeneous equilibria and write its reaction quotient, Q_c:
(a) $H_2O(l) + SO_3(g) \rightleftharpoons H_2SO_4(aq)$
(b) $KNO_3(s) \rightleftharpoons KNO_2(s) + O_2(g)$
(c) $S_8(s) + F_2(g) \rightleftharpoons SF_6(g)$

17.17 Write Q_c for each of the following:
(a) Hydrogen chloride gas reacts with oxygen gas to produce chlorine gas and water vapor.
(b) Solid diarsenic trioxide reacts with fluorine gas to produce liquid arsenic pentafluoride and oxygen gas.
(c) Gaseous sulfur tetrafluoride reacts with liquid water to produce gaseous sulfur dioxide and hydrogen fluoride gas.
(d) Solid molybdenum(VI) oxide reacts with gaseous xenon difluoride to form liquid molybdenum(VI) fluoride, xenon gas, and oxygen gas.

17.18 The interhalogen ClF_3 is prepared in a two-step fluorination of chlorine gas:

$$Cl_2(g) + F_2(g) \rightleftharpoons ClF(g)$$
$$ClF(g) + F_2(g) \rightleftharpoons ClF_3(g)$$

(a) Balance each step and write the overall equation.
(b) Show that the overall Q_c equals the product of the Q_c's for the individual steps.

Expressing Equilibria with Pressure Terms: Relation Between K_c and K_p
(Sample Problem 17.2)

17.19 Guldberg and Waage proposed the definition of the equilibrium constant as a certain ratio of *concentrations*. What relationship allows us to use a particular ratio of *partial pressures* (for a gaseous reaction) to express an equilibrium constant? Explain.

17.20 When are K_c and K_p equal, and when are they not?

17.21 A certain reaction at equilibrium has more moles of gaseous products than of gaseous reactants.
(a) Is K_c larger or smaller than K_p?
(b) Write a general statement about the relative sizes of K_c and K_p for any gaseous equilibrium.

17.22 Determine Δn_{gas} for each of the following reactions:
(a) $2KClO_3(s) \rightleftharpoons 2KCl(s) + 3O_2(g)$
(b) $2PbO(s) + O_2(g) \rightleftharpoons 2PbO_2(s)$
(c) $I_2(s) + 3XeF_2(s) \rightleftharpoons 2IF_3(s) + 3Xe(g)$

17.23 Determine Δn_{gas} for each of the following reactions:
(a) $MgCO_3(s) \rightleftharpoons MgO(s) + CO_2(g)$
(b) $2H_2(g) + O_2(g) \rightleftharpoons 2H_2O(l)$
(c) $HNO_3(l) + ClF(g) \rightleftharpoons ClONO_2(g) + HF(g)$

17.24 Calculate K_c for each of the following equilibria:
(a) $CO(g) + Cl_2(g) \rightleftharpoons COCl_2(g)$; $K_p = 3.9 \times 10^{-2}$ at 1000. K
(b) $S_2(g) + C(s) \rightleftharpoons CS_2(g)$; $K_p = 28.5$ at 500. K

17.25 Calculate K_c for each of the following equilibria:
(a) $H_2(g) + I_2(g) \rightleftharpoons 2HI(g)$; $K_p = 49$ at 730. K
(b) $2SO_2(g) + O_2(g) \rightleftharpoons 2SO_3(g)$; $K_p = 2.5 \times 10^{10}$ at 500. K

Reaction Direction: Comparing Q and K
(Sample Problem 17.3)

17.26 When the numerical value of Q is less than K, in which direction does the reaction proceed to reach equilibrium? Explain.

17.27 At 425°C, $K_p = 4.18 \times 10^{-9}$ for the reaction
$$2HBr(g) \rightleftharpoons H_2(g) + Br_2(g)$$
In one experiment, 0.20 atm of HBr(g), 0.010 atm of $H_2(g)$, and 0.010 atm of $Br_2(g)$ are introduced into a container. Is the reaction at equilibrium? If not, in which direction will it proceed?

17.28 At 100°C, $K_p = 60.6$ for the reaction
$$2NOBr(g) \rightleftharpoons 2NO(g) + Br_2(g)$$
In a given experiment, 0.10 atm of each component is placed in a container. Is the system at equilibrium? If not, in which direction will the reaction proceed?

17.29 The water-gas shift reaction plays a central role in the chemical methods for obtaining cleaner fuels from coal:
$$CO(g) + H_2O(g) \rightleftharpoons CO_2(g) + H_2(g)$$
At a given temperature, $K_p = 2.7$. If 0.13 mol of CO, 0.56 mol of H_2O, 0.62 mol of CO_2, and 0.43 mol of H_2 are introduced into a 2.0-L flask, in which direction must the reaction proceed to reach equilibrium?

How to Solve Equilibrium Problems
(Sample Problems 17.4 to 17.8)

17.30 For a problem involving the catalyzed reaction of methane and steam, the following reaction table was prepared:

Pressure (atm)	$CH_4(g)$	+	$2H_2O(g)$	$\rightleftharpoons$	$CO_2(g)$	+	$4H_2(g)$
Initial	0.30		0.40		0		0
Change	$-x$		$-2x$		$+x$		$+4x$
Equilibrium	$0.30 - x$		$0.40 - 2x$		x		$4x$

Explain the entries in the "Change" and "Equilibrium" rows.

17.31 (a) What is the basis of the approximation that avoids using the quadratic formula to find an equilibrium concentration? (b) When should this approximation *not* be made?

17.32 In an experiment to study the formation of HI(g),
$$H_2(g) + I_2(g) \rightleftharpoons 2HI(g)$$
$H_2(g)$ and $I_2(g)$ were placed in a sealed container at a certain temperature. At equilibrium, $[H_2] = 6.50 \times 10^{-5}$ M, $[I_2] = 1.06 \times 10^{-3}$ M, and $[HI] = 1.87 \times 10^{-3}$ M. Calculate K_c for the reaction at this temperature.

17.33 Gaseous ammonia was introduced into a sealed container and heated to a certain temperature:
$$2NH_3(g) \rightleftharpoons N_2(g) + 3H_2(g)$$
At equilibrium, $[NH_3] = 0.0225$ M, $[N_2] = 0.114$ M, and $[H_2] = 0.342$ M. Calculate K_c for the reaction at this temperature.

17.34 Gaseous PCl_5 decomposes according to the reaction
$$PCl_5(g) \rightleftharpoons PCl_3(g) + Cl_2(g)$$
In one experiment, 0.15 mol of $PCl_5(g)$ was introduced into a 2.0-L container. Construct the reaction table for this process.

17.35 Hydrogen fluoride, HF, can be made from the reaction
$$H_2(g) + F_2(g) \rightleftharpoons 2HF(g)$$
In one experiment, 0.10 mol of $H_2(g)$ and 0.050 mol of $F_2(g)$ are added to a 0.50-L flask. Write a reaction table for this process.

17.36 For the following reaction, $K_p = 6.5 \times 10^4$ at 308 K:
$$2NO(g) + Cl_2(g) \rightleftharpoons 2NOCl(g)$$
At equilibrium, $P_{NO} = 0.35$ atm and $P_{Cl_2} = 0.10$ atm. What is the equilibrium partial pressure of NOCl(g)?

17.37 For the following reaction, $K_p = 0.262$ at 1000°C:
$$C(s) + 2H_2(g) \rightleftharpoons CH_4(g)$$
At equilibrium, P_{H_2} is 1.22 atm. What is the equilibrium partial pressure of $CH_4(g)$?

17.38 Ammonium hydrogen sulfide decomposes according to the following reaction, for which $K_p = 0.11$ at 250°C:
$$NH_4HS(s) \rightleftharpoons H_2S(g) + NH_3(g)$$
If 55.0 g of $NH_4HS(s)$ is placed in a sealed 5.0-L container, what is the partial pressure of $NH_3(g)$ at equilibrium?

17.39 Hydrogen sulfide decomposes according to the following reaction, for which $K_c = 9.30 \times 10^{-8}$ at 700°C:
$$2H_2S(g) \rightleftharpoons 2H_2(g) + S_2(g)$$
If 0.45 mol of H_2S is placed in a 3.0-L container, what is the equilibrium concentration of $H_2(g)$ at 700°C?

17.40 Even at high T, the formation of nitric oxide is not favored:
$$N_2(g) + O_2(g) \rightleftharpoons 2NO(g) \qquad K_c = 4.10 \times 10^{-4} \text{ at } 2000°C$$
What is [NO] when a mixture of 0.20 mol of $N_2(g)$ and 0.15 mol of $O_2(g)$ reach equilibrium in a 1.0-L container at 2000°C?

17.41 Nitrogen dioxide decomposes according to the reaction
$$2NO_2(g) \rightleftharpoons 2NO(g) + O_2(g)$$
where $K_p = 4.48 \times 10^{-13}$ at a certain temperature. A pressure of 0.75 atm of NO_2 is introduced into a container and allowed to come to equilibrium. What are the equilibrium partial pressures of NO(g) and $O_2(g)$?

17.42 In an analysis of interhalogen reactivity, 0.500 mol of ICl was placed in a 5.00-L flask, where it decomposed at a high T: $2ICl(g) \rightleftharpoons I_2(g) + Cl_2(g)$. Calculate the equilibrium concentrations of I_2, Cl_2, and ICl ($K_c = 0.110$ at this temperature).

17.43 A United Nations toxicologist studying the properties of mustard gas, $S(CH_2CH_2Cl)_2$, a blistering agent used in warfare, prepares a mixture of 0.675 M SCl_2 and 0.973 M C_2H_4 and allows it to react at room temperature (20.0°C):
$$SCl_2(g) + 2C_2H_4(g) \rightleftharpoons S(CH_2CH_2Cl)_2(g)$$
At equilibrium, $[S(CH_2CH_2Cl)_2] = 0.350$ M. Calculate K_p.

17.44 The first step in industrial production of nitric acid is the catalyzed oxidation of ammonia. Without a catalyst, a different reaction predominates:
$$4NH_3(g) + 3O_2(g) \rightleftharpoons 2N_2(g) + 6H_2O(g)$$
When 0.0150 mol of $NH_3(g)$ and 0.0150 mol of $O_2(g)$ are placed in a 1.00-L container at a certain temperature, the N_2 concentration at equilibrium is 1.96×10^{-3} M. Calculate K_c.

17.45 A key step in the extraction of iron from its ore is
$$FeO(s) + CO(g) \rightleftharpoons Fe(s) + CO_2(g) \quad K_p = 0.403 \text{ at } 1000°C$$
This step occurs in the 700°C to 1200°C zone within a blast furnace. What are the equilibrium partial pressures of CO(g) and $CO_2(g)$ when 1.00 atm of CO(g) and excess FeO(s) react in a sealed container at 1000°C?

Reaction Conditions and the Equilibrium State: Le Châtelier's Principle

(Sample Problems 17.9 to 17.12)

17.46 What is the difference between the equilibrium position and the equilibrium constant of a reaction? Which changes as a result of a change in reactant concentration?

17.47 Scenes A, B, and C depict this reaction at three temperatures:

$$NH_4Cl(s) \rightleftharpoons NH_3(g) + HCl(g) \quad \Delta H^\circ_{rxn} = 176 \text{ kJ}$$

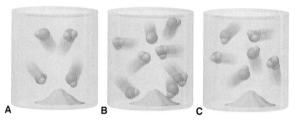

A   **B**   **C**

(a) Which best represents the reaction mixture at the highest temperature? Explain.

(b) Which best represents the reaction mixture at the lowest temperature? Explain.

17.48 What is implied by the word "constant" in the term *equilibrium constant*? Give two reaction parameters that can be changed without changing the value of an equilibrium constant.

17.49 Le Châtelier's principle is related ultimately to the rates of the forward and reverse steps in a reaction. Explain (a) why an increase in reactant concentration shifts the equilibrium position to the right but does not change K; (b) why a decrease in V shifts the equilibrium position toward fewer moles of gas but does not change K; and (c) why a rise in T shifts the equilibrium position of an exothermic reaction toward reactants and also changes K.

17.50 Le Châtelier's principle predicts that a rise in the temperature of an endothermic reaction from T_1 to T_2 results in K_2 being larger than K_1. Explain.

17.51 Consider this equilibrium system:

$$CO(g) + Fe_3O_4(s) \rightleftharpoons CO_2(g) + 3FeO(s)$$

How does the equilibrium position shift as a result of each of the following disturbances?
(a) CO is added.
(b) CO_2 is removed by adding solid NaOH.
(c) Additional $Fe_3O_4(s)$ is added to the system.
(d) Dry ice is added at constant temperature.

17.52 Sodium bicarbonate undergoes thermal decomposition according to the reaction

$$2NaHCO_3(s) \rightleftharpoons Na_2CO_3(s) + CO_2(g) + H_2O(g)$$

How does the equilibrium position shift as a result of each of the following disturbances?
(a) 0.20 atm of argon gas is added.
(b) $NaHCO_3(s)$ is added.
(c) $Mg(ClO_4)_2(s)$ is added as a drying agent to remove H_2O.
(d) Dry ice is added at constant temperature.

17.53 Predict the effect of *increasing* the container volume on the amounts of each reactant and product in the following reactions:
(a) $F_2(g) \rightleftharpoons 2F(g)$
(b) $2CH_4(g) \rightleftharpoons C_2H_2(g) + 3H_2(g)$

17.54 Predict the effect of *decreasing* the container volume on the amounts of each reactant and product in the following reactions:

(a) $C_3H_8(g) + 5O_2(g) \rightleftharpoons 3CO_2(g) + 4H_2O(l)$
(b) $4NH_3(g) + 3O_2(g) \rightleftharpoons 2N_2(g) + 6H_2O(g)$

17.55 How would you adjust the *volume* of the reaction vessel in order to maximize product yield in each of the following reactions?
(a) $Fe_3O_4(s) + 4H_2(g) \rightleftharpoons 3Fe(s) + 4H_2O(g)$
(b) $2C(s) + O_2(g) \rightleftharpoons 2CO(g)$

17.56 How would you adjust the *volume* of the reaction vessel in order to maximize product yield in each of the following reactions?
(a) $Na_2O_2(s) \rightleftharpoons 2Na(l) + O_2(g)$
(b) $C_2H_2(g) + 2H_2(g) \rightleftharpoons C_2H_6(g)$

17.57 Predict the effect of *increasing* the temperature on the amounts of products in the following reactions:
(a) $CO(g) + 2H_2(g) \rightleftharpoons CH_3OH(g) \quad \Delta H^\circ_{rxn} = -90.7 \text{ kJ}$
(b) $C(s) + H_2O(g) \rightleftharpoons CO(g) + H_2(g) \quad \Delta H^\circ_{rxn} = 131 \text{ kJ}$
(c) $2NO_2(g) \rightleftharpoons 2NO(g) + O_2(g)$ (endothermic)
(d) $2C(s) + O_2(g) \rightleftharpoons 2CO(g)$ (exothermic)

17.58 Predict the effect of *decreasing* the temperature on the amounts of reactants in the following reactions:
(a) $C_2H_2(g) + H_2O(g) \rightleftharpoons CH_3CHO(g) \quad \Delta H^\circ_{rxn} = -151 \text{ kJ}$
(b) $CH_3CH_2OH(l) + O_2(g) \rightleftharpoons CH_3CO_2H(l) + H_2O(g)$
$$\Delta H^\circ_{rxn} = -451 \text{ kJ}$$
(c) $2C_2H_4(g) + O_2(g) \rightleftharpoons 2CH_3CHO(g)$ (exothermic)
(d) $N_2O_4(g) \rightleftharpoons 2NO_2(g)$ (endothermic)

17.59 The minerals hematite (Fe_2O_3) and magnetite (Fe_3O_4) exist in equilibrium with atmospheric oxygen:

$$4Fe_3O_4(s) + O_2(g) \rightleftharpoons 6Fe_2O_3(s) \quad K_p = 2.5 \times 10^{87} \text{ at } 298 \text{ K}$$

(a) Determine P_{O_2} at equilibrium. (b) Given that P_{O_2} in air is 0.21 atm, in which direction will the reaction proceed to reach equilibrium? (c) Calculate K_c at 298 K.

17.60 The oxidation of SO_2 is the key step in H_2SO_4 production:

$$SO_2(g) + \tfrac{1}{2}O_2(g) \rightleftharpoons SO_3(g) \quad \Delta H^\circ_{rxn} = -99.2 \text{ kJ}$$

(a) What qualitative combination of T and P maximizes SO_3 yield?
(b) How does addition of O_2 affect Q? K?
(c) Why is catalysis used for this reaction?

Comprehensive Problems

Problems with an asterisk (*) are more challenging.

17.61 What three criteria characterize a chemical system at equilibrium?

17.62 The "filmstrip" represents five molecular-level scenes of a gaseous mixture as it reaches equilibrium over time:

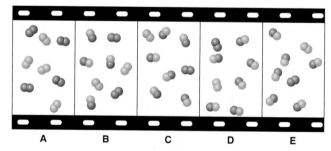

A   **B**   **C**   **D**   **E**

X is purple and Y is orange: $X_2(g) + Y_2(g) \rightleftharpoons 2XY(g)$.
(a) Write the reaction quotient, Q, for this reaction.
(b) If each particle represents 0.1 mol of particles, calculate Q for each scene.

(c) If $K > 1$, is time progressing to the right or to the left? Explain.

(d) Calculate K at this temperature.

(e) If $\Delta H^\circ_{rxn} < 0$, which scene, if any, best represents the mixture at a higher temperature? Explain.

(f) Which scene, if any, best represents the mixture at a higher pressure (lower volume)? Explain.

17.63 A study of the water-gas shift reaction (see Problem 17.29) was made in which equilibrium was reached with $[CO] = [H_2O] = [H_2] = 0.10$ M and $[CO_2] = 0.40$ M. After 0.60 mol of H_2 is added to the 2.0-L container and equilibrium is reestablished, what are the new concentrations of all the components?

17.64 Isolation of Group 8B(10) elements, used as industrial catalysts, involves a series of steps. For nickel, the sulfide ore is roasted in air: $Ni_3S_2(s) + O_2(g) \rightleftharpoons NiO(s) + SO_2(g)$. The metal oxide is reduced by the H_2 in water gas ($CO + H_2$) to impure Ni: $NiO(s) + H_2(g) \rightleftharpoons Ni(s) + H_2O(g)$. The CO in water gas then reacts with the metal in the Mond process to form gaseous nickel carbonyl, $Ni(s) + CO(g) \rightleftharpoons Ni(CO)_4(g)$, which is subsequently decomposed to the metal.

(a) Balance each of the three steps, and obtain an overall balanced equation for the conversion of Ni_3S_2 to $Ni(CO)_4$.

(b) Show that the overall Q_c is the product of the Q_c's for the individual reactions.

17.65 One of the most important industrial sources of ethanol is the reaction of steam with ethylene derived from crude oil:

$$C_2H_4(g) + H_2O(g) \rightleftharpoons C_2H_5OH(g)$$
$$\Delta H^\circ_{rxn} = -47.8 \text{ kJ} \qquad K_c = 9 \times 10^3 \text{ at } 600. \text{ K}$$

(a) At equilibrium, $P_{C_2H_5OH} = 200.$ atm and $P_{H_2O} = 400.$ atm. Calculate $P_{C_2H_4}$.

(b) Is the highest yield of ethanol obtained at high or low pressures? High or low temperatures?

(c) In ammonia manufacture, the yield is increased by condensing the NH_3 to a liquid and removing it from the vessel. Would condensing the C_2H_5OH have the same effect for ethanol production? Explain.

17.66 An industrial chemist introduces 2.0 atm of H_2 and 2.0 atm of CO_2 into a 1.00-L container at 25.0°C and then raises the temperature to 700.°C, at which $K_c = 0.534$:

$$H_2(g) + CO_2(g) \rightleftharpoons H_2O(g) + CO(g)$$

How many grams of H_2 are present at equilibrium?

17.67 As an EPA scientist studying catalytic converters and urban smog, you want to find K_c for the following reaction:

$$2NO_2(g) \rightleftharpoons N_2(g) + 2O_2(g) \qquad K_c = ?$$

Use the following data to find the unknown K_c:

$$\tfrac{1}{2}N_2(g) + \tfrac{1}{2}O_2(g) \rightleftharpoons NO(g) \qquad\qquad K_c = 4.8 \times 10^{-10}$$
$$2NO_2(g) \rightleftharpoons 2NO(g) + O_2(g) \qquad K_c = 1.1 \times 10^{-5}$$

17.68 An engineer examining the oxidation of SO_2 in the manufacture of sulfuric acid determines that $K_c = 1.7 \times 10^8$ at 600. K:

$$2SO_2(g) + O_2(g) \rightleftharpoons 2SO_3(g)$$

(a) At equilibrium, $P_{SO_3} = 300.$ atm and $P_{O_2} = 100.$ atm. Calculate P_{SO_2}.

(b) The engineer places a mixture of 0.0040 mol of $SO_2(g)$ and 0.0028 mol of $O_2(g)$ in a 1.0-L container and raises the temperature to 1000 K. At equilibrium, 0.0020 mol of $SO_3(g)$ is present. Calculate K_c and P_{SO_2} for this reaction at 1000. K.

17.69 When 0.100 mol of $CaCO_3(s)$ and 0.100 mol of $CaO(s)$ are placed in an evacuated sealed 10.0-L container and heated to 385 K, $P_{CO_2} = 0.220$ atm after equilibrium is established:

$$CaCO_3(s) \rightleftharpoons CaO(s) + CO_2(g)$$

An additional 0.300 atm of $CO_2(g)$ is then pumped into the container. What is the total mass (in g) of $CaCO_3$ after equilibrium is re-established?

17.70 Use each of the following reaction quotients to write the balanced equation:

(a) $Q = \dfrac{[CO_2]^2[H_2O]^2}{[C_2H_4][O_2]^3}$ (b) $Q = \dfrac{[NH_3]^4[O_2]^7}{[NO_2]^4[H_2O]^6}$

17.71 In combustion studies of H_2 as an alternative fuel, you find evidence that the hydroxyl radical (HO) is formed in flames by the reaction $H(g) + \tfrac{1}{2}O_2(g) \rightleftharpoons HO(g)$. Use the following data to calculate K_c for the reaction:

$$\tfrac{1}{2}H_2(g) + \tfrac{1}{2}O_2(g) \rightleftharpoons HO(g) \qquad K_c = 0.58$$
$$\tfrac{1}{2}H_2(g) \rightleftharpoons H(g) \qquad\qquad K_c = 1.6 \times 10^{-3}$$

17.72 Aluminum is one of the most versatile metals. It is produced by the Hall-Heroult process, in which molten cryolite, Na_3AlF_6, is used as a solvent for the aluminum ore. Cryolite undergoes very slight decomposition with heat to produce a tiny amount of F_2, which escapes into the atmosphere above the solvent. K_c is 2×10^{-104} at 1300 K for the reaction:

$$Na_3AlF_6(l) \rightleftharpoons 3Na(l) + Al(l) + 3F_2(g)$$

What is the concentration of F_2 over a bath of molten cryolite at this temperature?

17.73 An equilibrium mixture of car exhaust gases consisting of 10.0 volumes of CO_2, 1.00 volume of unreacted O_2, and 50.0 volumes of unreacted N_2 leaves the engine at 4.0 atm and 800. K.

(a) Given this equilibrium, what is the partial pressure of CO?

$$2CO_2(g) \rightleftharpoons 2CO(g) + O_2(g) \qquad K_p = 1.4 \times 10^{-28} \text{ at } 800. \text{ K}$$

(b) Assuming the mixture has enough time to reach equilibrium, what is the concentration in picograms per liter (pg/L) of CO in the exhaust gas? (The actual concentration of CO in car exhaust is much higher because the gases do *not* reach equilibrium in the short transit time through the engine and exhaust system.)

17.74 Consider the following reaction:

$$3Fe(s) + 4H_2O(g) \rightleftharpoons Fe_3O_4(s) + 4H_2(g)$$

(a) What are the apparent oxidation states of Fe and of O in Fe_3O_4?

(b) Fe_3O_4 is a compound of iron in which Fe occurs in two oxidation states. What are the oxidation states of Fe in Fe_3O_4?

(c) At 900°C, K_c for the reaction is 5.1. If 0.050 mol of $H_2O(g)$ and 0.100 mol of $Fe(s)$ are placed in a 1.0-L container at 900°C, how many grams of Fe_3O_4 are present at equilibrium?

Note: The synthesis of ammonia is a major process throughout the industrialized world. Problems 17.75 to 17.79 refer to various aspects of this all-important reaction:

$$N_2(g) + 3H_2(g) \rightleftharpoons 2NH_3(g) \qquad \Delta H^\circ_{rxn} = -91.8 \text{ kJ}$$

* **17.75** When ammonia is made industrially, the mixture of N_2, H_2, and NH_3 that emerges from the reaction chamber is far from equilibrium. Why does the plant supervisor use reaction conditions that produce less than the maximum yield of ammonia?

17.76 The following reaction is sometimes used to produce the H_2 needed for the synthesis of ammonia:

$$CH_4(g) + CO_2(g) \rightleftharpoons 2CO(g) + 2H_2(g)$$

(a) What is the percent yield of H_2 when an equimolar mixture of CH_4 and CO_2 with a total pressure of 20.0 atm reaches equilibrium at 1200. K, at which $K_p = 3.548 \times 10^6$?
(b) What is the percent yield of H_2 for this system at 1300. K, at which $K_p = 2.626 \times 10^7$?

17.77 Using CH_4 and steam as a source of H_2 for NH_3 synthesis requires high temperatures. Rather than burning CH_4 separately to heat the mixture, it is more efficient to inject some oxygen into the reaction mixture. All of the H_2 is thus released for the synthesis, and the heat of reaction for the combustion of CH_4 helps maintain the required temperature. Imagine the reaction occurring in two steps:

$$2CH_4(g) + O_2(g) \rightleftharpoons 2CO(g) + 4H_2(g)$$
$$K_p = 9.34 \times 10^{28} \text{ at } 1000. \text{ K}$$

$$CO(g) + H_2O(g) \rightleftharpoons CO_2(g) + H_2(g)$$
$$K_p = 1.374 \text{ at } 1000. \text{ K}$$

(a) Write the overall equation for the reaction of methane, steam, and oxygen to form carbon dioxide and hydrogen.
(b) What is K_p for the overall reaction?
(c) What is K_c for the overall reaction?
(d) A mixture of 2.0 mol of CH_4, 1.0 mol of O_2, and 2.0 mol of steam with a total pressure of 30. atm reacts at 1000. K at constant volume. Assuming that the reaction is complete and the ideal gas law is a valid approximation, what is the final pressure?

17.78 One mechanism for the synthesis of ammonia proposes that N_2 and H_2 molecules catalytically dissociate into atoms:

$$N_2(g) \rightleftharpoons 2N(g) \quad \log K_p = -43.10$$
$$H_2(g) \rightleftharpoons 2H(g) \quad \log K_p = -17.30$$

(a) Find the partial pressure of N in N_2 at 1000. K and 200. atm.
(b) Find the partial pressure of H in H_2 at 1000. K and 600. atm.
(c) How many N atoms and H atoms are present per liter?
(d) Based on these answers, which of the following is a more reasonable step to continue the mechanism after the catalytic dissociation? Explain.

$$N(g) + H(g) \longrightarrow NH(g)$$
$$N_2(g) + H(g) \longrightarrow NH(g) + N(g)$$

* 17.79 You are a member of a research team of chemists discussing the plans to operate an ammonia processing plant:

$$N_2(g) + 3H_2(g) \rightleftharpoons 2NH_3(g)$$

(a) The plant operates at close to 700 K, at which K_p is 1.00×10^{-4}, and employs the stoichiometric 1:3 ratio of N_2:H_2.

At equilibrium, the partial pressure of NH_3 is 50. atm. Calculate the partial pressures of each reactant and P_{total}.
(b) One member of the team makes the following suggestion: since the partial pressure of H_2 is cubed in the reaction quotient, the plant could produce the same amount of NH_3 if the reactants were in a 1:6 ratio of N_2:H_2 and could do so at a lower pressure, which would cut operating costs. Calculate the partial pressure of each reactant and P_{total} under these conditions, assuming an unchanged partial pressure of 50. atm for NH_3. Is the team member's argument valid?

17.80 The two most abundant atmospheric gases react to a tiny extent at 298 K in the presence of a catalyst:

$$N_2(g) + O_2(g) \rightleftharpoons 2NO(g) \qquad K_p = 4.35 \times 10^{-31}$$

(a) What are the equilibrium pressures of the three components when the atmospheric partial pressures of O_2 (0.210 atm) and of N_2 (0.780 atm) are put into an evacuated 1.00-L flask at 298 K with catalyst?
(b) What is P_{total} in the container?
(c) Find K_c for this reaction at 298 K.

17.81 Isopentyl alcohol reacts with pure acetic acid to form isopentyl acetate, the essence of banana oil:

$$C_5H_{11}OH + CH_3COOH \rightleftharpoons CH_3COOC_5H_{11} + H_2O$$

A student adds a drying agent to remove H_2O and thus increase the yield of banana oil. Is this approach reasonable? Explain.

* 17.82 For the equilibrium

$$2H_2S(g) \rightleftharpoons 2H_2(g) + S_2(g) \qquad K_c = 9.0 \times 10^{-8} \text{ at } 700°C$$

the initial concentrations of the three gases are 0.300 M H_2S, 0.300 M H_2, and 0.150 M S_2. Determine the equilibrium concentrations of the gases.

17.83 Glauber's salt, $Na_2SO_4 \cdot 10H_2O$, was used by J. R. Glauber in the 17th century as a medicinal agent. At 25°C, $K_p = 4.08 \times 10^{-25}$ for the loss of water of hydration from Glauber's salt:

$$Na_2SO_4 \cdot 10H_2O(s) \rightleftharpoons Na_2SO_4(s) + 10H_2O(g)$$

(a) What is the vapor pressure of water at 25°C in a closed container holding a sample of $Na_2SO_4 \cdot 10H_2O(s)$?
(b) How do the following changes affect the ratio (higher, lower, same) of hydrated form to anhydrous form for the system above?
(1) Add more $Na_2SO_4(s)$ (2) Reduce the container volume
(3) Add more water vapor (4) Add N_2 gas

* 17.84 In a study of synthetic fuels, 0.100 mol of CO and 0.100 mol of water vapor are added to a 20.00-L container at 900.°C, and they react to form CO_2 and H_2. At equilibrium, [CO] is 2.24×10^{-3} M. (a) Calculate K_c at this temperature. (b) Calculate P_{total} in the flask at equilibrium. (c) How many moles of CO must be added to double this pressure? (d) After P_{total} is doubled and the system reattains equilibrium, what is $[CO]_{eq}$?

CHAPTER EIGHTEEN

Acid-Base Equilibria

Cooking with Acids and Bases The acids in lemon react with the bases in salmon in a culinary neutralization. In this chapter, you'll examine the equilibrium nature and the expanding definitions of acid-base reactions.

Key Principles

◆ In aqueous systems, the proton always exists as a *hydronium ion, H_3O^+*.

◆ In the *Arrhenius acid-base definition*, an acid is an *H-containing substance* that yields H_3O^+ in water, a base is an *OH-containing substance* that yields OH^- in water, and an acid-base (*neutralization*) reaction occurs when H_3O^+ and OH^- form H_2O.

◆ The dissociation of a weak acid, HA, in water is a reversible process associated with an equilibrium constant called the *acid-dissociation constant, K_a*. The stronger the acid, the higher its K_a: weak acids typically have K_a values that are several orders of magnitude less than 1.

◆ Water molecules dissociate (*autoionize*) to a very small extent into H_3O^+ and OH^-, a process associated with the *ion-product constant for water, K_w*. Acidity or basicity is determined by the relative magnitudes of $[H_3O^+]$ and $[OH^-]$: in a neutral solution (or pure water), $[H_3O^+] = [OH^-]$, in an acidic solution, $[H_3O^+] > [OH^-]$, and in a basic solution, $[H_3O^+] < [OH^-]$.

◆ The *pH*, or the negative log of $[H_3O^+]$, is a measure of a solution's acidity: pH < 7 means the solution is acidic, and pH > 7 means it is basic. Because K_w, the product of $[H_3O^+]$ and $[OH^-]$, is a constant, the values for $[H_3O^+]$, $[OH^-]$, pH, and pOH are *interconvertible*.

◆ In the *Brønsted-Lowry acid-base definition*, a base is any species that *accepts a proton;* thus, there are many more Brønsted-Lowry bases than Arrhenius bases. When base B accepts a proton from acid HA, the species BH^+ and A^- form. HA and A^- are a *conjugate acid-base pair,* as are BH^+ and B. Thus, an acid-base reaction is a *proton-transfer process* between two conjugate acid-base pairs, with the stronger acid and base forming the weaker acid and base.

◆ *Polyprotic acids* have more than one ionizable proton, but in solution, essentially all the H_3O^+ comes from the first dissociation.

◆ Ammonia, amines, and anions of weak acids behave as weak bases in a process associated with a *base-dissociation constant, K_b*. The reaction of HA with H_2O added to the reaction of A^- with H_2O gives the reaction for the autoionization of water; thus, $K_a \times K_b = K_w$.

◆ The electronegativity of atoms and the polarity, length, and energy of bonds determine acid strength.

◆ In a salt solution, the ion that reacts with water to a greater extent (higher K) determines the solution's acidity or basicity.

◆ In the *Lewis acid-base definition*, an acid is any species that *accepts a lone pair* to form a new bond in an *adduct*. Thus, there are many more Lewis acids than other types. Lewis acids include molecules with electron-deficient atoms, molecules with polar multiple bonds, and metal cations.

Outline

Acids and bases have been used as laboratory chemicals for centuries, and they remain indispensable, not only in academic and industrial labs, but in the home as well. Common household acids include acetic acid (CH_3COOH, vinegar), citric acid ($H_3C_6H_5O_7$, in citrus fruits), and phosphoric acid (H_3PO_4, a flavoring agent in many carbonated beverages as well as a rust remover). Sodium hydroxide (NaOH, drain cleaner), ammonia (NH_3, glass cleaner), and sodium hydrogen carbonate ($NaHCO_3$, baking soda) are some familiar bases.

Some acids (e.g., acetic and citric) have a sour taste. In fact, sourness had been a defining property since the 17th century: an acid was any substance that had a sour taste; reacted with active metals, such as aluminum and zinc, to produce hydrogen gas; and turned certain organic compounds characteristic colors. (We discuss *indicators* later and in Chapter 19.) A base was any substance that had a bitter taste and slippery feel and turned the same organic compounds different characteristic colors. (Please remember *NEVER* to taste or touch laboratory chemicals; instead, try some acetic acid in the form of vinegar on your next salad.) Moreover, it was known that *when acids and bases react, each cancels the properties of the other in a process called neutralization.* But definitions in science evolve because, as descriptions become too limited, they must be replaced by broader ones. Although the early definitions of acids and bases described distinctive properties, they inevitably gave way to definitions based on molecular behavior.

In this chapter, we develop three definitions of acids and bases that allow us to understand ever-increasing numbers of reactions. In the process, we apply the principles of chemical equilibrium to this essential group of substances.

18.1 ACIDS AND BASES IN WATER

Although water is not an essential participant in all modern acid-base definitions, most laboratory work with acids and bases involves water, as do most environmental, biological, and industrial applications. Recall from our discussion in Chapter 4 that *water is a product in all reactions between strong acids and strong bases:*

$$HCl(aq) + NaOH(aq) \longrightarrow NaCl(aq) + H_2O(l)$$

Indeed, as the net ionic equation of this reaction shows, water is *the* product:

$$H^+(aq) + OH^-(aq) \longrightarrow H_2O(l)$$

Furthermore, whenever an acid dissociates in water, solvent molecules participate in the reaction:

$$HA(g \text{ or } l) + H_2O(l) \longrightarrow A^-(aq) + H_3O^+(aq)$$

Water molecules surround the proton to form species with the general formula $H(H_2O)_n^+$. Because the proton is so small, its charge density is very high, so its attraction to water is especially strong. The proton bonds covalently to one of the lone electron pairs of a water molecule's O atom to form a **hydronium ion, H_3O^+,** or $H(H_2O)^+$, which forms H bonds to several other water molecules. For example, $H_7O_3^+$, or $H(H_2O)_3^+$, is shown in Figure 18.1. To emphasize the active role of water and the nature of the proton-water interaction, the hydrated proton is usually shown in the text as $H_3O^+(aq)$, although in some cases this hydrated species is shown more simply as $H^+(aq)$.

Release of H^+ or OH^- and the Arrhenius Acid-Base Definition

The earliest and simplest definition of acids and bases that reflects their molecular nature was suggested by Svante Arrhenius, whose work on reaction rate we

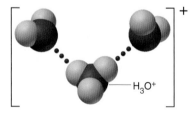

Figure 18.1 The hydrated proton. The charge of the H^+ ion is highly concentrated because the ion is so small. In aqueous solution, it forms a covalent bond to a water molecule, yielding an H_3O^+ ion that associates tightly with other H_2O molecules. Here, the $H_7O_3^+$ ion is shown.

encountered in Chapter 16. In the **Arrhenius acid-base definition,** acids and bases are classified in terms of their formulas and their behavior *in water:*

- An *acid* is a substance that has H in its formula and dissociates in water to yield H_3O^+.
- A *base* is a substance that has OH in its formula and dissociates in water to yield OH^-.

Some typical Arrhenius acids are HCl, HNO_3, and HCN, and some typical bases are NaOH, KOH, and $Ba(OH)_2$. Although Arrhenius bases contain discrete OH^- ions in their structures, Arrhenius acids *never* contain H^+ ions. On the contrary, these acids contain *covalently bonded H atoms that ionize in water.*

When an acid and a base react, they undergo **neutralization.** The meaning of acid-base reactions has changed along with the definitions of acid and base, but in the Arrhenius sense, neutralization occurs when *the H^+ ion from the acid and the OH^- ion from the base combine to form H_2O.* This description explains why all neutralization reactions between strong acids and strong bases (those that dissociate completely in water) have the same heat of reaction. No matter which strong acid and base react, and no matter which salt forms, ΔH°_{rxn} is about -56 kJ per mole of water formed because the actual reaction is always the same—a hydrogen ion and a hydroxide ion form water:

$$H^+(aq) + OH^-(aq) \longrightarrow H_2O(l) \qquad \Delta H^\circ_{rxn} = -55.9 \text{ kJ}$$

The dissolved salt that forms along with the water does not affect the ΔH°_{rxn} but exists as hydrated spectator ions.

Despite its importance at the time, limitations in the Arrhenius definition soon became apparent. Arrhenius and many others realized that even though some substances do *not* have OH in their formulas, they still behave as bases. For example, NH_3 and K_2CO_3 also yield OH^- in water. As you'll see shortly, broader acid-base definitions are required to include these species.

Variation in Acid Strength: The Acid-Dissociation Constant (K_a)

Acids and bases differ greatly in their *strength* in water, that is, in the amount of H_3O^+ or OH^- produced per mole of substance dissolved. We generally classify acids and bases as either strong or weak, according to the extent of their dissociation into ions in water (see Table 4.2). Remember, however, that a *gradation* in strength exists, as we'll examine quantitatively in a moment. Acids and bases are electrolytes in water, so this classification of acid and base strength correlates with our earlier classification of electrolyte strength: *strong electrolytes dissociate completely, and weak electrolytes dissociate partially.*

Animation: Dissociation of Strong and Weak Acids
Online Learning Center

- *Strong acids dissociate completely into ions in water* (Figure 18.2 on the next page):

$$HA(g \text{ or } l) + H_2O(l) \longrightarrow H_3O^+(aq) + A^-(aq)$$

In a dilute solution of a strong acid, *virtually no HA molecules are present;* that is, $[H_3O^+] \approx [HA]_{init}$. In other words, $[HA]_{eq} \approx 0$, so the value of K_c is extremely large:

$$Q_c = \frac{[H_3O^+][A^-]}{[HA][H_2O]} \qquad (\text{at equilibrium, } Q_c = K_c \gg 1)$$

Because the reaction is essentially complete, it is not very useful to express it as an equilibrium process. In a dilute aqueous nitric acid solution, for example, there are virtually no undissociated nitric acid molecules:

$$HNO_3(l) + H_2O(l) \longrightarrow H_3O^+(aq) + NO_3^-(aq)$$

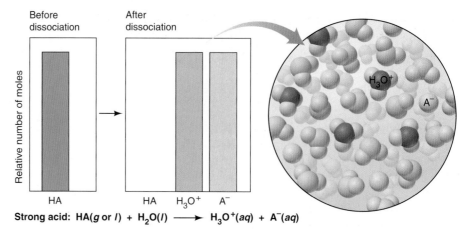

Figure 18.2 The extent of dissociation for strong acids. The bar graphs show the relative numbers of moles of species before (*left*) and after (*right*) acid dissociation occurs. When a strong acid dissolves in water, it dissociates completely, yielding $H_3O^+(aq)$ and $A^-(aq)$ ions; virtually no HA molecules are present.

- *Weak acids dissociate very slightly into ions in water* (Figure 18.3):

$$HA(aq) + H_2O(l) \rightleftharpoons H_3O^+(aq) + A^-(aq)$$

In a dilute solution of a weak acid, *the great majority of HA molecules are undissociated.* Thus, $[H_3O^+] \ll [HA]_{init}$. In other words, $[HA]_{eq} \approx [HA]_{init}$, so the value of K_c is very small. Hydrocyanic acid is an example of a weak acid:

$$HCN(aq) + H_2O(l) \rightleftharpoons H_3O^+(aq) + CN^-(aq)$$

$$Q_c = \frac{[H_3O^+][CN^-]}{[HCN][H_2O]} \quad \text{(at equilibrium, } Q_c = K_c \ll 1)$$

(In this chapter, we are dealing with systems *at equilibrium,* so instead of writing Q and stating that Q equals K at equilibrium, from here on, we'll express K directly as a collection of equilibrium concentration terms.)

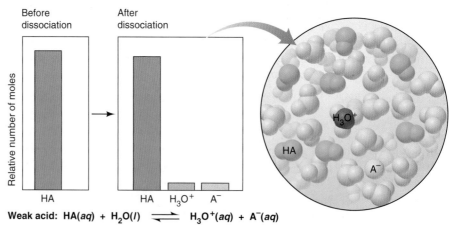

Figure 18.3 The extent of dissociation for weak acids. In contrast to a strong acid in water (see Figure 18.2), a weak acid dissociates very little, remaining mostly as intact acid molecules and, thus, yielding relatively few $H_3O^+(aq)$ and $A^-(aq)$ ions.

The Meaning of K_a There is a *specific* equilibrium constant for acid dissociation that highlights only those species whose concentrations change to any significant extent. The equilibrium expression for the dissociation of a general *weak acid, HA,* in water is

$$K_c = \frac{[H_3O^+][A^-]}{[HA][H_2O]}$$

In general, the concentration of water, $[H_2O]$, is so much larger than $[HA]$ that it changes negligibly when HA dissociates; thus, it is treated as a constant. Therefore, as you saw for solids in Section 17.2, we simplify the equilibrium expression by multiplying $[H_2O]$ by K_c to define a new equilibrium constant, the **acid-dissociation constant** (or **acid-ionization constant**), K_a:

$$K_c[H_2O] = K_a = \frac{[H_3O^+][A^-]}{[HA]} \qquad (18.1)$$

Like any equilibrium constant, K_a is a number whose magnitude is temperature dependent and tells how far to the right the reaction has proceeded to reach equilibrium. Thus, *the stronger the acid, the higher the $[H_3O^+]$ at equilibrium, and the larger the K_a:*

$$\text{Stronger acid} \Longrightarrow \text{higher } [H_3O^+] \Longrightarrow \text{larger } K_a$$

The range of values for the acid-dissociation constants of weak acids extends over many orders of magnitude. Listed below are some benchmark K_a values for typical weak acids to give you a general idea of the fraction of HA molecules that dissociate into ions:

- For a weak acid with a relatively high K_a ($\sim 10^{-2}$), a 1 M solution has $\sim 10\%$ of the HA molecules dissociated. The K_a of chlorous acid ($HClO_2$) is 1.1×10^{-2}, and 1 M $HClO_2$ is 10.% dissociated.
- For a weak acid with a moderate K_a ($\sim 10^{-5}$), a 1 M solution has $\sim 0.3\%$ of the HA molecules dissociated. The K_a of acetic acid (CH_3COOH) is 1.8×10^{-5}, and 1 M CH_3COOH is 0.42% dissociated.
- For a weak acid with a relatively low K_a ($\sim 10^{-10}$), a 1 M solution has $\sim 0.001\%$ of the HA molecules dissociated. The K_a of HCN is 6.2×10^{-10}, and 1 M HCN is 0.0025% dissociated.

Thus, for solutions of the same initial HA concentration, *the smaller the K_a, the lower the percent dissociation of HA:*

$$\text{Smaller } K_a \Longrightarrow \text{lower \% dissociation of HA} \Longrightarrow \text{weaker acid}$$

A list of K_a values for some common acids appears in Appendix C.

Classifying the Relative Strengths of Acids and Bases

Using a table of acid-dissociation constants is the surest way to quantify relative strengths of weak acids, but you can often classify acids and bases qualitatively as strong or weak just from their formulas:

- *Strong acids.* Two types of strong acids, with examples that *you should memorize,* are
  1. The hydrohalic acids HCl, HBr, and HI
  2. Oxoacids in which the number of O atoms exceeds the number of ionizable protons by two or more, such as HNO_3, H_2SO_4, and $HClO_4$; for example, in H_2SO_4, 4 O's $-$ 2 H's $= 2$

- *Weak acids.* There are many *more* weak acids than strong ones. Four types, with examples, are
  1. The hydrohalic acid HF
  2. Acids in which H is not bonded to O or to a halogen, such as HCN and H_2S
  3. Oxoacids in which the number of O atoms equals or exceeds by one the number of ionizable protons, such as HClO, HNO_2, and H_3PO_4
  4. Carboxylic acids (general formula RCOOH, with the ionizable proton shown in red), such as CH_3COOH and C_6H_5COOH

- *Strong bases.* Water-soluble compounds containing O^{2-} or OH^- ions are strong bases. The cations are usually those of the most active metals:
  1. M_2O or MOH, where M = Group 1A(1) metal (Li, Na, K, Rb, Cs)
  2. MO or $M(OH)_2$, where M = Group 2A(2) metal (Ca, Sr, Ba)
     [MgO and $Mg(OH)_2$ are only slightly soluble in water, but the soluble portion dissociates completely.]

- *Weak bases.* Many compounds with an electron-rich nitrogen atom are weak bases (none are Arrhenius bases). The common structural feature is an N atom with a lone electron pair (shown here in blue):
  1. Ammonia ($\ddot{N}H_3$)
  2. Amines (general formula $R\ddot{N}H_2$, $R_2\ddot{N}H$, or $R_3\ddot{N}$), such as $CH_3CH_2\ddot{N}H_2$, $(CH_3)_2\ddot{N}H$, and $(C_3H_7)_3\ddot{N}$

SAMPLE PROBLEM 18.1 Classifying Acid and Base Strength from the Chemical Formula

Problem Classify each of the following compounds as a strong acid, weak acid, strong base, or weak base:
(a) H_2SeO_4 (b) $(CH_3)_2CHCOOH$
(c) KOH (d) $(CH_3)_2CHNH_2$

Plan We examine the formula and classify each acid or base, using the text descriptions. Particular points to note for acids are the numbers of O atoms relative to H atoms and the presence of the —COOH group. For bases, note the nature of the cation or the presence of an N atom that has a lone pair.

Solution (a) Strong acid: H_2SeO_4 is an oxoacid in which the number of O atoms exceeds the number of ionizable protons by two.

(b) Weak acid: $(CH_3)_2CHCOOH$ is a carboxylic acid, as indicated by the —COOH group.

(c) Strong base: KOH is one of the Group 1A(1) hydroxides.

(d) Weak base: $(CH_3)_2CHNH_2$ has a lone pair on the N and is an amine.

FOLLOW-UP PROBLEM 18.1 Which member of each pair is the stronger acid or base?
(a) HClO or $HClO_3$
(b) HCl or CH_3COOH
(c) NaOH or CH_3NH_2

SECTION SUMMARY

Acids and bases are essential substances in home, industry, and the environment. In aqueous solution, water combines with the proton released from an acid to form the hydrated species represented by $H_3O^+(aq)$. In the Arrhenius definition, acids contain H and yield H_3O^+ in water, bases contain OH and yield OH^- in water, and an acid-base reaction (neutralization) is the reaction of H^+ and OH^- to form H_2O. Acid strength depends on $[H_3O^+]$ relative to [HA] in aqueous solution. Strong acids dissociate completely and weak acids slightly. The extent of dissociation is expressed by the acid-dissociation constant, K_a. Weak acids have K_a values ranging from about 10^{-1} to 10^{-12}. Many acids and bases can be classified qualitatively as strong or weak based on their formulas.

18.2 AUTOIONIZATION OF WATER AND THE pH SCALE

Before we discuss the next major definition of acid-base behavior, let's examine a crucial property of water that enables us to quantify $[H_3O^+]$ in any aqueous system: *water is an extremely weak electrolyte.* The electrical conductivity of tap water is due almost entirely to dissolved ions, but even water that has been repeatedly distilled and deionized exhibits a tiny conductance. The reason is that water itself dissociates into ions very slightly in an equilibrium process known as **autoionization** (or self-ionization):

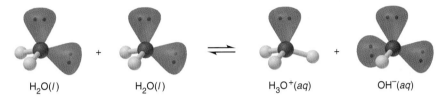

$$H_2O(l) \qquad H_2O(l) \qquad\qquad H_3O^+(aq) \qquad OH^-(aq)$$

The Equilibrium Nature of Autoionization: The Ion-Product Constant for Water (K_w)

Like any equilibrium process, the autoionization of water is described quantitatively by an equilibrium constant:

$$K_c = \frac{[H_3O^+][OH^-]}{[H_2O]^2}$$

Because the concentration of H_2O is essentially constant here, we simplify this equilibrium expression by including the constant $[H_2O]^2$ term with the value of K_c to obtain a new equilibrium constant, the **ion-product constant for water, K_w:**

$$K_c[H_2O]^2 = K_w = [H_3O^+][OH^-] = 1.0\times10^{-14} \text{ (at 25°C)} \qquad \text{(18.2)}$$

Notice that *one H_3O^+ ion and one OH^- ion appear for each H_2O molecule that dissociates.* Therefore, in pure water, we find that

$$[H_3O^+] = [OH^-] = \sqrt{1.0\times10^{-14}} = 1.0\times10^{-7} \, M \text{ (at 25°C)}$$

Pure water has a concentration of about 55.5 M $\left(\text{that is, } \dfrac{1000 \text{ g/L}}{18.02 \text{ g/mol}}\right)$, so these equilibrium concentrations are attained when only 1 in 555 million water molecules dissociates reversibly into ions!

Autoionization of water has two major consequences for aqueous acid-base chemistry:

1. *A change in $[H_3O^+]$ causes an inverse change in $[OH^-]$,* and vice versa:

$$\text{Higher } [H_3O^+] \Longrightarrow \text{lower } [OH^-] \qquad \text{and} \qquad \text{Higher } [OH^-] \Longrightarrow \text{lower } [H_3O^+]$$

Recall from our discussion of Le Châtelier's principle (Section 17.6) that a change in concentration of either ion shifts the equilibrium position, but it does *not* change the equilibrium constant. Therefore, if some acid is added, $[H_3O^+]$ increases, and so $[OH^-]$ must decrease; if some base is added, $[OH^-]$ increases, and so $[H_3O^+]$ must decrease. However, the addition of H_3O^+ or OH^- merely leads to the formation of H_2O, so the value of K_w is maintained.

2. *Both ions are present in all aqueous systems.* Thus, all acidic solutions contain a low concentration of OH^- ions, and all basic solutions contain a low concentration of H_3O^+ ions. The equilibrium nature of autoionization allows us to define "acidic" and "basic" solutions in terms of relative magnitudes of $[H_3O^+]$ and $[OH^-]$:

$$\text{In an } \textit{acidic} \text{ solution,} \qquad [H_3O^+] > [OH^-]$$
$$\text{In a } \textit{neutral} \text{ solution,} \qquad [H_3O^+] = [OH^-]$$
$$\text{In a } \textit{basic} \text{ solution,} \qquad [H_3O^+] < [OH^-]$$

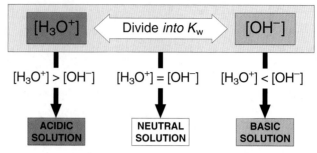

Figure 18.4 The relationship between $[H_3O^+]$ and $[OH^-]$ and the relative acidity of solutions.

Figure 18.4 summarizes these relationships and the relative solution acidity. Moreover, if you know the value of K_w at a particular temperature and the concentration of one of these ions, you can easily calculate the concentration of the other ion by solving for it from the K_w expression:

$$[H_3O^+] = \frac{K_w}{[OH^-]} \quad \text{or} \quad [OH^-] = \frac{K_w}{[H_3O^+]}$$

SAMPLE PROBLEM 18.2   Calculating $[H_3O^+]$ and $[OH^-]$ in Aqueous Solution

Problem A research chemist adds a measured amount of HCl gas to pure water at 25°C and obtains a solution with $[H_3O^+] = 3.0 \times 10^{-4} M$. Calculate $[OH^-]$. Is the solution neutral, acidic, or basic?

Plan We use the known value of K_w at 25°C (1.0×10^{-14}) and the given $[H_3O^+]$ ($3.0 \times 10^{-4} M$) to solve for $[OH^-]$. Then, we compare $[H_3O^+]$ with $[OH^-]$ to determine whether the solution is acidic, basic, or neutral (see Figure 18.4).

Solution Calculating $[OH^-]$:

$$[OH^-] = \frac{K_w}{[H_3O^+]} = \frac{1.0 \times 10^{-14}}{3.0 \times 10^{-4}}$$

$$= 3.3 \times 10^{-11} M$$

Because $[H_3O^+] > [OH^-]$, the solution is **acidic**.

Check It makes sense that adding an acid to water results in an acidic solution. Moreover, because $[H_3O^+]$ is greater than $10^{-7} M$, $[OH^-]$ must be less than $10^{-7} M$ to give a constant K_w.

FOLLOW-UP PROBLEM 18.2 Calculate $[H_3O^+]$ in a solution that is at 25°C and has $[OH^-] = 6.7 \times 10^{-2} M$. Is the solution neutral, acidic, or basic?

Expressing the Hydronium Ion Concentration: The pH Scale

In aqueous solutions, $[H_3O^+]$ can vary from about $10 M$ to $10^{-15} M$. To handle numbers with negative exponents more conveniently in calculations, we convert them to positive numbers using a numerical system called a *p-scale,* the negative of the common (base-10) logarithm of the number. Applying this numerical system to $[H_3O^+]$ gives **pH,** the negative logarithm of $[H^+]$ (or $[H_3O^+]$):

$$pH = -\log [H_3O^+] \tag{18.3}$$

What is the pH of $10^{-12} M$ H_3O^+ solution?

$$pH = -\log [H_3O^+] = -\log 10^{-12} = (-1)(-12) = 12$$

Similarly, a $10^{-3} M$ H_3O^+ solution has a pH of 3, and a $5.4 \times 10^{-4} M$ H_3O^+ solution has a pH of 3.27:

$$pH = -\log [H_3O^+] = (-1)(\log 5.4 + \log 10^{-4}) = 3.27$$

As with any measurement, the number of significant figures in a pH value reflects the precision with which the concentration is known. However, it is a logarithm, so the number of significant figures in the concentration equals the number of digits *to the right of the decimal point in the logarithm* (see Appendix A). In the preceding example, 5.4×10^{-4} *M* has two significant figures, so its negative logarithm, 3.27, has two digits to the right of the decimal point.

Note in particular that *the higher the pH, the lower the $[H_3O^+]$*. Therefore, *an acidic solution has a lower pH (higher $[H_3O^+]$) than a basic solution*. At 25°C in pure water, $[H_3O^+]$ is 1.0×10^{-7} *M*, so

$$\begin{aligned} \text{pH of a neutral solution} &= 7.00 \\ \text{pH of an acidic solution} &< 7.00 \\ \text{pH of a basic solution} &> 7.00 \end{aligned}$$

Figure 18.5 shows that the pH values of some familiar aqueous solutions fall within a range of 0 to 14.

Another important point arises when we compare $[H_3O^+]$ in different solutions. Because the pH scale is logarithmic, a solution of pH 1.0 has an $[H_3O^+]$ that is 10 times higher than that of a pH 2.0 solution, 100 times higher than that of a pH 3.0 solution, and so forth. To find the $[H_3O^+]$ from the pH, you perform the opposite arithmetic process; that is, you find the negative antilog of pH:

$$[H_3O^+] = 10^{-pH}$$

A p-scale is used to express other quantities as well:

- Hydroxide ion concentration can be expressed as pOH:

$$pOH = -\log [OH^-]$$

Acidic solutions have a higher pOH (lower $[OH^-]$) than basic solutions.
- Equilibrium constants can be expressed as p*K*:

$$pK = -\log K$$

A low pK corresponds to a high K. A reaction that reaches equilibrium with mostly products present (that proceeds far to the right) has a low p*K* (high *K*), whereas one that has mostly reactants present at equilibrium has a high p*K* (low *K*). Table 18.1 shows this relationship for aqueous equilibria of some weak acids.

The Relations Among pH, pOH, and pKw

Taking the negative log of both sides of the K_w expression gives a very useful relationship among pK_w, pH, and pOH:

$$K_w = [H_3O^+][OH^-] = 1.0 \times 10^{-14} \text{ (at 25°C)}$$
$$-\log K_w = (-\log [H_3O^+]) + (-\log [OH^-]) = -\log (1.0 \times 10^{-14})$$
$$\boxed{pK_w = pH + pOH = 14.00 \quad \text{(at 25°C)}} \quad (18.4)$$

Thus, the sum of pH and pOH is 14.00 in any aqueous solution at 25°C. With pH, pOH, $[H_3O^+]$, and $[OH^-]$ interrelated through K_w, knowing any one of the values allows us to determine the others (Figure 18.6 on the next page).

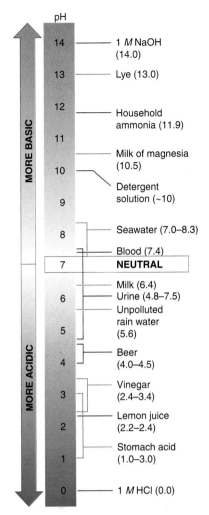

Figure 18.5 The pH values of some familiar aqueous solutions.

pH scale (from the figure):

- 14 — 1 *M* NaOH (14.0)
- 13 — Lye (13.0)
- 12 —
- Household ammonia (11.9)
- 11 —
- Milk of magnesia (10.5)
- 10 —
- Detergent solution (~10)
- 9 —
- 8 — Seawater (7.0–8.3)
- Blood (7.4)
- 7 — NEUTRAL
- 6 — Milk (6.4), Urine (4.8–7.5)
- Unpolluted rain water (5.6)
- 5 —
- Beer (4.0–4.5)
- 4 —
- Vinegar (2.4–3.4)
- 3 —
- Lemon juice (2.2–2.4)
- 2 —
- Stomach acid (1.0–3.0)
- 1 —
- 0 — 1 *M* HCl (0.0)

MORE BASIC / MORE ACIDIC

Table 18.1 The Relationship Between K_a and pK_a		
Acid Name (Formula)	K_a at 25°C	pK_a
Hydrogen sulfate ion (HSO_4^-)	1.0×10^{-2}	1.99
Nitrous acid (HNO_2)	7.1×10^{-4}	3.15
Acetic acid (CH_3COOH)	1.8×10^{-5}	4.74
Hypobromous acid ($HBrO$)	2.3×10^{-9}	8.64
Phenol (C_6H_5OH)	1.0×10^{-10}	10.00

Figure 18.6 The relations among [H₃O⁺], pH, [OH⁻], and pOH. Because K_w is constant, $[H_3O^+]$ and $[OH^-]$ are interdependent, and they change in opposite directions as the acidity or basicity of the aqueous solution increases. The pH and pOH are interdependent in the same way. Note that at 25°C, the product of $[H_3O^+]$ and $[OH^-]$ is 1.0×10^{-14}, and the sum of pH and pOH is 14.00.

	$[H_3O^+]$	pH	$[OH^-]$	pOH
	1.0×10^{-15}	15.00	1.0×10^{1}	-1.00
	1.0×10^{-14}	14.00	1.0×10^{0}	0.00
	1.0×10^{-13}	13.00	1.0×10^{-1}	1.00
BASIC	1.0×10^{-12}	12.00	1.0×10^{-2}	2.00
	1.0×10^{-11}	11.00	1.0×10^{-3}	3.00
	1.0×10^{-10}	10.00	1.0×10^{-4}	4.00
	1.0×10^{-9}	9.00	1.0×10^{-5}	5.00
	1.0×10^{-8}	8.00	1.0×10^{-6}	6.00
NEUTRAL	1.0×10^{-7}	7.00	1.0×10^{-7}	7.00
	1.0×10^{-6}	6.00	1.0×10^{-8}	8.00
	1.0×10^{-5}	5.00	1.0×10^{-9}	9.00
	1.0×10^{-4}	4.00	1.0×10^{-10}	10.00
	1.0×10^{-3}	3.00	1.0×10^{-11}	11.00
ACIDIC	1.0×10^{-2}	2.00	1.0×10^{-12}	12.00
	1.0×10^{-1}	1.00	1.0×10^{-13}	13.00
	1.0×10^{0}	0.00	1.0×10^{-14}	14.00
	1.0×10^{1}	-1.00	1.0×10^{-15}	15.00

MORE BASIC / MORE ACIDIC

SAMPLE PROBLEM 18.3   Calculating [H₃O⁺], pH, [OH⁻], and pOH

Problem In an art restoration project, a conservator prepares copper-plate etching solutions by diluting concentrated HNO_3 to 2.0 M, 0.30 M, and 0.0063 M HNO_3. Calculate $[H_3O^+]$, pH, $[OH^-]$, and pOH of the three solutions at 25°C.

Plan We know from its formula that HNO_3 is a strong acid, so it dissociates completely; thus, $[H_3O^+] = [HNO_3]_{init}$. We use the given concentrations and the value of K_w at 25°C (1.0×10^{-14}) to find $[H_3O^+]$ and $[OH^-]$ and then use them to calculate pH and pOH.

Solution Calculating the values for 2.0 M HNO_3:

$$[H_3O^+] = 2.0\ M$$

$$pH = -\log [H_3O^+] = -\log 2.0 = -0.30$$

$$[OH^-] = \frac{K_w}{[H_3O^+]} = \frac{1.0 \times 10^{-14}}{2.0} = 5.0 \times 10^{-15}\ M$$

$$pOH - -\log (5.0 \times 10^{-15}) = 14.30$$

Calculating the values for 0.30 M HNO_3:

$$[H_3O^+] = 0.30\ M$$

$$pH = -\log [H_3O^+] = -\log 0.30 = 0.52$$

$$[OH^-] = \frac{K_w}{[H_3O^+]} = \frac{1.0 \times 10^{-14}}{0.30} = 3.3 \times 10^{-14}\ M$$

$$pOH = -\log (3.3 \times 10^{-14}) = 13.48$$

Calculating the values for 0.0063 M HNO_3:

$$[H_3O^+] = 6.3 \times 10^{-3}\ M$$

$$pH = -\log [H_3O^+] = -\log (6.3 \times 10^{-3}) = 2.20$$

$$[OH^-] = \frac{K_w}{[H_3O^+]} = \frac{1.0 \times 10^{-14}}{6.3 \times 10^{-3}} = 1.6 \times 10^{-12}\ M$$

$$pOH = -\log (1.6 \times 10^{-12}) = 11.80$$

Check As the solution becomes more dilute, $[H_3O^+]$ decreases, so pH increases, as we expect. An $[H_3O^+]$ greater than 1.0 M, as in 2.0 M HNO_3, gives a positive log, so it results in a negative pH. The arithmetic seems correct because pH + pOH = 14.00 in each case.

Comment On most calculators, finding the pH requires several keystrokes. For example, to find the pH of 6.3×10^{-3} M HNO_3 solution, you enter: 6.3, EXP, 3, +/−, log, +/−.

FOLLOW-UP PROBLEM 18.3 A solution of NaOH has a pH of 9.52. What is its pOH, $[H_3O^+]$, and $[OH^-]$ at 25°C?

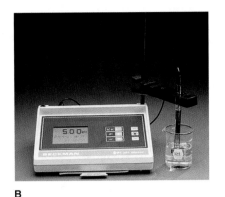

A

B

Figure 18.7 **Methods for measuring the pH of an aqueous solution. A,** A few drops of the solution are placed on a strip of pH paper, and the color is compared with the color chart. **B,** The electrodes of a pH meter immersed in the test solution measure $[H_3O^+]$. (In this instrument, the two electrodes are housed in one probe.)

Measuring pH In the laboratory, pH values are usually obtained with an acid-base indicator or, more precisely, with an instrument called a pH meter. **Acid-base indicators** are organic molecules whose colors depend on the acidity or basicity of the solution in which they are dissolved. The pH of a solution is estimated quickly with *pH paper,* a paper strip impregnated with one or a mixture of indicators. A drop of test solution is placed on the paper strip, and the color of the strip is compared with a color chart, as shown in Figure 18.7A.

The *pH meter* measures $[H_3O^+]$ by means of two electrodes immersed in the test solution. One electrode provides a stable reference voltage; the other has an extremely thin, conducting, glass membrane that separates a known internal $[H_3O^+]$ from the unknown external $[H_3O^+]$. The difference in $[H_3O^+]$ creates a voltage difference across the membrane, which is measured and displayed in pH units (Figure 18.7B). We examine this process further in Chapter 21.

SECTION SUMMARY
Pure water has a low conductivity because it autoionizes to a small extent. This process is described by an equilibrium reaction whose equilibrium constant is the ion-product constant for water, K_w (1.0×10^{-14} at 25°C). Thus, $[H_3O^+]$ and $[OH^-]$ are inversely related: in acidic solution, $[H_3O^+]$ is greater than $[OH^-]$; the reverse is true in basic solution; and the two are equal in neutral solution. To express small values of $[H_3O^+]$ more simply, we use the pH scale (pH $= -\log [H_3O^+]$). A high pH represents a low $[H_3O^+]$. Similarly, pOH $= -\log [OH^-]$, and p$K = -\log K$. In acidic solutions, pH < 7.00; in basic solutions, pH > 7.00; and in neutral solutions, pH $= 7.00$. The sum of pH and pOH equals pK_w (14.00 at 25°C).

18.3 PROTON TRANSFER AND THE BRØNSTED-LOWRY ACID-BASE DEFINITION

Earlier we noted a major shortcoming of the Arrhenius acid-base definition: many substances that yield OH^- ions when they dissolve in water do not contain OH in their formulas. Examples include ammonia, the amines, and many salts of weak acids, such as NaF. Another limitation of the Arrhenius definition was that water had to be the solvent for acid-base reactions. In the early 20th century, J. N. Brønsted and T. M. Lowry suggested definitions that remove these limitations. (Recall that we discussed their ideas briefly in Section 4.4.) According to the **Brønsted-Lowry acid-base definition,**

- *An acid is a **proton donor,** any species that donates an H^+ ion.* An acid must contain H in its formula; HNO_3 and $H_2PO_4^-$ are two of many examples. All Arrhenius acids are Brønsted-Lowry acids.

• A base is a **proton acceptor,** any species that accepts an H^+ ion. A base must contain a lone pair of electrons to bind the H^+ ion; a few examples are NH_3, CO_3^{2-}, and F^-, as well as OH^- itself. Brønsted-Lowry bases are not Arrhenius bases, but all Arrhenius bases contain the Brønsted-Lowry base OH^-.

From the Brønsted-Lowry perspective, the only requirement for an acid-base reaction is that *one species donates a proton and another species accepts it: an acid-base reaction is a proton-transfer process.* Acid-base reactions can occur between gases, in nonaqueous solutions, and in heterogeneous mixtures, as well as in aqueous solutions.

An acid and a base always work together in the transfer of a proton. In other words, one species behaves as an acid only if another species *simultaneously* behaves as a base, and vice versa. Even when an acid or a base merely dissolves in water, an acid-base reaction occurs because water acts as the other partner. Consider two typical acidic and basic solutions:

1. *Acid donates a proton to water* (Figure 18.8A). When HCl dissolves in water, an H^+ ion (a proton) is transferred from HCl to H_2O, where it becomes attached to a lone pair of electrons on the O atom, forming H_3O^+. In effect, HCl (the acid) has *donated* the H^+, and H_2O (the base) has *accepted* it:

$$HCl(g) + H_2\ddot{O}(l) \longrightarrow Cl^-(aq) + H_3\ddot{O}^+(aq)$$

2. *Base accepts a proton from water* (Figure 18.8B). In an aqueous solution of ammonia, proton transfer also occurs. An H^+ from H_2O attaches to the N atom's lone pair, forming NH_4^+. Having transferred an H^+, the H_2O becomes an OH^- ion:

$$\ddot{N}H_3(aq) + H_2O(l) \rightleftharpoons NH_4^+(aq) + OH^-(aq)$$

In this case, H_2O (the acid) has *donated* the H^+, and NH_3 (the base) has *accepted* it. Thus, H_2O is *amphoteric:* it acts as a base in one case and as an acid in the other. As you'll see, many other species are amphoteric as well.

Figure 18.8 Proton transfer as the essential feature of a Brønsted-Lowry acid-base reaction. A, When HCl dissolves in water, it acts as an acid by donating a proton to water, which acts as a base by accepting it. **B,** In aqueous solution, NH_3 acts as a base by accepting a proton from water, which acts as an acid by donating it. Thus, in the Brønsted-Lowry sense, an acid-base reaction occurs in both cases.

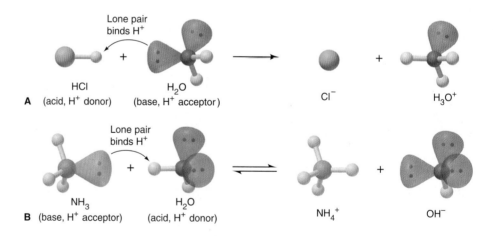

The Conjugate Acid-Base Pair

The Brønsted-Lowry definition provides a new way to look at acid-base reactions because it focuses on the reactants *and* the products. For example, let's examine the reaction between hydrogen sulfide and ammonia:

$$H_2S + NH_3 \rightleftharpoons HS^- + NH_4^+$$

In the forward reaction, H_2S acts as an acid by donating an H^+ to NH_3, which acts as a base by accepting it. The reverse reaction involves another acid-base pair. The ammonium ion, NH_4^+, acts as an acid by donating an H^+ to the hydrogen sulfide ion, HS⁻, which acts as a base. Notice that the acid, H_2S, becomes a base, HS⁻, and the base, NH_3, becomes an acid, NH_4^+.

In Brønsted-Lowry terminology, H_2S and HS⁻ are a **conjugate acid-base pair**: HS⁻ is the conjugate base of the acid H_2S. Similarly, NH_3 and NH_4^+ form a conjugate acid-base pair: NH_4^+ is the conjugate acid of the base NH_3. *Every acid has a conjugate base, and every base has a conjugate acid.* Note that, for any conjugate acid-base pair,

- The conjugate base has one *fewer* H and one *more* minus charge than the acid.
- The conjugate acid has one *more* H and one *fewer* minus charge than the base.

A Brønsted-Lowry acid-base reaction occurs when *an acid and a base react to form their conjugate base and conjugate acid, respectively:*

$$acid_1 + base_2 \rightleftharpoons base_1 + acid_2$$

Table 18.2 shows some Brønsted-Lowry acid-base reactions. Notice that

- Each reaction has an acid and a base as reactants *and* as products, and these comprise two conjugate acid-base pairs.
- Acids and bases can be neutral, cationic, or anionic.
- The same species can be an acid or a base, depending on the other species reacting. Water behaves this way in reactions 1 and 4, and HPO_4^{2-} does so in reactions 4 and 6.

Table 18.2 The Conjugate Pairs in Some Acid-Base Reactions

	Acid	+	Base	⇌	Base	+	Acid
Reaction 1	HF	+	H_2O	⇌	F^-	+	H_3O^+
Reaction 2	HCOOH	+	CN^-	⇌	$HCOO^-$	+	HCN
Reaction 3	NH_4^+	+	CO_3^{2-}	⇌	NH_3	+	HCO_3^-
Reaction 4	$H_2PO_4^-$	+	OH^-	⇌	HPO_4^{2-}	+	H_2O
Reaction 5	H_2SO_4	+	$N_2H_5^+$	⇌	HSO_4^-	+	$N_2H_6^{2+}$
Reaction 6	HPO_4^{2-}	+	SO_3^{2-}	⇌	PO_4^{3-}	+	HSO_3^-

SAMPLE PROBLEM 18.4 Identifying Conjugate Acid-Base Pairs

Problem The following reactions are important environmental processes. Identify the conjugate acid-base pairs.
(a) $H_2PO_4^-(aq) + CO_3^{2-}(aq) \rightleftharpoons HCO_3^-(aq) + HPO_4^{2-}(aq)$
(b) $H_2O(l) + SO_3^{2-}(aq) \rightleftharpoons OH^-(aq) + HSO_3^-(aq)$
Plan To find the conjugate pairs, we find the species that donated an H^+ (acid) and the species that accepted it (base). The acid (or base) on the left becomes its conjugate base (or conjugate acid) on the right. Remember, the conjugate acid has one more H and one fewer minus charge than its conjugate base.
Solution (a) $H_2PO_4^-$ has one more H^+ than HPO_4^{2-}; CO_3^{2-} has one fewer H^+ than HCO_3^-. Therefore, $H_2PO_4^-$ and HCO_3^- are the acids, and HPO_4^{2-} and CO_3^{2-} are the bases. The conjugate acid-base pairs are $H_2PO_4^-/HPO_4^{2-}$ and HCO_3^-/CO_3^{2-}.
(b) H_2O has one more H^+ than OH^-; SO_3^{2-} has one fewer H^+ than HSO_3^-. The acids are H_2O and HSO_3^-; the bases are OH^- and SO_3^{2-}. The conjugate acid-base pairs are H_2O/OH^- and HSO_3^-/SO_3^{2-}.

FOLLOW-UP PROBLEM 18.4 Identify the conjugate acid-base pairs:
(a) $CH_3COOH(aq) + H_2O(l) \rightleftharpoons CH_3COO^-(aq) + H_3O^+(aq)$
(b) $H_2O(l) + F^-(aq) \rightleftharpoons OH^-(aq) + HF(aq)$

Relative Acid-Base Strength and the Net Direction of Reaction

The net *direction* of an acid-base reaction depends on the relative strengths of the acids and bases involved. *A reaction proceeds to the greater extent in the direction in which a stronger acid and stronger base form a weaker acid and weaker base.* If the stronger acid and base are written on the left, the net direction is to the right, so $K_c > 1$. You can think of the process as a *competition for the proton between the two bases,* in which the stronger base wins.

In the same sense, the extent of acid (HA) dissociation in water depends on a competition for the proton between the two bases, A^- and H_2O. Strong acids and weak acids give very different results. When the acid HNO_3 dissolves in water, it transfers an H^+ to the base, H_2O, forming the conjugate base of HNO_3, which is NO_3^-, and the conjugate acid of H_2O, which is H_3O^+:

$$HNO_3 \ + \ H_2O \ \longrightarrow \ NO_3^- \ + \ H_3O^+$$

stronger acid + stronger base $\longrightarrow$ weaker base + weaker acid

(In this case, the net direction is so far to the right that it would be inappropriate to show an equilibrium arrow.) HNO_3 is a stronger acid than H_3O^+, and H_2O is a stronger base than NO_3^-. Thus, with strong acids such as HNO_3, the H_2O wins the competition for the proton because A^- (NO_3^-) is a much weaker base. On the other hand, with weak acids such as HF, the A^- (F^-) wins because it is a stronger base than H_2O:

$$HF \ + \ H_2O \ \rightleftharpoons \ F^- \ + \ H_3O^+$$

weaker acid + weaker base $\longleftarrow$ stronger base + stronger acid

Based on evidence gathered from the results of many such reactions, we can rank conjugate pairs in terms of the ability of the acid to transfer its proton (Figure 18.9). Note, especially, that *a weaker acid has a stronger conjugate base.* This makes perfect sense: the acid gives up its proton less readily because its conjugate base holds it more strongly. We can use this list to predict the direction of a reaction between any two pairs, that is, whether the equilibrium position lies predominantly to the right ($K_c > 1$) or to the left ($K_c < 1$). *An acid-base reaction proceeds to the right if the acid reacts with a base that is lower on the list* because this combination produces a weaker conjugate base and a weaker conjugate acid.

SAMPLE PROBLEM 18.5 Predicting the Net Direction of an Acid-Base Reaction

Problem Predict the net direction and whether K_c is greater or less than 1 for the following reaction (assume equal initial concentrations of all species):

$$H_2PO_4^-(aq) + NH_3(aq) \rightleftharpoons NH_4^+(aq) + HPO_4^{2-}(aq)$$

Plan We first identify the conjugate acid-base pairs. To predict the direction, we consult Figure 18.9 to see which acid and base are stronger. The stronger acid and base form the weaker acid and base, so the reaction proceeds in that net direction. If the reaction *as written* proceeds to the right, then [products] is higher than [reactants], and $K_c > 1$.

Solution The conjugate pairs are $H_2PO_4^-/HPO_4^{2-}$ and NH_4^+/NH_3. $H_2PO_4^-$ is higher on the list of acids, so it is stronger than NH_4^+; and NH_3 is lower on the list of bases, so it is stronger than HPO_4^{2-}. Therefore,

$$H_2PO_4^-(aq) + NH_3(aq) \rightleftharpoons NH_4^+(aq) + HPO_4^{2-}(aq)$$

stronger acid + stronger base $\longrightarrow$ weaker acid + weaker base

The net direction is to the right, so $K_c > 1$.

FOLLOW-UP PROBLEM 18.5 Predict the net direction and whether K_c is greater or less than 1 for the following reaction:

$$H_2O(l) \ + \ HS^-(aq) \rightleftharpoons OH^-(aq) \ + \ H_2S(aq)$$

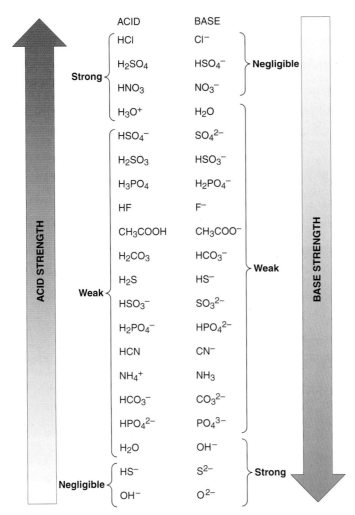

Figure 18.9 Strengths of conjugate acid-base pairs. The stronger the acid is, the weaker its conjugate base. The strongest acid appears at top left and the strongest base at bottom right. When an acid reacts with a base farther down the list, the reaction proceeds to the right ($K_c > 1$).

SECTION SUMMARY

The Brønsted-Lowry acid-base definition does not require that bases contain OH or that acid-base reactions occur in aqueous solution. It defines an acid as a species that donates a proton and a base as one that accepts it. An acid and a base act together in proton transfer. When an acid donates a proton, it becomes the conjugate base; when a base accepts a proton, it becomes the conjugate acid. In an acid-base reaction, acids and bases form their conjugates. A stronger acid has a weaker conjugate base, and vice versa. Thus, the reaction proceeds in the net direction in which a stronger acid and base form a weaker base and acid.

18.4 SOLVING PROBLEMS INVOLVING WEAK-ACID EQUILIBRIA

Just as you saw in Chapter 17 for equilibrium problems in general, there are two types of equilibrium problems involving weak acids and their conjugate bases:

1. Given equilibrium concentrations, find K_a.
2. Given K_a and some concentration information, find the other equilibrium concentrations.

For all of these problems, we'll apply the same problem-solving approach, notation system, and assumptions:

- *The problem-solving approach.* As always, start with what is given in the problem statement and move logically toward what you want to find. Make a habit of applying the following steps:
  1. Write the balanced equation and K_a expression; these will tell you what to find.
  2. Define x as the unknown change in concentration that occurs during the reaction. Frequently, $x = [HA]_{dissoc}$, the concentration of HA that dissociates, which, through the use of certain assumptions, also equals $[H_3O^+]$ and $[A^-]$ at equilibrium.
  3. Construct a reaction table that incorporates the unknown.
  4. Make assumptions that simplify the calculations, usually that x is very small relative to the initial concentration.
  5. Substitute the values into the K_a expression, and solve for x.
  6. Check that the assumptions are justified. (Apply the 5% test that was first used in Sample Problem 17.7.) If they are not justified, use the quadratic formula to find x.

- *The notation system.* As always, the molar concentration of each species is shown with brackets. A subscript refers to where the species comes from or when it occurs in the reaction process. For example, $[H_3O^+]_{from\ HA}$ is the molar concentration of H_3O^+ that comes from the dissociation of HA; $[HA]_{init}$ is the initial molar concentration of HA, that is, before the dissociation occurs; $[HA]_{dissoc}$ is the molar concentration of HA that dissociates; and so forth. Recall from Chapter 17 that brackets with no subscript indicate the molar concentration of the species *at equilibrium*.

- *The assumptions.* We make two assumptions to simplify the arithmetic:
  1. The $[H_3O^+]$ from the autoionization of water is negligible. In fact, it is so much smaller than the $[H_3O^+]$ from the dissociation of HA that we can neglect it:

  $$[H_3O^+] = [H_3O^+]_{from\ HA} + [H_3O^+]_{from\ H_2O} \approx [H_3O^+]_{from\ HA}$$

  Indeed, Le Châtelier's principle tells us that $[H_3O^+]_{from\ HA}$ *decreases* the extent of autoionization of water, so $[H_3O^+]_{from\ H_2O}$ in the HA solution is even less than $[H_3O^+]$ in pure water. Note that each molecule of HA that dissociates forms one H_3O^+ and one A^-, so $[A^-] = [H_3O^+]$.
  2. A weak acid has a small K_a. Therefore, it dissociates to such a small extent that we can neglect the change in its concentration to find its equilibrium concentration:

  $$[HA] = [HA]_{init} - [HA]_{dissoc} \approx [HA]_{init}$$

Finding K_a Given Concentrations

This type of problem involves finding K_a of a weak acid from the concentration of one of the species in solution, usually H_3O^+ from a given pH:

$$HA(aq) + H_2O(l) \rightleftharpoons H_3O^+(aq) + A^-(aq)$$

$$K_a = \frac{[H_3O^+][A^-]}{[HA]}$$

A common approach is to prepare an aqueous solution of HA and measure its pH. You prepared the solution, so you know $[HA]_{init}$. You can calculate $[H_3O^+]$ from the measured pH and then determine $[A^-]$ and $[HA]$ at equilibrium. Then, you substitute these values into the K_a expression and solve for K_a. We'll show how to make the assumptions discussed above and go through the entire approach in Sample Problem 18.6. We simplify later problems by omitting some of the recurring steps.

SAMPLE PROBLEM 18.6   Finding K_a of a Weak Acid from pH of Its Solution

Problem Phenylacetic acid ($C_6H_5CH_2COOH$, simplified here to HPAc) builds up in the blood of persons with phenylketonuria, an inherited disorder that, if untreated, causes mental retardation and death. A study of the acid shows that the pH of 0.12 M HPAc is 2.62. What is the K_a of phenylacetic acid?

Plan We are given $[HPAc]_{init}$ (0.12 M) and the pH (2.62) and must find K_a. We first write the equation for HPAc dissociation and the expression for K_a to see which values we need to find:

$$HPAc(aq) + H_2O(l) \rightleftharpoons H_3O^+(aq) + PAc^-(aq) \qquad K_a = \frac{[H_3O^+][PAc^-]}{[HPAc]}$$

- To find $[H_3O^+]$: We know the pH, so we can find $[H_3O^+]$. Because a pH of 2.62 is more than four pH units (10^4-fold) *lower* than the pH of pure water itself (pH = 7.0), we can assume that $[H_3O^+]_{from\ HPAc} \gg [H_3O^+]_{from\ H_2O}$. Therefore, $[H_3O^+]_{from\ HPAc} + [H_3O^+]_{from\ H_2O} \approx [H_3O^+]_{from\ HPAc} \approx [H_3O^+]$.
- To find $[PAc^-]$: Because each HPAc that dissociates forms one H_3O^+ and one PAc^-, $[H_3O^+] \approx [PAc^-]$.
- To find $[HPAc]$: We know $[HPAc]_{init}$. Because HPAc is a weak acid, we assume that very little dissociates, so $[HPAc]_{init} - [HPAc]_{dissoc} = [HPAc] \approx [HPAc]_{init}$.

We set up a reaction table, make the assumptions, substitute the equilibrium values, solve for K_a, and then check the assumptions.

Solution Calculating $[H_3O^+]$:

$$[H_3O^+] = 10^{-pH} = 10^{-2.62} = 2.4 \times 10^{-3}\ M$$

Setting up the reaction table, with $x = [HPAc]_{dissoc} = [H_3O^+]_{from\ HPAc} = [PAc^-] \approx [H_3O^+]$:

Concentration (M)	HPAc(aq)	+	H$_2$O(l)	$\rightleftharpoons$	H$_3$O$^+$(aq)	+	PAc$^-$(aq)
Initial	0.12		—		1×10^{-7}		0
Change	$-x$		—		$+x$		$+x$
Equilibrium	$0.12 - x$		—		$x + (<1 \times 10^{-7})$		x

Making the assumptions:
1. The calculated $[H_3O^+]$ ($2.4 \times 10^{-3}\ M$) $\gg [H_3O^+]_{from\ H_2O}$ ($<1 \times 10^{-7}\ M$), so we assume that $[H_3O^+] \approx [H_3O^+]_{from\ HPAc} = x$ (the change in $[HPAc]$).
2. HPAc is a weak acid, so we assume that $[HPAc] = 0.12\ M - x \approx 0.12\ M$.

Solving for the equilibrium concentrations:

$$x \approx [H_3O^+] = [PAc^-] = 2.4 \times 10^{-3}\ M$$
$$[HPAc] = 0.12\ M - x = 0.12\ M - (2.4 \times 10^{-3}\ M) \approx 0.12\ M\ (\text{to 2 sf})$$

Substituting these values into K_a:

$$K_a = \frac{[H_3O^+][PAc^-]}{[HPAc]} \approx \frac{(2.4 \times 10^{-3})(2.4 \times 10^{-3})}{0.12} = 4.8 \times 10^{-5}$$

Checking the assumptions by finding the percent error in concentration:
1. For $[H_3O^+]_{from\ H_2O}$: $\dfrac{1 \times 10^{-7}\ M}{2.4 \times 10^{-3}\ M} \times 100 = 4 \times 10^{-3}\%$ (<5%; assumption is justified).
2. For $[HPAc]_{dissoc}$: $\dfrac{2.4 \times 10^{-3}\ M}{0.12\ M} \times 100 = 2.0\%$ (<5%; assumption is justified). We had already shown above that, to two significant figures, the concentration had not changed, so this check was not really necessary.

Check The $[H_3O^+]$ makes sense: pH 2.62 should give $[H_3O^+]$ between 10^{-2} and $10^{-3}\ M$. The K_a calculation also seems in the correct range: $(10^{-3})^2/10^{-1} = 10^{-5}$, and this value seems reasonable for a weak acid.

Comment $[H_3O^+]_{from\ H_2O}$ is so small relative to $[H_3O^+]_{from\ HA}$ that, from here on, we will disregard it and enter it as zero in reaction tables.

FOLLOW-UP PROBLEM 18.6 The conjugate acid of ammonia is NH_4^+, a weak acid. If a 0.2 M NH_4Cl solution has a pH of 5.0, what is the K_a of NH_4^+?

Finding Concentrations Given K_a

The second type of equilibrium problem involving weak acids gives some concentration data and the K_a value and asks for the equilibrium concentration of some component. Such problems are very similar to those we solved in Chapter 17 in which a substance with a given initial concentration reacted to an unknown extent (see Sample Problems 17.6 to 17.8).

SAMPLE PROBLEM 18.7 Determining Concentrations from K_a and Initial [HA]

Problem Propanoic acid (CH_3CH_2COOH, which we simplify as HPr) is a carboxylic acid whose salts are used to retard mold growth in foods. What is the $[H_3O^+]$ of 0.10 M HPr ($K_a = 1.3 \times 10^{-5}$)?

Plan We know the initial concentration (0.10 M) and K_a (1.3×10^{-5}) of HPr, and we need to find $[H_3O^+]$. First, we write the balanced equation and the expression for K_a:

$$HPr(aq) + H_2O(l) \rightleftharpoons H_3O^+(aq) + Pr^-(aq) \qquad K_a = \frac{[H_3O^+][Pr^-]}{[HPr]} = 1.3 \times 10^{-5}$$

We know $[HPr]_{init}$ but not $[HPr]$. If we let $x = [HPr]_{dissoc}$, x is also $[H_3O^+]_{from\ HPr}$ and $[Pr^-]$ because each HPr that dissociates yields one H_3O^+ and one Pr^-. With this information, we can set up a reaction table. In solving for x, we assume that, because HPr has a small K_a, it dissociates very little; therefore, $[HPr]_{init} - x = [HPr] \approx [HPr]_{init}$. After we find x, we check the assumption.

Solution Setting up a reaction table, with $x = [HPr]_{dissoc} = [H_3O^+]_{from\ HPr} = [Pr^-] = [H_3O^+]$:

Concentration (M)	HPr(aq)	+	H_2O(l)	$\rightleftharpoons$	H_3O^+(aq)	+	Pr^-(aq)
Initial	0.10		—		0		0
Change	$-x$		—		$+x$		$+x$
Equilibrium	$0.10 - x$		—		x		x

Making the assumption: K_a is small, so x is small compared with $[HPr]_{init}$; therefore, $0.10\ M - x \approx 0.10\ M$. Substituting into the K_a expression and solving for x:

$$K_a - \frac{[H_3O^+][Pr^-]}{[HPr]} = 1.3 \times 10^{-5} \approx \frac{(x)(x)}{0.10}$$

$$x \approx \sqrt{(0.10)(1.3 \times 10^{-5})} = 1.1 \times 10^{-3}\ M = [H_3O^+]$$

Checking the assumption:

For $[HPr]_{dissoc}$: $\dfrac{1.1 \times 10^{-3}\ M}{0.10\ M} \times 100 = 1.1\%$ (<5%; assumption is justified).

Check The $[H_3O^+]$ seems reasonable for a dilute solution of a weak acid with a moderate K_a. By reversing the calculation, we can check the math: $(1.1 \times 10^{-3})^2/0.10 = 1.2 \times 10^{-5}$, which is within rounding of the given K_a.

Comment In these problems, we assume that the concentration of HA that dissociates ($[HA]_{dissoc} = x$) can be neglected because K_a is relatively small. However, this is true only if $[HA]_{init}$ is relatively large. Here's a benchmark to see if the assumption is justified:

- If $\dfrac{[HA]_{init}}{K_a} > 400$, the assumption is justified: neglecting x introduces an error < 5%.

- If $\dfrac{[HA]_{init}}{K_a} < 400$, the assumption is *not* justified; neglecting x introduces an error > 5%,

  so we solve a quadratic equation to find x.

The latter situation occurs in the follow-up problem.

FOLLOW-UP PROBLEM 18.7 Cyanic acid (HOCN) is an extremely acrid, unstable substance. What is the $[H_3O^+]$ and pH of 0.10 M HOCN ($K_a = 3.5 \times 10^{-4}$)?

The Effect of Concentration on the Extent of Acid Dissociation

If we repeat the calculation in Sample Problem 18.7, but start with a lower [HPr], we observe a very interesting fact about the extent of dissociation of a weak acid. Suppose the initial concentration of HPr is one-tenth as much, 0.010 M rather than 0.10 M. After filling in the reaction table and making the same assumptions, we find that

$$x = [HPr]_{dissoc} = 3.6 \times 10^{-4} M$$

Now, let's compare the percentages of HPr molecules dissociated at the two different initial acid concentrations, using the relationship

$$\text{Percent HA dissociated} = \frac{[HA]_{dissoc}}{[HA]_{init}} \times 100 \qquad (18.5)$$

Case 1: $[HPr]_{init} = 0.10 \ M$

$$\text{Percent dissociated} = \frac{1.1 \times 10^{-3} \ M}{1.0 \times 10^{-1} \ M} \times 100 = 1.1\%$$

Case 2: $[HPr]_{init} = 0.010 \ M$

$$\text{Percent dissociated} = \frac{3.6 \times 10^{-4} \ M}{1.0 \times 10^{-2} \ M} \times 100 = 3.6\%$$

As the initial acid concentration decreases, the percent dissociation of the acid increases. Don't confuse the *concentration* of HA dissociated with the *percent* HA dissociated. The concentration ($[HA]_{dissoc}$) is lower in the diluted HA solution because the actual *number* of dissociated HA molecules is smaller. It is the *fraction* (and thus the *percent*) of dissociated HA molecules that increases with dilution.

The Behavior of Polyprotic Acids

Acids with more than one ionizable proton are **polyprotic acids.** In a solution of a polyprotic acid, one proton at a time dissociates from the acid molecule, and each dissociation step has a different K_a. For example, phosphoric acid is a triprotic acid (three ionizable protons), so it has three K_a values:

$$H_3PO_4(aq) + H_2O(l) \rightleftharpoons H_2PO_4^-(aq) + H_3O^+(aq)$$
$$K_{a1} = \frac{[H_2PO_4^-][H_3O^+]}{[H_3PO_4]} = 7.2 \times 10^{-3}$$

$$H_2PO_4^-(aq) + H_2O(l) \rightleftharpoons HPO_4^{2-}(aq) + H_3O^+(aq)$$
$$K_{a2} = \frac{[HPO_4^{2-}][H_3O^+]}{[H_2PO_4^-]} = 6.3 \times 10^{-8}$$

$$HPO_4^{2-}(aq) + H_2O(l) \rightleftharpoons PO_4^{3-}(aq) + H_3O^+(aq)$$
$$K_{a3} = \frac{[PO_4^{3-}][H_3O^+]}{[HPO_4^{2-}]} = 4.2 \times 10^{-13}$$

As you can see from the relative K_a values, H_3PO_4 is a much stronger acid than $H_2PO_4^-$, which is much stronger than HPO_4^{2-}. (Appendix C lists several polyprotic acids and their multiple K_a values.) Notice that, for H_3PO_4 and every polyprotic acid, the first proton comes off to a much greater extent than the second and, where applicable, the second does to a much greater extent than the third:

$$K_{a1} > K_{a2} > K_{a3}$$

This trend makes sense: it is more difficult for an H^+ ion to leave a singly charged anion (such as $H_2PO_4^-$) than to leave a neutral molecule (such as H_3PO_4), and more difficult still for it to leave a doubly charged anion (HPO_4^{2-}). Successive acid-dissociation constants typically differ by several orders of magnitude. This fact greatly simplifies pH calculations involving polyprotic acids because *we can usually neglect the H_3O^+ coming from the subsequent dissociations.*

SECTION SUMMARY
Two common types of weak-acid equilibrium problems involve finding K_a from a concentration and finding a concentration from K_a. We summarize the information in a reaction table, and we simplify the arithmetic by assuming (1) $[H_3O^+]_{from\ H_2O}$ is so small relative to $[H_3O^+]_{from\ HA}$ that it can be neglected, and (2) weak acids dissociate so little that $[HA]_{init} \approx [HA]$ at equilibrium. The *fraction* of weak acid molecules that dissociates is greater in a more dilute solution, even though the total $[H_3O^+]$ is less. Polyprotic acids have more than one ionizable proton, but we assume that the first dissociation provides virtually all the H_3O^+.

18.5   WEAK BASES AND THEIR RELATION TO WEAK ACIDS

By focusing on where the proton comes from and goes to, the Brønsted-Lowry concept expands the definition of a base to encompass a host of species that the Arrhenius definition excludes: a base is any species that accepts a proton; to do so, *the base must have a lone electron pair*. (The lone electron pair also plays the central role in the Lewis acid-base definition, as you'll see later in this chapter.)

Now let's examine the equilibrium system of a weak base and focus, as we did for weak acids, on the base (B) dissolving in water. When B dissolves, it accepts a proton from H_2O, which acts as an acid, leaving behind an OH^- ion:

$$B(aq) + H_2O(aq) \rightleftharpoons BH^+(aq) + OH^-(aq)$$

This general reaction for a base in water is described by the following equilibrium expression:

$$K_c = \frac{[BH^+][OH^-]}{[B][H_2O]}$$

Based on our earlier reasoning that $[H_2O]$ is treated as a constant in aqueous reactions, we incorporate $[H_2O]$ in the value of K_c and obtain the **base-dissociation constant** (or **base-ionization constant**), K_b:

$$K_b = \frac{[BH^+][OH^-]}{[B]} \tag{18.6}$$

Despite the name "base-dissociation constant," *no base dissociates in the process*, as you can see from the reaction.

As with the relation between pK_a and K_a, we know that pK_b, the negative logarithm of the base-dissociation constant, decreases with increasing K_b (that is, increasing base strength). In aqueous solution, the two large classes of weak bases are (1) ammonia and the amines and (2) the anions of weak acids.

Molecules as Weak Bases: Ammonia and the Amines

Ammonia is the simplest nitrogen-containing compound that acts as a weak base in water:

$$NH_3(aq) + H_2O(l) \rightleftharpoons NH_4^+(aq) + OH^-(aq) \qquad K_b = 1.76 \times 10^{-5} \text{ (at 25°C)}$$

Despite labels on reagent bottles that read "ammonium hydroxide," an aqueous solution of ammonia consists largely of *unprotonated* NH_3 molecules, as its small K_b indicates. In a 1.0 M NH_3 solution, for example, $[OH^-] = [NH_4^+] = 4.2 \times 10^{-3}$ M, so about 99.58% of the NH_3 is not ionized. A list of several bases with their K_b values appears in Appendix C.

If one or more of the H atoms in NH_3 is replaced by an organic group (designated as R), an *amine* results: RNH_2, R_2NH, or R_3N (Section 15.4; see Figure 15.11). The key structural feature of these organic compounds, as in all Brønsted-Lowry bases, is *a lone pair of electrons that can bind the proton donated by the acid*. Figure 18.10 depicts this process for methylamine, the simplest amine.

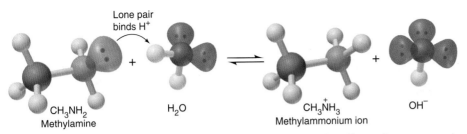

Figure 18.10 **Abstraction of a proton from water by methylamine.** The amines are organic derivatives of ammonia. Methylamine, the simplest amine, acts as a base in water by accepting a proton, thereby increasing [OH^-].

Finding the pH of a solution of a molecular weak base uses an approach very similar to that for a weak acid. We write the equilibrium expression, set up a reaction table to find [base]$_{reacting}$, make the usual assumptions, and then solve for [OH^-]. The main difference is that we must convert [OH^-] to [H_3O^+] in order to calculate pH.

SAMPLE PROBLEM 18.8 Determining pH from K_b and Initial [B]

Problem Dimethylamine, $(CH_3)_2NH$ (see margin), a key intermediate in detergent manufacture, has a K_b of 5.9×10^{-4}. What is the pH of 1.5 M $(CH_3)_2NH$?

Plan We know the initial concentration (1.5 M) and K_b (5.9×10^{-4}) of $(CH_3)_2NH$ and have to find the pH. The amine reacts with water to form OH^-, so we have to find [OH^-] and then calculate [H_3O^+] and pH. The balanced equation and K_b expression are

$$(CH_3)_2NH(aq) + H_2O(l) \rightleftharpoons (CH_3)_2NH_2^+(aq) + OH^-(aq)$$

$$K_b = \frac{[(CH_3)_2NH_2^+][OH^-]}{[(CH_3)_2NH]}$$

Dimethylamine

Because $K_b \gg K_w$, the [OH^-] from the autoionization of water is negligible, and we disregard it. Therefore,

$$[OH^-]_{from\ base} = [(CH_3)_2NH_2^+] = [OH^-]$$

Because K_b is small, we assume that the amount of amine reacting is small, so

$$[(CH_3)_2NH]_{init} - [(CH_3)_2NH]_{reacting} = [(CH_3)_2NH] \approx [(CH_3)_2NH]_{init}$$

We proceed as usual, setting up a reaction table, making the assumption, and solving for x. Then we check the assumption and convert [OH^-] to [H_3O^+] using K_w; finally, we calculate pH.

Solution Setting up the reaction table, with

$$x = [(CH_3)_2NH]_{reacting} = [(CH_3)_2NH_2^+] = [OH^-]$$

Concentration (M)	$(CH_3)_2NH(aq)$	+	$H_2O(l)$	$\rightleftharpoons$	$(CH_3)_2NH_2^+(aq)$	+	$OH^-(aq)$
Initial	1.5		—		0		0
Change	$-x$		—		$+x$		$+x$
Equilibrium	$1.5 - x$		—		x		x

Making the assumption:
K_b is small, so $[(CH_3)_2NH]_{init} \approx [(CH_3)_2NH]$; thus, 1.5 M $- x \approx 1.5$ M.

Substituting into the K_b expression and solving for x:

$$K_b = \frac{[(CH_3)_2NH_2^+][OH^-]}{[(CH_3)_2NH]}$$

$$= 5.9 \times 10^{-4} \approx \frac{x^2}{1.5}$$

$$x = [OH^-] \approx 3.0 \times 10^{-2}\ M$$

Checking the assumption:

$$\frac{3.0\times10^{-2}\,M}{1.5\,M} \times 100 = 2.0\%\ (<5\%;\ \text{assumption is justified})$$

Note that the Comment in Sample Problem 18.7 applies to weak bases as well:

$$\frac{[\text{B}]_{\text{init}}}{K_b} = \frac{1.5}{5.9\times10^{-4}} = 2.5\times10^3 > 400$$

Calculating pH:

$$[\text{H}_3\text{O}^+] = \frac{K_w}{[\text{OH}^-]} = \frac{1.0\times10^{-14}}{3.0\times10^{-2}} = 3.3\times10^{-13}\,M$$
$$\text{pH} = -\log(3.3\times10^{-13})$$
$$= 12.48$$

Check The value of x seems reasonable: $\sqrt{(\sim6\times10^{-4})(1.5)} = \sqrt{9\times10^{-4}} = 3\times10^{-2}$. Because $(\text{CH}_3)_2\text{NH}$ is a weak base, the pH should be several pH units greater than 7.

FOLLOW-UP PROBLEM 18.8 Pyridine ($\text{C}_5\text{H}_5\text{N}$, see margin) plays a major role in organic syntheses as a solvent and base. It has a pK_b of 8.77. What is the pH of 0.10 M pyridine?

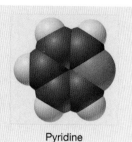

Pyridine

Anions of Weak Acids as Weak Bases

The other large group of Brønsted-Lowry bases consists of anions of weak acids:

$$\text{A}^-(aq) + \text{H}_2\text{O}(l) \rightleftharpoons \text{HA}(aq) + \text{OH}^-(aq) \qquad K_b = \frac{[\text{HA}][\text{OH}^-]}{[\text{A}^-]}$$

For example, F^-, the anion of the weak acid HF, acts as a weak base:

$$\text{F}^-(aq) + \text{H}_2\text{O}(l) \rightleftharpoons \text{HF}(aq) + \text{OH}^-(aq) \qquad K_b = \frac{[\text{HF}][\text{OH}^-]}{[\text{F}^-]}$$

Why is a solution of HA acidic and a solution of A^- basic? Let's approach the question by examining the relative concentrations of species present in 1 M HF and in 1 M NaF:

1. *The acidity of HA(aq).* Because HF is a weak acid, most of it is undissociated. The small fraction of HF that does dissociate yields small concentrations of H_3O^+ and F^-. The equilibrium position of the system lies far to the left:

$$\text{HF}(aq) + \text{H}_2\text{O}(l) \overset{\longleftarrow}{\rightleftharpoons} \text{H}_3\text{O}^+(aq) + \text{F}^-(aq)$$

Water molecules also dissociate to contribute minute amounts of H_3O^+ and OH^-, but these concentrations are extremely small:

$$2\text{H}_2\text{O}(l) \overset{\longleftarrow}{\rightleftharpoons} \text{H}_3\text{O}^+(aq) + \text{OH}^-(aq)$$

Of all the species present—HF, H_2O, H_3O^+, F^-, and OH^-—the two that can influence the acidity of the solution are H_3O^+, predominantly from HF, and OH^- from water. The solution is acidic because $[\text{H}_3\text{O}^+]_{\text{from HF}} \gg [\text{OH}^-]_{\text{from H}_2\text{O}}$.

2. *The basicity of A^-(aq).* Now, consider the species present in 1 M NaF. The salt dissociates completely to yield a stoichiometric concentration of F^-. The Na^+ ion behaves as a spectator, and some F^- reacts as a weak base to produce very small amounts of HF and OH^-:

$$\text{F}^-(aq) + \text{H}_2\text{O}(l) \overset{\longleftarrow}{\rightleftharpoons} \text{HF}(aq) + \text{OH}^-(aq)$$

As before, dissociation of water contributes minute amounts of H_3O^+ and OH^-. Thus, in addition to the Na^+ ion, the species present are the same as in the HF solution: HF, H_2O, H_3O^+, F^-, and OH^-. The two species that affect the acidity are OH^-, predominantly from the F^- reaction with water, and H_3O^+ from water. In this case, $[\text{OH}^-]_{\text{from F}^-} \gg [\text{H}_3\text{O}^+]_{\text{from H}_2\text{O}}$, so the solution is basic.

To summarize, *the relative concentrations of HA and A⁻ determine the acidity or basicity of the solution:*

- In an HA solution, $[HA] \gg [A^-]$ and $[H_3O^+]_{\text{from HA}} \gg [OH^-]_{\text{from H}_2\text{O}}$, so the solution is acidic.
- In an A⁻ solution, $[A^-] \gg [HA]$ and $[OH^-]_{\text{from A}^-} \gg [H_3O^+]_{\text{from H}_2\text{O}}$, so the solution is basic.

The Relation Between K_a and K_b of a Conjugate Acid-Base Pair

An important relationship exists between the K_a of HA and the K_b of A⁻, which we can see by treating the two dissociation reactions as a reaction sequence and adding them together:

$$HA + H_2O \rightleftharpoons H_3O^+ + A^-$$
$$A^- + H_2O \rightleftharpoons HA + OH^-$$
$$\overline{2H_2O \rightleftharpoons H_3O^+ + OH^-}$$

The sum of the two dissociation reactions is the autoionization of water. Recall from Chapter 17 that, for a reaction that is the *sum* of two or more reactions, the overall equilibrium constant is the *product* of the individual equilibrium constants. Therefore, writing the expressions for each reaction gives

$$\frac{[H_3O^+][A^-]}{[HA]} \times \frac{[HA][OH^-]}{[A^-]} = [H_3O^+][OH^-]$$

or
$$K_a \quad \times \quad K_b \quad = \quad K_w \qquad (18.7)$$

This relationship allows us to find K_a of the acid in a conjugate pair given K_b of the base, and vice versa. Reference tables typically have K_a and K_b values for *molecular species only*. The K_b for F⁻ or the K_a for $CH_3NH_3^+$, for example, does not appear in standard tables, but you can calculate either value simply by looking up the value for the molecular conjugate species and relating it to K_w. To find the K_b value for F⁻, for instance, we look up the K_a value for HF and relate it to K_w:

$$K_a \text{ of HF} = 6.8 \times 10^{-4} \text{ (from Appendix C)}$$

So, we have
$$K_a \text{ of HF} \times K_b \text{ of F}^- = K_w$$

or
$$K_b \text{ of F}^- = \frac{K_w}{K_a \text{ of HF}} = \frac{1.0 \times 10^{-14}}{6.8 \times 10^{-4}} = 1.5 \times 10^{-11}$$

We can use this calculated K_b value to finish solving the problem.

SAMPLE PROBLEM 18.9 Determining the pH of a Solution of A⁻

Problem Sodium acetate (CH_3COONa, or NaAc for this problem) has applications in photographic development and textile dyeing. What is the pH of 0.25 M NaAc? K_a of acetic acid (HAc) is 1.8×10^{-5}.

Plan From the formula (NaAc) and the fact that all sodium salts are water soluble, we know that the initial concentration of acetate ion, Ac⁻, is 0.25 M. We also know the K_a of the parent acid, HAc (1.8×10^{-5}). We have to find the pH of the solution of Ac⁻, which acts as a base in water:

$$Ac^-(aq) + H_2O(l) \rightleftharpoons HAc(aq) + OH^-(aq) \qquad K_b = \frac{[HAc][OH^-]}{[Ac^-]}$$

If we calculate [OH⁻], we can find $[H_3O^+]$ and convert it to pH. To solve for [OH⁻], we need the K_b of Ac⁻, which we obtain from the K_a of HAc and K_w. All sodium salts are soluble, so we know that $[Ac^-] = 0.25$ M. Our usual assumption is that $[Ac^-]_{\text{init}} \approx [Ac^-]$.

Solution Setting up the reaction table, with $x = [Ac^-]_{reacting} = [HAc] = [OH^-]$:

Concentration (M)	$Ac^-(aq)$	+	$H_2O(l)$	$\rightleftharpoons$	$HAc(aq)$	+	$OH^-(aq)$
Initial	0.25		—		0		0
Change	$-x$		—		$+x$		$+x$
Equilibrium	$0.25 - x$		—		x		x

Solving for K_b:

$$K_b = \frac{K_w}{K_a} = \frac{1.0 \times 10^{-14}}{1.8 \times 10^{-5}} = 5.6 \times 10^{-10}$$

Making the assumption: Because K_b is small, $0.25\,M - x \approx 0.25\,M$.
Substituting into the expression for K_b and solving for x:

$$K_b = \frac{[HAc][OH^-]}{[Ac^-]} = 5.6 \times 10^{-10} \approx \frac{x^2}{0.25} \qquad x = [OH^-] \approx 1.2 \times 10^{-5}\,M$$

Checking the assumption:

$$\frac{1.2 \times 10^{-5}\,M}{0.25\,M} \times 100 = 4.8 \times 10^{-3}\% \ (<5\%; \text{assumption is justified})$$

Note that $\dfrac{0.25}{5.6 \times 10^{-10}} = 4.5 \times 10^8 > 400$

Solving for pH:

$$[H_3O^+] = \frac{K_w}{[OH^-]} = \frac{1.0 \times 10^{-14}}{1.2 \times 10^{-5}} = 8.3 \times 10^{-10}\,M$$

$$pH = -\log(8.3 \times 10^{-10}) = 9.08$$

Check The K_b calculation seems reasonable: $\sim 10 \times 10^{-15}/2 \times 10^{-5} = 5 \times 10^{-10}$. Because Ac^- is a weak base, $[OH^-] > [H_3O^+]$; thus, pH > 7, which makes sense.

FOLLOW-UP PROBLEM 18.9 Sodium hypochlorite (NaClO) is the active ingredient in household laundry bleach. What is the pH of 0.20 M NaClO?

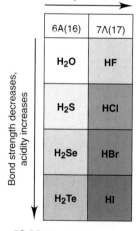

Electronegativity increases, acidity increases

6A(16)	7A(17)
H_2O	HF
H_2S	HCl
H_2Se	HBr
H_2Te	HI

Bond strength decreases, acidity increases

Figure 18.11 The effect of atomic and molecular properties on nonmetal hydride acidity. As the electronegativity of the nonmetal (E) bonded to the ionizable proton increases (*left to right*), the acidity increases. As the length of the E—H bond increases (*top to bottom*), the bond strength decreases, so the acidity increases. (In water, HCl, HBr, and HI are equally strong.)

SECTION SUMMARY

The extent to which a weak base accepts a proton from water to form OH^- is expressed by a base-dissociation constant, K_b. Brønsted-Lowry bases include NH_3 and amines and the anions of weak acids. All produce basic solutions by accepting H^+ from water, which yields OH^- and thus makes $[H_3O^+] < [OH^-]$. A solution of HA is acidic because $[HA] >> [A^-]$, so $[H_3O^+] > [OH^-]$. A solution of A^- is basic because $[A^-] >> [HA]$, so $[OH^-] > [H_3O^+]$. By multiplying the expressions for K_a of HA and K_b of A^-, we obtain K_w. This relationship allows us to calculate either K_a of BH^+, the cationic conjugate acid of a molecular weak base B, or K_b of A^-, the anionic conjugate base of a molecular weak acid HA.

18.6 MOLECULAR PROPERTIES AND ACID STRENGTH

The strength of an acid depends on its ability to donate a proton, which depends in turn on the strength of the bond to the acidic proton. In this section, we apply trends in atomic and bond properties to determine the trends in acid strength of nonmetal hydrides and oxoacids and discuss the acidity of hydrated metal ions.

Trends in Acid Strength of Nonmetal Hydrides

Two factors determine how easily a proton is released from a nonmetal hydride: the electronegativity of the central nonmetal (E) and the strength of the E—H bond. Figure 18.11 displays two periodic trends:

1. *Across a period, nonmetal hydride acid strength increases.* Across a period, the electronegativity of the nonmetal E determines the trend. As E becomes more electronegative, electron density around H is withdrawn, and the E—H bond becomes more polar. As a result, H^+ is released more easily to a surrounding water molecule. In aqueous solution, the hydrides of Groups 3A(13) to 5A(15) do not behave as acids, but an increase in acid strength is seen in Groups 6A(16) and 7A(17). Thus, HCl is a stronger acid than H_2S because Cl is more electronegative (EN = 3.0) than S (EN = 2.5). The same relationship holds across each period.

2. *Down a group, nonmetal hydride acid strength increases.* Down a group, E—H bond strength determines the trend. As E becomes larger, the E—H bond becomes longer and weaker, so H^+ comes off more easily. Thus, in Group 6A(16), for example,

$$H_2O < H_2S < H_2Se < H_2Te$$

Trends in Acid Strength of Oxoacids

All oxoacids have the acidic H atom bonded to an O atom, so bond strength (length) is not a factor in their acidity, as it is with the nonmetal hydrides. Rather, two factors determine the acid strength of oxoacids: the electronegativity of the central nonmetal (E) and the number of O atoms.

1. *For oxoacids with the **same** number of oxygens around E, acid strength increases with the electronegativity of E.* Consider the hypohalous acids (written here as HOE, where E is a halogen atom). The more electronegative E is, the more electron density it pulls from the O—H bond; the more polar the O—H bond becomes, the more easily H^+ is lost (Figure 18.12A). Electronegativity decreases down the group, so we predict that acid strength decreases: HOCl > HOBr > HOI. Our prediction is confirmed by the K_a values:

$$K_a \text{ of HOCl} = 2.9 \times 10^{-8} \qquad K_a \text{ of HOBr} = 2.3 \times 10^{-9} \qquad K_a \text{ of HOI} = 2.3 \times 10^{-11}$$

We also predict (correctly) that in Group 6A(16), H_2SO_4 is stronger than H_2SeO_4; in Group 5A(15), H_3PO_4 is stronger than H_3AsO_4, and so forth.

2. *For oxoacids with **different** numbers of oxygens around a given E, acid strength increases with number of O atoms.* The electronegative O atoms pull electron density away from E, which makes the O—H bond more polar. The more O atoms present, the greater the shift in electron density, and the more easily the H^+ ion comes off (Figure 18.12B). Therefore, we predict, for instance, that

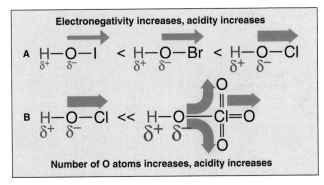

Figure 18.12 The relative strengths of oxoacids. A, Among these hypohalous acids, HOCl is the strongest and HOI the weakest. Because Cl is the most electronegative of the halogens shown here, it withdraws electron density (indicated by thickness of green arrow) from the O—H bond most effectively, making that bond most polar in HOCl (indicated by the relative sizes of the δ symbols). **B,** Among the chlorine oxoacids, the additional O atoms in $HOClO_3$ pull electron density from the O—H bond, making the bond much more polar than that in HOCl.

chlorine oxoacids (written here as $HOClO_n$, with n from 0 to 3) increase in strength in the order $HOCl < HOClO < HOClO_2 < HOClO_3$:

$$K_a \text{ of HOCl (hypochlorous acid)} = 2.9 \times 10^{-8}$$
$$K_a \text{ of HOClO (chlorous acid)} = 1.12 \times 10^{-2}$$
$$K_a \text{ of HOClO}_2 \text{ (chloric acid)} \approx 1$$
$$K_a \text{ of HOClO}_3 \text{ (perchloric acid)} = >10^7$$

It follows from this that HNO_3 is stronger than HNO_2, that H_2SO_4 is stronger than H_2SO_3, and so forth.

Acidity of Hydrated Metal Ions

The aqueous solutions of certain metal ions are acidic because the *hydrated* metal ion transfers an H^+ ion to water. Consider a general metal nitrate, $M(NO_3)_n$, as it dissolves in water. The ions separate and become bonded to a specific number of surrounding H_2O molecules. This equation shows the hydration of the cation (M^{n+}) with H_2O molecules; hydration of the anion (NO_3^-) is indicated by (aq):

$$M(NO_3)_n(s) + xH_2O(l) \longrightarrow M(H_2O)_x^{n+}(aq) + nNO_3^-(aq)$$

If the metal ion, M^{n+}, is *small and highly charged,* it has a high charge density and withdraws sufficient electron density from the O—H bonds of these bonded water molecules for a proton to be released. That is, the hydrated cation, $M(H_2O)_x^{n+}$, acts as a typical Brønsted-Lowry acid. In the process, the bound H_2O molecule that releases the proton becomes a bound OH^- ion:

$$M(H_2O)_x^{n+}(aq) + H_2O(l) \rightleftharpoons M(H_2O)_{x-1}OH^{(n-1)+}(aq) + H_3O^+(aq)$$

Each type of hydrated metal ion that releases a proton has a characteristic K_a value. Some common examples appear in Appendix C.

Aluminum ion, for example, has the small size and high positive charge needed to produce an acidic solution. When an aluminum salt, such as $Al(NO_3)_3$, dissolves in water, the following steps occur:

$$Al(NO_3)_3(s) + 6H_2O(l) \longrightarrow Al(H_2O)_6^{3+}(aq) + 3NO_3^-(aq)$$

[dissolution and hydration]

$$Al(H_2O)_6^{3+}(aq) + H_2O(l) \rightleftharpoons Al(H_2O)_5OH^{2+}(aq) + H_3O^+(aq)$$

[dissociation of weak acid]

Note the formulas of the hydrated metal ions in the last step. When H^+ is released, the number of bound H_2O molecules decreases by 1 (from 6 to 5) and the number of bound OH^- ions increases by 1 (from 0 to 1), which reduces the ion's positive charge by 1 (from 3 to 2) (Figure 18.13). Salts of most M^{2+} and M^{3+} ions yield acidic aqueous solutions.

Figure 18.13 The acidic behavior of the hydrated Al^{3+} ion. When a metal ion enters water, it is hydrated as water molecules bond to it. If the ion is small and multiply charged, as is the Al^{3+} ion, it pulls sufficient electron density from the O—H bonds of the attached water molecules to make the bonds more polar, and an H^+ ion is transferred to a nearby water molecule.

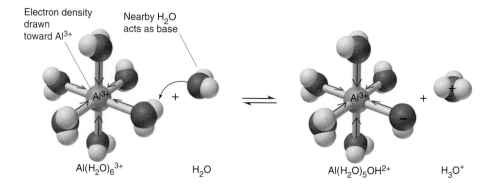

Electron density drawn toward Al^{3+}

Nearby H_2O acts as base

$Al(H_2O)_6^{3+}$ H_2O $Al(H_2O)_5OH^{2+}$ H_3O^+

SECTION SUMMARY
The strength of an acid depends on the ease with which the ionizable proton is released. For nonmetal hydrides, acid strength increases across a period, with the electronegativity of the nonmetal (E), and down a group, with the length of the E—H bond. For oxoacids with the same number of O atoms, acid strength increases with electronegativity of E; for oxoacids with the same E, acid strength increases with number of O atoms. Small, highly charged metal ions are acidic in water because they withdraw electron density from the O—H bonds of bound H_2O molecules, releasing an H^+ ion to the solution.

18.7 ACID-BASE PROPERTIES OF SALT SOLUTIONS

Up to now you've seen that cations of weak bases (such as NH_4^+) are acidic, anions of weak acids (such as CN^-) are basic, anions of polyprotic acids (such as $H_2PO_4^-$) are often acidic, and small, highly charged metal cations (such as Al^{3+}) are acidic. Therefore, when salts containing these ions dissolve in water, the pH of the solution is affected. You can predict the relative acidity of a salt solution from the relative ability of the cation and/or anion to react with water.

Salts That Yield Neutral Solutions

A salt consisting of the anion of a strong acid and the cation of a strong base yields a neutral solution because the ions do not react with water. To see why the ions don't react, let's consider the dissociation of the parent acid and base. When a strong acid such as HNO_3 dissolves, complete dissociation takes place:

$$HNO_3(l) + H_2O(l) \longrightarrow NO_3^-(aq) + H_3O^+(aq)$$

H_2O is a much stronger base than NO_3^-, so the reaction proceeds essentially to completion. The same argument can be made for any strong acid: *the anion of a strong acid is a much weaker base than water.* Therefore, a strong acid anion is hydrated, but nothing further happens.

Now consider the dissociation of a strong base, such as NaOH:

$$NaOH(s) \xrightarrow{H_2O} Na^+(aq) + OH^-(aq)$$

When Na^+ enters water, it becomes hydrated but nothing further happens. *The cations of all strong bases behave this way.*

The anions of strong acids are the halide ions, except F^-, and those of strong oxoacids, such as NO_3^- and ClO_4^-. The cations of strong bases are those from Group 1A(1) and Ca^{2+}, Sr^{2+}, and Ba^{2+} from Group 2A(2). Salts containing only these ions, such as NaCl and $Ba(NO_3)_2$, *yield neutral solutions because no reaction takes place between the ions and water.*

Salts That Yield Acidic Solutions

A salt consisting of the anion of a strong acid and the cation of a weak base yields an acidic solution because the cation acts as a weak acid, and the anion does not react. For example, NH_4Cl produces an acidic solution because the NH_4^+ ion, the cation that forms from the weak base NH_3, is a weak acid, and the Cl^- ion, the anion of a strong acid, does not react:

$$NH_4Cl(s) \xrightarrow{H_2O} NH_4^+(aq) + Cl^-(aq) \qquad \text{[dissolution and hydration]}$$
$$NH_4^+(aq) + H_2O(l) \rightleftharpoons NH_3(aq) + H_3O^+(aq) \qquad \text{[dissociation of weak acid]}$$

As you saw earlier, *small, highly charged metal ions* make up another group of cations that yield H_3O^+ in solution. For example, $Fe(NO_3)_3$ produces an acidic solution because the hydrated Fe^{3+} ion acts as a weak acid, whereas the NO_3^- ion, the anion of a strong acid, does not react:

$$Fe(NO_3)_3(s) + 6H_2O(l) \xrightarrow{H_2O} Fe(H_2O)_6^{3+}(aq) + 3NO_3^-(aq)$$

[dissolution and hydration]

$$Fe(H_2O)_6^{3+}(aq) + H_2O(l) \rightleftharpoons Fe(H_2O)_5OH^{2+}(aq) + H_3O^+(aq)$$

[dissociation of weak acid]

A third group of salts that yield H_3O^+ ions in solutions consists of *cations of strong bases and anions of polyprotic acids that still have one or more ionizable protons.* For example, NaH_2PO_4 yields an acidic solution because Na^+, the cation of a strong base, does not react, while $H_2PO_4^-$, the first anion of the weak polyprotic acid H_3PO_4, is also a weak acid:

$$NaH_2PO_4(s) \xrightarrow{H_2O} Na^+(aq) + H_2PO_4^-(aq) \qquad \text{[dissolution and hydration]}$$
$$H_2PO_4^-(aq) + H_2O(l) \rightleftharpoons HPO_4^{2-}(aq) + H_3O^+(aq) \qquad \text{[dissociation of weak acid]}$$

Salts That Yield Basic Solutions

A salt consisting of the anion of a weak acid and the cation of a strong base yields a basic solution in water because the anion acts as a weak base, and the cation does not react. The anion of a weak acid accepts a proton from water to yield OH^- ion. Sodium acetate, for example, yields a basic solution because the Na^+ ion, the cation of a strong base, does not react with water, and the CH_3COO^- ion, the anion of the weak acid CH_3COOH, acts as a weak base:

$$CH_3COONa(s) \xrightarrow{H_2O} Na^+(aq) + CH_3COO^-(aq) \qquad \text{[dissolution and hydration]}$$
$$CH_3COO^-(aq) + H_2O(l) \rightleftharpoons CH_3COOH(aq) + OH^-(aq) \qquad \text{[reaction of weak base]}$$

Table 18.3 displays the acid-base behavior of the various types of salts in water.

SAMPLE PROBLEM 18.10   Predicting Relative Acidity of Salt Solutions

Problem Predict whether aqueous solutions of the following are acidic, basic, or neutral, and write an equation for the reaction of any ion with water:
(a) Potassium perchlorate, $KClO_4$ **(b)** Sodium benzoate, C_6H_5COONa
(c) Chromium trichloride, $CrCl_3$ **(d)** Sodium hydrogen sulfate, $NaHSO_4$
Plan We examine the formulas to determine the cations and anions. Depending on the nature of these ions, the solution will be neutral (strong-acid anion and strong-base cation), acidic (weak-base cation and strong-acid anion, highly charged metal cation, or first anion of a polyprotic acid), or basic (weak-acid anion and strong-base cation).
Solution (a) Neutral. The ions are K^+ and ClO_4^-. The K^+ ion is from the strong base KOH, and the ClO_4^- anion is from the strong acid $HClO_4$. Neither ion reacts with water.
(b) Basic. The ions are Na^+ and $C_6H_5COO^-$. Na^+ is the cation of the strong base $NaOH$ and does not react with water. The benzoate ion, $C_6H_5COO^-$, is from the weak acid benzoic acid, so it reacts with water to produce OH^- ion:

$$C_6H_5COO^-(aq) + H_2O(l) \rightleftharpoons C_6H_5COOH(aq) + OH^-(aq)$$

(c) Acidic. The ions are Cr^{3+} and Cl^-. Cl^- is the anion of the strong acid HCl, so it does not react with water. Cr^{3+} is a small metal ion with a high positive charge, so the hydrated ion, $Cr(H_2O)_6^{3+}$, reacts with water to produce H_3O^+:

$$Cr(H_2O)_6^{3+}(aq) + H_2O(l) \rightleftharpoons Cr(H_2O)_5OH^{2+}(aq) + H_3O^+(aq)$$

(d) Acidic. The ions are Na^+ and HSO_4^-. Na^+ is the cation of the strong base $NaOH$, so it does not react with water. HSO_4^- is the first anion of the diprotic acid H_2SO_4, and it reacts with water to produce H_3O^+:

$$HSO_4^-(aq) + H_2O(l) \rightleftharpoons SO_4^{2-}(aq) + H_3O^+(aq)$$

FOLLOW-UP PROBLEM 18.10 Write equations to predict whether solutions of the following salts are acidic, basic, or neutral: **(a)** $KClO_2$; **(b)** $CH_3NH_3NO_3$; **(c)** CsI.

Table 18.3	**The Behavior of Salts in Water**				
Salt Solution	**Examples**	**pH**	**Nature of Ions**		**Ion That Reacts with Water**
Neutral	$NaCl$, KBr, $Ba(NO_3)_2$	7.0	Cation of strong base Anion of strong acid		None
Acidic	NH_4Cl, NH_4NO_3, CH_3NH_3Br	<7.0	Cation of weak base Anion of strong acid		Cation
Acidic	$Al(NO_3)_3$, $CrCl_3$, $FeBr_3$	<7.0	Small, highly charged cation Anion of strong base		Cation
Acidic	NaH_2PO_4, $KHSO_4$, $NaHSO_3$	<7.0	Cation of strong base First anion of polyprotic acid		Anion
Basic	CH_3COONa, KF, Na_2CO_3	>7.0	Cation of strong base Anion of weak acid		Anion

Salts of Weakly Acidic Cations and Weakly Basic Anions

The only salts left to consider are those consisting of a cation that acts as a weak acid *and* an anion that acts as a weak base. In these cases, both ions react with water, so the overall acidity of the solution depends on the relative acid strength (K_a) or base strength (K_b) of the separated ions.

For example, will an aqueous solution of ammonium cyanide, NH_4CN, be acidic or basic? First, we write equations for any reactions that occur between the separated ions and water. Ammonium ion is the conjugate acid of a weak base, so it acts as a weak acid:

$$NH_4^+(aq) + H_2O(l) \rightleftharpoons NH_3(aq) + H_3O^+(aq)$$

Cyanide ion is the anion of the weak acid HCN and it acts as a weak base:

$$CN^-(aq) + H_2O(l) \rightleftharpoons HCN(aq) + OH^-(aq)$$

The reaction that goes farther to the right will have the greater influence on the pH of the solution, so we must compare the K_a of NH_4^+ with the K_b of CN^-. But only molecular compounds are listed in K_a and K_b tables, so we first have to calculate these values for the ions:

$$K_a \text{ of } NH_4^+ = \frac{K_w}{K_b \text{ of } NH_3} = \frac{1.0 \times 10^{-14}}{1.76 \times 10^{-5}} = 5.7 \times 10^{-10}$$

$$K_b \text{ of } CN^- = \frac{K_w}{K_a \text{ of } HCN} = \frac{1.0 \times 10^{-14}}{6.2 \times 10^{-10}} = 1.6 \times 10^{-5}$$

The difference in magnitude of the equilibrium constants ($K_b \approx 3 \times 10^4 K_a$) tells us that the acceptance of a proton from H_2O by CN^- proceeds much further than the donation of a proton to H_2O by NH_4^+. In other words, because K_b of $CN^- > K_a$ of NH_4^+, the NH_4CN solution is basic.

SAMPLE PROBLEM 18.11 Predicting the Relative Acidity of a Salt Solution from K_a and K_b of the Ions

Problem Determine whether an aqueous solution of zinc formate, $Zn(HCOO)_2$, is acidic, basic, or neutral.

Plan The formula consists of the small, highly charged, and therefore weakly acidic, Zn^{2+} cation and the weakly basic $HCOO^-$ anion of the weak acid HCOOH. To determine the relative acidity of the solution, we write equations that show the reactions of the ions with water, and then find K_a of Zn^{2+} (from Appendix C) and calculate K_b of $HCOO^-$ (from K_a of HCOOH in Appendix C) to see which ion reacts to a greater extent.

Solution Writing the reactions with water:

$$Zn(H_2O)_6^{2+}(aq) + H_2O(l) \rightleftharpoons Zn(H_2O)_5OH^+(aq) + H_3O^+(aq)$$
$$HCOO^-(aq) + H_2O(l) \rightleftharpoons HCOOH(aq) + OH^-(aq)$$

Obtaining K_a and K_b of the ions: The K_a of $Zn(H_2O)_6^{2+}(aq)$ is 1×10^{-9}. We obtain K_a of HCOOH and solve for K_b of HCOO⁻:

$$K_b \text{ of HCOO}^- = \frac{K_w}{K_a \text{ of HCOOH}} = \frac{1.0 \times 10^{-14}}{1.8 \times 10^{-4}} = 5.6 \times 10^{-11}$$

K_a of $Zn(H_2O)_6^{2+} > K_b$ of HCOO⁻, so the solution is acidic.

FOLLOW-UP PROBLEM 18.11 Determine whether solutions of the following salts are acidic, basic, or neutral: (a) $Cu(CH_3COO)_2$; (b) NH_4F.

SECTION SUMMARY

Salts that yield a neutral solution consist of ions that do not react with water. Salts that yield an acidic solution contain an unreactive anion and a cation that releases a proton to water. Salts that yield a basic solution contain an unreactive cation and an anion that accepts a proton from water. If both cation and anion react with water, the ion that reacts to the greater extent (higher K) determines the acidity or basicity of the salt solution.

18.8 ELECTRON-PAIR DONATION AND THE LEWIS ACID-BASE DEFINITION

The final acid-base concept we consider was developed by Gilbert N. Lewis, whose contribution to understanding the importance of valence electron pairs in molecular bonding we discussed in Chapter 9. Whereas the Brønsted-Lowry concept focuses on the proton in defining a species as an acid or a base, the Lewis concept highlights the role of the *electron pair*. The **Lewis acid-base definition** holds that

- A *base* is any species that *donates* an electron pair.
- An *acid* is any species that *accepts* an electron pair.

The Lewis definition, like the Brønsted-Lowry definition, requires that a base have an electron pair to donate, so it does not expand the classes of bases. However, *it greatly expands the classes of acids.* Many species, such as CO_2 and Cu^{2+}, that do not contain H in their formula (and thus cannot be Brønsted-Lowry acids) function as Lewis acids by accepting an electron pair in their reactions. Lewis stated his objection to the proton as the defining feature of an acid this way: "To restrict the group of acids to those substances which contain hydrogen interferes as seriously with the systematic understanding of chemistry as would the restriction of the term oxidizing agent to those substances containing oxygen." Moreover, in the Lewis sense, the proton itself functions as an acid because it accepts the electron pair donated by a base:

$$B\!:\,+\,H^+ \rightleftharpoons B\!-\!H^+$$

Thus, *all Brønsted-Lowry acids donate H⁺, a Lewis acid.*

The product of any Lewis acid-base reaction is called an **adduct,** *a single species that contains a **new** covalent bond:*

$$A\,+\,:B \rightleftharpoons A\!-\!B \text{ (adduct)}$$

Thus, the Lewis concept radically broadens the idea of acid-base reactions. What to Arrhenius was the formation of H_2O from H^+ and OH^- became, to Brønsted and Lowry, the transfer of a proton from a stronger acid to a stronger base to form a weaker base and weaker acid. To Lewis, the same process became *the donation and acceptance of an electron pair to form a covalent bond in an adduct.*

As we've seen, the key feature of a *Lewis base is a lone pair of electrons to donate.* The key feature of a *Lewis acid is a vacant orbital* (or the ability to rearrange its bonds to form one) to accept that lone pair and form a new bond. In the upcoming discussion, you'll encounter a variety of neutral molecules and positively charged ions that satisfy this requirement.

Molecules as Lewis Acids

Many neutral molecules function as Lewis acids. In every case, the atom that accepts the electron pair is low in electron density because of either an electron deficiency or a polar multiple bond.

Lewis Acids with Electron-Deficient Atoms Some molecular Lewis acids contain a central atom that is *electron deficient,* one surrounded by fewer than eight valence electrons. The most important of these acids are covalent compounds of the Group 3A(13) elements boron and aluminum. As noted in Chapters 10 and 14, these compounds react to complete their octet. For example, boron trifluoride accepts an electron pair from ammonia to form a covalent bond in a gaseous Lewis acid-base reaction:

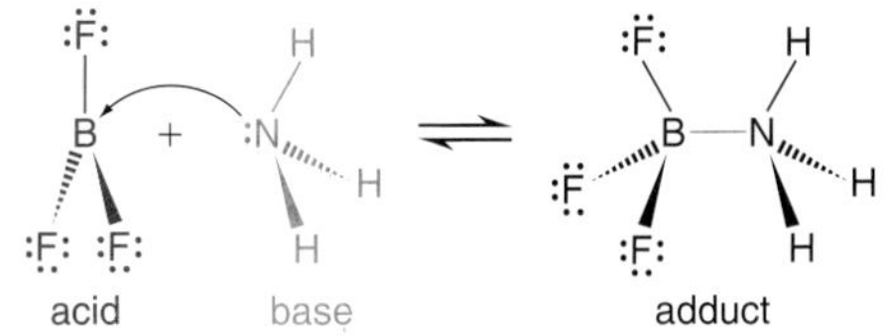

Unexpected solubility behavior is sometimes due to adduct formation. Aluminum chloride, for instance, dissolves freely in relatively nonpolar diethyl ether because of a Lewis acid-base reaction, in which the ether's O atom donates an electron pair to Al to form a covalent bond:

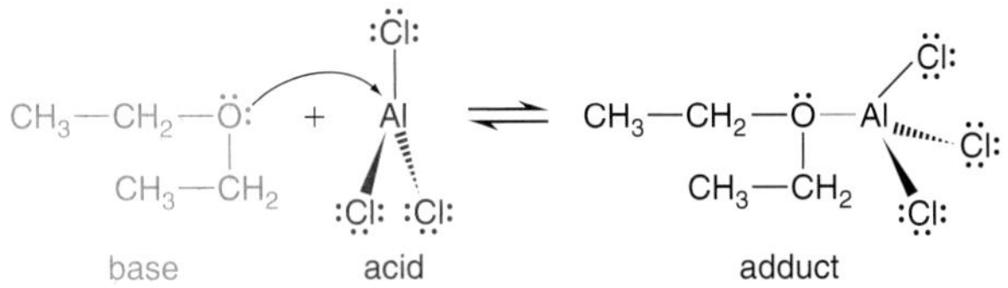

Lewis Acids with Polar Multiple Bonds Molecules that contain a polar double bond also function as Lewis acids. As the electron pair on the Lewis base approaches the partially positive end of the double bond, one of the bonds breaks to form the new bond in the adduct. For example, consider the reaction that occurs when SO_2 dissolves in water. The electronegative O atoms in SO_2 withdraw electron density from the central S, so it is partially positive. The O atom of water donates a lone pair to the S, breaking one of the π bonds and forming an S—O bond, and a proton is transferred from water to that O. The resulting adduct is sulfurous acid, and the overall process is

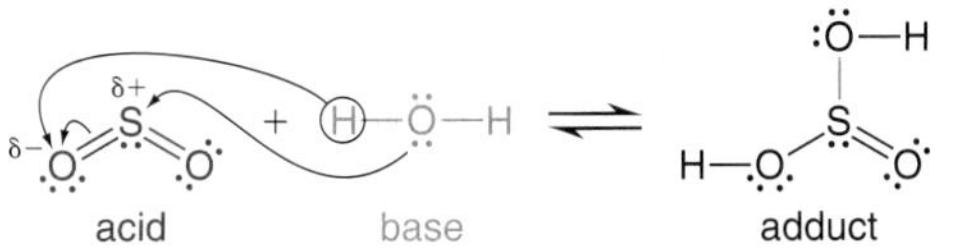

The formation of carbonates from a metal oxide and carbon dioxide is an analogous reaction that occurs in a nonaqueous heterogeneous system. The O^{2-} ion (shown below from CaO) donates an electron pair to the partially positive C in CO_2, a π bond breaks, and the CO_3^{2-} ion forms as the adduct:

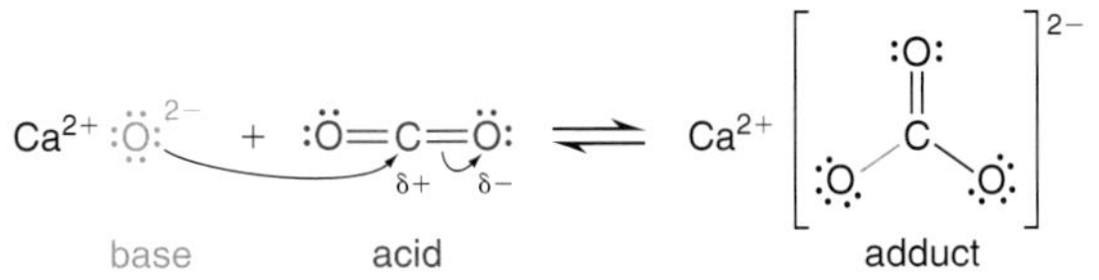

Metal Cations as Lewis Acids

Earlier we saw that certain hydrated metal ions act as Brønsted-Lowry acids. In the Lewis sense, the hydration process itself is an acid-base reaction. The hydrated cation is the adduct, as lone electron pairs on the O atoms of water form covalent bonds to the positively charged ion; thus, *any metal ion acts as a Lewis acid when it dissolves in water:*

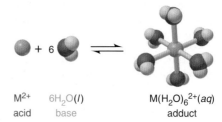

$$M^{2+} \qquad 6H_2O(l) \qquad\qquad M(H_2O)_6{}^{2+}(aq)$$
$$\text{acid} \qquad \text{base} \qquad\qquad\qquad \text{adduct}$$

Ammonia is a stronger Lewis base than water because it displaces H_2O from a hydrated ion when aqueous NH_3 is added:

$$\underset{\text{hydrated adduct}}{Ni(H_2O)_6{}^{2+}(aq)} + \underset{\text{base}}{6NH_3(aq)} \rightleftharpoons \underset{\text{ammoniated adduct}}{Ni(NH_3)_6{}^{2+}(aq)} + 6H_2O(l)$$

We discuss the equilibrium nature of these acid-base reactions in greater detail in Chapter 19, and we investigate the structures of these ions in Chapter 22.

    Many biomolecules are Lewis adducts with central metal ions. Most often, O and N atoms of organic groups, with their lone pairs, serve as the Lewis bases. Chlorophyll is a Lewis adduct of a Mg^{2+} ion and four N atoms in an organic ring system. Vitamin B_{12} has a similar structure with a central Co^{3+}, and so does heme, but with a central Fe^{2+}. Several other metal ions, such as Zn^{2+}, Mo^{2+}, and Cu^{2+}, are bound at the active sites of enzymes and function as Lewis acids in the enzymes' catalytic action.

SAMPLE PROBLEM 18.12   Identifying Lewis Acids and Bases

Problem Identify the Lewis acids and Lewis bases in the following reactions:
(a) $H^+ + OH^- \rightleftharpoons H_2O$
(b) $Cl^- + BCl_3 \rightleftharpoons BCl_4{}^-$
(c) $K^+ + 6H_2O \rightleftharpoons K(H_2O)_6{}^+$
Plan We examine the formulas to see which species accepts the electron pair (Lewis acid) and which donates it (Lewis base) in forming the adduct.
Solution (a) The H^+ ion accepts an electron pair from the OH^- ion in forming a bond. H^+ is the acid and OH^- is the base.
(b) The Cl^- ion has four lone pairs and uses one to form a new bond to the central B. Therefore, BCl_3 is the acid and Cl^- is the base.
(c) The K^+ ion does not have any valence electrons to provide, so the bond is formed when electron pairs from O atoms of water enter empty orbitals on K^+. Thus, K^+ is the acid and H_2O is the base.
Check The Lewis acids (H^+, BCl_3, and K^+) each have an unfilled valence shell that can accept an electron pair from the Lewis bases (OH^-, Cl^-, and H_2O).

FOLLOW-UP PROBLEM 18.12 Identify the Lewis acids and Lewis bases in the following reactions:
(a) $OH^- + Al(OH)_3 \rightleftharpoons Al(OH)_4{}^-$
(b) $SO_3 + H_2O \rightleftharpoons H_2SO_4$
(c) $Co^{3+} + 6NH_3 \rightleftharpoons Co(NH_3)_6{}^{3+}$

SECTION SUMMARY

The Lewis acid-base definition focuses on the donation or acceptance of an electron pair to form a new covalent bond in an adduct, the product of an acid-base reaction. Lewis bases donate the electron pair, and Lewis acids accept it. Thus, many species that do not contain H are Lewis acids. Molecules with polar double bonds act as Lewis acids, as do those with electron-deficient atoms. Metal ions act as Lewis acids when they dissolve in water, which acts as a Lewis base, to form an adduct, a hydrated cation. Many metal ions function as Lewis acids in biomolecules.

For Review and Reference (Numbers in parentheses refer to pages, unless noted otherwise.)

Learning Objectives

To help you review these learning objectives, the numbers of related sections (§), sample problems (SP), and upcoming end-of-chapter problems (EP) are listed in parentheses.

1. Understand the nature of the hydrated proton, the Arrhenius definition of an acid and a base, and why all strong acid–strong base reactions have the same ΔH°_{rxn}; describe how acid strength is expressed by K_a; classify strong and weak acids and bases from their formulas (§ 18.1) (SP 18.1) (EPs 18.1–18.12)
2. Describe the autoionization of water and the meaning of K_w and pH; understand why $[H_3O^+]$ is inversely related to $[OH^-]$ and how their relative magnitudes define the acidity of a solution; interconvert pH, pOH, $[H_3O^+]$, and $[OH^-]$ (§ 18.2) (SPs 18.2, 18.3) (EPs 18.13–18.23)
3. Understand the Brønsted-Lowry definitions of an acid and a base; discuss how water can act as a base or as an acid and how an acid-base reaction is a proton-transfer process involving two conjugate acid-base pairs, with the stronger acid and base forming the weaker base and acid (§ 18.3) (SPs 18.4, 18.5) (EPs 18.24–18.39)
4. Calculate K_a from the pH of an HA solution or pH from the K_a and initial [HA]; explain why the percent dissociation of HA in-creases as [HA] decreases and know how to find percent dissociation (§ 18.4) (SPs 18.6, 18.7) (EPs 18.40–18.54)
5. Understand the meaning of K_b, and why ammonia and amines and weak-acid anions are bases; discuss how relative [HA] and $[A^-]$ determine the acidity or basicity of a solution, and show the relationship between K_a and K_b of a conjugate acid-base pair and K_w; calculate pH from K_b and $[B]_{init}$, and find K_b of A^- from K_a of HA (§ 18.5) (SPs 18.8, 18.9) (EPs 18.55–18.67)
6. Understand how electronegativity and bond length, polarity, and strength affect the acid strength of nonmetal hydrides and oxoacids, and explain why certain metal ions form acidic solutions (§ 18.6) (EPs 18.68–18.75)
7. Understand how combinations of cations and anions lead to acidic, basic, or neutral solutions (§ 18.7) (SPs 18.10, 18.11) (EPs 18.76–18.82)
8. Describe how a Lewis acid-base reaction involves formation of a new covalent bond, and identify Lewis acids and bases (§ 18.9) (SP 18.12) (EPs 18.83–18.93)

Key Terms

Section 18.1
hydronium ion, H_3O^+ (578)
Arrhenius acid-base
    definition (579)
neutralization (579)
acid-dissociation (acid-
    ionization) constant
    (K_a) (581)

Section 18.2
autoionization (583)
ion-product constant for water
    (K_w) (583)
pH (584)
acid-base indicator (587)

Section 18.3
Brønsted-Lowry acid-base
    definition (587)

proton donor (587)
proton acceptor (588)
conjugate acid-base pair (589)

Section 18.4
polyprotic acid (595)

Section 18.5
base-dissociation (base-
    ionization) constant
    (K_b) (596)

Section 18.8
Lewis acid-base
    definition (606)
adduct (606)

Key Equations and Relationships

18.1 Defining the acid-dissociation constant (581):
$$K_a = \frac{[H_3O^+][A^-]}{[HA]}$$

18.2 Defining the ion-product constant for water (583):
$$K_w = [H_3O^+][OH^-] = 1.0\times10^{-14} \text{ (at 25°C)}$$

18.3 Defining pH (584):
$$pH = -\log [H_3O^+]$$

18.4 Relating pK_w to pH and pOH (585):
$$pK_w = pH + pOH = 14.00 \text{ (at 25°C)}$$

18.5 Finding the percent dissociation of HA (595):
$$\text{Percent HA dissociated} = \frac{[HA]_{dissoc}}{[HA]_{init}} \times 100$$

18.6 Defining the base-dissociation constant (596):
$$K_b = \frac{[BH^+][OH^-]}{[B]}$$

18.7 Expressing the relationship among K_a, K_b, and K_w (599):
$$K_a \times K_b = K_w$$

Brief Solutions to Follow-up Problems

18.1 (a) $HClO_3$; (b) HCl; (c) NaOH

18.2 $[H_3O^+] = \dfrac{1.0\times10^{-14}}{6.7\times10^{-2}} = 1.5\times10^{-13} \; M$; basic

18.3 $pOH = 14.00 - 9.52 = 4.48$
$[H_3O^+] = 10^{-9.52} = 3.0\times10^{-10} \; M$
$[OH^-] = \dfrac{1.0\times10^{-14}}{3.0\times10^{-10}} = 3.3\times10^{-5} \; M$

18.4 (a) CH_3COOH/CH_3COO^- and H_3O^+/H_2O
(b) H_2O/OH^- and HF/F^-
18.5 H_2S is higher on list of acids; OH^- is lower on list of bases. Thus, we have
$$H_2O(l) + HS^-(aq) \rightleftharpoons OH^-(aq) + H_2S(aq)$$
weaker acid + weaker base ← stronger base + stronger acid
The net direction is left, so $K_c < 1$.

18.6 $NH_4^+(aq) + H_2O(l) \rightleftharpoons NH_3(aq) + H_3O^+(aq)$

$[H_3O^+] = 10^{-pH} = 10^{-5.0} = 1 \times 10^{-5}\ M = [NH_3]$

From reaction table, $K_a = \dfrac{[NH_3][H_3O^+]}{[NH_4^+]} = \dfrac{(x)(x)}{0.2 - x}$

$\approx \dfrac{(1 \times 10^{-5})^2}{0.2} = 5 \times 10^{-10}$

18.7 $K_a = \dfrac{[H_3O^+][OCN^-]}{[HOCN]} = \dfrac{(x)(x)}{0.10 - x} = 3.5 \times 10^{-4}$

Since $\dfrac{[HOCN]_{init}}{K_a} = \dfrac{0.10}{3.5 \times 10^{-4}} = 286 < 400$, you must solve a quadratic equation: $x^2 + (3.5 \times 10^{-4})x - (3.5 \times 10^{-5}) = 0$

$x = [H_3O^+] = 5.7 \times 10^{-3}\ M;\ pH = 2.24$

18.8 $K_b = \dfrac{[C_5H_5NH^+][OH^-]}{[C_5H_5N]} = 10^{-8.77} = 1.7 \times 10^{-9}$

Assuming $0.10\ M - x \approx 0.10\ M$, $K_b = 1.7 \times 10^{-9} \approx \dfrac{(x)(x)}{0.10}$;

$x = [OH^-] \approx 1.3 \times 10^{-5}\ M;\ [H_3O^+] = 7.7 \times 10^{-10}\ M$;

$pH = 9.11$

18.9 K_b of $ClO^- = \dfrac{K_w}{K_a \text{ of HClO}} = \dfrac{1.0 \times 10^{-14}}{2.9 \times 10^{-8}} = 3.4 \times 10^{-7}$

Assuming $0.20\ M - x \approx 0.20\ M$,

$K_b = 3.4 \times 10^{-7} = \dfrac{[HClO][OH^-]}{[ClO^-]} \approx \dfrac{x^2}{0.20}$;

$x = [OH^-] \approx 2.6 \times 10^{-4}\ M;\ [H_3O^+] = 3.8 \times 10^{-11}\ M;\ pH = 10.42$

18.10 (a) Basic:

     $ClO_2^-(aq) + H_2O(l) \rightleftharpoons HClO_2(aq) + OH^-(aq)$

K^+ is from strong base KOH.

(b) Acidic:

     $CH_3NH_3^+(aq) + H_2O(l) \rightleftharpoons CH_3NH_2(aq) + H_3O^+(aq)$

NO_3^- is from strong acid HNO_3.

(c) Neutral: Cs^+ is from strong base CsOH; I^- is from strong acid HI.

18.11 (a) K_a of $Cu(H_2O)_6^{2+} = 3 \times 10^{-8}$

K_b of $CH_3COO^- = \dfrac{K_w}{K_a \text{ of } CH_3COOH} = 5.6 \times 10^{-10}$

Since $K_a > K_b$, $Cu(CH_3COO)_2(aq)$ is acidic.

(b) K_a of $NH_4^+ = \dfrac{K_w}{K_b \text{ of } NH_3} = 5.7 \times 10^{-10}$

K_b of $F^- = \dfrac{K_w}{K_a \text{ of HF}} = 1.5 \times 10^{-11}$

Since $K_a > K_b$, $NH_4F(aq)$ is acidic.

18.12 (a) OH^- is the Lewis base; $Al(OH)_3$ is the Lewis acid.

(b) H_2O is the Lewis base; SO_3 is the Lewis acid.

(c) NH_3 is the Lewis base; Co^{3+} is the Lewis acid.

Problems

Problems with **colored** numbers are answered in Appendix E. Sections match the text and provide the numbers of relevant sample problems. Bracketed problems are grouped in pairs (indicated by a short rule) that cover the same concept. Comprehensive Problems are based on material from any section or previous chapter.

Note: Unless stated otherwise, all problems refer to aqueous solutions at 298 K (25°C).

Acids and Bases in Water

(Sample Problem 18.1)

18.1 Describe the role of water according to the Arrhenius acid-base definition.

18.2 What characteristics do all Arrhenius acids have in common? What characteristics do all Arrhenius bases have in common? Explain neutralization in terms of the Arrhenius acid-base definition. What quantitative finding led Arrhenius to propose this idea of neutralization?

18.3 Why is the Arrhenius acid-base definition considered too limited? Give an example of a case in which the Arrhenius definition does not apply.

18.4 What is meant by the words "strong" and "weak" in terms of acids and bases? Weak acids have K_a values that vary over more than 10 orders of magnitude. What do they have in common that classifies them as "weak"?

18.5 Which of the following are Arrhenius acids?

(a) H_2O (b) $Ca(OH)_2$ (c) H_3PO_3 (d) HI

18.6 Which of the following are Arrhenius bases?

(a) CH_3COOH (b) HOH   (c) CH_3OH (d) H_2NNH_2

18.7 Write the K_a expression for each of the following in water:

(a) HNO_2 (b) CH_3COOH (c) $HBrO_2$

18.8 Write the K_a expression for each of the following in water:

(a) $H_2PO_4^-$ (b) H_3PO_2 (c) HSO_4^-

18.9 Use Appendix C to rank the following in order of *increasing* acid strength: HIO_3, HI, CH_3COOH, HF.

18.10 Use Appendix C to rank the following in order of *decreasing* acid strength: HClO, HCl, HCN, HNO_2.

18.11 Classify each as a strong or weak acid or base:

(a) H_3AsO_4 (b) $Sr(OH)_2$ (c) HIO   (d) $HClO_4$

18.12 Classify each as a strong or weak acid or base:

(a) CH_3NH_2 (b) K_2O (c) HI   (d) HCOOH

Autoionization of Water and the pH Scale

(Sample Problems 18.2 and 18.3)

18.13 What is an autoionization reaction? Write equations for the autoionization reactions of H_2O and of H_2SO_4.

18.14 (a) What is the change in pH when $[OH^-]$ increases by a factor of 10?

(b) What is the change in $[H_3O^+]$ when the pH decreases by 2 units?

18.15 Which solution has the higher pH? Explain.

(a) A 0.1 M solution of an acid with $K_a = 1 \times 10^{-4}$ or one with $K_a = 4 \times 10^{-5}$

(b) A 0.1 M solution of an acid with $pK_a = 3.0$ or one with $pK_a = 3.5$

(c) A 0.1 M solution of a weak acid or a 0.01 M solution of the same acid

(d) A 0.1 M solution of a weak acid or a 0.1 M solution of a strong acid

(e) A 0.1 M solution of an acid or a 0.1 M solution of a base

(f) A solution of pOH 6.0 or one of pOH 8.0

18.16 (a) What is the pH of 0.0111 M NaOH? Is the solution neutral, acidic, or basic?

(b) What is the pOH of 1.23×10^{-3} M HCl? Is the solution neutral, acidic, or basic?

18.17 (a) What is the pH of 0.0333 M HNO_3? Is the solution neutral, acidic, or basic?

(b) What is the pOH of 0.0347 M KOH? Is the solution neutral, acidic, or basic?

18.18 (a) What are $[H_3O^+]$, $[OH^-]$, and pOH in a solution with a pH of 9.78?

(b) What are $[H_3O^+]$, $[OH^-]$, and pH in a solution with a pOH of 10.43?

18.19 (a) What are $[H_3O^+]$, $[OH^-]$, and pOH in a solution with a pH of 3.47?

(b) What are $[H_3O^+]$, $[OH^-]$, and pH in a solution with a pOH of 4.33?

18.20 How many moles of H_3O^+ or OH^- must you add to 6.5 L of HA solution to adjust its pH from 4.82 to 5.22? Assume a negligible volume change.

18.21 How many moles of H_3O^+ or OH^- must you add to 87.5 mL of HA solution to adjust its pH from 8.92 to 6.33? Assume a negligible volume change.

18.22 Although the text asserts that water is an extremely weak electrolyte, parents commonly warn their children of the danger of swimming in a pool or lake during a lightning storm. Explain.

18.23 Like any equilibrium constant, K_w changes with temperature.

(a) Given that autoionization is an endothermic process, does K_w increase or decrease with rising temperature? Explain with an equation that includes heat as reactant or product.

(b) In many medical applications, the value of K_w at 37°C (body temperature) may be more appropriate than the value at 25°C, 1.0×10^{-14}. The pH of pure water at 37°C is 6.80. Calculate K_w, pOH, and $[OH^-]$ at this temperature.

Proton Transfer and the Brønsted-Lowry Acid-Base Definition
(Sample Problems 18.4 and 18.5)

18.24 How do the Arrhenius and Brønsted-Lowry definitions of an acid and a base differ? How are they similar? Name two Brønsted-Lowry bases that are not considered Arrhenius bases. Can you do the same for acids? Explain.

18.25 What is a conjugate acid-base pair? What is the relationship between the two members of the pair?

18.26 A Brønsted-Lowry acid-base reaction proceeds in the net direction in which a stronger acid and stronger base form a weaker acid and weaker base. Explain.

18.27 What is an amphoteric species? Name one and write balanced equations that show why it is amphoteric.

18.28 Give the formula of the conjugate base:
(a) HCl   (b) H_2CO_3 (c) H_2O

18.29 Give the formula of the conjugate base:
(a) HPO_4^{2-} (b) NH_4^+ (c) HS^-

18.30 Give the formula of the conjugate acid:
(a) NH_3 (b) NH_2^- (c) nicotine, $C_{10}H_{14}N_2$

18.31 Give the formula of the conjugate acid:
(a) O^{2-} (b) SO_4^{2-} (c) H_2O

18.32 In each equation, label the acids, bases, and conjugate acid-base pairs:
(a) $NH_3 + H_3PO_4 \rightleftharpoons NH_4^+ + H_2PO_4^-$
(b) $CH_3O^- + NH_3 \rightleftharpoons CH_3OH + NH_2^-$
(c) $HPO_4^{2-} + HSO_4^- \rightleftharpoons H_2PO_4^- + SO_4^{2-}$

18.33 In each equation, label the acids, bases, and conjugate acid-base pairs:
(a) $NH_4^+ + CN^- \rightleftharpoons NH_3 + HCN$
(b) $H_2O + HS^- \rightleftharpoons OH^- + H_2S$
(c) $HSO_3^- + CH_3NH_2 \rightleftharpoons SO_3^{2-} + CH_3NH_3^+$

18.34 Write balanced net ionic equations for the following reactions, and label the conjugate acid-base pairs:
(a) $NaOH(aq) + NaH_2PO_4(aq) \rightleftharpoons H_2O(l) + Na_2HPO_4(aq)$
(b) $KHSO_4(aq) + K_2CO_3(aq) \rightleftharpoons K_2SO_4(aq) + KHCO_3(aq)$

18.35 Write balanced net ionic equations for the following reactions, and label the conjugate acid-base pairs:
(a) $HNO_3(aq) + Li_2CO_3(aq) \rightleftharpoons LiNO_3(aq) + LiHCO_3(aq)$
(b) $2NH_4Cl(aq) + Ba(OH)_2(aq) \rightleftharpoons$
$$2H_2O(l) + BaCl_2(aq) + 2NH_3(aq)$$

18.36 The following aqueous species constitute two conjugate acid-base pairs. Use them to write one acid-base reaction with $K_c > 1$ and another with $K_c < 1$: HS^-, Cl^-, HCl, H_2S.

18.37 The following aqueous species constitute two conjugate acid-base pairs. Use them to write one acid-base reaction with $K_c > 1$ and another with $K_c < 1$: NO_3^-, F^-, HF, HNO_3.

18.38 Use Figure 18.9 to determine whether $K_c > 1$ for
(a) $HCl + NH_3 \rightleftharpoons NH_4^+ + Cl^-$
(b) $H_2SO_3 + NH_3 \rightleftharpoons HSO_3^- + NH_4^+$

18.39 Use Figure 18.9 to determine whether $K_c < 1$ for
(a) $H_2PO_4^- + F^- \rightleftharpoons HPO_4^{2-} + HF$
(b) $CH_3COO^- + HSO_4^- \rightleftharpoons CH_3COOH + SO_4^{2-}$

Solving Problems Involving Weak-Acid Equilibria
(Sample Problems 18.6 and 18.7)

18.40 In each of the following cases, would you expect the concentration of acid before and after dissociation to be nearly the same or very different? Explain your reasoning.
(a) A concentrated solution of a strong acid
(b) A concentrated solution of a weak acid
(c) A dilute solution of a weak acid
(d) A dilute solution of a strong acid

18.41 In which of the following solutions will $[H_3O^+]$ be approximately equal to $[CH_3COO^-]$: (a) 0.1 M CH_3COOH; (b) 1×10^{-7} M CH_3COOH; (c) a solution containing both 0.1 M CH_3COOH and 0.1 M CH_3COONa? Explain.

18.42 Why do successive K_a's decrease for all polyprotic acids?

18.43 A 0.15 M solution of butanoic acid, $CH_3CH_2CH_2COOH$, contains 1.51×10^{-3} M H_3O^+. What is the K_a of butanoic acid?

18.44 A 0.035 M solution of a weak acid (HA) has a pH of 4.88. What is the K_a of the acid?

18.45 Nitrous acid, HNO_2, has a K_a of 7.1×10^{-4}. What are $[H_3O^+]$, $[NO_2^-]$, and $[OH^-]$ in 0.50 M HNO_2?

18.46 Hydrofluoric acid, HF, has a K_a of 6.8×10^{-4}. What are $[H_3O^+]$, $[F^-]$, and $[OH^-]$ in 0.75 M HF?

18.47 Chloroacetic acid, $ClCH_2COOH$, has a pK_a of 2.87. What are $[H_3O^+]$, pH, $[ClCH_2COO^-]$, and $[ClCH_2COOH]$ in 1.05 M $ClCH_2COOH$?

18.48 Hypochlorous acid, HClO, has a pK_a of 7.54. What are [H_3O^+], pH, [ClO^-], and [HClO] in 0.115 M HClO?

18.49 In a 0.25 M solution, a weak acid is 3.0% dissociated.
(a) Calculate the [H_3O^+], pH, [OH^-], and pOH of the solution.
(b) Calculate K_a of the acid.

18.50 In a 0.735 M solution, a weak acid is 12.5% dissociated.
(a) Calculate the [H_3O^+], pH, [OH^-], and pOH of the solution.
(b) Calculate K_a of the acid.

18.51 The weak acid HZ has a K_a of 1.55×10^{-4}.
(a) Calculate the pH of 0.075 M HZ.
(b) Calculate the pOH of 0.045 M HZ.

18.52 The weak acid HQ has a pK_a of 4.89.
(a) Calculate the [H_3O^+] of 3.5×10^{-2} M HQ.
(b) Calculate the [OH^-] of 0.65 M HQ.

18.53 Acetylsalicylic acid (aspirin), $HC_9H_7O_4$, is the most widely used pain reliever and fever reducer. Find the pH of 0.018 M aqueous aspirin at body temperature (K_a at 37°C = 3.6×10^{-4}).

18.54 Formic acid, HCOOH, the simplest carboxylic acid, has many uses in the textile and rubber industries. It is an extremely caustic liquid that is secreted as a defense by many species of ants (family *Formicidae*). Calculate the percent dissociation of 0.50 M HCOOH.

Weak Bases and Their Relation to Weak Acids
(Sample Problems 18.8 and 18.9)

18.55 What is the key structural feature of all Brønsted-Lowry bases? How does this feature function in an acid-base reaction?

18.56 Why are most anions basic in H_2O? Give formulas of four anions that are not basic.

18.57 Except for the Na^+ spectator ion, aqueous solutions of CH_3COOH and CH_3COONa contain the same species. (a) What are the species (other than H_2O)? (b) Why is 0.1 M CH_3COOH acidic and 0.1 M CH_3COONa basic?

18.58 Write balanced equations and K_b expressions for these Brønsted-Lowry bases in water:
(a) Pyridine, C_5H_5N (b) CO_3^{2-}

18.59 Write balanced equations and K_b expressions for these Brønsted-Lowry bases in water:
(a) Benzoate ion, $C_6H_5COO^-$ (b) $(CH_3)_3N$

18.60 What is the pH of 0.050 M dimethylamine?

18.61 What is the pH of 0.12 M diethylamine?

18.62 (a) What is the pK_b of ClO_2^-?
(b) What is the pK_a of the dimethylammonium ion, $(CH_3)_2NH_2^+$?

18.63 (a) What is the pK_b of NO_2^-?
(b) What is the pK_a of the hydrazinium ion, $H_2N-NH_3^+$ (K_b of hydrazine = 8.5×10^{-7})?

18.64 (a) What is the pH of 0.050 M KCN?
(b) What is the pH of 0.30 M triethylammonium chloride, $(CH_3CH_2)_3NHCl$?

18.65 (a) What is the pH of 0.100 M sodium phenolate, C_6H_5ONa, the sodium salt of phenol?
(b) What is the pH of 0.15 M methylammonium bromide, CH_3NH_3Br (K_b of $CH_3NH_2 = 4.4\times10^{-4}$)?

18.66 Sodium hypochlorite solution, sold as "chlorine bleach," is recognized as a potentially dangerous household product. The dangers arise from its basicity and from ClO^-, the active bleaching ingredient. What is [OH^-] in an aqueous solution that is

5.0% NaClO by mass? What is the pH of the solution? (Assume *d* of solution = 1.0 g/mL.)

18.67 Codeine ($C_{18}H_{21}NO_3$) is a narcotic pain reliever that forms a salt with HCl. What is the pH of 0.050 M codeine hydrochloride (pK_b of codeine = 5.80)?

Molecular Properties and Acid Strength

18.68 Across a period, how does the electronegativity of a nonmetal affect the acidity of its binary hydride?

18.69 How does the atomic size of a nonmetal affect the acidity of its binary hydride?

18.70 A strong acid has a weak bond to its acidic proton, whereas a weak acid has a strong bond to its acidic proton. Explain.

18.71 Perchloric acid, $HClO_4$, is the strongest of the halogen oxoacids, and hypoiodous acid, HIO, is the weakest. What two factors govern this difference in acid strength?

18.72 Choose the *stronger* acid in each of the following pairs:
(a) H_2Se or H_3As (b) $B(OH)_3$ or $Al(OH)_3$ (c) $HBrO_2$ or HBrO

18.73 Choose the *weaker* acid in each of the following pairs:
(a) HI or HBr   (b) H_3AsO_4 or H_2SeO_4 (c) HNO_3 or HNO_2

18.74 Use Appendix C to choose the solution with the *lower* pH:
(a) 0.1 M $CuSO_4$ or 0.05 M $Al_2(SO_4)_3$
(b) 0.1 M $ZnCl_2$ or 0.1 M $PbCl_2$

18.75 Use Appendix C to choose the solution with the *higher* pH:
(a) 0.1 M $NiCl_2$ or 0.1 M NaCl
(b) 0.1 M $Sn(NO_3)_2$ or 0.1 M $Co(NO_3)_2$

Acid-Base Properties of Salt Solutions
(Sample Problems 18.10 and 18.11)

18.76 What determines whether an aqueous solution of a salt will be acidic, basic, or neutral? Give an example of each type of salt.

18.77 Why is aqueous NaF basic but aqueous NaCl neutral?

18.78 The NH_4^+ ion forms acidic solutions, and the CH_3COO^- ion forms basic solutions. However, a solution of ammonium acetate is almost neutral. Do all of the ammonium salts of weak acids form neutral solutions? Explain your answer.

18.79 Explain with equations and calculations, when necessary, whether an aqueous solution of each of these salts is acidic, basic, or neutral: (a) KBr; (b) NH_4I; (c) KCN.

18.80 Explain with equations and calculations, when necessary, whether an aqueous solution of each of these salts is acidic, basic, or neutral: (a) $Cr(NO_3)_3$; (b) NaHS; (c) $Zn(CH_3COO)_2$.

18.81 Rank the following salts in order of *increasing* pH of their 0.1 M aqueous solutions:
(a) KNO_3, K_2SO_3, K_2S
(b) NH_4NO_3, $NaHSO_4$, $NaHCO_3$, Na_2CO_3

18.82 Rank the following salts in order of *decreasing* pH of their 0.1 M aqueous solutions:
(a) NH_4Cl, $MgCl_2$, $KClO_2$
(b) NH_4Br, $NaBrO_2$, NaBr, $NaClO_2$

Electron-Pair Donation and the Lewis Acid-Base Definition
(Sample Problem 18.12)

18.83 What feature must a molecule or ion have for it to act as a Lewis base? A Lewis acid? Explain the roles of these features.

18.84 How do Lewis acids differ from Brønsted-Lowry acids? How are they similar? Do Lewis bases differ from Brønsted-Lowry bases? Explain.

18.85 (a) Is a weak Brønsted-Lowry base necessarily a weak Lewis base? Explain with an example.
(b) Identify the Lewis bases in the following reaction:

$$Cu(H_2O)_4^{2+}(aq) + 4CN^-(aq) \rightleftharpoons$$
$$Cu(CN)_4^{2-}(aq) + 4H_2O(l)$$

(c) Given that $K_c > 1$ for the reaction in part (b), which Lewis base is stronger?

18.86 In which of the three concepts of acid-base behavior discussed in the text can water be a product of an acid-base reaction? In which is it the only product?

18.87 (a) Give an example of a *substance* that is a base in two of the three acid-base definitions, but not in the third.
(b) Give an example of a *substance* that is an acid in one of the three acid-base definitions, but not in the other two.

18.88 Which are Lewis acids and which are Lewis bases?
(a) Cu^{2+} (b) Cl^- (c) $SnCl_2$ (d) OF_2

18.89 Which are Lewis acids and which are Lewis bases?
(a) Na^+ (b) NH_3 (c) CN^- (d) BF_3

18.90 Identify the Lewis acid and Lewis base in each equation:
(a) $Na^+ + 6H_2O \rightleftharpoons Na(H_2O)_6^+$
(b) $CO_2 + H_2O \rightleftharpoons H_2CO_3$
(c) $F^- + BF_3 \rightleftharpoons BF_4^-$

18.91 Identify the Lewis acid and Lewis base in each equation:
(a) $Fe^{3+} + 2H_2O \rightleftharpoons FeOH^{2+} + H_3O^+$
(b) $H_2O + H^- \rightleftharpoons OH^- + H_2$
(c) $4CO + Ni \rightleftharpoons Ni(CO)_4$

18.92 Classify the following as Arrhenius, Brønsted-Lowry, or Lewis acid-base reactions. A reaction may fit all, two, one, or none of the categories:
(a) $Ag^+ + 2NH_3 \rightleftharpoons Ag(NH_3)_2^+$
(b) $H_2SO_4 + NH_3 \rightleftharpoons HSO_4^- + NH_4^+$
(c) $2HCl \rightleftharpoons H_2 + Cl_2$
(d) $AlCl_3 + Cl^- \rightleftharpoons AlCl_4^-$

18.93 Classify the following as Arrhenius, Brønsted-Lowry, or Lewis acid-base reactions. A reaction may fit all, two, one, or none of the categories:
(a) $Cu^{2+} + 4Cl^- \rightleftharpoons CuCl_4^{2-}$
(b) $Al(OH)_3 + 3HNO_3 \rightleftharpoons Al^{3+} + 3H_2O + 3NO_3^-$
(c) $N_2 + 3H_2 \rightleftharpoons 2NH_3$
(d) $CN^- + H_2O \rightleftharpoons HCN + OH^-$

Comprehensive Problems

Problems with an asterisk (*) are more challenging.

18.94 Bodily processes in humans maintain the pH of blood within a narrow range. In fact, a condition called *acidosis* occurs if the blood pH goes below 7.35, and another called *alkalosis* occurs if the pH goes above 7.45. Given that the pK_w of blood is 13.63 at 37°C (body temperature), what is the normal range of $[H_3O^+]$ and of $[OH^-]$ in blood?

18.95 When carbon dioxide dissolves in water, it undergoes a multistep equilibrium process, with $K_{overall} = 4.5 \times 10^{-7}$, which is simplified to the following:

$$CO_2(g) + H_2O(l) \rightleftharpoons H_2CO_3(aq)$$
$$H_2CO_3(aq) + H_2O(l) \rightleftharpoons HCO_3^-(aq) + H_3O^+(aq)$$

(a) Classify each step as a Lewis or a Brønsted-Lowry reaction.
(b) What is the pH of nonpolluted rainwater in equilibrium with clean air (P_{CO_2} in clean air $= 3.2 \times 10^{-4}$ atm; Henry's law constant for CO_2 at 25°C is 0.033 mol/L·atm)?

(c) What is $[CO_3^{2-}]$ in rainwater (K_a of $HCO_3^- = 4.7 \times 10^{-11}$)?
(d) If the partial pressure of CO_2 in clean air doubles in the next few decades, what will the pH of rainwater become?

* 18.96 Use Appendix C to calculate $[H_2C_2O_4]$, $[HC_2O_4^-]$, $[C_2O_4^{2-}]$, $[H_3O^+]$, pH, $[OH^-]$, and pOH in a 0.200 M solution of the diprotic acid oxalic acid. (*Hint:* Assume all the $[H_3O^+]$ comes from the first dissociation.)

18.97 Many molecules with central atoms from Period 3 or higher take part in Lewis acid-base reactions in which the central atom expands its valence shell. $SnCl_4$ reacts with $(CH_3)_3N$ as follows:

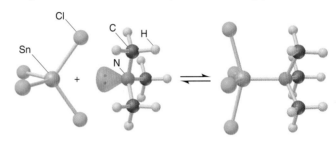

(a) Identify the Lewis acid and the Lewis base in the reaction.
(b) Give the nl designation of the sublevel of the central atom in the acid that accepts the lone pair.

18.98 A chemist makes four successive 1:10 dilutions of 1.0×10^{-5} M HCl. Calculate the pH of the original solution and of each diluted solution (through 1.0×10^{-9} M HCl).

18.99 Hydrogen peroxide, H_2O_2 ($pK_a = 11.75$), is commonly used as a bleaching agent and an antiseptic. The product sold in stores is 3% H_2O_2 by mass and contains 0.001% phosphoric acid by mass to stabilize the solution. Which contributes more H_3O^+ to this commercial solution, the H_2O_2 or the H_3PO_4?

18.100 Esters, RCOOR′, are formed by the reaction of carboxylic acids, RCOOH, and alcohols, R′OH, where R and R′ are hydrocarbon groups. Many esters are responsible for the odors of fruit and, thus, have important uses in the food and cosmetics industries. The first two steps in the mechanism of ester formation are

(1) $R{-}\overset{\displaystyle :O:}{\overset{\|}{C}}{-}\ddot{O}H \ + \ H^+ \ \rightleftharpoons \ R{-}\overset{\displaystyle :\ddot{O}H}{\underset{+}{C}}{-}\ddot{O}H$

(2) $R{-}\overset{\displaystyle :\ddot{O}H}{\underset{+}{C}}{-}\ddot{O}H \ + \ R'{-}\ddot{O}H \ \rightleftharpoons \ R{-}\overset{\displaystyle :\ddot{O}H}{\underset{\displaystyle \underset{H}{\overset{+}{:}\overset{}{O}{-}R'}}{C}}{-}\ddot{O}H$

Identify the Lewis acids and Lewis bases in these two steps.

18.101 Thiamine hydrochloride ($C_{12}H_{18}ON_4SCl_2$) is a water-soluble form of thiamine (vitamin B_1; $K_a = 3.37 \times 10^{-7}$). How many grams of the hydrochloride must be dissolved in 10.00 mL of water to give a pH of 3.50?

18.102 When Fe^{3+} salts are dissolved in water, the solution becomes acidic due to formation of $Fe(H_2O)_5OH^{2+}$ and H_3O^+. The overall process involves both Lewis and Brønsted-Lowry acid-base reactions. Write the equations for the process.

18.103 At 50°C and 1 atm, $K_w = 5.19 \times 10^{-14}$. Calculate parts (a)–(c) under these conditions:
(a) $[H_3O^+]$ in pure water
(b) $[H_3O^+]$ in 0.010 M NaOH

(c) $[OH^-]$ in 0.0010 M $HClO_4$

(d) Calculate $[H_3O^+]$ in 0.0100 M KOH at 100°C and 1000 atm pressure ($K_w = 1.10 \times 10^{-12}$).

(e) Calculate the pH of pure water at 100°C and 1000 atm.

* 18.104 A 1.000 m solution of chloroacetic acid ($ClCH_2COOH$) freezes at -1.93°C. Use these data to find the K_a of chloroacetic acid. (Assume the molarities equal the molalities.)

18.105 Calcium propionate [$Ca(CH_3CH_2COO)_2$] is a mold inhibitor used in food, tobacco, and pharmaceuticals. (a) Use balanced equations to show whether aqueous calcium propionate is acidic, basic, or neutral. (b) Use Appendix C to find the pH of a solution made by dissolving 7.05 g of $Ca(CH_3CH_2COO)_2$ in water to give 0.500 L of solution.

18.106 Carbon dioxide is less soluble in dilute HCl than in dilute NaOH. Explain.

18.107 (a) If $K_w = 1.139 \times 10^{-15}$ at 0°C and 5.474×10^{-14} at 50°C, find $[H_3O^+]$ and pH of water at 0°C and 50°C.

(b) The autoionization constant for heavy water (deuterium oxide, D_2O) is 3.64×10^{-16} at 0°C and 7.89×10^{-15} at 50°C. Find $[D_3O^+]$ and pD of heavy water at 0°C and 50°C.

(c) Suggest a reason for these differences.

* 18.108 HX ($\mathcal{M} = 150.$ g/mol) and HY ($\mathcal{M} = 50.0$ g/mol) are weak acids. A solution of 12.0 g/L of HX has the same pH as one containing 6.00 g/L of HY. Which is the stronger acid? Why?

* 18.109 Nitrogen is discharged from wastewater treatment facilities into rivers and streams, usually as NH_3 and NH_4^+:

$$NH_3(aq) + H_2O(l) \rightleftharpoons NH_4^+(aq) + OH^-(aq) \qquad K_b = 1.76 \times 10^{-5}$$

One strategy for removing it is to raise the pH and "strip" the NH_3 from solution by bubbling air through the water. (a) At pH 7.00, what fraction of the total nitrogen in solution is NH_3, defined as $[NH_3]/([NH_3] + [NH_4^+])$? (b) What is the fraction at pH 10.00? (c) Explain the basis of ammonia stripping.

18.110 Polymers and other large molecules are not very soluble in water, but their solubility increases if they have charged groups. (a) Casein is a protein in milk that contains many carboxylic acid groups on its side chains. Explain how the solubility of casein in water varies with pH.

(b) Histones are proteins that are essential to the proper function of DNA. They are weakly basic due to the presence of side chains with $-NH_2$ and $=NH$ groups. Explain how the solubility of histones in water varies with pH.

18.111 Hemoglobin (Hb) transports oxygen in the blood:

$$HbH^+(aq) + O_2(aq) + H_2O(l) \longrightarrow HbO_2(aq) + H_3O^+(aq)$$

In blood, $[H_3O^+]$ is held nearly constant at 4×10^{-8} M.

(a) How does the equilibrium position change in the lungs?

(b) How does it change in O_2-deficient cells?

(c) Excessive vomiting may lead to metabolic *alkalosis,* in which $[H_3O^+]$ in blood *decreases.* How does this condition affect the ability of Hb to transport O_2?

(d) Diabetes mellitus may lead to metabolic *acidosis,* in which $[H_3O^+]$ in blood *increases.* How does this condition affect the ability of Hb to transport O_2?

18.112 Vitamin C (ascorbic acid, $H_2C_6H_6O_6$) is a weak diprotic acid. It is essential for the synthesis of collagen, the major protein in connective tissue. If the pH of a 5.0% (w/v) solution of vitamin C in water is 2.77, calculate the K_{a1} of vitamin C.

* 18.113 Quinine ($C_{20}H_{24}N_2O_2$; *see below*) is a natural product with antimalarial properties that saved thousands of lives during construction of the Panama Canal. It stands as a classic example of the medicinal wealth of tropical forests. Both N atoms are basic, but the N (colored) of the 3° amine group is far more basic ($pK_b = 5.1$) than the N within the aromatic ring system ($pK_b = 9.7$).

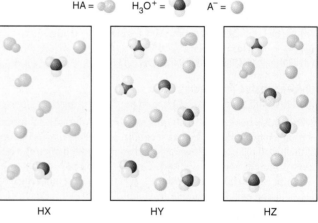

(a) Quinine is not very soluble in water: a saturated solution is only 1.6×10^{-3} M. What is the pH of this solution?

(b) Show that the aromatic N contributes negligibly to the pH of the solution.

(c) Because of its low solubility as a free base in water, quinine is given as an amine salt. For instance, quinine hydrochloride ($C_{20}H_{24}N_2O_2 \cdot HCl$) is about 120 times more soluble in water than quinine. What is the pH of 0.53 M quinine hydrochloride?

(d) An antimalarial concentration in water is 1.5% quinine hydrochloride by mass ($d = 1.0$ g/mL). What is the pH?

* 18.114 Drinking water is often disinfected with chlorine gas, which hydrolyzes to form hypochlorous acid (HClO), a weak acid but powerful disinfectant:

$$Cl_2(aq) + 2H_2O(l) \longrightarrow HClO(aq) + H_3O^+(aq) + Cl^-(aq)$$

The fraction of HClO in solution is defined as

$$\frac{[HClO]}{[HClO] + [ClO^-]}$$

(a) What is the fraction of HClO at pH 7.00 (K_a of HClO = 2.9×10^{-8})? (b) What is the fraction at pH 10.00?

* 18.115 The following scenes represent three weak acids HA (where A = X, Y, or Z) dissolved in water (H_2O is not shown):

HA = ⬤⬤ H_3O^+ = ⬤⬤ A^- = ⬤

HX   HY   HZ

(a) Rank the acids in order of increasing K_a.

(b) Rank the acids in order of increasing pK_a.

(c) Rank the conjugate bases in order of increasing pK_b.

(d) What is the percent dissociation of HX?

(e) If equimolar amounts of the sodium salts of the acids (NaX, NaY, and NaZ) were dissolved in water, which solution would have the highest pOH? The lowest pH?

CHAPTER NINETEEN

Ionic Equilibria in Aqueous Systems

Creating Coral Many remarkable natural formations, such as caves and reefs, arise through the subtle interplay of the aqueous ionic equilibria that you'll study in this chapter.

Key Principles

◆ For an aqueous ionic system at equilibrium, Le Châtelier's principle predicts that a substance dissociates less if one of the ions in the substance is already present in the solution. As a result of this *common-ion effect*, HA dissociates more in water than in a solution containing A^-. The A^- already present combines with some H_3O^+ and thus *lowers the acidity* of the HA solution.

◆ An *acid-base buffer* is a solution containing high concentrations of a conjugate acid-base pair. It resists changes in pH because of the common-ion effect: the conjugate base (or acid) component reacts with added H_3O^+ from a strong acid (or OH^- from a strong base) to keep pH relatively constant.

◆ A concentrated buffer has more *capacity* to resist a pH change than a dilute buffer. Buffers have a *range* of about 2 pH units, which corresponds to a value from 10 to 0.1 for the ratio $[A^-]/[HA]$.

◆ The *equivalence point* of a titration occurs when the moles of acid equal the moles of base. The pH at the equivalence point depends on the acid-base properties of the cation and the anion present: in a strong acid–strong base titration, the equivalence point is at pH 7; in a weak acid–strong base titration, pH > 7, and in a weak base–strong acid titration, pH < 7.

◆ The dissolution in water of a *slightly soluble ionic compound* reaches an equilibrium characterized by a *solubility-product constant, K_{sp},* that is much less than 1. Addition of a common ion lowers such a compound's solubility. Lowering the pH (adding H_3O^+) increases the solubility if the anion of the ionic compound is that of a weak acid.

◆ A *complex ion* consists of a central metal ion bonded to molecules or anions called *ligands*. Complex ions form in a stepwise process characterized by a *formation constant, K_f,* that is much greater than 1. Adding a ligand increases the solubility of a slightly soluble ionic compound if the ligand forms a complex ion with the ionic compound's cation.

Outline

Europa, one of Jupiter's moons, has an icy surface with hints of vast oceans of liquid water beneath. Is there life on Europa? If so, perhaps some Europan astronomer viewing Earth would be asking a similar question, because liquid water is essential for the aqueous systems that maintain life. Every astronaut has felt awe at seeing our "beautiful blue orb" from space. A biologist peering at the fabulous watery world of a living cell probably feels the same way. A chemist is awed by the principles of equilibrium and their universal application to aqueous solutions wherever they occur.

Consider just a few cases of aqueous equilibria. The magnificent formations in limestone caves and the vast expanses of oceanic coral reefs result from subtle shifts in carbonate solubility equilibria. Carbonates also influence soil pH and prevent acidification of lakes by acid rain. Equilibria involving carbon dioxide and phosphates help organisms maintain cellular pH within narrow limits. Equilibria involving clays in soils control the availability of ionic nutrients for plants. The principles of ionic equilibrium also govern how water is softened, how substances are purified by precipitation of unwanted ions, and even how the weak acids in wine and vinegar influence the delicate taste of a fine French sauce. In this chapter, we explore three aqueous ionic equilibrium systems: acid-base buffers, slightly soluble salts, and complex ions.

19.1 EQUILIBRIA OF ACID-BASE BUFFER SYSTEMS

Why do some lakes become acidic when showered by acid rain, while others remain unaffected? How does blood maintain a constant pH in contact with countless cellular acid-base reactions? How can a chemist sustain a nearly constant $[H_3O^+]$ in reactions that consume or produce H_3O^+ or OH^-? The answer in each case depends on the action of a buffer.

In everyday language, a buffer is something that lessens the impact of an external force. An **acid-base buffer** is a solution that *lessens the impact on pH from the addition of acid or base.* Figure 19.1 shows that a small amount of H_3O^+ or OH^- added to an unbuffered solution (or just water) changes the pH by several units. Note that, because of the logarithmic nature of pH, *this change is several orders of magnitude larger* than the change that results from the same addition to a buffered solution, shown in Figure 19.2. To withstand the addition of strong acid or strong base without significantly changing its pH, a buffer must contain an acidic component that can react with the added OH^- ion *and* a basic component that can react with added H_3O^+ ion. However, these buffer components cannot be just any acid and base because they would neutralize each other.

Figure 19.1 The effect of addition of acid or base to an unbuffered solution. A, A 100-mL sample of dilute HCl is adjusted to pH 5.00.

B, After the addition of 1 mL of 1 *M* HCl (*left*) or of 1 *M* NaOH (*right*), the pH changes by several units.

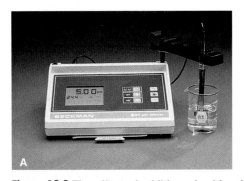

Figure 19.2 The effect of addition of acid or base to a buffered solution. A, A 100-mL sample of a buffered solution, made by mixing 1 M CH$_3$COOH with 1 M CH$_3$COONa, is adjusted to pH 5.00.

B, After the addition of 1 mL of 1 M HCl (*left*) or of 1 M NaOH (*right*), the pH change is negligibly small. Compare these changes with those in Figure 19.1.

Most commonly, *the components of a buffer are the conjugate acid-base pair of a weak acid.* The buffer used in Figure 19.2, for example, is a mixture of acetic acid (CH_3COOH) and acetate ion (CH_3COO^-).

How a Buffer Works: The Common-Ion Effect

Buffers work through a phenomenon known as the **common-ion effect.** An example of this effect occurs when acetic acid dissociates in water and some sodium acetate is added. As you know, acetic acid dissociates only slightly in water:

$$CH_3COOH(aq) + H_2O(l) \rightleftharpoons H_3O^+(aq) + CH_3COO^-(aq)$$

From Le Châtelier's principle (Section 17.6), we know that if some CH_3COO^- ion is added (from the soluble sodium acetate), the equilibrium position shifts to the left; thus, $[H_3O^+]$ decreases, in effect lowering the extent of acid dissociation:

$$CH_3COOH(aq) + H_2O(l) \overset{\longleftarrow}{\rightleftharpoons} H_3O^+(aq) + CH_3COO^-(aq; \text{ added})$$

Similarly, if we dissolve acetic acid *in* a sodium acetate solution, acetate ion and H_3O^+ ion from the acid enter the solution. The acetate ion already present in the solution acts to suppress as much acid from dissociating, which lowers $[H_3O^+]$. Thus, the effect again is to lower the acid dissociation. Acetate ion is called *the common ion* in this case because it is "common" to both the acetic acid and sodium acetate solutions; that is, acetate ion from the acid enters a solution in which it is already present. *The common-ion effect occurs when a given ion is added to an equilibrium mixture that already contains that ion, and the position of equilibrium shifts away from forming more of it.*

Table 19.1 shows the percent dissociation and the pH of an acetic acid solution containing various concentrations of acetate ion (supplied from solid sodium acetate). Note that the *common ion, CH_3COO^-, suppresses the dissociation of CH_3COOH,* which makes the solution less acidic (higher pH).

Table 19.1	**The Effect of Added Acetate Ion on the Dissociation of Acetic Acid**		
$[CH_3COOH]_{init}$	**$[CH_3COO^-]_{added}$**	**% Dissociation***	**pH**
0.10	0.00	1.3	2.89
0.10	0.050	0.036	4.44
0.10	0.10	0.018	4.74
0.10	0.15	0.012	4.92

*% Dissociation $= \dfrac{[CH_3COOH]_{dissoc}}{[CH_3COOH]_{init}} \times 100$

The Essential Feature of a Buffer In the previous example, we prepared a buffer by mixing a weak acid (CH_3COOH) and its conjugate base (CH_3COO^-). *How does this solution resist pH changes when H_3O^+ or OH^- is added?* The essential feature of a buffer is that *it consists of high concentrations of the acidic (HA) and basic (A^-) components.* When small amounts of H_3O^+ or OH^- ions are added to the buffer, they cause *a small amount of one buffer component to convert into the other,* which changes the relative concentrations of the two components. As long as the amount of H_3O^+ or OH^- added is much smaller than the amounts of HA and A^- present originally, *the added ions have little effect on the pH because they are consumed by one or the other buffer component:* a large excess of A^- ions react with any added H_3O^+, and a large excess of HA molecules react with any added OH^-.

Consider what happens to a solution containing high [CH_3COOH] and high [CH_3COO^-] when we add small amounts of strong acid or base. The expression for HA dissociation at equilibrium is

$$K_a = \frac{[CH_3COO^-][H_3O^+]}{[CH_3COOH]}$$

Solving for [H_3O^+] gives

$$[H_3O^+] = K_a \times \frac{[CH_3COOH]}{[CH_3COO^-]}$$

Note that because K_a is constant, *the [H_3O^+] of the solution depends directly on the buffer-component concentration ratio,* $\dfrac{[CH_3COOH]}{[CH_3COO^-]}$:

- If the ratio [HA]/[A^-] goes up, [H_3O^+] goes up.
- If the ratio [HA]/[A^-] goes down, [H_3O^+] goes down.

When we add a small amount of strong acid, the increased amount of H_3O^+ ion reacts with a *stoichiometric amount* of acetate ion from the buffer to form more acetic acid:

$$H_3O^+(aq; \text{ added}) + CH_3COO^-(aq; \text{ from buffer}) \longrightarrow CH_3COOH(aq) + H_2O(l)$$

As a result, [CH_3COO^-] goes down by that amount and [CH_3COOH] goes up by that amount, which increases the buffer-component concentration ratio, as you can see in Figure 19.3. The [H_3O^+] also increases but only very slightly.

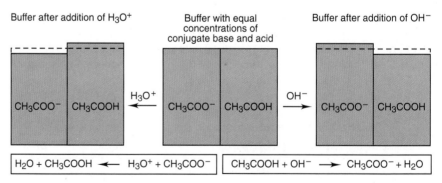

Figure 19.3 How a buffer works. A buffer consists of high concentrations of a conjugate acid-base pair, in this case, acetic acid (CH_3COOH) and acetate ion (CH_3COO^-). When a small amount of H_3O^+ is added (*left*), that same amount of CH_3COO^- combines with it, which increases the amount of CH_3COOH slightly. Similarly, when a small amount of OH^- is added (*right*), that amount of CH_3COOH combines with it, which increases the amount of CH_3COO^- slightly. In both cases, the relative changes in amounts of the buffer components are small, so their concentration ratio, and therefore the pH, changes very little.

Adding a small amount of strong base produces the opposite result. It supplies OH^- ions, which react with a *stoichiometric amount* of CH_3COOH from the buffer, forming that much more CH_3COO^-:

$$CH_3COOH(aq;\ from\ buffer) + OH^-(aq;\ added) \longrightarrow CH_3COO^-(aq) + H_2O(l)$$

The buffer-component concentration ratio decreases, which decreases $[H_3O^+]$, but once again, the change is very slight.

Thus, the buffer components consume virtually all the added H_3O^+ or OH^-. To reiterate, as long as the amount of added H_3O^+ or OH^- is small compared with the amounts of the buffer components, *the conversion of one component into the other produces a small change in the buffer-component concentration ratio and, consequently, a small change in [H₃O⁺] and in pH.* Sample Problem 19.1 demonstrates how small these pH changes typically are. Note that the latter two parts of the problem combine a stoichiometry portion, like the problems in Chapter 3, and a weak-acid dissociation portion, like those in Chapter 18.

SAMPLE PROBLEM 19.1 Calculating the Effect of Added H_3O^+ or OH^- on Buffer pH

Problem Calculate the pH:
(a) Of a buffer solution consisting of 0.50 M CH_3COOH and 0.50 M CH_3COONa
(b) After adding 0.020 mol of solid NaOH to 1.0 L of the buffer solution in part (a)
(c) After adding 0.020 mol of HCl to 1.0 L of the buffer solution in part (a)
K_a of $CH_3COOH = 1.8 \times 10^{-5}$. (Assume the additions cause negligible volume changes.)
Plan In each case, we know, or can find, $[CH_3COOH]_{init}$ and $[CH_3COO^-]_{init}$ and the K_a of CH_3COOH (1.8×10^{-5}), and we need to find $[H_3O^+]$ at equilibrium and convert it to pH. In (a), we use the given concentrations of buffer components (each 0.50 M) as the initial values. As in earlier problems, we assume that x, the $[CH_3COOH]$ that dissociates, which equals $[H_3O^+]$, is so small relative to $[CH_3COOH]_{init}$ that it can be neglected. We set up a reaction table, solve for x, and check the assumption. In (b) and (c), we assume that the added OH^- or H_3O^+ reacts completely with the buffer components to yield new $[CH_3COOH]_{init}$ and $[CH_3COO^-]_{init}$, which then dissociate to an unknown extent. We set up two reaction tables. The first summarizes the stoichiometry of adding strong base (0.020 mol) or acid (0.020 mol). The second summarizes the dissociation of the new HA concentrations, so we proceed as in part (a) to find the new $[H_3O^+]$.
Solution (a) The original pH: $[H_3O^+]$ in the original buffer.
Setting up a reaction table with $x = [CH_3COOH]_{dissoc} = [H_3O^+]$ (as in Chapter 18, we assume that $[H_3O^+]$ from H_2O is negligible and disregard it):

Concentration (M)	$CH_3COOH(aq)$	+	$H_2O(l)$	$\rightleftharpoons$	$CH_3COO^-(aq)$	+	$H_3O^+(aq)$
Initial	0.50		—		0.50		0
Change	$-x$		—		$+x$		$+x$
Equilibrium	$0.50 - x$		—		$0.50 + x$		x

Making the assumption and finding the equilibrium $[CH_3COOH]$ and $[CH_3COO^-]$: With K_a small, x is small, so we assume

$$[CH_3COOH] = 0.50\ M - x \approx 0.50\ M \quad and \quad [CH_3COO^-] = 0.50\ M + x \approx 0.50\ M$$

Solving for x ($[H_3O^+]$ at equilibrium):

$$x = [H_3O^+] = K_a \times \frac{[CH_3COOH]}{[CH_3COO^-]} \approx (1.8 \times 10^{-5}) \times \frac{0.50}{0.50} = 1.8 \times 10^{-5}\ M$$

Checking the assumption:

$$\frac{1.8 \times 10^{-5}\ M}{0.50\ M} \times 100 = 3.6 \times 10^{-3}\% < 5\%$$

The assumption is justified, and we will use the same assumption in parts (b) and (c).
Calculating pH:

$$pH = -\log [H_3O^+] = -\log (1.8 \times 10^{-5}) = 4.74$$

(b) The pH after adding base (0.020 mol of NaOH to 1.0 L of buffer). Finding $[OH^-]_{added}$:

$$[OH^-]_{added} = \frac{0.020 \text{ mol } OH^-}{1.0 \text{ L soln}} = 0.020 \text{ } M \text{ } OH^-$$

Setting up a reaction table for the *stoichiometry* of adding OH^- to CH_3COOH:

Concentration (M)	$CH_3COOH(aq)$	+	$OH^-(aq)$	$\longrightarrow$	$CH_3COO^-(aq)$	+	$H_2O(aq)$
Before addition	0.50		—		0.50		—
Addition	—		0.020		—		—
After addition	0.48		0		0.52		—

Setting up a reaction table for the *acid dissociation,* using these new initial concentrations. As in part (a), $x = [CH_3COOH]_{dissoc} = [H_3O^+]$:

Concentration (M)	$CH_3COOH(aq)$	+	$H_2O(l)$	$\rightleftharpoons$	$CH_3COO^-(aq)$	+	$H_3O^+(aq)$
Initial	0.48		—		0.52		0
Change	$-x$		—		$+x$		$+x$
Equilibrium	$0.48 - x$		—		$0.52 + x$		x

Making the assumption that x is small, and solving for x:

$$[CH_3COOH] = 0.48 \text{ } M - x \approx 0.48 \text{ } M \quad \text{and} \quad [CH_3COO^-] = 0.52 \text{ } M + x \approx 0.52 \text{ } M$$

$$x = [H_3O^+] = K_a \times \frac{[CH_3COOH]}{[CH_3COO^-]} \approx (1.8 \times 10^{-5}) \times \frac{0.48}{0.52} = 1.7 \times 10^{-5} \text{ } M$$

Calculating the pH:

$$pH = -\log [H_3O^+] = -\log (1.7 \times 10^{-5}) = 4.77$$

The addition of strong base increased the concentration of the basic buffer component at the expense of the acidic buffer component. Note especially that the pH *increased only slightly,* from 4.74 to 4.77.

(c) The pH after adding acid (0.020 mol of HCl to 1.0 L of buffer). Finding $[H_3O^+]_{added}$:

$$[H_3O^+]_{added} = \frac{0.020 \text{ mol } H_3O^+}{1.0 \text{ L soln}} = 0.020 \text{ } M \text{ } H_3O^+$$

Now we proceed as in part (b), by first setting up a reaction table for the *stoichiometry* of adding H_3O^+ to CH_3COO^-:

Concentration (M)	$CH_3COO^-(aq)$	+	$H_3O^+(aq)$	$\longrightarrow$	$CH_3COOH(aq)$	+	$H_2O(l)$
Before addition	0.50		—		0.50		—
Addition	—		0.020		—		—
After addition	0.48		0		0.52		—

The reaction table for the acid dissociation, with $x = [CH_3COOH]_{dissoc} = [H_3O^+]$ is

Concentration (M)	$CH_3COOH(aq)$	+	$H_2O(l)$	$\rightleftharpoons$	$CH_3COO^-(aq)$	+	$H_3O^+(aq)$
Initial	0.52		—		0.48		0
Change	$-x$		—		$+x$		$+x$
Equilibrium	$0.52 - x$		—		$0.48 + x$		x

Making the assumption that x is small, and solving for x:

$$[CH_3COOH] = 0.52 \text{ } M - x \approx 0.52 \text{ } M \quad \text{and} \quad [CH_3COO^-] = 0.48 \text{ } M + x \approx 0.48 \text{ } M$$

$$x = [H_3O^+] = K_a \times \frac{[CH_3COOH]}{[CH_3COO^-]} \approx (1.8 \times 10^{-5}) \times \frac{0.52}{0.48} = 2.0 \times 10^{-5} \text{ } M$$

Calculating the pH:

$$pH = -\log [H_3O^+] = -\log (2.0 \times 10^{-5}) = 4.70$$

The addition of strong acid increased the concentration of the acidic buffer component at the expense of the basic buffer component and *lowered* the pH only slightly, from 4.74 to 4.70.

Check The changes in $[CH_3COOH]$ and $[CH_3COO^-]$ occur in opposite directions in parts (b) and (c), which makes sense. The additions were of equal amounts, so the pH increase in (b) should equal the pH decrease in (c), within rounding.

Comment In part (a), we justified our assumption that x can be neglected. Therefore, in parts (b) and (c), we could have used the "After addition" values from the last line of the stoichiometry tables directly for the ratio of buffer components; that would have allowed us to dispense with a reaction table for the dissociation. In subsequent problems in this chapter, we will follow this simplified approach.

FOLLOW-UP PROBLEM 19.1 Calculate the pH of a buffer consisting of 0.50 M HF and 0.45 M F^- **(a)** before and **(b)** after addition of 0.40 g of NaOH to 1.0 L of the buffer (K_a of HF = 6.8×10^{-4}).

The Henderson-Hasselbalch Equation

For any weak acid, HA, the dissociation equation and K_a expression are

$$HA + H_2O \rightleftharpoons H_3O^+ + A^- \qquad K_a = \frac{[H_3O^+][A^-]}{[HA]}$$

The key variable that determines $[H_3O^+]$ is the concentration *ratio* of acid species to base species, so rearranging to isolate $[H_3O^+]$ gives

$$[H_3O^+] = K_a \times \frac{[HA]}{[A^-]}$$

Taking the negative common logarithm (base 10) of both sides gives

$$-\log [H_3O^+] = -\log K_a - \log \left(\frac{[HA]}{[A^-]}\right)$$

from which we obtain

$$pH = pK_a + \log \left(\frac{[A^-]}{[HA]}\right)$$

(Note the inversion of the buffer-component concentration ratio when the sign of the logarithm is changed.) A key point we'll emphasize again later is that when $[A^-]$ = $[HA]$, their ratio becomes 1; the log term then becomes 0, and thus pH = pK_a.

Generalizing the previous equation for any conjugate acid-base pair gives the **Henderson-Hasselbalch equation:**

$$pH = pK_a + \log \left(\frac{[base]}{[acid]}\right) \qquad (19.1)$$

This relationship is very useful for two reasons. First, it allows us to solve directly for pH instead of having to calculate $[H_3O^+]$ first. For instance, by applying the Henderson-Hasselbalch equation in part (b) of Sample Problem 19.1, we could have found the pH of the buffer after the addition of NaOH as follows:

$$pH = pK_a + \log \left(\frac{[CH_3COO^-]}{[CH_3COOH]}\right) = 4.74 + \log \left(\frac{0.52}{0.48}\right) = 4.77$$

Second, as we'll see shortly, it allows us to prepare a buffer of a desired pH just by mixing the appropriate amounts of A^- and HA.

Buffer Capacity and Buffer Range

As you've seen, a buffer resists a pH change as long as the concentrations of buffer components are *large* compared with the amount of strong acid or base added. **Buffer capacity** is a measure of this ability to resist pH change and depends on both the absolute and relative component concentrations.

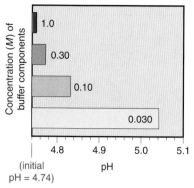

Figure 19.4 The relation between buffer capacity and pH change. The four bars in the graph represent CH_3COOH-CH_3COO^- buffers with the same initial pH (4.74) but different component concentrations (labeled on or near each bar). When a given amount of strong base is added to each buffer, the pH increases. The length of the bar corresponds to the pH increase. Note that the more concentrated the buffer, the greater its capacity, and the smaller the pH change.

In absolute terms, *the more concentrated the components of a buffer, the greater the buffer capacity.* In other words, you must add more H_3O^+ or OH^- to a high-capacity (concentrated) buffer than to a low-capacity (dilute) buffer to obtain the same pH change. Conversely, adding the same amount of H_3O^+ or OH^- to buffers of different capacities produces a smaller pH change in the higher capacity buffer (Figure 19.4). It's important to realize that *the pH of a buffer is distinct from its buffer capacity.* A buffer made of equal volumes of 1.0 M CH_3COOH and 1.0 M CH_3COO^- has the same pH (4.74) as a buffer made of equal volumes of 0.10 M CH_3COOH and 0.10 M CH_3COO^-, but the more concentrated buffer has a much larger capacity for resisting a pH change.

Buffer capacity is also affected by the *relative* concentrations of the buffer components. As a buffer functions, the concentration of one component increases relative to the other. Because the ratio of these concentrations determines the pH, the less the ratio changes, the less the pH changes. *For a given addition of acid or base, the buffer-component concentration ratio changes less when the concentrations are similar than when they are different.* Suppose that a buffer has $[HA] = [A^-] = 1.000$ M. When we add 0.010 mol of OH^- to 1.00 L of buffer, $[A^-]$ becomes 1.010 M and $[HA]$ becomes 0.990 M:

$$\frac{[A^-]_{init}}{[HA]_{init}} = \frac{1.000\ M}{1.000\ M} = 1.000$$

$$\frac{[A^-]_{final}}{[HA]_{final}} = \frac{1.010\ M}{0.990\ M} = 1.02$$

$$\text{Percent change} = \frac{1.02 - 1.000}{1.000} \times 100 = 2\%$$

Now suppose that the component concentrations are $[HA] = 0.250$ M and $[A^-] = 1.750$ M. The same addition of 0.010 mol of OH^- to 1.00 L of buffer gives $[HA] = 0.240$ M and $[A^-] = 1.760$ M, so the ratios are

$$\frac{[A^-]_{init}}{[HA]_{init}} = \frac{1.750\ M}{0.250\ M} = 7.00$$

$$\frac{[A^-]_{final}}{[HA]_{final}} = \frac{1.760\ M}{0.240\ M} = 7.33$$

$$\text{Percent change} = \frac{7.33 - 7.00}{7.00} \times 100 = 4.7\%$$

As you can see, the change in the buffer-component concentration ratio is much larger when the initial concentrations of the components are very different.

It follows that *a buffer has the highest capacity when the component concentrations are equal,* that is, when $[A^-]/[HA] = 1$:

$$pH = pK_a + \log\left(\frac{[A^-]}{[HA]}\right) = pK_a + \log 1 = pK_a + 0 = pK_a$$

Note this important result: for a given concentration, *a buffer whose pH is equal to or near the pK_a of its acid component has the highest buffer capacity.*

The **buffer range** is the pH range over which the buffer acts effectively, and it is related to the relative component concentrations. The further the buffer-component concentration ratio is from 1, the less effective the buffering action (that is, the lower the buffer capacity). In practice, if the $[A^-]/[HA]$ ratio is greater than 10 or less than 0.1—that is, if one component concentration is more than 10 times the other—buffering action is poor. Recalling that log 10 = +1 and log 0.1 = −1, we find that *buffers have a usable range within ±1 pH unit of the pK_a of the acid component:*

$$pH = pK_a + \log\left(\frac{10}{1}\right) = pK_a + 1 \qquad \text{and} \qquad pH = pK_a + \log\left(\frac{1}{10}\right) = pK_a - 1$$

Preparing a Buffer

Chemical supply houses sell buffers having a variety of pH values and concentrations, but chemists or lab technicians often have to prepare a buffer solution for a specific environmental or biomedical application. Several steps are required to prepare a buffer:

1. *Decide on the conjugate acid-base pair.* The choice is determined mostly by the desired pH. Remember that a buffer is most effective when the buffer-component concentration ratio is close to 1; in that case, the pH is close to the pK_a of the acid. Convert pK_a to K_a, choose the acid from a list, such as that in Appendix C, and use the sodium salt as the conjugate base.
2. *Find the ratio of [A⁻]/[HA] that gives the desired pH, using the Henderson-Hasselbalch equation.* Note that, because HA is a weak acid, and thus dissociates very little, the equilibrium concentrations are approximately equal to the initial concentrations; that is,

$$\text{pH} = pK_a + \log \left(\frac{[A^-]}{[HA]} \right) \approx pK_a + \log \left(\frac{[A^-]_{init}}{[HA]_{init}} \right)$$

Therefore, you can use the ratio directly in the next step.
3. *Choose the buffer concentration and calculate the amounts to mix.* Remember that the higher the concentration, the greater the buffer capacity. For most laboratory applications, concentrations from 0.05 *M* to 0.5 *M* are suitable. From a given amount (usually in the form of concentration and volume) of one component, find the amount of the other component using the buffer-component concentration ratio.
4. *Mix the amounts together and adjust the buffer pH to the desired value.* Add small amounts of strong acid or strong base, while monitoring the solution with a pH meter.

The following sample problem goes through steps 2 and 3.

SAMPLE PROBLEM 19.2 Preparing a Buffer

Problem An environmental chemist needs a carbonate buffer of pH 10.00 to study the effects of the acid rain on limestone-rich soils. How many grams of Na_2CO_3 must she add to 1.5 L of freshly prepared 0.20 *M* $NaHCO_3$ to make the buffer? K_a of HCO_3^- is 4.7×10^{-11}.
Plan The conjugate pair is already chosen, HCO_3^- (acid) and CO_3^{2-} (base), as are the volume (1.5 L) and concentration (0.20 *M*) of HCO_3^-, so we must find the buffer-component concentration ratio that gives pH 10.00 and the mass of Na_2CO_3 to dissolve. We find the amount (mol) of $NaHCO_3$ and use the ratio to find the amount (mol) of Na_2CO_3, which we convert to mass (g) using the molar mass.
Solution Solving for the buffer-component concentration ratio:

$$\text{pH} = pK_a + \log \left(\frac{[CO_3^{2-}]}{[HCO_3^-]} \right) \quad \text{or} \quad 10.00 = 10.33 + \log \left(\frac{[CO_3^{2-}]}{[HCO_3^-]} \right)$$

$$\log \left(\frac{[CO_3^{2-}]}{[HCO_3^-]} \right) = -0.33 \quad \text{so} \quad \frac{[CO_3^{2-}]}{[HCO_3^-]} = 0.47$$

Calculating the amount (mol) and mass (g) of Na_2CO_3 that will give the needed 0.47:1.0 ratio:

$$\text{Amount (mol) of } NaHCO_3 = 1.5 \text{ L} \times \frac{0.20 \text{ M } NaHCO_3}{1.0 \text{ L soln}} = 0.30 \text{ mol } NaHCO_3$$

$$\text{Amount (mol) of } Na_2CO_3 = 0.30 \text{ mol } NaHCO_3 \times \frac{0.47 \text{ mol } Na_2CO_3}{1 \text{ mol } NaHCO_3}$$

$$= 0.14 \text{ mol } Na_2CO_3$$

$$\text{Mass (g) of } Na_2CO_3 = 0.14 \text{ mol } Na_2CO_3 \times \frac{105.99 \text{ g } Na_2CO_3}{1 \text{ mol } Na_2CO_3} = 15 \text{ g } Na_2CO_3$$

We dissolve 15 g of Na_2CO_3 into approximately 1.3 L of 0.20 M $NaHCO_3$ and add 0.20 M $NaHCO_3$ to make 1.5 L. Using a pH meter, we adjust the pH to 10.00 with strong acid or base.

Check For a useful buffer range, the concentration of the acidic component, $[HCO_3^-]$, must be within a factor of 10 of the concentration of the basic component, $[CO_3^{2-}]$. We have 0.30 mol of HCO_3^- and 0.14 mol of CO_3^{2-}; 0.30:0.14 = 2.1, which seems fine. Make sure the relative amounts of components seem reasonable: we want a pH lower than the pK_a of HCO_3^- (10.33), so it makes sense that we have more of the acidic than the basic species.

Comment In the follow-up problem, we use an alternative calculation that does not rely on the Henderson-Hasselbalch equation.

FOLLOW-UP PROBLEM 19.2 How would you prepare a benzoic acid–benzoate buffer with pH = 4.25, starting with 5.0 L of 0.050 M sodium benzoate (C_6H_5COONa) solution and adding the acidic component? K_a of benzoic acid (C_6H_5COOH) is 6.3×10^{-5}.

Another way to prepare a buffer is to form one of the components during the final mixing step by *partial neutralization* of the other component. For example, suppose you need a $HCOOH$-$HCOO^-$ buffer. You can prepare it by mixing appropriate amounts of $HCOOH$ solution and $NaOH$ solution. As the OH^- ions react with the $HCOOH$ molecules, neutralization of part of the total $HCOOH$ present produces the $HCOO^-$ needed:

$HCOOH$ (HA total) + OH^- (amt added) $\longrightarrow$

$HCOOH$ (HA total $-$ OH^- amt added) + $HCOO^-$ (OH^- amt added) + H_2O

This method is based on the same chemical process that occurs when a weak acid is titrated with a strong base, as you'll see in the next section.

SECTION SUMMARY

A buffered solution exhibits a much smaller change in pH when H_3O^+ or OH^- is added than does an unbuffered solution. A buffer consists of relatively high concentrations of the components of a conjugate weak acid–base pair. The buffer-component concentration ratio determines the pH, and the ratio and pH are related by the Henderson-Hasselbalch equation. As H_3O^+ or OH^- is added, one buffer component reacts with it and is converted into the other component; therefore, the buffer-component concentration ratio, and consequently the $[H_3O^+]$ (and pH), changes only slightly. A concentrated buffer undergoes smaller changes in pH than a dilute buffer. When the buffer pH equals the pK_a of the acid component, the buffer has its highest capacity. A buffer has an effective range of $pK_a \pm 1$ pH unit. To prepare a buffer, you choose the conjugate acid-base pair, calculate the ratio of buffer components, determine buffer concentration, and adjust the final solution to the desired pH.

19.2 ACID-BASE TITRATION CURVES

In Chapter 4, we discussed the acid-base titration as an analytical method. Let's re-examine it, this time tracking the change in pH with an **acid-base titration curve,** a plot of pH vs. volume of titrant added. The behavior of an acid-base indicator and its role in the titration are described first. To better understand the titration process, we apply the principles of the acid-base behavior of salt solutions (Section 18.7) and, later in the section, the principles of buffer action.

Monitoring pH with Acid-Base Indicators

The two common devices for measuring pH in the laboratory are pH meters and acid-base indicators. (We discuss the operation of pH meters in Chapter 21.) An *acid-base indicator* is a weak organic acid (denoted here as HIn) that has a different color than its conjugate base (In^-), with the color change occurring over

pH

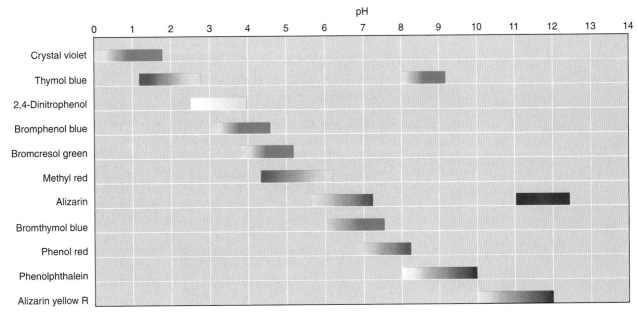

Figure 19.5 Colors and approximate pH range of some common acid-base indicators. Most have a range of about 2 pH units, in keeping with the useful buffer range of 2 pH units ($pK_a \pm 1$). (The pH range depends to some extent on the solvent used to prepare the indicator.)

a specific and relatively narrow pH range. Typically, one or both of the forms are intensely colored, so only a tiny amount of indicator is needed, far too little to affect the pH of the solution being studied.

Figure 19.5 shows the color changes and their pH ranges for some common acid-base indicators. Selecting an indicator requires that you know the approximate pH of the titration end point, which in turn requires that you know which ionic species are present. Because the indicator molecule is a weak acid, the ratio of the two forms is governed by the $[H_3O^+]$ of the test solution:

$$HIn(aq) + H_2O(l) \rightleftharpoons H_3O^+(aq) + In^-(aq) \qquad K_a \text{ of } HIn = \frac{[H_3O^+][In^-]}{[HIn]}$$

Therefore,
$$\frac{[HIn]}{[In^-]} = \frac{[H_3O^+]}{K_a}$$

How we perceive colors has a major influence on the use of indicators. Typically, the experimenter will see the HIn color if the $[HIn]/[In^-]$ ratio is 10:1 or greater and the In^- color if the $[HIn]/[In^-]$ ratio is 1:10 or less. Between these extremes, the colors of the two forms merge into an intermediate hue. Therefore, an indicator has a *color range* that reflects a 100-fold range in the $[HIn]/[In]$ ratio, which means that an *indicator changes color over a range of about 2 pH units.* For example, as you can see in Figure 19.5, bromthymol blue has a pH range of about 6.0 to 7.6 and, as Figure 19.6 shows, it is yellow below that range, blue above it, and greenish in between.

Figure 19.6 The color change of the indicator bromthymol blue. The acidic form of bromthymol blue is yellow (*left*) and the basic form is blue (*right*). Over the pH range in which the indicator is changing, both forms are present, so the mixture appears greenish (*center*).

Strong Acid–Strong Base Titration Curves

A typical curve for the titration of a strong acid with a strong base appears in Figure 19.7, along with the data used to construct it.

Features of the Curve   There are three distinct regions of the curve, which correspond to three major changes in slope:

1. The pH starts out low, reflecting the high $[H_3O^+]$ of the strong acid, and increases slowly as acid is gradually neutralized by the added base.
2. Suddenly, the pH rises steeply. This rise begins when the moles of OH^- that have been added nearly equal the moles of H_3O^+ originally present in the acid. An additional drop or two of base neutralizes the final tiny excess of acid and introduces a tiny excess of base, so the pH jumps 6 to 8 units.
3. Beyond this steep portion, the pH increases slowly as more base is added.

The **equivalence point,** which occurs within the nearly vertical portion of the curve, is the point at which the *number of moles of added OH^- equals the number of moles of H_3O^+* originally present. At the equivalence point of a strong acid–strong base titration, *the solution consists of the anion of the strong acid and the cation of the strong base.* Recall from Chapter 18 that *these ions do not react with water, so the solution is neutral: pH = 7.00.* The volume and concentration of base needed to reach the equivalence point allow us to calculate the amount of acid originally present (see Sample Problem 4.5).

Before the titration begins, we add a few drops of an appropriate indicator to the acid solution to signal when we reach the equivalence point. The **end point** of the titration occurs when the indicator changes color. *We choose an indicator with an end point close to the equivalence point,* one that changes color in the pH range on the steep vertical portion of the curve. Figure 19.7 shows the color changes for two indicators that are suitable for a strong acid–strong base titration. Methyl red changes from red at pH 4.2 to yellow at pH 6.3, whereas phenolphthalein changes from colorless at pH 8.3 to pink at pH 10.0. Even though neither color change occurs *at* the equivalence point (pH 7.00), both occur on the vertical portion of the curve, where a single drop of base causes a large pH change: when methyl red turns yellow, or when phenolphthalein turns pink, we know we are within a drop

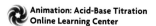

Animation: Acid-Base Titration
Online Learning Center

Figure 19.7 Curve for a strong acid–strong base titration. A, Data obtained from the titration of 40.00 mL of 0.1000 *M* HCl with 0.1000 *M* NaOH. **B,** Acid-base titration curve from data in part A. The pH increases gradually at first. When the amount (mol) of OH^- added is slightly less than the amount (mol) of H_3O^+ originally present, a large pH change accompanies a small addition of OH^-. The equivalence point occurs when amount (mol) of OH^- added = amount (mol) of H_3O^+ originally present. Note that, for a strong acid–strong base titration, pH = 7.00 at the equivalence point. Added before the titration begins, either methyl red or phenolphthalein is a suitable indicator in this case because each changes color on the steep portion of the curve, as shown by the color strips. Photos showing the color changes from 1–2 drops of indicator appear nearby. Beyond this point, added OH^- causes a gradual pH increase again.

Volume of NaOH added (mL)	pH
00.00	1.00
10.00	1.22
20.00	1.48
30.00	1.85
35.00	2.18
39.00	2.89
39.50	3.20
39.75	3.50
39.90	3.90
39.95	4.20
39.99	4.90
40.00	7.00
40.01	9.10
40.05	9.80
40.10	10.10
40.25	10.50
40.50	10.79
41.00	11.09
45.00	11.76
50.00	12.05
60.00	12.30
70.00	12.43
80.00	12.52

A

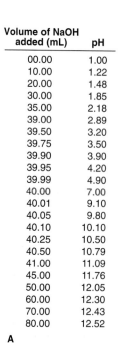

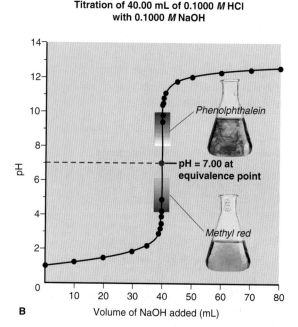

Titration of 40.00 mL of 0.1000 *M* HCl with 0.1000 *M* NaOH

Phenolphthalein

pH = 7.00 at equivalence point

Methyl red

B

Volume of NaOH added (mL)

or two of the equivalence point. For example, in going from 39.90 to 39.99 mL, one to two drops, the pH changes one whole unit. For all practical purposes, then, the *visible* change in color of the indicator (end point) signals the *invisible* point at which moles of added base equal the original moles of acid (equivalence point).

Calculating the pH By knowing the chemical species present during the titration, we can calculate the pH at various points along the way:

1. *Original solution of strong HA.* In Figure 19.7, 40.00 mL of 0.1000 M HCl is titrated with 0.1000 M NaOH. Because a strong acid is completely dissociated, $[HCl] = [H_3O^+] = 0.1000 \, M$. Therefore, the initial pH is*

$$pH = -\log [H_3O^+] = -\log (0.1000) = 1.00$$

2. *Before the equivalence point.* As soon as we start adding base, two changes occur that affect the pH calculations: (1) some acid is neutralized, and (2) the volume of solution increases. To find the pH at various points up to the equivalence point, we find the initial *amount (mol)* of H_3O^+ present, subtract the amount reacted, *which equals the amount (mol) of OH^- added,* and then use the change in volume to calculate the *concentration,* $[H_3O^+]$, and convert to pH. For example, after adding 20.00 mL of 0.1000 M NaOH:

- *Find the moles of H_3O^+ remaining.* Subtracting the number of moles of H_3O^+ reacted from the number initially present gives the number remaining. Moles of H_3O^+ reacted equals moles of OH^- added, so

$$\text{Initial moles of } H_3O^+ = 0.04000 \text{ L} \times 0.1000 \, M = 0.004000 \text{ mol } H_3O^+$$
$$\underline{-\text{Moles of } OH^- \text{ added} = 0.02000 \text{ L} \times 0.1000 \, M = 0.002000 \text{ mol } OH^-}$$
$$\text{Moles of } H_3O^+ \text{ remaining} = \qquad\qquad\qquad 0.002000 \text{ mol } H_3O^+$$

- *Calculate $[H_3O^+]$, taking the total volume into account.* To find the ion concentrations, we use the *total volume* because the water of one solution dilutes the ions of the other:

$$[H_3O^+] = \frac{\text{amount (mol) of } H_3O^+ \text{ remaining}}{\text{original volume of acid} + \text{volume of added base}}$$
$$= \frac{0.002000 \text{ mol } H_3O^+}{0.04000 \text{ L} + 0.02000 \text{ L}} = 0.03333 \, M \qquad pH = 1.48$$

Given the moles of OH^- added, we are halfway to the equivalence point; but we are still on the initial slow rise of the curve, so the pH is still very low. Similar calculations give values up to the equivalence point.

3. *At the equivalence point.* After 40.00 mL of 0.1000 M NaOH has been added, the equivalence point is reached. All the H_3O^+ from the acid has been neutralized, and the solution contains Na^+ and Cl^-, neither of which reacts with water. Because of the autoionization of water, however,

$$[H_3O^+] = 1.0 \times 10^{-7} \, M \qquad pH = 7.00$$

In this example, 0.004000 mol of OH^- reacted with 0.004000 mol of H_3O^+ to reach the equivalence point.

4. *After the equivalence point.* From the equivalence point on, the pH calculation is based on the moles of *excess* OH^- present. For example, after adding 50.00 mL of NaOH, we have

$$\text{Total moles of } OH^- \text{ added} = 0.05000 \text{ L} \times 0.1000 \, M = 0.005000 \text{ mol } OH^-$$
$$\underline{-\text{Moles of } H_3O^+ \text{ consumed} = 0.04000 \text{ L} \times 0.1000 \, M = 0.004000 \text{ mol } H_3O^+}$$
$$\text{Moles of excess } OH^- = \qquad\qquad\qquad 0.001000 \text{ mol } OH^-$$

$$[OH^-] = \frac{0.001000 \text{ mol } OH^-}{0.04000 \text{ L} + 0.05000 \text{ L}} = 0.01111 \, M \qquad pOH = 1.95$$
$$pH = pK_w - pOH = 14.00 - 1.95 = 12.05$$

---

*In acid-base titrations, volumes and concentrations are usually known to four significant figures, but pH is generally reported to no more than two digits to the right of the decimal point.

Weak Acid–Strong Base Titration Curves

Now let's turn to the titration of a weak acid with a strong base. Figure 19.8 shows the curve obtained when we use 0.1000 M NaOH to titrate 40.00 mL of 0.1000 M propanoic acid, a weak organic acid (CH_3CH_2COOH; $K_a = 1.3 \times 10^{-5}$). (We abbreviate the acid as HPr and the conjugate base, $CH_3CH_2COO^-$, as Pr^-.)

Features of the Curve When we compare this weak acid–strong base titration curve with the strong acid–strong base titration curve (dotted curve portion in Figure 19.8 corresponds to bottom half of curve in Figure 19.7), four key regions appear, and the first three differ from the strong acid case:

1. *The initial pH is higher.* Because the weak acid (HPr) dissociates slightly, less H_3O^+ is present than with the strong acid.
2. *A gradually rising portion of the curve, called the buffer region, appears before the steep rise to the equivalence point.* As HPr reacts with the strong base, more and more conjugate base (Pr^-) forms, which creates an HPr-Pr^- buffer. At the midpoint of the buffer region, *half the original HPr has reacted,* so [HPr] = [Pr^-], or [Pr^-]/[HPr] = 1. Therefore, *the pH equals the pK_a*:

$$pH = pK_a + \log\left(\frac{[Pr^-]}{[HPr]}\right) = pK_a + \log 1 = pK_a + 0 = pK_a$$

   Observing the pH at the midpoint of the buffer region is a common method for estimating the pK_a of an unknown acid.
3. *The pH at the equivalence point is greater than 7.00.* The solution contains the strong-base cation Na^+, which does not react with water, and the weak-acid anion Pr^-, which acts as a weak base to accept a proton from H_2O and yield OH^-.
4. Beyond the equivalence point, the pH increases slowly as *excess OH^-* is added.

   Our choice of indicator is more limited here than for a strong acid–strong base titration because the steep rise occurs over a smaller pH range. Phenolphthalein is suitable because its color change lies within this range (Figure 19.8). However, the figure shows that methyl red, our other choice for the strong acid–strong base titration, changes color earlier and slowly over a large volume (~10 mL) of titrant, thereby giving a vague and false indication of the equivalence point.

Figure 19.8 Curve for a weak acid–strong base titration. The curve for the titration of 40.00 mL of 0.1000 M CH_3CH_2COOH (HPr) with 0.1000 M NaOH is compared with that for the strong acid HCl (*dotted curve portion*). Phenolphthalein (*photo*) is a suitable indicator here.

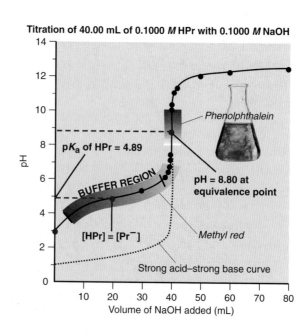

Titration of 40.00 mL of 0.1000 M HPr with 0.1000 M NaOH

pK_a of HPr = 4.89

BUFFER REGION

[HPr] = [Pr^-]

pH = 8.80 at equivalence point

Phenolphthalein

Methyl red

Strong acid–strong base curve

pH

Volume of NaOH added (mL)

Calculating the pH The calculation procedure for the weak acid–strong base titration is different from that for the strong acid–strong base titration because we have to consider the partial dissociation of the weak acid and the reaction of the conjugate base with water. There are four key regions of the titration curve, each of which requires a different type of calculation to find $[H_3O^+]$:

1. *Solution of HA.* Before base is added, the $[H_3O^+]$ is that of a weak-acid solution, so we find $[H_3O^+]$ as in Section 18.4: we set up a reaction table with $x = [HPr]_{dissoc}$, assume $[H_3O^+] = [HPr]_{dissoc} << [HPr]_{init}$, and solve for x:

$$K_a = \frac{[H_3O^+][Pr^-]}{[HPr]} \approx \frac{x^2}{[HPr]_{init}} \qquad \text{therefore,} \qquad x = [H_3O^+] \approx \sqrt{K_a \times [HPr]_{init}}$$

2. *Solution of HA and added base.* As soon as we add NaOH, it reacts with HPr to form Pr^-. This means that up to the equivalence point, we have a mixture of acid and conjugate base, and an HPr-Pr^- buffer solution exists over much of that interval. Therefore, we find $[H_3O^+]$ from the relationship

$$[H_3O^+] = K_a \times \frac{[HPr]}{[Pr^-]}$$

(Of course, we can find pH directly with the Henderson-Hasselbalch equation, which is just an alternative form of this relationship.) Note that in this calculation we do *not* have to consider the new total volume because *the volumes cancel in the ratio of concentrations.* That is, $[HPr]/[Pr^-]$ = moles of HPr/moles of Pr^-, so we need not calculate concentrations.

3. *Equivalent amounts of HA and added base.* At the equivalence point, the original amount of HPr has reacted, so the flask contains a solution of Pr^-, a weak base that reacts with water to form OH^-:

$$Pr^-(aq) + H_2O(l) \rightleftharpoons HPr(aq) + OH^-(aq)$$

Therefore, as mentioned previously, in a weak acid–strong base titration, the solution at the equivalence point is slightly basic, pH $>$ 7.00. We calculate $[H_3O^+]$ as in Section 18.5: we first find K_b of Pr^- from K_a of HPr, set up a reaction table (assume $[Pr^-] >> [Pr^-]_{reacting}$), and solve for $[OH^-]$. We need a single concentration, $[Pr^-]$, to solve for $[OH^-]$, so we *do* need the total volume. Then, we convert to $[H_3O^+]$. These two steps are

(1) $[OH^-] \approx \sqrt{K_b \times [Pr^-]}$,   where $K_b = \dfrac{K_w}{K_a}$ and $[Pr^-] = \dfrac{\text{moles of } HPr_{init}}{\text{total volume}}$

(2) $[H_3O^+] = \dfrac{K_w}{[OH^-]}$

Combining them into one step gives

$$[H_3O^+] \approx \frac{K_w}{\sqrt{K_b \times [Pr^-]}}$$

4. *Solution of excess added base.* Beyond the equivalence point, we are just adding excess OH^- ion, so the calculation is the same as for the strong acid–strong base titration:

$$[H_3O^+] = \frac{K_w}{[OH^-]}, \qquad \text{where } [OH^-] = \frac{\text{moles of excess } OH^-}{\text{total volume}}$$

Sample Problem 19.3 shows the overall approach.

SAMPLE PROBLEM 19.3   Finding the pH During a Weak Acid–Strong Base Titration

Problem Calculate the pH during the titration of 40.00 mL of 0.1000 *M* propanoic acid (HPr; $K_a = 1.3 \times 10^{-5}$) after adding the following volumes of 0.1000 *M* NaOH:
(a) 0.00 mL   **(b)** 30.00 mL   **(c)** 40.00 mL   **(d)** 50.00 mL
Plan (a) 0.00 mL: No base has been added yet, so this is a weak-acid solution. Thus, we calculate the pH as we did in Section 18.4. **(b)** 30.00 mL: A mixture of Pr^- and HPr is present. We find the amount (mol) of each, substitute into the K_a expression to solve for

$[H_3O^+]$, and convert to pH. **(c)** 40.00 mL: The amount (mol) of NaOH added equals the initial amount (mol) of HPr, so a solution of Na^+ and the weak base Pr^- exists. We calculate the pH as we did in Section 18.5, except that we need *total* volume to find $[Pr^-]$. **(d)** 50.00 mL: Excess NaOH is added, so we calculate the amount (mol) of excess OH^- in the total volume and convert to $[H_3O^+]$ and then pH.

Solution (a) 0.00 mL of 0.1000 *M* NaOH added. Following the approach used in Sample Problem 18.7 and just described in the text, we obtain

$$[H_3O^+] \approx \sqrt{K_a \times [HPr]_{init}} = \sqrt{(1.3 \times 10^{-5})(0.1000)} = 1.1 \times 10^{-3}\ M$$
$$pH = 2.96$$

(b) 30.00 mL of 0.1000 *M* NaOH added. Calculating the ratio of moles of HPr to Pr^-:

Original moles of HPr = 0.04000 L × 0.1000 *M* = 0.004000 mol HPr

Moles of NaOH added = 0.03000 L × 0.1000 *M* = 0.003000 mol OH^-

For 1 mol of NaOH that reacts, 1 mol of Pr^- forms, so we construct the following reaction table for the stoichiometry:

Amount (mol)	HPr(aq)	+	OH⁻(aq)	⟶	Pr⁻(aq)	+	H₂O(l)
Before addition	0.004000		—		0		—
Addition	—		0.003000		—		—
After addition	0.001000		0		0.003000		—

The last line of this table shows the new initial amounts of HPr and Pr^- that will react to attain a new equilibrium. However, with x very small, we assume that the $[HPr]/[Pr^-]$ ratio at equilibrium is essentially equal to the ratio of these new initial amounts (see Comment in Sample Problem 19.1). Thus,

$$\frac{[HPr]}{[Pr^-]} = \frac{0.001000\ \text{mol}}{0.003000\ \text{mol}} = 0.3333$$

Solving for $[H_3O^+]$: $[H_3O^+] = K_a \times \dfrac{[HPr]}{[Pr^-]} = (1.3 \times 10^{-5})(0.3333) = 4.3 \times 10^{-6}\ M$

$$pH = 5.37$$

(c) 40.00 mL of 0.1000 *M* NaOH added. Calculating $[Pr^-]$ after all HPr has reacted:

$$[Pr^-] = \frac{0.004000\ \text{mol}}{0.04000\ \text{L} + 0.04000\ \text{L}} = 0.05000\ M$$

Calculating K_b: $K_b = \dfrac{K_w}{K_a} = \dfrac{1.0 \times 10^{-14}}{1.3 \times 10^{-5}} = 7.7 \times 10^{-10}$

Solving for $[H_3O^+]$ as described in the text:

$$[H_3O^+] \approx \frac{K_w}{\sqrt{K_b \times [Pr^-]}} = \frac{1.0 \times 10^{-14}}{\sqrt{(7.7 \times 10^{-10})(0.05000)}} = 1.6 \times 10^{-9}\ M$$
$$pH = 8.80$$

(d) 50.00 mL of 0.1000 *M* NaOH added.

Moles of excess OH^- = (0.1000 *M*)(0.05000 L − 0.04000 L) = 0.001000 mol

$$[OH^-] = \frac{\text{moles of excess}\ OH^-}{\text{total volume}} = \frac{0.001000\ \text{mol}}{0.09000\ \text{L}} = 0.01111\ M$$

$$[H_3O^+] = \frac{K_w}{[OH^-]} = \frac{1.0 \times 10^{-14}}{0.01111} = 9.0 \times 10^{-13}\ M$$

$$pH = 12.05$$

Check As expected from the continuous addition of base, the pH increases through the four stages. Be sure to round off and check the arithmetic along the way.

FOLLOW-UP PROBLEM 19.3 A chemist titrates 20.00 mL of 0.2000 *M* HBrO ($K_a = 2.3 \times 10^{-9}$) with 0.1000 *M* NaOH. What is the pH **(a)** before any base is added; **(b)** when [HBrO] = [BrO⁻]; **(c)** at the equivalence point; **(d)** when the moles of OH^- added are twice the moles of HBrO originally present? **(e)** Sketch the titration curve.

Weak Base–Strong Acid Titration Curves

In the previous case, we titrated a weak acid with a strong base. The opposite process is the titration of a weak base (NH_3) with a strong acid (HCl), shown in Figure 19.9. Note that *the curve has the same shape as the weak acid–strong base curve* (Figure 19.8), *but it is inverted.* Thus, the regions of the curve have the same features, but *the pH decreases* throughout the process:

1. The initial solution is that of a weak base, so *the pH starts out above 7.00.*
2. The pH decreases gradually in the buffer region, where significant amounts of base (NH_3) and conjugate acid (NH_4^+) are present. At the midpoint of the buffer region, *the pH equals the pK_a of the ammonium ion.*
3. After the buffer region, the curve drops vertically to the equivalence point, at which all the NH_3 has reacted and the solution contains only NH_4^+ and Cl^-. Note that *the pH at the equivalence point is below 7.00* because Cl^- does not react with water and NH_4^+ is acidic:

$$NH_4^+(aq) + H_2O(l) \rightleftharpoons NH_3(aq) + H_3O^+(aq)$$

4. Beyond the equivalence point, the pH decreases slowly as *excess H_3O^+* is added.

For this titration also, we must be more careful in choosing the indicator than for a strong acid–strong base titration. Phenolphthalein changes color too soon and too slowly to indicate the equivalence point; on the other hand, methyl red lies on the steep portion of the curve and straddles the equivalence point, so it is a perfect choice.

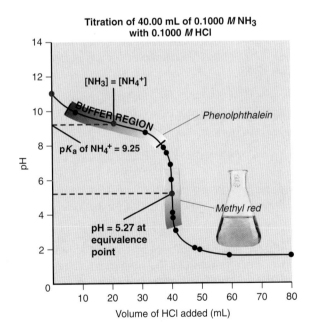

Titration of 40.00 mL of 0.1000 *M* NH_3 with 0.1000 *M* HCl

[NH_3] = [NH_4^+]

BUFFER REGION

pK_a of NH_4^+ = 9.25

Phenolphthalein

Methyl red

pH = 5.27 at equivalence point

pH

Volume of HCl added (mL)

Figure 19.9 Curve for a weak base–strong acid titration. Titrating 40.00 mL of 0.1000 *M* NH_3 with a solution of 0.1000 *M* HCl leads to a curve whose shape is the same as that of the weak acid–strong base curve in Figure 19.8 but inverted. The midpoint of the buffer region occurs when [NH_3] = [NH_4^+]; the pH at this point equals the pK_a of NH_4^+. Methyl red (*photo*) is a suitable indicator here.

SECTION SUMMARY

An acid-base (pH) indicator is a weak acid that has differently colored acidic and basic forms and changes color over about 2 pH units. In a strong acid–strong base titration, the pH starts out low, rises slowly, then shoots up near the equivalence point (pH = 7). In a weak acid–strong base titration, the pH starts out higher than in the strong acid titration, rises slowly in the buffer region (pH = pK_a at the midpoint), then rises more quickly near the equivalence point (pH > 7). A weak base–strong acid titration is the inverse of this, with the pH decreasing to the equivalence point (pH < 7).

19.3 EQUILIBRIA OF SLIGHTLY SOLUBLE IONIC COMPOUNDS

In this section, we explore the aqueous equilibria of slightly soluble ionic compounds, which up to now we've called "insoluble." In a saturated solution at a particular temperature, equilibrium exists between the undissolved and dissolved solute (Chapter 13). Slightly soluble ionic compounds have a relatively low solubility, so they reach equilibrium with relatively little solute dissolved. At this point, it would be a good idea for you to review the solubility rules listed in Table 4.1.

When a soluble ionic compound dissolves in water, it dissociates completely into ions. In this discussion, we will assume that the small amount of a slightly soluble ionic compound that does dissolve in water also dissociates completely into ions. In reality, however, this is not the case. Many slightly soluble salts, particularly those of transition metals and heavy main-group metals, have metal-nonmetal bonds with significant covalent character, and their solutions often contain other species that are partially dissociated or even undissociated. For example, when lead(II) chloride is thoroughly stirred in water, a small amount dissolves, and the solution contains not only the $Pb^{2+}(aq)$ and $Cl^-(aq)$ ions expected from complete dissociation, but also undissociated $PbCl_2(aq)$ molecules and $PbCl^+(aq)$ ions. In solutions of some other salts, such as $CaSO_4$, there are no molecules, but pairs of ions exist, such as $Ca^{2+}SO_4^{2-}(aq)$. These species increase the solubility above what we calculate assuming complete dissociation. For these reasons, it is best to treat the results of our calculations here as first approximations.

The Ion-Product Expression (Q_{sp}) and the Solubility-Product Constant (K_{sp})

If we make the assumption that there is complete dissociation of a slightly soluble ionic compound into its component ions, then *equilibrium exists between solid solute and aqueous ions*. Thus, for example, for a saturated solution of lead(II) sulfate in water, we have

$$PbSO_4(s) \rightleftharpoons Pb^{2+}(aq) + SO_4^{2-}(aq)$$

As with all the other equilibrium systems we've looked at, this one can be expressed by a reaction quotient:

$$Q_c = \frac{[Pb^{2+}][SO_4^{2-}]}{[PbSO_4]}$$

As in previous cases, we incorporate the constant concentration of the solid, $[PbSO_4]$, into the value of Q_c, which gives the *ion-product expression*, Q_{sp}:

$$Q_{sp} = Q_c[PbSO_4] = [Pb^{2+}][SO_4^{2-}]$$

And, when solid $PbSO_4$ attains equilibrium with Pb^{2+} and SO_4^{2-} ions, that is, when the solution reaches saturation, the numerical value of Q_{sp} attains a constant value, called the **solubility-product constant, K_{sp}.** The K_{sp} for $PbSO_4$ at 25°C, for example, is 1.6×10^{-8}.

As we've seen with other equilibrium constants, a given K_{sp} value depends only on the temperature, not on the individual ion concentrations. Suppose, for example, you add some lead(II) nitrate, a soluble lead salt, to increase the solution's $[Pb^{2+}]$. The equilibrium position shifts to the left, and $[SO_4^{2-}]$ goes down as more $PbSO_4$ precipitates; so the K_{sp} value is maintained.

The form of Q_{sp} is identical to that of the other reaction quotients we have written: each ion concentration is raised to an exponent equal to the coefficient in the balanced equation, which in this case also *equals the subscript of each ion in the compound's formula*. Thus, in general, for a saturated solution of a slightly soluble ionic compound, M_pX_q, composed of the ions M^{n+} and X^{z-}, the equilibrium condition is

$$Q_{sp} = [M^{n+}]^p[X^{z-}]^q = K_{sp} \tag{19.2}$$

At saturation, the concentration terms have their equilibrium values, so we can write the ion-product expression directly with the symbol K_{sp}. For example, the equation and ion-product expression that describe a saturated solution of $Cu(OH)_2$ are

$$Cu(OH)_2(s) \rightleftharpoons Cu^{2+}(aq) + 2OH^-(aq) \qquad K_{sp} = [Cu^{2+}][OH^-]^2$$

Insoluble metal sulfides present a slightly different case. The sulfide ion, S^{2-}, is so basic that it is not stable in water and reacts completely to form the hydrogen sulfide ion (HS^-) and the hydroxide ion (OH^-):

$$S^{2-}(aq) + H_2O(l) \longrightarrow HS^-(aq) + OH^-(aq)$$

For instance, when manganese(II) sulfide is shaken with water, the solution contains Mn^{2+}, HS^-, and OH^- ions. Although the sulfide ion does not exist as such in water, you can imagine the dissolution process as the sum of two steps, with S^{2-} occurring as an intermediate that is consumed immediately:

$$\begin{aligned} MnS(s) &\rightleftharpoons Mn^{2+}(aq) + \cancel{S^{2-}(aq)} \\ \cancel{S^{2-}(aq)} + H_2O(l) &\longrightarrow HS^-(aq) + OH^-(aq) \\ \hline MnS(s) + H_2O(l) &\rightleftharpoons Mn^{2+}(aq) + HS^-(aq) + OH^-(aq) \end{aligned}$$

Therefore, the ion-product expression is

$$K_{sp} = [Mn^{2+}][HS^-][OH^-]$$

SAMPLE PROBLEM 19.4 Writing Ion-Product Expressions for Slightly Soluble Ionic Compounds

Problem Write the ion-product expression for each of the following compounds:
(a) Magnesium carbonate   **(b)** Iron(II) hydroxide
(c) Calcium phosphate   **(d)** Silver sulfide

Plan We write an equation that describes a saturated solution and then write the ion-product expression, K_{sp}, according to Equation 19.2, noting the sulfide in part (d).

Solution **(a)** Magnesium carbonate:

$$MgCO_3(s) \rightleftharpoons Mg^{2+}(aq) + CO_3^{2-}(aq) \qquad K_{sp} = [Mg^{2+}][CO_3^{2-}]$$

(b) Iron(II) hydroxide:

$$Fe(OH)_2(s) \rightleftharpoons Fe^{2+}(aq) + 2OH^-(aq) \qquad K_{sp} = [Fe^{2+}][OH^-]^2$$

(c) Calcium phosphate:

$$Ca_3(PO_4)_2(s) \rightleftharpoons 3Ca^{2+}(aq) + 2PO_4^{3-}(aq) \qquad K_{sp} = [Ca^{2+}]^3[PO_4^{3-}]^2$$

(d) Silver sulfide:

$$\begin{aligned} Ag_2S(s) &\rightleftharpoons 2Ag^+(aq) + \cancel{S^{2-}(aq)} \\ \cancel{S^{2-}(aq)} + H_2O(l) &\longrightarrow HS^-(aq) + OH^-(aq) \\ \hline Ag_2S(s) + H_2O(l) &\rightleftharpoons 2Ag^+(aq) + HS^-(aq) + OH^-(aq) \quad K_{sp} = [Ag^+]^2[HS^-][OH^-] \end{aligned}$$

Check Except for part (d), you can check by reversing the process to see if you obtain the formula of the compound from K_{sp}.

Comment In part (d), we include H_2O as reactant to obtain a balanced equation.

FOLLOW-UP PROBLEM 19.4 Write the ion-product expression for each of the following compounds:
(a) Calcium sulfate   **(b)** Chromium(III) carbonate
(c) Magnesium hydroxide   **(d)** Arsenic(III) sulfide

The value of K_{sp} indicates how far to the right the dissolution proceeds at equilibrium (saturation). Table 19.2 presents some representative K_{sp} values. (Appendix C includes a much more extensive list.) Even though the values are all quite low, they range over many orders of magnitude.

Table 19.2 Solubility-Product Constants (K_{sp}) of Selected Ionic Compounds at 25°C

Name, Formula	K_{sp}
Aluminum hydroxide, $Al(OH)_3$	3×10^{-34}
Cobalt(II) carbonate, $CoCO_3$	1.0×10^{-10}
Iron(II) hydroxide, $Fe(OH)_2$	4.1×10^{-15}
Lead(II) fluoride, PbF_2	3.6×10^{-8}
Lead(II) sulfate, $PbSO_4$	1.6×10^{-8}
Mercury(I) iodide, Hg_2I_2	4.7×10^{-29}
Silver sulfide, Ag_2S	8×10^{-48}
Zinc iodate, $Zn(IO_3)_2$	3.9×10^{-6}

Calculations Involving the Solubility-Product Constant

In Chapters 17 and 18, we described two types of equilibrium problems. In one type, we use concentrations to find K, and in the other, we use K to find concentrations. Here we encounter the same two types.

Determining K_{sp} from Solubility The solubilities of ionic compounds are determined experimentally, and several chemical handbooks tabulate them. Most solubility values are given in units of grams of solute dissolved in 100 grams of H_2O. Because the mass of compound in solution is small, a negligible error is introduced if we assume that "100 g of water" is equal to "100 mL of solution." We then convert the solubility from grams of solute per 100 mL of solution to **molar solubility,** the amount (mol) of solute dissolved per liter of solution (that is, the molarity of the solute). Next, we use the equation for the dissolution of the solute to find the molarity of each ion and substitute into the ion-product expression to find the value of K_{sp}.

SAMPLE PROBLEM 19.5 Determining K_{sp} from Solubility

Problem (a) Lead(II) sulfate ($PbSO_4$) is a key component in lead-acid car batteries. Its solubility in water at 25°C is 4.25×10^{-3} g/100 mL solution. What is the K_{sp} of $PbSO_4$?
(b) When lead(II) fluoride (PbF_2) is shaken with pure water at 25°C, the solubility is found to be 0.64 g/L. Calculate the K_{sp} of PbF_2.
Plan We are given the solubilities in various units and must find K_{sp}. For each compound, we write an equation for its dissolution to see the number of moles of each ion, and then write the ion-product expression. We convert the solubility to molar solubility, find the molarity of each ion, and substitute into the ion-product expression to calculate K_{sp}.
Solution (a) For $PbSO_4$. Writing the equation and ion-product (K_{sp}) expression:

$$PbSO_4(s) \rightleftharpoons Pb^{2+}(aq) + SO_4{}^{2-}(aq) \qquad K_{sp} = [Pb^{2+}][SO_4{}^{2-}]$$

Converting solubility to molar solubility:

$$\text{Molar solubility of } PbSO_4 = \frac{0.00425 \text{ g } PbSO_4}{100 \text{ mL soln}} \times \frac{1000 \text{ mL}}{1 \text{ L}} \times \frac{1 \text{ mol } PbSO_4}{303.3 \text{ g } PbSO_4}$$
$$= 1.40\times10^{-4} \, M \, PbSO_4$$

Determining molarities of the ions: Because 1 mol of Pb^{2+} and 1 mol of $SO_4{}^{2-}$ form when 1 mol of $PbSO_4$ dissolves, $[Pb^{2+}] = [SO_4{}^{2-}] = 1.40\times10^{-4} \, M$.
Calculating K_{sp}:

$$K_{sp} = [Pb^{2+}][SO_4{}^{2-}] = (1.40\times10^{-4})^2 = 1.96\times10^{-8}$$

(b) For PbF_2. Writing the equation and K_{sp} expression:

$$PbF_2(s) \rightleftharpoons Pb^{2+}(aq) + 2F^-(aq) \qquad K_{sp} = [Pb^{2+}][F^-]^2$$

Converting solubility to molar solubility:

$$\text{Molar solubility of } PbF_2 = \frac{0.64 \text{ g } PbF_2}{1 \text{ L soln}} \times \frac{1 \text{ mol } PbF_2}{245.2 \text{ g } PbF_2} = 2.6\times10^{-3} \, M \, PbF_2$$

Determining molarities of the ions: 1 mol of Pb^{2+} and 2 mol of F^- form when 1 mol of PbF_2 dissolves, so we have

$$[Pb^{2+}] = 2.6\times10^{-3} \, M \qquad \text{and} \qquad [F^-] = 2(2.6\times10^{-3} \, M) = 5.2\times10^{-3} \, M$$

Calculating K_{sp}:

$$K_{sp} = [Pb^{2+}][F^-]^2 = (2.6\times10^{-3})(5.2\times10^{-3})^2 = 7.0\times10^{-8}$$

Check The low solubilities are consistent with K_{sp} values being small. In **(a)**, the molar solubility seems about right: $\sim\dfrac{4\times10^{-2} \text{ g/L}}{3\times10^2 \text{ g/mol}} \approx 1.3\times10^{-4} \, M$. Squaring this number gives 1.7×10^{-8}, close to the calculated K_{sp}. In **(b)**, we check the final step: $\sim(3\times10^{-3})(5\times10^{-3})^2 = 7.5\times10^{-8}$, close to the calculated K_{sp}.
Comment 1. In part (b), the formula PbF_2 means that $[F^-]$ is twice $[Pb^{2+}]$. Then we square this value of $[F^-]$. Always follow the ion-product expression explicitly.

2. The tabulated K_{sp} values for these compounds (Table 19.2) are lower than our calculated values. For PbF_2, for instance, the tabulated value is 3.6×10^{-8}, but we calculated 7.0×10^{-8} from solubility data. The discrepancy arises because we assumed that the PbF_2 in solution dissociates completely to Pb^{2+} and F^-. Here is an example of the complexity pointed out at the beginning of this section. Actually, about a third of the PbF_2 dissolves as $PbF^+(aq)$ and a small amount as undissociated $PbF_2(aq)$. The solubility (0.64 g/L) is determined experimentally and includes these other species, which we did not include in our simple calculation. This is why we treat such calculated K_{sp} values as approximations.

FOLLOW-UP PROBLEM 19.5 When powdered fluorite (CaF_2) is shaken with pure water at 18°C, 1.5×10^{-4} g dissolves for every 10.0 mL of solution. Calculate the K_{sp} of CaF_2 at 18°C.

Determining Solubility from K_{sp} The reverse of the previous type of problem involves finding the solubility of a compound based on its formula and K_{sp} value. An approach similar to the one we used for weak acids in Sample Problem 18.7 is to define the unknown amount dissolved—molar solubility—as S. Then we define the ion concentrations in terms of this unknown in a reaction table, and solve for S.

SAMPLE PROBLEM 19.6   Determining Solubility from K_{sp}

Problem Calcium hydroxide (slaked lime) is a major component of mortar, plaster, and cement, and solutions of $Ca(OH)_2$ are used in industry as a cheap, strong base. Calculate the solubility of $Ca(OH)_2$ in water if the K_{sp} is 6.5×10^{-6}.
Plan We write the dissolution equation and the ion-product expression. We know K_{sp} (6.5×10^{-6}); to find molar solubility (S), we set up a reaction table that expresses $[Ca^{2+}]$ and $[OH^-]$ in terms of S, substitute into the ion-product expression, and solve for S.
Solution Writing the equation and ion-product expression:

$$Ca(OH)_2(s) \rightleftharpoons Ca^{2+}(aq) + 2OH^-(aq) \qquad K_{sp} = [Ca^{2+}][OH^-]^2 = 6.5 \times 10^{-6}$$

Setting up a reaction table, with S = molar solubility:

Concentration (M)	$Ca(OH)_2(s)$	$\rightleftharpoons$	$Ca^{2+}(aq)$	+	$2OH^-(aq)$
Initial	—		0		0
Change	—		$+S$		$+2S$
Equilibrium	—		S		$2S$

Substituting into the ion-product expression and solving for S:

$$K_{sp} = [Ca^{2+}][OH^-]^2 = (S)(2S)^2 = (S)(4S^2) = 4S^3 = 6.5 \times 10^{-6}$$

$$S = \sqrt[3]{\frac{6.5 \times 10^{-6}}{4}} = 1.2 \times 10^{-2}\ M$$

Check We expect a low solubility from a slightly soluble salt. If we reverse the calculation, we should obtain the given K_{sp}: $4(1.2 \times 10^{-2})^3 = 6.9 \times 10^{-6}$, close to 6.5×10^{-6}.
Comment 1. Note that we did not double and *then* square $[OH^-]$. $2S$ *is* the $[OH^-]$, so we just squared it, as the ion-product expression required.
2. Once again, we assumed that the solid dissociates completely. Actually, the solubility is increased to about 2.0×10^{-2} M by the presence of $CaOH^+(aq)$ formed in the reaction $Ca(OH)_2(s) \rightleftharpoons CaOH^+(aq) + OH^-(aq)$. Our calculated answer is only approximate because we did not take this other species into account.

FOLLOW-UP PROBLEM 19.6 A suspension of $Mg(OH)_2$ in water is marketed as "milk of magnesia," which alleviates minor symptoms of indigestion by neutralizing stomach acid. The $[OH^-]$ is too low to harm the mouth and throat, but the suspension dissolves in the acidic stomach juices. What is the molar solubility of $Mg(OH)_2$ ($K_{sp} = 6.3 \times 10^{-10}$) in pure water?

No. of Ions	Formula	Cation:Anion	K_{sp}	Solubility (M)
2	$MgCO_3$	1:1	3.5×10^{-8}	1.9×10^{-4}
2	$PbSO_4$	1:1	1.6×10^{-8}	1.3×10^{-4}
2	$BaCrO_4$	1:1	2.1×10^{-10}	1.4×10^{-5}
3	$Ca(OH)_2$	1:2	6.5×10^{-6}	1.2×10^{-2}
3	BaF_2	1:2	1.5×10^{-6}	7.2×10^{-3}
3	CaF_2	1:2	3.2×10^{-11}	2.0×10^{-4}
3	Ag_2CrO_4	2:1	2.6×10^{-12}	8.7×10^{-5}

Table 19.3 **Relationship Between K_{sp} and Solubility at 25°C**

Using K_{sp} Values to Compare Solubilities The K_{sp} values provide a guide to *relative* solubility, as long as we compare compounds whose formulas contain the *same total number of ions*. In such cases, *the higher the K_{sp}, the greater the solubility*. Table 19.3 shows this point for several compounds. Note that for compounds that form three ions, the relationship holds whether the cation:anion ratio is 1:2 or 2:1, because the mathematical expression containing S is the same ($4S^3$) in the calculation (see Sample Problem 19.6).

The Effect of a Common Ion on Solubility

The presence of a common ion decreases the solubility of a slightly soluble ionic compound. As we saw in the case of acid-base systems, Le Châtelier's principle helps explain this effect. Let's examine the equilibrium condition for a saturated solution of lead(II) chromate:

$$PbCrO_4(s) \rightleftharpoons Pb^{2+}(aq) + CrO_4^{2-}(aq) \qquad K_{sp} = [Pb^{2+}][CrO_4^{2-}] = 2.3 \times 10^{-13}$$

At a given temperature, K_{sp} depends on the product of the ion concentrations. If the concentration of either ion goes up, the other must go down to maintain K_{sp}. Suppose we add Na_2CrO_4, a soluble salt, to the saturated $PbCrO_4$ solution. The concentration of the common ion, CrO_4^{2-}, increases, and some of it combines with Pb^{2+} ion to form more solid $PbCrO_4$ (Figure 19.10). The overall effect is a shift in the position of equilibrium to the left:

$$PbCrO_4(s) \overset{\longleftarrow}{\rightleftharpoons} Pb^{2+}(aq) + CrO_4^{2-}(aq; \text{added})$$

After the addition, $[CrO_4^{2-}]$ is higher, but $[Pb^{2+}]$ is lower. In this case, $[Pb^{2+}]$ represents the amount of $PbCrO_4$ dissolved; thus, in effect, the solubility of $PbCrO_4$ has decreased. The same result is obtained if we dissolve $PbCrO_4$ in a Na_2CrO_4 solution. We also obtain this result by adding a soluble lead(II) salt, such as $Pb(NO_3)_2$. The added Pb^{2+} ion combines with some $CrO_4^{2-}(aq)$, thereby lowering the amount of dissolved $PbCrO_4$.

Figure 19.10 The effect of a common ion on solubility. When a common ion is added to a saturated solution of an ionic compound, the solubility is lowered and more of the compound precipitates. **A,** Lead(II) chromate, a slightly soluble salt, forms a saturated aqueous solution. **B,** When Na_2CrO_4 solution is added, the amount of $PbCrO_4(s)$ increases. Thus, $PbCrO_4$ is less soluble in the presence of the common ion CrO_4^{2-}.

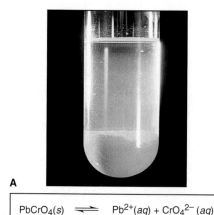

A

$$PbCrO_4(s) \rightleftharpoons Pb^{2+}(aq) + CrO_4^{2-}(aq)$$

B

$$PbCrO_4(s) \overset{\longleftarrow}{\rightleftharpoons} Pb^{2+}(aq) + CrO_4^{2-}(aq; \text{added})$$

SAMPLE PROBLEM 19.7 Calculating the Effect of a Common Ion on Solubility

Problem In Sample Problem 19.6, we calculated the solubility of $Ca(OH)_2$ in water. What is its solubility in 0.10 M $Ca(NO_3)_2$? K_{sp} of $Ca(OH)_2$ is 6.5×10^{-6}.

Plan Addition of Ca^{2+}, the common ion, should lower the solubility. We write the equation and ion-product expression and set up a reaction table, with $[Ca^{2+}]_{init}$ coming from $Ca(NO_3)_2$ and S equal to $[Ca^{2+}]_{from\ Ca(OH)_2}$. To simplify the math, we assume that, because K_{sp} is low, S is so small relative to $[Ca^{2+}]_{init}$ that it can be neglected. Then we solve for S and check the assumption.

Solution Writing the equation and ion-product expression:

$$Ca(OH)_2(s) \rightleftharpoons Ca^{2+}(aq) + 2OH^-(aq) \qquad K_{sp} = [Ca^{2+}][OH^-]^2 = 6.5\times10^{-6}$$

Setting up the reaction table, with $S = [Ca^{2+}]_{from\ Ca(OH)_2}$:

Concentration (M)	$Ca(OH)_2(s)$	$\rightleftharpoons$	$Ca^{2+}(aq)$	+	$2OH^-(aq)$
Initial	—		0.10		0
Change	—		+S		+2S
Equilibrium	—		0.10 + S		2S

Making the assumption: K_{sp} is small, so $S << 0.10$ M; thus, 0.10 M + $S \approx$ 0.10 M. Substituting into the ion-product expression and solving for S:

$$K_{sp} = [Ca^{2+}][OH^-]^2 = 6.5\times10^{-6} \approx (0.10)(2S)^2$$

Therefore,

$$4S^2 \approx \frac{6.5\times10^{-6}}{0.10} \qquad so \qquad S \approx \sqrt{\frac{6.5\times10^{-5}}{4}} = 4.0\times10^{-3}\ M$$

Checking the assumption:

$$\frac{4.0\times10^{-3}\ M}{0.10\ M} \times 100 = 4.0\% < 5\%$$

Check In Sample Problem 19.6, the solubility of $Ca(OH)_2$ was 0.012 M, but here, it is 0.0040 M, so the solubility *decreased* in the presence of added Ca^{2+}, the common ion, as expected.

FOLLOW-UP PROBLEM 19.7 To improve the quality of x-ray images used in the diagnosis of intestinal disorders, the patient drinks an aqueous suspension of $BaSO_4$ before the x-ray procedure. The Ba^{2+} in the suspension is opaque to x-rays, but it is also toxic; thus, the Ba^{2+} concentration is lowered by the addition of dilute Na_2SO_4. What is the solubility of $BaSO_4$ ($K_{sp} = 1.1\times10^{-10}$) in **(a)** pure water and in **(b)** 0.10 M Na_2SO_4?

The Effect of pH on Solubility

The hydronium ion concentration can have a profound effect on the solubility of an ionic compound. *If the compound contains the anion of a weak acid, addition of H_3O^+ (from a strong acid) increases its solubility.* Once again, Le Châtelier's principle explains why. An especially interesting case occurs with calcium carbonate. In a saturated solution of $CaCO_3$, we have

$$CaCO_3(s) \rightleftharpoons Ca^{2+}(aq) + CO_3^{2-}(aq)$$

Adding some strong acid introduces a large amount of H_3O^+, which immediately reacts with CO_3^{2-} to form the weak acid HCO_3^-:

$$CO_3^{2-}(aq) + H_3O^+(aq) \longrightarrow HCO_3^-(aq) + H_2O(l)$$

Thus, more $CaCO_3$ dissolves. In this particular case, the effect is increased by gas formation. If enough H_3O^+ is added, further reaction occurs to form carbonic acid, which decomposes immediately to H_2O and CO_2, and the gas escapes the container:

$$HCO_3^-(aq) + H_3O^+(aq) \longrightarrow H_2CO_3(aq) + H_2O(l) \longrightarrow CO_2(g) + 2H_2O(l)$$

Figure 19.11 Test for the presence of a carbonate. When a mineral that contains carbonate ion is treated with strong acid, the added H_3O^+ shifts the equilibrium position of the carbonate solubility. More carbonate dissolves, and the carbonic acid that is formed breaks down to water and gaseous CO_2.

As this sequence of changes shows, the net effect of added H_3O^+ is a shift in the equilibrium position to the right:

$$CaCO_3(s) \rightleftharpoons Ca^{2+} + CO_3^{2-} \xrightarrow{H_3O^+} HCO_3^- \xrightarrow{H_3O^+} H_2CO_3 \longrightarrow$$
$$CO_2(g) + H_2O + Ca^{2+}$$

In fact, this example illustrates a qualitative field test for carbonate minerals because the CO_2 bubbles vigorously (Figure 19.11).

In contrast, adding H_3O^+ to a saturated solution of a compound with a strong-acid anion, such as silver chloride, has no effect on the equilibrium position:

$$AgCl(s) \rightleftharpoons Ag^+(aq) + Cl^-(aq)$$

Because Cl^- ion is the conjugate base of a strong acid (HCl), it can coexist in solution with high $[H_3O^+]$. The Cl^- does not leave the system, so the equilibrium position is not affected.

SAMPLE PROBLEM 19.8   Predicting the Effect on Solubility of Adding Strong Acid

Problem Write balanced equations to explain whether addition of H_3O^+ from a strong acid affects the solubility of these ionic compounds:
(a) Lead(II) bromide   **(b)** Copper(II) hydroxide   **(c)** Iron(II) sulfide
Plan We write the balanced dissolution equation and note the anion: Weak-acid anions react with H_3O^+ and shift the equilibrium position toward more dissolution. Strong-acid anions do not react, so added H_3O^+ has no effect.
Solution (a) $PbBr_2(s) \rightleftharpoons Pb^{2+}(aq) + 2Br^-(aq)$
No effect. Br^- is the anion of HBr, a strong acid, so it does not react with H_3O^+.
(b) $Cu(OH)_2(s) \rightleftharpoons Cu^{2+}(aq) + 2OH^-(aq)$
Increases solubility. OH^- is the anion of H_2O, a very weak acid, so it reacts with the added H_3O^+:

$$OH^-(aq) + H_3O^+(aq) \longrightarrow 2H_2O(l)$$

(c) $FeS(s) + H_2O(l) \rightleftharpoons Fe^{2+}(aq) + HS^-(aq) + OH^-(aq)$
Increases solubility. We noted earlier that the S^{2-} ion reacts immediately with water to form HS^- and OH^-. The added H_3O^+ reacts with both of these weak-acid anions:

$$HS^-(aq) + H_3O^+(aq) \longrightarrow H_2S(aq) + H_2O(l)$$
$$OH^-(aq) + H_3O^+(aq) \longrightarrow 2H_2O(l)$$

FOLLOW-UP PROBLEM 19.8 Write balanced equations to show how addition of $HNO_3(aq)$ affects the solubility of these ionic compounds:
(a) Calcium fluoride   **(b)** Zinc sulfide   **(c)** Silver iodide

---

Predicting the Formation of a Precipitate: Q_{sp} vs. K_{sp}

In Chapter 17, we compared the values of Q and K to see if a reaction had reached equilibrium and, if not, in which net direction it would move until it did. In this discussion, we use the same approach to see if a precipitate will form and, if not, what changes in the concentrations of the component ions will cause it to do so. As you know, $Q_{sp} = K_{sp}$ when the solution is saturated. If Q_{sp} is greater than K_{sp}, the solution is momentarily supersaturated, and some solid precipitates until the remaining solution becomes saturated ($Q_{sp} = K_{sp}$). If Q_{sp} is less than K_{sp}, the solution is unsaturated, and no precipitate forms at that temperature (more solid can dissolve). To summarize,

- $Q_{sp} = K_{sp}$: solution is saturated and no change occurs.
- $Q_{sp} > K_{sp}$: precipitate forms until solution is saturated.
- $Q_{sp} < K_{sp}$: solution is unsaturated and no precipitate forms.

SAMPLE PROBLEM 19.9 Predicting Whether a Precipitate Will Form

Problem A common laboratory method for preparing a precipitate is to mix solutions containing the component ions. Does a precipitate form when 0.100 L of 0.30 M $Ca(NO_3)_2$ is mixed with 0.200 L of 0.060 M NaF?

Plan First, we must decide which slightly soluble salt could form and look up its K_{sp} value in Appendix C. To see whether mixing these solutions will form the precipitate, we find the ion concentrations by calculating the amount (mol) of each ion from its concentration and volume, and then dividing by the *total* volume because one solution dilutes the other. Finally, we write the ion-product expression, calculate Q_{sp}, and compare Q_{sp} with K_{sp}.

Solution The ions present are Ca^{2+}, Na^+, F^-, and NO_3^-. All sodium and all nitrate salts are soluble (Table 4.1), so the only possibility is CaF_2 ($K_{sp} = 3.2 \times 10^{-11}$). Calculating the ion concentrations:

$$\text{Moles of } Ca^{2+} = 0.30 \ M \ Ca^{2+} \times 0.100 \ L$$
$$= 0.030 \ \text{mol } Ca^{2+}$$
$$[Ca^{2+}] = \frac{0.030 \ \text{mol } Ca^{2+}}{0.100 \ L + 0.200 \ L}$$
$$= 0.10 \ M \ Ca^{2+}$$
$$\text{Moles of } F^- = 0.060 \ M \ F^- \times 0.200 \ L$$
$$= 0.012 \ \text{mol } F^-$$
$$[F^-] = \frac{0.012 \ \text{mol } F^-}{0.100 \ L + 0.200 \ L}$$
$$= 0.040 \ M \ F^-$$

Substituting into the ion-product expression and comparing Q_{sp} with K_{sp}:

$$Q_{sp} = [Ca^{2+}][F^-]^2 = (0.10)(0.040)^2 = 1.6 \times 10^{-4}$$

Because $Q_{sp} > K_{sp}$, CaF_2 will precipitate until $Q_{sp} = 3.2 \times 10^{-11}$.

Check Remember to round off and quickly check the math. For example, $Q_{sp} = (1 \times 10^{-1})(4 \times 10^{-2})^2 = 1.6 \times 10^{-4}$. With K_{sp} so low, CaF_2 must have a low solubility, and given the sizable concentrations being mixed, we would expect CaF_2 to precipitate.

FOLLOW-UP PROBLEM 19.9 As a result of mineral erosion and biological activity, phosphate ion is common in natural waters, where it often precipitates as insoluble salts, such as $Ca_3(PO_4)_2$. If $[Ca^{2+}] = [PO_4^{3-}] = 1.0 \times 10^{-9} \ M$ in a given river, will $Ca_3(PO_4)_2$ precipitate? K_{sp} of $Ca_3(PO_4)_2$ is 1.2×10^{-29}.

Applying Ionic Equilibria to the Acid-Rain Problem

The effect of industrial society on the environment is especially apparent in the problem of *acid rain;* the underlying chemistry applies several principles of ionic equilibria. Acidic precipitation—rain, snow, fog, or dry deposits on particles—has been recorded in all parts of North America, the Amazon basin, Europe, including Russia, much of Asia, and even at the North and South Poles. Three major substances are involved:

1. *Sulfurous acid.* Sulfur dioxide (SO_2) from the burning of high-sulfur coal forms sulfurous acid in contact with water. Oxidizing air pollutants, such as hydrogen peroxide, convert sulfurous acid to sulfuric acid:

$$H_2O_2(aq) + H_2SO_3(aq) \longrightarrow H_2SO_4(aq) + H_2O(l)$$

2. *Sulfuric acid.* Sulfur trioxide (SO_3) forms through the atmospheric oxidation of SO_2 and becomes H_2SO_4 in contact with water.

3. *Nitric acid.* Nitrogen oxides (denoted NO_x) form when N_2 and O_2 react. NO is produced during combustion in car engines and electric power plants, and then forms NO_2 and HNO_3. At night, NO_x are converted to N_2O_5, which hydrolyzes to HNO_3 in water.

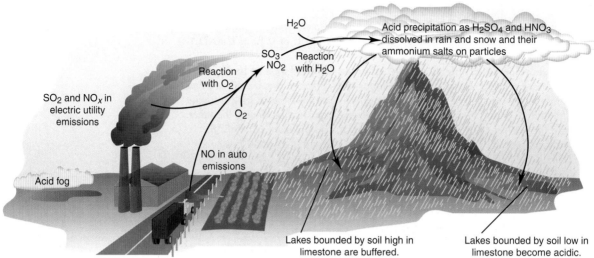

Figure 19.12 Formation of acidic precipitation. A complex interplay of human activities, atmospheric chemistry, and environmental distribution leads to acidic precipitation and its harmful effects. Car exhaust and electrical utility waste gases contain lower oxides of nitrogen and sulfur. These are oxidized in the atmosphere by O_2 (or O_3, not shown) to higher oxides (NO_2, SO_3), which react with moisture to form acidic rain, snow, and fog. In contact with acidic precipitation, many lakes become acidified, whereas limestone-bounded lakes form a carbonate buffer that prevents acidification.

Figure 19.12 illustrates some of the main sources of the two strong acids.

Unpolluted rainwater is weakly acidic (pH = 5.6) because it contains dissolved CO_2:

$$CO_2(g) + 2H_2O(l) \rightleftharpoons H_3O^+(aq) + HCO_3^-(aq)$$

In contrast, the average pH of rainfall in much of the United States was 4.2 as early as 1984; rain with a pH of 2.7 (about the same as vinegar) has been observed in Sweden and with a pH of 1.8 (between lemon juice and stomach acid) in West Virginia.

These 10- to 10,000-fold excesses of $[H_3O^+]$ are very destructive to fish (many species die at a pH below 5) and to forests. The aluminosilicates that make up most soils are nearly insoluble. But contact with H_3O^+ dissolves some of the bound Al^{3+}, which is extremely toxic to fish, and many ions that act as nutrients for plants and animals are dissolved and carried away.

Acid rain also dissolves the calcium carbonate in the marble and limestone of buildings and monuments. Ironically, the same process that destroys these structures saves lakes in limestone-rich soil. As we discussed previously, added H_3O^+ shifts the following equilibrium to the right to dissolve more limestone and form more bicarbonate:

$$CO_3^{2-}(aq) + H_3O^+(aq) \rightleftharpoons HCO_3^-(aq) + H_2O(l)$$

With time, limestone-bounded lakes become enormous HCO_3^-/CO_3^{2-} buffers that maintain a relatively stable pH as they absorb additional H_3O^+. Lakes in limestone-poor soils can be treated with limestone. Sweden spent tens of millions of dollars during the 1990s to add limestone to about 3000 lakes. This method provides only temporary improvement, however, and the lakes become re-acidified.

More effective approaches reduce SO_2 and NO_x at the source. SO_2 is removed from power-plant emissions with limestone or, in a newer method, it is partially reduced to H_2S and then converted to sulfur:

$$16H_2S(g) + 8SO_2(g) \longrightarrow 3S_8(s) + 16H_2O(l)$$

Coal can also be converted into gaseous and liquid low-sulfur fuels. The catalytic converter in an automobile exhaust system reduces NO_x to N_2 and NH_3, and in power plants, NO_x is removed from the hot stack gases with ammonia:

$$4NO(g) + 4NH_3(g) + O_2(g) \longrightarrow 4N_2(g) + 6H_2O(g)$$

Amendments to the Clean Air Act to further curb NO_x emissions and help states meet ozone standards will reduce HNO_3 in the process.

SECTION SUMMARY

As an approximation, the dissolved portion of a slightly soluble salt dissociates completely into ions. In a saturated solution, the ions are in equilibrium with the solid, and the product of the ion concentrations, each raised to the power of its subscript in the compound's formula, has a constant value ($Q_{sp} = K_{sp}$). The value of K_{sp} can be obtained from the solubility, and vice versa. Adding a common ion lowers an ionic compound's solubility. Adding H_3O^+ (lowering the pH) increases a compound's solubility if the anion of the compound is that of a weak acid. If $Q_{sp} > K_{sp}$ for an ionic compound, a precipitate forms when two solutions, each containing one of the compound's ions, are mixed. Lakes bounded by limestone-rich soils form buffer systems that prevent harmful acidification by acid rain.

19.4 EQUILIBRIA INVOLVING COMPLEX IONS

The final type of aqueous ionic equilibrium we consider involves a different kind of ion than we've examined up to now. A *simple ion,* such as Na^+ or SO_4^{2-}, consists of one or a few bound atoms, with an excess or deficit of electrons. A **complex ion** consists of a central metal ion covalently bonded to two or more anions or molecules, called **ligands.** Hydroxide, chloride, and cyanide ions are some ionic ligands; water, carbon monoxide, and ammonia are some molecular ligands. In the complex ion $Cr(NH_3)_6^{3+}$, for example, Cr^{3+} is the central metal ion and six NH_3 molecules are the ligands, giving an overall $3+$ charge (Figure 19.13).

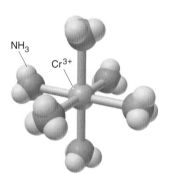

NH₃

Cr³⁺

Figure 19.13 $Cr(NH_3)_6^{3+}$, a typical complex ion. A complex ion consists of a central metal ion, such as Cr^{3+}, covalently bonded to a specific number of ligands, such as NH_3.

As we discussed in Section 18.8, *all complex ions are Lewis adducts.* The metal ion acts as a Lewis acid (accepts an electron pair) and the ligand acts as a Lewis base (donates an electron pair). The acidic hydrated metal ions that we discussed in Section 18.6 are complex ions with water molecules as ligands. In Chapter 22, we discuss the transition metals and the structures and properties of the numerous complex ions they form. Our focus here is on equilibria of hydrated ions with ligands other than water.

Formation of Complex Ions

Whenever a metal ion enters water, a complex ion forms, with water as the ligand. In many cases, when we treat this hydrated cation with a solution of another ligand, the bound water molecules exchange for the other ligand. For example, a hydrated M^{2+} ion, $M(H_2O)_4^{2+}$, forms $M(NH_3)_4^{2+}$ in aqueous NH_3:

$$M(H_2O)_4^{2+}(aq) + 4NH_3(aq) \rightleftharpoons M(NH_3)_4^{2+}(aq) + 4H_2O(l)$$

At equilibrium, this system is expressed by a ratio of concentration terms whose form follows that of any other equilibrium expression:

$$K_c = \frac{[M(NH_3)_4^{2+}][H_2O]^4}{[M(H_2O)_4^{2+}][NH_3]^4}$$

Once again, because the concentration of water is essentially constant in aqueous reactions, we incorporate it into K_c and obtain the expression for a new equilibrium constant, the **formation constant, K_f:**

$$K_f = \frac{K_c}{[H_2O]^4} = \frac{[M(NH_3)_4{}^{2+}]}{[M(H_2O)_4{}^{2+}][NH_3]^4}$$

At the molecular level (Figure 19.14), the actual process is stepwise, with ammonia molecules replacing water molecules one at a time to give a series of intermediate species, each with its own formation constant:

$$M(H_2O)_4{}^{2+}(aq) + NH_3(aq) \rightleftharpoons M(H_2O)_3(NH_3)^{2+}(aq) + H_2O(l)$$
$$K_{f1} = \frac{[M(H_2O)_3(NH_3)^{2+}]}{[M(H_2O)_4{}^{2+}][NH_3]}$$

$$M(H_2O)_3(NH_3)^{2+}(aq) + NH_3(aq) \rightleftharpoons M(H_2O)_2(NH_3)_2{}^{2+}(aq) + H_2O(l)$$
$$K_{f2} = \frac{[M(H_2O)_2(NH_3)_2{}^{2+}]}{[M(H_2O)_3(NH_3)^{2+}][NH_3]}$$

$$M(H_2O)_2(NH_3)_2{}^{2+}(aq) + NH_3(aq) \rightleftharpoons M(H_2O)(NH_3)_3{}^{2+}(aq) + H_2O(l)$$
$$K_{f3} = \frac{[M(H_2O)(NH_3)_3{}^{2+}]}{[M(H_2O)_2(NH_3)_2{}^{2+}][NH_3]}$$

$$M(H_2O)(NH_3)_3{}^{2+}(aq) + NH_3(aq) \rightleftharpoons M(NH_3)_4{}^{2+}(aq) + H_2O(l)$$
$$K_{f4} = \frac{[M(NH_3)_4{}^{2+}]}{[M(H_2O)(NH_3)_3{}^{2+}][NH_3]}$$

The *sum* of the equations gives the overall equation, so the *product* of the individual formation constants gives the overall formation constant:

$$K_f = K_{f1} \times K_{f2} \times K_{f3} \times K_{f4}$$

In this case, the K_f for each step is much larger than 1 because ammonia is a stronger Lewis base than water. Therefore, if we add excess ammonia to the $M(H_2O)_4{}^{2+}$ solution, the H_2O ligands are replaced and essentially all the M^{2+} ion exists as $M(NH_3)_4{}^{2+}$.

Appendix C shows the formation constants (K_f) of several complex ions; note that all are 10^6 or greater, which means that the equilibria of the formation reactions lie far to the right.

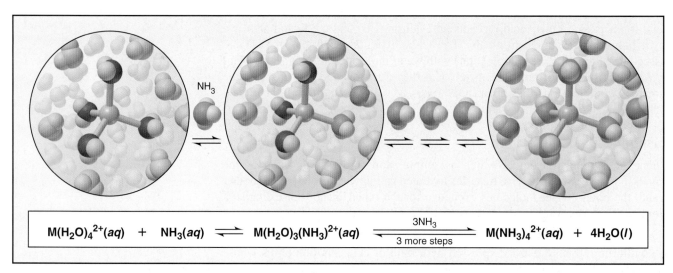

$$M(H_2O)_4{}^{2+}(aq) + NH_3(aq) \rightleftharpoons M(H_2O)_3(NH_3)^{2+}(aq) \xrightleftharpoons[\text{3 more steps}]{\text{3NH}_3} M(NH_3)_4{}^{2+}(aq) + 4H_2O(l)$$

Figure 19.14 The stepwise exchange of NH$_3$ for H$_2$O in M(H$_2$O)$_4$$^{2+}$. The ligands of a complex ion can exchange for other ligands. When ammonia is added to a solution of the hydrated M^{2+} ion, M(H$_2$O)$_4$$^{2+}$, NH$_3$ molecules replace the bound H$_2$O molecules one at a time to form the M(NH$_3$)$_4$$^{2+}$ ion. The molecular-scale views show the first exchange and the fully ammoniated ion.

Complex Ions and the Solubility of Precipitates

In Section 19.3, you saw that H_3O^+ increases the solubility of a slightly soluble ionic compound if its anion is that of a weak acid. Similarly, *a ligand increases the solubility of a slightly soluble ionic compound if it forms a complex ion with the cation.* For example, zinc sulfide is very slightly soluble:

$$ZnS(s) + H_2O(l) \rightleftharpoons Zn^{2+}(aq) + HS^-(aq) + OH^-(aq) \qquad K_{sp} = 2.0 \times 10^{-22}$$

When we add some 1.0 M NaCN, the CN^- ions act as ligands and react with the small amount of $Zn^{2+}(aq)$ to form the complex ion $Zn(CN)_4^{2-}$:

$$Zn^{2+}(aq) + 4CN^-(aq) \rightleftharpoons Zn(CN)_4^{2-}(aq) \qquad K_f = 4.2 \times 10^{19}$$

To see the effect of complex-ion formation on the solubility of ZnS, we add the equations and, therefore, multiply their equilibrium constants:

$$ZnS(s) + 4CN^-(aq) + H_2O(l) \rightleftharpoons Zn(CN)_4^{2-}(aq) + HS^-(aq) + OH^-(aq)$$
$$K_{overall} = K_{sp} \times K_f = (2.0 \times 10^{-22})(4.2 \times 10^{19}) = 8.4 \times 10^{-3}$$

The overall equilibrium constant increased by more than a factor of 10^{19} in the presence of the ligand; this reflects the increased amount of ZnS in solution.

SAMPLE PROBLEM 19.10   Calculating the Effect of Complex-Ion Formation on Solubility

Problem In black-and-white film developing, excess AgBr is removed from the film negative by "hypo," an aqueous solution of sodium thiosulfate ($Na_2S_2O_3$), which forms the complex ion $Ag(S_2O_3)_2^{3-}$. Calculate the solubility of AgBr in (a) H_2O; (b) 1.0 M hypo. K_f of $Ag(S_2O_3)_2^{3-}$ is 4.7×10^{13} and K_{sp} of AgBr is 5.0×10^{-13}.

Plan (a) After writing the equation and the ion-product expression, we use the given K_{sp} to solve for S, the molar solubility of AgBr. (b) In hypo, Ag^+ forms a complex ion with $S_2O_3^{2-}$, which shifts the equilibrium and dissolves more AgBr. We write the complex-ion equation and add it to the equation for dissolving AgBr to obtain the overall equation for dissolving AgBr in hypo. We multiply K_{sp} by K_f to find $K_{overall}$. To find the solubility of AgBr in hypo, we set up a reaction table, with $S = [Ag(S_2O_3)_2^{3-}]$, substitute into the expression for $K_{overall}$, and solve for S.

Solution (a) Solubility in water. Writing the equation for the saturated solution and the ion-product expression:

$$AgBr(s) \rightleftharpoons Ag^+(aq) + Br^-(aq) \qquad K_{sp} = [Ag^+][Br^-]$$

Solving for solubility (S) directly from the equation: We know that

$$S = [AgBr]_{dissolved} = [Ag^+] = [Br^-]$$

Thus,
$$K_{sp} = [Ag^+][Br^-] = S^2 = 5.0 \times 10^{-13}$$

so
$$S = \boxed{7.1 \times 10^{-7} \, M}$$

(b) Solubility in 1.0 M hypo. Writing the overall equation:

$$AgBr(s) \rightleftharpoons \cancel{Ag^+(aq)} + Br^-(aq)$$
$$\underline{\cancel{Ag^+(aq)} + 2S_2O_3^{2-}(aq) \rightleftharpoons Ag(S_2O_3)_2^{3-}(aq)}$$
$$AgBr(s) + 2S_2O_3^{2-}(aq) \rightleftharpoons Ag(S_2O_3)_2^{3-}(aq) + Br^-(aq)$$

Calculating $K_{overall}$:

$$K_{overall} = \frac{[Ag(S_2O_3)_2^{3-}][Br^-]}{[S_2O_3^{2-}]^2} = K_{sp} \times K_f = (5.0 \times 10^{-13})(4.7 \times 10^{13}) = 24$$

Setting up a reaction table, with $S = [AgBr]_{dissolved} = [Ag(S_2O_3)_2^{3-}]$:

Concentration (M)	AgBr(s)	+	$2S_2O_3^{2-}(aq)$	$\rightleftharpoons$	$Ag(S_2O_3)_2^{3-}(aq)$	+	$Br^-(aq)$
Initial	—		1.0		0		0
Change	—		$-2S$		$+S$		$+S$
Equilibrium	—		$1.0 - 2S$		S		S

Substituting the values into the expression for $K_{overall}$ and solving for S:

$$K_{overall} = \frac{[Ag(S_2O_3)_2{}^{3-}][Br^-]}{[S_2O_3{}^{2-}]^2} = \frac{S^2}{(1.0\ M - 2S)^2} = 24$$

Taking the square root of both sides gives

$$\frac{S}{1.0\ M - 2S} = \sqrt{24} = 4.9 \qquad [Ag(S_2O_3)_2{}^{3-}] = S = 0.45\ M$$

Check **(a)** From the number of ions in the formula of AgBr, we know that $S = \sqrt{K_{sp}}$, so the order of magnitude seems right: $\sim\sqrt{10^{-14}} = 10^{-7}$. **(b)** The $K_{overall}$ seems correct: the exponents cancel, and $5 \times 5 = 25$. Most importantly, the answer makes sense because the photographic process requires the remaining AgBr to be washed off the film and the large $K_{overall}$ confirms that. We can check S by rounding and working backward to find $K_{overall}$: from the reaction table, we find that

$$[(S_2O_3)^{2-}] = 1.0\ M - 2S = 1.0\ M - 2(0.45\ M) = 1.0\ M - 0.90\ M = 0.1\ M$$

so $K_{overall} \approx (0.45)^2/(0.1)^2 = 20$, within rounding of the calculated value.

FOLLOW-UP PROBLEM 19.10 How does the solubility of AgBr in $1.0\ M\ NH_3$ compare with its solubility in hypo? K_f of $Ag(NH_3)_2{}^+$ is 1.7×10^7.

SECTION SUMMARY

A complex ion consists of a central metal ion covalently bonded to two or more negatively charged or neutral ligands. Its formation is described by a formation constant, K_f. A hydrated metal ion is a complex ion with water molecules as ligands. Other ligands can displace the water in a stepwise process. In most cases, the K_f value of each step is large, so the fully substituted complex ion forms almost completely in the presence of excess ligand. Adding a solution containing a ligand increases the solubility of an ionic precipitate if the cation forms a complex ion with the ligand.

For Review and Reference (Numbers in parentheses refer to pages, unless noted otherwise.)

Learning Objectives

To help you review these learning objectives, the numbers of related sections (§), sample problems (SP), and upcoming end-of-chapter problems (EP) are listed in parentheses.

1. Explain how a common ion suppresses a reaction that forms it; describe buffer capacity and buffer range, and understand why the concentrations of buffer components must be high relative to the amount of added H_3O^+ or OH^-; calculate how to prepare a buffer (§ 19.1) (SPs 19.1, 19.2) (EPs 19.1–19.24)
2. Understand how an acid-base indicator works, how the equivalence point and end point in an acid-base titration differ, and how strong acid–strong base, weak acid–strong base, and strong acid–weak base titration curves differ; explain the significance of the pH at the midpoint of the buffer region; choose an appropriate indicator, and calculate the pH at any point in a titration (§ 19.2) (SP 19.3) (EPs 19.25–19.42)

3. Describe the equilibrium of a slightly soluble ionic compound in water, and explain the meaning of K_{sp}; understand how a common ion and pH affect solubility and how to predict precipitate formation from the values of Q_{sp} and K_{sp} (§ 19.3) (SPs 19.4–19.9) (EPs 19.43–19.62)
4. Describe the stepwise formation of a complex ion, and explain the meaning of K_f; calculate the effect of complex-ion formation on solubility (§ 19.4) (SP 19.10) (EPs 19.63–19.71)

Key Terms

Section 19.1
acid-base buffer (616)
common-ion effect (617)
Henderson-Hasselbalch equation (621)
buffer capacity (621)
buffer range (622)

Section 19.2
acid-base titration curve (624)
equivalence point (626)
end point (626)

Section 19.3
solubility-product constant (K_{sp}) (632)
molar solubility (634)

Section 19.4
complex ion (641)
ligand (641)
formation constant (K_f) (642)

Key Equations and Relationships

19.1 Finding the pH from known concentrations of a conjugate acid-base pair (Henderson-Hasselbalch equation) (621):

$$pH = pK_a + \log\left(\frac{[\text{base}]}{[\text{acid}]}\right)$$

19.2 Defining the equilibrium condition for a saturated solution of a slightly soluble compound, M_pX_q, composed of M^{n+} and X^{z-} ions (632):

$$Q_{sp} = [M^{n+}]^p[X^{z-}]^q = K_{sp}$$

Brief Solutions to Follow-up Problems

19.1 (a) Before addition:
Assuming x is small enough to be neglected,
$[HF] = 0.50\ M$ and $[F^-] = 0.45\ M$

$$[H_3O^+] = K_a \times \frac{[HF]}{[F^-]} \approx (6.8\times10^{-4})\left(\frac{0.50}{0.45}\right) = 7.6\times10^{-4}\ M$$

$pH = 3.12$

(b) After addition of 0.40 g of NaOH (0.010 mol of NaOH) to 1.0 L of buffer,
$[HF] = 0.49\ M$ and $[F^-] = 0.46\ M$

$$[H_3O^+] \approx (6.8\times10^{-4})\left(\frac{0.49}{0.46}\right) = 7.2\times10^{-4}\ M;\ pH = 3.14$$

19.2 $[H_3O^+] = 10^{-pH} = 10^{-4.25} = 5.6\times10^{-5}$

$$[C_6H_5COOH] = \frac{[H_3O^+][C_6H_5COO^-]}{K_a}$$

$$= \frac{(5.6\times10^{-5})(0.050)}{6.3\times10^{-5}} = 0.044\ M$$

Mass (g) of $C_6H_5COOH = 5.0$ L soln $\times \dfrac{0.044\ \text{mol } C_6H_5COOH}{1\ \text{L soln}}$

$$\times \frac{122.12\ \text{g } C_6H_5COOH}{1\ \text{mol } C_6H_5COOH}$$

$$= 27\ \text{g } C_6H_5COOH$$

Dissolve 27 g of C_6H_5COOH in 4.9 L of 0.050 M C_6H_5COONa and add solution to make 5.0 L. Adjust pH to 4.25 with strong acid or base.

19.3 (a) $[H_3O^+] \approx \sqrt{(2.3\times10^{-9})(0.2000)} = 2.1\times10^{-5}\ M$
$\qquad pH = 4.68$

(b) $[H_3O^+] = K_a \times \dfrac{[HBrO]}{[BrO^-]} = (2.3\times10^{-9})(1) = 2.3\times10^{-9}\ M$

$\qquad pH = 8.64$

(c) $[BrO^-] = \dfrac{\text{moles of } BrO^-}{\text{total volume}} = \dfrac{0.004000\ \text{mol}}{0.06000\ \text{L}} = 0.06667\ M$

K_b of $BrO^- = \dfrac{K_w}{K_a\ \text{of } HBrO} = 4.3\times10^{-6}$

$$[H_3O^+] = \frac{K_w}{\sqrt{K_b \times [BrO^-]}} \approx \frac{1.0\times10^{-14}}{\sqrt{(4.3\times10^{-6})(0.06667)}}$$

$$= 1.9\times10^{-11}\ M$$

$\qquad pH = 10.72$

(d) Moles of OH^- added $= 0.008000$ mol
Volume (L) of OH^- soln $= 0.08000$ L

$$[OH^-] = \frac{\text{moles of } OH^-\ \text{unreacted}}{\text{total volume}}$$

$$= \frac{0.008000\ \text{mol} - 0.004000\ \text{mol}}{(0.02000 + 0.08000)\ \text{L}} = 0.04000\ M$$

$$[H_3O^+] = \frac{K_w}{[OH^-]} = 2.5\times10^{-13}$$

$\qquad pH = 12.60$

(e)
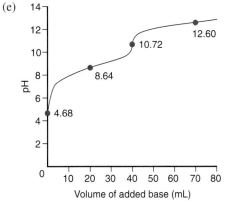

19.4 (a) $K_{sp} = [Ca^{2+}][SO_4^{2-}]$
(b) $K_{sp} = [Cr^{3+}]^2[CO_3^{2-}]^3$
(c) $K_{sp} = [Mg^{2+}][OH^-]^2$
(d) $K_{sp} = [As^{3+}]^2[HS^-]^3[OH^-]^3$

19.5 $[CaF_2] = \dfrac{1.5\times10^{-4}\ \text{g } CaF_2}{10.0\ \text{mL soln}} \times \dfrac{1000\ \text{mL}}{1\ \text{L}} \times \dfrac{1\ \text{mol } CaF_2}{78.08\ \text{g } CaF_2}$

$\qquad = 1.9\times10^{-4}\ M$

$CaF_2(s) \rightleftharpoons Ca^{2+}(aq) + 2F^-(aq)$
$[Ca^{2+}] = 1.9\times10^{-4}\ M$ and $[F^-] = 3.8\times10^{-4}\ M$
$K_{sp} = [Ca^{2+}][F^-]^2 = (1.9\times10^{-4})(3.8\times10^{-4})^2 = 2.7\times10^{-11}$

19.6 From the reaction table, $[Mg^{2+}] = S$ and $[OH^-] = 2S$
$K_{sp} = [Mg^{2+}][OH^-]^2 = 4S^3 = 6.3\times10^{-10}$; $S = 5.4\times10^{-4}\ M$

19.7 (a) In water: $K_{sp} = [Ba^{2+}][SO_4^{2-}] = S^2 = 1.1\times10^{-10}$;
$S = 1.0\times10^{-5}$

(b) In 0.10 M Na_2SO_4: $[SO_4^{2-}] = 0.10\ M$
$K_{sp} = 1.1\times10^{-10} \approx S \times 0.10$; $S = 1.1\times10^{-9}\ M$
S decreases in presence of the common ion SO_4^{2-}.

19.8 (a) Increases solubility.
$\qquad CaF_2(s) \rightleftharpoons Ca^{2+}(aq) + 2F^-(aq)$
$F^-(aq) + H_3O^+(aq) \longrightarrow HF(aq) + H_2O(l)$
(b) Increases solubility.
$\qquad ZnS(s) + H_2O(l) \rightleftharpoons Zn^{2+}(aq) + HS^-(aq) + OH^-(aq)$
$HS^-(aq) + H_3O^+(aq) \longrightarrow H_2S(aq) + H_2O(l)$
$OH^-(aq) + H_3O^+(aq) \longrightarrow 2H_2O(l)$
(c) No effect. $I^-(aq)$ is conjugate base of strong acid, HI.

19.9 $Ca_3(PO_4)_2(s) \rightleftharpoons 3Ca^{2+}(aq) + 2PO_4^{3-}(aq)$
$Q_{sp} = [Ca^{2+}]^3[PO_4^{3-}]^2 = (1.0\times10^{-9})^5 = 1.0\times10^{-45}$
$Q_{sp} < K_{sp}$, so $Ca_3(PO_4)_2$ will not precipitate.

19.10 $AgBr(s) + 2NH_3(aq) \rightleftharpoons Ag(NH_3)_2^+(aq) + Br^-(aq)$
$K_{overall} = K_{sp}$ of $AgBr \times K_f$ of $Ag(NH_3)_2^+ = 8.5\times10^{-6}$
From the reaction table,

$$\frac{S}{1.0 - 2S} = \sqrt{8.5\times10^{-6}} = 2.9\times10^{-3}$$

$$S = [Ag(NH_3)_2^+] = 2.9\times10^{-3}\ M$$

Solubility is greater in 1 M hypo than in 1 M NH_3.

Problems

Problems with **colored** numbers are answered in Appendix E. Sections match the text and provide the numbers of relevant sample problems. Bracketed problems are grouped in pairs (indicated by a short rule) that cover the same concept. Comprehensive Problems are based on material from any section or previous chapter.

Note: Unless stated otherwise, all of the problems for this chapter refer to aqueous solutions at 298 K (25°C).

Equilibria of Acid-Base Buffer Systems
(Sample Problems 19.1 and 19.2)

19.1 What is the purpose of an acid-base buffer?

19.2 How do the acid and base components of a buffer function? Why are they often a conjugate acid-base pair of a weak acid?

19.3 What is the common-ion effect? How is it related to Le Châtelier's principle? Explain with equations that include HF and NaF.

19.4 When a small amount of H_3O^+ is added to a buffer, does the pH remain constant? Explain.

19.5 What is the difference between buffers with high and low capacities? Will adding 0.01 mol of HCl produce a greater pH change in a buffer with a high or a low capacity? Explain.

19.6 Which of these factors influence buffer capacity? How?
(a) Conjugate acid-base pair   (b) pH of the buffer
(c) Concentration of buffer components   (d) Buffer range
(e) pK_a of the acid component

19.7 What is the relationship between the buffer range and the buffer-component concentration ratio?

19.8 A chemist needs a pH 3.5 buffer. Should she use NaOH with formic acid ($K_a = 1.8 \times 10^{-4}$) or with acetic acid ($K_a = 1.8 \times 10^{-5}$)? Why? What is the disadvantage of choosing the other acid? What is the role of the NaOH?

19.9 What are the $[H_3O^+]$ and the pH of a propanoic acid–propanoate buffer that consists of 0.25 M CH_3CH_2COONa and 0.15 M CH_3CH_2COOH (K_a of propanoic acid = 1.3×10^{-5})?

19.10 What are the $[H_3O^+]$ and the pH of a benzoic acid–benzoate buffer that consists of 0.33 M C_6H_5COOH and 0.28 M C_6H_5COONa (K_a of benzoic acid = 6.3×10^{-5})?

19.11 Find the pH of a buffer that consists of 1.0 M sodium phenolate (C_6H_5ONa) and 1.2 M phenol (C_6H_5OH) (pK_a of phenol = 10.00).

19.12 Find the pH of a buffer that consists of 0.12 M boric acid (H_3BO_3) and 0.82 M sodium borate (NaH_2BO_3) (pK_a of boric acid = 9.24).

19.13 Find the pH of a buffer that consists of 0.20 M NH_3 and 0.10 M NH_4Cl (pK_b of NH_3 = 4.75).

19.14 Find the pH of a buffer that consists of 0.50 M methylamine (CH_3NH_2) and 0.60 M CH_3NH_3Cl (pK_b of CH_3NH_2 = 3.35).

19.15 What is the buffer-component concentration ratio, $[Pr^-]/[HPr]$, of a buffer that has a pH of 5.11 (K_a of HPr = 1.3×10^{-5})?

19.16 What is the buffer-component concentration ratio, $[NO_2^-]/[HNO_2]$, of a buffer that has a pH of 2.95 (K_a of HNO_2 = 7.1×10^{-4})?

19.17 A buffer containing 0.2000 M of acid, HA, and 0.1500 M of its conjugate base, A^-, has a pH of 3.35. What is the pH after 0.0015 mol of NaOH is added to 0.5000 L of this solution?

19.18 A buffer that contains 0.40 M base, B, and 0.25 M of its conjugate acid, BH^+, has a pH of 8.88. What is the pH after 0.0020 mol of HCl is added to 0.25 L of this solution?

19.19 A buffer is prepared by mixing 184 mL of 0.442 M HCl and 0.500 L of 0.400 M sodium acetate. (See Appendix C.) (a) What is the pH? (b) How many grams of KOH must be added to 0.500 L of the buffer to change the pH by 0.15 units?

19.20 A buffer is prepared by mixing 50.0 mL of 0.050 M sodium bicarbonate and 10.7 mL of 0.10 M NaOH. (See Appendix C.) (a) What is the pH? (b) How many grams of HCl must be added to 25.0 mL of the buffer to change the pH by 0.07 units?

19.21 Choose specific acid-base conjugate pairs suitable for preparing the following buffers: (a) pH ≈ 4.0; (b) pH ≈ 7.0. (See Appendix C.)

19.22 Choose specific acid-base conjugate pairs suitable for preparing the following buffers: (a) $[H_3O^+]$ ≈ 1×10^{-9} M; (b) $[OH^-]$ ≈ 3×10^{-5} M. (See Appendix C.)

19.23 An industrial chemist studying the effect of pH on bleaching and sterilizing processes prepares several hypochlorite buffers. Calculate the pH of each of the following buffers:
(a) 0.100 M HClO and 0.100 M NaClO
(b) 0.100 M HClO and 0.150 M NaClO
(c) 0.150 M HClO and 0.100 M NaClO
(d) One liter of the solution in part (a) after 0.0050 mol of NaOH has been added

19.24 Oxoanions of phosphorus are buffer components in blood. For a KH_2PO_4-Na_2HPO_4 solution with pH = 7.40 (pH of normal arterial blood), what is the buffer-component concentration ratio?

Acid-Base Titration Curves
(Sample Problem 19.3)

19.25 How can you estimate the pH range of an indicator's color change? Why do some indicators have two separate pH ranges?

19.26 Why does the color change of an indicator take place over a range of about 2 pH units?

19.27 Why doesn't the addition of an acid-base indicator affect the pH of the test solution?

19.28 What is the difference between the end point of a titration and the equivalence point? Is the equivalence point always reached first? Explain.

19.29 Some automatic titrators measure the slope of a titration curve to determine the equivalence point. What happens to the slope that enables the instrument to recognize this point?

19.30 Explain how *strong acid*–strong base, *weak acid*–strong base, and *weak base*–strong acid titrations using the same concentrations differ in terms of (a) the initial pH and (b) the pH at the equivalence point. (The component in italics is in the flask.)

19.31 What species are in the buffer region of a weak acid–strong base titration? How are they different from the species at the equivalence point? How are they different from the species in the buffer region of a weak base–strong acid titration?

19.32 Why is the center of the buffer region of a weak acid–strong base titration significant?

19.33 The indicator cresol red has $K_a = 5.0 \times 10^{-9}$. Over what approximate pH range does it change color?

19.34 The indicator thymolphthalein has $K_a = 7.9 \times 10^{-11}$. Over what approximate pH range does it change color?

19.35 Use Figure 19.5 to find an indicator for these titrations:
(a) 0.10 M HCl with 0.10 M NaOH
(b) 0.10 M HCOOH (Appendix C) with 0.10 M NaOH

19.36 Use Figure 19.5 to find an indicator for these titrations:
(a) 0.10 M CH$_3$NH$_2$ (Appendix C) with 0.10 M HCl
(b) 0.50 M HI with 0.10 M KOH

19.37 Calculate the pH during the titration of 50.00 mL of 0.1000 M HCl with 0.1000 M NaOH solution after the following additions of base: (a) 0 mL; (b) 25.00 mL; (c) 49.00 mL; (d) 49.90 mL; (e) 50.00 mL; (f) 50.10 mL; (g) 60.00 mL.

19.38 Calculate the pH during the titration of 30.00 mL of 0.1000 M KOH with 0.1000 M HBr solution after the following additions of acid: (a) 0 mL; (b) 15.00 mL; (c) 29.00 mL; (d) 29.90 mL; (e) 30.00 mL; (f) 30.10 mL; (g) 40.00 mL.

19.39 Find the pH during the titration of 20.00 mL of 0.1000 M butanoic acid, CH$_3$CH$_2$CH$_2$COOH ($K_a = 1.54 \times 10^{-5}$), with 0.1000 M NaOH solution after the following additions of titrant: (a) 0 mL; (b) 10.00 mL; (c) 15.00 mL; (d) 19.00 mL; (e) 19.95 mL; (f) 20.00 mL; (g) 20.05 mL; (h) 25.00 mL.

19.40 Find the pH during the titration of 20.00 mL of 0.1000 M triethylamine, (CH$_3$CH$_2$)$_3$N ($K_b = 5.2 \times 10^{-4}$), with 0.1000 M HCl solution after the following additions of titrant: (a) 0 mL; (b) 10.00 mL; (c) 15.00 mL; (d) 19.00 mL; (e) 19.95 mL; (f) 20.00 mL; (g) 20.05 mL; (h) 25.00 mL.

19.41 Find the pH and volume (mL) of 0.0372 M NaOH needed to reach the equivalence point in titrations of
(a) 42.2 mL of 0.0520 M CH$_3$COOH
(b) 23.4 mL of 0.0390 M HNO$_2$

19.42 Find the pH and the volume (mL) of 0.135 M HCl needed to reach the equivalence point(s) in titrations of the following:
(a) 55.5 mL of 0.234 M NH$_3$
(b) 17.8 mL of 1.11 M CH$_3$NH$_2$

Equilibria of Slightly Soluble Ionic Compounds
(Sample Problems 19.4 to 19.9)

19.43 The molar solubility of M$_2$X is 5×10^{-5} M. What is the molarity of each ion? How do you set up the calculation to find K_{sp}? What assumption must you make about the dissociation of M$_2$X into ions? Why is the calculated K_{sp} higher than the actual value?

19.44 Why does pH affect the solubility of CaF$_2$ but not of CaCl$_2$?

19.45 In a gaseous equilibrium, the reverse reaction occurs when $Q_c > K_c$. What occurs in aqueous solution when $Q_{sp} > K_{sp}$?

19.46 Write the ion-product expressions for (a) silver carbonate; (b) barium fluoride; (c) copper(II) sulfide.

19.47 Write the ion-product expressions for (a) iron(III) hydroxide; (b) barium phosphate; (c) tin(II) sulfide.

19.48 The solubility of silver carbonate is 0.032 M at 20°C. Calculate its K_{sp}.

19.49 The solubility of zinc oxalate is 7.9×10^{-3} M at 18°C. Calculate its K_{sp}.

19.50 The solubility of silver dichromate at 15°C is 8.3×10^{-3} g/100 mL solution. Calculate its K_{sp}.

19.51 The solubility of calcium sulfate at 30°C is 0.209 g/100 mL solution. Calculate its K_{sp}.

19.52 Find the molar solubility of SrCO$_3$ ($K_{sp} = 5.4 \times 10^{-10}$) in (a) pure water and (b) 0.13 M Sr(NO$_3$)$_2$.

19.53 Find the molar solubility of BaCrO$_4$ ($K_{sp} = 2.1 \times 10^{-10}$) in (a) pure water and (b) 1.5×10^{-3} M Na$_2$CrO$_4$.

19.54 Calculate the molar solubility of Ca(IO$_3$)$_2$ in (a) 0.060 M Ca(NO$_3$)$_2$ and (b) 0.060 M NaIO$_3$. (See Appendix C.)

19.55 Calculate the molar solubility of Ag$_2$SO$_4$ in (a) 0.22 M AgNO$_3$ and (b) 0.22 M Na$_2$SO$_4$. (See Appendix C.)

19.56 Which compound in each pair is more soluble in water?
(a) Magnesium hydroxide or nickel(II) hydroxide
(b) Lead(II) sulfide or copper(II) sulfide
(c) Silver sulfate or magnesium fluoride

19.57 Which compound in each pair is more soluble in water?
(a) Strontium sulfate or barium chromate
(b) Calcium carbonate or copper(II) carbonate
(c) Barium iodate or silver chromate

19.58 Write equations to show whether the solubility of either of the following is affected by pH: (a) AgCl; (b) SrCO$_3$.

19.59 Write equations to show whether the solubility of either of the following is affected by pH: (a) CuBr; (b) Ca$_3$(PO$_4$)$_2$.

19.60 Does any solid Cu(OH)$_2$ form when 0.075 g of KOH is dissolved in 1.0 L of 1.0×10^{-3} M Cu(NO$_3$)$_2$?

19.61 Does any solid PbCl$_2$ form when 3.5 mg of NaCl is dissolved in 0.250 L of 0.12 M Pb(NO$_3$)$_2$?

19.62 When blood is donated, sodium oxalate solution is used to precipitate Ca^{2+}, which triggers clotting. A 104-mL sample of blood contains 9.7×10^{-5} g Ca^{2+}/mL. A technologist treats the sample with 100.0 mL of 0.1550 M Na$_2$C$_2$O$_4$. Calculate [Ca^{2+}] after the treatment. (See Appendix C for K_{sp} of CaC$_2$O$_4$·H$_2$O.)

Equilibria Involving Complex Ions
(Sample Problem 19.10)

19.63 How can a positive metal ion be at the center of a negative complex ion?

19.64 Write equations to show the stepwise reaction of Cd(H$_2$O)$_4^{2+}$ in an aqueous solution of KI to form CdI$_4^{2-}$. Show that $K_{f(overall)} = K_{f1} \times K_{f2} \times K_{f3} \times K_{f4}$.

19.65 Consider the dissolution of PbS in water:
$$PbS(s) + H_2O(l) \rightleftharpoons Pb^{2+}(aq) + HS^-(aq) + OH^-(aq)$$
Adding aqueous NaOH causes more PbS to dissolve. Does this violate Le Châtelier's principle? Explain.

19.66 Write a balanced equation for the reaction of Hg(H$_2$O)$_4^{2+}$ in aqueous KCN.

19.67 Write a balanced equation for the reaction of Zn(H$_2$O)$_4^{2+}$ in aqueous NaCN.

19.68 Write a balanced equation for the reaction of Ag(H$_2$O)$_2^+$ in aqueous Na$_2$S$_2$O$_3$.

19.69 Write a balanced equation for the reaction of Al(H$_2$O)$_6^{3+}$ in aqueous KF.

19.70 Find the solubility of AgI in 2.5 M NH$_3$ [K_{sp} of AgI = 8.3×10^{-17}; K_f of Ag(NH$_3$)$_2^+$ = 1.7×10^7].

19.71 Find the solubility of Cr(OH)$_3$ in a buffer of pH 13.0 [K_{sp} of Cr(OH)$_3$ = 6.3×10^{-31}; K_f of Cr(OH)$_4^-$ = 8.0×10^{29}].

Comprehensive Problems

Problems with an asterisk (*) are more challenging.

19.72 A microbiologist is preparing a medium on which to culture *E. coli* bacteria. She buffers the medium at pH 7.00 to minimize the effect of acid-producing fermentation. What volumes of equimolar aqueous solutions of K_2HPO_4 and KH_2PO_4 must she combine to make 100. mL of the pH 7.00 buffer?

19.73 Tris(hydroxymethyl)aminomethane [$(HOCH_2)_3CNH_2$, known as TRIS or THAM] is a weak base widely used in biochemical experiments to make buffer solutions in the pH range of 7 to 9. A certain TRIS buffer has a pH of 8.10 at 25°C and a pH of 7.80 at 37°C. Why does the pH change with temperature?

19.74 Gout is caused by an error in nucleic acid metabolism that leads to a buildup of uric acid in body fluids, which is deposited as slightly soluble sodium urate ($C_5H_3N_4O_3Na$) in the soft tissues of joints. If the extracellular [Na^+] is 0.15 *M* and the solubility in water of sodium urate is 0.085 g/100. mL, what is the minimum urate ion concentration (abbreviated [Ur^-]) that will cause a deposit of sodium urate?

19.75 Cadmium ion in solution is analyzed by being precipitated as the sulfide, a yellow compound used as a pigment in everything from artists' oil paints to glass and rubber. Calculate the molar solubility of cadmium sulfide at 25°C.

19.76 The solubility of KCl is 3.7 *M* at 20°C. Two beakers contain 100. mL of saturated KCl solution: 100. mL of 6.0 *M* HCl is added to the first beaker and 100. mL of 12 *M* HCl to the second. (a) Find the ion-product constant of KCl at 20°C. (b) What mass, if any, of KCl will precipitate from each beaker?

19.77 Manganese(II) sulfide is one of the compounds found in the nodules on the ocean floor that may eventually be a primary source of many transition metals. The solubility of MnS is 4.7×10^{-4} g/100 mL solution. Estimate the K_{sp} of MnS.

19.78 The normal pH of blood is 7.40 ± 0.05 and is controlled in part by the H_2CO_3-HCO_3^- buffer system.
(a) Assuming that the K_a value for carbonic acid at 25°C applies to blood, what is the [H_2CO_3]/[HCO_3^-] ratio in normal blood?
(b) In a condition called *acidosis*, the blood is too acidic. What is the [H_2CO_3]/[HCO_3^-] ratio in a patient whose blood pH is 7.20 (severe acidosis)?

19.79 Tooth enamel consists of hydroxyapatite, $Ca_5(PO_4)_3OH$ ($K_{sp} = 6.8 \times 10^{-37}$). Fluoride ion added to drinking water reacts with $Ca_5(PO_4)_3OH$ to form the more tooth decay–resistant fluorapatite, $Ca_5(PO_4)_3F$ ($K_{sp} = 1.0 \times 10^{-60}$). Fluoridated water has dramatically decreased cavities among children. Calculate the solubility of $Ca_5(PO_4)_3OH$ and of $Ca_5(PO_4)_3F$ in water.

19.80 The acid-base indicator ethyl orange turns from red to yellow over the pH range 3.4 to 4.8. Estimate K_a for ethyl orange.

* 19.81 Instrumental acid-base titrations use a pH meter to monitor the changes in pH and volume. The equivalence point is found from the volume at which the curve has the steepest slope.
(a) Use Figure 19.7 to calculate the slope $\Delta pH/\Delta V$ for all pairs of adjacent points and the average volume (V_{avg}) for each interval.

(b) Plot $\Delta pH/\Delta V$ vs. V_{avg} to find the steepest slope, and thus the volume at the equivalence point. (For example, the first pair of points gives $\Delta pH = 0.22$, $\Delta V = 10.00$ mL; hence, $\Delta pH/\Delta V = 0.022$ mL^{-1}, and $V_{avg} = 5.00$ mL.)

19.82 What is the pH of a solution of 6.5×10^{-9} mol of $Ca(OH)_2$ in 10.0 L of water [K_{sp} of $Ca(OH)_2 = 6.5 \times 10^{-6}$]?

19.83 A student wants to dissolve the maximum amount of CaF_2 ($K_{sp} = 3.2 \times 10^{-11}$) to make 1 L of aqueous solution.
(a) Into which of the following should she dissolve the salt?
(I) Pure water   (II) 0.01 *M* HF   (III) 0.01 *M* NaOH
(IV) 0.01 *M* HCl   (V) 0.01 *M* $Ca(OH)_2$
(b) Which would dissolve the least amount of salt?

19.84 The Henderson-Hasselbalch equation gives a relationship for obtaining the pH of a buffer solution consisting of HA and A^-. Derive an analogous relationship for obtaining the pOH of a buffer solution consisting of B and BH^+.

19.85 Calculate the molar solubility of $Hg_2C_2O_4$ ($K_{sp} = 1.75 \times 10^{-13}$) in 0.13 *M* $Hg_2(NO_3)_2$.

19.86 The well water in an area is "hard" because it is in equilibrium with $CaCO_3$ in the surrounding rocks. What is the concentration of Ca^{2+} in the well water (assuming the water's pH is such that the CO_3^{2-} ion is not hydrolyzed)? (See Appendix C for K_{sp} of $CaCO_3$.)

19.87 An environmental technician collects a sample of rainwater. A light on her portable pH meter indicates low battery power, so she uses indicator solutions to estimate the pH. A piece of litmus paper turns red, indicating acidity, so she divides the sample into thirds and obtains the following results: thymol blue turns yellow; bromphenol blue turns green; and methyl red turns red. Estimate the pH of the rainwater.

* 19.88 Quantitative analysis of Cl^- ion is often performed by a titration with silver nitrate, using sodium chromate as an indicator. As standardized $AgNO_3$ is added, both white AgCl and red Ag_2CrO_4 precipitate, but so long as some Cl^- remains, the Ag_2CrO_4 redissolves as the mixture is stirred. When the red color is permanent, the equivalence point has been reached.
(a) Calculate the equilibrium constant for the reaction

$$2AgCl(s) + CrO_4^{2-}(aq) \rightleftharpoons Ag_2CrO_4(s) + 2Cl^-(aq)$$

(b) Explain why the silver chromate redissolves.

19.89 Some kidney stones form by the precipitation of calcium oxalate monohydrate ($CaC_2O_4 \cdot H_2O$, $K_{sp} = 2.3 \times 10^{-9}$). The pH of urine varies from 5.5 to 7.0, and the average [Ca^{2+}] in urine is 2.6×10^{-3} *M*.
(a) If the concentration of oxalic acid in urine is 3.0×10^{-13} *M*, will kidney stones form at pH = 5.5?
(b) At pH = 7.0?
(c) Vegetarians have a urine pH above 7. Are they more or less likely to form kidney stones?

19.90 A 35.00-mL solution of 0.2500 *M* HF is titrated with a standardized 0.1532 *M* solution of NaOH at 25°C.
(a) What is the pH of the HF solution before titrant is added?
(b) How many milliliters of titrant are required to reach the equivalence point?
(c) What is the pH at 0.50 mL before the equivalence point?
(d) What is the pH at the equivalence point?
(e) What is the pH at 0.50 mL after the equivalence point?

* **19.91** Because of the toxicity of mercury compounds, mercury(I) chloride is used in antibacterial salves. The mercury(I) ion (Hg_2^{2+}) consists of two bound Hg^+ ions.
(a) What is the empirical formula of mercury(I) chloride?
(b) Calculate [Hg_2^{2+}] in a saturated solution of mercury(I) chloride ($K_{sp} = 1.5 \times 10^{-18}$).
(c) A seawater sample contains 0.20 lb of NaCl per gallon. Find [Hg_2^{2+}] if the seawater is saturated with mercury(I) chloride.
(d) How many grams of mercury(I) chloride are needed to saturate 4900 km^3 of water (the volume of Lake Michigan)?
(e) How many grams of mercury(I) chloride are needed to saturate 4900 km^3 of seawater?

* **19.92** A lake that has a surface area of 10.0 acres (1 acre = 4.840×10^3 yd^2) receives 1.00 in. of rain of pH 4.20. (Assume the acidity of the rain is due to a strong, monoprotic acid.)
(a) How many moles of H_3O^+ are in the rain falling on the lake?
(b) If the lake is unbuffered (pH = 7.00) and its average depth is 10.0 ft before the rain, find the pH after the rain has been mixed with lake water. (Ignore runoff from the surrounding land.)
(c) If the lake contains hydrogen carbonate ions (HCO_3^-), what mass of HCO_3^- would neutralize the acid in the rain?

19.93 Sodium chloride is purified for use as table salt by adding HCl to a saturated solution of NaCl (317 g/L). When 25.5 mL of 7.85 M HCl is added to 0.100 L of saturated solution, how many grams of purified NaCl precipitate?

19.94 Calcium ion present in water supplies is easily precipitated as calcite ($CaCO_3$):
$$Ca^{2+}(aq) + CO_3^{2-}(aq) \rightleftharpoons CaCO_3(s)$$
Because the K_{sp} decreases with temperature, heating hard water forms a calcite "scale," which clogs pipes and water heaters. Find the solubility of calcite in water (a) at 10°C ($K_{sp} = 4.4 \times 10^{-9}$) and (b) at 30°C ($K_{sp} = 3.1 \times 10^{-9}$).

19.95 Scenes A to D represent tiny portions of 0.10 M aqueous solutions of a weak acid HA ($K_a = 4.5 \times 10^{-5}$), its conjugate base A^-, or a mixture of the two (only these species are shown):

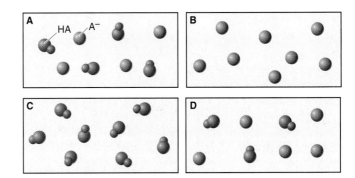

(a) Which scene(s) show(s) a buffer?
(b) What is the pH of each solution?
(c) Arrange the scenes in sequence, assuming that they represent stages in a weak acid–strong base titration.
(d) Which scene represents the titration at its equivalence point?

19.96 Scenes A to C represent aqueous solutions of the slightly soluble salt MZ (only the ions of this salt are shown):
$$MZ(s) \rightleftharpoons M^{2+}(aq) + Z^{2-}(aq)$$

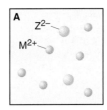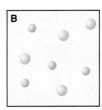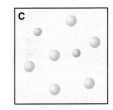

(a) Which scene represents the solution just after solid MZ is stirred thoroughly in distilled water?
(b) If each sphere represents 2.5×10^{-6} M of ions, what is the K_{sp} of MZ?
(c) Which scene represents the solution after $Na_2Z(aq)$ is added?
(d) If Z^{2-} is CO_3^{2-}, which scene represents the solution after the pH has been lowered?

CHAPTER TWENTY

Thermodynamics: Entropy, Free Energy, and the Direction of Chemical Reactions

Down's Easier than Up After an initial push, a sled goes down a hill by itself, but you must continually pull it up. In the same sense, as you'll learn in this chapter, some reactions occur by themselves once started, while others occur only through a continual input of energy.

Key Principles

◆ A process, such as a rock falling or a fuel burning, is called *spontaneous* because, once started, it continues by itself. A *nonspontaneous* process requires a continuous input of energy to occur. A *reaction proceeding toward equilibrium* is a spontaneous process.

◆ The total kinetic energy of a system is the sum of the translational, rotational, and vibrational energies of its particles, each of which is *quantized*. A *microstate* of the system is any specific combination of the quantized energy states of all the particles. The *entropy* of a system is directly related to the number of microstates over which the system *disperses its energy,* which is closely associated with the *freedom of motion* of the particles. A substance has more entropy in its gaseous state than in its liquid state, and more in its liquid state than in its solid state.

◆ We predict the direction of a spontaneous change from the *second law* of thermodynamics: a spontaneous change occurs in the direction that increases the entropy of the universe (system plus surroundings). In other words, a change occurs spontaneously if the energy of the universe becomes more dispersed.

◆ The *third law* of thermodynamics—the entropy of a perfect crystal is zero at 0 K—allows us to calculate *absolute* entropies. The *standard molar entropy* ($S°$) of a substance is influenced by temperature, physical state, dissolution, and atomic size or molecular complexity.

◆ Gases have such high entropy that if a reaction yields a net increase in moles of gas, the standard entropy change of the reaction is positive ($\Delta S°_{rxn} > 0$); if there is a net decrease in moles of gas, $\Delta S°_{rxn} < 0$.

◆ For a spontaneous process, $\Delta S_{univ} > 0$. Thus, the only way the entropy of the system can decrease is if the entropy of the surroundings increases even more.

◆ The *free energy change* (ΔG) of a process is a measure of its spontaneity. The free energy of a system decreases in a spontaneous process; that is, if $\Delta S_{univ} > 0$, $\Delta G_{sys} < 0$. In a *coupling of reactions,* a spontaneous step with a larger negative ΔG drives a nonspontaneous step with a smaller positive ΔG.

◆ ΔG is that portion of the total energy change available to do *work*. In any real process, however, some of the free energy is always *lost as heat*.

Outline

In the last few chapters, we've posed and answered some essential questions about chemical and physical change: How fast does the change occur, and how is this rate affected by concentration and temperature? How much product will be present when the net change ceases, and how is this yield affected by concentration and temperature?

Now it's time to stand back and ask the most profound question of all: why does a change occur in the first place? From everyday experience, it seems that some changes happen by themselves—that is, spontaneously—almost as if a force were driving them in one direction and not the other. Turn on a gas stove, for example, and the methane mixes with oxygen and burns immediately to yield carbon dioxide and water vapor. But those products will not remake methane and oxygen no matter how long they mix. A shiny steel shovel left outside slowly rusts, but put a rusty one outside and it won't become shiny. A cube of sugar dissolves in a cup of coffee after a few seconds of stirring, but stir for another century and the dissolved sugar won't reappear as a cube. In this chapter, we discuss the nature of such spontaneous changes. The principles of thermodynamics that we cover here apply, as far as we know, to every system in the universe!

Concepts & Skills to Review
Before You Study This Chapter

- internal energy, heat, and work (Section 6.1)
- state functions (Section 6.1) and standard states (Section 6.6)
- enthalpy, ΔH, and Hess's law (Sections 6.2 and 6.5)
- entropy and solution formation (Section 13.2)
- comparing Q and K to find reaction direction (Section 17.4)

20.1 THE SECOND LAW OF THERMODYNAMICS: PREDICTING SPONTANEOUS CHANGE

In a formal sense, a **spontaneous change** of a system, whether a chemical or physical change or just a change in location, is one that occurs by itself under specified conditions, without an ongoing input of energy from outside the system. The freezing of water, for example, is spontaneous at 1 atm and $-5°C$. A spontaneous process such as burning or falling may need a little "push" to get started—a spark to ignite gasoline, a shove to knock a book off your desk—but once the process begins, it keeps going without the need for any external input of energy.

In contrast, for a nonspontaneous change to occur, the surroundings must supply the system with a *continuous* input of energy. A book falls spontaneously, but it rises only if something else, such as a human hand (or a hurricane-force wind), supplies energy in the form of work. Under a given set of conditions, *if a change is spontaneous in one direction, it is **not** spontaneous in the other.*

Note that the term *spontaneous* does not mean *instantaneous* and has nothing to do with how long a process takes to occur; it means that, given enough time, the process will happen by itself. Many processes are spontaneous but slow—ripening, rusting, and (happily!) aging.

A chemical reaction proceeding toward equilibrium is an example of a spontaneous change. As you learned in Chapter 17, we can predict the net direction of the reaction—its spontaneous direction—by comparing the reaction quotient (Q) with the equilibrium constant (K). But *why* is there a drive to attain equilibrium? And what determines the value of the equilibrium constant? Can we tell the direction of a spontaneous change in cases that are not as obvious as burning gasoline or falling books? Because energy changes seem to be involved, let's begin by reviewing the idea of conservation of energy to see whether it can help uncover the criterion for spontaneity.

Limitations of the First Law of Thermodynamics

In Chapter 6, we discussed the first law of thermodynamics (the law of conservation of energy). It states that the internal energy (E) of a system, the sum of the kinetic and potential energy of all its particles, changes when heat (q) and/or work (w) are added or removed:

$$\Delta E = q + w$$

Whatever is not part of the system (sys) is part of the surroundings (surr), so the system and surroundings together constitute the universe (univ):

$$E_{univ} = E_{sys} + E_{surr}$$

Heat and/or work gained by the system is lost by the surroundings, and vice versa:

$$(q + w)_{sys} = -(q + w)_{surr}$$

It follows from these ideas that *the total energy of the universe is constant:**

$$\Delta E_{sys} = -\Delta E_{surr} \qquad \text{therefore} \qquad \Delta E_{sys} + \Delta E_{surr} = 0 = \Delta E_{univ}$$

Is the first law sufficient to explain why a natural process takes place as it does? It certainly accounts for the energy involved. When gasoline burns in your car's engine, the first law states that the potential energy difference between the chemical bonds in the fuel mixture and those in the exhaust gases is converted to the kinetic energy of the moving car and its parts plus the heat released to the environment. If you could measure the work and heat involved, you would find that energy is conserved as it is converted from one form to another.

However, the first law does not help us make sense of the *direction* of the change. Why doesn't the heat released in the car engine convert exhaust fumes back into gasoline and oxygen? This change would not violate the first law—energy would still be conserved—but it would never happen. The first law by itself tells nothing about the direction of a spontaneous change, so we must look elsewhere for a way to predict that direction.

The Sign of ΔH Cannot Predict Spontaneous Change

In the mid-19th century, some thought that the sign of the enthalpy change (ΔH), the heat added or removed at constant pressure (q_P), was the criterion for spontaneity. They thought that exothermic processes ($\Delta H < 0$) were spontaneous and endothermic ones ($\Delta H > 0$) were nonspontaneous. This hypothesis had a lot of support from observation; after all, many spontaneous processes *are* exothermic. All combustion reactions, such as methane burning, are spontaneous and exothermic:

$$CH_4(g) + 2O_2(g) \longrightarrow CO_2(g) + 2H_2O(g) \qquad \Delta H^\circ_{rxn} = -802 \text{ kJ}$$

Iron metal oxidizes spontaneously and exothermically:

$$2Fe(s) + \tfrac{3}{2}O_2(g) \longrightarrow Fe_2O_3(s) \qquad \Delta H^\circ_{rxn} = -826 \text{ kJ}$$

Ionic compounds, such as NaCl, form spontaneously and exothermically from their elements:

$$Na(s) + \tfrac{1}{2}Cl_2(g) \longrightarrow NaCl(s) \qquad \Delta H^\circ_{rxn} = -411 \text{ kJ}$$

However, in many other cases, the sign of ΔH is no help. An exothermic process occurs spontaneously under certain conditions, whereas the opposite, endothermic, process occurs spontaneously under other conditions. Consider the following examples of phase changes, dissolving salts, and chemical changes.

At ordinary pressure, water freezes below 0°C but melts above 0°C. Both changes are spontaneous, but the first is exothermic and the second endothermic:

$$H_2O(l) \longrightarrow H_2O(s) \qquad \Delta H^\circ_{rxn} = -6.02 \text{ kJ } (exothermic; \text{ spontaneous at } T < 0°C)$$
$$H_2O(s) \longrightarrow H_2O(l) \qquad \Delta H^\circ_{rxn} = +6.02 \text{ kJ } (endothermic; \text{ spontaneous at } T > 0°C)$$

At ordinary pressure and room temperature, liquid water vaporizes spontaneously in dry air, another endothermic change:

$$H_2O(l) \longrightarrow H_2O(g) \qquad \Delta H^\circ_{rxn} = +44.0 \text{ kJ}$$

In fact, all melting and vaporizing are endothermic changes that are spontaneous under proper conditions.

---

*Any modern statement of conservation of energy must take into account mass-energy equivalence and the processes in stars, which convert enormous amounts of matter into energy. These can be included by stating that the total *mass-energy* of the universe is constant.

Recall from Chapter 13 that most water-soluble salts have a positive $\Delta H^\circ_{\text{soln}}$ yet dissolve spontaneously:

$$NaCl(s) \xrightarrow{\text{H}_2\text{O}} Na^+(aq) + Cl^-(aq) \qquad \Delta H^\circ_{\text{soln}} = +3.9 \text{ kJ}$$

$$NH_4NO_3(s) \xrightarrow{\text{H}_2\text{O}} NH_4^+(aq) + NO_3^-(aq) \qquad \Delta H^\circ_{\text{soln}} = +25.7 \text{ kJ}$$

Some endothermic chemical changes are also spontaneous:

$$N_2O_5(s) \longrightarrow 2NO_2(g) + \tfrac{1}{2}O_2(g) \qquad \Delta H^\circ_{\text{rxn}} = +109.5 \text{ kJ}$$

$$Ba(OH)_2 \cdot 8H_2O(s) + 2NH_4NO_3(s) \longrightarrow$$
$$Ba^{2+}(aq) + 2NO_3^-(aq) + 2NH_3(aq) + 10H_2O(l) \qquad \Delta H^\circ_{\text{rxn}} = +62.3 \text{ kJ}$$

Freedom of Particle Motion and Dispersal of Particle Energy

What features common to the previous endothermic processes can help us see why they occur spontaneously? In each case, the particles that make up the matter have more freedom of motion after the change occurs. And this means that their energy of motion becomes more dispersed. As we'll see below, "dispersed" means spread over more quantized energy levels.

Phase changes lead from a solid, in which particle motion is restricted, to a liquid, in which the particles have more freedom to move around each other, to a gas, with its much greater freedom of particle motion. Along with this greater freedom of motion, the energy of the particles becomes dispersed over more levels. Dissolving a salt leads from a crystalline solid and pure liquid to ions and solvent molecules moving and interacting throughout the solution; their energy of motion, therefore, is much more dispersed. In the chemical reactions shown, *fewer* moles of crystalline solids produce *more* moles of gases and/or solvated ions. In these cases, there is not only more freedom of motion, but more particles to disperse their energy over more levels.

Thus, in each process, the particles have more freedom of motion and, therefore, their energy of motion has more levels over which to be dispersed:

less freedom of particle motion $\longrightarrow$ more freedom of particle motion
localized energy of motion $\longrightarrow$ dispersed energy of motion

Phase change: solid $\longrightarrow$ liquid $\longrightarrow$ gas
Dissolving of salt: crystalline solid + liquid $\longrightarrow$ ions in solution
Chemical change: crystalline solids $\longrightarrow$ gases + ions in solution

In thermodynamic terms, a change in the freedom of motion of particles in a system and in the dispersal of their energy of motion is a key factor determining the direction of a spontaneous process.

Entropy and the Number of Microstates

Let's see how freedom of motion and dispersal of energy relate to spontaneous change. Picture a system of, say, 1 mol of N_2 gas and focus on one molecule. At any instant, it is moving through space (translating) at a certain speed, it is rotating at a certain frequency, and its atoms are vibrating at a certain frequency. In the next instant, the molecule collides with another, and these motional (kinetic) energy states change. In our brief discussion of IR spectroscopy (Section 9.3), we mentioned that, just as the electronic energy states of molecules are quantized, so are their vibrational energy states. The complete quantum state of the molecule at any instant is given by a combination of its particular electronic, translational, rotational, and vibrational states. Clearly, many such combinations are possible for this single molecule, and the number of quantized energy states possible for the system of a mole of molecules is staggering—on the order of $10^{10^{23}}$. Each quantized state of the system is called a *microstate,* and every microstate has the same total energy at a given set of conditions. With each microstate equally possible for the system, the laws of probability say that, over time, all microstates are equally occupied. If we focus only on microstates associated with thermal energy, the number of microstates (W) of a system is the number of ways it can disperse its thermal energy among the various modes of motion of all its molecules.

In 1877, the Austrian mathematician and physicist Ludwig Boltzmann defined the **entropy** (S) of a system in terms of W:

$$S = k \ln W \qquad (20.1)$$

where k, the *Boltzmann constant,* is the universal gas constant (R) divided by Avogadro's number (N_A), or R/N_A, and equals 1.38×10^{-23} J/K. Because W is just a number of microstates and has no units, S has units of joules/kelvin (J/K). From this relationship, we conclude that

- A system with fewer microstates (smaller W) among which to spread its energy has *lower entropy (lower S).*
- A system with more microstates (larger W) among which to spread its energy has *higher entropy (higher S).*

Thus, for our earlier examples,

lower entropy (fewer microstates) $\longrightarrow$ higher entropy (more microstates)

Phase change: solid $\longrightarrow$ liquid $\longrightarrow$ gas

Dissolving of salt: crystalline solid + liquid $\longrightarrow$ ions in solution

Chemical change: crystalline solids $\longrightarrow$ gases + ions in solution

(In Chapter 13, we used some of these ideas to explain solution behavior.)

Changes in Entropy If the number of microstates increases during a physical or chemical change, there are more ways for the energy of the system to be dispersed among them. Thus, the entropy increases:

$$S_{\text{more microstates}} > S_{\text{fewer microstates}}$$

If the number of microstates decreases, the entropy decreases.

Like internal energy (E) and enthalpy (H), entropy is a thermodynamic state function, which means it depends only on the present state of the system, not on the path it took to arrive at that state (Chapter 6). Therefore, the change in entropy of the system (ΔS_{sys}) depends only on the difference between its final and initial values:

$$\Delta S_{\text{sys}} = S_{\text{final}} - S_{\text{initial}}$$

Like any state function, $\Delta S_{\text{sys}} > 0$ when its value increases during a change. For example, when dry ice sublimes to gaseous carbon dioxide, we have

$$CO_2(s) \longrightarrow CO_2(g) \qquad \Delta S_{\text{sys}} = S_{\text{gaseous } CO_2} - S_{\text{solid } CO_2} > 0$$

Similarly, $\Delta S_{\text{sys}} < 0$ when the entropy decreases during a change, as when water vapor condenses:

$$H_2O(g) \longrightarrow H_2O(l) \qquad \Delta S_{\text{sys}} = S_{\text{liquid } H_2O} - S_{\text{gaseous } H_2O} < 0$$

Or consider the decomposition of dinitrogen tetraoxide (written as $O_2N—NO_2$):

$$O_2N—NO_2(g) \longrightarrow 2NO_2(g)$$

When the N—N bond in 1 mol of dinitrogen tetraoxide molecules breaks, the 2 mol of NO_2 molecules have much more freedom of motion; thus, their energy is spread over more microstates:

$$\Delta S_{\text{sys}} = \Delta S_{\text{rxn}} = S_{\text{final}} - S_{\text{initial}} = S_{\text{products}} - S_{\text{reactants}} = 2S_{NO_2} - S_{N_2O_4} > 0$$

Quantitative Meaning of the Entropy Change Two approaches for quantifying an entropy change look different but give the same result. The first is a statistical approach based on the number of microstates possible for the particles in a system. The second is based on the heat absorbed (or released) by a system. We'll explore both in a simple case of 1 mol of an ideal gas, say neon, expanding from 10 L to 20 L at 298 K:

1 mol neon (initial: 10 L and 298 K) $\longrightarrow$ 1 mol neon (final: 20 L and 298 K)

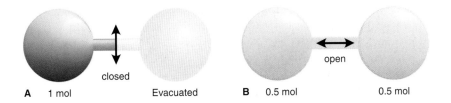

Figure 20.1 Spontaneous expansion of a gas The container consists of two identical flasks connected by a stopcock. **A,** With the stopcock closed, 1 mol of neon gas occupies one flask, and the other is evacuated. **B,** Open the stopcock, and the gas expands spontaneously until each flask contains 0.5 mol.

We use a statistical approach to find ΔS_{sys} by applying the definition of entropy expressed by Equation 20.1. Figure 20.1A shows a container consisting of two identical flasks connected by a stopcock, with 1 mol of neon in the left flask and an evacuated right flask. We know from experience that when we open the stopcock, the gas will expand to fill both flasks with 0.5 mol each—but *why*?

Let's start with one neon atom and think through what happens as we add more atoms and open the stopcock (Figure 20.2). One atom has some number of microstates (W) possible for it in the left flask and the same number possible in the right flask. Opening the stopcock increases the volume, which increases the number of possible particle locations and, thus, translational energy levels. As a result, the system has 2^1, or 2, times as many microstates possible when the atom moves through both flasks (final state, W_{final}) as when it is confined to one flask (initial state, $W_{initial}$).

With more atoms, different combinations of atoms can occupy various energy levels, and each combination represents a microstate. With 2 atoms, A and B, moving through both flasks, there are 2^2, or 4, times as many microstates as when they are confined initially to one flask—some number of microstates with A and B in the left, the same number with A in the left and B in the right, that number with B in the left and A in the right, and that number with A and B in the right. Add another atom and there are 2^3, or 8, times as many microstates when the

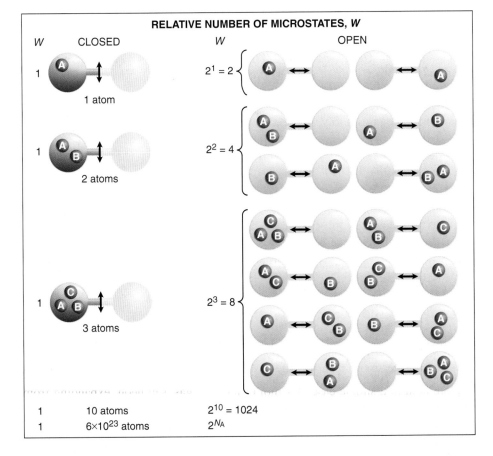

Figure 20.2 Expansion of a gas and the increase in number of microstates. When a gas confined to one flask is allowed to spread through two flasks, the energy of the particles is dispersed over more microstates, and so the entropy is higher. Each combination of particles in the available volume represents a different microstate. The increase in the number of possible microstates that occurs when the volume increases is given by 2^n, where n is the number of particles.

stopcock is open—some number with all three in the left, that number with A and B in the left and C in the right, that number with A and C in the left and B in the right, and so on. With 10 neon atoms, there are 2^{10}, or 1024, times as many microstates for the gas in both flasks. Finally, with 1 mol (N_A) of neon atoms, there are 2^{N_A} times as many microstates possible for the atoms in both flasks (W_{final}) as in one flask ($W_{initial}$). In other words, for 1 mol, we have

$$W_{final}/W_{initial} = 2^{N_A}$$

Now let's find ΔS_{sys} through the Boltzmann equation, $S = k \ln W$. From Appendix A, we know that $\ln A - \ln B = \ln A/B$. Thus,

$$\Delta S_{sys} = S_{final} - S_{initial} = k \ln W_{final} - k \ln W_{initial} = k \ln (W_{final}/W_{initial})$$

Also from Appendix A, $\ln A^y = y \ln A$; with $k = R/N_A$, we have

$$\Delta S_{sys} = R/N_A \ln 2^{N_A} = (R/N_A)N_A \ln 2 = R \ln 2 = (8.314 \text{ J/mol·K})(0.693)$$
$$= 5.76 \text{ J/mol·K}$$

or, for 1 mol, $\qquad\qquad\qquad \Delta S_{sys} = 5.76 \text{ J/K}$

The second approach for finding ΔS_{sys} is based on heat changes and relates closely to 19th-century attempts to understand the work done by steam engines. In such a process, the entropy change is defined by

$$\Delta S_{sys} = \frac{q_{rev}}{T} \qquad\qquad (20.2)$$

where T is the temperature at which the heat change occurs and q is the heat absorbed. The subscript "rev" refers to a *reversible process,* one that occurs slowly enough for equilibrium to be maintained continuously, so that the direction of the change can be reversed by an infinitesimal reversal of conditions.

A truly reversible expansion of an ideal gas can only be imagined, but we can approximate it by placing the 10-L neon sample in a piston-cylinder assembly surrounded by a heat reservoir maintained at 298 K, with a beaker of sand on the piston exerting the pressure. We remove one grain of sand (an "infinitesimal" change in pressure) with a pair of tweezers, and the gas expands a tiny amount, raising the piston and doing work on the surroundings, $-w$. If the neon behaves ideally, it absorbs from the reservoir a tiny increment of heat q, equivalent to $-w$. We remove another grain of sand, and the gas expands a tiny bit more and absorbs another tiny increment of heat. This expansion is very close to being reversible because we can reverse it at any point by putting a grain of sand back into the beaker, which causes a tiny compression of the gas and a tiny release of heat into the reservoir. If we continue this expansion process to 20 L and apply calculus to add together all the tiny increments of heat, we find q_{rev} is 1718 J. Thus, applying Equation 20.2, the entropy change is

$$\Delta S_{sys} = q_{rev}/T = 1718 \text{ J/298 K} = 5.76 \text{ J/K}$$

This is the same result we obtained by the statistical approach. That approach helps us visualize entropy changes in terms of the number of microstates over which the energy is dispersed, but the calculations are limited to simple systems like ideal gases. This approach, which involves incremental heat changes, is less easy to visualize but can be applied to liquids, solids, and solutions, as well as gases.

Entropy and the Second Law of Thermodynamics

Now back to our earlier question: what criterion determines the direction of a spontaneous change? The change in entropy is essential, but to evaluate it correctly, we have to consider more than just the system. After all, some changes, such as ice melting or a crystal dissolving, occur spontaneously and end up with higher entropy, whereas others, such as water freezing or a crystal forming, occur spontaneously and end up with lower entropy. If we consider changes in *both* the system *and* its surroundings, however, we find that *all real processes occur spontaneously in the direction that increases the entropy of the universe (system plus surroundings).* This is one way to state the **second law of thermodynamics.**

Notice that the second law places no limitations on the entropy change of the system *or* the surroundings: either may be negative; that is, either system *or* surroundings may have lower entropy after the process. The law does state, however, that for a spontaneous process, the *sum* of the entropy changes must be positive. If the entropy of the system decreases, the entropy of the surroundings increases even more to offset the system's decrease, and so the entropy of the universe (system *plus* surroundings) increases. A quantitative statement of the second law is that, for any real spontaneous process,

$$\Delta S_{univ} = \Delta S_{sys} + \Delta S_{surr} > 0 \qquad (20.3)$$

Standard Molar Entropies and the Third Law

Both entropy and enthalpy are state functions, but the nature of their values differs in a fundamental way. Recall that we cannot determine absolute *enthalpies* because we have no easily measurable starting point, no baseline value for the enthalpy of a substance. Therefore, we measure only enthalpy *changes*.

In contrast, we *can* determine the absolute *entropy* of a substance. To do so requires application of the **third law of thermodynamics,** which states that *a perfect crystal has zero entropy at a temperature of absolute zero:* $S_{sys} = 0$ at 0 K. "Perfect" means that all the particles are aligned flawlessly in the crystal structure, with no defects of any kind. At absolute zero, all particles in the crystal have the minimum energy, and there is only one way it can be dispersed: thus, in Equation 20.1, $W = 1$, so $S = k \ln 1 = 0$. When we warm the crystal, its total energy increases, so the particles' energy can be dispersed over more microstates (Figure 20.3). Thus, $W > 1$, $\ln W > 0$, and $S > 0$.

To obtain a value for S at a given temperature, we first cool a crystalline sample of the substance as close to 0 K as possible. Then we heat it in small increments, dividing q by T to get the increase in S for each increment, and add up all the entropy increases to the temperature of interest, usually 298 K. The entropy of a substance at a given temperature is therefore an *absolute* value that is equal to the entropy increase obtained when the substance is heated from 0 K to that temperature.

As with other thermodynamic variables, we usually compare entropy values for substances in their *standard states* at the temperature of interest: *1 atm for gases, 1 M for solutions, and the pure substance in its most stable form for solids or liquids.* Because entropy is an *extensive* property, that is, one that depends on the amount of substance, we are interested in the **standard molar entropy ($S°$)** in units of J/mol·K (or J·mol^{-1}·K^{-1}). The $S°$ values at 298 K for many elements, compounds, and ions appear, with other thermodynamic variables, in Appendix B.

Predicting Relative $S°$ Values of a System Based on an understanding of systems at the molecular level and the effects of heat absorbed, we can often predict how the entropy of a substance is affected by temperature, physical state, dissolution, and atomic or molecular complexity. (All $S°$ values in the following discussion have units of J/mol·K and, unless stated otherwise, refer to the system at 298 K.)

1. *Temperature changes.* For a given substance, $S°$ *increases as the temperature rises.* Consider these typical values for copper metal:

T (K):	273	295	298
$S°$:	31.0	32.9	33.2

The temperature increases as heat is absorbed ($q > 0$), which represents an increase in the average kinetic energy of the particles. Recall from Figure 5.12 that the kinetic energies of gas particles in a sample are distributed over a range, which becomes wider as the temperature rises. The same general behavior occurs for liquids and

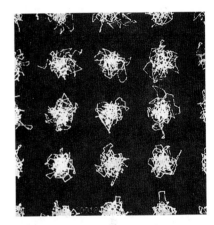

Figure 20.3 Random motion in a crystal. This computer simulation shows the paths of the particle centers in a crystalline solid. At any temperature greater than 0 K, each particle moves about its lattice position. The higher the temperature, the more vigorous the movement. Adding thermal energy increases the total energy, and the particle energies can be distributed over more microstates; thus, the entropy increases.

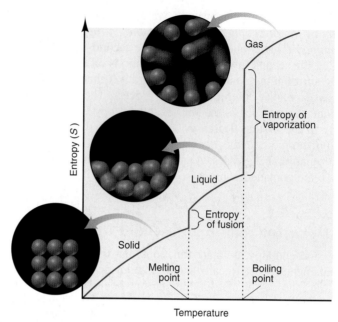

Temperature

Figure 20.4 The increase in entropy from solid to liquid to gas. A plot of entropy vs. temperature shows the gradual increase in entropy within a phase and the abrupt increase with a phase change. The molecular-scale views depict the increase in freedom of motion of the particles as the solid melts and, even more so, as the liquid vaporizes.

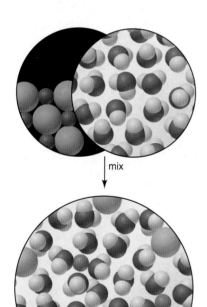

Figure 20.5 The entropy change accompanying the dissolution of a salt. When a crystalline salt and pure liquid water form a solution, the entropy change has two contributions: a positive contribution as the crystal separates into ions and the pure liquid disperses them, and a negative contribution as water molecules become organized around each ion. The relative magnitudes of these contributions determine the overall entropy change. The entropy of a salt solution is usually *greater* than that of the solid and water.

solids. With more microstates in which the energy can be dispersed, the entropy of the substance goes up. In other words, raising the temperature populates more microstates. Thus, $S°$ *increases for a substance as it is heated.*

2. *Physical states and phase changes.* For a phase change such as melting or vaporizing, heat is absorbed ($q > 0$). The particles have more freedom of motion and their energy is more dispersed, so the entropy change is positive. Thus, $S°$ *increases for a substance as it changes from a solid to a liquid to a gas:*

	Na	H_2O	C(graphite)
$S°(s \text{ or } l)$:	51.4(s)	69.9(l)	5.7(s)
$S°(g)$:	153.6	188.7	158.0

Figure 20.4 shows the entropy of a typical substance as it is heated and undergoes a phase change. Note the gradual increase within a phase as the temperature rises and the large, sudden increase at the phase change. The solid has the least energy dispersed within it and, thus, the lowest entropy. Its particles vibrate about their positions but, on average, remain fixed. As the temperature rises, the entropy gradually increases with the increase in the particles' kinetic energy. When the solid melts, the particles move much more freely between and around each other, so there is an abrupt increase in entropy. Further heating increases the speed of the particles in the liquid, and the entropy increases gradually. Finally, freed from intermolecular forces, the particles undergo another abrupt entropy increase and move chaotically as a gas. Note that *the increase in entropy from liquid to gas is much larger than that from solid to liquid:* $\Delta S°_{vap} >> \Delta S°_{fus}$.

3. *Dissolving a solid or liquid.* The entropy of a dissolved solid or liquid is usually *greater* than the entropy of the pure solute, but the nature of solute *and* solvent and the dissolving process affect the overall entropy change (Figure 20.5):

	NaCl	$AlCl_3$	CH_3OH
$S°(s \text{ or } l)$:	72.1(s)	167(s)	127(l)
$S°(aq)$:	115.1	−148	132

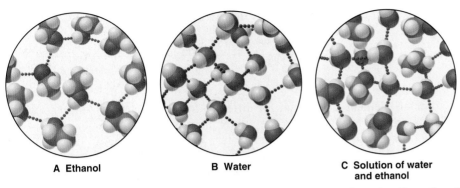

A Ethanol **B** Water **C** Solution of water and ethanol

Figure 20.6 The small increase in entropy when ethanol dissolves in water. Pure ethanol **(A)** and pure water **(B)** have many intermolecular H bonds. **C,** In a solution of these two substances, the molecules form H bonds to one another, so their freedom of motion does not change significantly. Thus, the entropy increase is relatively small and is due solely to random mixing.

When an ionic solid dissolves in water, the crystal breaks down, and the ions experience a great increase in freedom of motion as they become hydrated and separate, with their energy dispersed over more microstates. We expect the entropy of the ions themselves to be greater in the solution than in the crystal. However, some of the water molecules become organized around the ions, their motions restricted (see Figure 13.2), which makes a negative contribution to the overall entropy change. In fact, for small, multiply charged ions, the solvent becomes so attracted to the ions, making its energy localized rather than dispersed, that this negative contribution can dominate and lead to negative $S°$ values for the ions in solution. For example, the $Al^{3+}(aq)$ ion has such a negative $S°$ value (-313 J/mol·K) that when $AlCl_3$ dissolves in water, even though $S°$ of $Cl^-(aq)$ is positive, the overall entropy of aqueous $AlCl_3$ is lower than that of solid $AlCl_3$.*

For molecular solutes, the increase in entropy upon dissolving is typically much smaller than for ionic solutes. For a solid such as glucose, there is no separation into ions, and for a liquid such as ethanol, the breakdown of a crystal structure is absent as well. Furthermore, in pure ethanol and in pure water, the molecules form many H bonds, so there is relatively little change in their freedom of motion when they are mixed (Figure 20.6). The small increase in the entropy of dissolved ethanol arises from the random mixing of the molecules.

4. *Dissolving a gas.* The particles in a gas already have so much freedom of motion and such highly dispersed energy that they always lose freedom when they dissolve in a liquid or solid. Therefore, the entropy of a solution of a gas in a liquid or a solid is always *less* than the entropy of the gas. For instance, when gaseous O_2 [$S°(g) = 205.0$ J/mol·K] dissolves in water, its entropy decreases dramatically [$S°(aq) = 110.9$ J/mol·K] (Figure 20.7). When a gas dissolves in another gas, however, the entropy increases from the mixing of the molecules.

5. *Atomic size or molecular complexity.* In general, differences in entropy values for substances in the same phase are based on atomic size and molecular complexity. For elements within a periodic group, energy levels (microstates) become closer together for heavier atoms, so entropy increases down the group:

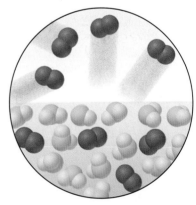

Figure 20.7 The large *decrease* in entropy of a gas when it dissolves in a liquid. The chaotic movement and high entropy of molecules of O_2 are reduced greatly when the gas dissolves in water.

	Li	Na	K	Rb	Cs
Atomic radius (pm):	152	186	227	248	265
Molar mass (g/mol):	6.941	22.99	39.10	85.47	132.9
$S°(s)$:	29.1	51.4	64.7	69.5	85.2

*An $S°$ value for a hydrated ion can be negative because it is relative to the $S°$ value for the hydrated proton, $H^+(aq)$, which is assigned a value of 0. In other words, $Al^{3+}(aq)$ has a lower entropy than $H^+(aq)$.

The same trend of increasing entropy down a group holds for similar compounds:

	HF	HCl	HBr	HI
Molar mass (g/mol):	20.01	36.46	80.91	127.9
$S°(g)$:	173.7	186.8	198.6	206.3

For an element that occurs in different forms (allotropes), the entropy is *higher* in the form that allows the atoms more freedom of motion, which disperses their energy over more microstates. For example, the $S°$ of graphite is 5.69 J/mol·K, whereas the $S°$ of diamond is 2.44 J/mol·K. In diamond, covalent bonds extend in three dimensions, allowing the atoms little movement; in graphite, covalent bonds extend only within a two-dimensional sheet, and motion of the sheets relative to each other is relatively easy.

For compounds, entropy increases with chemical complexity, that is, with the number of atoms in a formula unit or molecule of the compound. This trend holds for both ionic and covalent substances, as long as they are in the same phase:

	NaCl	$AlCl_3$	P_4O_{10}	NO	NO_2	N_2O_4
$S°(s)$:	72.1	167	229			
$S°(g)$:				211	240	304

The trend is based on the types of movement, and thus number of microstates, possible for the atoms (or ions) in each compound. For example, as Figure 20.8 shows, among the nitrogen oxides listed above, the two atoms of NO can vibrate only toward and away from each other. The three atoms of NO_2 have more vibrational motions, and the six atoms of N_2O_4 have even more.

For larger molecules, we also consider how one part of a molecule moves relative to other parts. A long hydrocarbon chain can rotate and vibrate in more ways than a short one, so entropy increases with chain length. A ring compound, such as cyclopentane (C_5H_{10}), has lower entropy than the corresponding chain compound, pentene (C_5H_{10}), because the ring structure restricts freedom of motion:

	$CH_4(g)$	$C_2H_6(g)$	$C_3H_8(g)$	$C_4H_{10}(g)$	$C_5H_{10}(g)$	$C_5H_{10}(cyclo, g)$	$C_2H_5OH(l)$
$S°$:	186	230	270	310	348	293	161

Remember, these trends hold only for *substances in the same physical state*. Gaseous methane (CH_4) has a greater entropy than liquid ethanol (C_2H_5OH), even though ethanol molecules are more complex. When gases are compared with liquids, *the effect of physical state usually dominates that of molecular complexity.*

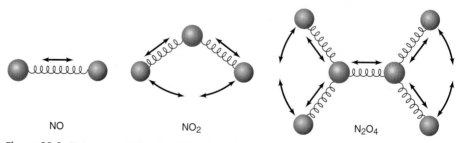

Figure 20.8 Entropy and vibrational motion. A diatomic molecule, such as NO, can vibrate in only one way. NO_2 can vibrate in more ways, and N_2O_4 in even more. Thus, as the number of atoms increases, a molecule can disperse its vibrational energy over more microstates, and so has higher entropy.

SAMPLE PROBLEM 20.1 Predicting Relative Entropy Values

Problem Choose the member with the higher entropy in each of the following pairs, and justify your choice [assume constant temperature, except in part (e)]:

(a) 1 mol of $SO_2(g)$ or 1 mol of $SO_3(g)$ (b) 1 mol of $CO_2(s)$ or 1 mol of $CO_2(g)$

(c) 3 mol of $O_2(g)$ or 2 mol of $O_3(g)$ (d) 1 mol of KBr(s) or 1 mol of KBr(aq)

(e) Seawater at 2°C or at 23°C   (f) 1 mol of $CF_4(g)$ or 1 mol of $CCl_4(g)$

Plan In general, we know that particles with more freedom of motion or more dispersed energy have higher entropy and that raising the temperature increases entropy. We apply the general categories described in the text to choose the member with the higher entropy.

Solution (a) 1 mol of $SO_3(g)$. For equal numbers of moles of substances with the same types of atoms in the same physical state, the more atoms in the molecule, the more types of motion available, and thus the higher the entropy.

(b) 1 mol of $CO_2(g)$. For a given substance, entropy increases in the sequence $s < l < g$.

(c) 3 mol of $O_2(g)$. The two samples contain the same number of oxygen atoms but different numbers of molecules. Despite the greater complexity of O_3, the greater number of molecules dominates in this case because there are many more microstates possible for three moles of particles than for two moles.

(d) 1 mol of KBr(aq). The two samples have the same number of ions, but their motion is more limited and their energy less dispersed in the solid than in the solution.

(e) Seawater at 23°C. Entropy increases with rising temperature.

(f) 1 mol of $CCl_4(g)$. For similar compounds, entropy increases with molar mass.

FOLLOW-UP PROBLEM 20.1 For 1 mol of substance at a given temperature, select the member in each pair with the higher entropy, and give the reason for your choice:

(a) $PCl_3(g)$ or $PCl_5(g)$ (b) $CaF_2(s)$ or $BaCl_2(s)$ (c) $Br_2(g)$ or $Br_2(l)$

SECTION SUMMARY

A change is spontaneous if it occurs in a given direction under specified conditions without a continuous input of energy. Neither the first law of thermodynamics nor the sign of ΔH predicts the direction. All spontaneous processes involve an increase in the dispersion of energy. Entropy is a state function that measures the extent of energy dispersal over the number of microstates possible for a system, which is related to the freedom of motion of its particles. The second law of thermodynamics states that, in a spontaneous process, the entropy of the universe (system plus surroundings) increases. Absolute entropy values can be found because perfect crystals have zero entropy at 0 K (third law). Standard molar entropy $S°$ (J/mol·K) is affected by temperature, phase changes, dissolution, and atomic size or molecular complexity.

20.2 CALCULATING THE CHANGE IN ENTROPY OF A REACTION

In addition to understanding trends in $S°$ values for different substances or for the same substance in different phases, chemists are especially interested in learning how to predict the sign *and* calculate the value of the change in entropy as a reaction occurs.

Entropy Changes in the System: Standard Entropy of Reaction ($\Delta S°_{rxn}$)

Based on the ideas we just discussed, we can often predict the sign of the **standard entropy of reaction, $\Delta S°_{rxn}$,** the entropy change that occurs when all reactants and products are in their standard states. A deciding event is usually a change in the number of moles of gas. Because gases have such great freedom of motion and thus high molar entropies, *if the number of moles of gas increases, $\Delta S°_{rxn}$ is usually positive; if the number decreases, $\Delta S°_{rxn}$ is usually negative.*

For example, when $H_2(g)$ and $I_2(s)$ form $HI(g)$, the total number of moles of *substance* stays the same, but we predict that the entropy increases because the number of moles of *gas* increases:

$$H_2(g) + I_2(s) \longrightarrow 2HI(g) \qquad \Delta S^\circ_{rxn} = S^\circ_{products} - S^\circ_{reactants} > 0$$

When ammonia forms from its elements, 4 mol of gas produces 2 mol of gas, so we predict that the entropy decreases:

$$N_2(g) + 3H_2(g) \rightleftharpoons 2NH_3(g) \qquad \Delta S^\circ_{rxn} = S^\circ_{products} - S^\circ_{reactants} < 0$$

In general *we cannot predict the sign of the entropy change unless the reaction involves a change in number of moles of gas.*

Recall that by applying Hess's law (Chapter 6), we can combine ΔH°_f values to find the standard heat of reaction, ΔH°_{rxn}. Similarly, we combine S° values to find the standard entropy of reaction, ΔS°_{rxn}:

$$\Delta S^\circ_{rxn} = \Sigma m S^\circ_{products} - \Sigma n S^\circ_{reactants} \qquad \text{(20.4)}$$

where m and n are the amounts (mol) of the individual species, given by their coefficients in the balanced equation. For the formation of ammonia, we have

$$\Delta S^\circ_{rxn} = [(2 \text{ mol } NH_3)(S^\circ \text{ of } NH_3)] - [(1 \text{ mol } N_2)(S^\circ \text{ of } N_2) + (3 \text{ mol } H_2)(S^\circ \text{ of } H_2)]$$

From Appendix B, we find the appropriate S° values:

$$\Delta S^\circ_{rxn} = [(2 \text{ mol})(193 \text{ J/mol·K})] - [(1 \text{ mol})(191.5 \text{ J/mol·K}) + (3 \text{ mol})(130.6 \text{ J/mol·K})]$$
$$= -197 \text{ J/K}$$

As we predicted from the decrease in number of moles of gas, $\Delta S^\circ_{rxn} < 0$.

SAMPLE PROBLEM 20.2   Calculating the Standard Entropy of Reaction, ΔS°_{rxn}

Problem Calculate ΔS°_{rxn} for the combustion of 1 mol of propane at 25°C:

$$C_3H_8(g) + 5O_2(g) \longrightarrow 3CO_2(g) + 4H_2O(l)$$

Plan To determine ΔS°_{rxn}, we apply Equation 20.4. We predict the sign of ΔS°_{rxn} from the change in the number of moles of gas: 6 mol of gas yields 3 mol of gas, so the entropy will probably decrease ($\Delta S^\circ_{rxn} < 0$).
Solution Calculating ΔS°_{rxn}. Using Appendix B values,

$$\Delta S^\circ_{rxn} = [(3 \text{ mol } CO_2)(S^\circ \text{ of } CO_2) + (4 \text{ mol } H_2O)(S^\circ \text{ of } H_2O)]$$
$$- [(1 \text{ mol } C_3H_8)(S^\circ \text{ of } C_3H_8) + (5 \text{ mol } O_2)(S^\circ \text{ of } O_2)]$$
$$= [(3 \text{ mol})(213.7 \text{ J/mol·K}) + (4 \text{ mol})(69.9 \text{ J/mol·K})]$$
$$- [(1 \text{ mol})(269.9 \text{ J/mol·K}) + (5 \text{ mol})(205.0 \text{ J/mol·K})]$$
$$= -374 \text{ J/K}$$

Check $\Delta S^\circ < 0$, so our prediction is correct. Rounding gives $[3(200) + 4(70)] - [270 + 5(200)] = 880 - 1270 = -390$, close to the calculated value.
Comment Remember that when there is no change in the amount (mol) of gas, you *cannot* confidently predict the sign of ΔS°_{rxn}.

FOLLOW-UP PROBLEM 20.2 Balance the following equations, predict the sign of ΔS°_{rxn} if possible, and calculate its value at 25°C:
(a) $NaOH(s) + CO_2(g) \longrightarrow Na_2CO_3(s) + H_2O(l)$
(b) $Fe(s) + H_2O(g) \longrightarrow Fe_2O_3(s) + H_2(g)$

Entropy Changes in the Surroundings: The Other Part of the Total

In many spontaneous reactions, such as the synthesis of ammonia and the combustion of propane, we see that the entropy of the reacting system decreases ($\Delta S^\circ_{rxn} < 0$). The second law dictates that *decreases in the entropy of the system can occur only if increases in the entropy of the surroundings outweigh them.* Let's examine the influence of the surroundings—in particular, the addition (or removal) of heat and the temperature at which this heat change occurs—on the *total* entropy change.

The essential role of the surroundings is to *either add heat to the system or remove heat from it.* In essence, the surroundings function as an enormous heat source or heat sink, one so large that its temperature remains constant, even though its entropy changes through the loss or gain of heat. The surroundings participate in the two possible types of enthalpy changes as follows:

1. *Exothermic change.* Heat lost by the system is gained by the surroundings. This heat gain increases the freedom of motion of particles in the surroundings, which disperses their energy; so the entropy of the surroundings increases:

   For an exothermic change: $q_{sys} < 0,$ $q_{surr} > 0,$ and $\Delta S_{surr} > 0$

2. *Endothermic change.* Heat gained by the system is lost by the surroundings. This heat loss reduces the freedom of motion of particles in the surroundings, which localizes their energy; so the entropy of the surroundings decreases:

   For an endothermic change: $q_{sys} > 0,$ $q_{surr} < 0,$ and $\Delta S_{surr} < 0$

The *temperature* of the surroundings at which the heat is transferred also affects ΔS_{surr}. Consider the effect of an exothermic reaction at a low and at a high temperature. At a low temperature, such as 20 K, there is very little random motion in the surroundings, that is, relatively little energy is dispersed there. Therefore, transferring heat to the surroundings has a large effect on how much energy is dispersed. At a higher temperature, such as 298 K, the surroundings already have a relatively large quantity of energy dispersed, so transferring the same amount of heat has a smaller effect on the total energy dispersed. In other words, the change in entropy of the surroundings is greater when heat is added at a lower temperature. Therefore, *the change in entropy of the surroundings is directly related to an opposite change in the heat of the system and inversely related to the temperature at which the heat is transferred.* Combining these relationships gives an equation that is closely related to Equation 20.2:

$$\Delta S_{surr} = -\frac{q_{sys}}{T}$$

Recall that for a process at *constant pressure*, the heat (q_P) is ΔH, so

$$\Delta S_{surr} = -\frac{\Delta H_{sys}}{T} \qquad (20.5)$$

This means that we can calculate ΔS_{surr} by measuring ΔH_{sys} and the temperature T at which the change takes place.

To restate the central point, if a spontaneous reaction has a negative ΔS_{sys} (energy dispersed over fewer microstates), ΔS_{surr} must be positive enough (energy dispersed over many more microstates) for ΔS_{univ} to be positive (energy dispersed over more microstates). Sample Problem 20.3 illustrates this situation for one of the reactions we considered earlier.

SAMPLE PROBLEM 20.3   Determining Reaction Spontaneity

Problem At 298 K, the formation of ammonia has a negative ΔS_{sys}°:

$$N_2(g) + 3H_2(g) \longrightarrow 2NH_3(g) \qquad \Delta S_{sys}^{\circ} = -197 \text{ J/K}$$

Calculate ΔS_{univ}, and state whether the reaction occurs spontaneously at this temperature.
Plan For the reaction to occur spontaneously, $\Delta S_{univ} > 0$, and so ΔS_{surr} must be greater than +197 J/K. To find ΔS_{surr}, we need ΔH_{sys}°, which is the same as ΔH_{rxn}°. We use ΔH_{f}° values from Appendix B to find ΔH_{rxn}°. Then, we use ΔH_{rxn}° and the given T (298 K) to find ΔS_{surr}. To find ΔS_{univ}, we add the calculated ΔS_{surr} to the given ΔS_{sys}° (-197 J/K).

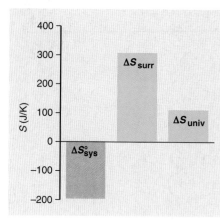

Solution Calculating ΔH°_{sys}:

$$\Delta H^{\circ}_{sys} = \Delta H^{\circ}_{rxn}$$
$$= [(2 \text{ mol NH}_3)(-45.9 \text{ kJ/mol})] - [(3 \text{ mol H}_2)(0 \text{ kJ/mol}) + (1 \text{ mol N}_2)(0 \text{ kJ/mol})]$$
$$= -91.8 \text{ kJ}$$

Calculating ΔS_{surr}:

$$\Delta S_{surr} = -\frac{\Delta H^{\circ}_{sys}}{T} = -\frac{-91.8 \text{ kJ} \times \dfrac{1000 \text{ J}}{1 \text{ kJ}}}{298 \text{ K}} = 308 \text{ J/K}$$

Determining ΔS_{univ}:

$$\Delta S_{univ} = \Delta S^{\circ}_{sys} + \Delta S_{surr} = -197 \text{ J/K} + 308 \text{ J/K} = 111 \text{ J/K}$$

$\Delta S_{univ} > 0$, so the reaction occurs spontaneously at 298 K (see figure in margin).

Check Rounding to check the math, we have

$$\Delta H^{\circ}_{rxn} \approx 2(-45 \text{ kJ}) = -90 \text{ kJ}$$
$$\Delta S_{surr} \approx -(-90,000 \text{ J})/300 \text{ K} = 300 \text{ J/K}$$
$$\Delta S_{univ} \approx -200 \text{ J/K} + 300 \text{ J/K} = 100 \text{ J/K}$$

Given the negative ΔH°_{rxn}, Le Châtelier's principle predicts that low temperature should favor NH_3 formation, and so the answer is reasonable.

Comments 1. Note that ΔH° has units of kJ, whereas ΔS has units of J/K. Don't forget to convert kJ to J, or you'll introduce a large error.

2. This example highlights the distinction between thermodynamic and kinetic considerations. Even though NH_3 forms spontaneously, it does so slowly; in the industrial production of ammonia by the Haber process (Section 17.6), a catalyst is used to form NH_3 at a practical rate.

FOLLOW-UP PROBLEM 20.3 Does the oxidation of $FeO(s)$ to $Fe_2O_3(s)$ occur spontaneously at 298 K?

The Entropy Change and the Equilibrium State

For a process spontaneously approaching equilibrium, $\Delta S_{univ} > 0$. When the process reaches equilibrium, there is no longer any force driving it to proceed further and, thus, no net change in either direction; that is, $\Delta S_{univ} = 0$. At that point, any entropy change in the system is exactly balanced by an opposite entropy change in the surroundings:

$$\text{At equilibrium:} \quad \Delta S_{univ} = \Delta S_{sys} + \Delta S_{surr} = 0 \quad \text{or} \quad \Delta S_{sys} = -\Delta S_{surr}$$

For example, let's calculate ΔS_{univ} for a phase change. For the vaporization-condensation of 1 mol of water at 100°C (373 K),

$$H_2O(l; 373 \text{ K}) \rightleftharpoons H_2O(g; 373 \text{ K})$$

First, we find ΔS_{univ} for the forward change (vaporization) by calculating ΔS°_{sys}:

$$\Delta S^{\circ}_{sys} = \Sigma m S^{\circ}_{products} - \Sigma n S^{\circ}_{reactants} = S^{\circ} \text{ of } H_2O(g; 373 \text{ K}) - S^{\circ} \text{ of } H_2O(l; 373 \text{ K})$$
$$= 195.9 \text{ J/K} - 86.8 \text{ J/K} = 109.1 \text{ J/K}$$

As we expect, the entropy of the system increases ($\Delta S^{\circ}_{sys} > 0$) as the liquid absorbs heat and changes to a gas.

For ΔS_{surr}, we have

$$\Delta S_{surr} = -\frac{\Delta H^{\circ}_{sys}}{T}$$

where $\Delta H^{\circ}_{sys} = \Delta H^{\circ}_{vap}$ at 373 K = 40.7 kJ/mol = 40.7×10^3 J/mol. For 1 mol of water, we have

$$\Delta S_{surr} = -\frac{\Delta H^{\circ}_{vap}}{T} = -\frac{40.7 \times 10^3 \text{ J}}{373 \text{ K}} = -109 \text{ J/K}$$

The surroundings lose heat, and the negative sign means that the entropy of the surroundings decreases. The two entropy changes have the same magnitude but opposite signs, so they cancel:

$$\Delta S_{univ} = 109 \text{ J/K} + (-109 \text{ J/K}) = 0$$

For the reverse change (condensation), ΔS_{univ} also equals zero, but ΔS°_{sys} and ΔS_{surr} have signs opposite those for vaporization. A similar treatment of a chemical change shows the same result: the entropy change of the forward reaction is *equal in magnitude but opposite in sign* to the entropy change of the reverse reaction. Thus, *when a system reaches equilibrium, neither the forward nor the reverse reaction is spontaneous,* and so there is no net reaction in either direction.

Spontaneous Exothermic and Endothermic Reactions: A Summary

We can now see why exothermic and endothermic spontaneous reactions occur. No matter what its *enthalpy* change, a reaction occurs because the total *entropy* of the reacting system *and* its surroundings increases. The two possibilities are

1. *For an exothermic reaction* ($\Delta H_{sys} < 0$), heat is released by the system, which increases the freedom of motion and energy dispersed and, thus, the entropy of the surroundings ($\Delta S_{surr} > 0$).

- If the reacting system yields products whose entropy is greater than that of the reactants ($\Delta S_{sys} > 0$), the total entropy change ($\Delta S_{sys} + \Delta S_{surr}$) will be positive (Figure 20.9A).
- If, on the other hand, the entropy of the system decreases as the reaction occurs ($\Delta S_{sys} < 0$), the entropy of the surroundings must increase even more ($\Delta S_{surr} >> 0$) to make the total ΔS positive (Figure 20.9B).

2. *For an endothermic reaction* ($\Delta H_{sys} > 0$), the heat lost by the surroundings decreases molecular freedom of motion and dispersal of energy and, thus, decreases the entropy of the surroundings ($\Delta S_{surr} < 0$). Therefore, the only way an endothermic reaction can occur spontaneously is if ΔS_{sys} is positive and large enough ($\Delta S_{sys} >> 0$) to outweigh the negative ΔS_{surr} (Figure 20.9C).

SECTION SUMMARY
The standard entropy of reaction, ΔS°_{rxn}, is calculated from S° values. When the amount (mol) of gas (Δn_{gas}) increases in a reaction, usually $\Delta S^{\circ}_{rxn} > 0$. The value of ΔS_{surr} is related directly to ΔH°_{sys} and inversely to the T at which the change occurs. In a spontaneous change, the entropy of the system can decrease only if the entropy of the surroundings increases even more. For a system at equilibrium, $\Delta S_{univ} = 0$.

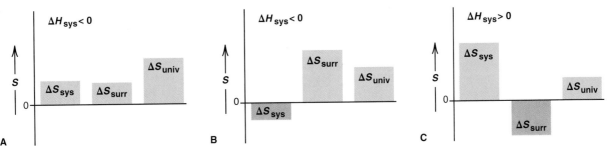

Figure 20.9 Components of ΔS_{univ} for spontaneous reactions. For a reaction to occur spontaneously, ΔS_{univ} must be positive. **A,** An exothermic reaction in which ΔS_{sys} increases; the size of ΔS_{surr} is not important. **B,** An exothermic reaction in which ΔS_{sys} decreases; ΔS_{surr} must be larger than ΔS_{sys}. **C,** An endothermic reaction in which ΔS_{sys} increases; ΔS_{surr} must be smaller than ΔS_{sys}.

20.3 ENTROPY, FREE ENERGY, AND WORK

By making *two* separate measurements, ΔS_{sys} and ΔS_{surr}, we can predict whether a reaction will be spontaneous at a particular temperature. It would be useful, however, to have *one* criterion for spontaneity that we can find by examining the system only. The Gibbs free energy, or simply **free energy (G),** is a function that combines the system's enthalpy and entropy:

$$G = H - TS$$

Named for Josiah Willard Gibbs (1839–1903), the physicist who proposed it and laid much of the foundation for chemical thermodynamics, this function provides the criterion for spontaneity we've been seeking.

Free Energy Change and Reaction Spontaneity

The free energy change (ΔG) is a measure of the spontaneity of a process and of the useful energy available from it. Let's see how the free energy change is derived from the second law of thermodynamics. Recall that by definition, the entropy change of the universe is the sum of the entropy changes of the system and the surroundings:

$$\Delta S_{univ} = \Delta S_{sys} + \Delta S_{surr}$$

At constant pressure,

$$\Delta S_{surr} = -\frac{\Delta H_{sys}}{T}$$

Substituting for ΔS_{surr} gives a relationship that lets us focus solely on the system:

$$\Delta S_{univ} = \Delta S_{sys} - \frac{\Delta H_{sys}}{T}$$

Multiplying both sides by $-T$ gives

$$-T\Delta S_{univ} = \Delta H_{sys} - T\Delta S_{sys}$$

Now we can introduce the new free energy quantity to replace the enthalpy and entropy terms. From $G = H - TS$, the *Gibbs equation* shows us the change in the free energy of the system (ΔG_{sys}) at constant temperature and pressure:

$$\Delta G_{sys} = \Delta H_{sys} - T\Delta S_{sys} \qquad (20.6)$$

Combining this equation with the previous one shows that

$$-T\Delta S_{univ} = \Delta H_{sys} - T\Delta S_{sys} = \Delta G_{sys}$$

*The **sign** of ΔG tells if a reaction is spontaneous.* The second law dictates

- $\Delta S_{univ} > 0$ for a spontaneous process
- $\Delta S_{univ} < 0$ for a nonspontaneous process
- $\Delta S_{univ} = 0$ for a process at equilibrium

Of course, absolute temperature is always positive, so

$$T\Delta S_{univ} > 0 \qquad \text{or} \qquad -T\Delta S_{univ} < 0 \text{ for a spontaneous process}$$

Because $\Delta G = -T\Delta S_{univ}$, we know that

- $\Delta G < 0$ for a spontaneous process
- $\Delta G > 0$ for a nonspontaneous process
- $\Delta G = 0$ for a process at equilibrium

An important point to keep in mind is that if a process is *nonspontaneous* in one direction ($\Delta G > 0$), it is *spontaneous* in the opposite direction ($\Delta G < 0$). By using ΔG, we have not incorporated any new ideas, but we can predict reaction spontaneity from one variable (ΔG_{sys}) rather than two (ΔS_{sys} and ΔS_{surr}).

Calculating Standard Free Energy Changes

Because free energy (G) combines three state functions, H, S, and T, it is also a state function. As with enthalpy, we focus on the free energy *change* (ΔG).

The Standard Free Energy Change As we did with the other thermodynamic variables, to compare the free energy changes of different reactions we calculate the *standard* **free energy change** ($\Delta G°$), which occurs when all components of the system are in their standard states. Adapting the Gibbs equation (20.6), we have

$$\Delta G°_{sys} = \Delta H°_{sys} - T\Delta S°_{sys} \tag{20.7}$$

This important relationship is used frequently to find any one of these three central thermodynamic variables, given the other two, as in this sample problem.

SAMPLE PROBLEM 20.4   Calculating $\Delta G°_{rxn}$ from Enthalpy and Entropy Values

Problem Potassium chlorate, a common oxidizing agent in fireworks and matchheads, undergoes a solid-state disproportionation reaction when heated:

$$4KClO_3(s) \xrightarrow{\Delta} 3KClO_4(s) + KCl(s)$$

Use $\Delta H°_f$ and $S°$ values to calculate $\Delta G°_{sys}$ ($\Delta G°_{rxn}$) at 25°C for this reaction.

Plan To solve for $\Delta G°$, we need values from Appendix B. We use $\Delta H°_f$ values to calculate $\Delta H°_{rxn}$ ($\Delta H°_{sys}$), use $S°$ values to calculate $\Delta S°_{rxn}$ ($\Delta S°_{sys}$), and then apply Equation 20.7.

Solution Calculating $\Delta H°_{sys}$ from $\Delta H°_f$ values (with Equation 6.8):

$$
\begin{aligned}
\Delta H°_{sys} = \Delta H°_{rxn} &= \Sigma m\Delta H°_{f(products)} - \Sigma n\Delta H°_{f(reactants)} \\
&= [(3 \text{ mol KClO}_4)(\Delta H°_f \text{ of KClO}_4) + (1 \text{ mol KCl})(\Delta H°_f \text{ of KCl})] \\
&\quad - [(4 \text{ mol KClO}_3)(\Delta H°_f \text{ of KClO}_3)] \\
&= [(3 \text{ mol})(-432.8 \text{ kJ/mol}) + (1 \text{ mol})(-436.7 \text{ kJ/mol})] \\
&\quad - [(4 \text{ mol})(-397.7 \text{ kJ/mol})] \\
&= -144 \text{ kJ}
\end{aligned}
$$

Calculating $\Delta S°_{sys}$ from $S°$ values (with Equation 20.4):

$$
\begin{aligned}
\Delta S°_{sys} = \Delta S°_{rxn} &= [(3 \text{ mol KClO}_4)(S° \text{ of KClO}_4) + (1 \text{ mol KCl})(S° \text{ of KCl})] \\
&\quad - [(4 \text{ mol KClO}_3)(S° \text{ of KClO}_3)] \\
&= [(3 \text{ mol})(151.0 \text{ J/mol·K}) + (1 \text{ mol})(82.6 \text{ J/mol·K})] \\
&\quad - [(4 \text{ mol})(143.1 \text{ J/mol·K})] \\
&= -36.8 \text{ J/K}
\end{aligned}
$$

Calculating $\Delta G°_{sys}$ at 298 K:

$$\Delta G°_{sys} = \Delta H°_{sys} - T\Delta S°_{sys} = -144 \text{ kJ} - \left[(298 \text{ K})(-36.8 \text{ J/K})\left(\frac{1 \text{ kJ}}{1000 \text{ J}}\right)\right] = -133 \text{ kJ}$$

Check Rounding to check the math:

$$\Delta H° \approx [3(-433 \text{ kJ}) + (-440 \text{ kJ})] - [4(-400 \text{ kJ})] = -1740 \text{ kJ} + 1600 \text{ kJ} = -140 \text{ kJ}$$
$$\Delta S° \approx [3(150 \text{ J/K}) + 85 \text{ J/K}] - [4(145 \text{ J/K})] = 535 \text{ J/K} - 580 \text{ J/K} = -45 \text{ J/K}$$
$$\Delta G° \approx -140 \text{ kJ} - 300 \text{ K}(-0.04 \text{ kJ/K}) = -140 \text{ kJ} + 12 \text{ kJ} = -128 \text{ kJ}$$

All values are close to the calculated ones.

Comments 1. For a spontaneous reaction under *any* conditions, the free energy change, ΔG, is negative. Under standard-state conditions, a spontaneous reaction has a negative *standard* free energy change; that is, $\Delta G° < 0$.
2. This reaction is spontaneous, but the rate is very low in the solid. When $KClO_3$ is heated slightly above its melting point, the ions are free to move and the reaction occurs readily.

FOLLOW-UP PROBLEM 20.4 Determine the standard free energy change at 298 K for the reaction $2NO(g) + O_2(g) \longrightarrow 2NO_2(g)$.

The Standard Free Energy of Formation Another way to calculate ΔG_{rxn}° is with values for the **standard free energy of formation** (ΔG_{f}°) of the components; ΔG_{f}° is the free energy change that occurs when 1 mol of compound is made *from its elements*, with all components in their standard states. Because free energy is a state function, we can combine ΔG_{f}° values of reactants and products to calculate ΔG_{rxn}° no matter how the reaction takes place:

$$\Delta G_{rxn}^{\circ} = \Sigma m \Delta G_{f(products)}^{\circ} - \Sigma n \Delta G_{f(reactants)}^{\circ} \tag{20.8}$$

ΔG_{f}° values have properties similar to ΔH_{f}° values:

- ΔG_{f}° of an element in its standard state is zero.
- An equation coefficient (*m* or *n* above) multiplies ΔG_{f}° by that number.
- Reversing a reaction changes the sign of ΔG_{f}°.

Many ΔG_{f}° values appear along with those for ΔH_{f}° and S° in Appendix B.

SAMPLE PROBLEM 20.5 Calculating ΔG_{rxn}° from ΔG_{f}° Values

Problem Use ΔG_{f}° values to calculate ΔG_{rxn}° for the reaction in Sample Problem 20.4:

$$4KClO_3(s) \longrightarrow 3KClO_4(s) + KCl(s)$$

Plan We apply Equation 20.8 to calculate ΔG_{rxn}°.

Solution
$$\Delta G_{rxn}^{\circ} = \Sigma m \Delta G_{f(products)}^{\circ} - \Sigma n \Delta G_{f(reactants)}^{\circ}$$
$$= [(3 \text{ mol } KClO_4)(\Delta G_{f}^{\circ} \text{ of } KClO_4) + (1 \text{ mol } KCl)(\Delta G_{f}^{\circ} \text{ of } KCl)]$$
$$- [(4 \text{ mol } KClO_3)(\Delta G_{f}^{\circ} \text{ of } KClO_3)]$$
$$= [(3 \text{ mol})(-303.2 \text{ kJ/mol}) + (1 \text{ mol})(-409.2 \text{ kJ/mol})]$$
$$- [(4 \text{ mol})(-296.3 \text{ kJ/mol})]$$
$$= -134 \text{ kJ}$$

Check Rounding to check the math:
$$\Delta G_{rxn}^{\circ} \approx [3(-300 \text{ kJ}) + 1(-400 \text{ kJ})] - 4(-300 \text{ kJ})$$
$$= -1300 \text{ kJ} + 1200 \text{ kJ} = -100 \text{ kJ}$$

Comment The slight discrepancy between this answer and that obtained in Sample Problem 20.4 is within experimental error. As you can see, when ΔG_{f}° values are available for a reaction taking place at 25°C, this method is simpler than that in Sample Problem 20.4.

FOLLOW-UP PROBLEM 20.5 Use ΔG_{f}° values to calculate the free energy change at 25°C for each of the following reactions:
(a) $2NO(g) + O_2(g) \longrightarrow 2NO_2(g)$ (from Follow-up Problem 20.4)
(b) $2C(graphite) + O_2(g) \longrightarrow 2CO(g)$

ΔG and the Work a System Can Do

The science of thermodynamics was born soon after the invention of the steam engine, and one of its most practical ideas relates the free energy change and the work a system can do:

- For a spontaneous process ($\Delta G < 0$) at constant T and P, ΔG is the *maximum useful work obtainable* **from** *the system* (−*w*) as the process takes place:

$$\Delta G = -w_{max} \tag{20.9}$$

- For a nonspontaneous process ($\Delta G > 0$) at constant T and P, ΔG is the *minimum work that must be done* **to** *the system to make the process take place.*

The free energy change is the maximum work a system can *possibly* do. But the work the system *actually* does depends on how the free energy is released. Suppose an expanding gas does work by lifting an object. The gas can do nearly the maximum work if the weight of the object can be adjusted in tiny increments and lifted in many small steps. An example would be the gas lifting a container of sand whose weight can be adjusted grain by grain (as described in Section 20.1

for calculating ΔS). In this way, the gas lifts the object in a very high number of steps. The maximum work could be done only in an infinite number of steps; that is, *the maximum work is done by a spontaneous process only if it is carried out reversibly*. In any *real* process, work is done irreversibly—in a finite number of steps—*so we can never obtain the maximum work*. The free energy not used for work is lost as heat.

Consider the work done by a *battery*, a packaged spontaneous redox reaction that releases free energy to the surroundings (flashlight, radio, motor, or other device). If we connect the battery terminals to each other through a short piece of wire, ΔG_{sys} is released all at once but does no work—it just heats the wire and battery and outside air, which increases the freedom of motion of the particles in the universe. If we connect the battery terminals to a motor, ΔG_{sys} is released more slowly, and much of it runs the motor; however, some is still lost as heat. Only if a battery could discharge infinitely slowly could we obtain the maximum work. This is the compromise that all engineers must face—*no real process uses all the available free energy to do work because some is always changed to heat.*

The Effect of Temperature on Reaction Spontaneity

In most cases, the enthalpy contribution (ΔH) to the free energy change (ΔG) is much *larger* than the entropy contribution ($T\Delta S$). For this reason, most exothermic reactions are spontaneous: the negative ΔH helps make ΔG negative. However, the *temperature of a reaction influences the magnitude of the $T\Delta S$ term*, so, for many reactions, the overall spontaneity depends on the temperature.

By scrutinizing the signs of ΔH and ΔS, we can predict the effect of temperature on the sign of ΔG. The values for the thermodynamic variables in this discussion are based on standard state values from Appendix B, but we show them without the degree sign to emphasize that the relationships among ΔG, ΔH, and ΔS are valid at any conditions. Also, we assume that ΔH and ΔS change little with temperature, which is true as long as no phase changes occur.

Let's examine the four combinations of positive and negative ΔH and ΔS; two are independent of temperature and two are dependent on temperature:

- *Temperature-independent cases.* When ΔH and ΔS have *opposite* signs, the reaction occurs spontaneously either at all temperatures or at none.
  1. *Reaction is spontaneous at all temperatures:* $\Delta H < 0$, $\Delta S > 0$. Both contributions favor the spontaneity of the reaction. ΔH is negative and ΔS is positive, so $-T\Delta S$ is negative; thus, ΔG is always negative. Most combustion reactions are in this category. The decomposition of hydrogen peroxide, a common disinfectant, is also spontaneous at all temperatures:
  $$2H_2O_2(l) \longrightarrow 2H_2O(l) + O_2(g) \qquad \Delta H = -196 \text{ kJ and } \Delta S = 125 \text{ J/K}$$
  2. *Reaction is nonspontaneous at all temperatures:* $\Delta H > 0$, $\Delta S < 0$. Both contributions oppose the spontaneity of the reaction. ΔH is positive and ΔS is negative, so $-T\Delta S$ is positive; thus, ΔG is always positive. The formation of ozone from oxygen is not spontaneous at any temperature:
  $$3O_2(g) \longrightarrow 2O_3(g) \qquad \Delta H = 286 \text{ kJ and } \Delta S = -137 \text{ J/K}$$
- *Temperature-dependent cases.* When ΔH and ΔS have the *same* sign, the relative magnitudes of the $-T\Delta S$ and ΔH terms determine the sign of ΔG. In these cases, the magnitude of T is crucial to reaction spontaneity.
  3. *Reaction is spontaneous at higher temperatures:* $\Delta H > 0$ and $\Delta S > 0$. Here, ΔS favors spontaneity ($-T\Delta S < 0$), but ΔH does not. For example,
  $$2N_2O(g) + O_2(g) \longrightarrow 4NO(g) \qquad \Delta H = 197.1 \text{ kJ and } \Delta S = 198.2 \text{ J/K}$$
  With a positive ΔH, the reaction will occur spontaneously only when $-T\Delta S$ is large enough to make ΔG negative, which will happen at higher temperatures. The oxidation of N_2O occurs spontaneously at $T > 994$ K.

4. *Reaction is spontaneous at lower temperatures:* $\Delta H < 0$ *and* $\Delta S < 0$. Now, ΔH favors spontaneity, but ΔS does not ($-T\Delta S > 0$). For example,

$$4Fe(s) + 3O_2(g) \longrightarrow 2Fe_2O_3(s) \qquad \Delta H = -1651 \text{ kJ and } \Delta S = -549.4 \text{ J/K}$$

With a negative ΔH, the reaction will occur spontaneously only if the $-T\Delta S$ term is smaller than the ΔH term, and this happens at lower temperatures. The production of iron(III) oxide occurs spontaneously at any $T < 3005$ K.

Table 20.1 summarizes these four possible combinations of ΔH and ΔS.

Table 20.1	**Reaction Spontaneity and the Signs of ΔH, ΔS, and ΔG**			
ΔH	ΔS	$-T\Delta S$	ΔG	Description
$-$	$+$	$-$	$-$	Spontaneous at all T
$+$	$-$	$+$	$+$	Nonspontaneous at all T
$+$	$+$	$-$	$+$ or $-$	Spontaneous at higher T; nonspontaneous at lower T
$-$	$-$	$+$	$+$ or $-$	Spontaneous at lower T; nonspontaneous at higher T

As you saw in Sample Problem 20.4, one way to calculate ΔG is from enthalpy and entropy changes. Because ΔH and ΔS usually change little with temperature if no phase changes occur, we can use their values at 298 K to examine the effect of temperature on ΔG and, thus, on reaction spontaneity.

SAMPLE PROBLEM 20.6 Determining the Effect of Temperature on ΔG

Problem A key step in the production of sulfuric acid is the oxidation of $SO_2(g)$ to $SO_3(g)$:

$$2SO_2(g) + O_2(g) \longrightarrow 2SO_3(g)$$

At 298 K, $\Delta G = -141.6$ kJ; $\Delta H = -198.4$ kJ; and $\Delta S = -187.9$ J/K.
(a) Use the data to decide if this reaction is spontaneous at 25°C, and predict how ΔG will change with increasing T.
(b) Assuming that ΔH and ΔS are constant with T, is the reaction spontaneous at 900.°C?
Plan (a) We note the sign of ΔG to see if the reaction is spontaneous and the signs of ΔH and ΔS to see the effect of T. **(b)** We use Equation 20.6 to calculate ΔG from the given ΔH and ΔS at the higher T (in K).
Solution (a) $\Delta G < 0$, so the reaction is spontaneous at 298 K: SO_2 and O_2 will form SO_3 spontaneously. With $\Delta S < 0$, the term $-T\Delta S > 0$, and this term will become more positive at higher T. Therefore,

ΔG will become less negative, and the reaction less spontaneous, with increasing T.
(b) Calculating ΔG at 900.°C ($T = 273 + 900. = 1173$ K):

$$\Delta G = \Delta H - T\Delta S = -198.4 \text{ kJ} - [(1173 \text{ K})(-187.9 \text{ J/K})(1 \text{ kJ/1000 J})] = 22.0 \text{ kJ}$$

$\Delta G > 0$, so the reaction is nonspontaneous at the higher T.
Check The answer in part (b) seems reasonable based on our prediction in part (a). The arithmetic seems correct, given considerable rounding:

$$\Delta G \approx -200 \text{ kJ} - [(1200 \text{ K})(-200 \text{ J/K})/1000 \text{ J}] = +40 \text{ kJ}$$

FOLLOW-UP PROBLEM 20.6 A reaction is nonspontaneous at room temperature but *is* spontaneous at −40°C. What can you say about the signs and relative magnitudes of ΔH, ΔS, and $-T\Delta S$?

The Temperature at Which a Reaction Becomes Spontaneous As you have just seen, when the signs of ΔH and ΔS are the same, some reactions that are nonspontaneous at one temperature become spontaneous at another, and vice versa. It would certainly be useful to know the temperature at which a reaction becomes spontaneous. This is the temperature at which a positive ΔG switches to a negative ΔG because of the changing magnitude of the $-T\Delta S$ term. We find this crossover temperature by setting ΔG equal to zero and solving for T:

$$\Delta G = \Delta H - T\Delta S = 0$$

Therefore,

$$\Delta H = T\Delta S \quad \text{and} \quad T = \frac{\Delta H}{\Delta S} \qquad (20.10)$$

Consider the reaction of copper(I) oxide with carbon, which does *not* occur at lower temperatures but is used at higher temperatures in a step during the extraction of copper metal from its ore:

$$Cu_2O(s) + C(s) \longrightarrow 2Cu(s) + CO(g)$$

We predict this reaction has a positive entropy change because the number of moles of gas increases; in fact, $\Delta S = 165$ J/K. Furthermore, because the reaction is *non*spontaneous at lower temperatures, it must have a positive ΔH (58.1 kJ). As the $-T\Delta S$ term becomes more negative at higher temperatures, it will eventually outweigh the positive ΔH term, and the reaction will occur spontaneously.

Let's calculate ΔG for this reaction at 25°C and then find the temperature above which the reaction is spontaneous. At 25°C (298 K),

$$\Delta G = \Delta H - T\Delta S = 58.1 \text{ kJ} - \left(298 \text{ K} \times 165 \text{ J/K} \times \frac{1 \text{ kJ}}{1000 \text{ J}}\right) = 8.9 \text{ kJ}$$

Because ΔG is positive, the reaction will not proceed on its own at 25°C. At the crossover temperature, $\Delta G = 0$, so

$$T = \frac{\Delta H}{\Delta S} = \frac{58.1 \text{ kJ} \times \dfrac{1000 \text{ J}}{1 \text{ kJ}}}{165 \text{ J/K}} = 352 \text{ K}$$

At any temperature above 352 K (79°C), a moderate one for recovering a metal from its ore, the reaction occurs spontaneously. Figure 20.10 depicts this result. The line for $T\Delta S$ increases steadily (and thus the $-T\Delta S$ term becomes more negative) with rising temperature. This line crosses the relatively constant ΔH line at 352 K. At any higher temperature, the $-T\Delta S$ term is greater than the ΔH term, so ΔG is negative.

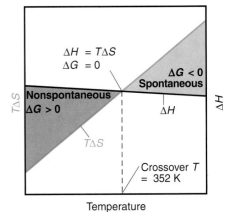

Figure 20.10 The effect of temperature on reaction spontaneity. The two terms that make up ΔG are plotted against T. The figure shows a relatively constant ΔH and a steadily increasing $T\Delta S$ (and thus more negative $-T\Delta S$) for the reaction between Cu_2O and C. At low T, the reaction is nonspontaneous ($\Delta G > 0$) because the positive ΔH term has a greater magnitude than the negative $T\Delta S$ term. At 352 K, $\Delta H = T\Delta S$, so $\Delta G = 0$. At any higher T, the reaction becomes spontaneous ($\Delta G < 0$) because the $-T\Delta S$ term dominates.

Coupling of Reactions to Drive a Nonspontaneous Change

When studying a multistep reaction, chemists often find that a nonspontaneous step is driven by a spontaneous step in a **coupling of reactions.** *One step supplies enough free energy for the other to occur,* as when the combustion of gasoline (spontaneous) supplies enough free energy to move a car (nonspontaneous).

Look again at the reaction of copper(I) oxide with carbon. Previously, we found that the *overall* reaction becomes spontaneous above 352 K. Dividing the reaction into two steps, we find that, even at the slightly higher temperature of 375 K, decomposition of copper(I) oxide to its elements is not spontaneous:

$$Cu_2O(s) \longrightarrow 2Cu(s) + \tfrac{1}{2}O_2(g) \qquad \Delta G_{375} = 140.0 \text{ kJ}$$

But the oxidation of carbon to CO is:

$$C(s) + \tfrac{1}{2}O_2(g) \longrightarrow CO(g) \qquad \Delta G_{375} = -143.8 \text{ kJ}$$

Coupling these reactions allows the reaction with the larger negative ΔG to "drive" the one with the smaller positive ΔG. Adding the reactions together gives

$$Cu_2O(s) + C(s) \longrightarrow 2Cu(s) + CO(g) \qquad \Delta G_{375} = -3.8 \text{ kJ}$$

Many biochemical processes—including the syntheses of proteins, nucleic acids, and fatty acids, the maintenance of ion balance, and the breakdown of nutrients—have nonspontaneous steps. Coupling these steps to spontaneous ones is a life-sustaining strategy that is common to *all organisms*. A key spontaneous biochemical reaction is the hydrolysis of a high-energy molecule called **adenosine triphosphate (ATP)** to adenosine diphosphate (ADP):

$$ATP^{4-} + H_2O \rightleftharpoons ADP^{3-} + HPO_4^{2-} + H^+ \qquad \Delta G^{\circ\prime} = -30.5 \text{ kJ}$$

(For biochemical systems, the standard-state concentration of H^+ is 10^{-7} M, not the usual 1 M, and the standard free energy change has the symbol $\Delta G^{\circ\prime}$.) In the metabolic breakdown of glucose, for example, the initial step, which is nonspontaneous, is the addition of a phosphate group to a glucose molecule:

$$\text{Glucose} + HPO_4^{2-} + H^+ \rightleftharpoons [\text{glucose phosphate}]^- + H_2O \qquad \Delta G^{\circ\prime} = 13.8 \text{ kJ}$$

Coupling this reaction to ATP hydrolysis makes the overall process spontaneous. If we add the two reactions, HPO_4^{2-}, H^+, and H_2O cancel, and we obtain

$$\text{Glucose} + ATP^{4-} \rightleftharpoons [\text{glucose phosphate}]^- + ADP^{3-} \qquad \Delta G^{\circ\prime} = -16.7 \text{ kJ}$$

Coupling of the two reactions is accomplished through an enzyme (Section 16.8) that simultaneously binds glucose and ATP and catalyzes the transfer of the phosphate group. The ADP is combined with phosphate to regenerate ATP in reactions catalyzed by other enzymes. Thus, there is a continuous cycling of ATP to ADP and back to ATP again to supply energy to cells (Figure 20.11).

Figure 20.11 The cycling of metabolic free energy through ATP. Processes that release free energy are coupled to the formation of ATP from ADP, whereas those that require free energy are coupled to the hydrolysis of ATP to ADP.

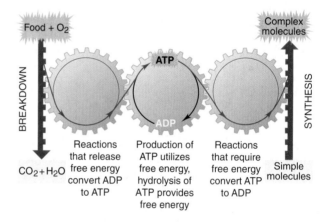

SECTION SUMMARY

The sign of the free energy change, $\Delta G = \Delta H - T\Delta S$, is directly related to reaction spontaneity: a negative ΔG corresponds to a positive ΔS_{univ}. We use the standard free energy change (ΔG°) to evaluate a reaction's spontaneity, and we use the standard free energy of formation (ΔG_f°) to calculate ΔG_{rxn}° at 25°C. The maximum work a system can do is never obtained from a real (irreversible) process because some free energy is always converted to heat. The magnitude of T influences the spontaneity of temperature-dependent reactions (same signs of ΔH and ΔS) by affecting the size of $T\Delta S$. For such reactions, the T at which the reaction becomes spontaneous can be estimated by setting $\Delta G = 0$. A nonspontaneous reaction ($\Delta G > 0$) can be coupled to a more spontaneous one ($\Delta G << 0$) to make it occur. In organisms, the hydrolysis of ATP drives many reactions with a positive ΔG.

20.4 FREE ENERGY, EQUILIBRIUM, AND REACTION DIRECTION

The sign of ΔG allows us to predict reaction direction, but you already know that it is not the only way to do so. In Chapter 17, we predicted direction by comparing the values of the reaction quotient (Q) and the equilibrium constant (K). Recall that

- If $Q < K$ ($Q/K < 1$), the reaction as written proceeds to the right.
- If $Q > K$ ($Q/K > 1$), the reaction as written proceeds to the left.
- If $Q = K$ ($Q/K = 1$), the reaction has reached equilibrium, and there is no net reaction in either direction.

As you might expect, these two ways of predicting reaction spontaneity—the sign of ΔG and the magnitude of Q/K—are related. Their relationship emerges when we compare the signs of $\ln Q/K$ with ΔG:

- If $Q/K < 1$, then $\ln Q/K < 0$: reaction proceeds to the right ($\Delta G < 0$).
- If $Q/K > 1$, then $\ln Q/K > 0$: reaction proceeds to the left ($\Delta G > 0$).
- If $Q/K = 1$, then $\ln Q/K = 0$: reaction is at equilibrium ($\Delta G = 0$).

Note that the signs of ΔG and $\ln Q/K$ are identical for a given reaction direction. In fact, ΔG and $\ln Q/K$ are proportional to each other and made equal through the constant RT:

$$\Delta G = RT \ln \frac{Q}{K} = RT \ln Q - RT \ln K \qquad (20.11)$$

What does this central relationship mean? As you know, Q represents the concentrations (or pressures) of a system's components at any time during the reaction, whereas K represents them when the reaction has reached equilibrium. Therefore, Equation 20.11 says that ΔG depends on how different the ratio of concentrations, Q, is from the equilibrium ratio, K.

The last term in Equation 20.11 is very important. By choosing standard-state values for Q, we obtain the standard free energy change ($\Delta G°$). When all concentrations are 1 M (or all pressures 1 atm), ΔG equals $\Delta G°$ and Q equals 1:

$$\Delta G° = RT \ln 1 - RT \ln K$$

We know that $\ln 1 = 0$, so the $RT \ln Q$ term drops out, and we have

$$\Delta G° = -RT \ln K \qquad (20.12)$$

This relationship allows us to calculate the standard free energy change of a reaction ($\Delta G°$) from its equilibrium constant, or vice versa. Because $\Delta G°$ is related logarithmically to K, even a small change in the value of $\Delta G°$ has a large effect on the value of K. Table 20.2 shows the K values that correspond to a range of $\Delta G°$ values. Note that as $\Delta G°$ becomes more positive, K becomes smaller, which means the reaction reaches equilibrium with less product and more reactant. Similarly, as $\Delta G°$ becomes more negative, K becomes larger. For example, if $\Delta G° = +10$ kJ, $K \approx 0.02$, which means that the product terms are about $\frac{1}{50}$ those of the reactant terms; whereas, if $\Delta G° = -10$ kJ, they are 50 times larger.

Of course, reactions do not usually begin with all components in their standard states. By substituting the relationship between $\Delta G°$ and K (Equation 20.12) into the expression for ΔG (Equation 20.11), we obtain a relationship that applies to any starting concentrations:

$$\Delta G = \Delta G° + RT \ln Q \qquad (20.13)$$

Sample Problem 20.7 illustrates how Equations 20.12 and 20.13 are applied.

Table 20.2	**The Relationship Between $\Delta G°$ and K at 298 K**	
$\Delta G°$ (kJ)	K	Significance
200	9×10^{-36}	Essentially no forward reaction; reverse reaction goes to completion
100	3×10^{-18}	
50	2×10^{-9}	
10	2×10^{-2}	
1	7×10^{-1}	Forward and reverse reactions proceed to same extent
0	1	
−1	1.5	
−10	5×10^{1}	
−50	6×10^{8}	
−100	3×10^{17}	
−200	1×10^{35}	Forward reaction goes to completion; essentially no reverse reaction

FORWARD REACTION / REVERSE REACTION

SAMPLE PROBLEM 20.7   Calculating ΔG at Nonstandard Conditions

Problem The oxidation of SO_2, which we considered in Sample Problem 20.6,

$$2SO_2(g) + O_2(g) \longrightarrow 2SO_3(g)$$

is too slow at 298 K to be useful in the manufacture of sulfuric acid. To overcome this low rate, the process is conducted at an elevated temperature.
(a) Calculate K at 298 K and at 973 K. (ΔG°_{298} = −141.6 kJ/mol for reaction as written; using ΔH° and ΔS° values at 973 K, ΔG°_{973} = −12.12 kJ/mol for reaction as written.)
(b) In experiments to determine the effect of temperature on reaction spontaneity, two sealed containers are filled with 0.500 atm of SO_2, 0.0100 atm of O_2, and 0.100 atm of SO_3 and kept at 25°C and at 700.°C. In which direction, if any, will the reaction proceed to reach equilibrium at each temperature?
(c) Calculate ΔG for the system in part (b) at each temperature.
Plan (a) We know ΔG°, T, and R, so we can calculate the K's from Equation 20.12.
(b) To determine if a net reaction will occur at the given pressures, we calculate Q with the given partial pressures and compare it with each K from part (a). **(c)** Because these are not standard-state pressures, we calculate ΔG at each T from Equation 20.13 with the values of ΔG° (given) and Q [found in part (b)].
Solution (a) Calculating K at the two temperatures:

$$\Delta G^\circ = -RT \ln K \qquad \text{so} \qquad K = e^{-(\Delta G^\circ/RT)}$$

At 298 K, the exponent is

$$-(\Delta G^\circ/RT) = -\left(\frac{-141.6 \text{ kJ/mol} \times \dfrac{1000 \text{ J}}{1 \text{ kJ}}}{8.314 \text{ J/mol·K} \times 298 \text{ K}} \right) = 57.2$$

So $K = e^{-(\Delta G^\circ/RT)} = e^{57.2} = 7 \times 10^{24}$

At 973 K, the exponent is

$$-(\Delta G^\circ/RT) = -\left(\frac{-12.12 \text{ kJ/mol} \times \dfrac{1000 \text{ J}}{1 \text{ kJ}}}{8.314 \text{ J/mol·K} \times 973 \text{ K}} \right) = 1.50$$

So $K = e^{-(\Delta G^\circ/RT)} = e^{1.50} = 4.5$

(b) Calculating the value of Q:

$$Q = \frac{P_{SO_3}^2}{P_{SO_2}^2 \times P_{O_2}} = \frac{0.100^2}{0.500^2 \times 0.0100} = 4.00$$

Because $Q < K$ at both temperatures, the denominator will decrease and the numerator increase—more SO_3 will form—until Q equals K. However, the reaction will go far to the right at 298 K while approaching equilibrium, whereas it will move only slightly to the right at 973 K.
(c) Calculating ΔG, the nonstandard free energy change, at 298 K:

$$\Delta G_{298} = \Delta G^\circ + RT \ln Q$$

$$= -141.6 \text{ kJ/mol} + \left(8.314 \text{ J/mol·K} \times \frac{1 \text{ kJ}}{1000 \text{ J}} \times 298 \text{ K} \times \ln 4.00 \right)$$

$$= -138.2 \text{ kJ/mol}$$

Calculating ΔG at 973 K:

$$\Delta G_{973} = \Delta G^\circ + RT \ln Q$$

$$= -12.12 \text{ kJ/mol} + \left(8.314 \text{ J/mol·K} \times \frac{1 \text{ kJ}}{1000 \text{ J}} \times 973 \text{ K} \times \ln 4.00 \right)$$

$$= -0.9 \text{ kJ/mol}$$

Check Note that in parts (a) and (c) we made the energy units in free energy changes (kJ) consistent with those in R (J). Based on the rules for significant figures in addition and subtraction, we retain one digit to the right of the decimal place in part (c).
Comment For these starting gas pressures at 973 K, the process is barely spontaneous (ΔG = −0.9 kJ/mol), so why use a higher temperature? Like the synthesis of NH_3 (Section 17.6), this process is carried out at a higher temperature *with a catalyst* to attain a higher *rate*, even though the *yield* is greater at a lower temperature.

FOLLOW-UP PROBLEM 20.7 At 298 K, hypobromous acid (HBrO) dissociates in water with a K_a of 2.3×10^{-9}.
(a) Calculate $\Delta G°$ for the dissociation of HBrO.
(b) Calculate ΔG if $[H_3O^+] = 6.0\times10^{-4}$ M, $[BrO^-] = 0.10$ M, and $[HBrO] = 0.20$ M.

Another Look at the Meaning of Spontaneity At this point, let's consider some terminology related to, but distinct from, the terms *spontaneous* and *nonspontaneous*. Consider the general reaction A $\rightleftharpoons$ B, for which $K = [B]/[A] > 1$; therefore, the reaction proceeds largely from left to right (Figure 20.12A). From pure A to the equilibrium point, $Q < K$ and the curved *green* arrow indicates the reaction is spontaneous ($\Delta G < 0$). From there on, the curved *red* arrow shows the reaction is nonspontaneous ($\Delta G > 0$). From pure B to the equilibrium point, $Q > K$ and the reaction is also spontaneous ($\Delta G < 0$), but not thereafter. In either case, *the free energy decreases as the reaction proceeds, until it reaches a minimum at the equilibrium mixture: $Q = K$ and $\Delta G = 0$.* For the overall reaction A $\rightleftharpoons$ B (starting with all components in their standard states), $G_B°$ is smaller than $G_A°$, so $\Delta G°$ is negative, which corresponds to $K > 1$. We call this a *product-favored* reaction because the final state of the system contains mostly product.

Now consider the opposite situation, a general reaction C $\rightleftharpoons$ D, for which $K = [D]/[C] < 1$: the reaction proceeds only slightly from left to right (Figure 20.12B). Here, too, whether we start with pure C or pure D, the reaction is spontaneous ($\Delta G < 0$) until the equilibrium point. But here, the equilibrium mixture contains mostly C (the reactant), so we say the reaction is *reactant favored*. In this case, $G_D°$ is *larger* than $G_C°$, so $\Delta G°$ is *positive*, which corresponds to $K < 1$. The point is that *spontaneous* refers to that portion of a reaction in which the free energy is decreasing, that is, from some starting mixture to the equilibrium mixture, whereas *product-favored* refers to a reaction that goes predominantly, but not necessarily completely, to product (see Table 20.2).

SECTION SUMMARY

Two ways of predicting reaction direction are from the value of ΔG and from the relation of Q to K. These variables represent different aspects of the same phenomenon and are related to each other by $\Delta G = RT \ln Q/K$. When $Q = K$, the system can release no more free energy. Beginning with Q at the standard state, the free energy change is $\Delta G°$, and it is related to the equilibrium constant by $\Delta G° = -RT \ln K$. For nonstandard conditions, ΔG has two components: $\Delta G°$ and $RT \ln Q$. Any nonequilibrium mixture of reactants and products moves spontaneously ($\Delta G < 0$) toward the equilibrium mixture. A product-favored reaction has $K > 1$ and, thus, $\Delta G° < 0$.

Figure 20.12 The relation between free energy and the extent of reaction. The free energy of the system is plotted against the extent of reaction. Each reaction proceeds spontaneously ($Q \neq K$ and $\Delta G < 0$; *curved green arrows*) from either pure reactants (A or C) or pure products (B or D) *to* the equilibrium mixture, at which point $\Delta G = 0$. The reaction *from* the equilibrium mixture to either pure reactants or products is nonspontaneous ($\Delta G > 0$; *curved red arrows*). **A,** For the product-favored reaction A $\rightleftharpoons$ B, $G_A° > G_B°$, so $\Delta G° < 0$ and $K > 1$. **B,** For the reactant-favored reaction C $\rightleftharpoons$ D, $G_D° > G_C°$, so $\Delta G° > 0$ and $K < 1$.

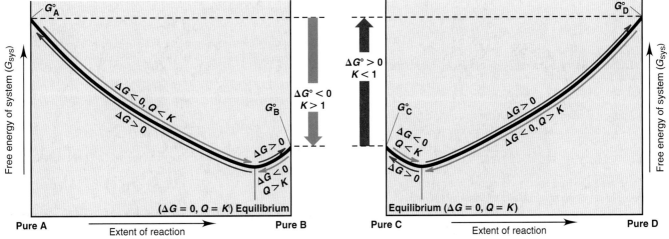

For Review and Reference (Numbers in parentheses refer to pages, unless noted otherwise.)

Learning Objectives

To help you review these learning objectives, the numbers of related sections (§), sample problems (SP), and upcoming end-of-chapter problems (EP) are listed in parentheses.

1. Discuss the meaning of a spontaneous change, and explain why the first law or the sign of $\Delta H°$ cannot predict its direction (§ 20.1) (EPs 20.1–20.3, 20.8, 20.9)
2. Understand the meaning of entropy (S) in terms of the number of microstates over which a system's energy is dispersed; describe how the second law provides the criterion for spontaneity, how the third law allows us to find absolute values of standard molar entropies ($S°$), and how conditions and properties of substances influence $S°$ (§ 20.1) (SP 20.1) (EPs 20.4–20.7, 20.10–20.23)
3. Calculate $\Delta S°_{rxn}$ from $S°$ of reactants and products, understand the influence of $\Delta S°_{surr}$ on $\Delta S°_{rxn}$, and describe the relationships be-

tween ΔS_{surr} and ΔH_{sys} and between ΔS_{univ} and K (§ 20.2) (SPs 20.2, 20.3) (EPs 20.24–20.34)
4. Derive the free energy change (ΔG) from the second law, and explain how ΔG is related to work; explain why temperature (T) affects the spontaneity of some reactions but not others; describe how a spontaneous change drives a nonspontaneous one; calculate $\Delta G°_{rxn}$ from $\Delta H°_f$ and $S°$ values or from $\Delta G°_f$ values and quantify the effect of T on $\Delta G°$; obtain the T at which a reaction becomes spontaneous (§ 20.3) (SPs 20.4–20.6) (EPs 20.35–20.50)
5. Know the relationships of ΔG to Q/K, $\Delta G°$ to K, and ΔG to $\Delta G°$ and Q, and understand why ΔG decreases as a reaction moves toward equilibrium (§ 20.4) (SP 20.7) (EPs 20.51–20.65)

Key Terms

Section 20.1
spontaneous change (651)
entropy (S) (654)
second law of thermodynamics (657)
third law of thermodynamics (657)

standard molar entropy ($S°$) (657)

Section 20.2
standard entropy of reaction ($\Delta S°_{rxn}$) (661)

Section 20.3
free energy (G) (666)
standard free energy change ($\Delta G°$) (667)
standard free energy of formation ($\Delta G°_f$) (668)

coupling of reactions (671)
adenosine triphosphate (ATP) (672)

Key Equations and Relationships

20.1 Defining entropy in terms of the number of microstates (W) over which the energy of a system can be distributed (654):
$$S = k \ln W$$

20.2 Quantifying the entropy change in terms of heat absorbed (or released) in a reversible process (656):
$$\Delta S_{sys} = \frac{q_{rev}}{T}$$

20.3 Stating the second law of thermodynamics, for a spontaneous process (657):
$$\Delta S_{univ} = \Delta S_{sys} + \Delta S_{surr} > 0$$

20.4 Calculating the standard entropy of reaction from the standard molar entropies of reactants and products (662):
$$\Delta S°_{rxn} = \Sigma m S°_{products} - \Sigma n S°_{reactants}$$

20.5 Relating the entropy change in the surroundings to the enthalpy change of the system and the temperature (663):
$$\Delta S_{surr} = -\frac{\Delta H_{sys}}{T}$$

20.6 Expressing the free energy change of the system in terms of its component enthalpy and entropy changes (Gibbs equation) (666):
$$\Delta G_{sys} = \Delta H_{sys} - T\Delta S_{sys}$$

20.7 Calculating the standard free energy change from standard enthalpy and entropy changes (667):
$$\Delta G°_{sys} = \Delta H°_{sys} - T\Delta S°_{sys}$$

20.8 Calculating the standard free energy change from the standard free energies of formation (668):
$$\Delta G°_{rxn} = \Sigma m \Delta G°_{f(products)} - \Sigma n \Delta G°_{f(reactants)}$$

20.9 Relating the free energy change to the maximum work a process can perform (668):
$$\Delta G = -w_{max}$$

20.10 Finding the temperature at which a reaction becomes spontaneous (671):
$$T = \frac{\Delta H}{\Delta S}$$

20.11 Expressing the free energy change in terms of Q and K (673):
$$\Delta G = RT \ln \frac{Q}{K} = RT \ln Q - RT \ln K$$

20.12 Expressing the free energy change when Q is evaluated at the standard state (673):
$$\Delta G° = -RT \ln K$$

20.13 Expressing the free energy change for nonstandard initial conditions (673):
$$\Delta G = \Delta G° + RT \ln Q$$

Brief Solutions to Follow-up Problems

20.1 (a) $PCl_5(g)$: higher molar mass and more complex molecule; (b) $BaCl_2(s)$: higher molar mass; (c) $Br_2(g)$: gases have more freedom of motion and dispersal of energy than liquids.

20.2 (a) $2NaOH(s) + CO_2(g) \longrightarrow Na_2CO_3(s) + H_2O(l)$
$\Delta n_{gas} = -1$, so $\Delta S^{\circ}_{rxn} < 0$
$\Delta S^{\circ}_{rxn} = [(1 \text{ mol } H_2O)(69.9 \text{ J/mol·K})$
$\qquad + (1 \text{ mol } Na_2CO_3)(139 \text{ J/mol·K})]$
$\qquad - [(1 \text{ mol } CO_2)(213.7 \text{ J/mol·K})$
$\qquad + (2 \text{ mol } NaOH)(64.5 \text{ J/mol·K})]$
$\qquad = -134 \text{ J/K}$
(b) $2Fe(s) + 3H_2O(g) \longrightarrow Fe_2O_3(s) + 3H_2(g)$
$\Delta n_{gas} = 0$, so cannot predict sign of ΔS°_{rxn}
$\Delta S^{\circ}_{rxn} = [(1 \text{ mol } Fe_2O_3)(87.4 \text{ J/mol·K})$
$\qquad + (3 \text{ mol } H_2)(130.6 \text{ J/mol·K})]$
$\qquad - [(2 \text{ mol } Fe)(27.3 \text{ J/mol·K})$
$\qquad + (3 \text{ mol } H_2O)(188.7 \text{ J/mol·K})]$
$\qquad = -141.5 \text{ J/K}$

20.3 $2FeO(s) + \frac{1}{2}O_2(g) \longrightarrow Fe_2O_3(s)$
$\Delta S^{\circ}_{sys} = (1 \text{ mol } Fe_2O_3)(87.4 \text{ J/mol·K})$
$\qquad - [(2 \text{ mol } FeO)(60.75 \text{ J/mol·K})$
$\qquad + (\frac{1}{2} \text{ mol } O_2)(205.0 \text{ J/mol·K})]$
$\qquad = -136.6 \text{ J/K}$
$\Delta H^{\circ}_{sys} = (1 \text{ mol } Fe_2O_3)(-825.5 \text{ kJ/mol})$
$\qquad - [(2 \text{ mol } FeO)(-272.0 \text{ kJ/mol})$
$\qquad + (\frac{1}{2} \text{ mol } O_2)(0 \text{ kJ/mol})]$
$\qquad = -281.5 \text{ kJ}$
$\Delta S_{surr} = -\dfrac{\Delta H^{\circ}_{sys}}{T} = -\dfrac{(-281.5 \text{ kJ} \times 1000 \text{ J/kJ})}{298 \text{ K}} = +945 \text{ J/K}$

$\Delta S_{univ} = \Delta S^{\circ}_{sys} + \Delta S_{surr} = -136.6 \text{ J/K} + 945 \text{ J/K}$
$\qquad = 808 \text{ J/K}$; reaction is spontaneous at 298 K.

20.4 Using ΔH°_f and S° values from Appendix B,
$\Delta H^{\circ}_{rxn} = -114.2 \text{ kJ}$ and $\Delta S^{\circ}_{rxn} = -146.5 \text{ J/K}$
$\Delta G^{\circ}_{rxn} = \Delta H^{\circ}_{rxn} - T\Delta S^{\circ}_{rxn} = -114.2 \text{ kJ}$
$\qquad - [(298 \text{ K})(-146.5 \text{ J/K})(1 \text{ kJ/1000 J})]$
$\qquad = -70.5 \text{ kJ}$

20.5 (a) $\Delta G^{\circ}_{rxn} = (2 \text{ mol } NO_2)(51 \text{ kJ/mol})$
$\qquad - [(2 \text{ mol } NO)(86.60 \text{ kJ/mol})$
$\qquad + (1 \text{ mol } O_2)(0 \text{ kJ/mol})]$
$\qquad = -71 \text{ kJ}$
(b) $\Delta G^{\circ}_{rxn} = (2 \text{ mol } CO)(-137.2 \text{ kJ/mol}) - [(2 \text{ mol } C)(0 \text{ kJ/mol})$
$\qquad + (1 \text{ mol } O_2)(0 \text{ kJ/mol})]$
$\qquad = -274.4 \text{ kJ}$

20.6 ΔG becomes negative at lower T, so $\Delta H < 0$, $\Delta S < 0$, and $-T\Delta S > 0$. At lower T, the negative ΔH value becomes larger than the positive $-T\Delta S$ value.

20.7 (a) $\Delta G^{\circ} = -RT \ln K = -8.314 \text{ J/mol·K} \times \dfrac{1 \text{ kJ}}{1000 \text{ J}} \times 298 \text{ K}$
$\qquad \times \ln (2.3 \times 10^{-9})$
$\qquad = 49 \text{ kJ/mol}$
(b) $Q = \dfrac{[H_3O^+][BrO^-]}{[HBrO]} = \dfrac{(6.0 \times 10^{-4})(0.10)}{0.20} = 3.0 \times 10^{-4}$
$\Delta G = \Delta G^{\circ} + RT \ln Q$
$\qquad = 49 \text{ kJ/mol}$
$\qquad + \left[8.314 \text{ J/mol·K} \times \dfrac{1 \text{ kJ}}{1000 \text{ J}} \times 298 \text{ K} \times \ln (3.0 \times 10^{-4}) \right]$
$\qquad = 29 \text{ kJ/mol}$

Problems

Problems with **colored** numbers are answered in Appendix E. Sections match the text and provide the numbers of relevant sample problems. Bracketed problems are grouped in pairs (indicated by a short rule) that cover the same concept. Comprehensive Problems are based on material from any section or previous chapter.

Note: Unless stated otherwise, problems refer to systems at 298 K (25°C). Solving these problems may require values from Appendix B.

The Second Law of Thermodynamics: Predicting Spontaneous Change

(Sample Problem 20.1)

20.1 Distinguish between the terms *spontaneous* and *instantaneous*. Give an example of a process that is spontaneous but very slow, and one that is very fast but not spontaneous.

20.2 Distinguish between the terms *spontaneous* and *nonspontaneous*. Can a nonspontaneous process occur? Explain.

20.3 State the first law of thermodynamics in terms of (a) the energy of the universe; (b) the creation or destruction of energy; (c) the energy change of system and surroundings. Does the first law reveal the direction of spontaneous change? Explain.

20.4 State qualitatively the relationship between entropy and freedom of particle motion. Use this idea to explain why you will probably never (a) be suffocated because all the air near you has moved to the other side of the room; (b) see half the water in your cup of tea freeze while the other half boils.

20.5 Why is ΔS_{vap} of a substance always larger than ΔS_{fus}?

20.6 How does the entropy of the surroundings change during an exothermic reaction? An endothermic reaction? Other than the examples cited in text, describe a spontaneous endothermic process.

20.7 (a) What is the entropy of a perfect crystal at 0 K?
(b) Does entropy increase or decrease as the temperature rises?
(c) Why is $\Delta H^{\circ}_f = 0$ but $S^{\circ} > 0$ for an element?
(d) Why does Appendix B list ΔH°_f values but not ΔS°_f values?

20.8 Which of these processes are spontaneous: (a) water evaporating from a puddle in summer; (b) a lion chasing an antelope; (c) an unstable isotope undergoing radioactive disintegration?

20.9 Which of these processes are nonspontaneous: (a) methane burning in air; (b) a teaspoonful of sugar dissolving in a cup of hot coffee; (c) a soft-boiled egg becoming raw?

20.10 Predict the sign of ΔS_{sys} for each process: (a) a piece of wax melting; (b) silver chloride precipitating from solution; (c) dew forming.

20.11 Predict the sign of ΔS_{sys} for each process: (a) alcohol evaporating; (b) a solid explosive converting to a gas; (c) perfume vapors diffusing through a room.

20.12 Without using Appendix B, predict the sign of ΔS° for
(a) $2K(s) + F_2(g) \longrightarrow 2KF(s)$
(b) $NH_3(g) + HBr(g) \longrightarrow NH_4Br(s)$
(c) $NaClO_3(s) \longrightarrow Na^+(aq) + ClO_3^-(aq)$

20.13 Without using Appendix B, predict the sign of ΔS° for
(a) $H_2S(g) + \frac{1}{2}O_2(g) \longrightarrow \frac{1}{8}S_8(s) + H_2O(g)$

(b) $HCl(aq) + NaOH(aq) \longrightarrow NaCl(aq) + H_2O(l)$
(c) $2NO_2(g) \longrightarrow N_2O_4(g)$

20.14 Without using Appendix B, predict the sign of $\Delta S°$ for
(a) $CaCO_3(s) + 2HCl(aq) \longrightarrow CaCl_2(aq) + H_2O(l) + CO_2(g)$
(b) $2NO(g) + O_2(g) \longrightarrow 2NO_2(g)$
(c) $2KClO_3(s) \longrightarrow 2KCl(s) + 3O_2(g)$

20.15 Without using Appendix B, predict the sign of $\Delta S°$ for
(a) $Ag^+(aq) + Cl^-(aq) \longrightarrow AgCl(s)$
(b) $KBr(s) \longrightarrow KBr(aq)$
(c) $CH_3CH{=}CH_2(g) \longrightarrow \overset{\displaystyle CH_2}{\underset{\displaystyle H_2C{-}CH_2(g)}{\diagup\diagdown}}$

20.16 Predict the sign of ΔS for each process:
(a) $C_2H_5OH(g)$ (350 K and 500 torr) $\longrightarrow$
$C_2H_5OH(g)$ (350 K and 250 torr)
(b) $N_2(g)$ (298 K and 1 atm) $\longrightarrow N_2(aq)$ (298 K and 1 atm)
(c) $O_2(aq)$ (303 K and 1 atm) $\longrightarrow O_2(g)$ (303 K and 1 atm)

20.17 Predict the sign of ΔS for each process:
(a) $O_2(g)$ (1.0 L at 1 atm) $\longrightarrow O_2(g)$ (0.10 L at 10 atm)
(b) $Cu(s)$ (350°C and 2.5 atm) $\longrightarrow Cu(s)$ (450°C and 2.5 atm)
(c) $Cl_2(g)$ (100°C and 1 atm) $\longrightarrow Cl_2(g)$ (10°C and 1 atm)

20.18 Predict which substance has greater molar entropy. Explain.
(a) Butane $CH_3CH_2CH_2CH_3(g)$ or 2-butene $CH_3CH{=}CHCH_3(g)$
(b) $Ne(g)$ or $Xe(g)$ (c) $CH_4(g)$ or $CCl_4(l)$

20.19 Predict which substance has greater molar entropy. Explain.
(a) $CH_3OH(l)$ or $C_2H_5OH(l)$ (b) $KClO_3(s)$ or $KClO_3(aq)$
(c) $Na(s)$ or $K(s)$

20.20 Without consulting Appendix B, arrange each group in order of *increasing* standard molar entropy ($S°$). Explain.
(a) Graphite, diamond, charcoal
(b) Ice, water vapor, liquid water   (c) O_2, O_3, O atoms

20.21 Without consulting Appendix B, arrange each group in order of *increasing* standard molar entropy ($S°$). Explain.
(a) Glucose ($C_6H_{12}O_6$), sucrose ($C_{12}H_{22}O_{11}$), ribose ($C_5H_{10}O_5$)
(b) $CaCO_3$, $Ca + C + \frac{3}{2}O_2$, $CaO + CO_2$
(c) $SF_6(g)$, $SF_4(g)$, $S_2F_{10}(g)$

20.22 Without consulting Appendix B, arrange each group in order of *decreasing* standard molar entropy ($S°$). Explain.
(a) $ClO_4^-(aq)$, $ClO_2^-(aq)$, $ClO_3^-(aq)$
(b) $NO_2(g)$, $NO(g)$, $N_2(g)$ (c) $Fe_2O_3(s)$, $Al_2O_3(s)$, $Fe_3O_4(s)$

20.23 Without consulting Appendix B, arrange each group in order of *decreasing* standard molar entropy ($S°$). Explain.
(a) Mg metal, Ca metal, Ba metal
(b) Hexane (C_6H_{14}), benzene (C_6H_6), cyclohexane (C_6H_{12})
(c) $PF_2Cl_3(g)$, $PF_5(g)$, $PF_3(g)$

Calculating the Change in Entropy of a Reaction
(Sample Problems 20.2 and 20.3)

20.24 What property of entropy allows Hess's law to be used in the calculation of entropy changes?

20.25 Describe the equilibrium condition in terms of the entropy changes of a system and its surroundings. What does this description mean about the entropy change of the universe?

20.26 For the reaction $H_2O(g) + Cl_2O(g) \longrightarrow 2HClO(g)$, you know $\Delta S°_{rxn}$ and $S°$ of $HClO(g)$ and of $H_2O(g)$. Write an expression to determine $S°$ of $Cl_2O(g)$.

20.27 For each reaction, predict the sign and find the value of $\Delta S°$:
(a) $3NO(g) \longrightarrow N_2O(g) + NO_2(g)$

(b) $3H_2(g) + Fe_2O_3(s) \longrightarrow 2Fe(s) + 3H_2O(g)$
(c) $P_4(s) + 5O_2(g) \longrightarrow P_4O_{10}(s)$

20.28 For each reaction, predict the sign and find the value of $\Delta S°$:
(a) $3NO_2(g) + H_2O(l) \longrightarrow 2HNO_3(l) + NO(g)$
(b) $N_2(g) + 3F_2(g) \longrightarrow 2NF_3(g)$
(c) $C_6H_{12}O_6(s) + 6O_2(g) \longrightarrow 6CO_2(g) + 6H_2O(g)$

20.29 Find $\Delta S°$ for the combustion of ethane (C_2H_6) to carbon dioxide and gaseous water. Is the sign of $\Delta S°$ as expected?

20.30 Find $\Delta S°$ for the reaction of nitric oxide with hydrogen to form ammonia and water vapor. Is the sign of $\Delta S°$ as expected?

20.31 Find $\Delta S°$ for the formation of $Cu_2O(s)$ from its elements.

20.32 Find $\Delta S°$ for the formation of $CH_3OH(l)$ from its elements.

20.33 Sulfur dioxide is released in the combustion of coal. Scrubbers use aqueous slurries of calcium hydroxide to remove the SO_2 from flue gases. Write a balanced equation for this reaction and calculate $\Delta S°$ at 298 K [$S°$ of $CaSO_3(s) = 101.4$ J/mol·K].

20.34 Oxyacetylene welding is used to repair metal structures, including bridges, buildings, and even the Statue of Liberty. Calculate $\Delta S°$ for the combustion of 1 mol of acetylene (C_2H_2).

Entropy, Free Energy, and Work
(Sample Problems 20.4 to 20.6)

20.35 What is the advantage of calculating free energy changes rather than entropy changes to determine reaction spontaneity?

20.36 Given that $\Delta G_{sys} = -T\Delta S_{univ}$, explain how the sign of ΔG_{sys} correlates with reaction spontaneity.

20.37 Is an endothermic reaction more likely to be spontaneous at higher temperatures or lower temperatures? Explain.

20.38 With its components in their standard states, a certain reaction is spontaneous only at high T. What do you know about the signs of $\Delta H°$ and $\Delta S°$? Describe a process for which this is true.

20.39 Calculate $\Delta G°$ for each reaction using $\Delta G°_f$ values:
(a) $2Mg(s) + O_2(g) \longrightarrow 2MgO(s)$
(b) $2CH_3OH(g) + 3O_2(g) \longrightarrow 2CO_2(g) + 4H_2O(g)$
(c) $BaO(s) + CO_2(g) \longrightarrow BaCO_3(s)$

20.40 Calculate $\Delta G°$ for each reaction using $\Delta G°_f$ values:
(a) $H_2(g) + I_2(s) \longrightarrow 2HI(g)$
(b) $MnO_2(s) + 2CO(g) \longrightarrow Mn(s) + 2CO_2(g)$
(c) $NH_4Cl(s) \longrightarrow NH_3(g) + HCl(g)$

20.41 Find $\Delta G°$ for the reactions in Problem 20.39 using $\Delta H°_f$ and $S°$ values.

20.42 Find $\Delta G°$ for the reactions in Problem 20.40 using $\Delta H°_f$ and $S°$ values.

20.43 Consider the oxidation of carbon monoxide:
$$CO(g) + \tfrac{1}{2}O_2(g) \longrightarrow CO_2(g)$$
(a) Predict the signs of $\Delta S°$ and $\Delta H°$. Explain.
(b) Calculate $\Delta G°$ by two different methods.

20.44 Consider the combustion of butane gas:
$$C_4H_{10}(g) + \tfrac{13}{2}O_2(g) \longrightarrow 4CO_2(g) + 5H_2O(g)$$
(a) Predict the signs of $\Delta S°$ and $\Delta H°$. Explain.
(b) Calculate $\Delta G°$ by two different methods.

20.45 One reaction used to produce small quantities of pure H_2 is
$$CH_3OH(g) \rightleftharpoons CO(g) + 2H_2(g)$$
(a) Determine $\Delta H°$ and $\Delta S°$ for the reaction at 298 K.
(b) Assuming that these values are relatively independent of temperature, calculate $\Delta G°$ at 38°C, 138°C, and 238°C.
(c) What is the significance of the different values of $\Delta G°$?

20.46 A reaction that occurs in the internal combustion engine is

$$N_2(g) + O_2(g) \rightleftharpoons 2NO(g)$$

(a) Determine $\Delta H°$ and $\Delta S°$ for the reaction at 298 K.
(b) Assuming that these values are relatively independent of temperature, calculate $\Delta G°$ at 100.°C, 2560.°C, and 3540.°C.
(c) What is the significance of the different values of $\Delta G°$?

20.47 Use $\Delta H°$ and $\Delta S°$ values for the following process at 1 atm to find the normal boiling point of Br_2: $Br_2(l) \rightleftharpoons Br_2(g)$

20.48 Use $\Delta H°$ and $\Delta S°$ values to find the temperature at which these sulfur allotropes reach equilibrium at 1 atm:

$$S(\text{rhombic}) \rightleftharpoons S(\text{monoclinic})$$

20.49 As a fuel, $H_2(g)$ produces only nonpolluting $H_2O(g)$ when it burns. Moreover, it combines with $O_2(g)$ in a fuel cell (Chapter 21) to provide electrical energy.
(a) Calculate $\Delta H°$, $\Delta S°$, and $\Delta G°$ per mol of H_2 at 298 K.
(b) Is the spontaneity of this reaction dependent on T? Explain.
(c) At what temperature does the reaction become spontaneous?

20.50 The United States requires a renewable component in automobile fuels. The fermentation of glucose from corn produces ethanol, which is added to gasoline to fulfill this requirement:

$$C_6H_{12}O_6(s) \longrightarrow 2C_2H_5OH(l) + 2CO_2(g)$$

Calculate $\Delta H°$, $\Delta S°$, and $\Delta G°$ for the reaction at 25°C. Is the spontaneity of this reaction dependent on T? Explain.

Free Energy, Equilibrium, and Reaction Direction
(Sample Problem 20.7)

20.51 (a) If $K \ll 1$ for a reaction, what do you know about the sign and magnitude of $\Delta G°$? (b) If $\Delta G° \ll 0$ for a reaction, what do you know about the magnitude of K? Of Q?

20.52 How is the free energy change of a process related to the work that can be obtained from the process? Is this quantity of work obtainable in practice? Explain.

20.53 What is the difference between $\Delta G°$ and ΔG? Under what circumstances does $\Delta G = \Delta G°$?

20.54 Calculate K at 298 K for each reaction:
(a) $NO(g) + \frac{1}{2}O_2(g) \rightleftharpoons NO_2(g)$
(b) $2HCl(g) \rightleftharpoons H_2(g) + Cl_2(g)$
(c) $2C(\text{graphite}) + O_2(g) \rightleftharpoons 2CO(g)$

20.55 Calculate K at 298 K for each reaction:
(a) $2H_2S(g) + 3O_2(g) \rightleftharpoons 2H_2O(g) + 2SO_2(g)$
(b) $H_2SO_4(l) \rightleftharpoons H_2O(l) + SO_3(g)$
(c) $HCN(aq) + NaOH(aq) \rightleftharpoons NaCN(aq) + H_2O(l)$

20.56 Use Appendix B to determine the K_{sp} of Ag_2S.

20.57 Use Appendix B to determine the K_{sp} of CaF_2.

20.58 For the reaction $I_2(g) + Cl_2(g) \rightleftharpoons 2ICl(g)$, calculate K_p at 25°C [$\Delta G_f°$ of $ICl(g) = -6.075$ kJ/mol].

20.59 For the reaction $CaCO_3(s) \rightleftharpoons CaO(s) + CO_2(g)$, calculate the equilibrium P_{CO_2} at 25°C.

20.60 The K_{sp} of $PbCl_2$ is 1.7×10^{-5} at 25°C. What is $\Delta G°$? Is it possible to prepare a solution that contains $Pb^{2+}(aq)$ and $Cl^-(aq)$, at their standard-state concentrations?

20.61 The K_{sp} of ZnF_2 is 3.0×10^{-2} at 25°C. What is $\Delta G°$? Is it possible to prepare a solution that contains $Zn^{2+}(aq)$ and $F^-(aq)$ at their standard-state concentrations?

20.62 The equilibrium constant for the reaction

$$2Fe^{3+}(aq) + Hg_2^{2+}(aq) \rightleftharpoons 2Fe^{2+}(aq) + 2Hg^{2+}(aq)$$

is $K_c = 9.1 \times 10^{-6}$ at 298 K.

(a) What is $\Delta G°$ at this temperature?
(b) If standard-state concentrations of the reactants and products are mixed, in which direction does the reaction proceed?
(c) Calculate ΔG when $[Fe^{3+}] = 0.20$ M, $[Hg_2^{2+}] = 0.010$ M, $[Fe^{2+}] = 0.010$ M, and $[Hg^{2+}] = 0.025$ M. In which direction will the reaction proceed to achieve equilibrium?

20.63 The formation constant for the reaction

$$Ni^{2+}(aq) + 6NH_3(aq) \rightleftharpoons Ni(NH_3)_6^{2+}(aq)$$

is $K_f = 5.6 \times 10^8$ at 25°C.
(a) What is $\Delta G°$ at this temperature?
(b) If standard-state concentrations of the reactants and products are mixed, in which direction does the reaction proceed?
(c) Determine ΔG when $[Ni(NH_3)_6^{2+}] = 0.010$ M, $[Ni^{2+}] = 0.0010$ M, and $[NH_3] = 0.0050$ M. In which direction will the reaction proceed to achieve equilibrium?

20.64 High levels of ozone (O_3) cause rubber to deteriorate, green plants to turn brown, and many people to have difficulty breathing.
(a) Is the formation of O_3 from O_2 favored at all T, no T, high T, or low T?
(b) Calculate $\Delta G°$ for this reaction at 298 K.
(c) Calculate ΔG at 298 K for this reaction in urban smog where $[O_2] = 0.21$ atm and $[O_3] = 5 \times 10^{-7}$ atm.

20.65 A $BaSO_4$ slurry is ingested before the gastrointestinal tract is x-rayed because it is opaque to x-rays and defines the contours of the tract. Ba^{2+} ion is toxic, but the compound is nearly insoluble. If $\Delta G°$ at 37°C (body temperature) is 59.1 kJ/mol for the process

$$BaSO_4(s) \rightleftharpoons Ba^{2+}(aq) + SO_4^{2-}(aq)$$

what is $[Ba^{2+}]$ in the intestinal tract? (Assume that the only source of SO_4^{2-} is the ingested slurry.)

Comprehensive Problems
Problems with an asterisk (*) are more challenging.

20.66 According to the advertisement, "a diamond is forever."
(a) Calculate $\Delta H°$, $\Delta S°$, and $\Delta G°$ at 298 K for the phase change

$$\text{Diamond} \longrightarrow \text{graphite}$$

(b) Given the conditions under which diamond jewelry is normally kept, argue for and against the statement in the ad.
(c) Given the answers in part (a), what would need to be done to make synthetic diamonds from graphite?
(d) Assuming $\Delta H°$ and $\Delta S°$ do not change with temperature, can graphite be converted to diamond spontaneously at 1 atm?

20.67 Replace each question mark with the correct information:

	ΔS_{rxn}	ΔH_{rxn}	ΔG_{rxn}	Comment
(a)	+	−	−	?
(b)	?	0	−	Spontaneous
(c)	−	+	?	Not spontaneous
(d)	0	?	−	Spontaneous
(e)	?	0	+	?
(f)	+	+	?	$T\Delta S > \Delta H$

20.68 What is the change in entropy when 0.200 mol of potassium freezes at 63.7°C ($\Delta H_{fus} = 2.39$ kJ/mol)?

* **20.69** Hemoglobin carries O_2 from the lungs to tissue cells, where the O_2 is released. The protein is represented as Hb in its unoxygenated form and as $Hb \cdot O_2$ in its oxygenated form. One reason CO is toxic is that it competes with O_2 in binding to Hb:

$$Hb \cdot O_2(aq) + CO(g) \rightleftharpoons Hb \cdot CO(aq) + O_2(g)$$

(a) If $\Delta G° \approx -14$ kJ at 37°C (body temperature), what is the ratio of [Hb·CO] to [Hb·O$_2$] at 37°C with [O$_2$] = [CO]?

(b) How is Le Châtelier's principle used to treat CO poisoning?

20.70 Magnesia (MgO) is used for fire brick, crucibles, and furnace linings because of its high melting point. It is produced by decomposing magnesite (MgCO$_3$) at around 1200°C.

(a) Write a balanced equation for magnesite decomposition.

(b) Use $\Delta H°$ and $S°$ values to find $\Delta G°$ at 298 K.

(c) Assuming $\Delta H°$ and $S°$ do not change with temperature, find the minimum temperature at which the reaction is spontaneous.

(d) Calculate the equilibrium P_{CO_2} above MgCO$_3$ at 298 K.

(e) Calculate the equilibrium P_{CO_2} above MgCO$_3$ at 1200 K.

20.71 Methanol, a major industrial feedstock, is made by several catalyzed reactions, such as $CO(g) + 2H_2(g) \longrightarrow CH_3OH(l)$.

(a) Show that this reaction is thermodynamically feasible.

(b) Is it favored at low or at high temperatures?

(c) One concern about using CH$_3$OH as an auto fuel is its oxidation in air to yield formaldehyde, CH$_2$O(g), which poses a health hazard. Calculate $\Delta G°$ at 100.°C for this oxidation.

20.72 (a) Write a balanced equation for the gaseous reaction between N$_2$O$_5$ and F$_2$ to form NF$_3$ and O$_2$. (b) Determine $\Delta G°_{rxn}$.
(c) Find ΔG_{rxn} at 298 K if $P_{N_2O_5} = P_{F_2} = 0.20$ atm, $P_{NF_3} = 0.25$ atm, and $P_{O_2} = 0.50$ atm.

20.73 Consider the following reaction:

$$2NOBr(g) \rightleftharpoons 2NO(g) + Br_2(g) \qquad K = 0.42 \text{ at } 373 \text{ K}$$

Given that $S°$ of NOBr(g) = 272.6 J/mol·K and that $\Delta S°_{rxn}$ and $\Delta H°_{rxn}$ are constant with temperature, find

(a) $\Delta S°_{rxn}$ at 298 K   (b) $\Delta G°_{rxn}$ at 373 K

(c) $\Delta H°_{rxn}$ at 373 K   (d) $\Delta H°_f$ of NOBr at 298 K

(e) $\Delta G°_{rxn}$ at 298 K   (f) $\Delta G°_f$ of NOBr at 298 K

20.74 Calculate the equilibrium constants for decomposition of the hydrogen halides at 298 K: $2HX(g) \rightleftharpoons H_2(g) + X_2(g)$ What do these values indicate about the extent of decomposition of HX at 298 K? Suggest a reason for this trend.

20.75 The key process in a blast furnace during the production of iron is the reaction of Fe$_2$O$_3$ and carbon to yield Fe and CO$_2$.

(a) Calculate $\Delta H°$ and $\Delta S°$. [Assume C(graphite).]

(b) Is the reaction spontaneous at low or at high T? Explain.

(c) Is the reaction spontaneous at 298 K?

(d) At what temperature does the reaction become spontaneous?

20.76 Bromine monochloride is formed from the elements:

$$Cl_2(g) + Br_2(g) \longrightarrow 2BrCl(g)$$
$$\Delta H°_{rxn} = -1.35 \text{ kJ/mol} \qquad \Delta G°_f = -0.88 \text{ kJ/mol}$$

Calculate (a) $\Delta H°_f$ and (b) $S°$ of BrCl(g).

20.77 Solid N$_2$O$_5$ reacts with water to form liquid HNO$_3$. Consider the reaction with all substances in their standard states.

(a) Is the reaction spontaneous at 25°C?

(b) The solid decomposes to NO$_2$ and O$_2$ at 25°C. Is the decomposition spontaneous at 25°C? At what T is it spontaneous?

(c) At what T does *gaseous* N$_2$O$_5$ decompose spontaneously? Explain the difference between this T and that in part (b).

20.78 Find K for (a) the hydrolysis of ATP, (b) the reaction of glucose with HPO$_4^{2-}$ to form glucose phosphate, and (c) the coupled reaction between ATP and glucose. (d) How does each K change when T changes from 25°C to 37°C?

* 20.79 Energy from ATP hydrolysis drives many nonspontaneous cell reactions:

$$ATP^{4-}(aq) + H_2O(l) \rightleftharpoons$$
$$ADP^{3-}(aq) + HPO_4^{2-}(aq) + H^+(aq) \qquad \Delta G°' = -30.5 \text{ kJ}$$

Energy for the reverse process comes ultimately from glucose metabolism:

$$C_6H_{12}O_6(s) + 6O_2(g) \longrightarrow 6CO_2(g) + 6H_2O(l)$$

(a) Find $\Delta G°'_{rxn}$ for metabolism of 1 mol of glucose. (b) How many moles of ATP can be produced by metabolism of 1 mol of glucose? (c) If 36 mol of ATP is formed, what is the actual yield?

20.80 A chemical reaction, such as HI forming from its elements, can reach equilibrium at many temperatures. In contrast, a phase change, such as ice melting, is in equilibrium at a given pressure only at the melting point. (a) Which graph depicts how G_{sys} changes for the formation of HI? Explain. (b) Which graph depicts how G_{sys} changes as ice melts at 1°C and 1 atm? Explain.

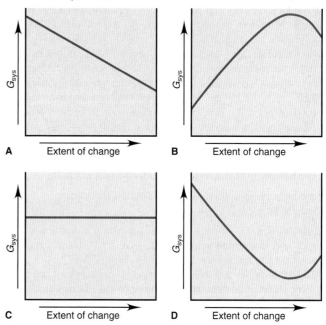

A Extent of change   **B** Extent of change

C Extent of change   **D** Extent of change

* 20.81 Consider the formation of ammonia:

$$N_2(g) + 3H_2(g) \rightleftharpoons 2NH_3(g)$$

(a) Assuming that $\Delta H°$ and $\Delta S°$ are constant with temperature, find the temperature at which $K_p = 1.00$. (b) Find K_p at 400.°C, a typical temperature for NH$_3$ production. (c) Given the lower K_p at the higher temperature, why are these conditions used industrially?

* 20.82 Kyanite, sillimanite, and andalusite all have the formula Al$_2$SiO$_5$. Each is stable under different conditions:

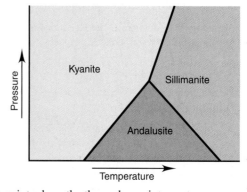

At the point where the three phases intersect:

(a) Which mineral, if any, has the lowest free energy?

(b) Which mineral, if any, has the lowest enthalpy?

(c) Which mineral, if any, has the highest entropy?

(d) Which mineral, if any, has the lowest density?

CHAPTER TWENTY-ONE

Electrochemistry: Chemical Change and Electrical Work

Plating It On Electroplating chromium onto this 1936 Packard hood ornament, both for beauty and protection against corrosion, requires energy. The other face of electrochemistry is represented by batteries, which supply energy. This chapter will reveal both faces.

Key Principles

◆ *Oxidation-reduction (redox)* reactions involve the movement of electrons. The *half-reaction method* of balancing a redox reaction separates the overall reaction into two half-reactions. This reflects the actual separation of the two *half-cells* in an *electrochemical cell*.

◆ In a *voltaic cell,* a spontaneous redox reaction ($\Delta G < 0$) is separated into an oxidation half-reaction (*anode half-cell*) and a reduction half-reaction (*cathode half-cell*). Electrons flow from anode to cathode through an external circuit, releasing electrical energy, and ions flow through a *salt bridge* to complete the circuit and balance the charge within the cell.

◆ The anode has a greater ability to give up electrons than the cathode, and the *cell potential,* or *voltage* (E_{cell}), is related to this difference. A negative ΔG correlates with a positive E_{cell}. Under standard-state conditions, each half-reaction is associated with a *standard electrode potential* ($E^{\circ}_{half\text{-}cell}$).

◆ The standard free energy change (ΔG°), the standard cell potential (E°_{cell}), and the equilibrium constant (K) are interrelated.

◆ Cell potential (E_{cell}) changes during operation of the cell. The *Nernst equation* shows that E_{cell} depends on E°_{cell} and a term for the potential at nonstandard-state concentrations. During the operation of a typical voltaic cell, reactant concentration starts out higher than product concentration, gradually becomes equal to it, and then less than it, until $Q = K$ and the cell can do no more work.

◆ In a *concentration cell,* each half-cell houses the same half-reaction, but the concentrations of reactants are different. During operation, the solution in the anode half-cell spontaneously becomes more concentrated and that in the cathode half-cell becomes less concentrated until their concentrations are equal.

◆ A *battery* is a group of voltaic cells arranged in series. In a *primary* battery, reactants become products until equilibrium is reached, at which point the battery is discarded. A *secondary* battery can be recharged by using an external energy source to convert the products back into reactants. In a *fuel cell,* reactants enter and products leave continually.

◆ The process of *corrosion* is a spontaneous electrochemical phenomenon similar to the operation of a voltaic cell. It is a major economic problem because the anode is typically a useful metal tool (e.g., shovel) or structure (e.g., bridge) that is harmed.

◆ In an *electrolytic cell,* an external energy source makes a *non*spontaneous redox reaction ($\Delta G > 0$) occur. In electrolysis of a molten binary ionic compound (salt), the cation is reduced to the metal and the anion is oxidized to the nonmetal. For an aqueous salt solution, the products depend on whether water or one of the ions of the salt requires less energy to be reduced or oxidized.

Outline

Concepts & Skills to Review Before You Study This Chapter

- redox terminology (Section 4.5)
- activity series of the metals (Section 4.6)
- free energy, work, and equilibrium (Sections 20.3 and 20.4)
- Q vs. K (Section 17.4) and ΔG vs. ΔG° (Section 20.4)

If you think thermodynamics relates mostly to expanding gases inside steam engines and has few practical, everyday applications, just look around. Some applications are probably within your reach right now, in the form of battery-operated devices—laptop computer, palm organizer, DVD remote, and, of course, wristwatch and calculator—or in the form of metal-plated jewelry or silverware. The operation and creation of these objects, and the many similar ones you use daily, involve the principles we cover in this chapter.

Electrochemistry, certainly one of the most important areas of applied thermodynamics, is the study of the relationship between chemical change and electrical work. It is typically investigated through the use of **electrochemical cells,** systems that incorporate a redox reaction to produce or utilize electrical energy.

21.1 REDOX REACTIONS AND ELECTROCHEMICAL CELLS

Whether an electrochemical process releases or absorbs free energy, it always involves the *movement of electrons from one chemical species to another* in an oxidation-reduction (redox) reaction. In this section, we review the redox process and describe the half-reaction method of balancing redox reactions. Then we see how such reactions are used in electrochemical cells.

A Quick Review of Oxidation-Reduction Concepts

In electrochemical reactions, as in any redox process, *oxidation* is the loss of electrons, and *reduction* is the gain of electrons. An *oxidizing agent* is the species that does the oxidizing, taking electrons from the substance being oxidized. A *reducing agent* is the species that does the reducing, giving electrons to the substance being reduced. After the reaction, the oxidized substance has a higher (more positive or less negative) oxidation number (O.N.), and the reduced substance has a lower (less positive or more negative) one. Keep in mind three key points:

- Oxidation (electron loss) always accompanies reduction (electron gain).
- The oxidizing agent is reduced, and the reducing agent is oxidized.
- The total number of electrons gained by the atoms/ions of the oxidizing agent always equals the total number lost by the atoms/ions of the reducing agent.

Figure 21.1 presents these ideas for the aqueous reaction between zinc metal and a strong acid. Be sure you can identify the oxidation and reduction parts of a redox process. If you're having trouble, see the full discussion in Chapter 4.

PROCESS	$Zn(s) + 2H^+(aq) \longrightarrow Zn^{2+}(aq) + H_2(g)$
OXIDATION • One reactant loses electrons. • Reducing agent is oxidized. • Oxidation number increases.	Zinc **loses** electrons. Zinc is the reducing agent and becomes **oxidized**. The oxidation number of Zn **increases** from 0 to +2.
REDUCTION • Other reactant gains electrons. • Oxidizing agent is reduced. • Oxidation number decreases.	Hydrogen ion **gains** electrons. Hydrogen ion is the oxidizing agent and becomes **reduced**. The oxidation number of H **decreases** from +1 to 0.

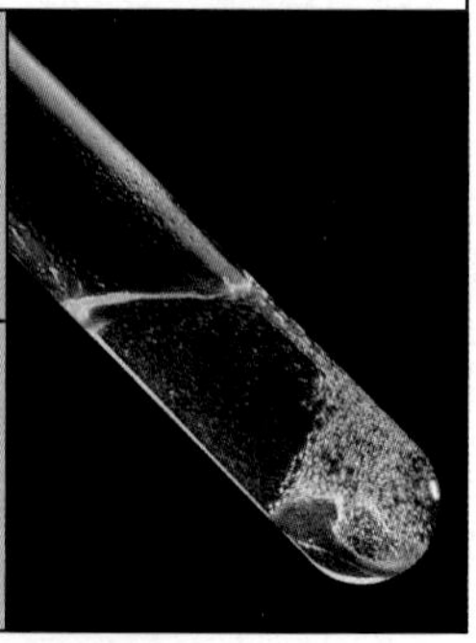

Figure 21.1 A summary of redox terminology. In the reaction between zinc and hydrogen ion, Zn is oxidized and H^+ is reduced.

Half-Reaction Method for Balancing Redox Reactions

The **half-reaction method** for balancing redox reactions *divides the overall redox reaction into oxidation and reduction half-reactions.* Each half-reaction is balanced for atoms and charge. Then, one or both are multiplied by some integer to make electrons gained equal electrons lost, and the half-reactions are recombined to give the balanced redox equation. The half-reaction method is commonly used for studying electrochemistry because

- It separates the oxidation and reduction steps, which reflects their actual physical separation in electrochemical cells.
- It is readily applied to redox reactions that take place in acidic or basic solution, which is common in these cells.
- It (usually) does *not* require assigning O.N.s. (In cases where the half-reactions are not obvious, we assign O.N.s to determine which atoms undergo a change and write half-reactions with the species that contain those atoms.)

In general, we begin with a "skeleton" ionic reaction, which shows only the species that are oxidized and reduced. *If the oxidized form of a species is on the left side of the skeleton reaction, the reduced form of that species must be on the right, and vice versa.* Unless H_2O, H^+, and OH^- are being oxidized or reduced, they do not appear in the skeleton reaction. The following steps are used in balancing a redox reaction by the half-reaction method:

Step 1. Divide the skeleton reaction into two half-reactions, each of which contains the oxidized and reduced forms of one of the species. (Which half-reaction is the oxidation and which is the reduction will become clear in the next step.)

Step 2. Balance the atoms and charges in each half-reaction.
- Atoms are balanced *in order:* atoms other than O and H, then O, and then H.
- Charge is balanced by *adding electrons* (e^-). They are added *to the left in the reduction half-reaction* because the reactant gains them; they are added *to the right in the oxidation half-reaction* because the reactant loses them.

Step 3. If necessary, multiply one or both half-reactions by an integer to make the number of e^- gained in the reduction equal the number lost in the oxidation.

Step 4. Add the balanced half-reactions, and include states of matter.

Step 5. Check that the atoms and charges are balanced.

We'll balance a redox reaction that occurs in acidic solution first and then go through Sample Problem 21.1 to balance one in basic solution.

Balancing Redox Reactions in Acidic Solution

When a redox reaction occurs in acidic solution, H_2O molecules and H^+ ions are available for balancing. Even though we've usually used H_3O^+ to indicate the proton in water, we use H^+ in this chapter because it makes the balanced equations less complex.

Let's balance the redox reaction between dichromate ion and iodide ion to form chromium(III) ion and solid iodine, which occurs in acidic solution (Figure 21.2). The skeleton ionic reaction shows only the oxidized and reduced species:

$$Cr_2O_7^{2-}(aq) + I^-(aq) \longrightarrow Cr^{3+}(aq) + I_2(s) \quad \text{[acidic solution]}$$

Step 1. Divide the reaction into half-reactions, each of which contains the oxidized and reduced forms of one species. The two chromium species make up one half-reaction, and the two iodine species make up the other:

$$Cr_2O_7^{2-} \longrightarrow Cr^{3+}$$
$$I^- \longrightarrow I_2$$

Figure 21.2 The redox reaction between dichromate ion and iodide ion. When $Cr_2O_7^{2-}$ (*left*) and I^- (*center*) are mixed in acid solution, they react to form Cr^{3+} and I_2 (*right*).

Step 2. Balance atoms and charges in each half-reaction. We use H_2O to balance O atoms, H^+ to balance H atoms, and e^- to balance positive charges.

- For the $Cr_2O_7^{2-}/Cr^{3+}$ half-reaction:

  a. *Balance atoms other than O and H.* We balance the two Cr on the left with a coefficient 2 on the right:

  $$Cr_2O_7^{2-} \longrightarrow 2Cr^{3+}$$

  b. *Balance O atoms by adding H_2O molecules.* Each H_2O has one O atom, so we add seven H_2O on the right to balance the seven O in $Cr_2O_7^{2-}$:

  $$Cr_2O_7^{2-} \longrightarrow 2Cr^{3+} + 7H_2O$$

  c. *Balance H atoms by adding H^+ ions.* Each H_2O contains two H, and we added seven H_2O, so we add 14 H^+ ions on the left:

  $$14H^+ + Cr_2O_7^{2-} \longrightarrow 2Cr^{3+} + 7H_2O$$

  d. *Balance charge by adding electrons.* Each H^+ ion has a 1+ charge, and 14 H^+ plus $Cr_2O_7^{2-}$ gives 12+ on the left. Two Cr^{3+} give 6+ on the right. There is an excess of 6+ on the left, so we add six e^- on the left:

  $$6e^- + 14H^+ + Cr_2O_7^{2-} \longrightarrow 2Cr^{3+} + 7H_2O$$

  This half-reaction is balanced, and we see it is the *reduction* because electrons appear on the *left, as reactants:* the reactant $Cr_2O_7^{2-}$ gains electrons (is reduced), so $Cr_2O_7^{2-}$ is the *oxidizing agent.* (Note that the O.N. of Cr decreases from +6 on the left to +3 on the right.)

- For the I^-/I_2 half-reaction:

  a. *Balance atoms other than O and H.* Two I atoms on the right require a coefficient 2 on the left:

  $$2I^- \longrightarrow I_2$$

  b. *Balance O atoms with H_2O.* Not needed; there are no O atoms.

  c. *Balance H atoms with H^+.* Not needed; there are no H atoms.

  d. *Balance charge with e^-.* To balance the 2− on the left, we add two e^- on the right:

  $$2I^- \longrightarrow I_2 + 2e^-$$

  This half-reaction is balanced, and it is the *oxidation* because electrons appear on the *right, as products:* the reactant I^- loses electrons (is oxidized), so I^- is the *reducing agent.* (Note that the O.N. of I increases from −1 to 0.)

Step 3. Multiply each half-reaction, if necessary, by an integer so that the number of e^- lost in the oxidation equals the number of e^- gained in the reduction. Two e^- are lost in the oxidation and six e^- are gained in the reduction, so we multiply the oxidation by 3:

$$3(2I^- \longrightarrow I_2 + 2e^-)$$
$$6I^- \longrightarrow 3I_2 + 6e^-$$

Step 4. Add the half-reactions together, canceling substances that appear on both sides, and include states of matter. In this example, only the electrons cancel:

$$\cancel{6e^-} + 14H^+ + Cr_2O_7^{2-} \longrightarrow 2Cr^{3+} + 7H_2O$$
$$6I^- \longrightarrow 3I_2 + \cancel{6e^-}$$

$$6I^-(aq) + 14H^+(aq) + Cr_2O_7^{2-}(aq) \longrightarrow 3I_2(s) + 7H_2O(l) + 2Cr^{3+}(aq)$$

Step 5. Check that atoms and charges balance:

Reactants (6I, 14H, 2Cr, 7O; 6+) $\longrightarrow$ products (6I, 14H, 2Cr, 7O; 6+)

Balancing Redox Reactions in Basic Solution

As you just saw, in acidic solution, H_2O molecules and H^+ ions are available for balancing. As Sample Problem 21.1 shows, in basic solution, H_2O molecules and OH^- ions are available. Only one additional step is needed to balance a redox equation that takes place in basic solution. It appears after both half-reactions have first been balanced *as if they took place in acidic solution* (steps 1 and 2), the e$^-$ lost have been made equal to the e$^-$ gained (step 3), and the half-reactions have been combined (step 4). At this point, *we add one OH$^-$ ion **to both sides of the equation** for every H$^+$ ion present.* (We label this step "4 Basic.") The H$^+$ ions on one side are combined with the added OH$^-$ ions to form H$_2$O, and OH$^-$ ions appear on the other side of the equation. Excess H$_2$O molecules are canceled, and states of matter are identified. Finally, we check that atoms and charges balance (step 5).

SAMPLE PROBLEM 21.1 Balancing Redox Reactions by the Half-Reaction Method

Problem Permanganate ion is a strong oxidizing agent, and its deep purple color makes it useful as an indicator in redox titrations. It reacts in basic solution with the oxalate ion to form carbonate ion and solid manganese dioxide. Balance the skeleton ionic equation for the reaction between NaMnO$_4$ and Na$_2$C$_2$O$_4$ in basic solution:

$$MnO_4^-(aq) + C_2O_4^{2-}(aq) \longrightarrow MnO_2(s) + CO_3^{2-}(aq) \quad \text{[basic solution]}$$

Plan We proceed through step 4 as if this took place in acidic solution. Then, we add the appropriate number of OH$^-$ ions and cancel excess H$_2$O molecules (step 4 Basic).

Solution

1. Divide into half-reactions.

$$MnO_4^- \longrightarrow MnO_2 \qquad\qquad C_2O_4^{2-} \longrightarrow CO_3^{2-}$$

2. Balance.

a. Atoms other than O and H,
　Not needed

b. O atoms with H$_2$O,
　$MnO_4^- \longrightarrow MnO_2 + 2H_2O$

c. H atoms with H$^+$,
　$4H^+ + MnO_4^- \longrightarrow MnO_2 + 2H_2O$

d. Charge with e$^-$,
　$3e^- + 4H^+ + MnO_4^- \longrightarrow MnO_2 + 2H_2O$
　　　　　　[reduction]

a. Atoms other than O and H,
　$C_2O_4^{2-} \longrightarrow 2CO_3^{2-}$

b. O atoms with H$_2$O,
　$2H_2O + C_2O_4^{2-} \longrightarrow 2CO_3^{2-}$

c. H atoms with H$^+$,
　$2H_2O + C_2O_4^{2-} \longrightarrow 2CO_3^{2-} + 4H^+$

d. Charge with e$^-$,
　$2H_2O + C_2O_4^{2-} \longrightarrow 2CO_3^{2-} + 4H^+ + 2e^-$
　　　　　　[oxidation]

3. Multiply each half-reaction, if necessary, by some integer to make e$^-$ lost equal e$^-$ gained.

$$2(3e^- + 4H^+ + MnO_4^- \longrightarrow MnO_2 + 2H_2O)$$
$$6e^- + 8H^+ + 2MnO_4^- \longrightarrow 2MnO_2 + 4H_2O$$

$$3(2H_2O + C_2O_4^{2-} \longrightarrow 2CO_3^{2-} + 4H^+ + 2e^-)$$
$$6H_2O + 3C_2O_4^{2-} \longrightarrow 6CO_3^{2-} + 12H^+ + 6e^-$$

4. Add half-reactions, and cancel substances appearing on both sides. The six e$^-$ cancel, eight H$^+$ cancel to leave four H$^+$ on the right, and four H$_2$O cancel to leave two H$_2$O on the left:

$$\cancel{6e^-} + \cancel{8}H^+ + 2MnO_4^- \longrightarrow 2MnO_2 + \cancel{4}H_2O$$
$$\underline{^2\cancel{6}H_2O + 3C_2O_4^{2-} \longrightarrow 6CO_3^{2-} + 4\cancel{12}H^+ + \cancel{6e^-}}$$
$$2MnO_4^- + 2H_2O + 3C_2O_4^{2-} \longrightarrow 2MnO_2 + 6CO_3^{2-} + 4H^+$$

4 Basic. Add OH$^-$ to both sides to neutralize H$^+$, and cancel H$_2$O. Adding four OH$^-$ to both sides forms four H$_2$O on the right, two of which cancel the two H$_2$O on the left, leaving two H$_2$O on the right:

$$2MnO_4^- + 2H_2O + 3C_2O_4^{2-} + 4OH^- \longrightarrow 2MnO_2 + 6CO_3^{2-} + [4H^+ + 4OH^-]$$
$$2MnO_4^- + \cancel{2}H_2O + 3C_2O_4^{2-} + 4OH^- \longrightarrow 2MnO_2 + 6CO_3^{2-} + ^2\cancel{4}H_2O$$

Including states of matter gives the final balanced equation:

$$2MnO_4^-(aq) + 3C_2O_4^{2-}(aq) + 4OH^-(aq) \longrightarrow 2MnO_2(s) + 6CO_3^{2-}(aq) + 2H_2O(l)$$

5. Check that atoms and charges balance.

$$(\text{2Mn, 24O, 6C, 4H; 12}-) \longrightarrow (\text{2Mn, 24O, 6C, 4H; 12}-)$$

Comment As a final step, we can obtain the balanced *molecular* equation for this reaction by noting the number of moles of each anion in the balanced ionic equation and adding the correct number of moles of spectator ions (in this case, Na$^+$) to obtain neutral compounds.

Thus, for instance, balancing the charge of 2 mol of MnO_4^- requires 2 mol of Na^+, so we have $2NaMnO_4$. The balanced molecular equation is

$$2NaMnO_4(aq) + 3Na_2C_2O_4(aq) + 4NaOH(aq) \longrightarrow$$
$$2MnO_2(s) + 6Na_2CO_3(aq) + 2H_2O(l)$$

FOLLOW-UP PROBLEM 21.1 Write a balanced molecular equation for the reaction between $KMnO_4$ and KI in basic solution. The skeleton ionic reaction is

$$MnO_4^-(aq) + I^-(aq) \longrightarrow MnO_4^{2-}(aq) + IO_3^-(aq) \qquad \text{[basic solution]}$$

The half-reaction method reveals a great deal about redox processes and is essential to understanding electrochemical cells. The major points are

- Any redox reaction can be treated as the sum of a reduction and an oxidation half-reaction.
- Atoms and charge are conserved in each half-reaction.
- Electrons lost in one half-reaction are gained in the other.
- Although the half-reactions are treated separately, electron loss and electron gain occur simultaneously.

An Overview of Electrochemical Cells

We distinguish two types of electrochemical cells based on the general thermodynamic nature of the reaction:

1. A **voltaic cell** (or **galvanic cell**) uses a spontaneous reaction ($\Delta G < 0$) to generate electrical energy. In the cell reaction, the difference in chemical potential energy between higher energy reactants and lower energy products is converted into electrical energy. This energy is used to operate the load—flashlight bulb, CD player, car starter motor, or other electrical device. In other words, *the system does work on the surroundings.* All batteries contain voltaic cells.
2. An **electrolytic cell** uses electrical energy to drive a nonspontaneous reaction ($\Delta G > 0$). In the cell reaction, electrical energy from an external power supply converts lower energy reactants into higher energy products. Thus, *the surroundings do work on the system.* Electroplating and recovering metals from ores involve electrolytic cells.

The two types of cell have certain design features in common (Figure 21.3). Two **electrodes,** which conduct the electricity between cell and surroundings, are dipped into an **electrolyte,** a mixture of ions (usually in aqueous solution) that are involved in the reaction or that carry the charge. An electrode is identified as either **anode** or **cathode** depending on the half-reaction that takes place there:

- *The oxidation half-reaction occurs at the anode.* Electrons are lost by the substance being oxidized (reducing agent) and *leave the cell* at the anode.
- *The reduction half-reaction occurs at the cathode.* Electrons are gained by the substance being reduced (oxidizing agent) and *enter the cell* at the cathode.

As shown in Figure 21.3, the relative charges of the electrodes are *opposite* in the two types of cell. As you'll see in the following sections, these opposite charges result from the different phenomena that cause the electrons to flow.

Here are some memory aids to help you remember which half-reaction occurs at which electrode:

1. The words *anode* and *oxidation* start with vowels; the words *cathode* and *reduction* start with consonants.
2. Alphabetically, the *A* in anode comes before the *C* in cathode, and the *O* in oxidation comes before the *R* in reduction.
3. Look at the first syllables and use your imagination:

ANode, OXidation; REDuction, CAThode $\Rightarrow$ AN OX and a RED CAT

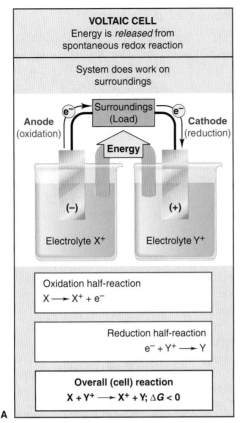

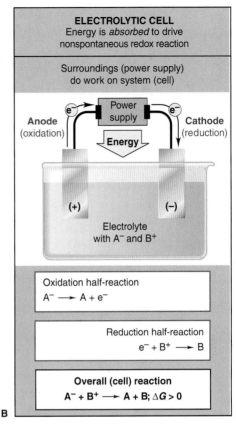

Figure 21.3 General characteristics of voltaic and electrolytic cells. A voltaic cell **(A)** generates energy from a spontaneous reaction ($\Delta G < 0$), whereas an electrolytic cell **(B)** requires energy to drive a nonspontaneous reaction ($\Delta G > 0$). In both types of cell, two electrodes dip into electrolyte solutions, and an external circuit provides the means for electrons to flow between them. Most important, notice that oxidation takes place at the anode and reduction takes place at the cathode, but the relative electrode charges are opposite in the two cells.

SECTION SUMMARY

An oxidation-reduction (redox) reaction involves the transfer of electrons from a reducing agent to an oxidizing agent. The half-reaction method of balancing divides the overall reaction into half-reactions that are balanced separately and then recombined. Both types of electrochemical cells are based on redox reactions. In a voltaic cell, a spontaneous reaction generates electricity and does work on the surroundings; in an electrolytic cell, the surroundings supply electricity that does work to drive a nonspontaneous reaction. In both types, two electrodes dip into electrolyte solutions; oxidation occurs at the anode, and reduction occurs at the cathode.

21.2 VOLTAIC CELLS: USING SPONTANEOUS REACTIONS TO GENERATE ELECTRICAL ENERGY

If you put a strip of zinc metal in a solution of Cu^{2+} ion, the blue color of the solution fades as a brown-black crust of Cu metal forms on the Zn strip (Figure 21.4). Judging from what we see, the reaction involves the reduction of Cu^{2+} ion to Cu metal, which must be accompanied by the oxidation of Zn metal to Zn^{2+} ion. The overall reaction consists of two half-reactions:

$$Cu^{2+}(aq) + 2e^- \longrightarrow Cu(s) \qquad \text{[reduction]}$$
$$\underline{Zn(s) \longrightarrow Zn^{2+}(aq) + 2e^- \qquad \text{[oxidation]}}$$
$$Zn(s) + Cu^{2+}(aq) \longrightarrow Zn^{2+}(aq) + Cu(s) \qquad \text{[overall reaction]}$$

In the remainder of this section, we examine this spontaneous reaction as the basis of a voltaic (galvanic) cell.

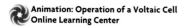

Construction and Operation of a Voltaic Cell

Electrons are being transferred in the Zn/Cu^{2+} reaction (Figure 21.4), but the system does not generate electrical energy because the oxidizing agent (Cu^{2+}) and the reducing agent (Zn) are in the same beaker. If, however, the half-reactions are physically separated and connected by an external circuit, the electrons are transferred by traveling through the circuit and an electric current is produced.

This separation of half-reactions is the essential idea behind a voltaic cell (Figure 21.5A). The components of each half-reaction are placed in a separate container, or **half-cell,** which consists of one electrode dipping into an electrolyte solution. The two half-cells are joined by the circuit, which consists of a wire and a salt bridge (the inverted U tube in the figure). In order to measure the voltage generated by the cell, a voltmeter is inserted in the path of the wire connecting the electrodes. A switch (not shown) closes (completes) or opens (breaks) the circuit. By convention, *the oxidation half-cell (anode compartment) is shown on the left and the reduction half-cell (cathode compartment) on the right.* Here are the key points about the Zn/Cu^{2+} voltaic cell:

1. *The oxidation half-cell.* In this case, the anode compartment consists of a zinc bar (the anode) immersed in a Zn^{2+} electrolyte (such as a solution of zinc sulfate, $ZnSO_4$). The zinc bar is the reactant in the oxidation half-reaction, and it conducts the released electrons *out* of its half-cell.

2. *The reduction half-cell.* In this case, the cathode compartment consists of a copper bar (the cathode) immersed in a Cu^{2+} electrolyte [such as a solution of copper(II) sulfate, $CuSO_4$]. Copper metal is the product in the reduction half-reaction, and the bar conducts electrons *into* its half-cell.

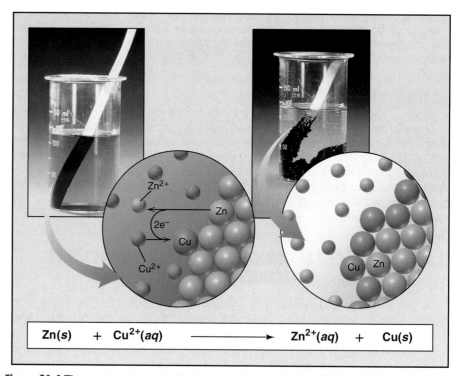

$$Zn(s) \;+\; Cu^{2+}(aq) \longrightarrow Zn^{2+}(aq) \;+\; Cu(s)$$

Figure 21.4 The spontaneous reaction between zinc and copper(II) ion. When a strip of zinc metal is placed in a solution of Cu^{2+} ion, a redox reaction begins (*left*), in which the zinc is oxidized to Zn^{2+} and the Cu^{2+} is reduced to copper metal. As the reaction proceeds (*right*), the deep blue color of the solution of hydrated Cu^{2+} ion lightens, and the Cu "plates out" on the Zn and falls off in chunks. (The Cu appears black because it is very finely divided.) At the atomic scale, each Zn atom loses two electrons, which are gained by a Cu^{2+} ion. The process is summarized with symbols in the balanced equation.

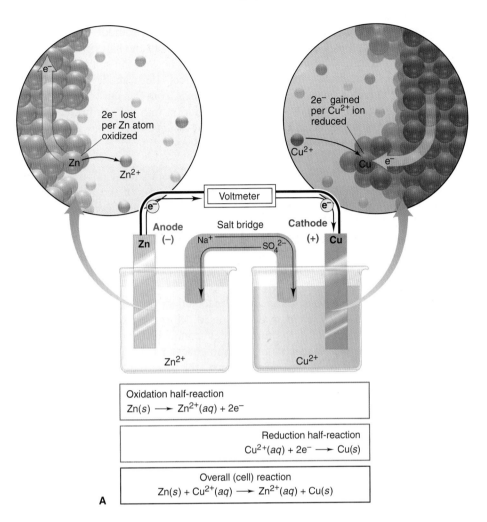

Oxidation half-reaction
Zn(s) ⟶ Zn^{2+}(aq) + 2e$^-$

Reduction half-reaction
Cu^{2+}(aq) + 2e$^-$ ⟶ Cu(s)

Overall (cell) reaction
Zn(s) + Cu^{2+}(aq) ⟶ Zn^{2+}(aq) + Cu(s)

A

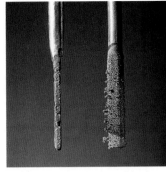

B

Figure 21.5 A voltaic cell based on the zinc-copper reaction. A, The anode half-cell (oxidation) consists of a Zn electrode dipping into a Zn^{2+} solution. The two electrons generated in the oxidation of each Zn atom move through the Zn bar and the wire, and into the Cu electrode, which dips into a Cu^{2+} solution in the cathode half-cell (reduction). There, the electrons reduce Cu^{2+} ions. Thus, electrons flow left to right through electrodes and wire. A salt bridge contains unreactive Na$^+$ and SO$_4^{2-}$ ions that maintain neutral charge in the electrolyte solutions: anions in the salt bridge flow to the left, and cations flow to the right. The voltmeter registers the electrical output of the cell. **B,** After the cell runs for several hours, the Zn anode weighs less because Zn atoms have been oxidized to aqueous Zn^{2+} ions, and the Cu cathode weighs more because aqueous Cu^{2+} ions have been reduced to Cu metal.

3. *Relative charges on the electrodes.* The electrode charges are determined by the *source of electrons* and the *direction of electron flow* through the circuit. In this cell, zinc atoms are oxidized at the anode to Zn^{2+} ions and electrons. The Zn^{2+} ions enter the solution, while the electrons enter the bar and then the wire. *The electrons flow left to right* through the wire to the cathode, where Cu^{2+} ions in the solution accept them and are reduced to Cu atoms. As the cell operates, electrons are continuously generated at the anode and consumed at the cathode. Therefore, the anode has an excess of electrons and a negative charge *relative* to the cathode. *In any* **voltaic** *cell, the anode is negative and the cathode is positive.*

4. *The purpose of the salt bridge.* The cell cannot operate unless the circuit is complete. The oxidation half-cell originally contains a neutral solution of Zn^{2+} and SO$_4^{2-}$ ions, but as Zn atoms in the bar lose electrons, the solution would develop a net positive charge from the Zn^{2+} ions entering. Similarly, in the reduction half-cell, the neutral solution of Cu^{2+} and SO$_4^{2-}$ ions would develop a net negative charge as Cu^{2+} ions leave the solution to form Cu atoms. A charge imbalance would arise and stop cell operation. To avoid this situation and enable the cell to operate, the two half-cells are joined by a **salt bridge,** which acts as a "liquid wire," allowing ions to flow through both compartments and complete the circuit. The salt bridge shown in Figure 21.5A is an inverted U tube containing a solution of the nonreacting ions Na$^+$ and SO$_4^{2-}$ in a gel. The solution cannot pour out, but ions can diffuse through it into or out of the half-cells.

To maintain neutrality in the reduction half-cell (right; cathode compartment) as Cu^{2+} ions change to Cu atoms, Na^+ ions move from the salt bridge into the solution (and some SO_4^{2-} ions move from the solution into the salt bridge). Similarly, to maintain neutrality in the oxidation half-cell (left; anode compartment) as Zn atoms change to Zn^{2+} ions, SO_4^{2-} ions move from the salt bridge into that solution (and some Zn^{2+} ions move from the solution into the salt bridge). Thus, as Figure 21.5A shows, the circuit is completed *as electrons move left to right through the wire, while anions move right to left and cations move left to right through the salt bridge.*

5. *Active vs. inactive electrodes.* The electrodes in the Zn/Cu^{2+} cell are *active* because the metal bars themselves are components of the half-reactions. As the cell operates, the mass of the zinc electrode gradually decreases, and the $[Zn^{2+}]$ in the anode half-cell increases. At the same time, the mass of the copper electrode increases, and the $[Cu^{2+}]$ in the cathode half-cell decreases; we say that the Cu^{2+} "plates out" on the electrode. Look at Figure 21.5B to see how the electrodes look, removed from their half-cells, after several hours of operation.

For many redox reactions, there are no reactants or products capable of serving as electrodes, so *inactive* electrodes are used. Most commonly, inactive electrodes are rods of *graphite* or *platinum*: they conduct electrons into or out of the half-cells but cannot take part in the half-reactions. In a voltaic cell based on the following half-reactions, for instance, the reacting species cannot act as electrodes:

$$2I^-(aq) \longrightarrow I_2(s) + 2e^- \qquad \text{[anode; oxidation]}$$
$$MnO_4^-(aq) + 8H^+(aq) + 5e^- \longrightarrow Mn^{2+}(aq) + 4H_2O(l) \qquad \text{[cathode; reduction]}$$

Therefore, each half-cell consists of inactive electrodes immersed in an electrolyte solution that contains *all the reactant species involved in that half-reaction* (Figure 21.6). In the anode half-cell, I^- ions are oxidized to solid I_2. The electrons that are released flow into the graphite anode, through the wire, and into the graphite cathode. From there, the electrons are consumed by MnO_4^- ions, which are reduced to Mn^{2+} ions. (A KNO_3 salt bridge is used.)

Notation for a Voltaic Cell

A useful shorthand notation describes the components of a voltaic cell. For example, the notation for the Zn/Cu^{2+} cell is

$$Zn(s) \,|\, Zn^{2+}(aq) \,\|\, Cu^{2+}(aq) \,|\, Cu(s)$$

Key parts of the notation are

- The components of the anode compartment (oxidation half-cell) are written *to the left* of the components of the cathode compartment (reduction half-cell).
- A single vertical line represents a phase boundary. For example, $Zn(s) \,|\, Zn^{2+}(aq)$ indicates that the *solid* Zn is a *different* phase from the *aqueous* Zn^{2+}. A comma separates the half-cell components that are in the *same* phase. For example, the notation for the voltaic cell housing the reaction between I^- and MnO_4^- shown in Figure 21.6 is

$$\text{graphite} \,|\, I^-(aq) \,|\, I_2(s) \,\|\, H^+(aq), MnO_4^-(aq), Mn^{2+}(aq) \,|\, \text{graphite}$$

That is, in the cathode compartment, H^+, MnO_4^-, and Mn^{2+} ions are all in aqueous solution with solid graphite immersed in it. Often, we specify the concentrations of dissolved components; for example, if the concentrations of Zn^{2+} and Cu^{2+} are 1 M, we write

$$Zn(s) \,|\, Zn^{2+}(1\ M) \,\|\, Cu^{2+}(1\ M) \,|\, Cu(s)$$

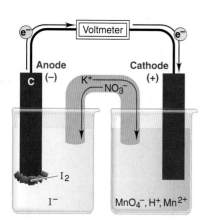

Oxidation half-reaction
$2I^-(aq) \longrightarrow I_2(s) + 2e^-$

Reduction half-reaction
$MnO_4^-(aq) + 8H^+(aq) + 5e^- \longrightarrow$ $Mn^{2+}(aq) + 4H_2O(l)$

Overall (cell) reaction
$2MnO_4^-(aq) + 16H^+(aq) + 10I^-(aq) \longrightarrow$ $2Mn^{2+}(aq) + 5I_2(s) + 8H_2O(l)$

Figure 21.6 A voltaic cell using inactive electrodes. The reaction between I^- and MnO_4^- in acidic solution does not have species that can be used as electrodes, so inactive graphite (C) electrodes are used.

- Half-cell components usually appear in the same order as in the half-reaction, and electrodes appear at the far left and far right of the notation.
- A double vertical line indicates the separated half-cells and represents the phase boundary on either side of the salt bridge (the ions in the salt bridge are omitted because they are not part of the reaction).

SAMPLE PROBLEM 21.2 Describing a Voltaic Cell with Diagram and Notation

Problem Draw a diagram, show balanced equations, and write the notation for a voltaic cell that consists of one half-cell with a Cr bar in a $Cr(NO_3)_3$ solution, another half-cell with an Ag bar in an $AgNO_3$ solution, and a KNO_3 salt bridge. Measurement indicates that the Cr electrode is negative relative to the Ag electrode.

Plan From the given contents of the half-cells, we can write the half-reactions. We must determine which is the anode compartment (oxidation) and which is the cathode (reduction). To do so, we must find the direction of the spontaneous redox reaction, which is given by the relative electrode charges. Electrons are released into the anode during oxidation, so it has a negative charge. We are told that Cr is negative, so it must be the anode; and, therefore, Ag is the cathode.

Solution Writing the balanced half-reactions. The Ag electrode is positive, so the half-reaction consumes e^-:

$$Ag^+(aq) + e^- \longrightarrow Ag(s) \quad \text{[reduction; cathode]}$$

The Cr electrode is negative, so the half-reaction releases e^-:

$$Cr(s) \longrightarrow Cr^{3+}(aq) + 3e^- \quad \text{[oxidation; anode]}$$

Writing the balanced overall cell reaction. We triple the reduction half-reaction to balance e^- and then combine the half-reactions to obtain the overall reaction:

$$Cr(s) + 3Ag^+(aq) \longrightarrow Cr^{3+}(aq) + 3Ag(s)$$

Determining direction of electron and ion flow. The released e^- in the Cr electrode (negative) flow through the external circuit to the Ag electrode (positive). As Cr^{3+} ions enter the anode electrolyte, NO_3^- ions enter from the salt bridge to maintain neutrality. As Ag^+ ions leave the cathode electrolyte and plate out on the Ag electrode, K^+ ions enter from the salt bridge to maintain neutrality. The diagram of this cell is shown in the margin. Writing the cell notation:

$$Cr(s) \mid Cr^{3+}(aq) \parallel Ag^+(aq) \mid Ag(s)$$

Check Always be sure that the half-reactions and cell reaction are balanced, the half-cells contain *all* components of the half-reactions, and the electron and ion flow are shown. You should be able to write the half-reactions from the cell notation as a check.

Comment The key to diagramming a voltaic cell is to use the direction of the spontaneous reaction to identify the oxidation (anode; negative) and reduction (cathode; positive) half-reactions.

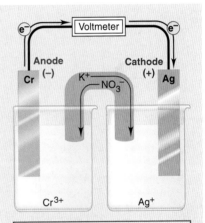

Oxidation half-reaction
$Cr(s) \longrightarrow Cr^{3+}(aq) + 3e^-$

Reduction half-reaction
$Ag^+(aq) + e^- \longrightarrow Ag(s)$

Overall (cell) reaction
$Cr(s) + 3Ag^+(aq) \longrightarrow Cr^{3+}(aq) + 3Ag(s)$

FOLLOW-UP PROBLEM 21.2 In one compartment of a voltaic cell, a graphite rod dips into an acidic solution of $K_2Cr_2O_7$ and $Cr(NO_3)_3$; in the other compartment, a tin bar dips into a $Sn(NO_3)_2$ solution. A KNO_3 salt bridge joins them. The tin electrode is negative relative to the graphite. Draw a diagram of the cell, show the balanced equations, and write the cell notation.

SECTION SUMMARY

A voltaic cell consists of oxidation (anode) and reduction (cathode) half-cells, connected by a wire to conduct electrons and a salt bridge to maintain charge neutrality as the cell operates. Electrons move from anode (left) to cathode (right), while cations move from the salt bridge into the cathode half-cell and anions from the salt bridge into the anode half-cell. The cell notation shows the species and their phases in each half-cell, as well as the direction of current flow.

21.3 CELL POTENTIAL: OUTPUT OF A VOLTAIC CELL

The purpose of a voltaic cell is to convert the free energy change of a spontaneous reaction into the kinetic energy of electrons moving through an external circuit (electrical energy). This electrical energy is proportional to the *difference in electrical potential between the two electrodes,* which is called the **cell potential** (E_{cell}), also the **voltage** of the cell or the **electromotive force (emf).**

Electrons flow spontaneously from the negative to the positive electrode, that is, toward the electrode with the more positive electrical potential. Thus, when the cell operates *spontaneously,* there is a *positive* cell potential:

$$E_{cell} > 0 \text{ for a spontaneous process} \tag{21.1}$$

The more positive E_{cell} is, the more work the cell can do, and the farther the reaction proceeds to the right as written. A *negative* cell potential, on the other hand, is associated with a *nonspontaneous* cell reaction. If $E_{cell} = 0$, the reaction has reached equilibrium and the cell can do no more work.

How are the units of cell potential related to those of energy available to do work? As you've seen, work is done when charge moves between electrode compartments that differ in electrical potential. The SI unit of electrical potential is the **volt (V),** and the SI unit of electrical charge is the **coulomb (C).** By definition, for two electrodes that differ by 1 volt of electrical potential, 1 joule of energy is released (that is, 1 joule of work can be done) for each coulomb of charge that moves between the electrodes. Thus,

$$1 \text{ V} = 1 \text{ J/C} \tag{21.2}$$

Table 21.1 lists the voltages of some commercial and natural voltaic cells. Next, we'll see how to measure cell potential.

Table 21.1   Voltages of Some Voltaic Cells

Voltaic Cell	Voltage (V)
Common alkaline flashlight battery	1.5
Lead-acid car battery (6 cells = 12 V)	2.0
Calculator battery (mercury)	1.3
Lithium-ion laptop battery	3.7
Electric eel (~5000 cells in 6-ft eel = 750 V)	0.15
Nerve of giant squid (across cell membrane)	0.070

Standard Cell Potentials

The measured potential of a voltaic cell is affected by changes in concentration as the reaction proceeds and by energy losses due to heating of the cell and the external circuit. Therefore, in order to compare the output of different cells, we obtain a **standard cell potential (E°_{cell}),** the potential measured at a specified temperature (usually 298 K) with no current flowing* and *all components in their standard states:* 1 atm for gases, 1 *M* for solutions, the pure solid for electrodes. When the zinc-copper cell that we diagrammed in Figure 21.5 begins operating

---

*The current required to operate modern digital voltmeters makes a negligible difference in the value of E°_{cell}.

under standard state conditions, that is, when $[Zn^{2+}] = [Cu^{2+}] = 1\ M$, the cell produces 1.10 V at 298 K (Figure 21.7):

$$Zn(s) + Cu^{2+}(aq;\ 1\ M) \longrightarrow Zn^{2+}(aq;\ 1\ M) + Cu(s) \qquad E^{\circ}_{cell} = 1.10\ V$$

Standard Electrode (Half-Cell) Potentials Just as each half-reaction makes up part of the overall reaction, the potential of each half-cell makes up a part of the overall cell potential. The **standard electrode potential ($E^{\circ}_{half\text{-}cell}$)** is the potential associated with a given half-reaction (electrode compartment) when all the components are in their standard states.

By convention, *a standard electrode potential always refers to the half-reaction written as a reduction.* For the zinc-copper reaction, for example, the standard electrode potentials for the zinc half-reaction (E°_{zinc}, anode compartment) and for the copper half-reaction (E°_{copper}, cathode compartment) refer to the processes written as reductions:

$$Zn^{2+}(aq) + 2e^- \longrightarrow Zn(s) \qquad E^{\circ}_{zinc}\ (E^{\circ}_{anode}) \qquad \text{[reduction]}$$
$$Cu^{2+}(aq) + 2e^- \longrightarrow Cu(s) \qquad E^{\circ}_{copper}\ (E^{\circ}_{cathode}) \qquad \text{[reduction]}$$

The overall cell reaction involves the *oxidation* of zinc at the anode, not the *reduction* of Zn^{2+}, so we reverse the zinc half-reaction:

$$Zn(s) \longrightarrow Zn^{2+}(aq) + 2e^- \qquad \text{[oxidation]}$$
$$Cu^{2+}(aq) + 2e^- \longrightarrow Cu(s) \qquad \text{[reduction]}$$

The overall redox reaction is the sum of these half-reactions:

$$Zn(s) + Cu^{2+}(aq) \longrightarrow Zn^{2+}(aq) + Cu(s)$$

Because electrons flow spontaneously toward the copper electrode (cathode), it must have a more positive $E^{\circ}_{half\text{-}cell}$ than the zinc electrode (anode). Therefore, to obtain a positive E°_{cell}, we subtract E°_{zinc} from E°_{copper}:

$$E^{\circ}_{cell} = E^{\circ}_{copper} - E^{\circ}_{zinc}$$

We can generalize this result for any voltaic cell: *the standard cell potential is the difference between the standard electrode potential of the cathode (reduction) half-cell and the standard electrode potential of the anode (oxidation) half-cell:*

$$E^{\circ}_{cell} = E^{\circ}_{cathode\ (reduction)} - E^{\circ}_{anode\ (oxidation)} \qquad (21.3)$$

For a spontaneous reaction at standard conditions, $E^{\circ}_{cell} > 0$.

Determining $E^{\circ}_{half\text{-}cell}$: The Standard Hydrogen Electrode What portion of E°_{cell} for the zinc-copper reaction is contributed by the anode half-cell (oxidation of Zn) and what portion by the cathode half-cell (reduction of Cu^{2+})? That is, how can we know half-cell potentials if we can only measure the potential of the complete cell? Half-cell potentials, such as E°_{zinc} and E°_{copper}, are not absolute quantities, but rather are values *relative* to that of a standard. *This standard reference half-cell has its standard electrode potential defined as zero* ($E^{\circ}_{reference} \equiv 0.00\ V$). The **standard reference half-cell** is a **standard hydrogen electrode,** which consists of a specially prepared platinum electrode immersed in a $1\ M$ aqueous solution of a strong acid, $H^+(aq)$ [or $H_3O^+(aq)$], through which H_2 gas at 1 atm is bubbled. Thus, the reference half-reaction is

$$2H^+(aq;\ 1\ M) + 2e^- \rightleftharpoons H_2(g;\ 1\ atm) \qquad E^{\circ}_{reference} = 0.00\ V$$

Now we can construct a voltaic cell consisting of this reference half-cell and another half-cell whose potential we want to determine. With $E^{\circ}_{reference}$ defined as zero, the overall E°_{cell} allows us to find the unknown standard electrode potential,

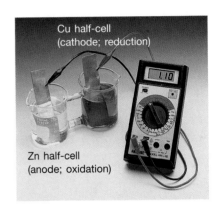

Cu half-cell
(cathode; reduction)

Zn half-cell
(anode; oxidation)

Figure 21.7 Measurement of a standard cell potential. The zinc-copper cell, operating at 298 K under standard-state conditions, produces a voltage of 1.10 V.

 **Animation: Galvanic Cell**
Online Learning Center

$E^\circ_{unknown}$. When H_2 is oxidized, the reference half-cell is the anode, and so reduction occurs at the unknown half-cell:

$$E^\circ_{cell} = E^\circ_{cathode} - E^\circ_{anode} = E^\circ_{unknown} - E^\circ_{reference} = E^\circ_{unknown} - 0.00 \text{ V} = E^\circ_{unknown}$$

When H^+ is reduced, the reference half-cell is the cathode, and so oxidation occurs at the unknown half-cell:

$$E^\circ_{cell} = E^\circ_{cathode} - E^\circ_{anode} = E^\circ_{reference} - E^\circ_{unknown} = 0.00 \text{ V} - E^\circ_{unknown} = -E^\circ_{unknown}$$

Figure 21.8 shows a voltaic cell that has the Zn/Zn^{2+} half-reaction in one compartment and the H^+/H_2 (or H_3O^+/H_2) half-reaction in the other. The zinc electrode is negative relative to the hydrogen electrode, so we know that the zinc is being oxidized and is the anode. The measured E°_{cell} is $+0.76$ V, and we use this value to find the unknown standard electrode potential, E°_{zinc}:

$$2H^+(aq) + 2e^- \longrightarrow H_2(g) \qquad E^\circ_{reference} = 0.00 \text{ V} \qquad \text{[cathode; reduction]}$$
$$\underline{Zn(s) \longrightarrow Zn^{2+}(aq) + 2e^- \qquad E^\circ_{zinc} = ? \text{ V} \qquad \text{[anode; oxidation]}}$$
$$Zn(s) + 2H^+(aq) \longrightarrow Zn^{2+}(aq) + H_2(g) \qquad E^\circ_{cell} = 0.76 \text{ V}$$
$$E^\circ_{cell} = E^\circ_{cathode} - E^\circ_{anode} = E^\circ_{reference} - E^\circ_{zinc}$$
$$E^\circ_{zinc} = E^\circ_{reference} - E^\circ_{cell} = 0.00 \text{ V} - 0.76 \text{ V} = -0.76 \text{ V}$$

Now let's return to the zinc-copper cell and use the measured value of E°_{cell} (1.10 V) and the value we just found for E°_{zinc} to calculate E°_{copper}:

$$E^\circ_{cell} = E^\circ_{cathode} - E^\circ_{anode} = E^\circ_{copper} - E^\circ_{zinc}$$
$$E^\circ_{copper} = E^\circ_{cell} + E^\circ_{zinc} = 1.10 \text{ V} + (-0.76 \text{ V}) = 0.34 \text{ V}$$

By continuing this process of constructing cells with one known and one unknown electrode potential, we can find many other standard electrode potentials.

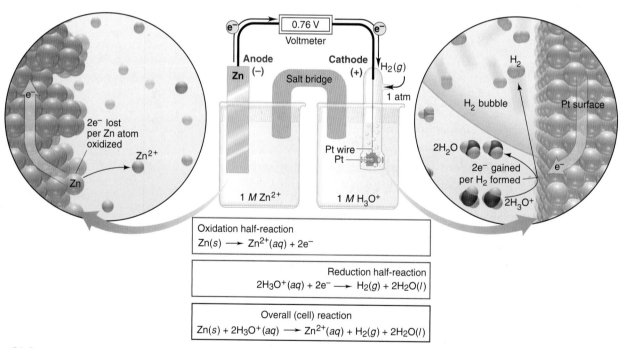

Figure 21.8 Determining an unknown $E^\circ_{half\text{-}cell}$ with the standard reference (hydrogen) electrode. A voltaic cell has the Zn half-reaction in one half-cell and the hydrogen reference half-reaction in the other. The magnified view of the hydrogen half-reaction shows two H_3O^+ ions being reduced to two H_2O molecules and an H_2 molecule, which enters the H_2 bubble. The Zn/Zn^{2+} half-cell potential is negative (anode), and the cell potential is 0.76 V. The potential of the standard reference electrode is defined as 0.00 V, so the cell potential equals the negative of the anode potential; that is,

$$0.76 \text{ V} = 0.00 \text{ V} - E^\circ_{zinc} \qquad \text{so} \qquad E^\circ_{zinc} = -0.76 \text{ V}$$

SAMPLE PROBLEM 21.3 Calculating an Unknown $E^{\circ}_{\text{half-cell}}$ from E°_{cell}

Problem A voltaic cell houses the reaction between aqueous bromine and zinc metal:

$$Br_2(aq) + Zn(s) \longrightarrow Zn^{2+}(aq) + 2Br^-(aq) \qquad E^{\circ}_{\text{cell}} = 1.83 \text{ V}$$

Calculate $E^{\circ}_{\text{bromine}}$, given $E^{\circ}_{\text{zinc}} = -0.76 \text{ V}$.

Plan E°_{cell} is positive, so the reaction is spontaneous as written. By dividing the reaction into half-reactions, we see that Br_2 is reduced and Zn is oxidized; thus, the zinc half-cell contains the anode. We use Equation 21.3 to find $E^{\circ}_{\text{unknown}}$ ($E^{\circ}_{\text{bromine}}$).

Solution Dividing the reaction into half-reactions:

$$Br_2(aq) + 2e^- \longrightarrow 2Br^-(aq) \qquad E^{\circ}_{\text{unknown}} = E^{\circ}_{\text{bromine}} = ? \text{ V}$$
$$Zn(s) \longrightarrow Zn^{2+}(aq) + 2e^- \qquad E^{\circ}_{\text{zinc}} = -0.76 \text{ V}$$

Calculating $E^{\circ}_{\text{bromine}}$:

$$E^{\circ}_{\text{cell}} = E^{\circ}_{\text{cathode}} - E^{\circ}_{\text{anode}} = E^{\circ}_{\text{bromine}} - E^{\circ}_{\text{zinc}}$$
$$E^{\circ}_{\text{bromine}} = E^{\circ}_{\text{cell}} + E^{\circ}_{\text{zinc}} = 1.83 \text{ V} + (-0.76 \text{ V})$$
$$= 1.07 \text{ V}$$

Check A good check is to make sure that calculating $E^{\circ}_{\text{bromine}} - E^{\circ}_{\text{zinc}}$ gives E°_{cell}: 1.07 V − (−0.76 V) = 1.83 V.

Comment Keep in mind that, whichever is the unknown half-cell, reduction is the cathode half-reaction and oxidation is the anode half-reaction. Always subtract E°_{anode} from $E^{\circ}_{\text{cathode}}$ to get E°_{cell}.

FOLLOW-UP PROBLEM 21.3 A voltaic cell based on the reaction between aqueous Br_2 and vanadium(III) ions has $E^{\circ}_{\text{cell}} = 1.39 \text{ V}$:

$$Br_2(aq) + 2V^{3+}(aq) + 2H_2O(l) \longrightarrow 2VO^{2+}(aq) + 4H^+(aq) + 2Br^-(aq)$$

What is $E^{\circ}_{\text{vanadium}}$, the standard electrode potential for the reduction of VO^{2+} to V^{3+}?

Relative Strengths of Oxidizing and Reducing Agents

One of the things we can learn from measuring potentials of voltaic cells is the relative strengths of the oxidizing and reducing agents involved. Three oxidizing agents present in the voltaic cell just discussed are Cu^{2+}, H^+, and Zn^{2+}. We can rank their relative oxidizing strengths by writing each half-reaction as a gain of electrons (reduction), with its corresponding standard electrode potential:

$$Cu^{2+}(aq) + 2e^- \longrightarrow Cu(s) \qquad E^{\circ} = 0.34 \text{ V}$$
$$2H^+(aq) + 2e^- \longrightarrow H_2(g) \qquad E^{\circ} = 0.00 \text{ V}$$
$$Zn^{2+}(aq) + 2e^- \longrightarrow Zn(s) \qquad E^{\circ} = -0.76 \text{ V}$$

The more positive the E° value, the more readily the reaction (as written) occurs; thus, Cu^{2+} gains two electrons more readily than H^+, which gains them more readily than Zn^{2+}. In terms of strength as an oxidizing agent, therefore, $Cu^{2+} > H^+ > Zn^{2+}$. Moreover, this listing also ranks the strengths of the reducing agents: $Zn > H_2 > Cu$. Notice that this list of half-reactions in order of *decreasing* half-cell potential shows, *from top to bottom,* the oxidizing agents (reactants) *decreasing* in strength and the reducing agents (products) *increasing* in strength; that is, Cu^{2+} (top left) is the strongest oxidizing agent, and Zn (bottom right) is the strongest reducing agent.

By combining many pairs of half-cells into voltaic cells, we can create a list of reduction half-reactions and arrange them in *decreasing* order of standard electrode potential (from most positive to most negative). Such a list, called an *emf series* or a *table of standard electrode potentials,* appears in Appendix D, with a few examples in Table 21.2 on the next page.

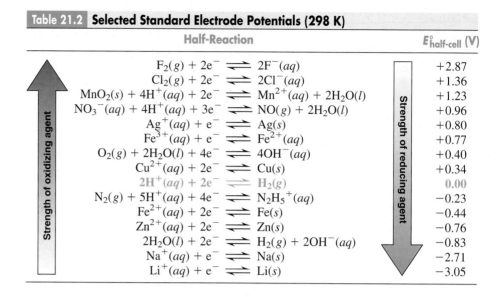

Table 21.2	Selected Standard Electrode Potentials (298 K)	
	Half-Reaction	$E^\circ_{\text{half-cell}}$ **(V)**
	$F_2(g) + 2e^- \rightleftharpoons 2F^-(aq)$	+2.87
	$Cl_2(g) + 2e^- \rightleftharpoons 2Cl^-(aq)$	+1.36
	$MnO_2(s) + 4H^+(aq) + 2e^- \rightleftharpoons Mn^{2+}(aq) + 2H_2O(l)$	+1.23
	$NO_3^-(aq) + 4H^+(aq) + 3e^- \rightleftharpoons NO(g) + 2H_2O(l)$	+0.96
	$Ag^+(aq) + e^- \rightleftharpoons Ag(s)$	+0.80
	$Fe^{3+}(aq) + e^- \rightleftharpoons Fe^{2+}(aq)$	+0.77
	$O_2(g) + 2H_2O(l) + 4e^- \rightleftharpoons 4OH^-(aq)$	+0.40
	$Cu^{2+}(aq) + 2e^- \rightleftharpoons Cu(s)$	+0.34
	$2H^+(aq) + 2e^- \rightleftharpoons H_2(g)$	0.00
	$N_2(g) + 5H^+(aq) + 4e^- \rightleftharpoons N_2H_5^+(aq)$	−0.23
	$Fe^{2+}(aq) + 2e^- \rightleftharpoons Fe(s)$	−0.44
	$Zn^{2+}(aq) + 2e^- \rightleftharpoons Zn(s)$	−0.76
	$2H_2O(l) + 2e^- \rightleftharpoons H_2(g) + 2OH^-(aq)$	−0.83
	$Na^+(aq) + e^- \rightleftharpoons Na(s)$	−2.71
	$Li^+(aq) + e^- \rightleftharpoons Li(s)$	−3.05

Strength of oxidizing agent (increasing upward) · Strength of reducing agent (increasing downward)

There are several key points to keep in mind:

- All values are relative to the standard hydrogen (reference) electrode:

$$2H^+(aq; 1\ M) + 2e^- \rightleftharpoons H_2(g; 1\ atm) \qquad E^\circ_{\text{reference}} = 0.00\ V$$

- By convention, the half-reactions are written as *reductions,* which means that *only reactants are oxidizing agents and only products are reducing agents.*
- The more positive the $E^\circ_{\text{half-cell}}$, the more readily the half-reaction occurs.
- Half-reactions are shown with an equilibrium arrow because each can occur as a reduction or an oxidation (that is, take place at the cathode or anode, respectively), depending on the $E^\circ_{\text{half-cell}}$ of the other half-reaction.
- As Appendix D (and Table 21.2) is arranged, the strength of the oxidizing agent (reactant) *increases going up (bottom to top),* and the strength of the reducing agent (product) *increases going down (top to bottom).*

Thus, $F_2(g)$ is the strongest oxidizing agent (has the largest positive E°), which means $F^-(aq)$ is the weakest reducing agent. Similarly, $Li^+(aq)$ is the weakest oxidizing agent (has the most negative E°), which means $Li(s)$ is the strongest reducing agent.

Writing Spontaneous Redox Reactions Appendix D can be used to write spontaneous redox reactions, which is useful for constructing voltaic cells.

Every redox reaction is the sum of two half-reactions, so there is a reducing agent and an oxidizing agent on each side. In the zinc-copper reaction, for instance, Zn and Cu are the reducing agents, and Cu^{2+} and Zn^{2+} are the oxidizing agents. The stronger oxidizing and reducing agents react spontaneously to form the weaker oxidizing and reducing agents:

$$\underset{\substack{\text{stronger}\\\text{reducing agent}}}{Zn(s)} + \underset{\substack{\text{stronger}\\\text{oxidizing agent}}}{Cu^{2+}(aq)} \longrightarrow \underset{\substack{\text{weaker}\\\text{oxidizing agent}}}{Zn^{2+}(aq)} + \underset{\substack{\text{weaker}\\\text{reducing agent}}}{Cu(s)}$$

Based on the order of the E° values in Appendix D, *the stronger oxidizing agent (species on the left) has a half-reaction with a larger (more positive or less negative) E° value, and the stronger reducing agent (species on the right) has a half-reaction with a smaller (less positive or more negative) E° value.* Therefore, a spontaneous reaction ($E^\circ_{\text{cell}} > 0$) will occur between an oxidizing agent and any reducing agent that lies *below* it in the list. For instance, Zn (right) lies below Cu^{2+} (left), and Cu^{2+} and Zn react spontaneously. In other words, *for a spontaneous reaction to occur, the half-reaction higher in the list proceeds at the cathode* *as written, and the half-reaction lower in the list proceeds at the anode in reverse.* This pairing ensures that the stronger oxidizing agent (higher on the left) and stronger reducing agent (lower on the right) will be the reactants.

However, if we know the electrode potentials, we can write a spontaneous redox reaction even if Appendix D is not available. Let's choose a pair of half-reactions from the appendix and, without referring to their relative positions in the list, use them to write a spontaneous redox reaction:

$$Ag^+(aq) + e^- \longrightarrow Ag(s) \qquad E^\circ_{silver} = 0.80 \text{ V}$$
$$Sn^{2+}(aq) + 2e^- \longrightarrow Sn(s) \qquad E^\circ_{tin} = -0.14 \text{ V}$$

There are two steps involved:

1. Reverse one of the half-reactions into an oxidation step such that the difference of the electrode potentials (cathode *minus* anode) gives a *positive* E°_{cell}. Note that when we reverse the half-reaction, we need *not* reverse the sign of $E^\circ_{half\text{-}cell}$ because the minus sign in Equation 21.3 ($E^\circ_{cell} = E^\circ_{cathode} - E^\circ_{anode}$) will do that.
2. Add the rearranged half-reactions to obtain a balanced overall equation. Be sure to multiply by coefficients so that e^- lost equals e^- gained and to cancel species common to both sides.

(You may be tempted in this particular case to add the two half-reactions as written, because you obtain a positive E°_{cell}, but you would then have two oxidizing agents forming two reducing agents, which cannot occur.)

We want to pair the stronger oxidizing and reducing agents as reactants. The larger (more positive) E° value for the silver half-reaction means that Ag^+ is a stronger oxidizing agent (gains electrons more readily) than Sn^{2+}, and the smaller (more negative) E° value for the tin half-reaction means that Sn is a stronger reducing agent (loses electrons more readily) than Ag. Therefore, we reverse the tin half-reaction (but *not* the sign of E°_{tin}):

$$Sn(s) \longrightarrow Sn^{2+}(aq) + 2e^- \qquad E^\circ_{tin} = -0.14 \text{ V}$$

Subtracting $E^\circ_{half\text{-}cell}$ of the tin half-reaction (anode, oxidation) from $E^\circ_{half\text{-}cell}$ of the silver half-reaction (cathode, reduction) gives a positive E°_{cell}; that is, 0.80 V $-$ (-0.14 V) $= 0.94$ V.

With the half-reactions written in the correct direction, we must next make sure that the *number of electrons lost in the oxidation equals the number gained in the reduction.* In this case, we double the silver (reduction) half-reaction. Adding the half-reactions and applying Equation 21.3 gives the balanced equation and E_{cell}:

$$
\begin{array}{lll}
2Ag^+(aq) + 2e^- \longrightarrow 2Ag(s) & E^\circ_{silver} = 0.80 \text{ V} & \text{[reduction]} \\
Sn(s) \longrightarrow Sn^{2+}(aq) + 2e^- & E^\circ_{tin} = -0.14 \text{ V} & \text{[oxidation]} \\
\hline
Sn(s) + 2Ag^+(aq) \longrightarrow Sn^{2+}(aq) + 2Ag(s) & E^\circ_{cell} = E^\circ_{silver} - E^\circ_{tin} = 0.94 \text{ V} &
\end{array}
$$

With the reaction spontaneous as written, the stronger oxidizing and reducing agents are reactants, which confirms that Sn is a stronger reducing agent than Ag, and Ag^+ is a stronger oxidizing agent than Sn^{2+}.

A very important point to note is that, when we doubled the coefficients of the silver half-reaction to balance the number of electrons, we did *not* double its E° value—it remained 0.80 V. That is, *changing the balancing coefficients of a half-reaction does **not** change the E° value.* The reason is that a standard electrode potential is an *intensive* property, one that does *not* depend on the amount of substance present. The potential is the *ratio* of energy to charge. When we change the coefficients, thus changing the amount of substance, the energy *and* the charge change proportionately, so their ratio stays the same. (Recall that density, which is also an intensive property, does not change with the amount of substance because the mass *and* the volume change proportionately.)

SAMPLE PROBLEM 21.4 Writing Spontaneous Redox Reactions

Problem Combine the following three half-reactions into three balanced equations (A, B, and C) for spontaneous reactions, and calculate $E°_{cell}$ for each.

$$(1) \quad NO_3^-(aq) + 4H^+(aq) + 3e^- \longrightarrow NO(g) + 2H_2O(l) \qquad E° = 0.96 \text{ V}$$
$$(2) \quad N_2(g) + 5H^+(aq) + 4e^- \longrightarrow N_2H_5^+(aq) \qquad E° = -0.23 \text{ V}$$
$$(3) \quad MnO_2(s) + 4H^+(aq) + 2e^- \longrightarrow Mn^{2+}(aq) + 2H_2O(l) \qquad E° = 1.23 \text{ V}$$

Plan To write the redox equations, we combine the possible pairs of half-reactions: (1) and (2), (1) and (3), and (2) and (3). They are all written as reductions, so the oxidizing agents appear as reactants and the reducing agents appear as products. In each pair, we reverse the reduction half-reaction that has the smaller (less positive or more negative) $E°$ value to an oxidation to obtain a positive $E°_{cell}$. We make e^- lost equal e^- gained, without changing the magnitude of the $E°$ value, add the half-reactions together, and then apply Equation 21.3 to find $E°_{cell}$.

Solution Combining half-reactions (1) and (2) gives equation (A). The $E°$ value for half-reaction (1) is larger (more positive) than that for (2), so we reverse (2) to obtain a positive $E°_{cell}$:

$$(1) \quad NO_3^-(aq) + 4H^+(aq) + 3e^- \longrightarrow NO(g) + 2H_2O(l) \qquad E° = 0.96 \text{ V}$$
$$(\text{rev } 2) \quad N_2H_5^+(aq) \longrightarrow N_2(g) + 5H^+(aq) + 4e^- \qquad E° = -0.23 \text{ V}$$

To make e^- lost equal e^- gained, we multiply (1) by four and the reversed (2) by three; then add half-reactions and cancel appropriate numbers of common species (H^+ and e^-):

$$4NO_3^-(aq) + 16H^+(aq) + 12e^- \longrightarrow 4NO(g) + 8H_2O(l) \qquad E° = 0.96 \text{ V}$$
$$3N_2H_5^+(aq) \longrightarrow 3N_2(g) + 15H^+(aq) + 12e^- \qquad E° = -0.23 \text{ V}$$
___

$$(A) \quad 3N_2H_5^+(aq) + 4NO_3^-(aq) + H^+(aq) \longrightarrow 3N_2(g) + 4NO(g) + 8H_2O(l)$$

$$E°_{cell} = 0.96 \text{ V} - (-0.23 \text{ V}) = 1.19 \text{ V}$$

Combining half-reactions (1) and (3) gives equation (B). Half-reaction (1) must be reversed:

$$(\text{rev } 1) \quad NO(g) + 2H_2O(l) \longrightarrow NO_3^-(aq) + 4H^+(aq) + 3e^- \qquad E° = 0.96 \text{ V}$$
$$(3) \quad MnO_2(s) + 4H^+(aq) + 2e^- \longrightarrow Mn^{2+}(aq) + 2H_2O(l) \qquad E° = 1.23 \text{ V}$$

We multiply reversed (1) by two and (3) by three, then add and cancel:

$$2NO(g) + 4H_2O(l) \longrightarrow 2NO_3^-(aq) + 8H^+(aq) + 6e^- \qquad E° = 0.96 \text{ V}$$
$$3MnO_2(s) + 12H^+(aq) + 6e^- \longrightarrow 3Mn^{2+}(aq) + 6H_2O(l) \qquad E° = 1.23 \text{ V}$$
___

$$(B) \quad 3MnO_2(s) + 4H^+(aq) + 2NO(g) \longrightarrow 3Mn^{2+}(aq) + 2H_2O(l) + 2NO_3^-(aq)$$

$$E°_{cell} = 1.23 \text{ V} - 0.96 \text{ V} = 0.27 \text{ V}$$

Combining half-reactions (2) and (3) gives equation (C). Half-reaction (2) must be reversed:

$$(\text{rev } 2) \quad N_2H_5^+(aq) \longrightarrow N_2(g) + 5H^+(aq) + 4e^- \qquad E° = -0.23 \text{ V}$$
$$(3) \quad MnO_2(s) + 4H^+(aq) + 2e^- \longrightarrow Mn^{2+}(aq) + 2H_2O(l) \qquad E° = 1.23 \text{ V}$$

We multiply reaction (3) by two, add the half-reactions, and cancel:

$$N_2H_5^+(aq) \longrightarrow N_2(g) + 5H^+(aq) + 4e^- \qquad E° = -0.23 \text{ V}$$
$$2MnO_2(s) + 8H^+(aq) + 4e^- \longrightarrow 2Mn^{2+}(aq) + 4H_2O(l) \qquad E° = 1.23 \text{ V}$$
___

$$(C) \quad N_2H_5^+(aq) + 2MnO_2(s) + 3H^+(aq) \longrightarrow N_2(g) + 2Mn^{2+}(aq) + 4H_2O(l)$$

$$E°_{cell} = 1.23 \text{ V} - (-0.23 \text{ V}) = 1.46 \text{ V}$$

Check As always, check that atoms and charge balance on each side of the equation. A good way to check that the reactions are spontaneous is to list the given half-reactions in order of decreasing $E°$ value:

$$MnO_2(s) + 4H^+(aq) + 2e^- \longrightarrow Mn^{2+}(aq) + 2H_2O(l) \qquad E° = 1.23 \text{ V}$$
$$NO_3^-(aq) + 4H^+(aq) + 3e^- \longrightarrow NO(g) + 2H_2O(l) \qquad E° = 0.96 \text{ V}$$
$$N_2(g) + 5H^+(aq) + 4e^- \longrightarrow N_2H_5^+(aq) \qquad E° = -0.23 \text{ V}$$

Then the oxidizing agents (reactants) decrease in strength going down the list, so the reducing agents (products) decrease in strength going up. Each of the three spontaneous reactions (A, B, and C) should combine a reactant with a product that is lower down on this list.

FOLLOW-UP PROBLEM 21.4 Is the following reaction spontaneous as written?

$$3Fe^{2+}(aq) \longrightarrow Fe(s) + 2Fe^{3+}(aq)$$

If not, write the equation for the spontaneous reaction, calculate $E°_{cell}$, and rank the three species of iron in order of decreasing reducing strength.

Relative Reactivities of Metals In Chapter 4, we discussed the activity series of the metals (see Figure 4.14), which ranks metals by their ability to "displace" one another from aqueous solution. Now you'll see *why* this displacement occurs, as well as why many, but not all, metals react with acid to form H_2, and why a few metals form H_2 even in water.

1. *Metals that can displace H_2 from acid.* The standard hydrogen half-reaction represents the reduction of H^+ ions from an acid to H_2:

$$2H^+(aq) + 2e^- \longrightarrow H_2(g) \qquad E° = 0.00 \text{ V}$$

To see which metals reduce H^+ (referred to as "displacing H_2") from acids, choose a metal, write its half-reaction as an oxidation, combine this half-reaction with the hydrogen half-reaction, and see if $E°_{cell}$ is positive. What you find is that the metals Li through Pb, those that lie *below* the standard hydrogen (reference) half-reaction in Appendix D, give a positive $E°_{cell}$ when reducing H^+. Iron, for example, reduces H^+ from an acid to H_2:

$$\begin{aligned} Fe(s) &\longrightarrow Fe^{2+}(aq) + 2e^- & E° = -0.44 \text{ V} \quad \text{[anode; oxidation]} \\ 2H^+(aq) + 2e^- &\longrightarrow H_2(g) & E° = 0.00 \text{ V} \quad \text{[cathode; reduction]} \\ \hline Fe(s) + 2H^+(aq) &\longrightarrow H_2(g) + Fe^{2+}(aq) & E°_{cell} = 0.00 \text{ V} - (-0.44 \text{ V}) = 0.44 \text{ V} \end{aligned}$$

The lower the metal in the list, the stronger it is as a reducing agent; therefore, the more positive its half-cell potential when the half-reaction is reversed, and the higher the $E°_{cell}$ for its reduction of H^+ to H_2. *If $E°_{cell}$ for the reduction of H^+ is more positive for metal A than it is for metal B, metal A is a stronger reducing agent than metal B and a more **active** metal.*

2. *Metals that cannot displace H_2 from acid.* Metals that are *above* the standard hydrogen (reference) half-reaction *cannot* reduce H^+ from acids. When we reverse the metal half-reaction, the $E°_{cell}$ is negative, so the reaction does not occur. For example, the coinage metals—copper, silver, and gold, which are in Group 1B(11)—are not strong enough reducing agents to reduce H^+ from acids:

$$\begin{aligned} Ag(s) &\longrightarrow Ag^+(aq) + e^- & E° = 0.80 \text{ V} \quad \text{[anode; oxidation]} \\ 2H^+(aq) + 2e^- &\longrightarrow H_2(g) & E° = 0.00 \text{ V} \quad \text{[cathode; reduction]} \\ \hline 2Ag(s) + 2H^+(aq) &\longrightarrow 2Ag^+(aq) + H_2(g) & E°_{cell} = 0.00 \text{ V} - 0.80 \text{ V} = -0.80 \text{ V} \end{aligned}$$

The higher the metal in the list, the more negative is its $E°_{cell}$ for the reduction of H^+ to H_2, the lower is its reducing strength, and the less active it is. Thus, gold is less active than silver, which is less active than copper.

3. *Metals that can displace H_2 from water.* Metals active enough to reduce H_2O lie *below that half-reaction:*

$$2H_2O(l) + 2e^- \longrightarrow H_2(g) + 2OH^-(aq) \qquad E = -0.42 \text{ V}$$

(The value shown here is the *nonstandard* electrode potential because, in pure water, $[OH^-]$ is 1.0×10^{-7} M, not the standard-state value of 1 M.) For example,

Oxidation half-reaction
$Ca(s) \longrightarrow Ca^{2+}(aq) + 2e^-$

Reduction half-reaction
$2H_2O(l) + 2e^- \longrightarrow H_2(g) + 2OH^-(aq)$

Overall (cell) reaction
$Ca(s) + 2H_2O(l) \longrightarrow Ca(OH)_2(aq) + H_2(g)$

Figure 21.9 The reaction of calcium in water. Calcium is one of the metals active enough to displace H_2 from H_2O.

consider the reaction of sodium in water (with the Na half-reaction reversed and doubled):

$$2Na(s) \longrightarrow 2Na^+(aq) + 2e^- \qquad E° = -2.71 \text{ V} \quad \text{[anode; oxidation]}$$
$$\underline{2H_2O(l) + 2e^- \longrightarrow H_2(g) + 2OH^-(aq) \qquad E = -0.42 \text{ V} \quad \text{[cathode; reduction]}}$$
$$2Na(s) + 2H_2O(l) \longrightarrow 2Na^+(aq) + H_2(g) + 2OH^-(aq)$$
$$E_{cell} = -0.42 \text{ V} - (-2.71 \text{ V}) = 2.29 \text{ V}$$

The alkali metals [Group 1A(1)] and the larger alkaline earth metals [Group 2A(2)] can reduce water, or "displace" H_2 from H_2O (Figure 21.9).

4. *Metals that can displace other metals from solution.* We can also predict whether one metal can reduce the aqueous ion of another metal. Any metal that is lower in the list in Appendix D can reduce the ion of a metal that is higher up, and thus displace that metal from solution. For example, zinc can displace iron from solution:

$$Zn(s) \longrightarrow Zn^{2+}(aq) + 2e^- \qquad E° = -0.76 \text{ V} \quad \text{[anode; oxidation]}$$
$$\underline{Fe^{2+}(aq) + 2e^- \longrightarrow Fe(s) \qquad\qquad E° = -0.44 \text{ V} \quad \text{[cathode; reduction]}}$$
$$Zn(s) + Fe^{2+}(aq) \longrightarrow Zn^{2+}(aq) + Fe(s) \qquad E°_{cell} = -0.44 \text{ V} - (-0.76 \text{ V}) = 0.32 \text{ V}$$

This particular reaction has tremendous economic importance in protecting iron from rusting, as you'll see shortly.

SECTION SUMMARY

The output of a cell is called the cell potential (E_{cell}) and is measured in volts (1 V = 1 J/C). When all substances are in their standard states, the output is the standard cell potential ($E°_{cell}$). $E°_{cell} > 0$ for a spontaneous reaction at standard-state conditions. By convention, a standard electrode potential ($E°_{half-cell}$) refers to the *reduction* half-reaction. $E°_{cell}$ equals $E°_{half-cell}$ of the cathode *minus* $E°_{half-cell}$ of the anode. Using a standard hydrogen (reference) electrode, other $E°_{half-cell}$ values can be measured and used to rank oxidizing (or reducing) agents (see Appendix D). Spontaneous redox reactions combine stronger oxidizing and reducing agents to form weaker ones. A metal can reduce another species (H^+, H_2O, or an ion of another metal) if $E°_{cell}$ for the reaction is positive.

21.4 FREE ENERGY AND ELECTRICAL WORK

In Chapter 20, we discussed the relationship of useful work, free energy, and the equilibrium constant. In this section, we examine this relationship in the context of electrochemical cells and see the effect of concentration on cell potential.

Standard Cell Potential and the Equilibrium Constant

As you know from Section 20.3, a spontaneous reaction has a *negative* free energy change ($\Delta G < 0$), and you've just seen that a spontaneous electrochemical reaction has a *positive* cell potential ($E_{cell} > 0$). Note that *the signs of ΔG and E_{cell} are opposite for a spontaneous reaction.* These two indications of spontaneity are proportional to each other:

$$\Delta G \propto -E_{cell}$$

Let's determine this proportionality constant by focusing on the electrical work done (w, in joules), which is the product of the potential (E_{cell}, in volts) and the amount of charge that flows (in coulombs):

$$w = E_{cell} \times \text{charge}$$

The value used for E_{cell} is measured with no current flowing and, therefore, no energy lost to heating the cell. Thus, E_{cell} is the maximum voltage the cell can generate, that is, the maximum work the system can do *on* the surroundings. Recall from Chapter 20 that *only a reversible process can do maximum work.* For no current to flow and the process to be reversible, E_{cell} must be opposed by an equal potential in the measuring circuit. (In this case, a reversible process means that, if the opposing potential is infinitesimally smaller, the cell reaction goes forward; if it is infinitesimally larger, the reaction goes backward.) Equation 20.9 shows that the maximum work done *on* the surroundings is $-\Delta G$:

$$w_{max} = E_{cell} \times \text{charge} = -\Delta G \qquad \text{or} \qquad \Delta G = -E_{cell} \times \text{charge}$$

The charge that flows through the cell equals the number of moles of electrons (*n*) transferred times the charge of 1 mol of electrons (symbol *F*):

$$\text{Charge} = \text{moles of e}^- \times \frac{\text{charge}}{\text{mol e}^-} \qquad \text{or} \qquad \text{charge} = nF$$

The charge of 1 mol of electrons is the **Faraday constant (*F*),** named in honor of Michael Faraday, the 19th-century British scientist who pioneered the study of electrochemistry:

$$F = \frac{96{,}485 \text{ C}}{\text{mol e}^-}$$

Because 1 V = 1 J/C, we have 1 C = 1 J/V, and

$$F = 9.65 \times 10^4 \frac{\text{J}}{\text{V·mol e}^-} \qquad \text{(3 sf)} \qquad \textbf{(21.4)}$$

Substituting for charge, the proportionality constant is *nF*:

$$\Delta G = -nFE_{cell} \qquad \textbf{(21.5)}$$

When all of the components are in their standard states, we have

$$\Delta G° = -nFE°_{cell} \qquad \textbf{(21.6)}$$

Using this relationship, we can relate the standard cell potential to the equilibrium constant of the redox reaction. Recall from Equation 20.12 that

$$\Delta G° = -RT \ln K$$

Substituting for $\Delta G°$ from Equation 21.6 gives

$$-nFE°_{cell} = -RT \ln K$$

Solving for $E°_{cell}$ gives

$$E°_{cell} = \frac{RT}{nF} \ln K \qquad \textbf{(21.7)}$$

Figure 21.10 summarizes the interconnections among the standard free energy change, the equilibrium constant, and the standard cell potential. The procedures

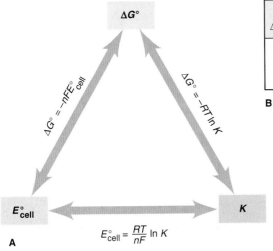

$\Delta G°$	K	$E°_{cell}$	Reaction at standard-state conditions
<0	>1	>0	Spontaneous
0	1	0	At equilibrium
>0	<1	<0	Nonspontaneous

B

Figure 21.10 The interrelationship of $\Delta G°$, $E°_{cell}$, and K. A, Any one of these three central thermodynamic parameters can be used to find the other two. **B,** The signs of $\Delta G°$ and $E°_{cell}$ determine the reaction direction at standard-state conditions.

presented in Chapter 20 for determining K required that we know $\Delta G°$, either from $\Delta H°$ and $\Delta S°$ values or from $\Delta G_f°$ values. For redox reactions, we now have a direct experimental method for determining K *and* $\Delta G°$: measure $E°_{cell}$.

It is common practice to simplify Equation 21.7 in calculations by

- Substituting the known value of 8.314 J/(mol rxn·K) for the constant R
- Substituting the known value of 9.65×10^4 J/(V·mol e⁻) for the constant F
- Substituting the standard temperature of 298.15 K for T, but keeping in mind that the cell can run at other temperatures.
- Multiplying by 2.303 to convert from natural to common (base-10) logarithms. This conversion shows that a *10-fold change in K makes $E°_{cell}$ change by 1.*

Thus, when n moles of e⁻ are transferred per mole of reaction in the balanced equation, this simplified relation between $E°_{cell}$ and K gives

$$E°_{cell} = \frac{RT}{nF} \ln K = 2.303 \frac{RT}{nF} \log K = 2.303 \times \frac{8.314 \frac{\text{J}}{\text{mol rxn·K}} \times 298.15 \text{ K}}{\frac{n \text{ mol e⁻}}{\text{mol rxn}} \left(9.65 \times 10^4 \frac{\text{J}}{\text{V·mol e⁻}}\right)} \log K$$

And, we have

$$E°_{cell} = \frac{0.0592 \text{ V}}{n} \log K \qquad \text{or} \qquad \log K = \frac{nE°_{cell}}{0.0592 \text{ V}} \qquad \text{(at 298.15 K)} \qquad \textbf{(21.8)}$$

SAMPLE PROBLEM 21.5 Calculating K and $\Delta G°$ from $E°_{cell}$

Problem Lead can displace silver from solution:

$$Pb(s) + 2Ag^+(aq) \longrightarrow Pb^{2+}(aq) + 2Ag(s)$$

As a consequence, silver is a valuable by-product in the industrial extraction of lead from its ore. Calculate K and $\Delta G°$ at 298.15 K for this reaction.

Plan We divide the spontaneous redox equation into the half-reactions and use values from Appendix D to calculate $E°_{cell}$. Then, we substitute this result into Equation 21.8 to find K and into Equation 21.6 to find $\Delta G°$.

Solution Writing the half-reactions and their $E°$ values:

(1) $Ag^+(aq) + e^- \longrightarrow Ag(s)$ $E° = 0.80$ V
(2) $Pb^{2+}(aq) + 2e^- \longrightarrow Pb(s)$ $E° = -0.13$ V

Calculating $E°_{cell}$: We double (1), reverse (2), add the half-reactions, and subtract $E°_{lead}$ from $E°_{silver}$:

$$2Ag^+(aq) + 2e^- \longrightarrow 2Ag(s) \qquad\qquad\qquad E° = 0.80 \text{ V}$$
$$\underline{\qquad\qquad Pb(s) \longrightarrow Pb^{2+}(aq) + 2e^- \qquad\qquad E° = -0.13 \text{ V}}$$
$$Pb(s) + 2Ag^+(aq) \longrightarrow Pb^{2+}(aq) + 2Ag(s) \qquad E°_{cell} = 0.80 \text{ V} - (-0.13 \text{ V}) = 0.93 \text{ V}$$

Calculating K with Equations 21.7 and 21.8:

$$E°_{cell} = \frac{RT}{nF} \ln K = 2.303 \frac{RT}{nF} \log K$$

The adjusted half-reactions show that 2 mol of e⁻ are transferred per mole of reaction as written, so $n = 2$. Then, performing the substitutions for R and F that we just discussed with the cell running at 25°C (298.15 K), we have

$$E°_{cell} = \frac{0.0592 \text{ V}}{2} \log K = 0.93 \text{ V}$$

So, $$\log K = \frac{0.93 \text{ V} \times 2}{0.0592 \text{ V}} = 31.42 \qquad \text{and} \qquad K = 2.6 \times 10^{31}$$

Calculating $\Delta G°$ (Equation 21.6):

$$\Delta G° = -nFE°_{cell} = -\frac{2 \text{ mol e}^-}{\text{mol rxn}} \times \frac{96.5 \text{ kJ}}{\text{V·mol e}^-} \times 0.93 \text{ V} = -1.8 \times 10^2 \text{ kJ/mol rxn}$$

Check The three variables are consistent with the reaction being spontaneous at standard-state conditions: $E°_{cell} > 0$, $\Delta G° < 0$, and $K > 1$. Be sure to round and check the order of magnitude: in the $\Delta G°$ calculation, for instance, $\Delta G° \approx -2 \times 100 \times 1 = -200$, so the overall math seems right. Another check would be to obtain $\Delta G°$ directly from its relation with K:

$$\Delta G° = -RT \ln K = -8.314 \text{ J/mol rxn·K} \times 298.15 \text{ K} \times \ln (2.6 \times 10^{31})$$
$$= -1.8 \times 10^5 \text{ J/mol rxn} = -1.8 \times 10^2 \text{ kJ/mol rxn}$$

FOLLOW-UP PROBLEM 21.5 When cadmium metal reduces Cu^{2+} in solution, Cd^{2+} forms in addition to copper metal. Given that $\Delta G° = -143$ kJ, calculate K at 25°C. What is $E°_{cell}$ of a voltaic cell that uses this reaction?

The Effect of Concentration on Cell Potential

So far, we've considered cells with all components in their standard states. But most cells don't start at those conditions, and even if they did, the concentrations change after a few moments of operation. Moreover, in all practical voltaic cells, such as batteries, reactant concentrations are far from standard-state values. Clearly, we must be able to determine E_{cell}, the cell potential under nonstandard conditions.

To do so, let's derive an expression for the relation between cell potential and concentration based on the relation between free energy and concentration. Recall from Chapter 20 (Equation 20.13) that ΔG equals $\Delta G°$ (the free energy change when the system moves from standard-state concentrations to equilibrium) *plus* $RT \ln Q$ (the free energy change when the system moves from nonstandard-state to standard-state concentrations):

$$\Delta G = \Delta G° + RT \ln Q$$

ΔG is related to E_{cell} and $\Delta G°$ to $E°_{cell}$ (Equations 21.5 and 21.6), so we substitute for them and get

$$-nFE_{cell} = -nFE°_{cell} + RT \ln Q$$

Dividing both sides by $-nF$, we obtain the **Nernst equation,** developed by the German chemist Walther Hermann Nernst in 1889:

$$E_{cell} = E°_{cell} - \frac{RT}{nF} \ln Q \tag{21.9}$$

The Nernst equation says that a cell potential under any conditions depends on the potential at standard-state concentrations *and* a term for the potential at nonstandard-state concentrations. How do changes in Q affect cell potential? From Equation 21.9, we see that

- When $Q < 1$ and thus [reactant] > [product], $\ln Q < 0$, so $E_{cell} > E°_{cell}$.
- When $Q = 1$ and thus [reactant] = [product], $\ln Q = 0$, so $E_{cell} = E°_{cell}$.
- When $Q > 1$ and thus [reactant] < [product], $\ln Q > 0$, so $E_{cell} < E°_{cell}$.

As before, to obtain a simplified form of the Nernst equation for use in calculations, let's substitute known values of R and F, operate the cell at 298.15 K, and convert to common (base-10) logarithms:

$$E_{cell} = E°_{cell} - \frac{RT}{nF} \ln Q = E°_{cell} - 2.303 \frac{RT}{nF} \log Q$$

$$= E°_{cell} - 2.303 \times \frac{8.314 \frac{\cancel{J}}{\text{mol } \cancel{rxn}\text{·K}} \times 298.15 \text{ K}}{\frac{n \cancel{\text{mol } e^-}}{\text{mol } \cancel{rxn}} \left(9.65 \times 10^4 \frac{\cancel{J}}{\text{V·}\cancel{\text{mol } e^-}}\right)} \log Q$$

And we obtain:

$$E_{cell} = E°_{cell} - \frac{0.0592\text{ V}}{n} \log Q \quad \text{(at 298.15 K)} \qquad \textbf{(21.10)}$$

Remember that the expression for Q *contains only those species with concentrations (and/or pressures) that can vary;* thus, solids do not appear, even when they are the electrodes. For example, in the reaction between cadmium and silver ion, the Cd and Ag electrodes do not appear in the expression for Q:

$$Cd(s) + 2Ag^+(aq) \longrightarrow Cd^{2+}(aq) + 2Ag(s) \qquad Q = \frac{[Cd^{2+}]}{[Ag^+]^2}$$

SAMPLE PROBLEM 21.6 Using the Nernst Equation to Calculate E_{cell}

Problem In a test of a new reference electrode, a chemist constructs a voltaic cell consisting of a Zn/Zn^{2+} half-cell and an H_2/H^+ half-cell under the following conditions:

$$[Zn^{2+}] = 0.010\ M \qquad [H^+] = 2.5\ M \qquad P_{H_2} = 0.30\text{ atm}$$

Calculate E_{cell} at 298.15 K.

Plan To apply the Nernst equation and determine E_{cell}, we must know $E°_{cell}$ and Q. We write the spontaneous reaction, calculate $E°_{cell}$ from standard electrode potentials (Appendix D), and use the given pressure and concentrations to find Q. (Recall that the ideal gas law allows us to use P at constant T as another way of writing concentration, n/V.) Then we substitute into Equation 21.10.

Solution Determining the cell reaction and $E°_{cell}$:

$2H^+(aq) + 2e^- \longrightarrow H_2(g)$	$E° = 0.00$ V
$Zn(s) \longrightarrow Zn^{2+}(aq) + 2e^-$	$E° = -0.76$ V
$2H^+(aq) + Zn(s) \longrightarrow H_2(g) + Zn^{2+}(aq)$	$E°_{cell} = 0.00$ V $- (-0.76$ V$) = 0.76$ V

Calculating Q:

$$Q = \frac{P_{H_2} \times [Zn^{2+}]}{[H^+]^2} = \frac{0.30 \times 0.010}{2.5^2} = 4.8 \times 10^{-4}$$

Solving for E_{cell} at 25°C (298.15 K), with $n = 2$:

$$E_{cell} = E°_{cell} - 2.303\frac{RT}{nF} \log Q = E°_{cell} - \frac{0.0592\text{ V}}{n} \log Q$$

$$= 0.76\text{ V} - \left[\frac{0.0592\text{ V}}{2} \log (4.8 \times 10^{-4})\right] = 0.76\text{ V} - (-0.0982\text{ V}) = \boxed{0.86\text{ V}}$$

Check After you check the arithmetic, reason through the answer: $E_{cell} > E°_{cell}$ ($0.86 > 0.76$) because the log Q term was negative, which is consistent with $Q < 1$; that is, the amounts of products, P_{H_2} and $[Zn^{2+}]$, are smaller than the amount of reactant, $[H^+]$.

FOLLOW-UP PROBLEM 21.6 Consider a voltaic cell based on the following reaction: $Fe(s) + Cu^{2+}(aq) \rightleftharpoons Fe^{2+}(aq) + Cu(s)$. If $[Cu^{2+}] = 0.30\ M$, what must $[Fe^{2+}]$ be to increase E_{cell} by 0.25 V above $E°_{cell}$ at 25°C?

Changes in Potential During Cell Operation

The potential of the zinc-copper cell changes as concentrations change during cell operation. The only concentrations that change are [reactant] $= [Cu^{2+}]$ and [product] $= [Zn^{2+}]$:

$$Zn(s) + Cu^{2+}(aq) \rightleftharpoons Zn^{2+}(aq) + Cu(s) \qquad Q = \frac{[Zn^{2+}]}{[Cu^{2+}]}$$

The positive $E°_{cell}$ value (1.10 V) means that this reaction proceeds *spontaneously* from the standard-state conditions, at which $[Zn^{2+}] = [Cu^{2+}] = 1\ M$ ($Q = 1$), to some point at which $[Zn^{2+}] > [Cu^{2+}]$ ($Q > 1$). If we start the cell when

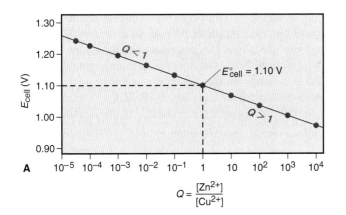

$$Q = \frac{[Zn^{2+}]}{[Cu^{2+}]}$$

Changes in E_{cell} and Concentration			
Stage in cell operation	Q	Relative [P] and [R]	$\frac{0.0592\ V}{n} \log Q$
1. $E > E°$	<1	[P] < [R]	<0
2. $E = E°$	=1	[P] = [R]	=0
3. $E < E°$	>1	[P] > [R]	>0
4. $E = 0$	=K	[P] ≫ [R]	=E°

Figure 21.11 The relation between E_{cell} and log Q for the zinc-copper cell. A, A plot of E_{cell} vs. Q (on a logarithmic scale) for the zinc-copper cell shows a linear decrease. When $Q < 1$ (*left*), [reactant] is relatively high, and the cell can do relatively more work. When $Q = 1$, $E_{cell} = E°_{cell}$. When $Q > 1$ (*right*), [product] is relatively high, and the cell can do relatively less work. **B,** A summary of the changes in E_{cell} as the cell operates, including the changes in $[Zn^{2+}]$, denoted [P] for [product], and $[Cu^{2+}]$, denoted [R] for [reactant].

$[Zn^{2+}] < [Cu^{2+}]$ ($Q < 1$), then the cell potential is *higher* than the standard cell potential.

As the cell operates, $[Zn^{2+}]$ increases (as the Zn electrode deteriorates) and $[Cu^{2+}]$ decreases (as Cu plates out on the Cu electrode). Although the changes during this process occur smoothly, we can identify four general stages of operation. Figure 21.11A shows the first three. The main point to note is *as the cell operates, its potential decreases:*

Stage 1. $E_{cell} > E°_{cell}$ when $Q < 1$: When the cell begins operation, $[Cu^{2+}] > [Zn^{2+}]$, so $E_{cell} > E°_{cell}$.

As cell operation continues, $[Zn^{2+}]$ increases and $[Cu^{2+}]$ decreases; thus, Q becomes larger, and E_{cell} decreases.

Stage 2. $E_{cell} = E°_{cell}$ when $Q = 1$: At the point when $[Cu^{2+}] = [Zn^{2+}]$, $Q = 1$, so $E_{cell} = E°_{cell}$.

Stage 3. $E_{cell} < E°_{cell}$ when $Q > 1$: As $[Cu^{2+}]$ falls below $[Zn^{2+}]$, $E_{cell} < E°_{cell}$.

Stage 4. $E_{cell} = 0$ when $Q = K$: Eventually, E_{cell} is zero. This occurs *when the system reaches **equilibrium**: no more free energy is released, so the cell can do no more work.* At this point, we say that a battery is "dead."

Figure 21.11B summarizes these four key stages in the operation of a voltaic cell.

Let's find K for the zinc-copper cell. At equilibrium, Equation 21.10 becomes

$$0 = E°_{cell} - \left(\frac{0.0592\ V}{n}\right) \log K, \text{ which rearranges to } E°_{cell} = \frac{0.0592\ V}{n} \log K$$

Note that this result is identical to Equation 21.8, which we obtained from $\Delta G°$. Solving for K of the zinc-copper cell ($E°_{cell} = 1.10$ V),

$$\log K = \frac{2 \times E°_{cell}}{0.0592\ V}, \text{ so } K = 10^{(2 \times 1.10\ V)/0.0592\ V} = 10^{37.16} = 1.4 \times 10^{37}$$

Thus, the zinc-copper cell does work until the $[Zn^{2+}]/[Cu^{2+}]$ ratio is *very* high.

Concentration Cells

If you mix a concentrated solution of a salt with a dilute solution of the salt, the final concentration equals some intermediate value. A **concentration cell** employs this phenomenon to generate electrical energy. The two solutions are in separate half-cells, so they do not mix; rather, their concentrations become equal as the cell operates.

How a Concentration Cell Works Suppose that both compartments of a voltaic cell house the Cu/Cu^{2+} half-reaction. The cell reaction is the sum of identical half-reactions, written in opposite directions, so the *standard* half-cell potentials cancel ($E°_{copper} - E°_{copper}$) and $E°_{cell}$ is zero. This occurs because standard electrode potentials are based on concentrations of 1 *M*. *In a concentration cell, however, the half-reactions are the same but the* **concentrations** *are different.* As a result, even though $E°_{cell}$ equals zero, the *nonstandard* cell potential, E_{cell}, does *not* equal zero because it depends on the *ratio of concentrations.*

In Figure 21.12A, a concentration cell has 0.10 *M* Cu^{2+} in the anode half-cell and 1.0 *M* Cu^{2+}, a 10-fold higher concentration, in the cathode half-cell:

$$Cu(s) \longrightarrow Cu^{2+}(aq; 0.10\ M) + 2e^- \qquad \text{[anode; oxidation]}$$
$$Cu^{2+}(aq; 1.0\ M) + 2e^- \longrightarrow Cu(s) \qquad \text{[cathode; reduction]}$$

The overall cell reaction is the sum of the half-reactions:

$$Cu^{2+}(aq; 1.0\ M) \longrightarrow Cu^{2+}(aq; 0.10\ M) \qquad E_{cell} = ?$$

The cell potential at the initial concentrations of 0.10 *M* (dilute) and 1.0 *M* (concentrated), with $n = 2$, is obtained from the Nernst equation:

$$E_{cell} = E°_{cell} - \frac{0.0592\ V}{2} \log \frac{[Cu^{2+}]_{dil}}{[Cu^{2+}]_{conc}} = 0\ V - \left(\frac{0.0592\ V}{2} \log \frac{0.10\ M}{1.0\ M} \right)$$

$$= 0\ V - \left[\frac{0.0592\ V}{2} (-1.00) \right] = 0.0296\ V$$

As you can see, because $E°_{cell}$ for a concentration cell equals zero, E_{cell} for nonstandard conditions depends entirely on the $[(0.0592\ V/n) \log Q]$ term.

What is actually going on as this cell operates? In the half-cell with dilute electrolyte (anode), the Cu atoms in the electrode give up electrons and become Cu^{2+} ions, which enter the solution and make it *more* concentrated. The electrons released at the anode flow to the cathode compartment. There, Cu^{2+} ions in the concentrated solution pick up the electrons and become Cu atoms, which plate out on the electrode, so that solution becomes *less* concentrated. As in any voltaic cell, E_{cell} decreases until equilibrium is attained, which happens when $[Cu^{2+}]$ is the same in both half-cells (Figure 21.12B). The same final concentration would result if we mixed the two solutions, but no electrical work would be done.

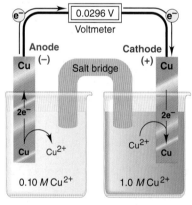

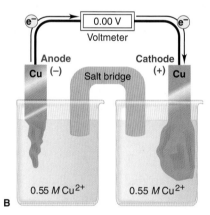

B

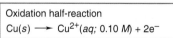

A

Figure 21.12 A concentration cell based on the Cu/Cu^{2+} half-reaction. A, The half-reactions are the same, so $E°_{cell} = 0$. The cell operates because the half-cell concentrations are different, which makes $E_{cell} > 0$ in this case. **B,** The cell operates until the half-cell concentrations are equal. Note the change in electrodes (exaggerated here for clarity) and the identical color of the solutions.

SAMPLE PROBLEM 21.7   Calculating the Potential of a Concentration Cell

Problem A concentration cell consists of two Ag/Ag^+ half-cells. In half-cell A, electrode A dips into 0.010 M $AgNO_3$; in half-cell B, electrode B dips into 4.0×10^{-4} M $AgNO_3$. What is the cell potential at 298.15 K? Which electrode has a positive charge?

Plan The standard half-cell reactions are identical, so E°_{cell} is zero, and we calculate E_{cell} from the Nernst equation. Because half-cell A has a higher $[Ag^+]$, Ag^+ ions will be reduced and plate out on electrode A. In half-cell B, Ag will be oxidized and Ag^+ ions will enter the solution. As in all voltaic cells, reduction occurs at the cathode, which is positive.

Solution Writing the spontaneous reaction: The $[Ag^+]$ decreases in half-cell A and increases in half-cell B, so the spontaneous reaction is

$$Ag^+(aq; 0.010\ M)\ [\text{half-cell A}] \longrightarrow Ag^+(aq; 4.0\times10^{-4}\ M)\ [\text{half-cell B}]$$

Calculating E_{cell}, with $n = 1$:

$$E_{cell} = E^\circ_{cell} - \frac{0.0592\ V}{1}\log\frac{[Ag^+]_{dil}}{[Ag^+]_{conc}} = 0\ V - \left(0.0592\ V\log\frac{4.0\times10^{-4}}{0.010}\right)$$

$$= 0.0828\ V$$

Reduction occurs at the cathode, electrode A: $Ag^+(aq; 0.010\ M) + e^- \longrightarrow Ag(s)$. Thus, electrode A has a positive charge due to a relative electron deficiency.

FOLLOW-UP PROBLEM 21.7 A concentration cell is built using two Au/Au^{3+} half-cells. In half-cell A, $[Au^{3+}] = 7.0\times10^{-4}$ M, and in half-cell B, $[Au^{3+}] = 2.5\times10^{-2}$ M. What is E_{cell}, and which electrode is negative?

Applications of Concentration Cells Chemists, biologists, and environmental scientists apply the principle of a concentration cell in a host of applications. The most important is the measurement of unknown ion concentrations in materials from various sources, such as blood, soil, natural waters, and industrial waste water. The most common laboratory application is measurement of $[H^+]$ to determine pH. Suppose we construct a concentration cell based on the H_2/H^+ half-reaction, in which the cathode compartment houses the standard hydrogen electrode and the anode compartment has the same apparatus dipping into an unknown $[H^+]$ in solution. The half-reactions and overall reaction are

$$H_2(g; 1\ atm) \longrightarrow 2H^+(aq; \text{unknown}) + 2e^- \qquad [\text{anode; oxidation}]$$
$$\underline{2H^+(aq; 1\ M) + 2e^- \longrightarrow H_2(g; 1\ atm) \qquad\qquad [\text{cathode; reduction}]}$$
$$2H^+(aq; 1\ M) \longrightarrow 2H^+(aq; \text{unknown}) \qquad E_{cell} = ?$$

As for the Cu/Cu^{2+} concentration cell, E°_{cell} is zero; however, the half-cells differ in $[H^+]$, so E_{cell} is *not* zero. From the Nernst equation, with $n = 2$, we obtain

$$E_{cell} = E^\circ_{cell} - \frac{0.0592\ V}{2}\log\frac{[H^+]^2_{unknown}}{[H^+]^2_{standard}}$$

Substituting 1 M for $[H^+]_{standard}$ and 0 V for E°_{cell} gives

$$E_{cell} = 0\ V - \frac{0.0592\ V}{2}\log\frac{[H^+]^2_{unknown}}{1^2} = -\frac{0.0592\ V}{2}\log[H^+]^2_{unknown}$$

Because $\log x^2 = 2\log x$ (see Appendix A), we obtain

$$E_{cell} = -\left[\frac{0.0592\ V}{2}(2\log[H^+]_{unknown})\right] = -0.0592\ V \times \log[H^+]_{unknown}$$

Substituting $-\log[H^+] = pH$, we have

$$E_{cell} = 0.0592\ V \times pH$$

Thus, by measuring E_{cell}, we can find the pH.

In the routine measurement of pH, a concentration cell incorporating two hydrogen electrodes is too bulky and difficult to maintain. Instead, as was pointed

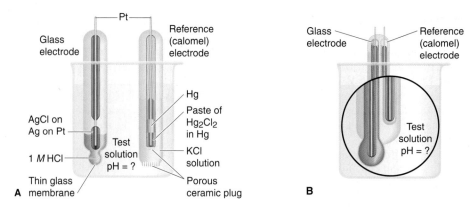

Figure 21.13 The laboratory measurement of pH. A, The glass electrode (*left*) is a self-contained Ag/AgCl half-cell immersed in an HCl solution of known concentration and enclosed by a thin glass membrane. It monitors the external [H$^+$] in the solution relative to its fixed internal [H$^+$]. The saturated calomel electrode (*right*) acts as a reference. **B,** Most modern laboratories use a combination electrode, which houses both the glass and reference electrodes in one tube.

out in Chapter 18, a pH meter is used. As shown in Figure 21.13A, two separate electrodes dip into the solution being tested. One of them is a *glass electrode,* which consists of an Ag/AgCl half-reaction immersed in an HCl solution of fixed concentration (usually 1.000 *M*) and enclosed by a thin (~0.05 mm) membrane made of a special glass that is highly sensitive to the presence of H$^+$ ions. The other electrode is a reference electrode, typically a *saturated calomel electrode.* It consists of a platinum wire immersed in a paste of Hg$_2$Cl$_2$ (calomel), liquid Hg, and saturated KCl solution. The glass electrode monitors the solution's [H$^+$] relative to its own fixed internal [H$^+$], and the instrument converts the potential difference between the glass and reference electrodes into a measure of pH. In modern instruments, a combination electrode is used, which houses both electrodes in one tube (Figure 21.13B).

SECTION SUMMARY
A spontaneous process is indicated by a negative ΔG or a positive E_{cell}, which are related: $\Delta G = -nFE_{cell}$. The ΔG of the cell reaction represents the maximum amount of electrical work the cell can do. Because the standard free energy change, $\Delta G°$, is related to $E°_{cell}$ and to K, we can use $E°_{cell}$ to determine K. At nonstandard conditions, the Nernst equation shows that E_{cell} depends on $E°_{cell}$ and a correction term based on Q. E_{cell} is high when Q is small (high [reactant]), and it decreases as the cell operates. At equilibrium, ΔG and E_{cell} are zero, which means that $Q = K$. Concentration cells have identical half-reactions, but solutions of differing concentration; thus they generate electrical energy as the concentrations become equal. The pH electrode measures the concentration of H$^+$ ions.

21.5 ELECTROCHEMICAL PROCESSES IN BATTERIES

Batteries are ingeniously engineered devices that house rather unusual half-reactions and half-cells, but they operate through the same electrochemical principles we've been discussing. Strictly speaking, a **battery** is a self-contained group of voltaic cells arranged in series (plus-to-minus-to-plus, and so on), so that their individual voltages are added together. In everyday speech, however, the term may also be applied to a single voltaic cell. In this section, we examine the three categories of batteries—primary, secondary, and fuel cells—and note important examples, including some newer designs, of each.

Figure 21.14 Alkaline battery.

- Positive button
- Steel case
- MnO₂ in KOH paste
- Zn (anode)
- Graphite rod (cathode)
- Absorbent/separator
- Negative end cap

Primary (Nonrechargeable) Batteries

A *primary battery* cannot be recharged, so it is discarded when the components have reached their equilibrium concentrations, that is, when the cell is "dead." We'll discuss the alkaline battery and the mercury and silver "button" batteries.

Alkaline Battery Today's ubiquitous alkaline battery has a zinc anode case that houses a mixture of MnO_2 and an alkaline paste of KOH and water. The cathode is an inactive graphite rod (Figure 21.14). The half-reactions are

Anode (oxidation): $\quad\quad\quad\quad Zn(s) + 2OH^-(aq) \longrightarrow ZnO(s) + H_2O(l) + 2e^-$

Cathode (reduction): $\quad MnO_2(s) + 2H_2O(l) + 2e^- \longrightarrow Mn(OH)_2(s) + 2OH^-(aq)$

Overall (cell) reaction:

$\quad\quad Zn(s) + MnO_2(s) + H_2O(l) \longrightarrow ZnO(s) + Mn(OH)_2(s) \quad\quad E_{cell} = 1.5 \text{ V}$

The alkaline battery powers portable radios, toys, flashlights, and so on, is safe, has a long shelf life, and comes in many sizes.

Mercury and Silver (Button) Batteries Mercury and silver batteries are quite similar. Both use a zinc container as the anode (reducing agent) in a basic medium. The mercury battery employs HgO as the oxidizing agent, the silver uses Ag_2O, and both use a steel can around the cathode. The solid reactants are compacted with KOH and separated with moist paper. The half-reactions are

Anode (oxidation): $\quad\quad\quad\quad\quad\quad Zn(s) + 2OH^-(aq) \longrightarrow ZnO(s) + H_2O(l) + 2e^-$

Cathode (reduction) (mercury): $\quad HgO(s) + H_2O(l) + 2e^- \longrightarrow Hg(l) + 2OH^-(aq)$

Cathode (reduction) (silver): $\quad Ag_2O(s) + H_2O(l) + 2e^- \longrightarrow 2Ag(s) + 2OH^-(aq)$

Overall (cell) reaction (mercury):

$\quad\quad\quad Zn(s) + HgO(s) \longrightarrow ZnO(s) + Hg(l) \quad\quad\quad E_{cell} = 1.3 \text{ V}$

Overall (cell) reaction (silver):

$\quad\quad\quad Zn(s) + Ag_2O(s) \longrightarrow ZnO(s) + 2Ag(s) \quad\quad\quad E_{cell} = 1.6 \text{ V}$

Both cells are manufactured as small button-sized batteries. The mercury cell is used in calculators (Figure 21.15). Because of its very steady output, the silver cell is used in watches, cameras, heart pacemakers, and hearing aids. Their major disadvantages are toxicity of discarded mercury and high cost of silver cells.

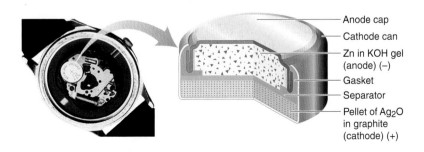

Figure 21.15 Silver button battery.

- Anode cap
- Cathode can
- Zn in KOH gel (anode) (–)
- Gasket
- Separator
- Pellet of Ag₂O in graphite (cathode) (+)

Secondary (Rechargeable) Batteries

In contrast to primary batteries, a *secondary*, or *rechargeable*, *battery* is recharged when it runs down *by supplying electrical energy to reverse the cell reaction* and re-form reactant. In other words, in this type of battery, the voltaic cells are periodically converted to electrolytic cells to restore *nonequilibrium* concentrations of the cell components. By far the most widely used secondary battery is the common car battery. Two newer types are the nickel–metal hydride battery and the lithium-ion battery.

Lead-Acid Battery A typical lead-acid car battery has six cells connected in series, each of which delivers about 2.1 V for a total of about 12 V. Each cell contains two lead grids loaded with the electrode materials: high-surface-area Pb in the anode and high-surface-area PbO_2 in the cathode. The grids are immersed in an electrolyte solution of ~4.5 M H_2SO_4. Fiberglass sheets between the grids prevent shorting due to physical contact (Figure 21.16). When the cell discharges, it generates electrical energy as a voltaic cell:

Anode (oxidation): $\quad Pb(s) + HSO_4^-(aq) \longrightarrow PbSO_4(s) + H^+(aq) + 2e^-$

Cathode (reduction):

$$PbO_2(s) + 3H^+(aq) + HSO_4^-(aq) + 2e^- \longrightarrow PbSO_4(s) + 2H_2O(l)$$

Notice that both half-reactions produce Pb^{2+} ions, one through the oxidation of Pb, the other through the reduction of PbO_2. The Pb^{2+} forms $PbSO_4(s)$ at both electrodes by reacting with HSO_4^-.

Overall (cell) reaction (discharge):

$$PbO_2(s) + Pb(s) + 2H_2SO_4(aq) \longrightarrow 2PbSO_4(s) + 2H_2O(l) \qquad E_{cell} = 2.1 \text{ V}$$

When the cell is recharged, it uses electrical energy as an electrolytic cell, and the half-cell and overall reactions are reversed.

Overall (cell) reaction (recharge):

$$2PbSO_4(s) + 2H_2O(l) \longrightarrow PbO_2(s) + Pb(s) + 2H_2SO_4(aq)$$

For more than a century, car and truck owners have relied on the lead-acid battery to provide the large burst of current to the starter motor needed to start the engine—and to do so for years in both hot and cold weather. Nevertheless, there are problems with the lead-acid battery, mainly loss of capacity and safety concerns. Loss of capacity arises from several factors, including corrosion of the positive (Pb) grid, detachment of active material as a result of normal mechanical bumping, and the formation of large crystals of $PbSO_4$, which make recharging more difficult.

Most of the safety concerns have been remedied in modern batteries. Older batteries had a cap on each cell for monitoring electrolyte density and replacing water lost on overcharging. During recharging, some water could be electrolyzed to H_2 and O_2, which could explode if sparked, and splatter H_2SO_4. Modern batteries are sealed, so they don't require addition of water during normal operation, and they use flame attenuators to reduce the explosion hazard.

Nickel–Metal Hydride (Ni-MH) Battery Concerns about the toxicity of the nickel-cadmium (nicad) battery are leading to its replacement by the nickel–metal hydride battery. The anode half-reaction oxidizes the hydrogen absorbed within a metal alloy (designated M; e.g., $LaNi_5$) in a basic (KOH) electrolyte, while nickel(III) in the form of NiO(OH) is reduced at the cathode (Figure 21.17):

Anode (oxidation): $\quad MH(s) + OH^-(aq) \longrightarrow M(s) + H_2O(l) + e^-$

Cathode (reduction): $NiO(OH)(s) + H_2O(l) + e^- \longrightarrow Ni(OH)_2(s) + OH^-(aq)$

Overall (cell) reaction: $\quad MH(s) + NiO(OH)(s) \longrightarrow M(s) + Ni(OH)_2(s)$

$$E_{cell} = 1.4 \text{ V}$$

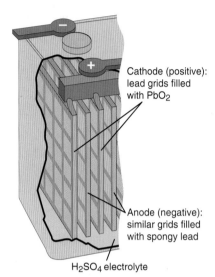

Cathode (positive): lead grids filled with PbO_2

Anode (negative): similar grids filled with spongy lead

H_2SO_4 electrolyte

Figure 21.16 Lead-acid battery.

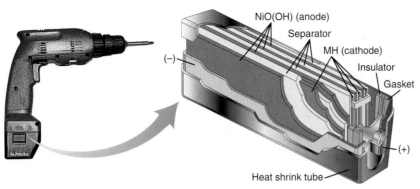

Figure 21.17 Nickel–metal hydride battery.

The cell reaction is reversed during recharging. The Ni-MH battery is common in cordless razors, camera flash units, and power tools. It is lightweight, has high power, and is nontoxic, but it may discharge excessively during long-term storage.

Lithium-Ion Battery The secondary lithium-ion battery has an anode of Li atoms that lie between sheets of graphite (designated Li_xC_6). The cathode is a lithium metal oxide, such as $LiMn_2O_4$ or $LiCoO_2$, and a typical electrolyte is $1\ M$ $LiPF_6$ in an organic solvent, such as a mixture of dimethyl carbonate and methylethyl carbonate. Electrons flow through the circuit, while solvated Li^+ ions flow from anode to cathode within the cell (Figure 21.18). The cell reactions are

Anode (oxidation): $\qquad\qquad\qquad Li_xC_6 \longrightarrow xLi^+ + xe^- + C_6(s)$

Cathode (reduction): $Li_{1-x}Mn_2O_4(s) + xLi^+ + xe^- \longrightarrow LiMn_2O_4(s)$

Overall (cell) reaction: $\qquad Li_xC_6 + Li_{1-x}Mn_2O_4(s) \longrightarrow LiMn_2O_4(s) + C_6(s)$

$$E_{cell} = 3.7\ V$$

The cell reaction is reversed during recharging. The lithium-ion battery powers countless laptop computers, cell phones, and camcorders. Its key drawbacks are expense and the flammability of the organic solvent.

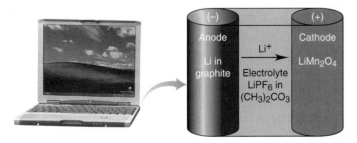

Figure 21.18 Lithium-ion battery.

Fuel Cells

In contrast to primary and secondary batteries, a **fuel cell,** sometimes called a *flow battery,* is not self-contained. The reactants (usually a combustible fuel and oxygen) enter the cell, and the products leave, generating electricity through the controlled oxidation of the fuel. In other words, *fuel cells use combustion to produce electricity.* The fuel does not burn because, as in other batteries, the half-reactions are separated, and the electrons are transferred through an external circuit.

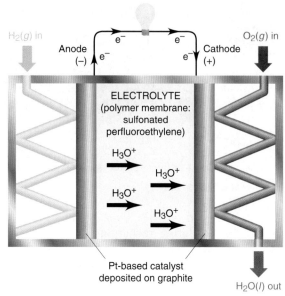

Figure 21.19 Hydrogen fuel cell.

The most common fuel cell being developed for use in cars is the *proton exchange membrane (PEM) cell,* which uses H_2 as the fuel and has an operating temperature of around 80°C (Figure 21.19). The cell reactions are

Anode (oxidation):	$2H_2(g) \longrightarrow 4H^+(aq) + 4e^-$
Cathode (reduction):	$O_2(g) + 4H^+(aq) + 4e^- \longrightarrow 2H_2O(g)$
Overall (cell) reaction:	$2H_2(g) + O_2(g) \longrightarrow 2H_2O(g) \qquad E_{cell} = 1.2 \text{ V}$

The reactions in fuel cells have much lower rates than those in other batteries, so they require an *electrocatalyst* to decrease the activation energy (Section 16.8). The PEM cell electrodes are composites consisting of nanoparticles of a Pt-based catalyst deposited on graphite. These are embedded in a polymer electrolyte membrane having a perfluoroethylene backbone ($-[F_2C-CF_2]_n-$) with attached sulfonic acid groups (RSO_3^-) that play a key role in ferrying protons from anode to cathode.

Hydrogen fuel cells have been used for years to provide electricity and pure water during space flights. In the very near future, similar ones will supply electric power for transportation, residential, and commercial needs. Already, every major car manufacturer has a fuel-cell prototype. By themselves, these cells produce no pollutants, and they convert about 75% of the fuel's bond energy into useable power, in contrast to 40% for a coal-fired power plant and 25% for a gasoline-powered car engine. Of course, their overall environmental impact will depend on how the H_2 is obtained; for example, electrolyzing water with solar power will have a positive impact, whereas electrolyzing it with electricity from a coal-fired plant will have a very negative one. Despite steady progress, current fuel-cell research remains focused on lowering costs by improving membrane conductivity and developing more efficient electrocatalysts.

SECTION SUMMARY

Batteries contain several voltaic cells in series and are classified as primary (e.g., alkaline, mercury, and silver), secondary (e.g., lead-acid, nickel–metal hydride, and lithium-ion), or fuel cell. Supplying electricity to a rechargeable (secondary) battery reverses the redox reaction, forming more reactant for further use. Fuel cells generate a current through the controlled oxidation of a fuel such as H_2.

21.6 CORROSION: A CASE OF ENVIRONMENTAL ELECTROCHEMISTRY

By now, you may be thinking that spontaneous electrochemical processes are always beneficial, but consider the problem of **corrosion,** the natural redox process that oxidizes metals to their oxides and sulfides. In chemical terms, corrosion is the reverse of isolating a metal from its oxide or sulfide ore; in electrochemical terms, the process shares many similarities with the operation of a voltaic cell. Damage from corrosion to cars, ships, buildings, and bridges runs into tens of billions of dollars annually, so it is a major problem in much of the world. We focus here on the corrosion of iron, but many other metals, such as copper and silver, also corrode.

The Corrosion of Iron

The most common and economically destructive form of corrosion is the rusting of iron. About 25% of the steel produced in the United States is made just to replace steel already in use that has corroded. Contrary to the simplified equation shown earlier in the text, rust is *not* a direct product of the reaction between iron and oxygen but arises through a complex electrochemical process. Let's look at the facts of iron corrosion and then use the features of a voltaic cell to explain them:

1. Iron does not rust in dry air: moisture must be present.
2. Iron does not rust in air-free water: oxygen must be present.
3. The loss of iron and the depositing of rust often occur at *different* places on the *same* object.
4. Iron rusts more quickly at low pH (high $[H^+]$).
5. Iron rusts more quickly in contact with ionic solutions.
6. Iron rusts more quickly in contact with a less active metal (such as Cu) and more slowly in contact with a more active metal (such as Zn).

Picture the magnified surface of a piece of iron or steel (Figure 21.20). Strains, ridges, and dents in contact with water are typically the sites of iron loss (fact 1). These sites are called *anodic regions* because the following half-reaction occurs there:

$$Fe(s) \longrightarrow Fe^{2+}(aq) + 2e^- \qquad \text{[anodic region; oxidation]}$$

Once the iron atoms lose electrons, the damage to the object has been done, and a pit forms where the iron is lost.

The freed electrons move through the external circuit—the piece of iron itself—until they reach a region of relatively high O_2 concentration (fact 2), near the surface of a surrounding water droplet, for instance. At this *cathodic region,* the electrons released from the iron atoms reduce O_2 molecules:

$$O_2(g) + 4H^+(aq) + 4e^- \longrightarrow 2H_2O(l) \qquad \text{[cathodic region; reduction]}$$

Figure 21.20 The corrosion of iron. A, A close-up view of an iron surface. Corrosion usually occurs at a surface irregularity. **B,** A schematic depiction of a small area of the surface, showing the steps in the corrosion process.

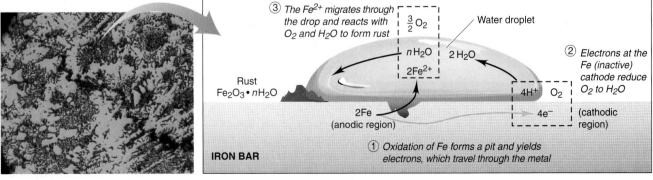

③ The Fe^{2+} migrates through the drop and reacts with O_2 and H_2O to form rust

$\frac{3}{2}O_2$

Water droplet

nH_2O · $2H_2O$

$2Fe^{2+}$

② Electrons at the Fe (inactive) cathode reduce O_2 to H_2O

Rust $Fe_2O_3 \cdot nH_2O$

$2Fe$ (anodic region)

$4H^+$ · O_2

$4e^-$ (cathodic region)

① Oxidation of Fe forms a pit and yields electrons, which travel through the metal

IRON BAR

A   **B**

Notice that this overall redox process is complete; thus, the iron loss has occurred without any rust forming:

$$2Fe(s) + O_2(g) + 4H^+(aq) \longrightarrow 2Fe^{2+}(aq) + 2H_2O(l)$$

Rust forms through another redox reaction in which the reactants make direct contact. The Fe^{2+} ions formed originally at the anodic region disperse through the surrounding water and react with O_2, often at some distance from the pit (fact 3). The overall reaction for this step is

$$2Fe^{2+}(aq) + \tfrac{1}{2}O_2(g) + (2+n)H_2O(l) \longrightarrow Fe_2O_3{\cdot}nH_2O(s) + 4H^+(aq)$$

[The inexact coefficient n for H_2O in the above equation appears because rust, $Fe_2O_3{\cdot}nH_2O$, is a form of iron(III) oxide with a variable number of waters of hydration.] The rust deposit is really incidental to the damage caused by loss of iron—a chemical insult added to the original injury.

Adding the previous two equations together shows the overall equation for the rusting of iron:

$$2Fe(s) + \tfrac{3}{2}O_2(g) + nH_2O(l) + \cancel{4H^+(aq)} \longrightarrow Fe_2O_3{\cdot}nH_2O(s) + \cancel{4H^+(aq)}$$

The canceled H^+ ions are shown to emphasize that they act as a catalyst; that is, they speed the process as they are used up in one step of the overall reaction and created in another. As a result of this action, rusting is faster at low pH (high $[H^+]$) (fact 4). Ionic solutions speed rusting by improving the conductivity of the aqueous medium near the anodic and cathodic regions (fact 5). The effect of ions is especially evident on ocean-going vessels and on the underbodies and around the wheel wells of cars driven in cold climates, where salts are used to melt ice on slippery roads.

The components of the corrosion process resemble those of a voltaic cell:

- Anodic and cathodic regions are separated in space.
- The regions are connected via an external circuit through which the electrons travel.
- In the anodic region, iron behaves like an active electrode, whereas in the cathodic region, it is inactive.
- The moisture surrounding the pit functions somewhat like a salt bridge, a means for ions to ferry back and forth and keep the solution neutral.

Protecting Against the Corrosion of Iron

A common approach to preventing or limiting corrosion is to eliminate contact with the corrosive factors. The simple act of washing off road salt removes the ionic solution from auto bodies. Iron objects are frequently painted to keep out O_2 and moisture, but if the paint layer chips, rusting proceeds. More permanent coatings include chromium plated on plumbing fixtures.

The only fact regarding corrosion that we have not yet addressed concerns the relative activity of other metals in contact with iron (fact 6), which leads to the most effective way to prevent corrosion. The essential idea is that *iron functions as both anode and cathode in the rusting process, but it is lost only at the anode.* Therefore, anything that makes iron behave more like the anode increases corrosion. As you can see in Figure 21.21A, when iron is in contact with a *less* active metal (weaker reducing agent), such as copper, its anodic function is enhanced. As a result, when iron plumbing is connected directly to copper plumbing, the iron pipe corrodes rapidly.

On the other hand, anything that makes iron behave more like the cathode prevents corrosion. In *cathodic protection,* iron makes contact with a *more* active metal (stronger reducing agent), such as zinc. The iron becomes cathodic and remains intact, while the zinc acts as the anode and loses electrons (Figure 21.21B). Coating steel with a "sacrificial" layer of zinc is the basis of the

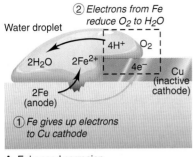

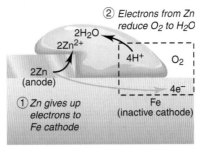

A Enhanced corrosion

B Cathodic protection

Figure 21.21 The effect of metal-metal contact on the corrosion of iron. A, When iron is in contact with a less active metal, such as copper, the iron loses electrons more readily (is more anodic), so it corrodes faster. **B,** When iron is in contact with a more active metal, such as zinc, the zinc acts as the anode and loses electrons. Therefore, the iron is cathodic, so it does not corrode. The process is known as *cathodic protection.*

galvanizing process. In addition to blocking physical contact with H_2O and O_2, the zinc is "sacrificed" (oxidized) instead of the iron.

Sacrificial anodes are employed to protect iron and steel structures (pipes, tanks, oil rigs, and so on) in marine and moist underground environments. The metals most frequently used for this purpose are magnesium and aluminum because they are much more active than iron. As a result, they act as the anode while iron acts as the cathode (Figure 21.22).

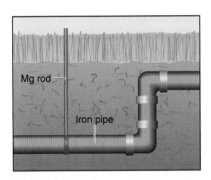

Figure 21.22 The use of sacrificial anodes to prevent iron corrosion. In cathodic protection, an active metal, such as magnesium or aluminum, is connected to underground iron pipes to prevent their corrosion. The active metal is sacrificed instead of the iron.

SECTION SUMMARY

Corrosion damages metal structures through a natural electrochemical change. Iron corrosion occurs in the presence of oxygen and moisture and is increased by high $[H^+]$, high [ion], or contact with a less active metal, such as Cu. Fe is oxidized and O_2 is reduced in one redox reaction, while rust (hydrated form of Fe_2O_3) is formed in another reaction that often takes place at a different location. Because Fe functions as both anode and cathode in the process, an iron or steel object can be protected by physically covering its surface or joining it to a more active metal (such as Zn, Mg, or Al), which acts as the anode in place of the Fe.

21.7 ELECTROLYTIC CELLS: USING ELECTRICAL ENERGY TO DRIVE NONSPONTANEOUS REACTIONS

Up to now, we've been considering voltaic cells, those that generate electrical energy from a spontaneous redox reaction. The principle of an electrolytic cell is exactly the opposite: *electrical energy from an external source drives a nonspontaneous reaction.*

Construction and Operation of an Electrolytic Cell

Let's examine the operation of an electrolytic cell by constructing one from a voltaic cell. Consider the tin-copper voltaic cell in Figure 21.23A. The Sn anode will gradually become oxidized to Sn^{2+} ions, and the Cu^{2+} ions will gradually be reduced and plate out on the Cu cathode because the cell reaction is spontaneous in that direction:

For the voltaic cell

$$Sn(s) \longrightarrow Sn^{2+}(aq) + 2e^- \qquad \text{[anode; oxidation]}$$
$$Cu^{2+}(aq) + 2e^- \longrightarrow Cu(s) \qquad \text{[cathode; reduction]}$$
$$Sn(s) + Cu^{2+}(aq) \longrightarrow Sn^{2+}(aq) + Cu(s) \qquad E°_{cell} = 0.48 \text{ V and } \Delta G° = -93 \text{ kJ}$$

Therefore, the *reverse* cell reaction is *non*spontaneous and never happens of its own accord, as the negative $E°_{cell}$ and positive $\Delta G°$ indicate:

$$Cu(s) + Sn^{2+}(aq) \longrightarrow Cu^{2+}(aq) + Sn(s) \qquad E°_{cell} = -0.48 \text{ V and } \Delta G° = 93 \text{ kJ}$$

However, we can make this process happen by supplying from an external source an electric potential *greater than* $E°_{cell}$. In effect, we have converted the voltaic cell into an electrolytic cell and changed the nature of the electrodes—anode is now cathode, and cathode is now anode (Figure 21.23B):

For the electrolytic cell

$$Cu(s) \longrightarrow Cu^{2+}(aq) + 2e^- \qquad \text{[anode; oxidation]}$$
$$Sn^{2+}(aq) + 2e^- \longrightarrow Sn(s) \qquad \text{[cathode; reduction]}$$
$$Cu(s) + Sn^{2+}(aq) \longrightarrow Cu^{2+}(aq) + Sn(s) \qquad \text{[overall (cell) reaction]}$$

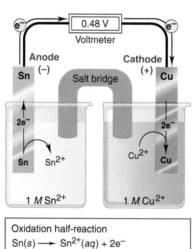
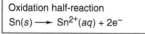

A Voltaic cell

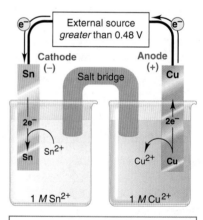
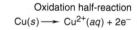

B Electrolytic cell

Figure 21.23 The tin-copper reaction as the basis of a voltaic and an electrolytic cell. A, At standard conditions, the spontaneous reaction between Sn and Cu^{2+} generates 0.48 V in a voltaic cell. **B,** If more than 0.48 V is supplied, the same apparatus and components become an electrolytic cell, and the nonspontaneous reaction between Cu and Sn^{2+} occurs. Note the changes in electrode charges and direction of electron flow.

Note that in an electrolytic cell, as in a voltaic cell, *oxidation takes place at the anode and reduction takes place at the cathode, but the direction of electron flow and the signs of the electrodes are reversed.*

To understand these changes, keep in mind the *cause* of the electron flow:

- In a voltaic cell, electrons are generated at the anode, so it is negative, and electrons are consumed at the cathode, so it is positive.
- In an electrolytic cell, the electrons come from the external power source, which *supplies* them *to* the cathode, so it is negative, and *removes* them *from* the anode, so it is positive.

Table 21.3 summarizes the processes and signs in the two types of electrochemical cells.

Table 21.3   Comparison of Voltaic and Electrolytic Cells

| Cell Type | ΔG | E_{cell} | Electrode | | |
			Name	**Process**	**Sign**
Voltaic	<0	>0	Anode	Oxidation	−
Voltaic	<0	>0	Cathode	Reduction	+
Electrolytic	>0	<0	Anode	Oxidation	+
Electrolytic	>0	<0	Cathode	Reduction	−

Predicting the Products of Electrolysis

Electrolysis, the splitting (lysing) of a substance by the input of electrical energy, is often used to decompose a compound into its elements. Electrolytic cells are involved in key industrial production steps for some of the most commercially important elements, including chlorine, copper, and aluminum. The first laboratory electrolysis of H_2O to H_2 and O_2 was performed in 1800, and the process is still used to produce these gases in ultrahigh purity. The electrolyte in an electrolytic cell can be the pure compound (such as H_2O or a molten salt), a mixture of molten salts, or an aqueous solution of a salt. The products obtained depend on several factors, so let's examine some actual cases.

Electrolysis of Molten Salts and the Industrial Production of Sodium   Many electrolytic applications involve isolating a metal or nonmetal from a molten binary ionic compound (salt). Predicting the product at each electrode is simple if the salt is pure because *the cation will be reduced and the anion oxidized.* The electrolyte is the molten salt itself, and the ions move through the cell because they are attracted by the oppositely charged electrodes.

Consider the electrolysis of molten (fused) calcium chloride. The two species present are Ca^{2+} and Cl^-, so Ca^{2+} ion is reduced and Cl^- ion is oxidized:

$$2Cl^-(l) \longrightarrow Cl_2(g) + 2e^- \qquad \text{[anode; oxidation]}$$
$$\underline{Ca^{2+}(l) + 2e^- \longrightarrow Ca(s) \qquad \text{[cathode; reduction]}}$$
$$Ca^{2+}(l) + 2Cl^-(l) \longrightarrow Ca(s) + Cl_2(g) \qquad \text{[overall]}$$

Metallic calcium is prepared industrially this way, as are several other active metals as well as the halogens Cl_2 and Br_2.

Another important application is the industrial production of sodium, which involves electrolysis of molten NaCl. The sodium ore is *halite* (largely NaCl), which is obtained either by evaporation of concentrated salt solutions (brines) or by mining vast salt deposits formed from the evaporation of prehistoric seas.

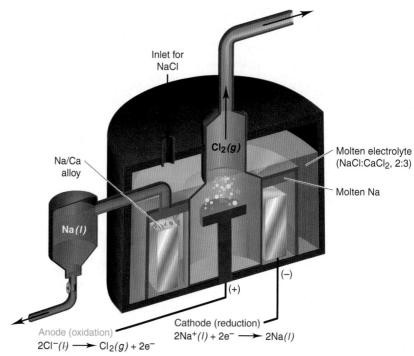

Inlet for
NaCl

Na/Ca
alloy

$Cl_2(g)$

Molten electrolyte
(NaCl:CaCl$_2$, 2:3)

Molten Na

Na(l)

(+)

(−)

Anode (oxidation)
$2Cl^-(l) \longrightarrow Cl_2(g) + 2e^-$

Cathode (reduction)
$2Na^+(l) + 2e^- \longrightarrow 2Na(l)$

Figure 21.24 The Downs cell for production of sodium. The mixture of solid NaCl and CaCl$_2$ forms the molten electrolyte. Sodium and calcium are formed at the cathode and float, but an Na/Ca alloy solidifies and falls back into the bath while liquid Na is separated. Chlorine gas forms at the anode.

Oxidation half-reaction
$2H_2O(l) \longrightarrow O_2(g) + 4H^+(aq) + 4e^-$

Reduction half-reaction
$2H_2O(l) + 2e^- \longrightarrow H_2(g) + 2OH^-(aq)$

Overall (cell) reaction
$2H_2O(l) \longrightarrow 2H_2(g) + O_2(g)$

Figure 21.25 The electrolysis of water. A certain volume of oxygen forms through the oxidation of H$_2$O at the anode (right), and twice that volume of hydrogen forms through the reduction of H$_2$O at the cathode (left).

The dry solid is crushed and fused (melted) in an electrolytic apparatus called the **Downs cell** (Figure 21.24). To reduce heating costs, the NaCl (mp = 801°C) is mixed with $1\frac{1}{2}$ parts CaCl$_2$ to form a mixture that melts at only 580°C. Reduction of the metal ions to Na and Ca takes place at a cylindrical steel cathode, with the molten metals floating on the denser molten salt mixture. As they rise through a short collecting pipe, the liquid Na is siphoned off, while a higher melting Na/Ca alloy solidifies and falls back into the molten electrolyte. Chloride ions are oxidized to Cl$_2$ gas at a large anode within an inverted cone-shaped chamber. The cell design separates the metals from the Cl$_2$ to prevent their explosive recombination. The Cl$_2$ gas is collected, purified, and sold as a valuable by-product.

Electrolysis of Water and Nonstandard Half-Cell Potentials Before we can analyze the electrolysis products of aqueous salt solutions, we must examine the electrolysis of water itself. Extremely pure water is difficult to electrolyze because very few ions are present to conduct a current. If we add a small amount of a salt that cannot be electrolyzed in water (such as Na$_2$SO$_4$), however, electrolysis proceeds rapidly. A glass electrolytic cell with separated gas compartments is used to keep the H$_2$ and O$_2$ gases from mixing (Figure 21.25). At the anode, water is oxidized as the O.N. of O changes from −2 to 0:

$$2H_2O(l) \longrightarrow O_2(g) + 4H^+(aq) + 4e^- \qquad E = 0.82 \text{ V} \qquad \text{[anode; oxidation]}$$

At the cathode, water is reduced as the O.N. of H changes from +1 to 0:

$$2H_2O(l) + 2e^- \longrightarrow H_2(g) + 2OH^-(aq) \qquad E = -0.42 \text{ V} \qquad \text{[cathode; reduction]}$$

Adding the half-reactions (which involves combining the H$^+$ and OH$^-$ into H$_2$O and canceling e$^-$ and excess H$_2$O), and calculating E_{cell}, the overall reaction is

$$2H_2O(l) \longrightarrow 2H_2(g) + O_2(g) \qquad E_{cell} = -0.42 \text{ V} - 0.82 \text{ V} = -1.24 \text{ V} \qquad \text{[overall]}$$

Notice that these electrode potentials are not written with a degree sign because they are *not* standard electrode potentials. The [H$^+$] and [OH$^-$] are 1.0×10^{-7} M rather than the standard-state value of 1 M. These E values are obtained by applying the Nernst equation. For example, the calculation for the anode potential (with $n = 4$) is

$$E_{cell} = E°_{cell} - \frac{0.0592\ V}{4} \log (P_{O_2} \times [H^+]^4)$$

The standard potential for the *oxidation* of water is -1.23 V (from Appendix D) and $P_{O_2} \approx 1$ atm in the half-cell, so we have

$$E_{cell} = -1.23\ V - \left\{ \frac{0.0592\ V}{4} \times [\log 1 + 4 \log (1.0 \times 10^{-7})] \right\} = -0.82\ V$$

In aqueous ionic solutions, $[H^+]$ and $[OH^-]$ are approximately 10^{-7} M also, so we use these nonstandard E_{cell} values to predict electrode products.

Electrolysis of Aqueous Salt Solutions; Overvoltage and the Chlor-Alkali Process

Aqueous salt solutions are mixtures of ions *and* water, so we compare electrode potentials to predict the products. When two half-reactions are possible,

- *The reduction with the less negative (more positive) electrode potential occurs.*
- *The oxidation with the less positive (more negative) electrode potential occurs.*

What happens, for instance, when a solution of KI is electrolyzed? The possible oxidizing agents are K^+ and H_2O, and their reduction half-reactions are

$$K^+(aq) + e^- \longrightarrow K(s) \qquad\qquad E° = -2.93\ V$$
$$2H_2O(l) + 2e^- \longrightarrow H_2(g) + 2OH^-(aq) \qquad E = -0.42\ V \qquad \text{[reduction]}$$

The less *negative* electrode potential for water means that it is much easier to reduce than K^+, so H_2 forms at the cathode. The possible reducing agents are I^- and H_2O, and their oxidation half-reactions are

$$2I^-(aq) \longrightarrow I_2(s) + 2e^- \qquad\qquad E° = 0.53\ V \qquad \text{[oxidation]}$$
$$2H_2O(l) \longrightarrow O_2(g) + 4H^+(aq) + 4e^- \qquad E = 0.82\ V$$

The less *positive* electrode potential for I^- means that a lower potential is needed to oxidize it than to oxidize H_2O, and so I_2 forms at the anode.

Unfortunately, the products predicted from electrode potentials are not always the products that form. For gases such as $H_2(g)$ and $O_2(g)$ to form at metal electrodes, an additional voltage is required. This increment above the expected required voltage is called the **overvoltage.** The phenomenon of overvoltage has major practical significance in the **chlor-alkali process** for the industrial production of chlorine and several other chemicals, which is based on the *electrolytic oxidation of Cl⁻ ion from concentrated aqueous NaCl solutions.* Chlorine ranks among the top 10 chemicals produced in the United States.

Because of overvoltage, electrolysis of NaCl solutions does not yield both of the component elements. Water is easier to reduce than Na^+, so H_2 forms at the cathode, even *with* an overvoltage of 0.6 V:

$$Na^+(aq) + e^- \longrightarrow Na(s) \qquad\qquad E° = -2.71\ V$$
$$2H_2O(l) + 2e^- \longrightarrow H_2(g) + 2OH^-(aq) \qquad E = -0.42\ V\ (\approx -1\ V\ \text{with overvoltage})$$
$$\text{[reduction]}$$

But Cl_2 *does* form at the anode, even though a comparison of electrode potentials would lead us to predict that O_2 should form:

$$2H_2O(l) \longrightarrow O_2(g) + 4H^+(aq) + 4e^- \qquad E = 0.82\ V \quad (\sim 1.4\ V\ \text{with overvoltage})$$
$$2Cl^-(aq) \longrightarrow Cl_2(g) + 2e^- \qquad\qquad E° = 1.36\ V \qquad \text{[oxidation]}$$

An overvoltage of ~ 0.6 V makes the potential for forming O_2 slightly above that for Cl_2. Therefore, the half-reactions for electrolysis of aqueous NaCl are

$$2Cl^-(aq) \longrightarrow Cl_2(g) + 2e^- \qquad\qquad E° = 1.36\ V \quad \text{[anode; oxidation]}$$
$$\underline{2H_2O(l) + 2e^- \longrightarrow 2OH^-(aq) + H_2(g) \qquad\qquad E = -1.0\ V \ \text{[cathode; reduction]}}$$
$$2Cl^-(aq) + 2H_2O(l) \longrightarrow 2OH^-(aq) + H_2(g) + Cl_2(g) \qquad E_{cell} = -1.0\ V - 1.36\ V = -2.4\ V$$

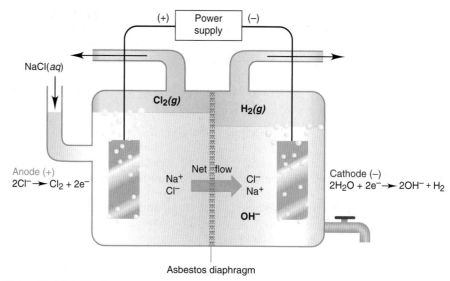

Figure 21.26 A diaphragm cell for the chlor-alkali process. This process uses concentrated aqueous NaCl to make NaOH, Cl_2, and H_2 in an electrolytic cell. The difference in liquid level between compartments keeps a net movement of solution into the cathode compartment, which prevents reaction between OH^- and Cl_2. The cathode electrolyte is concentrated and fractionally crystallized to give industrial-grade NaOH.

To obtain commercial amounts of Cl_2, a voltage almost twice this value and a current in excess of 3×10^4 A are used.

When we include the spectator ion Na^+, the total ionic equation shows another important product:

$$2Na^+(aq) + 2Cl^-(aq) + 2H_2O(l) \longrightarrow 2Na^+(aq) + 2OH^-(aq) + H_2(g) + Cl_2(g)$$

As Figure 21.26 shows, the sodium salts in the cathode compartment exist as an aqueous mixture of NaCl and NaOH; the NaCl is removed by fractional crystallization, which separates the compounds by differences in solubility. Thus, in this version of the chlor-alkali process, which uses an *asbestos diaphragm* to separate the anode and cathode compartments, the products are Cl_2, H_2, and industrial-grade NaOH, an important base.

Like other reactive products, H_2 and Cl_2 are kept apart to prevent explosive recombination. Note the higher liquid level in the anode compartment. This slight hydrostatic pressure difference minimizes backflow of NaOH, which prevents the disproportionation (self–oxidation-reduction) reactions of Cl_2 that occur in the presence of OH^-, such as

$$Cl_2(g) + 2OH^-(aq) \longrightarrow Cl^-(aq) + ClO^-(aq) + H_2O(l)$$

A newer chlor-alkali *membrane-cell* process, in which the diaphragm is replaced by a polymeric membrane to separate the cell compartments, has been adopted in much of the industrialized world. The membrane allows only cations to move through it and only from anode to cathode compartments. Thus, as Cl^- ions are removed at the anode through oxidation to Cl_2, Na^+ ions in the anode compartment move through the membrane to the cathode compartment and form an NaOH solution. In addition to forming purer NaOH than the older diaphragm-cell method, the membrane-cell process uses less electricity.

Based on many studies, we can determine which elements can be prepared electrolytically from aqueous solutions of their salts:

1. Cations of less active metals *are* reduced to the metal, including gold, silver, copper, chromium, platinum, and cadmium.

2. Cations of more active metals *are not* reduced, including those in Groups 1A(1) and 2A(2), and Al from 3A(13). Water is reduced to H_2 and OH^- instead.

3. Anions that *are* oxidized, because of overvoltage from O_2 formation, include the halides ($[Cl^-]$ must be high), except for F^-.

4. Anions that *are not* oxidized include F^- and common oxoanions, such as SO_4^{2-}, CO_3^{2-}, NO_3^-, and PO_4^{3-}, because the central nonmetal in these oxoanions is already in its highest oxidation state. Water is oxidized to O_2 and H^+ instead.

SAMPLE PROBLEM 21.8   Predicting the Electrolysis Products of Aqueous Ionic Solutions

Problem What products form during electrolysis of aqueous solutions of the following salts: **(a)** KBr; **(b)** $AgNO_3$; **(c)** $MgSO_4$?

Plan We identify the reacting ions and compare their electrode potentials with those of water, taking the 0.4 to 0.6 V overvoltage into consideration. The reduction half-reaction with the less negative electrode potential occurs at the cathode, and the oxidation half-reaction with the less positive electrode potential occurs at the anode.

Solution

(a)
$$K^+(aq) + e^- \longrightarrow K(s) \qquad E° = -2.93 \text{ V}$$
$$2H_2O(l) + 2e^- \longrightarrow H_2(g) + 2OH^-(aq) \qquad E = -0.42 \text{ V}$$

Despite the overvoltage, which makes E for the reduction of water between -0.8 and -1.0 V, H_2O is still easier to reduce than K^+, so $H_2(g)$ forms at the cathode.

$$2Br^-(aq) \longrightarrow Br_2(l) + 2e^- \qquad E° = 1.07 \text{ V}$$
$$2H_2O(l) \longrightarrow O_2(g) + 4H^+(aq) + 4e^- \qquad E = 0.82 \text{ V}$$

Because of the overvoltage, which makes E for the oxidation of water between 1.2 and 1.4 V, Br^- is easier to oxidize than water, so $Br_2(l)$ forms at the anode (see photo).

(b)
$$Ag^+(aq) + e^- \longrightarrow Ag(s) \qquad E° = 0.80 \text{ V}$$
$$2H_2O(l) + 2e^- \longrightarrow H_2(g) + 2OH^-(aq) \qquad E = -0.42 \text{ V}$$

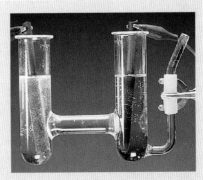
Electrolysis of aqueous KBr.

As the cation of an inactive metal, Ag^+ is a better oxidizing agent than H_2O, so Ag forms at the cathode. NO_3^- cannot be oxidized, because N is already in its highest $(+5)$ oxidation state. Thus, O_2 forms at the anode:

$$2H_2O(l) \longrightarrow O_2(g) + 4H^+(aq) + 4e^-$$

(c)
$$Mg^{2+}(aq) + 2e^- \longrightarrow Mg(s) \qquad E° = -2.37 \text{ V}$$

Like K^+ in part (a), Mg^{2+} cannot be reduced in the presence of water, so H_2 forms at the cathode. The SO_4^{2-} ion cannot be oxidized because S is in its highest $(+6)$ oxidation state. Thus, H_2O is oxidized, and O_2 forms at the anode:

$$2H_2O(l) \longrightarrow O_2(g) + 4H^+(aq) + 4e^-$$

FOLLOW-UP PROBLEM 21.8 Write half-reactions showing the products you predict will form in the electrolysis of aqueous $AuBr_3$.

Industrial Electrochemistry: Purifying Copper and Isolating Aluminum

In addition to the Downs cell for sodium production and the chlor-alkali process for chlorine manufacture, industrial methods based on voltaic and electrolytic cells are used commonly to obtain metals and nonmetals from their ores or to purify them for later use. Here we focus on two key electrochemical processes.

The Electrorefining of Copper The most common copper ore is chalcopyrite, $CuFeS_2$, a mixed sulfide of FeS and CuS. Most remaining deposits contain less than 0.5% Cu by mass. To "win" this small amount of copper from the ore

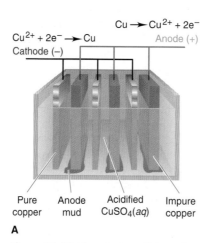

Figure 21.27 The electrorefining of copper. A, Copper is refined electrolytically, using impure slabs of copper as anodes and sheets of pure copper as cathodes. The Cu^{2+} ions released from the anode are reduced to Cu metal and plate out at the cathode. The "anode mud" contains valuable metal by-products. **B,** A small section of an industrial facility for electrorefining copper.

requires several steps, including a final refining to achieve the purity needed for electrical wiring, copper's most important application. More than 2.5 billion pounds of copper is produced in the United States annually.

After removing the iron(II) sulfide and reducing the copper(II) sulfide, the copper obtained is usable for plumbing, but it must be purified for electrical applications by removing unwanted impurities (Fe and Ni) as well as valuable ones (Ag, Au, and Pt). Purification is accomplished by *electrorefining,* which involves the oxidation of Cu to form Cu^{2+} ions in solution, followed by their reduction and the plating out of Cu metal (Figure 21.27). To do this, impure copper is cast into plates to be used as anodes, and cathodes are made from already purified copper. The electrodes are immersed in acidified $CuSO_4$ solution, and a controlled voltage is applied that accomplishes two tasks simultaneously:

1. Copper and the more active impurities (Fe, Ni) are oxidized to their cations, while the less active ones (Ag, Au, Pt) are not. As the anode slabs react, these unoxidized metals fall off as a valuable "anode mud" and are purified separately. Sale of the precious metals in the anode mud nearly offsets the cost of electricity to operate the cell, making Cu wire inexpensive.
2. Because Cu is much less active than the Fe and Ni impurities, Cu^{2+} ions are reduced at the cathode, but Fe^{2+} and Ni^{2+} ions remain in solution:

$$Cu^{2+}(aq) + 2e^- \longrightarrow Cu(s) \qquad E^\circ = 0.34 \text{ V}$$
$$Ni^{2+}(aq) + 2e^- \longrightarrow Ni(s) \qquad E^\circ = -0.25 \text{ V}$$
$$Fe^{2+}(aq) + 2e^- \longrightarrow Fe(s) \qquad E^\circ = -0.44 \text{ V}$$

The copper obtained by electrorefining is over 99.99% pure.

The Isolation of Aluminum Aluminum, the most abundant metal in Earth's crust by mass, is found in numerous aluminosilicate minerals. Through eons of weathering, certain of these became *bauxite,* a mixed oxide-hydroxide that is the major ore of aluminum. In general terms, the isolation of aluminum is a two-step process that combines several physical and chemical separations. In the first, the mineral oxide, Al_2O_3, is separated from bauxite; in the second, which we focus on here, the oxide is converted to the metal.

Aluminum is much too strong a reducing agent to be formed at the cathode from aqueous solution, so the oxide itself must be electrolyzed. However, the melting point of Al_2O_3 is very high (2030°C), so major energy (and cost) savings are realized by dissolving the oxide in molten *cryolite* (Na_3AlF_6) to give a mixture that is electrolyzed at ~1000°C. The electrolytic step, called the *Hall-Heroult process,* takes place in a graphite-lined furnace, with the lining itself acting as the cathode. Anodes of graphite dip into the molten Al_2O_3-Na_3AlF_6 mixture (Figure 21.28). The cell typically operates at a moderate voltage of 4.5 V, but with an enormous current flow of 1.0×10^5 to 2.5×10^5 A.

Because the process is complex and still not entirely known, the following reactions are chosen from among several other possibilities. Molten cryolite contains several ions (including AlF_6^{3-}, AlF_4^-, and F^-), which react with Al_2O_3 to form fluoro-oxy ions (including $AlOF_3^{2-}$, $Al_2OF_6^{2-}$, and $Al_2O_2F_4^{2-}$) that dissolve in the mixture. For example,

$$2Al_2O_3(s) + 2AlF_6^{3-}(l) \longrightarrow 3Al_2O_2F_4^{2-}(l)$$

Al forms at the cathode (reduction), shown here with AlF_6^{3-} as reactant:

$$AlF_6^{3-}(l) + 3e^- \longrightarrow Al(l) + 6F^-(l) \qquad \text{[cathode; reduction]}$$

The graphite anodes are oxidized and form carbon dioxide gas. Using one of the fluoro-oxy species as an example, the anode reaction is

$$Al_2O_2F_4^{2-}(l) + 8F^-(l) + C(graphite) \longrightarrow 2AlF_6^{3-}(l) + CO_2(g) + 4e^-$$

$$\text{[anode; oxidation]}$$

The anodes are consumed in this half-reaction and must be replaced frequently.

Combining the three previous equations gives the overall reaction:

$$2Al_2O_3(\text{in } Na_3AlF_6) + 3C(graphite) \longrightarrow 4Al(l) + 3CO_2(g) \qquad \text{[overall (cell) reaction]}$$

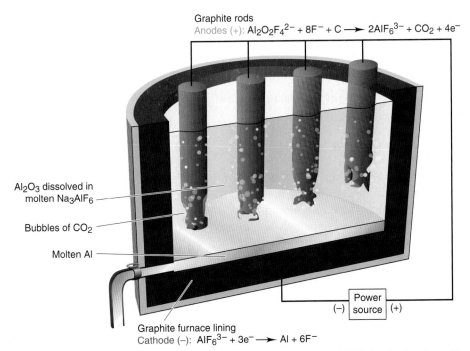

Figure 21.28 The electrolytic cell in aluminum manufacture. Purified Al_2O_3 is mixed with cryolite (Na_3AlF_6) and melted. Reduction at the graphite furnace lining (cathode) gives molten Al. Oxidation at the graphite rods (anodes) slowly converts them to CO_2, so they must be replaced periodically.

Aluminum production accounts for more than 5% of total U.S. electrical usage! Recent estimates for the entire manufacturing process (including mining, maintaining operating conditions, and so forth) show that aluminum recycling requires less than 1% as much energy as manufacturing it from the ore, which explains why recycling has become so widespread.

The Stoichiometry of Electrolysis: The Relation Between Amounts of Charge and Product

As you've seen, the charge flowing through an electrolytic cell yields products at the electrodes. In the electrolysis of molten NaCl, for example, the power source supplies electrons to the cathode, where Na^+ ions migrate to pick them up and become Na metal. At the same time, the power source pulls from the anode the electrons that Cl^- ions release as they become Cl_2 gas. It follows that the more electrons picked up by Na^+ ions and released by Cl^- ions, the greater the amounts of Na and Cl_2 that form. This relationship was first determined experimentally by Michael Faraday and is referred to as *Faraday's law of electrolysis: the amount of substance produced at each electrode is directly proportional to the quantity of charge flowing through the cell.*

Each balanced half-reaction shows the amounts (mol) of reactant, electrons, and product involved in the change, so it contains the information we need to answer such questions as "How much material will form as a result of a given quantity of charge?" or, conversely, "How much charge is needed to produce a given amount of material?" To apply Faraday's law,

1. Balance the half-reaction to find the number of moles of electrons needed per mole of product.
2. Use the Faraday constant ($F = 9.65 \times 10^4$ C/mol e^-) to find the corresponding charge.
3. Use the molar mass to find the charge needed for a given mass of product.

In practice, to supply the correct amount of electricity, we need some means of finding the charge flowing through the cell. We cannot measure charge directly, but we *can* measure current, the charge flowing per unit time. The SI unit of current is the **ampere (A),** which is defined as a charge of 1 coulomb flowing through a conductor in 1 second:

$$1 \text{ ampere} = 1 \text{ coulomb/second} \qquad \text{or} \qquad 1 \text{ A} = 1 \text{ C/s} \qquad \textbf{(21.11)}$$

Thus, the current multiplied by the time gives the charge:

$$\text{Current} \times \text{time} = \text{charge} \qquad \text{or} \qquad A \times s = \frac{C}{s} \times s = C$$

Therefore, we find the charge by measuring the current *and* the time during which the current flows. This, in turn, relates to the amount of product formed. Figure 21.29 summarizes these relationships.

Problems based on Faraday's law often ask you to calculate current, mass of material, or time. The electrode half-reaction provides the key to solving these problems because it is related to the mass for a certain quantity of charge.

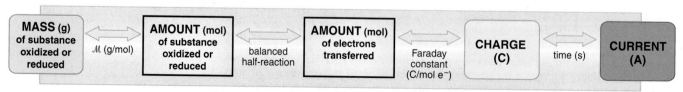

Figure 21.29 A summary diagram for the stoichiometry of electrolysis.

As an example, let's consider a typical problem in practical electrolysis: how long does it take to produce 3.0 g of $Cl_2(g)$ by electrolysis of aqueous NaCl using a power supply with a current of 12 A? The problem asks for the time needed to produce a certain mass, so let's first relate mass to number of moles of electrons to find the charge needed. Then, we'll relate the charge to the current to find the time.

We know the mass of Cl_2 produced, so we can find the amount (mol) of Cl_2. The half-reaction tells us that the loss of 2 mol of electrons produces 1 mol of chlorine gas:

$$2Cl^-(aq) \longrightarrow Cl_2(g) + 2e^-$$

We use this relationship as a conversion factor, and multiplying by the Faraday constant gives us the total charge:

$$\text{Charge (C)} = 3.0 \text{ g } Cl_2 \times \frac{1 \text{ mol } Cl_2}{70.90 \text{ g } Cl_2} \times \frac{2 \text{ mol } e^-}{1 \text{ mol } Cl_2} \times \frac{9.65 \times 10^4 \text{ C}}{1 \text{ mol } e^-} = 8.2 \times 10^3 \text{ C}$$

Now we use the relationship between charge and current to find the time needed:

$$\text{Time (s)} = \frac{\text{charge (C)}}{\text{current (A, or C/s)}} = 8.2 \times 10^3 \text{ C} \times \frac{1 \text{ s}}{12 \text{ C}} = 6.8 \times 10^2 \text{ s } (\sim 11 \text{ min})$$

Note that the entire calculation follows Figure 21.29 until the last step:

$$\text{grams of } Cl_2 \Rightarrow \text{moles of } Cl_2 \Rightarrow \text{moles of } e^- \Rightarrow \text{coulombs} \Rightarrow \text{seconds}$$

Sample Problem 21.9 demonstrates the steps as they appear in Figure 21.29.

SAMPLE PROBLEM 21.9 Applying the Relationship Among Current, Time, and Amount of Substance

Problem A technician is plating a faucet with 0.86 g of chromium from an electrolytic bath containing aqueous $Cr_2(SO_4)_3$. If 12.5 min is allowed for the plating, what current is needed?

Plan To find the current, we divide the charge by the time; therefore, we need to find the charge. First we write the half-reaction for Cr^{3+} reduction. From it, we know the number of moles of e^- required per mole of Cr. As the roadmap shows, to find the charge, we convert the mass of Cr needed (0.86 g) to amount (mol) of Cr. The balanced half-reaction gives the amount (mol) of e^- transferred. Then, we use the Faraday constant (9.65×10^4 C/mol e^-) to find the charge and divide by the time (12.5 min, converted to seconds) to obtain the current.

Solution Writing the balanced half-reaction:

$$Cr^{3+}(aq) + 3e^- \longrightarrow Cr(s)$$

Combining steps to find amount (mol) of e^- transferred for mass of Cr needed:

$$\text{Moles of } e^- \text{ transferred} = 0.86 \text{ g Cr} \times \frac{1 \text{ mol Cr}}{52.00 \text{ g Cr}} \times \frac{3 \text{ mol } e^-}{1 \text{ mol Cr}} = 0.050 \text{ mol } e^-$$

Calculating the charge:

$$\text{Charge (C)} = 0.050 \text{ mol } e^- \times \frac{9.65 \times 10^4 \text{ C}}{1 \text{ mol } e^-} = 4.8 \times 10^3 \text{ C}$$

Calculating the current:

$$\text{Current (A)} = \frac{\text{charge (C)}}{\text{time (s)}} = \frac{4.8 \times 10^3 \text{ C}}{12.5 \text{ min}} \times \frac{1 \text{ min}}{60 \text{ s}} = 6.4 \text{ C/s} = 6.4 \text{ A}$$

Check Rounding gives

$$(\sim 0.9 \text{ g})(1 \text{ mol Cr/50 g})(3 \text{ mol } e^-/1 \text{ mol Cr}) = 5 \times 10^{-2} \text{ mol } e^-$$

then

$$(5 \times 10^{-2} \text{ mol } e^-)(\sim 1 \times 10^5 \text{ C/mol } e^-) = 5 \times 10^3 \text{ C}$$

and

$$(5 \times 10^3 \text{ C/12 min})(1 \text{ min/60 s}) = 7 \text{ A}$$

Mass (g) of Cr needed

divide by $\mathcal{M}$ (g/mol)

Amount (mol) of Cr needed

3 mol e^- = 1 mol Cr

Amount (mol) of e^- transferred

1 mol e^- = 9.65×10^4 C

Charge (C)

divide by time (convert min to s)

Current (A)

Comment For the sake of introducing Faraday's law, the details of the electroplating process have been simplified here. Actually, electroplating chromium is only 30% to 40% efficient and must be run at a particular temperature range for the plate to appear bright. Nearly 10,000 metric tons (2×10^8 mol) of chromium is used annually for electroplating.

FOLLOW-UP PROBLEM 21.9 Using a current of 4.75 A, how many minutes does it take to plate onto a sculpture 1.50 g of Cu from a $CuSO_4$ solution?

SECTION SUMMARY

An electrolytic cell uses electrical energy to drive a nonspontaneous reaction. Oxidation occurs at the anode and reduction at the cathode, but the direction of electron flow and the charges of the electrodes are opposite those in voltaic cells. When two products can form at each electrode, the more easily oxidized substance reacts at the anode and the more easily reduced at the cathode. The reduction or oxidation of water takes place at nonstandard conditions. Overvoltage causes the actual voltage to be unexpectedly high and can affect the electrode product that forms. The industrial production of many elements, such as sodium, chlorine, copper, and aluminum, utilizes electrolytic cells. The amount of product that forms depends on the quantity of charge flowing through the cell, which is related to the magnitude of the current and the time it flows.

For Review and Reference (Numbers in parentheses refer to pages, unless noted otherwise.)

Learning Objectives

To help you review these learning objectives, the numbers of related sections (§), sample problems (SP), and upcoming end-of-chapter problems (EP) are listed in parentheses.

1. Understand the meanings of *oxidation, reduction, oxidizing agent,* and *reducing agent;* balance redox reactions by the half-reaction method; distinguish between voltaic and electrolytic cells in terms of the sign of ΔG (§ 21.1) (SP 21.1) (EPs 21.1–21.12)
2. Describe the physical makeup of a voltaic cell, and explain the direction of electron flow; draw a diagram and write the notation for a voltaic cell (§ 21.2) (SP 21.2) (EPs 21.13–21.23)
3. Describe how standard electrode potentials ($E^\circ_{\text{half-cell}}$ values) are combined to give E°_{cell} and how the standard reference electrode is used to find an unknown $E^\circ_{\text{half-cell}}$; explain how the reactivity of a metal is related to its $E^\circ_{\text{half-cell}}$; write spontaneous redox reactions using an emf series like that in Appendix D (§ 21.3) (SPs 21.3, 21.4) (EPs 21.24–21.40)
4. Understand how E_{cell} is related to ΔG and the charge flowing through the cell; use the interrelationship of ΔG°, E°_{cell}, and K to calculate any one of these variables; explain how E_{cell} changes as

the cell operates (Q changes), and use the Nernst equation to find E_{cell}; describe how a concentration cell works and calculate its E_{cell} (§ 21.4) (SPs 21.5–21.7) (EPs 21.41–21.56)
5. Understand how a battery operates, and describe the components of primary and secondary batteries and fuel cells (§ 21.5) (EPs 21.57–21.59)
6. Explain how corrosion occurs and is prevented (§ 21.6) (EPs 21.60–21.62)
7. Understand the basis of an electrolytic cell; describe the Downs cell for the production of Na, the chlor-alkali process and the importance of overvoltage for the production of Cl_2, the electrorefining of Cu, and the use of cryolite in the production of Al; know how water influences the products at the electrodes during electrolysis of aqueous salt solutions (§ 21.7) (SP 21.8) (EPs 21.63–21.75, 21.82)
8. Understand the relationship between charge and amount of product, and calculate the current (or time) needed to produce a given amount of product or vice versa (§ 21.7) (SP 21.9) (EPs 21.76–21.81, 21.83, 21.84)

Key Terms

electrochemistry (682)
electrochemical cell (682)

Section 21.1
half-reaction method (683)
voltaic (galvanic) cell (686)
electrolytic cell (686)
electrode (686)
electrolyte (686)
anode (686)
cathode (686)

Section 21.2
half-cell (688)
salt bridge (689)

Section 21.3
cell potential (E_{cell}) (692)
voltage (692)
electromotive force (emf) (692)
volt (V) (692)
coulomb (C) (692)

standard cell potential (E°_{cell}) (692)
standard electrode (half-cell) potential ($E^\circ_{\text{half-cell}}$) (693)
standard reference half-cell (standard hydrogen electrode) (693)

Section 21.4
Faraday constant (F) (701)
Nernst equation (703)
concentration cell (705)

Section 21.5
battery (708)
fuel cell (711)

Section 21.6
corrosion (713)

Section 21.7
electrolysis (717)
Downs cell (718)
overvoltage (719)
chlor-alkali process (719)
ampere (A) (724)

Key Equations and Relationships

21.1 Relating a spontaneous process to the sign of the cell potential (692):

$$E_{cell} > 0 \text{ for a spontaneous process}$$

21.2 Relating electric potential to energy and charge in SI units (692):

$$\text{Potential} = \text{energy/charge} \quad \text{or} \quad 1 \text{ V} = 1 \text{ J/C}$$

21.3 Relating standard cell potential to standard electrode potentials in a voltaic cell (693):

$$E°_{cell} = E°_{cathode \, (reduction)} - E°_{anode \, (oxidation)}$$

21.4 Defining the Faraday constant (701):

$$F = 9.65 \times 10^4 \frac{J}{V \cdot mol \, e^-} \quad (3 \text{ sf})$$

21.5 Relating the free energy change to electrical work and cell potential (701):

$$\Delta G = -w_{max} = -nFE_{cell}$$

21.6 Finding the standard free energy change from the standard cell potential (701):

$$\Delta G° = -nFE°_{cell}$$

21.7 Finding the equilibrium constant from the standard cell potential (701):

$$E°_{cell} = \frac{RT}{nF} \ln K$$

21.8 Substituting known values of R, F, and T into Equation 21.7 and converting to common logarithms (702):

$$E°_{cell} = \frac{0.0592 \text{ V}}{n} \log K \quad \text{or} \quad \log K = \frac{nE°_{cell}}{0.0592 \text{ V}} \quad (\text{at } 298.15 \text{ K})$$

21.9 Calculating the nonstandard cell potential (Nernst equation) (703):

$$E_{cell} = E°_{cell} - \frac{RT}{nF} \ln Q$$

21.10 Substituting known values of R, F, and T into the Nernst equation and converting to common logarithms (704):

$$E_{cell} = E°_{cell} - \frac{0.0592 \text{ V}}{n} \log Q \quad (\text{at } 298.15 \text{ K})$$

21.11 Relating current to charge and time (724):

$$\text{Current} = \text{charge/time} \quad \text{or} \quad 1 \text{ A} = 1 \text{ C/s}$$

Brief Solutions to Follow-up Problems

21.1 $6KMnO_4(aq) + 6KOH(aq) + KI(aq) \longrightarrow$
$\qquad\qquad 6K_2MnO_4(aq) + KIO_3(aq) + 3H_2O(l)$

21.2 $\qquad\qquad Sn(s) \longrightarrow Sn^{2+}(aq) + 2e^-$
$\qquad\qquad\qquad\qquad\qquad\qquad\qquad$ [anode; oxidation]

$6e^- + 14H^+(aq) + Cr_2O_7^{2-}(aq) \longrightarrow 2Cr^{3+}(aq) + 7H_2O(l)$
$\qquad\qquad\qquad\qquad\qquad\qquad\qquad$ [cathode; reduction]

$3Sn(s) + Cr_2O_7^{2-}(aq) + 14H^+(aq) \longrightarrow$
$\qquad\qquad 3Sn^{2+}(aq) + 2Cr^{3+}(aq) + 7H_2O(l)$ [overall]

Cell notation:
$Sn(s) \mid Sn^{2+}(aq) \parallel H^+(aq), Cr_2O_7^{2-}(aq), Cr^{3+}(aq) \mid graphite$

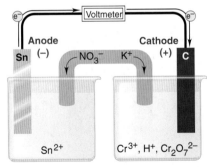

21.3 $Br_2(aq) + 2e^- \longrightarrow 2Br^-(aq) \qquad E°_{bromine} = 1.07 \text{ V}$
$\qquad\qquad\qquad\qquad\qquad\qquad\qquad\qquad\qquad\qquad$ [cathode]
$2V^{3+}(aq) + 2H_2O(l) \longrightarrow 2VO^{2+}(aq) + 4H^+(aq) + 2e^-$
$\qquad\qquad\qquad\qquad\qquad E°_{vanadium} = ? \qquad$ [anode]

$E°_{vanadium} = E°_{bromine} - E°_{cell} = 1.07 \text{ V} - 1.39 \text{ V} = -0.32 \text{ V}$

21.4 $Fe^{2+}(aq) + 2e^- \longrightarrow Fe(s) \qquad\qquad E° = -0.44 \text{ V}$
$\quad\quad 2[Fe^{2+}(aq) \longrightarrow Fe^{3+}(aq) + e^-] \qquad E° = 0.77 \text{ V}$

$3Fe^{2+}(aq) \longrightarrow 2Fe^{3+}(aq) + Fe(s)$

$E°_{cell} = -0.44 \text{ V} - 0.77 \text{ V} = -1.21 \text{ V}$

The reaction is nonspontaneous. The spontaneous reaction is
$2Fe^{3+}(aq) + Fe(s) \longrightarrow 3Fe^{2+}(aq) \qquad E°_{cell} = 1.21 \text{ V}$
$Fe > Fe^{2+} > Fe^{3+}$

21.5 $Cd(s) + Cu^{2+}(aq) \longrightarrow Cd^{2+}(aq) + Cu(s)$
$\Delta G° = -RT \ln K = -8.314 \text{ J/mol} \cdot K \times 298 \text{ K} \times \ln K$
$\qquad = -143 \text{ kJ}; K = 1.2 \times 10^{25}$

$E°_{cell} = \frac{0.0592 \text{ V}}{2} \log (1.2 \times 10^{25}) = 0.742 \text{ V}$

21.6 $\quad\quad Fe(s) \longrightarrow Fe^{2+}(aq) + 2e^- \qquad E° = -0.44 \text{ V}$
$\quad\quad Cu^{2+}(aq) + 2e^- \longrightarrow Cu(s) \qquad E° = 0.34 \text{ V}$

$Fe(s) + Cu^{2+}(aq) \longrightarrow Fe^{2+}(aq) + Cu(s) \quad E°_{cell} = 0.78 \text{ V}$

So $E_{cell} = 0.78 \text{ V} + 0.25 \text{ V} = 1.03 \text{ V}$

$1.03 \text{ V} = 0.78 \text{ V} - \frac{0.0592 \text{ V}}{2} \log \frac{[Fe^{2+}]}{[Cu^{2+}]}$

$\frac{[Fe^{2+}]}{[Cu^{2+}]} = 3.6 \times 10^{-9}$

$[Fe^{2+}] = 3.6 \times 10^{-9} \times 0.30 \text{ M} = 1.1 \times 10^{-9} \text{ M}$

21.7 $Au^{3+}(aq; 2.5 \times 10^{-2} \text{ M}) \text{ [B]} \longrightarrow$
$\qquad\qquad\qquad\qquad\qquad Au^{3+}(aq; 7.0 \times 10^{-4} \text{ M}) \text{ [A]}$

$E_{cell} = 0 \text{ V} - \left(\frac{0.0592 \text{ V}}{3} \times \log \frac{7.0 \times 10^{-4}}{2.5 \times 10^{-2}} \right) = 0.0306 \text{ V}$

The electrode in A is negative, so it is the anode.

21.8 The reduction with the more positive electrode potential is
$Au^{3+}(aq) + 3e^- \longrightarrow Au(s); E° = 1.50 \text{ V}$
$\qquad\qquad\qquad\qquad\qquad\qquad\qquad$ [cathode; reduction]
Because of overvoltage, O_2 will not form at the anode, so Br_2 will form:
$2Br^-(aq) \longrightarrow Br_2(l) + 2e^-; E° = 1.07 \text{ V}$
$\qquad\qquad\qquad\qquad\qquad\qquad\qquad$ [anode; oxidation]

21.9 $Cu^{2+}(aq) + 2e^- \longrightarrow Cu(s)$; therefore,
2 mol e^-/1 mol Cu = 2 mol e^-/63.55 g Cu

Time (min) = $1.50 \text{ g Cu} \times \frac{2 \text{ mol } e^-}{63.55 \text{ g Cu}}$

$\times \frac{9.65 \times 10^4 \text{ C}}{1 \text{ mol } e^-} \times \frac{1 \text{ s}}{4.75 \text{ C}} \times \frac{1 \text{ min}}{60 \text{ s}} = 16.0 \text{ min}$

Problems

Problems with **colored** numbers are answered in Appendix E. Sections match the text and provide the numbers of relevant sample problems. Bracketed problems are grouped in pairs (indicated by a short rule) that cover the same concept. Comprehensive Problems are based on material from any section or previous chapter.

Note: Unless stated otherwise, all problems refer to systems at 298.15 K (25°C).

Redox Reactions and Electrochemical Cells
(Sample Problem 21.1)

21.1 Define *oxidation* and *reduction* in terms of electron transfer and change in oxidation number.

21.2 Can one half-reaction in a redox process take place independently of the other? Explain.

21.3 Which type of electrochemical cell has $\Delta G_{sys} < 0$? Which type shows an increase in free energy?

21.4 Which statements are true? Correct any that are false.
(a) In a voltaic cell, the anode is negative relative to the cathode.
(b) Oxidation occurs at the anode of a voltaic or an electrolytic cell.
(c) Electrons flow into the cathode of an electrolytic cell.
(d) In a voltaic cell, the surroundings do work on the system.
(e) A metal that plates out of an electrolytic cell appears on the cathode.
(f) The cell electrolyte provides a solution of mobile electrons.

21.5 Consider the following balanced redox reaction:
$$16H^+(aq) + 2MnO_4^-(aq) + 10Cl^-(aq) \longrightarrow$$
$$2Mn^{2+}(aq) + 5Cl_2(g) + 8H_2O(l)$$
(a) Which species is being oxidized?
(b) Which species is being reduced?
(c) Which species is the oxidizing agent?
(d) Which species is the reducing agent?
(e) From which species to which does electron transfer occur?
(f) Write the balanced molecular equation, with K^+ and SO_4^{2-} as the spectator ions.

21.6 Consider the following balanced redox reaction:
$$2CrO_2^-(aq) + 2H_2O(l) + 6ClO^-(aq) \longrightarrow$$
$$2CrO_4^{2-}(aq) + 3Cl_2(g) + 4OH^-(aq)$$
(a) Which species is being oxidized?
(b) Which species is being reduced?
(c) Which species is the oxidizing agent?
(d) Which species is the reducing agent?
(e) From which species to which does electron transfer occur?
(f) Write the balanced molecular equation, with Na^+ as the spectator ion.

21.7 Balance the following skeleton reactions and identify the oxidizing and reducing agents:
(a) $ClO_3^-(aq) + I^-(aq) \longrightarrow I_2(s) + Cl^-(aq)$ [acidic]
(b) $MnO_4^-(aq) + SO_3^{2-}(aq) \longrightarrow$
$$MnO_2(s) + SO_4^{2-}(aq)$$ [basic]
(c) $MnO_4^-(aq) + H_2O_2(aq) \longrightarrow Mn^{2+}(aq) + O_2(g)$ [acidic]

21.8 Balance the following skeleton reactions and identify the oxidizing and reducing agents:
(a) $O_2(g) + NO(g) \longrightarrow NO_3^-(aq)$ [acidic]
(b) $CrO_4^{2-}(aq) + Cu(s) \longrightarrow$
$$Cr(OH)_3(s) + Cu(OH)_2(s)$$ [basic]
(c) $AsO_4^{3-}(aq) + NO_2^-(aq) \longrightarrow$
$$AsO_2^-(aq) + NO_3^-(aq)$$ [basic]

21.9 Balance the following skeleton reactions and identify the oxidizing and reducing agents:
(a) $BH_4^-(aq) + ClO_3^-(aq) \longrightarrow$
$$H_2BO_3^-(aq) + Cl^-(aq)$$ [basic]
(b) $CrO_4^{2-}(aq) + N_2O(g) \longrightarrow Cr^{3+}(aq) + NO(g)$ [acidic]
(c) $Br_2(l) \longrightarrow BrO_3^-(aq) + Br^-(aq)$ [basic]

21.10 Balance the following skeleton reactions and identify the oxidizing and reducing agents:
(a) $Sb(s) + NO_3^-(aq) \longrightarrow Sb_4O_6(s) + NO(g)$ [acidic]
(b) $Mn^{2+}(aq) + BiO_3^-(aq) \longrightarrow$
$$MnO_4^-(aq) + Bi^{3+}(aq)$$ [acidic]
(c) $Fe(OH)_2(s) + Pb(OH)_3^-(aq) \longrightarrow$
$$Fe(OH)_3(s) + Pb(s)$$ [basic]

21.11 In many residential water systems, the aqueous Fe^{3+} concentration is high enough to stain sinks and turn drinking water light brown. The iron content is analyzed by first reducing the Fe^{3+} to Fe^{2+} and then titrating with MnO_4^- in acidic solution. Balance the skeleton reaction of the titration step:
$$Fe^{2+}(aq) + MnO_4^-(aq) \longrightarrow Mn^{2+}(aq) + Fe^{3+}(aq)$$

21.12 *Aqua regia,* a mixture of concentrated HNO_3 and HCl, was developed by alchemists as a means to "dissolve" gold. The process is actually a redox reaction with the following simplified skeleton reaction:
$$Au(s) + NO_3^-(aq) + Cl^-(aq) \longrightarrow AuCl_4^-(aq) + NO_2(g)$$
(a) Balance the reaction by the half-reaction method.
(b) What are the oxidizing and reducing agents?
(c) What is the function of HCl in aqua regia?

Voltaic Cells: Using Spontaneous Reactions to Generate Electrical Energy
(Sample Problem 21.2)

21.13 Consider the following general voltaic cell:

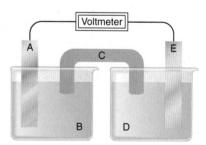

Identify the (a) anode, (b) cathode, (c) salt bridge, (d) electrode at which e^- leave the cell, (e) electrode with a positive charge, and (f) electrode that gains mass as the cell operates (assuming that a metal plates out).

21.14 Why does a voltaic cell not operate unless the two compartments are connected through an external circuit?

21.15 What purpose does the salt bridge serve in a voltaic cell, and how does it accomplish this purpose?

21.16 What is the difference between an active and an inactive electrode? Why are inactive electrodes used? Name two substances commonly used for inactive electrodes.

21.17 When a piece of metal A is placed in a solution containing ions of metal B, metal B plates out on the piece of A.
(a) Which metal is being oxidized?
(b) Which metal is being displaced?
(c) Which metal would you use as the anode in a voltaic cell incorporating these two metals?
(d) If bubbles of H_2 form when B is placed in acid, will they form if A is placed in acid? Explain.

21.18 A voltaic cell is constructed with an Sn/Sn^{2+} half-cell and a Zn/Zn^{2+} half-cell. The zinc electrode is negative.
(a) Write balanced half-reactions and the overall reaction.
(b) Draw a diagram of the cell, labeling electrodes with their charges and showing the directions of electron flow in the circuit and of cation and anion flow in the salt bridge.

21.19 A voltaic cell is constructed with an Ag/Ag^+ half-cell and a Pb/Pb^{2+} half-cell. The silver electrode is positive.
(a) Write balanced half-reactions and the overall reaction.
(b) Draw a diagram of the cell, labeling electrodes with their charges and showing the directions of electron flow in the circuit and of cation and anion flow in the salt bridge.

21.20 Consider the following voltaic cell:

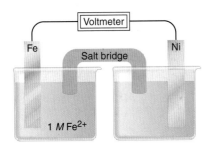

(a) In which direction do electrons flow in the external circuit?
(b) In which half-cell does oxidation occur?
(c) In which half-cell do electrons enter the cell?
(d) At which electrode are electrons consumed?
(e) Which electrode is negatively charged?
(f) Which electrode decreases in mass during cell operation?
(g) Suggest a solution for the cathode electrolyte.
(h) Suggest a pair of ions for the salt bridge.
(i) For which electrode could you use an inactive material?
(j) In which direction do anions within the salt bridge move to maintain charge neutrality?
(k) Write balanced half-reactions and an overall cell reaction.

21.21 Consider the following voltaic cell:

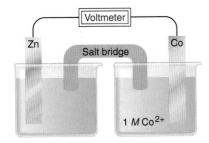

(a) In which direction do electrons flow in the external circuit?
(b) In which half-cell does reduction occur?
(c) In which half-cell do electrons leave the cell?
(d) At which electrode are electrons generated?
(e) Which electrode is positively charged?
(f) Which electrode increases in mass during cell operation?
(g) Suggest a solution for the anode electrolyte.
(h) Suggest a pair of ions for the salt bridge.
(i) For which electrode could you use an inactive material?
(j) In which direction do cations within the salt bridge move to maintain charge neutrality?
(k) Write balanced half-reactions and an overall cell reaction.

21.22 Write the cell notation for the voltaic cell that incorporates each of the following redox reactions:
(a) $Al(s) + Cr^{3+}(aq) \longrightarrow Al^{3+}(aq) + Cr(s)$
(b) $Cu^{2+}(aq) + SO_2(g) + 2H_2O(l) \longrightarrow$
$$Cu(s) + SO_4^{2-}(aq) + 4H^+(aq)$$

21.23 Write a balanced equation from each cell notation:
(a) $Mn(s) \mid Mn^{2+}(aq) \parallel Cd^{2+}(aq) \mid Cd(s)$
(b) $Fe(s) \mid Fe^{2+}(aq) \parallel NO_3^-(aq) \mid NO(g) \mid Pt(s)$

Cell Potential: Output of a Voltaic Cell
(Sample Problems 21.3 and 21.4)

21.24 How is a standard reference electrode used to determine unknown $E^\circ_{half-cell}$ values?

21.25 What does a negative E°_{cell} indicate about a redox reaction? What does a negative E°_{cell} indicate about the reverse reaction?

21.26 The standard cell potential is a thermodynamic state function. How are E° values treated similarly to ΔH°, ΔG°, and S° values? How are they treated differently?

21.27 In basic solution, Se^{2-} and SO_3^{2-} ions react spontaneously:
$$2Se^{2-}(aq) + 2SO_3^{2-}(aq) + 3H_2O(l) \longrightarrow$$
$$2Se(s) + 6OH^-(aq) + S_2O_3^{2-}(aq) \qquad E^\circ_{cell} = 0.35 \text{ V}$$
(a) Write balanced half-reactions for the process.
(b) If $E^\circ_{sulfite}$ is -0.57 V, calculate $E^\circ_{selenium}$.

21.28 In acidic solution, O_3 and Mn^{2+} ion react spontaneously:
$$O_3(g) + Mn^{2+}(aq) + H_2O(l) \longrightarrow$$
$$O_2(g) + MnO_2(s) + 2H^+(aq) \qquad E^\circ_{cell} = 0.84 \text{ V}$$
(a) Write the balanced half-reactions.
(b) Using Appendix D to find E°_{ozone}, calculate $E^\circ_{manganese}$.

21.29 Use the emf series (Appendix D) to arrange the species.
(a) In order of *decreasing* strength as *oxidizing* agents: Fe^{3+}, Br_2, Cu^{2+}
(b) In order of *increasing* strength as *oxidizing* agents: Ca^{2+}, $Cr_2O_7^{2-}$, Ag^+

21.30 Use the emf series (Appendix D) to arrange the species.
(a) In order of *decreasing* strength as *reducing* agents: SO_2, $PbSO_4$, MnO_2
(b) In order of *increasing* strength as *reducing* agents: Hg, Fe, Sn

21.31 Balance each skeleton reaction, calculate E°_{cell}, and state whether the reaction is spontaneous:
(a) $Co(s) + H^+(aq) \longrightarrow Co^{2+}(aq) + H_2(g)$
(b) $Mn^{2+}(aq) + Br_2(l) \longrightarrow MnO_4^-(aq) + Br^-(aq)$ [acidic]
(c) $Hg_2^{2+}(aq) \longrightarrow Hg^{2+}(aq) + Hg(l)$

21.32 Balance each skeleton reaction, calculate $E°_{cell}$, and state whether the reaction is spontaneous:
(a) $Cl_2(g) + Fe^{2+}(aq) \longrightarrow Cl^-(aq) + Fe^{3+}(aq)$
(b) $Mn^{2+}(aq) + Co^{3+}(aq) \longrightarrow MnO_2(s) + Co^{2+}(aq)$ [acidic]
(c) $AgCl(s) + NO(g) \longrightarrow$
$\qquad Ag(s) + Cl^-(aq) + NO_3^-(aq)$ [acidic]

21.33 Balance each skeleton reaction, calculate $E°_{cell}$, and state whether the reaction is spontaneous:
(a) $Ag(s) + Cu^{2+}(aq) \longrightarrow Ag^+(aq) + Cu(s)$
(b) $Cd(s) + Cr_2O_7^{2-}(aq) \longrightarrow Cd^{2+}(aq) + Cr^{3+}(aq)$
(c) $Ni^{2+}(aq) + Pb(s) \longrightarrow Ni(s) + Pb^{2+}(aq)$

21.34 Balance each skeleton reaction, calculate $E°_{cell}$, and state whether the reaction is spontaneous:
(a) $Cu^+(aq) + PbO_2(s) + SO_4^{2-}(aq) \longrightarrow$
$\qquad PbSO_4(s) + Cu^{2+}(aq)$ [acidic]
(b) $H_2O_2(aq) + Ni^{2+}(aq) \longrightarrow O_2(g) + Ni(s)$ [acidic]
(c) $MnO_2(s) + Ag^+(aq) \longrightarrow MnO_4^-(aq) + Ag(s)$ [basic]

21.35 Use the following half-reactions to write three spontaneous reactions and calculate $E°_{cell}$ for each reaction:
(1) $Al^{3+}(aq) + 3e^- \longrightarrow Al(s)$ $E° = -1.66$ V
(2) $N_2O_4(g) + 2e^- \longrightarrow 2NO_2^-(aq)$ $E° = 0.867$ V
(3) $SO_4^{2-}(aq) + H_2O(l) + 2e^- \longrightarrow SO_3^{2-}(aq) + 2OH^-(aq)$
$\qquad\qquad\qquad\qquad\qquad\qquad E° = 0.93$ V

21.36 Use the following half-reactions to write three spontaneous reactions and calculate $E°_{cell}$ for each reaction:
(1) $Au^+(aq) + e^- \longrightarrow Au(s)$ $E° = 1.69$ V
(2) $N_2O(g) + 2H^+(aq) + 2e^- \longrightarrow N_2(g) + H_2O(l)$
$\qquad\qquad\qquad\qquad\qquad\qquad E° = 1.77$ V
(3) $Cr^{3+}(aq) + 3e^- \longrightarrow Cr(s)$ $E° = -0.74$ V

21.37 Use the following half-reactions to write three spontaneous reactions and calculate $E°_{cell}$ for each reaction:
(1) $2HClO(aq) + 2H^+(aq) + 2e^- \longrightarrow Cl_2(g) + 2H_2O(l)$
$\qquad\qquad\qquad\qquad\qquad\qquad E° = 1.63$ V
(2) $Pt^{2+}(aq) + 2e^- \longrightarrow Pt(s)$ $E° = 1.20$ V
(3) $PbSO_4(s) + 2e^- \longrightarrow Pb(s) + SO_4^{2-}(aq)$ $E° = -0.31$ V

21.38 Use the following half-reactions to write three spontaneous reactions and calculate $E°_{cell}$ for each reaction:
(1) $I_2(s) + 2e^- \longrightarrow 2I^-(aq)$ $E° = 0.53$ V
(2) $S_2O_8^{2-}(aq) + 2e^- \longrightarrow 2SO_4^{2-}(aq)$ $E° = 2.01$ V
(3) $Cr_2O_7^{2-}(aq) + 14H^+(aq) + 6e^- \longrightarrow$
$\qquad 2Cr^{3+}(aq) + 7H_2O(l)$ $E° = 1.33$ V

21.39 When metal A is placed in a solution of a salt of metal B, the surface of metal A changes color. When metal B is placed in acid solution, gas bubbles form on the surface of the metal. When metal A is placed in a solution of a salt of metal C, no change is observed in the solution or on the metal A surface. Will metal C cause formation of H_2 when placed in acid solution? Rank metals A, B, and C in order of *decreasing* reducing strength.

21.40 When a clean iron nail is placed in an aqueous solution of copper(II) sulfate, the nail becomes coated with a brownish black material.
(a) What is the material coating the iron?
(b) What are the oxidizing and reducing agents?
(c) Can this reaction be made into a voltaic cell?
(d) Write the balanced equation for the reaction.
(e) Calculate $E°_{cell}$ for the process.

Free Energy and Electrical Work
(Sample Problems 21.5 to 21.7)

21.41 (a) How do the relative magnitudes of Q and K relate to the signs of ΔG and E_{cell}? Explain.
(b) Can a cell do work when $Q/K > 1$ or $Q/K < 1$? Explain.

21.42 A voltaic cell consists of a metal A/A^+ electrode and a metal B/B^+ electrode, with the A/A^+ electrode negative. The initial $[A^+]/[B^+]$ is such that $E_{cell} > E°_{cell}$.
(a) How do $[A^+]$ and $[B^+]$ change as the cell operates?
(b) How does E_{cell} change as the cell operates?
(c) What is $[A^+]/[B^+]$ when $E_{cell} = E°_{cell}$? Explain.
(d) Is it possible for E_{cell} to be less than $E°_{cell}$? Explain.

21.43 Explain whether E_{cell} of a voltaic cell will increase or decrease with each of the following changes:
(a) Decrease in cell temperature
(b) Increase in [active ion] in the anode compartment
(c) Increase in [active ion] in the cathode compartment
(d) Increase in pressure of a gaseous reactant in the cathode compartment

21.44 In a concentration cell, is the more concentrated electrolyte in the cathode or the anode compartment? Explain.

21.45 What is the value of the equilibrium constant for the reaction between each pair at 25°C?
(a) $Ni(s)$ and $Ag^+(aq)$ (b) $Fe(s)$ and $Cr^{3+}(aq)$

21.46 What is the value of the equilibrium constant for the reaction between each pair at 25°C?
(a) $Al(s)$ and $Cd^{2+}(aq)$ (b) $I_2(s)$ and $Br^-(aq)$

21.47 Calculate $\Delta G°$ for each of the reactions in Problem 21.45.
21.48 Calculate $\Delta G°$ for each of the reactions in Problem 21.46.

21.49 What are $E°_{cell}$ and $\Delta G°$ of a redox reaction at 25°C for which $n = 1$ and $K = 5.0 \times 10^3$?
21.50 What are $E°_{cell}$ and $\Delta G°$ of a redox reaction at 25°C for which $n = 2$ and $K = 0.075$?

21.51 A voltaic cell consists of a standard hydrogen electrode in one half-cell and a Cu/Cu^{2+} half-cell. Calculate $[Cu^{2+}]$ when E_{cell} is 0.25 V.

21.52 A voltaic cell consists of an Mn/Mn^{2+} half-cell and a Pb/Pb^{2+} half-cell. Calculate $[Pb^{2+}]$ when $[Mn^{2+}]$ is 1.3 M and E_{cell} is 0.42 V.

21.53 A voltaic cell with Ni/Ni^{2+} and Co/Co^{2+} half-cells has the following initial concentrations: $[Ni^{2+}] = 0.80$ M; $[Co^{2+}] = 0.20$ M.
(a) What is the initial E_{cell}?
(b) What is $[Ni^{2+}]$ when E_{cell} reaches 0.03 V?
(c) What are the equilibrium concentrations of the ions?

21.54 A voltaic cell with Mn/Mn^{2+} and Cd/Cd^{2+} half-cells has the following initial concentrations: $[Mn^{2+}] = 0.090$ M; $[Cd^{2+}] = 0.060$ M.
(a) What is the initial E_{cell}?
(b) What is E_{cell} when $[Cd^{2+}]$ reaches 0.050 M?
(c) What is $[Mn^{2+}]$ when E_{cell} reaches 0.055 V?
(d) What are the equilibrium concentrations of the ions?

21.55 A concentration cell consists of two H_2/H^+ half-cells. Half-cell A has H_2 at 0.90 atm bubbling into 0.10 M HCl. Half-cell B has H_2 at 0.50 atm bubbling into 2.0 M HCl. Which half-cell houses the anode? What is the voltage of the cell?

21.56 A concentration cell consists of two Sn/Sn^{2+} half-cells. The electrolyte in compartment A is 0.13 M $Sn(NO_3)_2$. The electrolyte in B is 0.87 M $Sn(NO_3)_2$. Which half-cell houses the cathode? What is the voltage of the cell?

Electrochemical Processes in Batteries

21.57 What is the direction of electron flow with respect to the anode and the cathode in a battery? Explain.

21.58 Both a D-sized and an AAA-sized alkaline battery have an output of 1.5 V. What property of the cell potential allows this to occur? What is different about these two batteries?

21.59 Many common electrical devices require the use of more than one battery.
(a) How many alkaline batteries must be placed in series to light a flashlight with a 6.0-V bulb?
(b) What is the voltage requirement of a camera that uses six silver batteries?
(c) How many volts can a car battery deliver if two of its anode/cathode cells are shorted?

Corrosion: A Case of Environmental Electrochemistry

21.60 During reconstruction of the Statue of Liberty, Teflon spacers were placed between the iron skeleton and the copper plates that cover the statue. What purpose do these spacers serve?

21.61 Why do steel bridge-supports rust at the waterline but not above or below it?

21.62 Which of the following metals are suitable for use as sacrificial anodes to protect against corrosion of underground iron pipes? If any are not suitable, explain why:
(a) Aluminum   (b) Magnesium
(c) Sodium   (d) Lead
(e) Nickel   (f) Zinc
(g) Chromium

Electrolytic Cells: Using Electrical Energy to Drive Nonspontaneous Reactions
(Sample Problems 21.8 and 21.9)

Note: Unless stated otherwise, assume that the electrolytic cells in the following problems operate at 100% efficiency.

21.63 Consider the following general electrolytic cell:

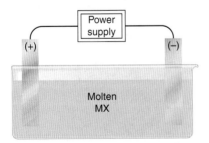

(a) At which electrode does oxidation occur?
(b) At which electrode does elemental M form?
(c) At which electrode are electrons being released by ions?
(d) At which electrode are electrons entering the cell?

21.64 A voltaic cell consists of Cr/Cr^{3+} and Cd/Cd^{2+} half-cells with all components in their standard states. After 10 minutes of operation, a thin coating of cadmium metal has plated out on the cathode. Describe what will happen if you attach the negative terminal of a dry cell (1.5 V) to the cell cathode and the positive terminal to the cell anode.

21.65 Why are $E_{\text{half-cell}}$ values for the oxidation and reduction of water different from $E^{\circ}_{\text{half-cell}}$ values for the same processes?

21.66 In an aqueous electrolytic cell, nitrate ions never react at the anode, but nitrite ions do. Explain.

21.67 How does overvoltage influence the products in the electrolysis of aqueous salts?

21.68 What property allows copper to be purified in the presence of iron and nickel impurities? Explain.

21.69 What is the practical reason for using cryolite in the electrolysis of aluminum oxide?

21.70 In the electrolysis of molten NaBr,
(a) What product forms at the anode?
(b) What product forms at the cathode?

21.71 In the electrolysis of molten BaI_2,
(a) What product forms at the negative electrode?
(b) What product forms at the positive electrode?

21.72 Identify those elements that can be prepared by electrolysis of their aqueous salts: copper, barium, aluminum, bromine.

21.73 Identify those elements that can be prepared by electrolysis of their aqueous salts: strontium, gold, tin, chlorine.

21.74 What product forms at each electrode in the aqueous electrolysis of the following salts: (a) LiF; (b) $SnSO_4$?

21.75 What product forms at each electrode in the aqueous electrolysis of the following salts: (a) $Cr(NO_3)_3$; (b) $MnCl_2$?

21.76 Electrolysis of molten $MgCl_2$ is the final production step in the isolation of magnesium from seawater. Assuming that 35.6 g of Mg metal forms,
(a) How many moles of electrons are required?
(b) How many coulombs are required?
(c) How many amps will produce this amount in 2.50 h?

21.77 Electrolysis of molten NaCl in a Downs cell is the major isolation step in the production of sodium metal. Assuming that 215 g of Na metal forms,
(a) How many moles of electrons are required?
(b) How many coulombs are required?
(c) How many amps will produce this amount in 9.50 h?

21.78 How many grams of radium can form by passing 215 C through an electrolytic cell containing a molten radium salt?

21.79 How many grams of aluminum can form by passing 305 C through an electrolytic cell containing a molten aluminum salt?

21.80 How many seconds does it take to deposit 85.5 g of Zn on a steel gate when 23.0 A is passed through a $ZnSO_4$ solution?

21.81 How many seconds does it take to deposit 1.63 g of Ni on a decorative drawer handle when 13.7 A is passed through a $Ni(NO_3)_2$ solution?

21.82 A professor adds Na_2SO_4 to water to facilitate its electrolysis in a lecture demonstration. (a) What is the purpose of the Na_2SO_4? (b) Why is the water electrolyzed instead of the salt?

21.83 A Downs cell operating at 75.0 A produces 30.0 kg of Na.
(a) What volume of $Cl_2(g)$ is produced at 1.0 atm and 580.°C?
(b) How many coulombs were passed through the cell?
(c) How long did the cell operate?

21.84 Zinc plating (galvanizing) is an important means of corrosion protection. Although the process is done customarily by dipping the object into molten zinc, the metal can also be electroplated from aqueous solutions. How many grams of zinc can be deposited on a steel tank from a $ZnSO_4$ solution when a 0.755-A current flows for 2.00 days?

Comprehensive Problems

Problems with an asterisk (*) are more challenging.

21.85 The MnO_2 used in alkaline batteries can be produced by an electrochemical process of which one half-reaction is

$$Mn^{2+}(aq) + 2H_2O(l) \longrightarrow MnO_2(s) + 4H^+(aq) + 2e^-$$

If a current of 25.0 A is used, how many hours are needed to produce 1.00 kg of MnO_2? At which electrode is the MnO_2 formed?

21.86 The overall cell reaction occurring in an alkaline battery is

$$Zn(s) + MnO_2(s) + H_2O(l) \longrightarrow ZnO(s) + Mn(OH)_2(s)$$

(a) How many moles of electrons flow per mole of reaction?
(b) If 2.50 g of zinc is oxidized, how many grams of manganese dioxide and of water are consumed?
(c) What is the total mass of reactants consumed in part (b)?
(d) How many coulombs are produced in part (b)?
(e) In practice, voltaic cells of a given capacity (coulombs) are heavier than the calculation in part (c) indicates. Explain.

* **21.87** Brass, an alloy of copper and zinc, can be produced by simultaneously electroplating the two metals from a solution containing their 2+ ions. If exactly 70.0% of the total current is used to plate copper, while 30.0% goes to plating zinc, what is the mass percent of copper in the brass?

21.88 Compare and contrast a voltaic cell and an electrolytic cell with respect to each of the following:
(a) Sign of the free energy change
(b) Nature of the half-reaction at the anode
(c) Nature of the half-reaction at the cathode
(d) Charge on the electrode labeled "anode"
(e) Electrode from which electrons leave the cell

* **21.89** A thin circular-disk earring 5.00 cm in diameter is plated with a coating of gold 0.20 mm thick from an Au^{3+} bath.
(a) How many days does it take to deposit the gold on one side of this earring if the current is 0.010 A (d of gold = 19.3 g/cm³)?
(b) How many days does it take to deposit the gold on both sides of a pair of these earrings?
(c) If the price of gold is $320 per troy ounce (31.10 g), what is the total cost of the gold plating?

21.90 (a) How many minutes does it take to form 10.0 L of O_2 measured at 99.8 kPa and 28°C from water if a current of 1.3 A passes through the electrolytic cell? (b) What mass of H_2 forms?

21.91 A silver button battery used in a watch contains 16.0 g of zinc and can run until 80% of the zinc is consumed. (a) How many days can the battery run at a current of 4.8 milliamps? (b) When the battery dies, 95% of the Ag_2O has been consumed. How many grams of Ag was used to make the battery? (c) If Ag costs $5.50 per troy ounce (31.10 g), what is the cost of the Ag consumed each day the watch runs?

21.92 If a chlor-alkali cell used a current of 3×10^4 A, how many pounds of Cl_2 would be produced in a typical 8-h operating day?

21.93 To improve conductivity in the electroplating of automobile bumpers, a thin coating of copper separates the steel from a heavy coating of chromium.

(a) What mass of Cu is deposited on an automobile trim piece if plating continues for 1.25 h at a current of 5.0 A?
(b) If the area of the trim piece is 50.0 cm², what is the thickness of the Cu coating (d of Cu = 8.95 g/cm³)?

21.94 Commercial electrolytic cells for producing aluminum operate at 5.0 V and 100,000 A.
(a) How long does it take to produce exactly 1 metric ton (1000 kg) of aluminum?
(b) How much electrical power (in kilowatt-hours, kW·h) is used [1 W = 1 J/s; 1 kW·h = 3.6×10^3 kJ]?
(c) If electricity costs 0.90¢ per kW·h and cell efficiency is 90.%, what is the cost of producing exactly 1 lb of aluminum?

21.95 Magnesium bars are connected electrically to underground iron pipes to serve as sacrificial anodes.
(a) Do electrons flow from the bar to the pipe or the reverse?
(b) A 12-kg Mg bar is attached to an iron pipe, and it takes 8.5 yr for the Mg to be consumed. What is the average current flowing between the Mg and the Fe during this period?

21.96 Bubbles of H_2 form when metal D is placed in hot H_2O. No reaction occurs when D is placed in a solution of a salt of metal E, but D is discolored and coated immediately when placed in a solution of a salt of metal F. What happens if E is placed in a solution of a salt of metal F? Rank metals D, E, and F in order of *increasing* reducing strength.

* **21.97** The following reactions are used in batteries:

I $\quad 2H_2(g) + O_2(g) \longrightarrow 2H_2O(l) \qquad E_{cell} = 1.23$ V

II $\quad Pb(s) + PbO_2(s) + 2H_2SO_4(aq) \longrightarrow$
$$2PbSO_4(s) + 2H_2O(l) \qquad E_{cell} = 2.04 \text{ V}$$

III $2Na(l) + FeCl_2(s) \longrightarrow 2NaCl(s) + Fe(s)$
$$E_{cell} = 2.35 \text{ V}$$

Reaction I is used in fuel cells, II in the automobile lead-acid battery, and III in an experimental high-temperature battery for powering electric vehicles. The aim is to obtain as much work as possible from a cell, while keeping its weight to a minimum.
(a) In each cell, find the moles of electrons transferred and ΔG.
(b) Calculate the ratio, in kJ/g, of w_{max} to mass of reactants for each of the cells. Which has the highest ratio, which the lowest, and why? (*Note:* For simplicity, ignore the masses of cell components that do not appear in the cell as reactants, including electrode materials, electrolytes, separators, cell casing, wiring, etc.)

21.98 From the skeleton reactions below, create a list of balanced half-reactions in which the strongest oxidizing agent is on top and the weakest is on the bottom:

$$U^{3+}(aq) + Cr^{3+}(aq) \longrightarrow Cr^{2+}(aq) + U^{4+}(aq)$$
$$Fe(s) + Sn^{2+}(aq) \longrightarrow Sn(s) + Fe^{2+}(aq)$$
$$Fe(s) + U^{4+}(aq) \longrightarrow \text{no reaction}$$
$$Cr^{3+}(aq) + Fe(s) \longrightarrow Cr^{2+}(aq) + Fe^{2+}(aq)$$
$$Cr^{2+}(aq) + Sn^{2+}(aq) \longrightarrow Sn(s) + Cr^{3+}(aq)$$

21.99 Use Appendix D to calculate the K_{sp} of AgCl.

21.100 Calculate the K_f of $Ag(NH_3)_2^+$ from

$$Ag^+(aq) + e^- \rightleftharpoons Ag(s) \qquad E° = 0.80 \text{ V}$$
$$Ag(NH_3)_2^+(aq) + e^- \rightleftharpoons Ag(s) + 2NH_3(aq) \qquad E° = 0.37 \text{ V}$$

21.101 Use Appendix D to create an activity series of Mn, Fe, Ag, Sn, Cr, Cu, Ba, Al, Na, Hg, Ni, Li, Au, Zn, and Pb. Rank these metals in order of decreasing reducing strength, and divide them into three groups: those that displace H_2 from water, those that displace H_2 from acid, and those that cannot displace H_2.

21.102 The overall cell reaction for aluminum production is

$$2Al_2O_3(\text{in } Na_3AlF_6) + 3C(\text{graphite}) \longrightarrow 4Al(l) + 3CO_2(g)$$

(a) Assuming 100% efficiency, how many metric tons (t) of Al_2O_3 are consumed per metric ton of Al produced?

(b) Assuming 100% efficiency, how many metric tons of the graphite anode are consumed per metric ton of Al produced?

(c) Actual conditions in an aluminum plant require 1.89 t of Al_2O_3 and 0.45 t of graphite per metric ton of Al. What is the percent yield of Al with respect to Al_2O_3?

(d) What is the percent yield of Al with respect to graphite?

(e) What volume of CO_2 (in m^3) is produced per metric ton of Al at operating conditions of 960.°C and exactly 1 atm?

* 21.103 Two concentration cells are prepared, both with 90.0 mL of 0.0100 M $Cu(NO_3)_2$ and a Cu bar in each half-cell.

(a) In the first concentration cell, 10.0 mL of 0.500 M NH_3 is added to one half-cell; the complex ion $Cu(NH_3)_4^{2+}$ forms, and E_{cell} is 0.129 V. Calculate K_f for the formation of the complex ion.

(b) Calculate E_{cell} when an additional 10.0 mL of 0.500 M NH_3 is added.

(c) In the second concentration cell, 10.0 mL of 0.500 M NaOH is added to one half-cell; the precipitate $Cu(OH)_2$ forms ($K_{sp} = 2.2 \times 10^{-20}$). Calculate E_{cell}°.

(d) What would the molarity of NaOH have to be for the addition of 10.0 mL to result in an E_{cell}° of 0.340 V?

21.104 A voltaic cell has one half-cell with a Cu bar in a 1.00 M Cu^{2+} salt solution, and the other half-cell with a Cd bar in the same volume of a 1.00 M Cd^{2+} salt solution.

(a) Find E_{cell}°, ΔG°, and K.

(b) As the cell operates, $[Cd^{2+}]$ increases; find E_{cell} and ΔG when $[Cd^{2+}]$ is 1.95 M.

(c) Find E_{cell}, ΔG, and $[Cu^{2+}]$ at equilibrium.

21.105 If the E_{cell} of the following cell is 0.915 V, what is the pH in the anode compartment?

$$Pt(s) \mid H_2(1.00 \text{ atm}) \mid H^+(aq) \parallel Ag^+(0.100 \ M) \mid Ag(s)$$

CHAPTER TWENTY-TWO

The Transition Elements and Their Coordination Compounds

Exploring the Center of the Table Many transition elements, like the iron and chromium in this stainless steel sculpture, are among the most useful metals known. In this chapter, you'll see why transition elements and their compounds differ so markedly from main-group elements.

Key Principles

◆ In the *transition elements* (*d* block), inner atomic orbitals are being filled, which causes horizontal and vertical trends in atomic properties that *differ* markedly from those of the main-group elements.

◆ Because the *ns* electrons are close in energy to the (*n* − 1)*d* electrons, transition elements can use different numbers of their electrons in bonding. Thus, transition elements have *multiple oxidation states,* and the lower states display more metallic behavior (ionic bonding and basic oxides). The compounds of ions with a partially filled *d* sublevel are *colored* and *paramagnetic.*

◆ Transition elements typically form *coordination compounds,* which consist of a *complex ion* and counter ions. A complex ion has a *central metal ion* and surrounding molecular or anionic *ligands.* The

number of ligands that the metal binds to determines the *shape* of the complex ion. Different positions and bonding arrangements of ligands lead to various types of *isomers.*

◆ According to *valence bond theory,* the shapes of complex ions arise from *hybridization* of different combinations of *d, s,* and *p* orbitals.

◆ According to *crystal field theory,* ligands approaching a metal ion *split its d-orbital energies,* creating two sets of orbitals. Each type of ligand causes a characteristic difference (*splitting energy,* Δ) in *d*-orbital energies, which allows us to rank ligands in a *spectrochemical series.* The *electron occupancy* of the *d* orbitals determines the magnetic behavior of the complex ion.

Outline

Our exploration of the elements to this point is far from complete; in fact, we have skirted the majority of them and some of the most familiar. Whereas most important uses of the main-group elements involve their compounds, the transition elements are remarkably useful in their uncombined form. Figure 22.1 shows that the **transition elements** (*transition metals*) make up the *d* block (B groups) and *f* block (*inner transition elements*).

In addition to copper, whose importance in plumbing and wiring we noted in Chapter 21, many other transition elements have essential uses: chromium in automobile parts, gold and silver in jewelry, tungsten in lightbulb filaments, platinum in automobile catalytic converters, titanium in bicycle frames and aircraft parts, and zinc in batteries, to mention just a few of the better known elements. You may be less aware of zirconium in nuclear-reactor liners, vanadium in axles and crankshafts, molybdenum in boiler plates, nickel in coins, tantalum in organ-replacement parts, palladium in telephone-relay contacts—the list goes on and on. As ions, many of these elements also play vital roles in living organisms.

In this chapter, we cover the *d*-block elements only. We first discuss some properties of the elements and then focus on the most distinctive feature of their chemistry, the formation of coordination compounds—substances that contain complex ions.

Concepts & Skills to Review
Before You Study This Chapter

- properties of light (Section 7.1)
- electron shielding of nuclear charge (Section 8.2)
- electron configuration, ionic size, and magnetic behavior (Sections 8.3 to 8.5)
- valence bond theory (Section 11.1)
- constitutional, geometric, and optical isomerism (Section 15.2)
- Lewis acid-base concepts (Section 18.8)
- complex-ion formation (Section 19.4)
- redox behavior and standard electrode potentials (Section 21.3)

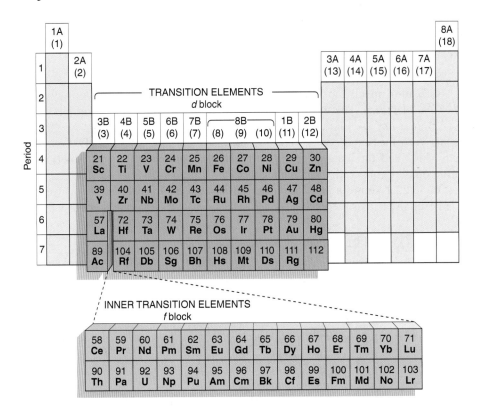

Figure 22.1 The transition elements (*d* block) and inner transition elements (*f* block) in the periodic table.

22.1 PROPERTIES OF THE TRANSITION ELEMENTS

The transition elements differ considerably in physical and chemical behavior from the main-group elements. In some ways, they are more uniform: main-group elements in each period change from metal to nonmetal, but *all transition elements are metals*. In other ways, the transition elements are more diverse: most main-group ionic compounds are colorless and diamagnetic, but *many transition metal compounds are highly colored and paramagnetic*. We first discuss electron configurations of the atoms and ions, and then examine certain key properties of transition elements, with an occasional comparison to the main-group elements.

Scandium, Sc; 3B(3)

Titanium, Ti; 4B(4)

Vanadium, V; 5B(5)

Chromium, Cr; 6B(6)

Manganese, Mn; 7B(7)

Figure 22.2 The Period 4 transition metals. Samples of all ten elements appear as pure metals, in chunk or powder form, in periodic-table order on this and the facing page.

Electron Configurations of the Transition Metals and Their Ions

The *d*-block (B-group) transition elements occur in four series that lie within Periods 4 through 7. Each transition series represents the filling of five *d* orbitals and, thus, contains ten elements. The first of these series occurs in Period 4 and consists of scandium (Sc) through zinc (Zn) (Figure 22.2). Lying between the first and second members of the *d*-block transition series in Periods 6 and 7 are the inner transition elements, whose *f* orbitals are being filled.

Even though there are several exceptions, in general, the *condensed* ground-state electron configuration for the elements in each *d*-block series is

$$[\text{noble gas}]\ ns^2(n-1)d^x, \text{ with } n = 4 \text{ to } 7 \text{ and } x = 1 \text{ to } 10$$

In Periods 6 and 7, the condensed configuration includes the *f* sublevel:

$$[\text{noble gas}]\ ns^2(n-2)f^{14}(n-1)d^x, \text{ with } n = 6 \text{ or } 7$$

The *partial* (valence-level) electron configuration for the *d*-block elements excludes the noble gas core and the filled inner *f* sublevel:

$$ns^2(n-1)d^x$$

Transition metal ions form through *the loss of the ns electrons before the (n − 1)d electrons.* Therefore, as one example, the electron configuration of Ti^{2+} is [Ar] $3d^2$, *not* [Ar] $4s^2$, and Ti^{2+} is referred to as a d^2 ion. Ions of different metals with the same configuration often have similar properties. For example, both Mn^{2+} and Fe^{3+} are d^5 ions; both have pale colors in aqueous solution and form complex ions with similar magnetic properties.

Table 22.1 shows a general pattern in number of unpaired electrons (or half-filled orbitals) across the Period 4 transition series. Note that the number increases

Table 22.1	Orbital Occupancy of the Period 4 Transition Metals		
Element	**Partial Orbital Diagram**		**Unpaired Electrons**
	4s · · · · · 3d · · · · · 4p		
Sc	↑↓ · ↑ □ □ □ □ · □ □ □		1
Ti	↑↓ · ↑ ↑ □ □ □ · □ □ □		2
V	↑↓ · ↑ ↑ ↑ □ □ · □ □ □		3
Cr	↑ · ↑ ↑ ↑ ↑ ↑ · □ □ □		6
Mn	↑↓ · ↑ ↑ ↑ ↑ ↑ · □ □ □		5
Fe	↑↓ · ↑↓ ↑ ↑ ↑ ↑ · □ □ □		4
Co	↑↓ · ↑↓ ↑↓ ↑ ↑ ↑ · □ □ □		3
Ni	↑↓ · ↑↓ ↑↓ ↑↓ ↑ ↑ · □ □ □		2
Cu	↑ · ↑↓ ↑↓ ↑↓ ↑↓ ↑↓ · □ □ □		1
Zn	↑↓ · ↑↓ ↑↓ ↑↓ ↑↓ ↑↓ · □ □ □		0

Iron, Fe; 8B(8)

Cobalt, Co; 8B(9)

Nickel, Ni; 8B(10)

Copper, Cu; 1B(11)

Zinc, Zn; 2B(12)

in the first half of the series and, when pairing begins, decreases through the second half. As you'll see, it is the electron configuration of the transition metal *atom* that correlates with physical properties of the *element,* such as density and magnetic behavior, whereas it is the electron configuration of the *ion* that determines the properties of the *compounds.*

SAMPLE PROBLEM 22.1 Writing Electron Configurations of Transition Metal Atoms and Ions

Problem Write *condensed* electron configurations for the following: **(a)** Zr; **(b)** V^{3+}; **(c)** Mo^{3+}. (Assume that elements in higher periods behave like those in Period 4.)

Plan We locate the element in the periodic table and count its position in the respective transition series. These elements are in Periods 4 and 5, so the general configuration is [noble gas] $ns^2(n-1)d^x$. For the ions, we recall that ns electrons are lost first.

Solution **(a)** Zr is the second element in the $4d$ series: [Kr] $5s^24d^2$.

(b) V is the third element in the $3d$ series: [Ar] $4s^23d^3$. In forming V^{3+}, three electrons are lost (two $4s$ and one $3d$), so V^{3+} is a d^2 ion: [Ar] $3d^2$.

(c) Mo lies below Cr in Group 6B(6), so we expect the same exception as for Cr. Thus, Mo is [Kr] $5s^14d^5$. To form the ion, Mo loses the one $5s$ and two of the $4d$ electrons, so Mo^{3+} is a d^3 ion: [Kr] $4d^3$.

Check Figure 8.5 shows we're correct for the atoms. Be sure that charge plus number of d electrons in the ion equals the sum of outer s and d electrons in the atom.

FOLLOW-UP PROBLEM 22.1 Write *partial* electron configurations for the following: **(a)** Ag^+; **(b)** Cd^{2+}; **(c)** Ir^{3+}.

Atomic and Physical Properties of the Transition Elements

The atomic properties of the transition elements contrast in several ways with those of a comparable set of main-group elements (Section 8.4).

Trends Across a Period Consider the variations in atomic size, electronegativity, and ionization energy across Period 4 (Figure 22.3):

- *Atomic size.* Atomic size decreases overall across the period (Figure 22.3A). However, there is a smooth, steady decrease across the main groups because

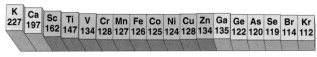

A Atomic radius (pm)

B Electronegativity

C First ionization energy (kJ/mol)

Figure 22.3 Horizontal trends in key atomic properties of the Period 4 elements. The atomic radius (**A**), electronegativity (**B**), and first ionization energy (**C**) of the elements in Period 4 are shown as posts of different heights, with darker shades for the transition series. The transition elements exhibit smaller, less regular changes for these properties than do the main-group elements.

the electrons are added to *outer* orbitals, which shield the increasing nuclear charge poorly. This steady decrease is suspended throughout the transition series, where *atomic size decreases at first but then remains fairly constant.* Recall that the *d* electrons fill *inner* orbitals, so they shield outer electrons from the increasing nuclear charge more effectively. As a result, the outer 4*s* electrons are not pulled closer.

• *Electronegativity.* Electronegativity generally increases across the period but, once again, the transition elements exhibit a relatively *small change in electronegativity* (Figure 22.3B), consistent with the relatively small change in size. In contrast, the main groups show a steady, much steeper increase between the metal potassium (0.8) and the nonmetal bromine (2.8). The transition elements all have intermediate electronegativity values.

• *Ionization energy.* The ionization energies of the Period 4 main-group elements rise steeply from left to right, more than tripling from potassium (419 kJ/mol) to krypton (1351 kJ/mol), as electrons become more difficult to remove from the poorly shielded, increasing nuclear charge. In the transition metals, however, the *first ionization energies increase relatively little* because the inner 3*d* electrons shield effectively (Figure 22.3C); thus, the outer 4*s* electron experiences only a slightly higher effective nuclear charge.

Trends Within a Group Vertical trends for transition elements are also different from those for the main groups.

• *Atomic size.* As expected, atomic size of transition elements increases from Period 4 to 5, as it does for the main-group elements, but there is virtually *no size increase from Period 5 to 6* (Figure 22.4A). The lanthanides (*Z* = 58 to 71), with their buried 4*f* sublevel, appear between the 4*d* (Period 5) and 5*d* (Period 6)

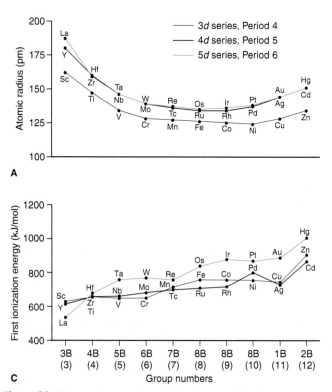

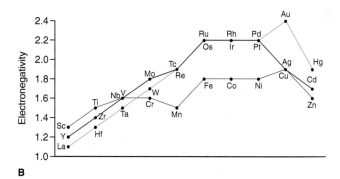

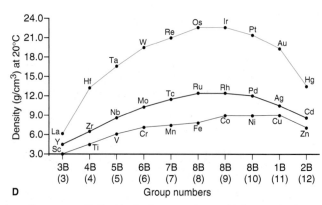

Figure 22.4 Vertical trends in key properties within the transition elements. The trends are unlike those for the main-group elements in several ways: **A,** The second and third members of a transition metal group are nearly the same size. **B,** Electronegativity increases down a transition group. **C,** First ionization energies are highest at the bottom of a transition group. **D,** Densities increase down a transition group because mass increases faster than volume.

series. Therefore, an element in Period 6 is separated from the one above it in Period 5 by 32 elements (ten $4d$, six $5p$, two $6s$, and fourteen $4f$ orbitals) instead of just 18. The extra shrinkage that results from the increase in nuclear charge due to the addition of 14 protons is called the **lanthanide contraction.** By coincidence, this *decrease* is about equal to the normal *increase* between periods, so the Periods 5 and 6 transition elements have about the same atomic sizes.

- *Electronegativity.* The vertical trend in electronegativity seen in most transition groups is opposite the trend in main groups. Here, we see an *increase* in electronegativity from Period 4 to Period 5, but then no further increase in Period 6 (Figure 22.4B). The heavier elements, especially gold (EN = 2.4), become quite electronegative, with values exceeding those of most metalloids and even some nonmetals (e.g., EN of Te and of P = 2.1).

- *Ionization energy.* The relatively small increase in size combined with the relatively large increase in nuclear charge also explains why *the first ionization energy generally increases* down a transition group (Figure 22.4C). This trend also runs counter to the pattern in the main groups.

- *Density.* Atomic size, and therefore volume, is inversely related to density. Across a period, densities increase, then level off, and finally dip a bit at the end of a series (Figure 22.4D). Down a transition group, densities increase dramatically because atomic volumes change little from Period 5 to 6, but atomic masses increase significantly. As a result, the Period 6 series contains some of the densest elements: tungsten, rhenium, osmium, iridium, platinum, and gold have densities about 20 times that of water and twice that of lead.

Chemical Properties of the Transition Metals

Like their atomic and physical properties, the chemical properties of the transition elements are very different from those of the main-group elements. Let's examine the key properties in the Period 4 transition series.

Oxidation States One of the most characteristic chemical properties of the transition metals is the occurrence of *multiple oxidation states.* For example, in their compounds, vanadium exhibits two common positive oxidation states, chromium three, and manganese three (Figure 22.5A), and many other oxidation states are seen less often. The ns and $(n - 1)d$ electrons are so close in energy that transition elements can use all or most of these electrons in bonding. This behavior is markedly different from that of the main-group metals, which display one or, at most, two oxidation states in their compounds.

The highest oxidation state of elements in Groups 3B(3) through 7B(7) is equal to the group number, as shown in Table 22.2. These oxidation states are seen when the elements combine with highly electronegative oxygen or fluorine. For instance, in the oxoanion solutions shown in Figure 22.5B, vanadium occurs as the vanadate ion (VO_4^{3-}; O.N. of V = +5), chromium occurs as the dichromate ion ($Cr_2O_7^{2-}$; O.N. of Cr = +6), and manganese occurs as the permanganate ion (MnO_4^-; O.N. of Mn = +7). In contrast, elements in Groups 8B(8), 8B(9), and 8B(10) exhibit fewer oxidation states, and the highest state is less common and never equal to the group number. For example, we never encounter iron in the +8 state and only rarely in the +6 state. The +2 and +3 states are the most common ones for iron and cobalt, and the +2 state is most common for nickel, copper, and zinc. *The +2 oxidation state is common because ns^2 electrons are readily lost.*

Metallic Behavior and Reducing Strength Atomic size and oxidation state have a major effect on the nature of bonding in transition metal compounds. Like the metals in Groups 3A(13), 4A(14), and 5A(15), the transition elements in their *lower* oxidation states behave chemically more like metals. That is, *ionic bonding is more prevalent for the lower oxidation states, and covalent bonding is more*

A

B

Figure 22.5 Aqueous oxoanions of transition elements. A, Often, a given transition element has multiple oxidation states. Here, Mn is shown in the +2 (Mn^{2+}, *left*), the +6 (MnO_4^{2-}, *middle*), and the +7 states (MnO_4^-, *right*). **B,** The highest possible oxidation state equals the group number in these oxoanions: VO_4^{3-} (*left*), $Cr_2O_7^{2-}$ (*middle*), and MnO_4^- (*right*).

Table 22.2 Oxidation States and d-Orbital Occupancy of the Period 4 Transition Metals*

Oxidation State	3B (3) Sc	4B (4) Ti	5B (5) V	6B (6) Cr	7B (7) Mn	8B (8) Fe	8B (9) Co	8B (10) Ni	1B (11) Cu	2B (12) Zn
0	d^1	d^2	d^3	d^5	d^5	d^6	d^7	d^8	d^{10}	d^{10}
+1			d^3	d^5	d^5	d^6	d^7	d^8	d^{10}	
+2		d^2	d^3	d^4	d^5	d^6	d^7	d^8	d^9	d^{10}
+3	d^0	d^1	d^2	d^3	d^4	d^5	d^6	d^7	d^8	
+4		d^0	d^1	d^2	d^3	d^4	d^5	d^6		
+5			d^0	d^1	d^2		d^4			
+6				d^0	d^1	d^2				
+7					d^0					

*The most important orbital occupancies are in color.

Table 22.3 Standard Electrode Potentials of Period 4 M²⁺ Ions

Half-Reaction	$E°$ (V)
$Ti^{2+}(aq) + 2e^- \rightleftharpoons Ti(s)$	−1.63
$V^{2+}(aq) + 2e^- \rightleftharpoons V(s)$	−1.19
$Cr^{2+}(aq) + 2e^- \rightleftharpoons Cr(s)$	−0.91
$Mn^{2+}(aq) + 2e^- \rightleftharpoons Mn(s)$	−1.18
$Fe^{2+}(aq) + 2e^- \rightleftharpoons Fe(s)$	−0.44
$Co^{2+}(aq) + 2e^- \rightleftharpoons Co(s)$	−0.28
$Ni^{2+}(aq) + 2e^- \rightleftharpoons Ni(s)$	−0.25
$Cu^{2+}(aq) + 2e^- \rightleftharpoons Cu(s)$	0.34
$Zn^{2+}(aq) + 2e^- \rightleftharpoons Zn(s)$	−0.76

prevalent for the higher states. For example, at room temperature, $TiCl_2$ is an ionic solid, whereas $TiCl_4$ is a molecular liquid. In the higher oxidation states, the atoms have higher charge densities, so they polarize the electron clouds of the nonmetal ions more strongly and the bonding is more covalent. For the same reason, the oxides become less basic as the oxidation state increases: TiO is weakly basic in water, but TiO_2 is amphoteric (reacts with both acid and base).

Table 22.3 shows the standard electrode potentials of the Period 4 transition metals in their +2 oxidation state in acid solution. Note that, in general, reducing strength decreases across the series. All the Period 4 transition metals, except copper, are active enough to reduce H^+ from aqueous acid to form hydrogen gas. In contrast to the rapid reaction at room temperature of the Group 1A(1) and 2A(2) metals with water, however, the transition metals have an oxide coating that allows rapid reaction only with hot water or steam.

Color and Magnetism of Compounds *Most main-group ionic compounds are colorless* because the metal ion has a filled outer level (noble gas electron configuration). With only much higher energy orbitals available to receive an excited electron, the ion does not absorb visible light. In contrast, electrons in a partially filled *d* sublevel can absorb visible wavelengths and move to slightly higher energy *d* orbitals. As a result, *many transition metal compounds have striking colors.* Exceptions are the compounds of scandium, titanium(IV), and zinc, which are colorless because their metal ions have either an empty *d* sublevel (Sc^{3+} or Ti^{4+}: [Ar] $3d^0$) or a filled one (Zn^{2+}: [Ar] $3d^{10}$) (Figure 22.6).

Figure 22.6 Colors of representative compounds of the Period 4 transition metals. Staggered from left to right, the compounds are scandium oxide (*white*), titanium(IV) oxide (*white*), vanadyl sulfate dihydrate (*light blue*), sodium chromate (*yellow*), manganese(II) chloride tetrahydrate (*light pink*), potassium ferricyanide (*red-orange*), cobalt(II) chloride hexahydrate (*violet*), nickel(II) nitrate hexahydrate (*green*), copper(II) sulfate pentahydrate (*blue*), and zinc sulfate heptahydrate (*white*).

Magnetic properties are also related to sublevel occupancy (Section 8.5). Recall that a *paramagnetic* substance has atoms or ions with unpaired electrons, which cause it to be attracted to an external magnetic field. A *diamagnetic* substance has only paired electrons, so it is unaffected (or slightly repelled) by a magnetic field. *Most main-group metal ions are diamagnetic* for the same reason they are colorless: all their electrons are paired. In contrast, *many transition metal compounds are paramagnetic because of their unpaired d electrons.* For example, $MnSO_4$ is paramagnetic, but $CaSO_4$ is diamagnetic. The Ca^{2+} ion has the electron configuration of argon, whereas Mn^{2+} has a d^5 configuration. Transition metal ions with a d^0 or d^{10} configuration are diamagnetic and colorless.

SECTION SUMMARY
All transition elements are metals. Atoms of the *d*-block elements have $(n - 1)d$ orbitals being filled, and their ions have an empty *ns* orbital. Unlike the trends in the main-group elements, atomic size, electronegativity, and first ionization energy change relatively little across a transition series. Because of the lanthanide contraction, atomic size changes little from Period 5 to 6 in a transition metal group; thus, electronegativity, first ionization energy, and density *increase* down the group. Transition metals typically have several oxidation states, with the +2 state most common. The elements exhibit more metallic behavior in their lower states. Most Period 4 transition metals are active enough to reduce hydrogen ion from acid solution. Many transition metal compounds are colored and paramagnetic because the metal ion has unpaired *d* electrons.

22.2 COORDINATION COMPOUNDS

The most distinctive aspect of transition metal chemistry is the formation of **coordination compounds** (also called *complexes*). These are substances that contain at least one **complex ion,** a species consisting of *a central metal cation (either a transition metal or a main-group metal) that is bonded to molecules and/or anions called* **ligands.** In order to maintain charge neutrality in the coordination compound, the complex ion is typically associated with other ions, called **counter ions.**

A typical coordination compound appears in Figure 22.7A: the coordination compound is $[Co(NH_3)_6]Cl_3$, the complex ion (always enclosed in square brackets) is $[Co(NH_3)_6]^{3+}$, the six NH_3 molecules bonded to the central Co^{3+} are ligands,

Figure 22.7 Components of a coordination compound. Coordination compounds, shown here as models (*top*), perspective drawings (*middle*), and chemical formulas (*bottom*), typically consist of a complex ion and counter ions to neutralize the charge. The complex ion has a central metal ion surrounded by ligands. **A,** When solid $[Co(NH_3)_6]Cl_3$ dissolves, the complex ion and the counter ions separate, but the ligands remain bound to the metal ion. Six ligands around the metal ion give the complex ion an octahedral geometry. **B,** Complex ions with a central d^8 metal ion have four ligands and a square planar geometry.

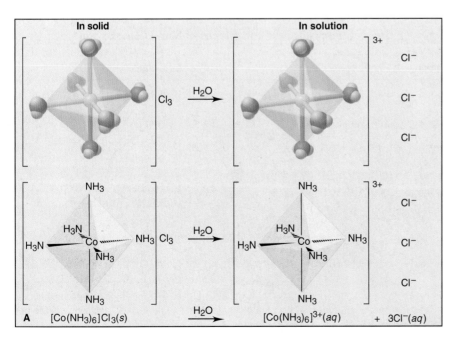

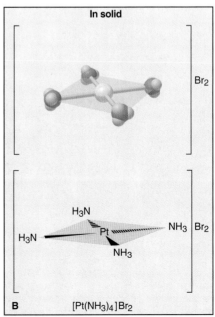

and the three Cl^- ions are counter ions. *A coordination compound behaves like an electrolyte in water:* the complex ion and counter ions separate from each other. But the complex ion behaves like a polyatomic ion: *the ligands and central metal ion remain attached.* Thus, as Figure 22.7A shows, 1 mol of $[Co(NH_3)_6]Cl_3$ yields 1 mol of $[Co(NH_3)_6]^{3+}$ ions and 3 mol of Cl^- ions.

We discussed the Lewis acid-base properties of hydrated metal ions, which are a type of complex ion, in Section 18.8, and we examined complex-ion equilibria in Section 19.4. In this section, we consider the bonding, structure, and properties of complex ions.

Complex Ions: Coordination Numbers, Geometries, and Ligands

A complex ion is described by the metal ion and the number and types of ligands attached to it. Its structure has three key characteristics—coordination number, geometry, and number of donor atoms per ligand:

- *Coordination number.* The **coordination number** is the *number of ligand atoms* bonded directly to the central metal ion and is *specific* for a given metal ion in a particular oxidation state and compound. In general, *the most common coordination number in complex ions is 6,* but 2 and 4 are often seen, and some higher ones are also known.
- *Geometry. The geometry (shape) of a complex ion depends on the coordination number and nature of the metal ion.* Table 22.4 shows the geometries associated with the coordination numbers 2, 4, and 6, with some examples of each. A complex ion whose metal ion has a coordination number of 2 is *linear.* The coordination number 4 gives rise to either of two geometries—square planar or tetrahedral. Most d^8 metal ions form *square planar* complex ions (Figure 22.7B). The d^{10} ions are among those that form *tetrahedral* complex ions. A coordination number of 6 results in an *octahedral* geometry, as shown by $[Co(NH_3)_6]^{3+}$ in Figure 22.7A. Note the similarity with some of the molecular shapes in VSEPR theory (Section 10.2).
- *Donor atoms per ligand.* The ligands of complex ions are *molecules or anions* with one or more **donor atoms** that each *donate a lone pair of electrons* to the metal ion to form a covalent bond. Because they have at least one lone pair, donor atoms often come from Group 5A(15), 6A(16), or 7A(17).

Table 22.4 **Coordination Numbers and Shapes of Some Complex Ions**

Coordination Number	Shape		Examples
2	Linear		$[CuCl_2]^-$, $[Ag(NH_3)_2]^+$, $[AuCl_2]^-$
4	Square planar		$[Ni(CN)_4]^{2-}$, $[PdCl_4]^{2-}$, $[Pt(NH_3)_4]^{2+}$, $[Cu(NH_3)_4]^{2+}$
4	Tetrahedral	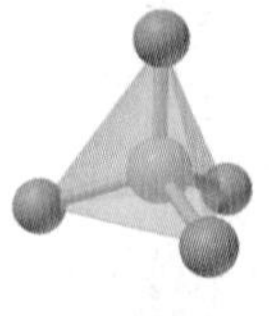	$[Cu(CN)_4]^{3-}$, $[Zn(NH_3)_4]^{2+}$, $[CdCl_4]^{2-}$, $[MnCl_4]^{2-}$
6	Octahedral	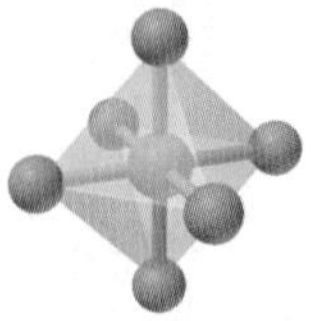	$[Ti(H_2O)_6]^{3+}$, $[V(CN)_6]^{4-}$, $[Cr(NH_3)_4Cl_2]^+$, $[Mn(H_2O)_6]^{2+}$, $[FeCl_6]^{3-}$, $[Co(en)_3]^{3+}$

Table 22.5 | **Some Common Ligands in Coordination Compounds**

Ligand Type	Examples
Monodentate	$H_2\ddot{O}:$ water $:\ddot{F}:^-$ fluoride ion $[:C\equiv N:]^-$ cyanide ion $[:\ddot{O}-H]^-$ hydroxide ion $:NH_3$ ammonia $:\ddot{Cl}:^-$ chloride ion $[:\ddot{S}=C=\ddot{N}:]^-$ thiocyanate ion (or) $[:\ddot{O}-N=\ddot{O}:]^-$ nitrite ion (or)

Bidentate

H_2C-CH_2 / H_2N NH_2 ethylenediamine (en)

oxalate ion $\left[\begin{array}{c}:\ddot{O}\quad\quad:\ddot{O}.\\ \backslash C-C\diagup\\ :\ddot{O}\diagup\quad\quad\diagdown\ddot{O}:\end{array}\right]^{2-}$

Polydentate

H_2C-CH_2 CH_2-CH_2 / H_2N NH NH_2
diethylenetriamine

triphosphate ion $\left[:\ddot{O}-\overset{:O:}{\underset{:O:}{\overset{\|}{P}}}-\ddot{O}-\overset{:O:}{\underset{:O:}{\overset{\|}{P}}}-\ddot{O}-\overset{:O:}{\underset{:O:}{\overset{\|}{P}}}-\ddot{O}:\right]^{5-}$

ethylenediaminetetraacetate ion ($EDTA^{4-}$)

Ligands are classified in terms of the number of donor atoms, or "teeth," that each uses to bond to the central metal ion. *Monodentate* (Latin, "one-toothed") ligands use a single donor atom. *Bidentate* ligands have two donor atoms, each of which bonds to the metal ion. *Polydentate* ligands have more than two donor atoms. Table 22.5 shows some common ligands in coordination compounds (with donor atoms in colored type). Bidentate and polydentate ligands give rise to *rings* in the complex ion. For instance, ethylenediamine (abbreviated *en* in formulas) has a chain of four atoms ($:N-C-C-N:$); the two electron-donating N atoms bond to the metal atom, which forms a five-membered ring. Such ligands seem to grab the metal ion like claws, so a complex ion that contains them is also called a **chelate** (pronounced "KEY-late"; Greek *chela,* "crab's claw").

Formulas and Names of Coordination Compounds

There are three important rules for writing the formulas of coordination compounds, the first two being the same rules for writing the formula of any ionic compound:

1. *The cation is written before the anion.*
2. *The charge of the cation(s) is balanced by the charge of the anion(s).*
3. *For the complex ion, neutral ligands are written before anionic ligands, and the formula for the whole ion is placed in brackets.*

Let's apply these rules as we examine the combinations of ions in several coordination compounds. *The whole complex ion may be a cation or an anion.* A complex cation has anionic counter ions, and a complex anion has cationic counter ions. It's easy to find the charge of the central metal ion. For example, in $K_2[Co(NH_3)_2Cl_4]$, two K^+ counter ions balance the charge of the complex anion $[Co(NH_3)_2Cl_4]^{2-}$, which contains two NH_3 molecules and four Cl^- ions as ligands. The two NH_3 are neutral, the four Cl^- have a total charge of $4-$, and the entire complex ion has a charge of $2-$, so the central metal ion must be Co^{2+}:

$$\text{Charge of complex ion} = \text{Charge of metal ion} + \text{total charge of ligands}$$
$$2- = \text{Charge of metal ion} + [(2\times 0) + (4\times 1-)]$$

So,   Charge of metal ion $= (2-) - (4-) = 2+$

In the compound $[Co(NH_3)_4Cl_2]Cl$, the complex ion is $[Co(NH_3)_4Cl_2]^+$ and one Cl^- is the counter ion. The four NH_3 ligands are neutral, the two Cl^- ligands have a total charge of $2-$, and the complex cation has a charge of $1+$, so the central metal ion must be Co^{3+} [that is, $1+ = (3+) + (2-)$]. Some coordination compounds have a complex cation *and* a complex anion, as in $[Co(NH_3)_5Br]_2[Fe(CN)_6]$. In this compound, the complex cation is $[Co(NH_3)_5Br]^{2+}$, with Co^{3+}, and the complex anion is $[Fe(CN)_6]^{4-}$, with Fe^{2+}.

Coordination compounds, originally named after the person who first prepared them or from their color, are now named systematically using a set of rules:

1. *The cation is named before the anion.* In naming $[Co(NH_3)_4Cl_2]Cl$, for example, we name the $[Co(NH_3)_4Cl_2]^+$ ion before the Cl^- ion. Thus, the name is

   tetraamminedichlorocobalt(III) chloride

   The only space in the name appears between the cation and the anion.

2. *Within the complex ion, the ligands are named, in alphabetical order,* **before** *the metal ion.* Note that in the $[Co(NH_3)_4Cl_2]^+$ ion of the compound named in rule 1, the four NH_3 and two Cl^- are named before the Co^{3+}.

3. *Neutral ligands generally have the molecule name,* but there are a few exceptions (Table 22.6). *Anionic ligands drop the -ide and add -o after the root name;* thus, the name *fluoride* for the F^- ion becomes the ligand name *fluoro.* The two ligands in $[Co(NH_3)_4Cl_2]^+$ are *ammine* (NH_3) and *chloro* (Cl^-) with *ammine* coming before *chloro* alphabetically.

Table 22.6   Names of Some Neutral and Anionic Ligands

Neutral		Anionic	
Name	**Formula**	**Name**	**Formula**
Aqua	H_2O	Fluoro	F^-
Ammine	NH_3	Chloro	Cl^-
Carbonyl	CO	Bromo	Br^-
Nitrosyl	NO	Iodo	I^-
		Hydroxo	OH^-
		Cyano	CN^-

4. *A numerical prefix indicates the number of ligands of a particular type.* For example, *tetra*ammine denotes *four* NH_3, and *di*chloro denotes *two* Cl^-. Other prefixes are *tri-, penta-,* and *hexa-.* These prefixes do *not* affect the alphabetical order; thus, *tetraammine* comes before *dichloro.* Because some ligand names already contain a numerical prefix (such as ethylene*di*amine), we use *bis* (2), *tris* (3), or *tetrakis* (4) to indicate the number of such ligands, followed by the ligand name in parentheses. Therefore, a complex ion that has two ethylenediamine ligands has *bis(ethylenediamine)* in its name.

5. *The oxidation state of the central metal ion is given by a Roman numeral (in parentheses) only* if the metal ion can have more than one state, as in the compound named in rule 1.

6. *If the complex ion is an anion, we drop the ending of the metal name and add -ate.* Thus, the name for $K[Pt(NH_3)Cl_5]$ is

   potassium amminepentachloroplatinate(IV)

   (Note that there is one K^+ counter ion, so the complex anion has a charge of $1-$. The five Cl^- ligands have a total charge of $5-$, so Pt must be in the $+4$ oxidation state.) For some metals, we use the Latin root with the *-ate* ending, as shown in Table 22.7. For example, the name for $Na_4[FeBr_6]$ is

   sodium hexabromoferrate(II)

Table 22.7   Names of Some Metal Ions in Complex Anions

Metal	Name in Anion
Iron	Ferrate
Copper	Cuprate
Lead	Plumbate
Silver	Argentate
Gold	Aurate
Tin	Stannate

SAMPLE PROBLEM 22.2 Writing Names and Formulas of
Coordination Compounds

Problem (a) What is the systematic name of $Na_3[AlF_6]$?
(b) What is the systematic name of $[Co(en)_2Cl_2]NO_3$?
(c) What is the formula of tetraamminebromochloroplatinum(IV) chloride?
(d) What is the formula of hexaamminecobalt(III) tetrachloroferrate(III)?

Plan We use the rules that were presented above and refer to Tables 22.6 and 22.7.

Solution (a) The complex ion is $[AlF_6]^{3-}$. There are six (*hexa-*) F^- ions (*fluoro*) as ligands, so we have *hexafluoro*. The complex ion is an anion, so the ending of the metal ion (aluminum) must be changed to *-ate:* hexafluoroaluminate. Aluminum has only the $+3$ oxidation state, so we do *not* use a Roman numeral. The positive counter ion is named first and separated from the anion by a space: sodium hexafluoroaluminate.

(b) Listed alphabetically, there are two Cl^- (*dichloro*) and two en [*bis(ethylenediamine)*] as ligands. The complex ion is a cation, so the metal name is unchanged, but we specify its oxidation state because cobalt can have several. One NO_3^- balances the $1+$ cation charge: with $2-$ for two Cl^- and 0 for two en, the metal must be *cobalt(III)*. The word *nitrate* follows a space: dichlorobis(ethylenediamine)cobalt(III) nitrate.

(c) The central metal ion is written first, followed by the neutral ligands and then (in alphabetical order) by the negative ligands. *Tetraammine* is four NH_3, *bromo* is one Br^-, *chloro* is one Cl^-, and *platinate(IV)* is Pt^{4+}, so the complex ion is $[Pt(NH_3)_4BrCl]^{2+}$. Its $2+$ charge is the sum of $4+$ for Pt^{4+}, 0 for four NH_3, $1-$ for one Br^-, and $1-$ for one Cl^-. To balance the $2+$ charge, we need two Cl^- counter ions: $[Pt(NH_3)_4BrCl]Cl_2$.

(d) This compound consists of two different complex ions. In the cation, *hexaammine* is six NH_3 and *cobalt(III)* is Co^{3+}, so the cation is $[Co(NH_3)_6]^{3+}$. The $3+$ charge is the sum of $3+$ for Co^{3+} and 0 for six NH_3. In the anion, *tetrachloro* is four Cl^-, and *ferrate(III)* is Fe^{3+}, so the anion is $[FeCl_4]^-$. The $1-$ charge is the sum of $3+$ for Fe^{3+} and $4-$ for four Cl^-. In the neutral compound, one $3+$ cation is balanced by three $1-$ anions: $[Co(NH_3)_6][FeCl_4]_3$.

Check Reverse the process to be sure you obtain the name or formula asked for in the problem.

FOLLOW-UP PROBLEM 22.2 (a) What is the name of $[Cr(H_2O)_5Br]Cl_2$?
(b) What is the formula of barium hexacyanocobaltate(III)?

Isomerism in Coordination Compounds

Isomers are compounds with the same chemical formula but different properties. We discussed many aspects of isomerism in the context of organic compounds in Section 15.2; it may be helpful to review that section now. Figure 22.8 presents an overview of the most common types of isomerism that occur in coordination compounds.

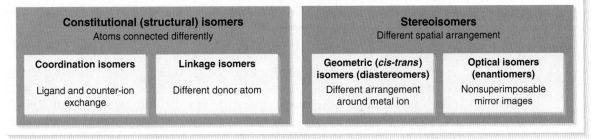

Figure 22.8 Important types of isomerism in coordination compounds.

A *Nitro* isomer, [Co(NH₃)₅(NO₂)]Cl₂

B *Nitrito* isomer, [Co(NH₃)₅(ONO)]Cl₂

Figure 22.9 Linkage isomers of a complex ion.

Constitutional Isomers: Same Atoms Connected Differently Two compounds with the same formula, but with the atoms connected differently, are called **constitutional (structural) isomers.** Coordination compounds exhibit two types of constitutional isomers: one involves a difference in the composition of the complex ion, the other in the donor atom of the ligand.

1. **Coordination isomers** occur when the composition of the complex ion changes but not that of the compound. One way this type of isomerism occurs is when ligand and counter ion exchange positions, as in $[Pt(NH_3)_4Cl_2](NO_2)_2$ and $[Pt(NH_3)_4(NO_2)_2]Cl_2$. In the first compound, the Cl^- ions are the ligands, and the NO_2^- ions are counter ions; in the second, the roles are reversed. Another way that this type of isomerism occurs is in compounds of two complex ions in which the two sets of ligands in one compound are reversed in the other, as in $[Cr(NH_3)_6][Co(CN)_6]$ and $[Co(NH_3)_6][Cr(CN)_6]$; note that NH_3 is a ligand of Cr^{3+} in one compound and of Co^{3+} in the other.

2. **Linkage isomers** occur when the composition of the complex ion remains the same but the attachment of the ligand donor atom changes. Some ligands can bind to the metal ion through *either of two donor atoms*. For example, the nitrite ion can bind through a lone pair on either the N atom (*nitro,* O₂N:) or one of the O atoms (*nitrito,* ONO:) to give linkage isomers, as in the orange compound pentaammine*nitro*cobalt(III) chloride $[Co(NH_3)_5(NO_2)]Cl_2$ (Figure 22.9A) and its red linkage isomer pentaammine*nitrito*cobalt(III) chloride $[Co(NH_3)_5(ONO)]Cl_2$ (Figure 22.9B). Another example of a ligand with two different donor atoms is the cyanate ion, which can attach via a lone pair on the O atom (*cyanato,* NCO:) or the N atom (*isocyanato,* OCN:); the thiocyanate ion behaves similarly, attaching via the S atom or the N atom:

$$\left[\begin{array}{c} \ddot{\text{O}} \\ \diagdown \text{N:} \\ \diagup \\ \ddot{\text{O}} \end{array} \right]^- \qquad [\ddot{\text{O}}=\text{C}=\ddot{\text{N}}:]^- \qquad [\ddot{\text{S}}=\text{C}=\ddot{\text{N}}:]^-$$

nitrite   cyanate   thiocyanate

Stereoisomers: Different Spatial Arrangements of Atoms In the case of **stereoisomers,** the atoms have the same connections but different spatial arrangements. The two types we discussed for organic compounds, called *geometric* and *optical* isomers, occur with coordination compounds as well:

1. **Geometric isomers** (also called *cis-trans* isomers and, sometimes, *diastereomers*) occur when atoms or groups of atoms are arranged differently in space relative to the central metal ion. For example, the square planar $[Pt(NH_3)_2Cl_2]$ has two arrangements (Figure 22.10A). The isomer with identical ligands *next* to each other is *cis*-diamminedichloroplatinum(II), and the one with identical ligands *across* from each other is *trans*-diamminedichloroplatinum(II); the *cis* isomer has striking antitumor activity, but the *trans* isomer has none! Octahedral complexes also exhibit *cis-trans* isomerism (Figure 22.10B). The *cis* isomer of the

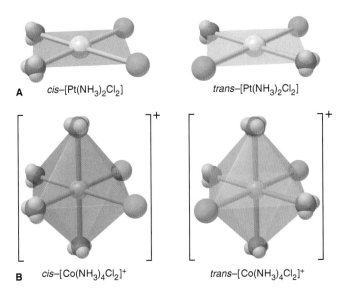

Figure 22.10 Geometric (*cis-trans*) isomerism. A, The *cis* and *trans* isomers of the square planar coordination compound [Pt(NH₃)₂Cl₂]. **B,** The *cis* and *trans* isomers of the octahedral complex ion [Co(NH₃)₄Cl₂]⁺. The colored shapes represent the actual colors of the species.

A *cis*–[Pt(NH₃)₂Cl₂]   *trans*–[Pt(NH₃)₂Cl₂]

B *cis*–[Co(NH₃)₄Cl₂]⁺   *trans*–[Co(NH₃)₄Cl₂]⁺

[Co(NH₃)₄Cl₂]⁺ ion has the two Cl⁻ ligands next to each other and is violet, whereas the *trans* isomer has these two ligands across from each other and is green.

2. **Optical isomers** (also called *enantiomers*) occur when a molecule and its mirror image cannot be superimposed (see Figures 15.8 and 15.9). Octahedral complex ions show many examples of optical isomerism, which we can observe by rotating one isomer and seeing if it is superimposable on the other isomer (its mirror image). For example, as you can see in Figure 22.11A, the two structures (I and II) of [Co(en)₂Cl₂]⁺, the *cis*-dichlorobis(ethylenediamine)cobalt(III) ion, are mirror images of each other. Rotate structure I 180° around a vertical axis, and you obtain III. The Cl⁻ ligands of III match those of II, but the en ligands do not: II and III (rotated I) are not superimposable; therefore, they are

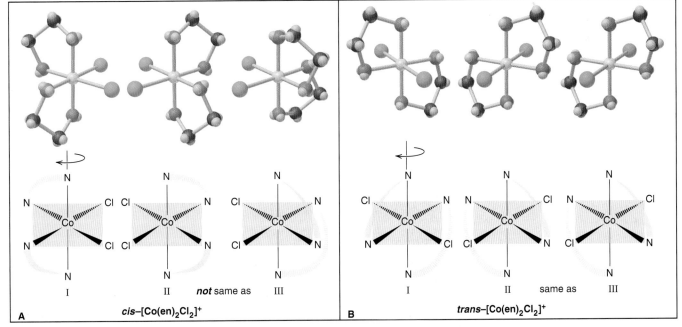

Figure 22.11 Optical isomerism in an octahedral complex ion. A, Structure I and its mirror image, structure II, are optical isomers of *cis*-[Co(en)₂Cl₂]⁺. Rotating structure I gives structure III, which is *not* the same as structure II. (The curved wedges represent the bidentate ligand ethylenediamine, H₂N—CH₂—CH₂—NH₂.) **B,** The *trans* isomer does *not* have optical isomers. Rotating structure I gives III, which is *identical* to II, the mirror image of I.

optical isomers. Unlike other types of isomers, which have distinct physical properties, optical isomers are physically identical in all ways but one: *the direction in which they rotate the plane of polarized light.* One isomer is designated *d*-[Co(en)$_2$Cl$_2$]$^+$ and the other is *l*-[Co(en)$_2$Cl$_2$]$^+$, depending on whether it rotates the plane of polarized light to the right (*d*- for "dextro-") or to the left (*l*- for "levo-"). (The *d*- or *l*- designation can only be determined experimentally, *not* by examination of the structure.) In contrast, as shown in Figure 22.11B, the two structures of the *trans*-dichlorobis(ethylenediamine)cobalt(III) ion are *not* optical isomers: rotate I 90° around a vertical axis and you obtain III, which *is* superimposable on II.

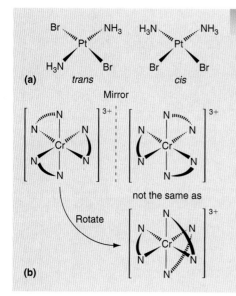

(a)   *trans*   *cis*

Mirror

not the same as

Rotate

(b)

SAMPLE PROBLEM 22.3   Determining the Type of Stereoisomerism

Problem Draw all stereoisomers for each of the following and state the type of isomerism:
(a) [Pt(NH$_3$)$_2$Br$_2$] (square planar)   **(b)** [Cr(en)$_3$]$^{3+}$ (en = H$_2$N̈CH$_2$CH$_2$N̈H$_2$)
Plan We first determine the geometry around each metal ion and the nature of the ligands. If there are two different ligands that can be placed in different positions relative to each other, geometric (*cis-trans*) isomerism occurs. Then, we see whether the mirror image of an isomer is superimposable on the original. If it is *not*, optical isomerism occurs.
Solution (a) The Pt(II) complex is square planar, and there are two different monodentate ligands. Each pair of ligands can lie next to or across from each other (see structures in margin). Thus, geometric isomerism occurs. Each isomer *is* superimposable on its mirror image, so there is no optical isomerism.
(b) Ethylenediamine (en) is a bidentate ligand. Cr^{3+} has a coordination number of 6 and an octahedral geometry, like Co^{3+}. The three bidentate ligands are identical, so there is no geometric isomerism. However, the complex ion has a nonsuperimposable mirror image (see structures in margin). Thus, optical isomerism occurs.

FOLLOW-UP PROBLEM 22.3 What stereoisomers, if any, are possible for the [Co(NH$_3$)$_2$(en)Cl$_2$]$^+$ ion?

SECTION SUMMARY
Coordination compounds consist of a complex ion and charge-balancing counter ions. The complex ion has a central metal ion bonded to neutral and/or anionic ligands, which have one or more donor atoms that each provide a lone pair of electrons. The most common geometry is octahedral (six ligand atoms bonding). Formulas and names of coordination compounds follow systematic rules. These compounds can exhibit constitutional isomerism (coordination and linkage) and stereoisomerism (geometric and optical).

22.3 THEORETICAL BASIS FOR THE BONDING AND PROPERTIES OF COMPLEXES

In this section, we consider two models that address, in different ways, several key features of complexes: how metal-ligand bonds form, why certain geometries are preferred, and why these complexes are brightly colored and often paramagnetic.

Application of Valence Bond Theory to Complex Ions

Valence bond (VB) theory, which helps explain bonding and structure in main-group compounds (Section 11.1), is also used to describe bonding in complex ions. In the formation of a complex ion, the filled ligand orbital overlaps the empty metal-ion orbital. *The ligand (Lewis base) donates the electron pair, and the metal ion (Lewis acid) accepts it to form one of the covalent bonds of the complex ion (Lewis adduct)* (Section 18.8). Such a bond, in which one atom in the bond contributes both electrons, is called a **coordinate covalent bond,** although, once formed, it is identical to any covalent single bond. Recall that the

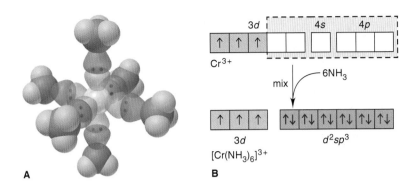

A

B

Figure 22.12 Hybrid orbitals and bonding in the octahedral $[Cr(NH_3)_6]^{3+}$ ion. A, VB depiction of the $Cr(NH_3)_6{}^{3+}$ ion. **B,** The partial orbital diagrams depict the mixing of two 3d, one 4s, and three 4p orbitals in Cr^{3+} to form six d^2sp^3 hybrid orbitals, which are filled with six NH_3 lone pairs (*red*).

VB concept of hybridization proposes the mixing of particular combinations of *s*, *p*, and *d* orbitals to give sets of hybrid orbitals, which have specific geometries. Similarly, for coordination compounds, the model proposes that *the number and type of metal-ion hybrid orbitals occupied by ligand lone pairs determine the geometry of the complex ion.* Let's discuss the orbital combinations that lead to octahedral, square planar, and tetrahedral geometries.

Octahedral Complexes The hexaamminechromium(III) ion, $[Cr(NH_3)_6]^{3+}$, illustrates the application of VB theory to an *octahedral complex* (Figure 22.12). The six lowest energy empty orbitals of the Cr^{3+} ion—two 3d, one 4s, and three 4p—mix and become six equivalent d^2sp^3 hybrid orbitals that point toward the corners of an octahedron.* Six NH_3 molecules donate lone pairs from their nitrogens to form six metal-ligand bonds. The three unpaired 3d electrons of the central Cr^{3+} ion ([Ar] $3d^3$), which make the complex ion paramagnetic, remain in unhybridized orbitals.

Square Planar Complexes Metal ions with a d^8 configuration usually form *square planar complexes* (Figure 22.13). In the $[Ni(CN)_4]^{2-}$ ion, for example, the model proposes that one 3d, one 4s, and two 4p orbitals of Ni^{2+} mix and form four dsp^2 hybrid orbitals, which point to the corners of a square and accept one electron pair from each of four CN^- ligands.

A look at the ground-state electron configuration of the Ni^{2+} ion, however, raises a key question: how can the Ni^{2+} ion ([Ar] $3d^8$) offer an empty 3d orbital for accepting a lone pair, if its eight 3d electrons lie in three filled and two half-filled orbitals? Apparently, in the d^8 configuration of Ni^{2+}, electrons in the half-filled orbitals pair up and leave one 3d orbital empty. This explanation is consistent with the fact that the complex is diamagnetic (no unpaired electrons).

---

*Note the distinction between the hybrid-orbital designation here and that for octahedral molecules like SF_6. The designation gives the orbitals in energy order within a given *n* value. In the $[Cr(NH_3)_6]^{3+}$ complex ion, the d orbitals have a *lower* n value than the s and p orbitals, so the hybrid orbitals are d^2sp^3. For the orbitals in SF_6, the d orbitals have the *same* n value as the s and p, so the hybrid orbitals are sp^3d^2.

A

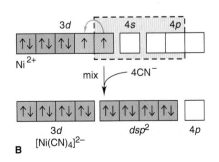

B

Figure 22.13 Hybrid orbitals and bonding in the square planar $[Ni(CN)_4]^{2-}$ ion. A, VB depiction of $[Ni(CN)_4]^{2-}$. **B,** Two lone 3d electrons pair up and free one 3d orbital for hybridization with the 4s and two of the 4p orbitals to form four dsp^2 orbitals, which become occupied with lone pairs (*red*) from four CN^- ligands.

Figure 22.14 Hybrid orbitals and bonding in the tetrahedral [Zn(OH)₄]²⁻ ion. A, VB depiction of [Zn(OH)₄]²⁻. **B,** Mixing one 4s and three 4p orbitals gives four sp³ hybrid orbitals available for accepting lone pairs (red) from OH⁻ ligands.

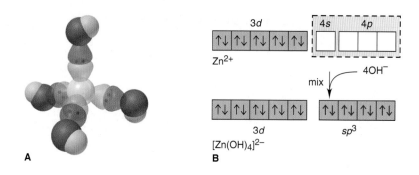

Moreover, it requires that the energy *gained* by using a 3d orbital for bonding in the hybrid orbital is greater than the energy *required* to overcome repulsions from pairing the 3d electrons.

Tetrahedral Complexes Metal ions that have a filled *d* sublevel, such as Zn^{2+} ([Ar] $3d^{10}$), often form *tetrahedral complexes* (Figure 22.14). For the complex ion $[Zn(OH)_4]^{2-}$, for example, VB theory proposes that the lowest available Zn^{2+} orbitals—one 4s and three 4p—mix to become four sp^3 hybrid orbitals that point to the corners of a tetrahedron and are occupied by four lone pairs, one from each of four OH⁻ ligands.

Crystal Field Theory

The VB model is easy to picture and rationalizes bonding and shape, but it treats the orbitals as little more than empty "slots" for accepting electron pairs. Consequently, it gives no insight into the colors of coordination compounds and sometimes predicts their magnetic properties incorrectly. In contrast to the VB approach, **crystal field theory** provides little insight about metal-ligand bonding but explains color and magnetism clearly. To do so, it highlights the *effects on the d-orbital energies of the metal ion as the ligands approach.* Before we discuss this theory, let's consider what causes a substance to be colored.

What Is Color? White light is electromagnetic radiation consisting of all wavelengths (λ) in the visible range (Section 7.1). It can be dispersed into a spectrum of colors, each of which has a narrower range of wavelengths. Objects appear colored in white light because they absorb certain wavelengths and reflect or transmit others: an opaque object *reflects* light, whereas a clear one *transmits* it. The reflected or transmitted light enters the eye and the brain perceives a color. If an object *absorbs* all visible wavelengths, it appears black; if it *reflects* all, it appears white.

Each color has a *complementary* color. For example, green and red are complementary colors. A mixture of complementary colors absorbs all visible wavelengths and appears black. Figure 22.15 shows these relationships on an artist's color wheel, where complementary colors appear as wedges opposite each other.

An object has a particular color for one of two reasons:

- It reflects (or transmits) light of *that* color. Thus, if an object absorbs all wavelengths *except* green, the reflected (or transmitted) light enters our eyes and is interpreted as green.
- It absorbs light of the *complementary* color. Thus, if the object absorbs only red, the *complement* of green, the remaining mixture of reflected (or transmitted) wavelengths enters our eyes and is interpreted as green also.

Table 22.8 lists the color absorbed and the resulting color perceived.

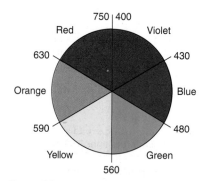

Figure 22.15 An artist's wheel. Colors, with approximate wavelength ranges (in nm), are shown as wedges. Complementary colors, such as red and green, lie opposite each other.

Table 22.8 | **Relation Between Absorbed and Observed Colors**

Absorbed Color	λ (nm)	Observed Color	λ (nm)
Violet	400	Green-yellow	560
Blue	450	Yellow	600
Blue-green	490	Red	620
Yellow-green	570	Violet	410
Yellow	580	Dark blue	430
Orange	600	Blue	450
Red	650	Green	520

Splitting of d Orbitals in an Octahedral Field of Ligands The crystal field model explains that the properties of complexes result from the splitting of d-orbital energies, which arises from electrostatic interactions between metal ion and ligands. The model assumes that a complex ion forms as a result of *electrostatic attractions between the metal cation and the negative charge of the ligands.* This negative charge is either partial, as in a polar neutral ligand like NH_3, or full, as in an anionic ligand like Cl^-. The ligands approach the metal ion along the mutually perpendicular x, y, and z axes, which minimizes the overall energy of the system.

Picture what happens as the ligands approach. Figure 22.16A shows six ligands moving toward a metal ion to form an octahedral complex. Let's see how the various d orbitals of the metal ion are affected as the complex forms. As the ligands approach, their electron pairs repel electrons in the five d orbitals. In the isolated metal ion, the d orbitals have equal energies despite their different orientations. In the electrostatic field of ligands, however, the d electrons are *repelled unequally because their orbitals have different orientations.* Because the ligands move along the x, y, and z axes, they approach *directly toward* the lobes of the $d_{x^2-y^2}$ and d_{z^2} orbitals (Figure 22.16B and C) but *between* the lobes of the d_{xy}, d_{xz}, and d_{yz} orbitals (Figure 22.16D to F). Thus, electrons in the $d_{x^2-y^2}$ and d_{z^2} orbitals experience *stronger* repulsions than those in the d_{xy}, d_{xz}, and d_{yz} orbitals.

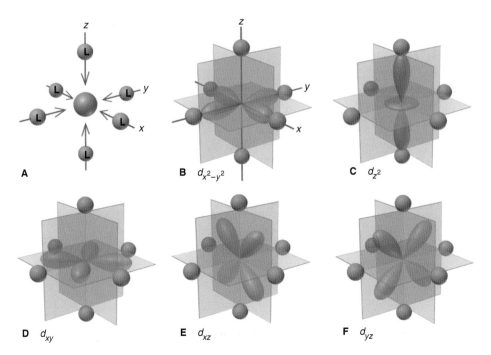

A **B** $d_{x^2-y^2}$ **C** d_{z^2}

D d_{xy} **E** d_{xz} **F** d_{yz}

Figure 22.16 The five d orbitals in an octahedral field of ligands. The direction of ligand approach influences the strength of repulsions of electrons in the five metal d orbitals. **A,** We assume that ligands approach a metal ion along the three linear axes in an octahedral orientation. **B** and **C,** Lobes of the $d_{x^2-y^2}$ and d_{z^2} orbitals lie *directly in line* with the approaching ligands, so repulsions are stronger. **D** to **F,** Lobes of the d_{xy}, d_{xz}, and d_{yz} orbitals lie *between* the approaching ligands, so repulsions are weaker.

Figure 22.17 Splitting of *d*-orbital energies by an octahedral field of ligands. Electrons in the *d* orbitals of the free metal ion experience an *average* net repulsion in the negative ligand field that increases all *d*-orbital energies. Electrons in the t_{2g} set are repelled less than those in the e_g set. The energy difference between these two sets is the crystal field splitting energy, Δ.

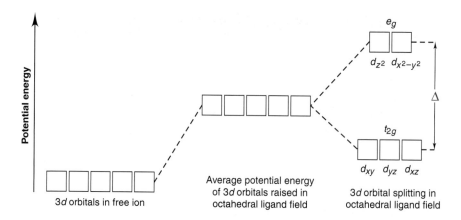

3*d* orbitals in free ion

Average potential energy of 3*d* orbitals raised in octahedral ligand field

3*d* orbital splitting in octahedral ligand field

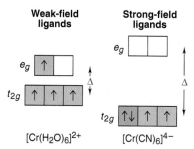

Weak-field ligands

$[Cr(H_2O)_6]^{2+}$

Strong-field ligands

$[Cr(CN)_6]^{4-}$

Figure 22.18 The effect of the ligand on splitting energy. Ligands interacting strongly with metal-ion *d* orbitals, such as CN^-, produce a larger Δ than those interacting weakly, such as H_2O.

An energy diagram of the orbitals shows that all five *d* orbitals are higher in energy in the forming complex than in the free metal ion because of repulsions from the approaching ligands, but *the orbital energies split, with two d orbitals higher in energy than the other three* (Figure 22.17). The two higher energy orbitals are called e_g **orbitals,** and the three lower energy ones are t_{2g} **orbitals.** (These designations refer to features of the orbitals that need not concern us here.)

The splitting of orbital energies is called the *crystal field effect,* and the difference in energy between the e_g and t_{2g} sets of orbitals is the **crystal field splitting energy (Δ).** Different ligands create crystal fields of different strength and, thus, cause the *d*-orbital energies to split to different extents. **Strong-field ligands** lead to a *larger* splitting energy (larger Δ); **weak-field ligands** lead to a *smaller* splitting energy (smaller Δ). For instance, H_2O is a weak-field ligand, and CN^- is a strong-field ligand (Figure 22.18). The magnitude of Δ relates directly to the color and magnetic properties of a complex.

Explaining the Colors of Transition Metal Compounds The remarkably diverse colors of coordination compounds are determined by the energy difference (Δ) between the t_{2g} and e_g orbital sets in their complex ions. When the ion absorbs light in the visible range, electrons are excited ("jump") from the lower energy t_{2g} level to the higher e_g level. In Chapter 7, you saw that the *difference* between two electronic energy levels in the ion is equal to the energy (and inversely related to the wavelength) of the absorbed photon:

$$\Delta E_{electron} = E_{photon} = h\nu = hc/\lambda$$

The substance has a color because only certain wavelengths of the incoming white light are absorbed.

Consider the $[Ti(H_2O)_6]^{3+}$ ion, which appears purple in aqueous solution (Figure 22.19). Hydrated Ti^{3+} is a d^1 ion, with the *d* electron in one of the three

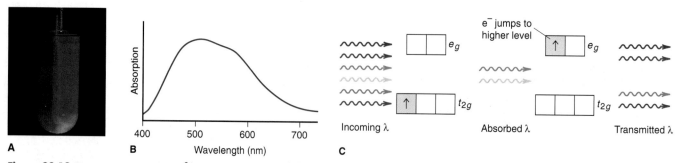

Figure 22.19 The color of $[Ti(H_2O)_6]^{3+}$. A, The hydrated Ti^{3+} ion is purple in aqueous solution. **B,** An absorption spectrum shows that incoming wavelengths corresponding to green and yellow light are absorbed, whereas other wavelengths are transmitted. **C,** A partial orbital diagram depicts the colors absorbed in the excitation of the *d* electron to the higher level.

lower energy t_{2g} orbitals. The energy difference (Δ) between the t_{2g} and e_g orbitals in this ion corresponds to the energy of photons spanning the green and yellow range. When white light shines on the solution, these colors of light are absorbed, and the electron jumps to one of the e_g orbitals. Red, blue, and violet light are transmitted, so the solution appears purple.

Absorption spectra show the wavelengths absorbed by a given metal ion with different ligands and by different metal ions with the same ligand. From such data, we relate the energy of the absorbed light to the Δ values, and two important observations emerge:

1. *For a given ligand, the color depends on the oxidation state of the metal ion.* For example, as shown in Figure 22.20A, a solution of $[V(H_2O)_6]^{2+}$ ion is violet, and a solution of $[V(H_2O)_6]^{3+}$ ion is yellow.
2. *For a given metal ion, the color depends on the ligand.* Even a single ligand substitution can have a major effect on the wavelengths absorbed and, thus, the color, as you can see for the two Cr^{3+} complex ions that are shown in Figure 22.20B.

The second observation allows us to rank ligands into a **spectrochemical series** with regard to their ability to split *d*-orbital energies. An abbreviated series, moving from weak-field ligands (small splitting, small Δ) to strong-field ligands (large splitting, large Δ), is shown in Figure 22.21.

Figure 22.20 Effects of the metal oxidation state and of ligand identity on color. A, Solutions of $[V(H_2O)_6]^{2+}$ (*left*) and $[V(H_2O)_6]^{3+}$ (*right*) ions have different colors. **B,** A change in even a single ligand can influence the color. The $[Cr(NH_3)_6]^{3+}$ ion is yellow-orange (*left*); the $[Cr(NH_3)_5Cl]^{2+}$ ion is purple (*right*).

Animation: Vanadium Reduction Online Learning Center

Figure 22.21 The spectrochemical series. As the crystal field strength of the ligand increases, the splitting energy (Δ) increases, so shorter wavelengths (λ) of light must be absorbed to excite electrons. Water is usually a weak-field ligand.

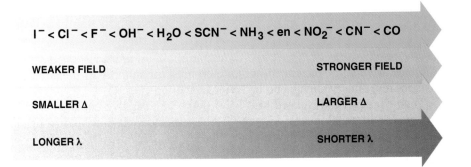

$I^- < Cl^- < F^- < OH^- < H_2O < SCN^- < NH_3 < en < NO_2^- < CN^- < CO$

WEAKER FIELD — STRONGER FIELD
SMALLER Δ — LARGER Δ
LONGER λ — SHORTER λ

Using this series, we can predict the *relative* size of Δ for a series of octahedral complexes of the same metal ion. Although it is difficult to predict the actual color of a given complex, we can determine whether a complex will absorb longer or shorter wavelengths than other complexes in the series.

SAMPLE PROBLEM 22.4 Ranking Crystal Field Splitting Energies for Complex Ions of a Given Metal

Problem Rank the ions $[Ti(H_2O)_6]^{3+}$, $[Ti(NH_3)_6]^{3+}$, and $[Ti(CN)_6]^{3-}$ in terms of the relative value of Δ and of the energy of visible light absorbed.
Plan The formulas show that titanium's oxidation state is +3 in the three ions. From the information given in Figure 22.21, we rank the ligands in terms of crystal field strength: the stronger the ligand, the greater the splitting, and the higher the energy of light absorbed.
Solution The ligand field strength is in the order $CN^- > NH_3 > H_2O$, so the relative size of Δ and energy of light absorbed is

$$[Ti(CN)_6]^{3-} > [Ti(NH_3)_6]^{3+} > [Ti(H_2O)_6]^{3+}$$

FOLLOW-UP PROBLEM 22.4 Which complex ion absorbs visible light of higher energy, $[V(H_2O)_6]^{3+}$ or $[V(NH_3)_6]^{3+}$?

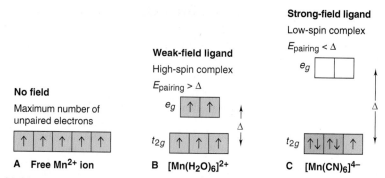

Figure 22.22 High-spin and low-spin complex ions of Mn²⁺. A, The free Mn²⁺ ion has five unpaired electrons. **B,** Bonded to weak-field ligands (smaller Δ), Mn²⁺ still has five unpaired electrons (high-spin complex). **C,** Bonded to strong-field ligands (larger Δ), Mn²⁺ has only one unpaired electron (low-spin complex).

Explaining the Magnetic Properties of Transition Metal Complexes

The splitting of energy levels influences magnetic properties by affecting the number of *unpaired* electrons in the metal ion's *d* orbitals. Based on Hund's rule (Section 8.3), electrons occupy orbitals one at a time as long as orbitals of equal energy are available. When all lower energy orbitals are half-filled, the next electron can

- enter a half-filled orbital and pair up by overcoming a repulsive *pairing energy* ($E_{pairing}$), or
- enter an empty, higher energy orbital by overcoming the crystal field splitting energy (Δ).

Thus, *the relative sizes of $E_{pairing}$ and Δ determine the occupancy of the d orbitals.* The orbital occupancy pattern, in turn, determines the number of unpaired electrons and, thus, the paramagnetic behavior of the ion.

As an example, the isolated Mn²⁺ ion ([Ar] $3d^5$) has five unpaired electrons in $3d$ orbitals of equal energy (Figure 22.22A). In an octahedral field of ligands, the orbital energies split. The orbital occupancy is affected by the ligand in one of two ways:

- *Weak-field ligands and high-spin complexes.* Weak-field ligands, such as H_2O in $[Mn(H_2O)_6]^{2+}$, cause a *small* splitting energy, so it takes *less* energy for *d* electrons to jump to the e_g set than to pair up in the t_{2g} set. As a result, the *d* electrons remain unpaired (Figure 22.22B). With weak-field ligands, the pairing energy is *greater* than the splitting energy ($E_{pairing} > Δ$); therefore, *the number of unpaired electrons in the complex ion is the **same** as in the free ion.* Weak-field ligands create **high-spin complexes,** those with the *maximum* number of unpaired electrons.
- *Strong-field ligands and low-spin complexes.* In contrast, strong-field ligands, such as CN^- in $[Mn(CN)_6]^{4-}$, cause a *large* splitting of the *d*-orbital energies, so it takes *more* energy for electrons to jump to the e_g set than to pair up in the t_{2g} set (Figure 22.22C). With strong-field ligands, the pairing energy is *smaller* than the splitting energy ($E_{pairing} < Δ$); therefore, *the number of unpaired electrons in the complex ion is **less** than in the free ion.* Strong-field ligands create **low-spin complexes,** those with *fewer* unpaired electrons.

Orbital diagrams for the d^1 through d^9 ions in octahedral complexes show that both high-spin and low-spin options are possible only for d^4, d^5, d^6, and d^7 ions (Figure 22.23). With three lower energy t_{2g} orbitals available, the d^1, d^2, and d^3 ions always form high-spin complexes because there is no need to pair up. Similarly, d^8 and d^9 ions always form high-spin complexes: because the t_{2g} set is filled with six electrons, the two e_g orbitals *must* have either two (d^8) or one (d^9) unpaired electron.

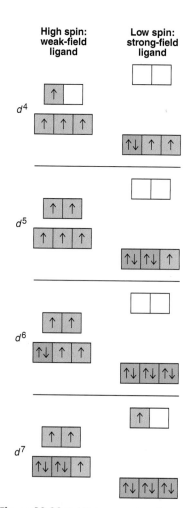

Figure 22.23 Orbital occupancy for high-spin and low-spin complexes of d^4 through d^7 metal ions.

SAMPLE PROBLEM 22.5 Identifying Complex Ions as High Spin or Low Spin

Problem Iron(II) forms an essential complex in hemoglobin. For each of the two octahedral complex ions $[Fe(H_2O)_6]^{2+}$ and $[Fe(CN)_6]^{4-}$, draw an orbital splitting diagram, predict the number of unpaired electrons, and identify the ion as low spin or high spin.

Plan The Fe^{2+} electron configuration gives us the number of d electrons, and the spectrochemical series in Figure 22.21 shows the relative strengths of the two ligands. We draw the diagrams, separating the t_{2g} and e_g sets by a greater distance for the strong-field ligand. Then we add electrons, noting that a weak-field ligand gives the *maximum* number of unpaired electrons and a high-spin complex, whereas a strong-field ligand leads to electron pairing and a low-spin complex.

Solution Fe^{2+} has the [Ar] $3d^6$ configuration. According to Figure 22.21, H_2O produces smaller splitting than CN^-. The diagrams are shown in the margin. The $[Fe(H_2O)_6]^{2+}$ ion has four unpaired electrons (high spin), and the $[Fe(CN)_6]^{4-}$ ion has no unpaired electrons (low spin).

Comments 1. H_2O is a weak-field ligand, so it almost always forms high-spin complexes. **2.** These results are correct, but we cannot confidently predict the spin of a complex without having actual values for Δ and $E_{pairing}$. **3.** Cyanide ions and carbon monoxide are highly toxic because they interact with the iron cations in essential proteins.

FOLLOW-UP PROBLEM 22.5 How many unpaired electrons do you expect for $[Mn(CN)_6]^{3-}$? Is this ion a high-spin or low-spin complex?

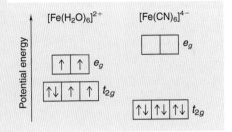

Crystal Field Splitting in Tetrahedral and Square Planar Complexes

Four ligands around a metal ion also cause d-orbital splitting, but the magnitude and pattern of the splitting depend on whether the ligands are in a tetrahedral or a square planar arrangement.

• *Tetrahedral complexes.* With the ligands approaching from the corners of a tetrahedron, none of the five d orbitals is directly in their paths (Figure 22.24). Thus, splitting of d-orbital energies is *less* in a tetrahedral complex than in an octahedral complex having the same ligands:

$$\Delta_{tetrahedral} < \Delta_{octahedral}$$

Minimal repulsions arise if the ligands approach the d_{xy}, d_{yz}, and d_{xz} orbitals closer than they approach the d_{z^2} and $d_{x^2-y^2}$ orbitals. This situation is the *opposite of the octahedral case,* and the relative d-orbital energies are reversed: the d_{xy}, d_{yz}, and d_{xz} orbitals become *higher* in energy than the d_{z^2} and $d_{x^2-y^2}$ orbitals. *Only high-spin tetrahedral complexes are known* because the magnitude of Δ is so small.

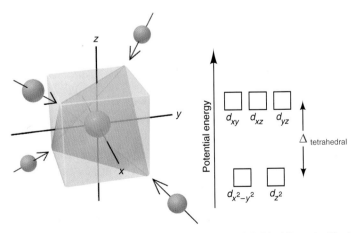

Figure 22.24 Splitting of d-orbital energies by a tetrahedral field of ligands. Electrons in d_{xy}, d_{yz}, and d_{xz} orbitals experience greater repulsions than those in $d_{x^2-y^2}$ and d_{z^2}, so the tetrahedral splitting pattern is the opposite of the octahedral pattern.

- *Square planar complexes.* The effects of the ligand field in the square planar case are easier to picture if we imagine starting with an octahedral geometry and then remove the two ligands along the *z*-axis, as depicted in Figure 22.25. With no *z*-axis interactions present, the d_{z^2} orbital energy decreases greatly, and the energies of the other orbitals with a *z*-axis component, the d_{xz} and d_{yz}, also decrease. As a result, the two *d* orbitals in the *xy* plane interact most strongly with the ligands, and because the $d_{x^2-y^2}$ orbital has its lobes *on* the axes, its energy is highest. As a consequence of this splitting pattern, square planar complexes with d^8 metal ions, such as $[PdCl_4]^{2-}$, are diamagnetic, with four pairs of *d* electrons filling the four lowest energy orbitals. Thus, as a general rule, *square planar complexes are low spin.*

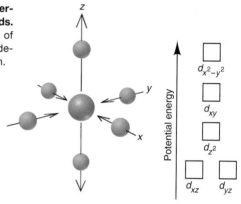

Figure 22.25 Splitting of *d*-orbital energies by a square planar field of ligands. In a square planar field, the energies of d_{xz}, d_{yz}, and especially d_{z^2} orbitals decrease relative to the octahedral pattern.

At this point, a final word about bonding theories may be helpful. As you have seen, the VB approach offers a simple picture of bond formation but does not explain color. The crystal field model predicts color and magnetic behavior but offers no insight about the covalent nature of metal-ligand bonding. Chemists now rely on *ligand field–molecular orbital theory,* which combines aspects of the previous two models with MO theory (Section 11.3). It yields information on bond properties that result from orbital overlap and on the spectral and magnetic properties that result from the splitting of a metal ion's *d* orbitals.

Transition Metal Complexes in Biological Systems

In addition to four *building-block elements* (C, O, H, and N) and seven elements known as *macro*nutrients (Na, Mg, P, S, Cl, K, and Ca), organisms contain a large number of *trace elements,* most of which are transition metals. With the exception of Sc, Ti, and Ni (in most species), the Period 4 transition elements are essential to many organisms (Table 22.9), and plants require Mo (from Period 5) as well. The principles of bonding and *d*-orbital splitting are the same in complex biomolecules containing transition metals as in simple inorganic systems. We focus here on an iron-containing complex.

Iron plays a crucial role in oxygen transport in all vertebrates. The O_2-transporting protein hemoglobin (Figure 22.26A) consists of four folded chains, each cradling the Fe-containing complex *heme.* Heme consists of iron(II) bonded to four N lone pairs of a tetradentate ring ligand known as a *porphyrin* to give a square planar complex. (Porphyrins are common biological ligands that are also found in chlorophyll, with Mg^{2+} at the center, and in vitamin B$_{12}$, with Co^{3+} at the center.) In hemoglobin (Figure 22.26B), the complex is *octahedral,* with the

Table 22.9 Essential Transition Metals in Humans	
Element	**Function**
Vanadium	Fat metabolism
Chromium	Glucose utilization
Manganese	Cell respiration
Iron	Oxygen transport; ATP formation
Cobalt	Component of vitamin B$_{12}$; development of red blood cells
Copper	Hemoglobin synthesis; ATP formation
Zinc	Elimination of CO_2; protein digestion

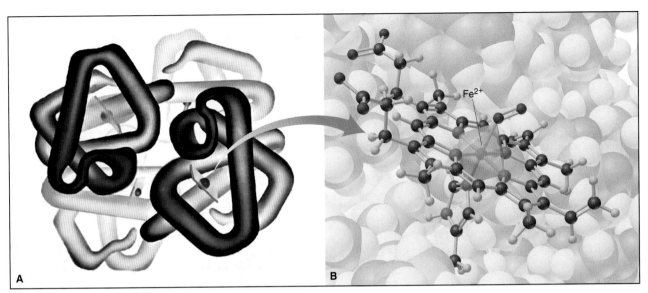

Figure 22.26 Hemoglobin and the octahedral complex in heme. A, Hemoglobin consists of four protein chains, each with a bound heme complex. (Illustration by Irving Geis. Rights owned by Howard Hughes Medical Institute. Not to be used without permission.) **B,** In the oxygenated form of hemoglobin, the octahedral complex in heme has iron(II) at the center surrounded by the four N atoms of the porphyrin ring, a fifth N from histidine (*below*), and an O_2 molecule (*above*).

fifth ligand of iron(II) being an N atom from a nearby amino acid (histidine), and the sixth an O atom from either an O_2 (shown) or an H_2O molecule.

Hemoglobin exists in two forms. In the arteries and lungs, the Fe^{2+} ion in heme binds to O_2; in the veins and tissues, O_2 is replaced by H_2O. Because H_2O is a weak-field ligand, the d^6 Fe^{2+} ion is part of a high-spin complex, and the relatively small d-orbital splitting makes venous blood absorb light at the red (low-energy) end of the spectrum and look purplish blue. On the other hand, O_2 is a strong-field ligand, so it increases the splitting energy and gives a low-spin complex. Thus, arterial blood absorbs at the blue (high-energy) end of the spectrum, which accounts for its bright red color.

Carbon monoxide is toxic because it binds to Fe^{2+} ion in heme about 200 times more strongly than O_2, which prevents the heme group from functioning:

$$\text{heme}-\text{CO} + O_2 \rightleftharpoons \text{heme}-O_2 + \text{CO}$$

Like O_2, CO is a strong-field ligand, which results in a bright red color of the blood. Because the binding is an equilibrium process, breathing extremely high concentrations of O_2 displaces CO from the heme and reverses CO poisoning.

SECTION SUMMARY

Valence bond theory pictures bonding in complex ions as arising from coordinate covalent bonding between Lewis bases (ligands) and Lewis acids (metal ions). Ligand lone pairs occupy hybridized metal-ion orbitals to form complex ions with characteristic shapes.

Crystal field theory explains the color and magnetism of complexes. As the result of a surrounding field of ligands, the d-orbital energies of the metal ion split. The magnitude of this crystal field splitting energy (Δ) depends on the charge of the metal ion and the crystal field strength of the ligand. In turn, Δ influences the energy of the photon absorbed (color) and the number of unpaired d electrons (paramagnetism). Strong-field ligands create a large Δ and produce low-spin complexes that absorb light of higher energy (shorter λ); the reverse is true of weak-field ligands. Several transition metals form complexes within proteins and are therefore important in living systems.

For Review and Reference (Numbers in parentheses refer to pages, unless noted otherwise.)

Learning Objectives

To help you review these learning objectives, the numbers of related sections (§), sample problems (SP), and upcoming end-of-chapter problems (EP) are listed in parentheses.

1. Write electron configurations of transition metal atoms and ions; compare periodic trends in atomic properties of transition elements with those of main-group elements; explain why transition elements have multiple oxidation states, how their metallic behavior (type of bonding and oxide acidity) changes with oxidation state, and why many of their compounds are colored and paramagnetic (§ 22.1) (SP 22.1) (EPs 22.1–22.17)
2. Be familiar with the coordination numbers, geometries, and ligands of complex ions; name and write formulas for coordination compounds; describe the types of constitutional and stereoisomerism they exhibit (§ 22.2) (SPs 22.2, 22.3) (EPs 22.18–22.39)
3. Correlate the shape of a complex ion with the number and type of hybrid orbitals of the central metal ion (§ 22.3) (EPs 22.40, 22.41, 22.47, 22.48)
4. Describe how approaching ligands cause d-orbital energies to split and give rise to octahedral, tetrahedral, and square-planar complexes; explain splitting energy (Δ) and how it accounts for the colors of complexes; explain how the relative magnitudes of pairing energy and Δ determine the magnetic properties of complexes; use a spectrochemical series to rank complex ions in terms of Δ, and determine if a complex is high spin or low spin (§ 22.3) (SPs 22.4, 22.5) (EPs 22.42–22.46, 22.49–22.57)

Key Terms

transition elements (735)

Section 22.1
lanthanide contraction (739)

Section 22.2
coordination compound (741)
complex ion (741)
ligand (741)
counter ion (741)
coordination number (742)

donor atom (742)
chelate (743)
isomer (745)
constitutional (structural) isomers (746)
coordination isomers (746)
linkage isomers (746)
stereoisomers (746)

geometric (*cis-trans*) isomers (746)
optical isomers (747)

Section 22.3
coordinate covalent bond (748)
crystal field theory (750)
e_g orbital (752)

t_{2g} orbital (752)
crystal field splitting energy (Δ) (752)
strong-field ligand (752)
weak-field ligand (752)
spectrochemical series (753)
high-spin complex (754)
low-spin complex (754)

Brief Solutions to Follow-up Problems

22.1 (a) Ag^+: $4d^{10}$
(b) Cd^{2+}: $4d^{10}$
(c) Ir^{3+}: $5d^6$
22.2 (a) Pentaaquabromochromium(III) chloride
(b) $Ba_3[Co(CN)_6]_2$
22.3 Two sets of *cis-trans* isomers, and the two *cis* isomers are optical isomers.

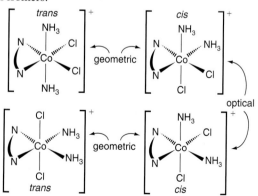

22.4 Both metal ions are V^{3+}; in terms of ligand field energy, $NH_3 > H_2O$, so $[V(NH_3)_6]^{3+}$ absorbs light of higher energy.
22.5 The metal ion is Mn^{3+}: [Ar] $3d^4$.

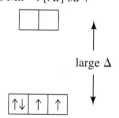

Two unpaired d electrons; low-spin complex

Problems

Problems with **colored** numbers are answered in Appendix E. Sections match the text and provide the numbers of relevant sample problems. Bracketed problems are grouped in pairs (indicated by a short rule) that cover the same concept. Comprehensive Problems are based on material from any section or previous chapter. *Note:* In these problems, the term *electron configuration* refers to the condensed, ground-state electron configuration.

Properties of the Transition Elements
(Sample Problem 22.1)

22.1 (a) Write the general electron configuration of a transition element in Period 5.
(b) Write the general electron configuration of a transition element in Period 6.

22.2 What is the general rule concerning the order in which electrons are removed from a transition metal atom to form an ion? Give an example from Group 5B(5). Name two types of measurements used to study electron configurations of ions.

22.3 How does the variation in atomic size across a transition series contrast with the change across the main-group elements of the same period? Why?

22.4 (a) What is the lanthanide contraction?
(b) How does it affect atomic size down a group of transition elements?
(c) How does it influence the densities of the Period 6 transition elements?

22.5 (a) What is the range in electronegativity values across the first ($3d$) transition series?
(b) What is the range across Period 4 of main-group elements?
(c) Explain the difference between the two ranges.

22.6 (a) Explain the major difference between the number of oxidation states of most transition elements and that of most main-group elements.
(b) Why is the +2 oxidation state so common among transition elements?

22.7 (a) What difference in behavior distinguishes a paramagnetic substance from a diamagnetic one?
(b) Why are paramagnetic ions common among the transition elements but not the main-group elements?
(c) Why are colored solutions of metal ions common among the transition elements but not the main-group elements?

22.8 Using the periodic table to locate each element, write the electron configuration of (a) V; (b) Y; (c) Hg.

22.9 Using the periodic table to locate each element, write the electron configuration of (a) Ru; (b) Cu; (c) Ni.

22.10 Give the electron configuration and the number of unpaired electrons for each of the following ions: (a) Sc^{3+}; (b) Cu^{2+}; (c) Fe^{3+}; (d) Nb^{3+}.

22.11 Give the electron configuration and the number of unpaired electrons for each of the following ions: (a) Cr^{3+}; (b) Ti^{4+}; (c) Co^{3+}; (d) Ta^{2+}.

22.12 Which transition metals have a maximum oxidation state of +6?

22.13 Which transition metals have a maximum oxidation state of +4?

22.14 In which compound does Cr exhibit greater metallic behavior, CrF_2 or CrF_6? Explain.

22.15 VF_5 is a liquid that boils at 48°C, whereas VF_3 is a solid that melts above 800°C. Explain this difference in properties.

22.16 Which oxide, CrO_3 or CrO, forms a more acidic aqueous solution? Explain.

22.17 Which oxide, Mn_2O_3 or Mn_2O_7, displays more basic behavior? Explain.

Coordination Compounds
(Sample Problems 22.2 and 22.3)

22.18 Describe the makeup of a complex ion, including the nature of the ligands and their interaction with the central metal ion. Explain how a complex ion can be positive or negative and how it occurs as part of a neutral coordination compound.

22.19 What is the coordination number of a metal ion in a complex ion? How does it differ from oxidation number?

22.20 What structural feature is characteristic of a complex described as a chelate?

22.21 What geometries are associated with the coordination numbers 2, 4, and 6?

22.22 In what sense is a complex ion the adduct of a Lewis acid-base reaction?

22.23 Is a linkage isomer a type of constitutional isomer or stereoisomer? Explain.

22.24 Give systematic names for the following formulas:
(a) $[Ni(H_2O)_6]Cl_2$ (b) $[Cr(en)_3](ClO_4)_3$ (c) $K_4[Mn(CN)_6]$

22.25 Give systematic names for the following formulas:
(a) $[Co(NH_3)_4(NO_2)_2]Cl$ (b) $[Cr(NH_3)_6][Cr(CN)_6]$
(c) $K_2[CuCl_4]$

22.26 What are the charge and coordination number of the central metal ion(s) in each compound of Problem 22.24?

22.27 What are the charge and coordination number of the central metal ion(s) in each compound of Problem 22.25?

22.28 Give systematic names for the following formulas:
(a) $K[Ag(CN)_2]$ (b) $Na_2[CdCl_4]$ (c) $[Co(NH_3)_4(H_2O)Br]Br_2$

22.29 Give systematic names for the following formulas:
(a) $K[Pt(NH_3)Cl_5]$ (b) $[Cu(en)(NH_3)_2][Co(en)Cl_4]$
(c) $[Pt(en)_2Br_2](ClO_4)_2$

22.30 Give formulas corresponding to the following names:
(a) Tetraamminezinc sulfate
(b) Pentaamminechlorochromium(III) chloride
(c) Sodium bis(thiosulfato)argentate(I)

22.31 Give formulas corresponding to the following names:
(a) Dibromobis(ethylenediamine)cobalt(III) sulfate
(b) Hexaamminechromium(III) tetrachlorocuprate(II)
(c) Potassium hexacyanoferrate(II)

22.32 What is the coordination number of the metal ion and the number of individual ions per formula unit in each of the compounds in Problem 22.30?

22.33 What is the coordination number of the metal ion and the number of individual ions per formula unit in each of the compounds in Problem 22.31?

22.34 Which of these ligands can give rise to linkage isomerism: (a) NO_2^-; (b) SO_2; (c) NO_3^-? Explain with Lewis structures.

22.35 Which of these ligands can give rise to linkage isomerism: (a) SCN^-; (b) $S_2O_3^{2-}$ (thiosulfate); (c) HS^-? Explain with Lewis structures.

22.36 For any of the following that can exist as isomers, state the type of isomerism and draw the structures:
(a) $[Pt(CH_3NH_2)_2Br_2]$ (b) $[Pt(NH_3)_2FCl]$
(c) $[Pt(H_2O)(NH_3)FCl]$

22.37 For any of the following that can exist as isomers, state the type of isomerism and draw the structures:
(a) $[Zn(en)F_2]$ (b) $[Zn(H_2O)(NH_3)FCl]$
(c) $[Pd(CN)_2(OH)_2]^{2-}$

22.38 For any of the following that can exist as isomers, state the type of isomerism and draw the structures:
(a) $[PtCl_2Br_2]^{2-}$ (b) $[Cr(NH_3)_5(NO_2)]^{2+}$
(c) $[Pt(NH_3)_4I_2]^{2+}$

22.39 For any of the following that can exist as isomers, state the type of isomerism and draw the structures:
(a) $[Co(NH_3)_5Cl]Br_2$ (b) $[Pt(CH_3NH_2)_3Cl]Br$
(c) $[Fe(H_2O)_4(NH_3)_2]^{2+}$

Theoretical Basis for the Bonding and Properties of Complexes
(Sample Problems 22.4 and 22.5)

22.40 According to valence bond theory, what set of orbitals is used by a Period 4 metal ion in forming (a) a square planar complex; (b) a tetrahedral complex?

22.41 A metal ion is described as using a d^2sp^3 set of orbitals when forming a complex. What is the coordination number of the metal ion and the shape of the complex?

22.42 A complex in solution absorbs green light. What is the color of the solution?

22.43 (a) What is the crystal field splitting energy (Δ)?
(b) How does it arise for an octahedral field of ligands?
(c) How is it different for a tetrahedral field of ligands?

22.44 What is the distinction between a weak-field ligand and a strong-field ligand? Give an example of each.

22.45 How do the relative magnitudes of $E_{pairing}$ and Δ affect the paramagnetism of a complex?

22.46 Why are there both high-spin and low-spin octahedral complexes but only high-spin tetrahedral complexes?

22.47 Give the number of d electrons (n of d^n) for the central metal ion in each of these species: (a) $[TiCl_6]^{2-}$; (b) $K[AuCl_4]$; (c) $[RhCl_6]^{3-}$.

22.48 Give the number of d electrons (n of d^n) for the central metal ion in each of these species: (a) $[Cr(H_2O)_6](ClO_3)_2$; (b) $[Mn(CN)_6]^{2-}$; (c) $[Ru(NO)(en)_2Cl]Br$.

22.49 Which of these ions *cannot* form both high- and low-spin octahedral complexes: (a) Ti^{3+}; (b) Co^{2+}; (c) Fe^{2+}; (d) Cu^{2+}?

22.50 Which of these ions *cannot* form both high- and low-spin octahedral complexes: (a) Mn^{3+}; (b) Nb^{3+}; (c) Ru^{3+}; (d) Ni^{2+}?

22.51 Draw orbital-energy splitting diagrams and use the spectrochemical series to show the orbital occupancy for each of the following (assuming that H_2O is a weak-field ligand):
(a) $[Cr(H_2O)_6]^{3+}$ (b) $[Cu(H_2O)_4]^{2+}$ (c) $[FeF_6]^{3-}$

22.52 Draw orbital-energy splitting diagrams and use the spectrochemical series to show the orbital occupancy for each of the following:
(a) $[Cr(CN)_6]^{3-}$ (b) $[Rh(CO)_6]^{3+}$ (c) $[Co(OH)_6]^{4-}$

22.53 Rank the following complex ions in order of *increasing* Δ and energy of visible light absorbed: $[Cr(NH_3)_6]^{3+}$, $[Cr(H_2O)_6]^{3+}$, $[Cr(NO_2)_6]^{3-}$.

22.54 Rank the following complex ions in order of *decreasing* Δ and energy of visible light absorbed: $[Cr(en)_3]^{3+}$, $[Cr(CN)_6]^{3-}$, $[CrCl_6]^{3-}$.

22.55 A complex, ML_6^{2+}, is violet. The same metal forms a complex with another ligand, Q, that creates a weaker field. What color might MQ_6^{2+} be expected to show? Explain.

22.56 $[Cr(H_2O)_6]^{2+}$ is violet. Another CrL_6 complex is green. Can ligand L be CN^-? Can it be Cl^-? Explain.

22.57 Three of the complex ions that are formed by Co^{3+} are $[Co(H_2O)_6]^{3+}$, $[Co(NH_3)_6]^{3+}$, and $[CoF_6]^{3-}$. These ions have the observed colors (listed in arbitrary order) yellow-orange, green, and blue. Match each complex with its color. Explain.

Comprehensive Problems
Problems with an asterisk (*) are more challenging.

* 22.58 How many different formulas are there for octahedral complexes with a metal M and four ligands A, B, C, and D? Give the number of isomers for each formula and describe the isomers.

22.59 Correct each name that has an error:
(a) $Na[FeBr_4]$, sodium tetrabromoferrate(II)
(b) $[Ni(NH_3)_6]^{2+}$, nickel hexaammine ion
(c) $[Co(NH_3)_3I_3]$, triamminetriiodocobalt(III)
(d) $[V(CN)_6]^{3-}$, hexacyanovanadium(III) ion
(e) $K[FeCl_4]$, potassium tetrachloroiron(III)

* 22.60 For the compound $[Co(en)_2Cl_2]Cl$, give:
(a) The coordination number of the metal ion
(b) The oxidation number of the central metal ion
(c) The number of individual ions per formula unit
(d) The moles of AgCl that precipitate immediately when 1 mol of compound is dissolved in water and treated with $AgNO_3$

22.61 Hexafluorocobaltate(III) ion is a high-spin complex. Draw the orbital-energy splitting diagram for its d orbitals.

22.62 A salt of each of the ions in Table 22.3 is dissolved in water. A Pt electrode is immersed in each solution and connected to a 0.38-V battery. All of the electrolytic cells are run for the same amount of time with the same current.
(a) In which cell(s) will a metal plate out? Explain.
(b) Which cell will plate out the least mass of metal? Explain.

22.63 In many species, a transition metal has an unusually high or low oxidation state. Write balanced equations for the following and find the oxidation state of the transition metal in the product:
(a) Iron(III) ion reacts with hypochlorite ion in basic solution to form ferrate ion (FeO_4^{2-}), Cl^-, and water.
(b) Heating sodium superoxide, NaO_2, with Co_3O_4 produces Na_4CoO_4 and O_2 gas.
(c) Heating cesium tetrafluorocuprate(II) with F_2 gas under pressure gives Cs_2CuF_6.
(d) Potassium tetracyanonickelate(II) reacts with hydrazine (N_2H_4) in basic solution to form $K_4[Ni_2(CN)_6]$ and N_2 gas.

22.64 An octahedral complex with three different ligands (A, B, and C) can have formulas with three different ratios of the ligands:

$[MA_4BC]^{n+}$,   such as $[Co(NH_3)_4(H_2O)Cl]^{2+}$

$[MA_3B_2C]^{n+}$,   such as $[Cr(H_2O)_3Br_2Cl]$

$[MA_2B_2C_2]^{n+}$,   such as $[Cr(NH_3)_2(H_2O)_2Br_2]^+$

For each example, give the name, state the type(s) of isomerism present, and draw all isomers.

22.65 In $[Cr(NH_3)_6]Cl_3$, the $[Cr(NH_3)_6]^{3+}$ ion absorbs visible light in the blue-violet range, and the compound is yellow-orange. In $[Cr(H_2O)_6]Br_3$, the $[Cr(H_2O)_6]^{3+}$ ion absorbs visible light in the red range, and the compound is blue-gray. Explain these differences in light absorbed and colors of the compounds.

22.66 Ionic liquids have many new applications in engineering and materials science. The dissolution of the metavanadate ion in chloroaluminate ionic liquids has been studied:

$$VO_3^- + AlCl_4^- \longrightarrow VO_2Cl_2^- + AlOCl_2^-$$

(a) What is the oxidation number of V and Al in each ion?

(b) In reactions with V_2O_5, acid concentration affects the product. At low acid concentration, $VO_2Cl_2^-$ and VO_3^- form:

$$V_2O_5 + HCl \longrightarrow VO_2Cl_2^- + VO_3^- + H^+$$

At high acid concentration, $VOCl_3$ forms:

$$V_2O_5 + HCl \longrightarrow VOCl_3 + H_2O$$

Balance each equation, and state which, if either, involves a redox process.

(c) What mass of $VO_2Cl_2^-$ or $VOCl_3$ can form from 10.0 g of V_2O_5 and the appropriate concentration of acid?

22.67 Several coordination isomers, with both Co and Cr as $3+$ ions, have the molecular formula $CoCrC_6H_{18}N_{12}$.

(a) Give the name and formula of the isomer in which the Co complex ion has six NH_3 groups.

(b) Give the name and formula of the isomer in which the Co complex ion has one CN and five NH_3 groups.

* **22.68** The enzyme carbonic anhydrase has zinc in a tetrahedral complex at its active site. Suggest a structural reason why carbonic anhydrase synthesized with Ni^{2+}, Fe^{2+}, or Mn^{2+} in place of Zn^{2+} gives an enzyme with less catalytic efficiency.

* **22.69** The effect of entropy on reactions is evident in the stabilities of certain complexes.

(a) Using the criterion of number of product particles, predict which of the following will be favored in terms of ΔS°_{rxn}:

$$[Cu(NH_3)_4]^{2+}(aq) + 4H_2O(l) \longrightarrow$$
$$[Cu(H_2O)_4]^{2+}(aq) + 4NH_3(aq)$$

$$[Cu(H_2NCH_2CH_2NH_2)_2]^{2+}(aq) + 4H_2O(l) \longrightarrow$$
$$[Cu(H_2O)_4]^{2+}(aq) + 2en(aq)$$

(b) Given that the Cu—N bond strength is approximately the same in both complexes, which complex will be more stable in water (less likely to exchange their ligands for H_2O molecules)? Explain.

22.70 The extent of crystal field splitting is often determined from spectra.

(a) Given the wavelength (λ) of maximum absorption, find the crystal field splitting energy (Δ), in kJ/mol, for each of the following complex ions:

Ion	λ (nm)	Ion	λ (nm)
$[Cr(H_2O)_6]^{3+}$	562	$[Fe(H_2O)_6]^{2+}$	966
$[Cr(CN)_6]^{3-}$	381	$[Fe(H_2O)_6]^{3+}$	730
$[CrCl_6]^{3-}$	735	$[Co(NH_3)_6]^{3+}$	405
$[Cr(NH_3)_6]^{3+}$	462	$[Rh(NH_3)_6]^{3+}$	295
$[Ir(NH_3)_6]^{3+}$	244		

(b) Construct a spectrochemical series for the ligands in the Cr complexes.

(c) Use the Fe data to state how oxidation state affects Δ.

(d) Use the Co, Rh, and Ir data to state how period number affects Δ.

CHAPTER TWENTY-THREE

Nuclear Reactions and Their Applications

From Infinitesimal to Infinite In a bubble chamber, subatomic particles leave distinct spiral tracks as they boil the liquid in their wake. At the other end of the size spectrum, in stars, interactions of subatomic particles create the chemical elements. In this chapter, you'll discover the astonishing world of the atomic nucleus.

Key Principles

◆ *Nuclear reactions* differ from chemical reactions in several important ways: (1) Element identity *does* change. (2) Nuclear particles, rather than electrons, participate. (3) The reactions release so much energy that mass *does* change. (4) Reaction rates are *not* affected by temperature, catalysts, or the chemical nature of the substance.

◆ The great majority of nuclei are unstable and undergo various types of radioactive decay: α decay, β decay, positron emission, electron capture, and γ emission. In nuclear reactions, total mass number (*A*) and total charge (*Z*) must be balanced. A plot of number of neutrons (*N*) versus number of protons (*Z*) for all nuclei shows a narrow *band of stable nuclei.* To become more stable; the type of decay can often be predicted from the *N/Z* ratio of the unstable nucleus. Certain heavy nuclei undergo a *series of decays* to reach stability.

◆ Radioactive decay is a *first-order process,* so the decay rate (*activity*) depends *only* on the number of nuclei. Therefore, the *half-life* does *not* depend on the number of nuclei. In *radiocarbon dating,* the age of an object is determined by comparing its ^{14}C activity with that of currently living things.

◆ *Particle accelerators* change one element into another (*nuclear transmutation*) by bombarding nuclei with high-energy particles.

◆ *Ionizing radiation* causes chemical changes in matter, while the much weaker (and less harmful) *nonionizing radiation* causes matter to heat up or emit photons.

◆ *Isotopes* of an element have nearly identical chemical properties. Therefore, a small amount of a radioactive isotope can act as a *tracer* for studying reaction mechanisms, physical movements of substances, and medical problems.

◆ The mass of a nucleus is *less* than the sum of the masses of the *nucleons* (protons and neutrons) that make it up. Einstein's equation shows that this *mass defect* is equivalent to the *nuclear binding energy.* The *binding energy per nucleon* is a measure of nuclear stability. Heavy nuclei split (*fission*) and light nuclei join (*fusion*) to increase the binding energy per nucleon. Both fission and fusion release enormous quantities of energy.

Outline

Far below the outer fringes of the cloud of electrons lies the atom's tiny, dense core, held together by the strongest force in the universe. For nearly the entire text, we have focused on an atom's electrons, treating the nucleus as their electrostatic anchor, examining the effect of its positive charge on atomic properties and, ultimately, chemical behavior. But, for the scientists probing the structure and behavior of the nucleus itself, that is the scene of real action, one that holds enormous potential benefit and great mystery and wonder.

Society is ambivalent about the applications of nuclear research, however. The promise of abundant energy and treatments for disease comes hand-in-hand with the threat of nuclear waste contamination, reactor accidents, and unimaginable destruction from nuclear war or terrorism. Can the power of the nucleus be harnessed for our benefit, or are the risks too great? In this chapter, we discuss the principles that can help you answer this vital question.

The changes that occur in atomic nuclei are strikingly different from chemical changes. In chemical reactions, electrons are shared or transferred to form *compounds,* while nuclei sit by passively, never changing their identities. In nuclear reactions, the roles are reversed: electrons in their orbitals are usually bystanders as the nuclei undergo changes that, in nearly every case, form different *elements.* Nuclear reactions are often accompanied by energy changes a million times greater than those in chemical reactions, energy changes so great that changes in mass *are* detectable. Moreover, nuclear reaction yields and rates are typically *not* subject to the effects of pressure, temperature, and catalysis. Table 23.1 summarizes these general differences.

Concepts & Skills to Review
Before You Study This Chapter

- discovery of the atomic nucleus (Section 2.4)
- protons, neutrons, mass number, and the $^A_Z X$ notation (Section 2.5)
- half-life and first-order reaction rate (Section 16.4)

Table 23.1 **Comparison of Chemical and Nuclear Reactions**	
Chemical Reactions	**Nuclear Reactions**
1. One substance is converted into another, but atoms never change identity.	1. Atoms of one element typically are converted into atoms of another element.
2. Orbital electrons are involved as bonds break and form; nuclear particles do not take part.	2. Protons, neutrons, and other particles are involved; orbital electrons rarely take part.
3. Reactions are accompanied by relatively small changes in energy and no measurable changes in mass.	3. Reactions are accompanied by relatively large changes in energy and measurable changes in mass.
4. Reaction rates are influenced by temperature, concentration, catalysts, and the compound in which an element occurs.	4. Reaction rates are affected by number of nuclei, but not by temperature, catalysts, or, normally, the compound in which an element occurs.

23.1   RADIOACTIVE DECAY AND NUCLEAR STABILITY

A stable nucleus remains intact indefinitely, but *the great majority of nuclei are unstable.* An unstable nucleus exhibits **radioactivity:** it spontaneously disintegrates, or *decays,* by emitting radiation. In the next section, you'll see that each type of unstable nucleus has its own characteristic *rate* of radioactive decay, which can range from a fraction of a second to billions of years. This section introduces important terms and notation for nuclei, describes the common types of radioactive emissions and decay, and discusses how to predict whether and how a given nucleus will decay.

The Components of the Nucleus: Terms and Notation

Recall from Chapter 2 that the nucleus contains essentially all the atom's mass but is only about 10^{-4} times its diameter (or 10^{-12} times its volume). Obviously, the nucleus is incredibly dense: about 10^{14} g/mL. *Protons* and *neutrons,* the elementary particles that make up the nucleus, are called **nucleons.** The term

nuclide refers to a nucleus with a particular composition, that is, with specific numbers of the two types of nucleons. Most elements occur in nature as a mixture of **isotopes,** atoms with the characteristic number of protons of the element but different numbers of neutrons. Therefore, each isotope of an element has a particular nuclide that we identify by the numbers of protons and neutrons it contains. The nuclide of the most abundant isotope of oxygen, for example, contains eight protons and eight neutrons, whereas the nuclide of the least abundant isotope contains eight protons and ten neutrons.

The relative mass and charge of a particle—a nucleon, another kind of elementary particle, or a nuclide—is described by the notation $^A_Z X$, where X is the *symbol* for the particle, *A* is the *mass number,* or the total number of nucleons, and *Z* is the *charge* of the particle; for nuclides, *A* is the *sum of protons and neutrons* and *Z* is the *number of protons* (atomic number). In this notation, the three subatomic elementary particles are

$$_{-1}^{0}e \text{ (electron), } _{1}^{1}p \text{ (proton), and } _{0}^{1}n \text{ (neutron)}$$

(In nuclear notation, the element symbol refers to the nucleus only, so a proton is also sometimes represented as $_1^1 H$.) The number of neutrons (*N*) in a nucleus is the mass number (*A*) minus the atomic number (*Z*): $N = A - Z$. The two naturally occurring isotopes of chlorine, for example, have 17 protons ($Z = 17$), but one has 18 neutrons ($_{17}^{35}Cl$, also written ^{35}Cl) and the other has 20 ($_{17}^{37}Cl$, or ^{37}Cl). Nuclides can also be designated with the element name followed by the mass number, for example, chlorine-35 and chlorine-37. Despite some small variations, in *naturally occurring* samples of an element or its compounds, the isotopes of the element are present in particular, fixed proportions. Thus, in a sample of sodium chloride (or any Cl-containing substance), 75.77% of the Cl atoms are chlorine-35 and the remaining 24.23% are chlorine-37.

To understand this chapter, it's very important that you are comfortable with nuclear notations, so please take a moment to review Sample Problem 2.2 and Problems 2.23 to 2.30 at the end of Chapter 2.

Types of Radioactive Emissions and Decay; Balancing Nuclear Equations

When a nuclide decays, it forms a nuclide of lower energy, and the excess energy is carried off by the emitted radiation. The three most common types of radioactive emission are:

- **Alpha particles** (symbolized α or $_2^4 He$) are dense, positively charged particles identical to helium nuclei.
- **Beta particles** (symbolized β, β⁻, or more usually $_{-1}^{0}\beta$) are negatively charged particles identified as high-speed electrons. (The emission of electrons from the nucleus may seem strange, but as you'll see shortly, β particles arise as a result of a nuclear reaction.)
- **Gamma rays** (symbolized as γ, or sometimes $_0^0\gamma$) are very high-energy photons, about 10^5 times as energetic as visible light.

The behavior of these three emissions in an electric field is shown in Figure 23.1. Note that α particles curve to a small extent toward the negative plate, β particles curve to a great extent toward the positive plate, and γ rays are not affected by the electric field.

The decaying, or reactant, nuclide is called the *parent;* the product nuclide is called the *daughter.* Nuclides can decay in several ways. As we discuss the major types of decay, which are summarized in Table 23.2, note the principle used to balance nuclear reactions: *the total Z (charge, number of protons)*

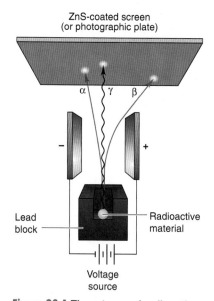

Figure 23.1 Three types of radioactive emissions in an electric field. Positively charged α particles curve toward the negative plate; negatively charged β particles curve toward the positive plate. The curvature is greater for β particles because they have much lower mass. The γ rays, uncharged high-energy photons, are unaffected by the field.

Table 23.2 Modes of Radioactive Decay*

Mode	Emission	Decay Process	Change in A	Z	N
α Decay	α ($^{4}_{2}$He)	Reactant (parent) → Product (daughter) + α expelled	−4	−2	−2
β Decay†	$^{0}_{-1}\beta$	$^{1}_{0}$n in nucleus → $^{1}_{1}$p in nucleus + $^{0}_{-1}\beta$ β expelled	0	+1	−1
Positron emission†	$^{0}_{1}\beta$	$h\nu$ + high-energy photon, nucleus with xp$^+$ and yn^0 → nucleus with (x − 1)p$^+$ and (y + 1)n^0 + $^{0}_{1}\beta$ positron expelled	0	−1	+1
Electron capture†	x-ray photon	$^{0}_{-1}$e absorbed from low-energy orbital + $^{1}_{1}$p in nucleus → $^{1}_{0}$n in nucleus	0	−1	+1
γ Emission	$^{0}_{0}\gamma$	excited nucleus → stable nucleus + $^{0}_{0}\gamma$ γ photon radiated	0	0	0

*Neutrinos (ν) are involved in several of these processes but are not shown.
†Nuclear chemists consider β decay to be a more general process that includes three decay modes: negatron emission (which the text calls "β decay"), positron emission, and electron capture.

and the total A (sum of protons and neutrons) of the reactants equal those of the products:

$$\text{Total } A_{\text{Total } Z} \text{ Reactants} = \text{Total } A_{\text{Total } Z} \text{ Products} \quad (23.1)$$

1. **Alpha decay** involves the loss of an α particle ($^{4}_{2}$He) from a nucleus. For each α particle emitted by the parent nucleus, A *decreases by 4 and Z decreases by 2*. Every element that is heavier than lead (Pb; Z = 82), as well as a few lighter ones, exhibits α decay. For example, radium undergoes α decay to yield radon (Rn; Z = 86):

$$^{226}_{88}\text{Ra} \longrightarrow ^{222}_{86}\text{Rn} + ^{4}_{2}\text{He}$$

Note that the A value for Ra equals the sum of the A values for Rn and He (226 = 222 + 4), and that the Z value for Ra equals the sum of the Z values for Rn and He (88 = 86 + 2).

2. **Beta decay** involves the ejection of a β particle ($^{0}_{-1}\beta$) from the nucleus.* This change does not involve the expulsion of a β particle that was actually in

*In formal nuclear chemistry terminology, *β decay* indicates a more general phenomenon (see footnote to Table 23.2).

the nucleus, but rather the *conversion of a neutron into a proton, which remains in the nucleus, and a β particle, which is expelled immediately:*

$$_{0}^{1}\text{n} \longrightarrow {}_{1}^{1}\text{p} + {}_{-1}^{0}\beta$$

As always, the totals of the *A* and the *Z* values for reactant and products are equal. Radioactive nickel-63 becomes stable copper-63 through β decay:

$$_{28}^{63}\text{Ni} \longrightarrow {}_{29}^{63}\text{Cu} + {}_{-1}^{0}\beta$$

Another example is the β decay of carbon-14, applied in radiocarbon dating:

$$_{6}^{14}\text{C} \longrightarrow {}_{7}^{14}\text{N} + {}_{-1}^{0}\beta$$

Note that *β decay results in a product nuclide with the same A but with Z one higher (one more proton) than in the reactant nuclide.* In other words, an atom of the element with the next *higher* atomic number is formed.

3. **Positron emission** involves the emission of a positron from the nucleus. A key idea of modern physics is that every fundamental particle has a corresponding *antiparticle* with the same mass but opposite charge. The **positron** (symbolized $_{1}^{0}\beta$; note the positive *Z*) is the antiparticle of the electron. Positron emission occurs through a process in which *a proton in the nucleus is converted into a neutron, and a positron is expelled.** *Positron emission has the opposite effect of β decay, resulting in a daughter nuclide with the same A but with Z one lower (one fewer proton) than the parent;* thus, an atom of the element with the next *lower* atomic number forms. Carbon-11, a synthetic radioisotope, decays to a stable boron isotope through emission of a positron:

$$_{6}^{11}\text{C} \longrightarrow {}_{5}^{11}\text{B} + {}_{1}^{0}\beta$$

4. **Electron capture** occurs when the nucleus of an atom draws in an electron from an orbital of the lowest energy level. The net effect is that *a nuclear proton is transformed into a neutron:*

$$_{1}^{1}\text{p} + {}_{-1}^{0}\text{e} \longrightarrow {}_{0}^{1}\text{n}$$

(We use the symbol $_{-1}^{0}\text{e}$ to distinguish an orbital electron from a beta particle, symbol $_{-1}^{0}\beta$.) The orbital vacancy is quickly filled by an electron that moves down from a higher energy level, and that energy difference appears as an *x-ray photon.* Radioactive iron forms stable manganese through electron capture:

$$_{26}^{55}\text{Fe} + {}_{-1}^{0}\text{e} \longrightarrow {}_{25}^{55}\text{Mn} + h\nu \text{ (x-ray)}$$

Electron capture has the same net effect as positron emission (Z lower by 1, A unchanged), even though the processes are entirely different.

5. **Gamma emission** involves the radiation of high-energy γ photons from an excited nucleus. Recall that an atom in an excited *electronic* state reduces its energy by emitting photons, usually in the UV and visible ranges. Similarly, a nucleus in an excited state lowers its energy by emitting γ photons, which are of much higher energy (much shorter wavelength) than UV photons. Many nuclear processes leave the nucleus in an excited state, so *γ emission accompanies most other types of decay.* Several γ photons (γ rays) of different frequencies can be emitted from an excited nucleus as it returns to the ground state. The decay of U-238 involves the release of γ rays:

$$_{92}^{238}\text{U} \longrightarrow {}_{90}^{234}\text{Th} + {}_{2}^{4}\text{He} + 2{}_{0}^{0}\gamma$$

Because γ rays have no mass or charge, *γ emission does not change A or Z.* Gamma rays also result when a particle and an antiparticle annihilate each other, as when an emitted positron meets an orbital electron:

$$_{1}^{0}\beta \text{ (from nucleus)} + {}_{-1}^{0}\text{e} \text{ (outside nucleus)} \longrightarrow 2{}_{0}^{0}\gamma$$

---

*The process, called *pair production,* involves a transformation of energy into matter. A high-energy ($>1.63\times10^{-13}$ J) photon becomes an electron and a positron simultaneously. The electron and a proton in the nucleus form a neutron, while the positron is expelled.

SAMPLE PROBLEM 23.1 Writing Equations for Nuclear Reactions

Problem Write balanced equations for the following nuclear reactions:
(a) Naturally occurring thorium-232 undergoes α decay.
(b) Chlorine-36 undergoes electron capture.
Plan We first write a skeleton equation that includes the mass numbers, atomic numbers, and symbols of all the particles, showing the unknown particles as $^A_Z X$. Then, because the total of mass numbers and the total of charges on the left side and the right side must be equal, we solve for A and Z, and use Z to determine X from the periodic table.
Solution (a) Writing the skeleton equation:

$$^{232}_{90}\text{Th} \longrightarrow\ ^A_Z X\ +\ ^4_2\text{He}$$

Solving for A and Z and balancing the equation: For A, $232 = A + 4$, so $A = 228$. For Z, $90 = Z + 2$, so $Z = 88$. From the periodic table, we see that the element with $Z = 88$ is radium (Ra). Thus, the balanced equation is

$$^{232}_{90}\text{Th} \longrightarrow\ ^{228}_{88}\text{Ra}\ +\ ^4_2\text{He}$$

(b) Writing the skeleton equation:

$$^{36}_{17}\text{Cl}\ +\ ^{\ \ 0}_{-1}\text{e} \longrightarrow\ ^A_Z X$$

Solving for A and Z and balancing the equation: For A, $36 + 0 = A$, so $A = 36$. For Z, $17 + (-1) = Z$, so $Z = 16$. The element with $Z = 16$ is sulfur (S), so we have

$$^{36}_{17}\text{Cl}\ +\ ^{\ \ 0}_{-1}\text{e} \longrightarrow\ ^{36}_{16}\text{S}$$

Check Always read across superscripts and then across subscripts, with the yield arrow as an equal sign, to check your arithmetic. In part (a), for example, $232 = 228 + 4$, and $90 = 88 + 2$.

FOLLOW-UP PROBLEM 23.1 Write a balanced equation for the reaction in which a nuclide undergoes β decay and produces cesium-133.

Nuclear Stability and the Mode of Decay

There are several ways that an unstable nuclide *might* decay, but can we predict how it *will* decay? Indeed, can we predict *if* a given nuclide will decay at all? Our knowledge of the nucleus is much less than that of the atom as a whole, but some patterns emerge from observation of the naturally occurring nuclides.

The Band of Stability and the Neutron-to-Proton (N/Z) Ratio A key factor that determines the stability of a nuclide is the ratio of the number of neutrons to the number of protons, the **N/Z ratio,** which we calculate from $(A - Z)/Z$. For lighter nuclides, one neutron for each proton ($N/Z \approx 1$) is enough to provide stability. However, for heavier nuclides to be stable, the number of neutrons must exceed the number of protons, and often by quite a lot. But, if the N/Z ratio is either too high or not high enough, the nuclide is unstable and decays.

Figure 23.2A on the next page is a plot of number of neutrons vs. number of protons for the *stable* nuclides. The nuclides form a narrow **band of stability** that gradually increases from an N/Z ratio of 1, near $Z = 10$, to an N/Z ratio slightly greater than 1.5, near $Z = 83$ for ^{209}Bi. Several key points are as follows:

- Very few stable nuclides exist with $N/Z < 1$; the only two are $^1_1$H and $^3_2$He. For lighter nuclides, $N/Z \approx 1$: for example, $^4_2$He, $^{12}_6$C, $^{16}_8$O, and $^{20}_{10}$Ne are particularly stable.
- The N/Z ratio of stable nuclides gradually increases as Z increases. No stable nuclide exists with $N/Z = 1$ for $Z > 20$. Thus, for $^{56}_{26}$Fe, $N/Z = 1.15$; for $^{107}_{47}$Ag, $N/Z = 1.28$; and for $^{184}_{74}$W, $N/Z = 1.49$.
- All nuclides with $Z > 83$ are unstable. Bismuth-209 is the heaviest stable nuclide. Therefore, the largest members of Groups 1A(1), 2A(2), 4A(14), 6A(16), 7A(17), and 8A(18) are radioactive, as are all the actinides (the 5*f* inner-transition elements) and the elements of the fourth *d*-block transition series (Period 7).

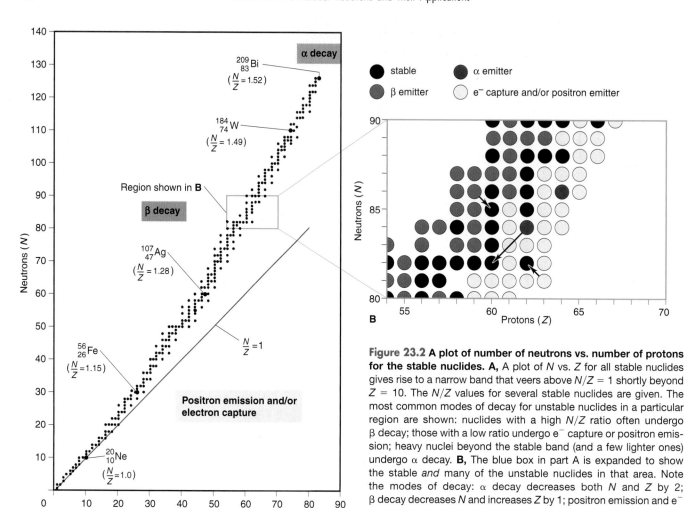

Figure 23.2 A plot of number of neutrons vs. number of protons for the stable nuclides. A, A plot of N vs. Z for all stable nuclides gives rise to a narrow band that veers above $N/Z = 1$ shortly beyond $Z = 10$. The N/Z values for several stable nuclides are given. The most common modes of decay for unstable nuclides in a particular region are shown: nuclides with a high N/Z ratio often undergo β decay; those with a low ratio undergo e^- capture or positron emission; heavy nuclei beyond the stable band (and a few lighter ones) undergo α decay. **B,** The blue box in part A is expanded to show the stable *and* many of the unstable nuclides in that area. Note the modes of decay: α decay decreases both N and Z by 2; β decay decreases N and increases Z by 1; positron emission and e^- capture increase N and decrease Z by 1.

Table 23.3 Number of Stable Nuclides for Elements 48 to 54*		
Element	**Atomic No. (Z)**	**No. of Nuclides**
Cd	**48**	**8**
In	49	2
Sn	**50**	**10**
Sb	51	2
Te	**52**	**8**
I	53	1
Xe	**54**	**9**

*Even Z shown in boldface.

Stability and Nuclear Structure The oddness or evenness of N and Z values is related to some important patterns of nuclear stability. Two interesting points become apparent when we classify the known stable nuclides:

- Elements with an even Z (number of protons) usually have a larger number of stable nuclides than elements with an odd Z. Table 23.3 demonstrates this point for cadmium ($Z = 48$) through xenon ($Z = 54$).
- Well over half the stable nuclides have *both* even N and even Z. Only seven nuclides with odd N and odd Z are either stable—$^2_1$H, $^6_3$Li, $^{10}_5$B, $^{14}_7$N—or decay so slowly that their amounts have changed little since Earth formed—$^{50}_{23}$V, $^{138}_{57}$La, and $^{176}_{71}$Lu.

One model of nuclear structure that attempts to explain these findings postulates that protons and neutrons lie in *nucleon shells,* or energy levels, and that stability results from the *pairing* of like nucleons. This arrangement leads to the stability of even values of N and Z. (The analogy to electron energy levels and the stability that arises from electron pairing is striking.)

Just as noble gases—with 2, 10, 18, 36, 54, and 86 electrons—are exceptionally stable because of their filled *electron* shells, nuclides with N or Z values of 2, 8, 20, 28, 50, 82 (and $N = 126$) are exceptionally stable. These so-called *magic numbers* are thought to correspond to the numbers of protons or neutrons in filled *nucleon* shells. A few examples are $^{50}_{22}$Ti ($N = 28$), $^{88}_{38}$Sr ($N = 50$), and the ten stable nuclides of tin ($Z = 50$). Some extremely stable nuclides have double magic numbers: $^4_2$He, $^{16}_8$O, $^{40}_{20}$Ca, and $^{208}_{82}$Pb ($N = 126$).

SAMPLE PROBLEM 23.2   Predicting Nuclear Stability

Problem Which of the following nuclides would you predict to be stable and which radioactive: (a) $^{18}_{10}Ne$; (b) $^{32}_{16}S$; (c) $^{236}_{90}Th$; (d) $^{123}_{56}Ba$? Explain.

Plan In order to evaluate the stability of each nuclide, we determine the N/Z ratio from $(A - Z)/Z$, the value of Z, stable N/Z ratios (from Figure 23.2), and whether Z and N are even or odd.

Solution (a) Radioactive. The ratio $N/Z = \dfrac{18 - 10}{10} = 0.8$. The minimum ratio for stability is 1.0; so, despite even N and Z, this nuclide has too few neutrons to be stable.
(b) Stable. This nuclide has $N/Z = 1.0$ and $Z < 20$, with even N and Z. Thus, it is most likely stable.
(c) Radioactive. Every nuclide with $Z > 83$ is radioactive.
(d) Radioactive. The ratio $N/Z = 1.20$. For Z from 55 to 60, Figure 23.2A shows $N/Z \geq 1.3$, so this nuclide probably has too few neutrons to be stable.

Check By consulting a table of isotopes, such as the one in the *CRC Handbook of Chemistry and Physics,* we find that our predictions are correct.

FOLLOW-UP PROBLEM 23.2 Why is $^{31}_{15}P$ stable but $^{30}_{15}P$ unstable?

Predicting the Mode of Decay An unstable nuclide generally decays in a mode that shifts its N/Z ratio toward the band of stability. This fact is illustrated in Figure 23.2B, which expands a small region of Figure 23.2A to show all of the stable *and* many of the radioactive nuclides in that region, as well as their modes of decay. Note the following points, and then we'll apply them in a sample problem:

1. *Neutron-rich nuclides.* Nuclides with too many neutrons for stability (a high N/Z) lie above the band of stability. They undergo *β decay,* which converts a neutron into a proton, thus reducing the value of N/Z.
2. *Neutron-poor nuclides.* Nuclides with too few neutrons for stability (a low N/Z) lie below the band. They undergo *positron emission* or *electron capture,* both of which convert a proton into a neutron, thus increasing the value of N/Z.
3. *Heavy nuclides.* Nuclides with $Z > 83$ are too heavy to lie within the band and undergo *α decay,* which reduces their Z and N values by two units per emission. (Several lighter nuclides also exhibit α decay.)

SAMPLE PROBLEM 23.3   Predicting the Mode of Nuclear Decay

Problem Predict the nature of the nuclear change(s) each of the following radioactive nuclides is likely to undergo: (a) $^{12}_{5}B$; (b) $^{234}_{92}U$; (c) $^{74}_{33}As$; (d) $^{127}_{57}La$.

Plan We use the N/Z ratio to decide where the nuclide lies relative to the band of stability and how its ratio compares with others in the nearby region of the band. Then, we predict which of the decay modes just discussed will yield a product nuclide that is closer to the band.

Solution (a) This nuclide has an N/Z ratio of 1.4, which is too high for this region of the band. It will probably undergo β decay, increasing Z to 6 and lowering the N/Z ratio to 1.
(b) This nuclide is heavier than those close to it in the band of stability. It will probably undergo α decay and decrease its total mass.
(c) This nuclide, with an N/Z ratio of 1.24, lies in the band of stability, so it will probably undergo either β decay or positron emission.
(d) This nuclide has an N/Z ratio of 1.23, which is too low for this region of the band, so it will decrease Z by either positron emission or electron capture.

Comment Both possible modes of decay are observed for the nuclides in parts (c) and (d).

FOLLOW-UP PROBLEM 23.3 What mode of decay would you expect for (a) $^{61}_{26}Fe$; (b) $^{241}_{95}Am$?

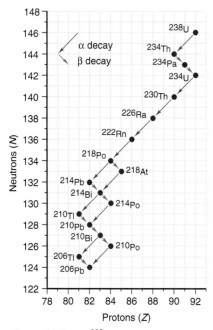

Figure 23.3 The ^{238}U decay series. Uranium-238 (*top right*) decays through a series of emissions of α or β particles to lead-206 (*bottom left*) in 14 steps.

Decay Series A parent nuclide may undergo a series of decay steps before a stable daughter nuclide forms. The succession of steps is called a **decay series,** or **disintegration series,** and is typically depicted on a gridlike display. Figure 23.3 shows the decay series from uranium-238 to lead-206. Numbers of neutrons (N) are plotted against numbers of protons (Z) to form the grid, which displays a series of α and β decays. The zigzag pattern is typical and occurs because α decay decreases both N and Z, whereas β decay decreases N but increases Z. Note that it is quite common for a given nuclide to undergo both types of decay. (Gamma emission accompanies many of these steps, but it does not affect the mass or type of the nuclide.) This decay series is one of three that occur in nature. All end with isotopes of lead whose nuclides all have one ($Z = 82$) or two ($N = 126$, $Z = 82$) magic numbers. A second series begins with uranium-235 and ends with lead-207, and a third begins with thorium-232 and ends with lead-208. (Neptunium-237 began a fourth series, but its half-life is so much less than the age of Earth that only traces of it remain today.)

SECTION SUMMARY

Nuclear reactions are not affected by reaction conditions or chemical composition and release much more energy than chemical reactions. A radioactive nuclide is unstable and may emit α particles ($_2^4$He nuclei), β particles ($_{-1}^0$β; high-speed electrons), positrons ($_1^0$β), or γ rays ($_0^0$γ; high-energy photons) or may capture an orbital electron ($_{-1}^0$e). A narrow band of neutron-to-proton ratios (N/Z) includes those of all the stable nuclides. Radioactive decay allows an unstable nuclide to achieve a more stable N/Z ratio. Certain "magic numbers" of neutrons and protons are associated with very stable nuclides. By comparing a nuclide's N/Z ratio with those in the band of stability, we can predict that, in general, heavy nuclides undergo α decay, neutron-rich nuclides undergo β decay, and proton-rich nuclides undergo positron emission or electron capture. Three naturally occurring decay series all end in isotopes of lead.

23.2 THE KINETICS OF RADIOACTIVE DECAY

Chemical and nuclear systems both tend toward maximum stability. Just as the concentrations in a chemical system change in a predictable direction to give a stable equilibrium ratio, the type and number of nucleons in an unstable nucleus change in a predictable direction to give a stable N/Z ratio. As you know, however, the tendency of a chemical system to become more stable tells nothing about how long that process will take, and the same holds true for nuclear systems. In this section, we examine the kinetics of nuclear change; later, we'll examine the energetics of nuclear change.

The Rate of Radioactive Decay

Radioactive nuclei decay at a characteristic rate, regardless of the chemical substance in which they occur. The *decay rate,* or **activity ($\mathscr{A}$),** of a radioactive sample is the change in number of nuclei ($\mathscr{N}$) divided by the change in time (t). As we saw with chemical reaction rates, because the number of nuclei is *decreasing,* a minus sign precedes the expression for the decay rate:

$$\text{Decay rate } (\mathscr{A}) = -\frac{\Delta\mathscr{N}}{\Delta t}$$

The SI unit of radioactivity is the **becquerel (Bq);** it is defined as one disintegration per second (d/s): 1 Bq = 1 d/s. A much larger and more common unit of radioactivity is the **curie (Ci):** 1 curie equals the number of nuclei disintegrating each second in 1 g of radium-226:

$$1 \text{ Ci} = 3.70\times10^{10} \text{ d/s} \qquad (23.2)$$

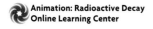
Animation: Radioactive Decay Online Learning Center

Because the curie is so large, the millicurie (mCi) and microcurie (μCi) are commonly used. We often express the radioactivity of a sample in terms of *specific activity,* the decay rate per gram.

An activity is meaningful only when we consider the large number of nuclei in a macroscopic sample. Suppose there are 1×10^{15} radioactive nuclei of a particular type in a sample and they decay at a rate of 10% per hour. Although any particular nucleus in the sample might decay in a microsecond or in a million hours, the *average* of all decays results in 10% of the entire collection of nuclei disintegrating each hour. During the first hour, 10% of the *original* number, or 1×10^{14} nuclei, will decay. During the next hour, 10% of the remaining 9×10^{14} nuclei, or 9×10^{13} nuclei, will decay. During the next hour, 10% of those remaining will decay, and so forth. Thus, for a large collection of radioactive nuclei, *the number decaying per unit time is proportional to the number present:*

$$\text{Decay rate } (\mathscr{A}) \propto \mathscr{N} \qquad \text{or} \qquad \mathscr{A} = k\mathscr{N}$$

where k is called the **decay constant** and is characteristic of each type of nuclide. The larger the value of k, the higher is the decay rate.

Combining the two rate expressions just given, we obtain

$$\mathscr{A} = -\frac{\Delta \mathscr{N}}{\Delta t} = k\mathscr{N} \qquad \textbf{(23.3)}$$

Note that the activity depends only on $\mathscr{N}$ raised to the first power (and on the constant value of k). Therefore, *radioactive decay is a first-order process* (see Section 16.3). The only difference in the case of nuclear decay is that we consider the *number* of nuclei rather than their concentration.

Half-Life of Radioactive Decay Decay rates are also commonly expressed in terms of the fraction of nuclei that decay over a given time interval. The **half-life** ($t_{1/2}$) of a nuclide is the time it takes for half the nuclei present in a sample to decay. *The number of nuclei remaining is halved after each half-life.* Thus, half-life has the same meaning for a nuclear change as for a chemical change (Section 16.4). Figure 23.4 shows the decay of carbon-14, which has a half-life of 5730 years, in terms of number of ^{14}C nuclei remaining:

$$^{14}_{6}\text{C} \longrightarrow {}^{14}_{7}\text{N} + {}^{\ 0}_{-1}\beta$$

We can also consider the half-life in terms of mass of substance. As ^{14}C decays to the product ^{14}N, its mass decreases. If we start with 1.0 g of carbon-14, half that mass of ^{14}C (0.50 g) will be left after 5730 years, half of that mass (0.25 g) after another 5730 years, and so on. The activity depends on the number of nuclei present, so the activity is halved after each succeeding half-life as well.

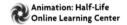

Animation: Half-Life
Online Learning Center

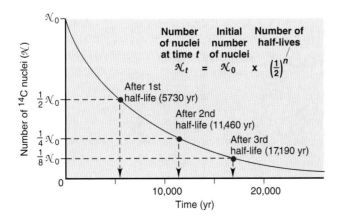

Figure 23.4 Decrease in number of ^{14}C nuclei over time. A plot of number of ^{14}C nuclei vs. time gives a decreasing curve. In each half-life (5730 yr), half the ^{14}C nuclei present undergo decay. A plot of mass of ^{14}C vs. time is identical.

We determine the half-life of a nuclear reaction from its rate constant. Rearranging Equation 23.3 and integrating over time gives

$$\ln \frac{\mathcal{N}_t}{\mathcal{N}_0} = -kt \qquad \text{or} \qquad \ln \frac{\mathcal{N}_0}{\mathcal{N}_t} = kt \qquad (23.4)$$

where $\mathcal{N}_0$ is the number of nuclei at $t = 0$, and $\mathcal{N}_t$ is the number of nuclei remaining at any time t. (Note the similarity to Equation 16.4.) To calculate the half-life ($t_{1/2}$), we set $\mathcal{N}_t$ equal to $\frac{1}{2}\mathcal{N}_0$ and solve for $t_{1/2}$:

$$\ln \frac{\mathcal{N}_0}{\frac{1}{2}\mathcal{N}_0} = kt_{1/2} \qquad \text{so} \qquad t_{1/2} = \frac{\ln 2}{k} \qquad (23.5)$$

Exactly analogous to the half-life of a first-order chemical change, *this half-life is **not** dependent on the number of nuclei and is inversely related to the decay constant:*

$$\text{large } k \Rightarrow \text{short } t_{1/2} \qquad \text{and} \qquad \text{small } k \Rightarrow \text{long } t_{1/2}$$

The decay constants and half-lives of radioactive nuclides vary over a very wide range, even those for the nuclides of a given element (Table 23.4).

Table 23.4 Decay Constants (k) and Half-Lives ($t_{1/2}$) of Beryllium Isotopes

Nuclide	k	$t_{1/2}$
^7_4Be	1.30×10^{-2}/day	53.3 days
^8_4Be	1.0×10^{16}/s	6.7×10^{-17} s
^9_4Be	Stable	
$^{10}_4\text{Be}$	4.3×10^{-7}/yr	1.6×10^6 yr
$^{11}_4\text{Be}$	5.02×10^{-2}/s	13.8 s

SAMPLE PROBLEM 23.4 Finding the Number of Radioactive Nuclei

Problem Strontium-90 is a radioactive by-product of nuclear reactors that behaves biologically like calcium, the element above it in Group 2A(2). When ^{90}Sr is ingested by mammals, it is found in their milk and eventually in the bones of those drinking the milk. If a sample of ^{90}Sr has an activity of 1.2×10^{12} d/s, what are the activity and the fraction of nuclei that have decayed after 59 yr ($t_{1/2}$ of $^{90}\text{Sr} = 29$ yr)?

Plan The fraction of nuclei that have decayed is the change in number of nuclei, expressed as a fraction of the starting number. The activity of the sample ($\mathcal{A}$) is proportional to the number of nuclei ($\mathcal{N}$), so we know that

$$\text{Fraction decayed} = \frac{\mathcal{N}_0 - \mathcal{N}_t}{\mathcal{N}_0} = \frac{\mathcal{A}_0 - \mathcal{A}_t}{\mathcal{A}_0}$$

We are given $\mathcal{A}_0$ (1.2×10^{12} d/s), so we find $\mathcal{A}_t$ from the integrated form of the first-order rate equation (Equation 23.4), in which t is 59 yr. To solve that equation, we need k, which we can calculate from the given $t_{1/2}$ (29 yr).

Solution Calculating the decay constant k:

$$t_{1/2} = \frac{\ln 2}{k} \qquad \text{so} \qquad k = \frac{\ln 2}{t_{1/2}} = \frac{0.693}{29 \text{ yr}} = 0.024 \text{ yr}^{-1}$$

Applying Equation 23.4 to calculate $\mathcal{A}_t$, the activity remaining at time t:

$$\ln \frac{\mathcal{N}_0}{\mathcal{N}_t} = \ln \frac{\mathcal{A}_0}{\mathcal{A}_t} = kt \qquad \text{or} \qquad \ln \mathcal{A}_0 - \ln \mathcal{A}_t = kt$$

So,
$$\ln \mathcal{A}_t = -kt + \ln \mathcal{A}_0 = -(0.024 \text{ yr}^{-1} \times 59 \text{ yr}) + \ln (1.2\times10^{12} \text{ d/s})$$
$$\ln \mathcal{A}_t = -1.4 + 27.81 = 26.4$$
$$\mathcal{A}_t = 2.9\times10^{11} \text{ d/s}$$

(All the data contain two significant figures, so we retained two in the answer.) Calculating the fraction decayed:

$$\text{Fraction decayed} = \frac{\mathcal{A}_0 - \mathcal{A}_t}{\mathcal{A}_0} = \frac{1.2\times10^{12} \text{ d/s} - 2.9\times10^{11} \text{ d/s}}{1.2\times10^{12} \text{ d/s}} = 0.76$$

Check The answer is reasonable: t is about 2 half-lives, so $\mathcal{A}_t$ should be about $\frac{1}{4}\mathcal{A}_0$, or about 0.3×10^{12}; therefore, the activity should have decreased by about $\frac{3}{4}$.

Comment An *alternative approach* is to use the number of half-lives ($t/t_{1/2}$) to find the fraction of activity (or nuclei) remaining. By combining Equations 23.4 and 23.5 and substituting (ln 2)/$t_{1/2}$ for k, we obtain

$$\ln \frac{\mathcal{N}_0}{\mathcal{N}_t} = \left(\frac{\ln 2}{t_{1/2}}\right)t = \frac{t}{t_{1/2}} \ln 2 = \ln 2^{t/t_{1/2}}$$

Thus,
$$\ln \frac{\mathcal{N}_t}{\mathcal{N}_0} = \ln \left(\frac{1}{2}\right)^{t/t_{1/2}}$$

Taking the antilog gives

$$\text{Fraction remaining} = \frac{\mathcal{N}_t}{\mathcal{N}_0} = \left(\frac{1}{2}\right)^{t/t_{1/2}} = \left(\frac{1}{2}\right)^{59/29} = 0.24$$

So,
$$\text{Fraction decayed} = 1.00 - 0.24 = 0.76$$

FOLLOW-UP PROBLEM 23.4 Sodium-24 ($t_{1/2} = 15$ h) is used to study blood circulation. If a patient is injected with a ^{24}NaCl solution whose activity is 2.5×10^9 d/s, how much of the activity is present in the patient's body and excreted fluids after 4.0 days?

Radioisotopic Dating

The historical record fades rapidly with time and virtually disappears for events of more than a few thousand years ago. Much of our understanding of prehistory comes from a technique called **radioisotopic dating**, which uses **radioisotopes** to determine the age of an object. The method supplies data in fields as diverse as art history, archeology, geology, and paleontology.

The technique of *radiocarbon dating*, for which the American chemist Willard F. Libby won the Nobel Prize in chemistry in 1960, is based on measuring the amounts of ^{14}C and ^{12}C in materials of biological origin. The accuracy of the method falls off after about six half-lives of ^{14}C ($t_{1/2} = 5730$ yr), so it is used to date objects up to about 36,000 years old.

Here is how the method works. High-energy neutrons resulting from cosmic ray collisions reach Earth continually from outer space. They enter the atmosphere and cause the slow formation of ^{14}C by bombarding ordinary ^{14}N atoms:

$$^{14}_{7}\text{N} + ^{1}_{0}\text{n} \longrightarrow ^{14}_{6}\text{C} + ^{1}_{1}\text{p}$$

Through the processes of formation and radioactive decay, the amount of ^{14}C in the atmosphere has remained nearly constant.*

The ^{14}C atoms combine with O_2, diffuse throughout the lower atmosphere, and enter the total carbon pool as gaseous $^{14}CO_2$ and aqueous $H^{14}CO_3^{-}$. They mix with ordinary $^{12}CO_2$ and $H^{12}CO_3^{-}$, reaching a constant ^{12}C:^{14}C ratio of about 10^{12}:1. The CO_2 is taken up by plants during photosynthesis, and then taken up and excreted by animals that eat the plants. Thus, the ^{12}C:^{14}C ratio of a living organism has the same constant value as the environment. When an organism dies, however, it no longer takes in ^{14}C, so the ^{12}C:^{14}C ratio steadily increases because the amount of *^{14}C decreases as it decays*:

$$^{14}_{6}\text{C} \longrightarrow ^{14}_{7}\text{N} + ^{0}_{-1}\beta$$

The difference between the ^{12}C:^{14}C ratio in a dead organism and the ratio in living organisms reflects the time elapsed since the organism died.

As you saw in Sample Problem 23.4, the first-order rate equation can be expressed in terms of a ratio of activities:

$$\ln \frac{\mathcal{N}_0}{\mathcal{N}_t} = \ln \frac{\mathcal{A}_0}{\mathcal{A}_t} = kt$$

We use this expression in radiocarbon dating, where $\mathcal{A}_0$ is the activity in a living organism and $\mathcal{A}_t$ is the activity in the object whose age is unknown. Solving for t gives the age of the object:

$$t = \frac{1}{k} \ln \frac{\mathcal{A}_0}{\mathcal{A}_t} \tag{23.6}$$

---

*Cosmic ray intensity does vary slightly with time, which affects the amount of atmospheric ^{14}C. From ^{14}C activity in ancient trees, we know the amount fell slightly about 3000 years ago to current levels. Recently, nuclear testing and fossil fuel combustion have also altered the fraction of ^{14}C slightly. Taking these factors into account improves the accuracy of the dating method.

To determine the ages of more ancient objects or of objects that do not contain carbon, different radioisotopes are measured. For example, by comparing the ratio of ^{238}U to its final decay product, ^{206}Pb, geochemists found that the oldest known surface rocks on Earth—granite in western Greenland—are about 3.7 billion years old, and determining this ratio in samples from meteorites gives 4.65 billion years for the age of the Solar System, and thus Earth.

SAMPLE PROBLEM 23.5 Applying Radiocarbon Dating

Problem The charred bones of a sloth in a cave in Chile represent the earliest evidence of human presence at the southern tip of South America. A sample of the bone has a specific activity of 5.22 disintegrations per minute per gram of carbon (d/min·g). If the ratio of ^{12}C:^{14}C in living organisms results in a specific activity of 15.3 d/min·g, how old are the bones ($t_{1/2}$ of ^{14}C = 5730 yr)?

Plan We first calculate k from the given $t_{1/2}$ (5730 yr). Then we apply Equation 23.6 to find the age (t) of the bones, using the given activities of the bones ($\mathscr{A}_t$ = 5.22 d/min·g) and of a living organism ($\mathscr{A}_0$ = 15.3 d/min·g).

Solution Calculating k for ^{14}C decay:

$$k = \frac{\ln 2}{t_{1/2}} = \frac{0.693}{5730 \text{ yr}} = 1.21 \times 10^{-4} \text{ yr}^{-1}$$

Calculating the age (t) of the bones:

$$t = \frac{1}{k} \ln \frac{\mathscr{A}_0}{\mathscr{A}_t} = \frac{1}{1.21 \times 10^{-4} \text{ yr}^{-1}} \ln \left(\frac{15.3 \text{ d/min·g}}{5.22 \text{ d/min·g}} \right) = 8.89 \times 10^3 \text{ yr}$$

The bones are about 8900 years old.

Check The activity of the bones is between $\frac{1}{2}$ and $\frac{1}{4}$ the activity of a living organism, so the age should be between one and two half-lives (5730 to 11,460 yr).

FOLLOW-UP PROBLEM 23.5 A sample of wood from an Egyptian mummy case has a specific activity of 9.41 d/min·g. How old is the case?

SECTION SUMMARY

The decay rate (activity) of a sample is proportional to the number of radioactive nuclei. Nuclear decay is a first-order process, so the half-life does not depend on the number of nuclei. Radioisotopic methods, such as ^{14}C dating, determine the ages of objects by measuring the ratio of specific isotopes in the sample.

23.3 NUCLEAR TRANSMUTATION: INDUCED CHANGES IN NUCLEI

The alchemists' dream of changing base metals into gold was never realized, but in the early 20th century, atomic physicists found that they *could* change one element into another. Research into **nuclear transmutation,** the *induced* conversion of one nucleus into another, was closely linked with research into atomic structure and led to the discovery of the neutron and to the production of artificial radioisotopes. Later, high-energy bombardment of nuclei in particle accelerators began a scientific endeavor, which continues to this day, of creating many new nuclides and a growing number of new elements. In fact, the majority of the nearly 1000 known radionuclides have been produced artificially.

During the 1930s and 1940s, researchers probing the nucleus bombarded elements with neutrons, α particles, protons, and **deuterons** (nuclei of the stable hydrogen isotope deuterium, ^{2}H). Neutrons are especially useful as projectiles because they have no charge and thus are not repelled as they approach a target nucleus. The other types of particles are all positive, so they must be given enough energy to overcome their repulsion by the target nuclei. **Particle accelerators**

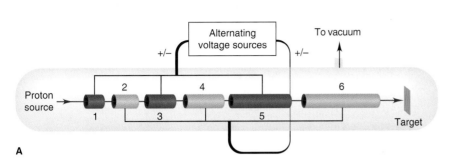

A

B

Figure 23.5 A linear accelerator. A, The voltage of each tubular section is alternated, so that the positively charged particle (a proton here) is repelled from the section it is leaving and attracted to the section it is entering. As a result, the particle's speed is continually increased. **B,** The linear accelerator operated by Stanford University in California.

were invented to impart high kinetic energies to particles by placing them in an electric field, usually in combination with a magnetic field.

A major advance in nuclear research was the *linear accelerator,* a series of separated tubes of increasing length that, through a source of alternating voltage, change their charge from positive to negative in synchrony with the movement of the particle through them (Figure 23.5A). A proton, for example, exits the first tube just when that tube becomes positive and the next tube negative. Repelled by the first tube and attracted by the second, the proton accelerates across the gap between them. Today, the Stanford Linear Accelerator (Figure 23.5B) and others like it accelerate particles much heavier than protons, such as B, C, O, and Ne nuclei. The *cyclotron* (Figure 23.6), invented by E. O. Lawrence in 1930, applies the principle of the linear accelerator but uses electromagnets to give the particle a spiral path, thus saving space.

Accelerators have many applications, from producing radioisotopes used in medical applications to studying the fundamental nature of matter. Perhaps their

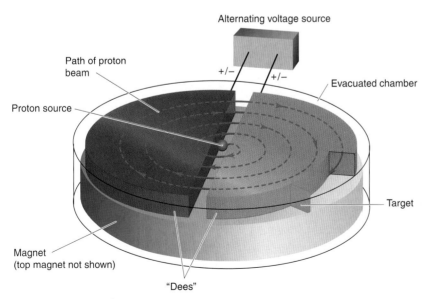

Figure 23.6 The cyclotron accelerator. When the positively charged particle reaches the gap between the two D-shaped electrodes ("dees"), it is repelled by one dee and attracted by the other. The particles move in a spiral path, so the cyclotron can be much smaller than a linear accelerator.

Table 23.5 Formation of Some Transuranium Nuclides	
Reaction	**Half-life of Product**
${}^{239}_{94}\text{Pu} + {}^{4}_{2}\text{He} \longrightarrow {}^{240}_{95}\text{Am} + {}^{1}_{1}\text{H} + 2{}^{1}_{0}\text{n}$	50.9 h
${}^{239}_{94}\text{Pu} + {}^{4}_{2}\text{He} \longrightarrow {}^{242}_{96}\text{Cm} + {}^{1}_{0}\text{n}$	163 days
${}^{244}_{96}\text{Cm} + {}^{4}_{2}\text{He} \longrightarrow {}^{245}_{97}\text{Bk} + {}^{1}_{1}\text{H} + 2{}^{1}_{0}\text{n}$	4.94 days
${}^{238}_{92}\text{U} + {}^{12}_{6}\text{C} \longrightarrow {}^{246}_{98}\text{Cf} + 4{}^{1}_{0}\text{n}$	36 h
${}^{253}_{99}\text{Es} + {}^{4}_{2}\text{He} \longrightarrow {}^{256}_{101}\text{Md} + {}^{1}_{0}\text{n}$	76 min
${}^{252}_{98}\text{Cf} + {}^{10}_{5}\text{B} \longrightarrow {}^{256}_{103}\text{Lr} + 6{}^{1}_{0}\text{n}$	28 s

most specific application for chemists is the synthesis of **transuranium elements,** those with atomic numbers higher than uranium, the heaviest naturally occurring element. Some reactions that were used to form several of these elements appear in Table 23.5.

SECTION SUMMARY
One nucleus can be transmuted to another through bombardment with high-energy particles. Accelerators increase the kinetic energy of particles and are used to produce radioisotopes for medical use and transuranium elements.

23.4 THE EFFECTS OF NUCLEAR RADIATION ON MATTER

In 1986, an accident at the Chernobyl nuclear facility in the former Soviet Union released radioactivity that is estimated to have already caused thousands of cancer deaths. In the same year, isotopes used in medical treatment emitted radioactivity that prevented thousands of cancer deaths. In this section and the next, we examine the harmful and beneficial effects of radioactivity.

The key to both of these outcomes is that *nuclear changes cause chemical changes in surrounding matter.* In other words, even though the nucleus of an atom undergoes a reaction with little or no involvement of the atom's electrons, the emissions from that reaction *do* affect the electrons of nearby atoms.

The Effects of Radioactive Emissions: Excitation and Ionization

Radioactive emissions interact with matter in two ways, depending on their energies:

- *Excitation.* In the process of **excitation,** radiation of relatively low energy interacts with an atom of a substance, which absorbs some of the energy and then re-emits it. Because electrons are not lost from the atom, the radiation that causes excitation is called **nonionizing radiation.** If the absorbed energy causes the atoms to move, vibrate, or rotate more rapidly, the material becomes hotter. For example, concentrated aqueous solutions of plutonium salts boil because the emissions excite the surrounding water molecules. Particles of somewhat higher energy excite electrons in other atoms to higher energy levels. As the atoms return to their ground state, they emit photons, often in the blue or UV region.

- *Ionization.* In the process of **ionization,** radiation collides with an atom energetically enough to dislodge an electron:

$$\text{Atom} \xrightarrow{\text{ionizing radiation}} \text{ion}^{+} + e^{-}$$

A cation and a free electron result, and the number of such *cation-electron pairs* that are produced is directly related to the energy of the incoming radiation. The high-energy radiation that gives rise to this effect is called **ionizing radiation.** The free electron of the pair often collides with another atom and ejects a second electron.

Effects of Ionizing Radiation on Living Matter

Whereas nonionizing radiation is relatively harmless, ionizing radiation has a destructive effect on living tissue. When the atom that was ionized is part of a biological macromolecule or membrane component, the results can be devastating.

Units of Radiation Dose and Its Effects To measure the effects of ionizing radiation, we need a unit for radiation dose. Units of radioactive decay, such as the becquerel and curie, measure the number of decay events in a given time but not their energy or absorption by matter. The number of cation-electron pairs produced in a given amount of living tissue is a measure of the energy absorbed by the tissue. The SI unit for such energy absorption is the **gray (Gy);** it is equal to 1 joule of energy absorbed per kilogram of body tissue: 1 Gy = 1 J/kg. A more widely used unit is the **rad (radiation-absorbed dose),** which is equal to 0.01 Gy:

$$1 \text{ rad} = 0.01 \text{ J/kg} = 0.01 \text{ Gy}$$

To measure actual tissue damage, we must account for differences in the strength of the radiation, the exposure time, and the type of tissue. To do this, we multiply the number of rads by a *relative biological effectiveness* (RBE) factor, which depends on the effect of a given type of radiation on a given tissue or body part. The product is the **rem (roentgen equivalent for man),** the unit of radiation dosage equivalent to a given amount of tissue damage in a human:

$$\text{no. of rems} = \text{no. of rads} \times \text{RBE}$$

Doses are often expressed in millirems (10^{-3} rem). The SI unit for dosage equivalent is the **sievert (Sv).** It is defined in the same way as the rem but with absorbed dose in grays; thus, 1 rem = 0.01 Sv.

Penetrating Power of Emissions The effect on living tissue of a radiation dose depends on the penetrating power *and* ionizing ability of the radiation. Figure 23.7 depicts the differences in penetrating power of the three common emissions. Note, in general, that *penetrating power is inversely related to the mass and charge of the emission.* In other words, if a particle interacts strongly with matter, it penetrates only slightly, and vice versa:

- *α Particles.* Alpha particles are massive and highly charged, which means that they interact with matter most strongly of the three common types of emissions. As a result, they penetrate so little that a piece of paper, light clothing, or the outer layer of skin can stop α radiation from an external source. However, if ingested, an α emitter such as plutonium-239 causes grave localized damage through extensive ionization.
- *β Particles and positrons.* Beta particles and positrons have less charge and much less mass than α particles, so they interact less strongly with matter. Even though a given particle has less chance of causing ionization, a β (or positron) emitter is a more destructive external source because the particles penetrate deeper. Specialized heavy clothing or a thick (0.5 cm) piece of metal is required to stop these particles.
- *γ Rays.* Neutral, massless γ rays interact least with matter and, thus, penetrate most. A block of lead several inches thick is needed to stop them. Therefore, an external γ ray source is the most dangerous because the energy can ionize many layers of living tissue.

Sources of Ionizing Radiation We are continuously exposed to ionizing radiation from natural and artificial sources (Table 23.6, next page). Indeed, life evolved in the presence of natural ionizing radiation, called **background radiation.** The same radiation that can alter bonds in DNA and cause harmful mutations also causes beneficial mutations that, over time, allow organisms to adapt and species to change.

α (~0.03 mm)

β (~2 mm)

γ (~10 cm)

Figure 23.7 Penetrating power of radioactive emissions. Penetrating power is often measured in terms of the depth of water that stops 50% of the incoming radiation. (Water is the main component of living tissue.) Alpha particles, with the highest mass and charge, have the lowest penetrating power, and γ rays have the highest. (Average values of actual penetrating distances are shown.)

Table 23.6 Typical Radiation Doses from Natural and Artificial Sources

Source of Radiation	Average Adult Exposure
Natural	
Cosmic radiation	30–50 mrem/yr
Radiation from the ground	
From clay soil and rocks	~25–170 mrem/yr
In wooden houses	10–20 mrem/yr
In brick houses	60–70 mrem/yr
In concrete (cinder block) houses	60–160 mrem/yr
Radiation from the air (mainly radon)	
Outdoors, average value	20 mrem/yr
In wooden houses	70 mrem/yr
In brick houses	130 mrem/yr
In concrete (cinder block) houses	260 mrem/yr
Internal radiation from minerals in tap water and daily intake of food (^{40}K, ^{14}C, Ra)	~40 mrem/yr
Artificial	
Diagnostic x-ray methods	
Lung (local)	0.04–0.2 rad/film
Kidney (local)	1.5–3 rad/film
Dental (dose to the skin)	≤1 rad/film
Therapeutic radiation treatment	Locally ≤ 10,000 rad
Other sources	
Jet flight (4 h)	~1 mrem
Nuclear testing	<4 mrem/yr
Nuclear power industry	<1 mrem/yr
TOTAL AVERAGE VALUE	100–200 mrem/yr

Background radiation has several sources. One source is *cosmic radiation,* which increases with altitude because of decreased absorption by the atmosphere. Thus, people in Denver absorb twice as much cosmic radiation as people in Los Angeles; even a jet flight involves measurable absorption. The sources of most background radiation are thorium and uranium minerals present in rocks and soil. Radon, the heaviest noble gas [Group 8A(18)], is a radioactive product of uranium and thorium decay, and its concentration in the air we breathe varies with type of local soil and rocks. About 150 g of K$^+$ ions is dissolved in the water in the tissues of an average adult, and 0.0118% of that amount is radioactive ^{40}K. The presence of these substances and of atmospheric ^{14}CO$_2$ means that all food, water, clothing, and building materials are slightly radioactive.

The largest artificial source of radiation, and the easiest to control, is associated with medical diagnostic techniques, especially x-rays. The radiation dosage from nuclear testing and radioactive waste disposal is miniscule for most people, but exposures for those living near test sites, nuclear energy facilities, or disposal areas may be many times higher.

SECTION SUMMARY

Relatively low-energy emissions cause excitation of atoms in surrounding matter, whereas high-energy emissions cause ionization. The effect of ionizing radiation on living matter depends on the quantity of energy absorbed and the extent of ionization in a given type of tissue. Radiation dose for the human body is measured in rem. All organisms are exposed to varying quantities of natural ionizing radiation.

23.5 APPLICATIONS OF RADIOISOTOPES

Our ability to detect minute amounts of radioisotopes makes them powerful tools for studying processes in biochemistry, medicine, materials science, environmental studies, and many other scientific and industrial fields. Such uses depend on the fact that *isotopes of an element exhibit* **very** *similar chemical and physical behavior.* In other words, except for having a less stable nucleus, a radioisotope has nearly the same chemical properties as a nonradioactive isotope of a given element. For example, the fact that $^{14}CO_2$ is utilized by a plant in the same way as $^{12}CO_2$ forms the basis of radiocarbon dating.

Radioactive Tracers: Applications of Nonionizing Radiation

A tiny amount of a radioisotope mixed with a large amount of the stable isotope can act as a **tracer,** a chemical "beacon" emitting nonionizing radiation that signals the presence of the substance.

Reaction Pathways Tracers help us choose from among possible reaction pathways. One well-studied example is the formation of an organic ester and water from a carboxylic acid and alcohol. Which portions of the reactants end up in the ester and which in the water? Figure 23.8 shows how ^{18}O-tracers answer the question: an ^{18}O-alcohol gives an ^{18}O-ester, but an ^{18}O-acid gives ^{18}O-water.

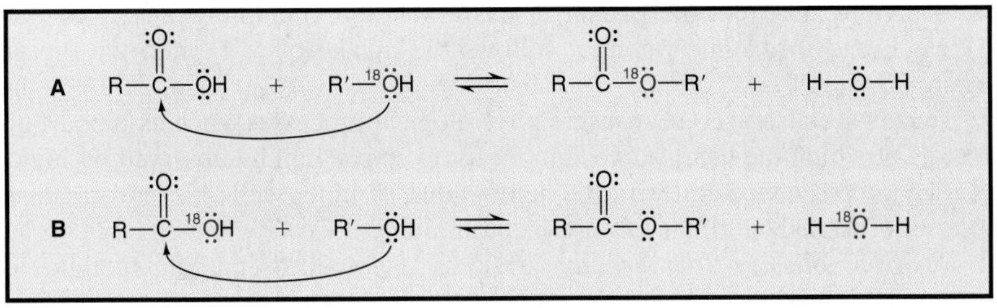

Figure 23.8 Which reactant contributes which group to the ester? An ester forms when a carboxylic acid reacts with an alcohol (see Section 15.4). To determine which reactant supplies the O atom in the —OR' part of the ester group, acid and alcohol were labeled with the ^{18}O and used as tracers. **A,** When R^{18}OH reacts with the unlabeled acid, the ester contains ^{18}O but the water doesn't. **B,** When RCO^{18}OH reacts with the unlabeled alcohol, the water contains ^{18}O. Thus, the alcohol supplies the —OR' part of the ester, and the acid supplies the RC=O part.

Far more complex reaction pathways can be followed with tracers as well. The photosynthetic pathway, the most essential and widespread metabolic process on Earth, in which energy from sunlight is used to form the chemical bonds of glucose, has an overall reaction that looks quite simple:

$$6CO_2(g) + 6H_2O(l) \xrightarrow[\text{chlorophyll}]{\text{light}} C_6H_{12}O_6(s) + 6O_2(g)$$

However, the actual process is extremely complex: 13 enzyme-catalyzed steps are required to incorporate each molecule of CO_2, so the six CO_2 molecules incorporated to form one molecule of $C_6H_{12}O_6$ require six repetitions of the pathway. Melvin Calvin and his coworkers took seven years to determine the pathway, using ^{14}C in CO_2 as the tracer and painstakingly separating the products formed after different times of light exposure. Calvin won the Nobel Prize in chemistry in 1961 for this remarkable achievement.

Material Flow Tracers are also used in studies of the flow of materials. Hydrologic engineers use tracers to study the volume and flow of large bodies of water. By following radionuclides formed during atmospheric nuclear bomb tests (^3H in H_2O, $^{90}Sr^{2+}$, and $^{137}Cs^+$), scientists have mapped the flow of water from land to lakes and streams to oceans. Surface and deep ocean currents that circulate around the globe are also studied, as are the mechanisms of hurricane formation and the mixing of the troposphere and stratosphere. Industries employ tracers to study material flow during manufacturing processes, such as the flow of ore pellets in smelting kilns, the paths of wood chips and bleach in paper mills, the diffusion of fungicide into lumber, and in a particularly important application, the porosity and leakage of oil and gas wells in geological formations.

Activation Analysis Another use of tracers is in *neutron activation analysis* (NAA). In this method, neutrons bombard a nonradioactive sample, converting a small fraction of its atoms to radioisotopes, which exhibit characteristic decay patterns, such as γ-ray spectra, that reveal the elements present. Unlike chemical analysis, NAA leaves the sample virtually intact, so the method can be used to determine the composition of a valuable object or a very small sample. For example, a painting thought to be a 16th-century Dutch masterpiece was shown through NAA to be a 20th-century forgery, because a microgram-sized sample of its pigment contained much less silver and antimony than the pigments used by the Dutch masters. Forensic chemists use NAA to detect traces of ammunition on a suspect's hand or traces of arsenic in the hair of a victim of poisoning. In 2004, space scientists used automated NAA in the *Spirit* and *Opportunity* robot vehicles to analyze the composition of Martian soils and rocks.

Medical Diagnosis The largest use of radioisotopes is in medical science. In fact, over 25% of U.S. hospital admissions are for diagnoses based on data from radioisotopes. Tracers with half-lives of a few minutes to a few days are employed to observe specific organs and body parts. For example, a healthy thyroid gland incorporates dietary I^- into iodine-containing hormones at a known rate. To assess thyroid function, the patient drinks a solution containing a trace amount of $Na^{131}I$, and a scanning monitor follows the uptake of $^{131}I^-$ into the thyroid (Figure 23.9).

Tracers are also used to measure physiological processes, such as blood flow. The rate at which the heart pumps blood, for example, can be observed by injecting ^{59}Fe, which concentrates in the hemoglobin of blood cells. Several radioisotopes used in medical diagnosis are listed in Table 23.7.

Positron-emission tomography (PET) is a powerful imaging method for observing brain structure and function. A biological substance is synthesized with one of its atoms replaced by an isotope that emits positrons. The substance is injected into a patient's bloodstream, from which it is taken up into the brain. The isotope emits positrons, each of which annihilates a nearby electron. In the annihilation process, two γ photons are emitted simultaneously 180° from each other:

$$^{0}_{1}\beta + ^{0}_{-1}e \longrightarrow 2^{0}_{0}\gamma$$

An array of detectors around the patient's head pinpoints the sites of γ emission, and the image is analyzed by computer. Two of the isotopes used are ^{15}O, injected as $H_2^{15}O$ to measure blood flow, and ^{18}F bonded to a glucose analog to measure glucose uptake, which is a marker for energy metabolism. Among many fascinating PET findings are those that show how changes in blood flow and glucose uptake accompany normal or abnormal brain activity (Figure 23.10).

Applications of Ionizing Radiation

To be used as a tracer, a radioisotope need emit only low-energy detectable radiation. Many other uses of radioisotopes, however, depend on the damage that high-energy, ionizing radiation can inflict on living systems.

Cancer cells divide more rapidly than normal cells, so radioisotopes that interfere with the cell-division process kill more cancer cells than normal ones. Implants of ^{198}Au or of a mixture of ^{90}Sr and ^{90}Y have been used to destroy

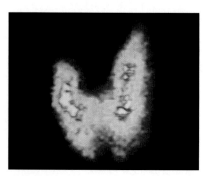

Figure 23.9 The use of a radioisotope to image the thyroid gland. Thyroid scanning is used to assess nutritional deficiencies, inflammation, tumor growth, and other thyroid-related ailments. In ^{131}I scanning, the thyroid gland absorbs $^{131}I^-$ ions whose β emissions expose a photographic film. The asymmetric shape revealed in the image indicates disease.

Animation: Nuclear Medicine
Online Learning Center

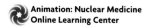

Isotope	Body Part or Process
^{11}C, ^{18}F, ^{13}N, ^{15}O	PET studies of brain, heart
^{60}Co, ^{192}Ir	Cancer therapy
^{64}Cu	Metabolism of copper
^{59}Fe	Blood flow, spleen
^{67}Ga	Tumor imaging
^{123}I, ^{131}I	Thyroid
^{111}In	Brain, colon
^{42}K	Blood flow
^{99}Tc	Heart, thyroid, liver, lung, bone
^{201}Tl	Heart muscle
^{90}Y	Cancer, arthritis

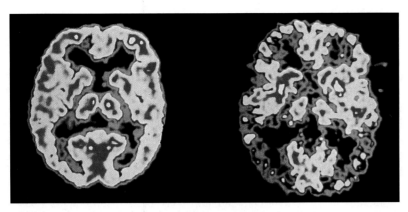

Figure 23.10 PET and brain activity. These PET scans show brain activity in a normal person (*left*) and in a patient with Alzheimer's disease (*right*). Red and yellow indicate relatively high activity within a region.

pituitary and breast tumor cells, and γ rays from ^{60}Co have been used to destroy brain tumors.

Irradiation of food increases shelf life by killing microorganisms that cause rotting or spoilage (Figure 23.11), but the practice is quite controversial. Advocates point to the benefits of preserving fresh foods, grains, and seeds for long periods, whereas opponents suggest that irradiation might lower the food's nutritional content or produce harmful by-products. The United Nations has approved irradiation for potatoes, wheat, chicken, and strawberries, and the United States allows irradiation of chicken.

Ionizing radiation has been used to control harmful insects. Captured males are sterilized by radiation and released to mate, thereby reducing the number of offspring. This method has been used to control the Mediterranean fruit fly in California and disease-causing insects, such as the tsetse fly and malarial mosquito, in other parts of the world.

Nonirradiated   Irradiated

Figure 23.11 The increased shelf life of irradiated food.

SECTION SUMMARY
Radioisotopic tracers emit nonionizing radiation and have been used to study reaction mechanisms, material flow, elemental composition, and medical conditions. Ionizing radiation has been used to destroy cancerous tissue, kill organisms that spoil food, and control insect populations.

23.6 THE INTERCONVERSION OF MASS AND ENERGY

Most of the nuclear processes we've considered so far have involved radioactive decay, in which a nucleus emits one or a few small particles or photons to become a slightly lighter nucleus. Two other nuclear processes cause much greater changes. In nuclear **fission,** a heavy nucleus splits into two much lighter nuclei, emitting several small particles at the same time. In nuclear **fusion,** the opposite process occurs: two lighter nuclei combine to form a heavier one. Both fission and fusion release enormous quantities of energy. Let's take a look at the origins of this energy by first examining the change in mass that accompanies the breakup of a nucleus into its nucleons and then considering the energy that is equivalent to this mass change.

The Mass Defect

We have known for most of the 20th century that mass and energy are interconvertible. The traditional mass and energy conservation laws have been combined to state that *the total quantity of mass-energy in the universe is constant.* Therefore, when *any* reacting system releases or absorbs energy, there must be an accompanying loss or gain in mass.

This relation between mass and energy is not important for chemical reactions because the energy changes involved in breaking or forming bonds are so small that the mass changes are negligible. When 1 mol of water breaks up into its atoms, for example, heat is absorbed:

$$H_2O(g) \longrightarrow 2H(g) + O(g) \qquad \Delta H^\circ_{rxn} = 2 \times BE \text{ of } O\text{—}H = 934 \text{ kJ}$$

We find the mass that is equivalent to this energy from *Einstein's equation:*

$$E = mc^2 \quad \text{or} \quad \Delta E = \Delta mc^2 \quad \text{so} \quad \Delta m = \frac{\Delta E}{c^2} \qquad (23.7)$$

where Δm is the change in mass between the reactants and the products. Substituting the heat of reaction (in J/mol) for ΔE and the numerical value for c (2.9979×10^8 m/s), we obtain

$$\Delta m = \frac{9.34 \times 10^5 \text{ J/mol}}{(2.9979 \times 10^8 \text{ m/s})^2}$$
$$= 1.04 \times 10^{-11} \text{ kg/mol}$$
$$= 1.04 \times 10^{-8} \text{ g/mol}$$

(Units of kg/mol are obtained because the joule includes the kilogram: 1 J = 1 kg·m²/s².) The mass of 1 mol of H_2O (reactant) is about 10 ng *less* than the combined masses of 2 mol of H and 1 mol of O (products), a change too small to measure with even the most sophisticated balance. Such minute mass changes when bonds break or form allow us to assume that mass is conserved in *chemical* reactions.

The much larger mass change that accompanies a *nuclear* process is related to the enormous energy required to bind the nucleus together or break it apart. Consider, for example, the change in mass that occurs when one ^{12}C nucleus breaks up into its nucleons: six protons and six neutrons. We calculate this change in mass by combining the mass of six H *atoms* and six neutrons and then subtracting the mass of one ^{12}C *atom*. This procedure cancels the masses of the electrons [six e⁻ (in six ^1H atoms) cancel six e⁻ (in one ^{12}C atom)]. The mass of one ^1H atom is 1.007825 amu, and the mass of one neutron is 1.008665 amu, so

Mass of six ^1H atoms =	6.046950 amu
Mass of six neutrons =	6.051990 amu
Total mass =	12.098940 amu

The mass of one ^{12}C atom is 12 amu (exactly). The difference in mass (Δm) is the total mass of the nucleons minus the mass of the nucleus:

$$\Delta m = 12.098940 \text{ amu} - 12.000000 \text{ amu}$$
$$= 0.098940 \text{ amu}/^{12}C$$
$$= 0.098940 \text{ g/mol } ^{12}C$$

Note that *the mass of the nucleus is **less** than the combined masses of its nucleons.* The mass decrease that occurs when nucleons are united into a nucleus is called the **mass defect.** The size of this mass change (9.89×10^{-2} g/mol) is nearly 10 million times that of the previous bond breakage (10.4×10^{-9} g/mol) and is easily observed on any laboratory balance.

Nuclear Binding Energy

Einstein's equation for the relation between mass and energy also allows us to find the energy equivalent of a mass defect. For ^{12}C, after converting grams to kilograms, we have

$$\Delta E = \Delta mc^2 = (9.8940 \times 10^{-5} \text{ kg/mol})(2.9979 \times 10^8 \text{ m/s})^2$$
$$= 8.8921 \times 10^{12} \text{ J/mol} = 8.8921 \times 10^9 \text{ kJ/mol}$$

This quantity of energy is called the **nuclear binding energy** for carbon-12. In general, the nuclear binding energy is the quantity of energy required to *break up 1 mol of nuclei into their individual nucleons:*

$$\text{Nucleus} + \text{nuclear binding energy} \longrightarrow \text{nucleons}$$

Thus, the nuclear binding energy is qualitatively analogous to the sum of bond energies of a covalent compound or the lattice energy of a salt. But, quantitatively, nuclear binding energies are typically several million times greater.

We use joules to express the binding energy per mole of nuclei, but the joule is much too large a unit to express the binding energy of a single nucleus. Instead, nuclear scientists use the **electron volt (eV),** the energy an electron acquires when it moves through a potential difference of 1 volt:

$$1 \text{ eV} = 1.602 \times 10^{-19} \text{ J}$$

Binding energies are commonly expressed in millions of electron volts, that is, in *mega–electron volts* (MeV):

$$1 \text{ MeV} = 10^6 \text{ eV} = 1.602 \times 10^{-13} \text{ J}$$

A particularly useful factor converts a given mass defect in atomic mass units to its energy equivalent in electron volts:

$$1 \text{ amu} = 931.5 \times 10^6 \text{ eV} = 931.5 \text{ MeV} \tag{23.8}$$

Earlier we found the mass defect of the ^{12}C nucleus to be 0.098940 amu. Therefore, the binding energy per ^{12}C nucleus, expressed in MeV, is

$$\frac{\text{Binding energy}}{^{12}C \text{ nucleus}} = 0.098940 \text{ amu} \times \frac{931.5 \text{ MeV}}{1 \text{ amu}} = 92.16 \text{ MeV}$$

We can compare the stability of nuclides of different elements by determining the *binding energy per nucleon.* For ^{12}C, we have

$$\text{Binding energy per nucleon} = \frac{\text{binding energy}}{\text{no. of nucleons}} = \frac{92.16 \text{ MeV}}{12 \text{ nucleons}} = 7.680 \text{ MeV/nucleon}$$

SAMPLE PROBLEM 23.6   Calculating the Binding Energy per Nucleon

Problem Iron-56 is an extremely stable nuclide. Compute the binding energy per nucleon for ^{56}Fe and compare it with that for ^{12}C (mass of ^{56}Fe atom = 55.934939 amu; mass of ^1H atom = 1.007825 amu; mass of neutron = 1.008665 amu).

Plan Iron-56 has 26 protons and 30 neutrons. We calculate the mass defect by finding the sum of the masses of 26 ^1H atoms and 30 neutrons and subtracting the given mass of 1 ^{56}Fe atom. Then we multiply Δm by the equivalent in MeV (931.5 MeV/amu) and divide by 56 (no. of nucleons) to obtain the binding energy per nucleon.

Solution Calculating the mass defect:

$$\text{Mass defect} = [(26 \times \text{mass } ^1H \text{ atom}) + (30 \times \text{mass neutron})] - \text{mass } ^{56}Fe \text{ atom}$$
$$= [(26)(1.007825 \text{ amu}) + (30)(1.008665 \text{ amu})] - 55.934939 \text{ amu}$$
$$= 0.52846 \text{ amu}$$

Calculating the binding energy per nucleon:

$$\text{Binding energy per nucleon} = \frac{0.52846 \text{ amu} \times 931.5 \text{ MeV/amu}}{56 \text{ nucleons}} = 8.790 \text{ MeV/nucleon}$$

An ^{56}Fe nucleus would require more energy to break up into its nucleons than would ^{12}C (7.680 MeV/nucleon), so ^{56}Fe is more stable than ^{12}C.

Check The answer is consistent with the great stability of ^{56}Fe. Given the number of decimal places in the values, rounding to check the math is useful only to find a *major* error. The number of nucleons (56) is an exact number, so we retain four significant figures.

FOLLOW-UP PROBLEM 23.6 Uranium-235 is the essential component of the fuel in nuclear power plants. Calculate the binding energy per nucleon for ^{235}U. Is this nuclide more or less stable than ^{12}C (mass of ^{235}U atom = 235.043924 amu)?

Fission or Fusion: Means of Increasing the Binding Energy Per Nucleon Calculations similar to Sample Problem 23.6 for other nuclides show that the binding energy per nucleon varies considerably. The essential point is that *the greater the binding energy per nucleon, the more stable the nuclide.*

Figure 23.12 shows a plot of the binding energy per nucleon vs. mass number. It provides information about nuclide stability and the two possible processes nuclides can undergo to form more stable nuclides. Most nuclides with fewer than 10 nucleons have a relatively small binding energy per nucleon. The ^{4}He nucleus is an exception, which is why it is emitted intact as an α particle. Above $A = 12$, the binding energy per nucleon varies from about 7.6 to 8.8 MeV.

The most important observation is that *the binding energy per nucleon peaks for elements with $A \approx 60$.* In other words, nuclides become more stable with increasing mass number up to around 60 nucleons and then become less stable with higher numbers of nucleons. The existence of a peak of stability suggests that there are two ways nuclides can increase their binding energy per nucleon:

- *Fission.* A heavier nucleus can *split into lighter ones (closer to $A \approx 60$)* by undergoing fission. The product nuclei have greater binding energy per nucleon (are more stable) than the reactant nucleus, and the difference in *energy is released.* Nuclear power plants generate energy through fission, as do atomic bombs.
- *Fusion.* Lighter nuclei, on the other hand, can *combine to form a heavier one (closer to $A \approx 60$)* by undergoing fusion. Once again, the product is more stable than the reactants, and *energy is released.* The Sun and other stars generate energy through fusion, as do hydrogen bombs. In these examples and in all current research efforts for developing fusion as a useful energy source, hydrogen nuclei fuse to form the very stable helium-4 nucleus.

Figure 23.12 The variation in binding energy per nucleon. A plot of the binding energy per nucleon vs. mass number shows that nuclear stability is greatest in the region near ^{56}Fe. Lighter nuclei may undergo fusion to become more stable; heavier ones may undergo fission. Note the exceptional stability of ^{4}He among extremely light nuclei.

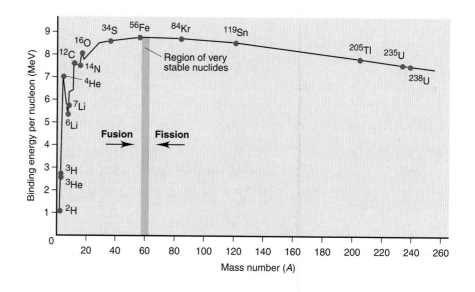

In the next section, we examine fission and fusion and the industrial energy facilities designed to utilize them.

SECTION SUMMARY
The mass of a nucleus is less than the sum of the masses of its nucleons by an amount called the mass defect. The energy equivalent to the mass defect is the nuclear binding energy, usually expressed in units of MeV. The binding energy per nucleon is a measure of nuclide stability and varies with the number of nucleons in a nuclide. Nuclides with $A \approx 60$ are most stable. Lighter nuclei can join (fusion) or heavier nuclei can split (fission) to become more stable.

23.7 APPLICATIONS OF FISSION AND FUSION

Of the many beneficial applications of nuclear reactions, the greatest is the potential for almost limitless amounts of energy. Our experience with nuclear energy from power plants, however, has shown that we must improve ways to tap this energy source safely and economically. In this section, we discuss how nuclear fission and fusion occur and how we are applying them.

The Process of Nuclear Fission

In the late 1930s, the Austrian physicist Lise Meitner and her nephew Otto Frisch proposed that the uranium-235 nucleus could be split into *smaller* nuclei, a process that they called *fission* because of its similarity to the fission a biological cell undergoes during reproduction.

The ^{235}U nucleus can split in many different ways, giving rise to various daughter nuclei, but all routes have the same general features. Figure 23.13 depicts one of these fission patterns. Neutron bombardment results in a highly excited ^{236}U nucleus, which splits apart in 10^{-14} s. The products are two nuclei of unequal mass, two or three neutrons (average of 2.4), and a large quantity of energy. A single ^{235}U nucleus releases 3.5×10^{-11} J when it splits; 1 mol of ^{235}U (about $\frac{1}{2}$ lb) releases 2.1×10^{13} J—a billion times as much energy as burning $\frac{1}{2}$ lb of coal (about 2×10^4 J)!

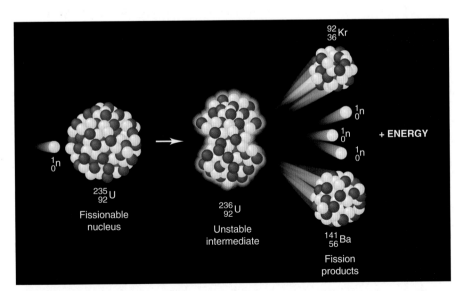

Figure 23.13 Induced fission of ^{235}U. A neutron bombarding a ^{235}U nucleus results in an extremely unstable ^{236}U nucleus, which becomes distorted in the act of splitting. In this case, which shows one of many possible splitting patterns, the products are ^{92}Kr and ^{141}Ba. Three neutrons and a great deal of energy are released also.

Figure 23.14 A chain reaction of ^{235}U. If a sample exceeds the critical mass, neutrons produced by the first fission event collide with other nuclei, causing their fission and the production of more neutrons to continue the process. Note that various product nuclei form. The vertical dashed lines identify succeeding "generations" of neutrons.

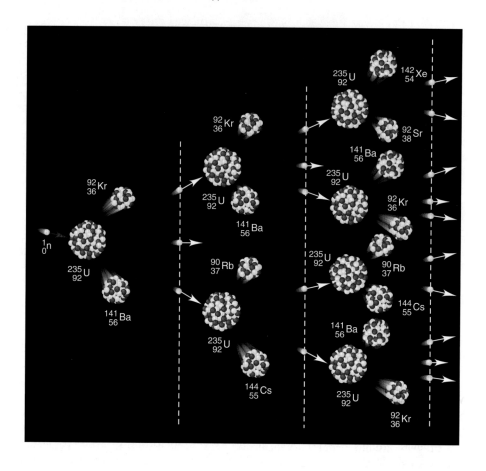

We harness the energy of nuclear fission, much of which appears as heat, by means of a **chain reaction,** illustrated in Figure 23.14: the two to three neutrons that are released by the fission of one nucleus collide with other fissionable nuclei and cause them to split, releasing more neutrons, which then collide with other nuclei, and so on, in a self-sustaining process. In this manner, the energy released increases rapidly because each fission event in a chain reaction releases two to three times as much energy as the preceding one.

Whether a chain reaction occurs depends on the mass (and thus the volume) of the fissionable sample. If the piece of uranium is large enough, the product neutrons strike another fissionable nucleus *before* flying out of the sample, and a chain reaction takes place. The mass required to achieve a chain reaction is called the **critical mass.** If the sample has less than the critical mass (a *subcritical mass*), too many product neutrons leave the sample before they collide with and cause the fission of another ^{235}U nucleus, and thus a chain reaction does not occur.

Nuclear Energy Reactors An uncontrolled fission chain reaction can be adapted to make an atomic bomb, but controlled fission can produce electric power more cleanly than can the combustion of coal. Like a coal-fired power plant, *a nuclear power plant generates heat to produce steam, which turns a turbine attached to an electric generator.*

Heat generation takes place in the **reactor core** of a nuclear plant (Figure 23.15). The core contains the *fuel rods,* which consist of fuel enclosed in tubes of a corrosion-resistant zirconium alloy. The fuel is uranium (IV) oxide (UO_2) that has been *enriched* from 0.7% ^{235}U, the natural abundance of this fissionable isotope, to the 3% to 4% ^{235}U required to sustain a chain reaction. Sandwiched between the fuel rods are movable *control rods* made of cadmium or boron (or, in nuclear submarines, hafnium), substances that absorb neutrons very efficiently.

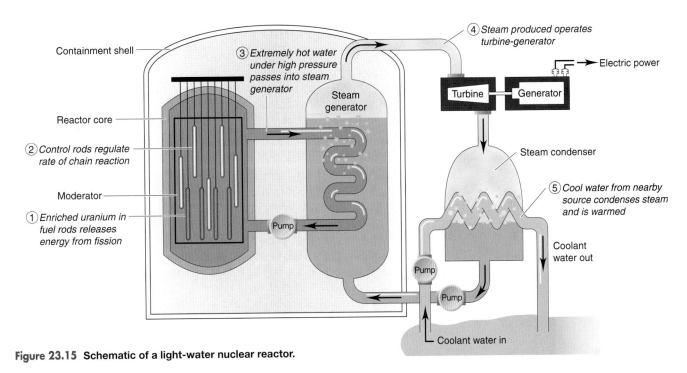

Figure 23.15 Schematic of a light-water nuclear reactor.

When the control rods are moved between the fuel rods, the chain reaction slows because fewer neutrons are available to bombard uranium atoms; when the control rods are removed, the chain reaction speeds up.

Flowing around the fuel and control rods in the reactor core is the *moderator,* a substance that slows the neutrons, making them much better at causing fission than the fast ones emerging directly from the fission event. In most modern reactors, the moderator also acts as the *coolant,* the fluid that transfers the released heat to the steam-producing region. Because ^1H absorbs neutrons, *light-water reactors* use H_2O as the moderator; in heavy-water reactors, D_2O is used. The advantage of D_2O is that it absorbs very few neutrons, leaving more available for fission, so heavy-water reactors can use *unenriched* uranium. As the coolant flows around the encased fuel, pumps circulate it through coils that transfer its heat to the water reservoir. Steam formed in the reservoir turns the turbine that runs the generator. The steam is then condensed in large cooling towers that use water taken from a lake or river; the condensed steam is returned to the water reservoir.

Some major accidents at nuclear power plants, such as the leak of radioactivity from the Three-Mile Island facility in Pennsylvania in 1979 and the far more serious one from the Chernobyl plant in Ukraine in 1986, have resulted in high levels of mistrust and fear in many people. Yet, despite these problems, nuclear energy remains an important source of electricity. In the late 1990s, nearly every European country employed nuclear fission in power plants, and such plants provide the majority of electricity in Sweden and France. Today, the United States obtains about 20% of its electricity from nuclear power plants, and Canada slightly less.

Even a smoothly operating nuclear power plant has certain inherent problems. Thermal pollution, resulting from the use of nearby natural waters to cool reactor parts, is a problem common to all power plants (Section 13.3). More serious is the problem of *nuclear waste disposal.* Many of the fission products formed in nuclear reactors have long half-lives, and proposals to bury containers of this waste in deep bedrock cannot be field-tested for the thousands of years that the material will remain harmful. It remains to be seen whether we can operate fission reactors and dispose of the waste safely and economically.

The Promise of Nuclear Fusion

Nuclear fusion is the ultimate source of nearly all the energy on Earth because almost all other sources depend, directly or indirectly, on the energy produced by nuclear fusion in the Sun. And much research is being devoted to making nuclear fusion practical as a direct source of energy on Earth. To understand the advantages of fusion, let's consider one of the most discussed fusion reactions, in which deuterium and tritium react:

$$^2_1H + ^3_1H \longrightarrow {}^4_2He + {}^1_0n$$

This reaction produces 1.7×10^9 kJ/mol, an enormous quantity of energy with no radioactive by-products. Moreover, the reactant nuclei are relatively easy to come by. Thus, fusion seems very promising, at least in principle. However, some extremely difficult technical problems remain. Fusion requires enormous energy in the form of heat to give the positively charged nuclei enough kinetic energy to force themselves together. The fusion of deuterium and tritium, for example, occurs at practical rates at about 10^8 K, hotter than the Sun's core! How can such temperatures be achieved?

Two research approaches are being used to achieve the necessary heat. In one, atoms are stripped of their electrons at high temperatures, which results in a gaseous *plasma,* a neutral mixture of positive nuclei and electrons. Because of the extreme temperatures needed for fusion, no *material* can contain the plasma. The most successful approach to date has been to enclose the plasma within a magnetic field. The *tokamak* design has a donut-shaped container in which a helical magnetic field confines the plasma and prevents it from contacting the walls (Figure 23.16). Scientists at the Princeton University Plasma Physics facility have achieved some success in generating energy from fusion this way. In another approach, the high temperature is reached by using many focused lasers to compress and heat the fusion reactants. In any event, as a practical, everyday source of energy, nuclear fusion still seems to be a long way off.

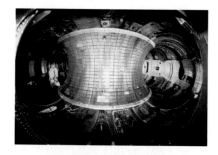

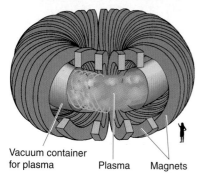

Vacuum container
for plasma   Plasma   Magnets

Figure 23.16 The tokamak design for magnetic containment of a fusion plasma. The donut-shaped chamber of the tokamak (photo, *top;* schematic, *bottom*) contains the plasma within a helical magnetic field.

SECTION SUMMARY
In nuclear fission, neutron bombardment causes a nucleus to split, releasing neutrons that split other nuclei to produce a chain reaction. A nuclear power plant controls the rate of the chain reaction to produce heat that creates steam, which is used to generate electricity. Potential hazards, such as radiation leaks, thermal pollution, and disposal of nuclear waste, remain current concerns. Nuclear fusion holds great promise as a source of clean abundant energy, but it requires extremely high temperatures and is not yet practical.

For Review and Reference (Numbers in parentheses refer to pages, unless noted otherwise.)

Learning Objectives

To help you review these learning objectives, the numbers of related sections (§), sample problems (SP), and upcoming end-of-chapter problems (EP) are listed in parentheses.

1. Describe the differences between nuclear and chemical changes; identify the three types of radioactive emissions and the types of radioactive decay, and know how each changes A and Z; explain how a decay series leads to a stable nuclide; write and balance nuclear equations; use the N/Z ratio to predict nuclear stability and the type of decay a nuclide undergoes (§ 23.1) (SPs 23.1–23.3) (EPs 23.1–23.16)

2. Understand why radioactive decay is a first-order process and the meaning of half-life; convert among units of radioactivity, and calculate specific activity, decay constant, half-life, and number of nuclei; estimate the age of an object from its specific activity (§ 23.2) (SPs 23.4, 23.5) (EPs 23.17–23.30)

3. Describe how particle accelerators are used to synthesize new nuclides and write balanced equations for nuclear transmutations (§ 23.3) (EPs 23.31–23.35)

4. Distinguish between excitation and ionization, and describe their effects on matter; convert among units of radiation dose, and

understand the penetrating power of emissions and how ionizing radiation is used beneficially (§ 23.4) (EPs 23.36–23.42)

5. Describe how radioisotopes are used in research, elemental analysis, and diagnosis (§ 23.5) (EPs 23.43–23.45)

6. Explain the mass defect and how it is related to nuclear binding energy; understand how nuclear stability is related to binding energy per nucleon and why unstable nuclides undergo either fission

or fusion; use Einstein's equation to find mass-energy equivalence in J and eV; compare nuclide stability from binding energy per nucleon (§ 23.6) (SP 23.6) (EPs 23.46–23.52)

7. Discuss the pros and cons of power generation by nuclear fission, and evaluate the potential of nuclear fusion (§ 23.7) (EPs 23.53–23.58)

Key Terms

Section 23.1
radioactivity (763)
nucleon (763)
nuclide (764)
isotope (764)
alpha (α) particle (764)
beta (β) particle (764)
gamma (γ) ray (764)
alpha decay (765)
beta decay (765)
positron emission (766)
positron (766)
electron capture (766)
gamma emission (766)
N/Z ratio (767)

band of stability (767)
decay (disintegration) series (770)

Section 23.2
activity ($\mathscr{A}$) (770)
becquerel (Bq) (770)
curie (Ci) (770)
decay constant (771)
half-life ($t_{1/2}$) (771)
radioisotopic dating (773)
radioisotope (773)

Section 23.3
nuclear transmutation (774)
deuteron (774)

particle accelerator (774)
transuranium element (776)

Section 23.4
excitation (776)
nonionizing radiation (776)
ionization (776)
ionizing radiation (776)
gray (Gy) (777)
rad (*radiation-absorbed dose*) (777)
rem (*roentgen equivalent for man*) (777)
sievert (Sv) (777)
background radiation (777)

Section 23.5
tracer (779)

Section 23.6
fission (781)
fusion (781)
mass defect (782)
nuclear binding energy (783)
electron volt (eV) (783)

Section 23.7
chain reaction (786)
critical mass (786)
reactor core (786)

Key Equations and Relationships

23.1 Balancing a nuclear equation (765):
$$\substack{\text{Total } A \\ \text{Total } Z} \text{Reactants} = \substack{\text{Total } A \\ \text{Total } Z} \text{Products}$$

23.2 Defining the unit of radioactivity (curie, Ci) (770):
$$1 \text{ Ci} = 3.70 \times 10^{10} \text{ disintegrations per second (d/s)}$$

23.3 Expressing the decay rate (activity) for radioactive nuclei (771):
$$\text{Decay rate } (\mathscr{A}) = -\frac{\Delta \mathscr{N}}{\Delta t} = k\mathscr{N}$$

23.4 Finding the number of nuclei remaining after a given time, $\mathscr{N}_t$ (772):
$$\ln \frac{\mathscr{N}_0}{\mathscr{N}_t} = kt$$

23.5 Finding the half-life of a radioactive nuclide (772):
$$t_{1/2} = \frac{\ln 2}{k}$$

23.6 Calculating the time to reach a given specific activity (age of an object in radioisotopic dating) (773):
$$t = \frac{1}{k} \ln \frac{\mathscr{A}_0}{\mathscr{A}_t}$$

23.7 Using Einstein's equation and the mass defect to calculate the nuclear binding energy (782):
$$\Delta E = \Delta mc^2$$

23.8 Relating the atomic mass unit to its energy equivalent in MeV (783):
$$1 \text{ amu} = 931.5 \times 10^6 \text{ eV} = 931.5 \text{ MeV}$$

Brief Solutions to Follow-up Problems

23.1 $^{133}_{54}\text{Xe} \longrightarrow ^{133}_{55}\text{Cs} + ^{0}_{-1}\beta$

23.2 Phosphorus-31 has a slightly higher N/Z ratio and an even N (16).

23.3 (a) $N/Z = 1.35$; too high for this region of band: β decay
(b) Mass too high for stability: α decay

23.4 $\ln \mathscr{A}_t = -kt + \ln \mathscr{A}_0$
$$= -\left(\frac{\ln 2}{15 \text{ h}} \times 4.0 \text{ days} \times \frac{24 \text{ h}}{1 \text{ day}}\right) + \ln (2.5 \times 10^9)$$
$$= 17.20$$
$$\mathscr{A}_t = 3.0 \times 10^7 \text{ d/s}$$

23.5 $t = \frac{1}{k} \ln \frac{\mathscr{A}_0}{\mathscr{A}_t} = \frac{5730 \text{ yr}}{\ln 2} \ln \left(\frac{15.3 \text{ d/min·g}}{9.41 \text{ d/min·g}}\right) = 4.02 \times 10^3 \text{ yr}$

The mummy case is about 4000 years old.

23.6 ^{235}U has 92 $^{1}_{1}\text{p}$ and 143 $^{1}_{0}\text{n}$.
$\Delta m = [(92 \times 1.007825 \text{ amu}) + (143 \times 1.008665 \text{ amu})]$
$$- 235.043924 \text{ amu} = 1.9151 \text{ amu}$$

$$\frac{\text{Binding energy}}{\text{nucleon}} = \frac{1.9151 \text{ amu} \times \dfrac{931.5 \text{ MeV}}{1 \text{ amu}}}{235 \text{ nucleons}}$$
$$= 7.591 \text{ MeV/nucleon}$$

Therefore, ^{235}U is less stable than ^{12}C.

Problems

Problems with **colored** numbers are answered in Appendix E. Sections match the text and provide the numbers of relevant sample problems. Bracketed problems are grouped in pairs (indicated by a short rule) that cover the same concept. Comprehensive Problems are based on material from any section or previous chapter.

Radioactive Decay and Nuclear Stability
(Sample Problems 23.1 to 23.3)

23.1 How do chemical and nuclear reactions differ in
(a) Magnitude of the energy change?
(b) Effect on rate of increasing temperature?
(c) Effect on rate of higher reactant concentration?
(d) Effect on yield of higher reactant concentration?

23.2 Which of the following types of radioactive decay produce an atom of a *different* element: (a) alpha; (b) beta; (c) gamma; (d) positron; (e) electron capture? Show how Z and N change, if at all, with each type.

23.3 Why is $_2^3$He stable but $_2^2$He so unstable that it has never been detected?

23.4 How do the modes of decay differ for a neutron-rich nuclide and a proton-rich nuclide?

23.5 Why can't you use the position of a nuclide's N/Z ratio relative to the band of stability to predict whether it is more likely to decay by positron emission or by electron capture?

23.6 Write balanced nuclear equations for the following:
(a) Alpha decay of $_{92}^{234}$U
(b) Electron capture by neptunium-232
(c) Positron emission by $_7^{12}$N

23.7 Write balanced nuclear equations for the following:
(a) Beta decay of sodium-26
(b) Beta decay of francium-223
(c) Alpha decay of $_{83}^{212}$Bi

23.8 Write balanced nuclear equations for the following:
(a) Formation of $_{22}^{48}$Ti through positron emission
(b) Formation of silver-107 through electron capture
(c) Formation of polonium-206 through α decay

23.9 Write balanced nuclear equations for the following:
(a) Production of $_{95}^{241}$Am through β decay
(b) Formation of $_{89}^{228}$Ac through β decay
(c) Formation of $_{83}^{203}$Bi through α decay

23.10 Which nuclide(s) would you predict to be stable? Why?
(a) $_8^{20}$O   (b) $_{27}^{59}$Co   (c) $_3^9$Li

23.11 Which nuclide(s) would you predict to be stable? Why?
(a) $_{60}^{146}$Nd   (b) $_{48}^{114}$Cd   (c) $_{42}^{88}$Mo

23.12 What is the most likely mode of decay for each?
(a) $_{92}^{238}$U   (b) $_{24}^{48}$Cr   (c) $_{25}^{50}$Mn

23.13 What is the most likely mode of decay for each?
(a) $_{26}^{61}$Fe   (b) $_{17}^{41}$Cl   (c) $_{44}^{110}$Ru

23.14 Why is $_{24}^{52}$Cr the most stable isotope of chromium?

23.15 Why is $_{20}^{40}$Ca the most stable isotope of calcium?

23.16 Neptunium-237 is the parent nuclide of a decay series that starts with α emission, followed by β emission, and then two more α emissions. Write a balanced nuclear equation for each step.

The Kinetics of Radioactive Decay
(Sample Problems 23.4 and 23.5)

23.17 What is the reaction order of radioactive decay? Explain.

23.18 After 1 minute, half the radioactive nuclei remain from an original sample of six nuclei. Is it valid to conclude that $t_{1/2}$ equals 1 minute? Would this conclusion be valid if the original sample contained 6×10^{12} nuclei? Explain.

23.19 Radioisotopic dating depends on the constant rate of decay and formation of various nuclides in a sample. How is the proportion of ^{14}C kept relatively constant in living organisms?

23.20 What is the specific activity (in Ci/g) if 1.55 mg of an isotope emits 1.66×10^6 α particles per second?

23.21 What is the specific activity (in Bq/g) if 8.58 μg of an isotope emits 7.4×10^4 α particles per minute?

23.22 If 1.00×10^{-12} mol of ^{135}Cs emits 1.39×10^5 β particles in 1.00 yr, what is the decay constant?

23.23 If 6.40×10^{-9} mol of ^{176}W emits 1.07×10^{15} positrons in 1.00 h, what is the decay constant?

23.24 The isotope $_{83}^{212}$Bi has a half-life of 1.01 yr. What mass (in mg) of a 2.00-mg sample will not have decayed after 3.75×10^3 h?

23.25 The half-life of radium-226 is 1.60×10^3 yr. How many hours will it take for a 2.50-g sample to decay to the point where 0.185 g of the isotope remains?

23.26 A rock contains 270 μmol of ^{238}U ($t_{1/2} = 4.5 \times 10^9$ yr) and 110 μmol of ^{206}Pb. Assuming that all the ^{206}Pb comes from decay of the ^{238}U, estimate the rock's age.

23.27 A fabric remnant from a burial site has a ^{14}C:^{12}C ratio of 0.735 of the original value. How old is the fabric?

23.28 Due to decay of ^{40}K, cow's milk has a specific activity of about 6×10^{-11} mCi per milliliter. How many disintegrations of ^{40}K nuclei are there per minute in 1.0 qt of milk?

23.29 Plutonium-239 ($t_{1/2} = 2.41 \times 10^4$ yr) represents a serious nuclear waste disposal problem. If seven half-lives are required to reach a tolerable level of radioactivity, how long must ^{239}Pu be stored?

23.30 A volcanic eruption melts a large area of rock, and all gases are expelled. After cooling, $_{18}^{40}$Ar accumulates from the ongoing decay of $_{19}^{40}$K in the rock ($t_{1/2} = 1.25 \times 10^9$ yr). When a piece of rock is analyzed, it is found to contain 1.38 mmol of ^{40}K and 1.14 mmol of ^{40}Ar. How long ago did the rock cool?

Nuclear Transmutation: Induced Changes in Nuclei

23.31 Why must the electrical polarity of the tubes in a linear accelerator be reversed at very short time intervals?

23.32 Why does bombardment with protons usually require higher energies than bombardment with neutrons?

23.33 Name the unidentified species in each transmutation, and write a full nuclear equation:
(a) Bombardment of ^{10}B with an α particle yields a neutron and a nuclide.
(b) Bombardment of ^{28}Si with ^{2}H yields ^{29}P and another particle.
(c) Bombardment of a nuclide with an α particle yields two neutrons and ^{244}Cf.

23.34 Name the unidentified species in each transmutation, and write a full nuclear equation:
(a) Bombardment of a nuclide with a γ photon yields a proton, a neutron, and ^{29}Si.
(b) Bombardment of ^{252}Cf with ^{10}B yields five neutrons and a nuclide.
(c) Bombardment of ^{238}U with a particle yields three neutrons and ^{239}Pu.

23.35 Elements 104, 105, and 106 have been named rutherfordium (Rf), dubnium (Db), and seaborgium (Sg), respectively. These elements are synthesized from californium-249 by bombarding with carbon-12, nitrogen-15, and oxygen-18 nuclei, respectively. Four neutrons are formed in each reaction as well. Write balanced nuclear equations for the formation of these elements.

The Effects of Nuclear Radiation on Matter

23.36 Gamma radiation and UV radiation cause different processes in matter. What are they and how do they differ?

23.37 Why is ionizing radiation more dangerous to children than to adults?

23.38 A 135-lb person absorbs 3.3×10^{-7} J of energy from radioactive emissions. (a) How many rads does she receive? (b) How many grays (Gy) does she receive?

23.39 A 3.6-kg laboratory animal receives a single dose of 8.92×10^{-4} Gy. (a) How many rads did the animal receive? (b) How many joules did the animal absorb?

23.40 A 70.-kg person exposed to ^{90}Sr absorbs 6.0×10^5 β particles, each with an energy of 8.74×10^{-14} J. (a) How many grays does the person receive? (b) If the RBE is 1.0, how many millirems is this? (c) What is the equivalent dose in sieverts (Sv)?

23.41 A laboratory rat weighs 265 g and absorbs 1.77×10^{10} β particles, each with an energy of 2.20×10^{-13} J. (a) How many rads does the animal receive? (b) What is this dose in Gy? (c) If the RBE is 0.75, what is the equivalent dose in Sv?

23.42 A small region of a cancer patient's brain is exposed for 27.0 min to 475 Bq of radioactivity from ^{60}Co for treatment of a tumor. If the brain mass exposed is 1.588 g and each β particle emitted has an energy of 5.05×10^{-14} J, what is the dose in rads?

Applications of Radioisotopes

23.43 Describe two ways that radioactive tracers are used in organisms.

23.44 Why is neutron activation analysis (NAA) useful to art historians and criminologists?

23.45 The oxidation of methanol to formaldehyde can be accomplished by reaction with chromic acid:

$$6H^+(aq) + 3CH_3OH(aq) + 2H_2CrO_4(aq) \longrightarrow$$
$$3CH_2O(aq) + 2Cr^{3+}(aq) + 8H_2O(l)$$

The reaction can be studied with the stable isotope tracer ^{18}O and mass spectrometry. When a small amount of $CH_3{}^{18}OH$ is present in the alcohol reactant, $H_2C^{18}O$ forms. When a small amount of $H_2Cr^{18}O_4$ is present, $H_2{}^{18}O$ forms. Does chromic acid or methanol supply the O atom to the aldehyde? Explain.

The Interconversion of Mass and Energy
(Sample Problem 23.6)

Note: Use the following data to solve the problems in this section: mass of ^1H atom = 1.007825 amu; mass of neutron = 1.008665 amu.

23.46 What is a mass defect, and how does it arise?

23.47 What is the binding energy per nucleon? Why is the binding energy per nucleon, rather than per nuclide, used to compare nuclide stability?

23.48 A ^3H nucleus decays with an energy of 0.01861 MeV. Convert this energy into (a) electron volts; (b) joules.

23.49 Arsenic-84 decays with an energy of 1.57×10^{-15} kJ per nucleus. Convert this energy into (a) eV; (b) MeV.

23.50 Cobalt-59 is the only stable isotope of this transition metal. One ^{59}Co atom has a mass of 58.933198 amu. Calculate the binding energy (a) per nucleon in MeV; (b) per atom in MeV; (c) per mole in kJ.

23.51 Iodine-131 is one of the most important isotopes used in the diagnosis of thyroid cancer. One atom has a mass of 130.906114 amu. Calculate the binding energy (a) per nucleon in MeV; (b) per atom in MeV; (c) per mole in kJ.

23.52 The ^{80}Br nuclide decays either by β decay or by electron capture. (a) What is the product of each process? (b) Which process releases more energy? (Masses of atoms: ^{80}Br = 79.918528 amu; ^{80}Kr = 79.916380 amu; ^{80}Se = 79.916520 amu; neglect the mass of the electron involved.)

Applications of Fission and Fusion

23.53 In what main way is fission different from radioactive decay? Are all fission events in a chain reaction identical? Explain.

23.54 What is the purpose of enrichment in the preparation of fuel rods?

23.55 Describe the nature and purpose of these components of a nuclear reactor: (a) control rods; (b) moderator.

23.56 State an advantage and a disadvantage of heavy-water reactors compared to light-water reactors.

23.57 What are the expected advantages of fusion reactors over fission reactors?

23.58 The reaction that will probably power the first commercial fusion reactor is $^3_1H + {}^2_1H \longrightarrow {}^4_2He + {}^1_0n$. How much energy would be produced per mole of reaction? (Masses of atoms: ^3_1H = 3.01605 amu; ^2_1H = 2.0140 amu; ^4_2He = 4.00260 amu; mass of ^1_0n = 1.008665 amu.)

Comprehensive Problems

Problems with an asterisk (*) are more challenging.

23.59 Some $^{243}_{95}Am$ was present when Earth formed, but it all decayed in the next billion years. The first three steps in this decay series are emission of an α particle, a β particle, and another α particle. What other isotopes were present on the young Earth in a rock that contained some $^{243}_{95}Am$?

23.60 Curium-243 undergoes α decay to plutonium-239:

$$^{243}Cm \longrightarrow {}^{239}Pu + {}^4He$$

(a) Calculate the change in mass, Δm (in kg). (Masses: ^{243}Cm = 243.0614 amu; ^{239}Pu = 239.0522 amu; ^4He = 4.0026 amu; 1 amu = 1.661×10^{-24} g.)

(b) Calculate the energy released in joules.

(c) Calculate the energy released in kJ/mol of reaction, and comment on the difference between this value and a typical heat of reaction for a chemical change of a few hundred kJ/mol.

23.61 Plutonium "triggers" for nuclear weapons were manufactured at the Rocky Flats plant in Colorado. An 85-kg worker inhaled a dust particle containing 1.00 μg of $^{239}_{94}Pu$, which resided in his body for 16 h ($t_{1/2}$ of $^{239}Pu = 2.41 \times 10^4$ yr; each disintegration released 5.15 MeV). (a) How many rads did he receive? (b) How many grays?

23.62 Archeologists removed some charcoal from a Native American campfire, burned it in O_2, and bubbled the CO_2 formed into $Ca(OH)_2$ solution (limewater). The $CaCO_3$ that precipitated was filtered and dried. If 4.38 g of the $CaCO_3$ had a radioactivity of 3.2 d/min, how long ago was the campfire?

23.63 A 5.4-μg sample of $^{226}RaCl_2$ has a radioactivity of 1.5×10^5 Bq. Calculate $t_{1/2}$ of ^{226}Ra.

23.64 The major reaction taking place during hydrogen fusion in a young star is $4^1_1H \longrightarrow {}^4_2He + 2^0_1\beta + 2^0_0\gamma + $ energy. How much energy (in MeV) is released per He nucleus formed? Per mole of He? (Masses: ^1_1H atom $= 1.007825$ amu; ^4_2He atom $= 4.00260$ amu; positron $= 5.48580 \times 10^{-4}$ amu.)

23.65 A sample of AgCl emits 175 nCi/g. A saturated solution prepared from the solid emits 1.25×10^{-2} Bq/mL due to radioactive Ag^+ ions. What is the molar solubility of AgCl?

* 23.66 Technetium-99m is a metastable nuclide used in numerous cancer diagnostic and treatment programs. It is prepared just before use because it decays rapidly through γ emission:

$$^{99m}Tc \longrightarrow {}^{99}Tc + \gamma$$

Use the data below to determine (a) the half-life of ^{99m}Tc, and (b) the percentage of the isotope that is lost if it takes 2.0 h to prepare and administer the dose.

Time (h)	γ Emission (photons/s)
0	5000.
4	3150.
8	2000.
12	1250.
16	788
20	495

23.67 In the 1950s, radioactive material was spread over the land from aboveground nuclear tests. A woman drinks some contaminated milk and ingests 0.0500 g of ^{90}Sr, which is taken up by bones and teeth and not eliminated. (a) How much ^{90}Sr ($t_{1/2} = 29$ yr) is present in her body after 10 yr? (b) How long will it take for 99.9% of the ^{90}Sr to decay?

* 23.68 What volume of radon will be produced per hour at STP from 1.000 g of ^{226}Ra ($t_{1/2} = 1599$ yr; 1 yr $= 8766$ h; mass of one ^{226}Ra atom $= 226.025402$ amu)?

23.69 Which isotope in each pair would you predict to be more stable? Why?
(a) $^{140}_{55}Cs$ or $^{133}_{55}Cs$ (b) $^{79}_{35}Br$ or $^{78}_{35}Br$
(c) $^{28}_{12}Mg$ or $^{24}_{12}Mg$ (d) $^{14}_{7}N$ or $^{18}_{7}N$

23.70 The 23rd-century starship *Enterprise* uses a substance called "dilithium crystals" as its fuel.
(a) Assuming this material is the result of fusion, what is the product of the fusion of two ^6Li nuclei?
(b) How much energy is released per kilogram of dilithium formed? (Mass of one ^6Li atom is 6.015121 amu.)
(c) When four ^1H atoms fuse to form ^4He, how many positrons are released?
(d) To determine the energy potential of the fusion processes in parts (b) and (c), compare the changes in mass per kilogram of dilithium and of ^4He.
(e) Compare the change in mass per kilogram in part (b) to that for the formation of ^4He by the method used in current fusion reactors (Section 23.7). (For masses, see Problem 23.58.)
(f) Using early 21st-century fusion technology, how much tritium can be produced per kilogram of ^6Li in the following reaction: $^6_3Li + {}^1_0n \longrightarrow {}^4_2He + {}^3_1H$? When this amount of tritium is fused with deuterium, what is the change in mass? How does this quantity compare with that for dilithium in part (b)?

23.71 Nuclear disarmament could be accomplished if weapons were not "replenished." The tritium in warheads decays to helium with a half-life of 12.26 yr and must be replaced or the weapon is useless. What fraction of the tritium is lost in 5.50 yr?

23.72 A decay series starts with the synthetic isotope $^{239}_{92}U$. The first four steps are emissions of a β particle, another β, an α particle, and another α. Write a balanced nuclear equation for each step. Which natural series could be started by this sequence?

23.73 The approximate date of a San Francisco earthquake is to be found by measuring the ^{14}C activity ($t_{1/2} = 5730$ yr) of parts of a tree uprooted during the event. The tree parts have an activity of 12.9 d/min·g C, and a living tree has an activity of 15.3 d/min·g C. How long ago did the earthquake occur?

23.74 Carbon from the most recent remains of an extinct Australian marsupial, called *Diprotodon*, has a specific activity of 0.61 pCi/g. Modern carbon has a specific activity of 6.89 pCi/g. How long ago did the *Diprotodon* apparently become extinct?

* 23.75 With our early 21st-century technology, hydrogen fusion requires temperatures around 10^8 K, but lower temperatures can be used if the hydrogen is compressed. In the late 24th century, the starship *Leinad* uses such methods to fuse hydrogen at 10^6 K.
(a) What is the kinetic energy of an H atom at 1.00×10^6 K?
(b) How many H atoms are heated to 1.00×10^6 K from the energy of one H and one anti-H atom annihilating each other?
(c) If these H atoms fuse into ^4He atoms (with the loss of two positrons per ^4He formed), how much energy (in J) is generated?
(d) How much more energy is generated by the fusion in (c) than by the hydrogen-antihydrogen collision in (b)?
(e) Should the captain of the *Leinad* change the technology and produce ^3He (mass $= 3.01603$ amu) instead of ^4He?

23.76 Seaborgium-263 (Sg; $Z = 106$) was the first isotope of this element synthesized. It was made, together with four neutrons, by bombarding californium-249 with oxygen-18. It then decayed by three α emissions. Write balanced equations for the synthesis and three decay steps of ^{263}Sg.

COMMON MATHEMATICAL OPERATIONS IN CHEMISTRY

In addition to basic arithmetic and algebra, four mathematical operations are used frequently in general chemistry: manipulating logarithms, using exponential notation, solving quadratic equations, and graphing data. Each is discussed briefly below.

MANIPULATING LOGARITHMS

Meaning and Properties of Logarithms

A *logarithm* is an exponent. Specifically, if $x^n = A$, we can say that the logarithm to the base x of the number A is n, and we can denote it as

$$\log_x A = n$$

Because logarithms are exponents, they have the following properties:

$$\log_x 1 = 0$$
$$\log_x (A \times B) = \log_x A + \log_x B$$
$$\log_x \frac{A}{B} = \log_x A - \log_x B$$
$$\log_x A^y = y \log_x A$$

Types of Logarithms

Common and natural logarithms are used frequently in chemistry and the other sciences. For *common* logarithms, the base (x in the examples above) is 10, but they are written without specifying the base; that is, $\log_{10} A$ is written more simply as $\log A$. The common logarithm of 1000 is 3; in other words, you must raise 10 to the 3rd power to obtain 1000:

$$\log 1000 = 3 \qquad \text{or} \qquad 10^3 = 1000$$

Similarly, we have

$$\log 10 = 1 \qquad \text{or} \qquad 10^1 = 10$$
$$\log 1{,}000{,}000 = 6 \qquad \text{or} \qquad 10^6 = 1{,}000{,}000$$
$$\log 0.001 = -3 \qquad \text{or} \qquad 10^{-3} = 0.001$$
$$\log 853 = 2.931 \qquad \text{or} \qquad 10^{2.931} = 853$$

The last example illustrates an important point about significant figures with all logarithms: the number of significant figures in the number equals the number of digits to the right of the decimal point in the logarithm. That is, the number 853 has three significant figures, and the logarithm 2.931 has three digits to the right of the decimal point.

To find a common logarithm with an electronic calculator, you simply enter the number and press the LOG button.

For *natural* logarithms, the base is the number e, which is $2.71828 \ldots$, and $\log_e A$ is written $\ln A$. The relationship between the common and natural logarithms is easily obtained: because

$$\log 10 = 1 \qquad \text{and} \qquad \ln 10 = 2.303$$

Therefore, we have

$$\ln A = 2.303 \log A$$

To find a natural logarithm with an electronic calculator, you simply enter the number and press the LN button. If your calculator does not have an LN button, enter the number, press the LOG button, and multiply by 2.303.

Antilogarithms

The *antilogarithm* is the number you obtain when you raise the base to the logarithm:

antilogarithm (antilog) of n is 10^n

Using two of the earlier examples, the antilog of 3 is 1000, and the antilog of 2.931 is 853. To obtain the antilog with a calculator, you enter the number and press the 10^x button. Similarly, to obtain the natural antilogarithm, you enter the number and press the e^x button. [On some calculators, you enter the number and first press INV and then the LOG (or LN) button.]

USING EXPONENTIAL (SCIENTIFIC) NOTATION

Many quantities in chemistry are very large or very small. For example, in the conventional way of writing numbers, the number of gold atoms in 1 gram of gold is

59,060,000,000,000,000,000,000 atoms (to four significant figures)

As another example, the mass in grams of one gold atom is

0.000000000000000000003272 g (to four significant figures)

Exponential (scientific) notation provides a much more practical way of writing such numbers. In exponential notation, we express numbers in the form

$$A \times 10^n$$

where A (the coefficient) is greater than or equal to 1 and less than 10 (that is, $1 \le A < 10$), and n (the exponent) is an integer.

If the number we want to express in exponential notation is larger than 1, the exponent is positive ($n > 0$); if the number is smaller than 1, the exponent is negative ($n < 0$). The size of n tells the number of places the decimal point (in conventional notation) must be moved to obtain a coefficient A greater than or equal to 1 and less than 10 (in exponential notation). In exponential notation, 1 gram of gold contains 5.906×10^{22} atoms, and each gold atom has a mass of 3.272×10^{-22} g.

Changing Between Conventional and Exponential Notation

In order to use exponential notation, you must be able to convert to it from conventional notation, and vice versa.

1. To change a number from conventional to exponential notation, move the decimal point to the left for numbers equal to or greater than 10 and to the right for numbers between 0 and 1:

    75,000,000 changes to 7.5×10^7 (decimal point 7 places to the left)
    0.006042 changes to 6.042×10^{-3} (decimal point 3 places to the right)

2. To change a number from exponential to conventional notation, move the decimal point the number of places indicated by the exponent to the right for numbers with positive exponents and to the left for numbers with negative exponents:

    1.38×10^5 changes to 138,000 (decimal point 5 places to the right)
    8.41×10^{-6} changes to 0.00000841 (decimal point 6 places to the left)

3. An exponential number with a coefficient greater than 10 or less than 1 can be changed to the standard exponential form by converting the coefficient to the standard form and adding the exponents:

$$582.3 \times 10^6 \text{ changes to } 5.823 \times 10^2 \times 10^6 = 5.823 \times 10^{(2+6)} = 5.823 \times 10^8$$
$$0.0043 \times 10^{-4} \text{ changes to } 4.3 \times 10^{-3} \times 10^{-4} = 4.3 \times 10^{[(-3)+(-4)]} = 4.3 \times 10^{-7}$$

Using Exponential Notation in Calculations

In calculations, you can treat the coefficient and exponents separately and apply the properties of exponents (see earlier section on logarithms).

1. To multiply exponential numbers, multiply the coefficients, add the exponents, and reconstruct the number in standard exponential notation:

$$(5.5 \times 10^3)(3.1 \times 10^5) = (5.5 \times 3.1) \times 10^{(3+5)} = 17 \times 10^8 = 1.7 \times 10^9$$
$$(9.7 \times 10^{14})(4.3 \times 10^{-20}) = (9.7 \times 4.3) \times 10^{[14+(-20)]} = 42 \times 10^{-6} = 4.2 \times 10^{-5}$$

2. To divide exponential numbers, divide the coefficients, subtract the exponents, and reconstruct the number in standard exponential notation:

$$\frac{2.6 \times 10^6}{5.8 \times 10^2} = \frac{2.6}{5.8} \times 10^{(6-2)} = 0.45 \times 10^4 = 4.5 \times 10^3$$
$$\frac{1.7 \times 10^{-5}}{8.2 \times 10^{-8}} = \frac{1.7}{8.2} \times 10^{[(-5)-(-8)]} = 0.21 \times 10^3 = 2.1 \times 10^2$$

3. To add or subtract exponential numbers, change all numbers so that they have the same exponent, then add or subtract the coefficients:

$$(1.45 \times 10^4) + (3.2 \times 10^3) = (1.45 \times 10^4) + (0.32 \times 10^4) = 1.77 \times 10^4$$
$$(3.22 \times 10^5) - (9.02 \times 10^4) = (3.22 \times 10^5) - (0.902 \times 10^5) = 2.32 \times 10^5$$

SOLVING QUADRATIC EQUATIONS

A *quadratic equation* is one in which the highest power of x is 2. The general form of a quadratic equation is

$$ax^2 + bx + c = 0$$

where a, b, and c are numbers. For given values of a, b, and c, the values of x that satisfy the equation are called *solutions* of the equation. We calculate x with the quadratic formula:

$$x = \frac{-b \pm \sqrt{b^2 - 4ac}}{2a}$$

We commonly require the quadratic formula when solving for some concentration in an equilibrium problem. For example, we might have an expression that is rearranged into the quadratic equation

$$4.3x^2 + 0.65x - 8.7 = 0$$

Applying the quadratic formula, with $a = 4.3$, $b = 0.65$, and $c = -8.7$, gives

$$x = \frac{-0.65 \pm \sqrt{(0.65)^2 - 4(4.3)(-8.7)}}{2(4.3)}$$

The "plus or minus" sign ($\pm$) indicates that there are always two possible values for x. In this case, they are

$$x = 1.3 \quad \text{and} \quad x = -1.5$$

In any real physical system, however, only one of the values will have any meaning. For example, if x were $[H_3O^+]$, the negative value would mean a negative concentration, which has no meaning.

GRAPHING DATA IN THE FORM OF A STRAIGHT LINE

Visualizing changes in variables by means of a graph is a very useful technique in science. In many cases, it is most useful if the data can be graphed in the form of a straight line. Any equation will appear as a straight line if it has, or can be rearranged to have, the following general form:

$$y = mx + b$$

where y is the dependent variable (typically plotted along the vertical axis), x is the independent variable (typically plotted along the horizontal axis), m is the slope of the line, and b is the intercept of the line on the y axis. The intercept is the value of y when $x = 0$:

$$y = m(0) + b = b$$

The slope of the line is the change in y for a given change in x:

$$\text{Slope } (m) = \frac{y_2 - y_1}{x_2 - x_1} = \frac{\Delta y}{\Delta x}$$

The *sign* of the slope tells the *direction* of the line. If y increases as x increases, m is positive, and the line slopes upward with higher values of x; if y decreases as x increases, m is negative, and the line slopes downward with higher values of x. The *magnitude* of the slope indicates the *steepness* of the line. A line with $m = 3$ is three times as steep (y changes three times as much for a given change in x) as a line with $m = 1$.

Consider the linear equation $y = 2x + 1$. A graph of this equation is shown in Figure A.1. In practice, you can find the slope by drawing a right triangle to the line, using the line as the hypotenuse. Then, one leg gives Δy, and the other gives Δx. In the figure, $\Delta y = 8$ and $\Delta x = 4$.

At several places in the text, an equation is rearranged into the form of a straight line in order to determine information from the slope and/or the intercept. For example, in Chapter 16, we obtained the following expression:

$$\ln \frac{[A]_0}{[A]_t} = kt$$

Based on the properties of logarithms, we have

$$\ln [A]_0 - \ln [A]_t = kt$$

Rearranging into the form of an equation for a straight line gives

$$\ln [A]_t = -kt + \ln [A]_0$$
$$y \quad\;\; = \;\; mx + \quad b$$

Thus, a plot of $\ln [A]_t$ vs. t is a straight line, from which you can see that the slope is $-k$ (the negative of the rate constant) and the intercept is $\ln [A]_0$ (the natural logarithm of the initial concentration of A).

At many other places in the text, linear relationships occur that were not shown in graphical terms. For example, the conversion between temperature scales in Chapter 1 can also be expressed in the form of a straight line:

$$°F = \tfrac{9}{5}(°C) + 32$$
$$y \;\; = \;\; mx \;\; + \;\; b$$

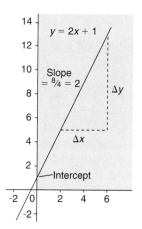

Figure A.1

Appendix B

STANDARD THERMODYNAMIC VALUES FOR SELECTED SUBSTANCES AT 298 K

Substance or Ion	ΔH_f° (kJ/mol)	ΔG_f° (kJ/mol)	S° (J/mol·K)	Substance or Ion	ΔH_f° (kJ/mol)	ΔG_f° (kJ/mol)	S° (J/mol·K)
$e^-(g)$	0	0	20.87	$CaCO_3(s)$	−1206.9	−1128.8	92.9
Aluminum				$CaO(s)$	−635.1	−603.5	38.2
$Al(s)$	0	0	28.3	$Ca(OH)_2(s)$	−986.09	−898.56	83.39
$Al^{3+}(aq)$	−524.7	−481.2	−313	$Ca_3(PO_4)_2(s)$	−4138	−3899	263
$AlCl_3(s)$	−704.2	−628.9	110.7	$CaSO_4(s)$	−1432.7	−1320.3	107
$Al_2O_3(s)$	−1676	−1582	50.94	Carbon			
Barium				$C(graphite)$	0	0	5.686
$Ba(s)$	0	0	62.5	$C(diamond)$	1.896	2.866	2.439
$Ba(g)$	175.6	144.8	170.28	$C(g)$	715.0	669.6	158.0
$Ba^{2+}(g)$	1649.9	—	—	$CO(g)$	−110.5	−137.2	197.5
$Ba^{2+}(aq)$	−538.36	−560.7	13	$CO_2(g)$	−393.5	−394.4	213.7
$BaCl_2(s)$	−806.06	−810.9	126	$CO_2(aq)$	−412.9	−386.2	121
$BaCO_3(s)$	−1219	−1139	112	$CO_3^{2-}(aq)$	−676.26	−528.10	−53.1
$BaO(s)$	−548.1	−520.4	72.07	$HCO_3^-(aq)$	−691.11	587.06	95.0
$BaSO_4(s)$	−1465	−1353	132	$H_2CO_3(aq)$	−698.7	−623.42	191
Boron				$CH_4(g)$	−74.87	−50.81	186.1
$B(\beta\text{-rhombo-}$	0	0	5.87	$C_2H_2(g)$	227	209	200.85
$\quad$hedral)				$C_2H_4(g)$	52.47	68.36	219.22
$BF_3(g)$	−1137.0	−1120.3	254.0	$C_2H_6(g)$	−84.667	−32.89	229.5
$BCl_3(g)$	−403.8	−388.7	290.0	$C_3H_8(g)$	−105	−24.5	269.9
$B_2H_6(g)$	35	86.6	232.0	$C_4H_{10}(g)$	−126	−16.7	310
$B_2O_3(s)$	−1272	−1193	53.8	$C_6H_6(l)$	49.0	124.5	172.8
$H_3BO_3(s)$	−1094.3	−969.01	88.83	$CH_3OH(g)$	−201.2	−161.9	238
Bromine				$CH_3OH(l)$	−238.6	−166.2	127
$Br_2(l)$	0	0	152.23	$HCHO(g)$	−116	−110	219
$Br_2(g)$	30.91	3.13	245.38	$HCOO^-(aq)$	−410	−335	91.6
$Br(g)$	111.9	82.40	174.90	$HCOOH(l)$	−409	−346	129.0
$Br^-(g)$	−218.9	—	—	$HCOOH(aq)$	−410	−356	164
$Br^-(aq)$	−120.9	−102.82	80.71	$C_2H_5OH(g)$	−235.1	−168.6	282.6
$HBr(g)$	−36.3	−53.5	198.59	$C_2H_5OH(l)$	−277.63	−174.8	161
Cadmium				$CH_3CHO(g)$	−166	−133.7	266
$Cd(s)$	0	0	51.5	$CH_3COOH(l)$	−487.0	−392	160
$Cd(g)$	112.8	78.20	167.64	$C_6H_{12}O_6(s)$	−1273.3	−910.56	212.1
$Cd^{2+}(aq)$	−72.38	−77.74	−61.1	$C_{12}H_{22}O_{11}(s)$	−2221.7	−1544.3	360.24
$CdS(s)$	−144	−141	71	$CN^-(aq)$	151	166	118
Calcium				$HCN(g)$	135	125	201.7
$Ca(s)$	0	0	41.6	$HCN(l)$	105	121	112.8
$Ca(g)$	192.6	158.9	154.78	$HCN(aq)$	105	112	129
$Ca^{2+}(g)$	1934.1	—	—	$CS_2(g)$	117	66.9	237.79
$Ca^{2+}(aq)$	−542.96	−553.04	−55.2	$CS_2(l)$	87.9	63.6	151.0
$CaF_2(s)$	−1215	−1162	68.87	$CH_3Cl(g)$	−83.7	−60.2	234
$CaCl_2(s)$	−795.0	−750.2	114	$CH_2Cl_2(l)$	−117	−63.2	179

(Continued)

Substance or Ion	ΔH_f° (kJ/mol)	ΔG_f° (kJ/mol)	S° (J/mol·K)	Substance or Ion	ΔH_f° (kJ/mol)	ΔG_f° (kJ/mol)	S° (J/mol·K)
$CHCl_3(l)$	−132	−71.5	203	$Fe^{2+}(aq)$	−87.9	−84.94	113
$CCl_4(g)$	−96.0	−53.7	309.7	$FeCl_2(s)$	−341.8	−302.3	117.9
$CCl_4(l)$	−139	−68.6	214.4	$FeCl_3(s)$	−399.5	−334.1	142
$COCl_2(g)$	−220	−206	283.74	$FeO(s)$	−272.0	−251.4	60.75
Cesium				$Fe_2O_3(s)$	−825.5	−743.6	87.400
$Cs(s)$	0	0	85.15	$Fe_3O_4(s)$	−1121	−1018	145.3
$Cs(g)$	76.7	49.7	175.5	Lead			
$Cs^+(g)$	458.5	427.1	169.72	$Pb(s)$	0	0	64.785
$Cs^+(aq)$	−248	−282.0	133	$Pb^{2+}(aq)$	1.6	−24.3	21
$CsF(s)$	−554.7	−525.4	88	$PbCl_2(s)$	−359	−314	136
$CsCl(s)$	−442.8	−414	101.18	$PbO(s)$	−218	−198	68.70
$CsBr(s)$	−395	−383	121	$PbO_2(s)$	−276.6	−219.0	76.6
$CsI(s)$	−337	−333	130	$PbS(s)$	−98.3	−96.7	91.3
Chlorine				$PbSO_4(s)$	−918.39	−811.24	147
$Cl_2(g)$	0	0	223.0	Lithium			
$Cl(g)$	121.0	105.0	165.1	$Li(s)$	0	0	29.10
$Cl^-(g)$	−234	−240	153.25	$Li(g)$	161	128	138.67
$Cl^-(aq)$	−167.46	−131.17	55.10	$Li^+(g)$	687.163	649.989	132.91
$HCl(g)$	−92.31	−95.30	186.79	$Li^+(aq)$	−278.46	−293.8	14
$HCl(aq)$	−167.46	−131.17	55.06	$LiF(s)$	−616.9	−588.7	35.66
$ClO_2(g)$	102	120	256.7	$LiCl(s)$	−408	−384	59.30
$Cl_2O(g)$	80.3	97.9	266.1	$LiBr(s)$	−351	−342	74.1
Chromium				$LiI(s)$	−270	−270	85.8
$Cr(s)$	0	0	23.8	Magnesium			
$Cr^{3+}(aq)$	−1971	—	—	$Mg(s)$	0	0	32.69
$CrO_4^{2-}(aq)$	−863.2	−706.3	38	$Mg(g)$	150	115	148.55
$Cr_2O_7^{2-}(aq)$	−1461	−1257	214	$Mg^{2+}(g)$	2351	—	—
Copper				$Mg^{2+}(aq)$	−461.96	−456.01	118
$Cu(s)$	0	0	33.1	$MgCl_2(s)$	−641.6	−592.1	89.630
$Cu(g)$	341.1	301.4	166.29	$MgCO_3(s)$	−1112	−1028	65.86
$Cu^+(aq)$	51.9	50.2	−26	$MgO(s)$	−601.2	−569.0	26.9
$Cu^{2+}(aq)$	64.39	64.98	−98.7	$Mg_3N_2(s)$	−461	−401	88
$Cu_2O(s)$	−168.6	−146.0	93.1	Manganese			
$CuO(s)$	−157.3	−130	42.63	$Mn(s, \alpha)$	0	0	31.8
$Cu_2S(s)$	−79.5	−86.2	120.9	$Mn^{2+}(aq)$	−219	−223	−84
$CuS(s)$	−53.1	−53.6	66.5	$MnO_2(s)$	−520.9	−466.1	53.1
Fluorine				$MnO_4^-(aq)$	−518.4	−425.1	190
$F_2(g)$	0	0	202.7	Mercury			
$F(g)$	78.9	61.8	158.64	$Hg(l)$	0	0	76.027
$F^-(g)$	−255.6	−262.5	145.47	$Hg(g)$	61.30	31.8	174.87
$F^-(aq)$	−329.1	−276.5	−9.6	$Hg^{2+}(aq)$	171	164.4	−32
$HF(g)$	−273	−275	173.67	$Hg_2^{2+}(aq)$	172	153.6	84.5
Hydrogen				$HgCl_2(s)$	−230	−184	144
$H_2(g)$	0	0	130.6	$Hg_2Cl_2(s)$	−264.9	−210.66	196
$H(g)$	218.0	203.30	114.60	$HgO(s)$	−90.79	−58.50	70.27
$H^+(aq)$	0	0	0	Nitrogen			
$H^+(g)$	1536.3	1517.1	108.83	$N_2(g)$	0	0	191.5
Iodine				$N(g)$	473	456	153.2
$I_2(s)$	0	0	116.14	$N_2O(g)$	82.05	104.2	219.7
$I_2(g)$	62.442	19.38	260.58	$NO(g)$	90.29	86.60	210.65
$I(g)$	106.8	70.21	180.67	$NO_2(g)$	33.2	51	239.9
$I^-(g)$	−194.7	—	—	$N_2O_4(g)$	9.16	97.7	304.3
$I^-(aq)$	−55.94	−51.67	109.4	$N_2O_5(g)$	11	118	346
$HI(g)$	25.9	1.3	206.33	$N_2O_5(s)$	−43.1	114	178
Iron				$NH_3(g)$	−45.9	−16	193
$Fe(s)$	0	0	27.3	$NH_3(aq)$	−80.83	26.7	110
$Fe^{3+}(aq)$	−47.7	−10.5	−293	$N_2H_4(l)$	50.63	149.2	121.2

(Continued)

Substance or Ion	ΔH_f° (kJ/mol)	ΔG_f° (kJ/mol)	S° (J/mol·K)	Substance or Ion	ΔH_f° (kJ/mol)	ΔG_f° (kJ/mol)	S° (J/mol·K)
$NO_3^-(aq)$	−206.57	−110.5	146	$Ag(g)$	289.2	250.4	172.892
$HNO_3(l)$	−173.23	−79.914	155.6	$Ag^+(aq)$	105.9	77.111	73.93
$HNO_3(aq)$	−206.57	−110.5	146	$AgF(s)$	−203	−185	84
$NF_3(g)$	−125	−83.3	260.6	$AgCl(s)$	−127.03	−109.72	96.11
$NOCl(g)$	51.71	66.07	261.6	$AgBr(s)$	−99.51	−95.939	107.1
$NH_4Cl(s)$	−314.4	−203.0	94.6	$AgI(s)$	−62.38	−66.32	114
Oxygen				$AgNO_3(s)$	−45.06	19.1	128.2
$O_2(g)$	0	0	205.0	$Ag_2S(s)$	−31.8	−40.3	146
$O(g)$	249.2	231.7	160.95	Sodium			
$O_3(g)$	143	163	238.82	$Na(s)$	0	0	51.446
$OH^-(aq)$	−229.94	−157.30	−10.54	$Na(g)$	107.76	77.299	153.61
$H_2O(g)$	−241.826	−228.60	188.72	$Na^+(g)$	609.839	574.877	147.85
$H_2O(l)$	−285.840	−237.192	69.940	$Na^+(aq)$	−239.66	−261.87	60.2
$H_2O_2(l)$	−187.8	−120.4	110	$NaF(s)$	−575.4	−545.1	51.21
$H_2O_2(aq)$	−191.2	−134.1	144	$NaCl(s)$	−411.1	−384.0	72.12
Phosphorus				$NaBr(s)$	−361	−349	86.82
$P_4(s, white)$	0	0	41.1	$NaOH(s)$	−425.609	−379.53	64.454
$P(g)$	314.6	278.3	163.1	$Na_2CO_3(s)$	−1130.8	−1048.1	139
$P(s, red)$	−17.6	−12.1	22.8	$NaHCO_3(s)$	−947.7	−851.9	102
$P_2(g)$	144	104	218	$NaI(s)$	−288	−285	98.5
$P_4(g)$	58.9	24.5	280	Strontium			
$PCl_3(g)$	−287	−268	312	$Sr(s)$	0	0	54.4
$PCl_3(l)$	−320	−272	217	$Sr(g)$	164	110	164.54
$PCl_5(g)$	−402	−323	353	$Sr^{2+}(g)$	1784	—	—
$PCl_5(s)$	−443.5	—	—	$Sr^{2+}(aq)$	−545.51	−557.3	−39
$P_4O_{10}(s)$	−2984	−2698	229	$SrCl_2(s)$	−828.4	−781.2	117
$PO_4^{3-}(aq)$	−1266	−1013	−218	$SrCO_3(s)$	−1218	−1138	97.1
$HPO_4^{2-}(aq)$	−1281	−1082	−36	$SrO(s)$	−592.0	−562.4	55.5
$H_2PO_4^-(aq)$	−1285	−1135	89.1	$SrSO_4(s)$	−1445	−1334	122
$H_3PO_4(aq)$	−1277	−1019	228	Sulfur			
Potassium				$S(rhombic)$	0	0	31.9
$K(s)$	0	0	64.672	$S(monoclinic)$	0.3	0.096	32.6
$K(g)$	89.2	60.7	160.23	$S(g)$	279	239	168
$K^+(g)$	514.197	481.202	154.47	$S_2(g)$	129	80.1	228.1
$K^+(aq)$	−251.2	−282.28	103	$S_8(g)$	101	49.1	430.211
$KF(s)$	−568.6	−538.9	66.55	$S^{2-}(aq)$	41.8	83.7	22
$KCl(s)$	−436.7	−409.2	82.59	$HS^-(aq)$	−17.7	12.6	61.1
$KBr(s)$	−394	−380	95.94	$H_2S(g)$	−20.2	−33	205.6
$KI(s)$	−328	−323	106.39	$H_2S(aq)$	−39	−27.4	122
$KOH(s)$	−424.8	−379.1	78.87	$SO_2(g)$	−296.8	−300.2	248.1
$KClO_3(s)$	−397.7	−296.3	143.1	$SO_3(g)$	−396	−371	256.66
$KClO_4(s)$	−432.75	−303.2	151.0	$SO_4^{2-}(aq)$	−907.51	−741.99	17
Rubidium				$HSO_4^-(aq)$	−885.75	−752.87	126.9
$Rb(s)$	0	0	69.5	$H_2SO_4(l)$	−813.989	−690.059	156.90
$Rb(g)$	85.81	55.86	169.99	$H_2SO_4(aq)$	−907.51	−741.99	17
$Rb^+(g)$	495.04	—	—	Tin			
$Rb^+(aq)$	−246	−282.2	124	$Sn(white)$	0	0	51.5
$RbF(s)$	−549.28	—	—	$Sn(gray)$	3	4.6	44.8
$RbCl(s)$	−435.35	−407.8	95.90	$SnCl_4(l)$	−545.2	−474.0	259
$RbBr(s)$	−389.2	−378.1	108.3	$SnO_2(s)$	−580.7	−519.7	52.3
$RbI(s)$	−328	−326	118.0	Zinc			
Silicon				$Zn(s)$	0	0	41.6
$Si(s)$	0	0	18.0	$Zn(g)$	130.5	94.93	160.9
$SiF_4(g)$	−1614.9	−1572.7	282.4	$Zn^{2+}(aq)$	−152.4	−147.21	−106.5
$SiO_2(s)$	−910.9	−856.5	41.5	$ZnO(s)$	−348.0	−318.2	43.9
Silver				$ZnS(s, zinc$	−203	−198	57.7
$Ag(s)$	0	0	42.702	blende)			

Appendix C

EQUILIBRIUM CONSTANTS AT 298 K

Dissociation (Ionization) Constants (K_a) of Selected Acids

Name and Formula	Lewis Structure*	K_{a1}	K_{a2}	K_{a3}
Acetic acid CH_3COOH		1.8×10^{-5}		
Acetylsalicylic acid[†] $CH_3COOC_6H_4COOH$		3.6×10^{-4}		
Adipic acid $HOOC(CH_2)_4COOH$		3.8×10^{-5}	3.8×10^{-6}	
Arsenic acid H_3AsO_4		6×10^{-3}	1.1×10^{-7}	3×10^{-12}
Ascorbic acid $H_2C_6H_6O_6$		1.0×10^{-5}	5×10^{-12}	
Benzoic acid C_6H_5COOH		6.3×10^{-5}		
Carbonic acid H_2CO_3		4.5×10^{-7}	4.7×10^{-11}	
Chloroacetic acid $ClCH_2COOH$		1.4×10^{-3}		
Chlorous acid $HClO_2$		1.1×10^{-2}		

*Acidic (ionizable) proton(s) shown in red.

[†]At 37°C in 0.15 M NaCl.

(*Continued*)

Dissociation (Ionization) Constants (K_a) of Selected Acids

Name and Formula	Lewis Structure*	K_{a1}	K_{a2}	K_{a3}
Citric acid $HOOCCH_2C(OH)(COOH)CH_2COOH$		7.4×10^{-4}	1.7×10^{-5}	4.0×10^{-7}
Formic acid $HCOOH$		1.8×10^{-4}		
Glyceric acid $HOCH_2CH(OH)COOH$		2.9×10^{-4}		
Glycolic acid $HOCH_2COOH$		1.5×10^{-4}		
Glyoxylic acid $HC(O)COOH$		3.5×10^{-4}		
Hydrocyanic acid HCN		6.2×10^{-10}		
Hydrofluoric acid HF		6.8×10^{-4}		
Hydrosulfuric acid H_2S		9×10^{-8}	1×10^{-17}	
Hypobromous acid $HBrO$		2.3×10^{-9}		
Hypochlorous acid $HClO$		2.9×10^{-8}		
Hypoiodous acid HIO		2.3×10^{-11}		
Iodic acid HIO_3		1.6×10^{-1}		
Lactic acid $CH_3CH(OH)COOH$		1.4×10^{-4}		
Maleic acid $HOOCCH=CHCOOH$		1.2×10^{-2}	4.7×10^{-7}	

(Continued)

Dissociation (Ionization) Constants (K_a) of Selected Acids (*continued*)

Name and Formula	Lewis Structure*	K_{a1}	K_{a2}	K_{a3}
Malonic acid $HOOCCH_2COOH$		1.4×10^{-3}	2.0×10^{-6}	
Nitrous acid HNO_2		7.1×10^{-4}		
Oxalic acid $HOOCCOOH$		5.6×10^{-2}	5.4×10^{-5}	
Phenol C_6H_5OH		1.0×10^{-10}		
Phenylacetic acid $C_6H_5CH_2COOH$		4.9×10^{-5}		
Phosphoric acid H_3PO_4		7.2×10^{-3}	6.3×10^{-8}	4.2×10^{-13}
Phosphorous acid $HPO(OH)_2$		3×10^{-2}	1.7×10^{-7}	
Propanoic acid CH_3CH_2COOH		1.3×10^{-5}		
Pyruvic acid $CH_3C(O)COOH$		2.8×10^{-3}		
Succinic acid $HOOCCH_2CH_2COOH$		6.2×10^{-5}	2.3×10^{-6}	
Sulfuric acid H_2SO_4		Very large	1.0×10^{-2}	
Sulfurous acid H_2SO_3		1.4×10^{-2}	6.5×10^{-8}	

Dissociation (Ionization) Constants (K_b) of Selected Amine Bases

Name and Formula	Lewis Structure*	K_{b1}	K_{b2}
Ammonia NH_3		1.76×10^{-5}	
Aniline $C_6H_5NH_2$		4.0×10^{-10}	
Diethylamine $(CH_3CH_2)_2NH$		8.6×10^{-4}	
Dimethylamine $(CH_3)_2NH$		5.9×10^{-4}	
Ethanolamine $HOCH_2CH_2NH_2$		3.2×10^{-5}	
Ethylamine $CH_3CH_2NH_2$		4.3×10^{-4}	
Ethylenediamine $H_2NCH_2CH_2NH_2$		8.5×10^{-5}	7.1×10^{-8}
Methylamine CH_3NH_2		4.4×10^{-4}	
tert-Butylamine $(CH_3)_3CNH_2$		4.8×10^{-4}	
Piperidine $C_5H_{10}NH$		1.3×10^{-3}	
n-Propylamine $CH_3CH_2CH_2NH_2$		3.5×10^{-4}	

*Blue type indicates the basic nitrogen and its lone pair.

(*Continued*)

Dissociation (Ionization) Constants (K_b) of Selected Amine Bases (*continued*)

Name and Formula	Lewis Structure*	K_{b1}	K_{b2}
Isopropylamine $(CH_3)_2CHNH_2$		4.7×10^{-4}	
Propylenediamine $H_2NCH_2CH_2CH_2NH_2$		3.1×10^{-4}	3.0×10^{-6}
Pyridine C_5H_5N		1.7×10^{-9}	
Triethylamine $(CH_3CH_2)_3N$		5.2×10^{-4}	
Trimethylamine $(CH_3)_3N$		6.3×10^{-5}	

Dissociation (Ionization) Constants (K_a) of Some Hydrated Metal Ions

Free Ion	Hydrated Ion	K_a
Fe^{3+}	$Fe(H_2O)_6^{3+}(aq)$	6×10^{-3}
Sn^{2+}	$Sn(H_2O)_6^{2+}(aq)$	4×10^{-4}
Cr^{3+}	$Cr(H_2O)_6^{3+}(aq)$	1×10^{-4}
Al^{3+}	$Al(H_2O)_6^{3+}(aq)$	1×10^{-5}
Cu^{2+}	$Cu(H_2O)_6^{2+}(aq)$	3×10^{-8}
Pb^{2+}	$Pb(H_2O)_6^{2+}(aq)$	3×10^{-8}
Zn^{2+}	$Zn(H_2O)_6^{2+}(aq)$	1×10^{-9}
Co^{2+}	$Co(H_2O)_6^{2+}(aq)$	2×10^{-10}
Ni^{2+}	$Ni(H_2O)_6^{2+}(aq)$	1×10^{-10}

Formation Constants (K_f) of Some Complex Ions

Complex Ion	K_f
$Ag(CN)_2^-$	3.0×10^{20}
$Ag(NH_3)_2^+$	1.7×10^7
$Ag(S_2O_3)_2^{3-}$	4.7×10^{13}
AlF_6^{3-}	4×10^{19}
$Al(OH)_4^-$	3×10^{33}
$Be(OH)_4^{2-}$	4×10^{18}
CdI_4^{2-}	1×10^6
$Co(OH)_4^{2-}$	5×10^9
$Cr(OH)_4^-$	8.0×10^{29}
$Cu(NH_3)_4^{2+}$	5.6×10^{11}
$Fe(CN)_6^{4-}$	3×10^{35}
$Fe(CN)_6^{3-}$	4.0×10^{43}
$Hg(CN)_4^{2-}$	9.3×10^{38}
$Ni(NH_3)_6^{2+}$	2.0×10^8
$Pb(OH)_3^-$	8×10^{13}
$Sn(OH)_3^-$	3×10^{25}
$Zn(CN)_4^{2-}$	4.2×10^{19}
$Zn(NH_3)_4^{2+}$	7.8×10^8
$Zn(OH)_4^{2-}$	3×10^{15}

Solubility-Product Constants (K_{sp}) of Slightly Soluble Ionic Compounds

Name, Formula	K_{sp}	Name, Formula	K_{sp}
Carbonates		Cobalt(II) hydroxide, $Co(OH)_2$	1.3×10^{-15}
Barium carbonate, $BaCO_3$	2.0×10^{-9}	Copper(II) hydroxide, $Cu(OH)_2$	2.2×10^{-20}
Cadmium carbonate, $CdCO_3$	1.8×10^{-14}	Iron(II) hydroxide, $Fe(OH)_2$	4.1×10^{-15}
Calcium carbonate, $CaCO_3$	3.3×10^{-9}	Iron(III) hydroxide, $Fe(OH)_3$	1.6×10^{-39}
Cobalt(II) carbonate, $CoCO_3$	1.0×10^{-10}	Magnesium hydroxide, $Mg(OH)_2$	6.3×10^{-10}
Copper(II) carbonate, $CuCO_3$	3×10^{-12}	Manganese(II) hydroxide, $Mn(OH)_2$	1.6×10^{-13}
Lead(II) carbonate, $PbCO_3$	7.4×10^{-14}	Nickel(II) hydroxide, $Ni(OH)_2$	6×10^{-16}
Magnesium carbonate, $MgCO_3$	3.5×10^{-8}	Zinc hydroxide, $Zn(OH)_2$	3×10^{-16}
Mercury(I) carbonate, Hg_2CO_3	8.9×10^{-17}	**Iodates**	
Nickel(II) carbonate, $NiCO_3$	1.3×10^{-7}	Barium iodate, $Ba(IO_3)_2$	1.5×10^{-9}
Strontium carbonate, $SrCO_3$	5.4×10^{-10}	Calcium iodate, $Ca(IO_3)_2$	7.1×10^{-7}
Zinc carbonate, $ZnCO_3$	1.0×10^{-10}	Lead(II) iodate, $Pb(IO_3)_2$	2.5×10^{-13}
Chromates		Silver iodate, $AgIO_3$	3.1×10^{-8}
Barium chromate, $BaCrO_4$	2.1×10^{-10}	Strontium iodate, $Sr(IO_3)_2$	3.3×10^{-7}
Calcium chromate, $CaCrO_4$	1×10^{-8}	Zinc iodate, $Zn(IO_3)_2$	3.9×10^{-6}
Lead(II) chromate, $PbCrO_4$	2.3×10^{-13}	**Oxalates**	
Silver chromate, Ag_2CrO_4	2.6×10^{-12}	Barium oxalate dihydrate, $BaC_2O_4 \cdot 2H_2O$	1.1×10^{-7}
Cyanides		Calcium oxalate monohydrate, $CaC_2O_4 \cdot H_2O$	2.3×10^{-9}
Mercury(I) cyanide, $Hg_2(CN)_2$	5×10^{-40}	Strontium oxalate monohydrate,	
Silver cyanide, $AgCN$	2.2×10^{-16}	$SrC_2O_4 \cdot H_2O$	5.6×10^{-8}
Halides		**Phosphates**	
Fluorides		Calcium phosphate, $Ca_3(PO_4)_2$	1.2×10^{-29}
Barium fluoride, BaF_2	1.5×10^{-6}	Magnesium phosphate, $Mg_3(PO_4)_2$	5.2×10^{-24}
Calcium fluoride, CaF_2	3.2×10^{-11}	Silver phosphate, Ag_3PO_4	2.6×10^{-18}
Lead(II) fluoride, PbF_2	3.6×10^{-8}	**Sulfates**	
Magnesium fluoride, MgF_2	7.4×10^{-9}	Barium sulfate, $BaSO_4$	1.1×10^{-10}
Strontium fluoride, SrF_2	2.6×10^{-9}	Calcium sulfate, $CaSO_4$	2.4×10^{-5}
Chlorides		Lead(II) sulfate, $PbSO_4$	1.6×10^{-8}
Copper(I) chloride, $CuCl$	1.9×10^{-7}	Radium sulfate, $RaSO_4$	2×10^{-11}
Lead(II) chloride, $PbCl_2$	1.7×10^{-5}	Silver sulfate, Ag_2SO_4	1.5×10^{-5}
Silver chloride, $AgCl$	1.8×10^{-10}	Strontium sulfate, $SrSO_4$	3.2×10^{-7}
Bromides		**Sulfides**	
Copper(I) bromide, $CuBr$	5×10^{-9}	Cadmium sulfide, CdS	1.0×10^{-24}
Silver bromide, $AgBr$	5.0×10^{-13}	Copper(II) sulfide, CuS	8×10^{-34}
Iodides		Iron(II) sulfide, FeS	8×10^{-16}
Copper(I) iodide, CuI	1×10^{-12}	Lead(II) sulfide, PbS	3×10^{-25}
Lead(II) iodide, PbI_2	7.9×10^{-9}	Manganese(II) sulfide, MnS	3×10^{-11}
Mercury(I) iodide, Hg_2I_2	4.7×10^{-29}	Mercury(II) sulfide, HgS	2×10^{-50}
Silver iodide, AgI	8.3×10^{-17}	Nickel(II) sulfide, NiS	3×10^{-16}
Hydroxides		Silver sulfide, Ag_2S	8×10^{-48}
Aluminum hydroxide, $Al(OH)_3$	3×10^{-34}	Tin(II) sulfide, SnS	1.3×10^{-23}
Cadmium hydroxide, $Cd(OH)_2$	7.2×10^{-15}	Zinc sulfide, ZnS	2.0×10^{-22}
Calcium hydroxide, $Ca(OH)_2$	6.5×10^{-6}		

Appendix D

STANDARD ELECTRODE (HALF-CELL) POTENTIALS AT 298 K*

Half-Reaction	$E°$ (V)
$F_2(g) + 2e^- \rightleftharpoons 2F^-(aq)$	+2.87
$O_3(g) + 2H^+(aq) + 2e^- \rightleftharpoons O_2(g) + H_2O(l)$	+2.07
$Co^{3+}(aq) + e^- \rightleftharpoons Co^{2+}(aq)$	+1.82
$H_2O_2(aq) + 2H^+(aq) + 2e^- \rightleftharpoons 2H_2O(l)$	+1.77
$PbO_2(s) + 3H^+(aq) + HSO_4^-(aq) + 2e^- \rightleftharpoons PbSO_4(s) + 2H_2O(l)$	+1.70
$Ce^{4+}(aq) + e^- \rightleftharpoons Ce^{3+}(aq)$	+1.61
$MnO_4^-(aq) + 8H^+(aq) + 5e^- \rightleftharpoons Mn^{2+}(aq) + 4H_2O(l)$	+1.51
$Au^{3+}(aq) + 3e^- \rightleftharpoons Au(s)$	+1.50
$Cl_2(g) + 2e^- \rightleftharpoons 2Cl^-(aq)$	+1.36
$Cr_2O_7^{2-}(aq) + 14H^+(aq) + 6e^- \rightleftharpoons 2Cr^{3+}(aq) + 7H_2O(l)$	+1.33
$MnO_2(s) + 4H^+(aq) + 2e^- \rightleftharpoons Mn^{2+}(aq) + 2H_2O(l)$	+1.23
$O_2(g) + 4H^+(aq) + 4e^- \rightleftharpoons 2H_2O(l)$	+1.23
$Br_2(l) + 2e^- \rightleftharpoons 2Br^-(aq)$	+1.07
$NO_3^-(aq) + 4H^+(aq) + 3e^- \rightleftharpoons NO(g) + 2H_2O(l)$	+0.96
$2Hg^{2+}(aq) + 2e^- \rightleftharpoons Hg_2^{2+}(aq)$	+0.92
$Hg_2^{2+}(aq) + 2e^- \rightleftharpoons 2Hg(l)$	+0.85
$Ag^+(aq) + e^- \rightleftharpoons Ag(s)$	+0.80
$Fe^{3+}(aq) + e^- \rightleftharpoons Fe^{2+}(aq)$	+0.77
$O_2(g) + 2H^+(aq) + 2e^- \rightleftharpoons H_2O_2(aq)$	+0.68
$MnO_4^-(aq) + 2H_2O(l) + 3e^- \rightleftharpoons MnO_2(s) + 4OH^-(aq)$	+0.59
$I_2(s) + 2e^- \rightleftharpoons 2I^-(aq)$	+0.53
$O_2(g) + 2H_2O(l) + 4e^- \rightleftharpoons 4OH^-(aq)$	+0.40
$Cu^{2+}(aq) + 2e^- \rightleftharpoons Cu(s)$	+0.34
$AgCl(s) + e^- \rightleftharpoons Ag(s) + Cl^-(aq)$	+0.22
$SO_4^{2-}(aq) + 4H^+(aq) + 2e^- \rightleftharpoons SO_2(g) + 2H_2O(l)$	+0.20
$Cu^{2+}(aq) + e^- \rightleftharpoons Cu^+(aq)$	+0.15
$Sn^{4+}(aq) + 2e^- \rightleftharpoons Sn^{2+}(aq)$	+0.13
$2H^+(aq) + 2e^- \rightleftharpoons H_2(g)$	0.00
$Pb^{2+}(aq) + 2e^- \rightleftharpoons Pb(s)$	−0.13
$Sn^{2+}(aq) + 2e^- \rightleftharpoons Sn(s)$	−0.14
$N_2(g) + 5H^+(aq) + 4e^- \rightleftharpoons N_2H_5^+(aq)$	−0.23
$Ni^{2+}(aq) + 2e^- \rightleftharpoons Ni(s)$	−0.25
$Co^{2+}(aq) + 2e^- \rightleftharpoons Co(s)$	−0.28
$PbSO_4(s) + H^+(aq) + 2e^- \rightleftharpoons Pb(s) + HSO_4^-(aq)$	−0.31
$Cd^{2+}(aq) + 2e^- \rightleftharpoons Cd(s)$	−0.40
$Fe^{2+}(aq) + 2e^- \rightleftharpoons Fe(s)$	−0.44
$Cr^{3+}(aq) + 3e^- \rightleftharpoons Cr(s)$	−0.74
$Zn^{2+}(aq) + 2e^- \rightleftharpoons Zn(s)$	−0.76
$2H_2O(l) + 2e^- \rightleftharpoons H_2(g) + 2OH^-(aq)$	−0.83
$Mn^{2+}(aq) + 2e^- \rightleftharpoons Mn(s)$	−1.18
$Al^{3+}(aq) + 3e^- \rightleftharpoons Al(s)$	−1.66
$Mg^{2+}(aq) + 2e^- \rightleftharpoons Mg(s)$	−2.37
$Na^+(aq) + e^- \rightleftharpoons Na(s)$	−2.71
$Ca^{2+}(aq) + 2e^- \rightleftharpoons Ca(s)$	−2.87
$Sr^{2+}(aq) + 2e^- \rightleftharpoons Sr(s)$	−2.89
$Ba^{2+}(aq) + 2e^- \rightleftharpoons Ba(s)$	−2.90
$K^+(aq) + e^- \rightleftharpoons K(s)$	−2.93
$Li^+(aq) + e^- \rightleftharpoons Li(s)$	−3.05

*Written as reductions; $E°$ value refers to all components in their standard states: 1 M for dissolved species; 1 atm pressure for the gas behaving ideally; the pure substance for solids and liquids.

Appendix E

ANSWERS TO SELECTED PROBLEMS

Chapter 1

1.2 Gas molecules fill the entire container; the volume of a gas is the volume of the container. Solids and liquids have a definite volume. The volume of the container does not affect the volume of a solid or liquid.   (a) gas   (b) liquid   (c) liquid   **1.3** Physical property: a characteristic shown by a substance itself, without any interaction with or change into other substances. Chemical property: a characteristic of a substance that appears as it interacts with, or transforms into, other substances.   (a) Color (yellow-green and silvery to white) and physical state (gas and metal to crystals) are physical properties. The interaction between chlorine gas and sodium metal is a chemical property.
(b) Color and magnetism are physical properties. No chemical changes.   **1.5**(a) Physical change; there is only a temperature change.   (b) Chemical change; the change in appearance indicates an irreversible chemical change.   (c) Physical change; there is only a change in size, not composition.   (d) Chemical change; the wood (and air) become different substances with different compositions.   **1.7**(a) fuel   (b) wood   **1.11** A well-designed experiment must have the following essential features:
(1) There must be at least two variables that are expected to be related; (2) there must be a way to control all the variables, so that only one at a time may be changed; (3) the results must be reproducible.   **1.14**(a) $(2.54 \text{ cm})^2/(1 \text{ in})^2$ (b) $(1000 \text{ m})^2/(1 \text{ km})^2$
(c) $(1.609 \times 10^3 \text{ m/mi})$ and $(1 \text{ h}/3600 \text{ s})$ (d) $(1000 \text{ g}/2.205 \text{ lb})$ and $(3.531 \times 10^{-5} \text{ ft}^3/\text{cm}^3)$ **1.16** An extensive property depends on the amount of material present. An intensive property is the same regardless of how much material is present.   (a) extensive property   (b) intensive property   (c) extensive property
(d) intensive property   **1.18**(a) increases   (b) remains the same
(c) decreases   (d) increases   (e) remains the same   **1.21** 1.43 nm
1.23(a) $1.77 \times 10^{-9} \text{ km}^2$ (b) 8.92 **1.25**(a) $5.52 \times 10^3 \text{ kg/m}^3$
(b) 345 lb/ft^3 **1.27**(a) $1.72 \times 10^{-9} \text{ mm}^3$ (b) 10^{-10} L
1.29(a) 9.626 cm^3 (b) 64.92 g **1.31** 2.70 g/cm^3 **1.33**(a) $22°C$; 295 K (b) 109 K; $-263°F$ (c) $-273°C$; $-460.°F$
1.36(a) $2.55 \times 10^{-7} \text{ m}$ (b) 6830 Å **1.41** Initial zeros are never significant; internal zeros are always significant; terminal zeros to the right of a decimal point are significant; terminal zeros to the left of a decimal point are significant only if they were measured.   **1.42**(a) none   (b) none   (c) $0.039\underline{0}$ (d) $3.0\underline{900} \times 10^4$
1.44(a) 1.34 m (b) $3.350 \times 10^3 \text{ cm}^3$ (c) 443 cm
1.46(a) 1.310000×10^5 (b) 4.7×10^{-4} (c) 2.10006×10^5
(d) 2.1605×10^3 **1.48**(a) 5550 (b) $10,070$ (c) 0.000000885
(d) 0.003004 **1.50**(a) $4.06 \times 10^{-19} \text{ J}$ (b) 1.56×10^{24} molecules
(c) $1.82 \times 10^5 \text{ J/mol}$ **1.52**(a) Height measured, not exact.
(b) Planets counted, exact.   (c) Number of grams in a pound is not a unit definition, not exact.   (d) Definition of "millimeter," exact.   **1.54** $7.50 \pm 0.05 \text{ cm}$ **1.56**(a) $I_{avg} = 8.72 \text{ g}$; $II_{avg} = 8.72 \text{ g}$; $III_{avg} = 8.50 \text{ g}$; $IV_{avg} = 8.56 \text{ g}$; sets I and II are most accurate.
(b) Set III is the most precise, but is the least accurate.   (c) Set I has the best combination of high accuracy and high precision.

(d) Set IV has both low accuracy and low precision.
1.59(a)

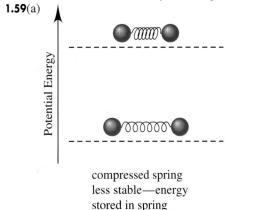

compressed spring
less stable—energy
stored in spring

(b)

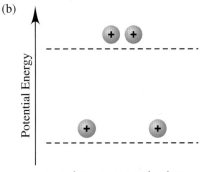

two charges near each other
less stable—repulsion of like
charges

1.62(a) $19.40 before price increase; $33.90 after price increase
(b) 51.7 coins   (c) 21.0 coins   **1.64**(a) 0.21 g/L, will float
(b) CO_2 is denser than air, will sink   (c) 0.30 g/L, will float
(d) O_2 is denser than air, will sink   (e) 1.38 g/L, will sink
(f) 0.50 g **1.67**(a) $8.0 \times 10^{12} \text{ g}$ (b) $4.1 \times 10^5 \text{ m}^3$ (c) 9.5×10^{13} dollars   **1.69**(a) $-195.79°C$ (b) $-320.42°F$ (c) 5.05 L
1.72 7.1×10^7 microparticles in the room; 1.2×10^3 microparticles in a breath   **1.73** $2.3 \times 10^{25} \text{ g}$ oxygen; $1.4 \times 10^{25} \text{ g}$ silicon; $5 \times 10^{15} \text{ g}$ ruthenium (and rhodium)   **1.75** freezing point = $-3.7°X$; boiling point = $63.3°X$

Chapter 2

2.1 Compounds contain different types of atoms; there is only one type of atom in an element.   **2.4**(a) The presence of more than one element makes pure calcium chloride a compound.
(b) There is only one kind of atom, so sulfur is an element.
(c) The presence of more than one compound makes baking powder a mixture.   (d) The presence of more than one type of atom means cytosine cannot be an element. The specific, not

variable, arrangement means it is a compound.   **2.6**(a) elements, compounds, and mixtures   (b) compounds   (c) compounds **2.7**(a) Law of definite composition: the composition is the same regardless of its source.   (b) Law of mass conservation: the total quantity of matter does not change.   (c) Law of multiple proportions: two elements can combine to form two different compounds that have different proportions of those elements.   **2.8**(a) No, the percent by mass of each element in a compound is fixed. (b) Yes, the *mass* of each element in a compound depends on the mass of the compound.   (c) No, the percent by mass of each element in a compound is fixed.   **2.9** The two experiments demonstrate the law of definite composition. The unknown compound decomposes the same way both times. The experiments also demonstrate the law of conservation of mass since the total mass before reaction equals the total mass after reaction. **2.11**(a) 1.34 g F   (b) 0.514 Ca; 0.486 F   (c) 51.4 mass % Ca; 48.6 mass % F   **2.13** 3.498×10^6 g Cu; 1.766×10^6 g S   **2.15** compound 1: 0.904 S/Cl; compound 2: 0.451 S/Cl; ratio: 2.00/1.00 **2.18** Coal A   **2.19** Dalton postulated that atoms of an element are identical and that compounds result from the chemical combination of specific ratios of different elements.   **2.20**(a) If you know the ratio of any two quantities and the value of one of them, the other can always be calculated; in this case, the charge and the charge-to-mass ratio were known.   (b) The charge on each oil droplet has a common factor of -1.602×10^{-19} C, the charge of the electron.   **2.23** All three isotopes have 18 protons and 18 electrons. Their respective mass numbers are 36, 38, and 40, with the respective numbers of neutrons being 18, 20, and 22. **2.25**(a) These have the same number of protons and electrons, but different numbers of neutrons; same Z.   (b) These have the same number of neutrons, but different numbers of protons and electrons; same N.   (c) These have different numbers of protons, neutrons, and electrons; same A.   **2.27**(a) $^{38}_{18}$Ar   (b) $^{55}_{25}$Mn (c) $^{109}_{47}$Ag   **2.29**(a) $^{48}_{22}$Ti   (b) $^{79}_{34}$Se   (c) $^{11}_{5}$B

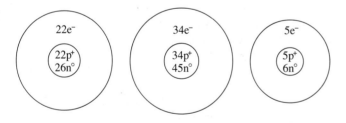

2.31 69.72 amu   **2.33** ^{35}Cl = 75.774%, ^{37}Cl = 24.226% **2.35**(a) In the modern periodic table, the elements are arranged in order of increasing atomic number.   (b) Elements in a group (or family) have similar chemical properties.   (c) Elements can be classified as metals, metalloids, or nonmetals.   **2.38**(a) germanium; Ge; 4A(14); metalloid   (b) sulfur; S; 6A(16); nonmetal (c) helium; He; 8A(18); nonmetal   (d) lithium; Li; 1A(1); metal (e) molybdenum; Mo; 6B(6); metal   **2.40**(a) Ra; 88   (b) As; 33 (c) Cu; 63.55 amu   (d) Br; 79.90 amu   **2.42** Atoms of these two kinds of substances will form ionic bonds, in which one or more electrons are transferred from the metal atom to the nonmetal atom to form a cation and an anion, respectively. **2.44** Coulomb's law states the energy of attraction in an ionic bond is directly proportional to the product of charges and inversely proportional to the distance between charges. The product of charges in MgO ($2+ \times 2-$) is greater than the product of

charges in LiF ($1+ \times 1-$). Thus, MgO has stronger ionic bonding.   **2.47** K^+; I^- **2.49**(a) oxygen; 17; 6A(16); 2   (b) fluorine; 19; 7A(17); 2   (c) calcium; 40; 2A(2); 4   **2.51** Lithium forms the Li^+ ion; oxygen forms the O^{2-} ion. Number of O^{2-} ions = 2.6×10^{20} O^{2-} ions.   **2.53** NaCl   **2.55** An empirical formula shows the simplest ratio of atoms of each element present in a compound, whereas a molecular formula describes the type and actual number of atoms of each element in a molecule of the compound. The empirical formula and the molecular formula can be the same. The molecular formula is always a whole-number multiple of the empirical formula.   **2.56** The two samples are similar in that both contain 20 billion oxygen atoms and 20 billion hydrogen atoms. They differ in that they contain different types of molecules: H_2O_2 molecules in the hydrogen peroxide sample, and H_2 and O_2 molecules in the mixture. In addition, the mixture contains 20 billion molecules (10 billion H_2 and 10 billion O_2), while the hydrogen peroxide sample contains 10 billion molecules.   **2.57**(a) NH_2 (b) CH_2O **2.59**(a) Li_3N, lithium nitride (b) SrO, strontium oxide   (c) $AlCl_3$, aluminum chloride **2.61**(a) MgF_2, magnesium fluoride   (b) Na_2S, sodium sulfide (c) $SrCl_2$, strontium chloride   **2.63**(a) $SnCl_4$ (b) iron(III) bromide   (c) CuBr   (d) manganese(III) oxide   **2.65**(a) BaO (b) $Fe(NO_3)_2$ (c) MgS   **2.67**(a) H_2SO_4; sulfuric acid (b) HIO_3; iodic acid   (c) HCN; hydrocyanic acid   (d) H_2S; hydrosulfuric acid   **2.69** Disulfur tetrafluoride, S_2F_4 **2.71**(a) 12 oxygen atoms; 342.2 amu   (b) 9 hydrogen atoms; 132.06 amu   (c) 8 oxygen atoms; 344.6 amu   **2.73**(a) $(NH_4)_2SO_4$; 132.15 amu   (b) NaH_2PO_4; 119.98 amu   (c) $KHCO_3$; 100.12 amu **2.75** disulfur dichloride; SCl; 135.04 amu   **2.77**(a) SO_3, sulfur trioxide, 80.07 amu   (b) N_2O, dinitrogen monoxide, 44.02 amu **2.79** Separating the components of a mixture requires physical methods only; that is, no chemical changes (no changes in composition) take place, and the components maintain their chemical identities and properties throughout. Separating the components of a compound requires a chemical change (change in composition). **2.82**(a) compound   (b) homogeneous mixture   (c) heterogeneous mixture   (d) homogeneous mixture   (e) homogeneous mixture **2.84**(a) fraction of volume = 5.2×10^{-13} (b) mass of nucleus = 6.64466×10^{-24} g; fraction of mass = 0.999726 **2.86**(a) I = NO; II = N_2O_3; III = N_2O_5 (b) I has 1.14 g O per 1.00 g N; II, 1.71 g O; III, 2.86 g O   **2.88**(a) Cl^-, 1.898 mass %; Na^+, 1.056 mass %; SO_4^{2-}, 0.265 mass %; Mg^{2+}, 0.127 mass %; Ca^{2+}, 0.04 mass %; K^+, 0.038 mass %; HCO_3^-, 0.014 mass % (b) 30.72%   (c) Alkaline earth metal ions, total mass % = 0.17%; alkali metal ions, total mass % = 1.094%   (d) Anions (2.177 mass %) make up a larger mass fraction than cations (1.26 mass %).   **2.90** Molecular, $C_4H_6O_4$; empirical, $C_2H_3O_2$; molecular mass, 118.09 amu; 40.68% by mass C; 5.122% by mass H; 54.20% by mass O   **2.92** 58.091 amu   **2.94** 0.370 lb C; 0.0222 lb H; 0.423 lb O; 0.185 lb N   **2.104**(1) chemical change (2) physical change (3) chemical change   (4) chemical change (5) physical change

Chapter 3

3.2(a) 12 mol C atoms   (b) 7.226×10^{24} C atoms   **3.6**(a) left (b) left   (c) left   (d) neither   **3.7**(a) 121.64 g/mol (b) 44.02 g/mol   (c) 106.44 g/mol   (d) 152.00 g/mol **3.9**(a) 150.7 g/mol   (b) 175.3 g/mol   (c) 342.17 g/mol (d) 125.84 g/mol   **3.11**(a) 9.0×10^1 g $KMnO_4$ (b) 0.331 mol

O atoms   (c) 1.8×10^{20} O atoms   **3.13**(a) 97 g $MnSO_4$
(b) 4.46×10^{-2} mol $Fe(ClO_4)_3$ (c) 1.74×10^{24} N atoms
3.15(a) 1.57×10^3 g Cu_2CO_3 (b) 0.366 g N_2O_5 (c) 0.473 mol
$NaClO_4$; 2.85×10^{23} formula units $NaClO_4$ (d) 2.85×10^{23} Na^+
ions; 2.85×10^{23} ClO_4^- ions; 2.85×10^{23} Cl atoms; 1.14×10^{24} O
atoms   **3.17**(a) 6.375 mass % H   (b) 71.52 mass % O
3.19(a) 0.9507 mol cisplatin   (b) 3.5×10^{24} H atoms
3.21 $CO(NH_2)_2 > NH_4NO_3 > (NH_4)_2SO_4 > KNO_3$
3.22(a) 883 mol PbS   (b) 1.88×10^{25} Pb atoms   **3.24**(b) From
the mass percent, determine the empirical formula. Add up the
total number of atoms in the empirical formula, and divide that
number into the total number of atoms in the molecule. The result
is the multiplier that makes the empirical formula into the molecular formula.   (c) (mass % expressed directly in grams)(1/molar
mass) = moles of each element   (e) Count the numbers of
the various types of atoms in the structural formula and put
these into a molecular formula.   **3.25**(a) CH_2; 14.03 g/mol
(b) CH_3O; 31.03 g/mol   (c) N_2O_5; 108.02 g/mol   (d) $Ba_3(PO_4)_2$;
601.8 g/mol   (e) TeI_4; 635.2 g/mol   **3.27**(a) C_3H_6 (b) N_2H_4
(c) N_2O_4 (d) $C_5H_5N_5$ **3.29**(a) Cl_2O_7 (b) $SiCl_4$ (c) CO_2
3.31(a) 1.20 mol F   (b) 24.0 g M   (c) calcium   **3.33** $C_{21}H_{30}O_5$
3.34 $C_{10}H_{20}O$ **3.36** b
3.37(a) $16Cu(s) + S_8(s) \longrightarrow 8Cu_2S(s)$
  (b) $P_4O_{10}(s) + 6H_2O(l) \longrightarrow 4H_3PO_4(l)$
  (c) $B_2O_3(s) + 6NaOH(aq) \longrightarrow 2Na_3BO_3(aq) + 3H_2O(l)$
  (d) $4CH_3NH_2(g) + 9O_2(g) \longrightarrow$
               $4CO_2(g) + 10H_2O(g) + 2N_2(g)$
3.39(a) $4Ga(s) + 3O_2(g) \longrightarrow 2Ga_2O_3(s)$
  (b) $2C_6H_{14}(l) + 19O_2(g) \longrightarrow 12CO_2(g) + 14H_2O(g)$
  (c) $3CaCl_2(aq) + 2Na_3PO_4(aq) \longrightarrow$
               $Ca_3(PO_4)_2(s) + 6NaCl(aq)$
3.41(a) 2.22×10^3 mol KNO_3 (b) 2.24×10^5 g KNO_3
3.43 150.2 g H_3BO_3; 14.69 g H_2 **3.45** 2.03×10^3 g Cl_2
3.47(a) 0.105 mol CaO   (b) 0.175 mol CaO   (c) calcium
(d) 5.88 g CaO   **3.49** 1.47 mol HIO_3, 258 g HIO_3; 38.0 g H_2O in
excess   **3.51** 4.40 g CO_2; 4.80 g O_2 in excess   **3.53** 6.4 g
$Al(NO_2)_3$; no NH_4Cl; 45.4 g $AlCl_3$; 28.6 g N_2; 36.8 g H_2O
3.55 53%   **3.57** 98.2%   **3.59** 24.5 g CH_3Cl **3.61** 52.9 g CF_4
3.64(a) C   (b) B   (c) C   (d) B   **3.65**(a) 5.76 g $Ca(C_2H_3O_2)_2$
(b) 0.254 M KI   (c) 124 mol NaCN   **3.67**(a) 0.0617 M KCl
(b) 0.00363 M $(NH_4)_2SO_4$ (c) 0.0150 M Na^+ **3.69**(a) 987 g
HNO_3/L   (b) 15.7 M HNO_3 **3.71** 845 mL   **3.73** 0.87 g $BaSO_4$
3.75(a) Instructions: Be sure to wear goggles to protect your eyes!
Pour approximately 3.0 gal of water into the container. Add to the
water, slowly and with mixing, 1.5 gal of concentrated HCl. Dilute
to 5.0 gal with more water.   (b) 22.4 mL   **3.76** $X = 3$
3.77 ethane > propane > cetyl palmitate > ethanol > benzene
3.81 89.8%   **3.82**(a) $2AB_2 + B_2 \longrightarrow 2AB_3$ (b) AB_2
(c) 5.0 mol AB_3 (d) 0.5 mol B_2 **3.85**(a) C   (b) B   (c) D
3.89 0.071 M KBr   **3.92** 586 g CO_2 **3.93** 10:0.66:1.0
3.95 32.7 mass % C   **3.97**(a) 192.12 g/mol; $C_6H_8O_7$
(b) 0.549 mol   **3.99**(a) 0.027 g heme   (b) 4.4×10^{-5} mol heme
(c) 2.4×10^{-3} g Fe   (d) 2.9×10^{-2} g hemin   **3.102**(a) 46.65
mass % N in urea; 31.98 mass % N in arginine; 21.04 mass %
N in ornithine   (b) 30.13 g N   **3.106** 44.6%

Chapter 4

4.2 Ions must be present and they come from ionic compounds
or from electrolytes such as acids and bases.   **4.5** (2)

4.8 (a) Benzene is likely to be insoluble in water because it is
nonpolar and water is polar.   (b) Sodium hydroxide, an ionic
compound, is likely to be very soluble in water.   (c) Ethanol
(CH_3CH_2OH) is likely to be soluble in water because the
alcohol group (—OH) is polar.   (d) Potassium acetate, an ionic
compound, is likely to be very soluble in water.   **4.10**(a) Yes,
CsI is a soluble salt.   (b) Yes, HBr is a strong acid.
4.12(a) 3.3 mol   (b) 8.93×10^{-5} mol   (c) 8.17×10^{-3} mol
4.14(a) 0.245 mol Al^{3+}; 0.735 mol Cl^-; 1.48×10^{23} Al^{3+} ions;
4.43×10^{23} Cl^- ions   (b) 0.0848 mol Li^+; 0.0424 mol
SO_4^{2-}; 5.11×10^{22} Li^+ ions; 2.55×10^{22} SO_4^{2-} ions
(c) 3.78×10^{21} K^+ ions; 3.78×10^{21} Br^- ions; 6.28×10^{-3} mol K^+;
6.28×10^{-3} mol Br^- **4.16**(a) 0.35 mol H^+ (b) 1.3×10^{-3} mol
H^+ (c) 0.43 mol H^+ **4.23** Assuming that the left beaker
contains $AgNO_3$ (because it has gray Ag^+ ion), the right must
contain NaCl. Then, NO_3^- is blue, Na^+ is brown, and Cl^- is
green.
Molecular equation: $AgNO_3(aq) + NaCl(aq) \longrightarrow$
               $AgCl(s) + NaNO_3(aq)$
Total ionic equation: $Ag^+(aq) + NO_3^-(aq) + Na^+(aq) +$
         $Cl^-(aq) \longrightarrow AgCl(s) + Na^+(aq) + NO_3^-(aq)$
Net ionic equation: $Ag^+(aq) + Cl^-(aq) \longrightarrow AgCl(s)$
4.24(a) Molecular: $Hg_2(NO_3)_2(aq) + 2KI(aq) \longrightarrow$
               $Hg_2I_2(s) + 2KNO_3(aq)$
   Total ionic: $Hg_2^{2+}(aq) + 2NO_3^-(aq) + 2K^+(aq) +$
         $2I^-(aq) \longrightarrow Hg_2I_2(s) + 2K^+(aq) + 2NO_3^-(aq)$
   Net ionic: $Hg_2^{2+}(aq) + 2I^-(aq) \longrightarrow Hg_2I_2(s)$
   Spectator ions are K^+ and NO_3^-.
(b) Molecular: $FeSO_4(aq) + Ba(OH)_2(aq) \longrightarrow$
               $Fe(OH)_2(s) + BaSO_4(s)$
   Total ionic: $Fe^{2+}(aq) + SO_4^{2-}(aq) + Ba^{2+}(aq) +$
         $2OH^-(aq) \longrightarrow Fe(OH)_2(s) + BaSO_4(s)$
   Net ionic: This is the same as the total ionic equation, be-
   cause there are no spectator ions.
4.26(a) No precipitate will form.   (b) A precipitate will form because silver ions, Ag^+, and iodide ions, I^-, will combine to form
a solid salt, silver iodide, AgI. The ammonium and nitrate ions
do not form a precipitate.
Molecular: $NH_4I(aq) + AgNO_3(aq) \longrightarrow$
               $AgI(s) + NH_4NO_3(aq)$
Total ionic: $NH_4^+(aq) + I^-(aq) + Ag^+(aq) + NO_3^-(aq) \longrightarrow$
               $AgI(s) + NH_4^+(aq) + NO_3^-(aq)$
Net ionic: $Ag^+(aq) + I^-(aq) \longrightarrow AgI(s)$
4.28 0.0389 M Pb^{2+} **4.30** 1.80 mass % Cl
4.36(a) Molecular equation: $KOH(aq) + HI(aq) \longrightarrow$
               $KI(aq) + H_2O(l)$
   Total ionic equation: $K^+(aq) + OH^-(aq) + H^+(aq) +$
         $I^-(aq) \longrightarrow K^+(aq) + I^-(aq) + H_2O(l)$
   Net ionic equation: $OH^-(aq) + H^+(aq) \longrightarrow H_2O(l)$
   The spectator ions are $K^+(aq)$ and $I^-(aq)$.
(b) Molecular equation: $NH_3(aq) + HCl(aq) \longrightarrow$
               $NH_4Cl(aq)$
   Total ionic equation: $NH_3(aq) + H^+(aq) + Cl^-(aq) \longrightarrow$
               $NH_4^+(aq) + Cl^-(aq)$
NH_3, a weak base, is written in the molecular (undissociated) form. HCl, a strong acid, is written as dissociated
ions. NH_4Cl is a soluble compound, because all ammonium compounds are soluble.
Net ionic equation: $NH_3(aq) + H^+(aq) \longrightarrow NH_4^+(aq)$
Cl^- is the only spectator ion.

4.38 Total ionic equation: $CaCO_3(s) + 2H^+(aq) + 2Cl^-(aq) \longrightarrow$
$Ca^{2+}(aq) + 2Cl^-(aq) + H_2O(l) + CO_2(g)$
Net ionic equation: $CaCO_3(s) + 2H^+(aq) \longrightarrow$
$Ca^{2+}(aq) + H_2O(l) + CO_2(g)$
4.40 0.05839 M CH_3COOH **4.45**(a) S has O.N. $= +6$ in SO_4^{2-}
(i.e., H_2SO_4), and O.N. $= +4$ in SO_2, so S has been reduced
(and I^- oxidized); H_2SO_4 acts as an oxidizing agent. (b) The
oxidation numbers remain constant throughout; H_2SO_4 transfers
a proton to F^- to produce HF, so it acts as an acid. **4.46**(a) -1
(b) -2 (c) -3 (d) $+3$ **4.48**(a) -3 (b) $+5$ (c) $+3$
4.50(a) $+6$ (b) $+3$ (c) $+7$ **4.52**(a) MnO_4^- is the oxidizing
agent; $H_2C_2O_4$ is the reducing agent. (b) Cu is the reducing
agent; NO_3^- is the oxidizing agent. **4.54**(a) Oxidizing agent is
NO_3^-; reducing agent is Sn. (b) Oxidizing agent is MnO_4^-;
reducing agent is Cl^-. **4.56** S is in Group 6A(16), so its highest
possible O.N. is $+6$ and its lowest possible O.N. is $6 - 8 = -2$.
(a) $S = -2$. The S can only increase its O.N. (oxidize), so S^{2-} can
function only as a reducing agent. (b) $S = +6$. The S can only
decrease its O.N. (reduce), so SO_4^{2-} can function only as an
oxidizing agent. (c) $S = +4$. The S can increase or decrease
its O.N. Therefore, SO_2 can function as either an oxidizing or
reducing agent.
4.62(a) $2Sb(s) + 3Cl_2(g) \longrightarrow 2SbCl_3(s)$; combination
(b) $2AsH_3(g) \longrightarrow 2As(s) + 3H_2(g)$; decomposition
(c) $3Mn(s) + 2Fe(NO_3)_3(aq) \longrightarrow$
$3Mn(NO_3)_2(aq) + 2Fe(s)$; displacement
4.64(a) $N_2(g) + 3H_2(g) \longrightarrow 2NH_3(g)$
(b) $2NaClO_3(s) \xrightarrow{\Delta} 2NaCl(s) + 3O_2(g)$
(c) $Ba(s) + 2H_2O(l) \longrightarrow Ba(OH)_2(aq) + H_2(g)$
4.66(a) $2Cs(s) + I_2(s) \longrightarrow 2CsI(s)$
(b) $2Al(s) + 3MnSO_4(aq) \longrightarrow Al_2(SO_4)_3(aq) + 3Mn(s)$
(c) $2SO_2(g) + O_2(g) \longrightarrow 2SO_3(g)$
(d) $C_3H_8(g) + 5O_2(g) \longrightarrow 3CO_2(g) + 4H_2O(g)$
(e) $2Al(s) + 3Mn^{2+}(aq) \longrightarrow 2Al^{3+}(aq) + 3Mn(s)$
4.68 315 g O_2; 3.95 kg Hg **4.70**(a) O_2 is in excess.
(b) 0.117 mol Li_2O (c) 0 g Li, 3.49 g Li_2O, and 4.13 g O_2
4.73 0.223 kg Fe
4.74(a) $\quad Fe(s) + 2H^+(aq) \longrightarrow Fe^{2+}(aq) + H_2(g)$
$\quad$ O.N.: 0 $\quad\quad +1 \quad\quad\quad\quad +2 \quad\quad\quad 0$
(b) 3.1×10^{21} Fe^{2+} ions
4.75 5.11 g C_2H_5OH; 2.49 L CO_2
4.80(a) Step 1: oxidizing agent is O_2; reducing agent is NH_3.
Step 2: oxidizing agent is O_2; reducing agent is NO. Step 3: oxidizing agent is NO_2; reducing agent is NO_2. (b) 1.2×10^4 kg NH_3
4.82 627 L air **4.86** 3.0×10^{-3} mol CO_2 (b) 0.11 L CO_2
4.88(a) $C_7H_5O_4Bi$ (b) $C_{21}H_{15}O_{12}Bi_3$
(c) $Bi(OH)_3(s) + 3HC_7H_5O_3(aq) \longrightarrow Bi(C_7H_5O_3)_3(s) + 3H_2O(l)$
(d) 0.490 mg $Bi(OH)_3$
4.90(a) Ethanol: $C_2H_5OH(l) + 3O_2(g) \longrightarrow 2CO_2(g) + 3H_2O(l)$
Gasoline: $2C_8H_{18}(l) + 25O_2(g) \longrightarrow$
$16CO_2(g) + 18H_2O(g)$
(b) 2.50×10^3 g O_2 (c) 1.75×10^3 L O_2 (d) 8.38×10^3 L air
4.92 yes **4.93**(a) Reaction (2) is a redox process. (b) 2.00×10^5 g
Fe_2O_3; 4.06×10^5 g $FeCl_3$ (c) 2.09×10^5 g Fe; 4.75×10^5 g
$FeCl_2$ (d) 0.313

Chapter 5

5.1(a) The volume of the liquid remains constant, but the volume
of the gas increases to the volume of the larger container. (b) The
volume of the container holding the gas sample increases when
heated, but the volume of the container holding the liquid sample
remains essentially constant when heated. (c) The volume of
the liquid remains essentially constant, but the volume of the gas
is reduced. **5.6** 979 cmH_2O **5.8**(a) 566 mmHg (b) 1.32 bar
(c) 3.60 atm (d) 107 kPa **5.14** At constant temperature and
volume, the pressure of a gas is directly proportional to number
of moles of the gas. **5.16**(a) Volume decreases to one-third of
the original volume. (b) Volume increases by a factor of 2.5.
(c) Volume increases by a factor of 3. **5.18** $-42°C$ **5.20** 35.3 L
5.22 0.061 mol Cl_2 **5.24** 0.674 g ClF_3 **5.28** Beaker is inverted
for H_2 and upright for CO_2. The molar mass of CO_2 is greater
than the molar mass of air, which, in turn, has a greater molar
mass than H_2. **5.31** 5.86 g/L **5.33** 1.78×10^{-3} mol AsH_3;
3.48 g/L **5.35** 51.1 g/mol **5.37** 1.33 atm **5.41** C_5H_{12}
5.43(a) 0.90 mol (b) 6.76 torr **5.44** 39.3 g P_4 **5.46** 41.2 g
PH_3 **5.48** 0.0249 g Al **5.50** 286 mL SO_2 **5.51** 0.0997 atm
SiF_4 **5.54** At STP, the volume occupied by a mole of any gas is
the same. At the same temperature, all gases have the same average kinetic energy, resulting in the same pressure. **5.57**(a) $P_A >$
$P_B > P_C$ (b) $E_A = E_B = E_C$ (c) $\text{rate}_A > \text{rate}_B > \text{rate}_C$ (d)
total $E_A >$ total $E_B >$ total E_C (e) $d_A = d_B = d_C$ **5.58** 13.21
5.60(a) curve 1 (b) curve 1 (c) curve 1; fluorine and argon
have about the same molar mass **5.62** 14.0 min **5.64** 4 atoms
per molecule **5.66** negative deviations; $N_2 < Kr < CO_2$
5.68 at 1 atm; because the pressure is lower. **5.71** 6.81×10^4 g/mol
5.73(a) 22.5 atm (b) 21.2 atm **5.75**(a) 597 torr N_2; 159 torr O_2;
0.3 torr CO_2; 3.5 torr H_2O (b) 74.9 mol % N_2; 13.7 mol % O_2;
5.3 mol % CO_2; 6.2 mol % H_2O (c) 1.6×10^{21} molecules O_2
5.77(a) 4×10^2 mL (b) 0.013 mol N_2 **5.78** 35.7 L NO_2
5.82 Al_2Cl_6 **5.84** 1.62×10^{-2} mol **5.88**(a) 9 volumes of $O_2(g)$
(b) CH_5N **5.90** 4.86; 52.5 ft to a depth of 73 ft **5.97**(a) xenon
(b) water vapor (c) mercury (d) water vapor **5.100** 17.2 g
CO_2; 17.8 g Kr

5.103(a) $\frac{1}{2}m\overline{u^2} = \frac{3}{2}\left(\frac{R}{N_A}\right)T$

$m\overline{u^2} = 3\left(\frac{R}{N_A}\right)T$

$\overline{u^2} = \frac{3RT}{mN_A}$

$u_{rms} = \sqrt{\frac{3RT}{\mathcal{M}}}$ where $\mathcal{M} = mN_A$

(b) $\overline{E}_k = \frac{1}{2}m_1\overline{u_1^2} = \frac{1}{2}m_2\overline{u_2^2}$

$m_1\overline{u_1^2} = m_2\overline{u_2^2}$

$\frac{m_1}{m_2} = \frac{\overline{u_2^2}}{\overline{u_1^2}}$; so $\frac{\sqrt{m_1}}{\sqrt{m_2}} = \frac{\overline{u_2}}{\overline{u_1}}$

Substitute molar mass, $\mathcal{M}$, for m:

$\frac{\sqrt{\mathcal{M}_1}}{\sqrt{\mathcal{M}_2}} = \frac{\text{rate}_2}{\text{rate}_1}$

5.107(a) 14.8 L CO_2 (b) $P_{H_2O} = 42.2$ torr; 3.7×10^2 torr
$P_{O_2} = P_{CO_2}$ **5.111** 332 steps **5.112** 1.6

Chapter 6

6.5 0 J **6.7**(a) 3.3×10^7 kJ (b) 7.9×10^6 kcal (c) 3.1×10^7 Btu
6.10(a) exothermic (b) endothermic (c) exothermic

(d) exothermic   (e) endothermic   (f) endothermic
(g) exothermic

6.11

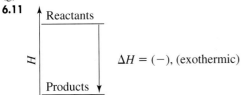

Reactants

H

Products

$\Delta H = (-)$, (exothermic)

6.13(a) Combustion of methane: $CH_4(g) + 2O_2(g) \longrightarrow$
$$CO_2(g) + 2H_2O(g) + \text{heat}$$

$CH_4 + 2O_2$ (initial)

H

$CO_2 + 2H_2O$ (final)

$\Delta H = (-)$, (exothermic)

(b) Freezing of water: $H_2O(l) \longrightarrow H_2O(s) + \text{heat}$

$H_2O(l)$ (initial)

H

$H_2O(s)$ (final)

$\Delta H = (-)$, (exothermic)

6.15(a) $C_2H_5OH(l) + 3O_2(g) \longrightarrow 2CO_2(g) + 3H_2O(g) + \text{heat}$

$C_2H_6O + 3O_2$ (initial)

H

$2CO_2 + 3H_2O$ (final)

$\Delta H = (-)$, (exothermic)

(b) $\frac{1}{2}N_2(g) + O_2(g) + \text{heat} \longrightarrow NO_2(g)$

NO_2 (final)

H

$\frac{1}{2}N_2 + O_2$ (initial)

$\Delta H = (+)$, (endothermic)

6.18 To determine the specific heat capacity of a substance, you need its mass, the heat added (or lost), and the change in temperature. **6.20** 4.0×10^3 J **6.22** 323°C **6.24** 77.5°C **6.26** 42°C
6.31 The reaction has a positive ΔH_{rxn}, because this reaction requires the input of energy to break the oxygen-oxygen bond.
6.32 ΔH is negative; it is opposite in sign and half of the value for the vaporization of 2 mol of H_2O. **6.33**(a) exothermic
(b) 20.2 kJ (c) -5.2×10^2 kJ (d) -12.6 kJ
6.35(a) $\frac{1}{2}N_2(g) + \frac{1}{2}O_2(g) \longrightarrow NO(g)$; $\Delta H = 90.29$ kJ
(b) -4.51 kJ
6.37 -2.11×10^6 kJ
6.39(a) $C_2H_4(g) + 3O_2(g) \longrightarrow 2CO_2(g) + 2H_2O(l)$;
$\Delta H_{rxn} = -1411$ kJ
(b) 1.39 g C_2H_4
6.42 -813.4 kJ
6.44 $N_2(g) + 2O_2(g) \longrightarrow 2NO_2(g)$; $\Delta H_{rxn} = +66.4$ kJ; A = 1, B = 2, C = 3 **6.47** The standard heat of reaction, ΔH°_{rxn}, is the enthalpy change for any reaction where all substances are in their standard states. The standard heat of formation, ΔH°_f, is the

enthalpy change that accompanies the formation of one mole of a compound in its standard state from elements in their standard states.
6.48(a) $\frac{1}{2}Cl_2(g) + Na(s) \longrightarrow NaCl(s)$
(b) $H_2(g) + \frac{1}{2}O_2(g) \longrightarrow H_2O(l)$
(c) no changes
6.49(a) $Ca(s) + Cl_2(g) \longrightarrow CaCl_2(s)$
(b) $Na(s) + \frac{1}{2}H_2(g) + C(graphite) + \frac{3}{2}O_2(g) \longrightarrow$
$$NaHCO_3(s)$$
(c) $C(graphite) + 2Cl_2(g) \longrightarrow CCl_4(l)$
(d) $\frac{1}{2}H_2(g) + \frac{1}{2}N_2(g) + \frac{3}{2}O_2(g) \longrightarrow HNO_3(l)$
6.51(a) -1036.8 kJ (b) -433 kJ **6.53** -157.3 kJ/mol
6.56(a) 503.9 kJ (b) $-\Delta H_1 + 2\Delta H_2 = 504$ kJ
6.57(a) $C_{18}H_{36}O_2(s) + 26O_2(g) \longrightarrow 18CO_2(g) + 18H_2O(g)$
(b) $-10,488$ kJ (c) -36.9 kJ; -8.81 kcal (d) 8.81 kcal/g $\times$ 11.0 g = 96.9 kcal **6.58**(a) 23.6 L/mol initial; 24.9 L/mol final
(b) 187 J (c) -1.2×10^2 J (d) 3.1×10^2 J (e) 310 J
(f) $\Delta H = \Delta E + P\Delta V = \Delta E - w = (q + w) - w = q_P$
6.66 721 kJ **6.73**(a) $\Delta H^{\circ}_{rxn1} = -657.0$ kJ; $\Delta H^{\circ}_{rxn2} = 32.9$ kJ
(b) -106.6 kJ **6.76**(a) 6.81×10^3 J (b) 243°C **6.77** -22.2 kJ
6.79(a) -1.25×10^3 kJ (b) 2.24×10^{3}°C

Chapter 7

7.2 (a) x-ray < ultraviolet < visible < infrared < microwave < radio waves (b) radio < microwave < infrared < visible < ultraviolet < x-ray (c) radio < microwave < infrared < visible < ultraviolet < x-ray **7.4** The energy of an atom is not continuous, but quantized. It exists only in certain fixed amounts called *quanta.* **7.6** 312 m; 3.12×10^{11} nm; 3.12×10^{12} Å
7.8 2.4×10^{-23} J **7.10** b < c < a **7.13**(a) 1.24×10^{15} s^{-1}; 8.21×10^{-19} J (b) 1.4×10^{15} s^{-1}; 9.0×10^{-19} J **7.16**(a) absorption (b) emission (c) emission (d) absorption
7.18 434.17 nm **7.20** -2.76×10^5 J/mol **7.22** d < a < c < b
7.24 $n = 4$ **7.27** Macroscopic objects do exhibit a wavelike motion, but the wavelength is too small for humans to perceive.
7.29 7.6×10^{-37} m **7.31** 2.2×10^{-26} m/s **7.33** 3.75×10^{-36} kg
7.37(a) principal determinant of the electron's energy or distance from the nucleus (b) determines the shape of the orbital
(c) determines the orientation of the orbital in three-dimensional space **7.38**(a) one (b) five (c) three (d) nine **7.40**(a) m_l: $-2, -1, 0, +1, +2$ (b) m_l: 0 (if $n = 1$, then $l = 0$) (c) m_l: $-3, -2, -1, 0, +1, +2, +3$

7.42

Sublevel	Allowable m_l	No. of orbitals
(a) d ($l = 2$)	$-2, -1, 0, +1, +2$	5
(b) p ($l = 1$)	$-1, 0, +1$	3
(c) f ($l = 3$)	$-3, -2, -1, 0, +1, +2, +3$	7

7.44(a) $n = 5$ and $l = 0$; one orbital (b) $n = 3$ and $l = 1$; three orbitals (c) $n = 4$ and $l = 3$; seven orbitals **7.46**(a) no; $n = 2$, $l = 1$, $m_l = -1$; $n = 2$, $l = 0$, $m_l = 0$ (b) allowed (c) allowed
(d) no; $n = 5$, $l = 3$, $m_l = +3$; $n = 5$, $l = 2$, $m_l = 0$ **7.48**(a) $E = -(2.180 \times 10^{-18}$ J)$(1/n^2)$. This is identical with the expression from Bohr's theory. (b) 3.028×10^{-19} J (c) 656.1 nm
7.49(a) The attraction of the nucleus for the electrons must be overcome. (b) The electrons in silver are more tightly held by the nucleus. (c) silver (d) Once the electron is freed from the atom, its energy increases in proportion to the frequency of the light.

7.52 Li^{2+} **7.55**(a) Ba; 462 nm (b) 278 to 292 nm
7.57(a) 2.7×10^2 s (b) 3.6×10^8 m **7.60** 6.4×10^{27} photons
7.63(a) 7.56×10^{-18} J; 2.63×10^{-8} m (b) 5.110×10^{-17} J;
3.890×10^{-9} m (c) 1.22×10^{-18} J; 1.63×10^{-7} m
7.65(a) red; green (b) 1.18 kJ (Sr); 1.17 kJ (Ba)
7.67(a) This is the wavelength of maximum absorbance, so it
gives the highest sensitivity. (b) ultraviolet region
(c) 1.93×10^{-2} g vitamin A/g oil **7.71** 1.0×10^{18} photons/s

Chapter 8

8.1 Elements are listed in the periodic table in an ordered, systematic way that correlates with a periodicity of their chemical
and physical properties. The theoretical basis for the table in
terms of atomic number and electron configuration does
not allow for a "new element" between Sn and Sb.
8.3(a) predicted atomic mass = 54.23 amu (b) predicted melting point = 6.3°C (c) predicted boiling point = −60.2°C
8.5 The quantum number m_s relates to just the electron;
all the others describe the orbital.
8.8 Shielding occurs when inner electrons protect, or shield,
outer electrons from the full nuclear attraction. The effective
nuclear charge is the nuclear charge an electron actually
experiences. As the number of inner electrons increases,
the effective nuclear charge decreases.
8.10(a) 6 (b) 10 (c) 2 **8.12**(a) 6 (b) 2 (c) 14
8.15 Electrons will occupy empty orbitals in the same sublevel
before filling half-filled orbitals so there is the maximum number
of unpaired electrons with parallel spins. N: $1s^22s^22p^3$.

$$\boxed{\uparrow\downarrow}\ \boxed{\uparrow\downarrow}\ \boxed{\uparrow}\boxed{\uparrow}\boxed{\uparrow}$$
$$1s\quad 2s\qquad 2p$$

8.17 Main-group elements from the same group have similar
outer electron configurations, and the (old) group number equals
the number of outer electrons. Outer electron configurations vary
in a periodic manner within a period, with each succeeding element having an additional electron.
8.18(a) $n = 5$, $l = 0$, $m_l = 0$, and $m_s = +\frac{1}{2}$ (b) $n = 3$, $l = 1$,
$m_l = +1$, and $m_s = -\frac{1}{2}$ (c) $n = 5$, $l = 0$, $m_l = 0$, and $m_s = +\frac{1}{2}$
(d) $n = 2$, $l = 1$, $m_l = +1$, and $m_s = -\frac{1}{2}$
8.20(a) Rb: $1s^22s^22p^63s^23p^64s^23d^{10}4p^65s^1$
(b) Ge: $1s^22s^22p^63s^23p^64s^23d^{10}4p^2$
(c) Ar: $1s^22s^22p^63s^23p^6$
8.22(a) Ti: [Ar] $4s^23d^2$

$$\boxed{\uparrow\downarrow}\ \boxed{\uparrow}\boxed{\uparrow}\boxed{\ }\boxed{\ }\boxed{\ }\ \boxed{\ }\boxed{\ }\boxed{\ }$$
$$4s\qquad 3d\qquad\qquad 4p$$

(b) Cl: [Ne] $3s^23p^5$

$$\boxed{\uparrow\downarrow}\ \boxed{\uparrow\downarrow}\boxed{\uparrow\downarrow}\boxed{\uparrow}$$
$$3s\qquad 3p$$

(c) V: [Ar] $4s^23d^3$

$$\boxed{\uparrow\downarrow}\ \boxed{\uparrow}\boxed{\uparrow}\boxed{\uparrow}\boxed{\ }\boxed{\ }\ \boxed{\ }\boxed{\ }\boxed{\ }$$
$$4s\qquad 3d\qquad\qquad 4p$$

8.24(a) O; Group 6A(16); Period 2

$$\boxed{\uparrow\downarrow}\ \boxed{\uparrow\downarrow}\boxed{\uparrow}\boxed{\uparrow}$$
$$2s\qquad 2p$$

(b) P; Group 5A(15); Period 3

$$\boxed{\uparrow\downarrow}\ \boxed{\uparrow}\boxed{\uparrow}\boxed{\uparrow}$$
$$3s\qquad 3p$$

8.26(a) [Ar] $4s^23d^{10}4p^1$; Group 3A(13) (b) [He] $2s^22p^6$;
Group 8A(18)

8.28

	Inner Electrons	Outer Electrons	Valence Electrons
(a) O	2	6	6
(b) Sn	46	4	4
(c) Ca	18	2	2
(d) Fe	18	2	8
(e) Se	28	6	6

8.30(a) B; Al, Ga, In, and Tl (b) S; O, Se, Te, and Po (c) La;
Sc, Y, and Ac **8.33** Atomic size increases down a group. Ionization energy decreases down a group. These trends result because
the outer electrons are more easily removed as the atom gets larger.
8.35 For a given element, successive ionization energies always
increase. As each successive electron is removed, the positive
charge on the ion increases, which results in a stronger attraction
between the leaving electron and the ion. When a large jump
between successive ionization energies is observed, the subsequent
electron must come from a lower energy level. **8.37** A high IE_1
and a very negative EA_1 suggest that the elements are halogens, in
Group 7A(17), which form 1− ions. **8.39**(a) K < Rb < Cs
(b) O < C < Be (c) Cl < S < K (d) Mg < Ca < K
8.41(a) Ba < Sr < Ca (b) B < N < Ne (c) Rb < Se < Br
(d) Sn < Sb < As **8.43** $1s^22s^22p^1$ (boron, B) **8.45**(a) Na
(b) Na (c) Be **8.47** (1) Metals conduct electricity, nonmetals
do not. (2) When they form stable ions, metal ions tend to have a
positive charge, nonmetal ions tend to have a negative charge.
(3) Metal oxides are ionic and act as bases in water; nonmetal oxides are covalent and act as acids in water. **8.48** Metallic character increases down a group and decreases toward the right across
a period. These trends are the same as those for atomic size and
opposite those for ionization energy. **8.52**(a) Rb (b) Ra (c) I
8.54 acidic solution; $SO_2(g) + H_2O(l) \longrightarrow H_2SO_3(aq)$
8.56(a) Cl^-: $1s^22s^22p^63s^23p^6$ (b) Na^+: $1s^22s^22p^6$
(c) Ca^{2+}: $1s^22s^22p^63s^23p^6$ **8.58**(a) 0 (b) 3 (c) 0 (d) 1
8.60(a) V^{3+}, [Ar] $3d^2$, paramagnetic (b) Cd^{2+}, [Kr] $4d^{10}$, diamagnetic (c) Co^{3+}, [Ar] $3d^6$, paramagnetic (d) Ag^+, [Kr] $4d^{10}$,
diamagnetic **8.62** For palladium to be diamagnetic, all of its
electrons must be paired. (a) You might first write the condensed
electron configuration for Pd as [Kr] $5s^24d^8$. However, the partial
orbital diagram is not consistent with diamagnetism.

$$\boxed{\uparrow\downarrow}\ \boxed{\uparrow\downarrow}\boxed{\uparrow\downarrow}\boxed{\uparrow\downarrow}\boxed{\uparrow}\boxed{\uparrow}\ \boxed{\ }\boxed{\ }\boxed{\ }$$
$$5s\qquad 4d\qquad\qquad 5p$$

(b) This is the only configuration that supports diamagnetism,
[Kr] $4d^{10}$.

$$\boxed{\ }\ \boxed{\uparrow\downarrow}\boxed{\uparrow\downarrow}\boxed{\uparrow\downarrow}\boxed{\uparrow\downarrow}\boxed{\uparrow\downarrow}\ \boxed{\ }\boxed{\ }\boxed{\ }$$
$$5s\qquad 4d\qquad\qquad 5p$$

(c) Promoting an s electron into the d sublevel still leaves two
electrons unpaired.

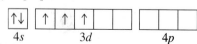
$$5s\qquad 4d\qquad\qquad 5p$$

8.64(a) $Li^+ < Na^+ < K^+$ (b) $Rb^+ < Br^- < Se^{2-}$ (c) $F^- <$
$O^{2-} < N^{3-}$ **8.67**(a) Cl_2O, dichlorine monoxide (b) Cl_2O_3,

dichlorine trioxide   (c) Cl_2O_5, dichlorine pentaoxide   (d) Cl_2O_7, dichlorine heptaoxide   (e) SO_3, sulfur trioxide   (f) SO_2, sulfur dioxide   (g) N_2O_5, dinitrogen pentaoxide   (h) N_2O_3, dinitrogen trioxide   (i) CO_2, carbon dioxide   (j) P_2O_5, diphosphorus pentaoxide   **8.69**(a) $SrBr_2$, strontium bromide   (b) CaS, calcium sulfide   (c) ZnF_2, zinc fluoride   (d) LiF, lithium fluoride   **8.70** All ions except Fe^{8+} and Fe^{14+} are paramagnetic; Fe^+ and Fe^{3+} would be most attracted.

Chapter 9

9.1(a) Greater ionization energy decreases metallic character. (b) Larger atomic radius increases metallic character.   (c) Higher number of outer electrons decreases metallic character.   (d) Larger effective nuclear charge decreases metallic character.
9.4(a) Cs   (b) Rb   (c) As   **9.6**(a) ionic   (b) covalent (c) metallic   **9.8**(a) Rb·   (b) ·Ṡi·   (c) :Ï·
9.10(a) 6A(16); [noble gas] ns^2np^4 (b) 3(A)13; [noble gas] ns^2np^1
9.13 Because the lattice energy is the result of electrostatic attractions between oppositely charged ions, its magnitude depends on several factors, including ionic size and ionic charge. For a particular arrangement of ions, the lattice energy increases as the charge on the ions increases and as their radii decrease.   **9.16**(a) Ba^{2+}, [Xe]; Cl^-, [Ne] $3s^23p^6$, :C̈l:⁻ ; $BaCl_2$ (b) Sr^{2+}, [Kr]; O^{2-}, [He] $2s^22p^6$, :Ö:²⁻; SrO   (c) Al^{3+}, [Ne]; F^-, [He] $2s^22p^6$, :F̈:⁻ ; AlF_3 (d) Rb^+, [Kr]; O^{2-}, [He] $2s^22p^6$, :Ö:²⁻; Rb_2O.
9.18(a) 3A(13)   (b) 2A(2)   (c) 6A(16)   **9.20**(a) BaS; Ba^{2+} is larger than Ca^{2+}.   (b) NaF; the charge on each ion is less than the charge on Mg and O.   **9.23** When two chlorine atoms are far apart, there is no interaction between them. As the atoms move closer together, the nucleus of each atom attracts the electrons of the other atom. The closer the atoms, the greater this attraction; however, the repulsions of the two nuclei and two electrons also increase at the same time. The final internuclear distance is the distance at which maximum attraction is achieved in spite of the repulsion.   **9.24** The bond energy is the energy required to break the bond between H atoms and Cl atoms in one mole of HCl molecules in the gaseous state. Energy is needed to break bonds, so bond energy is always endothermic and $\Delta H°_{\text{bond breaking}}$ is positive. The amount of energy needed to break the bond is released upon its formation, so $\Delta H°_{\text{bond forming}}$ has the same magnitude as $\Delta H°_{\text{bond breaking}}$ but is opposite in sign (always exothermic and negative).   **9.28**(a) I—I < Br—Br < Cl—Cl   (b) S—Br < S—Cl < S—H   (c) C—N < C=N < C≡N   **9.30**(a) C—O < C=O; the C=O bond (bond order = 2) is stronger than the C—O bond (bond order = 1).   (b) C—H < O—H; O is smaller than C so the O—H bond is shorter and stronger than the C—H bond.
9.33 Less energy is required to break weak bonds.   **9.35** Both are one-carbon molecules. Since methane contains fewer C—O bonds, it will have the greater heat of combustion per mole.   **9.37** −168 kJ
9.39 −22 kJ   **9.40** −59 kJ   **9.41** Electronegativity increases from left to right and increases from bottom to top within a group. Fluorine and oxygen are the two most electronegative elements. Cesium and francium are the two least electronegative elements.
9.43 The H—O bond in water is polar covalent. A nonpolar covalent bond occurs between two atoms with identical electronegativities. A polar covalent bond occurs when the atoms have differing electronegativities. Ionic bonds result from electron transfer between atoms.   **9.46**(a) Si < S < O   (b) Mg < As < P

9.48(a) N⃡—B   (b) N⃡—O   (c) C—S (none)
(d) S⃡—O   (e) N⃖—H   (f) Cl⃡—O
9.50 a, d, and e   **9.52**(a) nonpolar covalent   (b) ionic   (c) polar covalent   (d) polar covalent   (e) nonpolar covalent   (f) polar covalent; $SCl_2 < SF_2 < PF_3$
9.54(a) H⃡—I   <   H⃡—Br   <   H⃡—Cl

(b) H⃡—C   <   H⃡—O   <   H⃡—F

(c) S⃡—Cl   <   P⃡—Cl   <   Si—Cl

9.58(a) 800. kJ/mol, which is lower than the value in Table 9.2   (b) -2.417×10^4 kJ   (c) 1690. g CO_2 (d) 65.2 L O_2
9.59(a) −125 kJ   (b) yes, since $H°_f$ is negative   (c) −392 kJ (d) No, $H°_f$ for $MgCl_2$ is much more negative than that for MgCl.   **9.60**(a) 406 nm   (b) 2.93×10^{-19} J   (c) 1.87×10^4 m/s
9.63 C—Cl: 3.53×10^{-7} m; bond in O_2: 2.40×10^{-7} m
9.64 XeF_2: 132 kJ/mol; XeF_4: 150. kJ/mol; XeF_6: 146 kJ/mol
9.66(a) The presence of the very electronegative fluorine atoms bonded to one of the carbons makes the C—C bond polar. This polar bond will tend to undergo heterolytic rather than homolytic cleavage. More energy is required to achieve heterolytic cleavage.   (b) 1420 kJ
9.69 8.70×10^{14} s⁻¹; 3.45×10^{-7} m, which is in the ultraviolet region of the electromagnetic spectrum.   **9.71**(a) $CH_3OCH_3(g)$: −326 kJ; $CH_3CH_2OH(g)$: −369 kJ   (b) The formation of gaseous ethanol is more exothermic.   (c) 43 kJ

Chapter 10

10.1 He and H cannot serve as central atoms in a Lewis structure. Both can have no more than two valence electrons. Fluorine needs only one electron to complete its valence level, and it does not have d orbitals available to expand its valence level. Thus, it can bond to only one other atom.   **10.3** All the structures obey the octet rule except c and g.
10.5(a) :F̈—Si̇—F̈: (with F above and below)   (b) :C̈l—S̈e—C̈l: (c) :F̈—C—F̈: with =O below
10.7(a) :F̈—P̈—F̈: (with F below)   (b) H—Ö—C—Ö—H (with =O below)   (c) :S̈=C=S̈:
10.9(a) [Ö=N=Ö]⁺
(b) :F̈—N(=O)—Ö: ⟷ :F̈—N(=O)—Ö: (resonance)
10.11(a) [:N̈=N=N̈:]⁻ ⟷ [:N̈—N≡N:]⁻ ⟷ [:N≡N—N̈:]⁻
(b) [:Ö=N—Ö:]⁻ ⟷ [:Ö—N=Ö:]⁻
10.13(a) :F̈: with :F̈—I—F̈: and :F̈ ⟍ | ⟋ F̈:   formal charges: I = 0, F = 0

(b) [structure of AlH₄ with H—Al—H]   formal charges: H = 0, Al = −1

10.15(a) [structure $O=Br=O$ with $:O:$ below]   formal charges: Br = 0, doubly bonded O = 0, singly bonded O = −1   O.N.: Br = +5; O = −2

(b) [structure $:O—S—O:^{2-}$ with $:O:$ below]   formal charges: S = 0, singly bonded O = −1, doubly bonded O = 0   O.N.: S = +4; O = −2

10.17(a) BH_3 has 6 valence electrons.   (b) As has an expanded valence level with 10 electrons.   (c) Se has an expanded valence level with 10 electrons.

[structures (a) BH_3; (b) AsF_4^-; (c) $SeCl_2$... actually $SeCl_4$?]

(a)   (b)   (c)

10.19(a) Br expands its valence level to 10 electrons.   (b) I has an expanded valence level of 10 electrons.   (c) Be has only 4 valence shell electrons.

(a) $:F—Br—F:$ with $:F:$ below   (b) $[:Cl—I—Cl:]^-$ (c) $:F—Be—F:$

10.21

$:Cl—Be—Cl:$ + $[:Cl:]^-$ and $[:Cl:]^-$ → $[:Cl—Be—Cl:]^{2-}$ with $:Cl:$ top and $:Cl:$ bottom

10.24 structure A   **10.26** The molecular shape and the electron-group arrangement are the same when no lone pairs are present on the central atom.   **10.28** tetrahedral, AX_4; trigonal pyramidal, AX_3E; bent or V shaped, AX_2E_2

10.30(a) [bent $X—A—X$]   (b) [trigonal planar X—A—X]   (c) [trigonal pyramidal X—A with X's]   (d) [X—A tetrahedral]   (e) [A with X's]   (f) [trigonal bipyramidal A with X's]

10.32(a) trigonal planar, bent, 120°   (b) tetrahedral, trigonal pyramidal, 109.5°   (c) tetrahedral, trigonal pyramidal, 109.5°   **10.34**(a) trigonal planar, trigonal planar, 120°   (b) trigonal planar, bent, 120°   (c) tetrahedral, tetrahedral, 109.5°   **10.36**(a) trigonal planar, AX_3, 120°   (b) trigonal pyramidal, AX_3E, 109.5°   (c) trigonal bipyramidal, AX_5, 90° and 120°   **10.38**(a) bent, 109.5°, less than 109.5°   (b) trigonal bipyramidal, 90° and 120°, angles are ideal   (c) see-saw, 90° and 120°, less than ideal   (d) linear, 180°, angle is ideal   **10.40**(a) C: tetrahedral, 109.5°; O: bent, < 109.5°   (b) N: trigonal planar, 120°   **10.42**(a) C in CH_3: tetrahedral, 109.5°; C with C=O: trigonal planar, 120°; O with H: bent, < 109.5°   (b) O: bent, < 109.5°

10.44 $OF_2 < NF_3 < CF_4 < BF_3 < BeF_2$
10.46(a) The C and N each have three groups so the ideal angles are 120°, and the O has four groups so the ideal angle is 109.5°. The N and O have lone pairs so the angles are less than ideal. (b) All central atoms have four pairs, so the ideal angles are 109.5°. The lone pairs on the O reduce this value. (c) The B has three groups and an ideal bond angle of 120°. All the O's have four groups (ideal bond angles of 109.5°), two of which are lone pairs that reduce the angle.
10.49

[three structures: PCl_5 trigonal bipyramidal; PCl_4^+ tetrahedral; PCl_6^- octahedral]

In the gas phase, PCl_5 is AX_5, so the shape is trigonal bipyramidal, and the bond angles are 120° and 90°. The PCl_4^+ ion is AX_4, so the shape is tetrahedral, and the bond angles are 109.5°. The PCl_6^- ion is AX_6, so the shape is octahedral, and the bond angles are 90°.   **10.51**(a) CF_4 (b) BrCl and SCl_2
10.53(a) SO_2, because it is polar and SO_3 is not.   (b) IF has a greater electronegativity difference between its atoms.   (c) SF_4, because it is polar and SiF_4 is not.   (d) H_2O has a greater electronegativity difference between its atoms.

10.55

[three structures X, Y, Z of $C_2H_2Cl_2$ isomers]

X   Y   Z

Compound Y has a dipole moment

10.56 $H—N—N—H$ with H's (Hydrazine) $H—N=N—H$ (Diazene) $:N≡N:$ (Nitrogen)

(a) The single N—N bond (bond order = 1) is weaker and longer than the others. The triple bond (bond order = 3) is stronger and shorter than the others. The double bond (bond order = 2) has an intermediate strength and length.
(b) $\Delta H^\circ_{rxn} = −367$ kJ

$H—N—N=N—N—H$ with H's → $H—N—N—H$ with H's + $:N≡N:$

10.57(a) formal charges: Al = −1, end Cl = 0, bridging Cl = +1; I = −1, end Cl = 0, bridging Cl = +1   (b) The iodine atoms are each AX_4E_2 and the shape around each is square planar. Placing these square planar portions adjacent gives a planar molecule.   **10.65**(a) −1267 kJ/mol   (b) −1226 kJ/mol (c) −1234.8 kJ/mol. The two answers differ by less than 10 kJ/mol. This is very good agreement since average bond energies were used to calculate answers a and b.   (d) −37 kJ

10.67 $H—O—C—C—O—H$ with $:O:$ and $:O:$ below (double bonds)

10.68 (a) The OH species only has 7 valence electrons, which is less than an octet, and 1 electron is unpaired.   (b) 426 kJ (c) 508 kJ   **10.70**(a) The F atoms will substitute at the axial positions first.   (b) PF_5 and PCl_3F_2 **10.73** 22 kJ
10.75 Trigonal planar molecules are nonpolar, so AY_3 cannot be that shape. Trigonal pyramidal molecules and T-shaped molecules are polar, so either could represent AY_3.

Chapter 11

11.1(a) sp^2 (b) sp^3d^2 (c) sp (d) sp^3 (e) sp^3d

11.3 C has only $2s$ and $2p$ atomic orbitals, allowing for a maximum of four hybrid orbitals. Si has $3s$, $3p$, and $3d$ atomic orbitals, allowing it to form more than four hybrid orbitals.

11.5(a) six, sp^3d^2 (b) four, sp^3 **11.7**(a) sp^2 (b) sp^2 (c) sp^2

11.9 (a) sp^3 (b) sp^3 (c) sp^3 **11.11**(a) Si: one s and three p atomic orbitals form sp^3 hybrid orbitals. (b) C: one s and one p atomic orbitals form sp hybrid orbitals.   **11.13**(a) S: one s, three p, and one d atomic orbital mix to form sp^3d hybrid orbitals. (b) N: one s and three p atomic orbitals mix to form sp^3 hybrid orbitals.

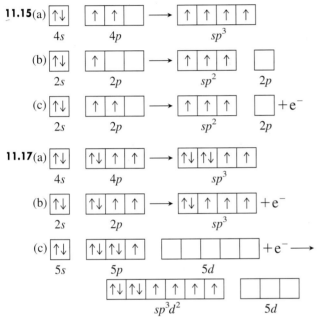

11.20(a) False. A double bond is one σ and one π bond. (b) False. A triple bond consists of one σ and two π bonds. (c) True   (d) True   (e) False. A π bond consists of a second pair of electrons after a σ bond has been previously formed. (f) False. End-to-end overlap results in a bond with electron density along the bond axis.   **11.21**(a) Nitrogen sp^2 with three σ bonds and one π bond   (b) Carbon sp with two σ bonds and two π bonds   (c) Carbon sp^2 with three σ bonds and one π bond

11.23(a) N: sp^2, forming 2 σ bonds and 1 π bond

:F̈—N̈=Ö:

(b) C: sp^2, forming 3 σ bonds and 1 π bond

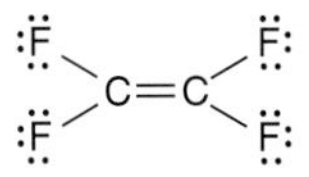

:F̈   F̈:
  \   /
   C=C
  /   \
:F̈   F̈:

(c) C: sp, forming 2 σ bonds and 2 π bonds

:N≡C—C≡N:

11.25 Four molecular orbitals form from the four p atomic orbitals. The total number of molecular orbitals must equal the number of atomic orbitals.   **11.27**(a) Bonding MOs have lower energy than antibonding MOs. Lower energy = more stable (b) Bonding MOs do not have a nodal plane perpendicular to the bond.   (c) Bonding MOs have higher electron density

between nuclei than antibonding MOs.   **11.29**(a) two (b) two   (c) four

11.31

a)   bonding $s + p$

antibonding $s - p$

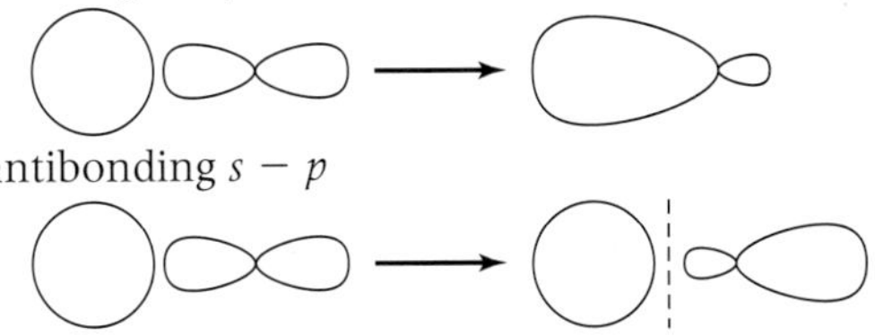

b)   bonding $p + p$

antibonding $p - p$

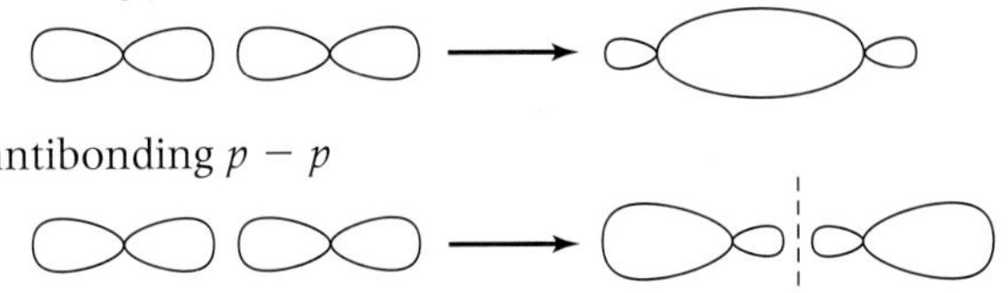

11.33(a) stable   (b) paramagnetic   (c) $(\sigma_{2s})^2(\sigma_{2s}^*)^1$

11.35(a) $C_2{}^+ < C_2 < C_2{}^-$ (b) $C_2{}^- < C_2 < C_2{}^+$ **11.39**(a) C (ring): sp^2; C (all others): sp^3; O (all): sp^3; N: sp^3 (b) 26 (c) 6   **11.41**(a) 17   (b) all carbons sp^2, the ring N is sp^2, the other N's are sp^3 **11.43**(a) B changes from sp^2 to sp^3.   (b) P changes from sp^3 to sp^3d.   (c) C changes from sp to sp^2. Two electron groups surround C in C_2H_2 and three electron groups surround C in C_2H_4.   (d) Si changes from sp^3 to sp^3d^2.   (e) no change for S   **11.44** P: tetrahedral, sp^3; N: trigonal pyramid, sp^3; C_1 and C_2: tetrahedral, sp^3; C_3: trigonal planar, sp^2 **11.48** The central C is sp hybridized, and the other two C atoms are sp^2 hybridized.

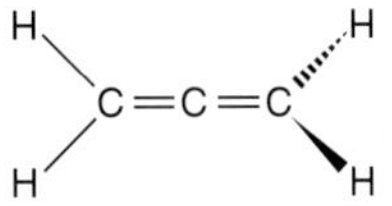

H   H
 \   ⋰
  C=C=C
 /   ↘
H   H

11.53(a) C in —CH_3: sp^3; all other C atoms: sp^2; O in two C—O bonds: sp^3; O in two C=O bonds: sp^2 (b) two (c) eight; one

Chapter 12

12.1 In a solid, the energy of attraction of the particles is greater than their energy of motion; in a gas, it is less. Gases have high compressibility and the ability to flow, while solids have neither.   **12.4**(a) Because the intermolecular forces are only partially overcome when fusion occurs but need to be totally overcome in vaporization.   (b) Because solids have greater intermolecular forces than liquids do.   (c) $\Delta H_{vap} = -\Delta H_{cond}$ **12.5**(a) condensation   (b) fusion   (c) vaporization   **12.7** The gas molecules slow down as the gas is compressed. Therefore, much of the kinetic energy lost by the propane molecules is released to the surroundings.   **12.11** At first, the vaporization of liquid molecules from the surface predominates, which increases the number of gas molecules and hence the vapor pressure. As more molecules enter the gas phase, gas molecules hit the surface of the liquid and "stick" more frequently, so the condensation rate increases. When the vaporization and condensation rates become equal, ' vapor pressure becomes constant.   **12.14** 4.16×10^3 J   **12.16** 0.692 atm

12.18

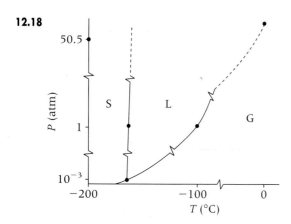

Solid ethylene is more dense than liquid ethylene.
12.20 42 atm   **12.24** O is smaller and more electronegative than Se; so the electron density on O is greater, which attracts H more strongly.   **12.25** All particles (atoms and molecules) exhibit dispersion forces, but the total force is weak in small molecules. Dipole–dipole forces in small polar molecules dominate the dispersion forces.   **12.28**(a) hydrogen bonding   (b) dispersion forces   (c) dispersion forces   **12.30**(a) dipole–dipole forces (b) dispersion forces   (c) hydrogen bonding

12.32(a)

$$CH_3-CH-\overset{\cdots}{\underset{|}{\underset{CH_3}{O}}}:\cdots\cdot H-\overset{\cdots}{\underset{|}{\underset{H}{O}}}-\overset{|}{\underset{CH_3}{CH}}-CH_3$$

(b)

$$H-\overset{\cdots}{\underset{}{F}}:\cdots\cdot H-\overset{\cdots}{\underset{}{F}}:\cdots\cdot H-\overset{\cdots}{\underset{}{F}}:$$

12.34(a) I^- (b) $CH_2{=}CH_2$ (c) H_2Se. In (a) and (c) the larger particle has the higher polarizability. In (b), the less tightly held π electron clouds are more easily distorted.   **12.36**(a) C_2H_6; it is a smaller molecule exhibiting weaker dispersion forces than C_4H_{10}.   (b) CH_3CH_2F; it has no H—F bonds, so it only exhibits dipole–dipole forces, which are weaker than the hydrogen bonds of CH_3CH_2OH.   (c) PH_3; it has weaker intermolecular forces (dipole–dipole) than NH_3 (hydrogen bonding).   **12.38**(a) Lithium chloride; it has ionic bonds versus dipole–dipole forces in HCl. (b) NH_3; it exhibits hydrogen bonding versus dipole–dipole forces in PH_3.   (c) I_2; its molecules are more polarizable than Xe atoms because of their larger size.   **12.40**(a) C_4H_8 (cyclobutane), because it is more compact than C_4H_{10}.   (b) PBr_3; the dipole–dipole forces in PBr_3 are weaker than the ionic bonds in NaBr.   (c) HBr; the dipole–dipole forces in HBr are weaker than the hydrogen bonds in water.   **12.45** The cohesive forces in water and mercury are stronger than the adhesive forces to the nonpolar wax on the floor. Weak adhesive forces result in spherical drops. The adhesive forces overcome the even weaker ·ohesive forces in the oil and so the oil drop spreads out.
·7 $CH_3CH_2CH_2OH < HOCH_2CH_2OH <$
$\,$·CH(OH)CH_2OH. More hydrogen bonding means more ·· between molecules so more energy is needed to ··ace area.   **12.51** Water is a good solvent for polar ···ances and a poor solvent for nonpolar substances. ··molecule and dissolves polar substances because ···ar forces are of similar strength.
···er molecule can form four H bonds. The two ···h form one H bond to oxygen atoms on

neighboring water molecules. The two lone pairs on the oxygen atom form H bonds with two hydrogen atoms on neighboring molecules.   **12.54** Water exhibits strong capillary action, which allows it to be easily absorbed by the narrow spaces in the plant's roots and transported upward to the leaves.   **12.60** A solid metal is a shiny solid that conducts heat, is malleable, and melts at high temperatures. (Other answers include relatively high boiling point and good conductor of electricity.)   **12.63** The energy gap is the energy difference between the highest filled energy level (valence band) and the lowest unfilled energy level (conduction band). In conductors and superconductors, the energy gap is zero because the valence band overlaps the conduction band. In semiconductors, the energy gap is small. In insulators, the gap is large.   **12.64**(a) face-centered cubic   (b) body-centered cubic (c) face-centered cubic   **12.66**(a) Tin, a metal, forms a metallic solid.   (b) Silicon is in the same group as carbon, so it forms network covalent bonding.   (c) Xenon is monatomic and forms an atomic solid.   **12.68** four   **12.70**(a) four Se^{2-} ions, four Zn^{2+} ions   (b) 577.48 amu   (c) $1.77{\times}10^{-22}\,cm^3$ (d) $5.61{\times}10^{-8}\,cm$
12.72(a) insulator   (b) conductor   (c) semiconductor
12.73(a) Conductivity increases.   (b) Conductivity increases. (c) Conductivity decreases.   **12.77** 259 K   **12.79**(a) simple (b) $3.99\,g/cm^3$
12.81(a) furfuryl alcohol

2-furoic acid

(b) furfuryl alcohol   2-furoic acid

12.85(a) $4r$ (b) $\sqrt{2}a$ (c) $a = 4r/\sqrt{3}$ (d) two   (e) 0.68017

Chapter 13

13.2 When a salt such as NaCl dissolves, ion-dipole forces cause the ions to separate, and many water molecules cluster around each ion in hydration shells.   **13.4**(a) KNO_3 is an ionic compound and is therefore more soluble in water.   **13.6**(a) ion–dipole forces   (b) hydrogen bonding   (c) dipole–induced dipole forces **13.8**(a) hydrogen bonding   (b) dipole–induced dipole forces (c) dispersion forces   **13.10**(a) HCl(g), because the molecular interactions (dipole–dipole forces) in ether are like those in HCl but not like the ionic bonding in NaCl.   (b) $CH_3CHO(l)$, because the molecular interactions with ether (dipole–dipole) can replace those between CH_3CHO, but not the H bonds in water. (c) $CH_3CH_2MgBr(s)$, because the molecular interactions (dispersion forces) are greater than between ether and the ions in $MgBr_2$.   **13.12** Gluconic acid is soluble in water due to extensive hydrogen bonding from its —OH groups attached to five of its carbons. The dispersion forces in the nonpolar tail of caproic acid are more similar to the dispersion forces in hexane; thus, caproic acid is soluble in hexane.   **13.17** Very soluble because a decrease in enthalpy and an increase in entropy both favor the formation of a solution.

13.18

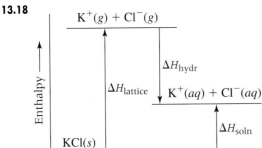

13.20(a) The volume of Na^+ is smaller, so it has the greater charge density.   (b) Sr^{2+} has a larger ionic charge and a smaller volume, so it has the greater charge density.   (c) Na^+ is smaller than Cl^-, so it has the greater charge density.   (d) O^{2-} has a larger ionic charge with a similar ion volume, so it has the greater charge density.   (e) OH^- has a smaller volume than SH^-, so it has the greater charge density.   **13.22**(a) Na^+ (b) Sr^{2+} (c) Na^+ (d) O^{2-} (e) OH^- **13.24**(a) -704 kJ/mol   (b) The K^+ ion contributes more because it is smaller and, therefore, has a greater charge density.   **13.26**(a) increases   (b) decreases   (c) increases   **13.29** Add a pinch of the solid solute to each solution. Addition of a "seed" crystal of solute to a supersaturated solution causes the excess solute to crystallize immediately, leaving behind a saturated solution. The solution in which the added solid solute dissolves is the unsaturated solution. The solution in which the added solid solute remains undissolved is the saturated solution.   **13.31**(a) increase   (b) decrease   **13.33**(a) 0.0819 g O_2 (b) 0.0171 g O_2 **13.36** 0.20 mol/L   **13.39** With just this information, you can convert between molality and molarity, but you need to know the molar mass of the solvent to convert to mole fraction.   **13.41**(a) 1.24 M $C_{12}H_{22}O_{11}$ (b) 0.158 M $LiNO_3$ **13.43**(a) 0.0750 M NaOH   (b) 0.31 M HNO_3 **13.45**(a) Add 4.22 g KH_2PO_4 to enough water to make 355 mL of aqueous solution.   (b) Add 107 mL of 1.25 M NaOH to enough water to make 425 mL of solution.   **13.47**(a) 0.942 m glycine   (b) 1.29 m glycerol   **13.49** 3.09 m C_6H_6 **13.51**(a) Add 2.13 g $C_2H_6O_2$ to 298 g H_2O.   (b) Add 0.0323 kg of 62.0% (w/w) HNO_3 to 0.968 kg H_2O to make 1.000 kg of 2.00% (w/w) HNO_3.   **13.53**(a) 0.27   (b) 56 mass %   (c) 21 m C_3H_7OH **13.55** 5.11 m NH_3; 4.53 M NH_3; mole fraction = 0.0843   **13.57** 2.2 ppm Ca^{2+}, 0.66 ppm Mg^{2+} **13.60** The boiling point is higher and the freezing point is lower for the solution compared to the solvent.   **13.64**(a) strong electrolyte   (b) strong electrolyte   (c) nonelectrolyte   (d) weak electrolyte   **13.66**(a) 0.4 mol of solute particles   (b) 0.14 mol   (c) 3×10^{-4} mol   (d) 0.07 mol   **13.68**(a) CH_3OH in H_2O (b) H_2O in CH_3OH solution   **13.70**(a) $\Pi_I = \Pi_{II} < \Pi_{III}$ (b) $bp_I = bp_{II} < bp_{III}$ (c) $fp_{III} < fp_I = fp_{II}$ (d) $vp_{III} < vp_I = vp_{II}$ **13.72** 23.36 torr   **13.74** $-0.206°C$ **13.76** 79.0°C   **13.78** 1.13×10^4 g $C_2H_6O_2$ **13.80**(a) NaCl: 0.173 m and $i = 1.84$ (b) CH_3COOH: 0.0837 m and $i = 1.02$ **13.83** 211 torr for CH_2Cl_2; 47.2 torr for CCl_4 **13.85** 3.5×10^9 L

13.90(a) 89.9 g/mol   (b) C_2H_5O; $C_4H_{10}O_2$.
(c) Forms H bonds

$$HO-\overset{\overset{H}{|}}{\underset{\underset{H}{|}}{C}}-\overset{\overset{H}{|}}{\underset{\underset{H}{|}}{C}}-\overset{\overset{H}{|}}{\underset{\underset{H}{|}}{C}}-\overset{\overset{H}{|}}{\underset{\underset{H}{|}}{C}}-OH$$

Does not form H bonds

$$H-\overset{\overset{H}{|}}{\underset{\underset{H}{|}}{C}}-O-\overset{\overset{H}{|}}{\underset{\underset{H}{|}}{C}}-O-\overset{\overset{H}{|}}{\underset{\underset{H}{|}}{C}}-\overset{\overset{H}{|}}{\underset{\underset{H}{|}}{C}}-H$$

13.92(a) 68 g/mol   (b) 2.1×10^2 g/mol   (c) The molar mass of CaN_2O_6 is 164.10 g/mol. This value is less than the 2.1×10^2 g/mol calculated when the compound is assumed to be a strong electrolyte and is greater than the 68 g/mol calculated when the compound is assumed to be a nonelectrolyte. Thus, the compound forms a non-ideal solution because the ions interact but do not dissociate completely in solution.   (d) 2.4   **13.95**(a) 1.82×10^4 g/mol   (b) $3.41\times10^{-5}°C$ **13.101**(a) CH_4N_2O. The empirical formula mass is 60.06 g/mol.   (b) 60. g/mol; CH_4N_2O.

Chapter 14

14.1 The outermost electron is attracted by a smaller effective nuclear charge in Li because of shielding by the inner electrons, and it is farther from the nucleus in Li. Both of these factors lead to a lower ionization energy.
14.2(a) $2Al(s) + 6HCl(aq) \longrightarrow 2AlCl_3(aq) + 3H_2(g)$
   (b) $LiH(s) + H_2O(l) \longrightarrow LiOH(aq) + H_2(g)$
14.4(a) $NaBH_4$: +1 for Na, +3 for B, -1 for H
   $Al(BH_4)_3$: +3 for Al, +3 for B, -1 for H
   $LiAlH_4$: +1 for Li, +3 for Al, -1 for H
   (b) tetrahedral

$$\left[\,H-\overset{\overset{H}{|}}{\underset{\underset{H}{|}}{B}}-H\,\right]^{-}$$

14.7(a) reducing agent   (b) Alkali metals have relatively low ionization energies, which means they easily lose the outermost electron.
(c) $2Na(s) + 2H_2O(l) \longrightarrow 2Na^+(aq) + 2OH^-(aq) + H_2(g)$
   $2Na(s) + Cl_2(g) \longrightarrow 2NaCl(s)$
14.9 Density and ionic size increase down a group; the other three properties decrease down a group.   **14.11** $2Na(s) + O_2(g) \longrightarrow Na_2O_2(s)$ **14.13** $K_2CO_3(s) + 2HI(aq) \longrightarrow 2KI(aq) + H_2O(l) + CO_2(g)$ **14.17** Group 2A(2) metals have an additional bonding electron to increase the strength of metallic bonding, which leads to higher melting points, higher boiling points, greater hardness, and greater density.   **14.18**(a) $CaO(s) + H_2O(l) \longrightarrow Ca(OH)_2(s)$ (b) $2Ca(s) + O_2(g) \longrightarrow 2CaO(s)$ **14.20**(a) $BeO(s) + H_2O(l) \longrightarrow$ no reaction   (b) $BeCl_2(l) + 2Cl^-(solvated) \longrightarrow BeCl_4{}^{2-}(solvated)$. This behaves like other Group 2A(2) elements.   **14.22** For Groups 1A(1) to 4A(14), the number of covalent bonds equals the (old) group number. For Groups 5A(15) to 7A(17), it equals 8 minus the (old) group number. There are exceptions in Period 3 to Period 6 because it is possible for the 3A(13) to 7A(17) elements to use d orbitals and form more bonds.   **14.25** The electron removed from Group 2A(2) atoms occupies the outer s orbital, whereas in

Group 3A(13) atoms, the electron occupies the outer p orbital. For example, the electron configuration for Be is $1s^2 2s^2$ and for B it is $1s^2 2s^2 2p^1$. It is easier to remove the p electron of B than an s electron of Be, because the energy of a p orbital is higher than that of the s orbital of the same level. Even though atomic size decreases because of increasing Z_{eff}, IE decreases from 2A(2) to 3A(13). **14.26**(a) Most atoms form stable compounds when they complete their outer shell (octet). Some compounds of Group 3A(13) elements, like boron, have only six electrons around the central atom. Having fewer than eight electrons is called an "electron deficiency."
(b) $BF_3(g) + NH_3(g) \longrightarrow F_3B{-}NH_3(g)$
$\quad B(OH)_3(aq) + OH^-(aq) \longrightarrow B(OH)_4^-(aq)$
14.28 $In_2O_3 < Ga_2O_3 < Al_2O_3$ **14.30** Apparent O.N., +3; actual O.N., +1. The anion I_3^- has the general formula AX_2E_3 and bond angles of 180°. $(Tl^{3+})(I^-)_3$ does not exist because of the low strength of the Tl—I bond. $\left[:\ddot{I}{-}\ddot{I}{-}\ddot{I}: \right]^-$

14.35 In general, network solids have very high melting and boiling points and are very hard, while molecular solids have low melting and boiling points and are soft. The properties of network solids reflect the necessity of breaking covalent bonds throughout the substances, whereas the properties of molecular solids reflect the weaker intermolecular forces between individual molecules.
14.36 Basicity in water is greater for the oxide of a metal. Tin(IV) oxide is more basic in water than carbon dioxide because tin has more metallic character than carbon. **14.38**(a) Ionization energy generally decreases down a group. (b) The deviations (increases) from the expected trend are due to the presence of the first transition series between Si and Ge and of the lanthanides between Sn and Pb. (c) Group 3A(13) **14.41** Atomic size increases down a group. As atomic size increases, ionization energy decreases and so it is easier to form a positive ion. An atom that is easier to ionize exhibits greater metallic character.
14.43(a)

$$\left[\begin{array}{c} :\ddot{O}: \quad\quad :\ddot{O}: \\ :\ddot{O}{-}Si{-}\ddot{O}{-}Si{-}\ddot{O}: \\ :\ddot{O}: \quad\quad :\ddot{O}: \\ :\ddot{O}{-}Si{-}\ddot{O}{-}Si{-}\ddot{O}: \\ :\ddot{O}: \quad\quad :\ddot{O}: \end{array} \right]^{8-}$$

(b)
$$\begin{array}{c} H \quad H \\ | \quad | \\ H{-}C{-}C{-}H \\ | \quad | \\ H{-}C{-}C{-}H \\ | \quad | \\ H \quad H \end{array}$$

14.46(a) diamond, C (b) calcium carbonate, $CaCO_3$ (c) carbon dioxide, CO_2 (d) carbon monoxide, CO (e) lead, Pb
14.50(a) -3 to $+5$ (b) For a group of nonmetals, the oxidation states range from the lowest, group number -8, or $5-8 = -3$ for Group 5A, to the highest, equal to the group number, or $+5$ for Group 5A. **14.51**(a) The greater the electronegativity of the element, the more covalent the bonding is in its oxide. (b) The more electronegative the element, the more acidic the oxide is.
14.54(a) $4As(s) + 5O_2(g) \longrightarrow 2As_2O_5(s)$ (b) $2Bi(s) + 3F_2(g) \longrightarrow 2BiF_3(s)$ (c) $Ca_3As_2(s) + 6H_2O(l) \longrightarrow 3Ca(OH)_2(s) + 2AsH_3(g)$ **14.60**(a) Boiling point and conductivity vary in similar ways down both groups. (b) Degree of metallic character and types of bonding vary in similar ways down both groups. (c) Both P and S have allotropes, and both bond covalently with almost every other nonmetal. (d) Both N and O are diatomic gases at normal temperatures and pressures. (e) O_2 is a reactive gas, whereas N_2 is not. Nitrogen can have any of six oxidation states, whereas oxygen has two.

14.62(a) $NaHSO_4(aq) + NaOH(aq) \longrightarrow Na_2SO_4(aq) + H_2O(l)$
(b) $S_8(s) + 24F_2(g) \longrightarrow 8SF_6(g)$
(c) $FeS(s) + 2HCl(aq) \longrightarrow H_2S(g) + FeCl_2(aq)$
14.64(a) acidic (b) acidic (c) basic (d) amphoteric
(e) basic **14.66**(a) O_3, ozone (b) SO_3, sulfur trioxide
(c) SO_2, sulfur dioxide **14.67** $S_2F_{10}(g) \longrightarrow SF_4(g) + SF_6(g)$; O.N. of S in S_2F_{10} is +5; O.N. of S in SF_4 is +4; O.N. of S in SF_6 is +6. **14.68**(a) Polarity is the molecular property that is responsible for the difference in boiling points between iodine monochloride (polar) and bromine (nonpolar). It arises from different EN values of the bonded atoms. (b) The boiling point of polar ICl is higher than the boiling point of Br_2. **14.70**(a) -1, $+1$, $+3$, $+5$, $+7$ (b) The electron configuration for Cl is [Ne] $3s^2 3p^5$. By gaining one electron, Cl achieves an octet. By forming covalent bonds, Cl completes or expands its valence level by maintaining electrons pairs in bonds or as lone pairs. (c) Fluorine has only the -1 oxidation state because its small size and absence of d orbitals prevent it from forming more than one covalent bond. **14.71**(a) Cl—Cl bond is stronger than Br—Br bond. (b) Br—Br bond is stronger than I—I bond.
(c) Cl—Cl bond is stronger than F—F bond. The fluorine atoms are so small that electron-electron repulsion of the lone pairs decreases the strength of the bond.
14.72(a) $3Br_2(l) + 6OH^-(aq) \longrightarrow$
$\quad\quad\quad\quad\quad 5Br^-(aq) + BrO_3^-(aq) + 3H_2O(l)$
14.73(a) $I_2(s) + H_2O(l) \longrightarrow HI(aq) + HIO(aq)$
(b) $Br_2(l) + 2I^-(aq) \longrightarrow I_2(s) + 2Br^-(aq)$
(c) $CaF_2(s) + H_2SO_4(l) \longrightarrow CaSO_4(s) + 2HF(g)$
14.76 helium; argon **14.77** Only dispersion forces hold atoms of noble gases together. **14.79** (a) Second ionization energies for alkali metals are so high because the electron being removed is from the next lower energy level and these are very tightly held by the nucleus.
(b) $2CsF_2(s) \longrightarrow 2CsF(s) + F_2(g)$; -405 kJ/mol.
14.81(a) hyponitrous acid, $H_2N_2O_2$; nitroxyl, HNO
(b) $H{-}\ddot{O}{-}\ddot{N}{=}\ddot{N}{-}\ddot{O}{-}H$ $\quad\quad H{-}\ddot{N}{=}\ddot{O}:$
(c) In both species the shape is bent about the N atoms.
(d)
$$\left[\begin{array}{c} \ddot{N}{=}N \\ / \quad \backslash \\ :\ddot{O}: \qu\quad :\ddot{O}: \end{array} \right]^{2-} \qu\quad \left[\begin{array}{c} :\ddot{O}: \\ | \\ \ddot{N}{=}N: \\ | \\ :\ddot{O}: \end{array} \right]^{2-}$$

14.84 In a disproportionation reaction, a substance acts as both a reducing agent and an oxidizing agent because atoms of an element within the substance attain both higher and lower oxidation states in the products. The disproportionation reactions are b, c, d, e, and f. **14.85**(a) Group 5A(15) (b) Group 7A(17) (c) Group 6A(16) (d) Group 1A(1) (e) Group 3A(13) (f) Group 8A(18) **14.86** 117.2 kJ
14.87
$$\left[\begin{array}{c} :\ddot{Cl}: \quad\quad :\ddot{Cl}: \\ :\ddot{Cl}{-}Al{-}\ddot{Cl}{-}Al{-}\ddot{Cl}: \\ :\ddot{Cl}: \quad\quad :\ddot{Cl}: \end{array} \right]^{-}$$

(a) sp^3 orbitals (b) tetrahedral (c) Since the ion is linear, the central Cl atom must be sp hybridized. (d) The sp hybridization means there are no lone pairs on the central Cl atom. Instead, the extra four electrons interact with the empty d orbitals on the

Al atoms to form double bonds between the chlorine and each aluminum atom.

14.88 $[\ddot{\text{O}}=\ddot{\text{N}}-\ddot{\text{O}}:]^- \longleftrightarrow [:\ddot{\text{O}}-\ddot{\text{N}}=\ddot{\text{O}}]^-$

$\ddot{\text{O}}=\ddot{\text{N}}-\ddot{\text{O}}: \longleftrightarrow :\ddot{\text{O}}-\ddot{\text{N}}=\ddot{\text{O}}$

$[\ddot{\text{O}}=\text{N}=\ddot{\text{O}}]^+$

The nitronium ion (NO_2^+) has a linear shape because the central N atom has two surrounding electron groups, which achieve maximum repulsion at 180°. The nitrite ion (NO_2^-) bond angle is more compressed than the nitrogen dioxide (NO_2) bond angle because the lone pair of electrons takes up more space than the lone electron.   **14.90**(a) 39.96 mass % in $CuHAsO_3$; As, 62.42 mass % in $(CH_3)_3As$ (b) 0.35 g $CuHAsO_3$

Chapter 15

15.1(a) Carbon's electronegativity is midway between the most metallic and nonmetallic elements of Period 2. To attain a filled outer level, carbon forms covalent bonds to other atoms in molecules, network covalent solids, and polyatomic ions. (b) Since carbon has four valence shell electrons, it forms four covalent bonds to attain an octet.   (c) To reach the He electron configuration, a carbon atom must lose four electrons, requiring too much energy to form the C^{4+} cation. To reach the Ne electron configuration, the carbon atom must gain four electrons, also requiring too much energy to form the C^{4-} anion.   (d) Carbon is able to bond to itself extensively because its small size allows for close approach and great orbital overlap. The extensive orbital overlap results in a strong, stable bond.   (e) The C—C bond is short enough to allow sideways overlap of unhybridized p orbitals of neighboring C atoms. The sideways overlap of p orbitals results in the π bonds that are part of double and triple bonds.   **15.2**(a) C, H, O, N, P, S, and halogens   (b) Heteroatoms are atoms of any element other than carbon and hydrogen. (c) More electronegative than C: N, O, F, Cl, and Br; less electronegative than C: H and P. Sulfur and iodine have the same electronegativity as carbon.   (d) Since carbon can bond to a wide variety of heteroatoms and to carbon atoms, it can form many different compounds.   **15.4** The C—H and C—C bonds are unreactive because electron density is shared equally between the two atoms. The C—I bond is somewhat reactive because it is long and weak. The C=O bond is reactive because oxygen is more electronegative than carbon and the electron-rich π bond makes it attract electron-poor atoms. The C—Li bond is also reactive because the bond polarity results in an electron-rich region around carbon and an electron-poor region around lithium.

15.5(a) An alkane and a cycloalkane are organic compounds that consist of carbon and hydrogen and have only single bonds. A cycloalkane has a ring of carbon atoms. An alkene is a hydrocarbon with at least one double bond. An alkyne is a hydrocarbon with at least one triple bond.   (b) alkane = C_nH_{2n+2}, cycloalkane = C_nH_{2n}, alkene = C_nH_{2n}, alkyne = C_nH_{2n-2} (c) Alkanes and cycloalkanes are saturated hydrocarbons.

15.8(a), (c), and (f)

15.9 (a)
```
C=C-C-C-C-C        C=C-C-C-C-C
    |                  |
    C                  C

C=C-C-C-C-C        C=C-C-C-C-C
      |                  |
      C                  C
                         |
                         C

C-C=C-C-C-C        C-C=C-C-C-C
    |                  |
    C                  C

C-C=C-C-C-C        C-C=C-C-C-C
      |                  |
      C                  C
                         |
                         C

C-C-C=C-C-C        C-C-C=C-C-C
    |                  |
    C                  C

C-C-C-C-C-C
      ‖
      C
```

(b)
```
C=C-C-C-C          C=C-C-C-C
    | |                |   |
    C C                C   C

C=C-C-C-C          C=C-C-C-C
    |                  | |
    C                  C C
    |
    C

        C
        |
C=C-C-C-C          C=C-C-C-C
      |                |
      C                C-C

C-C=C-C-C          C-C=C-C-C
    | |                |   |
    C C                C   C

                       C
                       |
C-C=C-C-C          C-C=C-C-C
      |                |
      C C              C
      |
(etc.)

C-C=C-C-C
      |
      C-C
```

(c)
```
  C   C            C                C
   \ /             |                |
    C              C                C
   / \ \          / \ \            / \ \
  C   C          C   C            C   C
  |   |          |   |   C        |   |
  C - C          C - C            C - C
                                      |
                                      C

     C    C-C
      \  /
   C---C
  / \
 C   C
 |   |
 C - C
```

15.11(a)
```
                CH₃
                |
CH₂=C-CH₂-CH₂-CH₂-CH₃        CH₂=CH-CH-CH₂-CH₂-CH₃
                                    |
                                    CH₃

CH₂=CH-CH₂-CH-CH₂-CH₃        CH₂=CH-CH₂-CH₂-CH-CH₃
           |                              |
           CH₃                            CH₃

CH₃-C=CH-CH₂-CH₂-CH₃        CH₃-CH=C-CH₂-CH₂-CH₃
    |                               |
    CH₃                             CH₃

CH₃-CH=CH-CH-CH₂-CH₃        CH₃-CH=CH-CH₂-CH-CH₃
              |                            |
              CH₃                          CH₃

CH₃-CH-CH=CH-CH₂-CH₃        CH₃-CH₂-C=CH-CH₂-CH₃
    |                               |
    CH₃                             CH₃

CH₃-CH₂-C-CH₂-CH₂-CH₃
        |
        CH₂
```

(b) $CH_2=C-CH-CH_2-CH_3$ (with CH_3, CH_3 branches)

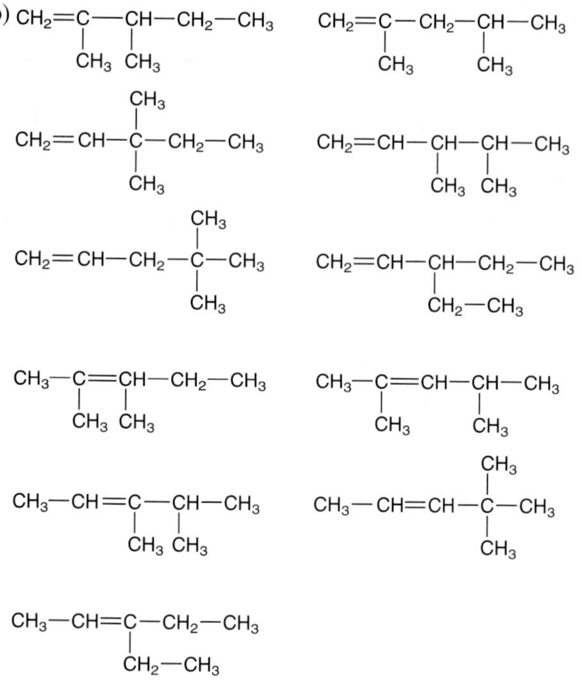

$CH_2=C-CH_2-CH-CH_3$ (with CH_3, CH_3 branches)

$CH_2=CH-C-CH_2-CH_3$ (with CH_3 above, CH_3 below)

$CH_2=CH-CH-CH-CH_3$ (with CH_3, CH_3 branches)

$CH_2=CH-CH_2-C-CH_3$ (with CH_3 above, CH_3 below)

$CH_2=CH-CH-CH_2-CH_3$ (with CH_2-CH_3 branch)

$CH_3-C=CH-CH_2-CH_3$ (with CH_3, CH_3 branches)

$CH_3-C=CH-CH-CH_3$ (with CH_3, CH_3 branches)

$CH_3-CH=C-CH-CH_3$ (with CH_3, CH_3 branches)

$CH_3-CH=CH-C-CH_3$ (with CH_3 above, CH_3 below)

$CH_3-CH=C-CH_2-CH_3$ (with CH_2-CH_3 branch)

(c)
CH_2 ring with $CH-CH_2-CH_3$ (cyclopentane, CH_2, H_2C-CH_2)

CH_2 ring with $CH-CH_3$ (CH_2, $H_2C-CH-CH_3$)

CH_2 ring with $CH-CH_3$ (H_2C, $H_3C-HC-CH_2$)

CH_2 ring with C—CH_3, CH_3 (H_2C, H_2C-CH_2)

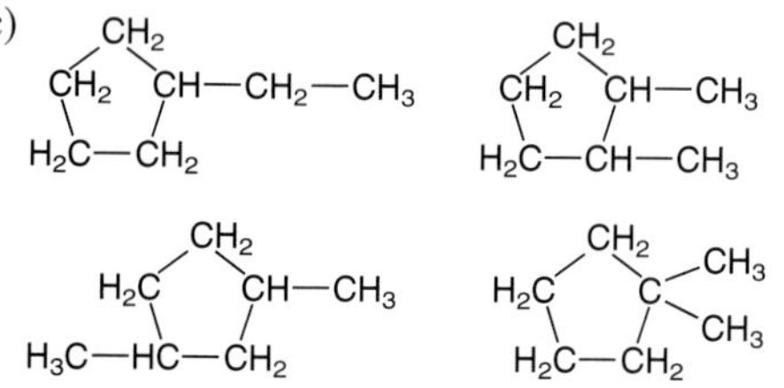

15.13(a)
$CH_3-C-CH_2-CH_3$ (with CH_3 above, CH_3 below)

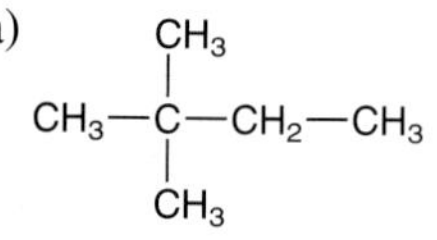

(b) $H_2C=CH-CH_2-CH_3$

(c) $HC\equiv C-CH-CH_3$ (with CH_2, CH_3 branch)

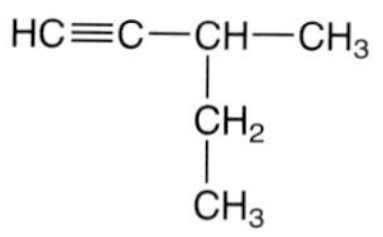

(d) Structure is correct.

15.15(a)
$CH_3-CH-CH-CH_2-CH_2-CH_2-CH_2-CH_3$ (with CH_3 above, CH_3 below)

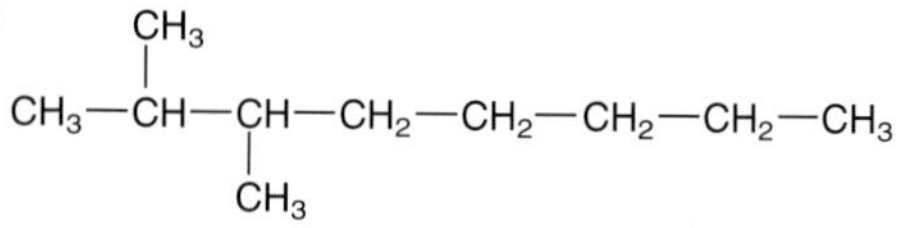

(b) cyclohexane ring with CH_2-CH_3 at position 1, CH_3 at position 4; positions numbered 1,2,3,4,5,6

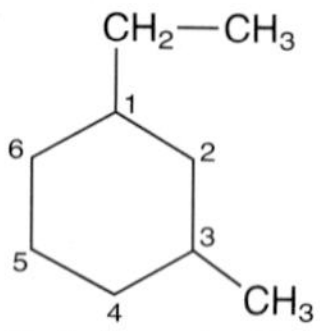

(c) 3,4-dimethylheptane   (d) 2,2-dimethylbutane

15.17(a) 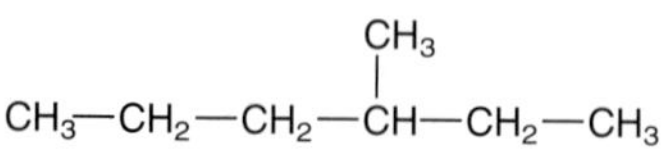
$CH_3-CH_2-CH_2-CH-CH_2-CH_3$ (with CH_3 branch)
Correct name is 3-methylhexane.

(b) 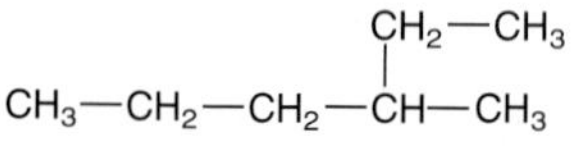
$CH_3-CH_2-CH_2-CH-CH_3$ (with CH_2-CH_3 branch)
Correct name is 3-methylhexane.

(c) cyclohexane ring with CH_3
Correct name is methylcyclohexane.

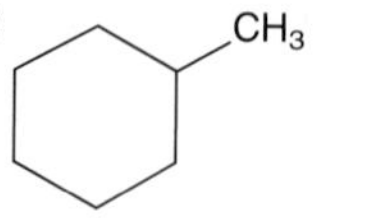

(d) 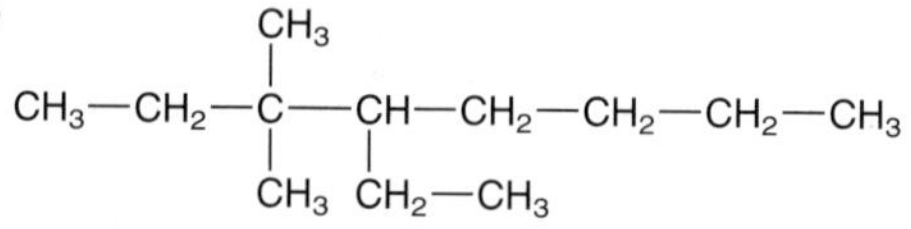
$CH_3-CH_2-C-CH-CH_2-CH_2-CH_2-CH_3$ (with CH_3 above, CH_3 below C; CH_2-CH_3 below CH)
Correct name is 4-ethyl-3,3-dimethyloctane.

15.19(a) three-membered ring with H, H at top carbon, $H-C-C-CH_3$, Cl H below

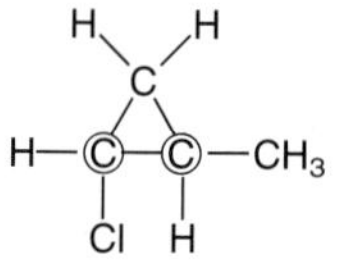

(b)
$CH_3-CH_2-C-C-CH_3$ (with H, CH_3 above; CH_3, H below)

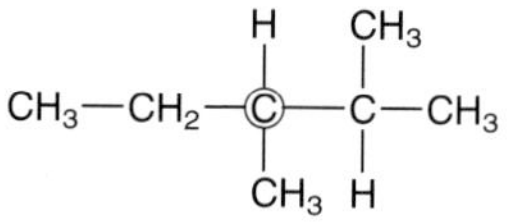

15.21(a) 3-Bromohexane is optically active.
$CH_3-CH_2-CH-CH_2-CH_2-CH_3$ (with Br below)

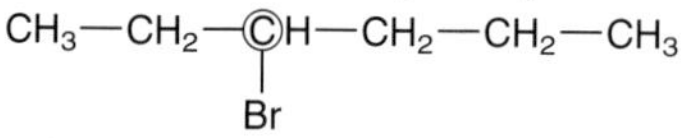

(b) 3-Chloro-3-methylpentane is not optically active.
$CH_3-CH_2-C-CH_2-CH_3$ (with CH_3 above, Cl below)

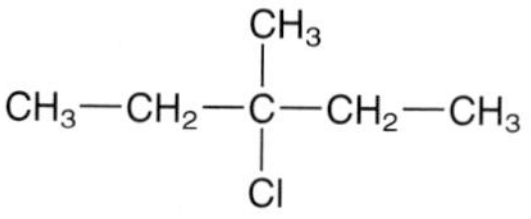

(c) 1,2-Dibromo-2-methylbutane is optically active.
$CH_2-C-CH_2-CH_3$ (with CH_3 above; Br, Br below)

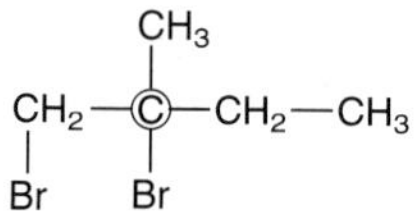

15.23(a)
$C=C$ with H, H top; CH_3CH_2, CH_3 bottom — *cis*-2-pentene

$C=C$ with H, CH_3 top; CH_3CH_2, H bottom — *trans*-2-pentene

(b) 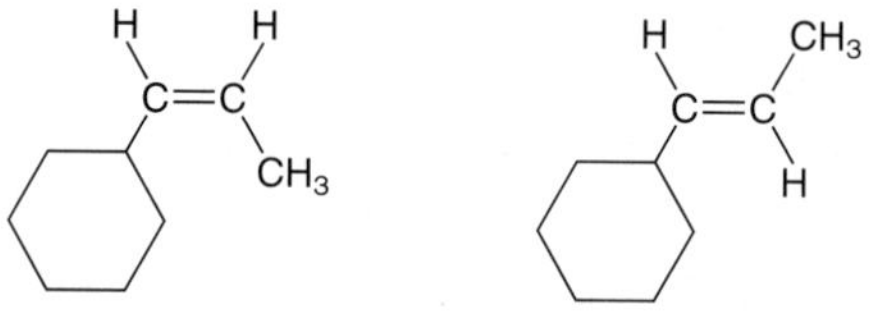
$C=C$ with H, H top; cyclohexyl, CH_3 bottom — *cis*-1-cyclohexylpropene

$C=C$ with H, CH_3 top; cyclohexyl, H bottom — *trans*-1-cyclohexylpropene

(c) no geometric isomers

15.25(a) no geometric isomers

(b) CH_3-CH_2 ... CH_2-CH_3, $C=C$ with H, H bottom — *cis*-3-hexene

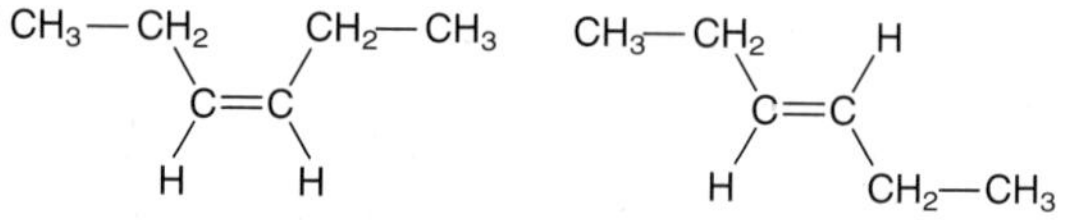

CH_3-CH_2 ... H, $C=C$ with H, CH_2-CH_3 bottom — *trans*-3-hexene

(c) no geometric isomers

(d)
$C=C$ with Cl, Cl top; H, H bottom — *cis*-1,2-dichloroethene

$C=C$ with Cl, H top; H, Cl bottom — *trans*-1,2-dichloroethene

15.27

1,2-dichlorobenzene   1,3-dichlorobenzene   1,4-dichlorobenzene
(o-dichlorobenzene)   (m-dichlorobenzene)   (p-dichlorobenzene)

15.29

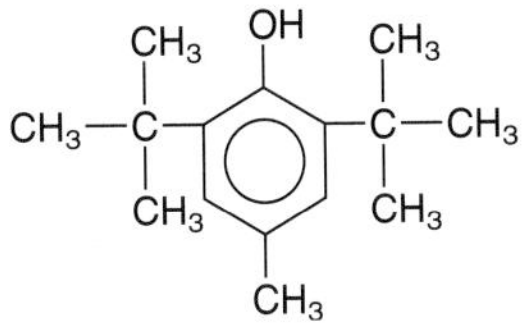

15.30

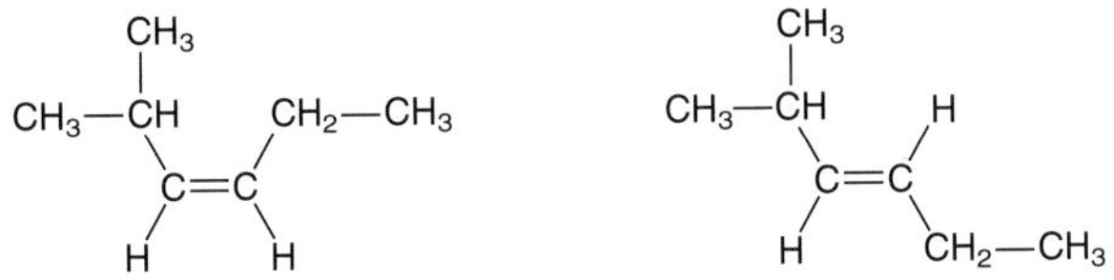

cis-2-methyl-3-hexene   trans-2-methyl-3-hexene

The compound 2-methyl-2-hexene does not have *cis-trans* isomers.   **15.32**(a) elimination   (b) addition
15.34(a) $CH_3CH_2CH{=}CHCH_2CH_3 + H_2O \xrightarrow{H^+}$
                    $CH_3CH_2CH_2CH(OH)CH_2CH_3$
   (b) $CH_3CHBrCH_3 + CH_3CH_2OK \longrightarrow$
                    $CH_3CH{=}CH_2 + CH_3CH_2OH + KBr$
   (c) $CH_3CH_3 + 2Cl_2 \xrightarrow{hv} CHCl_2CH_3 + 2HCl$
15.37(a) Methylethylamine is more soluble because it has the
ability to form H bonds with water molecules.   (b) 1-Butanol
has a higher melting point because it can form intermolecular H
bonds.   (c) Propylamine has a higher boiling point because it
contains N—H bonds that allow H bonding, and trimethylamine
cannot form H bonds.   **15.39** Both groups react by addition to
the π bond. The very polar C=O bond attracts the electron-rich
O of water to the partially positive C. There is no such polarity in
the alkene, so either C atom can be attacked.

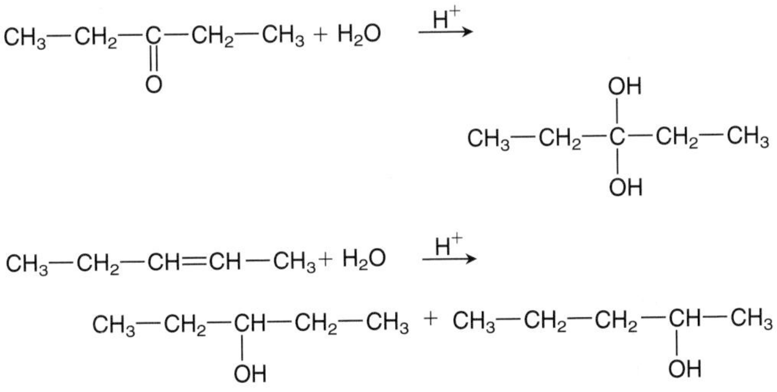

$CH_3-CH_2-C(O)-CH_2-CH_3 + H_2O \xrightarrow{H^+}$

$CH_3-CH_2-C(OH)(OH)-CH_2-CH_3$

$CH_3-CH_2-CH{=}CH-CH_3 + H_2O \xrightarrow{H^+}$

$CH_3-CH_2-CH(OH)-CH_2-CH_3$ + $CH_3-CH_2-CH_2-CH(OH)-CH_3$

15.41 Esters and acid anhydrides form through dehydration-
condensation reactions, and water is the other product.
15.43(a) alkyl halide   (b) nitrile   (c) carboxylic acid   (d) aldehyde
15.45(a) CH_3- CH=CH $-CH_2-OH$

alkene   alcohol

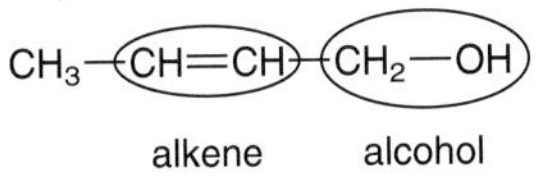

(b)

Cl—CH₂ ... C(O)—OH

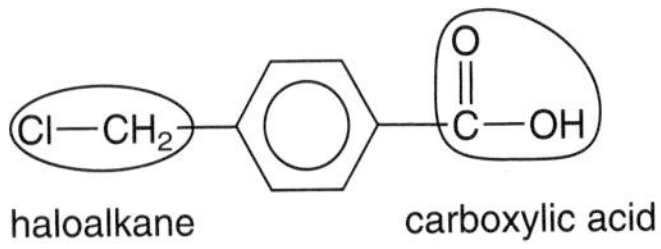

haloalkane   carboxylic acid

(c) amide / alkene

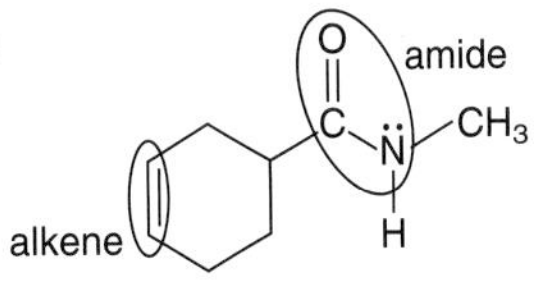

(d) nitrile / ketone

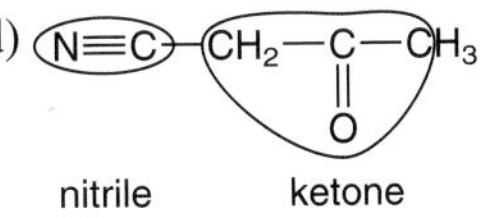

(e) ester

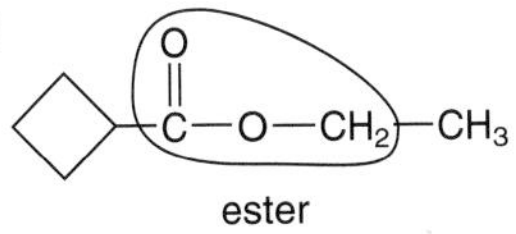

15.47 $H_3C-CH_2-CH_2-CH_2-CH_2-OH$
$H_3C-CH_2-CH_2-CH(OH)-CH_3$

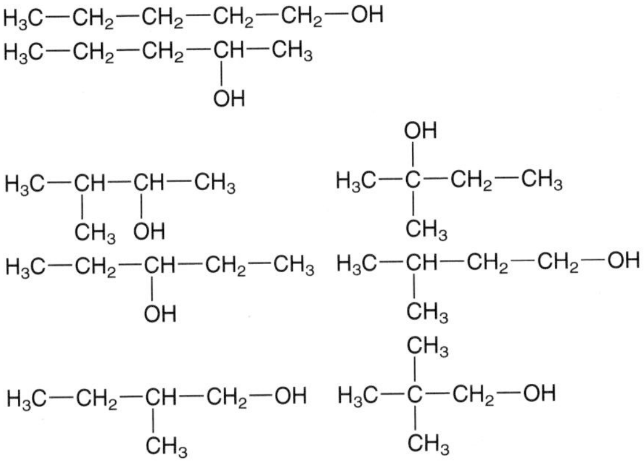

$H_3C-CH(CH_3)-CH(OH)-CH_3$ $H_3C-C(OH)(CH_3)-CH_2-CH_3$

$H_3C-CH_2-CH(OH)-CH_2-CH_3$ $H_3C-CH(CH_3)-CH_2-CH_2-OH$

$H_3C-CH_2-CH(CH_3)-CH_2-OH$ $H_3C-C(CH_3)_2-CH_2-OH$

15.49 $H_3C-CH_2-CH_2-CH_2-NH_2$ $H_3C-CH_2-CH(NH_2)-CH_3$

$H_3C-CH(CH_3)-CH_2-NH_2$ $H_3C-C(CH_3)_2-NH_2$

$H_3C-CH_2-N(H)-CH_2-CH_3$ $H_3C-CH_2-CH_2-N(H)-CH_3$

$H_3C-CH_2-N(CH_3)-CH_3$ $H_3C-CH(CH_3)-N(H)-CH_3$

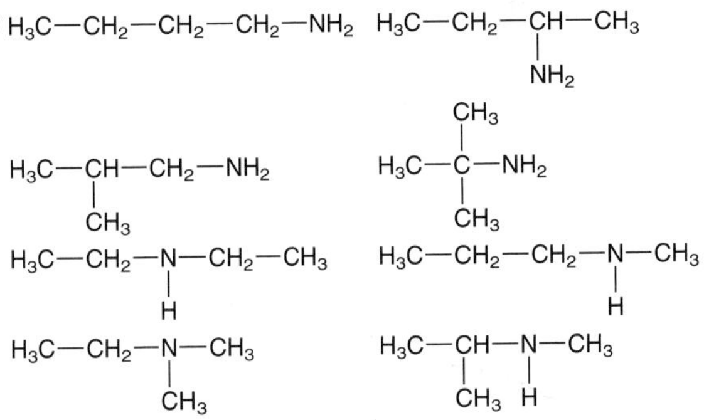

15.51(a)

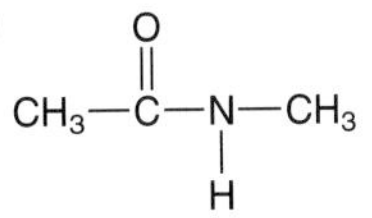

$CH_3-C(O)-N(H)-CH_3$

(b)

$CH_3-CH_2-CH_2-C(O)-O-CH(CH_3)-CH_3$

(c)

$H-C(O)-O-CH_2-CH(CH_3)-CH_3$

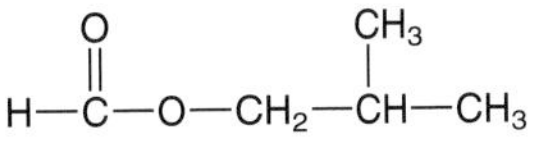

15.53(a)

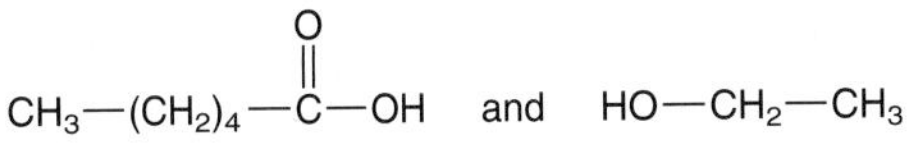

$CH_3-(CH_2)_4-C(O)-OH$ and $HO-CH_2-CH_3$

(b)

phenyl$-C(O)-OH$ and $HO-CH_2-CH_2-CH_3$

(c)

CH_3-CH_2-OH and $HO-C(O)-CH_2-CH_2-$phenyl

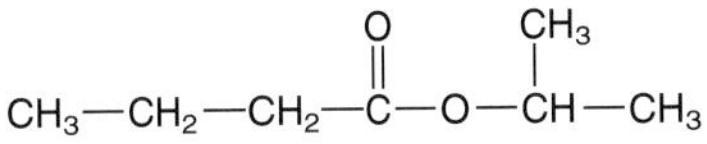

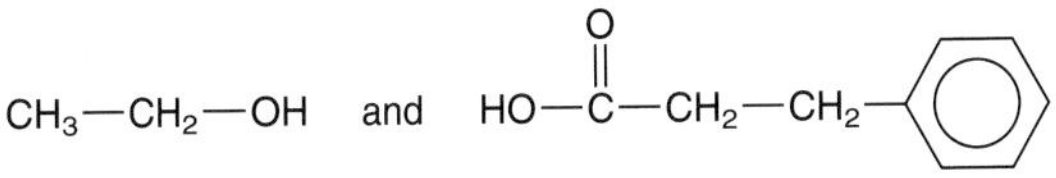

15.55(a) CH_3-CH_2-OH

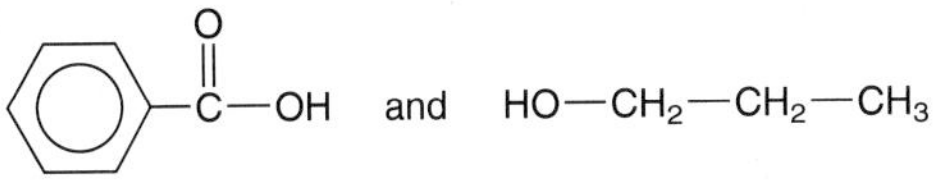

$CH_3-CH_2-C(O)-O-CH_2-CH_3$

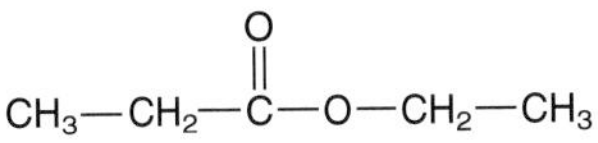

(b)

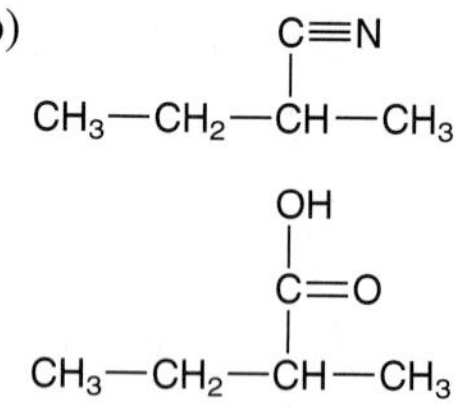

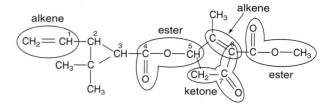

15.59 addition reactions and condensation reactions **15.61** Dispersion forces strongly attract the long, unbranched chains of high-density polyethylene (HDPE). Low-density polyethylene (LDPE) has branching in the chains that prevents packing and weakens the attractions. **15.63** An amine and a carboxylic acid react to form nylon; a carboxylic acid and an alcohol form a polyester.

15.64(a)   (b)

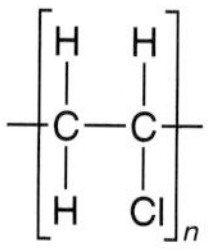

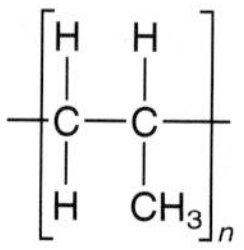

15.66

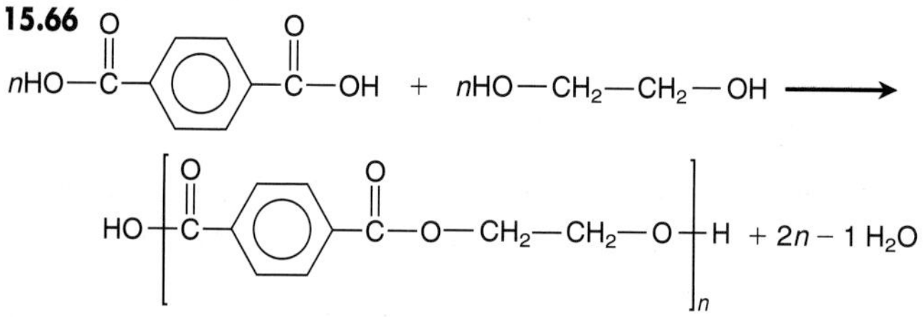

15.68(a) condensation   (b) addition   (c) condensation (d) condensation **15.70** The amino acid sequence in a protein determines its shape and structure, which determine its function. **15.72**(a)   (b) 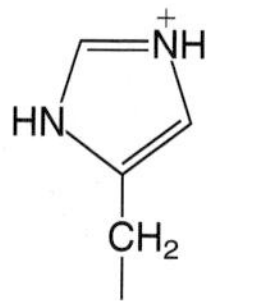   (c)

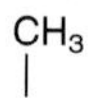

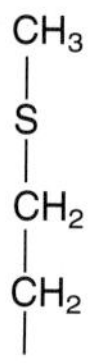

15.74(a)

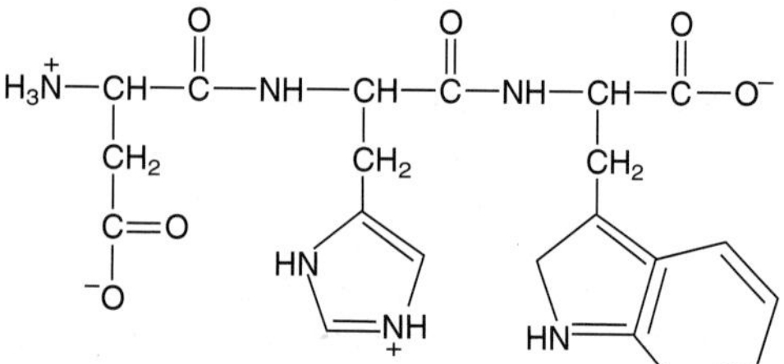

(b)

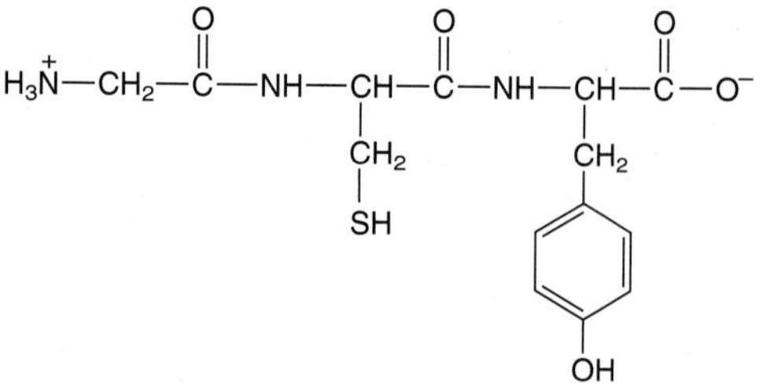

15.76(a) AATCGG   (b) TCTGTA **15.78**(a) Both R groups are from cysteine, which can form a disulfide bond (covalent bond). (b) Lysine and aspartic acid give a salt link.   (c) Asparagine and serine will hydrogen bond.   (d) Valine and phenylalanine interact through dispersion forces. **15.80** CH_3—CH=CH—CH_3

15.81(a)

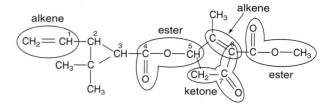

(b) Carbon 1 is sp^2 hybridized.   Carbon 2 is sp^3 hybridized. Carbon 3 is sp^3 hybridized.   Carbon 4 is sp^2 hybridized. Carbon 5 is sp^3 hybridized.   Carbons 6 and 7 are sp^2 hybridized. (c) Carbons 2, 3, and 5 are chiral centers, as they are each bonded to four different groups. **15.83**

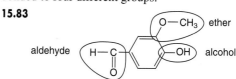

The shortest carbon-oxygen bond is the double bond in the aldehyde group.

Chapter 16

16.2 Reaction rate is proportional to concentration. An increase in pressure will increase the concentration, resulting in an increased reaction rate. **16.3** The addition of water will dilute the concentrations of all dissolved solutes, and the rate of the reaction will decrease. **16.5** An increase in temperature affects the rate of a reaction by increasing the number of collisions between particles, but more importantly, the energy of collisions increases. Both these factors increase the rate of reaction. **16.8**(a) The slope of the line joining any two points on a graph of concentration versus time gives the average rate between the two points. The closer the points, the closer the average rate will be to the instantaneous rate.   (b) The initial rate is the instantaneous rate at the point on the graph where time = 0, that is, when reactants are mixed.

16.10

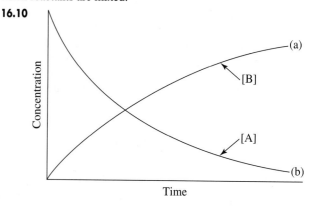

16.12(a) $rate = -\left(\dfrac{1}{2}\right)\dfrac{\Delta[AX_2]}{\Delta t}$

$= -\left(\dfrac{1}{2}\right)\dfrac{(0.0088\ M - 0.0500\ M)}{(20.0\ s - 0\ s)}$

$= 0.0010\ M/s$

(b)

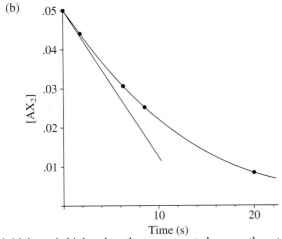

The initial rate is higher than the average rate because the rate will decrease as reactant concentration decreases.

16.14 rate $= -\dfrac{\Delta[A]}{\Delta t} = -\dfrac{1}{2}\dfrac{\Delta[B]}{\Delta t} = \dfrac{\Delta[C]}{\Delta t}$; 0.2 mol/L·s

16.16 $2N_2O_5(g) \longrightarrow 4NO_2(g) + O_2(g)$

16.19(a) rate $= -\dfrac{1}{3}\dfrac{\Delta[O_2]}{\Delta t} = \dfrac{1}{2}\dfrac{\Delta[O_3]}{\Delta t}$ (b) 1.45×10^{-5} mol/L·s

16.20(a) k is the rate constant, the proportionality constant in the rate law; it is reaction and temperature specific. (b) m represents the order of the reaction with respect to [A], and n represents the order of the reaction with respect to [B]. The order of a reactant does not necessarily equal its stoichiometric coefficient in the balanced equation. (c) L^2/mol²·min **16.21**(a) Rate doubles. (b) Rate decreases by a factor of four. (c) Rate increases by a factor of nine.
16.22 first order in BrO_3^-; first order in Br^-; second order in H^+; fourth order overall **16.24**(a) Rate doubles. (b) Rate is halved.
(c) The rate increases by a factor of 16. **16.26**(a) second order in A; first order in B (b) rate $= k[A]^2[B]$ (c) 5.00×10^3 L^2/mol²·min **16.29**(a) first order (b) second order (c) zero order **16.31** 7 s **16.33**(a) $k = 0.0660$ min^{-1} (b) 21.0 min
16.36 Measure the rate constant at a series of temperatures and plot ln k versus $1/T$. The slope of the line equals $-E_a/R$.
16.38 0.033 s^{-1} **16.42** No, other factors that affect the fraction of collisions that lead to reaction are the energy and orientation of the collisions. **16.45** At the same temperature, both reaction mixtures have the same average kinetic energy, but not the same velocity. The trimethylamine molecule has greater mass than the ammonia molecule, so trimethylamine molecules will collide less often with HCl. Moreover, the bulky groups bonded to nitrogen in trimethylamine mean that collisions with HCl having the correct orientation occur less frequently. Therefore, the rate of the first reaction is greater. **16.46** 12 unique collisions **16.48** 2.96×10^{-18}
16.50(a)

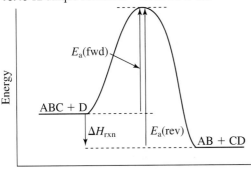

(b) 2.70×10^2 kJ/mol (c)

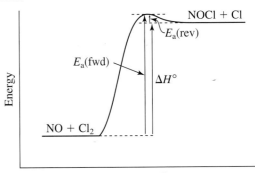

16.52(a) Because the enthalpy change is positive, the reaction is endothermic.

(b) 3 kJ (c) $:\!\ddot{C}l\!\cdots\!\cdots\!\ddot{C}l\cdots\!\cdots\!\overset{..}{N}\!\!\diagdown\!\!O:$

16.53 The rate of an overall reaction depends on the rate of the slowest step. The rate of the overall reaction will be slower than the average of the individual rates because the average includes faster rates as well. **16.57** The probability of three particles colliding with one another with the proper energy and orientation is much less than the probability for two particles. **16.58** No, the overall rate law must contain only reactants (no intermediates), and the overall rate is determined by the slow step.
16.59(a) $CO_2(aq) + 2OH^-(aq) \longrightarrow CO_3^{2-}(aq) + H_2O(l)$
  (b) $HCO_3^-(aq)$
  (c) (1) molecularity $= 2$; rate$_1 = k_1[CO_2][OH^-]$
      (2) molecularity $= 2$; rate$_2 = k_2[HCO_3^-][OH^-]$
  (d) Yes
16.60(a) $Cl_2(g) + 2NO_2(g) \longrightarrow 2NO_2Cl(g)$ (b) $Cl(g)$
  (c) (1) molecularity $= 2$; rate$_1 = k_1[Cl_2][NO_2]$
      (2) molecularity $= 2$; rate$_2 = k_2[Cl][NO_2]$ (d) Yes
16.61(a) $A(g) + B(g) + C(g) \longrightarrow D(g)$
  (b) X and Y are intermediates.

(c)

Step	Molecularity	Rate Law
$A(g) + B(g) \rightleftharpoons X(g)$	bimolecular	rate$_1 = k_1[A][B]$
$X(g) + C(g) \longrightarrow Y(g)$	bimolecular	rate$_2 = k_2[X][C]$
$Y(g) \longrightarrow D(g)$	unimolecular	rate$_3 = k_3[Y]$

(d) yes (e) yes **16.63** The proposed mechanism is valid because the individual steps are chemically reasonable add to give the overall equation, and the rate law for the mechanism matches the observed rate law. **16.66** No. A catalyst changes the mechanism of a reaction to one with lower activation energy. Lower activation energy means a faster reaction. An increase in temperature does not influence the activation energy, but increases the fraction of collisions with sufficient energy to equal or exceed the activation energy. **16.69** 4.61×10^4 J/mol **16.72**(a) Rate increases 2.5 times. (b) Rate is halved. (c) Rate decreases by a factor of 0.01. (d) Rate does not change. **16.75** 57 yr
16.77(a) 0.21 h^{-1}; 3.3 h (b) 6.6 h (c) If the concentration of sucrose is relatively low, the concentration of water remains nearly constant even with small changes in the amount of water. This gives an apparent zero-order reaction with respect to

water. Thus, the reaction is first order overall because the rate does not change with changes in the amount of water. **16.81** 7.3×10^3 J/mol　**16.82**(a) 2.4×10^{-15} M (b) 2.4×10^{-11} mol/L·s

Chapter 17

17.1 If the change is one of concentrations, it results temporarily in more products and less reactants. After equilibrium is reestablished, the K_c remains unchanged because the ratio of products and reactants remains the same. If the change is one of temperature, [product] and K_c increase and [reactant] decreases. **17.6** The equilibrium constant expression is $K = [O_2]$. If the temperature remains constant, K remains constant. If the initial amount of Li_2O_2 present is sufficient to reach equilibrium, the amount of O_2 obtained will be constant.

17.7(a) $Q = \dfrac{[HI]^2}{[H_2][I_2]}$

The value of Q increases as a function of time until it reaches the value of K.　(b) no　**17.10** Yes. If Q_1 is for the formation of 1 mol NH_3 from H_2 and N_2, and Q_2 is for the formation of NH_3 from H_2 and 1 mol of N_2, then $Q_2 = Q_1^2$.
17.11(a) $4NO(g) + O_2(g) \rightleftharpoons 2N_2O_3(g)$;

$Q_c = \dfrac{[N_2O_3]^2}{[NO]^4[O_2]}$

(b) $SF_6(g) + 2SO_3(g) \rightleftharpoons 3SO_2F_2(g)$;

$Q_c = \dfrac{[SO_2F_2]^3}{[SF_6][SO_3]^2}$

(c) $2SClF_5(g) + H_2(g) \rightleftharpoons S_2F_{10}(g) + 2HCl(g)$;

$Q_c = \dfrac{[S_2F_{10}][HCl]^2}{[SClF_5]^2[H_2]}$

17.13(a) 7.9　(b) 3.2×10^{-5}
17.15(a) $2Na_2O_2(s) + 2CO_2(g) \rightleftharpoons 2Na_2CO_3(s) + O_2(g)$;

$Q_c = \dfrac{[O_2]}{[CO_2]^2}$

(b) $H_2O(l) \rightleftharpoons H_2O(g)$;　　$Q_c = [H_2O(g)]$
(c) $NH_4Cl(s) \rightleftharpoons NH_3(g) + HCl(g)$;　　$Q_c = [NH_3][HCl]$
17.18(a) (1)　　$Cl_2(g) + F_2(g) \rightleftharpoons 2ClF(g)$
　　(2)　　$2ClF(g) + 2F_2(g) \rightleftharpoons 2ClF_3(g)$
　　overall: $Cl_2(g) + 3F_2(g) \rightleftharpoons 2ClF_3(g)$

(b) $Q_{overall} = Q_1Q_2 = \dfrac{[\cancel{ClF}]^2}{[Cl_2][F_2]} \times \dfrac{[ClF_3]^2}{[\cancel{ClF}]^2[F_2]^2}$

$= \dfrac{[ClF_3]^2}{[Cl_2][F_2]^3}$

17.20 K_c and K_p are equal when $\Delta n_{gas} = 0$.　**17.21**(a) smaller (b) Assuming that $RT > 1$ ($T > 12.2$ K), $K_p > K_c$ if there are more moles of products than reactants at equilibrium, and $K_p < K_c$ if there are more moles of reactants than products.

17.22(a) 3　(b) −1　(c) 3　**17.24**(a) 3.2　(b) 28.5　**17.26** The reaction quotient (Q) and equilibrium constant (K) are determined by the ratio [products]/[reactants]. When $Q < K$, the reaction proceeds to the right to form more products.　**17.27** No, to the left　**17.31**(a) The approximation applies when the change in concentration from initial concentration to equilibrium concentration is so small that it is insignificant; this occurs when K is small and initial concentration is large.　(b) This approximation should not be used when the change in concentration is greater than 5%. This can occur when [reactant]$_{initial}$ is very small or when change in [reactant] is relatively large due to a large K.　**17.32** 50.8
17.34

Concentration (M)	$PCl_5(g)$	$\rightleftharpoons$	$PCl_3(g) +$	$Cl_2(g)$
Initial	0.075		0	0
Change	$-x$		$+x$	$+x$
Equilibrium	$0.075 - x$		x	x

17.36 28 atm　**17.38** 0.33 atm　**17.40** 3.5×10^{-3} M
17.42 $[I_2]_{eq} = [Cl_2]_{eq} = 0.0200$ M; $[ICl]_{eq} = 0.060$ M
17.44 6.01×10^{-6}　**17.46** Equilibrium position refers to the specific concentrations or pressures of reactants and products that exist at equilibrium, whereas equilibrium constant is the overall ratio of equilibrium concentrations or pressures.　**17.47**(a) B, because the amount of product increases with temperature (b) A, because the lowest temperature will give the least product **17.50** A rise in temperature favors the forward direction of an endothermic reaction. The addition of heat makes K_2 larger than K_1. **17.51**(a) shifts toward products　(b) shifts toward products (c) does not shift　(d) shifts toward reactants　**17.53**(a) more F and less F_2　(b) more C_2H_2 and H_2 and less CH_4　**17.55**(a) no change　(b) increase volume　**17.57**(a) amount decreases (b) amount increases　(c) amount increases　(d) amount decreases　**17.60**(a) lower temperature; higher pressure (b) Q decreases; no change in K　(c) Reaction rates are lower at lower temperatures so a catalyst is used to speed up the reaction. **17.65**(a) 3×10^{-3} atm　(b) high pressure; low temperature (c) No, because water condenses at a higher temperature. **17.68**(a) 0.016 atm　(b) $K_c = 5.6 \times 10^2$; $P_{SO_2} = 0.16$ atm **17.69** 12.5 g $CaCO_3$　**17.73**(a) 3.0×10^{-14} atm (b) 0.013 pg CO/L　**17.76**(a) 98.0%　(b) 99.0% **17.77**(a) $2CH_4(g) + O_2(g) + 2H_2O(g) \rightleftharpoons 2CO_2(g) + 6H_2(g)$ (b) 1.76×10^{29}　(c) 3.19×10^{23}　(d) 48 atm **17.78**(a) 4.0×10^{-21} atm　(b) 5.5×10^{-8} atm　(c) 29 N atoms/L; 4.0×10^{14} H atoms/L　(d) $N_2(g) + H(g) \longrightarrow NH(g) + N(g)$ **17.79**(a) $P_{N_2} = 31$ atm; $P_{H_2} = 93$ atm; $P_{total} = 174$ atm; (b) $P_{N_2} = 18$ atm; $P_{H_2} = 111$ atm; $P_{total} = 179$ atm; not a valid argument　**17.80**(a) $P_{N_2} = 0.780$ atm; $P_{O_2} = 0.210$ atm; $P_{NO} = 2.67 \times 10^{-16}$ atm　(b) 0.990 atm　(c) $K_c = K_p = 4.35 \times 10^{-31}$ **17.84**(a) 1.52　(b) 0.9626 atm　(c) 0.2000 mol CO　(d) 0.01128 M

Chapter 18

18.2 All Arrhenius acids contain hydrogen in their formula and produce hydronium ion (H_3O^+) in aqueous solution. All Arrhenius bases produce hydroxide ion (OH^-) in aqueous solution. Neutralization occurs when each H_3O^+ ion combines with an OH^- ion to form two molecules of H_2O. Chemists found the reaction of any strong base with any strong acid always produced 56 kJ/mol ($\Delta H = -56$ kJ/mol), which was consistent with Arrhenius' hypothesis describing neutralization.　**18.4** Strong acids and bases dissociate completely into ions when dissolved in water. Weak acids and bases dissociate only partially. The characteristic property of all weak acids is that the great majority of acid molecules are undissociated. **18.5**(a), (c), and (d)

18.7(a) $K_a = \dfrac{[NO_2^-][H_3O^+]}{[HNO_2]}$

(b) $K_a = \dfrac{[CH_3COO^-][H_3O^+]}{[CH_3COOH]}$

(c) $K_a = \dfrac{[BrO_2^-][H_3O^+]}{[HBrO_2]}$

18.9 $CH_3COOH < HF < HIO_3 < HI$ **18.11**(a) weak acid (b) strong base (c) weak acid (d) strong acid **18.15**(a) The acid with the smaller K_a (4×10^{-5}) has the higher pH, because less dissociation yields fewer hydronium ions. (b) The acid with the larger pK_a (3.5) has the high pH, because a larger pK_a means a smaller K_a. (c) Lower concentration (0.01 M) gives fewer hydronium ions. (d) A 0.1 M weak acid solution gives fewer hydronium ions. (e) The 0.1 M base solution has a lower concentration of hydronium ions. (f) The pOH = 6.0 because pH = 14.0 − 6.0 = 8.0 **18.16**(a) 12.05; basic (b) 11.09; acidic **18.18**(a) $[H_3O^+] = 1.7\times10^{-10}$ M, pOH = 4.22, $[OH^-] = 6.0\times10^{-5}$ M (b) pH = 3.57, $[H_3O^+] = 2.7\times10^{-4}$ M, $[OH^-] = 3.7\times10^{-11}$ M **18.20** 6×10^{-5} mol OH^- **18.23**(a) Rising temperature increases the value of K_w. (b) $K_w = 2.5\times10^{-14}$; pOH = 6.80; $[OH^-] = 1.6\times10^{-7}$ M **18.24** The Brønsted-Lowry theory defines acids as proton donors and bases as proton acceptors, while the Arrhenius definition looks at acids as containing ionizable hydrogen atoms and at bases as containing hydroxide ions. In both definitions, an acid produces hydronium ions and a base produces hydroxide ions when added to water. Ammonia and carbonate ion are two Brønsted-Lowry bases that are not Arrhenius bases because they do not contain hydroxide ions. Brønsted-Lowry acids must contain an ionizable hydrogen atom in order to be proton donors, so a Brønsted-Lowry acid is also an Arrhenius acid. **18.27** An amphoteric species can act as either an acid or a base. The dihydrogen phosphate ion, $H_2PO_4^-$, is an example.

18.28(a) Cl^- (b) HCO_3^- (c) OH^- **18.30**(a) NH_4^+ (b) NH_3 (c) $C_{10}H_{14}N_2H^+$

18.32(a) $NH_3 + H_3PO_4 \rightleftharpoons NH_4^+ + H_2PO_4^-$
  base acid acid base
Conjugate acid-base pairs: $H_3PO_4/H_2PO_4^-$ and NH_4^+/NH_3
(b) $CH_3O^- + NH_3 \rightleftharpoons CH_3OH + NH_2^-$
  base acid acid base
Conjugate acid-base pairs: NH_3/NH_2^- and CH_3OH/CH_3O^-
(c) $HPO_4^{2-} + HSO_4^- \rightleftharpoons H_2PO_4^- + SO_4^{2-}$
  base acid acid base
Conjugate acid-base pairs: HSO_4^-/SO_4^{2-} and $H_2PO_4^-/HPO_4^{2-}$

18.34(a) $OH^-(aq) + H_2PO_4^-(aq) \rightleftharpoons H_2O(l) + HPO_4^{2-}(aq)$
Conjugate acid-base pairs: $H_2PO_4^-/HPO_4^{2-}$ and H_2O/OH^-
(b) $HSO_4^-(aq) + CO_3^{2-}(aq) \rightleftharpoons SO_4^{2-}(aq) + HCO_3^-(aq)$
Conjugate acid-base pairs: HSO_4^-/SO_4^{2-} and HCO_3^-/CO_3^{2-}

18.36 $K_c > 1$: $HS^- + HCl \rightleftharpoons H_2S + Cl^-$
$K_c < 1$: $H_2S + Cl^- \rightleftharpoons HS^- + HCl$

18.38 $K_c > 1$ for both (a) and (b) **18.40**(a) A strong acid is 100% dissociated, so the acid concentration will be very different after dissociation. (b) A weak acid dissociates to a very small extent, so the acid concentration before and after dissociation is nearly the same. (c) same as (b), but with the extent of dissociation greater. (d) same as (a) **18.43** 1.5×10^{-5} **18.45** $[H_3O^+] = [NO_2^-] = 1.9\times10^{-2}$ M; $[OH^-] = 5.3\times10^{-13}$ M **18.47** $[H_3O^+] = [ClCH_2COO^-] = 0.038$ M; $[ClCH_2COOH] = 1.01$ M; pH = 1.42 **18.49**(a) $[H_3O^+] = 7.5\times10^{-3}$ M; pH = 2.12; $[OH^-] = 1.3\times10^{-12}$ M; pOH = 11.88 (b) 2.3×10^{-4} **18.51**(a) 2.47 (b) 11.41 **18.54** 1.9% **18.55** All Brønsted-Lowry bases contain at least one lone pair of electrons, which binds an H^+ and allows the base to act as a proton acceptor. **18.58**(a) $C_5H_5N(aq) + H_2O(l) \rightleftharpoons OH^-(aq) + C_5H_5NH^+(aq)$;

$K_b = \dfrac{[C_5H_5NH^+][OH^-]}{[C_5H_5N]}$

(b) $CO_3^{2-}(aq) + H_2O(l) \rightleftharpoons OH^-(aq) + HCO_3^-(aq)$;

$K_b = \dfrac{[HCO_3^-][OH^-]}{[CO_3^{2-}]}$

18.60 11.71 **18.62** (a) 12.04 (b) 10.77 **18.64** (a) 10.95 (b) 5.62 **18.66** $[OH^-] = 4.8\times10^{-4}$ M; pH = 10.68 **18.68** As a nonmetal becomes more electronegative, the acidity of its binary hydride increases. The electronegative nonmetal attracts the electrons more strongly in the polar bond, shifting the electron density away from H, thus making the H^+ more easily transferred to a water molecule to form H_3O^+. **18.71** Chlorine is more electronegative than iodine, and $HClO_4$ has more oxygen atoms than HIO. **18.72** (a) H_2Se (b) $B(OH)_3$ (c) $HBrO_2$ **18.74**(a) 0.05 M $Al_2(SO_4)_3$ (b) 0.1 M $PbCl_2$ **18.77** NaF contains the anion of the weak acid HF, so F^- acts as a base. NaCl contains the anion of the strong acid HCl.

18.79(a) $KBr(s) + H_2O(l) \longrightarrow K^+(aq) + Br^-(aq)$; neutral
(b) $NH_4I(s) + H_2O(l) \longrightarrow NH_4^+(aq) + I^-(aq)$
$NH_4^+(aq) + H_2O(l) \rightleftharpoons NH_3(aq) + H_3O^+(aq)$; acidic
(c) $KCN(s) + H_2O(l) \longrightarrow K^+(aq) + CN^-(aq)$
$CN^-(aq) + H_2O(l) \rightleftharpoons HCN(aq) + OH^-(aq)$; basic

18.81(a) $KNO_3 < K_2SO_3 < Na_2CO_3$
(b) $NaHSO_4 < NH_4NO_3 < NaHCO_3 < Na_2CO_3$

18.84 A Lewis acid is an electron pair acceptor while a Brønsted-Lowry acid is a proton donor. The proton of a Brønsted-Lowry acid fits the definition of a Lewis acid because it accepts an electron pair when it bonds with a base. All Lewis acids are not Brønsted-Lowry acids. A Lewis base is an electron pair donor and a Brønsted-Lowry base is a proton acceptor. All Brønsted-Lowry bases can be Lewis bases, and vice versa. **18.85**(a) No, $Zn(H_2O)_6^{2+}(aq) + 6NH_3(aq) \rightleftharpoons Zn(NH_3)_6^{2+} + 6H_2O(l)$; NH_3 is a weak Brønsted-Lowry base, but a strong Lewis base. (b) cyanide ion and water (c) cyanide ion **18.88**(a) Lewis acid (b) Lewis base (c) Lewis acid (d) Lewis base **18.90**(a) Lewis acid: Na^+; Lewis base: H_2O (b) Lewis acid: CO_2; Lewis base: H_2O (c) Lewis acid: BF_3; Lewis base: F^- **18.92**(a) Lewis (b) Brønsted-Lowry and Lewis (c) none (d) Lewis **18.94** 3.5×10^{-8} to 4.5×10^{-8} M H_3O^+; 5.2×10^{-7} to 6.6×10^{-7} M OH^- **18.97**(a) $SnCl_4$ is the Lewis acid; $(CH_3)_3N$ is the Lewis base (b) $5d$ **18.98** pH = 5.00, 6.00, 6.79, 6.98, 7.00 **18.99** H_3PO_4 **18.104** 0.00147 **18.105**(a) Ca^{2+} does not react with water; $CH_3CH_2COO^-(aq) + H_2O(l) \rightleftharpoons CH_3CH_2COOH(aq) + OH^-(aq)$; basic (b) 9.03 **18.111**(a) The concentration of oxygen is higher in the lungs so the equilibrium shifts to the right.   (b) In an oxygen-deficient environment the equilibrium shifts to the left to release oxygen.   (c) A decrease in $[H_3O^+]$ shifts the equilibrium to the right. More oxygen is absorbed, but it will be more difficult to remove the O_2.   (d) An increase in $[H_3O^+]$ shifts the equilibrium to the left. Less oxygen is bound to Hb, but it will be easier to remove it.   **18.113**(a) 10.0 (b) The pK_b for the 3° amine group is much smaller than that for the aromatic ring, thus the K_b is significantly larger (yielding a much greater amount of OH^-).   (c) 4.6   (d) 5.1

Chapter 19

19.2 The acid component neutralizes added base and the base component neutralizes added acid so the pH of the buffer solution remains relatively constant. The components of a buffer do not neutralize one another because they are a conjugate acid-base pair.   **19.4** The pH of a buffer decreases only slightly with added H_3O^+.   **19.7** The buffer range, the pH over which the buffer acts effectively, is greatest when the buffer-component ratio is 1; the range decreases as the component ratio deviates from 1.
19.9 $[H_3O^+] = 7.8 \times 10^{-6}$ M; pH = 5.11   **19.11** 9.92   **19.13** 9.55   **19.15** 1.7   **19.17** 3.37   **19.19**(a) 4.91   (b) 0.66 g KOH   **19.21**(a) HCOOH/HCOO$^-$ or C$_6$H$_5$NH$_2$/C$_6$H$_5$NH$_3$$^+$ (b) $H_2PO_4^-$/HPO_4^{2-} or $H_2AsO_4^-$/$HAsO_4^{2-}$ **19.24** 1.6
19.26 To see a distinct color in a mixture of two colors, you need one to have about 10 times the intensity of the other. For this to be the case, the concentration ratio $[HIn]/[In^-]$ has to be greater than 10:1 or less than 1:10. This occurs when pH = $pK_a - 1$ or pH = $pK_a + 1$, respectively, giving a pH range of about two units.   **19.28** The equivalence point in a titration is the point at which the number of moles of base equals the number of moles of acid. The endpoint is the point at which the added indicator changes color. If an appropriate indicator is selected, the endpoint is close to the equivalence point, but they are not usually the same. The endpoint, or color change, may precede or follow the equivalence point, depending on the indicator chosen.
19.30(a) initial pH: *strong acid–strong base* < *weak acid–strong base* < *strong acid–weak base*   (b) equivalence point: strong acid–*weak base* < *strong acid–strong base* < *weak acid–strong base*   **19.32** At the center of the buffer region, the concentrations of weak acid and conjugate base are equal, so the pH = pK_a of the acid.   **19.33** pH range from 7.3 to 9.3   **19.35**(a) bromthymol blue   (b) thymol blue or phenolphthalein   **19.37**(a) 1.00 (b) 1.48   (c) 3.00   (d) 4.00   (e) 7.00   (f) 10.00   (g) 11.96
19.39(a) 2.91   (b) 4.81   (c) 5.29   (d) 6.09   (e) 7.40   (f) 8.76 (g) 10.10   (h) 12.05   **19.41**(a) 59.0 mL and 8.54   (b) 24.5 mL and 7.71   **19.44** Fluoride ion is the conjugate base of a weak acid and reacts with H_2O: $F^-(aq) + H_2O(l) \rightleftharpoons HF(aq) + OH^-(aq)$. As the pH increases, the equilibrium shifts to the left and $[F^-]$ increases. As the pH decreases, the equilibrium shifts to the right and $[F^-]$ decreases. The changes in $[F^-]$ influence the solubility of CaF_2. Chloride ion is the conjugate base of a strong acid so it does not react with water and its concentration is not influenced by pH.   **19.45** The compound precipitates.
19.46(a) $K_{sp} = [Ag^+]^2[CO_3^{2-}]$ (b) $K_{sp} = [Ba^{2+}][F^-]^2$ (c) $K_{sp} = [Cu^{2+}][HS^-][OH^-]$ **19.48** 1.3×10^{-4}
19.50 2.8×10^{-11} **19.52**(a) 2.3×10^{-5} M (b) 4.2×10^{-9} M
19.54(a) 1.7×10^{-3} M (b) 2.0×10^{-4} M **19.56**(a) Mg(OH)$_2$

(b) PbS   (c) Ag$_2$SO$_4$ **19.58**(a) AgCl(s) $\rightleftharpoons$ Ag$^+$(aq) + Cl$^-$(aq). The chloride ion is the anion of a strong acid, so it does not react with H_3O^+. No change with pH.   (b) SrCO$_3$(s) $\rightleftharpoons$ Sr^{2+}(aq) + CO$_3$$^{2-}$(aq). The strontium ion is the cation of a strong base so pH will not affect its solubility. The carbonate ion acts as a base: CO$_3$$^{2-}$(aq) + H$_2$O(l) $\rightleftharpoons$ HCO$_3$$^-$(aq) + OH$^-$(aq); also CO$_2$(g) forms and escapes: CO$_3$$^{2-}$(aq) + 2H$_3O^+$(aq) $\longrightarrow$ CO$_2$(g) + 2H$_2$O(l). Therefore, the solubility of SrCO$_3$ will increase with addition of H_3O^+ (decreasing pH).
19.60 yes   **19.65** No, because it indicates that a complex ion forms between the lead ion and hydroxide ions: Pb^{2+}(aq) + nOH$^-$(aq) $\rightleftharpoons$ Pb(OH)$_n$$^{2-n}$(aq)   **19.66** Hg(H$_2$O)$_4$$^{2+}$(aq) + 4CN$^-$(aq) $\rightleftharpoons$ Hg(CN)$_4$$^{2-}$(aq) + 4H$_2$O(l)   **19.68** Ag(H$_2$O)$_2$$^+$(aq) + 2S$_2O_3$$^{2-}$(aq) $\rightleftharpoons$ Ag(S$_2$O$_3$)$_2$$^{3-}$(aq) + 2H$_2$O(l)
19.70 9.4×10^{-5} M **19.74** 1.3×10^{-4} M **19.76**(a) 14 (b) 1 g   **19.78**(a) 0.088   (b) 0.14   **19.80** 8×10^{-5}
19.81(a)

V (mL)	pH	ΔpH/ΔV	V$_{average}$ (mL)
0.00	1.00		
10.00	1.22	0.022	5.00
20.00	1.48	0.026	15.00
30.00	1.85	0.037	25.00
35.00	2.18	0.066	32.50
39.00	2.89	0.18	37.00
39.50	3.20	0.62	39.25
39.75	3.50	1.2	39.63
39.90	3.90	2.7	39.83
39.95	4.20	6	39.93
39.99	4.90	18	39.97
40.00	7.00	200	40.00
40.01	9.40	200	40.01
40.05	9.80	10	40.03
40.10	10.40	10	40.08
40.25	10.50	0.67	40.18
40.50	10.79	1.2	40.38
41.00	11.09	0.60	40.75
45.00	11.76	0.17	43.00
50.00	12.05	0.058	47.50
60.00	12.30	0.025	55.00
70.00	12.43	0.013	65.00
80.00	12.52	0.009	75.00

(b)

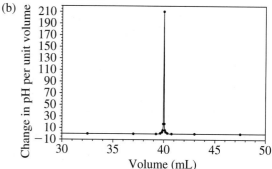

19.84 $K_b = \dfrac{[BH^+][OH^-]}{[B]}$

Rearranging to isolate $[OH^-]$:

$$[OH^-] = K_b \frac{[B]}{[BH^+]}$$

Taking the negative log:

$$-\log [OH^-] = -\log K_b - \log \frac{[B]}{[BH^+]}$$

Therefore, $pOH = pK_b + \log \frac{[BH^+]}{[B]}$

19.92(a) 65 mol   (b) 6.28   (c) 4.0×10^3 g   **19.93** No. NaCl will precipitate.   **19.95**(a) A and D   (b) $pH_A = 4.35$; $pH_B = 8.67$; $pH_C = 2.67$; $pH_D = 4.57$ (c) C, A, D, B   (d) B

Chapter 20

20.2 A spontaneous process occurs by itself, whereas a nonspontaneous process requires a continuous input of energy to make it happen. It is possible to cause a nonspontaneous process to occur, but the process stops once the energy source is removed. A reaction that is nonspontaneous under one set of conditions may be spontaneous under a different set of conditions.   **20.5** The transition from liquid to gas involves a greater increase in dispersal of energy and freedom of motion than does the transition from solid to liquid.   **20.6** In an exothermic reaction, $\Delta S_{surr} > 0$. In an endothermic reaction, $\Delta S_{surr} < 0$. A chemical cold pack for injuries is an example of an application using a spontaneous endothermic process.   **20.8**(a), (b), and (c)   **20.10**(a) positive (b) negative   (c) negative   **20.12**(a) negative   (b) negative (c) positive   **20.14**(a) positive   (b) negative   (c) positive **20.16**(a) positive   (b) negative   (c) positive   **20.18**(a) Butane. The double bond in 2-butene restricts freedom of rotation. (b) Xe(g) because it has the greater molar mass   (c) $CH_4(g)$. Gases have greater entropy than liquids.   **20.20**(a) diamond $<$ graphite $<$ charcoal. Freedom of motion is least in the network solid; more freedom between graphite sheets; most freedom in amorphous solid.   (b) ice $<$ liquid water $<$ water vapor. Entropy increases as a substance changes from solid to liquid to gas.   (c) O atoms $< O_2 < O_3$. Entropy increases with molecular complexity.   **20.22**(a) $ClO_4^-(aq) > ClO_3^-(aq) > ClO_2^-(aq)$; decreasing molecular complexity   (b) $NO_2(g) > NO(g) > N_2(g)$. N_2 has lower standard molar entropy because it consists of two of the same atoms; the other species have two different types of atoms. NO_2 is more complex than NO.   (c) $Fe_3O_4(s) > Fe_2O_3(s) > Al_2O_3(s)$. Fe_3O_4 is more complex and more massive. Fe_2O_3 is more massive than Al_2O_3.   **20.25** For a system at equilibrium, $\Delta S_{univ} = \Delta S_{sys} + \Delta S_{surr} = 0$. For a system moving to equilibrium, $\Delta S_{univ} > 0$.   **20.26** $S^\circ_{Cl_2O(g)} = 2S^\circ_{HClO(g)} - S^\circ_{H_2O(g)} - \Delta S^\circ_{rxn}$ **20.27**(a) negative; $\Delta S^\circ = -172.4$ J/K   (b) positive; $\Delta S^\circ = 141.6$ J/K   (c) negative; $\Delta S^\circ = -837$ J/K   **20.29** $\Delta S^\circ = 93.1$ J/K; yes, the positive sign of ΔS is expected because there is a net increase in the number of gas molecules.   **20.31** -75.6 J/K **20.34** -97.2 J/K   **20.36** A spontaneous process has $\Delta S_{univ} > 0$. Since the absolute temperature is always positive, ΔG_{sys} must be negative ($\Delta G_{sys} < 0$) for a spontaneous process.   **20.38** ΔH°_{rxn} is positive and ΔS°_{sys} is positive. Melting is an example. **20.39**(a) -1138.0 kJ   (b) -1379.4 kJ   (c) -224 kJ **20.41**(a) -1138 kJ   (b) -1379 kJ   (c) -226 kJ   **20.43**(a) Entropy decreases (ΔS° is negative) because the number of moles of gas decreases. The combustion of CO releases energy (ΔH° is negative).   (b) -257.2 kJ or 257.3 kJ, depending on the method   **20.45**(a) $\Delta H^\circ_{rxn} = 90.7$ kJ; $\Delta S^\circ_{rxn} = 221$ J/K   (b) at $38°C$, $\Delta G^\circ = 22.1$ kJ; at $138°C$, $\Delta G^\circ = 0.0$ kJ; at $238°C$, $\Delta G^\circ = -22.1$ kJ (c) For the substances in their standard states, the reaction is nonspontaneous at $38°C$, near equilibrium at $138°C$, and spontaneous at $238°C$.   **20.47** $\Delta H^\circ = 30910$ J, $\Delta S^\circ = 93.15$ J/K, $T = 331.8$ K   **20.49**(a) $\Delta H^\circ_{rxn} = -241.826$ kJ, $\Delta S^\circ_{rxn} = -44.4$ J/K, $\Delta G^\circ_{rxn} = -228.60$ kJ   (b) Yes. The reaction will become nonspontaneous at higher temperatures.   (c) The reaction is spontaneous below 5.45×10^3 K.   **20.51**(a) ΔG° is a relatively large positive value.   (b) $K \gg 1$. Q depends on initial conditions, not equilibrium conditions.   **20.53** The standard free energy change, ΔG°, applies when all components of the system are in their standard states; $\Delta G^\circ = \Delta G$.   **20.54**(a) 1.7×10^6 (b) 3.89×10^{-34} (c) 1.26×10^{48} **20.56** 4.89×10^{-51} **20.58** 3.36×10^5 **20.60** 2.7×10^4 J/mol; no   **20.62**(a) 2.9×10^4 J/mol   (b) The reverse direction, formation of reactants, is spontaneous so the reaction proceeds to the left.   (c) 7.0×10^3 J/mol; the reaction proceeds to the left to reach equilibrium. **20.64**(a) no T (b) 163 kJ   (c) 1×10^2 kJ/mol   **20.67**(a) spontaneous   (b) +   (c) +   (d) −   (e) −, not spontaneous   (f) − **20.69**(a) 2.3×10^2 (b) Administer oxygen-rich air to counteract the CO poisoning.   **20.72**(a) $2N_2O_5(g) + 6F_2(g) \longrightarrow 4NF_3(g) + 5O_2(g)$ (b) $\Delta G^\circ_{rxn} = -569$ kJ   (c) $\Delta G_{rxn} = -5.60 \times 10^2$ kJ/mol **20.75**(a) $\Delta H^\circ_{rxn} = 470.5$ kJ; $\Delta S^\circ_{rxn} = 558.4$ J/K   (b) The reaction will be spontaneous at high T, because the $-T\Delta S$ term will be larger in magnitude than ΔH.   (c) no   (d) 842.5 K   **20.77**(a) yes, negative Gibbs free energy   (b) Yes. It becomes spontaneous at 270.8 K.   (c) 234 K. The temperature is different because the ΔH and ΔS values for N_2O_5 vary with physical state. **20.81**(a) 465 K   (b) 6.59×10^{-4} (c) The reaction rate is higher at the higher temperature. The shorter time required (kinetics) overshadows the lower yield (thermodynamics).

Chapter 21

21.1 Oxidation is the loss of electrons and results in a higher oxidation number; reduction is the gain of electrons and results in a lower oxidation number.   **21.2** No, one half-reaction cannot take place independently because there is a transfer of electrons from one substance to another. If one substance loses electrons, another substance must gain them.   **21.3** Spontaneous reactions, $\Delta G_{sys} < 0$, take place in voltaic cells (also called galvanic cells). Nonspontaneous reactions, $\Delta G_{sys} > 0$, take place in electrolytic cells.   **21.5**(a) Cl^- (b) MnO_4^- (c) MnO_4^- (d) Cl^- (e) from Cl^- to MnO_4^-
(f) $8H_2SO_4(aq) + 2KMnO_4(aq) + 10KCl(aq) \longrightarrow$
$\qquad 2MnSO_4(aq) + 5Cl_2(g) + 8H_2O(l) + 6K_2SO_4(aq)$
21.7(a) $ClO_3^-(aq) + 6H^+(aq) + 6I^-(aq) \longrightarrow$
$\qquad Cl^-(aq) + 3H_2O(l) + 3I_2(s)$
Oxidizing agent is ClO_3^- and reducing agent is I^-.
(b) $2MnO_4^-(aq) + H_2O(l) + 3SO_3^{2-}(aq) \longrightarrow$
$\qquad 2MnO_2(s) + 3SO_4^{2-}(aq) + 2OH^-(aq)$
Oxidizing agent is MnO_4^- and reducing agent is SO_3^{2-}.
(c) $2MnO_4^-(aq) + 6H^+(aq) + 5H_2O_2(aq) \longrightarrow$
$\qquad 2Mn^{2+}(aq) + 8H_2O(l) + 5O_2(g)$
Oxidizing agent is MnO_4^- and reducing agent is H_2O_2.
21.10(a) $4NO_3^-(aq) + 4H^+(aq) + 4Sb(s) \longrightarrow$
$\qquad 4NO(g) + 2H_2O(l) + Sb_4O_6(s)$
Oxidizing agent is NO_3^- and reducing agent is Sb.
(b) $5BiO_3^-(aq) + 14H^+(aq) + 2Mn^{2+}(aq) \longrightarrow$
$\qquad 5Bi^{3+}(aq) + 7H_2O(l) + 2MnO_4^-(aq)$
Oxidizing agent BiO_3^- and reducing agent is Mn^{2+}.

(c) $Pb(OH)_3^-(aq) + 2Fe(OH)_2(s) \longrightarrow$
$$Pb(s) + 2Fe(OH)_3(s) + OH^-(aq)$$
Oxidizing agent is $Pb(OH)_3^-$ and reducing agent is $Fe(OH)_2$.
21.12(a) $Au(s) + 3NO_3^-(aq) + 4Cl^-(aq) + 6H^+(aq) \longrightarrow$
$AuCl_4^-(aq) + 3NO_2(g) + 3H_2O(l)$ (b) Oxidizing agent is NO_3^- and reducing agent is Au.   (c) HCl provides chloride ions that combine with the gold(III) ion to form the stable $AuCl_4^-$ ion.
21.13(a) A   (b) E   (c) C   (d) A   (e) E   (f) E   **21.16** An active electrode is a reactant or product in the cell reaction. An inactive electrode does not take part in the reaction and is present only to conduct a current. Platinum and graphite are commonly used as inactive electrodes.   **21.17**(a) A   (b) B   (c) A
(d) Hydrogen bubbles will form when metal A is placed in acid. Metal A is a better reducing agent than metal B, so if metal B reduces H^+ in acid, then metal A will also.
21.18(a) Oxidation: $Zn(s) \longrightarrow Zn^{2+}(aq) + 2e^-$
Reduction: $Sn^{2+}(aq) + 2e^- \longrightarrow Sn(s)$
Overall: $Zn(s) + Sn^{2+}(aq) \longrightarrow Zn^{2+}(aq) + Sn(s)$

(b)

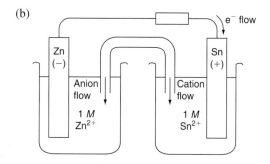

21.20(a) left to right   (b) left   (c) right   (d) Ni   (e) Fe   (f) Fe
(g) $1 M NiSO_4$ (h) K^+ and NO_3^- (i) Fe   (j) from right to left
(k) Oxidation: $Fe(s) \longrightarrow Fe^{2+}(aq) + 2e^-$
Reduction: $Ni^{2+}(aq) + 2e^- \longrightarrow Ni(s)$
Overall: $Fe(s) + Ni^{2+}(aq) \longrightarrow Fe^{2+}(aq) + Ni(s)$
21.22(a) $Al(s) \mid Al^{3+}(aq) \parallel Cr^{3+}(aq) \mid Cr(s)$
(b) $Pt(s) \mid SO_2(g) \mid SO_4^{2-}(aq), H^+(aq) \parallel Cu^{2+}(aq) \mid Cu(s)$
21.25 A negative E_{cell}° indicates that the redox reaction is not spontaneous, that is, $\Delta G^{\circ} > 0$. The reverse reaction is spontaneous with $E_{cell}^{\circ} > 0$.   **21.26** Similar to other state functions, E° changes sign when a reaction is reversed. Unlike ΔG°, ΔH°, and S°, E° (the ratio of energy to charge) is an intensive property. When the coefficients in a reaction are multiplied by a factor, the values of ΔG°, ΔH°, and S° are multiplied by that factor. However, E° does not change because both the energy and charge are multiplied by the factor and thus their ratio remains unchanged.
21.27(a) Oxidation: $Se^{2-}(aq) \longrightarrow Se(s) + 2e^-$
Reduction: $2SO_3^{2-}(aq) + 3H_2O(l) + 4e^- \longrightarrow$
$$S_2O_3^{2-}(aq) + 6OH^-(aq)$$
(b) $E_{anode}^{\circ} = E_{cathode}^{\circ} - E_{cell}^{\circ} = -0.57 V - 0.35 V = -0.92 V$
21.29(a) $Br_2 > Fe^{3+} > Cu^{2+}$ (b) $Ca^{2+} < Ag^+ < Cr_2O_7^{2-}$
21.31(a) $Co(s) + 2H^+(aq) \longrightarrow Co^{2+}(aq) + H_2(g)$
$E_{cell}^{\circ} = 0.28 V$; spontaneous
(b) $2Mn^{2+}(aq) + 5Br_2(l) + 8H_2O(l) \longrightarrow$
$$2MnO_4^-(aq) + 10Br^-(aq) + 16H^+(aq)$$
$E_{cell}^{\circ} = -0.44 V$; not spontaneous
(c) $Hg_2^{2+}(aq) \longrightarrow Hg^{2+}(aq) + Hg(l)$
$E_{cell}^{\circ} = -0.07 V$; not spontaneous

21.33(a) $2Ag(s) + Cu^{2+}(aq) \longrightarrow 2Ag^+(aq) + Cu(s)$
$E_{cell}^{\circ} = -0.46 V$; not spontaneous
(b) $Cr_2O_7^{2-}(aq) + 3Cd(s) + 14H^+(aq) \longrightarrow$
$$2Cr^{3+}(aq) + 3Cd^{2+}(aq) + 7H_2O(l)$$
$E_{cell}^{\circ} = 1.73 V$; spontaneous
(c) $Pb(s) + Ni^{2+}(aq) \longrightarrow Pb^{2+}(aq) + Ni(s)$
$E_{cell}^{\circ} = -0.12 V$; not spontaneous
21.35 $3N_2O_4(g) + 2Al(s) \longrightarrow 6NO_2^-(aq) + 2Al^{3+}(aq)$
$E_{cell}^{\circ} = 0.867 V - (-1.66 V) = 2.53 V$
$2Al(s) + 3SO_4^{2-}(aq) + 3H_2O(l) \longrightarrow$
$$2Al^{3+}(aq) + 3SO_3^{2-}(aq) + 6OH^-(aq)$$
$E_{cell}^{\circ} = 2.59 V$
$SO_4^{2-}(aq) + 2NO_2^-(aq) + H_2O(l) \longrightarrow$
$$SO_3^{2-}(aq) + N_2O_4(g) + 2OH^-(aq)$$
$E_{cell}^{\circ} = 0.06 V$
21.37 $2HClO(aq) + Pt(s) + 2H^+(aq) \longrightarrow$
$$Cl_2(g) + Pt^{2+}(aq) + 2H_2O(l)$$
$E_{cell}^{\circ} = 0.43 V$
$2HClO(aq) + Pb(s) + SO_4^{2-}(aq) + 2H^+(aq) \longrightarrow$
$$Cl_2(g) + PbSO_4(s) + 2H_2O(l)$$
$E_{cell}^{\circ} = 1.94 V$
$Pt^{2+}(aq) + Pb(s) + SO_4^{2-}(aq) \longrightarrow Pt(s) + PbSO_4(s)$
$E_{cell}^{\circ} = 1.51 V$
21.39 Yes; $C > A > B$ **21.42** $A(s) + B^+(aq) \longrightarrow A^+(aq) + B(s)$ with $Q = [A^+]/[B^+]$.   (a) $[A^+]$ increases and $[B^+]$ decreases.   (b) E_{cell} decreases.   (c) $E_{cell} = E_{cell}^{\circ} - (RT/nF)$ ln $([A^+]/[B^+])$; $E_{cell} = E_{cell}^{\circ}$ when (RT/nF) ln $([A^+]/[B^+]) = 0$. This occurs when ln $([A^+]/[B^+]) = 0$, that is, $[A^+]$ equals $[B^+]$. (d) Yes, when $[A^+] > [B^+]$.   **21.44** In a concentration cell, the overall reaction decreases the concentration of the more concentrated electrolyte because it is reduced in the cathode compartment.
21.45(a) 3×10^{35} (b) 4×10^{-31} **21.47**(a) -2.03×10^5 J
(b) 1.73×10^5 J   **21.49** $\Delta G^{\circ} = -2.1 \times 10^4$ J; $E^{\circ} = 0.22 V$
21.51 $9 \times 10^{-4} M$ **21.53**(a) 0.05 V   (b) 0.50 M   (c) $[Co^{2+}] = 0.91 M$; $[Ni^{2+}] = 0.09 M$ **21.55** A; 0.085 V   **21.57** Electrons flow from the anode, where oxidation occurs, to the cathode, where reduction occurs. The electrons always flow from the anode to the cathode no matter what type of battery.   **21.58** A D-sized alkaline battery is much larger than an AAA-sized one, so the D-sized battery contains greater amounts of the cell components. The cell potential is an intensive property and does not depend on the amounts of the cell components. The total charge, however, depends on the amount of cell components so the D-sized battery produces more charge than the AAA-sized battery.   **21.60** The Teflon spacers keep the two metals separated so the copper cannot conduct electrons that would promote the corrosion (rusting) of the iron skeleton.   **21.62** Sacrificial anodes are made of metals with E° less than that of iron, $-0.44 V$, so they are more easily oxidized than iron. Only (b), (f), and (g) will work for iron. (a) will form an oxide coating that prevents further oxidation. (c) would react with groundwater quickly.   **21.64** To reverse the reaction requires 0.34 V with the cell in its standard state. A 1.5 V cell supplies more than enough potential, so the cadmium metal is oxidized to Cd^{2+} and chromium plates out.   **21.66** The oxidation number of N in NO_3^- is $+5$, the maximum O.N. for N. In the nitrite ion, NO_2^-, the O.N. of N is $+3$, so nitrogen can be further oxidized.
21.68 Iron and nickel are more easily oxidized and less easily reduced than copper. They are separated from copper in the roasting step and converted to slag. In the electrorefining process, all three metals are in solution, but only Cu^{2+} ions are reduced at the

cathode to form Cu(s). **21.70**(a) Br_2 (b) Na **21.72** copper and bromine
21.74(a) Anode: $2H_2O(l) \longrightarrow O_2(g) + 4H^+(aq) + 4e^-$
Cathode: $2H_2O(l) + 2e^- \longrightarrow H_2(g) + 2OH^-(aq)$
(b) Anode: $2H_2O(l) \longrightarrow O_2(g) + 4H^+(aq) + 4e^-$
Cathode: $Sn^{2+}(aq) + 2e^- \longrightarrow Sn(s)$
21.76(a) 2.93 mol e$^-$ (b) 2.83×10^5 C (c) 31.4 A
21.78 0.282 g Ra **21.80** 1.10×10^4 s **21.82**(a) The sodium and sulfate ions make the water conductive so the current will flow through the water, facilitating electrolysis. Pure water, which contains very low (10^{-7} M) concentrations of H$^+$ and OH$^-$, conducts electricity very poorly. (b) The reduction of H_2O has a more positive half-potential than does the reduction of Na$^+$; the oxidation of H_2O is the only reaction possible because SO_4^{2-} cannot be oxidized. Thus, it is easier to reduce H_2O than Na$^+$ and easier to oxidize H_2O than SO_4^{2-}. **21.83**(a) 4.6×10^4 L
(b) 1.26×10^8 C (c) 1.68×10^6 s **21.84** 44.2 g Zn
21.87 69.4 mass % Cu **21.89**(a) 13 days (b) 5.2×10^1 days
(c) \$310 **21.91**(a) 91 days (b) 44 g (c) \$0.08/day
21.92 7×10^2 lb Cl_2 **21.94**(a) 1.073×10^5 s (b) 1.5×10^4 kW·h
(c) 6.8¢ **21.96** F < D < E. If metal E and a salt of metal F are mixed, the salt is reduced, producing metal F because E has the greatest reducing strength of the three metals.
21.97(a) Cell I: 4 mol electrons; $\Delta G = -4.75 \times 10^5$ J
Cell II: 2 mol electrons; $\Delta G = -3.94 \times 10^5$ J
Cell III: 2 mol electrons; $\Delta G = -4.53 \times 10^5$ J
(b) Cell I: -13.2 kJ/g
Cell II: -0.613 kJ/g
Cell III: -2.62 kJ/g
Cell I has the highest ratio (most energy released per gram) because the reactants have very low mass, while Cell II has the lowest ratio because the reactants have large masses.
21.98 $Sn^{2+}(aq) + 2e^- \longrightarrow Sn(s)$
$Cr^{3+}(aq) + e^- \longrightarrow Cr^{2+}(aq)$
$Fe^{2+}(aq) + 2e^- \longrightarrow Fe(s)$
$U^{4+}(aq) + e^- \longrightarrow U^{3+}(aq)$
21.101 Li > Ba > Na > Al > Mn > Zn > Cr > Fe > Ni > Sn > Pb > Cu > Ag > Hg > Au. Metals with potentials lower than that of water (-0.83 V) can displace H_2 from water: Li, Ba, Na, Al, and Mn. Metals with potentials lower than that of hydrogen (0.00 V) can displace H_2 from acid: Li, Ba, Na, Al, Mn, Zn, Cr, Fe, Ni, Sn, and Pb. Metals with potentials greater than that of hydrogen (0.00 V) cannot displace H_2: Cu, Ag, Hg, and Au.
21.102(a) 1.890 t Al_2O_3 (b) 0.3339 t C (c) 100% (d) 74%
(e) 2.813×10^3 m^3 **21.103**(a) 5.3×10^{-11} (b) 0.20 V
(c) 0.43 V (d) 8.2×10^{-4} M NaOH **21.105** 2.94

Chapter 22

22.1(a) $1s^2 2s^2 2p^6 3s^2 3p^6 4s^2 3d^{10} 4p^6 5s^2 4d^x$
(b) $1s^2 2s^2 2p^6 3s^2 3p^6 4s^2 3d^{10} 4p^6 5s^2 4d^{10} 5p^6 6s^2 4f^{14} 5d^x$
22.4(a) The elements should increase in size as they increase in mass from Period 5 to Period 6. Because 14 additional elements lie between Periods 5 and 6, the effective nuclear charge increases significantly; so the atomic size decreases, or "contracts." This effect is significant enough that Zr^{4+} and Hf^{4+} are almost the same size but differ greatly in atomic mass. (b) The atomic size increases from Period 4 to Period 5, but stays fairly constant

from Period 5 to Period 6. (c) Atomic mass increases significantly from Period 5 to Period 6, but atomic radius (and thus volume) increases slightly, so Period 6 elements are very dense.
22.7(a) A paramagnetic substance is attracted to a magnetic field, while a diamagnetic substance is slightly repelled by one.
(b) Ions of transition elements often have half-filled d orbitals whose unpaired electrons make the ions paramagnetic. Ions of main-group elements usually have a noble gas configuration with no partially filled levels. (c) Some d orbitals in the transition element ions are empty, which allows an electron from one d orbital to move to a slightly higher energy one. The energy required for this transition is small and falls in the visible wavelength range. All orbitals are filled in ions of main-group elements, so enough energy would have to be added to move an electron to the next principal energy level, not just another orbital within the same energy level. This amount of energy is very large and much greater than the visible range of wavelengths.
22.8(a) $1s^2 2s^2 2p^6 3s^2 3p^6 4s^2 3d^3$
(b) $1s^2 2s^2 2p^6 3s^2 3p^6 4s^2 3d^{10} 4p^6 5s^2 4d^1$ (c) [Xe] $6s^2 4f^{14} 5d^{10}$
22.10(a) [Ar], no unpaired electrons (b) [Ar] $3d^9$, one unpaired electron (c) [Ar] $3d^5$, five unpaired electrons (d) [Kr] $4d^2$, two unpaired electrons **22.12** Cr, Mo, and W **22.14** In CrF_2, because the chromium is in a lower oxidation state.
22.16 CrO_3, with Cr in a higher oxidation state, yields a more acidic aqueous solution. **22.19** The coordination number indicates the number of ligand atoms bonded to the metal ion. The oxidation number represents the number of electrons lost to form the ion. The coordination number is unrelated to the oxidation number. **22.21** 2, linear; 4, tetrahedral or square planar; 6, octahedral **22.24**(a) hexaaquanickel(II) chloride
(b) tris(ethylenediamine)chromium(III) perchlorate (c) potassium hexacyanomanganate(II) **22.26**(a) 2+, 6 (b) 3+, 6
(c) 2+, 6 **22.28**(a) potassium dicyanoargentate(I) (b) sodium tetrachlorocadmate(II) (c) tetraammineaquabromocobalt(III) bromide **22.30**(a) $[Zn(NH_3)_4]SO_4$ (b) $[Cr(NH_3)_5Cl]Cl_2$
(c) $Na_3[Ag(S_2O_3)_2]$ **22.32**(a) 4, two ions (b) 6, three ions
(c) 2, four ions **22.34**(a) The nitrite ion forms linkage isomers because it can bind to the metal ion through either the nitrogen or one of the oxygen atoms—both have a lone pair of electrons.

ligand lone pair

(b) Sulfur dioxide molecules form linkage isomers because both the sulfur and the oxygen atoms can bind the central metal ion.

ligand lone pair

(c) Nitrate ions have three oxygen atoms, all with a lone pair that can bond to the metal ion, but all of the oxygen atoms are equivalent, so there are no linkage isomers.

22.36(a) geometric isomerism

(b) geometric isomerism

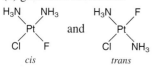

cis   *trans*

(c) geometric isomerism

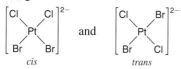

22.38(a) geometric isomerism

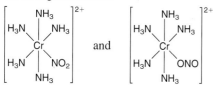

cis   *trans*

(b) linkage isomerism

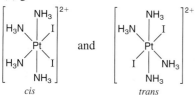

(c) geometric isomerism

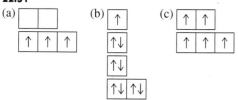

cis   *trans*

22.40(a) dsp^2 (b) sp^3 **22.43**(a) The crystal field splitting energy (Δ) is the energy difference between the two sets of *d* orbitals that result from the bonding of ligands to a central transition metal atom. (b) In an octahedral field of ligands, the ligands approach along the *x*-, *y*-, and *z*-axes. The $d_{x^2-y^2}$ and d_{z^2} orbitals are located *along* the *x*-, *y*-, and *z*-axes, so ligand interaction is higher in energy. The other orbital-ligand interactions are lower in energy because the d_{xy}, d_{yz}, and d_{xz} orbitals are located *between* the *x*-, *y*-, and *z*-axes. (c) In a tetrahedral field of ligands, the ligands do not approach along the *x*-, *y*-, and *z*-axes. The ligand interaction is greater for the d_{xy}, d_{yz}, and d_{xz} orbitals and lesser for the $d_{x^2-y^2}$ and d_{z^2} orbitals. Therefore, the crystal field splitting is reversed, and the d_{xy}, d_{yz}, and d_{xz} orbitals are higher in energy than the $d_{x^2-y^2}$ and d_{z^2} orbitals. **22.45** If Δ is greater than $E_{pairing}$, electrons will pair their spins in the lower energy *d* orbitals before adding as unpaired electrons to the higher energy *d* orbitals. If Δ is less than $E_{pairing}$, electrons will add as unpaired electrons to the higher energy *d* orbitals before pairing in the lower energy *d* orbitals. **22.47**(a) no *d* electrons (b) eight *d* electrons (c) six *d* electrons **22.49**(a) and (d)
22.51

(a) [ ][ ][ ] [↑][↑][↑]

(b) [↑] [↑↓] [↑↓] [↑↓][↑↓]

(c) [↑][↑] [↑][↑][↑]

22.53 $[Cr(H_2O)_6]^{3+} < [Cr(NH_3)_6]^{3+} < [Cr(NO_2)_6]^{3-}$
22.55 A violet complex absorbs yellow-green light. The light absorbed by a complex with a weaker field ligand would be at a lower energy and higher wavelength. Light of lower energy than yellow-green light is yellow, orange, or red. The color observed would be blue or green. **22.60**(a) 6 (b) +3 (c) two (d) 1 mol
22.64 $[Co(NH_3)_4(H_2O)Cl]^{2+}$

tetraammineaquachlorocobalt(III) ion

2 geometric isomers

cis   *trans*

$[Cr(H_2O)_3Br_2Cl]$ triaquadibromochlorochromium(III)

3 geometric isomers

Br's are *trans*   Br's are *cis*   Br's and H₂O's are *cis*
H₂O's are *cis* and *trans*

$[Cr(NH_3)_2(H_2O)_2Br_2]^+$
diamminediaquadibromochromium(III) ion

6 isomers (5 geometric)

all ligands are *trans*   only NH₃ is *trans*   only Br is *trans*

only H₂O is *trans*   optical isomers
all three ligands are *cis*

22.69(a) The first reaction shows no change in the number of particles. In the second reaction, the number of reactant particles is greater than the number of product particles. A decrease in the number of particles means a decrease in entropy. Based on entropy change only, the first reaction is favored. (b) The ethylenediamine complex will be more stable with respect to ligand exchange in water because the entropy change for that exchange is unfavorable.

Chapter 23

23.1(a) Chemical reactions are accompanied by relatively small changes in energy; nuclear reactions are accompanied by relatively large changes in energy. (b) Increasing temperature increases the rate of a chemical reaction but has no effect on a nuclear reaction. (c) Both chemical and nuclear reaction rates increase with higher reactant concentrations. (d) If the reactant is limiting in a chemical reaction, then more reactant produces more product and the yield increases. The presence of more radioactive reactant results in more decay product, so a higher reactant concentration increases the yield. **23.2**(a) *Z* down by 2, *N* down by 2 (b) *Z* up by 1, *N* down by 1 (c) no change in *Z* or *N*

(d) Z down by 1, N up by 1; a different element is produced in all cases except (c). **23.4** A neutron-rich nuclide decays by beta decay. A neutron-poor nuclide undergoes positron decay or electron capture. **23.6**(a) $^{234}_{92}U \longrightarrow \, ^{4}_{2}He + \, ^{230}_{90}Th$
(b) $^{232}_{93}Np + \, ^{0}_{-1}e \longrightarrow \, ^{232}_{92}U$ (c) $^{12}_{7}N \longrightarrow \, ^{0}_{1}\beta + \, ^{12}_{6}C$
23.8(a) $^{48}_{23}V \longrightarrow \, ^{48}_{22}Ti + \, ^{0}_{1}\beta$ (b) $^{107}_{48}Cd + \, ^{0}_{-1}e \longrightarrow \, ^{107}_{47}Ag$
(c) $^{210}_{86}Rn \longrightarrow \, ^{206}_{84}Po + \, ^{4}_{2}He$ **23.10**(a) Appears stable because its N and Z values are both magic numbers, but its N/Z ratio (1.50) is too high; it is unstable. (b) Appears unstable because its Z value is an odd number, but its N/Z ratio (1.19) is in the band of stability, so it is stable (c) Unstable because its N/Z ratio is too high. **23.12**(a) alpha decay (b) positron decay or electron capture (c) positron decay or electron capture **23.14** Stability results from a favorable N/Z ratio, even numbered N and/or Z, and the occurrence of magic numbers. The N/Z ratio of ^{52}Cr is 1.17, which is within the band of stability. The fact that Z is even does not account for the variation in stability because all isotopes of chromium have the same Z. However, ^{52}Cr has 28 neutrons, so N is both an even number and a magic number for this isotope only. **23.18** No, it is not valid to conclude that $t_{1/2}$ equals 1 min because the number of nuclei is so small. Decay rate is an average rate and is only meaningful when the sample is macroscopic and contains a large number of nuclei. For the sample containing 6×10^{12} nuclei, the conclusion is valid. **23.20** 2.89×10^{-2} Ci/g **23.22** 2.31×10^{-7} yr^{-1} **23.24** 1.49 mg **23.26** 2.2×10^{9} yr **23.28** 1×10^{2} dpm **23.32** Protons are repelled from the target nuclei due to interaction with like (positive) charges. Higher energy is required to overcome the repulsion. **23.33**(a) $^{13}_{7}N$; $^{10}_{5}B + \, ^{4}_{2}He \longrightarrow \, ^{1}_{0}n + \, ^{13}_{7}N$ (b) $^{1}_{0}n$; $^{28}_{14}Si + \, ^{2}_{1}H \longrightarrow$
$^{29}_{15}P + \, ^{1}_{0}n$ (c) $^{242}_{96}Cm$; $^{242}_{96}Cm + \, ^{4}_{2}He \longrightarrow 2 \, ^{1}_{0}n + \, ^{244}_{98}Cf$
23.37 Ionizing radiation is more dangerous to children because their rapidly dividing cells are more susceptible to radiation than an adult's slowly dividing cells. **23.38**(a) 5.4×10^{-7} rad (b) 5.4×10^{-9} Gy **23.40**(a) 7.5×10^{-10} Gy (b) 7.5×10^{-5} mrem (c) 7.5×10^{-10} Sv **23.42** 2.45×10^{-3} rad **23.44** NAA does not destroy the sample, while chemical analyses do. Neutrons bombard a nonradioactive sample, inducing some atoms within the sample to be radioactive. The radioisotopes decay by emitting radiation characteristic of each isotope. **23.45** The oxygen isotope in the methanol reactant appears in the formaldehyde product. The oxygen isotope in the chromic acid reactant appears in the water product. The isotope traces the oxygen in methanol to the oxygen in formaldehyde. **23.48**(a) 1.861×10^{4} eV (b) 2.981×10^{-15} J **23.50**(a) 8.768 MeV/nucleon (b) 517.3 MeV/atom (c) 4.99×10^{10} kJ/mol **23.53** Radioactive decay is a spontaneous process in which unstable nuclei emit radioactive particles and energy. Fission occurs as the result of high-energy bombardment of nuclei with small particles that cause the nuclei to break into smaller nuclides, radioactive particles, and energy. **23.56** The water serves to slow the neutrons so that they are better able to cause a fission reaction. Heavy water is a better moderator because it does not absorb neutrons as well as light water does, so more neutrons are available to initiate the fission process. However, D_2O does not occur naturally in great abundance, so its production adds to the cost of a heavy-water reactor. **23.60**(a) 1.1×10^{-29} kg (b) 9.8×10^{-13} J (c) 5.9×10^{8} kJ/mol. This is approximately 1 million times larger than a typical heat of reaction. **23.62** 7.6×10^{3} yr **23.65** 1.35×10^{-5} M **23.66**(a) 5.99 h (b) 21% **23.68** 4.904×10^{-9} L/h **23.75**(a) 2.07×10^{-17} J (b) 1.45×10^{7} H atoms (c) 1.4960×10^{-5} J (d) 1.4959×10^{-5} J (e) No, the Captain should continue using the current technology.

Glossary

Numbers in parentheses refer to the page(s) on which a term is introduced and/or discussed.

A

absolute scale (also *Kelvin scale*) The preferred temperature scale in scientific work, which has absolute zero (0 K, or −273.15°C) as the lowest temperature. (19) [See also *kelvin (K)*.]

absorption spectrum The spectrum produced when atoms absorb specific wavelengths of incoming light and become excited from lower to higher energy levels. (217)

accuracy The closeness of a measurement to the actual value. (24)

acid In common laboratory terms, any species that produces H^+ ions when dissolved in water. (117) [See also *Arrhenius, Brønsted-Lowry,* and *Lewis acid-base definitions.*]

acid anhydride A compound, sometimes formed by a dehydration-condensation reaction of an oxoacid, that yields two molecules of the acid when it reacts with water. (480)

acid-base buffer (also *buffer*) A solution that resists changes in pH when a small amount of either strong acid or strong base is added. (616)

acid-base indicator A species whose color is different in acid and in base, which is used to monitor the equivalence point of a titration or the pH of a solution. (587)

acid-base reaction Any reaction between an acid and a base. (117) (See also *neutralization reaction.*)

acid-base titration curve A plot of the pH of a solution of acid (or base) versus the volume of base (or acid) added to the solution. (624)

acid-dissociation (acid-ionization) constant (K_a) An equilibrium constant for the dissociation of an acid (HA) in H_2O to yield the conjugate base (A^-) and H_3O^+:

$$K_a = \frac{[H_3O^+][A^-]}{[HA]} \qquad (581)$$

actinides The Period 7 elements that constitute the second inner transition series ($5f$ block), which includes thorium (Th; $Z = 90$) through lawrencium (Lr; $Z = 103$). (247)

activated complex (See *transition state.*)

activation energy (E_a) The minimum energy with which molecules must collide to react. (516)

active site The region of an enzyme formed by specific amino acid side chains at which catalysis occurs. (531)

activity ($\mathscr{A}$) (also *decay rate*) The change in number of nuclei ($\mathscr{N}$) of a radioactive sample divided by the change in time (t). (770)

activity series of the metals A listing of metals arranged in order of their decreasing strength as reducing agents in aqueous reactions. (130)

actual yield The amount of product actually obtained in a chemical reaction. (93)

addition polymer (also *chain-reaction,* or *chain-growth, polymer*) A polymer formed when monomers (usually containing C=C) combine through an addition reaction. (483)

addition reaction A type of organic reaction in which atoms linked by a multiple bond become bonded to more atoms. (472)

adduct The product of a Lewis acid-base reaction characterized by the formation of a new covalent bond. (606)

adenosine triphosphate (ATP) A high-energy molecule that serves most commonly as a store and source of energy in organisms. (672)

alcohol An organic compound (ending, *-ol*) that contains a
—C—Ö—H functional group. (475)

aldehyde An organic compound (ending, *-al*) that contains the carbonyl functional group (C=Ö) in which the carbonyl C is also bonded to H. (478)

alkane A hydrocarbon that contains only single bonds (general formula, C_nH_{2n+2}). (463)

alkene A hydrocarbon that contains at least one C=C bond (general formula, C_nH_{2n}). (468)

alkyl group A saturated hydrocarbon chain with one bond available. (472)

alkyl halide (See *haloalkane.*)

alkyne A hydrocarbon that contains at least one C≡C bond (general formula, C_nH_{2n-2}). (469)

allotrope One of two or more crystalline or molecular forms of an element. In general, one allotrope is more stable than another at a particular pressure and temperature. (435)

alloy A mixture with metallic properties that consists of solid phases of two or more pure elements, a solid-solid solution, or distinct intermediate phases. (395)

alpha (α) decay A radioactive process in which an alpha particle is emitted from a nucleus. (765)

alpha particle (α or $^4_2$He) A positively charged particle, identical to a helium nucleus, that is one of the common types of radioactive emissions. (764)

amide An organic compound that contains the $-\overset{\overset{\displaystyle :O:}{\parallel}}{C}-\overset{\displaystyle |}{\underset{\displaystyle |}{N}}-$ functional group. (479)

amine An organic compound (general formula, $-\overset{\displaystyle |}{\underset{\displaystyle |}{C}}-\overset{\displaystyle |}{\underset{\displaystyle |}{N}}-$) derived structurally by replacing one or more H atoms of ammonia with alkyl groups; a weak organic base. (476)

amino acid An organic compound [general formula, $H_2N-CH(R)-COOH$] with at least one carboxyl and one amine group on the same molecule; the monomer unit of a protein. (487)

amorphous solid A solid that occurs in different shapes because it lacks extensive molecular-level ordering of its particles. (369)

ampere (A) The SI unit of electric current; 1 ampere of current results when 1 coulomb flows through a conductor in 1 second. (724)

amphoteric Able to act as either an acid or a base. (258)

amplitude The height of the crest (or depth of the trough) of a wave; related to the intensity of the energy. (207)

angular momentum quantum number (*l*) (also *orbital-shape quantum number*)   An integer from 0 to $n - 1$ that is related to the shape of an atomic orbital. (224)

anion   A negatively charged ion. (48)

anode   The electrode at which oxidation occurs in an electrochemical cell. Electrons are given up by the reducing agent and leave the cell at the anode. (686)

antibonding MO   A molecular orbital formed when wave functions are subtracted from each other, which decreases electron density between the nuclei and leaves a node. Electrons occupying such an orbital destabilize the molecule. (335)

aqueous solution   A solution in which water is the solvent. (60)

aromatic hydrocarbon   A compound of C and H with one or more rings of C atoms (often drawn with alternating C—C and C=C bonds), in which there is extensive delocalization of π electrons. (471)

Arrhenius acid-base definition   A model of acid-base behavior in which an acid is a substance that has H in its formula and produces H^+ in water, and a base is a substance that has OH in its formula and produces OH^- in water. (579)

Arrhenius equation   An equation that expresses the exponential relationship between temperature and the rate constant: $k = Ae^{-E_a/RT}$. (516)

atmosphere   (See *standard atmosphere.*)

atom   The smallest particle of an element that retains the chemical nature of the element. A neutral, spherical entity composed of a positively charged central nucleus surrounded by one or more negatively charged electrons. (36)

atomic mass (also *atomic weight*)   The average of the masses of the naturally occurring isotopes of an element weighted according to their abundances. (44)

atomic mass unit (amu) (also *dalton, Da*)   A mass exactly equal to $\frac{1}{12}$ the mass of a carbon-12 atom. (43)

atomic number (*Z*)   The unique number of protons in the nucleus of each atom of an element (equal to the number of electrons in the neutral atom). An integer that expresses the positive charge of a nucleus in multiples of the electronic charge. (42)

atomic orbital (also *wave function*)   A mathematical expression that describes the motion of the electron's matter-wave in terms of time and position in the region of the nucleus. The term is used qualitatively to mean the region of space in which there is a high probability of finding the electron. (222)

atomic size   A term referring to the atomic radius, one-half the distance between nuclei of identical bonded elements. (249) (See also *covalent radius* and *metallic radius.*)

atomic solid   A solid consisting of individual atoms held together by dispersion forces; the frozen noble gases are the only examples. (376)

atomic symbol (also *element symbol*)   A one- or two-letter abbreviation for the English, Latin, or Greek name of an element. (42)

ATP   (See *adenosine triphosphate.*)

aufbau principle (also *building-up principle*)   The conceptual basis of a process of building up atoms by adding one proton (and one or more neutrons) at a time to the nucleus and one electron around it to obtain the ground-state electron configurations of the elements. (240)

autoionization (also *self-ionization*)   A reaction in which two molecules of a substance react to give ions. The most important example is for water:

$$2H_2O(l) \rightleftharpoons H_3O^+(aq) + OH^-(aq) \qquad (583)$$

average rate   The change in concentration of reactants (or products) divided by a finite time period. (502)

Avogadro's law   The gas law stating that, at fixed temperature and pressure, equal volumes of any ideal gas contain equal numbers of particles, and, therefore, the volume of a gas is directly proportional to its amount (mol): $V \propto n$. (147)

Avogadro's number   A number (6.022×10^{23} to four significant figures) equal to the number of atoms in exactly 12 g of carbon-12; the number of atoms, molecules, or formula units in one mole of an element or compound. (70)

axial group   A group (or atom) that lies above or below the trigonal plane of a trigonal bipyramidal molecule, or a similar structural feature in a molecule. (311)

B

background radiation   Natural ionizing radiation, the most important form of which is cosmic radiation. (777)

balancing coefficient (also *stoichiometric coefficient*)   A numerical multiplier of all the atoms in the formula immediately following it in a chemical equation. (83)

band of stability   The narrow band of stable nuclides that appears on a plot of number of neutrons vs. number of protons for all nuclides. (767)

band theory   An extension of molecular orbital (MO) theory that explains many properties of metals, in particular, the differences in electrical conductivity of conductors, semiconductors, and insulators. (381)

barometer   A device used to measure atmospheric pressure. Most commonly, a tube open at one end, which is filled with mercury and inverted into a dish of mercury. (141)

base   In common laboratory terms, any species that produces OH^- ions when dissolved in water. (118) [See also *Brønsted-Lowry, classical (Arrhenius),* and *Lewis acid-base definitions.*]

base-dissociation (base-ionization) constant (*K*ₐb)   An equilibrium constant for the reaction of a base (B) with H_2O to yield the conjugate acid (BH^+) and OH^-:

$$K_b = \frac{[BH^+][OH^-]}{[B]} \qquad (596)$$

base pair   Two complementary bases in mononucleotides that are H bonded to each other; guanine (G) always pairs with cytosine (C), and adenine (A) always pairs with thymine (T) (or uracil, U). (491)

base unit (also *fundamental unit*)   A unit that defines the standard for one of the seven physical quantities in the International System of Units (SI). (13)

battery   A self-contained group of voltaic cells arranged in series. (708)

becquerel (Bq)   The SI unit of radioactivity; 1 Bq = 1 d/s (disintegration per second). (770)

bent shape (also *V shape*)   A molecular shape that arises when a central atom is bonded to two other atoms and has one or two lone pairs; occurs as the AX_2E shape class (bond angle < 120°) in the trigonal planar arrangement and as the AX_2E_2 shape class (bond angle < 109.5°) in the tetrahedral arrangement. (309)

beta (β) decay   A radioactive process in which a beta particle is emitted from a nucleus. (765)

beta particle (β, β⁻, or $_{-1}^{0}\beta$)   A negatively charged particle identified as a high-speed electron that is one of the common types of radioactive emissions. (764)

bimolecular reaction An elementary reaction involving the collision of two reactant species. (524)

binary covalent compound A compound that consists of atoms of two elements in which bonding occurs primarily through electron sharing. (57)

binary ionic compound A compound that consists of the oppositely charged ions of two elements. (48)

body-centered cubic unit cell A unit cell in which a particle lies at each corner and in the center of a cube. (370)

boiling point (bp or T_b) The temperature at which the vapor pressure of a gas equals the external (atmospheric) pressure. (355)

boiling point elevation (ΔT_b) The increase in the boiling point of a solvent caused by the presence of dissolved solute. (408)

bond angle The angle formed by the nuclei of two surrounding atoms with the nucleus of the central atom at the vertex. (307)

bond energy (BE) (also *bond strength*) The enthalpy change accompanying the breakage of a given bond in a mole of gaseous molecules. (279)

bond length The distance between the nuclei of two bonded atoms. (279)

bond order The number of electron pairs shared by two bonded atoms. (278)

bonding MO A molecular orbital formed when wave functions are added to each other, which increases electron density between the nuclei. Electrons occupying such an orbital stabilize the molecule. (335)

bonding pair (also *shared pair*) An electron pair shared by two nuclei; the mutual attraction between the nuclei and the electron pair forms a covalent bond. (278)

Boyle's law The gas law stating that, at constant temperature and amount of gas, the volume occupied by a gas is inversely proportional to the applied (external) pressure: $V \propto 1/P$. (144)

Brønsted-Lowry acid-base definition A model of acid-base behavior based on proton transfer, in which an acid and a base are defined, respectively, as species that donate and accept a proton. (587)

buffer (See *acid-base buffer.*)

buffer capacity A measure of the ability of a buffer to resist a change in pH; related to the total concentrations and relative proportions of buffer components. (621)

buffer range The pH range over which a buffer acts effectively; related to the relative component concentrations. (622)

C

calibration The process of correcting for systematic error of a measuring device by comparing it to a known standard. (25)

calorie (cal) A unit of energy defined as exactly 4.184 joules; originally defined as the heat needed to raise the temperature of 1 g of water 1°C (from 14.5°C to 15.5°C). (182)

calorimeter A device used to measure the heat released or absorbed by a physical or chemical process taking place within it. (188)

capillarity (or *capillary action*) A property that results in a liquid rising through a narrow space against the pull of gravity. (366)

carbonyl group The $C{=}O$ grouping of atoms. (478)

carboxylic acid An organic compound (ending, *-oic acid*)

that contains the $-\overset{\displaystyle :O:}{\underset{}{\overset{\|}{C}}}-\ddot{O}H$ group. (479)

catalyst A substance that increases the rate of a reaction without being used up in the process. (529)

cathode The electrode at which reduction occurs in an electrochemical cell. Electrons enter the cell and are acquired by the oxidizing agent at the cathode. (686)

cathode ray The ray of light emitted by the cathode (negative electrode) in a gas discharge tube; travels in straight lines, unless deflected by magnetic or electric fields. (37)

cation A positively charged ion. (48)

cell potential (E_{cell}) (also *electromotive force,* or *emf; cell voltage*) The potential difference between the electrodes of an electrochemical cell when no current flows. (692)

Celsius scale (formerly *centigrade scale*) A temperature scale in which the freezing and boiling points of water are defined as 0°C and 100°C, respectively. (19)

chain reaction In nuclear fission, a self-sustaining process in which neutrons released by splitting of one nucleus cause other nuclei to split, which releases more neutrons, and so on. (786)

change in enthalpy (ΔH) The change in internal energy plus the product of the constant pressure and the change in volume: $\Delta H = \Delta E + P\Delta V$; the heat lost or gained at constant pressure: $\Delta H = q_P$. (185)

charge density The ratio of the charge of an ion to its volume. (397)

Charles's law The gas law stating that at constant pressure, the volume occupied by a fixed amount of gas is directly proportional to its absolute temperature: $V \propto T$. (145)

chelate A complex ion in which the metal ion is bonded to a bidentate or polydentate ligand. (743)

chemical bond The force that holds two atoms together in a molecule (or formula unit). (48)

chemical change (also *chemical reaction*) A change in which a substance is converted into a substance with different composition and properties. (3)

chemical equation A statement that uses chemical formulas to express the identities and quantities of the substances involved in a chemical or physical change. (83)

chemical formula A notation of atomic symbols and numerical subscripts that shows the type and number of each atom in a molecule or formula unit of a substance. (51)

chemical kinetics The study of the rates and mechanisms of reactions. (499)

chemical property A characteristic of a substance that appears as it interacts with, or transforms into, other substances. (3)

chemical reaction (See *chemical change.*)

chemistry The scientific study of matter and the changes it undergoes. (2)

chiral molecule One that is not superimposable on its mirror image; an optically active molecule. In organic compounds, a chiral molecule typically contains a C atom bonded to four different groups (asymmetric C). (467)

chlor-alkali process An industrial method that electrolyzes concentrated aqueous NaCl and produces Cl_2, H_2, and NaOH. (719)

cis-trans isomers (See *geometric isomers.*)

Clausius-Clapeyron equation An equation that expresses the relationship between vapor pressure P of a liquid and temperature T:

$$\ln P = \frac{-\Delta H_{vap}}{R}\left(\frac{1}{T}\right) + C, \text{ where } C \text{ is a constant} \quad (354)$$

colligative property A property of a solution that depends on the number, not the identity, of solute particles. (407) (See also *boiling point elevation, freezing point depression, osmotic pressure,* and *vapor pressure lowering.*)

collision theory   A model that explains reaction rate as the result of particles colliding with a certain minimum energy. (518)

combustion analysis   A method for determining the formula of a compound from the amounts of its combustion products; used commonly for organic compounds. (80)

common-ion effect   The shift in the position of an ionic equilibrium away from formation of an ion that is caused by the addition (or presence) of that ion. (617)

complex   (See *coordination compound.*)

complex ion   An ion consisting of a central metal ion bonded covalently to molecules and/or anions called ligands. (641, 741)

composition   The types and amounts of simpler substances that make up a sample of matter. (2)

compound   A substance composed of two or more elements that are chemically combined in fixed proportions. (33)

concentration   A measure of the quantity of solute dissolved in a given quantity of solution. (95)

concentration cell   A voltaic cell in which both compartments contain the same components but at different concentrations. (705)

condensation   The process of a gas changing into a liquid. (349)

condensation polymer   A polymer formed by monomers with two functional groups that are linked together in a dehydration-condensation reaction. (485)

conduction band   In band theory, the empty, higher energy portion of the band of molecular orbitals into which electrons move when conducting heat and electricity. (381)

conductor   A substance (usually a metal) that conducts an electric current well. (382)

conjugate acid-base pair   Two species related to each other through the gain or loss of a proton; the acid has one more proton than its conjugate base. (589)

constitutional isomers (also *structural isomers*)   Compounds with the same molecular formula but different arrangements of atoms. (465, 746)

controlled experiment   An experiment that measures the effect of one variable at a time by keeping other variables constant. (9)

conversion factor   A ratio of equivalent quantities that is equal to 1 and used to convert the units of a quantity. (10)

coordinate covalent bond   A covalent bond formed when one atom donates both electrons to give the shared pair; once formed, it is identical to any covalent single bond. (748)

coordination compound (also *complex*)   A substance containing at least one complex ion. (741)

coordination isomers   Two or more coordination compounds with the same composition in which the complex ions have different ligand arrangements. (746)

coordination number   In a crystal, the number of nearest neighbors surrounding a particle. (370) In a complex, the number of ligand atoms bonded to the central metal ion. (742)

core electrons   (See *inner electrons.*)

corrosion   The natural redox process that results in unwanted oxidation of a metal. (713)

coulomb (C)   The SI unit of electric charge. One coulomb is the charge of 6.242×10^{18} electrons; one electron possesses a charge of 1.602×10^{-19} C. (692)

Coulomb's law   A law stating that the electrostatic force associated with two charges A and B is directly proportional to the product of their magnitudes and inversely proportional to the square of the distance between them:

$$\text{electrostatic force} \propto \frac{\text{charge A} \times \text{charge B}}{(\text{distance})^2} \qquad (274)$$

counter ion   A simple ion associated with a complex ion in a coordination compound. (741)

coupling of reactions   The pairing of reactions of which one releases enough free energy for the other to occur. (671)

covalent bond   A type of bond in which atoms are bonded through the sharing of two electrons; the mutual attraction of the nuclei and an electron pair that holds atoms together in a molecule. (50, 277)

covalent bonding   The idealized bonding type that is based on localized electron-pair sharing between two atoms with little difference in their tendencies to lose or gain electrons (most commonly nonmetals). (270)

covalent compound   A compound that consists of atoms bonded together by shared electron pairs. (47)

covalent radius   One-half the distance between nuclei of identical covalently bonded atoms. (249)

critical mass   The minimum mass needed to achieve a chain reaction. (786)

critical point   The point on a phase diagram above which the vapor cannot be condensed to a liquid; the end of the liquid-gas curve. (356)

crystal field splitting energy (Δ)   The difference in energy between two sets of metal-ion d orbitals that results from electrostatic interactions with the surrounding ligands. (752)

crystal field theory   A model that explains the color and magnetism of coordination compounds based on the effects of ligands on metal-ion d-orbital energies. (750)

crystalline solid   Solid with a well-defined shape because of the orderly arrangement of the atoms, molecules, or ions. (369)

cubic closest packing   A crystal structure based on the face-centered cubic unit cell in which the layers have an *abcabc . . .* pattern. (372)

cubic meter (m^3)   The SI derived unit of volume. (15)

curie (Ci)   The most common unit of radioactivity, defined as the number of nuclei disintegrating each second in 1 g of radium-226; $1 \text{ Ci} = 3.70 \times 10^{10}$ d/s (disintegrations per second). (770)

cyclic hydrocarbon   A hydrocarbon with one or more rings in its structure. (465)

D

***d* orbital**   An atomic orbital with $l = 2$. (228)

dalton (Da)   A unit of mass identical to *atomic mass unit.* (43)

Dalton's law of partial pressures   A gas law stating that, in a mixture of unreacting gases, the total pressure is the sum of the partial pressures of the individual gases: $P_{\text{total}} = P_1 + P_2 + \cdots$. (155)

data   Pieces of quantitative information obtained by observation. (8)

de Broglie wavelength   The wavelength of a moving particle obtained from the de Broglie equation: $\lambda = h/mu$. (219)

decay constant   The rate constant k for radioactive decay. (771)

decay rate   (See *activity.*)

decay series (also *disintegration series*)   The succession of steps a parent nucleus undergoes as it decays into a stable daughter nucleus. (770)

dehydration-condensation reaction   A reaction in which H and OH groups on two molecules react to form water as one of the products. (443)

delocalization   (See *electron-pair delocalization.*)

density (d)   An intensive physical property of a substance at a given temperature and pressure, defined as the ratio of the mass to the volume: $d = m/V$. (17)

deposition   The process of changing directly from gas to solid. (350)

derived unit   Any of various combinations of the seven SI base units. (13)

deuterons   Nuclei of the stable hydrogen isotope deuterium, ^{2}H. (774)

diagonal relationship   Physical and chemical similarities between a Period 2 element and one located diagonally down and to the right in Period 3. (428)

diamagnetism   The tendency of a species not to be attracted (or to be slightly repelled) by a magnetic field as a result of its electrons being paired. (261)

diffraction   The phenomenon in which a wave striking the edge of an object bends around it. A wave passing through a slit as wide as its wavelength forms a circular wave. (208)

diffusion   The movement of one fluid through another. (165)

dimensional analysis   (also *factor-label method*)   A calculation method in which arithmetic steps are accompanied by the appropriate canceling of units. (11)

dipole-dipole force   The intermolecular attraction between oppositely charged poles of nearby polar molecules. (360)

dipole–induced dipole force   The intermolecular attraction between a polar molecule and the oppositely charged pole it induces in a nearby molecule. (392)

dipole moment (μ)   A measure of molecular polarity; the magnitude of the partial charges on the ends of a molecule (in coulombs) times the distance between them (in meters). (316)

disaccharide   An organic compound formed by a dehydration-condensation reaction between two simple sugars (monosaccharides). (486)

disintegration series   (See *decay series.*)

dispersion force   (also *London force*)   The intermolecular attraction between all particles as a result of instantaneous polarizations of their electron clouds; the intermolecular force primarily responsible for the condensed states of nonpolar substances. (363)

disproportionation reaction   A reaction in which a given substance is both oxidized and reduced. (441)

donor atom   An atom that donates a lone pair of electrons to form a covalent bond, usually from ligand to metal ion in a complex. (742)

double bond   A covalent bond that consists of two bonding pairs; two atoms sharing four electrons in the form of one σ and one π bond. (278)

double-displacement reaction   (See *metathesis reaction.*)

double helix   The two intertwined polynucleotide strands held together by H bonds that form the structure of DNA (deoxyribonucleic acid). (491)

Downs cell   An industrial apparatus that electrolyzes molten NaCl to produce sodium and chlorine. (718)

dynamic equilibrium   In a chemical or physical change, the condition at which the forward and reverse processes are taking place at the same rate, so there is no net change in the amounts of reactants or products. (353)

E

e_g orbitals   The set of orbitals (composed of $d_{x^2-y^2}$ and d_{z^2}) that results when the energies of the metal-ion d orbitals are split by a ligand field. This set is higher in energy than the other (t_{2g}) set in an octahedral field of ligands and lower in energy in a tetrahedral field. (752)

effective collision   A collision in which the particles meet with sufficient energy and an orientation that allows them to react. (519)

effective nuclear charge (Z_{eff})   The nuclear charge an electron actually experiences as a result of shielding effects due to the presence of other electrons. (239)

effusion   The process by which a gas escapes from its container through a tiny hole into an evacuated space. (164)

electrochemical cell   A system that incorporates a redox reaction to produce or use electrical energy. (682)

electrochemistry   The study of the relationship between chemical change and electrical work. (682)

electrode   The part of an electrochemical cell that conducts the electricity between the cell and the surroundings. (686)

electrolysis   The nonspontaneous lysing (splitting) of a substance, often to its component elements, by supplying electrical energy. (717)

electrolyte   A substance that conducts a current when it dissolves in water. (110, 407) A mixture of ions, in which the electrodes of an electrochemical cell are immersed, that conducts a current. (686)

electrolytic cell   An electrochemical system that uses electrical energy to drive a nonspontaneous chemical reaction ($\Delta G > 0$). (686)

electromagnetic (EM) radiation   (also *electromagnetic energy* or *radiant energy*)   Oscillating, perpendicular electric and magnetic fields moving simultaneously through space as waves and manifested as visible light, x-rays, microwaves, radio waves, and so on. (206)

electromagnetic spectrum   The continuum of wavelengths of radiant energy. (207)

electromotive force (emf)   (See *cell potential.*)

electron (e⁻)   A subatomic particle that possesses a unit negative charge (1.60218×10^{-19} C) and occupies the space around the atomic nucleus. (41)

electron affinity (EA)   The energy change (in kJ) accompanying the addition of one mole of electrons to one mole of gaseous atoms or ions. (255)

electron capture   A type of radioactive decay in which a nucleus draws in an orbital electron, usually one from the lowest energy level, and releases energy. (766)

electron cloud   An imaginary representation of an electron's rapidly changing position around the nucleus over time. (222)

electron configuration   The distribution of electrons within the orbitals of the atoms of an element; also the notation for such a distribution. (236)

electron deficient   Referring to a bonded atom, such as Be or B, that has fewer than eight valence electrons. (303)

electron density diagram   (also *electron probability density diagram*)   The pictorial representation for a given energy sublevel of the quantity ψ^2 (the probability density of the electron lying within a particular tiny volume) as a function of r (distance from the nucleus). (222)

electron-pair delocalization (also *delocalization*) The process by which electron density is spread over several atoms rather than remaining between two. (301)

electron-sea model A qualitative description of metallic bonding proposing that metal atoms pool their valence electrons into a delocalized "sea" of electrons in which the metal cores (metal ions) are submerged in an orderly array. (380)

electron volt (eV) The energy (in joules, J) that an electron acquires when it moves through a potential difference of 1 volt; $1 \text{ eV} = 1.602 \times 10^{-19} \text{ J}$. (783)

electronegativity (EN) The relative ability of a bonded atom to attract shared electrons. (287)

electronegativity difference (ΔEN) The difference in electronegativities between the atoms in a bond. (289)

element The simplest type of substance with unique physical and chemical properties. An element consists of only one kind of atom, so it cannot be broken down into simpler substances. (32)

elementary reaction (also *elementary step*) A simple reaction that describes a single molecular event in a proposed reaction mechanism. (524)

elimination reaction A type of organic reaction in which C atoms are bonded to fewer atoms in the product than in the reactant, which leads to multiple bonding. (472)

emission spectrum The line spectrum produced when excited atoms return to lower energy levels and emit photons characteristic of the element. (217)

empirical formula A chemical formula that shows the lowest relative numbers of atoms of elements in a compound. (51)

enantiomers (See *optical isomers*.)

end point The point in a titration at which the indicator changes color. (120, 626)

endothermic process One occurring with an absorption of heat from the surroundings and therefore an increase in the enthalpy of the system ($\Delta H > 0$). (186)

energy The capacity to do work, that is, to move matter. (5) [See also *kinetic energy* (E_k) and *potential energy* (E_p).]

enthalpy (H) A thermodynamic quantity that is the sum of the internal energy plus the product of the pressure and volume. (185)

enthalpy diagram A graphic depiction of the enthalpy change of a system. (185)

enthalpy of hydration (ΔH_{hydr}) (See *heat of hydration*.)

enthalpy of solution (ΔH_{soln}) (See *heat of solution*.)

entropy (S) A thermodynamic quantity related to the number of ways the energy of a system can be dispersed through the motions of its particles. (398, 654)

enzyme A biological macromolecule (usually a protein) that acts as a catalyst. (531)

equatorial group A group (or atom) that lies in the trigonal plane of a trigonal bipyramidal molecule, or a similar structural feature in a molecule. (311)

equilibrium constant (K) The value obtained when equilibrium concentrations are substituted into the reaction quotient. (542)

equivalence point The point in a titration when the number of moles of the added species is stoichiometrically equivalent to the original number of moles of the other species. (120, 626)

ester An organic compound that contains the $-\overset{\overset{\displaystyle :\!O:}{\|}}{C}-\overset{..}{O}-\overset{|}{\underset{|}{C}}-$ group. (479)

exact number A quantity, usually obtained by counting or based on a unit definition, that has no uncertainty associated with it and, therefore, contains as many significant figures as a calculation requires. (23)

excitation The process by which a substance absorbs energy from low-energy radioactive particles, causing its electrons to move to higher energy levels. (776)

excited state Any electron configuration of an atom or molecule other than the lowest energy (ground) state. (214)

exclusion principle A principle developed by Wolfgang Pauli stating that no two electrons in an atom can have the same set of four quantum numbers. The principle arises from the fact that an orbital has a maximum occupancy of two electrons and their spins are paired. (238)

exothermic process One occurring with a release of heat to the surroundings and therefore a decrease in the enthalpy of the system ($\Delta H < 0$). (185)

expanded valence shell A valence level that can accommodate more than 8 electrons by using available d orbitals; occurs only for elements in Period 3 or higher. (305)

experiment A clear set of procedural steps that tests a hypothesis. (8)

extensive property A property, such as mass, that depends on the quantity of substance present. (17)

F

face-centered cubic unit cell A unit cell in which a particle occurs at each corner and in the center of each face of a cube. (370)

factor-label method (See *dimensional analysis*.)

Faraday constant (F) The physical constant representing the charge of 1 mol of electrons: $F = 96,485 \text{ C/mol e}^-$. (701)

fatty acid A carboxylic acid that has a long hydrocarbon chain and is derived from a natural source. (480)

first law of thermodynamics (See *law of conservation of energy*.)

fission The process by which a heavier nucleus splits into lighter nuclei with the release of energy. (781)

formal charge The hypothetical charge on an atom in a molecule or ion, equal to the number of valence electrons minus the sum of all the unshared and half the shared valence electrons. (302)

formation constant (K_f) An equilibrium constant for the formation of a complex ion from the hydrated metal ion and ligands. (642)

formation equation An equation in which 1 mole of a compound forms from its elements. (194)

formula mass The sum (in amu) of the atomic masses of a formula unit of an ionic compound. (58)

formula unit The chemical unit of a compound that contains the number and type of atoms (or ions) expressed in the chemical formula. (53)

fraction by mass (also *mass fraction*) The portion of a compound's mass contributed by an element; the mass of an element in a compound divided by the mass of the compound. (34)

free energy (G) A thermodynamic quantity that is the difference between the enthalpy and the product of the absolute temperature and the entropy: $G = H - TS$. (666)

free radical A molecular or atomic species with one or more unpaired electrons, which typically make it very reactive. (304)

freezing The process of cooling a liquid until it solidifies. (349)

freezing point depression (ΔT_f) A lowering of the freezing point of a solvent caused by the presence of dissolved solute particles. (410)

frequency (ν) The number of cycles a wave undergoes per second, expressed in units of 1/second, or s^{-1} [also called hertz (Hz)]. (206)

frequency factor (A) The product of the collision frequency Z and an orientation probability factor p that is specific for a reaction. (519)

fuel cell (also *flow battery*) A battery that is not self-contained and in which electricity is generated by the controlled oxidation of a fuel. (711)

functional group A specific combination of atoms, typically containing a carbon-carbon multiple bond and/or carbon-heteroatom bond, that reacts in a characteristic way no matter what molecule it occurs in. (460)

fundamental unit (See *base unit*.)

fusion (See *melting*.)

fusion (nuclear) The process by which light nuclei combine to form a heavier nucleus with the release of energy. (781)

G

galvanic cell (See *voltaic cell*.)

gamma emission The type of radioactive decay in which gamma rays are emitted from an excited nucleus. (766)

gamma (γ) ray A very high-energy photon. (764)

gas One of the three states of matter. A gas fills its container regardless of the shape. (3)

genetic code The set of three-base sequences that is translated into specific amino acids during the process of protein synthesis. (492)

geometric isomers (also *cis-trans isomers* or *diastereomers*) Stereoisomers in which the molecules have the same connections between atoms but differ in the spatial arrangements of the atoms. The *cis* isomer has similar groups on the same side of a structural feature; the *trans* isomer has them on opposite sides. (468, 746)

Graham's law of effusion A gas law stating that the rate of effusion of a gas is inversely proportional to the square root of its density (or molar mass):

$$\text{rate} \propto \frac{1}{\sqrt{\mathcal{M}}} \tag{164}$$

gray (Gy) The SI unit of absorbed radiation dose; 1 Gy = 1 J/kg tissue. (777)

ground state The electron configuration of an atom or ion that is lowest in energy. (214)

group A vertical column in the periodic table. (45)

H

H bond (See *hydrogen bond*.)

Haber process An industrial process used to form ammonia from its elements. (569)

half-cell A portion of an electrochemical cell in which a half-reaction takes place. (688)

half-life ($t_{1/2}$) In chemical processes, the time required for half the initial reactant concentration to be consumed. (771) In nuclear processes, the time required for half the initial number of nuclei in a sample to decay. (513)

half-reaction method A method of balancing redox reactions by treating the oxidation and reduction half-reactions separately. (683)

haloalkane (also *alkyl halide*) A hydrocarbon with one or more halogen atoms (X) in place of H; contains a group. (475)

heat (q) The energy transferred between objects because of differences in their temperatures only; thermal energy. (18, 179)

heat capacity The quantity of heat required to change the temperature of an object by 1 K. (187)

heat of fusion (ΔH°_{fus}) The enthalpy change occurring when 1 mol of a solid substance melts. (349)

heat of hydration (ΔH_{hydr}) (also *enthalpy of hydration*) The enthalpy change occurring when 1 mol of a gaseous species is hydrated. The sum of the enthalpies from separating water molecules and mixing the gaseous species with them. (397)

heat of reaction (ΔH°_{rxn}) The enthalpy change of a reaction. (185)

heat of solution (ΔH_{soln}) (also *enthalpy of solution*) The enthalpy change occurring when a solution forms from solute and solvent. The sum of the enthalpies from separating solute and solvent molecules and mixing them. (396)

heat of sublimation (ΔH°_{subl}) The enthalpy change occurring when 1 mol of a solid substance changes directly to a gas. The sum of the heats of fusion and vaporization. (350)

heat of vaporization (ΔH°_{vap}) The enthalpy change occurring when 1 mol of a liquid substance vaporizes. (349)

heating-cooling curve A plot of temperature vs. time for a substance when heat is absorbed or released by the system at a constant rate. (351)

Henderson-Hasselbalch equation An equation for calculating the pH of a buffer system:

$$\text{pH} = \text{p}K_a + \log\left(\frac{[\text{base}]}{[\text{acid}]}\right) \tag{621}$$

Henry's law A law stating that the solubility of a gas in a liquid is directly proportional to the partial pressure of the gas above the liquid: $S_{gas} = k_H \times P_{gas}$. (402)

Hess's law of heat summation A law stating that the enthalpy change of an overall process is the sum of the enthalpy changes of the individual steps of the process. (192)

heteroatom Any atom in an organic compound other than C or H. (459)

heterogeneous catalyst A catalyst that occurs in a different phase from the reactants, usually a solid interacting with gaseous or liquid reactants. (530)

heterogeneous mixture A mixture that has one or more visible boundaries among its components. (60)

hexagonal closest packing A crystal structure based on the hexagonal unit cell in which the layers have an *abab* . . . pattern. (372)

high-spin complex Complex ion that has the same number of unpaired electrons as in the isolated metal ion; contains weak-field ligands. (754)

homogeneous catalyst A catalyst (gas, liquid, or soluble solid) that exists in the same phase as the reactants. (530)

homogeneous mixture (also *solution*) A mixture that has no visible boundaries among its components. (60)

homologous series A series of organic compounds in which each member differs from the next by a —CH_2— (methylene) group. (463)

homonuclear diatomic molecule A molecule composed of two identical atoms. (337)

Hund's rule A principle stating that when orbitals of equal energy are available, the electron configuration of lowest energy has the maximum number of unpaired electrons with parallel spins. (241)

hybrid orbital An atomic orbital postulated to form during bonding by the mathematical mixing of specific combinations of nonequivalent orbitals in a given atom. (325)

hybridization A postulated process of orbital mixing to form hybrid orbitals. (325)

hydrate A compound in which a specific number of water molecules are associated with each formula unit. (55)

hydration Solvation in water. (397)

hydration shell The oriented cluster of water molecules that surrounds an ion in aqueous solution. (391)

hydrocarbon An organic compound that contains only H and C atoms. (460)

hydrogen bond (H bond) A type of dipole-dipole force that arises between molecules that have an H atom bonded to a small, highly electronegative atom with lone pairs, usually N, O, or F. (361)

hydrogenation The addition of hydrogen to a carbon-carbon multiple bond to form a carbon-carbon single bond. (531)

hydrolysis Cleaving a molecule by reaction with water, in which one part of the molecule bonds to the water —OH and the other to the water H. (480)

hydronium ion (H_3O^+) A proton covalently bonded to a water molecule. (578)

hypothesis A testable proposal made to explain an observation. If inconsistent with experimental results, a hypothesis is revised or discarded. (8)

I

ideal gas A hypothetical gas that exhibits linear relationships among volume, pressure, temperature, and amount (mol) at all conditions; approximated by simple gases at ordinary conditions. (143)

ideal gas law (also *ideal gas equation*) An equation that expresses the relationships among volume, pressure, temperature, and amount (mol) of an ideal gas: $PV = nRT$. (148)

ideal solution A solution whose vapor pressure equals the mole fraction of the solvent times the vapor pressure of the pure solvent; approximated only by very dilute solutions. (407) (See also *Raoult's law*.)

indicator (See *acid-base indicator*.)

infrared (IR) Radiation in the region of the electromagnetic spectrum between the microwave and visible regions. (207)

infrared (IR) spectroscopy An instrumental technique for determining the types of bonds in a covalent molecule by measuring the absorption of IR radiation. (282)

initial rate The instantaneous rate occurring as soon as the reactants are mixed, that is, at $t = 0$. (503)

inner electrons (also *core electrons*) Electrons that fill all the energy levels of an atom except the valence level; electrons also present in atoms of the previous noble gas and any completed transition series. (246)

inner transition elements The elements of the periodic table in which f orbitals are being filled; the lanthanides and actinides. (247)

instantaneous rate The reaction rate at a particular time, given by the slope of a tangent to a plot of reactant concentration vs. time. (502)

insulator A substance (usually a nonmetal) that does not conduct an electric current. (382)

integrated rate law A mathematical expression for reactant concentration as a function of time. (510)

intensive property A property, such as density, that does not depend on the quantity of substance present. (17)

intermolecular forces (also *interparticle forces*) The attractive and repulsive forces among the particles—molecules, atoms, or ions—in a sample of matter. (348)

internal energy (E) The sum of the kinetic and potential energies of all the particles in a system. (178)

ion A charged particle that forms from an atom (or covalently bonded group of atoms) when it gains or loses one or more electrons. (48)

ion-dipole force The intermolecular attractive force between an ion and a polar molecule (dipole). (360)

ion–induced dipole force The intermolecular attractive force between an ion and the dipole it induces in the electron cloud of a nearby particle. (392)

ion pair A pair of ions that form a gaseous ionic molecule; sometimes formed when a salt boils. (276)

ion-product constant for water (K_w) The equilibrium constant for the autoionization of water:

$$K_w = [H_3O^+][OH^-] \qquad (583)$$

ionic atmosphere A cluster of ions of net opposite charge surrounding a given ion in solution. (414)

ionic bonding The idealized type of bonding based on the attraction of oppositely charged ions that arise through electron transfer between atoms with large differences in their tendencies to lose or gain electrons (typically metals and nonmetals). (269)

ionic compound A compound that consists of oppositely charged ions. (47)

ionic radius The size of an ion as measured by the distance between the centers of adjacent ions in a crystalline ionic compound. (262)

ionic solid A solid whose unit cell contains cations and anions. (377)

ionization The process by which a substance absorbs energy from high-energy radioactive particles and loses an electron to become ionized. (776)

ionization energy (IE) The energy (in kJ) required to remove completely one mole of electrons from one mole of gaseous atoms or ions. (252)

ionizing radiation The high-energy radiation that forms ions in a substance by causing electron loss. (776)

isoelectronic Having the same number and configuration of electrons as another species. (259)

isomer One of two or more compounds with the same molecular formula but different properties, often as a result of different arrangements of atoms. (81, 465, 745)

isotopes Atoms of a given atomic number (that is, of a specific element) that have different numbers of neutrons and therefore different mass numbers. (42, 764)

isotopic mass The mass (in amu) of an isotope relative to the mass of carbon-12. (44)

J

joule (J)   The SI unit of energy; $1\ \text{J} = 1\ \text{kg·m}^2/\text{s}^2$. (182)

K

kelvin (K)   The SI base unit of temperature. The kelvin is the same size as the Celsius degree. (18)

Kelvin scale   (See *absolute scale*.)

ketone   An organic compound (ending, *-one*) that contains a carbonyl group bonded to two other C atoms, $-\overset{\displaystyle |}{\underset{\displaystyle |}{\text{C}}}-\overset{\displaystyle :\text{O}:}{\overset{\displaystyle \|}{\text{C}}}-\overset{\displaystyle |}{\underset{\displaystyle |}{\text{C}}}-$. (478)

kilogram (kg)   The SI base unit of mass. (16)

kinetic energy (E_k)   The energy an object has because of its motion. (5)

kinetic-molecular theory   The model that explains gas behavior in terms of particles in random motion whose volumes and interactions are negligible. (160)

L

lanthanide contraction   The additional decrease in atomic and ionic size, beyond the expected trend, caused by the poor shielding of the increasing nuclear charge by f electrons in the elements following the lanthanides. (739)

lanthanides (also *rare earths*)   The Period 6 (4f) series of inner transition elements, which includes cerium (Ce; $Z = 58$) through lutetium (Lu; $Z = 71$). (247)

lattice   The three-dimensional arrangement of points created by choosing each point to be at the same location within each particle of a crystal; thus, the lattice consists of all points with identical surroundings. (370)

lattice energy ($\Delta H^{\circ}_{\text{lattice}}$)   The enthalpy change (always positive) that occurs when 1 mol of an ionic compound separates into gaseous ions, with all components in their standard states. (274)

law   (See *natural law*.)

law of chemical equilibrium (also *law of mass action*)   The law stating that when a system reaches equilibrium at a given temperature, the ratio of quantities that make up the reaction quotient has a constant numerical value. (544)

law of conservation of energy (also *first law of thermodynamics*)   A basic observation that the total energy of the universe is constant: $\Delta E_{\text{universe}} = \Delta E_{\text{system}} + \Delta E_{\text{surroundings}} = 0$. (181)

law of definite (or constant) composition   A mass law stating that, no matter what its source, a particular compound is composed of the same elements in the same parts (fractions) by mass. (34)

law of mass action   (See *law of chemical equilibrium*.)

law of mass conservation   A mass law stating that the total mass of substances does not change during a chemical reaction. (34)

law of multiple proportions   A mass law stating that if elements A and B react to form two compounds, the different masses of B that combine with a fixed mass of A can be expressed as a ratio of small whole numbers. (35)

Le Châtelier's principle   A principle stating that if a system in a state of equilibrium is disturbed, it will undergo a change that shifts its equilibrium position in a direction that reduces the effect of the disturbance. (561)

level (also *shell*)   A specific energy state of an atom given by the principal quantum number n. (225)

Lewis acid-base definition   A model of acid-base behavior in which acids and bases are defined, respectively, as species that accept and donate an electron pair. (606)

Lewis electron-dot symbol   A notation in which the element symbol represents the nucleus and inner electrons, and surrounding dots represent the valence electrons. (271)

Lewis structure (also *Lewis formula*)   A structural formula consisting of electron-dot symbols, with lines as bonding pairs and dots as lone pairs. (297)

ligand   A molecule or anion bonded to a central metal ion in a complex ion. (641, 741)

like-dissolves-like rule   An empirical observation stating that substances having similar kinds of intermolecular forces dissolve in each other. (391)

limiting reactant (also *limiting reagent*)   The reactant that is consumed when a reaction occurs and therefore the one that determines the maximum amount of product that can form. (90)

line spectrum   A series of separated lines of different colors representing photons whose wavelengths are characteristic of an element. (212) (See also *emission spectrum*.)

linear arrangement   The geometric arrangement obtained when two electron groups maximize their separation around a central atom. (308)

linear shape   A molecular shape formed by three atoms lying in a straight line, with a bond angle of 180° (shape class AX_2 or AX_2E_3). (308)

linkage isomers   Coordination compounds with the same composition but with different ligand donor atoms linked to the central metal ion. (746)

lipid   Any of a class of biomolecules, including fats and oils, that are soluble in nonpolar solvents. (480)

liquid   One of the three states of matter. A liquid fills a container to the extent of its own volume and thus forms a surface. (3)

liter (L)   A non-SI unit of volume equivalent to 1 cubic decimeter ($0.001\ \text{m}^3$). (15)

London force   (See *dispersion force*.)

lone pair (also *unshared pair*)   An electron pair that is part of an atom's valence shell but not involved in covalent bonding. (278)

low-spin complex   Complex ion that has fewer unpaired electrons than in the free metal ion because of the presence of strong-field ligands. (754)

M

macromolecule   (See *polymer*.)

magnetic quantum number (m_l) (also *orbital-orientation quantum number*)   An integer from $-l$ through 0 to $+l$ that specifies the orientation of an atomic orbital in the three-dimensional space about the nucleus. (224)

mass   The quantity of matter an object contains. Balances are designed to measure mass. (16)

mass defect   The mass decrease that occurs when nucleons combine to form a nucleus. (782)

mass fraction   (See *fraction by mass*.)

mass number (A)   The total number of protons and neutrons in the nucleus of an atom. (42)

mass percent (also *mass %* or *percent by mass*)   The fraction by mass expressed as a percentage. A concentration term [% (w/w)] expressed as the mass in grams of solute dissolved per 100. g of solution. (34, 404)

mass spectrometry An instrumental method for measuring the relative masses of particles in a sample by creating charged particles and separating them according to their mass-charge ratio. (43)

matter Anything that possesses mass and occupies volume. (2)

melting (also *fusion*) The change of a substance from a solid to a liquid. (349)

melting point (mp or T_f) The temperature at which the solid and liquid forms of a substance are at equilibrium. (356)

metal A substance or mixture that is relatively shiny and malleable and is a good conductor of heat and electricity. In reactions, metals tend to transfer electrons to nonmetals and form ionic compounds. (46)

metallic bonding An idealized type of bonding based on the attraction between metal ions and their delocalized valence electrons. (270) (See also *electron-sea model*.)

metallic radius One-half the distance between the nuclei of adjacent individual atoms in a crystal of an element. (249)

metallic solid A solid whose individual atoms are held together by metallic bonding. (378)

metalloid (also *semimetal*) An element with properties between those of metals and nonmetals. (47)

metathesis reaction (also *double-displacement reaction*) A reaction in which atoms or ions of two compounds exchange bonding partners. (116)

meter (m) The SI base unit of length. The distance light travels in a vacuum in 1/299,792,458 second. (15)

milliliter (mL) A volume (0.001 L) equivalent to 1 cm^3. (15)

millimeter of mercury (mmHg) A unit of pressure based on the difference in the heights of mercury in a barometer or manometer. Renamed the *torr* in honor of Evangelista Torricelli. (142)

miscible Soluble in any proportion. (390)

mixture A group of two or more elements and/or compounds that are physically intermingled. (33)

MO bond order One-half the difference between the numbers of electrons in bonding and antibonding MOs. (336)

model (also *theory*) A simplified conceptual picture based on experiment that explains how an aspect of nature occurs. (9)

molality (m) A concentration term expressed as number of moles of solute dissolved in 1000 g (1 kg) of solvent. (403)

molar heat capacity (C) The quantity of heat required to change the temperature of 1 mol of a substance by 1 K. (187)

molar mass (M) (also *gram-molecular weight*) The mass of 1 mol of entities (atoms, molecules, or formula units) of a substance, in units of g/mol. (72)

molar solubility The solubility expressed in terms of amount (mol) of dissolved solute per liter of solution. (634)

molarity (M) A concentration term expressed as the moles of solute dissolved in 1 L of solution. (95)

mole (mol) The SI base unit for amount of a substance. The amount that contains a number of objects equal to the number of atoms in exactly 12 g of carbon-12. (70)

mole fraction (X) A concentration term expressed as the ratio of moles of one component of a mixture to the total moles present. (156, 404)

molecular equation A chemical equation showing a reaction in solution in which reactants and products appear as intact, undissociated compounds. (113)

molecular formula A formula that shows the actual number of atoms of each element in a molecule. (51)

molecular mass (also *molecular weight*) The sum (in amu) of the atomic masses of a formula unit of a compound. (58)

molecular orbital (MO) An orbital of given energy and shape that extends over a molecule and can be occupied by no more than two electrons. (334)

molecular orbital (MO) diagram A depiction of the relative energy and number of electrons in each MO, as well as the atomic orbitals from which the MOs form. (336)

molecular orbital (MO) theory A model that describes a molecule as a collection of nuclei and electrons in which the electrons occupy orbitals that extend over the entire molecule. (334)

molecular polarity The overall distribution of electronic charge in a molecule, determined by its shape and bond polarities. (315)

molecular shape The three-dimensional structure defined by the relative positions of the atomic nuclei in a molecule. (307)

molecular solid A solid held together by intermolecular forces between individual molecules. (377)

molecularity The number of reactant particles involved in an elementary step. (524)

molecule A structure consisting of two or more atoms that are chemically bound together and behave as an independent unit. (32)

monatomic ion An ion derived from a single atom. (48)

monomer A small molecule, linked covalently to others of the same or similar type to form a polymer, on which the repeat unit of the polymer is based. (483)

mononucleotide A monomer unit of a nucleic acid, consisting of an N-containing base, a sugar, and a phosphate group. (490)

monosaccharide A simple sugar; a polyhydroxy ketone or aldehyde with three to nine C atoms. (486)

N

natural law (also *law*) A summary, often in mathematical form, of a universal observation. (8)

Nernst equation An equation stating that the voltage of an electrochemical cell under any conditions depends on the standard cell voltage and the concentrations of the cell components:

$$E_{cell} = E°_{cell} - \frac{RT}{nF} \ln Q \qquad (703)$$

net ionic equation A chemical equation of a reaction in solution in which spectator ions have been eliminated to show the actual chemical change. (114)

network covalent solid A solid in which all the atoms are bonded covalently. (378)

neutralization In the Arrhenius acid-base definition, the combination of the H^+ ion from the acid and the OH^- ion from the base to form H_2O. (579)

neutralization reaction An acid-base reaction that yields water and a solution of a salt; when a strong acid reacts with a stoichiometrically equivalent amount of a strong base, the solution is neutral. (117)

neutron (n⁰) An uncharged subatomic particle found in the nucleus, with a mass slightly greater than that of a proton. (41)

nitrile An organic compound containing the $—C\equiv N$: group. (482)

node A region of an orbital where the probability of finding the electron is zero. (227)

nonelectrolyte A substance whose aqueous solution does not conduct an electric current. (112, 407)

nonionizing radiation Radiation that does not cause loss of electrons. (776)

nonmetal An element that lacks metallic properties. In reactions, nonmetals tend to bond with each other to form covalent compounds or accept electrons from metals to form ionic compounds. (47)

nonpolar covalent bond A covalent bond between identical atoms that share the bonding pair equally. (288)

nuclear binding energy The energy required to break 1 mol of nuclei of an element into individual nucleons. (783)

nuclear transmutation The induced conversion of one nucleus into another by bombardment with a particle. (774)

nucleic acid An unbranched polymer consisting of mononucleotides that occurs as two types, DNA and RNA (deoxyribonucleic and ribonucleic acids), which differ chemically in the nature of the sugar portion of the mononucleotides. (490)

nucleon A subatomic particle that makes up a nucleus; a proton or neutron. (763)

nucleus The tiny central region of the atom that contains all the positive charge and essentially all the mass. (40)

nuclide A nuclear species with specified numbers of protons and neutrons. (764)

N/Z ratio The ratio of the number of neutrons to the number of protons, a key factor that determines the stability of a nuclide. (767)

O

observation A fact obtained with the senses, often with the aid of instruments. Quantitative observations provide data that can be compared objectively. (8)

octahedral arrangement The geometric arrangement obtained when six electron groups maximize their space around a central atom; when all six groups are bonding groups, the molecular shape is octahedral (AX_6; ideal bond angle $= 90°$). (312)

octet rule The observation that when atoms bond, they often lose, gain, or share electrons to attain a filled outer shell of eight electrons. (272)

optical isomers (also *enantiomers*) A pair of stereoisomers consisting of a molecule and its mirror image that cannot be superimposed on each other. (467, 747)

optically active Able to rotate the plane of polarized light. (468)

orbital diagram A depiction of electron number and spin in an atom's orbitals by means of arrows in a series of small boxes, lines, or circles. (240)

organic compound A compound in which carbon is nearly always bonded to at least one other carbon, to hydrogen, and often to other elements. (458)

osmosis The process by which solvent flows through a semipermeable membrane from a dilute to a concentrated solution. (411)

osmotic pressure (Π) The pressure that results from the inability of solute particles to cross a semipermeable membrane. The pressure required to prevent the net movement of solvent across the membrane. (411)

outer electrons Electrons that occupy the highest energy level (highest n value) and are, on average, farthest from the nucleus. (246)

overvoltage The additional voltage, usually associated with gaseous products, that is required above the standard cell voltage to accomplish electrolysis. (719)

oxidation The loss of electrons by a species, accompanied by an increase in oxidation number. (124)

oxidation number (O.N.) (also *oxidation state*) A number equal to the magnitude of the charge an atom would have if its shared electrons were held completely by the atom that attracts them more strongly. (124)

oxidation-reduction reaction (also *redox reaction*) A process in which there is a net movement of electrons from one reactant (reducing agent) to another (oxidizing agent). (123)

oxidation state (See *oxidation number.*)

oxidizing agent The substance that accepts electrons in a redox reaction and undergoes a decrease in oxidation number. (124)

oxoanion An anion in which an element is bonded to one or more oxygen atoms. (55)

P

p orbital An atomic orbital with $l = 1$. (228)

packing efficiency The percentage of the available volume occupied by atoms, ions, or molecules in a unit cell. (372)

paramagnetism The tendency of a species with unpaired electrons to be attracted by an external magnetic field. (261)

partial ionic character An estimate of the actual charge separation in a bond (caused by the electronegativity difference of the bonded atoms) relative to complete separation. (289)

partial pressure The portion of the total pressure contributed by a gas in a mixture of gases. (155)

particle accelerator A device used to impart high kinetic energies to nuclear particles. (774)

pascal (Pa) The SI unit of pressure; $1 \text{ Pa} = 1 \text{ N/m}^2$. (141)

penetration The process by which an outer electron moves through the region occupied by the core electrons to spend part of its time closer to the nucleus; penetration increases the average effective nuclear charge for that electron. (239)

percent by mass (mass %) (See *mass percent.*)

percent yield (% yield) The actual yield of a reaction expressed as a percentage of the theoretical yield. (94)

period A horizontal row of the periodic table. (45)

periodic law A law stating that when the elements are arranged by atomic number, they exhibit a periodic recurrence of properties. (236)

periodic table of the elements A table in which the elements are arranged by atomic number into columns (groups) and rows (periods). (45)

pH The negative common logarithm of $[H_3O^+]$. (584)

phase A physically distinct portion of a system. (348)

phase change A physical change from one phase to another, usually referring to a change in physical state. (348)

phase diagram A diagram used to describe the stable phases and phase changes of a substance as a function of temperature and pressure. (356)

photoelectric effect The observation that when monochromatic light of sufficient frequency shines on a metal, an electric current is produced. (210)

photon A quantum of electromagnetic radiation. (211)

physical change A change in which the physical form (or state) of a substance, but not its composition, is altered. (3)

physical property A characteristic shown by a substance itself, without interacting with or changing into other substances. (2)

pi (π) bond   A covalent bond formed by sideways overlap of two atomic orbitals that has two regions of electron density, one above and one below the internuclear axis. (332)

pi (π) MO   A molecular orbital formed by combination of two atomic (usually p) orbitals whose orientations are perpendicular to the internuclear axis. (338)

Planck's constant (h)   A proportionality constant relating the energy and the frequency of a photon, equal to 6.626×10^{-34} J·s. (210)

polar covalent bond   A covalent bond in which the electron pair is shared unequally, so the bond has partially negative and partially positive poles. (288)

polar molecule   A molecule with an unequal distribution of charge as a result of its polar covalent bonds and shape. (109)

polarizability   The ease with which a particle's electron cloud can be distorted. (362)

polyatomic ion   An ion in which two or more atoms are bonded covalently. (51)

polymer (also *macromolecule*)   An extremely large molecule that results from the covalent linking of many simpler molecular units (monomers). (483)

polyprotic acid   An acid with more than one ionizable proton. (595)

polysaccharide   A macromolecule composed of many simple sugars linked covalently. (486)

positron ($_{1}^{0}\beta$)   The antiparticle of an electron. (766)

positron emission   A type of radioactive decay in which a positron is emitted from a nucleus. (766)

potential energy (E_p)   The energy an object has as a result of its position relative to other objects or because of its composition. (5)

precipitate   The insoluble product of a precipitation reaction. (115)

precipitation reaction   A reaction in which two soluble ionic compounds form an insoluble product, a precipitate. (115)

precision (also *reproducibility*)   The closeness of a measurement to other measurements of the same phenomenon in a series of experiments. (24)

pressure (P)   The force exerted per unit of surface area. (140)

pressure-volume work (PV work)   A type of work in which a volume change occurs against an external pressure. (181)

principal quantum number (n)   A positive integer that specifies the energy and relative size of an atomic orbital. (224)

probability contour   A shape that defines the volume around an atomic nucleus within which an electron spends a given percentage of its time. (223)

product   A substance formed in a chemical reaction. (83)

property   A characteristic that gives a substance its unique identity. (2)

protein   A natural, linear polymer composed of any of about 20 types of amino acid monomers linked together by peptide bonds. (487)

proton (p^+)   A subatomic particle found in the nucleus that has a unit positive charge (1.60218×10^{-19} C). (41)

proton acceptor   A substance that accepts an H^+ ion; a Brønsted-Lowry base. (588)

proton donor   A substance that donates an H^+ ion; a Brønsted-Lowry acid. (587)

pseudo–noble gas configuration   The $(n - 1)d^{10}$ configuration of a p-block metal atom that has emptied its outer energy level. (259)

pure substance   (See *substance*.)

Q

quantum   A packet of energy equal to $h\nu$. The smallest quantity of energy that can be emitted or absorbed. (210)

quantum mechanics   The branch of physics that examines the wave motion of objects on the atomic scale. (221)

quantum number   A number that specifies a property of an orbital or an electron. (210)

R

rad (radiation-absorbed dose)   The quantity of radiation that results in 0.01 J of energy being absorbed per kilogram of tissue; 1 rad = 0.01 J/kg tissue = 10^{-2} Gy. (777)

radial probability distribution plot   The graphic depiction of the total probability distribution (sum of ψ^2) of an electron in the region near the nucleus. (222)

radioactivity   The emissions resulting from the spontaneous disintegration of an unstable nucleus. (763)

radioisotope   An isotope with an unstable nucleus that decays through radioactive emissions. (773)

radioisotopic dating   A method for determining the age of an object based on the rate of decay of a particular radioactive nuclide. (773)

random error   Human error that occurs in all measurements and results in values *both* higher and lower than the actual value. (24)

Raoult's law   A law stating that the vapor pressure of a solution is directly proportional to the mole fraction of solvent: $P_{\text{solvent}} = X_{\text{solvent}} \times P^{\circ}_{\text{solvent}}$. (407)

rare earths   (See *lanthanides*.)

rate constant (k)   The proportionality constant that relates the reaction rate to reactant (and product) concentrations. (505)

rate-determining step (also *rate-limiting step*)   The slowest step in a reaction mechanism and therefore the step that limits the overall rate. (525)

rate law (also *rate equation*)   An equation that expresses the rate of a reaction as a function of reactant (and product) concentrations. (505)

reactant   A starting substance in a chemical reaction. (83)

reaction energy diagram   A graph that shows the potential energy of a reacting system as it progresses from reactants to products. (521)

reaction intermediate   A substance that is formed and used up during the overall reaction and therefore does not appear in the overall equation. (524)

reaction mechanism   A series of elementary steps that sum to the overall reaction and is consistent with the rate law. (523)

reaction order   The exponent of a reactant concentration in a rate law that shows how the rate is affected by changes in that concentration. (505)

reaction quotient (Q)   A ratio of terms for a given reaction consisting of product concentrations multiplied together and divided by reactant concentrations multiplied together, each raised to the power of their balancing coefficient. The value of Q changes until the system reaches equilibrium, at which point it equals K. (544)

reaction rate   The change in the concentrations of reactants (or products) with time. (501)

reactor core   The part of a nuclear reactor that contains the fuel rods and generates heat from fission. (786)

redox reaction   (See *oxidation-reduction reaction*.)

reducing agent   The substance that donates electrons in a redox reaction and undergoes an increase in oxidation number. (124)

reduction The gain of electrons by a species, accompanied by a decrease in oxidation number. (124)

refraction A phenomenon in which a wave changes its speed and therefore its direction as it passes through a phase boundary. (208)

rem (roentgen equivalent for *man*) The unit of radiation dosage for a human based on the product of the number of rads and a factor related to the biological tissue; 1 rem $= 10^{-2}$ Sv. (777)

reproducibility (See *precision*.)

resonance hybrid The weighted average of the resonance structures of a molecule. (300)

resonance structure (also *resonance form*) One of two or more Lewis structures for a molecule that cannot be adequately depicted by a single structure. Resonance structures differ only in the position of bonding and lone electron pairs. (300)

rms (root-mean-square) speed (u_{rms}) The speed of a molecule having the average kinetic energy; very close to the most probable speed. (163)

round off The process of removing digits based on a series of rules to obtain an answer with the proper number of significant figures (or decimal places). (22)

S

s orbital An atomic orbital with $l = 0$. (226)

salt An ionic compound that results from an Arrhenius acid-base reaction. (119)

salt bridge An inverted U tube containing a solution of nonreacting electrolyte that connects the compartments of a voltaic cell and maintains neutrality by allowing ions to flow between compartments. (689)

saturated hydrocarbon A hydrocarbon in which each C is bonded to four other atoms. (463)

saturated solution A solution that contains the maximum amount of dissolved solute at a given temperature in the presence of undissolved solute. (399)

Schrödinger equation An equation that describes how the electron matter-wave changes in space around the nucleus. Solutions of the equation provide allowable energy levels of the H atom. (221)

scientific method A process of creative thinking and testing aimed at objective, verifiable discoveries of the causes of natural events. (7)

second (s) The SI base unit of time. (20)

second law of thermodynamics A law stating that a process occurs spontaneously in the direction that increases the entropy of the universe. (657)

seesaw shape A molecular shape caused by the presence of one equatorial lone pair in a trigonal bipyramidal arrangement (AX_4E). (311)

self-ionization (See *autoionization*.)

semiconductor A substance whose electrical conductivity is poor at room temperature but increases significantly with rising temperature. (382)

semimetal (See *metalloid*.)

semipermeable membrane A membrane that allows solvent, but not solute, to pass through. (411)

shared pair (See *bonding pair*.)

shell (See *level*.)

shielding The ability of other electrons, especially inner ones, to lessen the nuclear attraction for an outer electron. (238)

SI unit A unit composed of one or more of the base units of the Système International d'Unités, a revised metric system. (13)

side reaction An undesired chemical reaction that consumes some of the reactant and reduces the overall yield of the desired product. (93)

sievert (Sv) The SI unit of human radiation dosage; 1 Sv $=$ 100 rem. (777)

sigma (σ) bond A type of covalent bond that arises through end-to-end orbital overlap and has most of its electron density along the bond axis. (331)

sigma (σ) MO A molecular orbital that is cylindrically symmetrical about an imaginary line that runs through the nuclei of the component atoms. (336)

significant figures The digits obtained in a measurement. The greater the number of significant figures, the greater the certainty of the measurement. (21)

silicate A type of compound found throughout rocks and soil and consisting of repeating $-Si-O$ groupings and, in most cases, metal cations. (437)

silicone A type of synthetic polymer containing $-Si-O$ repeat units, with organic groups and crosslinks. (437)

simple cubic unit cell A unit cell in which a particle occurs at each corner of a cube. (370)

single bond A bond that consists of one shared electron pair. (278)

solid One of the three states of matter. A solid has a fixed shape that does not conform to the container shape. (3)

solubility (S) The maximum amount of solute that dissolves in a fixed quantity of a particular solvent at a specified temperature when excess solute is present. (390)

solubility-product constant (K_{sp}) An equilibrium constant for the dissolving of a slightly soluble ionic compound in water. (632)

solute The substance that dissolves in the solvent. (95, 340)

solution (See *homogeneous mixture*.)

solvated Surrounded closely by solvent molecules. (110)

solvation The process of surrounding a solute particle with solvent particles. (397)

solvent The substance in which the solute(s) dissolve. (95, 390)

sp hybrid orbital An orbital formed by the mixing of one *s* and one *p* orbital of a central atom. (325)

sp^2 hybrid orbital An orbital formed by the mixing of one *s* and two *p* orbitals of a central atom. (327)

sp^3 hybrid orbital An orbital formed by the mixing of one *s* and three *p* orbitals of a central atom. (327)

sp^3d hybrid orbital An orbital formed by the mixing of one *s*, three *p*, and one *d* orbital of a central atom. (328)

sp^3d^2 hybrid orbital An orbital formed by the mixing of one *s*, three *p*, and two *d* orbitals of a central atom. (329)

specific heat capacity (c) The quantity of heat required to change the temperature of 1 gram of a substance by 1 K. (187)

spectator ion An ion that is present as part of a reactant but is not involved in the chemical change. (114)

spectrochemical series A ranking of ligands in terms of their ability to split *d*-orbital energies. (753)

spectrophotometry A group of instrumental techniques that create an electromagnetic spectrum to measure the atomic and molecular energy levels of a substance. (217)

speed of light (c) A fundamental constant giving the speed at which electromagnetic radiation travels in a vacuum: $c = 2.99792458 \times 10^8$ m/s. (207)

spin quantum number (m_s) A number, either $+\frac{1}{2}$ or $-\frac{1}{2}$, that indicates the direction of electron spin. (237)

spontaneous change A change that occurs by itself, that is, without an ongoing input of energy. (651)

square planar shape A molecular shape (AX_4E_2) caused by the presence of two axial lone pairs in an octahedral arrangement. (312)

square pyramidal shape A molecular shape (AX_5E) caused by the presence of one lone pair in an octahedral arrangement. (312)

standard atmosphere (atm) The average atmospheric pressure measured at sea level, defined as 1.01325×10^5 Pa. (141)

standard cell potential (E°_{cell}) The potential of a cell measured with all components in their standard states and no current flowing. (692)

standard electrode potential ($E^\circ_{half-cell}$) (also *standard half-cell potential*) The standard potential of a half-cell, with the half-reaction written as a reduction. (693)

standard entropy of reaction (ΔS°_{rxn}) The entropy change that occurs when all components are in their standard states. (661)

standard free energy change (ΔG°) The free energy change that occurs when all components are in their standard states. (667)

standard free energy of formation (ΔG°_f) The standard free energy change that occurs when 1 mol of a compound is made from its elements. (668)

standard half-cell potential (See *standard electrode potential*.)

standard heat of formation (ΔH°_f) The enthalpy change that occurs when 1 mol of a compound forms from its elements, with all substances in their standard states. (194)

standard heat of reaction (ΔH°_{rxn}) The enthalpy change that occurs during a reaction, with all substances in their standard states. (194)

standard hydrogen electrode (See *standard reference half-cell*.)

standard molar entropy (S°) The entropy of 1 mol of a substance in its standard state. (657)

standard molar volume The volume of 1 mol of an ideal gas at standard temperature and pressure: 22.4141 L. (147)

standard reference half-cell (also *standard hydrogen electrode*) A specially prepared platinum electrode immersed in 1 M $H^+(aq)$ through which H_2 gas at 1 atm is bubbled. $E^\circ_{half-cell}$ is defined as 0 V. (693)

standard states A set of specifications used to compare thermodynamic data: 1 atm for gases behaving ideally, 1 M for dissolved species, or the pure substance for liquids and solids. (194)

standard temperature and pressure (STP) The reference conditions for a gas:

$$0°C \ (273.15 \ K) \ \text{and} \ 1 \ \text{atm} \ (760 \ \text{torr}) \qquad (147)$$

state function A property of a system determined by its current state, regardless of how it arrived at that state. (183)

state of matter One of the three physical forms of matter: solid, liquid, or gas. (3)

stationary state In the Bohr model, one of the allowable energy levels of the atom in which it does not release or absorb energy. (213)

stereoisomers Molecules with the same connections of atoms but different orientations of groups in space. (467, 746) (See also *geometric isomers* and *optical isomers*.)

stoichiometric coefficient (See *balancing coefficient*.)

stoichiometry The study of the mass-mole-number relationships of chemical formulas and reactions. (70)

strong-field ligand A ligand that causes larger crystal field splitting energy and therefore is part of a low-spin complex. (752)

structural formula A formula that shows the actual numbers of atoms, their relative placement, and the bonds between them. (51)

structural isomers (See *constitutional isomers*.)

sublevel (also *subshell*) An energy substate of an atom within a level. Given by the n and l values, the sublevel designates the size and shape of the atomic orbitals. (225)

sublimation The process by which a solid changes directly into a gas. (350)

substance (also *pure substance*) A type of matter, either an element or a compound, that has a fixed composition. (32)

substitution reaction An organic reaction that occurs when an atom (or group) from one reactant substitutes for one in another reactant. (472)

superconductivity The ability to conduct a current with no loss of energy to resistive heating. (382)

supersaturated solution An unstable solution in which more solute is dissolved than in a saturated solution. (400)

surface tension The energy required to increase the surface area of a liquid by a given amount. (365)

surroundings All parts of the universe other than the system being considered. (178)

system The defined part of the universe under study. (178)

systematic error A type of error producing values that are all either higher or lower than the actual value, often caused by faulty equipment or a consistent fault in technique. (24)

T

t_{2g} orbitals The set of orbitals (composed of d_{xy}, d_{yz}, and d_{xz}) that results when the energies of the metal-ion d orbitals are split by a ligand field. This set is lower in energy than the other (e_g) set in an octahedral field and higher in energy in a tetrahedral field. (752)

T shape A molecular shape caused by the presence of two equatorial lone pairs in a trigonal bipyramidal arrangement (AX_3E_2). (311)

temperature (T) A measure of how hot or cold a substance is relative to another substance. (18)

tetrahedral arrangement The geometric arrangement formed when four electron groups maximize their separation around a central atom; when all four groups are bonding groups, the molecular shape is tetrahedral (AX_4; ideal bond angle 109.5°). (309)

theoretical yield The amount of product predicted by the stoichiometrically equivalent molar ratio in the balanced equation. (93)

theory (See *model*.)

thermochemical equation A chemical equation that shows the heat of reaction for the amounts of substances specified. (191)

thermochemistry The branch of thermodynamics that focuses on the heat involved in chemical reactions. (178)

thermodynamics The study of heat (thermal energy) and its interconversions. (178)

thermometer A device for measuring temperature that contains a fluid that expands or contracts within a graduated tube. (18)

third law of thermodynamics A law stating that the entropy of a perfect crystal is zero at 0 K. (657)

titration A method of determining the concentration of a solution by monitoring its reaction with a solution of known concentration. (119)

torr A unit of pressure identical to 1 mmHg. (142)

total ionic equation   A chemical equation for an aqueous reaction that shows all the soluble ionic substances dissociated into ions. (114)

tracer   A radioisotope that signals the presence of the species of interest by emitting nonionizing radiation. (779)

transition element (also *transition metal*)   An element that occupies the *d* block of the periodic table; one whose *d* orbitals are being filled. (244, 735)

transition state (also *activated complex*)   An unstable species formed in an effective collision of reactants that exists momentarily when the system is highest in energy and that can either form products or re-form reactants. (520)

transition state theory   A model that explains how the energy of reactant collision is used to form a high-energy transitional species that can change to reactant or product. (520)

transuranium element   An element with atomic number higher than that of uranium ($Z = 92$). (735, 776)

trigonal bipyramidal arrangement   The geometric arrangement formed when five electron groups maximize their separation around a central atom. When all five groups are bonding groups, the molecular shape is trigonal bipyramidal (AX_5; ideal bond angles, axial-center-equatorial $= 90°$ and equatorial-center-equatorial $= 120°$). (311)

trigonal planar arrangement   The geometric arrangement formed when three electron groups maximize their separation around a central atom. (308)

trigonal planar shape   A molecular shape (AX_3) formed when three atoms around a central atom lie at the corners of an equilateral triangle; ideal bond angle $= 120°$. (307)

trigonal pyramid   A molecular shape (AX_3E) caused by the presence of one lone pair in a tetrahedral arrangement. (310)

triple bond   A covalent bond that consists of three bonding pairs, two atoms sharing six electrons; one σ and two π bonds. (278)

triple point   The pressure and temperature at which three phases of a substance are in equilibrium. In a phase diagram, the point at which three phase-transition curves meet. (357)

U

ultraviolet (UV)   Radiation in the region of the electromagnetic spectrum between the visible and the x-ray regions. (207)

uncertainty   A characteristic of every measurement that results from the inexactness of the measuring device and the necessity of estimating when taking a reading. (21)

uncertainty principle   The principle stated by Werner Heisenberg that it is impossible to know simultaneously the exact position and velocity of a particle; the principle becomes important only for particles of very small mass. (221)

unimolecular reaction   An elementary reaction that involves the decomposition or rearrangement of a single particle. (524)

unit cell   The smallest portion of a crystal that, if repeated in all three directions, gives the crystal. (370)

universal gas constant (R)   A proportionality constant that relates the energy, amount of substance, and temperature of a system; $R = 0.0820578$ atm·L/mol·K $= 8.31447$ J/mol·K. (148)

unsaturated hydrocarbon   A hydrocarbon with at least one carbon-carbon multiple bond; one in which at least two C atoms are bonded to fewer than four atoms. (468)

unsaturated solution   A solution in which more solute can be dissolved at a given temperature. (400)

unshared pair   (See *lone pair.*)

V

V shape   (See *bent shape.*)

valence band   In band theory, the lower energy portion of the band of molecular orbitals, which is filled with valence electrons. (381)

valence bond (VB) theory   A model that attempts to reconcile the shapes of molecules with those of atomic orbitals through the concepts of orbital overlap and hybridization. (324)

valence electrons   The electrons involved in compound formation; in main-group elements, the electrons in the valence (outer) level. (246)

valence-shell electron-pair repulsion (VSEPR) theory   A model explaining that the shapes of molecules and ions result from minimizing electron-pair repulsions around a central atom. (306)

van der Waals equation   An equation that accounts for the behavior of real gases. (168)

van der Waals radius   One-half of the closest distance between the nuclei of identical nonbonded atoms. (359)

vapor pressure (also *equilibrium vapor pressure*)   The pressure exerted by a vapor at equilibrium with its liquid in a closed system. (353)

vapor pressure lowering (ΔP)   The lowering of the vapor pressure of a solvent caused by the presence of dissolved solute particles. (407)

vaporization   The process of changing from a liquid to a gas. (349)

variable   A quantity that can have more than a single value. (9) (See also *controlled experiment.*)

viscosity   A measure of the resistance of a liquid to flow. (367)

volt (V)   The SI unit of electric potential: 1 V $= 1$ J/C. (692)

voltage   (See *cell potential.*)

voltaic cell (also *galvanic cell*)   An electrochemical cell that uses a spontaneous reaction to generate electric energy. (686)

volume (V)   The space occupied by a sample of matter. (15)

volume percent [% (v/v)]   A concentration term defined as the volume of solute in 100. volumes of solution. (404)

W

wave function   (See *atomic orbital.*)

wave-particle duality   The principle stating that both matter and energy have wavelike and particle-like properties. (220)

wavelength (λ)   The distance between any point on a wave and the corresponding point on the next wave, that is, the distance a wave travels during one cycle. (206)

weak-field ligand   A ligand that causes smaller crystal field splitting energy and therefore is part of a high-spin complex. (752)

weight   The force exerted by a gravitational field on an object. (16)

work (w)   The energy transferred when an object is moved by a force. (179)

X

x-ray diffraction analysis   An instrumental technique used to determine spatial dimensions of a crystal structure by measuring the diffraction patterns caused by x-rays impinging on the crystal. (374)

Credits

Index

Page numbers followed by *f* indicate figures; *n*, footnotes; and *t*, tables.

A

Absolute temperature scale, 19, 145, G-1
Absolute zero, 19, 145
Absorption spectrum, 217, 217*f*, G-1
Accuracy, in measurement, 24–25, 25*f*, G-1
Acetaldehyde
  boiling point, 360, 360*f*
  dipole moment, 360, 360*f*
  molecular structure, 474*t*, 478*f*
Acetaminophen, molecular structure, 481*f*
Acetate ion, 54*t*, 617–621, 617*t*
Acetic acid
  boiling point elevation constant, 409*t*
  dissociation of, 122, 617–621, 617*t*
  freezing point depression constant, 409*t*
  K_a, 581
  molecular structure, 474*t*
Acetone
  bonds and orbitals, 334
  molecular shape, 315
  molecular structure, 474*t*, 478*f*
Acetonitrile
  boiling point, 360*f*
  dipole moment, 360*f*
  molecular structure, 474*t*
Acetylene, 157, 331, 332, 333*f*, 474*t*
Acid(s)
  acidic solutions, balancing redox reactions in, 683–684, 683*f*
  Arrhenius acids, 579
  Brønsted-Lowry acids, 121–122, 587–591, 598–600, 602
  buffers, 616–624
  concentration, finding from titration, 120–121
  conjugate acid-base pairs, 588–591, 589*t*, 591*f*, 599–600, 618*f*, 621, 623
  defined, 117, 121, 579, 587, G-1
  displacement of hydrogen from, 128–129
  dissociation, 112–113, 579–582, 580*f*, 595, 598–599, 617–621, 617*t*, A-8–A-10
  equilibria (*see* Acid-base equilibria)
  indicators, 120, 120*f*, 587, 587*f*, 624–625, 625*f*
  Lewis acids, 606–608, 641
  naming, 56–57
  pH (*see* pH)
  polyprotic, 595, 604
  strength (*see* Acid strength)
  strong acids, 118, 118*t*, 121–122, 122*f*, 579, 580*f*, 581, 590, 591*f*, 598–599, 603–604, 605*t*, 626–627, 631
  titrations, 119–121, 624–631
  in water, 578–582, 580*f*
  weak acids, 118, 118*t*, 122, 580–582, 580*f*, 590, 591–600, 591*f*, 603–605, 605*t*, 628–630
  weak bases and, 596–600
  weak-acid equilibrium, 591–596
Acid anhydride, 480, G-1
Acid rain, 86, 616, 639–640, 640*f*

Acid strength
  acid-dissociation constant and, 579–581
  classifying, 581–582
  conjugate acid-base pairs, 590, 591*f*
  direction of reaction and, 590, 591*f*
  hydrated metal ions, 602, 602*f*
  molecular properties and, 600–603
  of nonmetal hydrides, 600–601, 600*f*
  of oxoacids, 601–602, 601*f*
  strong acids, 118, 118*t*, 121–122, 122*f*, 579, 580*f*, 581, 590, 591*f*, 598–599, 603–604, 605*t*, 626–627, 631
  weak acids, 118, 118*t*, 122, 580–582, 580*f*, 590, 591–600, 591*f*, 603–605, 605*t*, 628–630
Acid-base behavior, of element oxides, 258
Acid-base buffer systems, 616–624, 616*f*–617*f*
  buffer capacity, 621–622, 622*f*
  buffer range, 622
  common-ion effect, 617–621
  conjugate base pair, 618*f*, 621, 623
  defined, 616, G-1
  essential features, 618–621
  Henderson-Hasselbalch equation, 621, 623
  preparing, 623–624
Acid-base equilibria, 577–608
  acid-base indicators, 120, 120*f*, 587, 587*f*, 624–625, 625*f*
  acid-dissociation constant (K_a), 579–582, 591–596, 599–600, A-8–A-10
    weak bases and weak acids, 599–600
    weak-acid equilibrium problem solving, 591–596
  autoionization of water, 583–587, 584*f*
  base-dissociation constant (K_b), 596–600, 597*f*, A-11–A-12
  Brønsted-Lowry acid-base definition, 587–591, 606
  conjugate acid-base pairs, 588–591, 589*t*, 591*f*, 599–600, 618*f*, 621
  hydrated metal ions, 602, 602*f*, A-12
  hydronium ion, 54*t*, 578–582, 578*f*, 580*f*, 584–587, 585*f*–587*f*
  ion-product constant for water (K_w), 583–587
  Lewis acid-base definition, 606–608, 641
  pH, 584–587, 585*f*–587*f*
  polyprotic acids, 595, 604
  salt solutions, 603–606
    of weakly acidic cations and weakly basic anions, 605–606
    yielding acidic solutions, 603–604, 605*t*
    yielding basic solutions, 604, 605*t*
    yielding neutral solutions, 603, 605*t*
  variation in acid strength, 579–582, 580*f*
  weak-acid equilibrium problem solving, 591–596
    acid-dissociation constant (K_a), 591–595
    concentrations, 592–595
    notation system, 592
    polyprotic acids, 595

Acid-base indicator
  acid-base titration, 120, 120*f*, 624–625, 625*f*
  color and pH range of common, 625*f*
  defined, 120, 587, 624, G-1
  pH paper, 587, 587*f*
Acid-base (neutralization) reactions, 117–122
  acid-base indicator, 120, 120*f*
  acid-base titrations, 119–121, 120*f*, 624–631
  acids, 117–118, 118*t*, 121
  bases, 117–118, 118*t*, 121
  defined, 117, G-1
  of element oxides, 258, 258*f*
  end point, 120
  equations, 118–119
  equivalence point, 120
  H_2O formation, 118–119
  key event, 118–119
  proton transfer, 121–122, 122*f*
  relative acid-base strength and the net direction of reaction, 590, 591*f*
  salt formation, 119, 122*f*
  strong acid with strong base, 121–122, 122*f*
  titrations, 119–121, 120*f*
  of weak acids, 118, 122
  of weak bases, 118
Acid-base titration, 119–121, 120*f*
Acid-base titration curves
  acid-base indicators, 624–625, 625*f*
  defined, 624, G-1
  end point, 120, 626
  equivalence point, 120, 626–629, 631
  strong acid–strong base, 626–627, 626*f*
  weak acid–strong base, 628–630, 628*f*
  weak base–strong acid, 631, 631*f*
Acid-dissociation constant (K_a)
  defined, 581, G-1
  of hydrated metal ions, A-12
  meaning of, 581
  relation with K_b in conjugate pair, 599–600
  tables of, A-8–A-10
  weak-acid equilibrium problem solving, 591–596
Acidic solutions, balancing redox reactions in, 683–684, 683*f*
Acid-ionization constant. *See* Acid-dissociation constant (K_a)
Acrylonitrile, 436*f*
Actinides
  defined, 247, G-1
  electron configuration, 246*f*, 247
  in periodic table, 46, 46*f*
Actinium, 247
Activated complex, 520
Activation analysis, 780
Activation energy (E_a), 516–521, 516*f*, 519*f*, 519*t*, 521*f*, 529–530, G-1
Active metal
  displacing hydrogen from water, 128*f*
  production by electrolysis, 128
  reactivity, 699
Active site, enzyme, 531–532, G-1
Activity (*A*), 770, G-1
Activity series of the halogens, 130

Activity series of the metals, 128–130, 130*f*, 699–700, G-1
Actual yield, 93, 94, 95, G-1
Addition polymers, 483–485, 484*t*, G-1
Addition reaction
  defined, 472, G-1
  organic, 472–473
Addition, significant figures and, 23
Adduct, 606–608, G-1
Adenine, 477*f*, 491
Adenosine triphosphate. *See* ATP
Air pollution
  acid rain, 86, 616, 639–640, 640*f*
  chlorofluorocarbons (CFCs), 532
  thermal pollution, 787
-al (suffix), 474*t*, 478
Alcohols
  defined, 475, G-1
  functional group, 460*f*, 474*t*, 475
  solubility, 392–393, 393*f*, 393*t*
Aldehydes, 474*t*, 478–479, 478*f*, G-1
Alizarin, 625*f*
Alkali metals
  alkaline earth metals compared, 428
  anomalous behavior of lithium, 427
  atomic radius, 252*f*
  electron configuration, 243, 425, 426
  first ionization energy, 253*f*
  important reactions, 426
  ion formation, 49
  melting point, 380
  properties, 425–427, 427*f*
  reactivity, 425, 427
Alkaline battery, 709, 709*f*
Alkaline earth metals, 428–430
  alkali metals compared, 428
  anomalous behavior of beryllium, 427
  atomic radius, 251
  diagonal relationships, 428, 428*f*, 430
  electron configuration, 428, 429
  important reactions, 429
  ion formation, 49
  melting point, 380
  properties, 428–430
Alkanes
  branched, 58
  chiral molecules, 467–468, 467*f*
  cycloalkanes, 465, 465*f*
  defined, 58, 463, G-1
  depicting with formulas and models, 464, 464*f*, 465*f*
  general formula, 463
  haloalkanes, 475–476
  homologous series, 463
  isomers
    constitutional, 465–466, 466*t*
    optical, 467–468, 467*f*
  naming, 58, 58*t*, 463–464, 463*t*, 464*t*, 470
  properties, 465–466, 466*t*, 467*f*
  saturated hydrocarbons, 463
  straight-chain, 58
Alkenes
  defined, 468, G-1
  functional group, 474*t*, 478
  general formula, 468
  isomerism, 468–469, 469*t*
  naming, 468, 470
  unsaturated hydrocarbons, 468

Fundamental Physical Constants (six significant figures)

Avogadro's number	N_A	$= 6.02214 \times 10^{23}$/mol
atomic mass unit	amu	$= 1.66054 \times 10^{-27}$ kg
charge of the electron (or proton)	e	$= 1.60218 \times 10^{-19}$ C
Faraday constant	F	$= 9.64853 \times 10^4$ C/mol
mass of the electron	m_e	$= 9.10939 \times 10^{-31}$ kg
mass of the neutron	m_n	$= 1.67493 \times 10^{-27}$ kg
mass of the proton	m_p	$= 1.67262 \times 10^{-27}$ kg
Planck's constant	h	$= 6.62607 \times 10^{-34}$ J·s
speed of light in a vacuum	c	$= 2.99792 \times 10^8$ m/s
standard acceleration of gravity	g	$= 9.80665$ m/s^2
universal gas constant	R	$= 8.31447$ J/(mol·K)
		$= 8.20578 \times 10^{-2}$ (atm·L)/(mol·K)

SI Unit Prefixes

p	n	μ	m	c	d	k	M	G
pico-	nano-	micro-	milli-	centi-	deci-	kilo-	mega-	giga-
10^{-12}	10^{-9}	10^{-6}	10^{-3}	10^{-2}	10^{-1}	10^3	10^6	10^9

Conversions and Relationships

Length
SI unit: meter, m

1 km	$= 1000$ m
	$= 0.62$ mile (mi)
1 inch (in)	$= 2.54$ cm
1 m	$= 1.094$ yards (yd)
1 pm	$= 10^{-12}$ m $= 0.01$ Å

Volume
SI unit: cubic meter, m^3

1 dm^3	$= 10^{-3}$ m^3
	$= 1$ liter (L)
	$= 1.057$ quarts (qt)
1 cm^3	$= 1$ mL
1 m^3	$= 35.3$ ft^3

Pressure
SI unit: pascal, Pa

1 Pa	$= 1$ N/m^2
	$= 1$ kg/m·s^2
1 atm	$= 1.01325 \times 10^5$ Pa
	$= 760$ torr
1 bar	$= 1 \times 10^5$ Pa

Mass
SI unit: kilogram, kg

1 kg	$= 10^3$ g
	$= 2.205$ lb
1 metric ton (t)	$= 10^3$ kg

Energy
SI unit: joule, J

1 J	$= 1$ kg·m^2/s^2
	$= 1$ coulomb·volt (1 C·V)
1 cal	$= 4.184$ J
1 eV	$= 1.602 \times 10^{-19}$ J

Math relationships

$$\pi = 3.1416$$
volume of sphere $= \frac{4}{3}\pi r^3$
volume of cylinder $= \pi r^2 h$

Temperature
SI unit: kelvin, K

0 K	$= -273.15$°C
mp of H_2O	$= 0$°C (273.15 K)
bp of H_2O	$= 100$°C (373.15 K)
T (K)	$= T$ (°C) $+ 273.15$
T (°C)	$= [T$ (°F) $- 32]\frac{5}{9}$
T (°F)	$= \frac{9}{5}T$ (°C) $+ 32$